Student Study Guide/ Solutions Manual

to accompany

Organic Chemistry

Fifth Edition

Janice Gorzynski Smith
University of Hawai'i at Mānoa

and

Erin R. Smith

STUDENT STUDY GUIDE/SOLUTIONS MANUAL TO ACCOMPANY
ORGANIC CHEMISTRY, FIFTH EDITION

Some ancillaries, including electronic and print components, may not be available to customers outside the United States.

This book is printed on acid-free paper.

1 2 3 4 5 6 7 8 9 0 RHR/RHR 1 0 9 8 7 6

ISBN 978-1-259-63706-3
MHID 1-259-63706-9

mheducation.com/highered

Contents

Chapter 1	Structure and Bonding	1-1
Chapter 2	Acids and Bases	2-1
Chapter 3	Introduction to Organic Molecules and Functional Groups	3-1
Chapter 4	Alkanes	4-1
Chapter 5	Stereochemistry	5-1
Chapter 6	Understanding Organic Reactions	6-1
Chapter 7	Alkyl Halides and Nucleophilic Substitution	7-1
Chapter 8	Alkyl Halides and Elimination Reactions	8-1
Chapter 9	Alcohols, Ethers, and Related Compounds	9-1
Chapter 10	Alkenes	10-1
Chapter 11	Alkynes	11-1
Chapter 12	Oxidation and Reduction	12-1
Chapter 13	Mass Spectrometry and Infrared Spectroscopy	13-1
Chapter 14	Nuclear Magnetic Resonance Spectroscopy	14-1
Chapter 15	Radical Reactions	15-1
Chapter 16	Conjugation, Resonance, and Dienes	16-1
Chapter 17	Benzene and Aromatic Compounds	17-1
Chapter 18	Reactions of Aromatic Compounds	18-1
Chapter 19	Carboxylic Acids and the Acidity of the O−H Bond	19-1
Chapter 20	Introduction to Carbonyl Chemistry; Organometallic Reagents; Oxidation and Reduction	20-1
Chapter 21	Aldehydes and Ketones—Nucleophilic Addition	21-1
Chapter 22	Carboxylic Acids and Their Derivatives—Nucleophilic Acyl Substitution	22-1
Chapter 23	Substitution Reactions of Carbonyl Compounds at the α Carbon	23-1
Chapter 24	Carbonyl Condensation Reactions	24-1
Chapter 25	Amines	25-1
Chapter 26	Carbon–Carbon Bond-Forming Reactions in Organic Synthesis	26-1
Chapter 27	Pericyclic Reactions	27-1
Chapter 28	Carbohydrates	28-1
Chapter 29	Amino Acids and Proteins	29-1
Chapter 30	Synthetic Polymers	30-1

Chapter 1: Structure and Bonding

Chapter Review

Important facts

- **The general rule of bonding:** Atoms strive to attain a complete outer shell of valence electrons (Section 1.2). H "wants" 2 electrons. Second-row elements "want" 8 electrons.

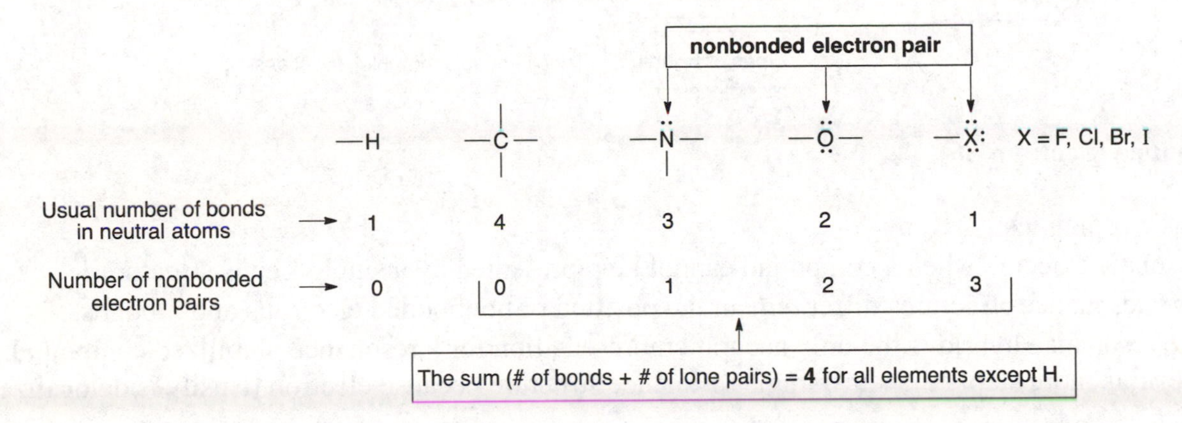

- **Formal charge** (FC) is the difference between the number of valence electrons on an atom and the number of electrons it "owns" (Section 1.3C). See Sample Problem 1.3 for a stepwise example.

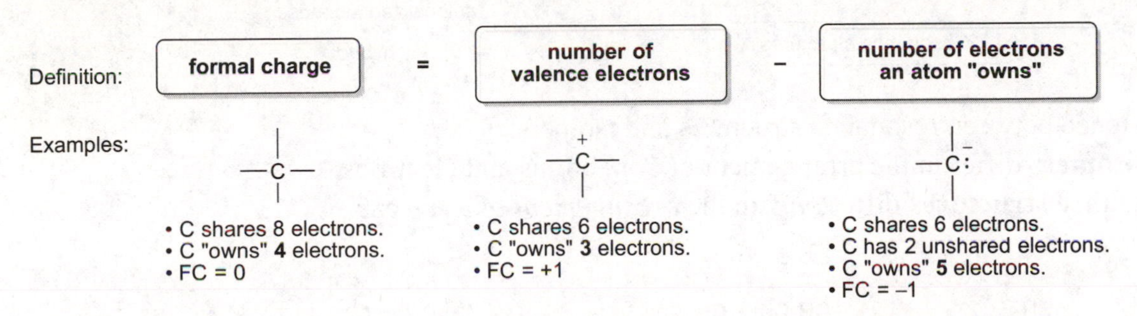

- **Curved arrow notation** shows the movement of an electron pair. The tail of the arrow always begins at an electron pair, either in a bond or a lone pair. The head points to where the electron pair "moves" (Section 1.6).

Use an electron pair to form a double bond.

Move an electron pair to O.

- **Electrostatic potential plots** are color-coded maps of electron density, indicating electron rich (red) and electron deficient (blue) regions (Section 1.12).

The importance of Lewis structures (Sections 1.3–1.5)

A properly drawn Lewis structure shows the number of bonds and lone pairs present around each atom in a molecule. In a valid Lewis structure, each H has two electrons, and each second-row element has no more than eight. This is the first step needed to determine many properties of a molecule.

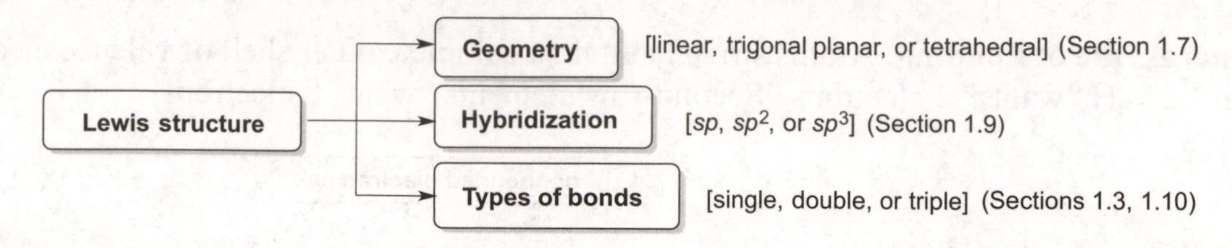

Resonance (Section 1.6)

The basic principles:

- Resonance occurs when a compound cannot be represented by a single Lewis structure.
- Two resonance structures differ *only* in the position of nonbonded electrons and π bonds.
- The resonance hybrid is the only accurate representation for a resonance-stabilized compound. A hybrid is more stable than any single resonance structure because electron density is delocalized.

$CH_3CH_2-C(=O)O^-$ ⟷ $CH_3CH_2-C(O^-)=O$ (resonance structures); hybrid: delocalized charges (δ^- on each O), delocalized π bonds

The difference between resonance structures and isomers:

- Two **isomers** differ in the arrangement of *both* atoms and electrons.
- **Resonance structures** differ *only* in the *arrangement of electrons*.

$CH_3-C(=O)-O-CH_3$ and $CH_3CH_2-C(=O)-O-H$ (isomers); $CH_3CH_2-C(=O)-O-H$ ⟷ $CH_3CH_2-C(O^-)=O^+-H$ (resonance structures)

Geometry and hybridization

The number of groups around an atom determines both its geometry (Section 1.7) and hybridization (Section 1.9).

Number of groups	Geometry	Bond angle (°)	Hybridization	Examples
2	linear	180	*sp*	BeH_2, $HC{\equiv}CH$
3	trigonal planar	120	*sp*^{2}	BF_3, $CH_2{=}CH_2$
4	tetrahedral	109.5	*sp*^{3}	CH_4, NH_3, H_2O

Drawing organic molecules (Section 1.8)

- Shorthand methods are used to abbreviate the structure of organic molecules.

skeletal structure = isooctane = condensed structure

$$CH_3-C(CH_3)(H)-C(H)(H)-C(CH_3)(CH_3)-CH_3 = (CH_3)_2CHCH_2C(CH_3)_3$$

- A carbon bonded to four atoms is tetrahedral in shape. The best way to represent a tetrahedron is to draw two bonds in the plane, one in front, and one behind.

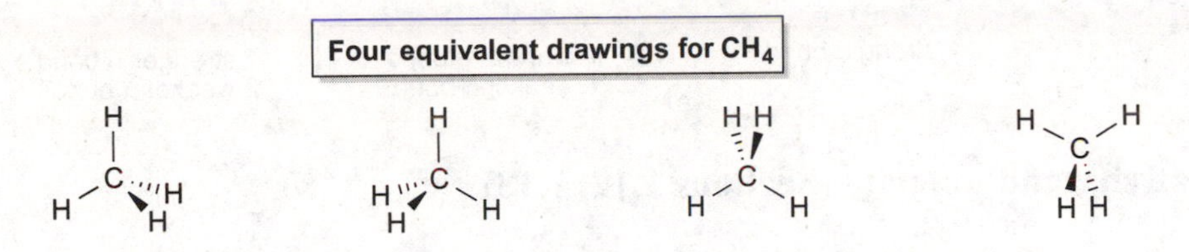

Each drawing has two solid lines, one wedge, and one dashed wedge.

Bond length

- Bond length decreases as you go from left to right across a row and increases down a column of the periodic table (Section 1.7A).

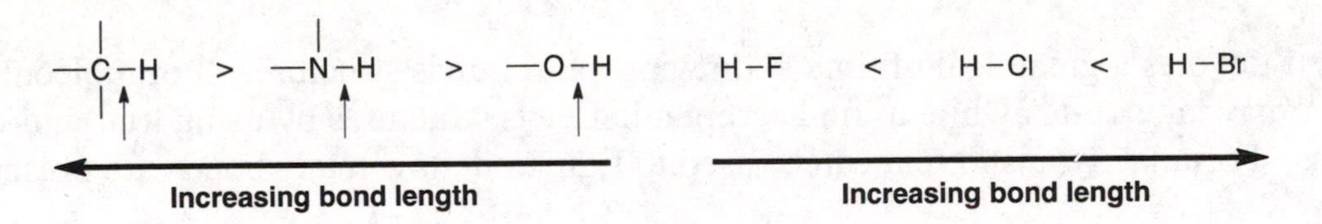

- Bond length decreases as the number of electrons between two nuclei increases (Section 1.11A).

CH_3-CH_3 < $CH_2=CH_2$ < $H-C\equiv C-H$

Increasing bond length

- Bond length increases as the percent *s*-character decreases (Section 1.11B).

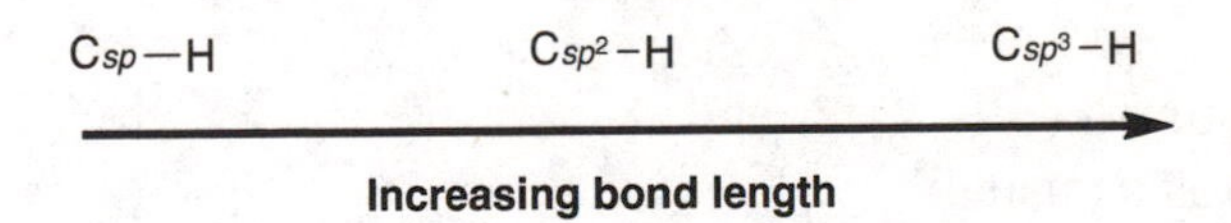

- Bond length and bond strength are inversely related. Shorter bonds are stronger bonds (Section 1.11).

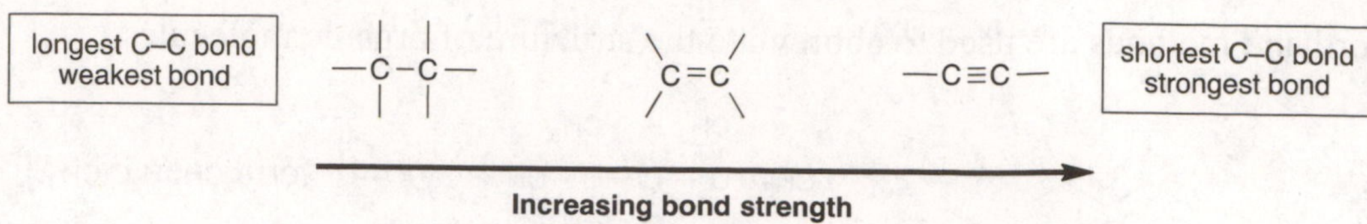

- Sigma (σ) bonds are generally stronger than π bonds (Section 1.10).

1 strong σ bond

1 stronger σ bond
1 weaker π bond

1 stronger σ bond
2 weaker π bonds

Electronegativity and polarity (Sections 1.12, 1.13)

- Electronegativity increases from left to right across a row and decreases down a column of the periodic table.
- A polar bond results when two atoms with different electronegativities are bonded together. Whenever C or H is bonded to N, O, or any halogen, the bond is polar.
- A polar molecule has either one polar bond, or two or more bond dipoles that reinforce.

Drawing Lewis structures: A shortcut

Chapter 1 devotes a great deal of time to drawing valid Lewis structures. For molecules with many bonds, it may take quite awhile to find acceptable Lewis structures by using trial-and-error to place electrons. Fortunately, a shortcut can be used to figure out how many bonds are present in a molecule.

Shortcut on drawing Lewis structures—Determining the number of bonds:

[1] Count up the number of valence electrons.

[2] Calculate how many electrons are needed if there are no bonds between atoms and every atom has a filled shell of valence electrons; that is, hydrogen gets two electrons, and second-row elements get eight.

[3] Subtract the number obtained in Step [1] from the sum obtained in Step [2]. **This difference tells how many electrons must be shared** to give every H two electrons and every second-row element eight. Because there are two electrons per bond, dividing this difference by two tells how many bonds are needed.

To draw the Lewis structure:

[1] Arrange the atoms as usual.

[2] Count up the number of valence electrons.

[3] Use the shortcut to determine how many bonds are present.

[4] Draw in the two-electron bonds to all the H's first. Then, draw the remaining bonds between other atoms, making sure that no second-row element gets more than eight electrons and that you use the total number of bonds determined previously.

[5] Finally, place unshared electron pairs on all atoms that do not have an octet of electrons, and calculate formal charge. You should have now used all the valence electrons determined in the first step.

Example: Draw all valid Lewis structures for CH_3NCO using the shortcut procedure.

[1] Arrange the atoms.

```
    H
H   C   N   C   O
    H
```

- In this case the arrangement of atoms is implied by the way the structure is drawn.

[2] Count up the number of valence electrons.

3 H's	x	1 electron per H	=	3 electrons
2 C's	x	4 electrons per C	=	8 electrons
1 N	x	5 electrons per N	=	5 electrons
1 O	x	6 electrons per O	=	+ 6 electrons
				22 electrons total

[3] Use the shortcut to figure out how many bonds are needed.

- Number of electrons needed if there were no bonds:

3 H's	x	2 electrons per H	=	6 electrons
4 second-row elements	x	8 electrons per element	=	+ 32 electrons
				38 electrons needed if there were no bonds

- Number of electrons that must be shared:

  38 electrons
  − 22 electrons
  16 electrons must be shared

- Every bond requires two electrons, so 16/2 = **8 bonds are needed.**

[4] Draw all possible Lewis structures.

- Draw the bonds to the H's first (three bonds). Then add five more bonds. Arrange them between the C's, N, and O, making sure that no atom gets more than eight electrons. There are three possible arrangements of bonds; that is, there are three resonance structures.
- Add additional electron pairs to give each atom an octet and check that all 22 electrons are used.

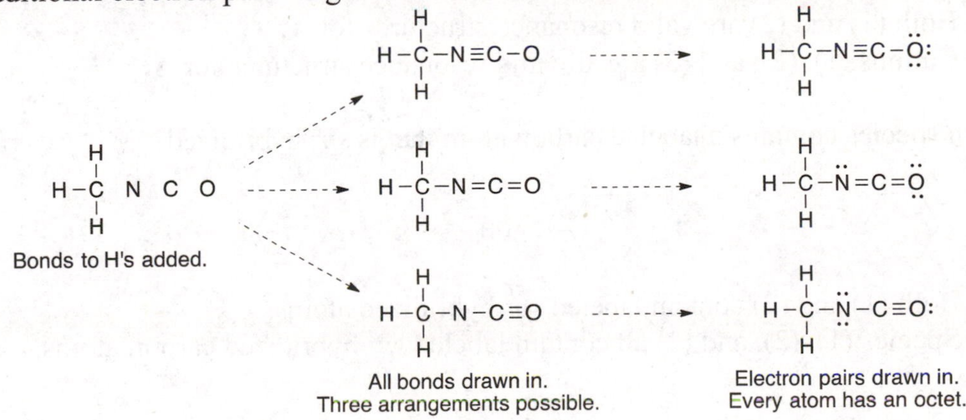
Bonds to H's added.
All bonds drawn in.
Three arrangements possible.
Electron pairs drawn in.
Every atom has an octet.

- Calculate the formal charge on each atom.

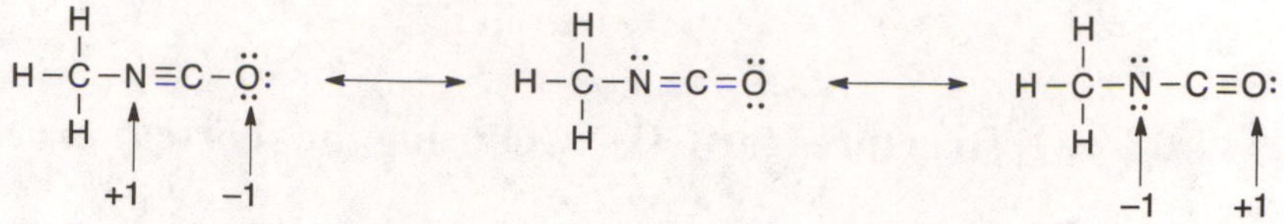

H–C(H)(H)–N≡C–Ö: ⟷ H–C(H)(H)–N̈=C=Ö ⟷ H–C(H)(H)–N̈–C≡O:

+1 −1 −1 +1

- You can evaluate the Lewis structures you have drawn. The middle structure is the best resonance structure, because it has no charged atoms.

Note: This method works for compounds that contain second-row elements in which every element gets an octet of electrons. It does NOT necessarily work for compounds with an atom that does not have an octet (such as BF_3), or compounds that have elements located in the third row and later in the periodic table.

Practice Test on Chapter Review

1.a. Which compound(s) contain a labeled atom with a +1 formal charge? All lone pairs of electrons have been drawn in.

1. 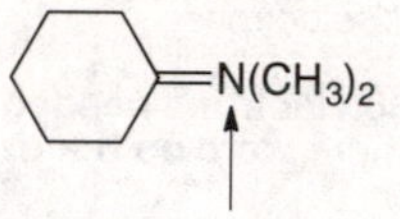 =N(CH3)2 2. Ö–CH3 3. CH_3–C̈–CH_3

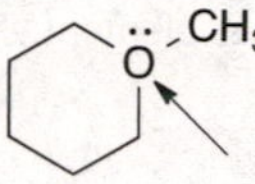

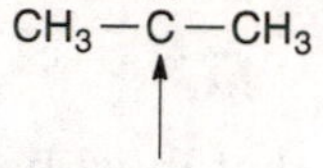

4. Both (1) and (2) have labeled atoms with a +1 charge.
5. Compounds (1), (2), and (3) all contain labeled atoms with a +1 formal charge.

b. Which of the following compounds is a valid resonance structure for **A?**

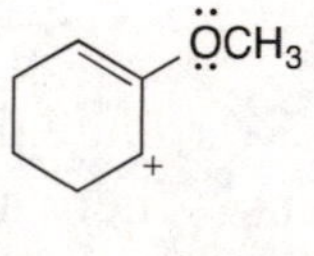

ÖCH3 **A** | 1. 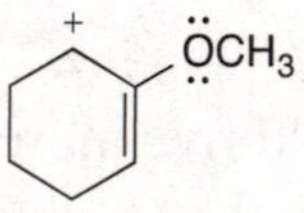 ÖCH3 2. 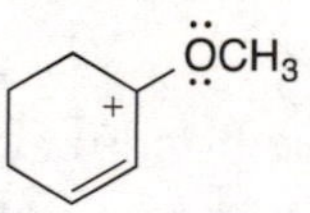 ÖCH3 3. 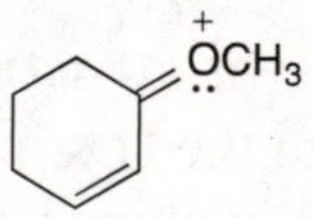 ÖCH3

4. Both (1) and (2) are valid resonance structures for **A.**
5. Cations (1), (2), and (3) are all valid resonance structures for **A.**

c. Which species contains a labeled carbon atom that is sp^2 hybridized?

1. 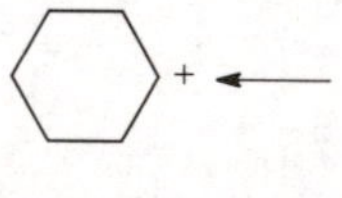 + 2. 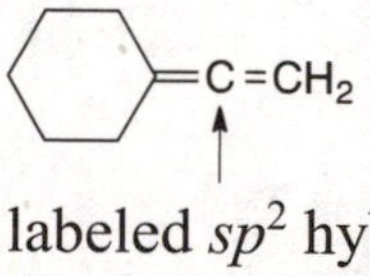 =C=CH2 3. 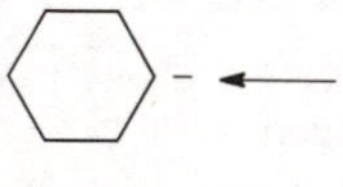 −

4. Both (1) and (2) contain labeled sp^2 hybridized atoms.
5. Species (1), (2), and (3) all contain labeled sp^2 hybridized carbon atoms.

d. Which of the following compounds has a net dipole?

1. $CH_3CH_2NHCH_2CH_3$
2. $CH_3CH_2CH_2OH$
3. $FCH_2CH_2CH_2F$
4. Compounds (1) and (2) both have net dipoles.
5. Compounds (1), (2), and (3) all have net dipoles.

2. Rank the labeled bonds in order of increasing bond length. Label the shortest bond as **1,** the longest bond as **4,** and the bonds of intermediate length as **2** and **3.**

B C A D HC≡C

3. Answer the following questions about compounds **A–D.**

NCH₂CH₃ [1] C=O [2]

A B C D

a. What is the hybridization of the labeled atom in **A?**
b. What is the molecular shape around the labeled atom in **B?**
c. In what type of orbital does the lone pair in **C** reside?
d. What orbitals are used to form bond [1] in **D?**
e. Which orbitals are used to form the carbon–oxygen double bond [2] in **D?**

4. Draw an acceptable Lewis structure for CH_3NO_3. Assume that the atoms are arranged as drawn.

H O
H C O N O
H

5. Follow the curved arrows and draw the product with all the needed charges and lone pairs.

Answers to Practice Test

1. a. 4
   b. 1
   c. 1
   d. 5

2. **A** – 1
   B – 4
   C – 2
   D – 3

3. a. sp^3
   b. trigonal planar
   c. sp^2
   d. Csp^3–Csp^2
   e. Csp–Osp^2, Cp–Op

4. One possibility:

5.

Answers to Problems

1.1 Two isotopes differ in the number of neutrons. Two isotopes have the same number of protons electrons, group number, and number of valence electrons. The **mass number** is the number of protons and neutrons. The **atomic number** is the number of protons and is the same for all isotopes.

	Nitrogen-14	Nitrogen-13
a. number of protons = atomic number for N = 7	7	7
b. number of neutrons = mass number – atomic number	7	6
c. number of electrons = number of protons	7	7
d. group number	5A	5A
e. number of valence electrons	5	5

1.2 **Ionic bonds** form when an element on the far left side of the periodic table transfers an electron to an element on the far right side of the periodic table. **Covalent bonds** result when two atoms *share* electrons.

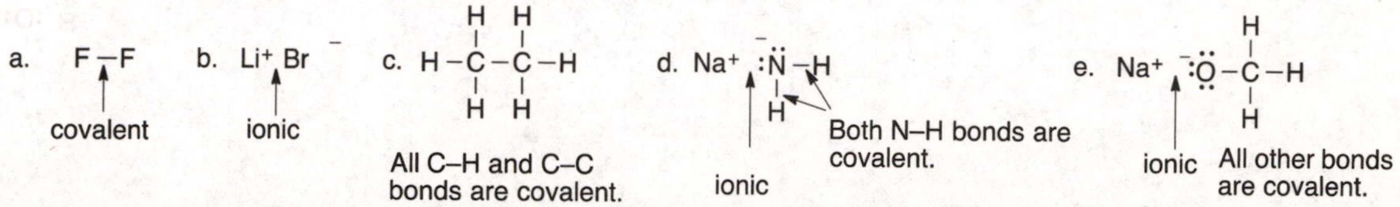

a. F–F (covalent)
b. Li^+ Br^- (ionic)
c. H–C(H)(H)–C(H)(H)–H — All C–H and C–C bonds are covalent.
d. Na^+ $^-$:N̈(H)–H — ionic; Both N–H bonds are covalent.
e. Na^+ $^-$:Ö–C(H)(H)–H — ionic; All other bonds are covalent.

1.3 Atoms with one, two, three, or four valence electrons form one, two, three, or four bonds, respectively. Atoms with five or more valence electrons form [8 – (number of valence electrons)] bonds.

a. O 8 – 6 valence e^- = 2 bonds
b. Al 3 valence e^- = 3 bonds
c. Br 8 – 7 valence e^- = 1 bond
d. Si 4 valence e^- = 4 bonds

1.4 [1] Arrange the atoms with the H's on the periphery.
[2] Count the valence electrons.
[3] Arrange the electrons around the atoms. Give the H's 2 electrons first, and then fill the octets of the other atoms.
[4] Assign formal charges (Section 1.3C).

a. [1] H C C H with H H above and H H below

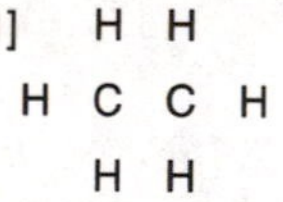

[2] Count valence e^-.
2 C x 4 e^- = 8
6 H x 1 e^- = 6
total e^- = 14

[3] H–C(H)(H)–C(H)(H)–H
All 14 e^- used.
All second-row elements have an octet.

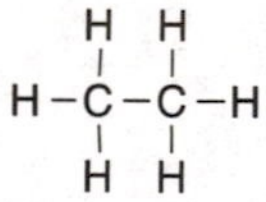

b. [1] H C N H with H above and H H below

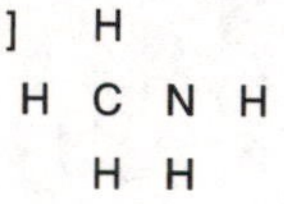

[2] Count valence e^-.
1 C x 4 e^- = 4
5 H x 1 e^- = 5
1 N x 5 e^- = 5
total e^- = 14

[3] H–C(H)(H)–N(H)–H → H–C(H)(H)–N̈(H)–H
12 e^- used.
N needs 2 more electrons for an octet.

c. [1]

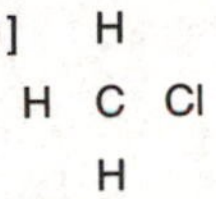

[2] Count valence e⁻.

1 C x 4 e⁻	= 4
3 H x 1 e⁻	= 3
1 Cl x 7 e⁻	= 7
total e⁻	= 14

[3]

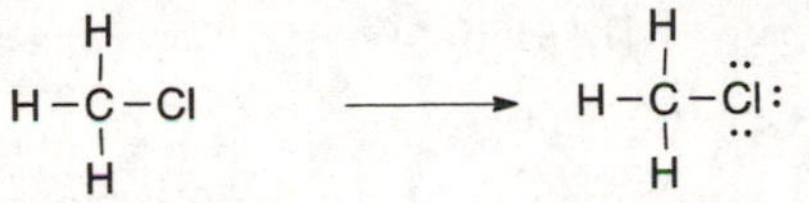

8 e⁻ used. Cl needs 6 more electrons for an octet.

Complete octet.

1.5 Follow the directions from Answer 1.4.

a. HCN H C N

Count valence e⁻.

1 C x 4 e⁻ =	4
1 H x 1 e⁻ =	1
1 N x 5 e⁻ =	5
total e⁻	= 10

4 e⁻ used.

Complete N and C octets.

b. H_2CO

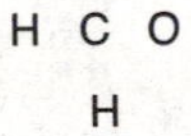

Count valence e⁻.

1 C x 4 e⁻ =	4
2 H x 1 e⁻ =	2
1 O x 6 e⁻ =	6
total e⁻	= 12

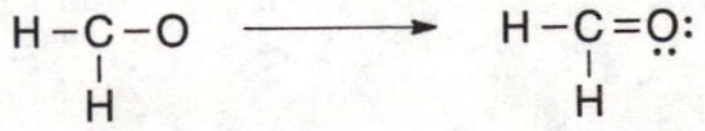

6 e⁻ used.

Complete O and C octets.

c. $HOCH_2CO_2H$

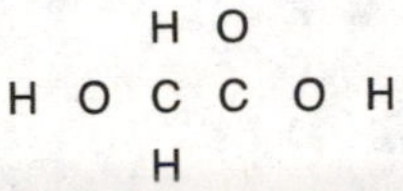

Count valence e⁻.

2 C x 4 e⁻ =	8
4 H x 1 e⁻ =	4
3 O x 6 e⁻ =	18
total e⁻	= 30

16 e⁻ used.

Complete octets.

1.6 **Formal charge** (FC) = number of valence electrons – [number of unshared electrons + (½)(number of shared electrons)]

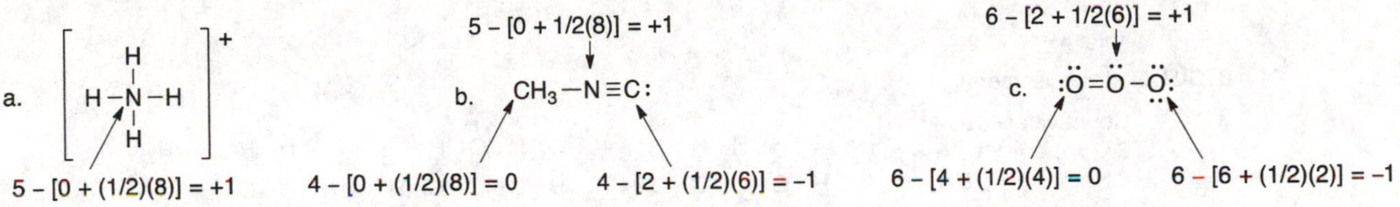

1.7

a. CH_3O^- [1]

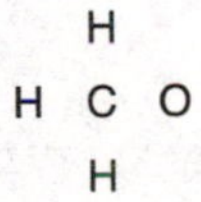

[2] Count valence e⁻.

1 C x 4 e⁻ =	4
3 H x 1 e⁻ =	3
1 O x 6 e⁻ =	6
total e⁻	= 13

Add 1 for (–) charge = 14

[3]

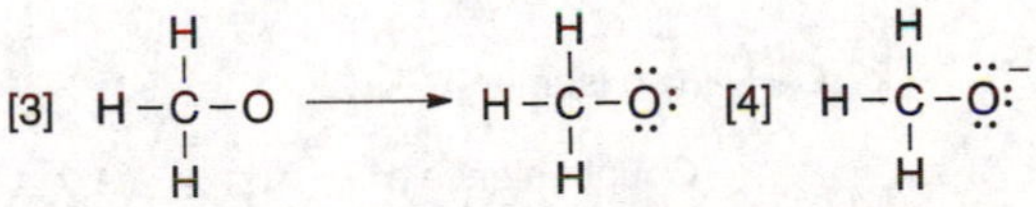

8 e⁻ used.

Assign charge.

b. HC_2^-

[1] H C C

[2] Count valence e⁻.
2 C x 4 e⁻ = 8
1 H x 1 e⁻ = 1
total e⁻ = 9
Add 1 for (–) charge = 10

[3] H–C–C ⟶ H–C≡C:
4 e⁻ used.

[4] H–C≡C:⁻
Assign charge.

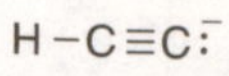

c. $(CH_3NH_3)^+$

[1] H H / H C N H / H H

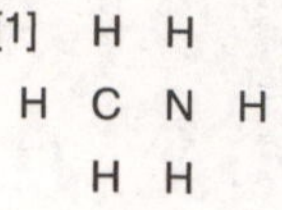

[2] Count valence e⁻.
1 C x 4 e⁻ = 4
6 H x 1 e⁻ = 6
1 N x 5 e⁻ = 5
total e⁻ = 15
Subtract 1 for (+) charge = 14

[3] 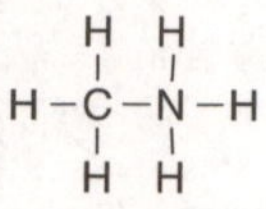 14 e⁻ used.

[4] 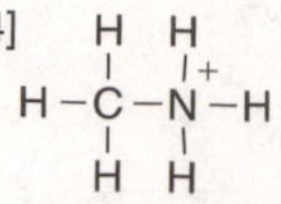 Assign charge.

d. $(CH_3NH)^-$

[1] H / H C N H / H

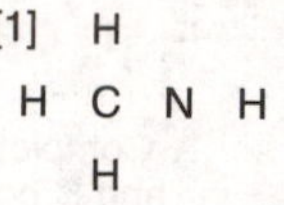

[2] Count valence e⁻.
1 C x 4 e⁻ = 4
4 H x 1 e⁻ = 4
1 N x 5 e⁻ = 5
total e⁻ = 13
Add 1 for (–) charge = 14

[3] 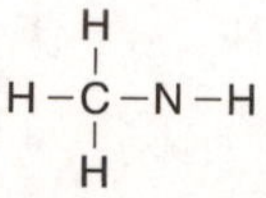 10 e⁻ used.

[4] 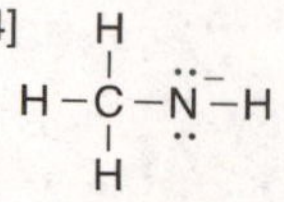 Complete octet and assign charge.

1.8

a. 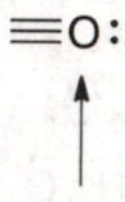 ≡O:
6 – [2 + (1/2)6] = +1

b. 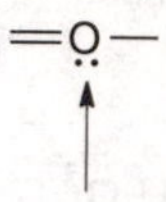 =O–
6 – [2 + (1/2)6] = +1

c. 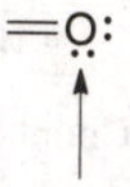 =O:
6 – [4 + (1/2)4] = 0

1.9

a. $C_2H_4Cl_2$ (two isomers)

Count valence e⁻.
2 C x 4 e⁻ = 8
4 H x 1 e⁻ = 4
2 Cl x 7 e⁻ = 14
total e⁻ = 26

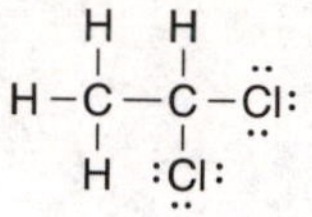

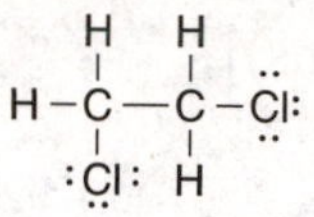

b. C_3H_8O (three isomers)

Count valence e⁻.
3 C x 4 e⁻ = 12
8 H x 1 e⁻ = 8
1 O x 6 e⁻ = 6
total e⁻ = 26

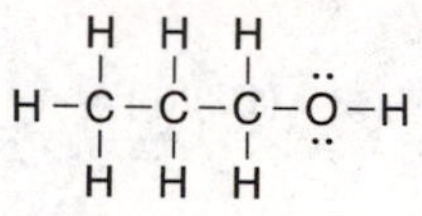

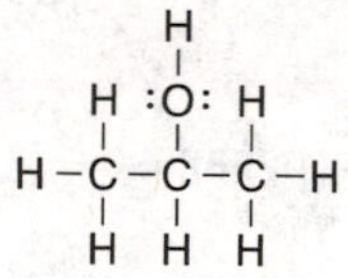

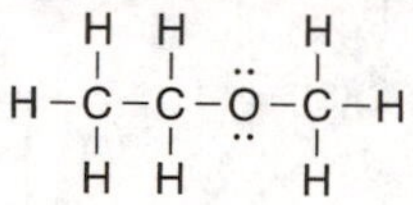

c. C_3H_6 (two isomers)

Count valence e⁻.
3 C x 4 e⁻ = 12
6 H x 1 e⁻ = 6
total e⁻ = 18

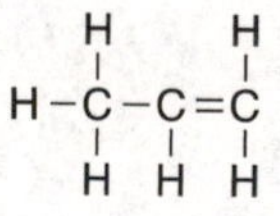

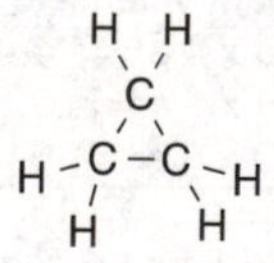

1.10 Two different definitions:

- **Isomers** have the same molecular formula and a *different* arrangement of atoms.
- **Resonance structures** have the same molecular formula and the *same* arrangement of atoms.

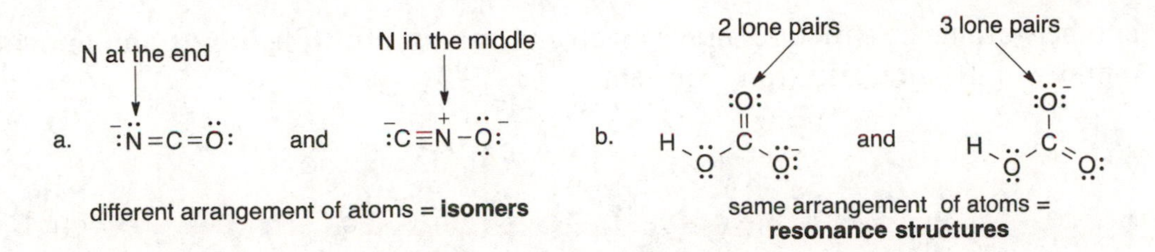

1.11 **Isomers** have the same molecular formula and a *different* arrangement of atoms.
Resonance structures have the same molecular formula and the *same* arrangement of atoms.

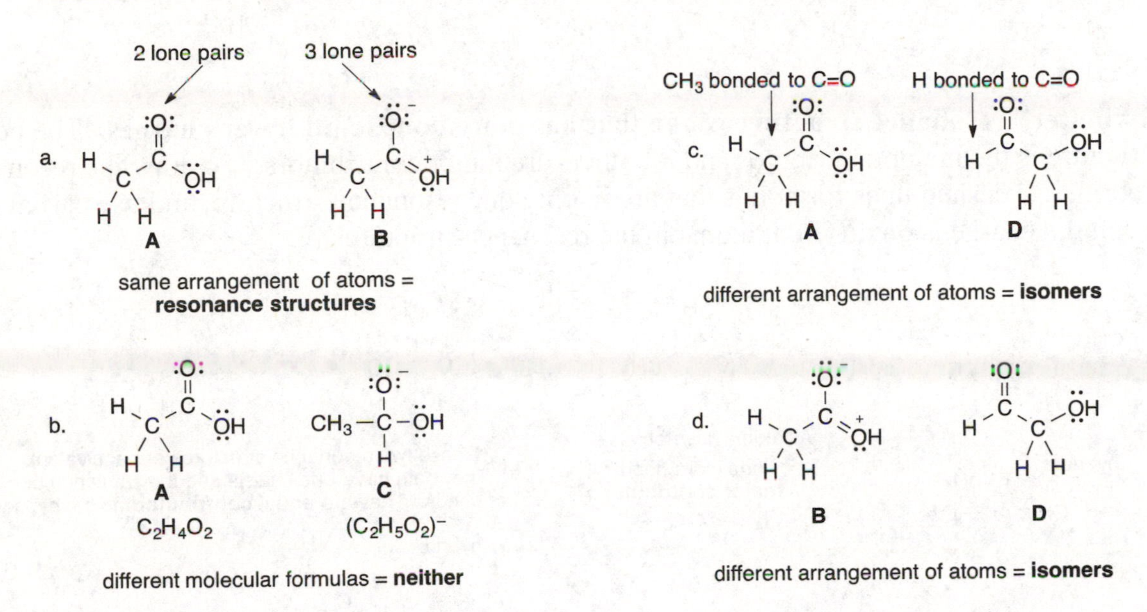

1.12 Curved arrow notation shows the movement of an electron pair. The tail begins at an electron pair (a bond or a lone pair) and the head points to where the electron pair moves.

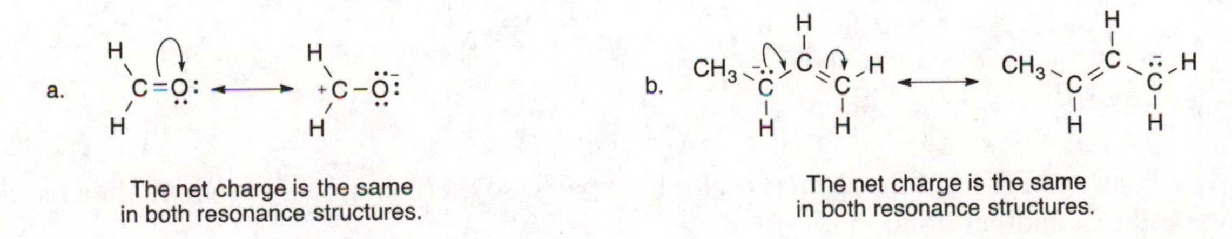

1.13 Compare the resonance structures to see what electrons have "moved." **Use one curved arrow to show the movement of each electron pair.**

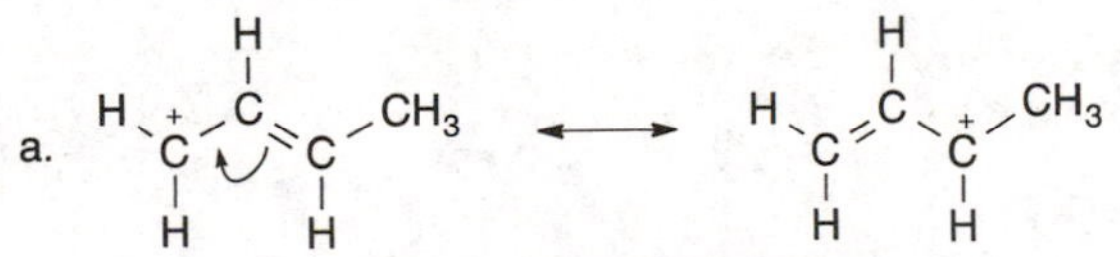

a.

One electron pair moves:
one curved arrow.

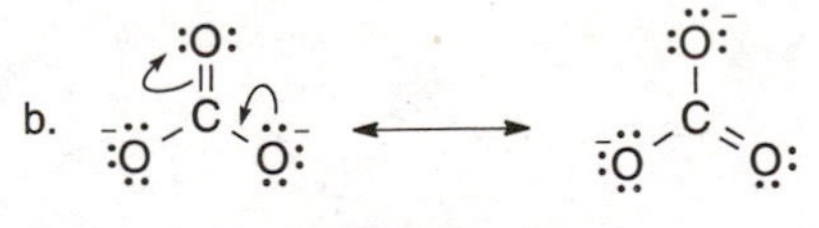

b.

Two electron pairs move:
two curved arrows.

1.14 To draw another resonance structure, **move electrons only in multiple bonds and lone pairs** and keep the number of unpaired electrons constant.

a.

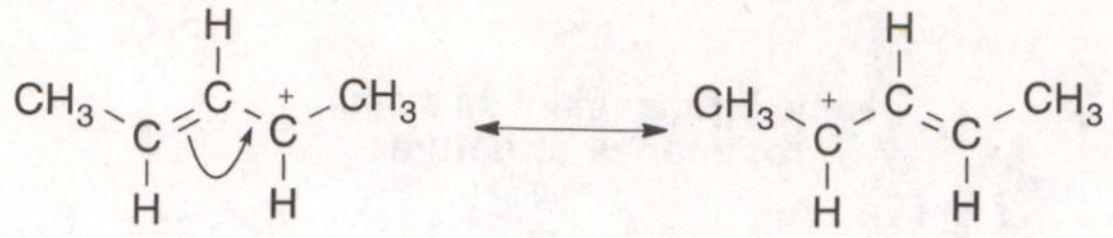

b.

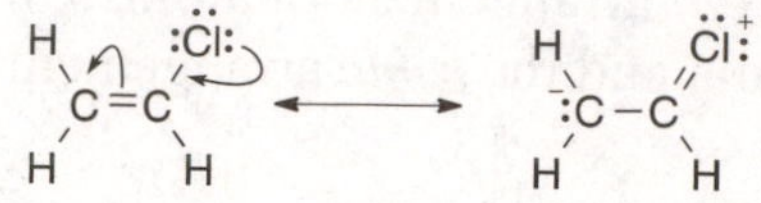

c.

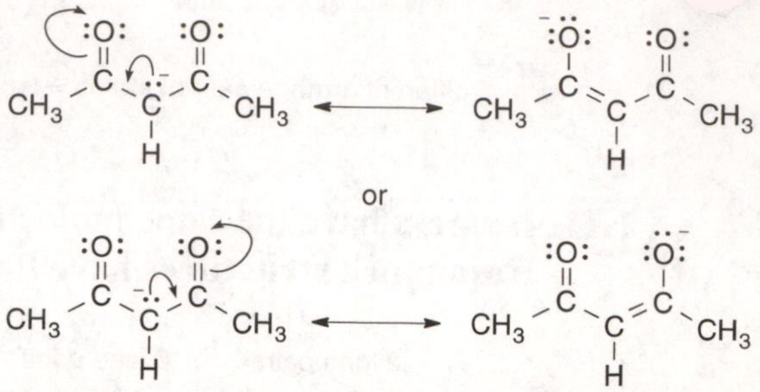

or

1.15 A "better" resonance structure is one that has more bonds and fewer charges. The better structure is the major contributor and all others are minor contributors. To draw the resonance hybrid, use dashed lines for bonds that are in only one resonance structure, and use partial charges when the charge is on different atoms in the resonance structures.

a.

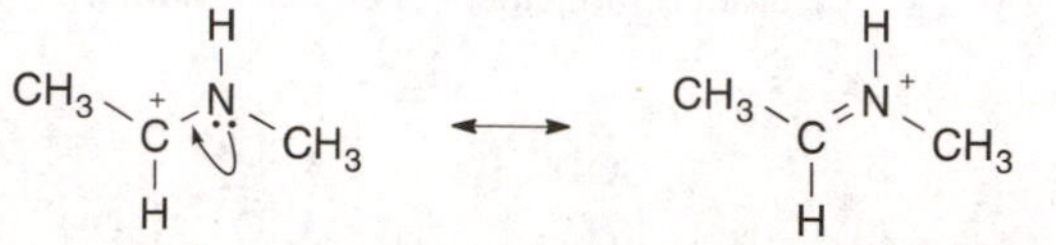

All atoms have octets.
one more bond
major contributor

hybrid:

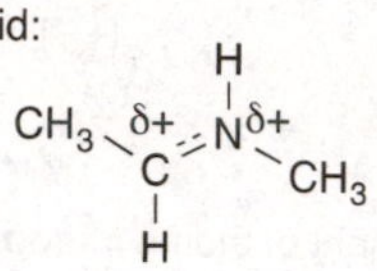

b.

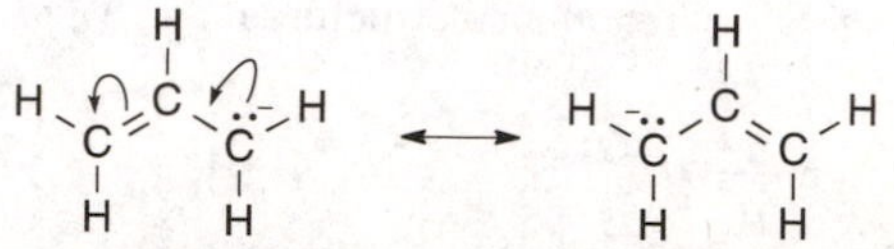

These two resonance structures are equivalent. They both have one charge and the same number of bonds. They are **equal contributors** to the hybrid.

hybrid:

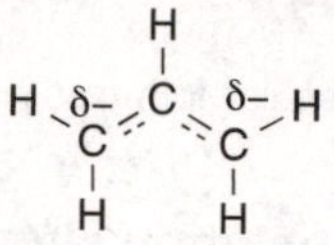

1.16

a.

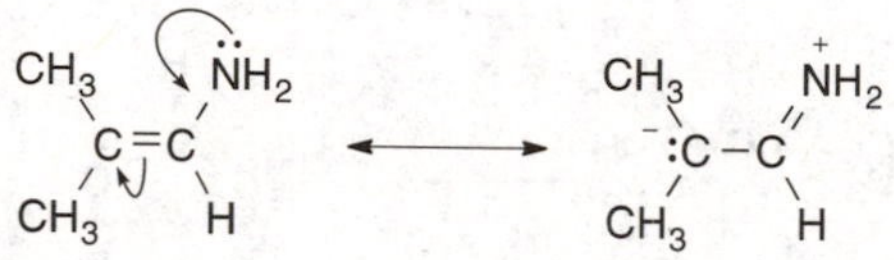

A

b. The N atom in **B** has four atoms and no lone pairs, so there is no way to move the electrons to give the N another bond.

1.17 To predict the geometry around an atom, **count the number of groups (atoms + lone pairs),** making sure to draw in any needed lone pairs or hydrogens: 2 groups = linear, 3 groups = trigonal planar, 4 groups = tetrahedral.

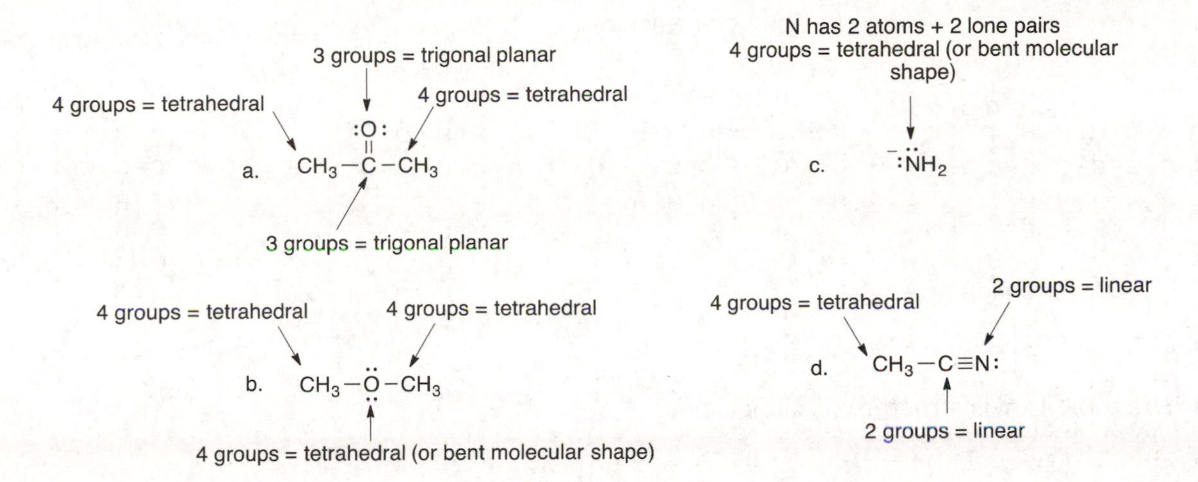

1.18 To predict the bond angle around an atom, **count the number of groups (atoms + lone pairs),** making sure to draw in any needed lone pairs or hydrogens: 2 groups = 180°, 3 groups = 120°, 4 groups = 109.5°.

2 groups = 180°

a. $CH_3–C≡C–\ddot{C}l:$

2 groups = 180°

This C has 3 groups, so both angles are 120°.

b. $CH_2=CH–\ddot{C}l:$

c. $CH_3–CH_2–\ddot{C}l:$

This C has 4 groups, so both angles are 109.5°.

1.19 To predict the geometry around an atom, use the rules in Answer 1.17.

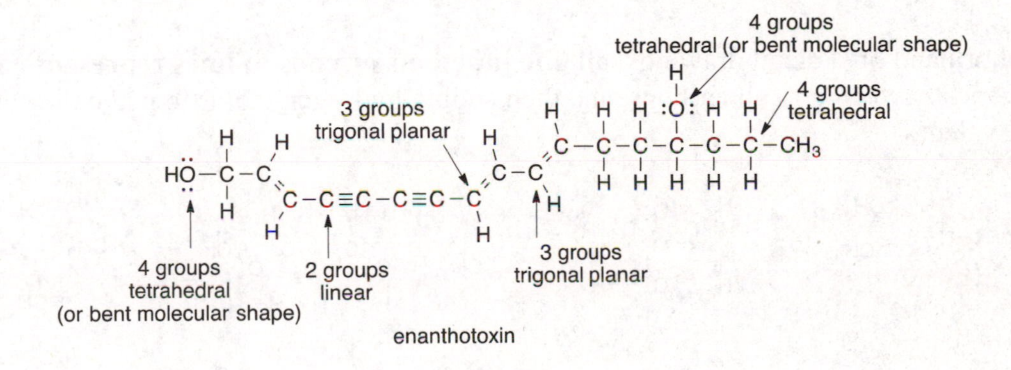

enanthotoxin

1.20 Reading from left-to-right, draw the molecule as a Lewis structure. Always check that carbon has four bonds and all heteroatoms have an octet by adding any needed lone pairs.

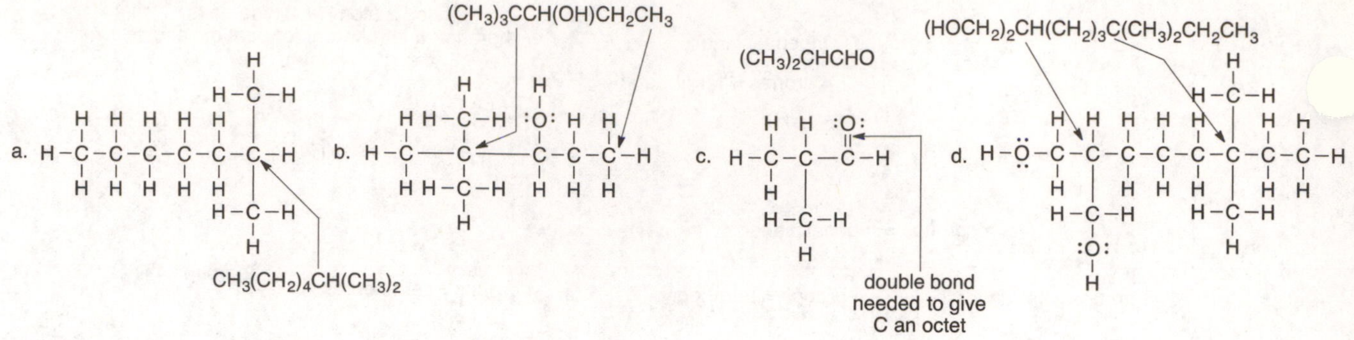

a. $CH_3(CH_2)_4CH(CH_3)_2$

b. $(CH_3)_3CCH(OH)CH_2CH_3$

c. $(CH_3)_2CHCHO$

double bond needed to give C an octet

d. $(HOCH_2)_2CH(CH_2)_3C(CH_3)_2CH_2CH_3$

1.21 Draw the Lewis structure of lactic acid.

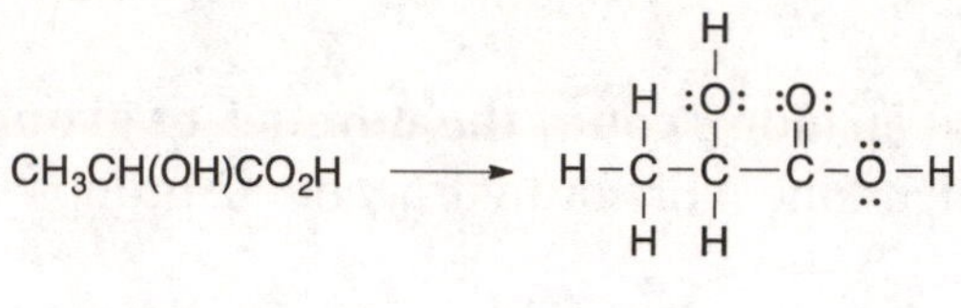

$CH_3CH(OH)CO_2H$ ⟶ H–C–C–C–Ö–H

1.22 In shorthand or skeletal drawings, **all line junctions or ends of lines represent carbon atoms.** The carbons are all tetravalent.

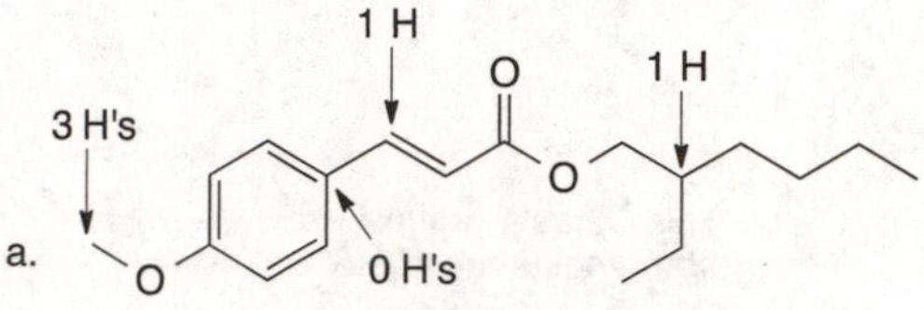

a. 3 H's, 1 H, 0 H's, 1 H

octinoxate
(2-ethylhexyl 4-methoxycinnamate)
$C_{18}H_{26}O_3$

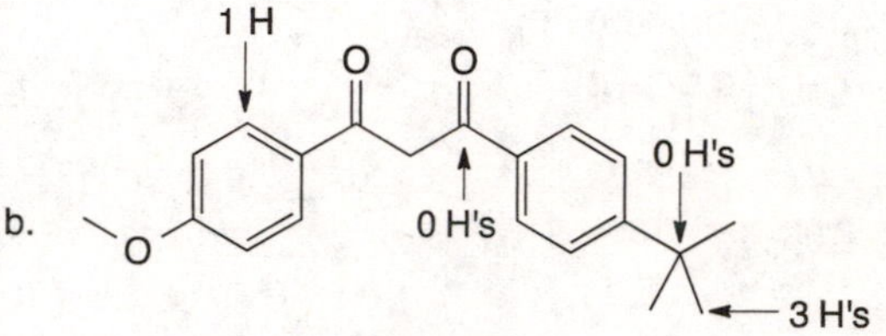

b. 1 H, 0 H's, 0 H's, 3 H's

avobenzone
$C_{20}H_{22}O_3$

1.23 In shorthand or skeletal drawings, **all line junctions or ends of lines represent carbon atoms.** Convert by writing in all carbons, and then adding hydrogen atoms to make the carbons tetravalent.

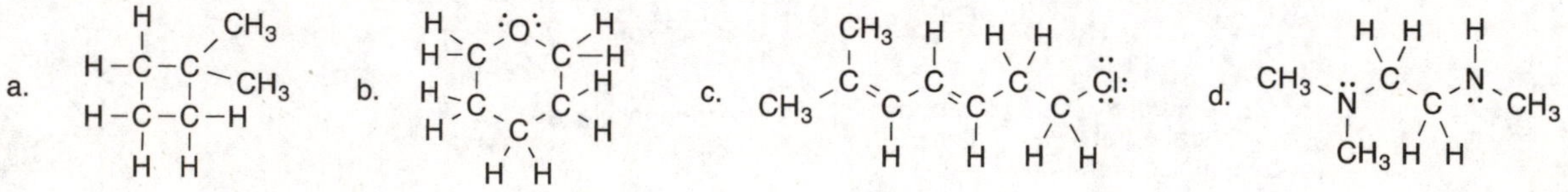

a. b. c. d.

1.24

HO, N, CH_3O, N

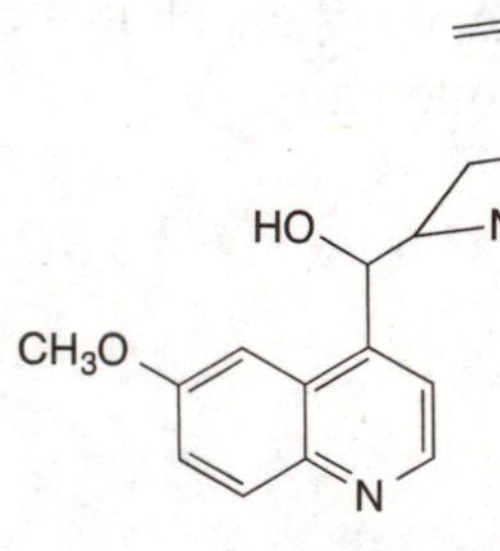

quinine

$C_{20}H_{24}N_2O_2$
molecular formula

1.25 A charge on a carbon atom takes the place of one hydrogen atom. **A negatively charged C has one lone pair, and a positively charged C has none.**

a. positive charge, no lone pairs, no H's needed

b. negative charge, one lone pair, one H needed

c. positive charge, no lone pairs, one H needed

d. negative charge, one lone pair, one H needed

1.26 Draw each indicated structure. Recall that in the skeletal drawings, a carbon atom is located at the intersection of any two lines and at the end of any line.

a. $(CH_3)_2C{=}CH(CH_2)_4CH_3$ =

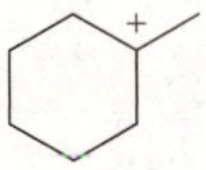

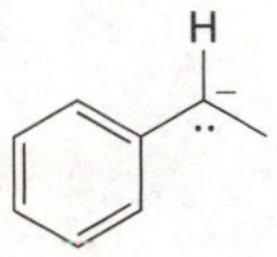

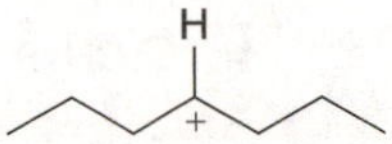

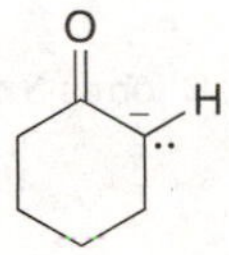

b. CH_3–CH(–)–CH(–)–CH_2CH_2Cl ; H_2N–CH–CH_2 (four-membered ring) =

c. = $(CH_3)_2CH(CH_2)_2CONHCH_3$

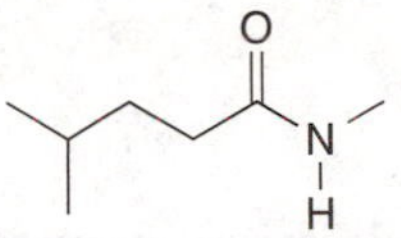

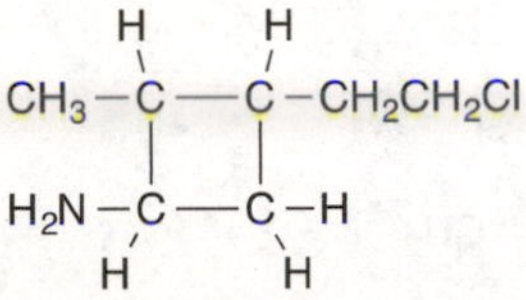

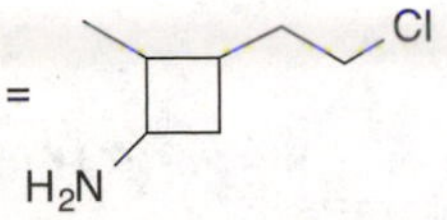

d. = $HO(CH_2)_2CH{=}CHCO_2CH(CH_3)_2$

1.27 To determine the orbitals used in bonding, **count the number of groups** (atoms + lone pairs): 4 groups = sp^3, 3 groups = sp^2, 2 groups = sp, H atom = 1*s* (no hybridization). All covalent single bonds are σ, and all double bonds contain one σ and one π bond.

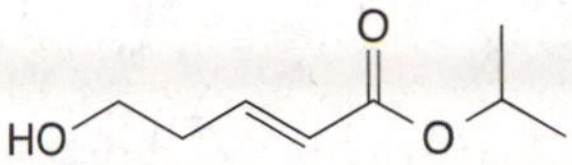

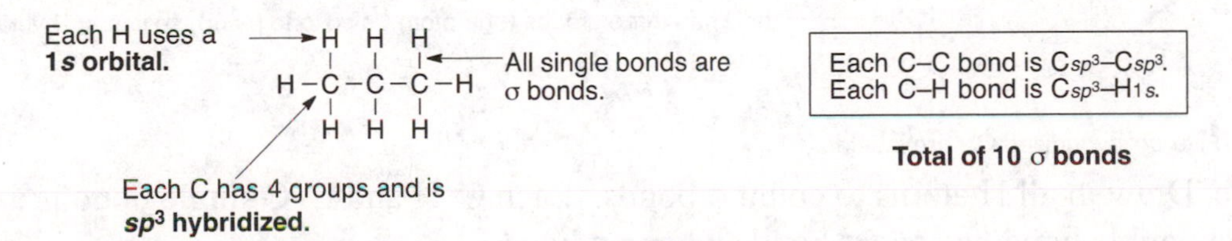

Each H uses a
1*s* orbital.
All single bonds are
σ bonds.
Each C has 4 groups and is
***sp*³ hybridized.**
Each C–C bond is C*sp*³–C*sp*³.
Each C–H bond is C*sp*³–H1*s*.
Total of 10 σ bonds

1.28 [1] Draw a valid Lewis structure for each molecule.
[2] **Count the number of groups** around each atom: 4 groups = sp^3, 3 groups = sp^2, 2 groups = sp, H atom = 1*s* (no hybridization).

Note: **Be and B** (Groups 2A and 3A) do not have enough valence e⁻ to form an octet, **and do not form an octet in neutral molecules.**

a. [1] H–C(H)(H)–Be–H
Be has 2 bonds.

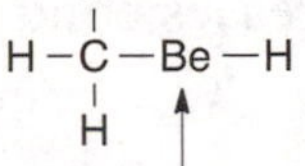

[2] Count groups around each atom:
H–C(H)(H)–Be–H
4 groups *sp*³
2 groups *sp*

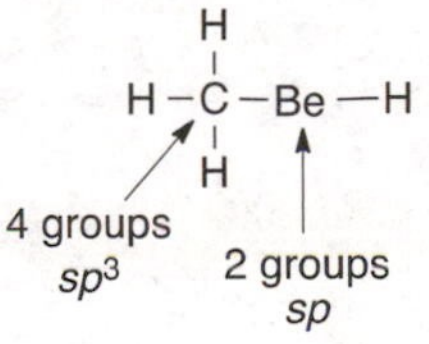

[3] All C–H bonds: C*sp*³–H1*s*
C–Be bond: C*sp*³–Be*sp*
Be–H bond: Be*sp*–H1*s*

b. [1] $CH_3-B(CH_3)-CH_3$

B forms 3 bonds.

[2] Count groups around each atom:

$CH_3-B(CH_3)-CH_3$

4 groups sp^3 3 groups sp^2

[3] All C–H bonds: $C_{sp^3}–H_{1s}$
C–B bonds: $C_{sp^3}–B_{sp^2}$

c. [1] $H_3C-\ddot{O}-CH_3$

[2] Count groups around each atom:

$H_3C-\ddot{O}-CH_3$

4 groups sp^3 4 groups sp^3

[3] All C–H bonds: $C_{sp^3}–H_{1s}$
C–O bonds: $C_{sp^3}–O_{sp^3}$

1.29 To determine the hybridization, **count the number of groups** around each atom: 4 groups = sp^3, 3 groups = sp^2, 2 groups = sp, H atom = $1s$ (no hybridization).

a. $CH_3-C{\equiv}CH$ — 4 groups sp^3; 2 groups sp

b. cyclohexylidene=N–CH_3 — 3 groups sp^2; 3 groups sp^2

c. $CH_2{=}C{=}CH_2$ — 3 groups sp^2; 2 groups sp

1.30

a. Five sp^2 hybridized C's are labeled.

b. O has three groups (one atom + two lone pairs), so it is sp^2 hybridized.

c. $C_{sp^2}–O_{sp^2}$ $C_p–O_p$

d, e. Draw in all H atoms to count σ bonds. Each C–H and C–C single bond is a σ bond. Each double bond has one σ bond and one π bond.

$H-CH_2-CH_2-CH{=}CH-CH_2-CH_2-CH{=}CH-CH{=}O$

23 σ bonds
3 π bonds

1.31 Single bonds are weaker and longer than double bonds, which are weaker and longer than triple bonds. Increasing percent s-character increases bond strength and decreases bond length.

a. 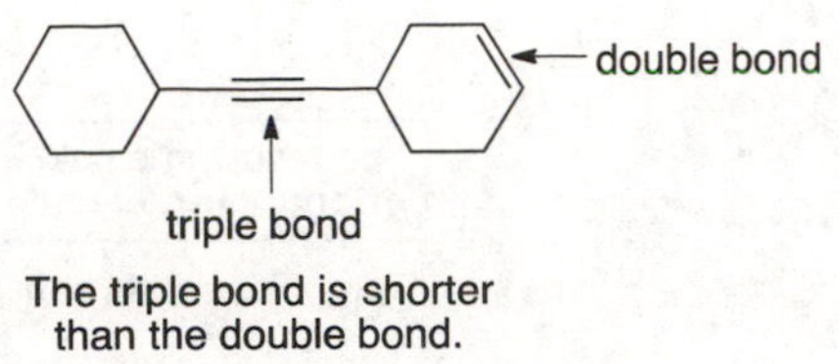 double bond

triple bond

The triple bond is shorter than the double bond.

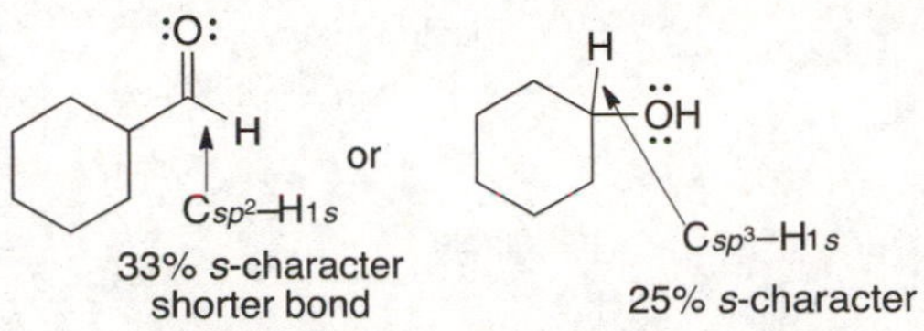

b. 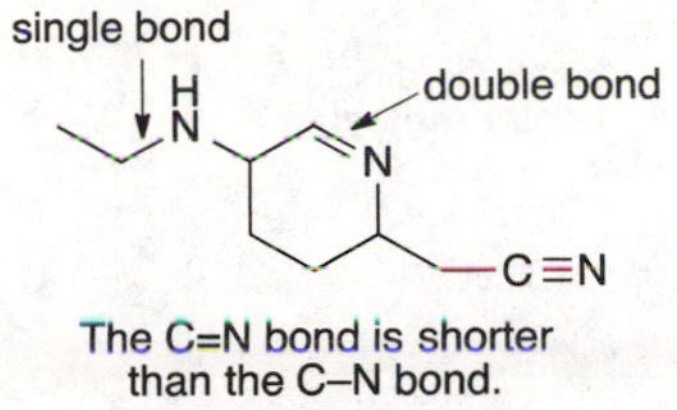 single bond

double bond

C≡N

The C=N bond is shorter than the C–N bond.

c. :O: H or H ÖH

C_{sp^2}–H_{1s}

33% *s*-character
shorter bond

C_{sp^3}–H_{1s}

25% *s*-character

d. 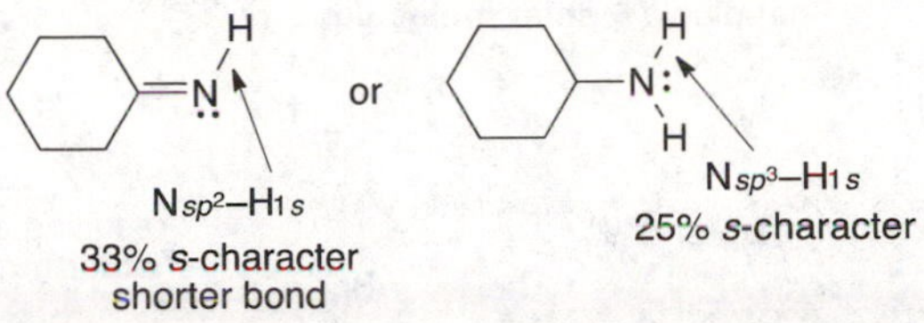 H =N or H N: H

N_{sp^2}–H_{1s}

33% *s*-character
shorter bond

N_{sp^3}–H_{1s}

25% *s*-character

1.32 **Electronegativity increases from left-to-right across a row of the periodic table and decreases down a column.** Look at the relative position of the atoms to determine their relative electronegativity.

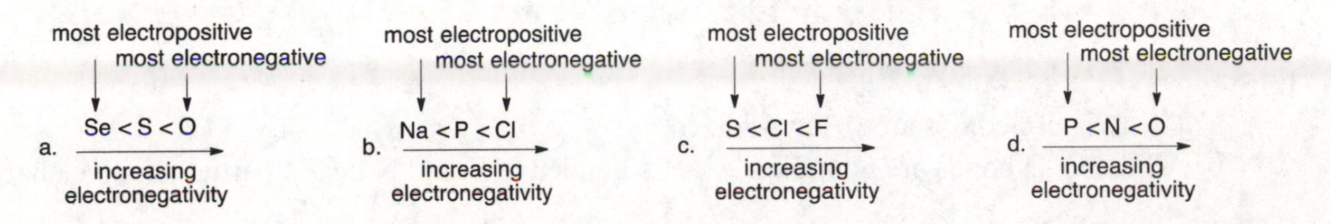

1.33 Dipoles result from unequal sharing of electrons in covalent bonds. More electronegative atoms "pull" electron density towards them, making a dipole. **Dipole arrows point towards the atom of higher electron density.**

a. 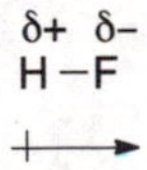 δ+ δ– H–F

b. 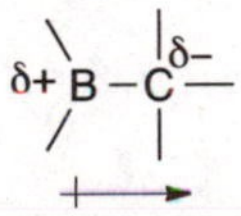 δ+ B–C δ–

c. δ– C–Li δ+

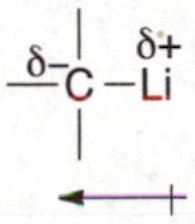

d. 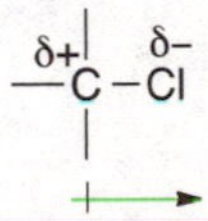 δ+ C–Cl δ–

1.34 Polar molecules result from a net dipole. To determine polarity, draw the molecule in three dimensions around any polar bonds, draw in the dipoles, and look to see whether the dipoles cancel or reinforce.

a.

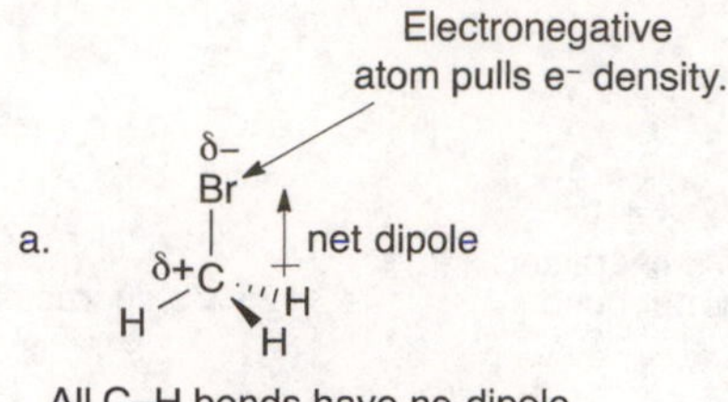

All C–H bonds have no dipole.
one polar bond
net dipole = **polar molecule**

c.

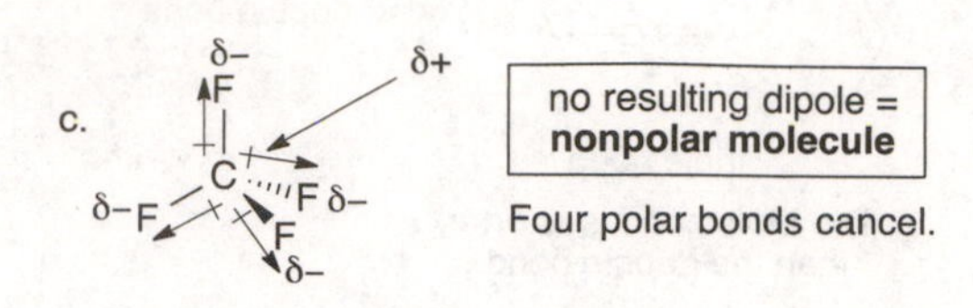

Four polar bonds cancel.

b. Br–C(H)(H)–Br re-draw → resulting dipole = **polar molecule**

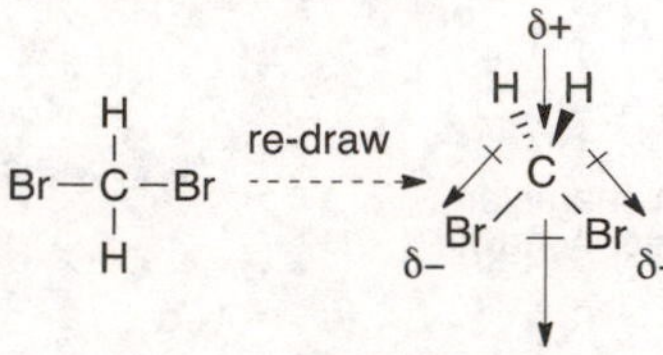

Note: You must draw the molecule in three dimensions to observe the net dipole. In the Lewis structure, it appears the dipoles would cancel out, when in fact they add to make a polar molecule.

d.

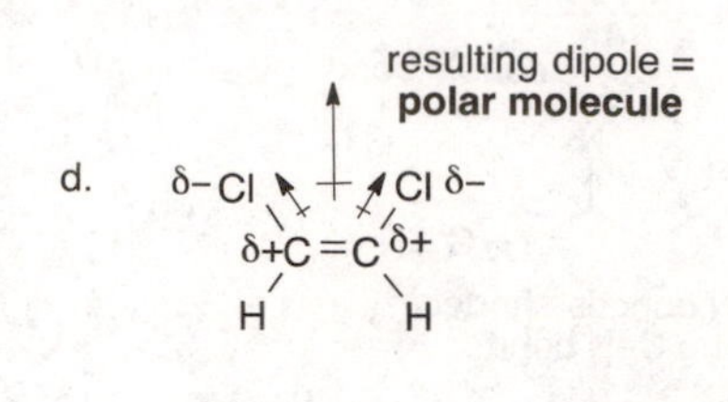

e. δ– Cl, H, δ+C=Cδ+, H, Cl δ–

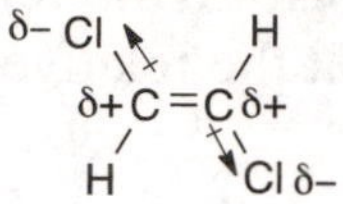

no resulting dipole =
nonpolar molecule

Two polar bonds are **equal and opposite** and cancel.

1.35

a. The two circled C's are sp^3 hybridized.

b. All the C–H bonds are nonpolar. All H's bonded to O and N bear a partial positive charge (δ+).

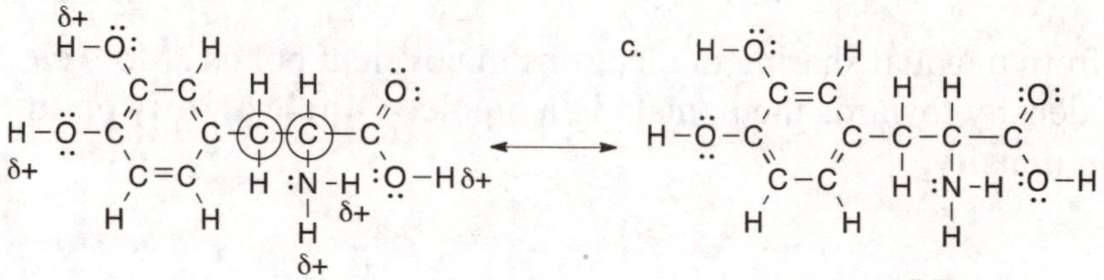

c. one possibility

1.36

a. skeletal structure

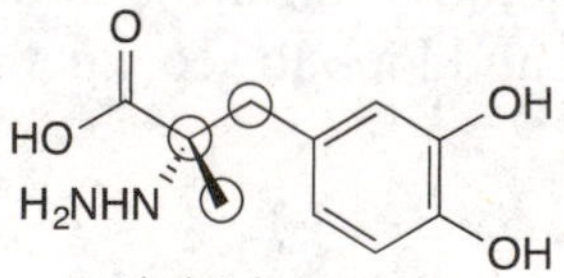

b. Circled carbons are sp^3 hybridized. All others are sp^2 hybridized.

c. Each N is surrounded by three atoms and a lone pair, making it sp^3 hybridized and trigonal pyramidal in molecular shape.

d. 11 polar bonds shown in bold

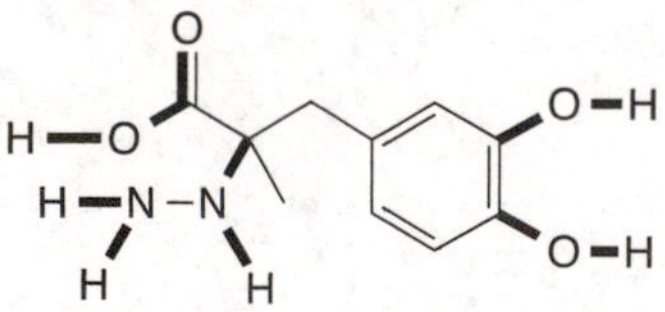

1.37

a, b, c.

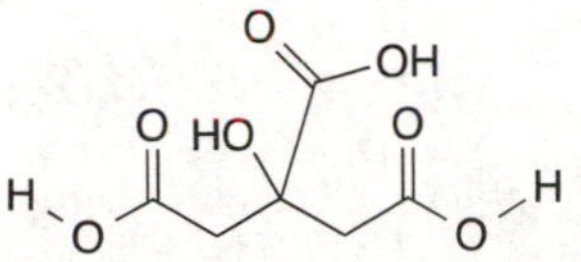

$C_6H_8O_7$

14 lone pairs, 2 lone pairs on each O

d. Each C that is part of a C=O is sp^2 hybridized, so there are three sp^2 C's.

e. Orbitals:

[1] C=O, Csp^2–Osp^2 and Cp–Op
[2] C–C, Csp^3–Csp^2
[3] O–H, Osp^3–$H1s$
[4] C–O, Csp^3–Osp^3

1.38

a, b, c.

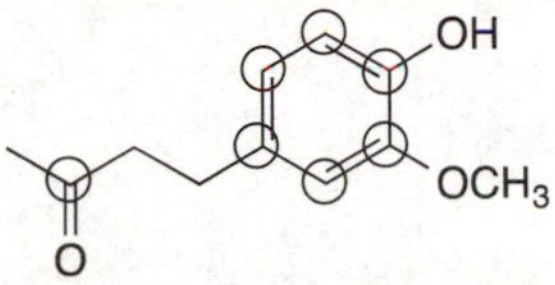

$C_{11}H_{14}O_3$

6 lone pairs, 2 lone pairs on each O

d. The sp^2 hybridized C's (seven) are labeled with circles.

e. Orbitals:

[1] C–C, Csp^3–Csp^2
[2] C–C, Csp^3–Csp^3
[3] C–H, Csp^3–$H1s$
[4] C–H, Csp^2– $H1s$

1.39 Formal charge (FC) = number of valence electrons – [number of unshared electrons + (½)(number of shared electrons)]. C is in group 4A.

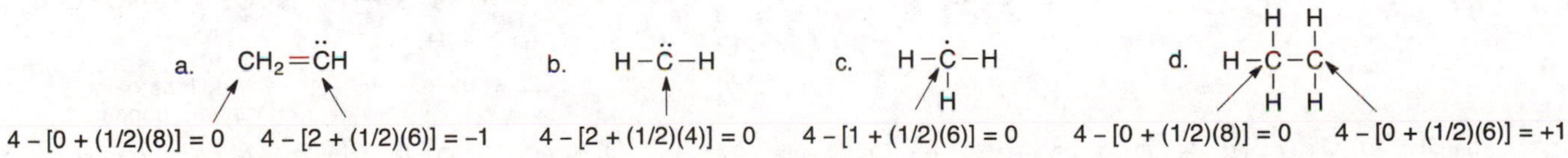

a. $CH_2=\ddot{C}H$: 4 – [0 + (1/2)(8)] = 0; 4 – [2 + (1/2)(6)] = –1

b. $H-\ddot{C}-H$: 4 – [2 + (1/2)(4)] = 0

c. $H-\dot{C}-H$ (with H): 4 – [1 + (1/2)(6)] = 0

d. H_3C-CH_2: 4 – [0 + (1/2)(8)] = 0; 4 – [0 + (1/2)(6)] = +1

1.40 Formal charge (FC) = number of valence electrons – [number of unshared electrons + (½)(number of shared electrons)]. N is in group 5A and O is in group 6A.

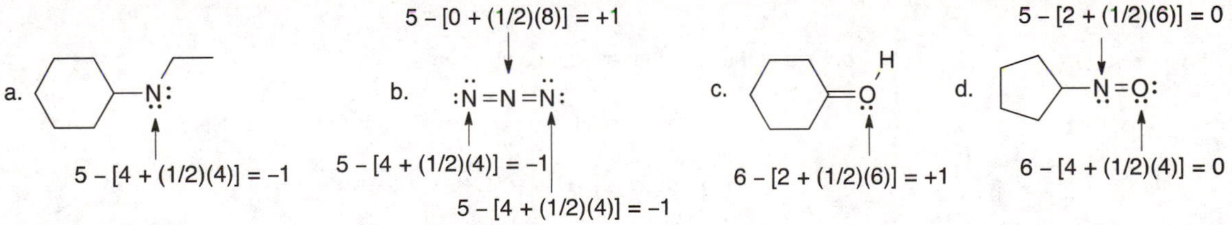

a. 5 – [4 + (1/2)(4)] = –1

b. $:\ddot{N}=N=\ddot{N}:$ 5 – [0 + (1/2)(8)] = +1; 5 – [4 + (1/2)(4)] = –1; 5 – [4 + (1/2)(4)] = –1

c. 6 – [2 + (1/2)(6)] = +1

d. 5 – [2 + (1/2)(6)] = 0; 6 – [4 + (1/2)(4)] = 0

1.41 Follow the steps in Answer 1.4 to draw Lewis structures.

a. CH_2N_2

valence e⁻	
1 C x 4 e⁻ =	4
2 H x 1 e⁻ =	2
2 N x 5 e⁻ =	10
total e⁻ =	16

```
      +  ..-
H–C=N=N:
  |
  H
       or
  ..-  +
H–C–N≡N:
  |
  H
```

c. CH_3CNO

valence e⁻	
2 C x 4 e⁻ =	8
3 H x 1 e⁻ =	3
1 N x 5 e⁻ =	5
1 O x 6 e⁻ =	6
total e⁻ =	22

```
  H         +  ..-              H   -    +
  |                             |
H–C–C≡N–O:       or           H–C–C=N=O:
  |            ..               |    ..
  H                             H

     H  2–
     |  ..   +  +
or H–C–C–N≡O:
     |  ..
     H
```

b. CH_3NO_2

valence e⁻	
1 C x 4 e⁻ =	4
3 H x 1 e⁻ =	3
1 N x 5 e⁻ =	5
2 O x 6 e⁻ =	12
total e⁻ =	24

```
  H   +  ..-            H   +
  |                     |
H–C–N–O:       or     H–C–N=Ö:
  |  ||  ..             |  |
  H :O:                 H :O:
                           ..-
```

d. $^{-}CH_2CN$

valence e⁻	
2 C x 4 e⁻	= 8
2 H x 1 e⁻	= 2
1 N x 5 e⁻	= 5
1 for (–) charge	= 1
total e⁻	= 16

```
        ..-              ..-
H–C=C=N:        or     H–C–C≡N:
  |                      |
  H                      H
```

1.42 Follow the steps in Answer 1.4 to draw Lewis structures.

a. $(CH_3CH_2)_2O$

[1]
```
  H H     H H
H C C O C C H
  H H     H H
```

[2] Count valence e⁻.

1 O x 6 e⁻ =	6
10 H x 1 e⁻ =	10
4 C x 4 e⁻ =	16
total e⁻ =	32

[3]
```
  H H     H H
  | |     | |
H–C–C–O–C–C–H
  | |     | |
  H H     H H
```
28 e⁻ used.

[4]
```
  H H     H H
  | |  .. | |
H–C–C–O–C–C–H
  | |  .. | |
  H H     H H
```
Add lone pairs.

b. CH_2CHCN

[1]
```
H H
C C C N
H
```

[2] Count valence e⁻.

1 N x 5 e⁻ =	5
3 H x 1 e⁻ =	3
3 C x 4 e⁻ =	12
total e⁻ =	20

[3]
```
H H
| |
C–C–C–N
|
H
```
12 e⁻ used.

[4]
```
H H
| |
C=C–C≡N:
|
H
```
Add lone pairs and π bonds.

c. $(HOCH_2)_2CO$

[1]
```
    H O H
H O C C C O H
    H   H
```

[2] Count valence e⁻.

3 O x 6 e⁻ =	18
6 H x 1 e⁻ =	6
3 C x 4 e⁻ =	12
total e⁻ =	36

[3]
```
    H O H
    | | |
H–O–C–C–C–O–H
    |   |
    H   H
```
22 e⁻ used.

[4]
```
     H :O: H
  ..  |  ||  |  ..
H–O–C–C–C–O–H
  ..  |      |  ..
     H      H
```
Add lone pairs and π bonds.

d. $(CH_3CO)_2O$

[1]
```
  H O     O H
H C C O C C H
  H         H
```

[2] Count valence e⁻.

3 O x 6 e⁻ =	18
6 H x 1 e⁻ =	6
4 C x 4 e⁻ =	16
total e⁻ =	40

[3]
```
  H O     O H
  | |     | |
H–C–C–O–C–C–H
  |         |
  H         H
```
24 e⁻ used.

[4]
```
  H :O:   :O: H
  |  ||  .. ||  |
H–C–C–O–C–C–H
  |      ..     |
  H             H
```
Add lone pairs and π bonds.

1.43

a.

```
     +
    NH2
    ||     ..        ..-
H2N    N          O:
 ..     |         ..
               :O:
```

b. Two of the possible resonance structures:

```
     ..                                   ..
    NH2                                  NH2
  +  |     ..    ..-                     |    +      ..-
H2N     N        O:     ⟷       H2N       N        O:
        |        ..               ..      |        ..
               :O:                                :O:
```

1.44 Isomers must have a different arrangement of atoms. Compounds are drawn as Lewis structures with no implied geometry.

a. Two isomers of molecular formula C_3H_7Cl

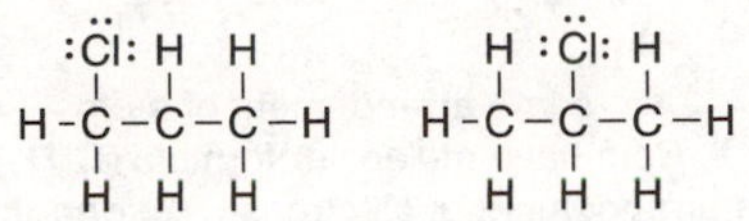

b. Three isomers of molecular formula C_2H_4O

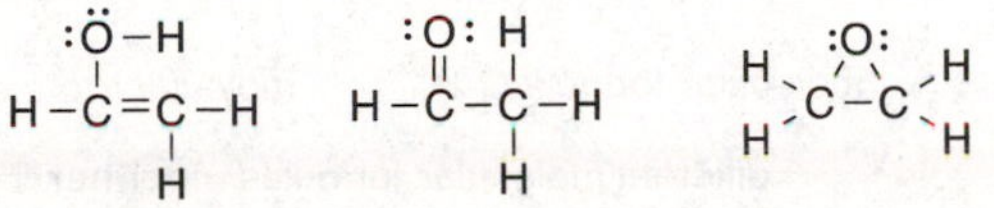

c. Four isomers of molecular formula C_3H_9N

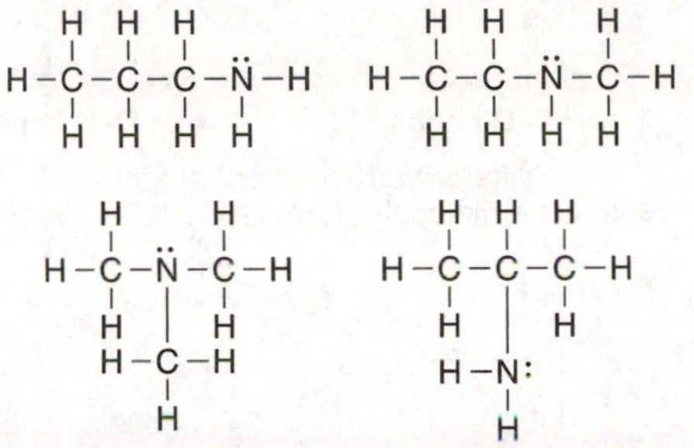

1.45

Nine isomers of C_3H_6O:

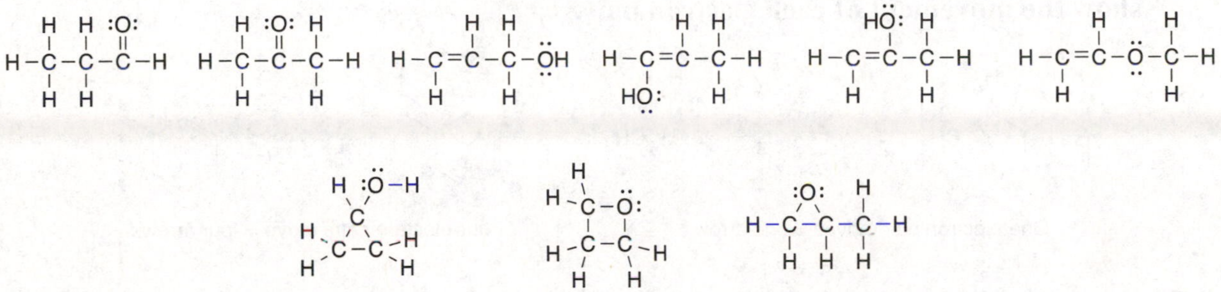

1.46 Use the definition of isomers and resonance structures in Answer 1.10.

C_6H_8O

A

a.

C_6H_8O
isomers

b.

$C_6H_{10}O$
different molecular formula
neither isomers nor resonance structures

c.

C_6H_8O
same arrangement of atoms
resonance structures

d.

5 membered ring

C_6H_8O
different arrangement of atoms
isomers

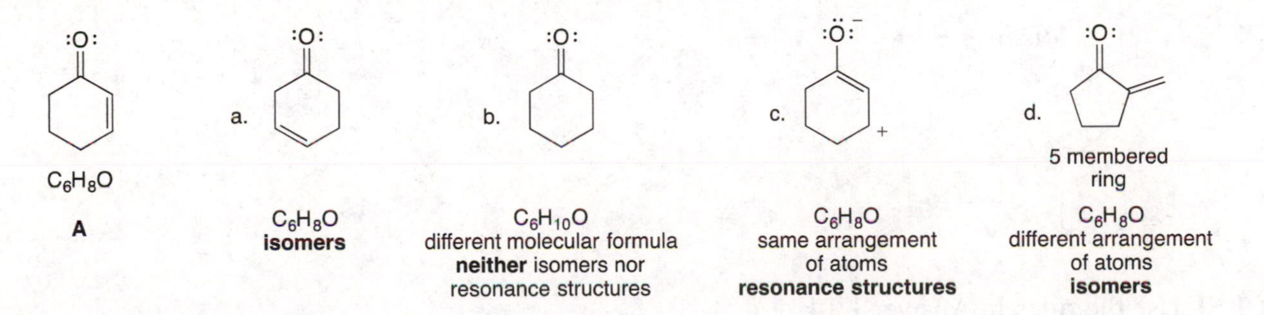

1.47 Use the definitions of isomers and resonance structures in Answer 1.9.

$(C_8H_{11})^+$
B

a.

$(C_8H_{11})^+$
different arrangement of atoms
isomers

b.

$(C_8H_{11})^+$
same arrangement of atoms
resonance structures

c.

$(C_8H_{11})^+$
different arrangement of atoms
isomers

d.

$(C_8H_9)^+$
different molecular formula
neither

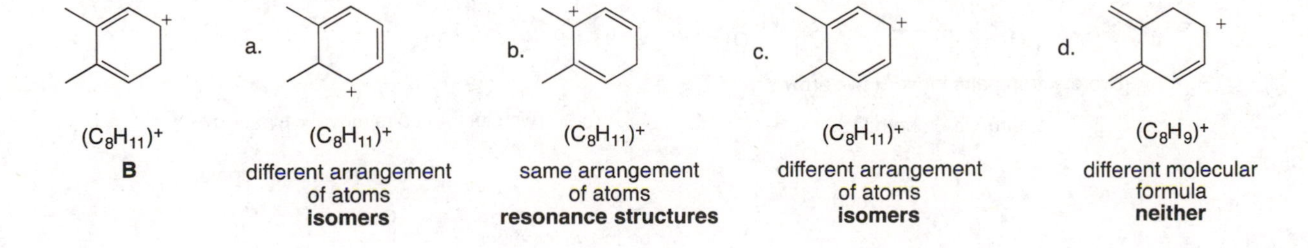

1.48 Use the definitions of isomers and resonance structures in Answer 1.9.

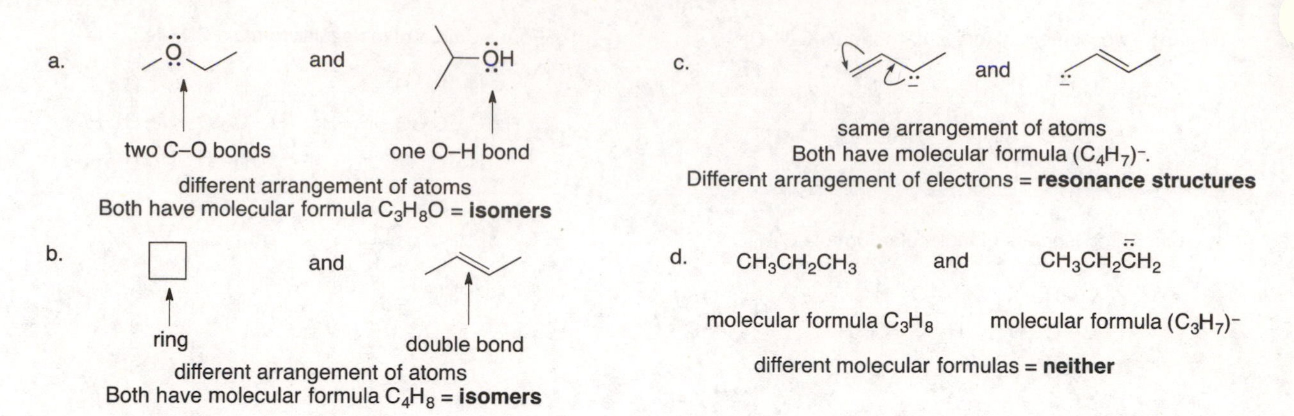

1.49 Compare the resonance structures to see what electrons have "moved." **Use one curved arrow to show the movement of each electron pair.**

a. One electron pair moves = one arrow

b. Four electron pairs move = four arrows

1.50 Curved arrow notation shows the movement of an electron pair. The tail begins at an electron pair (a bond or a lone pair) and the head points to where the electron pair moves.

a.

b.

c.

1.51 Use the rules in Answer 1.14.

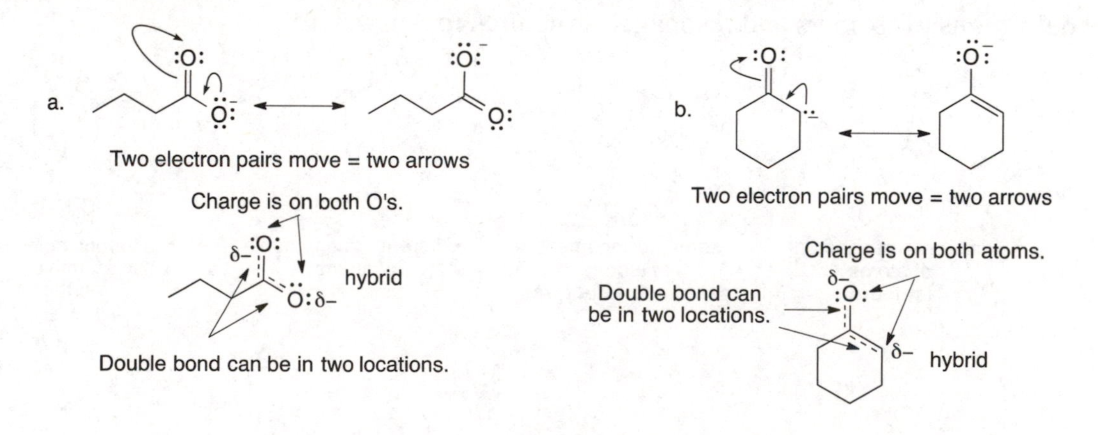

c. 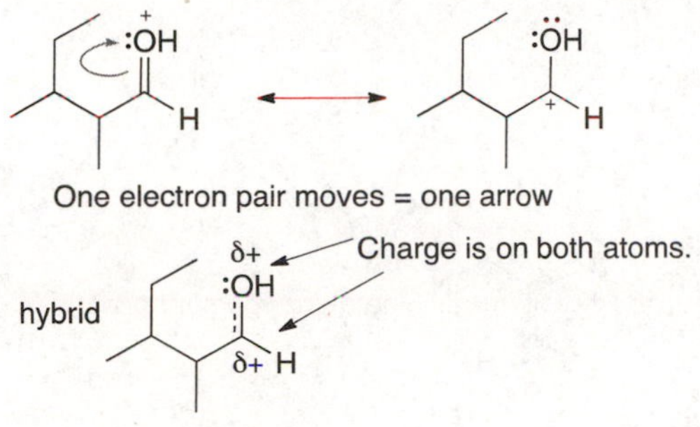 :OH H ⟷ :ÖH + H

One electron pair moves = one arrow

δ+ Charge is on both atoms.
:OH
hybrid
δ+ H

C–O bond has
partial double bond character.

1.52 For the compounds where the arrangement of atoms is not given, first draw a Lewis structure. Then use the rules in Answer 1.14.

a. O_3

Count valence e⁻.	
3 O x 6 e⁻ = 18	
total e⁻	= 18

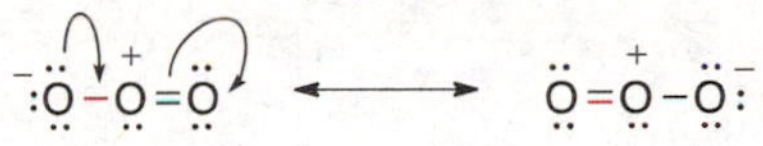

$^{-}$:Ö–Ö$^{+}$=Ö: ⟷ Ö=Ö$^{+}$–Ö:$^{-}$

b. NO_3^- (a central N atom)

Count valence e⁻.	
1 N x 5 e⁻	= 5
3 O x 6 e⁻	= 18
(–) charge	= 1
total e⁻	= 24

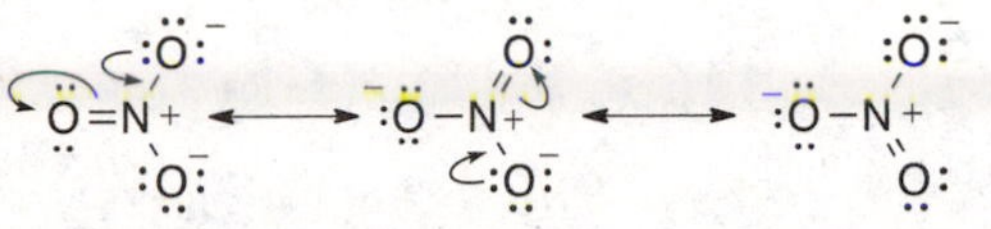

:Ö:$^{-}$ O=N+ :Ö:$^{-}$ ⟷ Ö: $^{-}$:Ö–N+ :Ö:$^{-}$ ⟷ :Ö:$^{-}$ $^{-}$:Ö–N+ Ö:

c. N_3^-

Count valence e⁻.	
3 N x 5 e⁻	= 15
(–) charge	= 1
total e⁻	= 16

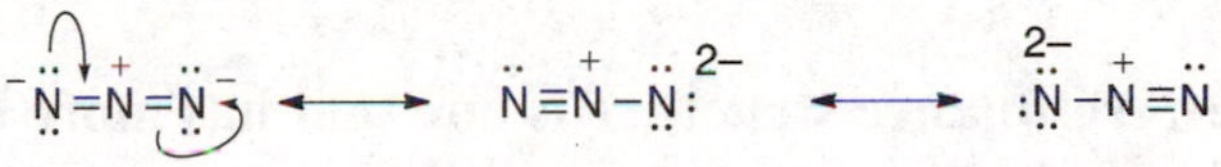

$^{-}$N̈=N^{+}=N̈$^{-}$ ⟷ N≡N^{+}–N̈:$^{2-}$ ⟷ $^{2-}$:N̈–N^{+}≡N

d. 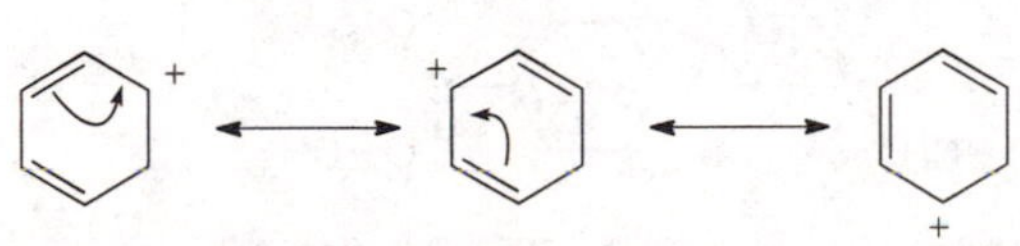 + ⟷ + ⟷ +

e. 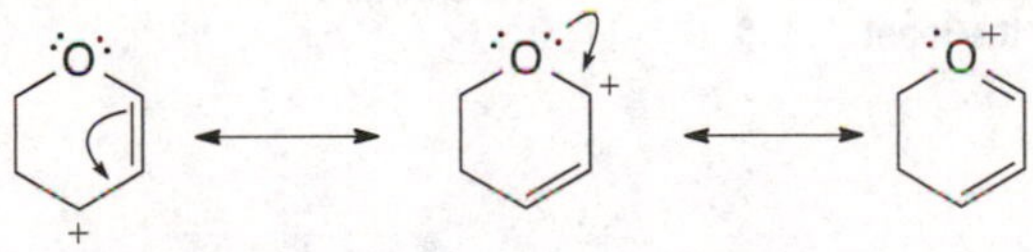 :Ö: + ⟷ :Ö: + ⟷ :O+

1.53 a. No additional Lewis structures can be drawn for **B**.

b. Additional Lewis structures can be drawn for **A, C,** and **D,** which are all examples of X=Y–Z* resonance.

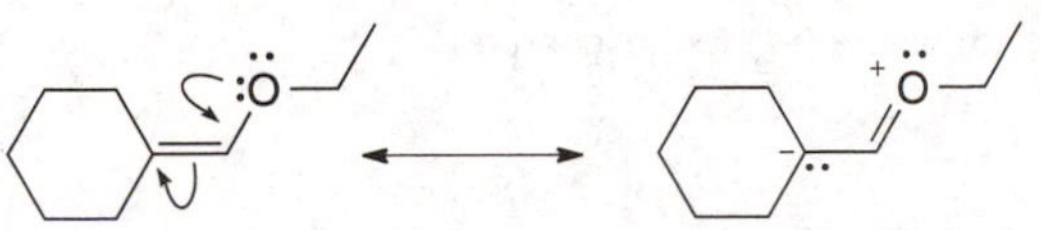

:Ö ⟷ +Ö ..

A

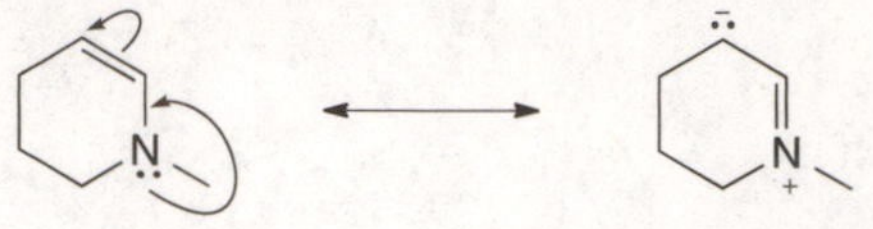

C

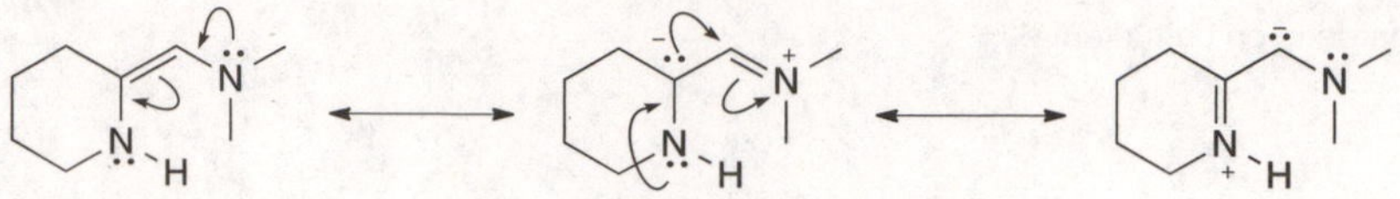

D

1.54 To draw the **resonance hybrid,** use the rules in Answer 1.15.

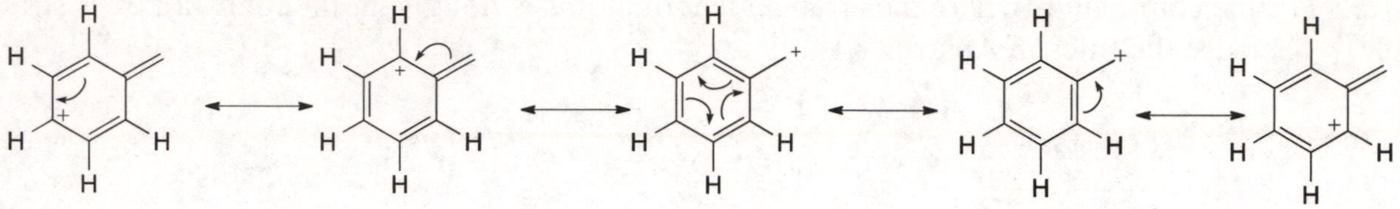

resonance hybrid

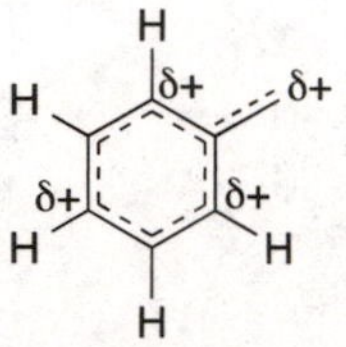

1.55 A "better" resonance structure is one that has more bonds and fewer charges. The better structure is the major contributor and all others are minor contributors.

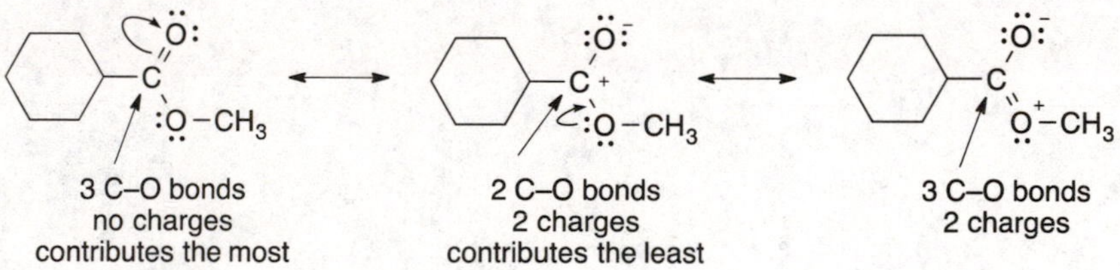

3 C–O bonds
no charges
contributes the most

2 C–O bonds
2 charges
contributes the least

3 C–O bonds
2 charges

1.56

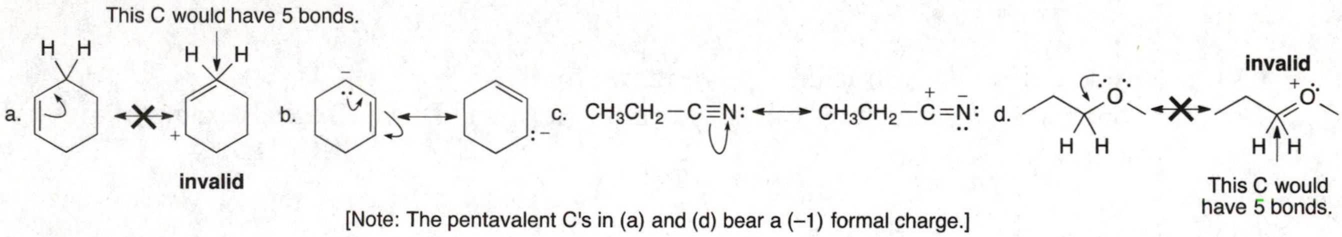

This C would have 5 bonds.

a. invalid

b.

c. $CH_3CH_2{-}C{\equiv}N{:} \longleftrightarrow CH_3CH_2{-}\overset{+}{C}{=}\overset{-}{N}{:}$

d. invalid

This C would have 5 bonds.

[Note: The pentavalent C's in (a) and (d) bear a (–1) formal charge.]

1.57 Use the rules in Answer 1.18.

a. CH_3Cl

4 groups = 109.5°

c. 120°

3 groups = 120°

4 groups = 109.5°

3 groups = 120°

e.

All C atoms have
3 groups = 120°.

b. H–N–O–H

4 groups = ~109.5°

4 groups = ~109.5°

d. HC≡C–C–OH

4 groups = 109.5°

both C's surrounded by
2 groups = 180°

4 groups = ~109.5°

1.58 To predict the geometry around an atom, use the rules in Answer 1.17.

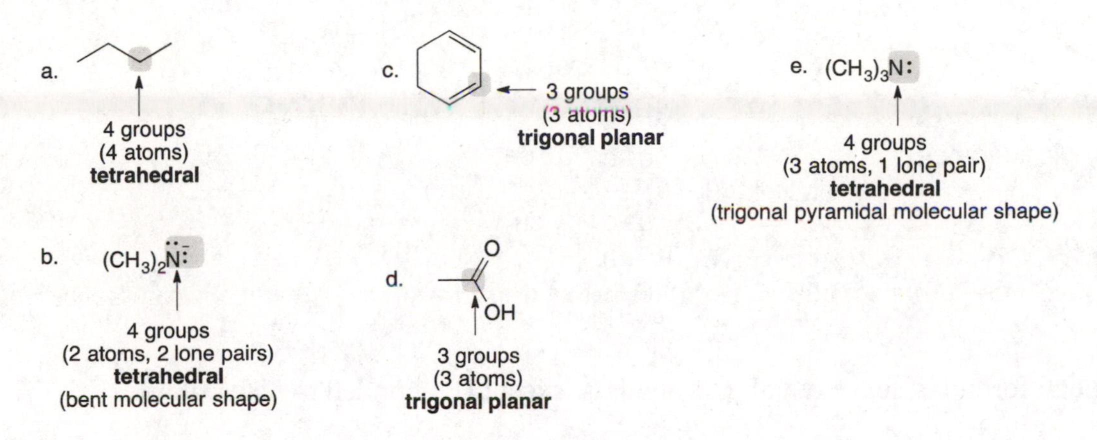

1.59 In shorthand or skeletal drawings, **all line junctions or ends of lines represent carbon atoms.** The C's are all tetravalent. All H's bonded to C's are drawn in the following structures. C's labeled with (*) have no H's bonded to them.

a.

b.

1.60 In shorthand or skeletal drawings, **all line junctions or ends of lines represent carbon atoms.** Convert by writing in all C's, and then adding H's to make the C's tetravalent.

a. 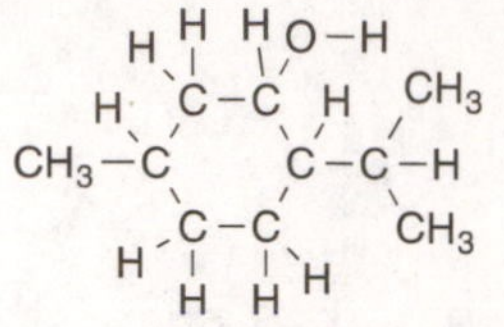 **menthol**
(isolated from peppermint oil)

b. 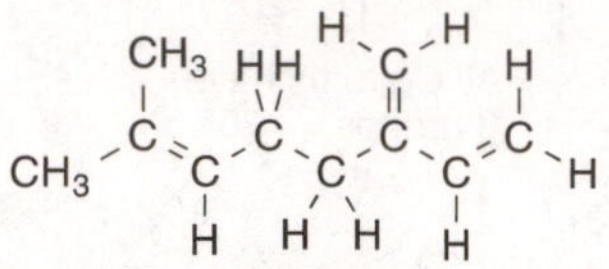 **myrcene**
(isolated from bayberry)

c. 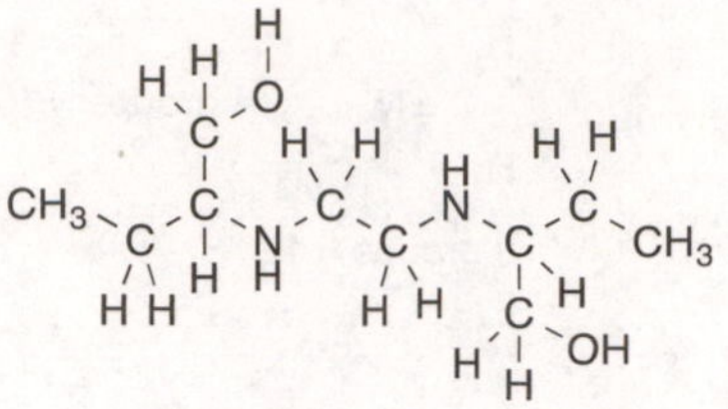 **ethambutol**
(drug used to treat tuberculosis)

d. 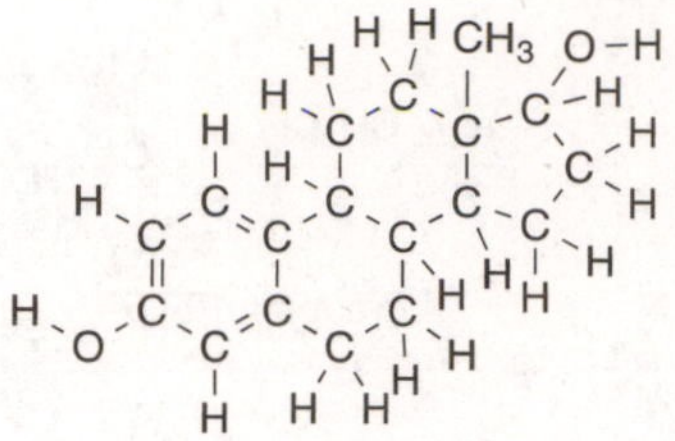 **estradiol**
(a female sex hormone)

1.61 In skeletal formulas, leave out all C's and H's, except H's bonded to heteroatoms.

a. $(CH_3)_2CHCH_2CH_2CH(CH_3)_2$

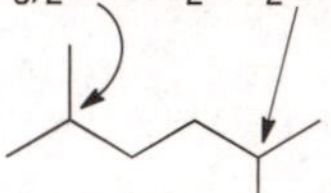

b. $CH_3CH(Cl)CH(OH)CH_3$

Cl

OH

c. $CH_3(CH_2)_2C(CH_3)_2CH(CH_3)CH(CH_3)CH(Br)CH_3$

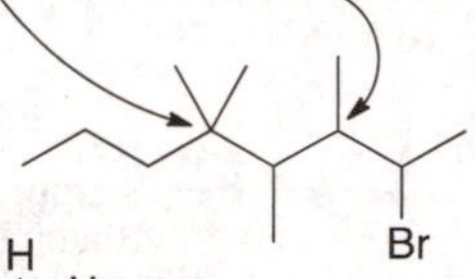

Br

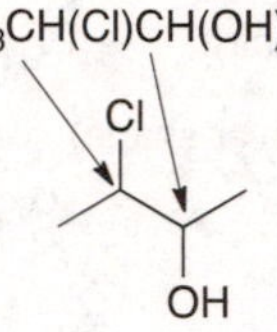

d. 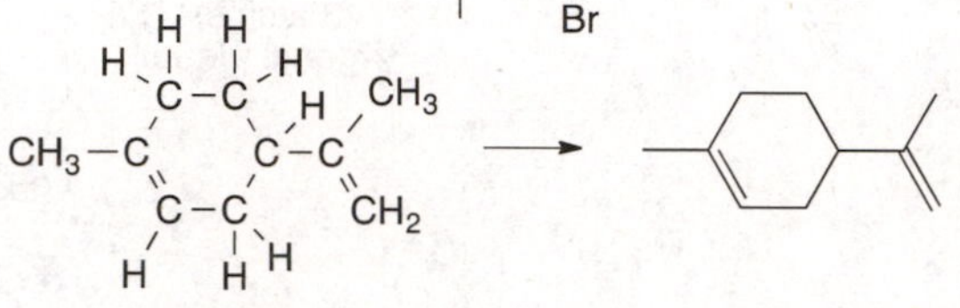 limonene
(oil of lemon)

1.62 In skeletal formulas, leave out all C's and H's, except H's bonded to heteroatoms.

a. $CH_3CONHCH_3$

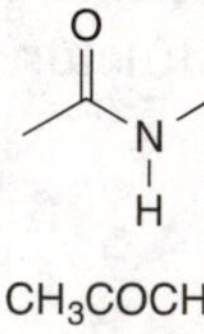

b. CH_3COCH_2Br

c. $(CH_3)_3COH$

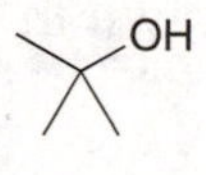

d. CH_3COCl

e. $CH_3COCH_2CO_2H$

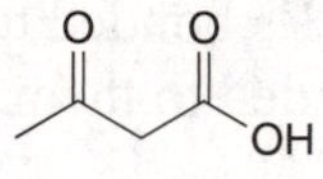

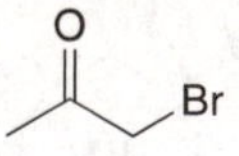

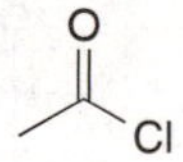

f. $HO_2CCH(OH)CO_2H$

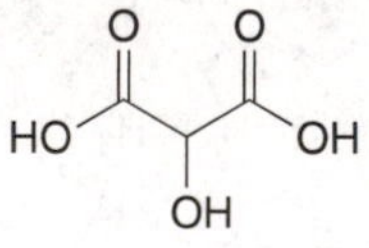

1.63 A charge on a C atom takes the place of one H atom. A negatively charged C has one lone pair, and a positively charged C has none.

a.

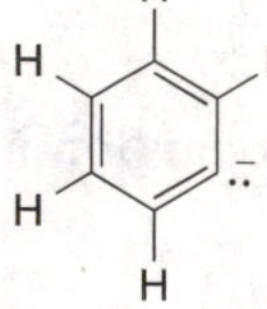

b.

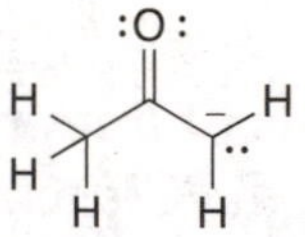

c.

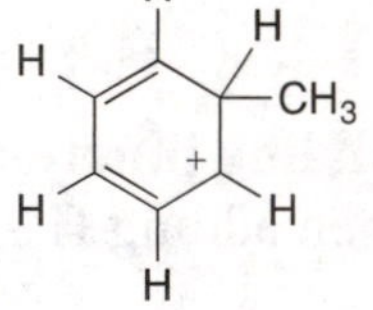

d. $CH_3-C\equiv\overset{+}{N}H$

e.

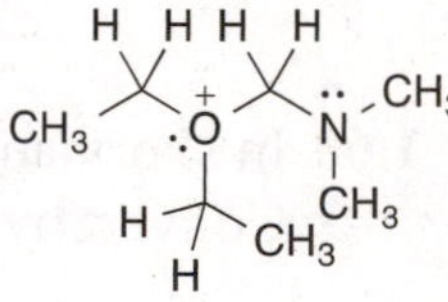

1.64 To determine the hybridization around the labeled atoms, use the procedure in Answer 1.29.

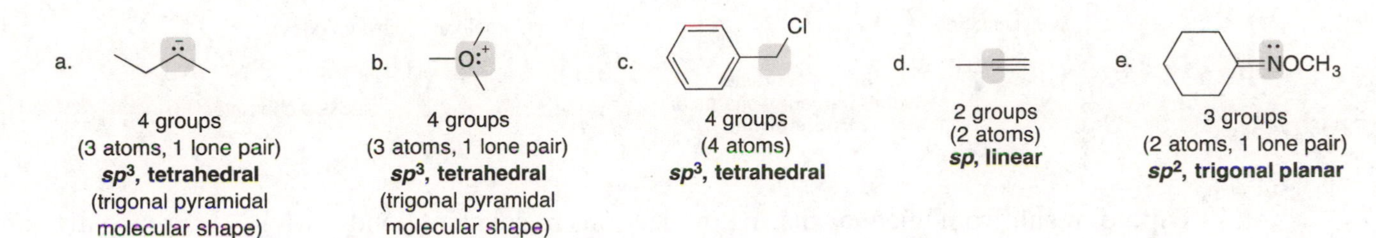

1.65 To determine what orbitals are involved in bonding, use the procedure in Answer 1.27.

a. C_{sp^2}–H_{1s}; σ: C_{sp^2}–C_{sp^2}, π: C_p–C_p; C_{sp^2}–C_{sp^3}

b. C_{sp^2}–C_{sp^3}; σ: C_{sp^2}–O_{sp^2}, π: C_p–O_p

c. H–C≡C–C(H)=N̈–CH_3: C_{sp}–C_{sp^2}; C_{sp}–H_{1s}; σ: C_{sp}–C_{sp}, π: C_p–C_p, π: C_p–C_p; σ: C_{sp^3}–N_{sp^2}

1.66

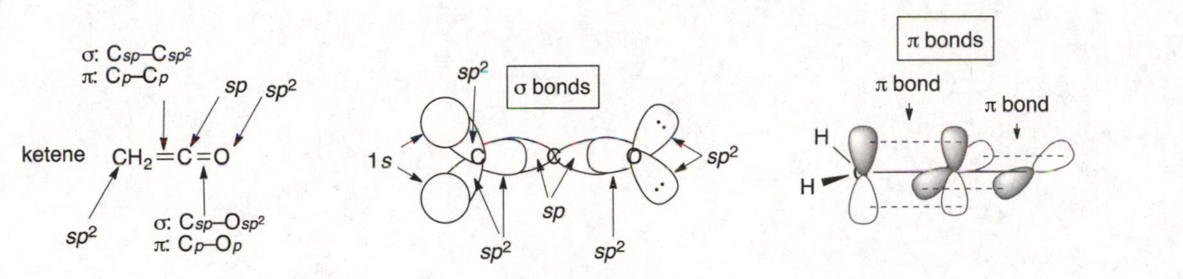

[For clarity, only the large bonding lobes of the hybrid orbitals are drawn.]

1.67 To determine relative bond length, use the rules in Answer 1.31. All C's and H's are drawn in for emphasis.

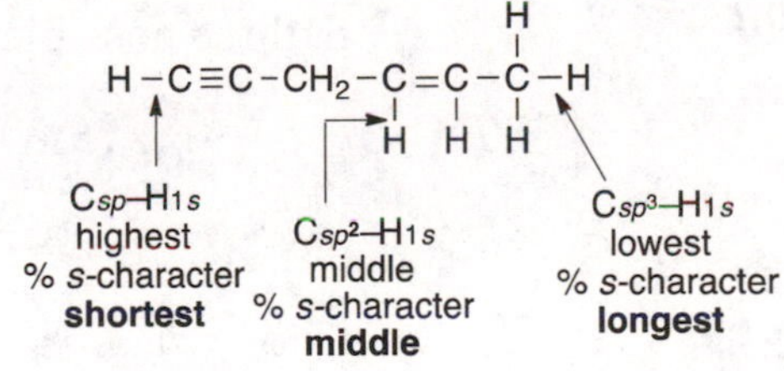

1.68

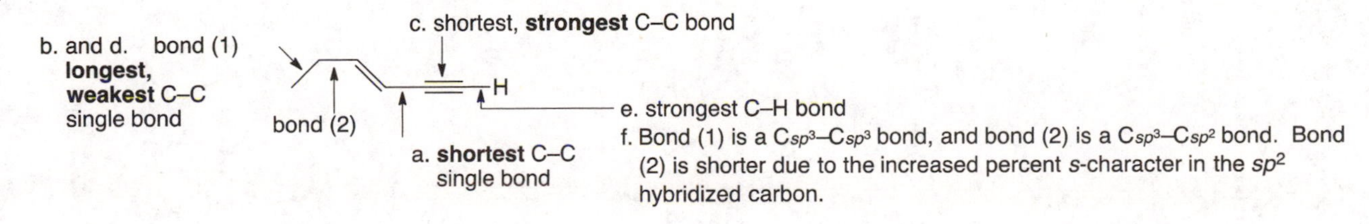

f. Bond (1) is a C_{sp^3}–C_{sp^3} bond, and bond (2) is a C_{sp^3}–C_{sp^2} bond. Bond (2) is shorter due to the increased percent *s*-character in the sp^2 hybridized carbon.

1.69 Percent *s*-character determines the strength of a bond. **The higher percent *s*-character of an orbital used to form a bond, the stronger the bond.**

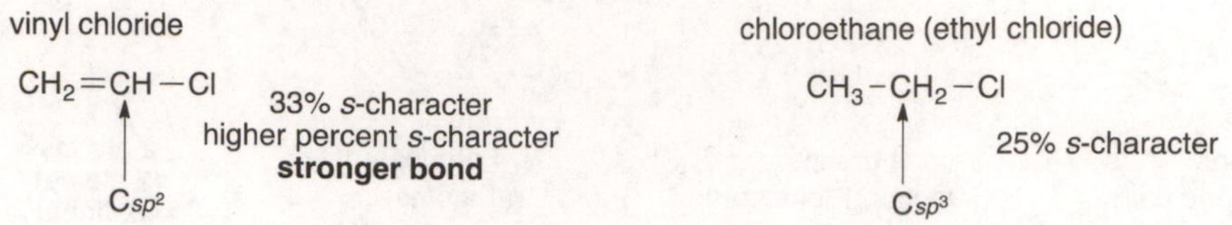

vinyl chloride

$CH_2=CH-Cl$

33% *s*-character
higher percent *s*-character
stronger bond

C_{sp^2}

chloroethane (ethyl chloride)

CH_3-CH_2-Cl

25% *s*-character

C_{sp^3}

1.70 Dipoles result from unequal sharing of electrons in covalent bonds. More electronegative atoms "pull" electron density towards them, making a dipole.

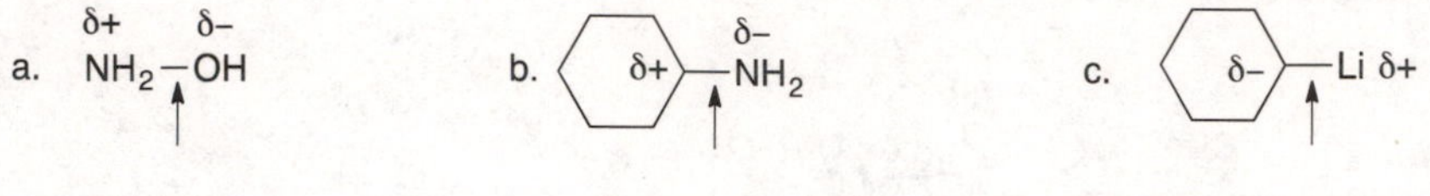

a. $\delta+$ $\delta-$ NH_2-OH

b. $\delta+$ $\delta-$ NH_2

c. $\delta-$ Li $\delta+$

1.71 Use the directions from Answer 1.34.

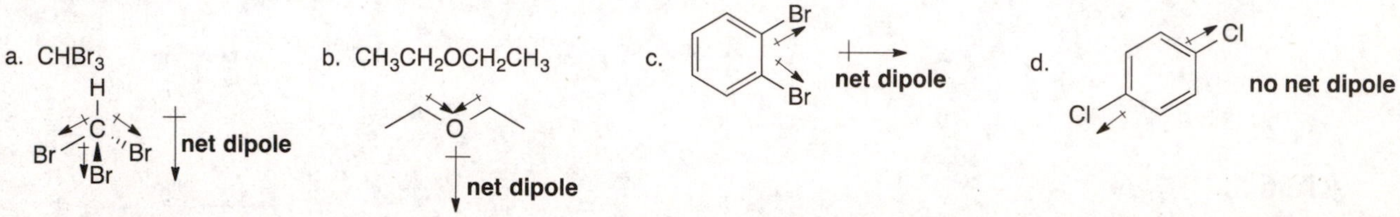

a. $CHBr_3$

H C Br Br Br net dipole

b. $CH_3CH_2OCH_2CH_3$

O net dipole

c. Br Br net dipole

d. Cl Cl no net dipole

1.72

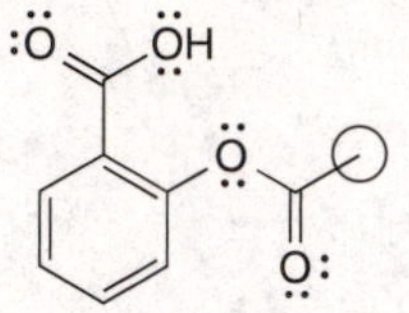

:Ö ÖH Ö: Ö:

aspirin

Ö: N N: :O N N

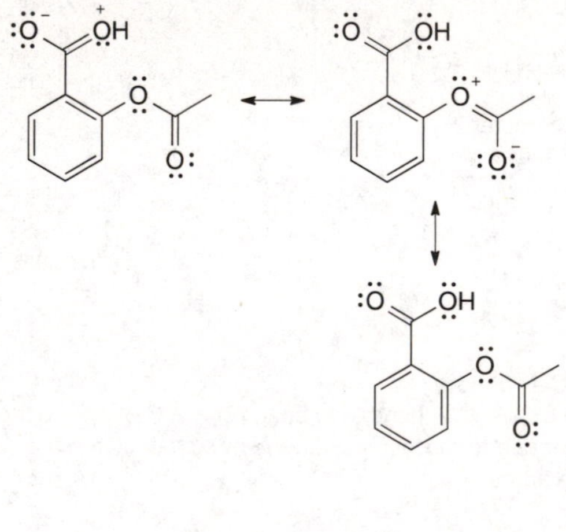

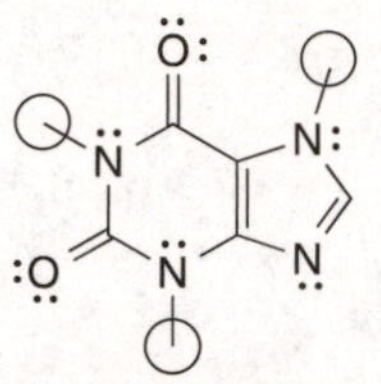

caffeine

a. molecular formula $C_9H_8O_4$
b. eight lone pairs total
c. C labeled with a circle is sp^3 hybridized. All other C's are sp^2 hybridized.
d. three possible resonance structures:

:Ö⁻ ÖH⁺ Ö: Ö: ⟷ :Ö ÖH Ö⁺ :Ö:⁻ ⟷ :Ö ÖH Ö: Ö:

a. molecular formula $C_8H_{10}N_4O_2$
b. eight lone pairs total
c. C's labeled with a circle are sp^3 hybridized. All other C's are sp^2 hybridized.
d. three possible resonance structures:

:Ö:⁻ N⁺ N: :O N N: ⟷ Ö: N N⁺ N: :O N N ⟷ Ö: N N⁺ :O N N⁻

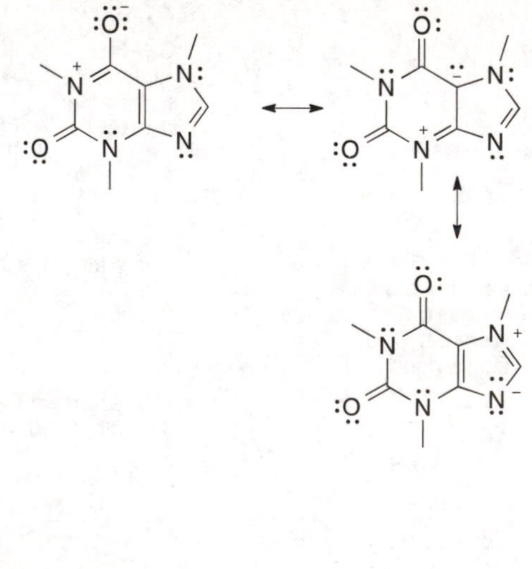

1.73

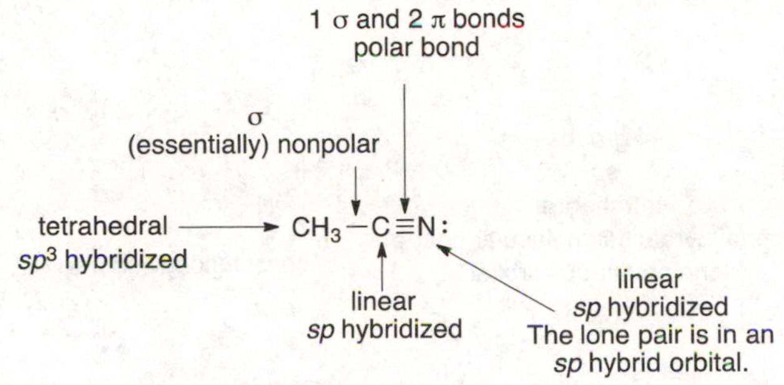

All C–H bonds are nonpolar σ bonds.
All H's use a 1*s* orbital in bonding.

1.74 a. sp^2

b. Each C is trigonal planar; the ring is flat, drawn as a hexagon.

c.

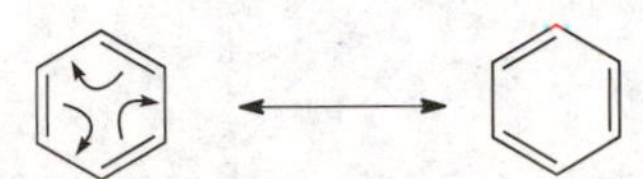

d. Benzene is stable because of its two resonance structures that contribute equally to the hybrid. (This is only part of the story. We'll learn more about benzene's unusual stability in Chapter 17.)

1.75

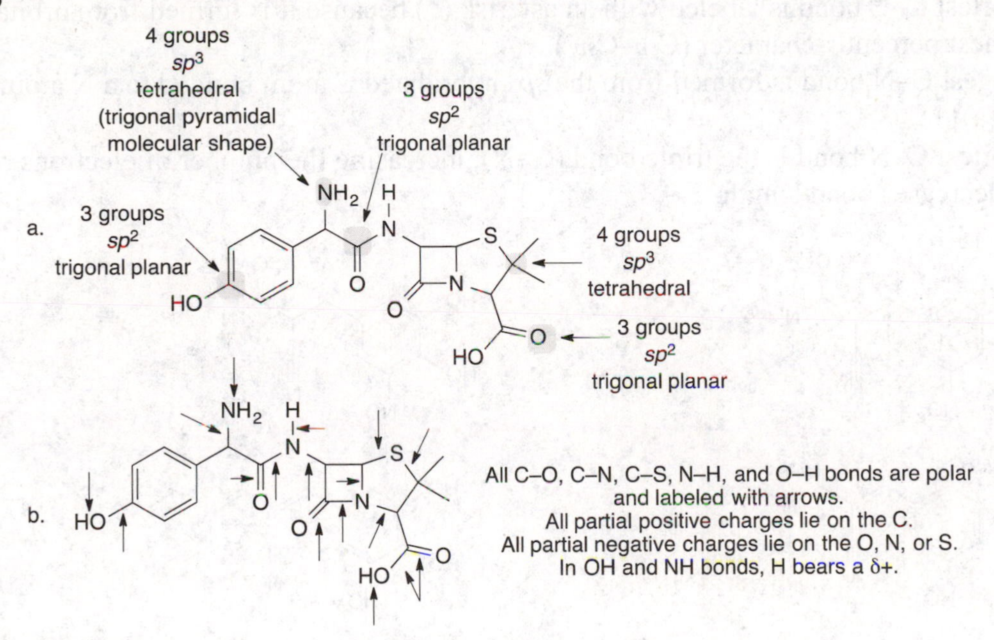

c.

6 π bonds

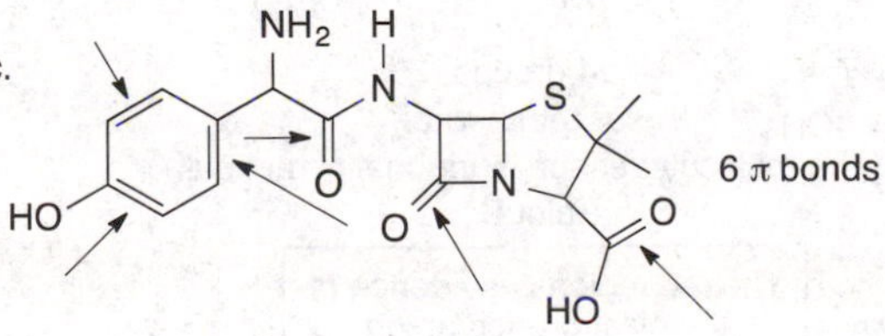

d. **33% *s*-character = *sp*² hybridized**

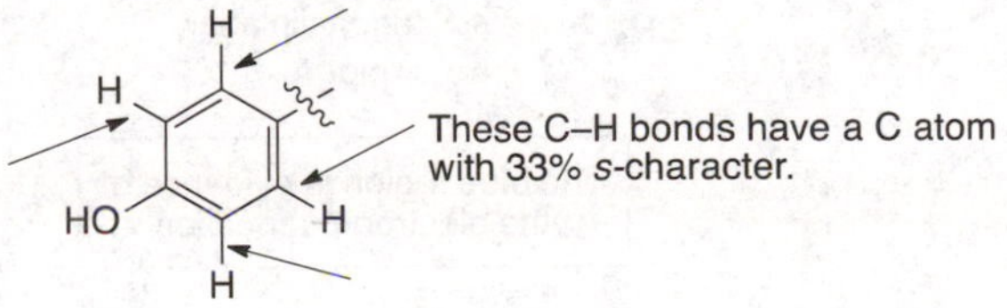

These C–H bonds have a C atom with 33% *s*-character.

1.76

a, b, c.

4 groups
sp^3
tetrahedral
(trigonal pyramidal molecular shape)
lone pair in sp^3 orbital

3 groups
sp^2
trigonal planar
lone pair in sp^2 orbital

d. constitutional isomer

e. resonance structure

1.77 a.

longest C–N bond

shortest C–N bond

longest C–C bond

b. The C–C bonds in the CH_2CH_3 groups are the longest because they are formed from sp^3 hybridized C's.

c. The shortest C–C bond is labeled with an asterisk (*) because it is formed from orbitals with the highest percent *s*-character (Csp–Csp^2).

d. The longest C–N bond is formed from the sp^3 hybridized C atom bonded to a N atom [labeled in part (a)].

e. The shortest C–N bond is the triple bond (C≡N); increasing the number of electrons between atoms decreases bond length.

f.

g.

1.78

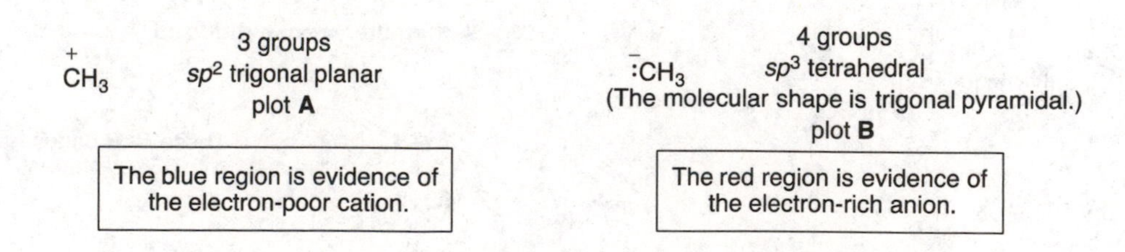

$\overset{+}{C}H_3$
3 groups
sp^2 trigonal planar
plot **A**
The blue region is evidence of the electron-poor cation.
$:\overset{-}{C}H_3$
4 groups
sp^3 tetrahedral
(The molecular shape is trigonal pyramidal.)
plot **B**
The red region is evidence of the electron-rich anion.

1.79 If the N atom is sp^2 hybridized, the lone pair occupies a p orbital, which can overlap with the π bond of the adjacent C=O. This allows electron density to delocalize, which is a stabilizing feature.

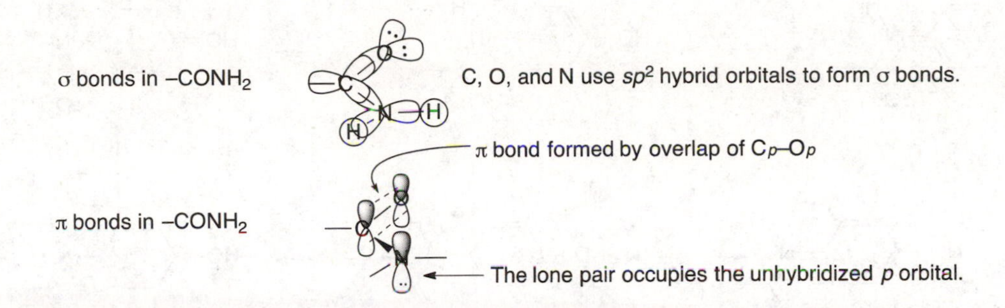

1.80

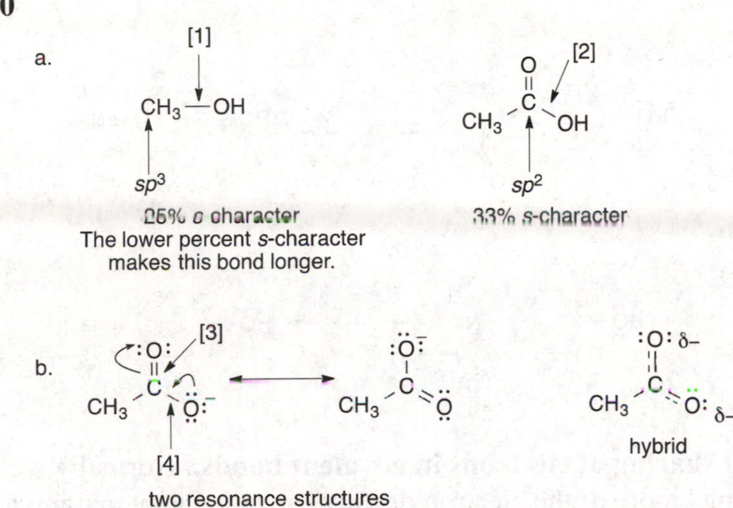

Bonds [3] and [4] are both equivalent in length, because the anion is resonance stabilized, and the C–O bond of the hybrid is a composite of one single bond and one double bond. Both resonance structures contribute equally to the hybrid. Since each C–O bond in the hybrid has partial double bond character, it is shorter than the C–O bond labeled [2].

1.81 Ten additional resonance structures are drawn. (There are more possibilities.)

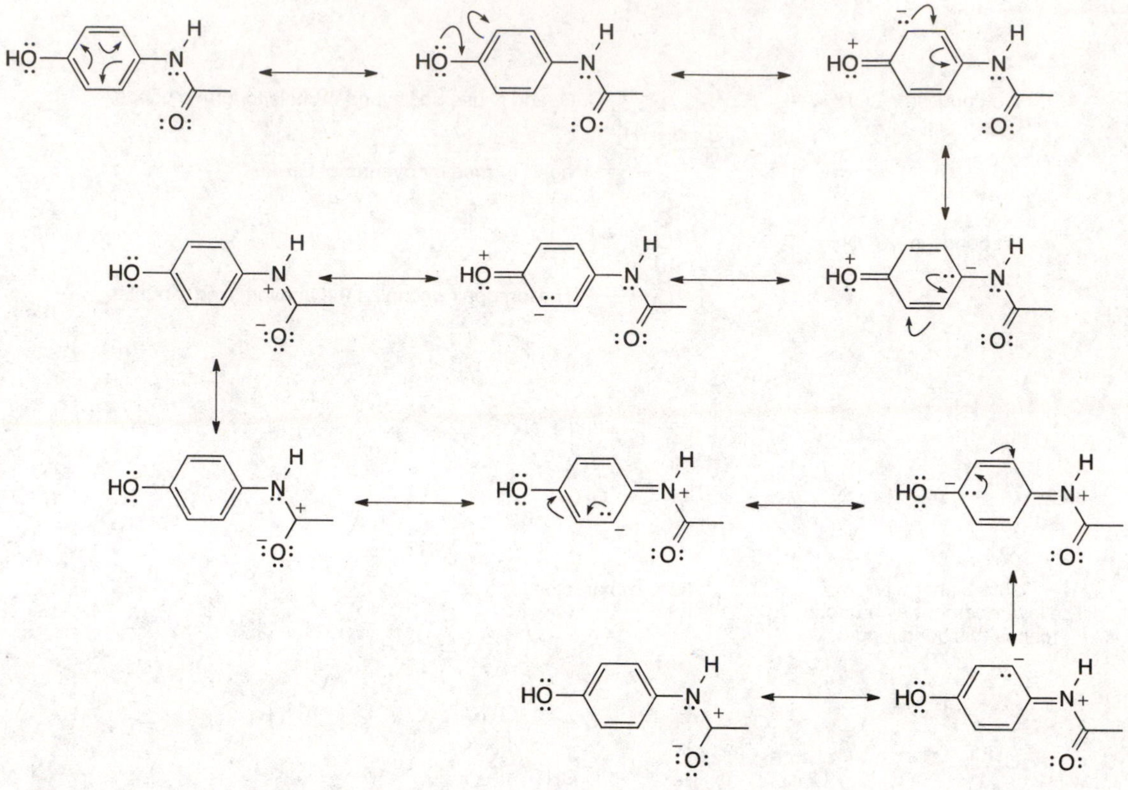

1.82 Polar bonds result from unequal sharing of electrons in covalent bonds. Normally we think c more electronegative atoms "pulling" more of the electron density towards them, making a dipole. In looking at a Csp^2–Csp^3 bond, the atom with a higher percent *s*-character will "pull" more of the electron density towards it, creating a small dipole.

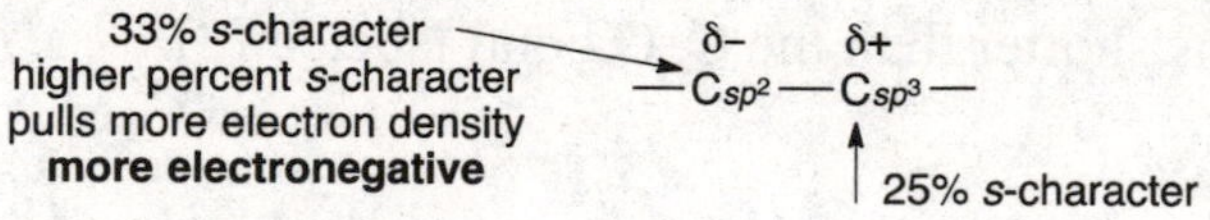

33% *s*-character
higher percent *s*-character
pulls more electron density
more electronegative

$\overset{\delta-}{C_{sp^2}}—\overset{\delta+}{C_{sp^3}}$

25% *s*-character

1.83

Isomers of C_4H_8:

These two compounds are different because of restricted rotation around the C=C (Section 8.2B).

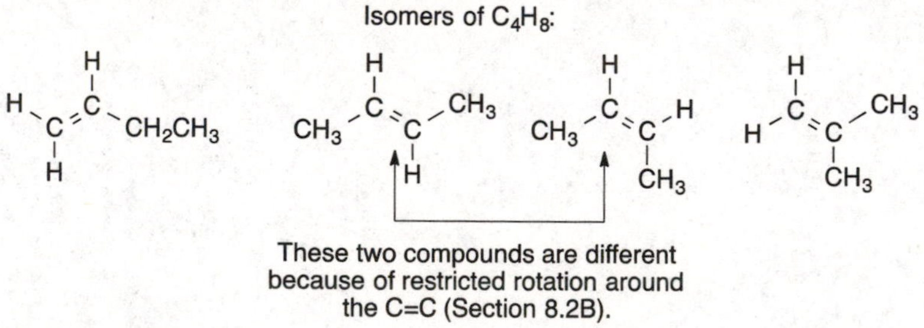

1.84 Carbocation **A** is more stable than carbocation **B** because resonance distributes the positive charge over two carbons. Delocalizing electron density is stabilizing. **B** has no possibility of resonance delocalization.

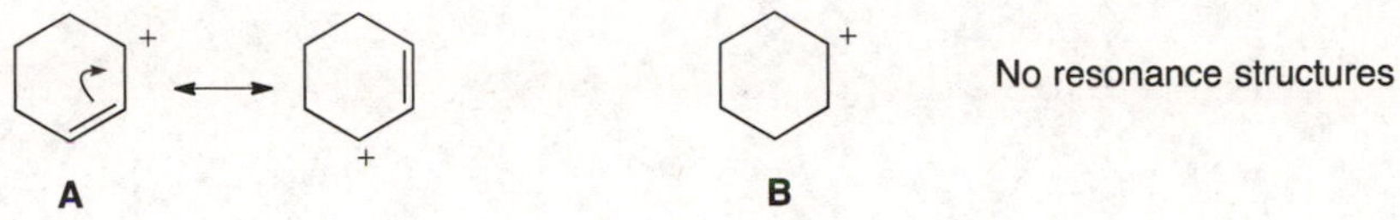

A

B

No resonance structures

1.85

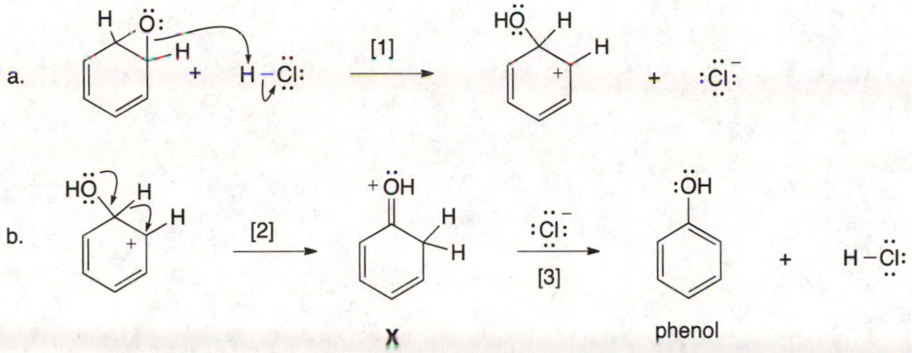

a. [1]

b. [2] [3]

X

phenol

Chapter 2: Acids and Bases

Chapter Review

A comparison of Brønsted–Lowry and Lewis acids and bases

Type	Definition	Structural feature	Examples
Brønsted–Lowry acid (2.1)	proton donor	a proton	HCl, H_2SO_4, H_2O, CH_3CO_2H, TsOH
Brønsted–Lowry base (2.1)	proton acceptor	a lone pair or a π bond	^-OH, $^-OCH_3$, H^-, $^-NH_2$, $CH_2{=}CH_2$
Lewis acid (2.8)	electron pair acceptor	a proton, or an unfilled valence shell, or a partial (+) charge	BF_3, $AlCl_3$, HCl, CH_3CO_2H, H_2O
Lewis base (2.8)	electron pair donor	a lone pair or a π bond	^-OH, $^-OCH_3$, H^-, $^-NH_2$, $CH_2{=}CH_2$

Acid–base reactions

[1] A Brønsted–Lowry acid donates a proton to a Brønsted–Lowry base (2.2).

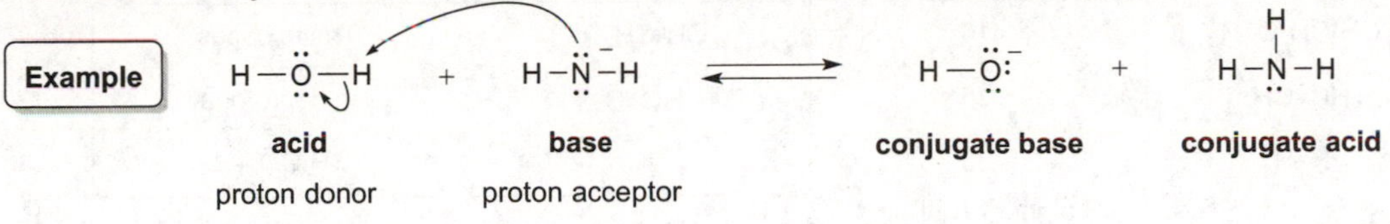

[2] A Lewis base donates an electron pair to a Lewis acid (2.8).

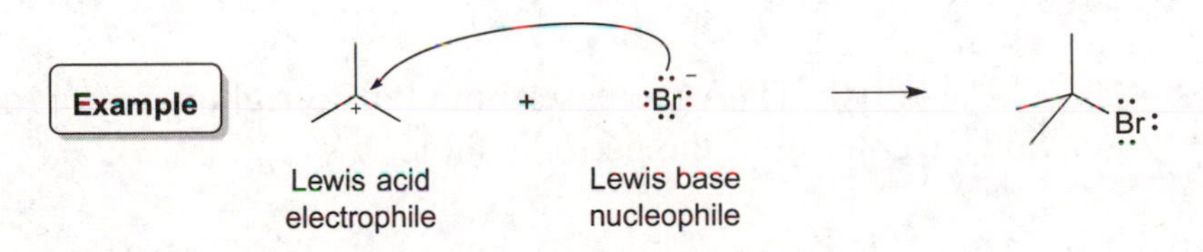

- Electron-rich species react with electron-poor ones.
- Nucleophiles react with electrophiles.

Important facts

- Definition: **pK_a = –log K_a.** The **lower the pK_a,** the **stronger** the acid (2.3).

NH_3
pK_a = 38

versus

H_2O
pK_a = 15.7
lower pK_a = stronger acid

- The stronger the acid, the weaker the conjugate base (2.3).

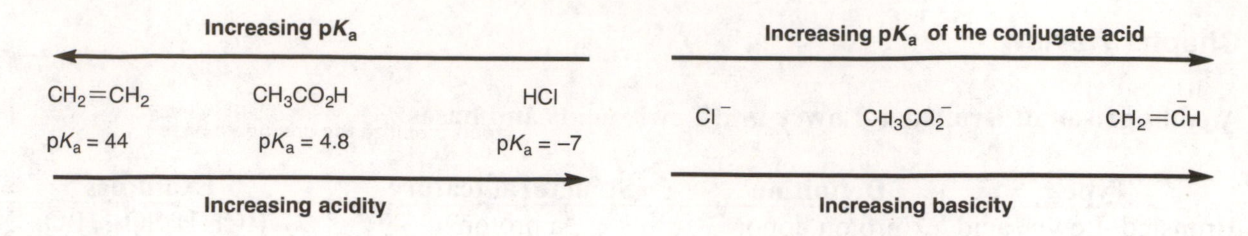

- In proton transfer reactions, equilibrium favors the weaker acid and the weaker base (2.4).

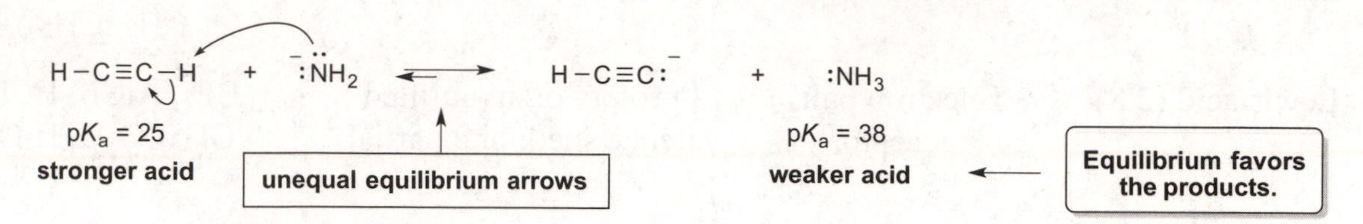

- An acid can be deprotonated by the conjugate base of any acid having a **higher pK_a** (2.4).

Acid	pK_a	Conjugate base	
CH_3CO_2–H	4.8	$CH_3CO_2^-$	
CH_3CH_2O–H	16	$CH_3CH_2O^-$	These bases
HC≡CH	25	HC≡C$^-$	can deprotonate
H–H	35	H^-	CH_3CO_2–H.
	higher pK_a than CH_3CO_2–H		

Factors that determine acidity (2.5)

[1] **Element effects** (2.5A) The acidity of H–A increases both left-to-right across a row and down a column of the periodic table.

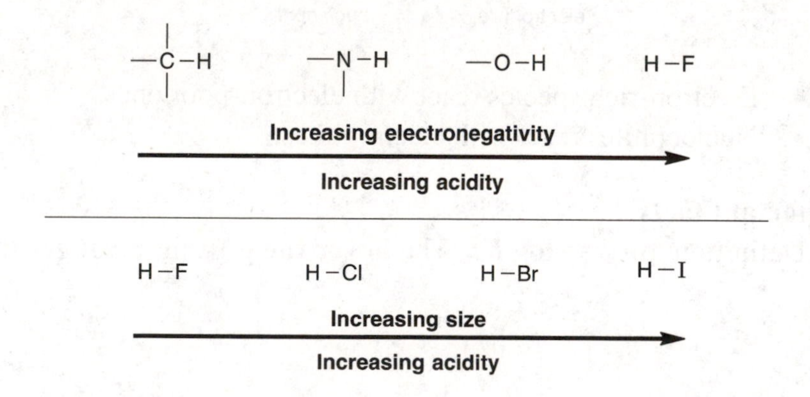

[2] Inductive effects (2.5B)

The acidity of H–A increases with the presence of electron-withdrawing groups in A.

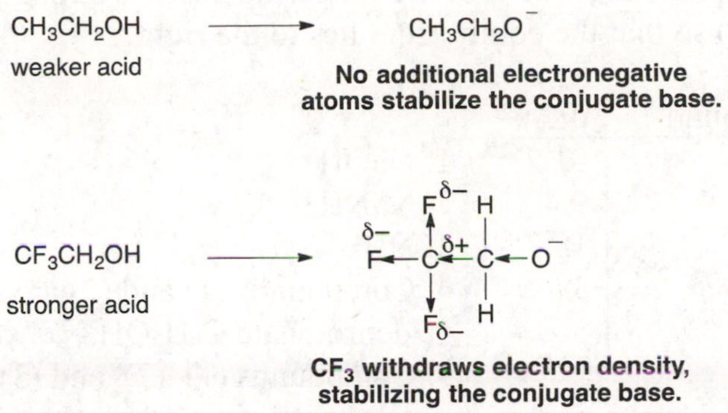

CH_3CH_2OH → $CH_3CH_2O^-$

weaker acid

No additional electronegative atoms stabilize the conjugate base.

CF_3CH_2OH → F_3C–CH_2–O^- (F δ−, C δ+)

stronger acid

CF_3 withdraws electron density, stabilizing the conjugate base.

[3] Resonance effects (2.5C)

The acidity of H–A increases when the conjugate base A:⁻ is resonance stabilized.

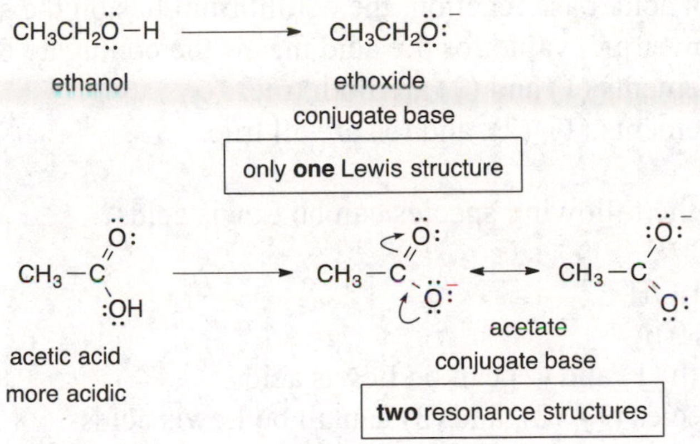

$CH_3CH_2\ddot{O}$–H → $CH_3CH_2\ddot{O}:^-$

ethanol

ethoxide conjugate base

only **one** Lewis structure

CH_3–C(=O)–OH → CH_3–C(=O)–O^- ⟷ CH_3–C(O^-)=O

acetic acid more acidic

acetate conjugate base

two resonance structures

[4] Hybridization effects (2.5D)

The acidity of H–A increases as the percent *s*-character of the A:⁻ increases.

CH_3CH_3	$CH_2{=}CH_2$	H–C≡C–H
ethane	ethylene	acetylene
$pK_a = 50$	$pK_a = 44$	$pK_a = 25$

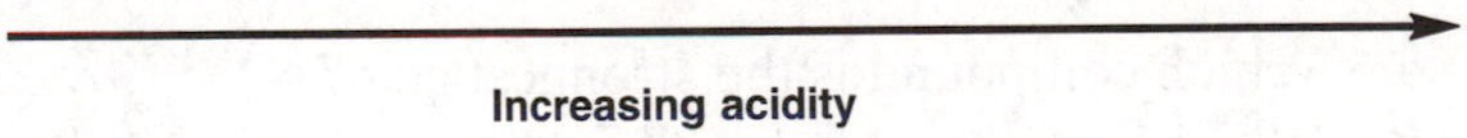

→ **Increasing acidity**

Practice Test on Chapter Review

1.a. Given the p*K*a data, which of the following bases is strong enough to deprotonate C_6H_5OH (p*K*a = 10) so that the equilibrium lies to the right?

Compound	p*K*a
H_3O^+	−1.7
NH_4^+	9.4
H_2O	15.7
NH_3	38

1. NaOH
2. $NaNH_2$
3. NH_3
4. Compounds (1) and (2) are strong enough to deprotonate C_6H_5OH.
5. Compounds (1), (2), and (3) are all strong enough to deprotonate C_6H_5OH.

b. Which of the following statements is true about p*K*a, acidity, and basicity?
1. A higher p*K*a means the acid is less acidic.
2. In an acid–base reaction, the equilibrium lies on the side of the acid with the higher p*K*a.
3. A lower p*K*a value for the acid means the conjugate base is more basic.
4. Statements (1) and (2) are both true.
5. Statements (1), (2), and (3) are all true.

c. Which of the following species can be Lewis acids?
1. BCl_3
2. CH_3OH
3. $(CH_3)_3C^+$
4. Both (1) and (2) can be Lewis acids.
5. Species (1), (2), and (3) can all be Lewis acids.

2. Answer the following questions about compounds **A–D**.

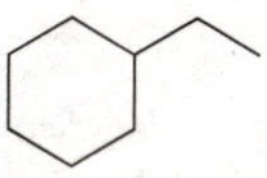

OH NH2 O OH

A 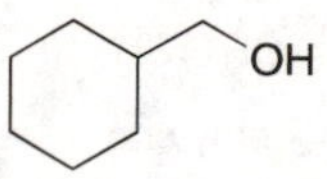 B 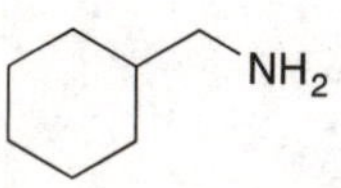 C 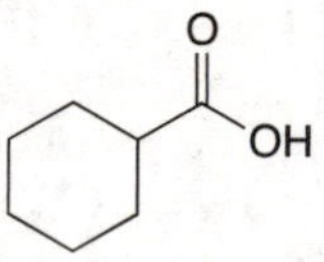 D

a. Which compound is the strongest acid?
b. Which compound forms the strongest conjugate base?
c. The conjugate base of **C** is strong enough to remove a proton on which compound(s), so that the equilibrium favors the products?

3. (a) Which compound is the strongest Brønsted–Lowry acid? (b) Which compound is the weakest Brønsted–Lowry acid?

Br NH2 Cl O Cl Cl OH OH

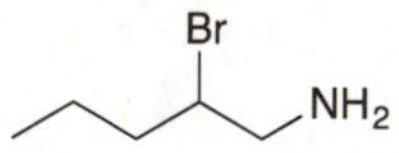

A 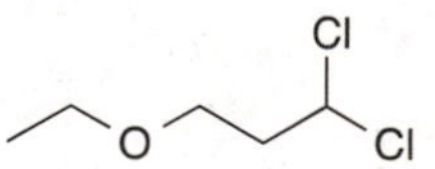 B 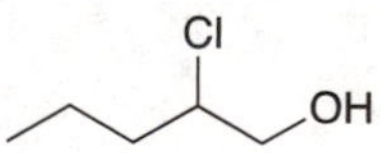 C 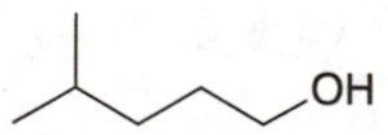 D

4. Draw all the products formed in the following reactions.

a. (benzene ring)–$CH_2CH(CH_3)NHCH_3$ + HBr ⟶

b. (benzene ring with HO, Cl, NO_2) + $Na^{+}\,{}^{-}NH_2$ ⟶

5. Draw the product(s) formed in the following Lewis acid–base reaction.

$(CH_3)_2\overset{+}{C}H$ + CH_3OH ⟶

Answers to Practice Test

1. a. 4 b. 4 c. 5
2. a. **D** b. **A** c. **B, D**
3. a. **C** b. **B**
4. a. (benzene ring)–$CH_2CH(CH_3)\overset{+}{N}H_2CH_3$ + Br^-
   b. (benzene ring with $Na^{+}\,{}^{-}O$, Cl, NO_2) + NH_3
5. $(CH_3)_2CH—\overset{+}{\ddot{O}}—CH_3$ (with H on O)

Answers to Problems

2.1 Brønsted–Lowry acids are **proton donors** and must contain a hydrogen atom.
Brønsted–Lowry bases are **proton acceptors** and must have an available electron pair (either a lone pair or a π bond).

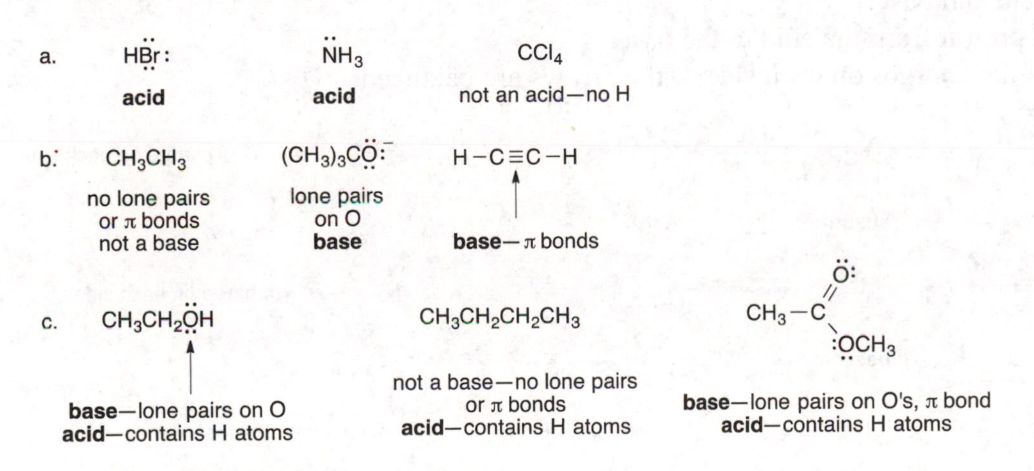

2.2 A Brønsted–Lowry base accepts a proton to form the conjugate acid. A Brønsted–Lowry acid loses a proton to form the conjugate base.

a. NH_3 ⟶ NH_4^+

Cl^- ⟶ HCl

$(CH_3)_2C{=}O$ ⟶ $(CH_3)_2C{=}\overset{+}{\ddot{O}}H$

b. HBr ⟶ Br^-

HSO_4^- ⟶ SO_4^{2-}

CH_3OH ⟶ CH_3O^-

2.3 a. True.

$CH_2{=}CH_2$ + H^+ ⟶ $CH_3CH_2^+$

base — conjugate acid

b. False. $CH_3CH_2^-$ cannot be the conjugate base of $CH_3CH_2^+$ because they both have the same number of H's and a conjugate base must have one fewer H.

c. False. $CH_2{=}CH_2$ and $CH_3CH_2^-$ differ by the presence of H^-, not H^+.

d. True.

$CH_2{=}CH_2$ —Remove H+⟶ $CH_2{=}\bar{C}H$

acid — conjugate base

e. True.

$CH_3CH_2^-$ + H^+ ⟶ CH_3CH_3

base — conjugate acid

2.4 The Brønsted–Lowry base accepts a proton to form the conjugate acid. The Brønsted–Lowry acid loses a proton to form the conjugate base. Use curved arrows to show the movement of electrons (***NOT protons***). Re-draw the starting materials if necessary to clarify the electron movement.

a. H–Cl: (acid) + diethyl ether (base) ⇌ :Cl:⁻ (conjugate base) + protonated diethyl ether (conjugate acid)

b. acetone (acid) + ethoxide (base) ⇌ enolate (conjugate base) + ethanol (conjugate acid)

2.5 To draw the products:

[1] Find the acid and base.

[2] Transfer a proton from the acid to the base.

[3] Check that the charges on each side of the arrows are balanced.

a. Cl_3CCOOH (acid) + $^-OCH_3$ (base) ⇌ Cl_3CCOO^- + HÖ–CH_3 — (–)1 charge on each side

b. cyclohexylacetylene (acid) + H^- (base) ⇌ cyclohexylacetylide + H_2 — (–)1 charge on each side

c. cyclopentyl–NH_2 (base) + H–Cl: (acid) ⇌ cyclopentyl–NH_3^+ + :Cl:⁻ — net neutral on each side

d. propanol (base) + H–OSO_3H (acid) ⇌ propyl–OH_2^+ + $^-OSO_3H$ — net neutral on each side

2.6 Draw the products in each reaction as in Answer 2.4.

a. (alcohol)–OH $\xrightarrow{HCl}$ (alcohol)–$\overset{+}{O}H_2$ + Cl^-

b. $CH_3CH_2OCH_2CH_3$ $\xrightarrow{HCl}$ $CH_3CH_2\overset{+}{O}(H)CH_2CH_3$ + Cl^-

c. (amine) N $\xrightarrow{HCl}$ (H)$\overset{+}{N}$ + Cl^-

d. NH (pyrrolidine) $\xrightarrow{HCl}$ $\overset{+}{N}H_2$ + Cl^-

2.7 The smaller the pK_a, the stronger the acid. The larger the K_a, the stronger the acid.

a. $CH_3CH_2CH_3$ ($pK_a = 50$) or CH_3CH_2OH ($pK_a = 16$)

smaller pK_a
stronger acid (CH₃CH₂OH)

b. phenol (OH) $K_a = 10^{-10}$ or toluene (CH₃) $K_a = 10^{-41}$

larger K_a
stronger acid (phenol)

2.8 To convert from K_a to pK_a, take (–) the log of the K_a; **pK_a = –log K_a.**
To convert pK_a to K_a, take the antilog of (–) the pK_a.

a. $K_a = 10^{-10}$ → $pK_a = 10$; $K_a = 10^{-21}$ → $pK_a = 21$; $K_a = 5.2 \times 10^{-5}$ → $pK_a = 4.3$

b. $pK_a = 7$ → $K_a = 10^{-7}$; $pK_a = 11$ → $K_a = 10^{-11}$; $pK_a = 3.2$ → $K_a = 6.3 \times 10^{-4}$

2.9 Since **strong acids form weak conjugate bases,** the basicity of conjugate bases increases with increasing pK_a of their acids. Find the pK_a of each acid from Table 2.1 and then rank the acids in order of increasing pK_a. This will also be the order of increasing basicity of their conjugate bases.

a. Increasing acidity (←)

	H_2O	NH_3	CH_4
pK_a =	15.7	38	50
conjugate bases:	^-OH	$^-NH_2$	$^-CH_3$

Increasing basicity (→)

b. Increasing acidity (←)

	HC≡CH	$CH_2{=}CH_2$	CH_4
pK_a =	25	44	50
conjugate bases:	$^-C{\equiv}CH$	$^-CH{=}CH_2$	$^-CH_3$

Increasing basicity (→)

2.10 Use the definitions in Answer 2.9 to compare the acids. The smaller the pK_a, the larger the K_a and the stronger the acid. When a stronger acid dissolves in water, the equilibrium lies further to the right.

HCO_2H
formic acid
$pK_a = 3.8$

$(CH_3)_3CCO_2H$
pivalic acid
$pK_a = 5.0$

a. smaller pK_a = larger K_a
b. smaller pK_a = stronger acid
c. weaker acid = stronger conjugate base
d. stronger acid = equilibrium further to the right

2.11 To estimate the pK_a of the indicated bond, find a similar bond in the pK_a table (H bonded to the same atom with the same hybridization).

a. For NH_3, pK_a is 38.
estimated pK_a = 38

b. For CH_3CH_2OH, pK_a is 16.
estimated pK_a = 16

c. For CH_3COOH, pK_a is 4.8.
estimated pK_a = 5

2.12 Label the acid and the base and then transfer a proton from the acid to the base. To determine if the reaction will proceed as written, compare the pK_a of the acid on the left with the conjugate acid on the right. **The equilibrium always favors the formation of the weaker acid and the weaker base.**

a. $CH_2{=}CH_2$ (acid, pK_a = 44, **weaker acid**) + $H:^-$ (base) ⇌ $CH_2{=}\ddot{C}H^-$ (conjugate base) + H_2 (conjugate acid, pK_a = 35)
Equilibrium favors the **starting materials.**

b. CH_4 (acid, pK_a = 50, **weaker acid**) + $:\ddot{O}H^-$ (base) ⇌ $:CH_3^-$ (conjugate base) + $H_2\ddot{O}:$ (conjugate acid, pK_a = 15.7)
Equilibrium favors the **starting materials.**

c. CH_3CO_2H (acid, pK_a = 4.8) + $CH_3CH_2\ddot{O}:^-$ (base) ⇌ $CH_3CO_2^-$ (conjugate base) + $CH_3CH_2\ddot{O}H$ (conjugate acid, pK_a = 16, **weaker acid**)
Equilibrium favors the **products.**

d. $:\ddot{Cl}:^-$ (base) + $CH_3CH_2\ddot{O}H$ (acid, pK_a = 16, **weaker acid**) ⇌ $H\ddot{Cl}:$ (conjugate acid, pK_a = −7) + $CH_3CH_2\ddot{O}:^-$ (conjugate base)
Equilibrium favors the **starting materials.**

2.13 An acid can be deprotonated by the conjugate base of any acid with a higher pK_a.

CH_3CN
pK_a = 25
Any base having a conjugate acid with a pK_a higher than 25 can deprotonate this acid.

Base	Conjugate acid	pK_a
NaH	H_2	35
Na_2CO_3	HCO_3^-	10.2
NaOH	H_2O	15.7
$NaNH_2$	NH_3	38
$NaHCO_3$	H_2CO_3	6.4

Only NaH and $NaNH_2$ are strong enough to deprotonate acetonitrile.

2.14 The acidity of H–Z **increases left-to-right across a row and down a column** of the periodic table.

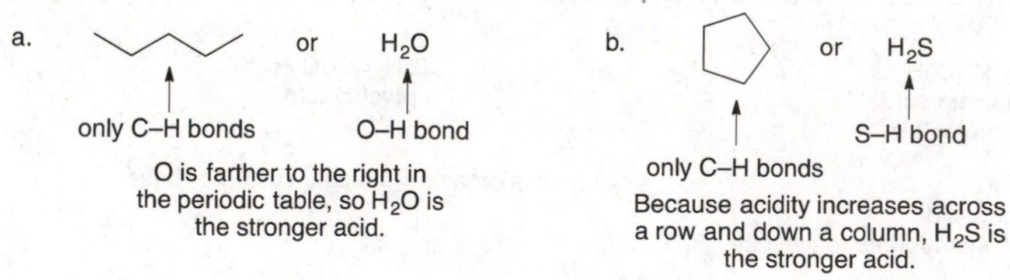
a.
or
H_2O
only C–H bonds
O–H bond
O is farther to the right in the periodic table, so H_2O is the stronger acid.
b.
or
H_2S
S–H bond
only C–H bonds
Because acidity increases across a row and down a column, H_2S is the stronger acid.

2.15

Because acidity increases across a row and down a column, the order of acidity is N–H < O–H < S–H.

2.16 Look at the element bonded to the acidic H and decide its acidity based on the periodic trends. **Farther to the right and down the periodic table is more acidic.**

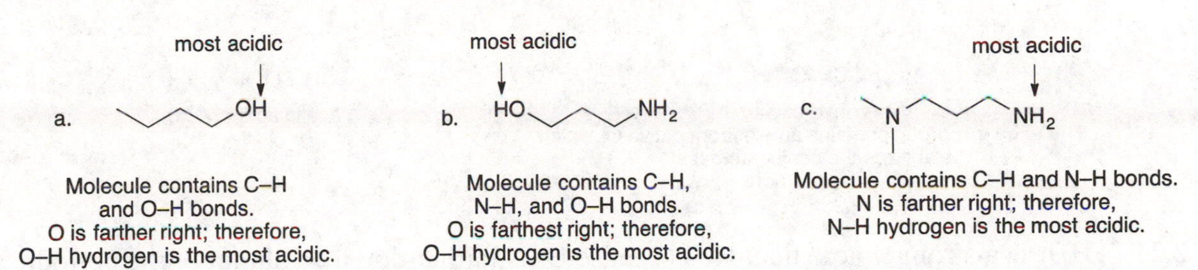

most acidic
a.
OH
Molecule contains C–H
and O–H bonds.
O is farther right; therefore,
O–H hydrogen is the most acidic.
most acidic
b.
HO
NH_2
Molecule contains C–H,
N–H, and O–H bonds.
O is farthest right; therefore,
O–H hydrogen is the most acidic.
most acidic
c.
N
NH_2
Molecule contains C–H and N–H bonds.
N is farther right; therefore,
N–H hydrogen is the most acidic.

2.17 The acidity of HA increases left-to-right across the periodic table. Pseudoephedrine contains C–H, N–H, and O–H bonds. The O–H bond is most acidic.

pseudoephedrine skeletal structure

2.18 Compare the most acidic protons in each compound to determine the stronger acid.

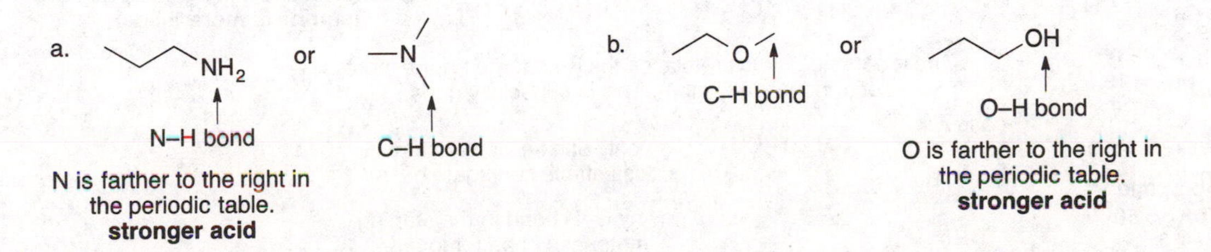

2.19 **More electronegative atoms stabilize the conjugate base, making the acid stronger.** Compare the electron-withdrawing groups on the acids below to decide which is a stronger acid **(more electronegative groups = more acidic).**

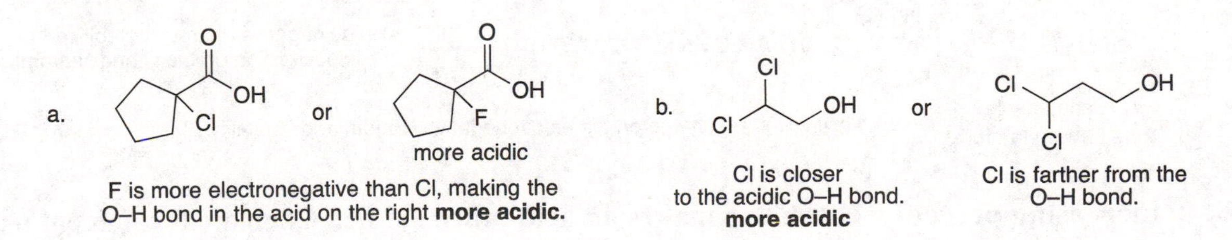

a.
O
OH
Cl
or
O
OH
F
more acidic
F is more electronegative than Cl, making the
O–H bond in the acid on the right more acidic.
b.
Cl
Cl
OH
or
Cl
OH
Cl
Cl is closer
to the acidic O–H bond.
more acidic
Cl is farther from the
O–H bond.

c. [structure] OH or [structure] OH, NO_2

more acidic

NO_2 is electron withdrawing, making the O–H bond in the acid on the right **more acidic.**

2.20 **More electronegative groups stabilize the conjugate base, making the acid stronger.**

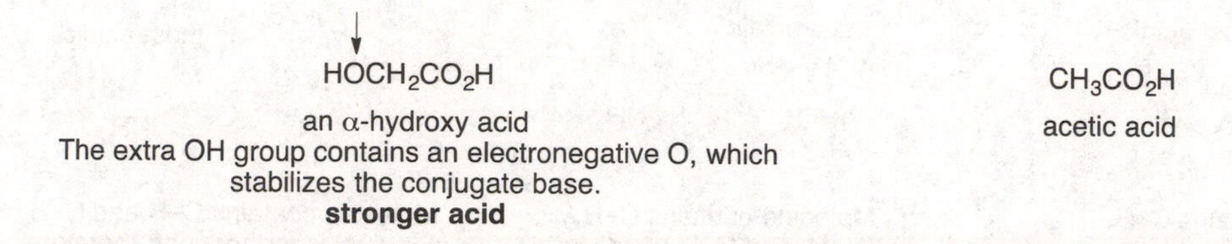

$HOCH_2CO_2H$
an α-hydroxy acid
The extra OH group contains an electronegative O, which stabilizes the conjugate base.
stronger acid
CH_3CO_2H
acetic acid

2.21 HBr is a stronger acid than HCl because Br is farther down a column of the periodic table, and the larger Br^- anion is more stable than the smaller Cl^- anion. In these acids the H is bonded directly to the halogen. In HOCl and HOBr, the H is bonded to O, and the halogens Cl and Br exert an inductive effect. In this case, the more electronegative Cl stabilizes ^-OCl more than the less electronegative Br stabilizes ^-OBr. Thus, HOCl forms the more stable conjugate base, making it the stronger acid.

2.22 The acidity of an acid increases when the conjugate base is resonance stabilized. Compare the conjugate bases of acetone and propane to explain why acetone is more acidic.

acetone, $pK_a = 19.2$ —base→ [two resonance structures of the conjugate base, $CH_3C(=O)CH_2^-$ ↔ $CH_3C(O^-)=CH_2$]

2 resonance structures
more stable conjugate base
Acetone is more acidic.

One resonance structure places the (–) charge on the more electronegative O atom. This is especially good.

$CH_3CH_2CH_3$ —base→ $CH_3CH_2\ddot{C}H_2$ (only one Lewis structure)

propane, $pK_a = 50$

only one Lewis structure
less stable conjugate base

(Any C–H bond in the starting material can be removed.)

2.23 The acidity of an acid increases when the conjugate base is resonance stabilized. Acetonitrile has a resonance-stabilized conjugate base, which accounts for its acidity.

H–C(H)(H)–C≡N: —base→ H–C̈(H)–C≡N: ↔ H–C(H)=C=N̈:⁻

acetonitrile
(one Lewis structure)

The negative charge is stabilized by delocalization on the C and N atoms.

Having the (–) charge on the electronegative N atom adds stability.

2.24 **Increasing percent *s*-character makes an acid more acidic.** Compare the percent *s*-character of the carbon atoms in each of the C–H bonds in question. A stronger acid has a weaker conjugate base.

a. [cyclopentyl–C≡C–H] sp hybridized C, 50% s-character, **more acidic** or [cyclopentyl–CH₂CH₂CH₃] sp^3 hybridized C, 25% s-character

base ↓ [cyclopentyl–C≡C:⁻] or [cyclopentyl–CH₂CH₂CH₂⁻] **stronger conjugate base**

b. [cyclohexene vinyl C–H] sp^2 hybridized C, 33% s-character, **more acidic** or [cyclohexane C–H] sp^3 hybridized C, 25% s-character

base ↓ [cyclohexenyl anion] or [cyclohexyl anion] **stronger conjugate base**

2.25 To compare the acids, first **look for element effects.** Then identify electron-withdrawing groups, resonance, or hybridization differences.

a. [butane] C is farthest left in the periodic table. **CH bond is least acidic.** [propylamine, NH_2] **intermediate acidity** [propanol, OH] O is farthest right in the periodic table. **OH bond is most acidic.**

c. [N,N-ethylmethylamine, N] C is farthest left in the periodic table. **CH bond is least acidic.** [butylamine, NH_2] **intermediate acidity** [butanol, OH] O is farthest right in the periodic table. **OH bond is most acidic.**

b. [propanol, OH] OH group **least acidic** [propanoic acid, OH] **intermediate acidity** [bromoacetic acid, Br, OH] Br is electron withdrawing and the conjugate base is resonance stabilized. **most acidic**

2.26 Look at the element bonded to the acidic H and decide its acidity based on the periodic trends. **Farther to the right and down the periodic table is more acidic.**

a. [THC structure, OH] **most acidic**

THC
tetrahydrocannabinol
The molecule contains C–H and O–H bonds.
O is farther right; therefore, O–H hydrogen is the most acidic.

b. [ketoprofen structure, COOH] **most acidic**

ketoprofen
The molecule contains C–H and O–H bonds.
O is farther right; therefore, O–H hydrogen is the most acidic.

2.27

:O:

A is resonance stabilized so **A** is the weaker base.

The negative charge is localized on O in **B,** so **B** is the stronger base.

A B

2.28 Draw the products of proton transfer from the acid to the base.

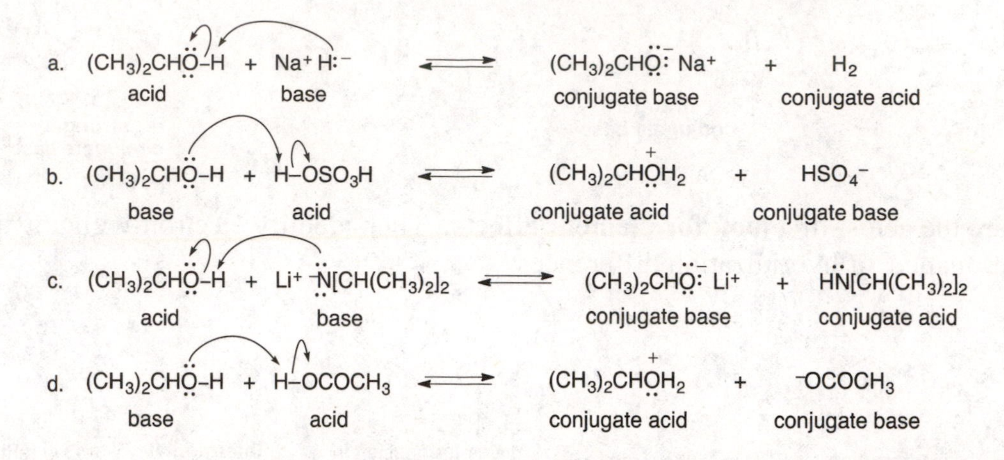

2.29 To cross a cell membrane, amphetamine must be in its neutral (not ionic) form.

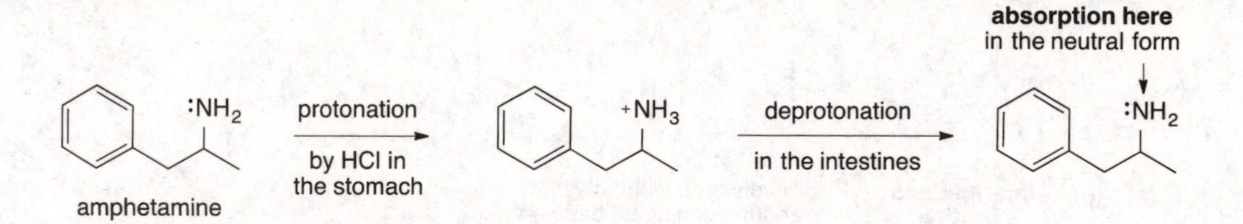

2.30 **Lewis bases are electron pair donors:** they contain a lone pair or a π bond.

a. $\ddot{N}H_3$ — **yes**—has lone pair

b. $CH_3CH_2CH_3$ — **no**—no lone pair or π bond

c. $H:^-$ — **yes**—has lone pair

d. $H{-}C{\equiv}C{-}H$ — **yes**—has 2 π bonds

2.31 **Lewis acids are electron pair acceptors.** Most Lewis acids contain a proton or an unfilled valence shell of electrons.

a. BBr_3 — **yes** unfilled valence shell on B

b. CH_3CH_2OH — **yes** contains a proton

c. $(CH_3)_3C^+$ — **yes** unfilled valence shell on C

d. Br^- — **no** no proton no unfilled valence shell

2.32 Label the Lewis acid and Lewis base and then draw the curved arrows.

a. BF_3 + CH_3OCH_3 → $F_3B^{-}–O^{+}(CH_3)_2$ (new bond)

Lewis acid unfilled valence shell on B

Lewis base lone pairs on O

b. carbocation + $^{-}$:ÖH → alcohol (:ÖH)

Lewis acid unfilled valence shell on C

Lewis base lone pairs on O

2.33 A Lewis acid is also called an **electrophile.** When a Lewis base reacts with an electrophile other than a proton, it is called a **nucleophile.** Label the electrophile and nucleophile in the starting materials and then draw the products.

a. diethyl ether + BBr_3 → $Br_3B^{-}–O^{+}$(diethyl)

Lewis base **nucleophile** lone pairs on O

Lewis acid **electrophile** unfilled valence shell on B

b. cyclohexanone + $AlCl_3$ → $Cl_3Al^{-}–O^{+}$=cyclohexane

Lewis base **nucleophile** lone pairs on O

Lewis acid **electrophile** unfilled valence shell on Al

2.34 Draw the product of each reaction by using an electron pair of the Lewis base to form a new bond to the Lewis acid.

a. $CH_3CH_2–\ddot{N}(CH_2CH_3)–CH_2CH_3$ + $B(CH_3)_3$ → $(CH_3)_3B^{-}–N^{+}(CH_2CH_3)_3$

Lewis base **nucleophile** lone pair on N

Lewis acid **electrophile** unfilled valence shell on B

b. $CH_3CH_2–\ddot{N}(CH_2CH_3)–CH_2CH_3$ + $^{+}C(CH_3)_3$ → $(CH_3)_3C–N^{+}(CH_2CH_3)_3$

Lewis base **nucleophile** lone pair on N

Lewis acid **electrophile** unfilled valence shell on C

c. $CH_3CH_2–\ddot{N}(CH_2CH_3)–CH_2CH_3$ + $AlCl_3$ → $Cl_3Al^{-}–N^{+}(CH_2CH_3)_3$

Lewis base **nucleophile** lone pair on N

Lewis acid **electrophile** unfilled valence shell on Al

2.35 Curved arrows begin at the Lewis base and point towards the Lewis acid.

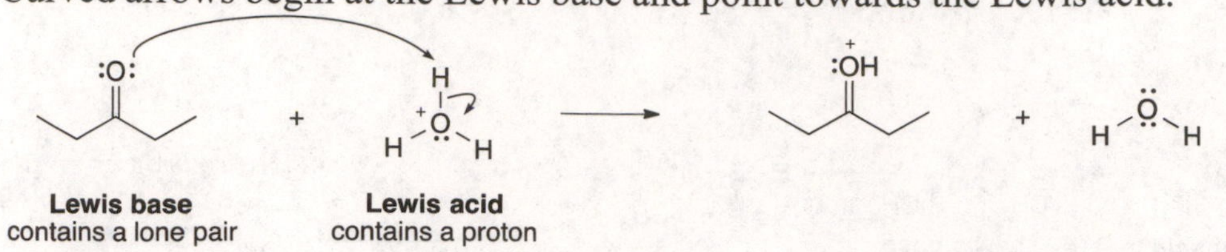

Lewis base
contains a lone pair

Lewis acid
contains a proton

2.36 a, b. Since acidity increases from left-to-right across a row of the periodic table and propranolol has C–H, N–H, and O–H bonds, the O–H bond is most acidic. NaH is a base that removes the most acidic OH proton.

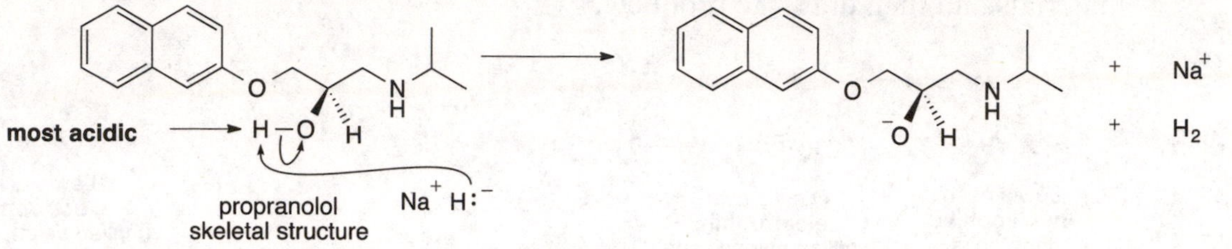

most acidic

propranolol
skeletal structure

Na^+ $H:^-$

+ Na^+

+ H_2

c, d. Of the atoms with lone pairs (N and O), N is to the left in the periodic table, making it the most basic site. HCl is an acid, which protonates the most basic site.

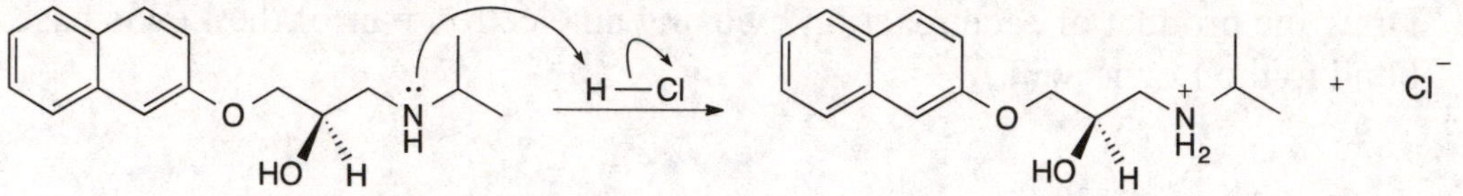

H—Cl

+ Cl^-

2.37 a, b. Using periodic trends, the N–H bond of amphetamine is most acidic. NaH is a base that removes an acidic proton on N, forming H_2 in the process.

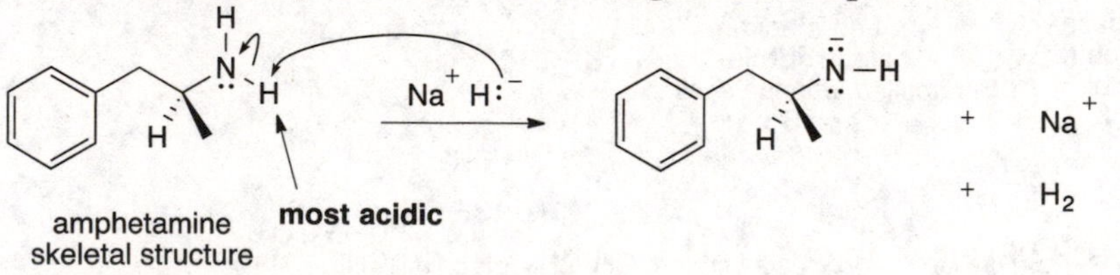

amphetamine
skeletal structure

most acidic

Na^+ $H:^-$

+ Na^+

+ H_2

c. HCl protonates the lone pair on N (the most basic site).

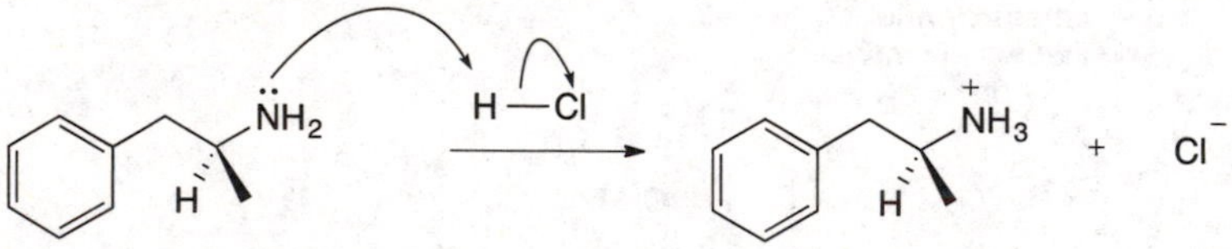

H—Cl

+ Cl^-

2.38 To draw the conjugate acid of a Brønsted–Lowry base, **add a proton to the base.**

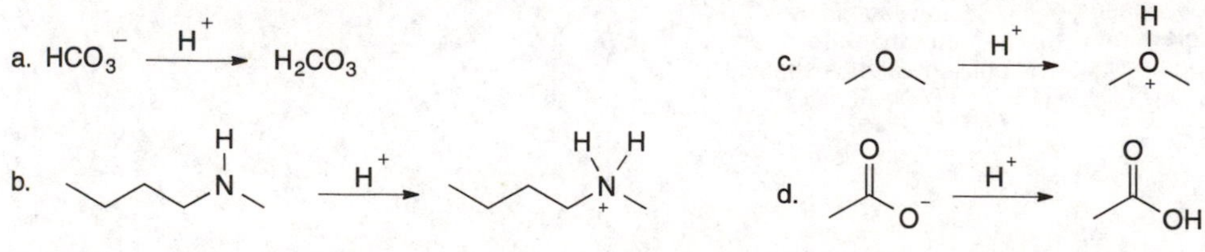

a. $HCO_3^- \xrightarrow{H^+} H_2CO_3$

b. $\xrightarrow{H^+}$

c. $\xrightarrow{H^+}$

d. $\xrightarrow{H^+}$

2.39 To draw the conjugate base of a Brønsted–Lowry acid, **remove a proton from the acid.**

a. HCO_3^- $\xrightarrow{-H^+}$ CO_3^{2-}

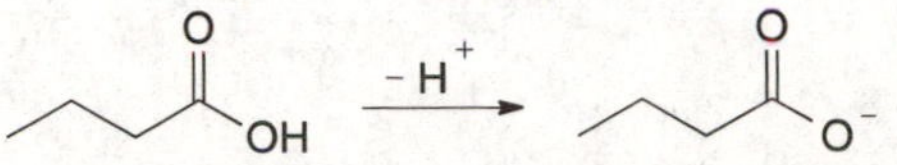

b. 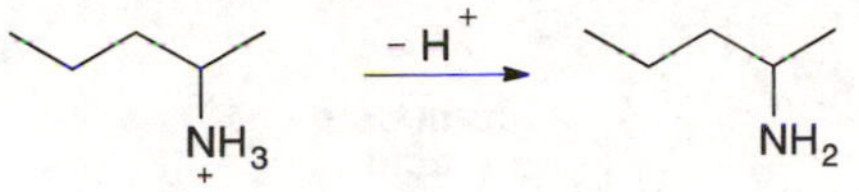 $\overset{+}{N}H_3$ $\xrightarrow{-H^+}$ NH_2

c. OH $\xrightarrow{-H^+}$ O^-

d. 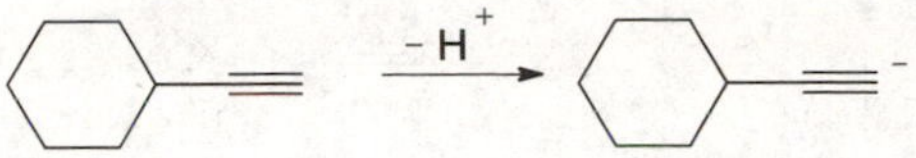 $\xrightarrow{-H^+}$

2.40 To draw the products of an acid–base reaction, transfer a proton from the acid (H_2SO_4 in this case) to the base.

a. 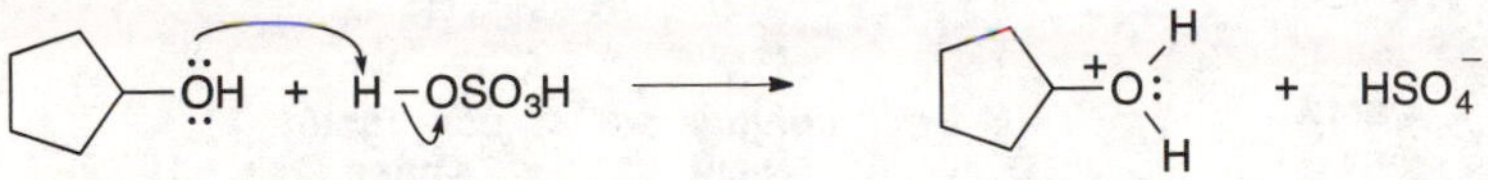 ÖH + H–OSO$_3$H ⟶ $\overset{+}{O}$(H)(H) + HSO_4^-

b. 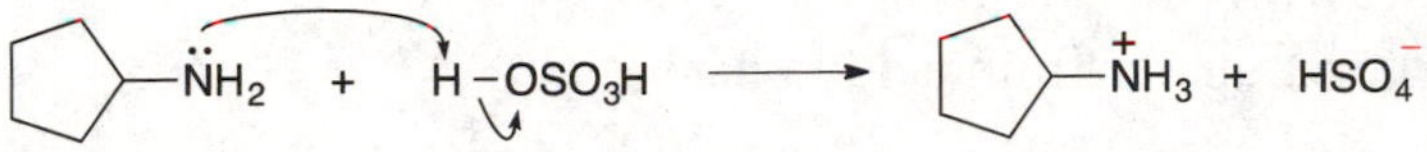 N̈H$_2$ + H–OSO$_3$H ⟶ $\overset{+}{N}H_3$ + HSO_4^-

c. 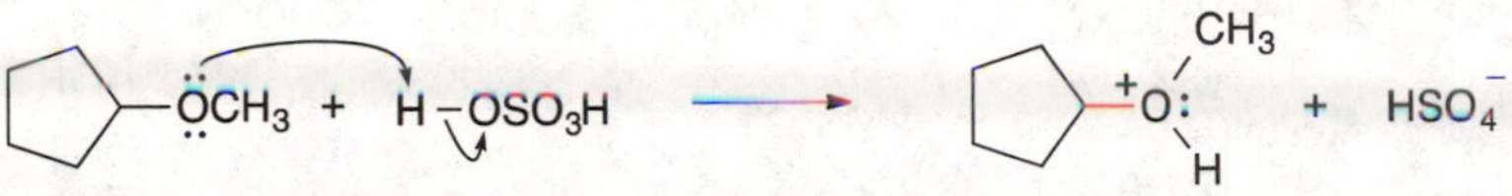 ÖCH$_3$ + H–OSO$_3$H ⟶ $\overset{+}{O}$(CH$_3$)(H) + HSO_4^-

d. 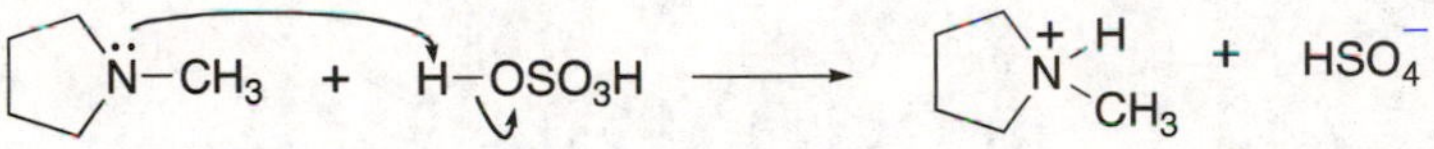 N̈–CH$_3$ + H–OSO$_3$H ⟶ $\overset{+}{N}$(H)(CH$_3$) + HSO_4^-

2.41 To draw the products of an acid–base reaction, transfer a proton from the acid to the base (^-OH in this case).

a.  Ö–H + K^+ $^-$:ÖH ⟶ Ö:$^-$ K^+ + H_2O

b. 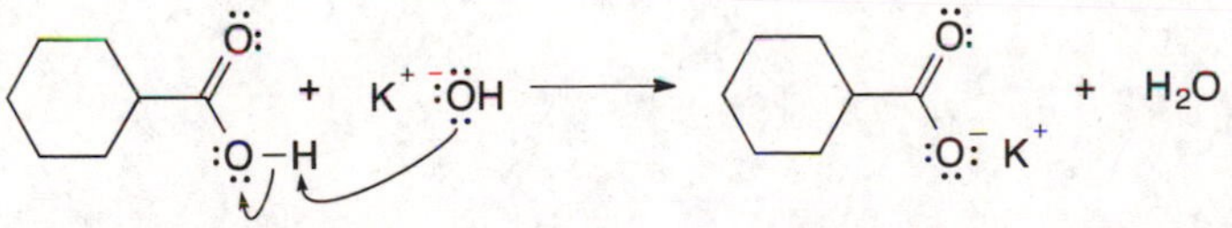 Ö: / :O–H + K^+ $^-$:ÖH ⟶ Ö: / :Ö:$^-$ K^+ + H_2O

c.  C≡C–H + K^+ $^-$:ÖH ⟶ C≡C:$^-$ K^+ + H_2O

d.  CH_3– –Ö–H + K^+ $^-$:ÖH ⟶ CH_3– –Ö:$^-$ K^+ + H_2O

2.42 Label the Brønsted–Lowry acid and Brønsted–Lowry base in the starting materials and **transfer a proton from the acid to the base** for the products.

a. 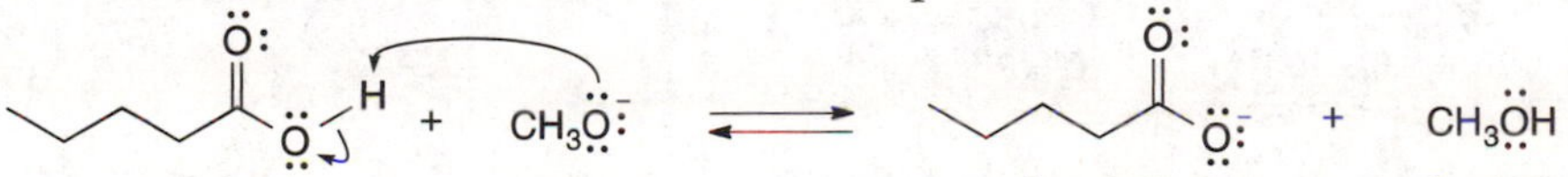 Ö: / Ö–H + $CH_3\ddot{O}$:$^-$ ⇌ Ö: / Ö:$^-$ + $CH_3\ddot{O}H$

acid **base** **conjugate base** **conjugate acid**

b. 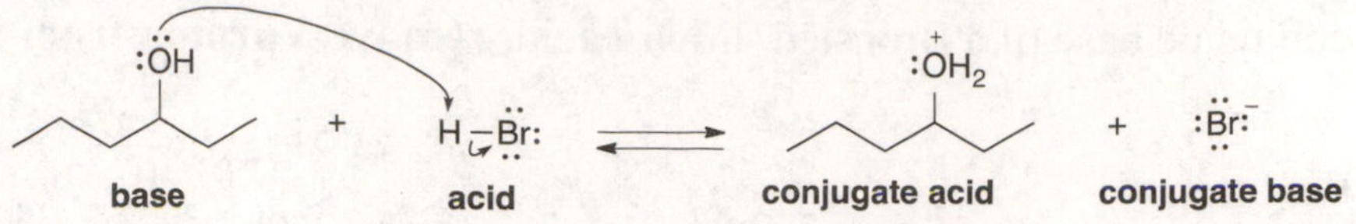 :ÖH + H–Br: ⇌ :ÖH$_2^+$ + :Br:$^-$

base acid conjugate acid conjugate base

c. 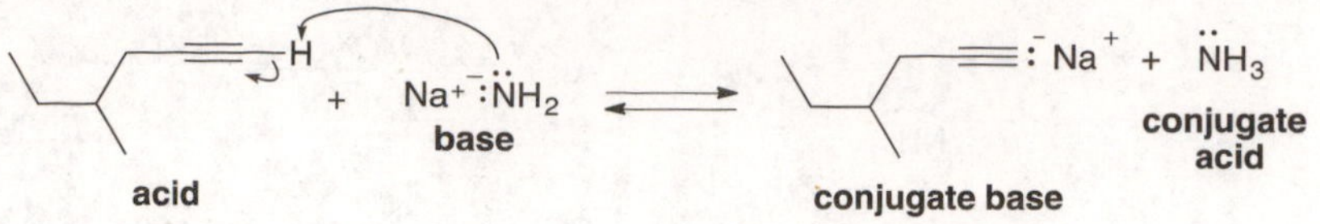 ≡–H + Na+ :NH$_2$ ⇌ ≡:$^-$ Na$^+$ + NH$_3$

acid base conjugate base conjugate acid

d. 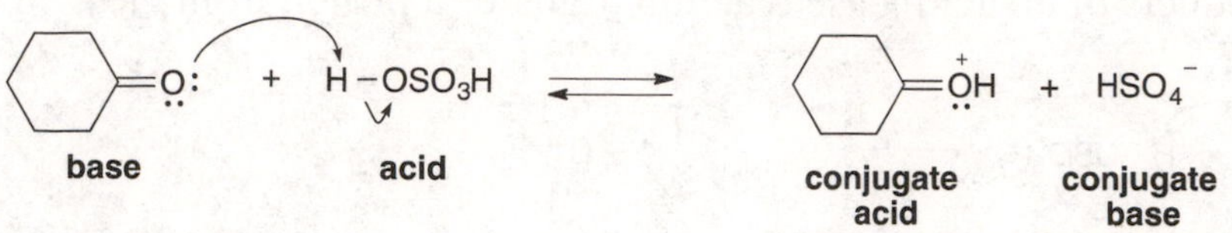 =O: + H–OSO$_3$H ⇌ =OH$^+$ + HSO$_4^-$

base acid conjugate acid conjugate base

2.43 Draw the products of proton transfer from acid to base.

a. 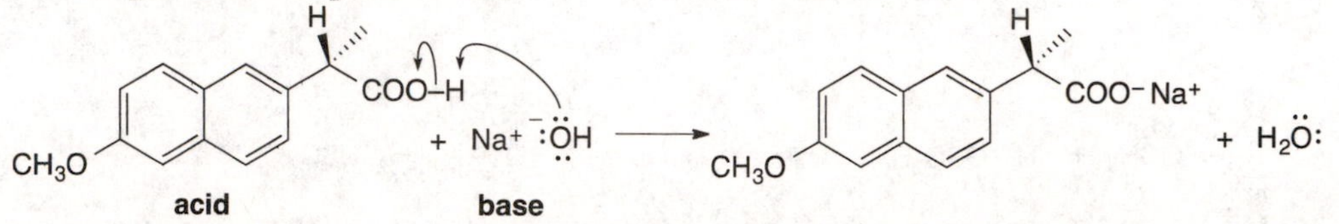 CH$_3$O, H, COO–H + Na+ :ÖH$^-$ ⟶ CH$_3$O, H, COO$^-$ Na+ + H$_2$Ö:

acid base

b. 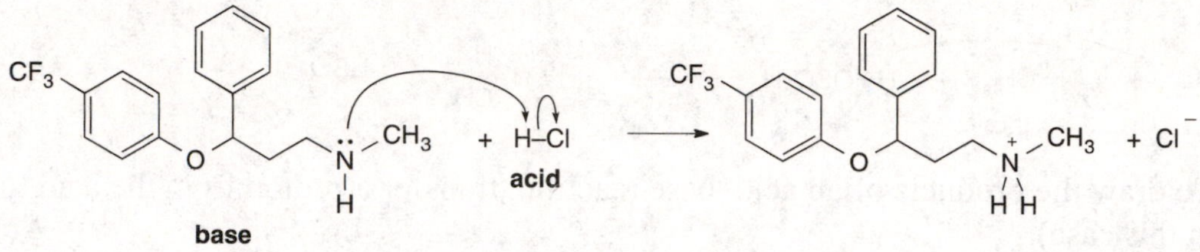 CF$_3$, O, N, CH$_3$, H + H–Cl ⟶ CF$_3$, O, N$^+$, CH$_3$, H H + Cl$^-$

base acid

2.44 Draw the products of proton transfer from acid to base.

CF$_3$, N, H + H, O, O ⟶ CF$_3$, N$^+$, H H + O, :Ö:$^-$

base acid

NH$_2$ + H, O, O ⟶ NH$_3^+$ + O, :Ö:$^-$

base acid

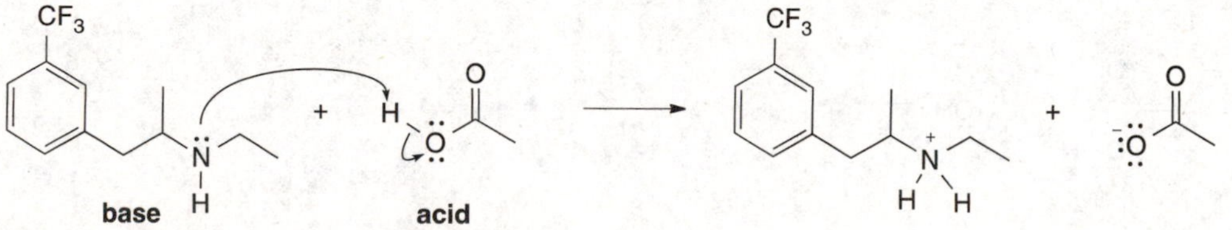

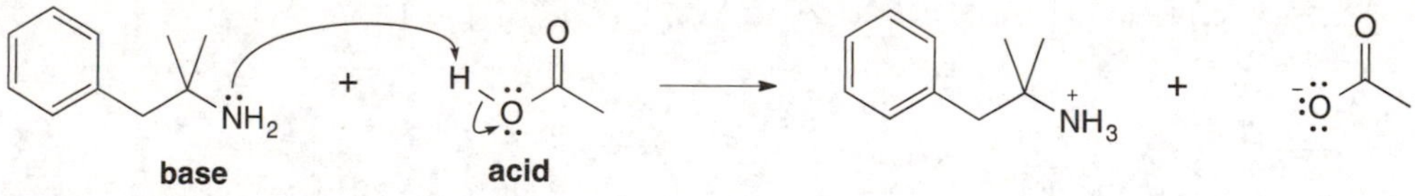

2.45 To convert pK_a to K_a, take the antilog of (–) the pK_a.

a. H_2S
pK_a = 7.0
$K_a = 10^{-7}$

b. $ClCH_2COOH$
pK_a = 2.8
$K_a = 1.6 \times 10^{-3}$

c. HCN
pK_a = 9.1
$K_a = 7.9 \times 10^{-10}$

2.46 To convert from K_a to pK_a, take (–) the log of the K_a; **pK_a = –log K_a.**

a. $C_6H_5CH_2\overset{+}{N}H_3$ — $K_a = 4.7 \times 10^{-10}$, p$K_a$ = 9.3

b. $C_6H_5\overset{+}{N}H_3$ — $K_a = 2.3 \times 10^{-5}$, pK_a = 4.6

c. CF_3COOH — $K_a = 5.9 \times 10^{-1}$, pK_a = 0.23

2.47 An acid can be deprotonated by the conjugate base of any acid with a higher pK_a.

$CH_3CH_2CH_2C{\equiv}CH$
pK_a = 25
Any base having a conjugate acid with a pK_a higher than 25 can deprotonate this acid.

Base	Conjugate acid	pK_a
H_2O	H_3O^+	–1.7
NaOH	H_2O	15.7
$NaNH_2$	NH_3	38
NH_3	NH_4^+	9.4
NaH	H_2	35
CH_3Li	CH_4	50

Only $NaNH_2$, NaH, and CH_3Li are strong enough to deprotonate the acid.

2.48 ^-OH can deprotonate any acid with a pK_a < 15.7.

a. HCOOH — pK_a = 3.8, stronger acid, deprotonated

b. H_2S — pK_a = 7.0, stronger acid, deprotonated

c. $C_6H_5CH_3$ — pK_a = 41, weaker acid

d. CH_3NH_2 — pK_a = 40, weaker acid

These acids are too weak to be deprotonated by ^-OH.

2.49 Draw the products and then compare the pK_a of the acid on the left and the conjugate acid on the right. **The equilibrium lies towards the side having the acid with a higher pK_a (weaker acid).**

a. $CH_3NH_2 + H{-}OSO_3H \rightleftharpoons CH_3\overset{+}{N}H_3 + HSO_4^-$ **products favored**
pK_a = –9 (left); pK_a = 10.7 (right)

b. CH_3CH_2COOH + $Na^+ \ :Cl:^-$ $\rightleftharpoons$ $CH_3CH_2COO^- \ Na^+$ + HCl: **starting material favored**
pK_a = ~5 (left); pK_a = –7 (right)

c. C_6H_5OH + $Na^+ \ HCO_3^-$ $\rightleftharpoons$ $C_6H_5O^- \ Na^+$ + H_2CO_3 **starting material favored**
pK_a = 10 (left); pK_a = 6.4 (right)

d. $H{-}C{\equiv}C{-}H$ + $CH_3CH_2^- \ Li^+$ $\rightleftharpoons$ $H{-}C{\equiv}C:^- \ Li^+$ + CH_3CH_3 **products favored**
pK_a = 25 (left); pK_a = 50 (right)

2.50 Compare element effects first and then resonance, hybridization, and electron-withdrawing groups to determine the relative strengths of the acids.

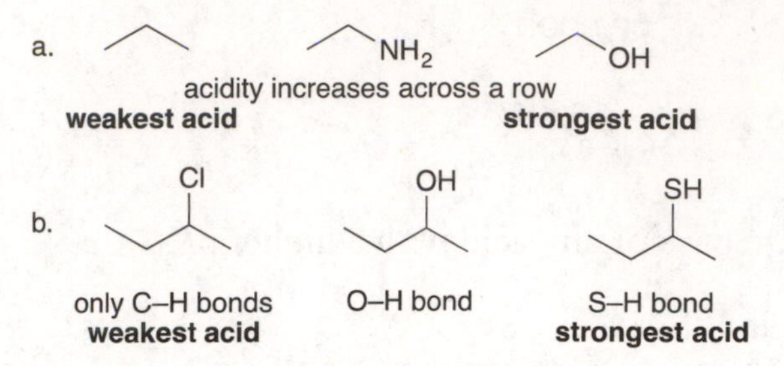

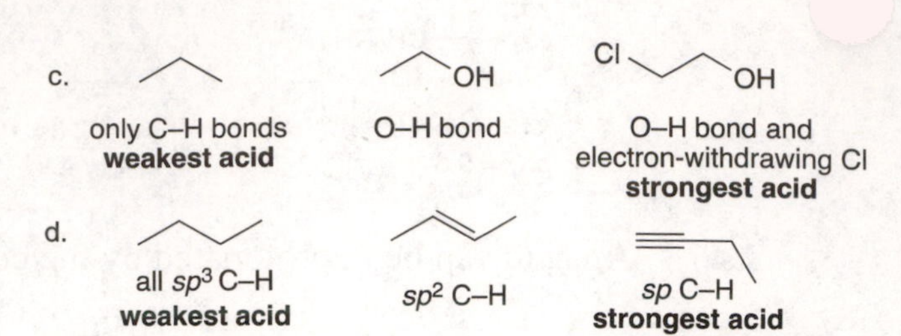

2.51 The strongest acid has the weakest conjugate base.

a. Draw the conjugate acid.
Increasing acidity of conjugate acids:
$CH_3CH_3 < CH_3NH_2 < CH_3OH$

increasing basicity: $CH_3O^- < CH_3NH^- < CH_3CH_2^-$

b. Draw the conjugate acid.
Increasing acidity of conjugate acids:
$CH_4 < H_2O < HBr$

increasing basicity: $Br^- < HO^- < CH_3^-$

c. Draw the conjugate acid.
Increasing acidity of conjugate acids:

OH < OH < Cl OH

increasing basicity:

Cl O⁻ < O⁻ < O⁻

d. Draw the conjugate acid.
Increasing acidity of conjugate acids:

< <

increasing basicity:

–C≡C: < <

2.52 **More electronegative atoms stabilize the conjugate base by an electron-withdrawing inductive effect, making the acid stronger.** Thus, an O atom increases the acidity of an acid.

$\overset{+}{N}H_2$

pK_a = 11.1

O $\overset{+}{N}H_2$

The O atom makes this cation the stronger acid.
pK_a = 8.33

2.53

H H_b O

H H_a

pentan-2-one

:B

resonance-stabilized conjugate base

H_a is more acidic than H_b because loss of H_a forms a resonance-stabilized conjugate base.

B: H H_b

only one Lewis structure

2.54

H
H H
pK_a = 50

H
H H
pK_a = 43

O
H
H
H
pK_a = 19.2

The negative charge on O is good. This makes this resonance structure especially good.

conjugate base:

one Lewis structure
weakest acid

two resonance structures
negative charge delocalized
on two carbons

two resonance structures
negative charge delocalized
on one O and one C
strongest acid

2.55 To draw the conjugate acid, look for the most basic site and protonate it. To draw the conjugate base, look for the most acidic site and remove a proton.

$\overset{+}{N}H_3$ — OH ← NH_2 — OH → NH_2 — O^-

conjugate acid | most basic site | **A** | most acidic proton | conjugate base

2.56 Estimate the pK_a of **B** as 16. A difference of 10^5 in acidity is a difference of 5 pK_a units.

a. HO — most acidic H, pK_a ~ 5

b. most acidic, pK_a ~ 25

c. OH — most acidic, pK_a ~ 16

2.57 Remove the most acidic proton to form the conjugate base. Protonate the most basic electron pair to form the conjugate acid.

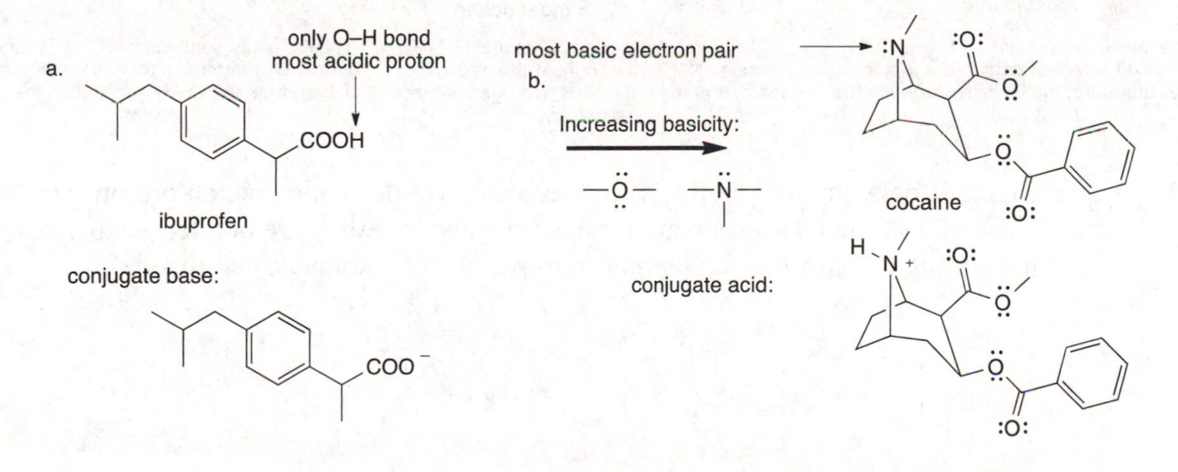

2.58 Compare the isomers.

dimethyl ether CH_3-O-CH_3

All H's are on C.

CH_3CH_2OH ethanol

One O–H bond
O–H bonds are more acidic than C–H bonds.
more acidic

2.59 Compare the Lewis structures of the conjugate bases when each H is removed. The more stable base makes the proton more acidic.

more acidic proton

The negative charge on N is stabilized by resonance. This conjugate base is more stable, so it is formed by removal of the more acidic H.

less acidic proton

no resonance stabilization formed from the weaker acid

2.60 Look at the element bonded to the acidic H and decide its acidity based on the periodic trends. **Farther to the right across a row and down a column of the periodic table is more acidic.**

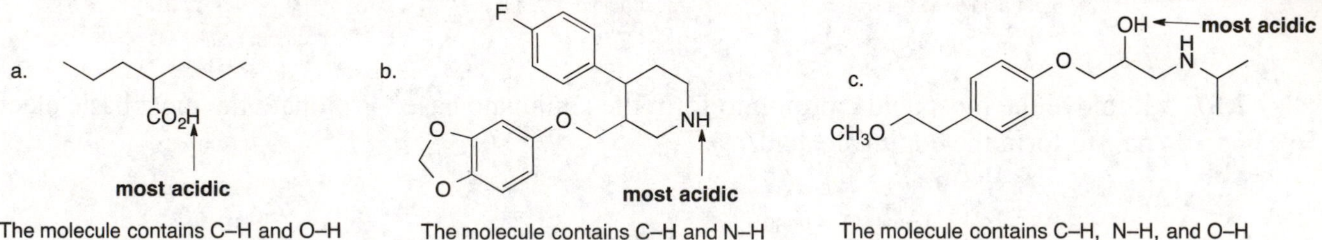

a. The molecule contains C–H and O–H bonds. O is farther right in the periodic table; therefore, the O–H hydrogen is the most acidic.

b. The molecule contains C–H and N–H bonds. N is farther right in the periodic table; therefore, the N–H hydrogen is the most acidic.

c. The molecule contains C–H, N–H, and O–H bonds. O is farthest right in the periodic table; therefore, the O–H hydrogen is the most acidic.

2.61 Use element effects, inductive effects, and resonance to determine which protons are the most acidic. The H's of the CH_3 group are least acidic, because they are bonded to an sp^3 hybridized C and the conjugate base formed by their removal is not resonance stabilized.

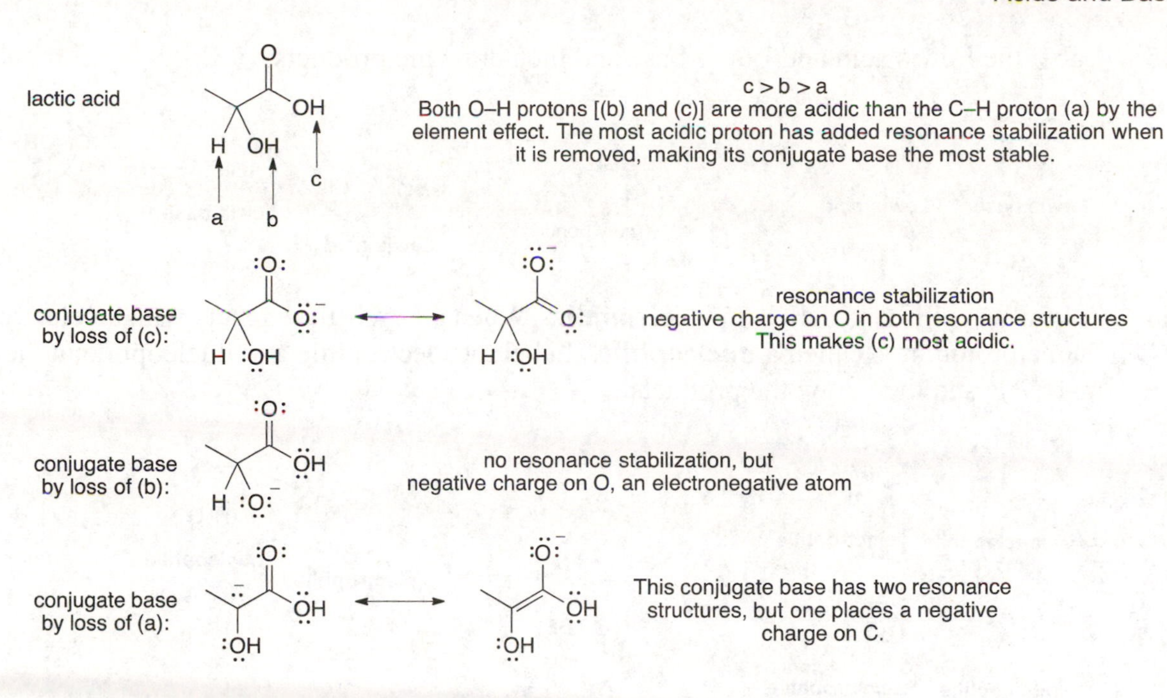

lactic acid
c > b > a
Both O–H protons [(b) and (c)] are more acidic than the C–H proton (a) by the element effect. The most acidic proton has added resonance stabilization when it is removed, making its conjugate base the most stable.
conjugate base by loss of (c):
resonance stabilization
negative charge on O in both resonance structures
This makes (c) most acidic.
conjugate base by loss of (b):
no resonance stabilization, but
negative charge on O, an electronegative atom
conjugate base by loss of (a):
This conjugate base has two resonance structures, but one places a negative charge on C.

2.62

This lone pair is localized on N, so it is more basic.

bupivacaine

This lone pair is delocalized by resonance, so it is less available to donate to an acid, making it less basic.

2.63 ***Lewis bases* are electron pair donors:** they contain a lone pair or a π bond. ***Brønsted–Lowry bases* are proton acceptors:** to accept a proton they need a lone pair or a π bond. This means Lewis bases are also Brønsted–Lowry bases.

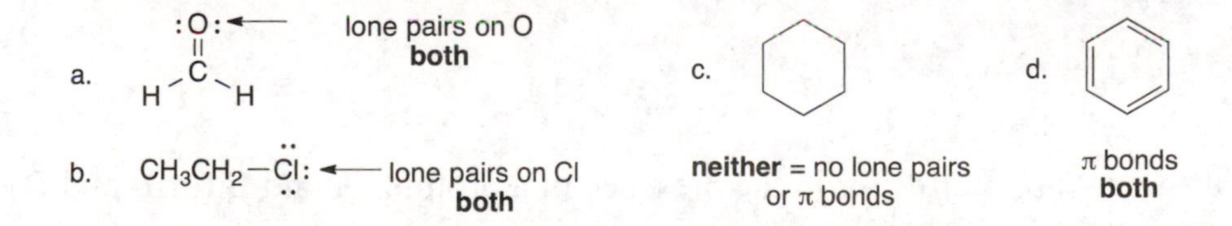

2.64 **A *Lewis acid* is an electron pair acceptor** and usually contains a proton or an unfilled valence shell of electrons. A ***Brønsted–Lowry acid* is a proton donor** and must contain a hydrogen atom. All Brønsted–Lowry acids are Lewis acids, though the reverse may not be true.

a. H_3O^+	b. Cl_3C^+	c. BCl_3	d. BF_4^-
both contains a H	**Lewis acid** unfilled valence shell on C	**Lewis acid** unfilled valence shell on B	**neither** no H or unfilled valence shell

2.65 Label the Lewis acid and Lewis base and then draw the products.

a. :Cl: + BCl_3 ⟶ Cl—B—Cl (with Cl above and below)
Lewis base Lewis acid new bond

b. + :OH ⟶ new bond
Lewis acid Lewis base

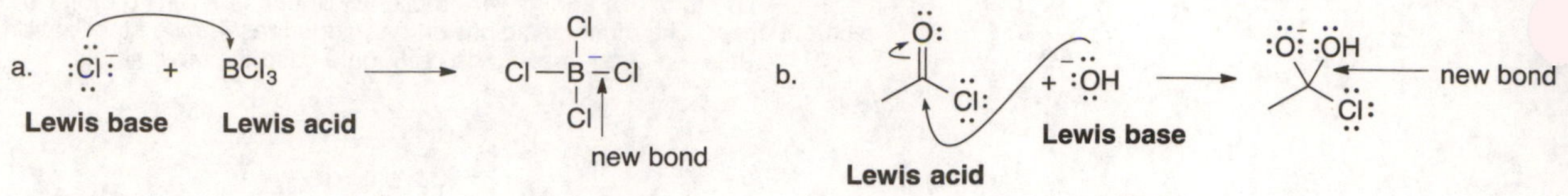

2.66 A Lewis acid is also called an **electrophile.** When a Lewis base reacts with an electrophile other than a proton, it is called a **nucleophile.** Label the electrophile and nucleophile in the starting materials and then draw the products.

a. + $AlCl_3$ ⟶ $^{-}AlCl_3$
nucleophile electrophile

b. + BF_3 ⟶ $O-BF_3$
nucleophile electrophile

c. + H_2O ⟶ $\overset{+}{O}H_2$
electrophile nucleophile

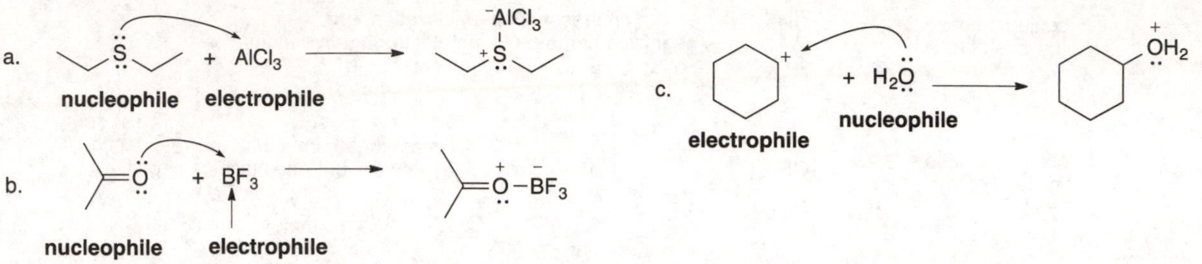

2.67 Draw the product of each reaction.

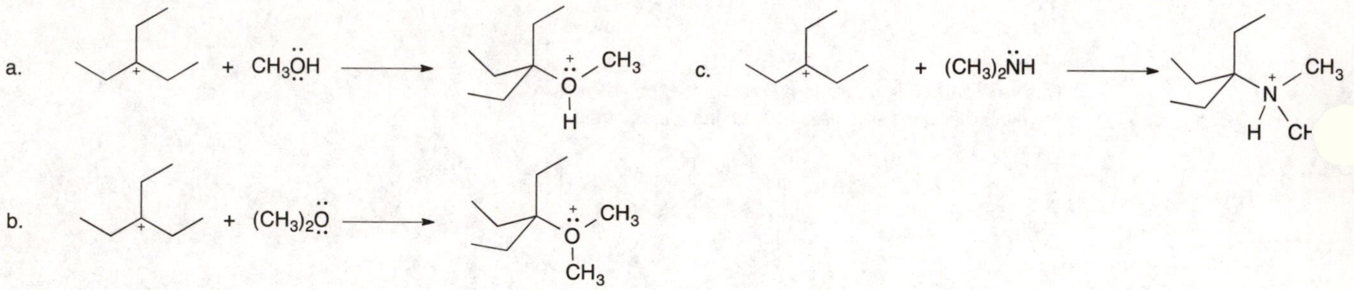

a. + CH_3OH ⟶ CH_3, O, H

b. + $(CH_3)_2O$ ⟶ CH_3, O, CH_3

c. + $(CH_3)_2NH$ ⟶ CH_3, N, H, CH

2.68

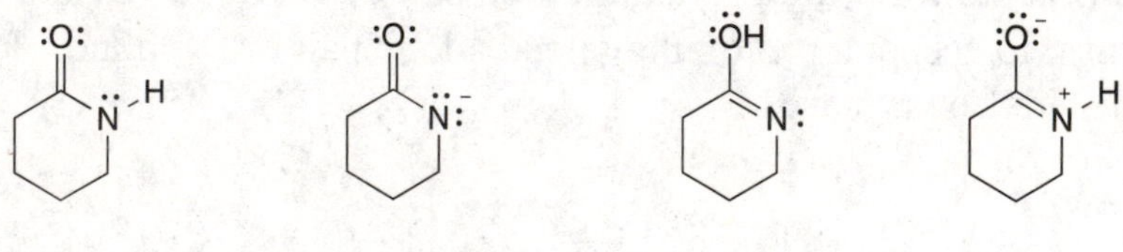

A B C D

a. A conjugate acid–base pair differ in the presence of a proton. **A** and **B** (or **B** and **D**) represent a conjugate acid–base pair.

b. **A** and **D** are resonance structures because the atom placement is the same, but the electron pairs are placed differently.

c. **A** and **C** (or **C** and **D**) are constitutional isomers because they have the same molecular formula (C_5H_9NO), but the bonds are different. **A** and **D** have an N–H bond, and **C** has an O–H bond.

2.69

a. :ÖH (isopropanol) + H–Br —proton transfer→ $\overset{+}{O}H_2$ + :Br:⁻ (nucleophile, electrophile) → 2-bromopropane + H_2O:

b. benzene (nucleophile) + Br^+ (electrophile) → carbocation + :Br:⁻ —proton transfer→ bromobenzene + HBr

2.70 Draw the products of each reaction. In part (a), ⁻OH pulls off a proton and thus acts as a Brønsted–Lowry base. In part (b), ⁻OH attacks a carbon and thus acts as a Lewis base.

a. carbocation + ⁻:ÖH → H_2O: + alkene

b. ⁻:ÖH + carbocation → tertiary alcohol (ÖH)

2.71 Answer each question about esmolol.

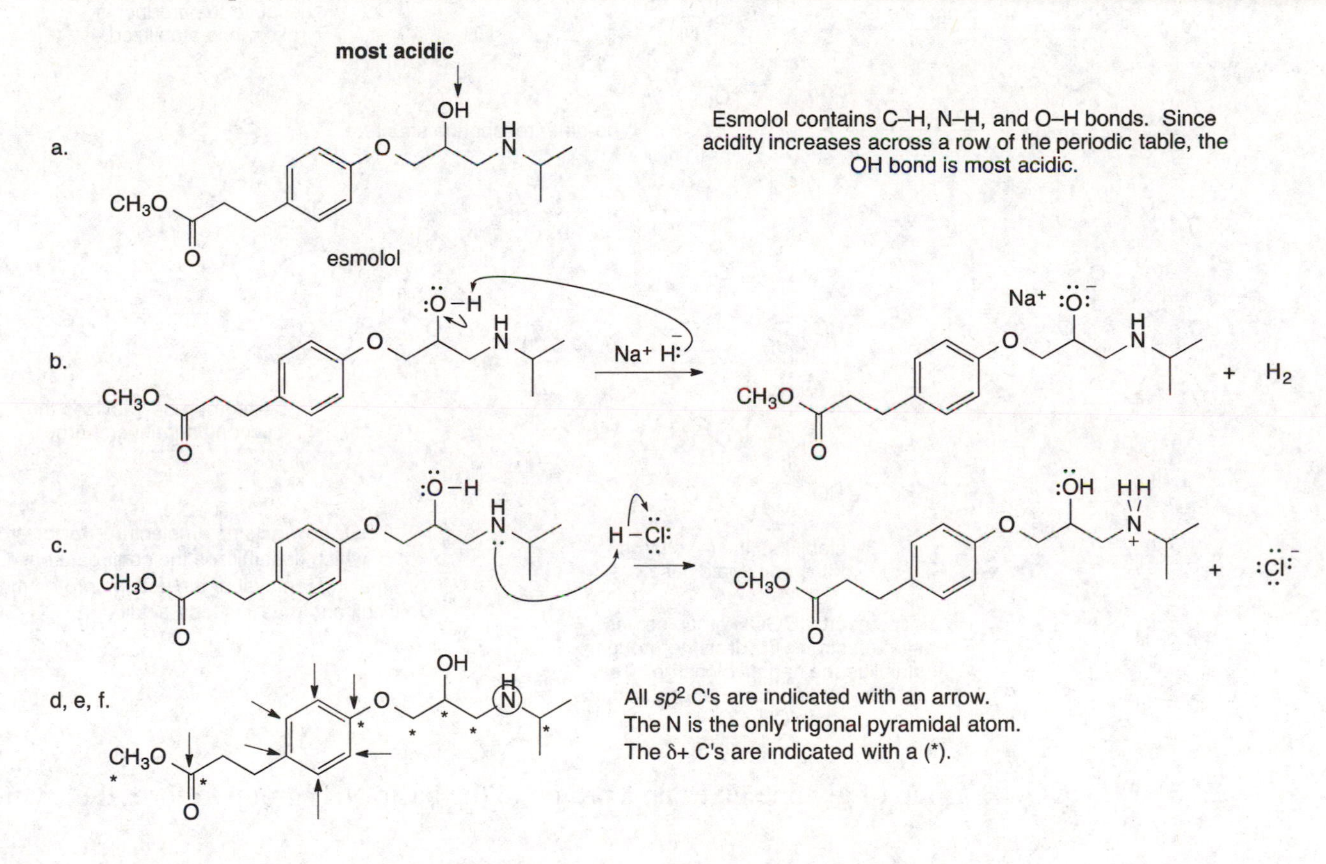
a.
most acidic
OH
H
N
O
CH3O
O
esmolol
Esmolol contains C–H, N–H, and O–H bonds. Since acidity increases across a row of the periodic table, the OH bond is most acidic.
b.
Na+ H:⁻
Na+ :O:⁻
+ H2
c.
H–Cl:
:ÖH HH
+ :Cl:⁻
d, e, f.
All sp2 C's are indicated with an arrow.
The N is the only trigonal pyramidal atom.
The δ+ C's are indicated with a (*).

2.72

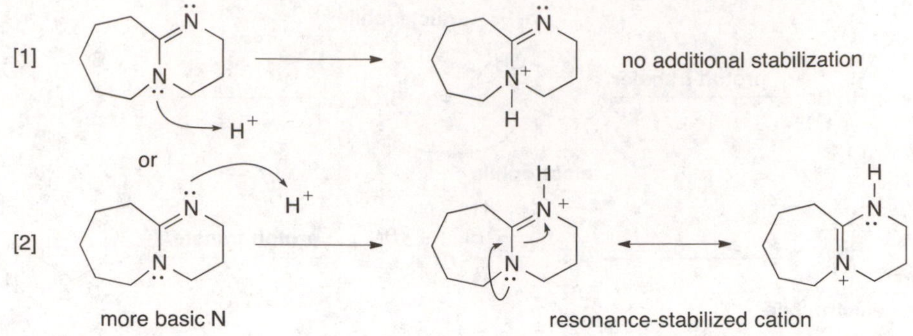

[1] no additional stabilization

or

[2] more basic N resonance-stabilized cation

Path [2] is favored because a resonance-stabilized conjugate acid is formed. The N that is part of the C=N is therefore more basic.

2.73 Draw the product of protonation of either O or N and compare the conjugate acids. When acetamide reacts with an acid, the O atom is protonated because it results in a resonance-stabilized conjugate acid.

acetamide

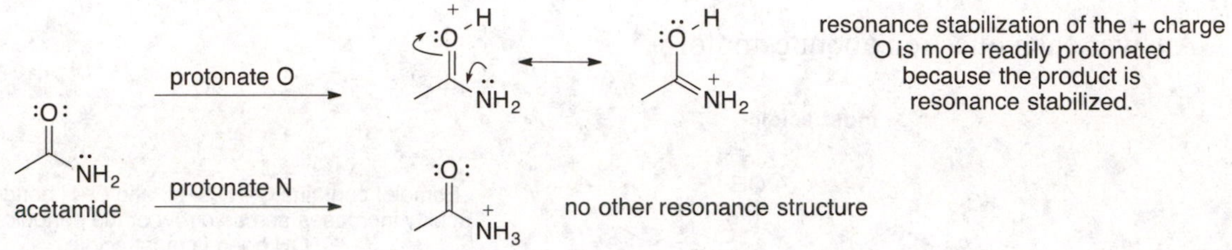

protonate O

resonance stabilization of the + charge
O is more readily protonated because the product is resonance stabilized.

protonate N

no other resonance structure

2.74

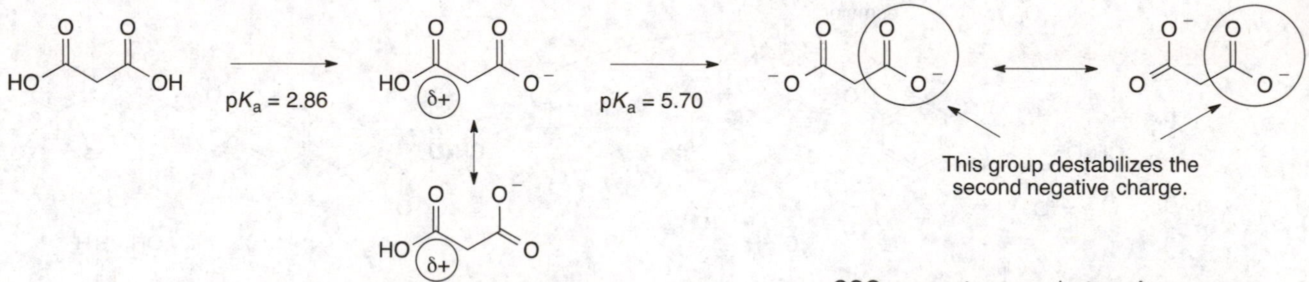

pK_a = 2.86

pK_a = 5.70

δ+ stabilizes the (–) charge of the conjugate base.

The nearby COOH group serves as an electron-**withdrawing** group to stabilize the negative charge. This makes the first proton **more** acidic than CH_3COOH.

This group destabilizes the second negative charge.

COO^- now acts as an electron-**donor** group which destabilizes the conjugate base, making removal of the second proton more difficult and thus it is **less** acidic than CH_3COOH.

2.75 The COOH group of glycine gives up a proton to the basic NH_2 group to form the zwitterion.

a. 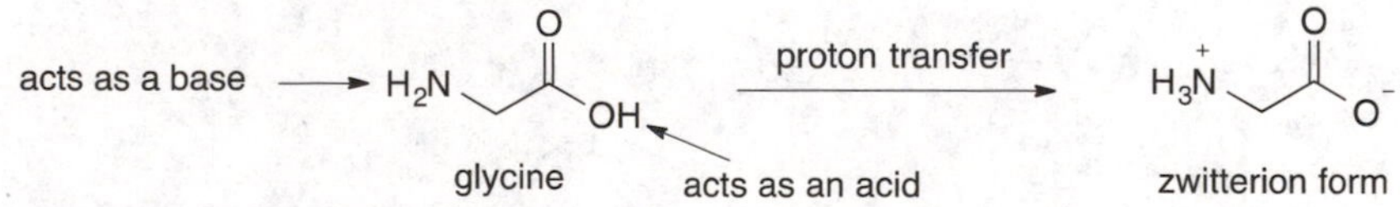 acts as a base → glycine ← acts as an acid

proton transfer

zwitterion form

b. $H_3\overset{+}{N}CH_2CO_2^-$ + H–Cl → $H_3\overset{+}{N}CH_2CO_2H$ + Cl^-

most basic site

c. $H_3\overset{+}{N}CH_2CO_2^-$ + Na^+ $^-$:ÖH → $H_2NCH_2CO_2^-$ + Na^+ + $H_2\ddot{O}$:

most acidic site

2.76 Use curved arrows to show how the reaction occurs.

[1]

[2]

Protonate the negative charge on this carbon to form the product.

2.77 Compare the OH bonds in vitamin C and decide which one is the most acidic.

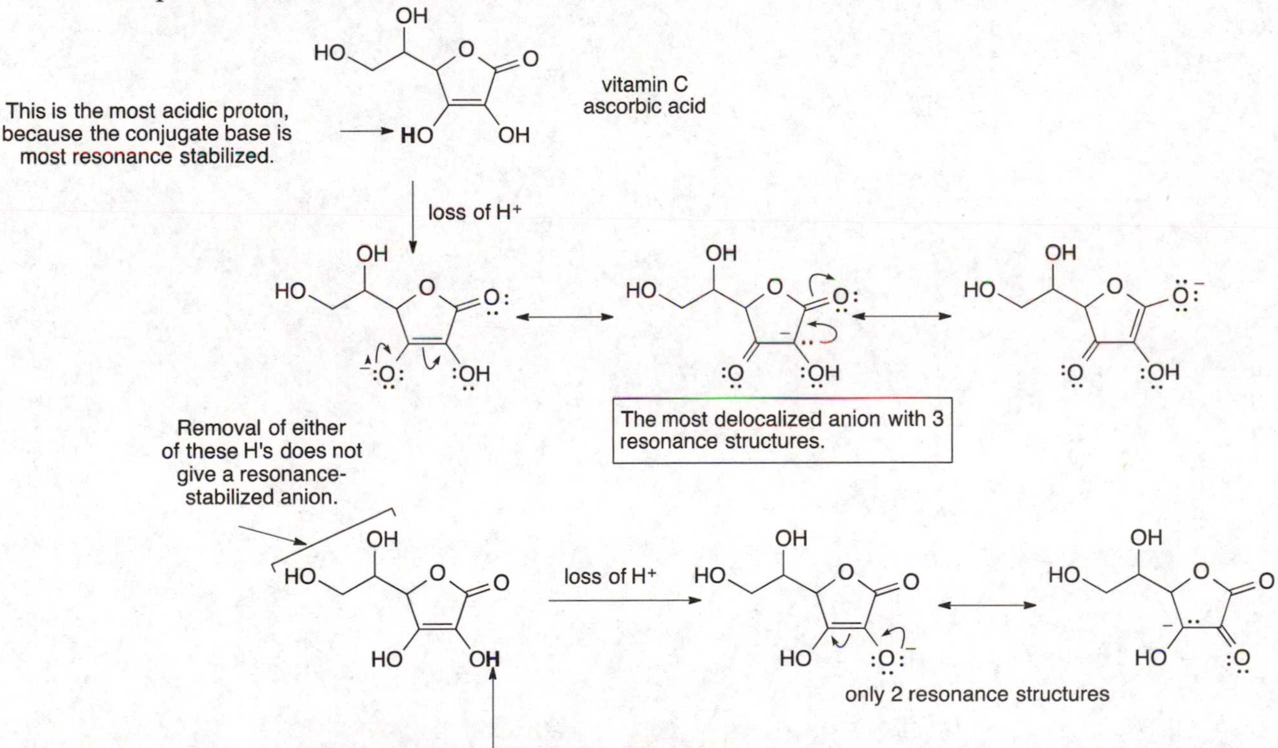
vitamin C
ascorbic acid
This is the most acidic proton, because the conjugate base is most resonance stabilized.
loss of H+
The most delocalized anion with 3 resonance structures.
Removal of either of these H's does not give a resonance-stabilized anion.
loss of H+
only 2 resonance structures
This proton is less acidic, because its conjugate base is less resonance stabilized.

2.78

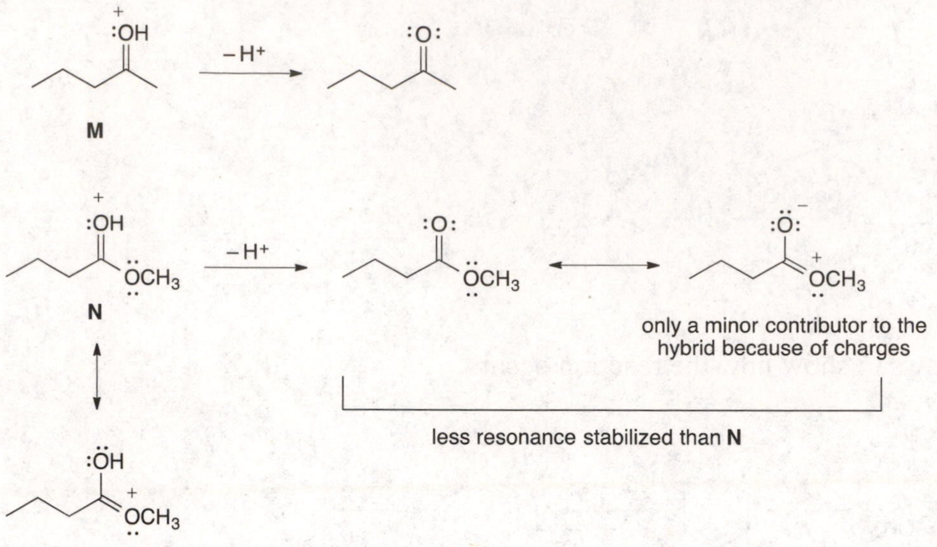

M $\xrightarrow{-H^+}$

N $\xrightarrow{-H^+}$

only a minor contributor to the hybrid because of charges

less resonance stabilized than **N**

N has two resonance structures with the same number of bonds and charges, so both contribute approximately equally to the hybrid. This makes **N** more resonance stabilized than its conjugate base, and less willing to give up a proton than **M,** which has no similar resonance stabilization. Thus **M** is a stronger acid than **N.** (Resonance structures that break the C=O bond are not drawn in this solution, because they are possible for both compounds.)

Chapter 3: Introduction to Organic Molecules and Functional Groups

Chapter Review

Classifying carbon atoms, hydrogen atoms, alcohols, alkyl halides, amines, and amides (3.2)

- Carbon atoms are classified by the number of carbons bonded to them; a 1° carbon is bonded to one other carbon, and so forth.
- Hydrogen atoms are classified by the type of carbon atom to which they are bonded; a 1° hydrogen is bonded to a 1° carbon, and so forth.
- Alkyl halides and alcohols are classified by the type of carbon to which the OH or X group is bonded; a 1° alcohol has an OH group bonded to a 1° carbon, and so forth.
- Amines and amides are classified by the number of carbons bonded to the nitrogen atom; a 1° amine has one carbon–nitrogen bond, and so forth.

Types of intermolecular forces (3.3)

(Increasing strength ↓)

Type of force	Cause	Examples
van der Waals (VDW)	Due to the interaction of temporary dipoles • Larger surface area, stronger forces • Larger, more polarizable atoms, stronger forces	All organic compounds
dipole–dipole (DD)	Due to the interaction of permanent dipoles	$(CH_3)_2C{=}O$, H_2O
hydrogen bonding (HB or H-bonding)	Due to the electrostatic interaction of a H atom in an O–H, N–H, or H–F bond with another N, O, or F atom.	H_2O
ion–ion	Due to the interaction of two ions	NaCl, LiF

Physical properties

Property	Observation
Boiling point (3.4A)	• For compounds of comparable molecular weight, the stronger the forces the higher the bp.

VDW
MW = 72
bp = 36 °C

VDW, DD
MW = 72
bp = 76 °C

VDW, DD, HB
MW = 74
bp = 118 °C

Increasing strength of intermolecular forces
Increasing boiling point

- For compounds with similar functional groups, the larger the surface area, the higher the bp.

bp = 0 °C bp = 36 °C

Increasing surface area
Increasing boiling point

- For compounds with similar functional groups, the more polarizable the atoms, the higher the bp.

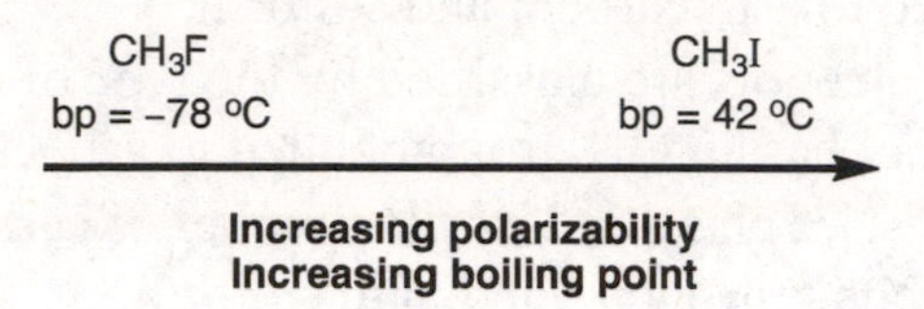

Melting point (3.4B)

- For compounds of comparable molecular weight, the stronger the forces the higher the mp.

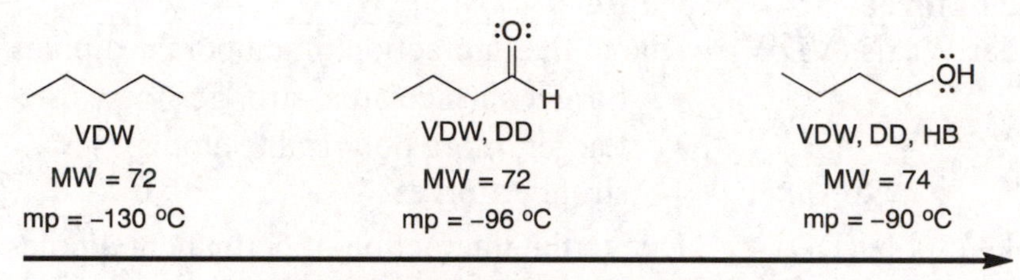

- For compounds with similar functional groups, the more symmetrical the compound, the higher the mp.

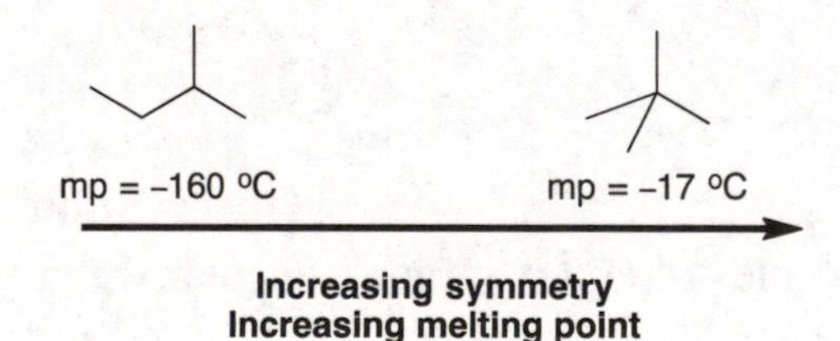

Solubility (3.4C)

Types of water-soluble compounds:
- Ionic compounds
- Organic compounds having ≤ 5 C's, and an O or N atom for hydrogen bonding (for a compound with one functional group)

Types of compounds soluble in organic solvents:
- Organic compounds regardless of size or functional group
- Examples:

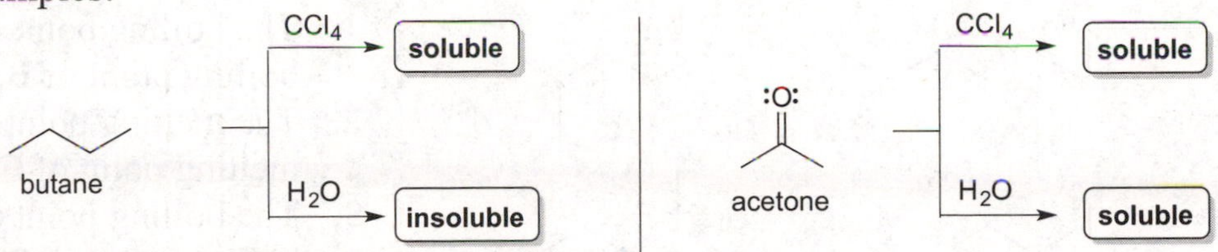

Key: VDW = van der Waals, DD = dipole–dipole, HB = hydrogen bonding
MW = molecular weight

Reactivity (3.8)

- **Nucleophiles react with electrophiles.**
- Electronegative heteroatoms create electrophilic carbon atoms that react with nucleophiles.
- Lone pairs and π bonds are nucleophilic sites that react with electrophiles.

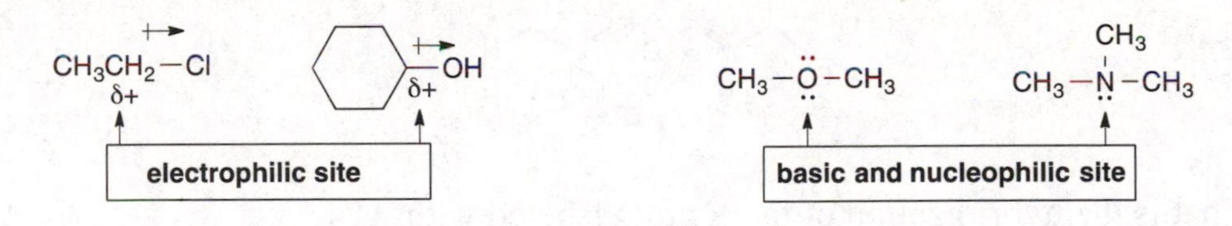

Practice Test on Chapter Review

1.a. Which of the following compounds exhibits dipole–dipole interactions?

1. CH_2Cl_2
2. $(CH_3)_2O$
3. $CH_3CH_2CH_2—OH$
4. Compounds (1) and (2) both exhibit dipole–dipole interactions.
5. Compounds (1), (2), and (3) all exhibit dipole–dipole interactions.

b. Which of the following compounds can hydrogen bond both to another molecule of itself and to water?

1. 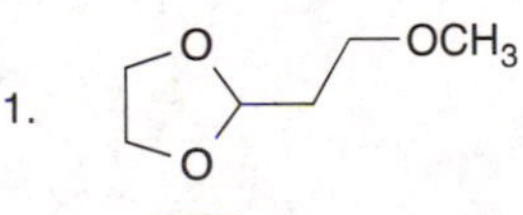

2. NH
3. $CH_3C{\equiv}N$
4. Compounds (1) and (2) can each hydrogen bond to another molecule like itself and water.
5. Each of the three compounds, (1), (2), and (3), can hydrogen bond to another molecule like itself and to water.

c. Which of the following compounds has the highest boiling point?

1. $CH_3CH_2CH_2OCH_2CH_3$
2. $CH_3(CH_2)_4CH_3$
3. $CH_3(CH_2)_4NH_2$
4. $(CH_3)_2CHCH(CH_3)NH_2$
5. $(CH_3)_2CHCH(CH_3)_2$

d. Which statement(s) is (are) true about the following compounds?

A C B D

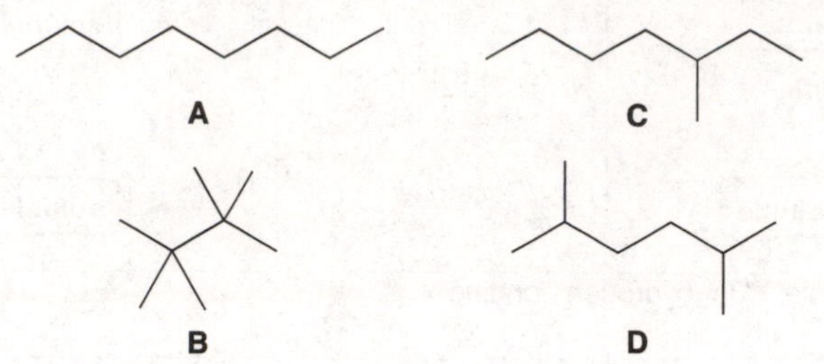

1. The boiling point of **A** is higher than the boiling point of **B.**
2. The melting point of **B** is higher than the melting point of **C.**
3. The boiling point of **A** is higher than the boiling point of **D.**
4. Statements (1) and (2) are both true.
5. Statements (1), (2), and (3) are all true.

2. Consider the anti-hypertensive agent atenolol drawn below.

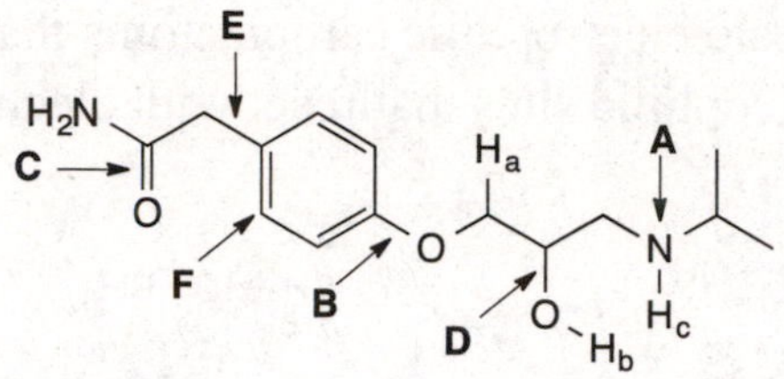

E H_2N C O H_a A O N F B D O H_c H_b

a. What is the hybridization of the N atom labeled with **A?**
b. What is the shape around the C atom labeled with **F?**
c. What orbitals are used to form the bond labeled with **E?**
d. Which bond (**B, C,** or **D**) is the longest bond?
e. Would you predict atenolol to be soluble in water?
f. Which of the labeled H atoms (H_a, H_b, or H_c) is most acidic?

Answers to Practice Test

1. a. 5
   b. 2
   c. 3
   d. 5

2. a. sp^3
   b. trigonal planar
   c. $C_{sp2}–C_{sp3}$
   d. **D**
   e. yes
   f. H_b

Answers to Problems

3.1

$CH_3CH_2-\ddot{O}H \xrightarrow{H-OSO_3H} CH_3CH_2-\overset{+}{O}H_2 + HSO_4^-$ $\qquad CH_3CH_3 \xrightarrow{H_2SO_4}$ no reaction

$CH_3CH_2-\ddot{O}-H \xrightarrow{Na^+\ H:^-} CH_3CH_2-\ddot{O}:^-\ Na^+ + H_2$ $\qquad CH_3CH_3 \xrightarrow{NaH}$ no reaction

3.2 To classify a carbon atom as 1°, 2°, 3°, or 4°, **determine how many carbon atoms it is bonded to (1° C = bonded to one other C, 2° C = bonded to two other C's, 3° C = bonded to three other C's, 4° C = bonded to four other C's).** Re-draw if necessary to see each carbon clearly.

To classify a hydrogen atom as 1°, 2°, or 3°, **determine if it is bonded to a 1°, 2°, or 3° C (a 1° H is bonded to a 1° C; a 2° H is bonded to a 2° C; a 3° H is bonded to a 3° C).** Re-draw if necessary.

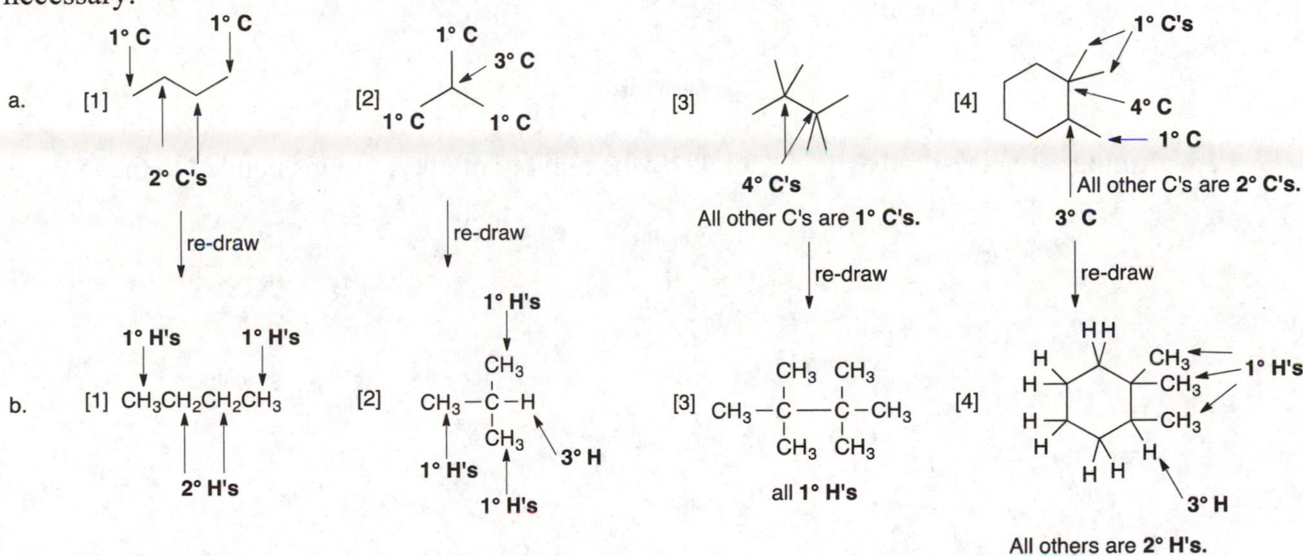

3.3 Use the definition of 1°, 2°, 3°, or 4° carbon atoms from Answer 3.2.

3.4 A 1° ROH or RX has the functional group bonded to a 1° C; a 2° ROH or RX has the functional group bonded to a 2° C; a 3° ROH or RX has the functional group bonded to a 3° C.

a. 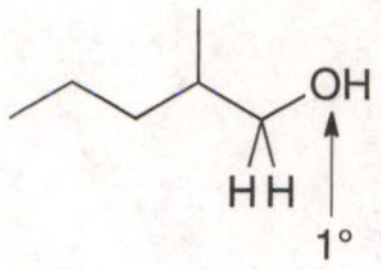 OH, HH, 1°   b. 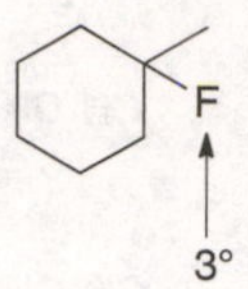 F, 3°   c. 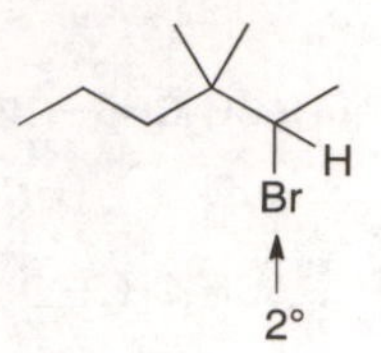 H, Br, 2°   d. 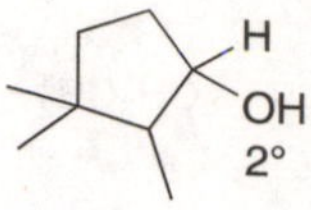 H, OH, 2°

3.5 Use the definitions from Answer 3.4.

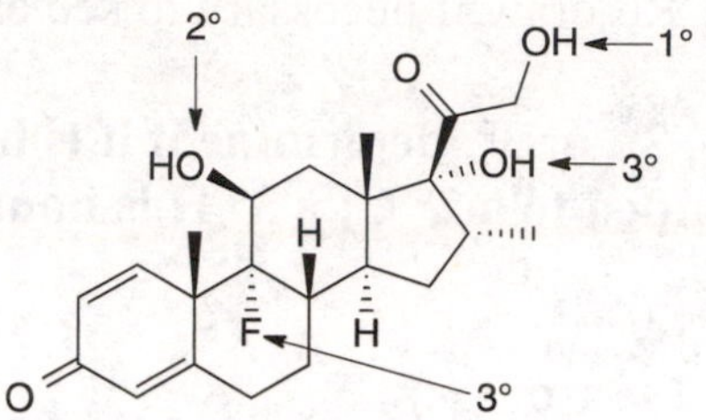

2°, HO, O, OH ← 1°, OH ← 3°, H, F, H, 3°

dexamethasone

3.6 Amines are classified as 1°, 2°, or 3° by the number of alkyl groups bonded to the *nitrogen* atom.

a. 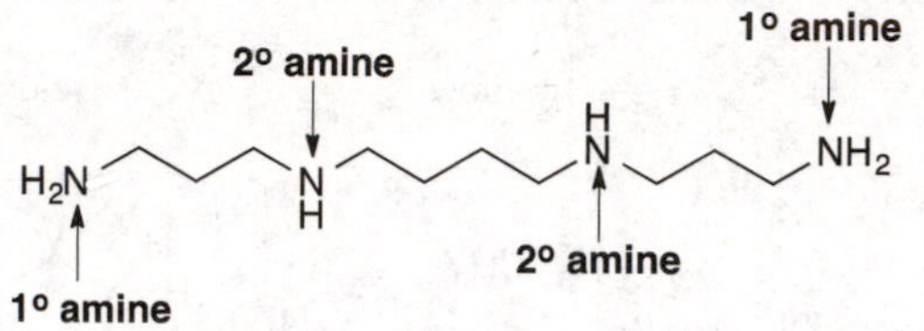 H_2N, 1° amine; N–H, 2° amine; N–H, 2° amine; NH_2, 1° amine   b. 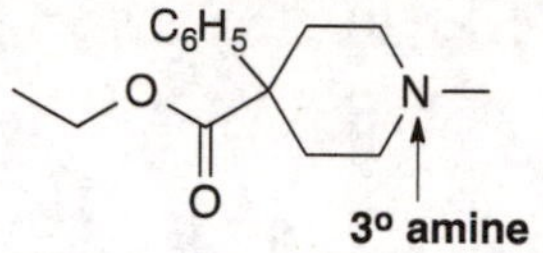 C_6H_5, O, O, N—, 3° amine

3.7

a. 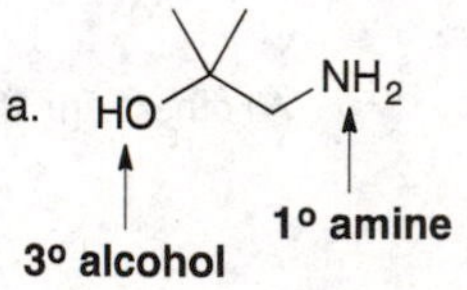 HO, 3° alcohol; NH_2, 1° amine   b. 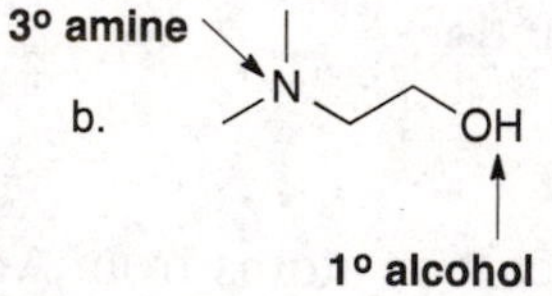 N, 3° amine; OH, 1° alcohol

3.8 A 1° amide has one carbon–nitrogen bond; a 2° amide has two carbon–nitrogen bonds; a 3° amide has three carbon–nitrogen bonds.

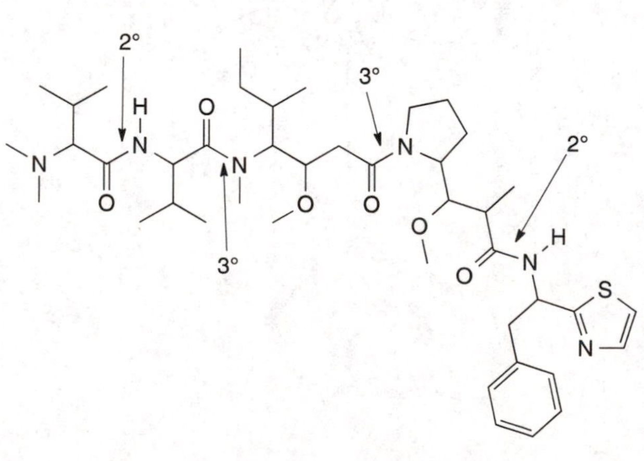

2°, H–N, O, N, 3°, 3°, N, O, O, O, O, 2°, H, N, O, S, N

3.9 Identify the functional groups based on Tables 3.1, 3.2, and 3.3.

alkene (double bond)
HO
CO_2H
HO
carboxylic acid
OH
shikimic acid
alcohols (hydroxy groups)

ether
O
alkene (double bond)
$CO_2CH_2CH_3$
O
N H
ester
NH_2
amide
oseltamivir
amine

3.10 One possible structure for each functional group:

a. aldehyde = O, C, R, H → O, H

c. carboxylic acid = O, C, R, OH → O, OH

b. ketone = O, C, R, R → O

d. ester = O, C, R, O, R → O, O

3.11 One possible structure for each description:

a. $C_5H_{10}O$

O, H
aldehyde

O
ketone

b. $C_6H_{10}O$

ketone → O
alkene

ketone → O
alkene

3.12 Summary of forces:

- **All compounds exhibit van der Waals forces (VDW).**
- **Polar molecules have dipole–dipole forces (DD).**
- **Hydrogen bonding (H-bonding)** can occur only when a **H is bonded to an O, N, or F.**

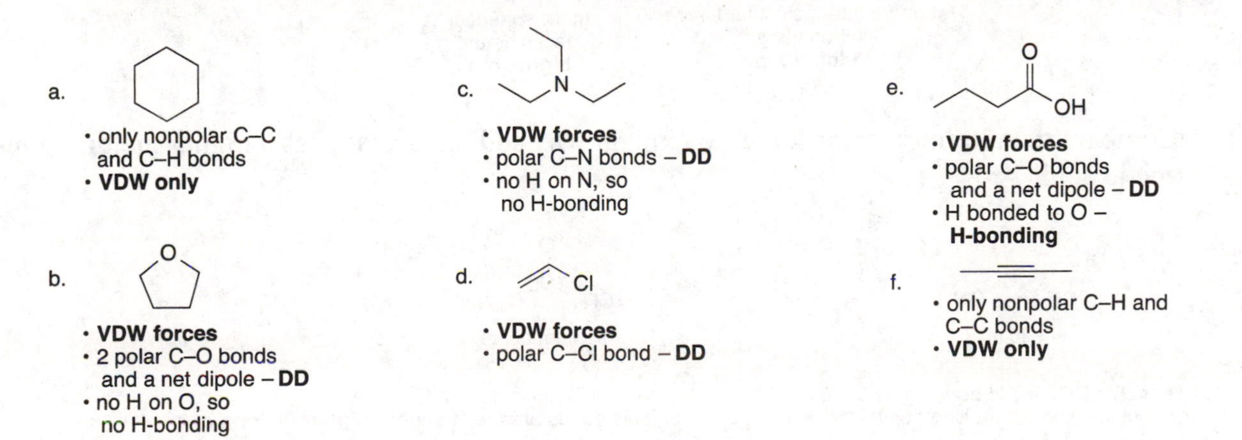

3.13 One principle governs boiling point:

- **Stronger intermolecular forces = higher bp.**
  Increasing intermolecular forces: van der Waals < dipole–dipole < hydrogen bonding

Two factors affect the strength of van der Waals forces, and thus affect bp:

- **Increasing surface area = increasing bp.**
  Longer molecules have a larger surface area. Any branching decreases the surface area of a molecule.
- **Increasing polarizability = increasing bp.**

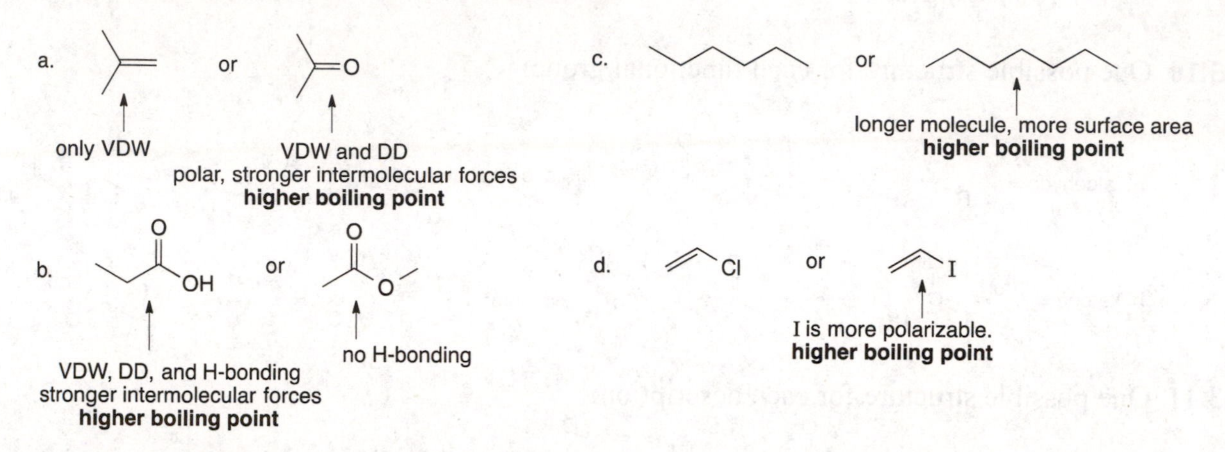

3.14 Increasing intermolecular forces: van der Waals < dipole–dipole < hydrogen bonding

3.15

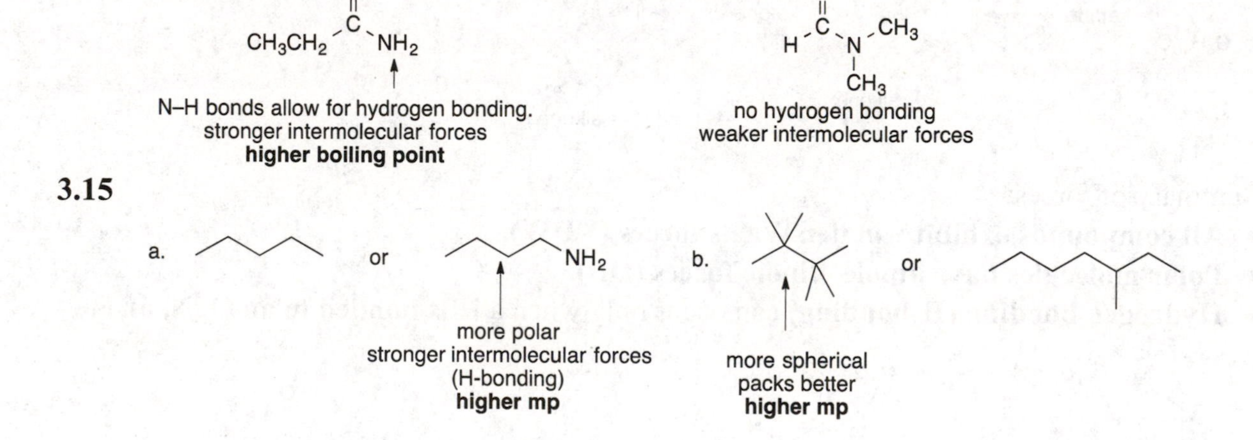

3.16 Compare the intermolecular forces to explain why sodium acetate has a higher melting point than acetic acid.

CH_3COOH — acetic acid

a. VDW, DD, and H-bonding
b. not ionic, lower melting point

$CH_3COO^{-}Na^{+}$ — sodium acetate

a. VDW, DD, ionic bonds
b. Ionic bonds are the strongest: **higher melting point.**

3.17 A compound is water soluble if it is ionic or if it has an O or N atom and ≤ 5 C's.

a. an O atom that can H-bond with water ≤ 5 C's **water soluble**

b. nonpolar **not water soluble**

c. an N atom that can H-bond to H_2O, but > 5 C's **not water soluble**

3.18 Hydrophobic portions will primarily be hydrocarbon chains. **Hydrophilic** portions will be polar. Circled regions are **hydrophilic** because they are polar.
All other regions are **hydrophobic** because they have only C and H.

can H-bond to itself

can H-bond to H_2O

can H-bond to H_2O

C norethindrone

D arachidonic acid

3.19 Like dissolves like.

- To be **soluble in water**, a molecule must be ionic, or have a polar functional group capable of H-bonding for every 5 C's.
- Organic compounds are generally **soluble in organic solvents** regardless of size or functional group.

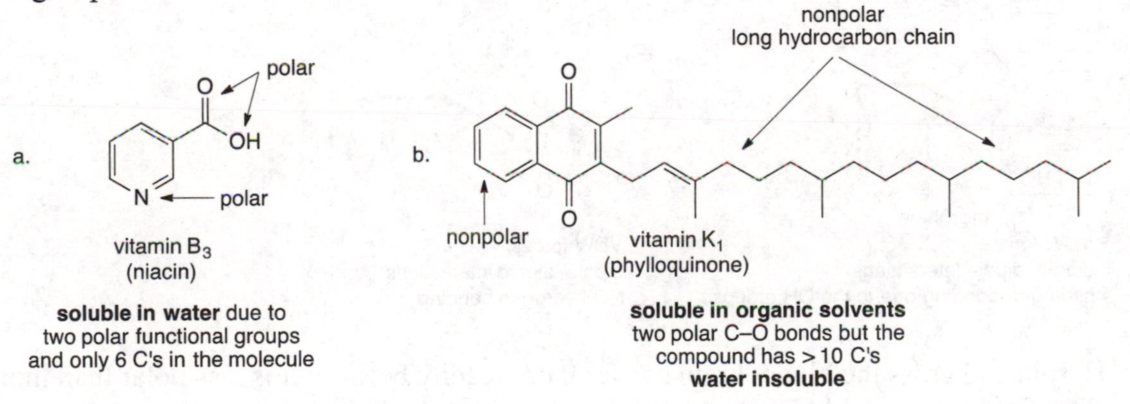

3.20 a.

alcohol → OH; HO; amide; carboxylic acid

b. The amide, carboxylic acid, and both alcohols can all hydrogen bond with water.

c. Since pantothenic acid has only nine carbons with four functional groups that can hydrogen bond, pantothenic acid is a water-soluble vitamin.

3.21 A soap contains both a long hydrocarbon chain and a carboxylic acid salt.

a. O⁻ Na+
short chain
carboxylic acid salt

c. OH
no salt

b. ionic salt
O⁻ Na+
long chain
This is a soap because it contains both a long chain and a carboxylic acid salt.

3.22 Detergents have a polar head consisting of oppositely charged ions, and a nonpolar tail consisting of C–C and C–H bonds, just like soaps do. Detergents clean by having the **hydrophobic ends of molecules surround grease**, while the **hydrophilic portion of the molecule interacts with the polar solvent** (usually water).

a detergent

SO_3^- Na+

nonpolar tail
hydrophobic
This end interacts with the grease to dissolve it.

polar head
ionic – hydrophilic
This end interacts with the water solvent to maintain the micelle's solubility in water.

3.23

a.

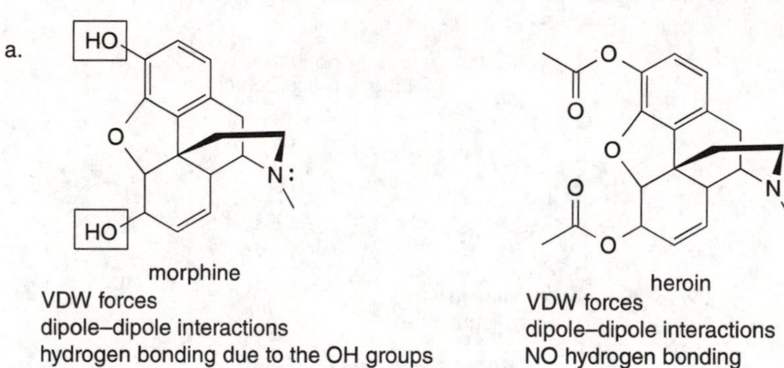

b. Heroin can cross the blood–brain barrier more readily because it is less polar than morphine, and therefore more soluble in the nonpolar interior of the cell membrane.

3.24 The noble gas xenon consists of uncharged atoms that exhibit only van der Waals interactions, so it is very soluble in the nonpolar interior of the cell membrane. It can thus cross the blood–brain barrier and act as an anesthetic.

3.25 Because the interior of a cell membrane is nonpolar, aspirin crosses a cell membrane as a neutral carboxylic acid, by the general rule that "like dissolves like."

3.26 Electronegative heteroatoms like N, O, or X make a carbon atom an *electrophile*.
A lone pair on a heteroatom makes it basic and nucleophilic.
Pi (π) bonds create *nucleophilic* sites and are more easily broken than σ bonds.

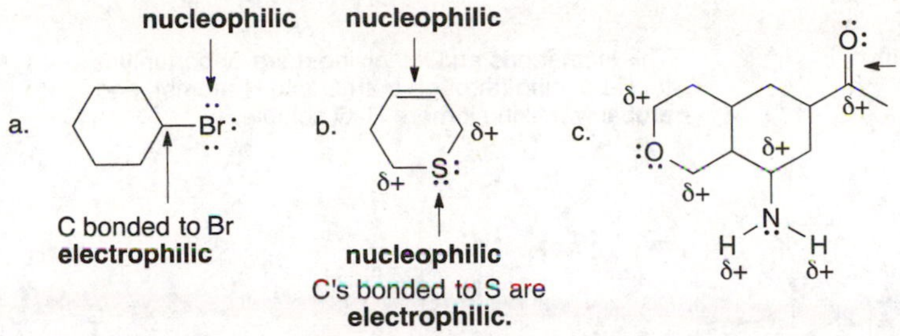

a. **nucleophilic** (Br:) — C bonded to Br **electrophilic**

b. **nucleophilic** (C=C), **nucleophilic** (S:) — C's bonded to S are **electrophilic.** δ+ δ+

c. ← **nucleophilic** (O:) δ+ δ+ δ+ δ+ δ+ δ+
All lone pairs—**nucleophilic** sites.
All atoms labeled with δ+ are **electrophilic** sites.

3.27 Electrophiles and nucleophiles react with each other.

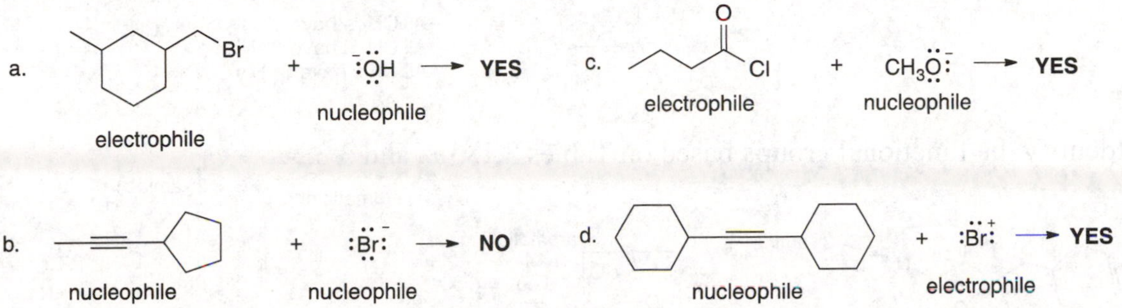

a. (Br) electrophile + :ÖH⁻ nucleophile → **YES**

b. nucleophile + :Br:⁻ nucleophile → **NO**

c. (Cl, O) electrophile + CH_3O^- nucleophile → **YES**

d. nucleophile + :Br:⁺ electrophile → **YES**

3.28

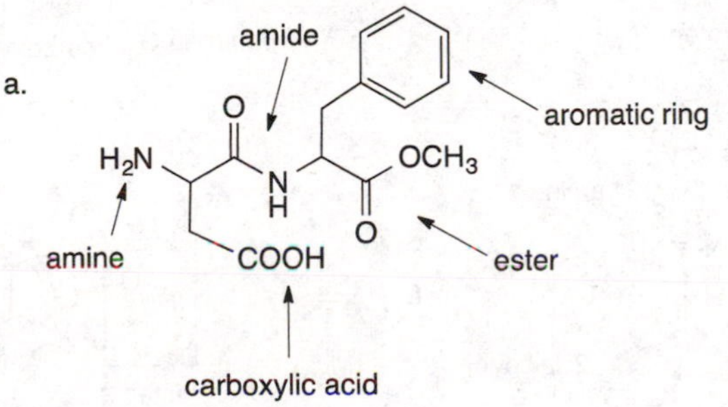

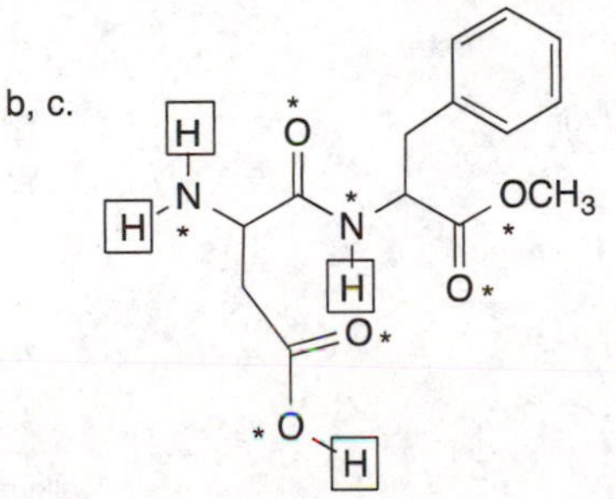

a. amide, aromatic ring, H_2N, OCH₃, amine, COOH, ester, carboxylic acid

b, c. H's that are boxed in can hydrogen bond to O of H_2O.
Atoms labeled with (*) can hydrogen bond to H of H_2O.

3.29

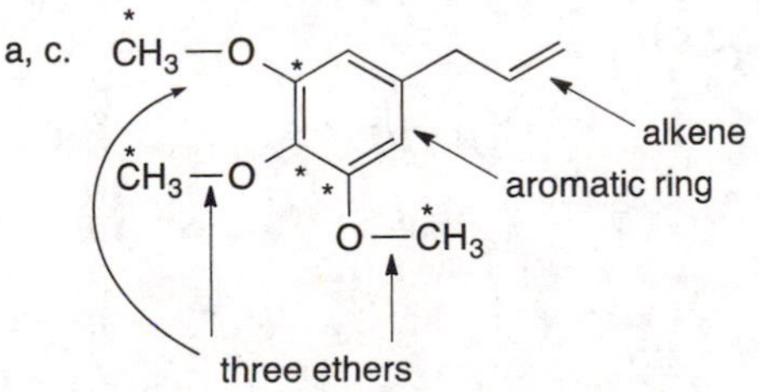

a, c. CH_3—O, CH_3—O, O—CH_3 — alkene, aromatic ring, three ethers
Electrophilic carbons are labeled with (*).

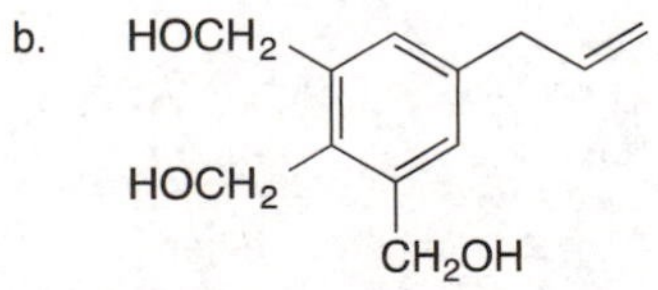

b. $HOCH_2$, $HOCH_2$, CH_2OH
Three OH's allow for H-bonding.
Stronger intermolecular forces mean a higher bp and mp.

3.30

a, c.

alkene

aldehyde

b.

OH

The most electrophilic C is labeled with *.

This isomer has an OH, giving more opportunities for H-bonding (through both O and H atoms), and probably making it more H_2O soluble.

3.31 Use the rules from Answer 3.2.

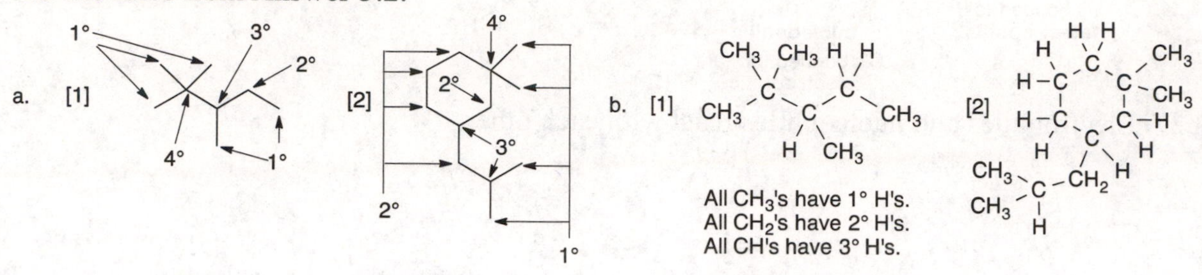

3.32 Identify the functional groups based on Tables 3.1, 3.2, and 3.3.

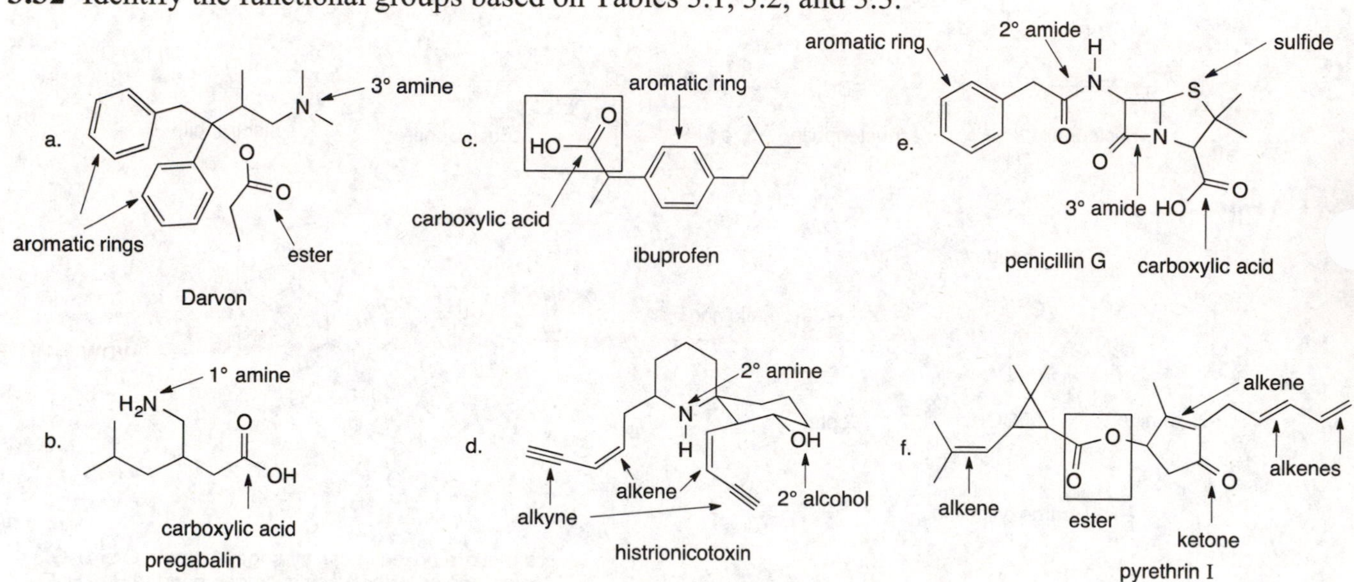

3.33 A cyclic ester is called a lactone. A cyclic amide is called a lactam.

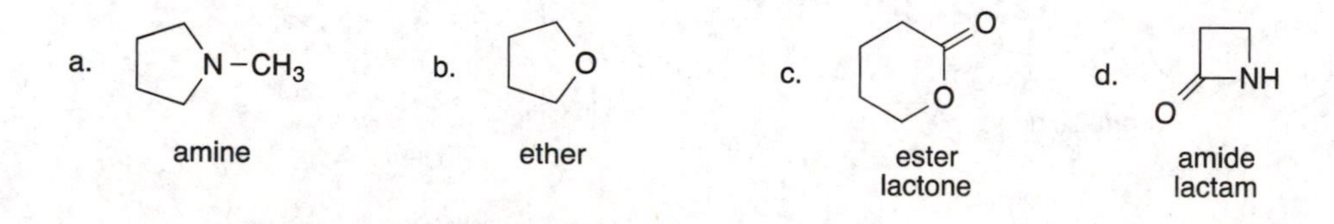

3.34

alkene
2° alcohol
OH
H
N
2° amide
O
ester
O
O
Cl
1° alkyl halide

salinosporamide A

3.35 Draw the constitutional isomers and identify the functional groups.

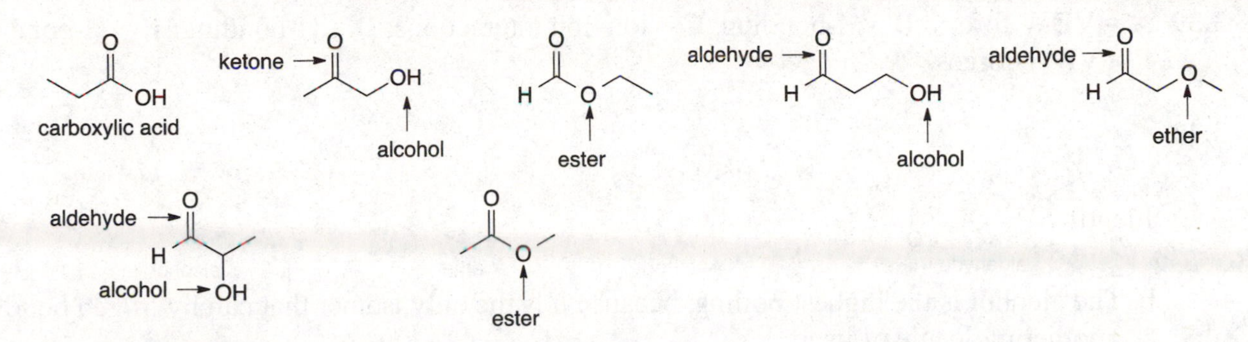

3.36 Use the rules from Answer 3.12.

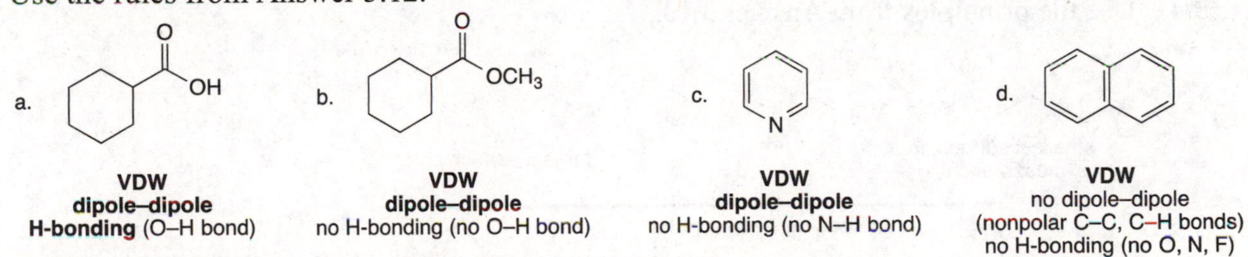

3.37 Increasing intermolecular forces: van der Waals < dipole–dipole < H-bonding

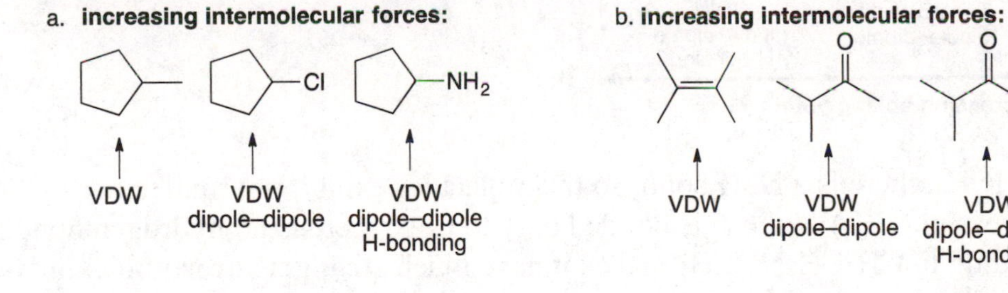

3.38

indinavir

Indinavir can hydrogen bond to another molecule of itself at boxed-in sites. Indinavir can hydrogen bond to water at both the boxed-in sites and the sites labeled with an arrow.

3.39 **A** = VDW forces; **B** = H-bonding; **C** = ion–ion interactions; **D** = H-bonding; **E** = H-bonding; **F** = VDW forces.

3.40

a. aldehyde ketone ether alcohol

b. The alcohol is the highest boiling, because it is the only isomer that can hydrogen bond with another molecule of itself.

3.41 Use the principles from Answer 3.13.

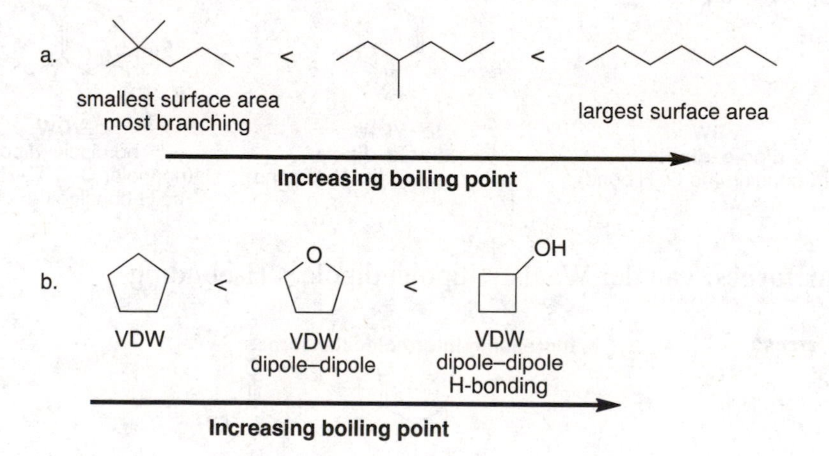

3.42 In $CH_3CH_2NHCH_3$, there is a N–H bond, so the molecules exhibit intermolecular hydrogen bonding, whereas in $(CH_3)_3N$ the N is bonded only to C, so there is no hydrogen bonding. The hydrogen bonding in $CH_3CH_2NHCH_3$ makes it have much **stronger intermolecular forces** than $(CH_3)_3N$. As intermolecular forces increase, the boiling point of a molecule of the same molecular weight increases.

3.43 Stronger forces, higher mp.

menthone
VDW
dipole–dipole
lower melting point

menthol
VDW
dipole–dipole
H-bonding
stronger forces
higher melting point

3.44 Stronger forces, higher mp.

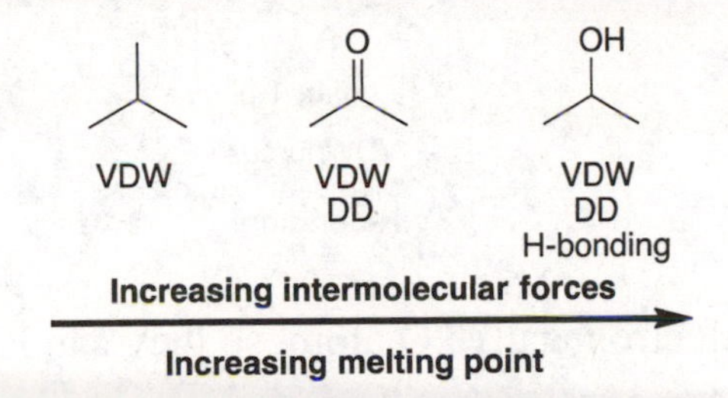

3.45 **Boiling point is determined solely by the strength of the intermolecular forces.** Because benzene has a smaller size, it has less surface area and weaker VDW interactions and therefore a lower boiling point than toluene. The increased melting point for benzene can be explained by symmetry: benzene is much more symmetrical than toluene. More symmetrical molecules can pack more tightly together, increasing their melting point. Symmetry has no effect on boiling point.

benzene
bp = 80 °C
mp = 5 °C

very symmetrical
closer packing in solid form
higher mp

and

toluene
bp = 111 °C
mp = –93 °C

less symmetrical
lower mp

3.46 Increasing polarity = increasing water solubility.

–Br
polar
no H-bonding

=O
polar
H-bonding to H_2O,
not itself

–OH
polar and
H-bonding
More opportunities
for H-bonding with its
O atom and its H on O.

3.47 Look for two things:

- To H-bond to another molecule of itself, the molecule must contain a **H bonded to O, N, or F**
- To H-bond with water, a molecule needs **only to contain an O, N, or F.**

Only (c) can H-bond to another molecule like itself.
c. $CH_3CH_2CONH_2$

These molecules can H-bond with water. All of these molecules have an O or N atom.
b. $(CH_3CH_2)_3N$
c. $CH_3CH_2CONH_2$
d. $CH_3CH_2COOCH_3$

3.48 Draw the molecules in question and look at the intermolecular forces involved.

no H bonded to O

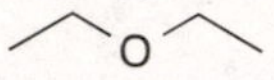

diethyl ether

VDW forces
dipole–dipole forces

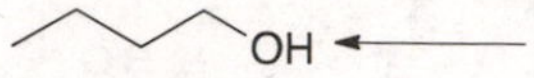

OH ← H bonded to O: hydrogen bonding

butan-1-ol

VDW forces
dipole–dipole forces
H-bonding

- Both have ≤ 5 C's and an electronegative O atom, so they can H-bond to water, making them soluble in water.
- Only butan-1-ol can H-bond to another molecule of itself, and this increases its boiling point.

3.49 Use the solubility rule from Answer 3.19.

a.

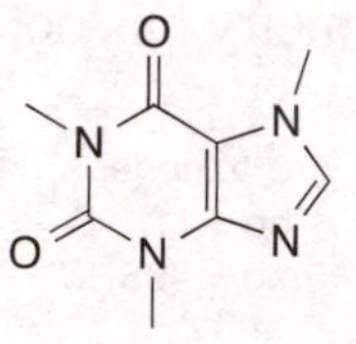

caffeine

many polar bonds with N and O atoms
many opportunities for H-bonding
water soluble

c.

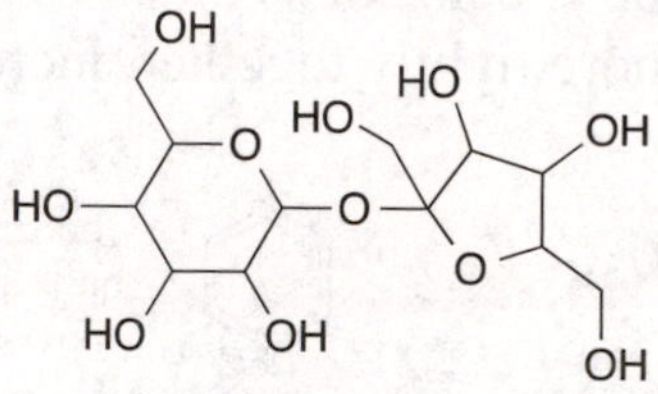

sucrose

many polar bonds with O
11 O's and 12 C's
many opportunities for H-bonding with H_2O
water soluble

b.

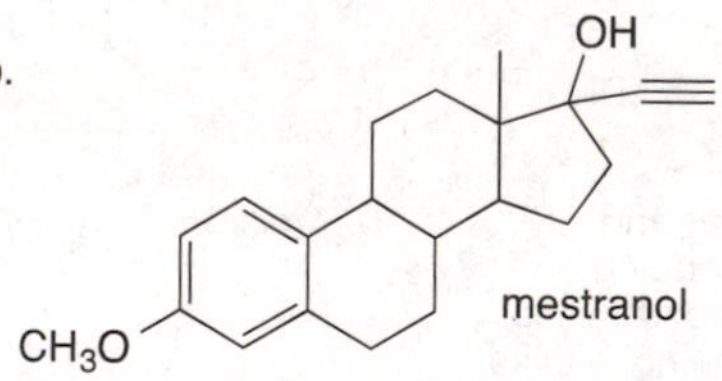

CH_3O … OH

mestranol

2 polar functional groups
but > 10 C's
not water soluble

d.

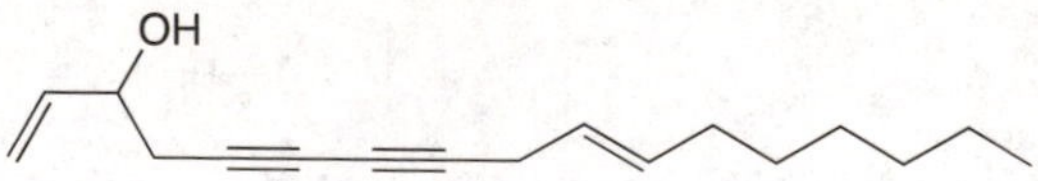

OH

carotatoxin

1 polar functional group
but > 10 C's
not water soluble

3.50 Water solubility is determined by polarity. Polar molecules are soluble in water, while nonpolar molecules are soluble in organic solvents.

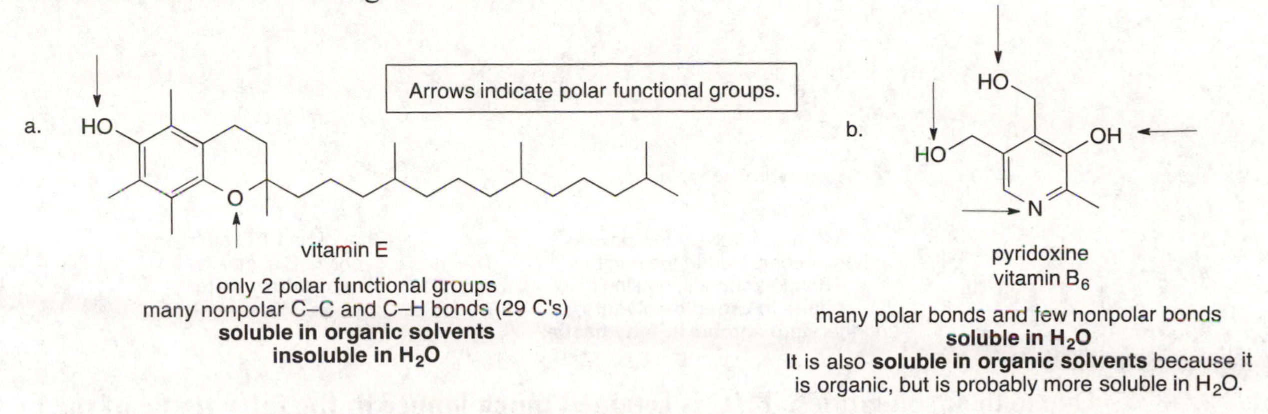

Arrows indicate polar functional groups.
a.
HO
O
vitamin E
only 2 polar functional groups
many nonpolar C–C and C–H bonds (29 C's)
soluble in organic solvents
insoluble in H_2O
b.
HO
HO
OH
N
pyridoxine
vitamin B_6
many polar bonds and few nonpolar bonds
soluble in H_2O
It is also soluble in organic solvents because it is organic, but is probably more soluble in H_2O.

3.51 Compare the functional groups in the two components of sunscreen. Dioxybenzone will most likely be washed off in water because it contains two hydroxy groups and is more water soluble.

avobenzone
two ketones
one ether

dioxybenzone
two hydroxy groups
one ketone
one ether
more water soluble

3.52 Because of the O atoms, PEG is capable of hydrogen bonding with water, which makes PEG water soluble and suitable for a product like shampoo. PVC cannot hydrogen bond to water, so PVC is water insoluble, even though it has many polar bonds. Because PVC is water insoluble, it can be used to transport and hold water.

H-bond

poly(ethylene glycol)
PEG
water soluble

no H-bonding

poly(vinyl chloride)
PVC
water insoluble

3.53 Molecules that dissolve in water are readily excreted from the body in urine, whereas less polar molecules that dissolve in organic solvents are soluble in fatty tissue and are retained for longer periods. Compare the solubility properties of THC and ethanol to determine why drug screenings can detect THC and not ethanol weeks after introduction to the body.

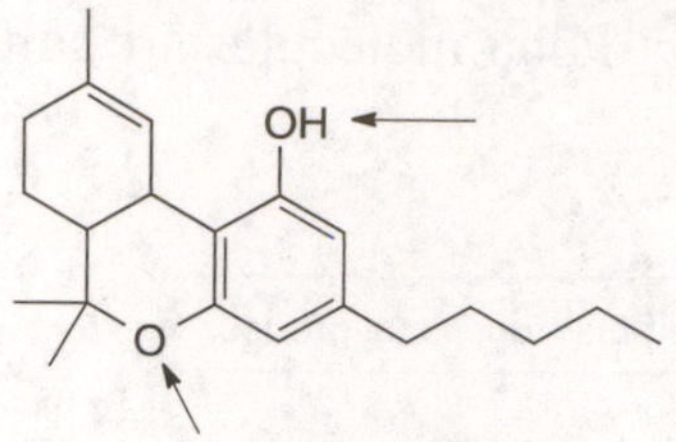

OH

tetrahydrocannabinol
THC

THC has relatively few polar bonds compared to the number of nonpolar bonds, making it **soluble in organic solvents** and therefore **soluble in fatty tissue.**

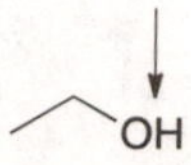

OH
ethanol

Ethanol has 1 O atom and only 2 C's, making it **soluble in water.**

Due to their solubilities, **THC is retained much longer in the fatty tissue of the body,** being slowly excreted over many weeks, while ethanol is excreted rapidly in urine after ingestion.

3.54 Compare the intermolecular forces of crack and cocaine hydrochloride. Stronger intermolecular forces increase both the boiling point and the water solubility.

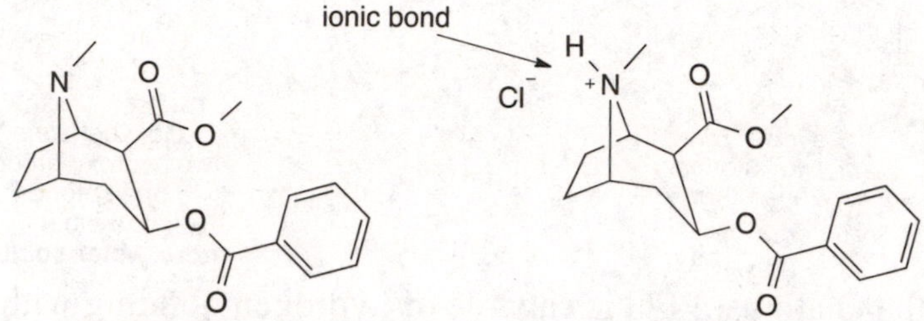

ionic bond

cocaine (crack)
neutral organic molecule

cocaine hydrochloride
a salt

The molecules are identical except for the ionic bond in cocaine hydrochloride. Ionic forces are extremely strong forces, and therefore the cocaine hydrochloride salt has a much **higher boiling point and is more water soluble.** Because the salt is highly water soluble, it can be injected directly into the bloodstream, where it dissolves. Crack is smoked because it can dissolve in the organic tissues of the nasal passages and lungs.

3.55

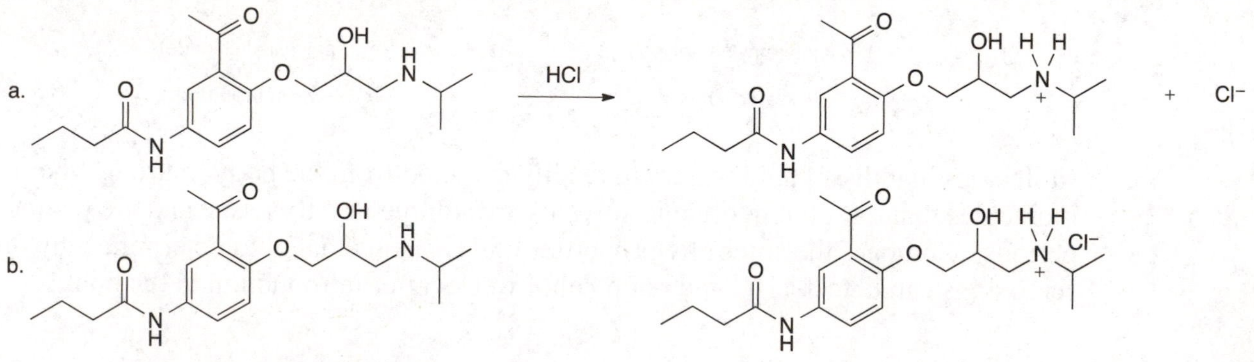

a. HCl → + Cl^-

b.

five functional groups that have many opportunities for H-bonding
water soluble

ionic salt
more water soluble

c. Because the hydrochloride salt is ionic and therefore more water soluble, it is more readily transported in the bloodstream.

3.56 Use the rules from Answer 3.26.

3.57

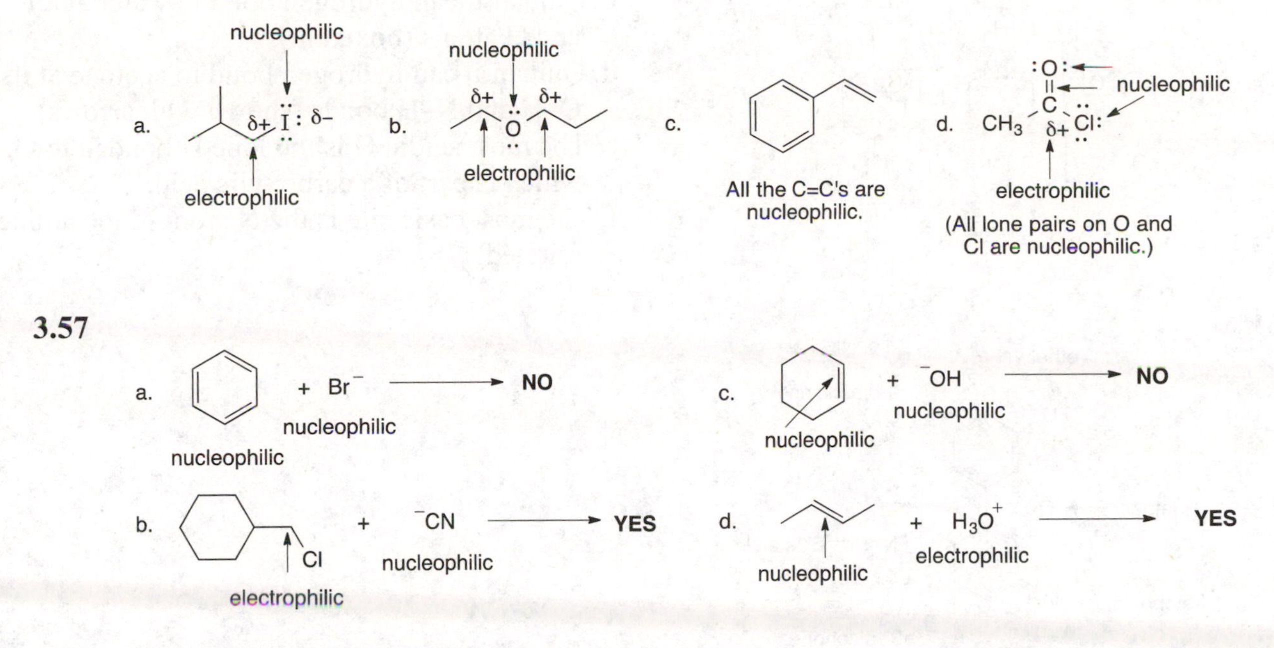

3.58 More rigid cell membranes have phospholipids with *fewer* C=C's. Each C=C introduces a bend in the molecule, making the phospholipids pack less tightly. Phospholipids without C=C's can pack very tightly, making the membrane less fluid and more rigid.

The double bonds introduce kinks in the chain, making packing of the hydrocarbon chains less efficient. This makes the cell membrane formed from them more fluid.

3.59 **B** is ionic, so it does not "dissolve" in the nonpolar interior of a cell membrane and it cannot cross the blood–brain barrier. **A** is a neutral organic molecule with polar bonds, so it can enter the nonpolar interior of the cell membrane and cross the blood–brain barrier.

3.60

a, b.

ester

2° amine

carboxylic acid

3° amide

aromatic ring

aromatic ring

quinapril

quinapril

c. Quinapril can hydrogen bond to water at all N and O atoms (boxed in).
d. Quinapril can hydrogen bond to acetone at its O–H and N–H bonds (shown with arrows).
e. The most acidic H is the lone H bonded to O, which is part of a carboxylic acid.
f. The most basic site is the N atom of the amine (circled).

3.61

a, b.

ether
aromatic ring
ether
3° amine
ketone
3° alcohol

oxycodone

c. The lone OH proton is the most acidic site.
d. The lone N atom is the most basic site.
e. The N atom is surrounded by three atoms and a lone pair (four groups), so it is sp^3 hybridized.
f. Oxycodone contains seven sp^2 hybridized C's labeled with gray circles.

3.62 Because the O atom in tetrahydrofuran is in a ring, the C atoms bonded to it are kept away from the lone pairs on O. This allows the O atom to more readily hydrogen bond with water, thus increasing its solubility in water.

Electron pairs are more exposed.
This facilitates H-bonding.

3.63

a. amide, amine, aromatic ring, aromatic ring

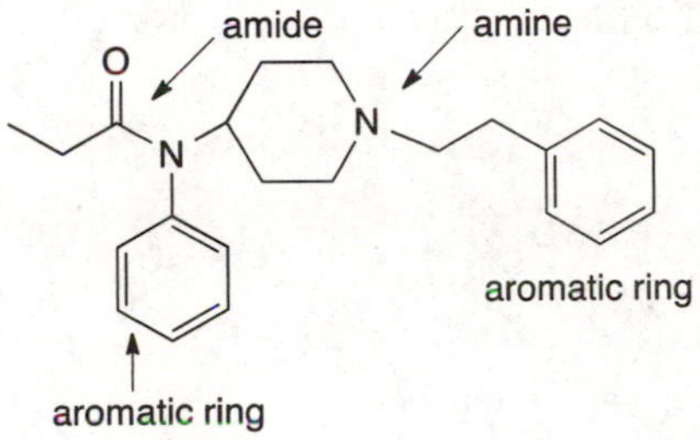

e. An isomer that can hydrogen bond should have a higher boiling point.

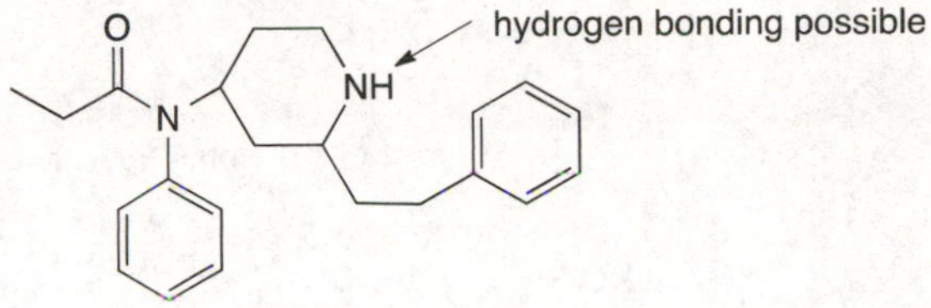

hydrogen bonding possible

b, c, f, g.

most basic atom

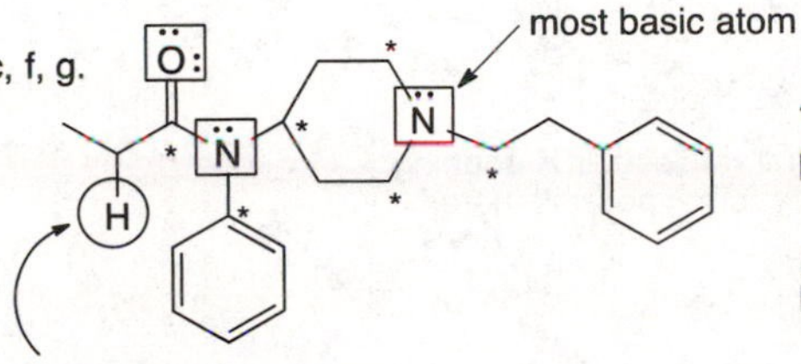

The N atom of the amide is less basic because the lone pair is part of resonance with the C=O.

Atoms that can hydrogen bond to H_2O are boxed in.
Electrophilic carbons are labeled with (*).

most acidic H, $pK_a \sim 25$
All H's are bonded to C's. Removal of the H on the C adjacent to the C=O results in a resonance-stabilized anion.

d. Fentanyl exhibits van der Waals and dipole–dipole interactions but no hydrogen bonding, because there is no H bonded to O or N.

3.64

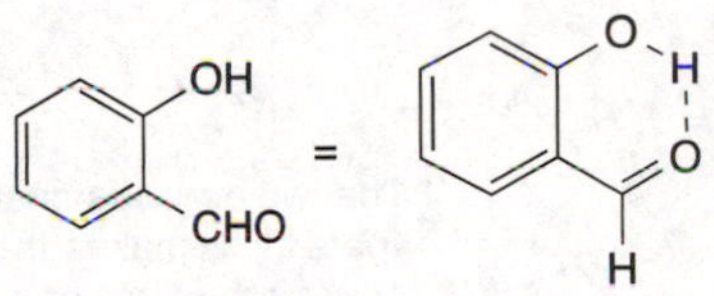

A

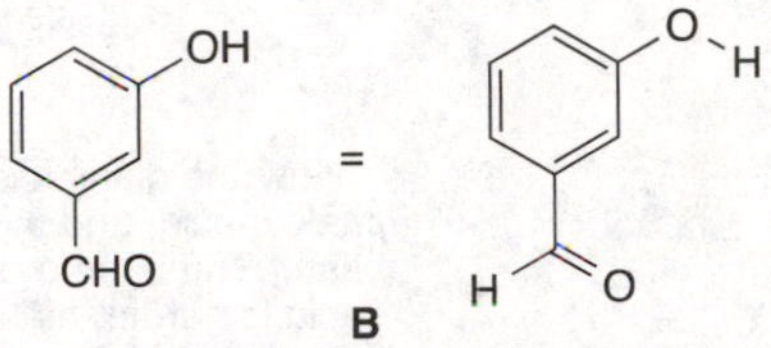

B

The OH and CHO groups are close enough that they can intramolecularly H-bond to each other. Because the two polar functional groups are involved in intramolecular H-bonding, they are less available for H-bonding to H_2O. This makes **A** less H_2O soluble than **B,** whose two functional groups are both available for H-bonding to the H_2O solvent.

The OH and the CHO are too far apart to intramolecularly H-bond to each other, leaving more opportunity to H-bond with solvent.

3.65

a. melting point

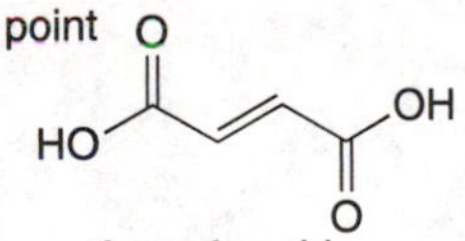

fumaric acid

Fumaric acid has its two larger COOH groups on opposite ends of the molecule, and in this way it can pack better in a lattice than maleic acid, giving it a **higher mp.**

b. solubility

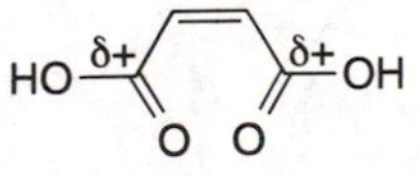

maleic acid

Maleic acid is more polar, giving it greater **H_2O solubility.** The bond dipoles in fumaric acid cancel.

c. removal of the first proton (pK_{a1})

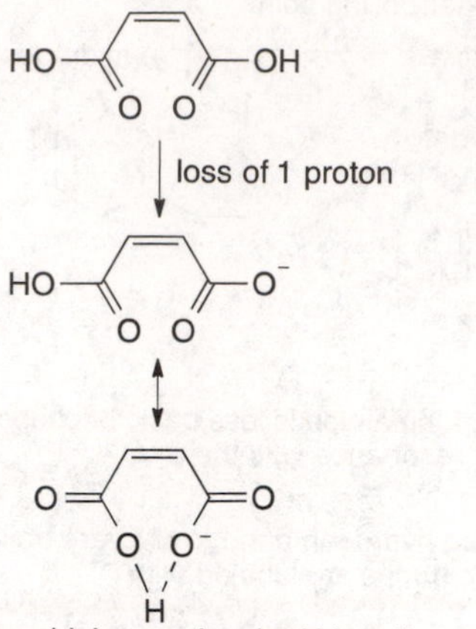

loss of 1 proton

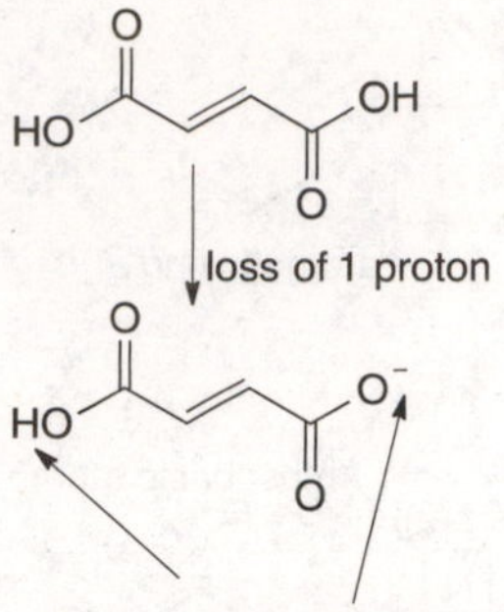

loss of 1 proton

Intramolecular H-bonding is not possible here.

In maleic acid, intramolecular H-bonding stabilizes the conjugate base after one H is removed, making maleic acid more acidic than fumaric acid.

d. removal of the second proton (pK_{a2})

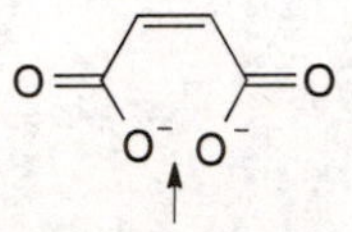

Now the dianion is held in close proximity in maleic acid, and this destabilizes the conjugate base. Thus, removing the second H in maleic acid is harder, making it a weaker acid than fumaric acid for removal of the second proton.

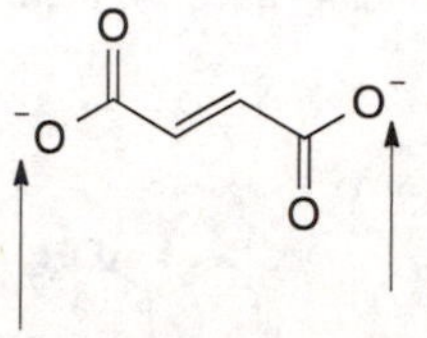

The two negative charges are much farther apart. This makes the dianion from fumaric acid more stable and thus pK_{a2} is lower for fumaric acid than maleic acid.

Chapter 4: Alkanes

Chapter Review

General facts about alkanes (4.1–4.3)

- Alkanes are composed of **tetrahedral,** ***sp*^{3}** hybridized C's.
- There are two types of alkanes: acyclic alkanes having molecular formula $\mathbf{C_nH_{2n+2}}$, and cycloalkanes having molecular formula $\mathbf{C_nH_{2n}}$.
- Alkanes have only **nonpolar C–C and C–H bonds** and no functional group, so they undergo few reactions.
- Alkanes are named with the suffix ***-ane.***

Names of alkyl groups (4.4A)

- CH_3— (methyl)
- CH_3CH_2— (ethyl)
- $CH_3CH_2CH_2$— (propyl)
- $(CH_3)_2CH$— (isopropyl)
- $CH_3CH_2CH_2CH_2$— (butyl)
- $CH_3CH_2CHCH_3$ (*sec*-butyl)
- $(CH_3)_2CHCH_2$— (isobutyl)
- $(CH_3)_3C$— (*tert*-butyl)

Conformations in acyclic alkanes (4.9, 4.10)

- Alkane conformations can be classified as **staggered, eclipsed, anti,** or **gauche** depending on the relative orientation of the groups on adjacent carbons.

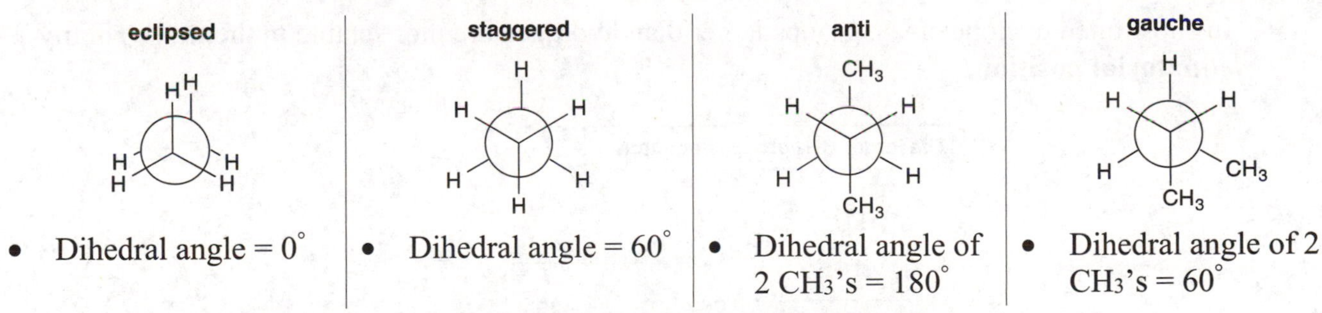

- Dihedral angle = 0°
- Dihedral angle = 60°
- Dihedral angle of 2 CH_3's = 180°
- Dihedral angle of 2 CH_3's = 60°

- A staggered conformation is **lower in energy** than an eclipsed conformation.
- An anti conformation is **lower in energy** than a gauche conformation.

Types of strain

- **Torsional strain**—an increase in energy due to eclipsing interactions (4.9).
- **Steric strain**—an increase in energy when atoms are forced too close to each other (4.10).
- **Angle strain**—an increase in energy when tetrahedral bond angles deviate from 109.5° (4.11).

Two types of isomers

[1] **Constitutional isomers**—isomers that differ in the way the atoms are connected to each other (4.1A).
[2] **Stereoisomers**—isomers that differ only in the way atoms are oriented in space (4.13B).

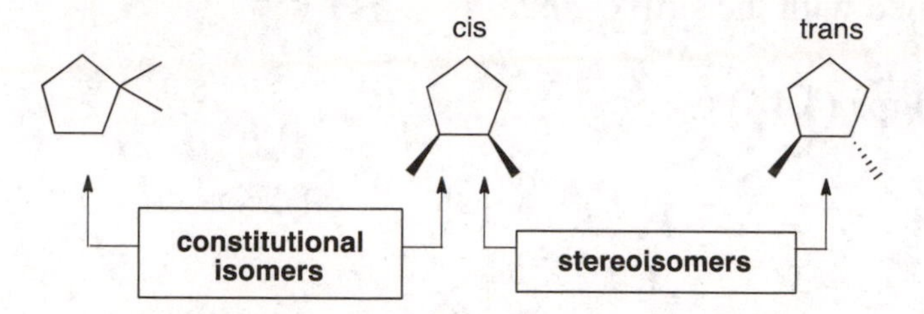

Conformations in cyclohexane (4.12, 4.13)

- Cyclohexane exists as **two chair conformations** in rapid equilibrium at room temperature.
- Each carbon atom on a cyclohexane ring has **one axial** and **one equatorial hydrogen.** Ring-flipping converts axial H's to equatorial H's, and vice versa.

An axial H flips equatorial.
H_{ax}
H_{eq}
Ring-flip.
H_{eq}
H_{ax}
An equatorial H flips axial.

- In substituted cyclohexanes, groups larger than hydrogen are more stable in the **more roomy equatorial position.**

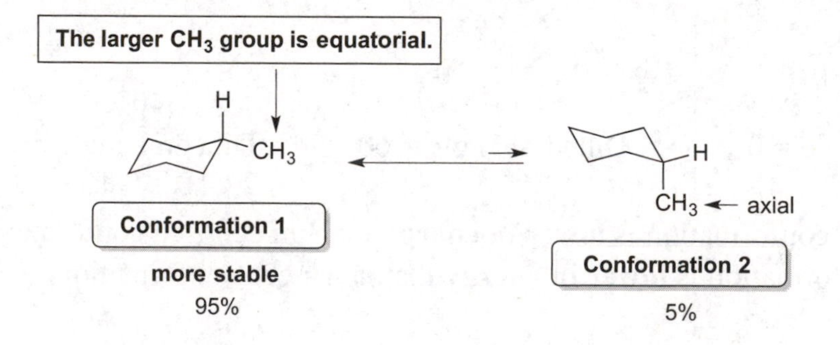

- Disubstituted cyclohexanes with substituents on different atoms exist as two possible stereoisomers.
  - The **cis** isomer has two groups on the **same side** of the ring, either both up or both down.
  - The **trans** isomer has two groups on **opposite sides** of the ring, one up and one down.

CH_3 CH_3 H H — **trans** isomer

CH_3 CH_3 H H — **cis** isomer

Oxidation–reduction reactions (4.14)

- **Oxidation** results in an **increase in the number of C–Z bonds** or a **decrease in the number of C–H bonds.**

$CH_3CH_2{-}OH$ (ethanol) ⟶ $CH_3C(=O)OH$ (acetic acid)

Increase in C–O bonds = **oxidation**

- **Reduction** results in a **decrease in the number of C–Z bonds** or an **increase in the number of C–H bonds.**

$H_2C{=}CH_2$ (ethylene) ⟶ $H_3C{-}CH_3$ (ethane)

Increase in C–H bonds = **reduction**

Practice Test on Chapter Review

1.a. Which statement is true about compounds **A**–**D** below?

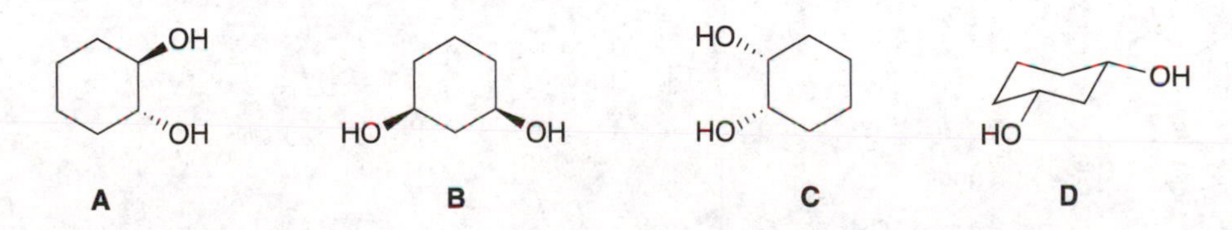

1. **A** and **C** are stereoisomers.
2. **B** and **D** are identical.
3. **A** and **B** are stereoisomers.
4. Statements (1) and (2) are both true.
5. Statements (1), (2), and (3) are all true.

b. Which of the following statements is true about *cis*-1-isopropyl-2-methylcyclohexane?

1. The more stable conformation has the isopropyl group in the equatorial position and the methyl group in the axial position.
2. The more stable conformation has both the methyl and isopropyl groups in the equatorial position.
3. *cis*-1-Isopropyl-2-methylcyclohexane is a stereoisomer of *trans*-1-isopopyl-3-methyl-cyclohexane.
4. Statements (1) and (2) are true.
5. Statements (1), (2), and (3) are all true.

c. Rank the following conformations in order of ***increasing energy.***

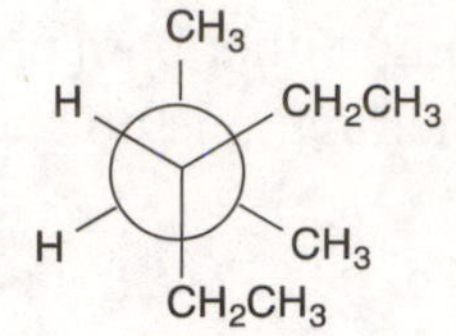

A 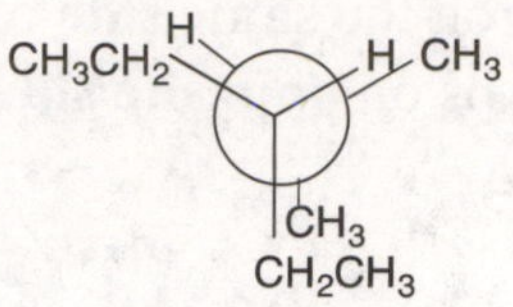 B 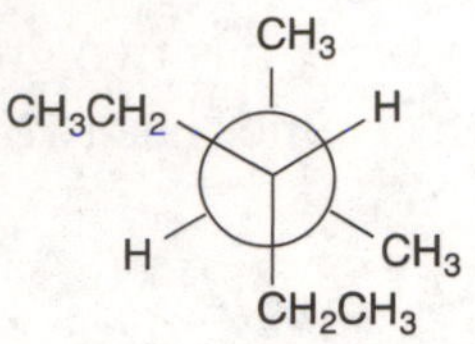 C

1. $C < B < A$
2. $C < A < B$
3. $A < C < B$
4. $B < A < C$
5. $A < B < C$

2. Give the IUPAC name for each of the following compounds.

a.

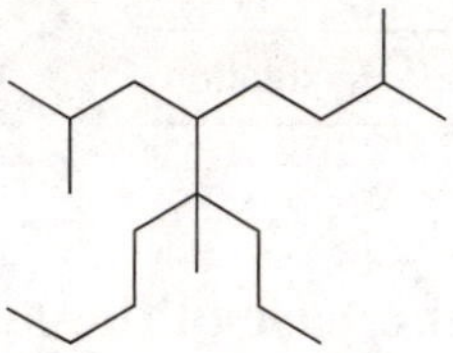

b.

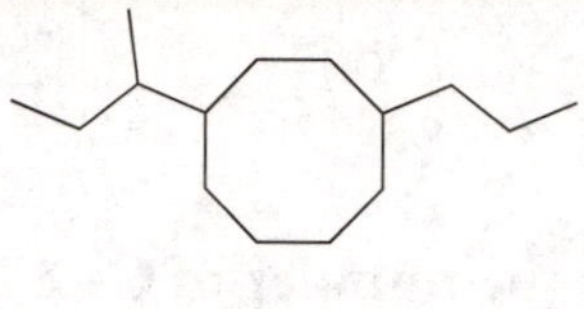

3. How are the molecules in each pair related? Are they constitutional isomers, stereoisomers, identical, or not isomers?

a. and

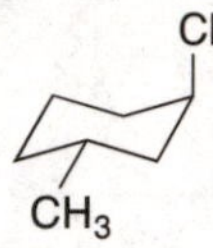

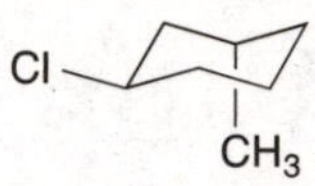

b. 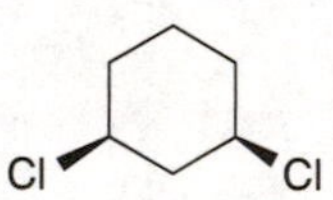 and

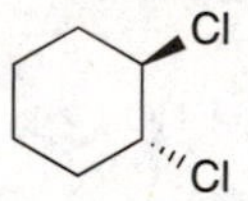

c. 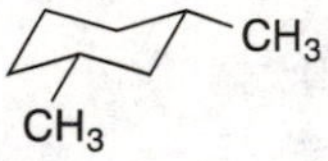 and

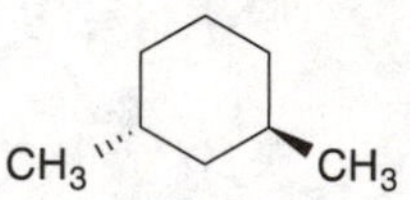

4. Rank the following conformations in order of increasing energy. Label the conformation of lowest energy as **1,** the highest energy as **4,** and the conformations of intermediate energy as **2** and **3.**

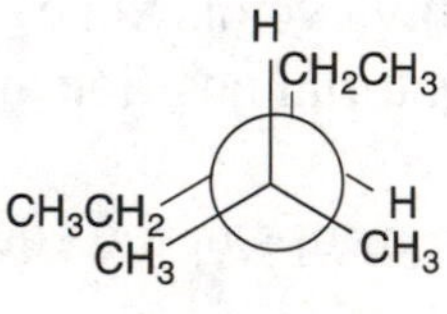

A 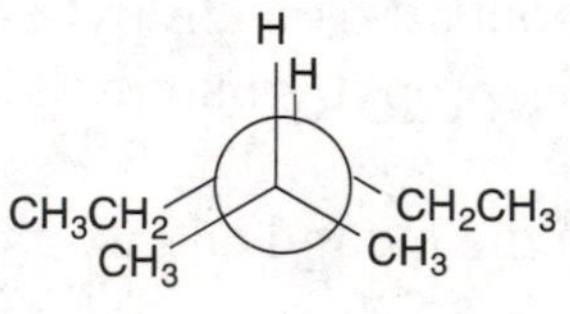 B 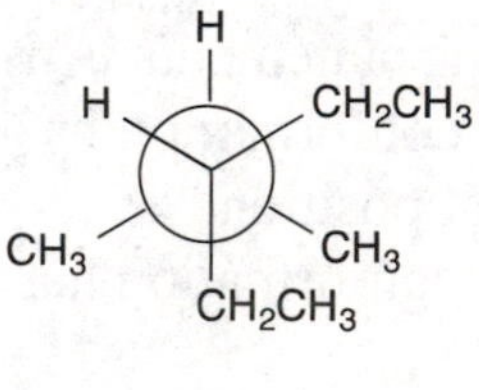 C 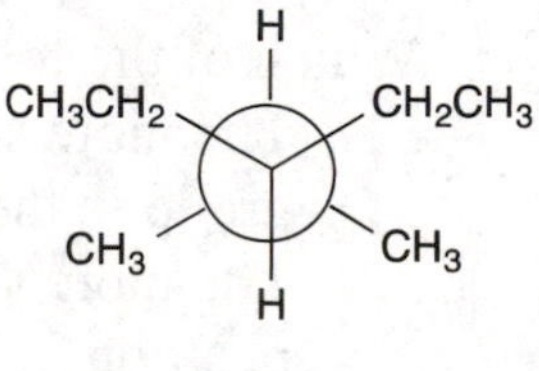 D

5. Consider the following disubstituted cyclohexane drawn below:

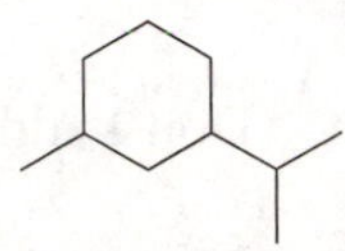

a. Draw the more stable chair conformation for the cis isomer.
b. Draw the more stable chair conformation for the trans isomer.

Answers to Practice Test

1. a. 4
   b. 1
   c. 2

2. a. 5-isobutyl-2,6-dimethyl-6-propyldecane
   b. 1-*sec*-butyl-4-propylcyclooctane

3. a. identical
   b. constitutional isomers
   c. stereoisomers

4. **A**–3
   B–4
   C–2
   D–1

5. a.

   b.

Answers to Problems

4.1 The general molecular formula for an acyclic alkane is $\mathbf{C_nH_{2n+2}}$.

Number of C atoms = *n*	**2*n* + 2**	Number of H atoms
23	2(23) + 2 =	48
25	2(25) + 2 =	52
27	2(27) + 2 =	56

4.2 2-Methylbutane has 4 C's in a row with a 1 C branch.

a. 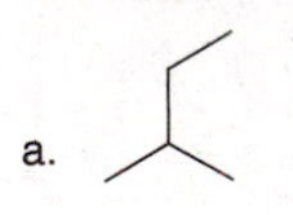

2-methylbutane

b.

2-methylbutane

c.

re-draw

2-methylbutane

d.

5 C's in a row
pentane

4.3 **Constitutional isomers differ in the way the atoms are connected to each other.** To draw all the constitutional isomers:

[1] Draw all of the C's in a long chain.

[2] Take off one C and use it as a substituent. (Don't add it to the end carbon: this re-makes the long chain.)

[3] Take off two C's and use these as substituents, etc.

Five **constitutional isomers** of molecular formula C_6H_{14}:

[1] long chain

[2] with one C as a substituent

[3] using two C's as substituents

4.4 Draw each alkane to satisfy the requirements.

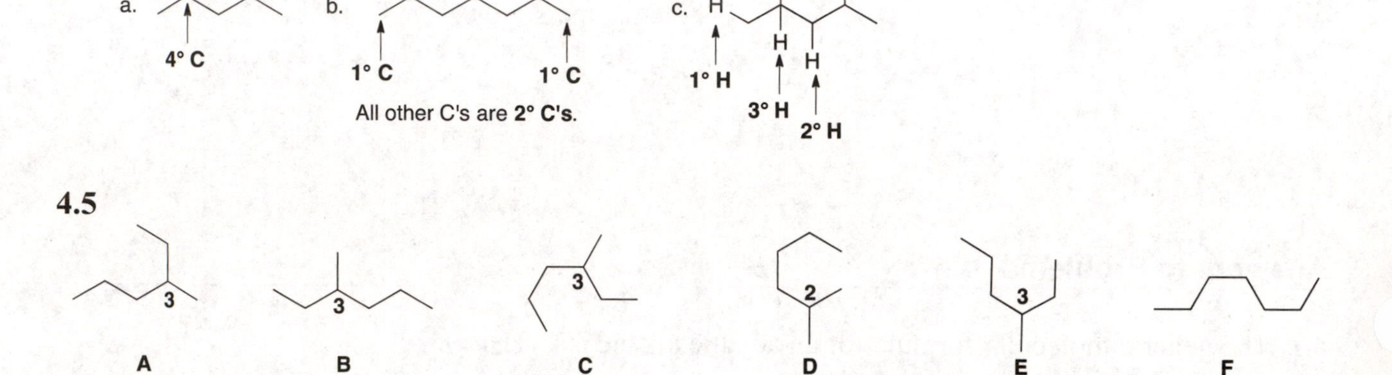

A	B	C	D	E	F
6 C chain CH_3 group on C3	identical	identical	isomer CH_3 bonded to C2	identical	isomer 7 C chain

4.6 Use the steps from Answer 4.3 to draw the constitutional isomers.

Five **constitutional isomers** of molecular formula C_5H_{10} having one ring:

[1] [2] [3]

4.7 Follow these steps to name an alkane:

[1] **Name the parent chain** by finding the longest C chain.

[2] **Number the chain** so that the first substituent gets the lower number. Then **name and number all substituents,** giving like substituents a prefix (di, tri, etc.).

[3] **Combine all parts,** alphabetizing the substituents, ignoring all prefixes except *iso*.

a. [1]

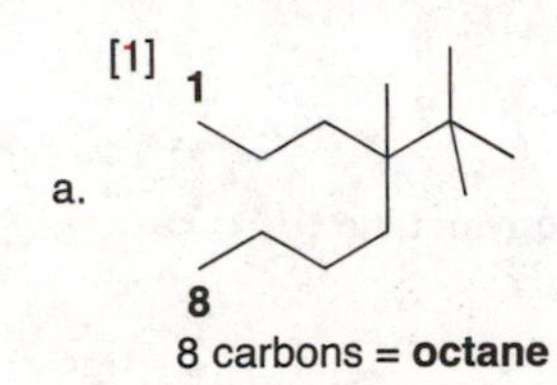

8 carbons = **octane**

[2]

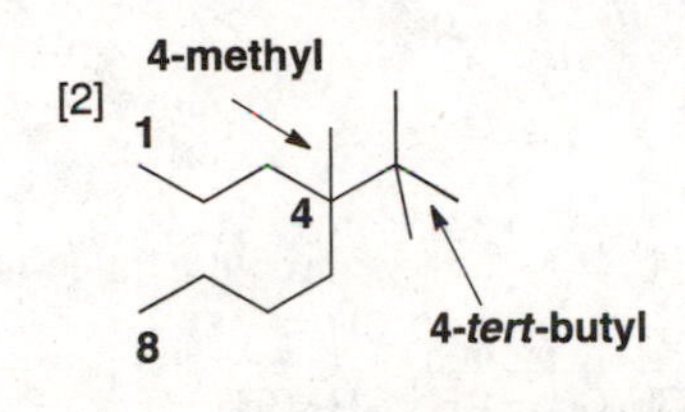

[3] **4-*tert*-butyl-4-methyloctane**

b. [1]

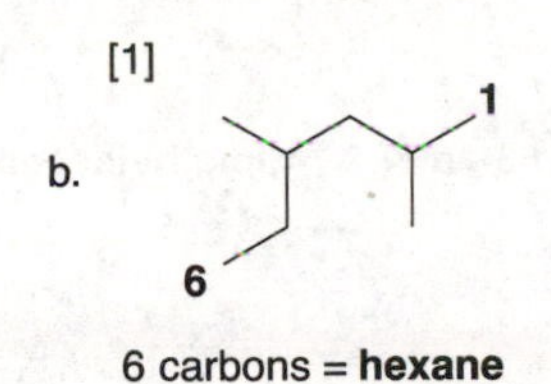

6 carbons = **hexane**

[2] **4-methyl** 1 6 **2-methyl**

[3] **2,4-dimethylhexane**

c. [1]

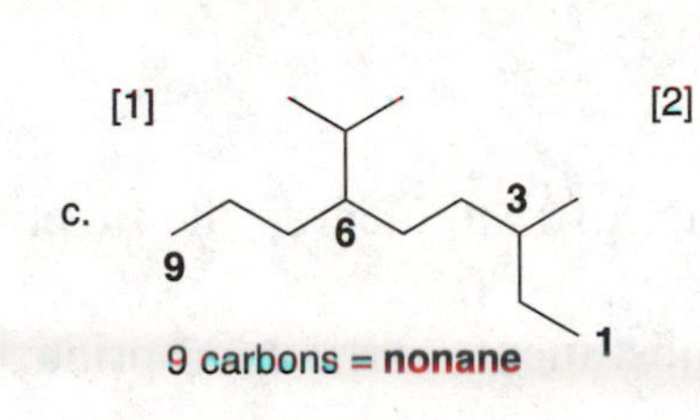

9 carbons = **nonane**

[2]

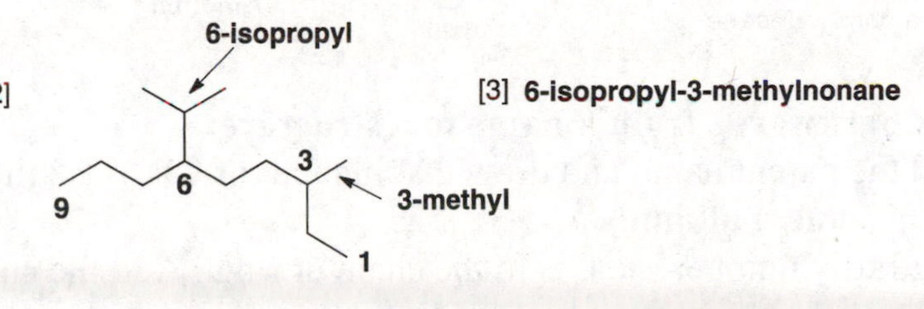

[3] **6-isopropyl-3-methylnonane**

d. [1]

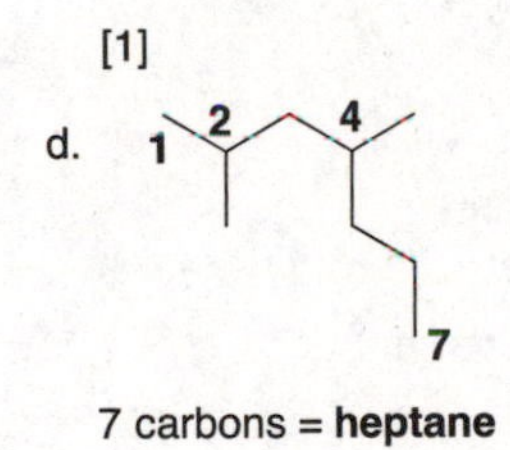

7 carbons = **heptane**

[2]

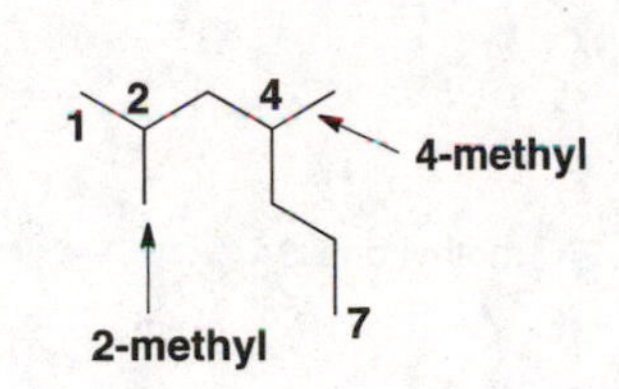

[3] **2,4-dimethylheptane**

4.8 Use the steps in Answer 4.7 to name each alkane.

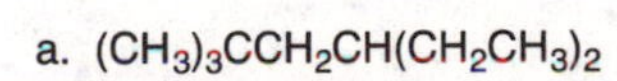

[1]

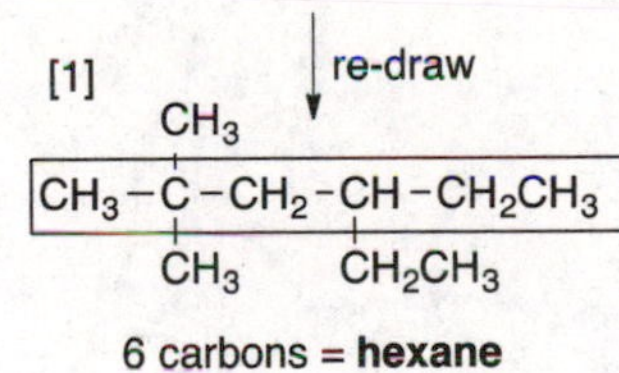

6 carbons = **hexane**

[2]

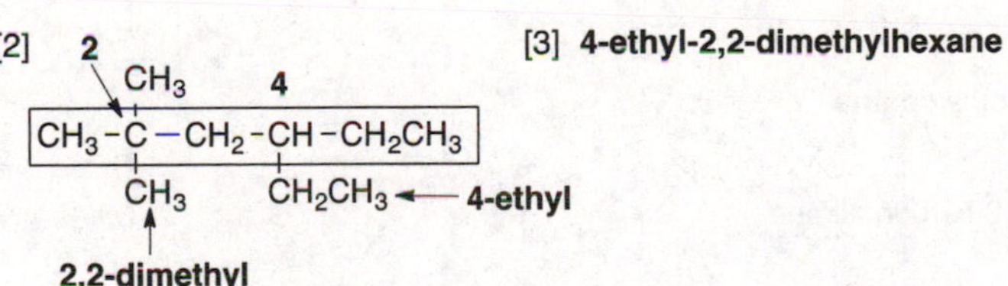

2,2-dimethyl

[3] **4-ethyl-2,2-dimethylhexane**

b. [1]

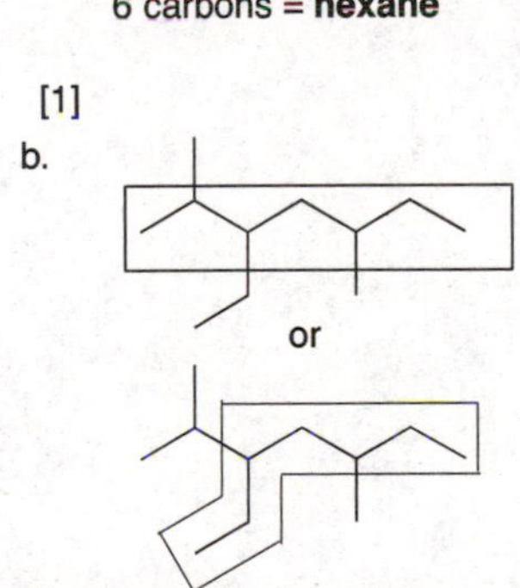

[2]

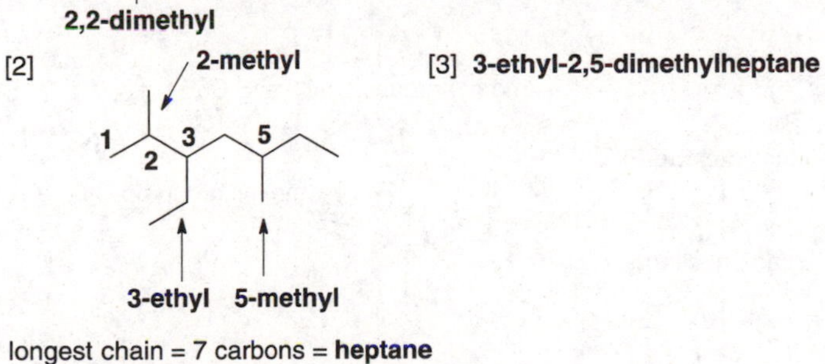

longest chain = 7 carbons = **heptane**
Number so there are more substituents.
Pick the upper option.

[3] **3-ethyl-2,5-dimethylheptane**

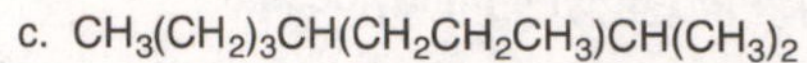

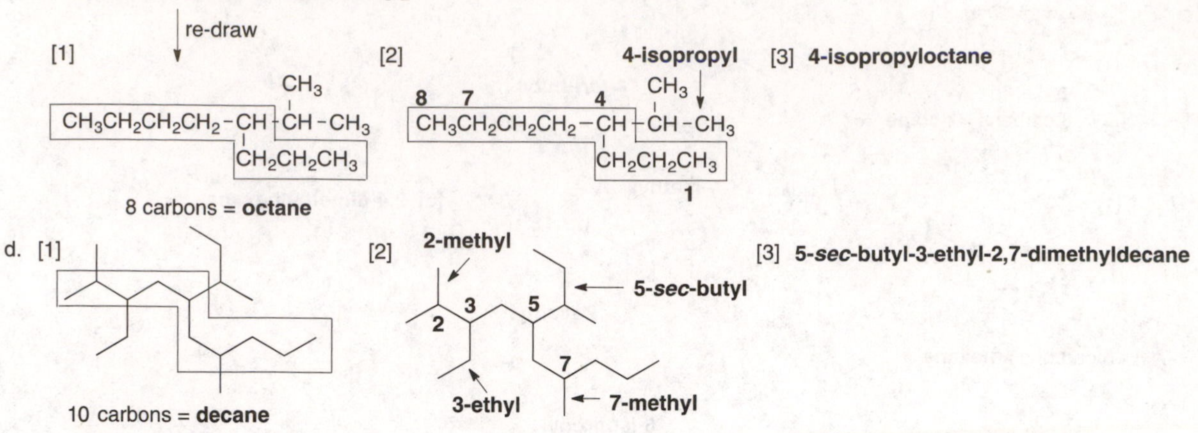

4.9 To work backwards from a name to a structure:

[1] Find the parent name and draw that number of C's. Use the suffix to identify the functional group **(-ane = alkane).**

[2] Arbitrarily number the C's in the chain or ring. Add the substituents to the appropriate C's.

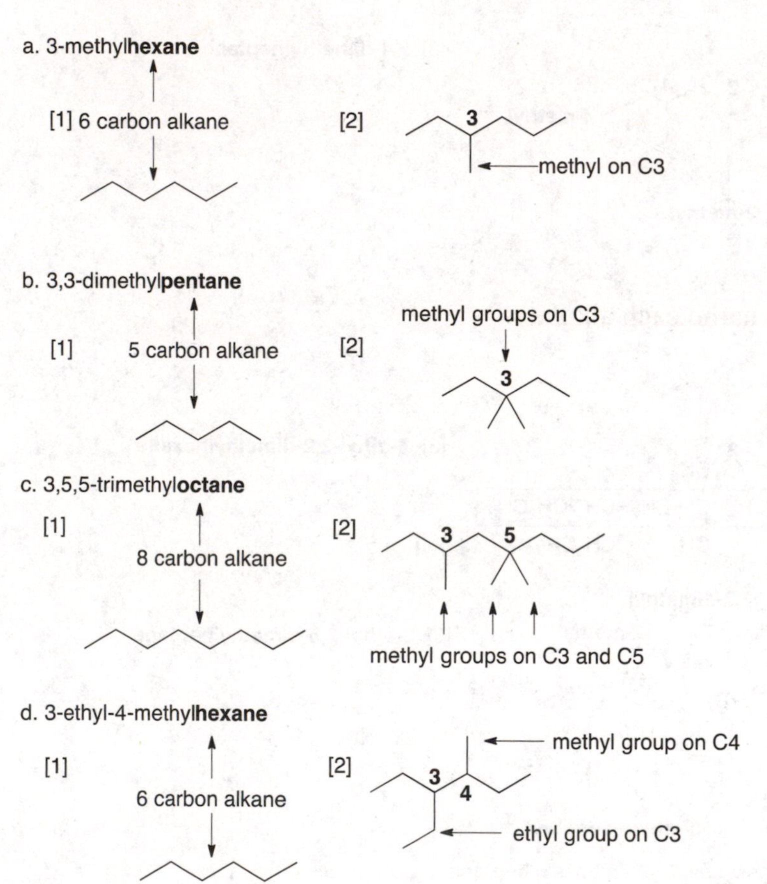

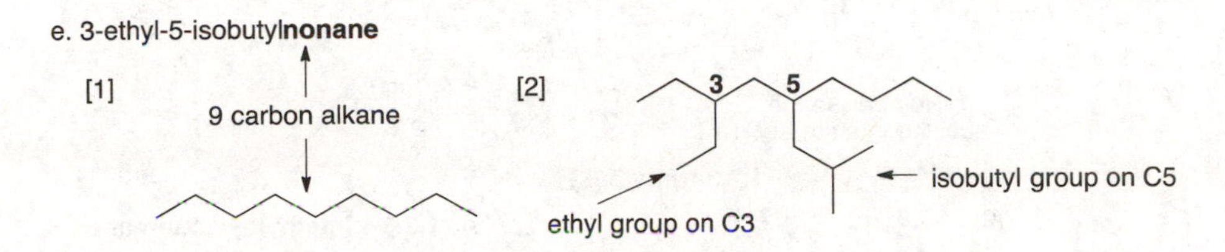

4.10 Use the steps in Answer 4.7 to name each alkane.

[1] 6 carbons = **hexane** [2] no substituents [3] **hexane**

[1] 5 carbons = **pentane** [2] **2-methyl** 2 [3] **2-methylpentane**

[1] 5 carbons = **pentane** [2] 3 **3-methyl** [3] **3-methylpentane**

[1] 4 carbons = **butane** [2] **2,2-dimethyl** 2 [3] **2,2-dimethylbutane**

[1] 4 carbons = **butane** [2] 2 3 **2,3-dimethyl** [3] **2,3-dimethylbutane**

4.11 Follow these steps to name a cycloalkane:

[1] **Name the parent cycloalkane** by counting the C's in the ring and adding cyclo-.

[2] **Numbering:**

a. **Number around the ring** beginning at a substituent and giving the second substituent the lower number.

b. **Number to assign the lower number to the substituents alphabetically.**

c. **Name and number all substituents,** giving like substituents a prefix (di, tri, etc.).

[3] **Combine all parts,** alphabetizing the substituents, ignoring all prefixes except *iso*. (Remember: If a carbon chain has more C's than the ring, the chain is the parent, and the ring is a substituent.)

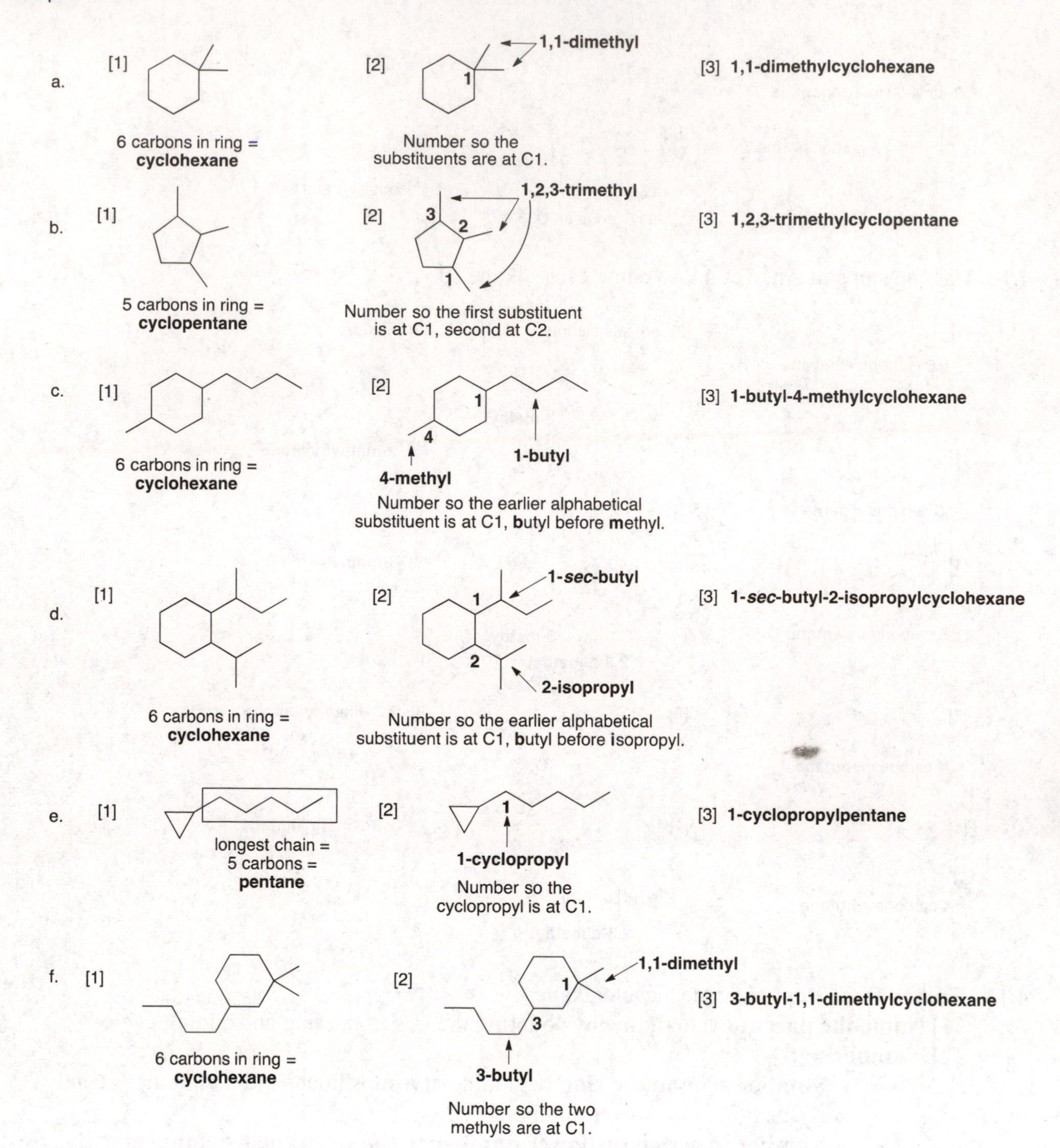
a.
[1]
6 carbons in ring =
cyclohexane
[2]
1,1-dimethyl
1
Number so the
substituents are at C1.
[3] 1,1-dimethylcyclohexane
b.
[1]
5 carbons in ring =
cyclopentane
[2]
1,2,3-trimethyl
3
2
1
Number so the first substituent
is at C1, second at C2.
[3] 1,2,3-trimethylcyclopentane
c.
[1]
6 carbons in ring =
cyclohexane
[2]
1
4
1-butyl
4-methyl
Number so the earlier alphabetical
substituent is at C1, butyl before methyl.
[3] 1-butyl-4-methylcyclohexane
d.
[1]
6 carbons in ring =
cyclohexane
[2]
1-sec-butyl
1
2
2-isopropyl
Number so the earlier alphabetical
substituent is at C1, butyl before isopropyl.
[3] 1-sec-butyl-2-isopropylcyclohexane
e.
[1]
longest chain =
5 carbons =
pentane
[2]
1
1-cyclopropyl
Number so the
cyclopropyl is at C1.
[3] 1-cyclopropylpentane
f.
[1]
6 carbons in ring =
cyclohexane
[2]
1,1-dimethyl
1
3
3-butyl
Number so the two
methyls are at C1.
[3] 3-butyl-1,1-dimethylcyclohexane

4.12 To draw the structures, use the steps in Answer 4.9.

a. 1,2-dimethyl**cyclobutane**

[1] 4 carbon cycloalkane

[2] methyl groups on C1 and C2

1
2

[3]

b. 1,1,2-trimethyl**cyclopropane**

[1] 3 carbon cycloalkane

[2] 3 CH_3's

2
1

[3]

c. 4-ethyl-1,2-dimethyl**cyclohexane**

[1] 6 carbon cycloalkane

[2] ethyl on C4

4 2
1

2 CH_3's

[3]

d. 1-*sec*-butyl-3-isopropyl**cyclopentane**

[1] 5 carbon cycloalkane

[2]

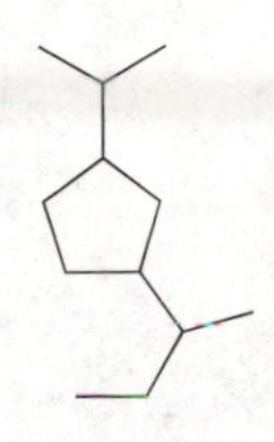

[3]

e. 1,1,2,3,4-pentamethyl**cycloheptane**

[1] 7 carbon cycloalkane

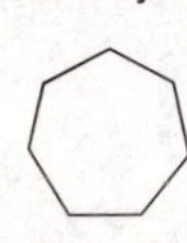

[2] 5 CH_3's

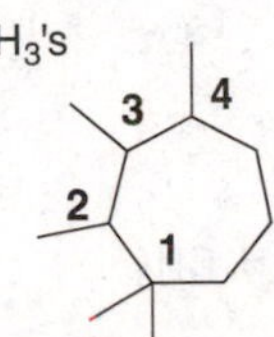

[3]

4.13 Compare the number of C's and surface area to determine relative boiling points. Rules:
[1] Increasing number of C's = increasing boiling point.
[2] Increasing surface area = increasing boiling point (branching decreases surface area).

8 C's
linear
largest number of C's
no branching
highest bp

7 C's
linear

7 C's
one branch

7 C's
three branches

increasing branching
decreasing surface area
decreasing bp

Increasing boiling point: $(CH_3)_3CCH(CH_3)_2$ < $CH_3CH_2CH_2CH_2CH(CH_3)_2$ < $CH_3(CH_2)_5CH_3$ < $CH_3(CH_2)_6CH_3$

4.14 To draw a Newman projection, visualize the carbons as one in front and one in back of each other. The C–C bond is not drawn. There is only one staggered and one eclipsed conformation.

rotation here
C in front
C behind
60°
1
staggered
2
eclipsed

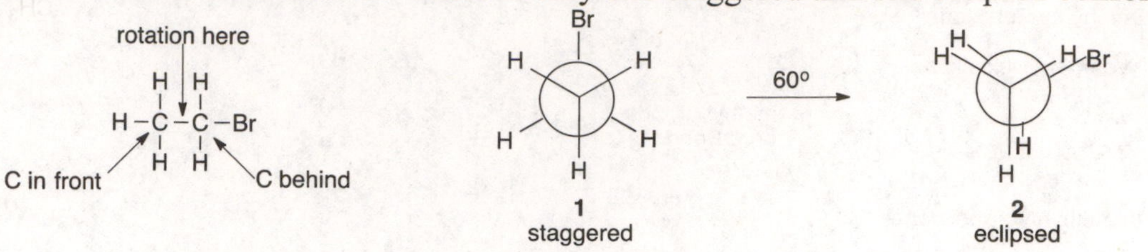

4.15 Re-draw each Newman projection as a skeletal structure, remembering that a Newman projection shows the substituents bonded to two C's but not the C's themselves.

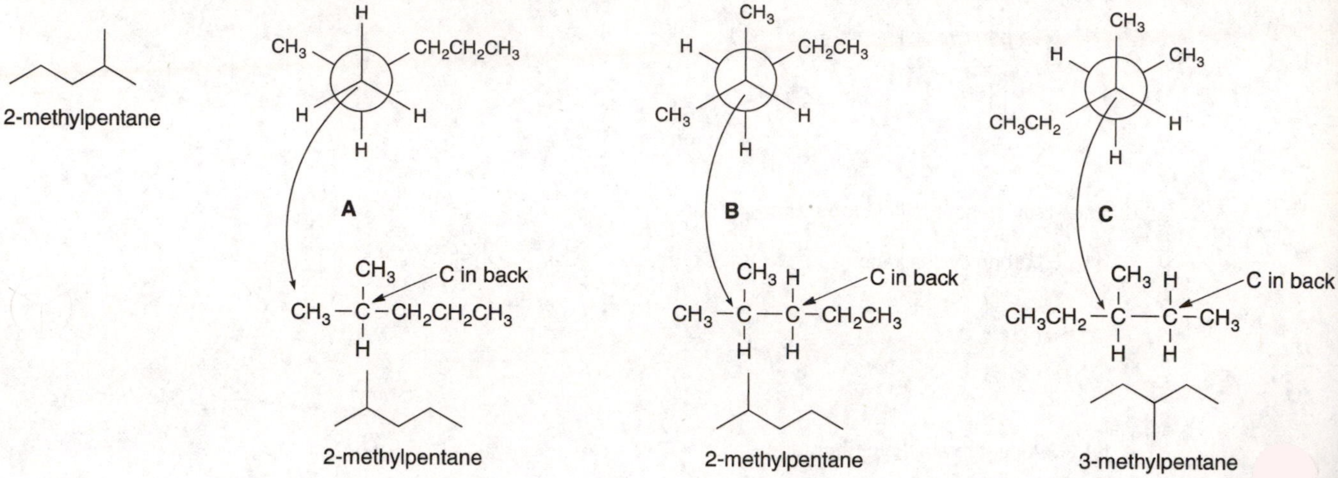

2-methylpentane
A
B
C
C in back
C in back
C in back
2-methylpentane
2-methylpentane
3-methylpentane

4.16

4.0 kJ/mol
H,H eclipsing
4.0 kJ/mol of destabilization
4.0 kJ/mol

To calculate H,CH_3 destabilization:

14 kJ/mol (total) –
8.0 kJ/mol for 2 H,H eclipsing interactions
= **6 kJ/mol** for one H,CH_3 eclipsing interaction

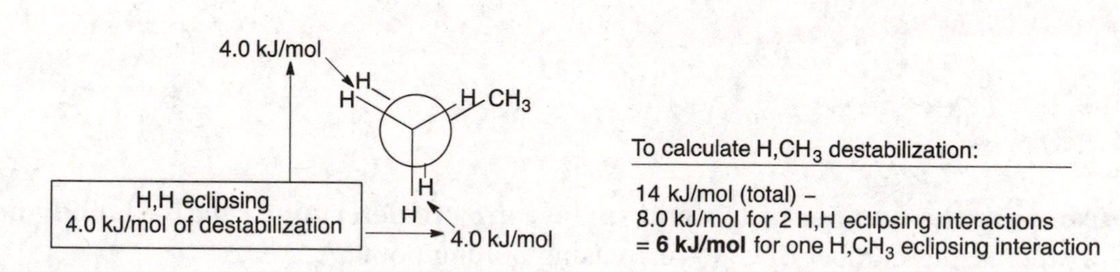

4.17 To determine the energy of conformations, keep two things in mind:
[1] Staggered conformations are more stable than eclipsed conformations.
[2] Minimize steric interactions: keep large groups away from each other.
The highest energy conformation is the eclipsed conformation in which the two largest groups are eclipsed. The lowest energy conformation is the staggered conformation in which the two largest groups are anti.

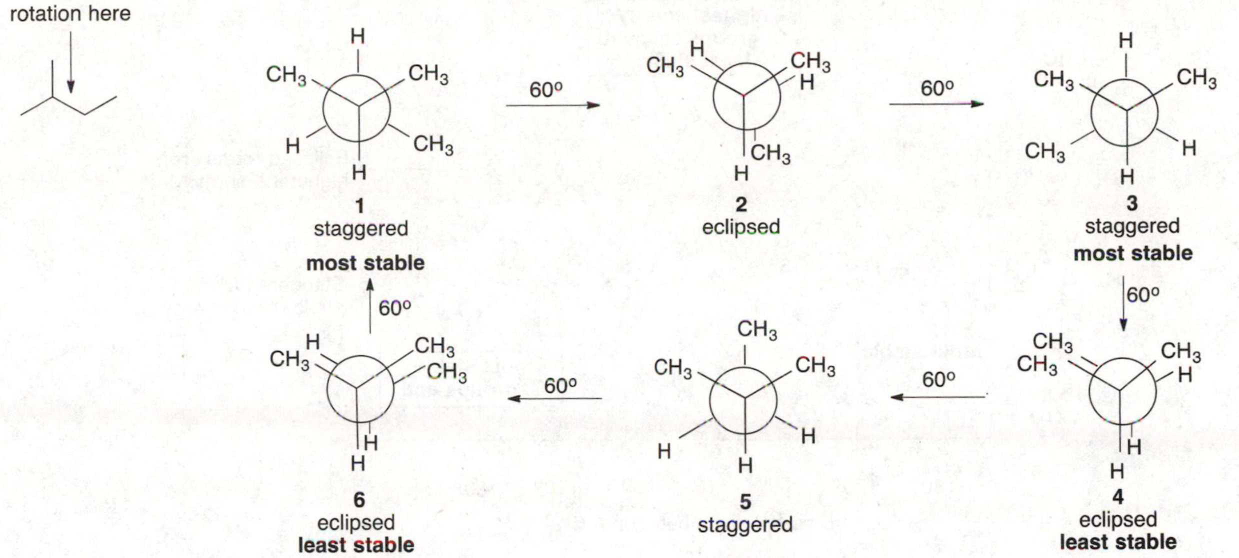

rotation here

60° 60° 60° 60° 60° 60°

1 staggered **most stable**

2 eclipsed

3 staggered **most stable**

4 eclipsed **least stable**

5 staggered

6 eclipsed **least stable**

4.18 Use the criteria in problem 4.17 to determine the relative energy.

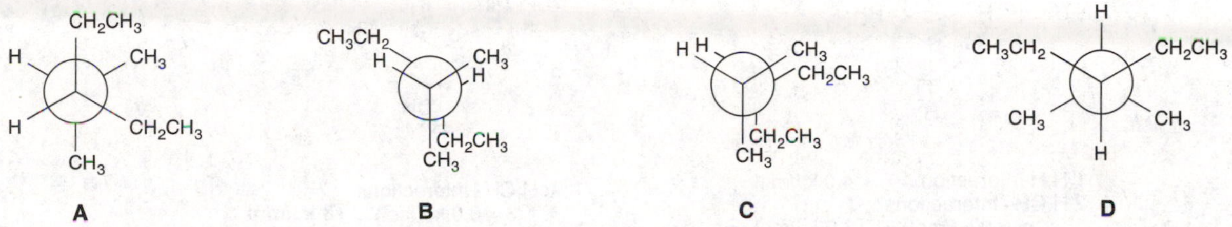

A **B** **C** **D**

Staggered conformations **A** and **D** are lower in energy than eclipsed conformations **B** and **C. D** is lower in energy than **A** because it has two gauche interactions and **A** has three. **C** is higher in energy than **B** because all the larger groups (CH_3 and CH_2CH_3) are eclipsed.
Ranking: **D < A < B < C**

4.19 To determine the most and least stable conformations, use the rules from Answer 4.17.

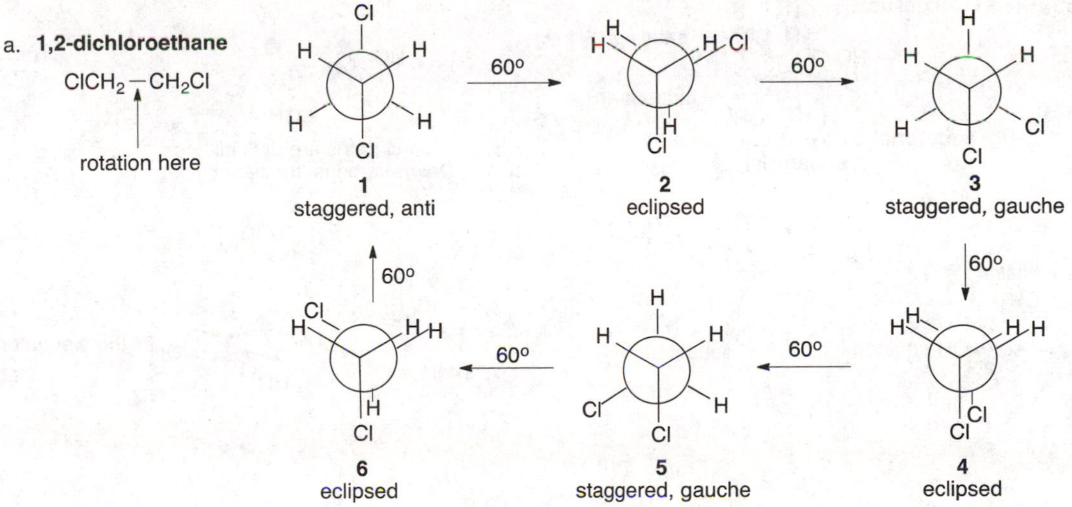

a. **1,2-dichloroethane**
$ClCH_2-CH_2Cl$
rotation here

60° 60° 60° 60° 60° 60°

1 staggered, anti

2 eclipsed

3 staggered, gauche

4 eclipsed

5 staggered, gauche

6 eclipsed

b.

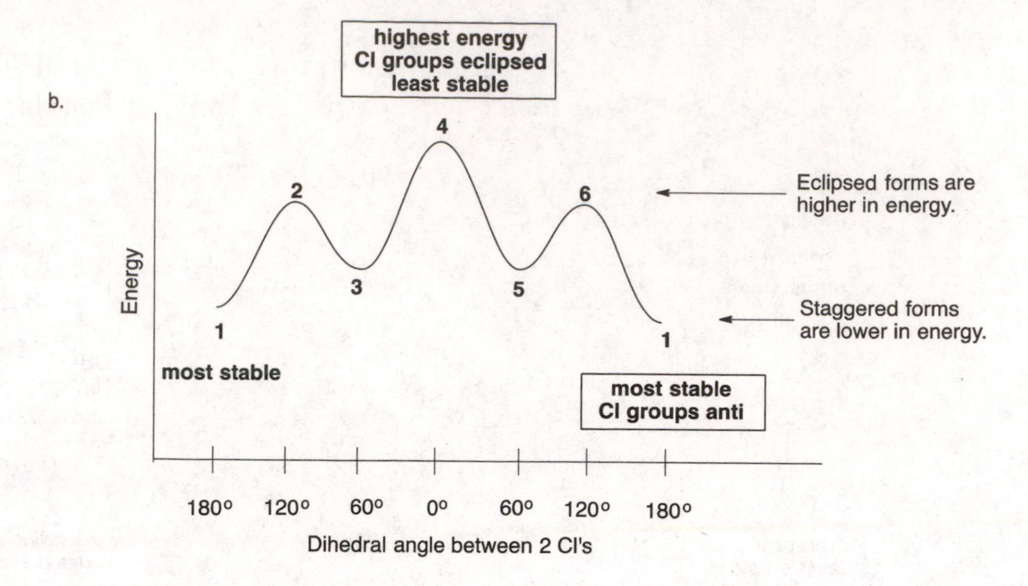

4.20 Add the energy increase for each eclipsing interaction to determine the destabilization.

a. H H H CH_3 H CH_3

1 H,H interaction = 4.0 kJ/mol
2 H,CH_3 interactions
(2 x 6.0 kJ/mol) = 12.0 kJ/mol

Total destabilization = 16 kJ/mol

b. CH_3 H H CH_3 CH_3 H

3 H,CH_3 interactions
(3 x 6.0 kJ/mol) = **18 kJ/mol**
Total destabilization

4.21 Two points:
- Axial bonds point up or down, while equatorial bonds point out.
- An *up* carbon has an axial *up* bond, and a *down* carbon has an axial *down* bond.

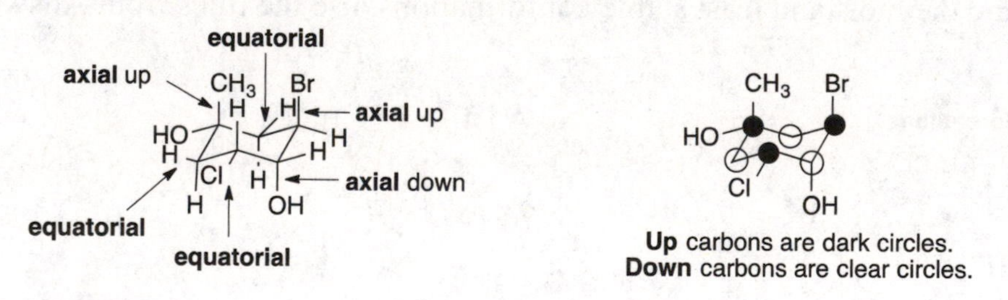

4.22

a. axial CH_3 — OH equatorial (1, 2)

b. equatorial CH_3 — OH axial (4, 6)

c. HO, OH, HO (5, 1, 2) three equatorial OH's

4.23 Draw the second chair conformation by flipping the ring.

- **The *up* carbons become *down* carbons, and the axial bonds become equatorial bonds.**
- **Axial bonds become equatorial, but *up* bonds stay *up*;** that is, an axial *up* bond becomes an equatorial *up* bond.
- The conformation with **larger groups equatorial is the more stable** conformation and is present in higher concentration at equilibrium.

a. Br

more stable
Br is equatorial.

Draw in the H
and label the C
as up or down.

axial
H eq
Br
Axial bond is up =
up carbon

Draw second conformation.
Up carbons switch
to down carbons.

H ← eq
Br
axial
Axial bond is down =
down carbon

b. Cl

Draw in the H
and label the C
as up or down.

axial
Cl
H
eq
Axial bond is up =
up carbon

Draw second conformation.
Up carbons switch
to down carbons.

eq
Cl
axial → H
Axial bond is down =
down carbon

more stable
Cl is equatorial.

c. CH_2CH_3

more stable
CH_2CH_3 is equatorial.

Draw in the H
and label the C
as up or down.

Axial bond is up =
up carbon
H
CH_2CH_3
eq

Draw second conformation.
Up carbons switch
to down carbons.

eq
H
CH_2CH_3
Axial bond is down =
down carbon

4.24

CH_3 ax
CH_2CH_3
eq
more stable
The larger ethyl group is in the
roomier equatorial position.

⇌

CH_3 eq
CH_2CH_3
ax

4.25 Wedges represent "up" groups in front of the page, whereas dashed wedges are "down" groups in back of the page. Cis groups are on the same side of the ring, whereas trans groups are on opposite sides of the ring.

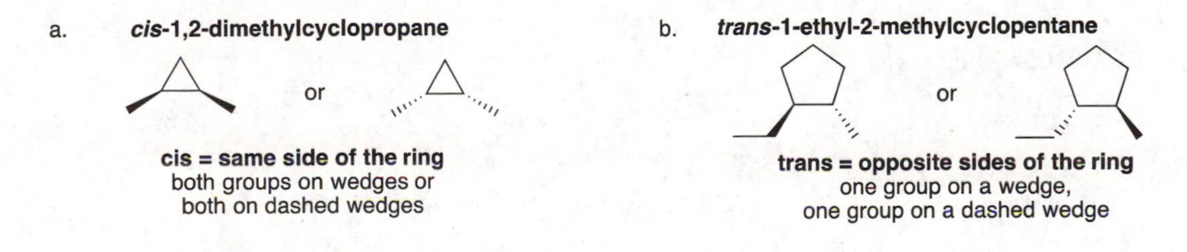

4.26 Cis and trans isomers are stereoisomers.

***cis*-1,3-diethylcyclobutane**

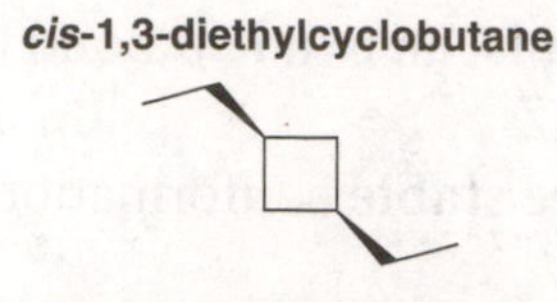

cis = same side of the ring
both groups on wedges or
both on dashed wedges

a. ***trans*-1,3-diethylcyclobutane**

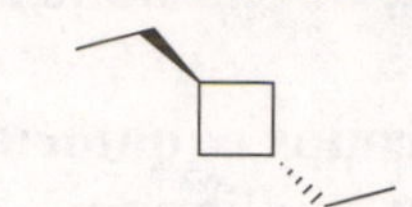

trans = opposite sides of the ring
one group on a wedge,
one group on a dashed wedge

b. ***cis*-1,2-diethylcyclobutane**

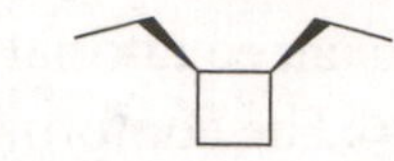

constitutional isomer
different arrangement of atoms

4.27 To classify a compound as a cis or trans isomer, **classify each non-hydrogen group as up or down. Groups on the same side = cis isomer; groups on opposite sides = trans isomer.**

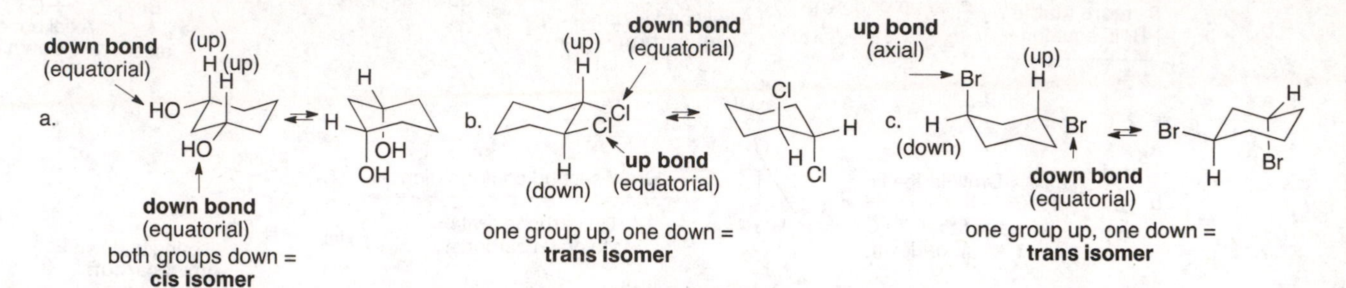

4.28

a.

groups on same side
cis isomer

groups on opposite sides
trans isomer
(one possibility)

b. cis: CH_3, H, CH_3, H ⇄ CH_3, CH_3, H, H

two chair conformations for the **cis isomer**

Same stability because they both have one equatorial, one axial CH_3 group.

c. trans: CH_3, H, H, CH_3 ⇄ H, CH_3, CH_3, H

both groups equatorial
more stable

two chair conformations for the **trans isomer**

d. The **trans isomer is more stable** because it can have both methyl groups in the more roomy **equatorial** position.

4.29

a. CH_2CH_3 ← axial CH_2CH_3; CH_3

1,1-disubstituted

b. CH_3 ← up (axial); H; CH_2CH_3 ← up; H

cis-1,2-disubstituted

c. up → CH_3CH_2; H; H; CH_3 ← down (equatorial)

trans-1,3-disubstituted

d. up (equatorial) → CH_3CH_2; H; CH_3 ← down; H

trans-1,4-disubstituted

4.30 *Oxidation* results in an *increase* in the number of C–Z bonds, or a *decrease* in the number of C–H bonds.
Reduction results in a *decrease* in the number of C–Z bonds, or an *increase* in the number of C–H bonds.

a. Decrease in the number of C–H bonds. Increase in the number of C–O bonds. **Oxidation**

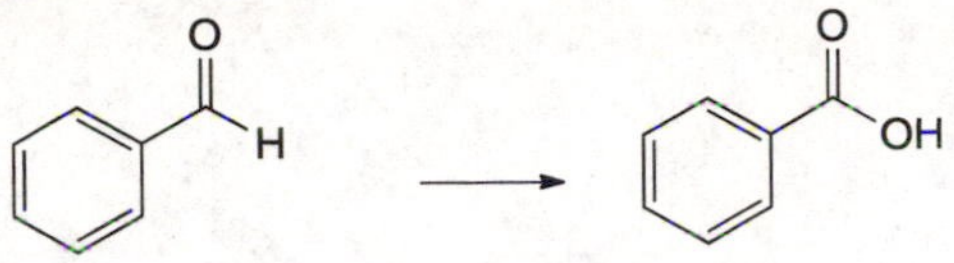

c. No change in the number of C–O or C–H bonds. **Neither**

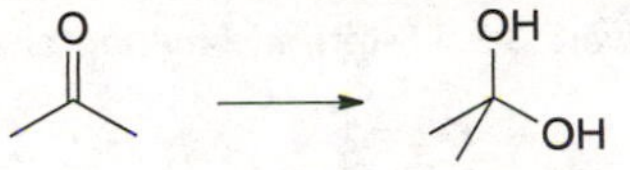

b. Decrease in the number of C–O bonds. Increase in the number of C–H bonds. **Reduction**

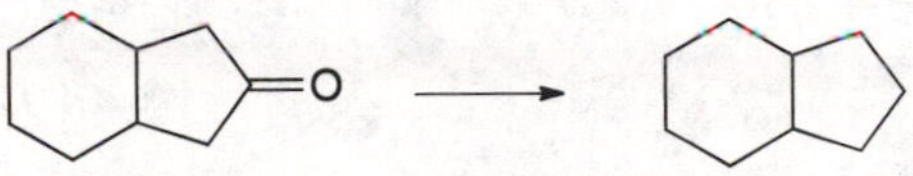

d. Decrease in the number of C–O bonds. Increase in the number of C–H bonds. **Reduction**

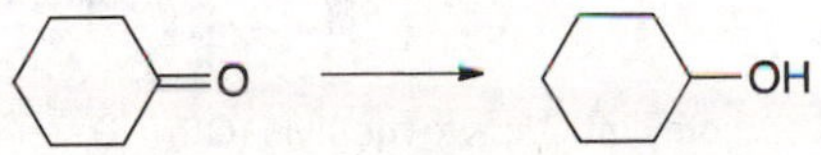

4.31 The products of a combustion reaction of a hydrocarbon are always the same: **CO_2 and H_2O.**

a. $CH_3CH_2CH_3 + 5\ O_2 \xrightarrow{\text{flamo}} 3\ CO_2 + 4\ H_2O + \text{heat}$

b.  cyclohexane $+\ 9\ O_2 \xrightarrow{\text{flame}} 6\ CO_2 + 6\ H_2O + \text{heat}$

4.32 "Like dissolves like." Beeswax is a lipid, so it will be more soluble in nonpolar solvents. H_2O is very polar, ethanol is slightly less polar, and chloroform is least polar. Beeswax is most soluble in the least polar solvent.

Increasing polarity ←

H_2O CH_3CH_2OH $CHCl_3$

→ Increasing solubility of beeswax

4.33 Re-draw the model as a skeletal structure. The longest chain has 15 C's, making it a derivative of pentadecane. Numbering from either direction gives the same numbers.

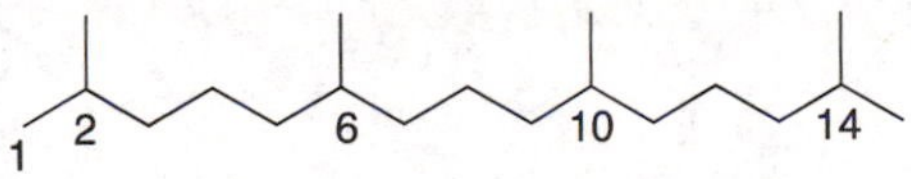

1 2 6 10 14

15 C's ----→ pentadecane

4 methyl groups at C2, C6, C10, and C14

2,6,10,14-tetramethylpentadecane

4.34 Re-draw each alkane as a skeletal structure, and use the steps in Answer 4.7 to name each compound. Classify C's as in Section 3.2.

a.

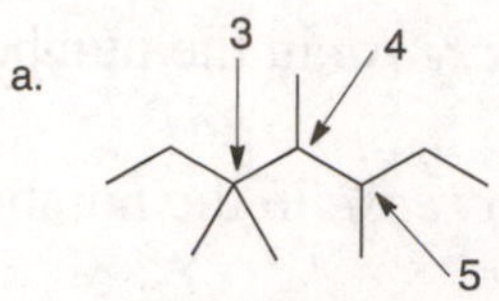

7 C's in longest chain = heptane
Number from left to right to put two substituents on C3.

● = 2° C
□ = 3° C
■ = 4° C
1° C's unlabeled

Four CH_3 groups ------> tetramethyl

Answer: **3,3,4,5-tetramethylheptane**

b.

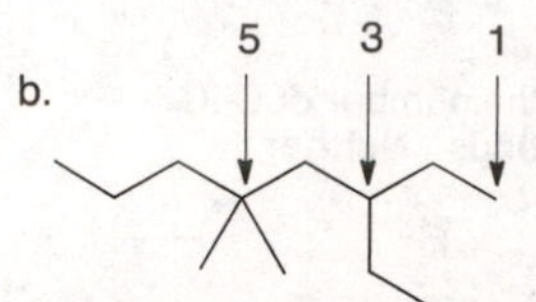

8 C's in longest chain = octane
Number from right to left to put first substituent at C3.

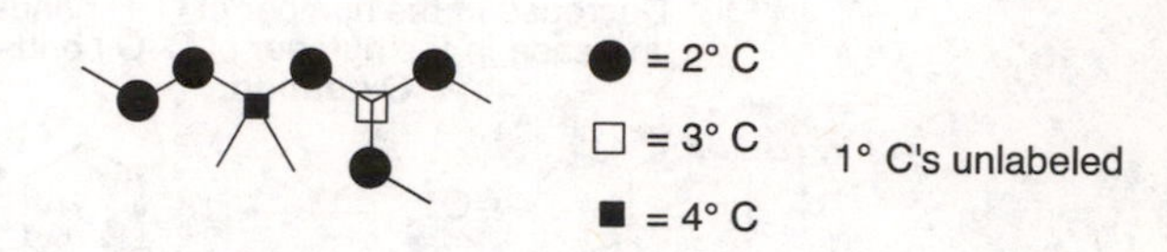

ethyl at C3; two methyls (C5) ------> dimethyl

Answer: **3-ethyl-5,5-dimethyloctane**

4.35 Re-draw the ball-and-stick model as a chair form.

a.

equatorial
H
H
4 $CH(CH_3)_2$
H
1
2
Br ← equatorial
CH_3
axial

b. CH_3 on C1 and Br on C2 are both down, making them cis.

c. Br on C2 and $CH(CH_3)_2$ on C4 are both down, making them cis.

d. Second chair form:

H
1
CH_3
H
4
H
2
Br
$CH(CH_3)_2$

This chair form is less stable because two groups [Br and $CH(CH_3)_2$] are in the more crowded axial position.

4.36

a.

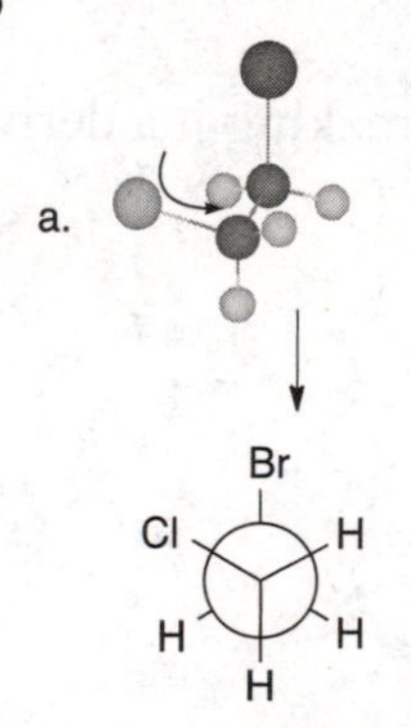

b.

Cl
H
H
H
H
CH_3

c.

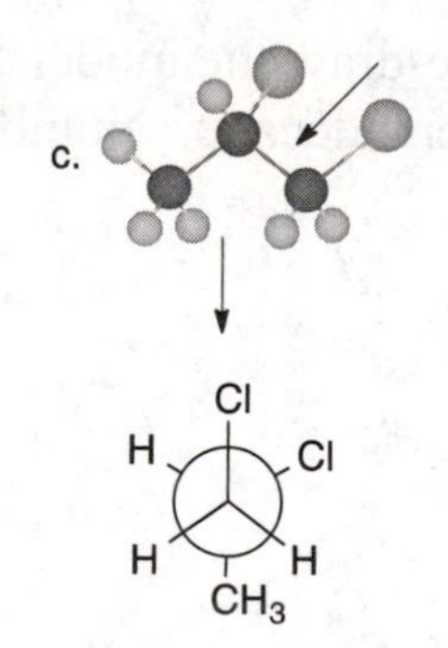

4.37

a. Five constitutional isomers of molecular formula C_4H_8:

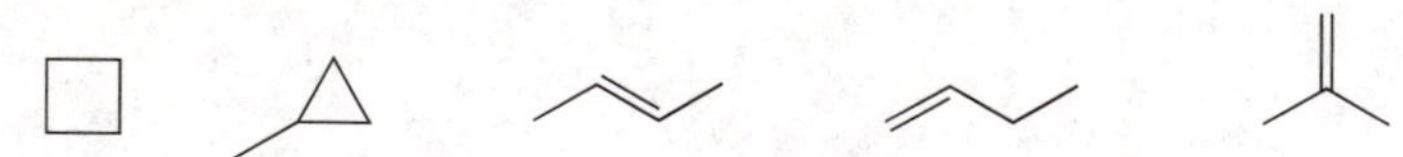

b. Nine constitutional isomers of molecular formula C_7H_{16}:

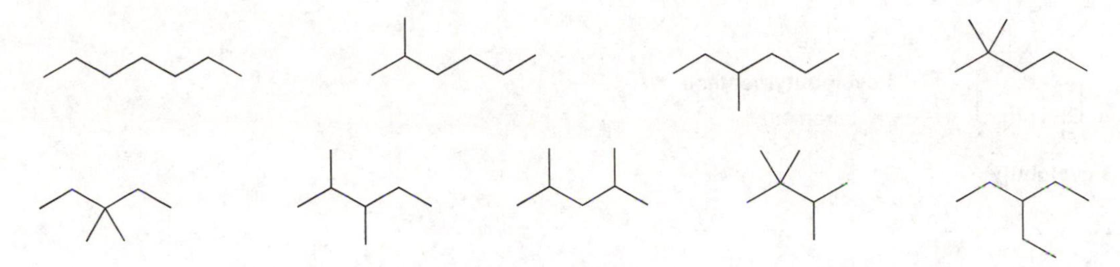

c. Twelve constitutional isomers of molecular formula C_6H_{12} containing one ring:

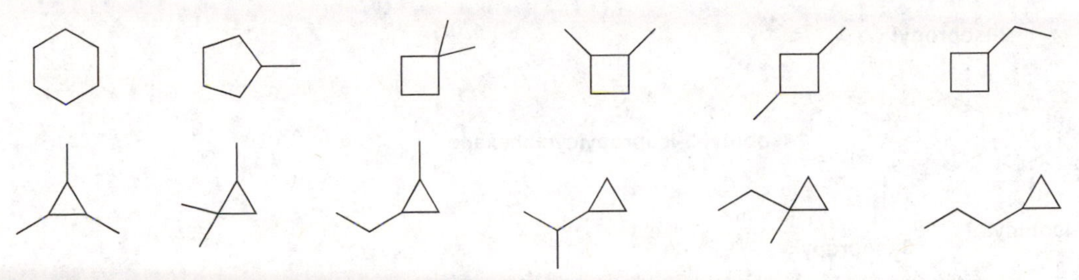

4.38 Use the steps in Answers 4.7 and 4.11 to name the alkanes.

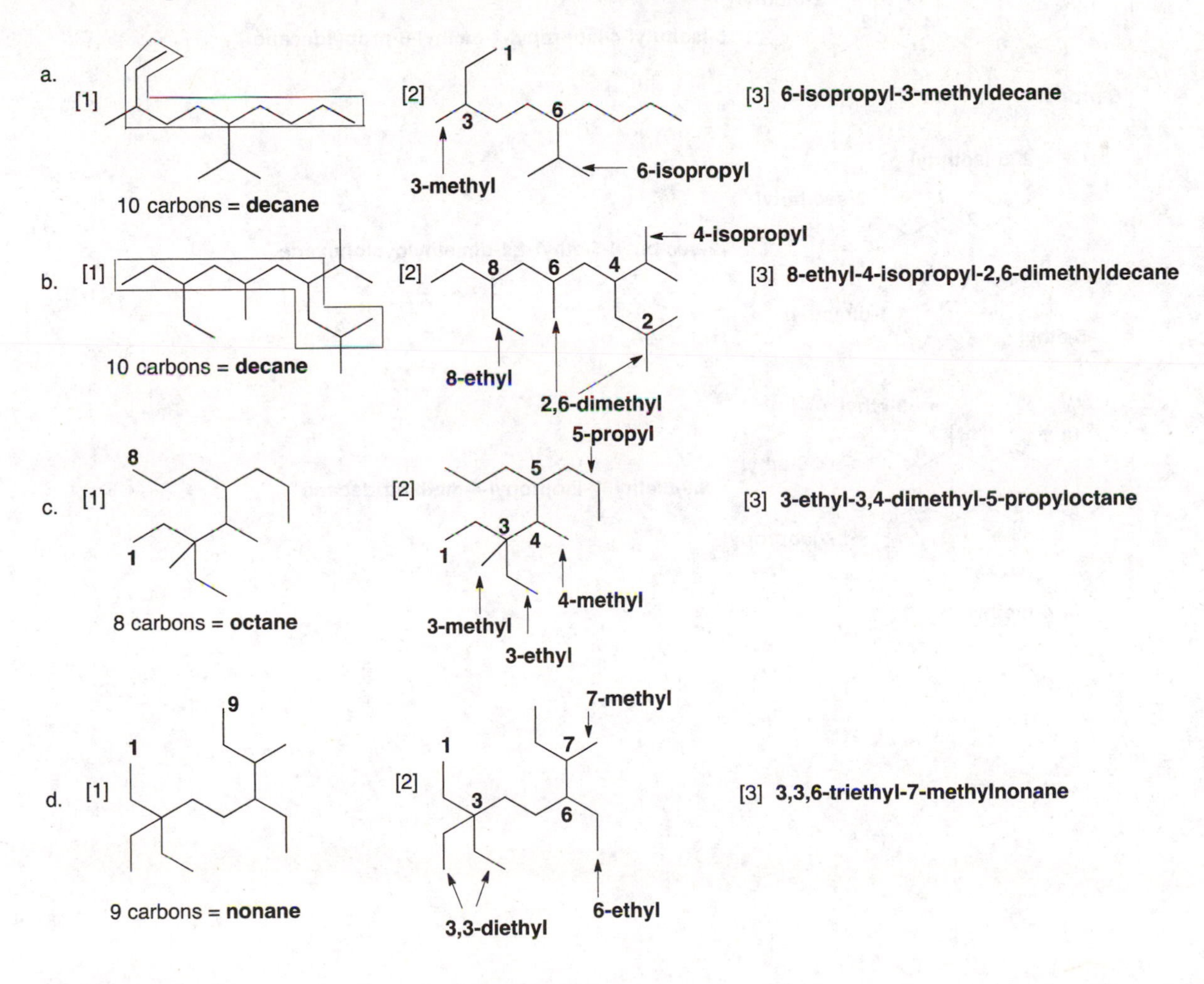

e. **3-cyclobutylpentane**

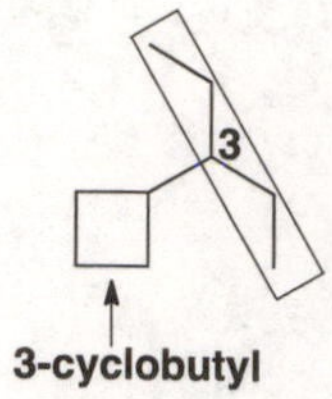

3-cyclobutyl

f. **1-*sec*-butyl-2-isopropylcyclopentane**

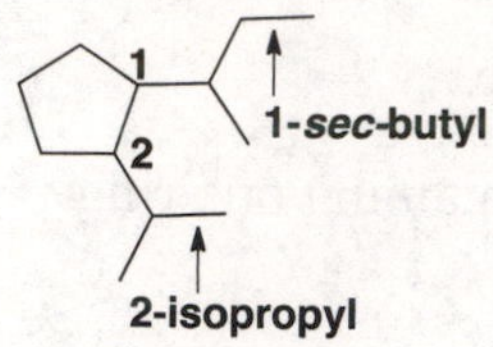

1-*sec*-butyl

2-isopropyl

g. **1-isobutyl-3-isopropylcyclohexane**

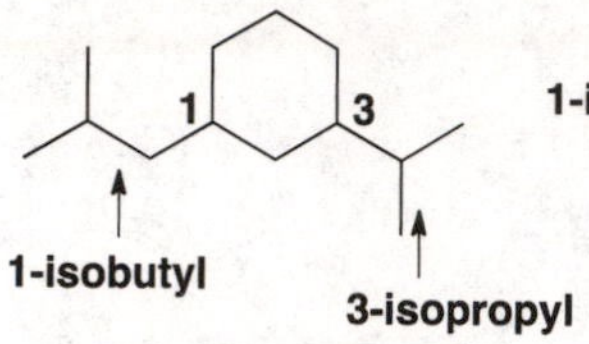

1-isobutyl

3-isopropyl

h. **5-isobutyl-4-isopropyl-2-methyl-6-propyldecane**

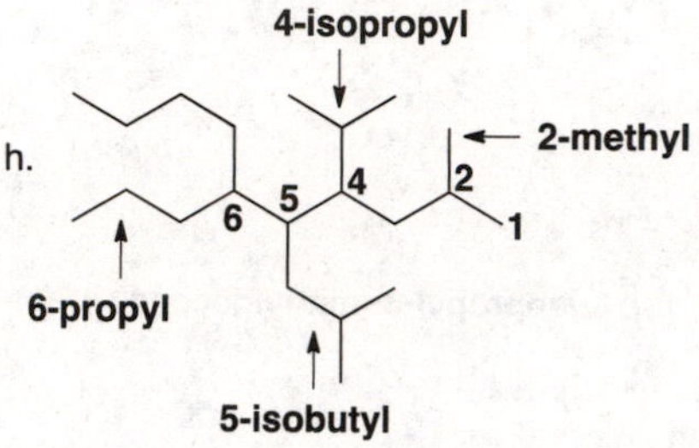

4-isopropyl

2-methyl

6-propyl

5-isobutyl

i. **2-*sec*-butyl-5-ethyl-1,1-dimethylcyclohexane**

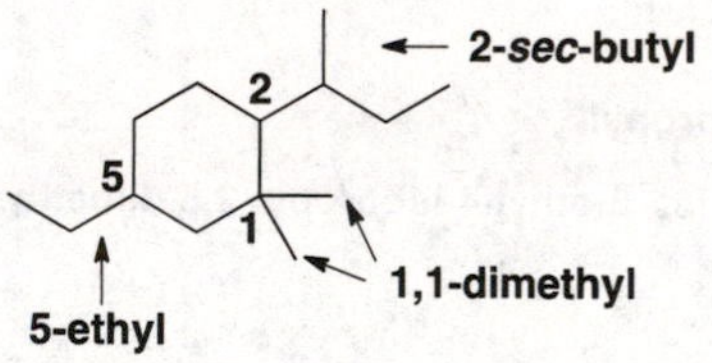

2-*sec*-butyl

1,1-dimethyl

5-ethyl

j. **8,9-diethyl-7-isopropyl-4-methyltridecane**

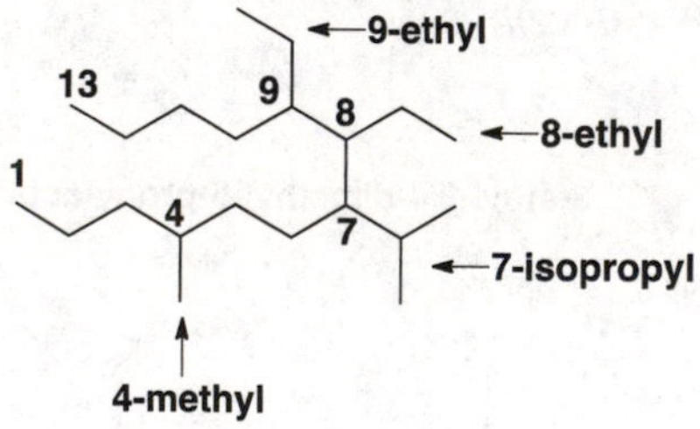

9-ethyl

8-ethyl

7-isopropyl

4-methyl

4.39

← 2,2-dimethyl
2

2,2-dimethylheptane

3
← 3,3-dimethyl

3,3-dimethylheptane

← 4,4-dimethyl
4

4,4-dimethylheptane

3 4

3,4-dimethylheptane

2 4

2,4-dimethylheptane

2 5

2,5-dimethylheptane

2 3

2,3-dimethylheptane

3 5

3,5-dimethylheptane

2 6

2,6-dimethylheptane

4.40 Use the steps in Answer 4.9 to draw the structures.

a. 3-ethyl-2-methyl**hexane**

[1] 6 C chain

[2]

methyl on C2

← ethyl on C3

b. *sec*-butyl**cyclopentane**

[1] 5 C ring

[2]

isopropyl on C4

c. 4-isopropyl-2,4,5-trimethyl**undecane**

[1] 11 C chain

[2]

methyls on C2, C4, and C5

d. cyclobutylcycloheptane

[1] 7 C cycloalkane

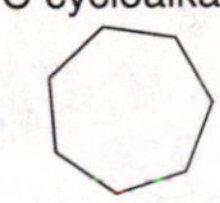

[2]

e. 3-ethyl-1,1-dimethyl**cyclohexane**

[1] 6 C cycloalkane

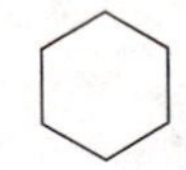

[2]

← 2 methyl
← groups on C1

ethyl on C3

2 ethyl groups

f. 4-butyl-1,1-diethyl**cyclooctane**

[1] 8 C cycloalkane

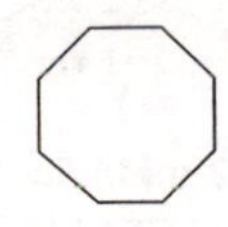

[2]

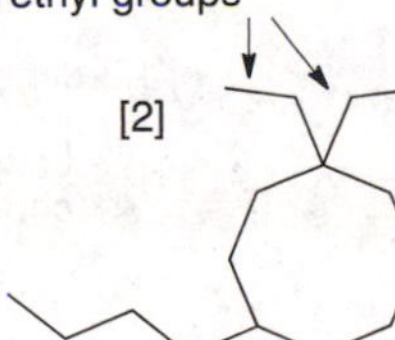

g. 6-isopropyl-2,3-dimethyl**dodecane**

[1] 12 C alkane

[2] methyl on C2, isopropyl on C6, methyl on C3

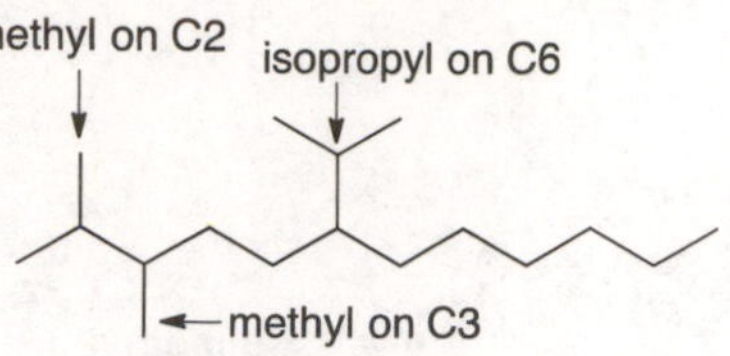

h. 2,2,6,6,7-pentamethyl**octane**

[1] 8 C alkane

[2] 5 methyl groups

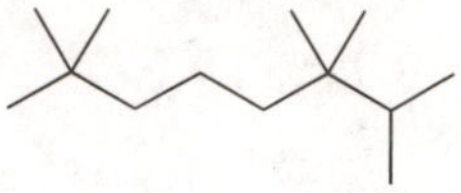

i. *cis*-1-ethyl-3-methyl**cyclopentane**

[1] 5 C ring

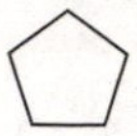

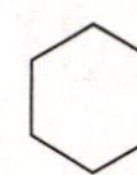

[2] 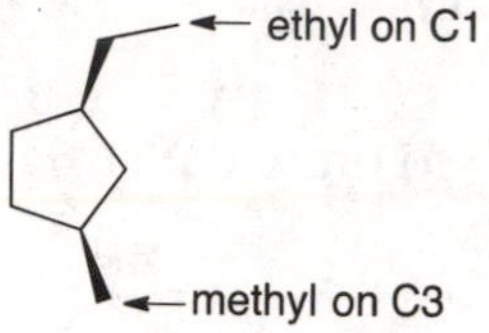 ethyl on C1, methyl on C3 or

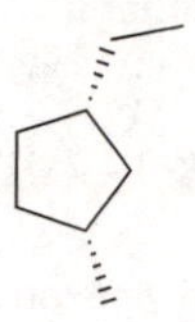

j. *trans*-1-*tert*-butyl-4-ethyl**cyclohexane**

[1] 6 C ring

[2]

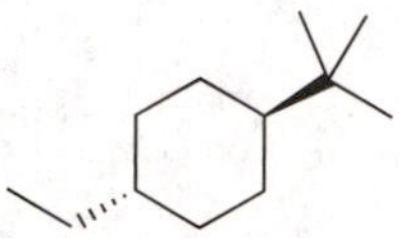

4.41 Draw the compounds.

a. 2,2-dimethyl-4-ethylheptane

alphabetized incorrectly
ethyl before **m**ethyl

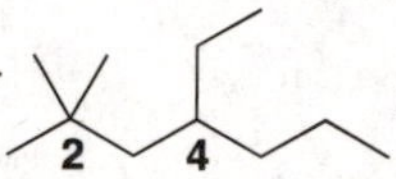

4-ethyl-2,2-dimethylheptane

b. 5-ethyl-2-methylhexane

Longest chain was not chosen = **heptane**

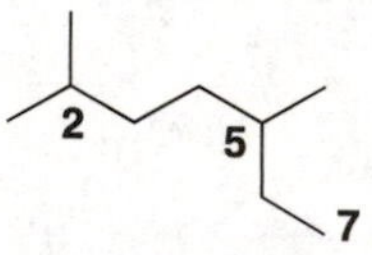

2,5-dimethylheptane

c. 2-methyl-2-isopropylheptane

Longest chain was not chosen = **octane**

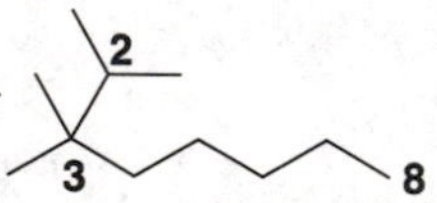

2,3,3-trimethyloctane

d. 1,5-dimethylcyclohexane

Numbered incorrectly. Re-number so methyls are at C1 and C3.

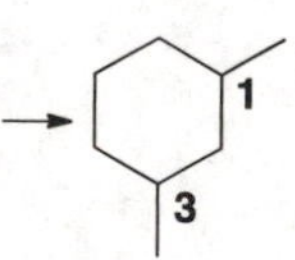

1,3-dimethylcyclohexane

e. 1-ethyl-2,6-dimethylcycloheptane

Numbered incorrectly. Re-number so methyls are at C1 and C4.

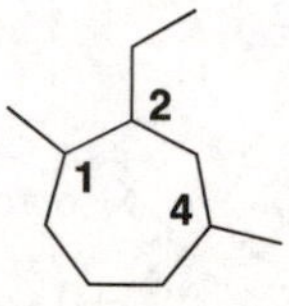

2-ethyl-1,4-dimethylcycloheptane

f. 5,5,6-trimethyloctane

Numbered incorrectly. Re-number so methyls are at C3 and C4.

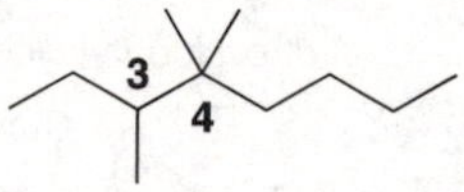

3,4,4-trimethyloctane

g. 3-butyl-2,2-dimethylhexane

Longest chain not chosen = **octane**

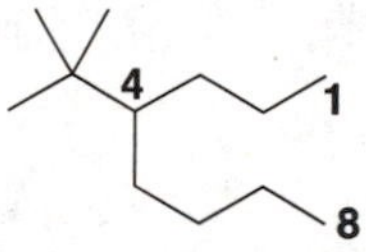

4-*tert*-butyloctane

h. 1,3-dimethylbutane

Longest chain not chosen = **pentane**

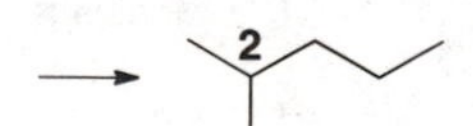

2-methylpentane

4.42

a. Newman projection: front carbon CH_3, H, $CH_2CH_2CH_3$; back carbon CH_3, H, $CH_2CH_2CH_3$

re-draw

4

4-isopropylheptane

b. Newman projection: front carbon CH_3, CH_3, CH_2CH_3; back carbon H, H, CH_2CH_3

re-draw

3

3-ethyl-3-methylpentane

c. Newman projection: front carbon CH_3, CH_3CH_2, $CH_2CH_2CH_3$; back carbon $CH_3CH_2CH_2$, H, CH_2CH_3

re-draw

4 5

4,4-diethyl-5-methyloctane

4.43 Use the rules from Answer 4.13.

most branching
lowest boiling point

least branching
highest boiling point

4.44 a.

$CH_3(CH_2)_6CH_3$
no branching = higher surface area
higher boiling point

$(CH_3)_3CC(CH_3)_3$
branching = lower surface area
lower boiling point
more spherical, better packing =
higher melting point

b. There is a 159 °C difference in the melting points, but only a 20 °C difference in the boiling points because the symmetry in $(CH_3)_3CC(CH_3)_3$ allows it to pack more tightly in the solid, thus requiring more energy to melt. In contrast, once the compounds are in the liquid state, symmetry is no longer a factor, the compounds are isomeric alkanes, and the boiling points are closer together.

4.45

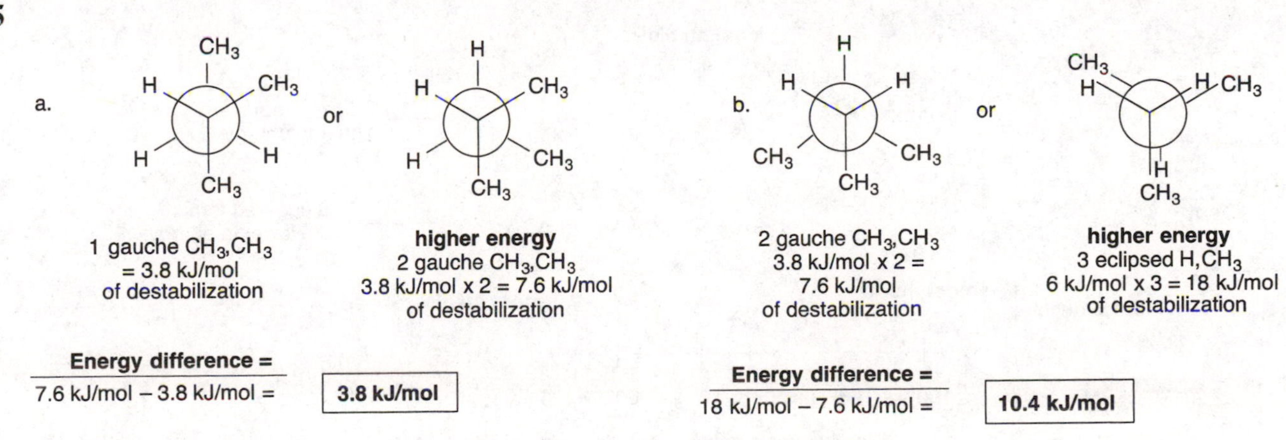

4.46 Use the rules from Answer 4.17 to determine the most and least stable conformations.

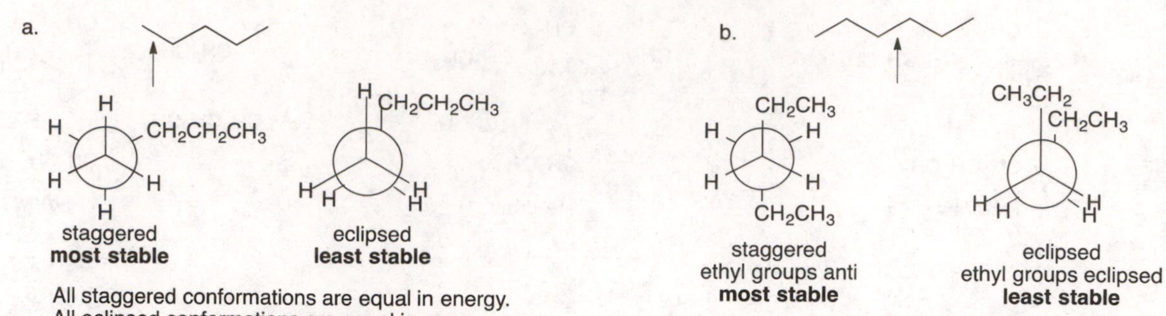

All staggered conformations are equal in energy.
All eclipsed conformations are equal in energy.

4.47

a.

b.

c.

4.48

[1]

Eclipsed forms are higher in energy.

Staggered forms are lower in energy.

least stable

most stable

most stable

Energy

180° 120° 60° 0° 60° 120° 180°

Dihedral angle between two alkyl groups

[2]

most stable
CH_3CH_2 H H H CH_3 CH_3
1

60°

H H H CH_2CH_3 CH_3 CH_3
2

60°

H H H CH_3 CH_2CH_3 CH_3
3

60°

H CH_3 H H CH_3 CH_2CH_3
least stable
4

60°

CH_3 H H CH_3CH_2 H CH_3
5

60°

CH_3CH_2 H H CH_3 CH_3 H
6

60°

least stable
4
2
6
3
5
1
most stable
1
most stable
Energy
Eclipsed forms are higher in energy.
Staggered forms are lower in energy.
180° 120° 60° 0° 60° 120° 180°
Dihedral angle
(between CH_3CH_2 in back and CH_3 in front)

4.49 Two types of strain:

- *Torsional strain* is due to eclipsed groups on adjacent carbon atoms.
- *Steric strain* is due to overlapping electron clouds of large groups (e.g. gauche interactions).

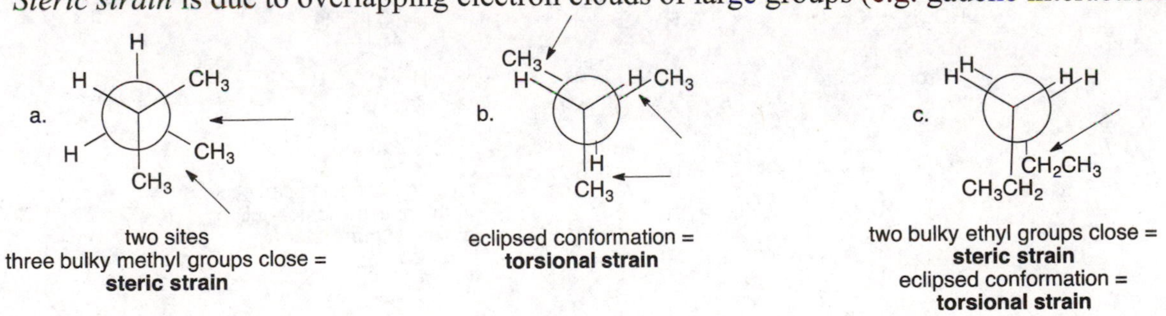

4.50 The barrier to rotation is equal to the difference in energy between the highest energy eclipsed and lowest energy staggered conformations of the molecule.

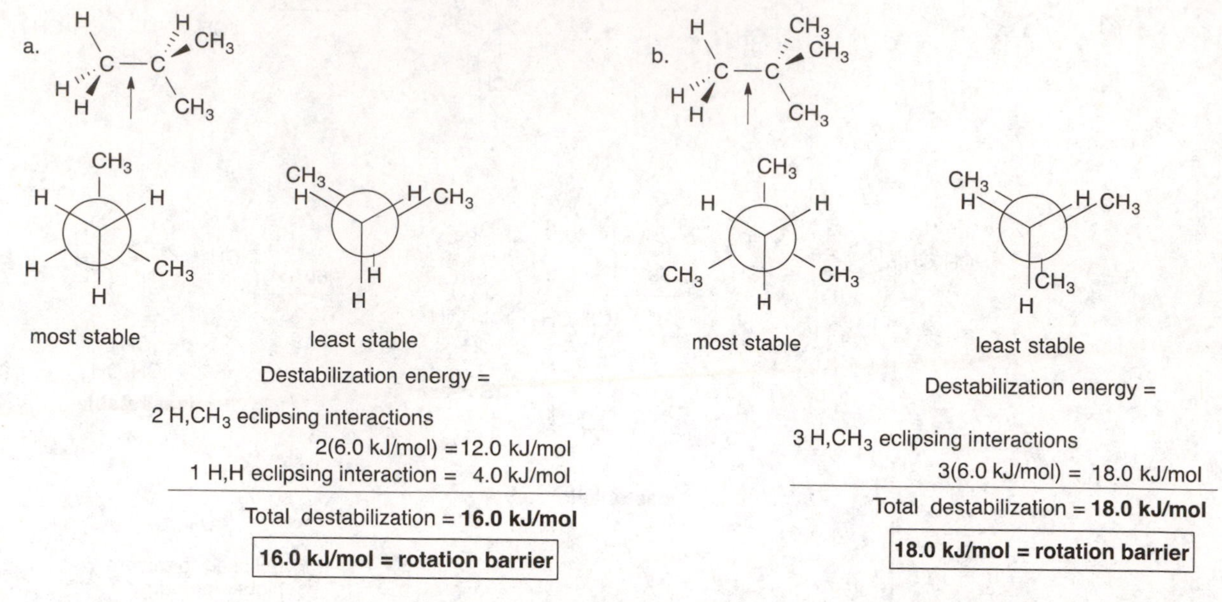

a.
most stable
least stable
Destabilization energy =
2 H,CH3 eclipsing interactions
2(6.0 kJ/mol) = 12.0 kJ/mol
1 H,H eclipsing interaction = 4.0 kJ/mol
Total destabilization = 16.0 kJ/mol
16.0 kJ/mol = rotation barrier
b.
most stable
least stable
Destabilization energy =
3 H,CH3 eclipsing interactions
3(6.0 kJ/mol) = 18.0 kJ/mol
Total destabilization = 18.0 kJ/mol
18.0 kJ/mol = rotation barrier

4.51

most stable   least stable

2 H,H eclipsing interactions = 2(4.0 kJ/mol) = 8.0 kJ/mol

Since the barrier to rotation is 15 kJ/mol, the difference between this value and the destabilization due to H,H eclipsing is the destabilization due to H,Cl eclipsing.

15.0 kJ/mol – 8.0 kJ/mol = 7.0 kJ/mol destabilization due to H,Cl eclipsing

4.52 The gauche conformation can intramolecularly hydrogen bond, making it the more stable conformation.

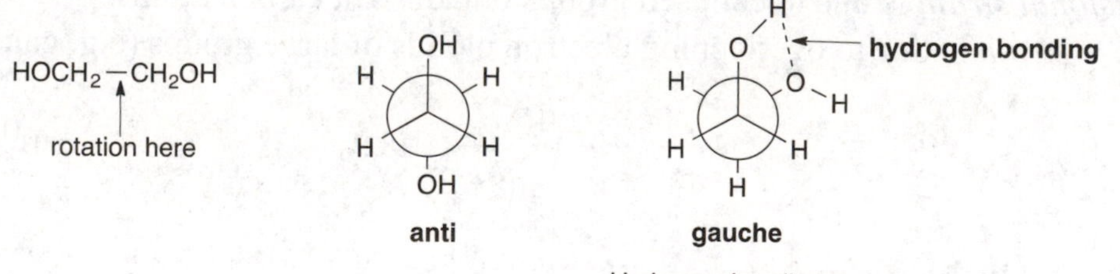

Hydrogen bonding can occur only in the gauche conformation, making it **more stable.**

4.53

[1] a. **axial** H, OH **axial**, HO **eq**, H **eq**

b. H **up** OH, **down** HO, H

one up, one down = **trans**

c. HO, OH

d. ax H ax OH, eq HO, H eq ⇌ H eq, eq HO, OH ax H ax

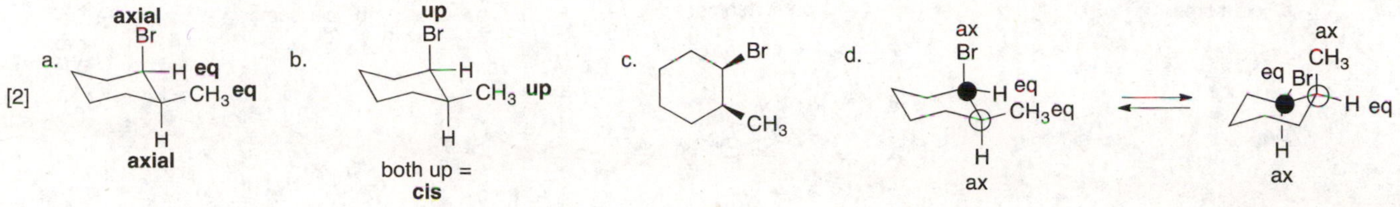

[2] a. axial Br, H eq, CH_3 eq, H axial
b. up Br, H, CH_3 up, H — both up = **cis**
c. Br, CH_3
d. ax Br, H eq, CH_3eq, H ax ⇌ ax CH_3, eq Br, H eq, H ax

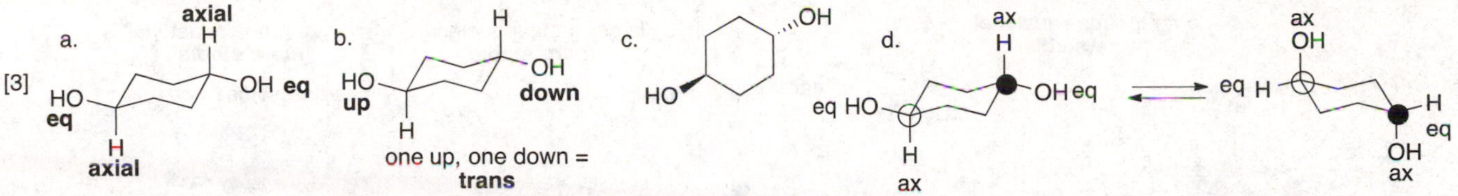

[3] a. axial H, OH eq, HO eq, H axial
b. H, OH down, HO up, H — one up, one down = **trans**
c. OH, HO
d. ax H, OH eq, eq HO, H ax ⇌ ax OH, eq H, H eq, OH ax

4.54

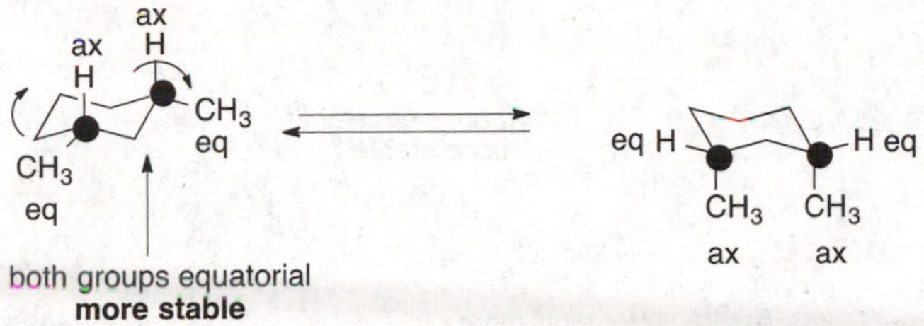

ax H, ax H, CH_3 eq, CH_3 eq ⇌ eq H, H eq, CH_3 ax, CH_3 ax

both groups equatorial
more stable

4.55 A **cis isomer** has two groups on the **same side** of the ring. The two groups can be drawn both up or both down. Only one possibility is drawn. A **trans isomer** has one group on one side of the ring and one group on the other side. Either group can be drawn on either side. Only one possibility is drawn.

[1] [2] [3]

a. **cis** **trans**

a. **cis** **trans**

a. **cis** **trans**

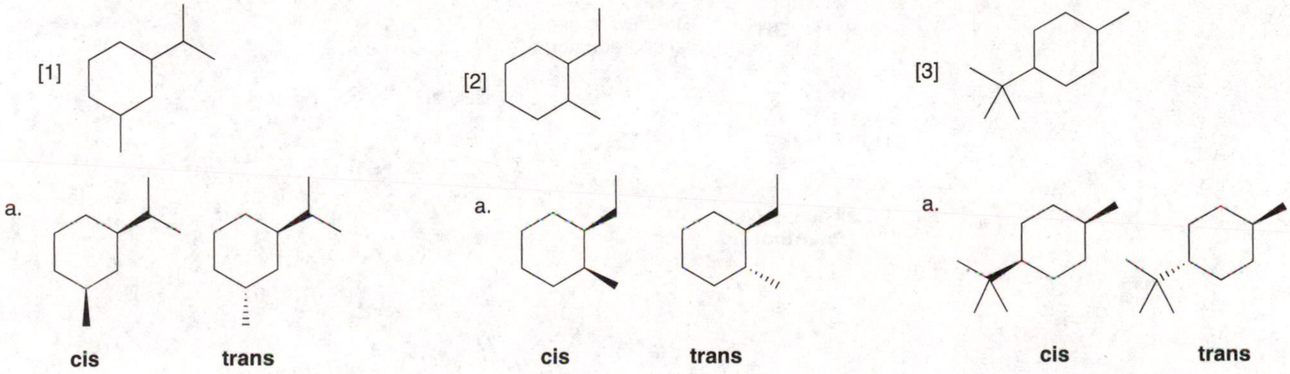

b. cis isomer

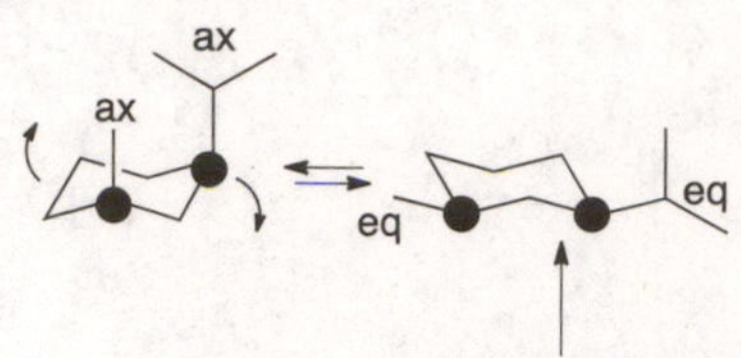

both groups equatorial
more stable

c. trans isomer

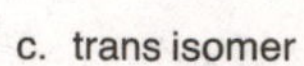

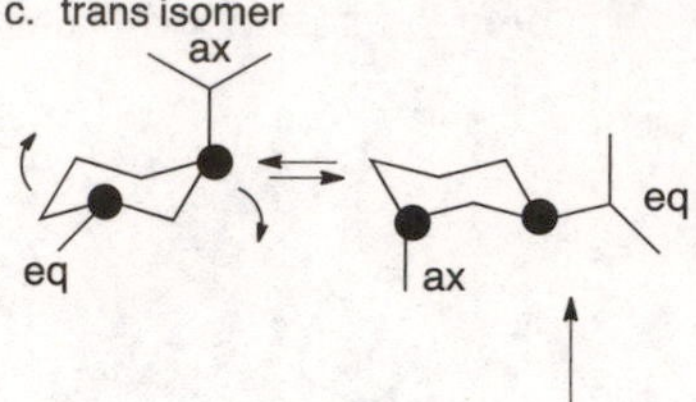

larger group equatorial
more stable

d.

The cis isomer is more stable than the trans because one conformation has both groups equatorial.

b. cis isomer

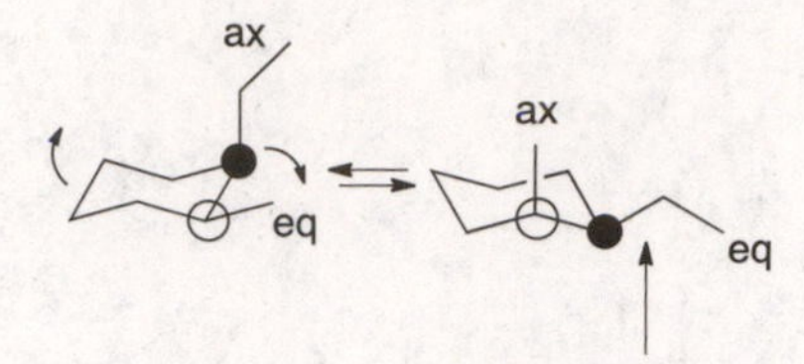

larger group equatorial
more stable

c. trans isomer

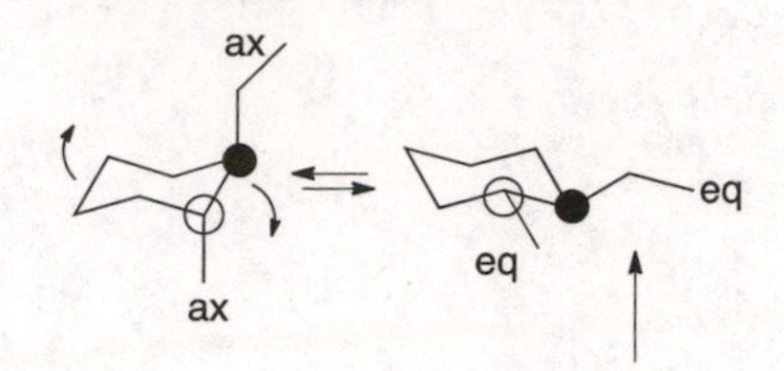

both groups equatorial
more stable

d.

The trans isomer is more stable than the cis because one conformation has both groups equatorial.

b. cis isomer

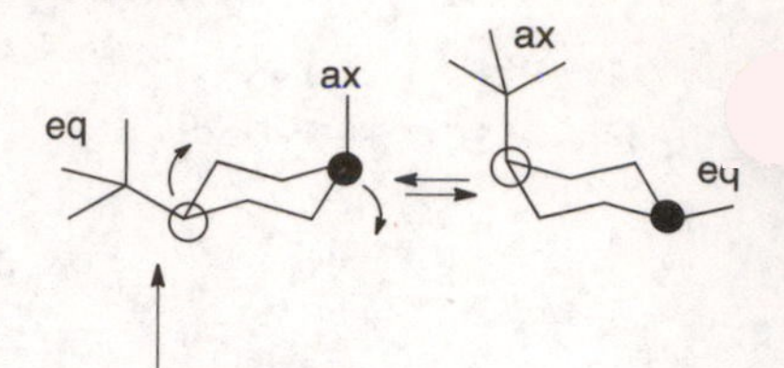

larger group equatorial
more stable

c. trans isomer

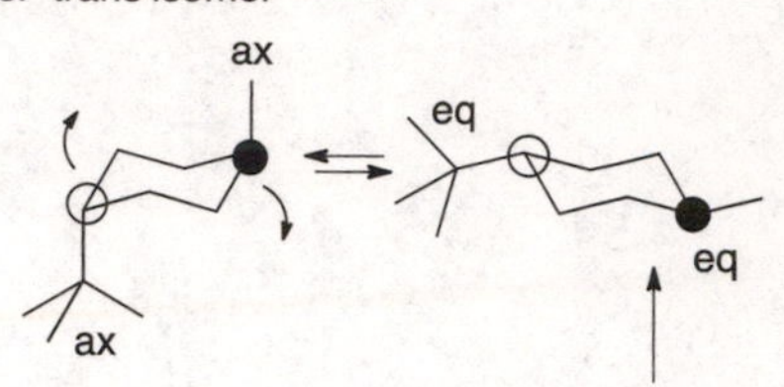

both groups equatorial
more stable

d.

The trans isomer is more stable than the cis because one conformation has both groups equatorial.

4.56

OH

Re-draw to see axial and equatorial.

OH

CH_3

HO

all equatorial
menthol

CH_3

HO

4.57 a.

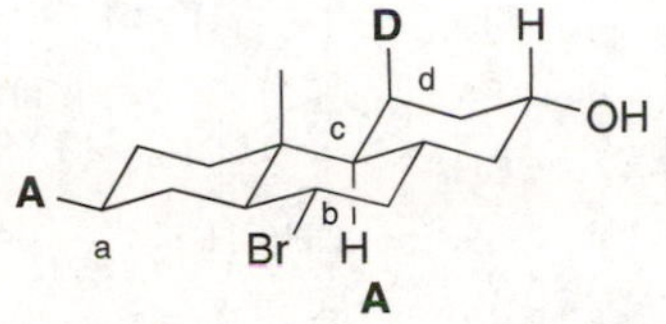

D H d c OH A a b Br H A

CH_3 is axial and up. OH is equatorial and down, so the two groups are trans.

b. A substituent (**A**) on C_a that is cis to the existing CH_3 must be up, so it is equatorial.
c. An equatorial Br on C_b is down, so it is cis to the OH group.
d. A H on C_c must be axial and down, making it cis to the OH group.
e. A substituent (**D**) on C_d that is trans to the OH must be up and axial.

4.58

a. HO HO O OH OH HO [or HO OH HO HO O OH]

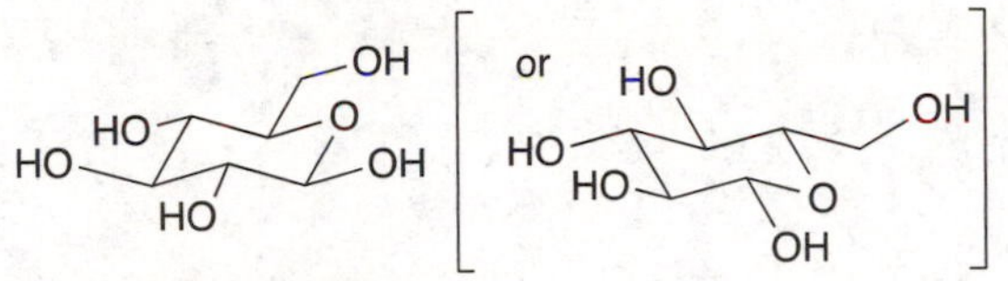

most stable
All groups are equatorial.

c. OH HO O OH OH HO

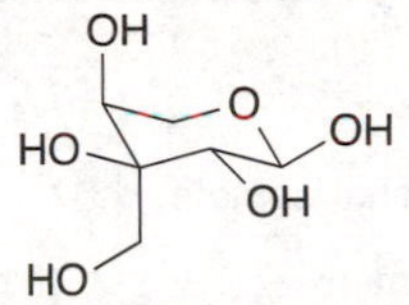

constitutional isomer

b. HO O HO OH HO OH [HO or O HO OH HO OH]

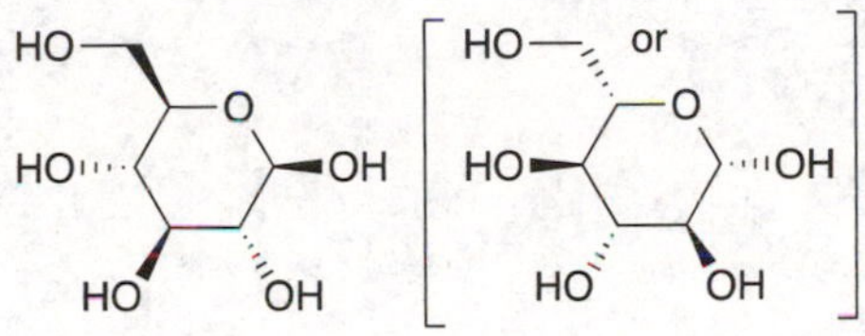

d. OH OH O HO OH OH

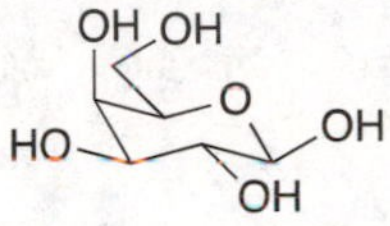

4.59

a. and H CH_3 CH_3 H =

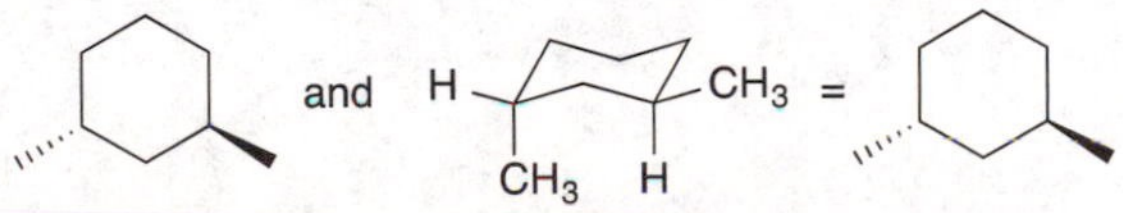

1 down, 1 up = **trans**

1 down, 1 up = **trans**

same arrangement in three dimensions
identical

b. and

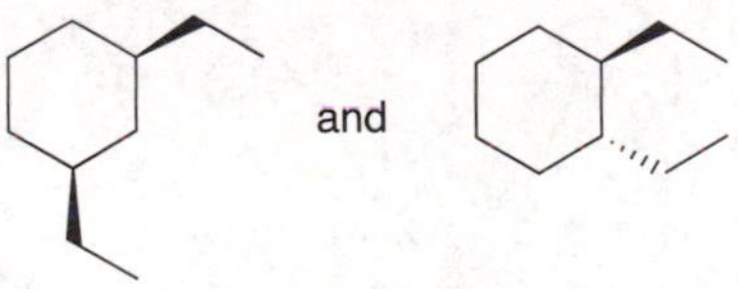

same molecular formula $C_{10}H_{20}$
different connectivity
constitutional isomers

c. and CH_2CH_3 CH_3CH_2

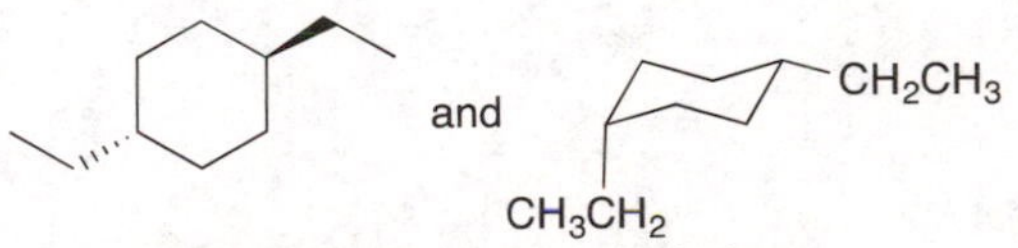

1 down, 1 up = **trans**

both down = **cis**

different arrangement in three dimensions
stereoisomers

d. **ax** CH_3 ← **up** **eq** CH_3CH_2 H H **up** and **ax** CH_2CH_3 **up** **eq** CH_3 H H **up**

both up = **cis**

both up = **cis**

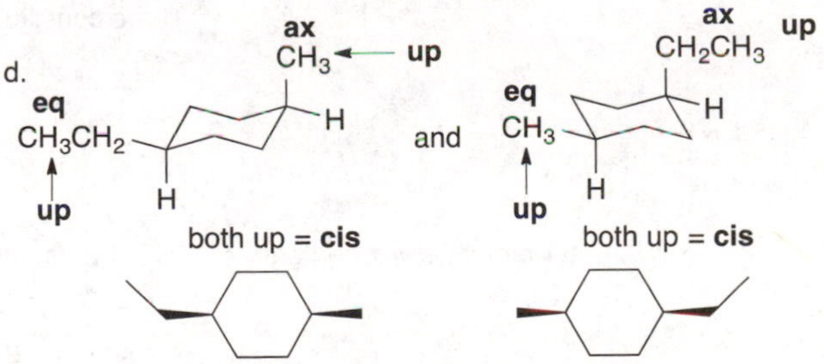

same arrangement in three dimensions
identical

4.60

a. $CH(CH_3)_2$, H, H, H, CH_2CH_3, CH_3 and CH_3CH_2, CH_3, H, H, CH_3, CH_2CH_3

re-draw

3-ethyl-2-methylpentane 3-ethyl-2-methylpentane

same molecular formula
same name
identical molecules

b. H, CH_3, H, CH_3, CH_3, $CH(CH_3)_2$ and

re-draw

same molecular formula
different arrangement of atoms
constitutional isomers

4.61

a. **A** and **B:** constitutional isomers
A and **C:** identical
B and **D:** stereoisomers

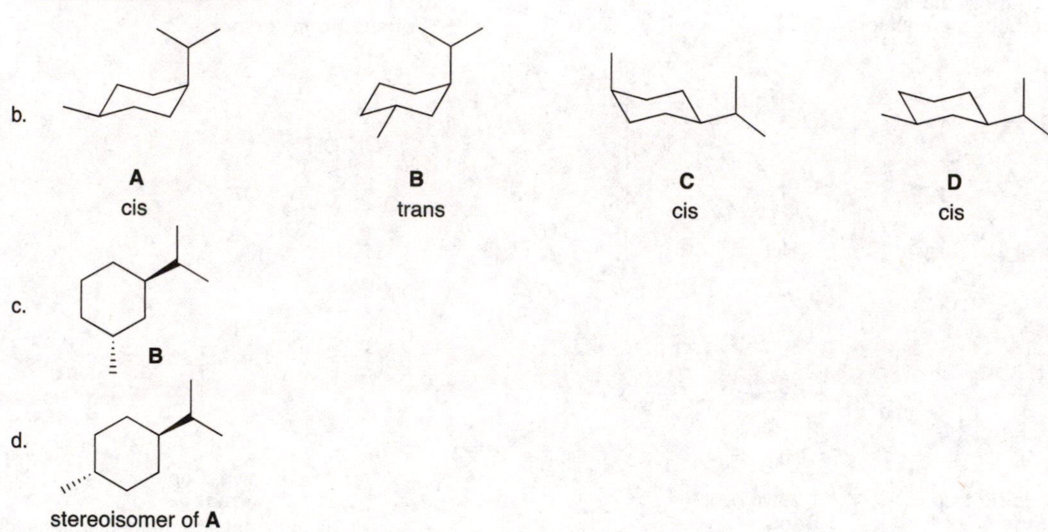

4.62

Three constitutional isomers of C_7H_{14}:

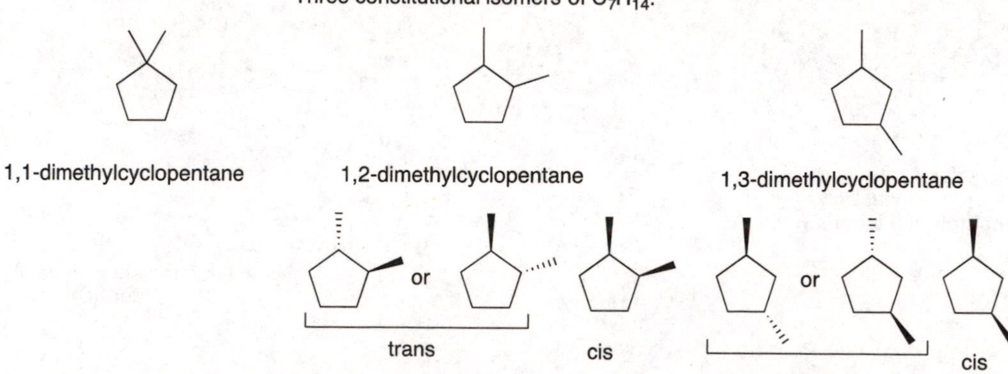

4.63 Use the definitions from Answer 4.30 to classify the reactions.

a. 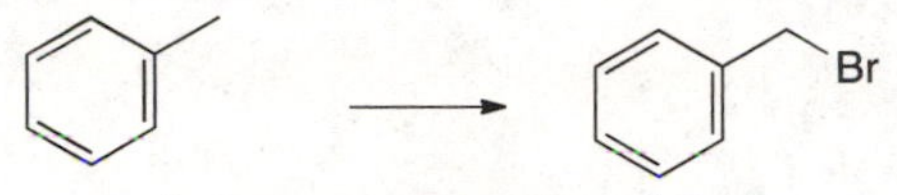 Increase in the number of C–Z bonds. **Oxidation**

b. 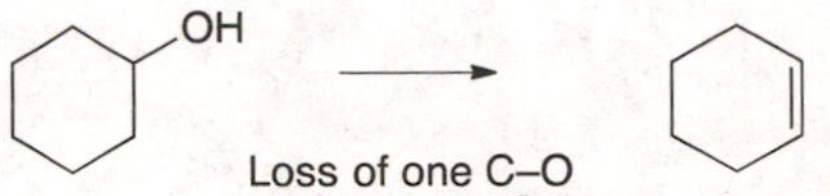 Loss of one C–O bond *and* one C–H bond. **Neither**

4.64 Use the rule from Answer 4.31.

a. 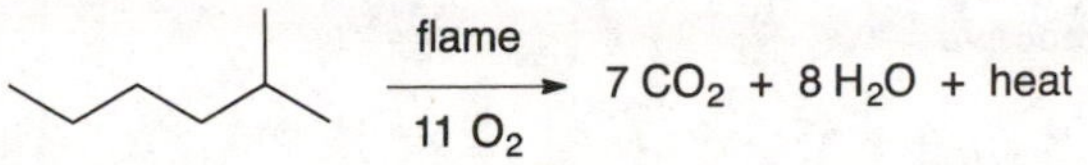 $\xrightarrow[11\ O_2]{\text{flame}}$ $7\ CO_2 + 8\ H_2O$ + heat

b. 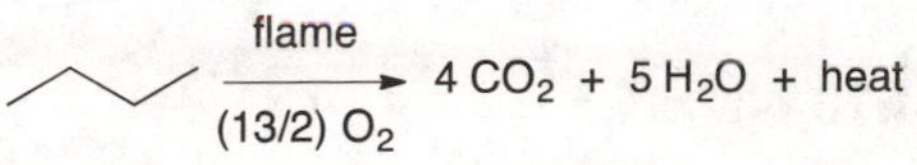 $\xrightarrow[(13/2)\ O_2]{\text{flame}}$ $4\ CO_2 + 5\ H_2O$ + heat

4.65

a. 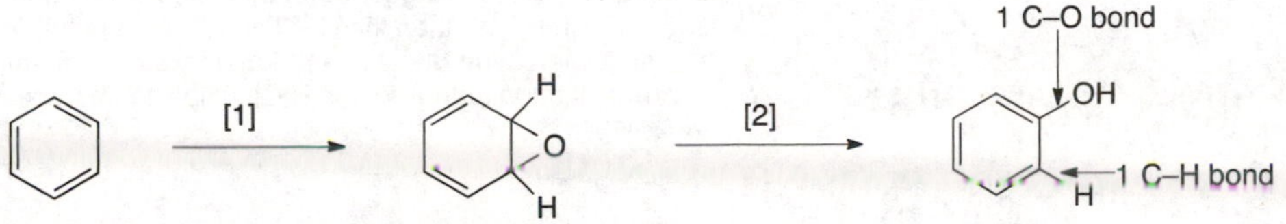 **benzene** →[1] **an arene oxide** (2 C–O bonds) →[2] **phenol** (1 C–O bond; 1 C–H bond)

[1] increase in C–O bonds **oxidation reaction**

[2] loss of 1 C–O bond, loss of 1 C–H bond **neither**

b. Phenol is more water soluble than benzene because it is **polar (contains an O–H group) and can hydrogen bond with water,** whereas benzene is nonpolar and cannot hydrogen bond.

4.66

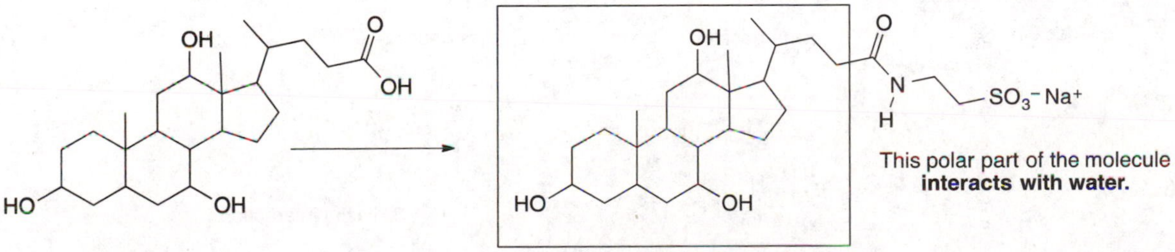

cholic acid
a bile acid

a bile salt

This polar part of the molecule **interacts with water.**

This nonpolar part of the molecule can **interact with lipids** to create micelles that allow for transport of lipids through aqueous environments.

4.67 The mineral oil can prevent the body's absorption of important fat-soluble vitamins. The vitamins dissolve in the mineral oil, and are thus not absorbed. Instead, they are expelled with the mineral oil.

4.68 Cyclopropane has larger angle strain than cyclobutane because the internal angles in the three-membered ring (60°) are smaller than they are in cyclobutane. Although cyclobutane is not flat, as shown in Figure 4.11, there are more C–H bonds than there are in cyclopropane, so there are more sites of torsional strain. Thus cyclopropane has more angle strain but less torsional strain. The result is that both cyclopropane and cyclobutane have roughly similar strain energies.

4.69 The amide in the four-membered ring has 90° bond angles giving it angle strain, which makes it more reactive.

amide

penicillin G

strained amide
more reactive

4.70

Example:

Although I is a much bigger atom than Cl, the C–I bond is also much longer than the C–Cl bond. As a result the eclipsing interaction of the H and I atoms is not very much different in magnitude from the H,Cl eclipsing interaction.

longer bond

4.71

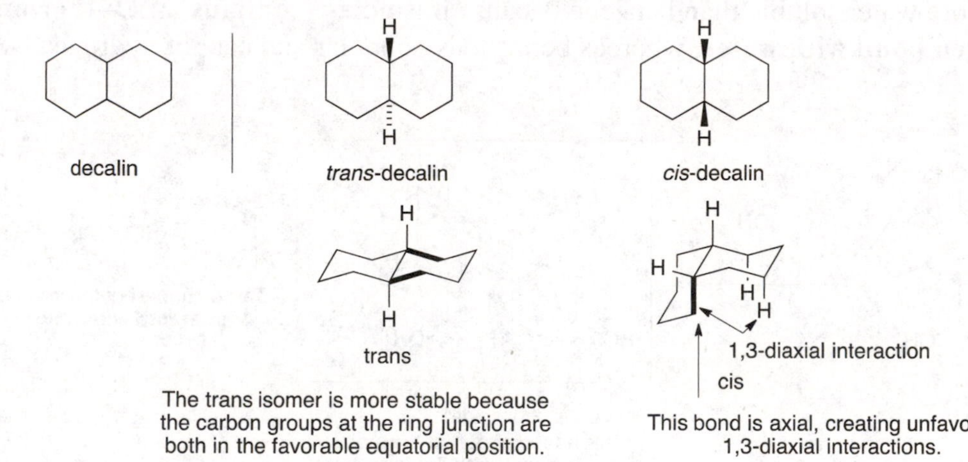

The trans isomer is more stable because the carbon groups at the ring junction are both in the favorable equatorial position.

This bond is axial, creating unfavorable 1,3-diaxial interactions.

4.72 Re-draw the ball-and-stick model using chair forms.

a.

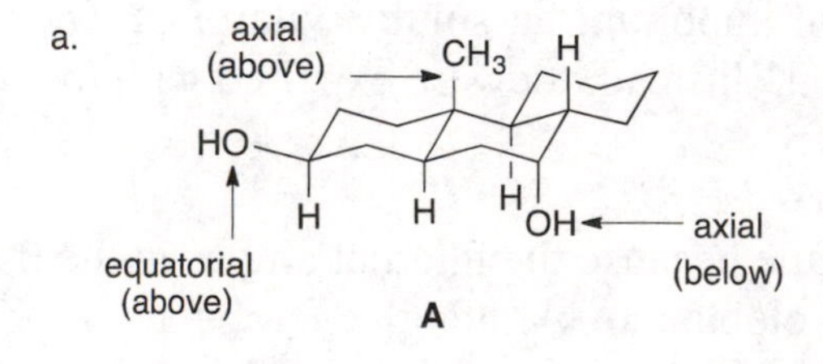

b. All bonds above the ring are on wedges and all bonds below the ring are on dashed wedges.

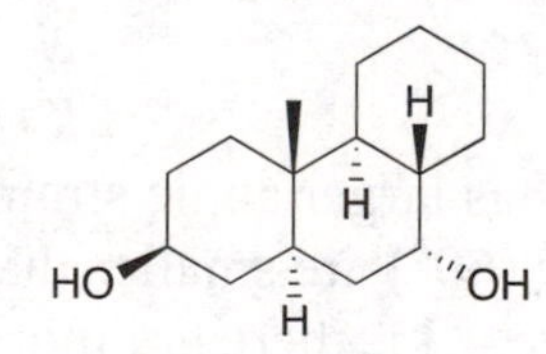

4.73

a, b. CH_3 ← axial; Cl ← axial; Cl ← equatorial

B

c. The circled H's at one ring fusion are cis. The boxed in CH_3 and H at the second ring fusion are trans.

4.74

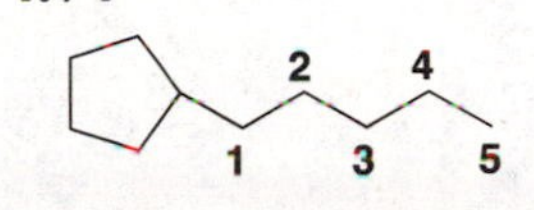

pentylcyclopentane

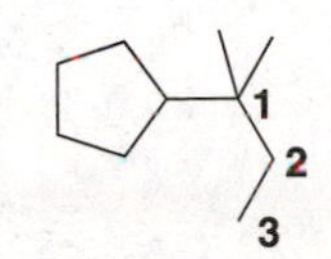

(1,1-dimethylpropyl)cyclopentane

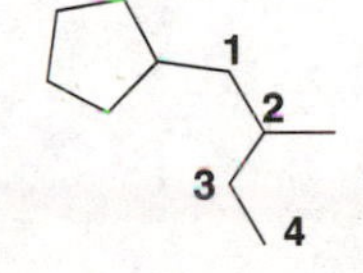

(2-methylbutyl)cyclopentane

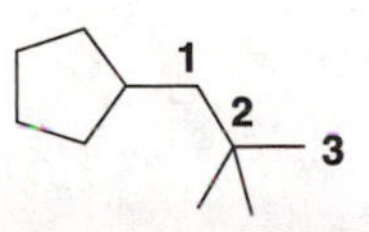

(2,2-dimethylpropyl)cyclopentane

(1-methylbutyl)cyclopentane

(1-ethylpropyl)cyclopentane

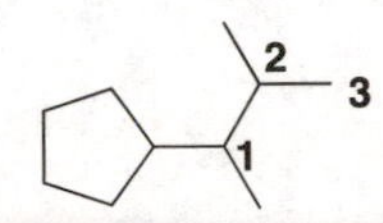

(1,2-dimethylpropyl)cyclopentane

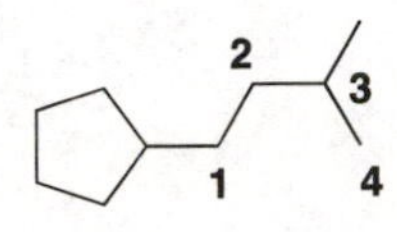

(3-methylbutyl)cyclopentane

4.75

a.

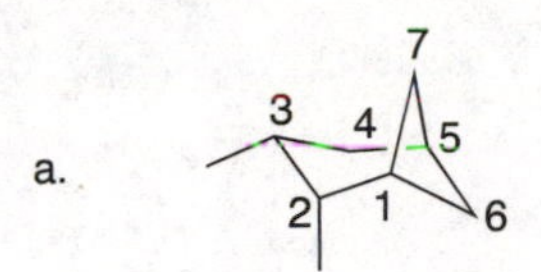

2,3-dimethylbicyclo[3.1.1]heptane

c.

1-methyl-7-propylbicyclo[3.2.1]octane

b.

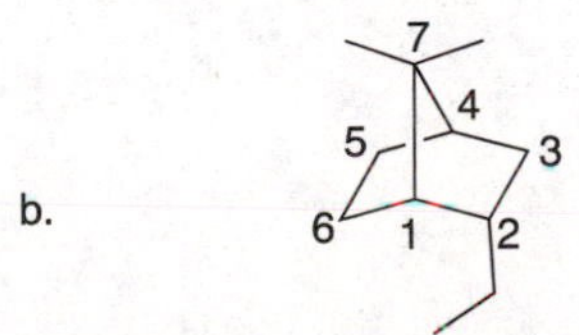

2-ethyl-7,7-dimethylbicyclo[2.2.1]heptane

d.

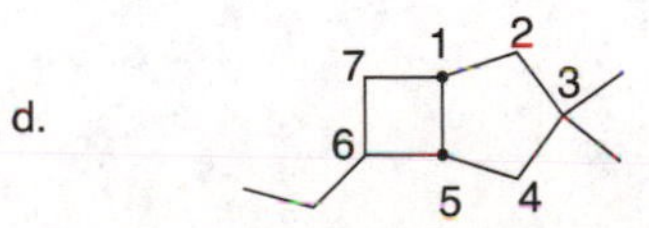

6-ethyl-3,3-dimethylbicyclo[3.2.0]heptane

Chapter 5: Stereochemistry

Chapter Review

Isomers are different compounds with the same molecular formula (5.2, 5.11).

[1] **Constitutional isomers**—isomers that differ in the way the atoms are connected to each other. They have:

- different IUPAC names
- the same or different functional groups
- different physical and chemical properties.

[2] **Stereoisomers**—isomers that differ only in the way atoms are oriented in space. They have the same functional group and the same IUPAC name except for prefixes such as cis, trans, *R*, and *S*.

- **Enantiomers**—stereoisomers that are nonsuperimposable mirror images of each other (5.4).
- **Diastereomers**—stereoisomers that are not mirror images of each other (5.7).

A B C D

enantiomers enantiomers

A and **B** are diastereomers of **C** and **D**.

Assigning priority (5.6)

- Assign priorities (1, 2, 3, or 4) to the atoms bonded directly to the stereogenic center in order of decreasing atomic number. The atom of *highest* atomic number gets the *highest* priority (1).
- If two atoms on a stereogenic center are the *same,* assign priority based on the atomic number of the atoms bonded to these atoms. *One* atom of higher atomic number determines a higher priority.
- If two isotopes are bonded to the stereogenic center, assign priorities in order of decreasing *mass* number.
- To assign a priority to an atom that is part of a multiple bond, treat a multiply bonded atom as an equivalent number of singly bonded atoms.

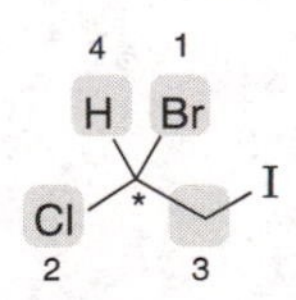

- The stereogenic center is bonded to Br, Cl, C, and H.
- The stereogenic center is *not* bonded directly to I.

- $CH(CH_3)_2$ gets the highest priority because the C is bonded to two other C's.

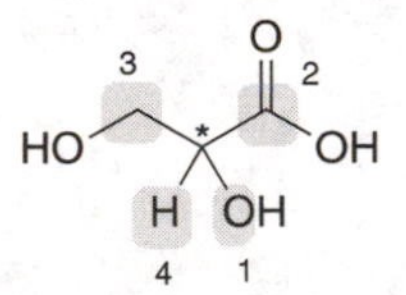

- OH gets the highest priority because O has the highest atomic number.
- CO_2H (three bonds to O) gets higher priority than CH_2OH (one bond to O).

Some basic principles

- When a compound and its mirror image are **superimposable,** they are **identical achiral compounds.** A plane of symmetry in one conformation makes a compound achiral (5.3).
- When a compound and its mirror image are **not superimposable,** they are **different chiral compounds** called **enantiomers.** A chiral compound has no plane of symmetry in any conformation (5.3).
- A **tetrahedral stereogenic center** is a carbon atom bonded to four different groups (5.4, 5.5).
- For ***n*** **stereogenic centers,** the maximum number of stereoisomers is $\mathbf{2^n}$ (5.7).

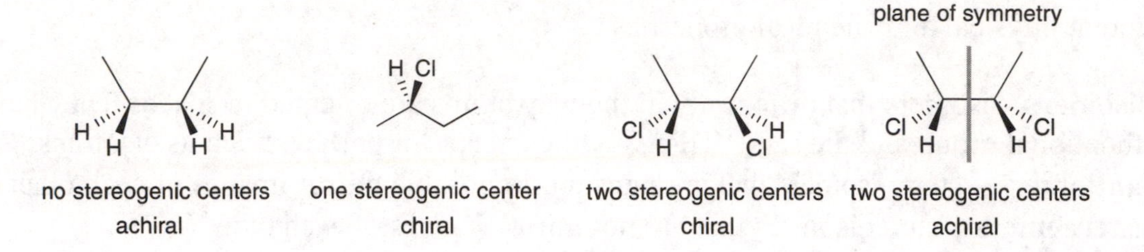

Optical activity is the ability of a compound to rotate plane-polarized light (5.12).

- An optically active solution contains a chiral compound.
- An optically inactive solution contains one of the following:
  - an achiral compound with no stereogenic centers.
  - a meso compound—an achiral compound with two or more stereogenic centers.
  - a racemic mixture—an equal amount of two enantiomers.

The prefixes *R* and *S* compared with *d* and *l*

The prefixes *R* and *S* are labels used in nomenclature. Rules on assigning *R,S* are found in Section 5.6.

- An enantiomer has every stereogenic center opposite in configuration.
- A diastereomer of this same compound has one stereogenic center with the same configuration and one that is opposite.

The prefixes *d* (or +) and *l* (or –) tell the direction a compound rotates plane-polarized light (5.12).

- *d* (or +) stands for dextrorotatory, rotating polarized light clockwise.
- *l* (or –) stands for levorotatory, rotating polarized light counterclockwise.

The physical properties of isomers compared (5.12)

Type of isomer	Physical properties
Constitutional isomers	Different
Enantiomers	Identical except the direction of rotation of polarized light
Diastereomers	Different
Racemic mixture	Possibly different from either enantiomer

Equations

- Specific rotation (5.12C):

$$\text{specific rotation} = [\alpha] = \frac{\alpha}{l \times c}$$

α = observed rotation (°)
l = length of sample tube (dm)
c = concentration (g/mL)

[dm = decimeter
1 dm = 10 cm]

- Enantiomeric excess (5.12D):

$$ee = \%\text{ of one enantiomer} - \%\text{ of other enantiomer}$$
$$= \frac{[\alpha]\text{ mixture}}{[\alpha]\text{ pure enantiomer}} \times 100\%$$

Practice Test on Chapter Review

1.a. Which of the following statements is true for compounds **A–D** below?

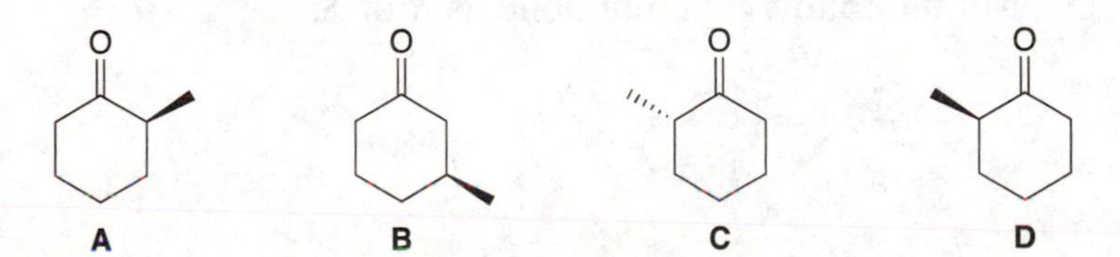

1. **A** and **B** are separable by physical methods such as distillation.
2. **A** and **C** are separable by physical methods such as distillation.
3. **A** and **D** are separable by physical methods such as distillation.
4. Statements (1) and (2) are both true.
5. Statements (1), (2), and (3) are all true.

b. Which of the following statements is true about compounds **A–C** below?

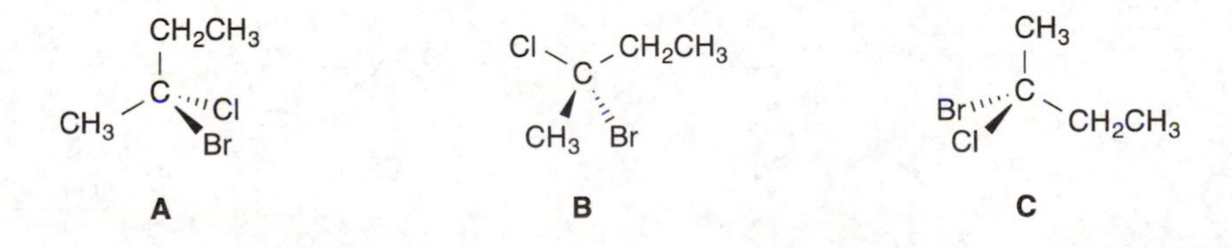

1. **A** and **B** are enantiomers.
2. **A** and **C** are enantiomers.
3. An equal mixture of **B** and **C** is optically active.
4. Statements (1) and (2) are true.
5. Statements (1), (2), and (3) are all true.

c. Which compound is a diastereomer of **A?**

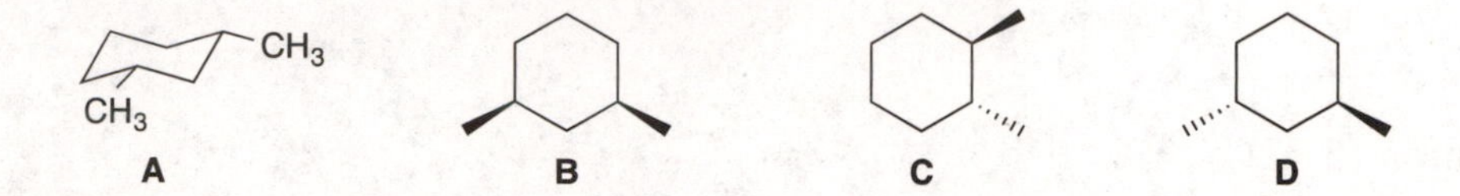

1. **B** only
2. **C** only
3. **D** only
4. Both **B** and **C**
5. Compounds **B, C,** and **D**

2. Rank the following four groups around a stereogenic center in order of decreasing priority. Rank the highest priority group as **1,** the lowest priority group as **4,** and the two groups of intermediate priority as **2** and **3.**

$-CH_2Cl$	$-CH_2CH_2Br$	$-COOH$	$-CH_2OH$
A	**B**	**C**	**D**

3. Label each stereogenic center in the following compound as *R* or *S*.

4. State how the compounds in each pair are related to each other. Choose from constitutional isomers, enantiomers, diastereomers, or identical compounds.

a. and

b. and

c. 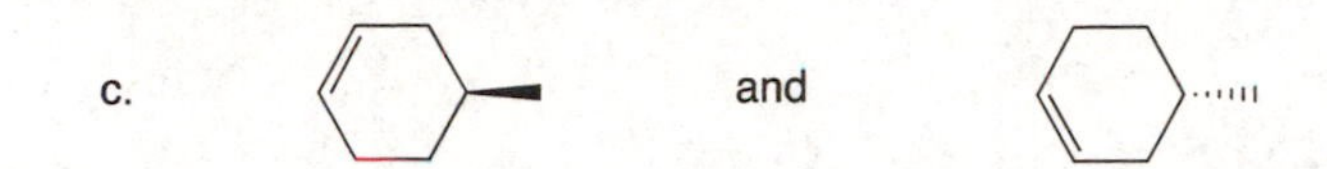

5. The enantiomeric excess of a mixture of **A** and **B** is 62% with **A** in excess. How much of **A** and **B** are present in the mixture?

Answers to Practice Test

1. a. 1	2. **A**–1	3. a. *S*	4. a. diastereomers	5. 81% **A**
b. 2	**B**–4	b. *S*	b. enantiomers	19% **B**
c. 3	**C**–2		c. identical	
	D–3			

Answers to Problems

5.1 Cellulose consists of long chains held together by intermolecular hydrogen bonds forming sheets that stack in extensive three-dimensional arrays. Most of the OH groups in cellulose are in the interior of this three-dimensional network, unavailable for hydrogen bonding to water. Thus, even though cellulose has many OH groups, its three-dimensional structure prevents many of the OH groups from hydrogen bonding with the solvent and this makes it water insoluble.

5.2 Constitutional isomers have atoms bonded to different atoms.
Stereoisomers differ only in the three-dimensional arrangement of atoms.

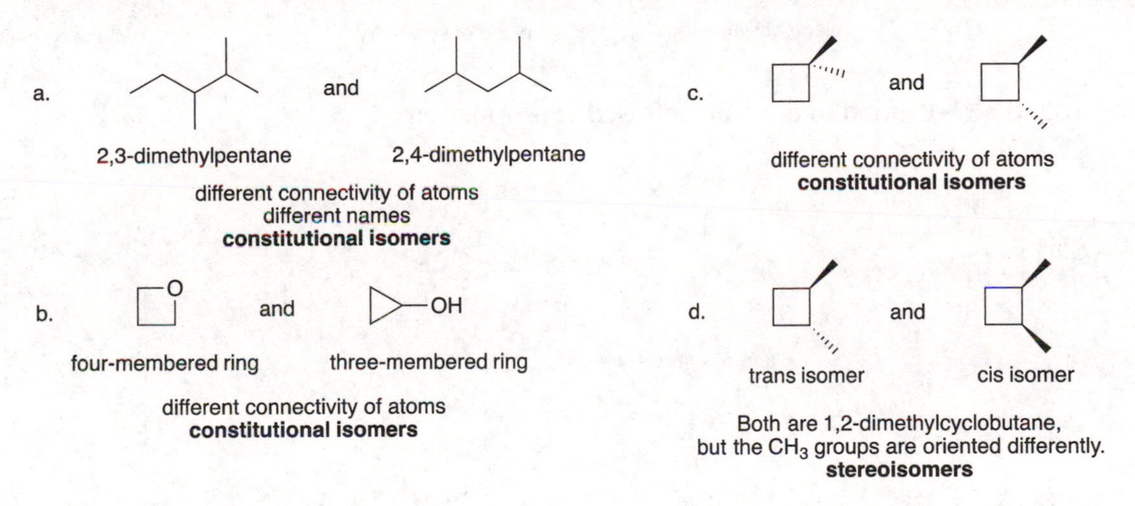

5.3 Draw the mirror image of each molecule by drawing a mirror plane and then drawing the molecule's reflection. **A chiral molecule is one that is not superimposable on its mirror image.** A molecule with one stereogenic center is always chiral. A molecule with zero stereogenic centers is not chiral (in general).

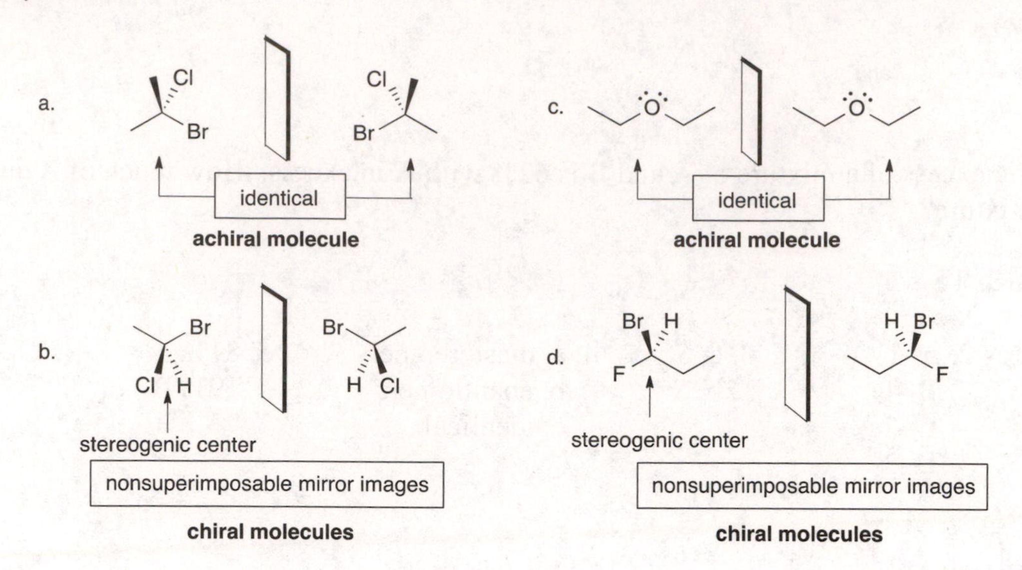

5.4 A plane of symmetry cuts the molecule into **two identical halves.**

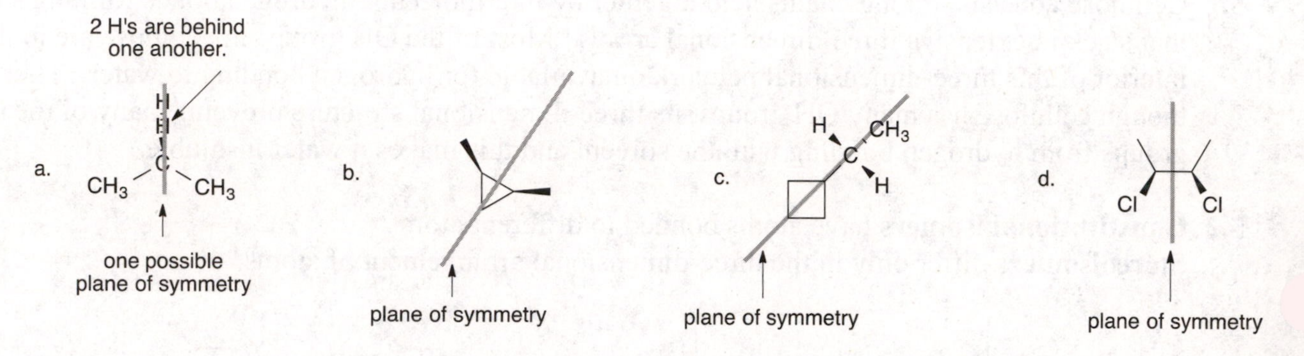

5.5 Rotate around a C–C bond to draw an eclipsed conformation.

a. CH_3, H, Br, C, C, Br, H, CH_3 — C2, C3 — rotate → H, H, Br, Br, C–C, CH_3, CH_3 — plane of symmetry

b. OH, OH — rotate → OH, OH — plane of symmetry

5.6 To locate a stereogenic center, omit all C's with two or more H's, all *sp* and sp^2 hybridized atoms, and all heteroatoms. [In Chapter 25, we will learn that the N atoms of ammonium salts ($R_4N^+ X^-$) can sometimes be stereogenic centers.] Then evaluate any remaining atoms. A tetrahedral stereogenic center has a carbon bonded to **four different groups.**

a. Cl

c. OH

e. OH OH OH

b. OH

d. O OH NH2

f. OH O OH OH

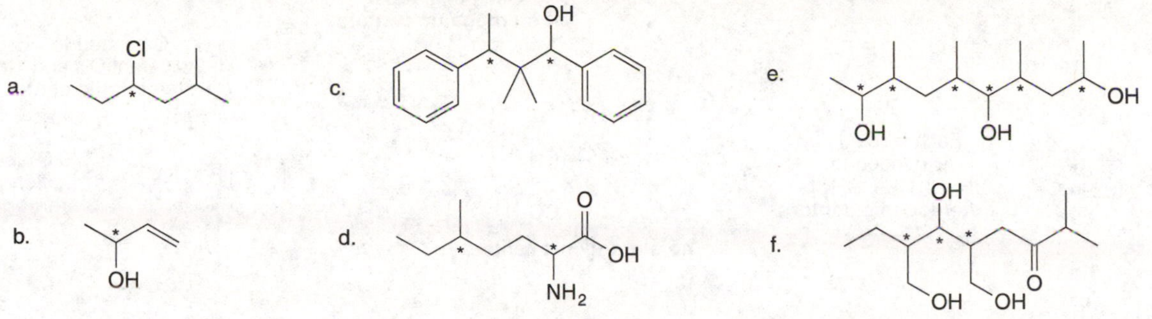

* = stereogenic center

5.7 Use the directions from Answer 5.6 to locate the stereogenic centers.

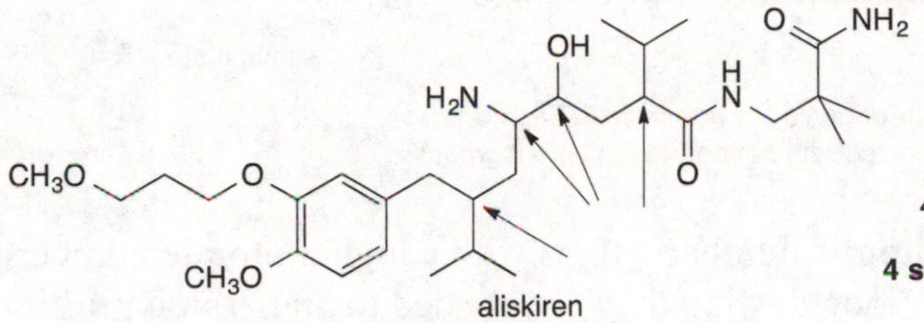

CH_3O O CH_3O H_2N OH H N O O NH_2

aliskiren

4 C's bonded to 4 different groups
4 stereogenic centers

5.8 Find the C bonded to four different groups in each molecule. At the stereogenic center, draw two bonds in the plane of the page, one in front (on a wedge), and one behind (on a dashed wedge). Then draw the mirror image (enantiomer).

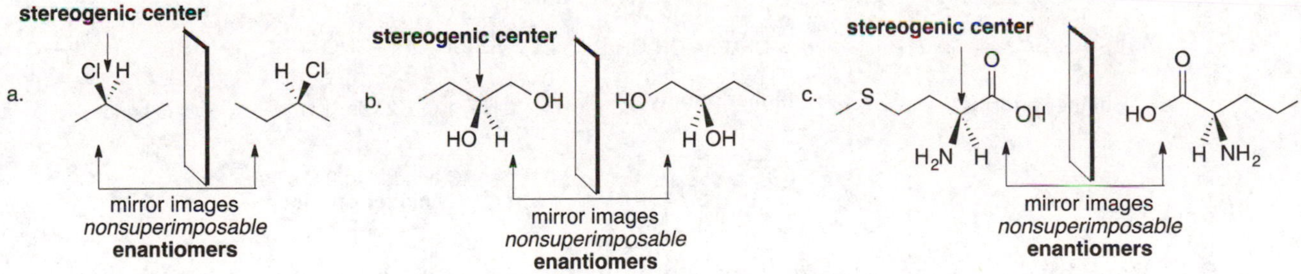

a. **stereogenic center** Cl H H Cl

mirror images
nonsuperimposable
enantiomers

b. **stereogenic center** OH HO HO H H OH

mirror images
nonsuperimposable
enantiomers

c. **stereogenic center** S O OH HO O S H_2N H H NH_2

mirror images
nonsuperimposable
enantiomers

5.9 Use the directions from Answer 5.6 to locate the stereogenic centers.

a. 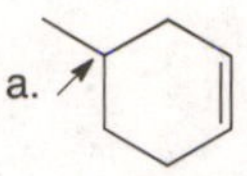 C bonded to H and 3 different C's: **1 stereogenic center**

c. 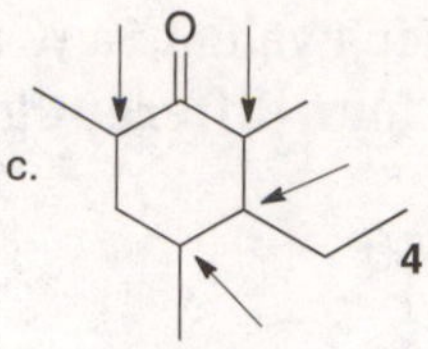 4 C's bonded to 4 different groups: **4 stereogenic centers**

e. 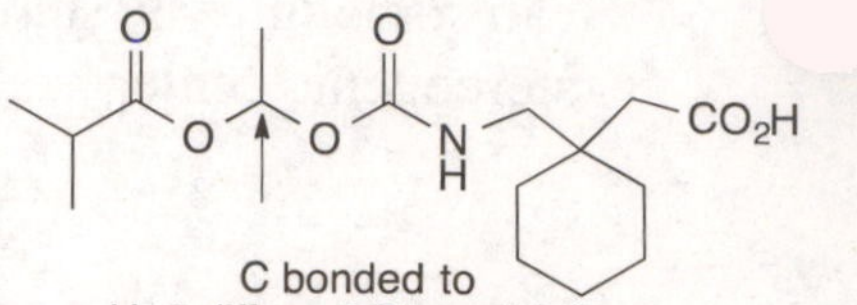 C bonded to H, 2 different O's and 1 C: **1 stereogenic center**

b. 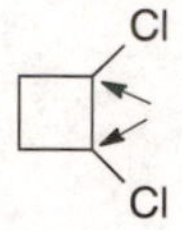 Each labeled C is bonded to: H, Cl, CH_2, CHCl: **2 stereogenic centers**

d. 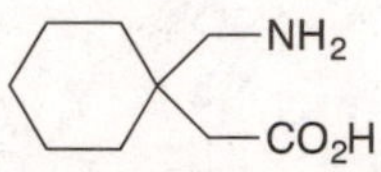 **no stereogenic centers**

f. 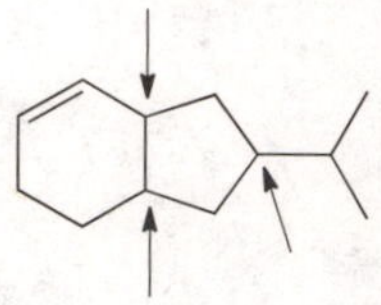 3 C's bonded to 4 different groups: **3 stereogenic centers**

5.10

a. 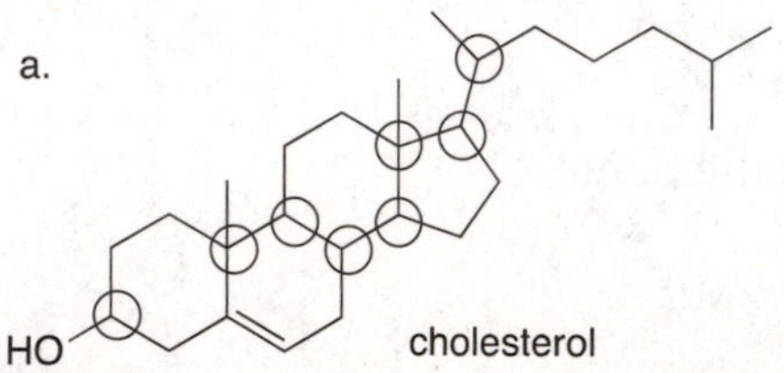 cholesterol

b. 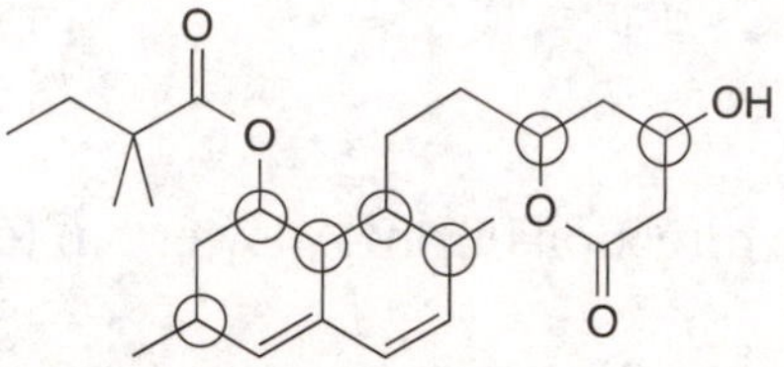 simvastatin

All stereogenic C's are circled. Each C is sp^3 hybridized and bonded to 4 different groups.

5.11 Assign priority based on atomic number: atoms with a higher atomic number get a higher priority. If two atoms are the same, look at what they are bonded to and assign priority based on the atomic number of these atoms.

a. $-CH_3$, $-CH_2CH_3$ ($-CH_2CH_3$: **higher priority**)

b. $-I$, $-Br$ ($-I$: **higher priority**)

c. $-H$, $-D$ ($-D$: higher mass, **higher priority**)

d. $-CH_2Br$, $-CH_2CH_2Br$ ($-CH_2Br$: **higher priority**)

e. $-CH_2CH_2Cl$, $-CH_2CH(CH_3)_2$ ($-CH_2CH_2Cl$: **higher priority**)

f. $-CH_2OH$, $-CHO$

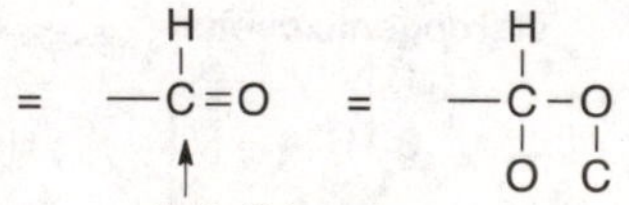

$-CH_2OH$: 2 H's, 1 O; $-CHO$: **2 O's**, 1 H (2 C–O bonds)

C bonded to 2 O's has **higher priority.**

5.12 Rank by decreasing priority. Lower atomic number = lower priority.

Highest priority = 1, Lowest priority = 4

a.		priority
$-COOH$	C = second lowest atomic number	3
$-H$	H = lowest atomic number	4
$-NH_2$	N = second highest atomic number	2
$-OH$	O = highest atomic number	1

decreasing priority: $-OH$, $-NH_2$, $-COOH$, $-H$

b.		priority
$-H$	H = lowest atomic number	4
$-CH_3$	C bonded to 3 H's	3
$-Cl$	Cl = highest atomic number	1
$-CH_2Cl$	C bonded to 2 H's + **1 Cl**	2

decreasing priority: $-Cl$, $-CH_2Cl$, $-CH_3$, $-H$

c.		priority
$-CH_2CH_3$	C bonded to 2 H's + **1 C**	2
$-CH_3$	C bonded to 3 H's	3
–H	H = lowest atomic number	4
$-CH(CH_3)_2$	C bonded to 1 H + **2 C's**	1

decreasing priority: $-CH(CH_3)_2$, $-CH_2CH_3$, $-CH_3$, –H

d.		priority
$-CH{=}CH_2$	C bonded to 1 H + **2 C's**	2
$-CH_3$	C bonded to 3 H's	3
$-C{\equiv}CH$	C bonded to **3 C's**	1
–H	H = lowest atomic number	4

decreasing priority: $-C{\equiv}CH$, $-CH{=}CH_2$, $-CH_3$, –H

5.13 To assign *R* or *S* to the molecule, first rank the groups. The lowest priority group must be oriented behind the page. If tracing a circle from (1) → (2) → (3) proceeds in the clockwise direction, then the stereogenic center is labeled *R;* if the circle is counterclockwise, then it is labeled *S.*

a. *S*

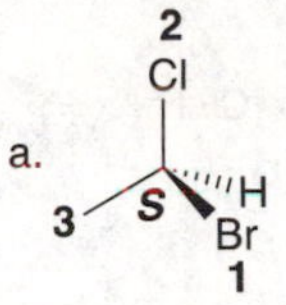

b. 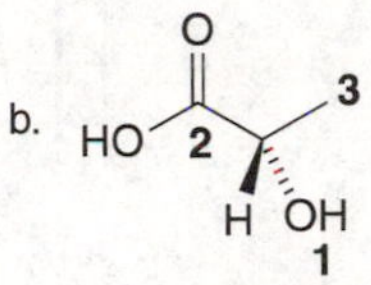 lowest priority forward
clockwise
It looks like *R*, but reverse answer
R* → *S

c. *R*

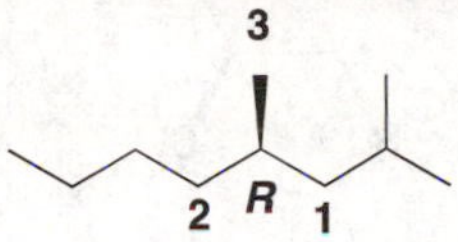

d. 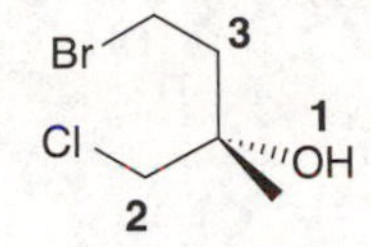 lowest priority forward
clockwise
It looks like *R*, but reverse answer
R* → *S

5.14

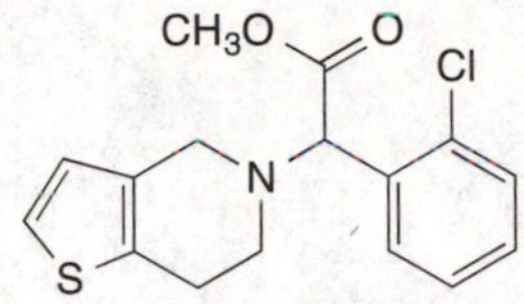

clopidogrel

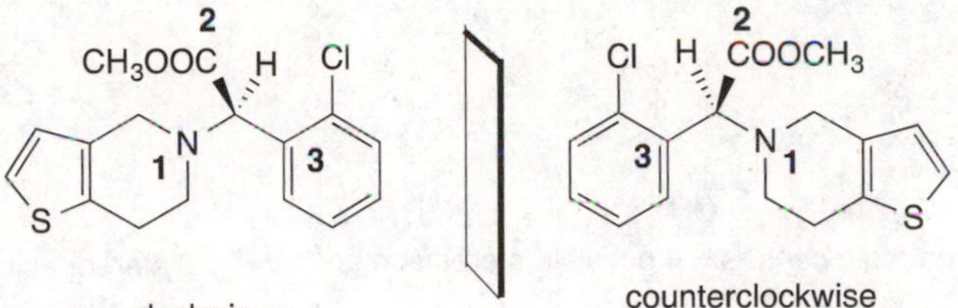

clockwise
R isomer

counterclockwise
S isomer
Plavix

5.15 a, b. Re-draw lisinopril as a skeletal structure, locate the stereogenic centers, and assign *R,S.*

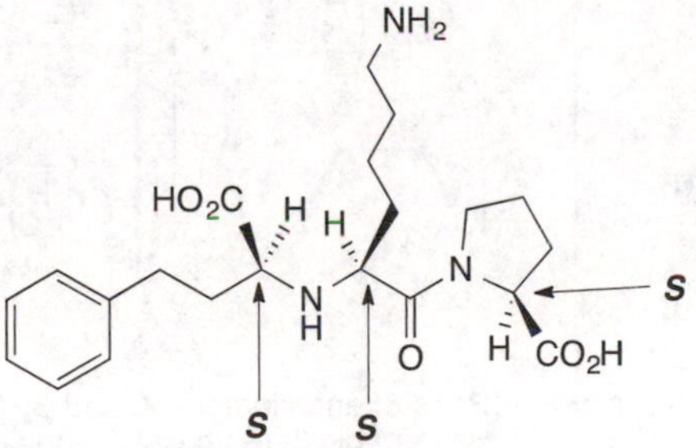

three stereogenic centers
All have the *S* configuration.

5.16 The maximum number of stereoisomers = 2^n where n = the number of stereogenic centers.

a. 3 stereogenic centers
2^3 = 8 stereoisomers

b. 8 stereogenic centers
2^8 = 256 stereoisomers

5.17

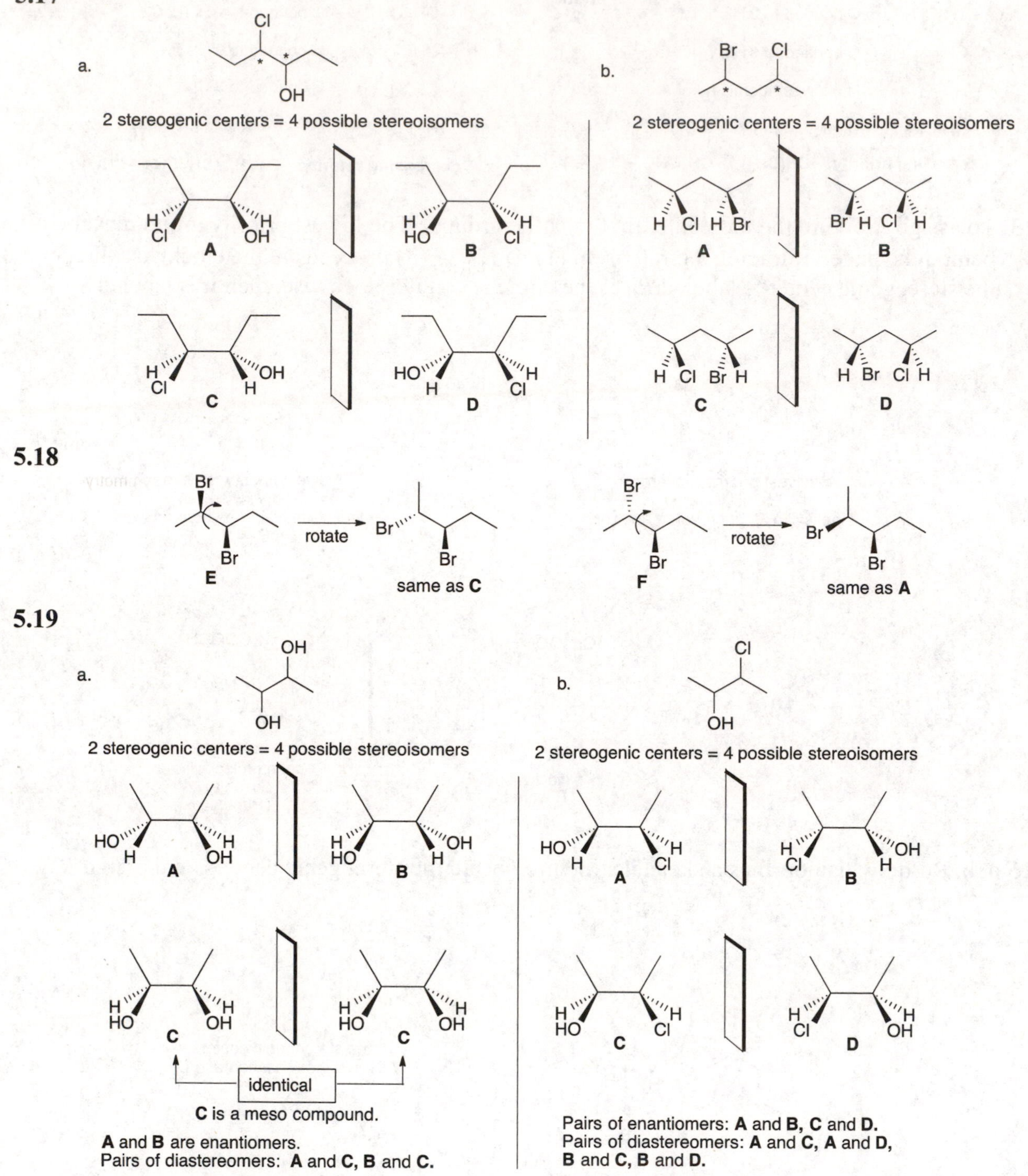

5.20 **A meso compound must have at least two stereogenic centers. Usually a meso compound has a plane of symmetry.** You may have to rotate around a C–C bond to see the plane of symmetry clearly.

a.

2 stereogenic centers
plane of symmetry
meso compound

b.

2 stereogenic centers
no plane of symmetry
not a meso compound

c. rotate

2 stereogenic centers
plane of symmetry
meso compound

5.21 Use the definition in Answer 5.20 to draw the meso compounds.

a. plane of symmetry

b. plane of symmetry

c. plane of symmetry

5.22 The enantiomer has the exact opposite *R,S* designations. Diastereomers with two stereogenic centers have one center the same and one different.

If a compound is ***R,S*:**

Its enantiomer is: ***S,R*** ← Exact opposite: *R* and *S* interchanged.

Its diastereomers are: ***R,R* and *S,S*** ← One designation remains the same, the other changes.

5.23 The enantiomer has the exact opposite *R,S* designations. For diastereomers, at least one of the *R,S* designations is the same, but not all of them.

a. (2*R*,3*S*)-hexane-2,3-diol and (2*R*,3*R*)-hexane-2,3-diol
One changes; one remains the same:
diastereomers

b. (2*R*,3*R*)-hexane-2,3-diol and (2*S*,3*S*)-hexane-2,3-diol
Both *R*'s change to *S*'s:
enantiomers

c. (2*R*,3*S*,4*R*)-hexane-2,3,4-triol and (2*S*,3*R*,4*R*)-hexane-2,3,4-triol
Two change; one remains the same:
diastereomers

5.24 The enantiomer must have the exact opposite *R,S* designations. For diastereomers, at least one of the *R,S* designations is the same, but not all of them.

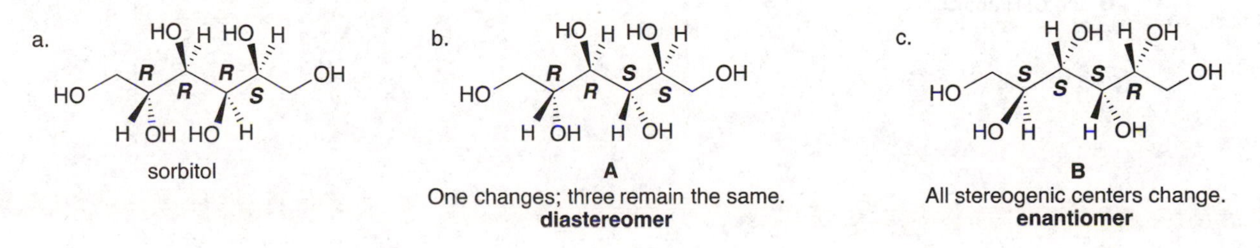

5.25 Meso compounds generally have a plane of symmetry. They cannot have just one stereogenic center.

a.

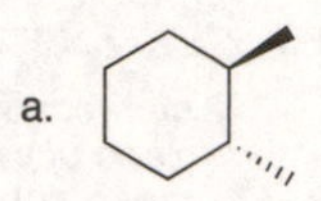

no plane of symmetry
not a meso compound

b.

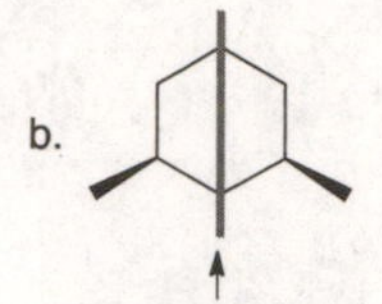

plane of symmetry
meso compound

c. Cl, OH

no plane of symmetry
not a meso compound

5.26

a. 2 stereogenic centers = 4 stereoisomers maximum

Draw the cis and trans isomers:

cis

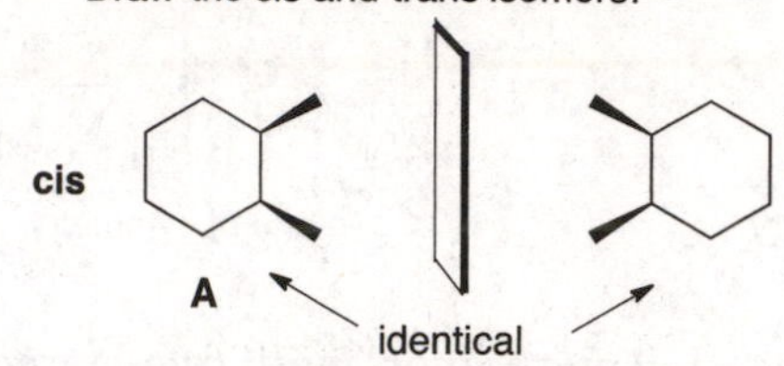

A identical

trans

B **C**

Pair of enantiomers: **B** and **C.**
Pairs of diastereomers: **A** and **B, A** and **C.**

Only 3 stereoisomers exist.

b. HO — 2 stereogenic centers = 4 stereoisomers maximum

Draw the cis and trans isomers:

cis

OH HO

A **B**

trans

OH HO

C **D**

Pairs of enantiomers: **A** and **B, C** and **D.**
Pairs of diastereomers: **A** and **C, A** and **D, B** and **C, B** and **D.**

All 4 stereoisomers exist.

c. Cl, Cl

Draw the cis and trans isomers:

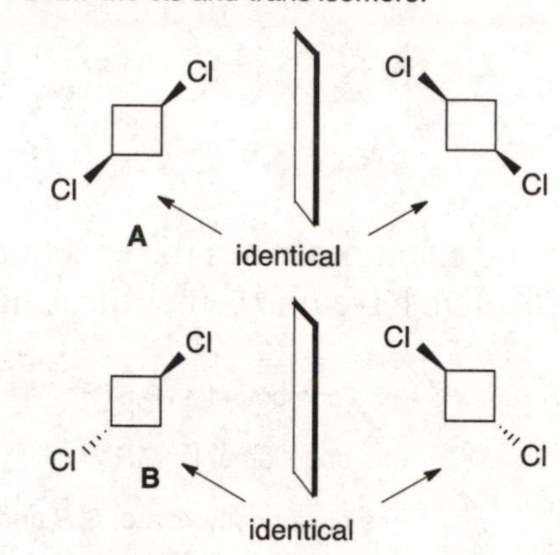

Pair of diastereomers: **A** and **B.**

Only 2 stereoisomers exist.

5.27 Four facts:

- **Enantiomers** are mirror image isomers.
- **Diastereomers** are stereoisomers that are not mirror images.
- **Constitutional isomers** have the same molecular formula but the atoms are bonded to different atoms.
- **Cis and trans isomers** are always diastereomers.

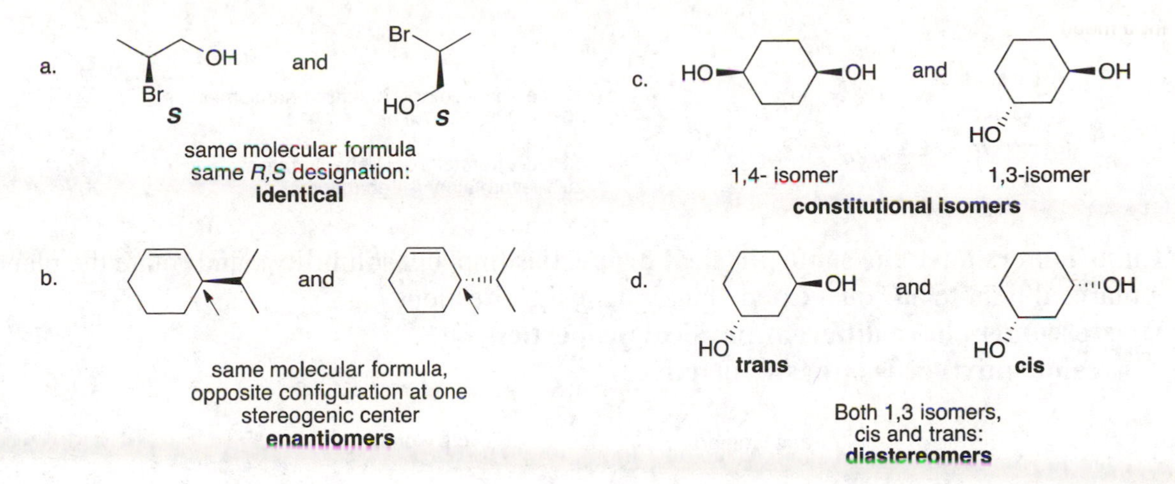

5.28

O
OH
NH_2
(*S*)-alanine
[α] = +8.5
mp = 297 °C

a. Mp = same as the *S* isomer.
b. The mp of a racemic mixture is often different from the melting point of the enantiomers.
c. –8.5, same as *S* but opposite sign
d. Zero. A racemic mixture is optically inactive.
e. Solution of pure (*S*)-alanine: **optically active**
Equal mixture of (*R*)- and (*S*)-alanine: **optically inactive**
75% (*S*)- and 25% (*R*)-alanine: **optically active**

5.29

$$[\alpha] = \frac{\alpha}{l \times c}$$

α = observed rotation
l = length of tube (dm)
c = concentration (g/mL)

$$[\alpha] = \frac{10°}{1 \text{ dm} \times (1 \text{ g}/10 \text{ mL})}$$ **= +100 = specific rotation**

5.30 Enantiomeric excess = *ee* = % of one enantiomer – % of other enantiomer.

a. 95% – 5% = **90% *ee***

b. 85% – 15% = **70% *ee***

5.31

a. 90% *ee* means 90% excess of **A** and 10% racemic mixture of **A** and **B** (5% each); therefore, **95% A and 5% B.**

b. 99% *ee* means 99% excess of **A** and 1% racemic mixture of **A** and **B** (0.5% each); therefore, **99.5% A and 0.5% B.**

c. 60% *ee* means 60% excess of **A** and 40% racemic mixture of **A** and **B** (20% each); therefore, **80% A and 20% B.**

5.32

$$ee = \frac{[\alpha]\ \text{mixture}}{[\alpha]\ \text{pure enantiomer}} \times 100\%$$

a. $\frac{+10}{+24} \times 100\% = 42\%\ ee$

b. $\frac{[\alpha]\ \text{solution}}{+24} \times 100\% = 80\%\ ee$

$[\alpha]$ solution = +19.2

5.33

a. $\frac{[\alpha]\ \text{mixture}}{+3.8} \times 100\% = 60\%\ ee$

$[\alpha]$ mixture = +2.3

b. % one enantiomer – % other enantiomer = *ee*
80% – 20% = 60% *ee*

80% dextrorotatory (+) enantiomer
20% levorotatory (–) enantiomer

5.34 • **Enantiomers have the same physical properties** (mp, bp, solubility), and rotate the plane of polarized light to an equal extent, but in opposite directions.
• **Diastereomers have different physical properties.**
• **A racemic mixture is optically inactive.**

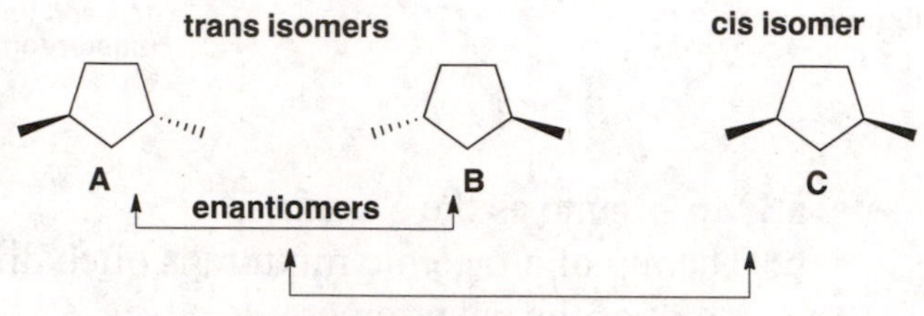

a. The bp's of **A** and **B** are the same. The bp's of **A** and **C** are different.
b. Pure **A:** optically active
Pure **B:** optically active
Pure **C:** optically inactive
Equal mixture of **A** and **B:** optically inactive
Equal mixture of **A** and **C:** optically active
c. There would be two fractions: one containing **A** and **B** (optically inactive), and one containing **C** (optically inactive).

5.35

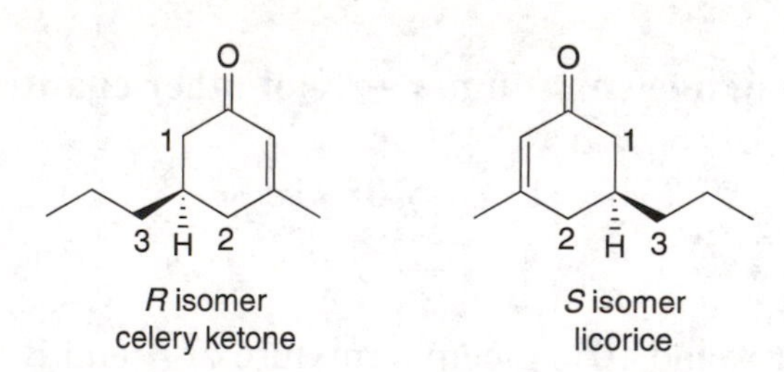

5.36

a, b.

three stereogenic centers

5.37

identical

a. **A** and **B,** same *R,S* assignment, identical
b. **A** and **C,** opposite *R,S* assignment, enantiomers
c. **A** and **D,** one stereogenic center different, diastereomers
d. **C** and **D,** one stereogenic center different, diastereomers

5.38 Use the definitions from Answer 5.2.

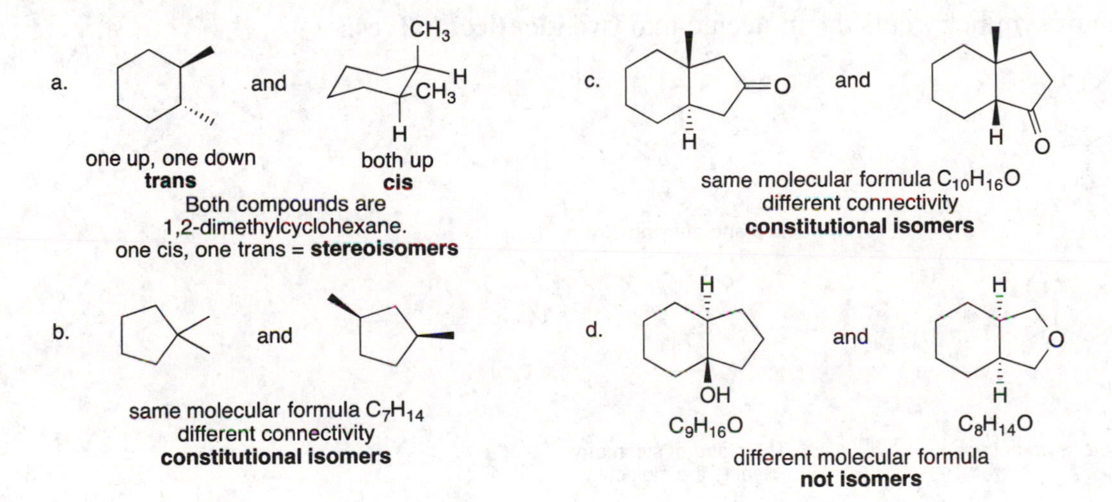

5.39 Use the definitions from Answer 5.3.

a. OH HO
H H
identical
achiral

c. Br Br
identical
achiral

b. HS OH HO SH
H_2N H H NH_2
O O
chiral

d. H OH OH HO OH H
O OH OH O
chiral

5.40

O H
H OH
R isomer

a. H OH H O
S
enantiomer

b. H H OH O
R
identical

c. H O H OH
S
enantiomer

5.41 A plane of symmetry cuts the molecule into **two identical halves.**

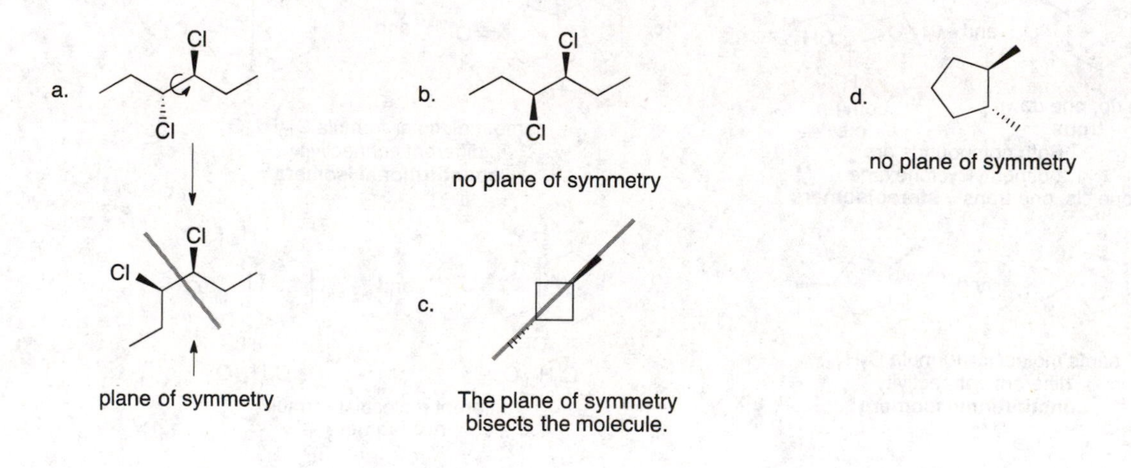

5.42 Use the directions from Answer 5.6 to locate the stereogenic centers.

a.

d.

f.

b.

e.

c.

5.43 Stereogenic centers are circled.

a. amoxicillin

b. norethindrone

c. heroin

5.44

a. amphetamine

b. ketoprofen

5.45 Assign priority based on the rules in Answer 5.11.

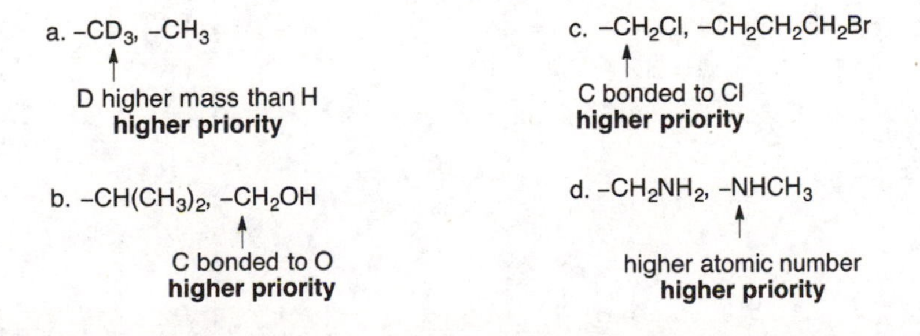

5.46 Assign priority based on the rules in Answer 5.11.

a. $-F > -OH > -NH_2 > -CH_3$

b. $-(CH_2)_3CH_3 > -CH_2CH_2CH_3 > -CH_2CH_3 > -CH_3$

c. $-NH_2 > -CH_2NHCH_3 > -CH_2NH_2 > -CH_3$

d. $-COOH > -CHO > -CH_2OH > -H$

e. $-Cl > -SH > -OH > -CH_3$

f. $-C{\equiv}CH > -CH{=}CH_2 > -CH(CH_3)_2 > -CH_2CH_3$

5.47 Use the rules in Answer 5.13 to assign *R* or *S* to each stereogenic center.

a. counterclockwise **S isomer**

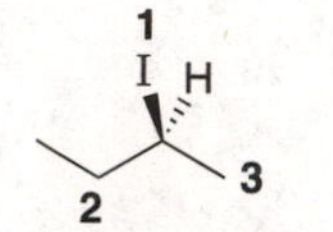

b. clockwise, but H in front **S isomer**

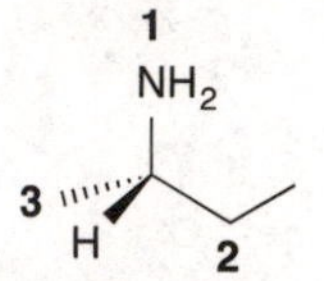

c. It looks like an *S* isomer, but we must reverse the answer, *S* to *R*. ***R* isomer**

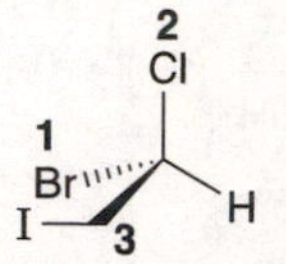

d. ***R, R***

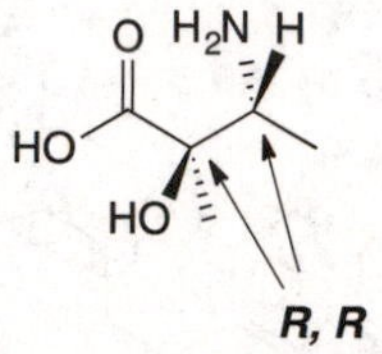

e. ***S***

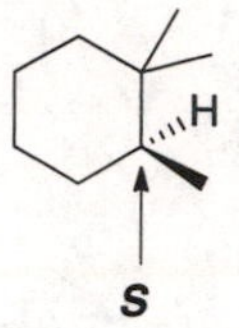

f. ***S***, ***S***

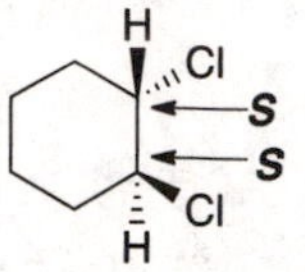

g. ***S***, ***S***

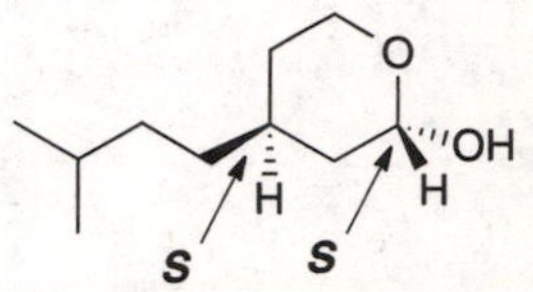

h. ***S***, ***S***

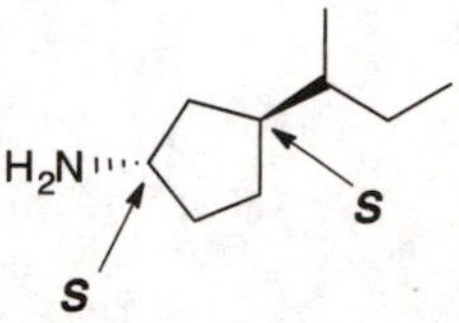

5.48

a. re-draw: ***R***, ***R***

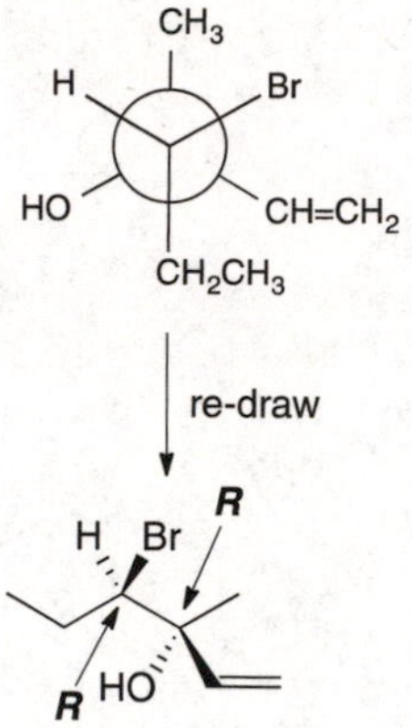

b. re-draw: ***S S***

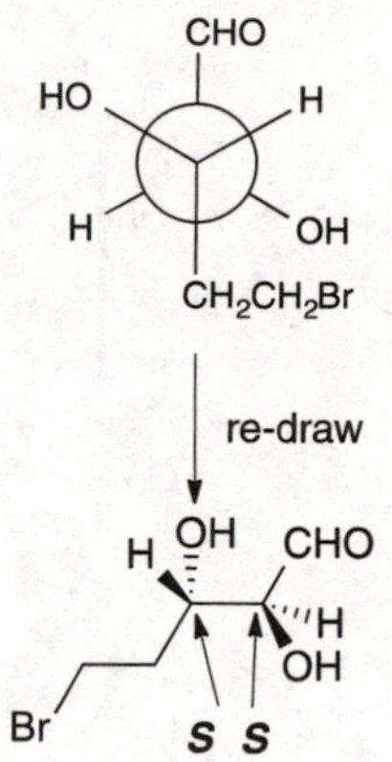

5.49

S, ***S***

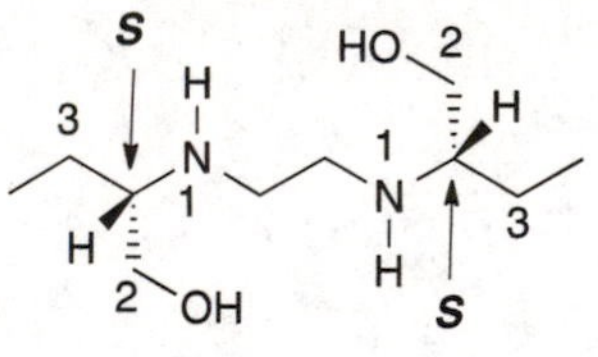

5.50

citalopram

***S* isomer**

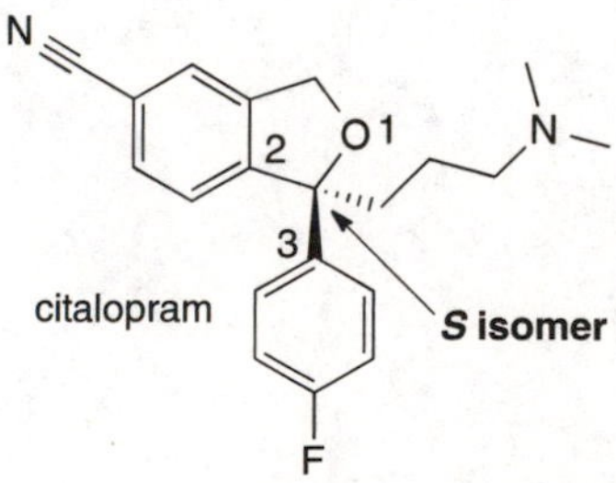

5.51

a. (*R*)-3-methylhexane

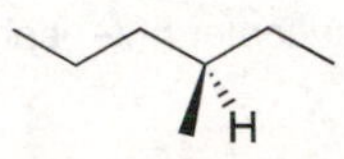

c. (3*R*,5*S*,6*R*)-5-ethyl-3,6-dimethylnonane

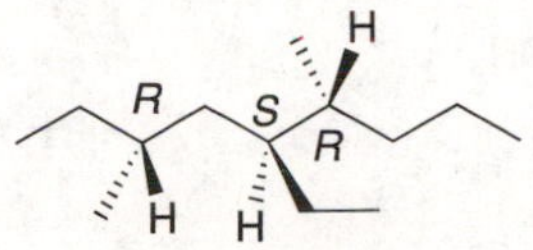

b. (4*R*,5*S*)-4,5-diethyloctane

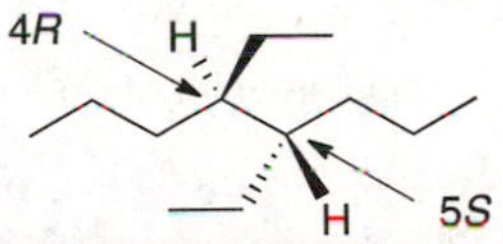

d. (3*S*,6*S*)-6-isopropyl-3-methyldecane

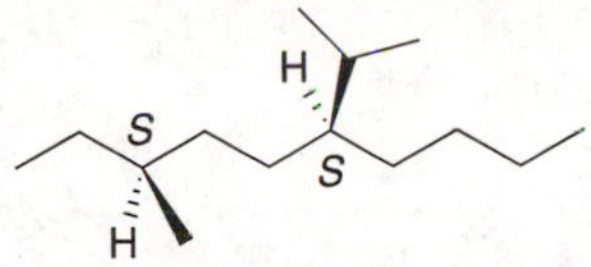

5.52

a. **(*S*)-3-methylhexane**

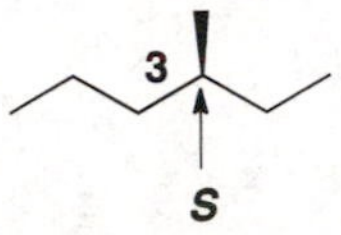

b. **(4*R*,6*R*)-4-ethyl-6-methyldecane**

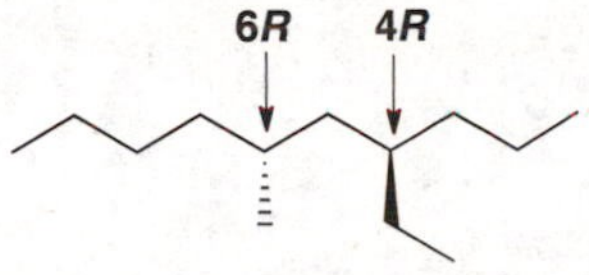

c. **(3*R*,5*S*,6*R*)-5-isobutyl-3,6-dimethylnonane**

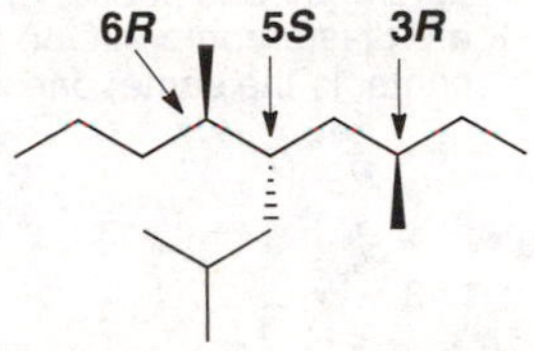

5.53

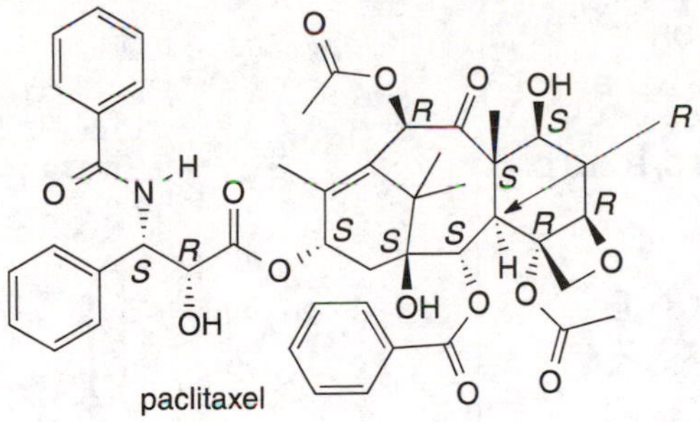

paclitaxel

5.54

a. 2 stereogenic centers
$2^2 = 4$ possible stereoisomers

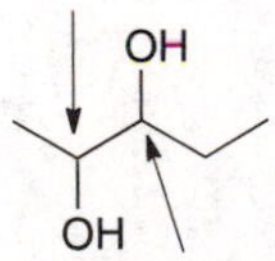

b. 0 stereogenic centers

c. 4 stereogenic centers
$2^4 = 16$ possible stereoisomers

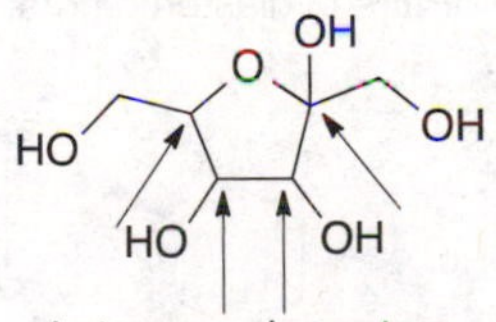

5.55

a. 1*R*,2*S*

C1 OH $NHCH_3$ C2

ephedrine

b. 1*S*,2*S*

C1 OH $NHCH_3$ C2

pseudoephedrine

c. Ephedrine and pseudoephedrine are diastereomers. One stereogenic center is the same; one is different.

d. OH *S* *R* $NHCH_3$ ← e. enantiomer of (–)-ephedrine

OH *R* *R* $NHCH_3$ ← diastereomer of (–)-ephedrine

5.56

a.

OH OH **A** | OH OH **B** | OH OH **C** | OH OH

identical

meso compound

Pair of enantiomers: **A** and **B.**
Pairs of diastereomers: **A** and **C, B** and **C.**

b.

Cl Br **A** | Br Cl **B**

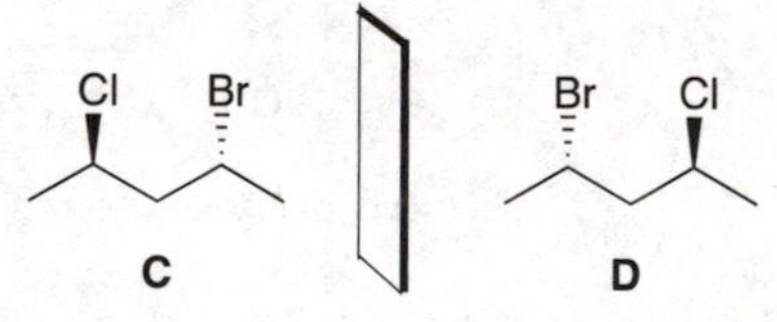

Br Cl **D**

Pairs of enantiomers: **A** and **B, C** and **D.**
Pairs of diastereomers: **A** and **C, A** and **D, B** and **C, B** and **D.**

c.

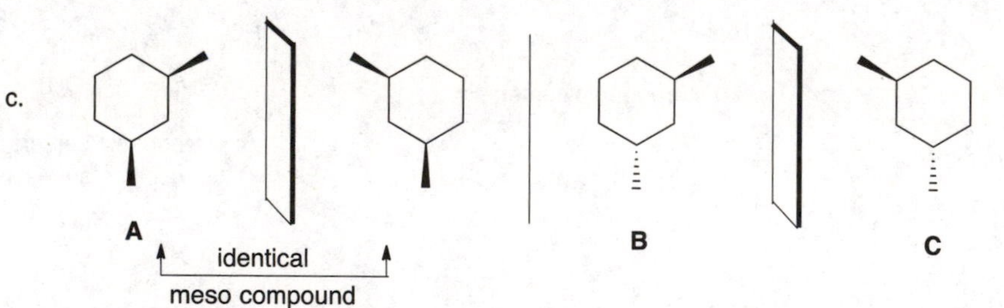

identical

meso compound

Pair of enantiomers: **B** and **C.**
Pairs of diastereomers: **A** and **B, A** and **C.**

d.

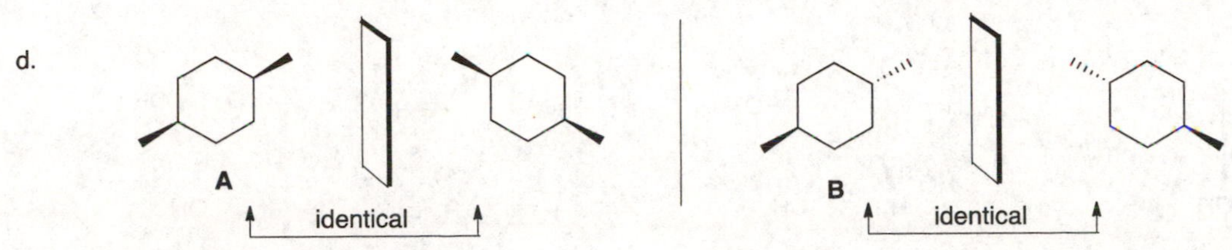

Pair of diastereomers: **A** and **B**.
Meso compounds: **A** and **B**.

5.57

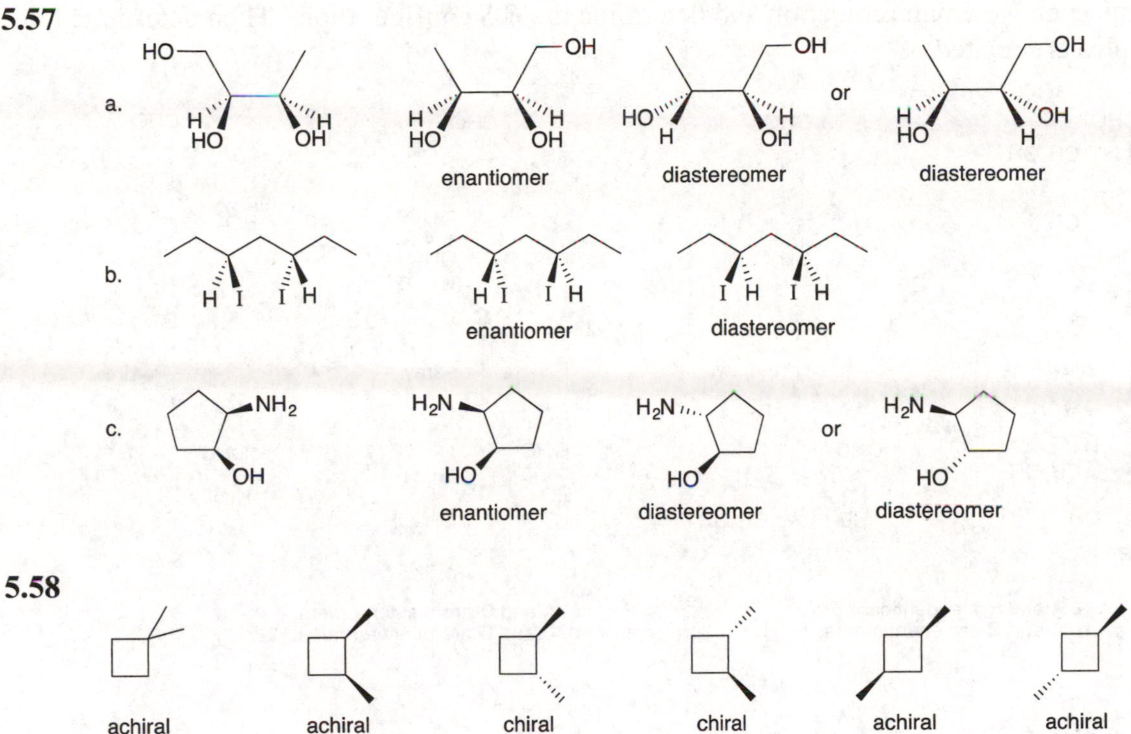

5.59 Explain each statement.

a. All molecules have a mirror image, but only chiral molecules have enantiomers. **A** is not chiral, and therefore, does not have an enantiomer.

b. **B** has one stereogenic center, and therefore, has an enantiomer. Only compounds with two or more stereogenic centers have diastereomers.

c. **C** is chiral and has two stereogenic centers, and therefore, has both an enantiomer and a diastereomer.

d.

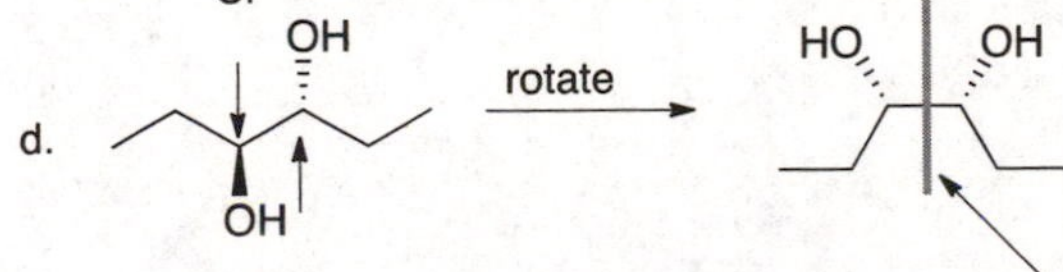

plane of symmetry

D has two stereogenic centers, but is a meso compound. Therefore, it has a diastereomer, but no enantiomer because it is achiral.

e. **E** has two stereogenic centers, but is a meso compound. Therefore, it has a diastereomer, but no enantiomer because it is achiral.

plane of symmetry

5.60

D-erythrose 2*R*,3*R*

a. 2*S*,3*R* diastereomer

b. 2*R*,3*R* identical

c. 2*S*,3*S* enantiomer

5.61 Re-draw each Newman projection and determine the *R*,*S* configuration. Then determine how the molecules are related.

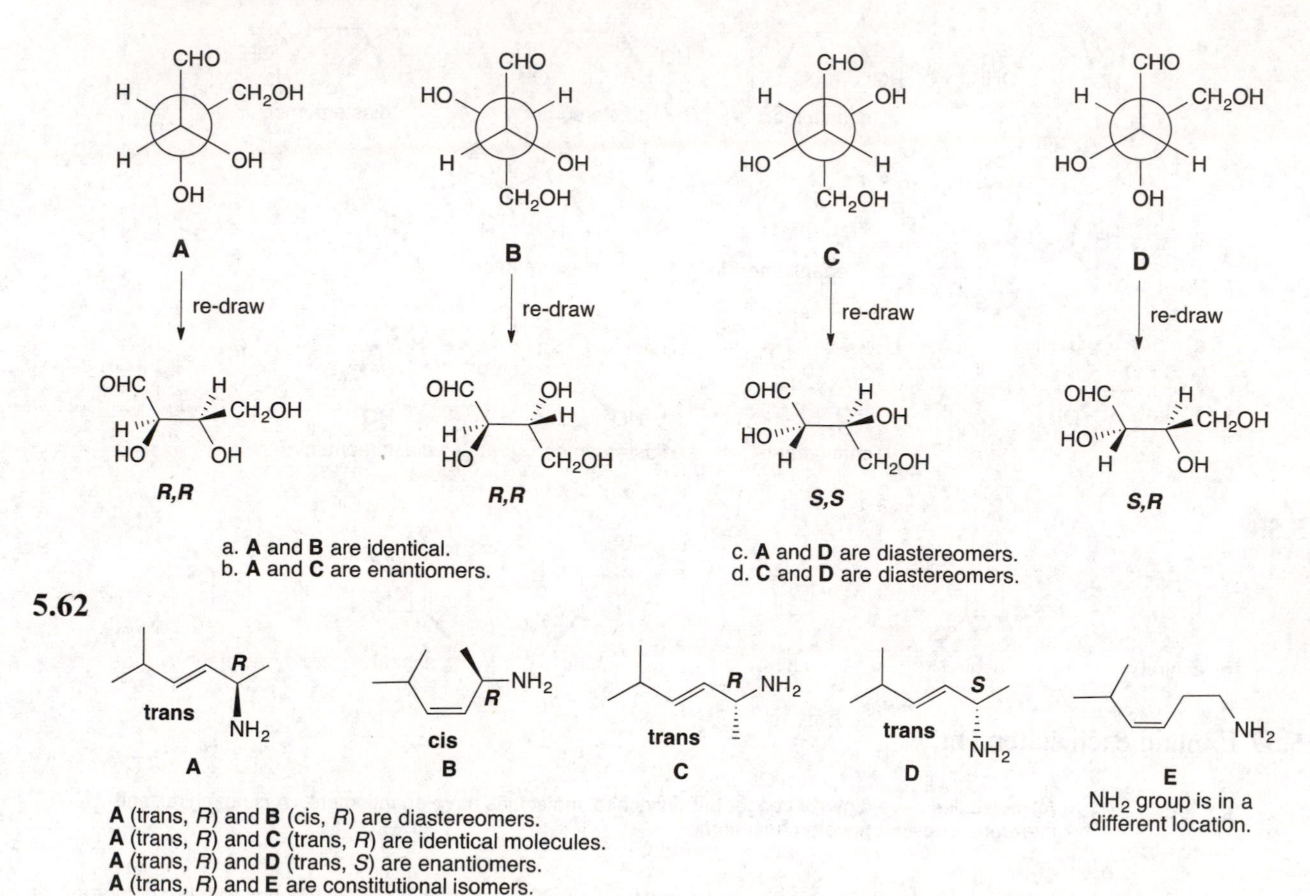

a. **A** and **B** are identical.
b. **A** and **C** are enantiomers.
c. **A** and **D** are diastereomers.
d. **C** and **D** are diastereomers.

5.62

A (trans, *R*) and **B** (cis, *R*) are diastereomers.
A (trans, *R*) and **C** (trans, *R*) are identical molecules.
A (trans, *R*) and **D** (trans, *S*) are enantiomers.
A (trans, *R*) and **E** are constitutional isomers.

5.63

a. and

enantiomers

b. 3*S* and

2*R* 2*R*,3*R*

one different configuration

diastereomers

c. and

mirror images

not superimposable

enantiomers

d. and

enantiomers

e. and

2*S*,3*S* 2*S*,3*S*

identical

f. and

1,4-trans 1,4-cis

diastereomers

5.64

a. **A** and **B** are constitutional isomers.
A and **C** are constitutional isomers.
B and **C** are diastereomers (cis and trans).
C and **D** are enantiomers.

b.

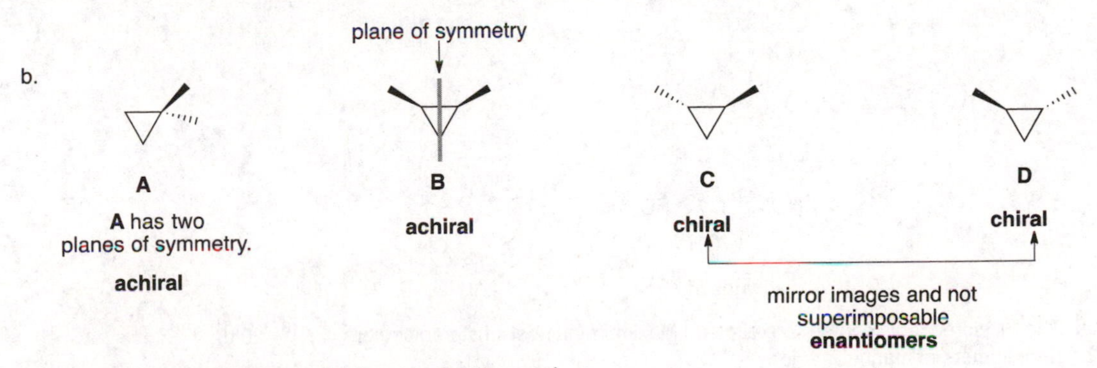

c. Alone, **C** and **D** would be optically active.

d. **A** and **B** have a plane of symmetry.

e. **A** and **B** have different boiling points.
B and **C** have different boiling points.
C and **D** have the same boiling point.

f. **B** is a meso compound.

g. An equal mixture of **C** and **D** is optically inactive because it is a racemic mixture.
An equal mixture of **B** and **C** would be optically active.

5.65

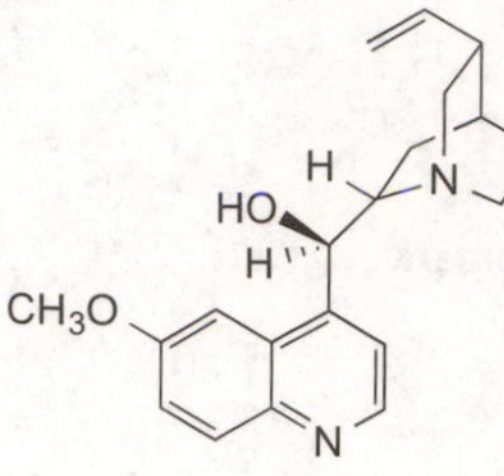

quinine

$$ee = \frac{[\alpha]\text{ mixture}}{[\alpha]\text{ pure enantiomer}} \times 100\%$$

quinine = **A**
quinine's enantiomer = **B**

a.

$$\frac{-50}{-165} \times 100\% = 30\%\ ee$$

$$\frac{-83}{-165} \times 100\% = 50\%\ ee$$

$$\frac{-120}{-165} \times 100\% = 73\%\ ee$$

b. 30% *ee* = 30% excess one compound (**A**)
remaining 70% = mixture of 2 compounds (35% each **A** and **B**)
Amount of **A** = 30 + 35 = **65%**
Amount of **B** = **35%**

50% *ee* = 50% excess one compound (**A**)
remaining 50% = mixture of 2 compounds (25% each **A** and **B**)
Amount of **A** = 50 + 25 = **75%**
Amount of **B** = **25%**

73% *ee* = 73% excess of one compound (**A**)
remaining 27% = mixture of 2 compounds (13.5% each **A** and **B**)
Amount of **A** = 73 + 13.5 = **86.5%**
Amount of **B** = **13.5%**

c. [α] = +165
d. 80% – 20% = 60% *ee*

e. $$60\% = \frac{[\alpha]\text{ mixture}}{-165} \times 100\%$$

[α] mixture = –99

5.66

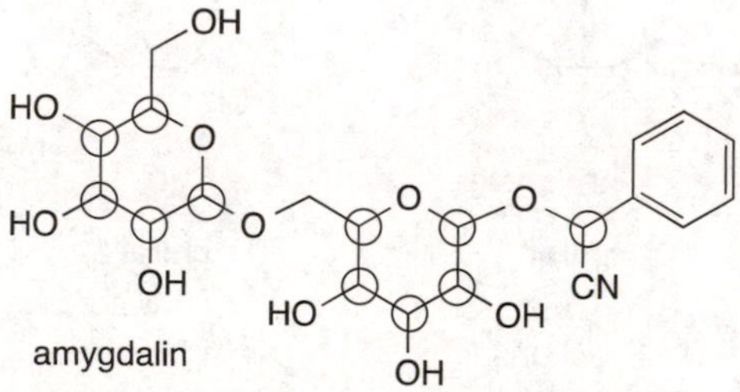

amygdalin

a. The 11 stereogenic centers are circled. Maximum number of stereoisomers = 2^{11} = 2048
b. Enantiomers of mandelic acid:

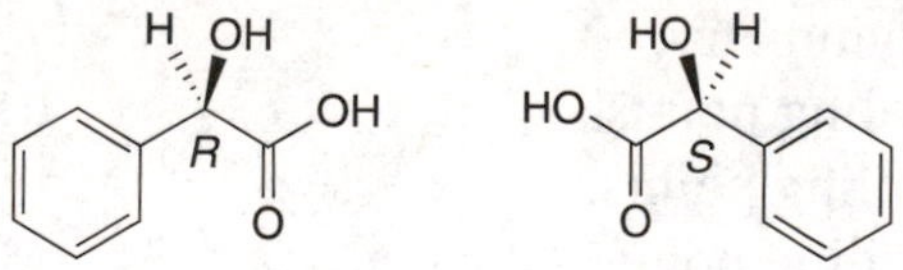

R   *S*

c. 60% – 40% = 20% *ee*
20% = [α] mixture/–154 x 100%
[α] mixture = –31

d. $$ee = \frac{+50}{+154} \times 100\% = 32\%\ ee$$

[α] for (*S*)-mandelic acid = +154

32% excess of the *S* enantiomer
68% of racemic *R* and *S* = 34% *S* and 34% *R*

S enantiomer: 32% + 34% = 66%
R enantiomer = 34%

5.67

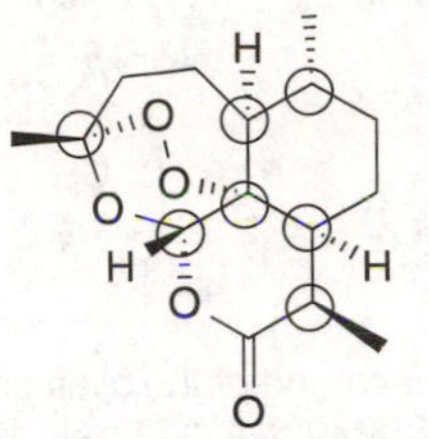

artemisinin

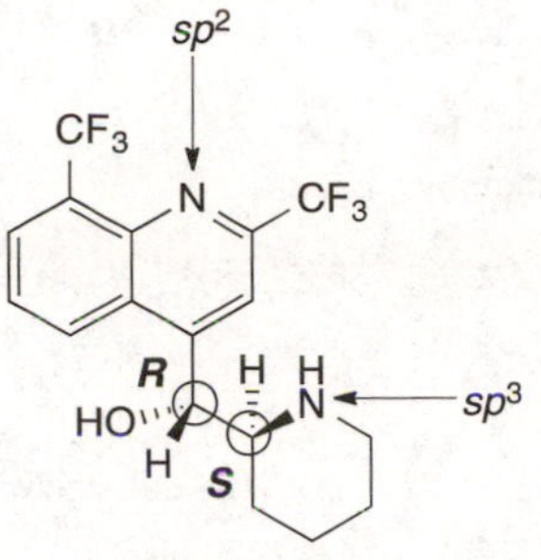

mefloquine

a. Each stereogenic center is circled.
b. The stereogenic centers in mefloquine are labeled.
c. Artemisinin has seven stereogenic centers. $2^n = 2^7 = 128$ possible stereoisomers
d. One N atom in mefloquine is sp^2 and one is sp^3.
e. Two molecules of artemisinin cannot intermolecularly H-bond because there are no O–H or N–H bonds.

f.

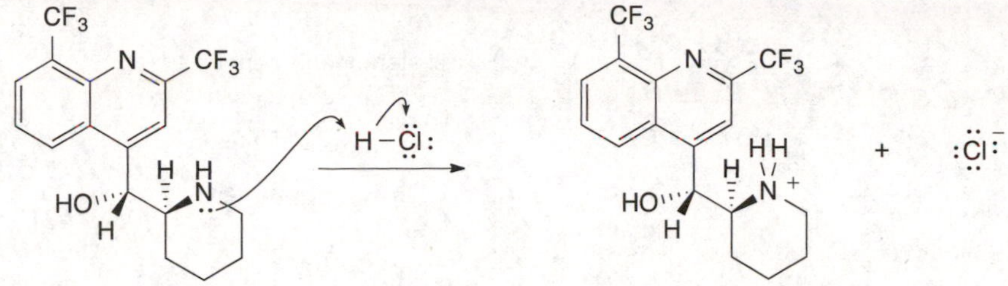

5.68

a. Each stereogenic center is circled.

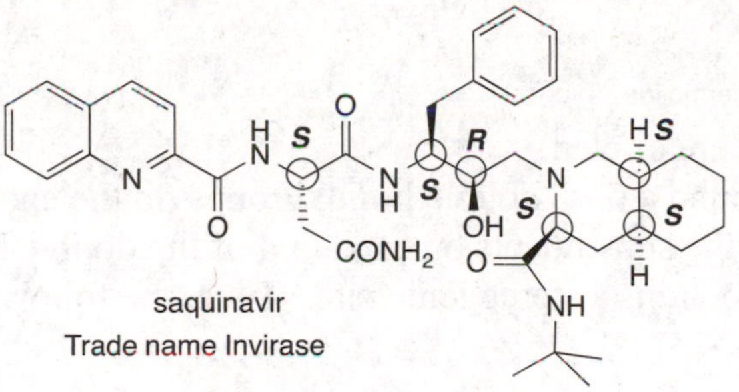

saquinavir
Trade name Invirase

b. enantiomer

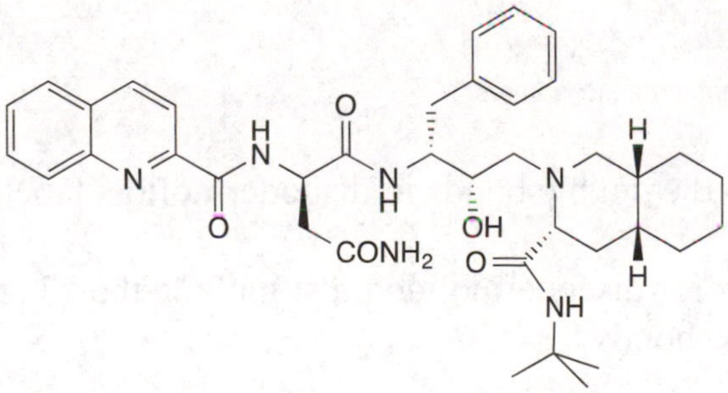

c. diastereomer

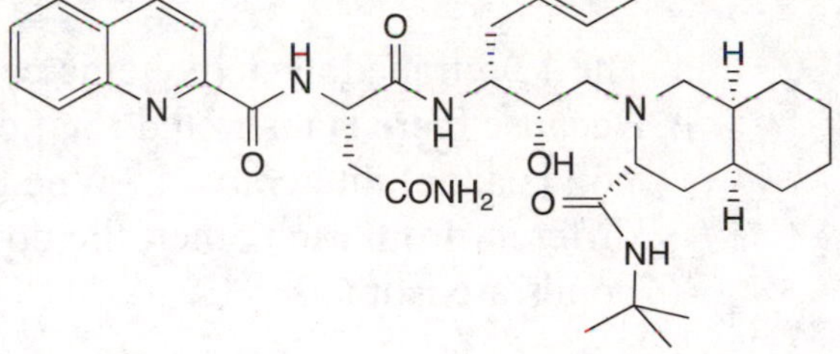

d. constitutional isomer

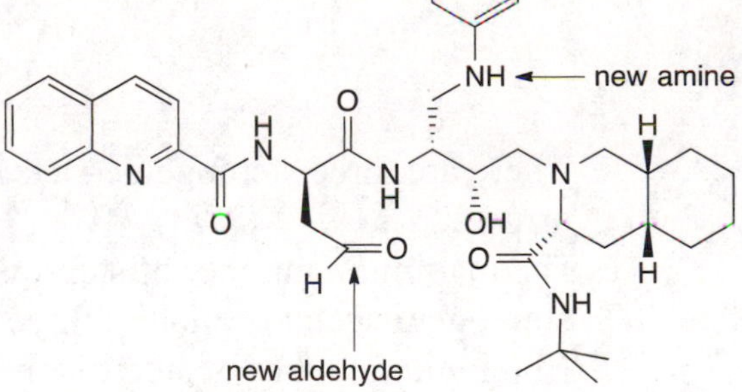

new amine

new aldehyde

5.69 Allenes contain an *sp* hybridized carbon atom doubly bonded to two other carbons. This makes the double bonds of an allene perpendicular to each other. When each end of the allene has two like substituents, the allene contains two planes of symmetry and it is achiral. When each end of the allene has two different groups, the allene has no plane of symmetry and it becomes chiral.

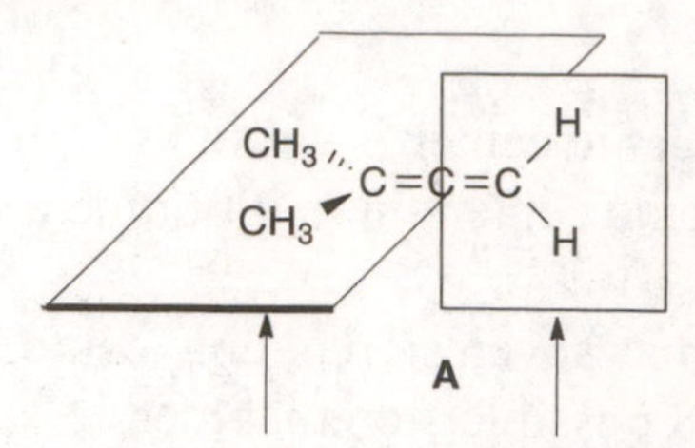

CH₃ C=C=C CH₃ H B H

no plane of symmetry
chiral

These two substituents are at 90° to these two substituents.
Allene **A** contains two planes of symmetry, making it **achiral.**

$HC{\equiv}C{-}C{\equiv}C{-}CH{=}C{=}CH{-}CH{=}CH{-}CH{=}CHCH_2CO_2H$

allene mycomycin

re-draw

$HC{\equiv}C{-}C{\equiv}C(H)C{=}C{=}C(H)CH{=}CH{-}CH{=}CHCH_2CO_2H$

The substituents on each end of the allene in mycomycin are different. Therefore, mycomycin is **chiral.**

5.70

H₂N, HO, O, *S*, *S*, *S*, O, *R*, N, NH₂, N, N, *S*, *R*, HO, OH

six stereogenic centers

5.71

HO, O, O, OH, OH, OH, O, NH₂, O

discodermolide

a. The 13 tetrahedral stereogenic centers are circled.

b. Because there is restricted rotation around a C–C double bond, groups on the end of the double bond cannot interconvert. Whenever the substituents on each end of the double bond are different from each other, the double bond is a stereogenic site. Thus, the following two double bonds are isomers:

R R C=C H H ⇹ R H C=C H R

These compounds are isomers.

There are three stereogenic sites due to the double bonds in discodermolide, labeled with arrows.

c. The maximum number of stereoisomers for discodermolide must include the 13 tetrahedral stereogenic centers and the three double bonds.
Maximum number of stereoisomers = 2^{16} = 65,536.

5.72 When the spiro compound has a plane of symmetry, it is achiral.

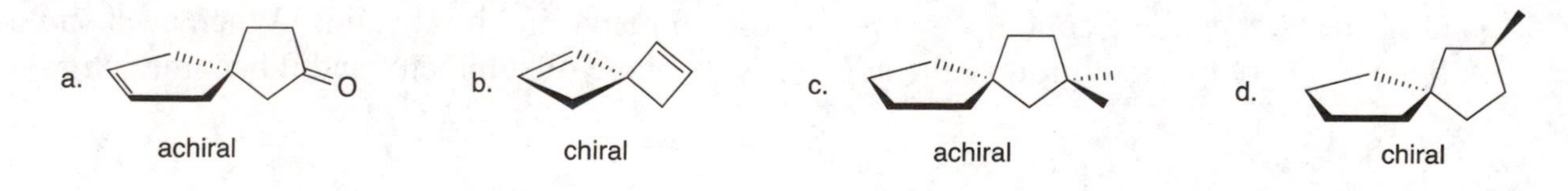

5.73

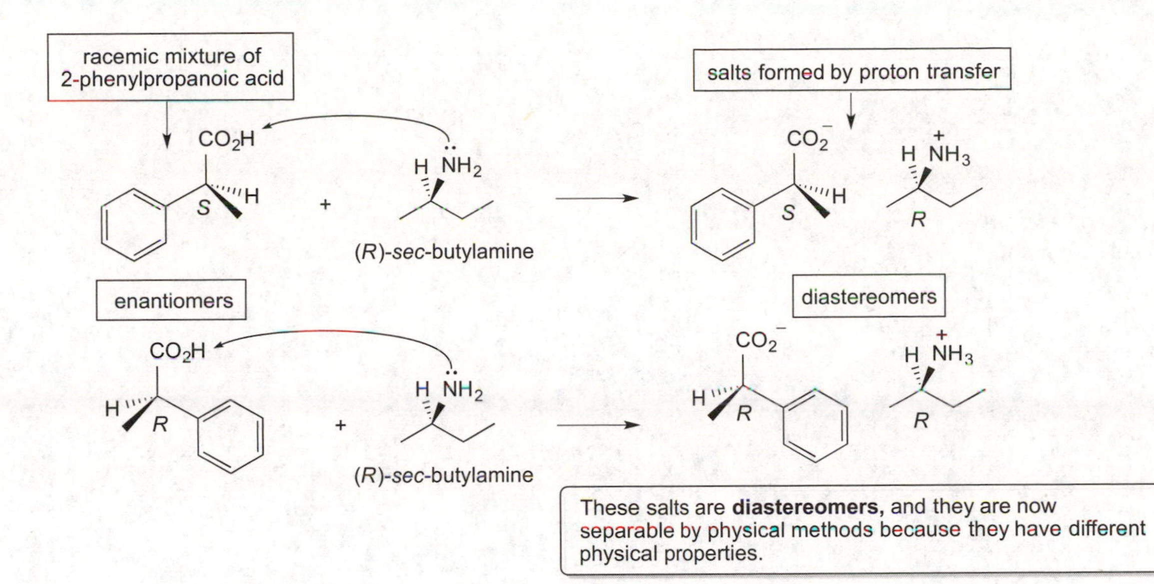

These salts are **diastereomers,** and they are now separable by physical methods because they have different physical properties.

Chapter 6: Understanding Organic Reactions

Chapter Review

Writing organic reactions (6.1)

- Use curved arrows to show the movement of electrons. Full-headed arrows are used for electron pairs and half-headed arrows are used for single electrons.

half-headed arrows

full-headed arrow

- Reagents can be drawn either on the left side of an equation or over an arrow. Catalysts are drawn over or under an arrow.

Types of reactions (6.2)

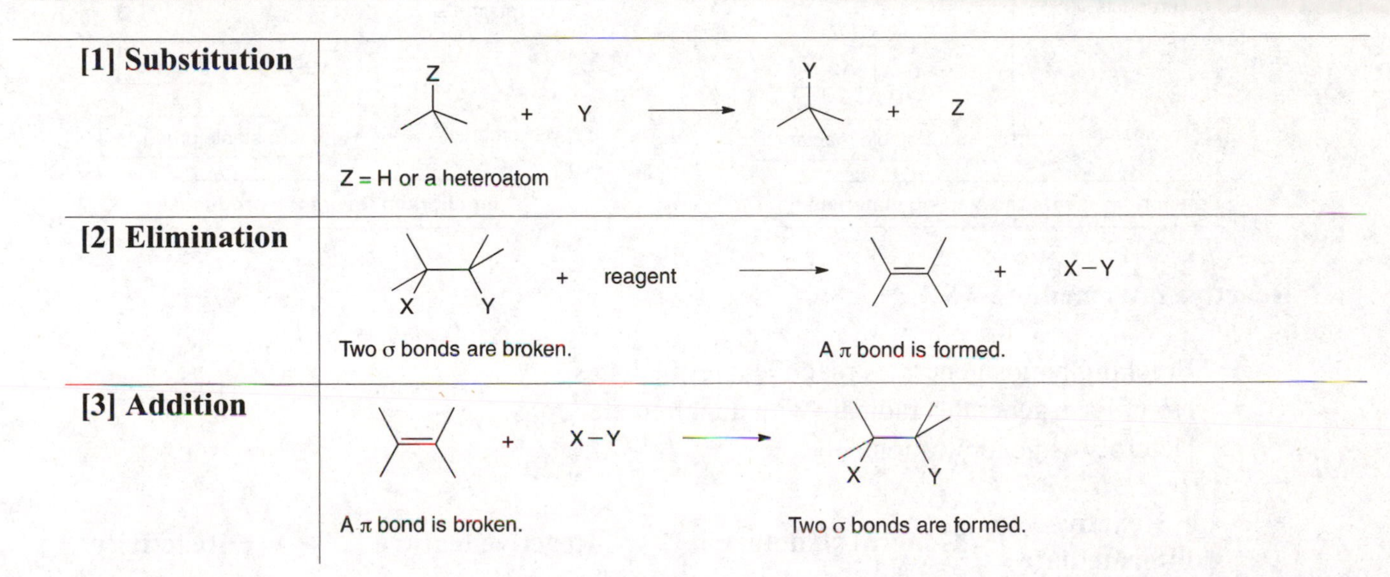

Important trends

Values compared	Trend
Bond dissociation energy and **bond strength**	The ***higher*** the bond dissociation energy, the ***stronger*** the bond (6.4). **Increasing size of the halogen** → CH_3-F: $\Delta H° = 456$ kJ/mol; CH_3-Cl: 351 kJ/mol; CH_3-Br: 293 kJ/mol; CH_3-I: 234 kJ/mol ← **Increasing bond strength**

E_a and reaction rate	The ***larger*** the energy of activation, the ***slower*** the reaction (6.9A). 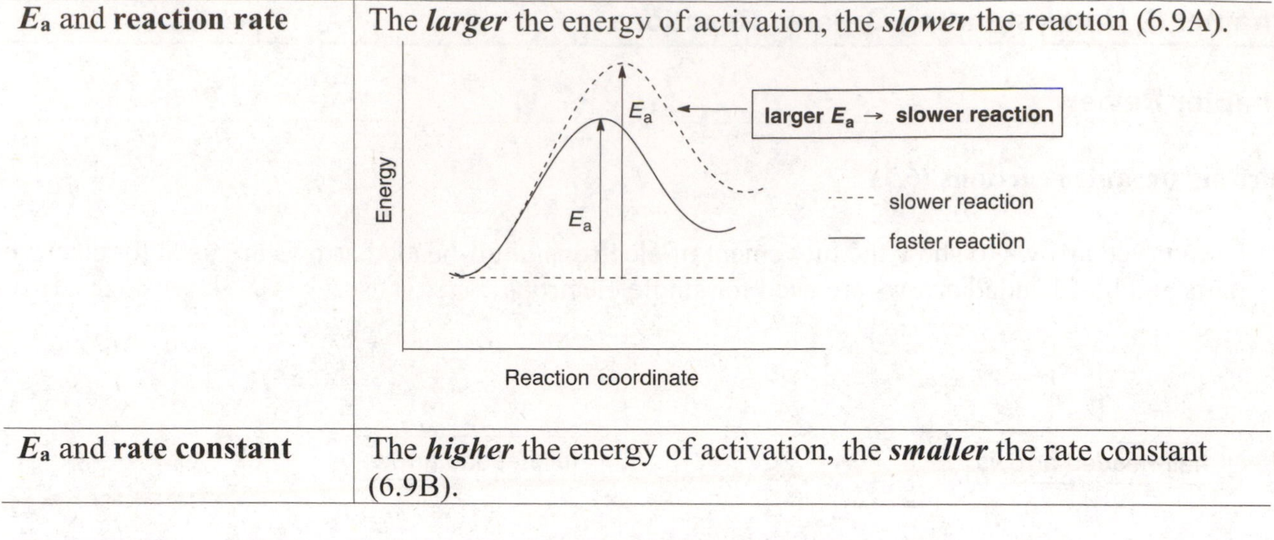
E_a and rate constant	The ***higher*** the energy of activation, the ***smaller*** the rate constant (6.9B).

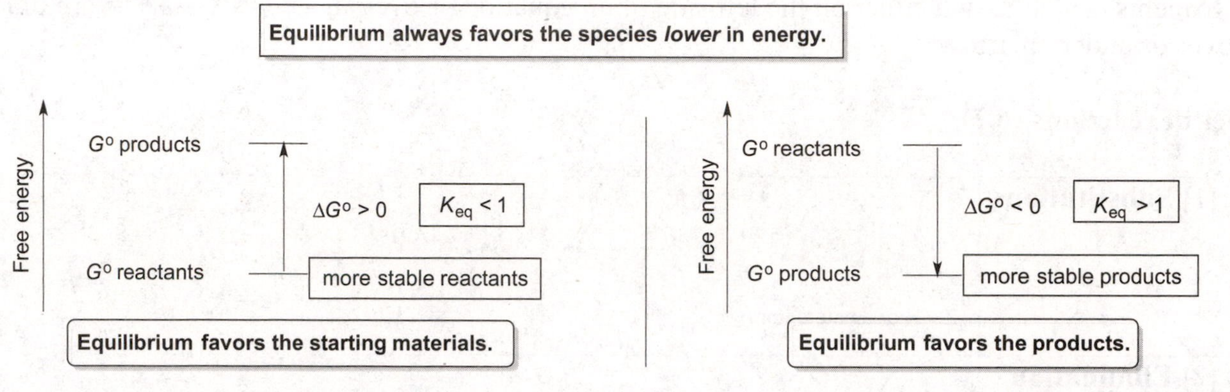

Reactive intermediates (6.3)

- Breaking bonds generates reactive intermediates.
- Homolysis generates radicals with unpaired electrons.
- Heterolysis generates ions.

Reactive intermediate	General structure	Reactive feature	Reactivity
radical		unpaired electron	electrophilic
carbocation		positive charge; only six electrons around C	electrophilic
carbanion		net negative charge; lone electron pair on C	nucleophilic

Energy diagrams (6.7, 6.8)

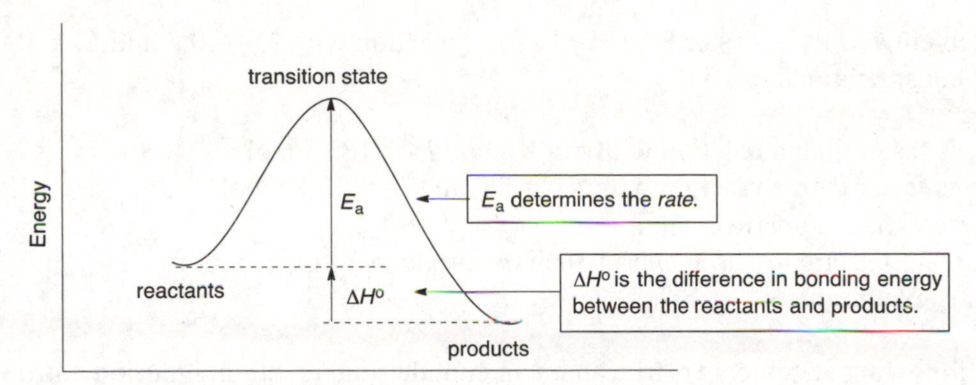

Conditions favoring product formation (6.5, 6.6)

Variable	Value	Meaning
K_{eq}	$K_{eq} > 1$	More product than starting material is present at equilibrium.
ΔG^o	$\Delta G^o < 0$	The energy of the products is **lower** than the energy of the reactants.
ΔH^o	$\Delta H^o < 0$	The bonds in the products are **stronger** than the bonds in the reactants.
ΔS^o	$\Delta S^o > 0$	The product is **more disordered** than the reactant.

Equations (6.5, 6.6)

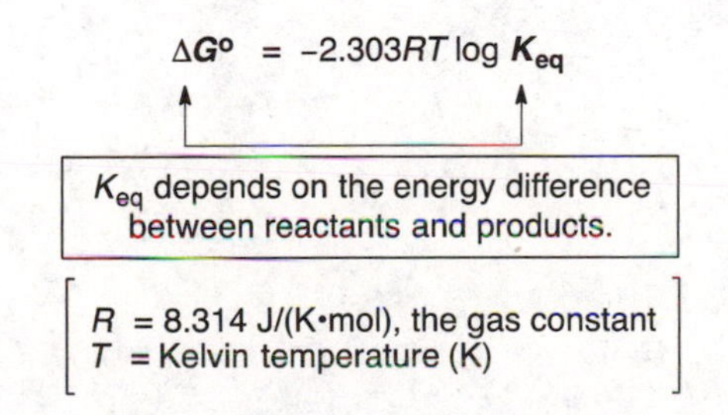

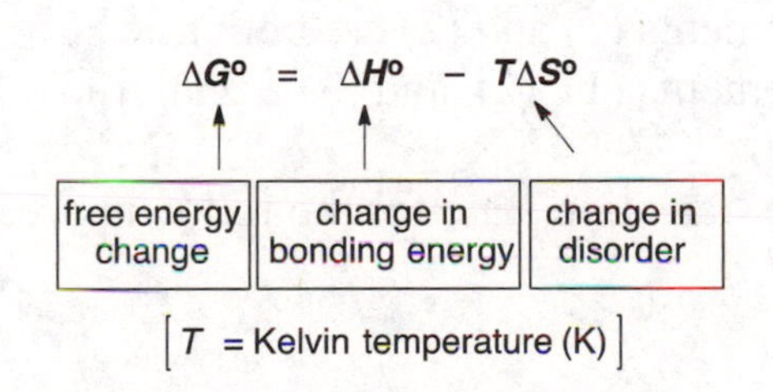

Factors affecting reaction rate (6.9)

Factor	Effect
energy of activation	higher $E_a \rightarrow$ slower reaction
concentration	higher concentration $\rightarrow$ faster reaction
temperature	higher temperature $\rightarrow$ faster reaction

Practice Test on Chapter Review

1. Label each statement as TRUE (T) or FALSE (F) for a reaction with $K_{eq} = 0.5$ and $E_a = 18$ kJ/mol. Ignore entropy considerations.

   a. The reaction is faster than a reaction with $K_{eq} = 8$ and $E_a = 18$ kJ/mol.
   b. The reaction is faster than a reaction with $K_{eq} = 0.5$ and $E_a = 12$ kJ/mol.
   c. $\Delta G°$ for the reaction is a positive value.
   d. The starting materials are lower in energy than the products of the reaction.
   e. The reaction is exothermic.

2.a. Which of the following statements is true about an endothermic reaction, ignoring entropy considerations?

   1. The bonds in the products are stronger than the bonds in the starting materials.
   2. $K_{eq} < 1$.
   3. A catalyst speeds up the rate of the reaction and gives a larger amount of product.
   4. Statements (1) and (2) are both true.
   5. Statements (1), (2), and (3) are all true.

b. Which of the following statements is true about a reaction with $K_{eq} = 10^3$ and $E_a = 2.5$ kJ/mol? Ignore entropy considerations.

   1. The reaction is faster than a reaction with $E_a = 4$ kJ/mol.
   2. The starting materials are higher in energy than the products of the reaction.
   3. $\Delta G°$ is positive.
   4. Statements (1) and (2) are both true.
   5. Statements (1), (2), and (3) are all true.

3.a. Draw the transition state for the following reaction.

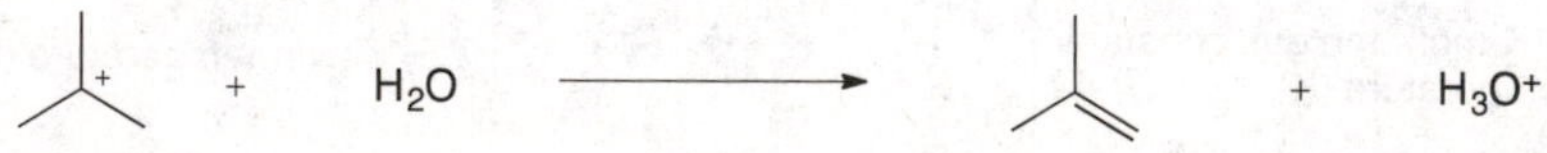

+ H_2O → + H_3O^+

b. Draw the transition state for the following one-step elimination reaction.

Br + $^-OCH_3$ → + CH_3OH + Br^-

Answers to Practice Test

1. a. F
   b. F
   c. T
   d. T
   e. F

2. a. 2
   b. 4

3. a. δ+ H--O:δ+ (H, H) ‡

   b. Br: δ− H ÖCH$_3$ δ− ‡

Answers to Problems

6.1 [1] In a **substitution reaction,** one group replaces another.
[2] In an **elimination reaction,** elements of the starting material are lost and a π bond is formed.
[3] In an **addition reaction,** elements are added to the starting material.

a. Br replaces OH = **substitution reaction**

b. addition of 2 H's **addition reaction**

c. Cl replaces H = **substitution reaction**

d. elements lost (H + OH) π bond formed **elimination reaction**

6.2 The elements of Cl and OH are added to a π bond, so the reaction is an addition.

A → B

6.3 **Heterolysis** means one atom gets both of the electrons when a bond is broken. A carbocation is a C with a positive charge, and a carbanion is a C with a negative charge.

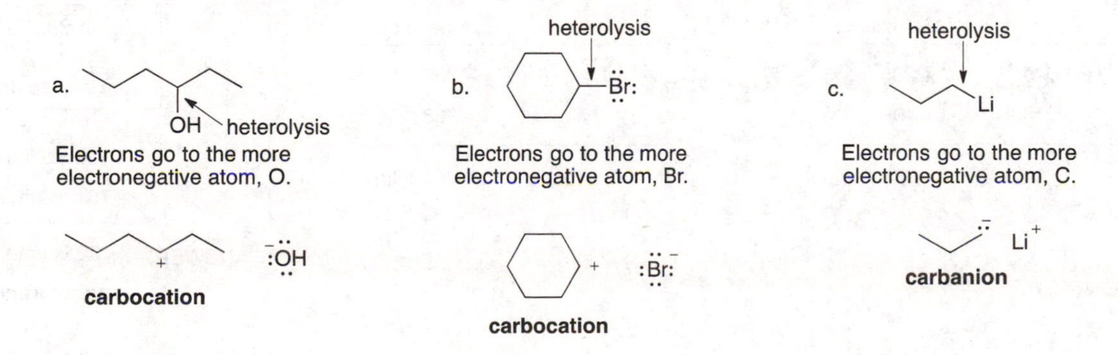

6.4 Use **full-headed arrows** to show the movement of electron pairs, and **half-headed arrows** to show the movement of single electrons.

a. + →

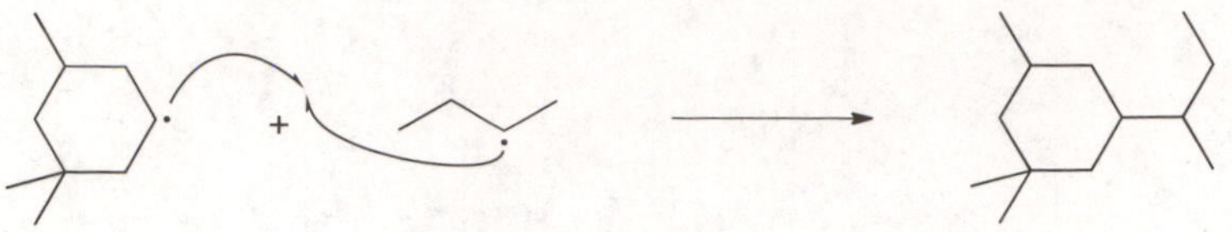

b. + :Br:⁻ → :Br:

6.5

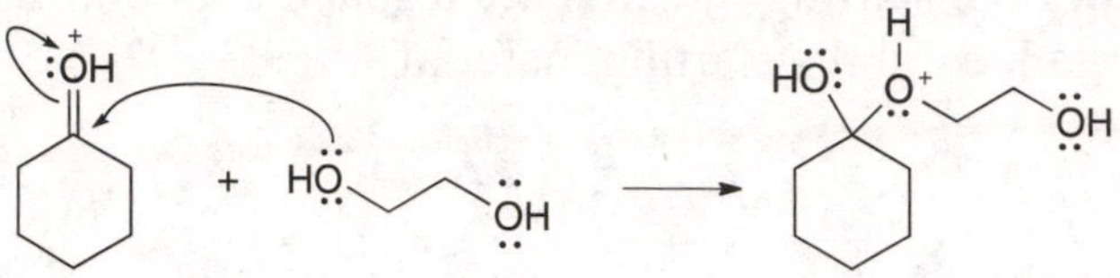

:OH (+) + HO OH → HO: H O+ OH

6.6 Increasing number of electrons between atoms = increasing bond strength = increasing bond dissociation energy = decreasing bond length.
Increasing size of an atom = increasing bond length = decreasing bond strength.

a. OH or SH

higher bond dissociation energy

S is larger than O. longer, weaker bond

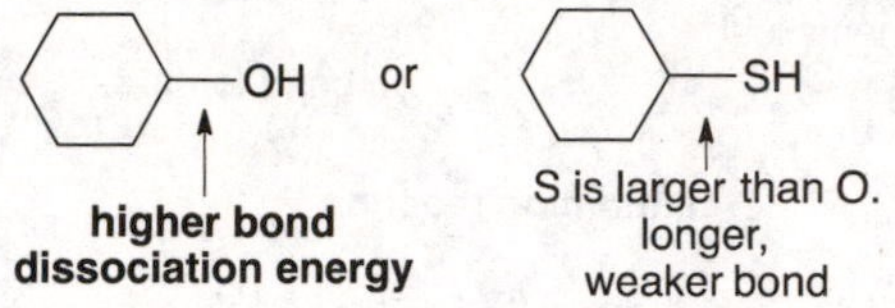

b. O or O

higher bond dissociation energy

single bond fewer electrons

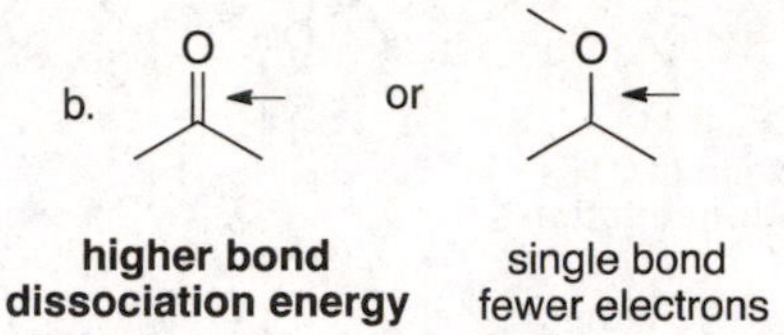

6.7 To determine Δ*H*° for a reaction:
[1] Add the bond dissociation energies for all bonds *broken* in the equation (+ values).
[2] Add the bond dissociation energies for all of the bonds *formed* in the equation (– values).
[3] *Add the energies together* to get the Δ*H*° for the reaction.
A positive Δ*H*° means the reaction is ***endothermic*****. A negative Δ*H*°** means the reaction is ***exothermic*****.**

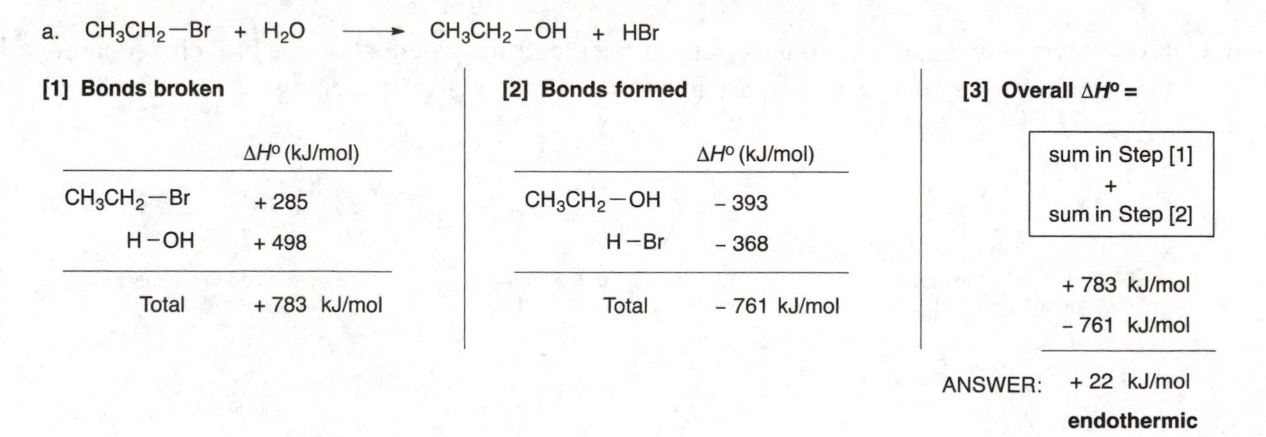

a. $CH_3CH_2-Br + H_2O \longrightarrow CH_3CH_2-OH + HBr$

[1] Bonds broken

	$\Delta H°$ (kJ/mol)
CH_3CH_2-Br	+ 285
H–OH	+ 498
Total	+ 783 kJ/mol

[2] Bonds formed

	$\Delta H°$ (kJ/mol)
CH_3CH_2-OH	- 393
H–Br	- 368
Total	- 761 kJ/mol

[3] Overall $\Delta H°$ =

sum in Step [1] + sum in Step [2]

+ 783 kJ/mol
- 761 kJ/mol

ANSWER: + 22 kJ/mol

endothermic

b. $CH_4 + Cl_2 \longrightarrow CH_3Cl + HCl$

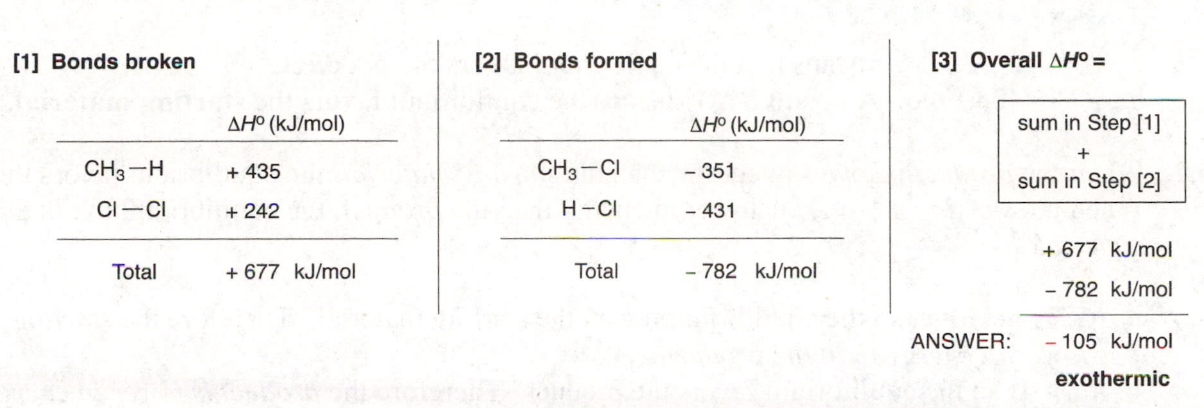

[1] Bonds broken	ΔH^o (kJ/mol)
CH_3-H	+ 435
$Cl-Cl$	+ 242
Total	+ 677 kJ/mol

[2] Bonds formed	ΔH^o (kJ/mol)
CH_3-Cl	– 351
$H-Cl$	– 431
Total	– 782 kJ/mol

[3] Overall ΔH^o = sum in Step [1] + sum in Step [2]

+ 677 kJ/mol
– 782 kJ/mol

ANSWER: – 105 kJ/mol
exothermic

6.8 Use the directions from Answer 6.7. In determining the number of bonds broken or formed, you must take into account the coefficients needed to balance an equation.

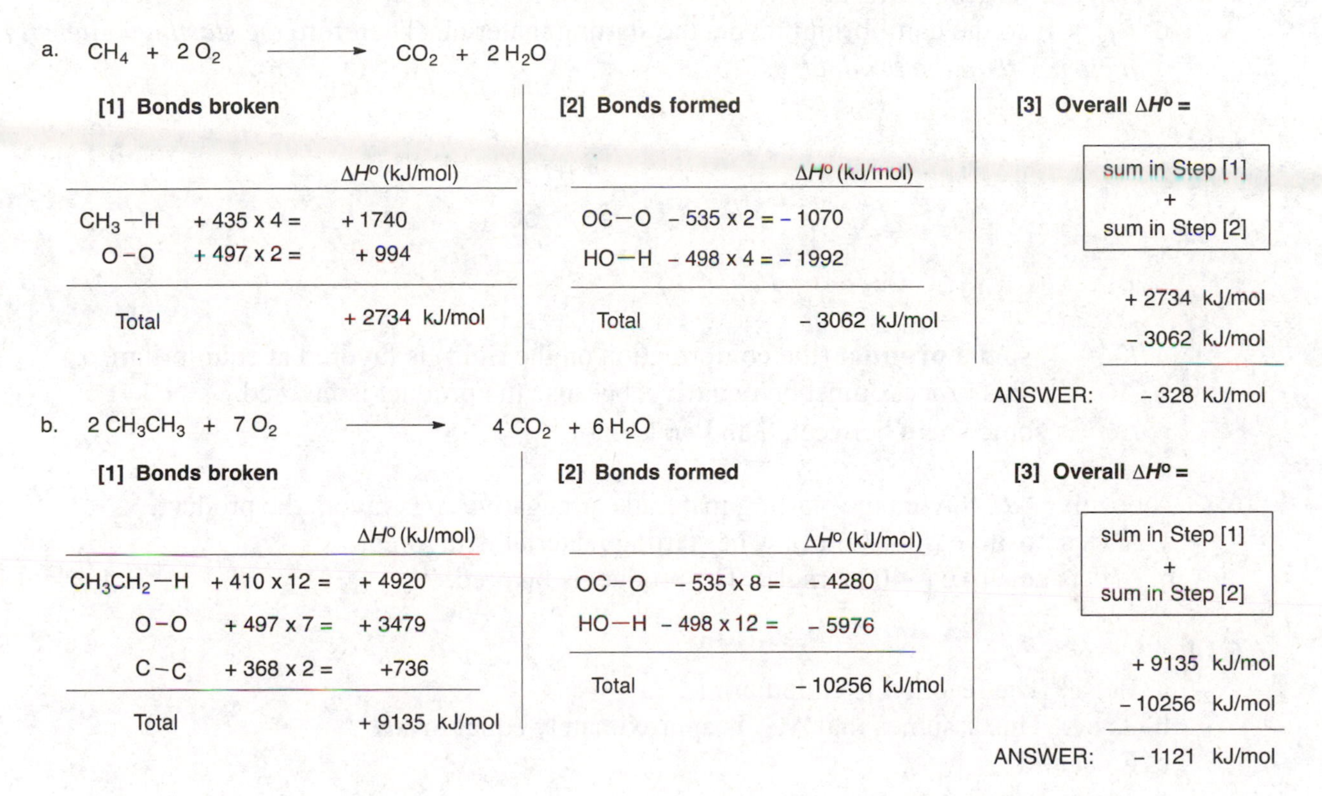

a. $CH_4 + 2\,O_2 \longrightarrow CO_2 + 2\,H_2O$

[1] Bonds broken		ΔH^o (kJ/mol)
CH_3-H	+ 435 x 4 =	+ 1740
$O-O$	+ 497 x 2 =	+ 994
Total		+ 2734 kJ/mol

[2] Bonds formed	ΔH^o (kJ/mol)
$OC-O$	– 535 x 2 = – 1070
$HO-H$	– 498 x 4 = – 1992
Total	– 3062 kJ/mol

[3] Overall ΔH^o = sum in Step [1] + sum in Step [2]

+ 2734 kJ/mol
– 3062 kJ/mol

ANSWER: – 328 kJ/mol

b. $2\,CH_3CH_3 + 7\,O_2 \longrightarrow 4\,CO_2 + 6\,H_2O$

[1] Bonds broken		ΔH^o (kJ/mol)
CH_3CH_2-H	+ 410 x 12 =	+ 4920
$O-O$	+ 497 x 7 =	+ 3479
$C-C$	+ 368 x 2 =	+736
Total		+ 9135 kJ/mol

[2] Bonds formed		ΔH^o (kJ/mol)
$OC-O$	– 535 x 8 =	– 4280
$HO-H$	– 498 x 12 =	– 5976
Total		– 10256 kJ/mol

[3] Overall ΔH^o = sum in Step [1] + sum in Step [2]

+ 9135 kJ/mol
– 10256 kJ/mol

ANSWER: – 1121 kJ/mol

6.9 Use the following relationships to answer the questions:
If $K_{eq} = 1$, then $\Delta G^\circ = 0$; if $K_{eq} > 1$, then $\Delta G^\circ < 0$; if $K_{eq} < 1$, then $\Delta G^\circ > 0$.

a. A negative value of ΔG° means the equilibrium favors the product and $K_{eq} > 1$. Therefore, $K_{eq} = 1000$ is the answer.

b. A lower value of ΔG° means a larger value of K_{eq}, and the products are more favored. $K_{eq} = 10^{-2}$ is larger than $K_{eq} = 10^{-5}$, so ΔG° is lower.

6.10 Use the relationships from Answer 6.9.

a. K_{eq} = 5.5. K_{eq} > 1 means that the equilibrium favors the **product.**
b. $\Delta G°$ = 40 kJ/mol. A positive $\Delta G°$ means the equilibrium favors the **starting material.**

6.11 When the *product* is lower in energy than the *starting material,* the equilibrium favors the *product.* When the *starting material* is lower in energy than the *product,* the equilibrium favors the *starting material.*

a. **$\Delta G°$ is positive,** so the equilibrium favors the starting material. Therefore the *starting material is lower in energy than the product.*
b. **K_{eq} > 1,** so the equilibrium favors the product. Therefore the *product is lower in energy than the starting material.*
c. **$\Delta G°$ is negative,** so the equilibrium favors the product. Therefore the *product is lower in energy than the starting material.*
d. **K_{eq} < 1,** so the equilibrium favors the starting material. Therefore *the starting material is lower in energy than the product.*

6.12

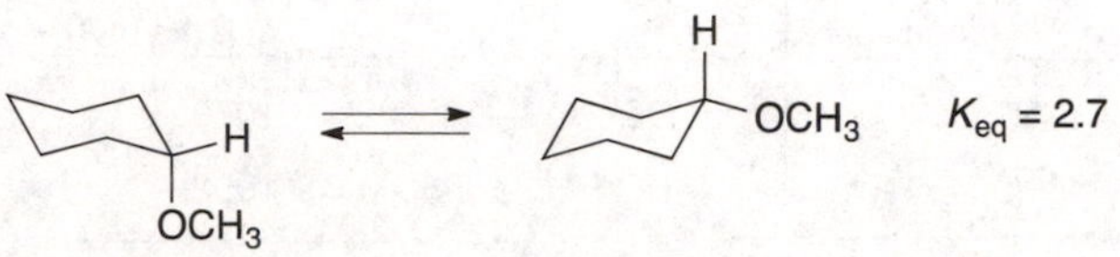

H
H
OCH_3
OCH_3
K_{eq} = 2.7

a. K_{eq} > 1, so the **product** (the conformation on the right) is favored at equilibrium.
b. $\Delta G°$ for this process must be **negative,** because the product is favored.
c. $\Delta G°$ is somewhere between 0 and –6 kJ/mol.

6.13 A positive $\Delta H°$ favors the starting material. A negative $\Delta H°$ favors the product.
a. $\Delta H°$ is positive (80 kJ/mol). The starting material is favored.
b. $\Delta H°$ is negative (–40 kJ/mol). The product is favored.

6.14

a. **False.** The reaction is endothermic.
b. **True.** This assumes that $\Delta G°$ is approximately equal to $\Delta H°$.
c. **False.** K_{eq} < 1.
d. **True.**
e. **False.** The starting material is favored at equilibrium.

6.15

a. **True.**
b. **False.** $\Delta G°$ for the reaction is negative.
c. **True.**
d. **False.** The bonds in the product are stronger than the bonds in the starting material.
e. **True.**

6.16

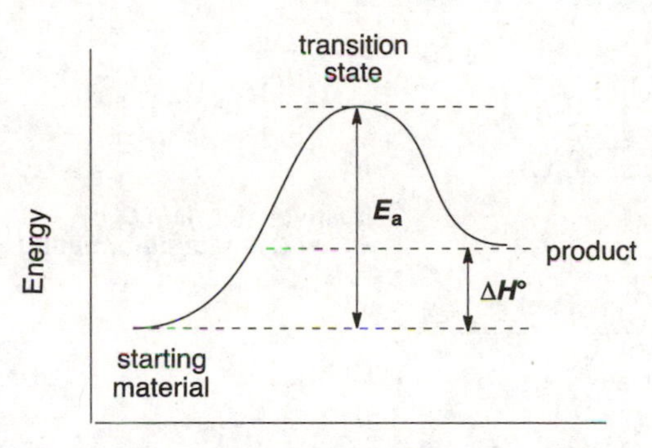

6.17 A transition state is drawn with dashed lines to indicate the partially broken and partially formed bonds. Any atom that gains or loses a charge contains a partial charge in the transition state.

a. $\overset{+}{OH_2}$ → + H_2O transition state: [δ+ :OH_2 δ+]‡

b. OH + $^-$OH → O^- + H_2O transition state: [δ– O H δ– OH]‡

6.18

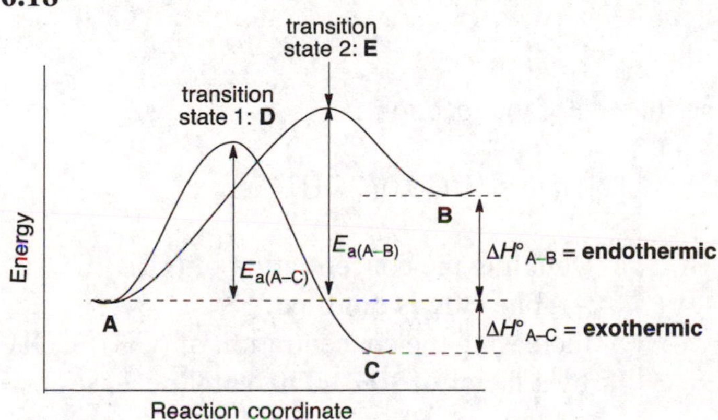

a. Reaction **A–C** is exothermic. Reaction **A–B** is endothermic.
b. Reaction **A–C** is faster.
c. Reaction **A–C** generates a lower-energy product.
d. See labels.
e. See labels.
f. See labels.

6.19

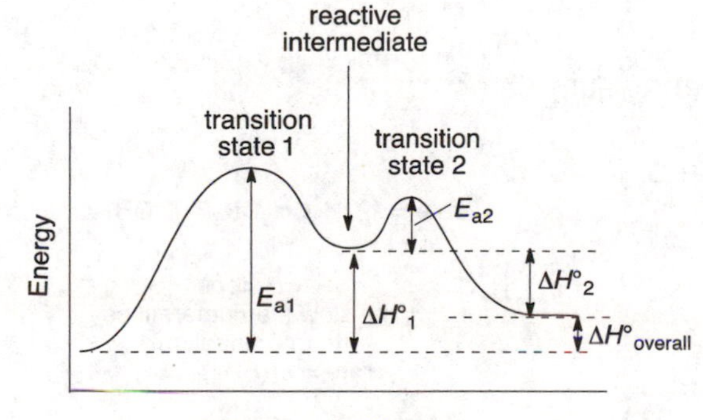

a. Two steps, because there are two energy barriers.
b. See labels.
c. See labels.
d. One reactive intermediate is formed (see label).
e. The first step is rate-determining, because its transition state is at higher energy.
f. The overall reaction is endothermic, because the energy of the products is higher than the energy of the reactants.

6.20

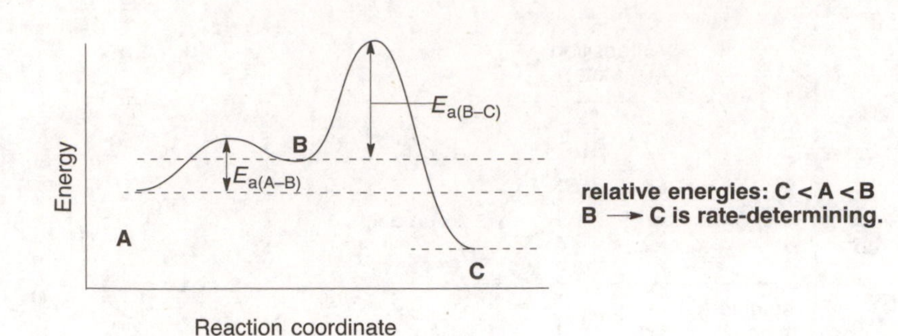

6.21 ***E*$_a$, concentration, and temperature affect reaction rate.** $\Delta H°$, $\Delta G°$, and K_{eq} do not affect reaction rate.

a. ***E*$_a$ = 4 kJ/mol** corresponds to a faster reaction rate.
b. A temperature of **25 °C** will have a faster reaction rate, because a higher temperature corresponds to a faster reaction.
c. **No change:** K_{eq} does not affect reaction rate.
d. **No change:** $\Delta H°$ does not affect reaction rate.

6.22

a. **False.** The reaction occurs at the same rate as a reaction with K_{eq} = 8 and E_a = 80 kJ/mol.
b. **False.** The reaction is slower than a reaction with K_{eq} = 0.8 and E_a = 40 kJ/mol.
c. **True.**
d. **True.**
e. **False.** The reaction is endothermic.

6.23 All reactants in the rate equation determine the rate of the reaction.

[1] rate = $k[CH_3CH_2Br][^{-}OH]$

a. Tripling the concentration of CH_3CH_2Br only → **The rate is tripled.**
b. Tripling the concentration of ^{-}OH only → **The rate is tripled.**
c. Tripling the concentration of both $CH_3CH_2CH_2Br$ and ^{-}OH → **The rate increases by a factor of 9 (3 × 3 = 9).**

[2] rate = $k[(CH_3)_3COH]$

a. Doubling the concentration of $(CH_3)_3COH$ → **The rate is doubled.**
b. Increasing the concentration of $(CH_3)_3COH$ by a factor of 10 → **The rate increases by a factor of 10.**

6.24 The rate equation is determined by the rate-determining step.

a. [1-bromopropane] + ^{-}OH → [propene] + H_2O + Br^-

one step
rate = $k[CH_3CH_2CH_2Br][^{-}OH]$

b. [tert-butyl bromide] —slow→ [tert-butyl carbocation] + Br^- —^{-}OH, fast→ [2-methylpropene] + H_2O

two steps
The slow step determines the rate equation.
rate = $k[(CH_3)_3CBr]$

6.25 A catalyst is not used up or changed in the reaction. It only speeds up the reaction rate.

OH and H are added to the starting material.

a. $CH_2{=}CH_2 \xrightarrow[H_2SO_4]{H_2O} CH_3CH_2OH$

H_2SO_4 is not used up = **catalyst.**

I^- not used up = **catalyst.**

b. $CH_3Cl \xrightarrow[^-OH]{I^-} CH_3OH$

^-OH substitutes for Cl^-.

6.26

a. cyclohexane (H, H) ⟶ cyclohexyl radical (H) + ·H

radical

b. CH_3CH_2OH ⟶ $CH_3\overset{+}{C}H_2$ + ^-OH

carbocation

6.27

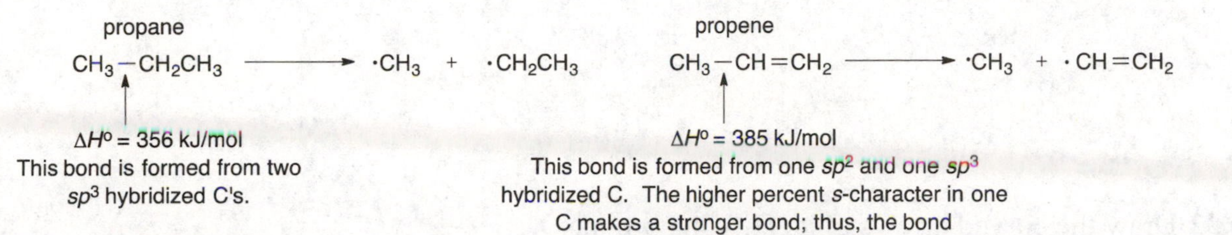

propane
CH3—CH2CH3 ⟶ ·CH3 + ·CH2CH3
ΔHº = 356 kJ/mol
This bond is formed from two sp3 hybridized C's.
propene
CH3—CH=CH2 ⟶ ·CH3 + ·CH=CH2
ΔHº = 385 kJ/mol
This bond is formed from one sp2 and one sp3 hybridized C. The higher percent s-character in one C makes a stronger bond; thus, the bond dissociation energy is higher.

6.28 Use the directions from Answer 6.1.

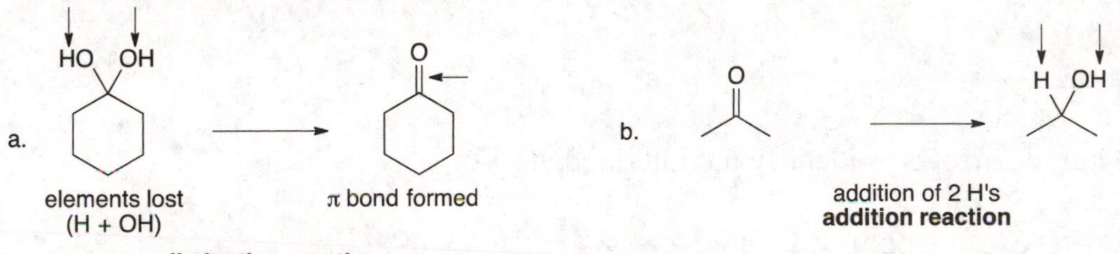

6.29 Use the rules in Answer 6.4 to draw the arrows.

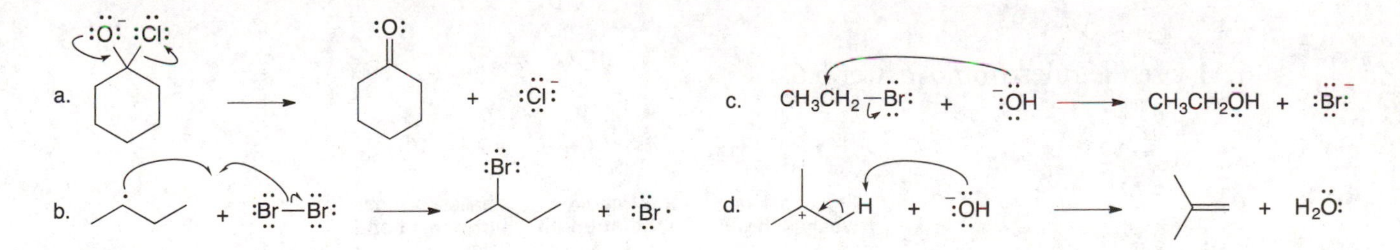

6.30

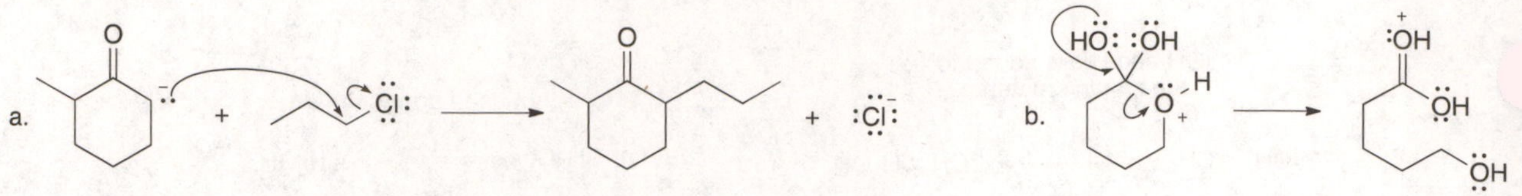

a. + $\rightarrow$ + :Cl:

b. HO: :OH H O $\rightarrow$ OH OH OH

6.31

a. :O–OH

A B C + :OH

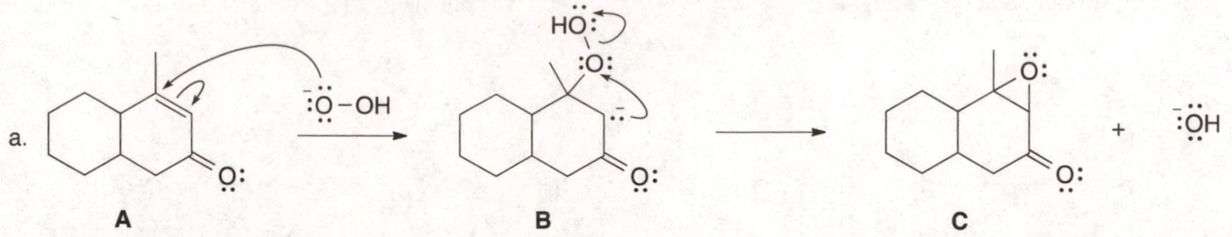

b. The conversion of **A** to **C** is an addition because a π bond is broken and an O atom is added.

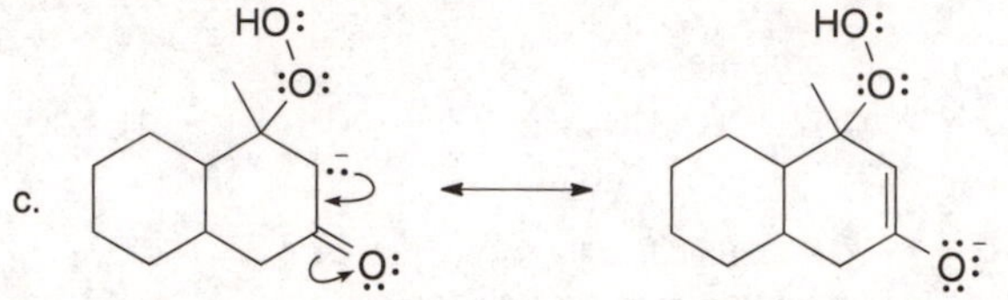

c. HO: :O: HO: :O:

6.32 Draw the curved arrows to identify the product **X**.

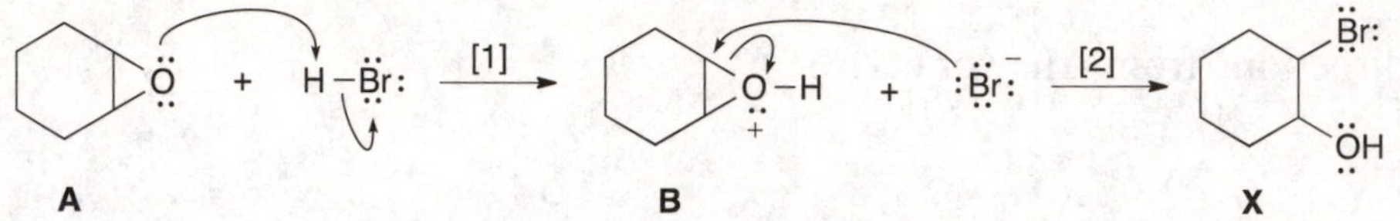

A + H–Br: [1] $\rightarrow$ B O–H + :Br: [2] $\rightarrow$ X Br: OH

6.33 Follow the curved arrows to identify the intermediate **Y**.

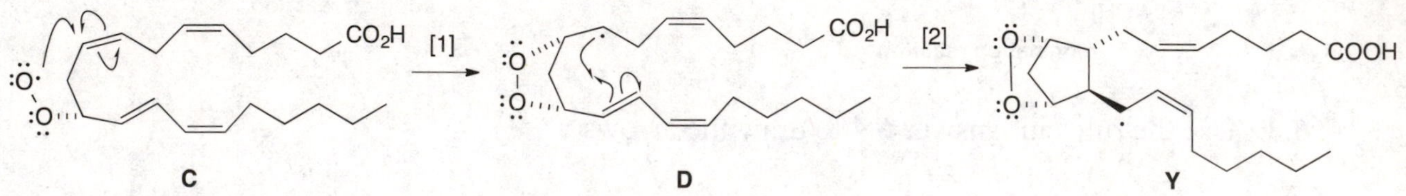

C CO_2H [1] $\rightarrow$ D CO_2H [2] $\rightarrow$ Y COOH

6.34 Use the rules from Answer 6.6.

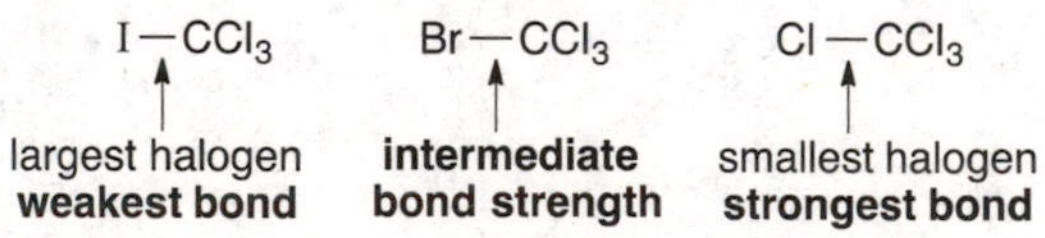

I–CCl_3	Br–CCl_3	Cl–CCl_3
largest halogen **weakest bond**	**intermediate bond strength**	smallest halogen **strongest bond**

6.35 Use the directions from Answer 6.7.

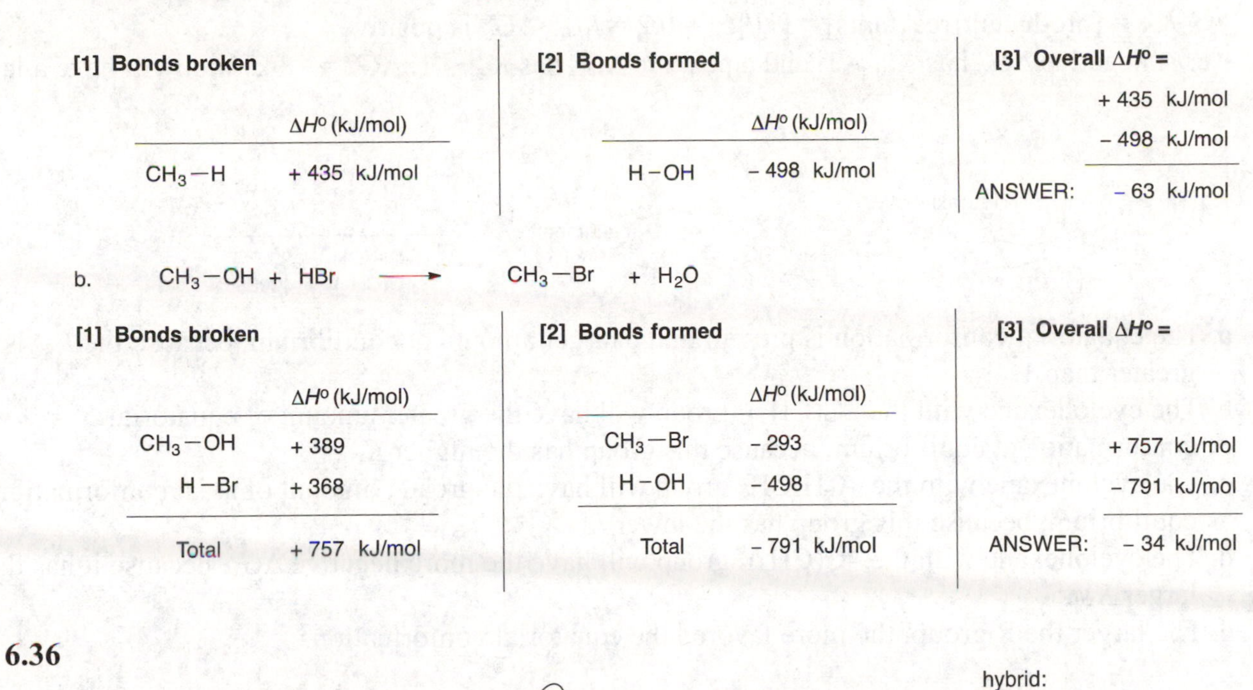

a. ·OH + CH4 → ·CH3 + H2O
[1] Bonds broken
ΔHº (kJ/mol)
CH3–H + 435 kJ/mol
[2] Bonds formed
ΔHº (kJ/mol)
H–OH – 498 kJ/mol
[3] Overall ΔHº =
+ 435 kJ/mol
– 498 kJ/mol
ANSWER: – 63 kJ/mol
b. CH3–OH + HBr → CH3–Br + H2O
[1] Bonds broken
ΔHº (kJ/mol)
CH3–OH + 389
H–Br + 368
Total + 757 kJ/mol
[2] Bonds formed
ΔHº (kJ/mol)
CH3–Br – 293
H–OH – 498
Total – 791 kJ/mol
[3] Overall ΔHº =
+ 757 kJ/mol
– 791 kJ/mol
ANSWER: – 34 kJ/mol

6.36

hybrid:

δ· δ·

6.37 The more stable radical is formed by a reaction with a smaller $\Delta H°$.

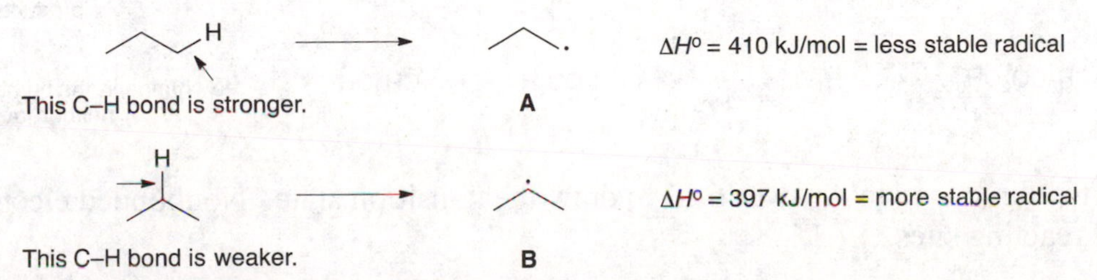

Because the bond dissociation for cleavage of the C–H bond to form radical **A** is higher, more energy must be added to form it. This makes **A** higher in energy and therefore less stable than **B**.

6.38 Use the rules from Answer 6.11.

a. $K_{eq} = 0.5$. K_{eq} is less than one, so the **starting material** is favored.
b. $\Delta G° = -100$ kJ/mol. $\Delta G°$ is less than 0, so the **product** is favored.
c. $\Delta H° = 8.0$ kJ/mol. $\Delta H°$ is positive, so the **starting material** is favored.
d. $K_{eq} = 16$. K_{eq} is greater than one, so the **product** is favored.
e. $\Delta G° = 2.0$ kJ/mol. $\Delta G°$ is greater than zero, so the **starting material** is favored.
f. $\Delta H° = 200$ kJ/mol. $\Delta H°$ is positive, so the **starting material** is favored.
g. $\Delta S° = 8$ J/(K•mol). $\Delta S°$ is greater than zero, so the **product** is more disordered and favored.
h. $\Delta S° = -8$ J/(K•mol). $\Delta S°$ is less than zero, so the **starting material** is more disordered and favored.

6.39

a. A negative $\Delta G°$ must have $K_{eq} > 1$. $K_{eq} = 10^2$.
b. K_{eq} = [products]/[reactants] = [1]/[5] = 0.2 = K_{eq}. $\Delta G°$ is positive.
c. A negative $\Delta G°$ has $K_{eq} > 1$, and a positive $\Delta G°$ has $K_{eq} < 1$. $\Delta G°$ = –8 kJ/mol will have a larger K_{eq}.

6.40

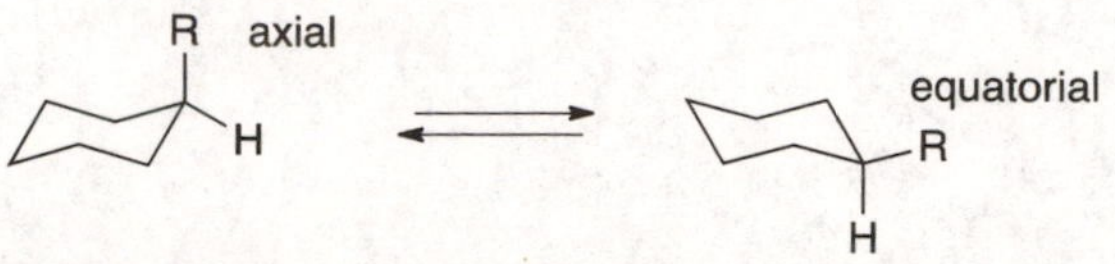

R axial ⇌ equatorial R

R	K_{eq}
$-CH_2CH_3$	23
$-C(CH_3)_3$	4000

a. The equatorial conformation is present in the larger amount at equilibrium, because the K_{eq} is greater than 1.
b. The cyclohexane with the $-C(CH_3)_3$ group will have the greater amount of equatorial conformation at equilibrium, because this group has the higher K_{eq}.
c. The cyclohexane with the $-CH_2CH_3$ group will have the greater amount of axial conformation at equilibrium, because this group has the lower K_{eq}.
d. The cyclohexane with the $-C(CH_3)_3$ group will have the more negative $\Delta G°$, because it has the larger K_{eq}.
e. The larger the R group, the more favored the equatorial conformation.

6.41 Reactions resulting in an increase in entropy are favored. When a single molecule forms two molecules, there is an increase in entropy.

a. [octane] → [propene] + [pentane]

increased number of molecules
$\Delta S°$ is positive.
products favored

b. $CH_3CO_2CH_3$ + H_2O → CH_3CO_2H + CH_3OH

no change in the number of molecules
neither favored

6.42 Use the directions in Answer 6.17 to draw the transition state. Nonbonded electron pairs are drawn in at reacting sites.

a. 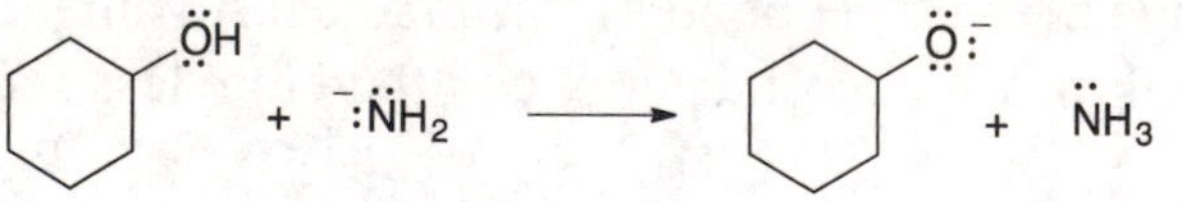 cyclohexanol + $^{-}$:$\ddot{N}H_2$ → cyclohexoxide + $\ddot{N}H_3$

transition state: [cyclohexyl–$\ddot{O}^{\delta-}$- -H- -$\ddot{N}H_2^{\delta-}$]‡

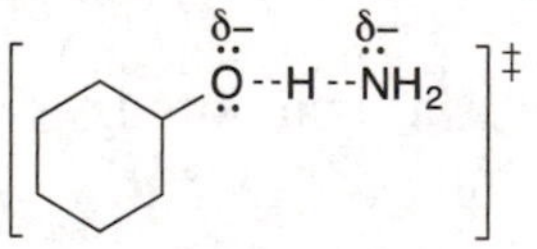

b. 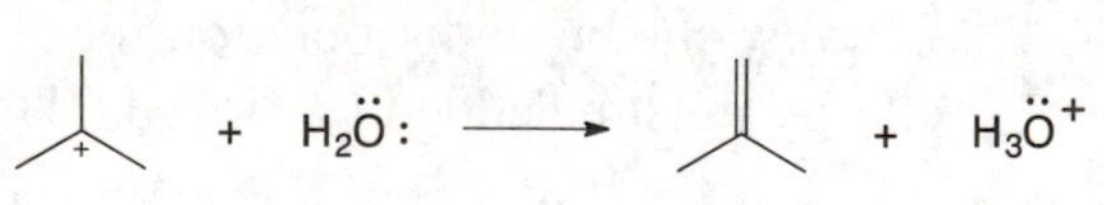 tert-butyl cation + $H_2\ddot{O}$: → 2-methylpropene + $H_3\ddot{O}^+$

transition state: [δ+ ... H ... $\ddot{O}H_2^{\delta+}$]‡

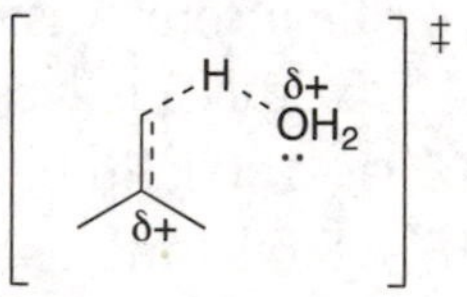

6.43

a.

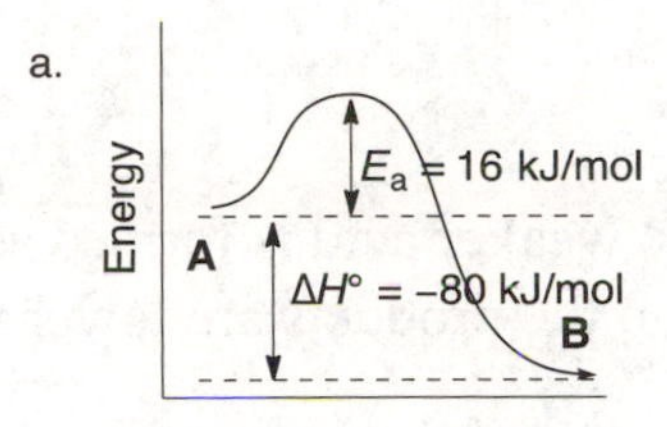

- one step **A** → **B**
- exothermic because **B** lower than **A**

b.

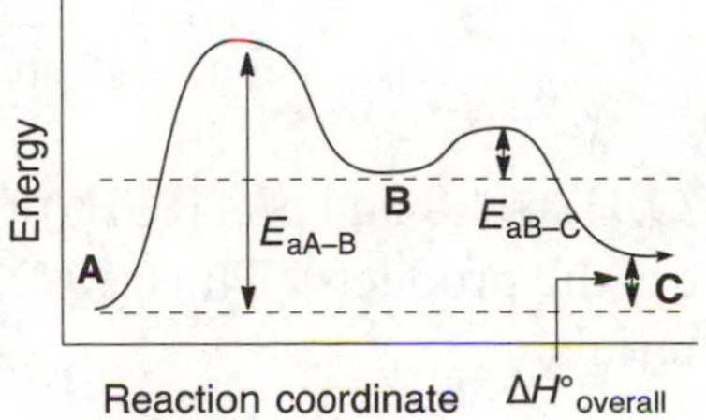

- two steps
- **A** lowest energy
- **B** highest energy
- $E_{a(A\text{-}B)}$ **is rate-determining,** because the transition state for Step [1] is higher in energy.

6.44

a. CH_3—H + ·$\ddot{Cl}$: → ·CH_3 + H$\ddot{Cl}$:

b. ·Cl + CH_4 → ·CH_3 + HCl

[1] Bonds broken

	$\Delta H°$ (kJ/mol)
CH_3—H	+ 435 kJ/mol

[2] Bonds formed

	$\Delta H°$ (kJ/mol)
H—Cl	– 431 kJ/mol

[3] Overall $\Delta H°$ =

+ 435 kJ/mol
– 431 kJ/mol

ANSWER: + 4 kJ/mol

c.

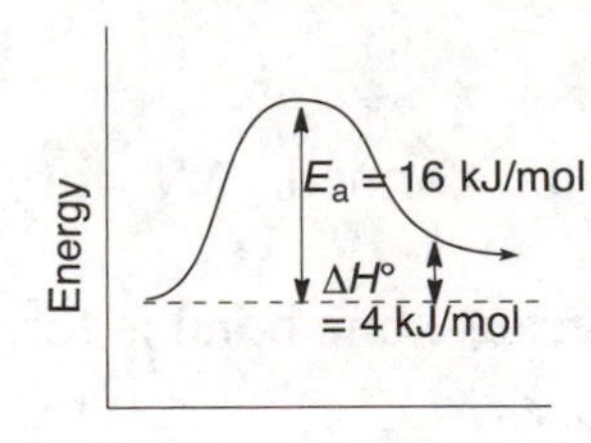

d. E_a for the reverse reaction is the difference in energy between the products and the transition state, 12 kJ/mol.

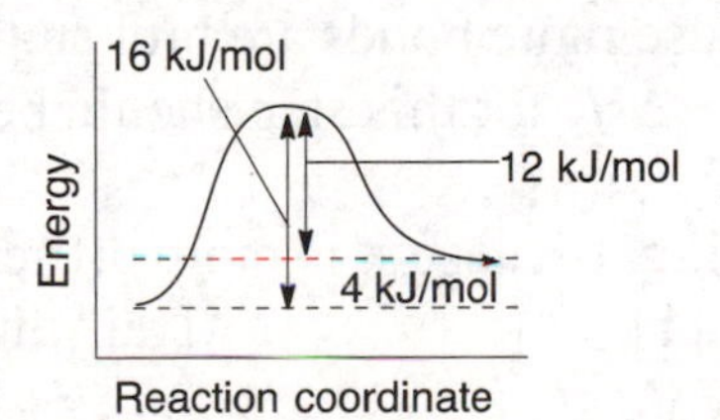

6.45

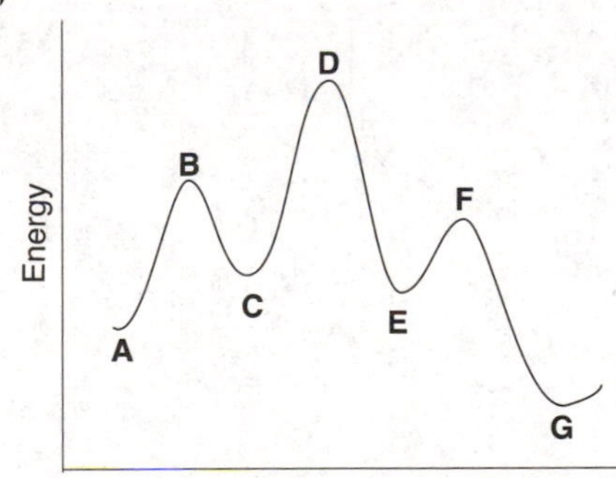

a. **B, D,** and **F** are transition states.
b. **C** and **E** are reactive intermediates.
c. The overall reaction has **three steps.**
d. **A–C** is endothermic.
C–E is exothermic.
E–G is exothermic.
e. The overall reaction is exothermic.

6.46

Because pK_a (CH_3CO_2H) = 4.8 and pK_a [$(CH_3)_3COH$] = 18, the weaker acid is formed as product, and equilibrium favors the products. Thus, $\Delta H°$ is negative, and the products are lower in energy than the starting materials.

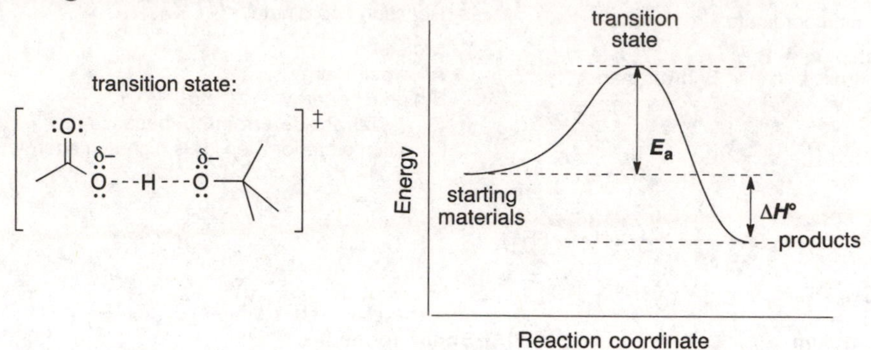

6.47

a. Step [1] breaks one π bond and the H–Cl bond, and one C–H bond is formed. $\Delta H°$ for this st should be positive, because more bonds are broken than formed.

b. Step [2] forms one bond. $\Delta H°$ for this step should be negative, because one bond is formed and none is broken.

c. Step [1] is rate-determining, because it is more difficult.

d. Transition state for Step [1]: Transition state for Step [2]:

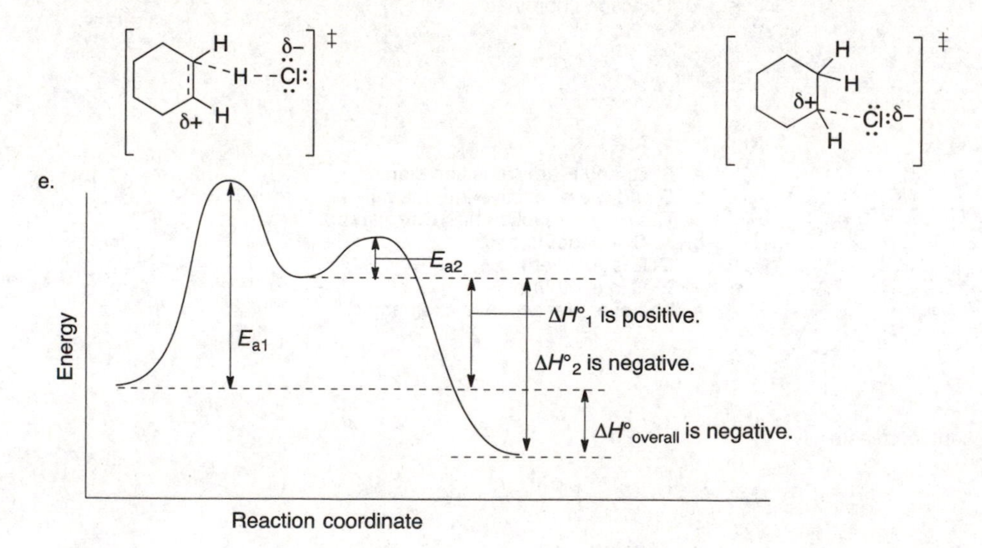

6.48 E_a, concentration, catalysts, rate constant, and temperature affect reaction rate so (c), (d), (e), (g), and (h) affect rate.

6.49

a. **rate = k[CH_3Br][NaCN]**
b. Double [CH_3Br] = **rate doubles.**
c. Halve [NaCN] = **rate halved.**
d. Increase both [CH_3Br] and [NaCN] by factor of 5 = [5][5] = **rate increases by a factor of 25.**

6.50

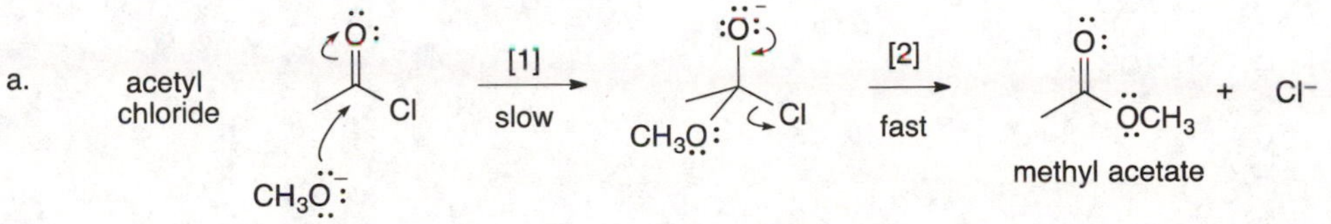

a. acetyl chloride + $CH_3\ddot{O}^-$ —[1] slow→ tetrahedral intermediate —[2] fast→ methyl acetate + Cl^-

b. Only the slow step is included in the rate equation: **Rate = k[CH_3O^-][CH_3COCl]**
c. CH_3O^- is in the rate equation. Increasing its concentration by 10 times would increase the rate by **10 times.**
d. When both reactant concentrations are increased by 10 times, the rate increases by **100 times (10 × 10 = 100).**
e. This is a **substitution reaction** (OCH_3 substitutes for Cl).

6.51

a. **True:** Increasing temperature increases reaction rate.
b. **True:** If a reaction is fast, then it has a large rate constant.
c. **False: Corrected**—There is no relationship between $\Delta G°$ and reaction rate.
d. **False: Corrected**—When the E_a is large, *the rate constant is small.*
e. **False: Corrected**—There is no relationship between K_{eq} and reaction rate.
f. **False: Corrected**—Increasing the concentration of a reactant increases the rate of a reaction *only if the reactant appears in the rate equation.*

6.52

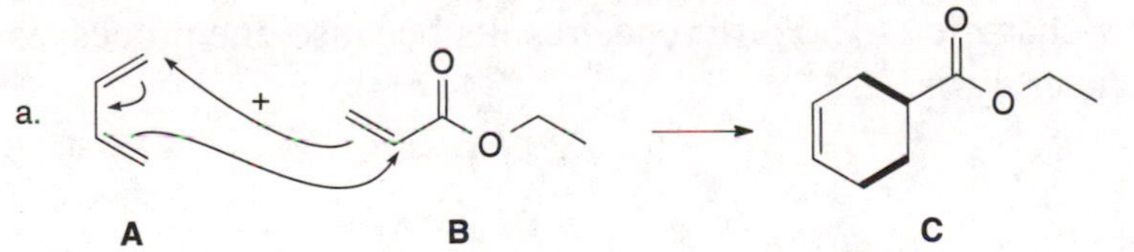

a. A + B → C

b. Two π bonds in **A** are broken and one π bond in **B** is broken. Two new σ bonds in **C** are formed (in bold), as well as a new π bond.
c. The reaction should be exothermic because more energy is released in forming two new C–C σ bonds than is required to break two C–C π bonds.
d. Entropy favors the reactants for two reasons. There are two molecules of reactant and only one product. The reactants are both acyclic and the product has a ring with fewer degrees of freedom.
e. The Diels–Alder reaction is an addition reaction because π bonds are broken and new σ bonds are formed.

6.53

a. The first mechanism has one step: **Rate = $k[(CH_3)_3CI][^-OH]$**

b. The second mechanism has two steps, but only the first step would be in the rate equation, because it is slow and therefore rate-determining: **Rate = $k[(CH_3)_3CI]$**

c. Possibility [1] is second order; possibility [2] is first order.

d. These rate equations can be used to show which mechanism is plausible by changing the concentration of ^-OH. If this affects the rate, then possibility [1] is reasonable. If it does not affect the rate, then possibility [2] is reasonable.

e.

f.

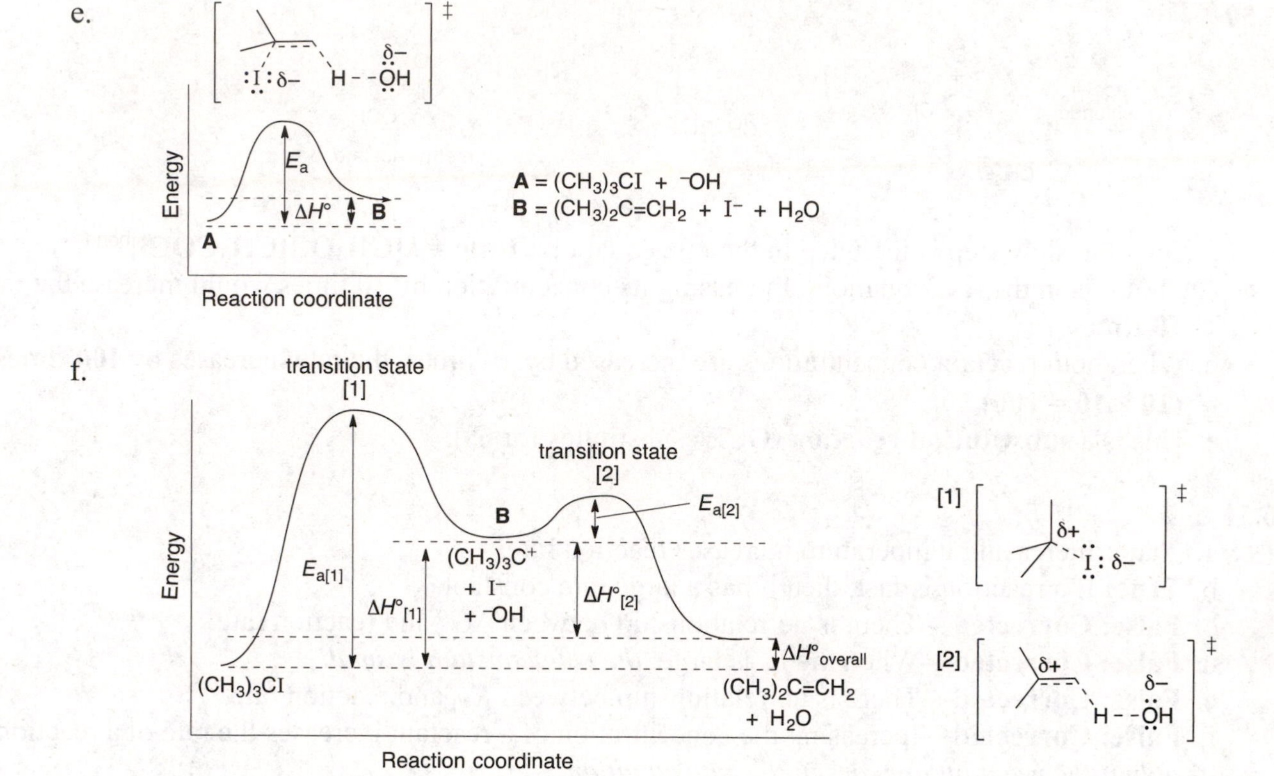

6.54 The difference in both the acidity and the bond dissociation energy of CH_3CH_3 versus HC≡CH is due to the same factor: percent *s*-character. The difference results because one process is based on homolysis and one is based on heterolysis.

Bond dissociation energy:

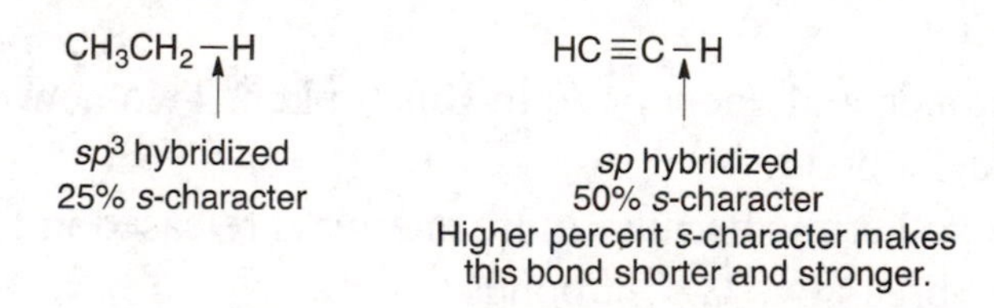

Acidity: To compare acidity, we must compare the stability of the conjugate bases:

$CH_3\bar{C}H_2$ — *sp*³ hybridized, 25% *s*-character

$HC{\equiv}C^-$ — *sp* hybridized, 50% *s*-character

Now a higher percent *s*-character stabilizes the conjugate base making the starting acid more acidic.

6.55 a. Re-draw **A** to see more clearly how cyclization occurs.

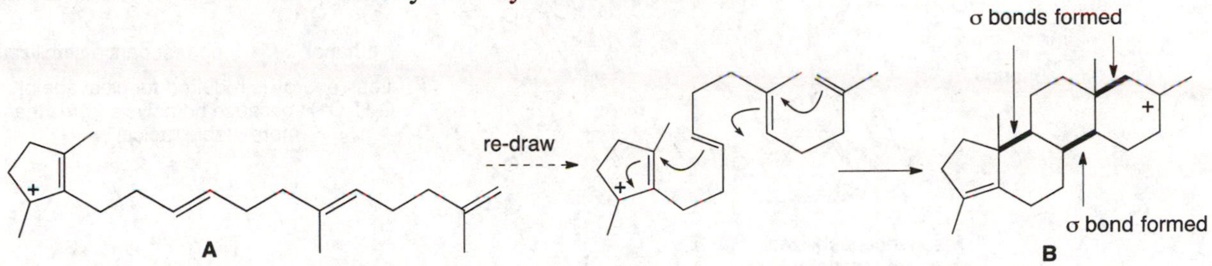

re-draw

σ bonds formed

σ bond formed

A **B**

b.

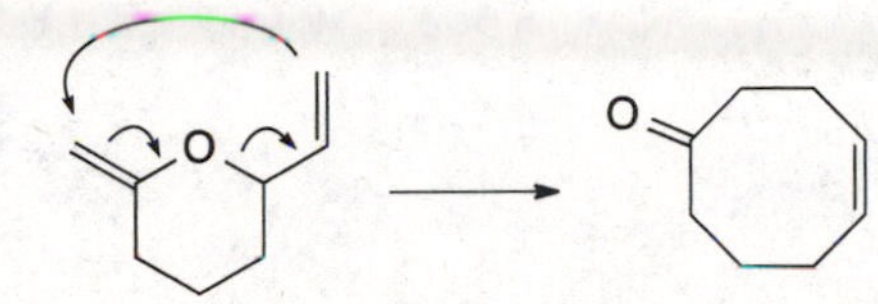

6.56

a. H_a H_b

b. H_a H_b

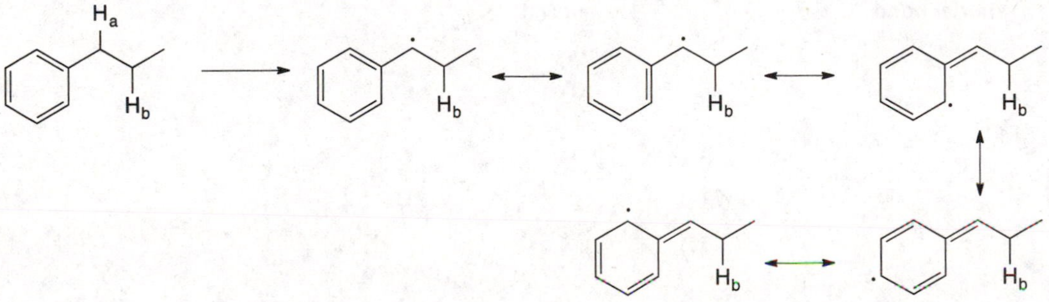

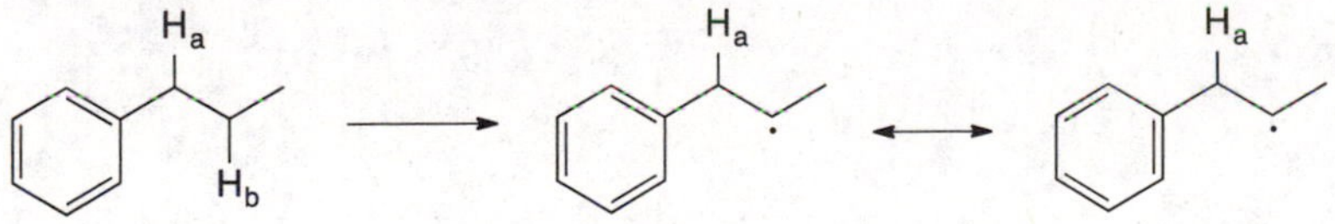

c. C–H_a is weaker than C–H_b because the carbon radical formed when the C–H_a bond is broken is highly resonance stabilized. This means the bond dissociation energy for C–H_a is lower.

6.57 In Reaction [1], the number of molecules of reactants and products stays the same, so entropy is not a factor. In Reaction [2], a single molecule of starting material forms two molecules of products, so entropy increases. This makes $\Delta G°$ more favorable, thus increasing K_{eq}.

6.58

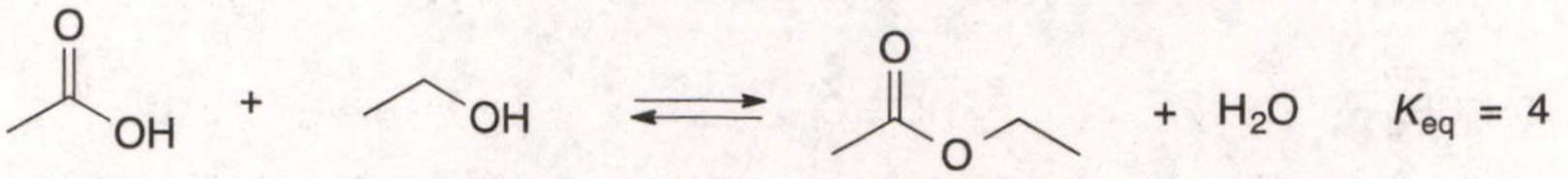

acetic acid + ethanol ⇌ ethyl acetate + H_2O $K_{eq} = 4$

ethyl acetate

To increase the yield of ethyl acetate, H_2O can be removed from the reaction mixture, or there can be a large excess of one of the starting materials.

6.59

a. phenol O–H →

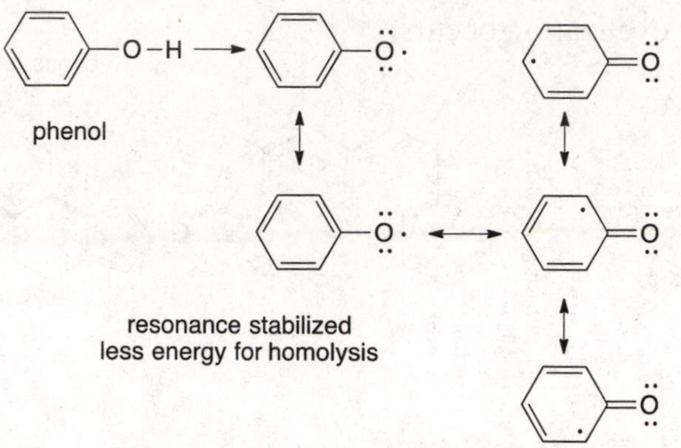

resonance stabilized
less energy for homolysis

ethanol → no resonance stabilization

Less energy is required for cleavage of C_6H_5O–H because homolysis forms the more stable radical.

b. O–H

Csp^2–O
higher % *s*-character
shorter bond

OH

Csp^3–O
lower % *s*-character
longer bond

Chapter 7 Alkyl Halides and Nucleophilic Substitution

Chapter Review

General facts about alkyl halides

- Alkyl halides contain a halogen atom X bonded to an sp^3 hybridized carbon (7.1).
- Alkyl halides are named as halo alkanes, with the halogen as a substituent (7.2).
- Alkyl halides have a polar C–X bond, so they exhibit dipole–dipole interactions but are incapable of intermolecular hydrogen bonding (7.3).
- The polar C–X bond containing an electrophilic carbon makes alkyl halides reactive towards nucleophiles and bases (7.5).

The central theme (7.6)

- Nucleophilic substitution is one of the two main reactions of alkyl halides. A nucleophile replaces a leaving group on an sp^3 hybridized carbon.

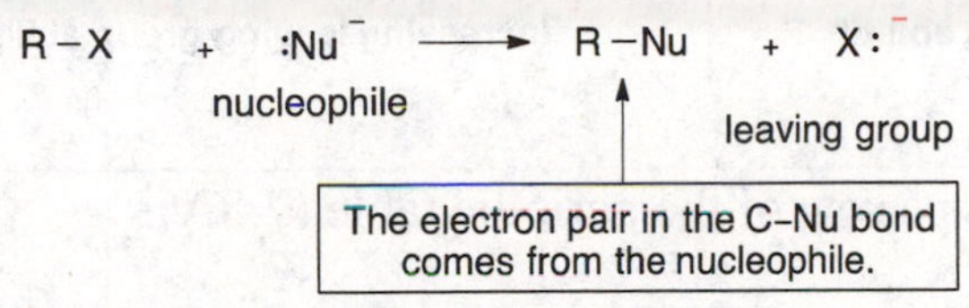

- One σ bond is broken and one σ bond is formed.
- There are two possible mechanisms: SN1 and SN2.

SN1 and SN2 mechanisms compared

	SN2 mechanism	SN1 mechanism
[1] Mechanism	• One step (7.11B)	• Two steps (7.12B)
[2] Alkyl halide	• Order of reactivity: CH_3X > RCH_2X > R_2CHX > R_3CX (7.11D)	• Order of reactivity: R_3CX > R_2CHX > RCH_2X > CH_3X (7.12D)
[3] Rate equation	• rate = k[RX][:Nu⁻] • second-order kinetics (7.11A)	• rate = k[RX] • first-order kinetics (7.12A)
[4] Stereochemistry	• backside attack of the nucleophile (7.11C) • inversion of configuration at a stereogenic center	• trigonal planar carbocation intermediate (7.12C) • racemization at a stereogenic center
[5] Nucleophile	• favored by stronger nucleophiles (7.15B)	• favored by weaker nucleophiles (7.15B)
[6] Leaving group	• better leaving group → faster reaction (7.15C)	• better leaving group → faster reaction (7.15C)
[7] Solvent	• favored by polar aprotic solvents (7.15D)	• favored by polar protic solvents (7.15D)

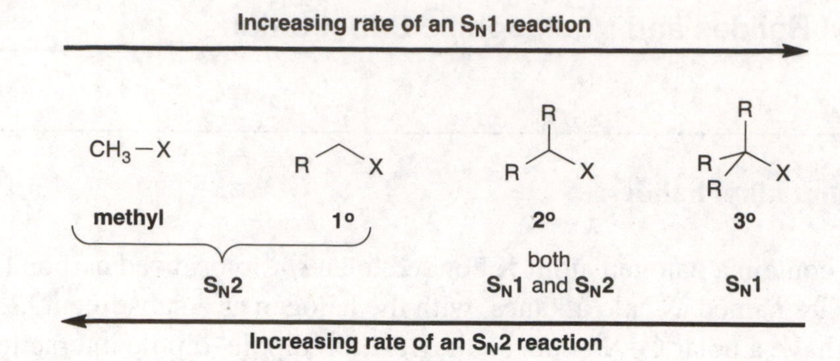

Important trends

- The best leaving group is the weakest base. Leaving group ability increases left-to-right across a row and down a column of the periodic table (7.7).

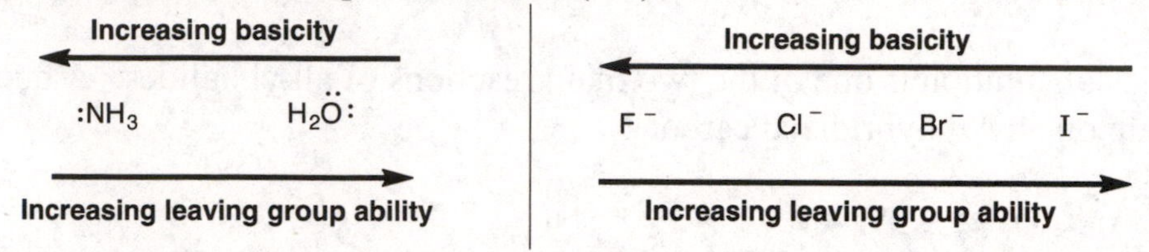

- Nucleophilicity decreases across a row of the periodic table (7.8A).

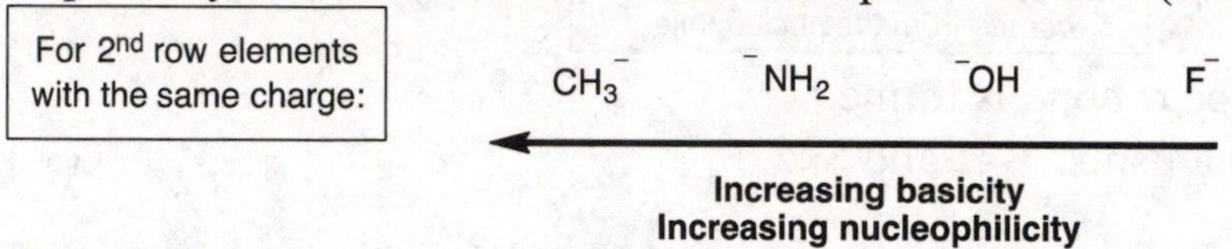

- Nucleophilicity decreases down a column of the periodic table in polar aprotic solvents (7.8C).

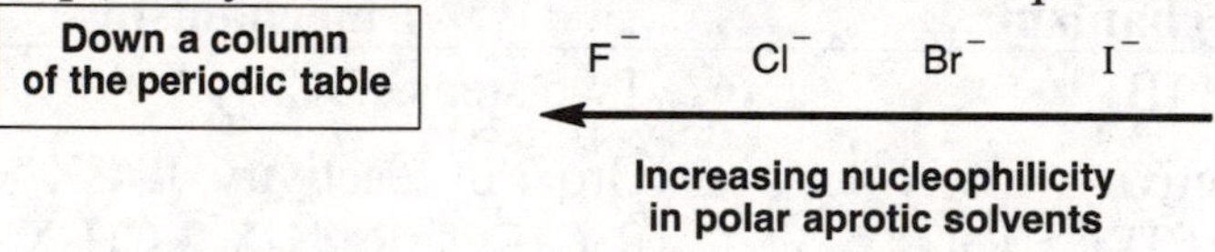

- Nucleophilicity increases down a column of the periodic table in polar protic solvents (7.8C).

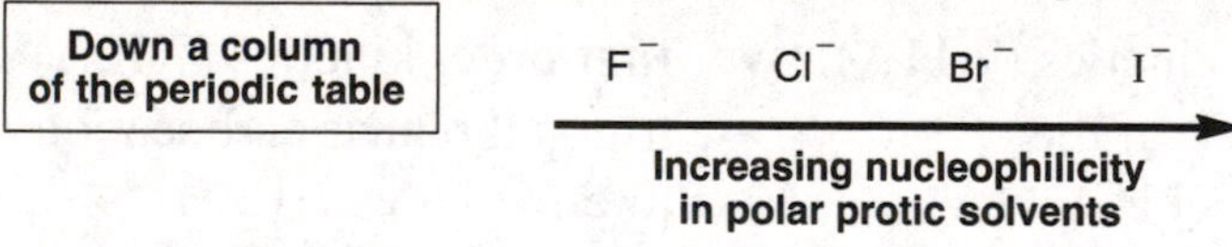

- The stability of a carbocation increases as the number of R groups bonded to the positively charged carbon increases (7.13).

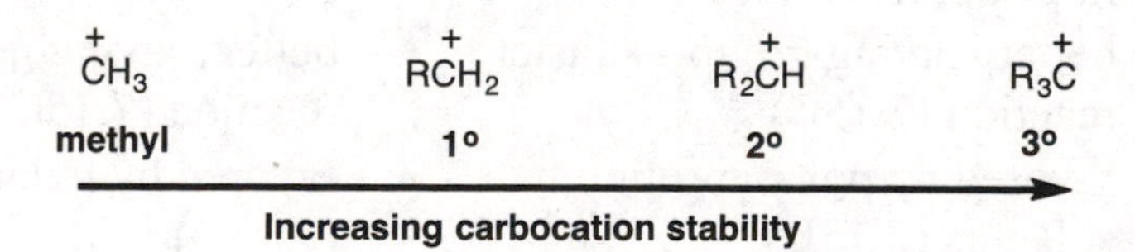

Important principles

Principle	Example
• Electron-donating groups (such as R groups) stabilize a positive charge (7.13A).	• 3° Carbocations (R_3C^+) are more stable than 2° carbocations (R_2CH^+), which are more stable than 1° carbocations (RCH_2^+).
• Steric hindrance decreases nucleophilicity but not basicity (7.8B).	• $(CH_3)_3CO^-$ is a stronger base but a weaker nucleophile than $CH_3CH_2O^-$.
• Hammond postulate: In an endothermic reaction, the more stable product is formed faster. In an exothermic reaction, this fact is not necessarily true (7.14).	• S_N1 reactions are faster when more stable (more substituted) carbocations are formed, because the rate-determining step is endothermic.
• Planar, *sp*^{2} hybridized atoms react with reagents from both sides of the plane (7.12C).	• A trigonal planar carbocation reacts with nucleophiles from both sides of the plane.

Practice Test on Chapter Review

1. Give the IUPAC name for the following compound, including the appropriate *R,S* prefix.

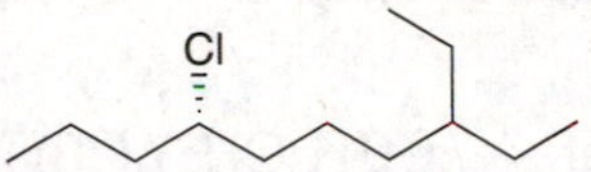

Cl

2.a. Which of the following carbocations is the most stable?

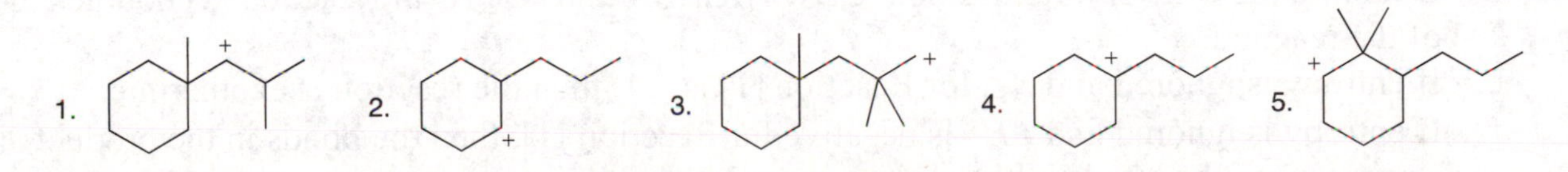

b. Which of the following anions is the best leaving group?

1. CH_3^- 2. ^-OH 3. H^- 4. $^-NH_2$ 5. Cl^-

c. Which species is the strongest nucleophile in polar protic solvents?

1. F^- 2. ^-OH 3. Cl^- 4. H_2O 5. ^-SH

d. Which of the following statements is true about the given reaction?

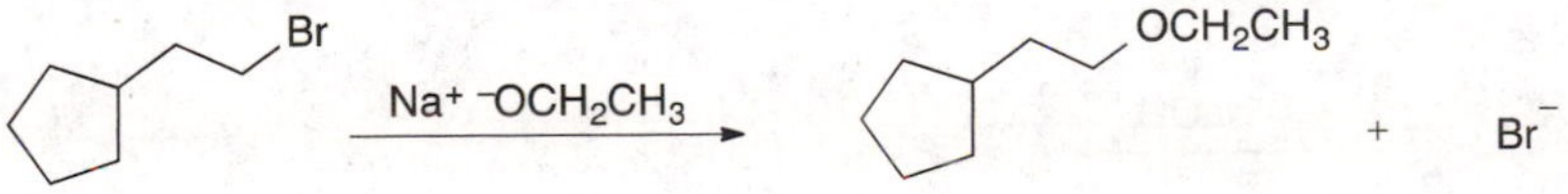

Br $\xrightarrow{Na^+\ ^-OCH_2CH_3}$ OCH_2CH_3 + Br^-

1. The reaction follows second-order kinetics.
2. The rate of the reaction increases when the solvent is changed from CH_3CH_2OH to DMSO.
3. The rate of the reaction increases when the leaving group is changes from Br to F.
4. Statements (1) and (2) are both true.
5. Statements (1), (2), and (3) are all true.

3. Rank the following compounds in order of ***increasing* reactivity in an S_N1** reaction. Rank the ***least reactive*** compound as **1,** the ***most reactive*** compound as **4,** and the compounds of intermediate reactivity as **2** and **3.**

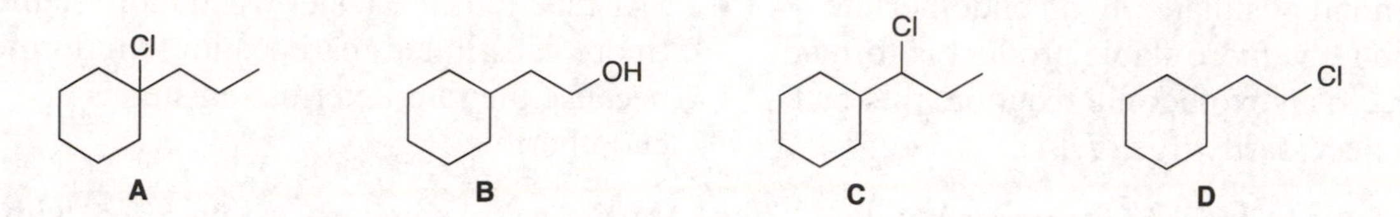

4. Consider the following two nucleophilic substitution reactions, labeled Reaction [1] and Reaction [2]. (Only the starting materials are drawn.) Then answer True (T) or False (F) to each of the following statements.

Reaction [1] $(CH_3CH_2)_3CBr$ + CH_3OH ⟶

Reaction [2] $CH_3CH_2CH_2Br$ + $^{-}OCH_3$ ⟶

a. The rate equation for Reaction [1] is rate = $k[(CH_3CH_2)_3CBr][CH_3OH]$.
b. Changing the leaving group from Br^- to Cl^- decreases the rate of both reactions.
c. Changing the solvent from CH_3OH to $(CH_3)_2S{=}O$ increases the rate of Reaction [2].
d. Doubling the concentration of both $CH_3CH_2CH_2Br$ and $^{-}OCH_3$ in Reaction [2] doubles the rate of the reaction.
e. If entropy is ignored and K_{eq} for Reaction [1] is < 1, then the reaction is exothermic.
f. If entropy is ignored and $\Delta H°$ is negative for Reaction [1], then the bonds in the product are stronger than the bonds in the starting materials.
g. The energy diagram for Reaction [2] exhibits only one energy barrier.

5. Draw the organic products formed in the following reactions. **Use wedges and dashed wedges to show stereochemistry in compounds with stereogenic centers.**

a. 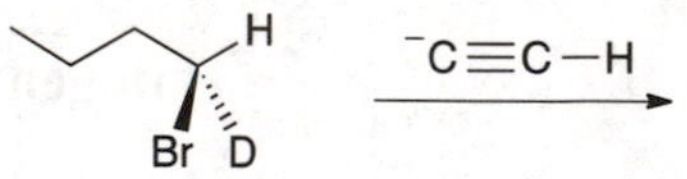 H, Br D ⟶ $^{-}C{\equiv}C{-}H$

b. 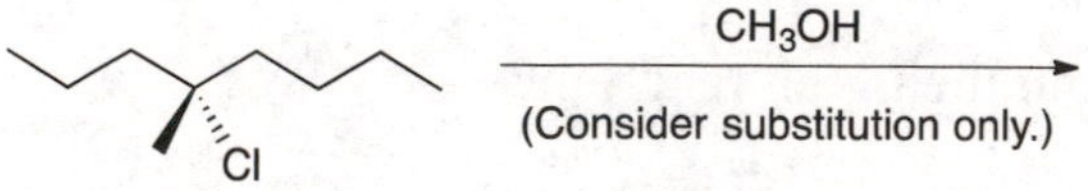 Cl ⟶ CH_3OH (Consider substitution only.)

c. Br ⟶ NaOH

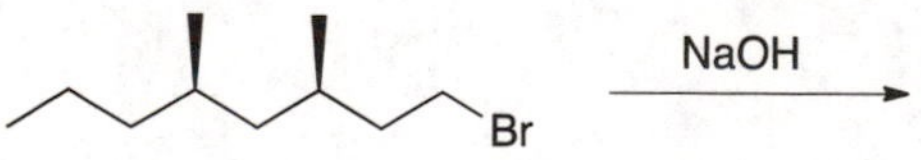

d. Br $\xrightarrow{CH_3CO_2^-}$

Answers to Practice Test

1. (*S*)-7-chloro-3-ethyldecane

2. a. 4
   b. 5
   c. 5
   d. 4

3. **A**–4
   B–1
   C–3
   D–2

4. a. F
   b. T
   c. T
   d. F
   e. F
   f. T
   g. T

5. a. H, D, C≡CH
   b. OCH_3 + CH_3O
   c. OH
   d. O_2CCH_3

Answers to Problems

7.1 Classify the alkyl halide as 1°, 2°, or 3° **by counting the number of carbons bonded directly to the carbon bonded to the halogen.**

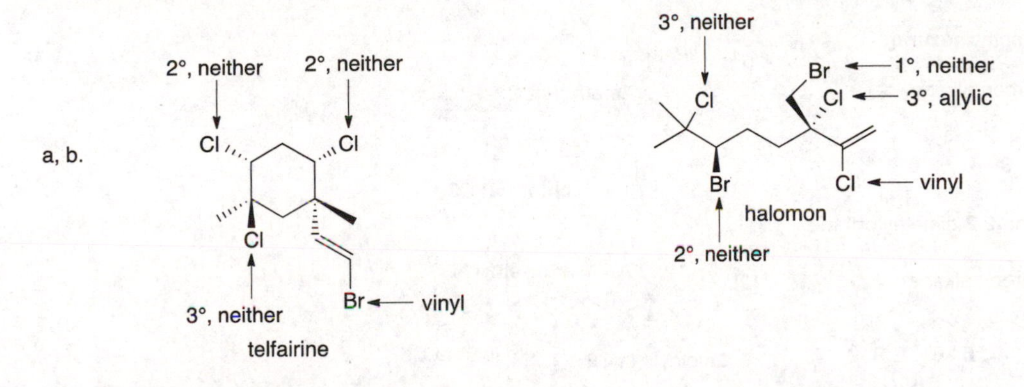

7.2 To name a compound with the IUPAC system:
[1] **Name the parent** chain by finding the longest carbon chain.
[2] **Number the chain** so the first substituent gets the lower number. Then **name and number all substituents,** giving like substituents a prefix (di, tri, etc.). **To name the halogen substituent, change the *-ine* ending to *-o*.**
[3] **Combine all parts,** alphabetizing substituents, and ignoring all prefixes except iso.

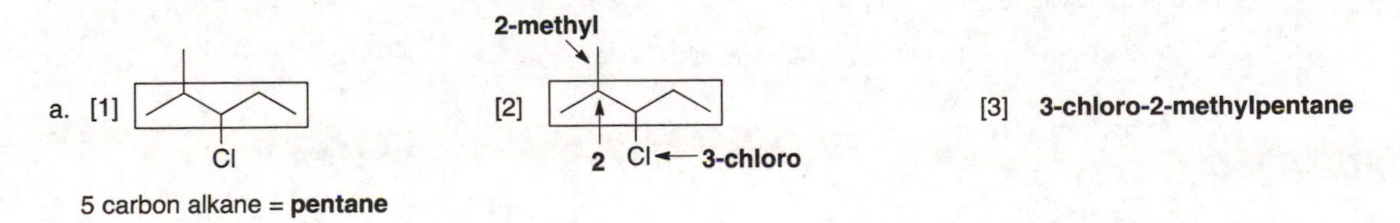

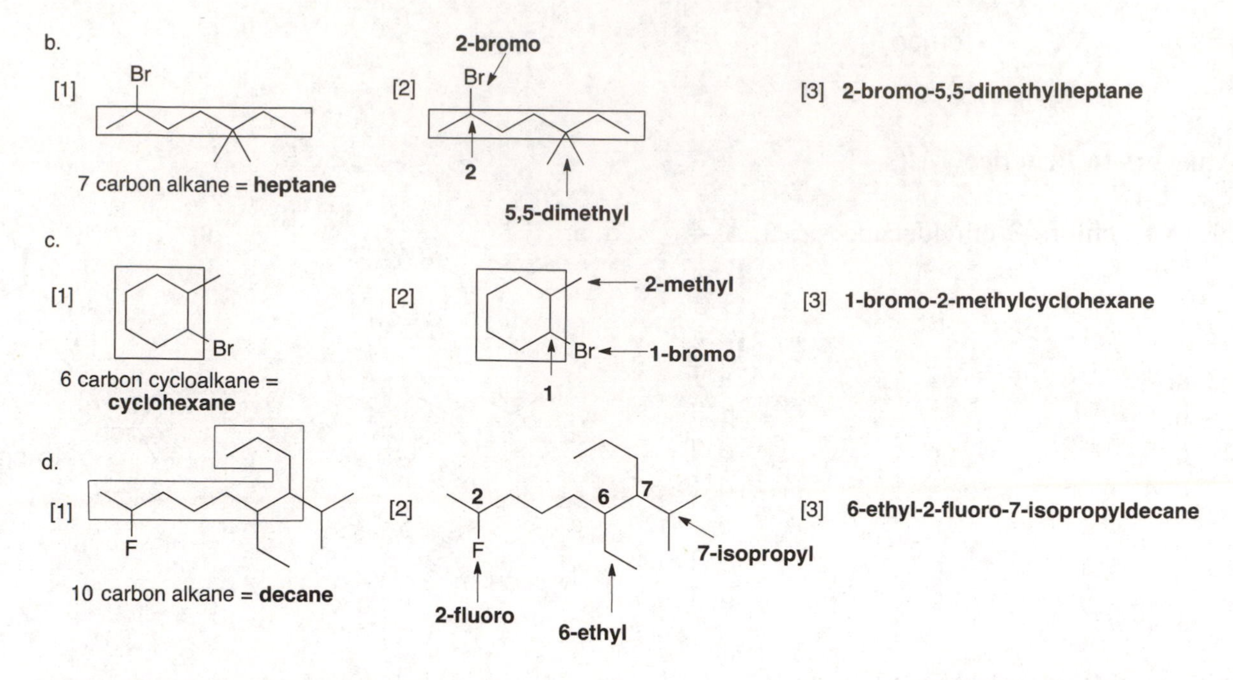

7.3 To work backwards from a name to a structure:

[1] Find the parent name and draw that number of carbons. Use the suffix to identify the functional group **(-ane = alkane).**

[2] Arbitrarily number the carbons in the chain. Add the substituents to the appropriate carbon.

a. 3-chloro-2-methyl**hexane**

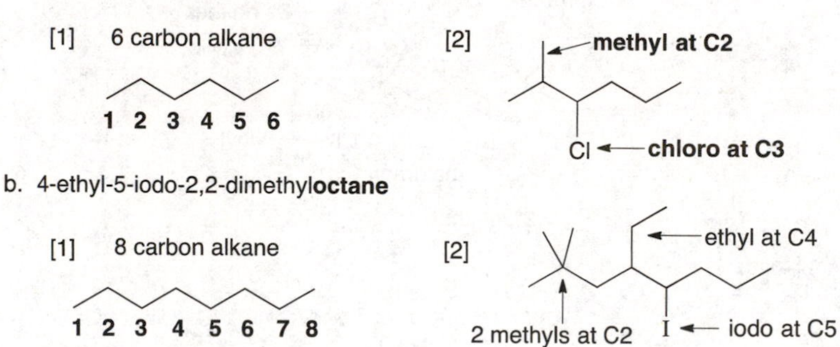

c. *cis*-1,3-dichloro**cyclopentane**

[1] 5 carbon cycloalkane

[2] chloro groups at C1 and C3, both on the same side

Cl Cl C1 C3

d. 1,1,3-tribromo**cyclohexane**

[1] 6 carbon cycloalkane

[2] 3 Br groups

Br C3 Br Br C1

e. ***sec*-butyl** bromide

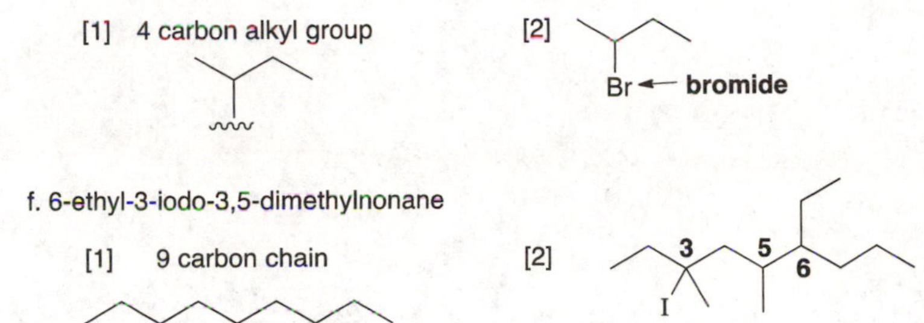

f. 6-ethyl-3-iodo-3,5-dimethylnonane

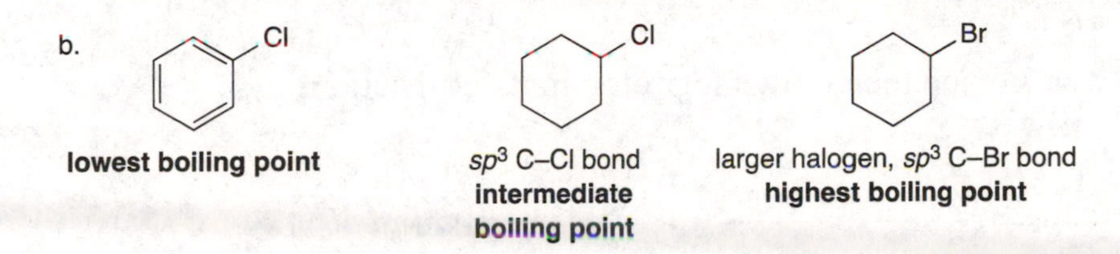

7.4 a. Because an sp^2 hybridized C has a higher percent *s*-character than an sp^3 hybridized C, it holds electron density closer to C. This pulls a little more electron density towards C, away from Cl, and thus a Csp^2–Cl bond is less polar than a Csp^3–Cl bond.

b.

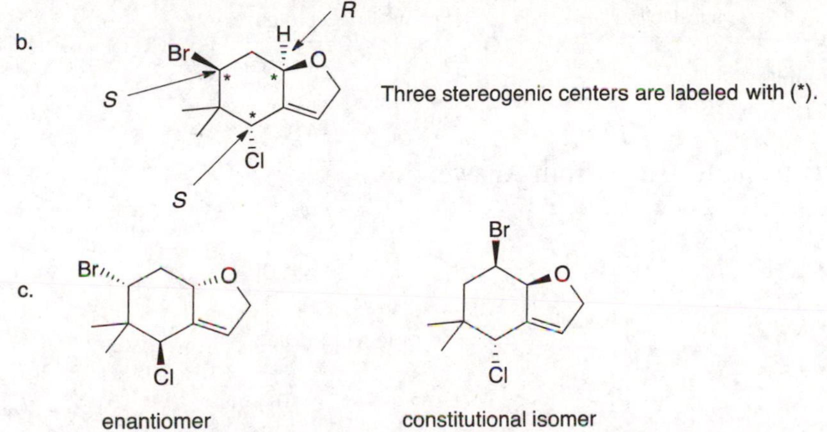

7.5 a. Since chondrocole A has 10 C's and only one functional group capable of hydrogen bonding to water (an ether), it is insoluble in H_2O. Because it is organic, it is soluble in CH_2Cl_2.

b. Three stereogenic centers are labeled with (*).

c.

enantiomer

constitutional isomer

7.6 To draw the products of a nucleophilic substitution reaction:
[1] **Find the *sp*³ hybridized electrophilic carbon** with a leaving group.
[2] **Find the nucleophile** with lone pairs or electrons in π bonds.
[3] **Substitute the nucleophile for the leaving group** on the electrophilic carbon.

a.

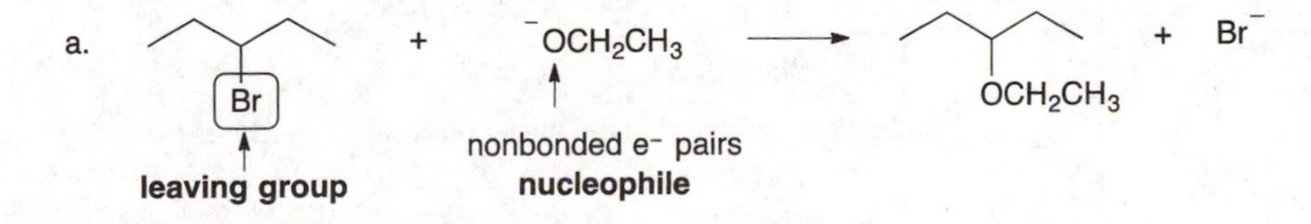

b. Cl + Na^+ ^-OH → OH + Na^+Cl^-

leaving group — nonbonded e^- pairs **nucleophile**

c. I + N_3^- → N_3 + I^-

leaving group — nonbonded e^- pairs **nucleophile**

d. Br + Na^+ ^-CN → CN + Na^+Br^-

leaving group — nonbonded e^- pairs **nucleophile**

7.7 Use the steps from Answer 7.6 and then draw the proton transfer reaction.

a. Br + $:N(CH_2CH_3)_3$ —substitution→ $\overset{+}{N}(CH_2CH_3)_3$ + Br^-

nucleophile — **leaving group**

b. Cl + $H_2\ddot{O}:$ —substitution→ H, $\ddot{O}:^+$, H + $:\ddot{Cl}:^-$ —proton transfer→ $\ddot{O}H$ + HCl

leaving group — **nucleophile**

7.8 Draw the structure of CPC using the steps from Answer 7.6.

N: + Cl

nucleophile — **leaving group**

substitution

$\overset{+}{N}$ + $:\ddot{Cl}:^-$

CPC

7.9

Cl, Cl → N, S, Cl

A — ticlopidine

These atoms come from the nucleophile.

NH, S

7.10 Compare the leaving groups based on these trends:
- Better leaving groups are weaker bases.
- A neutral leaving group is always better than its conjugate base.

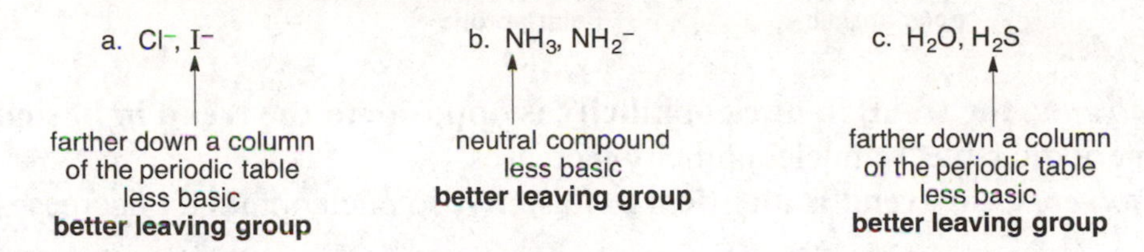

7.11 Good leaving groups include Cl^-, Br^-, I^-, and H_2O.

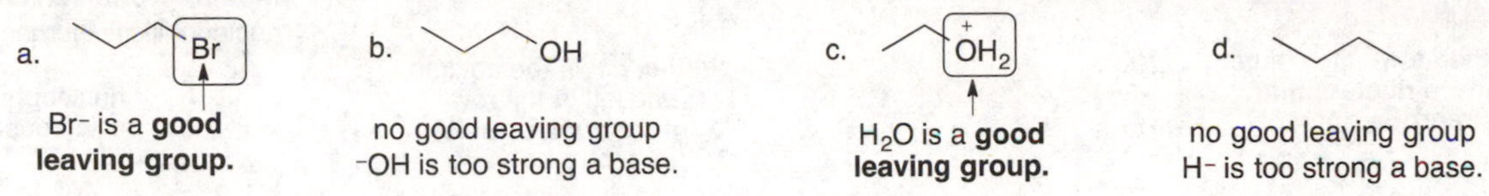

7.12 To decide whether the equilibrium favors the starting material or the products, **compare the nucleophile and the leaving group.** The reaction proceeds towards the weaker base.

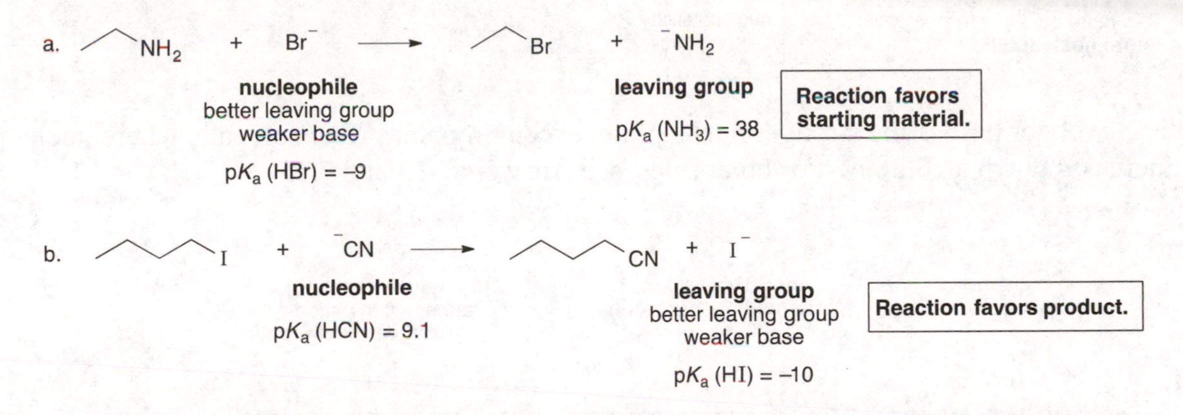

7.13 Use these three rules to find the stronger nucleophile in each pair:

[1] Comparing two nucleophiles having the ***same attacking atom***, **the stronger base is a stronger nucleophile.**

[2] **Negatively charged nucleophiles** are always **stronger than their conjugate acids.**

[3] **Nucleophilicity decreases left to right across a row of the periodic table,** when comparing species of similar charge.

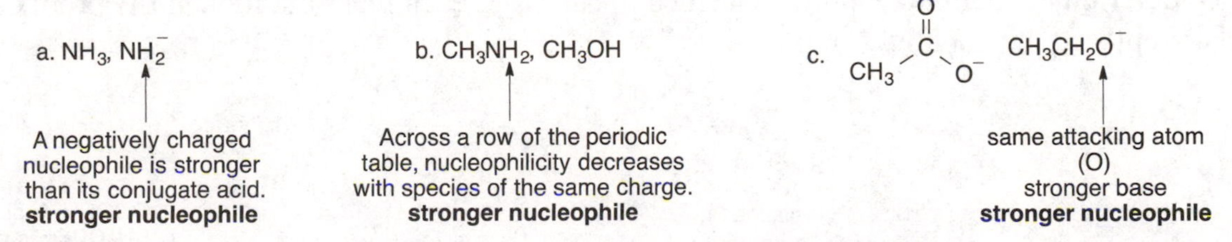

7.14 ***Polar protic solvents*** **are capable of hydrogen bonding,** and therefore must contain a **H bonded to an electronegative O or N.** ***Polar aprotic solvents*** **are incapable of hydrogen bonding,** and therefore do not contain any O–H or N–H bonds.

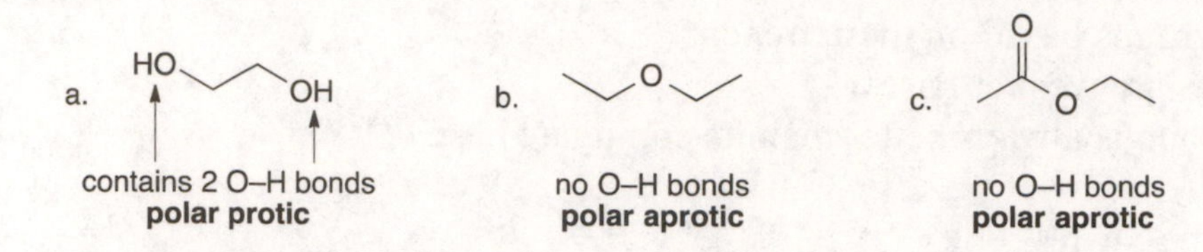

7.15 • In ***polar protic solvents,*** **the trend in nucleophilicity is opposite to the trend in basicity** down a column of the periodic table, so nucleophilicity increases.

• In ***polar aprotic solvents,*** **the trend is identical to basicity,** so nucleophilicity decreases down a column.

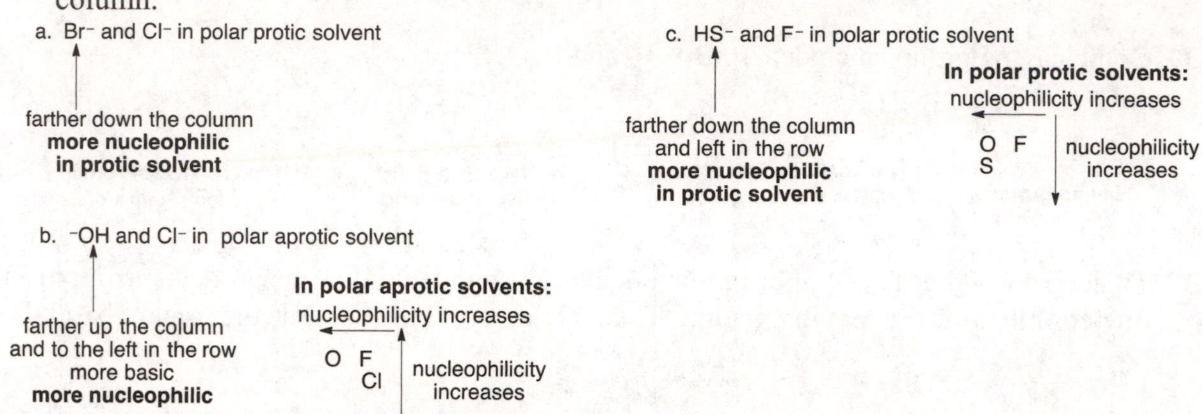

7.16 The stronger base is the stronger nucleophile, except in polar protic solvents, where nucleophilicity increases down a column. For other rules, see Answers 7.13 and 7.15.

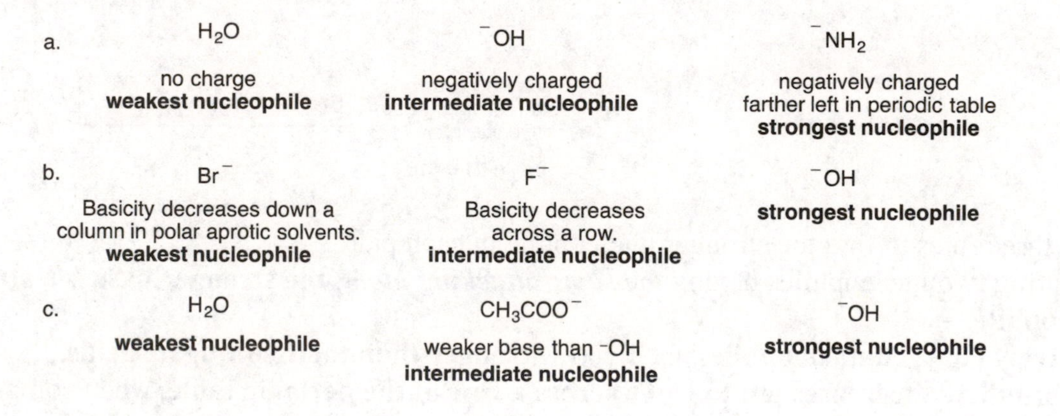

7.17 To determine what nucleophile is needed to carry out each reaction, look at the product to see what has replaced the leaving group.

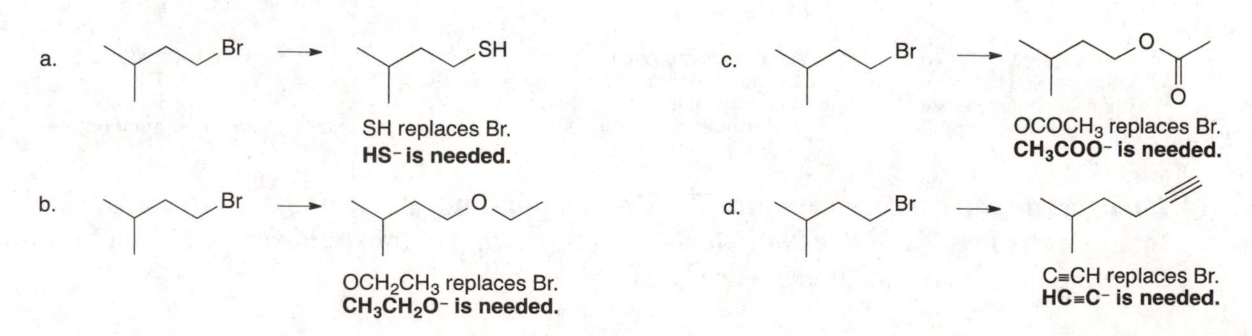

7.18 The general rate equation for an SN2 reaction is rate = k[RX][:Nu$^-$].

a. [RX] is tripled, and [:Nu$^-$] stays the same: **rate triples.**
b. Both [RX] and [:Nu$^-$] are tripled: **rate increases by a factor of 9 (3 × 3 = 9).**
c. [RX] is halved, and [:Nu$^-$] stays the same: **rate halved.**
d. [RX] is halved, and [:Nu$^-$] is doubled: **rate stays the same (1/2 × 2 = 1).**

7.19 All SN2 reactions have one step. The transition state in an SN2 reaction has **dashed bonds to both the leaving group and the nucleophile,** and must contain partial charges.

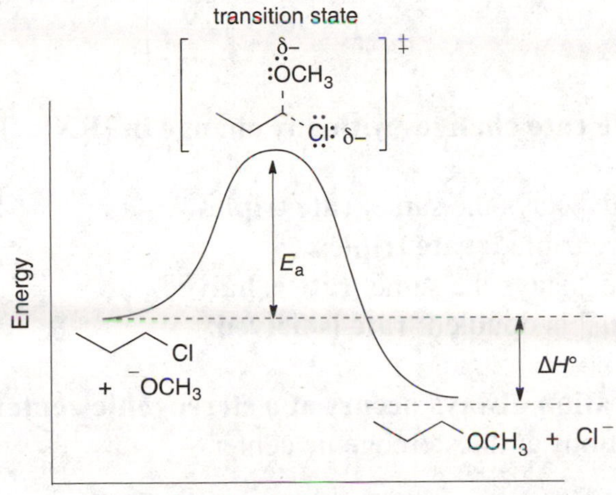

7.20 To draw the products of SN2 reactions, **replace the leaving group by the nucleophile, and then draw the stereochemistry with *inversion* at the stereogenic center.**

a. D, H, Br + :Ö⁻ ⟶ :O, D, H

b. I + ⁻:C≡N: ⟶ C≡N:

7.21 *Increasing* the number of R groups *increases* crowding of the transition state and *decreases* the rate of an SN2 reaction.

a. Cl or Cl

2° alkyl halide — 1° alkyl halide **faster reaction**

b. Br or Br

2° alkyl halide **faster reaction** — 3° alkyl halide

7.22

Br
H
N
O
C
Cl

Br
H
N
O
D
N
N
from the nucleophile
H
N
N

7.23 In a first-order reaction, **the rate changes with any change in [RX].** The rate is independent of any change in [:Nu⁻].

a. [RX] is tripled, and [:Nu⁻] stays the same: **rate triples.**
b. Both [RX] and [:Nu⁻] are tripled: **rate triples.**
c. [RX] is halved, and [:Nu⁻] stays the same: **rate is halved.**
d. [RX] is halved, and [:Nu⁻] is doubled: **rate is halved.**

7.24 In S_N1 reactions, racemization always occurs at a stereogenic center. Draw two products, with the two possible configurations at the stereogenic center.

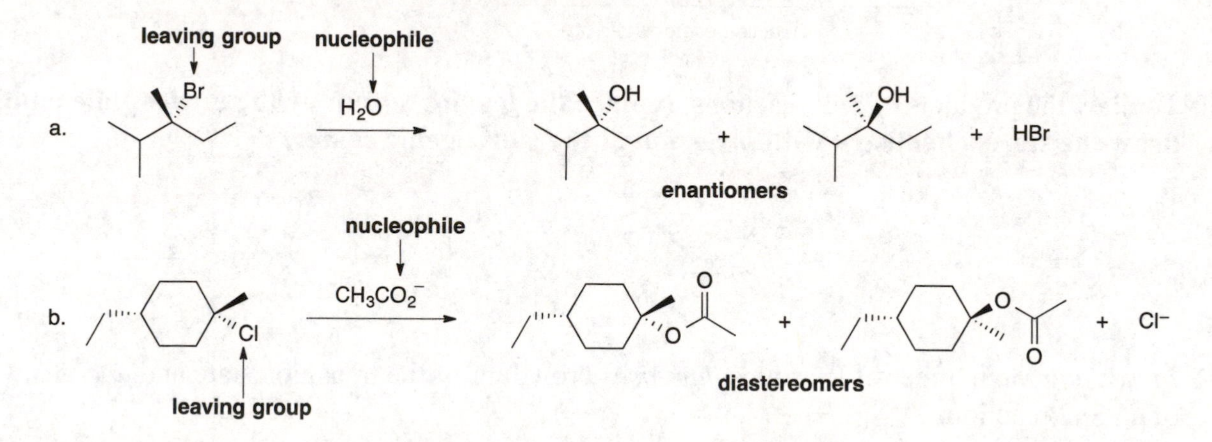

7.25 Carbocations are classified by the number of R groups bonded to the carbon: 0 R groups = methyl, 1 R group = 1°, 2 R groups = 2°, and 3 R groups = 3°.

a. 2 R groups **2° carbocation**

b. 1 R group **1° carbocation**

c. 3 R groups **3° carbocation**

d. 2 R groups **2° carbocation**

7.26 For carbocations: Increasing number of R groups = Increasing stability.

1° carbocation	2° carbocation	3° carbocation
least stable	**intermediate stability**	**most stable**

7.27 For carbocations: Increasing number of R groups = Increasing stability.

1° carbocation	2° carbocation	3° carbocation
least stable	**intermediate stability**	**most stable**

7.28 The rate of an S_N1 reaction increases with increasing alkyl substitution.

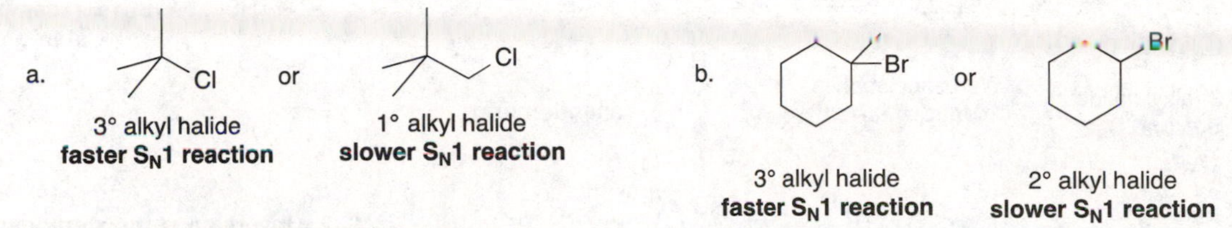

7.29 • For **methyl and 1° alkyl halides,** only S_N2 will occur.
• For **2° alkyl halides,** S_N1 and S_N2 will occur.
• For **3° alkyl halides,** only S_N1 will occur.

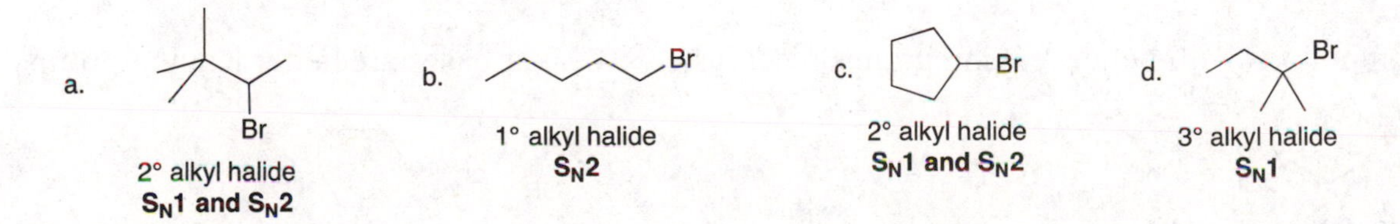

7.30 • Draw the product of nucleophilic substitution for each reaction.
• For **methyl and 1° alkyl halides,** only S_N2 will occur.
• For **2° alkyl halides,** S_N1 and S_N2 will occur and other factors determine which mechanism operates.
• For **3° alkyl halides,** only S_N1 will occur.

a. Cl + CH_3OH → OCH_3 + HCl

3° alkyl halide
only S_N1

b. Br + ^-SH → SH + Br^-

1° alkyl halide
only S_N2

c. I + $CH_3CH_2O^-$ → OCH_2CH_3 + I^-

Strong nucleophile
favors S_N2.

2° alkyl halide
Both S_N1 and S_N2 are possible.

d. Br + CH_3OH → OCH_3 + HBr

Weak nucleophile
favors S_N1.

2° alkyl halide
Both S_N1 and S_N2 are possible.

7.31 First decide whether the reaction will proceed via an S_N1 or S_N2 mechanism. Then draw the products with stereochemistry.

a. H Br + H_2O → H OH + HO H + HBr

2° alkyl halide
S_N1 and S_N2

Weak nucleophile
favors S_N1.

enantiomers

S_N1 = racemization at the stereogenic C

b. Cl, H D + $^-$:C≡C–H → HC≡C, D H + Cl^-

1° alkyl halide
S_N2 only

S_N2 = inversion at the stereogenic C

7.32 Compounds with better leaving groups react faster. Weaker bases are better leaving groups.

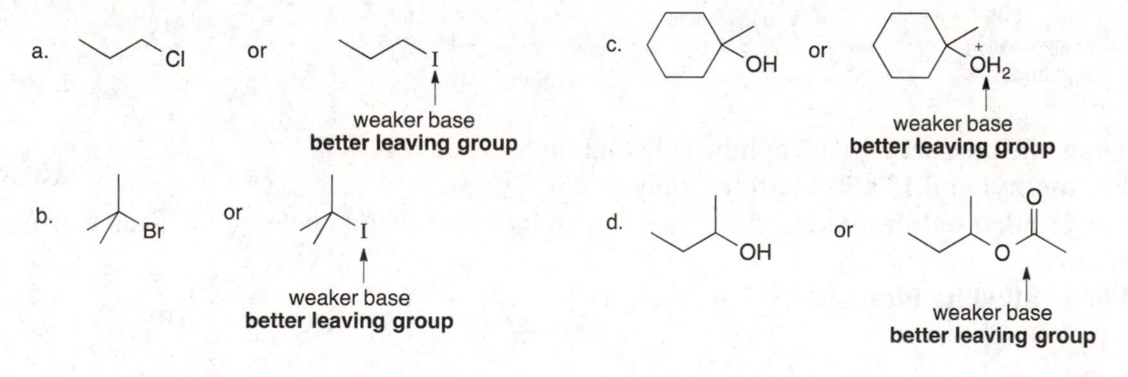

7.33
- **Polar protic solvents** favor the S_N1 mechanism by solvating the intermediate carbocation and halide.
- **Polar aprotic solvents** favor the S_N2 mechanism by making the nucleophile stronger.

a. CH_3CH_2OH
polar protic solvent
contains an O–H bond
favors S_N1

b. CH_3CN
polar aprotic solvent
no O–H or N–H bond
favors S_N2

c. CH_3COOH
polar protic solvent
contains an O–H bond
favors S_N1

d. $CH_3CH_2OCH_2CH_3$
polar aprotic solvent
no O–H or N–H bond
favors S_N2

7.34 Compare the solvents in the reactions below. **For the solvent to increase the reaction rate of an S_N1 reaction, the solvent must be *polar protic*. For the solvent to increase the reaction rate of an S_N2 reaction, the solvent must be *polar aprotic*.**

a. + CH_3OH → (CH_3OH or DMSO) + HCl

Cl; OCH_3

3° RX – S_N1 reaction

CH_3OH
Polar protic solvent increases the rate of an S_N1 reaction.

b. Br + ^-OH → (H_2O or DMF) OH + Br^-

1° RX – S_N2 reaction

DMF [$HCON(CH_3)_2$]
Polar aprotic solvent increases the rate of an S_N2 reaction.

c. + CH_3O^- → (CH_3OH or HMPA) + Cl^-

H Cl; H OCH_3

2° RX; strong nucleophile

S_N2 reaction

HMPA [$[(CH_3)_2N]_3P{=}O$]
Polar aprotic solvent increases the rate of an S_N2 reaction.

7.35 To predict whether the reaction follows an S_N1 or S_N2 mechanism:

[1] **Classify RX as a methyl, 1°, 2°, or 3° halide.** (Methyl, 1° = S_N2; 3° = S_N1; 2° = either.)
[2] **Classify the nucleophile as strong or weak.** (Strong favors S_N2; weak favors S_N1.)
[3] **Classify the solvent as polar protic or polar aprotic.** (Polar protic favors S_N1; polar aprotic favors S_N2.)

a. Br + O^- → O + Br^- S_N2 reaction

1° alkyl halide
S_N2

b. Br + N_3^- → N_3 + Br^-

2° alkyl halide
S_N1 or S_N2

Strong nucleophile
favors S_N2.

S_N2 reaction = ***inversion*** at the stereogenic center
The leaving group was "up."
The nucleophile attacks from below.

c. I + CH_3OH → OCH_3 + HI S_N1 reaction

3° alkyl halide
S_N1

Weak nucleophile
favors S_N1.

d. Cl + H_2O → OH + HO + HCl

3° alkyl halide
S_N1

Weak nucleophile
favors S_N1.

S_N1 reaction forms **two enantiomers.**

7.36

A: N, N:, H + CH_3–$\overset{+}{S}R_2$ → N, $\overset{+}{N}$, H, CH_3 + SR_2 → (loss of a proton) → N, N:, CH_3

A

nicotine

7.37 Vinyl carbocations are even less stable than 1° carbocations.

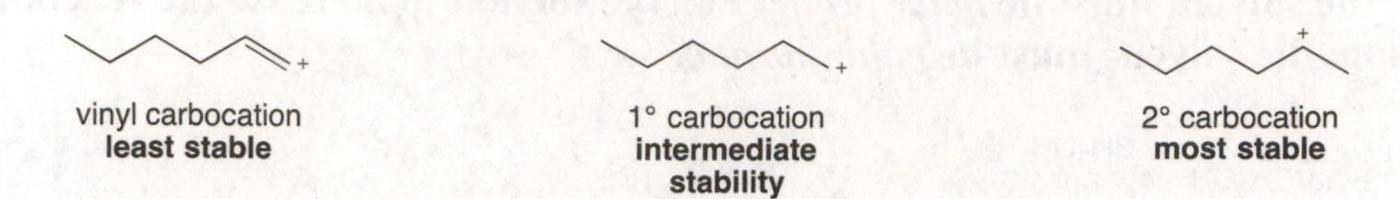

7.38 Convert each ball-and-stick model to a skeletal or condensed structure and draw the reactants.

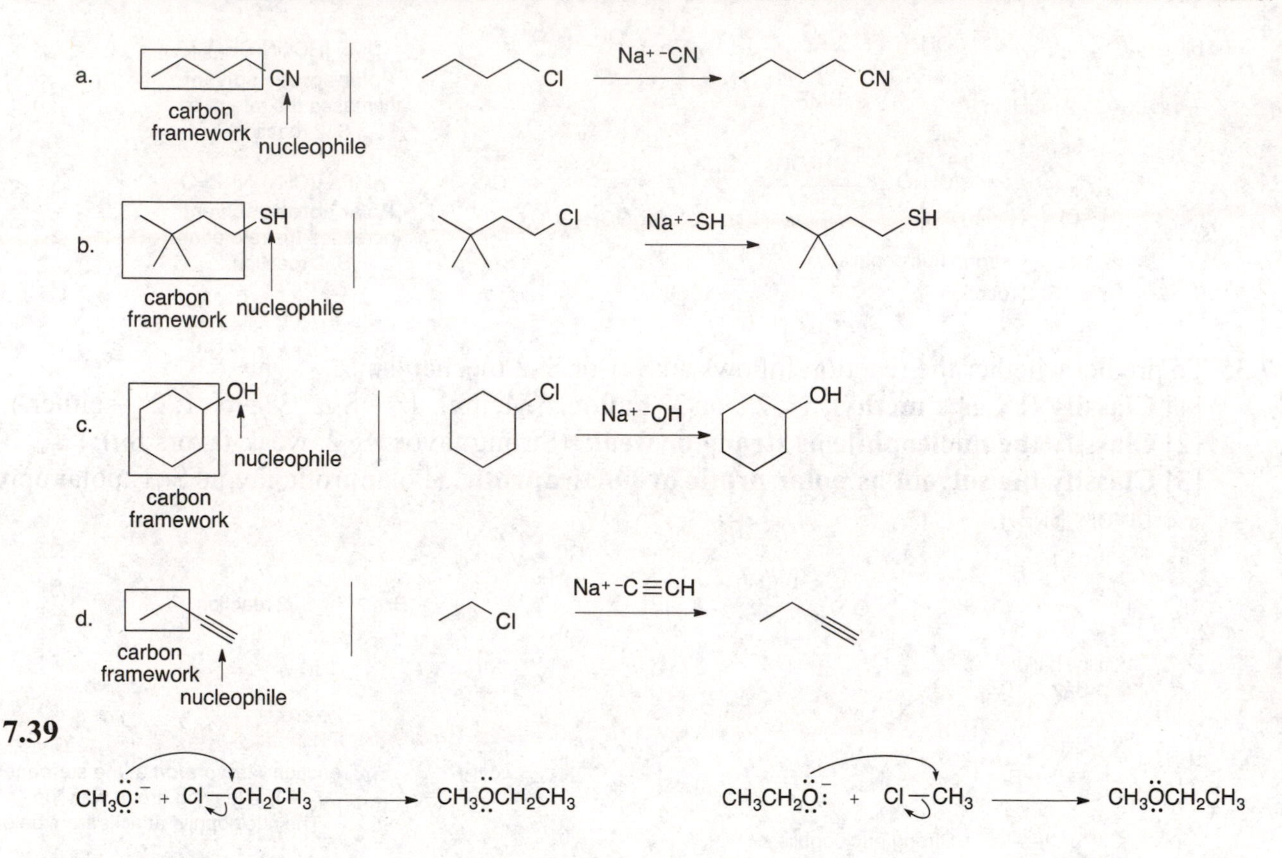

7.40 Use the directions from Answer 7.2 to name the compounds.

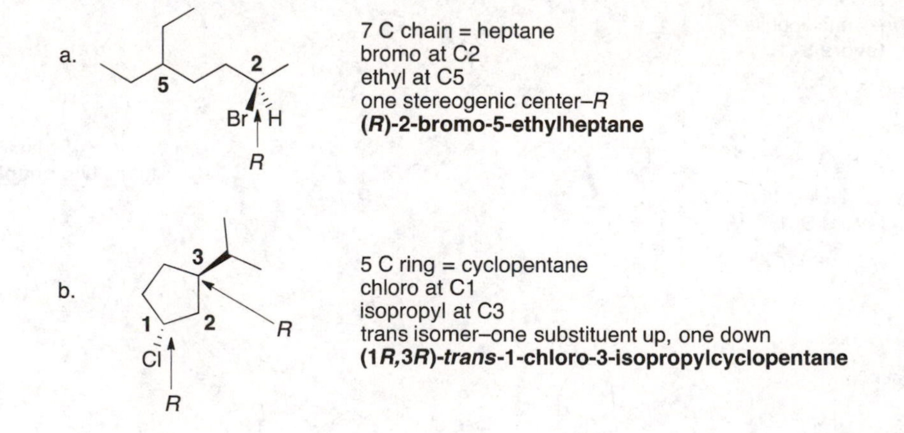

a.
5
2
Br
H
R
7 C chain = heptane
bromo at C2
ethyl at C5
one stereogenic center–R
(R)-2-bromo-5-ethylheptane
b.
3
1
2
Cl
R
R
5 C ring = cyclopentane
chloro at C1
isopropyl at C3
trans isomer–one substituent up, one down
(1R,3R)-trans-1-chloro-3-isopropylcyclopentane

7.41

a. H, Br **(eq)** —$^{-}$CN, acetone→ CN **(axial)**, H   **inversion** (equatorial to axial)

polar aprotic solvent
S_N2 reaction

Large *tert*-butyl group is in more roomy equatorial position.

b. Br **(axial)**, H —$^{-}$CN, acetone→ H, CN **(eq)**   **inversion** (axial to equatorial)

polar aprotic solvent
S_N2 reaction

7.42 Use the directions from Answer 7.2 to name the compounds.

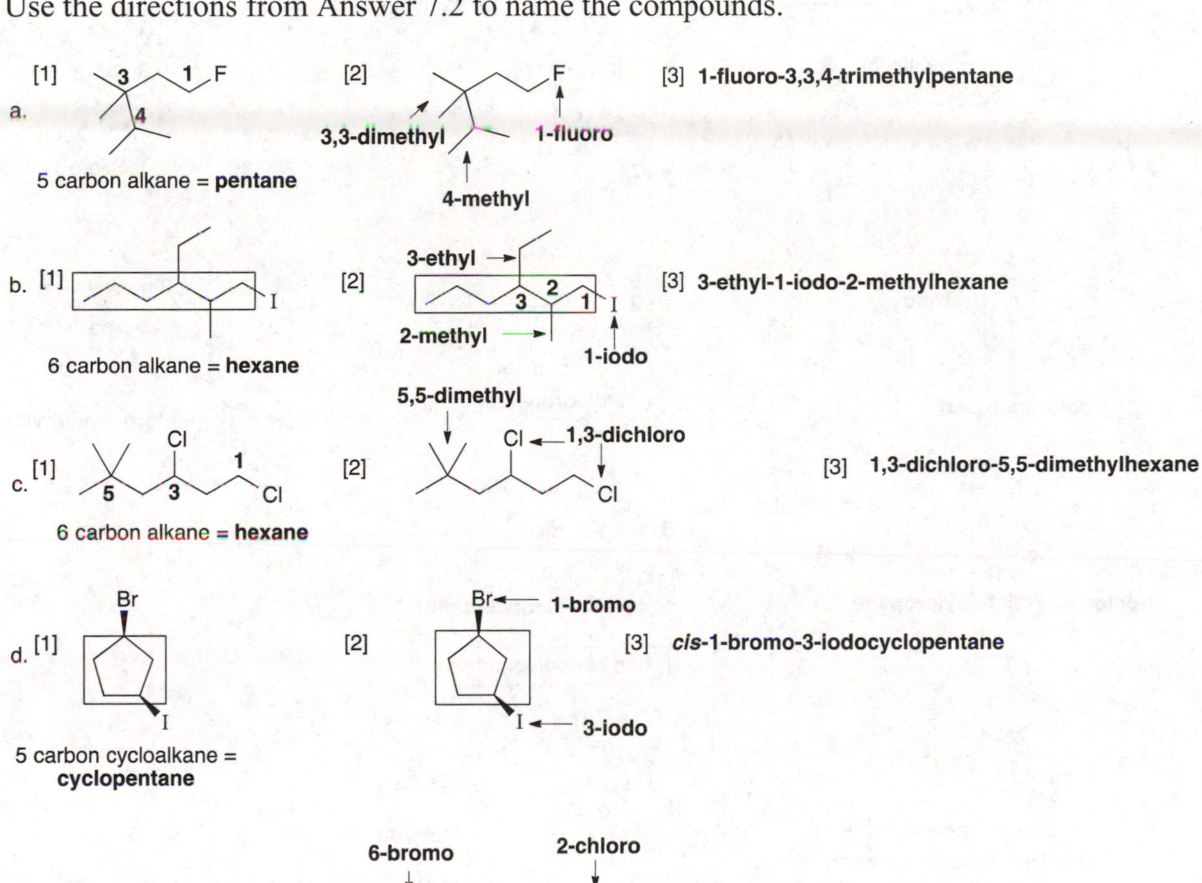

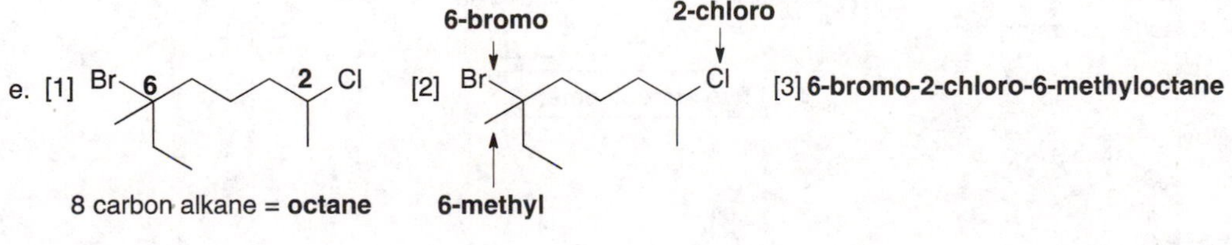

f. [1]

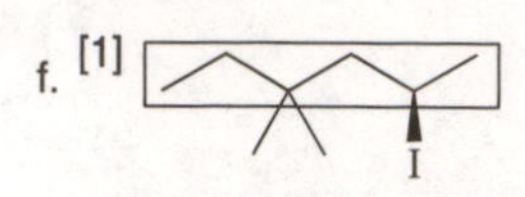

6 carbon alkane = **hexane**
(Indicate the *R,S* designation also)

[2]

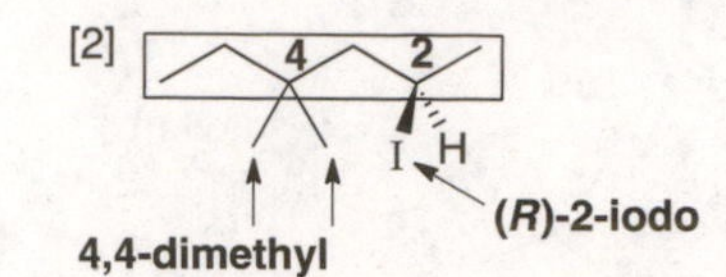

4,4-dimethyl

(*R*)-2-iodo

[3] **(*R*)-2-iodo-4,4-dimethylhexane**

2 3 1 H 4 I

Clockwise
R

7.43 To work backwards to a structure, use the directions in Answer 7.3.

a. 3-bromo-4-ethyl**heptane**

4-ethyl, 3, 4, Br ← 3-bromo

b. 1,1-dichloro-2-methyl**cyclohexane**

1 Cl ← Cl ← 1,1-dichloro; 2 ← 2-methyl

c. 1-bromo-4-ethyl-3-fluoro**octane**

Br, 1-bromo, 1, 3, 4, ← 4-ethyl, F ← 3-fluoro

d. (*S*)-3-iodo-2-methyl**nonane**

← 2-methyl, 3, 3*S*, I, H, 3-iodo

e. (1*R*,2*R*)-*trans*-1-bromo-2-chloro**cyclohexane**

1*R*, 1, Br ← 1-bromo, Cl ← 2-chloro, 2*R*

f. (*R*)-4,4,5-trichloro-3,3-dimethyl**decane**

5*R*, Cl, H, 3,3-dimethyl, 4, 5, 3, Cl, Cl, 4,4,5-trichloro

7.44

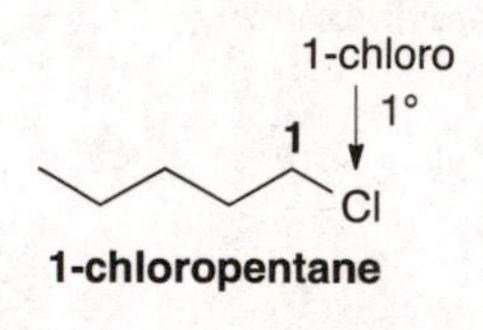

1-chloropentane

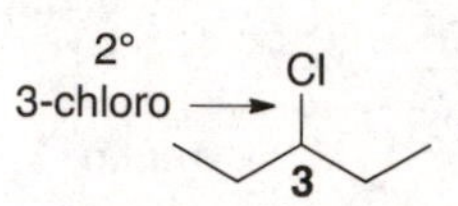

3-chloropentane

3-methyl →, Cl, 1, 3, 1°, 1-chloro

1-chloro-3-methylbutane

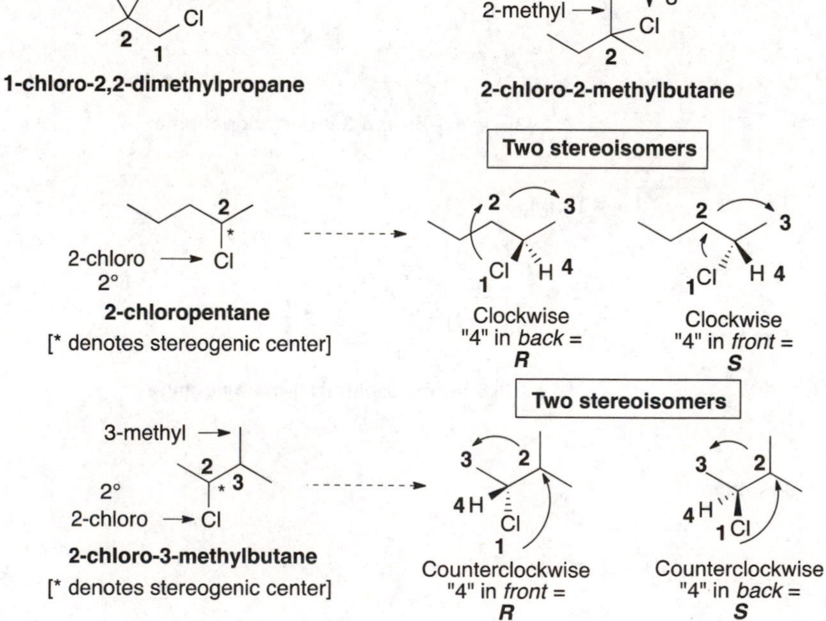

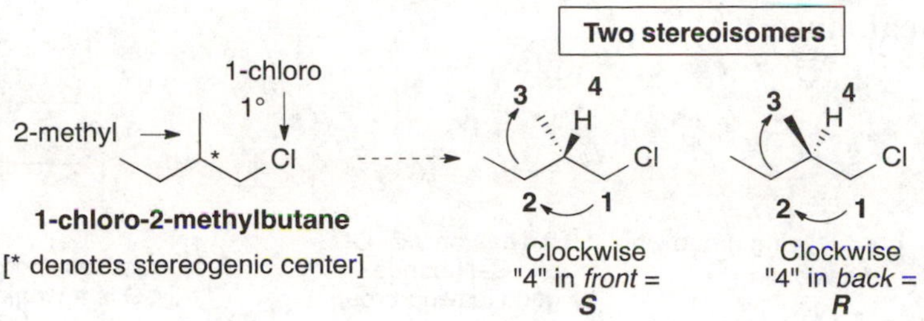

Two stereoisomers

1-chloro
1°
2-methyl
Cl

1-chloro-2-methylbutane
[* denotes stereogenic center]

Clockwise
"4" in *front* =
S

Clockwise
"4" in *back* =
R

7.45

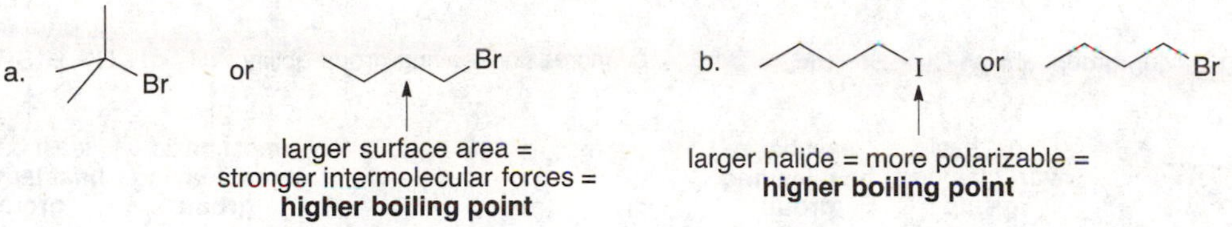

a. Br or Br

larger surface area =
stronger intermolecular forces =
higher boiling point

b. I or Br

larger halide = more polarizable =
higher boiling point

7.46 Use the steps from Answer 7.6 and then draw the proton transfer reaction, when necessary.

a. 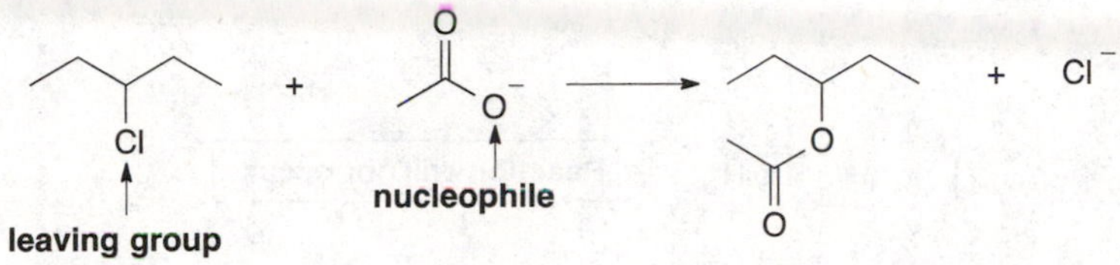 + → + Cl^-

leaving group **nucleophile**

b. 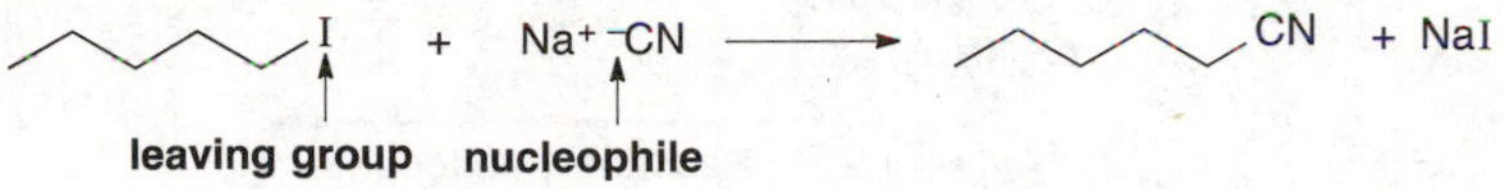 I + $Na^+ \ ^-CN$ → CN + NaI

leaving group **nucleophile**

c. 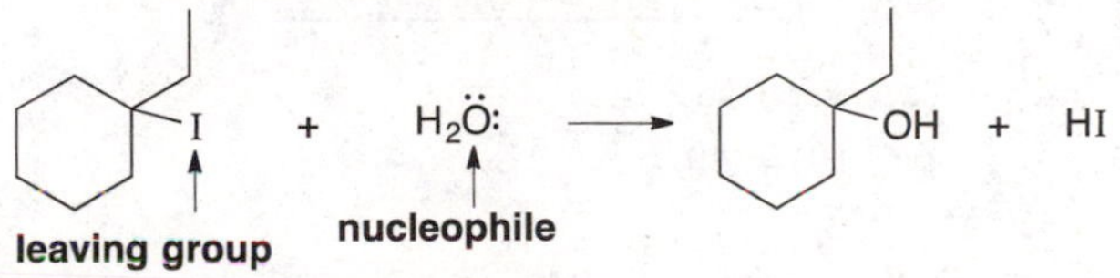 I + $H_2\ddot{O}$: → OH + HI

leaving group **nucleophile**

d. 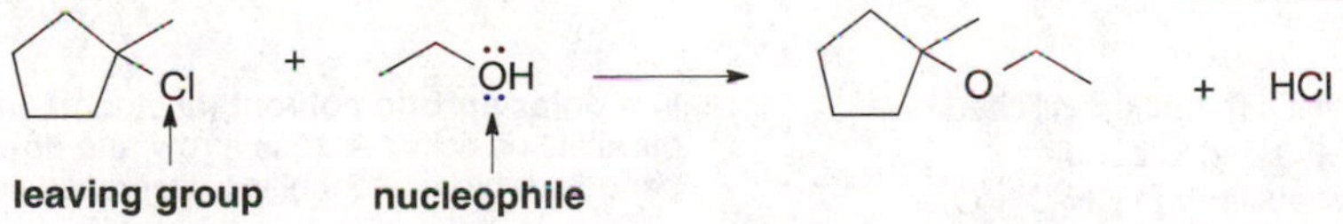 Cl + $\ddot{O}H$ → O + HCl

leaving group **nucleophile**

e. 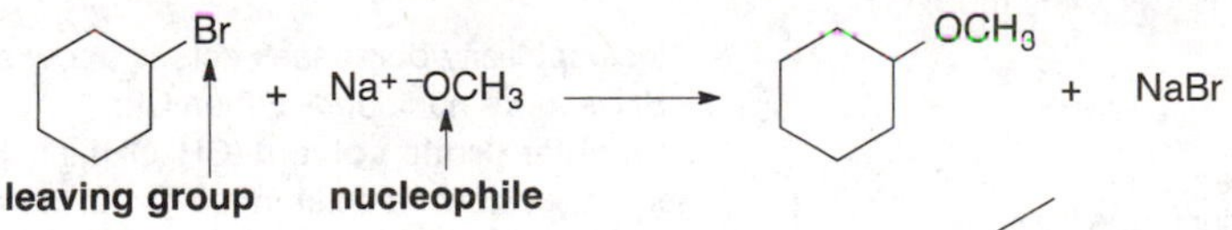 Br + $Na^+ \ ^-OCH_3$ → OCH_3 + NaBr

leaving group **nucleophile**

f. 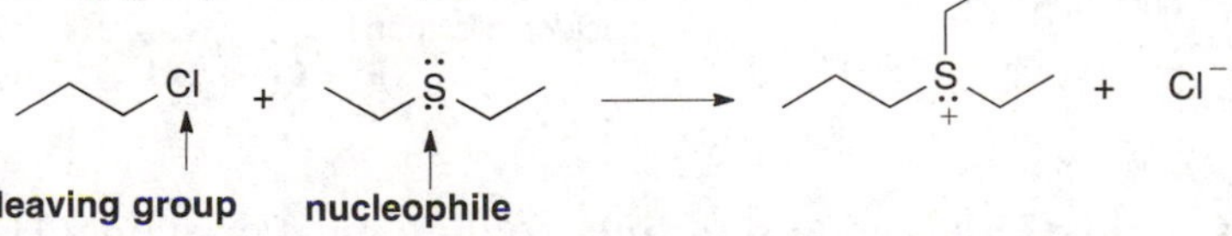 Cl + $\ddot{S}$ → S^+ + Cl^-

leaving group **nucleophile**

7.47 A good leaving group is a weak base.

a. OH — bad leaving group, ^{-}OH is a strong base.

b. Cl — Cl^{-} good leaving group, weak base

c. This has only C–C and C–H bonds. no good leaving group

d. $\overset{+}{O}H_2$ — good leaving group, H_2O is a weak base.

7.48 Use the rules from Answer 7.10.

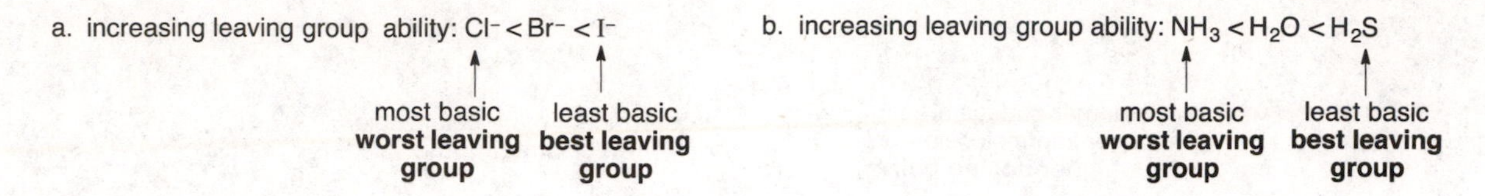

7.49 Compare the nucleophile and the leaving group in each reaction. The reaction will occur if it proceeds towards the weaker base. Remember that the stronger the acid (lower p*K*a), the weaker the conjugate base.

a. NH_2 + I^{-} ⟶ (✗) I + $^{-}NH_2$ **Reaction will not occur.**

weaker base, pK_a (HI) = –10; stronger base, pK_a (NH_3) = 38

b. I + $^{-}O{-}CH_3$ ⟶ O + I^{-} **Reaction will occur.**

stronger base, pK_a (CH_3OH) = 15.5; weaker base, pK_a (HI) = –10

7.50 Use the directions in Answer 7.13.

a. • In a **polar protic solvent** (CH_3OH), nucleophilicity *increases down a column* of the periodic table, so: $CH_3CH_2S^{-}$ is more nucleophilic than $CH_3CH_2O^{-}$.
• For two species with the same attacking atom, the more basic is the more nucleophilic, so $CH_3CH_2O^{-}$ is more nucleophilic than CH_3COO^{-}.

$CH_3COO^{-} < CH_3CH_2O^{-} < CH_3CH_2S^{-}$

b. Compare the nucleophilicity of N, S, and O. In a polar aprotic solvent (acetone), nucleophilicity parallels basicity.

$CH_3SH < CH_3OH < CH_3NH_2$

c. In a **polar aprotic solvent** (acetone), nucleophilicity parallels basicity. Across a row and down a column of the periodic table nucleophilicity decreases.

$Cl^{-} < F^{-} < {}^{-}OH$

d. Nucleophilicity decreases across a row so ^{-}SH is more nucleophilic than Cl^{-}. In a **polar protic solvent** (CH_3OH), nucleophilicity increases down a column, so Cl^{-} is more nucleophilic than F^{-}.

$F^{-} < Cl^{-} < {}^{-}SH$

7.51 ***Polar protic solvents* are capable of hydrogen bonding,** so they must contain a H bonded to an electronegative O or N. ***Polar aprotic solvents* are incapable of hydrogen bonding,** so they do not contain any O–H or N–H bonds.

a. $(CH_3)_2CHOH$

contains O–H bond
protic

b. CH_3NO_2

no O–H or N–H bond
aprotic

c. CH_2Cl_2

no O–H or N–H bond
aprotic

d. NH_3

contains N–H bond
protic

e. $N(CH_3)_3$

no O–H or N–H bond
aprotic

f. $HCONH_2$

contains an N–H bond
protic

7.52

The amine N is more nucleophilic because the electron pair is localized on the N.

The amide N is less nucleophilic because the electron pair is delocalized by resonance.

7.53

1° alkyl halide
S_N2 reaction

a. Mechanism: + :CN → (acetone) CN + Br^-

b. Energy diagram:

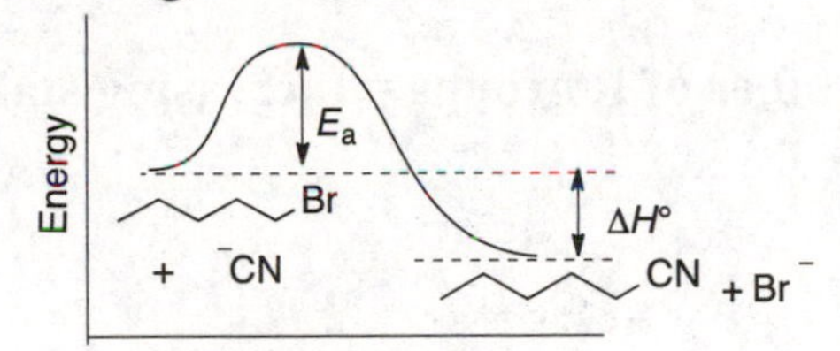

c. Transition state:

d. Rate equation: one-step reaction with both nucleophile and alkyl halide in the only step:
rate = k[R–Br][⁻CN]

e. [1] The leaving group is changed from Br^- to I^-:
Leaving group becomes less basic → a better leaving group → faster reaction.

[2] The solvent is changed from acetone to CH_3CH_2OH:
Solvent changed to polar protic → decreases reaction rate.

[3] The alkyl halide is changed from $CH_3(CH_2)_4Br$ to $CH_3CH_2CH_2CH(Br)CH_3$:
Changed from 1° to 2° alkyl halide → the alkyl halide gets more crowded and the reaction rate decreases.

[4] The concentration of ⁻CN is increased by a factor of 5.
Reaction rate will increase by a factor of 5.

[5] The concentration of both the alkyl halide and ⁻CN are increased by a factor of 5:
Reaction rate will increase by a factor of 25 (5 × 5 = 25).

7.54 a. CH_3CH_2Br reacts faster than CH_3CH_2Cl because Br^- is a better leaving group than Cl^-.

b. The S_N2 reaction with NaOH is faster than the S_N2 reaction with $NaOCOCH_3$ because ^-OH is a stronger nucleophile than $^-OCOCH_3$.

c. The S_N2 reaction is faster in the polar aprotic solvent DMSO because the nucleophile $^-OCH_3$ is stronger.

7.55 All S_N2 reactions proceed with backside attack of the nucleophile. When nucleophilic attack occurs at a stereogenic center, inversion of configuration occurs.

a. H, Cl, D + $^-OCH_3$ → H, D, O + Cl^- inversion of configuration

b. Cl + $^-OCH_2CH_3$ → OCH_2CH_3 + Cl^-

[No bond to the stereogenic center is broken, because the leaving group is not bonded to the stereogenic center.]

c. Br + ^-CN → CN + Br^- inversion of configuration

[* denotes a stereogenic center]

7.56 For carbocations: **Increasing number of R groups = Increasing stability.**

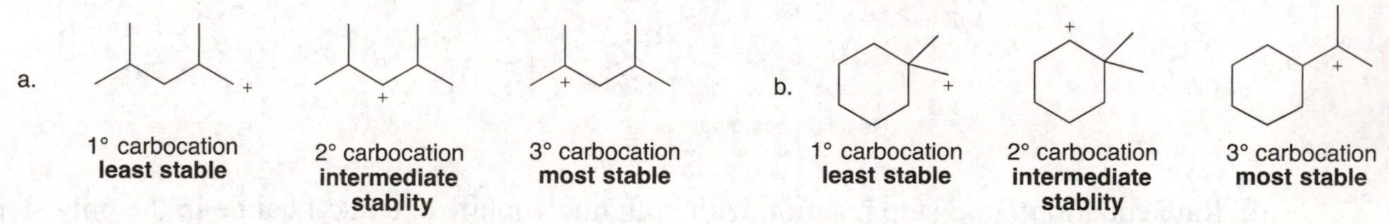

a. 1° carbocation **least stable**; 2° carbocation **intermediate stablity**; 3° carbocation **most stable**

b. 1° carbocation **least stable**; 2° carbocation **intermediate stablity**; 3° carbocation **most stable**

7.57 Both **A** and **B** are resonance stabilized, but the N atom in **B** is more basic and therefore more willing to donate its electron pair.

A O ⟷ O+

B N, H ⟷ N+, H

more basic N atom

This resonance form stabilizes the carbocation more than the equivalent resonance structure for **A**. Thus, **B** is more stable than **A**.

7.58

a. Mechanism: **S_N1 only**

A (I) + H_2O —Step [1]→ **B** (+) + I^- + H_2O —Step [2]→ **C** ($:\overset{+}{O}H_2$)

b. Energy diagram:

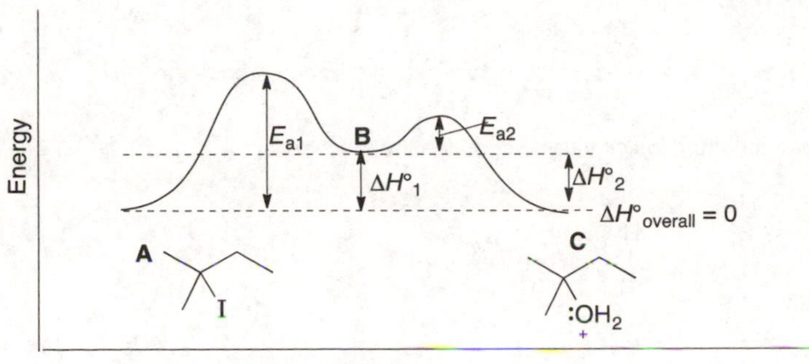

Reaction coordinate

c. Transition states:

$[\ \delta+ \ \ :I:\ \delta- \ ]^{‡}$ $[\ \delta+ \ \ :OH_2\ \delta+ \ ]^{‡}$

d. Rate equation: **rate = k[$(CH_3)_2CICH_2CH_3$]**

e. [1] Leaving group changed from I^- to Cl^-: **rate decreases** because I^- is a better leaving group.
   [2] Solvent changed from H_2O (polar protic) to DMF (polar aprotic):
   rate decreases because polar protic solvent favors S_N1.
   [3] Alkyl halide changed from 3° to 2°: **rate decreases** because 2° carbocations are less stable.
   [4] [H_2O] increased by factor of five: **no change in rate** because H_2O is not in rate equation.
   [5] [R–X] and [H_2O] increased by factor of five: **rate increases** by a factor of five. (Only the concentration of R–X affects the rate.)

7.59 a. S_N1 reaction of $(CH_3)_3CI$ is faster than S_N1 reaction with $(CH_3)_3CCl$ because I^- is a better leaving group than Cl^-.
b. S_N1 reaction is faster with the 3° alkyl halide $(CH_3)_3CBr$ than with the 1° alkyl halide $(CH_3)_2CHCH_2Br$.
c. S_N1 reaction is faster with the polar protic solvent H_2O rather than the aprotic solvent DMSO.

7.60

a. Br + CH_3CH_2OH ⟶ OCH_2CH_3 + CH_3CH_2O + HBr

b. Br + H_2O ⟶ OH + OH + HBr

7.61 The 1° alkyl halide is also allylic, so it forms a resonance-stabilized carbocation. Increasing the stability of the carbocation by resonance increases the rate of the S_N1 reaction.

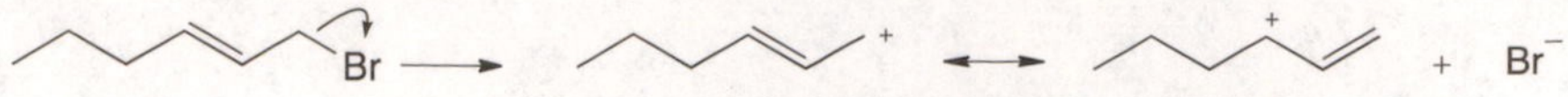

Br → + ⟷ + + Br^-

resonance-stabilized carbocation

Use each resonance structure individually to continue the mechanism:

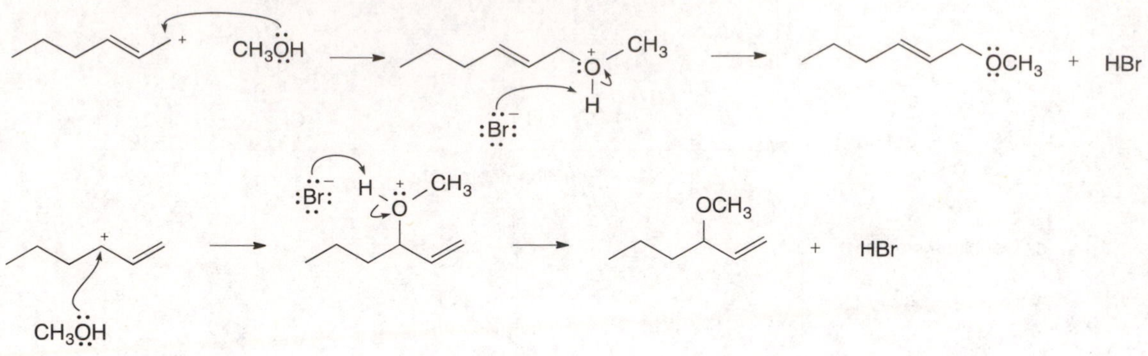

+ $CH_3\ddot{O}H$ → CH_3 O H :Br: → $\ddot{O}CH_3$ + HBr

:Br: H + CH_3 O → OCH_3 + HBr

$CH_3\ddot{O}H$

7.62

Br Br Br

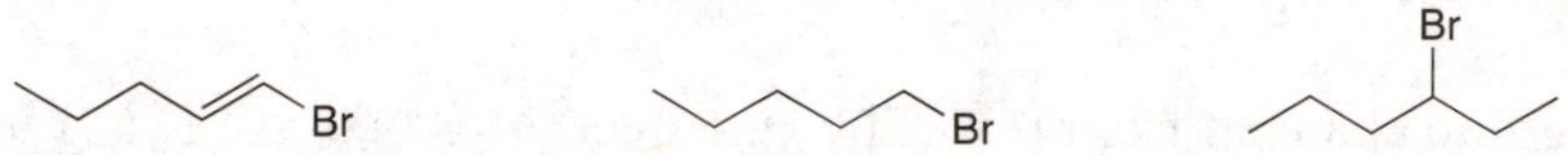

A **B** **C**

Vinyl halides like **A** do not react by either an S_N1 or S_N2 mechanism. With S_N2, the rate is 3° < 2° < 1°, and with S_N1, the rate is 1° < 2° < 3°.

a. S_N2 reactivity: **A < C < B**

b. S_N1 reactivity: **A < B < C**

7.63

a. Br + ^-CN →(acetone) CN + Br^-

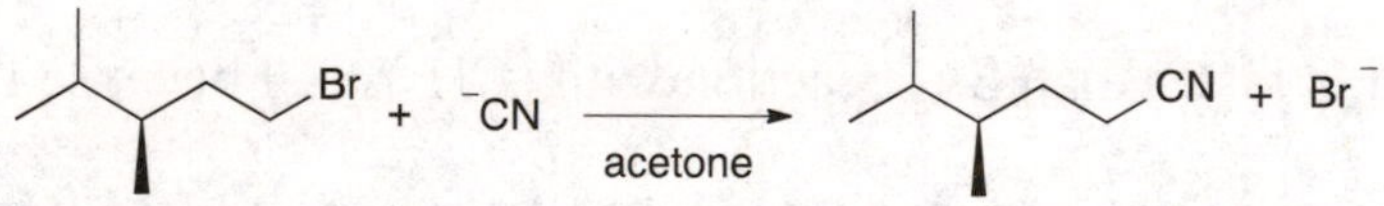

1° alkyl halide
S_N2 only

b. Br H + $^-OCH_3$ →(DMSO) H OCH_3 + Br^- reaction at a stereogenic center **inversion of configuration**

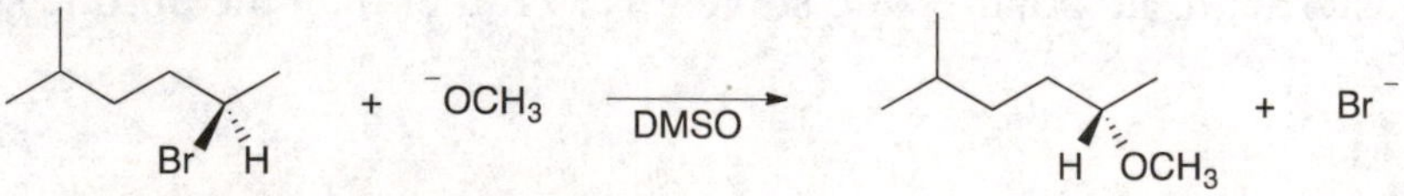

2° alkyl halide
S_N1 and S_N2

strong nucleophile
polar aprotic solvent
Both favor S_N2.

c. Br + CH_3OH → OCH_3 + HBr

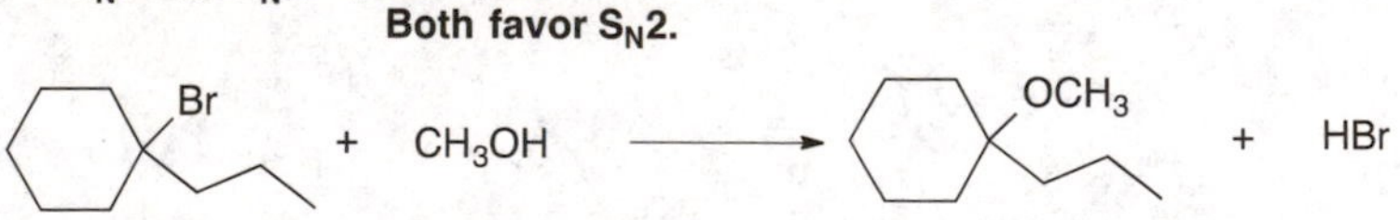

3° alkyl halide
S_N1 only

d. I + CH_3CO_2H → O_2CCH_3 + O_2CHCH + HI reaction at a stereogenic center **racemization of product**

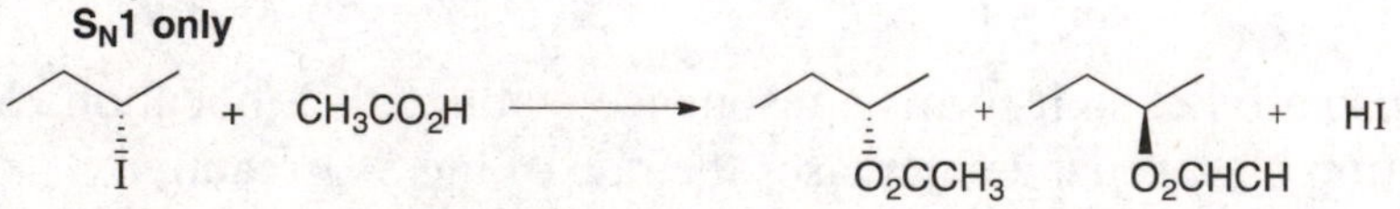

2° alkyl halide
S_N1 and S_N2

Weak nucleophile
favors S_N1.

e. Br + $^{-}OCH_2CH_3$ → (DMF) OCH_2CH_3 + Br^{-}

2° alkyl halide
S_N1 and S_N2

strong nucleophile
polar aprotic solvent
Both favor S_N2.

reaction at a stereogenic center
inversion of configuration

f. Cl + CH_3CH_2OH → OCH_2CH_3 + OCH_2CH_3 + HCl

2° alkyl halide
S_N1 and S_N2

Weak nucleophile
favors S_N1.

two products – **diastereomers**
Nucleophile attacks
from above and below.

7.64 The reaction follows an S_N2 mechanism.

CO_3^{2-} S–H HO O O H F H O F **A** → S⁻ Br–CH_2F + HCO_3^- → (S_N2) F S fluticasone + Br^-

7.65 An S_N1 mechanism means the reaction occurs in a stepwise fashion by way of a carbocation.

P–O O O–PP HO OH **A** → P–O O + HO OH + $^{-}$O–PP H N O N H O O⁻ **B** → P–O N⁺ H $^{-}$O–PP HO OH O O⁻ → P–O N HO OH **C** + HO–PP

7.66

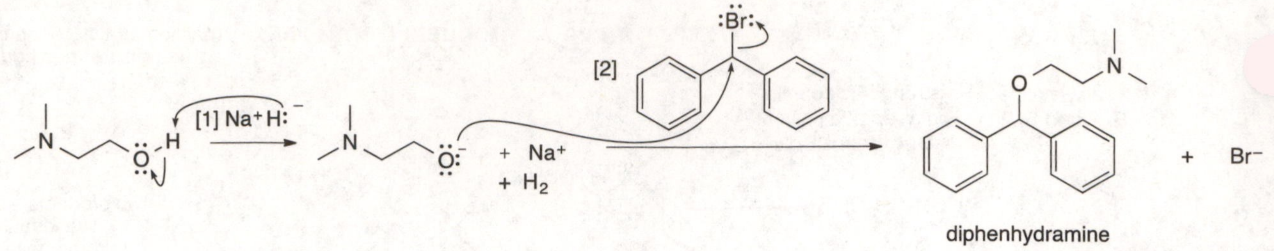

[1] $Na^+H:^-$

+ Na^+
+ H_2

[2]

:Br:

diphenhydramine

+ Br^-

7.67 First decide whether the reaction will proceed via an S_N1 or S_N2 mechanism (Answer 7.38), and then draw the mechanism.

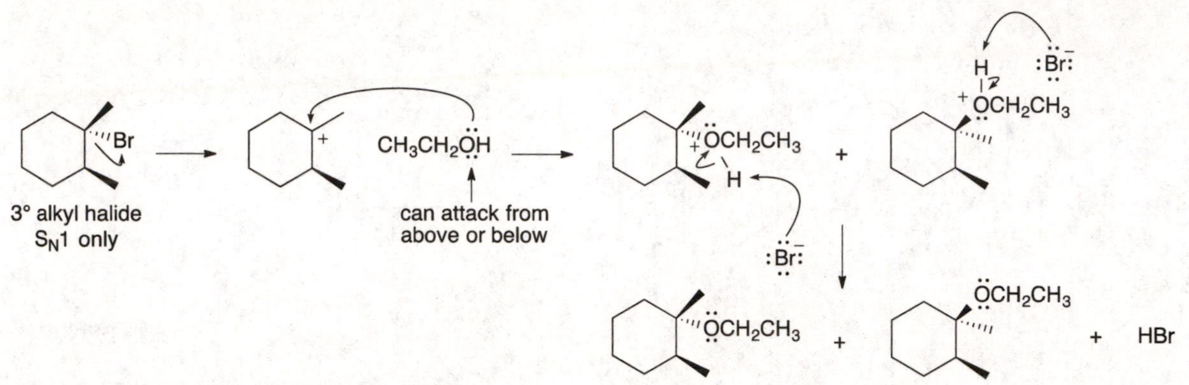

3° alkyl halide
S_N1 only

$CH_3CH_2\ddot{O}H$

can attack from
above or below

OCH_2CH_3 + OCH_2CH_3

:Br:$^-$

OCH_2CH_3 + OCH_2CH_3 + HBr

7.68

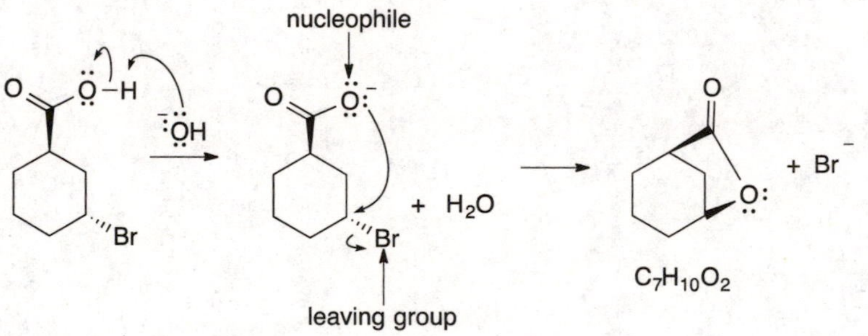

nucleophile

:$\ddot{O}H$

+ H_2O

leaving group

+ Br^-

$C_7H_{10}O_2$

7.69

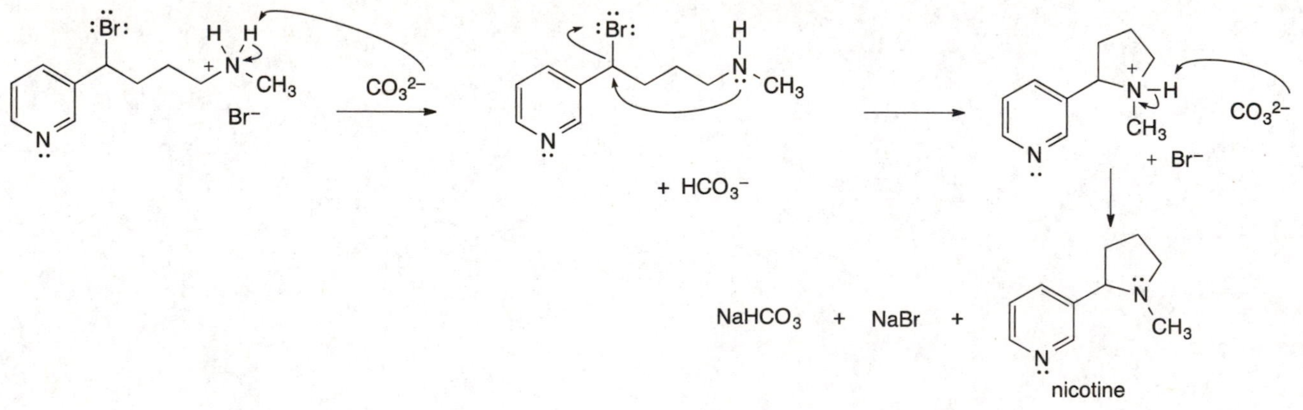

:Br:

CH_3

Br^-

CO_3^{2-}

+ HCO_3^-

CH_3

CO_3^{2-}

+ Br^-

$NaHCO_3$ + NaBr +

CH_3

nicotine

7.70

a. Two diastereomers (**C** and **D**) are formed as products from the two enantiomers of **A.**

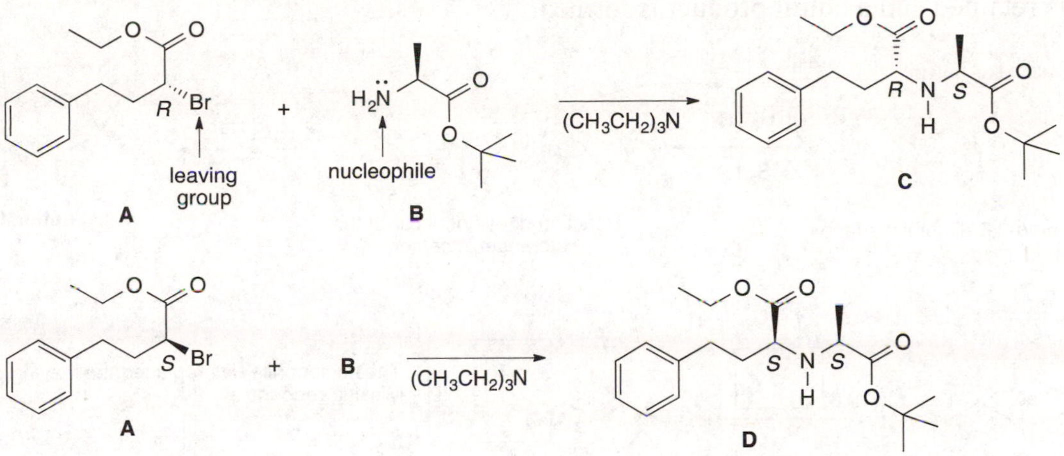

R Br
leaving group
A
+
H_2N
nucleophile
B
$(CH_3CH_2)_3N$
R N S
H
C

S Br
A
+ **B**
$(CH_3CH_2)_3N$
S N S
H
D

b.

S N S
H
N
OH
Both have the *S* configuration
quinapril

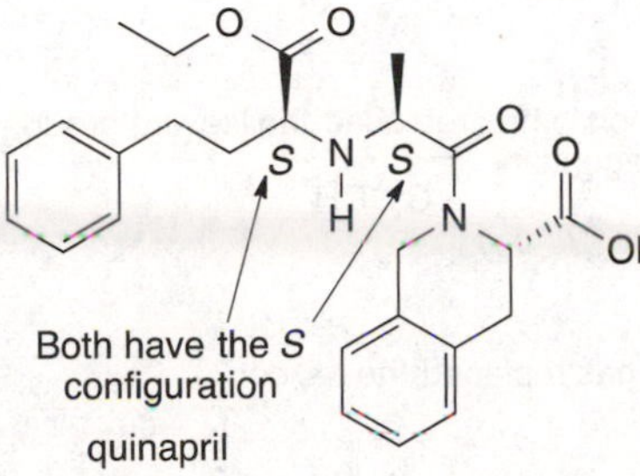

Since both stereogenic centers in **D** have the *S* configuration and the corresponding stereogenic centers in quinapril are also *S*, **D** is needed to synthesize the drug.

7.71

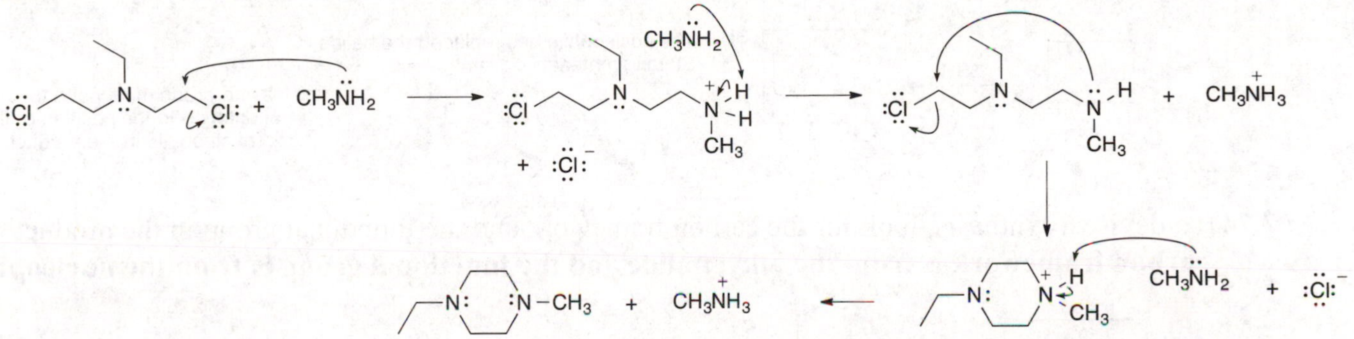

:Cl N Cl: + CH_3NH_2 → :Cl N N–H CH_3 + :Cl:$^-$ → :Cl N N H CH_3 + $CH_3\overset{+}{N}H_3$

→ N: N H CH_3 CH_3NH_2 + :Cl:$^-$ → N: N–CH_3 + $CH_3\overset{+}{N}H_3$

7.72 In the first reaction, substitution occurs at the stereogenic center. Because an achiral, planar carbocation is formed, the nucleophile can attack from either side, thus generating a racemic mixture.

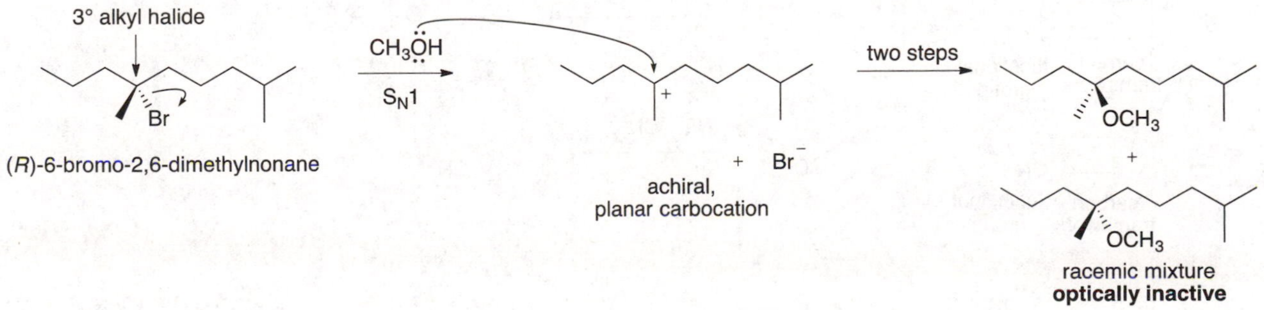

3° alkyl halide
Br
(*R*)-6-bromo-2,6-dimethylnonane
CH_3OH
S_N1
+ Br^-
achiral, planar carbocation
two steps
OCH_3
+
OCH_3
racemic mixture
optically inactive

In the second reaction, the starting material contains a stereogenic center, but the nucleophile does not attack at that carbon. Because a bond to the stereogenic center is not broken, the configuration is retained and a chiral product is formed.

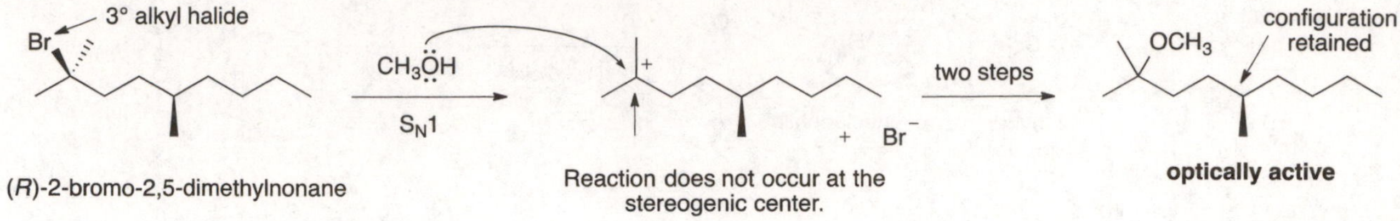

3° alkyl halide

Br

$CH_3\ddot{O}H$

S_N1

\+ Br^-

two steps

OCH_3

configuration retained

(*R*)-2-bromo-2,5-dimethylnonane

Reaction does not occur at the stereogenic center.

optically active

7.73

a. 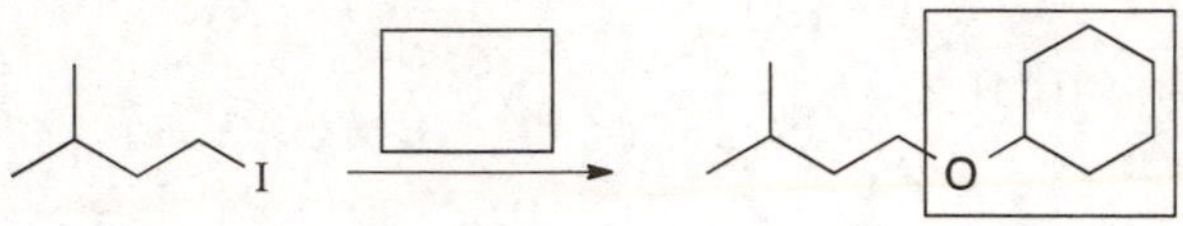 I

The nucleophile has replaced the leaving group.
Missing reagent:

^-O

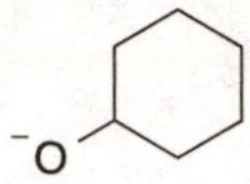

b. 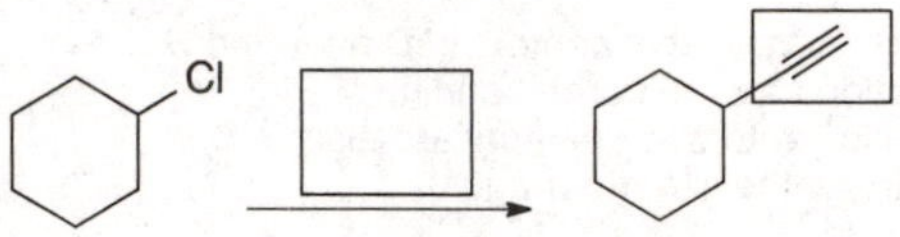 Cl

The nucleophile has replaced the leaving group.
Missing reagent:

$^-C{\equiv}CH$

c. 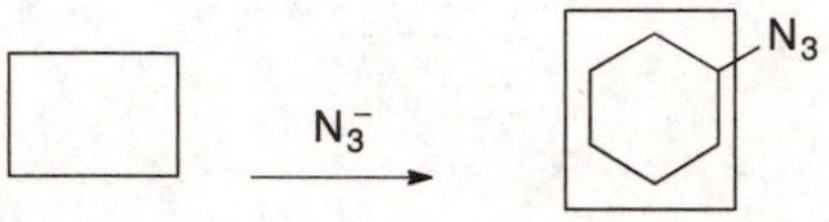 N_3^-

N_3

The nucleophile has replaced the halide.
Starting material:

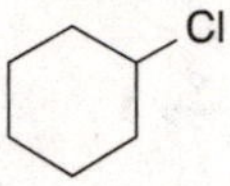

Cl

d. 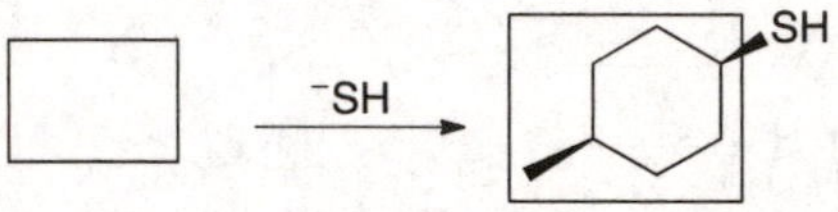 ^-SH

SH

The nucleophile has replaced the halide.
Starting material:

Cl

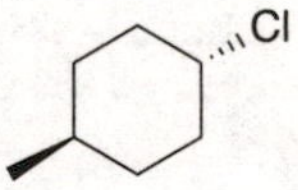

The leaving group must have the opposite orientation to the position of the nucleophile in the product.

7.74 To devise a synthesis, look for the carbon framework and the functional group in the product. **The carbon framework is from the alkyl halide and the functional group is from the nucleophile.**

a. 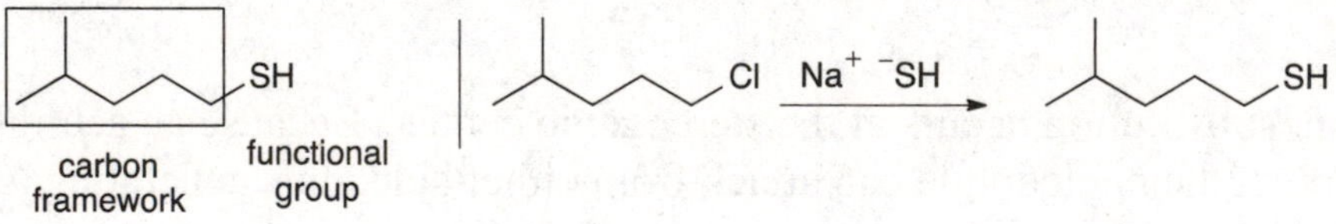 SH

carbon framework

functional group

Cl $Na^+\ ^-SH$ → SH

b. 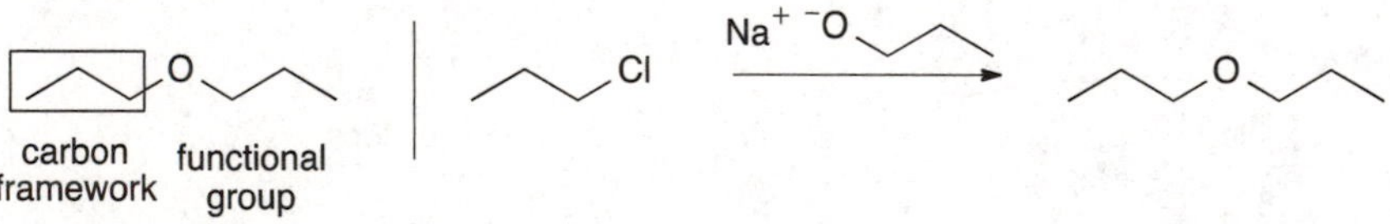 O

carbon framework

functional group

Cl $Na^+\ ^-O$ → O

c. 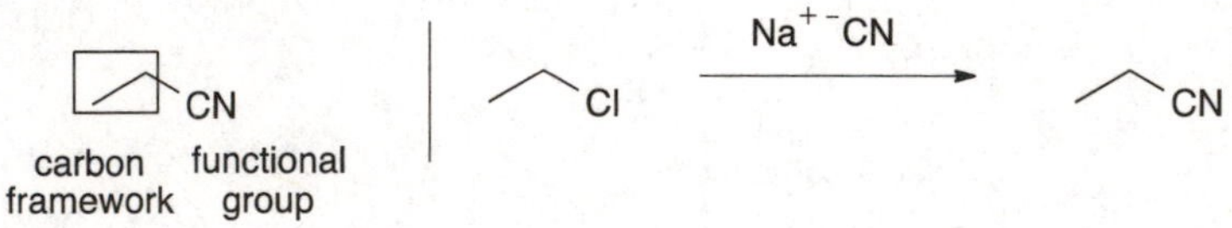 CN

carbon framework

functional group

Cl $Na^+\ ^-CN$ → CN

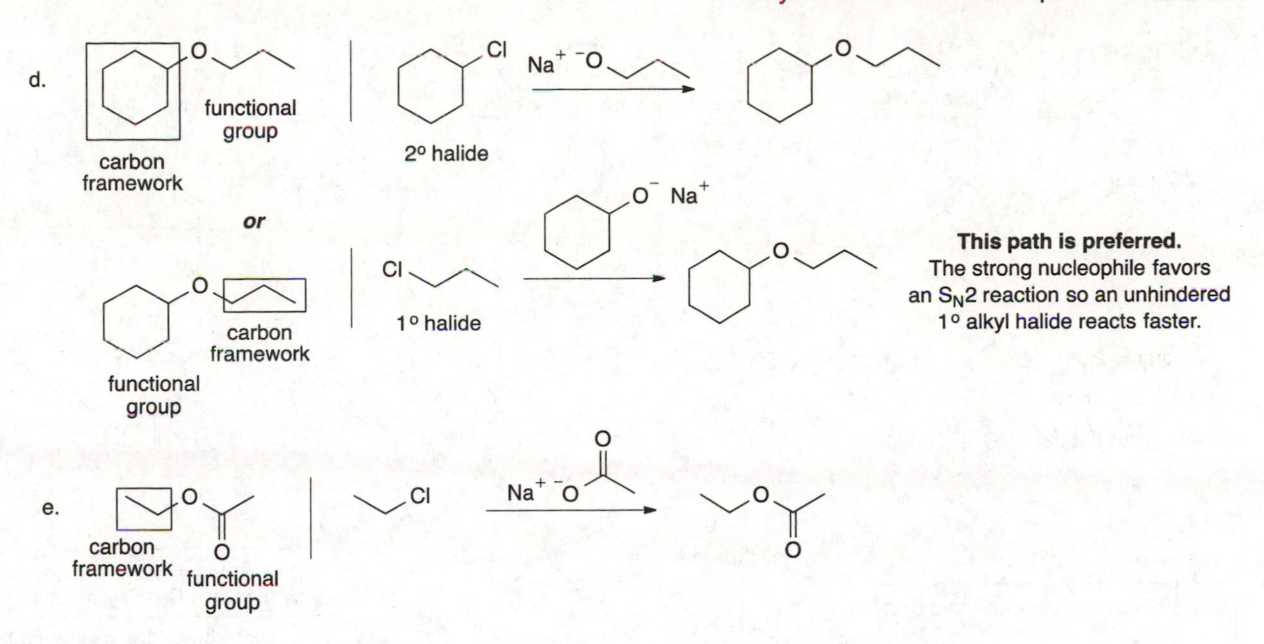

7.75

B (very crowded 3° halide) with I + $Na^{+}:\ddot{O}CH_3$ (C) → E ($\ddot{O}CH_3$)

D ($\ddot{O}:^{-}\ Na^{+}$) + CH_3I (A, unhindered methyl halide) → E ($\ddot{O}CH_3$)

preferred method
The strong nucleophile favors S_N2 reaction, so the alkyl halide should be unhindered for a faster reaction.

7.76

$H-C\equiv C-H \xrightarrow{NaH} H-C\equiv C^{-}$ (A) $+ H_2$ → (with decyl bromide, Br) B

B $\xrightarrow{NaH}$ C $+ H_2$

C → (with long-chain alkyl bromide, Br) D

D $\xrightarrow{\text{addition of } H_2 \text{ (1 equiv)}}$ muscalure

7.77

a. (Chapter 9) hexanol $\ddot{O}$–H $\xrightarrow{[1]\ Na^{+}H:^{-}}$ $\ddot{O}:^{-}$ $\xrightarrow{[2]\ CH_3-\ddot{B}r:}$ $\ddot{O}$–CH_3 + $Na^{+}\ Br^{-}$ + H_2

b. [1] $Na^+ :\ddot{N}H_2^-$ [2] $\ddot{Br}$: + $Na^+ Br^-$ + NH_3

(Chapter 11)

c. [1] $Na^+ :\ddot{O}CH_2CH_3^-$ [2] $\ddot{Br}$: + $Na^+ Br^-$ + CH_3CH_2OH

(Chapter 23)

7.78

The three alkyl groups are "tied back" in a ring, making the electron pair more available.

quinuclidine

triethylamine

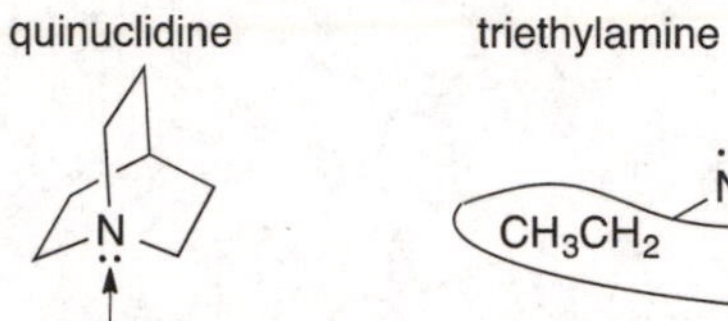

This electron pair is more hindered by the three CH_2CH_3 groups.

These bulky groups around the N cause steric hindrance and this decreases nucleophilicity.

This electron pair on quinuclidine is much more available than the one on triethylamine.

less steric hindrance
more nucleophilic

7.79

[1] + H_2

CH_3-Br [2] minor product

CH_3-Br [2] major product

+ NaBr

7.80

a. base → intramolecular S_N2

b. base → intramolecular S_N2

c.

Br

OH

base

Br

:O:⁻

3° alkyl halide
harder reaction

intramolecular
S_N2

O

d.

OH

Br

base

:O:⁻

Br

3° alkyl halide
harder reaction

intramolecular
S_N2

O

7.81

Cl bonded to *sp*² C
cannot undergo S_N1.

Cl bonded to *sp*³ C
no resonance stabilization possible for the
carbocation formed here

Cl bonded to *sp*³ C
Resonance-stabilized carbocation forms.
best for S_N1

J

CH_3OH
(1 equiv)

+ :Cl:⁻

re-draw

CH_3OH

$O-CH_3$

H

+ :Cl:⁻

+ HCl

OCH_3

K

7.82

a.

H OH

(*S*)-1-phenylpropan-1-ol
[α] = –48

H OH

(*R*)-1-phenylpropan-1-ol
[α] = +48

$$ee = \frac{[\alpha]\ \text{mixture}}{[\alpha]\ \text{pure enantiomer}} \times 100\%$$

$$= \frac{+5.0}{+48} \times 100\% = 10.\%\ \text{excess of } R \text{ isomer}$$

90% racemic mixture = 45% *R* and 45% *S*
Total *R* isomer = 45 + 10 = 55% *R* isomer

b. The *R* product is the product of inversion and it predominates.

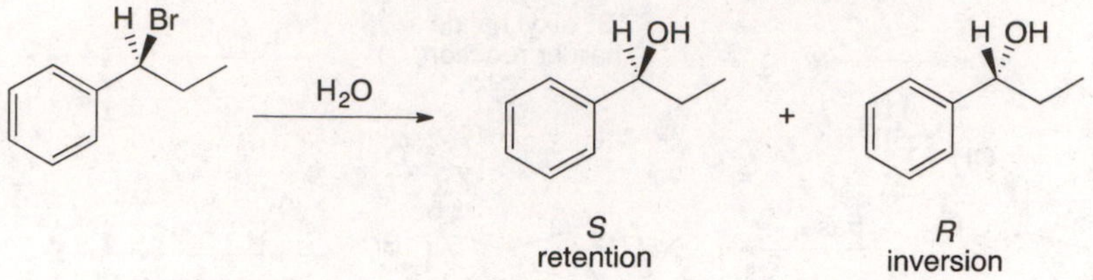

H Br

H_2O

H OH

\+

H OH

S
retention

R
inversion

c. The weak nucleophile favors an S_N1 reaction, which occurs by way of an intermediate carbocation. Perhaps there is more inversion than retention because H_2O attacks the intermediate carbocation while the Br^- leaving group is still in the vicinity of the carbocation. The Br^- would then shield one side of the carbocation and backside attack would be slightly favored.

Chapter 8 Alkyl Halides and Elimination Reactions

Chapter Review

A comparison between nucleophilic substitution and β-elimination

Nucleophilic substitution—A nucleophile attacks a carbon atom (7.6).

X + :Nu⁻ → Nu (substitution product) + :X⁻ (good leaving group)

β-Elimination—A base attacks a proton (8.1).

B: H … X → (elimination product) + H–B⁺ + :X⁻ (good leaving group)

Similarities	Differences
• In both reactions RX acts as an electrophile, reacting with an electron-rich reagent.	• In substitution, a nucleophile attacks a single carbon atom.
• Both reactions require a **good leaving group X:⁻** willing to accept the electron density in the C–X bond.	• In elimination, a Brønsted–Lowry base removes a proton to form a π bond, and two carbons are involved in the reaction.

The importance of the base in E2 and E1 reactions (8.9)

The strength of the base determines the mechanism of elimination.

- Strong bases favor E2 reactions.
- Weak bases favor E1 reactions.

Br → with $^{-}$:OH (strong base) → E2 product + H_2O: + :Br:⁻

Br → with H_2O: (weak base) → E1 product + H_3O^+ + :Br:⁻

E1 and E2 mechanisms compared

	E2 mechanism	E1 mechanism
[1] Mechanism	• one step (8.4B)	• two steps (8.6B)
[2] Alkyl halide	• rate: $R_3CX > R_2CHX > RCH_2X$ (8.4C)	• rate: $R_3CX > R_2CHX > RCH_2X$ (8.6C)
[3] Rate equation	• rate = k[RX][B:] • second-order kinetics (8.4A)	• rate = k[RX] • first-order kinetics (8.6A)
[4] Stereochemistry	• anti periplanar arrangement of H and X (8.8)	• trigonal planar carbocation intermediate (8.6B)
[5] Base	• favored by strong bases (8.4B)	• favored by weak bases (8.6C)
[6] Leaving group	• better leaving group → faster reaction (8.4B)	• better leaving group → faster reaction (Table 8.4)
[7] Solvent	• favored by polar aprotic solvents (8.4B)	• favored by polar protic solvents (Table 8.4)
[8] Product	• more substituted alkene favored (Zaitsev rule, 8.5)	• more substituted alkene favored (Zaitsev rule, 8.6C)

Summary chart on the four mechanisms: S_N1, S_N2, E1, and E2 (8.11)

Alkyl halide type	Conditions	Mechanism
1° RCH_2X	strong nucleophile	**S_N2**
	strong bulky base	**E2**
2° R_2CHX	strong base and nucleophile	**S_N2 + E2**
	strong bulky base	**E2**
	weak base and nucleophile	**S_N1 + E1**
3° R_3CX	weak base and nucleophile	**S_N1 + E1**
	strong base	**E2**

Zaitsev rule

- β-Elimination affords the more stable product having the more substituted double bond.
- Zaitsev products predominate in E2 reactions except when a cyclohexane ring prevents trans diaxial arrangement.

Practice Test on Chapter Review

1. Which of the following is true about an E1 reaction?
   1. The reaction is faster with better leaving groups.
   2. The reaction is fastest with 3° alkyl halides.
   3. The reaction is faster with stronger bases.
   4. Statements (1) and (2) are true.
   5. Statements (1), (2), and (3) are all true.

2. Consider the S_N2 and E1 reaction mechanisms. What effect on the rate of the reaction is observed when each of the following changes is made? Fill in each box of the table with one of the following phrases: **increases, decreases,** or **remains the same.**

Change	S_N2 mechanism	E1 mechanism
a. The alkyl halide is changed from $(CH_3)_3CBr$ to $CH_3CH_2CH_2CH_2Br$.		
b. The solvent is changed from $(CH_3)_2CO$ to CH_3CH_2OH.		
c. The nucleophile/base is changed from ^-OH to H_2O.		
d. The alkyl halide is changed from CH_3CH_2Cl to CH_3CH_2I.		
e. The concentration of the base/nucleophile is increased by a factor of five.		

3. Rank the following compounds in order of ***increasing*** **reactivity in an E2 elimination** reaction. Rank the ***most reactive*** compound as **3,** the ***least reactive*** compound as **1,** and the compound of intermediate reactivity as **2.**

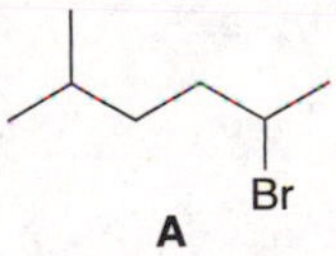

Br
A

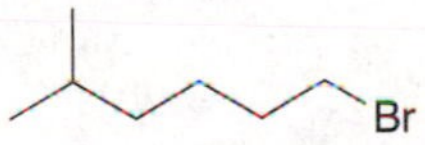

Br
B

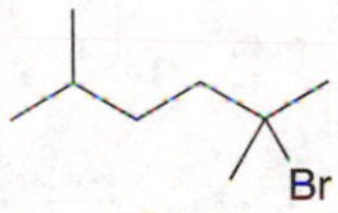

Br
C

4. Draw the organic products formed in the following reactions.

a. Br KOH →

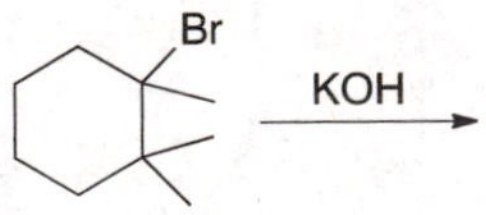

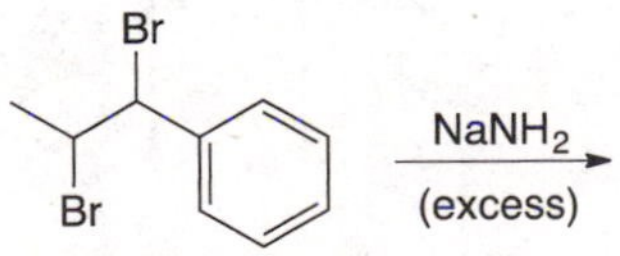

c. Cl $K^+ \ ^-OC(CH_3)_3$ →

b. Br Br $NaNH_2$ (excess) →

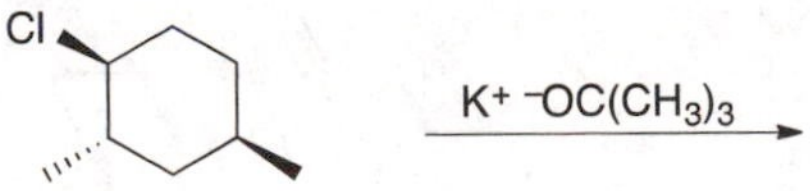

d. Br $NaOCH_3$ →

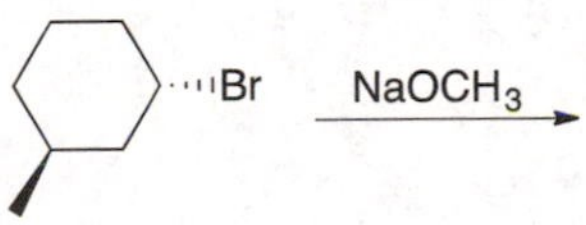

5.a. Fill in the appropriate alkyl halide needed to synthesize the following compound as a single product using the given reagents.

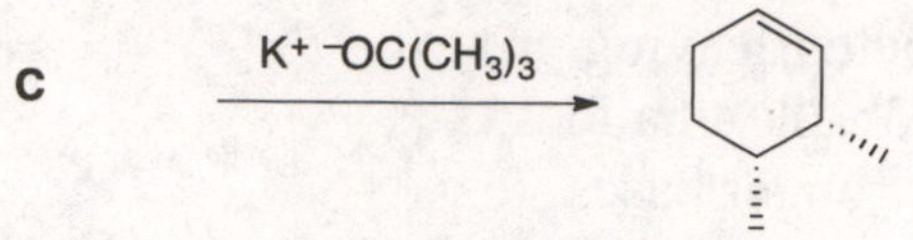

C $\xrightarrow{K^+ \ ^-OC(CH_3)_3}$

b. What starting material is needed for the following reaction? The starting material must yield product cleanly, in one step without any other organic side products.

D $\xrightarrow{K^+ \ ^-OC(CH_3)_3}$ $CH_3CH_2CH{=}CHCH_3$ (cis and trans mixture)

6. Draw all products formed in the following reaction.

Cl $\xrightarrow{CH_3OH}$

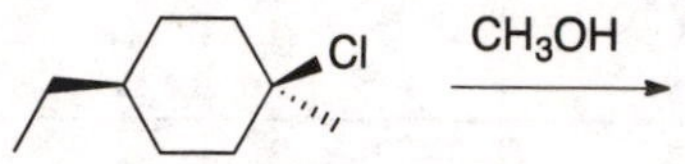

Answers to Practice Test

1. 4

2.

	SN2	E1
a.	increases	decreases
b.	decreases	increases
c.	decreases	same
d.	increases	increases
e.	increases	same

3. **A–2**
B–1
C–3

4.
a. 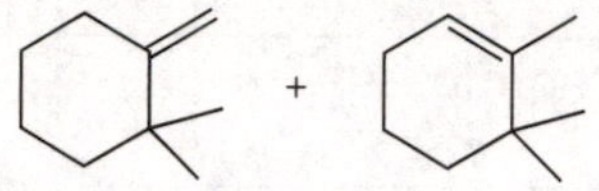 +
b. 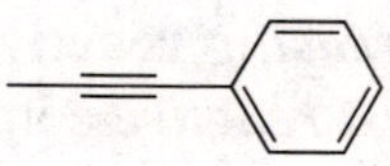
c. 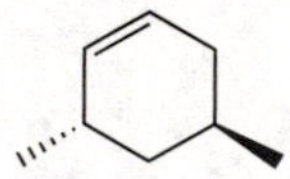
d. 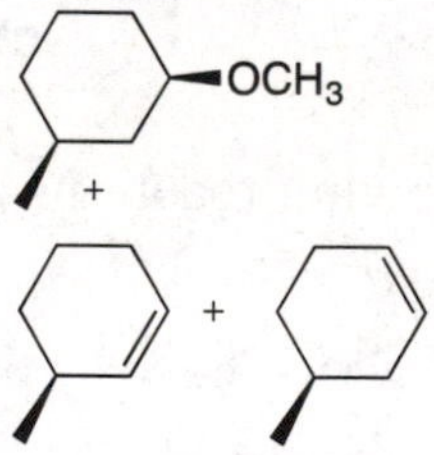 OCH_3 + +

5.
a. 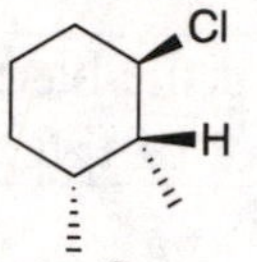 Cl H C
b. 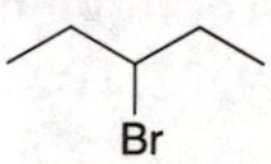 Br (or Cl or I) D

6. OCH_3 OCH_3

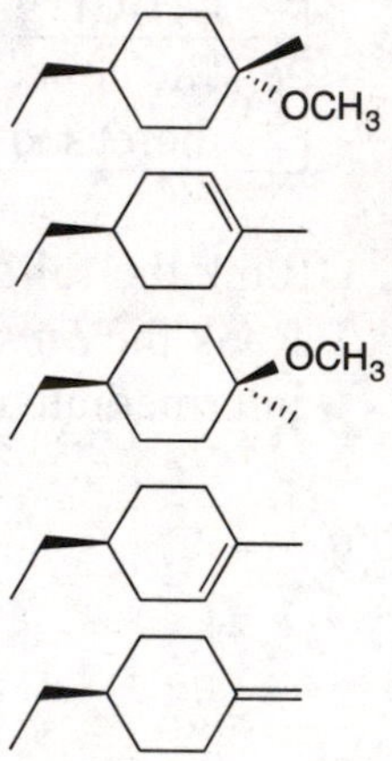

Answers to Problems

8.1 • The carbon bonded to the leaving group is the **α carbon.** Any carbon bonded to it is a **β carbon.**
• **To draw the products of an elimination reaction:** Remove the leaving group from the α carbon and a H from the β carbon and form a π bond.

a. 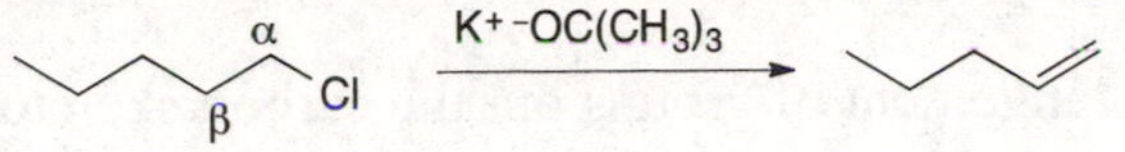 α, β, Cl; $K^{+}\,{}^{-}OC(CH_3)_3$

b. 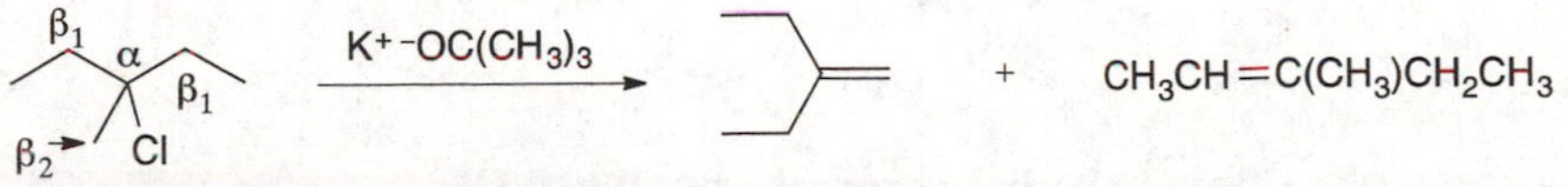 β_1, α, β_1, β_2, Cl; $K^{+}\,{}^{-}OC(CH_3)_3$; + $CH_3CH{=}C(CH_3)CH_2CH_3$

c. β_1, β_2, α, Br, β_1; $K^{+}\,{}^{-}OC(CH_3)_3$; +

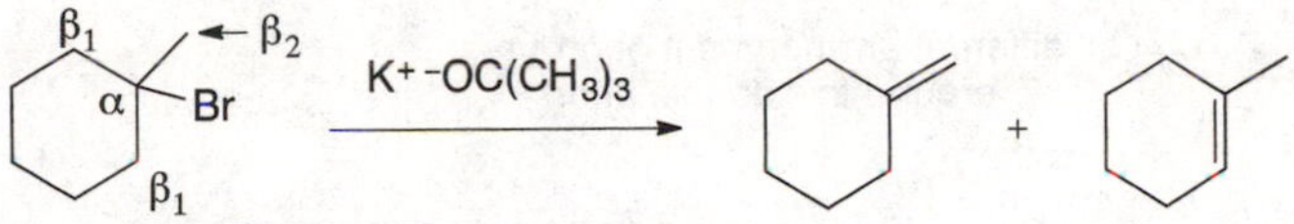

8.2 **Alkenes are classified by the number of carbon atoms bonded to the double bond.** A monosubstituted alkene has one carbon atom bonded to the double bond, a disubstituted alkene has two carbon atoms bonded to the double bond, etc.

a. 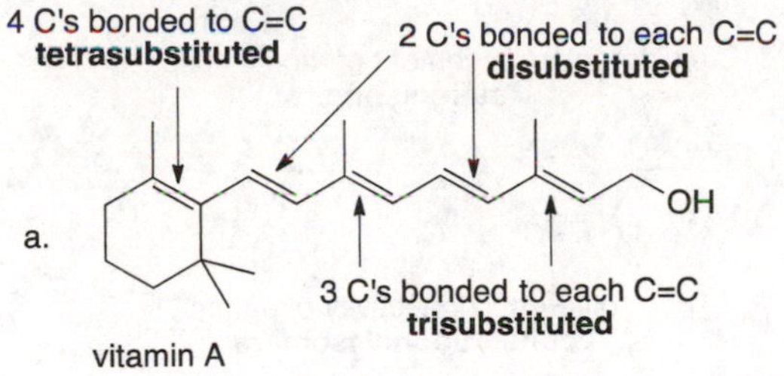 4 C's bonded to C=C **tetrasubstituted**; 2 C's bonded to each C=C **disubstituted**; OH; 3 C's bonded to each C=C **trisubstituted**; vitamin A

b. 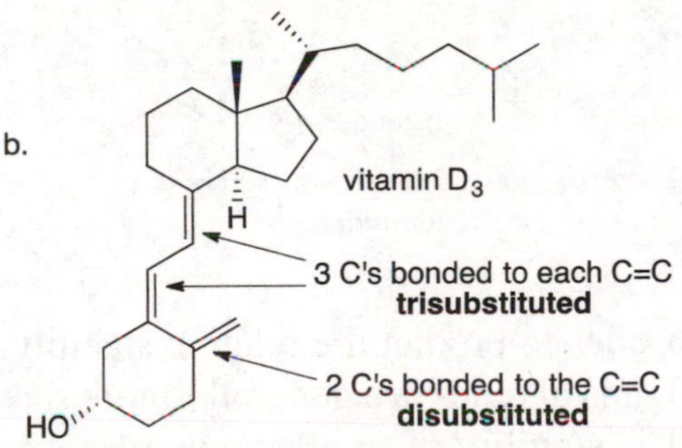 vitamin D_3; H; HO; 3 C's bonded to each C=C **trisubstituted**; 2 C's bonded to the C=C **disubstituted**

8.3 To have stereoisomers at a C=C, the two groups on each end of the double bond must be different from each other.

a. 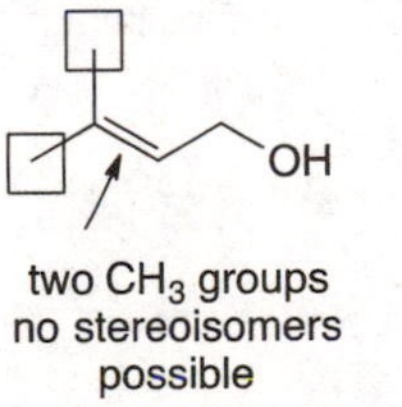 OH; two CH_3 groups no stereoisomers possible

b. 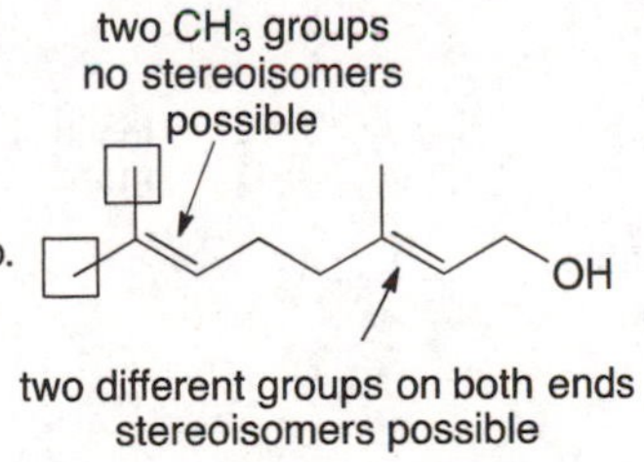 two CH_3 groups no stereoisomers possible; OH; two different groups on both ends stereoisomers possible

c. 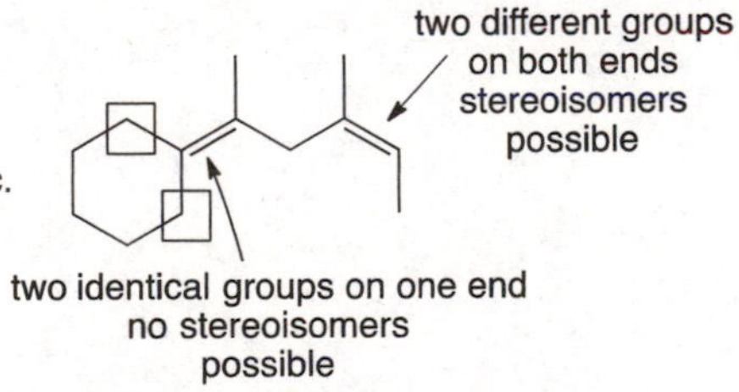 two different groups on both ends stereoisomers possible; two identical groups on one end no stereoisomers possible

8.4

a. 2 CH_3's on one end → ← 2 H's on one end

Only this C=C exhibits stereoisomerism.

b. A diastereomer has a different 3-D arrangement of groups but the carbon skeleton and the double bonds must stay in the original positions.

diastereomer — different arrangement of groups around this double bond

8.5 Two definitions:

- **Constitutional isomers** differ in the connectivity of the atoms.
- **Stereoisomers** differ only in the 3-D arrangement of the atoms in space.

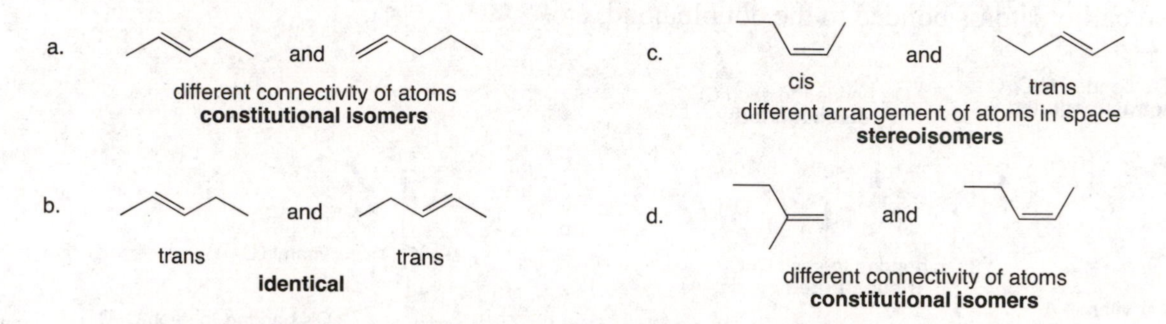

8.6 Two rules to predict the relative stability of alkenes:
[1] Trans alkenes are generally more stable than cis alkenes.
[2] The stability of an alkene increases as the number of R groups on the C=C increases.

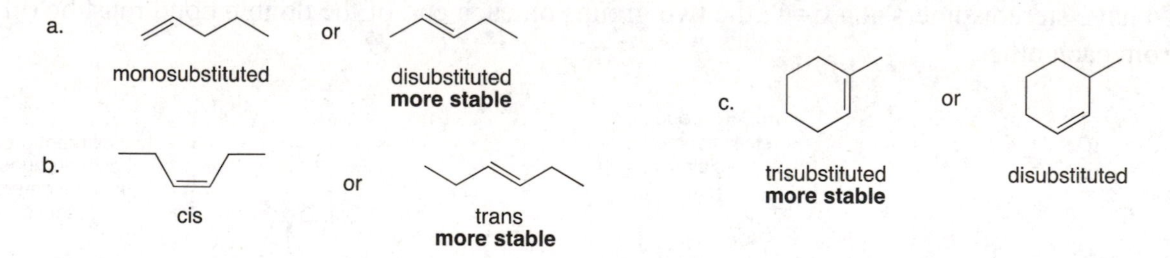

8.7

Alkene **A** is more stable than alkene **B** because the double bond in **A** is in a six-membered ring. The double bond in **B** is in a four-membered ring, which has considerable angle strain due to the small ring size.

8.8 In an E2 mechanism, four bonds are involved in the single step. Use curved arrows to show these simultaneous actions:
[1] The base attacks a hydrogen on a β carbon.
[2] A π bond forms.
[3] The leaving group comes off.

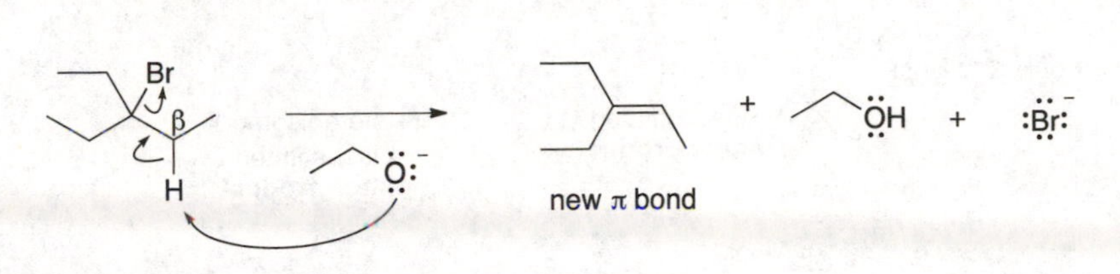

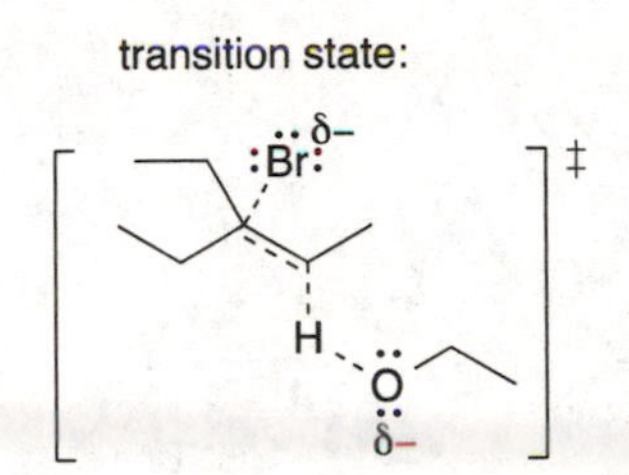

8.9 In both cases, the rate of elimination decreases.

a. stronger base **faster reaction**

$CH_3CH_2{-}Br + {}^{-}OC(CH_3)_3 \longrightarrow$

$CH_3CH_2{-}Br + {}^{-}OH \longrightarrow$

b. better leaving group **faster reaction**

$CH_3CH_2{-}Br + {}^{-}OC(CH_3)_3 \longrightarrow$

$CH_3CH_2{-}Cl + {}^{-}OC(CH_3)_3 \longrightarrow$

8.10 As the number of R groups on the carbon with the leaving group increases, the rate of an E2 reaction increases.

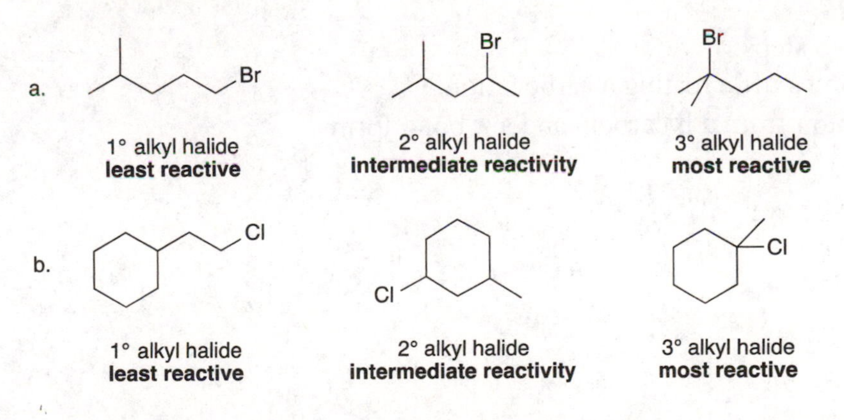

8.11 Use the following characteristics of an E2 reaction to answer the questions:
[1] E2 reactions are second order and one step.
[2] More substituted halides react faster.
[3] Reactions with strong bases or better leaving groups are faster.
[4] Reactions with polar aprotic solvents are faster.

Rate equation: rate = k[RX][Base]

a. tripling the concentration of the alkyl halide = **rate triples**
b. halving the concentration of the base = **rate is halved**
c. changing the solvent from CH_3OH to DMSO = **rate increases** (Polar aprotic solvent is better for E2.)
d. changing the leaving group from I^- to Br^- = **rate decreases** (I^- is a better leaving group.)
e. changing the base from ^-OH to H_2O = **rate decreases** (weaker base)
f. changing the alkyl halide from CH_3CH_2Br to $(CH_3)_2CHBr$ = **rate increases** (More substituted halide reacts faster.)

8.12 According to the Zaitsev rule, the major product in a β-elimination reaction has the *more* substituted double bond.

a. Br, α — loss of H and Br → trisubstituted **major product** + (+ stereoisomer) disubstituted **minor product**

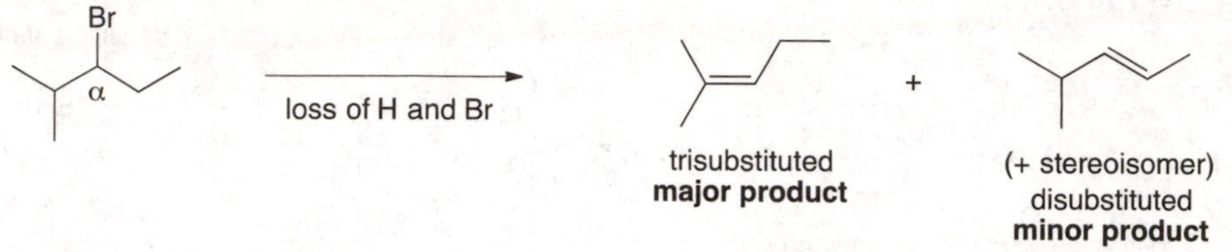

b. Br, α — loss of H and Br → trisubstituted **minor product** + tetrasubstituted **major product** + disubstituted **minor product**

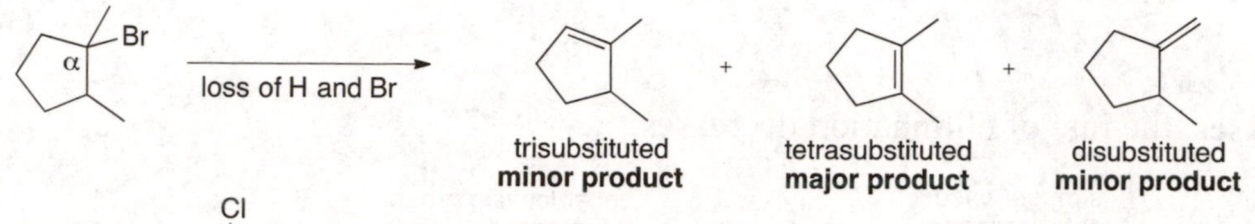

c. Cl, α — loss of H and Cl → monosubstituted **minor product** + (+ stereoisomer) disubstituted **major product**

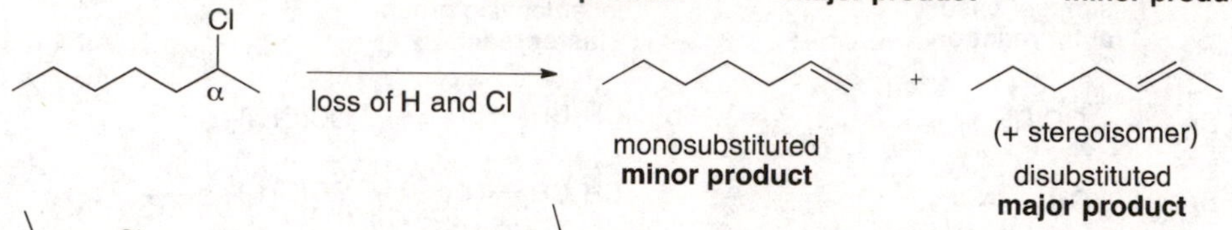

d. α Cl — loss of H and Cl → trisubstituted **ONLY product**

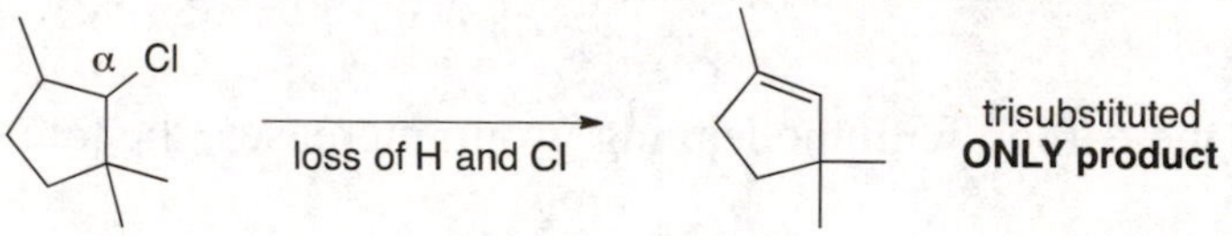

8.13 An E1 mechanism has two steps:
[1] The leaving group comes off, creating a carbocation.
[2] A base pulls off a proton from a β carbon, and a π bond forms.

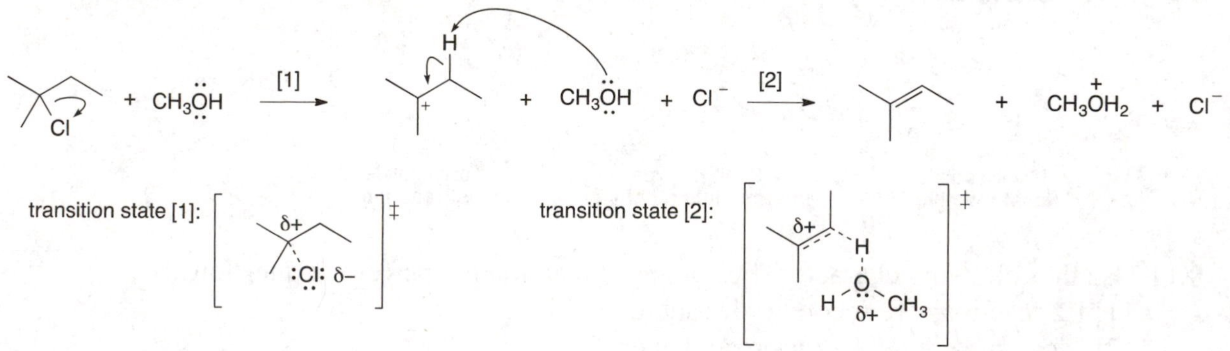

Cl + $CH_3\ddot{O}H$ —[1]→ H, + + $CH_3\ddot{O}H$ + Cl^- —[2]→ + $CH_3\overset{+}{O}H_2$ + Cl^-

transition state [1]: $[\ \delta+ \ \ :\ddot{Cl}:\ \delta-\ ]^{\ddagger}$

transition state [2]: $[\ \delta+ \ \ H \ \ \ddot{O} \ \ H \ \ CH_3 \ \ \delta+\ ]^{\ddagger}$

8.14 According to the Zaitsev rule, the major product in a β-elimination reaction has the *more* substituted double bond.

a. β2 Cl α β1 β1 —E1→ (+ stereoisomer) trisubstituted **major product** + disubstituted

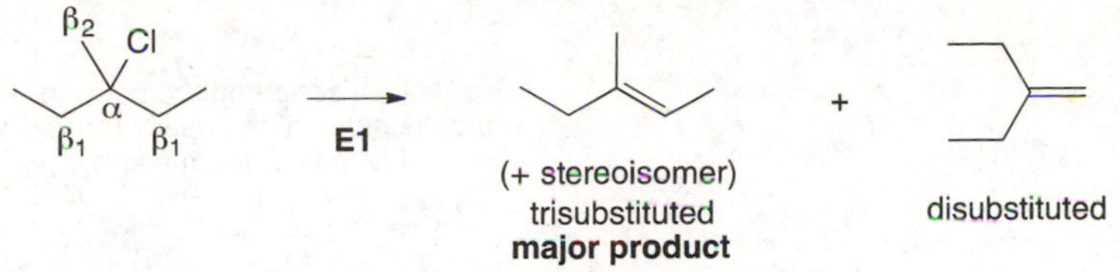

b. β2 I α β3 β1 —E1→ tetrasubstituted **major product** + disubstituted + trisubstituted

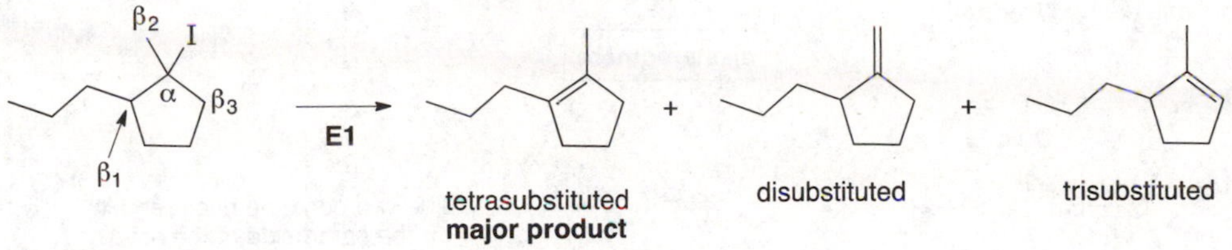

8.15 Use the following characteristics of an **E1 reaction** to answer the questions:
[1] E1 reactions are first order and two steps.
[2] More substituted halides react faster.
[3] Weaker bases are preferred.
[4] Reactions with better leaving groups are faster.
[5] Reactions in polar protic solvents are faster.

Rate equation: rate = *k*[RX]. The base doesn't affect rate.
a. doubling the concentration of the alkyl halide = **rate doubles**
b. doubling the concentration of the base = **no change** (The base is not in the rate equation.)
c. changing the alkyl halide from $(CH_3)_3CBr$ to $CH_3CH_2CH_2Br$ = **rate decreases** (More substituted halides react faster.)
d. changing the leaving group from Cl^- to Br^- = **rate increases** (better leaving group)
e. changing the solvent from DMSO to CH_3OH = **rate increases** (Polar protic solvent favors E1.)

8.16 Both S_N1 and E1 reactions occur by forming a carbocation. To draw the products:
[1] **For the S_N1 reaction,** substitute the nucleophile for the leaving group.
[2] **For the E1 reaction,** remove a proton from a β carbon and create a new π bond.

a. Br (leaving group) + H_2O (nucleophile and base) → OH (**S_N1 product**) + + (**E1 products**)

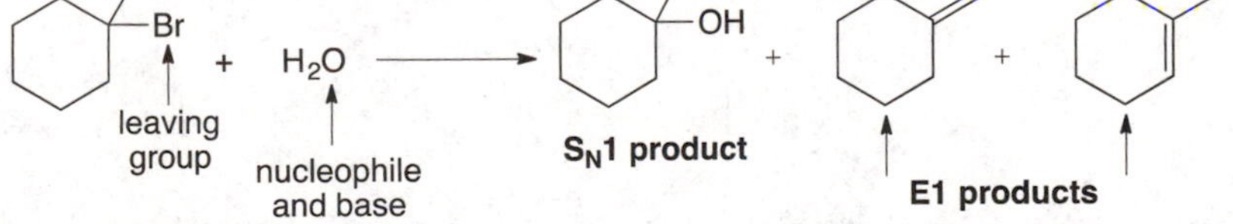

b. Cl (leaving group) + OH (nucleophile and base) → O (**S_N1 product**) + + (**E1 products**)

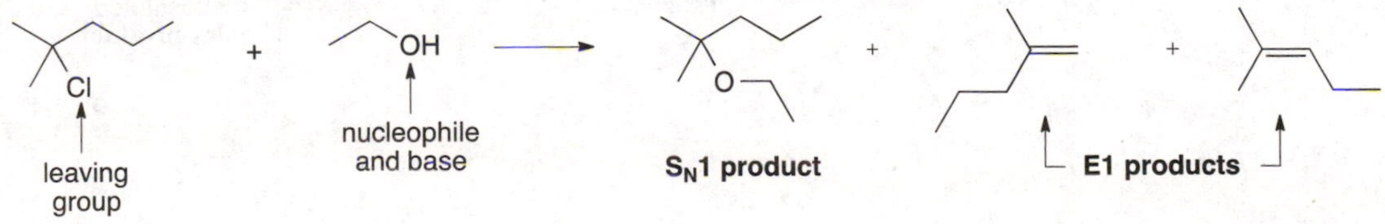

8.17 The E2 elimination reactions will occur in the anti periplanar orientation as drawn. To draw the product of elimination, maintain the orientation of the remaining groups around the C=C.

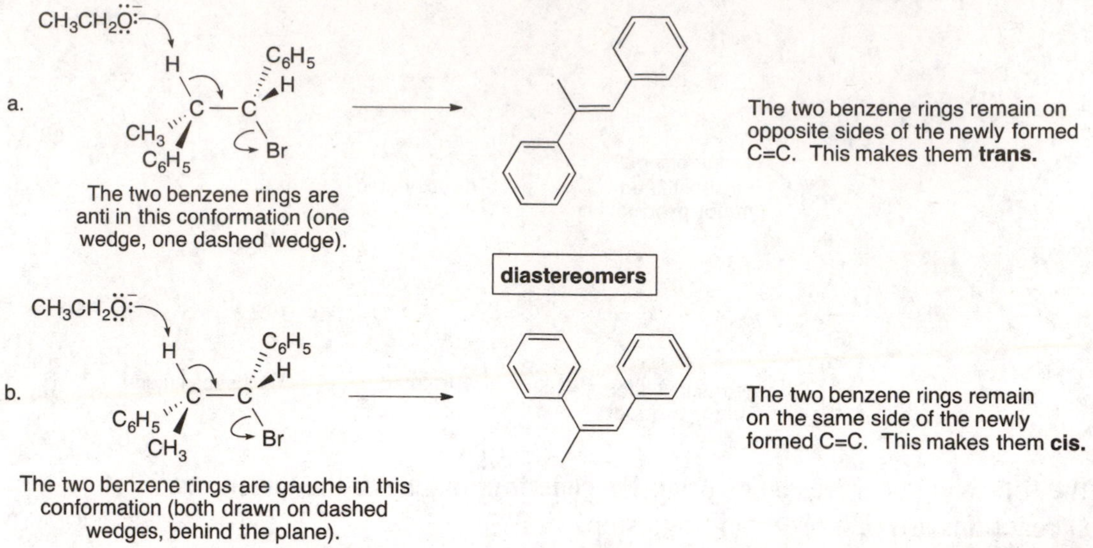

a. $CH_3CH_2\ddot{O}:^-$ H C–C C_6H_5 H CH_3 C_6H_5 Br

The two benzene rings are anti in this conformation (one wedge, one dashed wedge).

The two benzene rings remain on opposite sides of the newly formed C=C. This makes them **trans**.

diastereomers

b. $CH_3CH_2\ddot{O}:^-$ H C–C C_6H_5 H C_6H_5 CH_3 Br

The two benzene rings are gauche in this conformation (both drawn on dashed wedges, behind the plane).

The two benzene rings remain on the same side of the newly formed C=C. This makes them **cis**.

8.18 Note: The Zaitsev products predominate in E2 elimination *except* when substituents on a cyclohexane ring prevent a **trans diaxial** arrangement of H and X.

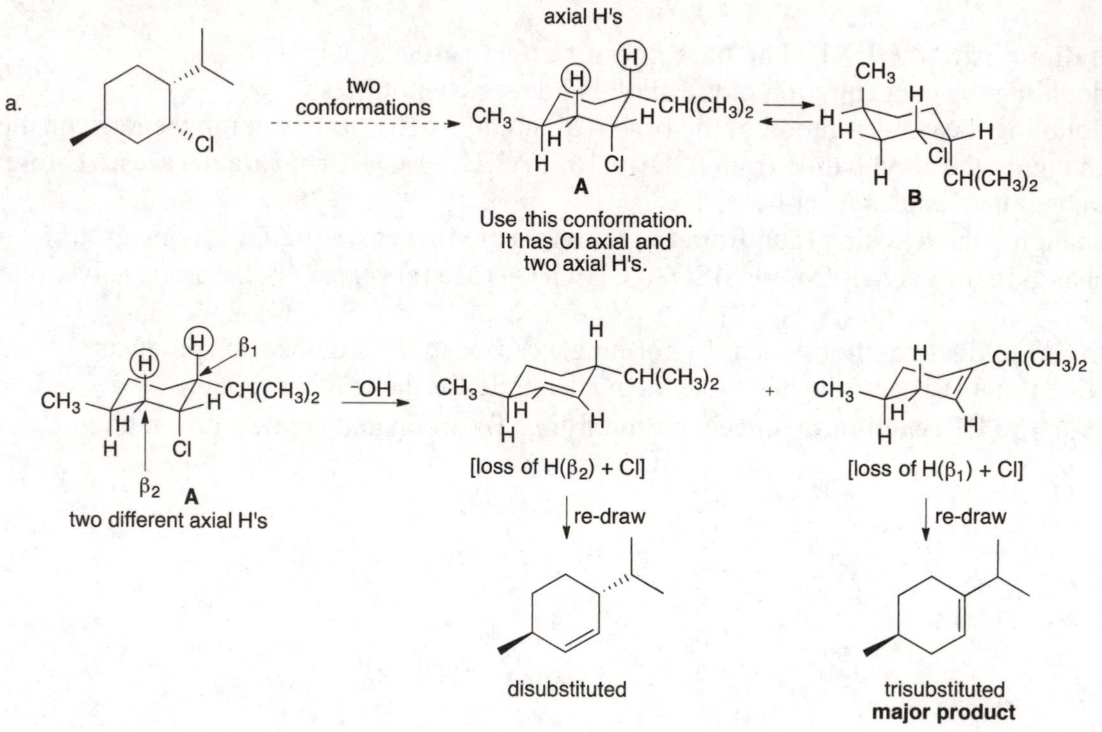

a. Cl

two conformations

axial H's

CH_3 H H H H Cl $CH(CH_3)_2$

A

Use this conformation. It has Cl axial and two axial H's.

CH_3 H H H H Cl H $CH(CH_3)_2$

B

CH_3 H H H H Cl $CH(CH_3)_2$ β_1 β_2 **A**

two different axial H's

^-OH

CH_3 H H H H $CH(CH_3)_2$

[loss of H(β_2) + Cl]

re-draw

disubstituted

+

CH_3 H H H H $CH(CH_3)_2$

[loss of H(β_1) + Cl]

re-draw

trisubstituted

major product

b. two conformations → A ⇌ B

Use this conformation.
It has Cl axial and
one axial H.

B: only one axial H on a β carbon

^-OH →

[loss of H(β_1) + Cl]

=

disubstituted
only product

8.19 Draw the chair conformations of *cis*-1-chloro-2-methylcyclohexane and its trans isomer. For E2 elimination reactions to occur, **there must be a H and X trans diaxial to each other.**

Two conformations of the cis isomer:

Two conformations of the trans isomer:

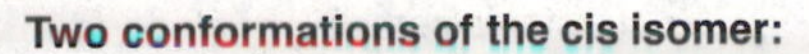

A
reacting conformation (axial Cl)

This reacting conformation has only one group axial, making it more stable and present in a higher concentration than **B.** This makes a **faster elimination reaction with the cis isomer.**

B
reacting conformation (axial Cl)

This conformation is less stable than **A,** because both CH_3 and Cl are axial. **This slows the rate of elimination from the trans isomer.**

8.20 E2 reactions are favored by strong negatively charged bases and occur with 1°, 2°, and 3° halides, with 3° being the most reactive.
E1 reactions are favored by weaker neutral bases and do not occur with 1° halides because they would have to form highly unstable carbocations.

a. + $^-OCH_3$ → strong negatively charged base **E2**

b. + H_2O → weak neutral base **E1**

c. + CH_3OH → weak neutral base **E1**

d. + $^-OC(CH_3)_3$ → strong negatively charged base **E2**

8.21 Draw the alkynes that result from removal of two equivalents of HX.

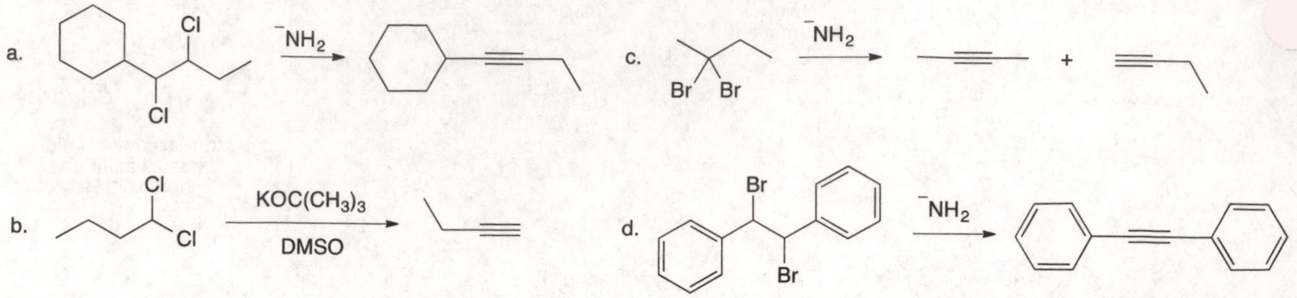

a. $^{-}NH_2$

b. $KOC(CH_3)_3$, DMSO

c. $^{-}NH_2$

d. $^{-}NH_2$

8.22

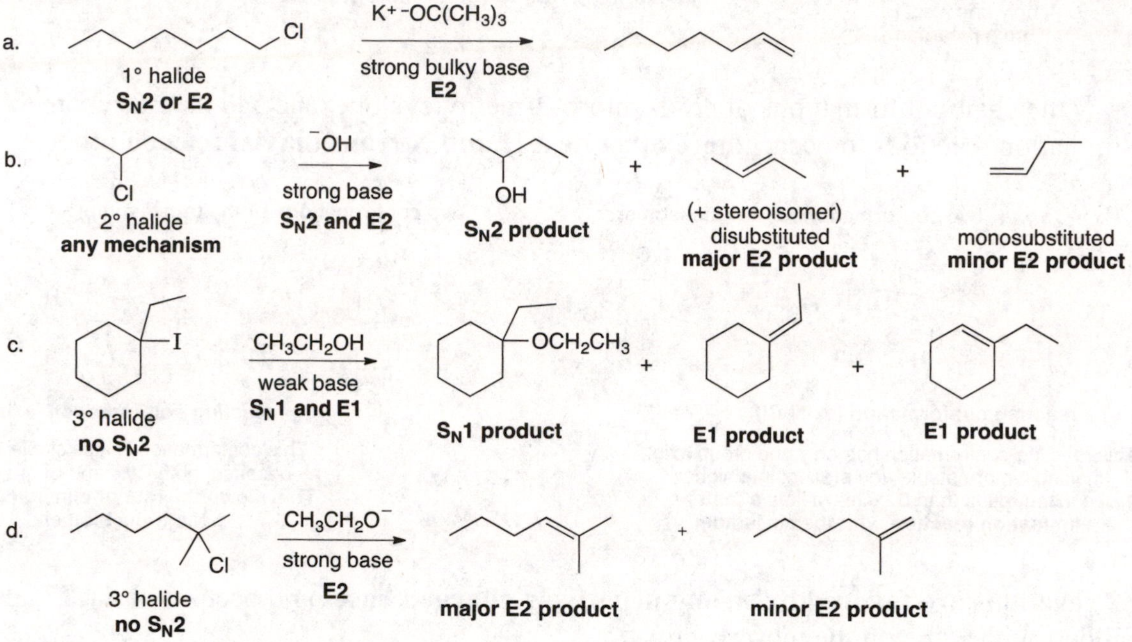

a. 1° halide, S_N2 or E2 — $K^{+}\ {}^{-}OC(CH_3)_3$, strong bulky base, E2

b. 2° halide, any mechanism — ^{-}OH, strong base, S_N2 and E2 → S_N2 product + (+ stereoisomer) disubstituted major E2 product + monosubstituted minor E2 product

c. 3° halide, no S_N2 — CH_3CH_2OH, weak base, S_N1 and E1 → OCH_2CH_3 S_N1 product + E1 product + E1 product

d. 3° halide, no S_N2 — $CH_3CH_2O^{-}$, strong base, E2 → major E2 product + minor E2 product

8.23

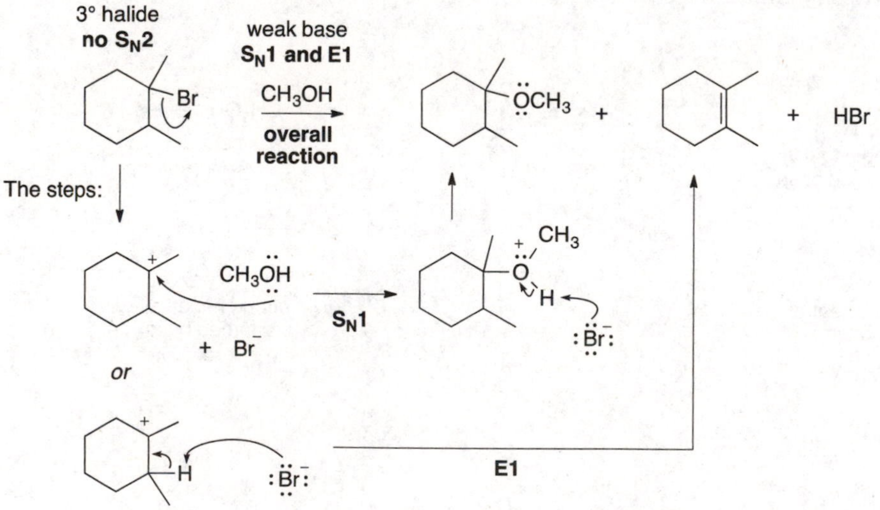

3° halide, no S_N2 — weak base, S_N1 and E1, CH_3OH, overall reaction → OCH_3 + alkene + HBr

The steps:

CH_3OH + Br^{-} → S_N1 → CH_3 ... H ... :Br:

or

:Br: → E1

8.24 More substituted alkenes are more stable. Trans alkenes are generally more stable than cis alkenes. Order of stability:

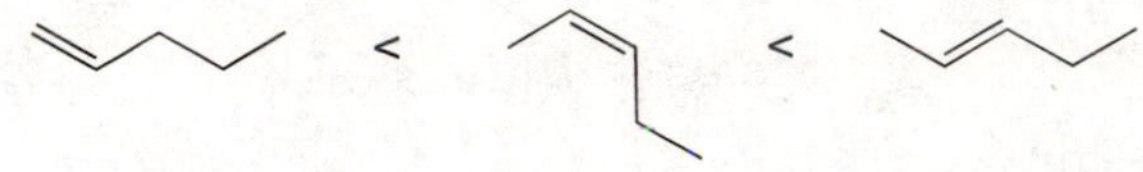

B < C < A

B least stable; A most stable

8.25 The trans isomer (**E**) reacts faster. During elimination, Br must be axial to give trans diaxial elimination. In the trans isomer, the more stable conformation has the bulky *tert*-butyl group in the more roomy equatorial position and Br in the axial position. In the cis isomer (**D**), elimination can occur only when both the *tert*-butyl and Br groups are axial, a conformation that is not energetically favorable.

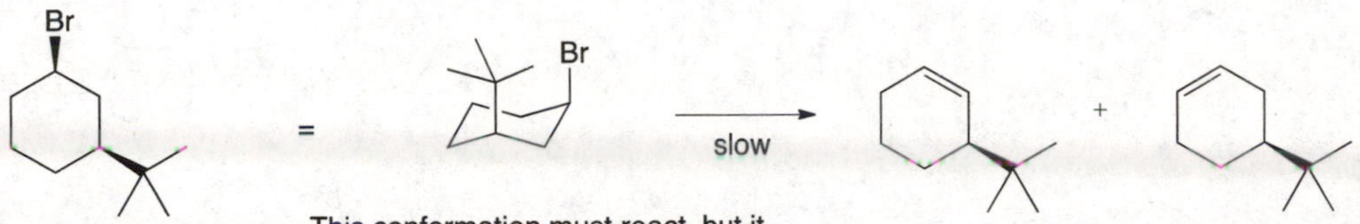

cis
D
cis-1-bromo-3-*tert*-butylcyclohexane

This conformation must react, but it contains two axial groups.

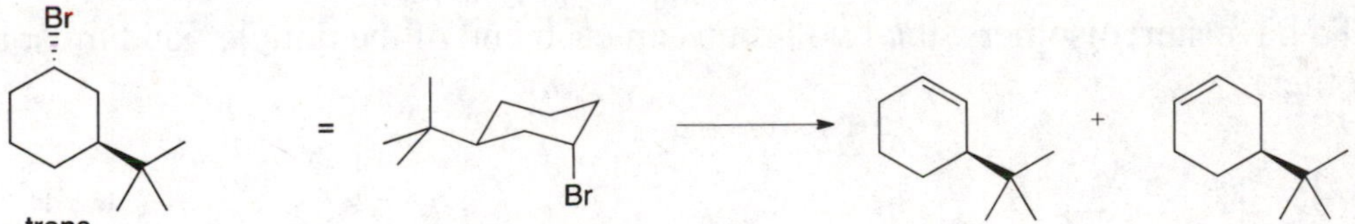

slow

trans
E
trans-1-bromo-3-*tert*-butylcyclohexane

preferred conformation

8.26 Translate each model to a structure and arrange H and Br to be anti periplanar.

a. rotate → E2 → $CH_3CH_2(CH_3)C{=}C(C_6H_5)H$

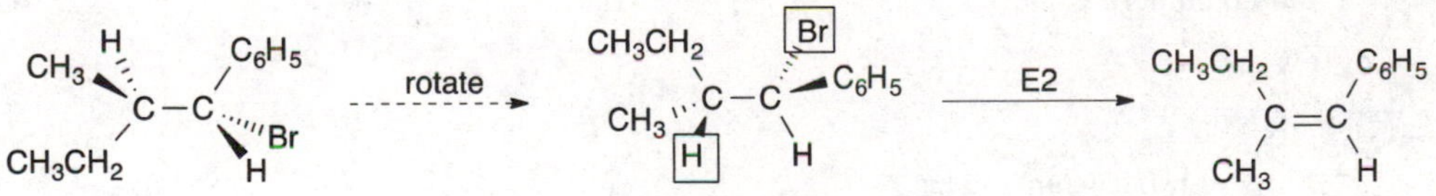

H and Br 180° away from each other

b. rotate → E2 → $CH_3CH_2(CH_3)C{=}C(H)C_6H_5$

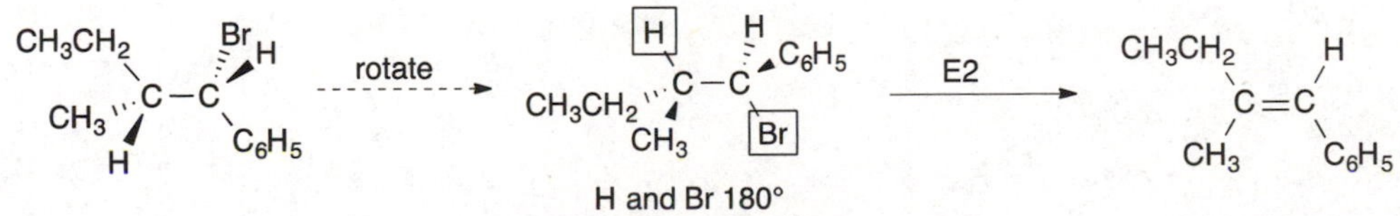

H and Br 180° away from each other

8.27

a.  (+ stereoisomer)

b. 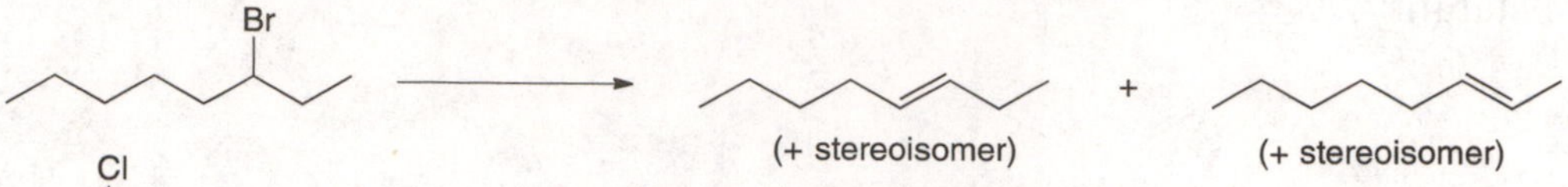 (+ stereoisomer) + (+ stereoisomer)

c. 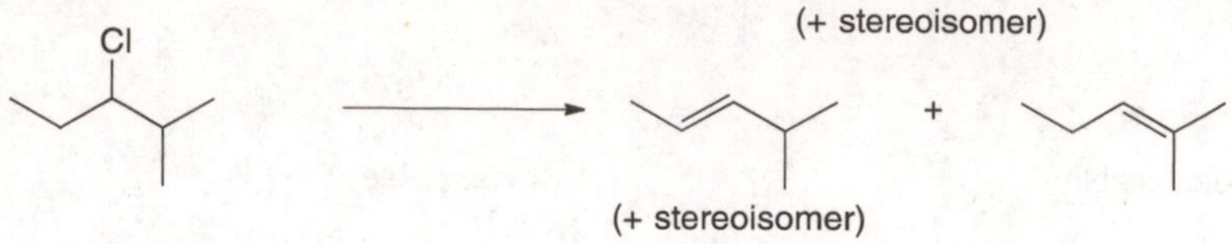 (+ stereoisomer) +

d.

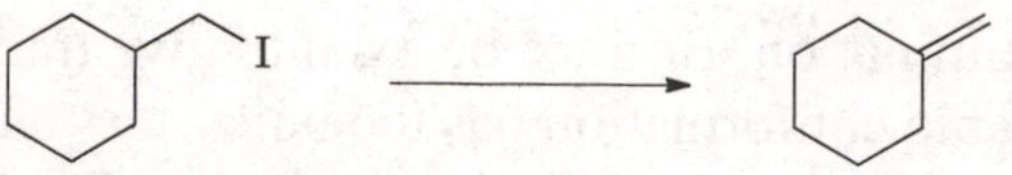

8.28 To give only one product in an elimination reaction, **the starting alkyl halide must have only one type of β carbon with H's.**

a.

b.

c. 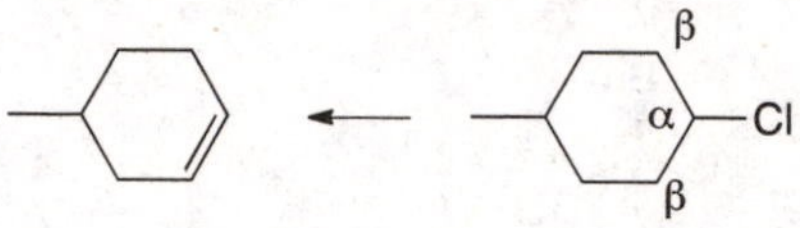 Two β carbons are identical.

d.  Two β carbons are identical.

8.29 To have stereoisomers, the two groups on each end of the double bond must be different from each other.

a. 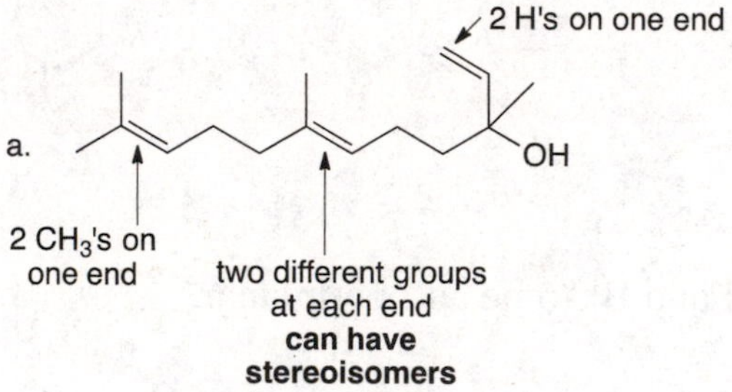 2 H's on one end; 2 CH_3's on one end; two different groups at each end **can have stereoisomers**

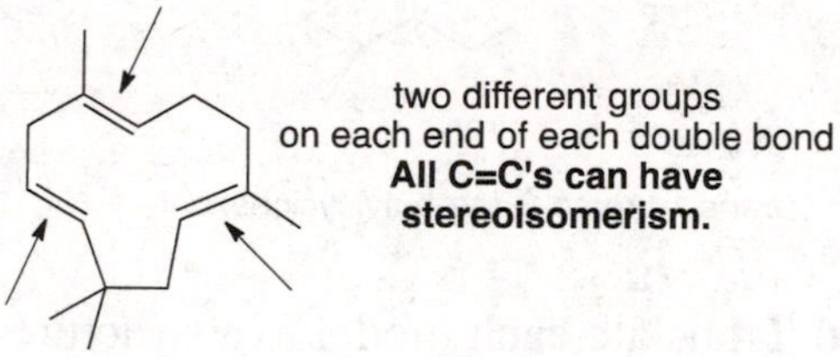

b. 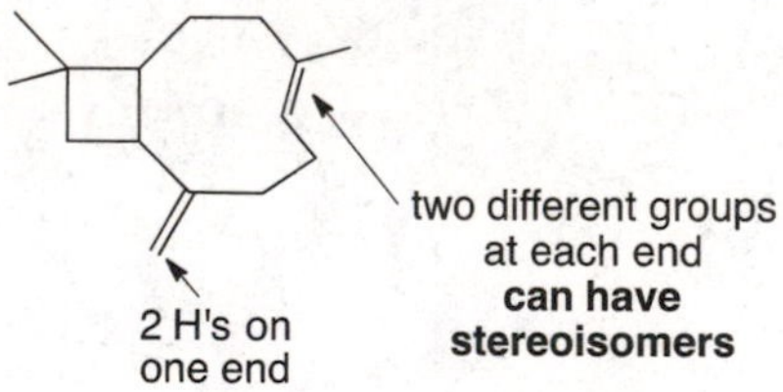 2 H's on one end; two different groups at each end **can have stereoisomers**

c. two different groups on each end of each double bond **All C=C's can have stereoisomerism.**

8.30 Use the definitions in Answer 8.5.

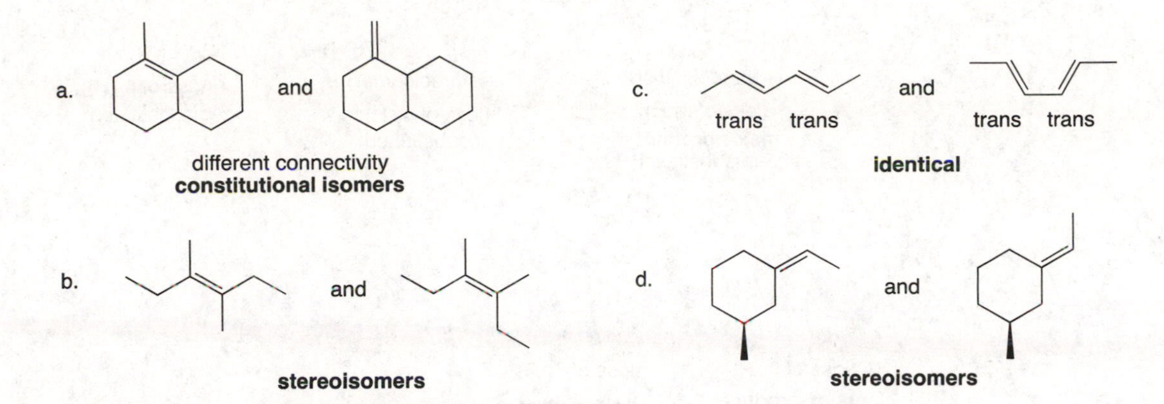

8.31 Use the rules from Answer 8.6 to rank the alkenes.

B	**A**	**C**
monosubstituted	disubstituted	trisubstituted
least stable	**intermediate stability**	**most stable**

8.32 A larger negative value for $\Delta H°$ means the reaction is more exothermic. Because both but-1-ene and *cis*-but-2-ene form the same product (butane), these data show that but-1-ene was higher in energy to begin with, **because more energy is released in the hydrogenation reaction.**

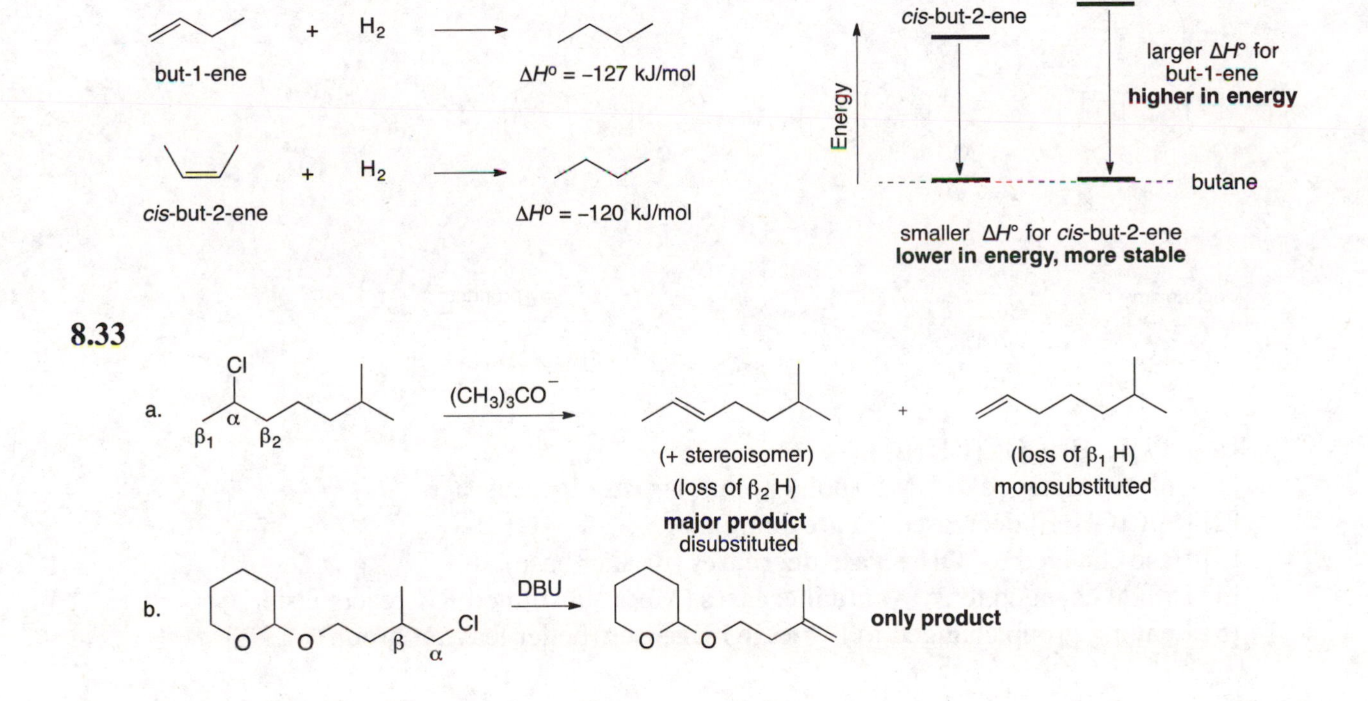

c. 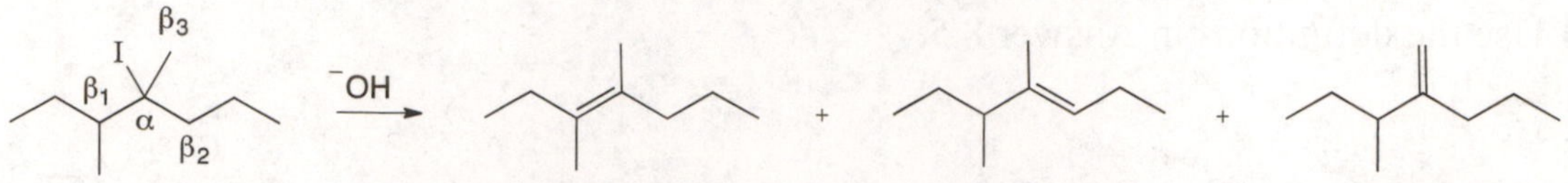 β_3 I β_1 α β_2 $\xrightarrow{^{-}OH}$ + +

(+ stereoisomer)
(loss of β_1 H)
major product
tetrasubstituted

(+ stereoisomer)
(loss of β_2 H)
trisubstituted

(loss of β_3 H)
disubstituted

d. 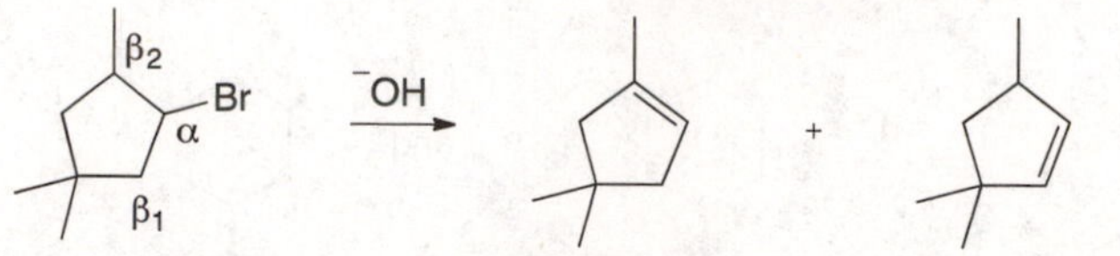 β_2 Br α β_1 $\xrightarrow{^{-}OH}$ +

(loss of β_2 H)
major product
trisubstituted

(loss of β_1 H)
disubstituted

8.34 To give only one alkene as the product of elimination, the alkyl halide must have either:
- only one β carbon with a hydrogen atom, or
- all identical β carbons, so the resulting elimination products are identical

a. 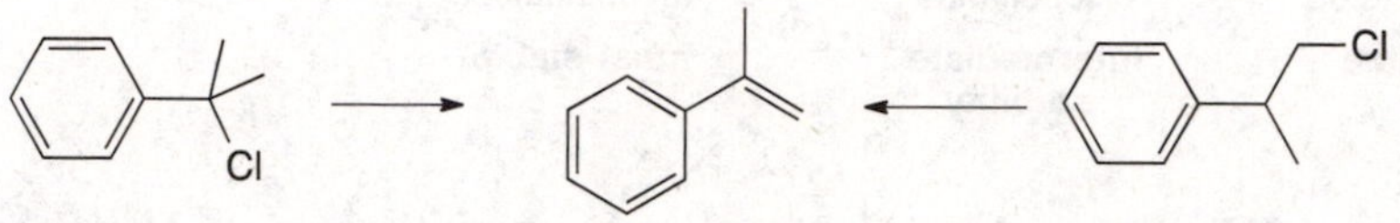 Cl → ← Cl

b. 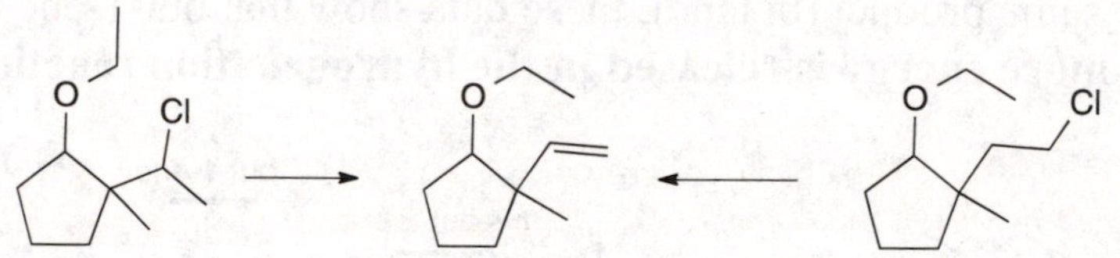 O Cl → O ← O Cl

c. 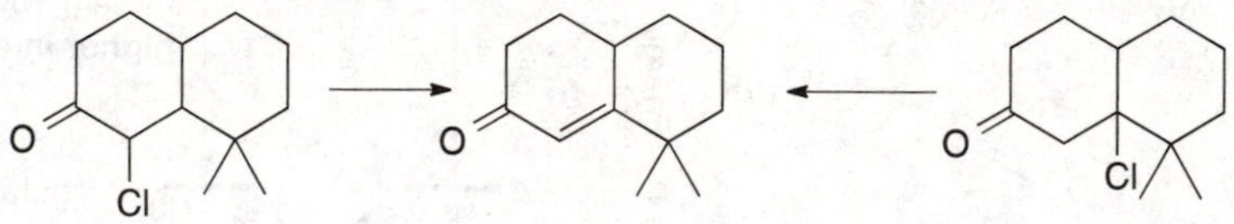 O Cl → O ← O Cl

8.35

a. Mechanism:

Br H $\xrightarrow[(CH_3)_3COH]{^{-}:\ddot{O}C(CH_3)_3}$ + $H\ddot{O}C(CH_3)_3$ + Br^- (by-products)

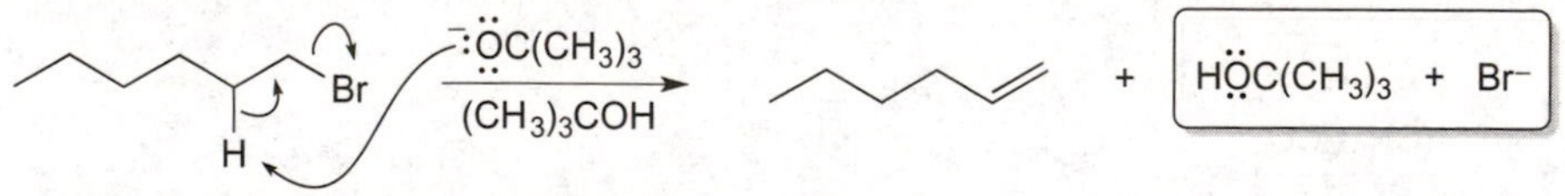

b. Rate = $k[R\text{–}Br][^{-}OC(CH_3)_3]$
[1] Solvent changed to DMF (polar aprotic) = **rate increases**
[2] $[^{-}OC(CH_3)_3]$ decreased = **rate decreases**
[3] Base changed to ^{-}OH = **rate decreases** (weaker base)
[4] Halide changed to 2° = **rate increases** (More substituted RX reacts faster.)
[5] Leaving group changed to I^- = **rate increases** (better leaving group)

8.36

$K^+ \ ^-OC(CH_3)_3$

Cl

1-chloro-1-methyl-cyclopropane

A + **B**

The dehydrohalogenation of an alkyl halide usually forms the more stable alkene. In this case, **A** is more stable than **B** even though **A** contains a disubstituted C=C whereas **B** contains a trisubstituted C=C. The double bond in **B** is part of a three-membered ring, and is less stable than **A** because of severe angle strain around both C's of the double bond.

8.37

a. 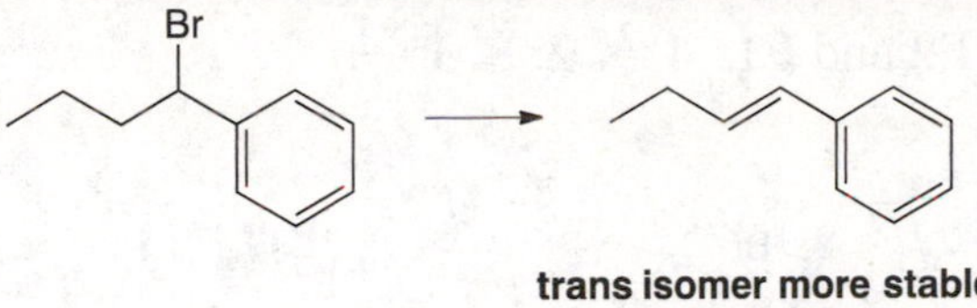 Br

trans isomer more stable
major product

b. 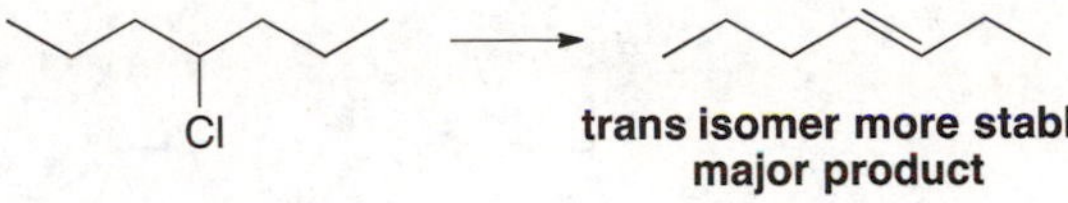 Cl

trans isomer more stable
major product

8.38

a. 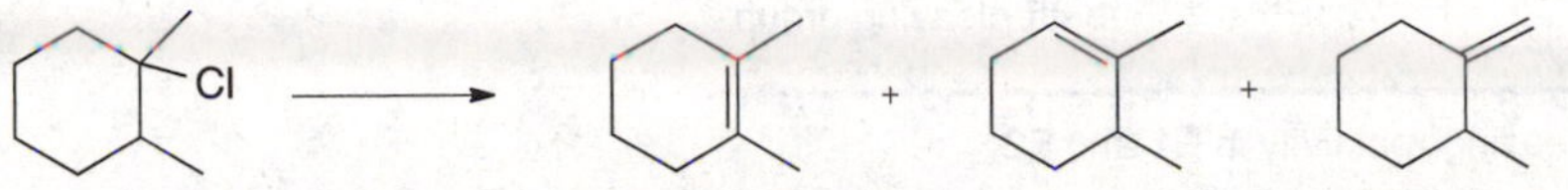 Cl

tetrasubstituted
major product

trisubstituted

disubstituted

b. 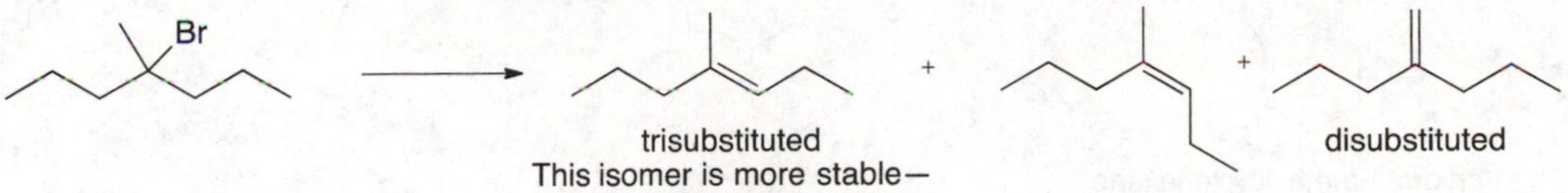 Br

trisubstituted
This isomer is more stable—
large groups farther away.
major product

trisubstituted

disubstituted

c. 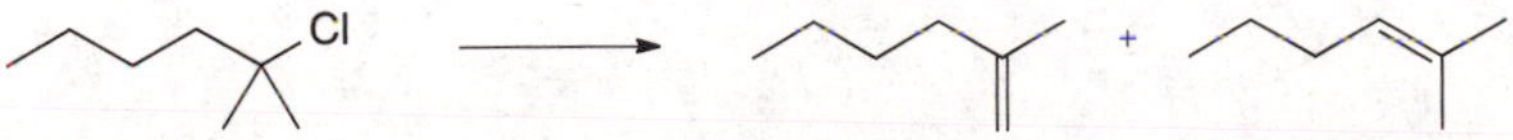 Cl

disubstituted

trisubstituted
major product

8.39 Use the rules from Answer 8.20.

a. 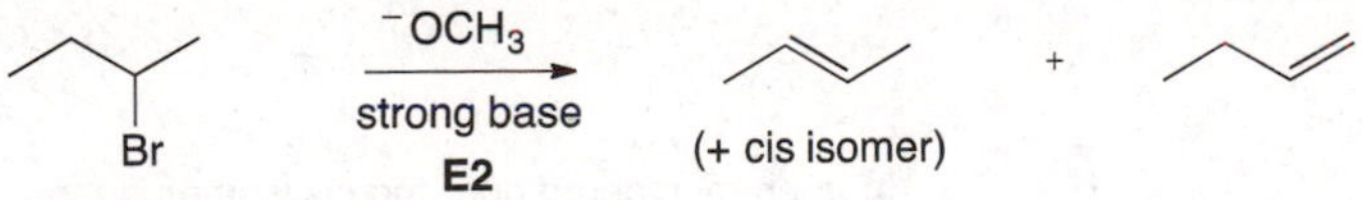 Br

2° halide

$^-OCH_3$
strong base
E2

(+ cis isomer)

b. 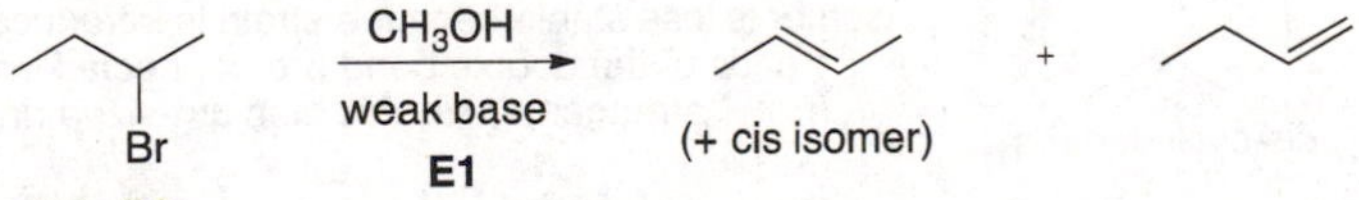 Br

2° halide

CH_3OH
weak base
E1

(+ cis isomer)

c. 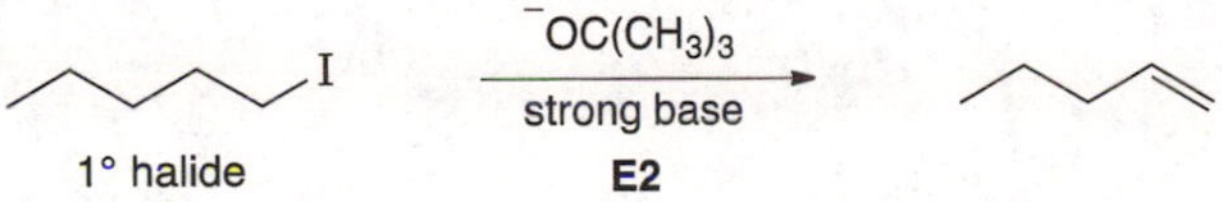 I

1° halide

$^-OC(CH_3)_3$
strong base
E2

d. Cl 3° halide $\xrightarrow[\text{weak base}]{H_2O}$ **E1** + +

e. Cl 2° halide $\xrightarrow[\text{strong base}]{^-OH}$ **E2**

f. 2° halide Cl $\xrightarrow[\text{strong base}]{^-OH}$ **E2** +

8.40 The order of reactivity is the same for both E2 and E1: 1° < 2° < 3°.

Cl Cl Br

2° halide — 3° halide — 3° halide + better leaving group

Increasing reactivity in E1 and E2

8.41

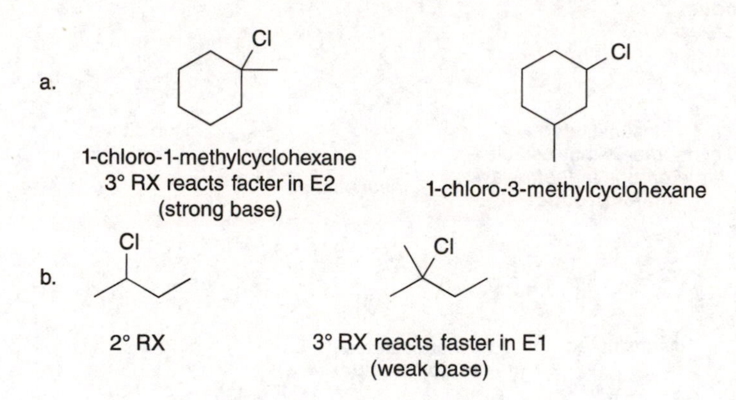

c. The mechanism is E2 because ^-OH is a strong base, so the reaction is faster in a polar aprotic solvent (DMSO).

8.42

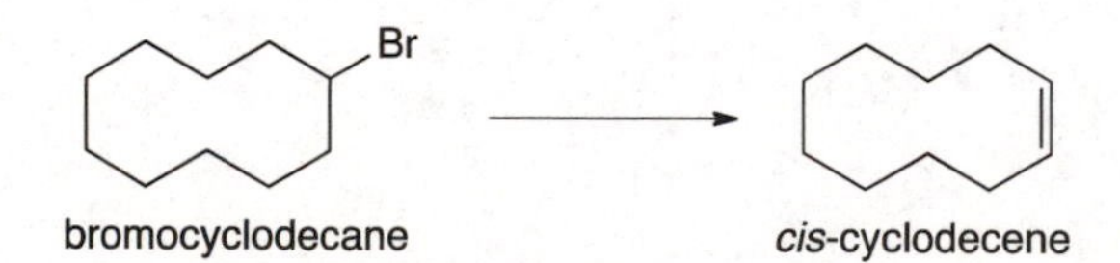

In a ten-membered ring, the cis isomer is more stable and, therefore, the preferred elimination product. The trans isomer is less stable because strain is introduced when two ends of the double bond are connected in a trans arrangement in this medium-sized ring.

8.43

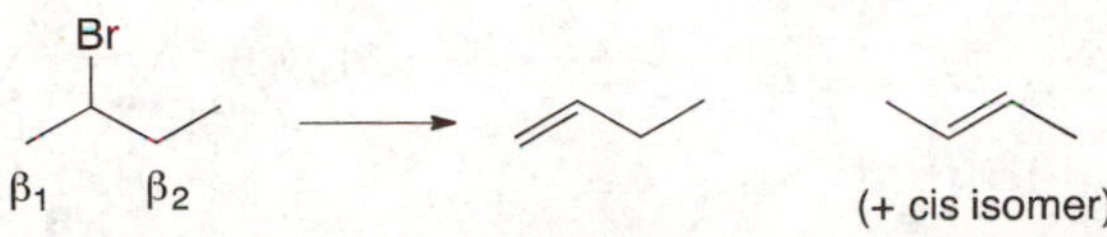

Br

β_1 β_2

(+ cis isomer)

	but-1-ene from loss of β_1 H	but-2-ene from loss of β_2 H
$Na^{+}\ ^{-}OCH_2CH_3$	19%	81%
$K^{+}\ ^{-}OC(CH_3)_3$	33%	67%

The H's on the CH_2 group of the β_2 carbon are more sterically hindered than the H's on the CH_3 group of the β_1 carbon. Because $K^{+}\ ^{-}OC(CH_3)_3$ is a much bulkier base than $Na^{+}\ ^{-}OCH_2CH_3$, it is easier to remove the more accessible H on β_1, giving $K^{+}\ ^{-}OC(CH_3)_3$ a higher percentage of but-1-ene than $Na^{+}\ ^{-}OCH_2CH_3$.

8.44 H and Br must be anti during the E2 elimination. Rotate if necessary to make them anti; then eliminate.

a. CH_3, C_6H_5, Br, H, CH_2CH_3, CH_3 —E2→ C_6H_5, CH_3, CH_3, CH_2CH_3

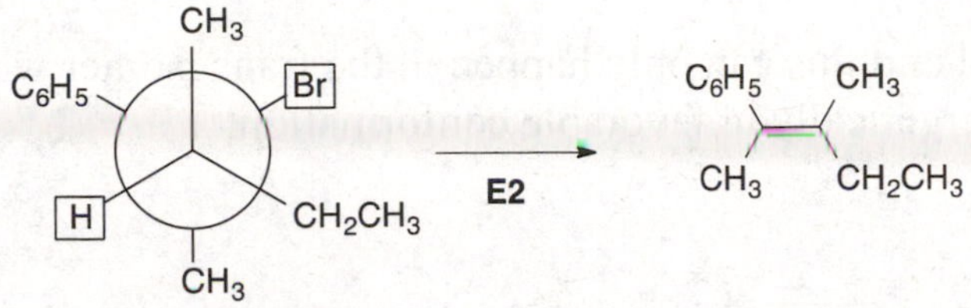

b. Br, H, CH_3, C_6H_5, CH_3, CH_2CH_3 —rotate→ Br, CH_3, CH_2CH_3, C_6H_5, CH_3, H —E2→ C_6H_5, CH_3, CH_3, CH_2CH_3

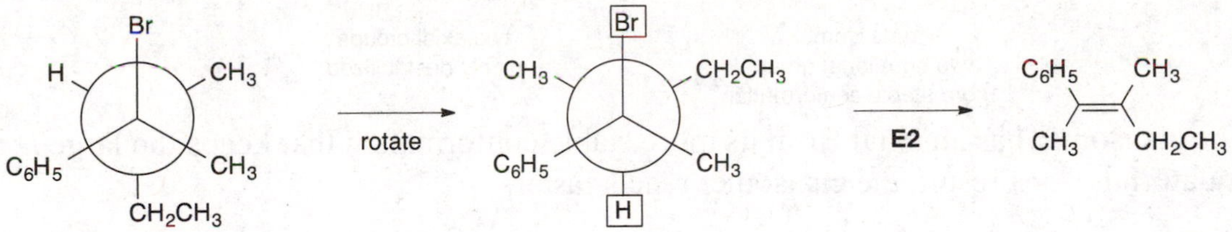

c. C_6H_5, CH_3, H, CH_3, Br, CH_2CH_3 —rotate→ C_6H_5, H, CH_2CH_3, CH_3, Br, CH_3 —E2→ C_6H_5, CH_2CH_3, CH_3, CH_3

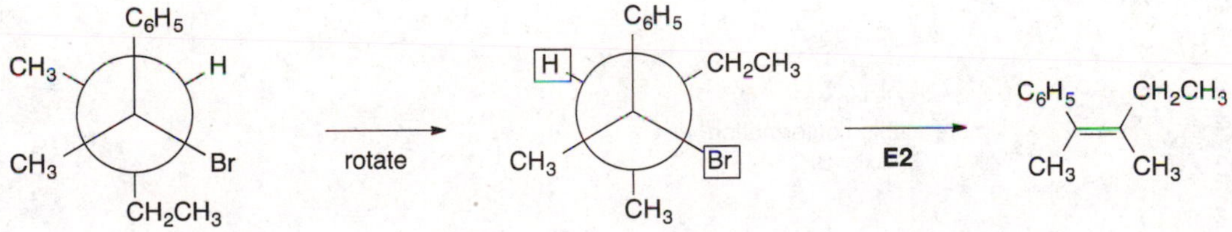

8.45

a. Cl —two chair conformations→ H, H, H, H, Cl, CH_3, H, $(CH_3)_2CH$ **A** ⇌ $(CH_3)_2CH$, H, **axial**, Cl, H, H, H, **B**, CH_3, H ← Choose this conformation. **axial Cl**

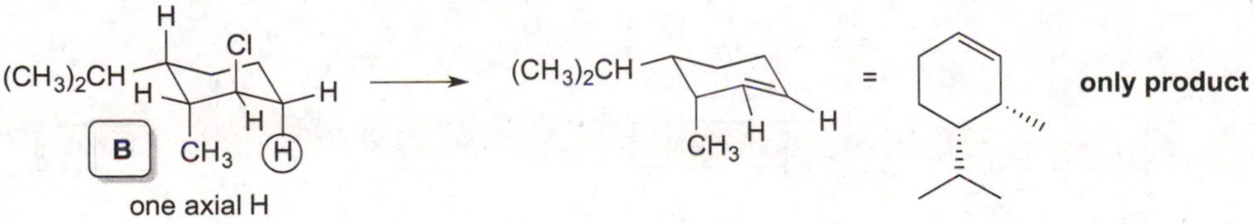

$(CH_3)_2CH$, H, Cl, H, H, H, **B**, CH_3, H, one axial H → $(CH_3)_2CH$, CH_3, H, H = **only product**

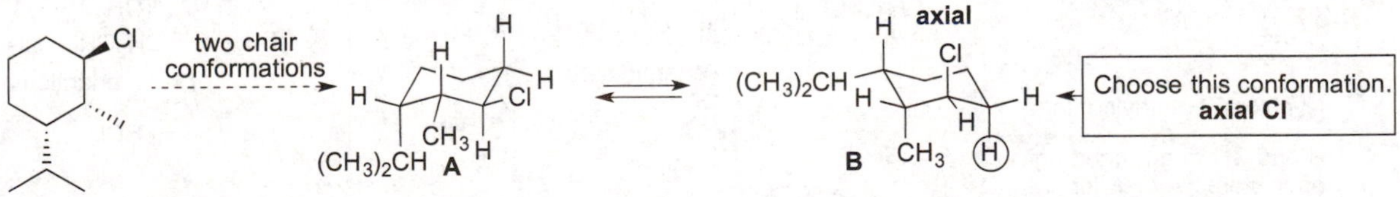

b.

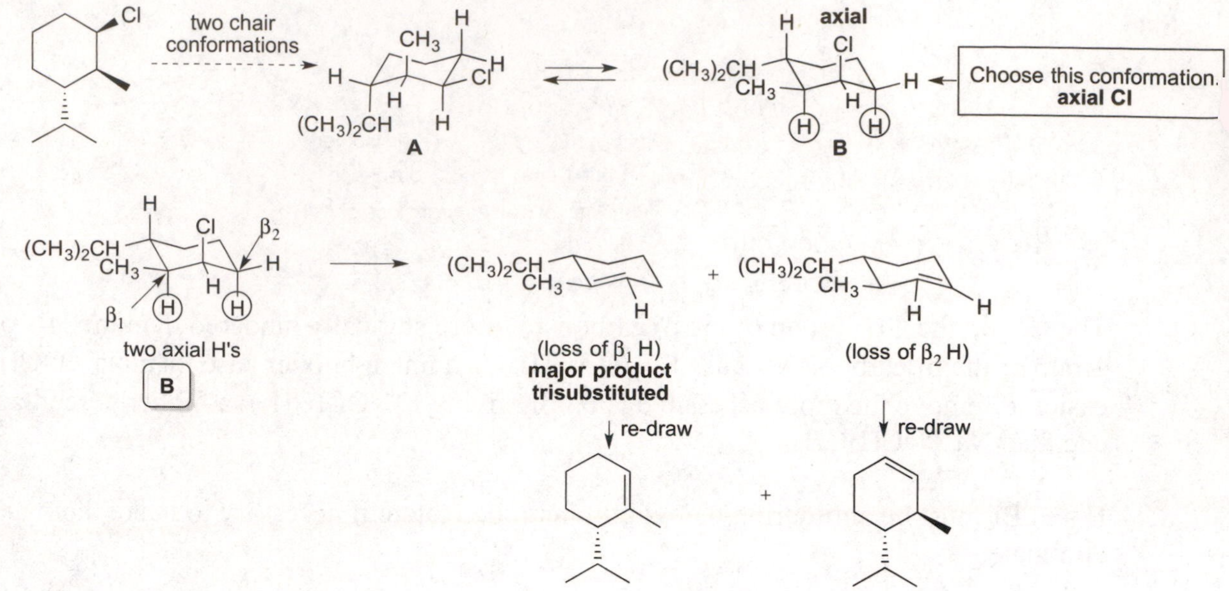
two chair conformations
A
axial
B
Choose this conformation. axial Cl
two axial H's
B
(loss of $β_1$ H)
major product
trisubstituted
(loss of $β_2$ H)
re-draw
re-draw

8.46 To react by E2, the Br must be axial and this can only happen in the trans isomer when the large *tert*-butyl group is also axial, an energetically unfavorable conformation.

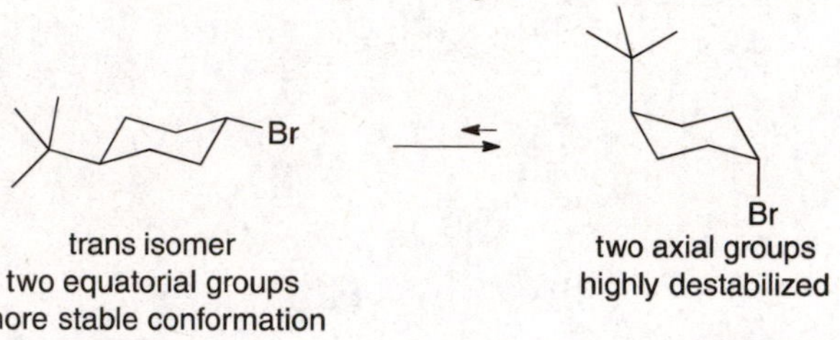
Br
Br
trans isomer
two equatorial groups
more stable conformation
two axial groups
highly destabilized

The cis isomer has an axial Br in its more stable conformation that keeps the large *tert*-butyl group equatorial. As a result, the cis isomer reacts faster.

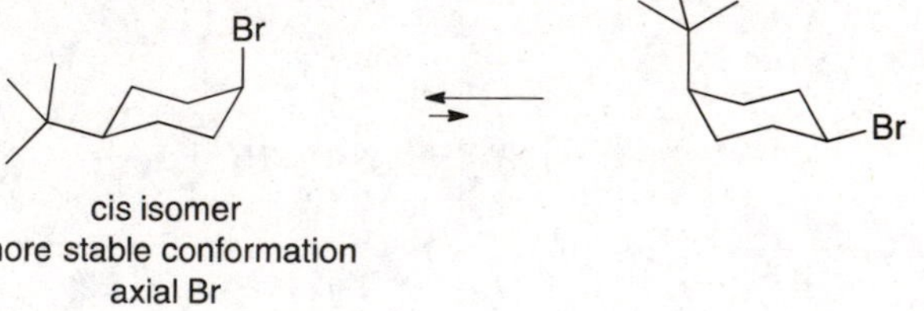
Br
Br
cis isomer
more stable conformation
axial Br

8.47

a.

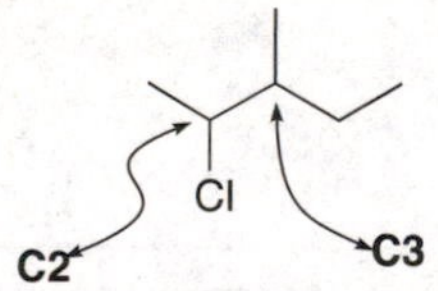
Cl
C2
C3

2-chloro-3-methylpentane

H and Cl are arranged anti in each stereoisomer, for anti periplanar elimination.

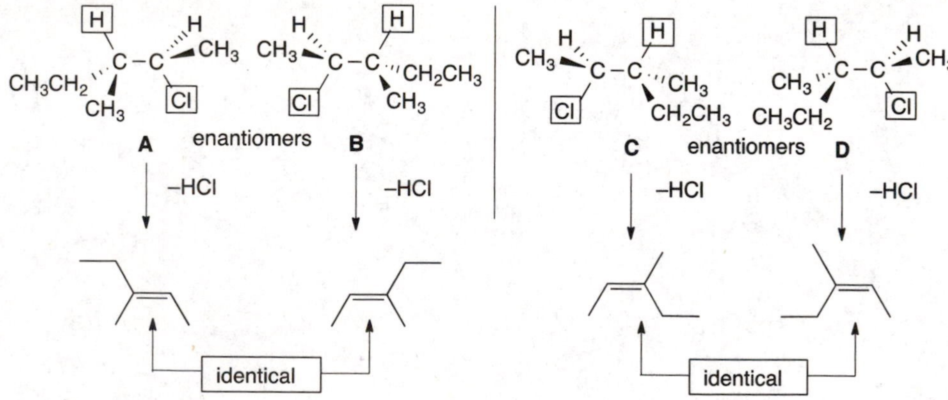
A
enantiomers
B
–HCl
–HCl
identical
C
enantiomers
D
–HCl
–HCl
identical

b. Two different alkenes are formed as products.

c. The products are diastereomers: Two enantiomers (**A** and **B**) give identical products. **A** and **B** are diastereomers of **C** and **D.** Each pair of enantiomers gives a single alkene. Thus, diastereomers give diastereomeric products.

8.48 To undergo an E2 reaction the Cl must be axial and the conformation with an axial Cl is highly unstable. Thus, no E2 reaction can occur because the needed conformation is too destabilized.

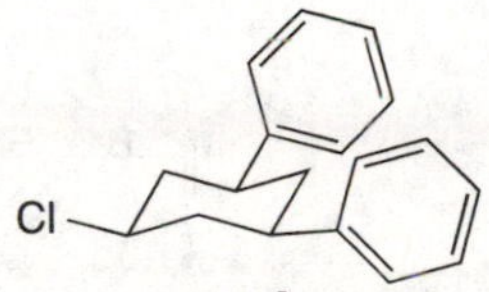

Cl

A

more stable conformation
All substituents are equatorial.

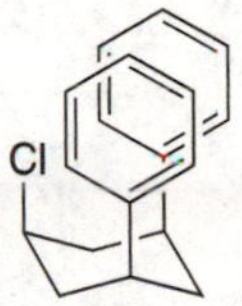

Cl

highly destabilized
All substituents are axial.

8.49 The alkyl chloride must have a Cl and H anti periplanar with all the substituents having the same arrangement that they have in the alkene product.

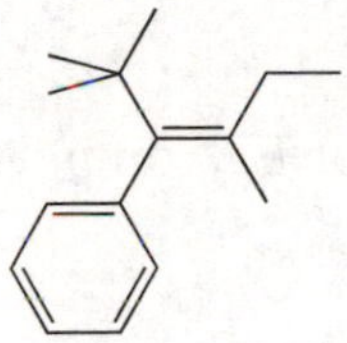

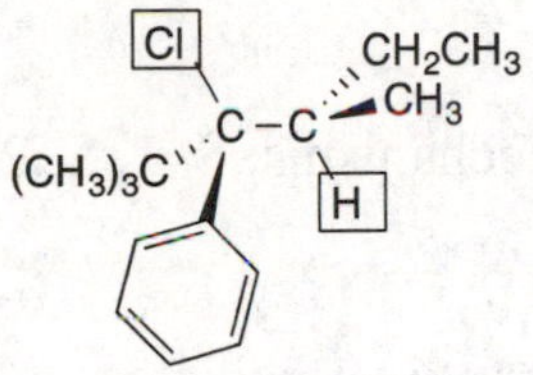

Cl
CH_2CH_3
CH_3
C–C
$(CH_3)_3C$
H

or

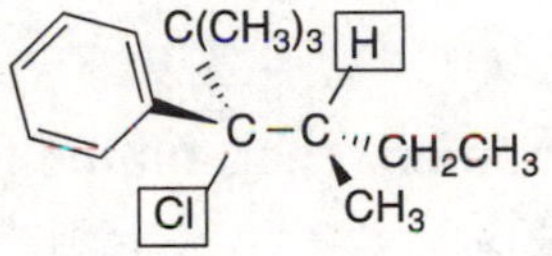

$C(CH_3)_3$ H
C–C
CH_2CH_3
CH_3
Cl

8.50

a. Cl, Cl — $NaNH_2$, excess →

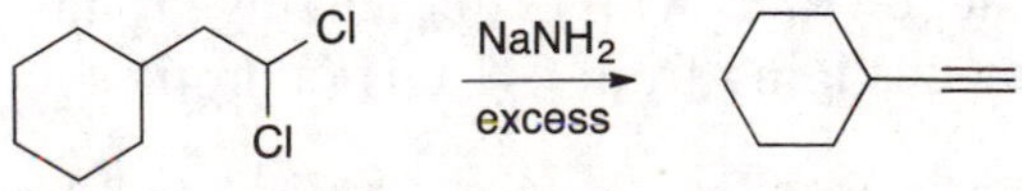

b. Br, Br — $NaNH_2$, excess →

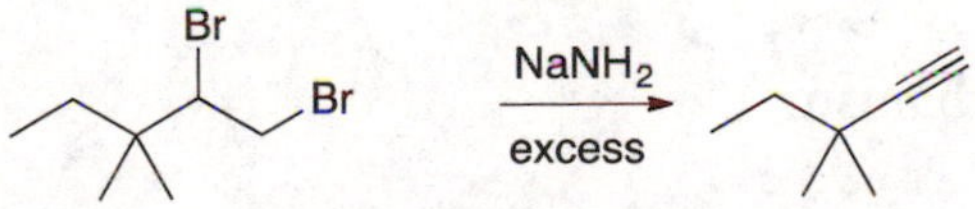

c. Cl Cl — $NaNH_2$, (excess) → +

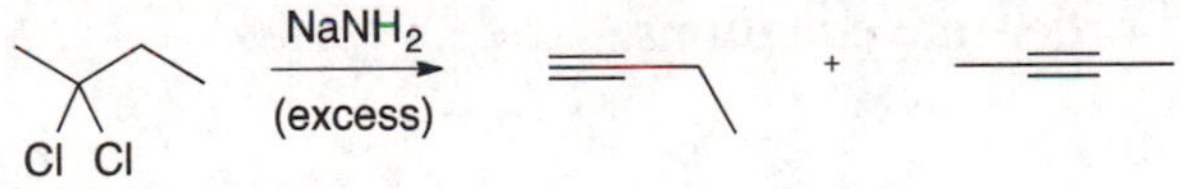

d. Cl, Cl — $NaNH_2$, excess →

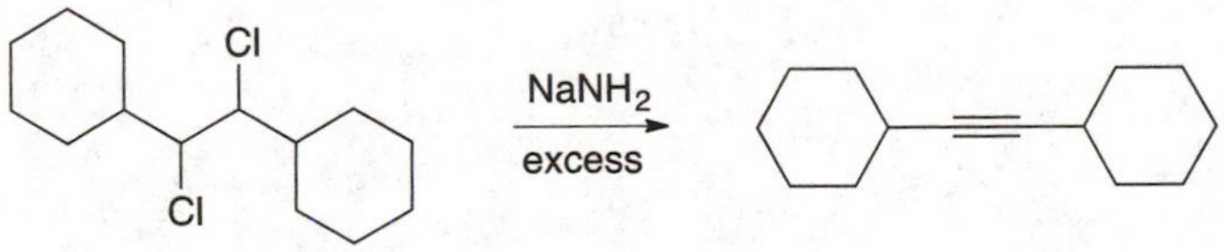

8.51

a. 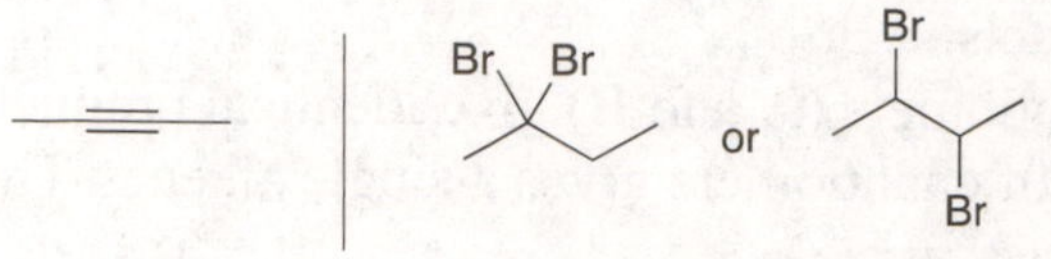 Br Br or Br Br

b. 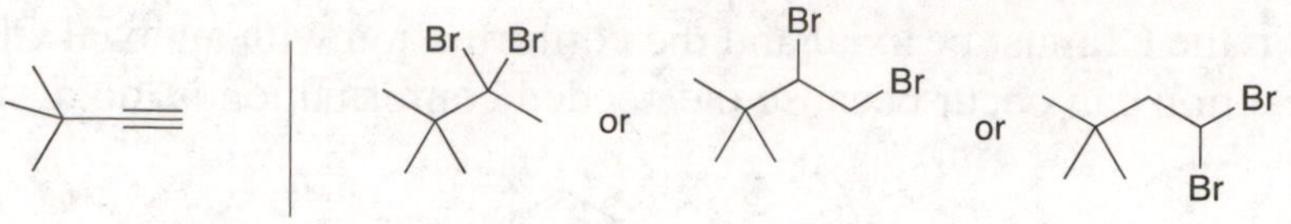 Br Br or Br Br or Br Br

c. 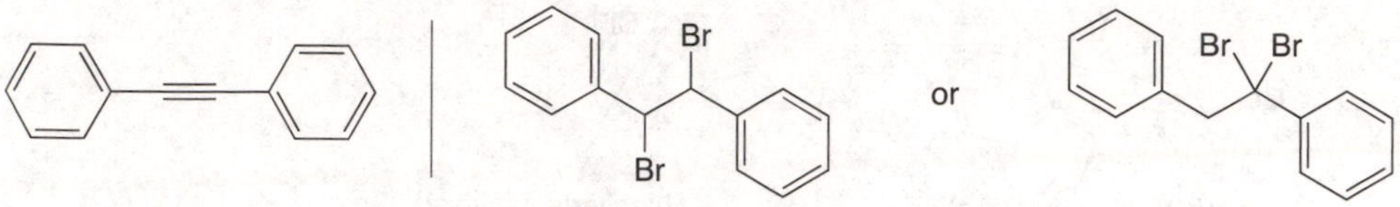 Br Br or Br Br

8.52

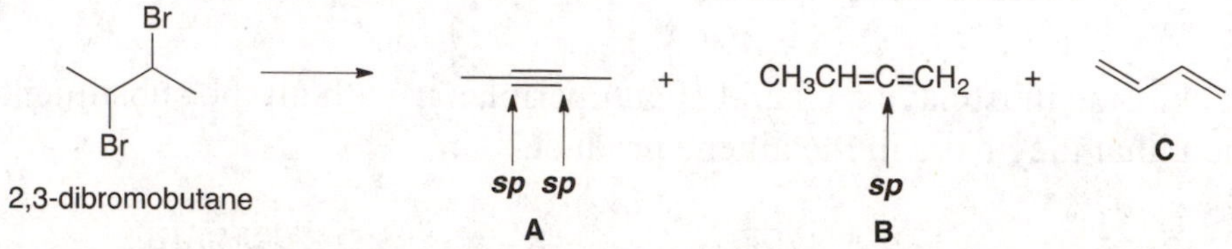

2,3-dibromobutane → A + $CH_3CH{=}C{=}CH_2$ + C

A: *sp sp*

B: *sp*

8.53 Use the "Summary chart on the four mechanisms: SN1, SN2, E1, or E2" on p. 8–2 to answer the questions.

a. Both SN1 and E1 involve carbocation intermediates.
b. Both SN1 and E1 have two steps.
c. SN1, SN2, E1, and E2 all have increased reaction rates with better leaving groups.
d. Both SN2 and E2 have increased rates when changing from CH_3OH (a protic solvent) to $(CH_3)_2SO$ (DMSO—an aprotic solvent).
e. In SN1 and E1 reactions, the rate depends on only the alkyl halide concentration.
f. Both SN2 and E2 are concerted reactions.
g. CH_3CH_2Br and NaOH react by an SN2 mechanism.
h. Racemization occurs in SN1 reactions.
i. In SN1, E1, and E2 mechanisms, 3° alkyl halides react faster than 1° or 2° halides.
j. E2 and SN2 reactions follow second-order rate equations.

8.54

a. 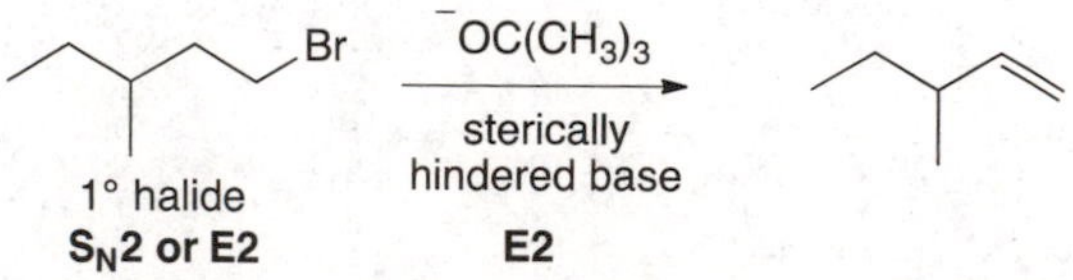 Br —$^-OC(CH_3)_3$, sterically hindered base→

1° halide
S_N2 or E2

E2

b. 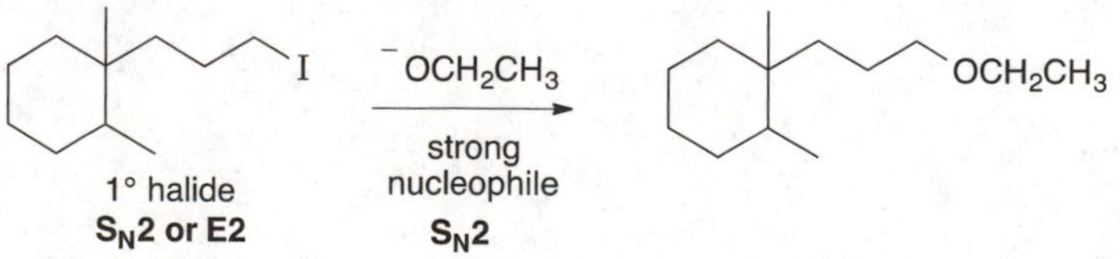 I —$^-OCH_2CH_3$, strong nucleophile→ OCH_2CH_3

1° halide
S_N2 or E2

S_N2

c. Cl Cl — dihalide — $^{-}NH_2$ (2 equiv) strong base →

d. Br — 1° halide **S_N2 or E2** — DBU sterically hindered base **E2** →

e. Br — 2° halide **S_N1, S_N2, E1, E2** — $^{-}OC(CH_3)_3$ sterically hindered base **E2** → **major product** +

f. Br — 3° halide **no S_N2** — CH_3CH_2OH weak base → OCH_2CH_3 **S_N1 product** + + **E1 products**

g. Br, Br — diahlide — 2 $NaNH_2$ →

h. Cl — 3° halide **no S_N2** — H_2O weak base → OH **S_N1 product** + (+ stereoisomer) **E1 product** + **E1 product**

8.55 [1] $NaOCOCH_3$ is a good nucleophile and weak base, and substitution is favored. [3] $KOC(CH_3)_3$ is a strong, bulky base that reacts by E2 elimination when there is a β hydrogen in the alkyl halide.

a. Cl
[1] $NaOCOCH_3$ → $OCOCH_3$
[2] $NaOCH_3$ → OCH_3
[3] $KOC(CH_3)_3$ →

b. Cl
[1] $NaOCOCH_3$ → $OCOCH_3$
[2] $NaOCH_3$ → OCH_3 (S_N2) + (E2)
[3] $KOC(CH_3)_3$ →

c. [1] $NaOCOCH_3$ → $OCOCH_3$

[2] $NaOCH_3$ →

[3] $KOC(CH_3)_3$ →

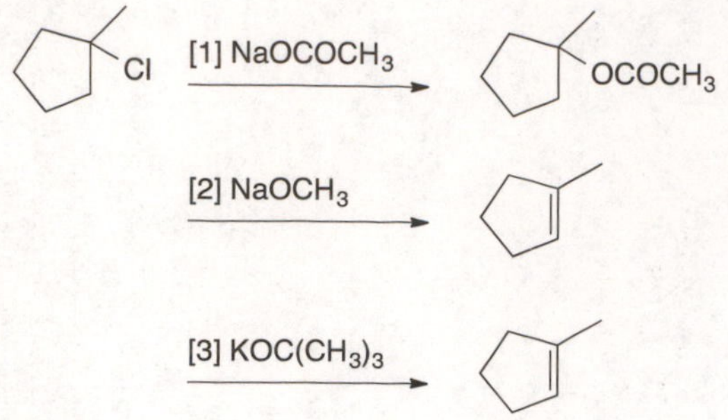

8.56

a. Cl H; ^{-}OH, strong base, **S_N2 and E2** → H OH + + +

2° halide
S_N1, S_N2, E1, E2

S_N2 product
inversion at
stereogenic center

major E2 product

minor E2 product

minor E2 product

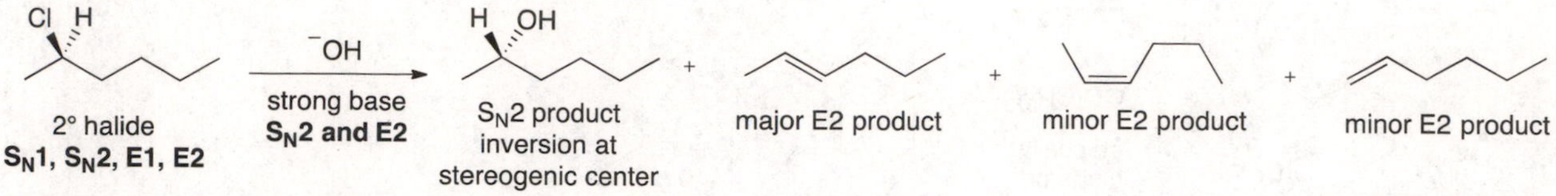

b. Cl H; H_2O, weak base, **S_N1 and E1** → HO H + H OH +

2° halide
S_N1, S_N2, E1, E2

S_N1 products

+ +

major E1 product

minor E1 product

minor E1 product

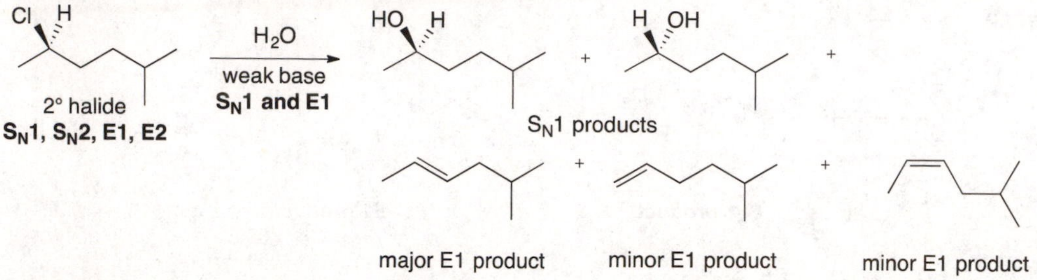

c. Cl, C_6H_5; CH_3OH, weak base, **S_N1 and E1** → CH_3O, C_6H_5 + C_6H_5, OCH_3 + C_6H_5

3° halide
no S_N2

S_N1 products

E1 product

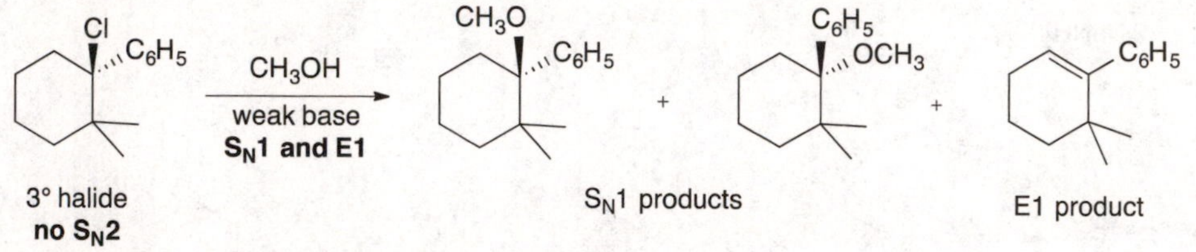

d. Br, D; KOH, strong base, **S_N2 and E2** → OH, D +

(trans diaxial elimination of D, Br)

2° halide
S_N1, S_N2, E1, E2

S_N2 product
inversion at
stereogenic center

E2 product

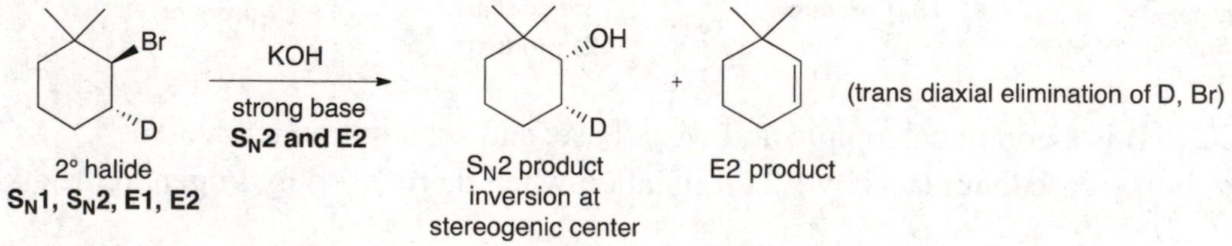

8.57

a. Cl, H; CH_3OH, weak base, **S_N1 and E1** → OCH_3, H + OCH_3, H + H + H

S_N1

S_N1

E1

E1

+ H

E1

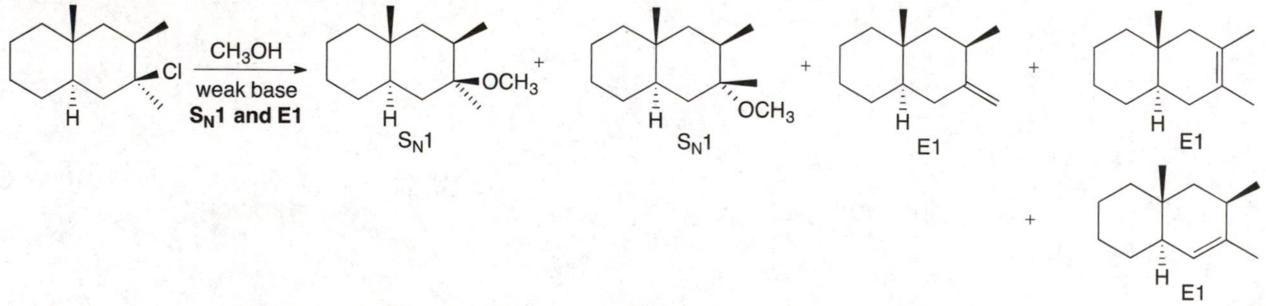

b. Cl, H; KOH, strong base, **E2** → H + H + H

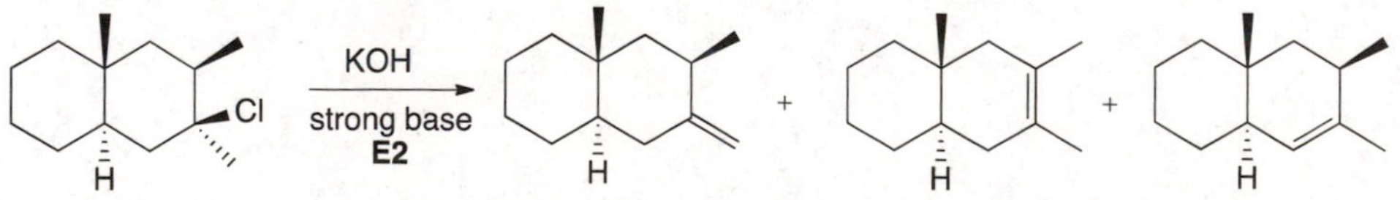

8.58

a.

3° halide + $^{-}OC(CH_3)_3$ — strong bulky base, **E2**

major product
more substituted alkene

No substitution occurs with a strong bulky base and a 3° RX. The C with the leaving group is too crowded for an S_N2 substitution to occur. Elimination occurs instead by an E2 mechanism.

b.

Br + $^{-}OCH_3$

1° halide — strong nucleophile, $\mathbf{S_N2}$ → OCH_3

All elimination reactions are slow with 1° halides.
The strong nucleophile reacts by an S_N2 mechanism instead.

c.

Cl — ^{-}OH → ← minor product only

3° halide — strong base, **E2** → More substituted alkene is favored.

d.

Cl — I⁻ → minor product only

2° halide — good nucleophile, weak base, $\mathbf{S_N2}$ **favored** → I, major product

The 2° halide can react by an E2 or S_N2 reaction with a negatively charged nucleophile or base. Since I⁻ is a weak base, substitution by an S_N2 mechanism is favored.

8.59

3° halide, weak base:
$\mathbf{S_N1}$ **and E1**

a.

Cl — OH (**overall reaction**) → O + + + HCl

The steps:

$\mathbf{S_N1}$ — OH, :O, H, :Cl:⁻ → + HCl

Any base (such as CH_3CH_2OH or Cl^-) can be used to remove a proton to form an alkene. If Cl^- is used, HCl is formed as a reaction by-product. If CH_3CH_2OH is used, $(CH_3CH_2OH_2)^+$ is formed instead.

or

H, :Cl:⁻ — **E1** → + HCl

or

H, :Cl:⁻ — **E1** → + HCl

b. 3° halide, strong base, **E2**

^{-}OH → **overall reaction** … + H_2O + Cl^-

Each product: one step → + $H_2\ddot{O}$ + Cl^- ($:\ddot{O}H^-$)

or

one step ($:\ddot{O}H^-$)

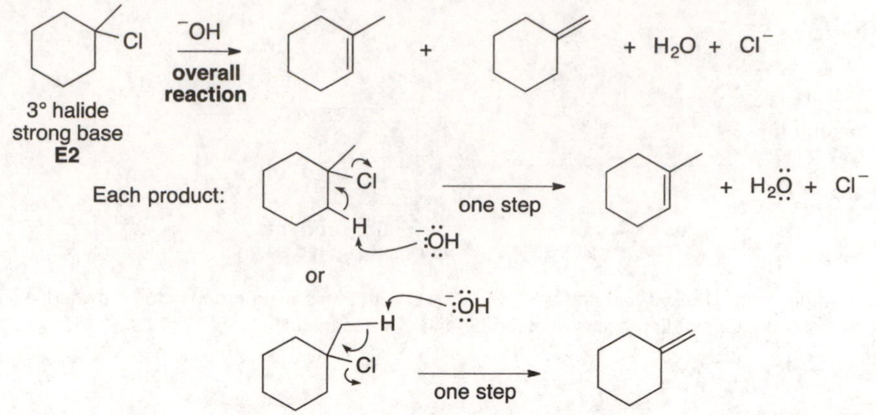

8.60 Draw the products of each reaction with the 1° alkyl halide.

a. Cl, H → $NaOCH_2CH_3$ (strong nucleophile, S_N2) → OCH_2CH_3, H

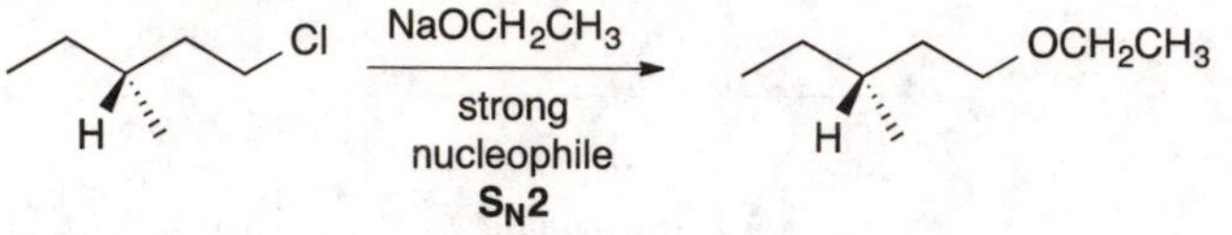

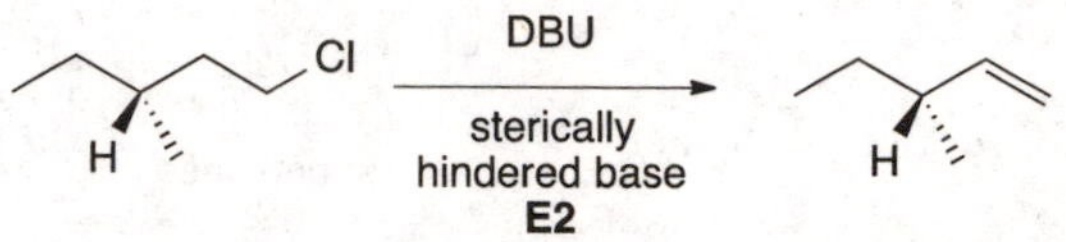

b. Cl, H → KCN (strong nucleophile, S_N2) → CN, H

c. Cl, H → DBU (sterically hindered base, **E2**) → H

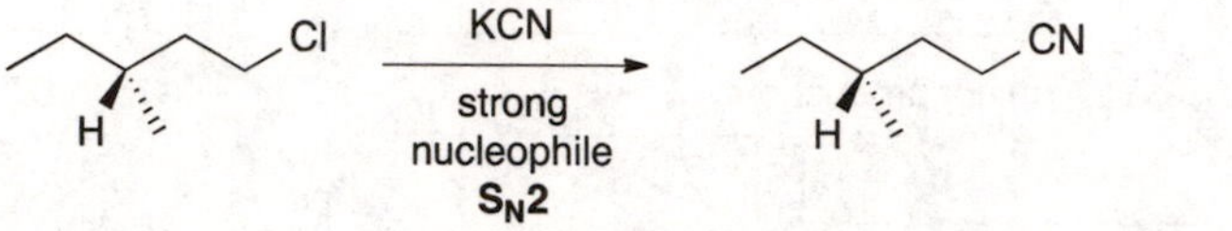

8.61

H, Br, H → $H_2\ddot{O}$: → H, +, H + Br^-

$H_2\ddot{O}$: above → H, $+\ddot{O}-H$, H, H ; $:\ddot{Br}:^-$ → H, $:\ddot{O}H$, H + HBr

$H_2\ddot{O}$: below → H, $+\ddot{O}-H$, H, H ; $:\ddot{Br}:^-$ → H, $:\ddot{O}H$, H + HBr

re-draw

H, +, H, H ; $:\ddot{Br}:^-$ → H, H + HBr

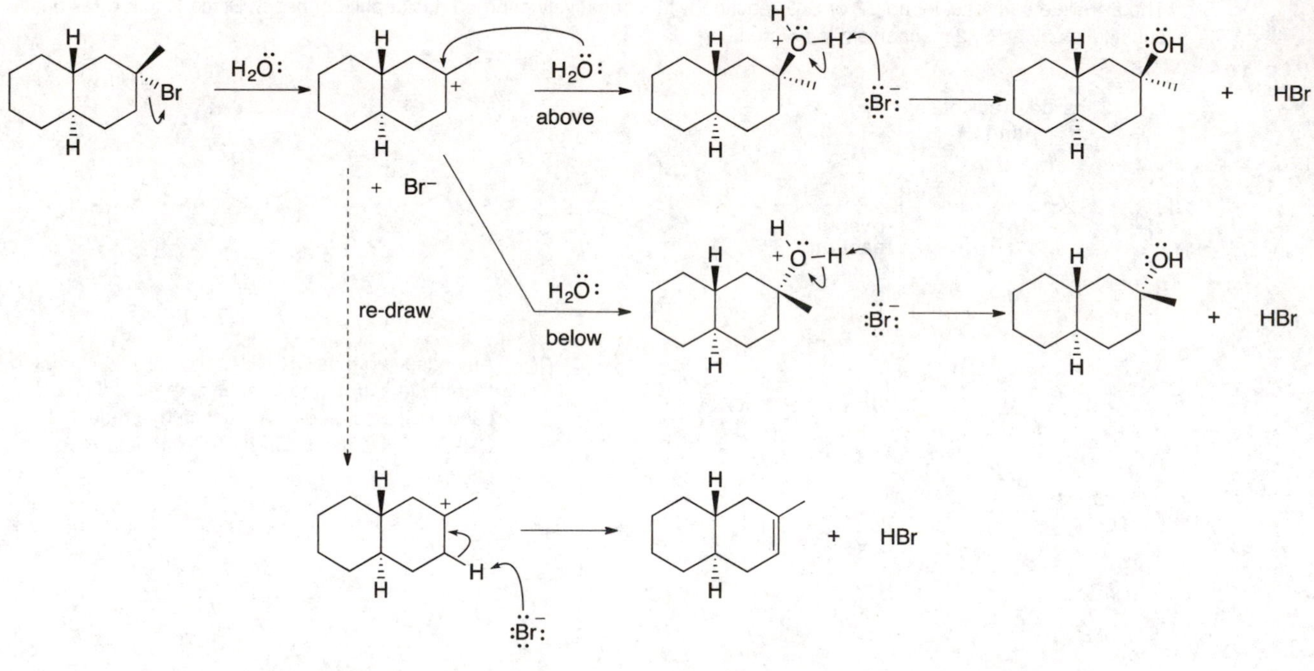

8.62

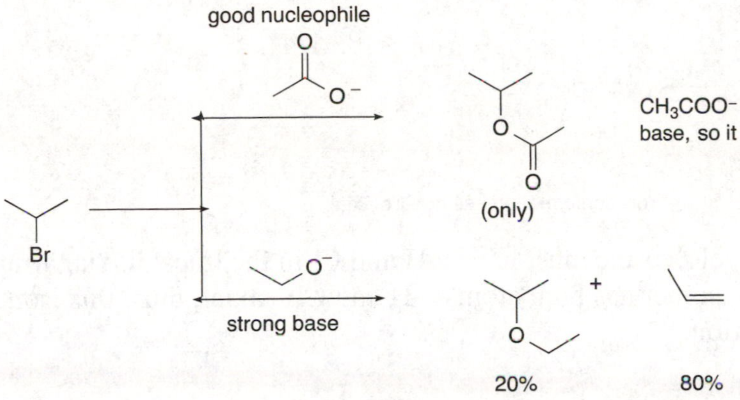
good nucleophile
O
O⁻
O
O
(only)
Br
O⁻
strong base
+
O
20%
80%

CH_3COO^- is a good nucleophile and a weak base, so it favors substitution by S_N2.

The strong base gives both S_N2 and E2 products, but because the 2° RX is somewhat hindered to substitution, the E2 product is favored.

8.63

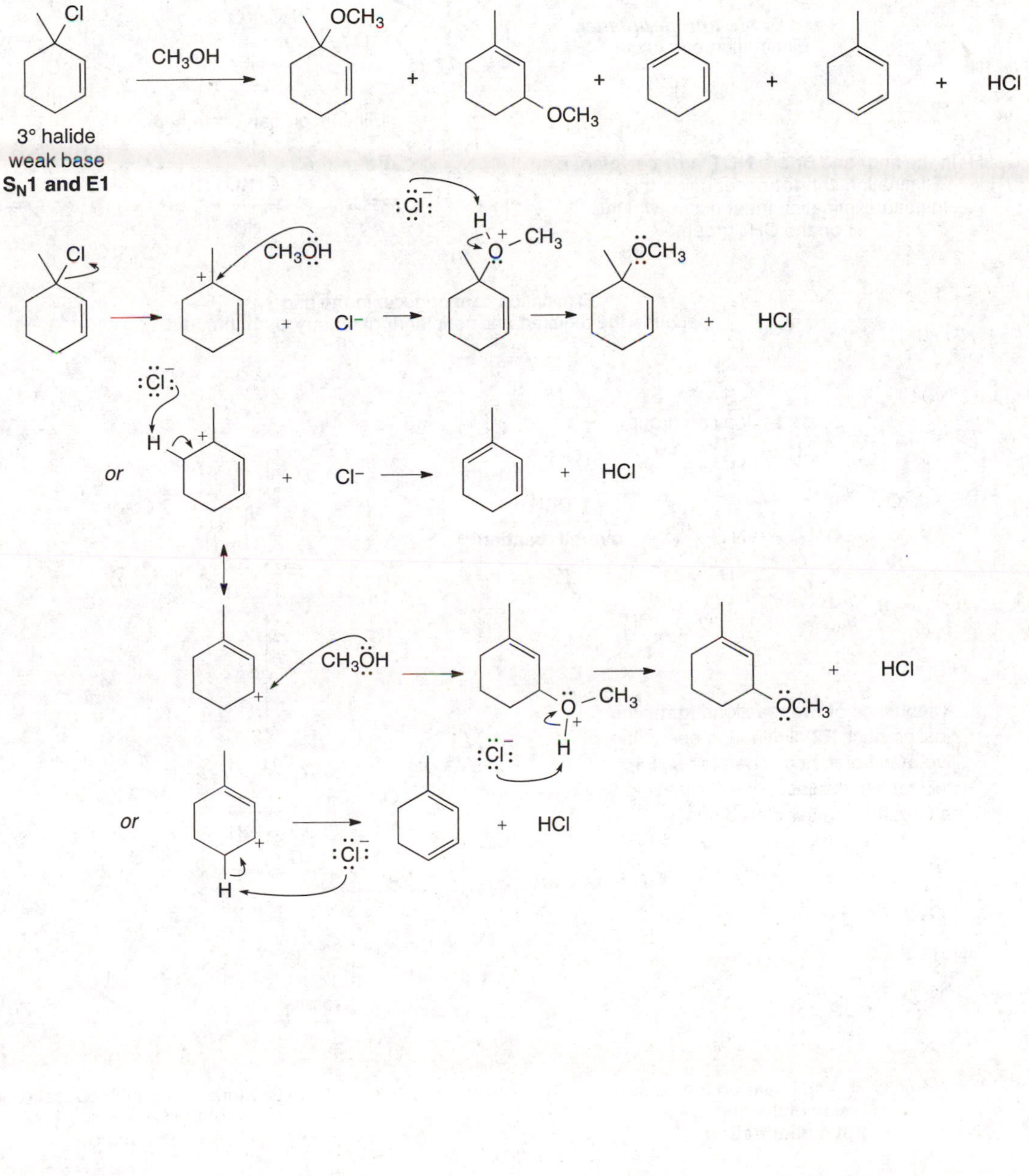
Cl
CH₃OH
OCH₃
+
OCH₃
+
+
+
HCl
3° halide
weak base
S_N1 and E1
Cl
CH₃ÖH
:Cl:⁻
H
+
O
CH₃
ÖCH₃
+
Cl⁻
+
HCl
:Cl:⁻
H
or
+
Cl⁻
+
HCl
CH₃ÖH
CH₃
H
:Cl:⁻
+
HCl
ÖCH₃
or
:Cl:
H
+
HCl

8.64 E2 elimination needs a leaving group and a hydrogen in the **trans diaxial** position.

Two different conformations:

This conformation has Cl's axial, but no H's axial. This conformation has no Cl's axial.

For elimination to occur, a cyclohexane must have a H and Cl in the trans diaxial arrangement. Neither conformation of this isomer has both atoms—H and Cl—axial; thus, this isomer only slowly loses HCl by elimination.

8.65

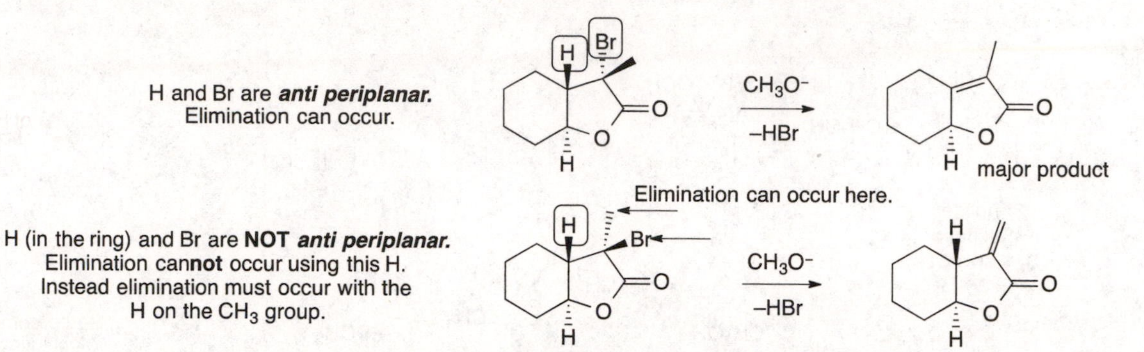

H and Br are ***anti periplanar.***
Elimination can occur.
CH_3O^-
–HBr
major product
Elimination can occur here.
H (in the ring) and Br are **NOT *anti periplanar.***
Elimination can**not** occur using this H.
Instead elimination must occur with the
H on the CH_3 group.
CH_3O^-
–HBr

Elimination cannot occur in the ring because the required anti periplanar geometry is not present.

8.66

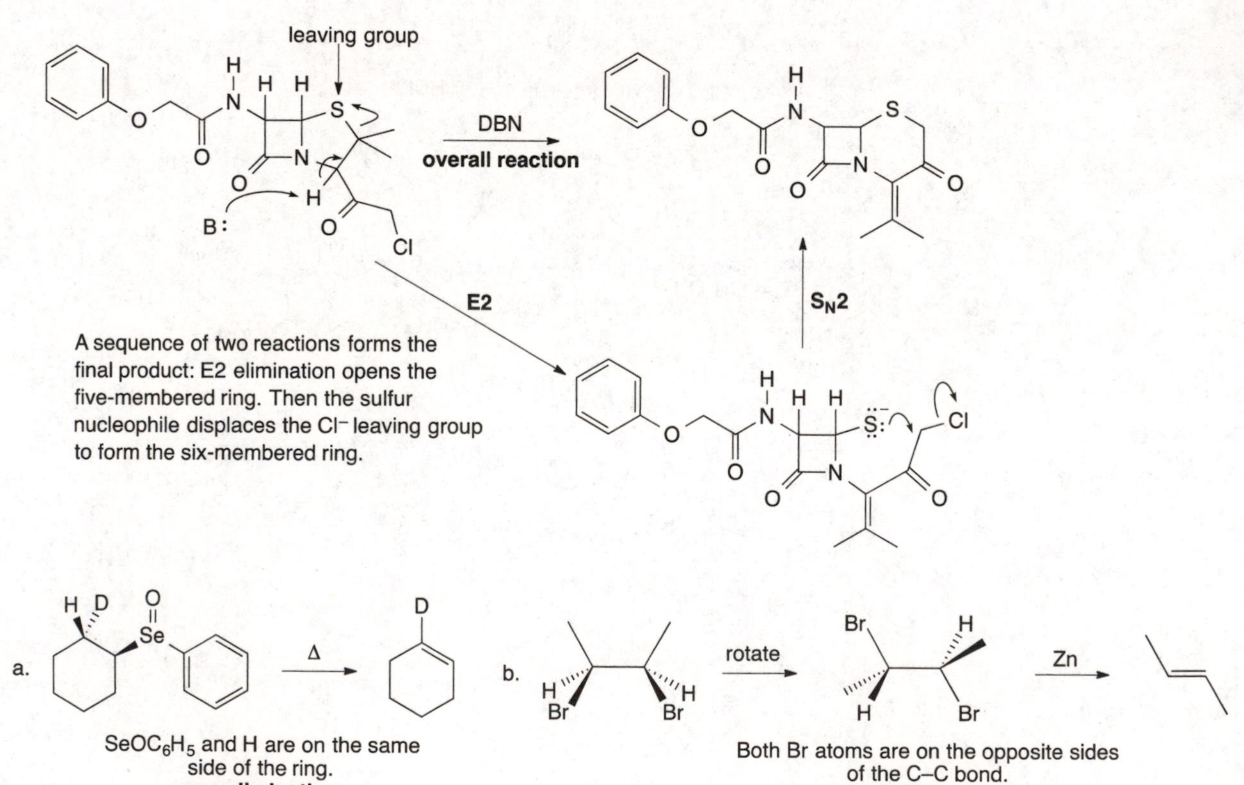

$SeOC_6H_5$ and H are on the same side of the ring.
syn elimination

Both Br atoms are on the opposite sides of the C–C bond.
anti elimination

8.68 One equivalent of $NaNH_2$ removes one mole of HBr in an anti periplanar fashion from each dibromide. Two modes of elimination are possible for each compound.

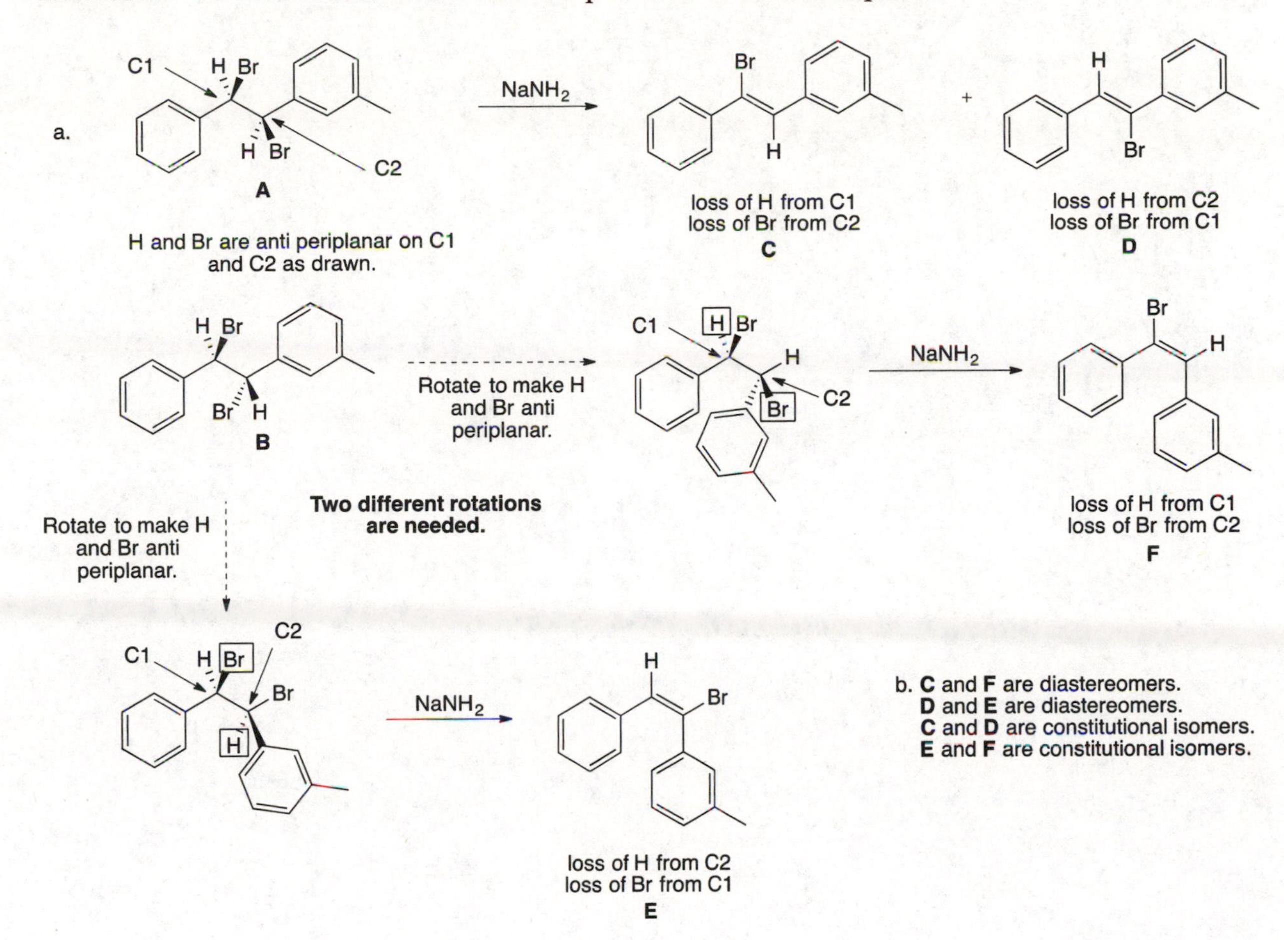

b. **C** and **F** are diastereomers.
D and **E** are diastereomers.
C and **D** are constitutional isomers.
E and **F** are constitutional isomers.

Chapter 9 Alcohols, Ethers, and Related Compounds

Chapter Review

General facts about ROH, ROR, and epoxides

- All three compounds contain an O atom that is sp^3 hybridized and tetrahedral (9.2).

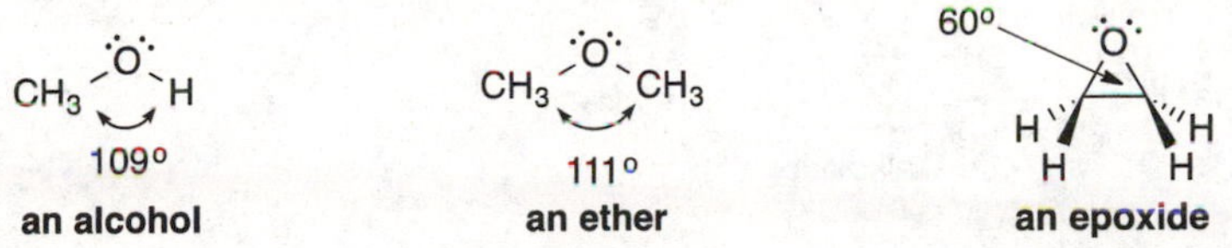

- All three compounds have polar C–O bonds, but only alcohols have an O–H bond for intermolecular hydrogen bonding (9.4).

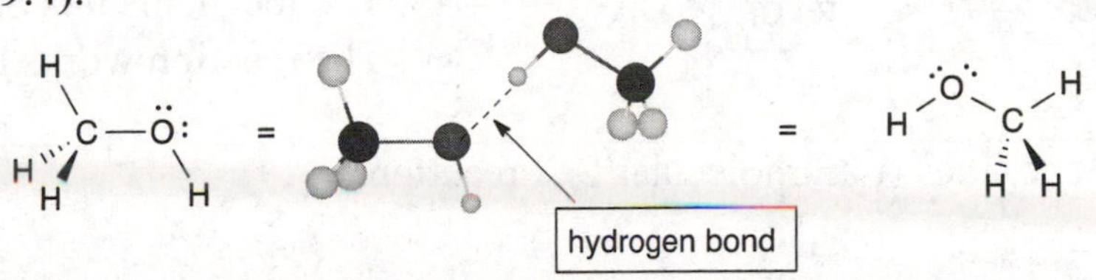

- Alcohols and ethers do not contain a good leaving group. Nucleophilic substitution can occur only after the OH (or OR) group is converted to a better leaving group (9.7A).

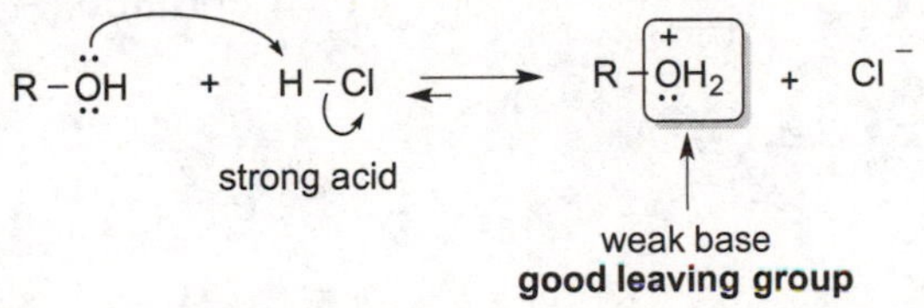

- Epoxides have a leaving group located in a strained three-membered ring, making them reactive to strong nucleophiles and acids HZ that contain a nucleophilic atom Z (9.16).

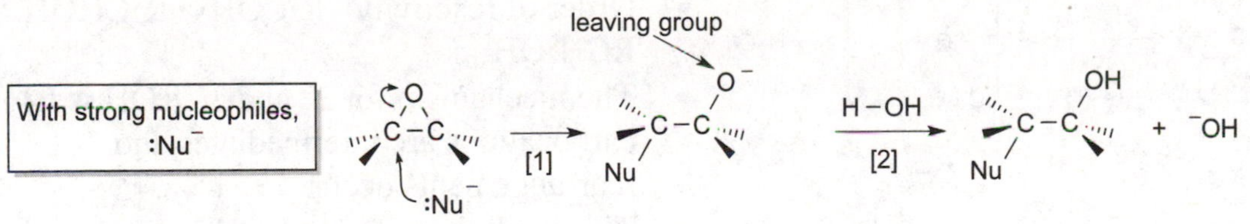

A new reaction of carbocations (9.9)

- Less stable carbocations rearrange to more stable carbocations by shift of a hydrogen atom or an alkyl group. Besides rearrangement, carbocations also react with nucleophiles (7.12) and bases (8.6).

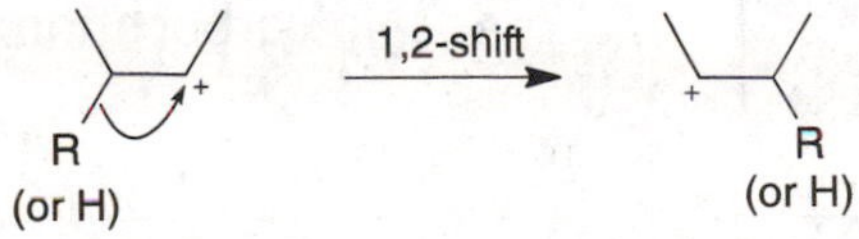

Preparation of alcohols, ethers, and epoxides (9.6)

[1] Preparation of alcohols

$$R{-}X + {}^{-}OH \longrightarrow R{-}OH + X^{-}$$

- The mechanism is S_N2.
- The reaction works best for CH_3X and 1° RX.

[2] Preparation of alkoxides (a Brønsted–Lowry acid–base reaction)

$$R{-}O{-}H + Na^{+}H^{-} \longrightarrow R{-}O^{-}\,Na^{+} + H_2$$

alkoxide

[3] Preparation of ethers (Williamson ether synthesis)

$$R{-}X + {}^{-}OR' \longrightarrow R{-}OR' + X^{-}$$

- The mechanism is S_N2.
- The reaction works best for CH_3X and 1° RX.

[4] Preparation of epoxides (intramolecular S_N2 reaction)

B: H–Ö: C–C X —[1]→ :Ö:⁻ C–C X + $H{-}B^{+}$ —[2]→ :O: C–C (epoxide) + X^{-}

halohydrin

- A two-step reaction sequence:
  [1] Removal of a proton with base forms an alkoxide.
  [2] Intramolecular **S_N2** reaction forms the epoxide.

Reactions of alcohols

[1] Dehydration to form alkenes

a. Using strong acid (9.8, 9.9)

$$-\overset{|}{\underset{H}{C}}-\overset{|}{\underset{OH}{C}}- \xrightarrow[\text{TsOH}]{H_2SO_4 \text{ or}} C{=}C + H_2O$$

- Order of reactivity: $R_3COH > R_2CHOH > RCH_2OH$.
- The mechanism for 2° and 3° ROH is E1; carbocations are intermediates and rearrangements occur.
- The mechanism for 1° ROH is E2.
- The Zaitsev rule is followed.

b. Using $POCl_3$ and pyridine (9.10)

$$-\overset{|}{\underset{H}{C}}-\overset{|}{\underset{OH}{C}}- \xrightarrow[\text{pyridine}]{POCl_3} C{=}C + H_2O$$

- The mechanism is E2.
- No carbocation rearrangements occur.

[2] Reaction with HX to form RX (9.11)

$$R{-}OH + H{-}X \longrightarrow \boxed{R{-}X} + H_2O$$

- Order of reactivity: $R_3COH > R_2CHOH > RCH_2OH$.
- The mechanism for 2° and 3° ROH is S_N1; carbocations are intermediates and rearrangements occur.
- The mechanism for CH_3OH and 1° ROH is S_N2.

---

[3] Reaction with other reagents to form RX (9.12)

$$R{-}OH + SOCl_2 \xrightarrow[\text{pyridine}]{} \boxed{R{-}Cl}$$

$$R{-}OH + PBr_3 \longrightarrow \boxed{R{-}Br}$$

- Reactions occur with CH_3OH and 1° and 2° ROH.
- The reactions follow an S_N2 mechanism.

---

[4] Reaction with tosyl chloride to form alkyl tosylates (9.13A)

$$R{-}OH + Cl{-}SO_2{-}C_6H_4{-}CH_3 \xrightarrow{\text{pyridine}} R{-}O{-}SO_2{-}C_6H_4{-}CH_3$$

$\boxed{R{-}OTs}$

- The C–O bond is not broken, so the configuration at a stereogenic center is retained.

Reactions of alkyl tosylates

Alkyl tosylates undergo either substitution or elimination depending on the reagent (9.13B).

$$-\overset{|}{\underset{H}{C}}-\overset{|}{\underset{OTs}{C}}- \xrightarrow{:Nu^-} \boxed{-\overset{|}{\underset{H}{C}}-\overset{|}{\underset{Nu}{C}}-} + {}^-OTs$$

$$-\overset{|}{\underset{H}{C}}-\overset{|}{\underset{OTs}{C}}- \xrightarrow{B:} \boxed{C{=}C} + HB^+ + {}^-OTs$$

- Substitution is carried out with strong $:Nu^-$, so the mechanism is S_N2.
- Elimination is carried out with strong bases, so the mechanism is E2.

Reactions of ethers

Only one reaction is useful: Cleavage with strong acids (9.14)

$$R{-}O{-}R' + \underset{\substack{(2\text{ equiv})\\(X = Br\text{ or }I)}}{H{-}X} \longrightarrow \boxed{R{-}X} + \boxed{R'{-}X} + H_2O$$

- With 2° and 3° R groups, the mechanism is S_N1.
- With CH_3 and 1° R groups, the mechanism is S_N2.

Reactions involving thiols and sulfides (9.15)

[1] Preparation of thiols

$$R{-}X \;+\; {}^{-}SH \longrightarrow \boxed{R{-}SH} \;+\; X^{-}$$

- The mechanism is S_N2.
- The reaction works best for CH_3X and 1° RX.

---

[2] Oxidation and reduction involving thiols

a. Oxidation of thiols to disulfides

$$R{-}SH \xrightarrow{Br_2 \text{ or } I_2} \boxed{RS{-}SR}$$

b. Reduction of disulfides to thiols

$$RS{-}SR \xrightarrow[HCl]{Zn} \boxed{R{-}SH}$$

---

[3] Preparation of sulfides

$$R{-}X \;+\; {}^{-}SR' \longrightarrow \boxed{R{-}SR'} \;+\; X^{-}$$

- The mechanism is S_N2.
- The reaction works best for CH_3X and 1° RX.

---

[4] Reaction of sulfides to form sulfonium ions

$$R'_2S \;+\; R{-}X \longrightarrow \boxed{R'_2\overset{+}{S}{-}R} \;+\; X^{-}$$

- The mechanism is S_N2.
- The reaction works best for CH_3X and 1° RX.

Reactions of epoxides

Epoxide rings are opened with nucleophiles $:Nu^{-}$ and acids HZ (9.16).

Epoxide $\xrightarrow[\text{or HZ}]{[1]\ :Nu^{-}\quad [2]\ H_2O}$ trans product with OH and Nu (Z) on adjacent carbons

- The reaction occurs with backside attack, resulting in trans or anti products.
- With $:Nu^{-}$, the mechanism is S_N2, and nucleophilic attack occurs at the *less* substituted C.
- With HZ, the mechanism is between S_N1 and S_N2, and attack of Z^{-} occurs at the *more* substituted C.

Practice Test on Chapter Review

1. Give the IUPAC name for each of the following compounds.

a. 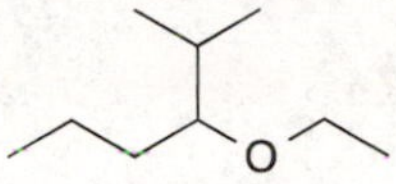 b. 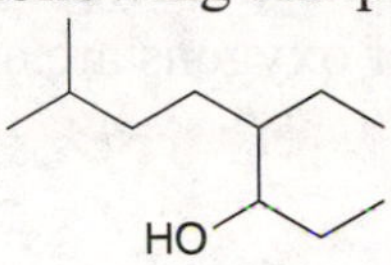 HO

2. Draw the organic products formed in each reaction. Draw all stereogenic centers using wedges and dashed wedges.

a. 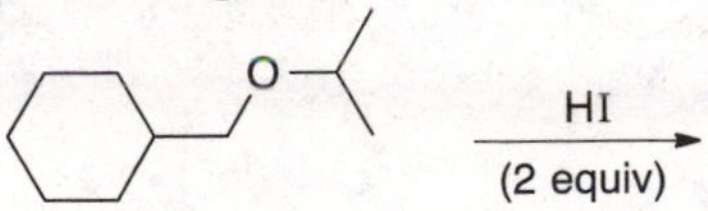 HI (2 equiv)

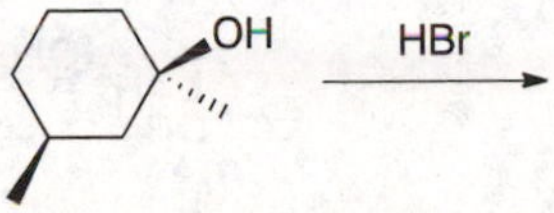

b. 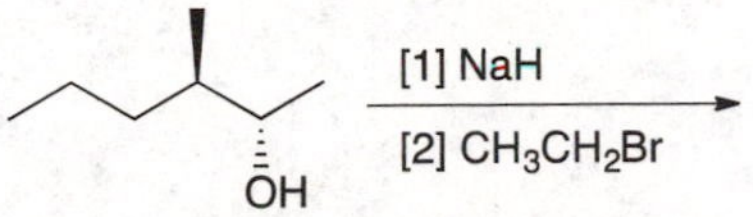 OH [1] NaH [2] CH_3CH_2Br

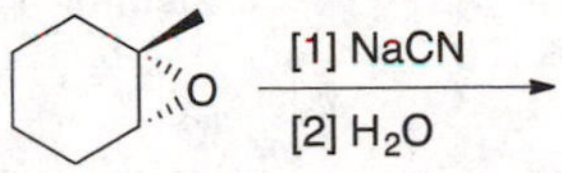

c. 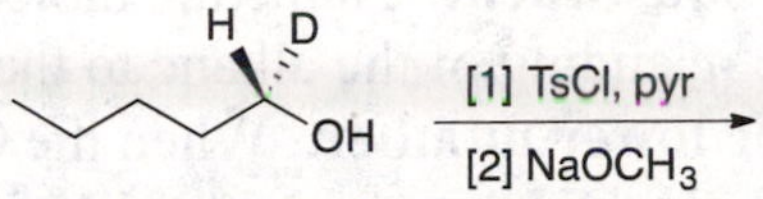 H D OH [1] TsCl, pyr [2] $NaOCH_3$

d. OH HBr

e. O [1] NaCN [2] H_2O

f. 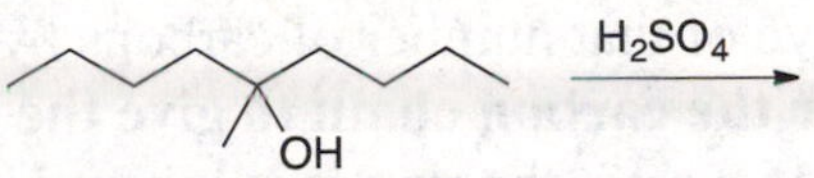 OH H_2SO_4

3. What starting material is needed for the following reaction?

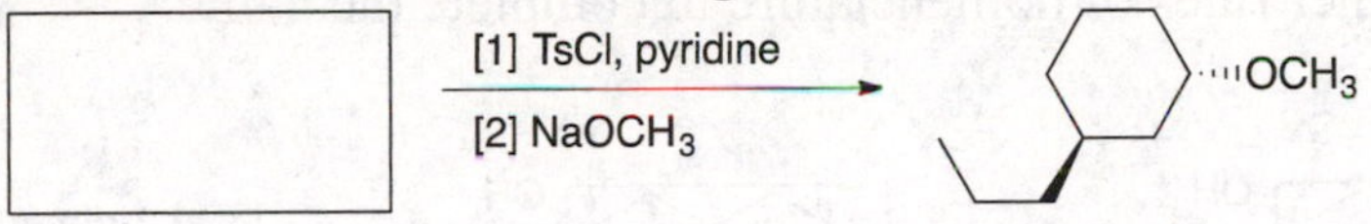

[1] TsCl, pyridine [2] $NaOCH_3$ → OCH_3

4. What alkoxide and alkyl halide are needed to make the following ether?

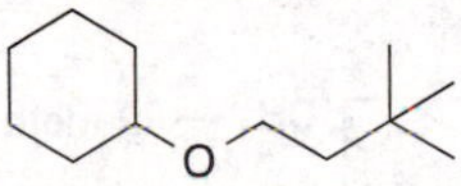

O

Answers to Practice Test

1.a. 3-ethoxy-2-methylhexane

b. 4-ethyl-7-methyl-octan-3-ol

2.

a. 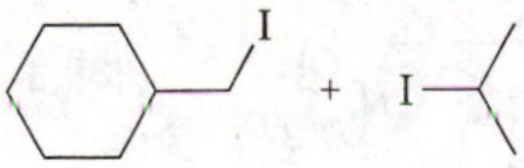 I + I

b. 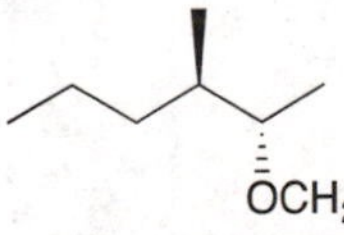 OCH_2CH_3

c. 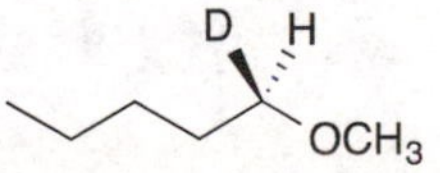 D H OCH_3

d. Br + Br

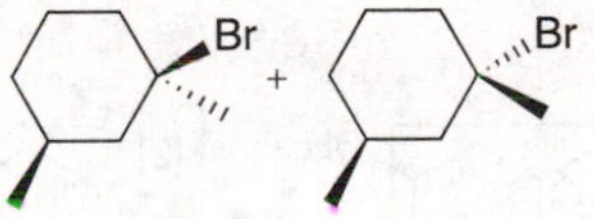

e. 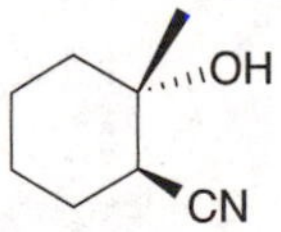 OH CN

f. 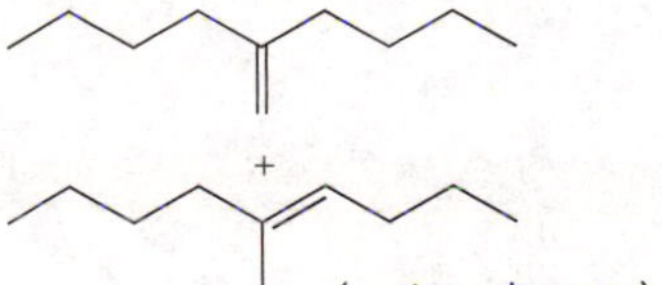 + (+ stereoisomer)

3.  OH

4. 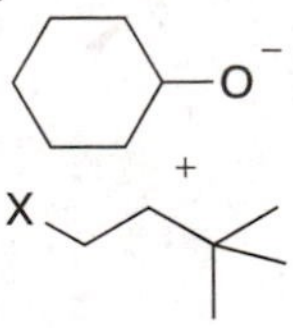 O^- + X

Answers to Problems

9.1 **Alcohols** are classified as 1°, 2°, or 3°, depending on the number of carbon atoms bonded to the carbon with the OH group. Five ether oxygens are circled.

9.2 To name an alcohol:

[1] **Find the longest chain that has the OH group as a substituent.** Name the molecule as a derivative of that number of carbons by changing the ***-e*** ending of the alkane to the suffix ***-ol.***

[2] **Number the carbon chain to give the OH group the lower number.** When the OH group is bonded to a ring, the ring is numbered beginning with the OH group, and the “1” is usually omitted.

[3] Apply the other rules of nomenclature to complete the name.

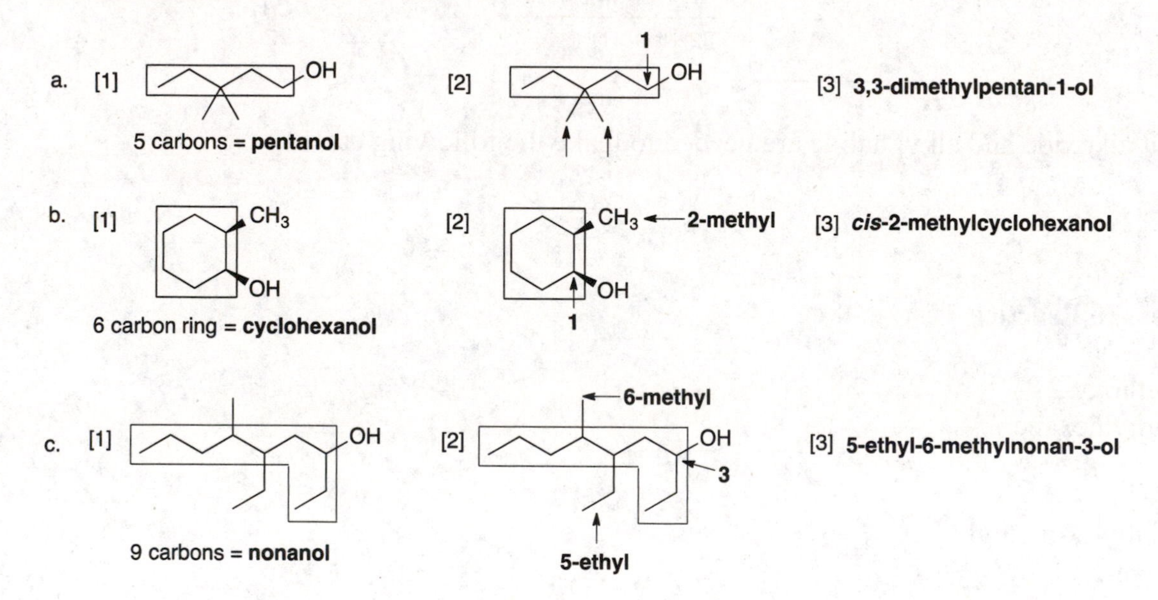

9.3 To work backwards from a name to a structure:

[1] Find the parent name and draw its structure.

[2] Add the substituents to the long chain.

a. 7,7-dimethyl**octan-4-ol**

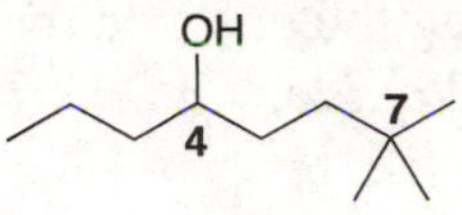

OH 4 7

c. 2-*tert*-butyl-3-methyl**cyclohexanol**

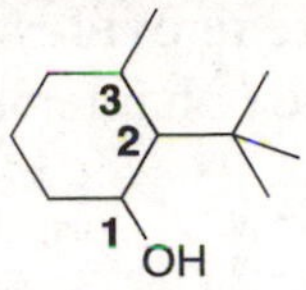

3 2 1 OH

b. 5-methyl-4-propyl**heptan-3-ol**

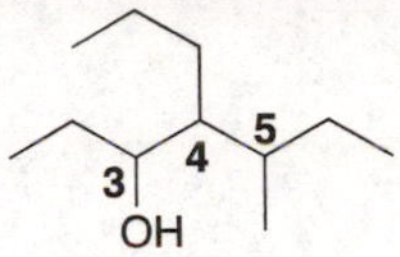

3 4 5 OH

d. *trans*-**cyclohexane-1,2-diol**

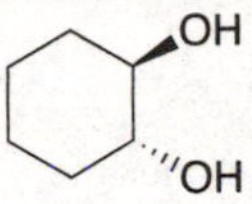

OH OH or OH OH

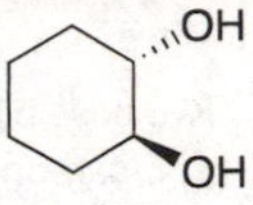

9.4 To name simple ethers:

[1] Name both alkyl groups bonded to the oxygen.

[2] Arrange these names alphabetically and add the word ***ether.*** For symmetrical ethers, name the alkyl group and add the prefix ***di.***

To name ethers using the IUPAC system:

[1] Find the two alkyl groups bonded to the ether oxygen. The smaller chain becomes the substituent, named as an alkoxy group.

[2] Number the chain to give the lower number to the first substituent.

a. **common name:**

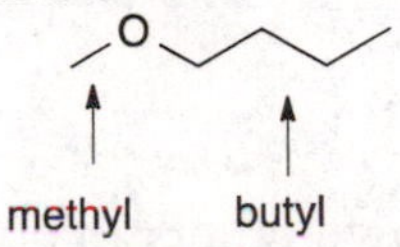

O

methyl butyl

butyl methyl ether

IUPAC name:

O 1 ← 4 C's, butane

substituent: methoxy

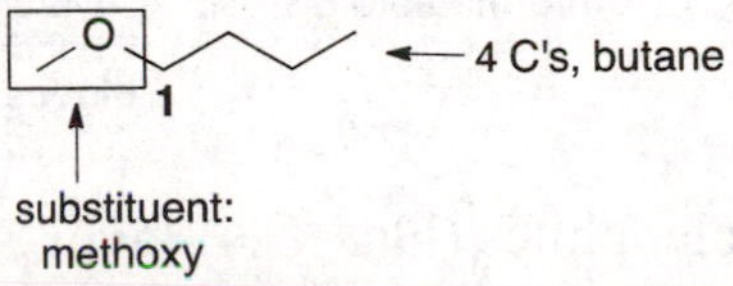

1-methoxybutane

b. **common name:**

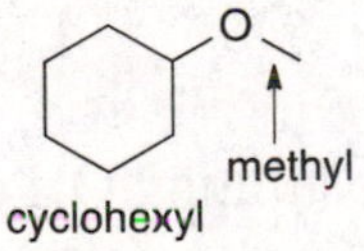

O

methyl

cyclohexyl

cyclohexyl methyl ether

IUPAC name:

O ← methoxy substituent

6 C's, cyclohexane

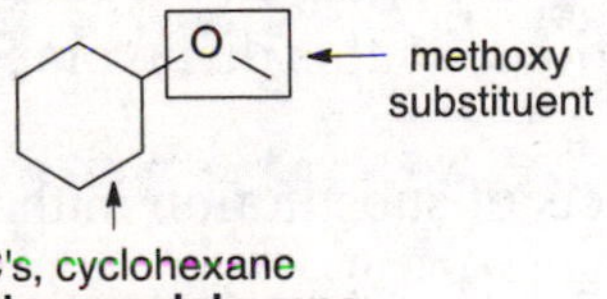

methoxycyclohexane

c. **common name:**

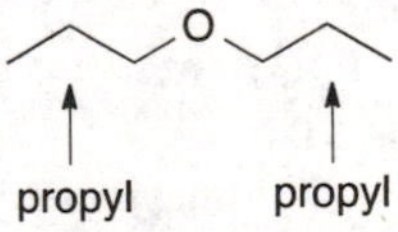

O

propyl propyl

dipropyl ether

IUPAC name:

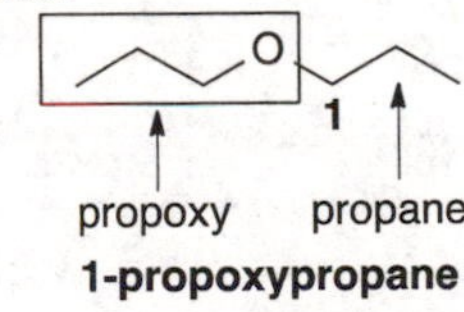

propoxy propane

1-propoxypropane

9.5 **Three ways to name epoxides:**

[1] Epoxides are named as derivatives of oxirane, the simplest epoxide.

[2] Epoxides can be named by considering the oxygen as a substituent called an **epoxy** group, bonded to a hydrocarbon chain or ring. Use two numbers to designate which two atoms the oxygen is bonded to.

[3] Epoxides can be named as **alkene oxides** by mentally replacing the epoxide oxygen by a double bond. Name the alkene (Chapter 10) and add the word ***oxide.***

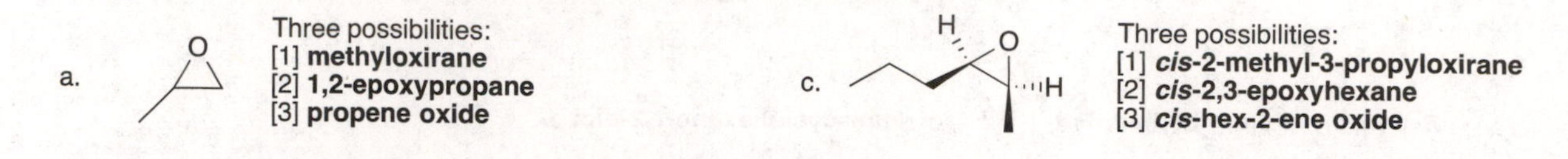

b. 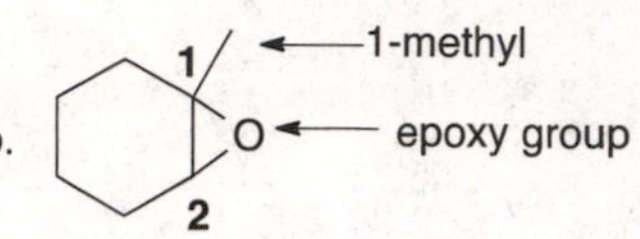 1 ← 1-methyl; O ← epoxy group; 2

Two possibilities:
[1] 6 carbons = cyclohexane
1,2-epoxy-1-methylcyclohexane
[2] **1-methylcyclohexene oxide**

9.6 Two rules for boiling point:

[1] **The stronger the intermolecular forces the higher the bp.**

[2] **Bp increases as the extent of the hydrogen bonding increases.** For alcohols with the same number of carbon atoms: 3° ROH < 2° ROH < 1° ROH.

a.

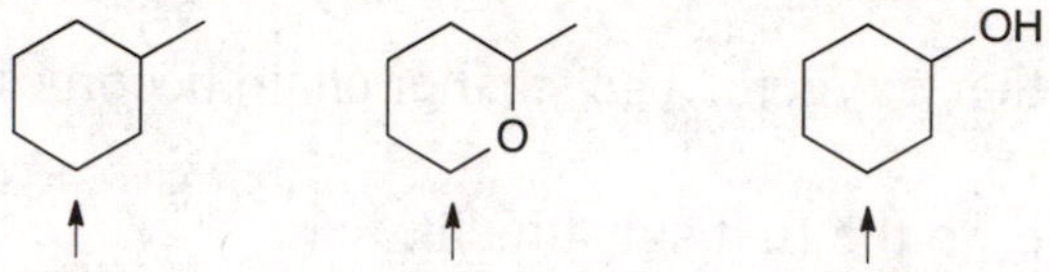

	O	OH
VDW **lowest bp**	VDW DD **intermediate bp**	VDW DD hydrogen bonding **highest bp**

b.

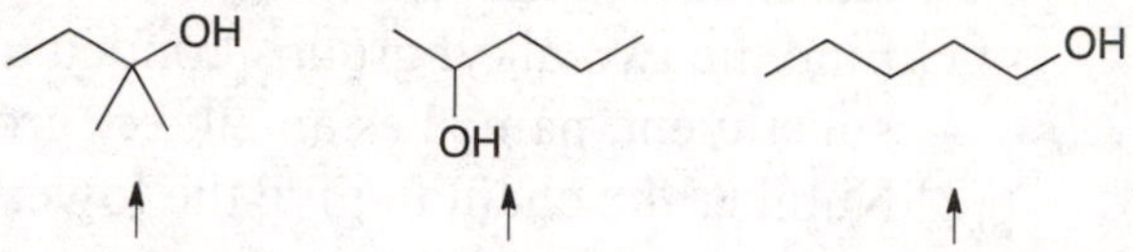

OH	OH	OH
3° ROH **lowest bp**	2° ROH **intermediate bp**	1° ROH **highest bp**

9.7 Strong nucleophiles (like $^{-}$CN) favor S_N2 reactions. The use of crown ethers in nonpolar solvents increases the nucleophilicity of the anion, and this increases the rate of the S_N2 reaction. The nucleophile does not appear in the rate equation for the S_N1 reaction. Nonpolar solvents cannot solvate carbocations, so this disfavors S_N1 reactions as well.

9.8 Draw the products of substitution in the following reactions by substituting OH or OR for X in the starting material.

a. 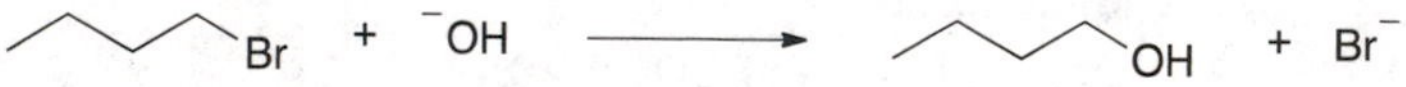 Br + $^{-}$OH ⟶ OH + Br^{-} **alcohol**

b. 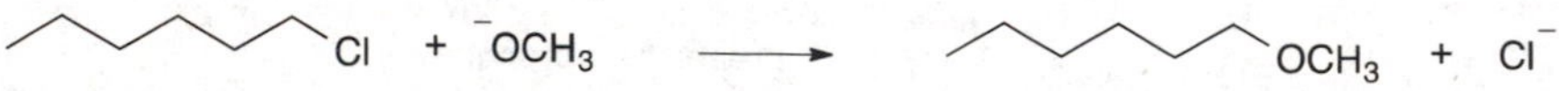 Cl + $^{-}OCH_3$ ⟶ OCH_3 + Cl^{-} **unsymmetrical ether**

c. 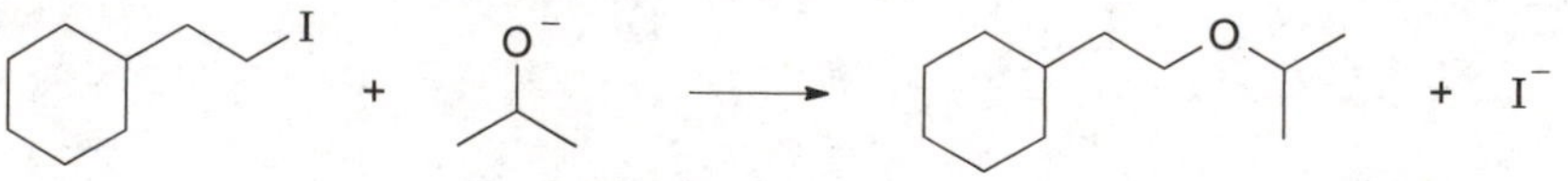 I + O^{-} ⟶ O + I^{-} **unsymmetrical ether**

d. 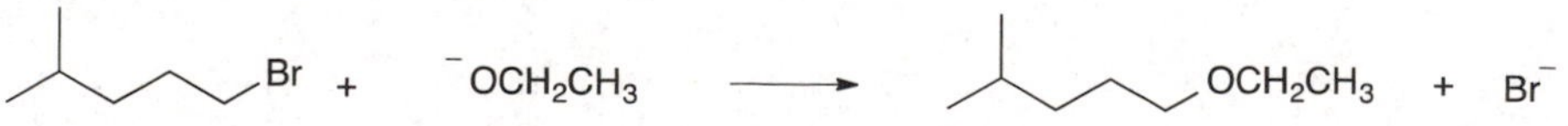 Br + $^{-}OCH_2CH_3$ ⟶ OCH_2CH_3 + Br^{-} **unsymmetrical ether**

9.9 Two possible routes to **X** are shown. Path [2] with a 1° alkyl halide is preferred. Path [1] cannot occur because the leaving group would be bonded to an sp^2 hybridized C, making it an unreactive aryl halide.

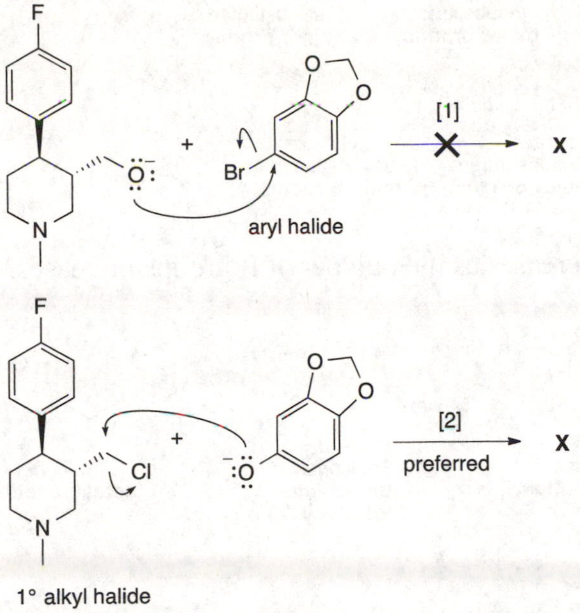

F, Ö:, +, Br, O, O, aryl halide, [1], X, N

F, Cl, +, :Ö, O, O, [2], preferred, X, N, 1° alkyl halide

9.10 NaH and NaNH₂ are strong bases that will remove a proton from an alcohol, creating a nucleophile.

a.  Ö–H + $Na^+H:^-$ → Ö: Na^+ + H_2

b. 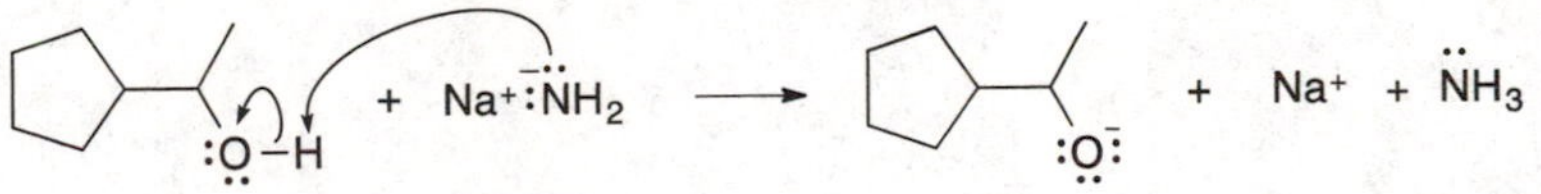 :Ö–H + $Na^+:\ddot{N}H_2$ → :Ö: + Na^+ + NH_3

c. 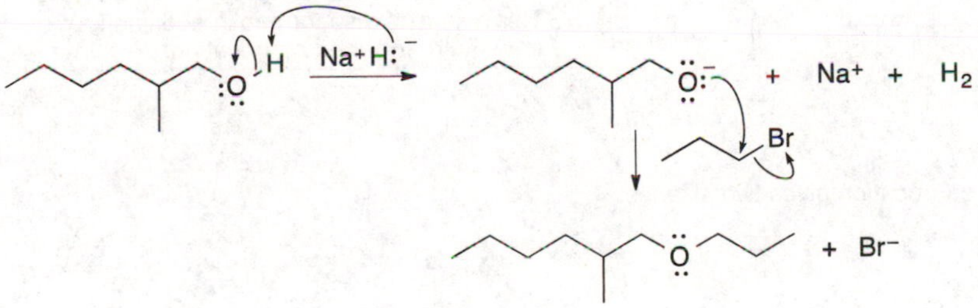 :Ö–H, $Na^+H:^-$ → $:\ddot{O}:^-$ + Na^+ + H_2; Br → Ö + Br^-

d. 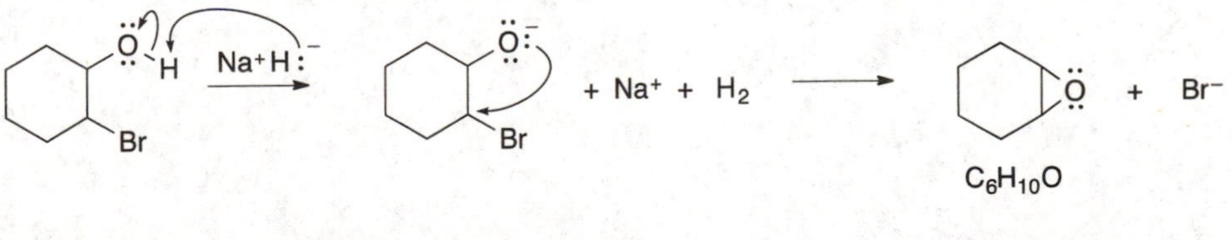 Ö–H, Br, $Na^+H:^-$ → $Ö:^-$, Br + Na^+ + H_2 → Ö + Br^-

$C_6H_{10}O$

9.11 Dehydration follows the Zaitsev rule, so the more stable, more substituted alkene is the major product.

a. 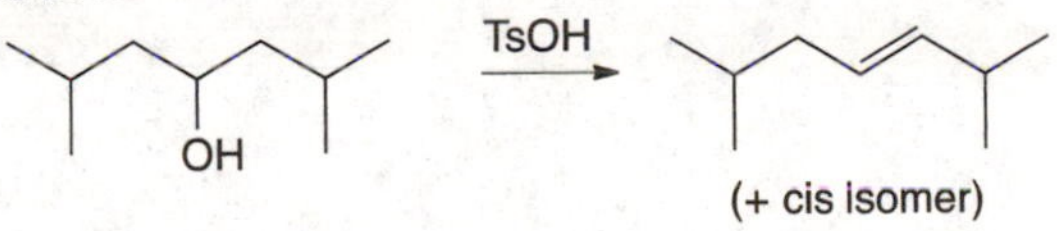 OH, TsOH → (+ cis isomer)

b. TsOH → (+ stereoisomer) + + H_2O

trisubstituted **major product**

disubstituted **minor product**

c. OH TsOH → + + H_2O

trisubstituted **major product**

disubstituted **minor product**

9.12 The rate of dehydration increases as the number of R groups increases.

1° alcohol **slowest reaction**

2° alcohol **intermediate reactivity**

3° alcohol **fastest reaction**

9.13

transition state [1]:

transition state [2]:

9.14

rearranged 3° carbocation

+ H_2SO_4

This alkene is also formed in addition to **Y** from the rearranged carbocation.

The initially formed 2° carbocation gives two alkenes:

or

9.15

a. 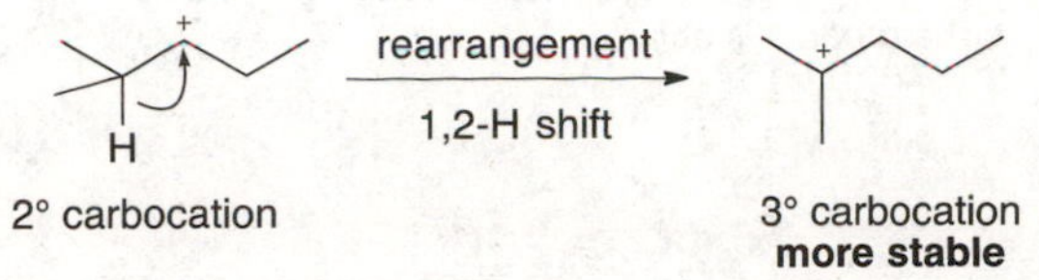 2° carbocation —rearrangement, 1,2-H shift→ 3° carbocation **more stable**

c. 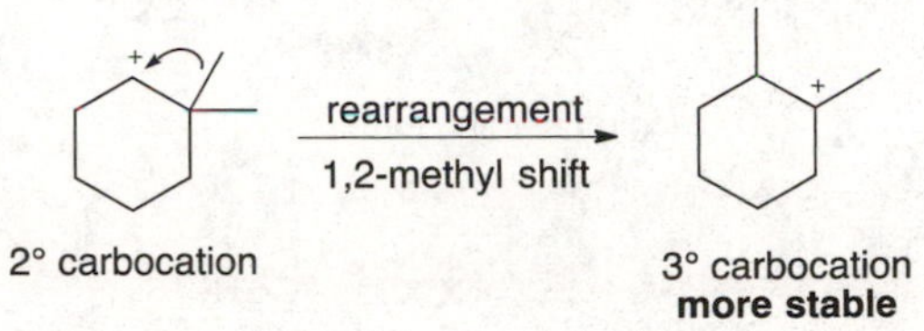 2° carbocation —rearrangement, 1,2-methyl shift→ 3° carbocation **more stable**

b. 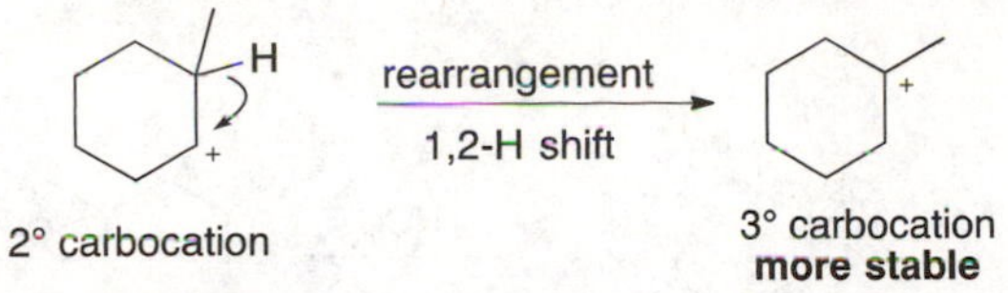 2° carbocation —rearrangement, 1,2-H shift→ 3° carbocation **more stable**

9.16

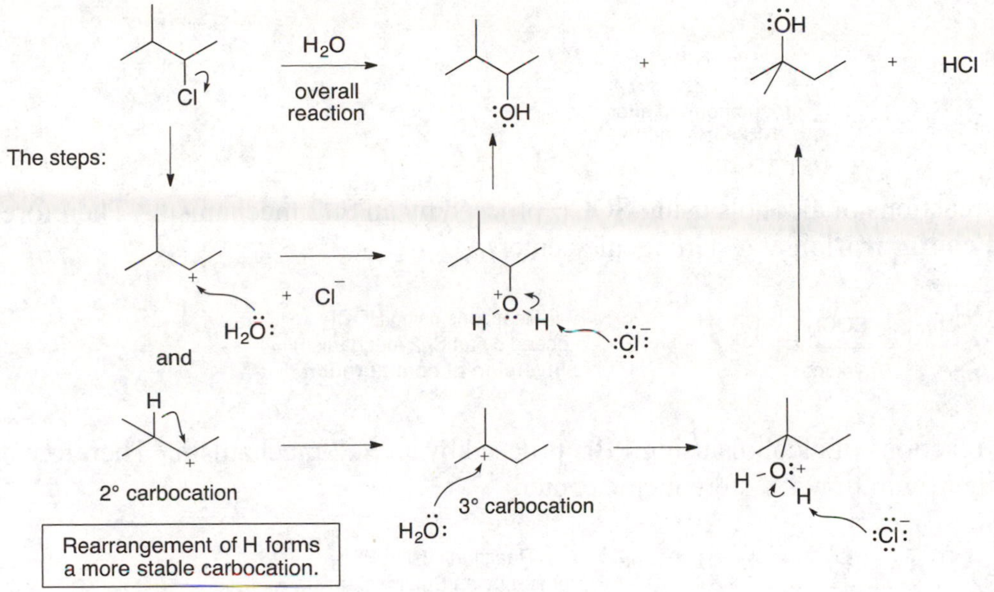

H_2O overall reaction → :OH + :OH + HCl

The steps:

+ Cl^-

$H_2\ddot{O}$:

:Cl:

and

2° carbocation

Rearrangement of H forms a more stable carbocation.

3° carbocation

$H_2\ddot{O}$:

:Cl:

9.17

a. 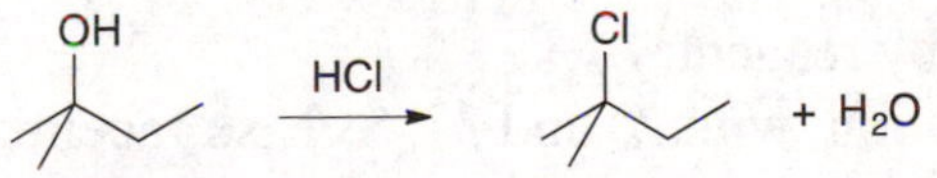 OH —HCl→ Cl + H_2O

c. 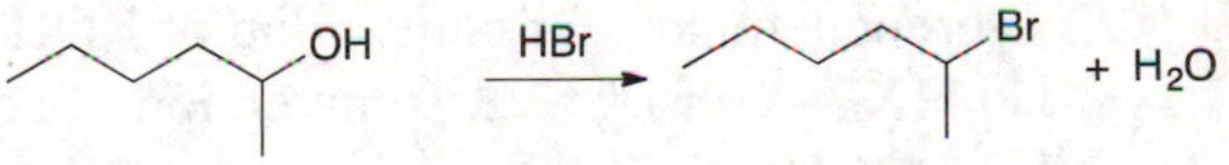 OH —HBr→ Br + H_2O

b. 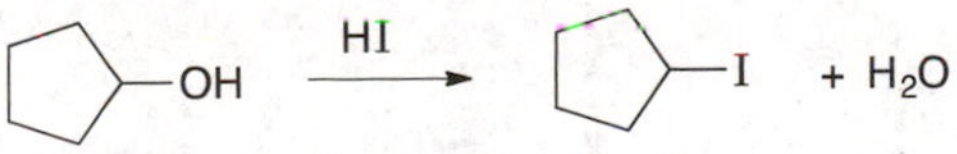 OH —HI→ I + H_2O

9.18 • **CH_3OH and 1° alcohols** follow an S_N2 mechanism, which results in inversion of configuration.
• **Secondary (2°) and 3° alcohols** follow an S_N1 mechanism, which results in racemization at a stereogenic center.

a. 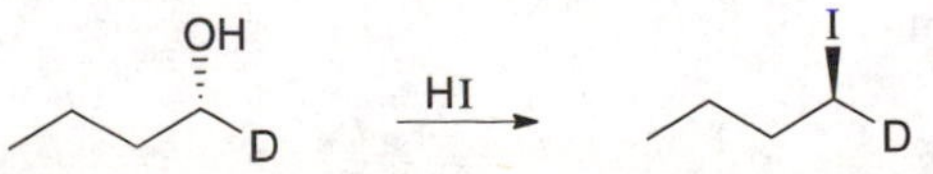 OH, D —HI→ I, D 1° alcohol, so **inversion of configuration**

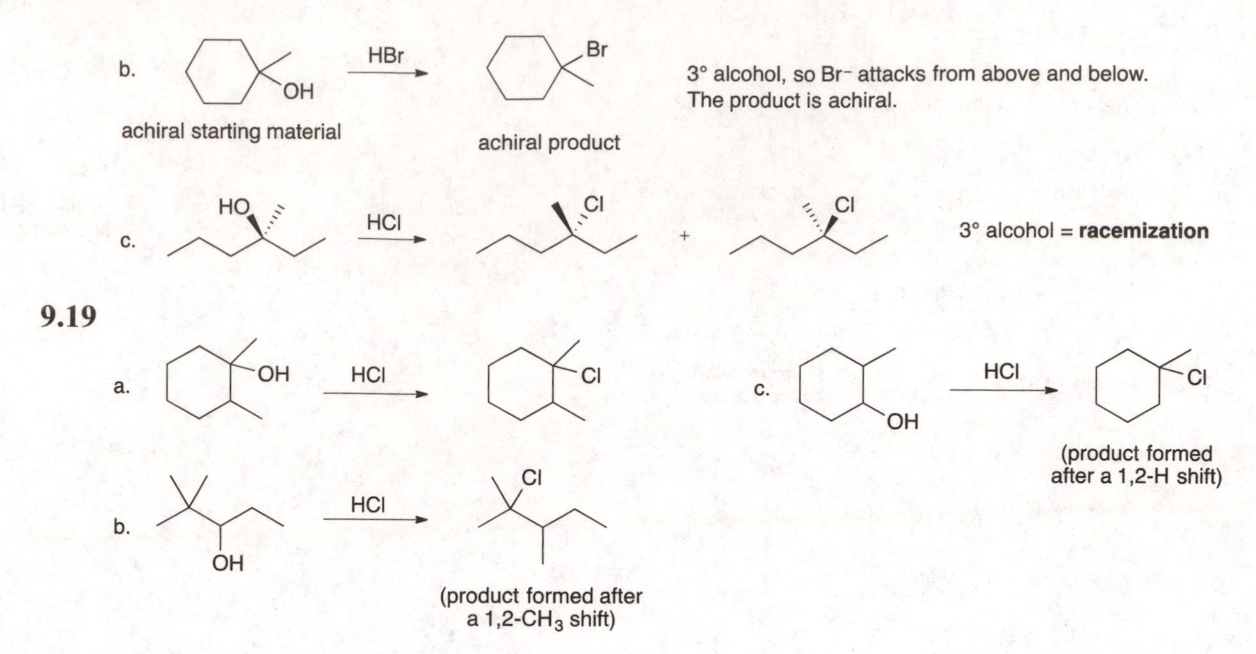

9.20 Substitution reactions of alcohols using $SOCl_2$ proceed by an S_N2 mechanism. Therefore, there is **inversion of configuration** at a stereogenic center.

H OH — $SOCl_2$, pyridine → Cl H

R → *S*

Reactions using $SOCl_2$ proceed by an S_N2 mechanism = **inversion of configuration.**

9.21 Substitution reactions of alcohols using PBr_3 proceed by an S_N2 mechanism. Therefore, there is inversion of configuration at a stereogenic center.

H OH — PBr_3 → Br H

R → *S*

Reactions using PBr_3 proceed by an S_N2 mechanism = **inversion of configuration.**

9.22 Stereochemistry for conversion of ROH to RX by reagent:

[1] **HX**—with 1°, S_N2, so inversion of configuration; with 2° and 3°, S_N1, so racemization.

[2] **$SOCl_2$**—S_N2, so inversion of configuration.

[3] **PBr_3**—S_N2, so inversion of configuration.

a. OH — $SOCl_2$, pyridine → Cl

b. OH — HI → I + I

3° alcohol, S_N1 = **racemization**

c. OH — PBr_3 → Br

S_N2 = **inversion**

9.23 To do a two-step synthesis with this starting material:

[1] Convert the OH group into a good leaving group (by using either PBr_3 or $SOCl_2$).

[2] Add the nucleophile for the S_N2 reaction.

OH → (PBr_3) → Br → (N_3^-) → N_3; → ($CH_3CH_2O^-$) → O

bad leaving group; good leaving group

9.24

a. OH + SO_2Cl → (pyridine) → O–S(=O)(=O)– + Cl^-

b. OH → (TsCl, pyridine) → OTs + Cl^-

9.25

a. OTs + ^-CN → (S_N2) → CN + ^-OTs

1° tosylate; strong nucleophile

b. OTs + K^+ $^-OC(CH_3)_3$ → (E2) → + K^+ ^-OTs + $HOC(CH_3)_3$

1° tosylate; strong bulky base

c. OTs + ^-SH → (S_N2) → SH

2° tosylate; strong nucleophile

(Substitution is favored over elimination.)

S_N2 product (inversion of configuration)

9.26

HO H (*S*) → (TsCl, pyridine) → TsO H (retention, *S*) → (NaOH, S_N2) → H OH (inversion, *R*)

one inversion from starting material to product

enantiomers

9.27 These reagents can be classified as follows:

[1] $SOCl_2$, PBr_3, HCl, and HBr replace OH with X by a substitution reaction.

[2] Tosyl chloride (TsCl) makes OH a better leaving group by converting it to OTs.

[3] Strong acids (H_2SO_4) and $POCl_3$ (pyridine) result in elimination by dehydration.

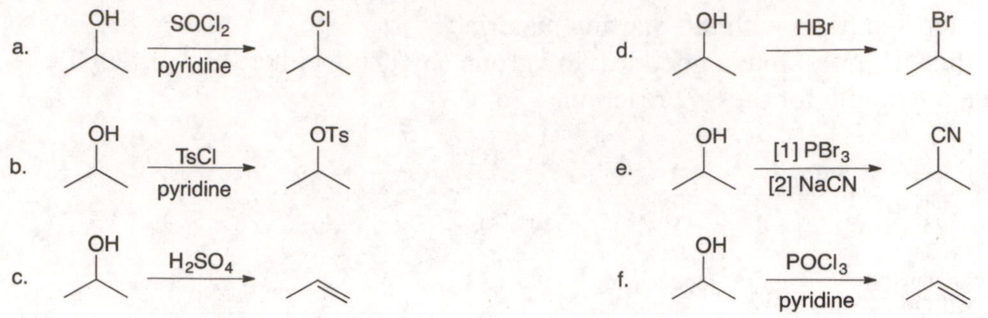

a. SOCl$_2$, pyridine
d. HBr
b. TsCl, pyridine
e. [1] PBr$_3$ [2] NaCN
c. H_2SO_4
f. POCl$_3$, pyridine

9.28

a. HBr → 2 + H_2O

b. HBr → + + H_2O

c. HBr → + + H_2O

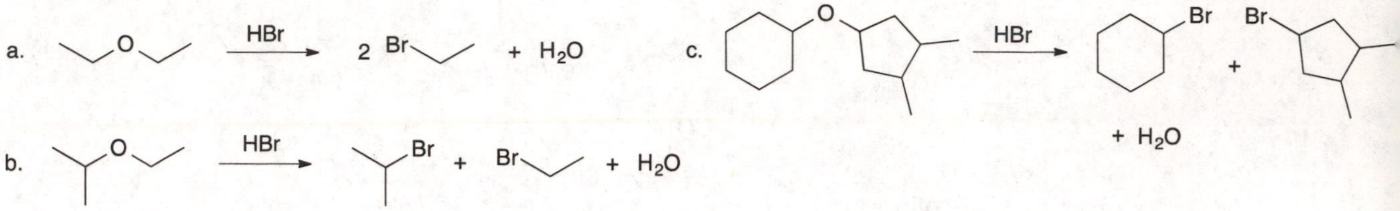

9.29 Ether cleavage can occur by either an SN1 or SN2 mechanism, but neither mechanism can occur when the ether O atom is bonded to an aromatic ring. An SN1 reaction would require formation of a highly unstable carbocation on a benzene ring, a process that does not occur. An SN2 reaction would require backside attack through the plane of the aromatic ring, which is also not possible. Thus, cleavage of the Ph–OCH$_3$ bond does not occur.

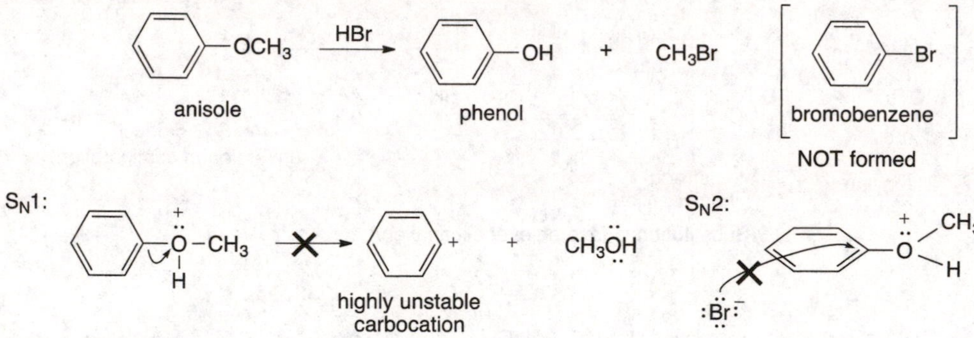

anisole —OCH$_3$ HBr → phenol —OH + CH_3Br [bromobenzene —Br NOT formed]

SN1: —O(H)$^+$–CH$_3$ ✕→ highly unstable carbocation + CH_3OH

SN2: :Br:$^-$ ✕ backside attack, O$^+$(CH$_3$)(H)

9.30

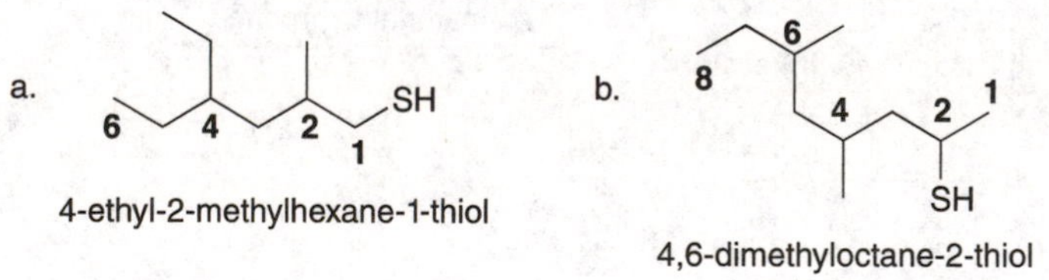

a. 4-ethyl-2-methylhexane-1-thiol

b. 4,6-dimethyloctane-2-thiol

9.31

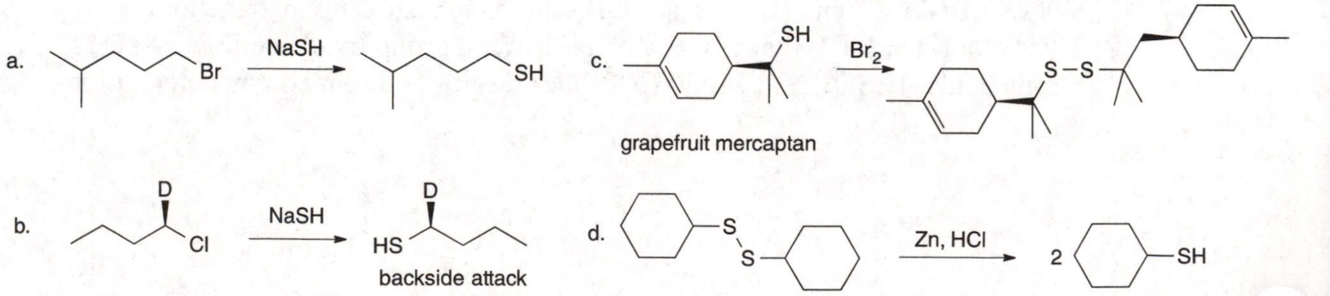

a. Br NaSH → SH

b. D, Cl NaSH → HS, D — backside attack

c. grapefruit mercaptan (SH) Br$_2$ → S–S

d. S–S Zn, HCl → 2 —SH

9.32

a. 1-ethylthiobutane or butyl ethyl sulfide

b. 2-methyl-1-methylthiocyclopentane

9.33

a. [1] NaH, [2] CH_3Br

b. + Cl^-

9.34 Two rules for the reaction of an epoxide:
[1] Nucleophiles attack from the **back side** of the epoxide.
[2] Negatively charged nucleophiles attack at the **less substituted carbon.**

a. [1] $CH_3CH_2O^-$, [2] H_2O → OH, OCH_2CH_3

Attack here:
less substituted C
backside attack

b. [1] $H{-}C{\equiv}C^-$, [2] H_2O

Attack here:
less substituted C
backside attack

9.35 In both isomers, ^-OH attacks from the back side at either C–O bond.

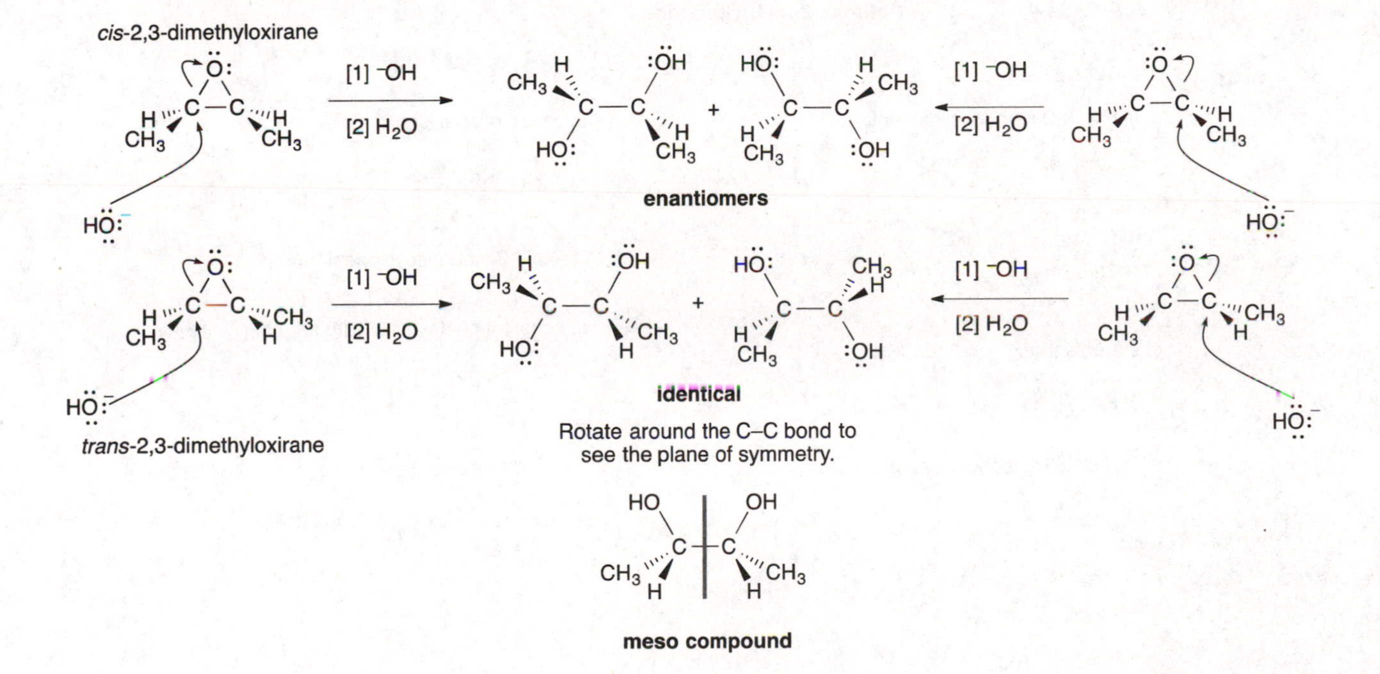

9.36 Remember the difference between negatively charged nucleophiles and neutral nucleophiles:
- **Negatively charged nucleophiles attack first,** followed by protonation, and the nucleophile attacks at the ***less*** **substituted carbon.**
- **Neutral nucleophiles have protonation first,** followed by nucleophilic attack at the ***more*** **substituted carbon.**

Trans or anti products are always formed, regardless of the nucleophile.

a.
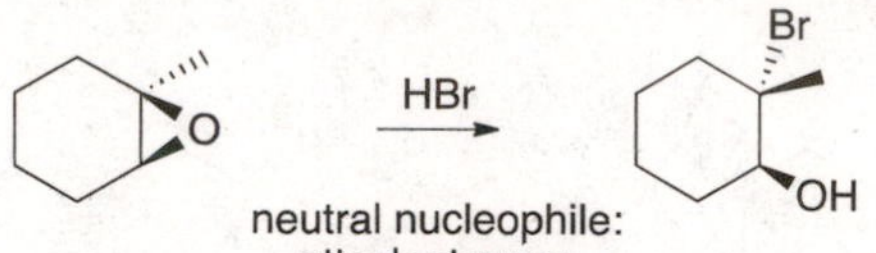
HBr
Br
OH
O
neutral nucleophile: attack at **more** substituted C

c.
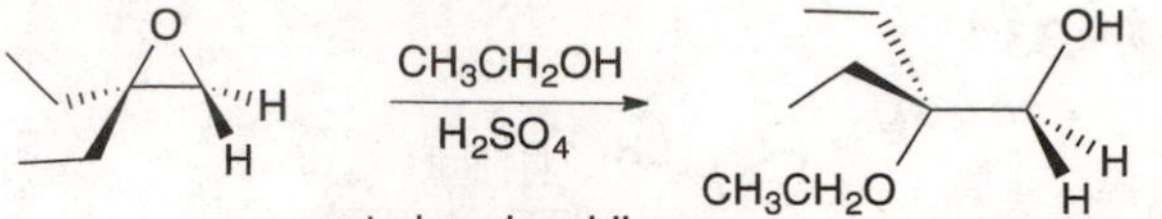
O
H
H
CH_3CH_2OH
H_2SO_4
OH
H
H
CH_3CH_2O
neutral nucleophile: attack at **more** substituted C

b.
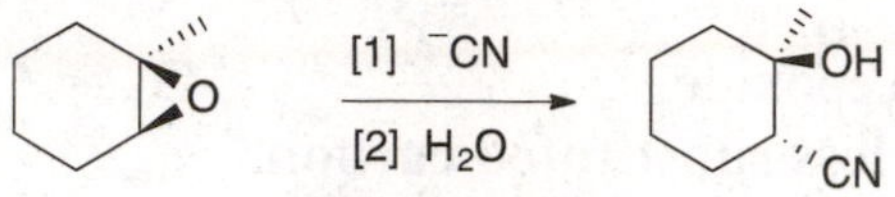
O
[1] ⁻CN
[2] H_2O
OH
CN
negatively charged nucleophile: attack at **less** substituted C

d.
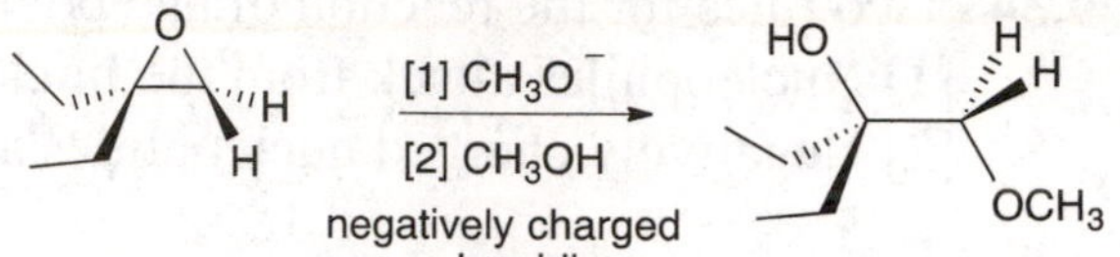
O
H
H
[1] CH_3O^-
[2] CH_3OH
HO
H
H
OCH_3
negatively charged nucleophile: attack at **less** substituted C

9.37

a.
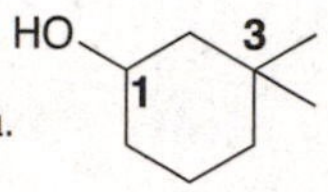
HO
1
3
6 C ring ⟶ cyclohexanol
2 CH_3's at C3
3,3-dimethylcyclohexanol

c.
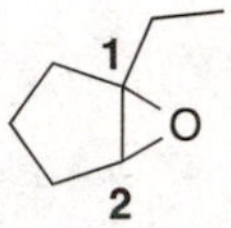
1
O
2
1,2-epoxy-1-ethylcyclopentane
or
1-ethylcyclopentene oxide

b.
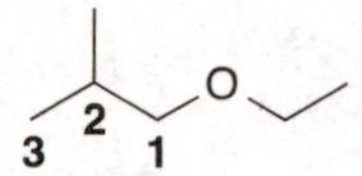
O
3
2
1
ethyl isobutyl ether
or
1-ethoxy-2-methylpropane

9.38

a. (1*R*,2*R*)-2-isobutylcyclopentanol

b. 2° alcohol
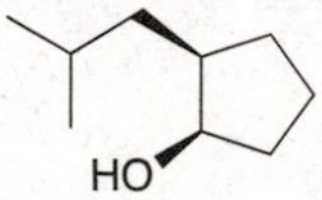
HO
A

c. stereoisomer
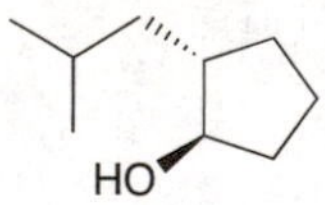
HO
(1*R*,2*S*)-2-isobutylcyclopentanol

d. constitutional isomer
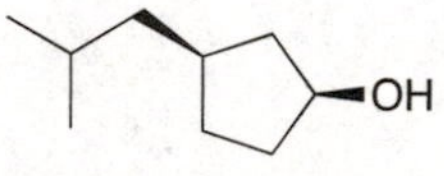
OH
(1*S*,3*S*)-3-isobutylcyclopentanol

e. constitutional isomer with an ether
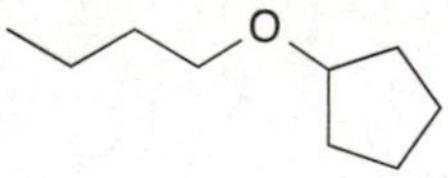
O
butoxycyclopentane

f.

[1] HO A — NaH → $^{-}$O

[2] HO — H_2SO_4 → +

[3] HO — $POCl_3$, pyridine → +

[4] HO — HCl → Cl

[5] HO — $SOCl_2$, pyridine → Cl

[6] HO — TsCl, pyridine → TsO

9.39

a. HO H — HBr → Br H (*S*) + H Br (*R*)   2° alcohol **S_N1 = racemization**

b. HO H — PBr_3 → H Br (*R*) PBr_3 follows **S_N2 = inversion.**

c. HO H — HCl → Cl H (*S*) + H Cl (*R*)   2° alcohol **S_N1 = racemization**

d. HO H — $SOCl_2$, pyridine → H Cl (*R*) $SOCl_2$ follows **S_N2 = inversion.**

9.40 Use the directions from Answer 9.2.

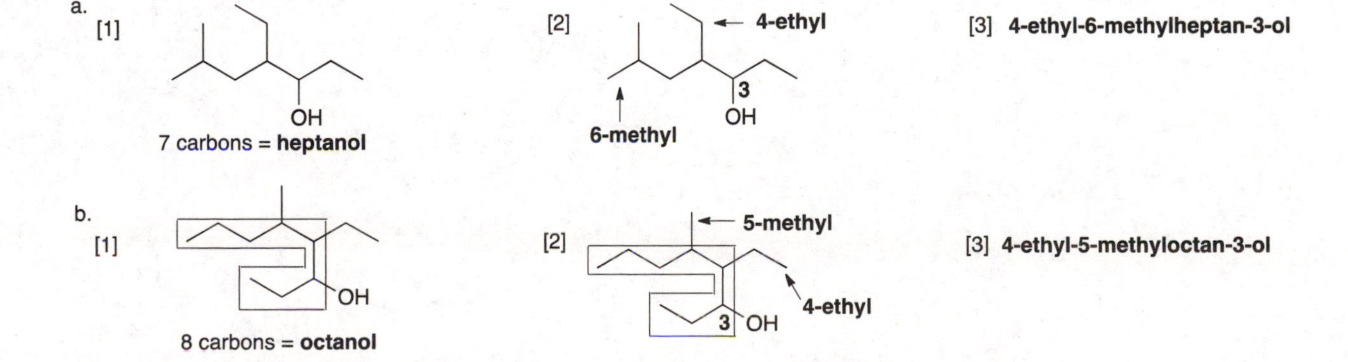

c.

[1] 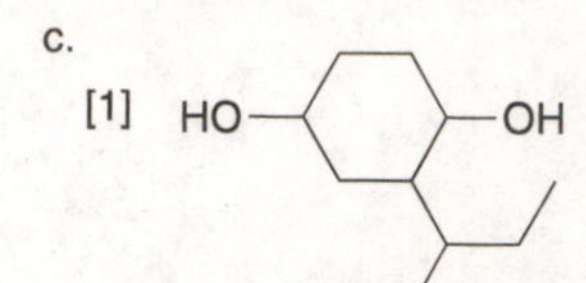

cyclohexanediol

[2] HO OH 4 1 3 2

[3] **2-*sec*-butylcyclohexane-1,4-diol**

d.

[1] 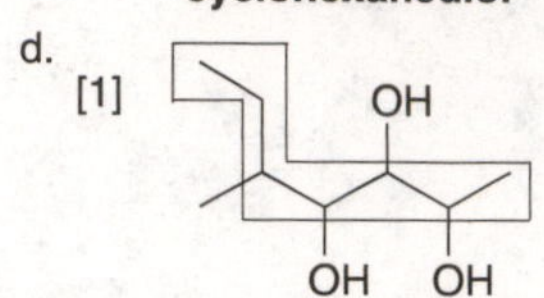

7 carbons = **heptanetriol**

[2] OH 4 3 2 OH OH

5-methyl

[3] **5-methylheptane-2,3,4-triol**

e.

[1] 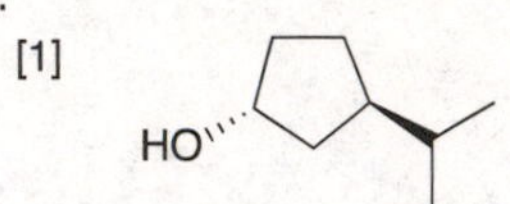

5 carbons = **cyclopentanol**

[2] HO 1

3-isopropyl

[3] ***trans*-3-isopropylcyclopentanol**

9.41 Use the rules from Answers 9.4 and 9.5.

a. O

dicyclohexyl ether

b. 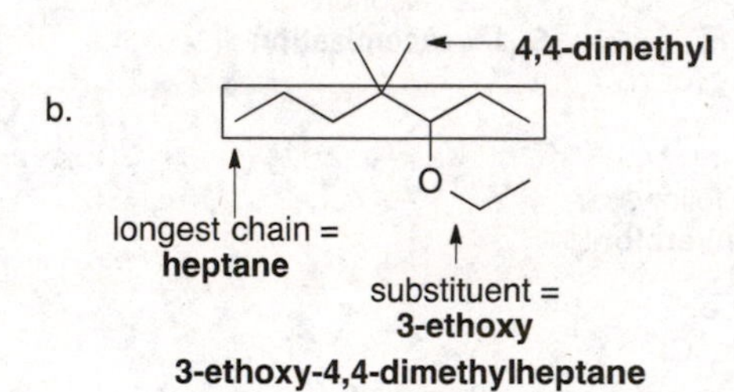

3-ethoxy-4,4-dimethylheptane

c. 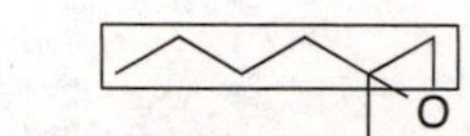

1,2-epoxy-2-methylhexane
or **2-butyl-2-methyloxirane**
or **2-methylhexene oxide**

d. 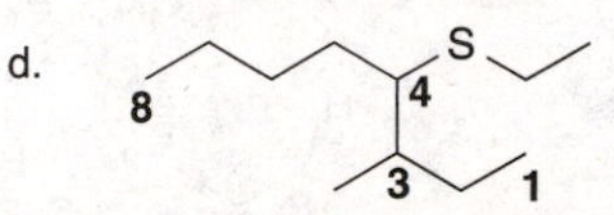

4-ethylthio-3-methyloctane

e. 2 4 1 SH

2,2,4-trimethylcyclopentanethiol

9.42 Use the directions from Answer 9.3.

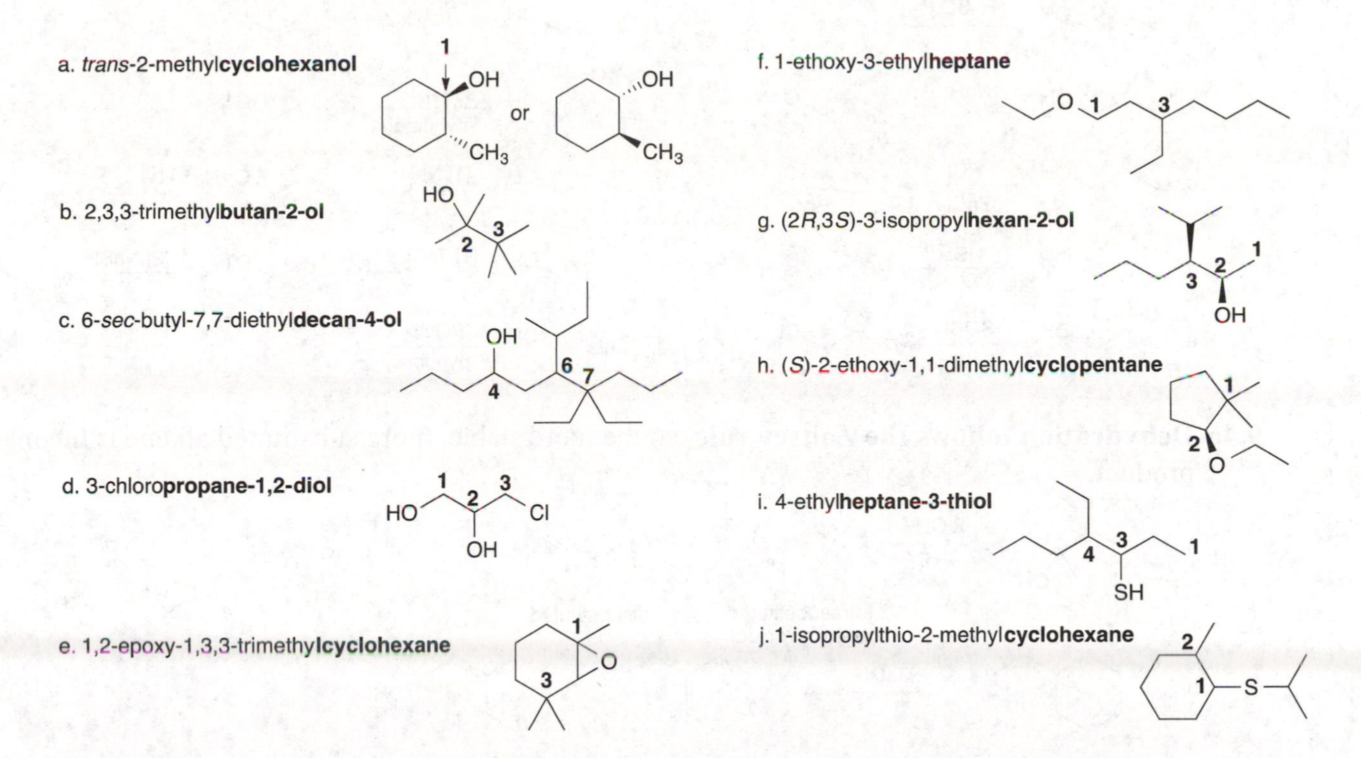

9.43 Melting points depend on intermolecular forces and symmetry. $(CH_3)_2CHCH_2OH$ has a lower melting point than $CH_3CH_2CH_2CH_2OH$ because branching decreases surface area and makes $(CH_3)_2CHCH_2OH$ less symmetrical, so it packs less well. Although $(CH_3)_3COH$ has the most branching and least surface area, it is the most symmetrical, so it packs best in a crystalline lattice, giving it the highest melting point.

OH	OH	OH
–108 °C lowest melting point	–90 °C intermediate melting point	26 °C highest melting point

9.44 Stronger intermolecular forces increase boiling point. All of the compounds can hydrogen bond, but both diols have more opportunity for hydrogen bonding because they have two OH groups, making their bp's higher than the bp of butan-1-ol. Propane-1,2-diol can also intramolecularly hydrogen bond. Intramolecular hydrogen bonding decreases the amount of intermolecular hydrogen bonding, so the bp of propane-1,2-diol is somewhat lower.

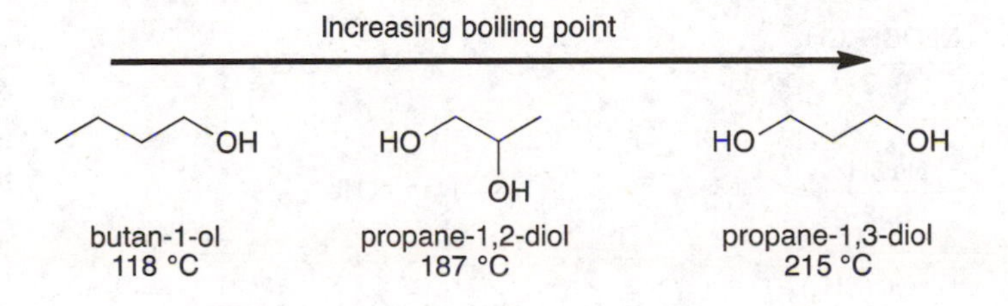

9.45

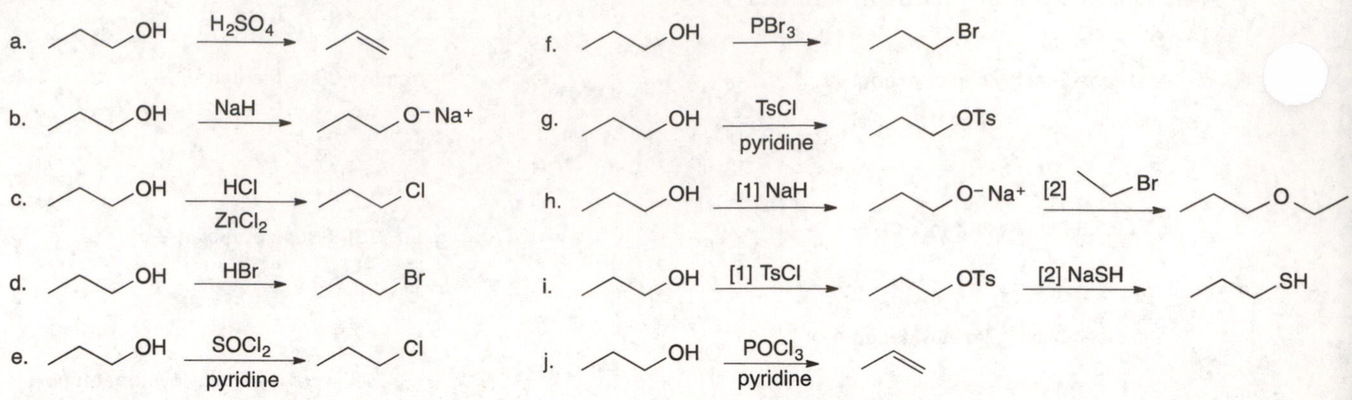

a. OH $\xrightarrow{H_2SO_4}$

b. OH $\xrightarrow{NaH}$ $O^- Na^+$

c. OH $\xrightarrow[ZnCl_2]{HCl}$ Cl

d. OH $\xrightarrow{HBr}$ Br

e. OH $\xrightarrow[pyridine]{SOCl_2}$ Cl

f. OH $\xrightarrow{PBr_3}$ Br

g. OH $\xrightarrow[pyridine]{TsCl}$ OTs

h. OH $\xrightarrow{[1]\ NaH}$ $O^- Na^+$ $\xrightarrow{[2]\ \text{Br}}$ O

i. OH $\xrightarrow{[1]\ TsCl}$ OTs $\xrightarrow{[2]\ NaSH}$ SH

j. OH $\xrightarrow[pyridine]{POCl_3}$

9.46 Dehydration follows the Zaitsev rule, so the more stable, more substituted alkene is the major product.

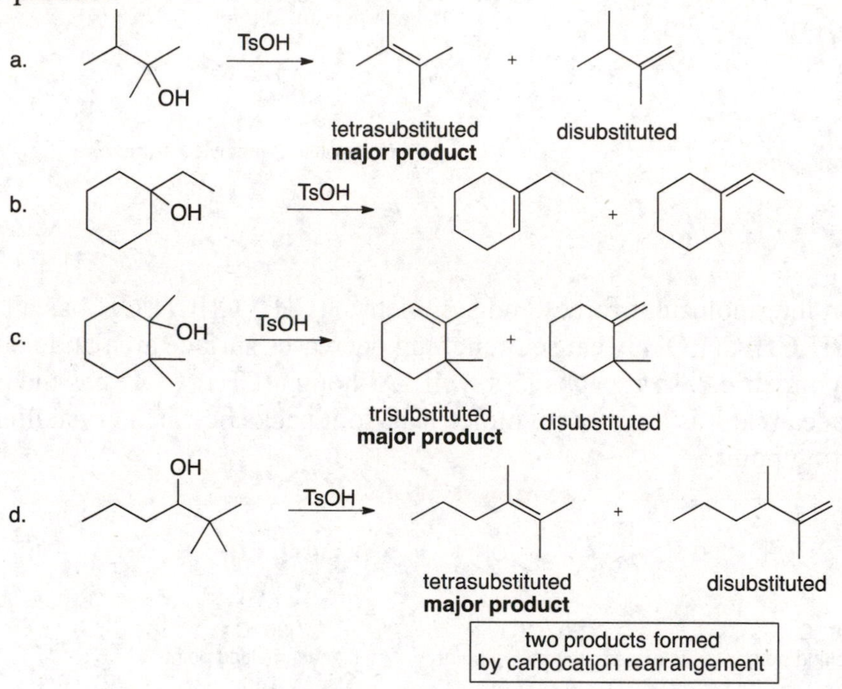

a. OH $\xrightarrow{TsOH}$ +

tetrasubstituted **major product**; disubstituted

b. OH $\xrightarrow{TsOH}$ +

c. OH $\xrightarrow{TsOH}$ +

trisubstituted **major product**; disubstituted

d. OH $\xrightarrow{TsOH}$ +

tetrasubstituted **major product**; disubstituted

two products formed by carbocation rearrangement

9.47 OTs is a good leaving group and will easily be replaced by a nucleophile. Draw the products by substituting the nucleophile in the reagent for OTs in the starting material.

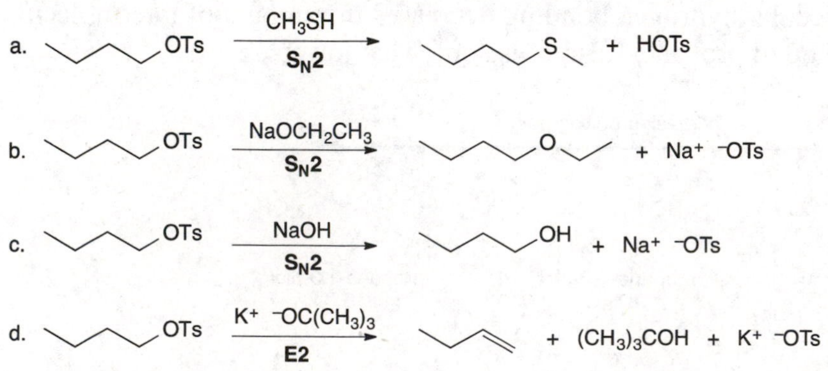

a. OTs $\xrightarrow[S_N2]{CH_3SH}$ S + HOTs

b. OTs $\xrightarrow[S_N2]{NaOCH_2CH_3}$ O + Na^+ ^-OTs

c. OTs $\xrightarrow[S_N2]{NaOH}$ OH + Na^+ ^-OTs

d. OTs $\xrightarrow[E2]{K^+\ ^-OC(CH_3)_3}$ + $(CH_3)_3COH$ + K^+ ^-OTs

9.48

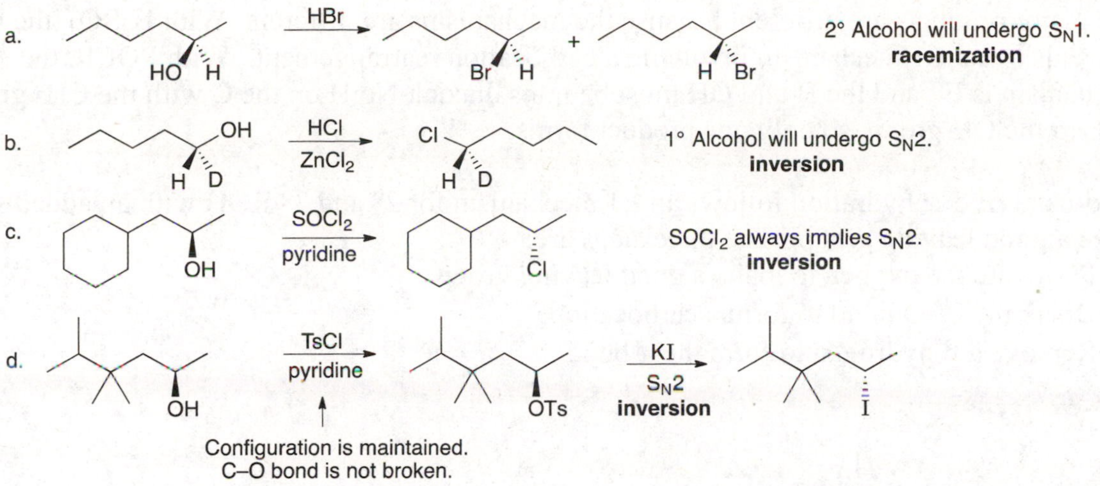

a. HO H → HBr → Br H + H Br

2° Alcohol will undergo S_N1.
racemization

b. OH, H D → HCl, $ZnCl_2$ → Cl, H D

1° Alcohol will undergo S_N2.
inversion

c. OH → $SOCl_2$, pyridine → Cl

$SOCl_2$ always implies S_N2.
inversion

d. OH → TsCl, pyridine → OTs → KI, S_N2, **inversion** → I

Configuration is maintained.
C–O bond is not broken.

9.49

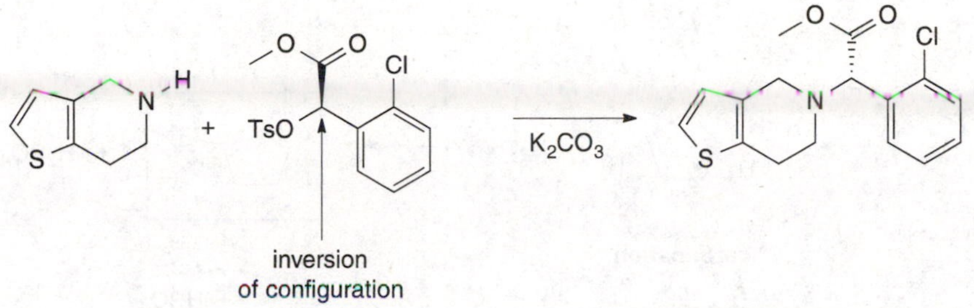

N–H + TsO (Cl) → K_2CO_3 → S, N, O, O, Cl

inversion
of configuration

9.50

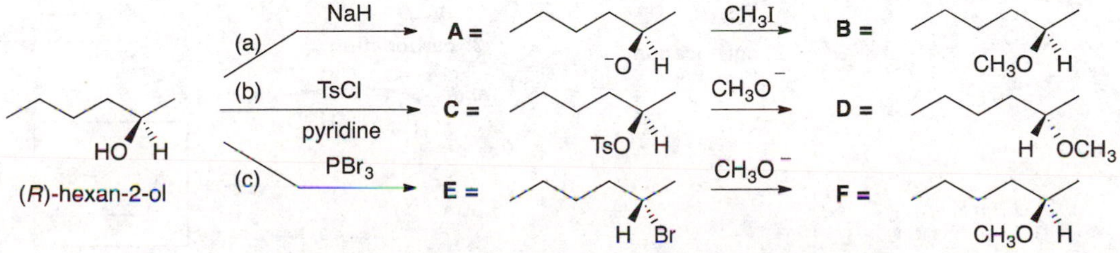

HO H
(*R*)-hexan-2-ol

(a) NaH → A = ⁻O H → CH_3I → B = CH_3O H

(b) TsCl, pyridine → C = TsO H → CH_3O^- → D = H OCH_3

(c) PBr_3 → E = H Br → CH_3O^- → F = CH_3O H

Routes (a) and (c) give identical products, labeled **B** and **F**.

9.51

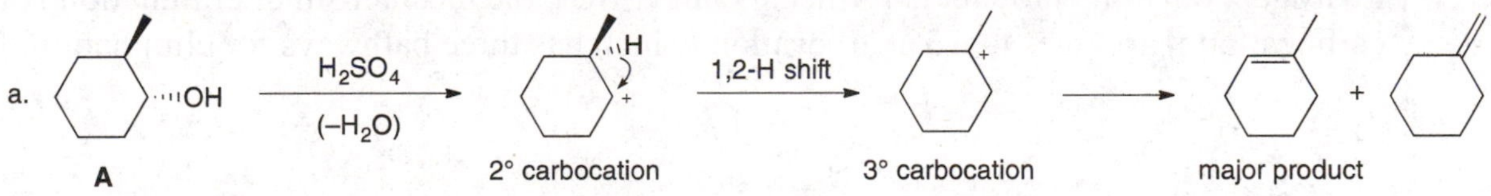

a. A (OH) → H_2SO_4, ($-H_2O$) → 2° carbocation (H) → 1,2-H shift → 3° carbocation → major product +

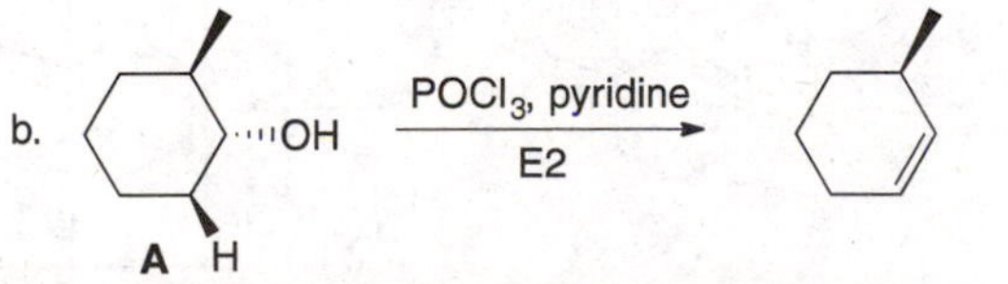

b. A (OH, H) → $POCl_3$, pyridine, E2 →

H and OH must be trans in the E2 reaction.

The major products are different because the mechanisms are different. With H_2SO_4, the reaction proceeds by an E1 mechanism involving a carbocation rearrangement. With $POCl_3$, the mechanism is E2 and the H and OH must be trans diaxial. No H on the C with the CH_3 group is trans to the OH group, so only one product forms.

9.52 Acid-catalyzed dehydration follows an E1 mechanism for 2° and 3° ROH with an added step to make a good leaving group. The three steps are:
[1] Protonate the oxygen to make a good leaving group.
[2] Break the C–O bond to form a carbocation.
[3] Remove a β hydrogen to form the π bond.

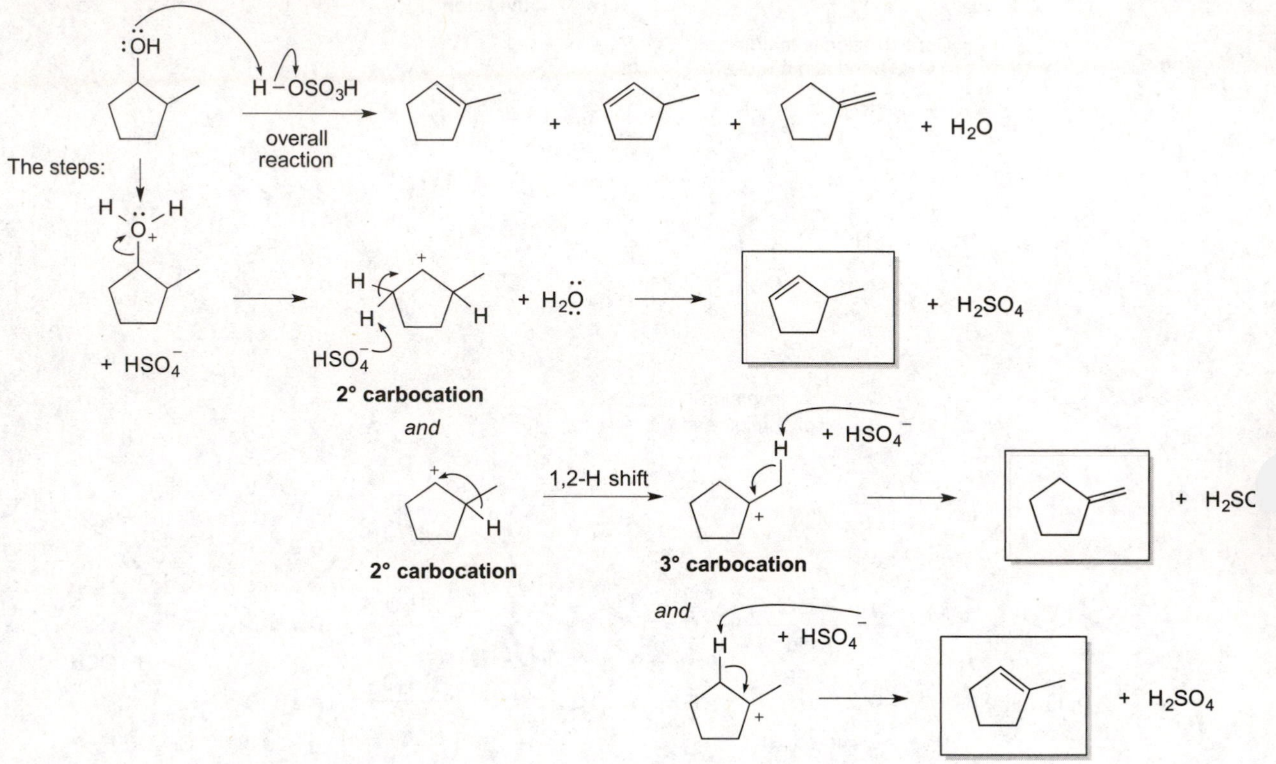

9.53 With $POCl_3$ (pyridine), elimination occurs by an E2 mechanism. Because only one carbon has a β hydrogen, only one product is formed. With H_2SO_4, the mechanism of elimination is E1. A 2° carbocation rearranges to a 3° carbocation, which has three pathways for elimination.

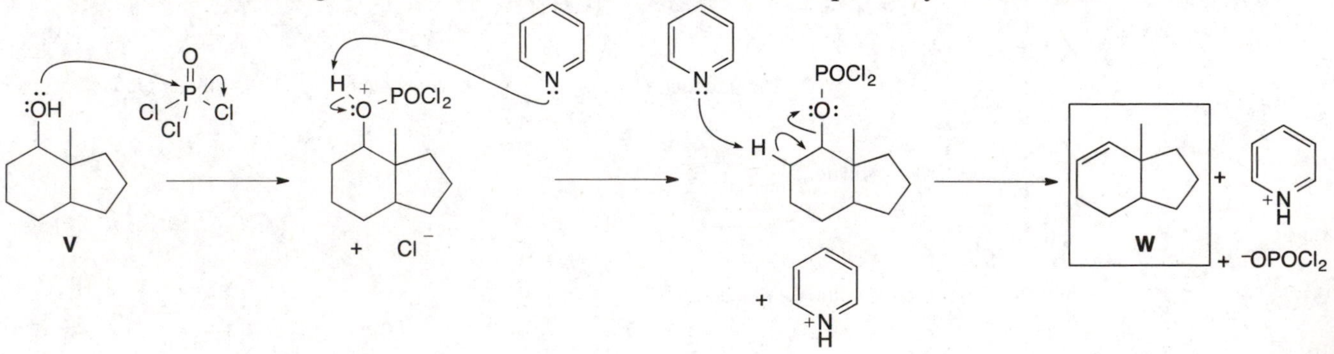

V

$+ HSO_4^-$

2° carbocation

$+ H_2O$

1,2–CH_3 shift

3° carbocation

3° carbocation

3° carbocation

3° carbocation

X

$+ H_2SO_4$

Y

$+ H_2SO_4$

Z

$+ H_2SO_4$

9.54 To draw the mechanism:
[1] Protonate the oxygen to make a good leaving group.
[2] Break the C–O bond to form a carbocation.
[3] Look for possible rearrangements to make a more stable carbocation.
[4] Remove a β hydrogen to form the π bond.

Dark and light circles are meant to show where the carbons in the starting material appear in the product.

$+ HSO_4^-$

2º carbocation

$+ H_2O$

$+ HSO_4^-$

3º carbocation

$+ H_2SO_4$

9.55

3-methylbutan-2-ol

[1] HBr
[2] $-H_2O$

2° carbocation

1,2-H shift

3° carbocation

The 2° alcohol reacts by an S_N1 mechanism to form a carbocation that rearranges.

2-methylpropan-1-ol

HBr

S_N2
no carbocation

$+ H_2O$

The 1° alcohol reacts with HBr by an S_N2 mechanism.
no carbocation intermediate = no rearrangement possible

9.56

A

H–Cl

$\ddot{O}H$

$\ddot{O}H_2^+$

+ $:\ddot{Cl}:^-$

two resonance structures for the carbocation

+ $H_2\ddot{O}$

$:\ddot{Cl}:^-$

B

Cl

$:\ddot{Cl}:^-$

Cl

C

9.57

$\ddot{O}H$

$H–OSO_3H$

$+\ddot{O}H_2$

+ HSO_4^-

+ $H_2\ddot{O}$

1,2-shift

H

+ HSO_4^-

+ H_2SO_4

9.58

$\ddot{O}H$

$H–OSO_3H$

$\ddot{O}H$

$\ddot{O}H_2^+$

+ HSO_4^-

+ $H_2\ddot{O}$

HSO_4^-

H

+ H_2SO_4

9.59

a.

O

O^- + Br

2° halide

Br + ^-O

1° halide

less hindered RX

preferred path

b.

O

O^-

+ Br

1° halide

less hindered RX

preferred path

O

Br

+ ^-O

2° halide

c.

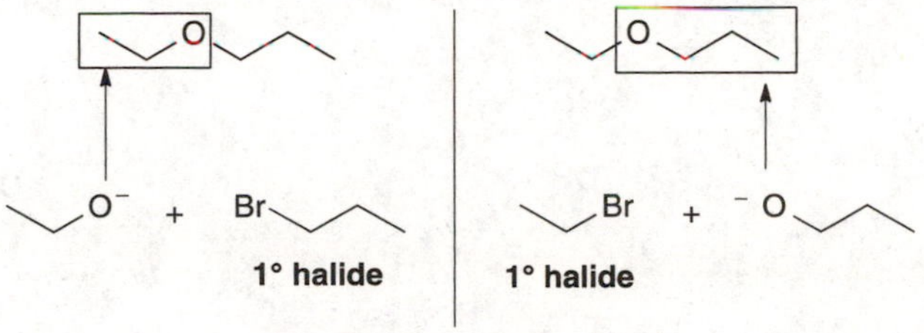
O
O
O⁻
Br
1° halide
Br
⁻O
1° halide

Neither path preferred.

9.60 A tertiary halide is too hindered and an aryl halide is too unreactive to undergo a Williamson ether synthesis.

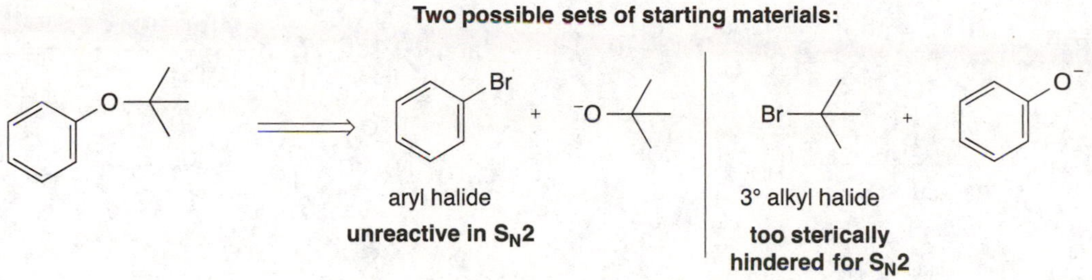
Two possible sets of starting materials:
O
Br
⁻O
Br
O⁻
aryl halide
unreactive in S_N2
3° alkyl halide
too sterically
hindered for S_N2

9.61

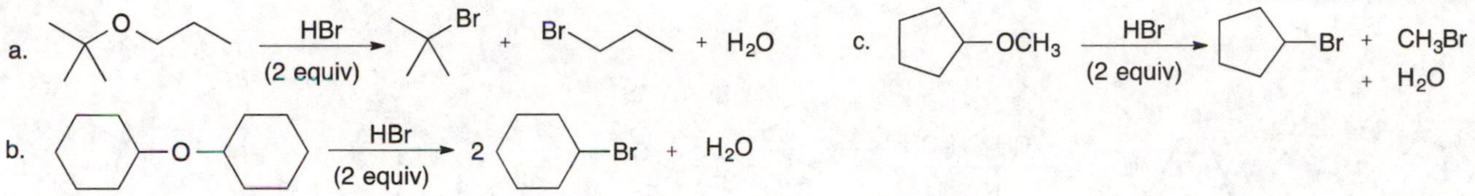
a.
O
HBr
(2 equiv)
Br
Br
+ H_2O
c.
OCH_3
HBr
(2 equiv)
Br
+ CH_3Br
+ H_2O
b.
O
HBr
(2 equiv)
2
Br
+ H_2O

9.62

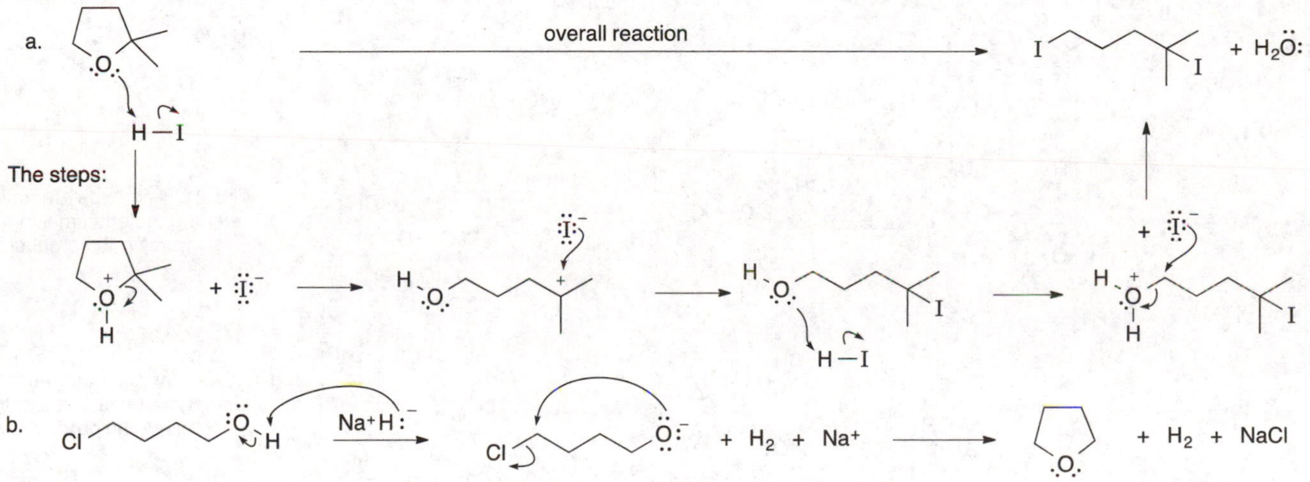
a.
overall reaction
I
I
+ H_2O:
H—I
The steps:
+ I⁻
H
O
I⁻
H
O
I
H—I
H
O
H
I
I⁻
b.
Cl
O
H
Na⁺H:⁻
Cl
O⁻
+ H_2 + Na⁺
O
+ H_2 + NaCl

9.63

$H{-}OCOCF_3$

$+\ CF_3CO_2^-$

$CF_3CO_2^-$

$+\ CF_3CO_2H$

9.64

a. HBr

b. H_2O, H_2SO_4

c. [1] $CH_3CH_2O^-$ [2] H_2O

d. [1] $HC{\equiv}C^-$ [2] H_2O

e. [1] ^-OH [2] H_2O

f. [1] CH_3S^- [2] H_2O

9.65

a. CH_3CH_2OH, H_2SO_4

b. [1] $CH_3CH_2O^-\ Na^+$ [2] H_2O

c. HBr

d. [1] NaCN [2] H_2O

9.66

a. $Na^+H{:}^-$ → C_4H_8O

The 2 CH_3 groups are anti in the starting material, making them trans in the product.

b. $Na^+H{:}^-$ → C_4H_8O

The 2 CH_3 groups are gauche in the starting material, making them cis in the product.

c. rotate; $Na^+H{:}^-$; backside attack → C_4H_8O

9.67 First, use the names to draw the structures of the starting material and both products. Because the product has two OH groups, one OH must come from the epoxide oxygen, and one must come from the nucleophile, either ^-OH or H_2O.

With ^-OH, the nucleophile attacks at the less substituted end of the epoxide to form the *R* isomer of the product.

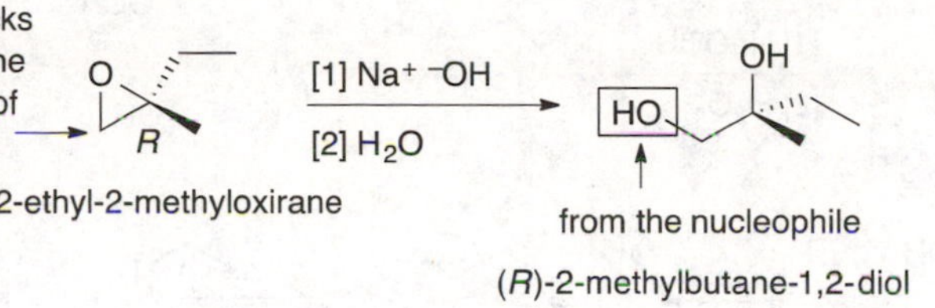

[1] $Na^+ \ ^-OH$
[2] H_2O

(*R*)-2-ethyl-2-methyloxirane → (*R*)-2-methylbutane-1,2-diol

HO — from the nucleophile

With H_2O and H_2SO_4, H_2O attacks at the more substituted end of the epoxide, from the back side, and the *S* isomer is formed.

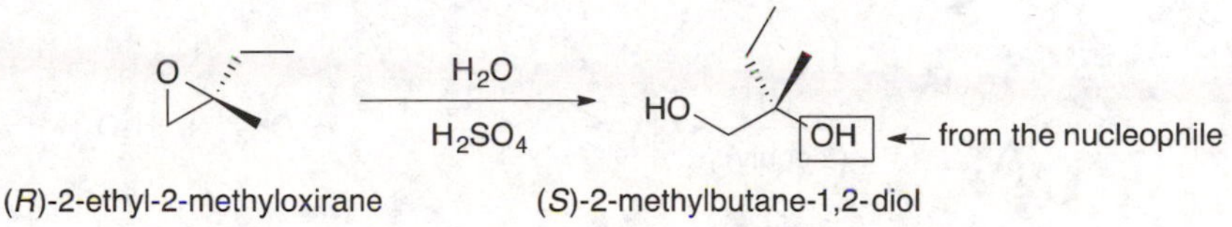

H_2O, H_2SO_4

(*R*)-2-ethyl-2-methyloxirane → (*S*)-2-methylbutane-1,2-diol

OH ← from the nucleophile

9.68

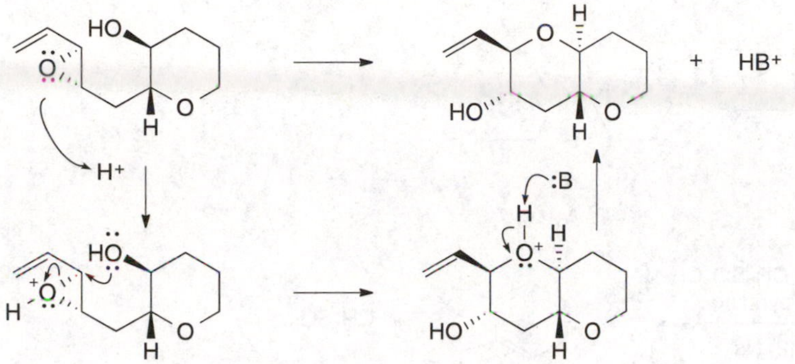

H^+ + HB^+ :B

9.69

a. Cyclopentyl–OTs —$KOC(CH_3)_3$→ cyclopentene. Bulky base favors E2.

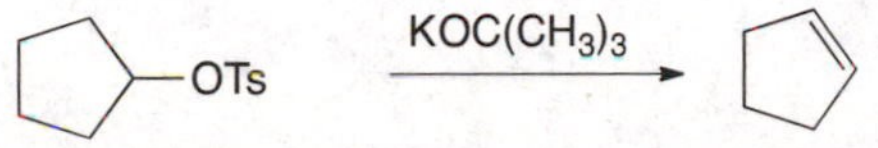

b. —HBr→ Br. Keep the stereochemistry at the stereogenic center [*] the same here because no bond to it is broken .

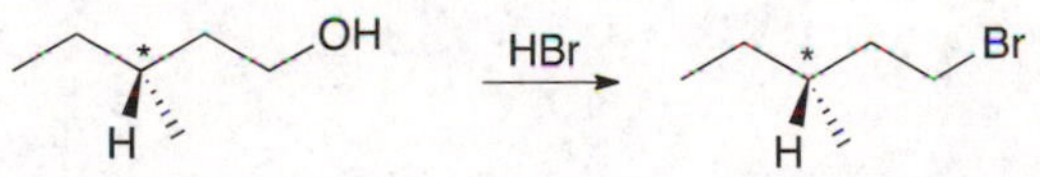

c. SH —Br_2→ S–S

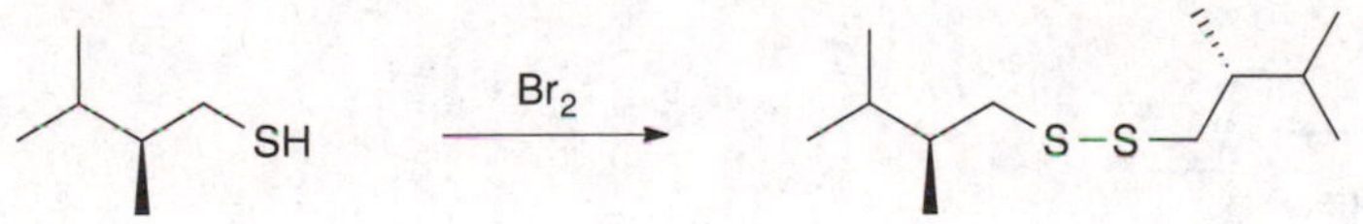

d. OTs —KSH→ SH

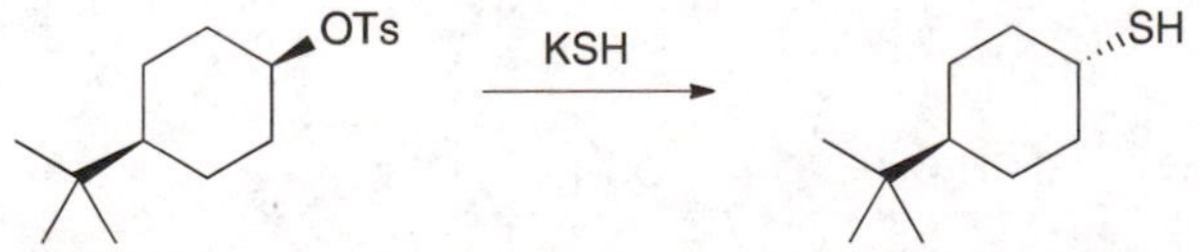

e. OH —PBr_3→ Br. S_N2 inversion

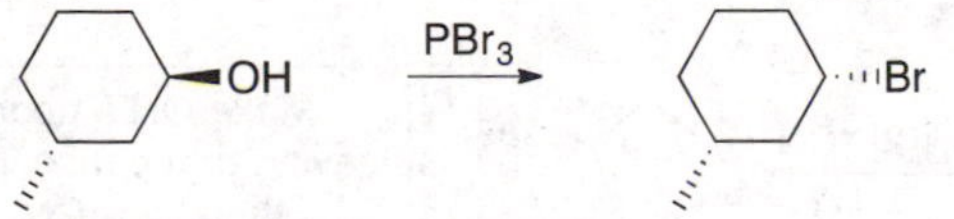

f. OH (H, D) —TsCl, pyridine→ OTs (H, D) —$CH_3CO_2^-$→ CH_3CO_2 (H, D)

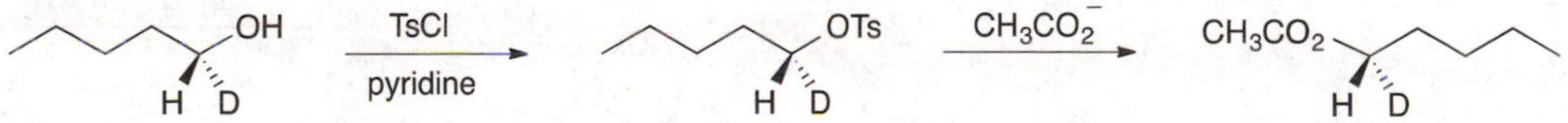

g. 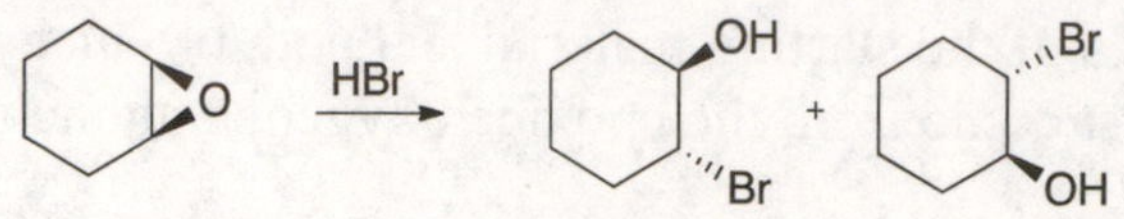 O; HBr; OH, Br + Br, OH

h. 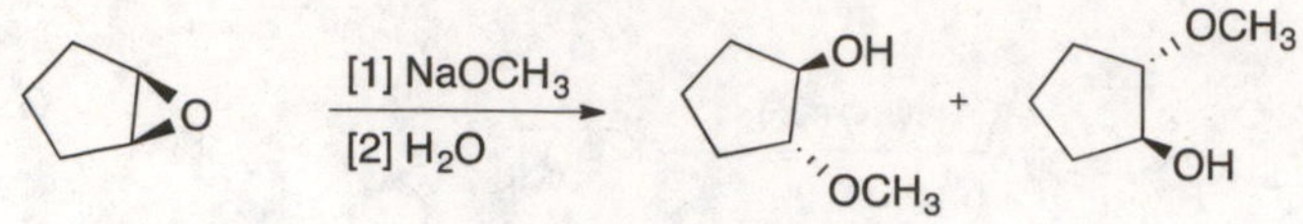 O; [1] $NaOCH_3$ [2] H_2O; OH, OCH_3 + OCH_3, OH

i. 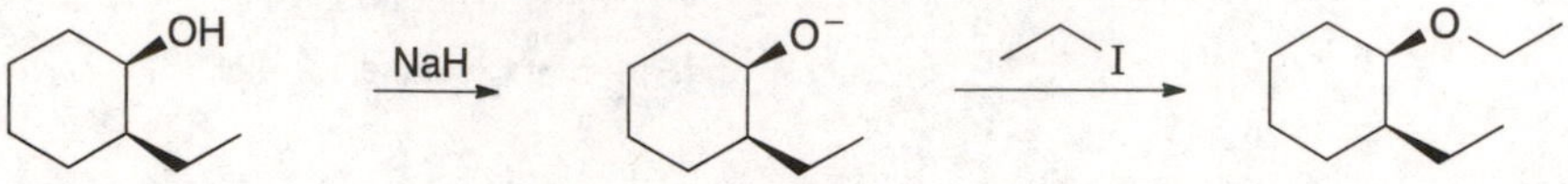 OH; NaH; O^-; I; O

j. 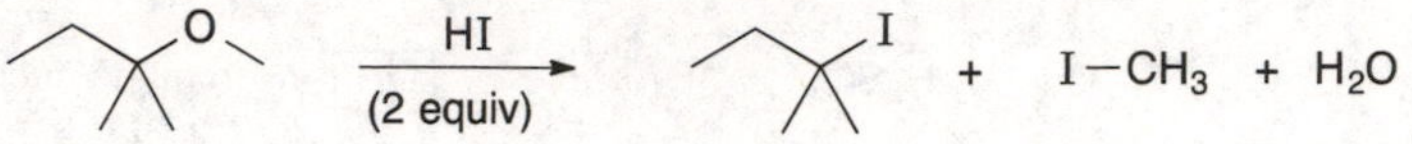 O; HI (2 equiv); I + $I-CH_3$ + H_2O

k. 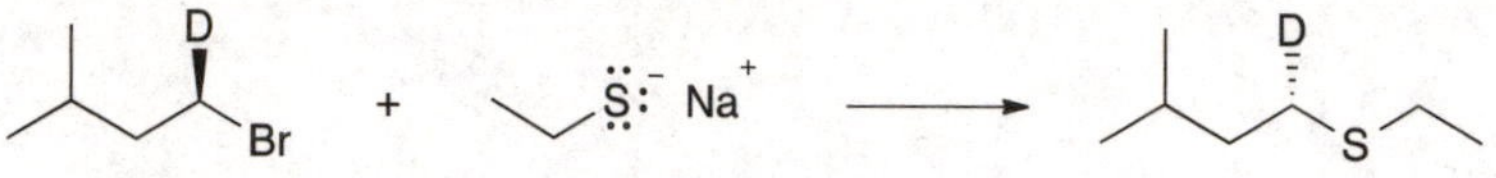 D, Br + $S:^-$ Na^+ → D, S

l. 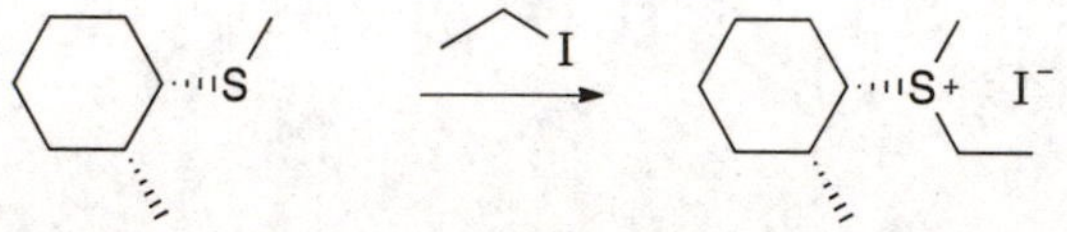 S; I; S^+ I^-

9.70

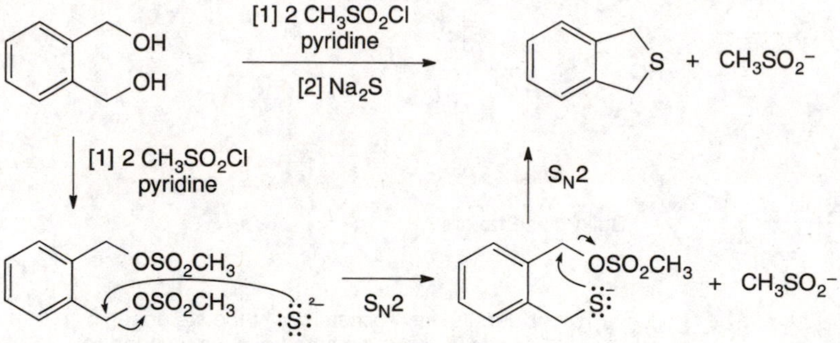

OH, OH; [1] 2 CH_3SO_2Cl pyridine [2] Na_2S; S + $CH_3SO_2^-$

[1] 2 CH_3SO_2Cl pyridine

OSO_2CH_3, OSO_2CH_3, $:S:^{2-}$; S_N2; OSO_2CH_3, $S:^-$ + $CH_3SO_2^-$; S_N2

9.71

a. 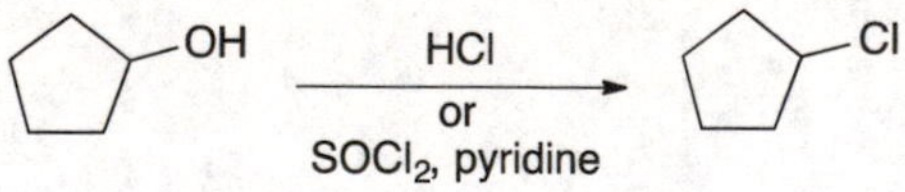 OH; HCl or $SOCl_2$, pyridine; Cl

b. 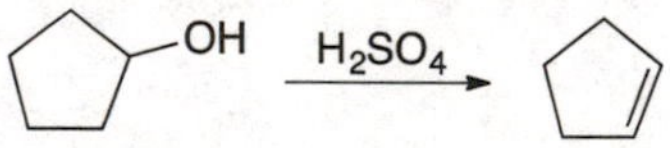 OH; H_2SO_4

c. 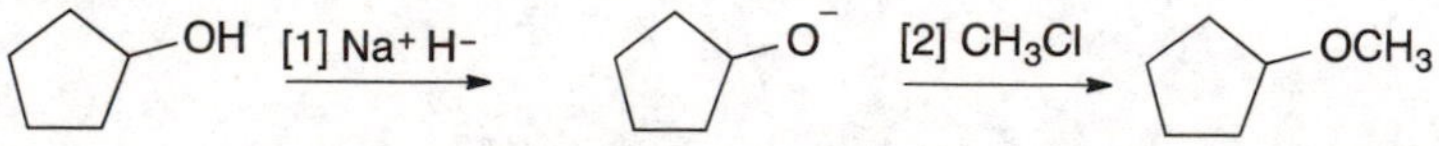 OH; [1] Na^+ H^-; O^-; [2] CH_3Cl; OCH_3

d. 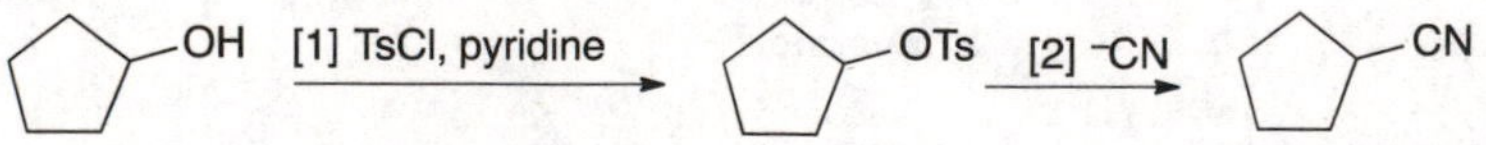 OH; [1] TsCl, pyridine; OTs; [2] ^-CN; CN

Make OH a good leaving group (use TsCl); then add ^-CN.

9.72

(a) = HBr or PBr_3 → Br

(b) = $KOC(CH_3)_3$ or other strong base

OH

(c) = TsCl, pyridine → OTs

(d) = $KOC(CH_3)_3$ or other strong base

(e) = H_2SO_4 or TsOH

NBS → Br

(f) = $KOC(CH_3)_3$ or other strong base

HOCl → OH, Cl

(g) = NaH → O

(h) = ^{-}OH, H_2O → OH, OH + enantiomer

9.73

OH

NaH

:Ö:$^{-}$

O

Cl

O

O

H

N

H

O

O^{-}

N

H H

proton transfer

O

N

H

OH

propranolol

9.74 With the cis isomer, ^{-}OH acts as a nucleophile to displace Br^{-} from the back side, forming a trans diol (**A**). With the trans isomer, the two functional groups are arranged in a manner that allows an intramolecular S_N2. ^{-}OH removes a proton to form an alkoxide, which can then displace Br^{-} by intramolecular backside attack to afford an ether (**B**). Such a reaction is not possible with the cis isomer because the nucleophile and leaving group are on the same side.

HO, Br

cis-4-bromocyclohexanol

$^{-}$:ÖH

S_N2

HO, OH

A

HO, Br

trans-4-bromocyclohexanol

$^{-}$:ÖH

proton transfer

:Ö:$^{-}$, Br

+ H_2O

S_N2

O

B

9.75 If the base is not bulky, it can react as a nucleophile and open the epoxide ring. The bulky base cannot act as a nucleophile, and will only remove the proton.

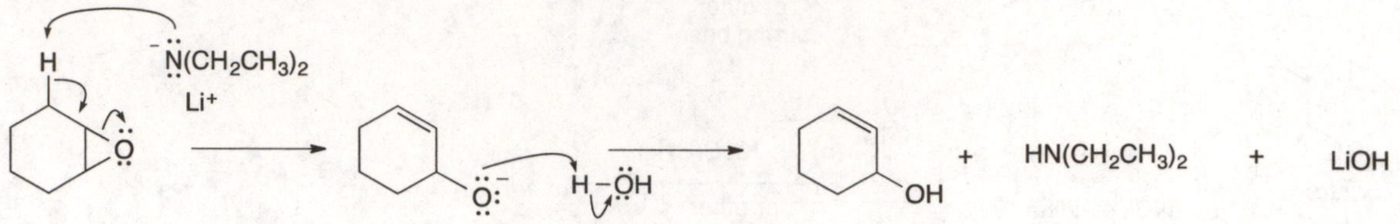

H $^-\ddot{N}(CH_2CH_3)_2$ Li^+ → $H-\ddot{O}H$ → OH + $HN(CH_2CH_3)_2$ + LiOH

9.76 First form the 2° carbocation. Then lose a proton to form each product.

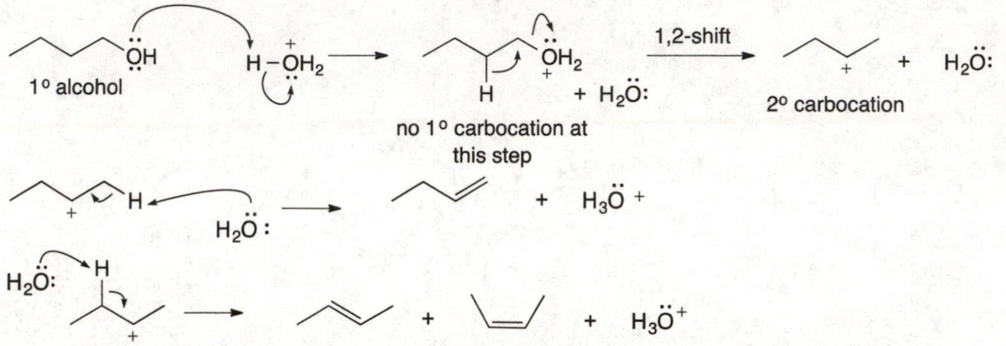

OH 1° alcohol; $H-\overset{+}{O}H_2$ → $\overset{+}{O}H_2$ H + $H_2\ddot{O}$: no 1° carbocation at this step; 1,2-shift → + $H_2\ddot{O}$: 2° carbocation

H $H_2\ddot{O}$: → + $H_3\ddot{O}$ +

$H_2\ddot{O}$: H → + + $H_3\ddot{O}^+$

9.77

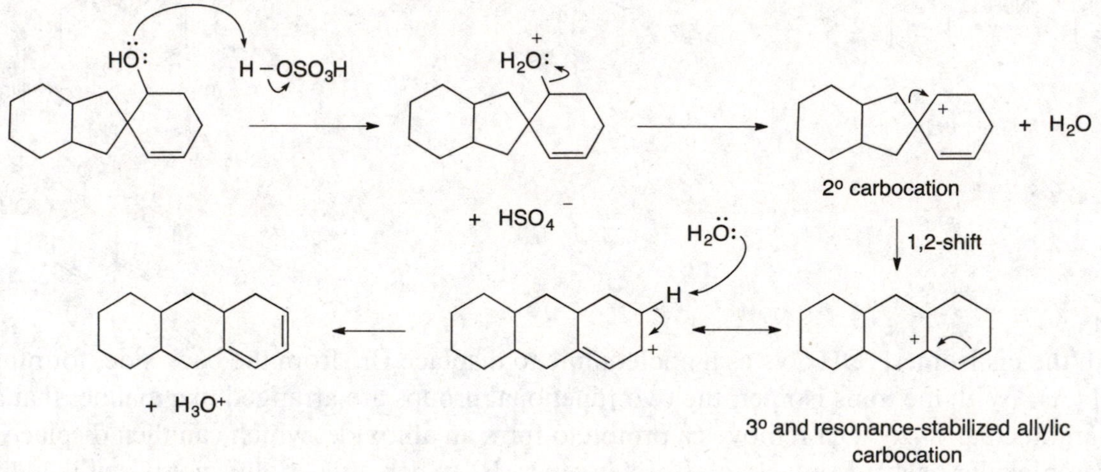

$H\ddot{O}$: $H-OSO_3H$ → $\overset{+}{H_2O}$: + HSO_4^- → + H_2O 2° carbocation

1,2-shift; 3° and resonance-stabilized allylic carbocation ⟷ $H_2\ddot{O}$: H → + H_3O^+

9.78

a.

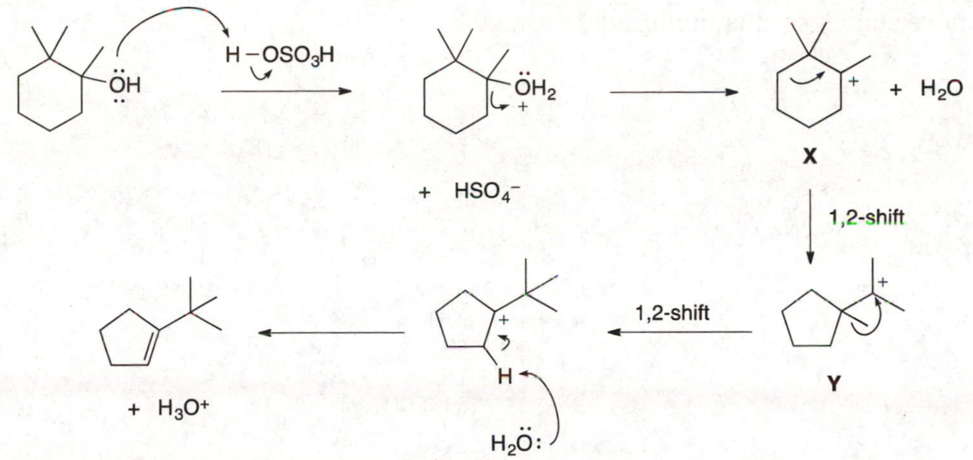

H–OSO_3H

+ HSO_4^-

+ H_2O

X

1,2-shift

Y

1,2-shift

+ H_3O^+

$H_2\ddot{O}$:

b. Other elimination products can form from carbocations **X** and **Y.**

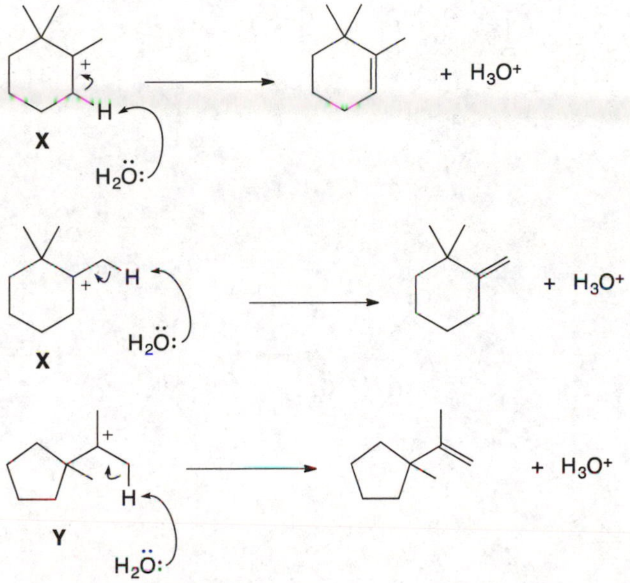

X

$H_2\ddot{O}$:

+ H_3O^+

X

$H_2\ddot{O}$:

+ H_3O^+

Y

$H_2\ddot{O}$:

+ H_3O^+

9.79

a.

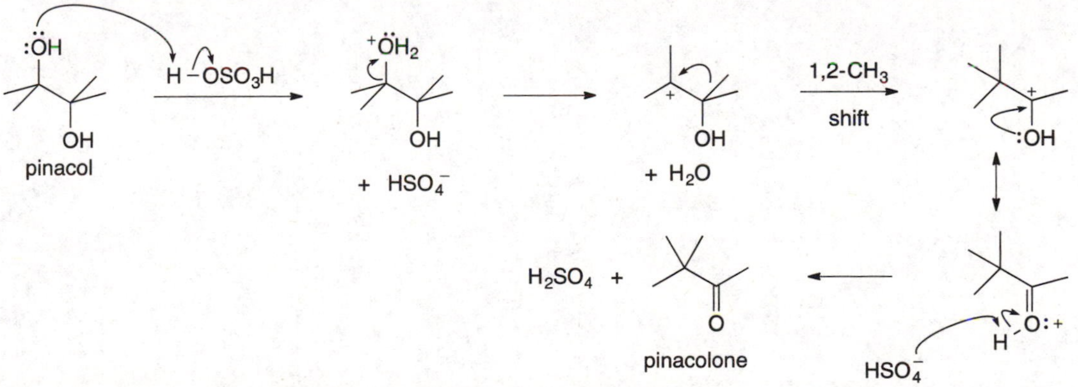

:ÖH

OH

pinacol

H–OSO_3H

$^+\ddot{O}H_2$

OH

+ HSO_4^-

OH

+ H_2O

1,2-CH_3

shift

:OH

H_2SO_4 +

O

pinacolone

H

:O:+

HSO_4^-

b. Two different carbocations can form. The carbocation with the (+) charge adjacent to the benzene rings (**A**) is more stable, so it is preferred.

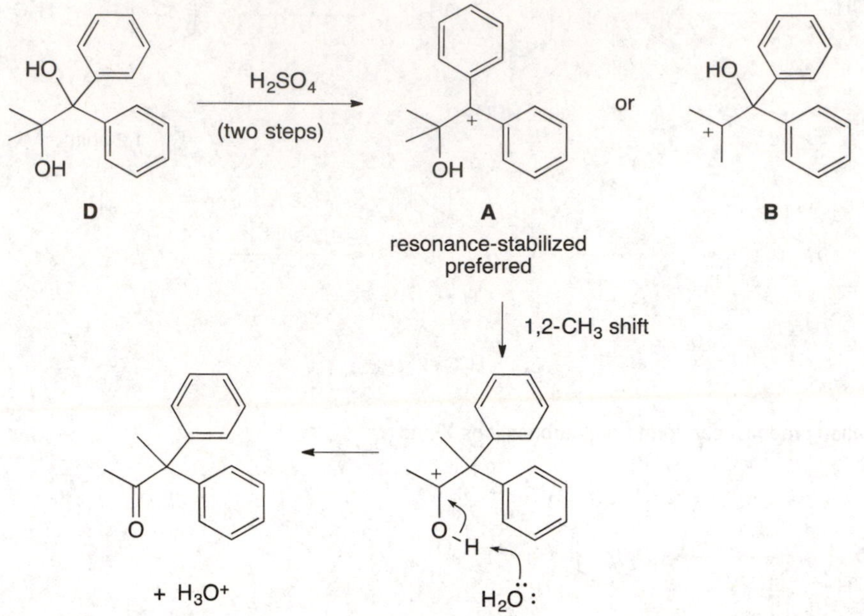

HO OH **D** $\xrightarrow[\text{(two steps)}]{H_2SO_4}$ OH **A** or HO **B**

A: resonance-stabilized preferred

1,2-CH_3 shift

+ H_3O^+ H_2O

9.80

Na+ :OH

+ I⁻

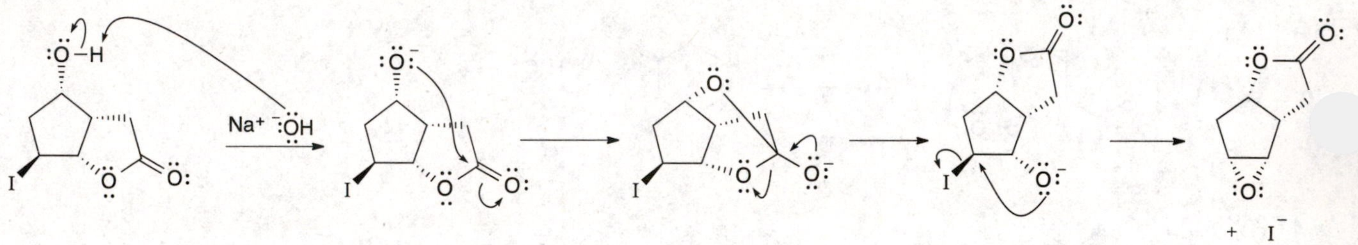

9.81 The conversion of **X** to **Y** requires two operations. **X** contains both a nucleophile (NH_2) and a leaving group (OSO_2CH_3), so an intramolecular S_N2 reaction forms an aziridine. Because the aziridine is strained, the amine nucleophile ($CH_2{=}CHCH_2NH_2$) opens the ring by backside attack, resulting in the trans stereochemistry of the two N's on the six-membered ring.

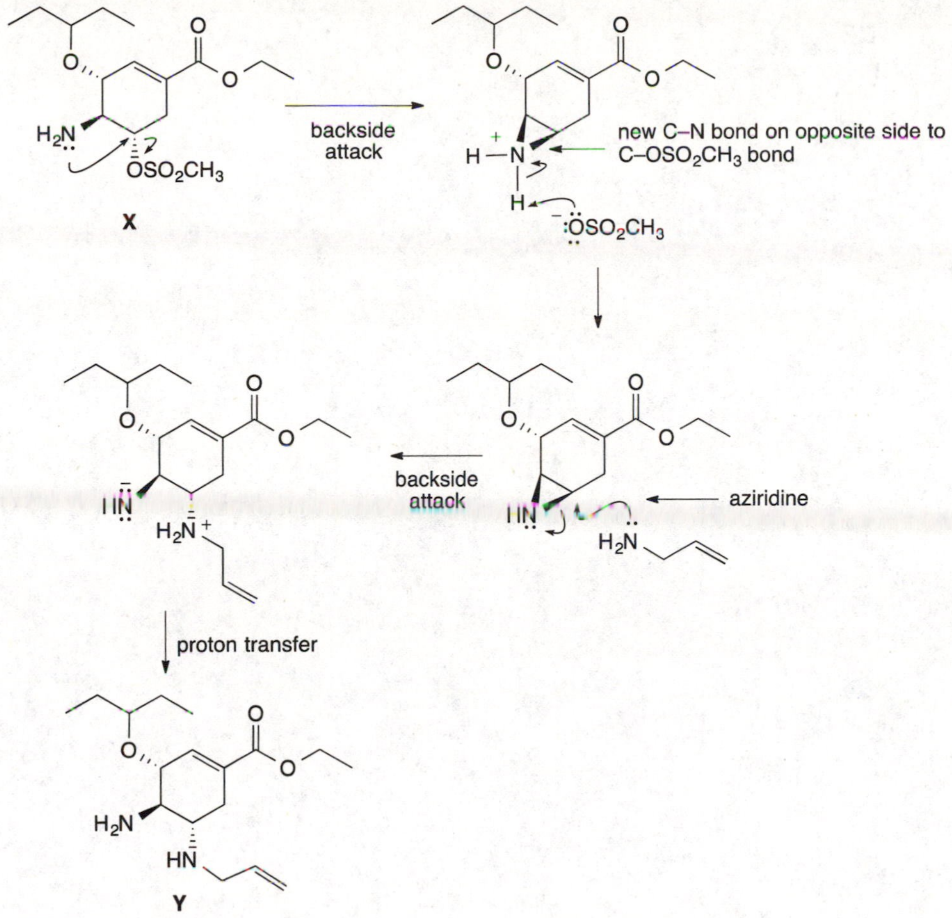

X

backside attack

new C–N bond on opposite side to C–OSO_2CH_3 bond

aziridine

backside attack

proton transfer

Y

Chapter 10 Alkenes

Chapter Review

General facts about alkenes

- Alkenes contain a carbon–carbon double bond consisting of a stronger σ bond and a weaker π bond. Each carbon is sp^2 hybridized and trigonal planar (10.1).
- Alkenes are named using the suffix ***-ene*** (10.3).
- Alkenes with different groups on each end of the double bond exist as a pair of diastereomers, identified by the prefixes *E* and *Z* (10.3B).

- Alkenes have weak intermolecular forces, giving them low mp's and bp's, and making them water insoluble. A cis alkene is more polar than a trans alkene, giving it a slightly higher boiling point (10.4).

cis-but-2-ene — **more polar isomer** — a small net dipole, **higher bp**

trans-but-2-ene — **less polar isomer** — **no net dipole**, **lower bp**

- A π bond is electron rich and much weaker than a σ bond, so alkenes undergo addition reactions with electrophiles (10.8).

Stereochemistry of alkene addition reactions (10.8)

A reagent XY adds to a double bond in one of three different ways:

- **Syn addition**—X and Y add from the same side.

  C=C —(H−BH_2)→ H and BH_2 on C−C (same side)

  - Syn addition occurs in **hydroboration.**

- **Anti addition**—X and Y add from opposite sides.

  C=C —(X_2 or X_2, H_2O)→ X and X(OH) on C−C (opposite sides)

  - Anti addition occurs in **halogenation** and **halohydrin formation.**

- **Both syn and anti addition** occur when carbocations are intermediates.

C=C —($H-X$ or H_2O, H^+)→ H–C–C–X(OH) and H–C–C–X(OH)

- Syn and anti addition occur in **hydrohalogenation** and **hydration.**

Addition reactions of alkenes

[1] **Hydrohalogenation**—Addition of HX (X = Cl, Br, I) (10.9–10.11)

R–CH=CH₂ + H–X → [R–CHX–CH₃] alkyl halide

- The mechanism has two steps.
- Carbocations are formed as intermediates.
- Carbocation rearrangements are possible.
- Markovnikov's rule is followed. H bonds to the less substituted C to form the more stable carbocation.
- Syn and anti addition occur.

[2] **Hydration** and related reactions—Addition of H_2O or ROH (10.12)

R–CH=CH₂ + H–OH —(H_2SO_4)→ [R–CH(OH)–CH₃] alcohol

R–CH=CH₂ + H–OR —(H_2SO_4)→ [R–CH(OR)–CH₃] ether

For both reactions:

- The mechanism has three steps.
- Carbocations are formed as intermediates.
- Carbocation rearrangements are possible.
- Markovnikov's rule is followed. H bonds to the less substituted C to form the more stable carbocation.
- Syn and anti addition occur.

[3] **Halogenation**—Addition of X_2 (X = Cl or Br) (10.13–10.14)

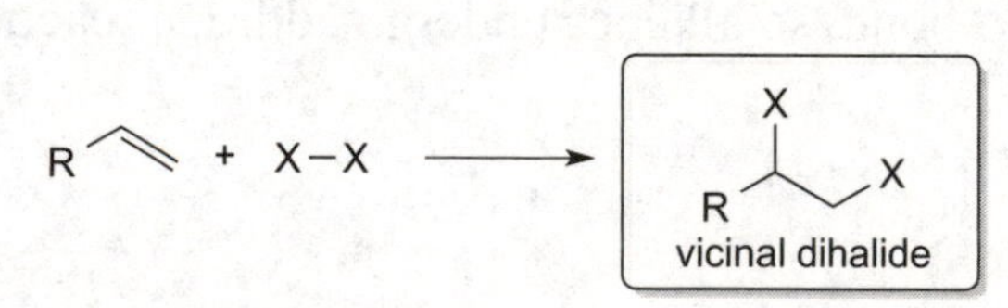

- The mechanism has two steps.
- Bridged halonium ions are formed as intermediates.
- No rearrangements occur.
- Anti addition occurs.

[4] **Halohydrin formation**—Addition of OH and X (X = Cl, Br) (10.15)

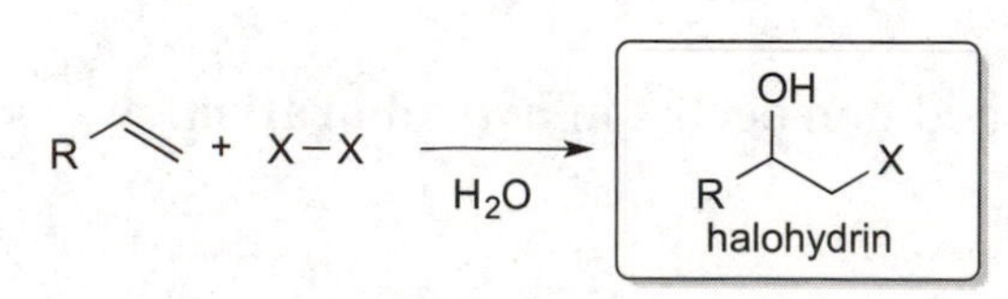

- The mechanism has three steps.
- Bridged halonium ions are formed as intermediates.
- No rearrangements occur.
- X bonds to the less substituted C.
- Anti addition occurs.
- NBS in DMSO and H_2O adds Br and OH in the same fashion.

[5] **Hydroboration–oxidation**—Addition of H_2O (10.16)

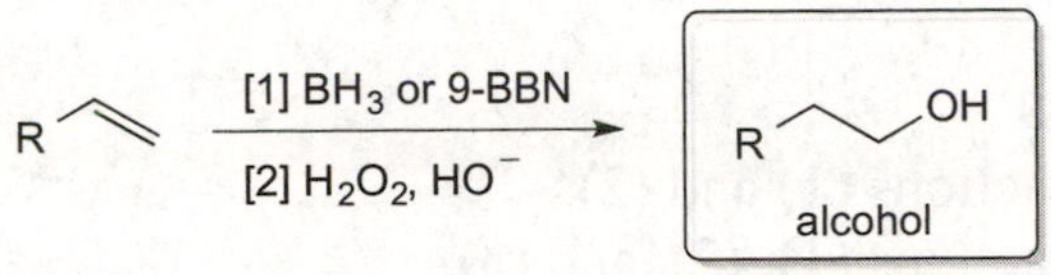

R–CH=CH₂ → [1] BH_3 or 9-BBN; [2] H_2O_2, HO^- → R–CH₂CH₂–OH (alcohol)

- Hydroboration has a one-step mechanism.
- No rearrangements occur.
- OH bonds to the less substituted C.
- Syn addition of H_2O results.

Practice Test on Chapter Review

1. Give the IUPAC name for the following compounds.

a.

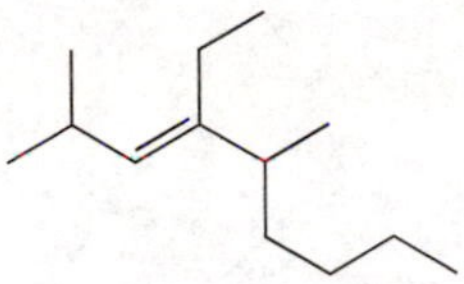

b.

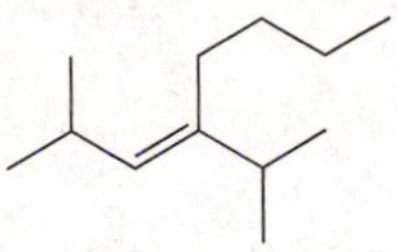

2. Draw the organic products formed in the following reactions. Draw all stereogenic centers using wedges and dashes.

a. 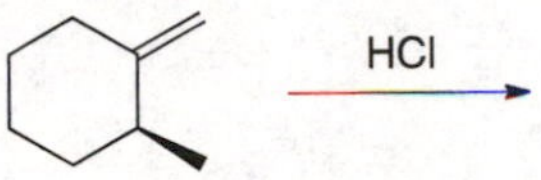 → HCl →

b. 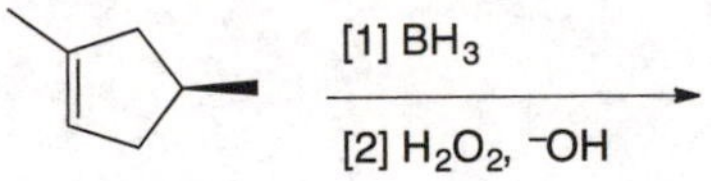 → [1] BH_3; [2] H_2O_2, ^-OH →

c. 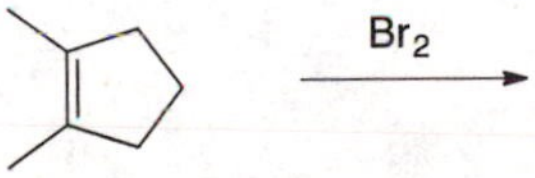 → Br_2 →

d. 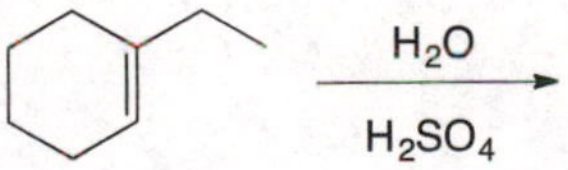 → H_2O, H_2SO_4 →

3. Fill in the table with the stereochemistry observed in the reaction of an alkene with each reagent. Choose from syn, anti, or both syn and anti.

Reagent	Stereochemistry
a. [1] 9-BBN; [2] H_2O_2, ^-OH	
b. H_2O, H_2SO_4	
c. Cl_2, H_2O	
d. HI	
e. Br_2	

4. a. In which of the following reactions are carbocation rearrangements observed?
   1. hydrohalogenation
   2. halohydrin formation
   3. hydroboration–oxidation
   4. Carbocation rearrangements are observed in reactions (1) and (2).
   5. Carbocation rearrangements are observed in reactions (1), (2), and (3).

   b. Which of the following products are formed when HCl is added to 3-methylpent-1-ene?
   1. 2-chloro-2-methylpentane
   2. 3-chloro-3-methylpentane
   3. 1-chloro-3-methylpentane
   4. Both (1) and (2) are formed.
   5. Products (1), (2), and (3) are all formed.

Answers to Practice Test

1. a. (*E*)-4-ethyl-2,5-dimethylnon-3-ene
   b. (*E*)-4-isopropyl-2-methyloct-3-ene

2. a. Cl + Cl
   b. HO + HO
   c. Br, Br + Br, Br
   d. OH

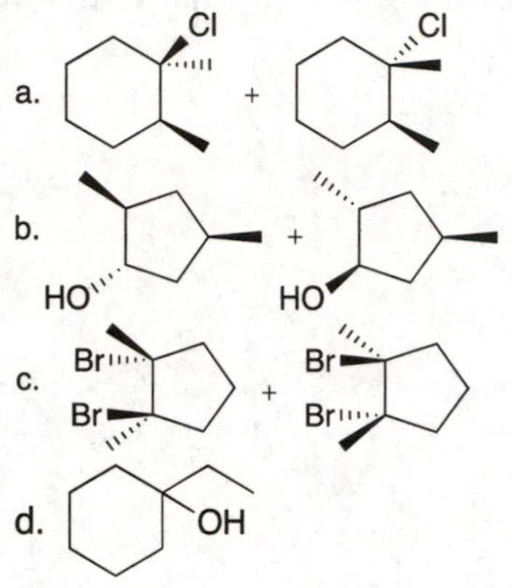

3. a. syn
   b. both
   c. anti
   d. both
   e. anti

4. a. 1
   b. 2

Answers to Problems

10.1

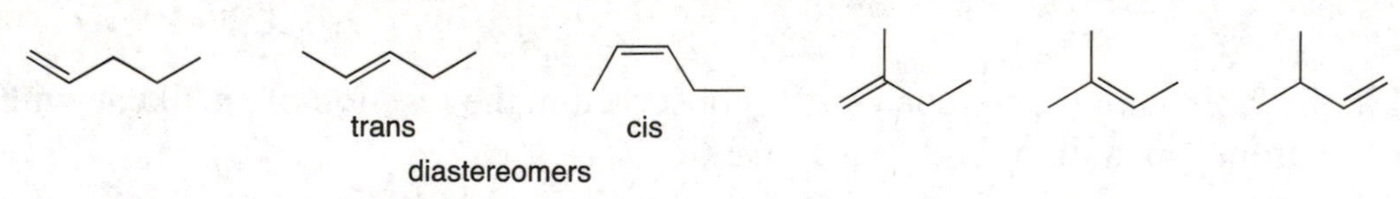

10.2 To determine the number of degrees of unsaturation:

[1] Calculate the maximum number of H's ($2n + 2$).
[2] Subtract the actual number of H's from the maximum number.
[3] Divide by two.

a. C_6H_6
[1] maximum number of H's = $2n + 2 = 2(6) + 2 = 14$
[2] subtract actual from maximum = $14 - 6 = 8$
[3] divide by two = $8/2 =$ **4 degrees of unsaturation**

b. C_8H_{18}
[1] maximum number of H's = $2n + 2 = 2(8) + 2 = 18$
[2] subtract actual from maximum = $18 - 18 = 0$
[3] divide by two = $0/2 =$ **0 degrees of unsaturation**

c. C_7H_8O
Ignore the O.
[1] maximum number of H's = $2n + 2 = 2(7) + 2 = 16$
[2] subtract actual from maximum = $16 - 8 = 8$
[3] divide by two = $8/2 =$ **4 degrees of unsaturation**

d. $C_7H_{11}Br$
Because of Br, add one more H (11 + 1 H = 12 H's).
[1] maximum number of H's = $2n + 2 = 2(7) + 2 = 16$
[2] subtract actual from maximum = $16 - 12 = 4$
[3] divide by two = $4/2 =$ **2 degrees of unsaturation**

e. C_5H_9N
Because of N, subtract one H (9 – 1 H = 8 H's).
[1] maximum number of H's = $2n + 2 = 2(5) + 2 = 12$
[2] subtract actual from maximum = $12 - 8 = 4$
[3] divide by two = $4/2 =$ **2 degrees of unsaturation**

10.3 Use the directions from Answer 10.2. Ignore O in calculating degrees of unsaturation. Add one more H for each halogen. Subtract one H for each N.

a. $C_{19}H_{21}N_3O$ is equivalent to $C_{19}H_{18}$ when calculating degrees of unsaturation.
For 19 C's, the maximum number of H's = $2n + 2 = 2(19) + 2 = 40$ H's
Subtract actual from maximum = $40 - 18 = 22$ H's fewer
Divide by two = $22/2 = 11$ degrees of unsaturation

b. $C_{17}H_{16}F_6N_2O$ is equivalent to $C_{17}H_{20}$ when calculating degrees of unsaturation.
For 17 C's, the maximum number of H's = $2n + 2 = 2(17) + 2 = 36$ H's
Subtract actual from maximum = $36 - 20 = 16$ H's fewer
Divide by two = $16/2 = 8$ degrees of unsaturation

10.4 To name an alkene:

[1] Find the longest chain that contains the double bond. Change the ending from *-ane* to ***-ene.***
[2] Number the chain to give the double bond the lower number. The alkene is named by the first number.
[3] Apply all other rules of nomenclature.

To name a cycloalkene:

[1] When a double bond is located in a ring, it is always located between C1 and C2. Omit the "1" in the name. Change the ending from *-ane* to ***-ene.***
[2] Number the ring clockwise or counterclockwise to give the first substituent the lower number.
[3] Apply all other rules of nomenclature.

a. [1] 5 C chain with double bond **pentene** [2] **1** **3-methyl** **pent-1-ene** [3] **3-methylpent-1-ene**

b. [1] 7 C chain with double bond **heptene** [2] **3** **hept-3-ene** **3-ethyl** [3] **3-ethylhept-3-ene**

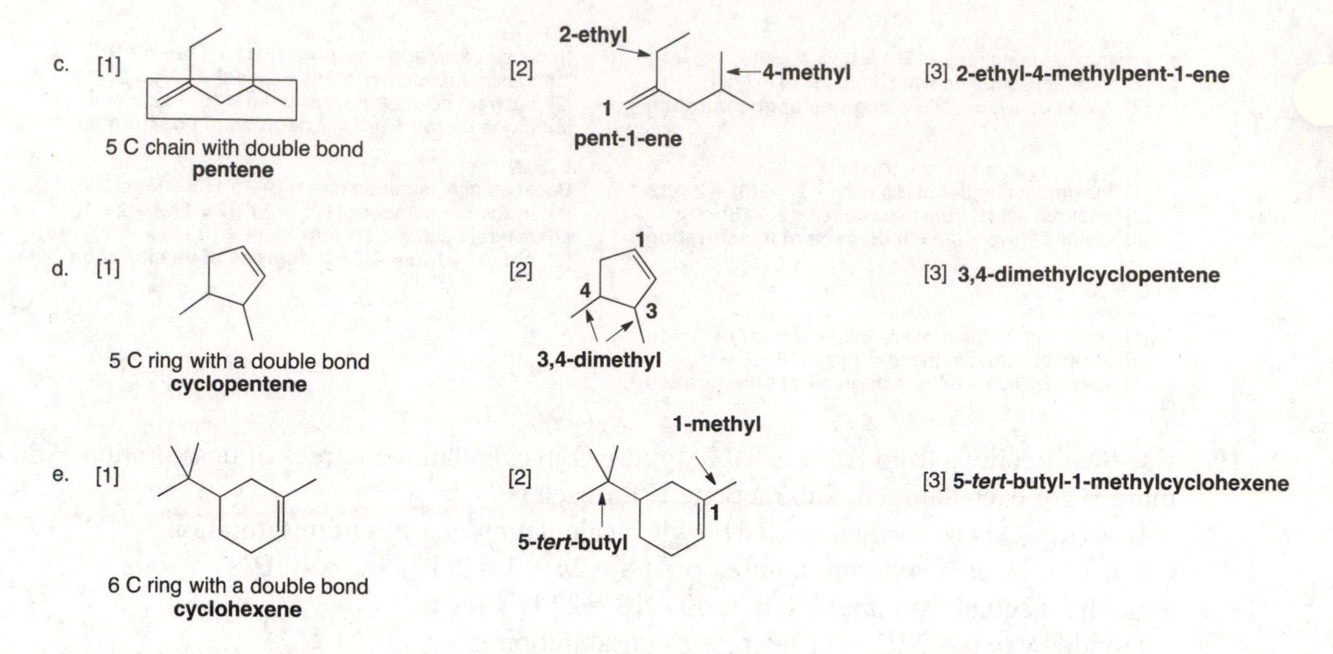

10.5 Use the rules from Answer 10.4 to name the compounds. Enols are named to give the OH the lower number. Compounds with two C=C's are named with the suffix ***-adiene.***

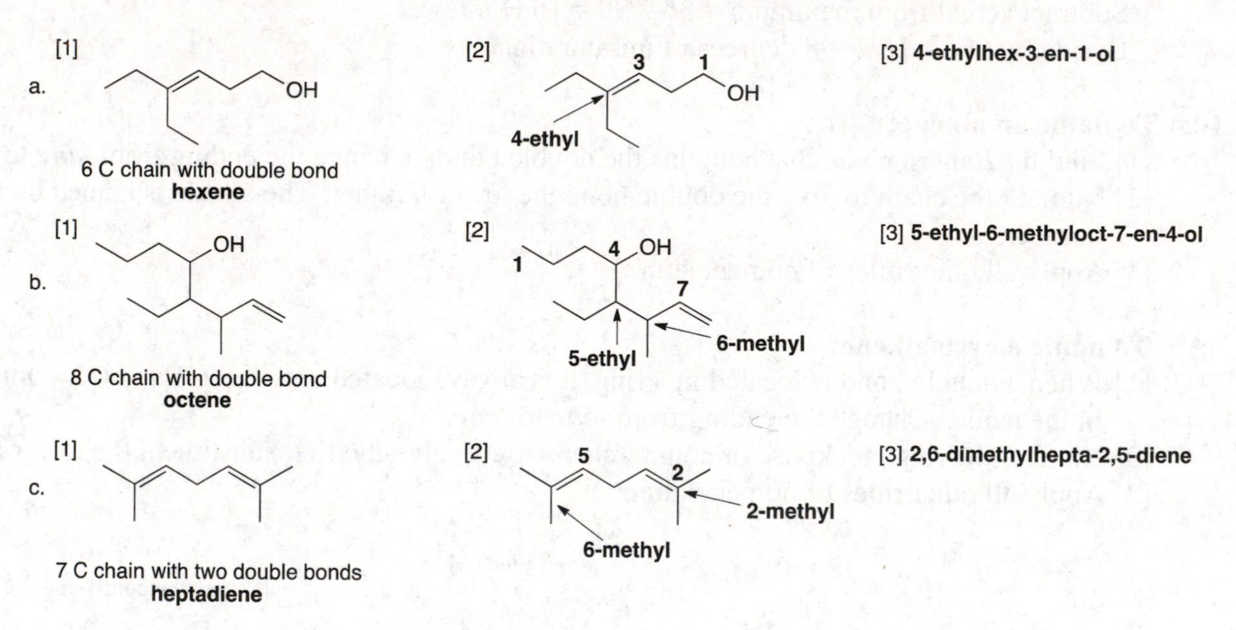

10.6 To label an alkene as *E* or *Z*:

[1] **Assign priorities** to the two substituents ***on each end*** using the rules for *R,S* nomenclature.

[2] **Assign *E* or *Z*** depending on the location of the two higher priority groups.

- The ***E*** prefix is used when the two higher priority groups are on **opposite sides** of the double bond.
- The ***Z*** prefix is used when the two higher priority groups are on the **same side** of the double bond.

a. higher priority → Cl

Br ← higher priority

Two higher priority groups are on opposite sides: ***E* isomer.**

b. higher priority →

← higher priority

Two higher priority groups are on the same side: ***Z* isomer.**

c. higher priority → OCH_3

H

← higher priority

H

higher priority → C_6H_5

O

H

higher priority

kavain

In both double bonds, the two higher priority groups are on opposite sides: ***E* isomers.**

10.7

E *E* *Z*

E

Z

CHO

skeletal structure of 11-*cis*-retinal

10.8 To work backwards from a name to a structure:

[1] Find the parent name and functional group and draw, remembering that the double bond is between C1 and C2 for cycloalkenes.

[2] Add the substituents to the appropriate carbons.

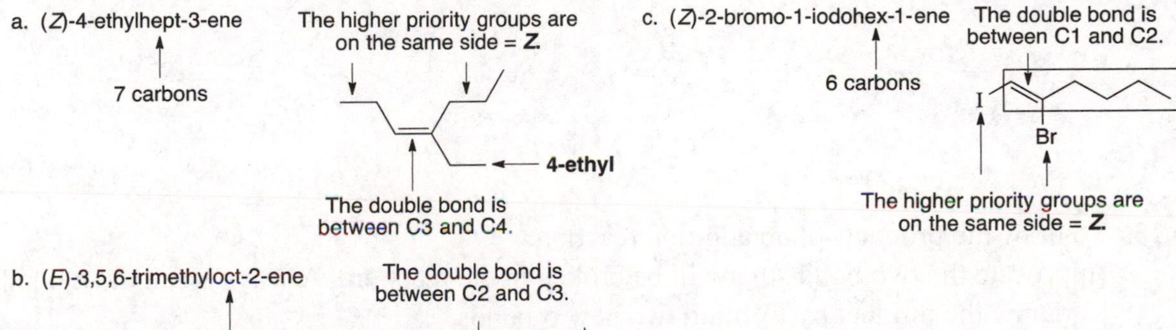
a. (*Z*)-4-ethylhept-3-ene
7 carbons
The higher priority groups are on the same side = *Z*.
4-ethyl
The double bond is between C3 and C4.
c. (*Z*)-2-bromo-1-iodohex-1-ene
6 carbons
The double bond is between C1 and C2.
I
Br
The higher priority groups are on the same side = *Z*.

b. (*E*)-3,5,6-trimethyloct-2-ene

8 carbons

The double bond is between C2 and C3.

The higher priority groups are on opposite sides = *E*.

3,5,6-trimethyl

10.9

(2*Z*,6*E*)-3-ethyl-7-methyldeca-2,6-dien-1-ol

10 *E* 6 *Z* 1 3 OH

10 C chain with OH at C1

2 C=C's beginning at C2, C6

10.10 To rank the isomers by increasing boiling point:
Look for polarity differences: *small net dipoles* make an alkene more polar, giving it a higher boiling point than an alkene with *no net dipole*. Cis isomers have a higher boiling point than their trans isomers.

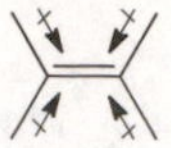

All dipoles cancel.
smallest surface area
no net dipole
lowest bp

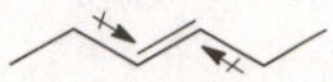

Two dipoles cancel.
no net dipole
trans isomer
intermediate bp

Two dipoles reinforce.
net dipole
cis isomer
highest bp

10.11 Increasing number of double bonds = decreasing melting point.

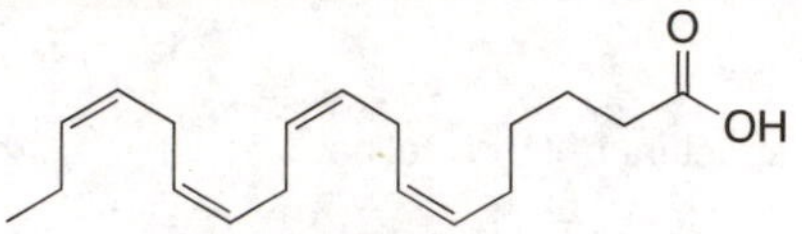

OH
stearidonic acid
4 double bonds
lowest melting point

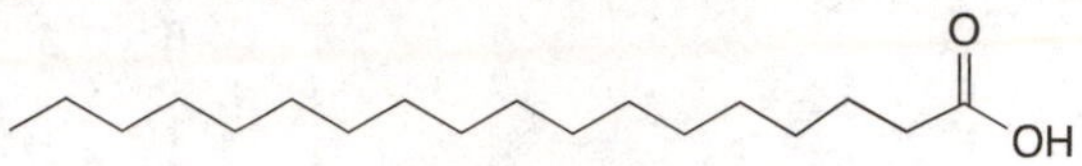

OH
stearic acid
no double bonds
highest melting point

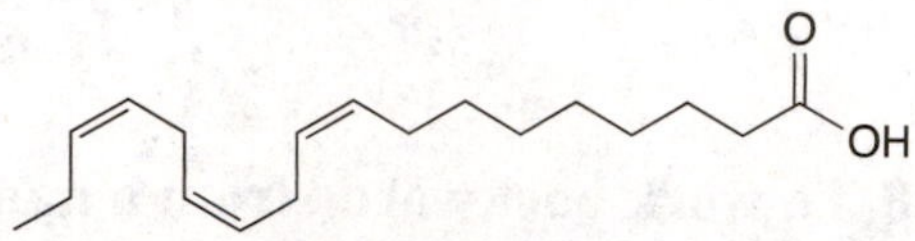

OH
linolenic acid
3 double bonds
intermediate melting point

10.12

a. OH $\xrightarrow{H_2SO_4}$ +

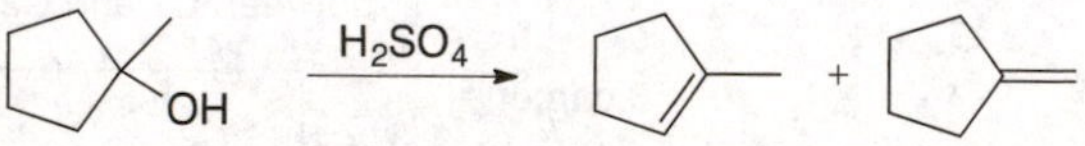

b. Br $\xrightarrow{NaOCH_2CH_3}$ + (+ cis isomer)

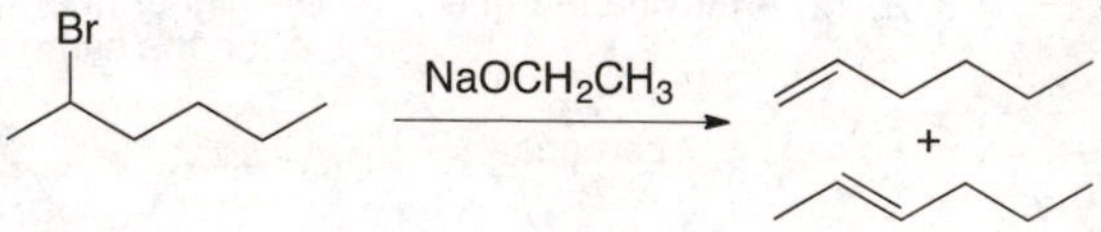

10.13 To draw the products of an addition reaction:
[1] Locate the two bonds that will be broken in the reaction. Always break the π bond.
[2] Draw the product by forming two new σ bonds.

a. $\xrightarrow{HCl}$ H, Cl — two new σ bonds

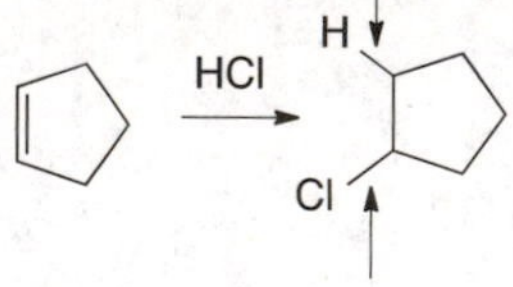

c. $\xrightarrow{HCl}$ H, Cl — two new σ bonds

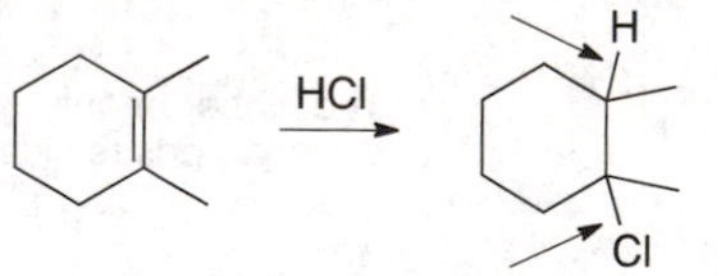

b. $\xrightarrow{HCl}$ H, Cl — two new σ bonds

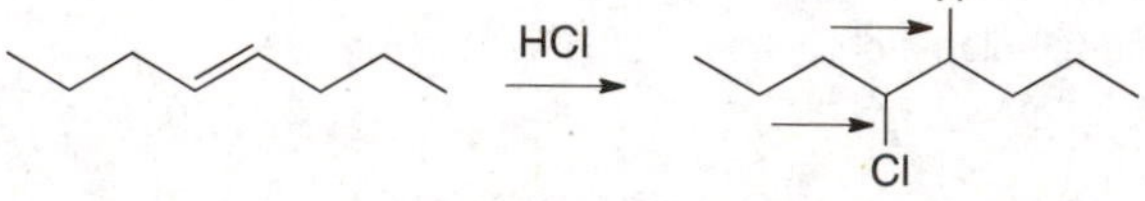

10.14 Addition reactions of HX occur in two steps:
[1] The double bond attacks the H atom of HX to form a carbocation.
[2] X^- attacks the carbocation to form a C–X bond.

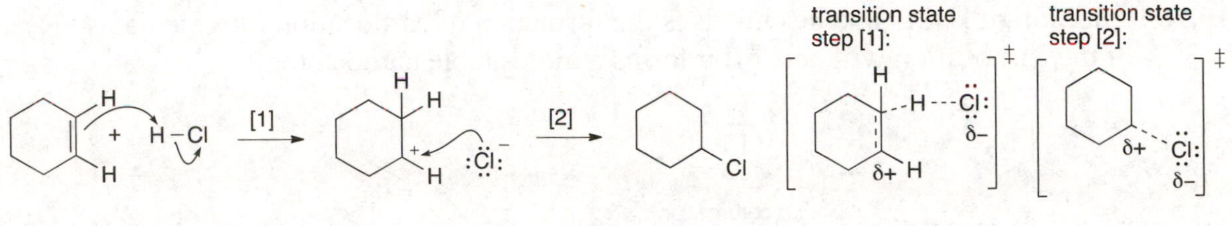

H
+
H−Cl
H
[1]
H
H
+
H
:Cl:−
[2]
Cl
transition state
step [1]:
H
H---Cl:
δ−
δ+ H
‡
transition state
step [2]:
δ+
Cl:
δ−
‡

10.15 Addition to alkenes follows Markovnikov's rule: When HX adds to an unsymmetrical alkene, the H bonds to the C that has more H's to begin with.

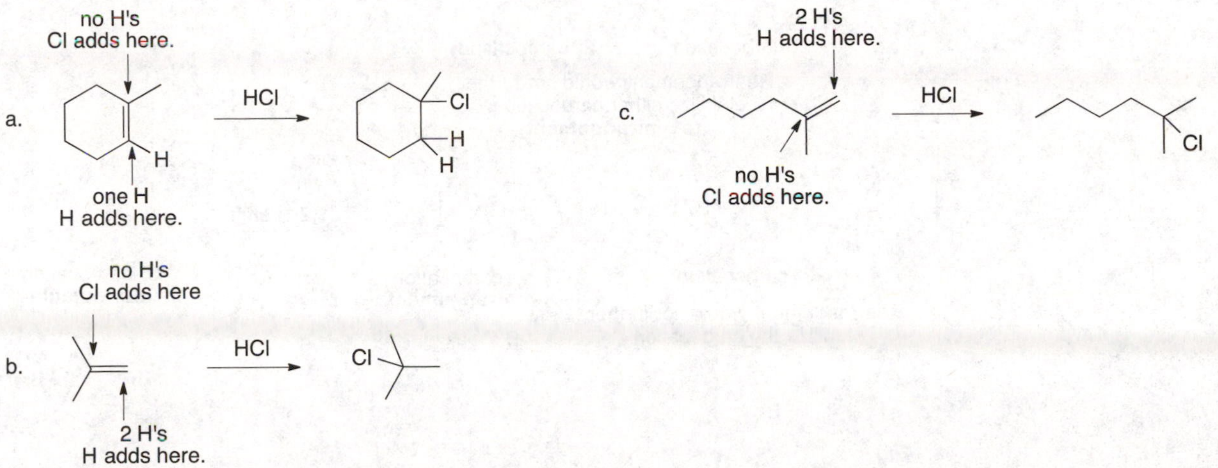

a.
no H's
Cl adds here.
H
one H
H adds here.
HCl
Cl
H
H
c.
2 H's
H adds here.
no H's
Cl adds here.
HCl
Cl
b.
no H's
Cl adds here
2 H's
H adds here.
HCl
Cl

10.16 To determine which alkene will react faster, draw the carbocation that forms in the rate-determining step. The more stable, more substituted the carbocation, the lower the E_a to form it and the faster the reaction.

H−X
+
3° carbocation
faster reaction
H−X
+
2° carbocation
slower reaction

10.17 Look for rearrangements of a carbocation intermediate to explain these results.

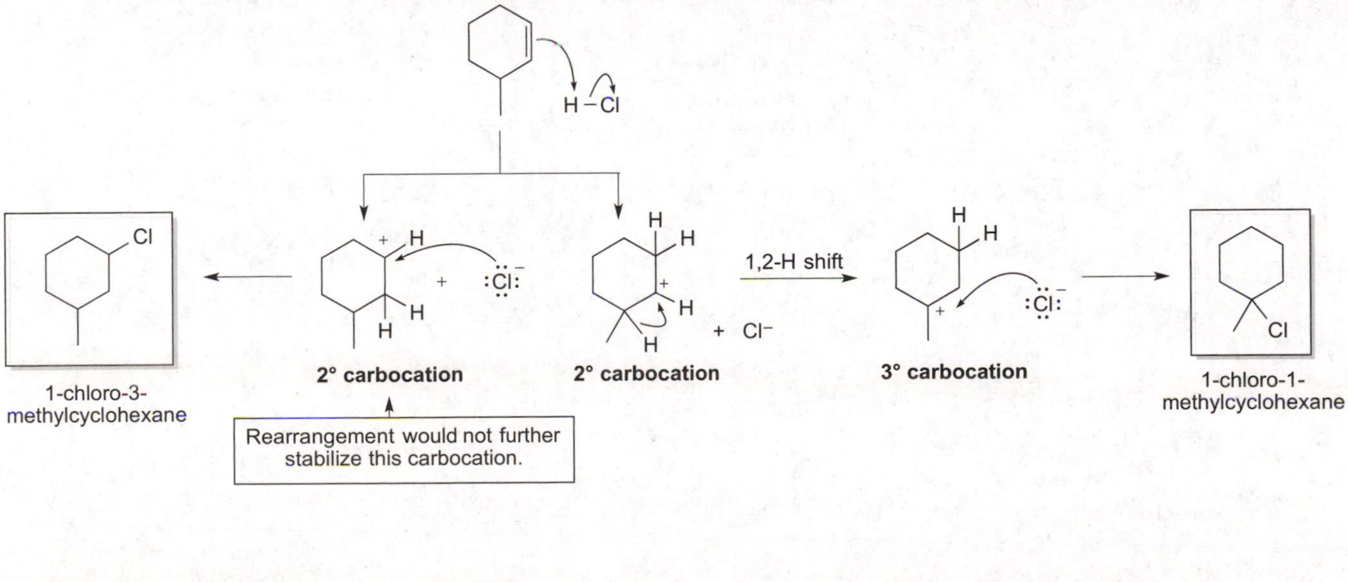

H−Cl
Cl
1-chloro-3-methylcyclohexane
H
H
H
+
:Cl:−
2° carbocation
Rearrangement would not further stabilize this carbocation.
H
H
H
H
+ Cl−
2° carbocation
1,2-H shift
H
H
:Cl:−
3° carbocation
Cl
1-chloro-1-methylcyclohexane

10.18 Addition of HX to alkenes involves the formation of carbocation intermediates. Rearrangement of the carbocation will occur if it forms a more stable carbocation.

a.

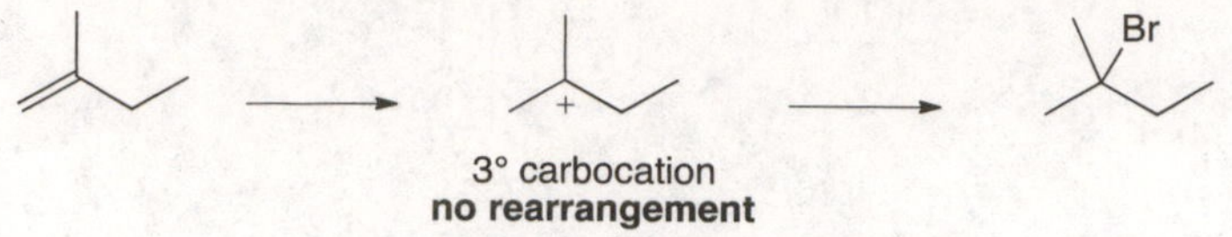

Br

3° carbocation
no rearrangement

b.

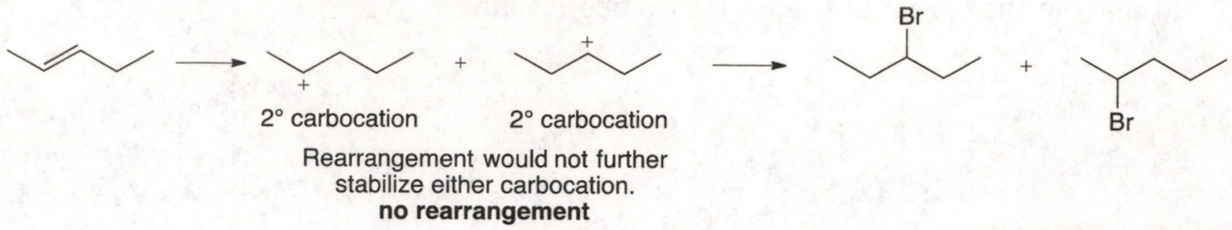

Br

2° carbocation 2° carbocation

Br

Rearrangement would not further stabilize either carbocation.
no rearrangement

c.

1,2-H shift

2° carbocation

2° carbocation
rearrangement

3° carbocation
more stable

Rearrangement would not further stabilize the carbocation.

Br

Br

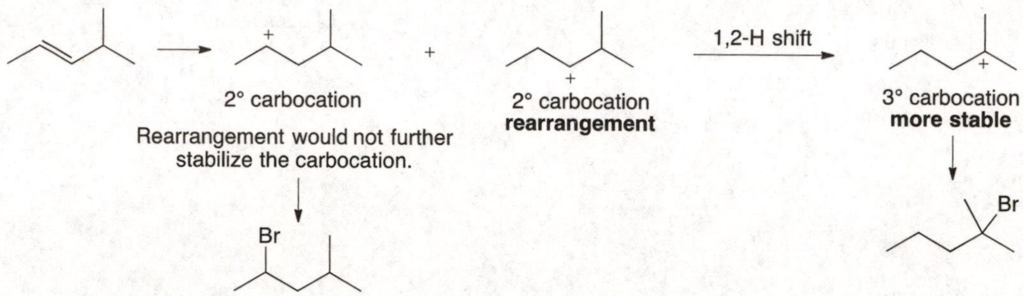

10.19 To draw the products, remember that addition of HX proceeds via a carbocation intermediate.

a.

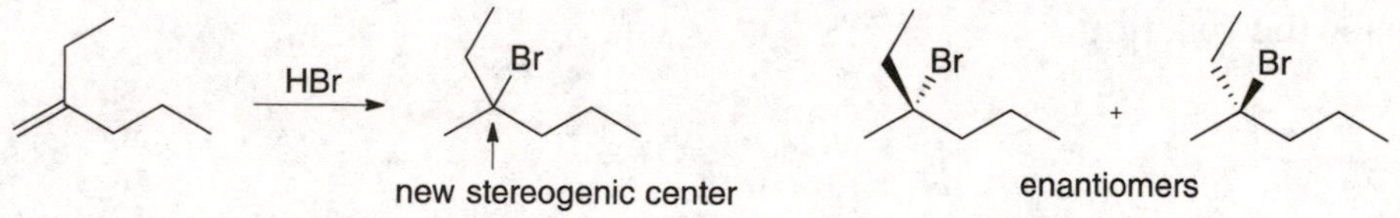

HBr

Br

Br Br

new stereogenic center

enantiomers

Addition of H^+ (from HCl) from above and below by Markovnikov's rule forms an achiral 3° carbocation.

Cl⁻ attacks from above and below.

b.

H H

Cl Cl

achiral, trigonal planar 3° carbocation

diastereomers

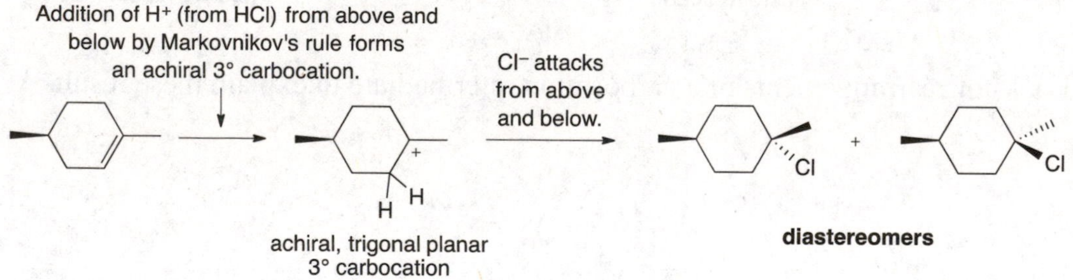

10.20 The product of syn addition will have H and Cl both up or down (both on wedges or both on dashed wedges), while the product of anti addition will have one up and one down (one wedge, one dashed wedge).

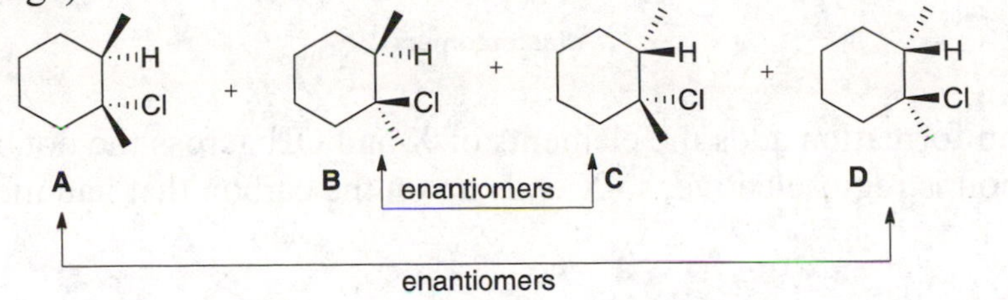

A 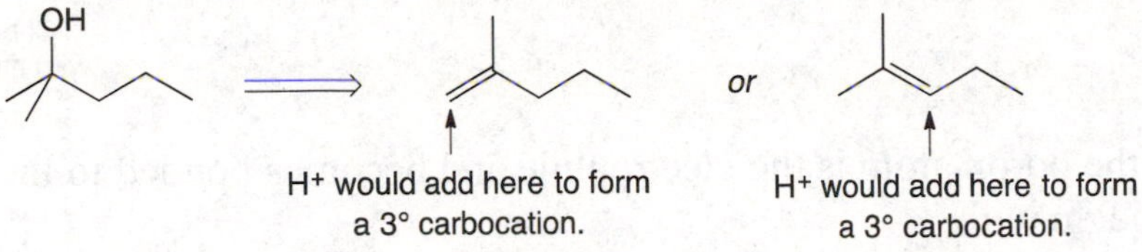 + B 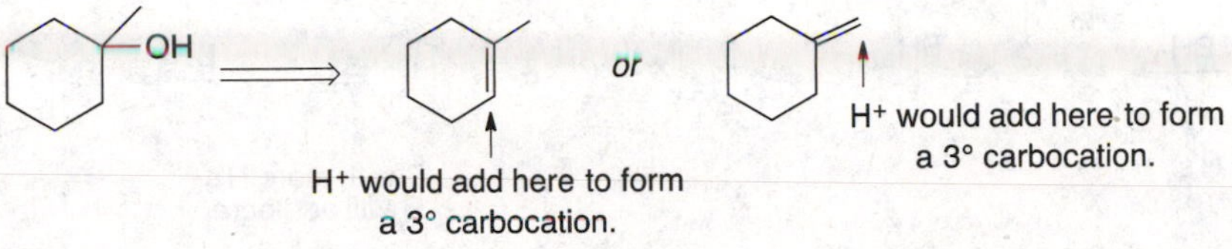 + C 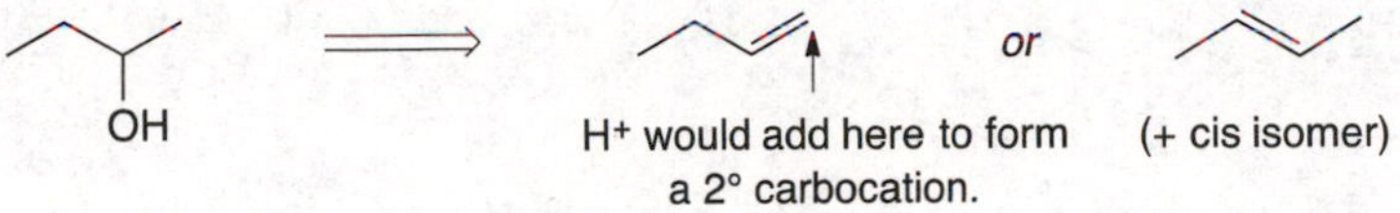 + D

B and C: enantiomers

A and D: enantiomers

10.21

a. ⟹ (alkene) *or* (alkene)

H+ would add here to form a 3° carbocation.

H+ would add here to form a 3° carbocation.

b. ⟹ (alkene) *or* (alkene)

H+ would add here to form a 3° carbocation.

H+ would add here to form a 3° carbocation.

c. ⟹ (alkene) *or* (alkene)

H+ would add here to form a 2° carbocation.

(+ cis isomer)

10.22

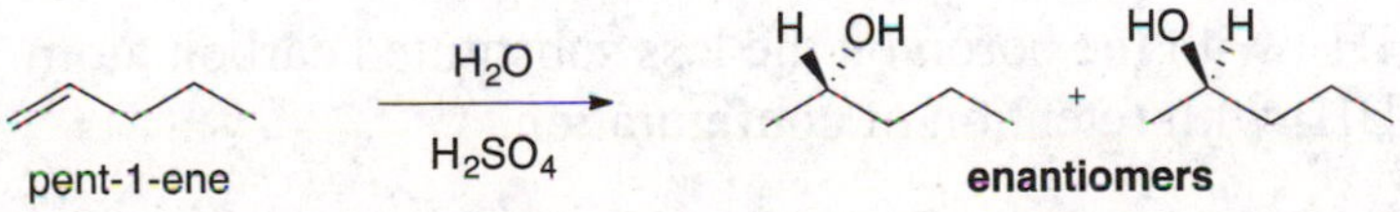

pent-1-ene $\xrightarrow[H_2SO_4]{H_2O}$ (H OH) + (HO H)

enantiomers

10.23 Halogenation of an alkene adds two elements of X in an anti fashion.

a. 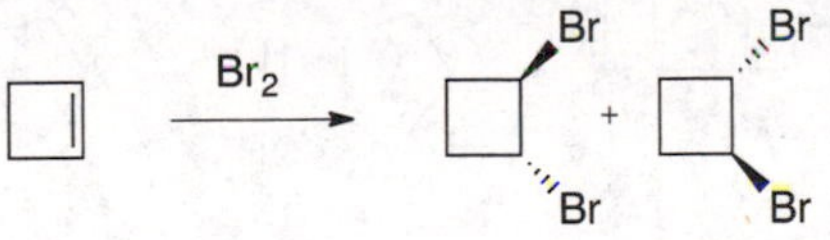 $\xrightarrow{Br_2}$ (Br, Br) + (Br, Br)

b. 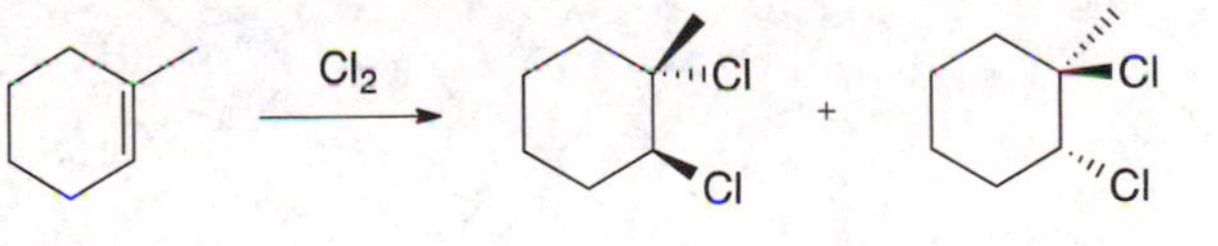 $\xrightarrow{Cl_2}$ (Cl, Cl) + (Cl, Cl)

10.24 To draw the products of halogenation of an alkene, remember that the halogen adds to both ends of the double bond but only anti addition occurs.

a. 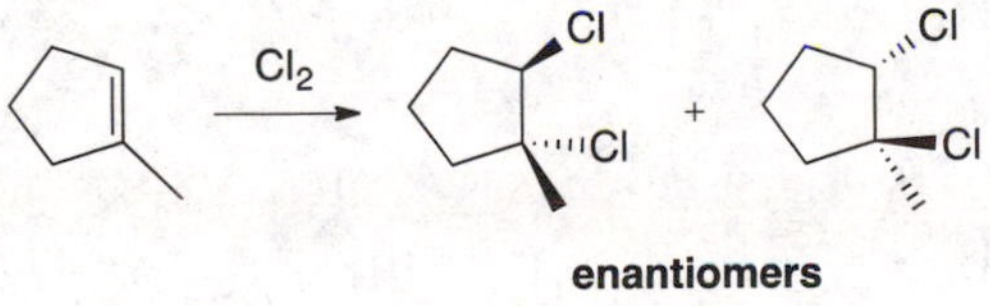 $\xrightarrow{Cl_2}$ (Cl, Cl) + (Cl, Cl)

enantiomers

b. 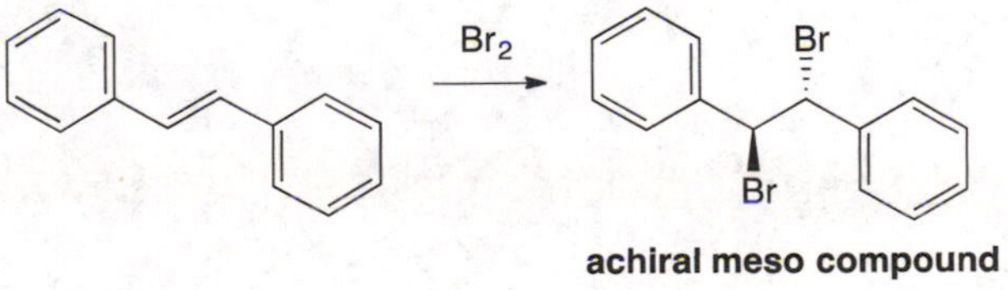 $\xrightarrow{Br_2}$ (Br, Br)

achiral meso compound

c. Br_2 → diastereomers

Br Br Br Br

diastereomers

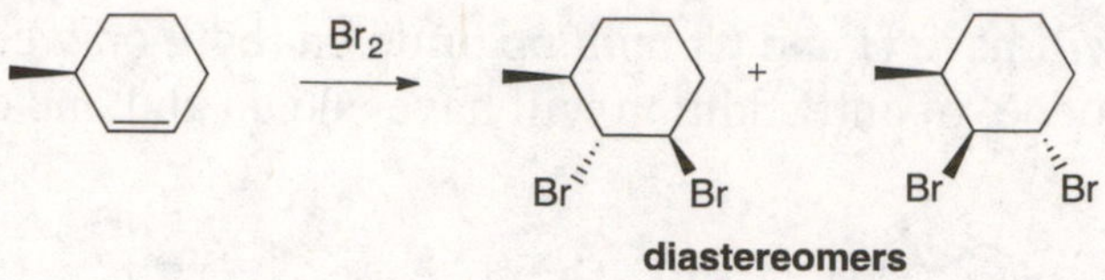

10.25 Halohydrin formation adds the elements of X and OH across the double bond in an anti fashion. The reaction is regioselective, so X ends up on the carbon that had more H's to begin with.

a. NBS, DMSO, H_2O → Br, HO + HO, Br

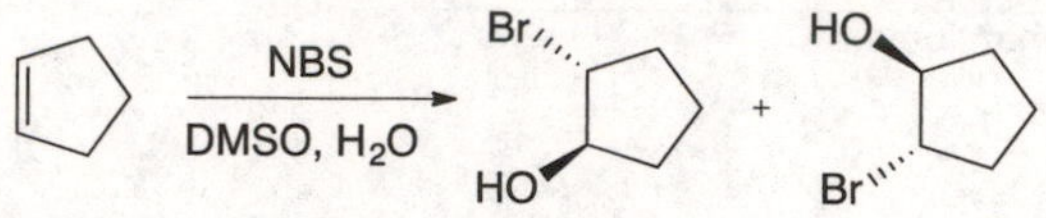

b. Cl_2, H_2O → OH, Cl + OH, Cl

Cl bonds to the carbon with more H's to begin with.

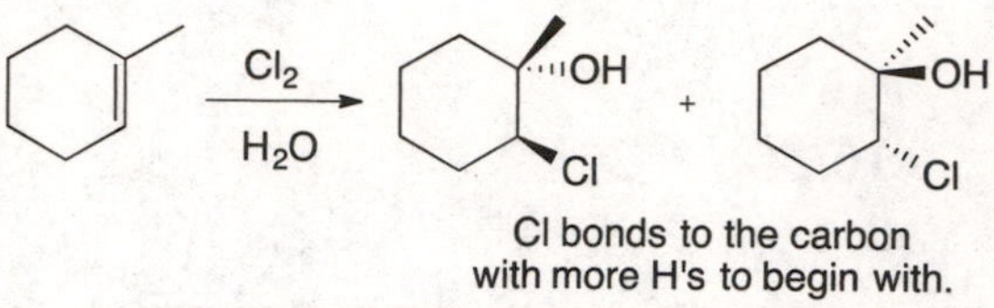

10.26 In hydroboration the boron atom is the electrophile and becomes bonded to the carbon atom that had more H's to begin with.

a. BH_3 → BH_2

C with more H's. B will add here.

c. BH_3 → BH_2

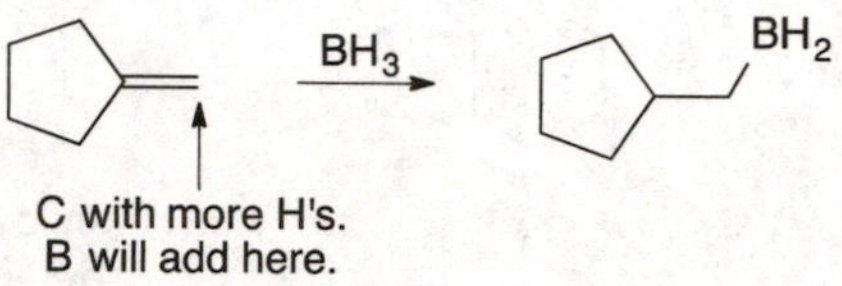

C with more H's. B will add here.

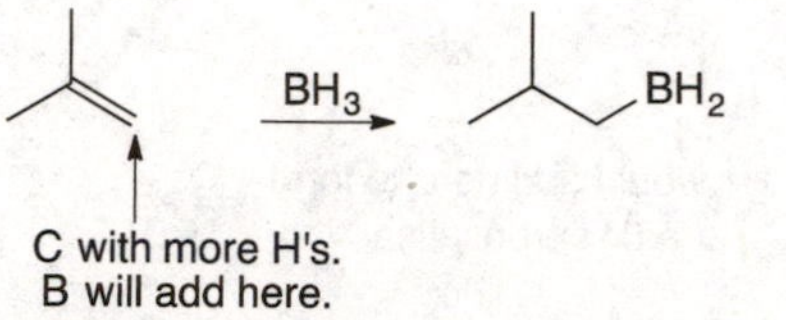

b. BH_3 → BH_2, H

C with more H's. B will add here.

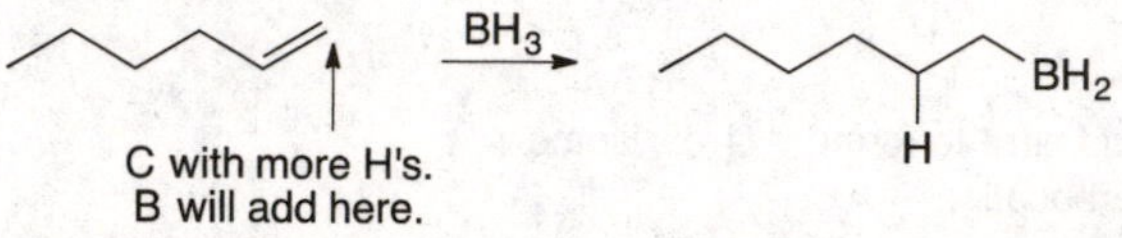

10.27 The hydroboration–oxidation reaction occurs in two steps:
[1] Syn addition of BH_3, with the boron on the less substituted carbon atom
[2] OH replaces the BH_2 with retention of configuration.

a. BH_3 → BH_2 → H_2O_2, ^-OH → OH

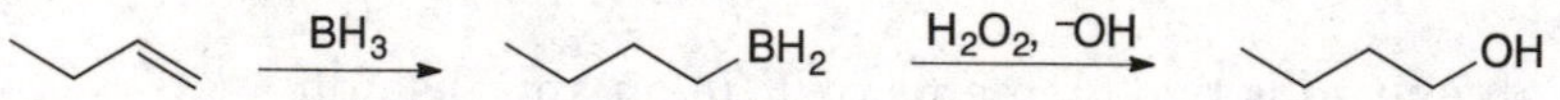

b. → H, BH_2 + H, BH_2 → H_2O_2, ^-OH → H, OH + H, OH

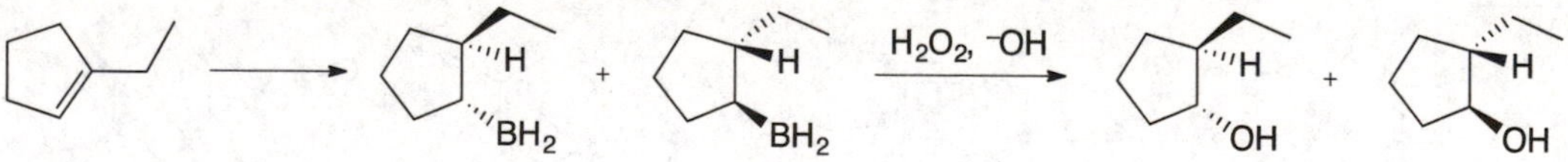

c. BH_3 → H, BH_2 + H, BH_2

H_2O_2, ^-OH

H, OH + H, OH

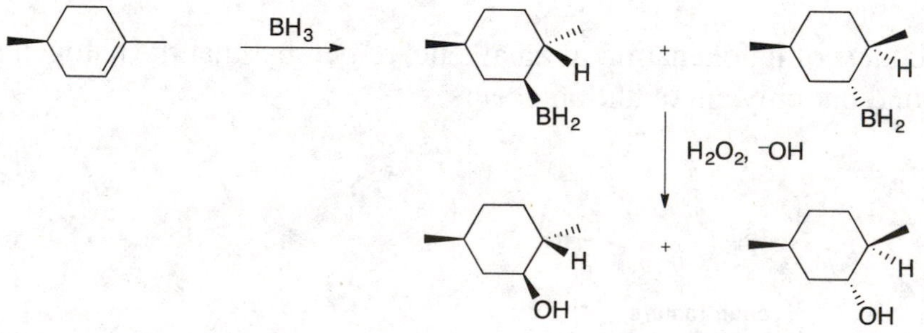

10.28 Remember that hydroboration results in addition of OH on the less substituted C.

a. HO

c. OH

b. OH

(*E* or *Z* isomer can be used.)

10.29

a. H_2O / H_2SO_4 — OH — Hydration places the OH on the more substituted carbon.

[1] BH_3 / [2] H_2O_2, HO^- — OH — Hydroboration–oxidation places the OH on the less substituted carbon.

b. H_2O / H_2SO_4 — OH — Hydration places the OH on the more substituted carbon.

[1] BH_3 / [2] H_2O_2, HO^- — OH — Hydroboration–oxidation places the OH on the less substituted carbon.

c. H_2O / H_2SO_4 — OH — Hydration places the OH on the more substituted carbon.

[1] BH_3 / [2] H_2O_2, HO^- — HO — Hydroboration–oxidation places the OH on the less substituted carbon.

10.30 There are always two steps in this kind of question:

[1] **Identify the functional group and decide what types of reactions it undergoes** (e.g., substitution, elimination, or addition).

[2] **Look at the reagent and determine if it is an electrophile, nucleophile, acid, or base.**

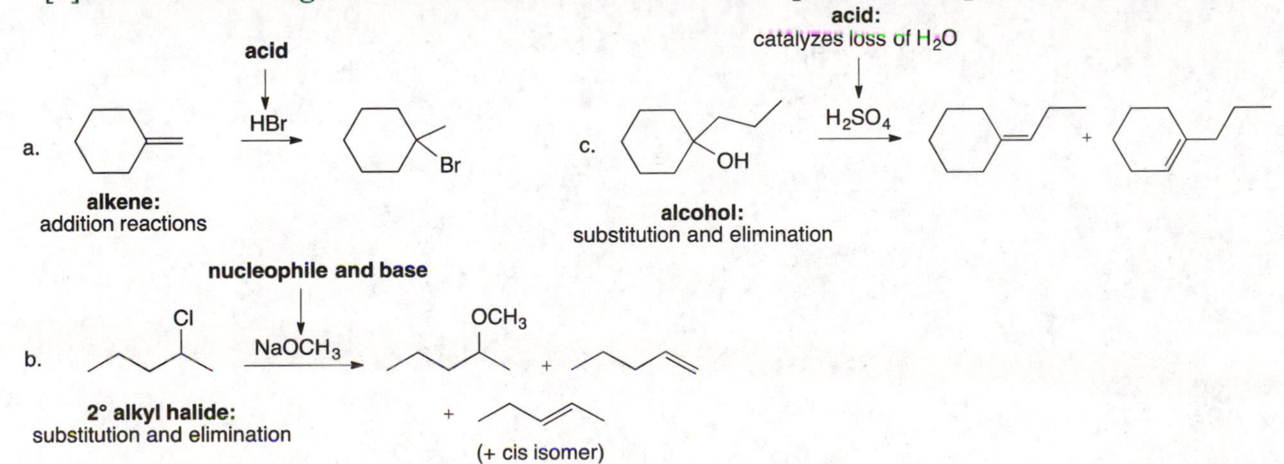

10.31 To devise a synthesis:
[1] Look at the starting material and decide what reactions it can undergo.
[2] Look at the product and decide what reactions could make it.

a.

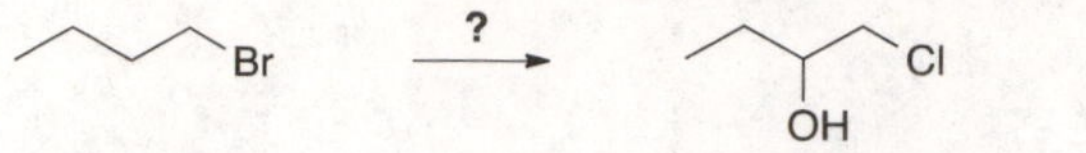

alkyl halide
Can undergo substitution and elimination.

halohydrin:
Can form from an alkene with Cl_2 and H_2O.

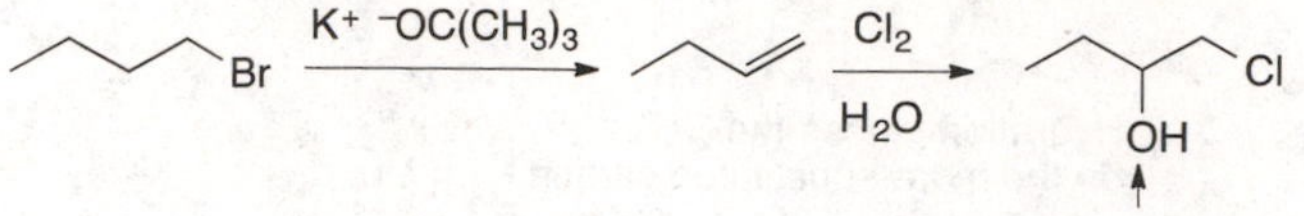

OH is added to the more substituted C.

b.

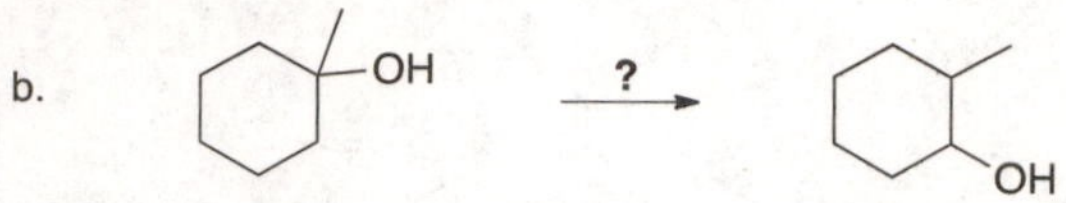

alcohol
Can undergo substitution and elimination.

alcohol
Can be formed by substitution and addition.

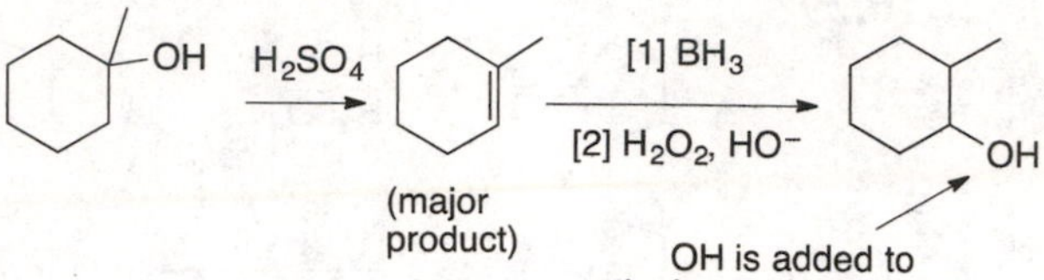

(major product)

OH is added to the less substituted C.

10.32 Convert each ball-and-stick model to a skeletal structure and then name the molecule.

a.

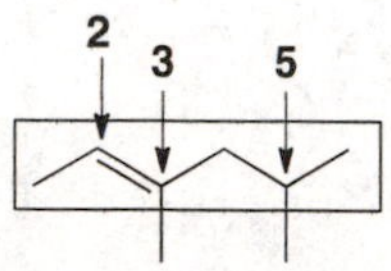

6 C chain with a C=C → hexene
C=C at C2 → hex-2-ene
2 CH_3's at C3 and C5
two higher priority groups on opposite sides → *E* isomer
Answer: ***(E)*-3,5-dimethylhex-2-ene**

b.

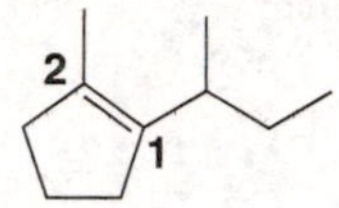

5 C ring with a C=C → cyclopentene
sec-butyl at C1
methyl at C2
Answer: **1-*sec*-butyl-2-methylcyclopentene**

10.33

a.

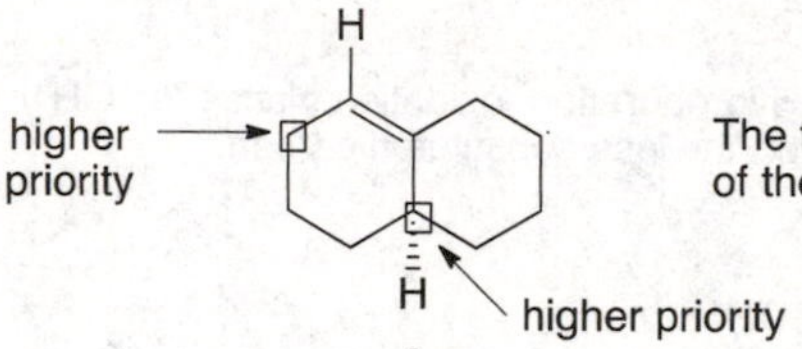

A

The two higher priority groups are on the same side of the C=C, making it a ***Z*** alkene.

b. Add H_2O in a Markovnikov fashion to form two products.

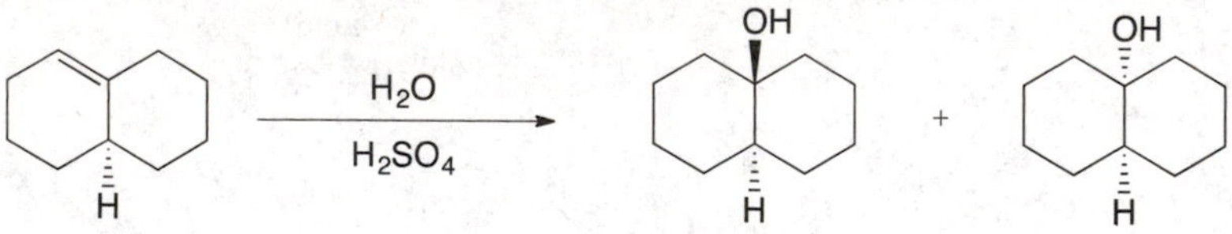

10.34

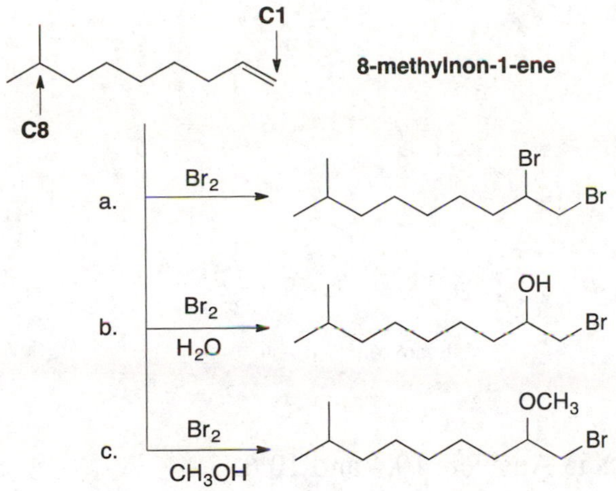

C1

8-methylnon-1-ene

C8

a. Br_2 → (Br, Br product)

b. Br_2, H_2O → (OH, Br product)

c. Br_2, CH_3OH → (OCH_3, Br product)

10.35 Use the directions from Answer 10.2 to calculate degrees of unsaturation.

a. C_6H_8
[1] maximum number of H's = 2*n* + 2 = 2(6) + 2 = 14
[2] subtract actual from maximum = 14 – 8 = 6
[3] divide by 2 = 6/2 = **3 degrees of unsaturation**

b. $C_{40}H_{56}$
[1] maximum number of H's = 2*n* + 2 = 2(40) + 2 = 82
[2] subtract actual from maximum = 82 – 56 = 26
[3] divide by 2 = 26/2 = **13 degrees of unsaturation**

c. $C_{10}H_{16}O_2$
Ignore both O's.
[1] maximum number of H's = 2*n* + 2 = 2(10) + 2 = 22
[2] subtract actual from maximum = 22 – 16 = 6
[3] divide by 2 = 6/2 = **3 degrees of unsaturation**

d. C_8H_9Br
Because of Br, add one H (9 + 1 = 10 H's).
[1] maximum number of H's = 2*n* + 2 = 2(8) + 2 = 18
[2] subtract actual from maximum = 18 – 10 = 8
[3] divide by 2 = 8/2 = **4 degrees of unsaturation**

e. C_8H_9ClO
Ignore the O; count Cl as one more H (9 + 1 = 10 H's).
[1] maximum number of H's = 2*n* + 2 = 2(8) + 2 = 18
[2] subtract actual from maximum = 18 – 10 = 8
[3] divide by 2 = 8/2 = **4 degrees of unsaturation**

f. $C_7H_{11}N$
Because of N, subtract one H (11 – 1 = 10 H's).
[1] maximum number of H's = 2*n* + 2 = 2(7) + 2 = 16
[2] subtract actual from maximum = 16 – 10 = 6
[3] divide by 2 = 6/2 = **3 degrees of unsaturation**

g. C_4H_8BrN
Because of Br, add one H, but subtract one for N
(8 + 1 – 1 = 8 H's).
[1] maximum number of H's = 2*n* + 2 = 2(4) + 2 = 10
[2] subtract actual from maximum = 10 – 8 = 2
[3] divide by 2 = 2/2 = **1 degree of unsaturation**

h. $C_{10}H_{18}ClNO$
Add one H because of Cl and subtract 1 H for N
(18 + 1 – 1 = 18) .
[1] maximum number of H's = 2*n* + 2 = 2(10) + 2 = 22
[2] subtract actual from maximum = 22 – 18 = 4
[3] divide by 2 = 4/2 = **2 degrees of unsaturation**

10.36 First determine the number of degrees of unsaturation in the compound. Then decide which combinations of rings and π bonds could exist.

$C_{10}H_{14}$
[1] maximum number of H's = 2*n* + 2 = 2(10) + 2 = 22
[2] subtract actual from maximum = 22 – 14 = 8
[3] divide by two = 8/2 = **4 degrees of unsaturation**

possibilities:
4 π bonds
3 π bonds + 1 ring
2 π bonds + 2 rings
1 π bond + 3 rings
4 rings

10.37

a. enclomiphene
E

b. tamoxifen
Z

c. clavulanic acid
Z

10.38 Name the alkenes using the rules in Answers 10.4 and 10.6.

a.

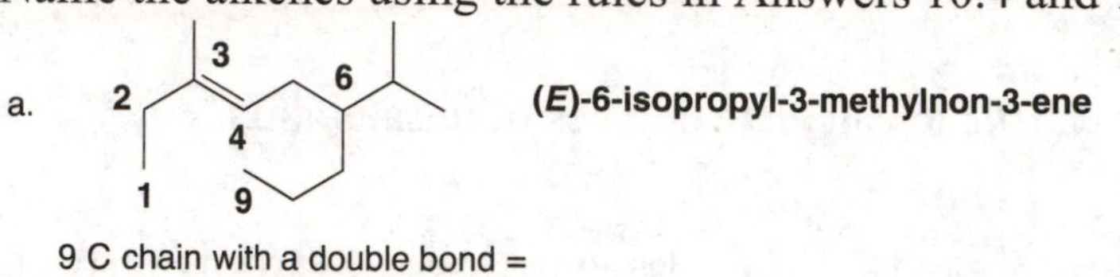

9 C chain with a double bond =
nonene

b.

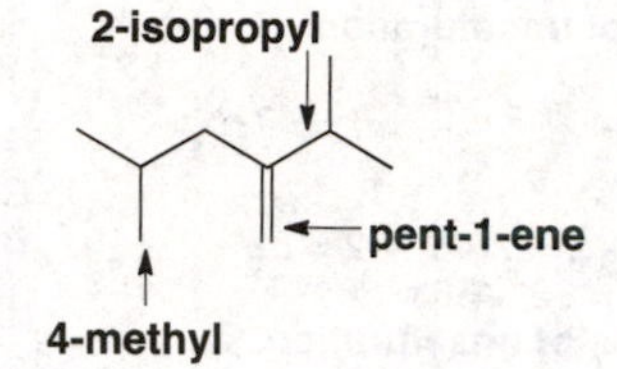

2-isopropyl-4-methylpent-1-ene

5 C chain with a double bond =
pentene

c. **5-*sec*-butyl-1,3,3-trimethylcyclohexene**

6 C ring with a double bond =
hexene

d. **4-isopropyl** → **3-ol** → **hept-4-ene**

***E* double bond** (higher priority groups on opposite sides with bold bonds)

(*E*)-4-isopropylhept-4-en-3-ol

7 C chain with a double bond =
heptene

e. **cyclohex-2-ene** **1-ol** **5-*sec*-butyl**

5-*sec*-butylcyclohex-2-enol

6 C ring with a double bond =
cyclohexene

f. **(*E*)-5-ethyl-3,4-dimethylnon-2-ene**

10.39 Use the directions from Answer 10.8.

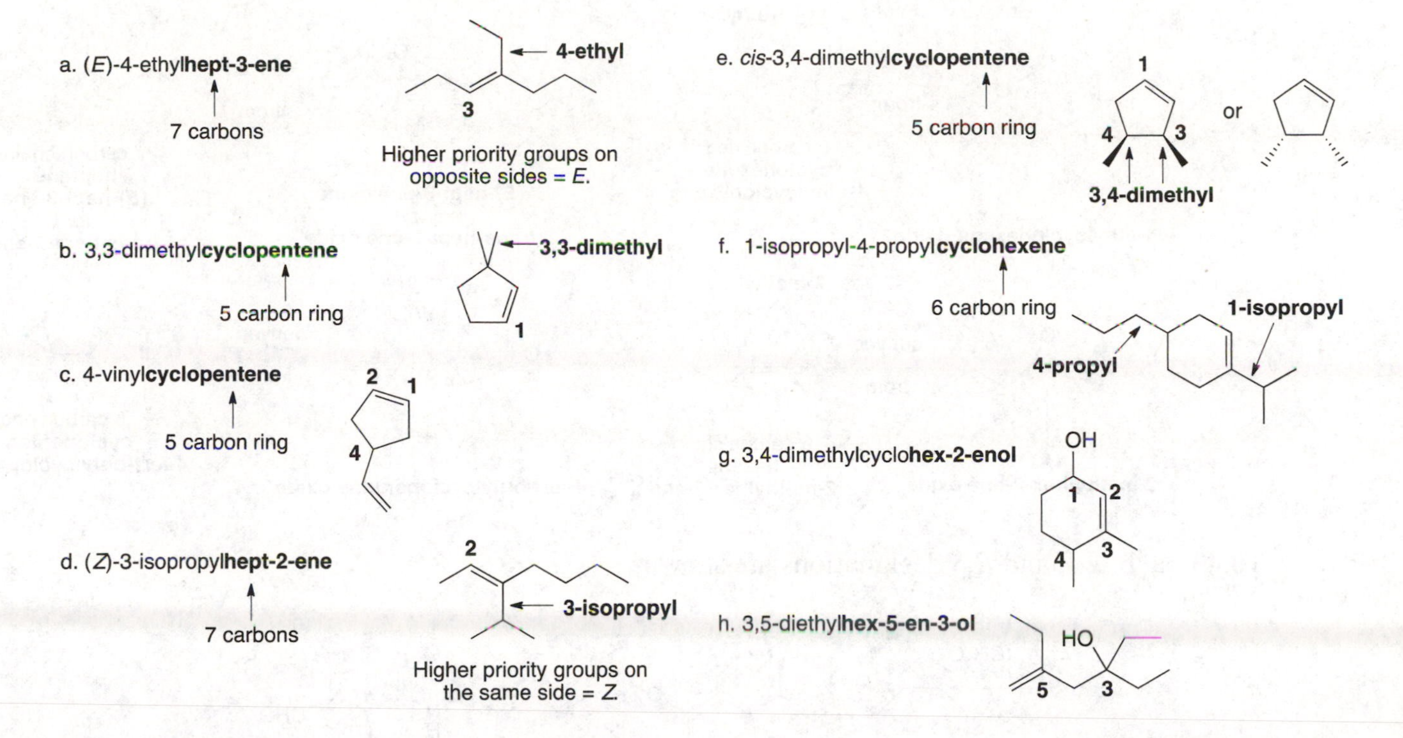

10.40

a.

(2*E*,4*S*)-4-methylnon-2-ene
A

(2*E*,4*R*)-4-methylnon-2-ene
B

(2*Z*,4*S*)-4-methylnon-2-ene
C

(2*Z*,4*R*)-4-methylnon-2-ene
D

b. **A** and **B** are enantiomers. **C** and **D** are enantiomers.
c. Pairs of diastereomers: **A** and **C**, **A** and **D**, **B** and **C**, **B** and **D**.

10.41

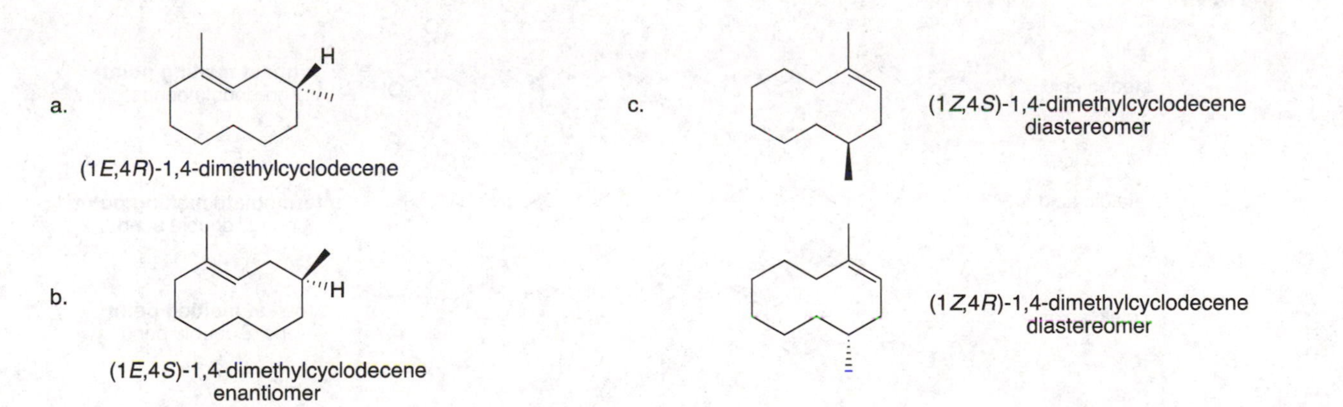

10.42 Name the alkene from which the epoxide can be derived and add the word *oxide*.

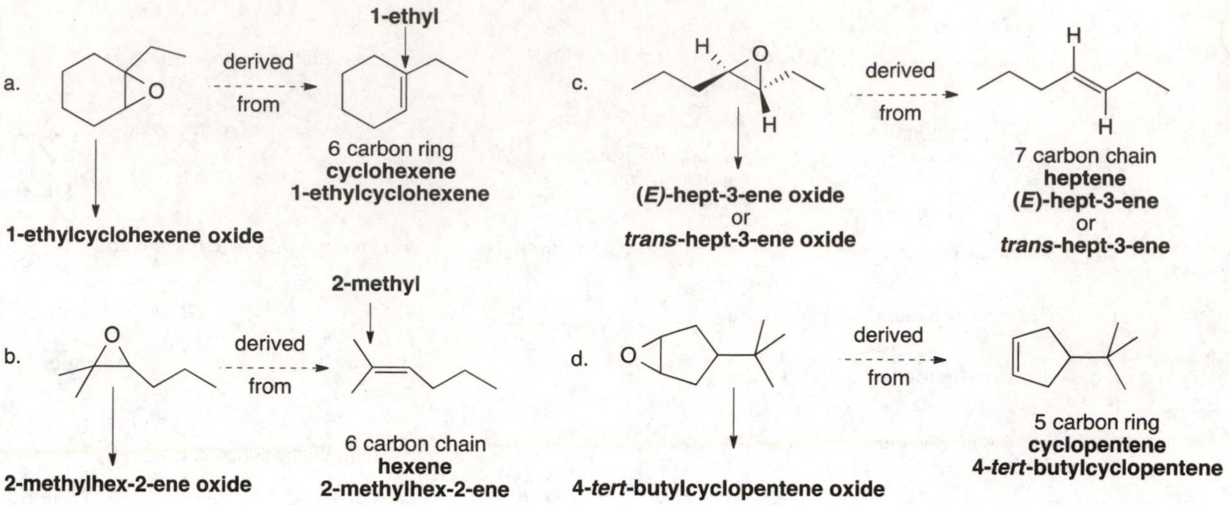

a. derived from — 1-ethyl

6 carbon ring
cyclohexene
1-ethylcyclohexene

1-ethylcyclohexene oxide

c. derived from

7 carbon chain
heptene
(*E*)-hept-3-ene
or
***trans*-hept-3-ene**

(*E*)-hept-3-ene oxide
or
***trans*-hept-3-ene oxide**

b. derived from — 2-methyl

6 carbon chain
hexene
2-methylhex-2-ene

2-methylhex-2-ene oxide

d. derived from

5 carbon ring
cyclopentene
4-*tert*-butylcyclopentene

4-*tert*-butylcyclopentene oxide

10.43 a, b. *E*,*Z* and *R*,*S* designations are shown.

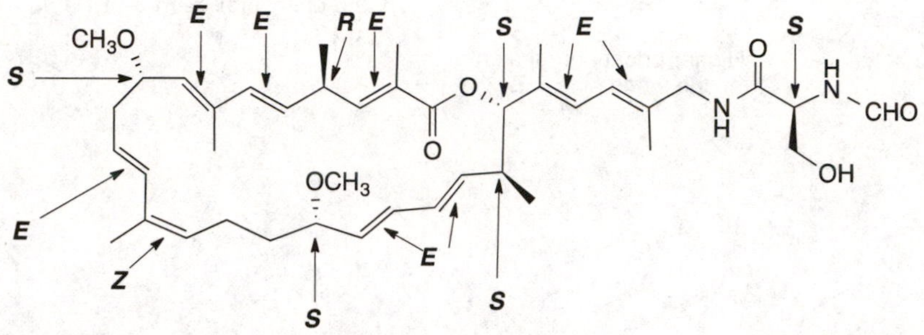

CH_3O, OCH_3, CHO, OH; designations: *S*, *E*, *E*, *R*, *E*, *S*, *E*, *S*, *E*, *Z*, *S*, *E*, *S*

c. Nine double bonds that can be *E* or *Z*
Six tetrahedral stereogenic centers
Maximum possible number of stereoisomers = 2^{15}

10.44

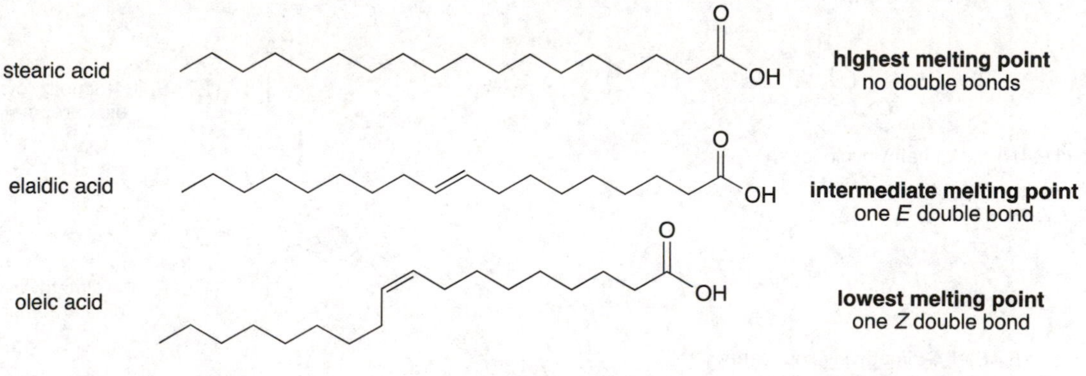

stearic acid		**highest melting point** no double bonds
elaidic acid		**intermediate melting point** one *E* double bond
oleic acid		**lowest melting point** one *Z* double bond

10.45

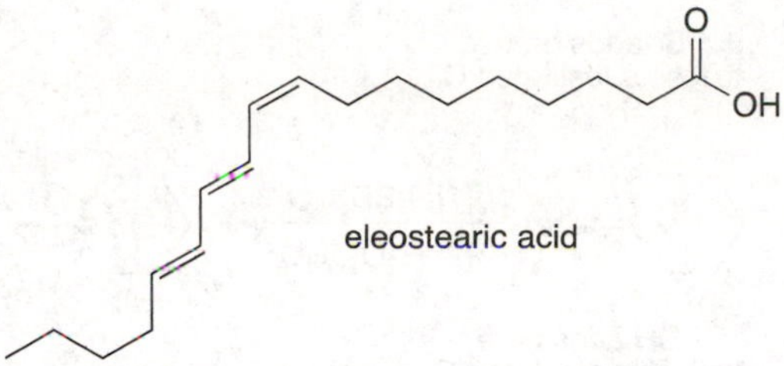
O
OH

eleostearic acid

a.

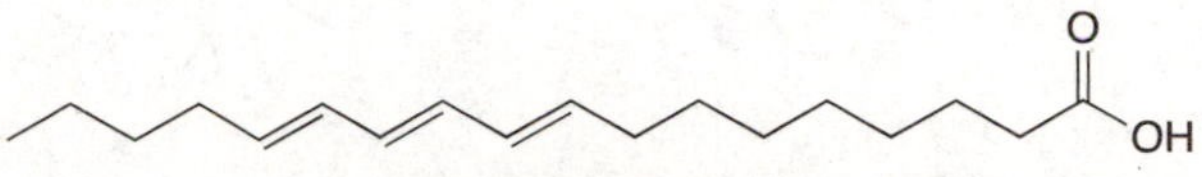
O
OH

all trans double bonds
higher melting point

b.

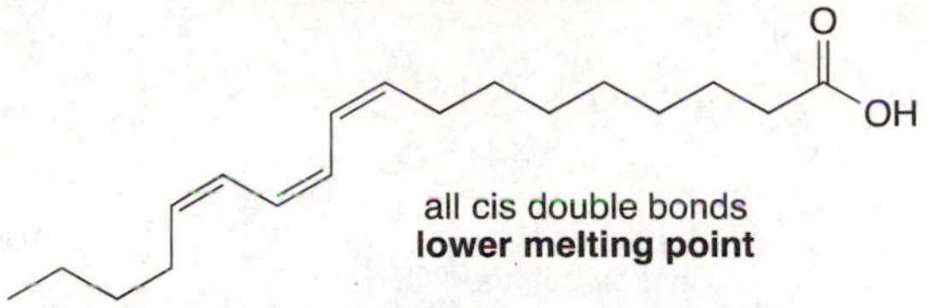
O
OH

all cis double bonds
lower melting point

10.46

a.

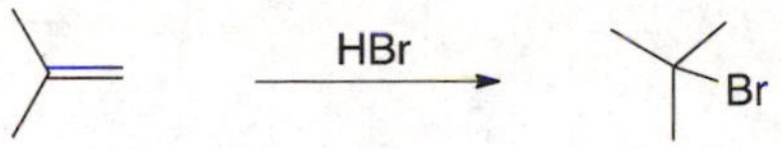
HBr
Br

b.

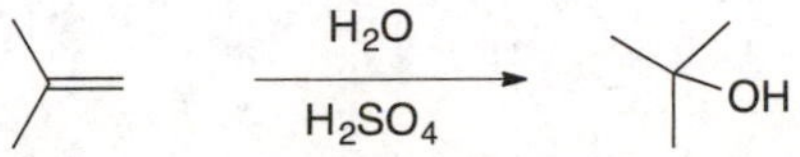
H2O
H2SO4
OH

c.

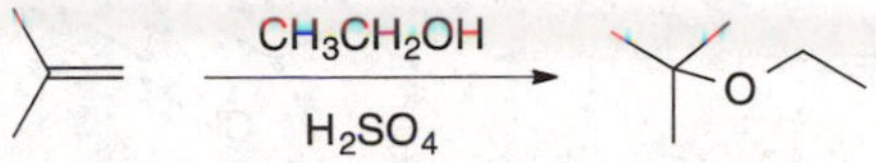
CH3CH2OH
H2SO4
O

d.

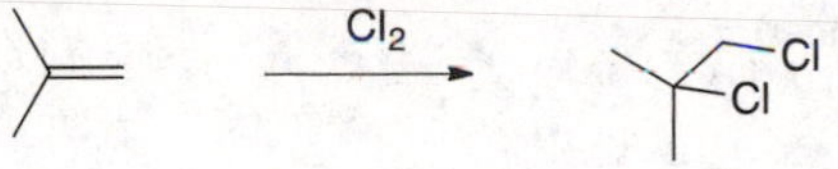
Cl2
Cl
Cl

e.

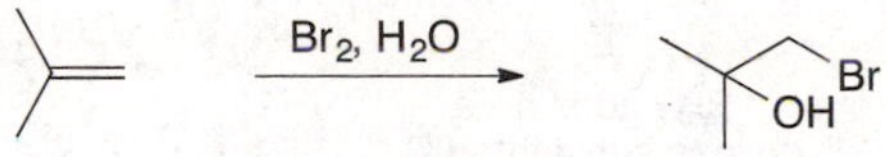
Br2, H2O
Br
OH

f.

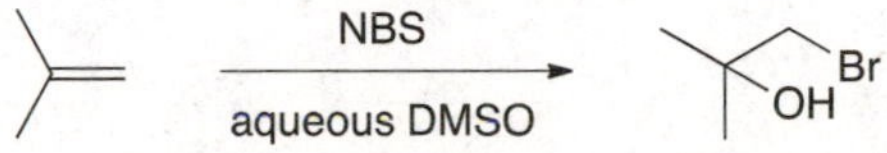
NBS
aqueous DMSO
Br
OH

g.

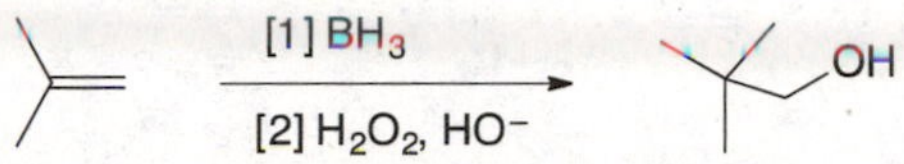
[1] BH3
[2] H2O2, HO−
OH

10.47

a.

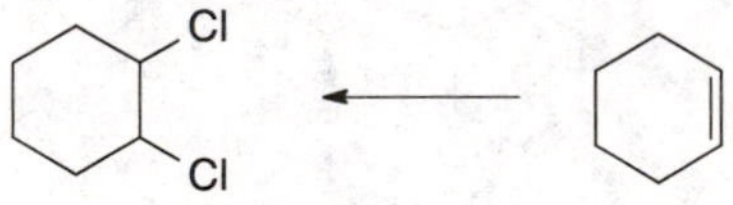
Cl
Cl

b.

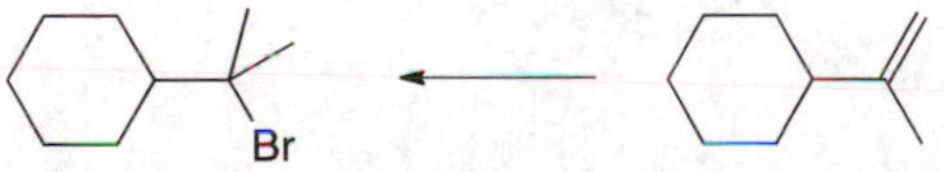
Br

c.

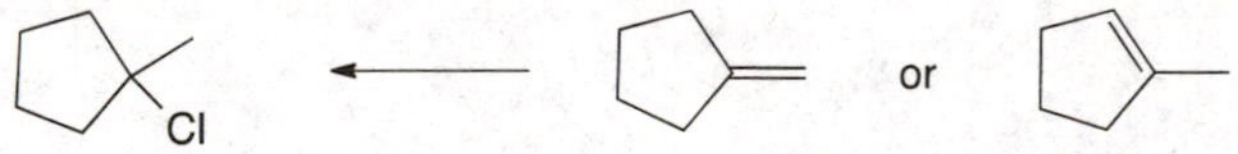
Cl
or

d.

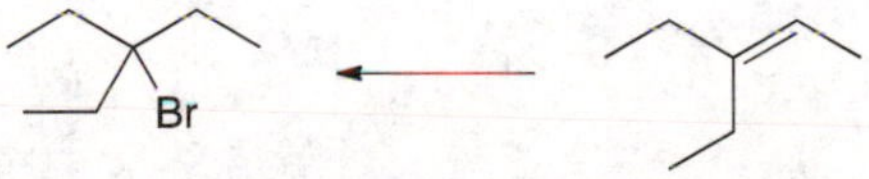
Br

10.48 Hydroboration–oxidation results in addition of an OH group on the *less* substituted carbon, whereas acid-catalyzed addition of H_2O results in the addition of an OH group on the *more* substituted carbon.

a.

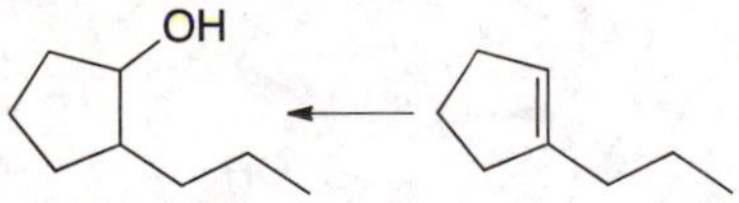
OH

hydroboration–oxidation

b.

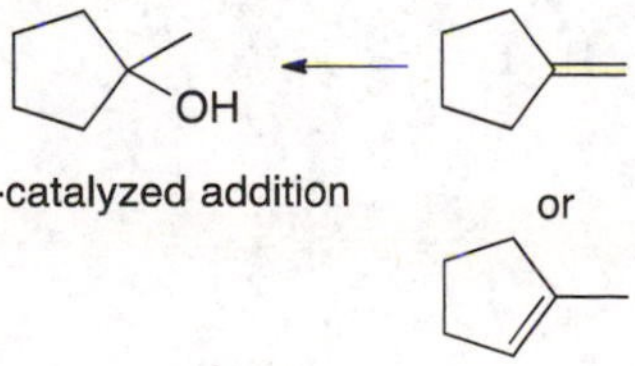
OH
or

acid-catalyzed addition

c.

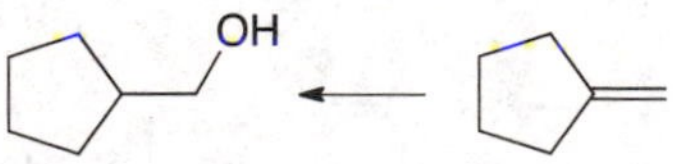
OH

hydroboration–oxidation

d.

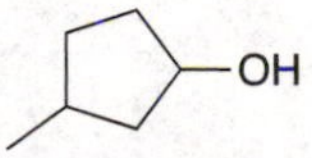
OH

Both methods would give product mixtures.

10.49

a. HCl

H adds here to less substituted C.

Cl

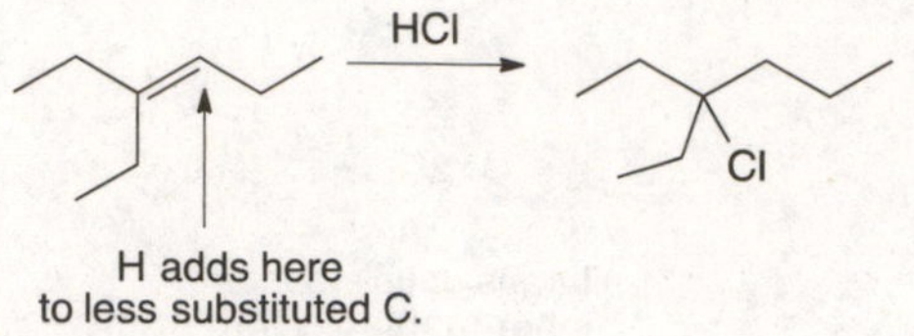

d. Br_2 / H_2O

Br

OH

Br adds here to less substituted C.

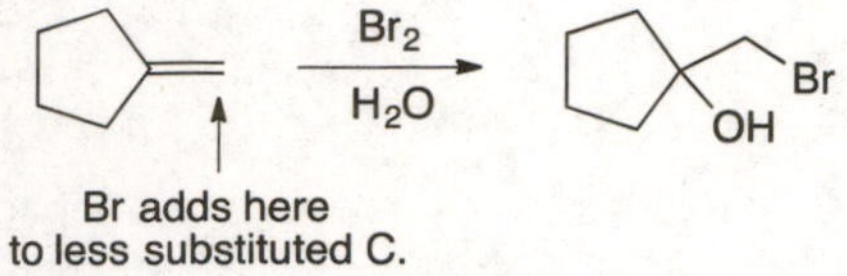

b. H_2O / H_2SO_4

OH

H adds here to less substituted C.

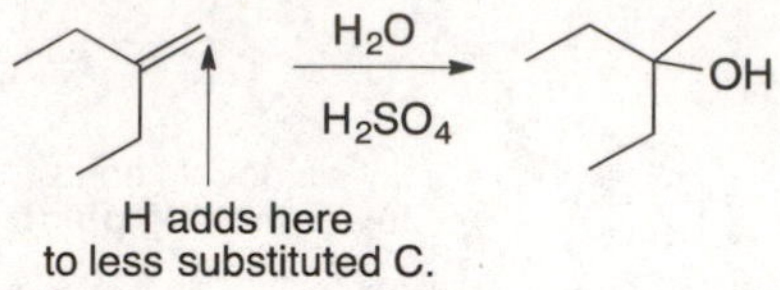

e. [1] 9-BBN / [2] H_2O_2, HO^-

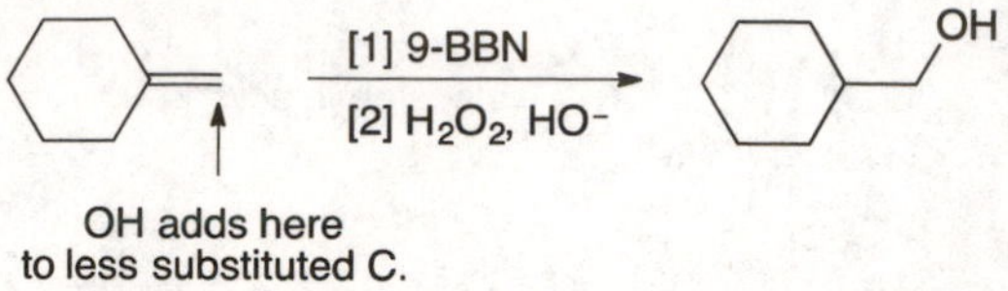

OH

OH adds here to less substituted C.

c. [1] BH_3 / [2] H_2O_2, HO^-

OH

BH_3 adds here to less substituted C.

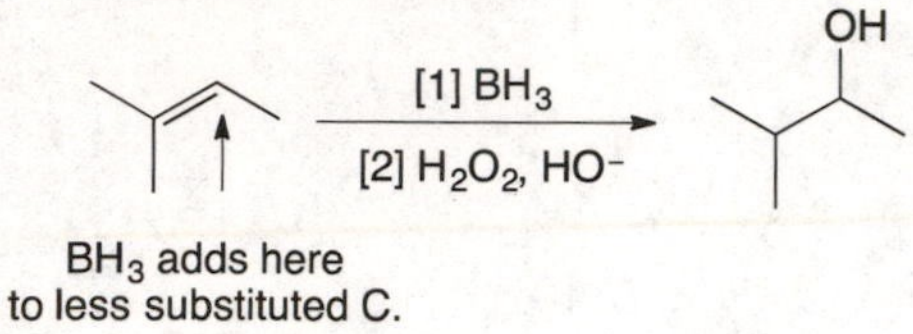

f. Br_2

Br Br

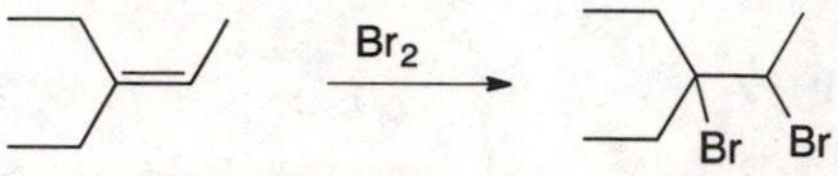

10.50

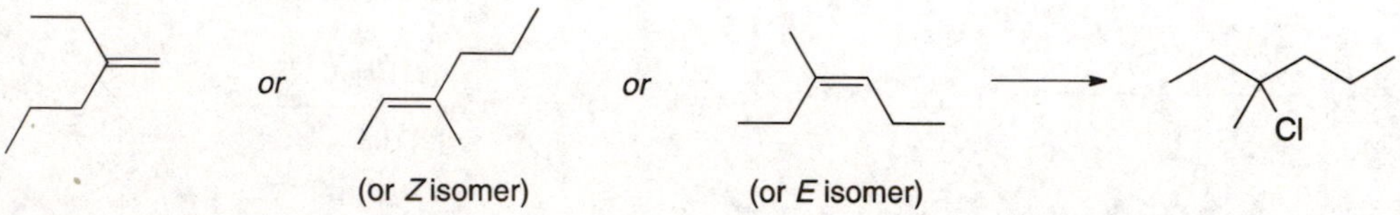

or (or *Z* isomer) *or* (or *E* isomer)

Cl

10.51

a. Br_2

Br Br + Br Br

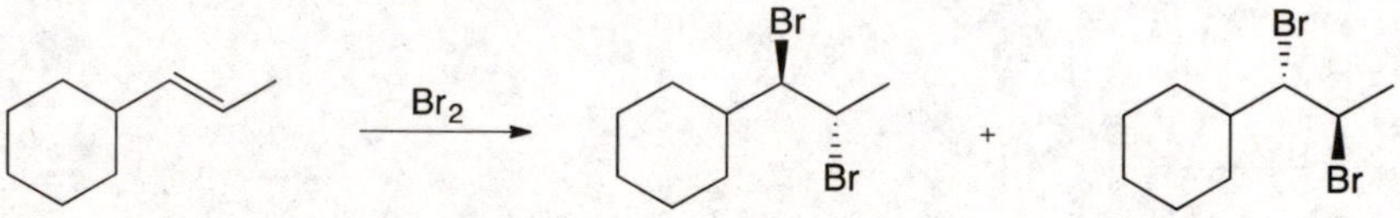

b. Cl_2 / H_2O

OH Cl + OH Cl

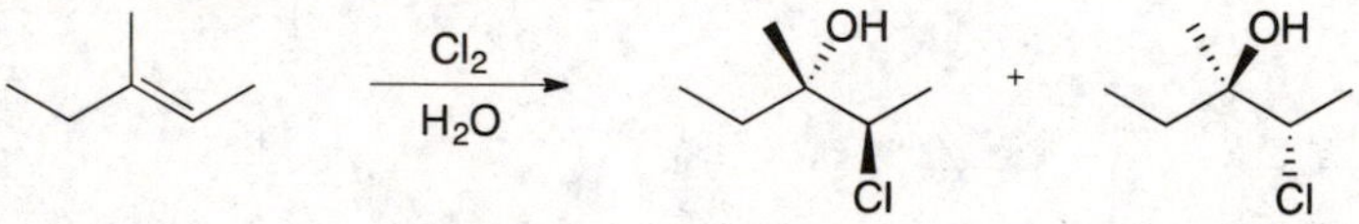

c. NBS / DMSO, H_2O

OH Br + OH Br

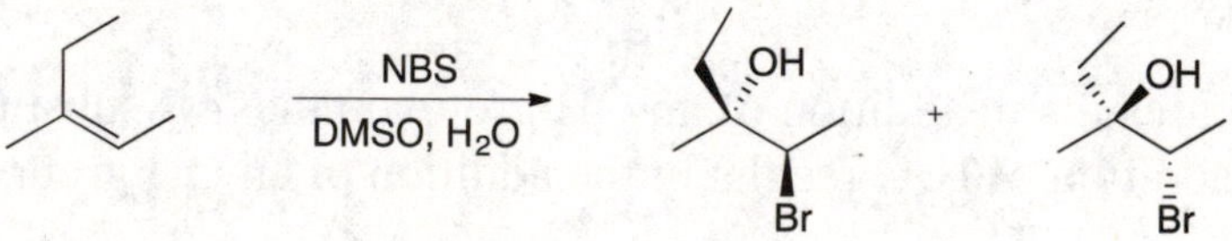

10.52

a. H_2O / H_2SO_4

OH + OH

b. HI

I H + H I

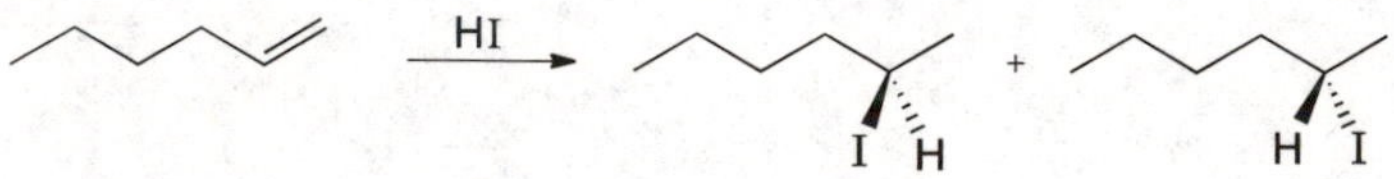

c. 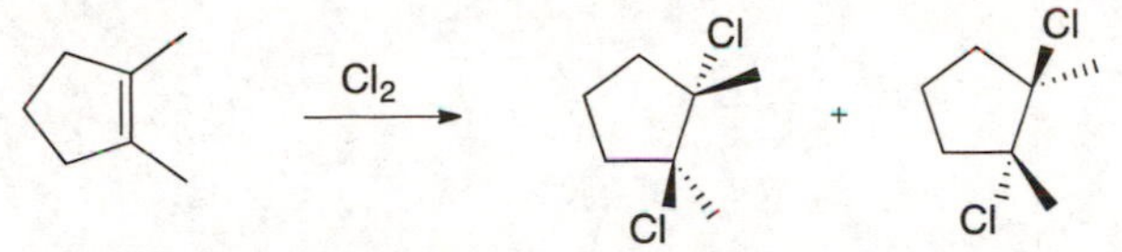 Cl_2 → only anti addition

d. 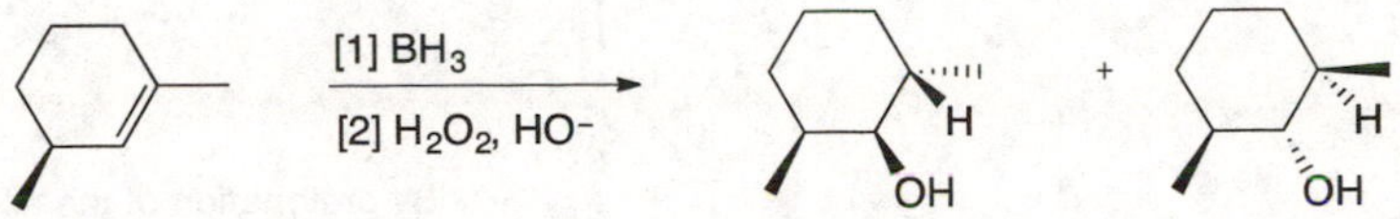 [1] BH_3 [2] H_2O_2, HO^- → only syn addition

e.  Cl_2, H_2O → only anti addition

f. 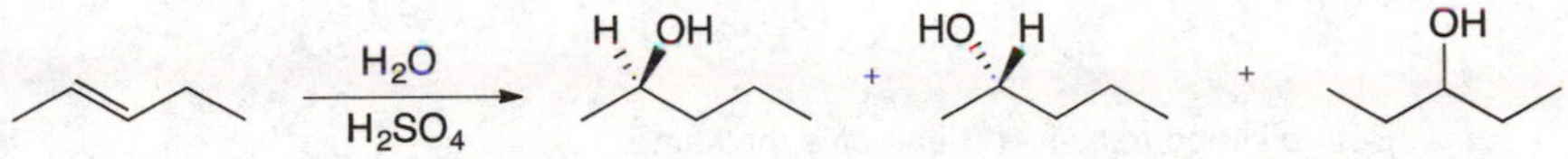 H_2O, H_2SO_4 →

10.53 The alkene that forms the more stable carbocation when H^+ is added reacts faster. Both alkenes form 2° carbocations, but the carbocation from **B** is also resonance stabilized by the benzene ring. As a result, **B** reacts faster.

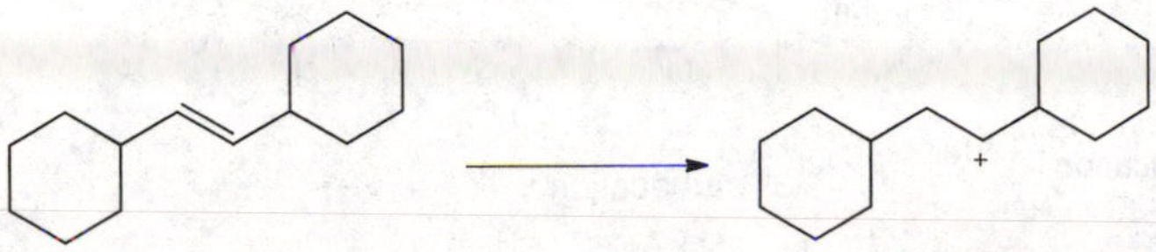

A → 2° carbocation

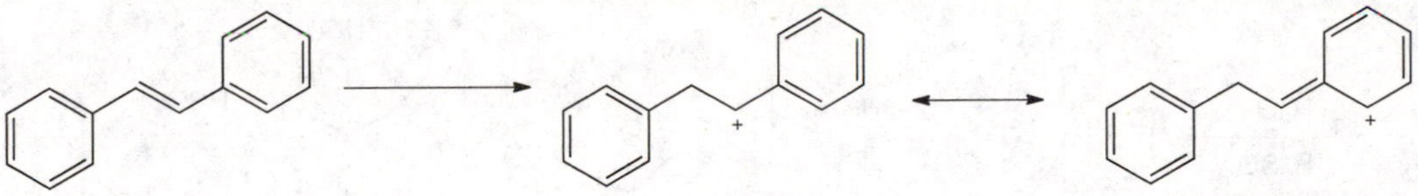

B → + 3 more resonance structures

10.54 Draw each reaction. (a) The cis isomer of oct-4-ene gives two enantiomers on addition of Br_2. (b) The trans isomer gives a meso compound.

a. 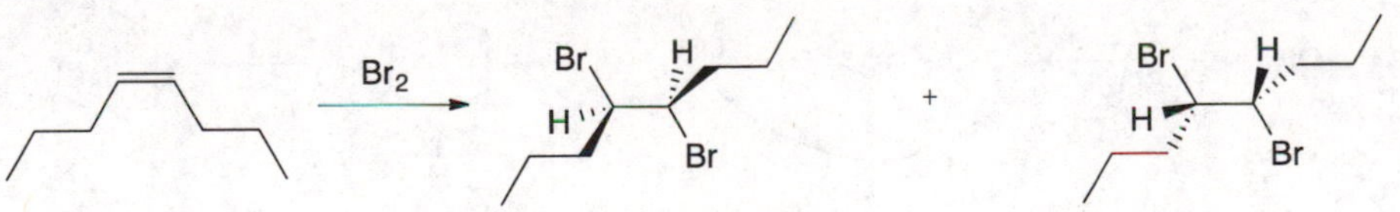 ***cis*-oct-4-ene** $\xrightarrow{Br_2}$ (4*R*,5*R*)-4,5-dibromooctane + (4*S*,5*S*)-4,5-dibromooctane

enantiomers

b. 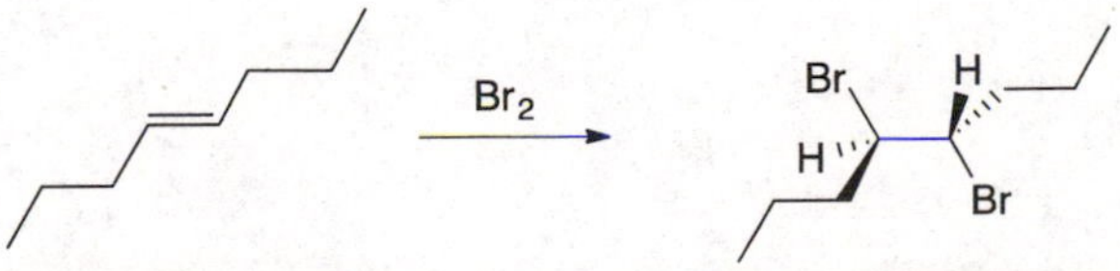 ***trans*-oct-4-ene** $\xrightarrow{Br_2}$ (4*R*,5*S*)-4,5-dibromooctane

meso compound

10.55

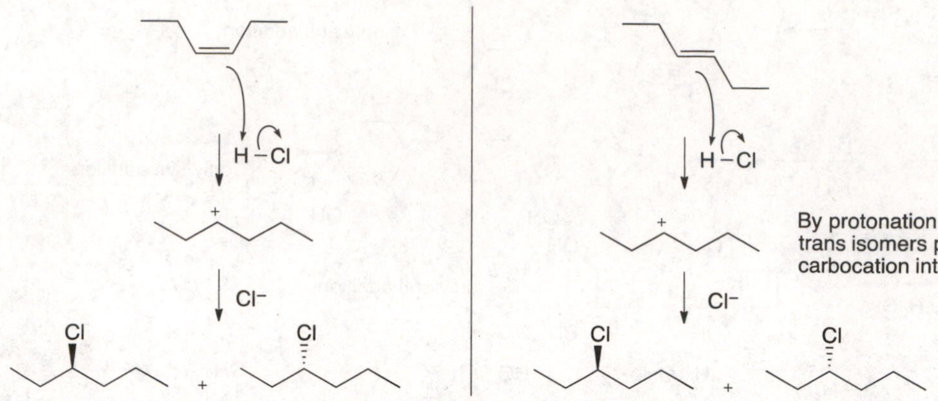

H–Cl

Cl⁻

By protonation of the alkene, the cis and trans isomers produce **identical** carbocation intermediates.

Cl + Cl

Both *cis-* and *trans*-hex-3-ene give the same racemic mixture of products, so the reaction is not stereospecific.

10.56

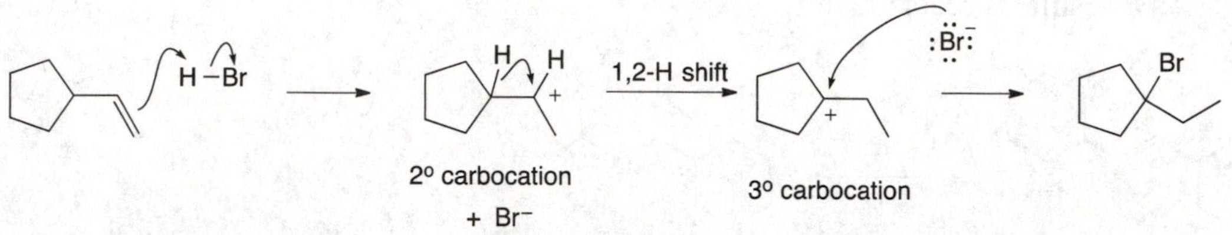

H–Br → 1,2-H shift → :Br: → Br

2° carbocation + Br^-

3° carbocation

10.57

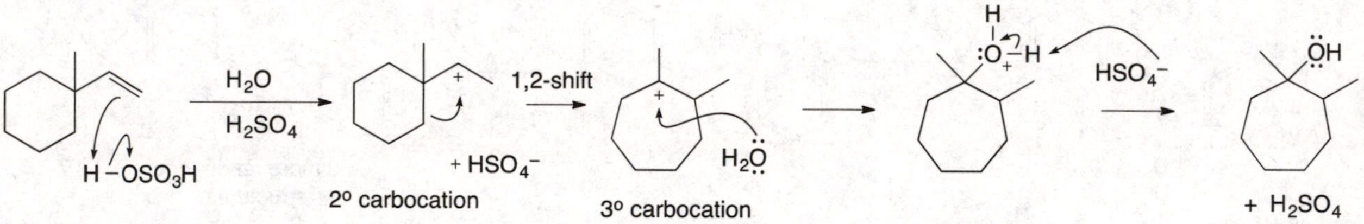

H_2O, H_2SO_4

$H-OSO_3H$

1,2-shift

+ HSO_4^-

2° carbocation

3° carbocation

$H_2\ddot{O}$

HSO_4^-

+ H_2SO_4

10.58

$H-OSO_3H$

H_2SO_4

+ HSO_4^-

$^-OSO_3H$

+ H_2SO_4

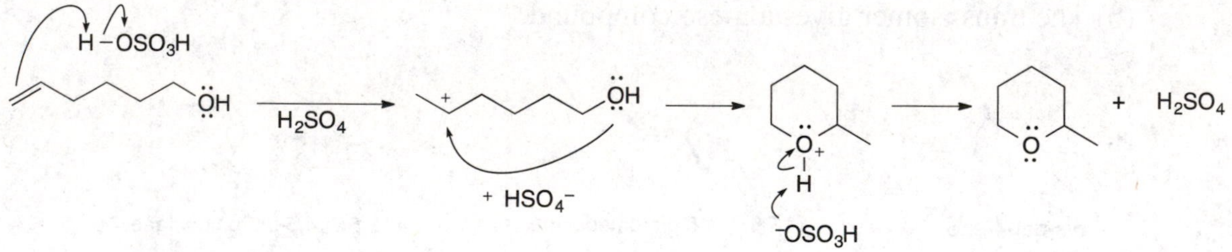

10.59

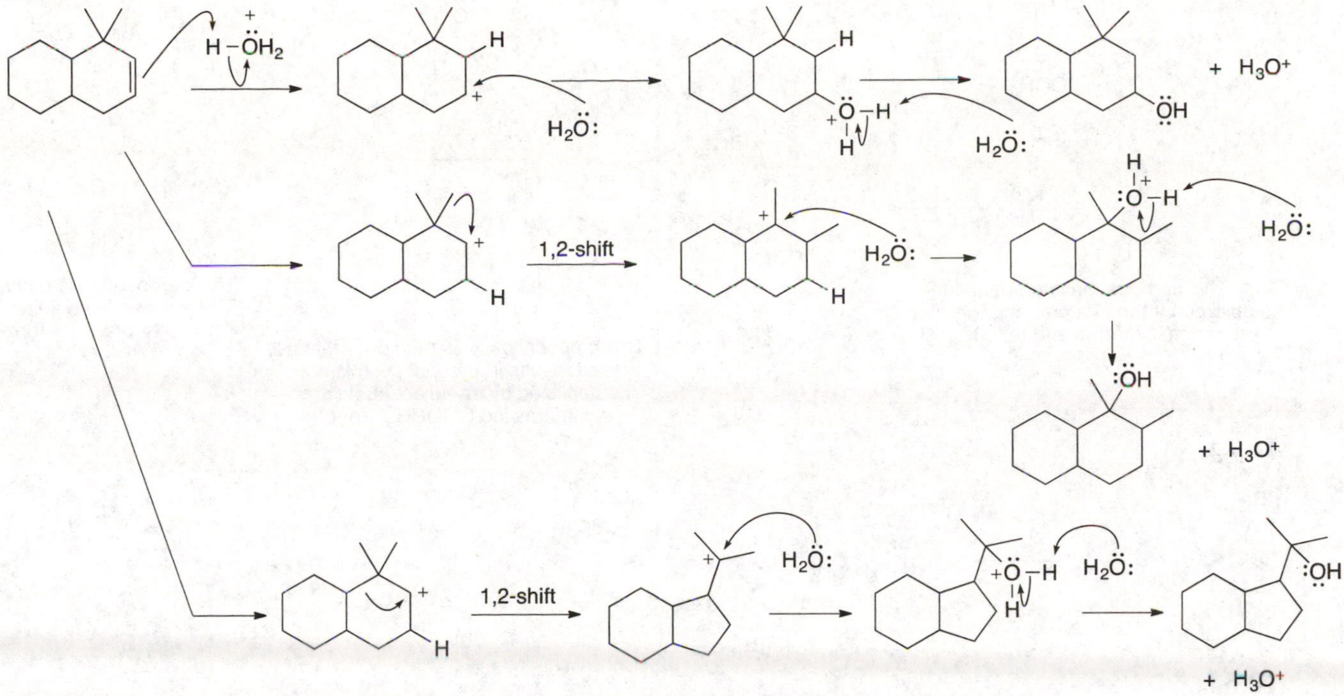

H–ÖH₂ ... 1,2-shift ... $H_2\ddot{O}$: ... + H_3O^+

10.60 The isomerization reaction occurs by protonation and deprotonation.

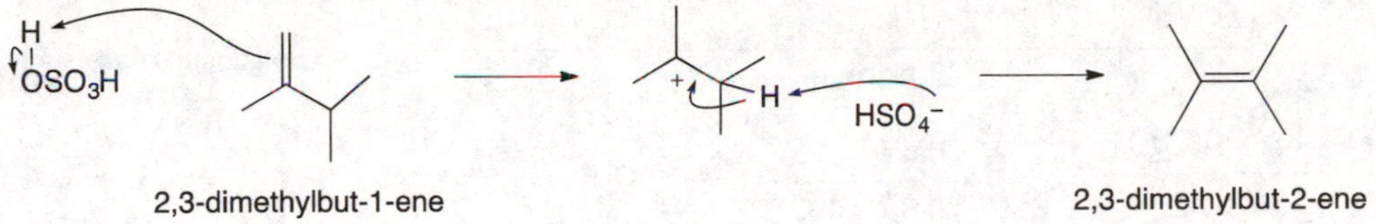

OSO_3H ... HSO_4^-

2,3-dimethylbut-1-ene

2,3-dimethylbut-2-ene

10.61

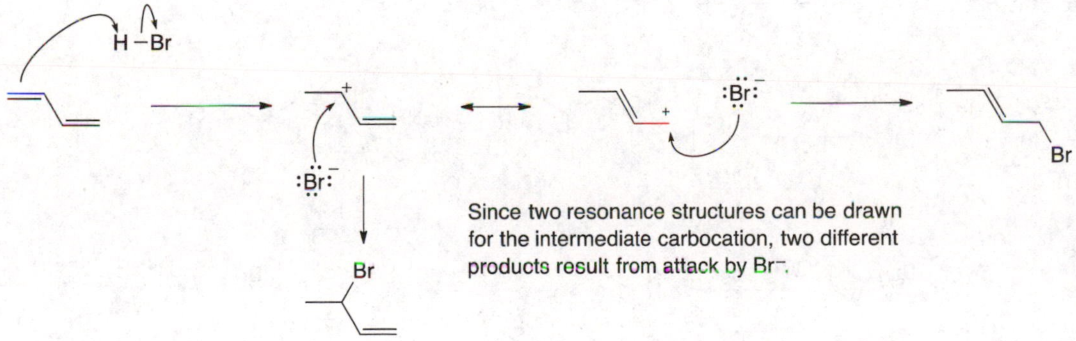

H–Br ... :Br:⁻ ... Br

Since two resonance structures can be drawn for the intermediate carbocation, two different products result from attack by Br⁻.

10.62

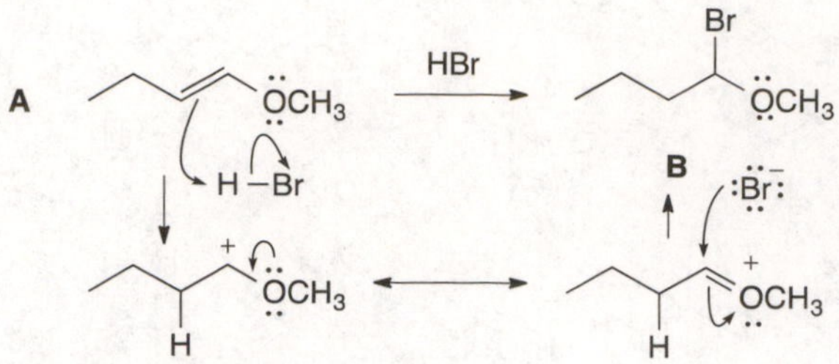

A → HBr → B

This carbocation is resonance stabilized by the O atom, and therefore preferentially forms and results in **B**.

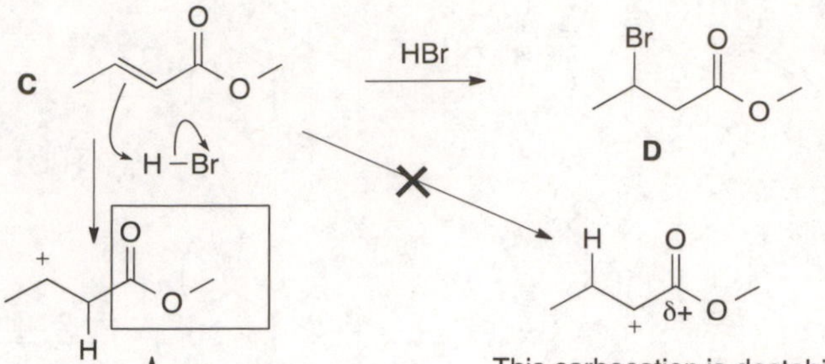

C → HBr → D

This carbocation is formed preferentially and results in product **D.** It is not destabilized by an adjacent electron-withdrawing $COOCH_3$ group.

This carbocation is destabilized by the δ+ on the adjacent C, so it does not form.

10.63

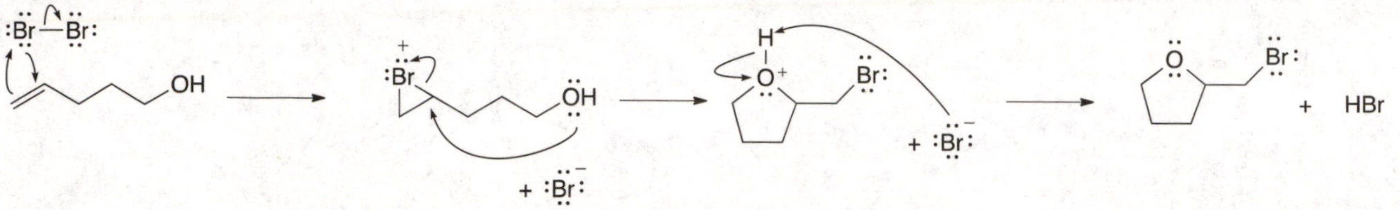

+ HBr

10.64

a. PBr_3; $K^+ {}^-OC(CH_3)_3$; H_2O + H_2SO_4; $POCl_3$, pyridine

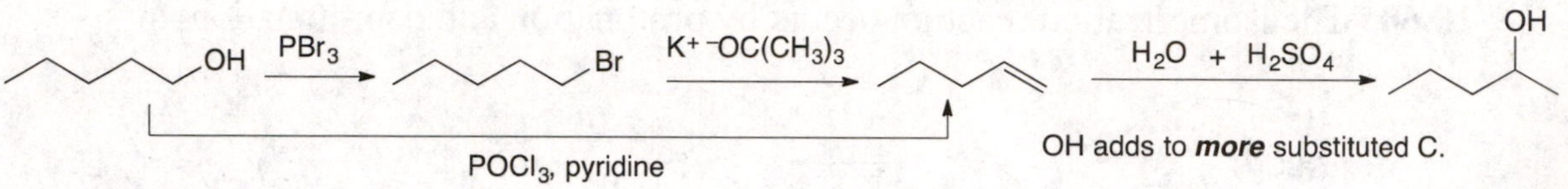

OH adds to ***more*** substituted C.

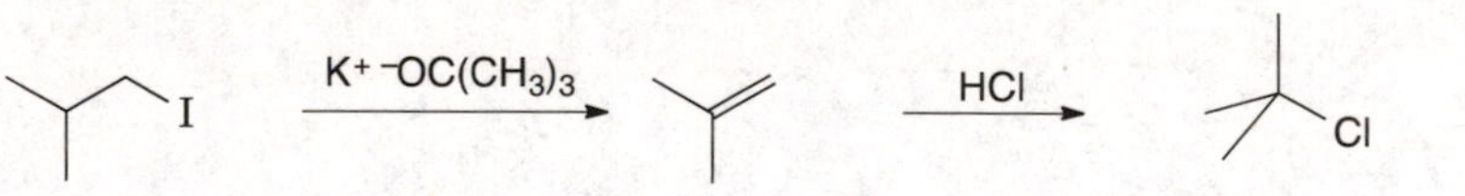

b. $K^+ {}^-OC(CH_3)_3$; [1] BH_3 [2] H_2O_2, HO^-; NaH; CH_3I; + H_2

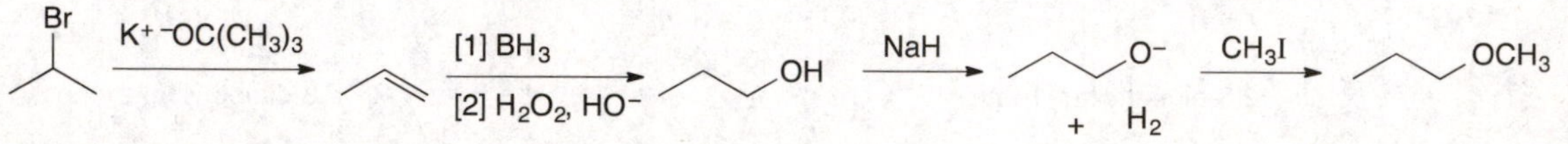

c. $K^+ {}^-OC(CH_3)_3$; HCl

d. Br_2; $NaNH_2$ excess

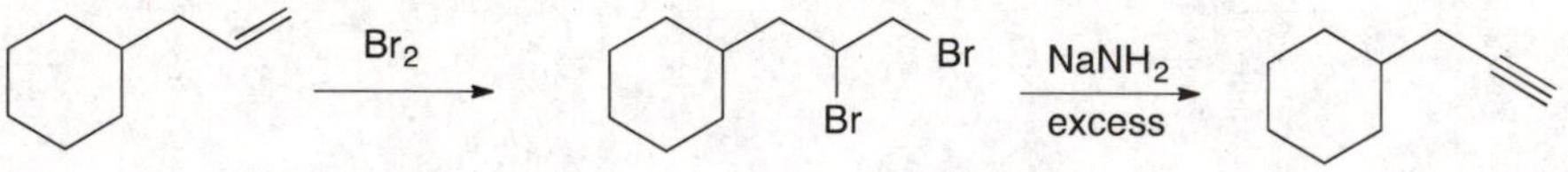

e. $K^+ {}^-OC(CH_3)_3$; Br_2; $NaNH_2$ excess

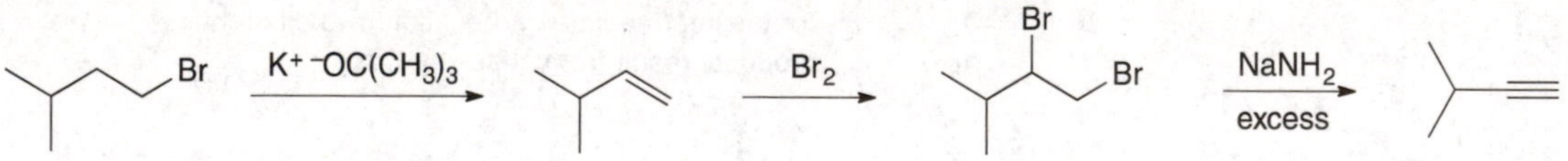

f. $K^+ {}^-OC(CH_3)_3$; Cl_2, H_2O; NaH

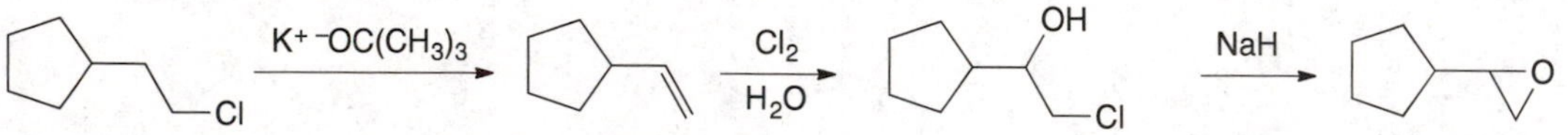

10.65

a. Cl_2, H_2O; $Na^+ H^-$; + H_2

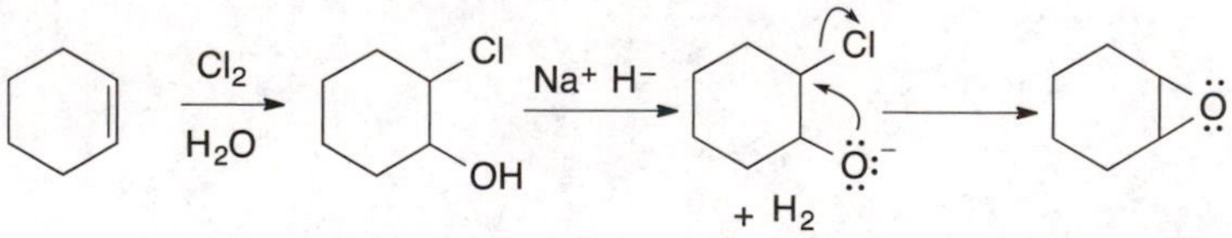

b. [cyclohexene] —HBr→ [bromocyclohexane] —NaCN→ [cyclohexanecarbonitrile]

c. [bromocyclohexane] (from b.) —NaOH→ [cyclohexanol] —NaH→ [cyclohexoxide] —I–propyl→ [cyclohexyl propyl ether]

d. [cyclohexene oxide] (from a.) —[1] ^-SH [2] H_2O→ [trans-2-mercaptocyclohexanol] + enantiomer

10.66 Having two rings joined together as in **A** and **B** creates a very rigid ring system and constrains bond angles. Evidently the bond angles around the C=C in **A** are close enough to the trigonal planar bond angle of 120°, so **A** is stable. With **B,** however, the C=C is located at a carbon shared by both rings and the bond angles around the C=C deviate greatly from the desired angle, so **B** is *not* a stable compound.

10.67 a. Br_2 adds in an anti fashion to form a meso dibromide. Rotate around the C–C bond to place H and Br anti periplanar in the second step. HBr can be eliminated in two ways, but both give the same product.

C_6H_5CH=CHC_6H_5 —Br_2, anti→ meso compound —rotate→ (H and Br must be anti.) —KOH (–HBr)→ (Br)(C_6H_5)C=CH(C_6H_5)

meso compound

H and Br must be anti.

Two higher priority groups are on opposite sides = ***E.***

b. Br_2 addition forms two enantiomers. Anti periplanar elimination of H and Br gives the same alkene from both compounds.

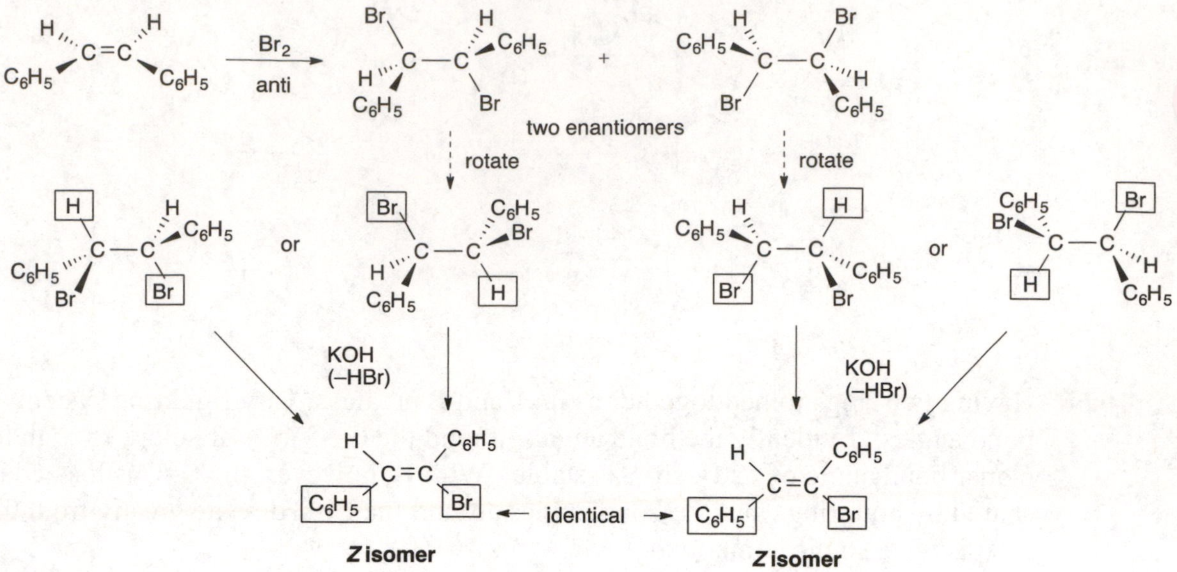

Br$_2$ anti

two enantiomers

rotate

or

KOH (–HBr)

identical

Z isomer

c. The products in (a) and (b) are diastereomers.

10.68

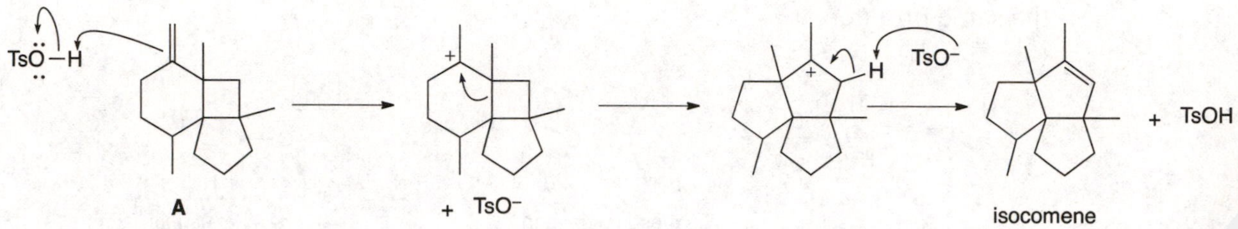

TsOH

A

+ TsO$^-$

isocomene

10.69

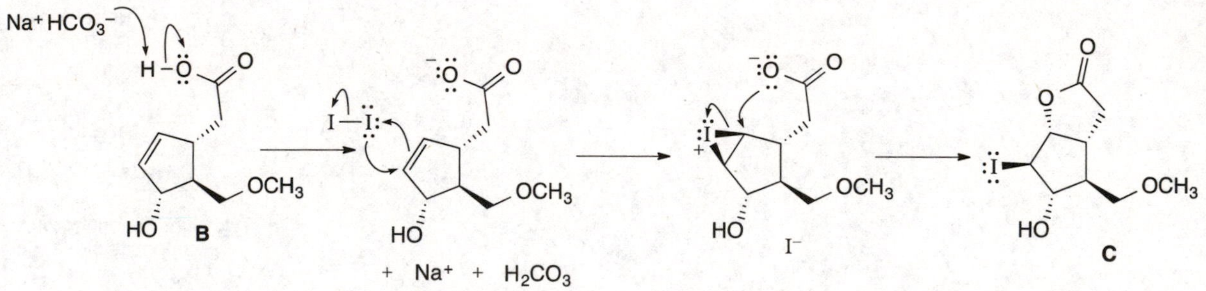

Na$^+$ HCO$_3^-$

B

+ Na$^+$ + H$_2$CO$_3$

I$^-$

C

10.70

+ HSO$_4^-$

2° carbocation

+ H$_2$O

1,2-H shift

3° carbocation

HSO$_4^-$

+ H$_2$SO$_4$

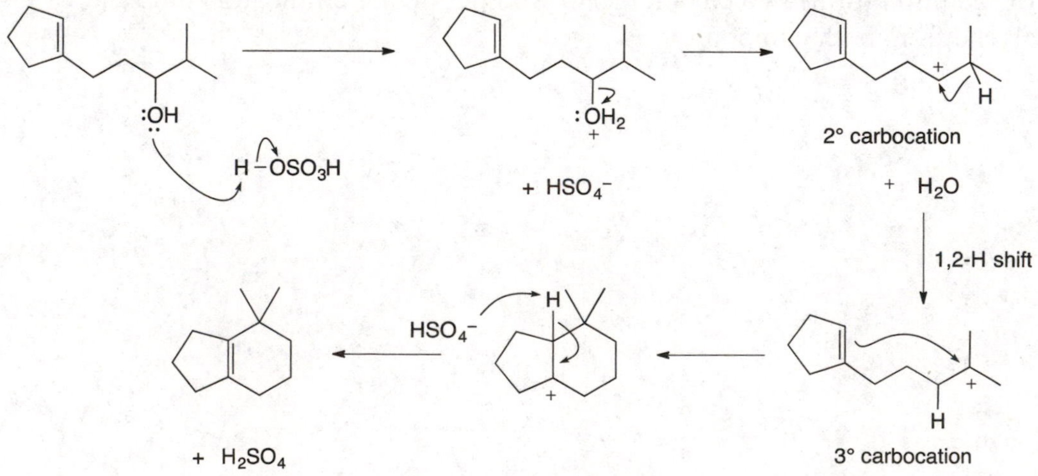

10.71

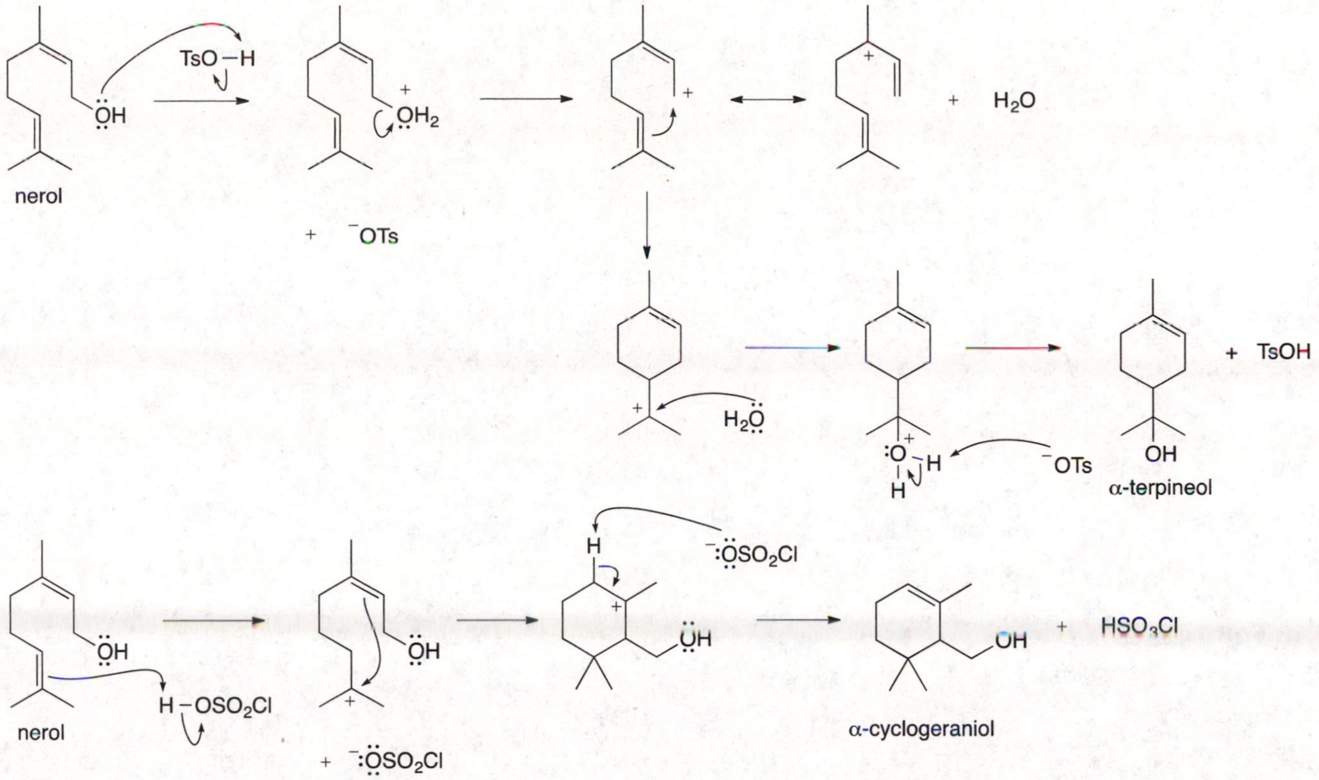

nerol

TsO–H

$\ddot{O}H$

$\overset{+}{O}H_2$

+ ^{-}OTs

+ H_2O

$H_2\ddot{O}$

:O–H

H

^{-}OTs

OH

α-terpineol

+ TsOH

nerol

$\ddot{O}H$

H–OSO_2Cl

+ $^{-}:\ddot{O}SO_2Cl$

H

$^{-}:\ddot{O}SO_2Cl$

$\ddot{O}H$

OH

α-cyclogeraniol

+ HSO_3Cl

10.72

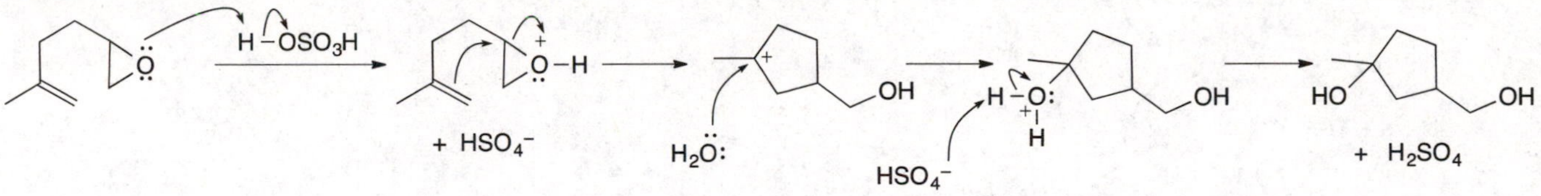

H–OSO_3H

$\ddot{O}$

O–H

+ HSO_4^-

OH

$H_2\ddot{O}$:

H–O:

H

HSO_4^-

OH

HO

OH

+ H_2SO_4

Chapter 11 Alkynes

Chapter Review

General facts about alkynes

- Alkynes contain a carbon–carbon triple bond consisting of a strong σ bond and two weak π bonds. Each carbon is *sp* hybridized and linear (11.1).

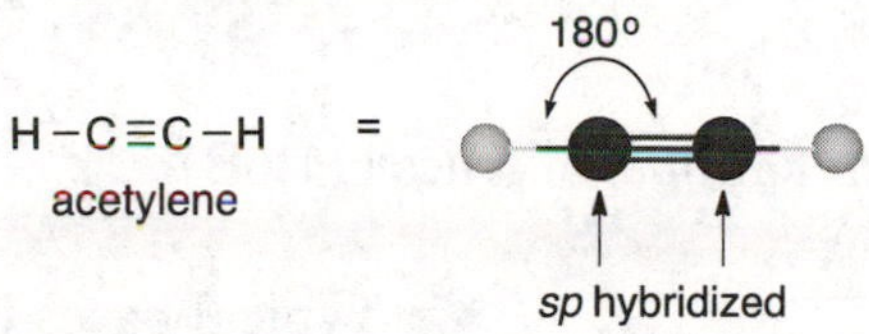
180°
H−C≡C−H
=
acetylene
sp hybridized

- Alkynes are named using the suffix ***-yne*** (11.2).
- Alkynes have weak intermolecular forces, giving them low mp's and low bp's, and making them water insoluble (11.3).
- Since its weaker π bonds make an alkyne electron rich, alkynes undergo addition reactions with electrophiles (11.6).

Addition reactions of alkynes

[1] Hydrohalogenation—Addition of HX (X = Cl, Br, or I) (11.7)

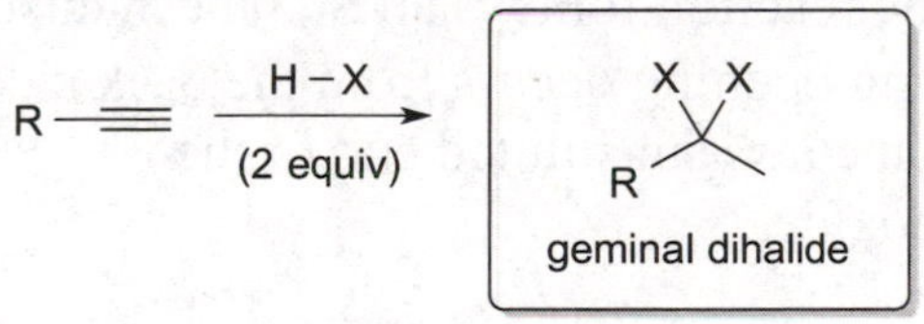
H−X
(2 equiv)
X X
R
geminal dihalide

- Markovnikov's rule is followed. H bonds to the *less* substituted C in order to form the more stable carbocation.

[2] Halogenation—Addition of X_2 (X = Cl or Br) (11.8)

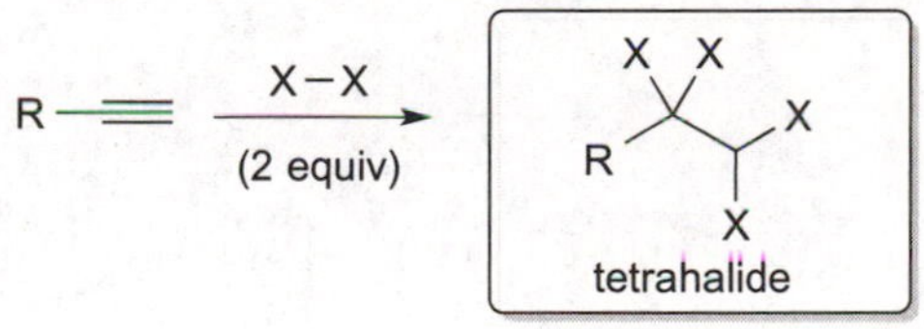
X−X
(2 equiv)
X X
X
R
X
tetrahalide

- Bridged halonium ions are formed as intermediates.
- Anti addition of X_2 occurs.

[3] Hydration—Addition of H_2O (11.9)

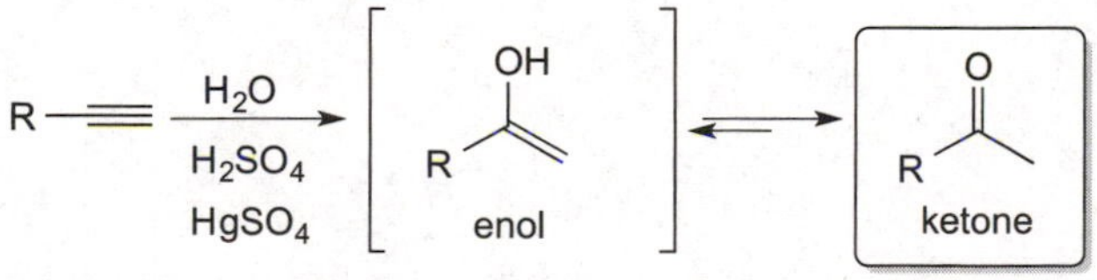
H2O
H2SO4
HgSO4
OH
R
enol
O
R
ketone

- Markovnikov's rule is followed. H bonds to the *less* substituted C in order to form the more stable carbocation.
- The unstable enol that is first formed rearranges to a carbonyl group.

[4] Hydroboration–oxidation—Addition of H_2O (11.10)

$$R{-}C{\equiv}CH \xrightarrow[\text{[2] } H_2O_2,\ HO^-]{\text{[1] } R_2BH} \left[\ \text{enol}\ \right] \rightleftarrows \boxed{\text{aldehyde}}$$

enol

aldehyde

- The unstable enol, first formed after oxidation, rearranges to a carbonyl group.

Reactions involving acetylide anions

[1] Formation of acetylide anions from terminal alkynes (11.6B)

$$R{-}C{\equiv}C{-}H \ + \ :B \rightleftarrows \boxed{R{-}C{\equiv}C{:}^-} \ + \ HB^+$$

- Typical bases used for the reaction are $NaNH_2$ and NaH.

---

[2] Reaction of acetylide anions with alkyl halides (11.11A)

$$H{-}C{\equiv}C{:}^- \ + \ R{-}X \longrightarrow H{-}C{\equiv}C{-}R \ + \ X^-$$

- The reaction follows an S_N2 mechanism.
- The reaction works best with CH_3X and RCH_2X.

---

[3] Reaction of acetylide anions with epoxides (11.11B)

$$H{-}C{\equiv}C{:}^- \xrightarrow[\text{[2] } H_2O]{\text{[1] epoxide}} \boxed{H{-}C{\equiv}C{-}CH_2CH_2OH}$$

- The reaction follows an S_N2 mechanism.
- Ring opening occurs from the back side at the *less* substituted end of the epoxide.

Practice Test on Chapter Review

1. Draw the structure for the compound with the following IUPAC name: 5-*tert*-butyl-6,6-dimethyl-non-3-yne.

2.a. Which of the following compounds is an enol tautomer of compound **A?**

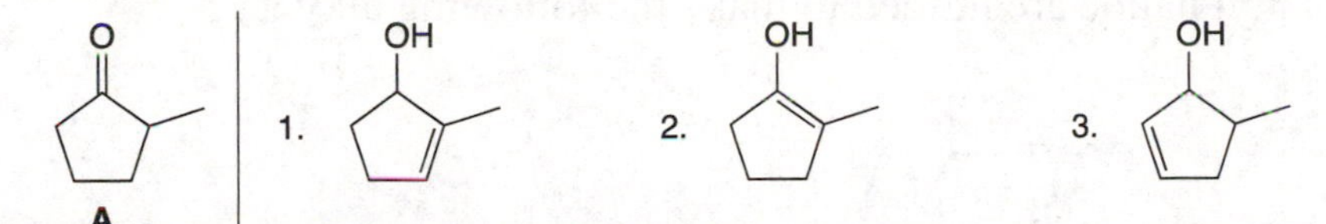

4. **A** can be a tautomer of both compounds (1) and (2).
5. **A** can be a tautomer of compounds (1), (2), and (3).

b. Which of the following bases is strong enough to deprotonate $CH_3C{\equiv}CH$ (propyne, pK_a = 25)? The pK_a's of the conjugate acids of the bases are given in parentheses.

1. CH_3Li (pK_a = 50)
2. $NaOCH_3$ (pK_a = 15.5)
3. $NaOCOCH_3$ (pK_a = 4.8)
4. The bases in (1) and (2) are both strong enough.
5. The bases in (1), (2), and (3) are all strong enough.

3. Draw the organic products formed in the following reactions.

a. [1] R_2BH ; [2] H_2O_2, ^-OH

b. [1] NaH ; [2] O (epoxide) ; [3] H_2O

c. HO H D ; [1] TsCl, pyridine ; [2] $^-C{\equiv}CH$

d. H_2O ; H_2SO_4 ; $HgSO_4$

e. 2 HBr

4. Draw two different enol tautomers for the following compound.

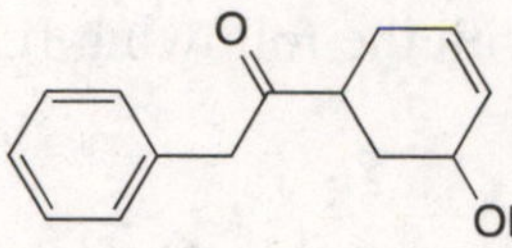

O

OH

5. What acetylide anion and alkyl halide are needed to make the following alkyne?

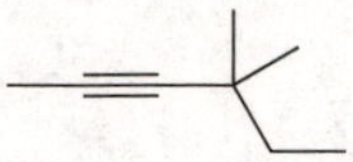

Answers to Practice Test

1.

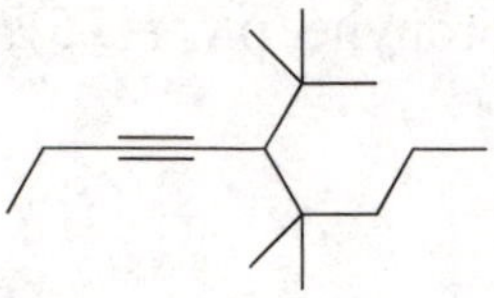

2.a. 2
b. 1

3.
a. H O
b. OH
c. HC≡C D H
d. O
e. Br Br

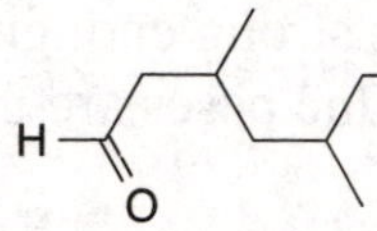

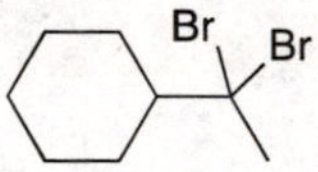

4.
HO OH (*E* + *Z*)

HO OH (*E* + *Z*)

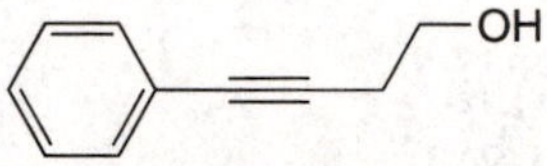

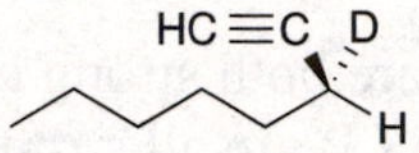

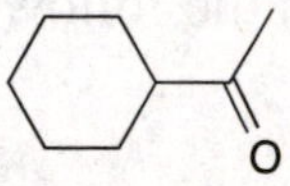

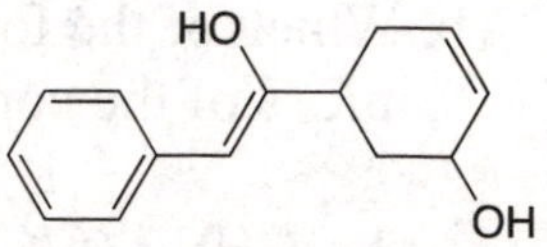

5.
CH_3X + $^-$

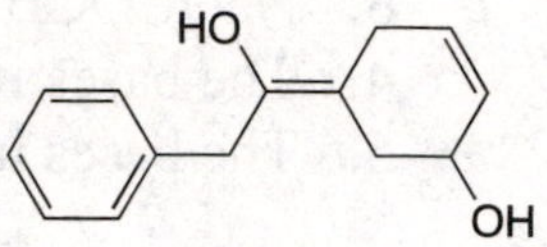

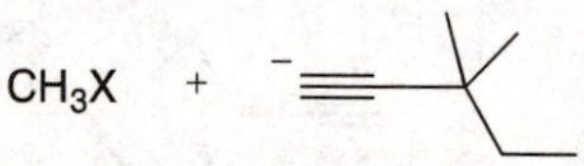

Answers to Problems

11.1

a. most acidic H

b.

nepheliosyne B

a. The most acidic H is part of the COOH group of the carboxylic acid.

b. The shortest C–C σ bond is formed from C atoms with a higher percent *s*-character. The labeled C–C bond is formed from an *sp* hybridized orbital and an sp^2 hybridized orbital.

c. Nepheliosyne B has four triple bonds and five double bonds = 13 degrees of unsaturation.

d. Six bonds (labeled with x) are formed from Csp–Csp^3.

e. All triple bonds are internal except the C≡C at the end of the chain, which is boxed in.

11.2 To name an alkyne:

[1] Find the longest chain that contains both atoms of the triple bond, change the *-ane* ending of the parent name to *-yne*, and number the chain to give the first carbon of the triple bond the lower number.

[2] Name all substituents following the other rules of nomenclature.

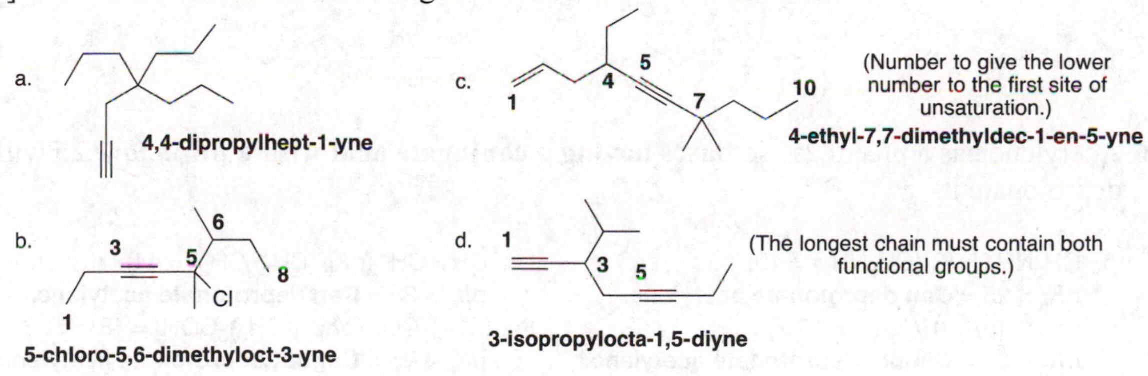

11.3 To work backwards from a name to a structure:

[1] Find the parent name and the functional group.

[2] Add the substituents to the appropriate carbon.

a. *trans*-2-ethynyl**cyclopentanol**

5 C ring with OH at C1

OH ← OH on C1

C≡CH ← ethynyl at C2

or

b. 4-*tert*-butyl**dec-5-yne**

10 C chain with a triple bond

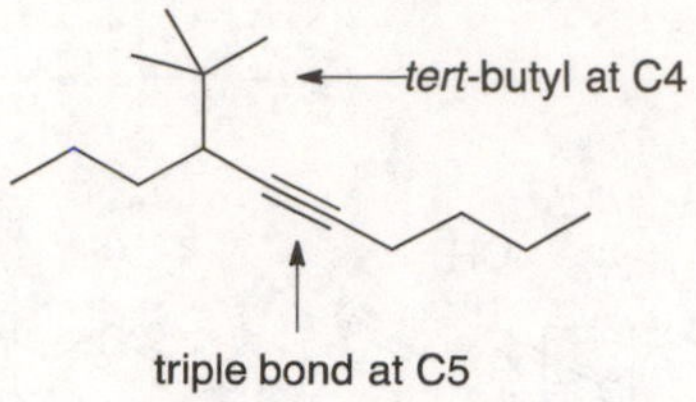

tert-butyl at C4

triple bond at C5

c. 3,3,5-trimethyl**cyclononyne**

9 C ring with a triple bond at C1

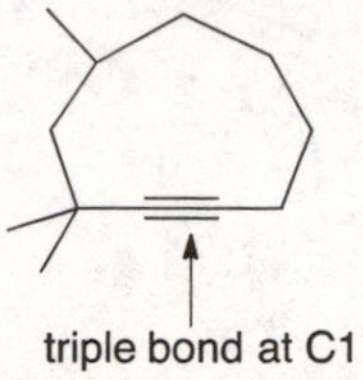

triple bond at C1

11.4 Two factors cause the boiling point increase. The linear *sp* hybridized C's of the alkyne allow for more van der Waals attraction between alkyne molecules. Also, a triple bond is more polarizable than a double bond, which increases the van der Waals forces between two alkyne molecules as well.

11.5 To convert an alkene to an alkyne:
[1] Make a vicinal dihalide from the alkene by addition of X_2.
[2] Add base to remove two equivalents of HX and form the alkyne.

a. Br, Br — $Na^+ \ ^-NH_2$ → [Br — not isolated] — $Na^+ \ ^-NH_2$ →

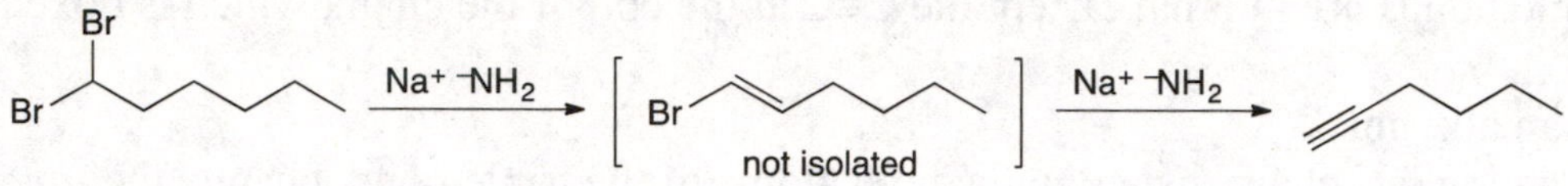

b. Cl — $Na^+ \ ^-NH_2$ →

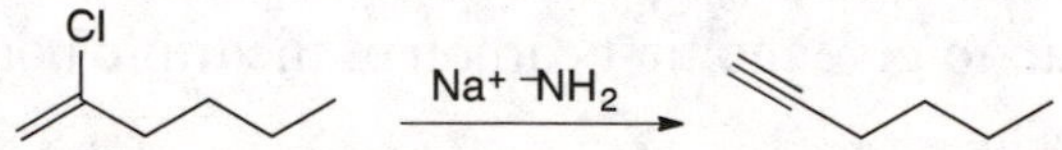

c. Cl_2 → Cl, Cl — $Na^+ \ ^-NH_2$ (2 equiv) →

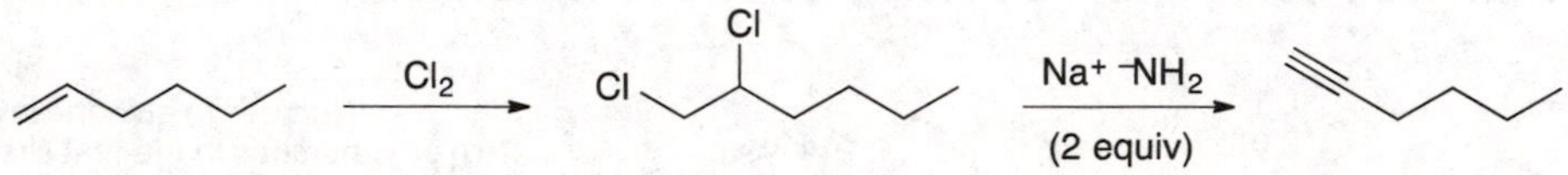

11.6 Acetylene has a pK_a of 25, so **bases having a conjugate acid with a pK_a *above* 25** will be able to deprotonate it.

a. CH_3NH^- [pK_a (CH_3NH_2) = 40]
pK_a > 25 = **Can deprotonate acetylene.**

b. CO_3^{2-} [pK_a (HCO_3^-) = 10.2]
pK_a < 25 = **Cannot deprotonate acetylene.**

c. $CH_2{=}CH^-$ [pK_a ($CH_2{=}CH_2$) = 44]
pK_a > 25 = **Can deprotonate acetylene.**

d. $(CH_3)_3CO^-$ {pK_a [$(CH_3)_3COH$] = 18}
pK_a < 25 = **Cannot deprotonate acetylene.**

11.7 To draw the products of reactions with HX:
- Add two moles of HX to the triple bond, following Markovnikov's rule.
- Both X's end up on the more substituted C.

a. 2 HBr → Br Br

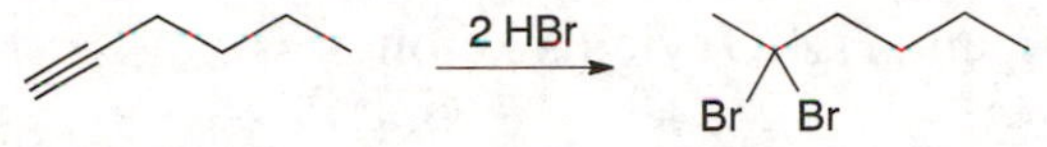

b. 2 HBr → Br Br + Br Br

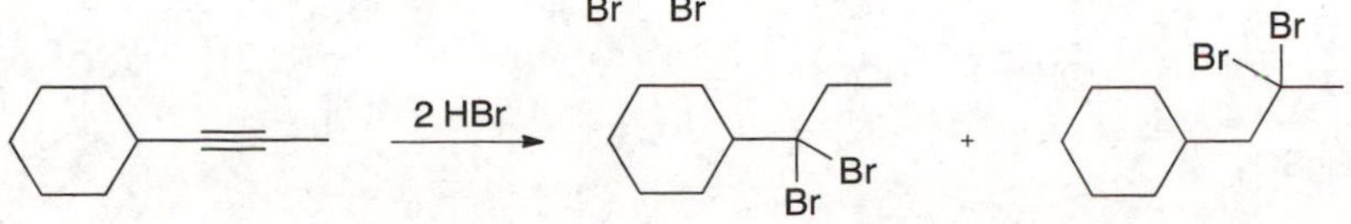

c. 2 HBr → Br Br

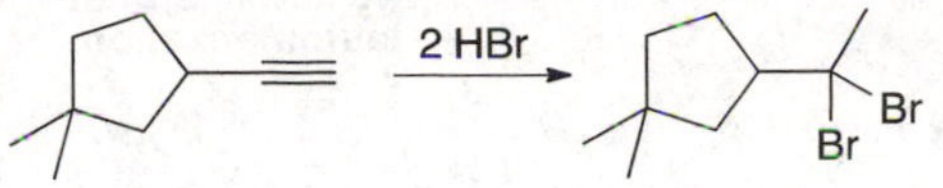

11.8

a. + Cl: ⟷ Cl:

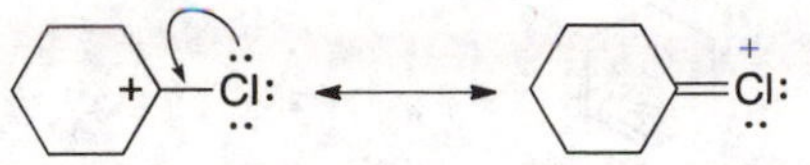

c. H N N H ⟷ H N N H ⟷ H N N H

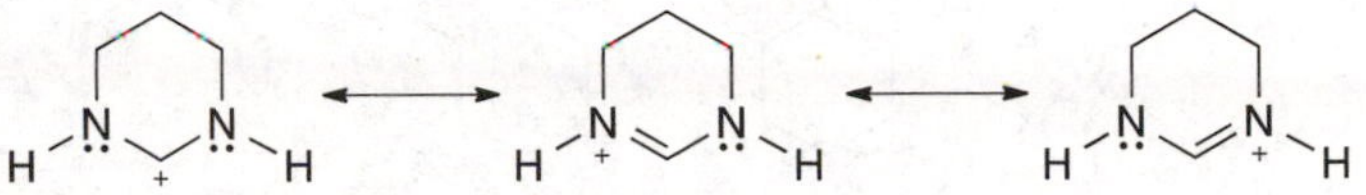

b. O ⟷ O ⟷ O

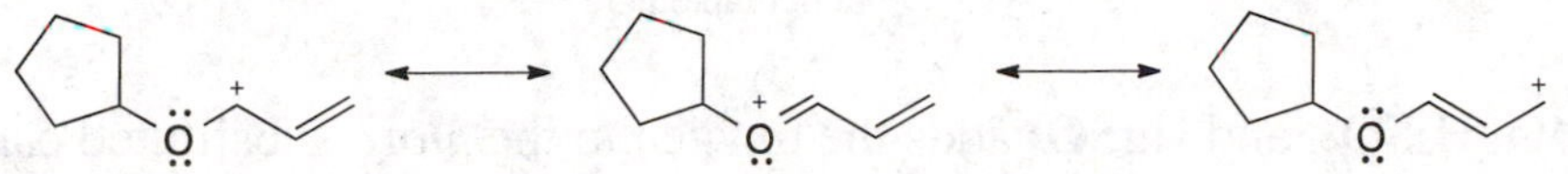

11.9 Addition of one equivalent of X_2 to alkynes forms trans dihalides. Addition of two equivalents of X_2 to alkynes forms tetrahalides.

a. 2 Br_2 → Br Br Br Br

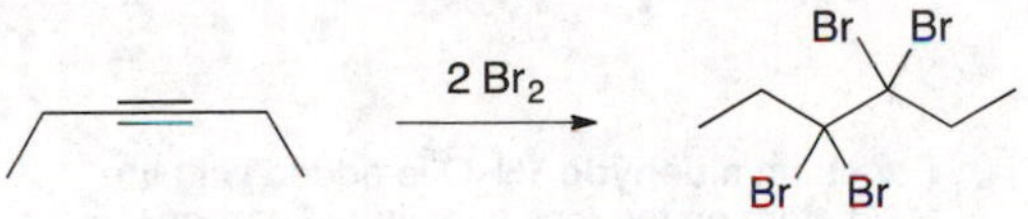

b. Cl_2 → Cl Cl **trans dihalide**

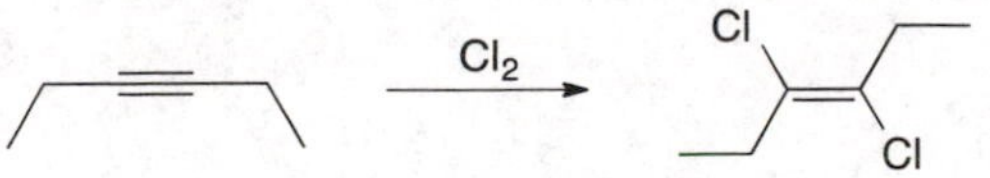

11.10

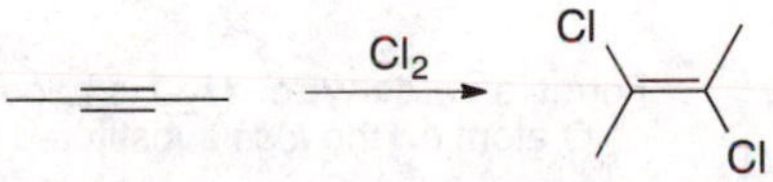

Cl_2 → Cl Cl

The two Cl atoms are electron withdrawing, making the π bond less electron rich and therefore less reactive with an electrophile.

11.11 To draw the keto form of each enol:

[1] Change the C–OH to a C=O at one end of the double bond.

[2] Add a proton at the other end of the double bond.

a. OH ⇌ O H new C–H bond

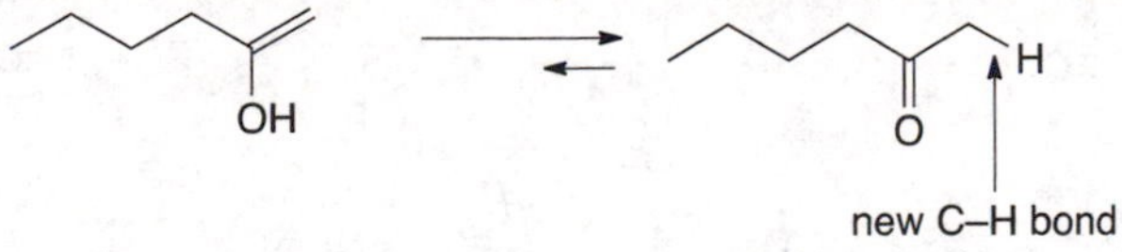

c. OH H ⇌ O H H new C–H bond

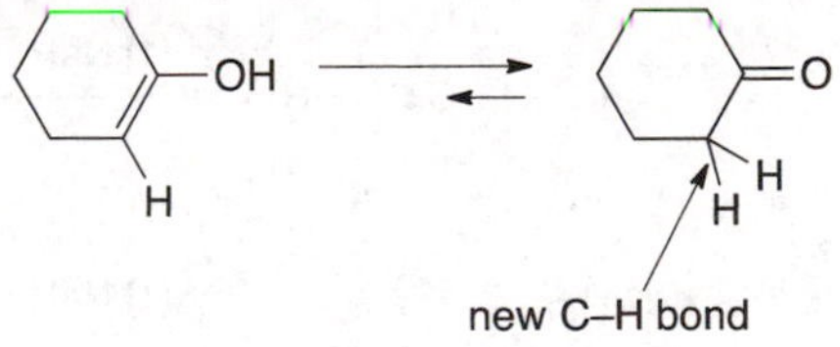

b. OH H ⇌ O H H new C–H bond

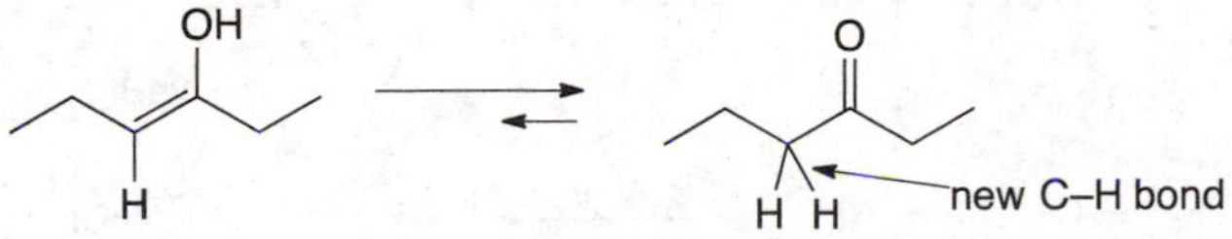

11.12 The treatment of alkynes with H_2O, H_2SO_4, and $HgSO_4$ yields ketones.

H_2O, H_2SO_4, $HgSO_4$

OH + HO

Two enols form. (*E* and *Z* isomers)

O + O

two ketones after tautomerization

11.13

a. OH OH

enol tautomers

b. OH OH

constitutional isomers, but not tautomers

11.14 Reaction with H_2O, H_2SO_4, and $HgSO_4$ adds the oxygen to the *more* substituted carbon. Reaction with [1] R_2BH, [2] H_2O_2, ^-OH adds the oxygen to the *less* substituted carbon.

a. H_2O, H_2SO_4, $HgSO_4$ → O

Forms a **ketone.** H_2O is added with the O atom on the *more* substituted carbon.

[1] R_2BH, [2] H_2O_2, HO^- → H, O

Forms an **aldehyde.** H_2O is added with the O atom on the *less* substituted carbon.

b. H_2O, H_2SO_4, $HgSO_4$ → O

Forms a **ketone.** H_2O is added with the O atom on the *more* substituted carbon.

[1] R_2BH, [2] H_2O_2, HO^- → O, H

Forms an **aldehyde.** H_2O is added with the O atom on the *less* substituted carbon.

11.15

a. $H-C\equiv C-H$ —[1] NaH→ $H-C\equiv C:^-$ + H_2 —[2] (Cl)→ + NaCl

b. $-C\equiv CH$ —[1] NaH→ $-C\equiv C:^-$ + H_2 —[2] (Br)→ + NaBr

1° alkyl halide **substitution product**

—[1] $NaNH_2$→ $-C\equiv C:^-$ + NH_3 —[2] (Cl)→ + NaCl +

3° alkyl halide **elimination product**

11.16

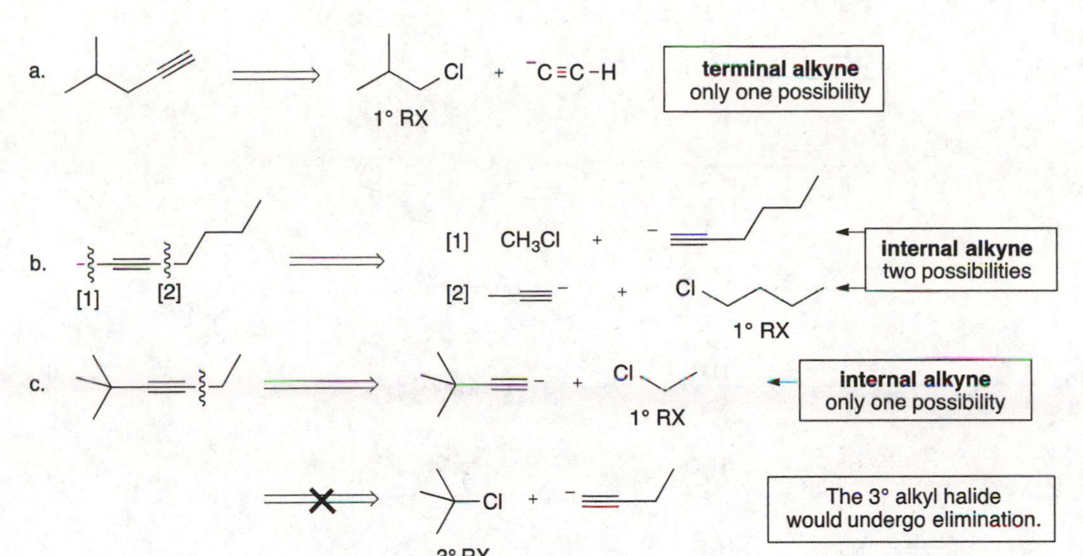

11.17

$HC{\equiv}C{-}H$ $\xrightarrow{Na^+H:^-}$ $HC{\equiv}C:^-$ + H_2 + Na^+

Br, + Na^+Br^-, $Na^+H:^-$, H, + H_2, Br, + Br^-

11.18

2,2,5,5-tetramethylhex-3-yne

X

The 3° alkyl halide is too crowded for nucleophilic substitution. Instead, it would undergo elimination with the acetylide anion.

11.19

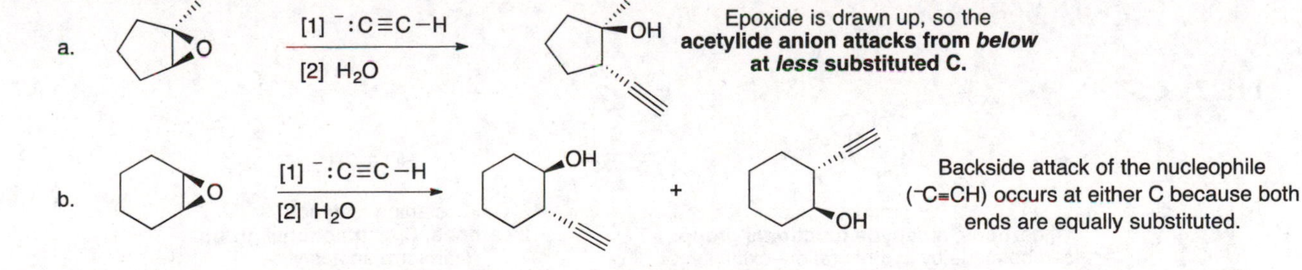

11.20

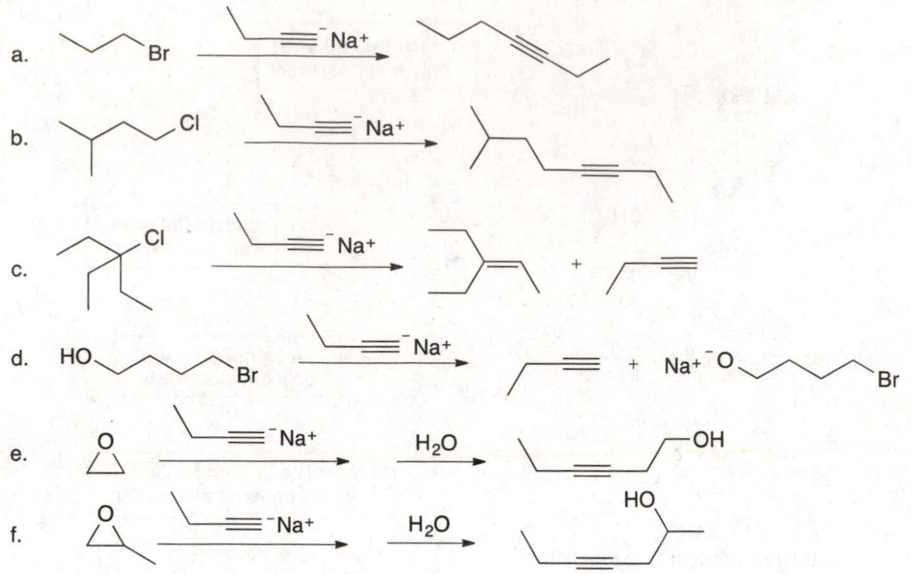

a. Br $\xrightarrow{\text{≡}^- Na^+}$

b. Cl $\xrightarrow{\text{≡}^- Na^+}$

c. Cl $\xrightarrow{\text{≡}^- Na^+}$ +

d. HO Br $\xrightarrow{\text{≡}^- Na^+}$ + $Na^+\ ^-O$ Br

e. O $\xrightarrow{\text{≡}^- Na^+}$ $\xrightarrow{H_2O}$ OH

f. O $\xrightarrow{\text{≡}^- Na^+}$ $\xrightarrow{H_2O}$ HO

11.21 To use a retrosynthetic analysis:

[1] **Count the number of carbon atoms** in the starting material and product.

[2] **Look at the functional groups** in the starting material and product.

- Determine what types of reactions can form the product.
- Determine what types of reactions the starting material can undergo.

[3] **Work backwards** from the product to make the starting material.

[4] Write out the synthesis in the synthetic direction.

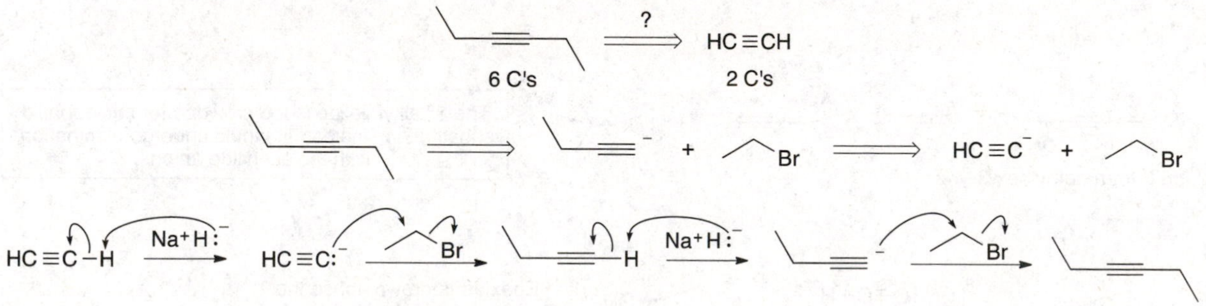

6 C's $\overset{?}{\Longrightarrow}$ HC≡CH 2 C's

$\Longrightarrow$ ≡$^-$ + Br $\Longrightarrow$ HC≡C$^-$ + Br

HC≡C–H $\xrightarrow{Na^+H:^-}$ HC≡C:$^-$ $\xrightarrow{\text{Br}}$ ≡–H $\xrightarrow{Na^+H:^-}$ ≡$^-$ $\xrightarrow{\text{Br}}$

11.22

O, H $\Longrightarrow$ HC≡CH

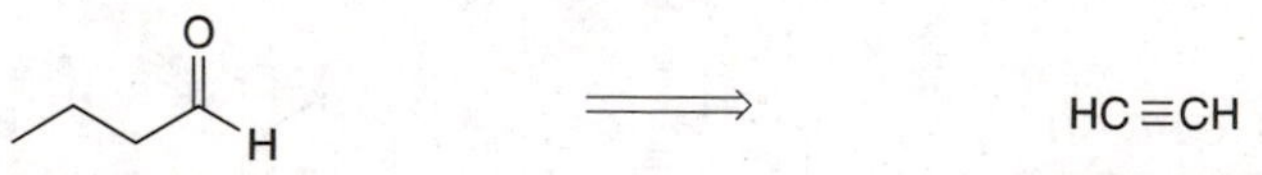

product:
4 carbons, aldehyde functional group
(can be made by hydroboration–oxidation of a terminal alkyne)

starting material:
2 carbons, C≡C functional group
(can form an acetylide anion by reaction with NaH)

Retrosynthetic analysis: O, H $\Longrightarrow$ ≡ $\Longrightarrow$ H–C≡C–H

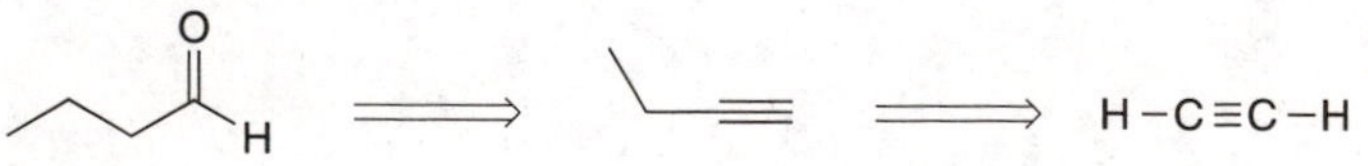

Forward direction: H–C≡C–H $\xrightarrow{Na^+H:^-}$ $^-$:C≡C–H $\xrightarrow{\text{Br}}$ ≡ $\xrightarrow[\text{[2] } H_2O_2,\ HO^-]{\text{[1] } R_2BH}$ O, H

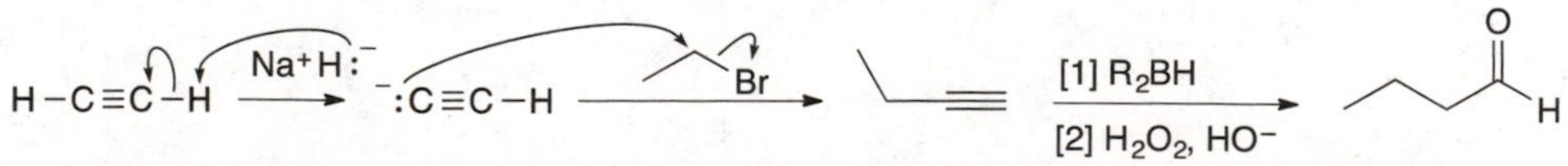

11.23

a.

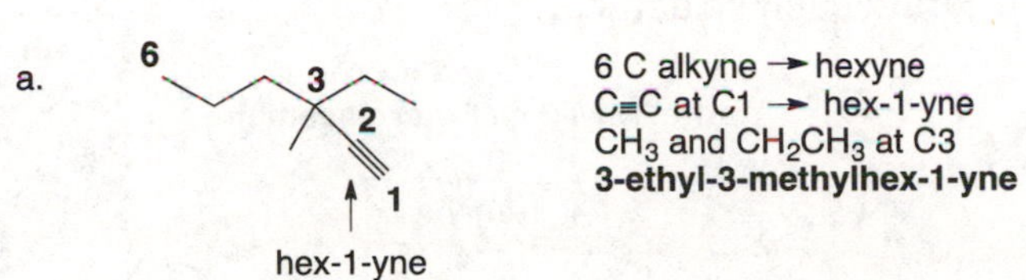

6 C alkyne → hexyne
C≡C at C1 → hex-1-yne
CH_3 and CH_2CH_3 at C3
3-ethyl-3-methylhex-1-yne

b. 3 5 1 11

12 C chain with 2 C≡C's → dodecadiyne
C≡C's at C3 and C5 → dodeca-3,5-diyne
CH_3 at C11
11-methyldodeca-3,5-diyne

11.24

a. O — keto form; OH — enol form

b. OH — enol form; O, H — keto form

11.25

a, b. O, O, O, O, N, N, HN, H

erlotinib

most acidic C–H proton

shortest C–C single bond
C_{sp}–C_{sp^2}

c. (1) (2) (3) O HO C*

d. * = *sp* hybridized C

e. 3 < 1 < 2

phomallenic acid C

most acidic proton

11.26 Use the rules from Answer 11.2 to name the alkynes.

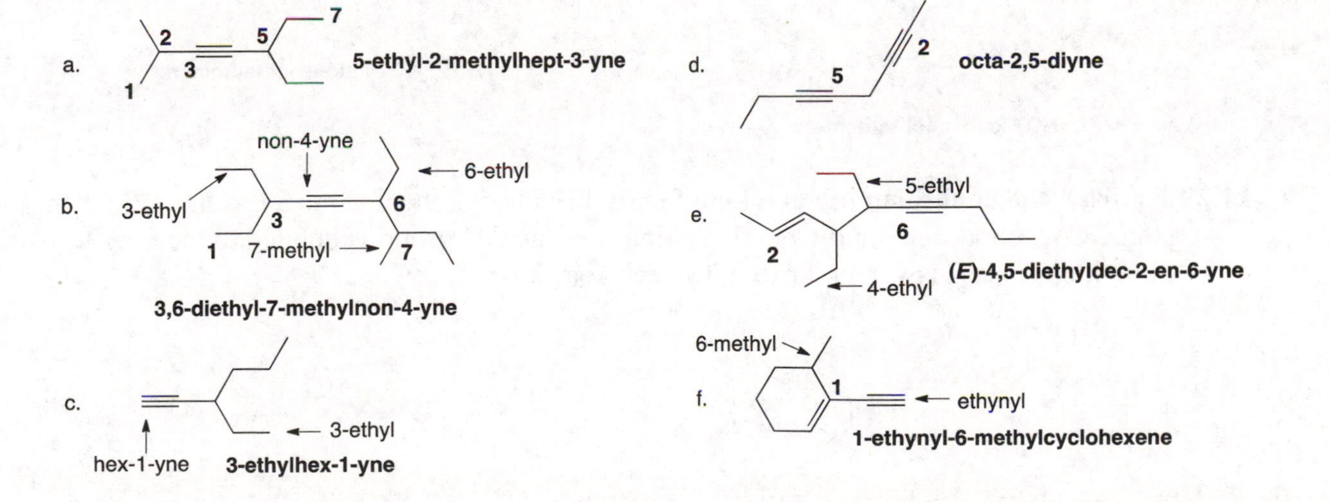

11.27 Use the directions from Answer 11.3 to draw each structure.

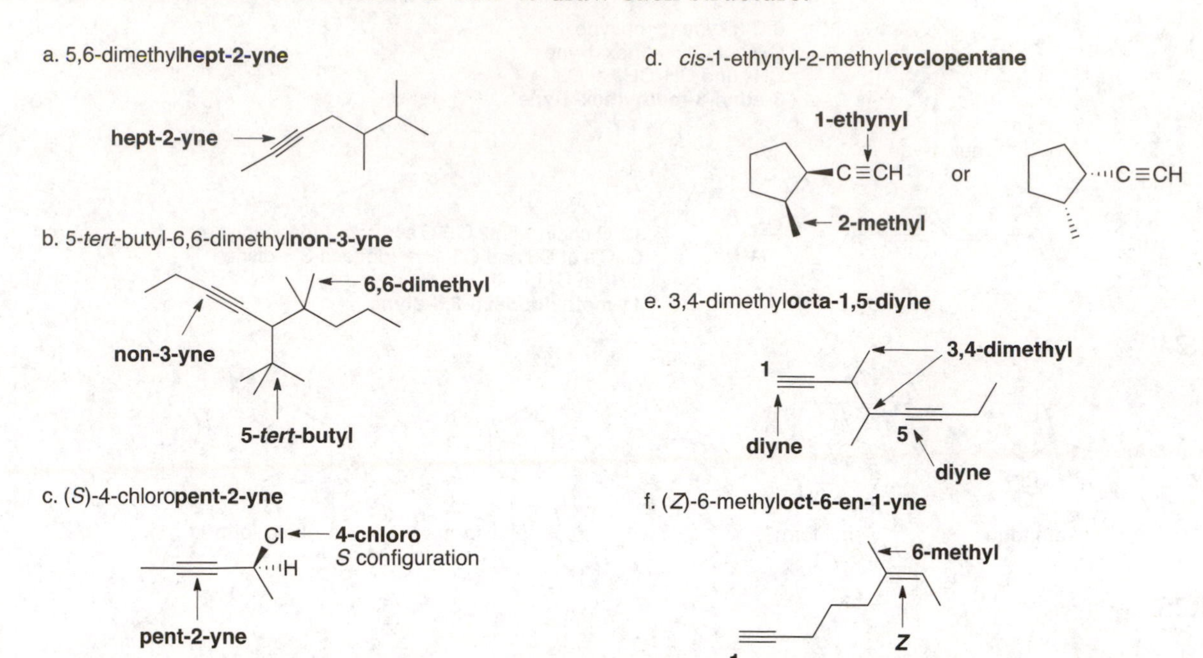

11.28 Keto–enol tautomers are constitutional isomers in equilibrium that differ in the location of a double bond and a hydrogen. The OH in an enol must be bonded to a C=C.

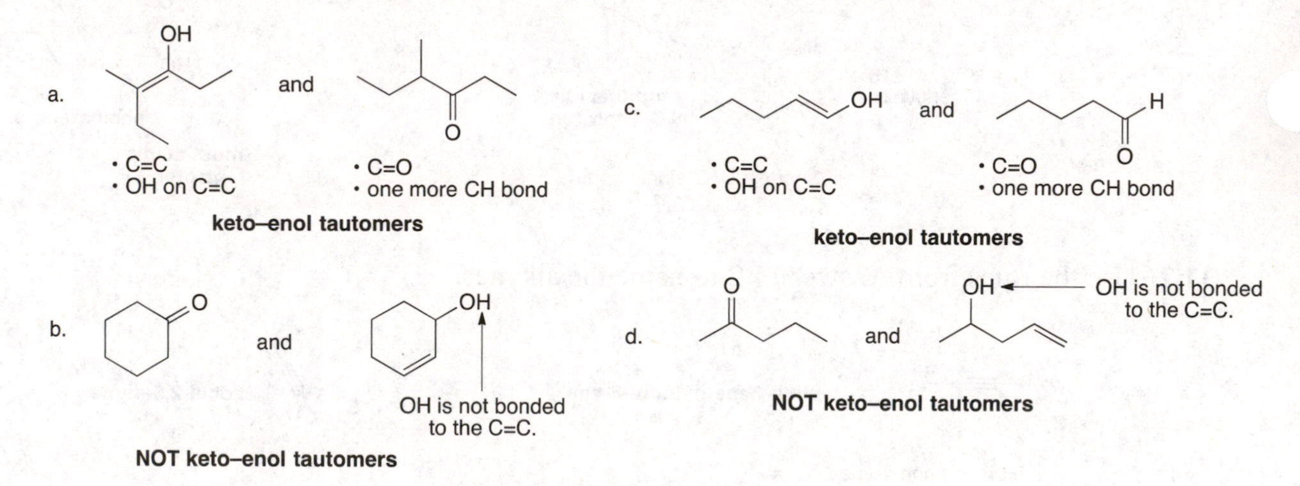

11.29 To draw the enol form of each keto form: [1] Change the C=O to a C–OH. [2] Change one single C–C bond to a double bond, making sure the OH group is bonded to the C=C. Use the directions from Answer 11.11 to draw each keto form.

a. O, H → OH, H

b. O → OH + OH

c. OH → O, H

d. OH → O

11.30 Tautomers are constitutional isomers that are in equilibrium and differ in the location of a double bond and a hydrogen atom.

A

a. tautomer

b. constitutional isomer

c. constitutional isomer

d. neither

11.31

2-butanone

C=C has one C bonded to it.

(*E* and *Z*)

C=C has two C's bonded to it. The more substituted double bond is **more stable.**

11.32

enamine X

$H_2\ddot{O}$–H

+ $H_2\ddot{O}$:

+ $H_2\ddot{O}$:

Y imine

+ $H_3\ddot{O}^+$

11.33

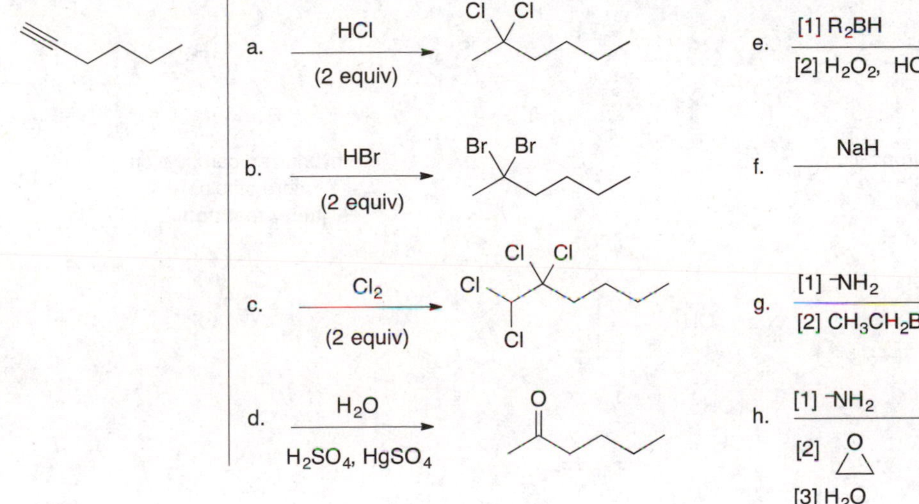

e. [1] R_2BH [2] H_2O_2, HO^- → H, O

f. NaH → Na^+

g. [1] $^-NH_2$ [2] CH_3CH_2Br

h. [1] $^-NH_2$ [2] (epoxide) [3] H_2O → HO

11.34

a. 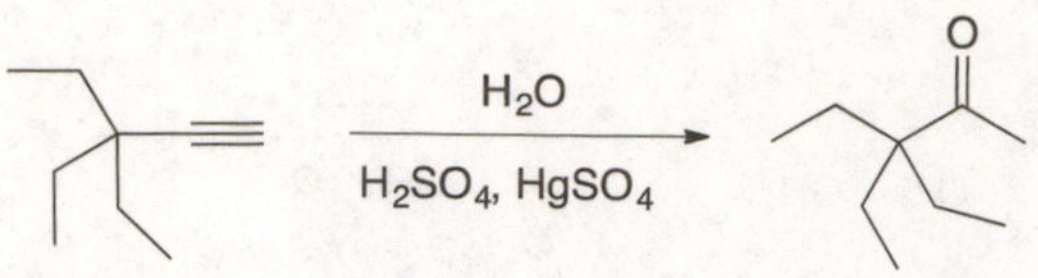 $\xrightarrow[H_2SO_4,\ HgSO_4]{H_2O}$

b. 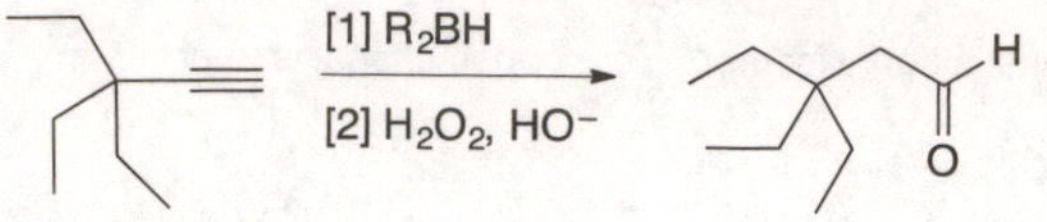 $\xrightarrow[\text{[2] } H_2O_2,\ HO^-]{\text{[1] } R_2BH}$ H O

c. 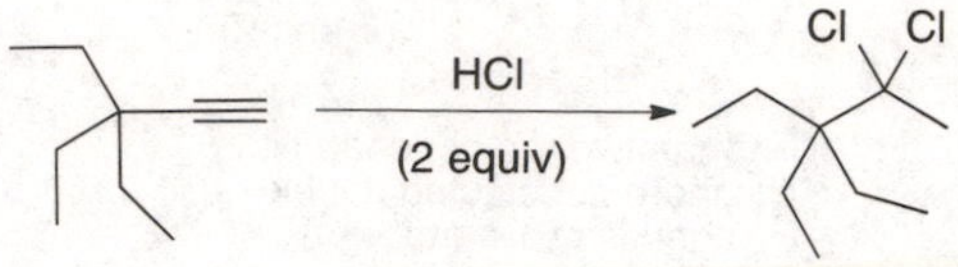 $\xrightarrow[\text{(2 equiv)}]{HCl}$ Cl Cl

d. 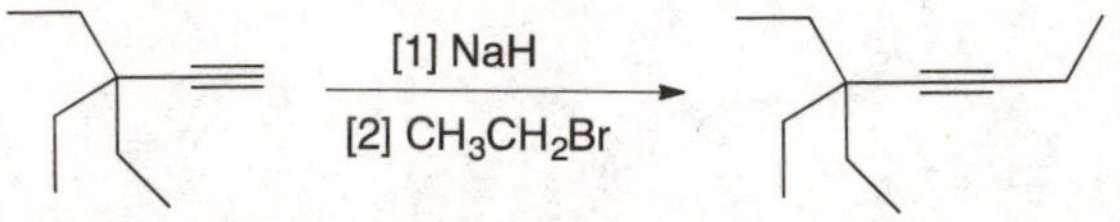 $\xrightarrow[\text{[2] } CH_3CH_2Br]{\text{[1] NaH}}$

11.35 Reaction rate (which is determined by E_a) and enthalpy ($\Delta H°$) are not related. More exothermic reactions are not necessarily faster. Because the addition of HX to an alkene forms a more stable carbocation in an endothermic, rate-determining step, this carbocation is formed faster by the Hammond postulate.

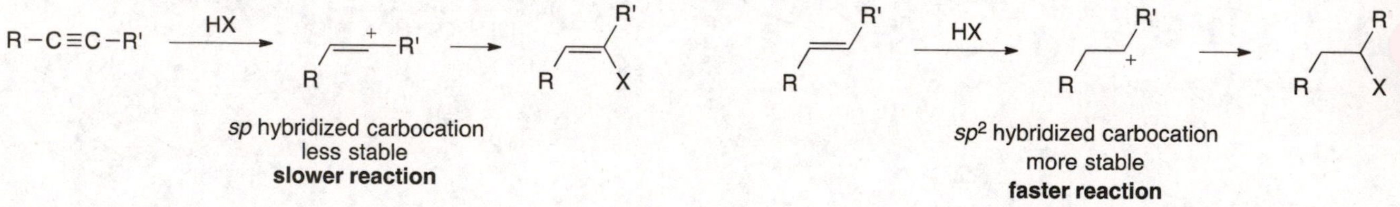

R–C≡C–R' $\xrightarrow{HX}$ R, +, R' → R, R', X

sp hybridized carbocation
less stable
slower reaction

R, R' $\xrightarrow{HX}$ R, R', + → R, R, X

sp[2] hybridized carbocation
more stable
faster reaction

11.36

a. 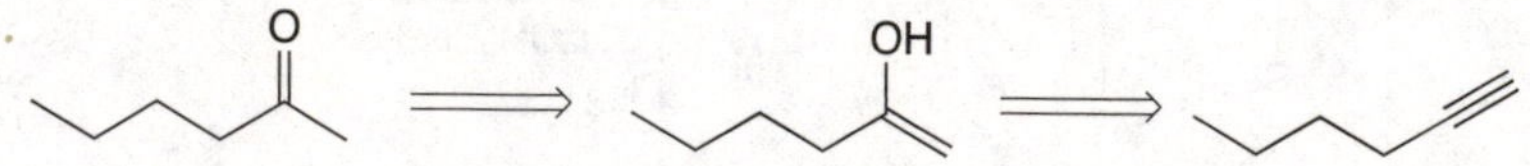 O ⟹ OH ⟹

b. 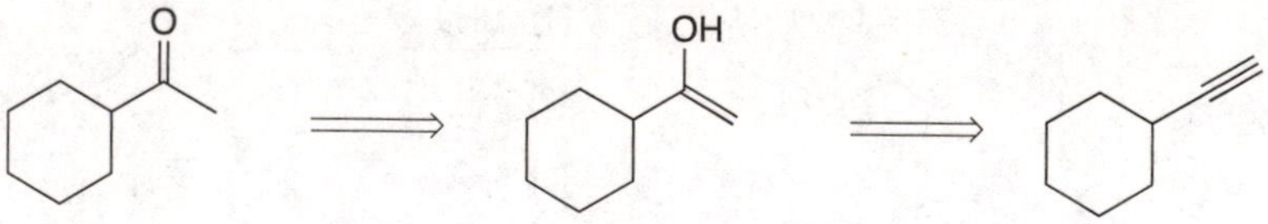 O ⟹ OH ⟹

c. 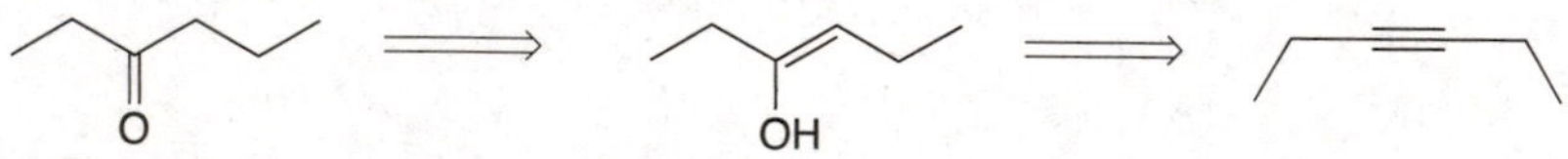 O ⟹ OH ⟹

d. 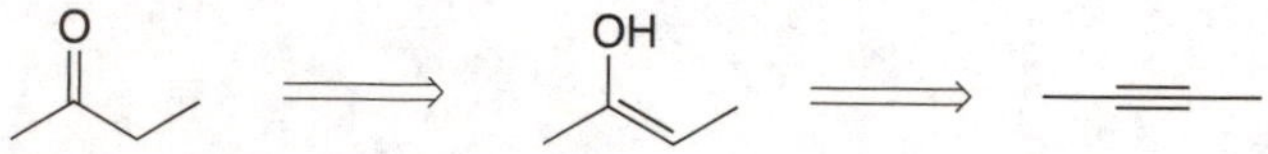 O ⟹ OH ⟹

11.37

a.  ⟹

b. 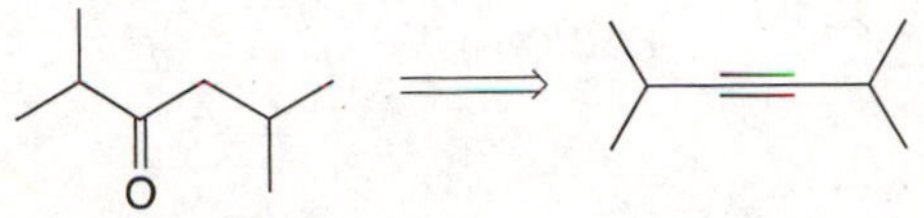 ⟹

11.38

a. 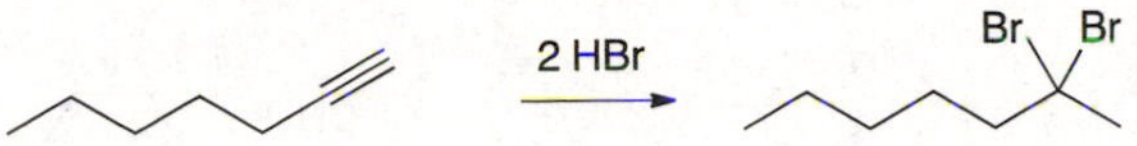 $\xrightarrow{2\ HBr}$ Br Br

b. 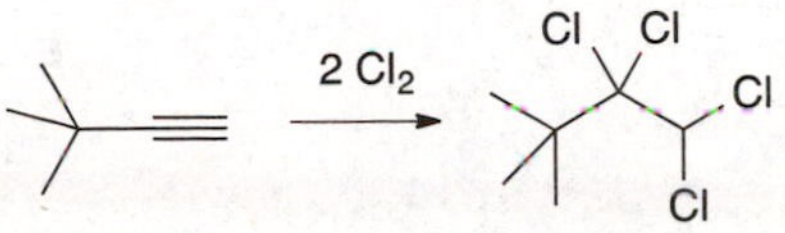 $\xrightarrow{2\ Cl_2}$ Cl Cl Cl Cl

c. 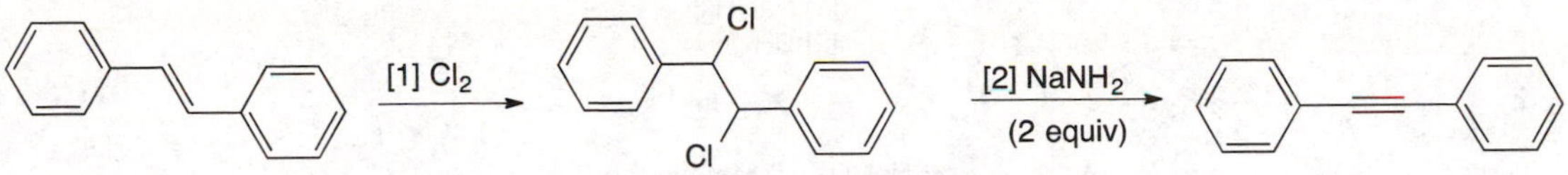 $\xrightarrow{[1]\ Cl_2}$ Cl Cl $\xrightarrow[(2\ equiv)]{[2]\ NaNH_2}$

d. 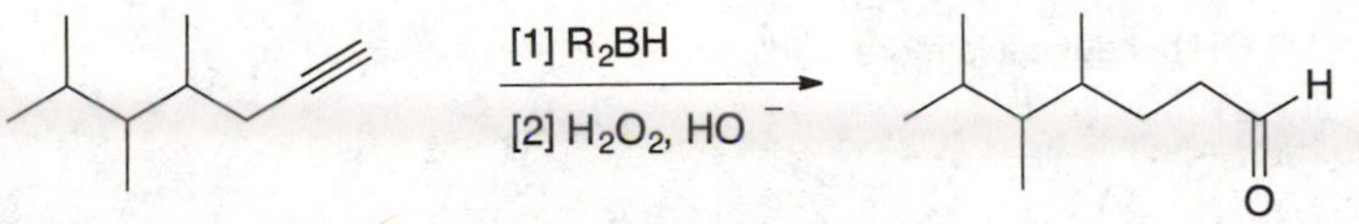 $\xrightarrow[[2]\ H_2O_2,\ HO^-]{[1]\ R_2BH}$ H O

e. $HC{\equiv}C^- + D_2O \longrightarrow HC{\equiv}CD + DO^-$

f. 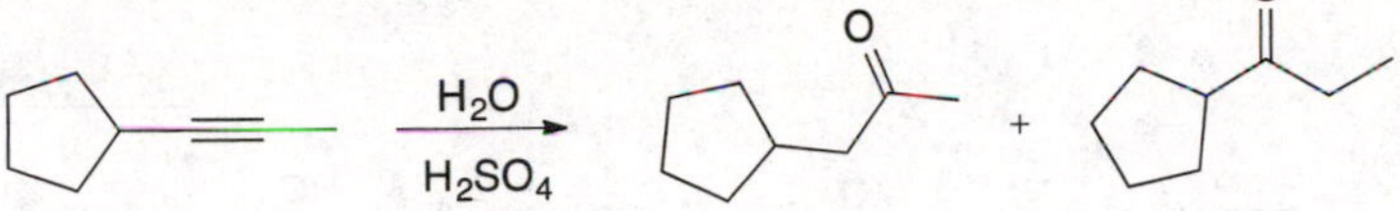 $\xrightarrow[H_2SO_4]{H_2O}$ O + O

g. 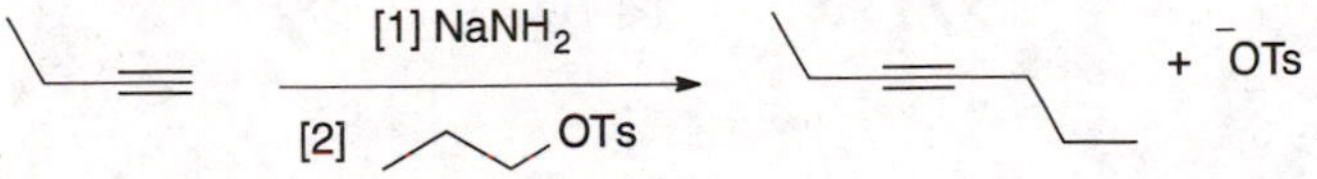 $\xrightarrow[[2]\ \text{OTs}]{[1]\ NaNH_2}$ + $^-$OTs

h. 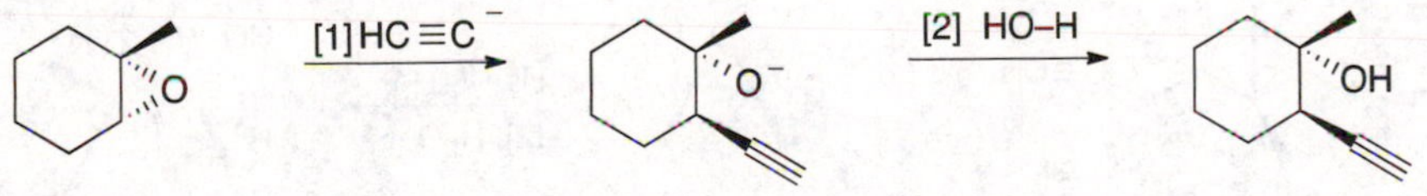 O $\xrightarrow{[1]\ HC{\equiv}C^-}$ O$^-$ $\xrightarrow{[2]\ \text{HO–H}}$ OH

i. $\xrightarrow{[1]\ NaNH_2}$ $^-$ $\xrightarrow{[2]}$ I

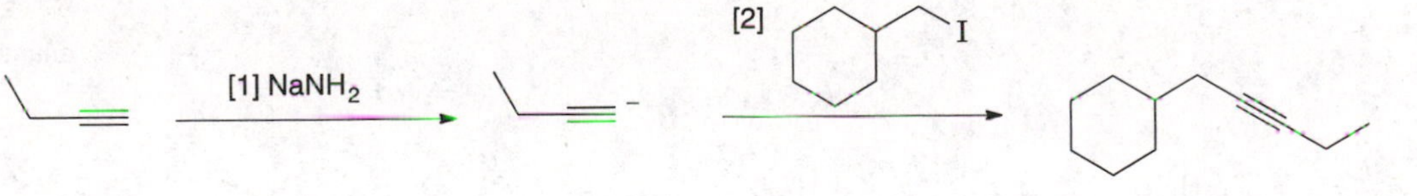

j. 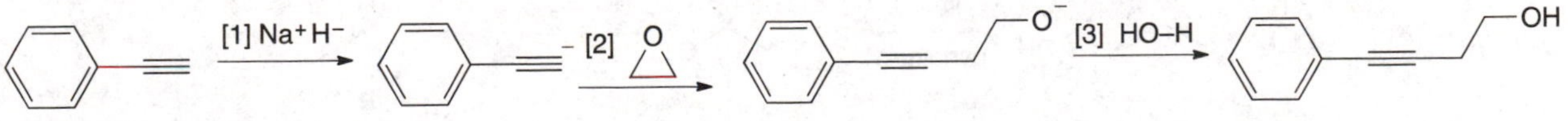 $\xrightarrow{[1]\ Na^+H^-}$ $^-$ $\xrightarrow{[2]\ O}$ O$^-$ $\xrightarrow{[3]\ \text{HO–H}}$ OH

11.39

Br

$KOC(CH_3)_3$

Br_2

Br

Br

$KOC(CH_3)_3$

DMSO

(2 equiv)

A

B

C

D

$NaNH_2$

CH_3I

E

11.40

most acidic H

[1] $NaNH_2$

[2] CH_3I

OH

OH

A

$NaNH_2$ will remove the proton from the OH because it is more acidic.

CH_3—I

= B

11.41

stereogenic center at the site of reaction

a.

Cl $HC{\equiv}C^-$

C≡CH

H D

D H

inversion

c.

H

H

[1] $HC{\equiv}C^-$

[2] H_2O

identical

HO

H

H

+

H

H

stereogenic center NOT at the site of reaction

Configuration is retained.

b.

Cl $HC{\equiv}C^-$

C≡CH

H

H

d.

[1] $HC{\equiv}C^-$

[2] H_2O

HO

H

H

+

H

OH

H

enantiomers

11.42

H D

OH

A

TsCl

pyridine

H D

OTs

$HC{\equiv}C^-$

S_N2

D H

B

retention

inversion

H D

OH

A

PBr_3

S_N2

D H

Br

$HC{\equiv}C^-$

S_N2

H D

C

inversion

inversion

11.43

a.

OH

PBr_3

Br

A

H—≡—OR'

CH_3^- Li^+

$^-$:≡—OR'

B

new C–C bond

OR'

C

b.

OR

[1] **D** = $HC{\equiv}C^-$

[2] H_2O

OH

OR

These 2 C's are added.

OH

OH

E = TsCl, pyridine

OH

OTs

OH

[1] NaH

[2] **F = CH_3CH_2Br**

[1] NaH

[2] CH_3I

new C–C bond

G

11.44

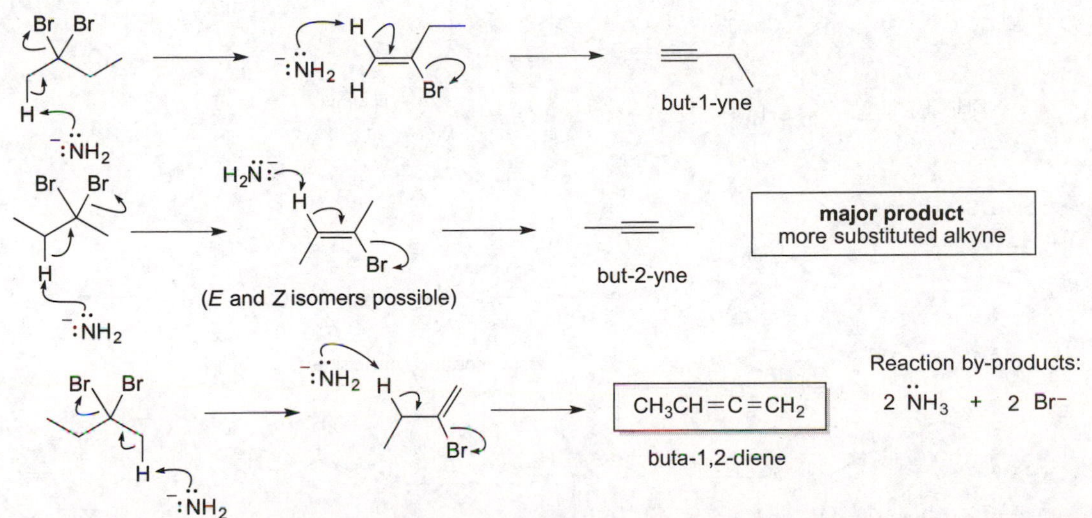

11.45 A carbanion is more stable when its lone pair is in an orbital with a higher percentage of the smaller *s* orbital. A carbocation is more stable when its positive charge is due to a vacant orbital with a lower percentage of the smaller *s* orbital. In $HC{\equiv}C^+$, the positively charged C uses two *p* orbitals to form two π bonds. If the σ bond is formed using an *sp* hybrid orbital, the second hybrid orbital would have to remain vacant, a highly unstable situation.

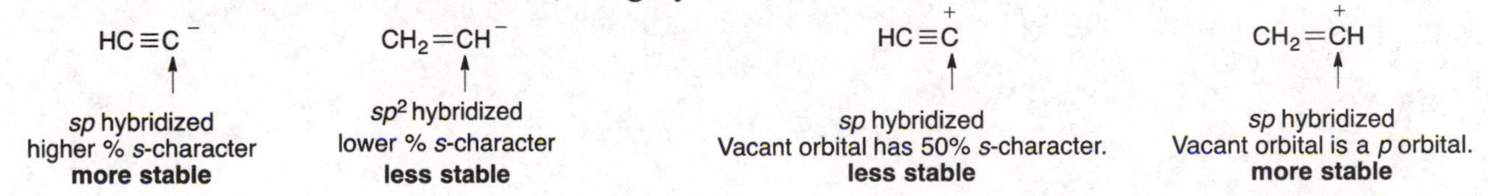

11.46

a.

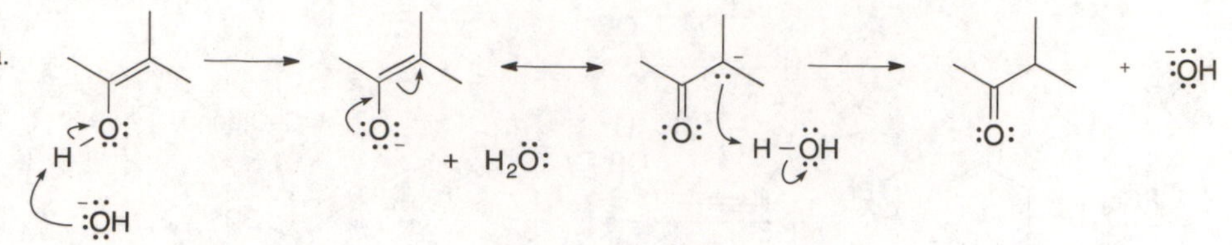

:O: H $^{-}$:OH → :O: + H_2O: ↔ :O: H–OH → :O: + :OH

b.

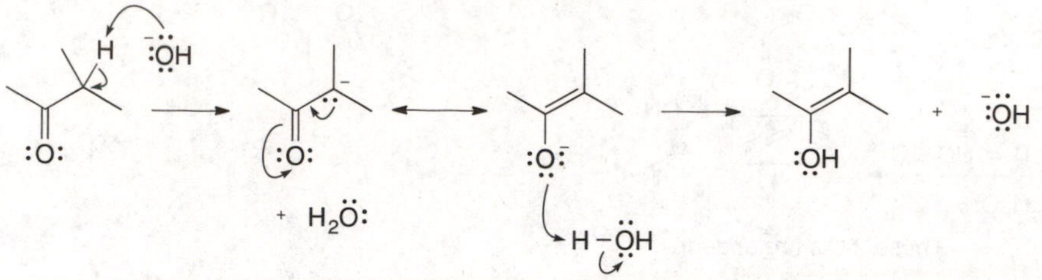

H :OH :O: → :O: + H_2O: ↔ :O: H–OH → :OH + :OH

11.47

a.

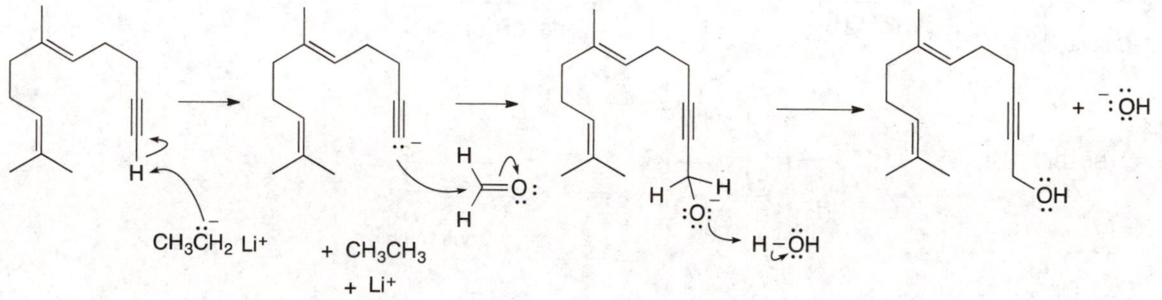

H $CH_3\overset{-}{C}H_2$ Li+ → + CH_3CH_3 + Li+ H H :O: → H H :O: H–OH → OH + :OH

b.

:OH H–OSO_3H → $\overset{+}{:}OH_2$ + HSO_4^- → resonance structures H_2O: → HSO_4^- H :O–H

:O–H H–OSO_3H → H H :O H ↔ H :O+ H HSO_4^- → H :O: + H_2SO_4

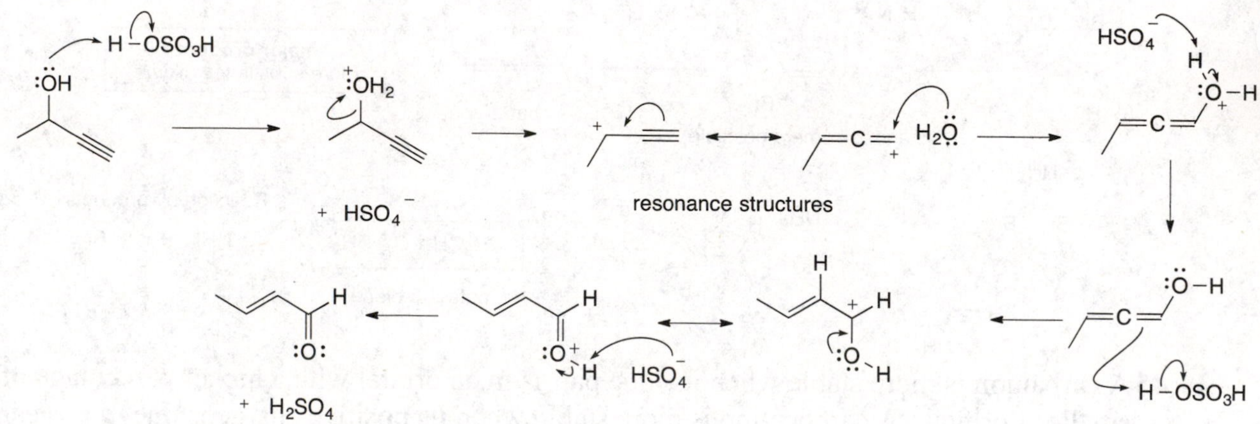

11.48

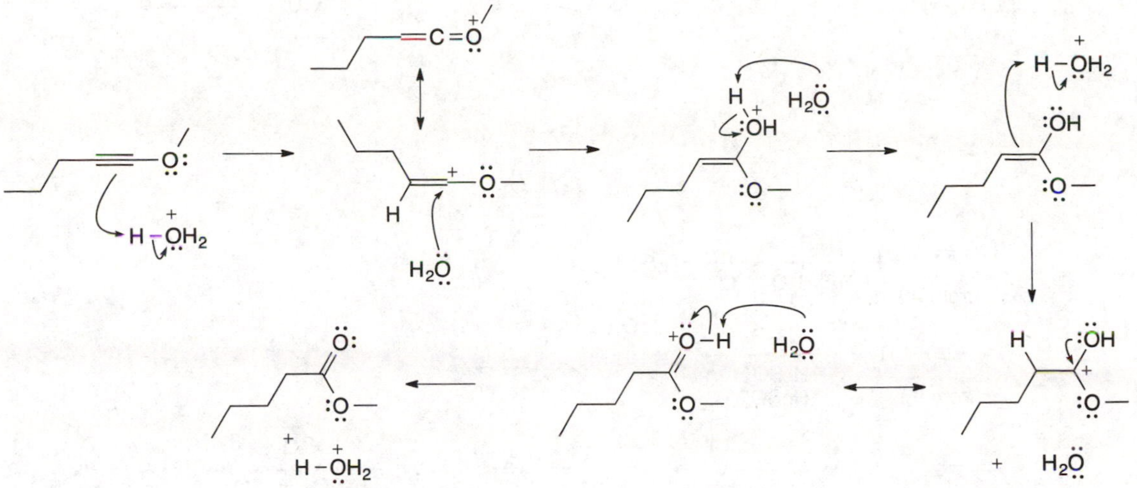

$H_2\ddot{O}$ $H-\overset{+}{O}H_2$:OH :O— H $+$ $H_2\ddot{O}$

11.49

a. 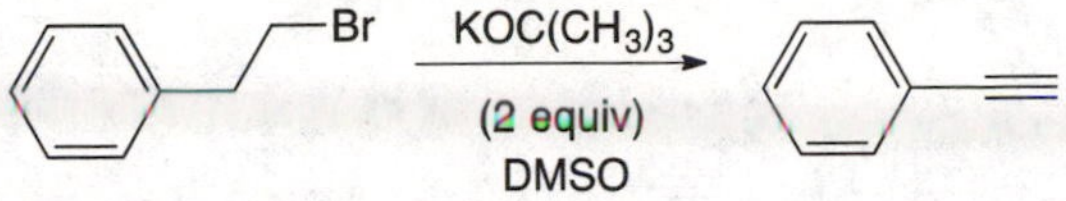 Br $\xrightarrow[\text{(2 equiv) DMSO}]{KOC(CH_3)_3}$

b. 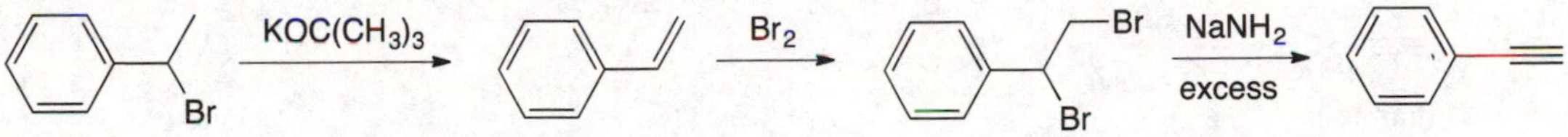 Br $\xrightarrow{KOC(CH_3)_3}$ $\xrightarrow{Br_2}$ Br, Br $\xrightarrow[\text{excess}]{NaNH_2}$

c. 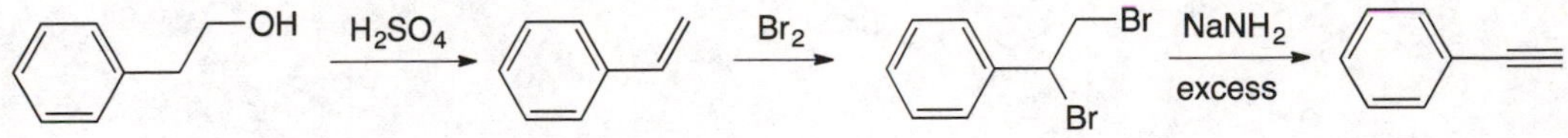 OH $\xrightarrow{H_2SO_4}$ $\xrightarrow{Br_2}$ Br, Br $\xrightarrow[\text{excess}]{NaNH_2}$

11.50 The alkyl halides must be methyl or 1°.

a. 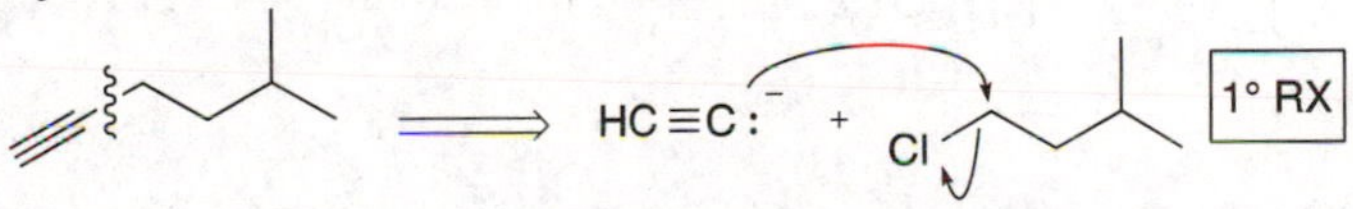 $\Longrightarrow$ $HC\equiv C:^-$ + Cl — 1° RX

b. 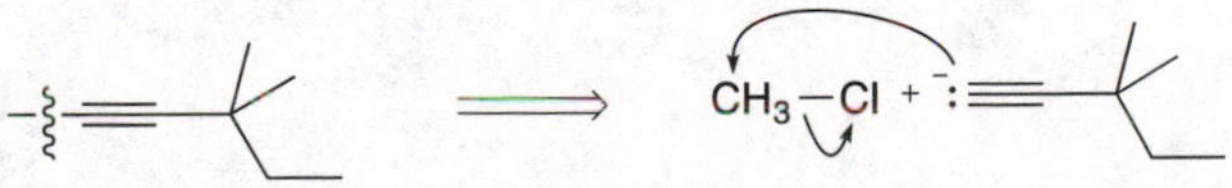 $\Longrightarrow$ CH_3-Cl + $^-:\equiv$

c. $\Longrightarrow$ $C\equiv C:^-$ Cl — 1° RX

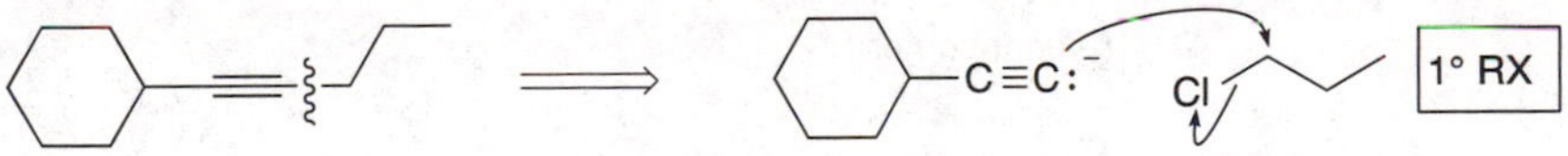

11.51

a. 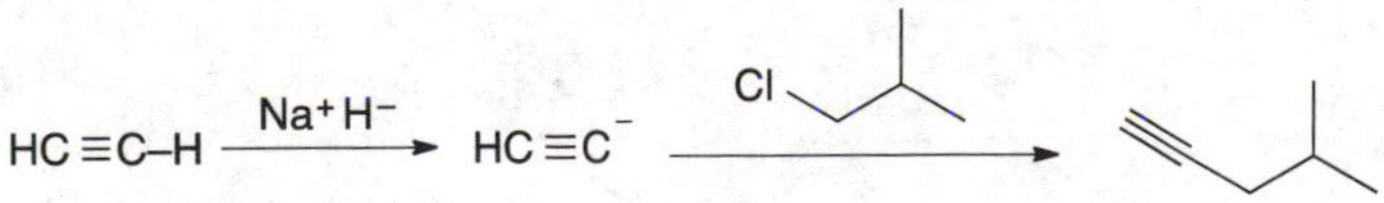 $HC\equiv C-H \xrightarrow{Na^+H^-} HC\equiv C^- \xrightarrow{Cl}$

b. 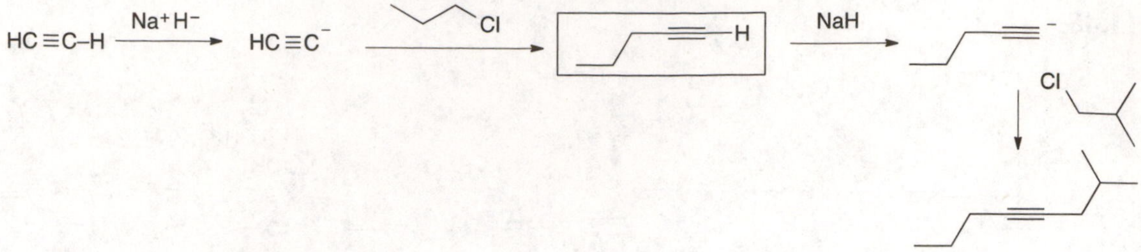 HC≡C–H —Na^+H^-→ HC≡C^- —Cl→ ≡–H —NaH→ ≡$^-$ —Cl→

c. ≡–H (from b.) —[1] R_2BH [2] H_2O_2, HO^-→ O H

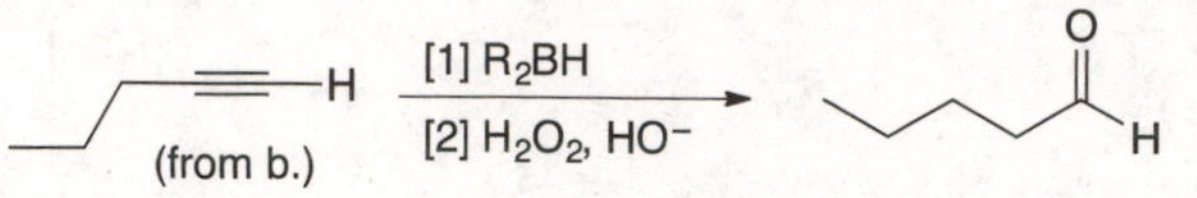

d. 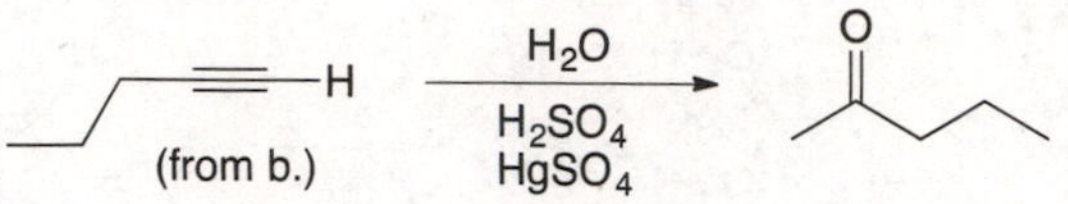 ≡–H (from b.) —H_2O H_2SO_4 $HgSO_4$→ O

e. 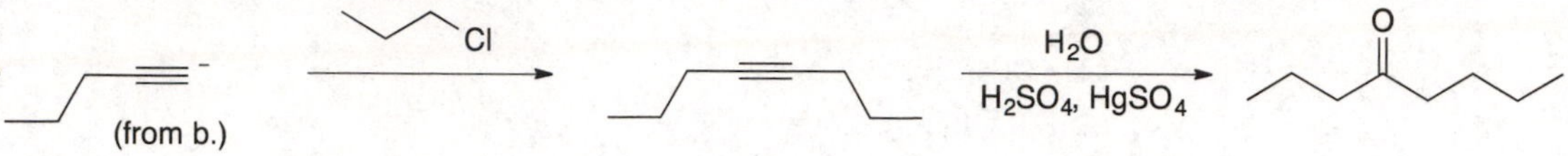 ≡$^-$ (from b.) —Cl→ —H_2O H_2SO_4, $HgSO_4$→ O

11.52

a. 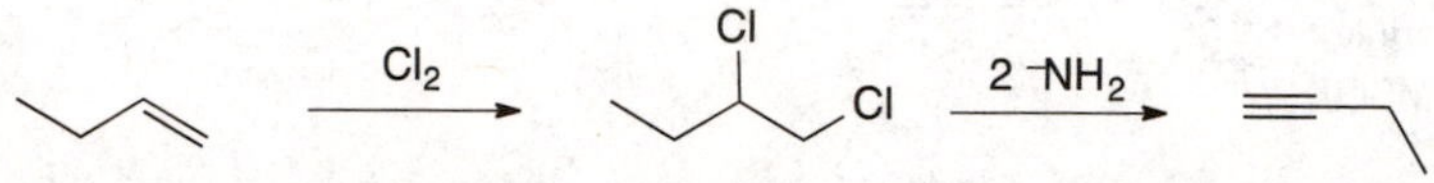 —Cl_2→ Cl Cl —$2\ ^-NH_2$→

b. 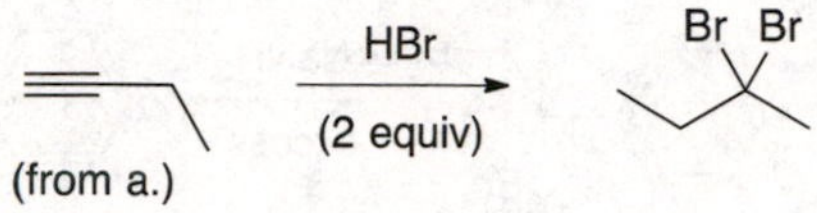 (from a.) —HBr (2 equiv)→ Br Br

c. 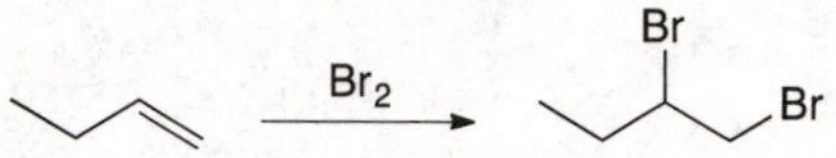 —Br_2→ Br Br

d. 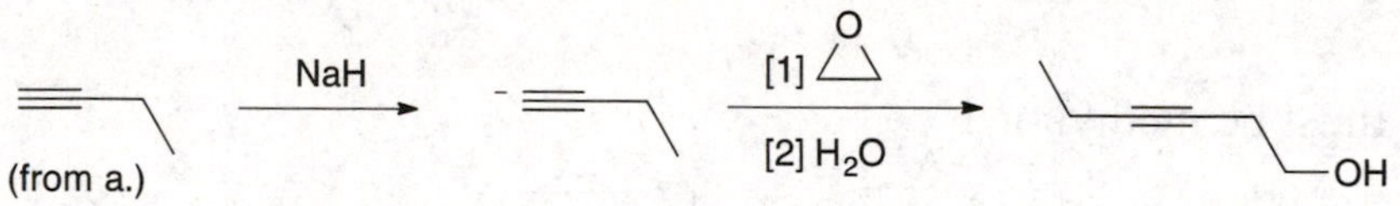 (from a.) —NaH→ $^-$ —[1] O [2] H_2O→ OH

e. 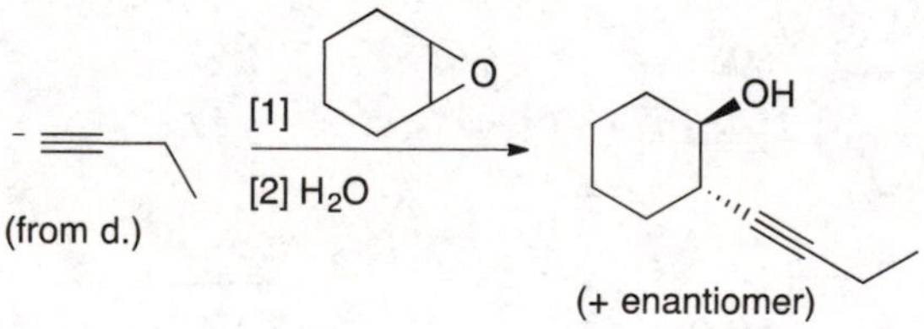 $^-$ (from d.) —[1] O [2] H_2O→ OH (+ enantiomer)

11.53

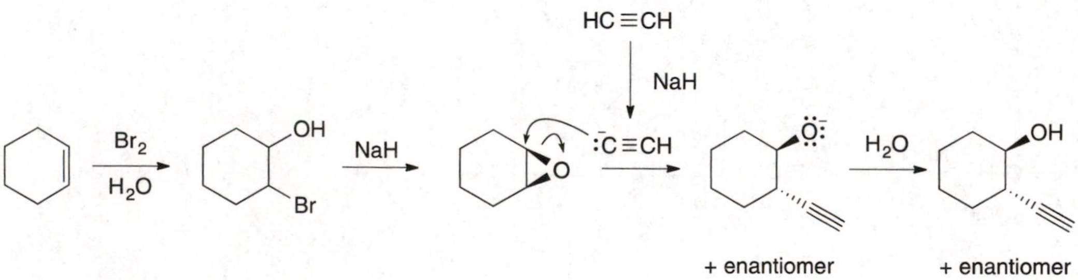

—Br_2 H_2O→ OH Br —NaH→ O —(HC≡CH, NaH, $:\bar{C}$≡CH)→ $\ddot{O}:^-$ + enantiomer —H_2O→ OH + enantiomer

11.54

a. $HC\equiv CH$ NaH $HC\equiv C^-$ Br NaH $^-$ Br

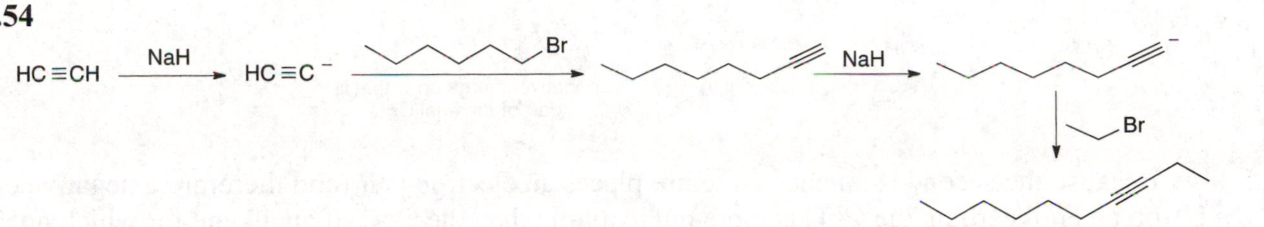

b. $^-$ (from a.) [1] O [2] H_2O OH [1] NaH [2] Br O

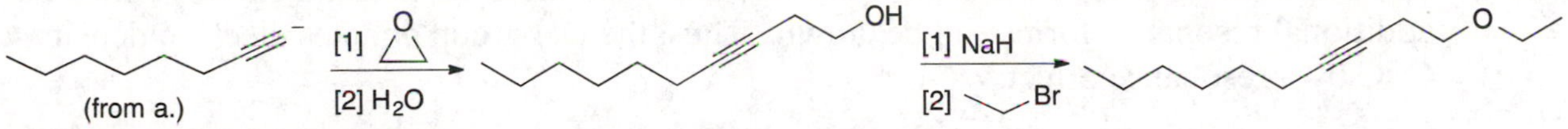

11.55

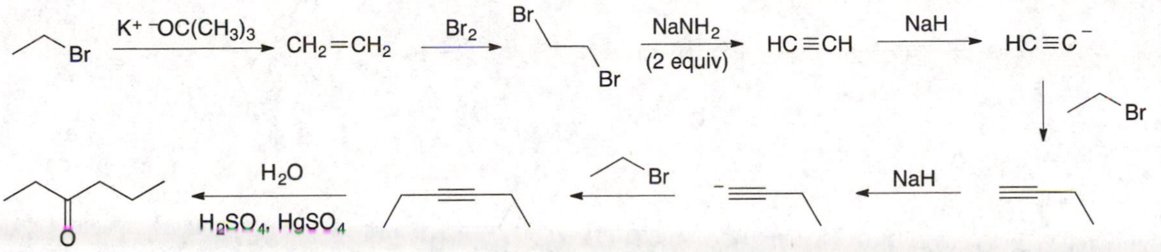

Br $K^+\ ^-OC(CH_3)_3$ $CH_2{=}CH_2$ Br_2 Br Br $NaNH_2$ (2 equiv) $HC\equiv CH$ NaH $HC\equiv C^-$ Br

O H_2O H_2SO_4, $HgSO_4$ Br $^-$ NaH

11.56

a. OH H_2SO_4 Br_2 Br Br $2\ ^-NH_2$ NaH $^-$

OH $SOCl_2$ Cl

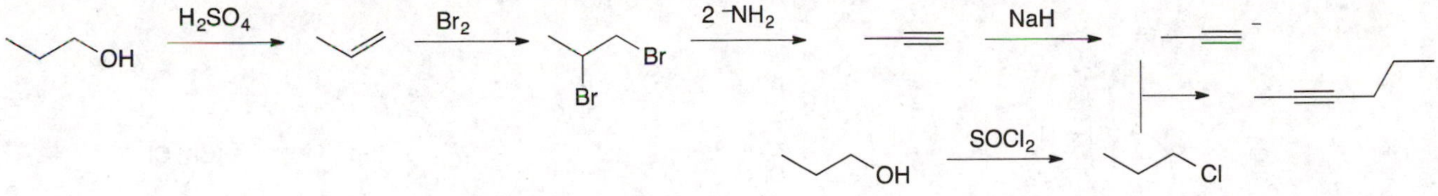

b. (from a.) Cl_2 H_2O OH Cl NaH O [1] $^-$ (from a.) [2] H_2O HO

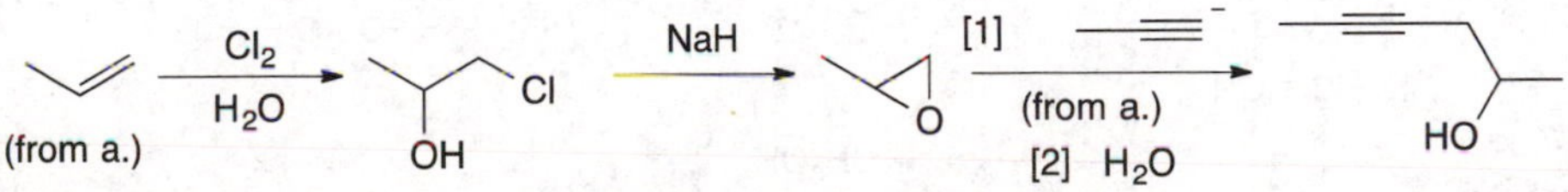

11.57

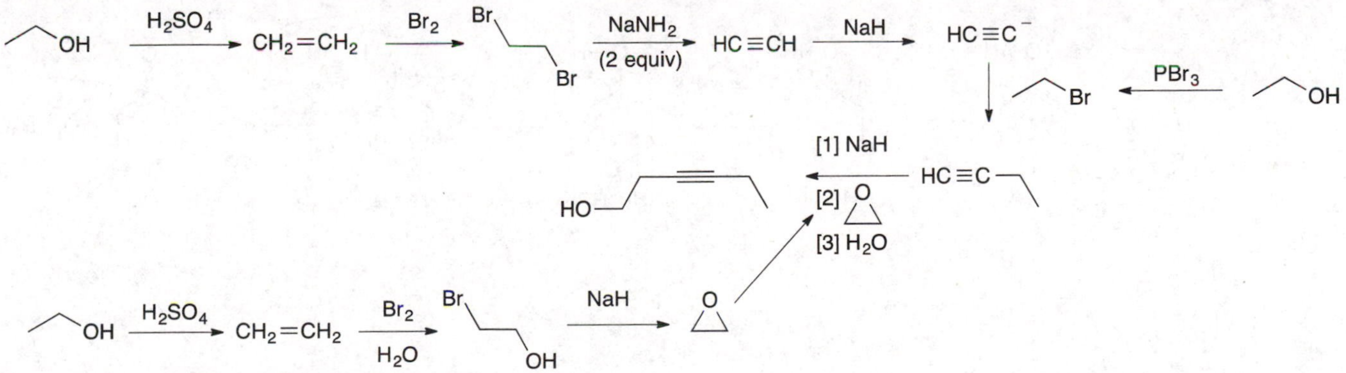

OH H_2SO_4 $CH_2{=}CH_2$ Br_2 Br Br $NaNH_2$ (2 equiv) $HC\equiv CH$ NaH $HC\equiv C^-$

Br PBr_3 OH

HO [1] NaH [2] O [3] H_2O $HC\equiv C$

OH H_2SO_4 $CH_2{=}CH_2$ Br_2 H_2O Br OH NaH O

11.58 Two resonance structures can be drawn for an enol.

:ÖH ⟷ + ÖH

negative charge on C that is part of the enol C=C

Because the second resonance structure places an electron pair (and therefore a negative charge) on an enol carbon, the C=C is more nucleophilic than the C=C of an alkene for which no additional resonance forms can be drawn. Thus, the OH group donates electron density to the C=C by a resonance effect.

11.59

but-2-yne + N–Cl $\xrightarrow{CH_3OH}$ –O O– Cl Cl

:Cl–N → :Cl: → :Cl: → :Cl:

$CH_3\ddot{O}H$ + :N: — –O: H — $CH_3\ddot{O}H$ — –O: + $CH_3\overset{+}{O}H_2$

:Cl–N Cl –O → :Cl: Cl –O → O :Cl: Cl –O: H → :Cl: Cl –O :O– + $CH_3\overset{+}{O}H_2$

$CH_3\ddot{O}H$ + :N: — $CH_3\ddot{O}H$

11.60

$H–\overset{+}{O}H_2$:Ö–H H → :Ö–H + H H ⟷ :Ö–H + H H ⟷ + Ö–H H H $H_2\ddot{O}$ → Ö: + H_3O^+

11.61

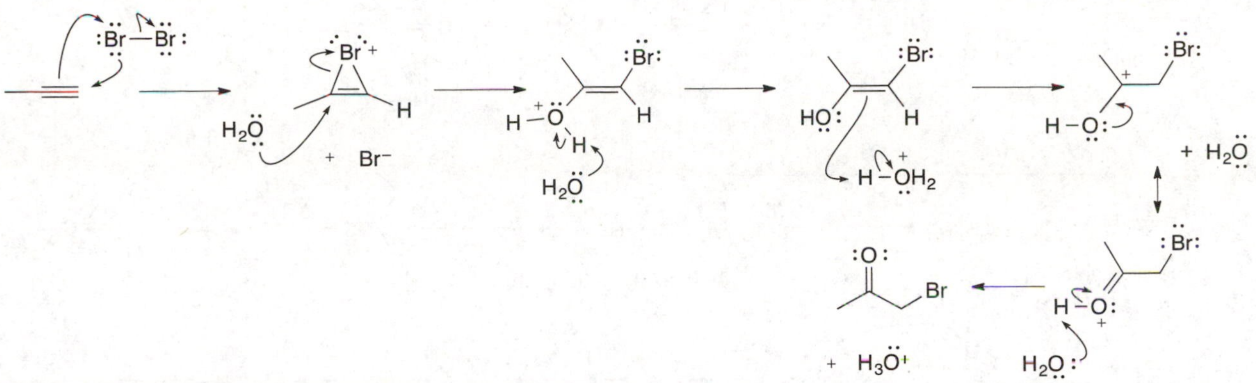

$:Br-Br:$ → Br^+ (bridged bromonium) + Br^-; H_2O attacks → $H-\overset{+}{O}$ … H_2O → HO … $H-\overset{+}{O}H_2$ → $H-O:$ (cation) + H_2O ↔ $H-\overset{+}{O}$ … $H_2O:$ → ketone with Br + H_3O^+

11.62

Only this carbocation forms because it is resonance stabilized. The positive charge is delocalized on oxygen.

TsOH, $TsÖ-H$ → + TsO^- ↔ [not resonance stabilized (not formed)]

re-draw

CH_3OH → $\overset{+}{O}CH_3$, H, CH_3OH → X NOT Y (OCH_3)

+ $CH_3\overset{+}{O}H_2$

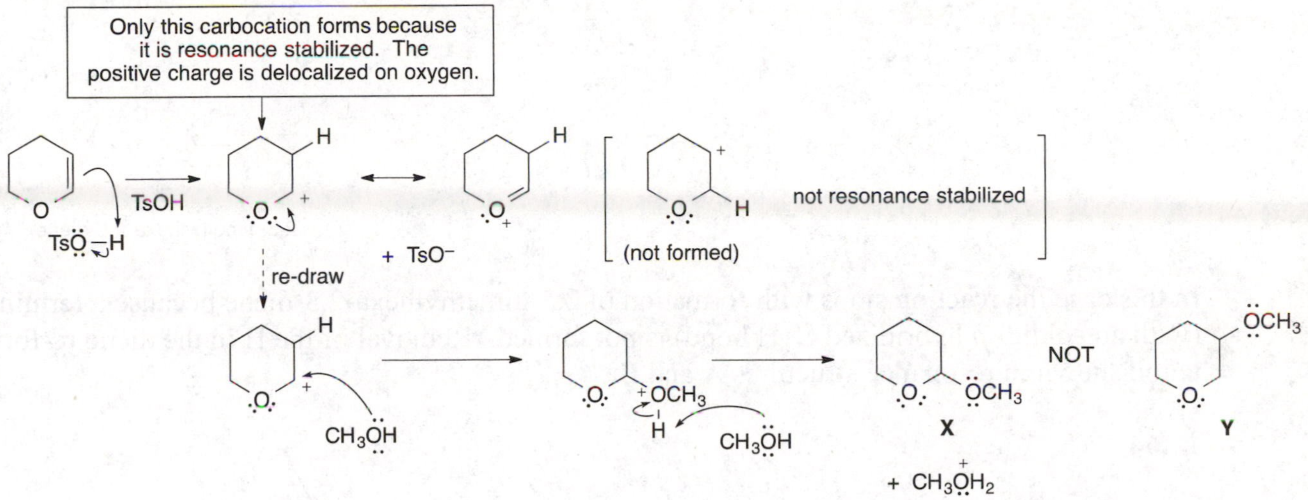

11.63 A more stable internal alkyne can be isomerized to a less stable terminal alkyne under these reaction conditions because when $CH_3CH_2C{\equiv}CH$ is first formed, it contains an *sp* hybridized C–H bond, which is more acidic than any proton in $CH_3–C{\equiv}C–CH_3$. Under the reaction conditions, this proton is removed with base. Formation of the resulting acetylide anion drives the equilibrium to favor its formation. Protonation of this acetylide anion gives the less stable terminal alkyne.

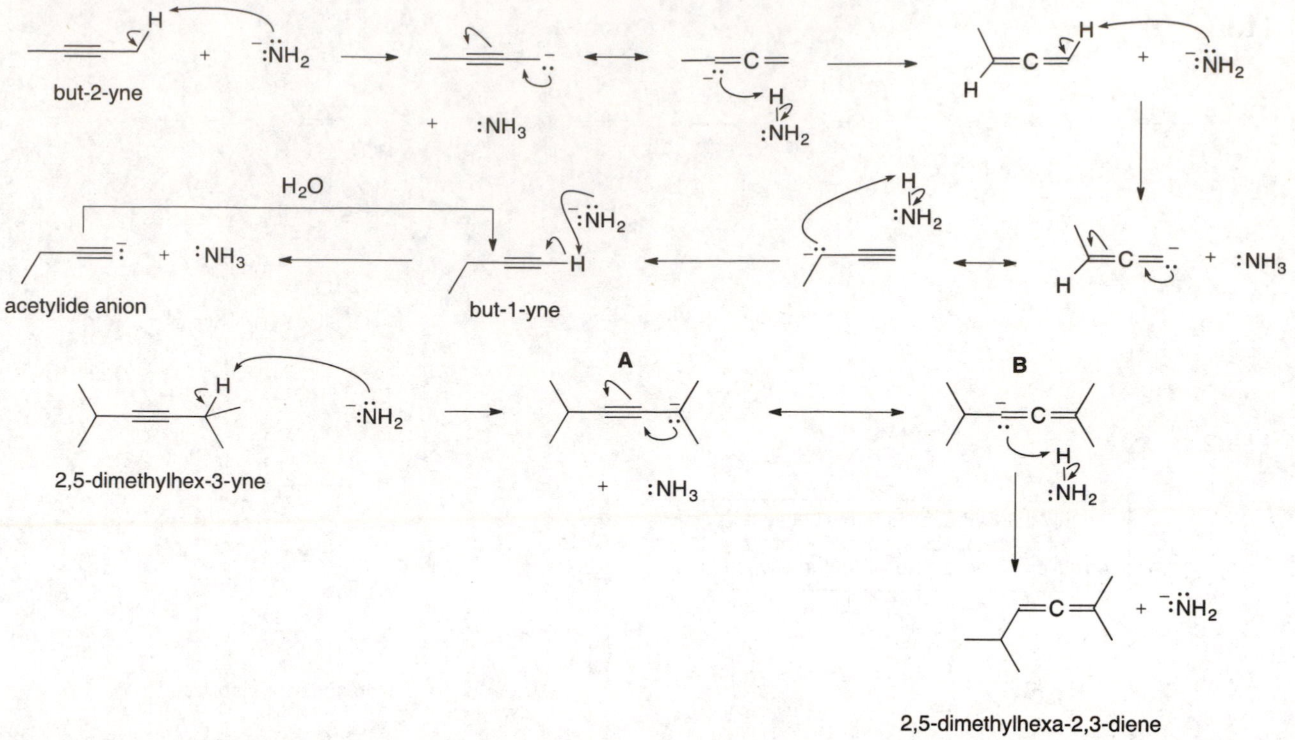

but-2-yne

acetylide anion

but-1-yne

2,5-dimethylhex-3-yne

A

B

2,5-dimethylhexa-2,3-diene

In this case the reaction stops with formation of 2,5-dimethylhexa-2,3-diene because a terminal alkyne (with an acidic *sp* hybridized C–H bond) is not formed. Removal of the H in the diene re-forms the anion shown in resonance structures **A** and **B.**

11.64

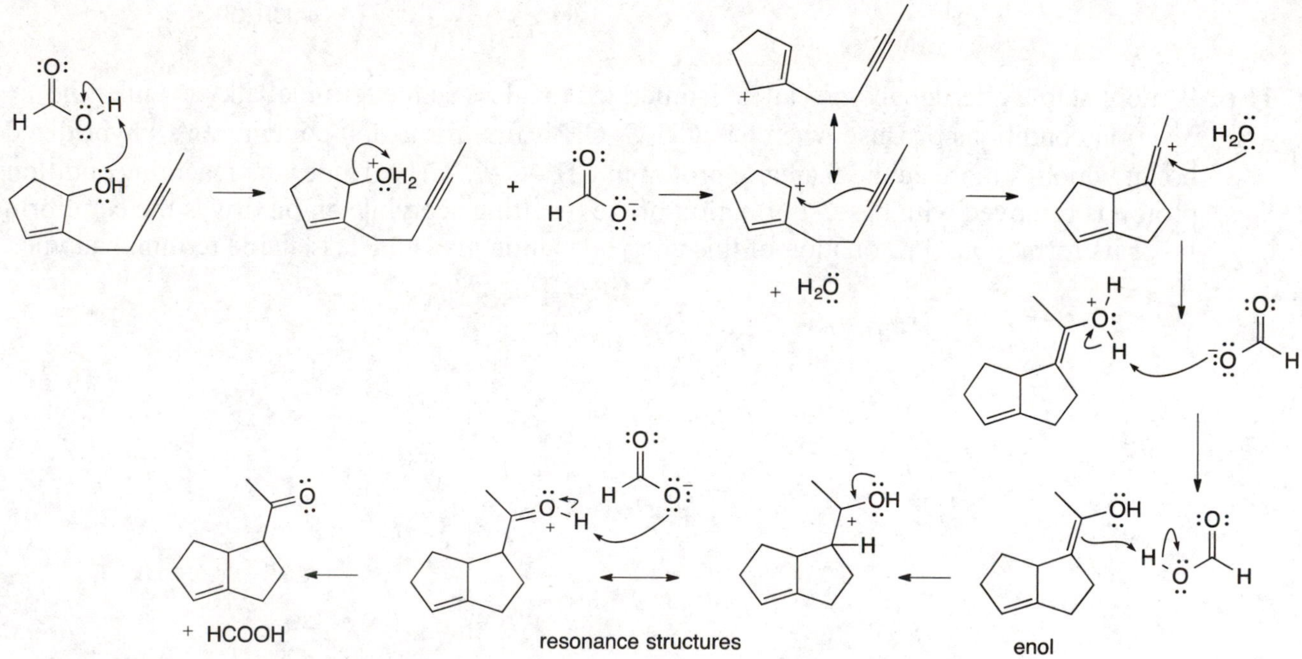

+ HCOOH

resonance structures

enol

11.65 In the presence of acid, (*R*)-α-methylbutyrophenone enolizes to form an achiral enol.

:O: H—$\overset{+}{O}H_2$:$\overset{+}{O}$H $H_2\ddot{O}$ OH

(*R*)-α-methylbutyrophenone

(+ 1 resonance structure)

(*E* and *Z* isomers) achiral enol

The achiral enol can then be protonated from above or below the plane to form a racemic mixture that is optically inactive.

:ÖH H—$\overset{+}{O}H_2$ below above :Ö–H $H_2\ddot{O}$: :O: + H_3O^+

racemic mixture

Chapter 12 Oxidation and Reduction

Chapter Review

Summary: Terms that describe reaction selectivity

- A **regioselective reaction** forms predominately or exclusively one constitutional isomer (Section 8.5).

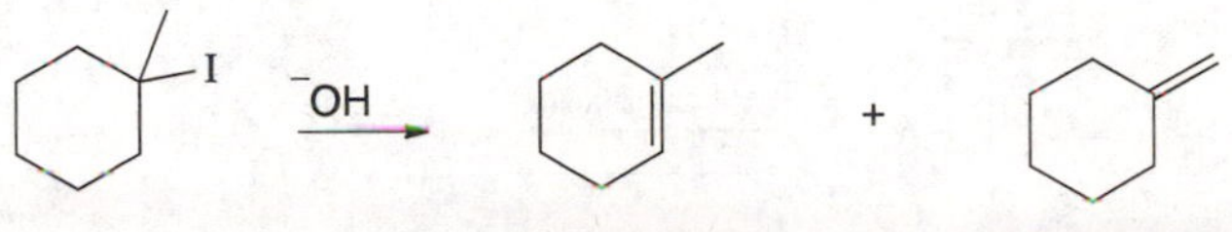

I $\xrightarrow{^-OH}$ major product trisubstituted alkene + minor product disubstituted alkene

- A **stereoselective reaction** forms predominately or exclusively one stereoisomer (Section 8.5).

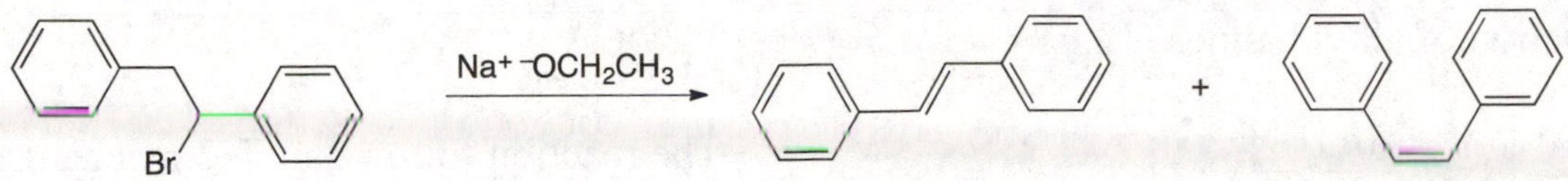

Br $\xrightarrow{Na^+ \ ^-OCH_2CH_3}$ trans alkene **major product** + cis alkene **minor product**

- An **enantioselective reaction** forms predominately or exclusively one enantiomer (Section 12.15).

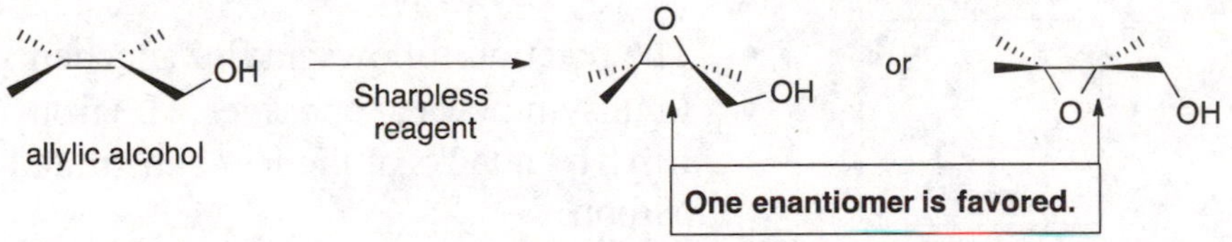

allylic alcohol (OH) → Sharpless reagent → O, OH or O, OH

One enantiomer is favored.

Definitions of oxidation and reduction

Oxidation reactions result in:
- an increase in the number of C–Z bonds, *or*
- a decrease in the number of C–H bonds.

Reduction reactions result in:
- a decrease in the number of C–Z bonds, *or*
- an increase in the number of C–H bonds.

[Z = an element more electronegative than C]

Reduction reactions

[1] Reduction of alkenes—Catalytic hydrogenation (12.3)

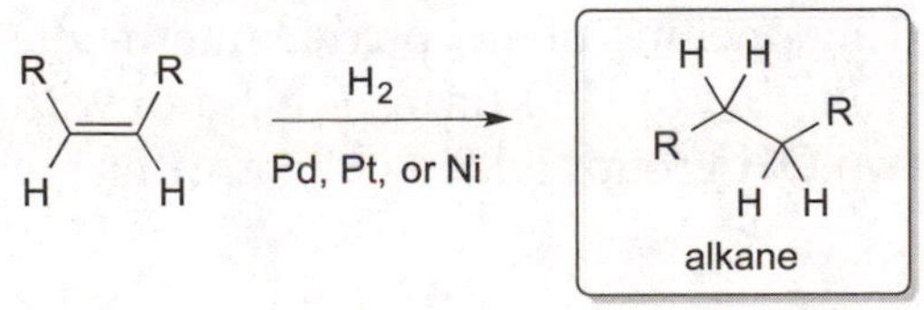

R, R, H, H $\xrightarrow[\text{Pd, Pt, or Ni}]{H_2}$ H H, R, R, H H — alkane

- **Syn addition** of H_2 occurs.
- Increasing alkyl substitution on the C=C decreases the rate of reaction.

[2] Reduction of alkynes

a. R—≡—R $\xrightarrow[\text{Pd-C}]{2\ H_2}$ alkane

- Two equivalents of H_2 are added and new C–H bonds are formed (12.5A).

b. R—≡—R $\xrightarrow[\text{Lindlar catalyst}]{H_2}$ cis alkene

- **Syn addition** of H_2 occurs, forming a **cis** alkene (12.5B).
- The Lindlar catalyst is deactivated so that reaction stops after one equivalent of H_2 has been added.

c. R—≡—R $\xrightarrow[NH_3]{Na}$ trans alkene

- **Anti addition** of H_2 occurs, forming a **trans** alkene (12.5C).

[3] Reduction of alkyl halides (12.6)

R–X $\xrightarrow[\text{[2] } H_2O]{\text{[1] } LiAlH_4}$ R–H (alkane)

- The reaction follows an S_N2 mechanism.
- CH_3X and RCH_2X react faster than more substituted RX.

[4] Reduction of epoxides (12.6)

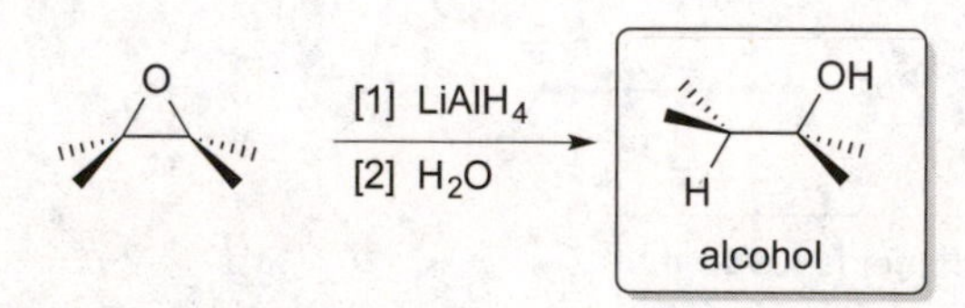

- The reaction follows an $\mathbf{S_N2}$ mechanism.
- In unsymmetrical epoxides, H^- (from $LiAlH_4$) attacks at the less substituted carbon.

Oxidation reactions

[1] Oxidation of alkenes

a. Epoxidation (12.8)

alkene + RCO_3H → epoxide

- The mechanism has **one step.**
- **Syn addition** of an O atom occurs.
- The reaction is stereospecific.

b. Anti dihydroxylation (12.9A)

alkene $\xrightarrow[\text{[2] } H_2O\ (H^+ \text{ or } HO^-)]{\text{[1] } RCO_3H}$ 1,2-diol

- Ring opening of an epoxide intermediate with ^-OH or H_2O forms a 1,2-diol with two OH groups added in an **anti** fashion.

c. Syn dihydroxylation (12.9B)

[1] OsO_4; [2] $NaHSO_3$, H_2O
or
[1] OsO_4, NMO; [2] $NaHSO_3$, H_2O
or
$KMnO_4$, H_2O, HO^-

HO, OH — 1,2-diol

- Each reagent adds two new C–O bonds to the C=C in a **syn** fashion.

d. Oxidative cleavage (12.10)

[1] O_3
[2] Zn, H_2O or CH_3SCH_3

ketone + aldehyde

- Both the σ and π bonds of the alkene are cleaved to form two carbonyl groups.

[2] Oxidative cleavage of alkynes (12.11)

a. R—≡—R' (internal alkyne) — [1] O_3; [2] H_2O → carboxylic acids

- The σ bond and both π bonds of the alkyne are cleaved.

b. R—≡—H (terminal alkyne) — [1] O_3; [2] H_2O → RCOOH + CO_2

[3] Oxidation of alcohols (12.12, 12.13)

a. 1° alcohol — PCC or $HCrO_4^-$–Amberlyst A-26 resin → aldehyde

- Oxidation of a 1° alcohol with PCC or $HCrO_4^-$ (Amberlyst A-26 resin) stops at the aldehyde stage. Only one C–H bond is replaced by a C–O bond.

b. 1° alcohol — CrO_3, H_2SO_4, H_2O → carboxylic acid

- Oxidation of a 1° alcohol under harsher reaction conditions—CrO_3 (or $Na_2Cr_2O_7$ or $K_2Cr_2O_7$) + H_2O + H_2SO_4—affords a RCO_2H. Two C–H bonds are replaced by two C–O bonds.

c.

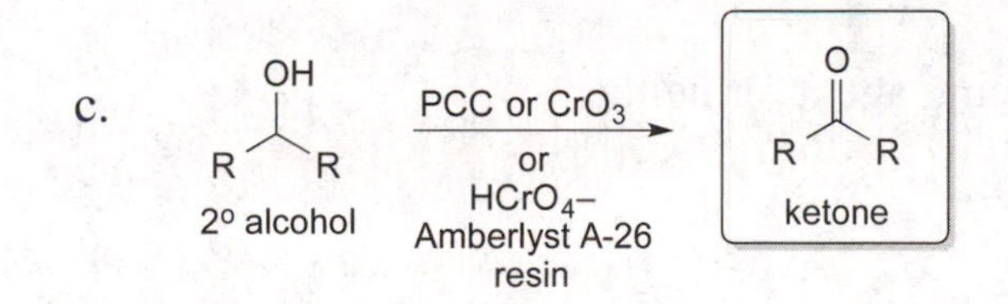

O, R R — ketone

- A 2° alcohol has only one C–H bond on the carbon bearing the OH group, so all Cr^{6+} reagents—PCC, CrO_3, $Na_2Cr_2O_7$, $K_2Cr_2O_7$, or $HCrO_4^-$ (Amberlyst A-26 resin)—oxidize a 2° alcohol to a ketone.

[4] Asymmetric epoxidation of allylic alcohols (12.15)

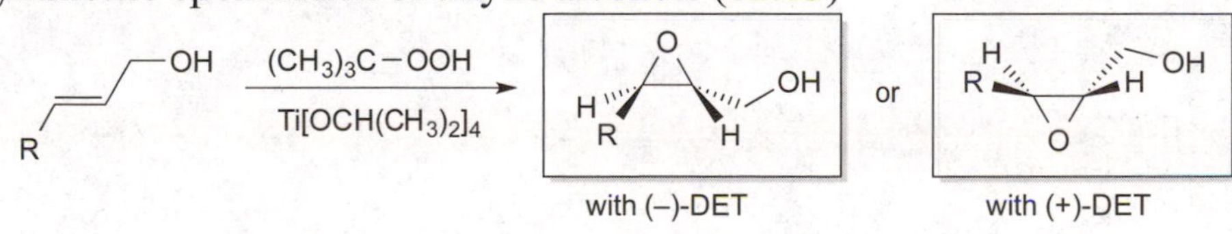

Practice Test on Chapter Review

1.a. Compound **X** has a molecular formula of C_9H_{12} and contains no triple bonds. **X** is hydrogenated to a compound of molecular formula C_9H_{14} with excess H_2 and a palladium catalyst. What can be said about **X?**

1. **X** has four rings.
2. **X** has three rings and one double bond.
3. **X** has two rings and two double bonds.
4. **X** has one ring and three double bonds.
5. **X** has four double bonds.

b. Syn addition to an alkene occurs exclusively with which reagents?

1. OsO_4
2. $KMnO_4$, H_2O, ^-OH
3. mCPBA, then H_2O, ^-OH
4. Both reagents (1) and (2) give syn addition exclusively.
5. Reagents (1), (2), and (3) give syn addition exclusively.

c. Which of the following reagents adds to an alkene exclusively in an anti fashion?

1. Br_2
2. H_2, Pd-C
3. BH_3, then H_2O_2, ^-OH
4. Reagents (1) and (2) both add in an anti fashion.
5. Reagents (1), (2), and (3) all add in an anti fashion.

2. Label each statement as True (T) or False (F).

a. PCC oxidizes 1° alcohols to aldehydes.
b. CrO_3 oxidizes 2° alcohols to ketones.
c. Treatment of hex-2-yne with Na in NH_3 forms *cis*-hex-2-ene.
d. Reduction of propene oxide with $LiAlH_4$ forms propan-1-ol.
e. Ozonolysis of 2-methyloct-2-ene forms one ketone and one aldehyde.
f. mCPBA is an oxidizing agent that converts alkenes to trans diols.
g. Oct-1-en-5-yne reacts with H_2 and Pd-C, but does not react with H_2 and Lindlar catalyst.
h. Treatment of cyclohexene with OsO_4 affords an optically inactive product mixture that contains two enantiomers.

3. Label each reagent as an oxidizing agent, reducing agent, or neither.

a. O_3
b. $LiAlH_4$
c. mCPBA
d. H_2O, H_2SO_4
e. PCC
f. Na, NH_3

4. Draw the organic products formed in each reaction and indicate stereochemistry when necessary.

a. Na / NH_3

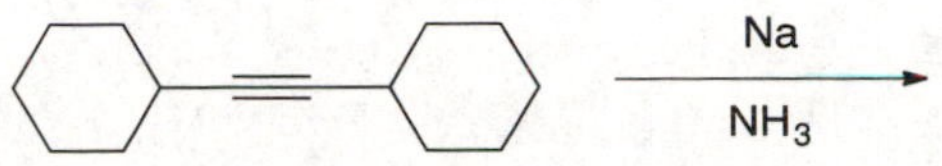

b. H_2 / Pd–C

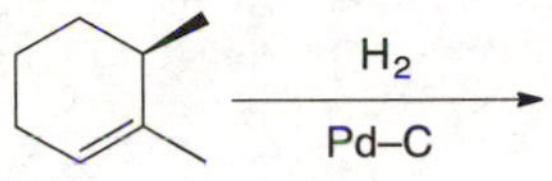

c. HO, OH; PCC

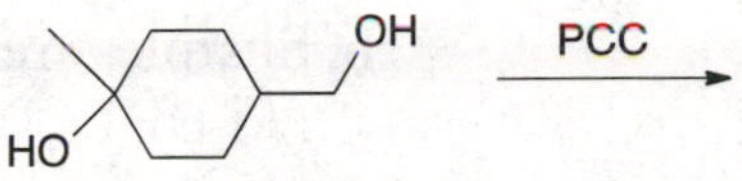

d. OH; [1] $SOCl_2$ [2] $LiAlH_4$

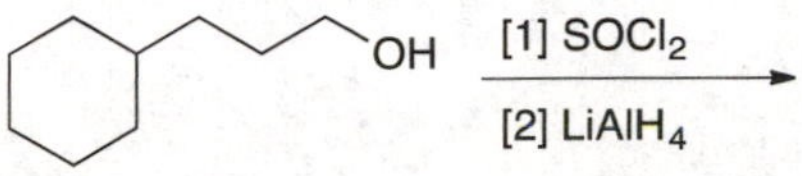

5 a. Fill in the appropriate starting material (including any needed stereochemistry) in the following reaction.

[1] OsO_4 [2] H_2O, $NaHSO_3$ → HO, H, C_6H_5, H, C_6H_5, OH + enantiomer

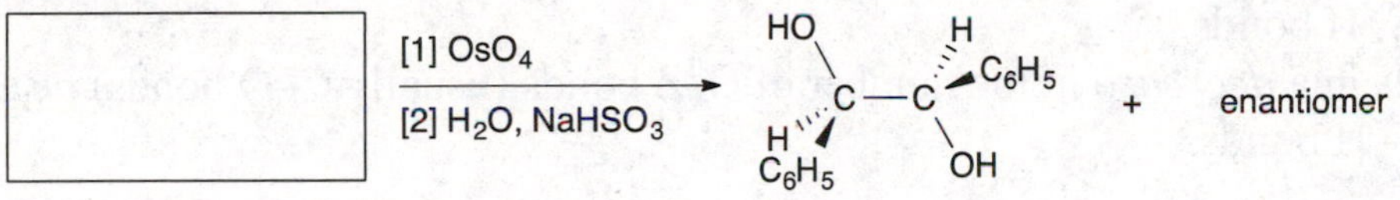

b. Fill in the appropriate reagent in the following reaction.

OH → O, H, OH

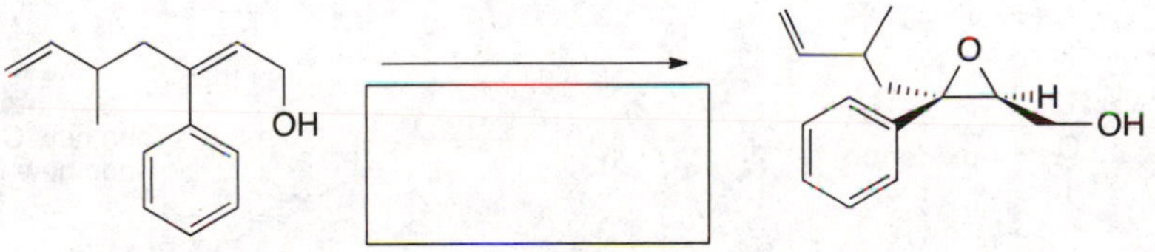

c. What starting material is needed for the following reaction?

[1] O_3 [2] $(CH_3)_2S$ → CHO, O

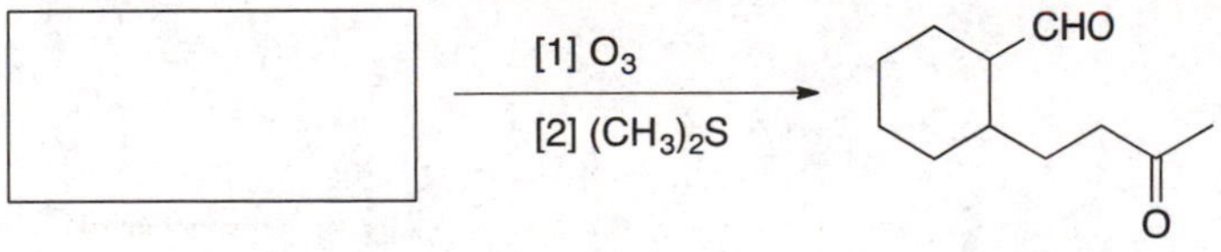

Answers to Practice Test

1.a. 2
b. 4
c. 1

2.a. T
b. T
c. F
d. F
e. T
f. F
g. F
h. F

3.a. oxidizing
b. reducing
c. oxidizing
d. neither
e. oxidizing
f. reducing

4.
a. 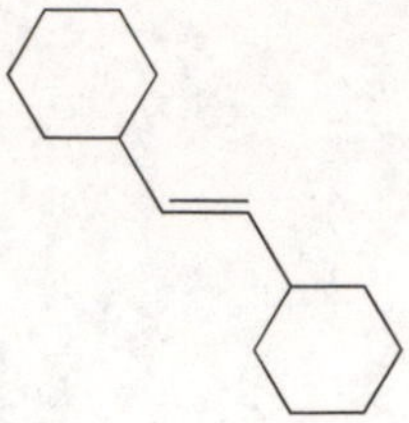
b. 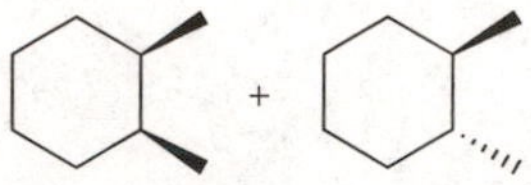 +
c. 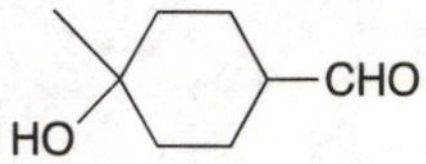 HO CHO
d.

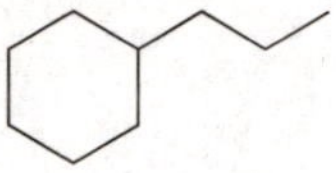

5.
a. 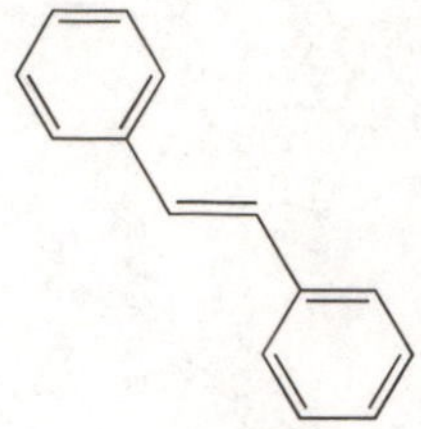
b. Sharpless reagent
(+)-DET
c.

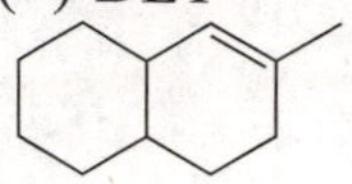

Chapter 12: Answers to Problems

12.1 *Oxidation* results in an *increase* in the number of C–Z bonds (usually C–O bonds) *or* a *decrease* in the number of C–H bonds.
Reduction results in a *decrease* in the number of C–Z bonds (usually C–O bonds) *or* an *increase* in the number of C–H bonds.

a. 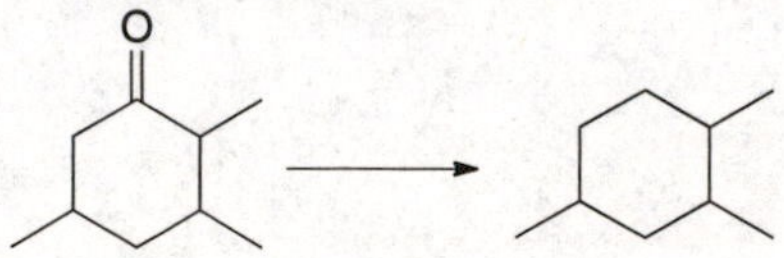 **reduction**

b. **oxidation**

c. 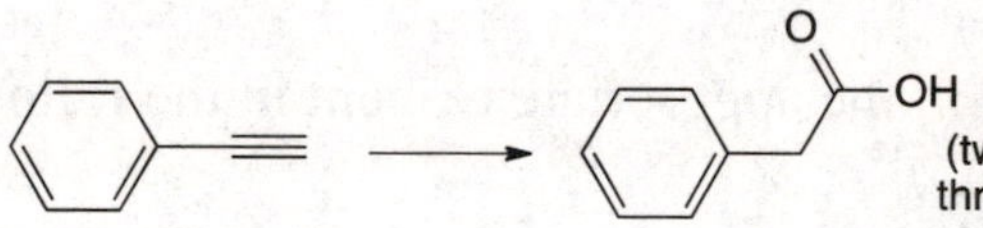 **oxidation** (two new C–H bonds, three new C–O bonds)

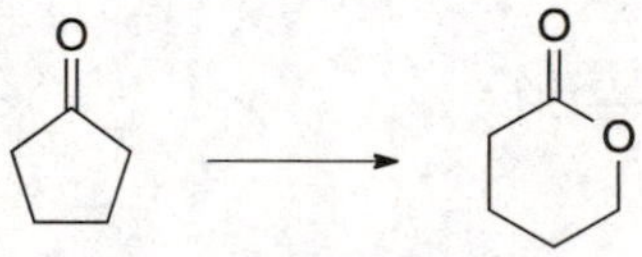

d. 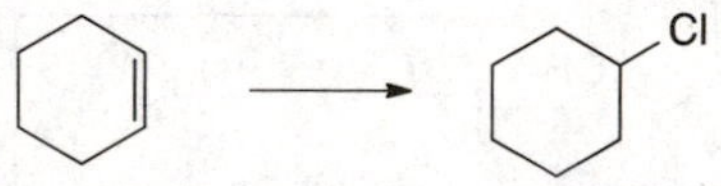 **neither** (one new C–H bond and one new C–Cl bond)

12.2 **Hydrogenation** is the addition of hydrogen. When alkenes are hydrogenated, they are *reduced* by the addition of H_2 to the π bond. To draw the alkane product, add a H to each C of the double bond.

a.

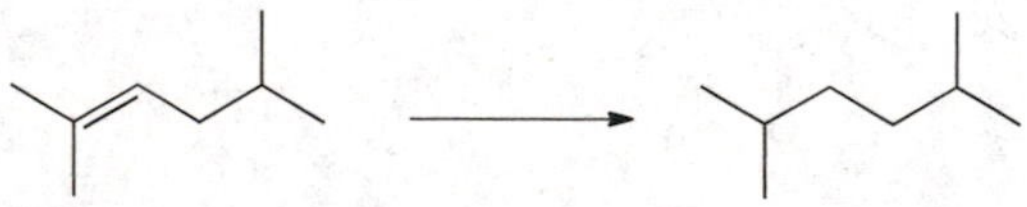

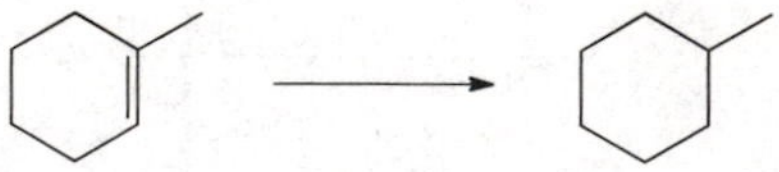

b.

c.

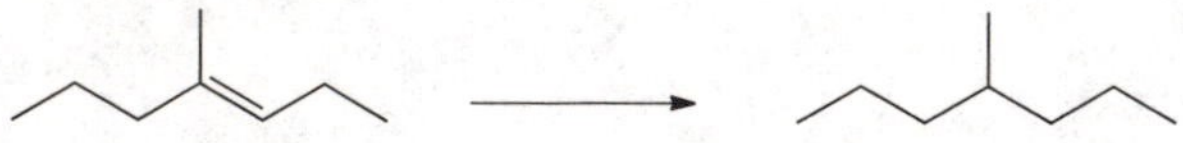

12.3 Draw the alkenes that form each alkane when hydrogenated.

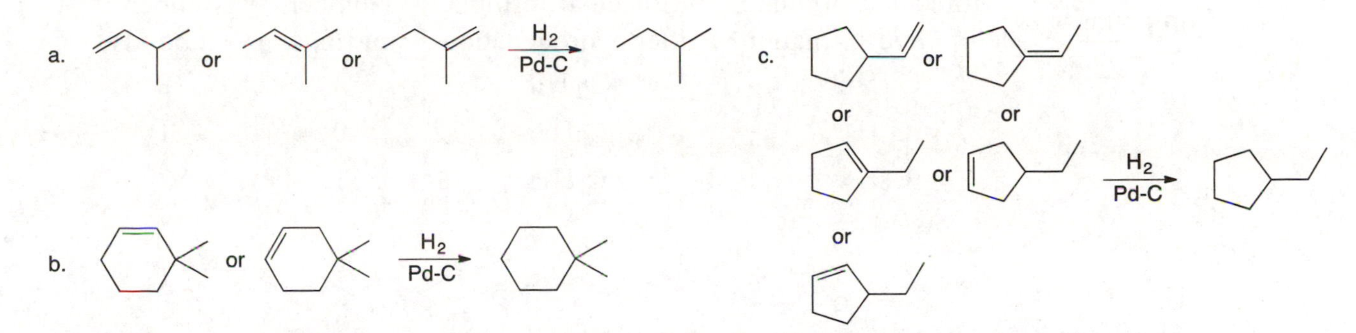

12.4 Cis alkenes are less stable than trans alkenes, so they have larger heats of hydrogenation. Increasing alkyl substitution increases the stability of a C=C, thus decreasing the heat of hydrogenation.

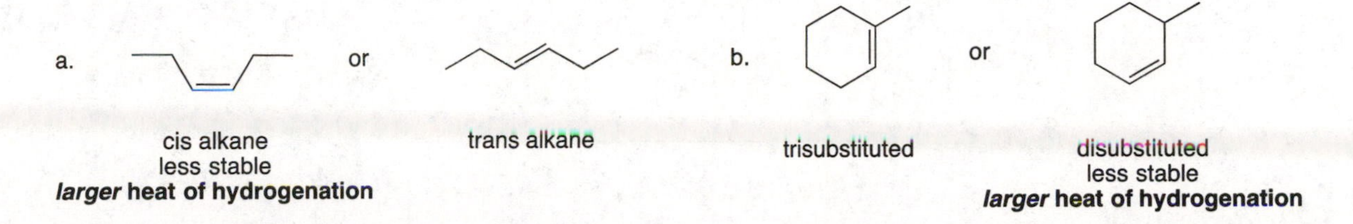

12.5 Hydrogenation products must be identical to use hydrogenation data to evaluate the relative stability of the starting materials.

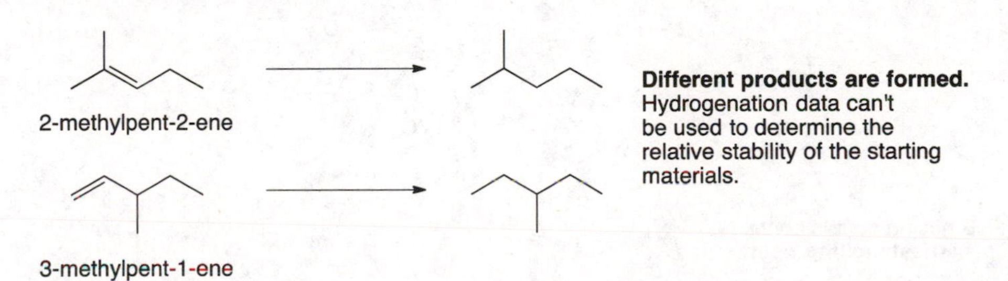

12.6

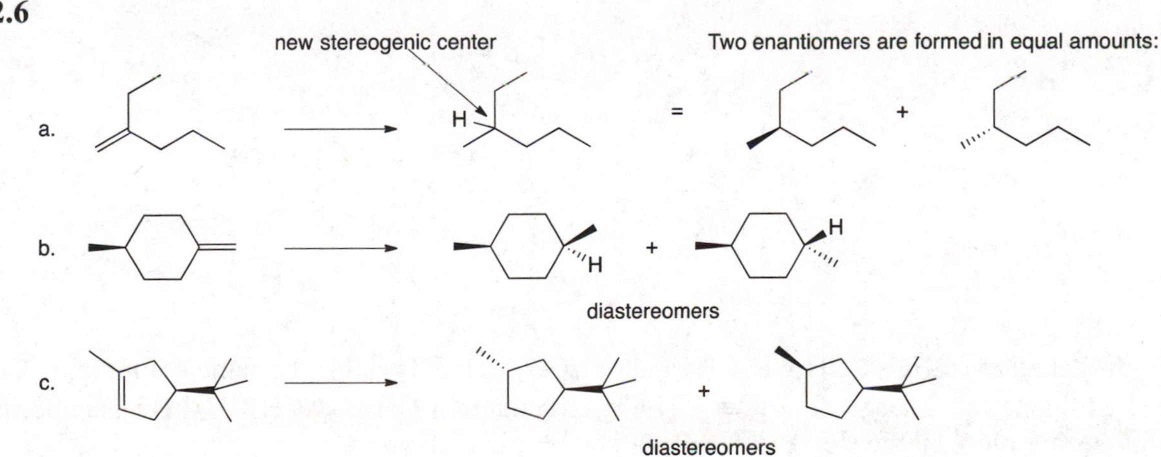

12.7

Compound	Molecular formula before hydrogenation	Molecular formula after hydrogenation	Number of rings	Number of π bonds
A	$C_{10}H_{12}$	$C_{10}H_{16}$	3	2
B	C_4H_8	C_4H_{10}	0	1
C	C_6H_8	C_6H_{12}	1	2

12.8

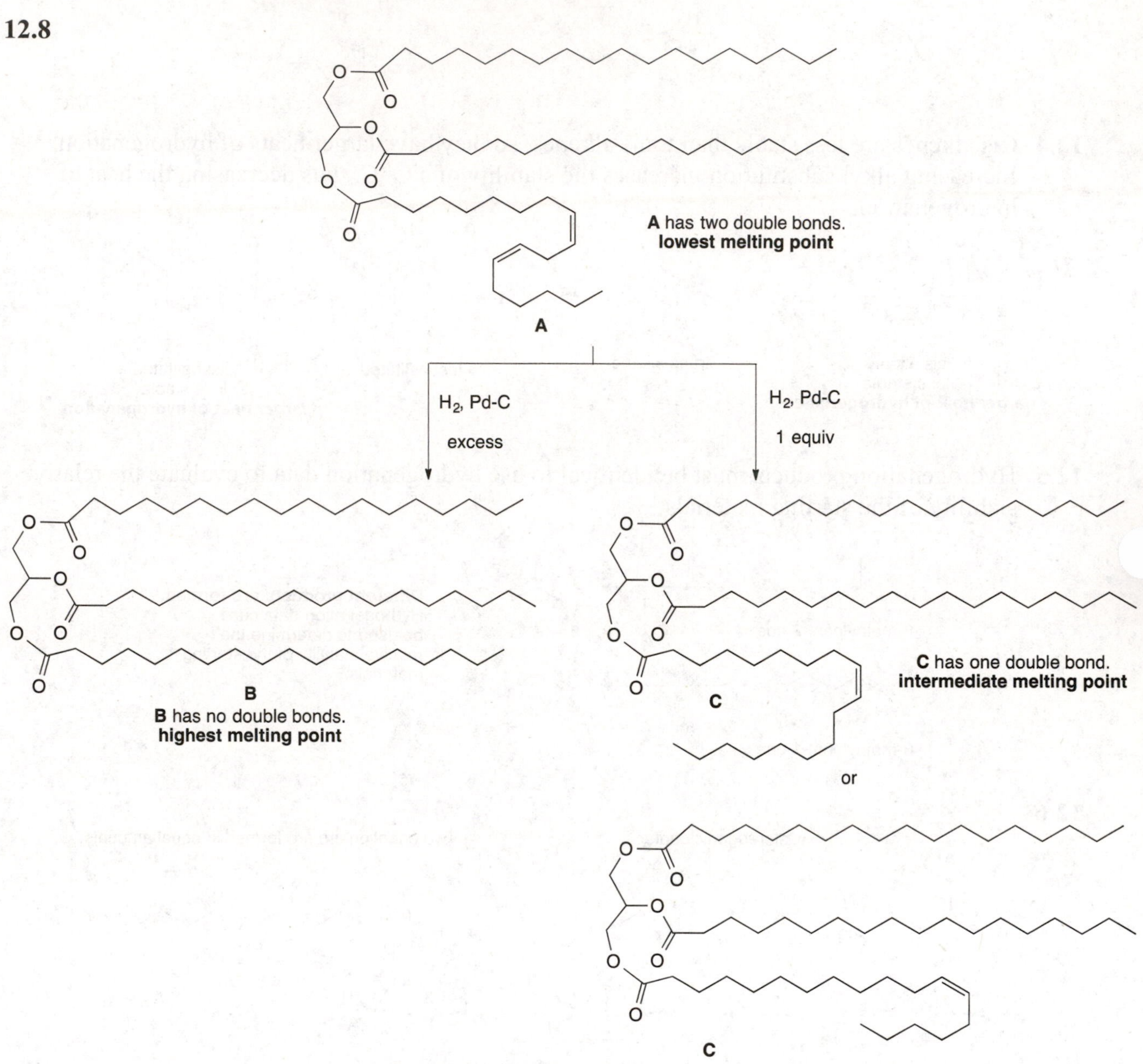

12.9 Hydrogenation of $HC{\equiv}CCH_2CH_2CH_3$ and $CH_3C{\equiv}CCH_2CH_3$ yields the same compound. The heat of hydrogenation is larger for $HC{\equiv}CCH_2CH_2CH_3$ than for $CH_3C{\equiv}CCH_2CH_3$ because internal alkynes are more stable (lower in energy) than terminal alkynes.

12.10 In the presence of the Lindlar catalyst, alkynes react with H_2 to form cis alkenes. Alkenes do not react with the Lindlar catalyst.

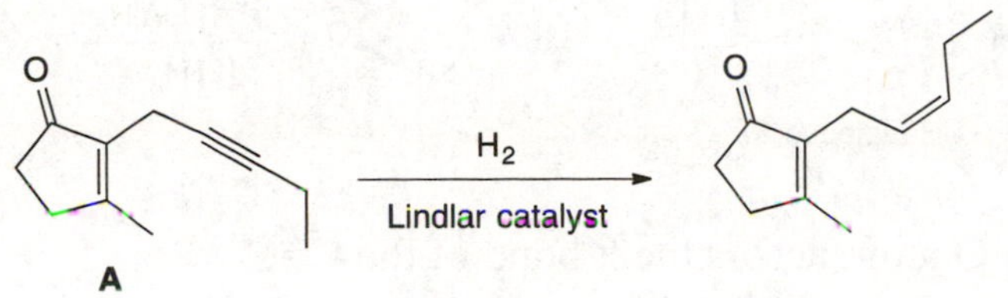

A → (H_2, Lindlar catalyst) → ***cis*-jasmone** (perfume component isolated from jasmine flowers)

12.11 In the presence of Pd-C, H_2 adds to alkenes and alkynes to form alkanes. In the presence of the Lindlar catalyst, alkynes react with H_2 to form cis alkenes. Alkenes do not react with the Lindlar catalyst.

a. C_6H_{10} + H_2 → Pd-C → cyclohexane; Lindlar catalyst → no reaction

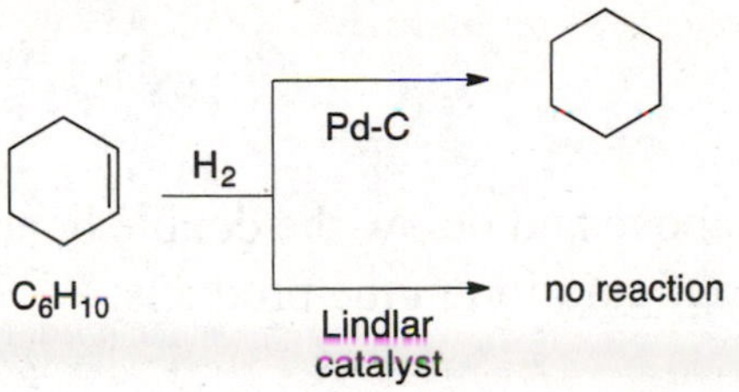

b. C_6H_{10} + H_2 → Pd-C → hexane; Lindlar catalyst → cis alkene

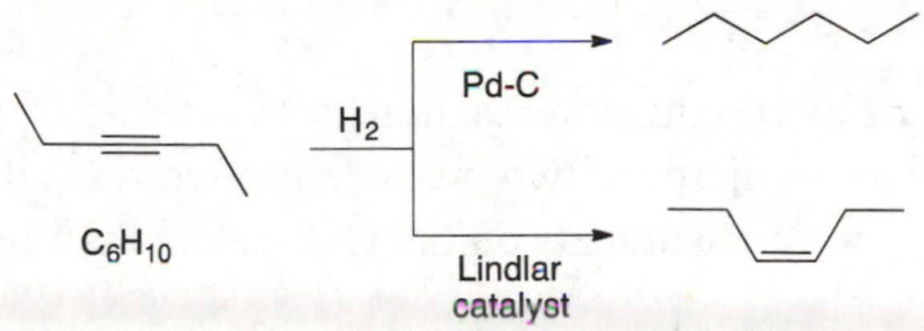

12.12 Use the directions from Answer 12.11.

a. H_2 (excess), Pd-C

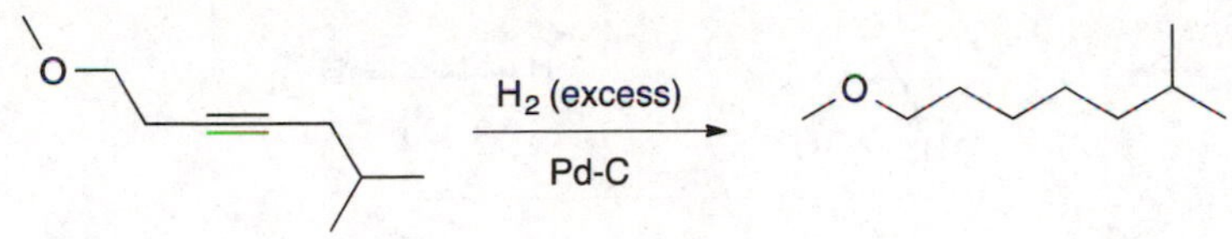

b. H_2 (1 equiv), Lindlar catalyst

c. H_2 (excess), Lindlar catalyst

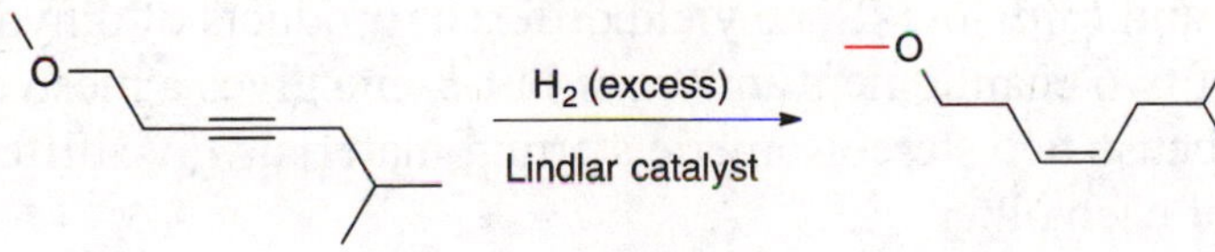

d. Na, NH_3

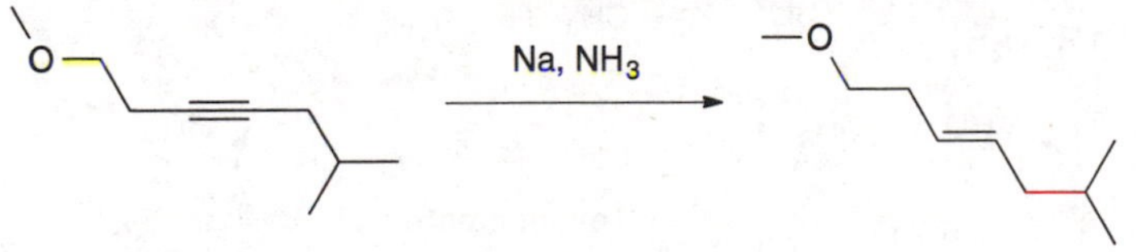

12.13

A → (H_2, Lindlar catalyst) → B (*R*)

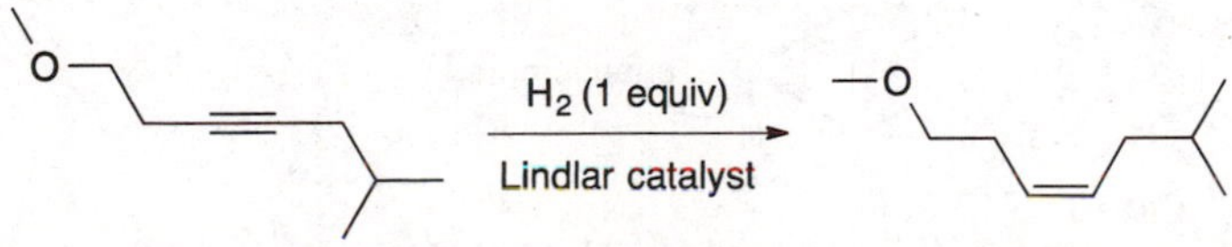

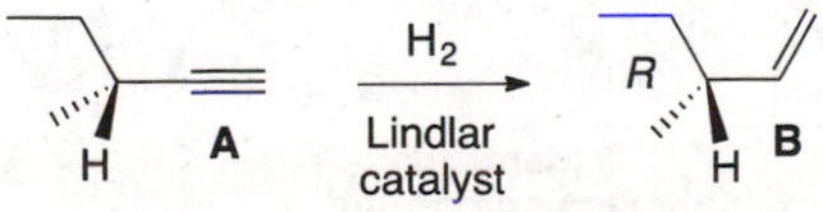

12.14 $LiAlH_4$ reduces alkyl halides to alkanes and epoxides to alcohols.

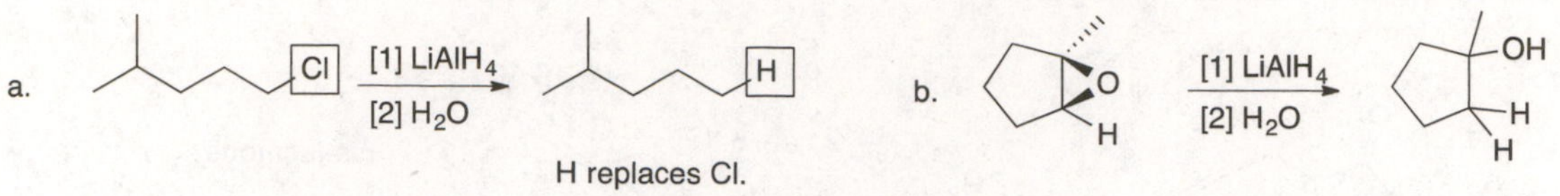

a. Cl — [1] $LiAlH_4$, [2] H_2O → H

H replaces Cl.

b. [1] $LiAlH_4$, [2] H_2O

12.15 To draw the product, add an O atom across the π bond of the C=C.

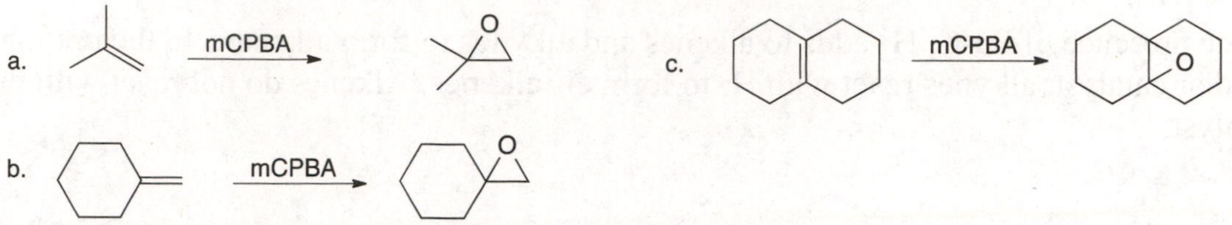

a. mCPBA

b. mCPBA

c. mCPBA

12.16 For epoxidation reactions:

- There are two possible products: O adds from above and below the double bond.
- Substituents on the C=C retain their original configuration in the products.

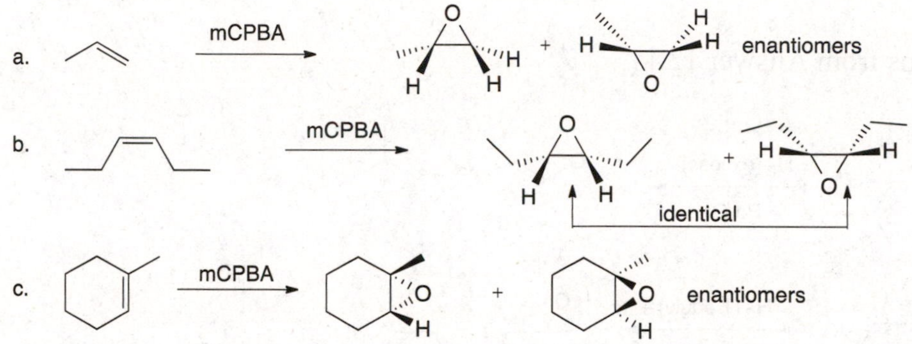

a. mCPBA → + enantiomers

b. mCPBA → + identical

c. mCPBA → + enantiomers

12.17 Treatment of an alkene with a peroxyacid followed by H_2O, HO^- adds two hydroxy groups in an **anti** fashion. *cis*-But-2-ene and *trans*-but-2-ene yield different products of dihydroxylation. *cis*-But-2-ene gives a mixture of two enantiomers and *trans*-but-2-ene gives a meso compound. The reaction is stereospecific because two stereoisomeric starting materials give different products that are also stereoisomers of each other.

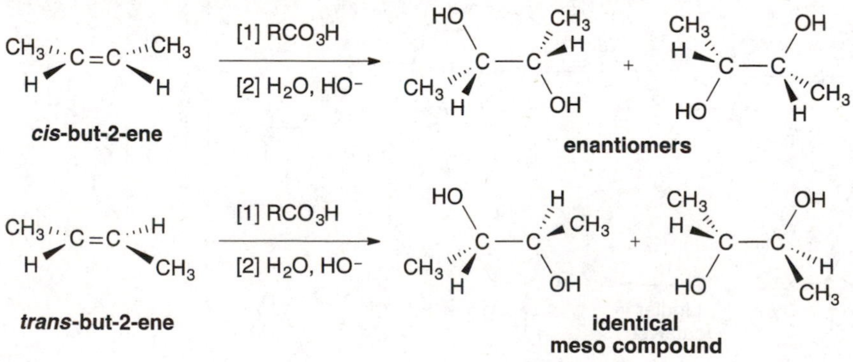

***cis*-but-2-ene** — [1] RCO_3H, [2] H_2O, HO^- → + **enantiomers**

***trans*-but-2-ene** — [1] RCO_3H, [2] H_2O, HO^- → + **identical meso compound**

12.18 Treatment of an alkene with OsO_4 adds two hydroxy groups in a **syn** fashion. *cis*-But-2-ene and *trans*-but-2-ene yield different stereoisomers in this dihydroxylation, so the reaction is stereospecific.

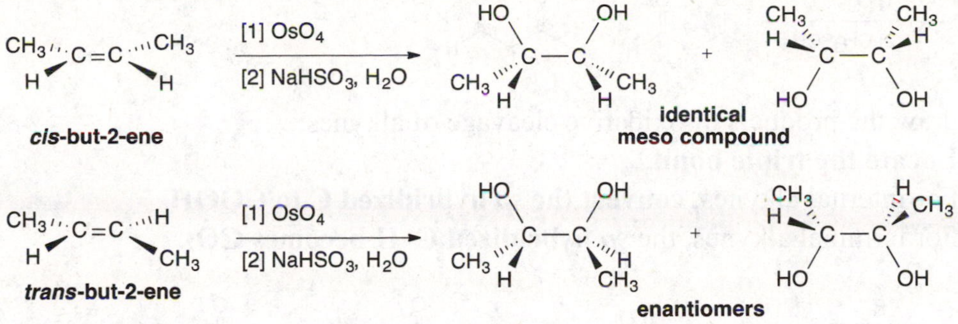

cis-but-2-ene → [1] OsO_4 [2] $NaHSO_3$, H_2O → identical **meso compound**

trans-but-2-ene → [1] OsO_4 [2] $NaHSO_3$, H_2O → **enantiomers**

12.19 To draw the oxidative cleavage products:

- **Locate all the π bonds** in the molecule.
- **Replace all C=C's with *two* C=O's.**

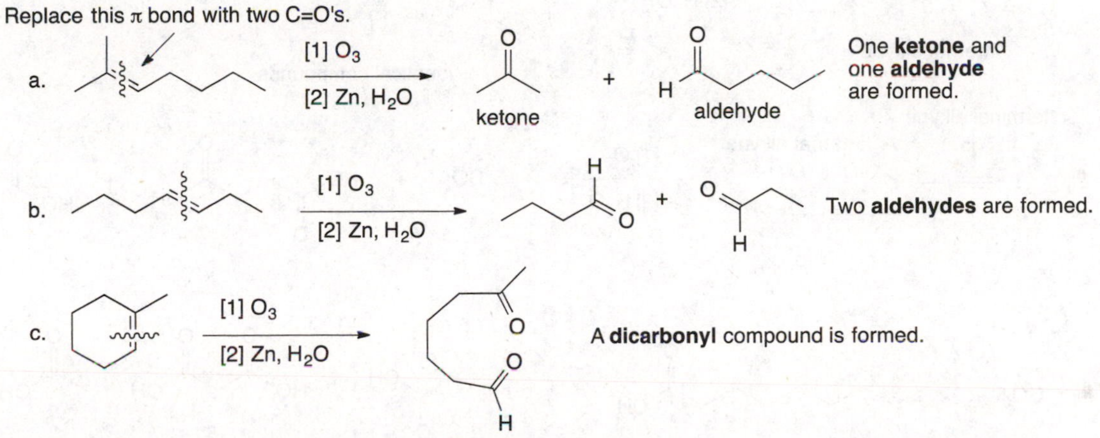

Replace this π bond with two C=O's.

a. [1] O_3 [2] Zn, H_2O → ketone + aldehyde — One **ketone** and one **aldehyde** are formed.

b. [1] O_3 [2] Zn, H_2O → Two **aldehydes** are formed.

c. [1] O_3 [2] Zn, H_2O → A **dicarbonyl** compound is formed.

12.20 To find the alkene that yields the oxidative cleavage products:

- **Find the two carbonyl groups** in the products.
- **Join the two carbonyl carbons** together with a double bond. This is the double bond that was broken during ozonolysis.

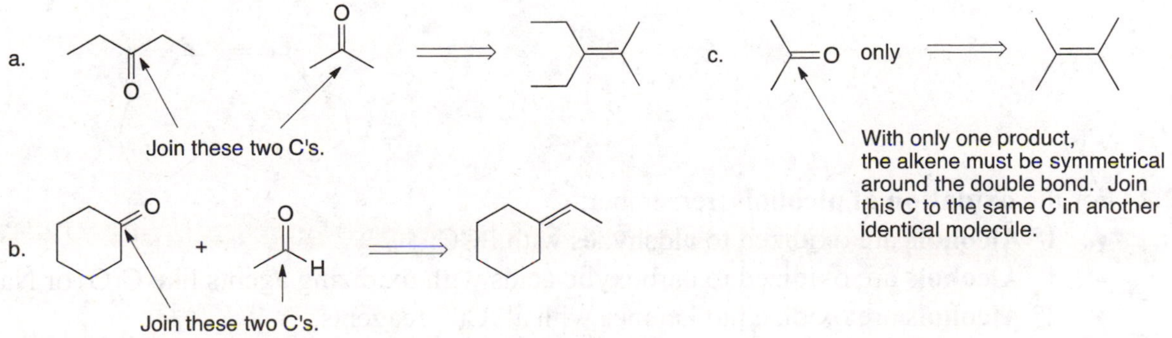

a. Join these two C's.

b. Join these two C's.

c. only — With only one product, the alkene must be symmetrical around the double bond. Join this C to the same C in another identical molecule.

12.21

a. 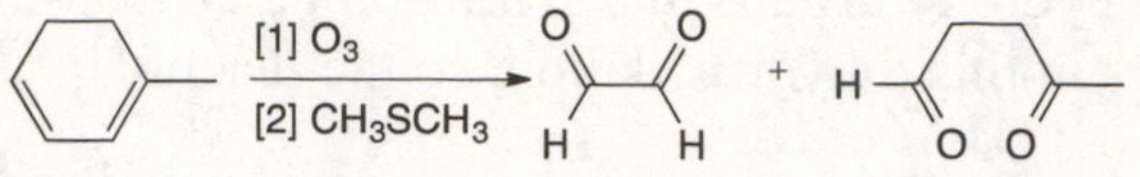 [1] O_3 [2] CH_3SCH_3

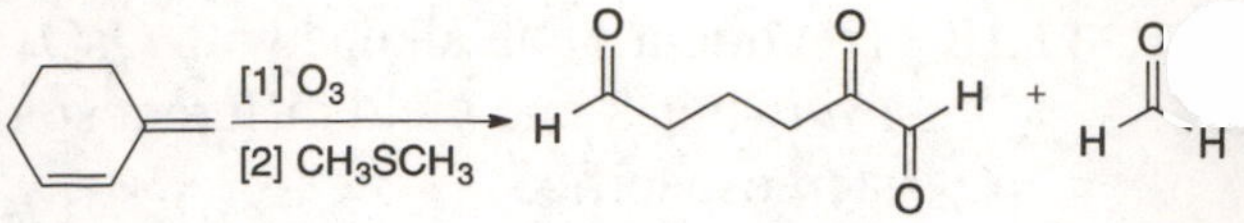

b. 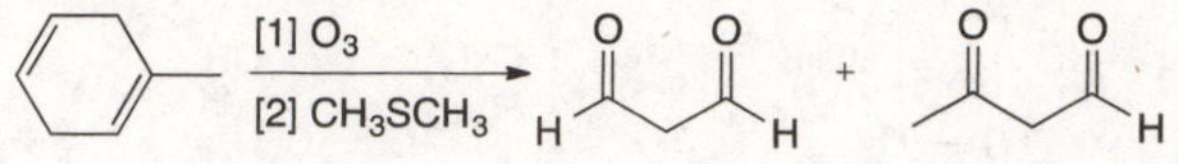 [1] O_3 [2] CH_3SCH_3

c. [1] O_3 [2] CH_3SCH_3

12.22 To draw the products of oxidative cleavage of alkynes:

- **Locate the triple bond.**
- For internal alkynes, **convert the *sp* hybridized C to COOH.**
- For terminal alkynes, the ***sp* hybridized C–H becomes CO_2.**

a. 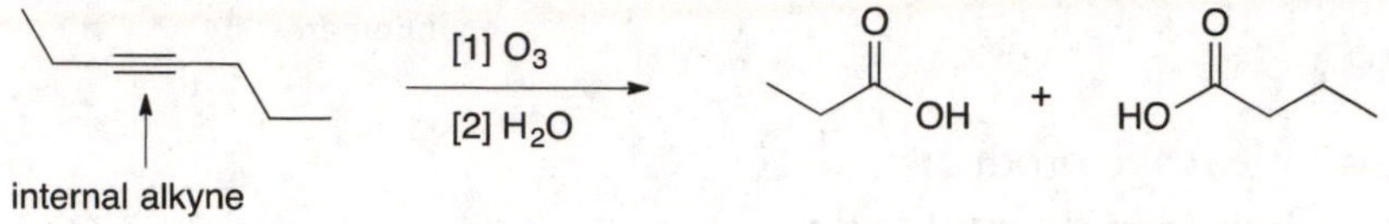 internal alkyne [1] O_3 [2] H_2O → OH + HO

b. 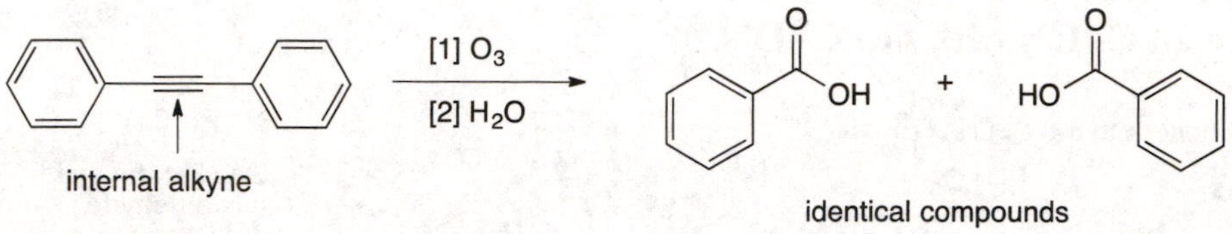 internal alkyne [1] O_3 [2] H_2O → OH + HO

identical compounds

c. 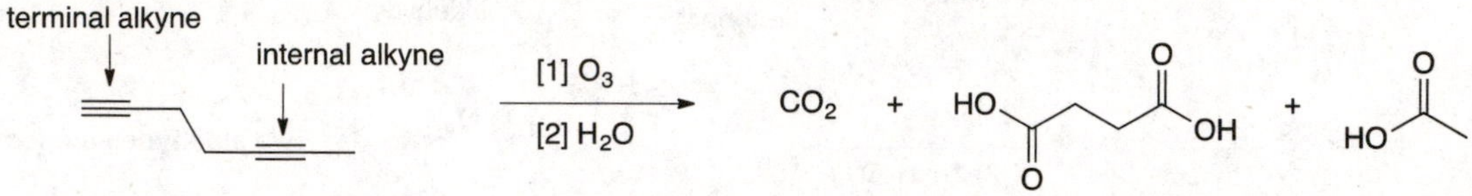 terminal alkyne, internal alkyne [1] O_3 [2] H_2O → CO_2 + HO OH + HO

12.23

a. 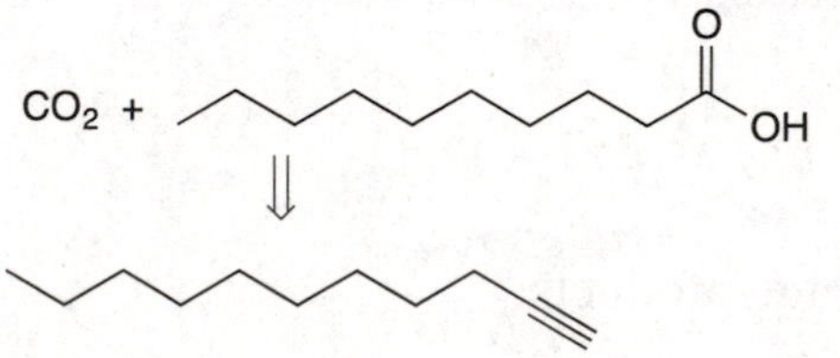 CO_2 + OH

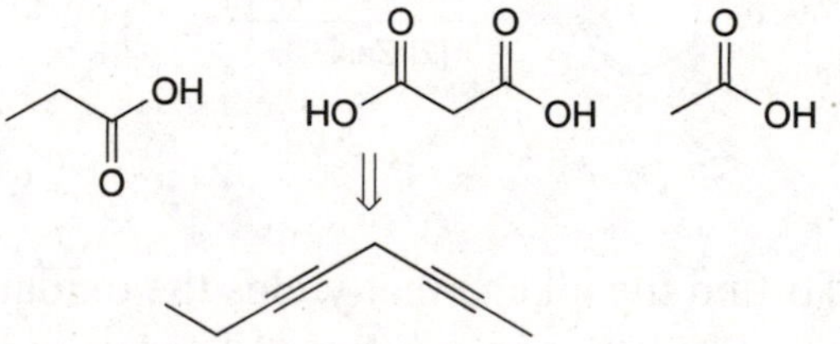

b. 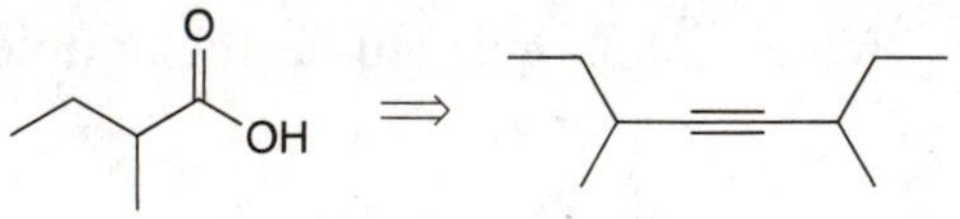 OH

c. OH HO OH OH

d. 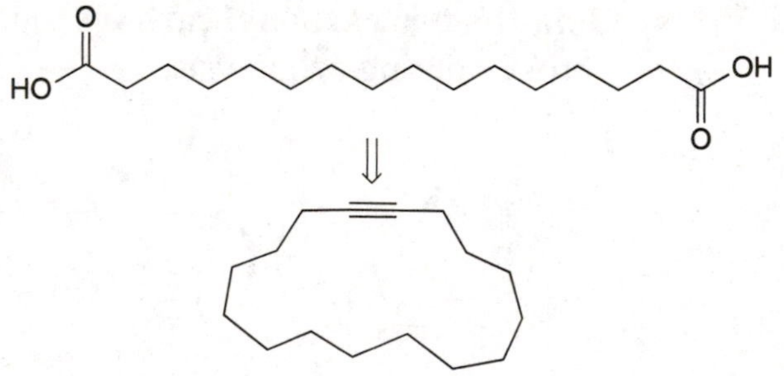 HO OH

12.24 For the **oxidation of alcohols,** remember:

- **1° Alcohols** are oxidized to aldehydes with PCC.
- **1° Alcohols** are oxidized to carboxylic acids with oxidizing agents like CrO_3 or $Na_2Cr_2O_7$.
- **2° Alcohols** are oxidized to ketones with all Cr^{6+} reagents.

a. 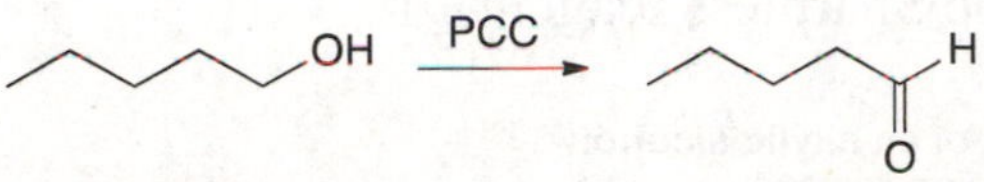 OH —PCC→ H, O

c. 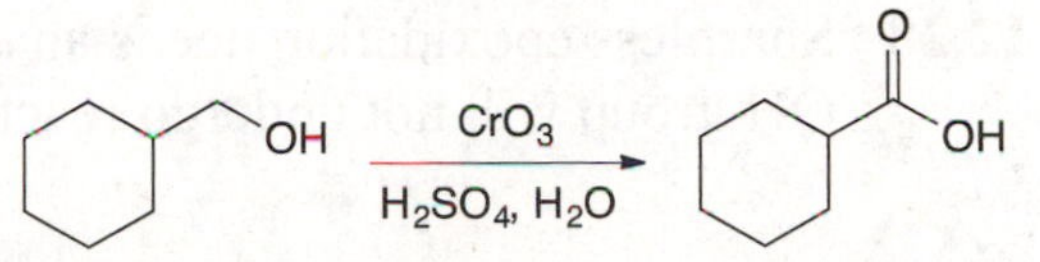 OH —CrO_3, H_2SO_4, H_2O→ O, OH

b. 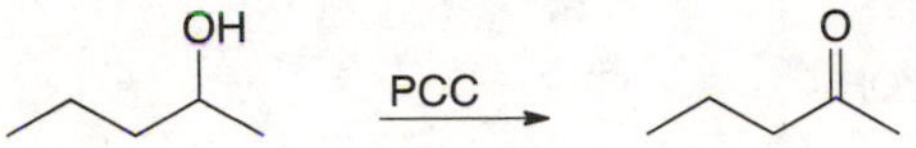 OH —PCC→ O

d. 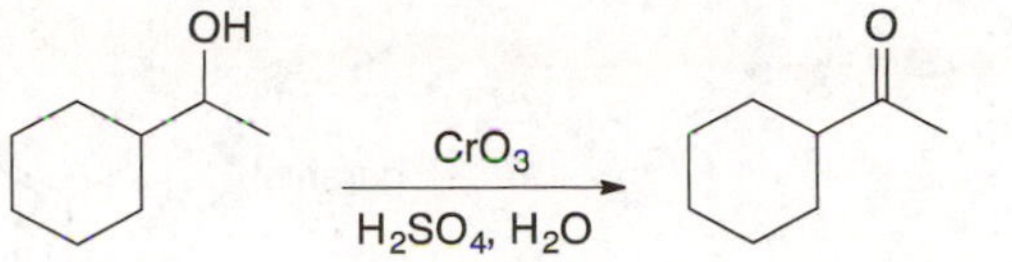 OH —CrO_3, H_2SO_4, H_2O→ O

12.25 Upon treatment with $HCrO_4^-$–Amberlyst A-26 resin:

- **1° Alcohols** are oxidized to aldehydes.
- **2° Alcohols** are oxidized to ketones.

a. 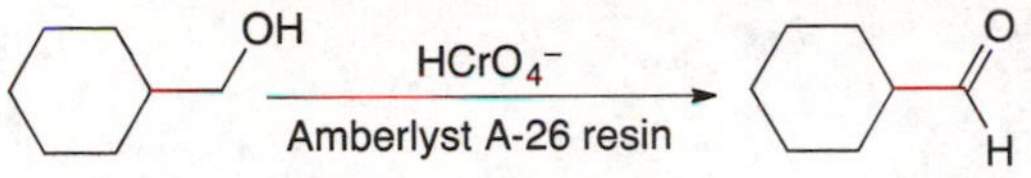 OH —$HCrO_4^-$, Amberlyst A-26 resin→ O, H

c. 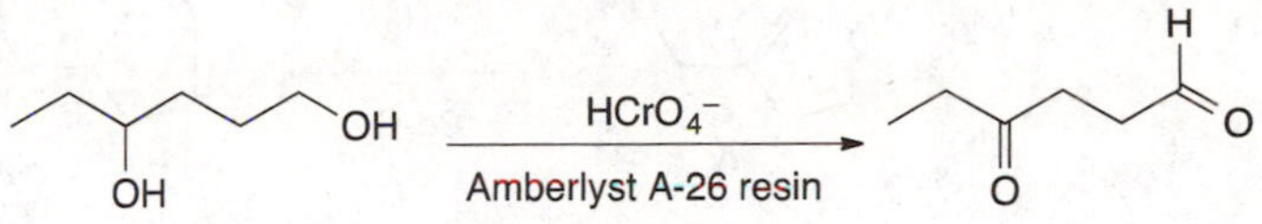 OH, OH —$HCrO_4^-$, Amberlyst A-26 resin→ H, O, O

b. 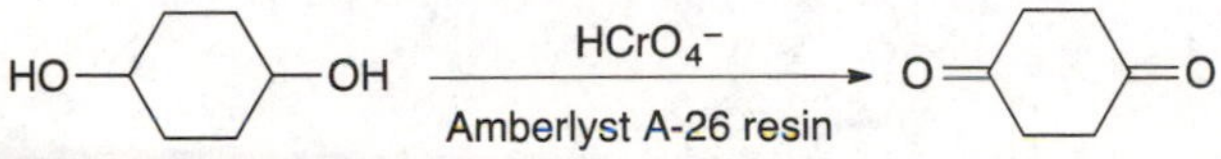 HO, OH —$HCrO_4^-$, Amberlyst A-26 resin→ O, O

12.26

a.  OH —NaOCl→ O + NaCl + H_2O

The by-products of the reaction with sodium hypochlorite are water and table salt (NaCl), as opposed to the by-products with $HCrO_4^-$–Amberlyst A-26 resin, which contain carcinogenic Cr^{3+} metal.

b. Oxidation with NaOCl has at least two advantages over oxidation with CrO_3, H_2SO_4 and H_2O. Because no Cr^{6+} is used as oxidant, there are no Cr by-products that must be disposed of. Also, CrO_3 oxidation is carried out in corrosive inorganic acids (H_2SO_4) and oxidation with NaOCl avoids this.

12.27 To draw the products of a **Sharpless epoxidation:**

- With the C=C horizontal, draw the allylic alcohol with the OH on the **top right** of the alkene.
- Add the new oxygen **above** the plane if (–)-DET is used and **below** the plane if (+)-DET is used.

a. 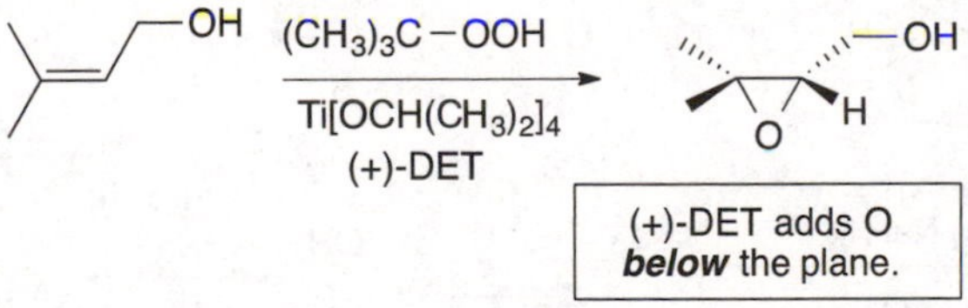 OH —$(CH_3)_3C-OOH$, $Ti[OCH(CH_3)_2]_4$, (+)-DET→ OH, H, O

(+)-DET adds O *below* the plane.

b. 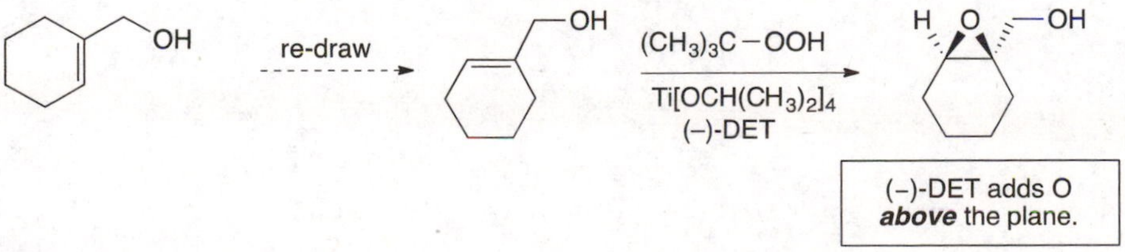 OH - - re-draw - -→ OH —$(CH_3)_3C-OOH$, $Ti[OCH(CH_3)_2]_4$, (–)-DET→ H, O, OH

(–)-DET adds O *above* the plane.

12.28 Sharpless epoxidation needs an ***allylic alcohol*** as the starting material. Alkenes with no allylic OH group will not undergo reaction with the Sharpless reagent.

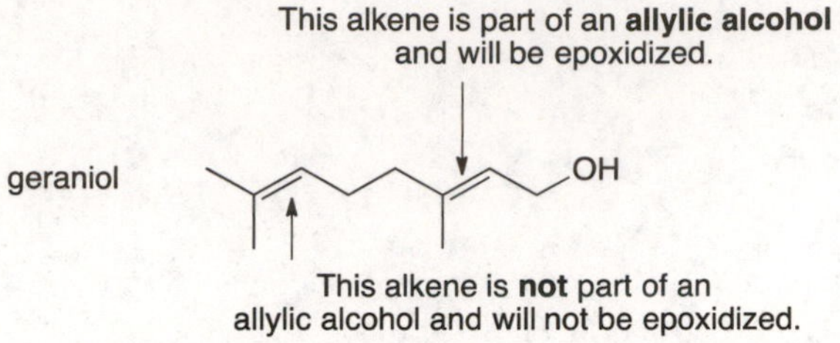

This alkene is part of an **allylic alcohol** and will be epoxidized.

geraniol

OH

This alkene is **not** part of an allylic alcohol and will not be epoxidized.

12.29

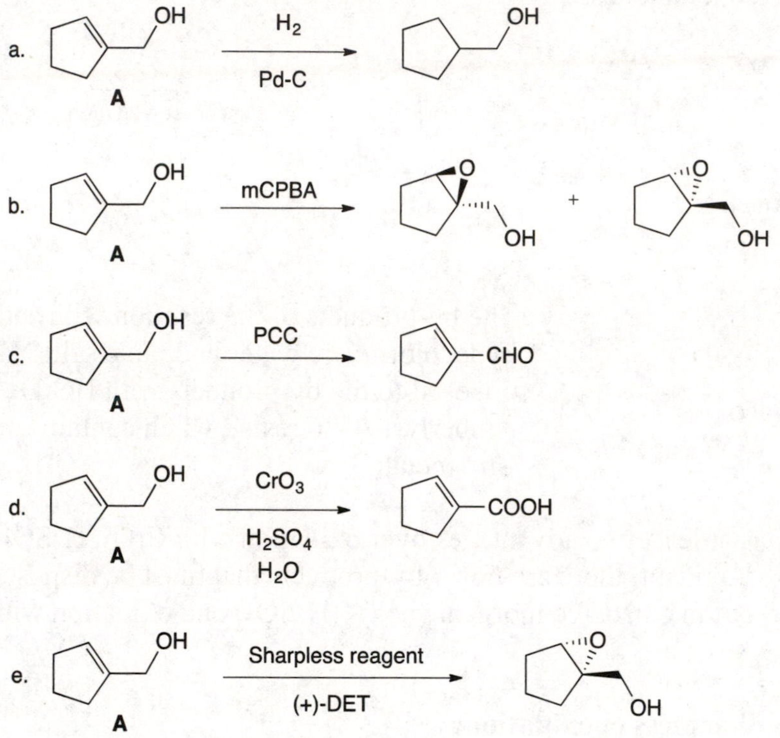

a. A (OH) $\xrightarrow[\text{Pd-C}]{H_2}$ (OH)

b. A (OH) $\xrightarrow{\text{mCPBA}}$ (O, OH) + (O, OH)

c. A (OH) $\xrightarrow{\text{PCC}}$ –CHO

d. A (OH) $\xrightarrow[H_2SO_4,\ H_2O]{CrO_3}$ –COOH

e. A (OH) $\xrightarrow[\text{(+)-DET}]{\text{Sharpless reagent}}$ (O, OH)

12.30

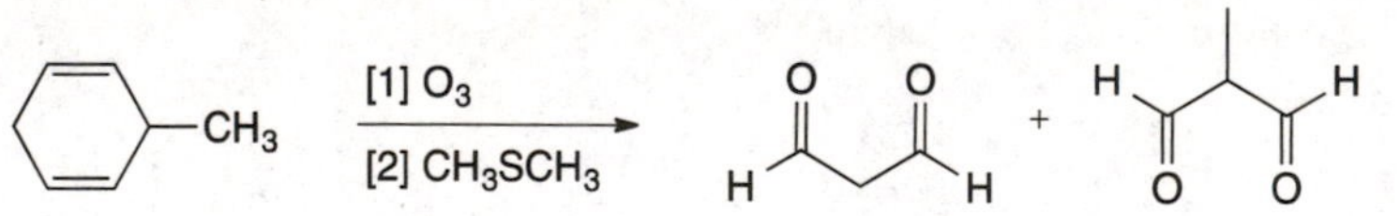

–CH_3 $\xrightarrow[\text{[2] } CH_3SCH_3]{\text{[1] } O_3}$ (O, O, H, H) + (H, H, O, O)

12.31

HC≡CH $\xrightarrow[\text{[2] } CH_3CH_2Br]{\text{[1] NaH}}$ $\xrightarrow[\text{[2] ethylene oxide, [3] } H_2O]{\text{[1] NaH}}$ –OH $\xrightarrow[NH_3]{\text{Na}}$ OH

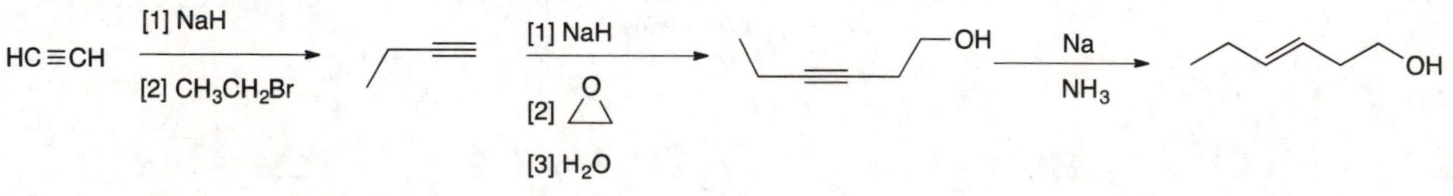

12.32 Use the rules from Answer 12.1.

a.  → reduction

b. HO–⟨⟩–OH → O=⟨⟩=O oxidation

c. 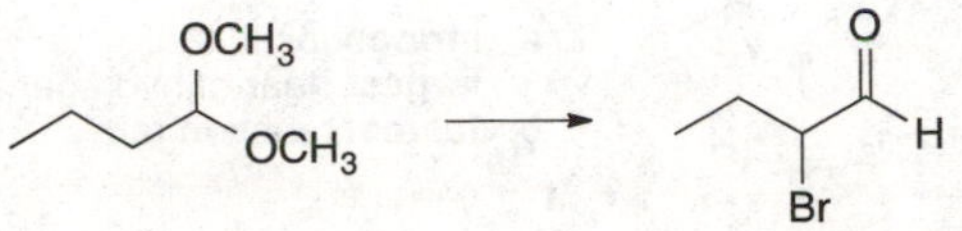 OCH_3 / OCH_3 → O / H / Br oxidation (one new C–Br bond)

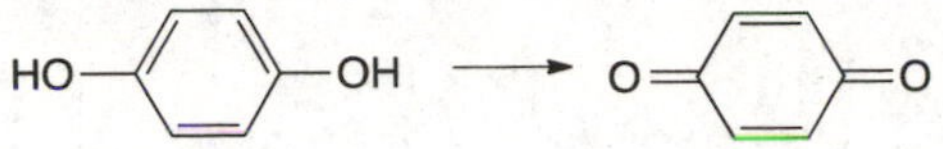

d. 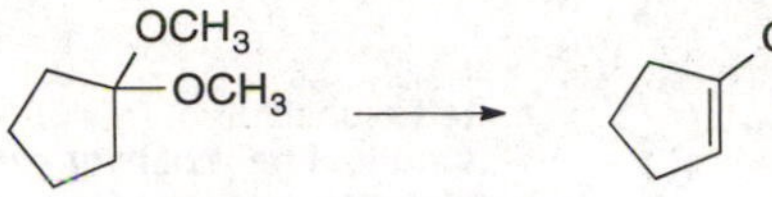 OCH_3 / OCH_3 → OCH_3 neither (one fewer C–O bond and one fewer C–H bond)

12.33 Use the principles from Answer 12.2 and draw the products of syn addition of H_2 from above and below the C=C.

a. 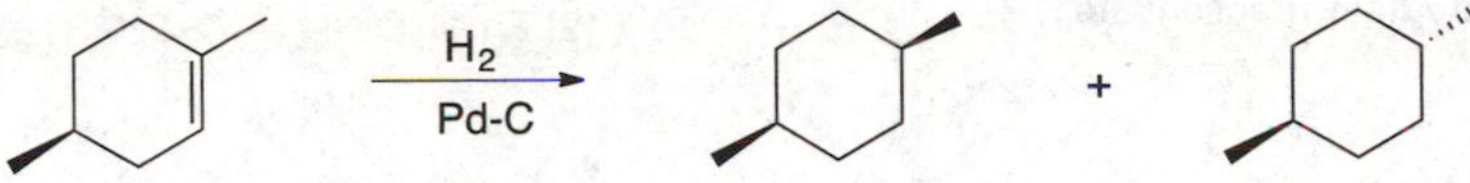 $\xrightarrow[\text{Pd-C}]{H_2}$ +

b. 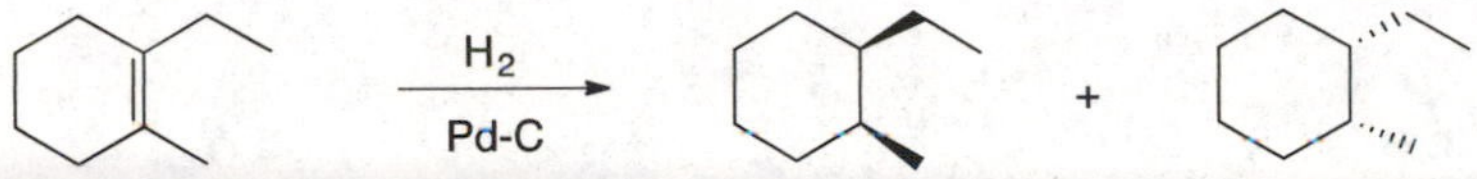 $\xrightarrow[\text{Pd-C}]{H_2}$ +

c. $\xrightarrow[\text{Pd-C}]{H_2}$ H + H

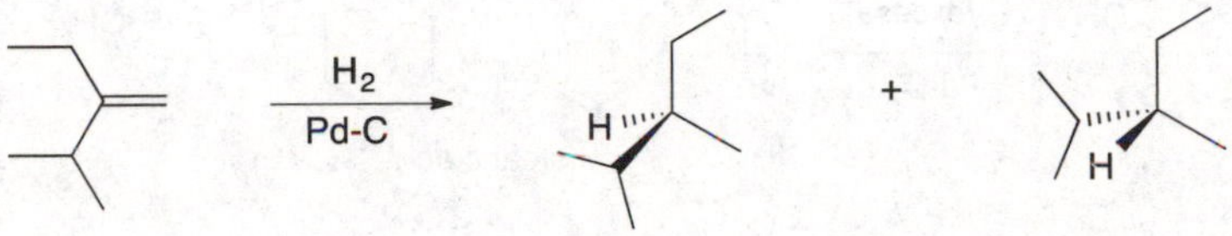

12.34 Increasing alkyl substitution increases alkene stability, thus decreasing the heat of hydrogenation.

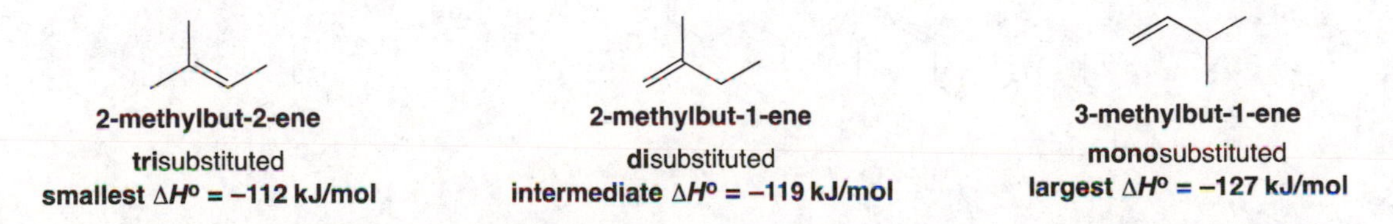

12.35

A possible structure:

a. Compound **A:** molecular formula C_5H_8: hydrogenated to C_5H_{10}.
2 degrees of unsaturation, 1 is hydrogenated.
1 ring and 1 π bond →

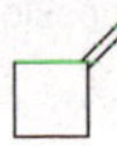

b. Compound **B:** molecular formula $C_{10}H_{16}$: hydrogenated to $C_{10}H_{18}$.
3 degrees of unsaturation, 1 is hydrogenated.
2 rings and 1 π bond →

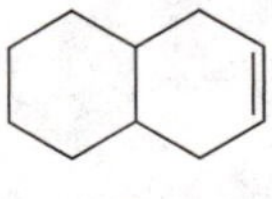

c. Compound **C:** molecular formula C_8H_8: hydrogenated to C_8H_{16}.
5 degrees of unsaturation, 4 are hydrogenated.
1 ring and 4 π bonds →

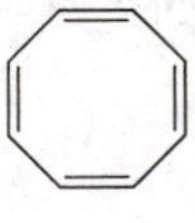

12.36

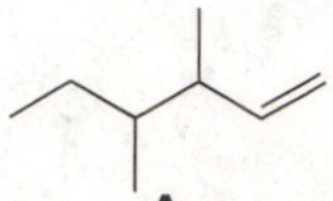

A

a. **mono**substituted
largest heat of hydrogenation
b. **fastest** reaction rate

c. [1] O_3 → [2] Zn, H_2O

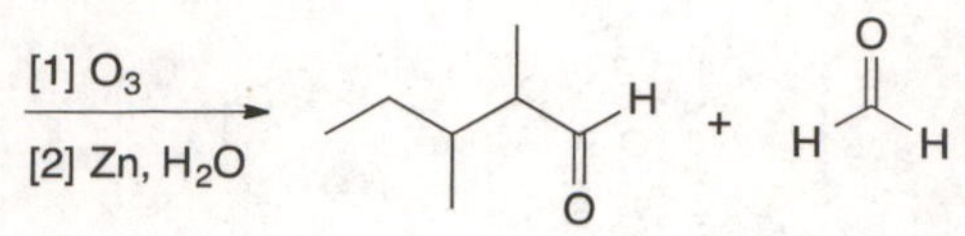

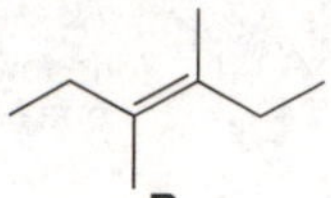

B

a. **tetra**substituted
smallest heat of hydrogenation
b. **slowest** reaction rate

c. [1] O_3 → [2] Zn, H_2O

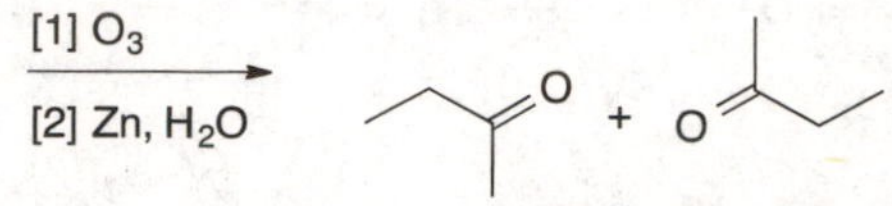

identical

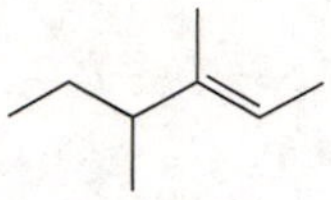

C

a. **tri**substituted
intermediate heat of hydrogenation
b. **intermediate** reaction rate

c. [1] O_3 → [2] Zn, H_2O

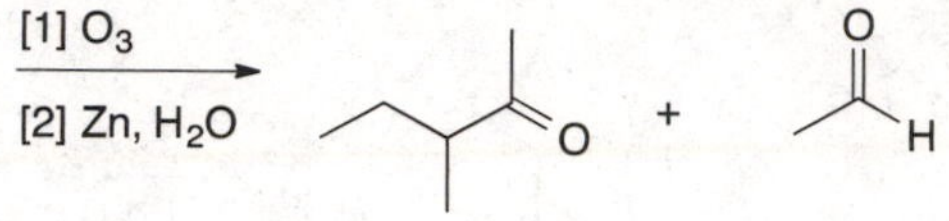

12.37

a. stearidonic acid —H_2 (excess), Pd-C→ stearic acid

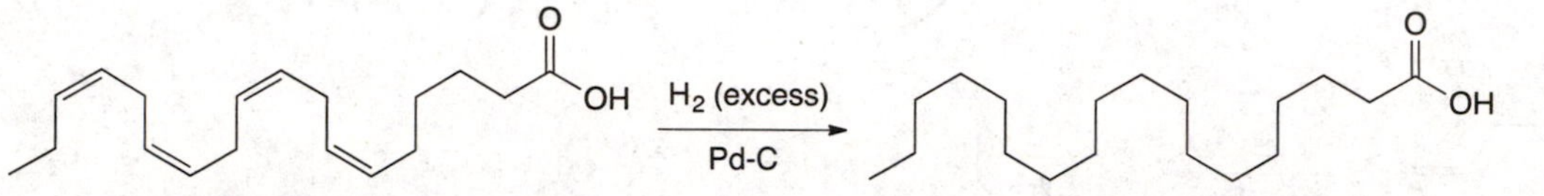

b. —H_2 (1 equiv), Pd-C→ + + +

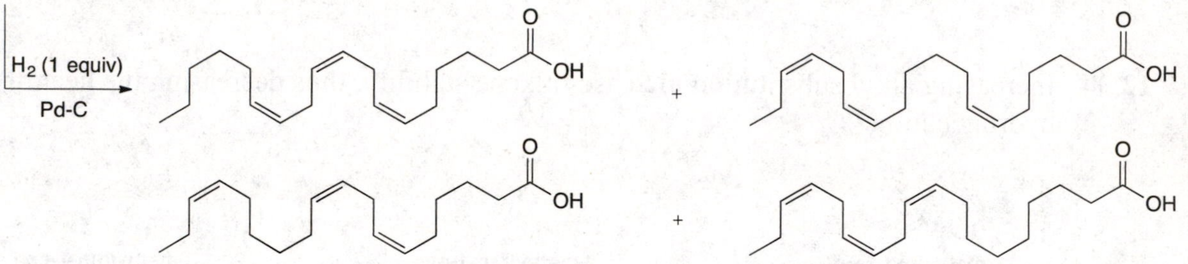

c. trans

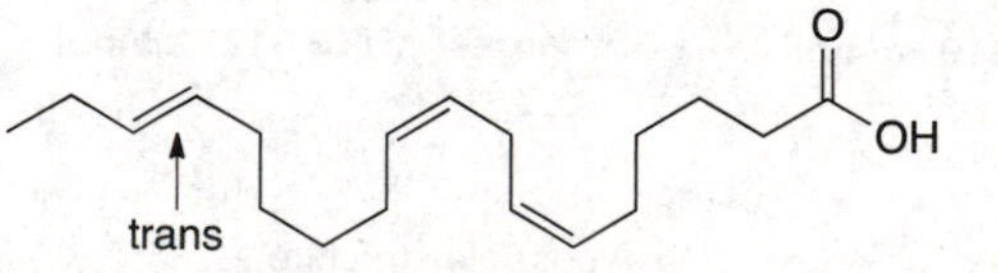

one possibility

d. stearidonic acid 4 cis C=C's < product in (b) 3 cis C=C's < product in (c) 2 cis and one trans C=C's

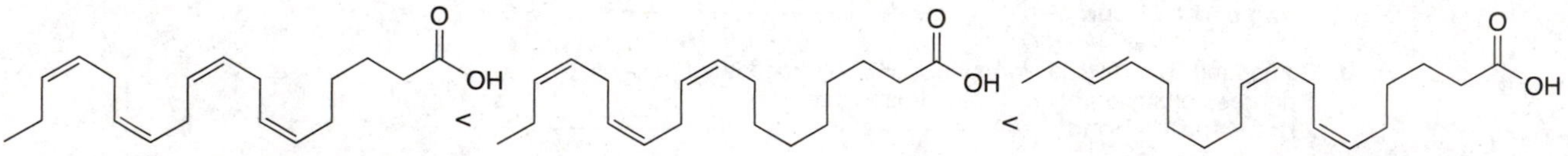

12.38

a. cyclopentene $\xrightarrow[\text{Pd-C}]{H_2}$ cyclopentane

b. cyclopentene $\xrightarrow[\text{Lindlar catalyst}]{H_2}$ no reaction

c. cyclopentene $\xrightarrow[NH_3]{Na}$ no reaction

d. cyclopentene $\xrightarrow{CH_3CO_3H}$ epoxide (O)

e. cyclopentene $\xrightarrow[\text{[2] } H_2O,\ HO^-]{\text{[1] } CH_3CO_3H}$ OH / OH + OH / OH

anti addition

f. cyclopentene $\xrightarrow[\text{[2] } NaHSO_3,\ H_2O]{\text{[1] } OsO_4 + NMO}$ OH / OH syn addition

g. cyclopentene $\xrightarrow[H_2O,\ HO^-]{KMnO_4}$ OH / OH syn addition

h. cyclopentene $\xrightarrow[\text{[2] } H_2O]{\text{[1] } LiAlH_4}$ no reaction

i. cyclopentene $\xrightarrow[\text{[2] } CH_3SCH_3]{\text{[1] } O_3}$ H(C=O)CH₂CH₂CH₂(C=O)H (O, O, H, H)

j. cyclopentene $\xrightarrow[\text{Ti[OCH(CH_3)_2]_4, (–)-DET}]{(CH_3)_3COOH}$ no reaction

k. cyclopentene $\xrightarrow{\text{mCPBA}}$ epoxide (O)

l. epoxide (O) $\xrightarrow[\text{[2] } H_2O]{\text{[1] } LiAlH_4}$ cyclopentanol (OH)

12.39

a. OH $\xrightarrow[\text{Pd-C}]{H_2}$ OH + OH

b. OH $\xrightarrow{\text{mCPBA}}$ O, H, OH + H, OH, O

c. OH $\xrightarrow{\text{PCC}}$ H, O

d. OH $\xrightarrow[H_2SO_4,\ H_2O]{CrO_3}$ OH, O

e. OH $\xrightarrow[\text{Ti[OCH(CH_3)_2]_4, (+)-DET}]{(CH_3)_3COOH}$ O, H, OH

f. OH $\xrightarrow[\text{Ti[OCH(CH_3)_2]_4, (–)-DET}]{(CH_3)_3COOH}$ H, OH, O

g. OH $\xrightarrow{\text{[1] } PBr_3}$ Br $\xrightarrow{\text{[2] } LiAlH_4,\ \text{[3] } H_2O}$ alkene

h. OH $\xrightarrow[\text{Amberlyst A-26}]{HCrO_4^-}$ H, O

12.40

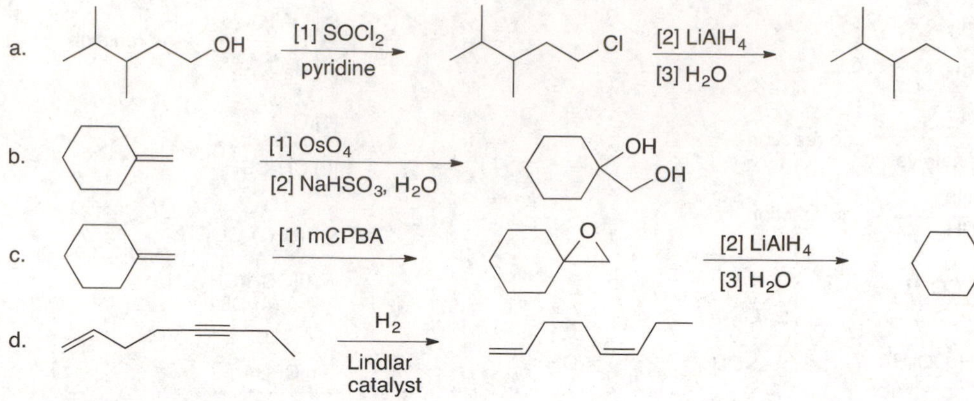

a. [1] $SOCl_2$, pyridine; Cl; [2] $LiAlH_4$, [3] H_2O

b. [1] OsO_4, [2] $NaHSO_3$, H_2O; OH, OH

c. [1] mCPBA; O; [2] $LiAlH_4$, [3] H_2O; OH

d. H_2, Lindlar catalyst

12.41

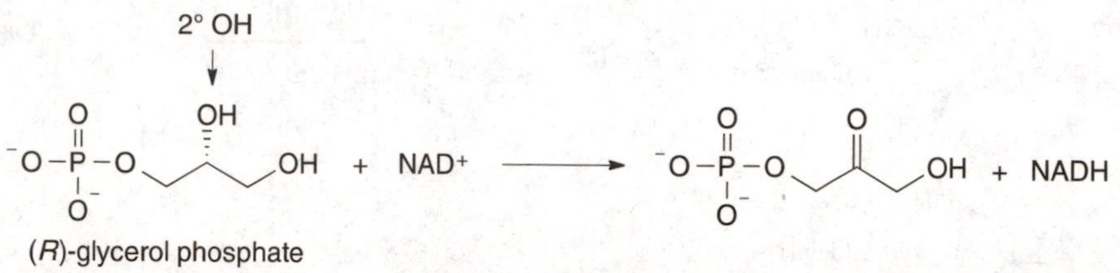

2° OH

$^{-}O-P(=O)(O^{-})-O$ ... OH (with OH) + NAD^+ ⟶ $^{-}O-P(=O)(O^{-})-O$ ... (C=O) ... OH + NADH

(*R*)-glycerol phosphate

With NAD^+ the 2° alcohol is oxidized to a ketone, and NAD^+ is reduced to NADH.

12.42 $LiAlH_4$ attacks at the less substituted end of an unsymmetrical epoxide to form an alcohol with an OH on the more substituted carbon.

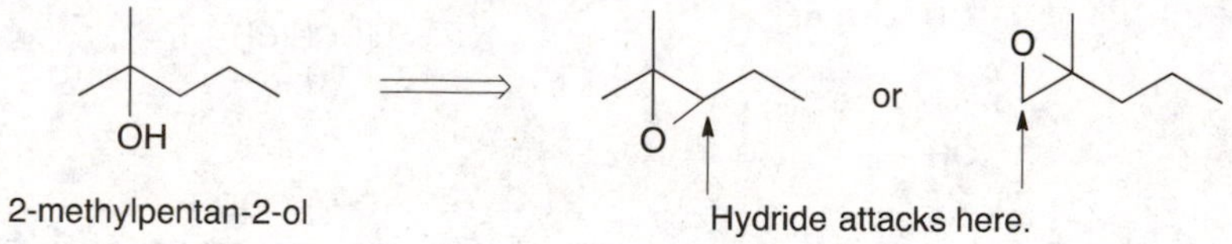

OH

2-methylpentan-2-ol ⟹ (epoxide, O) or (epoxide, O)

Hydride attacks here.

12.43 The two sides of the C=C of **A** are different. D_2 adds only from above, so this side must be less sterically hindered. Other reagents will also add from the same side.

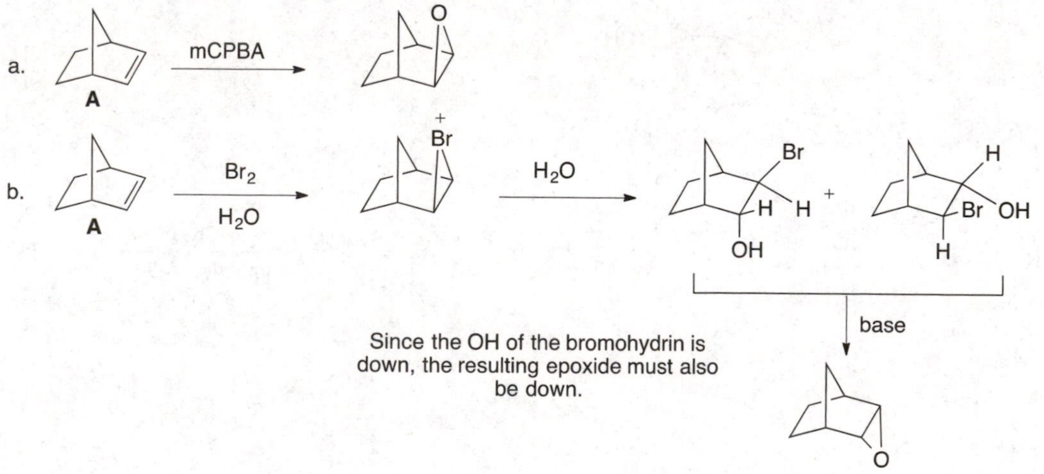

a. **A** —mCPBA→ (epoxide, O)

b. **A** —Br_2, H_2O→ (Br^+) —H_2O→ (Br, H, H, OH) + (H, Br, OH, H)

base

Since the OH of the bromohydrin is down, the resulting epoxide must also be down.

(epoxide, O)

12.44 Alkenes treated with [1] OsO_4 followed by $NaHSO_3$ in H_2O will undergo **syn** addition, whereas alkenes treated with [2] CH_3CO_3H followed by ^-OH in H_2O will undergo **anti** addition.

a. [1] [1] OsO_4 [2] $NaHSO_3$, H_2O — HO, OH

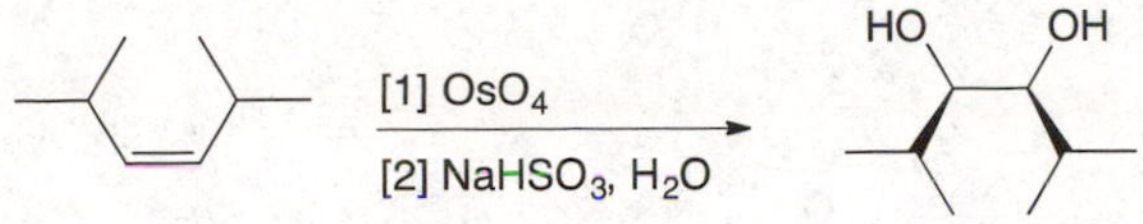

[2] [1] CH_3CO_3H [2] ^-OH, H_2O anti addition — HO, OH — rotate — HO, OH

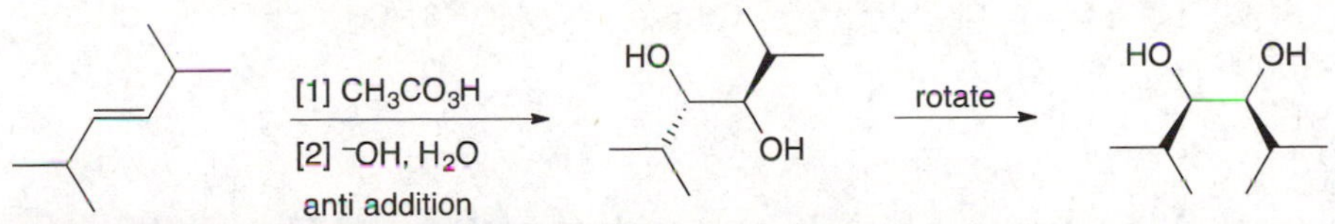

b. [1] [1] OsO_4 [2] $NaHSO_3$, H_2O syn addition — OH, OH — rotate — OH, OH + enantiomer

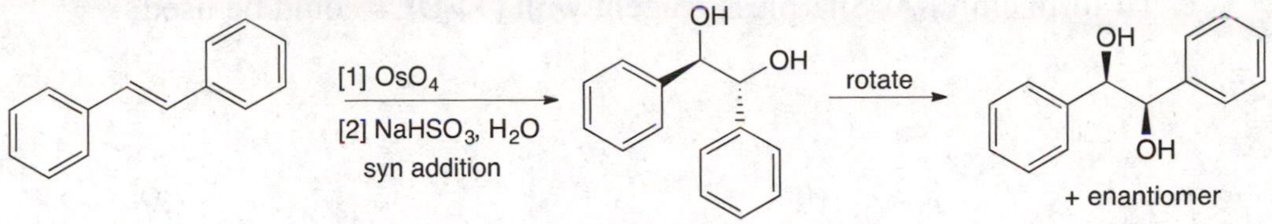

[2] [1] CH_3CO_3H [2] ^-OH, H_2O — OH, OH + enantiomer

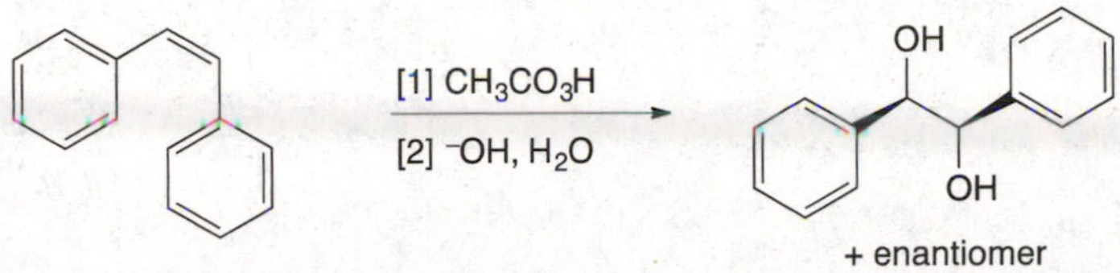

c. [1] [1] OsO_4 [2] $NaHSO_3$, H_2O syn addition — HO, OH — rotate — OH, $CH_3CH_2CH_2$, H, H, $CH_2CH_2CH_3$, OH

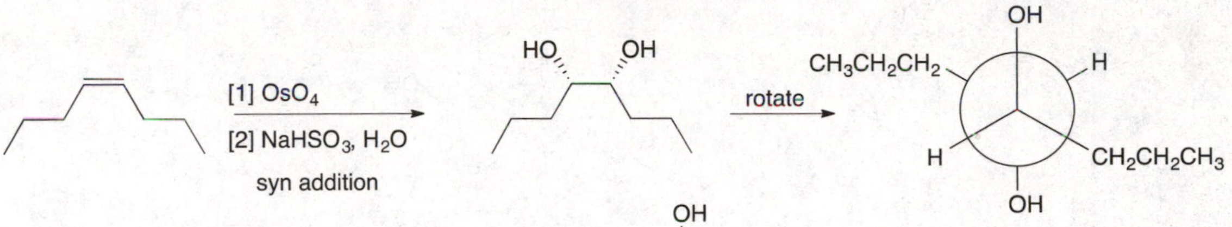

[2] [1] CH_3CO_3H [2] ^-OH, H_2O — OH, $CH_3CH_2CH_2$, H, H, $CH_2CH_2CH_3$, OH

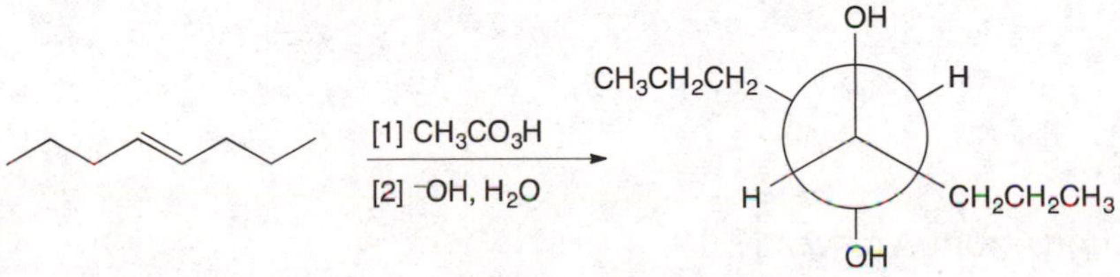

12.45

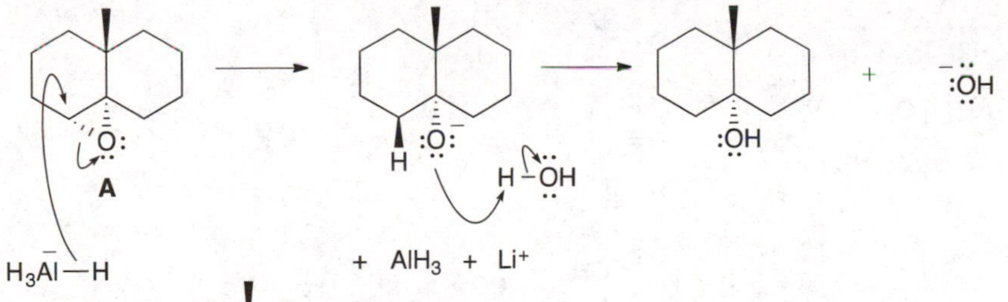

A — $H_3\bar{A}l$—H Li^+ → H :Ö:$^-$ H—ÖH + AlH_3 + Li^+ → :ÖH + $^-$:ÖH:

D :ÖH

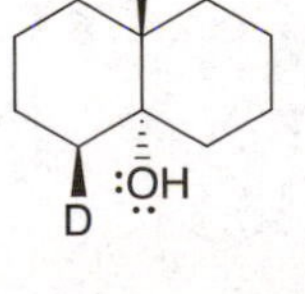

D^- (from $LiAlD_4$) opens the epoxide ring from the back side, so it is oriented on a wedge in the final product.

12.46

a. 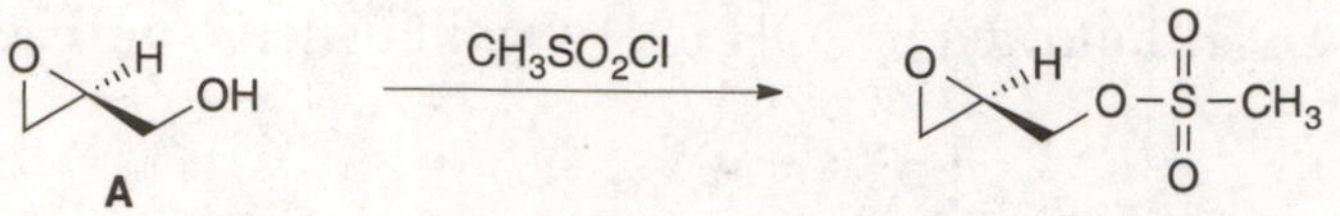 A + CH_3SO_2Cl → O–S(=O)(=O)–CH_3

b. CH_3S^- → CH_3S ... O–S(=O)(=O)–CH_3 → CH_3S = SCH_3 B

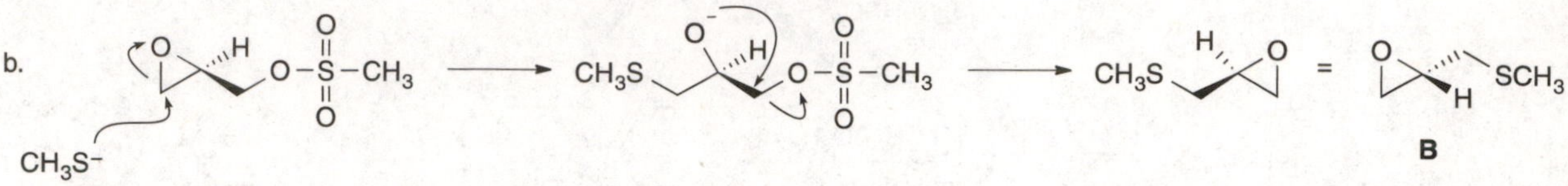

c. To form chiral **A,** Sharpless reagent with (+)-DET could be used.

12.47

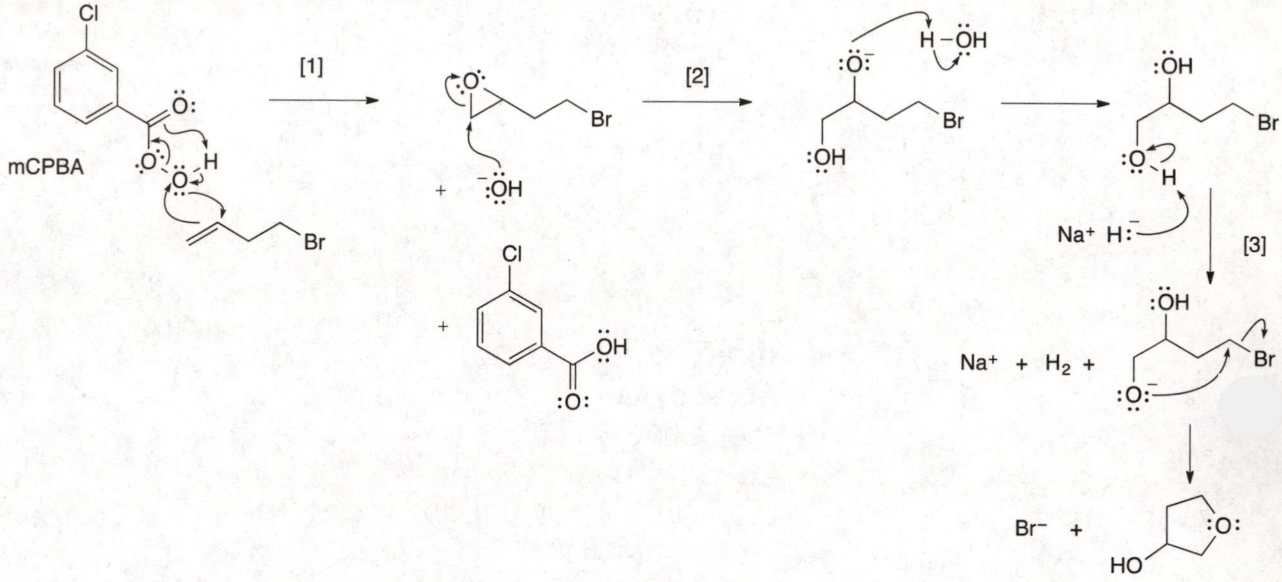

mCPBA, Cl, Br → [1] → Br + :ÖH + Cl, ÖH, :O: → [2] → :OH, H–ÖH, Br → :ÖH, Br, Na+ H: → [3] → Na+ + H_2 + :ÖH, Br → Br– + HO, :O:

12.48 Use the directions from Answer 12.19.

a. 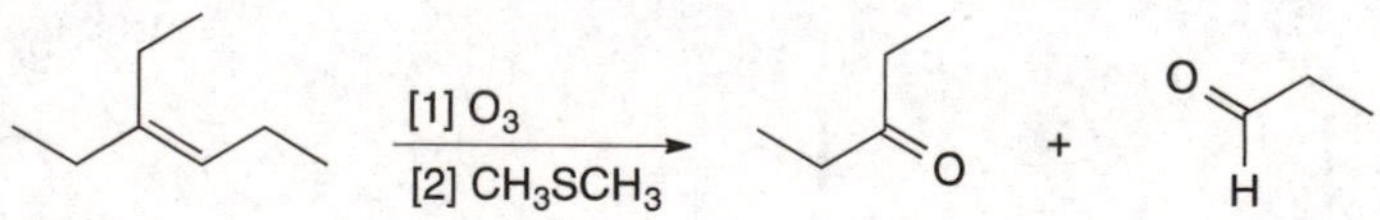 [1] O_3 [2] CH_3SCH_3 → O + O, H

b. 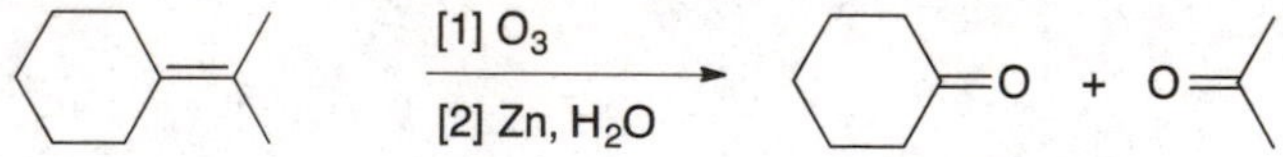 [1] O_3 [2] Zn, H_2O → O + O

c. 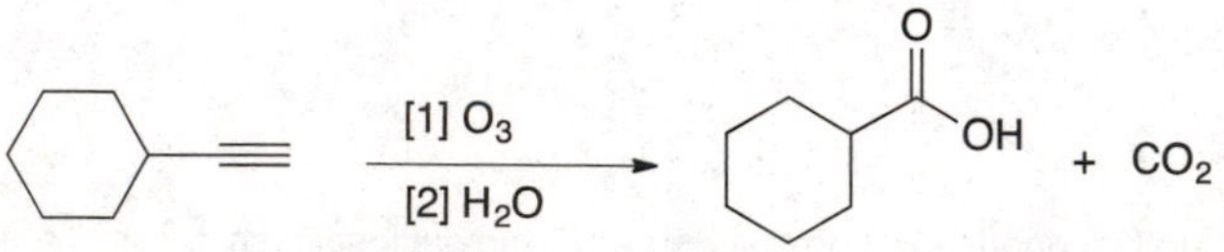 [1] O_3 [2] H_2O → O, OH + CO_2

d. 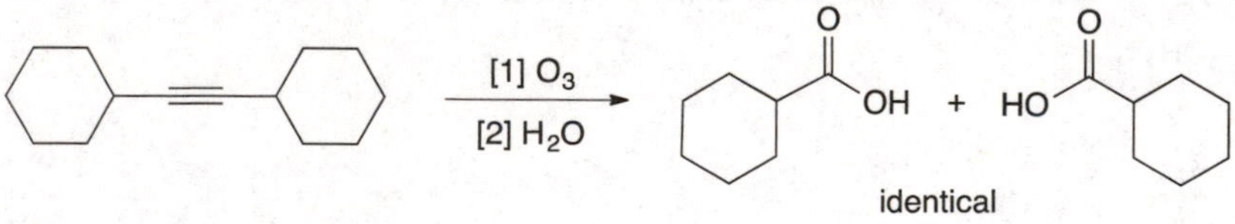 [1] O_3 [2] H_2O → O, OH + HO, O

identical

12.49

a. 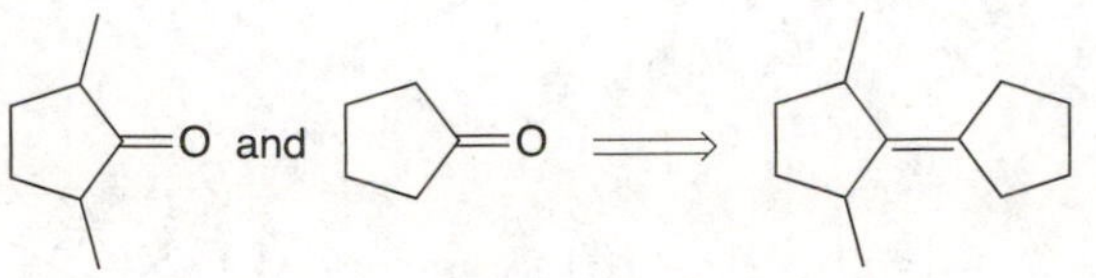 and $\Longrightarrow$

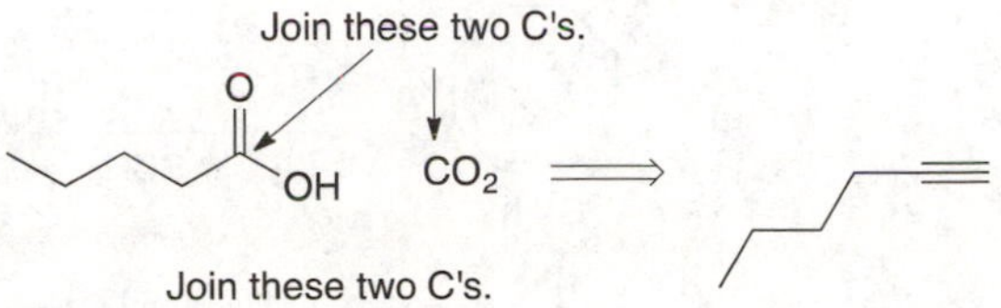

b. 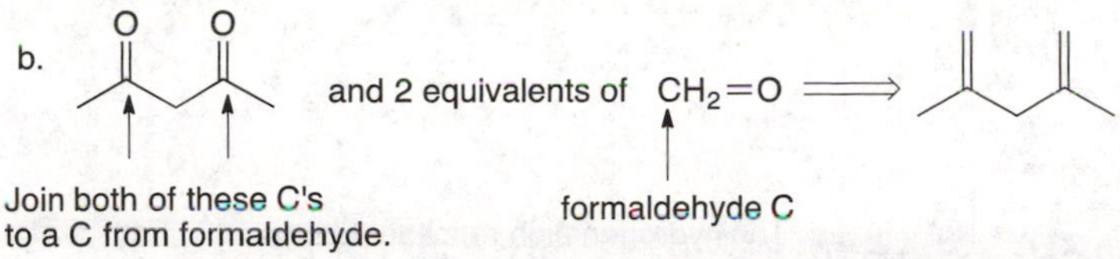 and 2 equivalents of $CH_2{=}O$ $\Longrightarrow$

Join both of these C's to a C from formaldehyde.

formaldehyde C

c. Join these two C's. OH CO_2 $\Longrightarrow$

d. 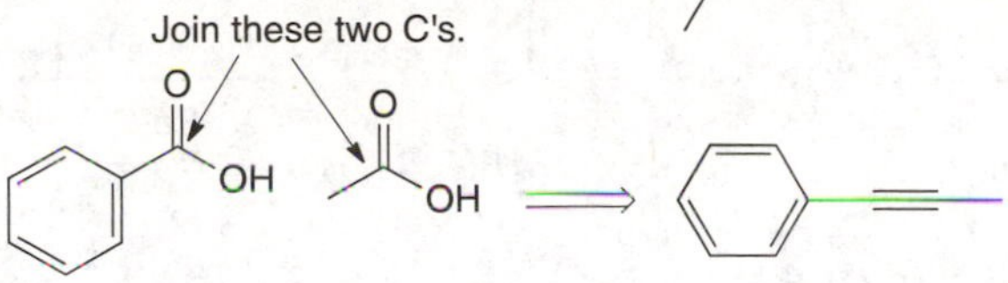 Join these two C's. OH OH $\Longrightarrow$

12.50 Use the directions from Answer 12.20.

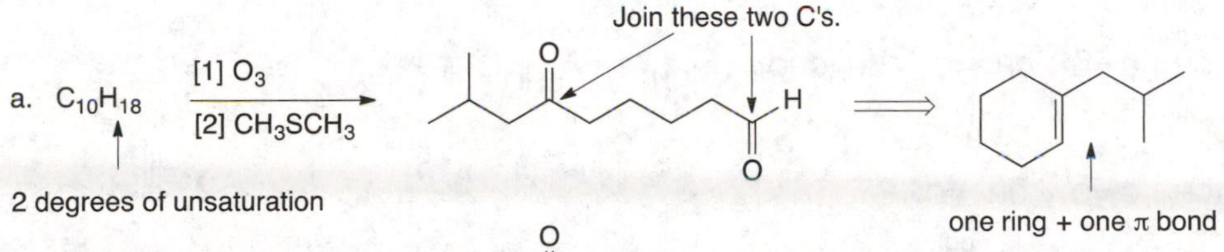

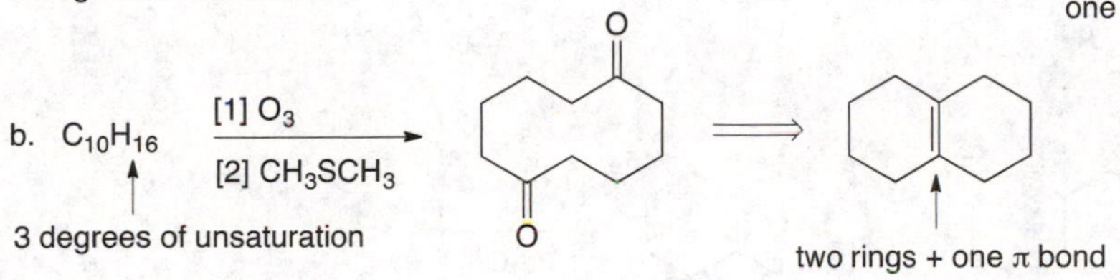

a. $C_{10}H_{18}$ $\xrightarrow[\text{[2] } CH_3SCH_3]{\text{[1] } O_3}$ $\Longrightarrow$

2 degrees of unsaturation

Join these two C's.

one ring + one π bond

b. $C_{10}H_{16}$ $\xrightarrow[\text{[2] } CH_3SCH_3]{\text{[1] } O_3}$ $\Longrightarrow$

3 degrees of unsaturation

two rings + one π bond

12.51

a. 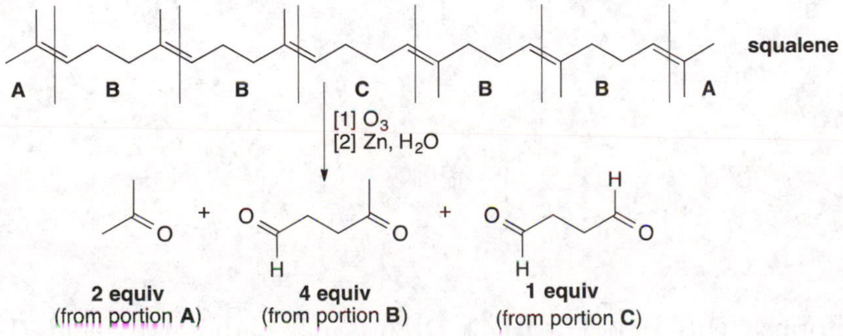 **squalene**

A B B C B B A

[1] O_3
[2] Zn, H_2O

2 equiv (from portion **A**) + **4 equiv** (from portion **B**) + **1 equiv** (from portion **C**)

b. 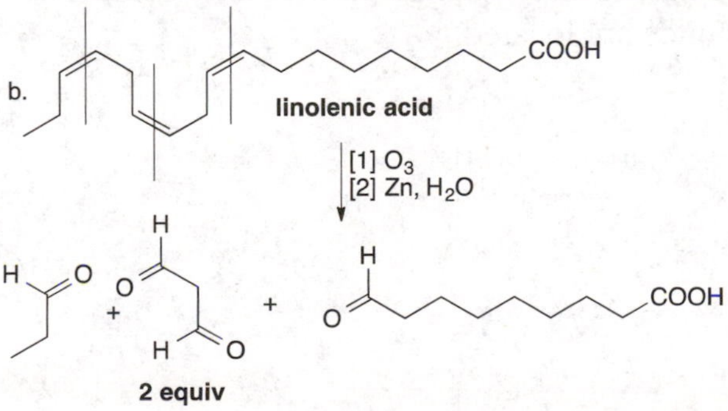 **linolenic acid** COOH

[1] O_3
[2] Zn, H_2O

+ **2 equiv** + COOH

c. **zingiberene** $\xrightarrow[\text{[2] Zn, } H_2O]{\text{[1] } O_3}$ + +

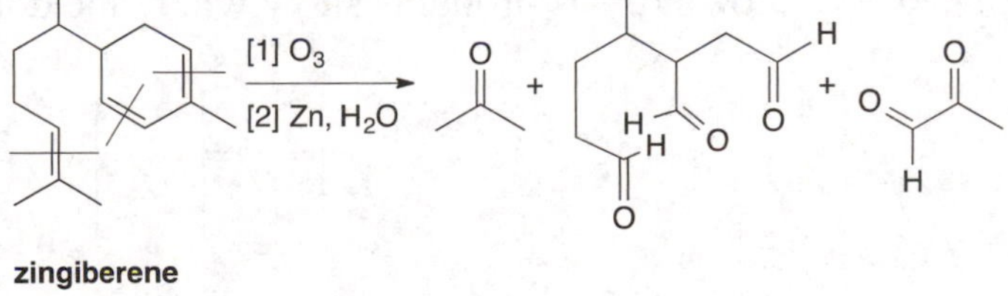

12.52

a. 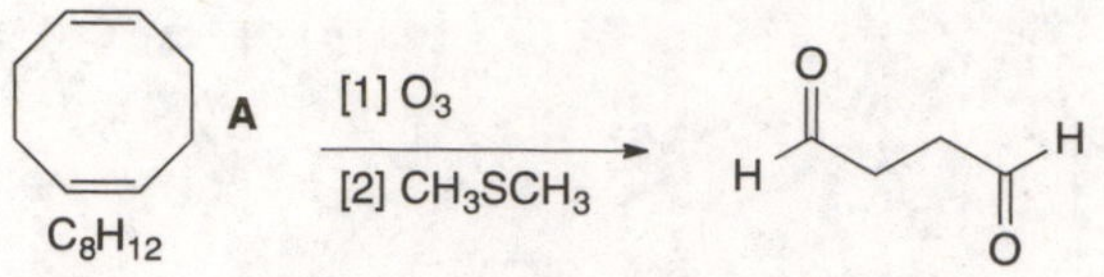 A C_8H_{12} → [1] O_3 [2] CH_3SCH_3 → H, O, O, H

b. 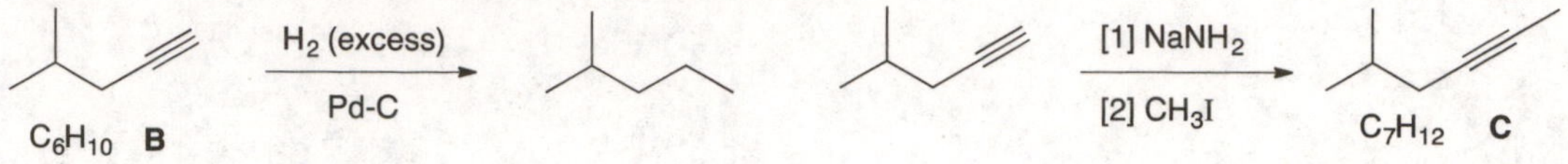 C_6H_{10} **B** → H_2 (excess), Pd-C → ; → [1] $NaNH_2$ [2] CH_3I → C_7H_{12} **C**

12.53

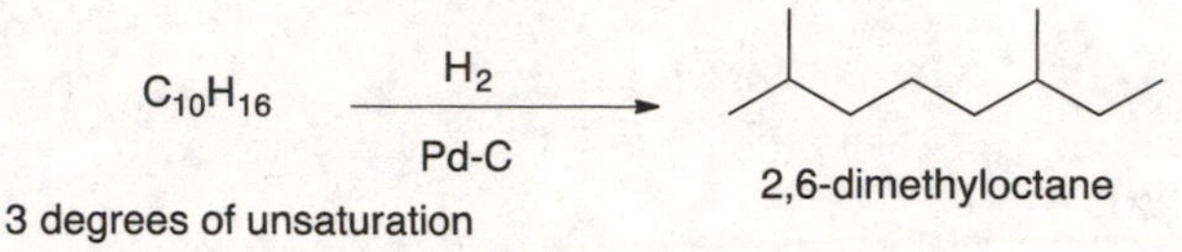

$C_{10}H_{16}$ → H_2, Pd-C → 2,6-dimethyloctane

3 degrees of unsaturation

The hydrogenation reaction tells you that both oximene and myrcene have 3 π bonds (and no rings). Use this carbon backbone and add in the double bonds based on the oxidative cleavage products.

Oximene: $(CH_3)_2C{=}O$ $CH_2{=}O$ $CH_2(CHO)_2$ CH_3COCHO ⟹

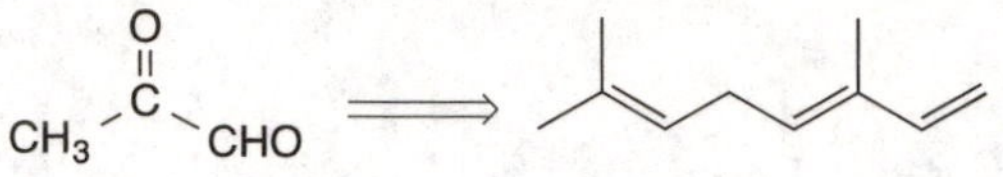

Myrcene: $(CH_3)_2C{=}O$ $CH_2{=}O$ (2 equiv) H, O, O, H, O ⟹

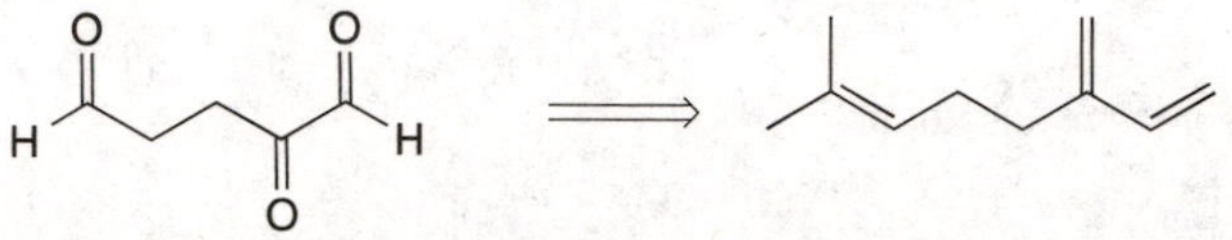

12.54

OH; $C_{10}H_{18}O$ **A** → H_2SO_4 → $C_{10}H_{16}$ **B** + $C_{10}H_{16}$ **C** → H_2, Pd-C → decalin

ozonolysis

D (O, CHO)

$C_{10}H_{16}O_2$ **E** (O, O)

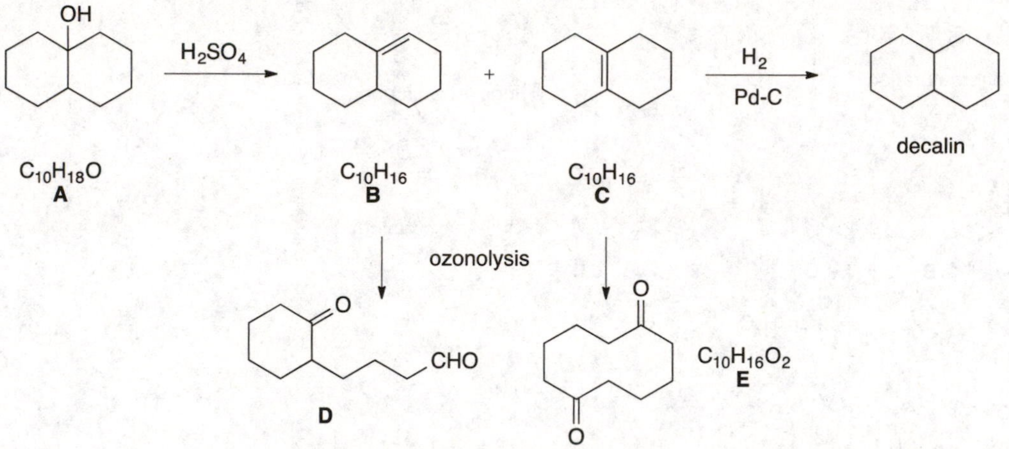

12.55 Hydrogenation of DHA forms $CH_3(CH_2)_{20}COOH$, so DHA is a 22-carbon fatty acid. The ozonolysis products show where the double bonds are located.

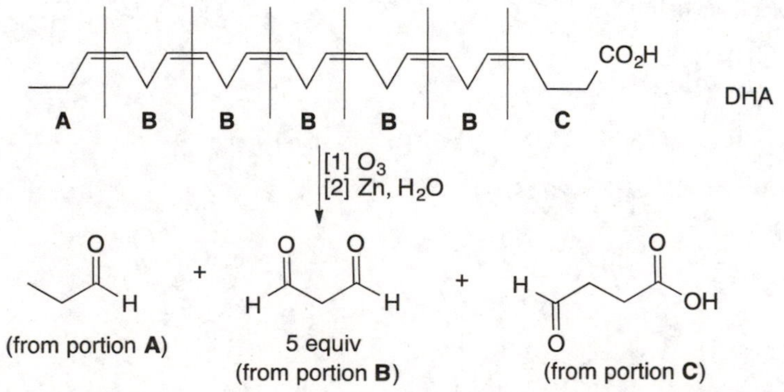

DHA (CO_2H); A B B B B B C

[1] O_3 [2] Zn, H_2O

O, H (from portion **A**) + O, O, H, H 5 equiv (from portion **B**) + H, O, O, OH (from portion **C**)

12.56 The stereogenic center (labeled with *) in both structures can be *R* or *S*.

ozonolysis

or

H_2

Pd-C

butylcyclopheptane

possible structures for **dictyopterene D'**

12.57

a. re-draw

$(CH_3)_3COOH$

$Ti[OC(CH_3)_2]_4$

(–)-DET

b. $(CH_3)_3COOH$

$Ti[OC(CH_3)_2]_4$

(+)-DET

12.58

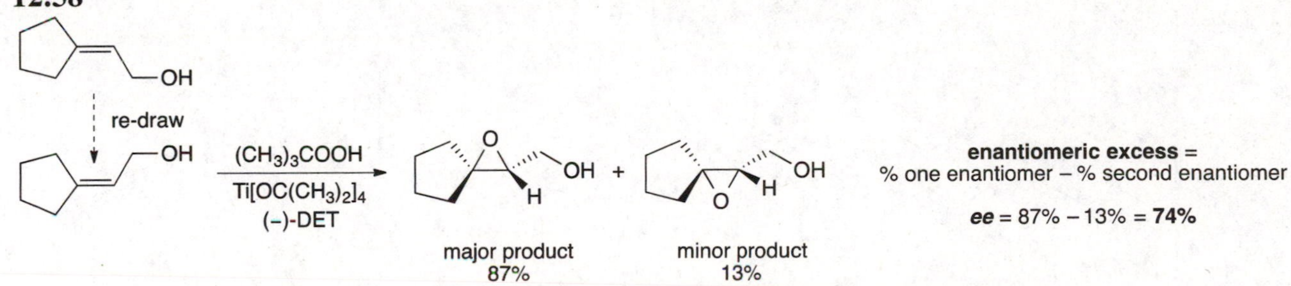

12.59

a. Replace this O to make an alkene.

(–)-DET

b. Replace this O to make an alkene.

(–)-DET

c. Replace this O to make an alkene.

(+)-DET

12.60

OH $\xrightarrow[\text{(–)-DET}]{\text{Sharpless reagent}}$ **A**

Na^+ :ÖH⁻ → H

Rotate around C–C bond.

OH

OH

B

B

$H{-}OCH_3$

12.61

$HC{\equiv}CH \xrightarrow{\text{NaH}} HC{\equiv}C^-$ → (Cl)

NaH

Cl

H_2, Lindlar catalyst

12.62

cis $\xrightarrow{Br_2}$ (Br, Br) $\xrightarrow[\text{(2 equiv)}]{^-NH_2}$ → $\xrightarrow[NH_3]{\text{Na}}$ trans

12.63

a. $HC{\equiv}CH \xrightarrow{\text{NaH}} {}^-C{\equiv}CH \xrightarrow{CH_3Cl}$ → $\xrightarrow{\text{NaH}}$ → $\xrightarrow{CH_3Cl}$ → $\xrightarrow[NH_3]{\text{Na}}$

b. 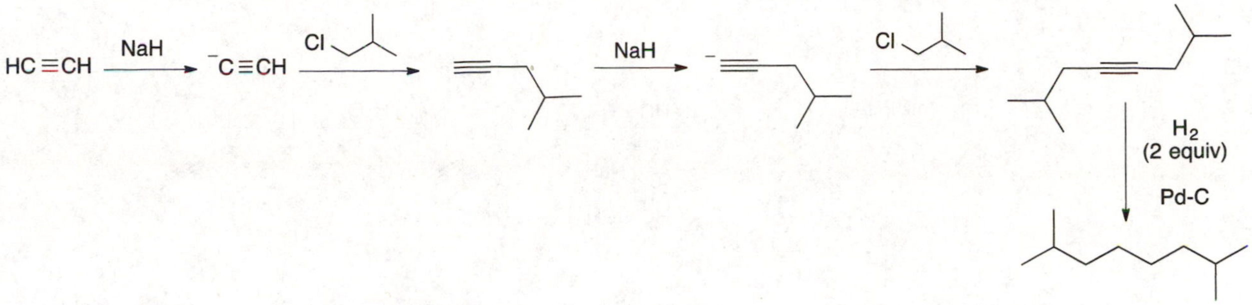 HC≡CH $\xrightarrow{NaH}$ $^{-}$C≡CH $\xrightarrow{Cl}$ $\xrightarrow{NaH}$ $\xrightarrow{Cl}$ $\xrightarrow[Pd\text{-}C]{H_2 \text{ (2 equiv)}}$

c. HC≡CH $\xrightarrow{NaH}$ HC≡C^{-} $\xrightarrow{CH_3Cl}$ $\xrightarrow{NaH}$ $\xrightarrow{CH_3Cl}$ $\xrightarrow{H_2 \text{ Lindlar catalyst}}$ $\xrightarrow{mCPBA}$ O, H, H

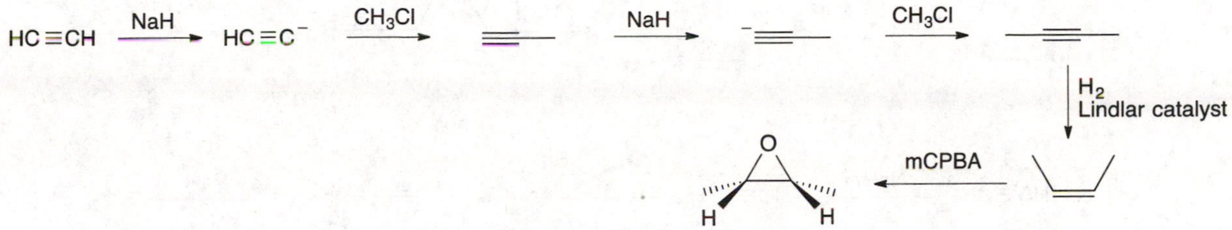

d. 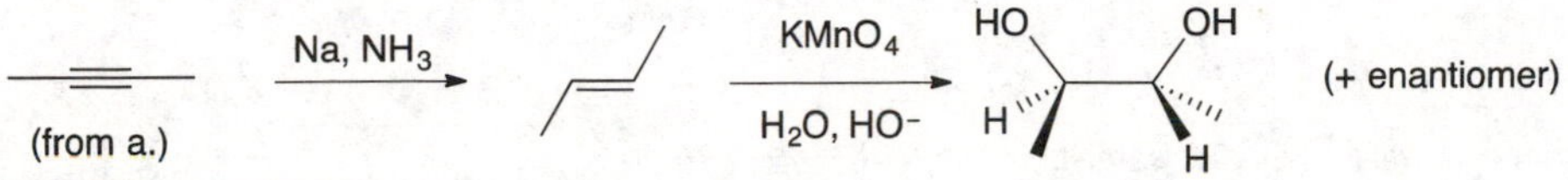 (from a.) $\xrightarrow{Na,\ NH_3}$ $\xrightarrow[H_2O,\ HO^-]{KMnO_4}$ HO, OH, H, H (+ enantiomer)

12.64

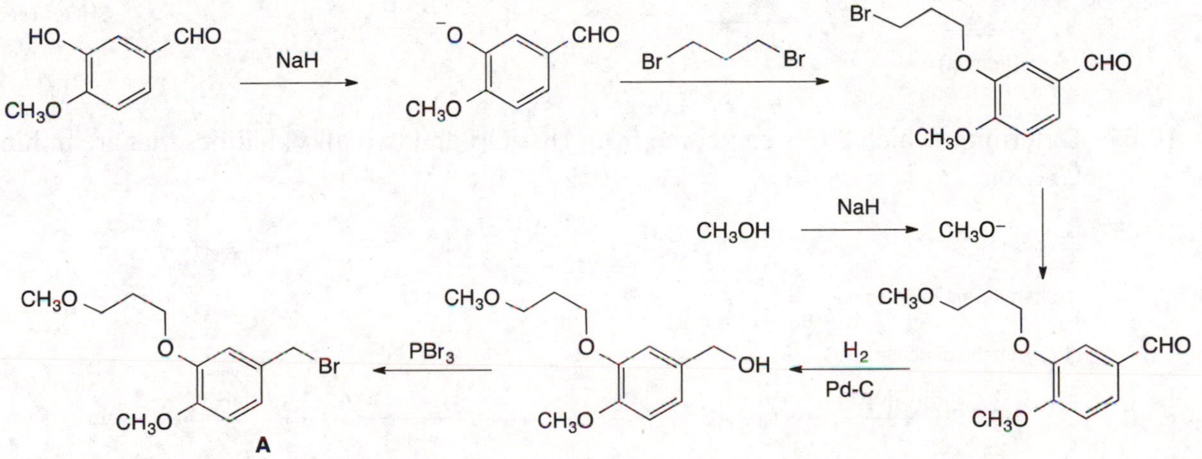

HO, CHO, CH_3O $\xrightarrow{NaH}$ $^{-}$O, CHO, CH_3O $\xrightarrow{Br\quad Br}$ Br, O, CHO, CH_3O

$CH_3OH \xrightarrow{NaH} CH_3O^-$

CH_3O, O, CHO, CH_3O $\xrightarrow[Pd\text{-}C]{H_2}$ CH_3O, O, OH, CH_3O $\xrightarrow{PBr_3}$ CH_3O, O, Br, CH_3O **A**

12.65

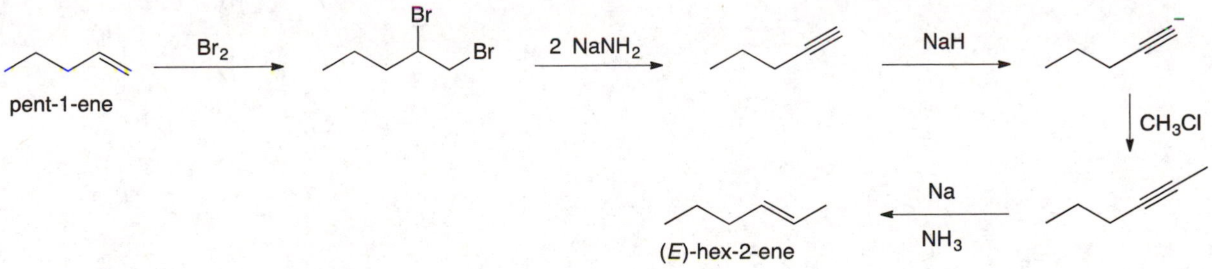

pent-1-ene $\xrightarrow{Br_2}$ Br, Br $\xrightarrow{2\ NaNH_2}$ $\xrightarrow{NaH}$ $\xrightarrow{CH_3Cl}$ $\xrightarrow[NH_3]{Na}$ (*E*)-hex-2-ene

12.66

a. mCPBA → HC≡C⁻ (NaH, HC≡CH) → H_2O → PCC →

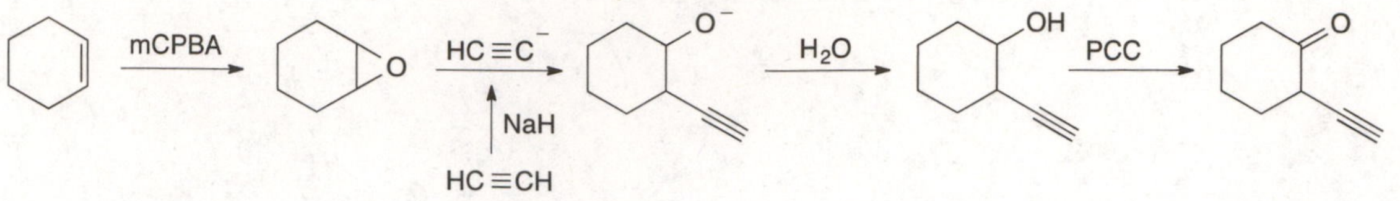

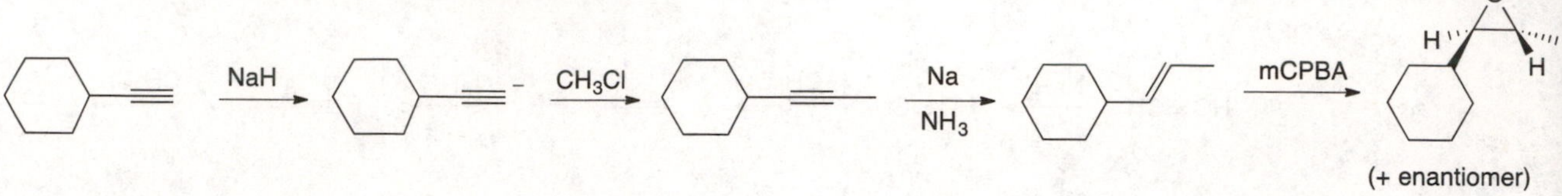

b. NaH → CH_3Cl → Na, NH_3 → mCPBA →

(+ enantiomer)

c. HC≡CH NaH → HC≡C⁻ Br → NaH → (epoxide) → H_2O → HO H_2, Lindlar catalyst → OH

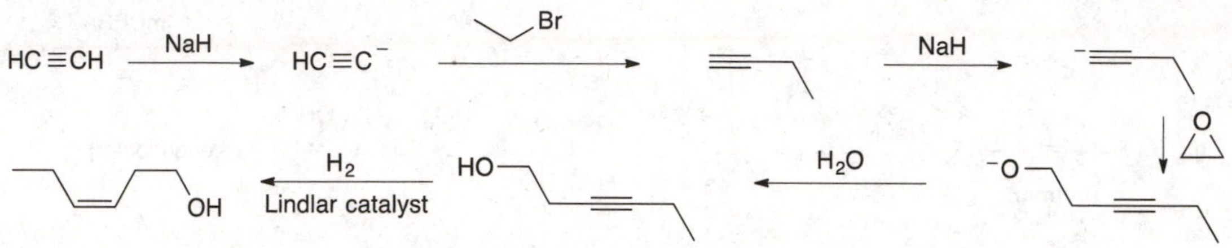

d. 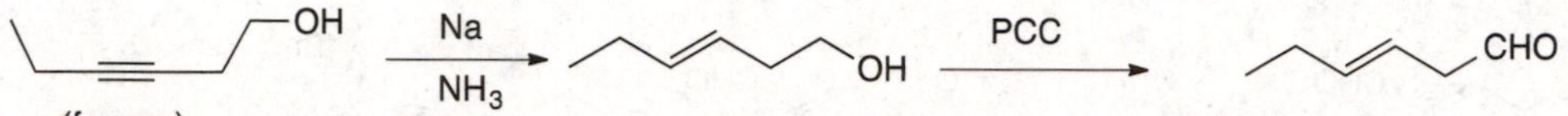 (from c.) OH Na, NH_3 → OH PCC → CHO

12.67 Determine which 2 C's can come from HC≡CH and two alkyl halides that are unhindered (1° or CH_3X).

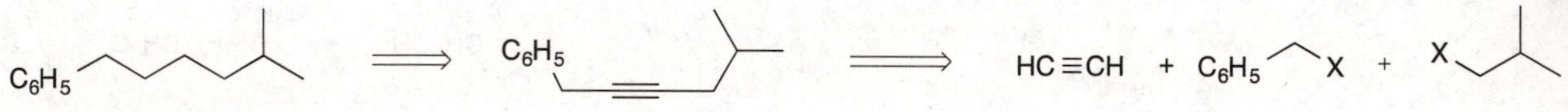

C_6H_5

1-phenyl-5-methylhexane ⟹ C_6H_5 ⟹ HC≡CH + C_6H_5 X + X

In synthetic direction:

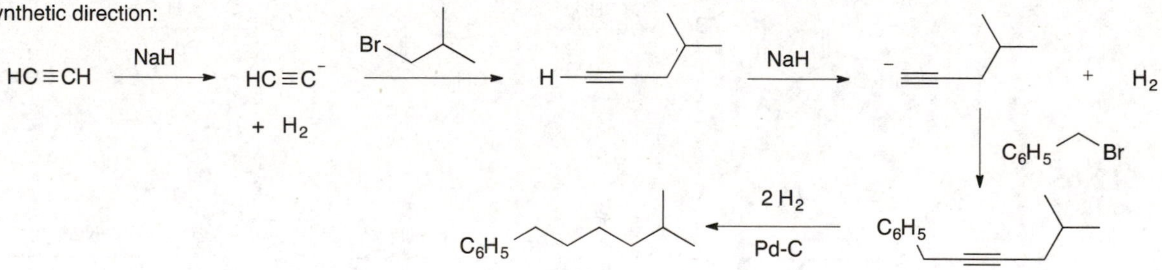

HC≡CH NaH → HC≡C⁻ + H_2 Br → H— NaH → ⁻ + H_2 → C_6H_5 Br → C_6H_5 → 2 H_2, Pd-C → C_6H_5

12.68

$HC\equiv CH \xrightarrow{NaH} {}^-C\equiv CH \xrightarrow{CH_3CH_2Br}$ 1-butyne $\xrightarrow{NaH}$ butynide $\xrightarrow{CH_3CH_2Br}$ 3-hexyne $\xrightarrow{Na,\ NH_3}$ trans-3-hexene $\xrightarrow{Cl_2,\ \text{anti addition}}$

(3*R*,4*S*)-3,4-dichlorohexane

12.69

a. $CH_3CH_2OH \xrightarrow{PBr_3} CH_3CH_2Br \xrightarrow{K^+\ {}^-OC(CH_3)_3} CH_2{=}CH_2 \xrightarrow{Br_2} BrCH_2CH_2Br \xrightarrow{2\ NaNH_2} HC\equiv CH$

$HC\equiv CH \xrightarrow{NaH} {}^-C\equiv CH \xrightarrow{CH_3CH_2Br}$ 1-butyne $\xrightarrow{NaH}$ butynide $\xrightarrow{CH_3CH_2Br}$ 3-hexyne $\xrightarrow{Na,\ NH_3}$ trans-3-hexene

b. 3-hexyne (from a.) $\xrightarrow[\text{Lindlar catalyst}]{H_2}$ cis-3-hexene $\xrightarrow{mCPBA}$ cis-epoxide

c. cis-3-hexene (from b.) $\xrightarrow[\text{[2] } NaHSO_3,\ H_2O]{\text{[1] } OsO_4}$ diol (HO, OH) $\xrightarrow{\text{rotate}}$ diol (OH, OH)

12.70

Anisole + Li· → radical anion + Li^+ ⟷ resonance form → (H–$\ddot{O}CH_2CH_3$) → H-added radical + $^-:\ddot{O}CH_2CH_3$

Radical + Li· → anion + Li^+ → (H–$\ddot{O}CH_2CH_3$) → $^-:\ddot{O}CH_2CH_3$ + 1-methoxy-1,4-cyclohexadiene ($-OCH_3$)

12.71

The favored conformation for both molecules places the *tert*-butyl group equatorial.

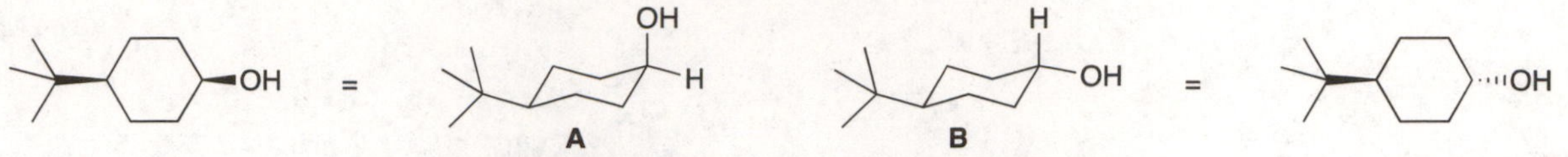

A

This OH is axial and will react **faster** because the OH group is more hindered.

B

This OH is equatorial and will react **more slowly** because the OH group is less hindered.

12.72

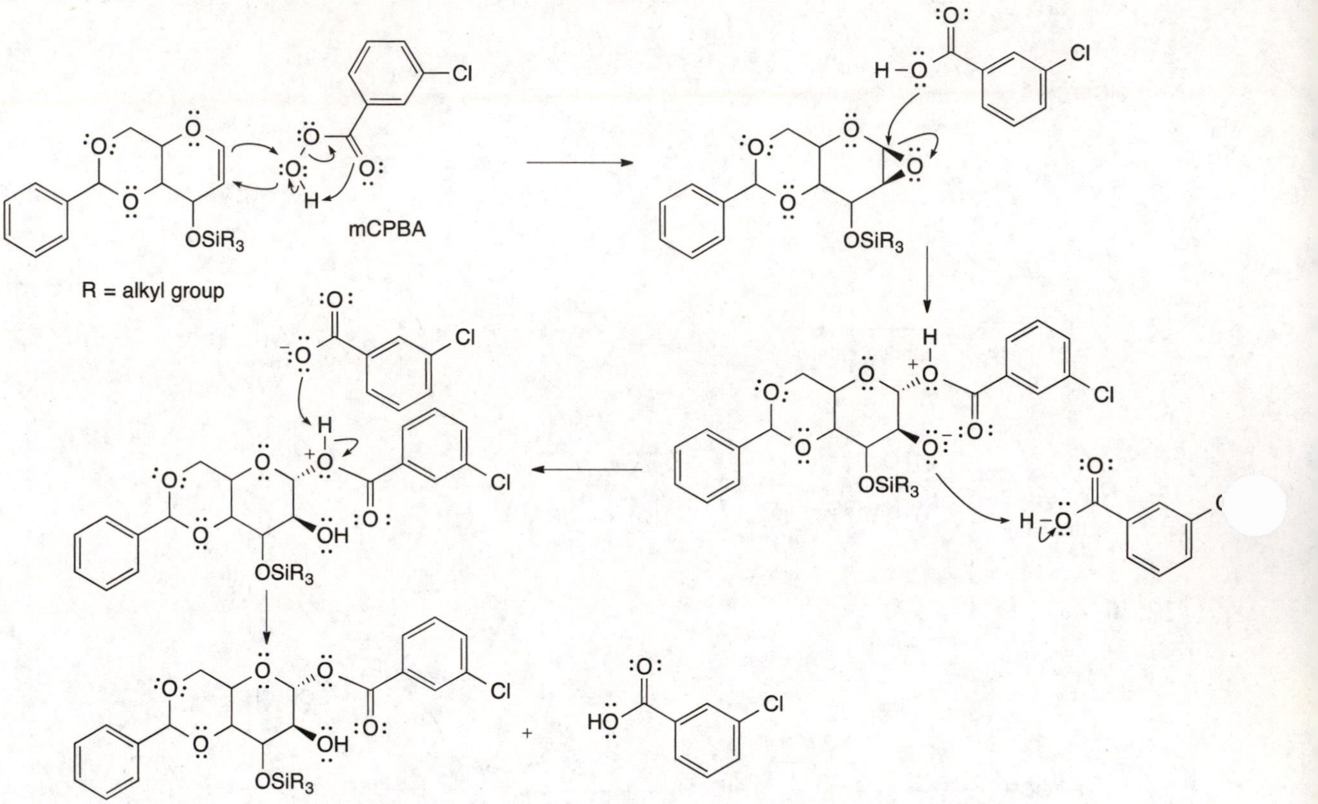

mCPBA

R = alkyl group

12.73 The two OH's are added to opposite faces of the C=C, so anti addition occurs.

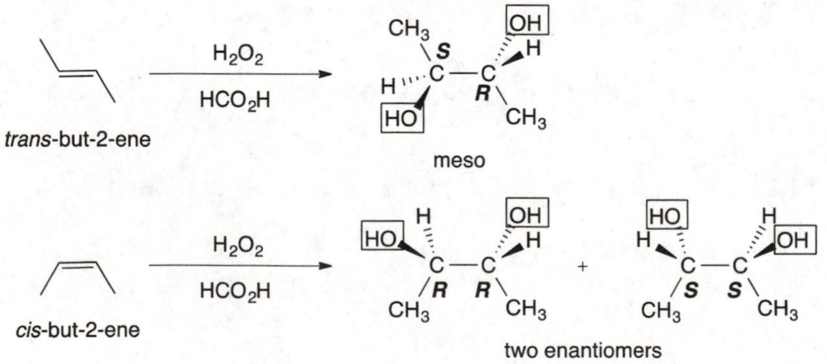

trans-but-2-ene $\xrightarrow[HCO_2H]{H_2O_2}$ meso

cis-but-2-ene $\xrightarrow[HCO_2H]{H_2O_2}$ two enantiomers

12.74

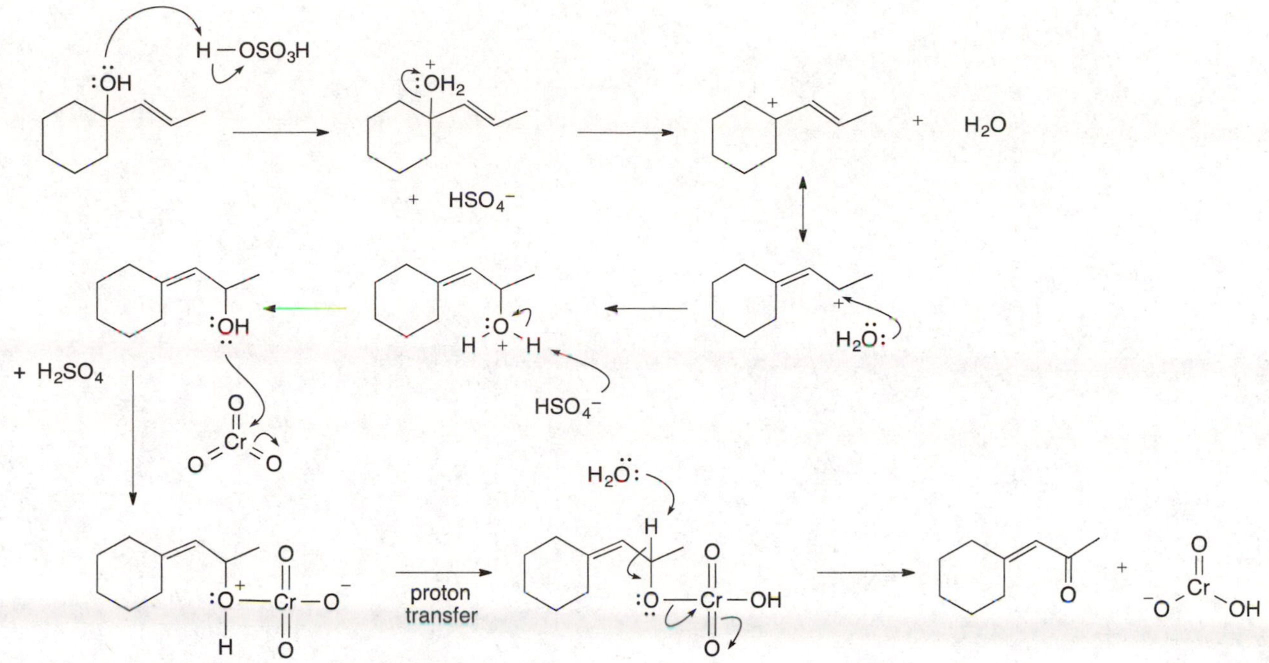

H–OSO$_3$H

:ÖH

:ÖH$_2$ (+)

\+ HSO$_4^-$

\+ H$_2$O

H$_2$Ö:

H + H

HSO$_4^-$

:ÖH

\+ H$_2$SO$_4$

H$_2$Ö:

H

proton transfer

:O–Cr–O$^-$

:O–Cr–OH

\+ $^-$O–Cr–OH

12.75

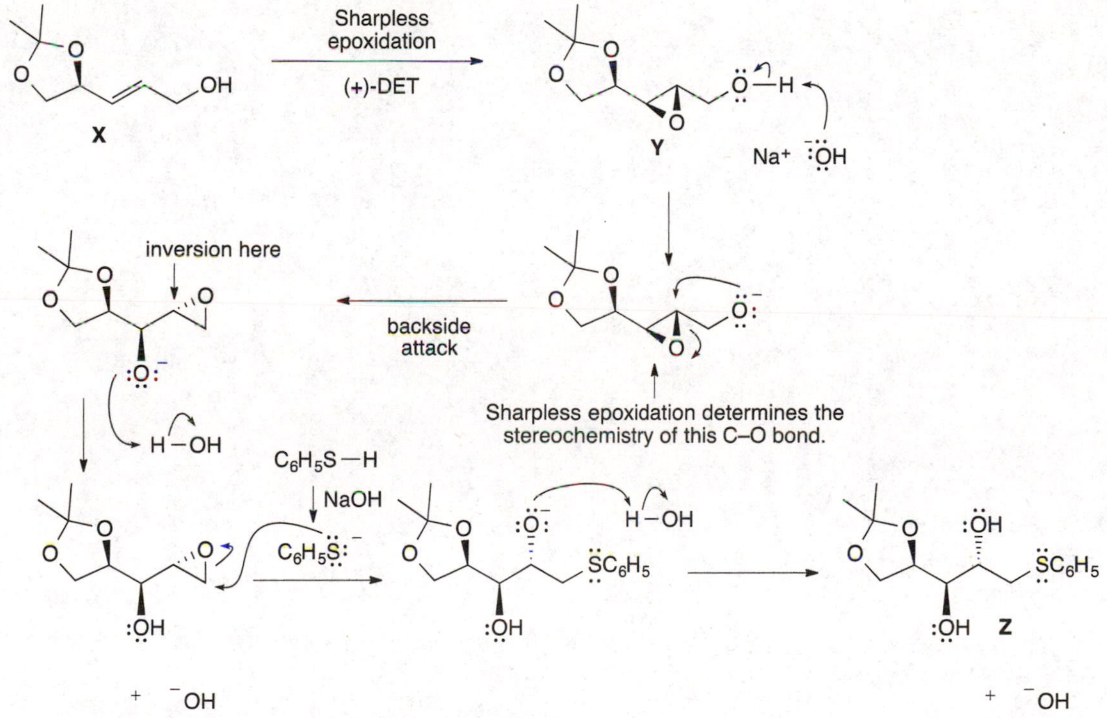

Sharpless epoxidation

(+)-DET

X

Y

Na$^+$ $^-$:ÖH

inversion here

backside attack

Sharpless epoxidation determines the stereochemistry of this C–O bond.

H–OH

C$_6$H$_5$S–H

NaOH

C$_6$H$_5$S:$^-$

:ÖH

SC$_6$H$_5$

H–OH

Z

\+ $^-$OH

\+ $^-$OH

Chapter 13 Mass Spectrometry and Infrared Spectroscopy

Chapter Review

Mass spectrometry (MS)

- Mass spectrometry measures the molecular weight of a compound (13.1A).
- The mass of the molecular ion (**M**) = the molecular weight of a compound. Except for isotope peaks at M + 1 and M + 2, the molecular ion has the highest mass in a mass spectrum (13.1A).
- The base peak is the tallest peak in a mass spectrum (13.1A).
- A compound with an odd number of N atoms gives an odd molecular ion. A compound with an even number of N atoms (including zero) gives an even molecular ion (13.1B).
- Organic chlorides show two peaks for the molecular ion (M and M + 2) in a 3:1 ratio (13.2).
- Organic bromides show two peaks for the molecular ion (M and M + 2) in a 1:1 ratio (13.2).
- The fragmentation of radical cations formed in a mass spectrometer gives lower molecular weight fragments, often characteristic of a functional group (13.3).
- High-resolution mass spectrometry gives the molecular formula of a compound (13.4A).

Electromagnetic radiation

- The wavelength (λ) and frequency (ν) of electromagnetic radiation are *inversely* related by the following equations, where c is the speed of light: $\boldsymbol{\lambda = c/\nu}$ or $\boldsymbol{\nu = c/\lambda}$ (13.5).
- The energy (E) of a photon is proportional to its frequency; the higher the frequency, the higher the energy (h = Planck's constant): $\boldsymbol{E = h\nu}$ (13.5).

Infrared spectroscopy (IR, 13.6 and 13.7)

- Infrared spectroscopy identifies functional groups.
- IR absorptions are reported in wavenumbers:

$$\textbf{wavenumber} = \tilde{\nu} = 1/\lambda$$

- The functional group region from **4000–1500 cm^{-1}** is the most useful region of an IR spectrum.
- C–H, O–H, and N–H bonds absorb at high frequency, $\geq$ 2500 cm^{-1}.
- As bond strength increases, the wavenumber of an absorption increases; thus triple bonds absorb at higher wavenumber than double bonds.

C=C ~1650 cm^{-1} → C≡C ~2250 cm^{-1}

Increasing bond strength
Increasing $\tilde{\nu}$

- The higher the percent *s*-character, the stronger the bond, and the higher the wavenumber of an IR absorption.

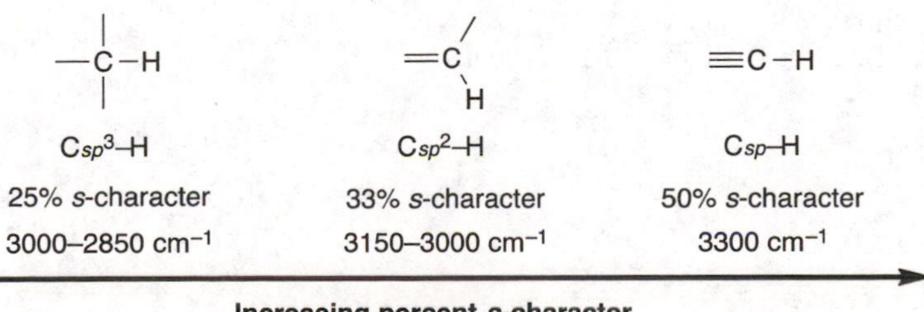

Practice Test on Chapter Review

1.a. Which compound has a molecular ion at 112 and a peak at 1720 cm^{-1} in its IR spectrum?

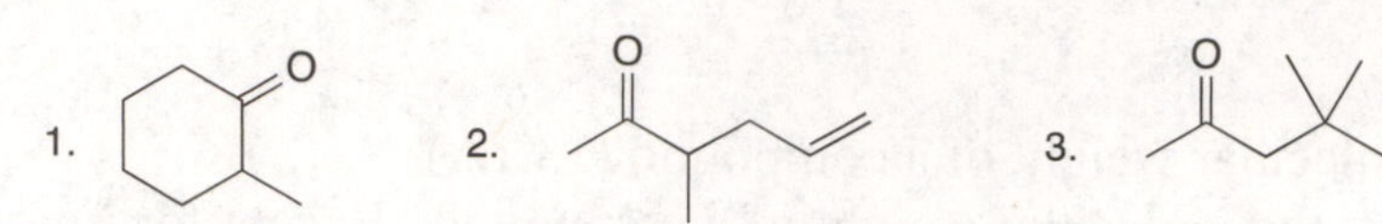

4. Compounds (1) and (2) both fit these criteria.
5. Compounds (1), (2), and (3) all fit these criteria.

b. Which of the following compounds has peaks at 3300, 3000, and 2250 cm^{-1} in its IR spectrum?

1. 2. N 3.

4. Compounds (1) and (2) have these peaks in their IR spectra.
5. Compounds (1), (2), and (3) all contain these peaks in their IR spectra.

c. What is the base peak in a mass spectrum?
1. the peak due to the radical cation formed when a molecule loses an electron
2. the tallest peak in the mass spectrum
3. the peak due to the fragment with the largest *m*/*z* ratio
4. Both (1) and (2) describe the base peak.
5. Statements (1), (2), and (3) all describe the base peak.

d. Which compounds are possible structures for a molecule that has a molecular ion at 150 in its mas spectrum?

1. O O 2. O 3. O H CH_3O OCH_3

4. Both (1) and (2) are possible structures.
5. Compounds (1), (2), and (3) are all possible structures.

e. Which compounds exhibit prominent M + 2 peaks in their mass spectra?

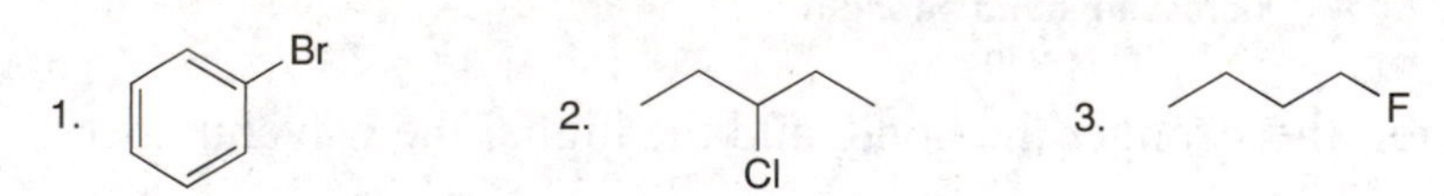

4. Both (1) and (2) show M + 2 peaks.
5. Compounds (1), (2), and (3) all show M + 2 peaks.

2. Answer each question with the number that corresponds to one of the following regions of an IR spectrum.

1. 4000–2500 cm^{-1}
2. 2500–2000 cm^{-1}
3. 2000–1500 cm^{-1}
4. < 1500 cm^{-1}

a. This region is called the fingerprint region of an IR spectrum.
b. The OH group of propan-1-ol absorbs in this region.
c. The C=N of $C_6H_5CH_2CH{=}NCH_3$ absorbs in this region.
d. An unsymmetrical C≡C absorbs in this region.
e. An *sp* hybridized C–H bond absorbs in this region.
f. Ethyl benzoate ($C_6H_5CO_2CH_2CH_3$) absorbs in all regions of the IR except this one.

3. Answer True (T) or False (F).

a. IR spectroscopy is useful for determining the molecular weight of a compound.
b. A C–H bond that absorbs at 3140 cm^{-1} is stronger than a C–H bond that absorbs at 2950 cm^{-1}.
c. A compound with a molecular ion at 109 contains a N atom.
d. A compound with a base peak at 57 must contain a N atom.
e. But-2-yne shows an IR absorption at 2250 cm^{-1}.
f. Propan-1-ol shows an IR absorption at 3200–3600 cm^{-1}.
g. An ether shows no IR absorptions at 3200–3600 or 1700 cm^{-1}.
h. In its mass spectrum, a compound that has a molecular ion with two peaks of approximately equal intensity at 124 and 126 contains chlorine.

Answers to Practice Test

1.a. 4
b. 1
c. 2
d. 4
e. 4

2.a. 4
b. 1
c. 3
d. 2
e. 1
f. 2

3.a. F
b. T
c. T
d. F
e. F
f. T
g. T
h. F

Answers to Problems

13.1 The molecular ion formed from each compound is equal to its molecular weight.

a. C_3H_6O
molecular weight = **58**
molecular ion (*m/z*) = **58**

b. $C_{10}H_{20}$
molecular weight = **140**
molecular ion (*m/z*) = **140**

c. $C_8H_8O_2$
molecular weight = **136**
molecular ion (*m/z*) = **136**

d. $C_{10}H_{15}N$
molecular weight = **149**
molecular ion (*m/z*) = **149**

13.2 Some possible formulas for each molecular ion:

a. Molecular ion at 72: C_5H_{12}, C_4H_8O, $C_3H_4O_2$
b. Molecular ion at 100: C_8H_4, C_7H_{16}, $C_6H_{12}O$, $C_5H_8O_2$
c. Molecular ion at 73: $C_4H_{11}N$, $C_2H_7N_3$

13.3 Use the molecular ion to propose molecular formulas that contain C, H, and O. Then determine which formula has five degrees of unsaturation.

M at 218: $C_{18}H_2$ (not enough H's)

$C_{17}H_{14} \xrightarrow[+\,O]{-\,CH_4} C_{16}H_{10}O$ (not enough H's)

$C_{16}H_{26} \xrightarrow[+\,O]{-\,CH_4} C_{15}H_{22}O$ (five degrees of unsaturation)
Answer

13.4 To calculate the molecular ions you would expect for compounds with Cl, calculate the molecular weight using each of the two most common isotopes of Cl (^{35}Cl and ^{37}Cl).

a. $C_4H_9{}^{35}Cl$ = **92**
$C_4H_9{}^{37}Cl$ = **94**
Two peaks in 3:1 ratio at *m/z* 92 and 94

b. C_3H_7F = **62**
One peak at *m/z* 62

c. $C_4H_{11}N$ = **73**
One peak at *m/z* 73

d. $C_4H_4N_2$ = **80**
One peak at *m/z* 80

13.5 Convert the ball-and-stick model to a skeletal structure and determine the molecular formula. Calculate the molecular weight using each of the two common isotopes for Br (^{79}Br and ^{81}Br).

Br

$C_6H_{11}Br$

$C_6H_{11}{}^{79}Br$ = **162**
$C_6H_{11}{}^{81}Br$ = **164**
Two peaks in a 1:1 ratio at *m/z* 162 and 164

13.6 After calculating the mass of the molecular ion, draw the structure and determine which C–C bond is broken to form fragments of the appropriate mass-to-charge ratio.

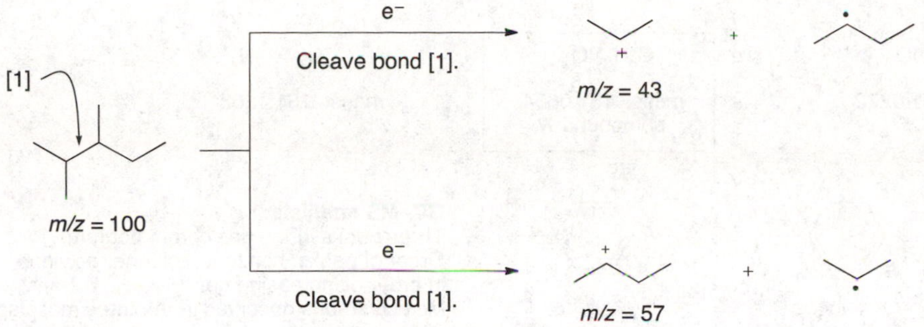

[1]

$m/z = 100$

e^- Cleave bond [1]. → $m/z = 43$ +

e^- Cleave bond [1]. → $m/z = 57$ +

13.7

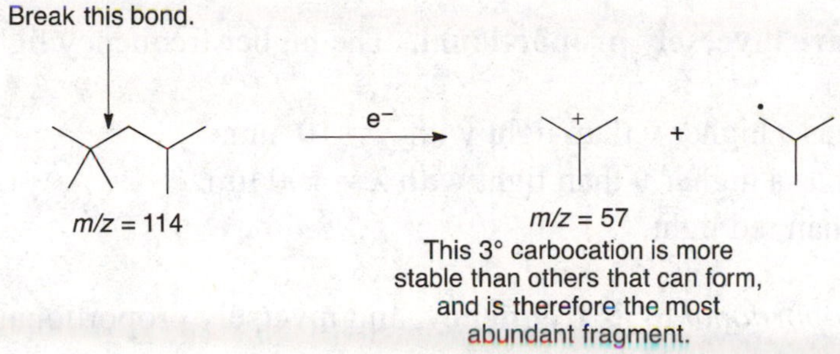

Break this bond.

$m/z = 114$ $\xrightarrow{e^-}$ $m/z = 57$ +

This 3° carbocation is more stable than others that can form, and is therefore the most abundant fragment.

13.8

a.

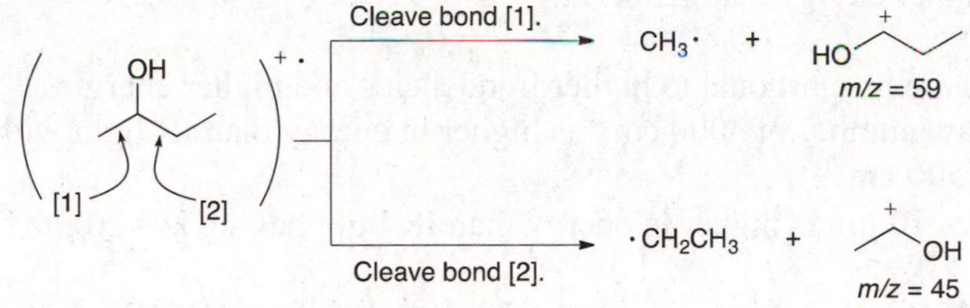

OH [1] [2]

Cleave bond [1]. → $CH_3\cdot$ + HO (+) $m/z = 59$

Cleave bond [2]. → $\cdot CH_2CH_3$ + (+) OH $m/z = 45$

b. OH $\xrightarrow{-H_2O}$ $m/z = 56$ + $m/z = 56$

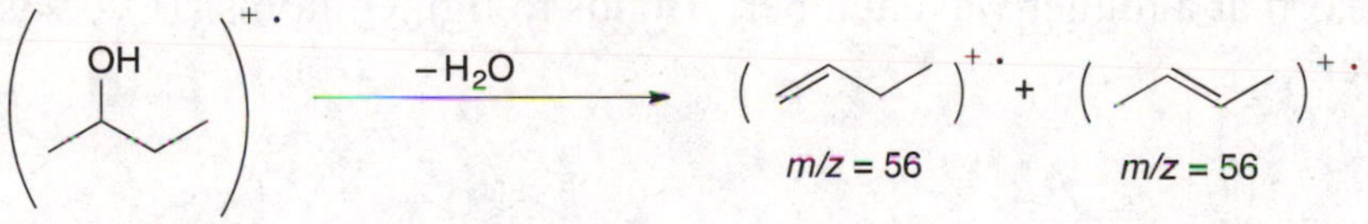

13.9

a. O [1] [2] → (from cleavage of bond [2]) + O= (from cleavage of bond [1])

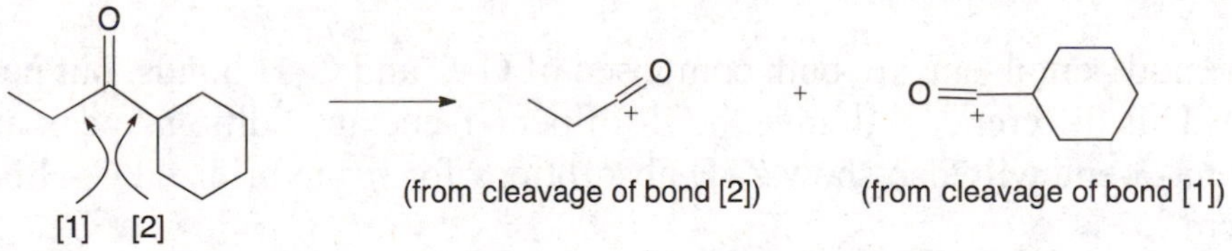

b. OH → $^+CH_2OH$

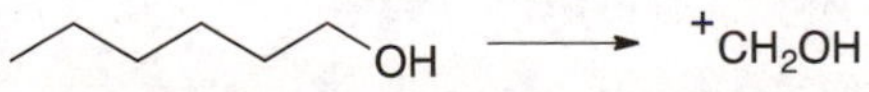

c. O H → + $H\overset{+}{C}=O$

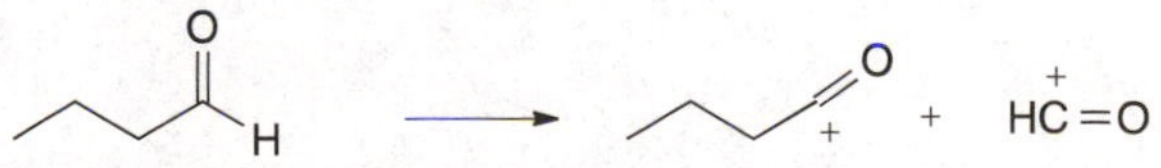

13.10 Use the exact mass values given in Table 13.1 to calculate the exact mass of each compound.

13.11

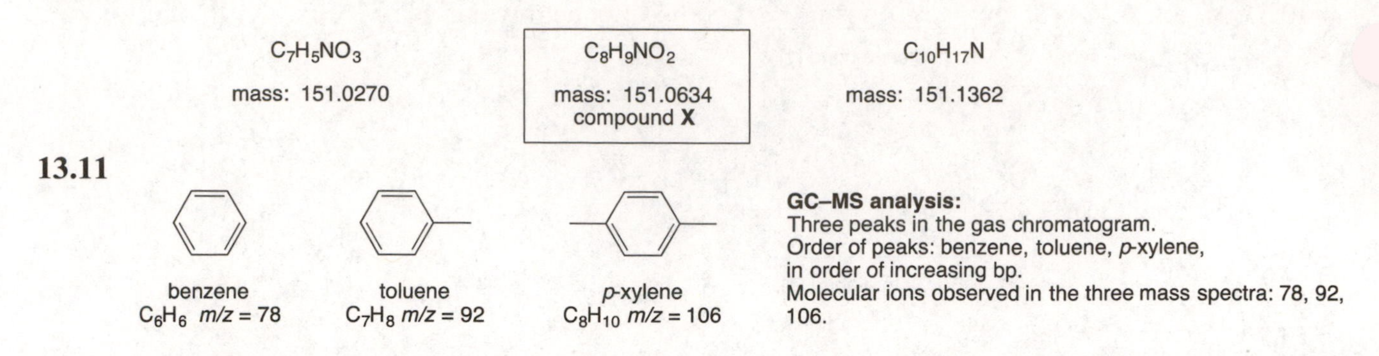

13.12 **Wavelength and frequency are inversely proportional.** The higher frequency light will have a shorter wavelength.

a. Light having $\lambda = 10^2$ nm has a higher ν than light with $\lambda = 10^4$ nm.
b. Light having $\lambda = 100$ nm has a higher ν than light with $\lambda = 100$ µm.
c. Blue light has a higher ν than red light.

13.13 The **energy of a photon** is *proportional* to its **frequency,** and inversely proportional to its wavelength.

a. Light having $\nu = 10^8$ Hz is of higher energy than light having $\nu = 10^4$ Hz.
b. Light having $\lambda = 10$ nm is of higher energy than light having $\lambda = 1000$ nm.
c. Blue light is of higher energy than red light.

13.14 Higher wavenumbers are proportional to higher frequencies and higher energies.

a. IR light with a wavenumber of 3000 cm^{-1} is higher in energy than IR light with a wavenumber of 1500 cm^{-1}.
b. IR light having $\lambda = 10$ µm is higher in energy than IR light having $\lambda = 20$ µm.

13.15 Stronger bonds absorb at a higher wavenumber. Bonds to lighter atoms (H versus D) absorb at higher wavenumber.

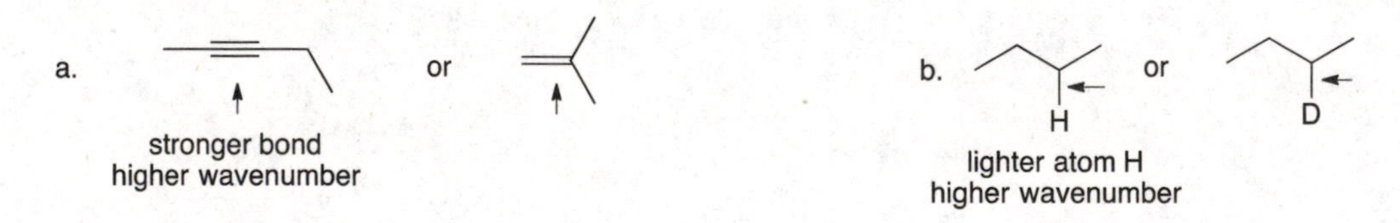

13.16 Cyclopentane and pent-1-ene are both composed of C–C and C–H bonds, but pent-1-ene also has a C=C bond. This difference will give the IR of pent-1-ene an additional peak at 1650 cm^{-1} (for the C=C). Pent-1-ene will also show C–H absorptions for sp^2 hybridized C–H bonds at 3150–3000 cm^{-1}.

13.17 Look at the functional groups in each compound below to explain how each IR is different.

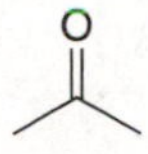

A: C=O peak at ~1700 cm^{-1}

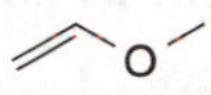

B: C=C peak at 1650 cm^{-1}; C_{sp^2}–H at 3150–3000 cm^{-1}

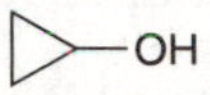

C: O–H peak at 3200–3600 cm^{-1}

13.18 a. Compound **A** has peaks at ~3150 (*sp*^{2} hybridized C–H), 3000–2850 (*sp*^{3} hybridized C–H), and 1650 (C=C) cm^{-1}.

b. Compound **B** has a peak at 3000–2850 (*sp*^{3} hybridized C–H) cm^{-1}.

13.19 All compounds show an absorption at 3000–2850 cm^{-1} due to the *sp*^{3} hybridized C–H bonds. Additional peaks in the functional group region for each compound are shown.

a. no additional peaks

b. O–H bond at 3600–3200 cm^{-1}

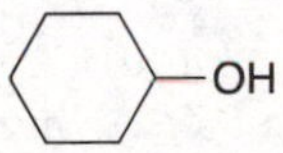

c. C_{sp^2}–H at 3150–3000 cm^{-1}; C=C bond at 1650 cm^{-1}

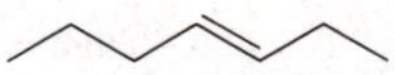

d. C=O bond at ~1700 cm^{-1}

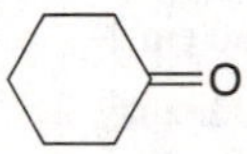

e. (CH_3O, HO, N–H, O)
O–H at 3600–3200 cm^{-1}
N–H at 3500–3200 cm^{-1}
C_{sp^2}–H at 3150–3000 cm^{-1}
C=O at ~1700 cm^{-1}
C=C at 1650 cm^{-1}
aromatic ring at 1600, 1500 cm^{-1}

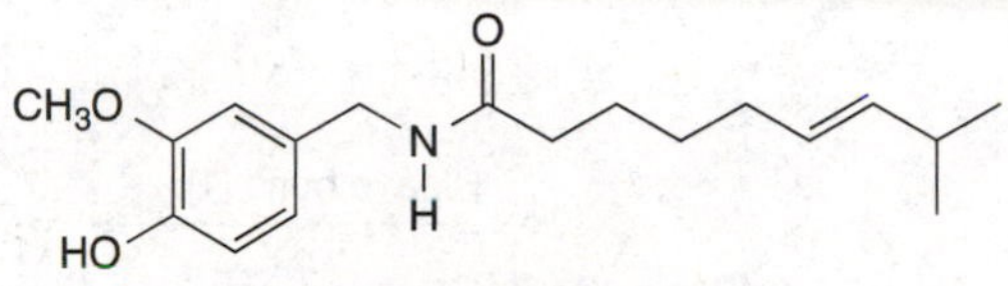

13.20

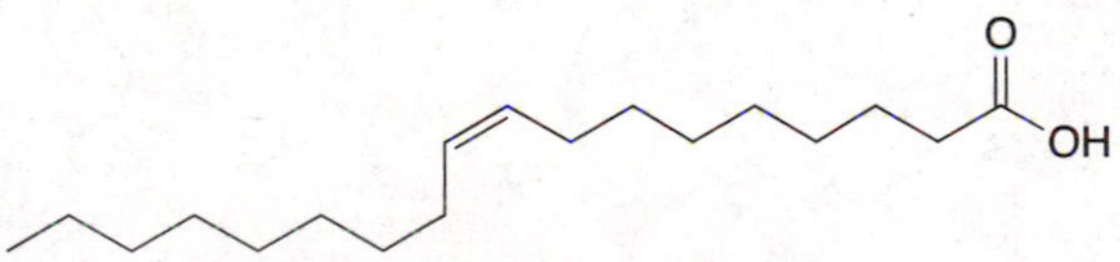

C_{sp^2}–H at 3150–3000 cm^{-1}
C_{sp^3}–H at 3000–2850 cm^{-1}
C=O at ~1700 cm^{-1}
C=C at 1650 cm^{-1}
O–H above 3000 cm^{-1}

[The OH of a COOH is much broader than the OH of an alcohol and occurs at 3500–2500 cm^{-1} (see Chapter 19).]

13.21 Possible structures are (a) $CH_3COOCH_2CH_3$ and (c) $CH_3CH_2COOCH_3$. Compounds (b) and (d) also have an OH group that would give a strong absorption at ~3600–3200 cm^{-1}, which is absent in the IR spectrum of **X,** thus excluding them as possibilities.

13.22

a. Hydrocarbon with a molecular ion at $m/z = 68$
IR absorptions at 3310 cm^{-1} = Csp–H bond
3000–2850 cm^{-1} = Csp^3–H bonds
2120 cm^{-1} = C≡C bond
Molecular formula: C_5H_8

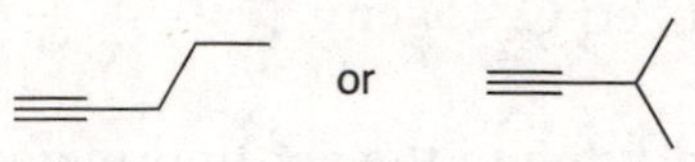

or

b. Compound with C, H, and O with a molecular ion at $m/z = 60$
IR absorptions at 3600–3200 cm^{-1} = O–H bond
3000–2850 cm^{-1} = Csp^3–H bonds
Molecular formula: C_3H_8O

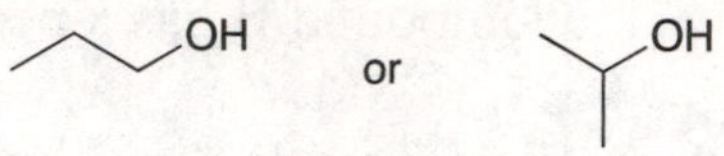

OH or OH

13.23

a. 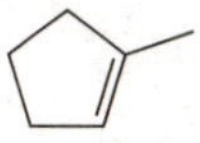
Csp^2–H at 3000–3150 cm^{-1}
Csp^3–H at 2850–3000 cm^{-1}
C=C at 1650 cm^{-1}

b. 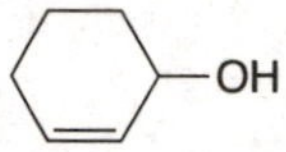 OH
O–H at 3200–3600 cm^{-1}
Csp^2–H at 3000–3150 cm^{-1}
Csp^3–H at 2850–3000 cm^{-1}
C=C at 1650 cm^{-1}

13.24

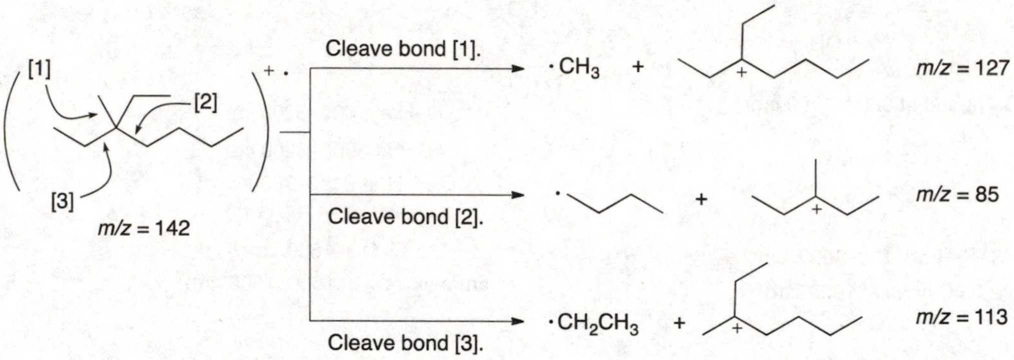

[1] [2] [3]
$m/z = 142$

Cleave bond [1]. → $\cdot CH_3$ + ... $m/z = 127$

Cleave bond [2]. → ... $m/z = 85$

Cleave bond [3]. → $\cdot CH_2CH_3$ + ... $m/z = 113$

13.25

a. 
molecular formula: C_6H_6
molecular ion (m/z): **78**

b. 
molecular formula: $C_{10}H_{16}$
molecular ion (m/z): **136**

c. 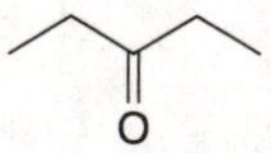 O
molecular formula: $C_5H_{10}O$
molecular ion (m/z): **86**

d. 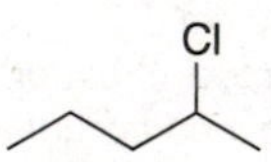 Cl
molecular formula: $C_5H_{11}Cl$
molecular ions (m/z): **106, 108**

e. Br
molecular formula: $C_8H_{17}Br$
molecular ions (m/z): **192, 194**

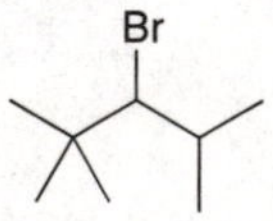

13.26

C_9H_{12}
molecular weight = 120

$C_9H_{10}O$
molecular weight = 134

$C_8H_{10}O$
molecular weight = 122

13.27 Examples are given for each molecular ion.

a. molecular ion 102: C_8H_6, $C_6H_{14}O$, $C_5H_{10}O_2$, $C_5H_{14}N_2$
b. molecular ion 98: C_8H_2, C_7H_{14}, $C_6H_{10}O$, $C_5H_6O_2$
c. molecular ion 119: C_8H_9N, $C_6H_5N_3$
d. molecular ion 74: C_6H_2, $C_4H_{10}O$, $C_3H_6O_2$

13.28 Likely molecular formula, C_8H_{16} (one degree of unsaturation—one ring or one π bond).

Four structures with *m/z* = 112

13.29 Use the molecular ion to propose molecular formulas for a compound with only C and H. Then determine which formula has four degrees of unsaturation.

M at 204: C_{17} (no H's)
$C_{16}H_{12}$ (not enough H's)
$C_{15}H_{24}$ (four degress of unsaturation)
Answer

13.30 Use the directions from Answer 13.3 and propose a formula with two degrees of unsaturation.

M at 154: $C_{12}H_{10}$ (not enough H's)

$$C_{11}H_{22} \xrightarrow[+\,O]{-\,CH_4} C_{10}H_{18}O$$ (two degrees of unsaturation)
Answer

13.31

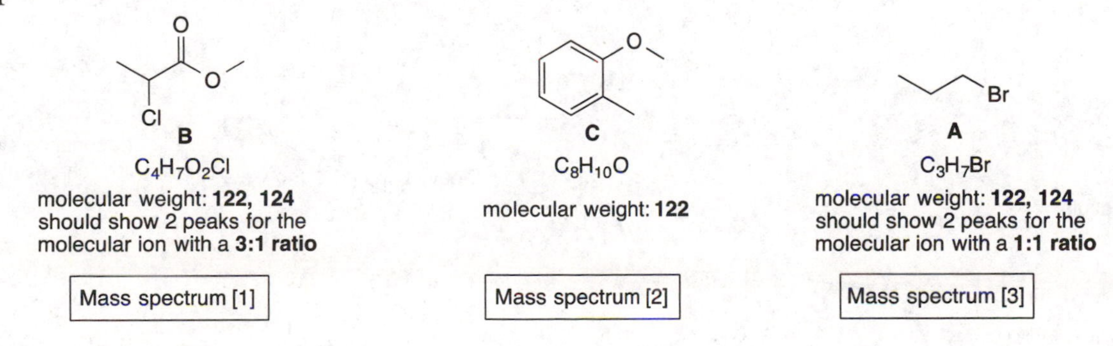

13.32

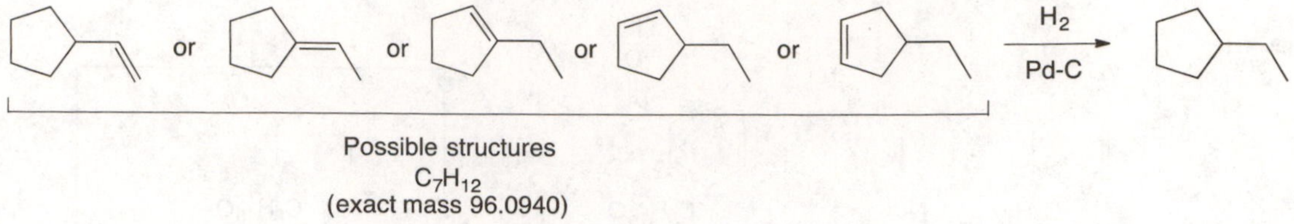

or or or or $\xrightarrow[\text{Pd-C}]{H_2}$

Possible structures
C_7H_{12}
(exact mass 96.0940)

13.33

a. [1] [2] OH → OH + OH

(from cleavage of bond [1]) (from cleavage of bond [2])

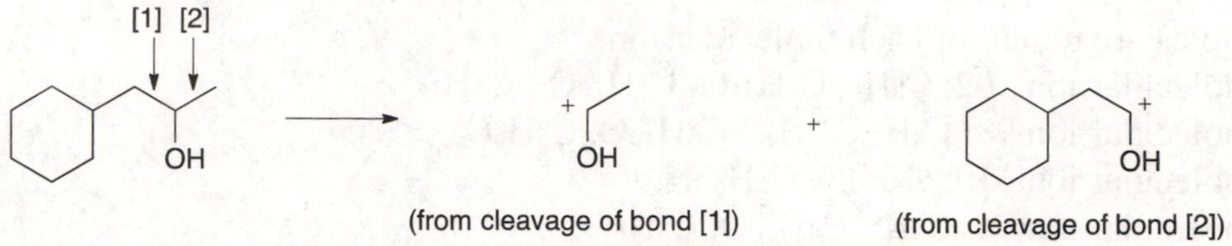

b. [1] O [2] → O + + O

(from cleavage of bond [1]) (from cleavage of bond [2])

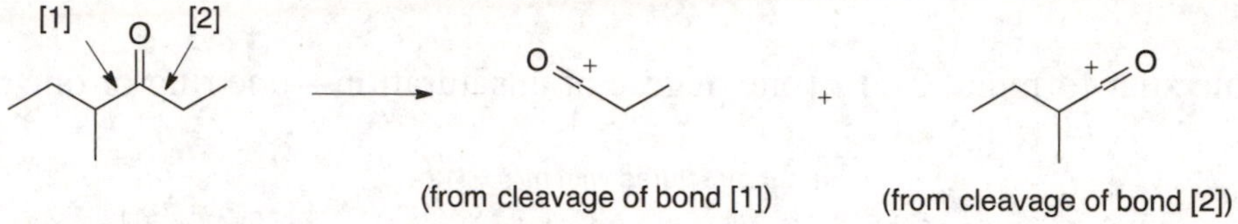

c. [1] [2] O → O + O

(from cleavage of bond [1]) (from cleavage of bond [2])

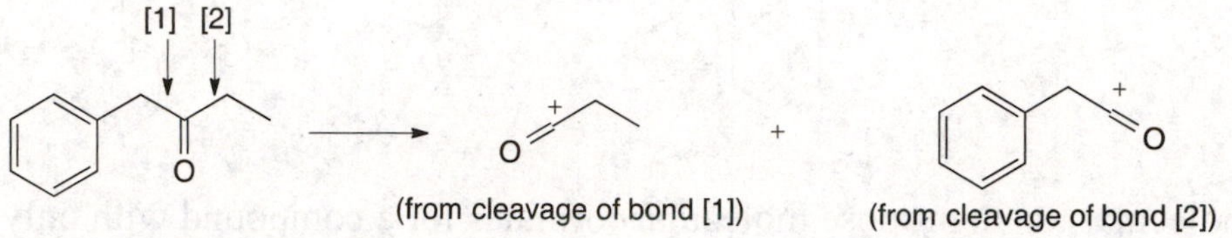

13.34 2,2-Dimethylbutane shows a peak at $m/z = 57$ in it s mass spectrum due to a 3° carbocation. 2,3-Dimethylbutane shows a peak at $m/z = 43$ due to a 2° carbocation.

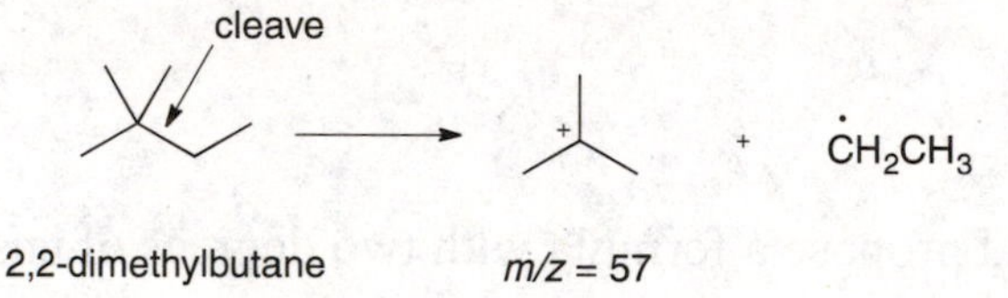

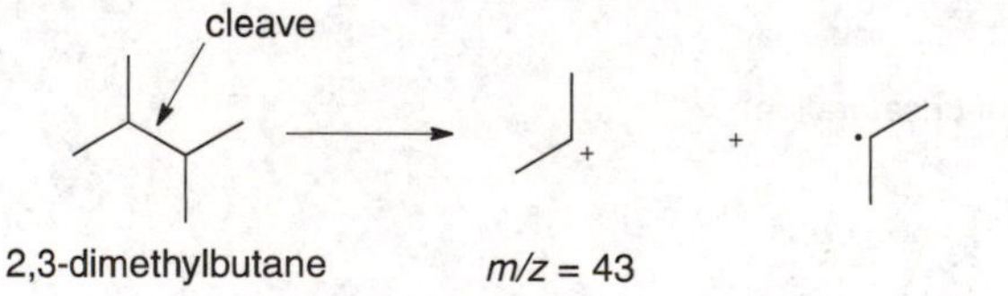

cleave

2,2-dimethylbutane → $m/z = 57$ + $\dot{C}H_2CH_3$

cleave

2,3-dimethylbutane → $m/z = 43$ +

13.35

a. $m/z = 104$ $\xleftarrow{-H_2O}$ [1] OH $m/z = 122$ $\xrightarrow[\text{Cleave bond [1].}]{e^-}$ $m/z = 91$ (resonance-stabilized carbocation)

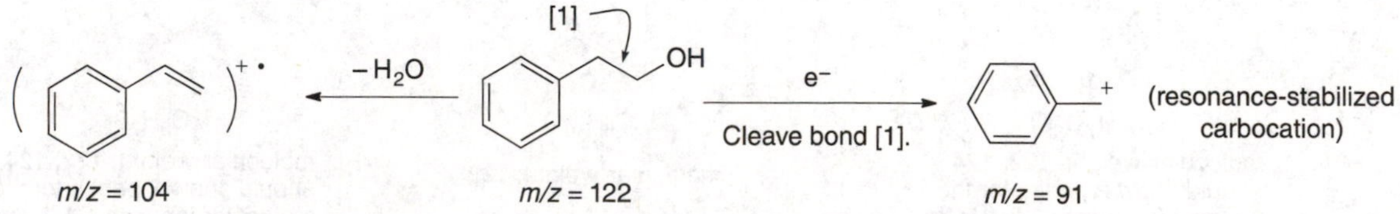

b.

$m/z = 68$ ← $-H_2O$ — $m/z = 86$ [1] [2] [3]

Cleave bond [1]. → $m/z = 71$

Cleave bond [2]. → $m/z = 41$

Cleave bond [3]. → $^+CH_2-OH$ $m/z = 31$

13.36

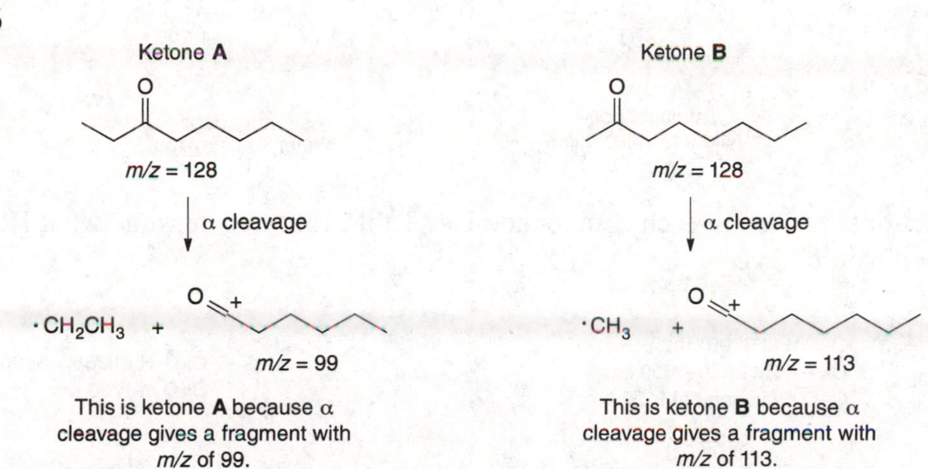

Ketone A
O
m/z = 128
α cleavage
·CH2CH3 +
O+
m/z = 99
This is ketone A because α cleavage gives a fragment with m/z of 99.
Ketone B
O
m/z = 128
α cleavage
·CH3 +
O+
m/z = 113
This is ketone B because α cleavage gives a fragment with m/z of 113.

13.37 One possible structure is drawn for each set of data:

a. a compound that contains a benzene ring and has a molecular ion at $m/z = 107$

NH_2

C_7H_9N

b. a hydrocarbon that contains only sp^3 hybridized carbons and a molecular ion at $m/z = 84$

C_6H_{12}

c. a compound that contains a carbonyl group and gives a molecular ion at $m/z = 114$

O

$C_7H_{14}O$

d. a compound that contains C, H, N, and O and has an exact mass for the molecular ion at 101.0841

O N H

$C_5H_{11}NO$

13.38 Use the values given in Table 13.1 to calculate the exact mass of each compound. $C_8H_{11}NO_2$ (exact mass 153.0790) is the correct molecular formula.

13.39 Alpha cleavage of a 1° alcohol (RCH_2OH) forms an alkyl radical (R•) and a resonance-stabilized carbocation with $m/z = 31$.

$^+CH_2OH \longleftrightarrow CH_2{=}\overset{+}{O}H$ $m/z = 31$

resonance-stabilized carbocation

13.40 An ether fragments by α cleavage because the resulting carbocation is resonance stabilized.

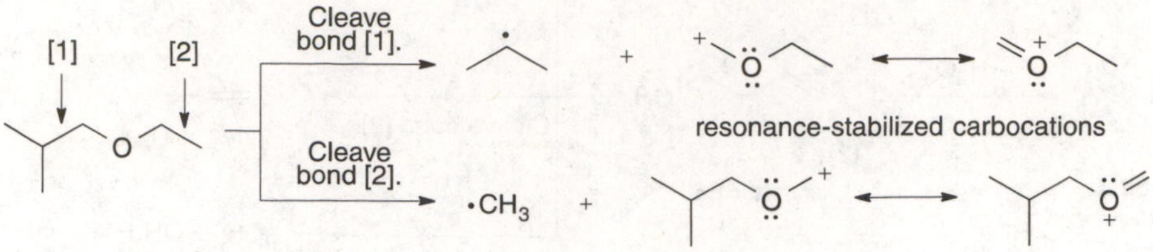

[1] [2]

Cleave bond [1].

Cleave bond [2].

$\cdot CH_3$ +

resonance-stabilized carbocations

13.41

a. 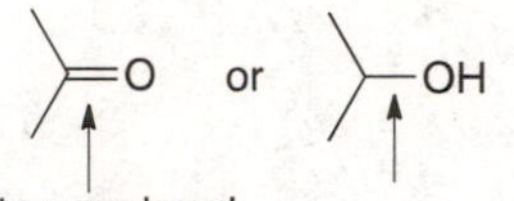 or

stronger bond
higher $\tilde{\nu}$ absorption

b. 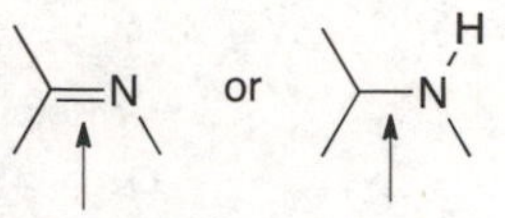 or

stronger bond
higher $\tilde{\nu}$ absorption

c. 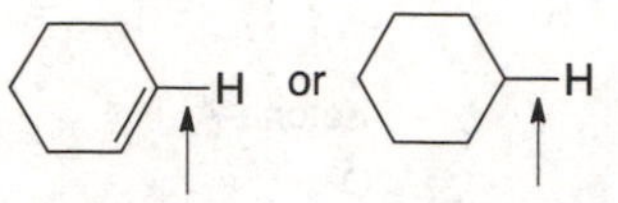 or

stronger bond
higher $\tilde{\nu}$ absorption

13.42 Locate the functional groups in each compound. Use Table 13.2 to determine what IR absorptions each would have.

a. 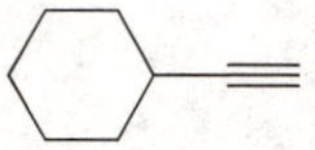
C_{sp}–H at 3300 cm^{-1}
C_{sp^3}–H at 2850–3000 cm^{-1}
C–C triple bond at 2250 cm^{-1}

b. 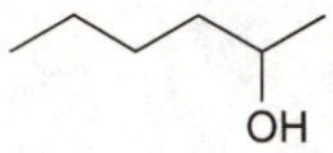
O–H at 3200–3600 cm^{-1}
C_{sp^3}–H at 2850–3000 cm^{-1}

c. 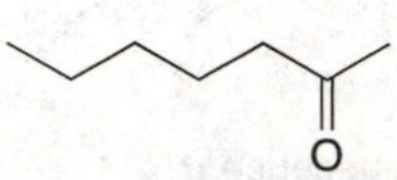
C_{sp^3}–H at 2850–3000 cm^{-1}
C=O at 1700 cm^{-1}

d. 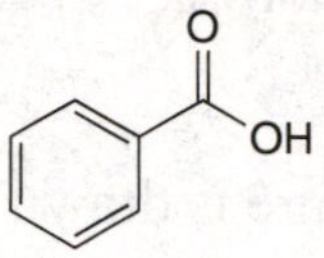
O–H at > 3000 cm^{-1}
C_{sp^2}–H at 3000–3150 cm^{-1}
C=O at ~1700 cm^{-1}
phenyl group at 1600, 1500 cm^{-1}

The OH of the RCOOH is even broader than the OH of an alcohol (3500–2500 cm^{-1}), as we will learn in Chapter 19.

13.43

a.  and

C=C bond
1650 cm^{-1}

C_{sp^2}–H at 3150–3000 cm^{-1}

C≡C bond
2250 cm^{-1}

C_{sp}–H at 3300 cm^{-1}

b. 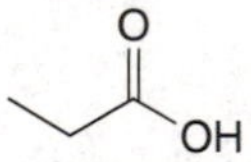 and

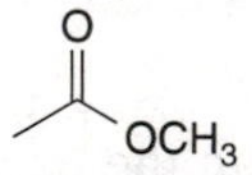

O–H bond
> 3000 cm^{-1}
[See note on OH in Answer 13.42.]

no O–H bond

c. 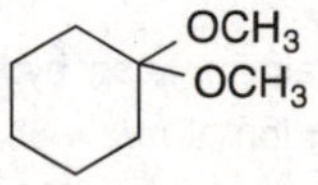 and

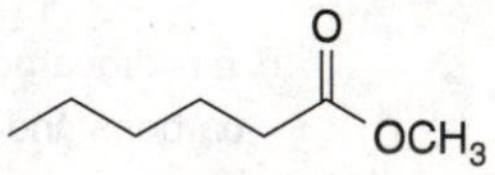

no C=O bond

C=O bond
~1700 cm^{-1}

d. 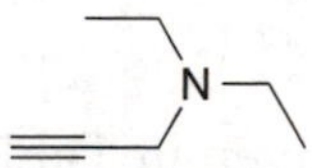 and

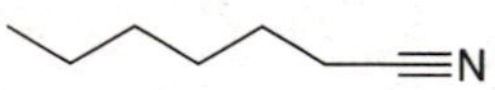

C_{sp}–H bond
3300 cm^{-1}

13.44 The IR absorptions above 1500 cm^{-1} are different for each of the narcotics.

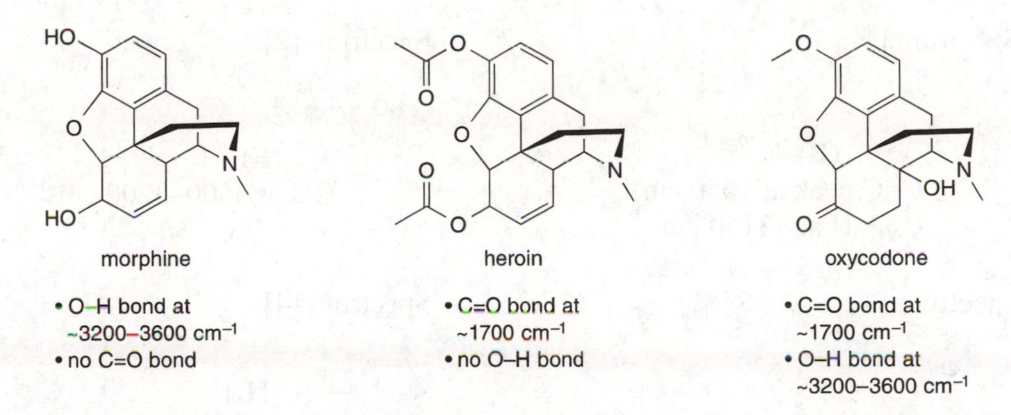

13.45 The three compounds show differences in their IR spectra.

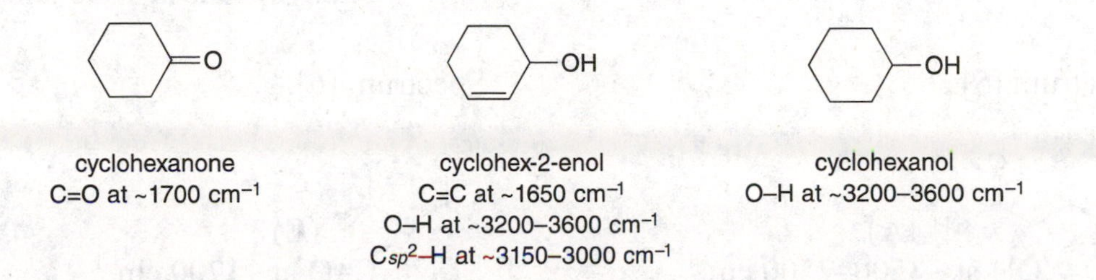

13.46 Look for a **change in functional groups** from starting material to product to see how IR could be used to determine when the reaction is complete.

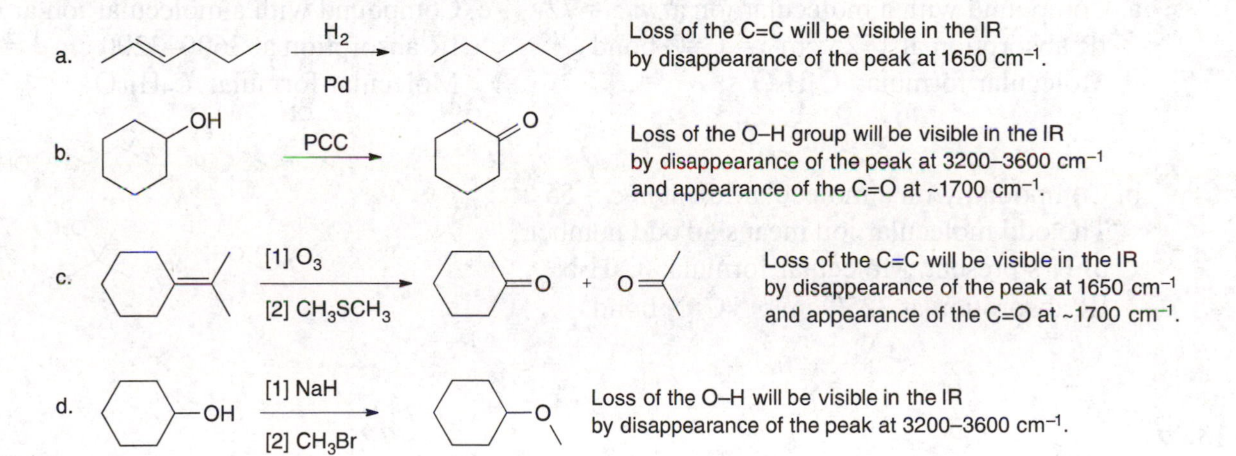

a.
H_2
Pd
Loss of the C=C will be visible in the IR
by disappearance of the peak at 1650 cm^{-1}.
b.
OH
PCC
O
Loss of the O–H group will be visible in the IR
by disappearance of the peak at 3200–3600 cm^{-1}
and appearance of the C=O at ~1700 cm^{-1}.
c.
[1] O_3
[2] CH_3SCH_3
O
+
O
Loss of the C=C will be visible in the IR
by disappearance of the peak at 1650 cm^{-1}
and appearance of the C=O at ~1700 cm^{-1}.
d.
OH
[1] NaH
[2] CH_3Br
O
Loss of the O–H will be visible in the IR
by disappearance of the peak at 3200–3600 cm^{-1}.

13.47 In addition to Csp^3–H at ~3000–2850 cm^{-1}:

Spectrum [1]:

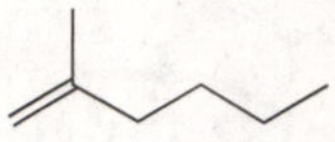

HO

(B)

C=C peak at 1650 cm^{-1}
Csp^2–H at ~3150 cm^{-1}

Spectrum [2]:

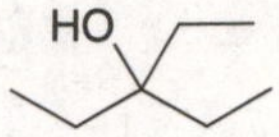

(F)

OH at 3600–3200 cm^{-1}

Spectrum [3]:

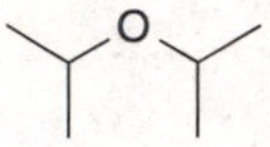

O

(D)

No other peaks above 1500 cm^{-1}

Spectrum [4]:

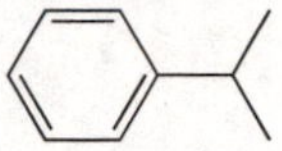

(C)

Csp^2–H at ~3150 cm^{-1}
Phenyl peaks at 1600 and 1500 cm^{-1}

Spectrum [5]:

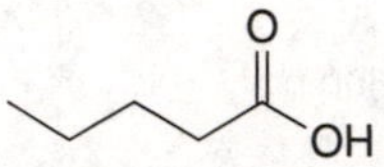

O

OH **(A)**

OH at ~3500–2500 cm^{-1}
C=O at ~1700 cm^{-1}

Spectrum [6]:

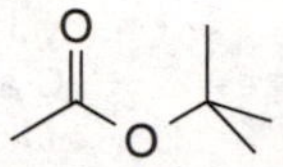

O

O **(E)**

C=O at ~1700 cm^{-1}

13.48

a. Compound with a molecular ion at $m/z = 72$
IR absorption at 1725 cm^{-1} = C=O bond
Molecular formula: C_4H_8O

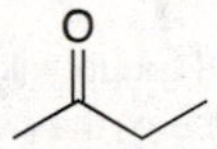

O

b. Compound with a molecular ion at $m/z = 55$
The odd molecular ion means an odd number of N's present. Molecular formula: C_3H_5N
IR absorption at 2250 cm^{-1} = C≡N bond

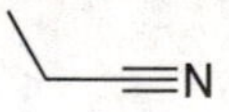

≡N

c. Compound with a molecular ion at $m/z = 74$
IR absorption at 3600–3200 cm^{-1} = O–H bond
Molecular formula: $C_4H_{10}O$

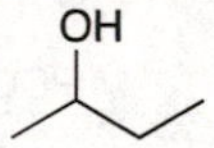

OH

or

OH

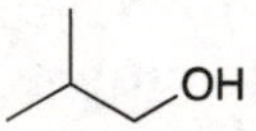

OH

or

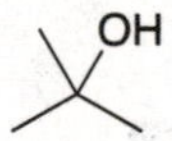

OH

13.49

Chiral hydrocarbon with a molecular ion at $m/z = 82$
Molecular formula: C_6H_{10}
IR absorptions at 3300 cm^{-1} = Csp–H bond
3000–2850 cm^{-1} = Csp^3–H bonds
2250 cm^{-1} = C≡C bond

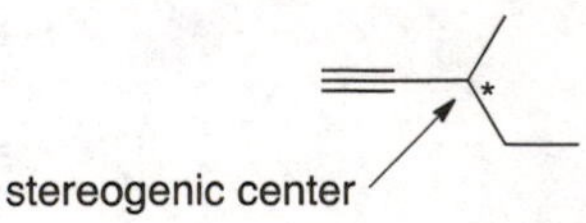

*

stereogenic center

Two possible enantiomers:

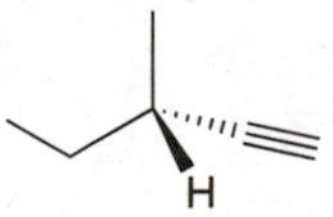

H

or

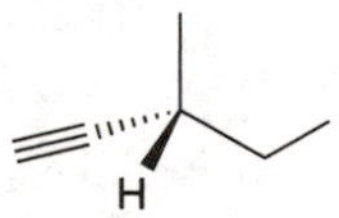

H

13.50 The chiral compound **Y** has a strong absorption at 2970–2840 cm^{-1} in its IR spectrum due to sp^3 hybridized C–H bonds. The two peaks of equal intensity at 136 and 138 indicate the presence of a Br atom. The molecular formula is C_4H_9Br. Only one constitutional isomer of this molecular formula has a stereogenic center:

Y = Br and Br

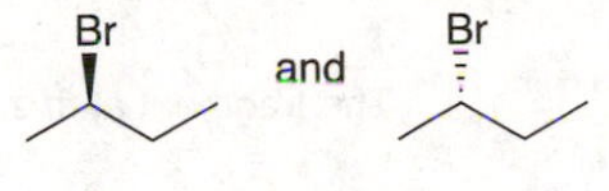

two possible enantiomers

13.51 The molecular ion of 192 suggests $C_{12}H_{16}O_2$ as a possible molecular formula. IR absorption at 1721 cm^{-1} is due to a C=O, and the absorptions around 3000 cm^{-1} are due to C_{sp^2}–H and C_{sp^3}–H. The compound is an ester, formed in the following manner.

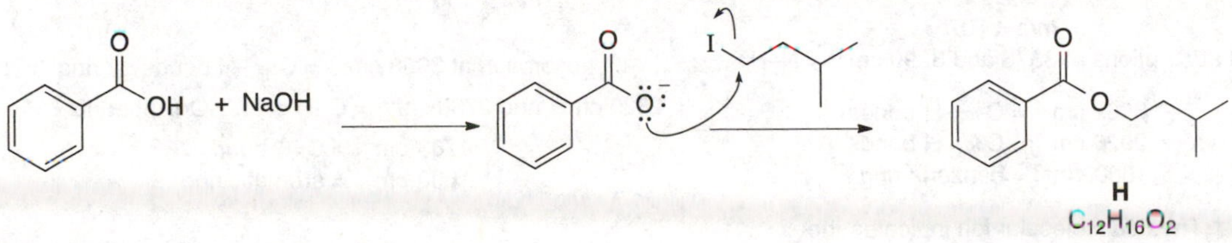

O OH + NaOH → O O⁻ I → O O

H

$C_{12}H_{16}O_2$

13.52

O H Zn(Hg) HCl → Z

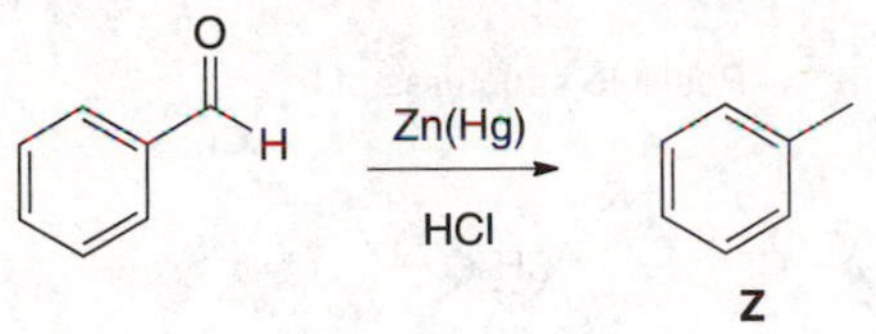

m/z = 92; molecular formula C_7H_8
IR absorptions at:
3150–2950 cm^{-1} = C_{sp^3}–H and C_{sp^2}–H bonds
1605 cm^{-1} and 1496 cm^{-1} due to phenyl group

13.53

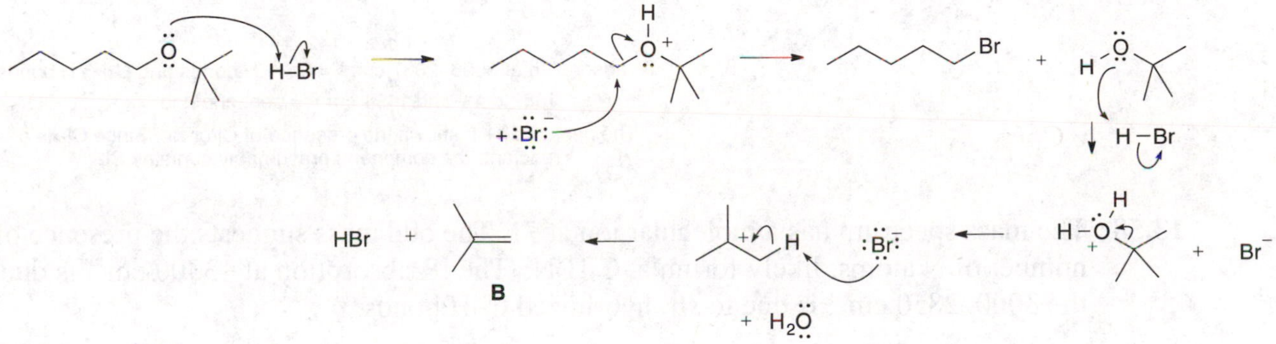

O H–Br → H O+ + :Br:⁻ → Br + H O H–Br

H H–O+ + Br⁻ → :Br:⁻ + H + H_2O → HBr + B

13.54 The molecular ion of 144 suggests $C_8H_{16}O_2$ as a possible molecular formula for **X.** The IR absorption at 1739 cm^{-1} is due to a C=O, and the absorptions at less than 3000 cm^{-1} are due to C_{sp^3}–H.

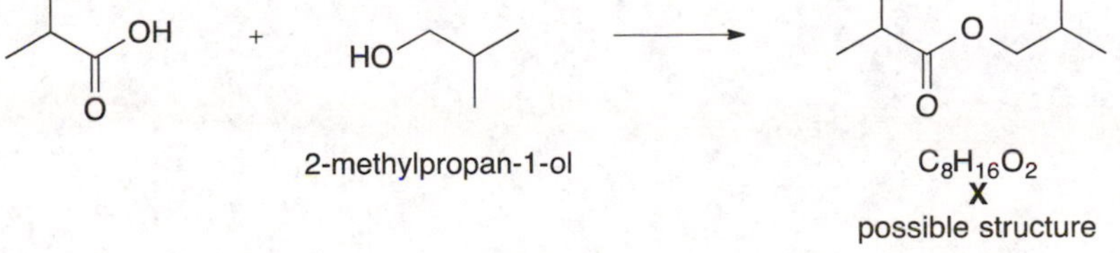

OH O + HO → O O

2-methylpropan-1-ol

$C_8H_{16}O_2$
X
possible structure

13.55

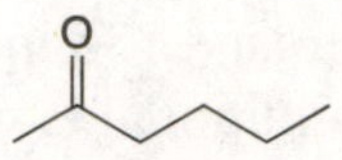

J

$C_6H_{12}O$
$m/z = 100$
IR absorption at 2962 cm^{-1} = Csp^3–H bonds
1718 cm^{-1} = C=O bond

fragments:

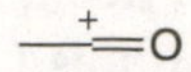

α cleavage product
$m/z = 43$

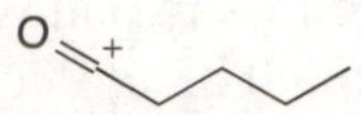

α cleavage product
$m/z = 85$

The fragment at $m/z = 57$ could be due to $(C_4H_9)^+$ or $(C_3H_5O)^+$.

13.56

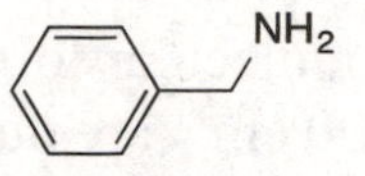

NH_2

K

C_7H_9N
$m/z = 107$
IR absorptions at 3373 and 3290 cm^{-1} = N–H

3062 cm^{-1} = Csp^2–H bonds
2920 cm^{-1} = Csp^3–H bonds
1600 cm^{-1} = benzene ring

The odd molecular ion indicates the presence of a N atom.

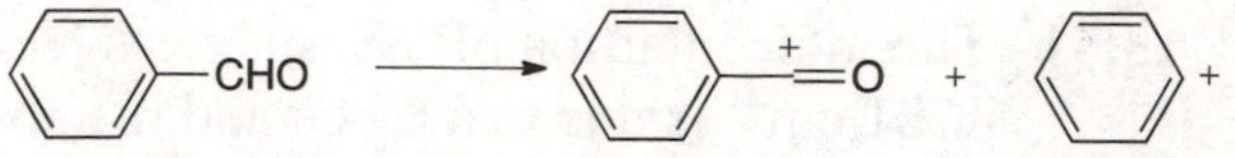

CHO

L
C_7H_6O
$m/z = 106$

$m/z = 105$

$m/z = 77$

IR absorption at 3068 cm^{-1} = Csp^2–H bonds on ring
2820 cm^{-1} and 2736 cm^{-1} = C–H of RCHO (Appendix E)
1703 cm^{-1} = C=O bond
1600 cm^{-1} = aromatic ring

13.57

Possible structures of **P**:

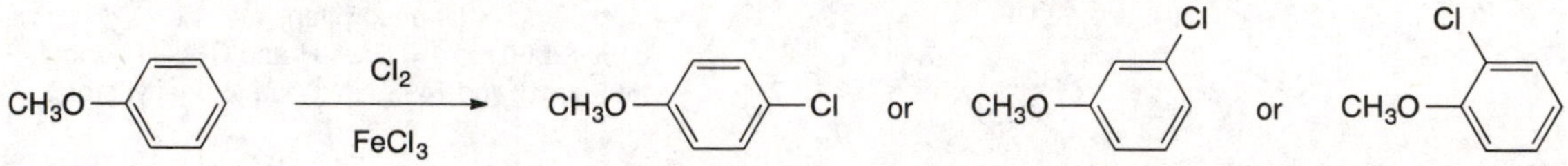

CH_3O $\xrightarrow[FeCl_3]{Cl_2}$ CH_3O Cl or CH_3O Cl or CH_3O Cl

C_7H_7ClO

$m/z = 142, 144$
IR absorption at 3096–2837 cm^{-1} = Csp^3–H bonds and Csp^2–H bonds
1582 cm^{-1} and 1494 cm^{-1} = benzene ring
The peak at M + 2 shows the presence of Cl or Br. Since Cl_2 is a reactant, the compound presumably contains Cl.

13.58 The mass spectrum has a molecular ion at 71. The odd mass suggests the presence of an odd number of N atoms; likely formula, C_4H_9N. The IR absorption at ~3300 cm^{-1} is due to N–H and the 3000–2850 cm^{-1} is due to sp^3 hybridized C–H bonds.

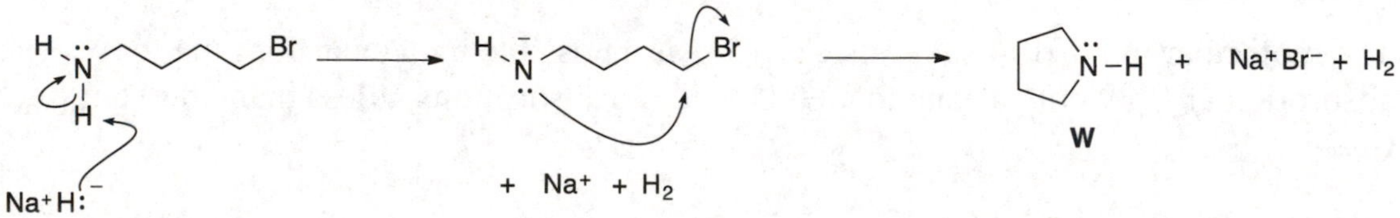

H N H Br Na^+H^- → H N Br + Na^+ + H_2 → N–H + Na^+Br^- + H_2

W

13.59 Because the carbonyl absorption of an amide is at lower wavenumber than the carbonyl absorption of an ester, the C=O of the amide must be weaker and have more single bond character. This can be explained by resonance. Although both an ester and amide are resonance stabilized, the N atom of the amide is more basic, making it more willing to donate its electron pair.

amide

ester

This form contributes more to the hybrid, so there is more single bond character.

This form contributes less to the hybrid.

Because the amide carbonyl has more single bond character, the bond is weaker and it absorbs at lower wavenumber.

13.60 The α,β-unsaturated carbonyl compound has three resonance structures, two of which place a single bond between the C and O atoms. This means that the C–O bond has partial single bond character, making it weaker than a regular C=O bond, and moving the absorption to lower wavenumber.

three resonance structures for cyclohex-2-enone

13.61 If a ketone carbonyl absorbs at lower wavenumber than an aldehyde carbonyl, the ketone carbonyl is weaker and has more single bond character. This can be explained by the fact that R groups are electron donating and stabilize an adjacent (+) charge.

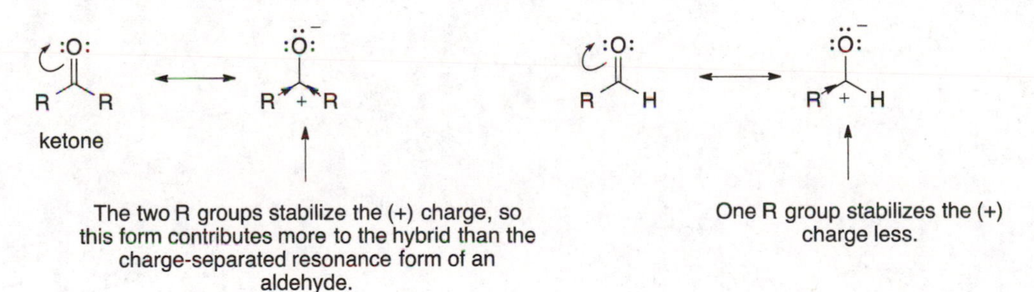

ketone
The two R groups stabilize the (+) charge, so this form contributes more to the hybrid than the charge-separated resonance form of an aldehyde.
One R group stabilizes the (+) charge less.

As a result, the charge-separated resonance form of a ketone, which contains a C–O single bond, contributes more to the hybrid of a ketone, making the C=O weaker and shifting the absorption to lower wavenumber.

13.62

a, b.

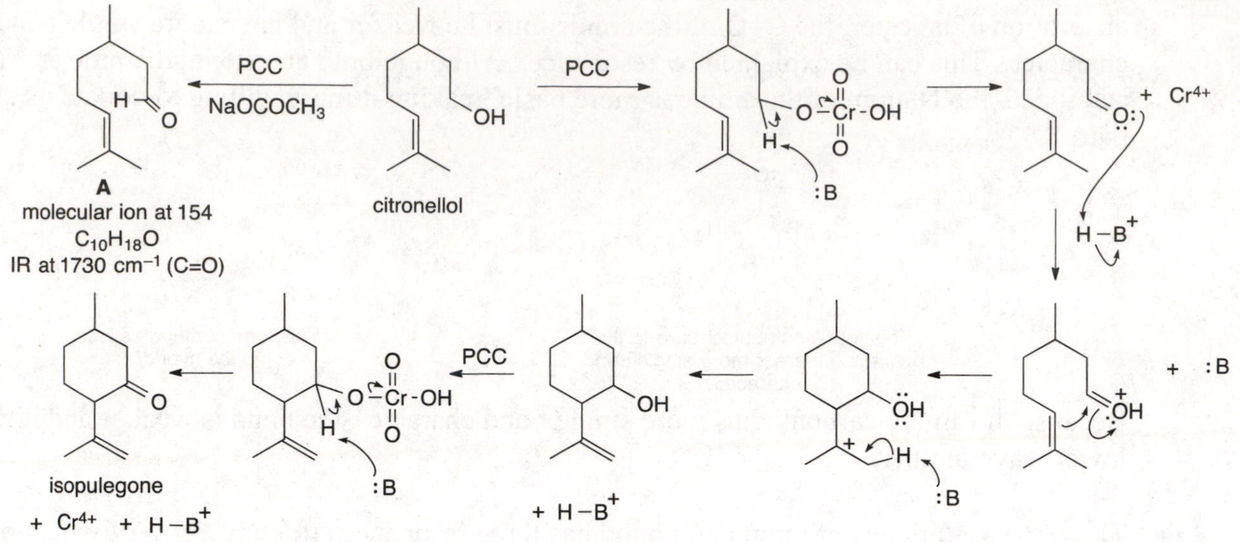

PCC

NaOCOCH$_3$

OH

PCC

O–Cr–OH

H

:B

+ Cr^{4+}

H–B$^+$

A

molecular ion at 154

$C_{10}H_{18}O$

IR at 1730 cm^{-1} (C=O)

citronellol

O

PCC

O–Cr–OH

H

:B

OH

+ H–B$^+$

ÖH

H

:B

ÖH

+ :B

isopulegone

+ Cr^{4+} + H–B$^+$

a, c.

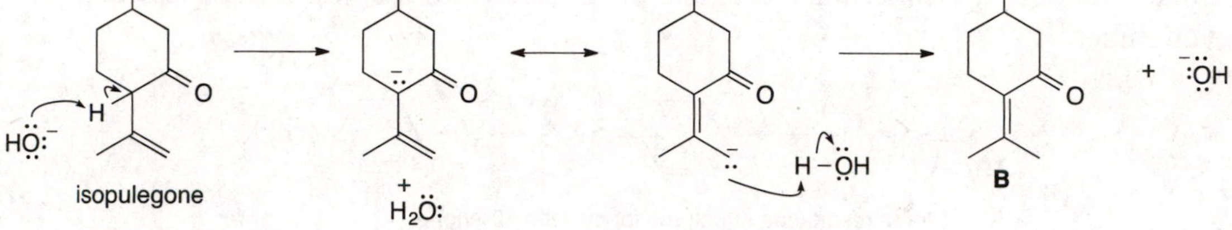

HÖ:$^-$

H

isopulegone

$H_2\ddot{O}$:

H–ÖH

B

+ :ÖH$^-$

Chapter 14 Nuclear Magnetic Resonance Spectroscopy

Chapter Review

^{1}H NMR spectroscopy

[1] The **number of signals** equals the number of different types of protons (14.2).

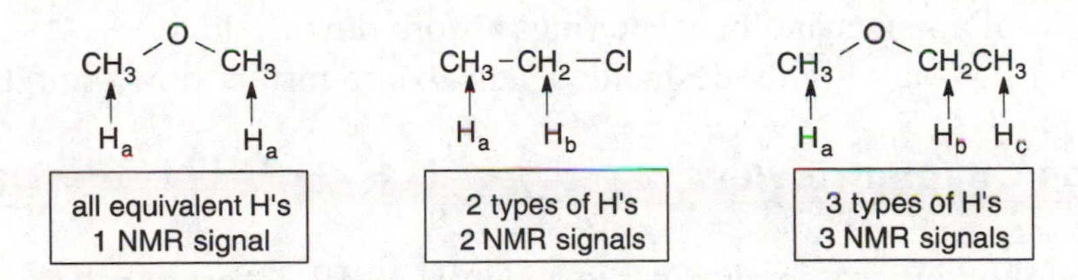

[2] The **position of a signal** (its chemical shift) is determined by shielding and deshielding effects.

- Shielding shifts an absorption upfield; deshielding shifts an absorption downfield.
- Electronegative atoms withdraw electron density, deshield a nucleus, and shift an absorption downfield (14.3).

H ← This proton is shielded. Its absorption is upfield, 0.9–2 ppm.

X, H ← This proton is deshielded. Its absorption is farther downfield, 2.5–4 ppm.

- Loosely held π electrons can either shield or deshield a nucleus. Protons on benzene rings and double bonds are deshielded and absorb downfield, whereas protons on triple bonds are shielded and absorb upfield (14.4).

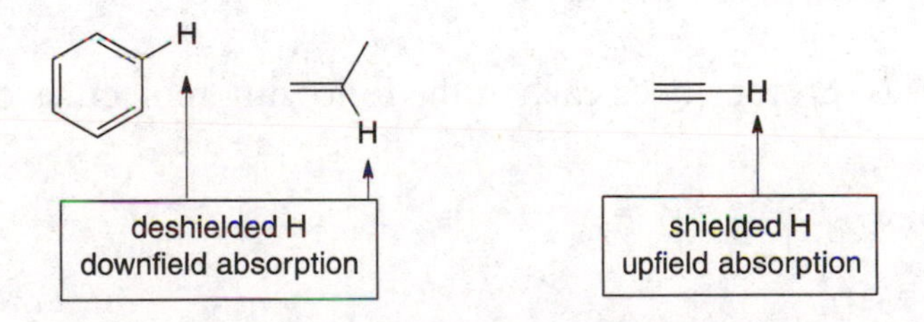

[3] The **area under an NMR signal** is proportional to the number of absorbing protons (14.5).

[4] **Spin–spin splitting** tells about nearby nonequivalent protons (14.6–14.8).

- Equivalent protons do not split each other's signals.
- A set of n nonequivalent protons on the same carbon or adjacent carbons split an NMR signal into $n + 1$ peaks.
- OH and NH protons do not cause splitting (14.9).
- When an absorbing proton has two sets of nearby nonequivalent protons that are equivalent to each other, use the $n + 1$ rule to determine splitting.
- When an absorbing proton has two sets of nearby nonequivalent protons that are not equivalent to each other, the number of peaks in the NMR signal = $(n + 1)(m + 1)$. In flexible alkyl chains, peak overlap often occurs, resulting in $n + m + 1$ peaks in an NMR signal.

^{13}C NMR spectroscopy (14.11)

[1] The number of signals equals the number of different types of carbon atoms. All signals are single lines.

[2] The relative position of ^{13}C signals is determined by shielding and deshielding effects.

- Carbons that are sp^3 hybridized are shielded and absorb upfield.
- Electronegative elements (N, O, and X) shift absorptions downfield.
- The carbons of alkenes and benzene rings absorb downfield.
- Carbonyl carbons are highly deshielded, and absorb farther downfield than other carbon types.

Practice Test on Chapter Review

1.a. Which of the following statements is true about ^{1}H NMR absorptions?

1. A signal that occurs at 1800 Hz on a 300 MHz NMR spectrometer occurs at 3000 Hz on a 500 MHz NMR spectrometer.
2. A signal that occurs at 3.3 ppm on a 60 MHz NMR absorbs at 198 Hz upfield from TMS.
3. A signal that occurs at 600 Hz is downfield from a signal that occurs at 800 Hz.
4. Statements (1) and (2) are both true.
5. Statements (1), (2), and (3) are all true.

b. Which of the following statements is true about ^{1}H NMR spectroscopy?

1. Electronegative elements shield a nucleus so an absorption shifts downfield.
2. A triplet is due to a proton that has four adjacent nonequivalent protons.
3. Circulating π electrons create a magnetic field that reinforces the applied field in the vicinity of the protons in benzene.
4. Statements (1) and (2) are both true.
5. Statements (1), (2), and (3) are all true.

2. How many different types of protons does each of the following molecules contain?

a. c. e. g. Cl Cl

b. O O Cl d. Cl Br Cl Br f. h. Cl H

3. Into how many peaks will each of the circled protons be split in a proton NMR spectrum?

a. 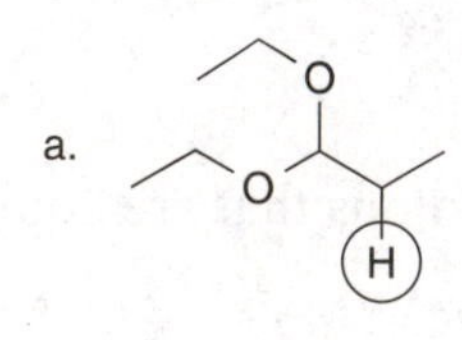

c. H

e. 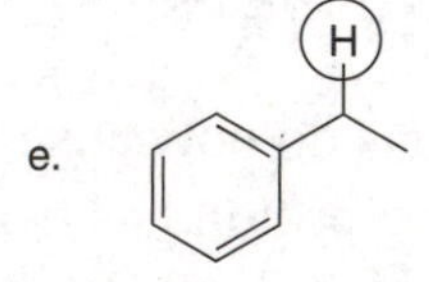

g. 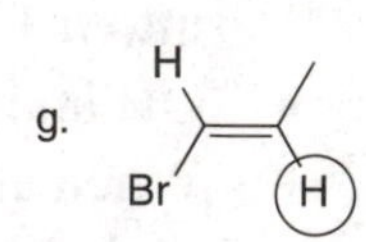

b. H

d. HO OH H

f. 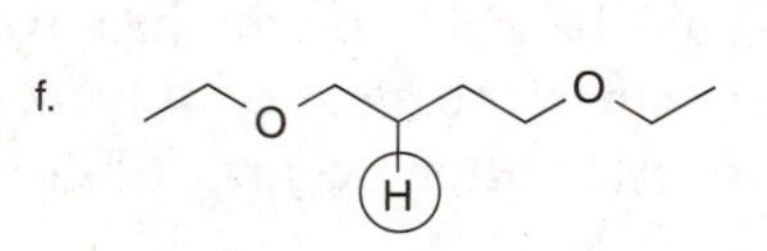

4. How many lines are presents in the ^{13}C NMR spectrum of each compound?

a. 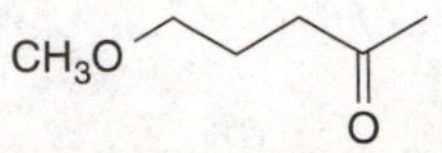 CH_3O ... O

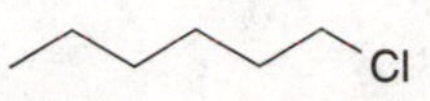

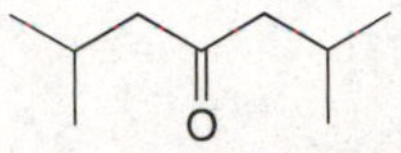

b.

c. Cl

d. 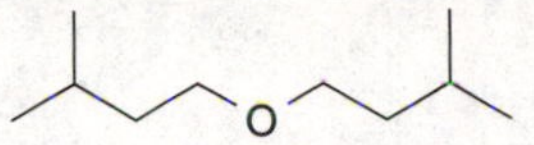 O

e. O

f.

5. With reference to the ^1H NMR absorptions in the following compound, (a) which proton absorbs farthest upfield; (b) which proton absorbs farthest downfield?

H_a H_b H_d O O H_c

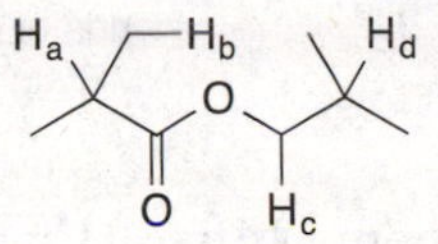

6. With reference to the ^{13}C NMR absorptions in the following compound, (a) which carbon absorbs farthest downfield; (b) which carbon absorbs farthest upfield?

C_b O C_a F C_c C_d

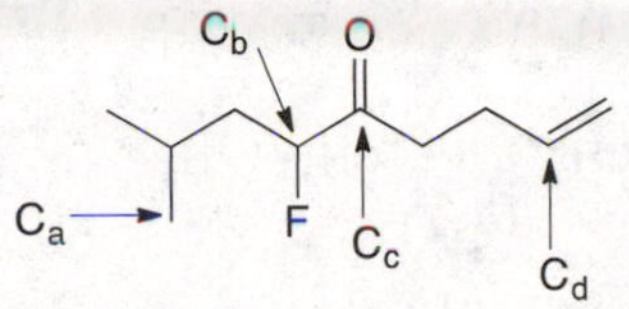

Answers to Practice Test

1.	2.	3.	4.	5.	6.
a. 1	a. 5	a. 8	a. 6	a. H_b	a. C_c
b. 3	b. 5	b. 3	b. 4	b. H_c	b. C_a
	c. 4	c. 7	c. 6		
	d. 5	d. 4	d. 4		
	e. 4	e. 4	e. 4		
	f. 5	f. 3	f. 5		
	g. 3	g. 8			
	h. 9				

Answers to Problems

14.1 Use the formula δ = [observed chemical shift (Hz)/ν of the NMR (MHz)] to calculate the chemical shifts.

a. **CH_3 protons:**
δ = [1715 Hz] / [500 MHz]
= **3.43 ppm**

OH proton:
δ = [1830 Hz] / [500 MHz]
= **3.66 ppm**

b. The positive direction of the δ scale is downfield from TMS. The CH_3 protons absorb upfield from the OH proton.

14.2 Calculate the chemical shifts as in Answer 14.1.

a. one signal:
δ = [1017 Hz] / [300 MHz]
= **3.39 ppm**

second signal:
δ = [1065 Hz] / [300 MHz]
= **3.55 ppm**

b. one signal:
3.39 = [*x* Hz] / [500 MHz]
x = **1695 Hz**

second signal:
3.55 = [*x* Hz] / [500 MHz]
x = **1775 Hz**

14.3 To determine if two H's are equivalent replace each by an atom X. If this yields the same compound or mirror images, the two H's are equivalent. Each kind of H will give one NMR signal.

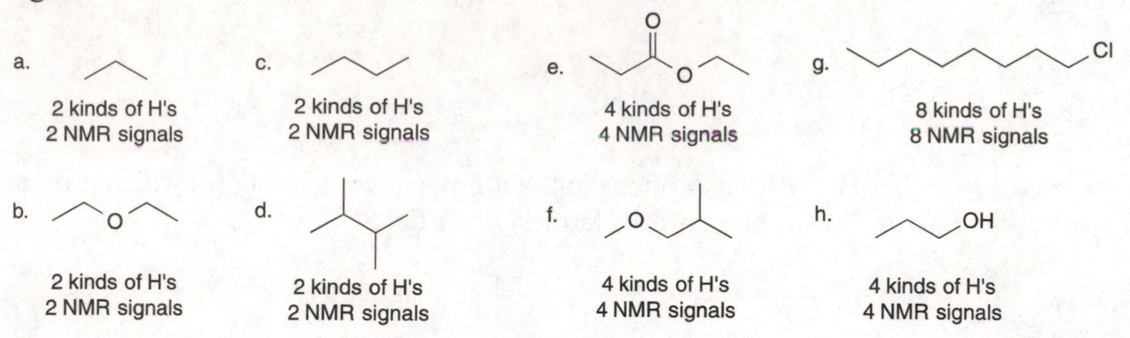

14.4 Draw in all of the H's and compare them. If two H's are cis and trans to the same group, they are equivalent.

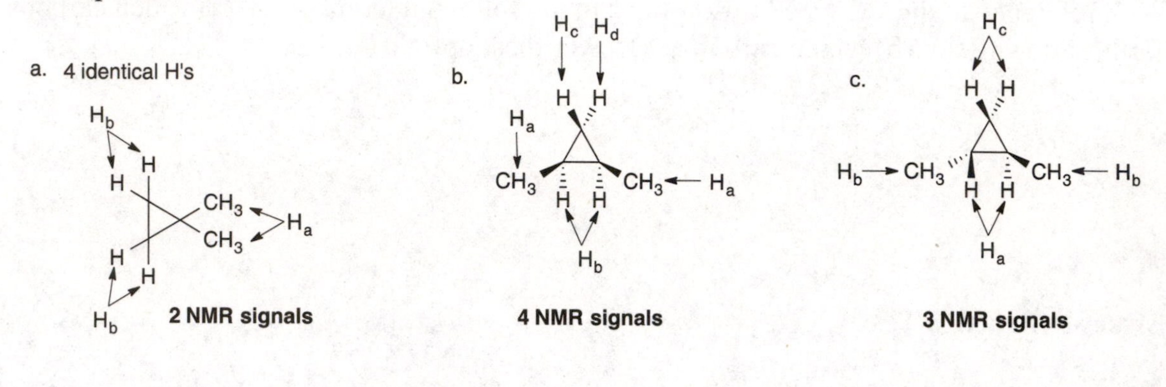

14.5

a. 3 NMR signals

b. 4 NMR signals

c. 6 NMR signals

d. 3 NMR signals

14.6 If replacement of H by X yields the same compound, the protons are **homotopic.**
If replacement of H with X yields enantiomers, the protons are **enantiotopic.**
If replacement of H with X yields diastereomers, the protons are **diastereotopic.** In general, if the compound has **one stereogenic center**, the protons in a CH_2 group are **diastereotopic.**

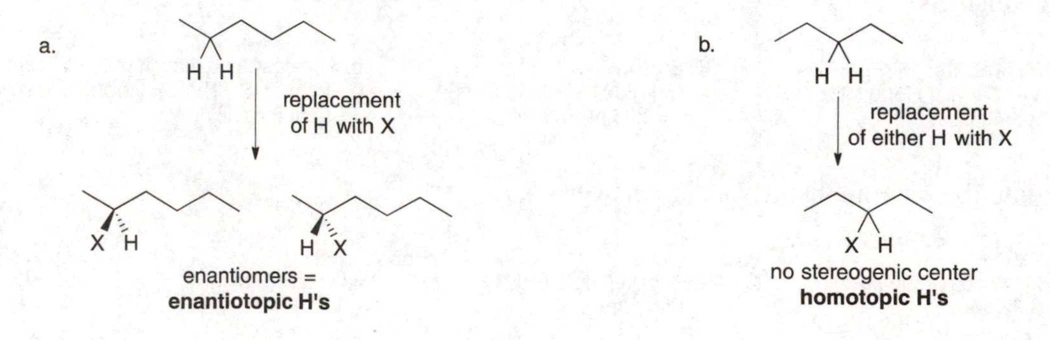

c. [structure: 2-butanol, OH] = CH_3, HO, H, C—C, H, H ← replacement of H with X, CH_2CH_3

Pick one configuration at the existing stereogenic center.

[structures: two products with H and X exchanged]

diastereomers =
diastereotopic H's

14.7 The two protons of a CH_2 group are different from each other if the compound has one stereogenic center. Replace one proton with X and compare the products.

a. The stereogenic center makes the H's in the CH_2 group diastereotopic and therefore different from each other.

stereogenic center, Cl, H_a, H_d, H_e, H_b and H_c

5 NMR signals

b. H_d and H_e, stereogenic center, H_c, O, H_a, Cl, H_b

5 NMR signals

c. stereogenic center, H_d and H_e, Br, H_f, H_g, H_a, H_b and H_c

7 NMR signals

14.8 Decreased electron density deshields a nucleus and the absorption goes downfield. Absorption also shifts downfield with increasing alkyl substitution.

a. $FC\underline{H_2}CH_2\underline{CH_2}Cl$
F is more electronegative than Cl. The CH_2 group adjacent to the F is more deshielded and the H's will absorb farther downfield.

b. $CH_3C\underline{H_2}CH_2\underline{CH_2}OCH_3$
The CH_2 group adjacent to the O will absorb farther downfield because it is closer to the electronegative O atom.

c. $C\underline{H_3}OC(C\underline{H_3})_3$
The CH_3 group bonded to the O atom will absorb farther downfield.

14.9

a. Cl, H_a, H_c, H_b, Br

3 types of protons:
$H_b < H_c < H_a$

b. O, O, H_a, H_b, H_c, H_c, H_c

3 types of protons:
$H_c < H_a < H_b$

c. O, H_a, H_b, H_c

3 types of protons:
$H_c < H_a < H_b$

14.10 a. False. When a nucleus is strongly shielded, the effective field is smaller than the applied field and the absorption shifts upfield.

b. True.

c. False. A nucleus that is strongly deshielded requires a higher field strength for resonance. Alternatively, a nucleus that is strongly *shielded* requires a lower field strength for resonance.

d. False. A nucleus that is strongly shielded absorbs at a smaller δ value. Alternatively, a nucleus that is strongly *deshielded* absorbs at a larger δ value.

14.11

a. —≡—H_a H_b H_c

H_c protons are shielded because they are bonded to an sp^3 C.
H_a is shielded because it is bonded to an sp C.
H_b protons are deshielded because they are bonded to an sp^2 C.

$H_c < H_a < H_b$

b. $CH_3C(=O)OCH_2CH_3$ (CH_3: H_a; OCH_2CH_3: H_b H_c)

H_c protons are shielded because they are bonded to an sp^3 C.
H_a protons are deshielded slightly because the CH_3 group is bonded to a C=O.
H_b protons are deshielded because the CH_2 group is bonded to an O atom.

$H_c < H_a < H_b$

14.12 An integration ratio of 2:3 means that there are two types of hydrogens in the compound, and that the ratio of one type to another type is 2:3.

a. (Cl) 2 types of H's 3:2 - YES

b. 2 types of H's 6:2 or 3:1 - no

c. (O) 2 types of H's 6:4 or 3:2 - YES

d. (O, O) 2 types of H's 6:4 or 3:2 - YES

14.13

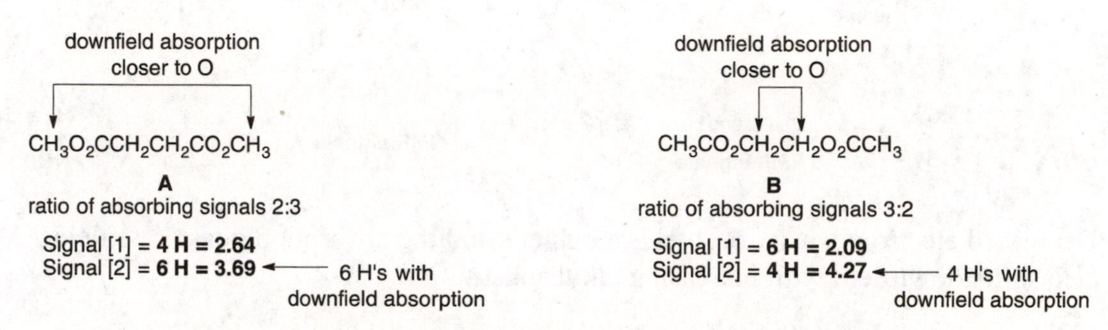

14.14 To determine the **splitting pattern** for a molecule:

- Determine the number of different kinds of protons.
- Nonequivalent protons on the same C or adjacent C's split each other.
- Apply the *n* + 1 rule.

a. $CH_3CH_2C(=O)Cl$ (H_a H_b)
H_a: 3 peaks - triplet
H_b: 4 peaks - quartet

b. CH_3CBr_2H (CH_3: H_a; H ← H_b)
H_a: 2 peaks - doublet
H_b: 4 peaks - quartet

c. $CH_3C(=O)CH_2CH_2Br$ (H_a; H_b H_c)
H_a: 1 peak - singlet
H_b: 3 peaks - triplet
H_c: 3 peaks - triplet

d. H_a → H, Cl, Br, H ← H_b
H_a: 2 peaks - doublet
H_b: 2 peaks - doublet

e. H ← H_a, H ← H_b
H_a: 2 peaks - doublet
H_b: 2 peaks - doublet

f. $ClCH_2CH(OCH_3)_2$ ($ClCH_2$: H_a; H: H_b)
H_a: 2 peaks - doublet
H_b: 3 peaks - triplet

14.15 Use the directions from Answer 14.14.

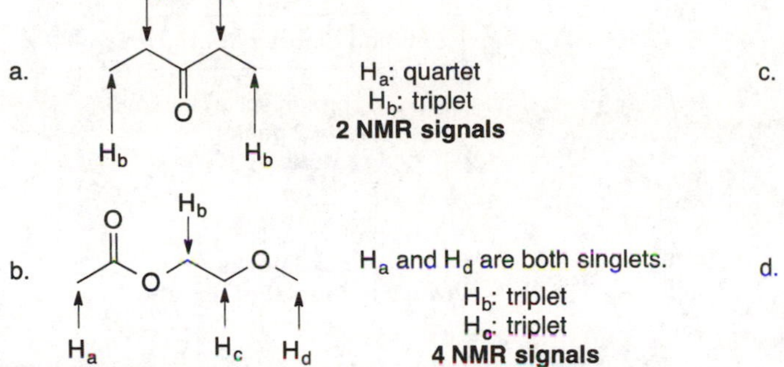

a. H_a: quartet
H_b: triplet
2 NMR signals

b. H_a and H_d are both singlets.
H_b: triplet
H_c: triplet
4 NMR signals

c. H_a: doublet
H_b: quartet
2 NMR signals

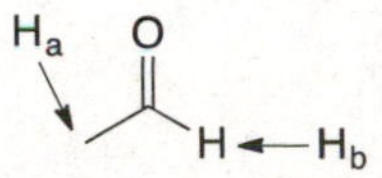

d. 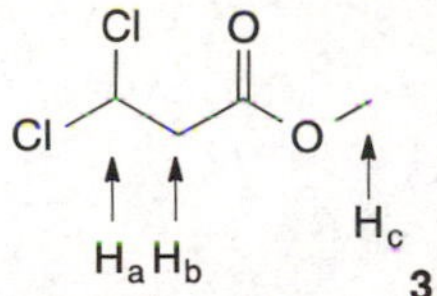 H_a: triplet
H_b: doublet
H_c: singlet
3 NMR signals

14.16 CH_3CH_2Cl

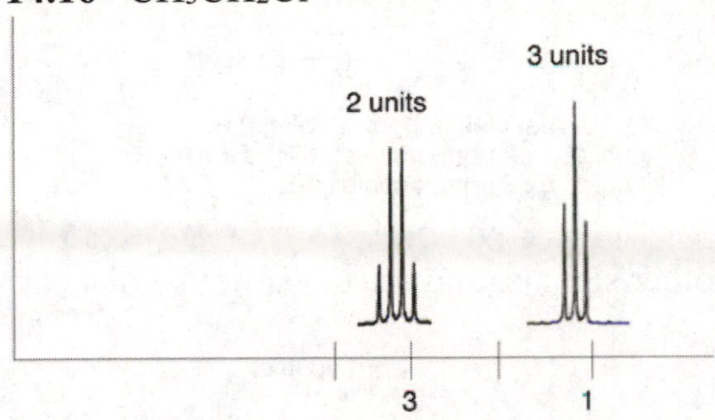

chemical shift (ppm)

There are two kinds of protons, and they can split each other. The CH_3 signal will be split by the CH_2 protons into 2 + 1 = 3 peaks. It will be upfield from the CH_2 protons because it is farther from the Cl. The CH_2 signal will be split by the CH_3 protons into 3 + 1 = 4 peaks. It will be downfield from the CH_3 protons because the CH_2 protons
are closer to the Cl. The ratio of integration units will be 3:2.

14.17

a.

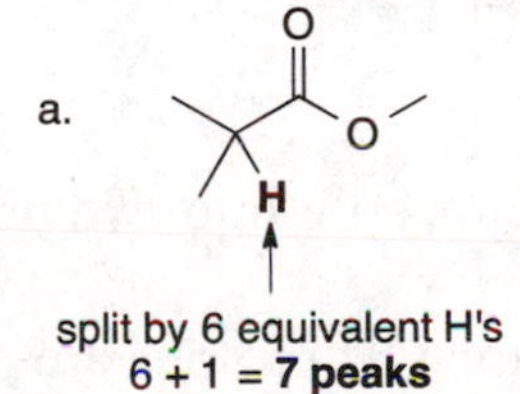

split by 6 equivalent H's
6 + 1 = **7 peaks**

b. 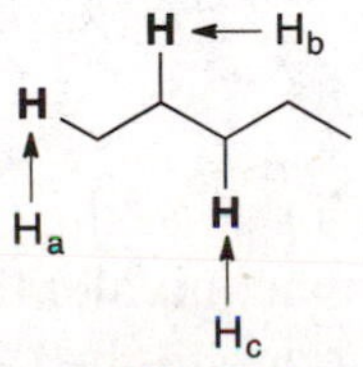 H_a: split by 2 H's
3 peaks
H_c: split by 4 equivalent H's
5 peaks
H_b: split by 2 sets of H's
(3 + 1)(2 + 1) = **12 peaks (maximum)**
Since this is a flexible alkyl chain, the signal due to H_b will have peak overlap, and
3 + 2 + 1 = **6 peaks** will likely be visible.

c. 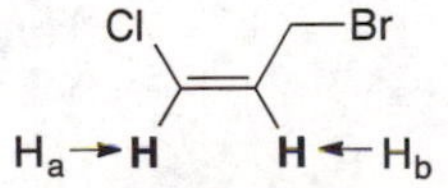 H_a: split by 1 H
2 peaks
H_b: split by 2 sets of H's
(1 + 1)(2 + 1) = **6 peaks**

d. 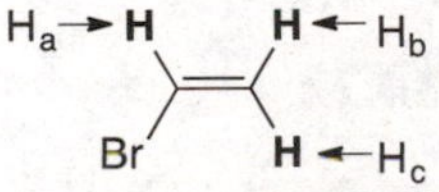 H_a: split by 2 different H's
(1+1)(1+1) = **4 peaks**
H_b: split by 2 different H's
(1+1)(1+1) = **4 peaks**
H_c: split by 2 different H's
(1+1)(1+1) = **4 peaks**

14.18

a. 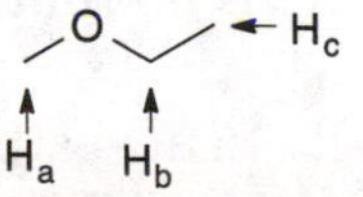 H_a: singlet at ~3 ppm
H_b: quartet at ~3.5 ppm
H_c: triplet at ~1 ppm

b. 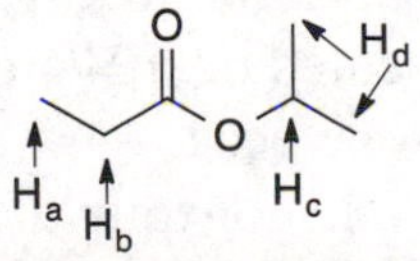 H_a: triplet at ~1 ppm
H_b: quartet at ~2 ppm
H_c: septet at ~3.5 ppm
H_d: doublet at ~1 ppm

c. 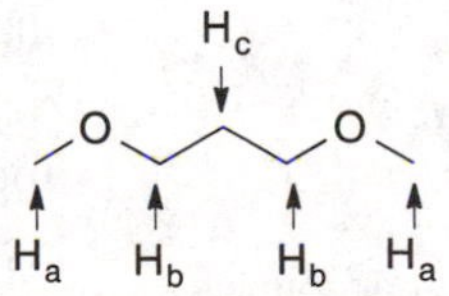 H_a: singlet at ~3 ppm
H_b: triplet at ~3.5 ppm
H_c: quintet at ~1.5 ppm

d. 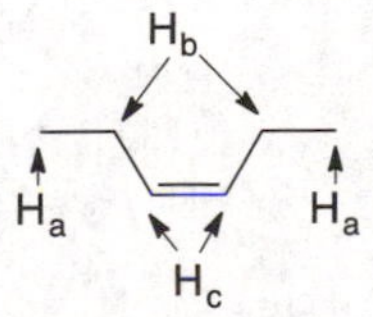 H_a: triplet at ~1 ppm
H_b: multiplet (8 peaks) at ~2.5 ppm
H_c: triplet at ~5 ppm

14.19

H_b
Cl Cl
H_c
H_a H_c 2 H_c protons

trans-1,3-dichloropropene

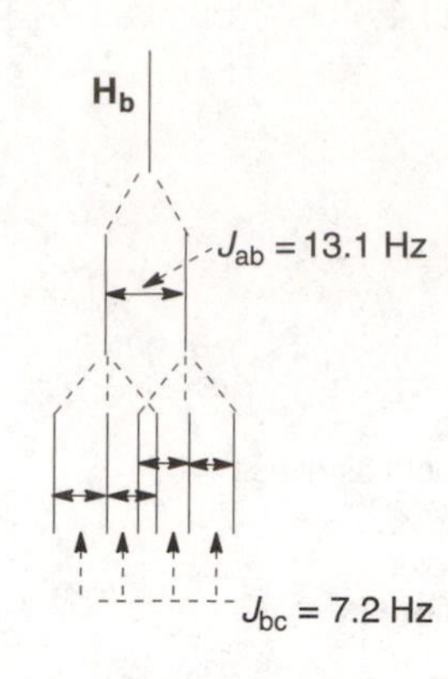

Splitting diagram for H_b

1 **trans** H_a proton splits H_b into
1 + 1 = 2 peaks
a doublet

2 H_c protons split H_b into
2 + 1 = 3 peaks
Now it's a doublet of triplets.

14.20

$C_3H_4Cl_2$

A

H_a: 1.75 ppm, doublet, 3 H, J = 6.9 Hz
H_b: 5.89 ppm, quartet, 1 H, J = 6.9 Hz

B

signal at 4.16 ppm, singlet, 2 H
signal at 5.42 ppm, doublet, 1 H, J = 1.9 Hz
signal at 5.59 ppm, doublet, 1 H, J = 1.9 Hz

14.21 Remember that OH (or NH) protons do not split other signals, and are not split by adjacent protons.

a.

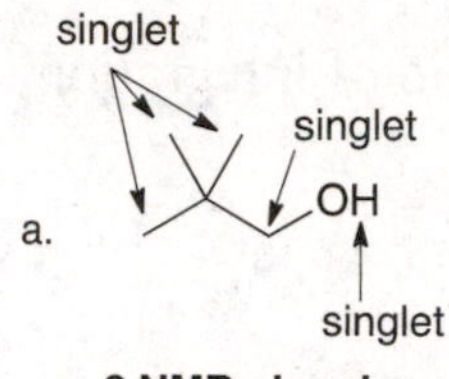

3 NMR signals

b.

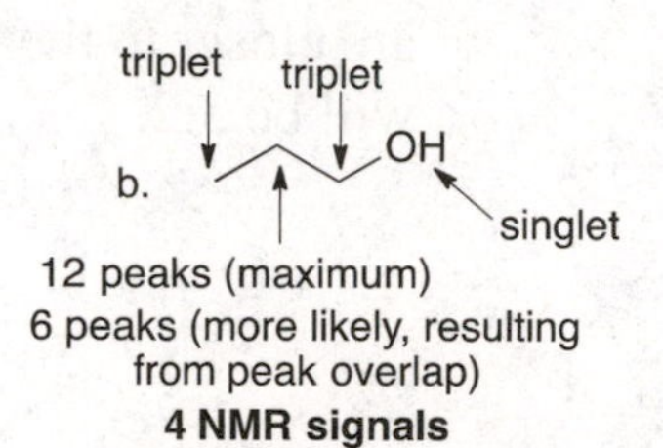

12 peaks (maximum)
6 peaks (more likely, resulting from peak overlap)
4 NMR signals

c.

doublet
singlet
NH_2
7 peaks
3 NMR signals

14.22

H_d
5 H's on benzene ring

A

H_a: doublet at ~1.4 due to the CH_3 group, split into two peaks by one adjacent nonequivalent H (H_c).
H_b: singlet at ~2.7 due to the OH group. OH protons are not split by nor do they split adjacent protons.
H_c: quartet at ~4.7 due to the CH group, split into four peaks by the adjacent CH_3 group.
H_d: multiplets at ~7.2–7.4 due to five protons on the benzene ring.

14.23

palau'amine

H_a: one adjacent nonequivalent H, so two peaks
H_b: one adjacent nonequivalent H, so two peaks
H_c: H_c is located on a N atom, so there is no splitting and it appears as one peak.
H_d: H_d has one nonequivalent H on the same carbon and one on an adjacent carbon, so it is split into (1 + 1)(1 + 1) = 4 peaks (a doublet of doublets).

14.24 Use these steps to propose a structure consistent with the molecular formula, IR, and NMR data.

- Calculate the **degrees of unsaturation.**
- Use the IR data to determine what types of **functional groups** are present.
- Determine the number of different **types of protons.**
- Calculate the **number of H's** giving rise to each signal.
- Analyze the **splitting pattern** and put the molecule together.
- Use the **chemical shift** information to check the structure.

- Molecular formula $C_7H_{14}O_2$

  $2n + 2 = 2(7) + 2 = 16$
  $16 - 14 = 2/2 =$ **1 degree of unsaturation**
  1 π bond or 1 ring

- IR peak at 1740 cm^{-1}

  C=O absorption is around 1700 cm^{-1} (causes the degree of unsaturation).
  No signal at 3200–3600 cm^{-1} means there is no O–H bond.

- NMR data:

absorption	ppm	relative area	
singlet	1.2	9	→ **9 H's**
triplet	1.3	3	→ **3 H's** (probably a CH_3 group)
quartet	4.1	2	→ **2 H's** (probably a CH_2 group)

  - 3 kinds of H's
  - number of H's per signal

    Because the sum of the relative areas equals the number of absorbing H's ($9 + 3 + 2 = 14$), the relative area shows the actual number of absorbing H's: 9 H's, 3 H's and 2 H's.
  - splitting pattern

    The singlet (9 H) is likely from a *tert*-butyl group: $—C(CH_3)_2—CH_3$

    The CH_3 and CH_2 groups split each other: $CH_3—CH_2—$

- Join the pieces together.

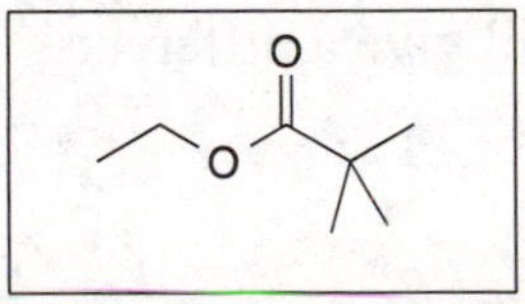

  or

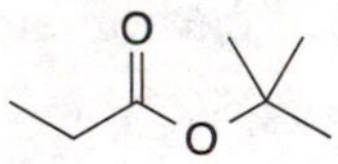

Pick this structure due to the chemical shift data.
The CH_2 group is shifted downfield (4 ppm), so it is close to the electron-withdrawing O.

14.25

- Molecular formula: C_3H_8O ➤ Calculate degrees of unsaturation
  $2n + 2 = 2(3) + 2 = 8$
  $8 - 8 =$ **0 degrees of unsaturation**
- IR peak at 3200–3600 cm^{-1} ➤ Peak at 3200–3600 cm^{-1} is due to an **O–H bond.**
- NMR data:
  - doublet at ~1.2 (6 H)
  - singlet at ~2.2 (1 H)
  - septet at ~4 (1 H)

3 types of H's
septet from 1 H ← split by 6 H's
singlet from 1 H ← from the O–H proton
doublet from 6 H's ← split by 1 H

➤ Put information together:

HO

14.26 Identify each compound from the ^1H NMR data.

a. $CH_2{=}CHCOCH_3$ —HCl→ A: CH3 (singlet); H H (triplet at 3.6); H H (triplet at 3.05); Cl

b. $(CH_3)_2C{=}O$ —base, H_2O→ B: singlet at 2.5; singlet at 1.3; OH singlet at 3.8; singlet at 2.2

14.27 Each different kind of carbon atom will give a different ^{13}C NMR signal.

a. C_a C_b C_b C_a
2 kinds of C's
2 ^{13}C NMR signals

b. Each C is different.
4 kinds of C's
4 ^{13}C NMR signals

c. C_a C_b C_c O C_c C_b C_a
same groups on both sides of O
3 kinds of C's
3 ^{13}C NMR signals

d. Each C is different.
4 kinds of C's
4 ^{13}C NMR signals

14.28

a.

Cl, Cl; H_d, H_c; 2 different H's H_a and H_b
4 ^1H NMR signals

Cl, Cl; H_b; H_a, H_a
2 ^1H NMR signals

H_b Cl, Cl; H_c, H_a
3 ^1H NMR signals

Cl Cl
all H's identical
1 ^1H NMR signal

b.

Cl, Cl
Each C is different.
3 kinds of C's
3 ^{13}C NMR signals

Cl, Cl; C_b; C_a, C_a
2 kinds of C's
2 ^{13}C NMR signals

Cl, Cl
Each C is different.
3 kinds of C's
3 ^{13}C NMR signals

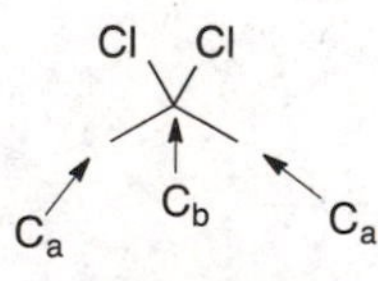

2 kinds of C's
2 ^{13}C NMR signals

c. Although the number of ^{13}C signals cannot be used to distinguish these isomers, each isomer exhibits a different number of signals in its ^{1}H NMR spectrum. As a result, the isomers are distinguishable by ^{1}H NMR spectroscopy.

14.29

These 2 C's are different because they are cis and trans to different groups.

OH

O

Every carbon is different so there are 10 lines for the 10 C atoms.

14.30 Electronegative elements shift absorptions downfield. The carbons of alkenes, benzene rings, and carbonyl groups are also shifted downfield.

a.

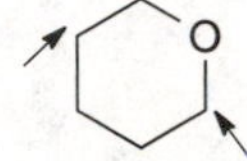

The C closer to the electronegative O will be farther downfield.

b.

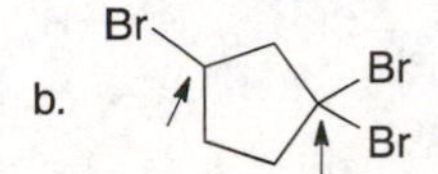

The C of the CBr_2 group has two bonds to electronegative Br atoms and will be farther downfield.

c.

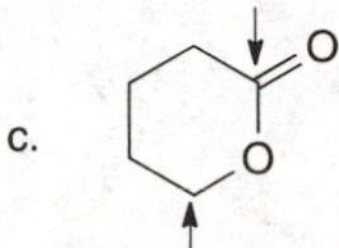

The carbonyl carbon is highly deshielded and will be farther downfield.

d.

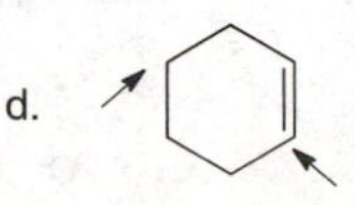

The C atom that is part of the double bond will be farther downfield.

14.31

a. In order of lowest to highest chemical shift:

OH

C_a C_b C_c C_d

$C_d < C_a < C_c < C_b$

b. In order of lowest to highest chemical shift:

O

C_a C_b C_c C_b C_a

$C_a < C_b < C_c$

14.32

- molecular formula $C_4H_8O_2$
  $2n + 2 = 2(4) + 2 = 10$
  $10 - 8 = 2/2 =$ **1 degree of unsaturation**
- no IR peaks at 3200–3600 or 1700 cm^{-1}
  no O–H or C=O
- ^{1}H NMR spectrum at 3.69 ppm
  only one kind of proton
- ^{13}C NMR spectrum at 67 ppm
  only one kind of carbon

O

O

This structure satisifies all the data. One ring is one degree of unsaturation. All carbons and protons are identical.

14.33

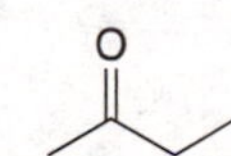

- molecular formula C_4H_8O
  $2n + 2 = 2(4) + 2 = 10$
  $10 - 8 = 2/2 =$ **1 degree of unsaturation**
- ^{13}C NMR signal at > 160 ppm due to C=O

OH

- molecular formula C_4H_8O
  $2n + 2 = 2(4) + 2 = 10$
  $10 - 8 = 2/2 =$ **1 degree of unsaturation**
- all ^{13}C NMR signals at < 160 ppm
  NO C=O

14.34

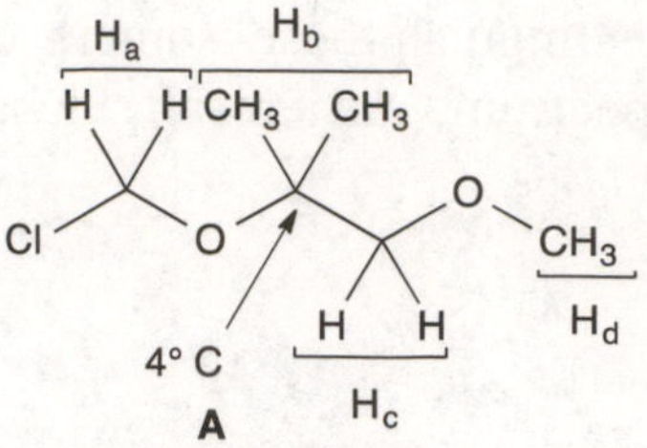

A

a. 4 ^1H NMR signals
b. 5 ^{13}C NMR signals (including the 4° C)

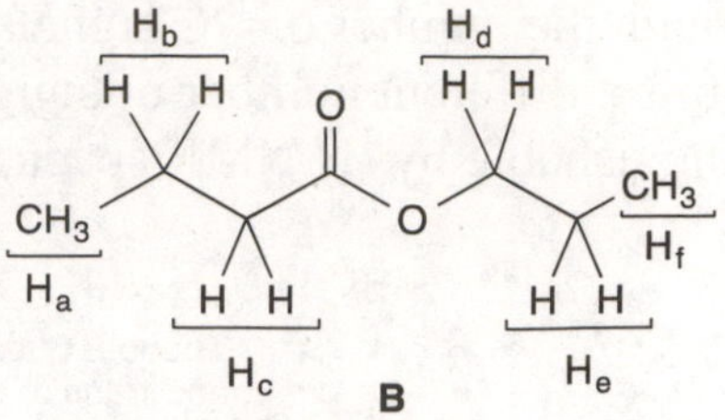

B

a. 6 ^1H NMR signals
b. 7 ^{13}C NMR signals (including the carbonyl C)

14.35

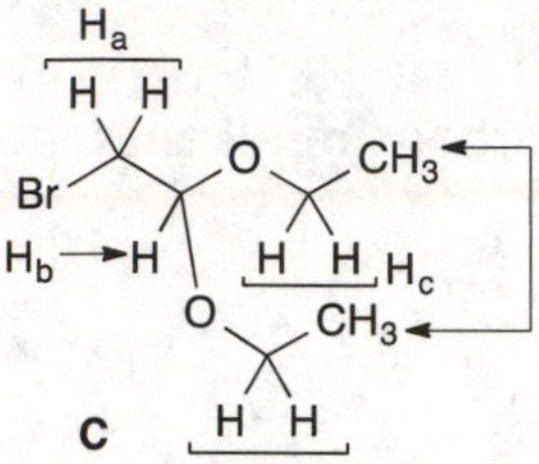

C

a. 4 ^1H NMR signals
b. H_a: 1 adjacent H, so 2 peaks
H_b: 2 adjacent H's, so 3 peaks
H_c: 3 adjacent H's, so 4 peaks
H_d: 2 adjacent H's, so 3 peaks

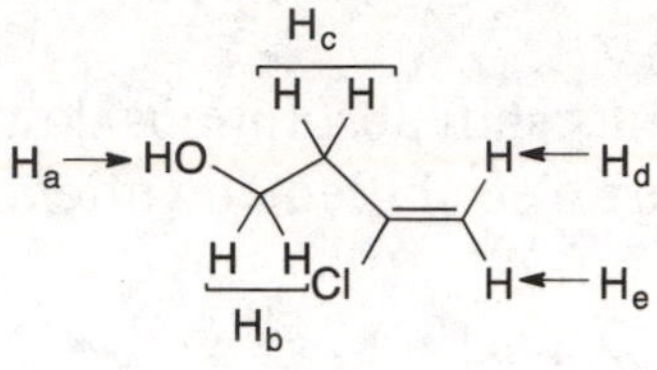

D

a. 5 ^1H NMR signals
b. H_a: singlet
H_b: 2 adjacent H's, so 3 peaks
H_c: 2 adjacent H's, so 3 peaks
H_d: 1 nonequivalent H on the same C, so 2 peaks
H_e: 1 nonequivalent H on the same C, so 2 peaks

14.36 Use the directions from Answer 14.3.

a.  **2 kinds** of H's

b. 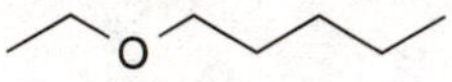 **7 kinds** of H's

c. 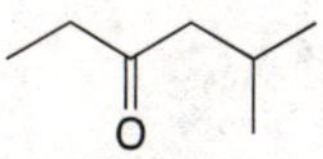 **5 kinds** of H's

d.  **3 kinds** of H's

e. 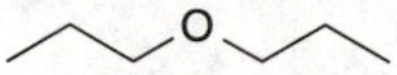 **3 kinds** of H's

f. 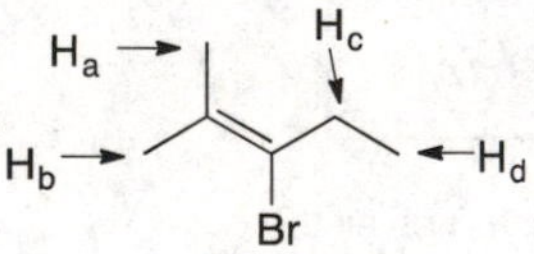 **4 kinds** of H's

g. 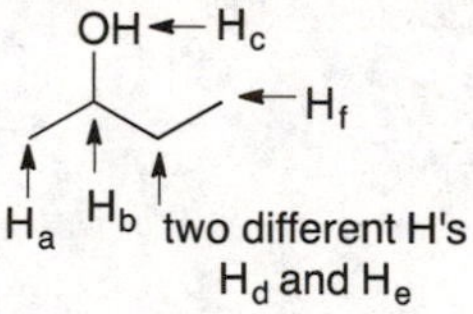 H_a, H_b, two different H's H_d and H_e, H_c (OH), H_f
6 kinds of H's

h. 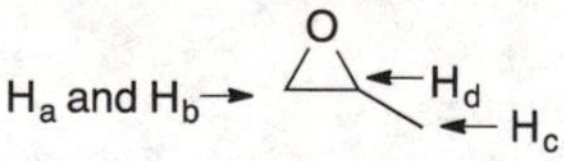 H_a and H_b, H_c, H_d
4 kinds of H's

i. 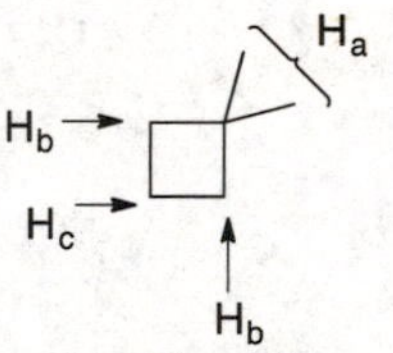 **3 kinds** of H's

j. 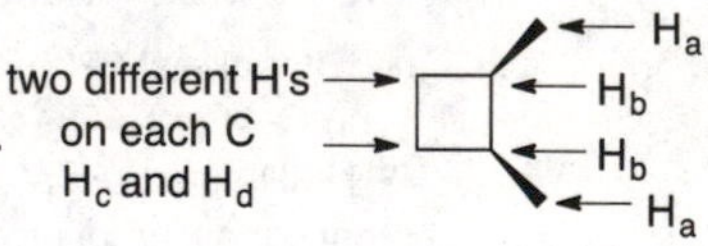 two different H's on each C H_c and H_d
4 kinds of H's

14.37

a. caffeine
4 NMR signals

b. vanillin
6 NMR signals

c. thymol
7 NMR signals

d. capsaicin
15 NMR signals

14.38

δ (in ppm) = [observed chemical shift (Hz)] / ν of the NMR (MHz)]

a. 2.5 = x Hz/300 MHz
x = **750 Hz**

b. ppm = 1200 Hz/300 MHz
= **4 ppm**

c. 2.0 = x Hz/300 MHz
x = **600 Hz**

14.39 a. The chemical shift in δ is independent of the operating frequency, so there is no change in δ when the ν is increased.

b. When the operating ν increases, the ν of an absorption increases as well, because the two quantities are proportional.

c. Coupling constants are independent of the operating ν, so the J value in Hz remains the same.

14.40 Use the directions from Answer 14.8.

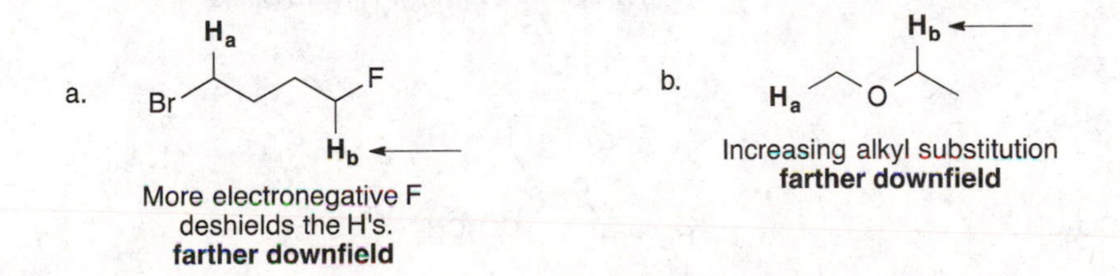

14.41

$H_a : H_b$ = 3:2

$H_a : H_b$ = 3:1

different ratio of peak areas

14.42

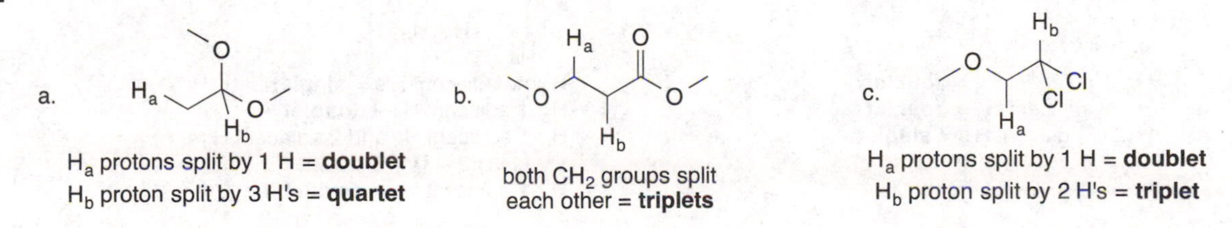

d. 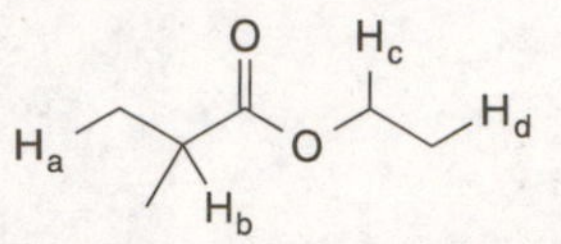 H_a O H_c O H_d H_b

H_a protons split by 1 H = **doublet**
H_b proton split by 6 H's = **septet**
H_c protons split by 3 H's = **quartet**
H_d protons split by 2 H's = **triplet**

g. H_a H_b O OH H_c

H_a protons split by 2 H's = **triplet**
H_c protons split by 2 H's = **triplet**
H_b protons split by CH_3 + CH_2 protons = **12 peaks** (maximum)
Since H_b is located in a flexible alkyl chain, it is likely that peak overlap occurs, so that only 3 + 2 + 1 = 6 peaks will be observed.

e. 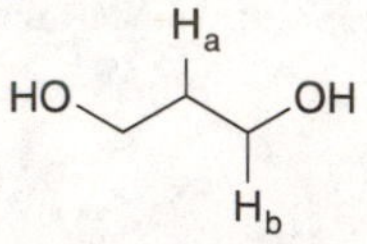 HO H_a OH H_b

H_a protons split by 2 CH_2 groups = **quintet**
H_b protons split by 2 H's = **triplet**

h. O H_b H_a

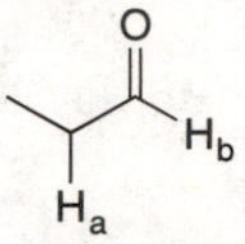

H_a: split by CH_3 group + H_b = **8 peaks** (maximum)
H_b: split by 2 H's = **triplet**

f. 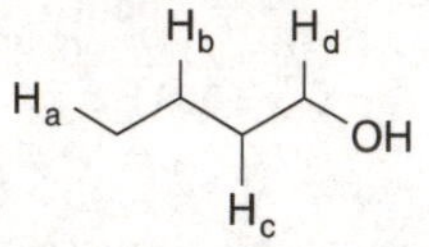 H_a H_b H_d OH H_c

H_a protons split by 2 H's = **triplet**
H_b protons split by CH_3 + CH_2 protons = **12 peaks** (maximum)
H_c protons split by 2 different CH_2 groups = **9 peaks** (maximum)
H_d protons split by 2 H's = **triplet**
Since H_b and H_c are located in a flexible alkyl chain, it is likely that peak overlap occurs, so that the following is observed: H_b (3 + 2 + 1 = 6 peaks), H_c (2 + 2 + 1 = 5 peaks).

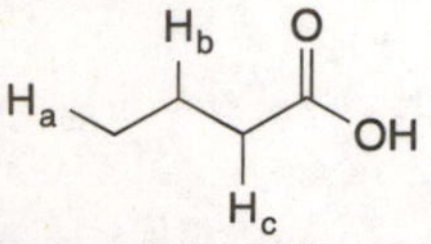

i. H_a H_b

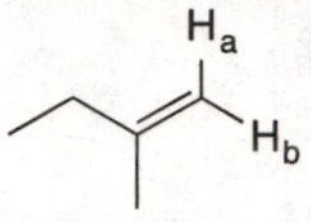

H_a: split by 1 H = **doublet**
H_b: split by 1 H = **doublet**

j. 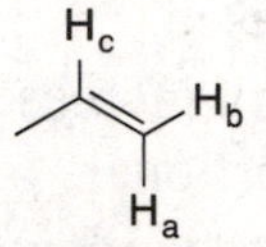 H_c H_b H_a

H_a: split by H_b + H_c = **doublet of doublets** (4 peaks)
H_b: split by H_a + H_c = **doublet of doublets** (4 peaks)
H_c: split by CH_3, H_a + H_b = **16 peaks**

14.43

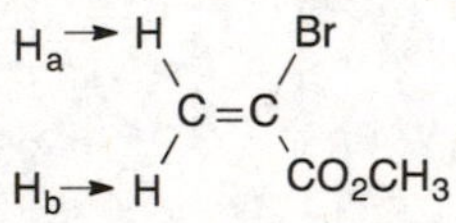

H_a → H, Br, C=C, H_b → H, CO_2CH_3

H_a: split by 1 H = **doublet**
H_b: split by 1 H = **doublet**
H_a and H_b are geminal.

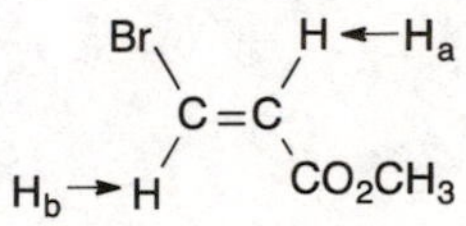

Br, H ← H_a, C=C, H_b → H, CO_2CH_3

H_a: split by 1 H = **doublet**
H_b: split by 1 H = **doublet**
H_a and H_b are trans.

Both compounds exhibit two doublets for the H's on the C=C, but the coupling constants ($J_{geminal}$ and J_{trans}) are different. $J_{geminal}$ is much smaller than J_{trans} (0–3 Hz versus 11–18 Hz).

14.44

a. 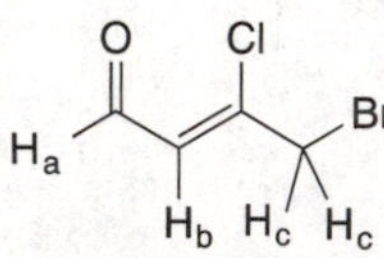 O Cl Br H_a H_b H_c H_c

H_a: 1 adjacent H_b = **doublet**
H_b: 1 adjacent H_a = **doublet**
H_c: no adjacent H's = **singlet**

b. 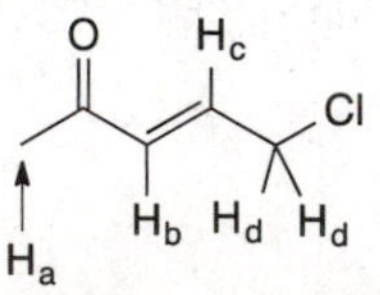 O H_c Cl H_a H_b H_d H_d

H_a: no adjacent H's = **singlet**
H_b: 1 adjacent H_c = **doublet**
H_c: 1 adjacent H_b and 2 adjacent H_d's (1+1)(2 + 1) = **6 peaks**
H_d: 1 adjacent H_c = **doublet**

14.45

H_b H_a C=C H_c CN

J_{ab} = 11.8 Hz
J_{bc} = 0.9 Hz
J_{ac} = 18 Hz

H_a: doublet of doublets at 5.7 ppm. Two large J values are seen for the H's cis (J_{ab} = 11.8 Hz) and trans (J_{ac} = 18 Hz) to H_a.
H_b: doublet of doublets at ~6.2 ppm. One large J value is seen for the cis H (J_{ab} = 11.8 Hz). The geminal coupling (J_{bc} = 0.9 Hz) is hard to see.
H_c: doublet of doublets at ~6.6 ppm. One large J value is seen for the trans H (J_{ac} = 18 Hz). The geminal coupling (J_{bc} = 0.9 Hz) is hard to see.

Splitting diagram for H_a

1 trans H_c proton splits H_a into
1 + 1 = 2 peaks
a doublet

1 cis H_b proton splits H_a into
1 + 1 = 2 peaks
Now it's a doublet of doublets.

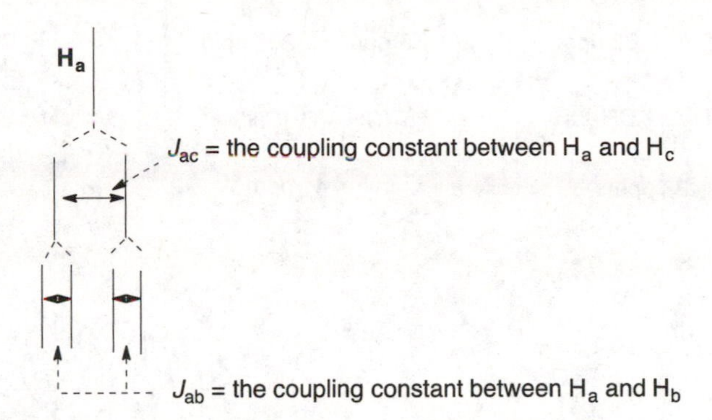

14.46

Four constitutional isomers of C_4H_9Br:

Br — 4 different C's
Br — 4 different C's
Br — 2 different C's
Br — 3 different C's

14.47 The O atom of an ester donates electron density, so the carbonyl carbon has less δ+, making it less deshielded than the carbonyl carbon of an aldehyde or ketone. Therefore, the carbonyl carbon of an aldehyde or ketone is more deshielded and absorbs farther downfield.

:O: R C OR' ⟷ :O:⁻ R C⁺ OR'

14.48

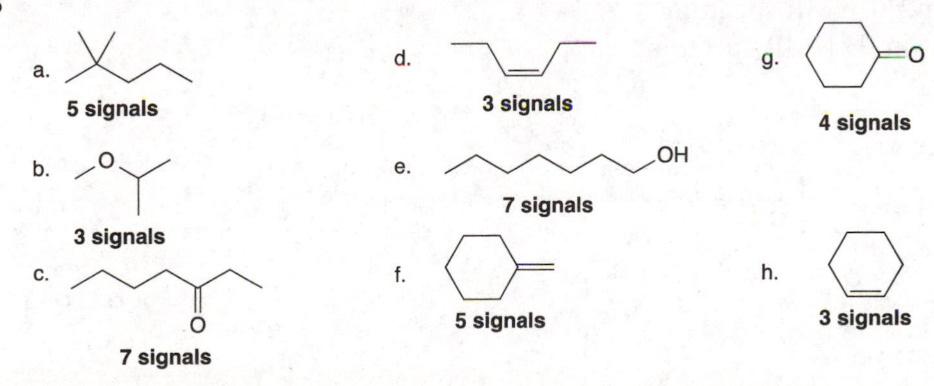

14.49

a. $C_a < C_b < C_c$

b. $C_b < C_c < C_a$

14.50

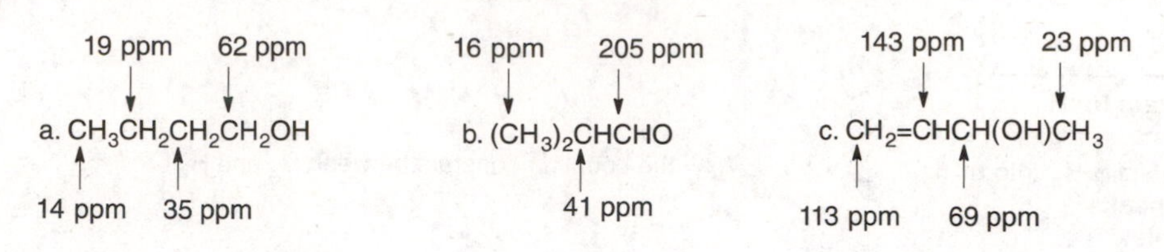

14.51

a. 3 different C's
3 signals

b. **3 signals**

4 signals

5 signals

14.52

1-hydroxybutan-2-one
A

4-hydroxybutan-2-one
B

The answers for parts (a)–(d) are the same for both compounds.

a. molecular ion for $C_4H_8O_2 = 88$

b. IR absorptions at 3200–3600 (OH), ~3000 (CH), and ~1700 (C=O) cm^{-1}.

c. Four lines in ^{13}C NMR spectrum

d. Four signals in ^{1}H NMR spectrum

e.

A: quartet at ~ 2.1 ppm; singlet at ~3.5 ppm; singlet anywhere in the 1–5 ppm region; triplet at ~1.0 ppm

B: triplet at ~3.5 ppm; singlet anywhere in the 1–5 ppm region; singlet at ~2.0 ppm; triplet at ~2.1 ppm

14.53 Use the directions from Answer 14.24.

a. **$C_4H_8Br_2$:** 0 degrees of unsaturation
IR peak at 3000–2850 cm^{-1}: **C*sp*^{3}–H bonds**
NMR: singlet at 1.87 ppm (6 H) (2 CH_3 groups)
singlet at 3.86 ppm (2 H) (CH_2 group)

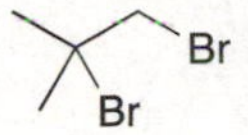

b. **$C_3H_6Br_2$:** 0 degrees of unsaturation
IR peak at 3000–2850 cm^{-1}: **C*sp*^{3}–H bonds**
NMR: quintet at 2.4 ppm (split by 2 CH_2 groups)
triplet at 3.5 ppm (split by 2 H's)

c. **$C_5H_{10}O_2$: 1 degree of unsaturation**
IR peak at 1740 cm^{-1}: **C=O**
NMR: triplet at 1.15 ppm (3 H) (CH_3 split by 2 H's)
triplet at 1.25 ppm (3 H) (CH_3 split by 2 H's)
quartet at 2.30 ppm (2 H) (CH_2 split by 3 H's)
quartet at 4.72 ppm (2 H) (CH_2 split by 3 H's)

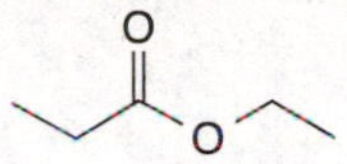

d. **$C_6H_{14}O$:** 0 degrees of unsaturation
IR peak at 3600–3200 cm^{-1}: **O–H**
NMR: triplet at 0.8 ppm (6 H) (2 CH_3 groups split by CH_2 groups)
singlet at 1.0 ppm (3 H) (CH_3)
quartet at 1.5 ppm (4 H) (2 CH_2 groups split by CH_3 groups)
singlet at 1.6 ppm (1 H) (O–H proton)

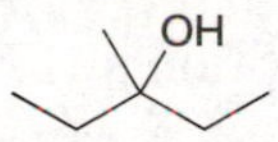

e. **$C_6H_{14}O$:** 0 degrees of unsaturation
IR peak at 3000–2850 cm^{-1}: **C*sp*^{3}–H bonds**
NMR: doublet at 1.10 ppm (relative area = 6) (from 12 H's)
septet at 3.60 ppm (relative area = 1) (from 2 H's)

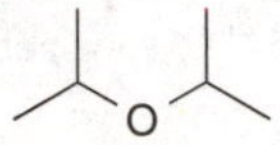

f. **C_3H_6O: 1 degree of unsaturation**
IR peak at 1730 cm^{-1}: **C=O**
NMR: triplet at 1.11 ppm
multiplet at 2.46 ppm
triplet at 9.79 ppm

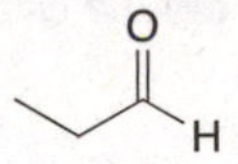

14.54

Two isomers of $C_9H_{10}O$: 5 degrees of unsaturation (benzene ring likely)

Compound A:
IR absorption at 1742 cm^{-1}: **C=O**
NMR data:
Absorptions:
singlet at 2.15 (3 H) (CH_3 group)
singlet at 3.70 (2 H) (CH_2 group)
broad singlet at 7.20 (5 H)
(likely a monosubstituted benzene ring)

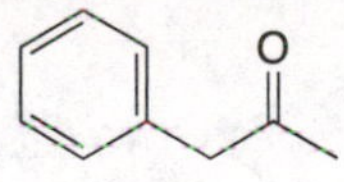

Compound B:
IR absorption at 1688 cm^{-1}: **C=O**
NMR data:
Absorptions:
triplet at 1.22 (3 H) (CH_3 group split by 2 H's)
quartet at 2.98 (2 H) (CH_2 group split by 3 H's)
multiplet at 7.28–7.95 (5 H)
(likely a monosubstituted benzene ring)

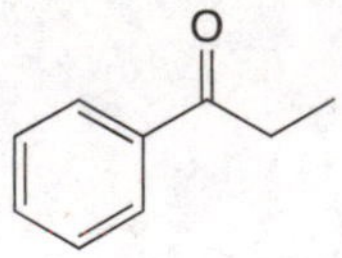

14.55 IR absorptions:

3088–2897 cm^{-1}: *sp*^{2} and *sp*^{3} hybridized C–H
1740 cm^{-1}: C=O
1606 cm^{-1}: benzene ring

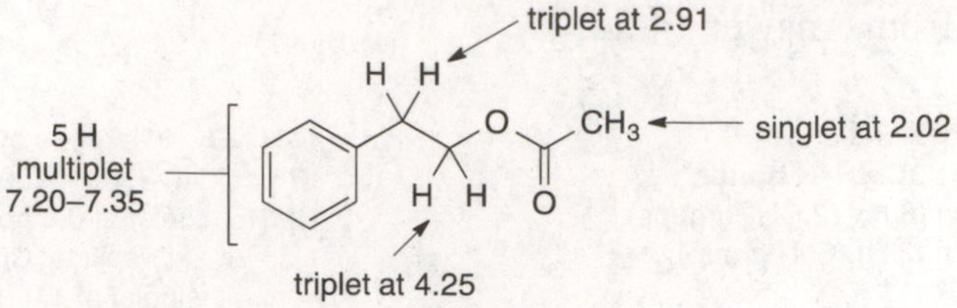

triplet at 2.91
5 H multiplet 7.20–7.35
singlet at 2.02
triplet at 4.25

$C_{10}H_{12}O_2$
W

14.56 IR absorption at 1713 cm^{-1} is due to C=O.

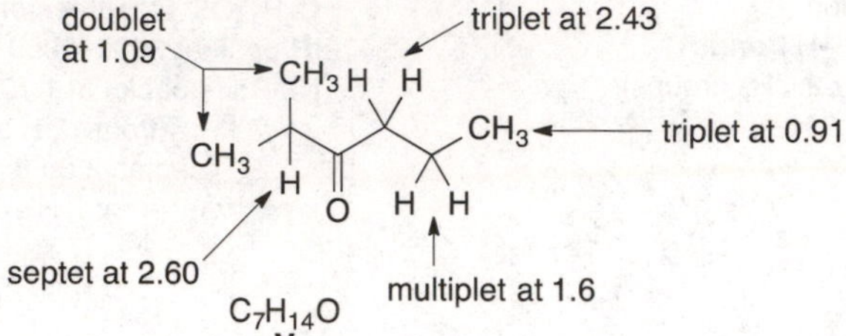

doublet at 1.09
triplet at 2.43
triplet at 0.91
septet at 2.60
multiplet at 1.6

$C_7H_{14}O$
V

14.57

Compound C:
molecular ion 146 (molecular formula $C_6H_{10}O_4$)
IR absorption at 1762 cm^{-1}: **C=O**
^{1}**H NMR** data:
Absorptions:
H_a: doublet at 1.47 (3 H) (CH_3 group adjacent to CH)
H_b: singlet at 2.07 (6 H) (2 CH_3 groups)
H_c: quartet at 6.84 (1 H adjacent to CH_3)

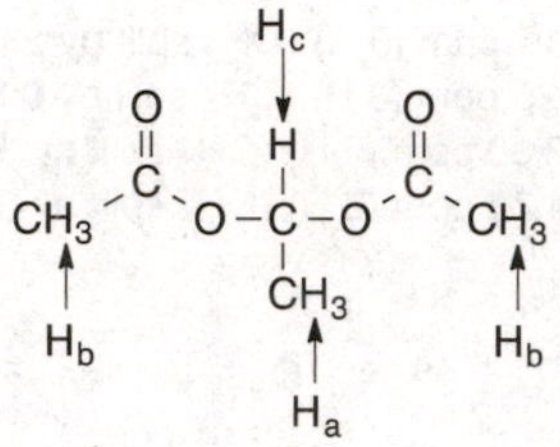

$CH_3C(=O)-O-CH(CH_3)-O-C(=O)CH_3$ (H_b on the acetyl CH_3 groups, H_c on CH, H_a on CH_3)

14.58

$(CH_3)_2C=O$ → [1] LiC≡CH, [2] H_2O → $CH_3-C(OH)(CH_3)-C≡CH$

(H_a on CH_3 groups, H_b on OH, H_c on ≡CH)

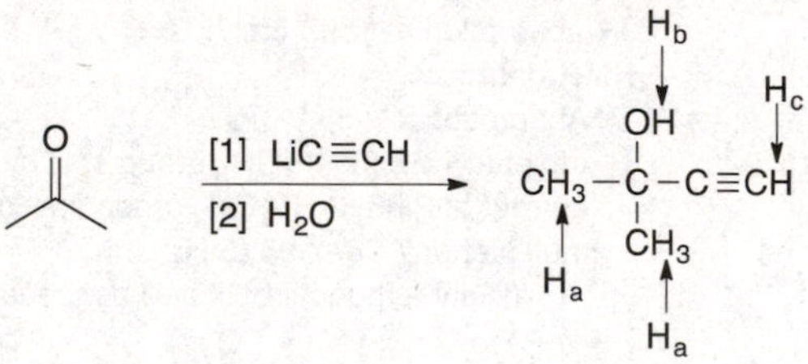

D

Compound D:
molecular ion 84 (molecular formula C_5H_8O)
IR absorptions at 3600–3200 cm^{-1}: OH
3303 cm^{-1}: C*sp*–H
2938 cm^{-1}: C*sp*^{3}–H
2120 cm^{-1}: C≡C
^{1}**H NMR** data:
Absorptions:
H_a: singlet at 1.53 (6 H) (2 CH_3 groups)
H_b: singlet at 2.37 (1 H)
H_c: singlet at 2.43 (1 H)
(H_b and H_c:) alkynyl CH and OH

14.59

Compound E:
$C_4H_8O_2$:
1 degree of unsaturation
IR absorption at 1743 cm^{-1}: **C=O**
NMR data:
H_a: quartet at 4.1 (**2 H**)
H_b: singlet at 2.0 (**3 H**)
H_c: triplet at 1.4 (**3 H**)

$CH_3C(=O)OCH_2CH_3$ — H_b = CH_3 (acetyl); H_a = OCH_2; H_c = CH_2CH_3

Compound F:
$C_4H_8O_2$:
1 degree of unsaturation
IR absorption at 1730 cm^{-1}: **C=O**
NMR data:
H_a: singlet at 4.1 (**2 H**)
H_b: singlet at 3.4 (**3 H**)
H_c: singlet at 2.1 (**3 H**)

$CH_3C(=O)CH_2OCH_3$ — H_c = CH_3 (acetyl); H_a = CH_2; H_b = OCH_3

14.60

Compound H:
$C_8H_{11}N$:
4 degrees of unsaturation
IR absorptions at 3365 cm^{-1}: N–H
3284 cm^{-1}: N–H
3026 cm^{-1}: Csp^2–H
2932 cm^{-1}: Csp^3–H
1603 cm^{-1}: due to benzene
1497 cm^{-1}: due to benzene
NMR data:
multiplet at 7.2–7.4 ppm, **5 H** on a benzene ring
H_a: triplet at 2.9 ppm, **2 H,** split by 2 H's
H_b: triplet at 2.8 ppm, **2 H,** split by 2 H's
H_c: singlet at 1.1 ppm, **2 H,** no splitting (on NH_2)

$C_6H_5CH_2CH_2NH_2$ — H_a, H_b, H_c (NH_2)

Compound I:
$C_8H_{11}N$:
4 degrees of unsaturation
IR absorptions at 3367 cm^{-1}: N–H
3286 cm^{-1}: N–H
3027 cm^{-1}: Csp^2–H
2962 cm^{-1}: Csp^3–H
1604 cm^{-1}: due to benzene
1492 cm^{-1}: due to benzene
NMR data:
multiplet at 7.2–7.4 ppm, **5 H** on a benzene ring
H_a: quartet at 4.1 ppm, **1 H,** split by 3H's
H_b: singlet at 1.45 ppm, **2 H,** no splitting (NH_2)
H_c: doublet at 1.4 ppm, **3 H,** split by 1 H

$C_6H_5CH(NH_2)CH_3$ — H_a = CH; H_b = NH_2; H_c = CH_3

14.61

a. **$C_9H_{10}O_2$:**
5 degrees of unsaturation
IR absorption at 1718 cm^{-1}: **C=O**
NMR data:
multiplet at 7.4–8.1 ppm, **5 H** on a benzene ring
quartet at 4.4 ppm, **2 H,** split by 3 H's
triplet at 1.3 ppm, **3 H,** split by 2 H's

$C_6H_5C(=O)OCH_2CH_3$ — OCH_2: downfield due to the O atom

b. **C_9H_{12}:**
4 degrees of unsaturation
IR absorption at 2850–3150 cm^{-1}:
C–H bonds
NMR data:
singlet at 7.1–7.4 ppm, **5 H,** benzene
septet at 2.8 ppm, **1 H,** split by 6 H's
doublet at 1.3 ppm, **6 H,** split by 1 H

$C_6H_5CH(CH_3)_2$

14.62 IR absorption at 1717 cm^{-1} is due to a C=O.

singlet at 1.02 — CH_3 CH_3 O, CH_3; CH_3 ← singlet at 2.13; H H ← singlet at 2.33

$C_7H_{14}O$
R

14.63 IR absorption at 1730 cm^{-1} is due to a C=O. Eight lines in the ^{13}C NMR spectrum means there are eight different types of C.

triplet at 4.20; H_a; H H; H_b; O; H ← singlet at 9.97; $(CH_3)_2N$; O; H_b; H H; H_a

singlet at 2.32; triplet at 3.05

H_a and H_b appear as two doublets.

14.64

a. Compound **J** has a molecular ion at 72: molecular formula **C_4H_8O**
1 degree of unsaturation
IR spectrum at 1710 cm^{-1}: **C=O**
^1H NMR data (ppm):
1.0 (triplet, 3 H), split by 2 H's
2.1 (singlet, 3 H)
2.4 (quartet, 2 H), split by 3 H's

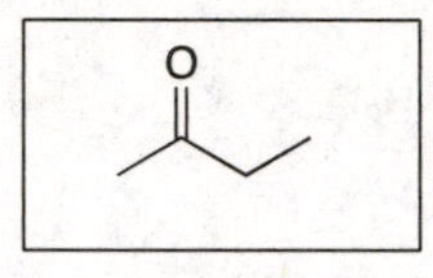

b. Compound **K** has a molecular ion at 88: molecular formula **$C_5H_{12}O$**
0 degrees of unsaturation
IR spectrum at 3600–3200 cm^{-1}: **O–H bond**
^1H NMR data (ppm):
0.9 (triplet, 3 H), split by 2 H's
1.2 (singlet, 6 H), due to 2 CH_3 groups
1.5 (quartet, 2 H), split by 3 H's
1.6 (singlet, 1 H), due to the OH proton

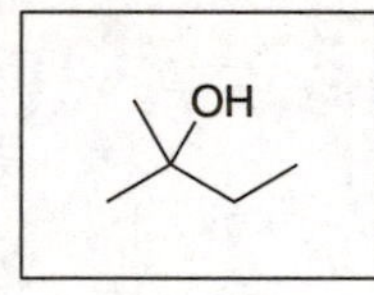

14.65

Compound **L** has a molecular ion at 90: molecular formula **$C_4H_{10}O_2$**
0 degrees of unsaturation
IR absorptions at 2992 and 2941 cm^{-1}: Csp^3–H
^1H NMR data (ppm):
H_a: 1.2 (doublet, 3 H), split by 1 H
H_b: 3.3 (singlet, 6 H), due to 2 CH_3 groups
H_c: 4.8 (quartet, 1 H), split by 3 adjacent H's

O, H + CH_3—OH $\xrightarrow{H^+}$ H_c; O ← H_b; H_a O ← H_b

L

14.66

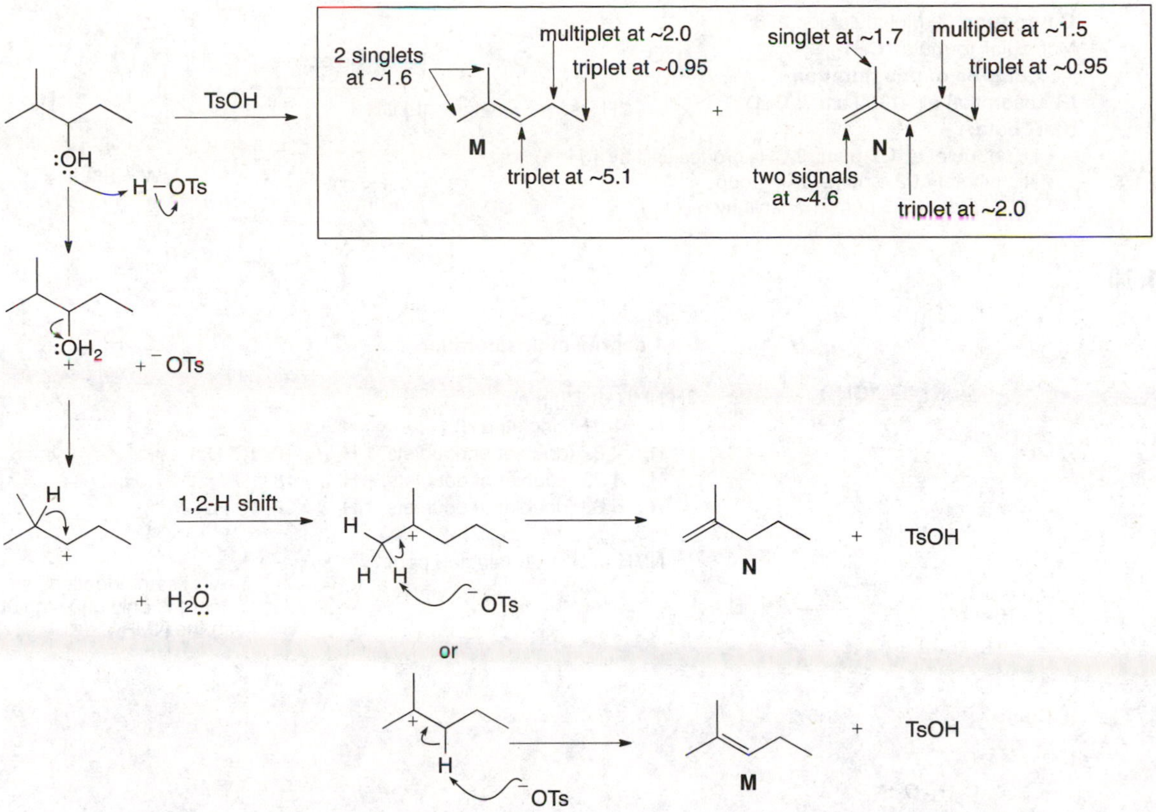

TsOH

M: 2 singlets at ~1.6; multiplet at ~2.0; triplet at ~0.95; triplet at ~5.1

N: singlet at ~1.7; multiplet at ~1.5; triplet at ~0.95; two signals at ~4.6; triplet at ~2.0

:OH + H–OTs

:OH_2 + $^-$OTs

+ $H_2\ddot{O}$

1,2-H shift

N + TsOH

$^-$OTs

or

M + TsOH

$^-$OTs

14.67

Compound **O** has a molecular formula **$C_{10}H_{12}O$.**

5 degrees of unsaturation

IR absorption at 1687 cm^{-1}

^{1}H NMR data (ppm):

- H_a: 1.0 (triplet, 3 H), due to CH_3 group, split by 2 adjacent H's
- H_b: 1.7 (sextet, 2 H), split by CH_3 and CH_2 groups
- H_c: 2.9 (triplet, 2 H), split by 2 H's
- 7.4–8.0 (multiplet, 5 H), benzene ring

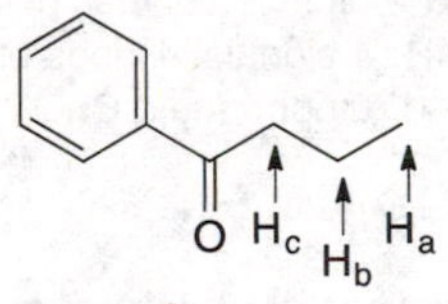

O H_c H_b H_a

O

14.68

Compound **P** has a molecular formula **$C_5H_9ClO_2$.**

1 degree of unsaturation

^{13}C NMR shows 5 different C's, including a C=O.

^{1}H NMR data (ppm):

- H_a: 1.3 (triplet, 3 H), split by 2 H's
- H_b: 2.8 (triplet, 2 H), split by 2 H's
- H_c: 3.7 (triplet, 2 H), split by 2 H's
- H_d: 4.2 (quartet, 2 H), split by CH_3 group

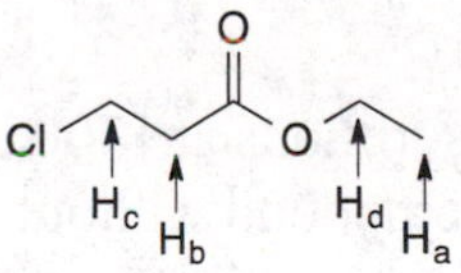

Cl O O H_c H_b H_d H_a

P

14.69

Compound Q: Molecular ion at 86.
Molecular formula: **$C_5H_{10}O$:**
1 degree of unsaturation
IR absorption at ~1700 cm^{-1}: **C=O**
NMR data:
H_a: doublet at 1.1 ppm, 2 CH_3 groups split by 1 H
H_b: singlet at 2.1 ppm, CH_3 group
H_c: septet at 2.6 ppm, 1 H split by 6 H's

[1] strong base
[2] CH_3I
H_b O H_c
H_a
H_a
= **Q**
MW = 86

14.70

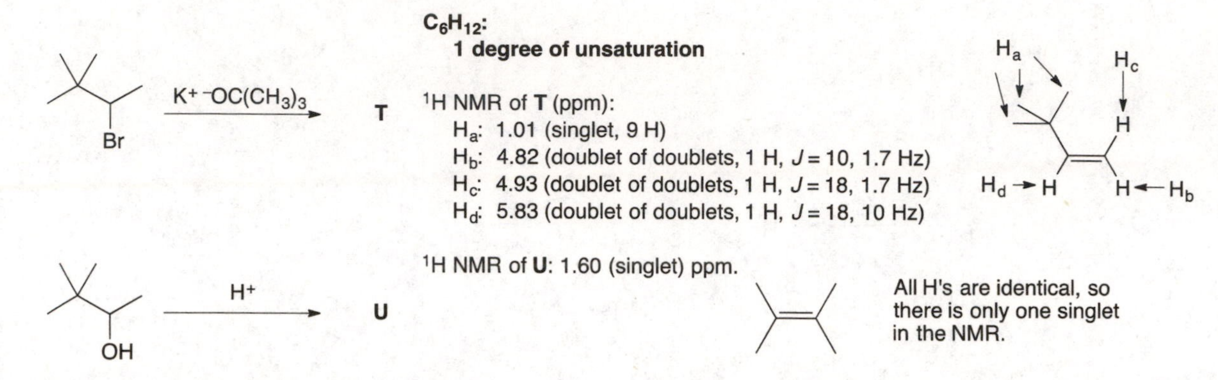

14.71

a.

$C_6H_{12}O_2$:
1 degree of unsaturation
IR peak at 1740 cm^{-1}: **C=O**
^{1}H NMR 2 signals: 2 types of H's
^{13}C NMR: 4 signals: 4 kinds of C's, including one at ~170 ppm due a C=O

b.

C_6H_{10}:
2 degrees of unsaturation
IR peak at 3000 cm^{-1}: **C*sp*^{3}–H bonds**
peak at 3300 cm^{-1}: **C_{sp}–H bond**
peak at ~2150 cm^{-1}: **C≡C bond**
^{13}C NMR: 4 signals: 4 kinds of C's

14.72

a. Because **A** has no absorptions at 1700 cm^{-1} or 3600–3200 cm^{-1}, it has no C=O or OH. An oxygen-containing compound without these functional groups is an ether (or an epoxide). Because **B** is formed from a reaction with HCl, **A** must contain an epoxide, because ethers are unreactive with HCl.

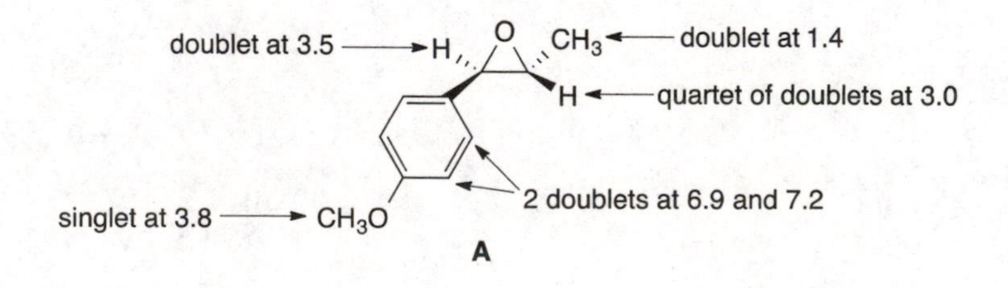

b. Epoxide **A** is equally substituted by R groups on both C's. With HCl, the epoxide is protonated first and then backside attack by Cl^- forms the chlorohydrin. Attack at the C adjacent to the benzene ring is preferred because the δ+ in the transition state at this carbon can be delocalized on the benzene ring.

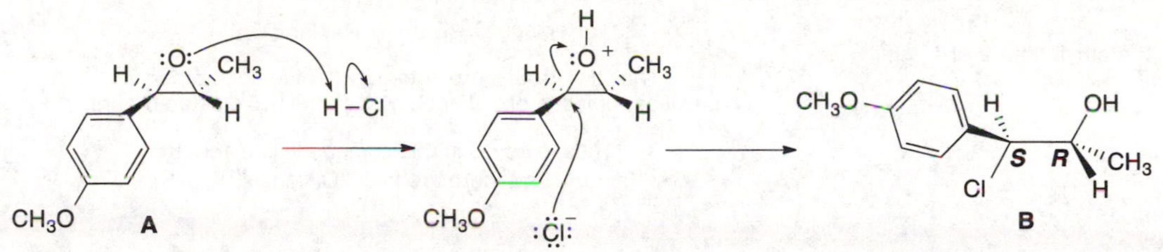

A   H–Cl   B

The benzene ring and CH_3 group must be trans in the epoxide to give the correct configuration at the two stereogenic centers in the product.

14.73 A second resonance structure for *N,N*-dimethylformamide places the two CH_3 groups in different environments. One CH_3 group is cis to the O atom, and one is cis to the H atom. This gives rise to two different absorptions for the CH_3 groups.

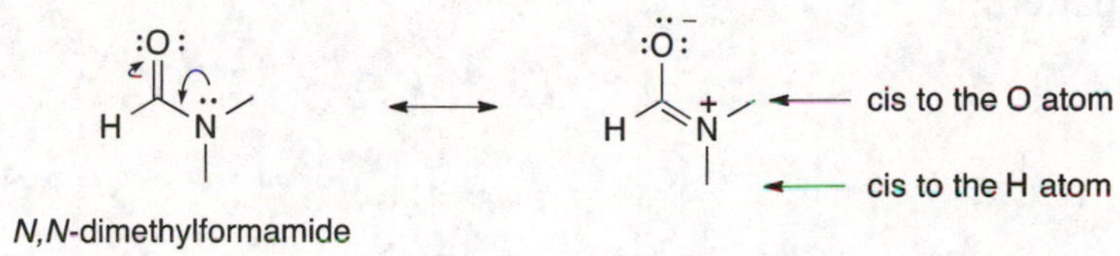

N,N-dimethylformamide

cis to the O atom

cis to the H atom

14.74

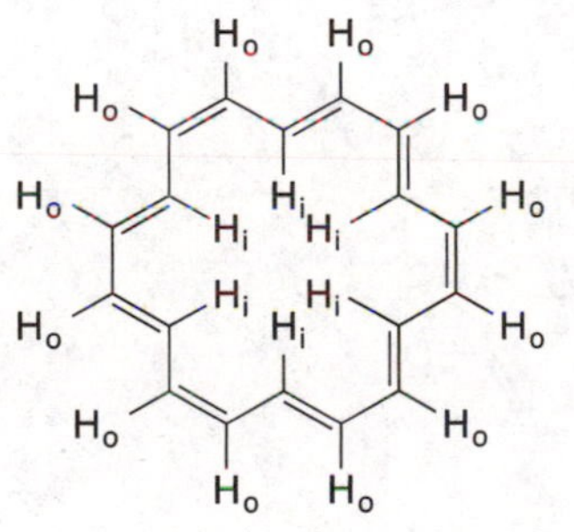

18-Annulene has 18 π electrons that create an **induced magnetic field** similar to the 6 π electrons of benzene. 18-Annulene has 12 protons that are oriented on the outside of the ring (labeled H_o), and 6 protons that are oriented inside the ring (labeled H_i). The induced magnetic field reinforces the external field in the vicinity of the protons on the outside of the ring. These H_o protons are deshielded, so they absorb downfield (8.9 ppm). In contrast, the induced magnetic field is opposite in direction to the applied magnetic field in the vicinity of the protons on the inside of the ring. This shields the H_i protons, so they absorb very far upfield, at –1.8 ppm, which is even higher than TMS.

14.75

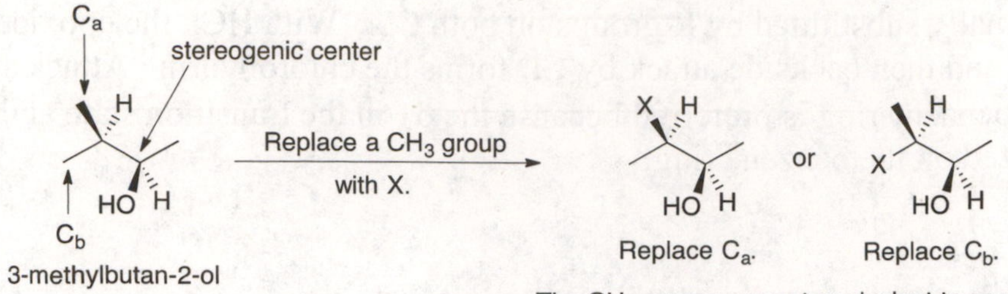

C_a

stereogenic center

H

HO H

C_b

3-methylbutan-2-ol

Replace a CH_3 group with X.

X H

HO H

Replace C_a.

or

H

X

HO H

Replace C_b.

The CH_3 groups are not equivalent to each other, because replacement of each by X forms two diastereomers.

Thus, every C in this compound is different and there are five ^{13}C signals.

14.76

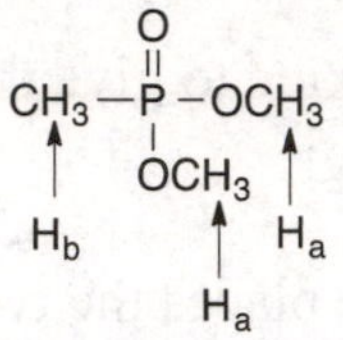

$CH_3-P(=O)(-OCH_3)(OCH_3)$

H_b H_a H_a

All 6 H_a protons are equivalent.

One P atom splits each nearby CH_3 into a doublet by the $n + 1$ rule, making two doublets.

14.77 a. Splitting pattern:

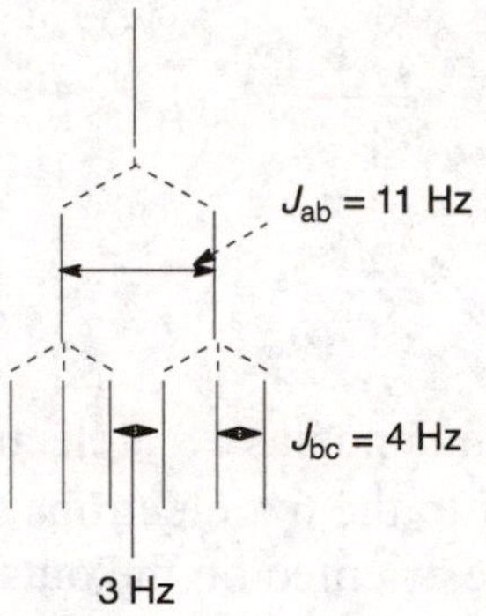

J_{ab} = 11 Hz

J_{bc} = 4 Hz

3 Hz

b. Three resonance structures can be drawn for cyclohex-2-enone.

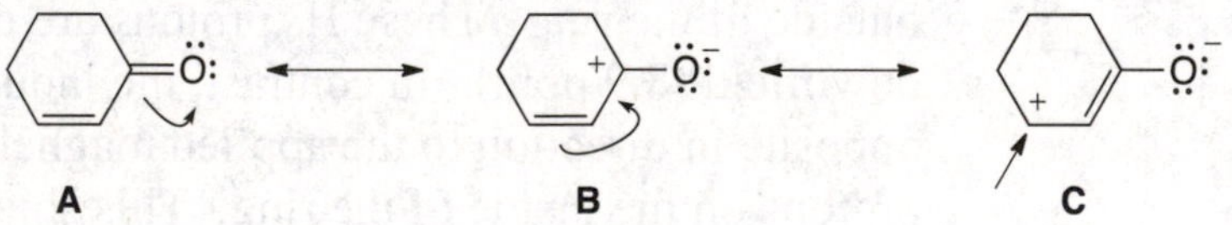

A B C

Resonance structure **C** places a (+) charge on one C of the C=C, deshielding the H attached to it and shifting the absorption downfield.

Chapter 15 Radical Reactions

Chapter Review

General features of radicals

- A radical is a reactive intermediate with an unpaired electron (15.1).
- A carbon radical is sp^2 hybridized and trigonal planar (15.1).
- The stability of a radical increases as the number of C's bonded to the radical carbon increases (15.1).

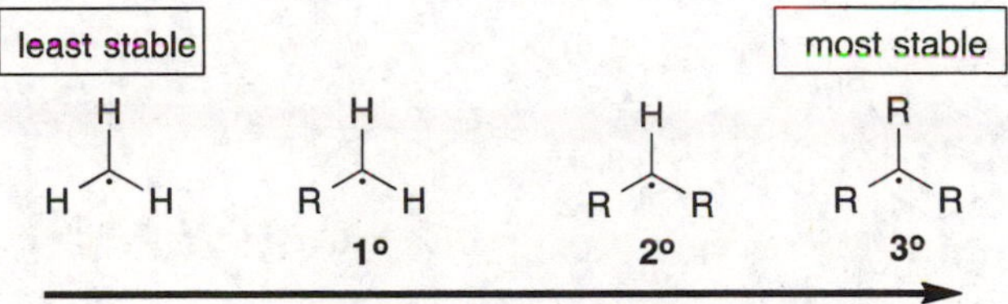

- Allylic radicals are stabilized by resonance, making them more stable than 3° radicals (15.10).

two resonance structures for the allyl radical

Radical reactions

[1] Halogenation of alkanes (15.4)

$$R{-}H \xrightarrow[h\nu \text{ or } \Delta]{X_2} R{-}X \text{ (alkyl halide)}$$

X = Cl or Br

- The reaction follows a radical chain mechanism.
- The weaker the C–H bond, the more readily the H is replaced by X.
- Chlorination is faster and less selective than bromination (15.6).
- Radical substitution results in racemization at a stereogenic center (15.8).

[2] Allylic halogenation (15.10)

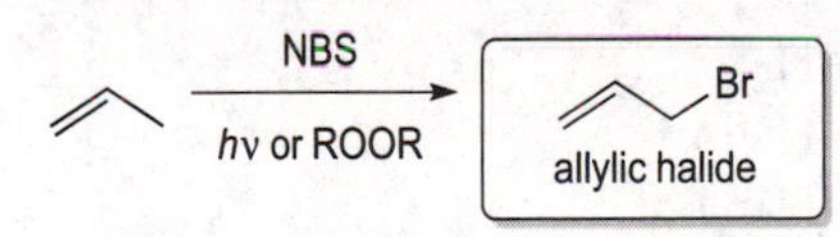

- The reaction follows a radical chain mechanism.

[3] Radical addition of HBr to an alkene (15.13)

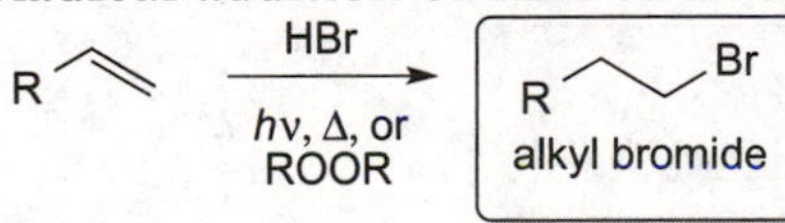

- A radical addition mechanism is followed.
- Br bonds to the less substituted carbon atom to form the more substituted, more stable radical.

[4] Radical polymerization of alkenes (15.14)

ROOR → polymer

- A radical addition mechanism is followed.

Practice Test on Chapter Review

1.a. Which alkyl halide(s) can be made in good yield by radical halogenation of an alkane?

1. 2. 3.

4. Both (1) and (2) can be made in good yield.
5. Compounds (1), (2), and (3) can all be made in good yield.

b. In which of the following reactions will rearrangement ***not*** occur?
1. halogenation of an alkane with Cl_2 and heat
2. addition of Cl_2 to an alkene
3. addition of HCl to an alkene
4. Rearrangements do not occur in reactions (1) and (2).
5. Rearrangements do not occur in reactions (1), (2), and (3).

c. Which labeled H is most easily abstracted in a radical halogenation reaction?

1. H_a
2. H_b
3. H_c
4. H_d
5. H_e

d. Which of the labeled C–H bonds in the following compound has the smallest bond dissociation energy?

1. C–H_a
2. C–H_b
3. C–H_c
4. C–H_d
5. C–H_e

2. (a) Which radical is the most stable? (b) Which radical is the least stable?

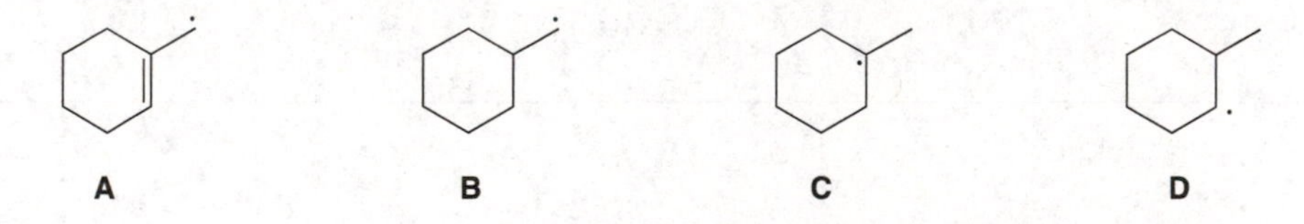

3. Draw all of the organic products formed in each reaction. Indicate stereochemistry in part (c).

a. 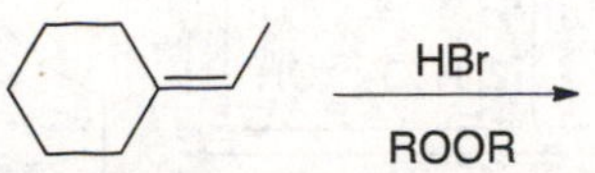 HBr, ROOR →

b.  NBS, $h\nu$ →

c. 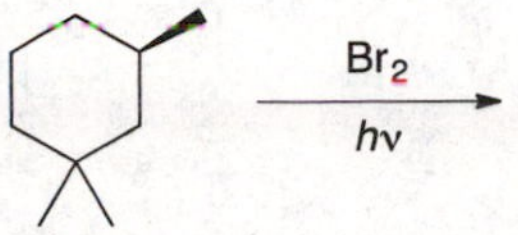 Br_2, $h\nu$ →

4. What monomer is needed to make the following polymer?

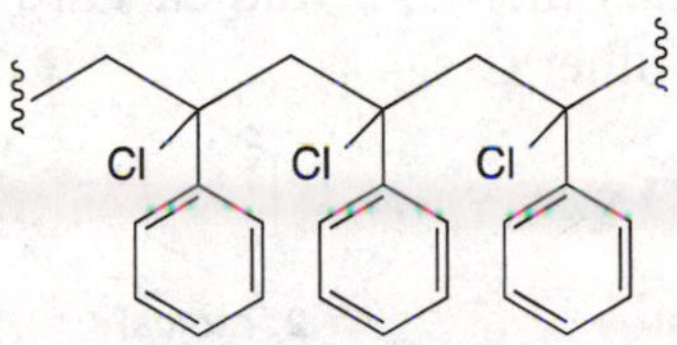

Cl Cl Cl

5. In each box, fill in the appropriate reagents needed to carry out the given reaction. This question involves reactions from Chapter 15, as well as previous chapters.

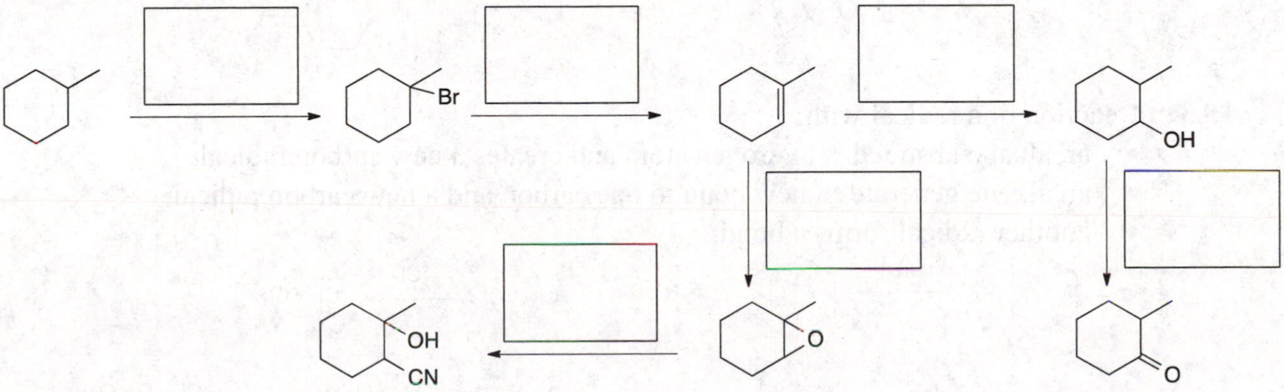

Br

OH

OH

CN

O

O

Answers to Practice Test

1.a. 2
b. 4
c. 2
d. 3

2.a. **A**
b. **B**

3.a.

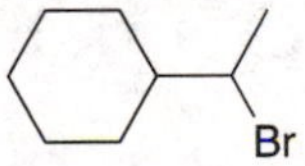

Br

3.b.

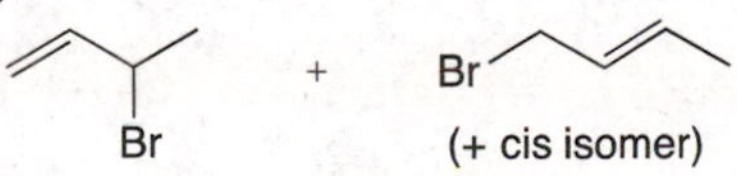

Br + Br (+ cis isomer)

3.c.

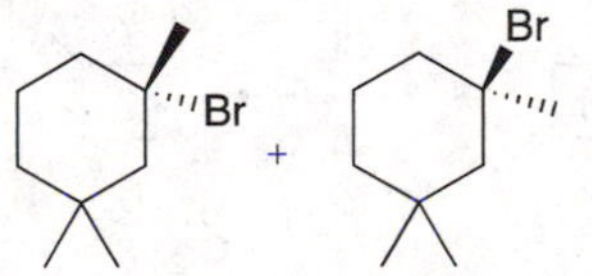

Br + Br

4.

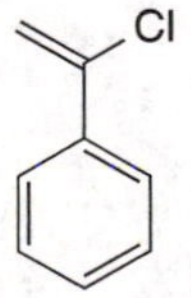

Cl

5.

Br_2 / $h\nu$ or Δ → Br → $KOC(CH_3)_3$ → [1] BH_3 [2] H_2O_2, ^-OH → OH

PCC or CrO_3 or $Na_2Cr_2O_7$ → O

mCPBA → O → [1] ^-CN [2] H_2O → OH CN

Answers to Problems

15.1 1° Radicals are on C's bonded to one other C; 2° radicals are on C's bonded to two other C's; 3° radicals are on C's bonded to three other C's.

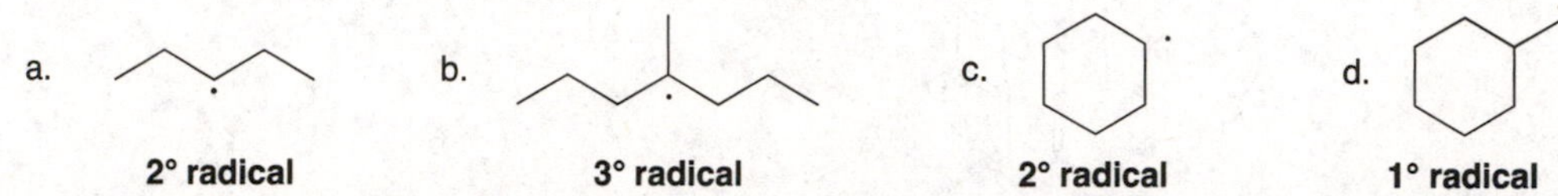

2° radical   3° radical   2° radical   1° radical

15.2 The stability of a radical increases as the number of alkyl groups bonded to the radical carbon increases. Draw the most stable radical.

a. b. c. d.

15.3 Reaction of a radical with:

- an alkane abstracts a hydrogen atom and creates a new carbon radical.
- an alkene generates a new bond to one carbon and a new carbon radical.
- another radical forms a bond.

a. :Cl· → + H–Cl:

c. :Cl· —(:Cl·)→ :Cl–Cl:

b. $CH_2{=}CH_2$ —(:Cl·)→ $\dot{C}H_2{-}CH_2$(:Cl:)

d. :Cl· + ·O–O· → :Cl–O–O·

15.4 **Monochlorination** is a radical substitution reaction in which a Cl replaces a H, thus generating an alkyl halide.

a. —(Cl_2, Δ)→ Cl

b. —(Cl_2, Δ)→ Cl + Cl + Cl

c. —(Cl_2, Δ)→ Cl + Cl

15.5

A [reaction: 2-methylpropane + Cl_2, Δ → 1-chloro-2,2-dimethyl product (Cl)]

B [reaction: pentane + Cl_2, Δ → 2-chloropentane + 3-chloropentane + 1-chloropentane]

15.6

Initiation: $:\ddot{Br}-\ddot{Br}:$ $\xrightarrow[\text{or } \Delta]{h\nu}$ $:\ddot{Br}\cdot$ + $\cdot\ddot{Br}:$

Propagation: CH_3-H + $\cdot\ddot{Br}:$ ⟶ $\dot{C}H_3$ + $H-\ddot{Br}:$

$\dot{C}H_3$ + $:\ddot{Br}-\ddot{Br}:$ ⟶ $CH_3-\ddot{Br}:$ + $\cdot\ddot{Br}:$

Termination: $:\ddot{Br}\cdot$ + $\cdot\ddot{Br}:$ ⟶ $:\ddot{Br}-\ddot{Br}:$

or

$\dot{C}H_3$ + $\dot{C}H_3$ ⟶ CH_3-CH_3

or

$\dot{C}H_3$ + $\cdot\ddot{Br}:$ ⟶ $CH_3-\ddot{Br}:$

15.7 The rate-determining step for halogenation reactions is formation of $CH_3\cdot$ + HX.

CH_3-H + $\cdot\ddot{I}:$ ⟶ $\cdot CH_3$ + $H-\ddot{I}:$ $\Delta H° = +138$ kJ/mol

1 bond broken +435 kJ/mol

1 bond formed −297 kJ/mol

This reaction is more endothermic and has a higher E_a than a similar reaction with Cl_2 or Br_2.

15.8 The **weakest C–H bond** in each alkane is the **most readily cleaved** during radical halogenation.

a. [structure with H indicated] 3° **most reactive**

b. [structure with H indicated] 3° **most reactive**

c. [structure with H indicated] 2° **most reactive**

15.9 To draw the product of bromination:

- Draw out the starting material and find the most reactive C–H bond (on the most substituted C).
- The major product is formed by **cleavage of the *weakest* C–H bond.**

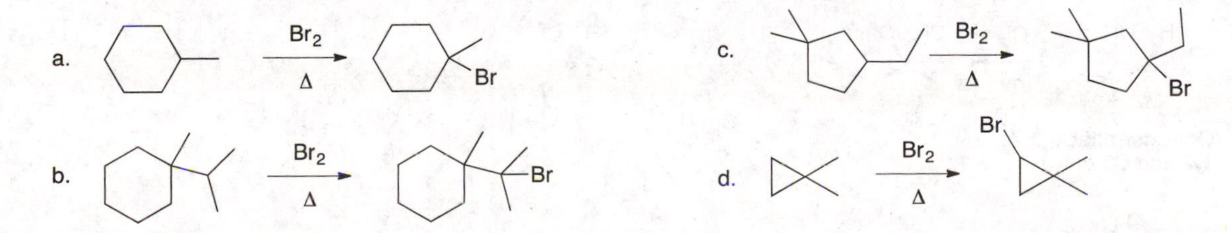

15.10 If 1° C–H and 3° C–H bonds were equally reactive there would be nine times as much $(CH_3)_2CHCH_2Cl$ as $(CH_3)_3CCl$ because the ratio of 1° to 3° H's is 9:1. The fact that the ratio is only 63:37 shows that the 1° C–H bond is less reactive than the 3° C–H bond. $(CH_3)_2CHCH_2Cl$ is still the major product, though, because there are nine 1° C–H bonds and only one 3° C–H bond.

15.11

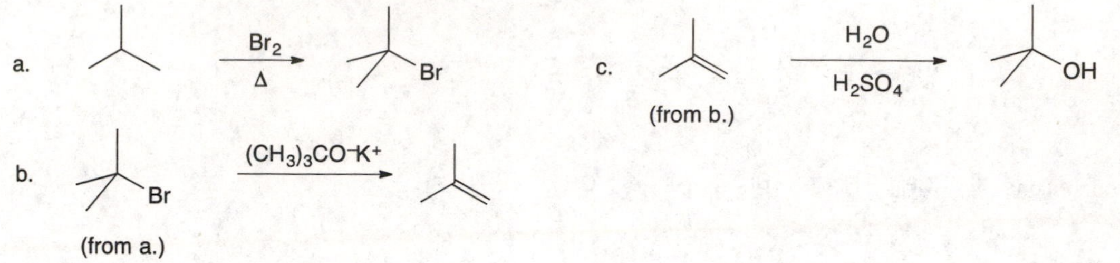

a. Br_2, Δ → Br

b. Br (from a.) $(CH_3)_3CO^-K^+$ →

c. (from b.) H_2O, H_2SO_4 → OH

15.12

a. Br_2, hν → Br; K^+ $^-OC(CH_3)_3$ → ; Br_2 → Br, Br + enantiomer

b. (from a.) mCPBA → O

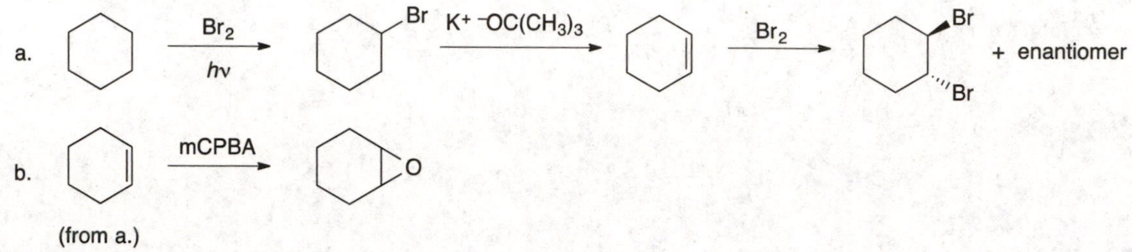

15.13 The reaction does not occur at the stereogenic center, so leave it as is.

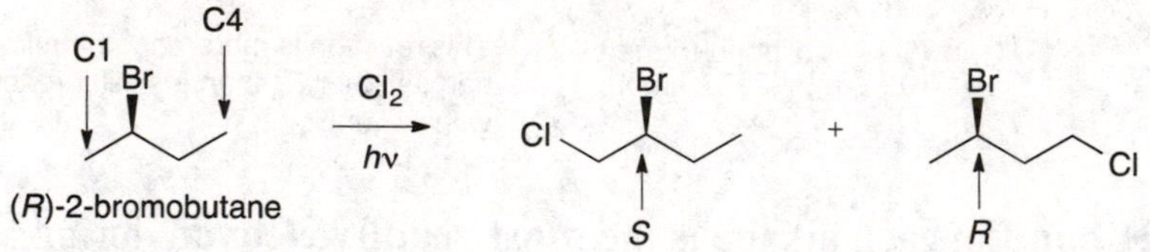

C1, C4, Br; (R)-2-bromobutane; Cl_2, hν → Cl, Br (S) + Br, Cl (R)

15.14

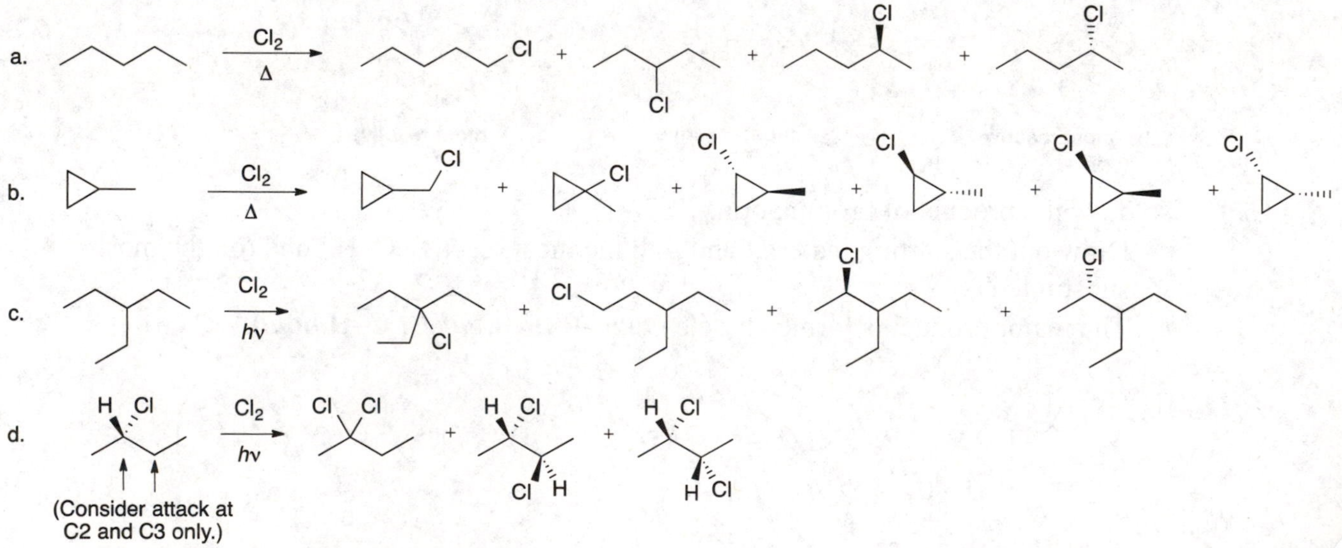

a. Cl_2, Δ → Cl + Cl + Cl + Cl

b. Cl_2, Δ → Cl + Cl + Cl + Cl + Cl + Cl

c. Cl_2, hν → Cl + Cl + Cl + Cl

d. H Cl (Consider attack at C2 and C3 only.) Cl_2, hν → Cl Cl + H Cl, Cl H + H Cl, H Cl

15.15

Chain propagation:

$:\ddot{O}=\ddot{N}\cdot + O_3 \longrightarrow :\ddot{O}=\ddot{N}-\ddot{O}\cdot + O_2$

$:\ddot{O}=\ddot{N}-\ddot{O}\cdot + \cdot\ddot{O}\cdot \longrightarrow :\ddot{O}=\ddot{N}\cdot + O_2$

The radical is re-formed.

15.16 Draw the resonance structure by moving the π bond and the unpaired electron. The hybrid is drawn with dashed lines for bonds that are in one resonance structure but not another. The symbol δ· is used on any atom that has an unpaired electron in any resonance structure.

a. 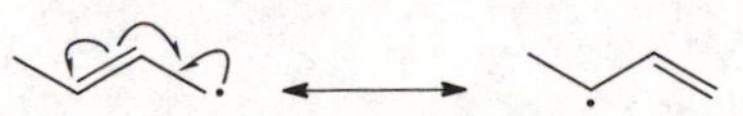 hybrid: δ· δ·

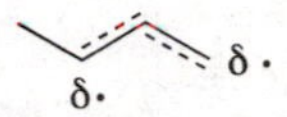

b. hybrid: δ· δ·

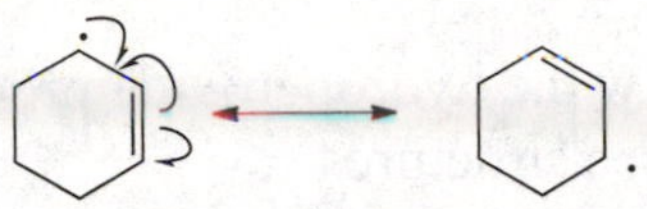

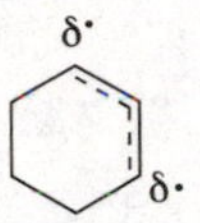

c. hybrid: δ· δ·

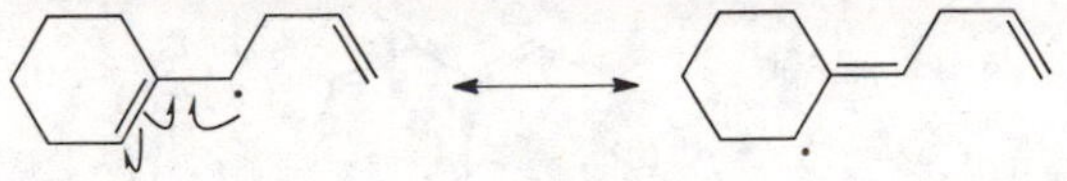

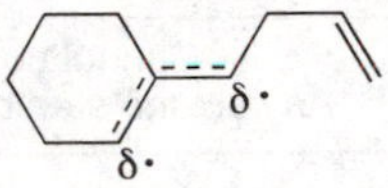

d. hybrid: δ· δ·

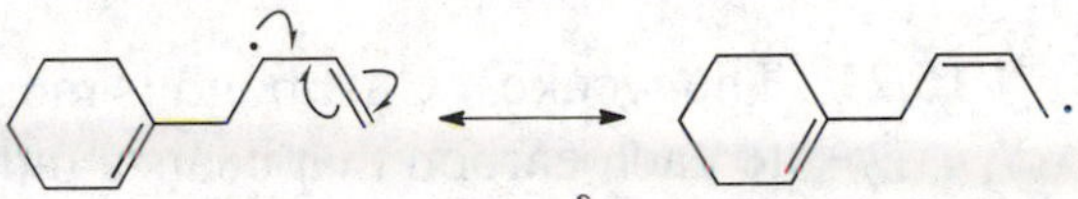

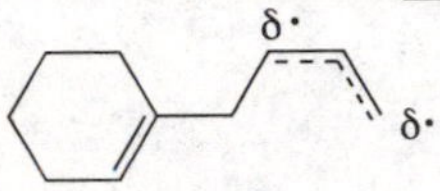

15.17 Reaction of an alkene with NBS or Br_2 + $h\nu$ yields allylic substitution products.

a. 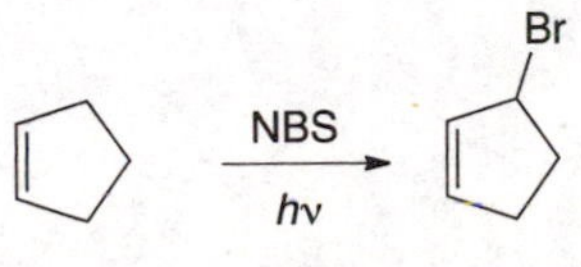 NBS, $h\nu$ → Br

b. 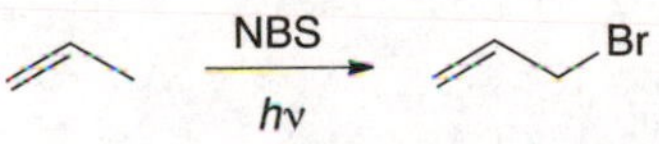 NBS, $h\nu$ → Br

c. 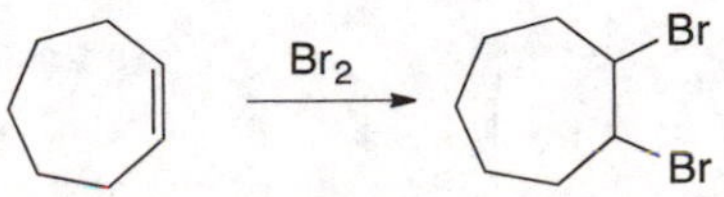 Br_2 → Br, Br

15.18

a. 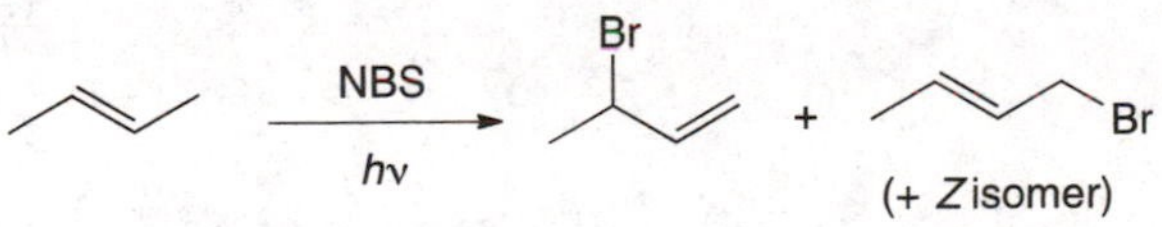 NBS, $h\nu$ → Br + Br (+ *Z* isomer)

b. 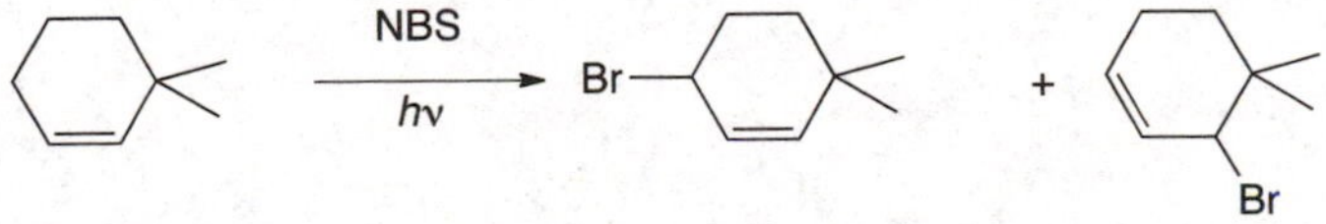 NBS, $h\nu$ → Br + Br

c. 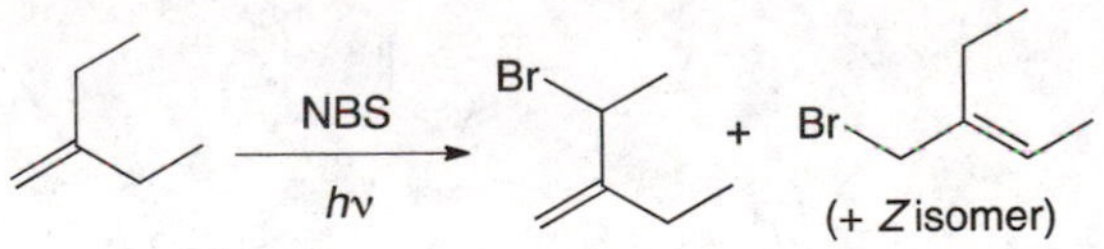 NBS, $h\nu$ → Br + Br (+ *Z* isomer)

15.19

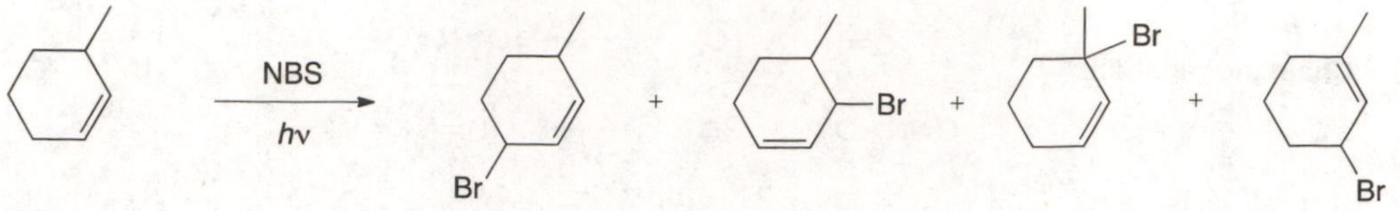

NBS
hν
Br
Br
Br
Br

15.20 Reaction of an alkene with NBS + *hν* yields allylic substitution products.

a. Br ⟹ one possible product: **high yield**

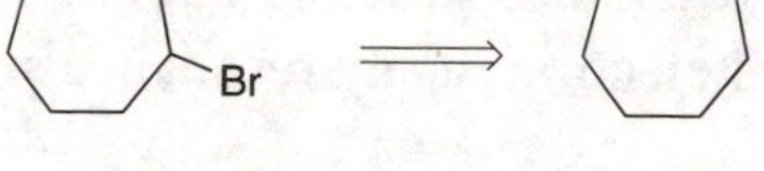

b. Br c. Br

Cannot be made in high yield by allylic halogenation.
Any alkene starting material would yield a mixture of allylic halides.

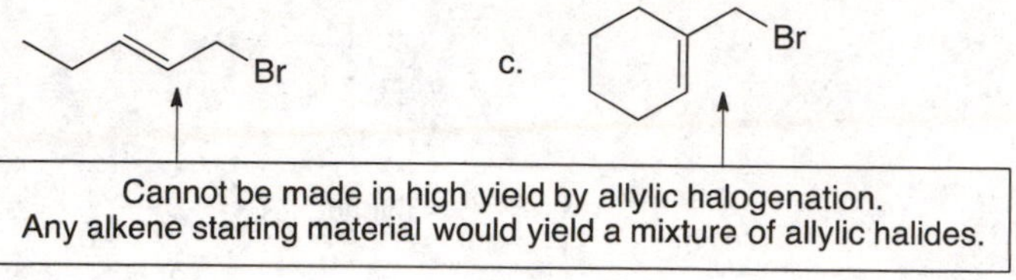

15.21 The weakest C–H bond is most readily cleaved. To draw the hydroperoxide products, add OOH to each carbon that bears a radical in one of the resonance structures.

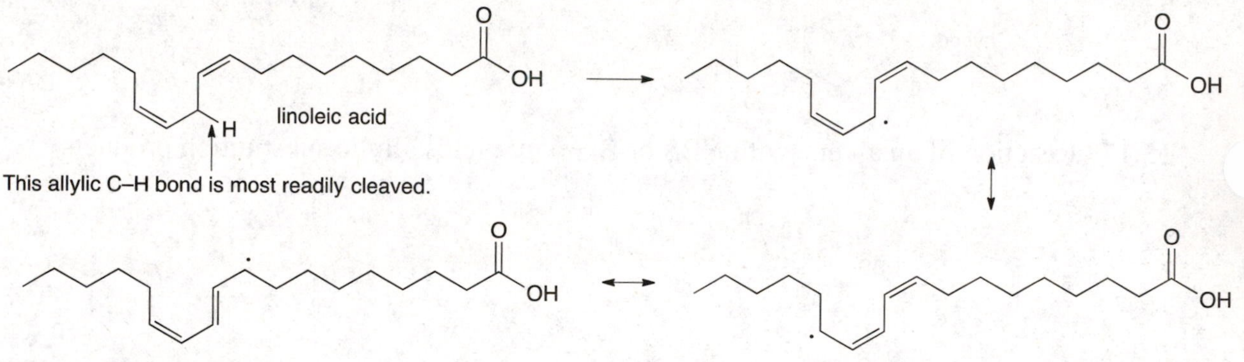

O OH linoleic acid H O OH

This allylic C–H bond is most readily cleaved.

O OH O OH

hydroperoxide products:

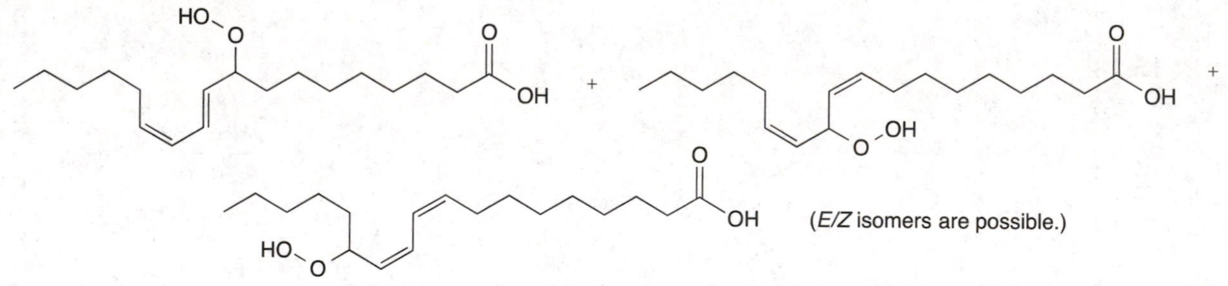

HO O O OH + O OH OH O +

O OH HO O (*E/Z* isomers are possible.)

15.22

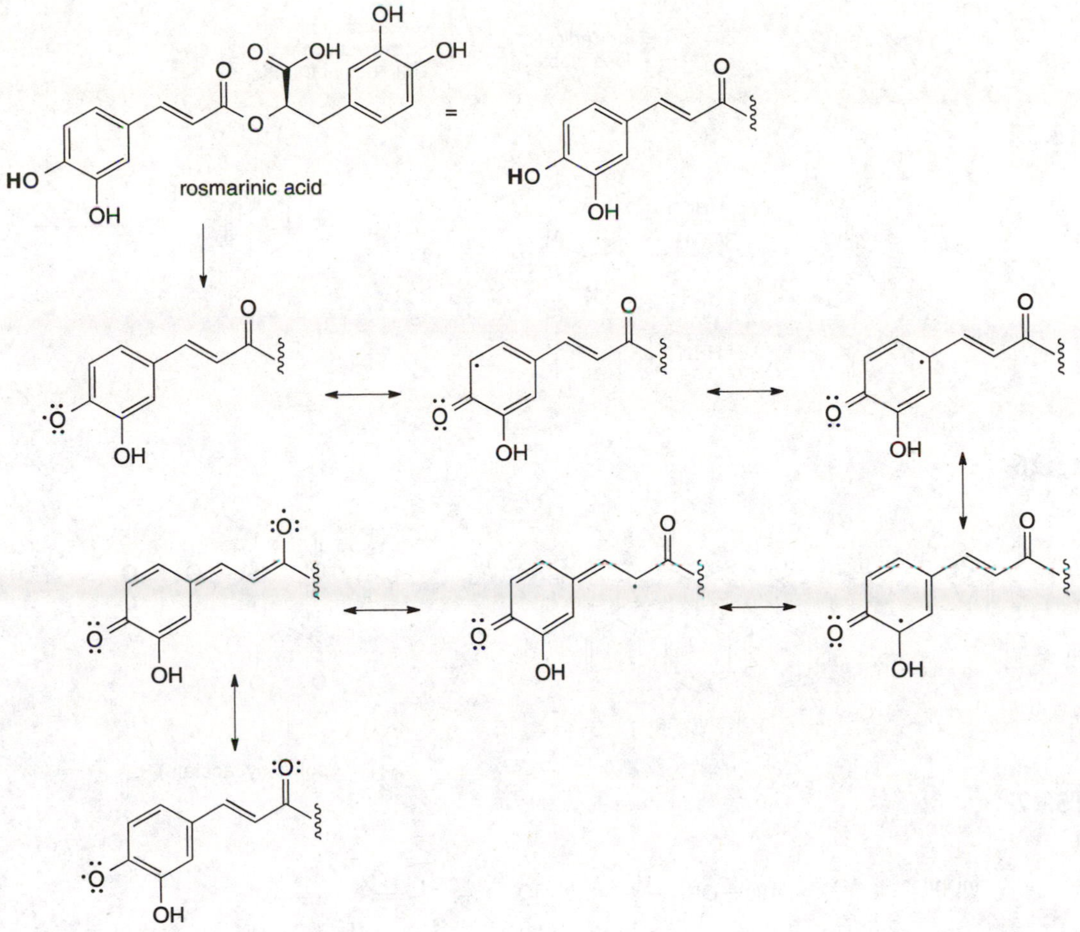

HO
OH
O
O
OH
O
OH
OH
rosmarinic acid
=
O
HO
OH

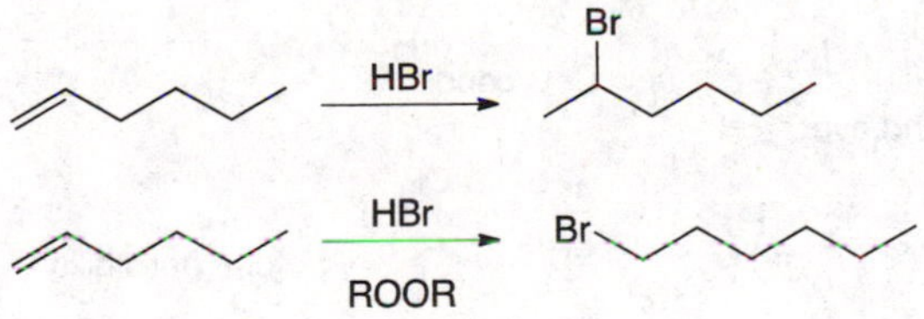

15.23

a. HBr → Br

HBr, ROOR → Br

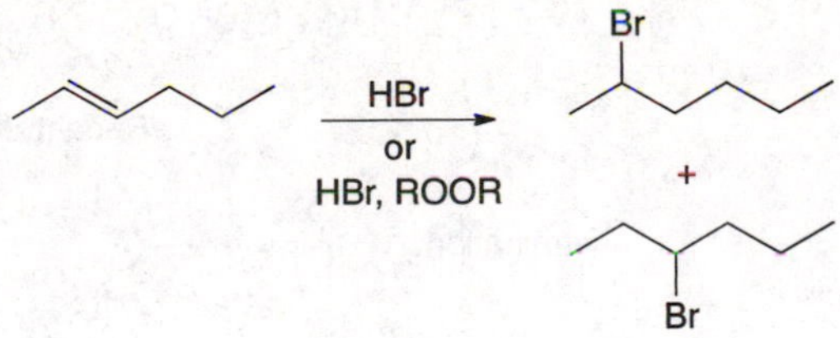

b. HBr → Br

HBr, ROOR → Br

c. HBr or HBr, ROOR → Br + Br

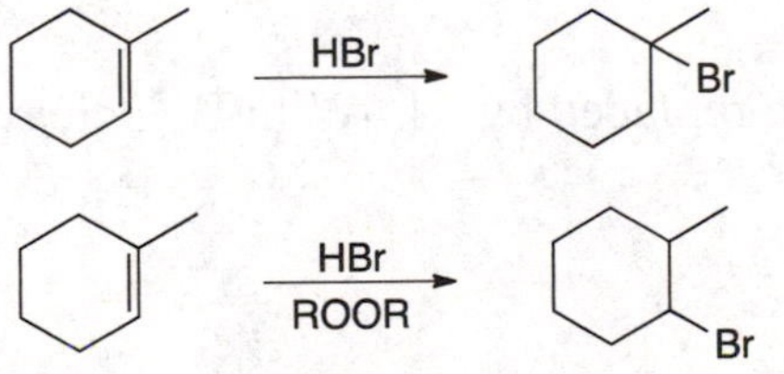

15.24 In addition of HBr under radical conditions:

- Br· adds first to form the more stable radical.
- Then H· is added to the carbon radical.

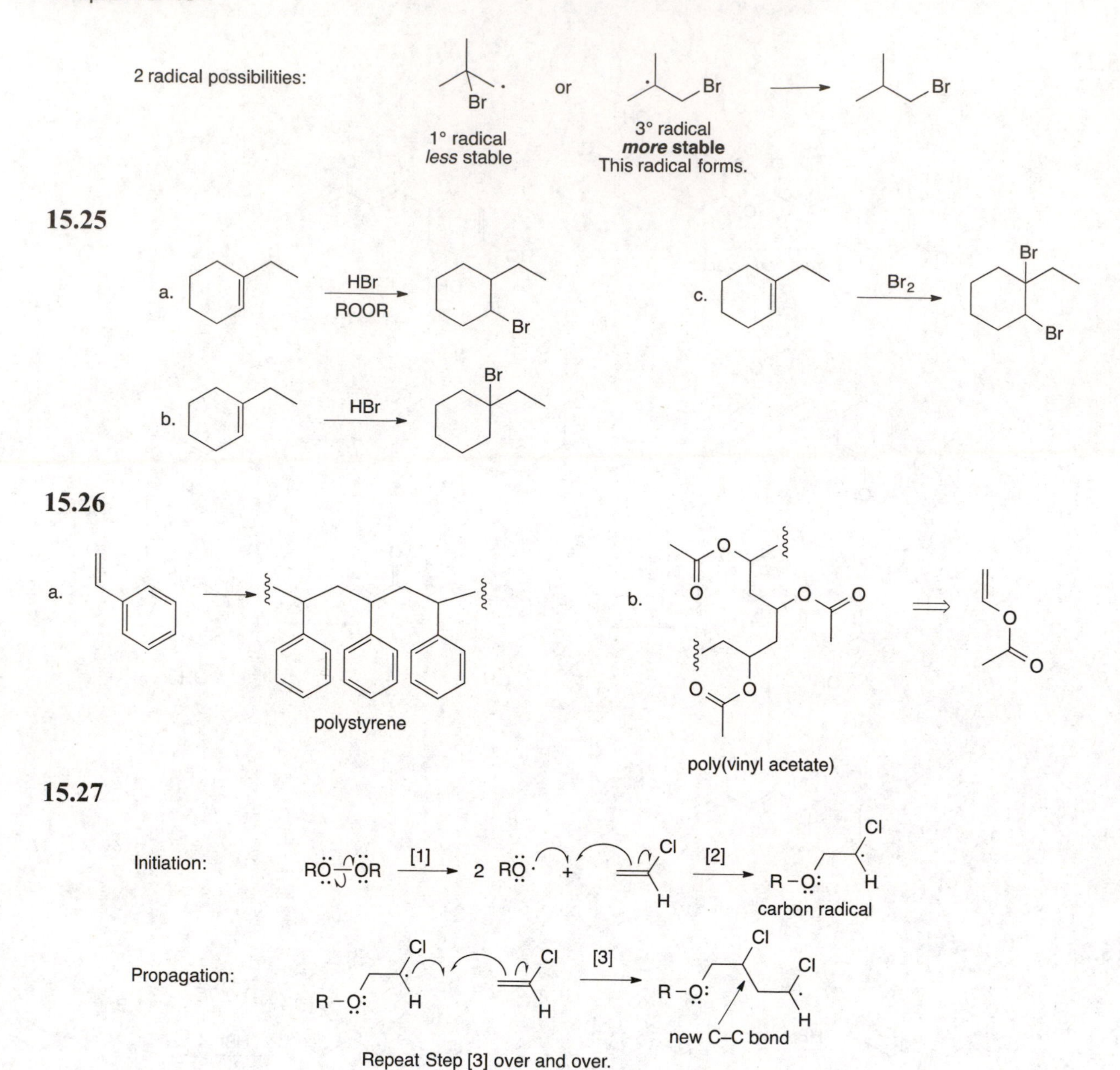

15.28 With Cl_2, each H of the starting material can be replaced by Cl. With Br_2, cleavage of the weakest C–H bond is preferred.

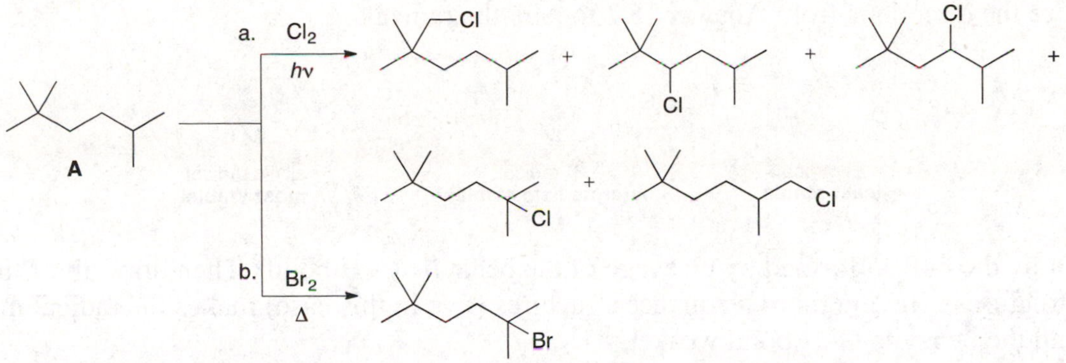

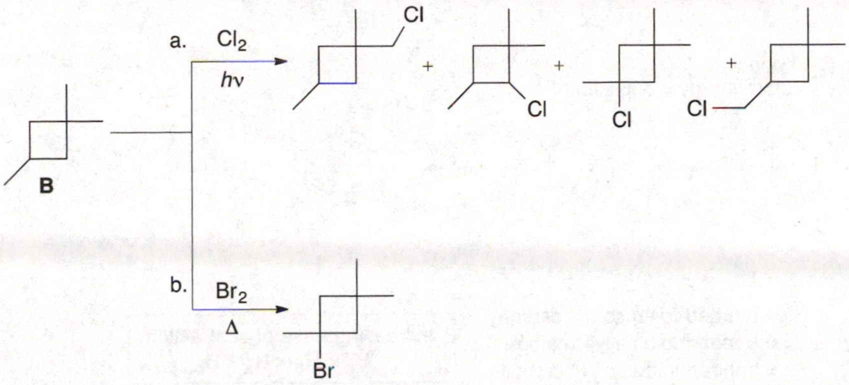

a. Cl_2 $h\nu$ → Cl + Cl + Cl + Cl + Cl

A

b. Br_2 Δ → Br

a. Cl_2 $h\nu$ → Cl + Cl + Cl + Cl

B

b. Br_2 Δ → Br

15.29

H–Ö–, –OCH_3

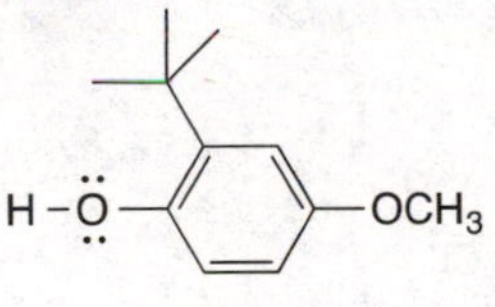

BHA

Abstraction of the phenol H
produces a resonance-stabilized radical.

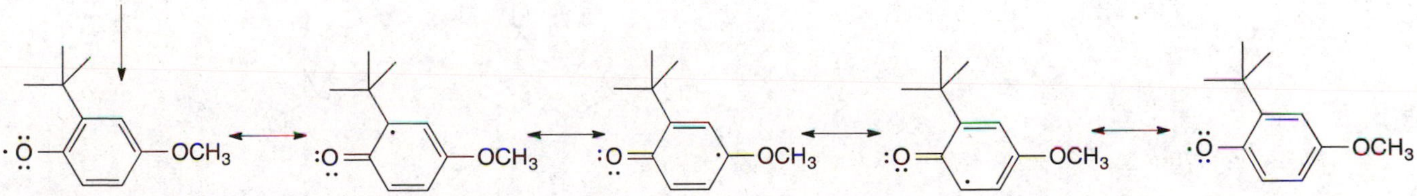

·Ö– –OCH_3 ↔ :O= –OCH_3 ↔ :O= –OCH_3 ↔ :O= –OCH_3 ↔ ·Ö– –OCH_3

15.30

a. increasing bond strength: 2 < 3 < 1

b. and c.

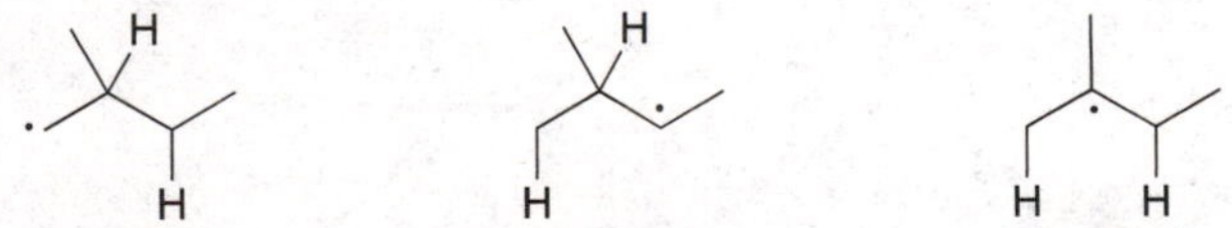

H H H H H H H

1° radical
least stable

2° radical
intermediate stability

3° radical
most stable

d. increasing ease of H abstraction: 1 < 3 < 2

15.31 Use the directions from Answer 15.2 to rank the radicals.

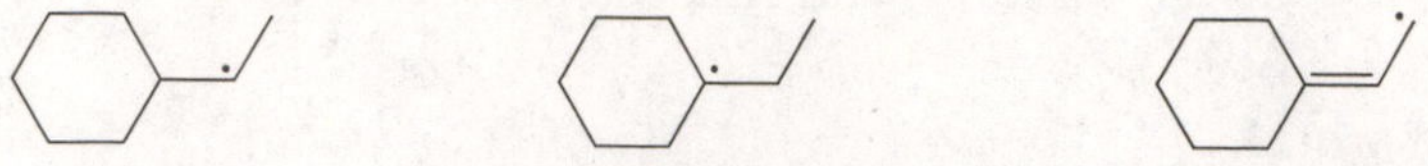

2° radical
least stable

3° radical
intermediate stability

allylic radical
most stable

15.32 Draw the radical formed by cleavage of the benzylic C–H bond. Then draw all of the resonance structures. Having more resonance structures (five in this case) makes the radical more stable, and the benzylic C–H bond weaker.

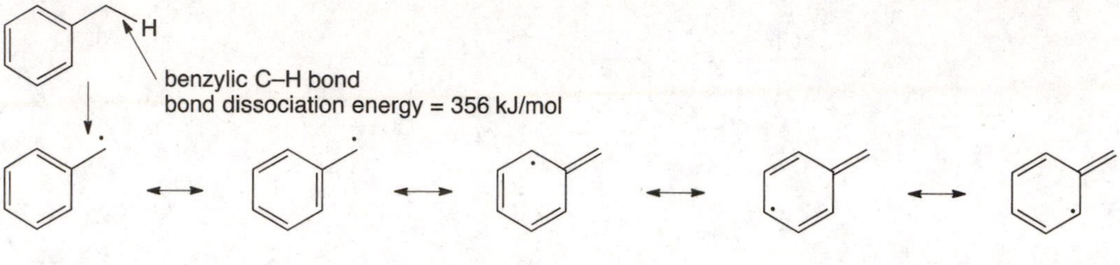

H

benzylic C–H bond
bond dissociation energy = 356 kJ/mol

15.33

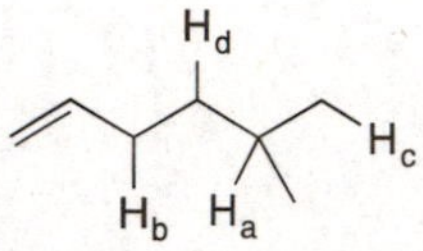

H_d H_c H_b H_a

H_a = bonded to an sp^3 3° carbon
H_b = bonded to an allylic carbon
H_c = bonded to an sp^3 1° carbon
H_d = bonded to an sp^3 2° carbon

Increasing ease of abstraction:
$H_c < H_d < H_a < H_b$

15.34

a. 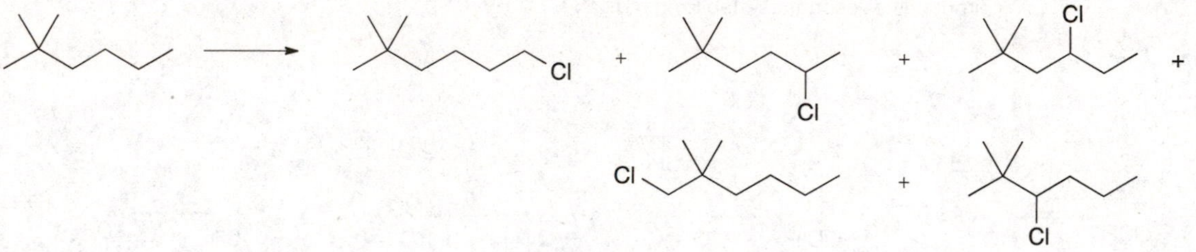 → Cl + Cl + Cl + Cl + Cl

b. 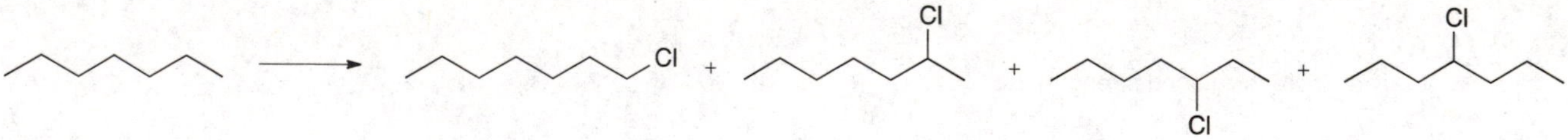 → Cl + Cl + Cl + Cl

c. 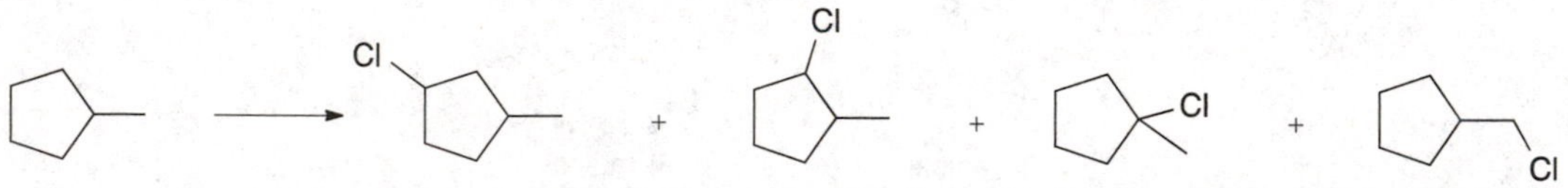 → Cl + Cl + Cl + Cl

15.35 To draw the product of bromination:

- Draw out the starting material and find the most reactive C–H bond (on the most substituted C).
- The major product is formed by **cleavage of the *weakest* C–H bond.**

a. 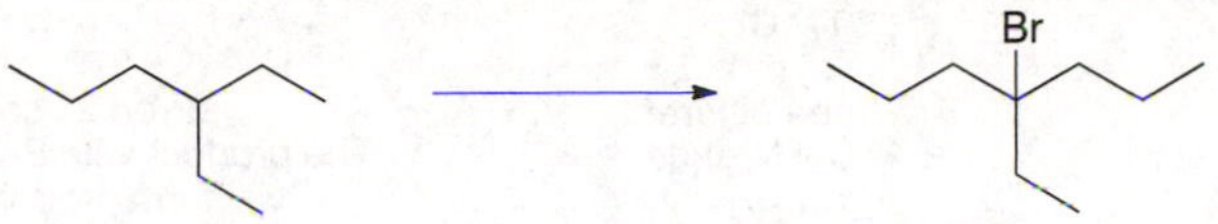 → Br

c. 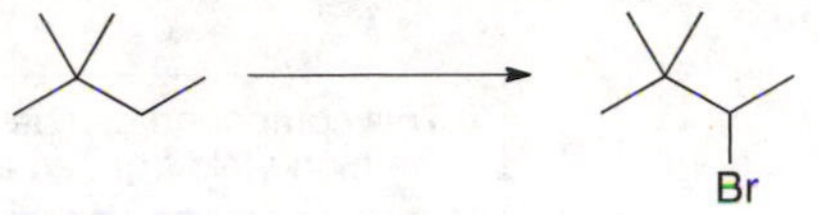 → Br

b. 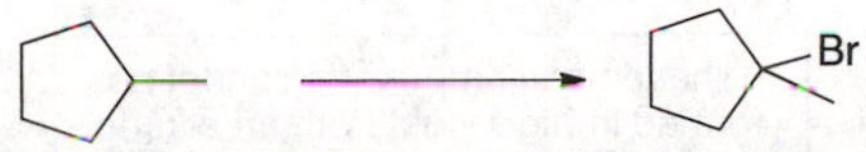 → Br

15.36 Draw all of the alkane isomers of C_6H_{14} and their products on chlorination. Then determine which letter corresponds to which alkane.

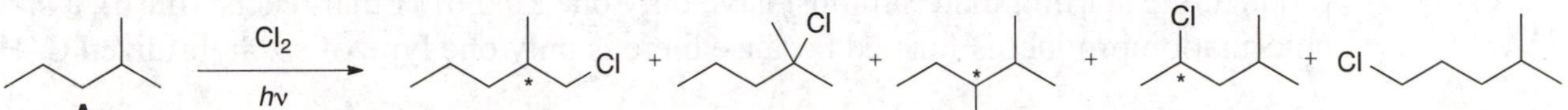

A $\xrightarrow[h\nu]{Cl_2}$ Cl + Cl + Cl + Cl + Cl

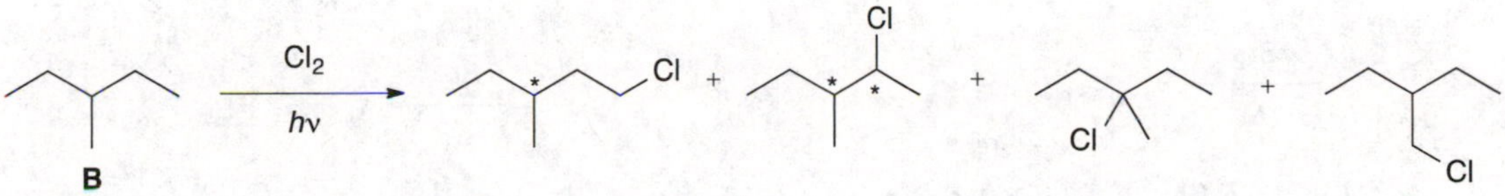

B $\xrightarrow[h\nu]{Cl_2}$ Cl + Cl + Cl + Cl

C $\xrightarrow[h\nu]{Cl_2}$ Cl + Cl

D $\xrightarrow[h\nu]{Cl_2}$ Cl + Cl + Cl

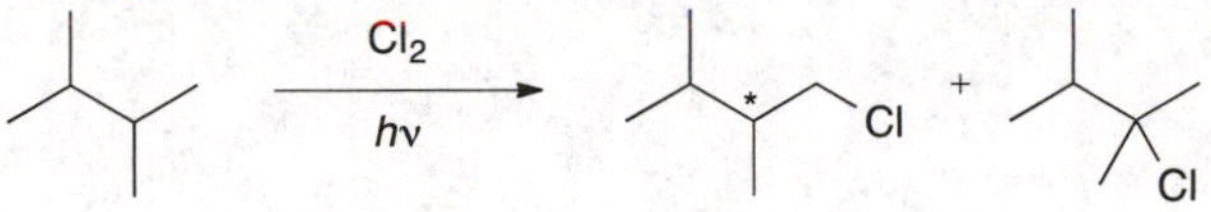

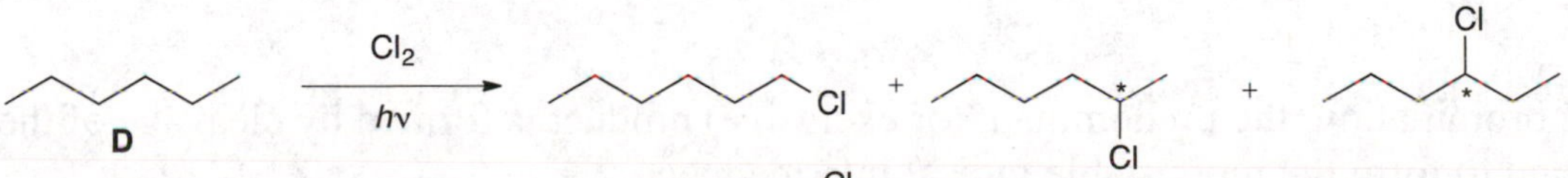

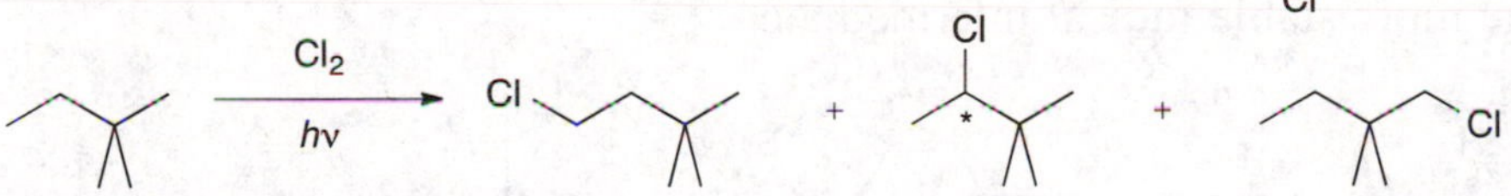

E $\xrightarrow[h\nu]{Cl_2}$ Cl + Cl + Cl

[* = stereogenic center]

15.37 Halogenation replaces a C–H bond with a C–X bond. To find the alkane needed to make each of the alkyl halides, replace the X with a H.

a. 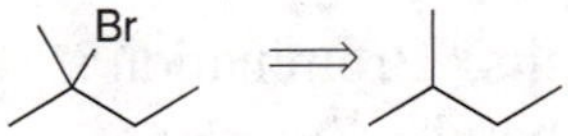 Br ⟹

b. 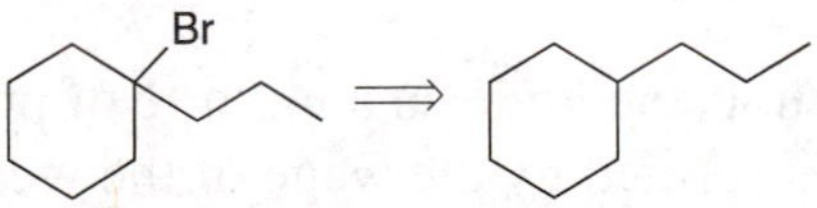 Br ⟹

c. 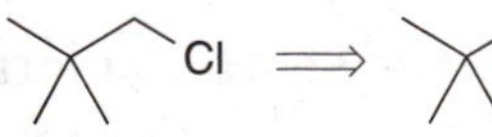 Cl ⟹

15.38 For an alkane to yield one major product on monohalogenation with Cl_2, all of the hydrogens must be identical in the starting material. For an alkane to yield one major product on bromination, it must have a more substituted carbon in the starting material.

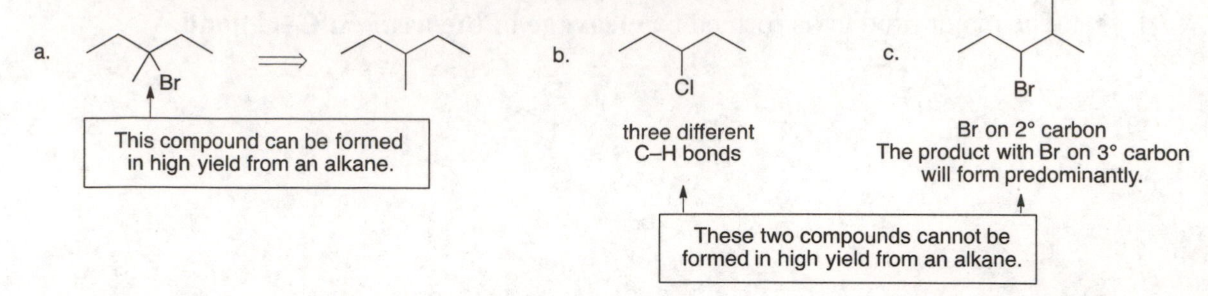
a.
Br
This compound can be formed in high yield from an alkane.
b.
Cl
three different C–H bonds
c.
Br
Br on 2° carbon
The product with Br on 3° carbon will form predominantly.
These two compounds cannot be formed in high yield from an alkane.

15.39 A single constitutional isomer is formed in both halogenations in (b) and (c). Bromination often forms a single product by cleavage of the weakest C–H bond. For chlorination to form a single product, the starting material must have only one kind of H that reacts. In (b), a single chlorination product is formed because there is only one type of sp^3 hybridized C–H bond.

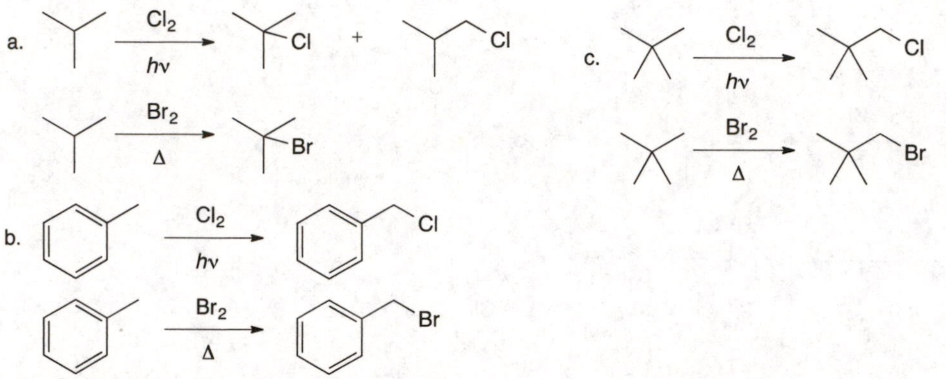

a. Cl_2, $h\nu$ → Cl + Cl

Br_2, Δ → Br

b. Cl_2, $h\nu$ → Cl

Br_2, Δ → Br

c. Cl_2, $h\nu$ → Cl

Br_2, Δ → Br

15.40 In bromination, the predominant (or exclusive) product is formed by cleavage of the weaker C–H bond to form the more stable radical intermediate.

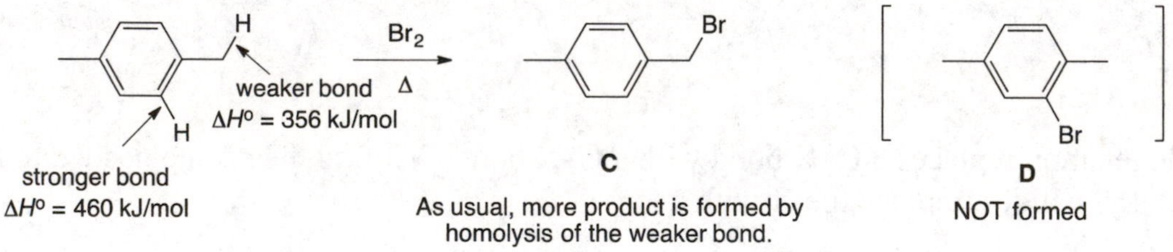

H — weaker bond, ΔH^o = 356 kJ/mol

H — stronger bond, ΔH^o = 460 kJ/mol

Br_2, Δ → Br

C

As usual, more product is formed by homolysis of the weaker bond.

Br

D

NOT formed

15.41 Chlorination is not selective, so a mixture of products results. Bromination is selective, and the major product is formed by cleavage of the weakest C–H bond.

a. **Y** — Cl_2, Δ → Cl + Cl + Cl + Cl + Cl + Cl

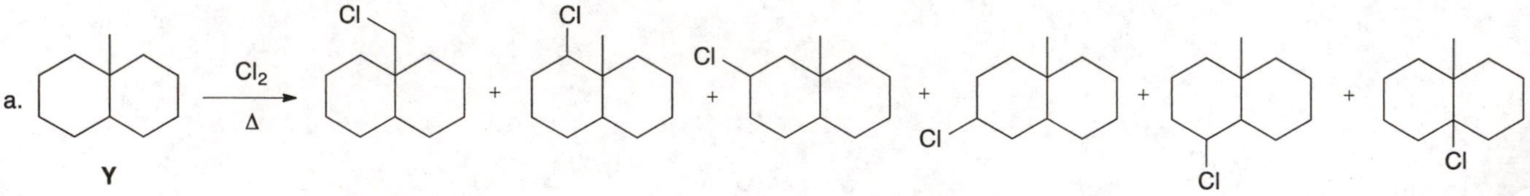

b. Y $\xrightarrow[\Delta]{Br_2}$ (Br product)

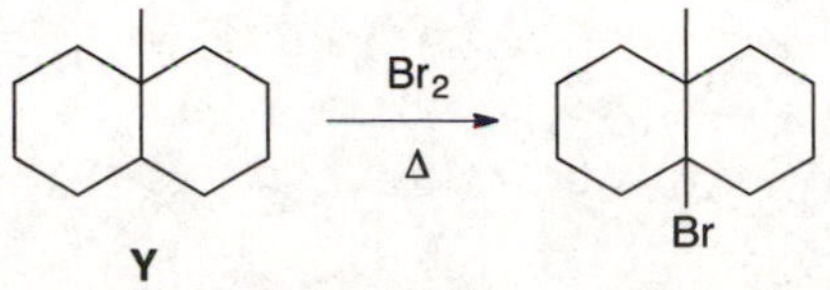

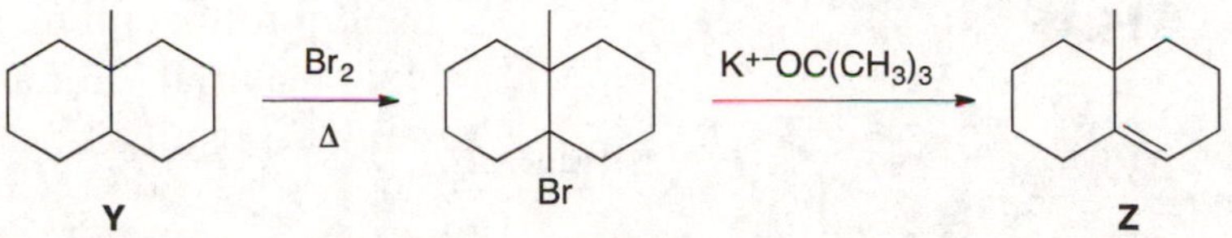

c. Y $\xrightarrow[\Delta]{Br_2}$ (Br product) $\xrightarrow{K^{+}\,{}^{-}OC(CH_3)_3}$ Z

15.42 Draw the resonance structures by moving the π bonds and the radical.

a.

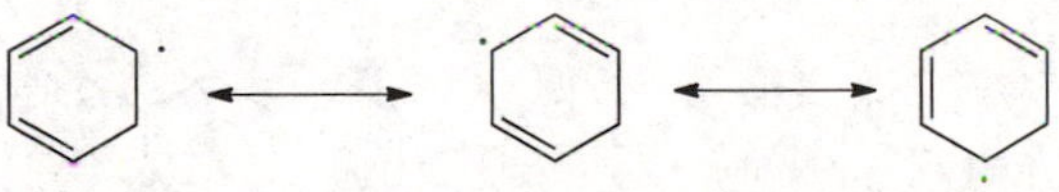

b.

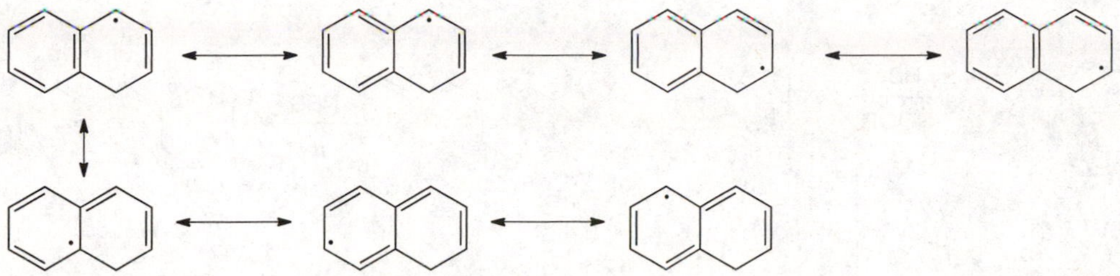

c.

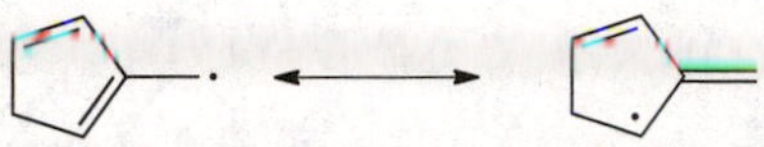

15.43 Reaction of an alkene with NBS + $h\nu$ yields allylic substitution products.

a. $\xrightarrow[h\nu]{NBS}$ Br

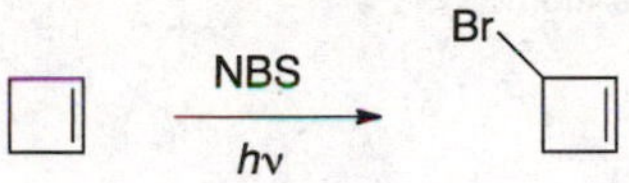

b. $\xrightarrow[h\nu]{NBS}$ Br + Br (+ *Z* isomer) + Br + Br

c. $\xrightarrow[h\nu]{NBS}$ Br + Br + Br + Br

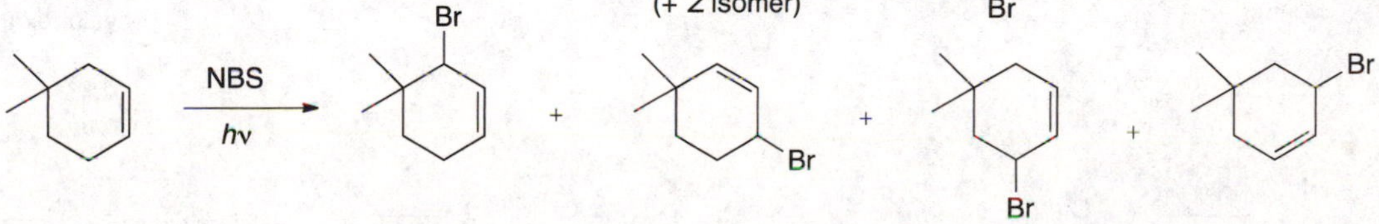

15.44

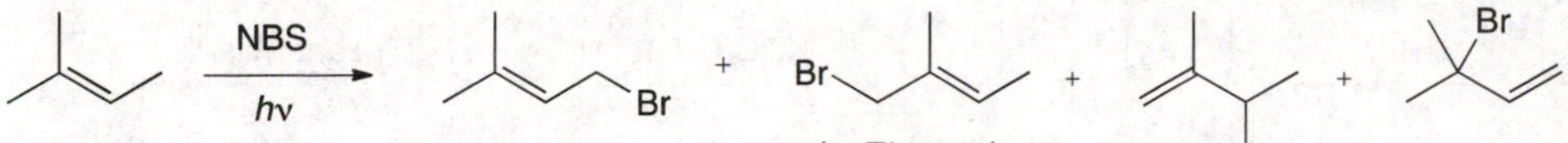

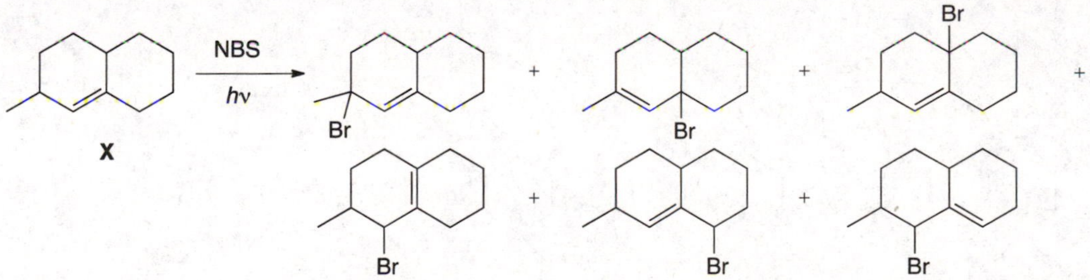

X $\xrightarrow[h\nu]{NBS}$ Br + Br + Br + Br + Br + Br

15.45

a. 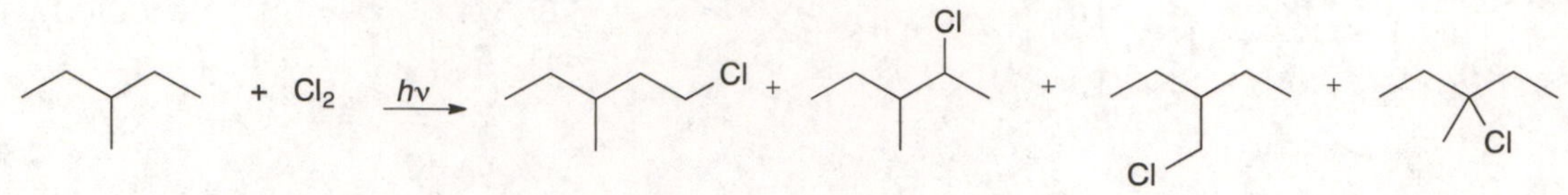 + Cl_2 $h\nu$ → Cl + Cl + Cl + Cl

b. 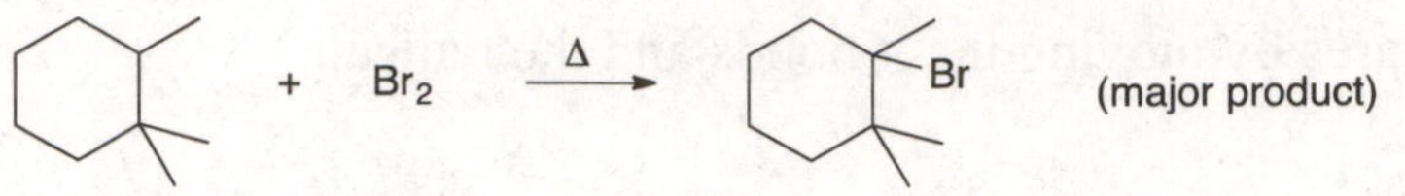 + Br_2 Δ → Br (major product)

c. 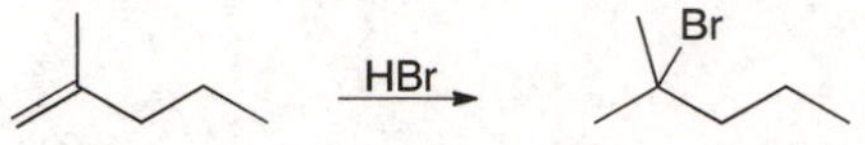 HBr → Br

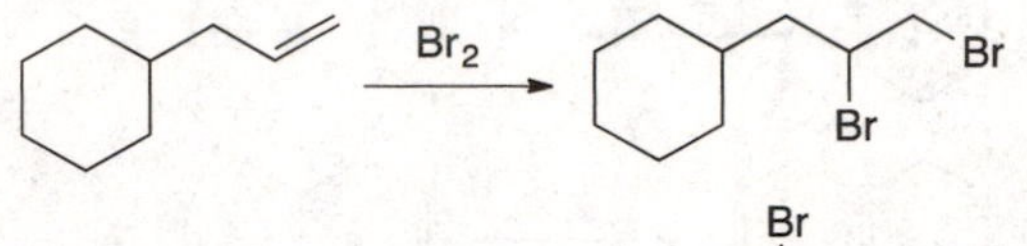

d. 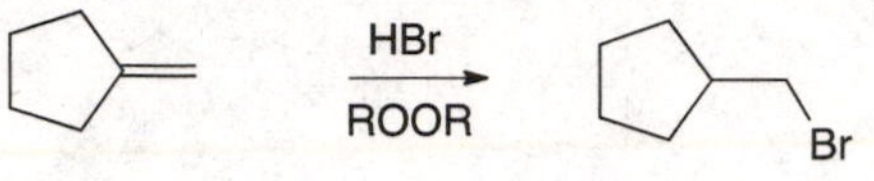 HBr ROOR → Br

e. Br_2 → Br Br

f. 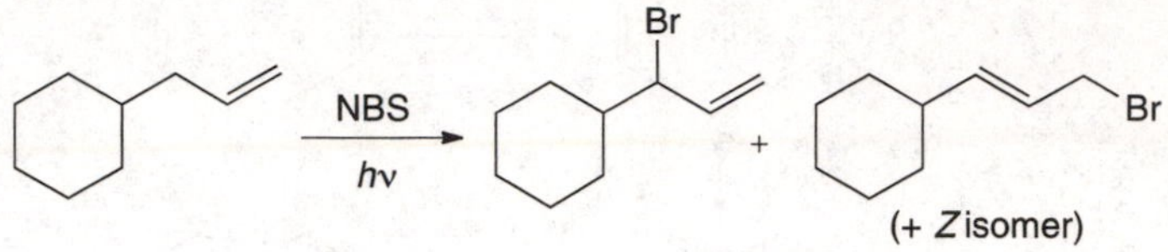 NBS $h\nu$ → Br + Br (+ *Z* isomer)

15.46

a. 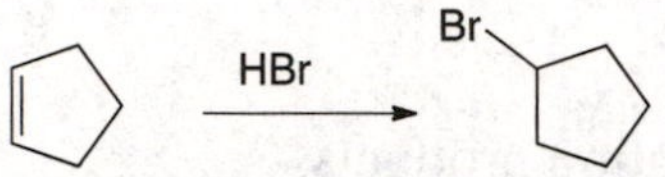 HBr → Br

b. 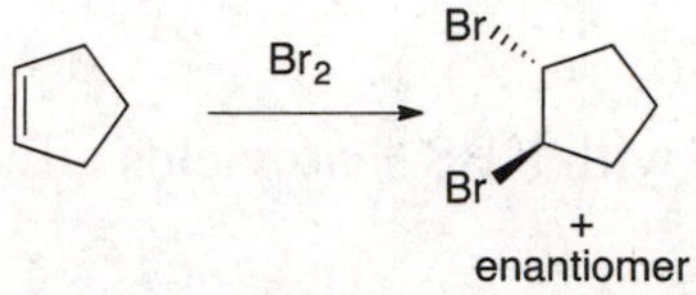 Br_2 → Br Br + enantiomer

c. 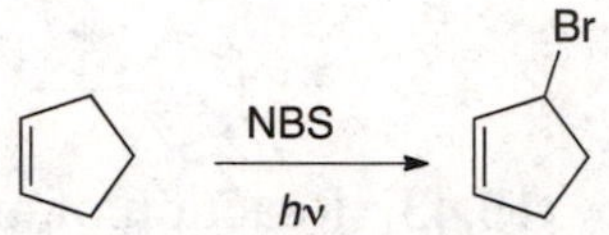 NBS $h\nu$ → Br

15.47

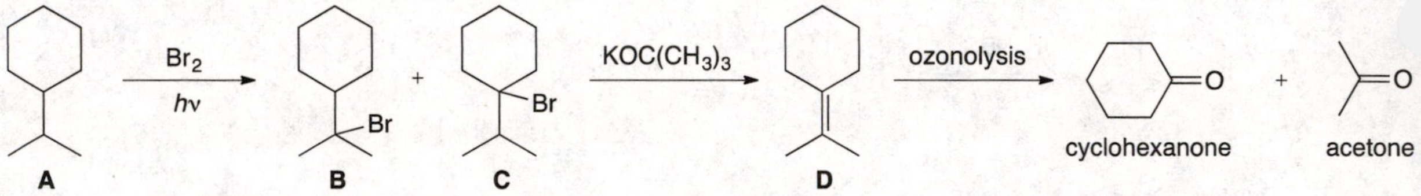

A Br_2 $h\nu$ → B Br + C Br $KOC(CH_3)_3$ → D ozonolysis → cyclohexanone O + acetone O

15.48

a. 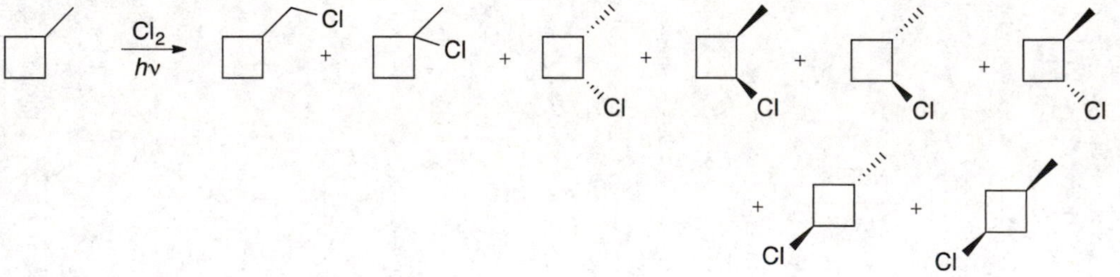 Cl_2 $h\nu$ → Cl + Cl + Cl + Cl + Cl + Cl + Cl + Cl

b. 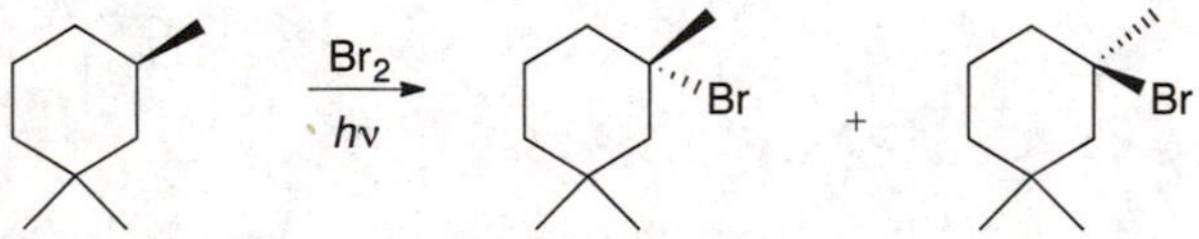 Br_2 $h\nu$ → Br + Br

c. 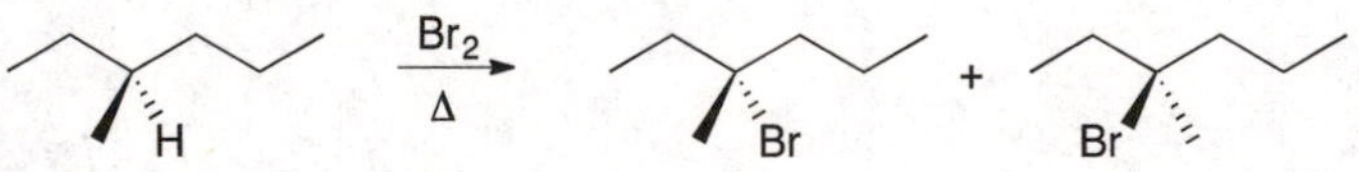 H Br_2 Δ → Br + Br

d. Cl Cl / Cl / F $\xrightarrow[\Delta]{Cl_2}$ Cl Cl Cl Cl F + Cl Cl Cl F Cl + Cl Cl Cl Cl F + Cl Cl Cl Cl F + Cl Cl Cl Cl F

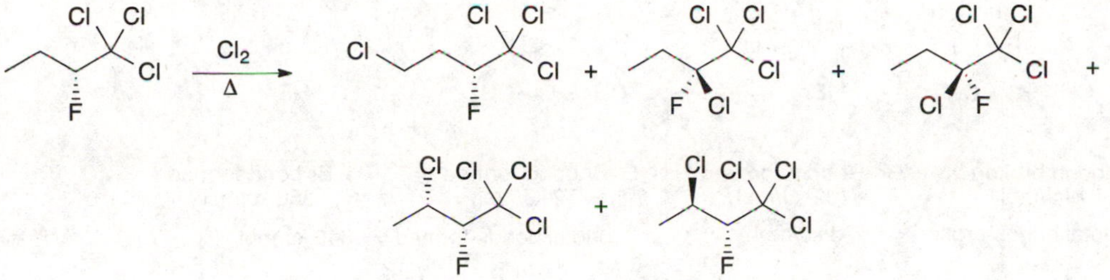

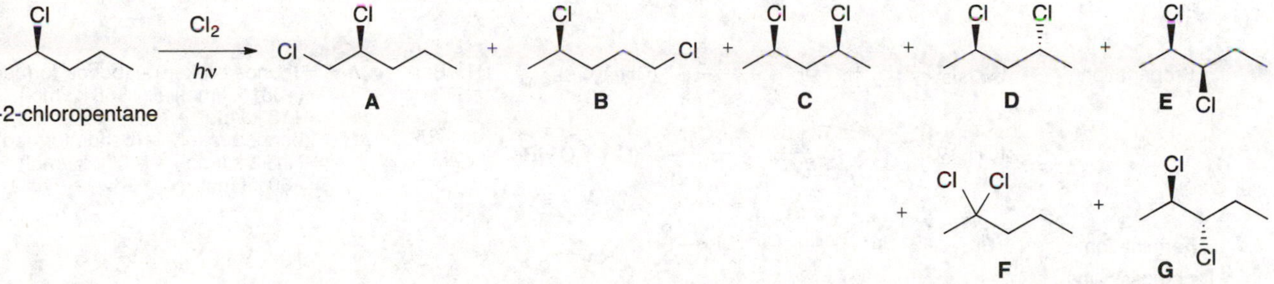

15.49

a. (*R*)-2-chloropentane $\xrightarrow[h\nu]{Cl_2}$ A + B + C + D + E + F + G

b. There would be seven fractions, because each molecule drawn has different physical properties.

c. Fractions **A, B, D, E,** and **G** would show optical activity.

15.50

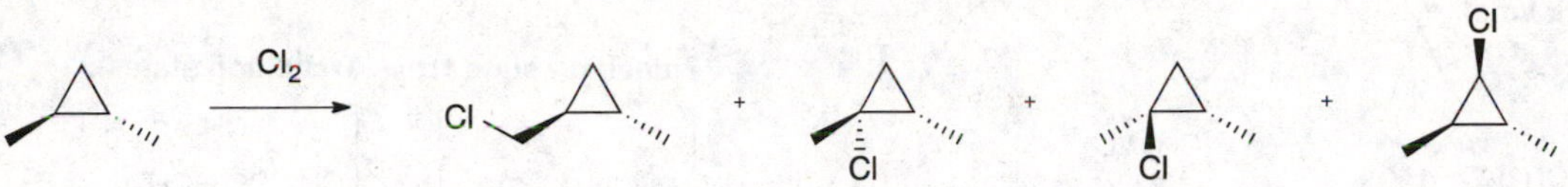

$\xrightarrow{Cl_2}$ Cl + Cl + Cl + Cl

15.51

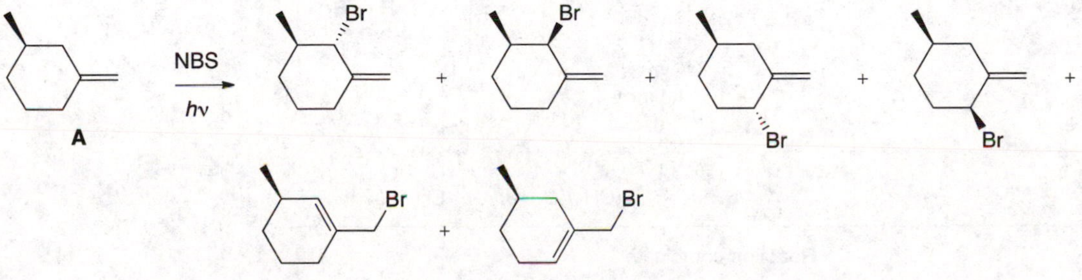

A $\xrightarrow[h\nu]{NBS}$ Br + Br + Br + Br + Br + Br

15.52

a. $\xrightarrow[ROOR]{HBr}$ Br H + Br H

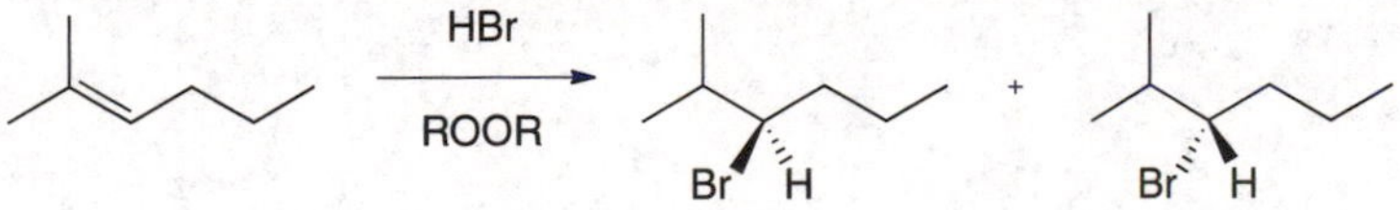

b. $\xrightarrow[ROOR]{HBr}$ Br H + Br H

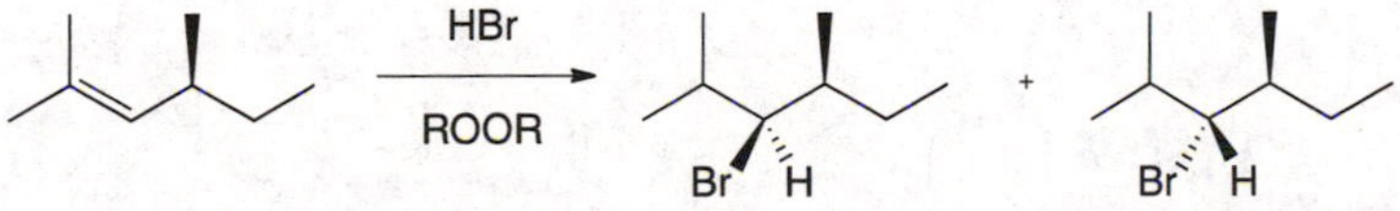

15.53

a. $CH_3-C(CH_3)(H)-CH_3 + Br_2 \xrightarrow{h\nu} CH_3-C(CH_3)(Br)-CH_3 + HBr$

C–H bond broken	Br–Br bond broken	C–Br bond formed	H–Br bond formed
+381 kJ/mol	+192 kJ/mol	–272 kJ/mol	–368 kJ/mol

total bonds broken = +573 kJ/mol total bonds formed = –640 kJ/mol $\Delta H°$ = –67 kJ/mol

b. Initiation:

$:\ddot{Br}-\ddot{Br}: \xrightarrow[\text{or } \Delta]{h\nu} :\ddot{Br}\cdot + \cdot\ddot{Br}:$

Propagation:

$(CH_3)_3C-H + \cdot\ddot{Br}: \longrightarrow (CH_3)_3C\cdot + H-\ddot{Br}:$

$(CH_3)_3C\cdot + :\ddot{Br}-\ddot{Br}: \longrightarrow (CH_3)_3C-Br + \cdot\ddot{Br}:$

c. $\Delta H°$ = (bonds broken) – (bonds formed)
= (+381 kJ/mol) + (–368 kJ/mol)
= +13 kJ/mol

$\Delta H°$ = (bonds broken) – (bonds formed)
= (+192 kJ/mol) + (–272 kJ/mol)
= –80 kJ/mol

Termination: (one possibility)

$:\ddot{Br}\cdot + \cdot\ddot{Br}: \longrightarrow :\ddot{Br}-\ddot{Br}:$

d, e.

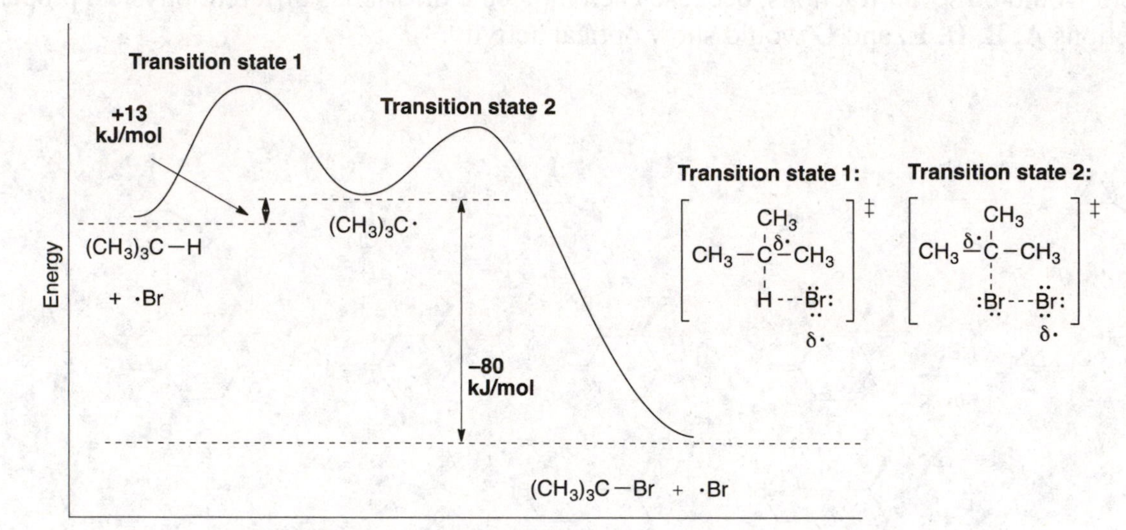

15.54

Initiation: NBS $\xrightarrow{h\nu}$ N· + ·Br:

Propagation: + :Br· → → + H–Br:

:Br–Br: (from NBS) :Br–Br: (from NBS)

Br + Br + ·Br:

Termination: (one possibility) :Br· + ·Br: → :Br–Br:

15.55 Calculate the $\Delta H°$ for the propagation steps of the reaction of CH_4 with I_2 to show why it does not occur at an appreciable rate.

CH_3–H + ·I: → $\dot{C}H_3$ + H–I: $\Delta H° = +138$ kJ/mol

+435 kJ/mol −297 kJ/mol

This step is highly endothermic, making it difficult for chain propagation to occur over and over again.

$\dot{C}H_3$ + :I–I: → CH_3–I + ·I: $\Delta H° = -83$ kJ/mol

+151 kJ/mol −234 kJ/mol

15.56

3,3-dimethylbut-1-ene $\xrightarrow{H-Br}$ 2° carbocation + Br^- $\xrightarrow{1,2\text{-}CH_3\text{ shift}}$ 3° carbocation $\xrightarrow{Br^-}$ 2-bromo-2,3-dimethylbutane

3,3-dimethylbut-1-ene $\xrightarrow{Br\cdot}$ (HBr + peroxide) The 2° radical does NOT rearrange. $\xrightarrow{HBr}$ 1-bromo-3,3-dimethylbutane + Br·

Addition of HBr without added peroxide occurs by an ionic mechanism and forms a 2° carbocation, which rearranges to a more stable 3° carbocation. The addition of H^+ occurs first, followed by Br^-. Addition of HBr with added peroxide occurs by a radical mechanism and forms a 2° radical that does not rearrange. In the radical mechanism, Br· adds first, followed by H·.

15.57

a. $\xrightarrow[\Delta]{Cl_2}$ –Cl $\xrightarrow{K^+\ {}^-OC(CH_3)_3}$ $\xrightarrow[H_2SO_4]{H_2O}$ –OH

b. (from a.) $\xrightarrow{\text{mCPBA}}$ O

c. (from a.) $\xrightarrow[\text{ROOR}]{\text{NBS}}$ –Br $\xrightarrow{Br_2}$ Br, Br, Br

d. –Br (from c.) $\xrightarrow{{}^-OH}$ –OH

e. O (from b.) $\xrightarrow[\text{[2] } H_2O]{\text{[1] NaCN}}$ OH, CN

15.58

$\xrightarrow[h\nu]{Br_2}$ Br $\xrightarrow{K^+\ {}^-OC(CH_3)_3}$ major product $\xrightarrow{\text{mCPBA}}$ O

1-methylcyclohexene oxide

15.59

$HC\equiv CH \xrightarrow{NaH} HC\equiv C^- \xrightarrow{CH_3CH_2Br}$ $\xrightarrow[\text{Lindlar catalyst}]{H_2}$ $\xrightarrow[\text{ROOR}]{HBr}$ Br

15.60

a. $CH_3-CH_3 \xrightarrow[h\nu]{Br_2} CH_3CH_2Br \xrightarrow{K^+\ {}^-OC(CH_3)_3} CH_2=CH_2 \xrightarrow{Br_2} BrCH_2-CH_2Br \xrightarrow{2\ NaNH_2} HC\equiv CH$

b. $CH_2=CH_2$ (from a.) $\xrightarrow{\text{mCPBA}}$ O $\xrightarrow[\text{[2] } H_2O]{\text{[1] } HC\equiv C^-}$ OH

$HC\equiv CH$ (from a.) $\xrightarrow{NaH}$ $HC\equiv C^-$

c. $HC\equiv CH$ (from a.) $\xrightarrow{NaH} HC\equiv C^- \xrightarrow[\text{(from a.)}]{CH_3CH_2Br}$ $\xrightarrow{NaH}$ $^-$ $\xrightarrow[\text{(from a.)}]{CH_3CH_2Br}$ $\xrightarrow[NH_3]{Na}$

d. (from c.) $\xrightarrow[\substack{H_2SO_4 \\ HgSO_4}]{H_2O}$ O

15.61

$\xrightarrow[h\nu]{Cl_2}$ Cl $\xrightarrow{K^+\ {}^-OC(CH_3)_3}$ $\xrightarrow[\text{[2] } (CH_3)_2S]{\text{[1] } O_3}$ OHC CHO

15.62

hexane-2,3-diol

OH, OH, Br, Br_2, $h\nu$, $K^+ \, ^-OC(CH_3)_3$, HBr, ROOR, Br, Br_2, Br, Br, $NaNH_2$ (excess), H, [1] NaH, [2] Br, H_2, Lindlar, [1] OsO_4, [2] $NaHSO_3$, H_2O, OH, OH

15.63

O_2 abstracts a H here.

H H, COOH, C_5H_{11}

arachidonic acid

$\cdot\ddot{O}-\ddot{O}\cdot$, COOH, C_5H_{11}, + $H\ddot{O}\ddot{O}\cdot$

$\cdot\ddot{O}-\ddot{O}\cdot$, COOH, C_5H_{11}

R–H, $:\ddot{O}-\ddot{O}:$, COOH, C_5H_{11}

R–H

another molecule of arachidonic acid

OOH, COOH, C_5H_{11}

5-HPETE

COOH, C_5H_{11}, +

This process is repeated.

15.64

O, H, + $\cdot\ddot{O}-\ddot{O}\cdot$, [1], O, + $H\ddot{O}\ddot{O}\cdot$, $\cdot\ddot{O}-\ddot{O}\cdot$, [2], O, $\ddot{O}-\ddot{O}\cdot$, H, O, [3], O, $\ddot{O}-\ddot{O}H$, +, O

Then, repeat Steps [2] and [3].

15.65

a. OOH, +, HOO, (+ cis isomer)

O$_2$ abstracts a H here.

b.

[1]

+ HOO·

[2a]

[3a]

+ R·   [RH = hex-1-ene]

OOH

[2b]

[3b]

HOO

+ R·

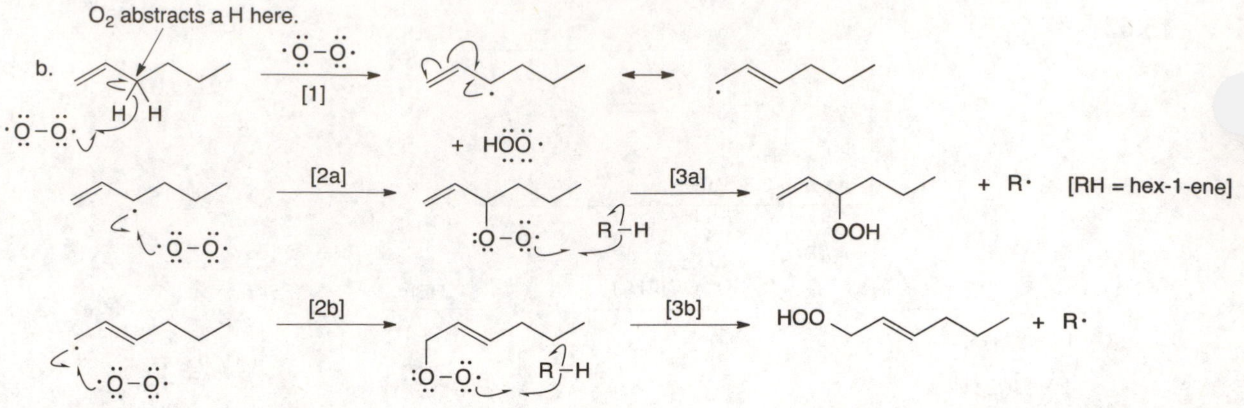

Repeat Steps [2] and [3] again and again.

15.66 For resonance structures **A–F**, an additional resonance form can be drawn that moves the position of the three π bonds in the ring bonded to two OH groups.

a.

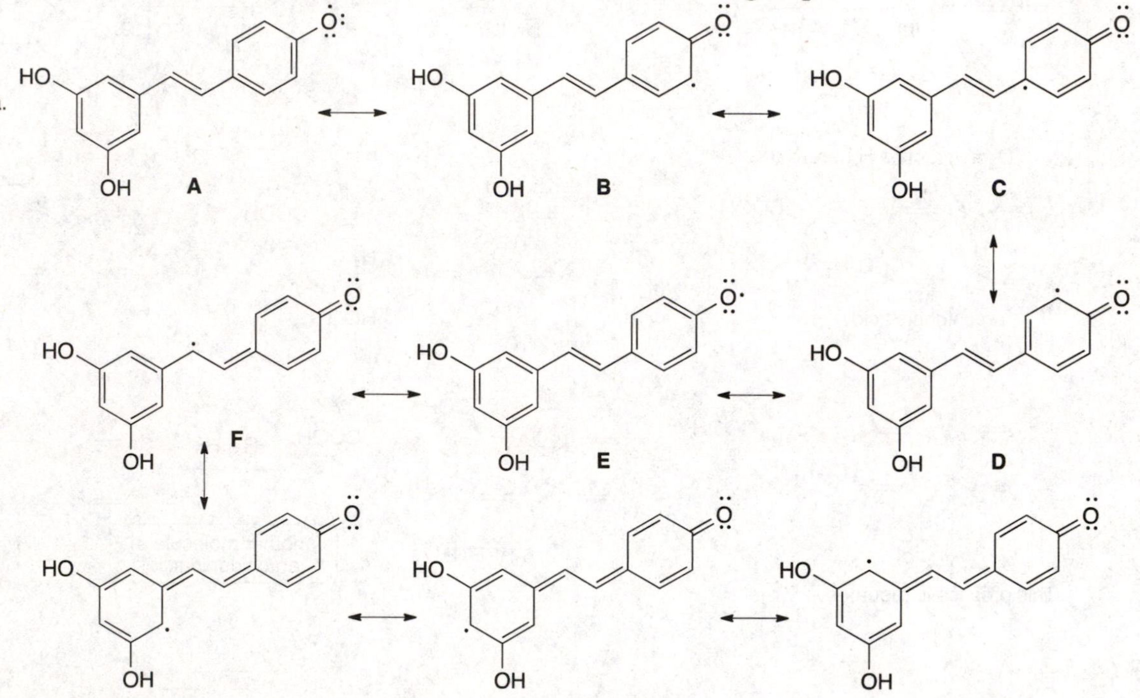

A B C D E F

b. Homolysis of the indicated OH bond is preferred because it allows the resulting radical to delocalize over both benzene rings. Cleavage of one of the other OH bonds gives a radical that delocalizes over only one of the benzene rings.

15.67 Abstraction of the labeled H forms a highly resonance-stabilized radical. Four of the possible resonance structures are drawn.

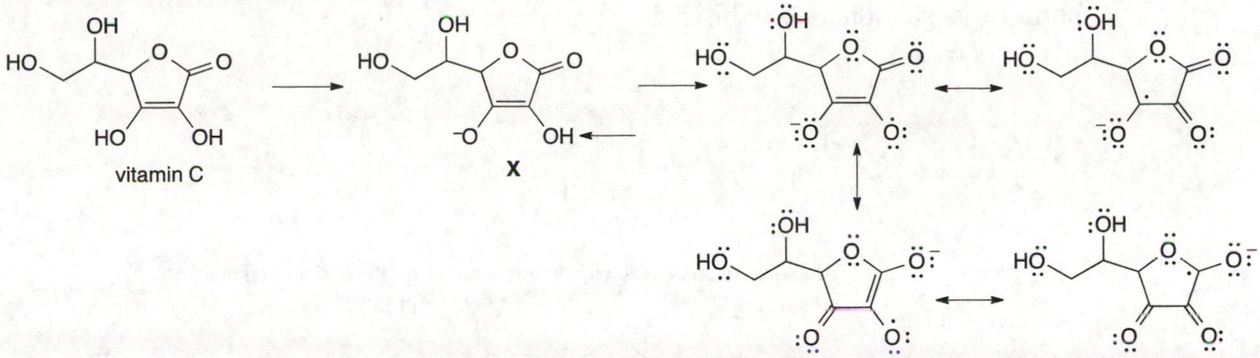

vitamin C

X

15.68 The monomers used in radical polymerization always contain double bonds.

a.

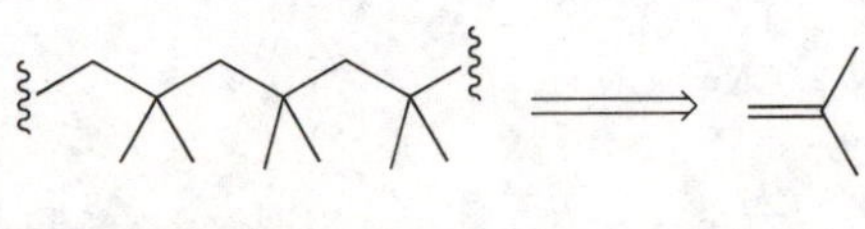

polyisobutylene

b.

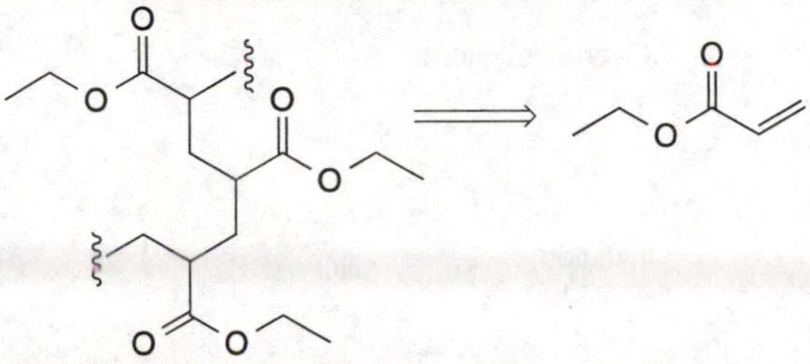

poly(ethyl acrylate)
(used in Latex paints)

15.69

a.

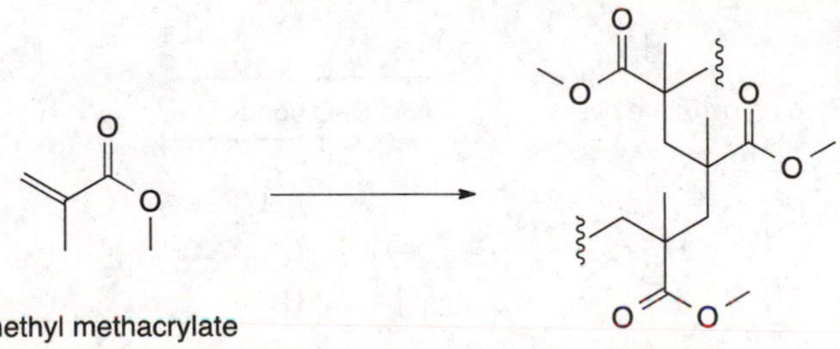

methyl methacrylate

PMMA

b.

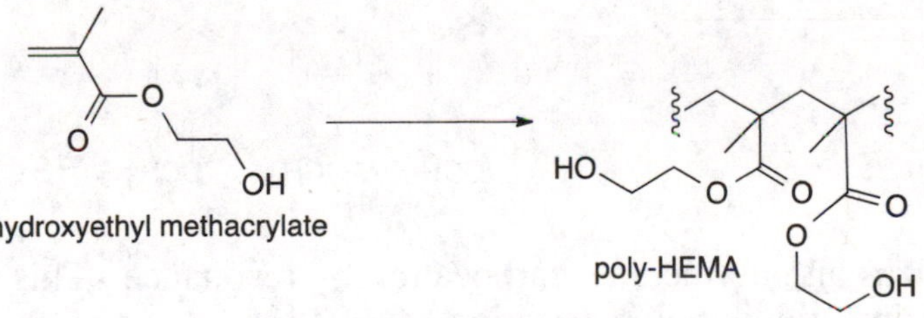

hydroxyethyl methacrylate

poly-HEMA

15.70 Polystyrene has H atoms bonded to benzylic carbons—that is, carbons bonded directly to a benzene ring. These C–H bonds are unusually weak because the radical that results from homolysis is resonance stabilized.

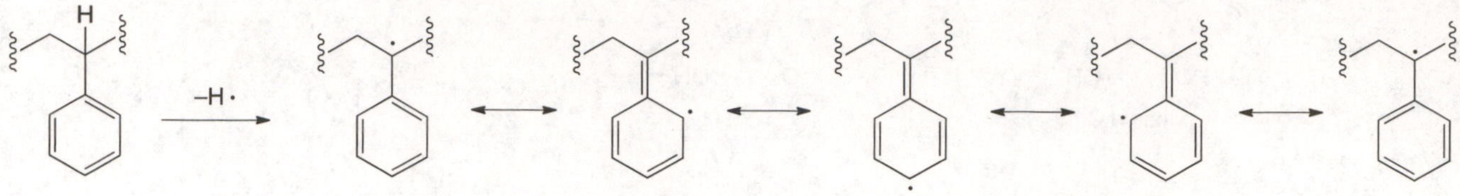

–H·

No such resonance stabilization is possible for the radical that results from C–H bond cleavage in polyethylene.

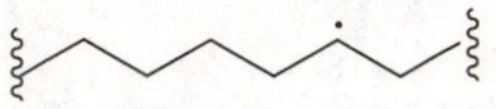

15.71

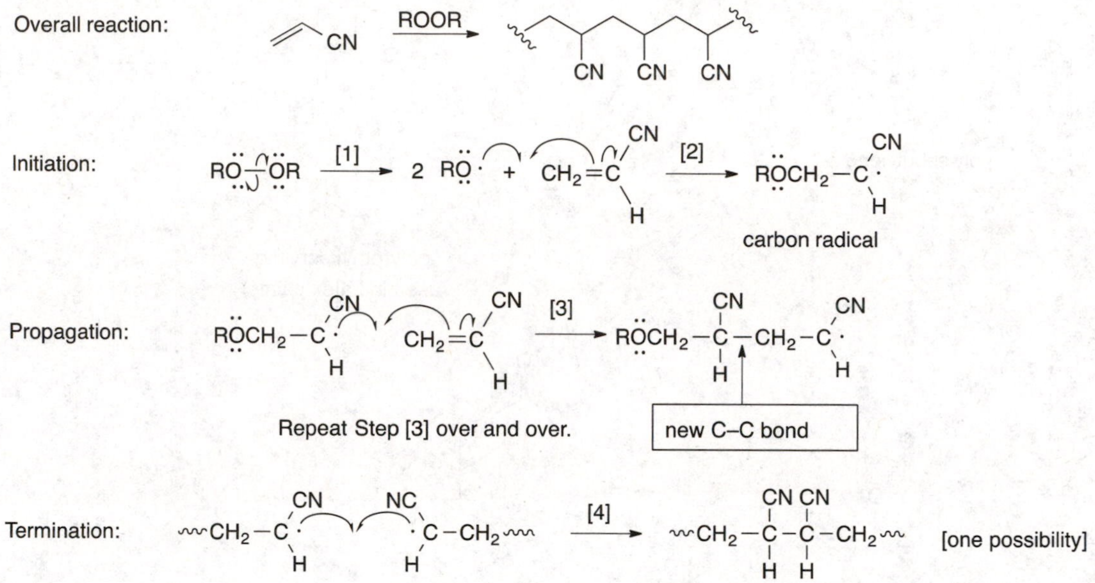

Overall reaction: CN —ROOR→ CN CN CN

Initiation: RO–OR —[1]→ 2 RO· + CH_2=C(CN)(H) —[2]→ $ROCH_2$–C·(CN)(H)

carbon radical

Propagation: $ROCH_2$–C·(CN)(H) + CH_2=C(CN)(H) —[3]→ $ROCH_2$–C(CN)(H)–CH_2–C·(CN)(H)

Repeat Step [3] over and over.

new C–C bond

Termination: ~CH_2–C·(CN)(H) + ·C(NC)(H)–CH_2~ —[4]→ ~CH_2–C(CN)(H)–C(CN)(H)–CH_2~ [one possibility]

15.72

a. CH_3O– (A) → OCH$_3$ OCH$_3$ OCH$_3$

A

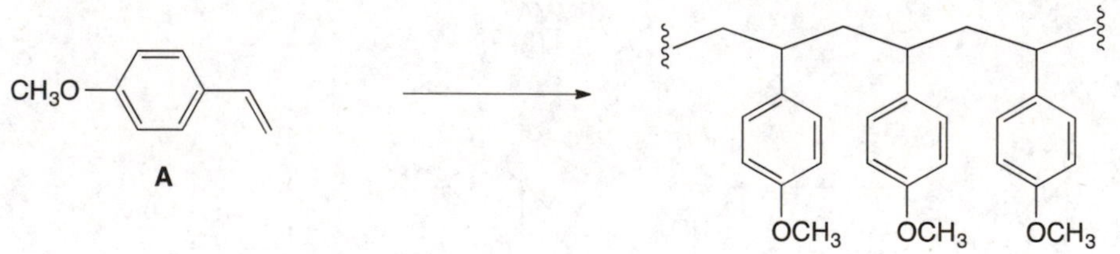

b. The OCH_3 group stabilizes an intermediate carbocation by resonance. This makes **A** react faster than styrene in cationic polymerization.

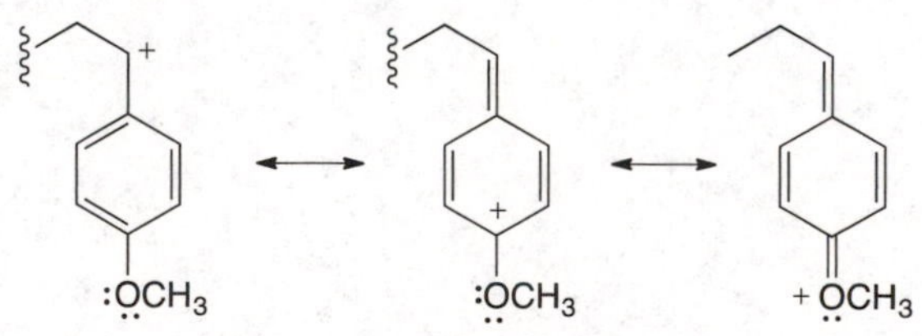

:$\ddot{O}CH_3$:$\ddot{O}CH_3$ + $\ddot{O}CH_3$

three of the possible resonance structures

15.73

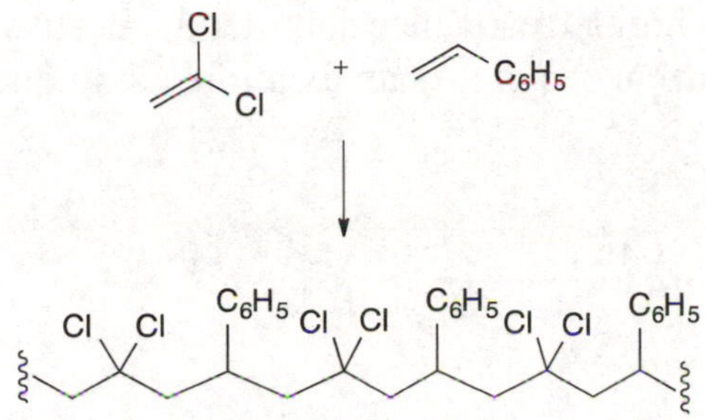

Cl

\+ C_6H_5

Cl

Cl Cl C_6H_5 Cl Cl C_6H_5 Cl Cl C_6H_5

alternating copolymer

15.74

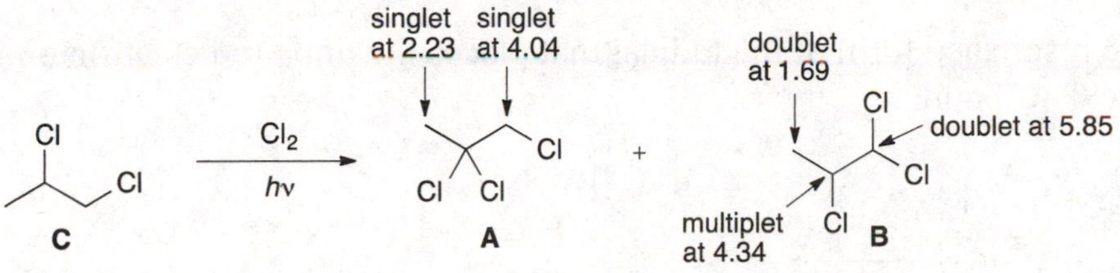

singlet at 2.23 singlet at 4.04

doublet at 1.69

Cl Cl C Cl_2 $h\nu$ Cl Cl Cl A + Cl doublet at 5.85 Cl

multiplet at 4.34 Cl B

15.75

triplet

Cl_2 $h\nu$ Cl Cl

quintet

minor product

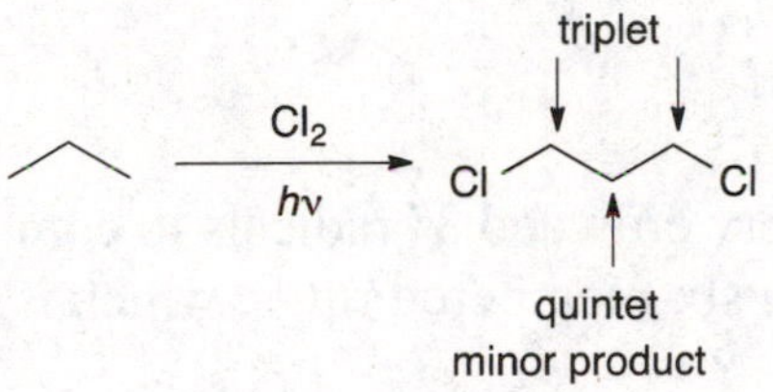

Molecular formula $C_3H_6Cl_2$

Integration: relative area 2:1

Because the compound has 6 H's and the sum of the relative areas is 3, each absorption is due to twice as many H's.

One signal is due to 4 H's.

The second signal is due to 2 H's (2 x 1).

^1H NMR data:

quintet at 2.2 (2 H's) split by 4 H's

triplet at 3.7 (4 H's) split by 2 H's

15.76

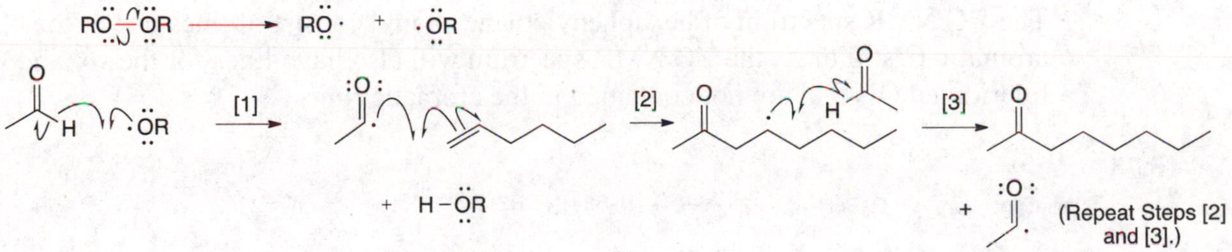

RO–OR → RO· + ·OR

O H ·OR [1] :O: [2] O H O [3] O

\+ H–OR

\+ :O: (Repeat Steps [2] and [3].)

15.77

a. The triphenylmethyl radical is highly resonance stabilized, because the radical can be delocalized on each of the benzene rings. As an example using one ring:

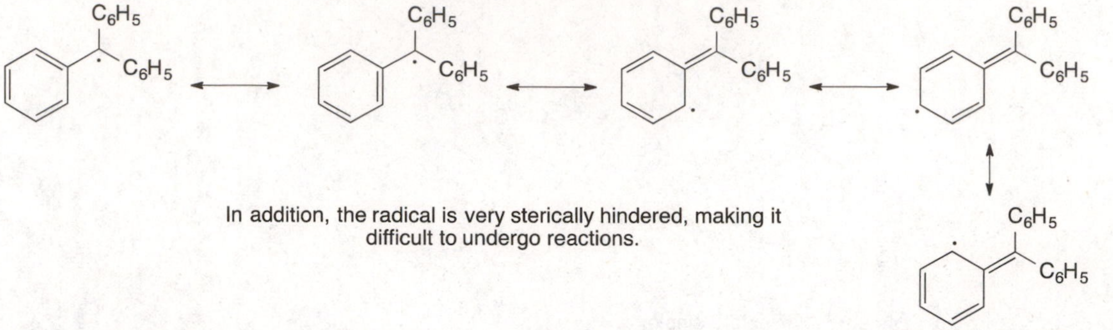

C_6H_5 C_6H_5 C_6H_5 C_6H_5 C_6H_5 C_6H_5 C_6H_5 C_6H_5 C_6H_5 C_6H_5

In addition, the radical is very sterically hindered, making it difficult to undergo reactions.

b. First, draw the resonance form of the radical that places the unpaired electron on the C that forms the new C–C bond.

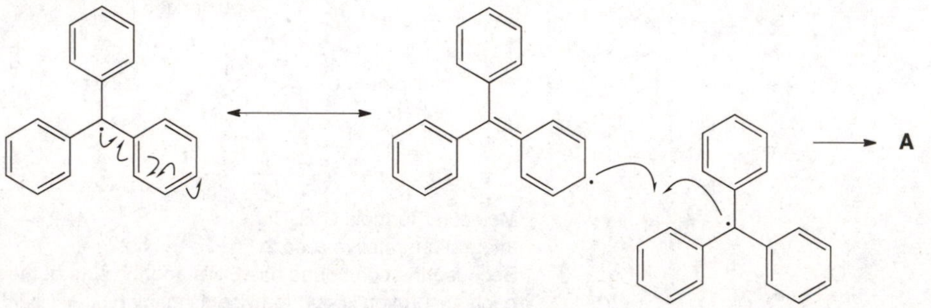

→ **A**

c. Hexaphenylethane formation would require two very crowded 3° radicals to combine. The formation of **A** results from a radical on one of the six-membered rings, which is much more accessible for reaction.

d. The ^{1}H NMR spectrum of hexaphenylethane should show signals only in the aromatic region (7–8 ppm), whereas the ^{1}H NMR spectrum of **A** will also have signals for the *sp*^{2} hybridized C–H bonds (4.5–6.0 ppm) of the alkenes, as well the single H on the *sp*^{3} hybridized carbon. The ^{13}C NMR spectrum of hexaphenylethane should consist of lines due to the 4° C's and the aromatic C's. For **A,** the ^{13}C NMR spectrum will also have lines for the *sp*^{3} and *sp*^{2} hybridized C's that are not contained in the aromatic rings.

15.78

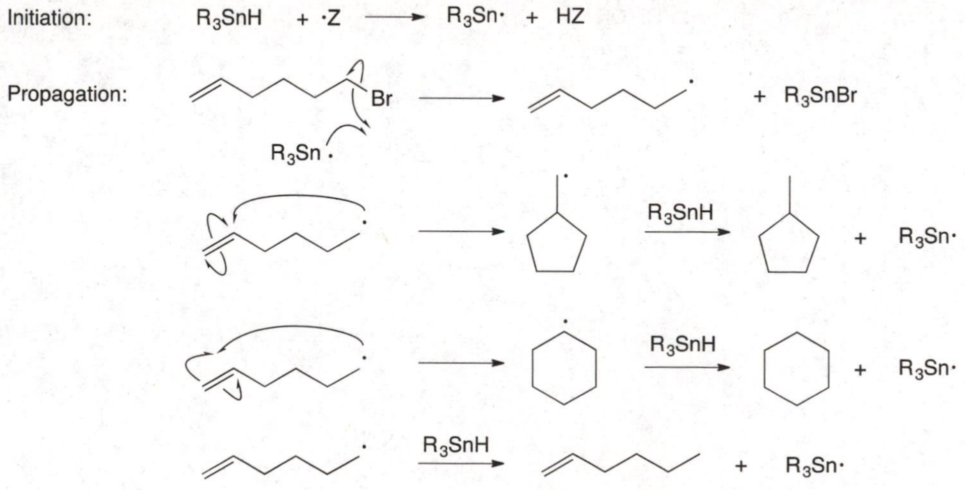

Initiation: R_3SnH + ·Z → R_3Sn· + HZ

Propagation: Br → + R_3SnBr

R_3Sn·

→ $\xrightarrow{R_3SnH}$ + R_3Sn·

→ $\xrightarrow{R_3SnH}$ + R_3Sn·

$\xrightarrow{R_3SnH}$ + R_3Sn·

15.79

O
OH
:Ö:
:O:
A
O
OH
O
O
O
OH
O
O
·Ö–Ö·
O
OH
O
O
:Ö–Ö·
R–H
O
OH
O
O
ÖOH
+ R·

Chapter 16: Conjugation, Resonance, and Dienes

Chapter Review

Conjugation and delocalization of electron density

- The overlap of *p* orbitals on three or more adjacent atoms allows electron density to delocalize, thus adding stability (16.1).
- An allyl carbocation ($CH_2{=}CHCH_2^+$) is more stable than a 1° carbocation because of *p* orbital overlap (16.2).
- In a system X=Y–Z:, Z is generally sp^2 hybridized to allow the lone pair to occupy a *p* orbital, making the system conjugated (16.5).

Four common examples of resonance (16.3)

[1] The three atom "allyl" system: X=Y–Z* ⟷ X*–Y=Z * = +, –, •, *or* ••

[2] Conjugated double bonds:

[3] Cations having a positive charge adjacent to a lone pair: $\ddot{X}$–$\overset{+}{Y}$ ⟷ $\overset{+}{X}$=Y

[4] Double bonds having one atom more electronegative than the other: X=Y ⟷ $\overset{+}{X}$–$\overset{-}{Y}$: [Electronegativity of Y > X]

Rules on evaluating the relative "stability" of resonance structures (16.4)

[1] Structures with more bonds and fewer charges are better.

all neutral atoms
one more bond
bettter resonance structure

[2] Structures in which every atom has an octet are better.

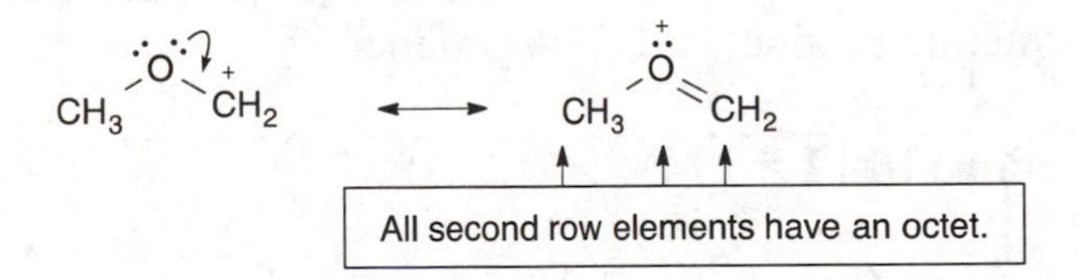

[3] Structures that place a negative charge on a more electronegative element are better.

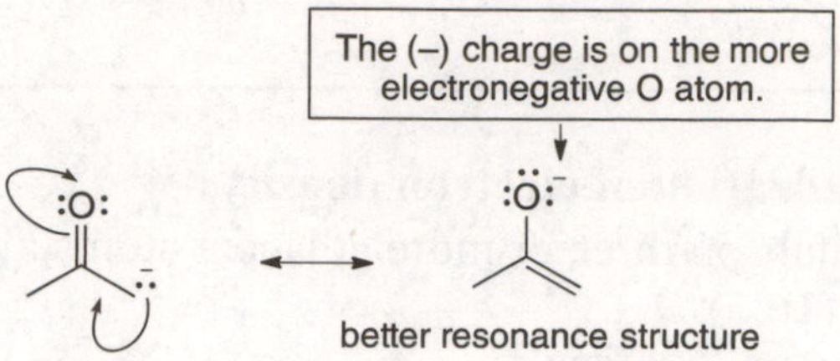

The (–) charge is on the more electronegative O atom.

better resonance structure

The unusual properties of conjugated dienes

[1] The C–C σ bond joining the two double bonds is unusually short (16.8).

[2] Conjugated dienes are more stable than similar isolated dienes. $\Delta H°$ of hydrogenation is smaller for a conjugated diene than for an isolated diene converted to the same product (16.9).

[3] The reactions are unusual:

- Electrophilic addition affords products of 1,2-addition and 1,4-addition (16.10, 16.11).
- Conjugated dienes undergo the Diels–Alder reaction, a reaction that does not occur with isolated dienes (16.12–16.14).

[4] Conjugated dienes absorb UV light in the 200–400 nm region. As the number of conjugated π bonds increases, the absorption shifts to longer wavelength (16.15).

Reactions of conjugated dienes

[1] Electrophilic addition of HX (X = halogen) (16.10, 16.11)

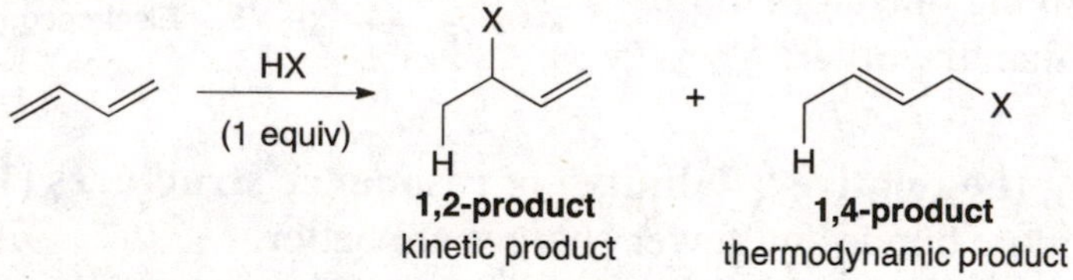

HX (1 equiv) → **1,2-product** kinetic product + **1,4-product** thermodynamic product

- The mechanism has two steps.
- Markovnikov's rule is followed. Addition of H^+ forms the *more* stable allylic carbocation.
- The 1,2-product is the kinetic product. When H^+ adds to the double bond, X^- adds to the end of the allylic carbocation to which it is closer (C2 not C4). The kinetic product is formed faster at low temperature.
- The thermodynamic product has the more substituted, more stable double bond. The thermodynamic product predominates at equilibrium. With buta-1,3-diene, the thermodynamic product is the 1,4-product.

[2] Diels–Alder reaction (16.12–16.14)

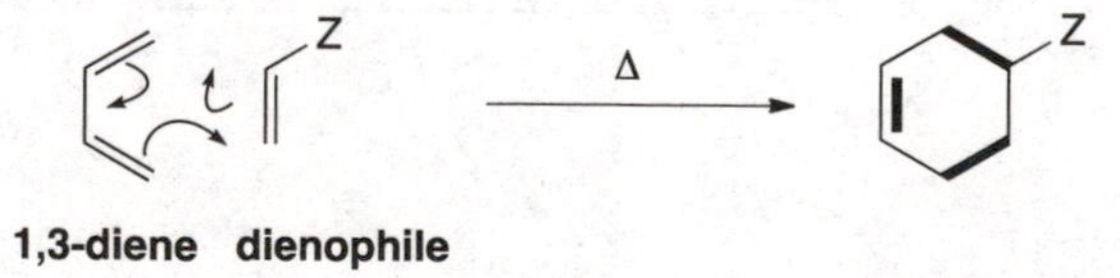

1,3-diene **dienophile** →(Δ)

The three new bonds are labeled in **bold**.

- The reaction forms two σ bonds and one π bond in a six-membered ring.
- The reaction is initiated by heat.
- The mechanism is concerted: all bonds are broken and formed in a single step.
- The diene must react in the *s*-cis conformation (16.13A).
- Electron-withdrawing groups in the dienophile increase the reaction rate (16.13B).
- The stereochemistry of the dienophile is retained in the product (16.13C).
- Endo products are preferred (16.13D).

Practice Test on Chapter Review

1.a. Which of the following statements is true about the Diels–Alder reaction?

1. The reaction is faster with electron-donating groups in the dienophile.
2. The reaction is endothermic.
3. The diene must adopt the *s*-cis conformation.
4. Statements (1) and (2) are true.
5. Statements (1), (2), and (3) are all true.

b. Which of the following statements is true about the absorption of ultraviolet light by unsaturated systems?

1. Penta-1,4-diene requires light having a wavelength < 200 nm for electron promotion.
2. Cyclohexa-1,3-diene absorbs ultraviolet light with a wavelength > 200 nm.
3. As the number of conjugated π bonds increases, the energy difference between the excited state and ground state decreases.
4. Statements (1) and (2) are true.
5. Statements (1), (2), and (3) are all true.

c. Which of the following compounds contains a labeled carbon atom that is sp^2 hybridized?

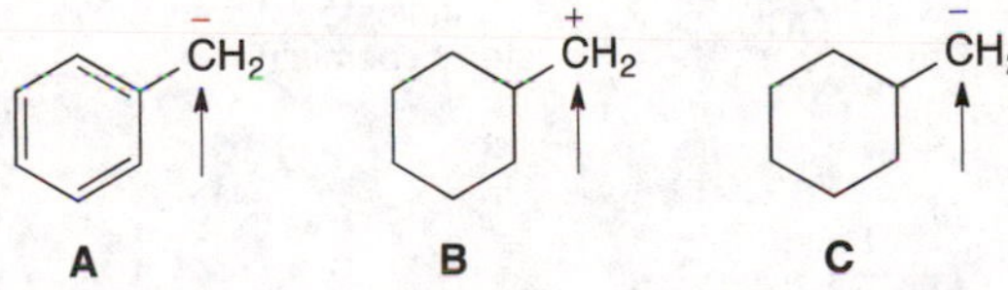

$\overset{-}{C}H_2$ $\overset{+}{C}H_2$ $\overset{-}{C}H_2$

A B C

1. **A** only
2. **B** only
3. **C** only
4. **A** and **B**
5. **A, B,** and **C**

d. Which of the following represent valid resonance structures for **A?**

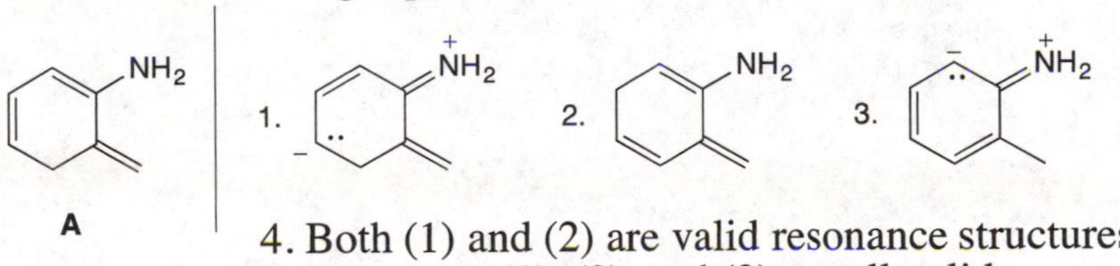

NH_2 1. $\overset{+}{N}H_2$ 2. NH_2 3. $\overset{+}{N}H_2$

A

4. Both (1) and (2) are valid resonance structures.
5. Structures (1), (2), and (3) are all valid.

2. Name the following compounds and indicate the conformation around the σ bond that joins the two double bonds.

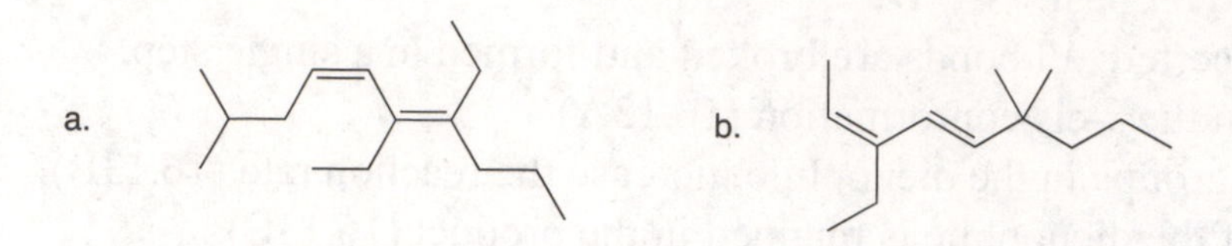

3.a. Consider the four hydrocarbons (**A–D**) drawn below. [1] Which compound absorbs the *shortest* wavelength of ultraviolet light? [2] Which compound absorbs the *longest* wavelength of ultraviolet light?

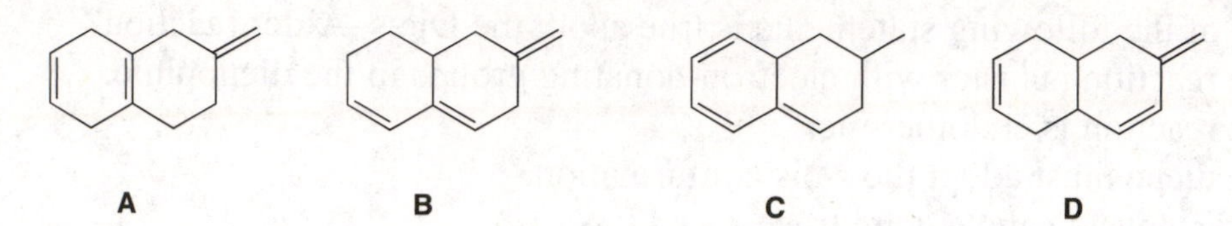

b. Consider the four dienes (**A–D**) drawn below. [1] Which diene is *most* reactive in the Diels–Alder reaction? [2] Which diene is the *least* reactive in the Diels–Alder reaction?

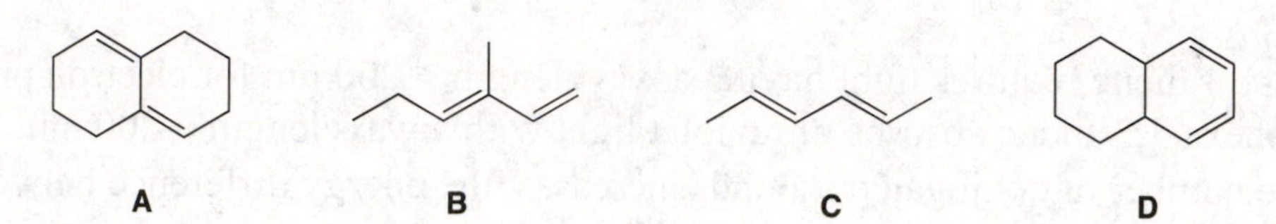

4. Draw the organic products formed in each reaction. In part (b), label the kinetic and thermodynamic products.

a. O= =O + OCH_3 —Δ, indicate stereochemistry→

b. —HBr (1 equiv)→

c. —[1] Δ; [2] CH_3O_2C CO_2CH_3, indicate stereochemistry→

5. What diene and dienophile are needed to synthesize the following Diels–Alder adducts?

a. CH_3O, CH_3O

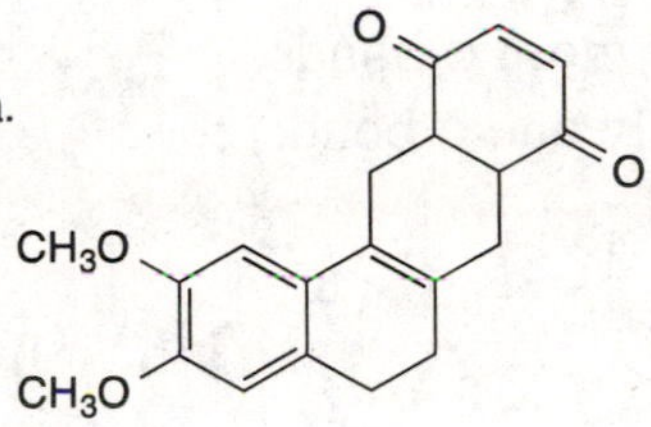

b. Cl, Cl, Cl, CN

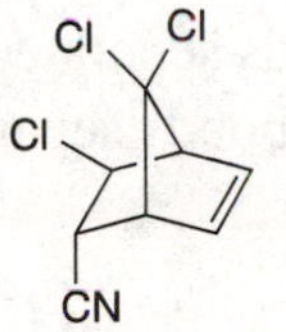

Answers to Practice Test

1.a. 3
  b. 5
  c. 4
  d. 1

2.a. (4*Z*,6*E*)-6,7-diethyl-2-methyldeca-4,6-diene, *s*-trans

  b. (2*Z*,4*E*)-3-ethyl-6,6-dimethylnona-2,4-diene, *s*-trans

3.a. [1] **A**; [2] **C**
  b. [1] **D**; [2] **A**

4.a. CH_3O, H, H, O, O

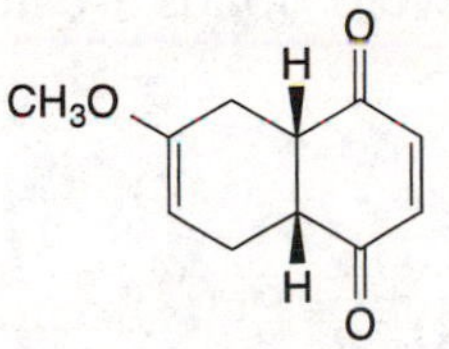

(both H's up or both H's down)

b. Br
kinetic

+

Br

thermodynamic

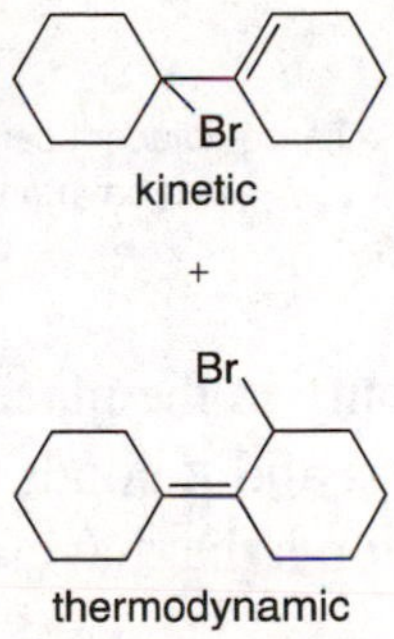

c. CO_2CH_3, CO_2CH_3

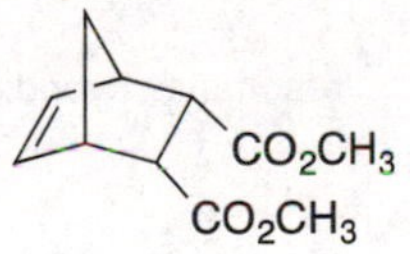

5.a. CH_3O, CH_3O

+

O, O

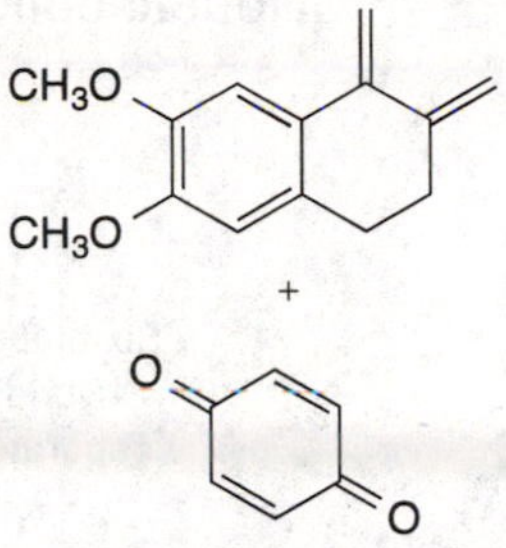

b. Cl, CN + Cl, Cl

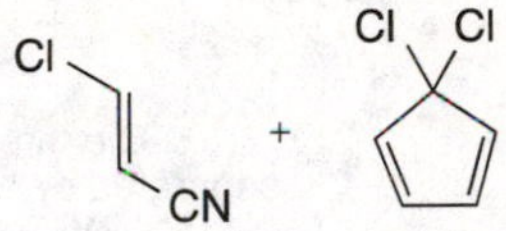

Answers to Problems

16.1 **Isolated dienes** have two double bonds separated by two or more σ bonds. **Conjugated dienes** have two double bonds separated by only one σ bond.

a. 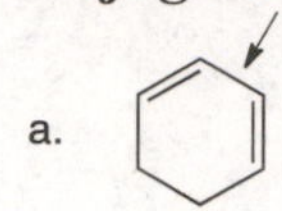

One σ bond separates two double bonds = **conjugated diene.**

b. Two σ bonds separate two double bonds = **isolated diene.**

c. One σ bond separates two double bonds = **conjugated diene.**

d. Four σ bonds separate two double bonds = **isolated diene.**

16.2 **Conjugation** occurs when there are overlapping *p* orbitals on three or more adjacent atoms. Double bonds separated by two σ bonds are not conjugated.

a. Six of the carbon atoms are sp^2 hybridized. Each π bond is separated by only one σ bond. **conjugated**

b. The three π bonds are separated by two or three σ bonds. **NOT conjugated**

c. The two π bonds are separated by only one σ bond. **conjugated**

d. Three adjacent carbon atoms are sp^2 hybridized and have an unhybridized *p* orbital. **conjugated**

e. This carbon is not sp^2 hybridized. **NOT conjugated**

16.3 Two resonance structures differ only in the placement of electrons. All σ bonds stay in the same place. Nonbonded electrons and π bonds can be moved. To draw the hybrid:

- Use a dashed line between atoms that have a π bond in one resonance structure and not the other.
- Use a δ symbol for atoms with a charge or radical in one structure but not the other.

a. resonance hybrid: δ+ δ+

The + charge is delocalized on two carbons.

b. resonance hybrid: δ+ δ+

The + charge is delocalized on two carbons.

c. resonance hybrid: δ+ δ+

The + charge is delocalized on two carbons.

16.4 S_N1 reactions proceed via a carbocation intermediate. Draw the carbocation formed on loss of Cl and compare. The more stable the carbocation, the faster the S_N1 reaction.

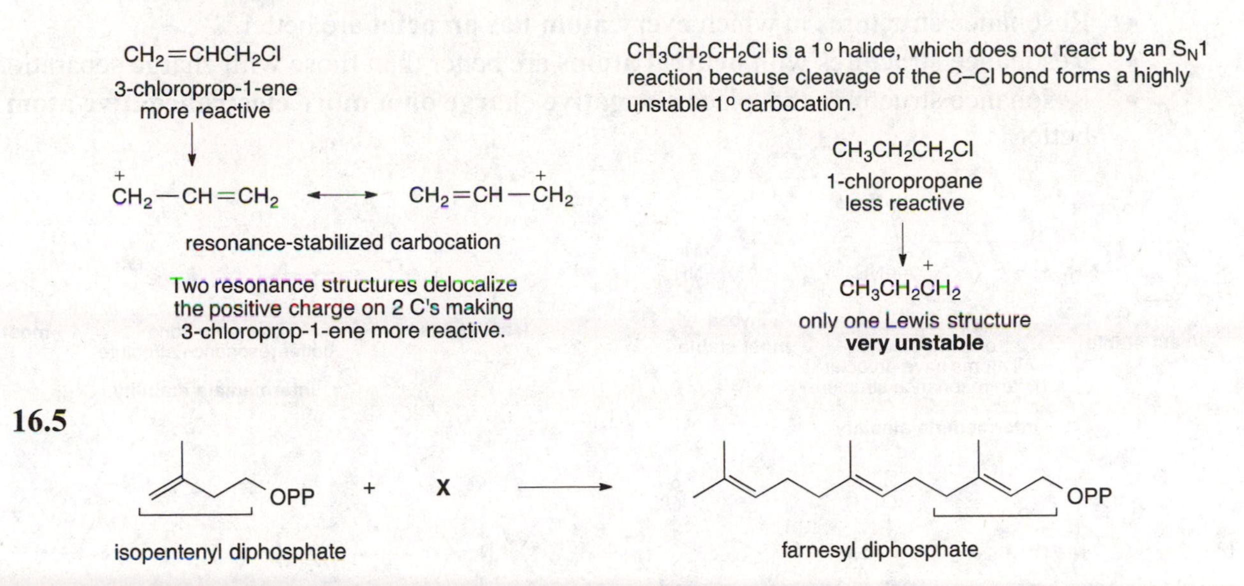
$CH_2{=}CHCH_2Cl$
3-chloroprop-1-ene
more reactive
$\overset{+}{C}H_2{-}CH{=}CH_2$ ⟷ $CH_2{=}CH{-}\overset{+}{C}H_2$
resonance-stabilized carbocation
Two resonance structures delocalize the positive charge on 2 C's making 3-chloroprop-1-ene more reactive.
$CH_3CH_2CH_2Cl$ is a 1° halide, which does not react by an S_N1 reaction because cleavage of the C–Cl bond forms a highly unstable 1° carbocation.
$CH_3CH_2CH_2Cl$
1-chloropropane
less reactive
$CH_3CH_2\overset{+}{C}H_2$
only one Lewis structure
very unstable
16.5
OPP
+
X
OPP
isopentenyl diphosphate
farnesyl diphosphate

Five C's of farnesyl diphosphate come from isopentenyl diphosphate, so the remaining 10 C's come from **X.** If the reaction is analogous to the formation of geranyl diphosphate, **X** must contain 10 C's and have an allylic diphosphate.

OPP

OPP = **X**

16.6

a. Move the charge and the double bond.

b. Move the charge and the double bond.

c. Move the lone pair.

d. Move the charge and the double bond.

16.7 To compare the resonance structures remember:

- Resonance structures with **more bonds** are better.
- Resonance structures in which **every atom has an octet** are better.
- Resonance structures with **neutral atoms** are better than those with charge separation.
- Resonance structures that place a **negative charge on a more electronegative atom** are better.

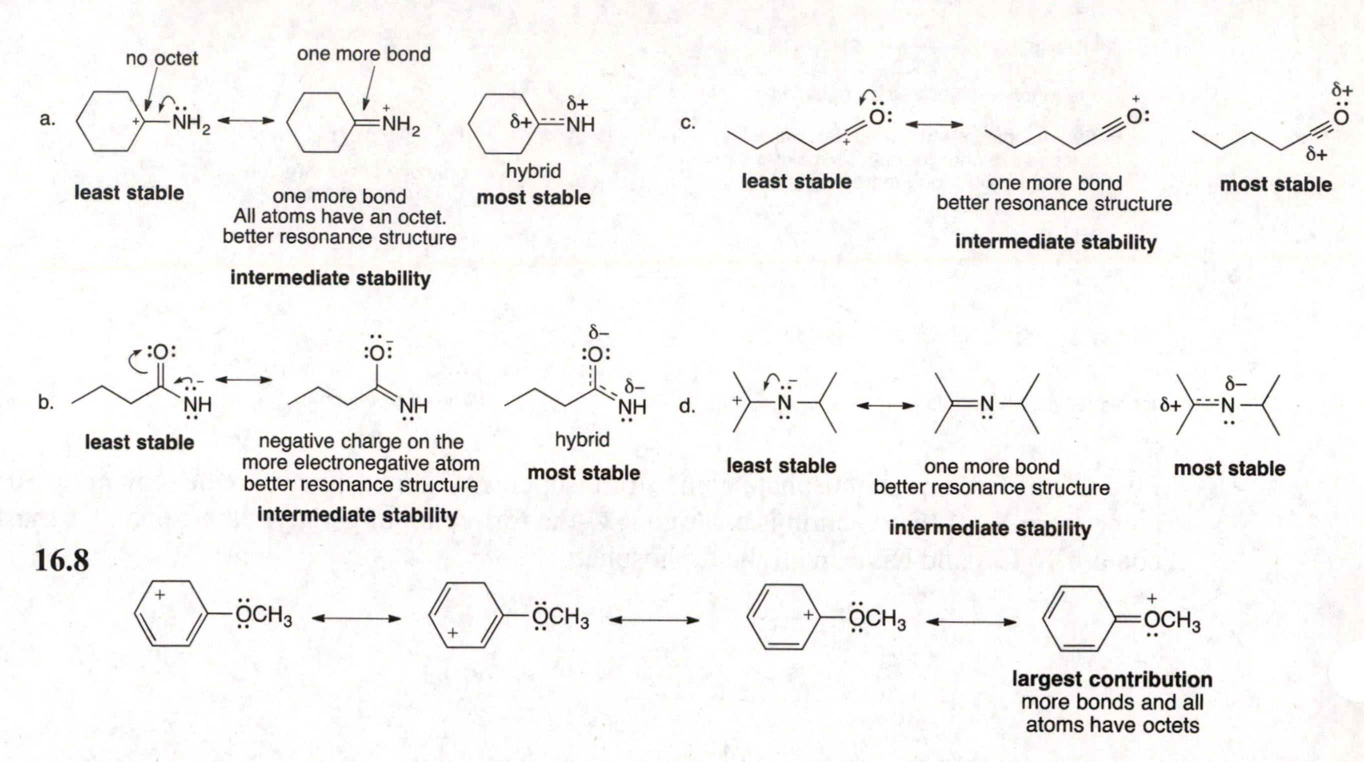

16.8

16.9 Remember that in any allyl system, there must be *p* orbitals to delocalize the lone pair.

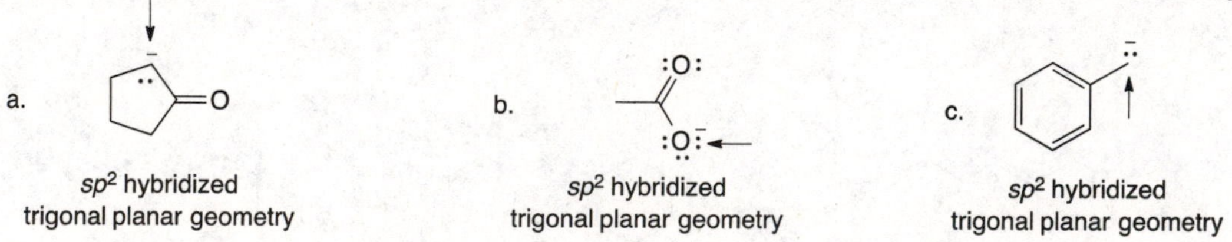

16.10 The **s-cis** conformation has two double bonds on the **same side** of the single bond. The **s-trans** conformation has two double bonds on **opposite sides** of the single bond.

a. (2*E*,4*E*)-**octa-2,4-diene** in the *s*-trans conformation

double bonds on opposite sides
***s*-trans**

b. (3*E*,5*Z*)- **nona-3,5-diene** in the *s*-cis conformation

Z
E

double bonds on the same side
***s*-cis**

c. (3*Z*,5*Z*)-4,5-dimethyl**deca-3,5-diene** in both the *s*-cis and *s*-trans conformations

Z
***s*-cis**
Z
Z
***s*-trans**
Z

16.11

***s*-cis**
conjugated
conjugated
***s*-trans**
conjugated
E E
Z
OH
isolated
isolated
Z
OH
HO
Z
Z
isolated
isolated

16.12 Bond length depends on hybridization and percent *s*-character. Bonds with a higher percent *s*-character have smaller orbitals and are shorter.

HC≡C–C≡CH
sp hybridized carbons
50% *s*-character
shortest bond

CH_2=CH–CH=CH_2
sp^2 hybridized carbons
33% *s*-character
intermediate length

CH_3–CH_3
sp^3 hybridized carbons
25% *s*-character
longest bond

16.13 Two equivalent resonance structures delocalize the π bond and the negative charge.

hybrid:
δ–
δ–
These bond lengths are equal because they are identical.

16.14 The **less stable** (higher energy) **diene** has the **larger heat of hydrogenation.** Isolated dienes are higher in energy than conjugated dienes, so they will have a larger heat of hydrogenation.

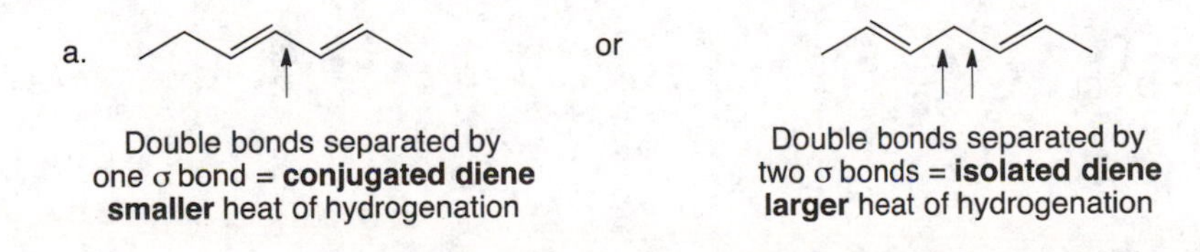

b.

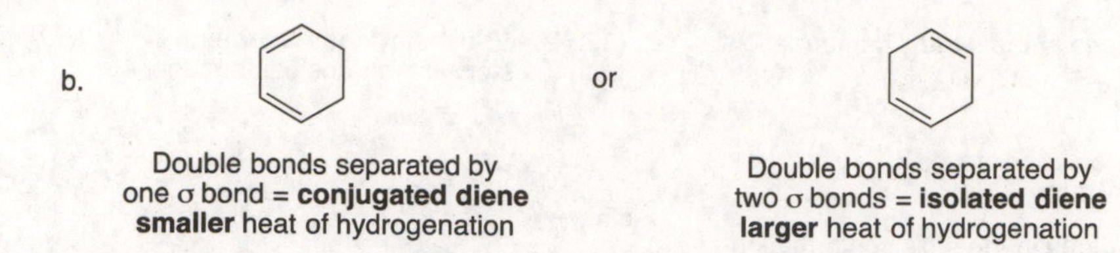

16.15 Isolated dienes are higher in energy than conjugated dienes. Compare the location of the double bonds in the compounds below.

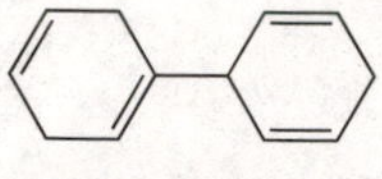

0 conjugated double bonds **least stable**

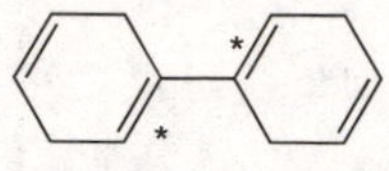

2 conjugated double bonds **intermediate stability**

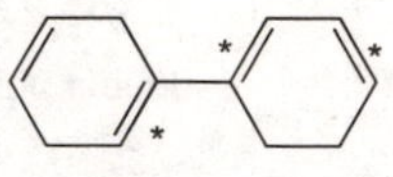

3 conjugated double bonds **most stable**

16.16 Conjugated dienes react with HX to form 1,2- and 1,4-products.

a. 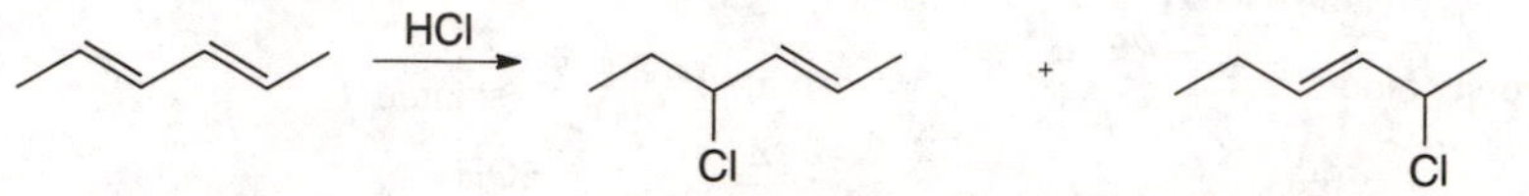 HCl → (+ *Z* isomer) **1,2-product** + (+ *Z* isomer) **1,4-product**

b. 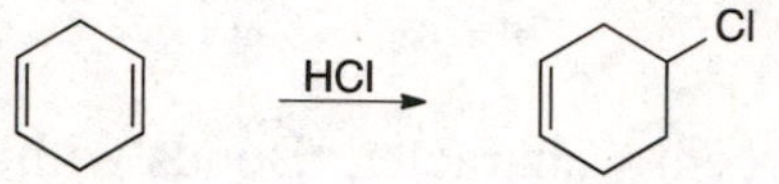 HCl → **isolated diene**

c. 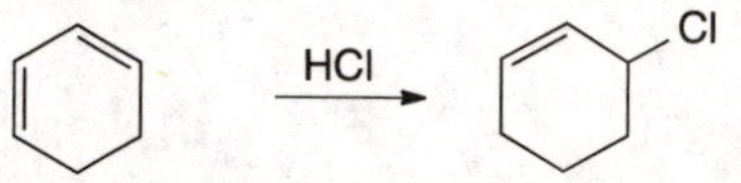 HCl →

d. 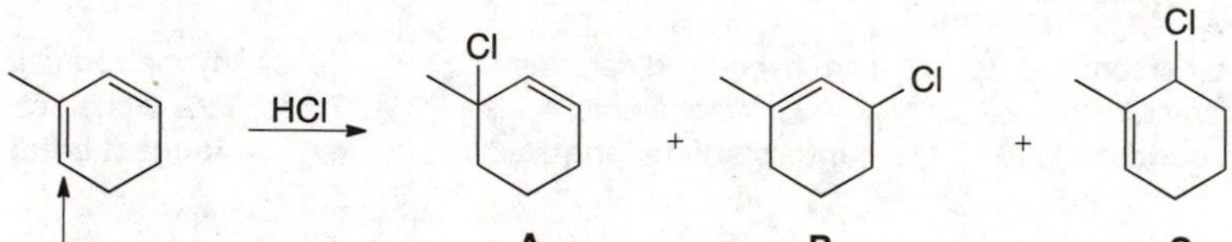 HCl → **A** + **B** + **C**

This double bond is more reactive, so **C** is probably a minor product because it results from HCl addition to the less reactive double bond.

16.17 The mechanism for addition of DCl has two steps:

[1] **Addition of D^+** forms a resonance-stabilized carbocation.

[2] **Nucleophilic attack of Cl^-** forms 1,2- and 1,4-products.

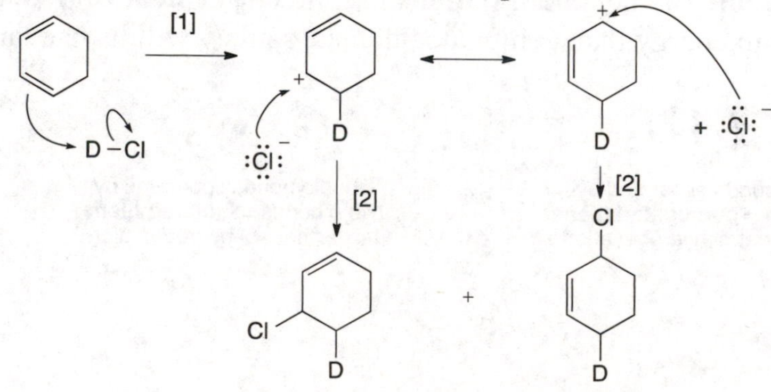

16.18 Label the products as 1,2- or 1,4-products. The 1,2-product is the kinetic product, and the 1,4-product, which has the more substituted double bond, is the thermodynamic product.

This is C1. This is C4. HCl
The H is added here. 1,2-product kinetic product
The H is added here. 1,4-product thermodynamic product

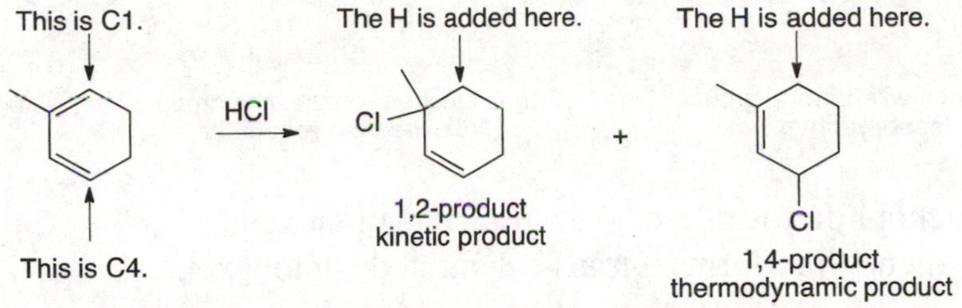

16.19 To draw the products of a Diels–Alder reaction:
[1] Find the 1,3-diene and the dienophile.
[2] Arrange them so the diene is on the left and the dienophile is on the right.
[3] Cleave three bonds and use arrows to show where the new bonds will be formed.

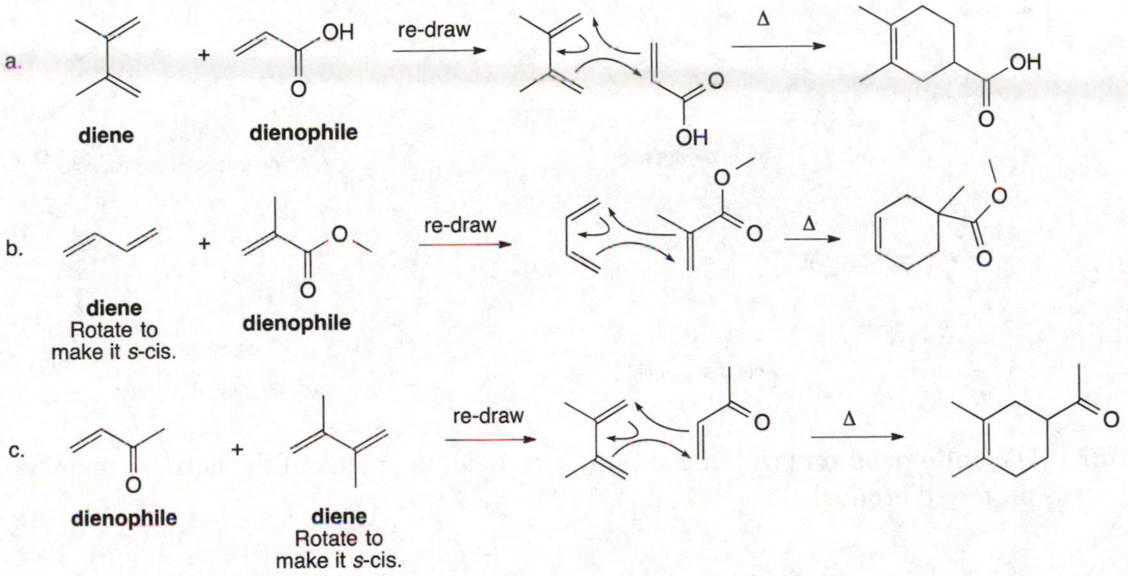

a. diene + dienophile —re-draw→ —Δ→

b. diene (Rotate to make it *s*-cis.) + dienophile —re-draw→ —Δ→

c. dienophile + diene (Rotate to make it *s*-cis.) —re-draw→ —Δ→

16.20 For a diene to be reactive in a Diels–Alder reaction, **a diene must be able to adopt an *s*-cis conformation.**

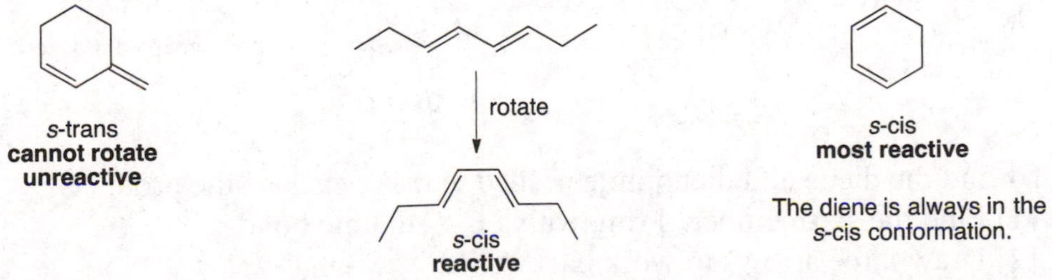

s-trans
cannot rotate
unreactive

rotate

s-cis
reactive

s-cis
most reactive
The diene is always in the *s*-cis conformation.

16.21 Zingiberene reacts much faster than β-sesquiphellandrene as a Diels–Alder diene because its diene is constrained in the *s*-cis conformation. The diene in β-sesquiphellandrene is constrained in the *s*-trans conformation, so it is unreactive in the Diels–Alder.

16.22 Electron-withdrawing substituents in the dienophile increase the reaction rate.

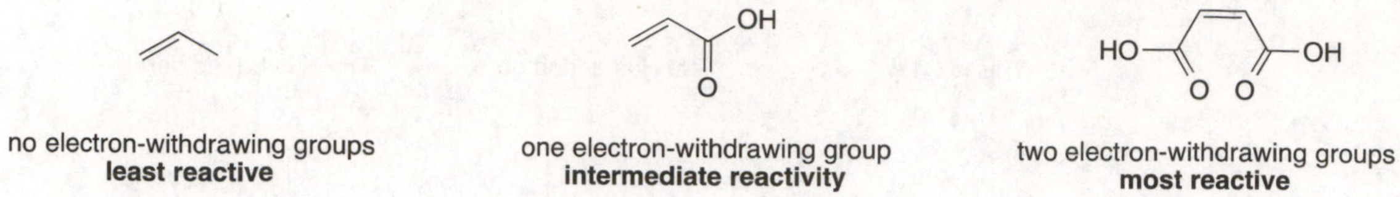

OH
HO OH
O O O

no electron-withdrawing groups
least reactive

one electron-withdrawing group
intermediate reactivity

two electron-withdrawing groups
most reactive

16.23 A cis dienophile forms a cis-substituted cyclohexene.
A trans dienophile forms a trans-substituted cyclohexene.

a. + Δ → +

cis dienophile

cis-substituted products

b. + Δ → +

trans dienophile

trans-substituted products

c. + Δ → +

cis dienophile

identical

cis-substituted product

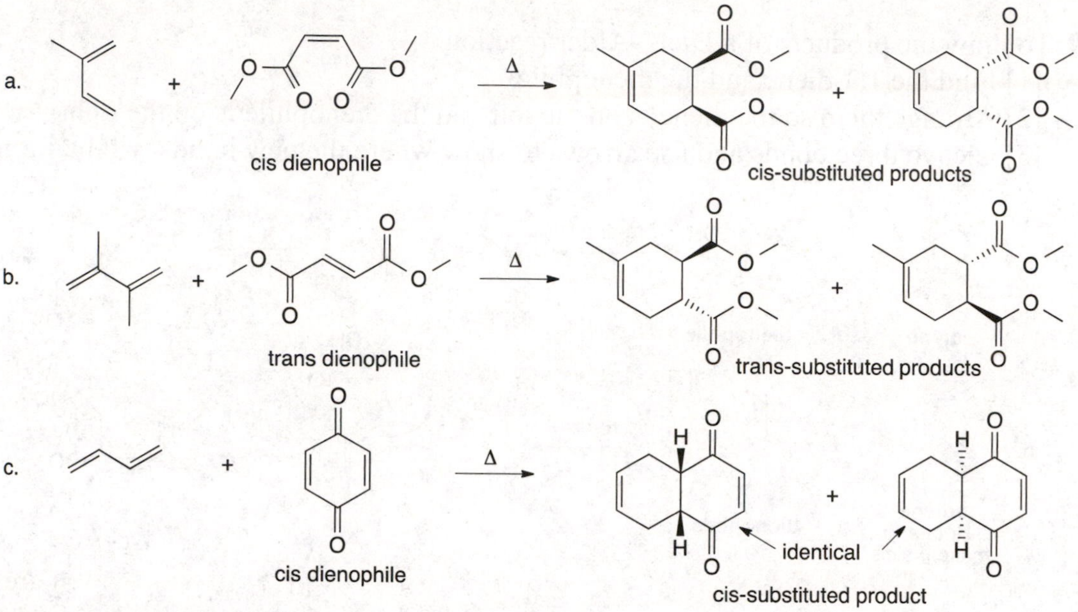

16.24 The **endo product** (with the substituents under the plane of the new six-membered ring) is the preferred product.

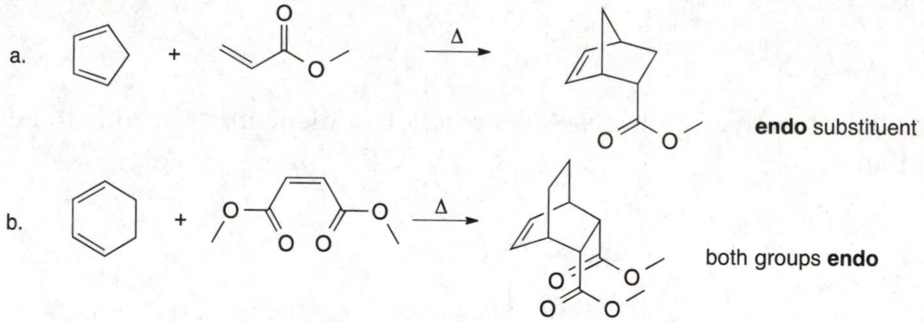

a. + Δ →

endo substituent

b. + Δ →

both groups **endo**

16.25 To find the diene and dienophile needed to make each of the products:
[1] Find the six-membered ring with a C–C double bond.
[2] Draw three arrows to work backwards.
[3] Follow the arrows to show the diene and dienophile.

a.

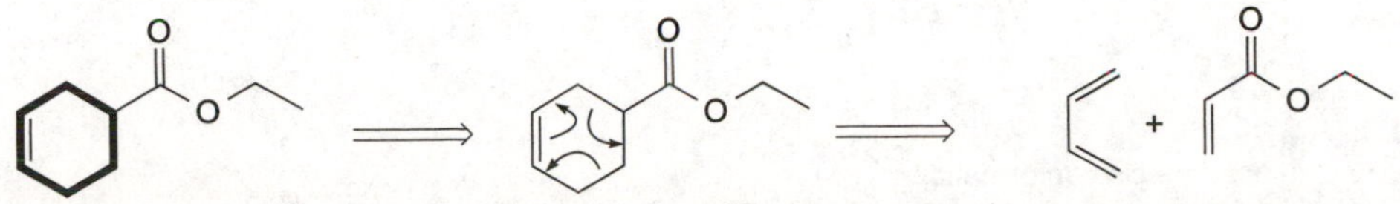

b.

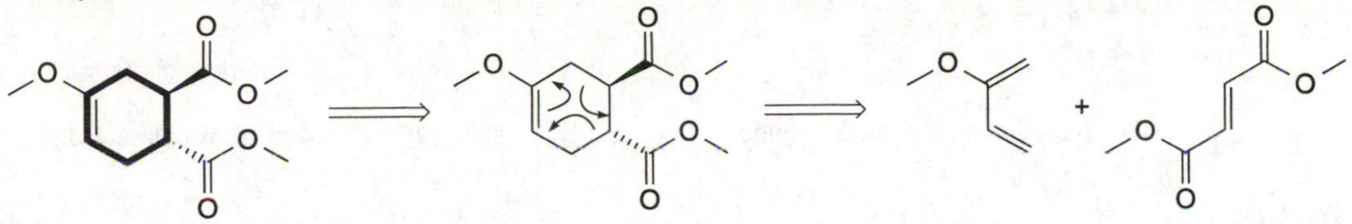

c.

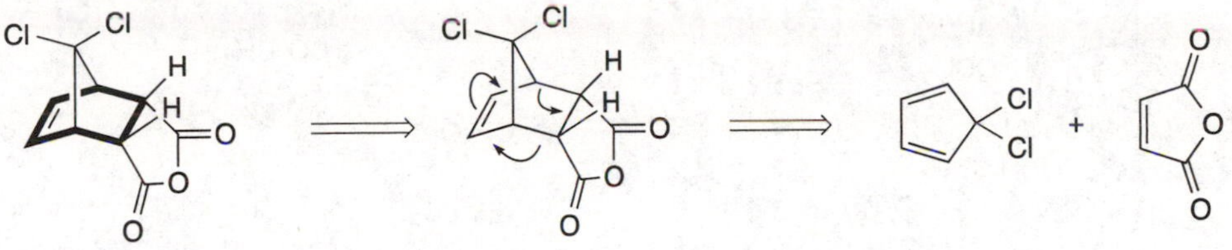

Cl Cl H H O O O ⟹ Cl Cl H H O O O ⟹ Cl Cl + O O O

16.26

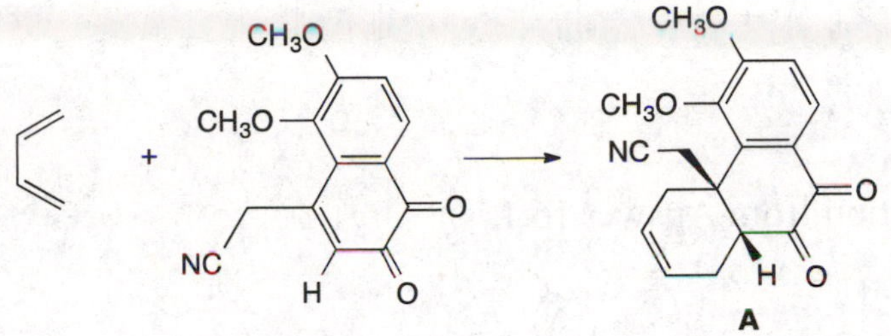

CH$_3$O CH$_3$O NC H O O + ⟶ CH$_3$O CH$_3$O NC H O O (+ enantiomer)

A

16.27 Conjugated molecules absorb light at a longer wavelength than molecules that are not conjugated.

a.  conjugated longer wavelength or not conjugated

b. 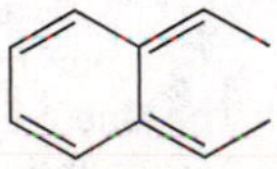 all double bonds conjugated longer wavelength or one set of conjugated dienes

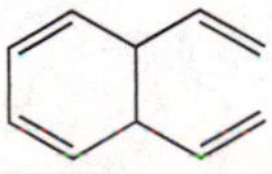

16.28 Sunscreens contain conjugated systems to absorb UV radiation from sunlight. Look for conjugated systems in the compounds below.

a. 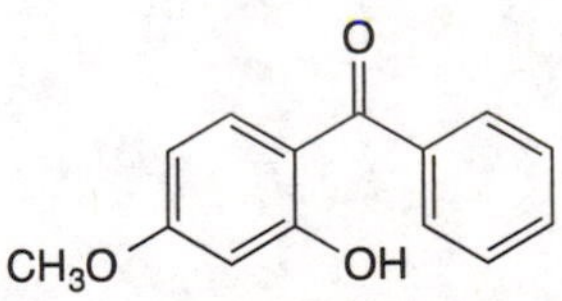 CH$_3$O OH **conjugated system** could be a sunscreen

b. 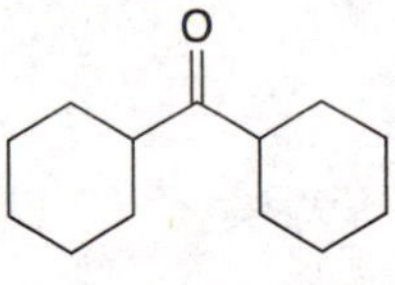 not a conjugated system

c. 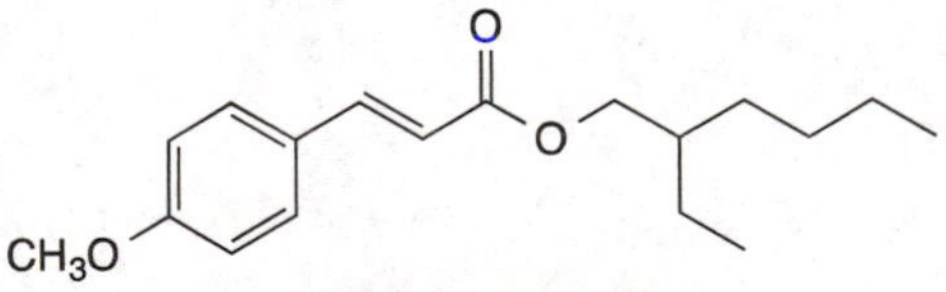 CH$_3$O **conjugated system** could be a sunscreen

16.29

a. 2, 3, 4 — *s*-cis conformation

(2*Z*,4*E*)-3,4-dimethylhepta-2,4-diene

b. 2, 3, 4 — *s*-trans conformation

(2*E*,4*Z*)-3,4-dimethylocta-2,4-diene

16.30

a. O, OCH_3, H, H, O, OCH_3 ⟹ + CO_2CH_3, CH_3O_2C

b. H, O, H, O ⟹ + O, O

16.31 Use the definition from Answer 16.1.

a. 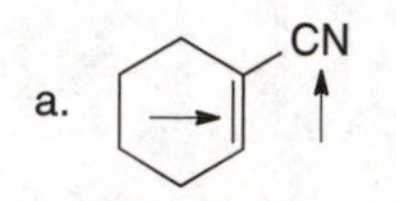

2 multiple bonds with only
1 σ bond between
conjugated

b. This C is *sp*[2].

1 π bond with
an adjacent
sp[2] hybridized atom
The lone pair occupies a *p* orbital,
so there are *p* orbitals on
three adjacent atoms.
conjugated

c. 1 π bond with
no adjacent
sp[2] hybridized atoms
NOT conjugated

d. O

1 π bond with
no adjacent
sp[2] hybridized atoms
NOT conjugated

16.32

a. ⟷ ⟷ ⟷

b. ⟷ ⟷

c. ⟷ ⟷

d. :O: :O: ⟷ :O: :Ö:⁻ ⟷ :Ö:⁻ :O:

e.

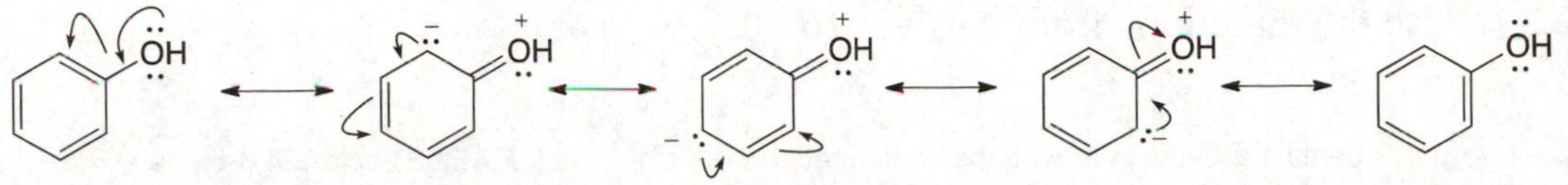

f.

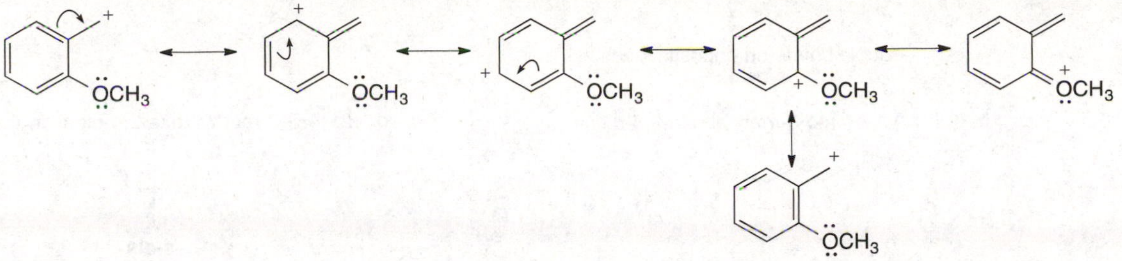

16.33 No additional resonance structures can be drawn for compounds (a) and (d).

b. 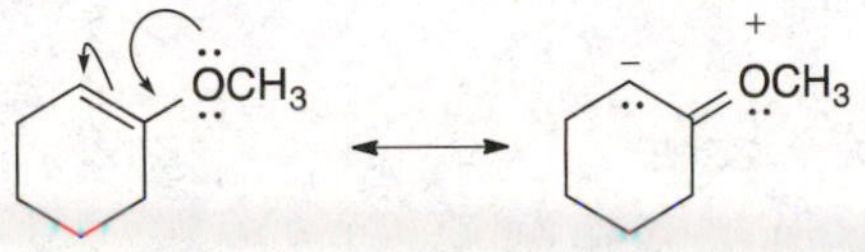 c.

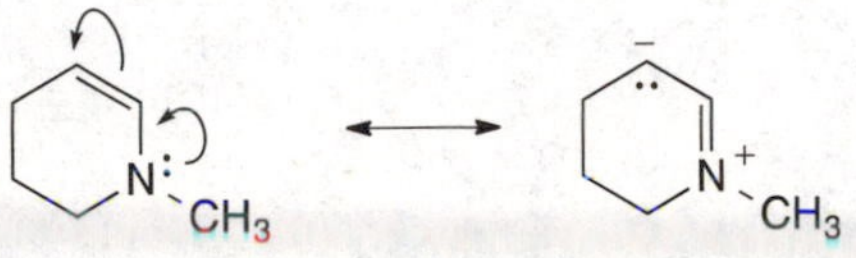

16.34

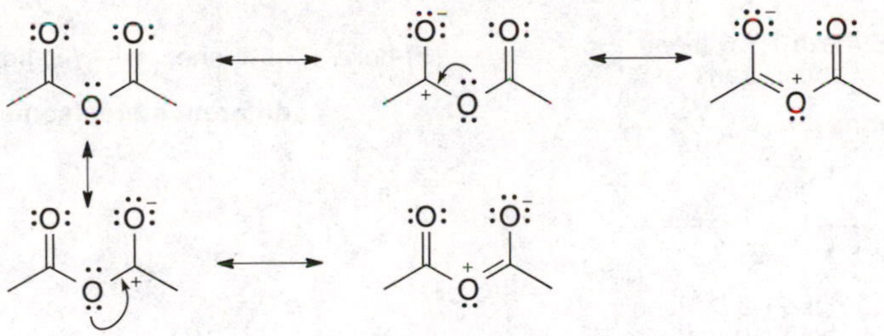

16.35

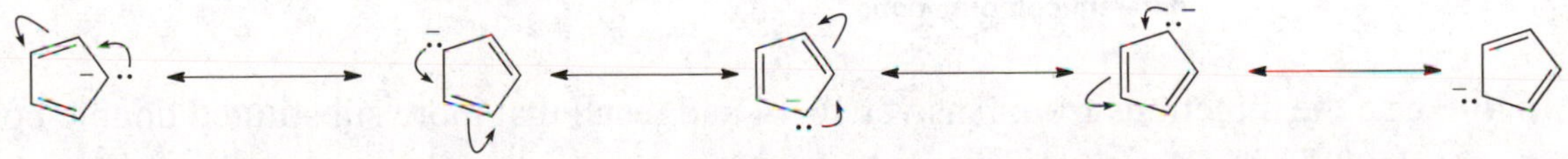

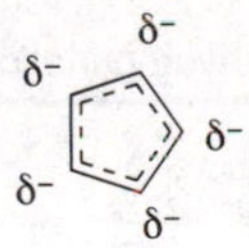

Five resonance structures delocalize the negative charge on five C's, making them all equivalent.

resonance hybrid:

All of the carbons are identical in the anion.

16.36

Draw the products of cleavage of the bond.

ethane CH_3—CH_3

$\cdot CH_3$ + $\cdot CH_3$

Two unstable radicals form.

but-1-ene

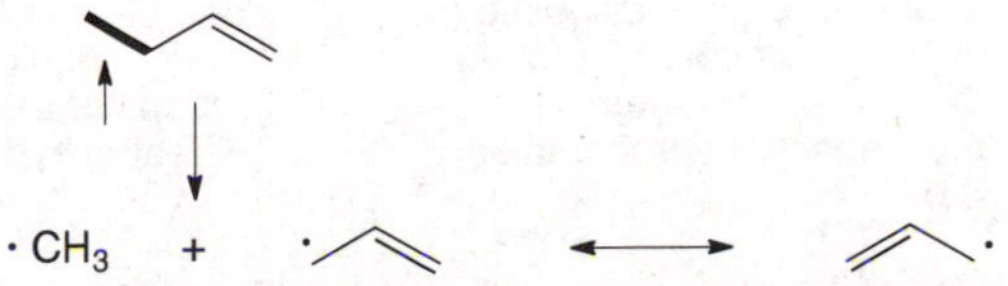

$\cdot CH_3$ +

One resonance-stabilized radical forms. This makes the bond dissociation energy lower because a more stable radical is formed.

16.37 Use the directions from Answer 16.10.

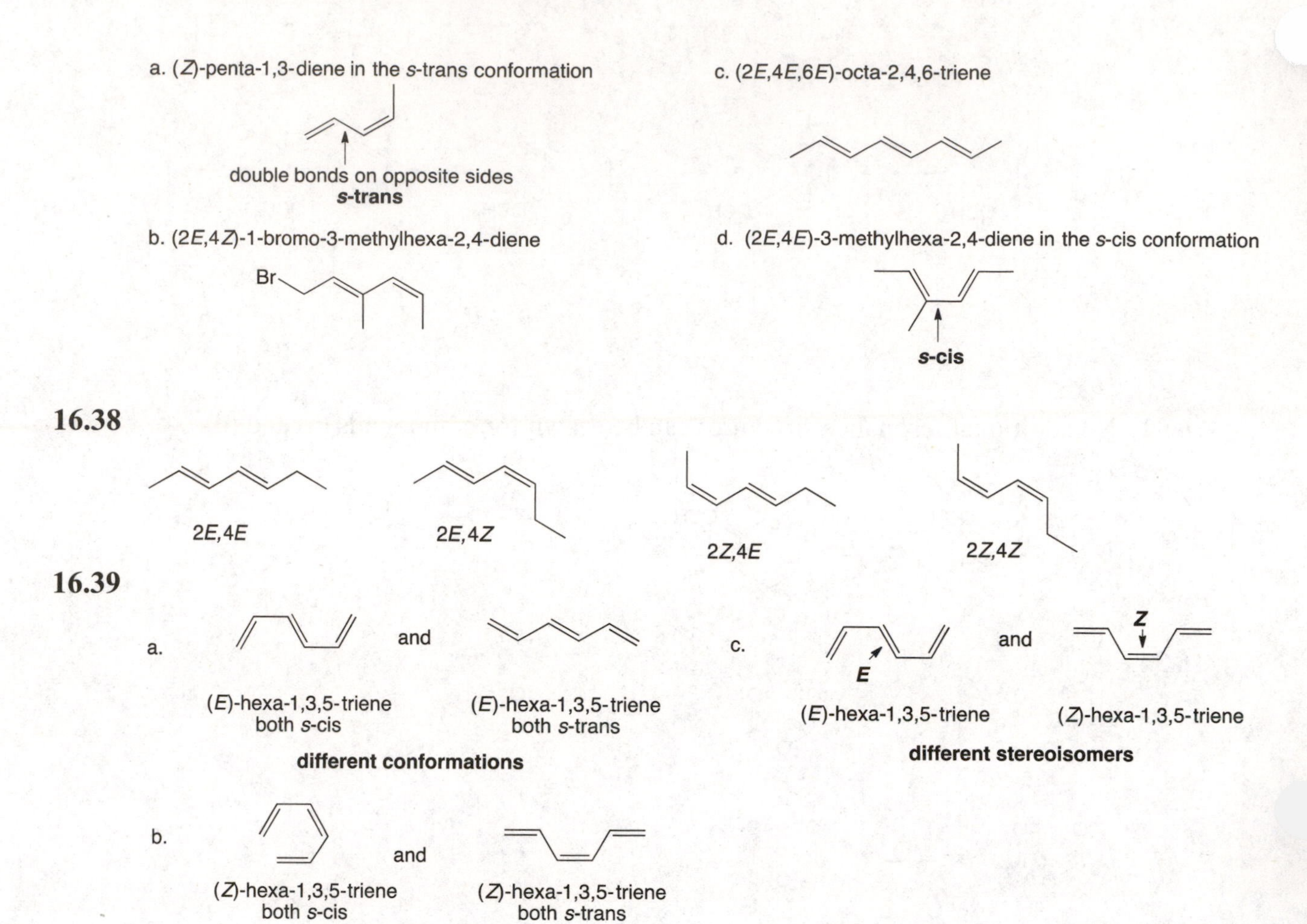

16.40 Use the directions from Answer 16.14 and recall that more substituted double bonds are more stable.

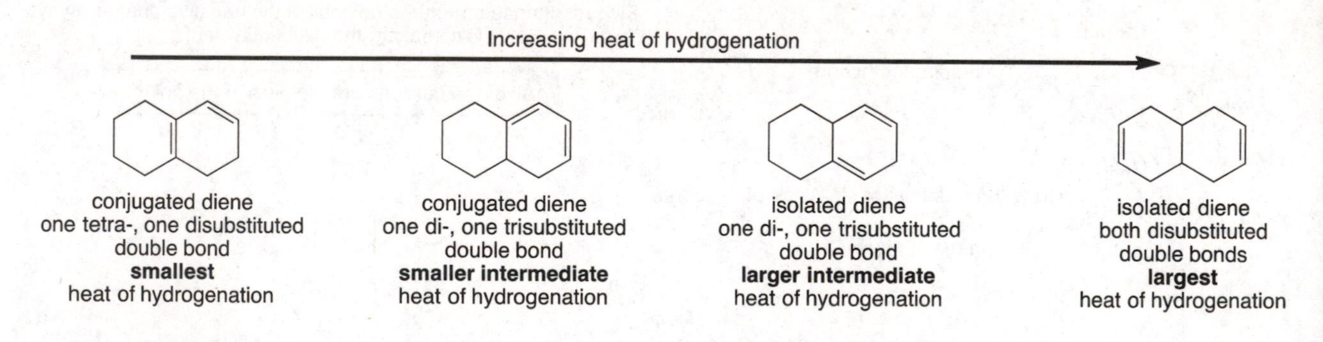

16.41 Conjugated dienes react with HX to form 1,2- and 1,4-products.

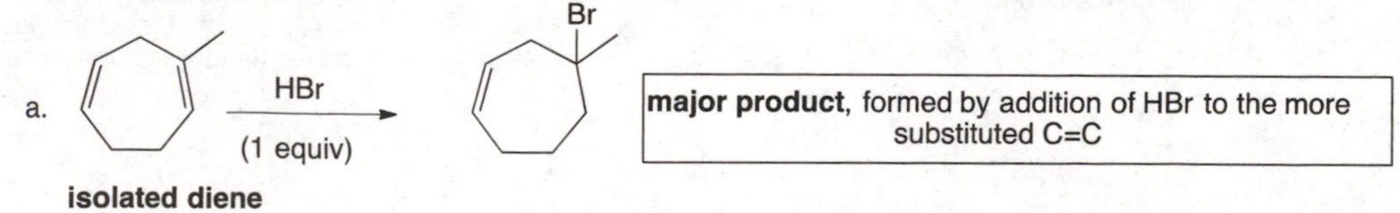

b. 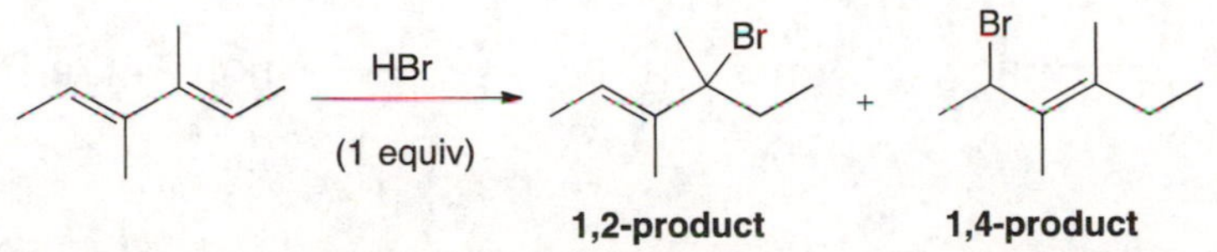 HBr (1 equiv) → 1,2-product + 1,4-product (*E* and *Z* isomers can form.)

c. 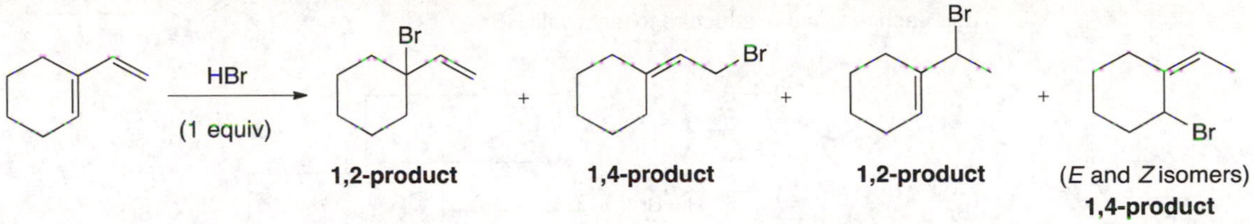 HBr (1 equiv) → 1,2-product + 1,4-product + 1,2-product + (*E* and *Z* isomers) 1,4-product

16.42

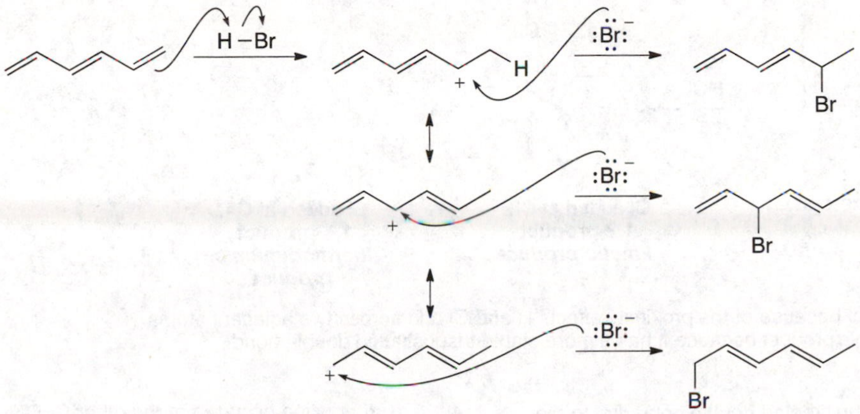

H–Br; :Br:⁻; Br

16.43

This cation forms because it is benzylic and resonance stabilized.

A; H–Br; :Br:⁻; Br; C

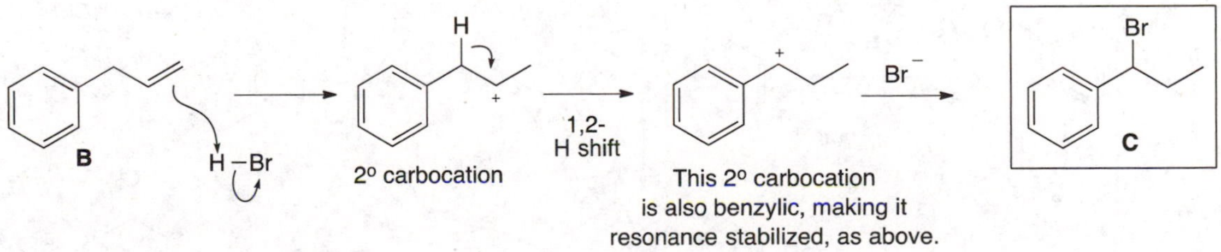

B; H–Br; 2° carbocation; 1,2- H shift; This 2° carbocation is also benzylic, making it resonance stabilized, as above.; Br^-; Br; C

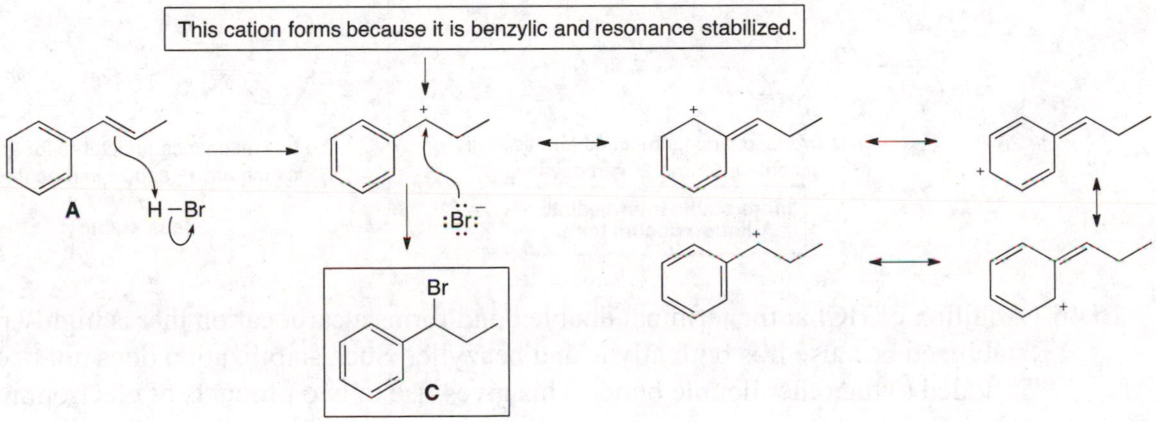

16.44 To draw the mechanism for the reaction of a diene with HBr and ROOR, recall from Chapter 15 that when an alkene is treated with HBr under these radical conditions, the Br ends up on the carbon with more H's to begin with.

RO–OR → RO· + ·OR + H–Br: → HOR + ·Br:

:Br (radical) ⟷ :Br (radical)

Use each resonance structure to react with HBr.

:Br + H–Br: → **:Br** + ·Br:

:Br + H–Br: → **:Br** + ·Br:

16.45

a, b.

HCl

X → Y + Z

C2

Cl

C4 →

Cl

H adds here at C1.

Cl added at C2.
1,2-product
kinetic product

Cl added at C4.
1,4-product
thermodynamic product

Y is the kinetic product because of the proximity effect. H and Cl add across two adjacent atoms.
Z is the thermodynamic product because it has a more stable trisubstituted double bond.

c.

Addition occurs at the labeled double bond due to the stability of the carbocation intermediate.

The two resonance structures for this allylic cation are 3° and 2° carbocations.

more stable intermediate
Addition occurs here.

If addition occurred at the other C=C, the following allylic carbocation would form:

The two resonance structures for this allylic cation are 1° and 2° carbocations.

less stable

16.46 Addition of HCl at the terminal double bond forms a carbocation that is highly resonance stabilized because it is both allylic and benzylic. Such stabilization does not occur when HCl is added to the other double bond. This gives rise to two products of electrophilic addition.

H–Cl

:Cl:⁻

Cl

1,2-product

:Cl:⁻

Cl

1,4-product

(+ three more resonance structures that delocalize the positive charge onto the benzene ring)

16.47 There are two possible products:

trisubstituted C=C
disubstituted C=C
Br
Br
1,2-product
1,4-product

The 1,2-product is always the kinetic product because of the proximity effect. In this case, it is also the thermodynamic (more stable) product because it contains a more highly substituted C=C (trisubstituted) than the 1,4-product (disubstituted). Thus, the 1,2-product is the major product at high and low temperature.

16.48 The electron pairs on O can be donated to the double bond through resonance. This increases the electron density of the double bond, making it less electrophilic and therefore less reactive in a Diels–Alder reaction.

methyl vinyl ether
This C now bears a net negative charge.

16.49 Use the directions from Answer 16.19.

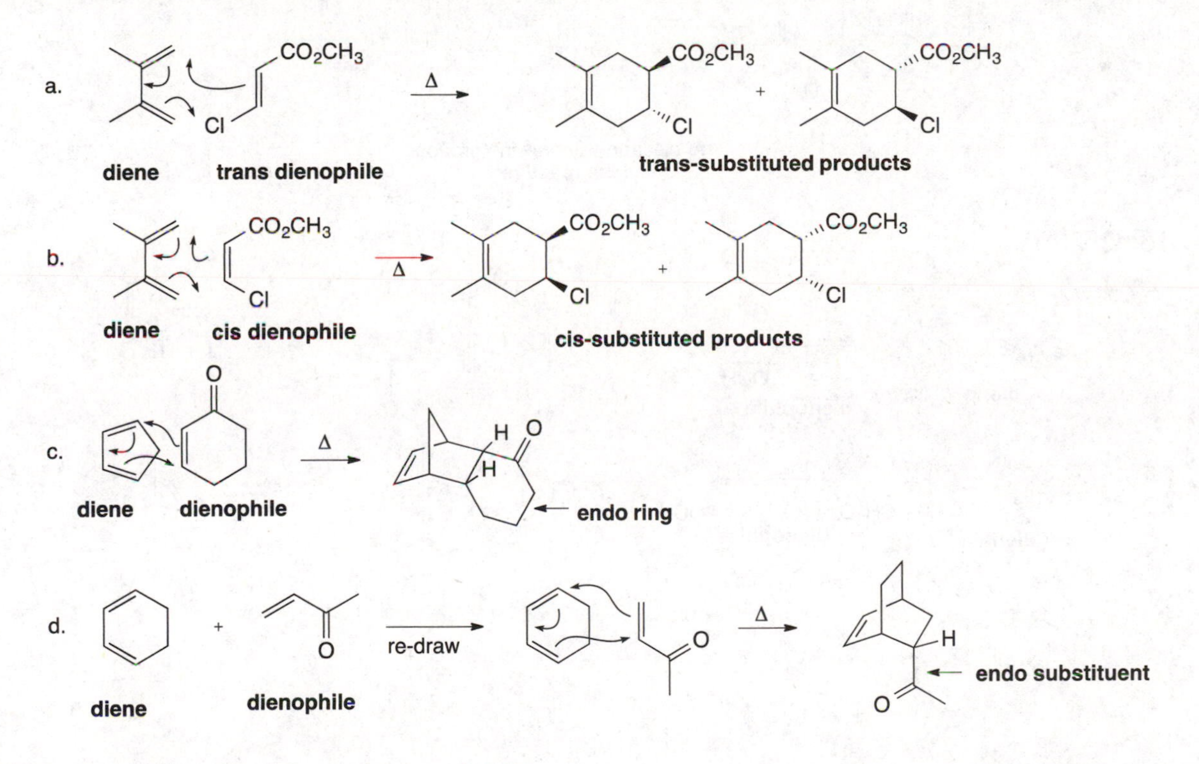

16.50 Use the directions from Answer 16.25.

a. CO_2CH_3 ⟹ CO_2CH_3 ⟹ + CO_2CH_3

b. 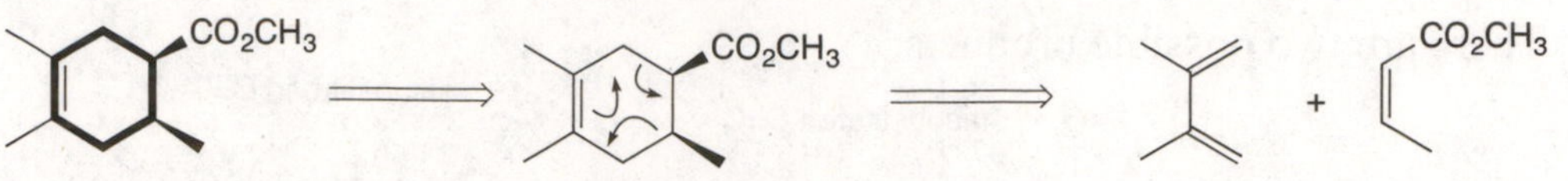 CO_2CH_3 ⟹ CO_2CH_3 ⟹ + CO_2CH_3

c. 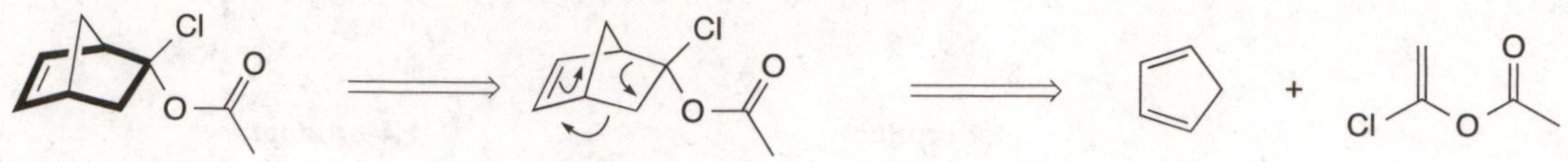 Cl ⟹ Cl ⟹ + Cl

d.

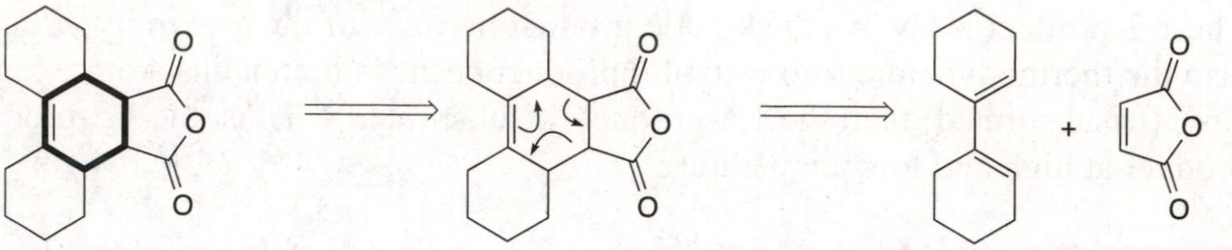

16.51

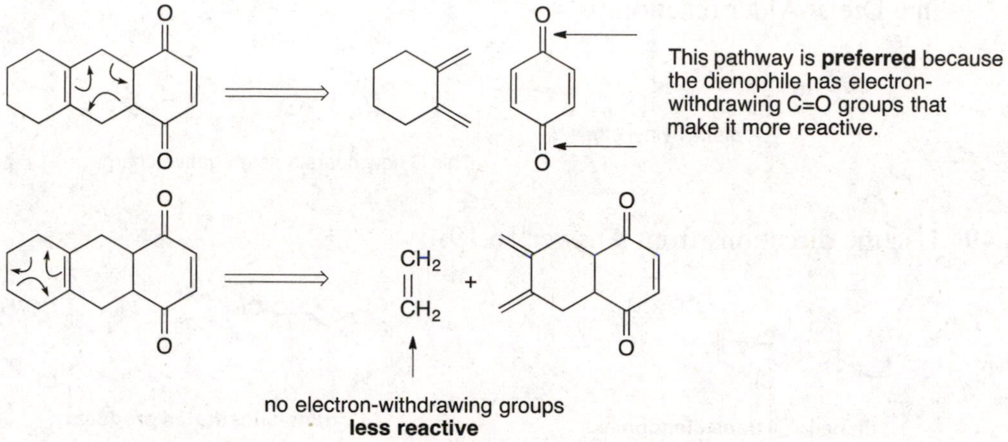

This pathway is **preferred** because the dienophile has electron-withdrawing C=O groups that make it more reactive.

CH_2 = CH_2 +

no electron-withdrawing groups
less reactive

16.52

a. 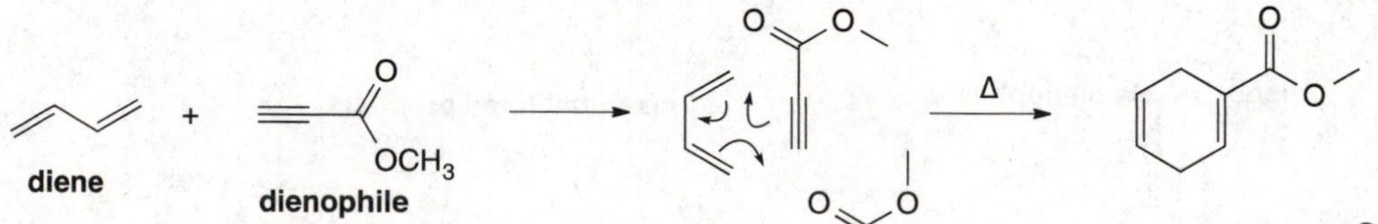 diene + dienophile (OCH_3) ⟶ Δ ⟶

diene **dienophile**

b. 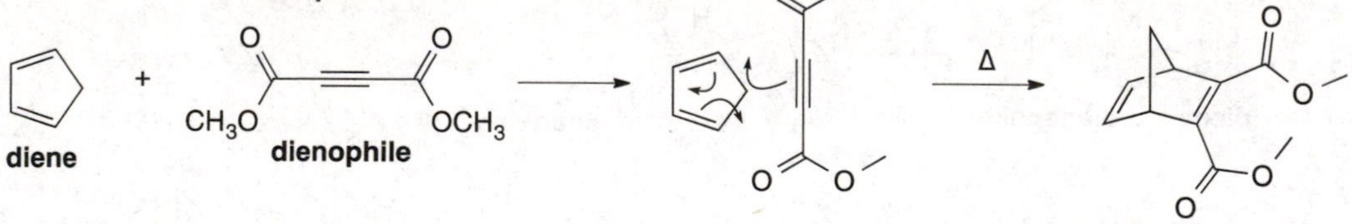 diene + CH_3O dienophile OCH_3 ⟶ Δ ⟶

diene **dienophile**

16.53

Δ

1,2-disubstituted product
major

1,3-disubstituted product
minor

The major product is formed when the circled carbons with a δ+ and δ− react.

resonance hybrids:

For the 1,2-product, carbons with unlike charges would react. This is favored because the electron-rich and the electron-poor C's can bond.

For the 1,3-product, there are no partial charges of opposite sign on reacting carbons. This arrangement is less attractive.

16.54

X

Y

aldrin

mCPBA
1 equiv

Z
dieldrin

This double bond is more electron rich, so it is epoxidized more readily.

16.55

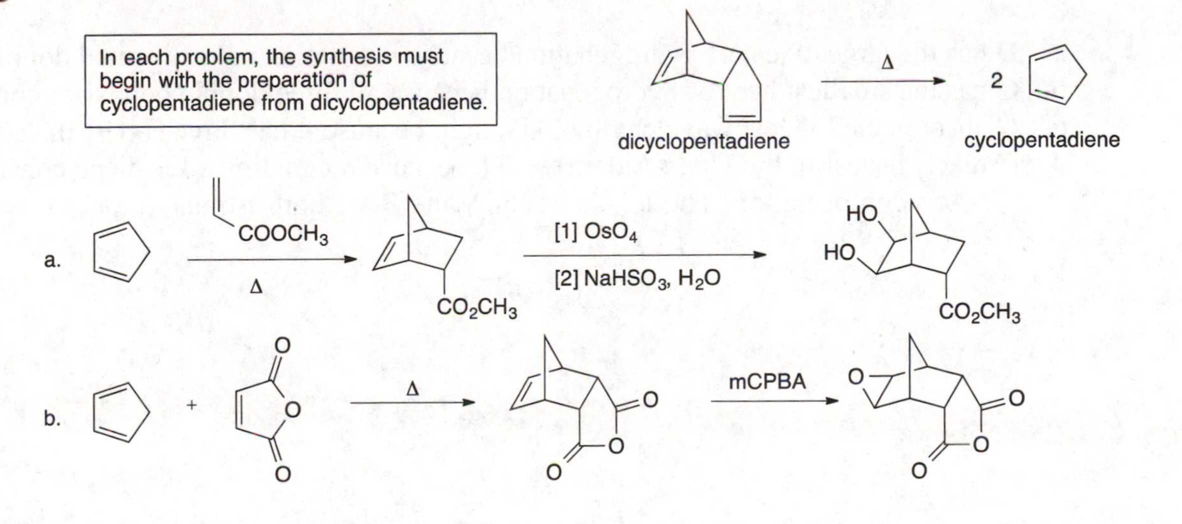

c. 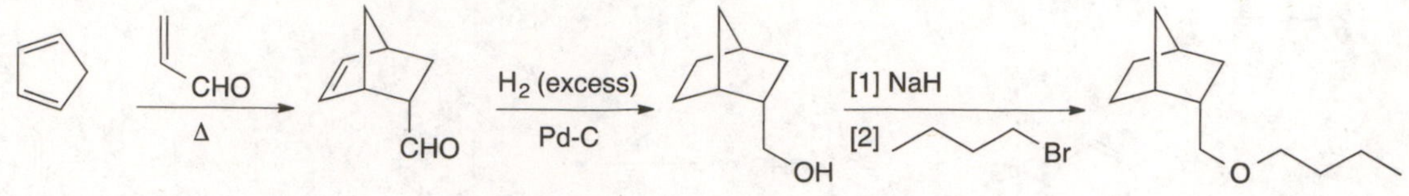 CHO Δ CHO H_2 (excess) Pd-C OH [1] NaH [2] Br O

16.56

a. 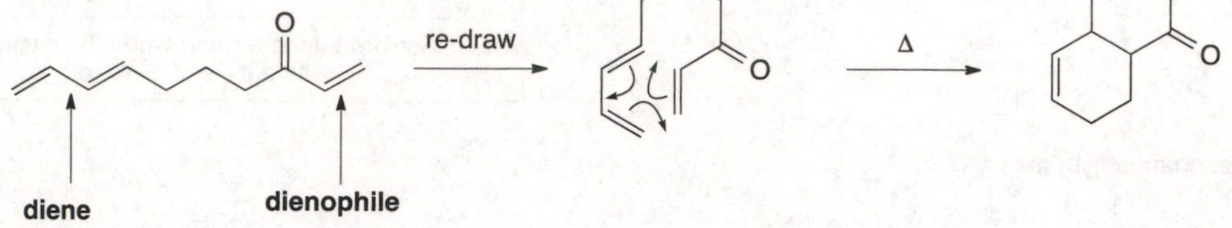 O re-draw O Δ O

diene dienophile

b. 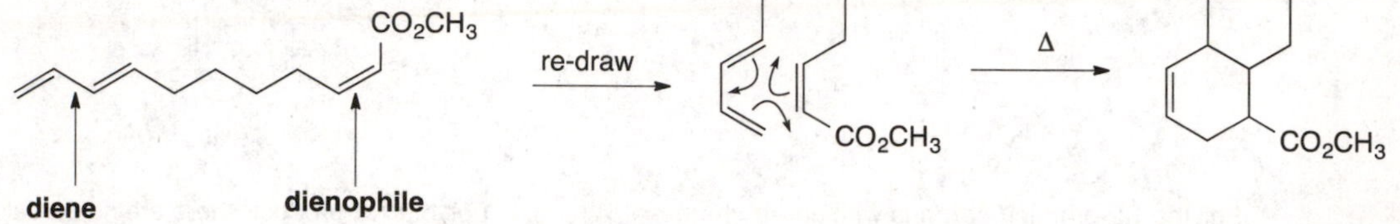 CO_2CH_3 re-draw CO_2CH_3 Δ CO_2CH_3

diene dienophile

16.57 A transannular Diels–Alder reaction forms a tricyclic product from a monocyclic starting material.

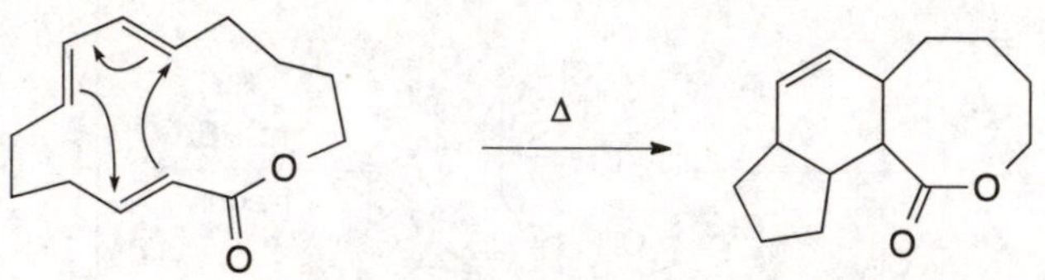

O O Δ O O

16.58

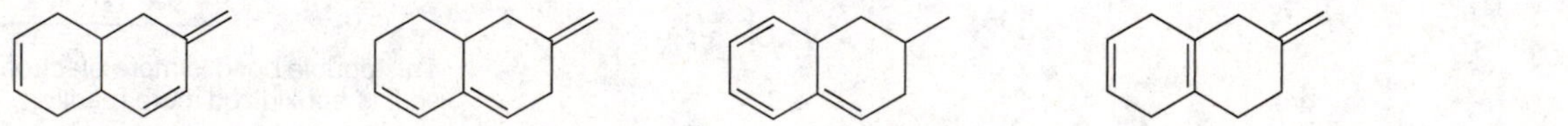

A B C D

a. **D** has the largest heat of hydrogenation because it contains no conjugated double bonds.
b. **C** has the smallest heat of hydrogenation because all three double bonds are conjugated.
c. **C** absorbs the longest wavelength of UV light because it has three C=C's in conjugation.
d. **C** reacts fastest in the Diels-Alder reaction because it contains a 1,3-diene constrained in the *s*-cis conformation. The 1,3-dienes in **A** and **B** are both *s*-trans.

16.59

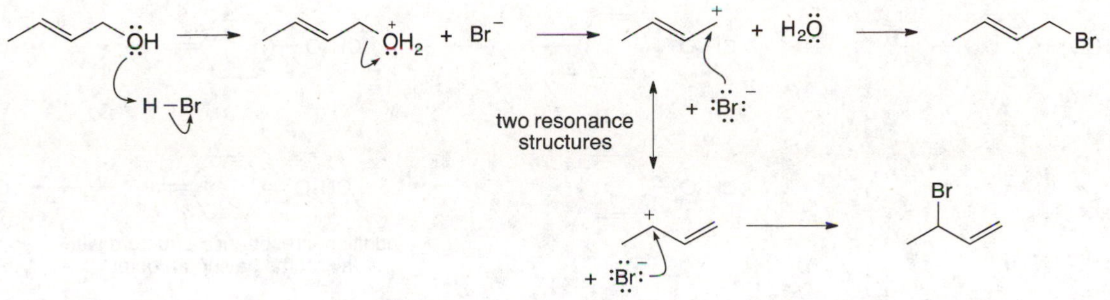

$\ddot{O}H$ → $\overset{+}{O}H_2$ + Br^- → + $H_2\ddot{O}$ → Br

H–Br

two resonance structures

+ :Br:⁻

+ :Br:⁻

Br

16.60

a. HI

conjugated diene | 1,2-product | 1,4-product

I + I

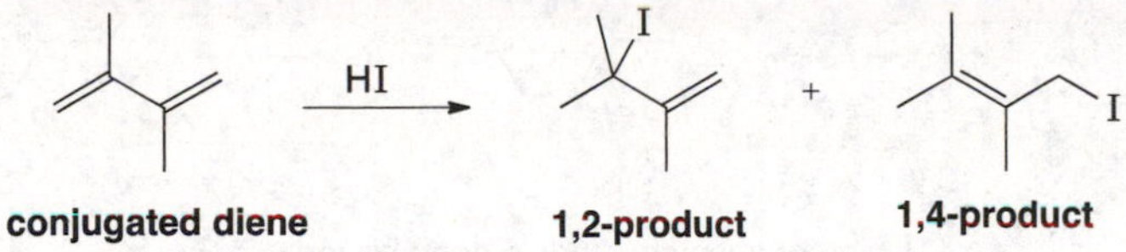

b. Δ

diene | dienophile

O, H, H, O, O

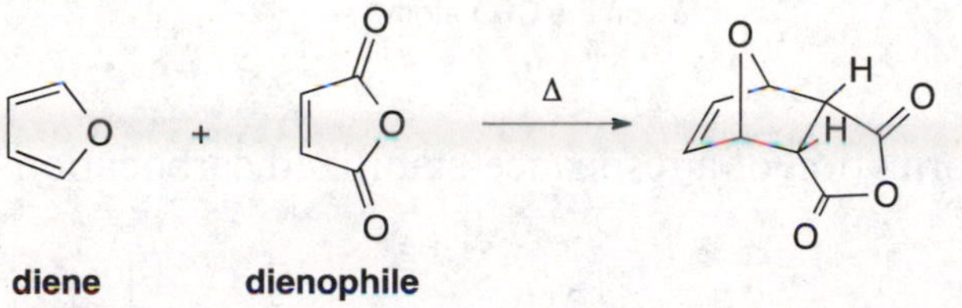

c. Δ

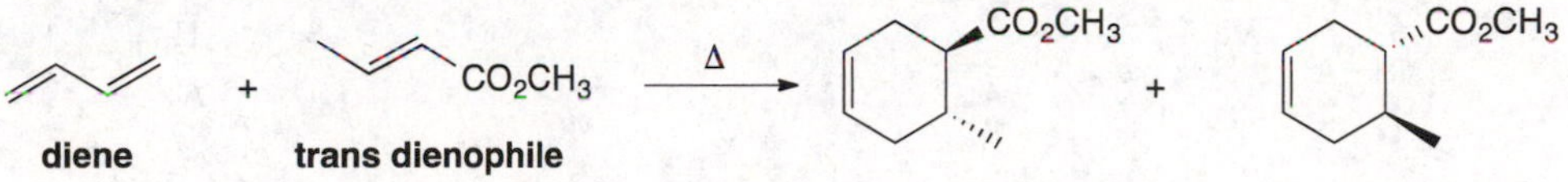

CO_2CH_3 → CO_2CH_3 + CO_2CH_3

diene | trans dienophile

d. HBr

Br + Br

conjugated diene | 1,2-product | 1,4-product

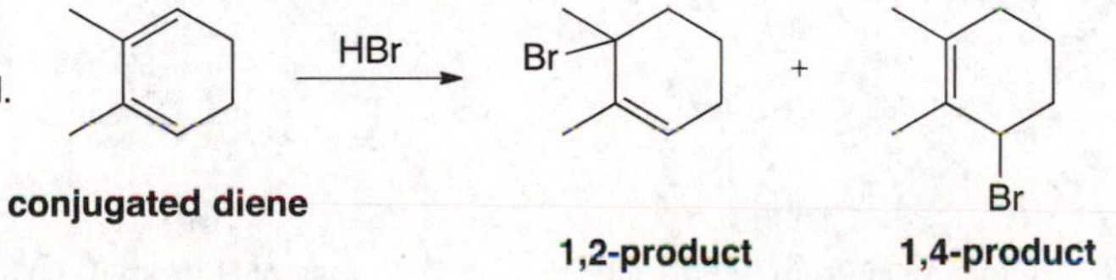

16.61

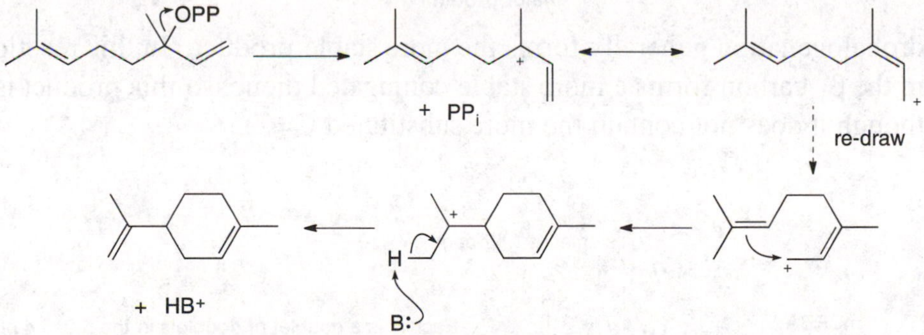

OPP → + PP_i ↔ +

re-draw

+ HB+ ← H, B: ←

16.62 The more stable the carbocation, the faster the SN1 reaction. The carbocation from **A** is more stable because the lone pairs on the O atom of the OCH_3 afford additional resonance stabilization.

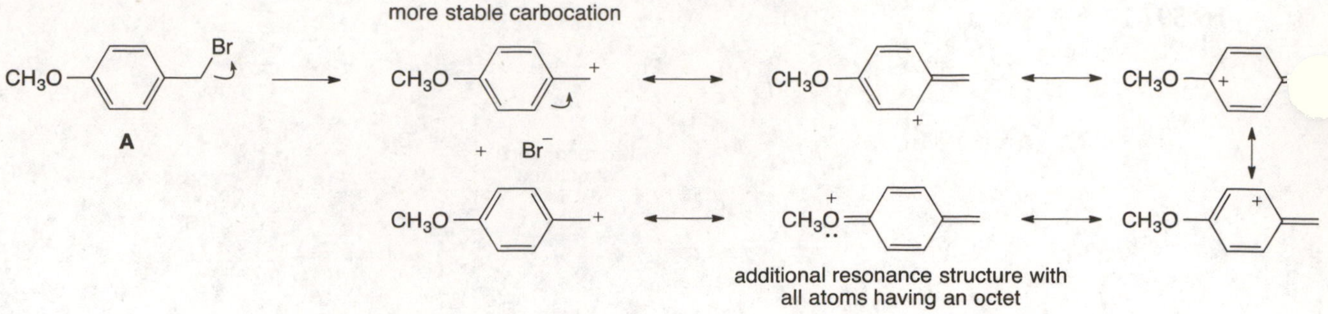

more stable carbocation

CH_3O — Br → CH_3O + ↔ CH_3O ↔ CH_3O

A

+ Br^-

CH_3O + ↔ $CH_3\overset{+}{O}$ ↔ CH_3O

additional resonance structure with
all atoms having an octet

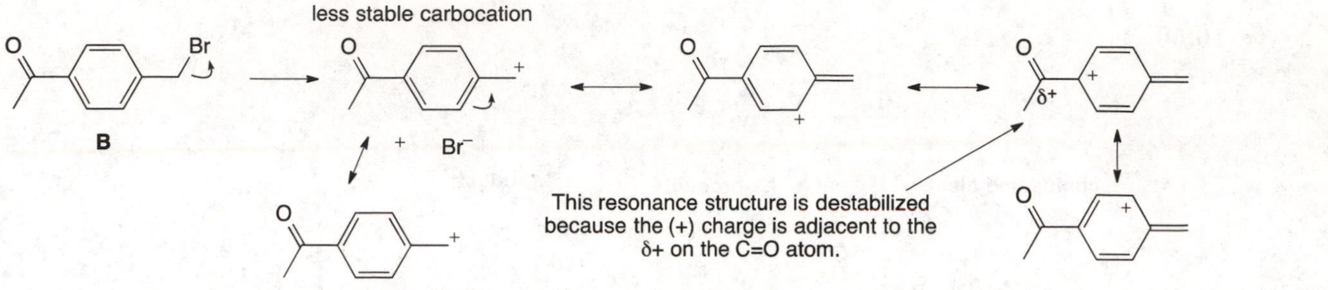

less stable carbocation

O Br → O + ↔ O ↔ O δ+

B + Br^-

O + O

This resonance structure is destabilized
because the (+) charge is adjacent to the
δ+ on the C=O atom.

16.63 The mechanism is E1, with formation of a resonance-stabilized carbocation.

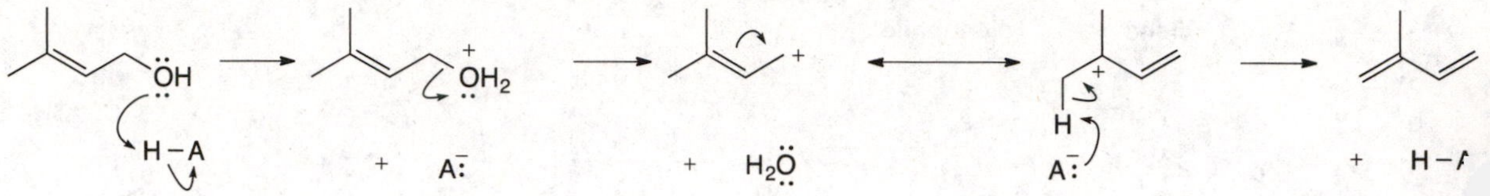

ÖH → $\overset{+}{O}H_2$ → + ↔ H + →

H–A + A:⁻ + $H_2\ddot{O}$ A:⁻ + H–A

16.64

a. Cl β_2 β_1 → +

loss of H (from β_1 C) + Cl
conjugated
more stable
major product

loss of H (from β_2 C) + Cl
more substituted

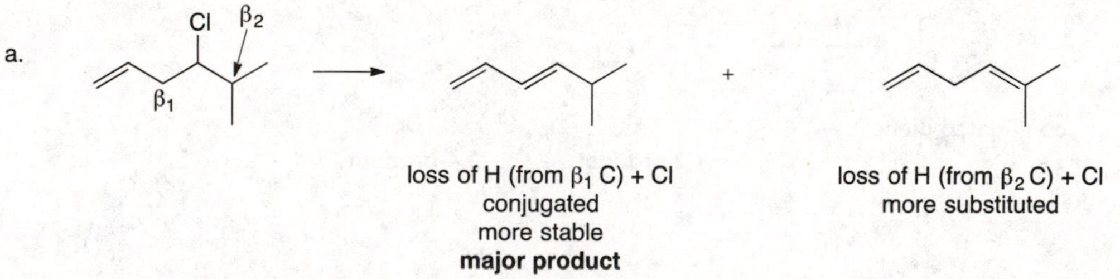

b. Dehydrohalogenation generally forms the more stable product. In this reaction, loss of H from the β_1 carbon forms a more stable conjugated diene, so this product is preferred even though it does not contain the more substituted C=C.

16.65

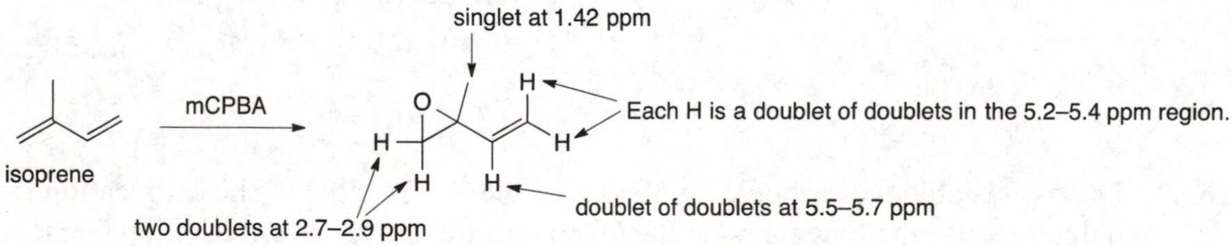

singlet at 1.42 ppm

isoprene mCPBA → O H H H H H

Each H is a doublet of doublets in the 5.2–5.4 ppm region.

doublet of doublets at 5.5–5.7 ppm

two doublets at 2.7–2.9 ppm

16.66

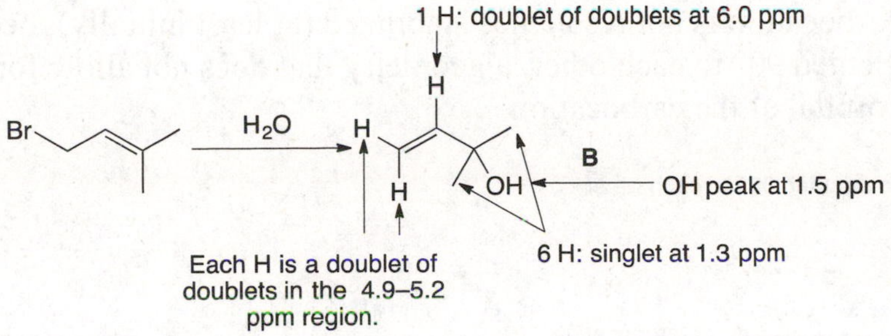

1 H: doublet of doublets at 6.0 ppm

Br ... $\xrightarrow{H_2O}$ **B**

OH peak at 1.5 ppm

6 H: singlet at 1.3 ppm

Each H is a doublet of doublets in the 4.9–5.2 ppm region.

The IR shows an OH absorption at 3200–3600 cm^{-1}.

16.67

A **B**

All double bonds in **A** are conjugated, whereas the double bonds in the circled ring in **B** are not conjugated. As a result, **A** absorbs at a longer wavelength of UV light.

16.68 **C** and **D** absorb UV light because they contain conjugated π systems.

O
C
conjugated
O
D
conjugated

16.69

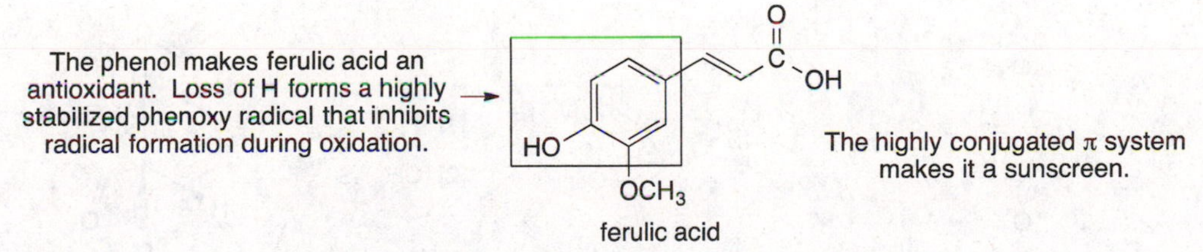
The phenol makes ferulic acid an antioxidant. Loss of H forms a highly stabilized phenoxy radical that inhibits radical formation during oxidation.
O
C
OH
HO
OCH3
ferulic acid
The highly conjugated π system makes it a sunscreen.

16.70 There are two possible modes of addition of HBr to allene. When H$^+$ adds to the terminal carbon, a 2° vinyl carbocation is formed, which affords 2-bromoprop-1-ene after nucleophilic attack.

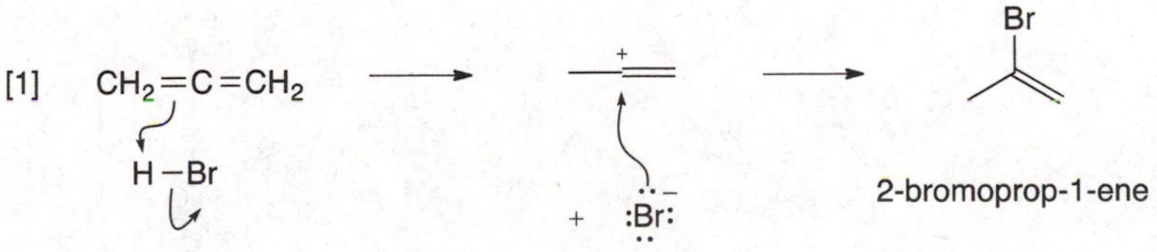

[1] $CH_2{=}C{=}CH_2$ + H−Br ⟶ 2° vinyl carbocation + :Br:$^-$ ⟶ 2-bromoprop-1-ene (Br)

When H^+ adds to the middle carbon, an intermediate carbocation with a (+) charge adjacent to the C=C is formed. This carbocation is not resonance stabilized (at least initially), because the two C=C's of allene are oriented 90° to each other, a geometry that does not allow for overlap the C=C with the empty *p* orbital of the carbocation.

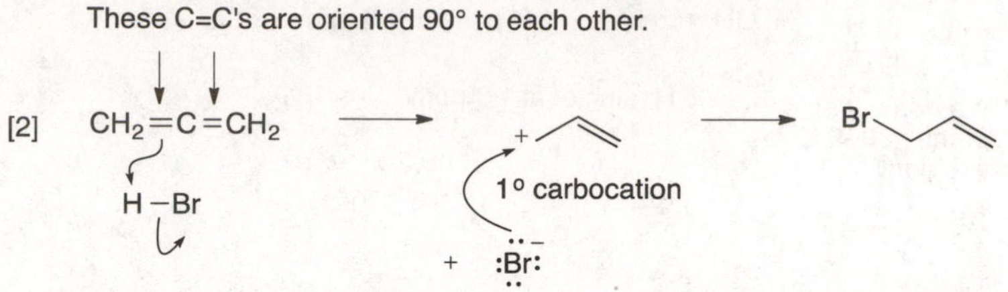

These C=C's are oriented 90° to each other.

[2] $CH_2=C=CH_2$ + H–Br → 1° carbocation + :Br:⁻ → Br product

As a result, path [1] forms a more stable carbocation and is preferred.

16.71

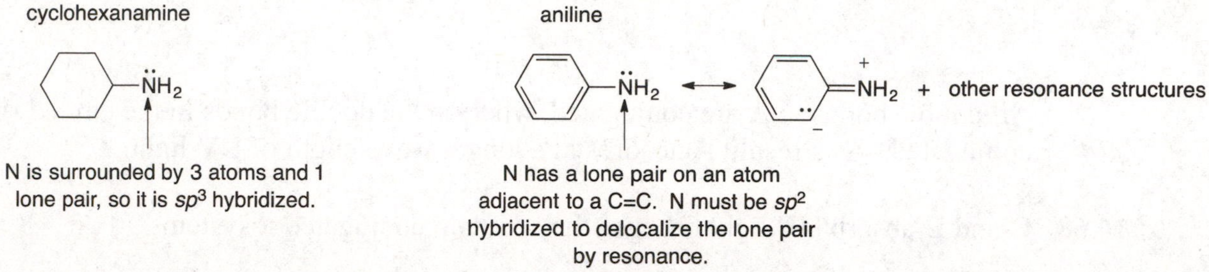

cyclohexanamine

N is surrounded by 3 atoms and 1 lone pair, so it is sp^3 hybridized.

aniline

NH_2 ⟷ $=NH_2^+$ + other resonance structures

N has a lone pair on an atom adjacent to a C=C. N must be sp^2 hybridized to delocalize the lone pair by resonance.

Basicity is a measure of how willing an atom is to donate an electron pair. The lone pair on the N in cyclohexanamine is localized on N, whereas the lone pair on the N in aniline is delocalized onto the benzene ring. As a result, the lone pair in cyclohexanamine is much more available for donation to a proton than the lone pair in aniline. This makes cyclohexanamine much more basic than aniline.

16.72

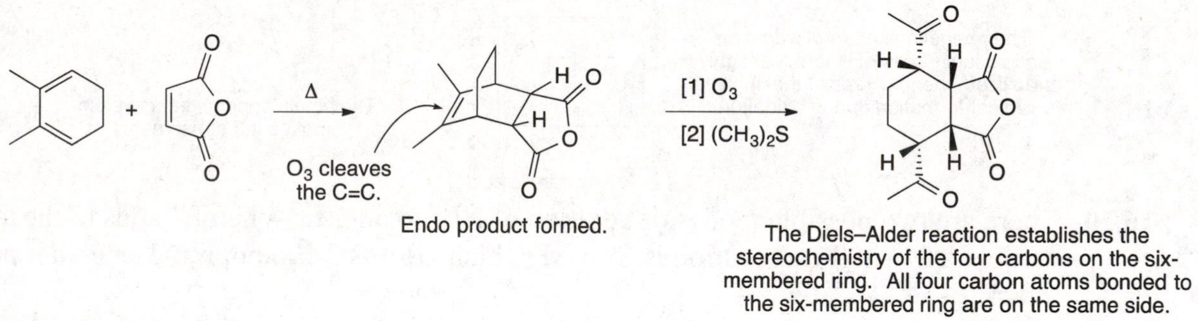

Δ → Endo product formed. O₃ cleaves the C=C. → [1] O_3 [2] $(CH_3)_2S$ →

The Diels–Alder reaction establishes the stereochemistry of the four carbons on the six-membered ring. All four carbon atoms bonded to the six-membered ring are on the same side.

16.73

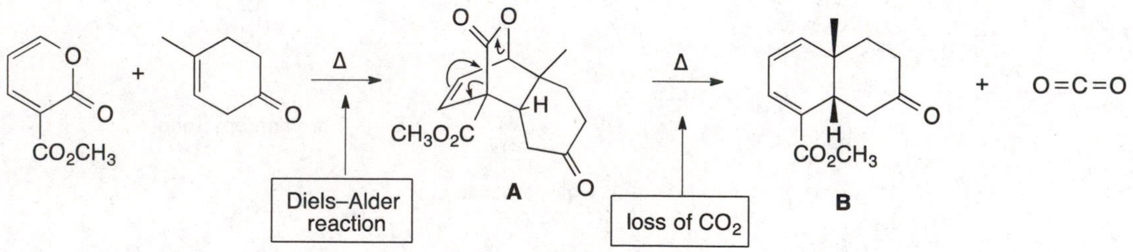

CO_2CH_3 + → Δ (Diels–Alder reaction) → **A** (CH_3O_2C) → Δ (loss of CO_2) → **B** (CO_2CH_3) + O=C=O

16.74

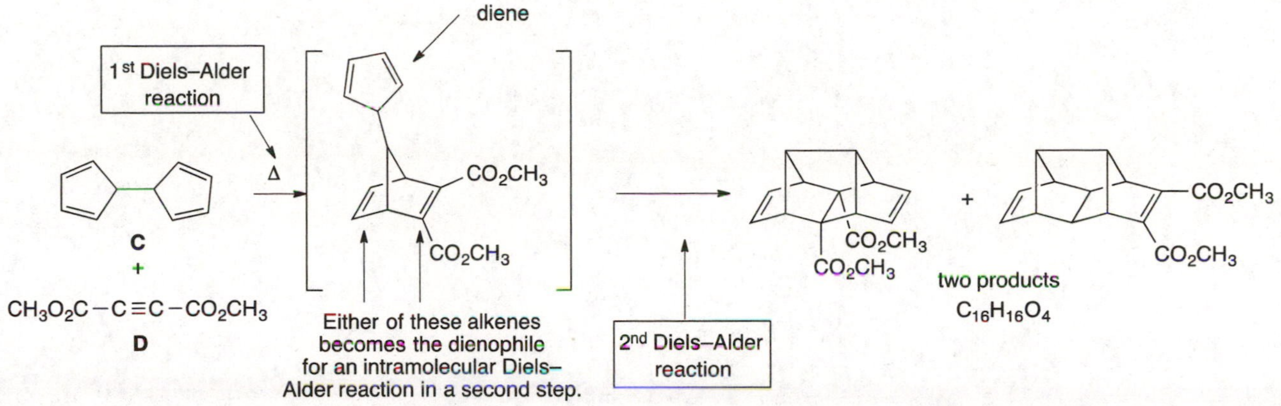

1st Diels–Alder reaction

diene

Δ

C + D

$CH_3O_2C-C{\equiv}C-CO_2CH_3$

CO_2CH_3

CO_2CH_3

Either of these alkenes becomes the dienophile for an intramolecular Diels–Alder reaction in a second step.

2nd Diels–Alder reaction

CO_2CH_3 CO_2CH_3 + CO_2CH_3 CO_2CH_3

two products

$C_{16}H_{16}O_4$

16.75 A retro Diels–Alder reaction forms a conjugated diene. An intramolecular Diels–Alder reaction then forms **N.**

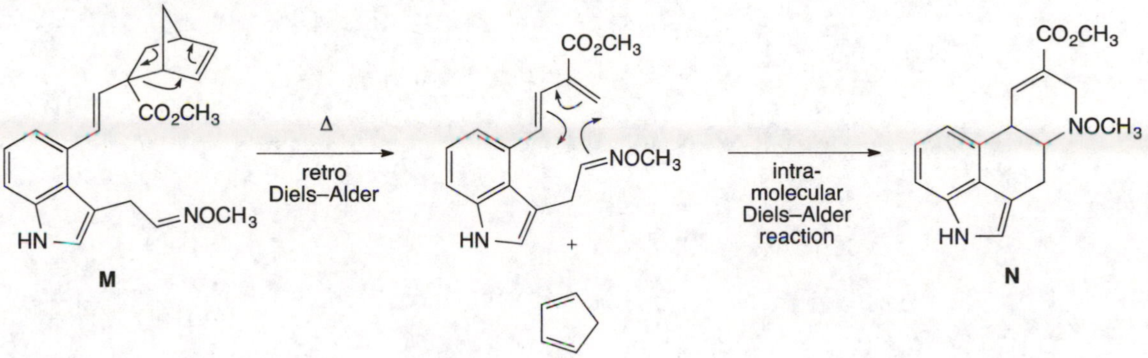

CO_2CH_3

HN

$=NOCH_3$

M

Δ

retro Diels–Alder

CO_2CH_3

$=NOCH_3$

HN

+

intra-molecular Diels–Alder reaction

CO_2CH_3

$NOCH_3$

HN

N

Chapter 17 Benzene and Aromatic Compounds

Chapter Review

Comparing aromatic, antiaromatic, and nonaromatic compounds (17.7)

- **Aromatic compound**
  - A cyclic, planar, completely conjugated compound that contains $4n + 2$ π electrons ($n = 0, 1, 2, 3$, and so forth).
  - An aromatic compound is more stable than a similar acyclic compound having the same number of π electrons.
- **Antiaromatic compound**
  - A cyclic, planar, completely conjugated compound that contains $4n$ π electrons ($n = 0, 1, 2, 3$, and so forth).
  - An antiaromatic compound is less stable than a similar acyclic compound having the same number of π electrons.
- **Nonaromatic compound**
  - A compound that lacks one (or more) of the requirements to be aromatic or antiaromatic.

Properties of aromatic compounds

- Every carbon in an aromatic ring has a *p* orbital to delocalize electron density (17.2).
- Aromatic compounds are unusually stable. $\Delta H°$ for hydrogenation is much less than expected, given the number of degrees of unsaturation (17.6).
- Aromatic compounds do not undergo the usual addition reactions of alkenes (17.6).
- ^{1}H NMR spectra show highly deshielded protons because of ring currents (17.4).

Examples of aromatic compounds with 6 π electrons (17.8)

benzene

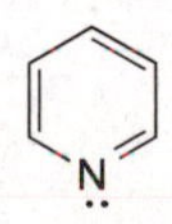

pyridine

pyrrole

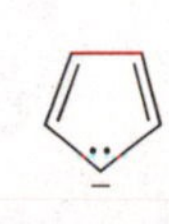
cyclopentadienyl anion

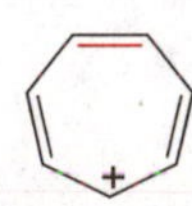

tropylium cation

Examples of compounds that are not aromatic (17.8)

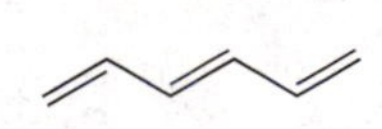
not cyclic

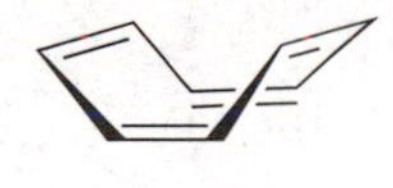
not planar

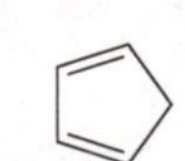
not completely conjugated

Practice Test on Chapter Review

1. Give the IUPAC name for each of the following compounds.

a. 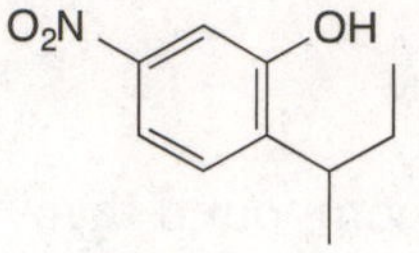 O_2N OH

b.

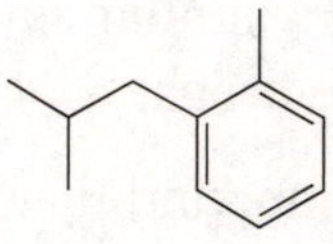

c. 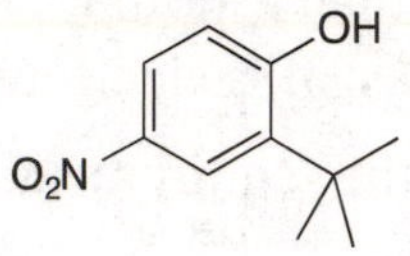 OH O_2N

2. Label each compound as aromatic, nonaromatic, or antiaromatic. Choose only **one** possibility. Assume all completely conjugated rings are planar.

a. 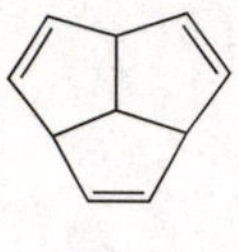 c. 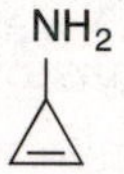 NH_2 e. 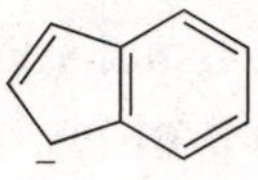 g. 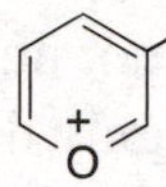 O + i. 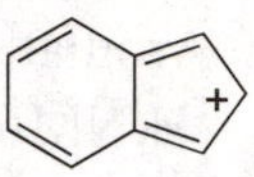 +

b. 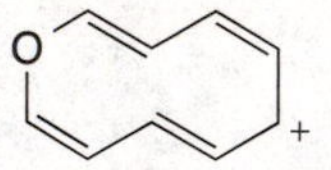 O + d. 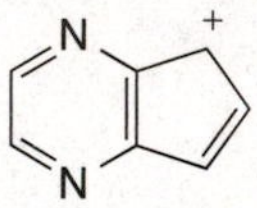 N N + f. 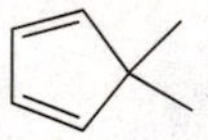 h. 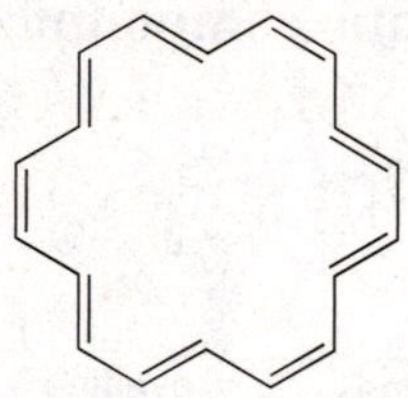 j. 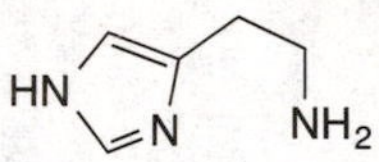 HN N NH_2

3. Answer the following questions about compounds **A–E** drawn below.

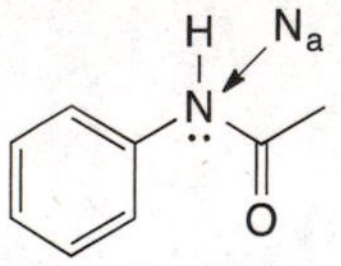

N_a H N O N_b N N [2] [1] [3] [4] H H_a H_b H_d H H_c

A 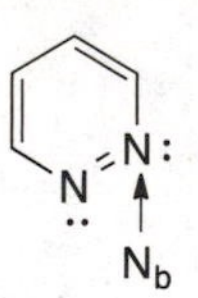 **B** 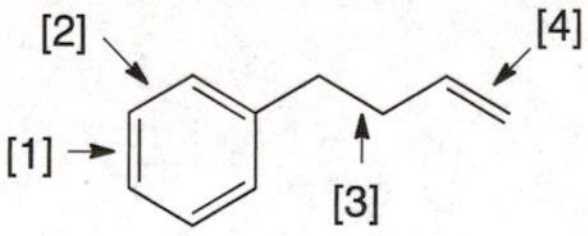 **C** 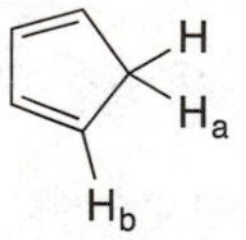 **D** 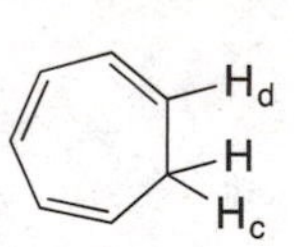 **E**

a. How is nitrogen N_a in compound **A** hybridized?
b. In what type of orbital does the lone pair on N_a reside?
c. How is nitrogen N_b in compound **B** hybridized?
d. In what type of orbital does the lone pair on N_b reside?
e. Which of the labeled bonds in compound **C** is the shortest?
f. Which of the labeled bonds in compound **C** is the longest?
g. When considering both compounds **D** and **E,** which of the labeled hydrogen atoms (H_a, H_b, H_c, or H_d) is the most acidic? Give only **one** answer.
h. When considering both compounds **D** and **E,** which of the labeled hydrogen atoms (H_a, H_b, H_c, or H_d) is the least acidic? Give only **one** answer.

Answers to Practice Test

1.a. 2-*sec*-butyl-5-nitrophenol
b. *o*-isobutyltoluene
c. 2-*tert*-butyl-4-nitrophenol

2.a. nonaromatic
b. aromatic
c. nonaromatic
d. antiaromatic
e. aromatic
f. nonaromatic
g. aromatic
h. aromatic
i. antiaromatic
j. aromatic

3.a. sp^2
b. p
c. sp^2
d. sp^2
e. 4
f. 3
g. H_a
h. H_c

Answers to Problems

17.1 Move the electrons in the π bonds to draw all major resonance structures.

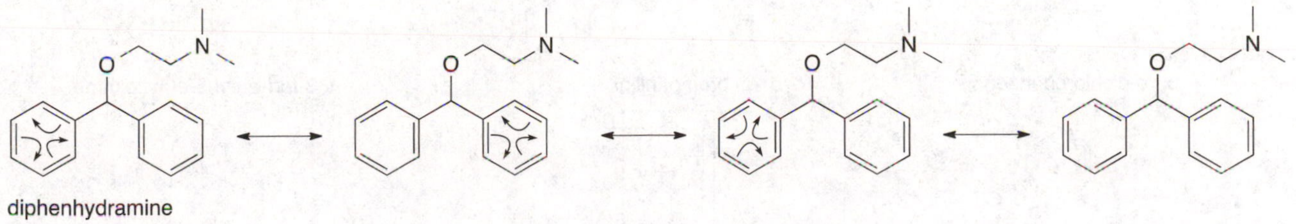

diphenhydramine

17.2 Look at the hybridization of the atoms involved in each bond. Carbons in a benzene ring are surrounded by three groups and are sp^2 hybridized.

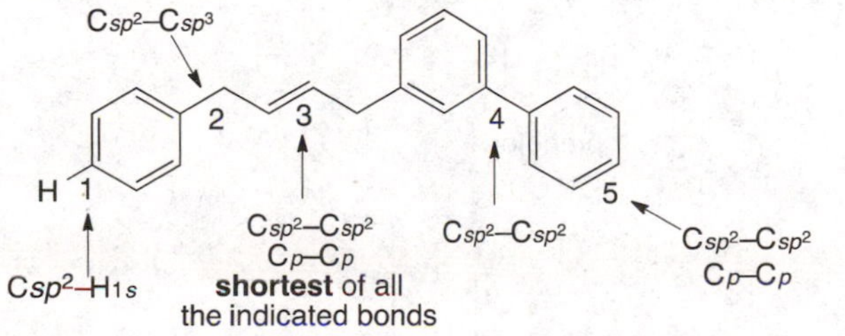

C_{sp^2}–C_{sp^3}

2 3 4 H 1 5

C_{sp^2}–H_{1s}

C_{sp^2}–C_{sp^2}
C_p–C_p
shortest of all the indicated bonds

C_{sp^2}–C_{sp^2}

C_{sp^2}–C_{sp^2}
C_p–C_p

17.3

- To name a benzene ring with **one substituent,** name the substituent and add the word *benzene*.
- To name a **disubstituted ring,** select the correct prefix (ortho = 1,2; meta = 1,3; para = 1,4) an alphabetize the substituents. Use a common name if it is a derivative of that monosubstituted benzene.
- To name a **polysubstituted ring,** number the ring to give the lowest possible numbers and then follow other rules of nomenclature.

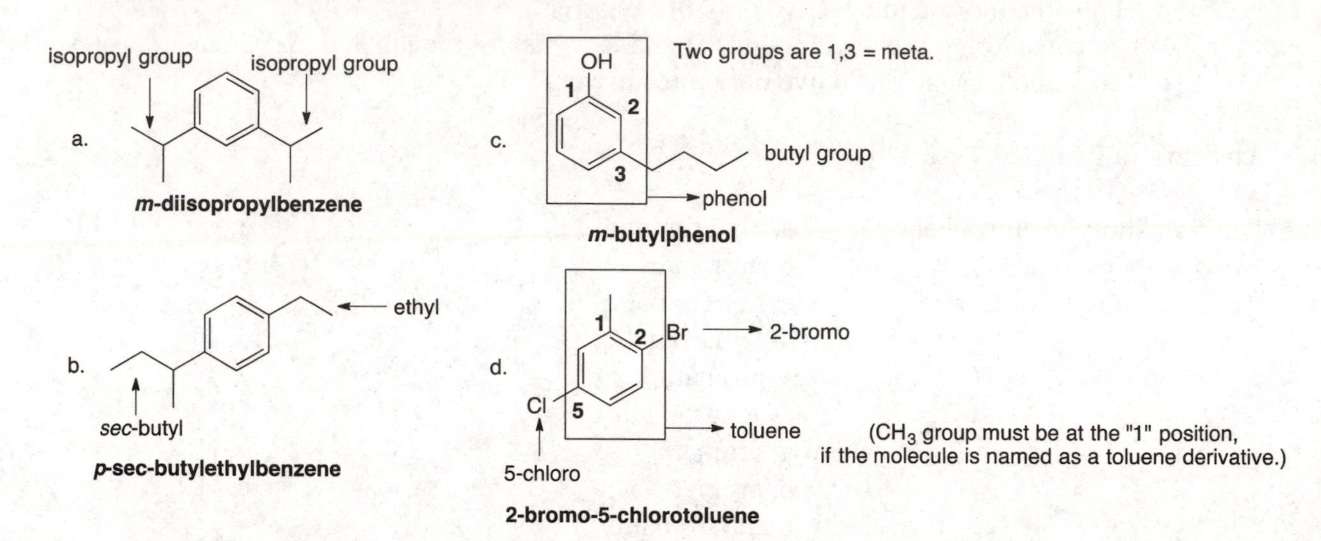

17.4 Work backwards to draw the structures from the names.

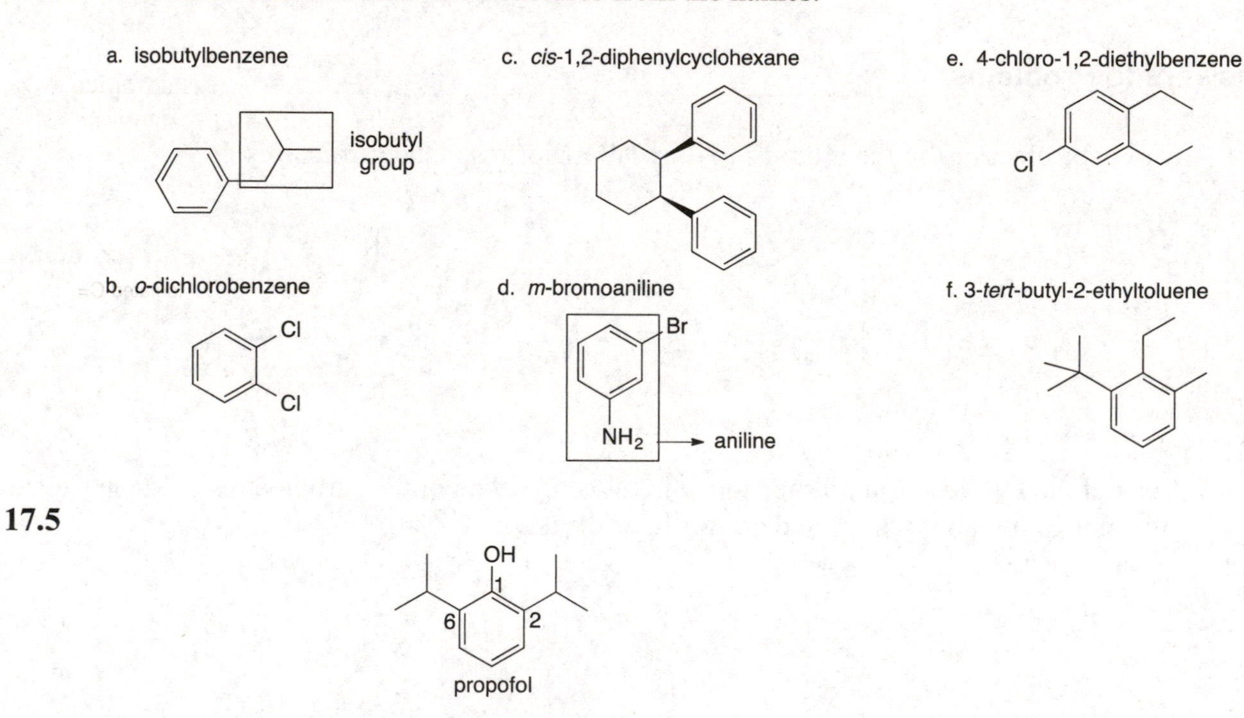

17.6

Molecular formula $C_{10}H_{14}O_2$: 4 degrees of unsaturation
IR absorption at 3150–2850 cm^{-1}: sp^2 and sp^3 hybridized C–H bonds
NMR absorptions (ppm):
- 1.4 (triplet, 6 H)
- 4.0 (quartet, 4 H)
- 6.8 (singlet, 4 H)

17.7 Count the different types of carbons to determine the number of ^{13}C NMR signals.

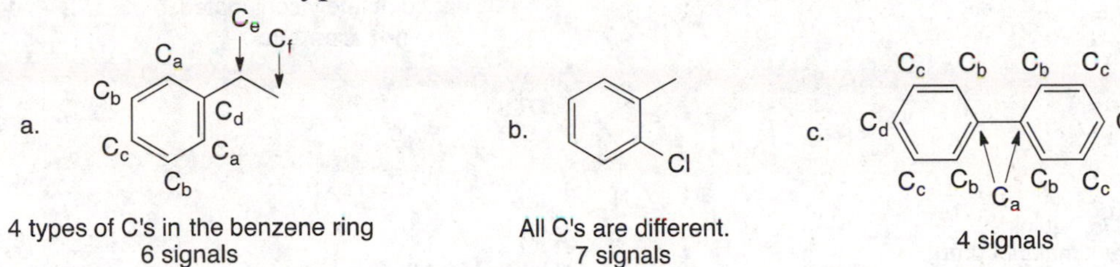

a. 4 types of C's in the benzene ring
6 signals

b. All C's are different.
7 signals

c. 4 signals

17.8 The less stable compound has a larger heat of hydrogenation.

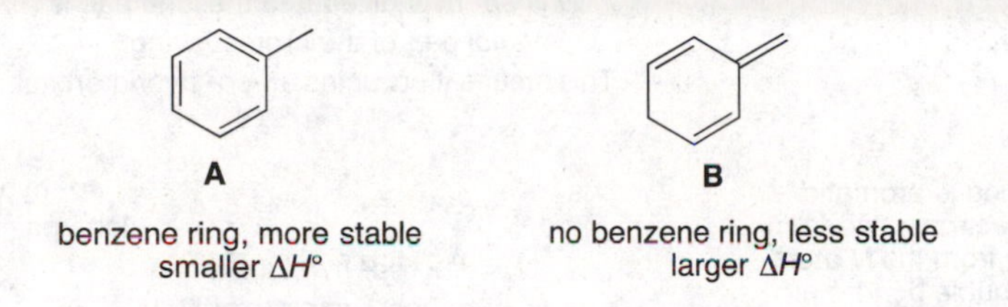

A: benzene ring, more stable
smaller $\Delta H°$

B: no benzene ring, less stable
larger $\Delta H°$

17.9 The protons on sp^2 hybridized carbons in aromatic hydrocarbons are highly deshielded and absorb at 6.5–8 ppm, whereas hydrocarbons that are not aromatic show an absorption at 4.5–6 ppm, typical of protons bonded to the C=C of an alkene.

a. aromatic ring H's ~6.5–8 ppm

b. not aromatic
alkene H's ~4.5–6 ppm

c. protons on isolated C=C ~4.5–6 ppm
protons on benzene ring ~6.5–8 ppm

17.10 To be aromatic, a ring must have $4n + 2$ π electrons.

16 π e⁻	20 π e⁻	22 π e⁻
$4n$	$4n$	$4n + 2$
4(4) = 16	4(5) = 20	4(5) + 2 = 22
antiaromatic	**antiaromatic**	**aromatic**

17.11

17.12 In determining if a heterocycle is aromatic, count a nonbonded electron pair if it makes the ring aromatic in calculating $4n + 2$. Lone pairs on atoms already part of a multiple bond cannot be delocalized in a ring, and so they are never counted in determining aromaticity.

a. Count one lone pair from O. $4n + 2 = 4(1) + 2 = 6$ **aromatic**

b. no lone pair from O $4n + 2 = 4(1) + 2 = 6$ **aromatic**

c. 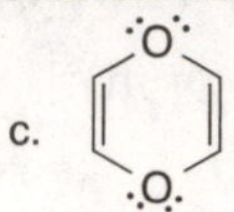
With one lone pair from each O there would be 8 π electrons. If O's are sp^3 hybridized, the ring is not completely conjugated. **not aromatic**

d. Both N atoms are part of a double bond, so the lone pairs cannot be counted: there are 6 π electrons. $4n + 2 = 4(1) + 2 = 6$ **aromatic**

17.13

quinine (antimalarial drug)

N is sp^3 hybridized and the lone pair is in an sp^3 hybrid orbital.

N is sp^2 hybridized and the lone pair is *not* part of the aromatic ring. This means it occupies an sp^2 hybrid orbital.

17.14

a. The five-membered ring is aromatic because it has 6 π electrons, two from each π bond and two from the N atom that is not part of a double bond.

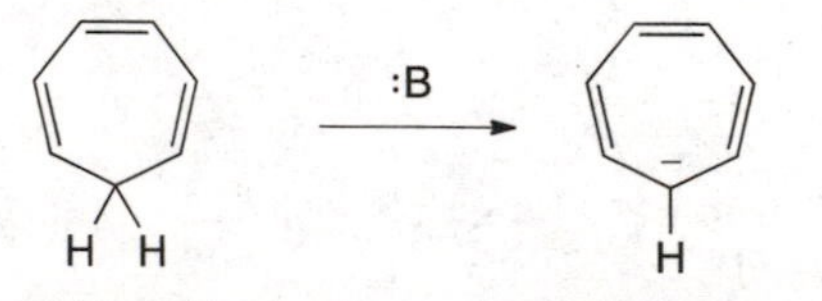

sitagliptin

b, c.

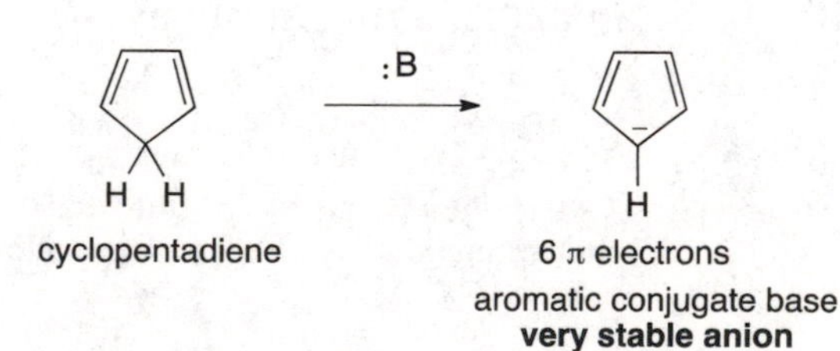

sp^2 hybridized N lone pair in *p* orbital

sp^2 hybridized N lone pair in *p* orbital

sp^2 hybridized N lone pair in sp^2 orbital

sp^2 hybridized N lone pair in sp^2 orbital

sp^3 hybridized N lone pair in sp^3 orbital

17.15 Compare the conjugate base of cyclohepta-1,3,5-triene with the conjugate base of cyclopentadiene. Remember that the compound with the more stable conjugate base will have a lower p*K*a.

cyclohepta-1,3,5-triene pK_a = 39 —:B→ 8 π Electrons make this conjugate base especially unstable (**antiaromatic**).

Because the conjugate base is unstable, the pK_a of cyclohepta-1,3,5-triene is **high.**

cyclopentadiene —:B→ 6 π electrons aromatic conjugate base **very stable anion**

Because the conjugate base is very stable, the pK_a of cyclopentadiene is much **lower.**

17.16 The compound with the most stable conjugate base is the most acidic.

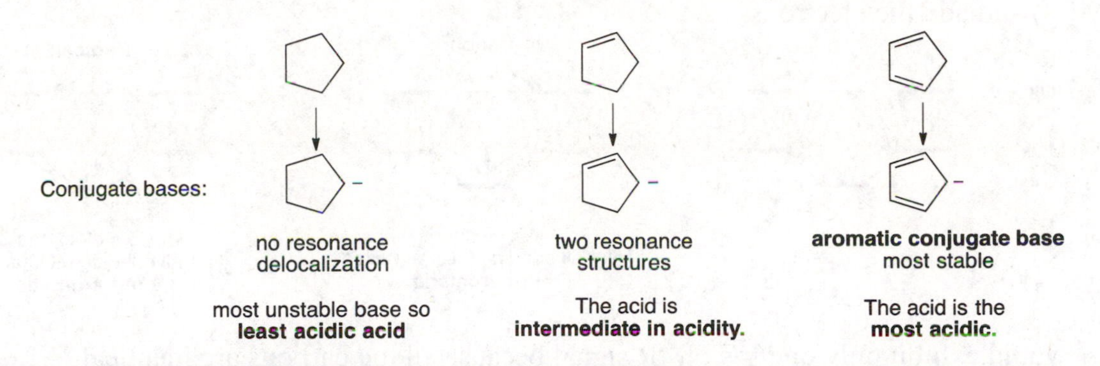

17.17

17.18 To be aromatic, the ions must have $4n + 2$ π electrons. Ions in (b) and (c) do not have the right number of π electrons to be aromatic.

a. 2 π electrons
$4(0) + 2 = 2$
aromatic

d. 10 π electrons
$4(2) + 2 = 10$
aromatic

17.19

A =

The NMR indicates that **A** is aromatic. (a) The C's of the triple bond are *sp* hybridized. (b) Each triple bond has one set of electrons in *p* orbitals that overlap with other *p* orbitals on adjacent atoms in the ring. This overlap allows electrons to delocalize. Each C of the triple bonds also has a *p* orbital in the plane of the ring. The electrons in these *p* orbitals are localized between the C's of the triple bond, not delocalized in the ring. (c) Although **A** has 24 π e^- total, only 18 e^- are delocalized around the ring.

17.20 In using the inscribed polygon method, always draw the vertex pointing down.

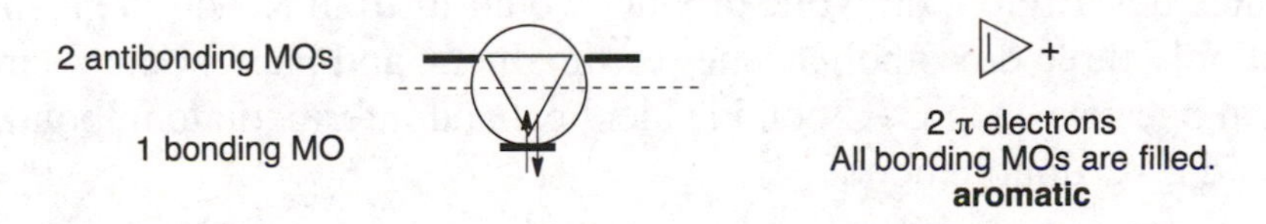

17.21 Draw the inscribed pentagons with the vertex pointing down. Then draw the molecular orbitals (MOs) and add the electrons.

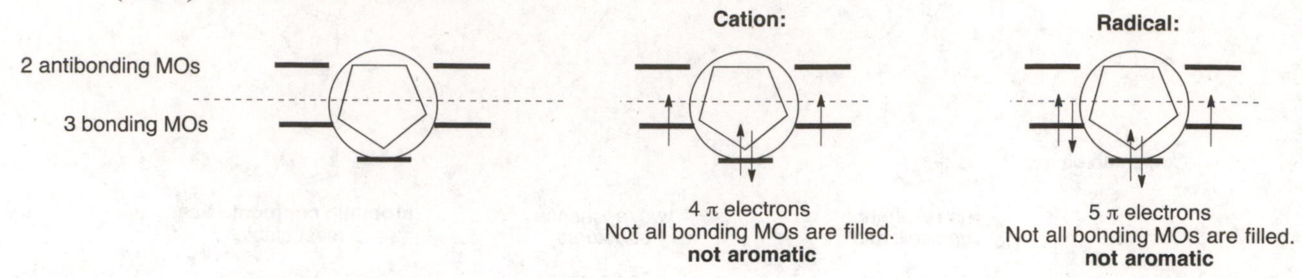

17.22 C_{60} would exhibit only one ^{13}C NMR signal because all the carbons are identical.

17.23

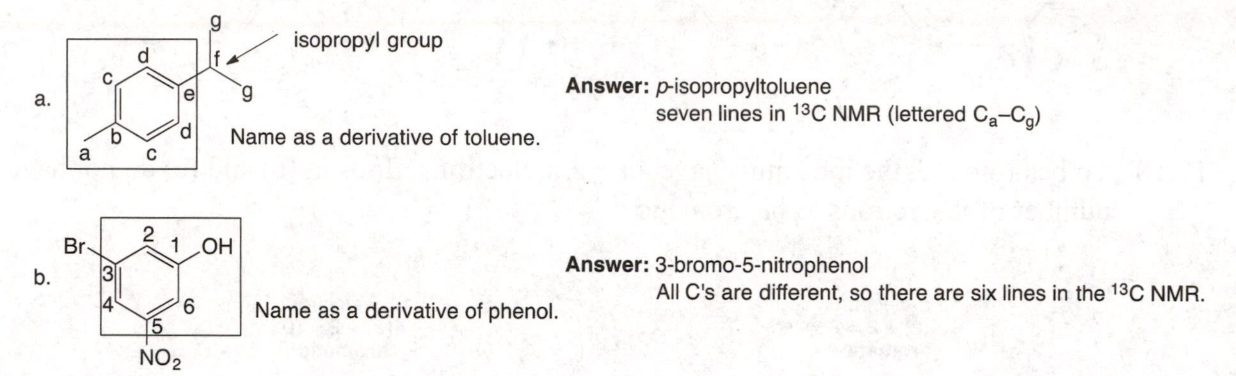

a.
isopropyl group
Name as a derivative of toluene.
Answer: p-isopropyltoluene
seven lines in ¹³C NMR (lettered Ca–Cg)
b.
Br
OH
NO2
Name as a derivative of phenol.
Answer: 3-bromo-5-nitrophenol
All C's are different, so there are six lines in the ¹³C NMR.

17.24

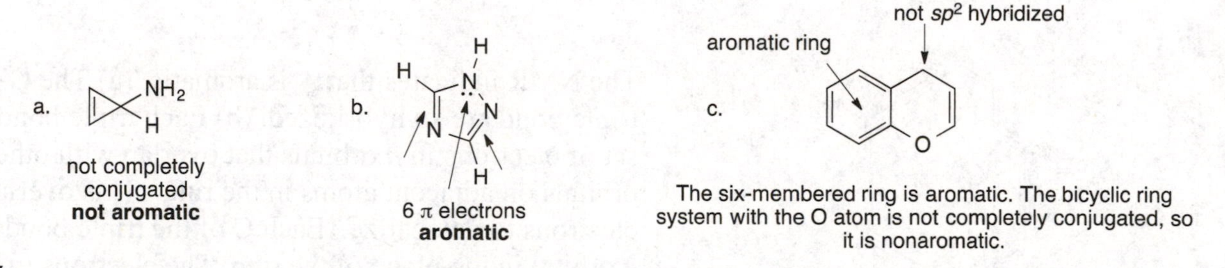

17.25

a. If the Kekulé description of benzene was accurate, only one product would form in Reaction [1], but there would be four (not three) dibromobenzenes (**A–D**), because adjacent C–C bonds are different—one is single and one is double. Thus, compounds **A** and **B** would *not* be identical. **A** has two Br's bonded to the same double bond, but **B** has two Br's on different double bonds.

b. In the resonance description, only one product would form in Reaction [1], because all C's are identical, but only three dibromobenzenes (ortho, meta, and para isomers) are possible. **A** and **B** are identical because each C–C bond is identical and intermediate in bond length between a C–C single and C–C double bond.

[1] → Br [2] → Br Br + Br Br + Br Br + Br Br

A **B** **C** **D**

17.26

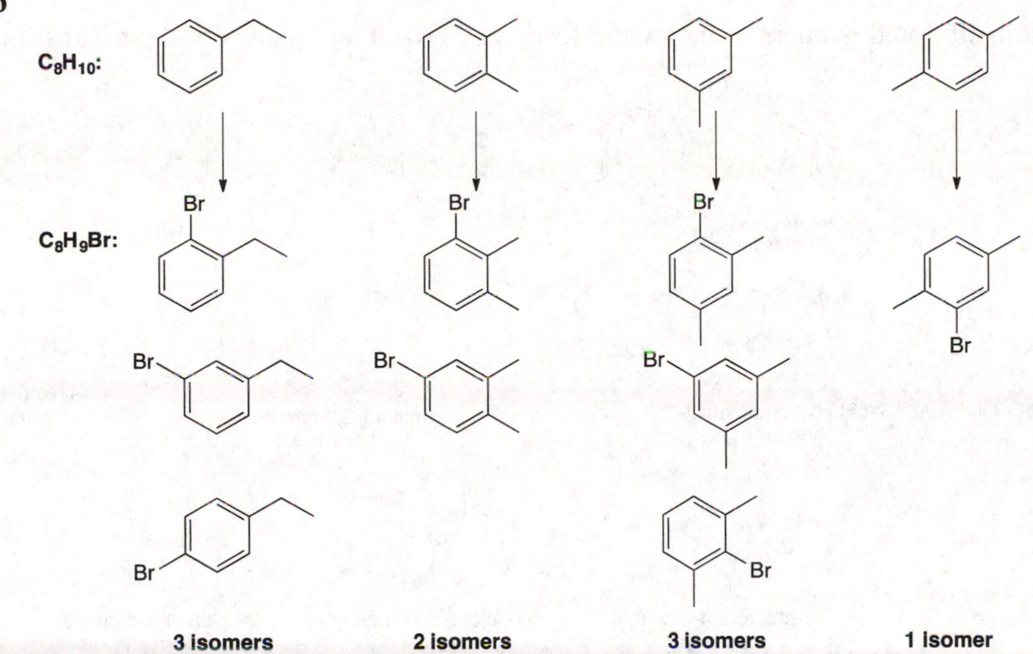

17.27 To name the compounds use the directions from Answer 17.3.

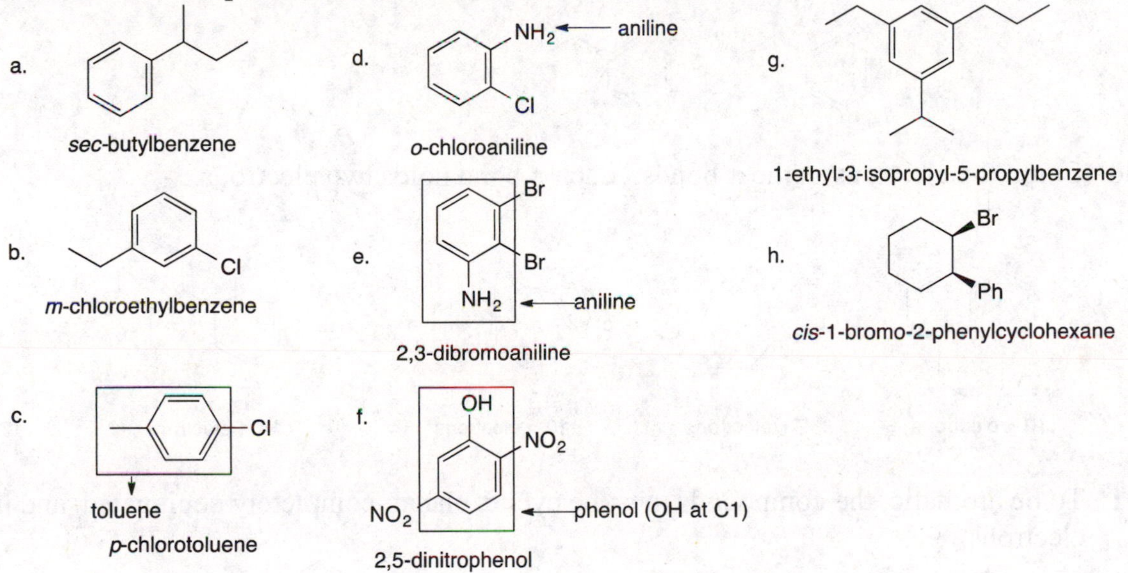

17.28

a. *p*-dichlorobenzene

c. *o*-bromonitrobenzene

e. 2-phenylprop-2-en-1-ol

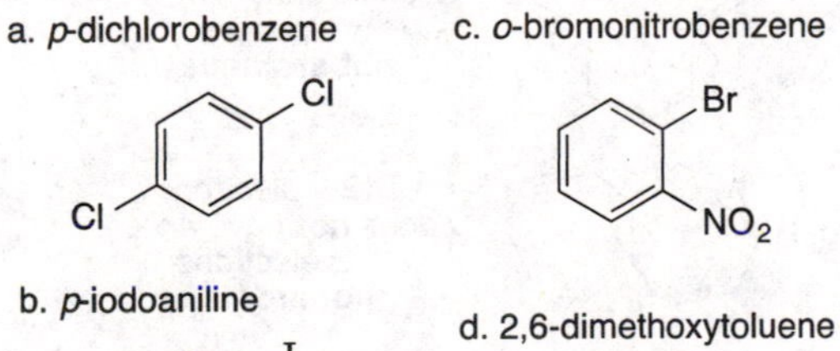

OH

b. *p*-iodoaniline

d. 2,6-dimethoxytoluene

f. *trans*-1-benzyl-3-phenylcyclopentane

I

H_2N

OCH_3

OCH_3

or

17.29

a, b. constitutional isomers of molecular formula $\mathbf{C_8H_9Cl}$ and names of the trisubstituted benz

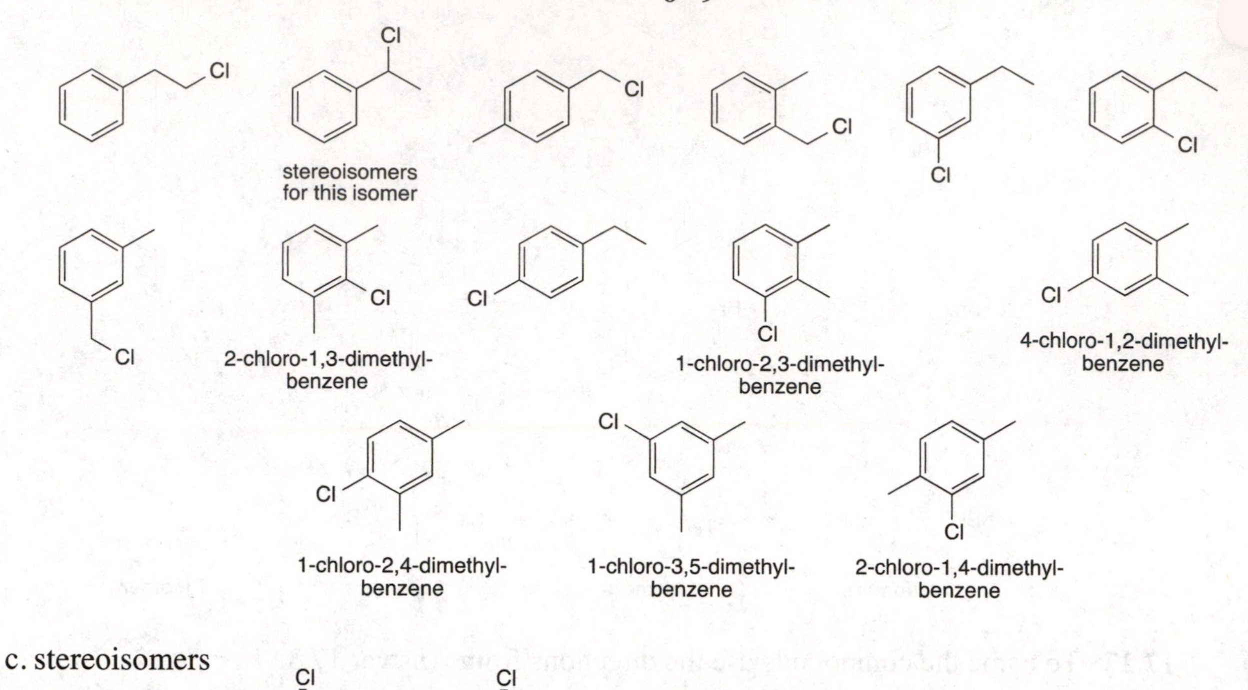

c. stereoisomers

17.30 Count the electrons in the π bonds. Each π bond holds two electrons.

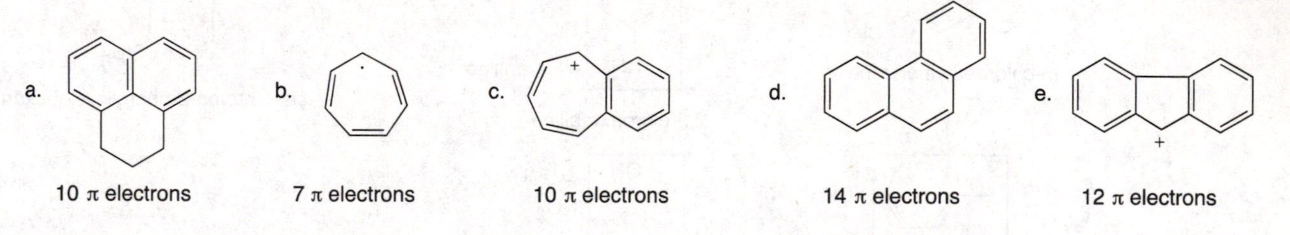

17.31 To be aromatic, the compounds must be cyclic, planar, completely conjugated, and have $4n + 2$ π electrons.

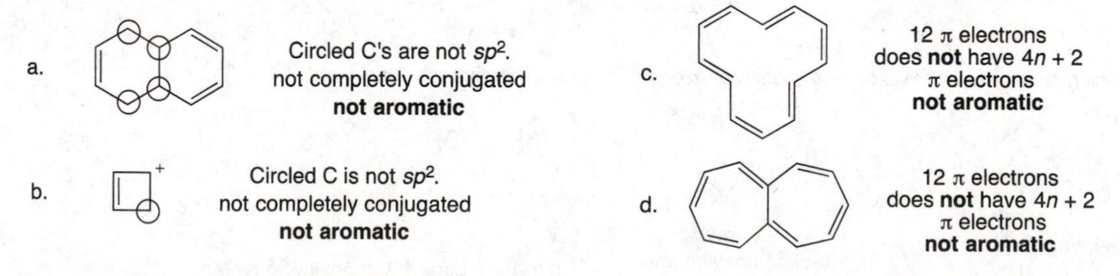

17.32 In determining if a heterocycle is aromatic, count a nonbonded electron pair if it makes the ring aromatic in calculating $4n + 2$. Lone pairs on atoms already part of a multiple bond cannot be delocalized in a ring, so they are never counted in determining aromaticity.

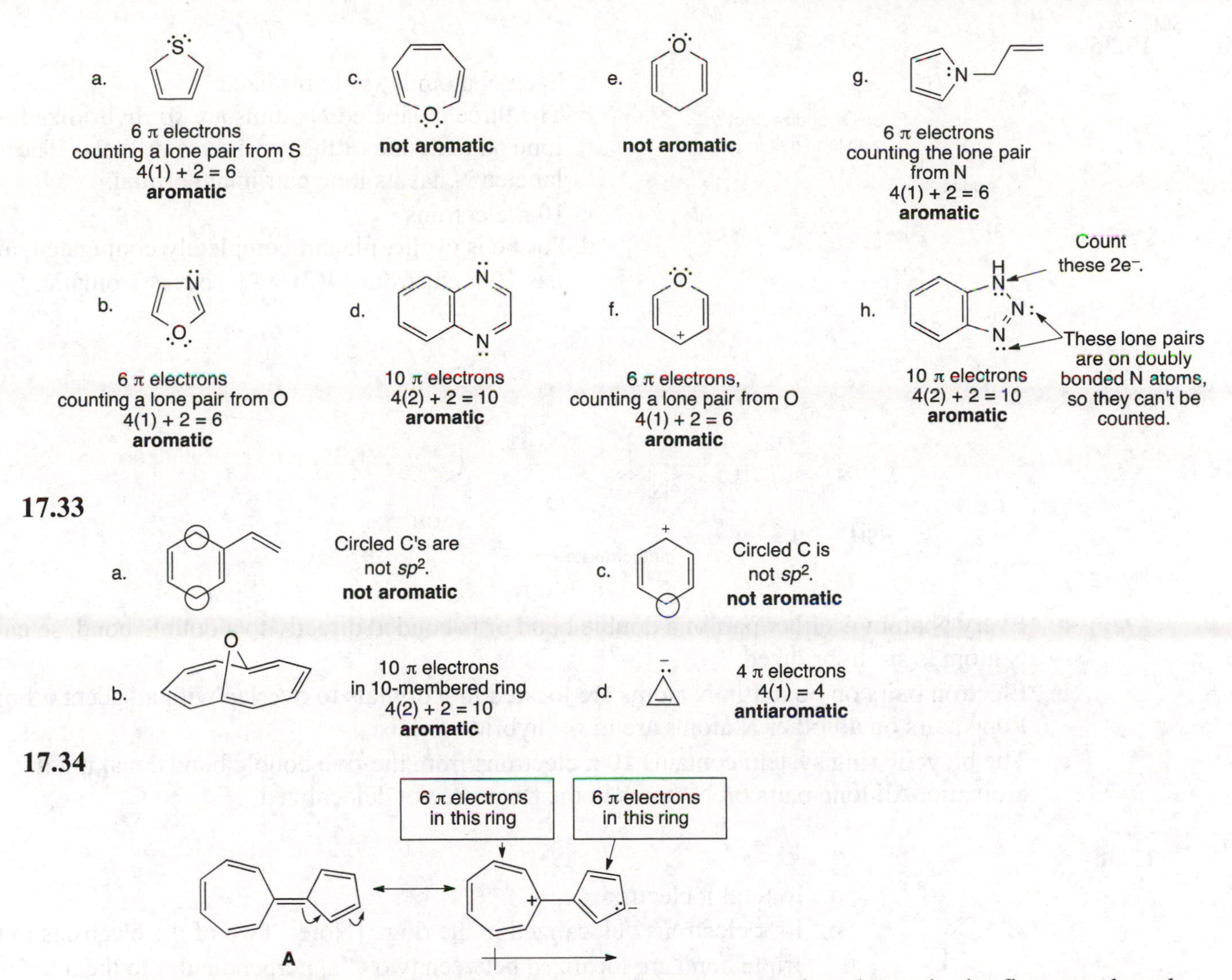

A resonance structure can be drawn for **A** that places a negative charge in the five-membered ring and a positive charge in the seven-membered ring. This resonance structure shows that each ring has six π electrons, making it aromatic. The molecule possesses a dipole such that the seven-membered ring is electron deficient and the five-membered ring is electron rich.

17.35 Each compound is completely conjugated. A compound with $4n + 2$ π electrons is especially stable, whereas a compound with $4n$ π electrons is especially unstable.

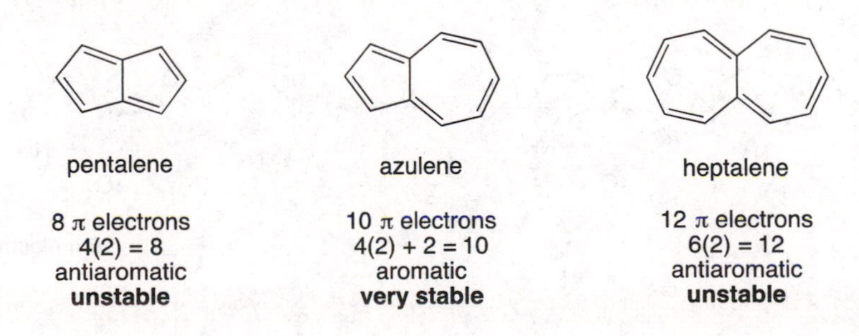

17.36

sp^2 hybridized but with lone pair in p orbital

purine

a. Each N atom is sp^2 hybridized.
b. The three unlabeled N atoms are sp^2 hybridized wi... lone pairs in one of the sp^2 hybrid orbitals. The labeled N has its lone pair in a p orbital.
c. 10 π electrons
d. Purine is cyclic, planar, completely conjugated, and has 10 π electrons [4(2) + 2], so it is aromatic.

17.37

methotrexate

a. Every N atom is either part of a double bond or is bonded directly to a double bond, so each N atom is sp^2 hybridized.
b. Electron pairs on boxed-in N atoms are located in p orbitals to overlap with adjacent π bonds. Lone pairs on all other N atoms are in sp^2 hybrid orbitals.
c. The bicyclic ring system contains 10 π electrons from the five double bonds, making it aromatic. All lone pairs on N atoms in the rings are not delocalized.

17.38

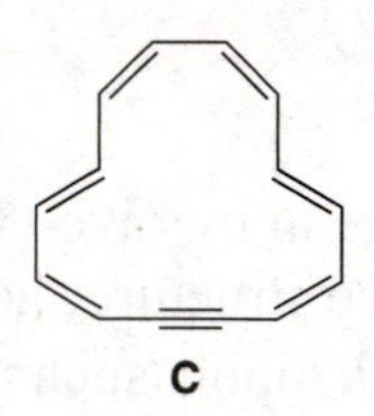

a. 16 total π electrons
b. 14 π electrons delocalized in the ring. [Note: Two of the electrons in the triple bond are localized between two C's, perpendicular to the π electrons delocalized in the ring.]
c. By having two of the p orbitals of the C–C triple bond co-planar with the p orbitals of all the C=C's, the total number of π electrons delocalized in the ring is 14. 4(3) + 2 = 14, so the ring is **aromatic.**

17.39 A second resonance structure can be drawn for the six-membered ring that gives it three π bonds, thus making it aromatic with six π electrons.

6 π electrons

17.40 The rate of an S_N1 reaction increases with increasing stability of the intermediate carbocation.

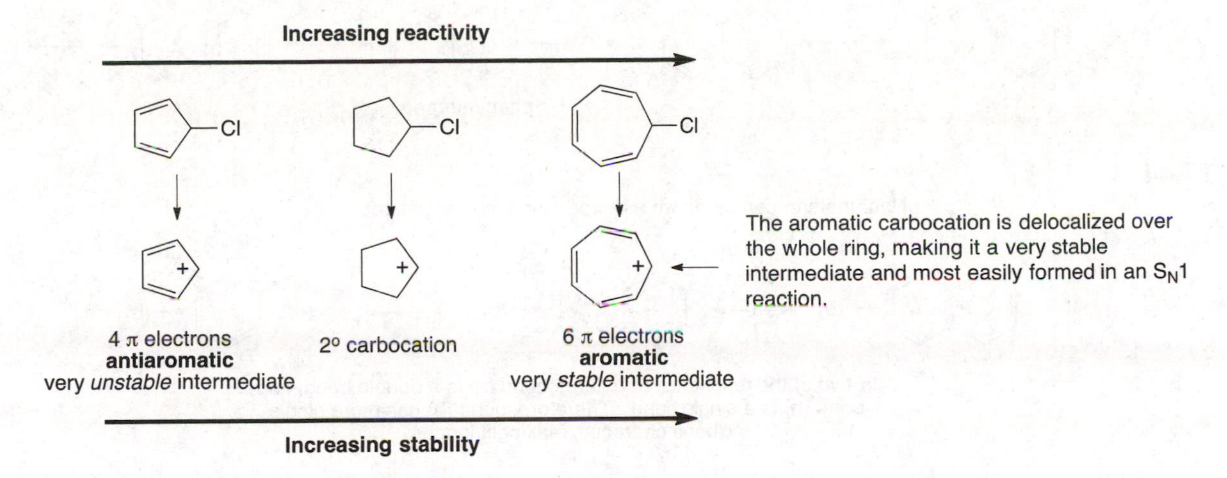
Increasing reactivity
Cl
Cl
Cl
+
+
+
The aromatic carbocation is delocalized over the whole ring, making it a very stable intermediate and most easily formed in an S_N1 reaction.
4 π electrons
antiaromatic
very *unstable* intermediate
2° carbocation
6 π electrons
aromatic
very *stable* intermediate
Increasing stability

17.41

Na+ H

H–OH

D

D

D

D

+ Na+ ⁻OH

+ H–D

HO–H

D

D

Two additional resonance structures are not drawn.

D

D

H–OH

17.42 α-Pyrone reacts like benzene because it is aromatic. A second resonance structure can be drawn showing how the ring has six π electrons. Thus, α-pyrone undergoes reactions characteristic of aromatic compounds—that is, substitution rather than addition.

α-pyrone

6 π electrons

17.43

a.

cyclopropenyl radical

b.

N–H

N+–H

N+–H

and

N–H

N+–H

N+–H

pyrrole

c.

phenanthrene

17.44

Naphthalene can be drawn as three resonance structures:

(a) (a) (a)

(b) (b) (b)

In two of the resonance structures bond (a) is a double bond, and bond (b) is a single bond. Therefore, bond (b) has more single bond character, making it longer.

17.45

a.

and

pyrrole

Pyrrole is less resonance stabilized than benzene because four of the resonance structures have charges, making them less good.

b.

and

furan

Furan is less resonance stabilized than pyrrole because its O atom is less basic, so it donates electron density less "willingly." Thus, charge-separated resonance forms are more minor contributors to the hybrid than the charge-separated resonance forms of pyrrole.

17.46 The compound with the more stable conjugate base is the stronger acid. Draw and compare the conjugate bases of each pair of compounds.

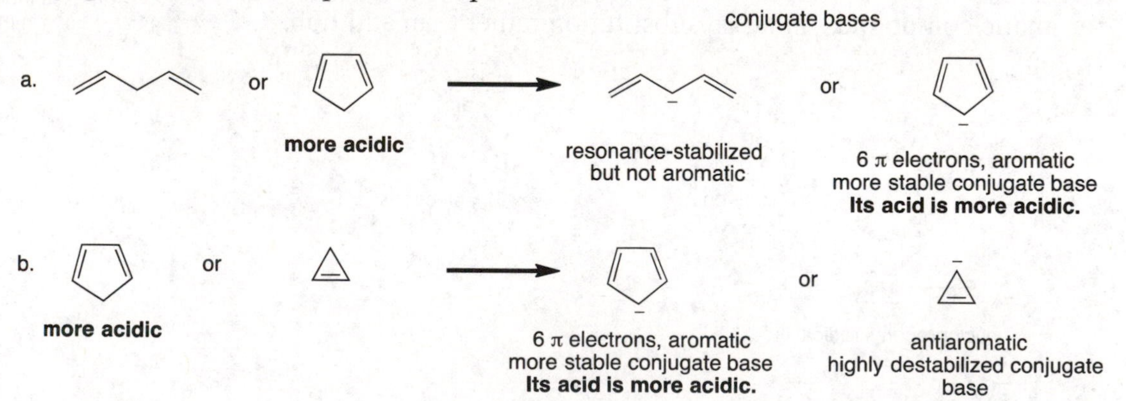

17.47

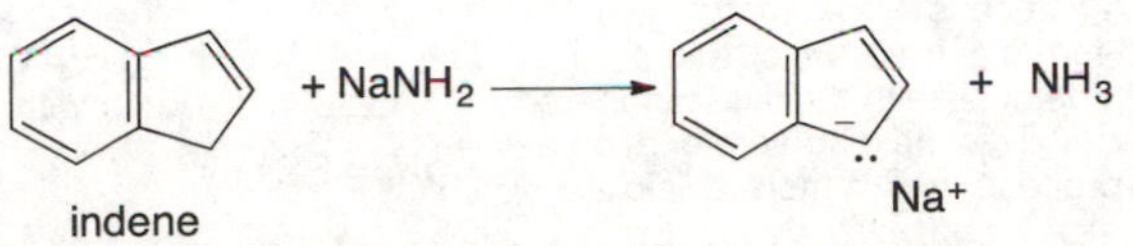

indene + $NaNH_2$ ⟶ (indenyl anion) Na^+ + NH_3

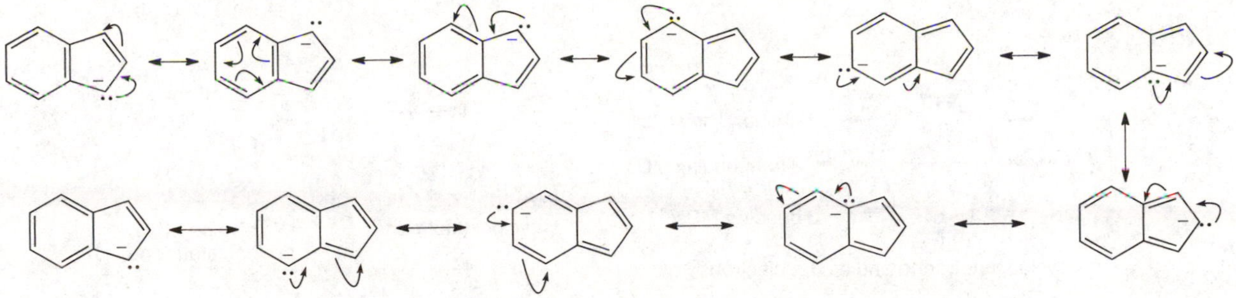

The conjugate base of indene has 10 π electrons, making it aromatic and very stable. Therefore, indene is more acidic than many hydrocarbons.

17.48

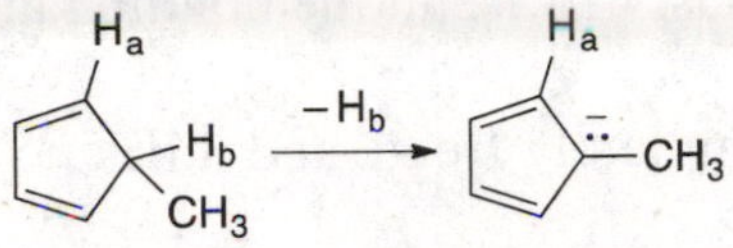

H_a, H_b, CH_3 — $-H_b$ ⟶ H_a, CH_3

A

H_b is most acidic because its conjugate base is aromatic (6 π electrons).

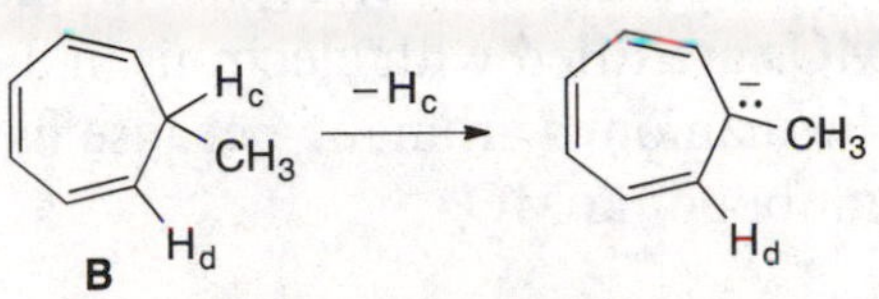

H_c, CH_3, H_d — $-H_c$ ⟶ CH_3, H_d

B

H_c is least acidic because its conjugate base has 8 π electrons, making it antiaromatic.

17.49

pyrrole ⟶ conjugate base

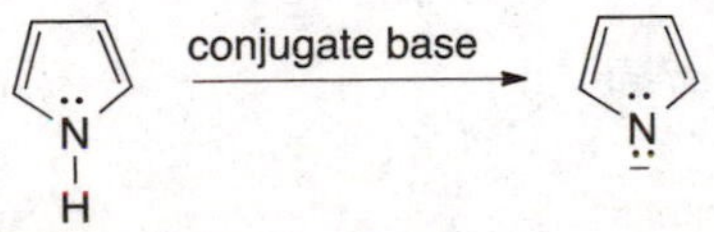

Both pyrrole and the conjugate base of pyrrole have 6 π electrons in the ring, making them both aromatic. Thus, deprotonation of pyrrole does not result in a gain of aromaticity because the starting material is aromatic to begin with.

cyclopentadiene ⟶ conjugate base

more acidic

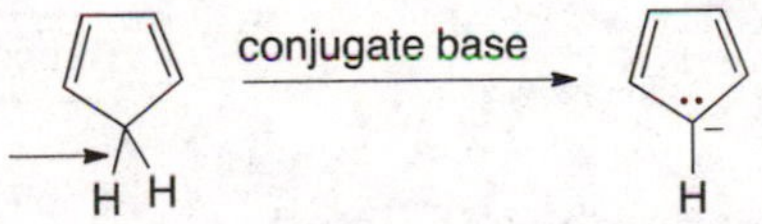

Cyclopentadiene is not aromatic, but the conjugate base has 6 π electrons and is therefore aromatic. This makes the C–H bond in cyclopentadiene more acidic than the N–H bond in pyrrole, because deprotonation of cyclopentadiene forms an aromatic conjugate base.

17.50

a. pyrrole + H^+ ⟶ **A** (C2) ⟷ ⟷

or

pyrrole + H^+ ⟶ **B**

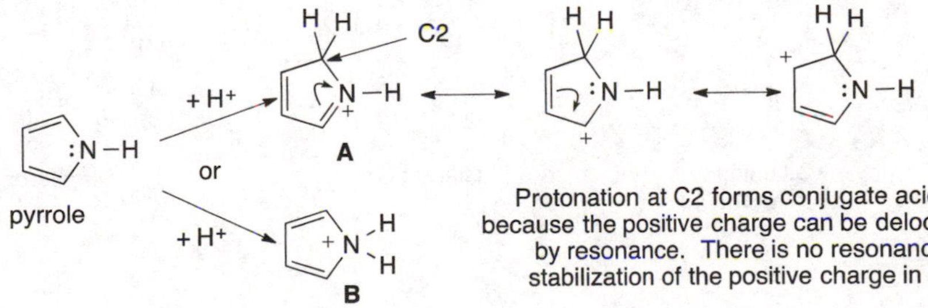

Protonation at C2 forms conjugate acid **A** because the positive charge can be delocalized by resonance. There is no resonance stabilization of the positive charge in **B**.

b.

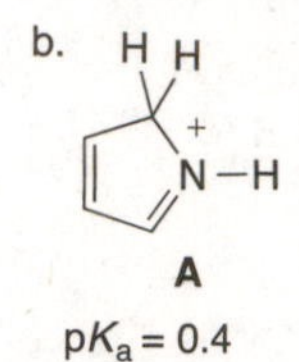

Loss of a proton from **A** (which is not aromatic) gives two electrons to N, and forms pyrrole, which has six π electrons that can then delocalize in the five-membered ring, making it aromatic. This makes deprotonation a highly favorable process, and **A** more acidic.

+N–H

pK_a = 5.2

C

Both **C** and its conjugate base pyridine are aromatic Since **C** has six π electrons, it is already aromatic tc begin with, so there is less to be gained by deprotonati and **C** is thus less acidic than **A**.

17.51

a.

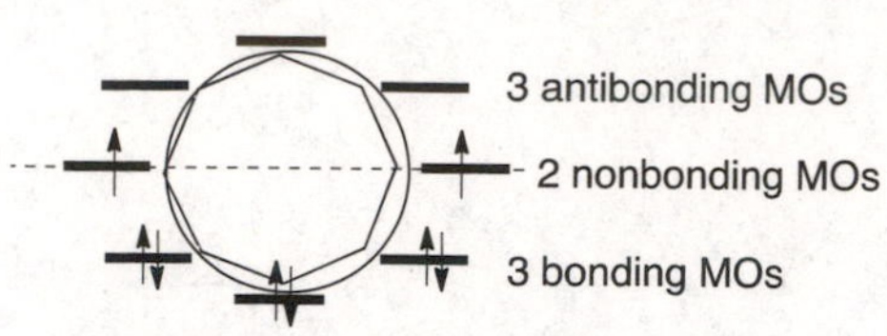

cyclooctatetraene and its 8 π electrons

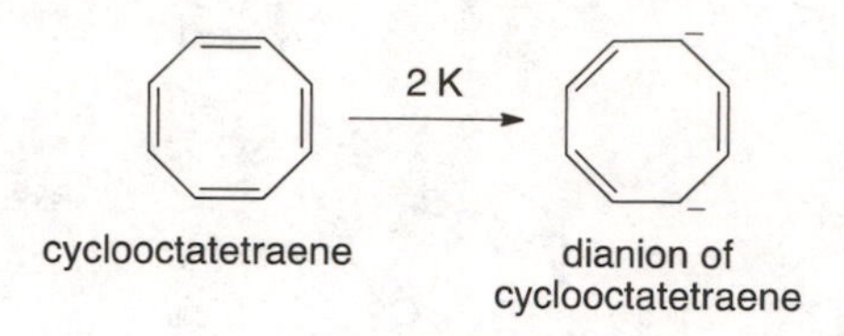

b. Even if cyclooctatetraene were flat, it has two unpaired electrons in its HOMOs (nonbonding MOs), so it cannot be aromatic.

c. The dianion has 10 π electrons.

d. The two additional electrons fill the nonbonding MOs; that is, all the bonding and nonbonding MOs are filled with electrons in the dianion.

e. The dianion is aromatic because its HOMOs are completely filled, and it has no electrons in antibonding MOs.

17.52

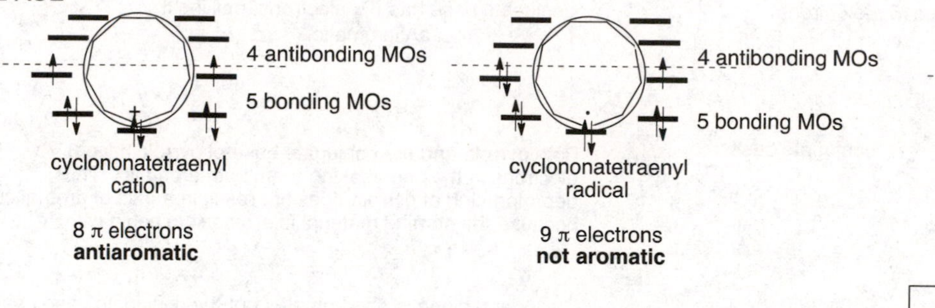

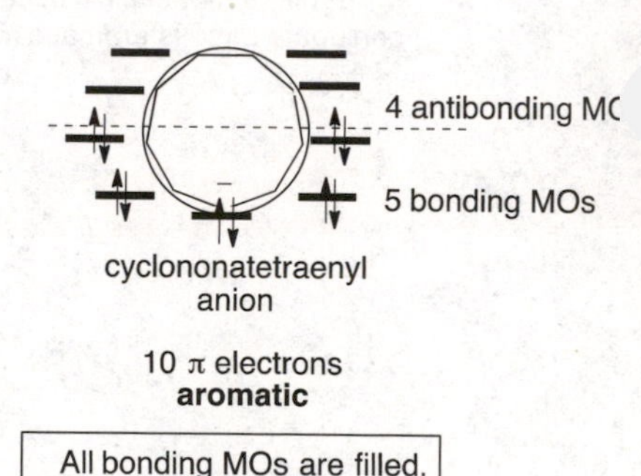

17.53 The number of different types of C's = the number of signals.

a.

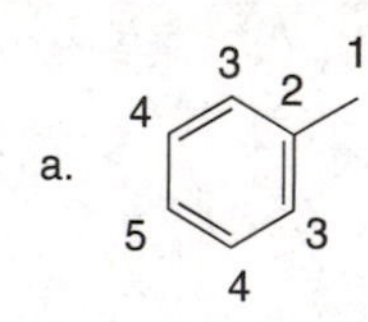

b.

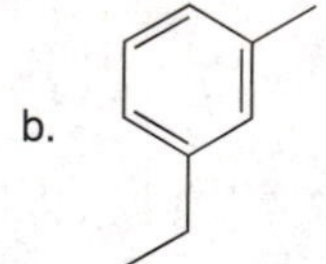

c.

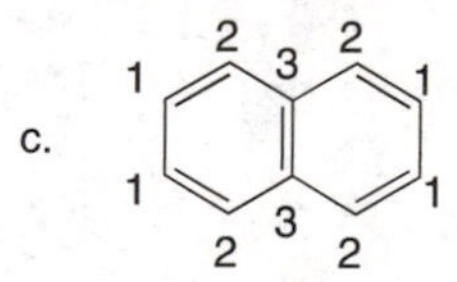

d.

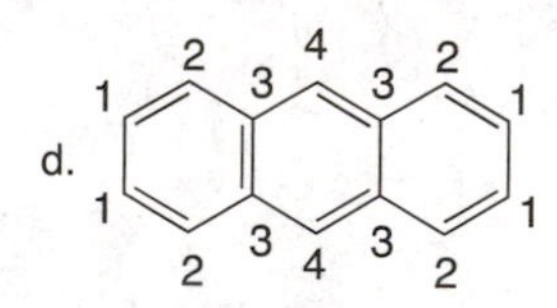

17.54 Draw the three isomers and count the different types of carbons in each. Then match the structures with the data.

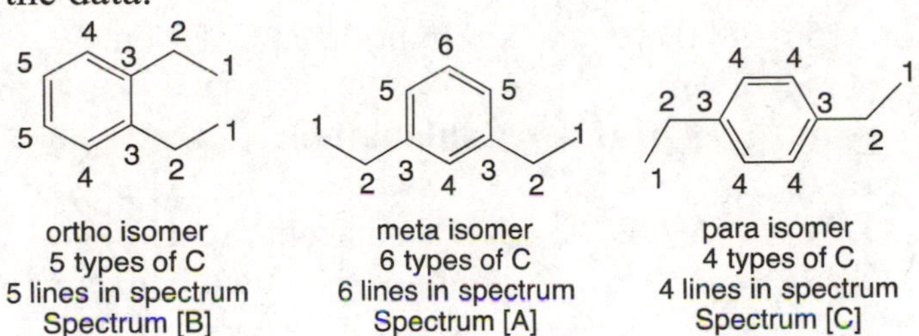

17.55

a. $C_{10}H_{14}$: IR absorptions at 3150–2850 (*sp*^{2} and *sp*^{3} hybridized C–H), 1600, and 1500 (due to a benzene ring) cm^{-1}
^{1}H NMR data:

Absorption	ppm	# of H's	Explanation
doublet	1.2	6	6 H's adjacent to 1 H
singlet	2.3	3	CH_3
septet	3.1	1	1 H adjacent to 6 H's
multiplet	7–7.4	4	a disubstituted benzene ring

$(CH_3)_2CH$ group

You can't tell from these data where the two groups are on the benzene ring. They are not para, because the para arrangement usually gives two sets of distinct peaks (resembling two doublets), so there are two possible structures—ortho and meta isomers.

or

b. C_9H_{12}: ^{13}C NMR signals at 21, 127, and 138 ppm → means three different types of C's.
^{1}H NMR shows two types of H's: 9 H's probably means 3 CH_3 groups; the other 3 H's are very deshielded so they are bonded to a benzene ring.
Only one possible structure fits:

c. C_8H_{10}: IR absorptions at 3108–2875 (*sp*^{2} and *sp*^{3} hybridized C–H), 1606, and 1496 (due to a benzene ring) cm^{-1}
^{1}H NMR data:

Absorption	ppm	# of H's	Explanation
triplet	1.3	3	3 H's adjacent to 2 H's
quartet	2.7	2	2 H's adjacent to 3 H's
multiplet	7.3	5	a monosubstituted benzene ring

Structure:

17.56

a. Compound **A:** Molecular formula $C_8H_{10}O$
IR absorption at 3150–2850 (sp^2 and sp^3 hybridized C–H) cm^{-1}
^1H NMR data:

Absorption	**ppm**	**# of H's**	**Explanation**
triplet	1.4	3	3 H's adjacent to 2 H's
quartet	3.95	2	2 H's adjacent to 3 H's
multiplet	6.8–7.3	5	a monosubstituted benzene ring

Structure:

O

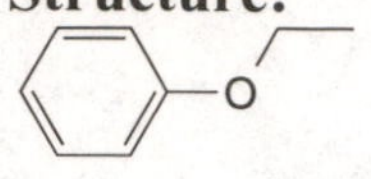

b. Compound **B:** Molecular formula $C_9H_{10}O_2$
IR absorption at 1669 (C=O) cm^{-1}
^1H NMR data:

Absorption	**ppm**	**# of H's**	**Explanation**
singlet	2.5	3	CH_3 group
singlet	3.8	3	CH_3 group
doublet	6.9	2	2 H's on a benzene ring
doublet	7.9	2	2 H's on a benzene ring

Structure:

O O

or

O O

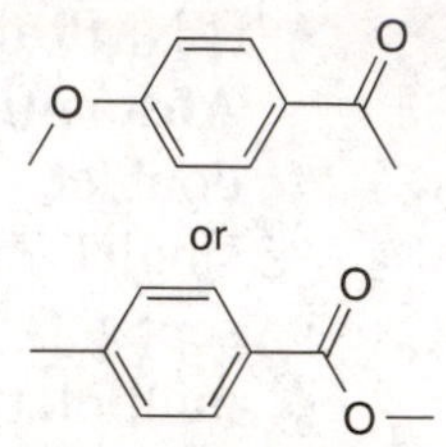

It would be hard to distinguish these two compounds with the given data.

17.57

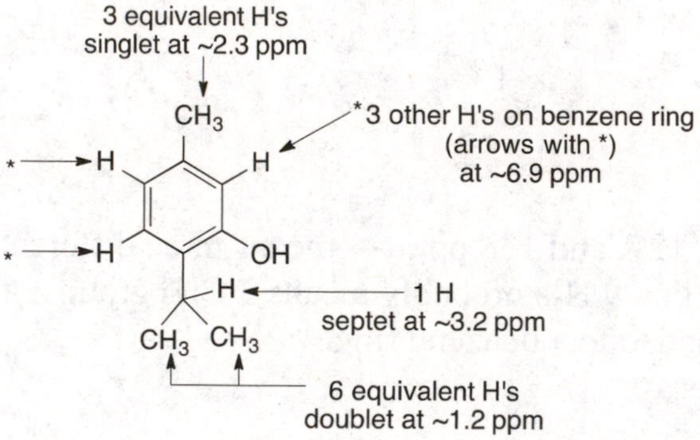

3 equivalent H's
singlet at ~2.3 ppm

CH_3

*3 other H's on benzene ring (arrows with *) at ~6.9 ppm

thymol

H H H OH H

1 H
septet at ~3.2 ppm

CH_3 CH_3

6 equivalent H's
doublet at ~1.2 ppm

IR absorptions:
3500–3200 cm^{-1} (O–H)
3150–2850 cm^{-1} (C–H bonds)
1621 and 1585 cm^{-1} (benzene ring)

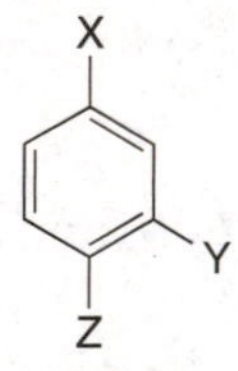

X Y Z

basic structure of thymol

Thymol must have this basic structure, given the NMR and IR data because it is a trisubstituted benzene ring with one singlet and two doublets in the NMR at ~6.9 ppm. However, which group [OH, CH_3, or $CH(CH_3)_2$] corresponds to X, Y, and Z is not readily distinguished with the given data. The correct structure for thymol is given.

17.58

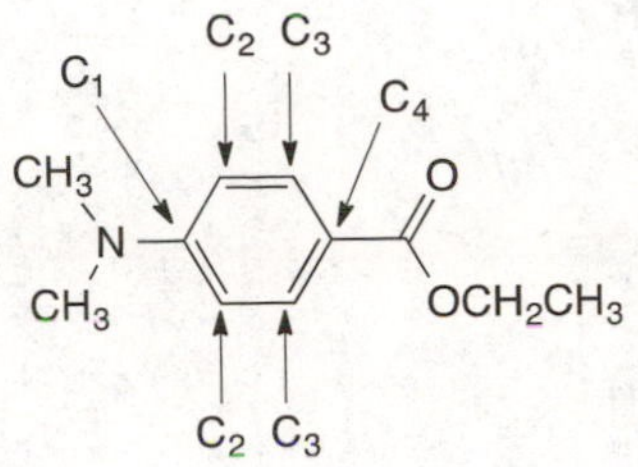

C_1 C_2 C_3 C_4 CH_3 N CH_3 O OCH_2CH_3 C_2 C_3

^{13}C NMR has four lines that are located in the aromatic region (~110–155 ppm), corresponding to the four different types of carbons in the aromatic ring of the para isomer. The ortho and meta isomers have six different C's, and so six lines would be expected for each of them.

17.59 Because tetrahydrofuran has a higher boiling point and is more water soluble, it must be more polar and have stronger intermolecular forces than furan. There are two contributing factors. One lone pair on furan's O atom is delocalized on the five-membered ring to make it aromatic. This makes it less available for H-bonding with water and other intermolecular interactions. Also, the C–O bonds in furan are less polar than the C–O bonds in tetrahydrofuran because of hybridization. The sp^2 hybridized C's of furan pull a little more electron density towards them than do the sp^3 hybridized C's of tetrahydrofuran. This counteracts the higher electronegativity of O compared to C to a small extent.

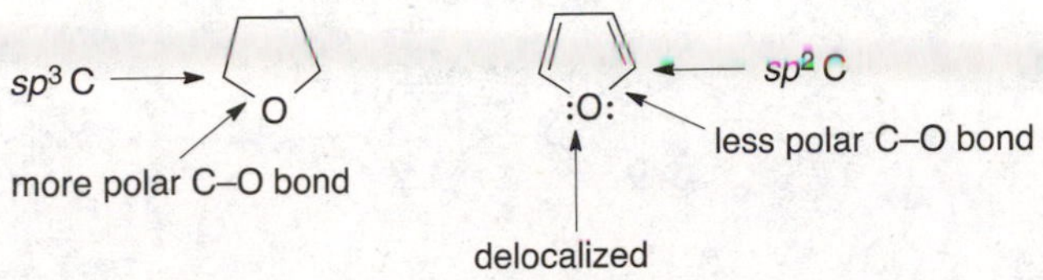

sp^3 C → O
more polar C–O bond

:O: ← sp^2 C
less polar C–O bond
delocalized

17.60

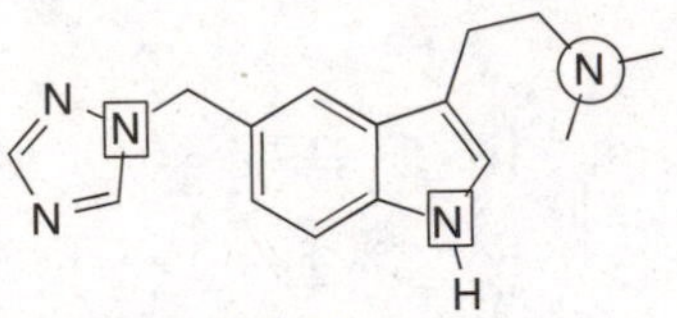

rizatriptan

a. Rizatriptan contains three aromatic rings.

b. The circled N atom is sp^3 hybridized. All other N atoms are either part of a π bond or bonded to a π bond, making them sp^2 hybridized.

c. The lone pair on the circled N atom occupies an sp^3 hybrid orbital. The lone pairs on the boxed-in N atoms are in a p orbital. The lone pairs on the unlabeled N atoms occupy an sp^2 hybrid orbital.

d.

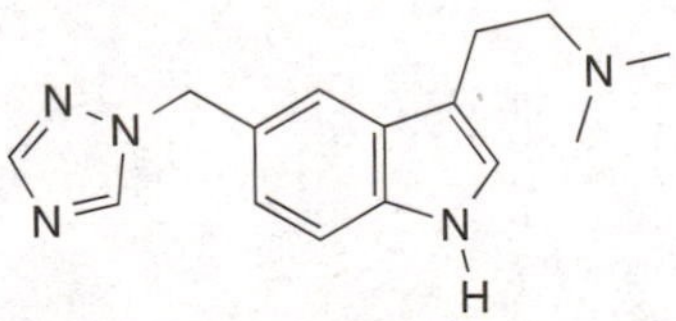

one additional resonance structure

e.

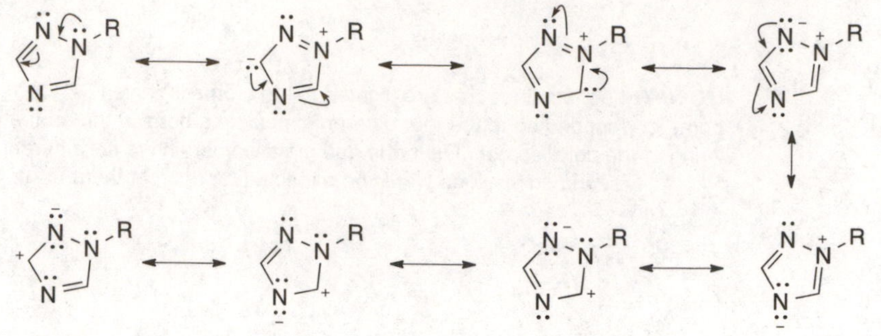

N N R ⟷ N N R ⟷ N N R ⟷ N N R

N N R ⟷ N N R ⟷ N N R ⟷ N N R

17.61

a, b.

electron pair in an sp^2 hybridized orbital

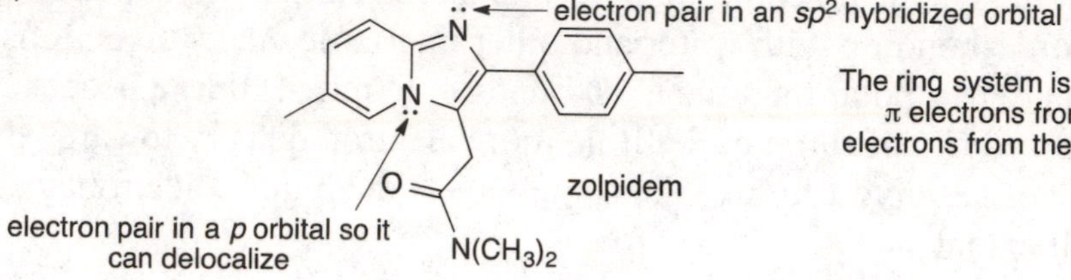

The ring system is aromatic with 10 π electrons, 8 π electrons from the double bonds and 2 π electrons from the N atom common to both rings.

zolpidem

electron pair in a *p* orbital so it can delocalize

O

$N(CH_3)_2$

c.

O $N(CH_3)_2$ ⟷ O $N(CH_3)_2$ ⟷ O $N(CH_3)_2$

O $N(CH_3)_2$ — O $N(CH_3)_2$ — O $N(CH_3)_2$

O $N(CH_3)_2$ ⟷ O $N(CH_3)_2$ ⟷ O $N(CH_3)_2$

O $N(CH_3)_2$ ⟷ O $N(CH_3)_2$

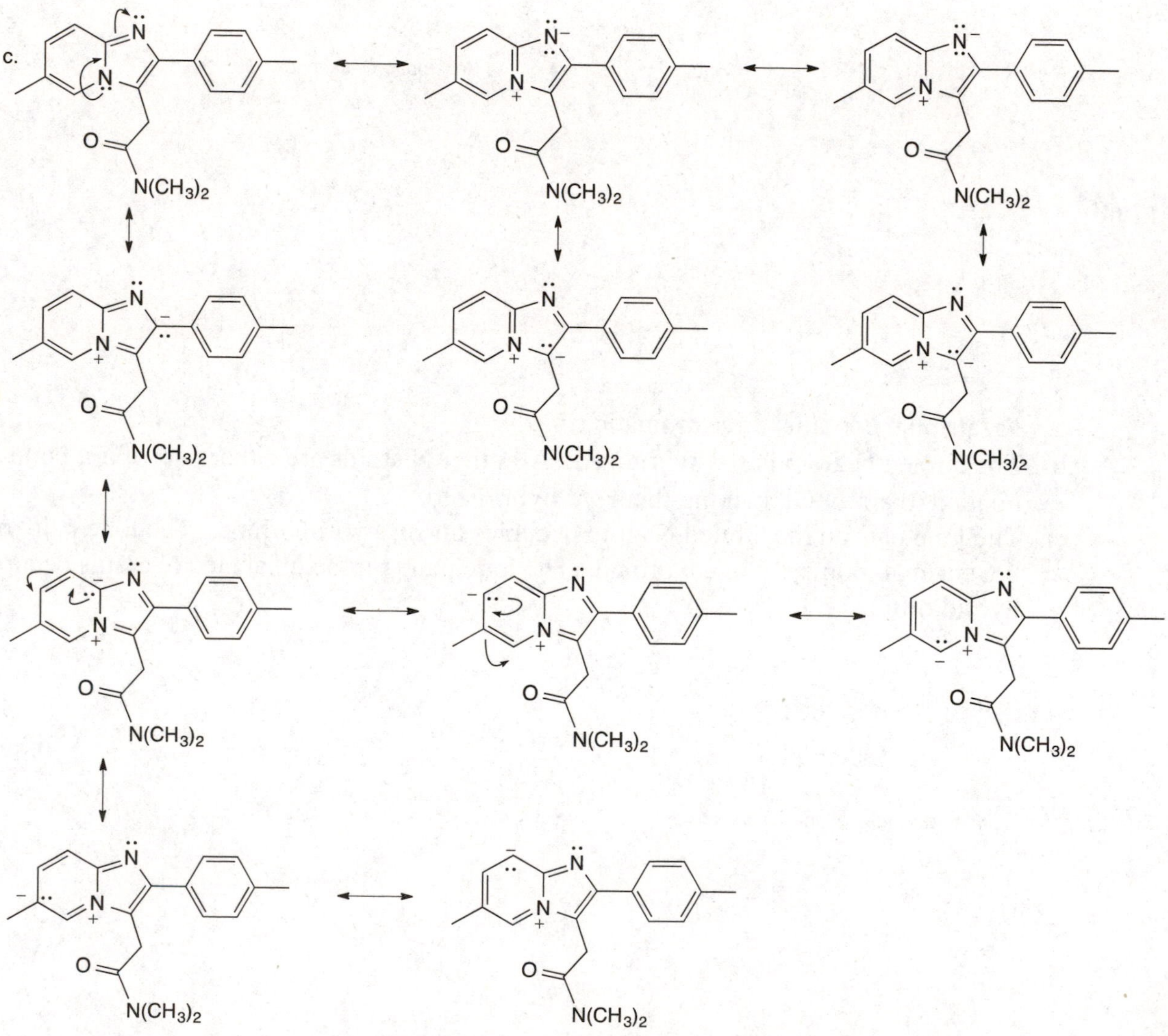

17.62

a.

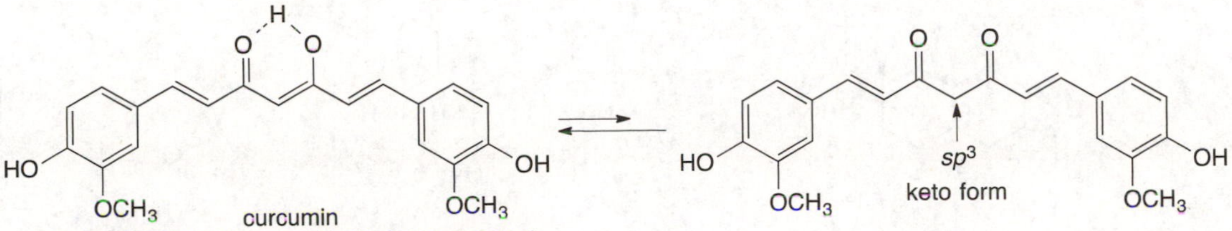

HO
OCH3
curcumin
OH
OCH3
HO
OCH3
sp3
keto form
OH
OCH3

The enol form is more stable because the enol double bond makes a highly conjugated system. The enol OH can also intramolecularly hydrogen bond to the nearby carbonyl O atom.

b.

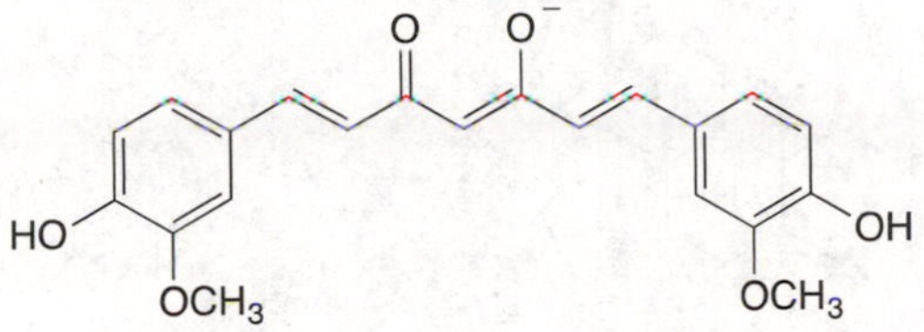

HO
OCH3
OH
OCH3

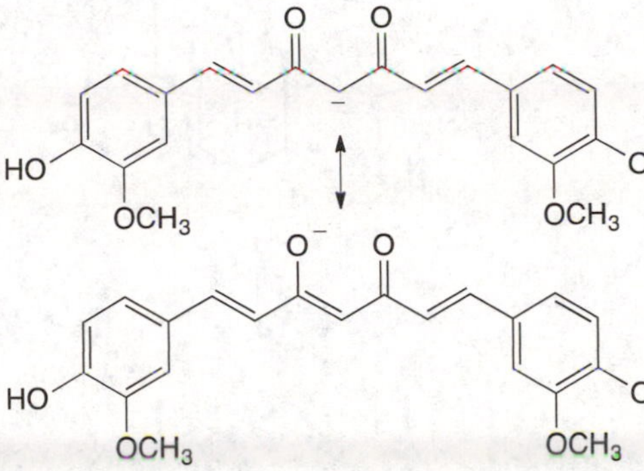

The enol O–H proton is more acidic than an alcohol O–H proton because the conjugate base is resonance stabilized.

c. Curcumin is colored because it has many conjugated π electrons, which shift absorption of light from the UV to the visible region.

d. Curcumin is an antioxidant because it contains a phenol. Homolytic cleavage affords a resonance-stabilized phenoxy radical, which can inhibit oxidation from occurring, much like vitamin E and BHT in Chapter 15.

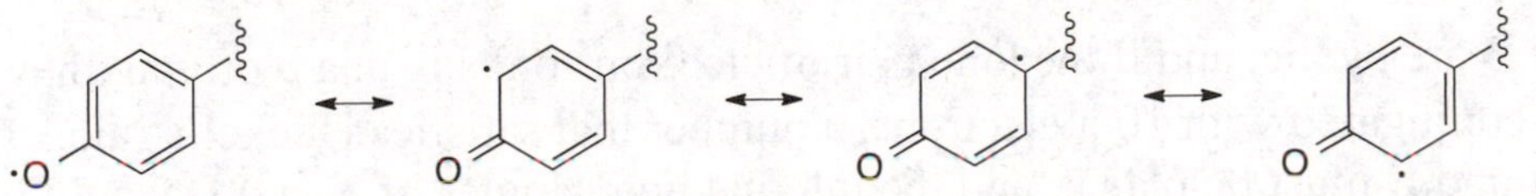

(+ other resonance structures)

phenoxy radical

Resonance delocalizes the radical on the ring and C chain of curcumin.

17.63

a. Pyrazole rings are aromatic because they have six π electrons—two from the lone pair on the N atom that is not part of the double bond and four from the double bonds.

b.

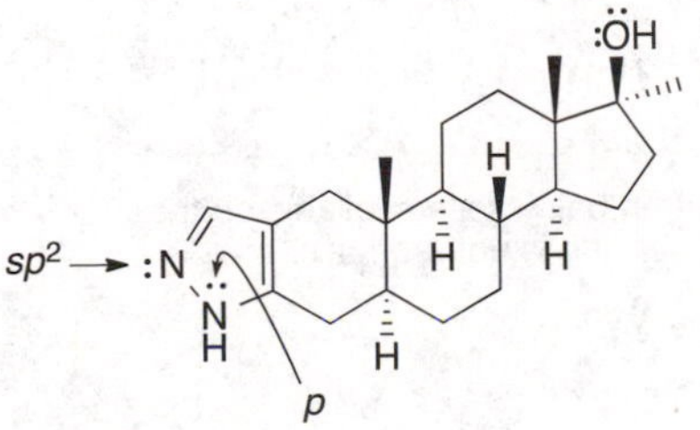

:ÖH
sp^2 → :N
N
H
p

c.

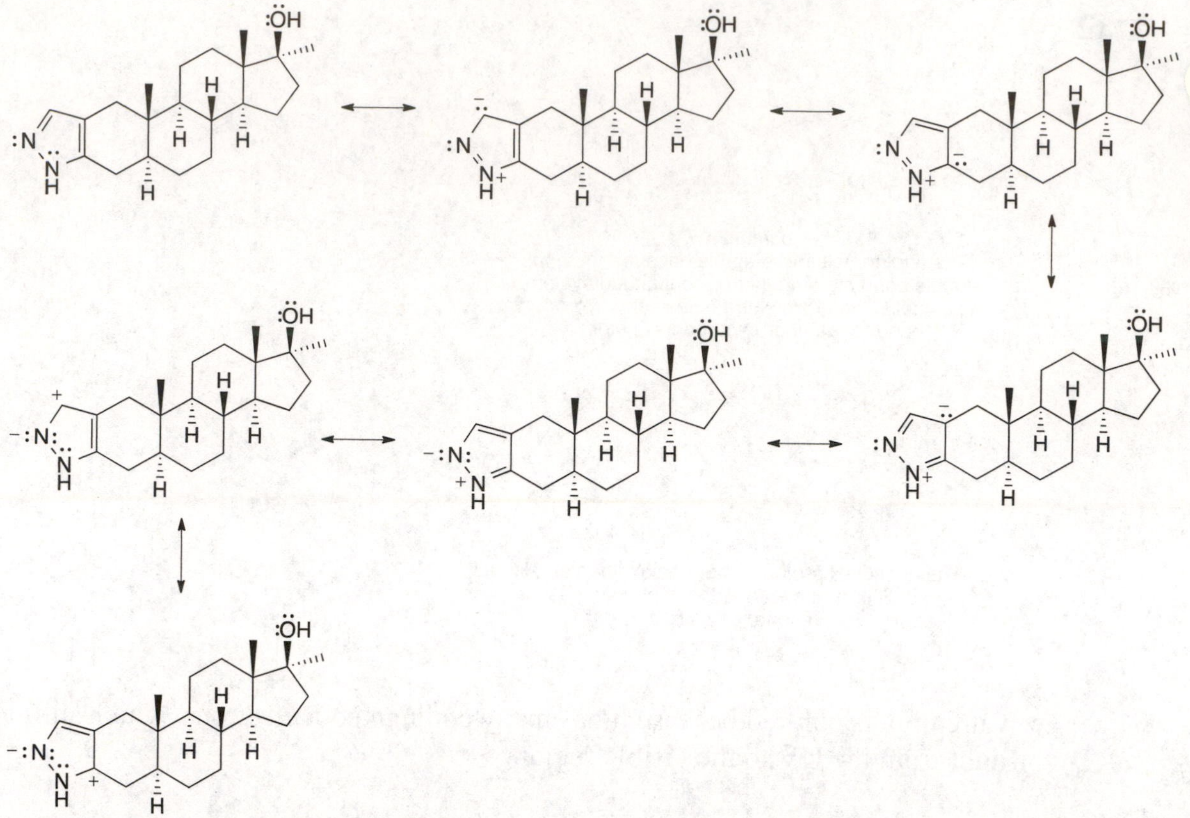

:ÖH :ÖH :ÖH
H H H
:N H H :N H H :N H H
N N+ N+
H H H H H H

:ÖH :ÖH :ÖH
H H H
+ −:N: H H −:N: H H :N H H
N +N N+
H H H H H H

:ÖH
H
−:N: H H
N +
H H

d. The N atom in the NH bond in the pyrazole ring is sp^2 hybridized with 33% *s*-character, increasing the acidity of the N–H bond. The N–H bond of CH_3NH_2 contains an sp^3 hybridized N atom (25% *s*-character).

17.64 Both **A** and **B** are cyclic, and if the lone pair of electrons on N is in a *p* orbital, they are completely conjugated with 10 π electrons, a number that satisfies Hückel's rule. To be aromatic, **A** and **B** must be planar, and the internal bond angles of **A** and **B** would be much larger than 120°, the theoretical bond angle of sp^2 hybridized C's. The fact that **A** is aromatic means that the lone pair on N occupies a *p* orbital, so it can delocalize onto the nine-membered ring. The stabilization gained by being aromatic is greater than any angle strain. With **B,** the lone pair on N is also delocalized onto the C=O, making it less available for donation to the ring, so the ring is not aromatic.

:NH
2 electrons in a *p* orbital
A
10 π electrons

:O: :O:⁻
:N OCH$_2$CH$_3$ ⟷ +N OCH$_2$CH$_3$
B
Electron pair is less available so the ring is not aromatic.

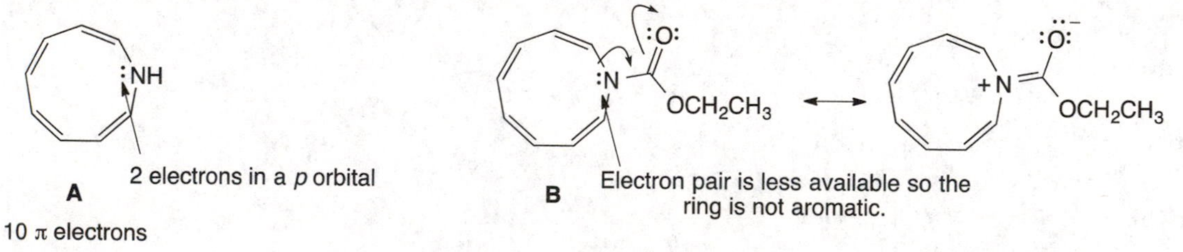

17.65 With 14 π electrons in the double bonds, the system is aromatic [4(3) + 2 = 14 π electrons]. The ring current generated by the circulating π electrons deshields the protons on the C=C's, so they absorb downfield (8.14–8.67 ppm). The CH_3 groups, however, are very shielded because they lie above and below the plane, so they absorb far upfield (–4.25 ppm). The dianion of **C** now has 16 π electrons, making it antiaromatic, so the position of the absorptions reverses. The C=C protons are now shielded (–3 ppm), and the CH_3 protons are now deshielded (21 ppm).

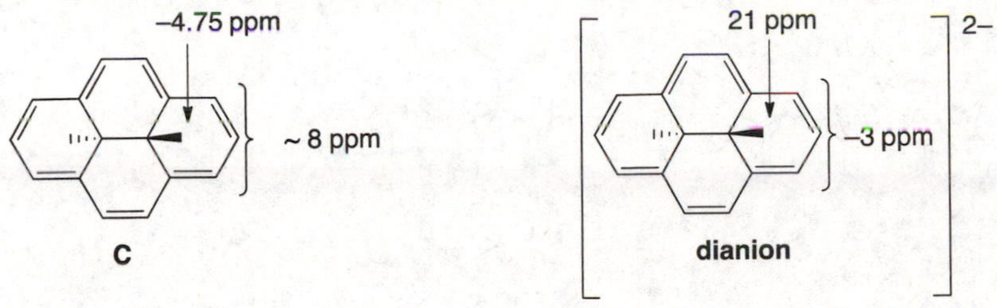

–4.75 ppm
~8 ppm
C

21 ppm
2–
–3 ppm
dianion

17.66 A second resonance structure for **A** shows that the ring is completely conjugated and has six π electrons, making it aromatic and especially stable. A similar charge-separated resonance structure for **B** makes the ring completely conjugated, but gives the ring four π electrons, making it antiaromatic and especially unstable.

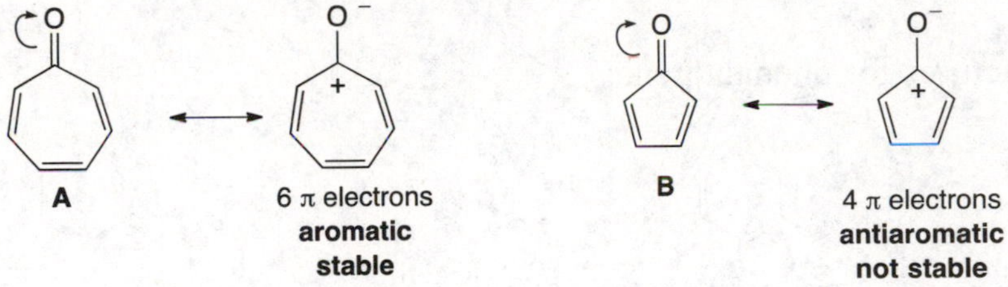

A

6 π electrons
aromatic
stable

B

4 π electrons
antiaromatic
not stable

17.67 The conversion of carvone to carvacrol involves acid-catalyzed isomerization of two double bonds and tautomerization of a ketone to an enol tautomer. In this case the enol form is part of an aromatic phenol. Each isomerization of a C=C involves Markovnikov addition of a proton, followed by deprotonation.

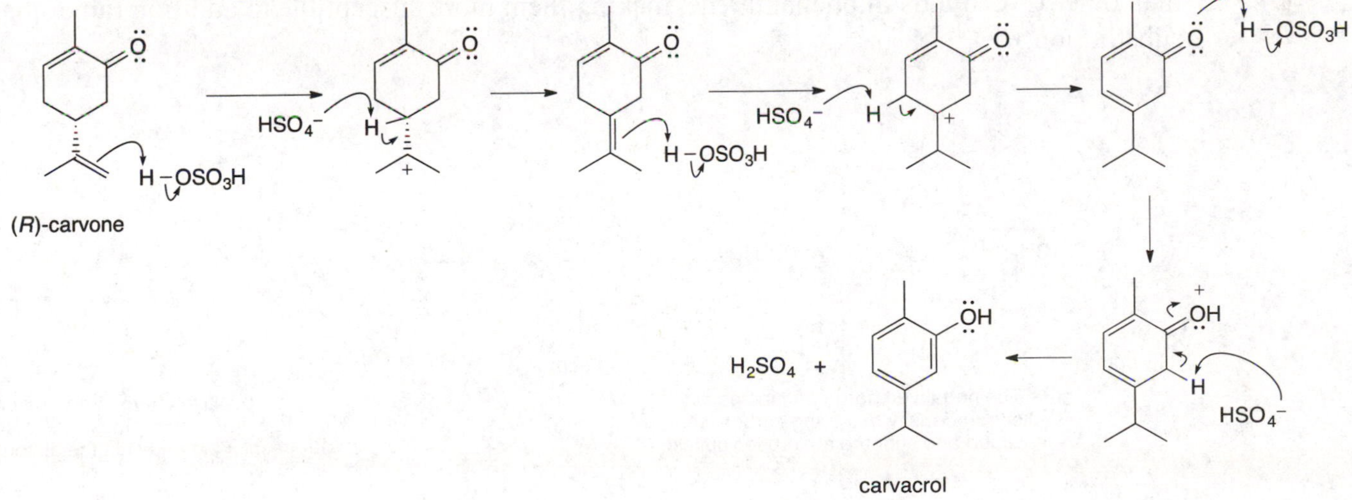

(*R*)-carvone

$H-OSO_3H$ HSO_4^- $H-OSO_3H$ HSO_4^- $H-OSO_3H$ HSO_4^-

H_2SO_4 +

carvacrol

17.68 Resonance structures for triphenylene:

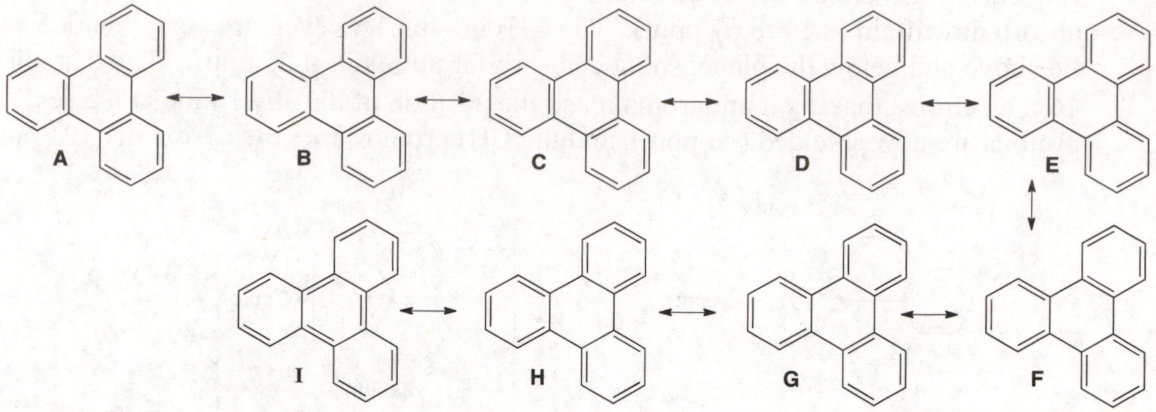

A ⟷ B ⟷ C ⟷ D ⟷ E ⟷ F ⟷ G ⟷ H ⟷ I

Resonance structures **A–H** all keep three double and three single bonds in the three six-membered rings on the periphery of the molecule. This means that each ring behaves like an isolated benzene ring undergoing substitution rather than addition because the π electron density is delocalized within each six-membered ring. Only resonance structure **I** does not have this form. Each C–C bond of triphenylene has four (or five) resonance structures in which it is a single bond and four (or five) resonance structures in which it is a double bond.

Resonance structures for phenanthrene:

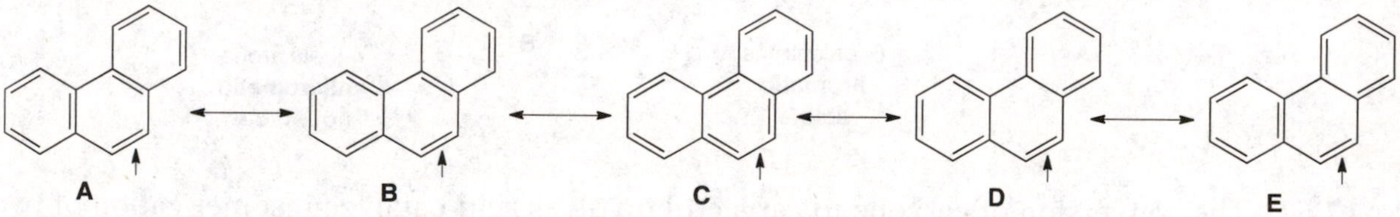

A ⟷ B ⟷ C ⟷ D ⟷ E

With phenanthrene, however, four of the five resonance structures keep a double bond at the labeled C's. (Only **C** does not.) This means that these two C's have more double bond character than other C–C bonds in phenanthrene, making them more susceptible to addition rather than substitution.

17.69

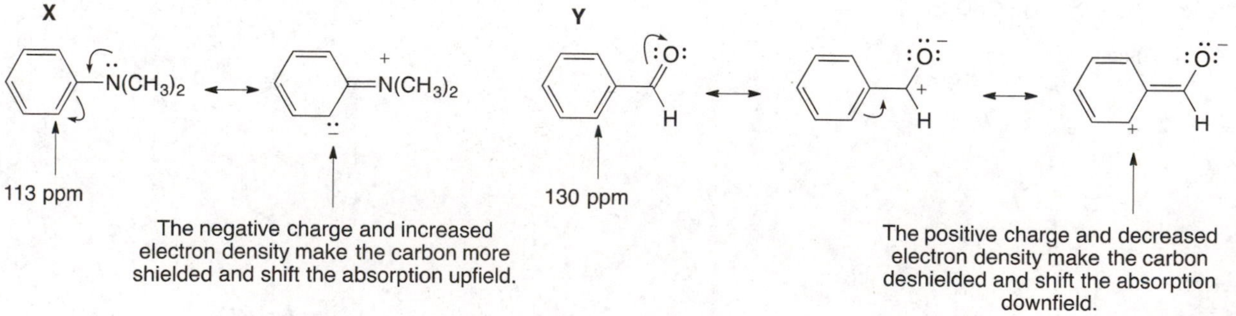

X

$-N(CH_3)_2$ ⟷ $=N(CH_3)_2$

113 ppm

The negative charge and increased electron density make the carbon more shielded and shift the absorption upfield.

Y

130 ppm

The positive charge and decreased electron density make the carbon deshielded and shift the absorption downfield.

Chapter 18 Reactions of Aromatic Compounds

Chapter Review

Mechanism of electrophilic aromatic substitution (18.2)

- Electrophilic aromatic substitution follows a two-step mechanism. Reaction of the aromatic ring with an electrophile forms a carbocation, and loss of a proton regenerates the aromatic ring.
- The first step is rate-determining.
- The intermediate carbocation is stabilized by resonance; a minimum of three resonance structures can be drawn. The positive charge is always located ortho or para to the new C–E bond.

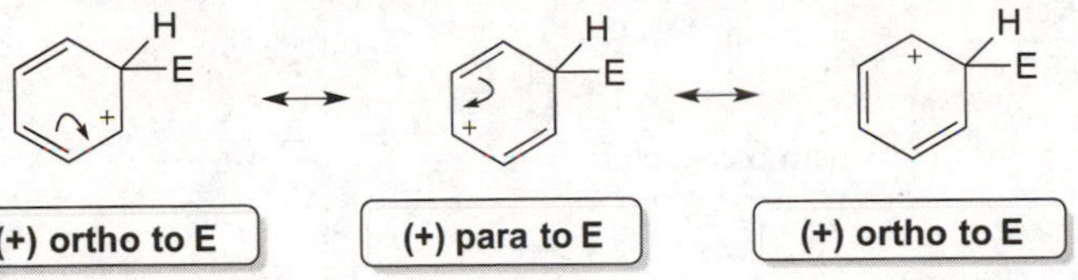

(+) ortho to E ⟷ (+) para to E ⟷ (+) ortho to E

Three rules describing the reactivity and directing effects of common substituents (18.7–18.9)

[1] All ortho, para directors except the halogens activate the benzene ring.
[2] All meta directors deactivate the benzene ring.
[3] The halogens deactivate the benzene ring.

Summary of substituent effects in electrophilic aromatic substitution (18.6–18.9)

	Substituent	Inductive effect	Resonance effect	Reactivity	Directing effect
[1]	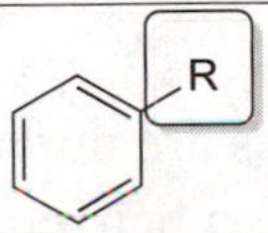 R (R = alkyl)	donating	none	activating	ortho, para
[2]	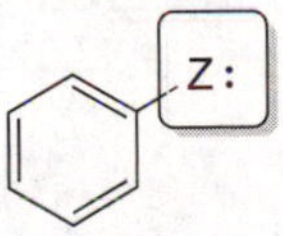 Z: (Z = N or O)	withdrawing	donating	activating	ortho, para
[3]	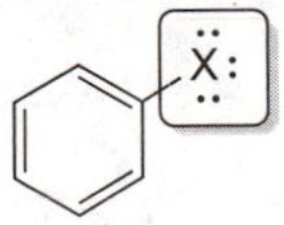 X: (X = halogen)	withdrawing	donating	deactivating	ortho, para
[4]	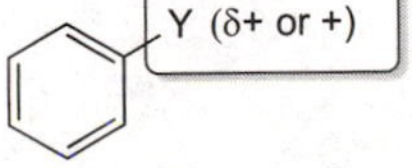 Y (δ+ or +)	withdrawing	withdrawing	deactivating	meta

Five examples of electrophilic aromatic substitution

[1] Halogenation—Replacement of H by Cl or Br (18.3)

H → (X_2, FeX_3) [X = Cl, Br] → Cl aryl chloride or Br aryl bromide

- Polyhalogenation occurs on benzene rings substituted by OH and NH_2 (and related substituents) (18.10A).

[2] Nitration—Replacement of H by NO_2 (18.4)

H → (HNO_3, H_2SO_4) → NO_2 nitro compound

[3] Sulfonation—Replacement of H by SO_3H (18.4)

H → (SO_3, H_2SO_4) → SO_3H benzenesulfonic acid

[4] Friedel–Crafts alkylation—Replacement of H by R (18.5)

H → (RCl, $AlCl_3$) → R alkyl benzene (arene)

- Rearrangements can occur.
- Vinyl halides and aryl halides are unreactive.
- The reaction does not occur on benzene rings substituted by meta deactivating groups or NH_2 groups (18.10B).
- Polyalkylation can occur.

Variations:

[1] with alcohols

H → (ROH, H_2SO_4) → R

[2] with alkenes

H → (CH2=CH–R, H_2SO_4) → R

[5] Friedel–Crafts acylation—Replacement of H by RCO (18.5)

RCOCl, $AlCl_3$ → ketone

- The reaction does not occur on benzene rings substituted by meta deactivating groups or NH_2 groups (18.10B).

Nucleophilic aromatic substitution (18.13)

[1] Nucleophilic substitution by an addition–elimination mechanism

:Nu⁻

X = F, Cl, Br, I
A = electron-withdrawing group

- The mechanism has two steps.
- Strong electron-withdrawing groups at the ortho or para position are required.
- Increasing the number of electron-withdrawing groups increases the rate.
- Increasing the electronegativity of the halogen increases the rate.

[2] Nucleophilic substitution by an elimination–addition mechanism

:Nu⁻

X = halogen

- Reaction conditions are harsh.
- Benzyne is formed as an intermediate.
- Product mixtures may result.

Other reactions of benzene derivatives

[1] Benzylic halogenation (18.14)

Br_2, $h\nu$ or Δ, or NBS, $h\nu$ or ROOR → benzylic bromide

[2] Oxidation of alkyl benzenes (18.15A)

$KMnO_4$ → benzoic acid (COOH)

- A benzylic C–H bond is needed for reaction.

[3] Reduction of ketones to alkyl benzenes (18.15B)

Zn(Hg), HCl
or
NH_2NH_2, ^-OH

alkyl benzene

[4] Reduction of nitro groups to amino groups (18.15C)

NO_2

H_2, Pd-C
or
Fe, HCl
or
Sn, HCl

NH_2

aniline

Practice Test on Chapter Review

1.a. Which of the following statements is true about an ethoxy substituent ($-OCH_2CH_3$) on a benzene ring?
1. OCH_2CH_3 increases the rates of both electrophilic substitution and nucleophilic substitution.
2. OCH_2CH_3 decreases the rates of both electrophilic substitution and nucleophilic substitution.
3. OCH_2CH_3 increases the rate of electrophilic substitution and decreases the rate of nucleophilic substitution.
4. OCH_2CH_3 decreases the rate of electrophilic substitution and increases the rate of nucleophilic substitution.
5. None of these statements is true.

b. Which of the following statements is true about a $-CO_2CH_3$ group on a benzene ring?
1. CO_2CH_3 increases the rates of both electrophilic substitution and nucleophilic substitution.
2. CO_2CH_3 decreases the rates of both electrophilic substitution and nucleophilic substitution.
3. CO_2CH_3 increases the rate of electrophilic substitution and decreases the rate of nucleophilic substitution.
4. CO_2CH_3 decreases the rate of electrophilic substitution and increases the rate of nucleophilic substitution.
5. None of these statements is true.

c. Which of the following is *not* a valid resonance structure for the carbocation that results from ortho attack of an electrophile on $C_6H_5C(CH_3){=}CH_2$?

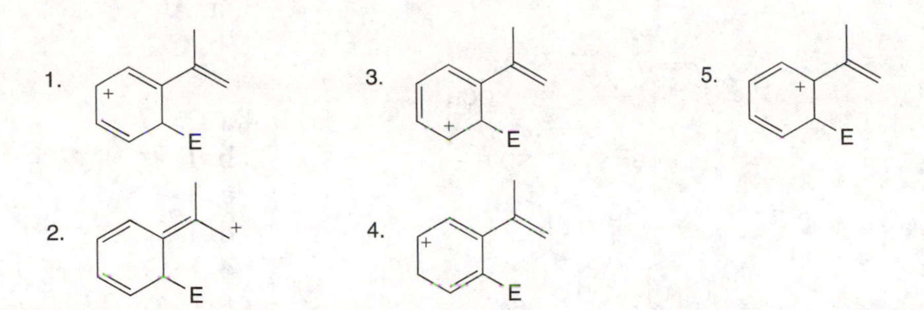

2. Draw the organic products formed in the following reactions.

a. [1] $CH_3CH_2CH_2COCl$, $AlCl_3$; [2] Zn(Hg), HCl

b. HNO_3, H_2SO_4; Fe, HCl

c. $KMnO_4$

d. CH_3O– ; HNO_3, H_2SO_4; H_2, Pd-C

e. OCH_3, Br; SO_3, H_2SO_4

f. Br_2 (1 equiv), $FeBr_3$

g. O_2N–, –Cl; $NaOCH_2CH_3$

h. –Cl; NaOH, Δ

3. (a) Considering the compound drawn below, which ring is *most* reactive in electrophilic aromatic substitution? (b) Which ring is the *least* reactive in electrophilic aromatic substitution?

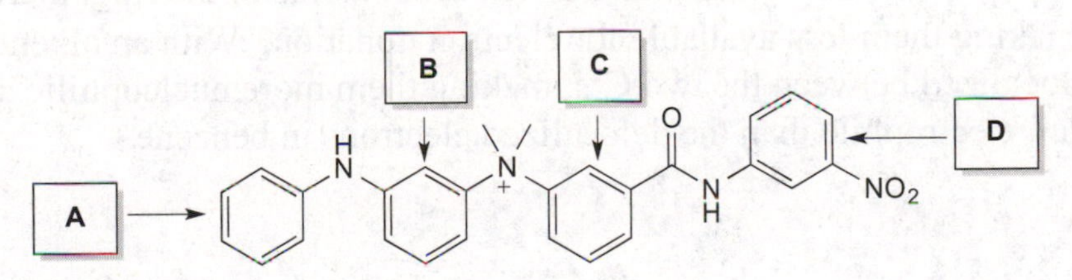

4. Classify each substituent as [1] ortho, para activating, [2] ortho, para deactivating, or [3] meta deactivating.

a. –Br
b. $–CH_2CH_2CH_2Br$
c. –COOH
d. $–NHCOCH_2CH_3$
e. $–N(CH_2)_6COCH_3$
f. $–CCl_3$

5. What reagents are needed to convert toluene ($C_6H_5CH_3$) to each compound?

a. C_6H_5COOH
b. $C_6H_5CH_2Br$
c. *p*-bromotoluene
d. *o*-nitrotoluene
e. *p*-ethyltoluene
f.

Answers to Practice Test

1.a. 3
b. 4
c. 4

2.

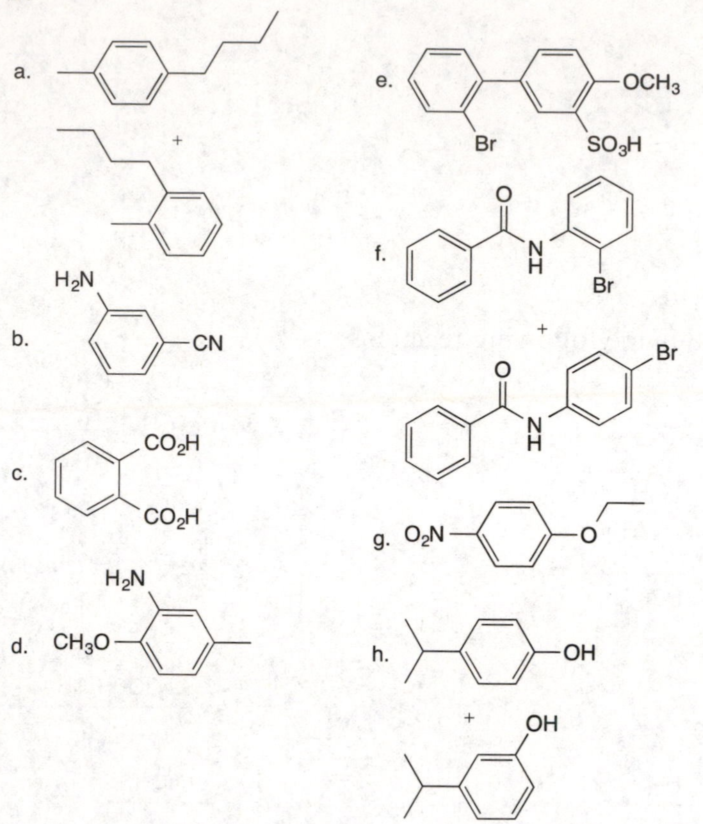

a.

+

b. H_2N, CN

c. CO_2H, CO_2H

d. CH_3O, H_2N

e. OCH_3, Br, SO_3H

f. O, N H, Br

+

O, N H, Br

g. O_2N, O

h. OH

+

OH

3. a. **A**
b. **C**

4.a. 2
b. 1
c. 3
d. 1
e. 1
f. 3

5.a. $KMnO_4$
b. Br_2, *hv*
c. Br_2, $FeBr_3$
d. HNO_3, H_2SO_4
e. CH_3CH_2Cl, $AlCl_3$
f. CH_3CH_2COCl, $AlCl_3$

Answers to Problems

18.1 The π electrons of benzene are delocalized over the six atoms of the ring, increasing benzene's stability and making them less available for electron donation. With an alkene, the two π electrons are localized between the two C's, making them more nucleophilic and thus more reactive with an electrophile than the delocalized electrons in benzene.

18.2

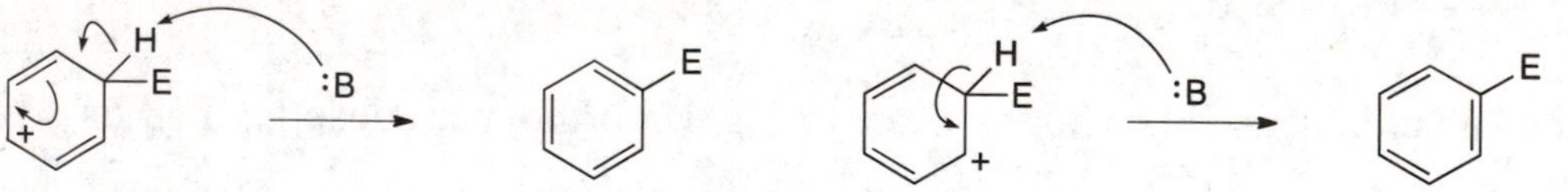

H, E, :B, E, H, E, :B, E

18.3 Reaction with Cl_2 and $FeCl_3$ as the catalyst occurs in two parts. First is the formation of an electrophile, followed by a two-step substitution reaction.

[1] :Cl–Cl: + $FeCl_3$ ⟶ :Cl–Cl(+)–$FeCl_3^-$

Lewis base Lewis acid electrophile

[2] benzene + :Cl–Cl(+)–$FeCl_3^-$ ⟶ resonance-stabilized carbocation + $FeCl_4^-$

resonance-stabilized carbocation

[3] carbocation + :Cl–$FeCl_3^-$ ⟶ chlorobenzene + HCl + $FeCl_3$

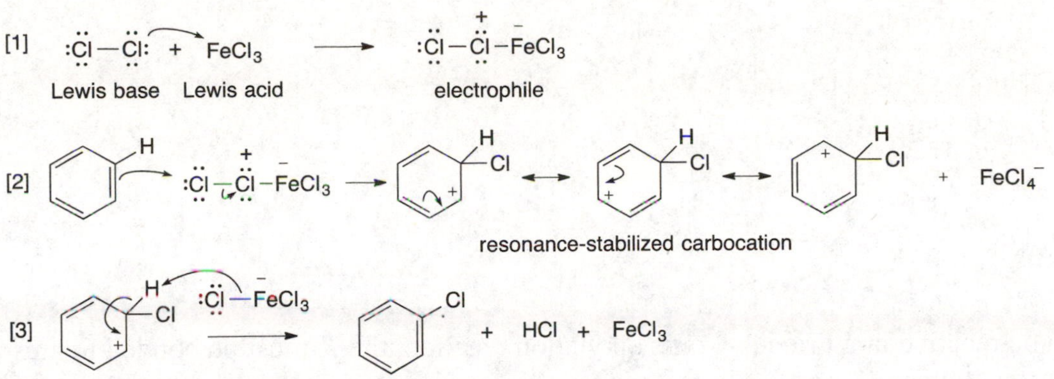

18.4 There are two parts in the mechanism. The first part is formation of an electrophile. The second part is a two-step substitution reaction.

A = R–C₆H₄–H

[1] SO_3 + H–OSO_3H ⟶ $^+SO_3H$ + HSO_4^-

electrophile

[2] R–C₆H₅ + $^+SO_3H$ ⟶ resonance-stabilized carbocation

resonance-stabilized carbocation

[3] carbocation + HSO_4^- ⟶ R–C₆H₄–SO_3H + H_2SO_4

B

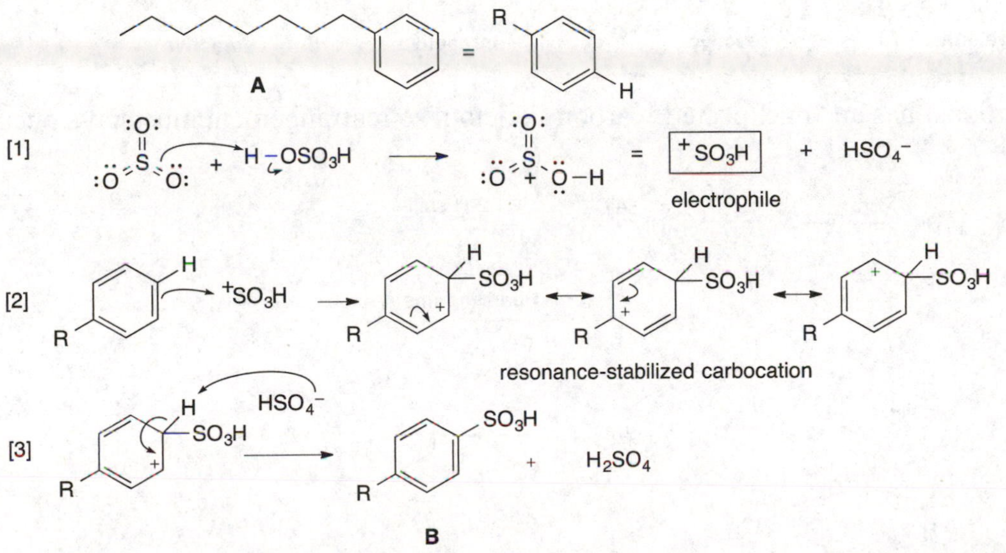

18.5 Friedel–Crafts alkylation results in the transfer of an alkyl group from a halogen to a benzene ring. In Friedel–Crafts acylation an acyl group is transferred from a halogen to a benzene ring.

a. benzene + 2-chloropropane —$AlCl_3$⟶ isopropylbenzene

b. benzene + chlorocyclohexane —$AlCl_3$⟶ cyclohexylbenzene

c. benzene + propanoyl chloride —$AlCl_3$⟶ propiophenone

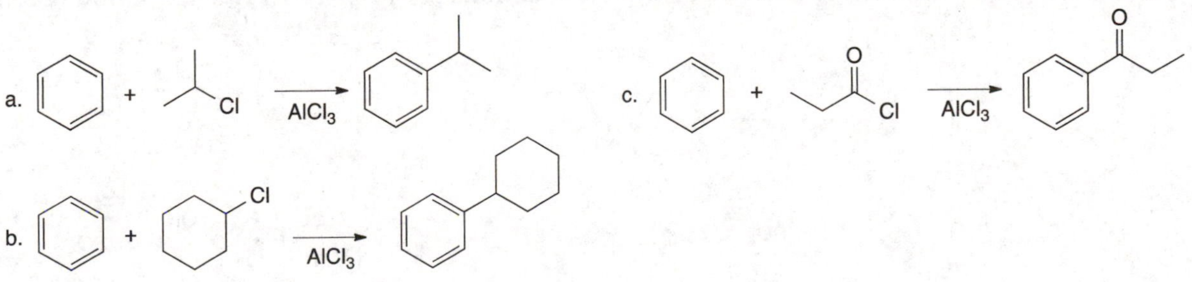

18.6 Remember that an acyl group is transferred from a Cl atom to a benzene ring. To draw the acid chloride, substitute a Cl for the benzene ring.

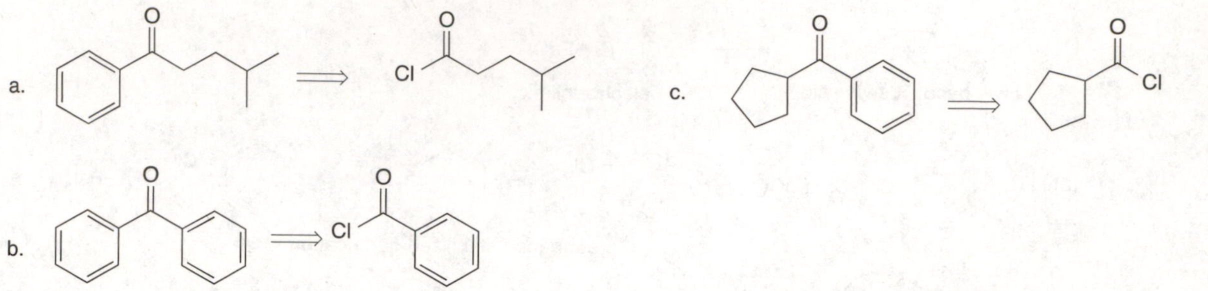

a. ⟹ Cl

c. ⟹ Cl

b. ⟹ Cl

18.7 To be reactive in a Friedel–Crafts alkylation reaction, the X must be bonded to an sp^3 hybridized carbon atom.

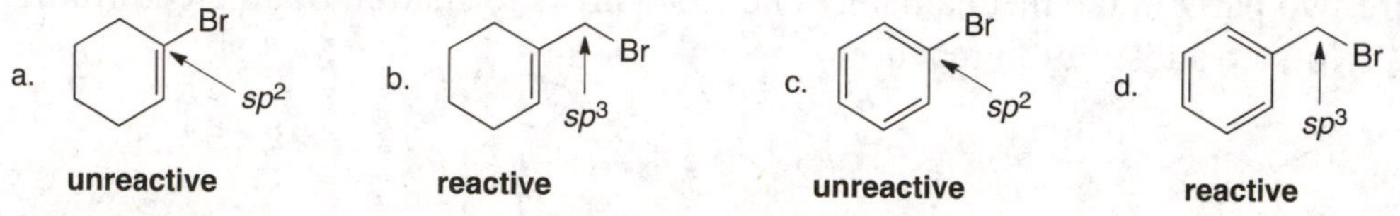

a. Br, sp^2 — **unreactive**

b. Br, sp^3 — **reactive**

c. Br, sp^2 — **unreactive**

d. Br, sp^3 — **reactive**

18.8 The product has an "unexpected" carbon skeleton, so rearrangement must have occurred.

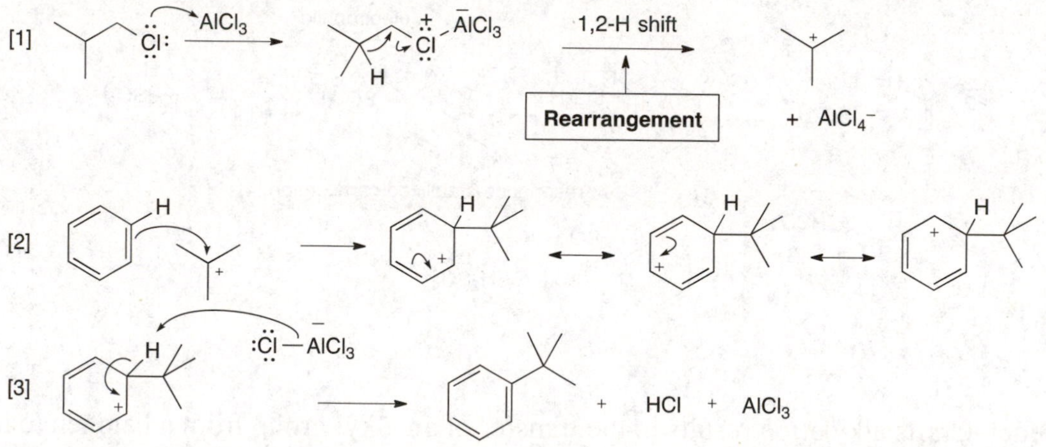

[1] Cl: + $AlCl_3$ → H, Cl, $AlCl_3$ —1,2-H shift→ (**Rearrangement**) + $AlCl_4^-$

[2] H → H ⟷ H ⟷ H

[3] H, :Cl—$AlCl_3$ → + HCl + $AlCl_3$

18.9 Both alkenes and alcohols can form carbocations for Friedel–Crafts alkylation reactions.

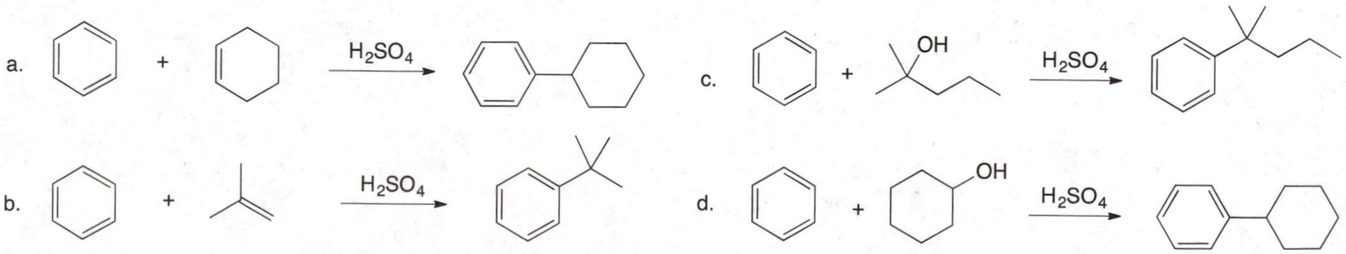

a. + —H_2SO_4→

b. + —H_2SO_4→

c. + OH —H_2SO_4→

d. + OH —H_2SO_4→

18.10

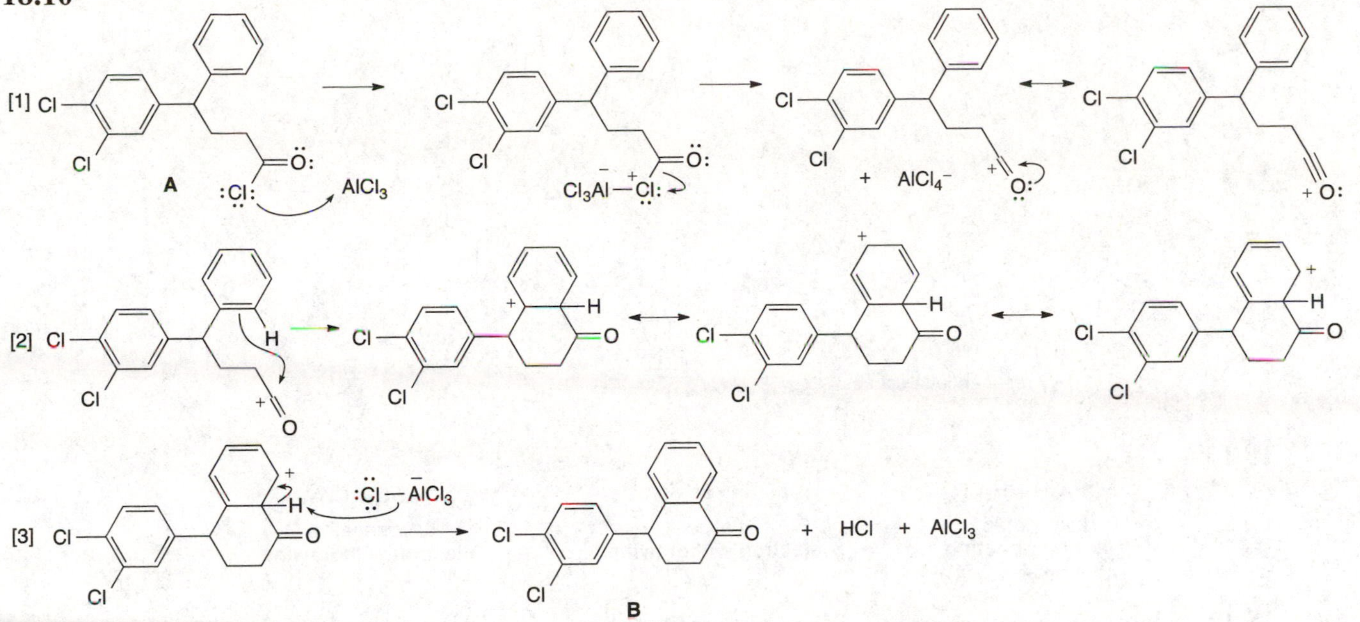

[1] Cl Cl A :Cl: Ö: $AlCl_3$ → Cl Cl Ö: $Cl_3\bar{A}l—\overset{+}{Cl}:$ → Cl Cl + $AlCl_4^-$ $\overset{+}{O}$: ↔ Cl Cl + O:

[2] Cl Cl H + O → Cl Cl + H O ↔ Cl Cl + H O ↔ Cl Cl + H O

[3] Cl Cl + H O :Cl—$\bar{A}lCl_3$ → Cl Cl O B + HCl + $AlCl_3$

18.11 In parts (b) and (c), a 1,2-shift occurs to afford a rearrangement product.

a. →

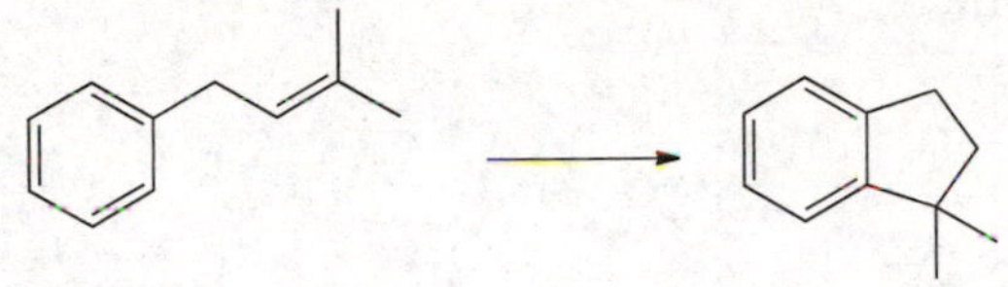

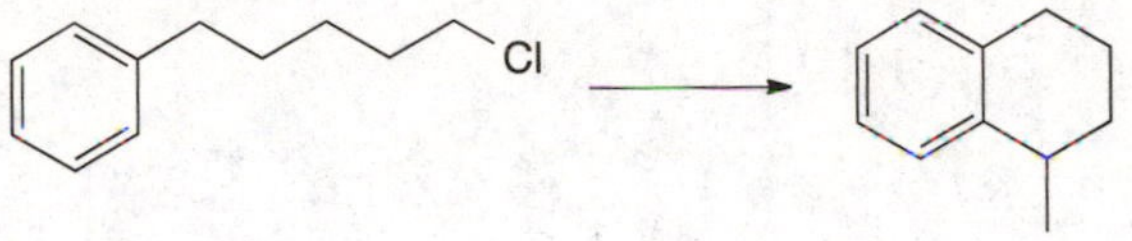

b. OH →

c. Cl →

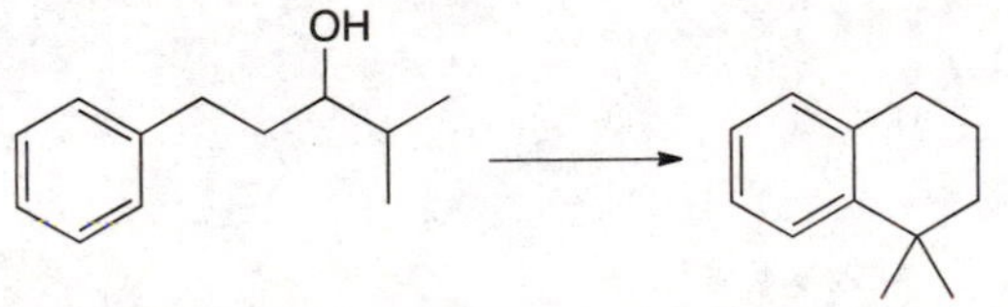

18.12

a. PPO → + ↕ +

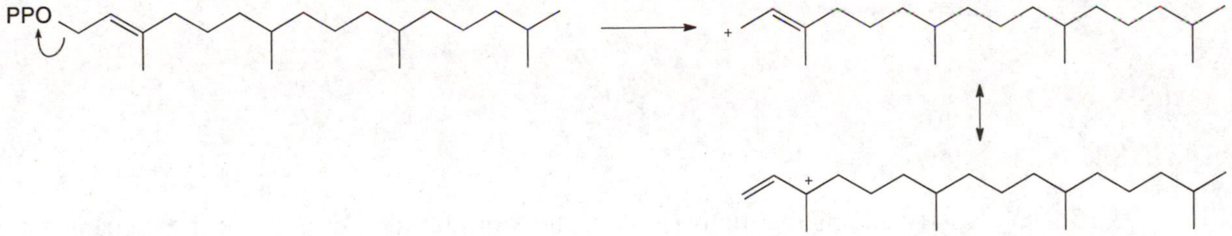

b.

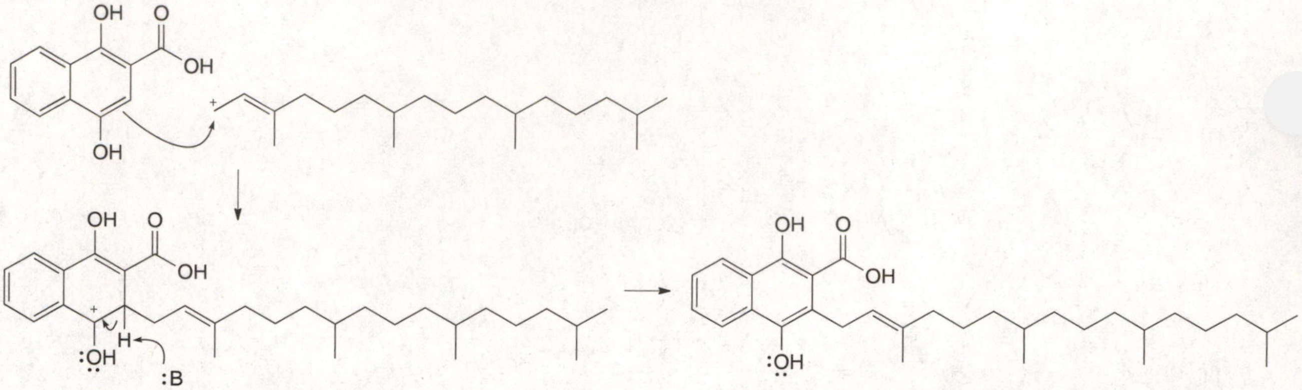

OH O OH OH

[+ 7 resonance structures]

:B

+ HB+ X

18.13

a. $-CH_2CH_2CH_2CH_3$
alkyl group
electron donating

b. $-Br$
halide
electron withdrawing

c. $-OCH_2CH_3$
electronegative O
electron withdrawing

18.14 Electron-donating groups place a negative charge in the benzene ring. Draw the resonance structures to show how $-OCH_3$ puts a negative charge in the ring. Electron-withdrawing groups place a positive charge in the benzene ring. Draw the resonance structures to show how $-COCH_3$ puts a positive charge in the ring.

a.

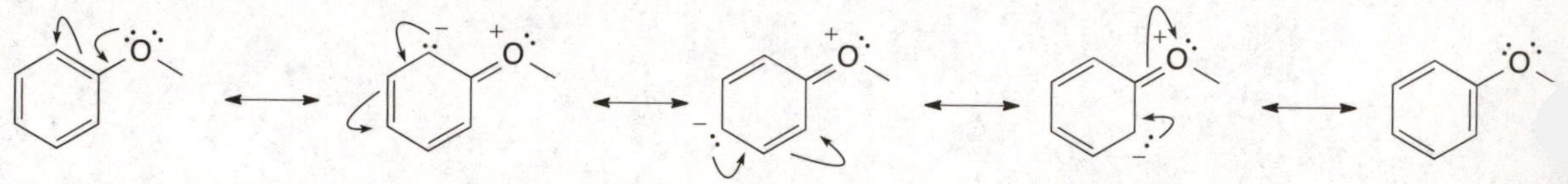

b.

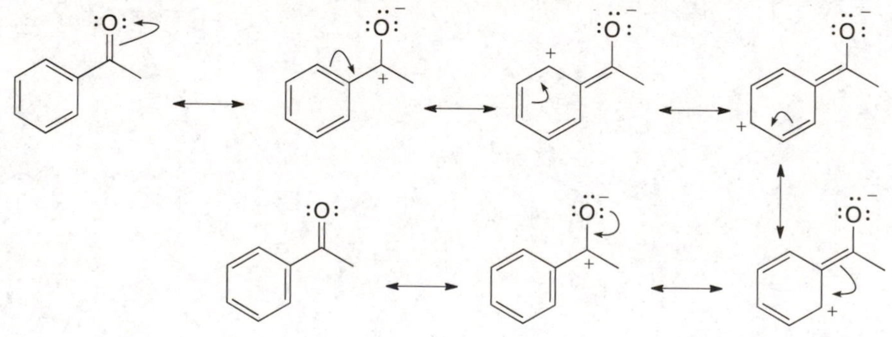

18.15 To classify each substituent, look at the atom bonded directly to the benzene ring. All R groups and Z groups (except halogens) are electron donating. All groups with a positive charge, δ+, or halogens are electron withdrawing.

a. 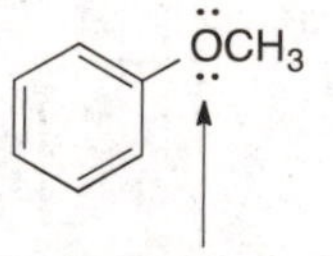 OCH_3
lone pair on O
electron donating

b. 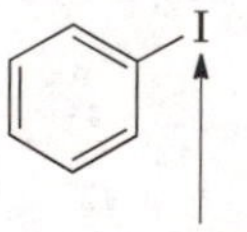 I
halogen
electron withdrawing

c. 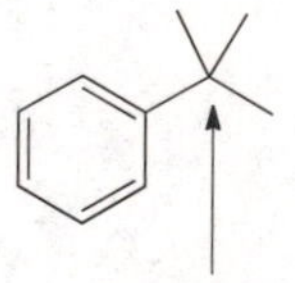
R group
electron donating

18.16 **Electron-donating groups** make the compound ***react faster*** than benzene in electrophilic aromatic substitution. **Electron-withdrawing groups** make the compound ***react more slowly*** than benzene in electrophilic aromatic substitution.

a. electron withdrawing
reacts slower

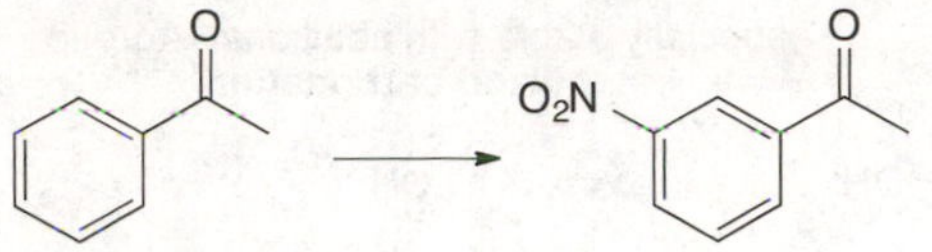

O_2N

b. CN

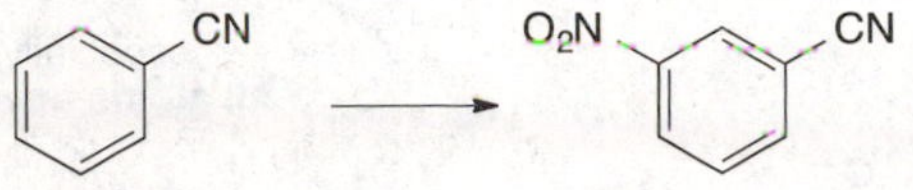

electron withdrawing
reacts slower

O_2N CN

c. OH

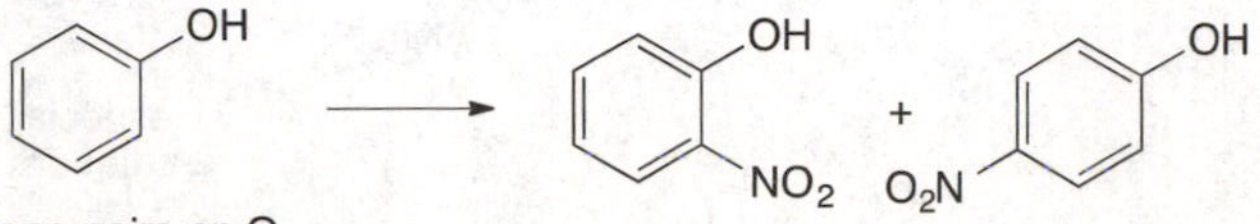

lone pairs on O
electron donating
reacts faster

OH NO_2 + OH O_2N

d. Cl

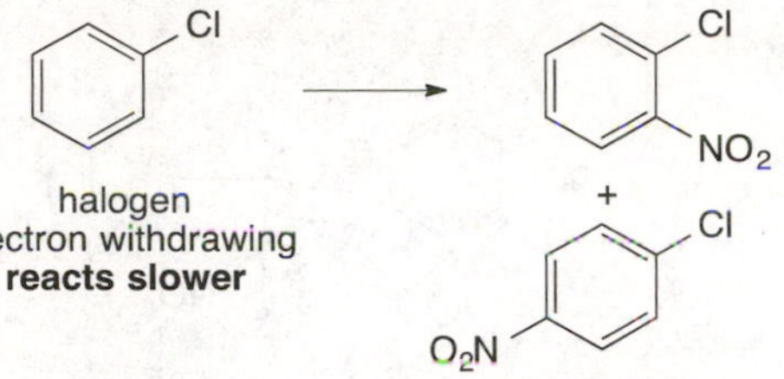

halogen
electron withdrawing
reacts slower

Cl NO_2 + Cl O_2N

e.

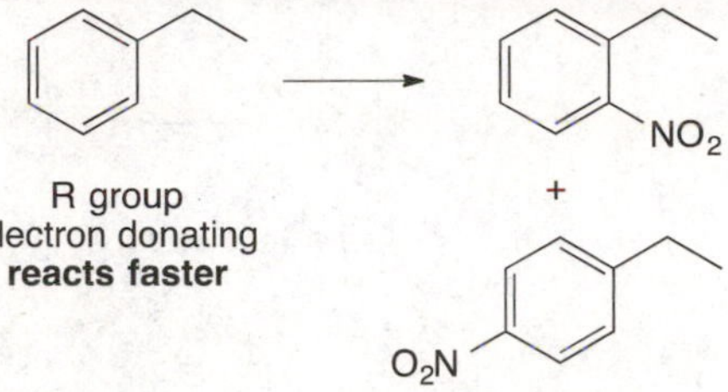

R group
electron donating
reacts faster

NO_2 + O_2N

18.17 **Electron-donating groups** make the compound ***more reactive*** than benzene in electrophilic aromatic substitution. **Electron-withdrawing groups** make the compound ***less reactive*** than benzene in electrophilic aromatic substitution.

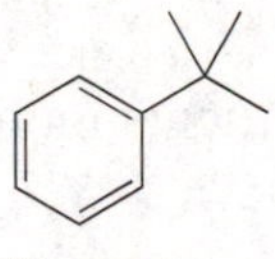

a. R group
electron donating
more reactive

b. OH OH
two OH's
electron donating
more reactive

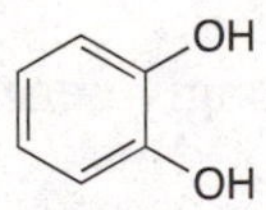

c. O O
C with 2 electronegative O's
electron withdrawing
less reactive

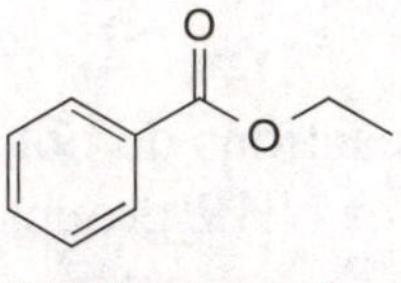

d. $\overset{+}{N}(CH_3)_3$
electron withdrawing
less reactive

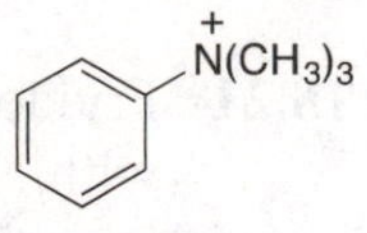

18.18 Chlorine inductively withdraws electron density and decreases the rate of electrophilic aromatic substitution. The closer the Cl is to the ring, the larger the effect it has. The larger the number of Cl's, the larger the effect.

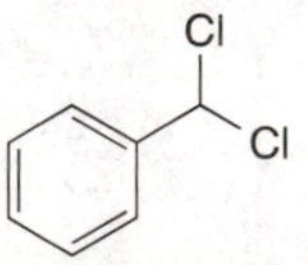

Cl Cl
least reactive

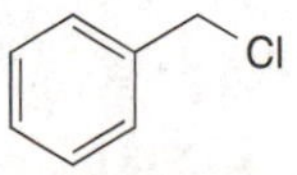

Cl
intermediate reactivity

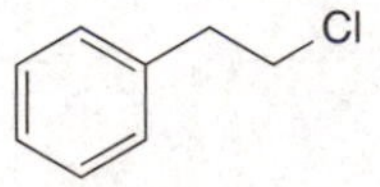

Cl
most reactive

18.19 Especially stable resonance structures have all atoms with an octet. Carbocations with additional electron donor R groups are also more stable structures. Especially unstable resonance structures have adjacent like charges.

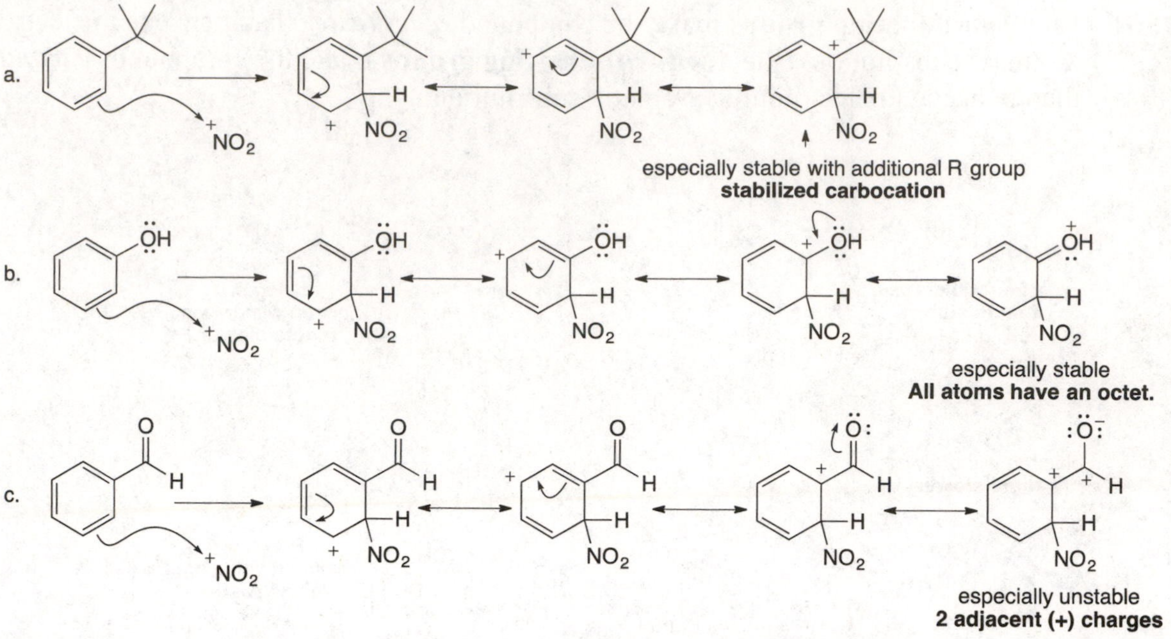

a. especially stable with additional R group
stabilized carbocation

b. especially stable
All atoms have an octet.

c. especially unstable
2 adjacent (+) charges

18.20 Polyhalogenation occurs with highly activated benzene rings containing OH, NH_2, and related groups with a catalyst.

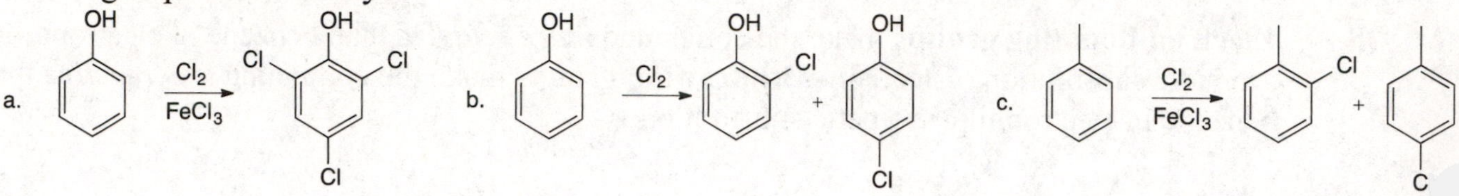

a. Cl_2, $FeCl_3$

b. Cl_2

c. Cl_2, $FeCl_3$

18.21 Friedel–Crafts reactions do not occur with strongly deactivating substituents including NO_2, or with NH_2, NR_2, or NHR groups.

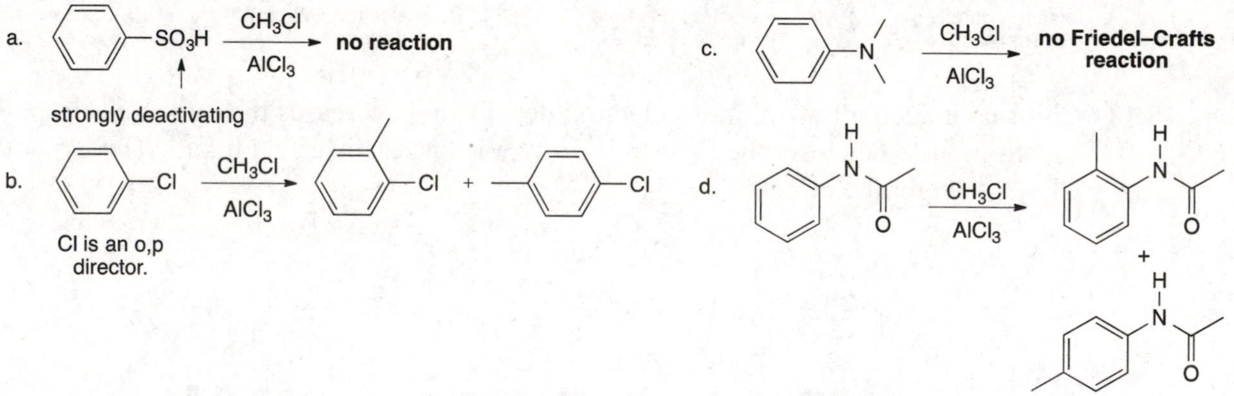

a. CH_3Cl, $AlCl_3$ → **no reaction**
strongly deactivating

b. CH_3Cl, $AlCl_3$
Cl is an o,p director.

c. CH_3Cl, $AlCl_3$ → **no Friedel–Crafts reaction**

d. CH_3Cl, $AlCl_3$

18.22 To draw the product of a reaction with these disubstituted benzene derivatives and HNO_3, H_2SO_4, remember the following:

- If the two directing effects reinforce each other, the new substituent will be on the position reinforced by both.
- If the directing effects oppose each other, the stronger activator wins.
- No substitution occurs between two meta substituents.

a. 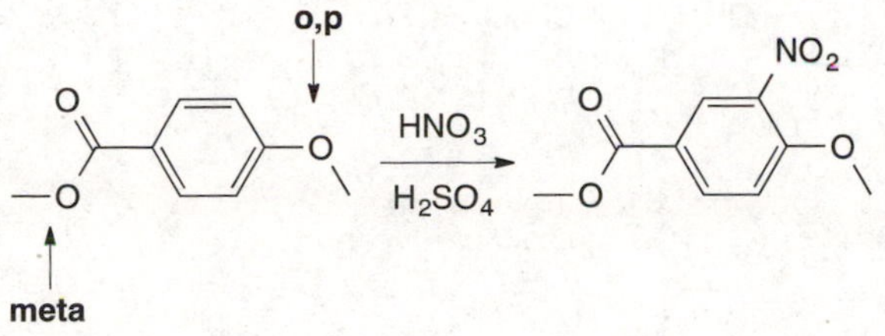 meta; o,p; HNO_3, H_2SO_4; NO_2

c. 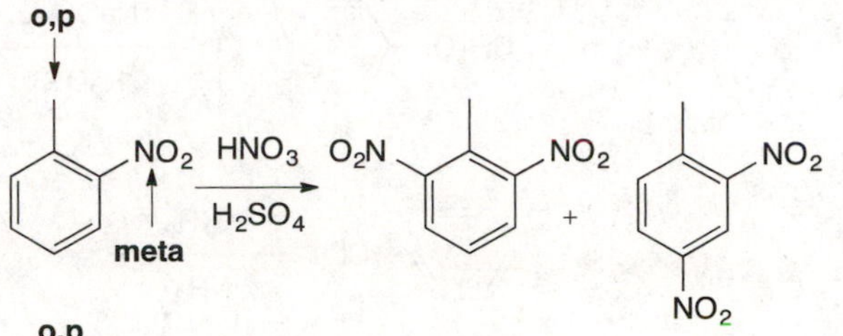 o,p; NO_2; meta; HNO_3, H_2SO_4; O_2N, NO_2 + NO_2, NO_2

b. 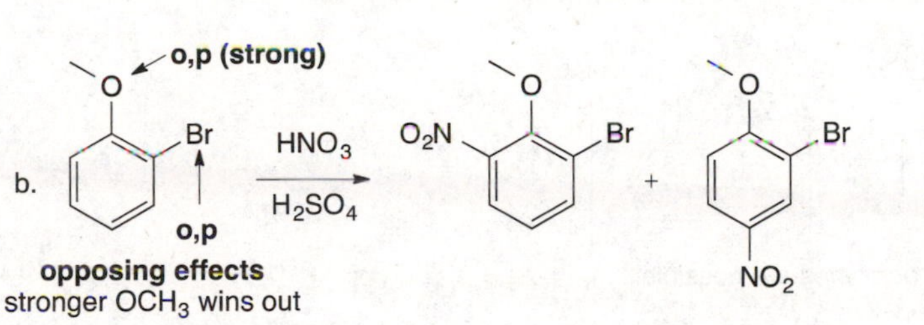 o,p (strong); Br; o,p; **opposing effects** stronger OCH_3 wins out; HNO_3, H_2SO_4; O_2N, Br + Br, NO_2

d. 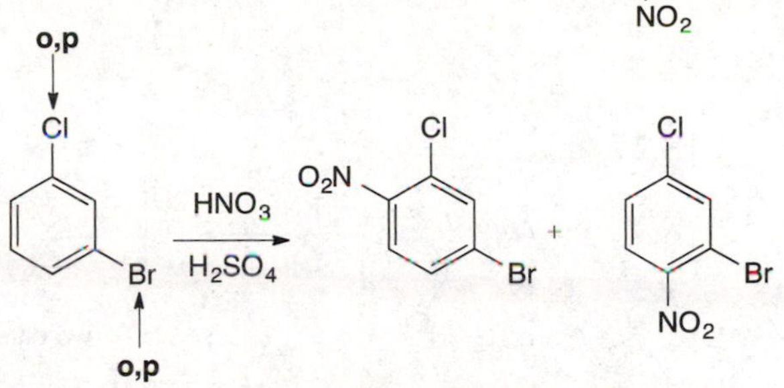 o,p; Cl; Br; o,p; HNO_3, H_2SO_4; O_2N, Cl, Br + Cl, Br, NO_2

18.23

a. 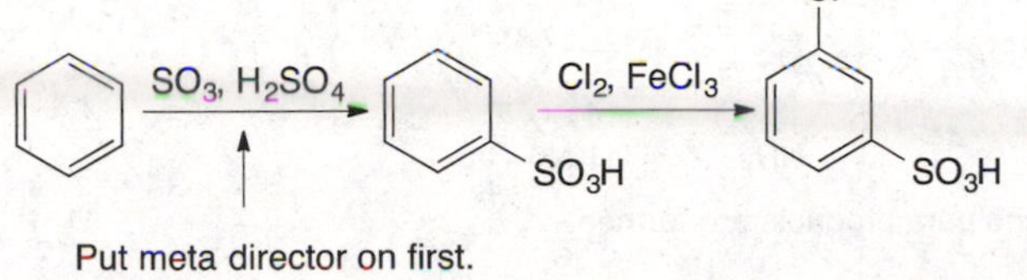 SO_3, H_2SO_4 → SO_3H; Cl_2, $FeCl_3$ → Cl, SO_3H

Put meta director on first.

c. 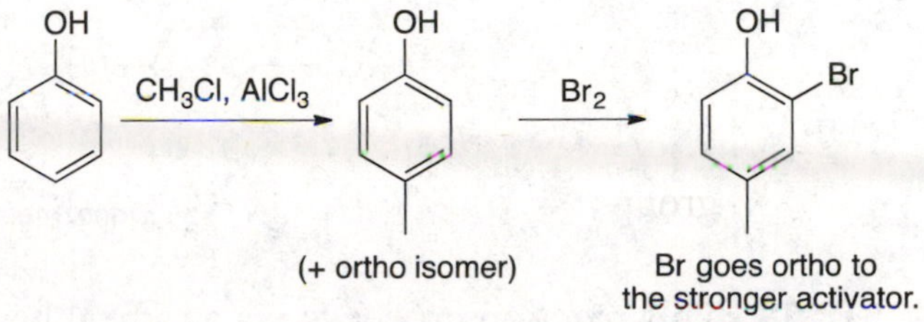 OH; CH_3Cl, $AlCl_3$ → OH (+ ortho isomer); Br_2 → OH, Br

Br goes ortho to the stronger activator.

b. 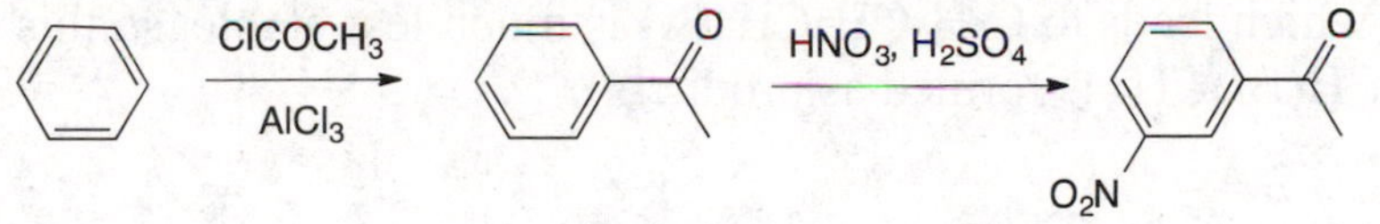 $ClCOCH_3$, $AlCl_3$ → O; HNO_3, H_2SO_4 → O, O_2N

18.24

a. 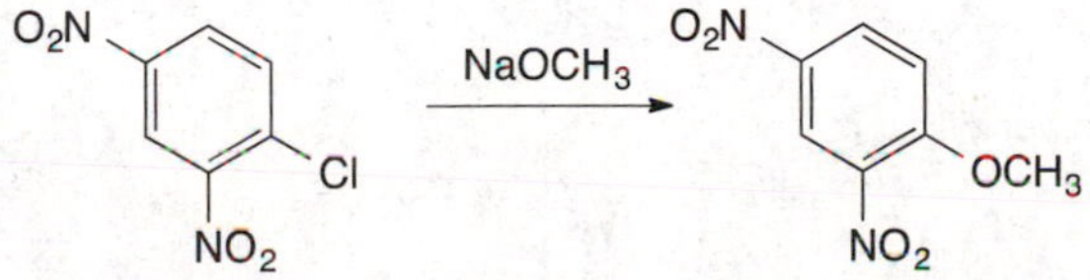 O_2N, Cl, NO_2; $NaOCH_3$ → O_2N, OCH_3, NO_2

b. 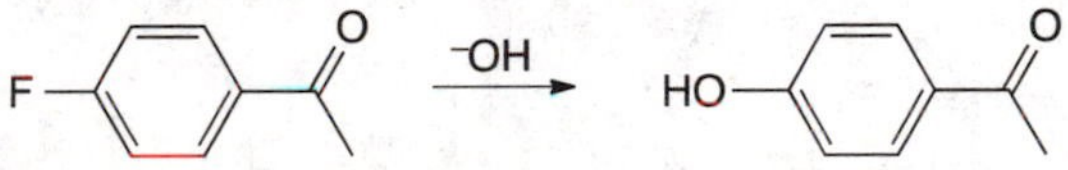 F; O; ^-OH → HO; O

18.25

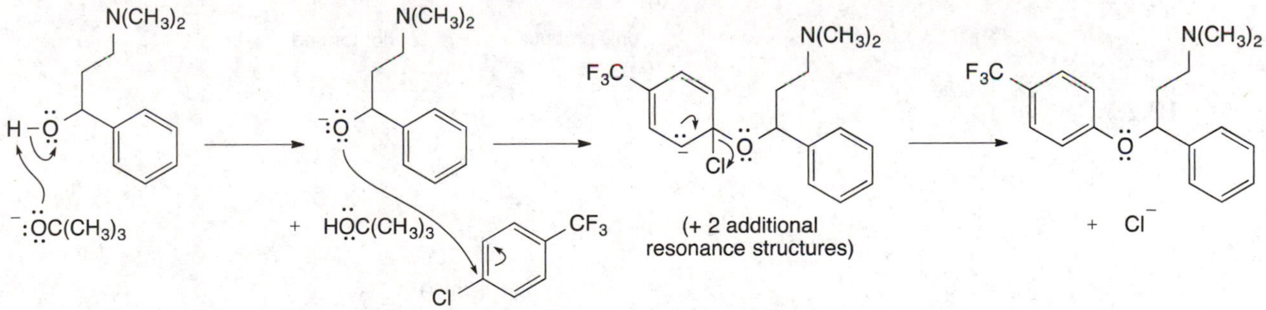

$N(CH_3)_2$; H–Ö; $^-$:ÖC$(CH_3)_3$ → $N(CH_3)_2$; $^-$:Ö + HÖC$(CH_3)_3$; Cl, CF_3 → F_3C, $N(CH_3)_2$, Cl (+ 2 additional resonance structures) → F_3C, $N(CH_3)_2$ + Cl^-

18.26

a. 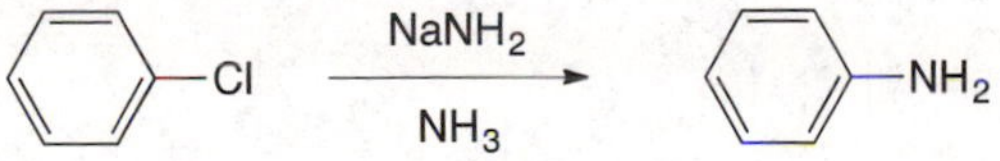 Cl; $NaNH_2$, NH_3 → NH_2

b. 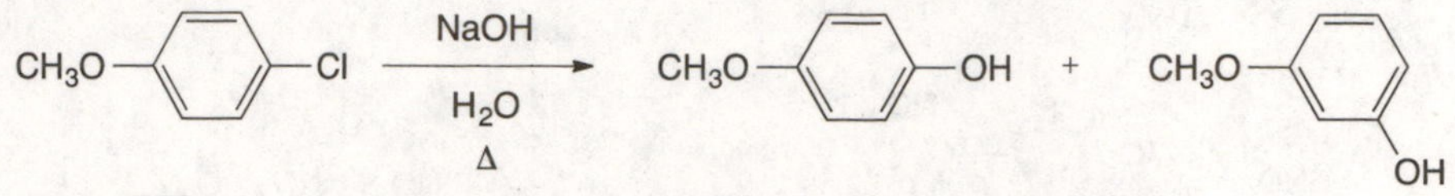 CH_3O–C₆H₄–Cl $\xrightarrow[H_2O, \Delta]{NaOH}$ CH_3O–C₆H₄–OH + CH_3O–C₆H₄–OH

c. 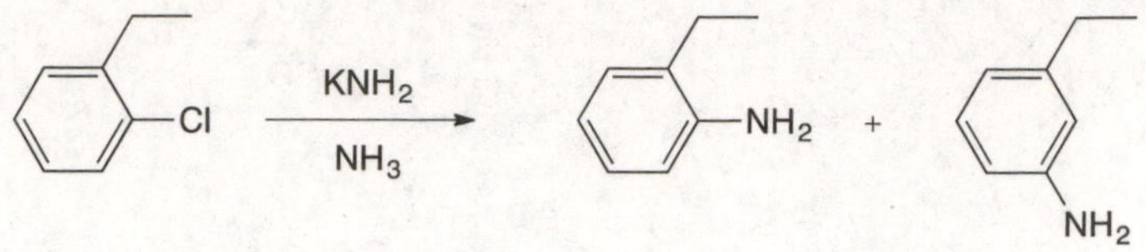 $\xrightarrow[NH_3]{KNH_2}$ NH_2 + NH_2

18.27

$\xrightarrow[NH_3]{KNH_2}$ [ ] + [ ]

Cl

Two different benzynes are possible.

NH_2 + +

NH_2 H_2N

Ortho, meta, and para products are formed.

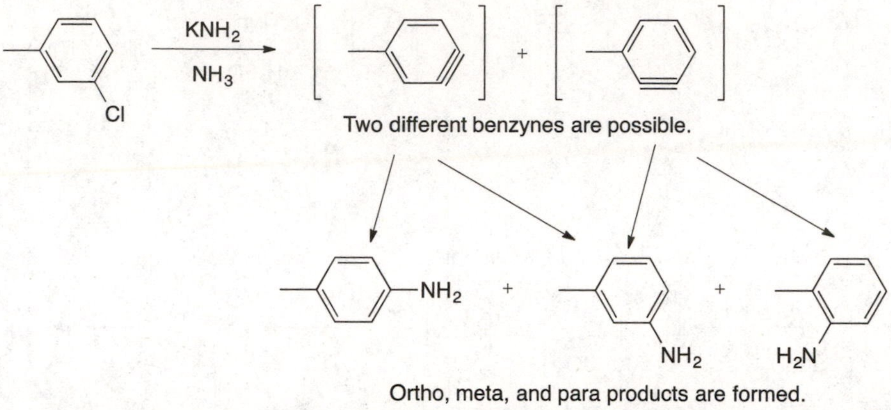

18.28 This reaction proceeds via a radical bromination mechanism and two radicals are possible: **A** (2° and benzylic) and **B** (1°). **B** (which leads to $C_6H_5CH_2CH_2Br$) is much less stable, so this radical is not formed and only $C_6H_5CH(Br)CH_3$ is formed as product.

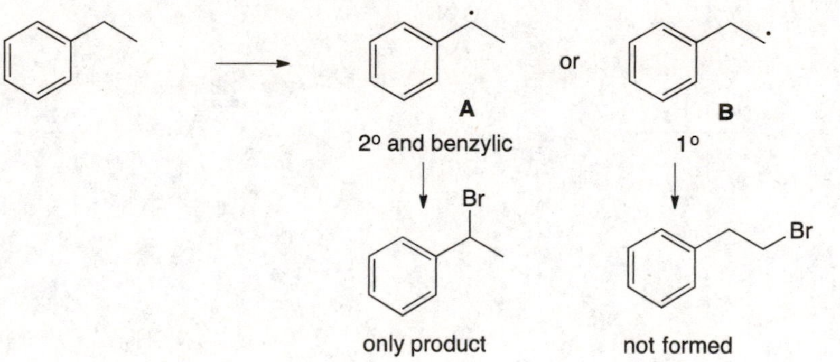

or

A **B**

2° and benzylic 1°

Br Br

only product not formed

18.29

a. 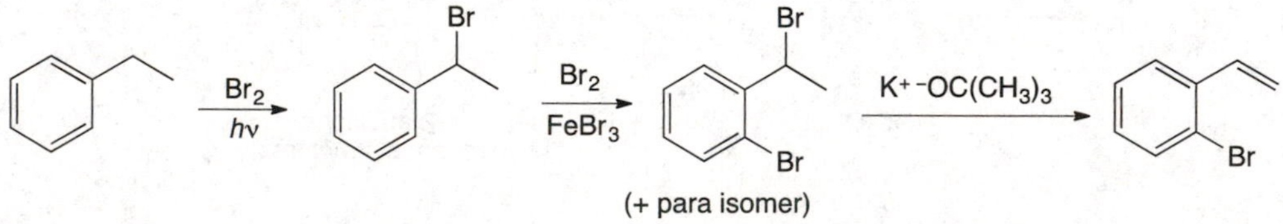 $\xrightarrow[h\nu]{Br_2}$ Br $\xrightarrow[FeBr_3]{Br_2}$ Br, Br (+ para isomer) $\xrightarrow{K^{+}\,{}^{-}OC(CH_3)_3}$ Br

b. 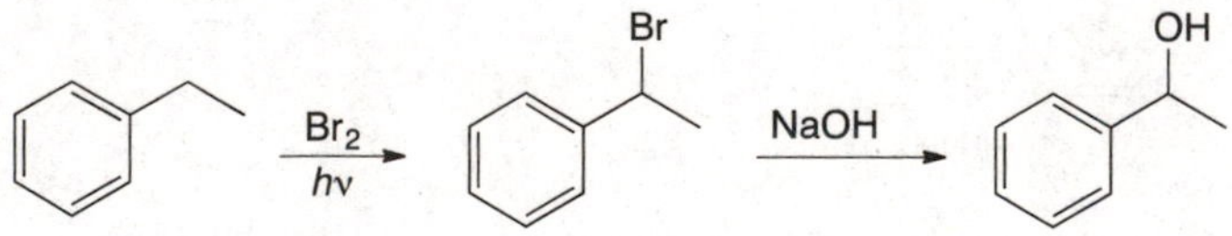 $\xrightarrow[h\nu]{Br_2}$ Br $\xrightarrow{NaOH}$ OH

c. $\xrightarrow[h\nu]{Br_2}$ Br $\xrightarrow{K^+ \,^-OC(CH_3)_3}$ $\xrightarrow{mCPBA}$ O

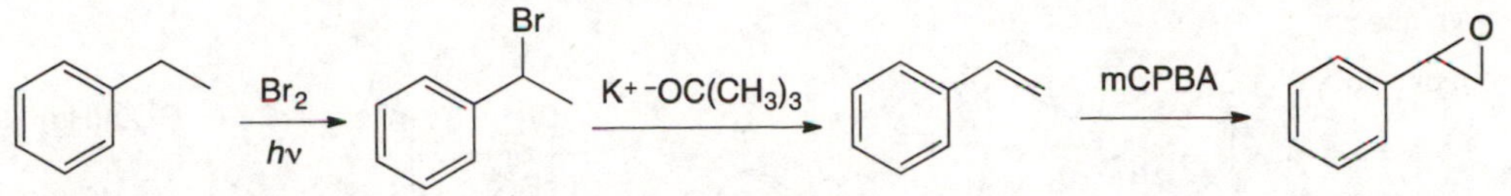

d. $\xrightarrow[[2]\ H_2O_2,\ HO^-]{[1]\ BH_3}$ OH

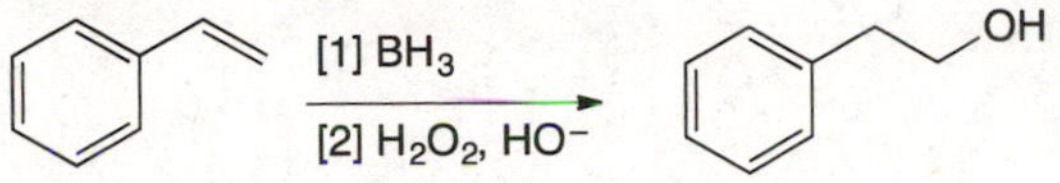

(from c.)

18.30 First use an acylation reaction, and then reduce the carbonyl group to form the alkyl benzenes.

a. O, Cl; $AlCl_3$; O; Zn(Hg) + HCl

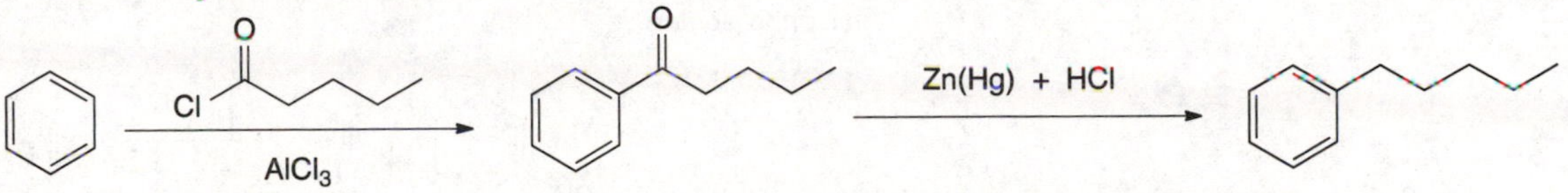

b. O, Cl; $AlCl_3$; O; Zn(Hg) + HCl

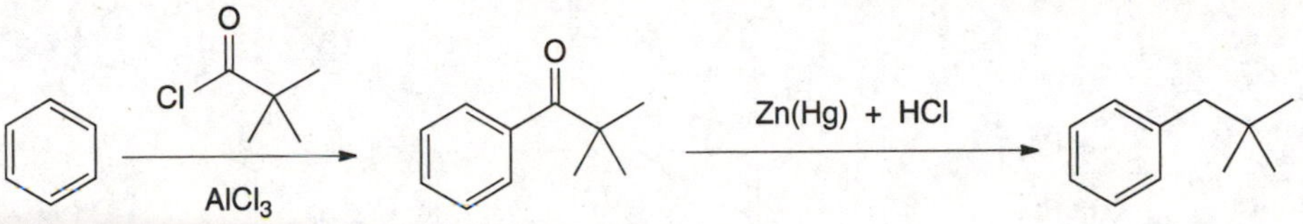

18.31

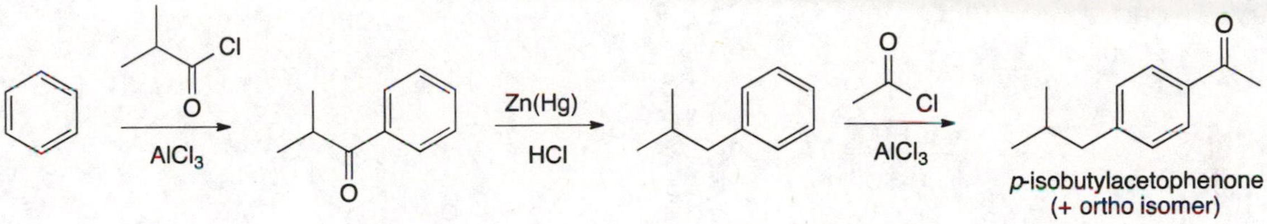

Cl, O; $AlCl_3$; O; Zn(Hg), HCl; O, Cl; $AlCl_3$; O

p-isobutylacetophenone
(+ ortho isomer)

18.32

a. $\xrightarrow[AlCl_3]{CH_3Cl}$ $\xrightarrow{KMnO_4}$ CO_2H

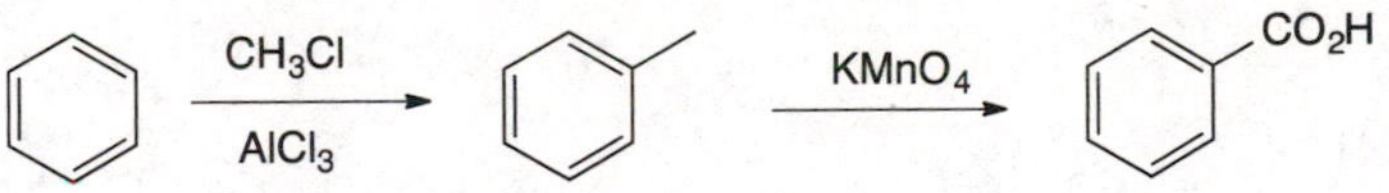

b. $\xrightarrow[H_2SO_4]{HNO_3}$ NO_2 $\xrightarrow[Pd\text{-}C]{H_2}$ NH_2

c. $\xrightarrow[AlCl_3]{CH_3Cl}$ $\xrightarrow[FeBr_3]{Br_2}$ Br $\xrightarrow{KMnO_4}$ CO_2H, Br

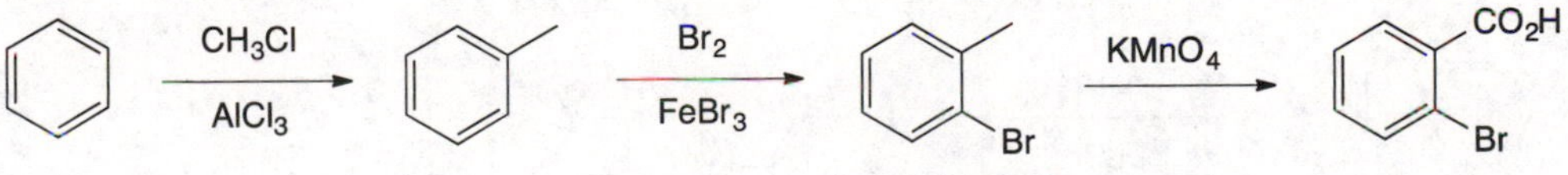

(+ para isomer)

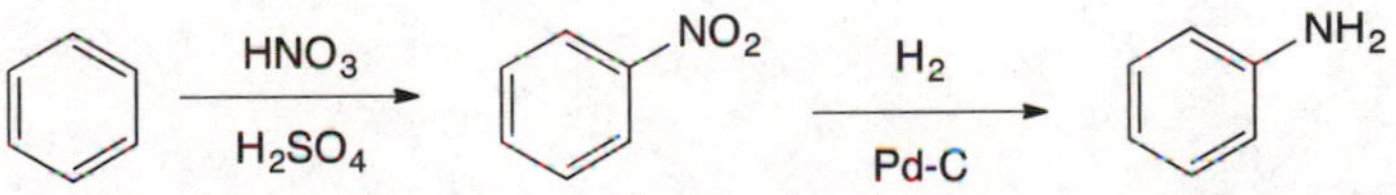

18.33

a. $\xrightarrow[AlCl_3]{ClCH_2CH_3}$ O, Cl, $AlCl_3$; O; $\xrightarrow[H_2SO_4]{SO_3}$ O, HO_3S

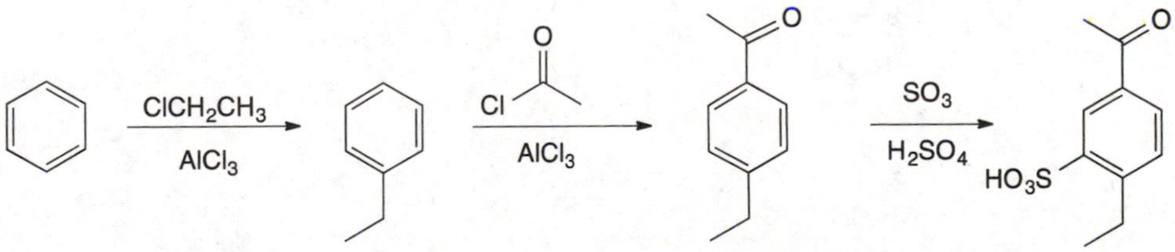

(+ ortho isomer)

b.

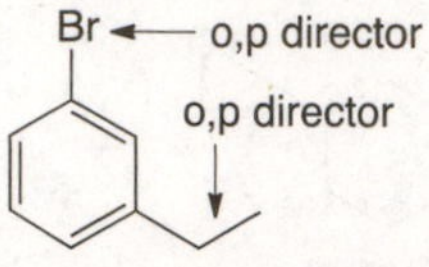

Br ← o,p director

o,p director

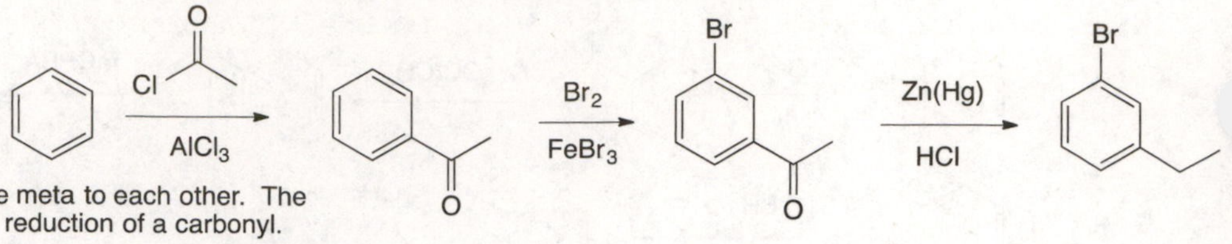

Cl

$AlCl_3$

Br_2

$FeBr_3$

Br

Zn(Hg)

HCl

Br

Both are o,p directors, but they are meta to each other. The alkyl group must be obtained by reduction of a carbonyl.

c.

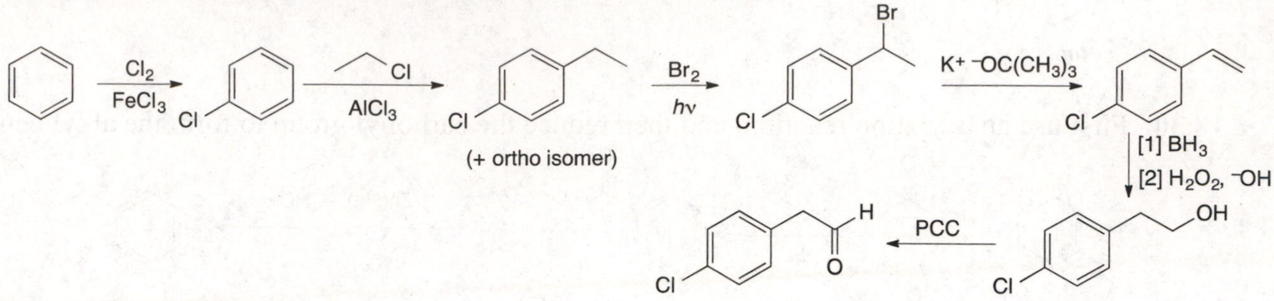

Cl_2

$FeCl_3$

Cl

Cl

$AlCl_3$

Cl

(+ ortho isomer)

Br_2

$h\nu$

Br

Cl

$K^+ \, ^-OC(CH_3)_3$

Cl

[1] BH_3

[2] H_2O_2, ^-OH

OH

Cl

PCC

H

O

Cl

18.34

A

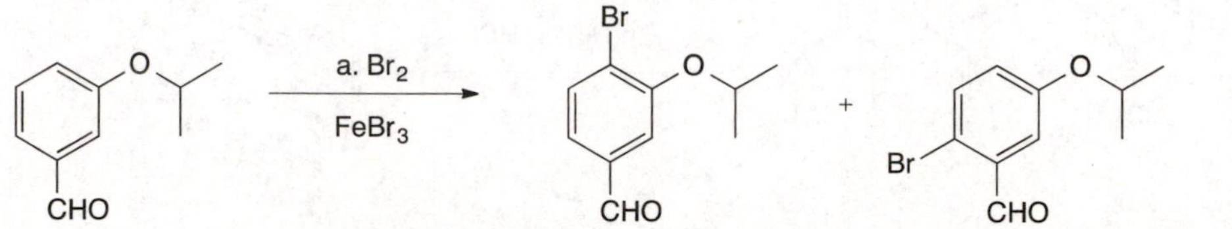

CHO

a. Br_2

$FeBr_3$

Br

CHO

\+

Br

CHO

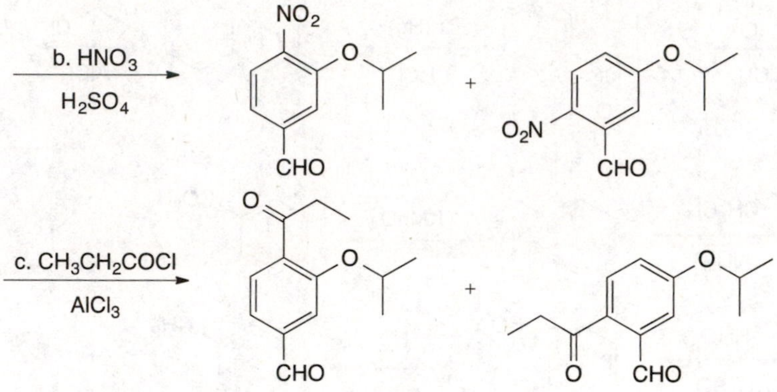

b. HNO_3

H_2SO_4

NO_2

CHO

\+

O_2N

CHO

c. CH_3CH_2COCl

$AlCl_3$

CHO

\+

CHO

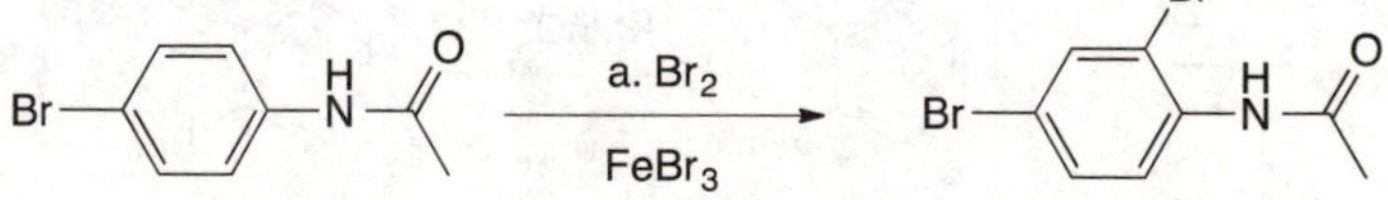

B

Br

H

N

a. Br_2

$FeBr_3$

Br

Br

H

N

b. HNO_3

H_2SO_4

NO_2

Br

H

N

c. CH_3CH_2COCl

$AlCl_3$

Br

H

N

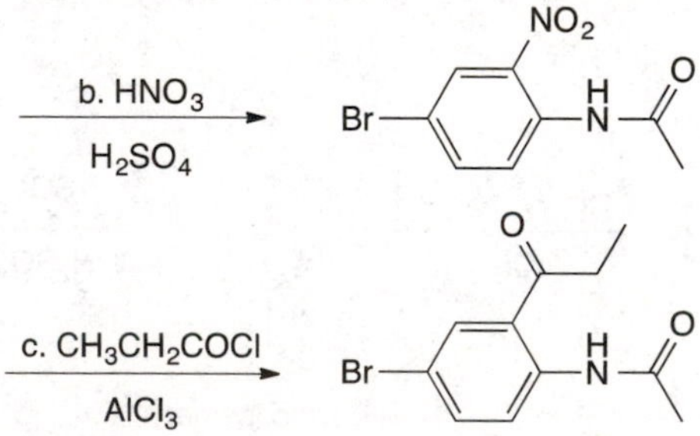

18.35 Intramolecular Friedel–Crafts acylation occurs on the more activated aromatic ring.

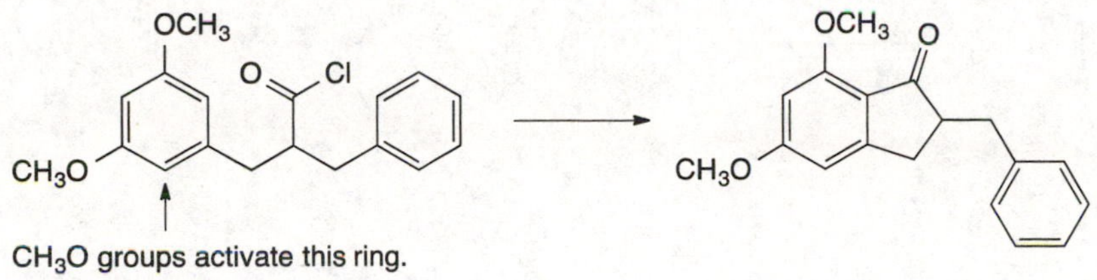

OCH_3 O   Cl CH_3O → OCH_3 O CH_3O

CH_3O groups activate this ring.

18.36 OH is an ortho, para director.

a. OH —[HNO_3 / H_2SO_4]→ OH, NO_2 + OH, NO_2 —[Sn / HCl]→ OH, NH_2 + OH, NH_2

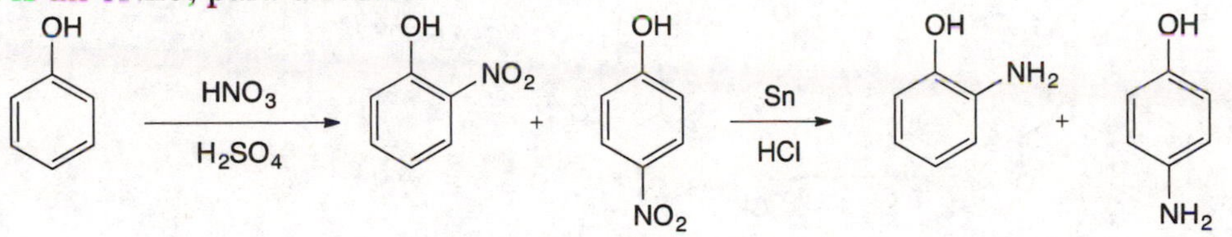

b. OH —[O, Cl / $AlCl_3$]→ OH O + OH, O —[Zn(Hg) / HCl]→ OH + OH

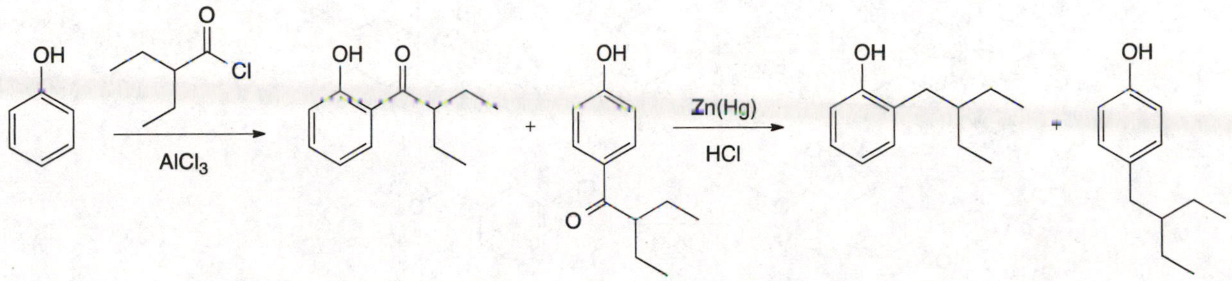

c. OH —[Cl / $AlCl_3$]→ OH + OH —[Br_2 / $h\nu$]→ OH Br + OH, Br

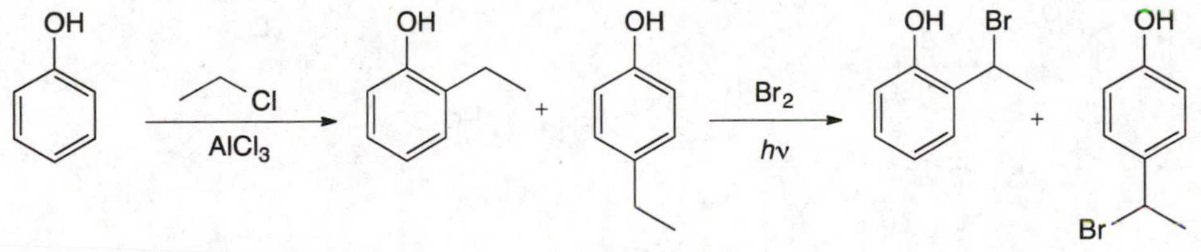

d. OH —[Cl / $AlCl_3$]→ OH + OH —[$KMnO_4$]→ OH O OH + OH, O OH

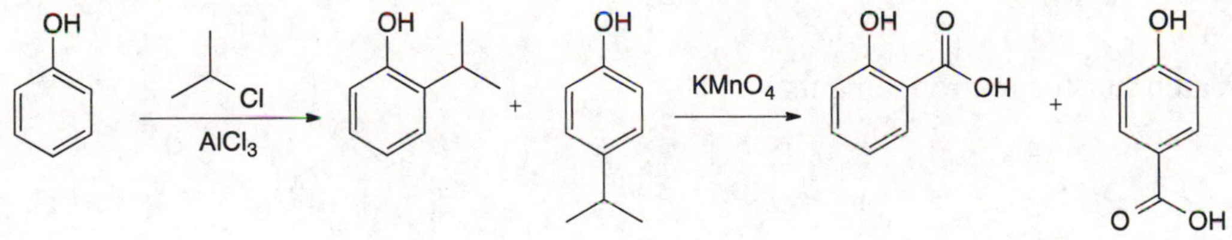

18.37

a. O —[O, Cl / $AlCl_3$]→ No reaction

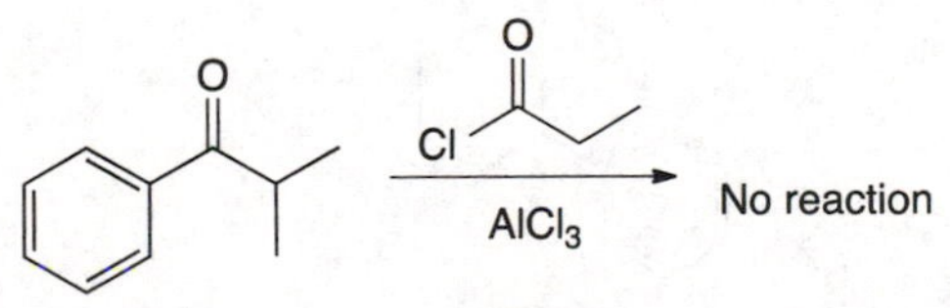

b. N —[O, Cl / $AlCl_3$]→ No Friedel–Crafts reaction

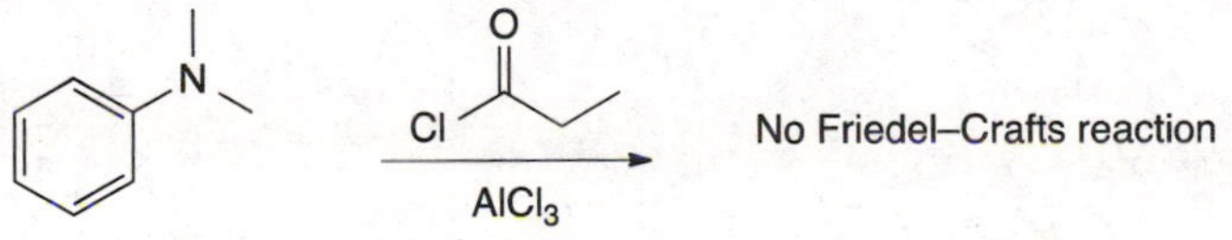

c. 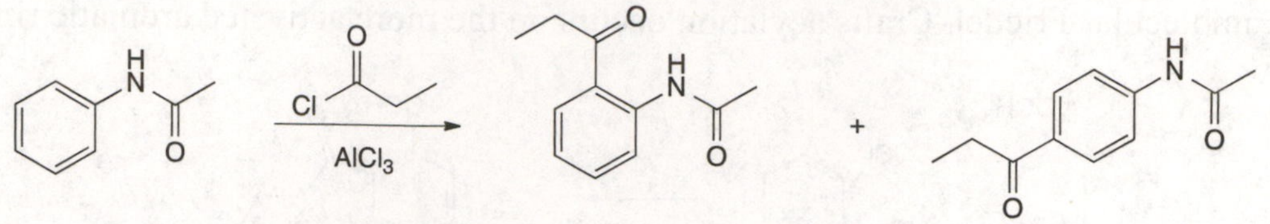 Cl (acyl chloride), $AlCl_3$ → +

18.38

a. 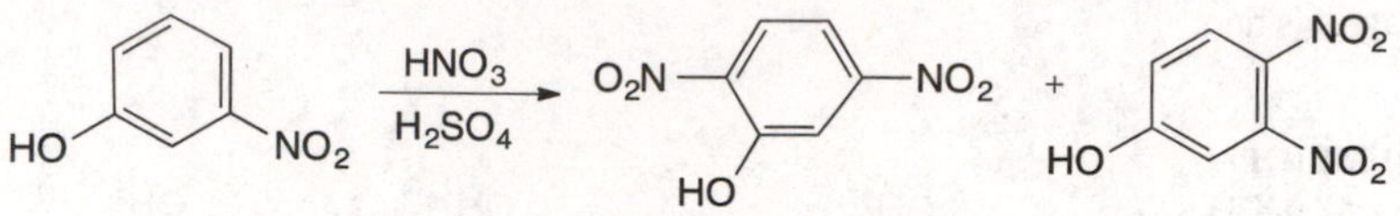 HO, NO_2 → HNO_3, H_2SO_4 → O_2N, NO_2, HO + NO_2, HO, NO_2

b. 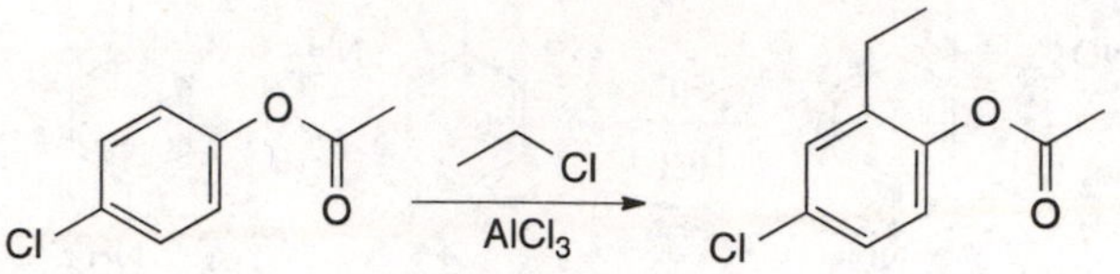 Cl → Cl, $AlCl_3$ → Cl

c. 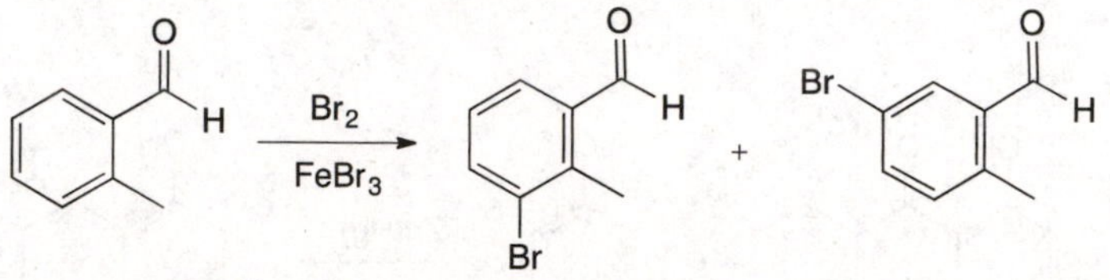 O, H → Br_2, $FeBr_3$ → O, H, Br + Br, O, H

d. 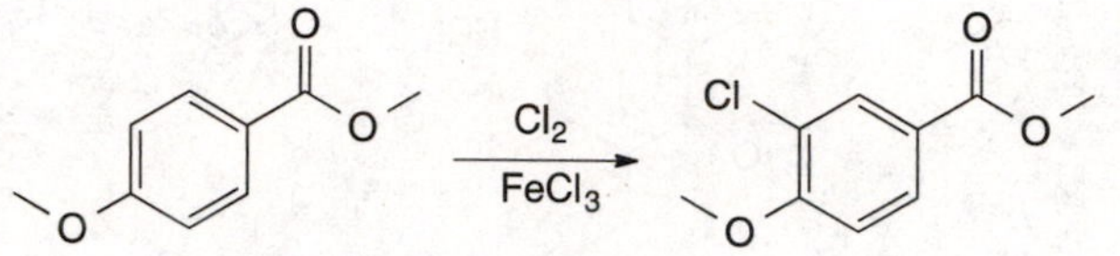 O, O, O → Cl_2, $FeCl_3$ → Cl, O, O, O

e. 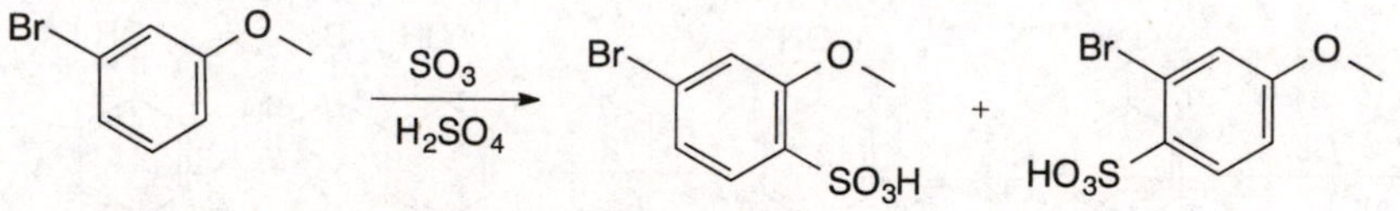 Br, O → SO_3, H_2SO_4 → Br, O, SO_3H + Br, O, HO_3S

f. 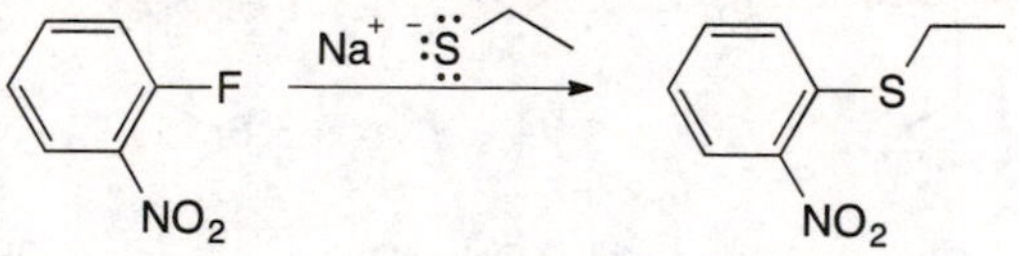 F, NO_2 → Na^+ $^-$:S: → S, NO_2

18.39 Watch out for rearrangements.

a. 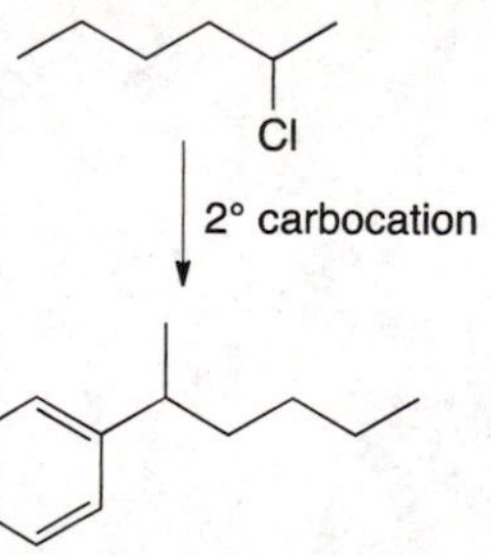 Cl → 2° carbocation

b. 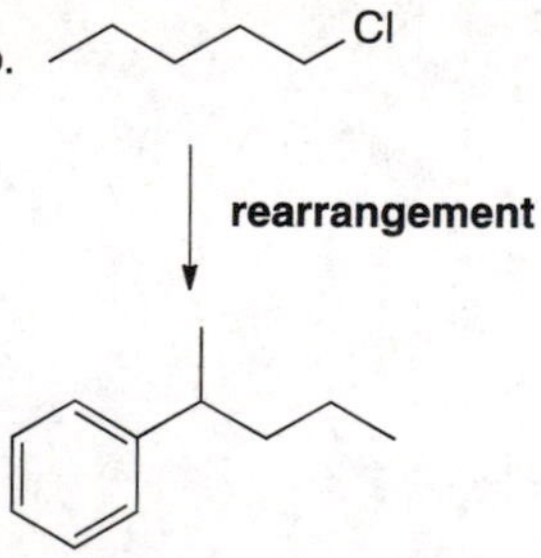 Cl → **rearrangement**

c. 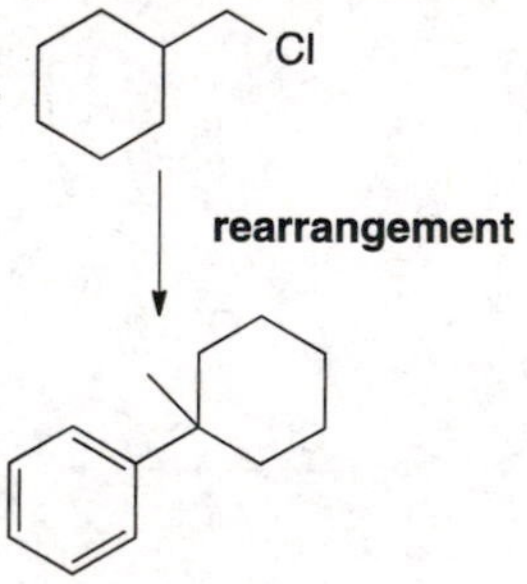 Cl → **rearrangement**

18.40

a. 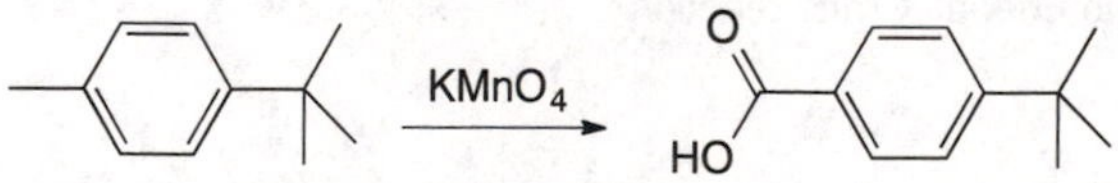 → $KMnO_4$ → O, HO

b. Br_2, $h\nu$; Br ; $KOC(CH_3)_3$; (+ cis isomer)

c. [1] Cl_2, $FeCl_3$; [2] Zn(Hg), HCl ; O ; Cl

d. NO_2 ; O_2N ; Cl ; NO_2 ; CH_3NH_2 ; NO_2 ; O_2N ; $NHCH_3$; NO_2 ; H_2 (excess) ; Pd-C ; NH_2 ; H_2N ; $NHCH_3$; NH_2

18.41

a. C bonded to 2 H's must use acylation followed by reduction. ; O ; Cl ; $AlCl_3$; O ; Zn(Hg) ; HCl

b. C bonded to 1 H can be added directly by alkylation. ; Cl ; $AlCl_3$

c. Ethyl group can be introduced by two methods. ; O ; Cl ; $AlCl_3$; O ; Zn(Hg) ; HCl ; Method [1] ; CH_3CH_2Cl ; $AlCl_3$; Method [2]

18.42

HO ; N ; F ; F ; N ; H ; KOH ; O ; N ; F ; F ; N ; H ; KOH ; O ; N ; F ; N ; H ; **A**

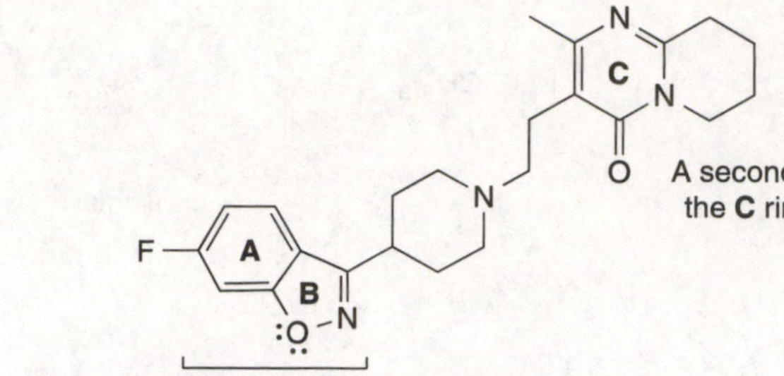

N
C
N
O
N
F
A
B
:O:
N

Rings **A** and **B** contain 10 π electrons, eight from the double bonds and two from O, making this ring system aromatic.

A second resonance structure can be drawn for the **C** ring, showing that it has six π electrons.

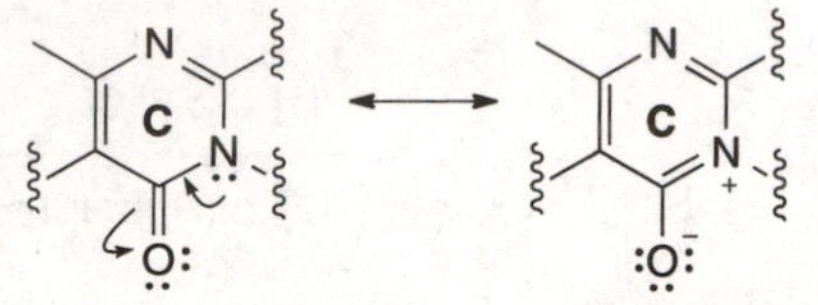

N
C
N
:O:
N
C
N
+
:O:⁻

6 π electrons

18.43

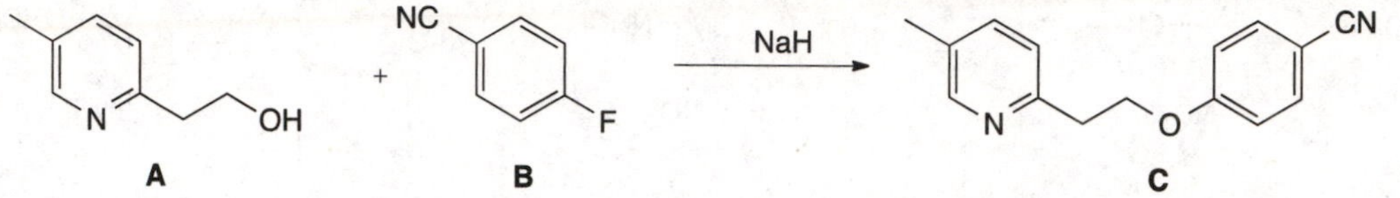

NC
N
OH
+
F
NaH
N
O
CN
A
B
C

18.44

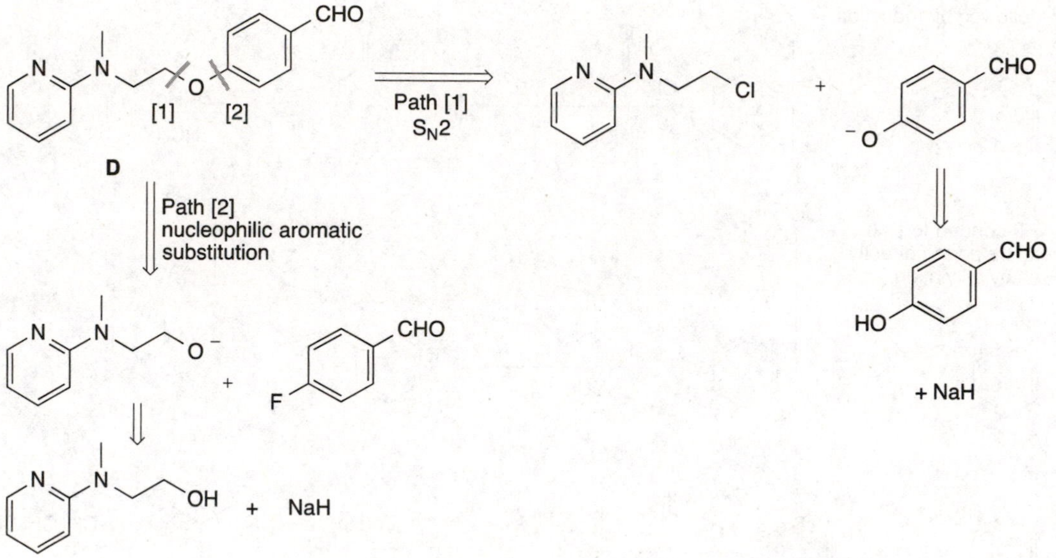

CHO
N
N
O
[1]
[2]
D
Path [1]
SN2
N
N
Cl
+
CHO
⁻O
CHO
HO
+ NaH
Path [2]
nucleophilic aromatic
substitution
N
N
O⁻
+
CHO
F
N
N
OH
+
NaH

18.45

a.

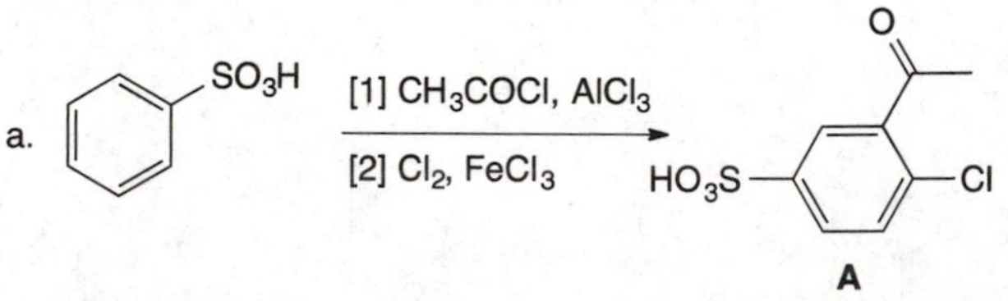

SO3H
[1] CH3COCl, AlCl3
[2] Cl2, FeCl3
O
HO3S
Cl
A

Step [1] won't work because a Friedel–Crafts reaction can't be done on a deactivated benzene ring, as is the case with the SO_3H substituent. Even if Step [1] did work, the second step would introduce Cl meta to SO_3H, not para as drawn.

Alternate synthesis:

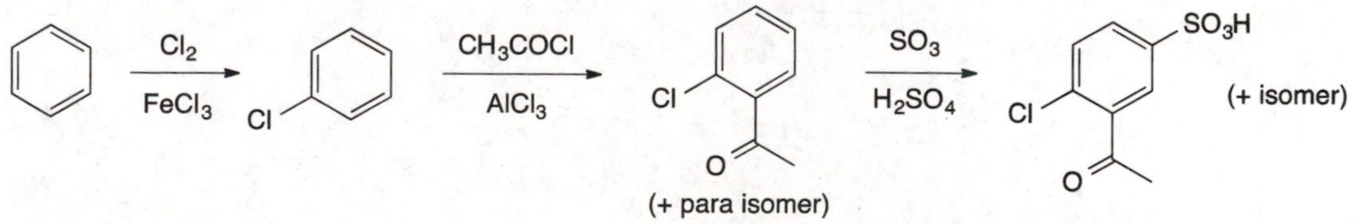

Cl2
FeCl3
Cl
CH3COCl
AlCl3
Cl
O
(+ para isomer)
SO3
H2SO4
Cl
SO3H
O
(+ isomer)

b. [1] $CH_3CH_2CH_2CH_2Cl$, $AlCl_3$; [2] HNO_3, H_2SO_4 → OCH_3, NO_2 = **B**

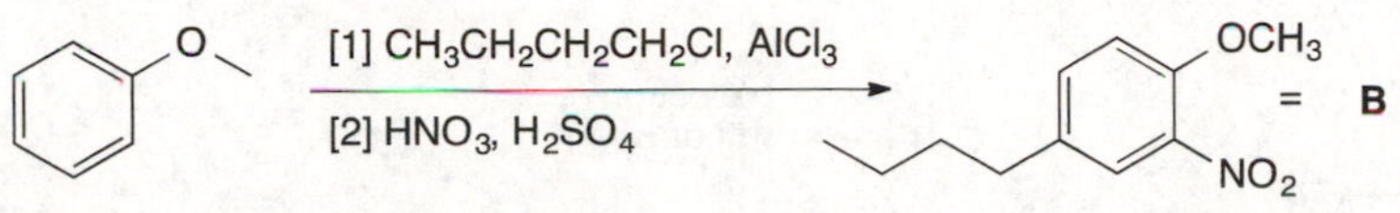

Step [1] involves a Friedel–Crafts alkylation using a 1° alkyl halide that will undergo rearrangement, so that a butyl group will not be introduced as a side chain.

Alternate synthesis:

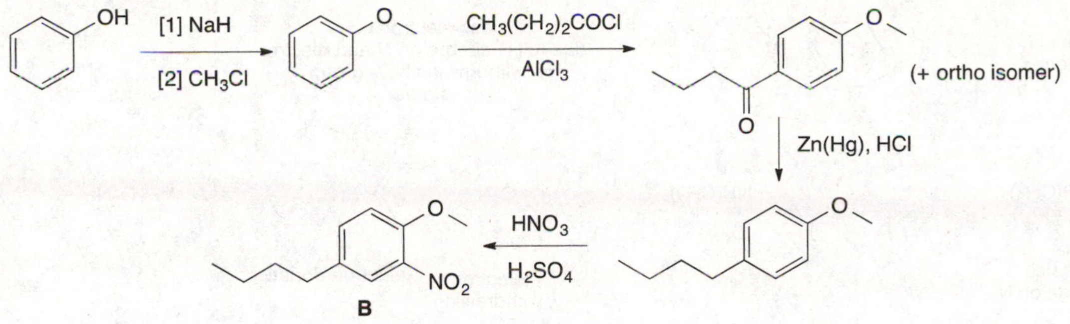

OH [1] NaH [2] CH_3Cl → $CH_3(CH_2)_2COCl$, $AlCl_3$ → (+ ortho isomer)

Zn(Hg), HCl

HNO_3, H_2SO_4 → NO_2 **B**

18.46 Use the directions from Answer 18.17 to rank the compounds.

a. 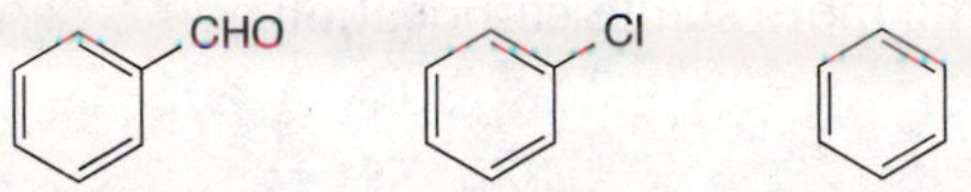 CHO, Cl

least reactive | intermediate reactivity | most reactive

b. 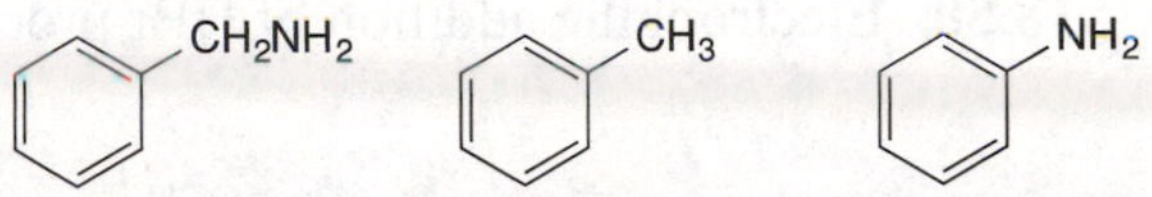 CH_2NH_2, CH_3, NH_2

least reactive | intermediate reactivity | most reactive

18.47

[1] Br

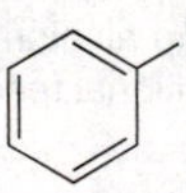

a. withdraw
b. donate
c. less
d. deactivate

[2] C≡N

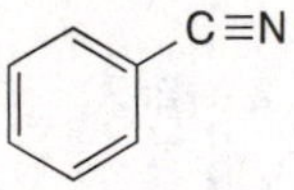

a. withdraw
b. withdraw
c. less
d. deactivate

[3]

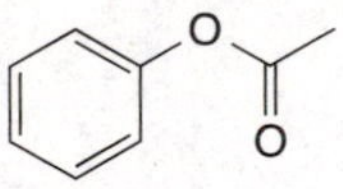

a. withdraw
b. donate
c. more
d. activate

18.48

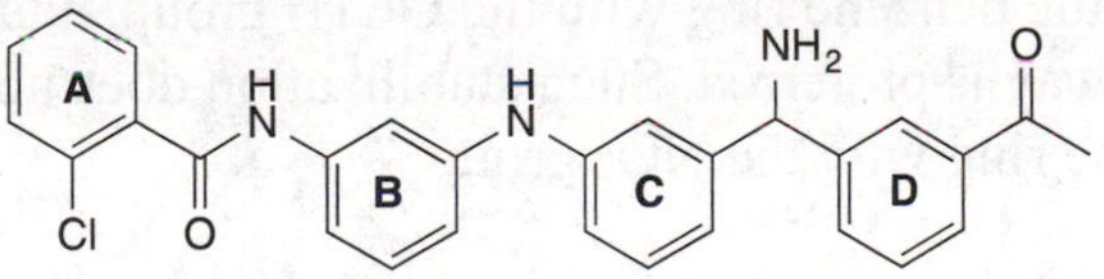

A, B, C, D, NH_2, H N, Cl, O

a. **B** b. **A** c.

Br

+

Br

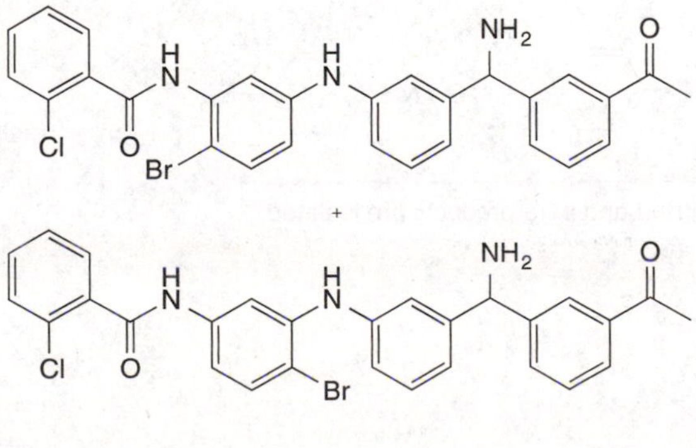

18.49

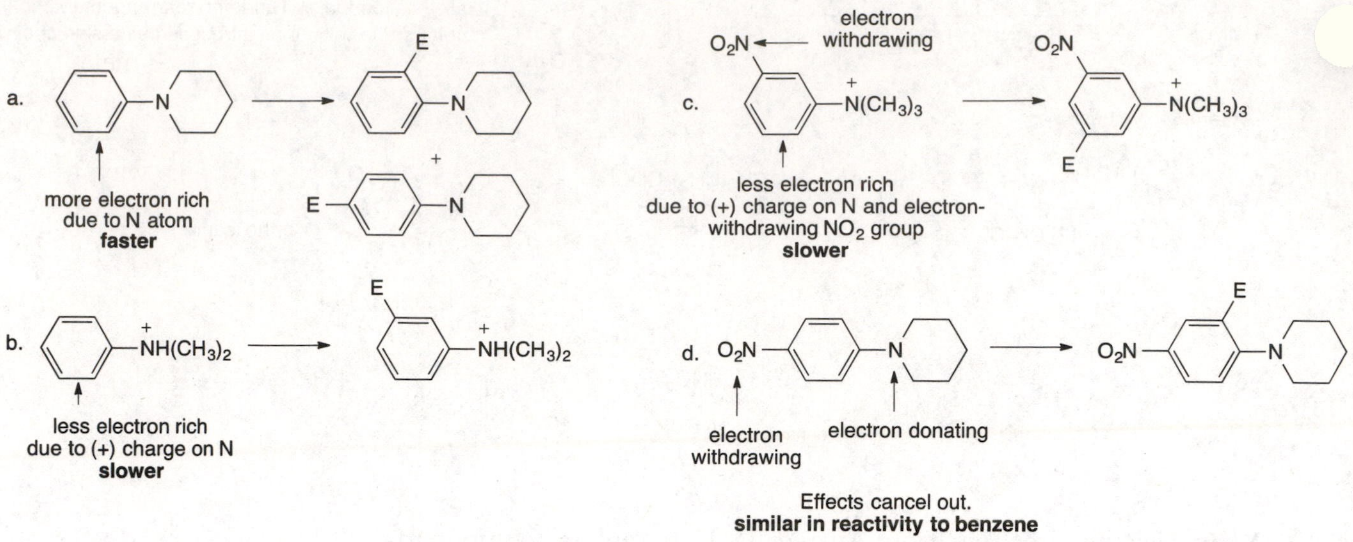

a. more electron rich due to N atom **faster**

b. less electron rich due to (+) charge on N **slower**

c. O_2N electron withdrawing; $\overset{+}{N}(CH_3)_3$

less electron rich due to (+) charge on N and electron-withdrawing NO_2 group **slower**

d. O_2N electron withdrawing; N electron donating

Effects cancel out.
similar in reactivity to benzene

18.50 Electrophilic addition of HBr proceeds by the more stable carbocation.

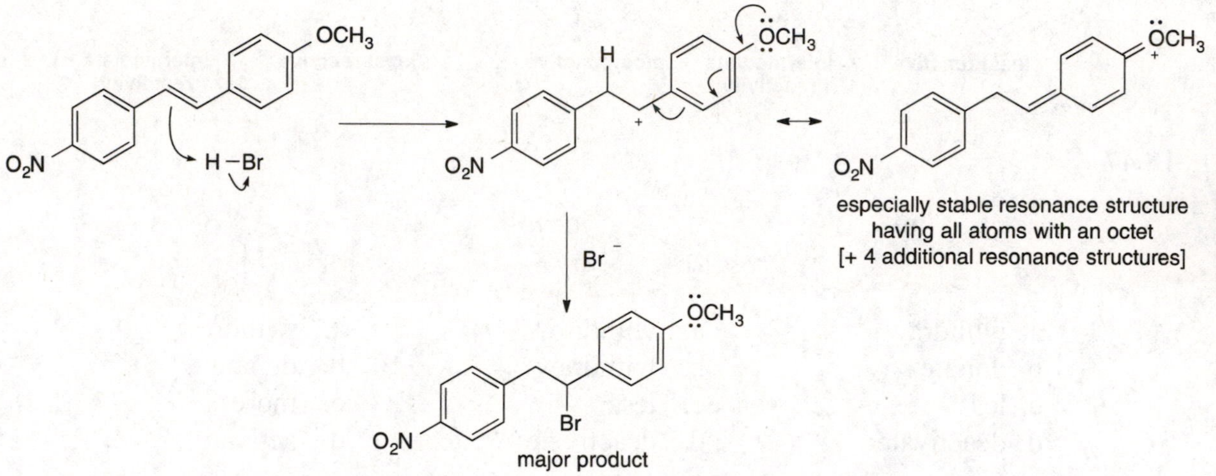

especially stable resonance structure having all atoms with an octet
[+ 4 additional resonance structures]

Br⁻

major product

When the (+) charge is benzylic to the benzene ring with the OCH_3 group, additional resonance stabilization is present, so this pathway is preferred. Such stabilization does not result when the (+) charge is benzylic to the benzene ring with the NO_2 group.

18.51

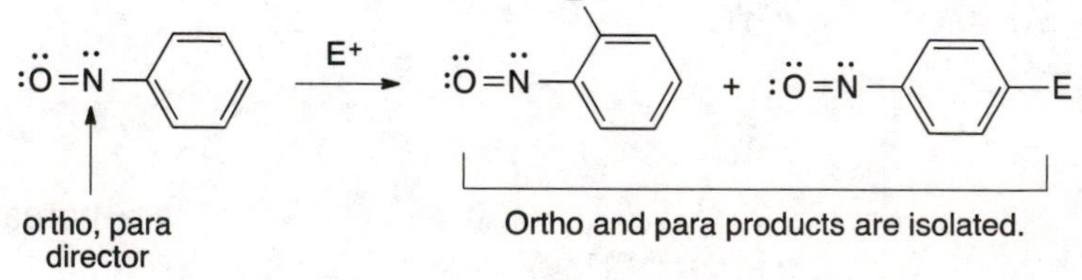

E^+

ortho, para director

Ortho and para products are isolated.

With ortho and para attack there is additional resonance stabilization that delocalizes the positive charge onto the nitroso group. Such additional stabilization is not possible with meta attack. This makes –NO an ortho, para director. Since the N atom bears a partial (+) charge (because it is bonded to a more electronegative O atom), the –NO group inductively withdraws electron density, thus deactivating the benzene ring towards electrophilic attack. In this way, the –NO group resembles the halogens. Thus, the electron-donating resonance effect makes –NO an o,p director, but the electron-withdrawing inductive effect makes it a deactivator.

Ortho attack:

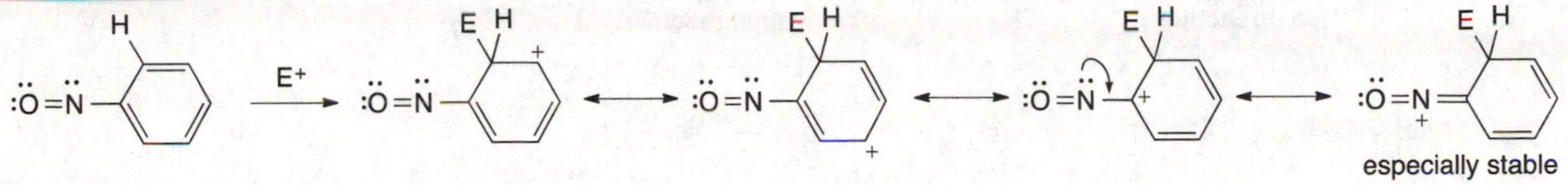

E^+

especially stable

Meta attack:

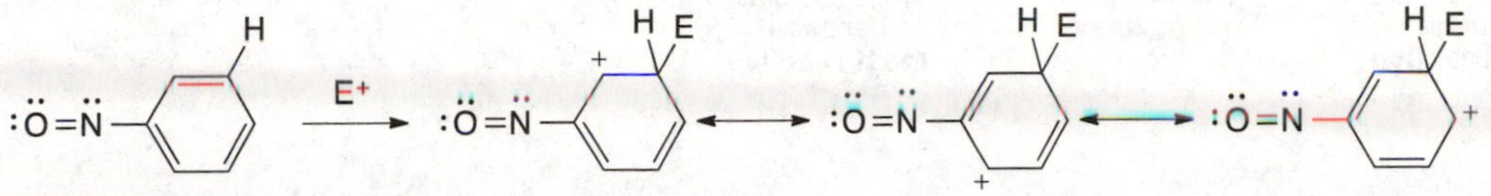

E^+

Para attack:

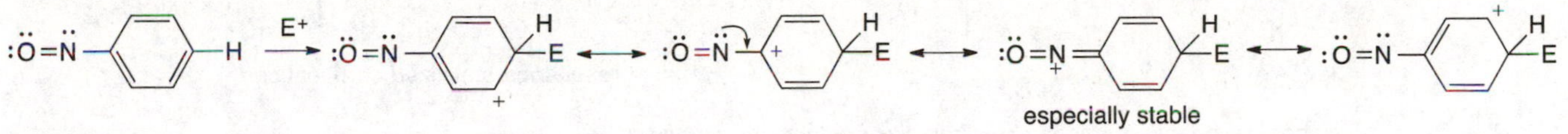

E^+

especially stable

18.52

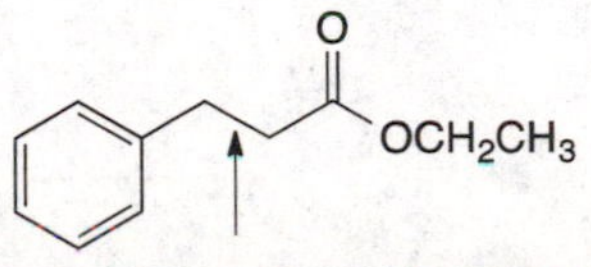

OCH_2CH_3

alkyl group on the benzene ring
R stabilizes (+) charges on the o,p positions by an electron-donating inductive effect. This group behaves like any other R group so that ortho and para products are formed in electrophilic aromatic substitution.

OCH_2CH_3 ⟷ OCH_2CH_3

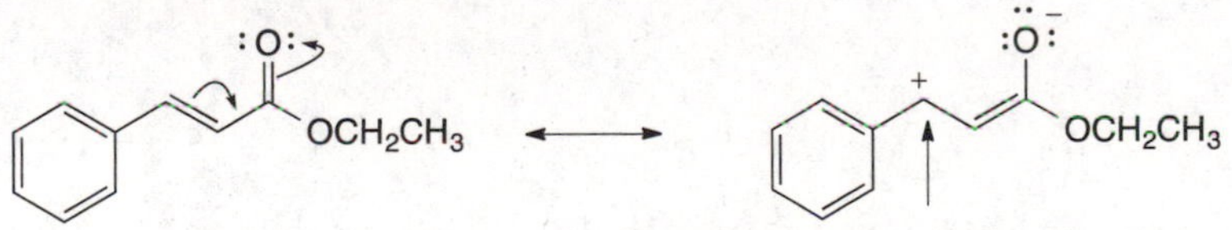

(+) charge on atom bonded to the benzene ring
Drawing resonance structures in electrophilic aromatic substitution results in especially unstable structures for attack at the o,p positions—two (+) charges on adjacent atoms. This doesn't happen with meta attack, so meta attack is preferred. This is identical to the situation observed with all meta directors.

18.53 Increasing the number of electron-withdrawing groups (especially at the ortho and para positions to the leaving group) increases the rate of nucleophilic aromatic substitution. Increasing the electronegativity of the halogen increases the rate.

a.

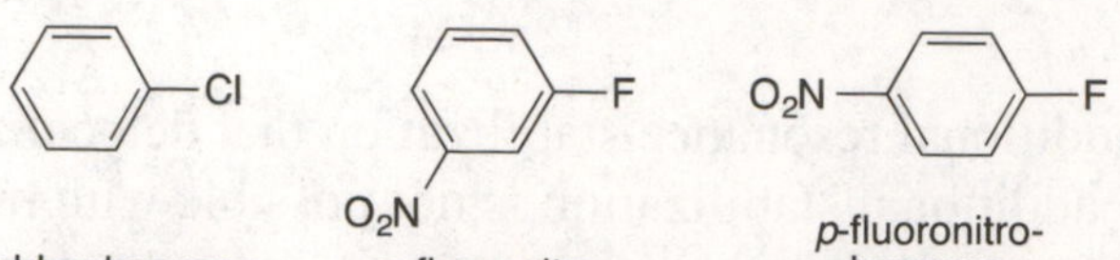

chlorobenzene **least reactive**

m-fluoronitrobenzene

p-fluoronitrobenzene **most reactive**

b.

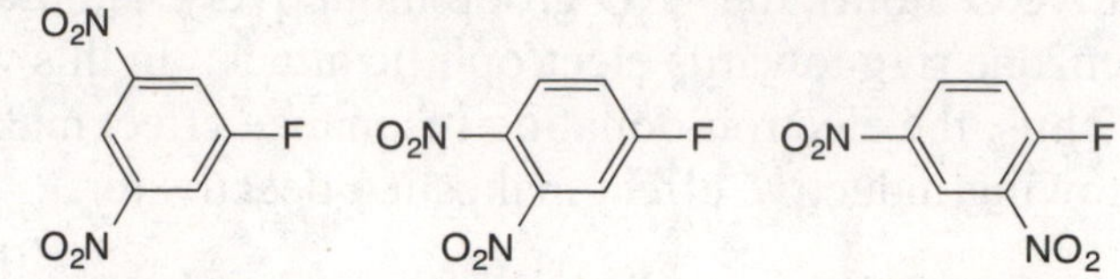

1-fluoro-3,5-dinitrobenzene **least reactive**

1-fluoro-3,4-dinitrobenzene

1-fluoro-2,4-dinitrobenzene **most reactive**

c.

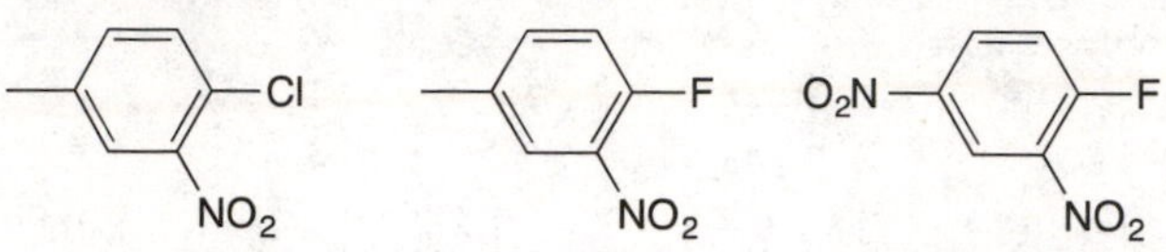

4-chloro-3-nitrotoluene **least reactive**

4-fluoro-3-nitrotoluene

1-fluoro-2,4-dinitrobenzene **most reactive**

18.54

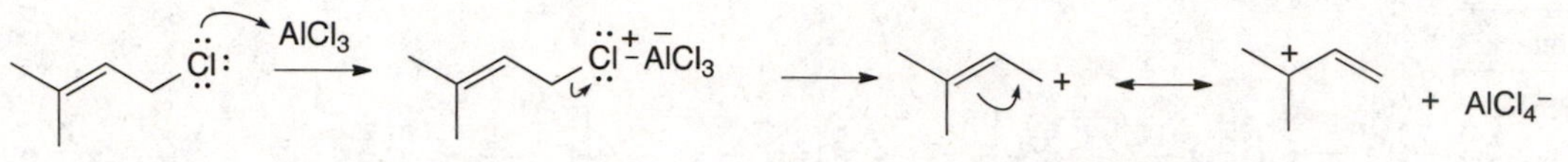

$AlCl_3$ → → resonance-stabilized carbocation + $AlCl_4^-$

Use both resonance forms to show how two products are formed.

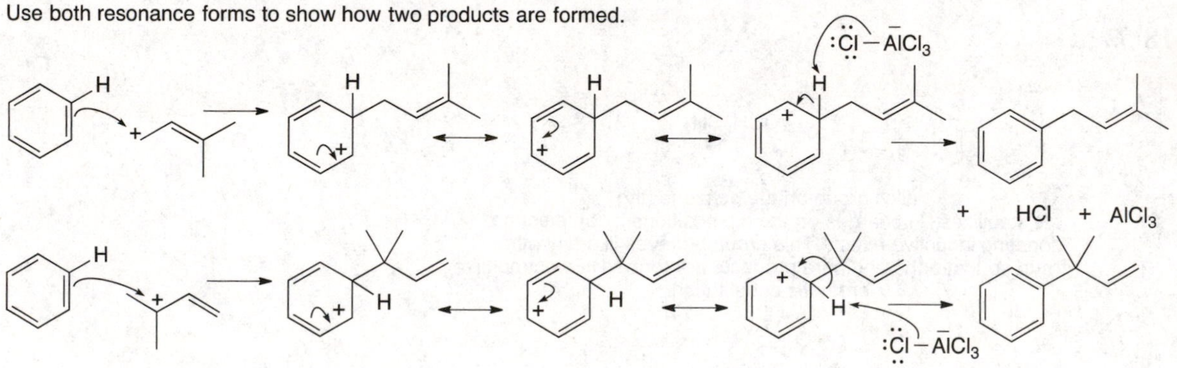

+ HCl + $AlCl_3$

18.55

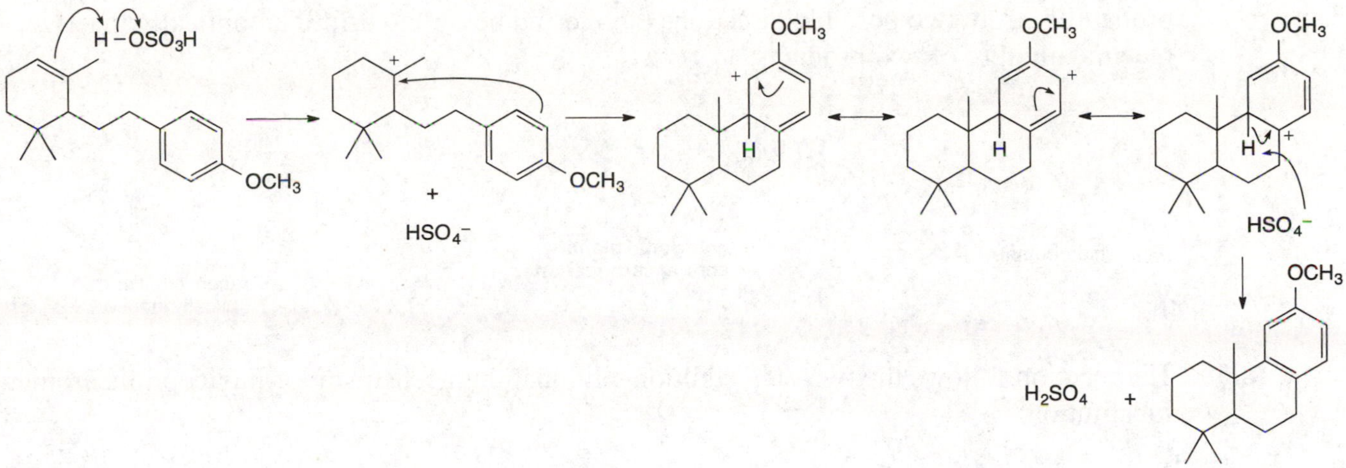

$H-\ddot{O}SO_3H$ $\quad$ OCH_3 $\quad$ + HSO_4^- $\quad$ OCH_3 $\quad$ OCH_3 $\quad$ OCH_3 $\quad$ HSO_4^-

H_2SO_4 + OCH_3

18.56

CH_3O $\quad$ $:\ddot{O}H$ $\quad$ $H-\ddot{O}SO_3H$ $\quad$ $CH_3\ddot{O}$ $\quad$ $\overset{+}{\ddot{O}}H_2$ $\quad$ + HSO_4^- $\quad$ $CH_3\ddot{O}$ $\quad$ + $H_2\ddot{O}$

1,2-CH_3 shift

$CH_3\ddot{O}$ $\quad$ $CH_3\ddot{O}$ $\quad$ $CH_3\ddot{O}$ $\quad$ $CH_3\overset{+}{O}$

$CH_3\ddot{O}$ $\quad$ HSO_4^- $\quad$ $CH_3\ddot{O}$ + H_2SO_4

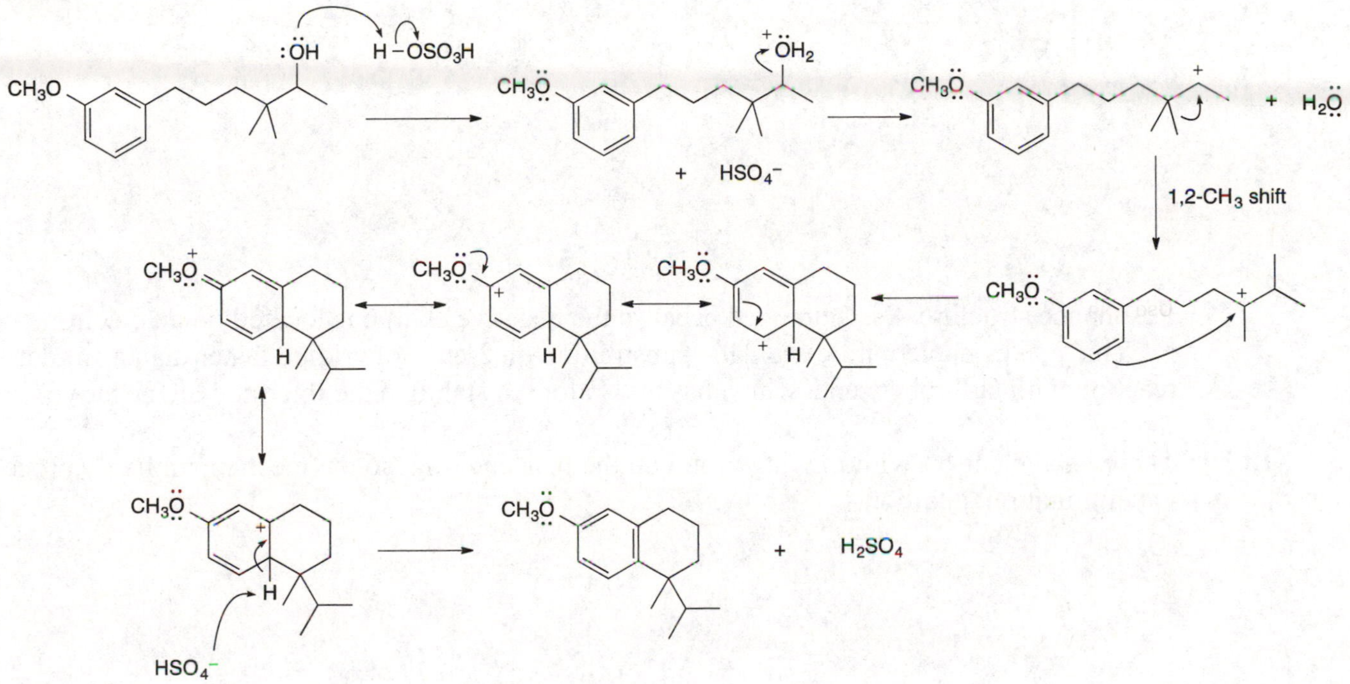

18.57

a. The product has one stereogenic center.

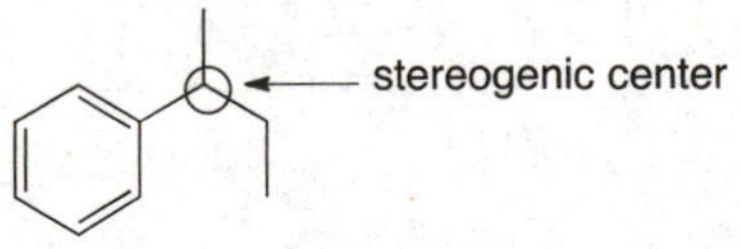

stereogenic center

b. The mechanism for Friedel–Crafts alkylation with this 2° halide involves formation of a trigonal planar carbocation. Because the carbocation is achiral, it reacts with benzene with equal probability from two possible directions (above and below) to afford an optically inactive, racemic mixture of two products.

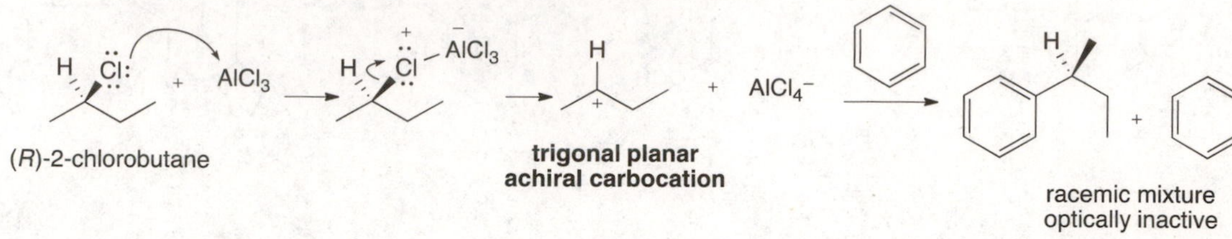

(R)-2-chlorobutane + $AlCl_3$ → … → **trigonal planar achiral carbocation** + $AlCl_4^-$ → racemic mixture, optically inactive

18.58 The reaction follows the two-step addition–elimination mechanism for nucleophilic aromatic substitution.

2-chloropyridine + $Na^+\ {}^-:\ddot{O}CH_3$ → A ↔ … ↔ … ; A → 2-methoxypyridine + Cl^-

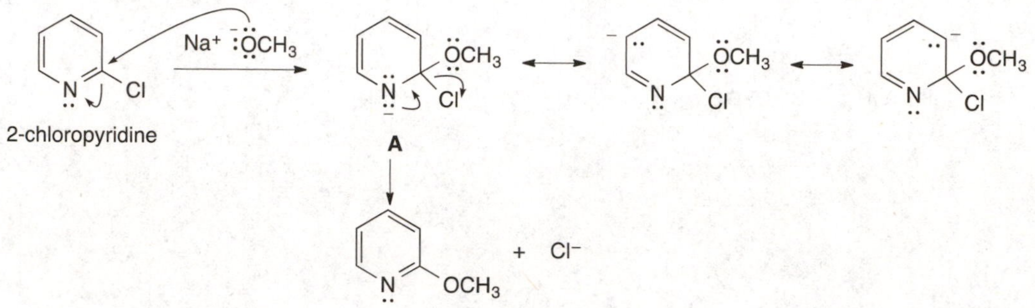

Resonance structure **A** is stabilized because the negative charge is located on an electronegative N. This makes nucleophilic aromatic substitution on 2-chloropyridine faster than a similar reaction with chlorobenzene, which has no N atom to stabilize the intermediate negative charge.

18.59 There is no electron-withdrawing group on the benzene ring, so the mechanism likely proceeds via elimination–addition.

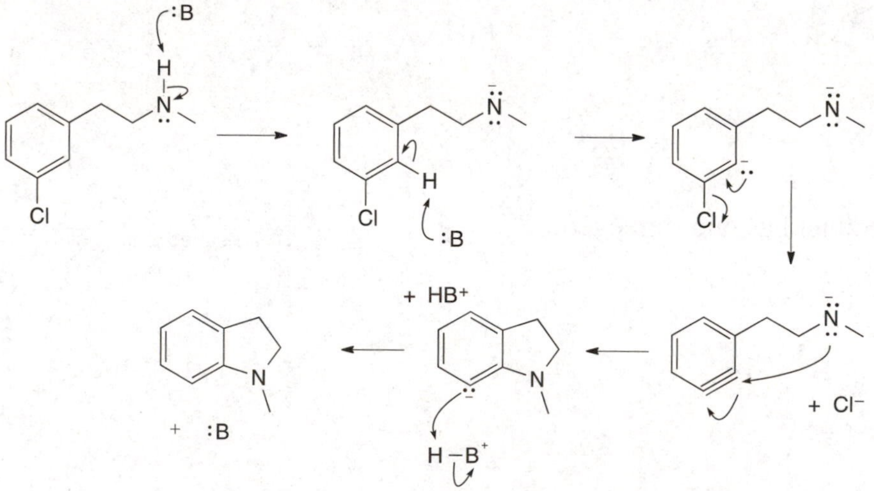

:B … + HB+ … + Cl^- … + :B

18.60

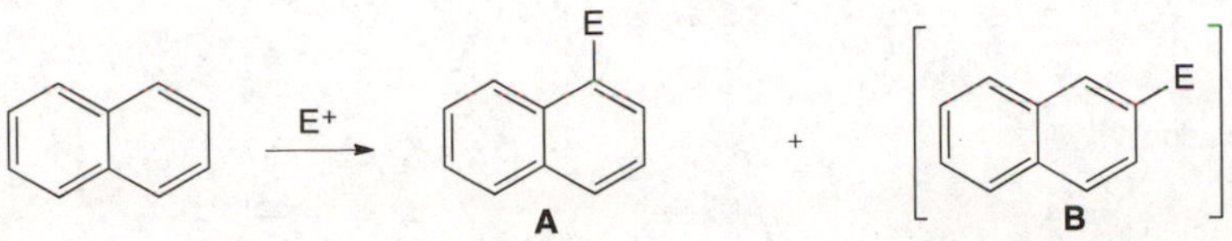

E⁺

A
This product is formed.

B
This product is *not* formed.

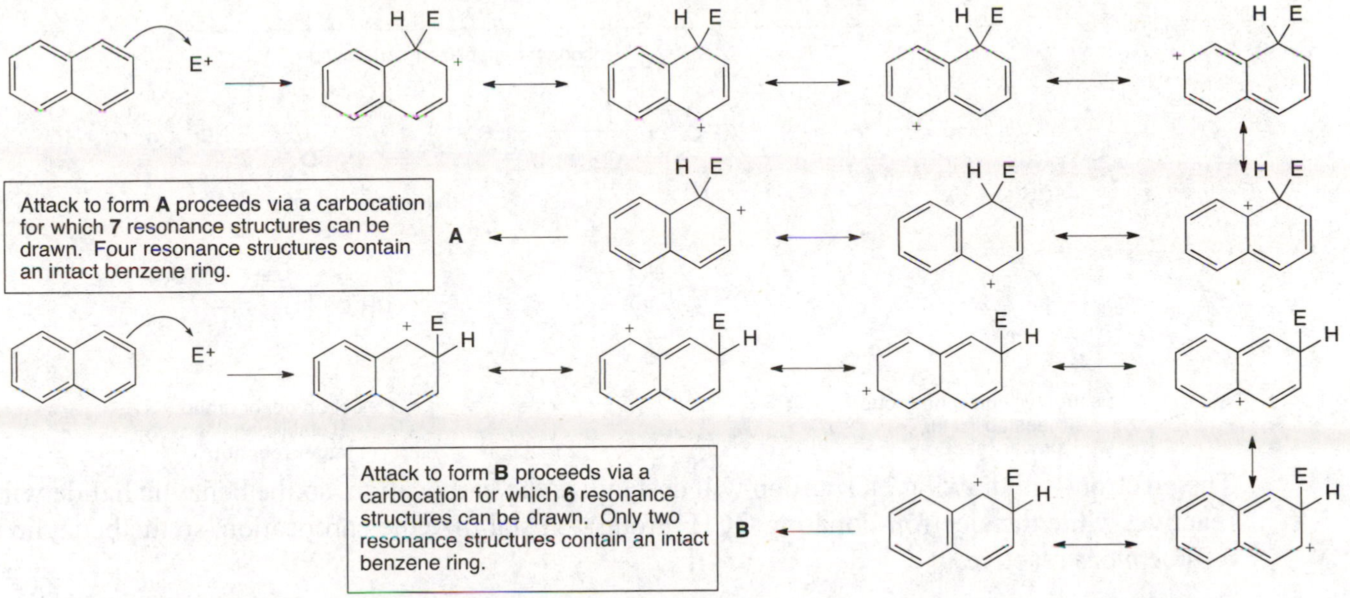

Attack to form **A** proceeds via a carbocation for which **7** resonance structures can be drawn. Four resonance structures contain an intact benzene ring.

Attack to form **B** proceeds via a carbocation for which **6** resonance structures can be drawn. Only two resonance structures contain an intact benzene ring.

A reaction that occurs by way of the more stable carbocation is preferred, so product **A** is formed.

18.61

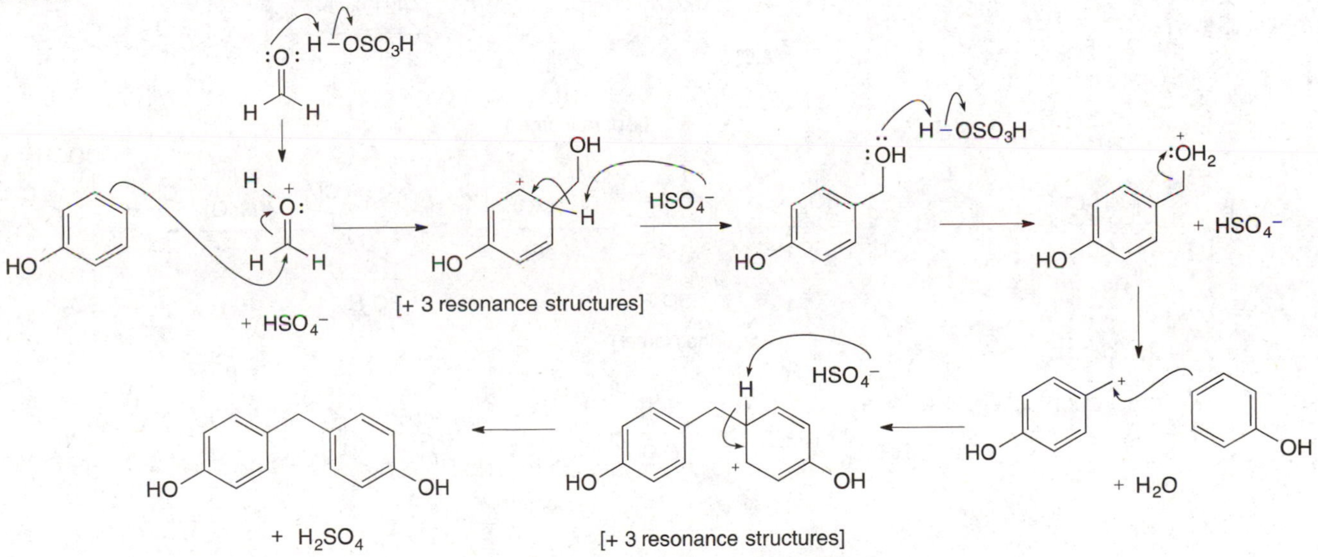

[+ 3 resonance structures]

$+ HSO_4^-$

$+ H_2O$

$+ H_2SO_4$

[+ 3 resonance structures]

18.62 Benzyl bromide forms a resonance-stabilized intermediate that allows it to react rapidly under SN1 conditions.

Formation of a resonance-stabilized carbocation:

benzyl bromide

resonance-stabilized carbocation

+ CH_3OH

+ :Br:$^-$

+ HBr

benzyl methyl ether

O_2N Br Br CH_3O Br

electron-withdrawing group
destabilizing
slower reaction

electron-donor group
stabilizing
faster reaction

The electron-withdrawing NO_2 group will destabilize the carbocation, so the benzylic halide will be *less* reactive, while the electron-donating OCH_3 group will stabilize the carbocation, so the benzylic halide will be *more* reactive.

18.63

a. Cl_2, $FeCl_3$ → Cl; Cl–C(=O)CH$_2$CH$_2$CH$_3$, $AlCl_3$ → (+ para isomer); Zn(Hg), HCl →

b. CH_3Cl, $AlCl_3$ → ; SO_3, H_2SO_4 → SO_3H (+ ortho isomer); Cl_2 (excess), $FeCl_3$ → Cl, Cl, SO_3H; $KMnO_4$ → COOH, Cl, Cl, SO_3H

c. $(CH_3)_2CHCl$, $AlCl_3$ → ; HNO_3, H_2SO_4 → NO_2 (+ para isomer); Cl_2, $FeCl_3$ → Cl, NO_2 (+ isomer); Br_2, $h\nu$ → Br, Cl, NO_2; K^+ $^-OC(CH_3)_3$ → Cl, NO_2

d. CH_3Cl, $AlCl_3$ → ; HNO_3, H_2SO_4 → NO_2 (+ ortho isomer); $KMnO_4$ → HOOC, NO_2; Sn, HCl → HOOC, NH_2

e. [Benzene → (CH_3CH_2COCl, $AlCl_3$) → propiophenone → (Br_2, $FeBr_3$) → Br → (NH_2NH_2, ^-OH) → Br → (HNO_3, H_2SO_4) → Br, NO_2 (+ isomer) → (H_2, Pd-C) → Br, NH_2]

f. [Benzene → (CH_3CH_2Cl, $AlCl_3$) → ethylbenzene → (CH_3CH_2COCl, $AlCl_3$) → (+ ortho isomer) → (Cl_2, $FeCl_3$) → Cl → (Br_2, $h\nu$) → Br, Cl → (K^+ $^-OC(CH_3)_3$) → Cl → (NH_2NH_2, ^-OH) → Cl]

g. [Benzene → (CH_3Cl, $AlCl_3$) → toluene → ($CH_3CH_2CH_2COCl$, $AlCl_3$) → (+ ortho isomer) → (Br_2, $FeBr_3$) → Br → (Br_2, $h\nu$) → Br, Br → (^-OCH₂CH₂CH₂CH₃, from HO-butyl + NaH) → Br, O-butyl]

h. [Benzene → (CH_3CH_2Cl, $AlCl_3$) → ethylbenzene → ($(CH_3)_2CHCOCl$, $AlCl_3$) → (+ ortho isomer) → (HNO_3, H_2SO_4) → O_2N → (Br_2, $h\nu$) → O_2N, Br → (K^+ $^-OC(CH_3)_3$) → O_2N → ([1] OsO_4; [2] $NaHSO_3$, H_2O) → O_2N, HO, OH]

18.64

a. [Toluene → (Br_2, $FeBr_3$) → Br (+ ortho isomer) → ($CH_3CH_2CH_2COCl$, $AlCl_3$) → Br, O → (Br_2, $h\nu$) → Br, Br, O → (^-OH) → Br, OH, O]

b.

$AlCl_3$ (+ ortho isomer) HNO_3 H_2SO_4 O_2N NH_2NH_2 ^-OH O_2N H_2 Pd-C H_2N

c.

$AlCl_3$ (+ ortho isomer) Cl_2 $FeCl_3$ Cl Br_2 $h\nu$ Br Cl ^-OH OH Cl PCC CHO Cl

18.65

a.

HO [1] NaH [2] CH_3CH_2Br CH_3Cl $AlCl_3$ (+ ortho isomer) Br_2 $h\nu$ Br ^-OH OH [1] NaH [2] CH_3I OCH_3

b.

HO [1] NaH [2] CH_3I CH_3Cl $AlCl_3$ (+ ortho isomer) HNO_3 H_2SO_4 (excess) O_2N NO_2 H_2 Pd-C H_2N NH_2

18.66

a.

HO $SOCl_2$ Cl $AlCl_3$ Br_2 $h\nu$ Br $K^+ {}^-OC(CH_3)_3$ Br_2 Br Br $2\ {}^-NH_2$

b. (from a.) $\xrightarrow[FeCl_3]{Cl_2}$ Cl (+ ortho isomer) $\xrightarrow[h\nu]{Br_2}$ Br, Cl $\xrightarrow{K^+ {}^-OC(CH_3)_3}$ Cl $\xrightarrow{Br_2}$ Br, Br, Cl $\xrightarrow{2\ {}^-NH_2}$ Cl $\xrightarrow{NaH}$ Cl $\xrightarrow{\text{ethylene oxide}}$ O^-, Cl $\xrightarrow{H_2O}$ OH, Cl

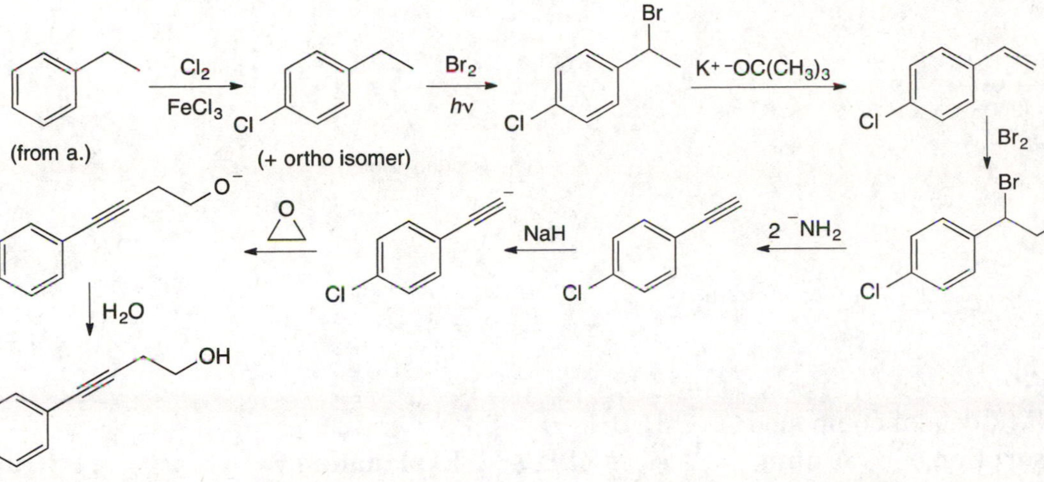

c. (from a.) $\xrightarrow[H_2SO_4]{HNO_3}$ O_2N (+ ortho isomer) $\xrightarrow[h\nu]{Br_2}$ O_2N, Br $\xrightarrow{K^+ {}^-OC(CH_3)_3}$ O_2N $\xrightarrow[ROOR]{HBr}$ O_2N, Br

$HC{\equiv}CH \xrightarrow{NaH} {}^-C{\equiv}CH$ → O_2N

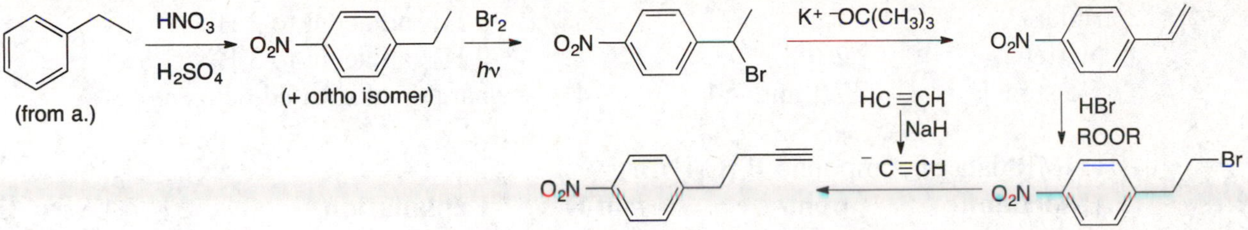

d. Cl (from b.) $\xrightarrow[H_2SO_4]{HNO_3}$ Cl, NO_2 $\xrightarrow[h\nu]{Br_2}$ Cl, Br, NO_2 $\xrightarrow{K^+ {}^-OC(CH_3)_3}$ Cl, NO_2 $\xrightarrow{mCPBA}$ Cl, O, NO_2

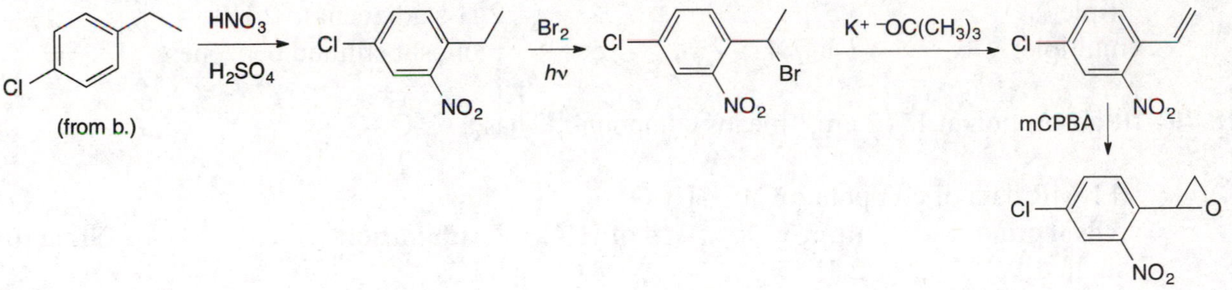

18.67 One possibility:

Benzene $\xrightarrow[AlCl_3]{\text{Cl (chloroethane)}}$ $\xrightarrow[AlCl_3]{\text{Cl, O (acid chloride)}}$ O (+ ortho isomer) $\xrightarrow[h\nu]{Br_2}$ Br, O $\xrightarrow{K^+ {}^-OC(CH_3)_3}$ O $\xrightarrow{[1]\ BH_3;\ [2]\ H_2O,\ {}^-OH}$ HO, O $\xrightarrow{Na_2Cr_2O_7,\ H_2O,\ H_2SO_4}$ HO, O, O $\xrightarrow{Zn(Hg),\ HCl}$ HO, O

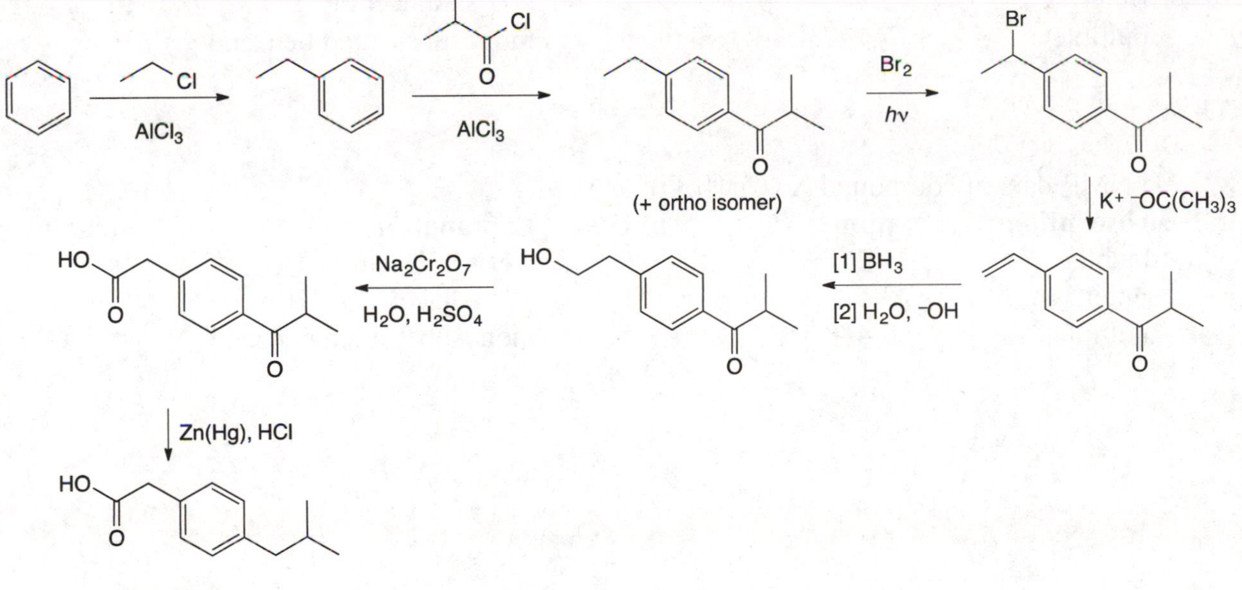

ibufenac

18.68

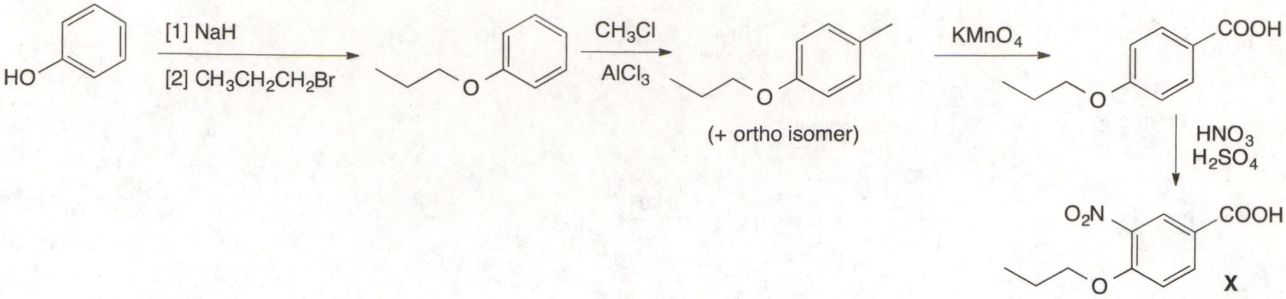

HO

[1] NaH

[2] $CH_3CH_2CH_2Br$

O

CH_3Cl

$AlCl_3$

O

(+ ortho isomer)

$KMnO_4$

COOH

O

HNO_3

H_2SO_4

O_2N

COOH

O

X

18.69

^{1}H NMR data of compound **A** (C_8H_9Br):

Absorption	ppm	# of H's	Explanation
triplet	1.2	3	3 H's adjacent to 2 H's
quartet	2.6	2	2 H's adjacent to 3 H's
two signals	7.1 and 7.4	4	para disubstituted benzene

Structure:

Br

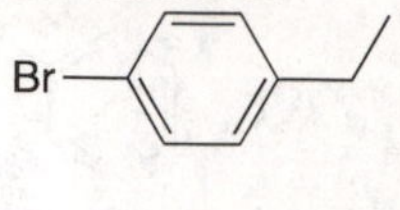

^{1}H NMR data of compound **B** (C_8H_9Br):

Absorption	ppm	# of H's	Explanation
triplet	3.1	2	2 H's adjacent to 2 H's
triplet	3.5	2	2 H's adjacent to 2 H's
multiplet	7.1–7.4	5	monosubstituted benzene

Structure:

Br

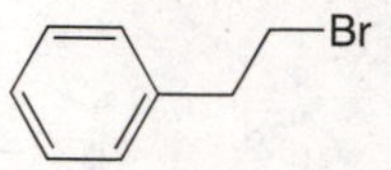

18.70 IR absorption at 1717 cm^{-1} means compound **C** has a C=O.

^{1}H NMR data of compound **C** ($C_{10}H_{12}O$):

Absorption	ppm	# of H's	Explanation
singlet	2.1	3	3 H's
triplet	2.8	2	2 H's adjacent to 2 H's
triplet	2.9	2	2 H's adjacent to 2 H's
multiplet	7.1–7.4	5	monosubstituted benzene

Structure:

O

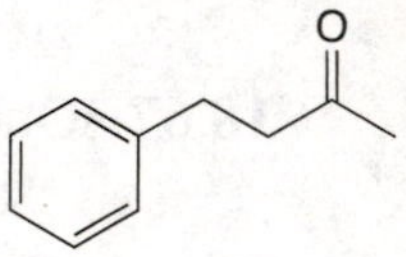

18.71

^{1}H NMR data of compound **X** ($C_{10}H_{12}O$):

Absorption	ppm	# of H's	Explanation
doublet	1.3	6	6 H's adjacent to 1 H
septet	3.5	1	1 H adjacent to 6 H's
multiplet	7.4–8.1	5	monosubstituted benzene

Structure:

O

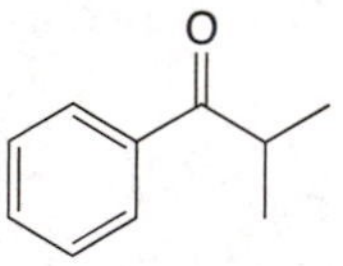

^{1}H NMR data of compound **Y** ($C_{10}H_{14}$):

Absorption	**ppm**	**# of H's**	**Explanation**	**Structure:**
doublet	0.9	6	6 H's adjacent to 1 H	
multiplet	1.8	1	1 H adjacent to many H's	
doublet	2.5	2	2 H's adjacent to 1 H	
multiplet	7.1–7.3	5	monosubstituted benzene	

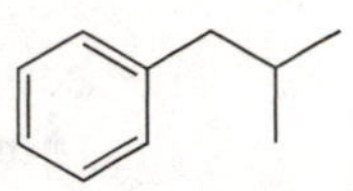

18.72

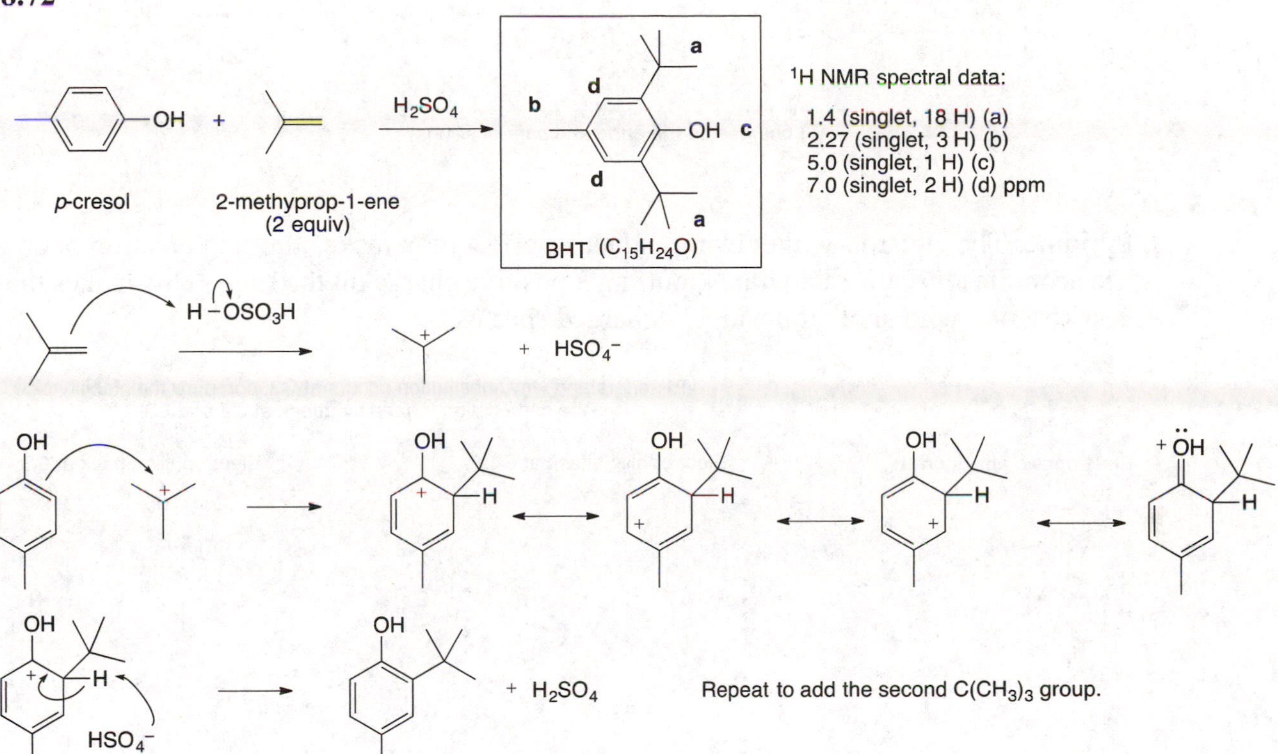

18.73

OH
Sharpless reagent (–)-DET
O
OH
TsCl
pyridine
OTs
CH_3NH_2
N–H
[1] $LiAlH_4$
[2] H_2O
OH
strong base
Cl
CF_3
O

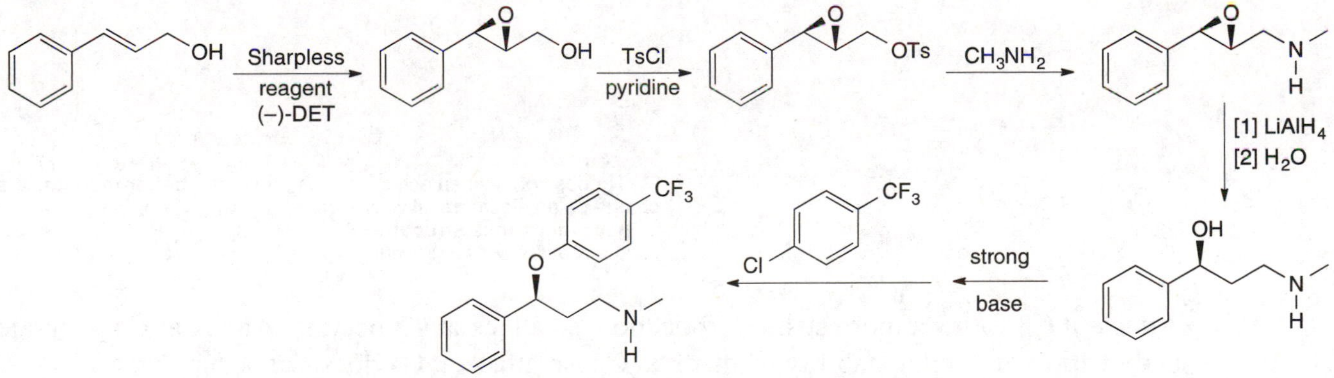

18.74 Five resonance structures can be drawn for phenol, three of which place a negative charge on the ortho and para carbons. These illustrate that the electron density at these positions is increased thus shielding the protons at these positions and shifting the absorptions to lower chemical shif... Similar resonance structures cannot be drawn with a negative charge at the meta position, so it is more deshielded and absorbs farther downfield, at higher chemical shift.

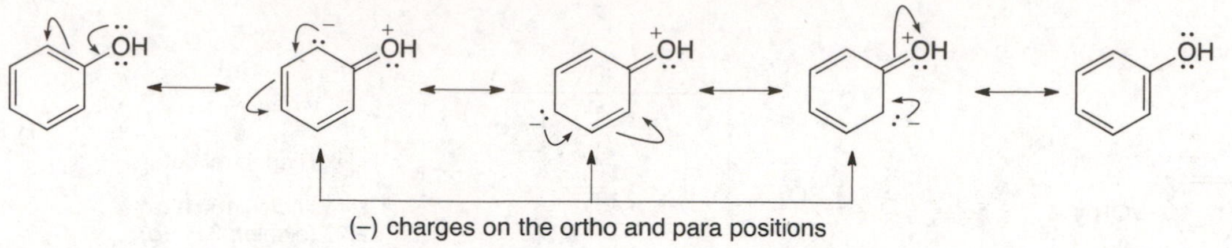

(–) charges on the ortho and para positions

18.75

a. Pyridine: The electron-withdrawing inductive effect of N makes the ring electron poor. Also, electrophiles E^+ can react with N, putting a positive charge on the ring. This makes the ring less reactive with another positively charged species.

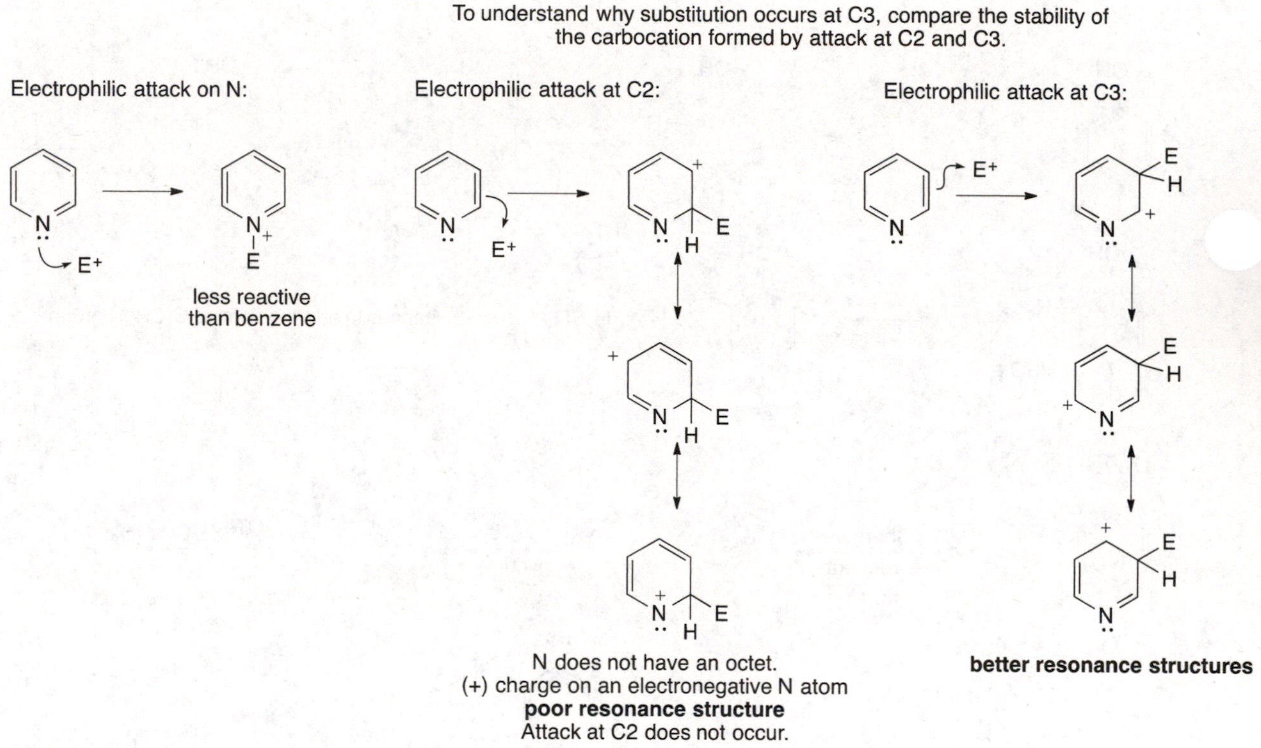

To understand why substitution occurs at C3, compare the stability of the carbocation formed by attack at C2 and C3.

Electrophilic attack on N:

less reactive than benzene

Electrophilic attack at C2:

N does not have an octet.
(+) charge on an electronegative N atom
poor resonance structure
Attack at C2 does not occur.

Electrophilic attack at C3:

better resonance structures

Attack at C3 forms a more stable carbocation, so attack at C3 occurs. Attack at C4 generates a carbocation of similar stability to attack at C2, so attack at C4 does *not* occur.

b. Pyrrole is more reactive than benzene because the C's are more electron rich. The lone pair on N has an electron-donating resonance effect.

more reactive than benzene

Attack at C2:

2-position

more resonance structures

attack at C2

Attack at C3:

3-position

fewer resonance structures

Attack at C3 does not occur.

Attack at C2 forms a more stable carbocation, so electrophilic substitution occurs at C2.

18.76

$H-OSO_3H$

$+ HSO_4^-$

1,2-CH_3 shift

HSO_4^-

18.77 Draw a stepwise mechanism for the following intramolecular reaction, which was used in the synthesis of the female sex hormone estrone.

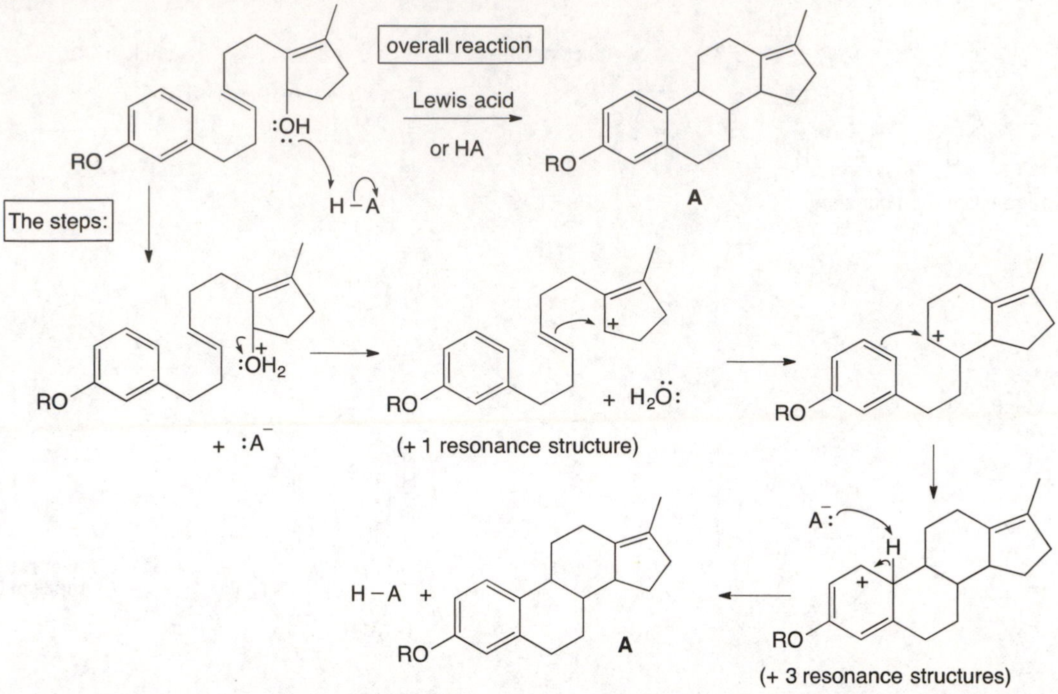

overall reaction

Lewis acid or HA

RO

:OH

H–A

A

The steps:

$:\ddot{O}H_2$

+ :A⁻

+ $H_2\ddot{O}:$

(+ 1 resonance structure)

A:⁻

H

H–A +

(+ 3 resonance structures)

18.78

a. In quinoline the lone pair on N occupies an *sp*² hybrid orbital, so it can never be donated to the ring by resonance. The N atom decreases the electron density of the ring in which it is locat by an electron-withdrawing inductive effect, so substitution occurs on the other ring. In indole, the N atom donates its electron pair (which is contained in a *p* orbital) to the five-membered ring, increasing its electron density, so substitution occurs on the five-membered ring with the N atom.

b. In the presence of acid, the N atom is protonated prior to electrophilic attack. For substitution to occur at C8 rather than C7, the carbocation that results from electrophilic addition at C8 must be more stable. Attack at C8 generates a carbocation with more resonance structures, four of which keep one ring aromatic (**1**–**4**). Attack at C7 generates a carbocation with fewer resonance structures and only two have an intact aromatic ring (**5** and **6**).

c. With indole, attack at C3 forms the more highly resonance-stabilized carbocation.

Attack at C3 forms resonance structures, all of which have an intact aromatic ring, and two of which have all atoms with octets. Attack at C2 forms a cation with more resonance structures, but only two have an intact aromatic ring, and only one has complete octets.

18.79

BF$_3$

+ HOBF$_3$

(+ 3 additional resonance structures)

1,2-shift

HO—BF$_3$

+ H_2O + BF$_3$

Chapter 19 Carboxylic Acids and the Acidity of the O–H Bond

Chapter Review

General facts

- Carboxylic acids contain a carboxy group (COOH). The central carbon is sp^2 hybridized and trigonal planar (19.1).
- Carboxylic acids are identified by the suffixes *-oic acid, carboxylic acid,* or *-ic acid* (19.2).
- Carboxylic acids are polar compounds that exhibit hydrogen bonding interactions (19.3).

Summary of spectroscopic absorptions (19.4)

IR absorptions	C=O	~1710 cm^{-1}
	O–H	3500–2500 cm^{-1} (very broad and strong)
^{1}H NMR absorptions	O–H	10–12 ppm (highly deshielded proton)
	C–H α to COOH	2–2.5 ppm (somewhat deshielded Csp^3–H)
^{13}C NMR absorption	C=O	170–210 ppm (highly deshielded carbon)

General acid–base reaction of carboxylic acids (19.9)

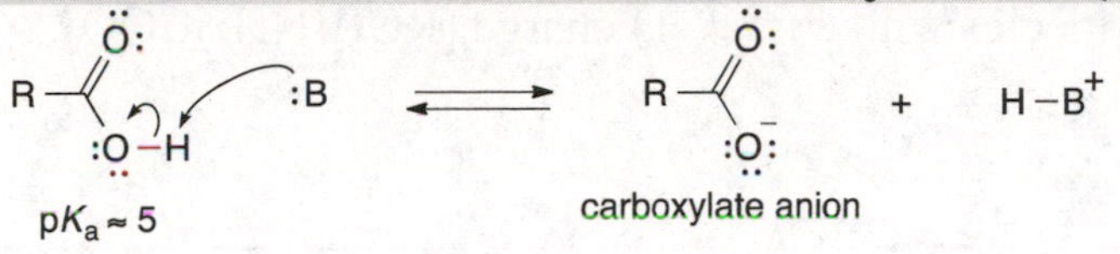

- Carboxylic acids are especially acidic because carboxylate anions are resonance stabilized.
- For equilibrium to favor the products, the base must have a conjugate acid with a pK_a > 5. Common bases are listed in Table 19.3.

Factors that affect acidity

Resonance effects. A carboxylic acid is more acidic than an alcohol or phenol because its conjugate base is more effectively stabilized by resonance (19.9).

ROH — pK_a = 16–18
OH (phenol) — pK_a = 10
RCOOH — pK_a ≈ 5

Increasing acidity

Inductive effects. Acidity increases with the presence of electron-withdrawing groups (like the electronegative halogens) and decreases with the presence of electron-donating groups (like polarizable alkyl groups) (19.10).

Substituted benzoic acids.

- Electron-donor groups (D) make a substituted benzoic acid *less* acidic than benzoic acid.
- Electron-withdrawing groups (W) make a substituted benzoic acid *more* acidic than benzo acid.

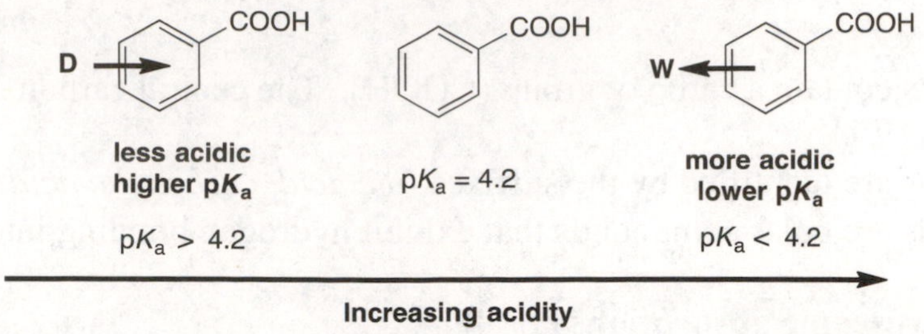

COOH COOH COOH

D W

less acidic
higher pK_a

pK_a = 4.2

more acidic
lower pK_a

pK_a > 4.2

pK_a < 4.2

Increasing acidity

Other facts

- Extraction is a useful technique for separating compounds having different solubility properties. Carboxylic acids can be separated from other organic compounds by extraction, because aqueous base converts a carboxylic acid into a water-soluble carboxylate anion (19.12).
- A sulfonic acid (RSO_3H) is a strong acid because it forms a weak, resonance-stabilized conjugate base on deprotonation (19.13).
- Amino acids have an amino group on the α carbon to the carboxy group [$RCH(NH_2)COOH$]. Amino acids exist as zwitterions at pH ≈ 6. Adding acid forms a species with a net (+1) charge [$RCH(NH_3)COOH]^+$. Adding base forms a species with a net (–1) charge [$RCH(NH_2)COO]^-$ (19.14).

Practice Test on Chapter Review

1.a. Give the IUPAC name for the following compound.

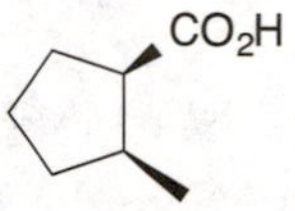

CO_2H

b. Draw the structure corresponding to the following name: sodium *m*-bromobenzoate.

2.a. Which of the labeled atoms is least acidic?

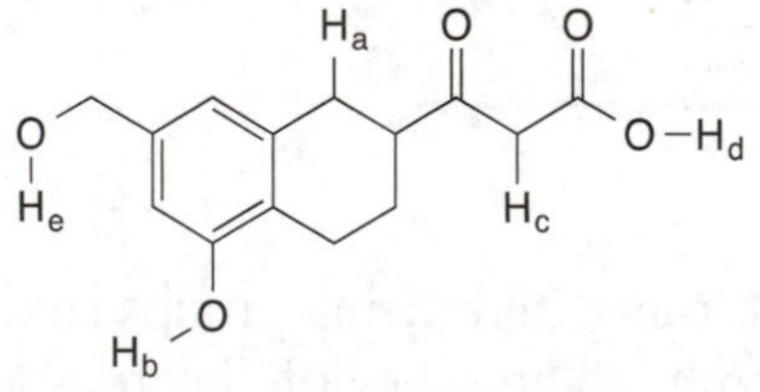

H_a O O

O $O-H_d$

H_e H_c

H_b O

1. H_a 2. H_b 3. H_c 4. H_d 5. H_e

b. Which of the following carboxylic acids has the lowest pK_a?

1. 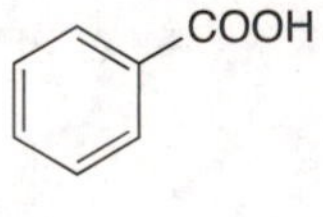 COOH
2. 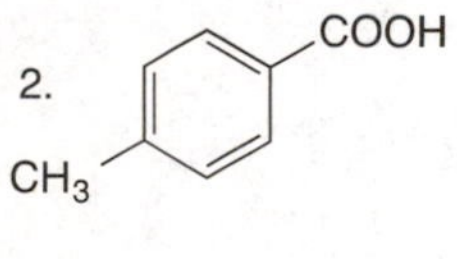 COOH, CH_3
3. 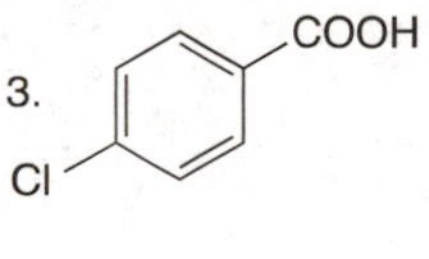 COOH, Cl
4. 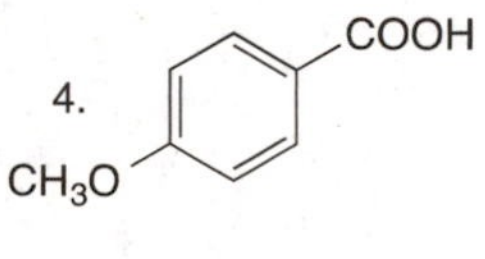 COOH, CH_3O
5. 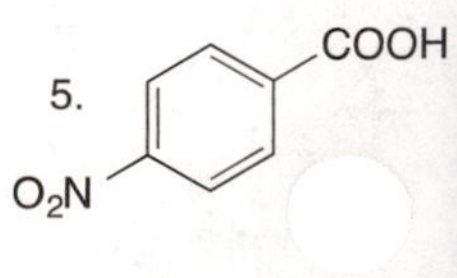 COOH, O_2N

c. Which compound(s) can be converted to **A** by an oxidation reaction?

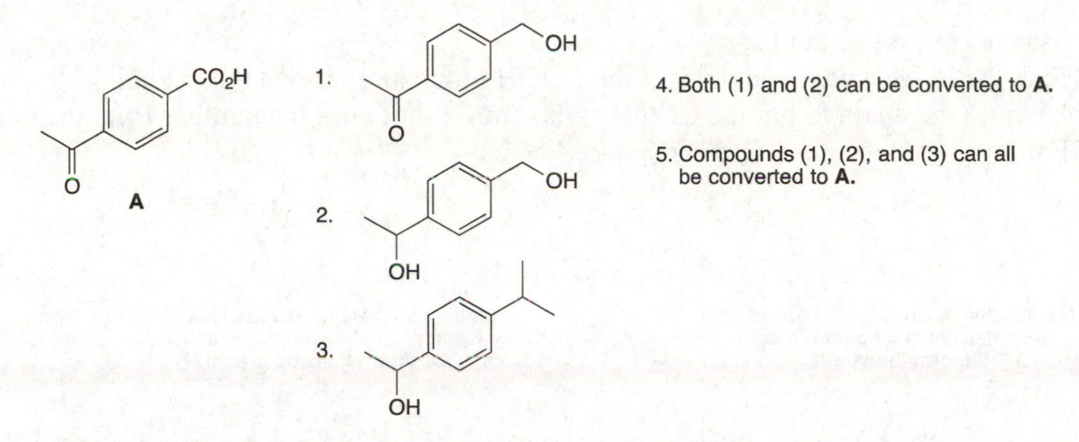

3. Rank the following compounds in order of increasing basicity. Label the ***least*** basic compound as **1,** the ***most*** basic compound as **3,** and the compound of ***intermediate*** basicity as **2.**

C_6H_5–COO⁻ (**A**) O_2N–C_6H_4–COO⁻ (**B**) C_6H_5–O⁻ (**C**)

4. Draw the organic products formed in each of the following reactions.

a. $CH_3CH(\overset{+}{N}H_3)COOH$ —NaOH (1 equiv)→

b. HO–C_6H_4–CH_2CH_2COOH —NaH (1 equiv)→

c. (2-hydroxyphenyl)–$CH_2CH_2N(CH_3)_2$ —[1] NaH; [2] CH_3I→

Answers to Practice Test

1.a. *cis*-2-methylcyclopentanecarboxylic acid

b. m-bromobenzoate: COO⁻ Na⁺, Br

2.a. 1
b. 5
c. 5

3. **A**–2
B–1
C–3

4.a. $CH_3CH(\overset{+}{N}H_3)COO^-$

b. HO–C_6H_4–CH_2CH_2COO⁻

c. (2-methoxyphenyl)–$CH_2CH_2N(CH_3)_2$ (OCH_3)

Answers to Problems

19.1 To name a carboxylic acid:
[1] Find the longest chain containing the COOH group and change the *-e* ending to *-oic acid.*
[2] Number the chain to put the COOH carbon at C1, but omit the number from the name.
[3] Follow all other rules of nomenclature.

a. Number the chain to put COOH at C1.
6 carbon chain = **hexanoic acid**
3,3-dimethylhexanoic acid

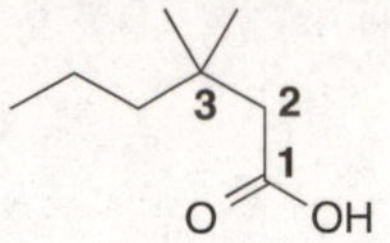

c. Number the chain to put COOH at C1.
6 carbon chain = **hexanoic acid**
2,4-diethylhexanoic acid

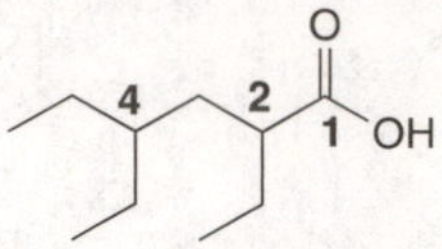

b. Number the chain to put COOH at C1.
5 carbon chain = **pentanoic acid**
3-bromo-4-chloro-2-fluoropentanoic acid

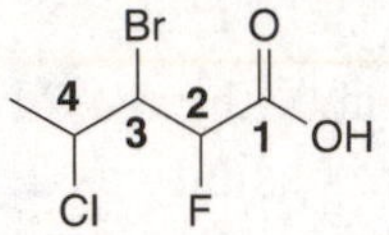

d. Number the chain to put COOH at C1.
9 carbon chain = **nonanoic acid**
4-isopropyl-6,8-dimethylnonanoic acid

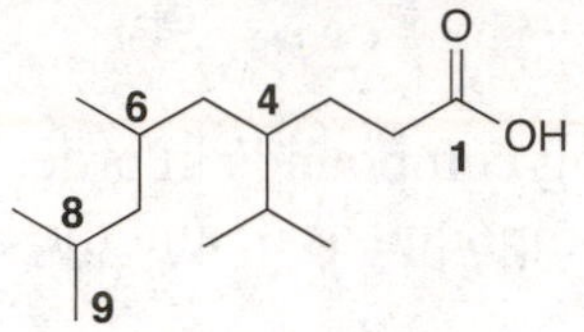

19.2

a. 2-bromo**butanoic acid**

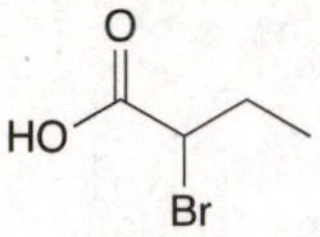

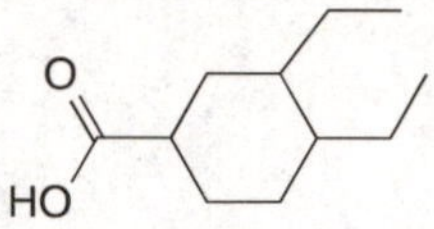

b. 2,3-dimethyl**pentanoic acid**

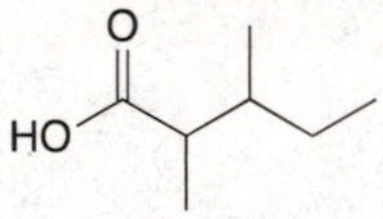

c. 3,3,4-trimethyl**heptanoic acid**

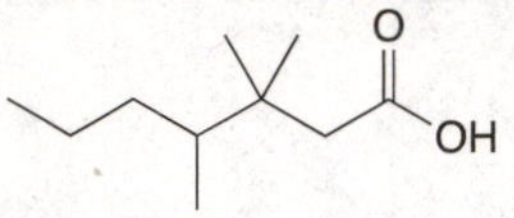

d. 2-*sec*-butyl-4,4-diethyl**nonanoic acid**

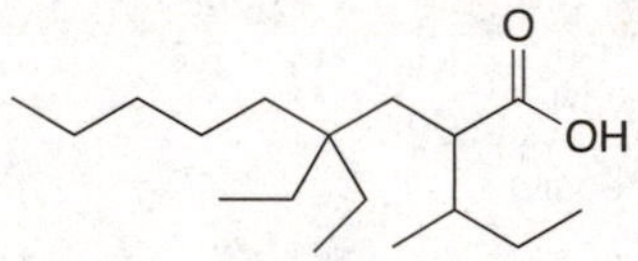

e. 3,4-diethyl**cyclohexanecarboxylic acid**

f. 1-isopropyl**cyclobutane-carboxylic acid**

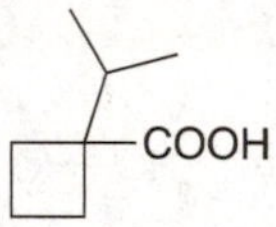

19.3

a. α-methoxy**valeric acid**

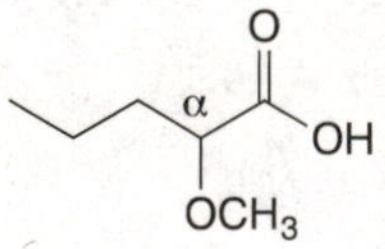

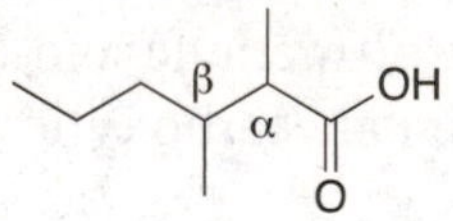

b. β-phenyl**propionic acid**

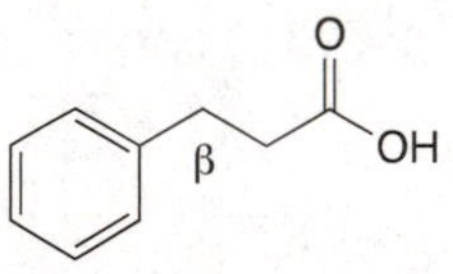

c. α,β-dimethyl**caproic acid**

d. α-chloro-β-methyl**butyric acid**

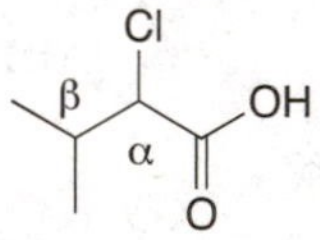

19.4

a. **lithium benzoate**

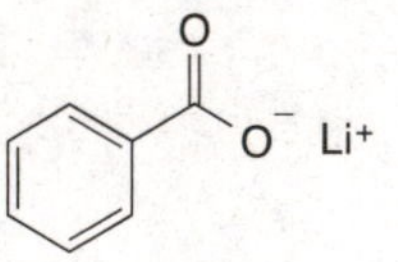

b. **sodium formate**

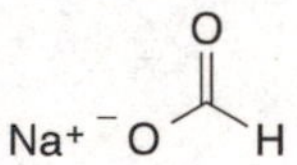

c. **potassium 2-methylpropanoate**

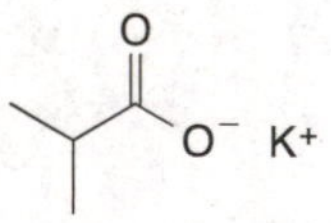

d. **sodium 4-bromo-6-ethyl-octanoate**

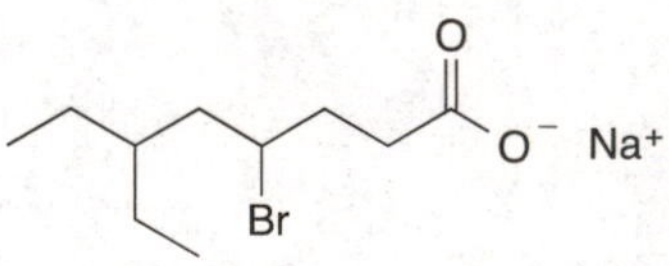

19.5

COOH
C2
2-propyl**pentanoic acid**

COO⁻ Na⁺
sodium 2-propyl**pentanoate**

19.6 More polar molecules have a higher boiling point and are more water soluble.

least polar
lowest boiling point
least H_2O soluble

intermediate polarity
intermediate boiling point

most polar
highest boiling point
most H_2O soluble

19.7 Look for functional group differences to distinguish the compounds by IR. Besides sp^3 hybridized C–H bonds at 3000–2850 cm^{-1} (which all three compounds have), the following functional group absorptions are seen:

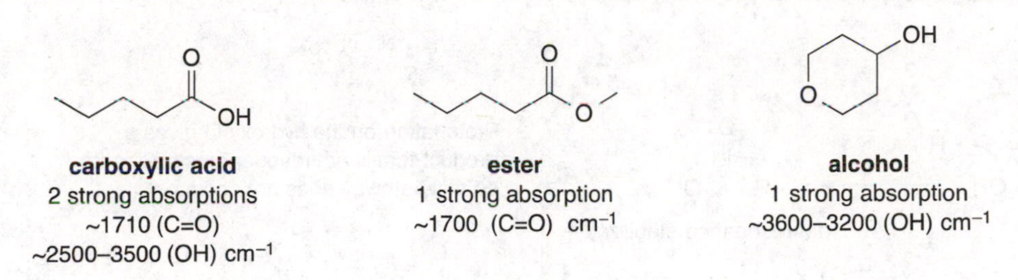

19.8

s-cis
s-trans
O
HO
Z
E
E
E
Z
s-trans
s-cis

19.9

HO
COOH
HO
OH
PGF$_{2\alpha}$
a prostaglandin

HO
COOH
HO
OH
enantiomer

There are five tetrahedral stereogenic centers labeled with (*). Both double bonds can exhibit cis–trans isomerism. Therefore, there are $2^7 = 128$ stereoisomers.

19.10 1° Alcohols are converted to carboxylic acids by oxidation reactions.

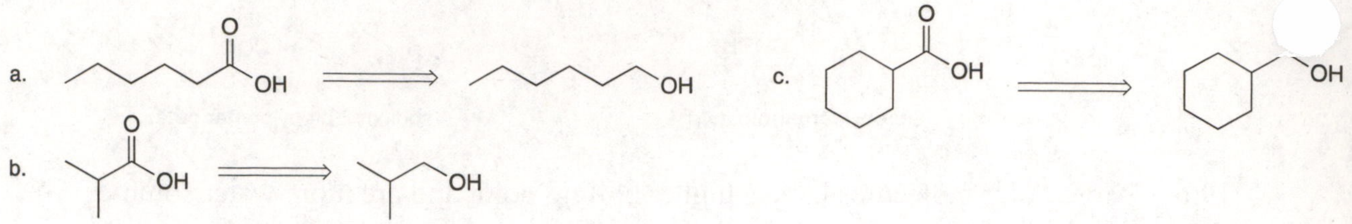

a. OH ⟹ OH

b. OH ⟹ OH

c. OH ⟹ OH

19.11

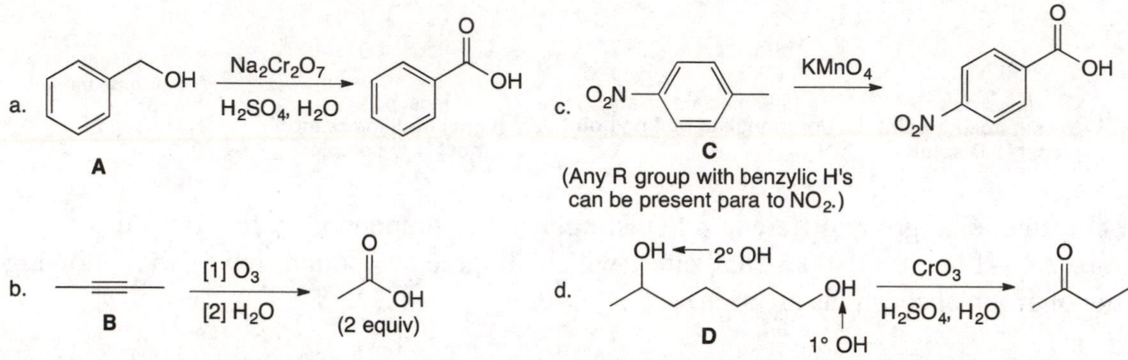

a. OH $\xrightarrow[H_2SO_4,\ H_2O]{Na_2Cr_2O_7}$ OH

A

b. B $\xrightarrow[[2]\ H_2O]{[1]\ O_3}$ OH (2 equiv)

c. O_2N — C $\xrightarrow{KMnO_4}$ O_2N — OH

(Any R group with benzylic H's can be present para to NO_2.)

d. OH ← 2° OH; OH ↑ 1° OH; D $\xrightarrow[H_2SO_4,\ H_2O]{CrO_3}$ OH

19.12

:O: R—ÖH + H–A ⟶ :O: R—$\overset{+}{O}H_2$ + A:

not resonance stabilized

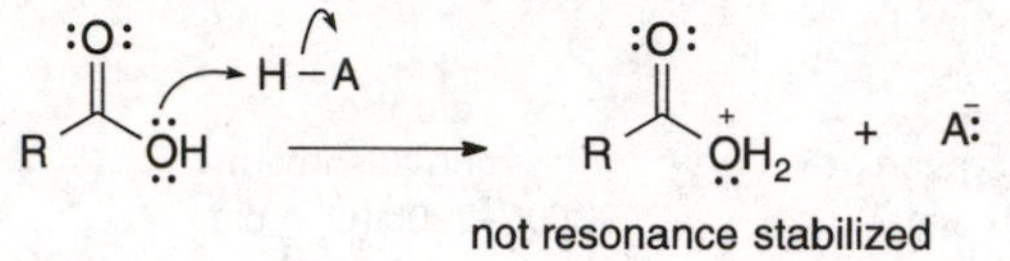

Protonation on the hydroxy O gives a product that is not resonance stabilized, so this pathway does not occur.

19.13

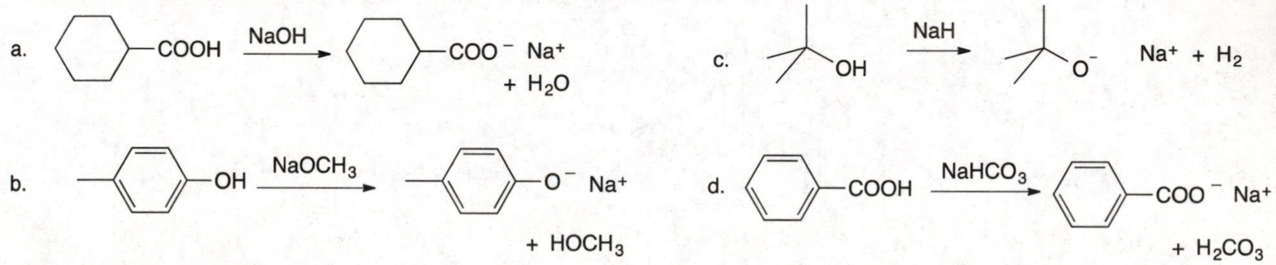

a. —COOH $\xrightarrow{NaOH}$ —COO^- Na^+ + H_2O

b. —OH $\xrightarrow{NaOCH_3}$ —O^- Na^+ + $HOCH_3$

c. OH $\xrightarrow{NaH}$ O^- Na^+ + H_2

d. —COOH $\xrightarrow{NaHCO_3}$ —COO^- Na^+ + H_2CO_3

19.14 CH_3COOH has a pK_a of 4.8. Any base having a conjugate acid with a pK_a higher than 4.8 can deprotonate it.

a. F^- pK_a (HF) = 3.2 **not strong enough**

b. $(CH_3)_3CO^-$ pK_a $[(CH_3)_3COH]$ = 18 **strong enough**

c. CH_3^- pK_a (CH_4) = 50 **strong enough**

d. $^-NH_2$ pK_a (NH_3) = 38 **strong enough**

e. Cl^- pK_a (HCl) = –7.0 **not strong enough**

19.15

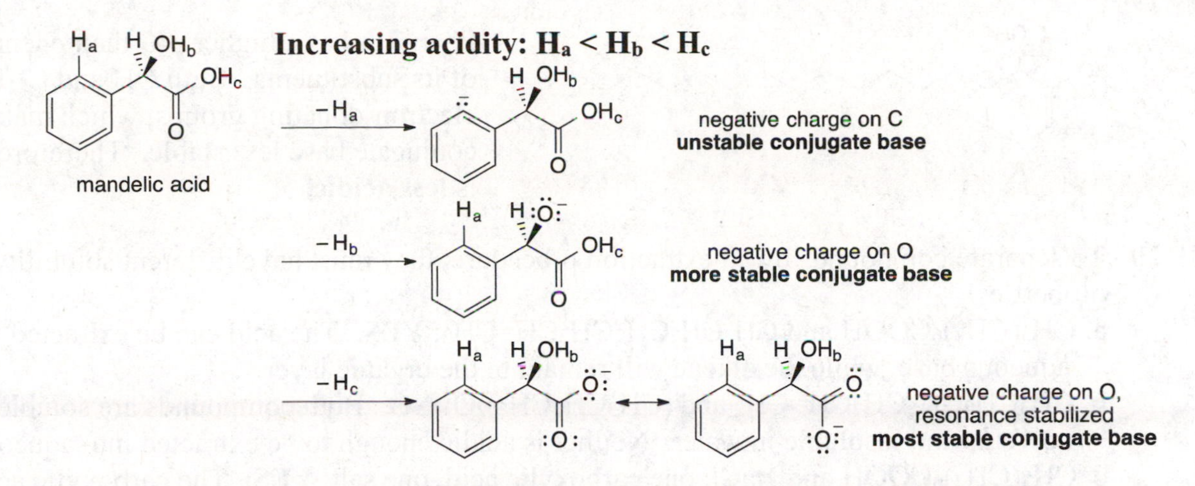

19.16 Electron-withdrawing groups make an acid more acidic, lowering its pK_a.

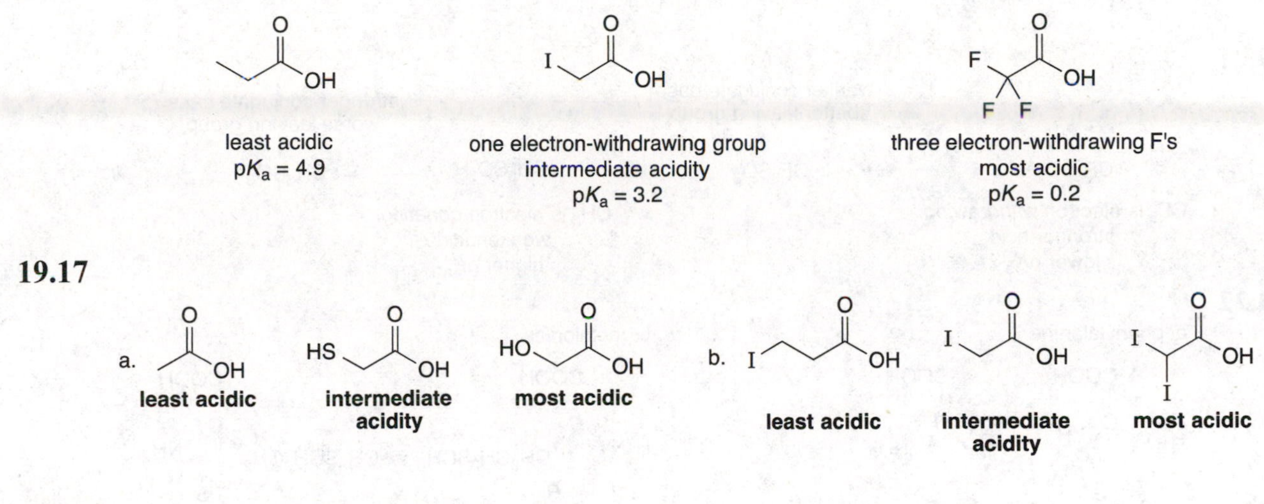

19.18

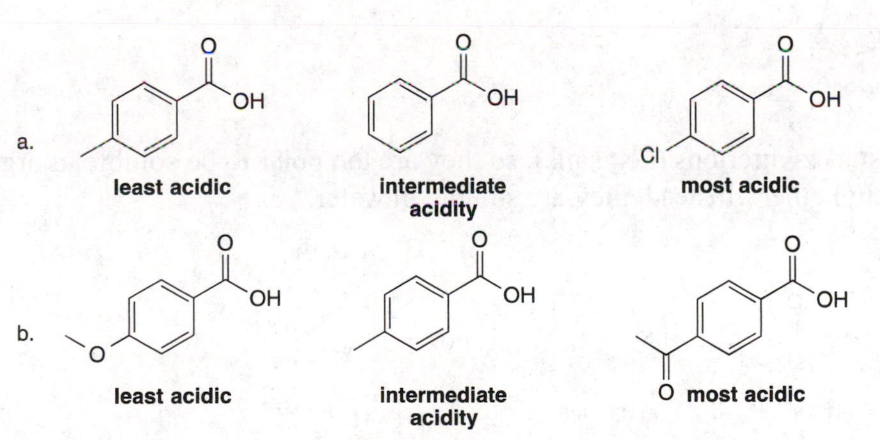

19.19

A

Phenol **A** has a higher pK_a than phenol because of its substituents. Both OH and $C_{15}H_{31}$ are electron-donating groups, which make the conjugate base less stable. Therefore, the acid is **less acidic.**

19.20 To separate compounds by an extraction procedure, they must have different solubility properties.

a. $CH_3(CH_2)_6COOH$ and $CH_3CH_2CH_2CH_2CH{=}CH_2$: **YES.** The acid can be extracted into aqueous base, while the alkene will remain in the organic layer.

b. $CH_3CH_2CH_2CH_2CH{=}CH_2$ and $(CH_3CH_2CH_2)_2O$: **NO.** Both compounds are soluble in organic solvents and insoluble in water. Neither is acidic enough to be extracted into aqueous base.

c. $CH_3(CH_2)_6COOH$ and NaCl: one carboxylic acid, one salt: **YES.** The carboxylic acid is soluble in an organic solvent while the salt is soluble in water.

d. NaCl and KCl: two salts: **NO.**

19.21

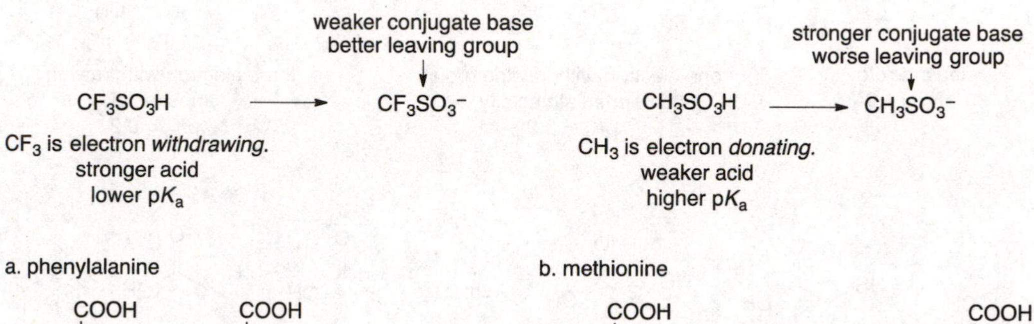

19.22

a. phenylalanine

R *S*

b. methionine

R *S*

19.23 Amino acids exist as zwitterions (i.e., salts), so they are too polar to be soluble in organic solvents like diethyl ether. Instead, they are soluble in water.

19.24

pH = 1

glycine
neutral form

pH = 11

19.25

$$pI = \frac{pK_a(COOH) + pK_a(NH_3^+)}{2} = \frac{(2.58) + (9.24)}{2} = 5.91$$

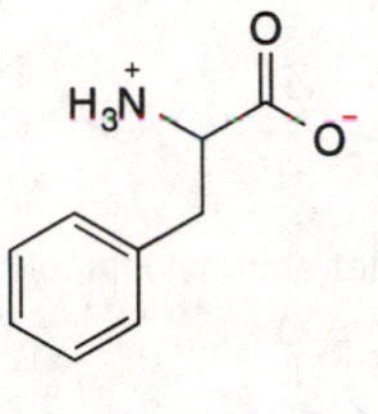

$H_3\overset{+}{N}$ O^-

19.26

$H_3\overset{+}{N}$ OH → $H_3\overset{+}{N}$ O^-

electron-withdrawing group

The nearby (+) stabilizes the conjugate base by an electron-withdrawing inductive effect, thus making the starting acid more acidic.

19.27

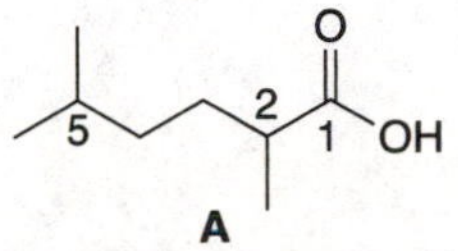

A

a. 2,5-dimethylhexanoic acid

NaOH

b. O^- Na^+

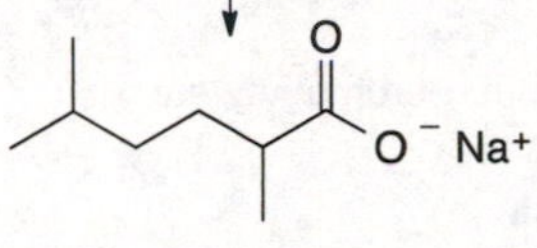

c. sodium 2,5-dimethylhexanoate

d. An alcohol or ether would have a much higher pK_a than a carboxylic acid.

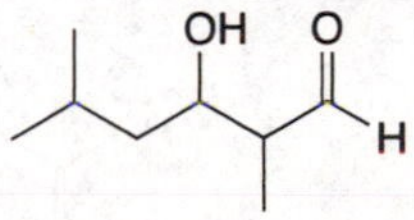

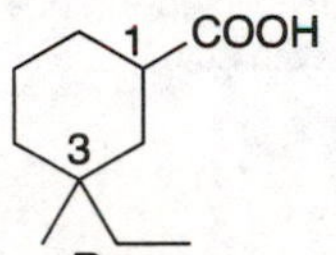

OH O H

B

COOH

a. 3-ethyl-3-methylcyclohexanecarboxylic acid

NaOH

b. COO^- Na^+

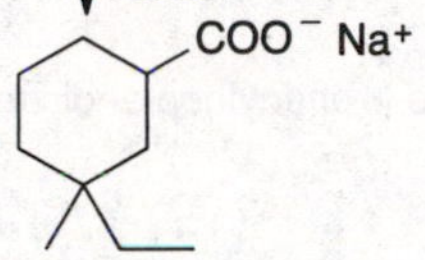

c. sodium 3-ethyl-3-methylcyclohexanecarboxylate

d. OH OH

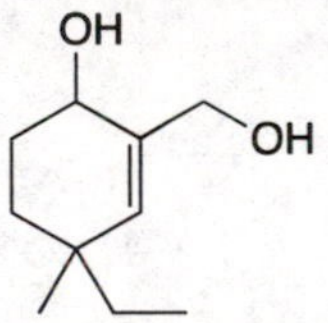

19.28

COOH COOH CF_3 COOH

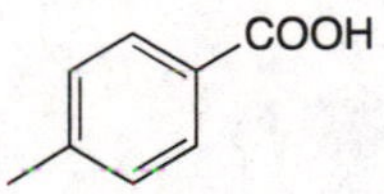

least acidic 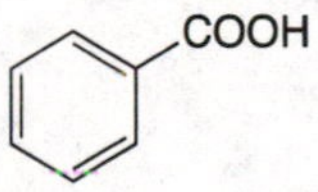 — intermediate acidity 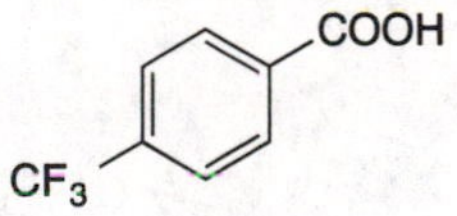 — most acidic

19.29 Use the directions from Answer 19.1 to name the compounds.

a.

4,4,5,5-tetramethyloctanoic acid

d. 5-bromo-4-ethyl-2-nitrobenzoic acid

b. lithium 2-ethylpentanoate

e. sodium 2-methylhexanoate

c. 1-ethyl-3-isobutylcyclopentane-carboxylic acid

f. 7-ethyl-5-isopropyl-3-methyldecanoic acid

19.30

a. 3,3-dimethylpentanoic acid

b. 4-chloro-3-phenylheptanoic acid

c. (*R*)-2-chloropropanoic acid

d. *m*-hydroxybenzoic acid

e. potassium acetate

f. sodium α-bromobutyrate

g. 2,2-dichloropentanedioic acid

h. 4-isopropyl-2-methyloctanedioic acid

19.31

a. C2 or α carbon

IUPAC: 2-hydroxy**propanoic acid**
common: α-hydroxy**propionic acid**

b. C5 or δ carbon; C3 or β carbon

IUPAC: 3,5-dihydroxy-3-methyl**pentanoic acid**
common: β,δ-dihydroxy-β-methyl**valeric acid**

19.32

lowest boiling point **intermediate boiling point** **highest boiling point**

19.33

a. (cyclohexylmethanol) $\xrightarrow[H_2SO_4,\ H_2O]{CrO_3}$ cyclohexanecarboxylic acid

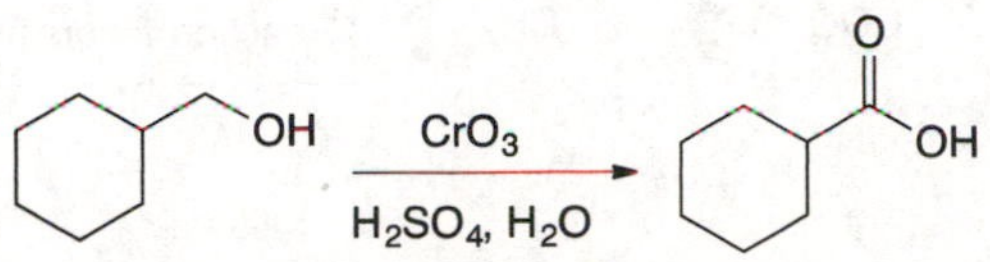

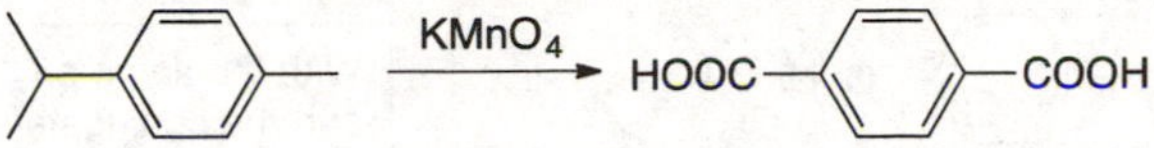

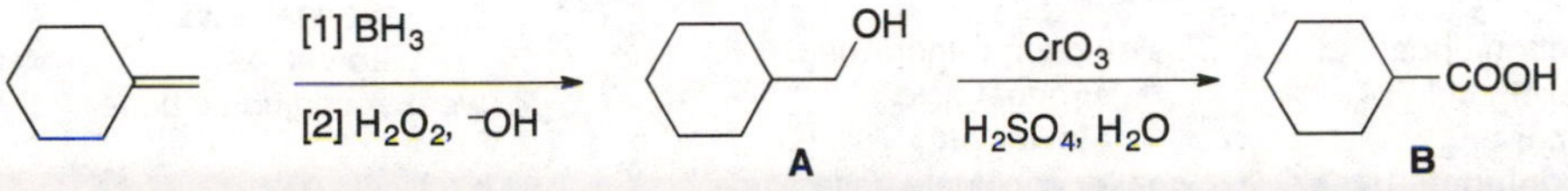

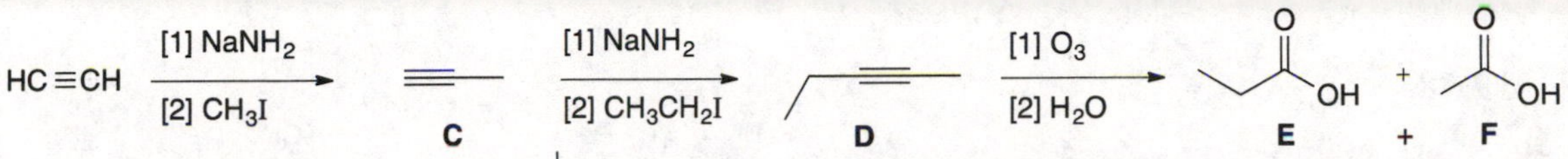

b. (p-isopropyltoluene) $\xrightarrow{KMnO_4}$ HOOC–C_6H_4–COOH

c. (ethynylcyclohexane) $\xrightarrow[[2]\ H_2O]{[1]\ O_3}$ C_6H_{11}–COOH + CO_2

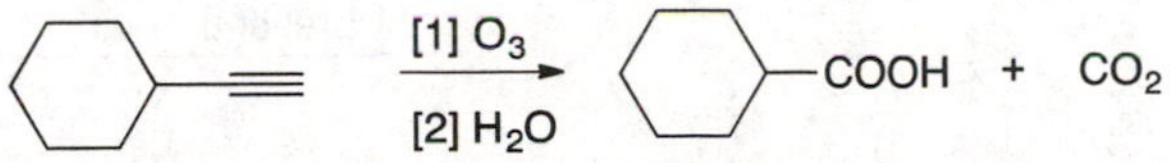

d. (octan-1-ol) OH $\xrightarrow[H_2SO_4,\ H_2O]{Na_2Cr_2O_7}$ (octanoic acid) OH

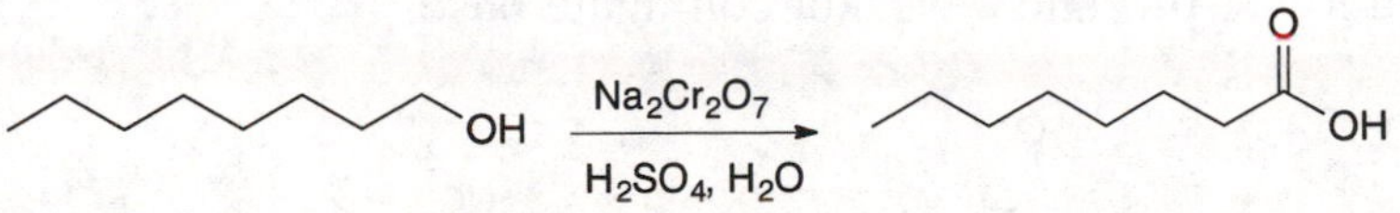

19.34

a. (methylenecyclohexane) $\xrightarrow[[2]\ H_2O_2,\ ^-OH]{[1]\ BH_3}$ (cyclohexylmethanol) OH **A** $\xrightarrow[H_2SO_4,\ H_2O]{CrO_3}$ C_6H_{11}–COOH **B**

b. HC≡CH $\xrightarrow[[2]\ CH_3I]{[1]\ NaNH_2}$ (propyne) **C** $\xrightarrow[[2]\ CH_3CH_2I]{[1]\ NaNH_2}$ (pent-2-yne) **D** $\xrightarrow[[2]\ H_2O]{[1]\ O_3}$ (propanoic acid) OH **E** + (acetic acid) OH **F**

c. (benzene) $\xrightarrow[AlCl_3]{(CH_3)_2CHCl}$ (isopropylbenzene) **G** $\xrightarrow{KMnO_4}$ (benzoic acid) COOH **H**

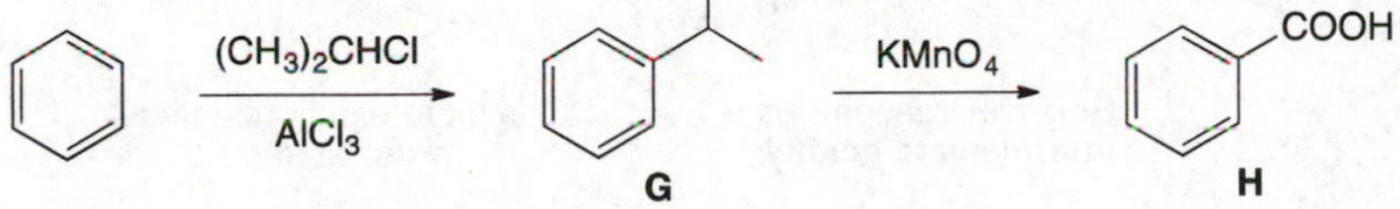

19.35

Bases: [1] ^-OH pK_a (H_2O) = 15.7; [2] $CH_3CH_2^-$ pK_a (CH_3CH_3) = 50; [3] $^-NH_2$ pK_a (NH_3) = 38; [4] NH_3 pK_a (NH_4^+) = 9.4; [5] HC≡C$^-$ pK_a (HC≡CH) = 25.

a. CH_3–C_6H_4–COOH
pK_a = 4.3
All of the bases can deprotonate this RCO_2H.

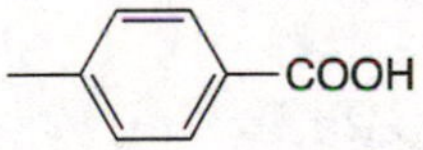

b. Cl–C_6H_4–OH
pK_a = 9.4
^-OH, $CH_3CH_2^-$, $^-NH_2$, and HC≡C$^-$ can deprotonate this phenol.

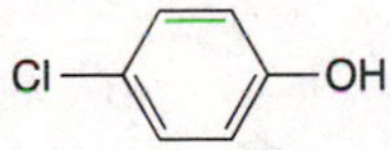

c. $(CH_3)_3COH$
pK_a = 18
$CH_3CH_2^-$, $^-NH_2$, and HC≡C$^-$ can deprotonate this ROH.

19.36

a. (pentan-1-ol) OH + NH_3 ⇌ (pentoxide) O^- + NH_4^+
pK_a ≈ 16 ; pK_a = 9.4
Reaction favors reactants.

b. C_6H_5–OH + Na^+ $^-NH_2$ ⇌ C_6H_5–O^- + NH_3 + Na^+
pK_a = 10 ; pK_a = 38
Reaction favors products.

c. (2-methylbenzoic acid) COOH + $CH_3^-Li^+$ ⇌ (2-methylbenzoate) $COO^-\ Li^+$ + CH_4

$pK_a \approx 4$ $\qquad$ $pK_a = 50$

Reaction favors products.

d. (4-methylphenol) OH + Na_2CO_3 ⇌ (4-methylphenoxide) O^-Na^+ + $Na^+\ HCO_3^-$

$pK_a = 10.2$ $\qquad$ $pK_a = 10.2$

With the same pK_a for the starting acid and the conjugate acid, an equal amount of starting materials and products is present.

19.37 The stronger acid has a lower pK_a and a weaker conjugate base.

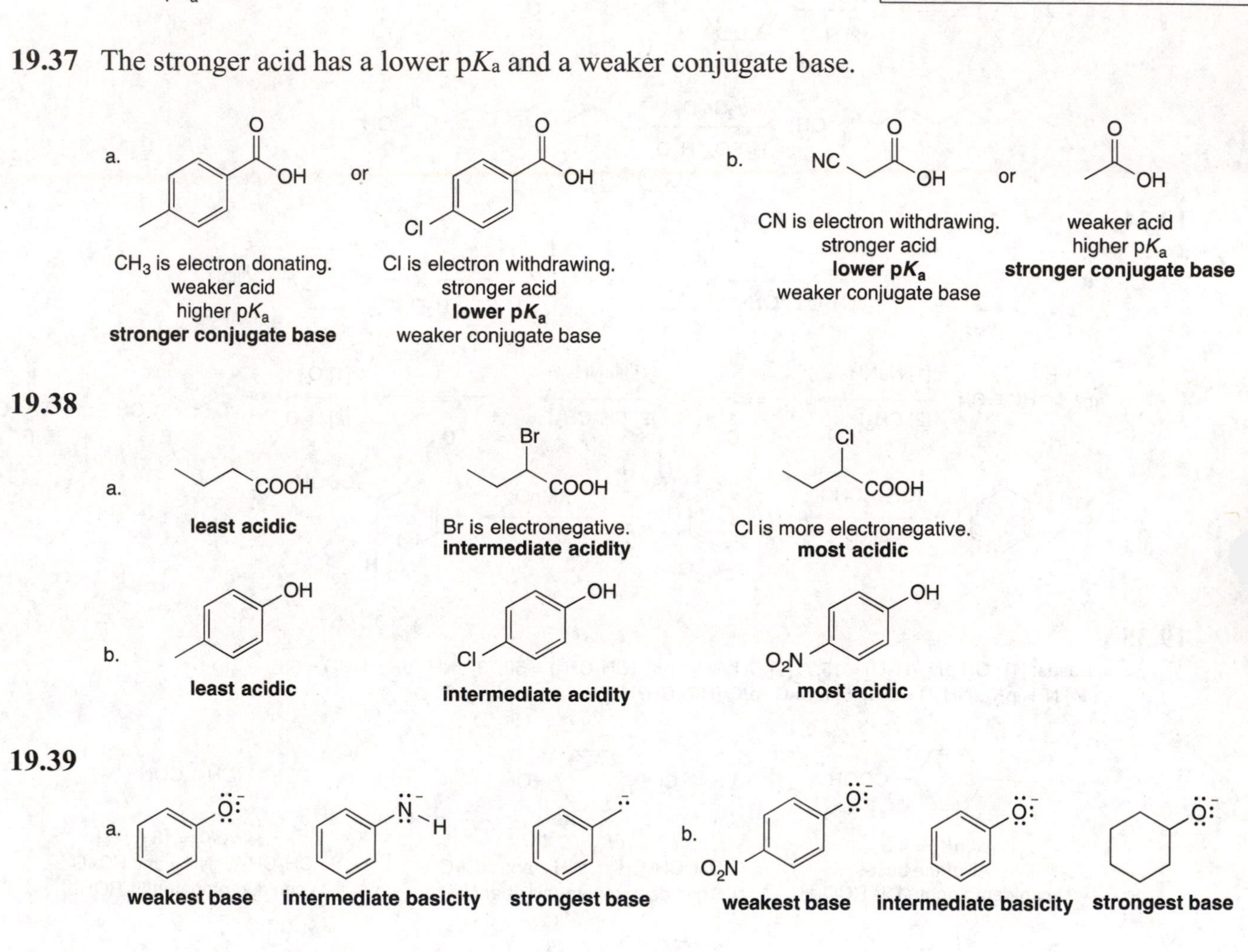

19.40

Increasing acidity →

pK_a values	ICH_2COOH	$BrCH_2COOH$	FCH_2COOH	$F_2CHCOOH$	F_3CCOOH
	least acidic 3.12	2.86	2.66	1.24	most acidic 0.28

19.41 The OH of the phenol group in morphine is more acidic than the OH of the alcohol (pK_a ≈ 10 versus pK_a ≈ 16). KOH is basic enough to remove the phenolic OH, the most acidic proton.

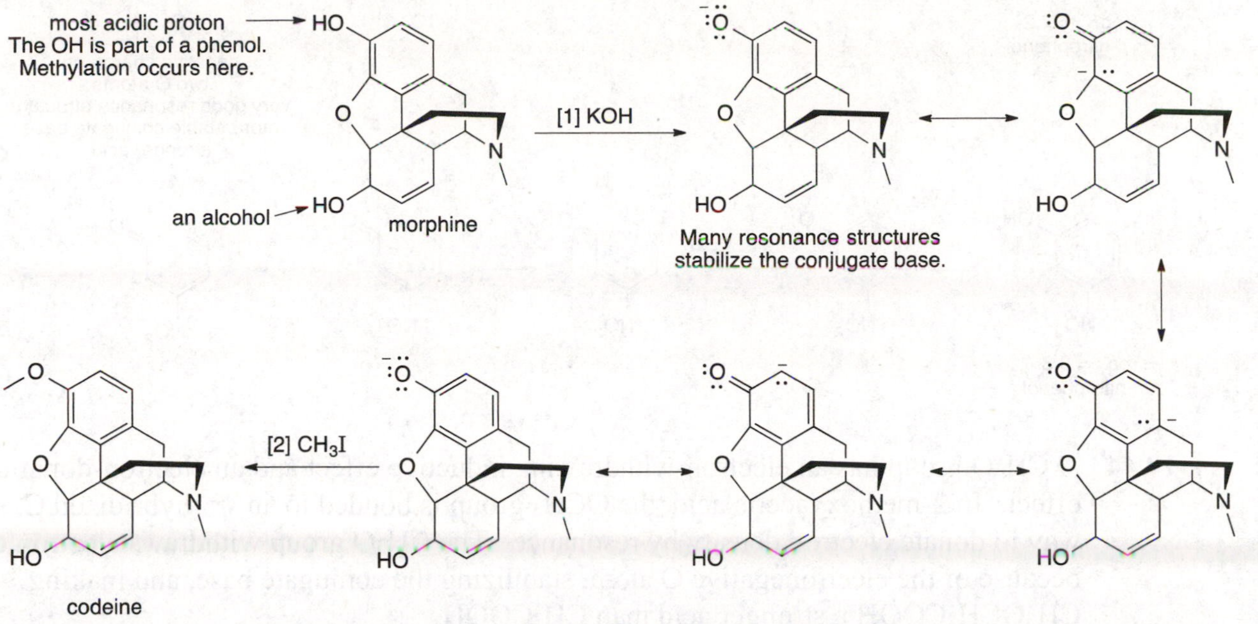

most acidic proton → HO
The OH is part of a phenol.
Methylation occurs here.
an alcohol → HO
morphine
[1] KOH
Many resonance structures stabilize the conjugate base.
[2] CH_3I
codeine

19.42 The closer the electron-withdrawing CH_3CO– group is to the carboxylic acid, the more it will stabilize the conjugate base, making the acid stronger.

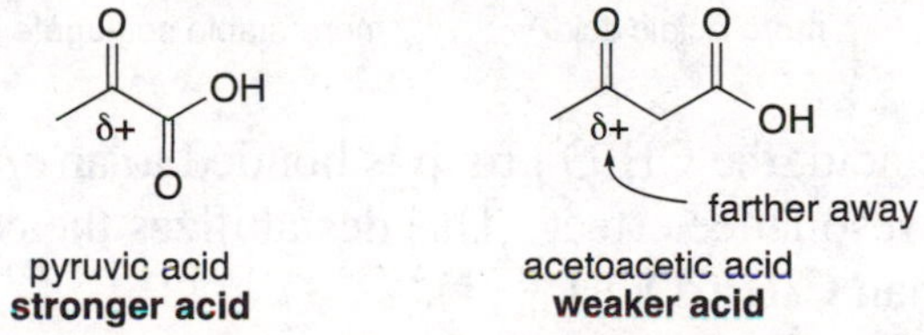

δ+
pyruvic acid
stronger acid
δ+
farther away
acetoacetic acid
weaker acid

19.43

a. The negative charge on the conjugate base of *p*-nitrophenol is delocalized on the NO_2 group, stabilizing the conjugate base, and making *p*-nitrophenol more acidic than phenol (where the negative charge is delocalized only around the benzene ring).

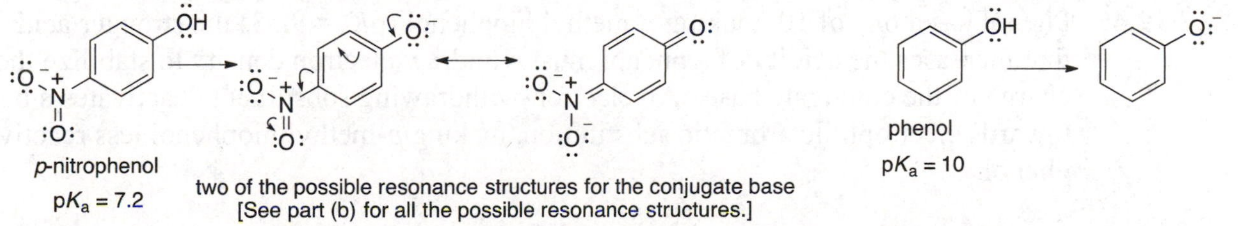

p-nitrophenol
pK_a = 7.2
two of the possible resonance structures for the conjugate base
[See part (b) for all the possible resonance structures.]
phenol
pK_a = 10

b. In the para isomer, the negative charge of the conjugate base is delocalized over both the benzene ring and onto the NO_2 group, whereas in the meta isomer it cannot be delocalized onto the NO_2 group. This makes the conjugate base from the para isomer more highly resonance stabilized, and the para-substituted phenol more acidic than its meta isomer.

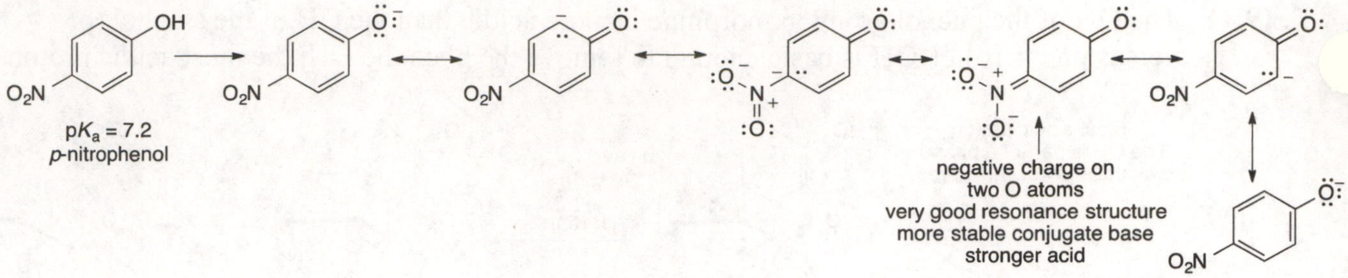

$pK_a = 7.2$
p-nitrophenol

negative charge on two O atoms
very good resonance structure
more stable conjugate base
stronger acid

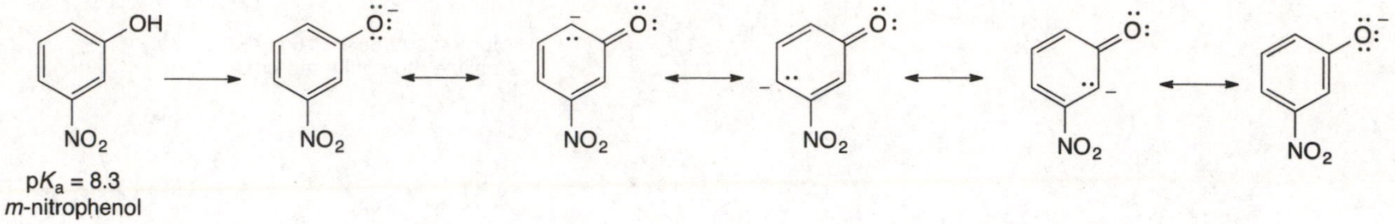

$pK_a = 8.3$
m-nitrophenol

19.44 A CH_3O group has an electron-withdrawing inductive effect and an electron-donating resonance effect. In 2-methoxyacetic acid, the OCH_3 group is bonded to an sp^3 hybridized C, so there is no way to donate electron density by resonance. The CH_3O group withdraws electron density because of the electronegative O atom, stabilizing the conjugate base, and making CH_3OCH_2COOH a stronger acid than CH_3COOH.

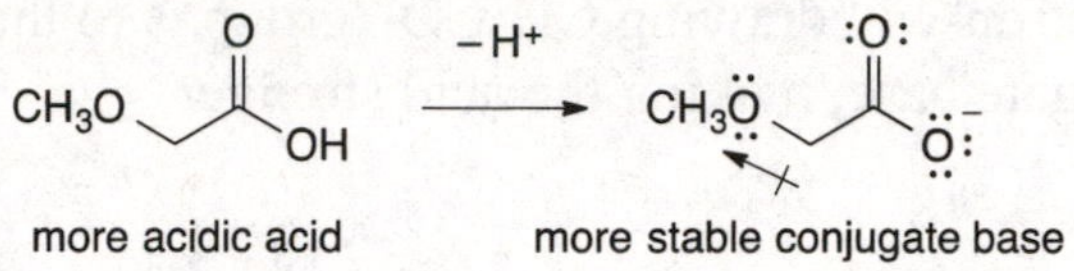

more acidic acid

more stable conjugate base

In *p*-methoxybenzoic acid, the CH_3O group is bonded to an sp^2 hybridized C, so it can donate electron density by a resonance effect. This destabilizes the conjugate base, making the starting material less acidic than C_6H_5COOH.

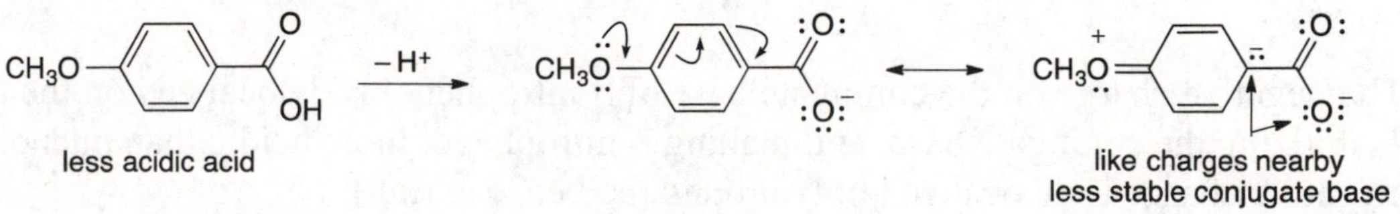

less acidic acid

like charges nearby
less stable conjugate base

19.45 Phenol has a pK_a of 10, making *p*-methylthiophenol (pK_a = 9.53) the stronger acid. A substituent that increases the acidity of a phenol must withdraw electron density to stabilize the negative charge of the conjugate base. An electron-withdrawing substituent deactivates a benzene ring towards electrophilic aromatic substitution, making *p*-methylthiophenol less reactive than phenol.

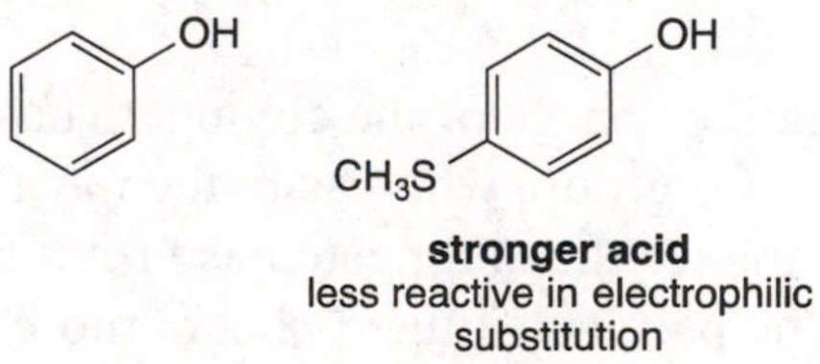

stronger acid
less reactive in electrophilic substitution

19.46

The O in **A** is more electronegative than the N in **C,** so there is a stronger electron-withdrawing inductive effect. This stabilizes the conjugate base of **A,** making **A** more acidic than **C.**

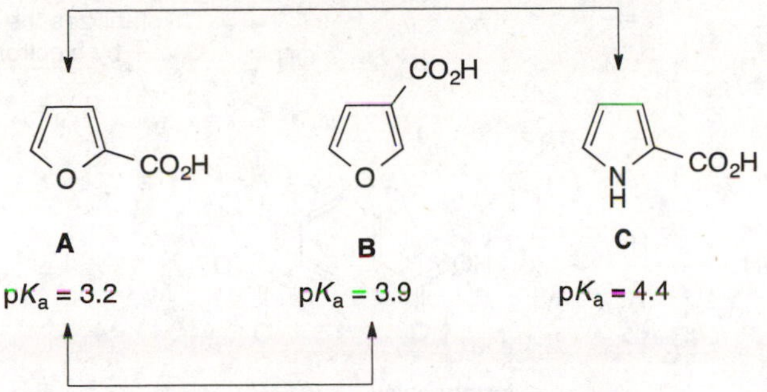

CO_2H CO_2H CO_2H

A **B** **C**

pK_a = 3.2 pK_a = 3.9 pK_a = 4.4

Since the O in **A** is closer to the COOH group than the O atom in **B,** there is a stronger electron-withdrawing inductive effect. This makes **A** more acidic than **B.**

19.47

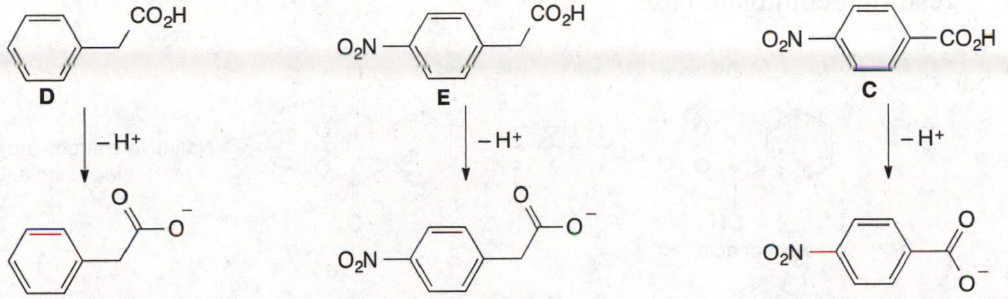

D **E** **C**

$-H^+$ $-H^+$ $-H^+$

Since the benzene ring is bonded to the α carbon (not the carbonyl carbon), this compound is not much different than any alkyl-substituted carboxylic acid.
least acidic

The electron-withdrawing inductive effect of the NO_2 group helps stabilize the COO^- group.
intermediate acidity

Since the NO_2 group is bonded to a benzene ring that is bonded directly to the carbonyl group, inductive effects and resonance effects stabilize the conjugate base. For example, a resonance structure can be drawn that places a (+) charge close to the COO^- group.
most acidic

Two of the resonance structures for the conjugate base of **C**:

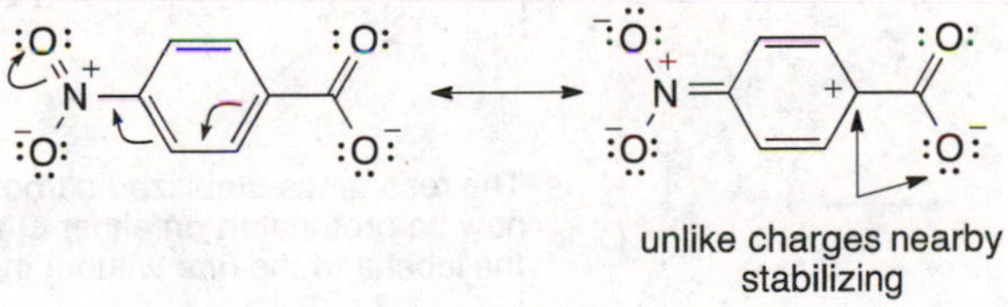

unlike charges nearby stabilizing

19.48 a. The pK_{a1} of phthalic acid is lower than the pK_{a1} of isophthalic acid because the electron-withdrawing CO_2H is closer to the negatively charged CO_2^- of the conjugate base.

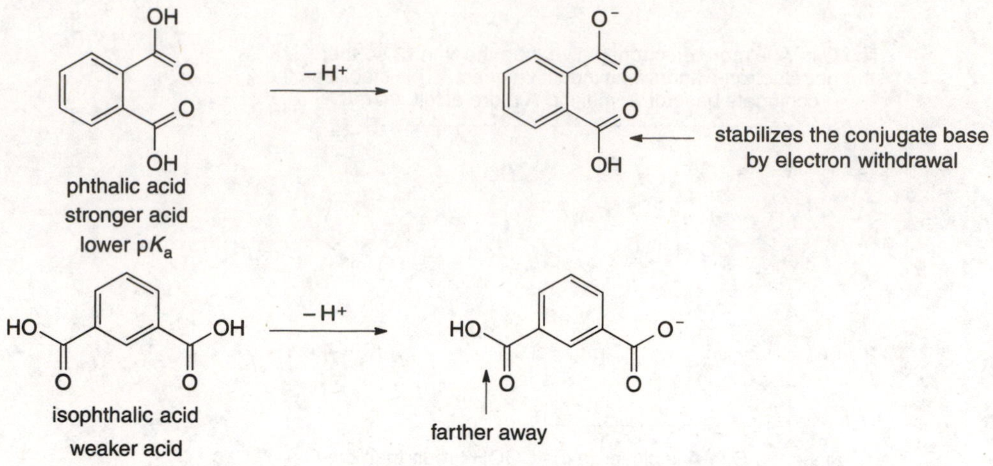

OH
O
O
OH
phthalic acid
stronger acid
lower pK_a

– H+

O⁻
O
O
OH
stabilizes the conjugate base by electron withdrawal

HO OH
O O
isophthalic acid
weaker acid

– H+

HO O⁻
O O
farther away

b. After loss of the first proton, each compound now has one CO_2H group and one CO_2^-. The CO_2^- is electron donating, so the closer it is located to the CO_2H, the more it destabilizes the resulting conjugate base.

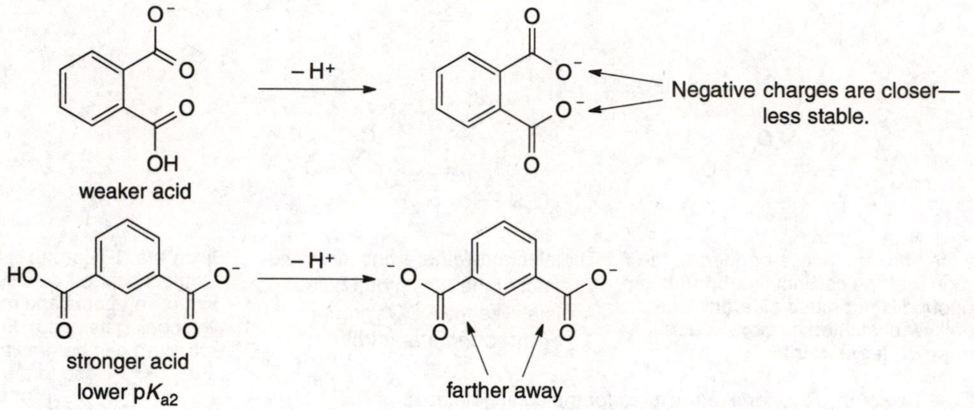

weaker acid

– H+

Negative charges are closer—less stable.

stronger acid
lower pK_{a2}

– H+

farther away

19.49

labeled O atom

NaOH

H_3O^+

The resonance-stabilized carboxylate anion can now be protonated on either O atom, the one with the label and the one without the label.

The label is now in two different locations.

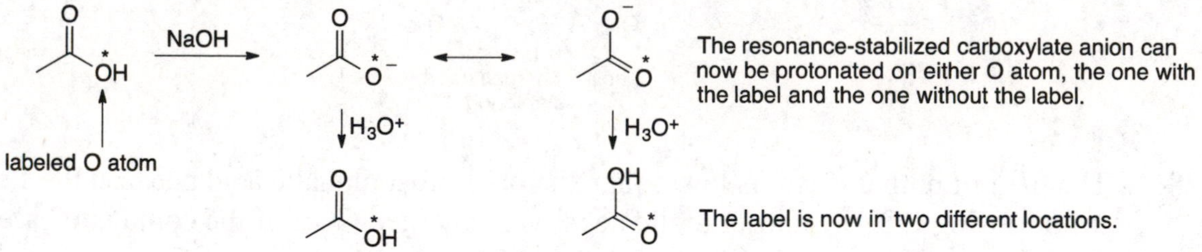

19.50

a.

cyclohexane-1,3-dione

increasing acidity: $H_b < H_a < H_c$

loss of H_b:

one Lewis structure

least stable conjugate base

The most acidic proton forms the most stable conjugate base.

loss of H_a:

2 resonance structures

intermediate stability

loss of H_c:

3 resonance structures

most stable conjugate base

b.

acetanilide

increasing acidity: $H_a < H_c < H_b$

loss of H_b:

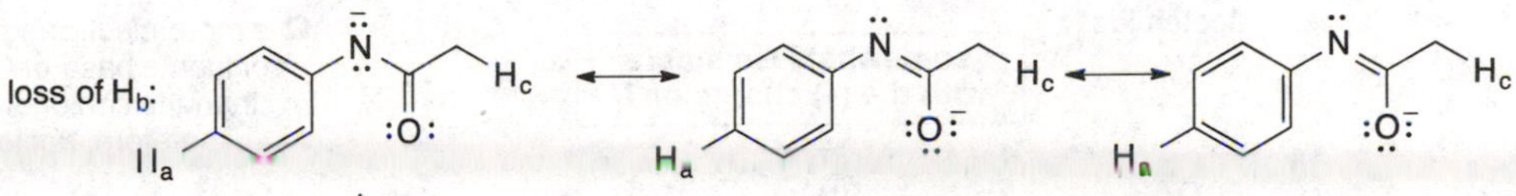

7 resonance structures

most stable conjugate base

loss of H_a:

one Lewis structure

least stable conjugate base

loss of H_c:

2 resonance structures that delocalize the negative charge

intermediate stability

19.51

The conjugate base is resonance stabilized. Two of the structures place a negative charge on an O atom.

weaker conjugate base

stronger acid

The conjugate base has only one Lewis structure.

stronger conjugate base

weaker acid

19.52

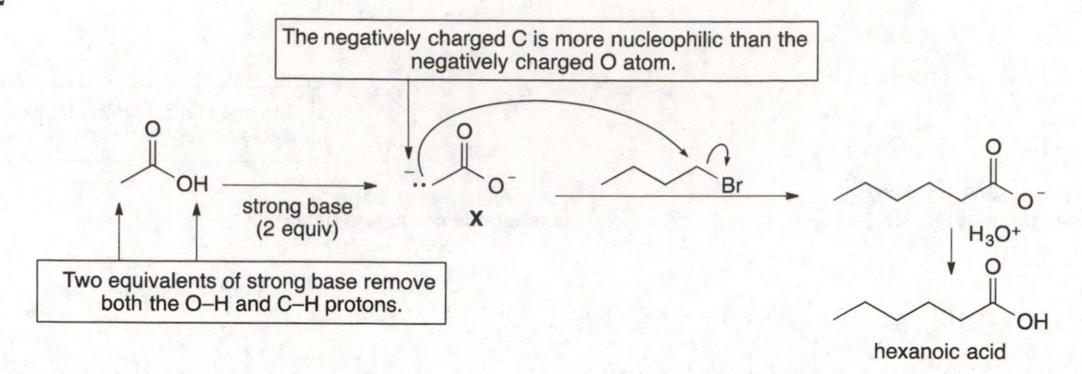

The negatively charged C is more nucleophilic than the negatively charged O atom.
OH
strong base
(2 equiv)
X
Br
H3O+
OH
hexanoic acid
Two equivalents of strong base remove both the O–H and C–H protons.

19.53

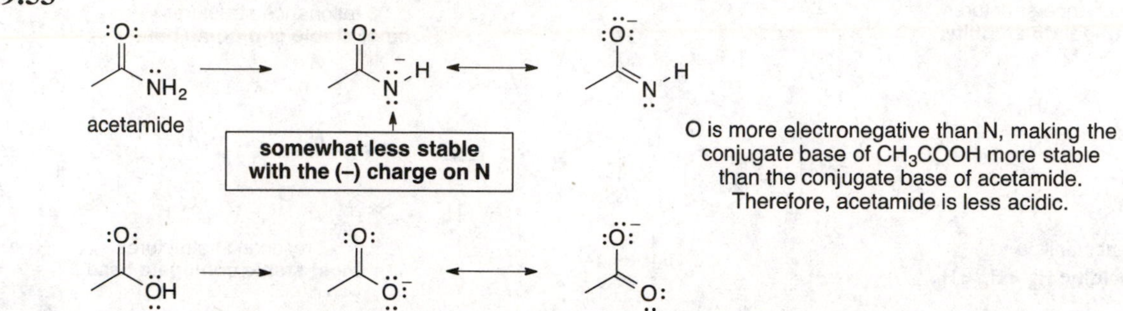

NH2
acetamide
somewhat less stable with the (–) charge on N
O is more electronegative than N, making the conjugate base of CH3COOH more stable than the conjugate base of acetamide. Therefore, acetamide is less acidic.
OH

19.54

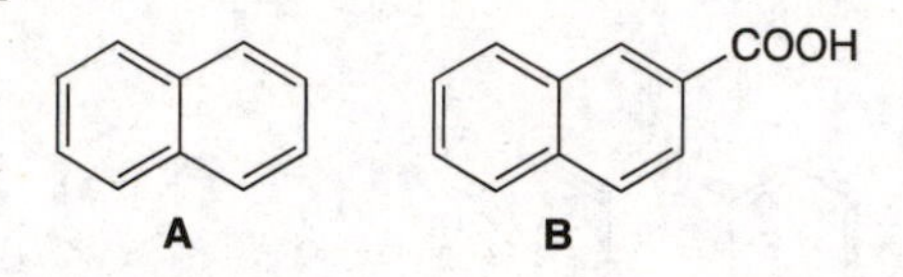

COOH
A
B

- Dissolve both compounds in CH_2Cl_2.
- Add 10% $NaHCO_3$ solution. This makes a carboxylate anion ($C_{10}H_7COO^-$) from **B,** which dissolves in the aqueous layer. The other compound (**A**) remains in the CH_2Cl_2.
- Separate the layers.

19.55

OH + O, OH + OH

CH_2Cl_2

10% $NaHCO_3$

aqueous → O, O^- Na+

CH_2Cl_2 → OH + OH

10% NaOH

aqueous → O^- Na+

CH_2Cl_2 → OH

19.56 To separate two compounds in an aqueous extraction, one must be water soluble (or be able to be converted into a water-soluble ionic compound by an acid–base reaction), and the other insoluble. Octan-1-ol has greater than 5 C's, making it insoluble in water. Octane is an alkane, so it, too, is insoluble in water. Neither compound is acidic enough to be deprotonated by a base in aqueous solution. Because their solubility properties are similar, they cannot be separated by an extraction procedure.

19.57

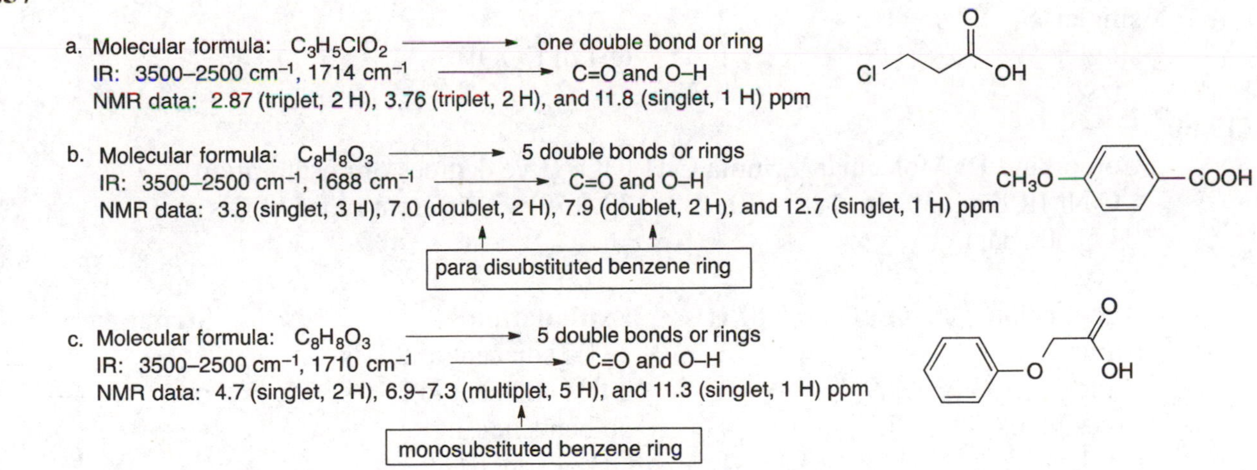

a. Molecular formula: $C_3H_5ClO_2$ → one double bond or ring
IR: 3500–2500 cm^{-1}, 1714 cm^{-1} → C=O and O–H
NMR data: 2.87 (triplet, 2 H), 3.76 (triplet, 2 H), and 11.8 (singlet, 1 H) ppm
Cl
O
OH
b. Molecular formula: $C_8H_8O_3$ → 5 double bonds or rings
IR: 3500–2500 cm^{-1}, 1688 cm^{-1} → C=O and O–H
NMR data: 3.8 (singlet, 3 H), 7.0 (doublet, 2 H), 7.9 (doublet, 2 H), and 12.7 (singlet, 1 H) ppm
para disubstituted benzene ring
CH_3O
COOH
c. Molecular formula: $C_8H_8O_3$ → 5 double bonds or rings
IR: 3500–2500 cm^{-1}, 1710 cm^{-1} → C=O and O–H
NMR data: 4.7 (singlet, 2 H), 6.9–7.3 (multiplet, 5 H), and 11.3 (singlet, 1 H) ppm
monosubstituted benzene ring
O
O
OH

19.58

Compound **A:** Molecular formula $C_4H_8O_2$ (one degree of unsaturation)
IR absorptions at 3600–3200 (O–H), 3000–2800 (C–H), and 1700 (C=O) cm^{-1}
^{1}H NMR data:

Absorption	ppm	# of H's	Explanation
singlet	2.2	3	a CH_3 group
singlet	2.55	1	1 H adjacent to none or OH
triplet	2.7	2	2 H's adjacent to 2 H's
triplet	3.8	2	2 H's adjacent to 2 H's

Structure:

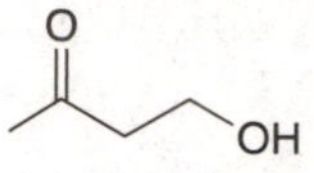

O OH A

Compound **B:** Molecular formula $C_4H_8O_2$ (one degree of unsaturation)
IR absorptions at 3500–2500 (O–H) and 1700 (C=O) cm^{-1}
^{1}H NMR data:

Absorption	ppm	# of H's	Explanation
doublet	1.0	6	6 H's adjacent to 1 H
septet	2.3	1	1 H adjacent to 6 H's
singlet (very broad)	10.6	1	OH of RCOOH

Structure:

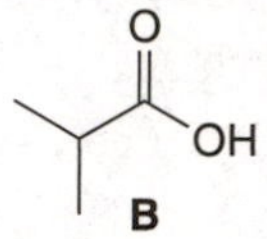

O OH B

19.59

Compound **C:** Molecular formula $C_4H_8O_3$ (one degree of unsaturation)
IR absorptions at 3600–2500 (O–H) and 1734 (C=O) cm^{-1}
^{1}H NMR data:

Absorption	ppm	# of H's	Explanation
triplet	1.3	3	a CH_3 group adjacent to 2 H's
quartet	3.6	2	2 H's adjacent to 3 H's
singlet	4.2	2	2 H's
singlet	11.3	1	OH of COOH

Structure:

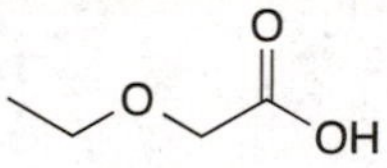

O O OH C

19.60

Compound **D:** Molecular formula $C_9H_9ClO_2$ (five degrees of unsaturation)
^{13}C NMR data: 30, 36, 128, 130, 133, 139, 179 = 7 different types of C's
^{1}H NMR data:

Absorption	ppm	# of H's	Explanation
triplet	2.7	2	2 H's adjacent to 2 H's
triplet	2.9	2	2 H's adjacent to 2 H's
two signals	7.2	4	on benzene ring
singlet	11.7	1	OH of COOH

Structure:

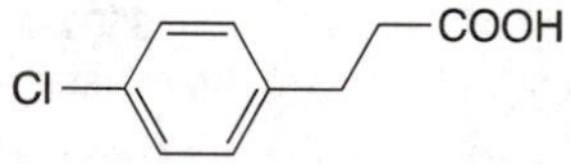

Cl COOH D

19.61

A COOH
3 different C's
Spectrum [2]: peaks at 27, 39, 186 ppm

B COOH
5 different C's
Spectrum [1]: peaks at 14, 22, 27, 34, 181 ppm

C O OH
4 different C's
Spectrum [3]: peaks at 22, 26, 43, 180 ppm

19.62

GBL: Molecular formula $C_4H_6O_2$ (two degrees of unsaturation)
IR absorption at 1770 (C=O) cm^{-1}
^{1}H NMR data:

Absorption	ppm	# of H's	Explanation
multiplet	2.28	2	2 H's adjacent to several H's
triplet	2.48	2	2 H's adjacent to 2 H's
triplet	4.35	2	2 H's adjacent to 2 H's

Structure:
O O
GBL

19.63

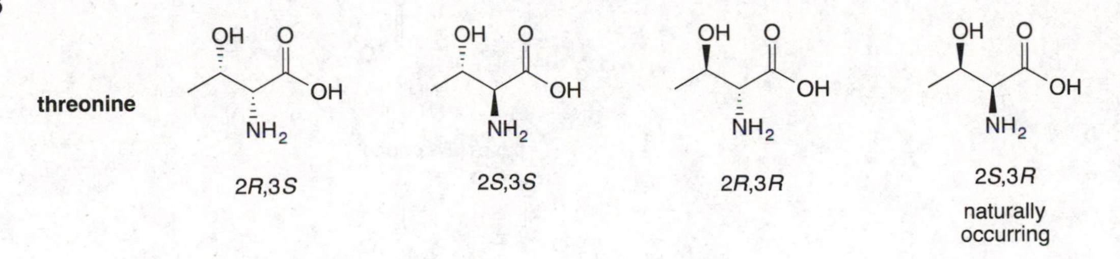

19.64

NH COOH **proline**
NH COOH enantiomer
$\overset{+}{NH_2}$ COO^- zwitterion

19.65

a. methionine

pH = 1 | pH = 6 form at isoelectric point | pH = 11

b. serine

pH = 1 | pH = 6 form at isoelectric point | pH = 11

19.66

a. cysteine $pI = \dfrac{pK_a(COOH) + pK_a(NH_3^+)}{2}$ = (2.05) + (10.25) / 2 = **6.15**

b. methionine $pI = \dfrac{pK_a(COOH) + pK_a(NH_3^+)}{2}$ = (2.28) + (9.21) / 2 = **5.75**

19.67

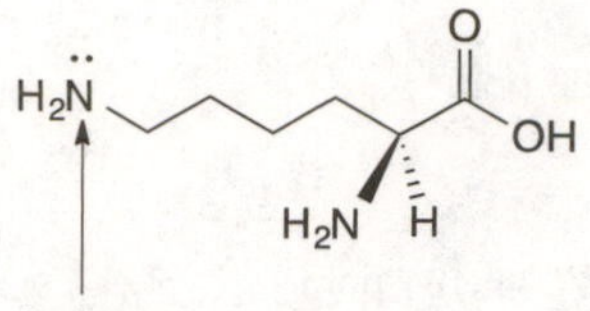

lysine
This lone pair is localized on the N atom, making it a base.

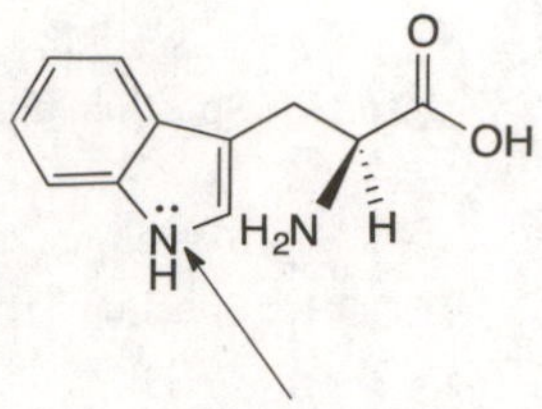

tryptophan
This lone pair is delocalized in the π system to give 10 π electrons, making it aromatic. This is similar to pyrrole (Chapter 17). Since these electrons are delocalized in the aromatic system, this N atom in tryptophan is not basic.

19.68

a. At pH = 1, the net charge is (+1).

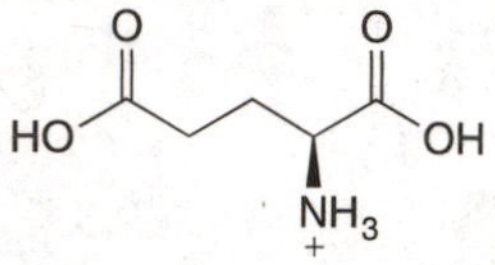

b. increasing pH: As base is added, the most acidic proton is removed first, then the next most acidic proton, and so forth.

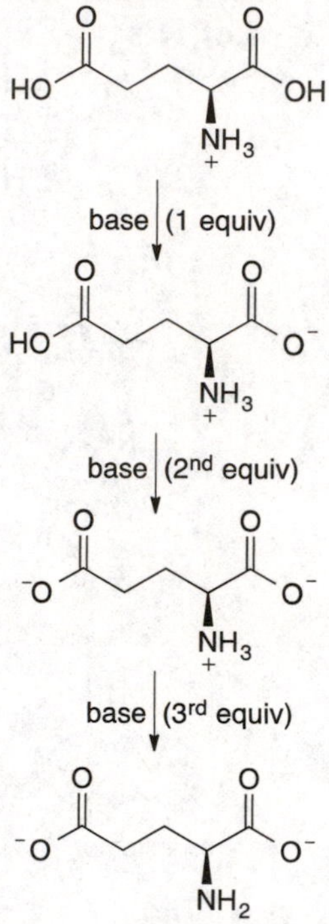

base (1 equiv)

base (2nd equiv)

base (3rd equiv)

c. monosodium glutamate

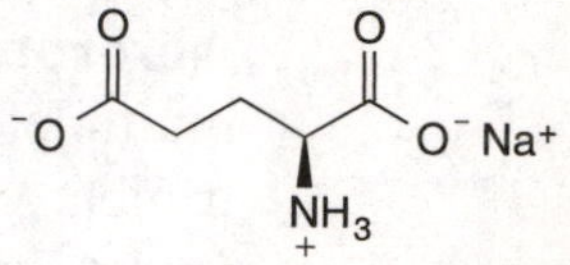

19.69 The first equivalent of NaH removes the most acidic proton—that is, the OH proton on the phenol. The resulting phenoxide can then act as a nucleophile to displace I to form a substitution product. With two equivalents, both OH protons are removed. In this case the more nucleophilic O atom is the stronger base—that is, the alkoxide derived from the alcohol (not the phenoxide), so this negatively charged O atom reacts first in a nucleophilic substitution reaction.

most acidic proton

HO, OH

NaH (1 equiv) → nucleophile → [1] CH_3I [2] H_2O

NaH (2 equiv) → nucleophile → [1] CH_3I [2] H_2O

HO

19.70

HO—C6H4—COOH

p-hydroxybenzoic acid
less acidic than benzoic acid

like charges on nearby atoms
destabilizing

The OH group donates electron density by its resonance effect and this destabilizes the conjugate base, making the acid less acidic than benzoic acid.

OH

COOH

o-hydroxybenzoic acid
more acidic than benzoic acid

Intramolecular hydrogen bonding stabilizes the conjugate base, making the acid more acidic than benzoic acid.

19.71

H_aO, H_d, OH_e, H_bO, H_c

2-hydroxybutanedioic acid
increasing acidity:
$H_d < H_c < H_b < H_e < H_a$

H_a and H_e must be the two most acidic protons because they are part of a carboxy group. Loss of a proton forms a resonance-stabilized carboxylate anion that has the negative charge delocalized on two O atoms. H_a is more acidic than H_e because the nearby OH group on the α carbon increases acidity by an electron-withdrawing inductive effect. H_b is the next most acidic proton because the conjugate base places a negative charge on the electronegative O atom, but it is not resonance stabilized.

The least acidic H's are H_c and H_d because these H's are bonded to C atoms. The electronegative O atom further acidifies H_c by an electron-withdrawing inductive effect.

19.72

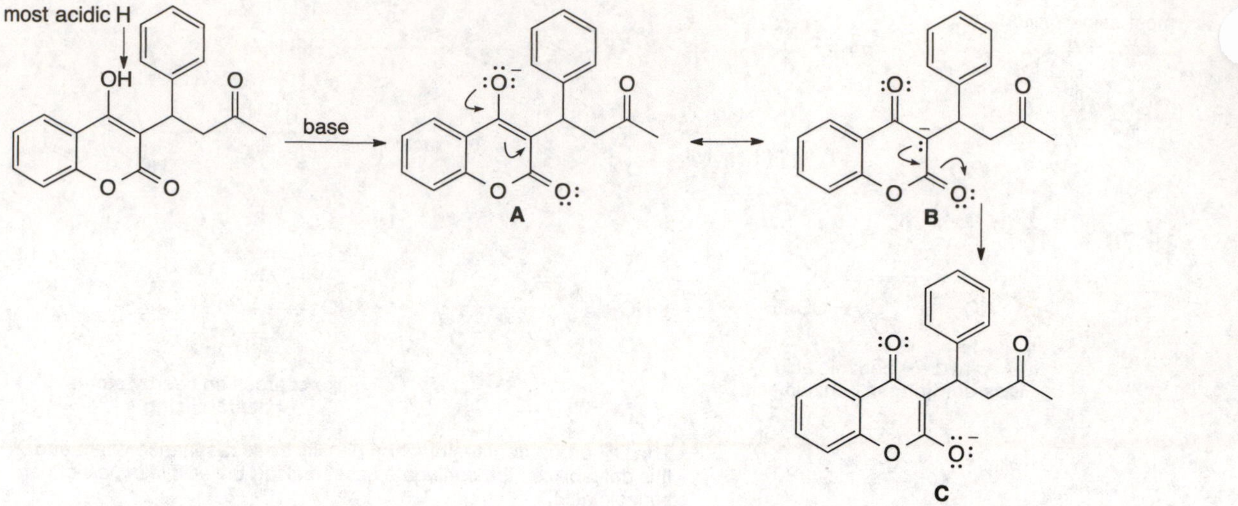

most acidic H

base

A

B

C

The conjugate base has three resonance structures, two of which place a negative charge on the oxygens. In this way the conjugate base resembles a carboxylate anion. In addition, the C=C's in **A** and **C** are conjugated.

Chapter 20 Introduction to Carbonyl Chemistry; Organometallic Reagents; Oxidation and Reduction

Chapter Review

Reduction reactions

[1] Reduction of aldehydes and ketones to 1° and 2° alcohols (20.4)

R' = H or alkyl

$NaBH_4$, CH_3OH
or
[1] $LiAlH_4$ [2] H_2O
or
H_2, Pd-C

1° or 2° alcohol

[2] Reduction of α,β-unsaturated aldehydes and ketones (20.4C)

$NaBH_4$, CH_3OH — • reduction of the C=O only

H_2 (1 equiv), Pd-C — • reduction of the C=C only

H_2 (excess), Pd-C — • reduction of both π bonds

[3] Enantioselective ketone reduction (20.6)

[1] (*S*)- or (*R*)-CBS reagent
[2] H_2O

(*R*) 2° alcohol or (*S*) 2° alcohol

- A single enantiomer is formed.

[4] Reduction of acid chlorides (20.7A)

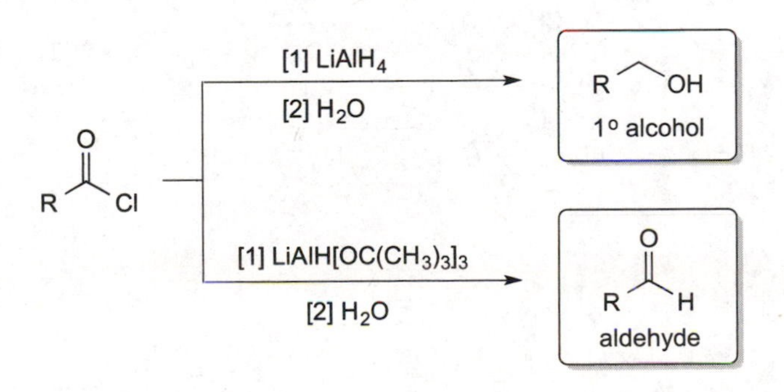

- $LiAlH_4$, a strong reducing agent, reduces an acid chloride to a 1° alcohol.
- With $LiAlH[OC(CH_3)_3]_3$, a milder reducing agent, reduction stops at the aldehyde stage.

[5] Reduction of esters (20.7A)

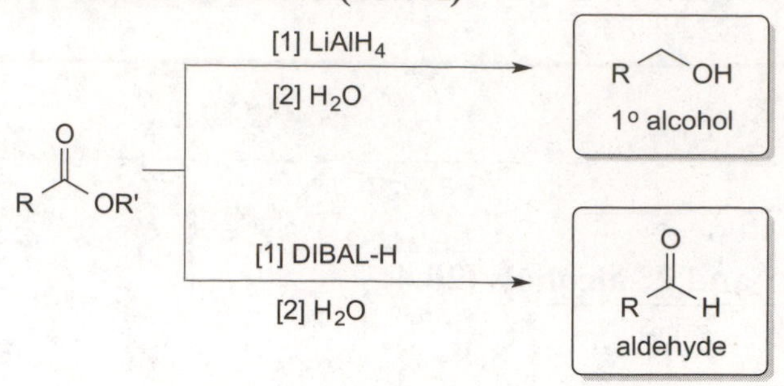

- $LiAlH_4$, a strong reducing agent, reduces an ester to a 1° alcohol.
- With DIBAL-H, a milder reducing agent, reduction stops at the aldehyde stage.

---

[6] Reduction of carboxylic acids to 1° alcohols (20.7B)

RCOOH —[1] $LiAlH_4$; [2] H_2O→ RCH_2OH (1° alcohol)

---

[7] Reduction of amides to amines (20.7B)

$RCON(CH_3)_2$ —[1] $LiAlH_4$; [2] H_2O→ $RCH_2N(CH_3)_2$ (amine)

Oxidation reactions

Oxidation of aldehydes to carboxylic acids (20.8)

RCHO —CrO_3, $Na_2Cr_2O_7$, $K_2Cr_2O_7$, $KMnO_4$ or Ag_2O, NH_4OH→ RCOOH (carboxylic acid)

- All Cr^{6+} reagents except PCC oxidize RCHO to RCOOH.
- Tollens reagent ($Ag_2O + NH_4OH$) oxidizes RCHO only. Primary (1°) and secondary (2°) alcohols do not react with Tollens reagent.

Preparation of organometallic reagents (20.9)

[1] Organolithium reagents:

$$R{-}X + 2\,Li \longrightarrow R{-}Li + LiX$$

[2] Grignard reagents:

$$R{-}X + Mg \xrightarrow{(CH_3CH_2)_2O} R{-}Mg{-}X$$

[3] Organocuprate reagents:

$$R{-}X + 2\,Li \longrightarrow R{-}Li + LiX$$

$$2\,R{-}Li + CuI \longrightarrow R_2Cu^{-}\,Li^{+} + LiI$$

[4] Lithium and sodium acetylides:

$$R{-}C{\equiv}C{-}H \xrightarrow{Na^+ \ {}^-NH_2} \boxed{R{-}C{\equiv}C^- \ Na^+} + NH_3$$

a sodium acetylide

$$R{-}C{\equiv}C{-}H \xrightarrow{R{-}Li} \boxed{R{-}C{\equiv}C{-}Li} + R{-}H$$

a lithium acetylide

Reactions with organometallic reagents

[1] Reaction as a base (20.9C)

$$R{-}M + H{-}\ddot{O}{-}R \longrightarrow R{-}H + M^+ + {:}\ddot{O}^-{-}R$$

- RM = RLi, RMgX, R_2CuLi
- This acid–base reaction occurs with H_2O, ROH, RNH_2, R_2NH, RSH, RCOOH, $RCONH_2$, and RCONHR.

[2] Reaction with aldehydes and ketones to form 1°, 2°, and 3° alcohols (20.10)

RCOR' (R' = H or alkyl) → [1] R"MgX or R"Li; [2] H_2O → RR'R"C–OH

1°, 2°, or 3° alcohol

[3] Reaction with esters to form 3° alcohols (20.13A)

RCOOR' → [1] R"Li or R"MgX (2 equiv); [2] H_2O → RR"R"C–OH

3° alcohol

[4] Reaction with acid chlorides (20.13)

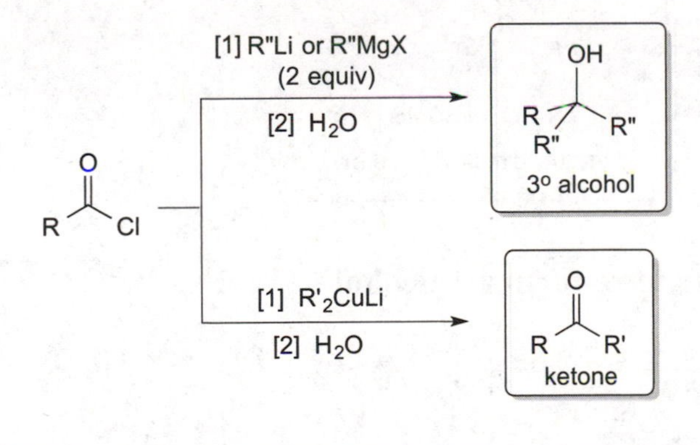

- More reactive organometallic reagents—R"Li and R"MgX—add two equivalents of R" to an acid chloride to form a 3° alcohol with two identical R" groups.
- Less reactive organometallic reagents—R'_2CuLi—add only one equivalent of R' to an acid chloride to form a ketone.

[5] Reaction with carbon dioxide—Carboxylation (20.14A)

R–MgX → [1] CO_2 [2] H_3O^+ → R–COOH

carboxylic acid

[6] Reaction with epoxides (20.14B)

epoxide → [1] RLi, RMgX, or R_2CuLi [2] H_2O → alcohol

[7] Reaction with α,β-unsaturated aldehydes and ketones (20.15B)

α,β-unsaturated carbonyl → [1] R'Li or R'MgX [2] H_2O → allylic alcohol

α,β-unsaturated carbonyl → [1] R'_2CuLi [2] H_2O → ketone

- More reactive organometallic reagents—R'Li and R'MgX—react with α,β-unsaturated carbonyls by 1,2-addition.
- Less reactive organometallic reagents—R'_2CuLi— react with α,β-unsaturated carbonyls by 1,4-addition.

Protecting groups (20.12)

[1] Protecting an alcohol as a *tert*-butyldimethylsilyl ether

R–O–H + Cl–Si [Cl–TBDMS] → (imidazole) → [R–O–TBDMS]

tert-butyldimethylsilyl ether

[2] Deprotecting a *tert*-butyldimethylsilyl ether to re-form an alcohol

[R–O–TBDMS] → $Bu_4N^+F^-$ → R–O–H + [F–TBDMS]

Practice Test on Chapter Review

1. Which compounds undergo nucleophilic addition and which undergo substitution?

a. 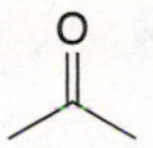 b. 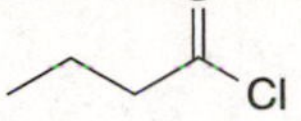 Cl c. 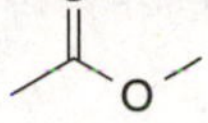 O d. 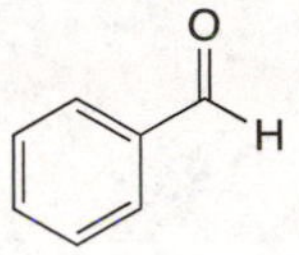 H

2. What product is formed when $CH_3CH_2CH_2Li$ reacts with each compound, followed by quenching with water and acid?

a. CH_3CH_2CHO
b. $(CH_3)_2CO$
c. $CH_3CH_2CO_2CH_3$
d. CH_3CH_2COCl
e. CO_2
f. $CH_2{=}CHCOCH_3$
g. ethylene oxide
h. CH_3COOH

3. What product is formed when $HO(CH_2)_4CHO$ is treated with each reagent?

a. $NaBH_4$, CH_3OH
b. PCC
c. Ag_2O, NH_4OH
d. $Na_2Cr_2O_7$, H_2SO_4, H_2O

4. What reagent is needed to convert $(CH_3CH_2)_2CHCOCl$ into each compound?

a. $(CH_3CH_2)_2CHCOCH_2CH_3$
b. $(CH_3CH_2)_2CHCHO$
c. $(CH_3CH_2)_2CHC(OH)(CH_2CH_3)_2$
d. $(CH_3CH_2)_2CHCH_2OH$

5. Draw the organic products formed in the following reactions.

a. 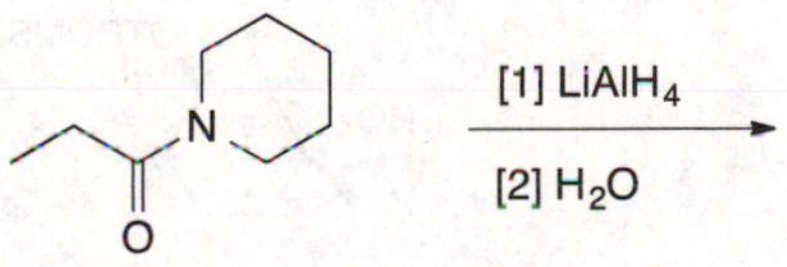 [1] $LiAlH_4$ [2] H_2O

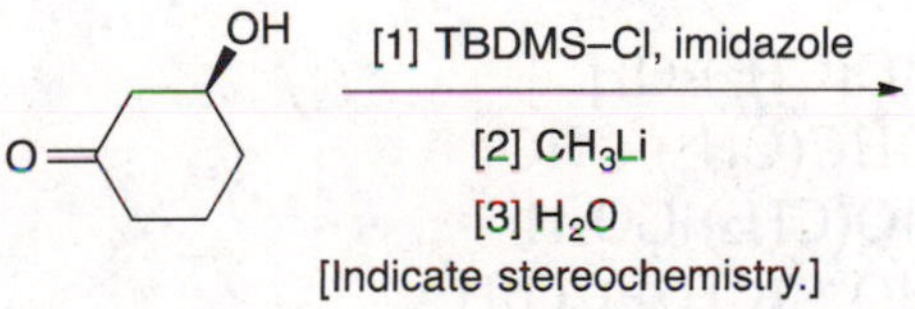

b. 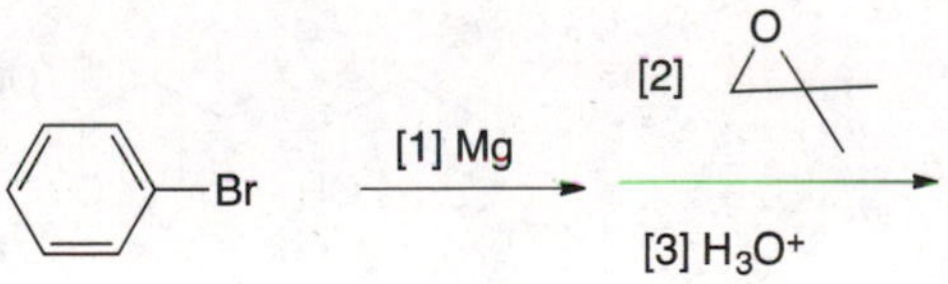 Br [1] Mg [2] (epoxide) [3] H_3O^+

c. OH [1] TBDMS–Cl, imidazole [2] CH_3Li [3] H_2O [Indicate stereochemistry.]

d. 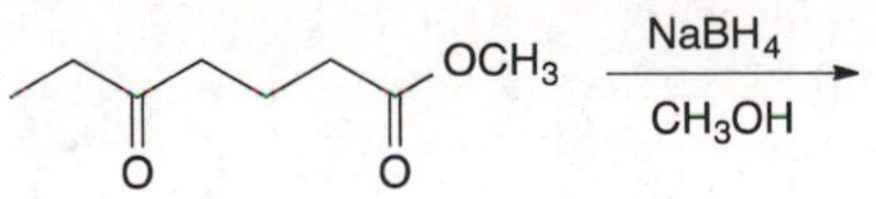 OCH_3 $NaBH_4$, CH_3OH

6. What starting materials are needed to synthesize each compound using the indicated reagent or functional group?

a. Synthesize: OH

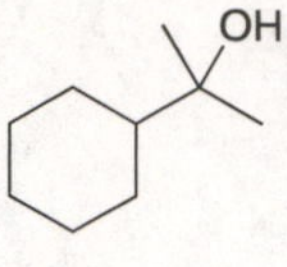

from an ester

c. Synthesize: OH

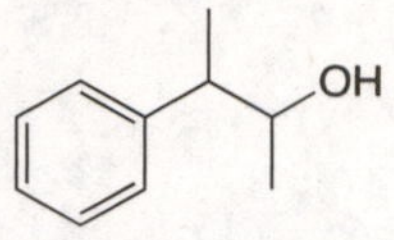

using a Grignard reagent

b. Synthesize: O

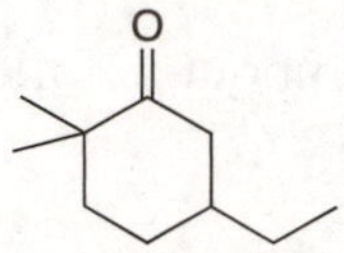

using an organocuprate reagent

Answers to Practice Test

1.a. addition
b. substitution
c. substitution
d. addition

2. a. $CH_3CH_2CH(OH)CH_2CH_2CH_3$
b. $(CH_3)_2C(OH)CH_2CH_2CH_3$
c. $CH_3CH_2C(OH)(CH_2CH_2CH_3)_2$
d. $CH_3CH_2C(OH)(CH_2CH_2CH_3)_2$
e. $CH_3CH_2CH_2CO_2H$
f. $CH_2{=}CHC(OH)(CH_3)CH_2CH_2CH_3$
g. $CH_3(CH_2)_4OH$
h. $CH_3CH_2CH_3$ + CH_3CO_2H

3.a. $HO(CH_2)_5OH$
b. $OHC(CH_2)_3CHO$
c. $HO(CH_2)_4CO_2H$
d. $HO_2C(CH_2)_3CO_2H$

4.a. $(CH_3CH_2)_2CuLi$
b. $LiAlH[OC(CH_3)_3]_3$
c. CH_3CH_2MgBr
d. $LiAlH_4$

5.
a. 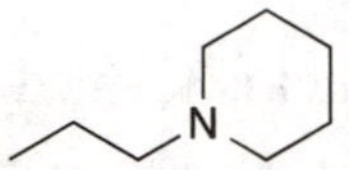 N
b. 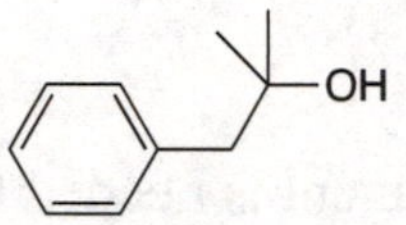 OH
c. 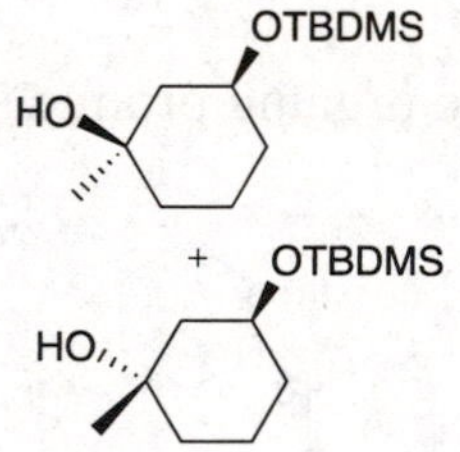 HO OTBDMS + HO OTBDMS
d. 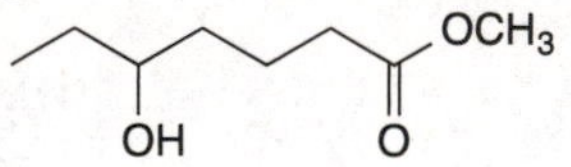 OCH3 OH O

6.
a. 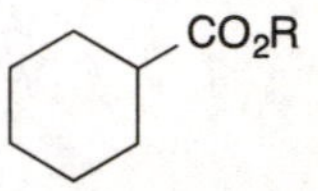 CO_2R + CH_3MgBr
b. 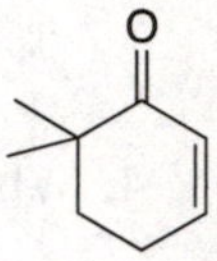 O +
c. 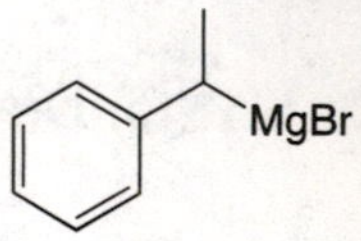 MgBr + O H

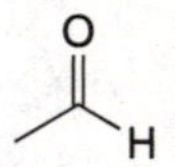

Answers to Problems

20.1

a. 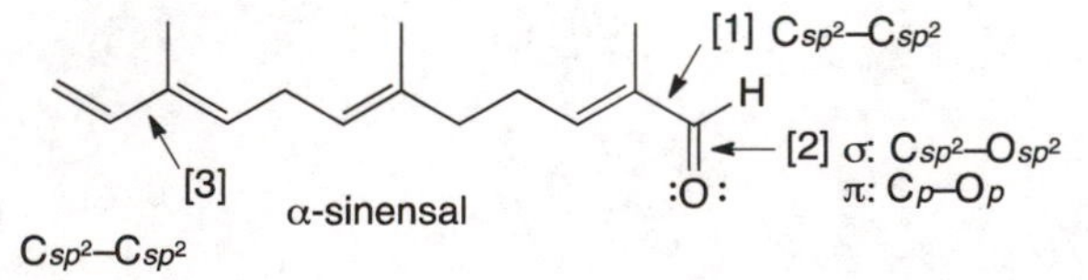 [1] C_{sp^2}–C_{sp^2} H [2] σ: C_{sp^2}–O_{sp^2} π: C_p–O_p :O: [3] α-sinensal C_{sp^2}–C_{sp^2}

b. The O is sp^2 hybridized.
Both lone pairs occupy sp^2 hybrid orbitals.

20.2 A carbonyl compound with a leaving group (NR_2 or OR bonded to the C=O) undergoes substitution reactions. Those without leaving groups undergo addition.

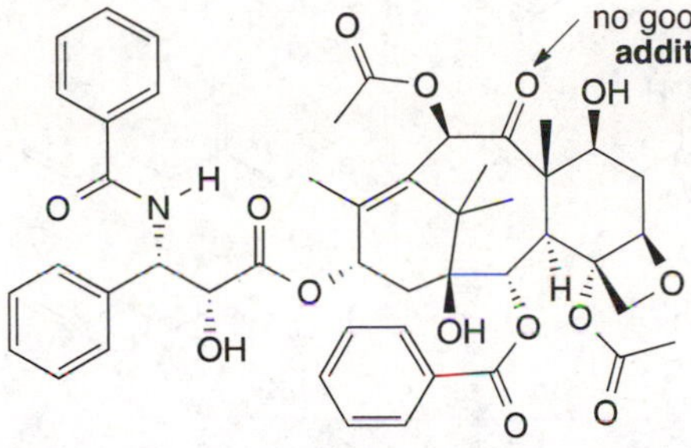

no good leaving group
addition reactions

All other C=O's have leaving groups.
substitution reactions

20.3 Aldehydes are more reactive than ketones. In carbonyl compounds with leaving groups, the better the leaving group, the more reactive the carbonyl compound.

a. 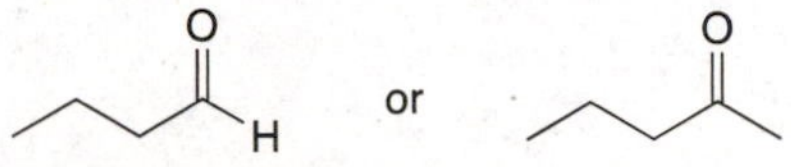 less hindered carbonyl **more reactive** (aldehyde) or (ketone)

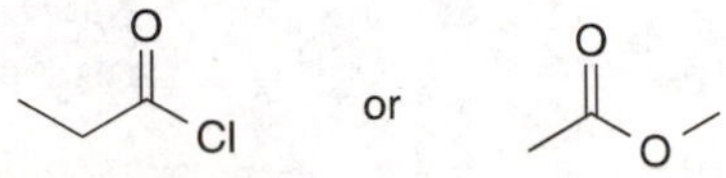

b. 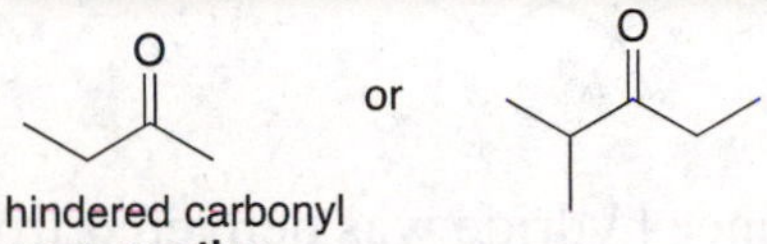 less hindered carbonyl **more reactive** or

c. better leaving group **more reactive** (Cl) or

d. 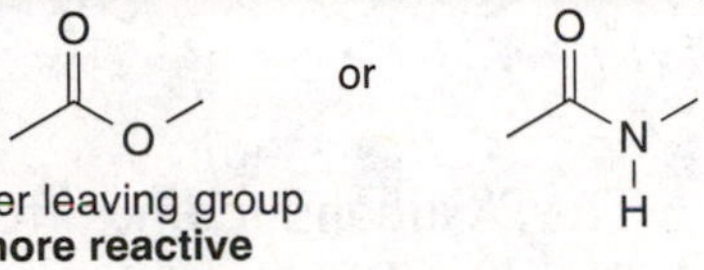 better leaving group **more reactive** or

20.4 $NaBH_4$ reduces aldehydes to 1° alcohols and ketones to 2° alcohols.

a. 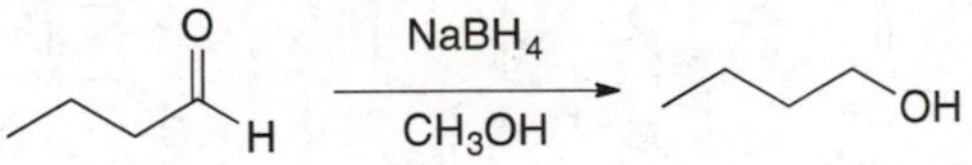 $\xrightarrow[CH_3OH]{NaBH_4}$ OH

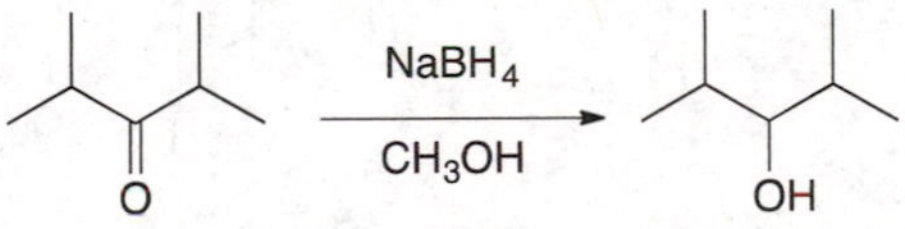

b. 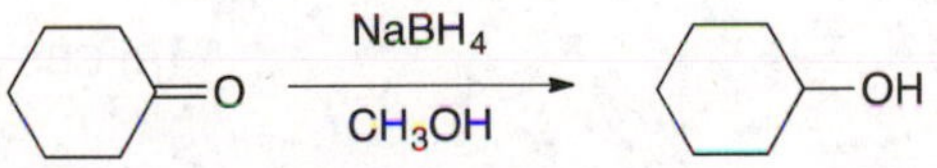 =O $\xrightarrow[CH_3OH]{NaBH_4}$ –OH

c. $\xrightarrow[CH_3OH]{NaBH_4}$ OH

20.5 1° Alcohols are prepared from aldehydes and 2° alcohols are from ketones.

a. 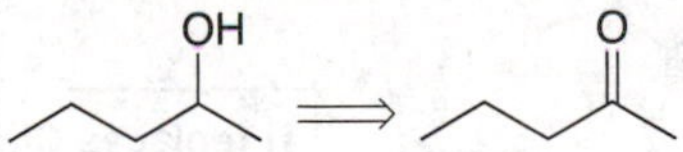 OH ⟹ O

b. 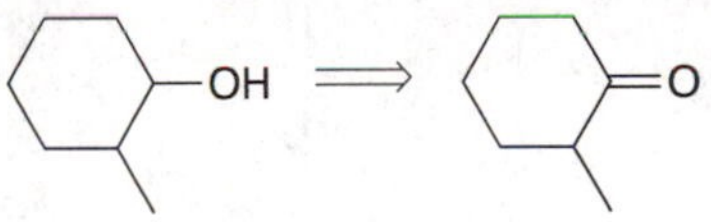 –OH ⟹ =O

c. 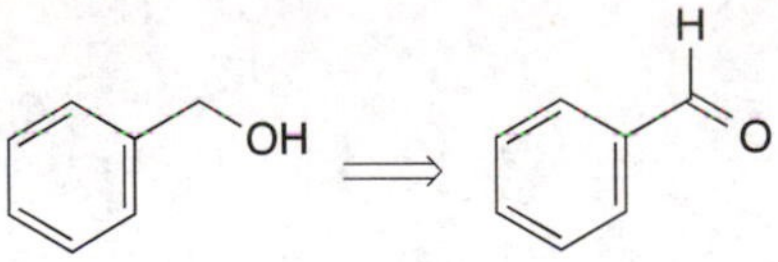 OH ⟹ H, O

20.6

–OH

1-methylcyclohexanol

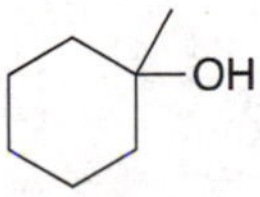

3° Alcohols cannot be made by reduction of a carbonyl group, because they do not contain a H on the C with the OH.

20.7

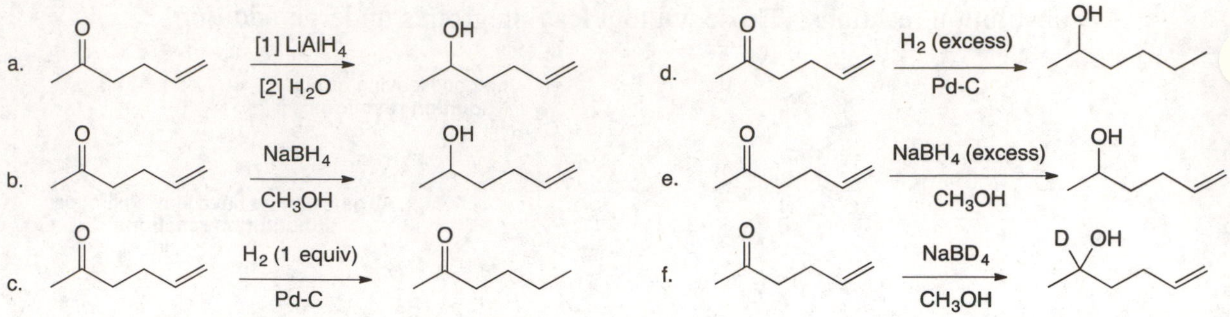

a. [1] $LiAlH_4$ / [2] H_2O

b. $NaBH_4$ / CH_3OH

c. H_2 (1 equiv) / Pd-C

d. H_2 (excess) / Pd-C

e. $NaBH_4$ (excess) / CH_3OH

f. $NaBD_4$ / CH_3OH

20.8

a. $NaBH_4$ / CH_3OH

b. $NaBH_4$ / CH_3OH

c. $NaBH_4$ / CH_3OH

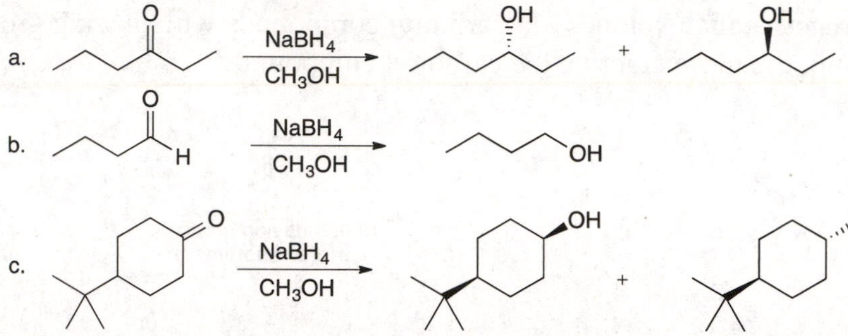

20.9 The 2° alcohol comes from a carbonyl group. Since hydride was delivered from the back side, the (*R*)-CBS reagent must be used.

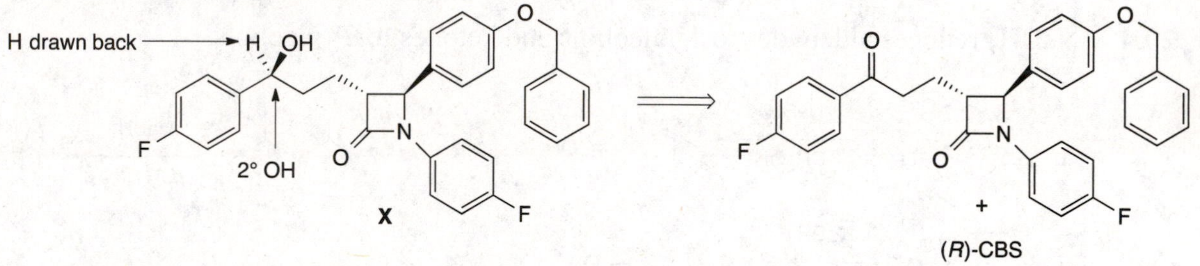

H drawn back → H OH

2° OH

X

(*R*)-CBS

20.10

Part [1]: Nucleophilic substitution of H for Cl

$H_3\bar{Al}$—H [1] + AlH_3 [2] aldehyde + Cl^-

H replaces Cl.

Part [2]: Nucleophilic addition of H^- to form an alcohol

$H_3\bar{Al}$—H [3] + AlH_3 H–OH [4] 1° alcohol + $:\ddot{O}H^-$

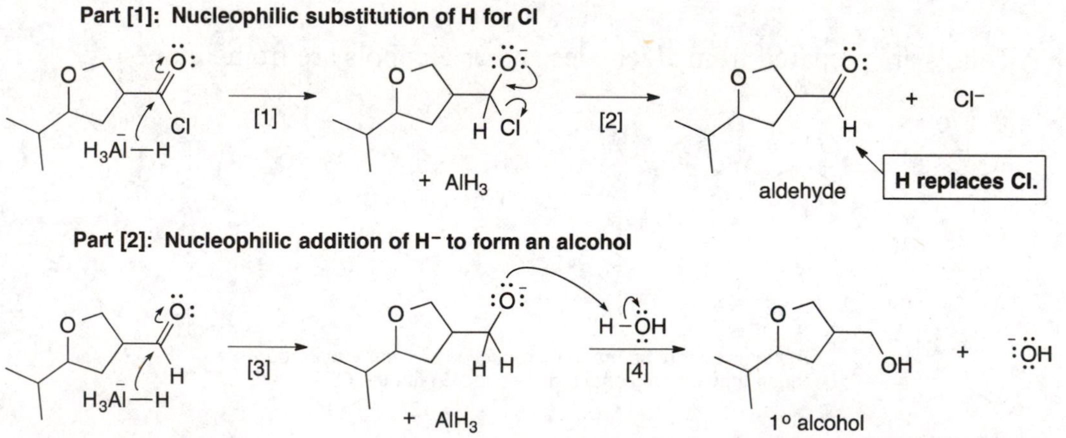

20.11 Acid chlorides and esters can be reduced to 1° alcohols. Keep the carbon skeleton the same in drawing an ester and acid chloride precursor.

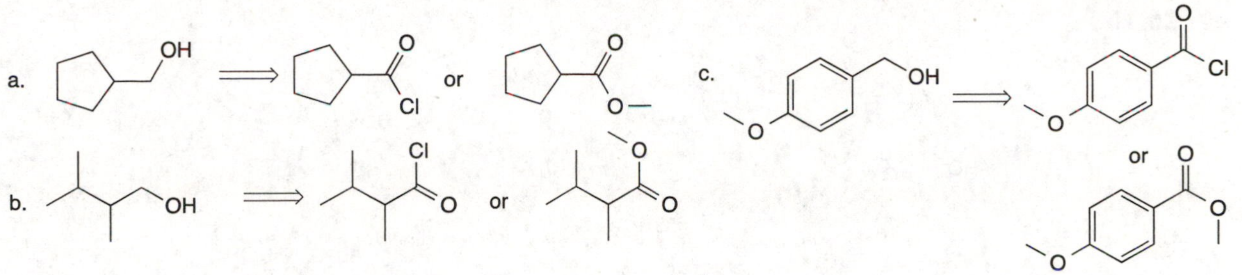

a. ⟹ or
b. OH ⟹ or
c. OH ⟹ Cl or

20.12

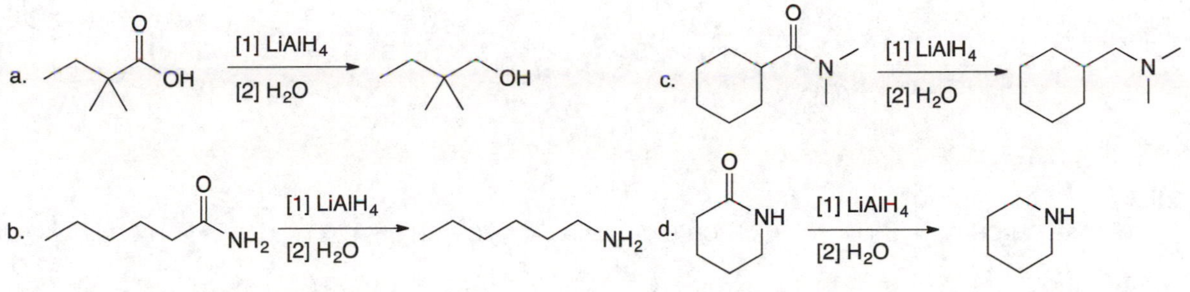

a. OH $\xrightarrow[\text{[2] } H_2O]{\text{[1] } LiAlH_4}$ OH
b. NH_2 $\xrightarrow[\text{[2] } H_2O]{\text{[1] } LiAlH_4}$ NH_2
c. N $\xrightarrow[\text{[2] } H_2O]{\text{[1] } LiAlH_4}$ N
d. NH $\xrightarrow[\text{[2] } H_2O]{\text{[1] } LiAlH_4}$ NH

20.13

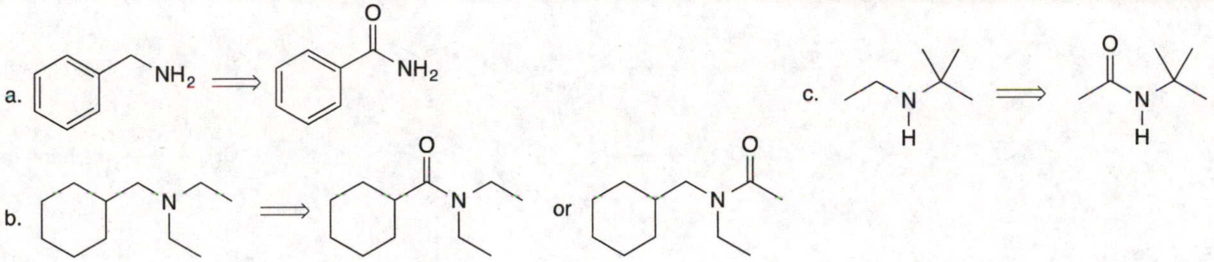

a. NH_2 ⟹ NH_2
b. N ⟹ N or N
c. N–H ⟹ N–H

20.14

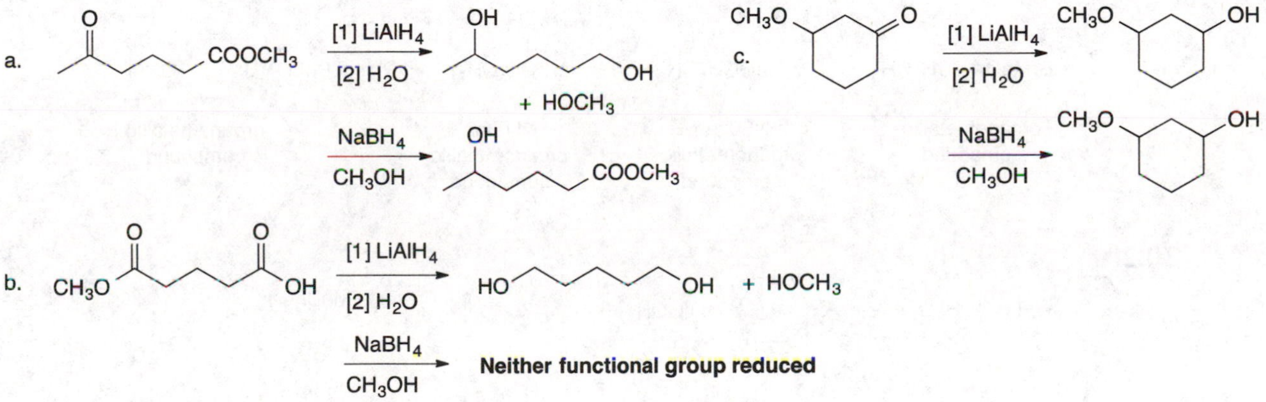

a. $COOCH_3$ $\xrightarrow[\text{[2] } H_2O]{\text{[1] } LiAlH_4}$ OH, OH + $HOCH_3$

$\xrightarrow[CH_3OH]{NaBH_4}$ OH, $COOCH_3$

b. CH_3O ... OH $\xrightarrow[\text{[2] } H_2O]{\text{[1] } LiAlH_4}$ HO ... OH + $HOCH_3$

$\xrightarrow[CH_3OH]{NaBH_4}$ **Neither functional group reduced**

c. CH_3O ... O $\xrightarrow[\text{[2] } H_2O]{\text{[1] } LiAlH_4}$ CH_3O ... OH

$\xrightarrow[CH_3OH]{NaBH_4}$ CH_3O ... OH

20.15 Tollens reagent reacts only with aldehydes.

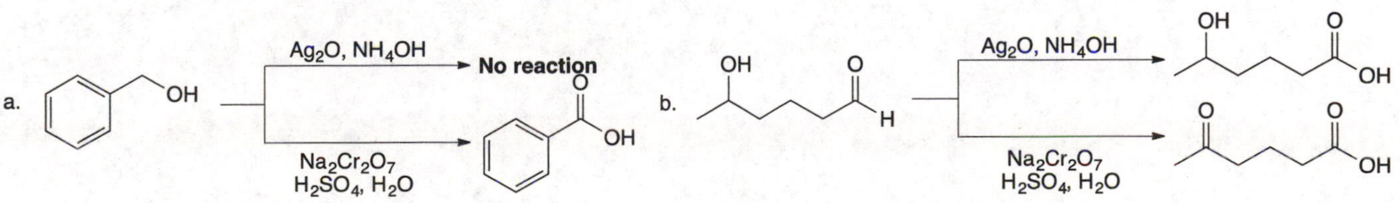

a. OH $\xrightarrow{Ag_2O,\ NH_4OH}$ **No reaction**

$\xrightarrow{Na_2Cr_2O_7,\ H_2SO_4,\ H_2O}$ OH

b. OH ... H $\xrightarrow{Ag_2O,\ NH_4OH}$ OH ... OH

$\xrightarrow{Na_2Cr_2O_7,\ H_2SO_4,\ H_2O}$ OH

20.16

B: HO–CH$_2$–(cyclohexane)–CH(OH)–CH$_2$CH$_2$CH$_2$–CHO

a. **B** $\xrightarrow[CH_3OH]{NaBH_4}$ HO–CH$_2$–(cyclohexane)–CH(OH)–CH$_2$CH$_2$CH$_2$CH$_2$–OH

b. **B** $\xrightarrow[[2]\ H_2O]{[1]\ LiAlH_4}$ HO–CH$_2$–(cyclohexane)–CH(OH)–CH$_2$CH$_2$CH$_2$CH$_2$–OH

c. **B** $\xrightarrow{PCC}$ OHC–(cyclohexane)–CO–CH$_2$CH$_2$CH$_2$–CHO

d. **B** $\xrightarrow{Ag_2O,\ NH_4OH}$ HO–CH$_2$–(cyclohexane)–CH(OH)–CH$_2$CH$_2$CH$_2$–COOH

e. **B** $\xrightarrow[H_2SO_4,\ H_2O]{CrO_3}$ HOOC–(cyclohexane)–CO–CH$_2$CH$_2$CH$_2$–COOH

20.17

a. $CH_3CH_2Br + 2\ Li \longrightarrow CH_3CH_2Li + LiBr$

b. $CH_3CH_2Br + Mg \longrightarrow CH_3CH_2MgBr$

c. $CH_3CH_2Br + 2\ Li \longrightarrow CH_3CH_2Li + LiBr$

$2\ CH_3CH_2Li + CuI \longrightarrow LiCu(CH_2CH_3)_2 + LiI$

20.18

HC≡C–(CH$_2$)$_5$CH$_3$ + NaH ⟶ Na^+ $^-$C≡C–(CH$_2$)$_5$CH$_3$ + H_2 ← hydrogen gas

HC≡C–(CH$_2$)$_5$CH$_3$ + CH_3MgBr ⟶ BrMg–C≡C–(CH$_2$)$_5$CH$_3$ + CH_4 ← methane gas

20.19 Organometallic compounds have a carbon–metal bond.

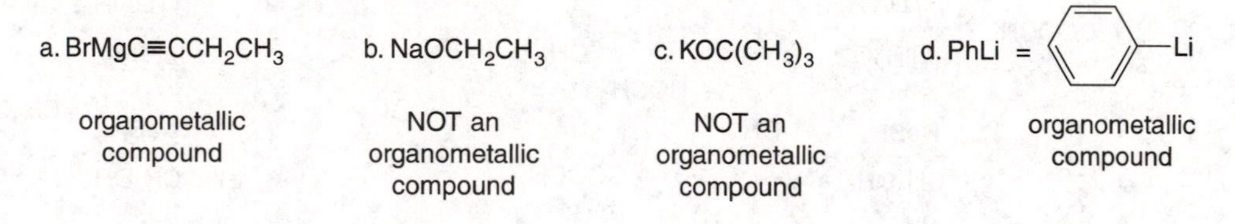

a. $BrMgC{\equiv}CCH_2CH_3$ — organometallic compound

b. $NaOCH_2CH_3$ — NOT an organometallic compound

c. $KOC(CH_3)_3$ — NOT an organometallic compound

d. PhLi = (C$_6$H$_5$–Li) — organometallic compound

20.20

a. (cyclohexyl)–Li + H_2O ⟶ cyclohexane + LiOH

b. $(CH_3)_3C$–MgBr + H_2O ⟶ $(CH_3)_3CH$ + HOMgBr

c. $C_6H_5CH_2$–MgBr + H_2O ⟶ $C_6H_5CH_3$ + HOMgBr

d. $CH_3CH_2C{\equiv}C$–Li + H_2O ⟶ $CH_3CH_2C{\equiv}CH$ + LiOH

20.21 To draw the products, add the alkyl or phenyl group to the carbonyl carbon and protonate the oxygen.

a. 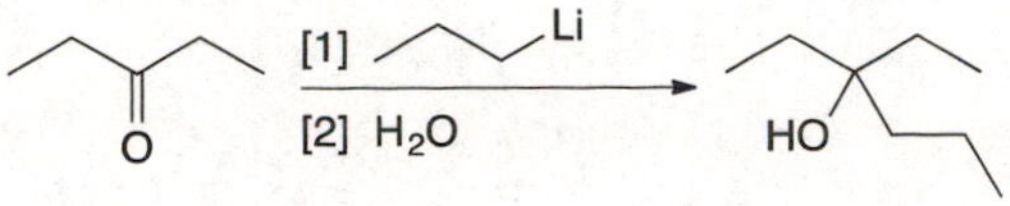 O; [1] Li; [2] H_2O; HO

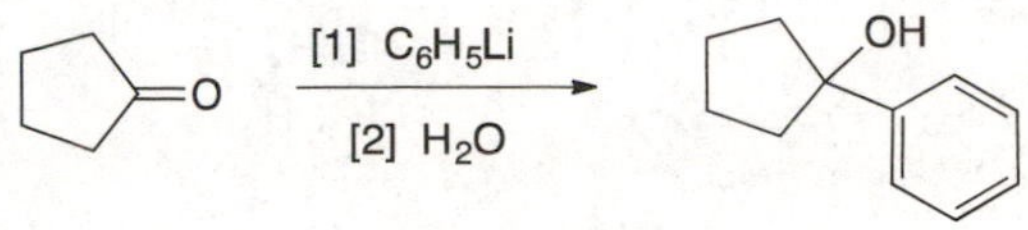

b. 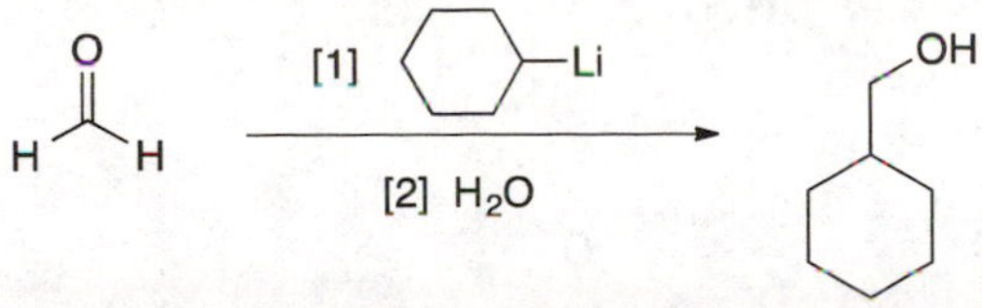 O; H H; [1] –Li; [2] H_2O; OH

c. O; [1] C_6H_5Li; [2] H_2O; OH

d. 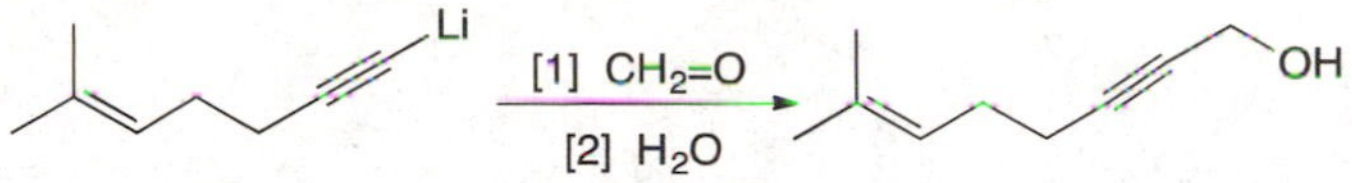 Li; [1] CH_2=O; [2] H_2O; OH

20.22 Addition of RM always occurs from above and below the plane of the molecule.

a. 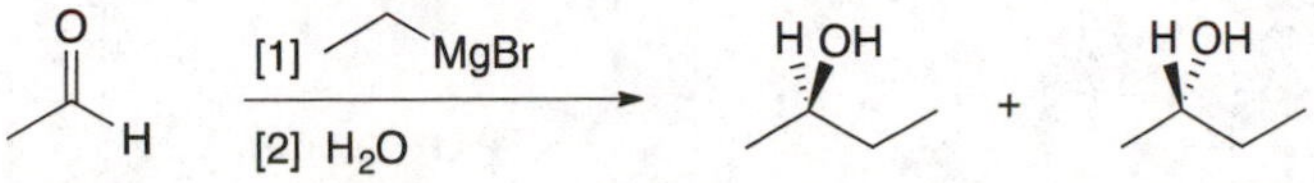 O; H; [1] MgBr; [2] H_2O; H OH + H OH

b.  O; [1] CH_3CH_2Li; [2] H_2O; OH + OH

20.23

a. 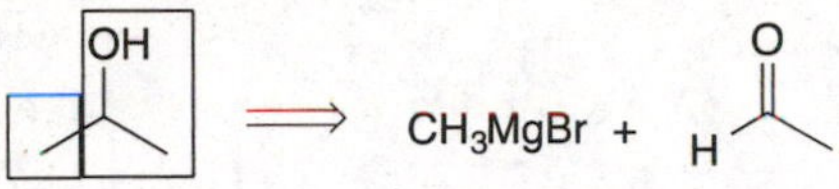 OH ⟹ CH_3MgBr + O; H

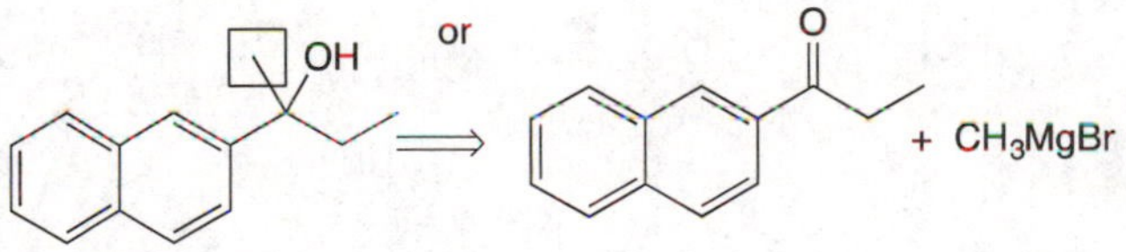

b. 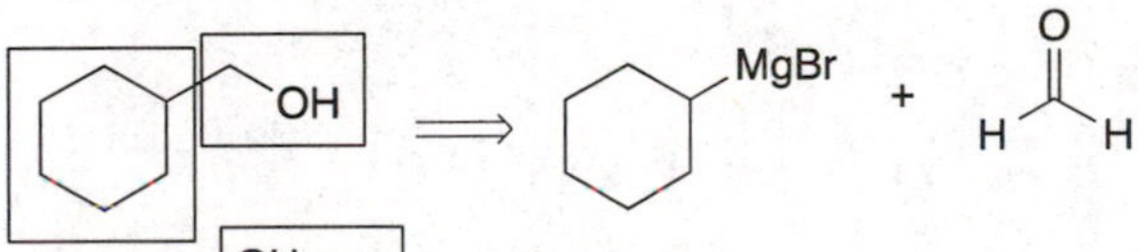 OH ⟹ MgBr + O; H H

c. 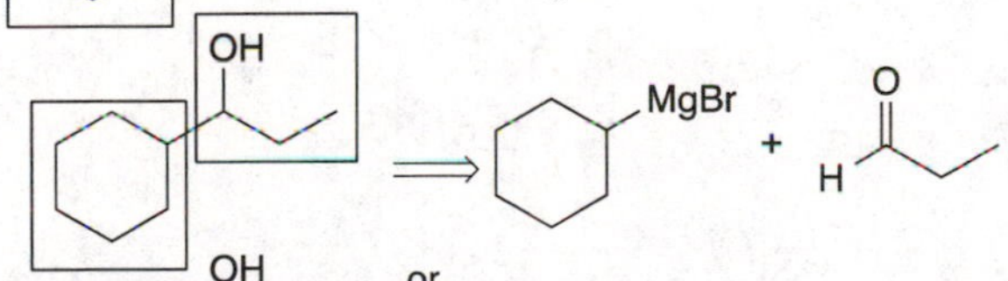 OH ⟹ MgBr + O; H

or

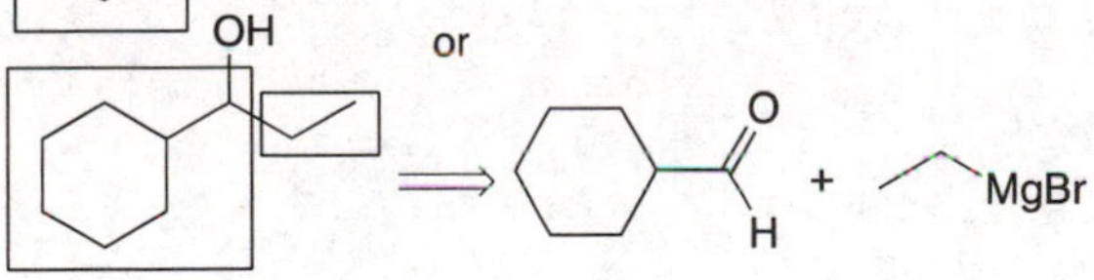

OH ⟹ O; H + MgBr

d. OH ⟹ O + MgBr

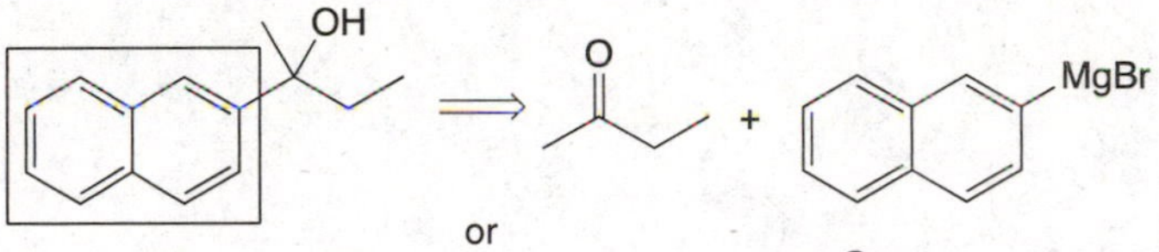

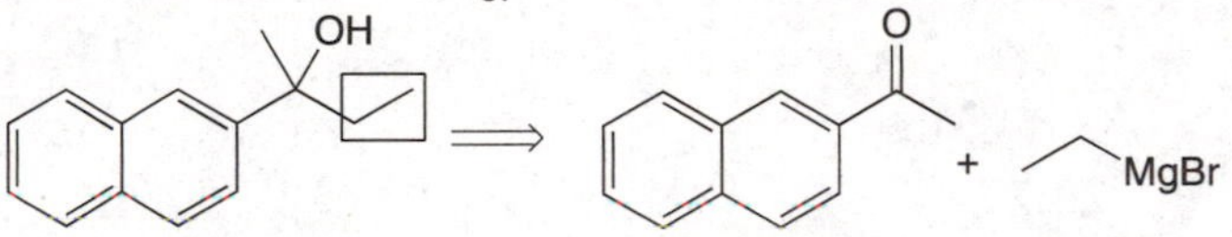

or

OH ⟹ O + MgBr

or

OH ⟹ O + CH_3MgBr

20.24

a. 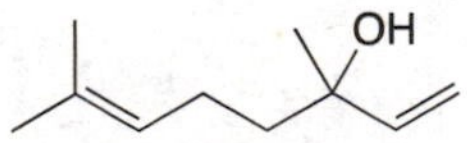 OH; linalool (three methods) ⟹ O + CH_3Li

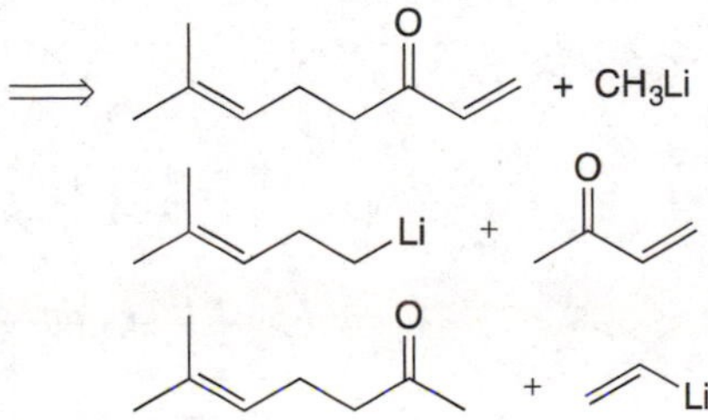

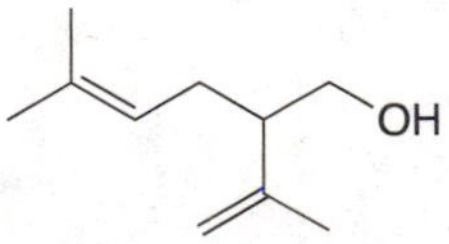

Li + O

O + Li

OH; lavandulol ⟹ Li + CH_2=O

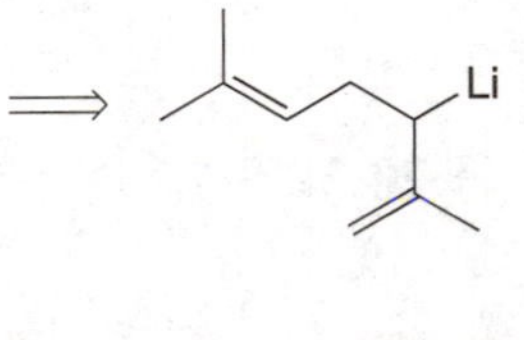

b. OH ⟹ CHO

$NaBH_4$
CH_3OH

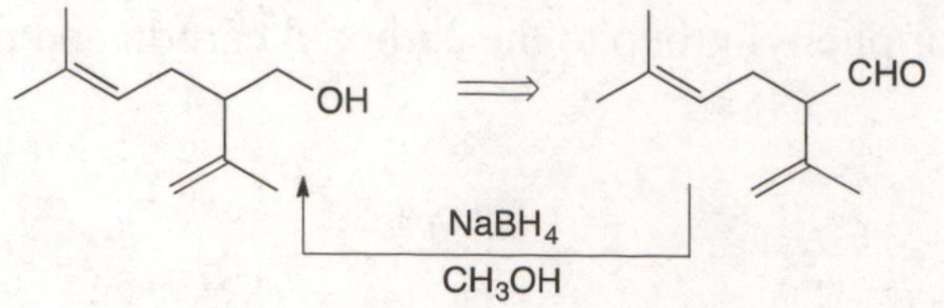

c. Linalool is a 3° ROH. Therefore, it has no H on the carbon with the OH group, and cannot be prepared by reduction of a carbonyl compound.

20.25

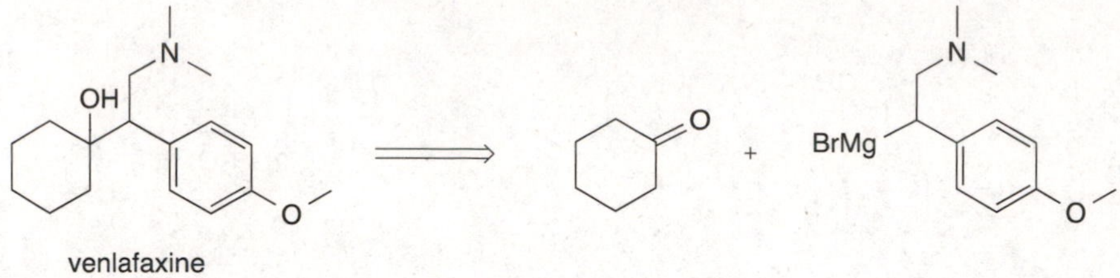

OH N

venlafaxine

⟹ O + BrMg N O

20.26

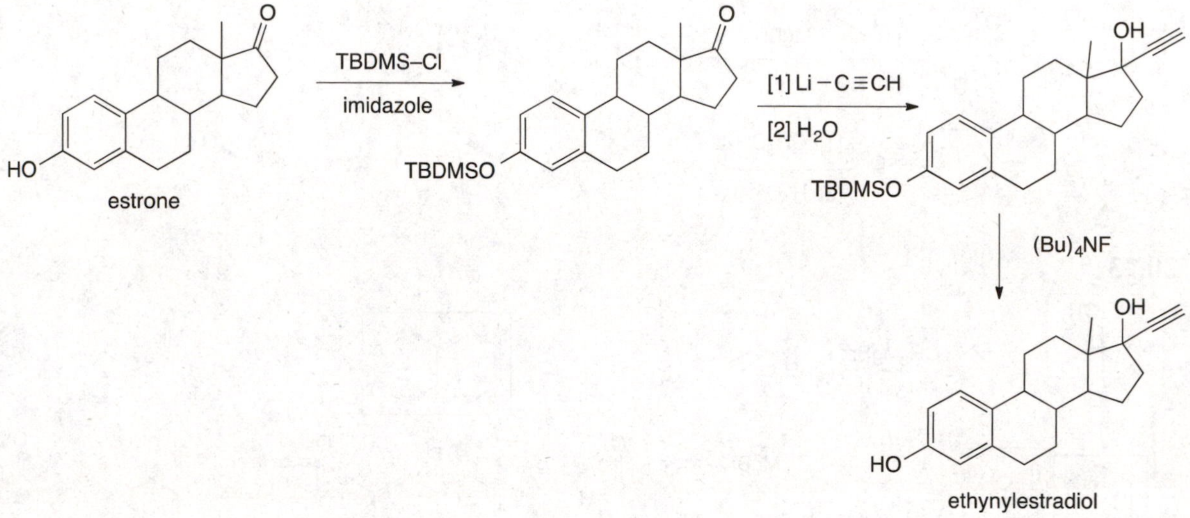

estrone

TBDMS–Cl
imidazole

TBDMSO

[1] Li–C≡CH
[2] H_2O

OH
TBDMSO

$(Bu)_4NF$

OH
HO

ethynylestradiol

20.27

a. O Cl

[1] MgBr (2 equiv)
[2] H_2O

HO

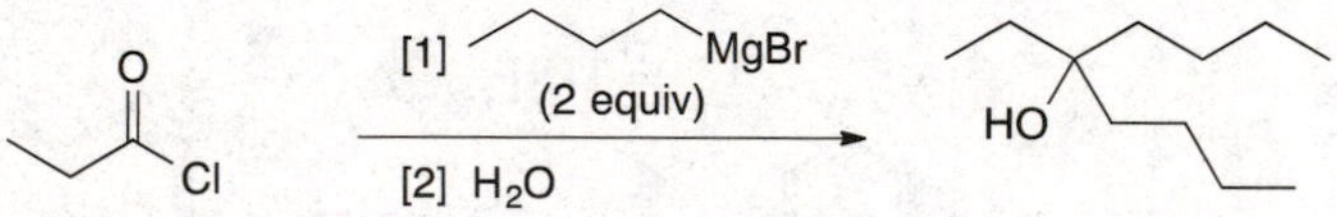

b. O O

[1] MgBr (2 equiv)
[2] H_2O

HO

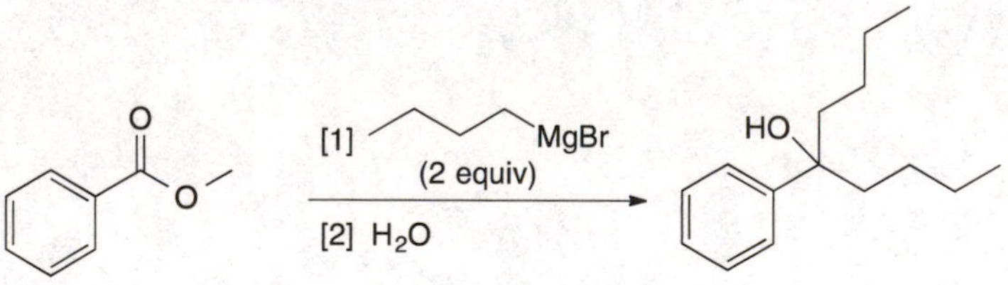

c. O O

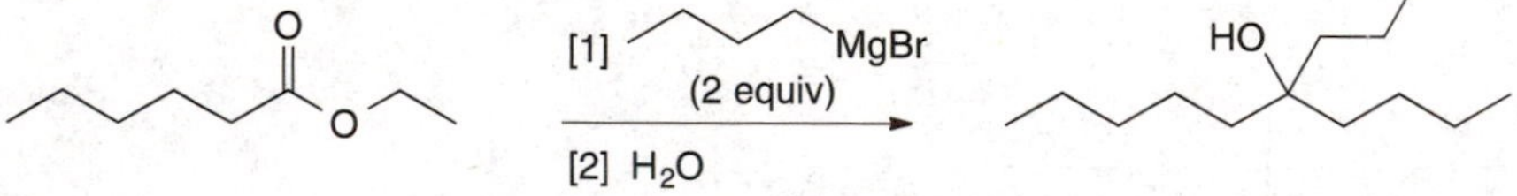

[1] MgBr (2 equiv)
[2] H_2O

HO

20.28

a. OH ⟹ O O + MgBr (2 equiv)

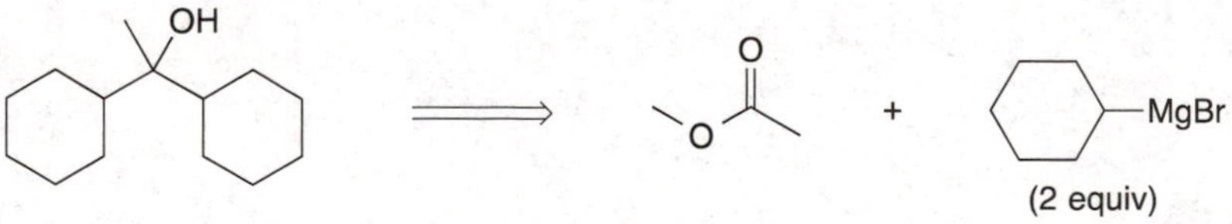

b. 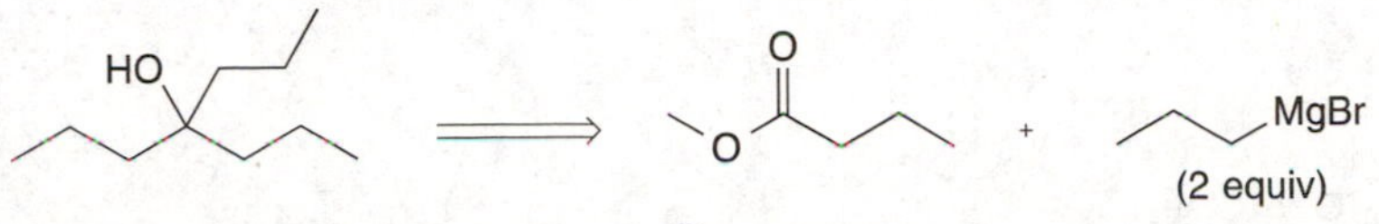 HO ⟹ O, O + MgBr (2 equiv)

c. 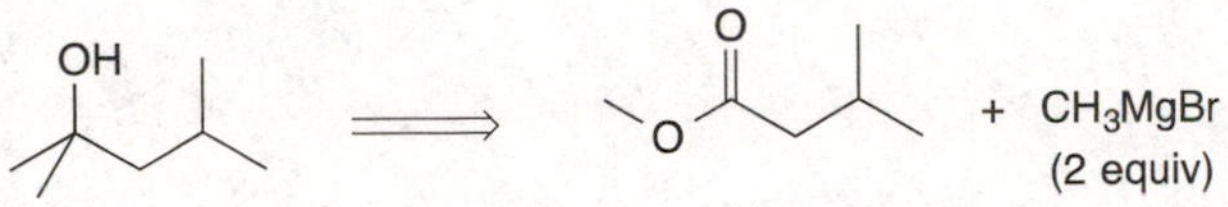 OH ⟹ O, O + CH_3MgBr (2 equiv)

20.29 The R group of the organocuprate has replaced the Cl on the acid chloride.

a. 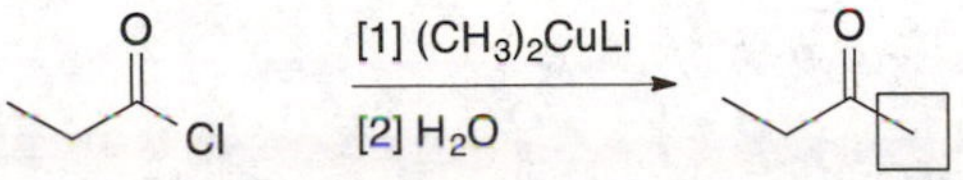 O, Cl —[1] $(CH_3)_2CuLi$, [2] H_2O→ O

c. 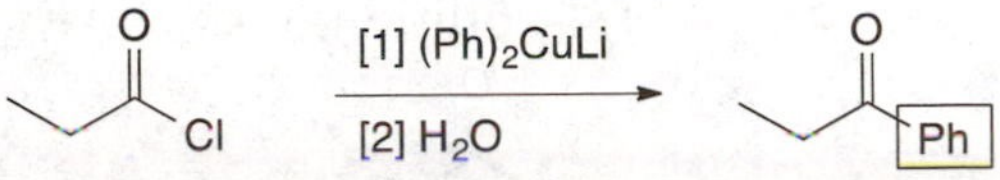 O, Cl —[1] $(Ph)_2CuLi$, [2] H_2O→ O, Ph

b. 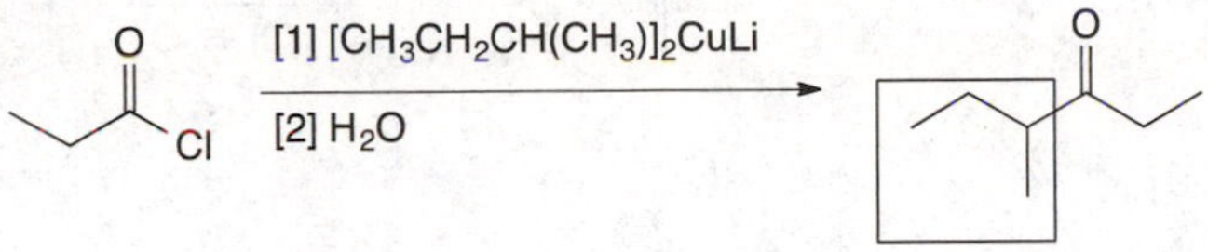 O, Cl —[1] $[CH_3CH_2CH(CH_3)]_2CuLi$, [2] H_2O→ O

20.30

a. 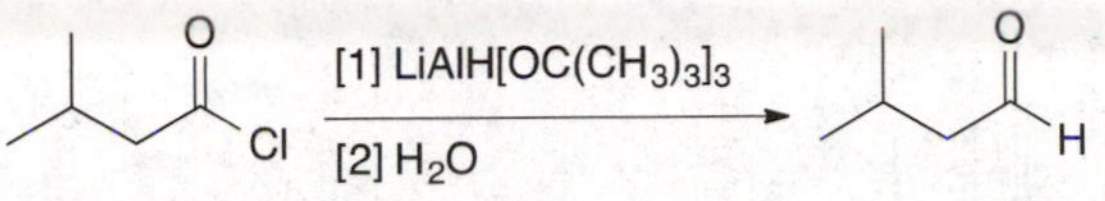 O, Cl —[1] $LiAlH[OC(CH_3)_3]_3$, [2] H_2O→ O, H

c. 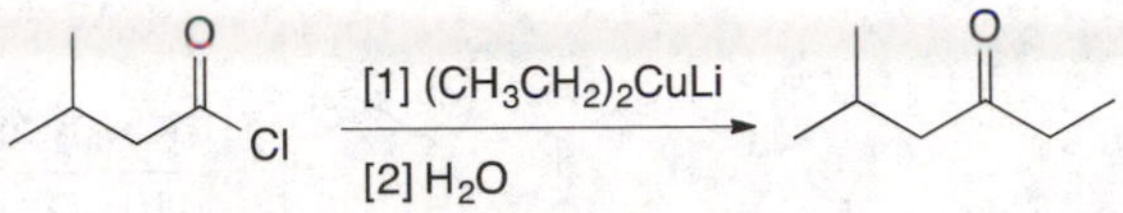 O, Cl —[1] $(CH_3CH_2)_2CuLi$, [2] H_2O→ O

b. 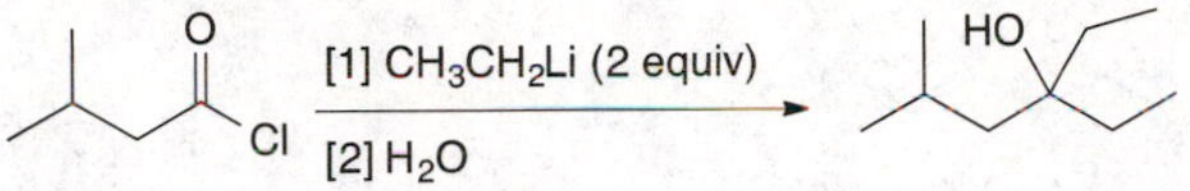 O, Cl —[1] CH_3CH_2Li (2 equiv), [2] H_2O→ HO

d. 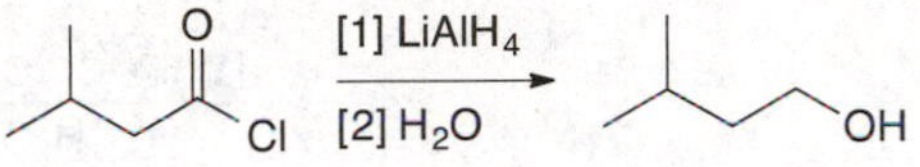 O, Cl —[1] $LiAlH_4$, [2] H_2O→ OH

20.31

a. 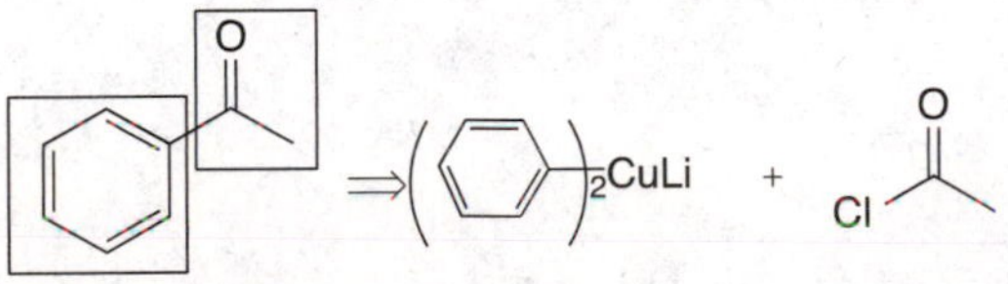 O ⟹ $(\text{Ph})_2CuLi$ + Cl, O

or

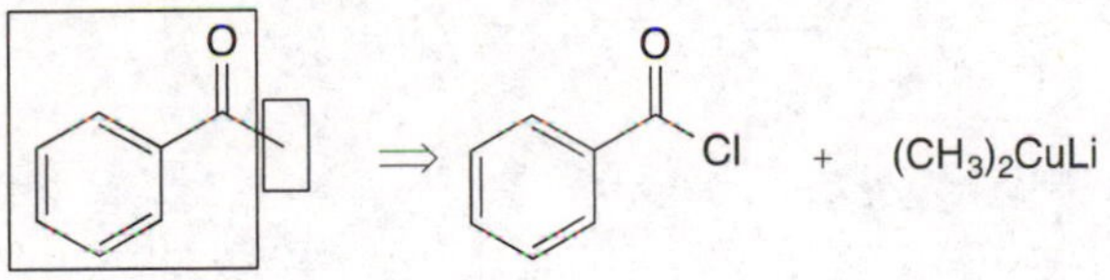

O ⟹ O, Cl + $(CH_3)_2CuLi$

b. 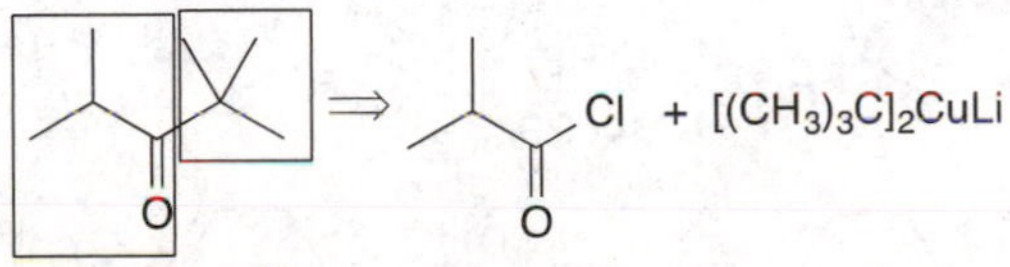 O ⟹ Cl, O + $[(CH_3)_3C]_2CuLi$

or

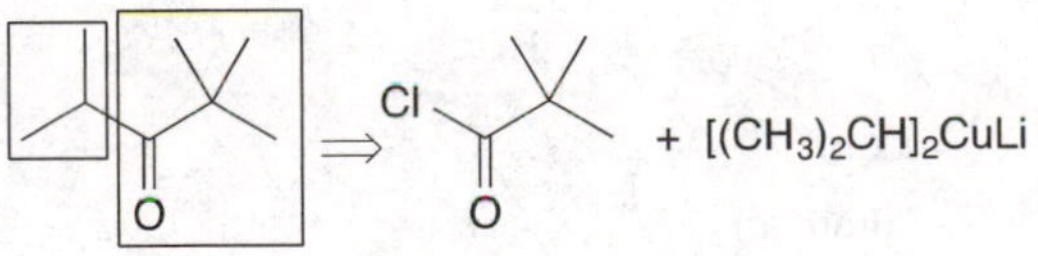

O ⟹ Cl, O + $[(CH_3)_2CH]_2CuLi$

20.32

a. 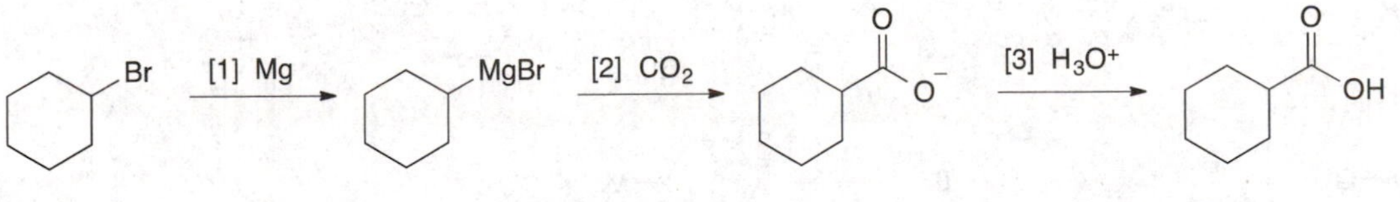 Br —[1] Mg→ MgBr —[2] CO_2→ O, O^- —[3] H_3O^+→ O, OH

b. 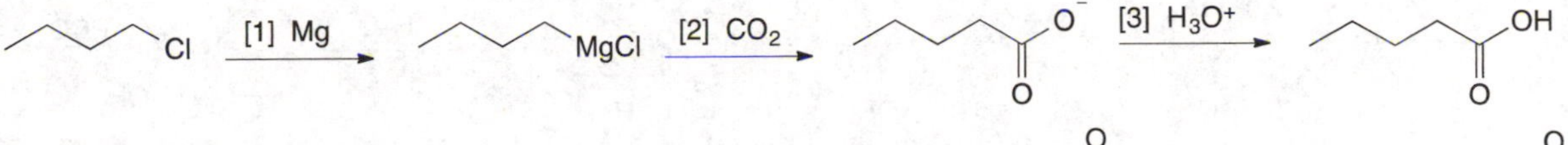 Cl —[1] Mg→ MgCl —[2] CO_2→ O^-, O —[3] H_3O^+→ OH, O

c. 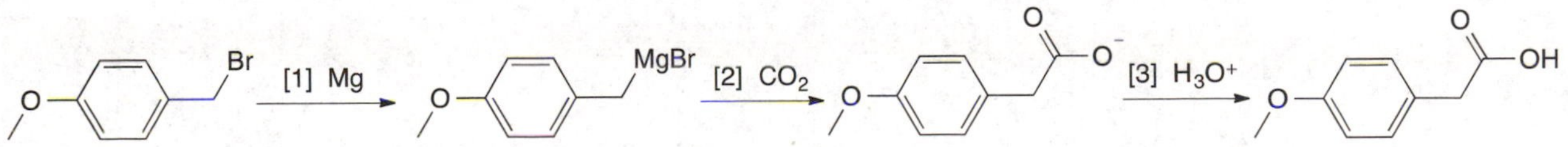 O, Br —[1] Mg→ O, MgBr —[2] CO_2→ O, O, O^- —[3] H_3O^+→ O, O, OH

20.33

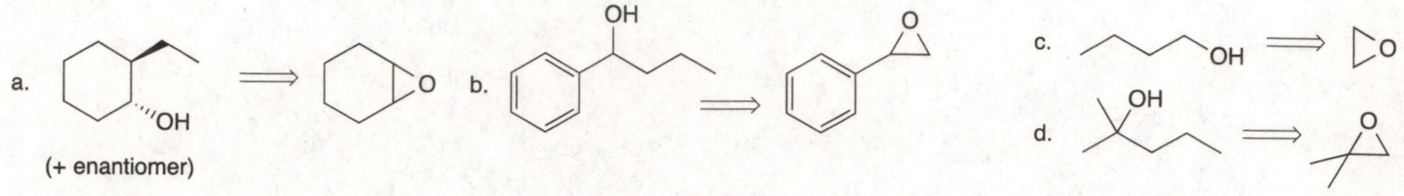

a. OH (+ enantiomer) ⟹ O

b. OH ⟹ O

c. OH ⟹ O

d. OH ⟹ O

20.34 The characteristic reaction of α,β-unsaturated carbonyl compounds is nucleophilic addition. Grignard and organolithium reagents react by 1,2-addition and organocuprate reagents react by 1,4-addition.

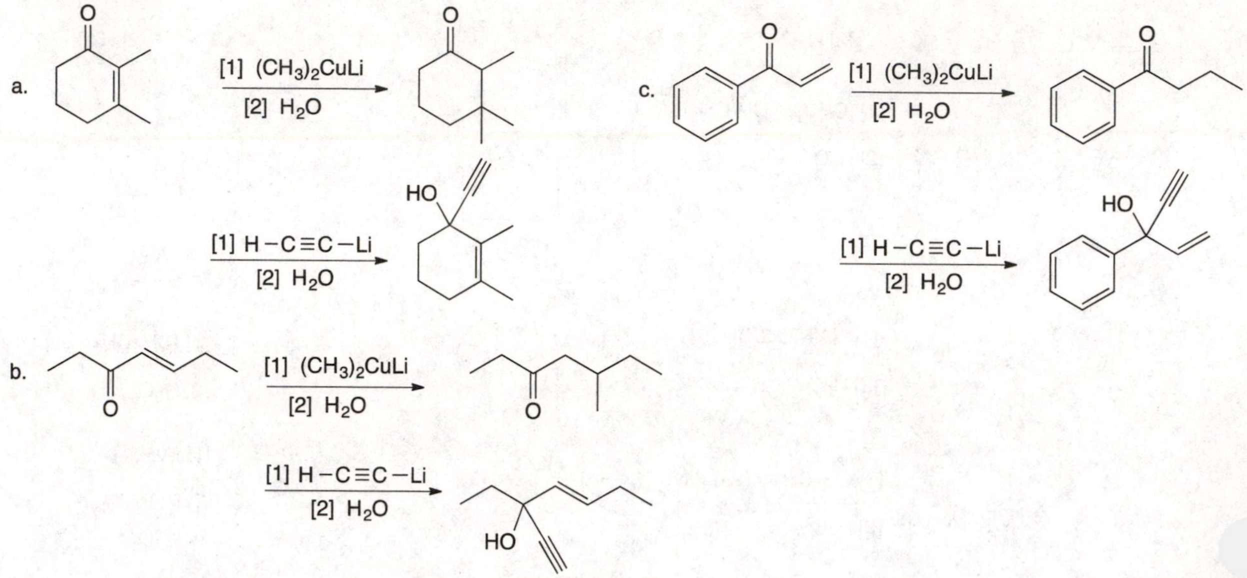

a. O — [1] $(CH_3)_2CuLi$, [2] H_2O → O

[1] H−C≡C−Li, [2] H_2O → HO

c. O — [1] $(CH_3)_2CuLi$, [2] H_2O → O

[1] H−C≡C−Li, [2] H_2O → HO

b. O — [1] $(CH_3)_2CuLi$, [2] H_2O → O

[1] H−C≡C−Li, [2] H_2O → HO

20.35

a. OH — HBr or PBr_3 → Br — Mg → MgBr

OH — PCC → =O — [1] MgBr, [2] H_2O → OH

b. OH (from a.) — HBr → Br

c. OH (from a.) — H_2SO_4 → + — H_2, Pd-C →

d. OH — HBr or PBr_3 → Br — Mg → MgBr

OH — PCC → O, H

} → H_2O → OH — PCC → O

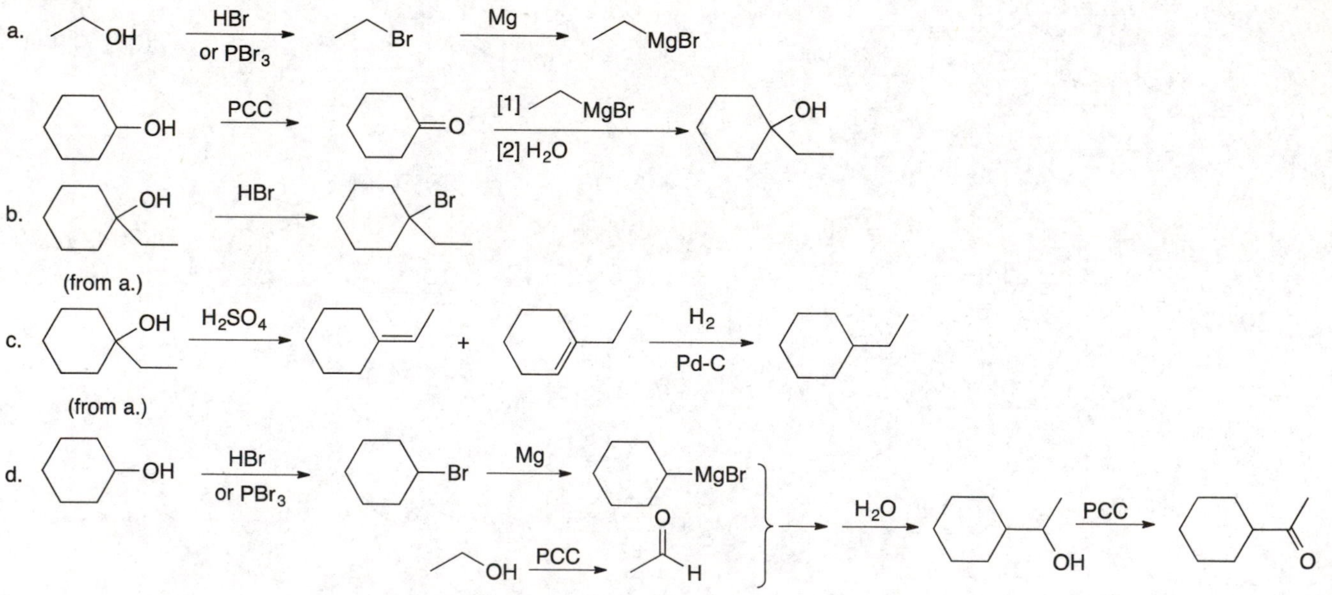

e. (cyclohexyl)MgBr (from d.) → [1] ethylene oxide, [2] H_2O → 2-cyclohexylethanol (OH)

OH (ethanol) $\xrightarrow{H_2SO_4}$ $CH_2{=}CH_2$ $\xrightarrow{mCPBA}$ ethylene oxide

20.36

A (2-pentanone)

a. $NaBH_4$, CH_3OH → 2-pentanol (OH)

b. [1] $LiAlH_4$ [2] H_2O → 2-pentanol (OH)

c. [1] CH_3MgBr (excess) [2] H_2O → 2-methyl-2-pentanol (OH)

d. [1] C_6H_5Li (excess) [2] H_2O → tertiary alcohol (HO, C_6H_5)

e. $Na_2Cr_2O_7$, H_2SO_4, H_2O → No reaction

B (CH_3O–CH_2, methyl-substituted cyclohexane with CO_2CH_3 (OCH_3))

a. $NaBH_4$, CH_3OH → No reaction

b. [1] $LiAlH_4$ [2] H_2O → CH_3O– ... –CH_2OH (OH)

c. [1] CH_3MgBr (excess) [2] H_2O → CH_3O– ... –$C(CH_3)_2OH$ (OH)

d. [1] C_6H_5Li (excess) [2] H_2O → CH_3O– ... –$C(C_6H_5)_2OH$ (C_6H_5 C_6H_5 OH)

e. $Na_2Cr_2O_7$, H_2SO_4, H_2O → No reaction

20.37

a. Ethanol (OH) $\xrightarrow{PBr_3}$ Br (bromoethane) $\xrightarrow{Mg}$ BrMg–CH_2CH_3

Ethanol (OH) $\xrightarrow{PCC}$ H–C(=O)–CH_3 (acetaldehyde) $\xrightarrow{[1]\ BrMg-CH_2CH_3,\ [2]\ H_2O}$ 2-butanol (OH)

b. (from a.) [1] CH_3MgBr (from CH_3OH → PBr_3 → CH_3Br → Mg), [2] H_2O; PCC; [1] CH_3MgBr, [2] H_2O

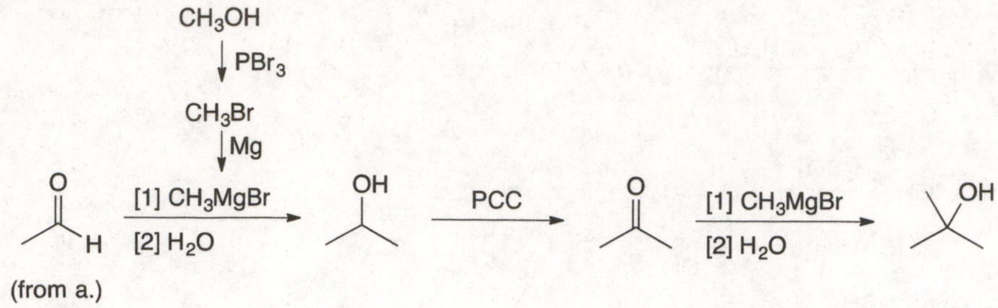

c. (from b.) PBr_3; [1] Mg, [2] $H_2C{=}O$, [3] H_2O (CH_3OH → PCC → $H_2C{=}O$)

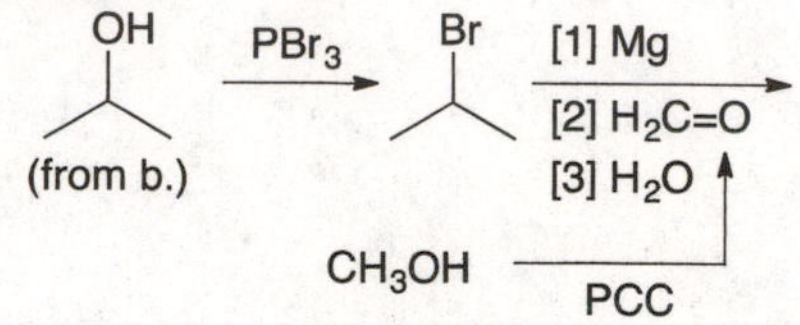

d. H_2SO_4 → $CH_2{=}CH_2$; mCPBA; [1] $BrMg{-}CH_2CH_3$ (from a.), [2] H_2O

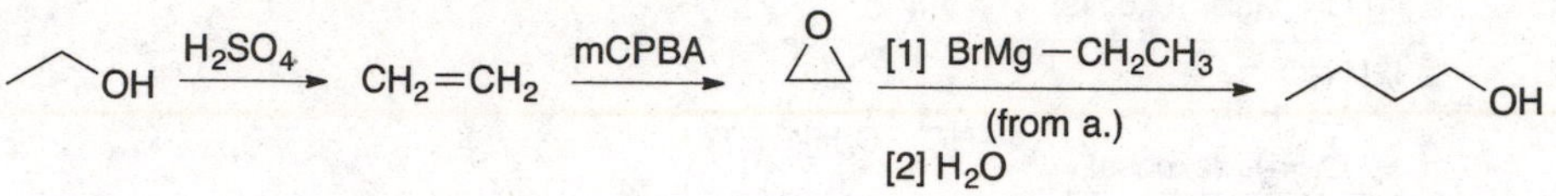

20.38

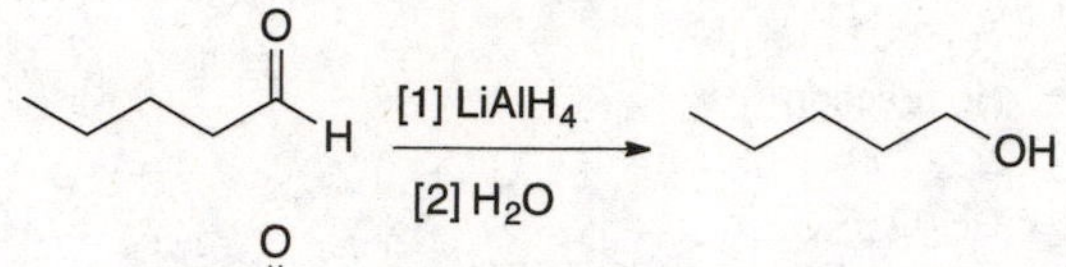

a. NaBH$_4$, CH_3OH

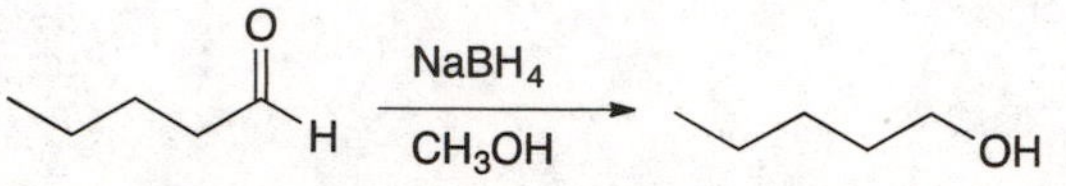

b. [1] $LiAlH_4$, [2] H_2O

c. H_2, Pd-C

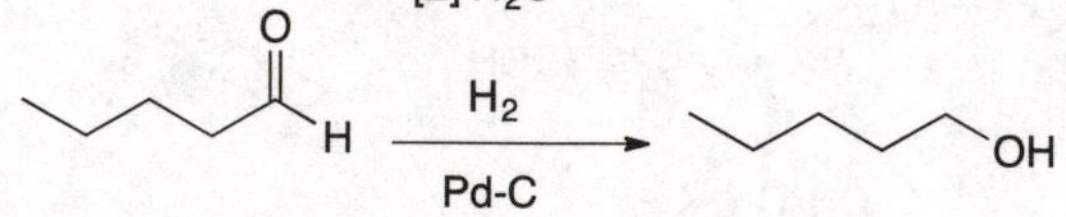

d. PCC → **No reaction**

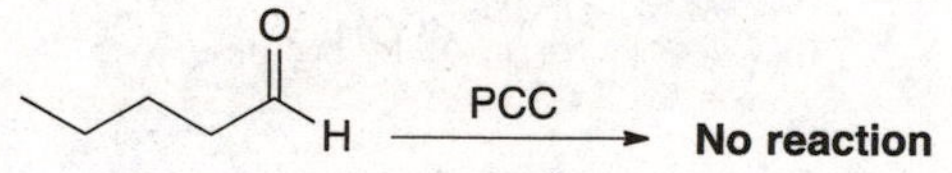

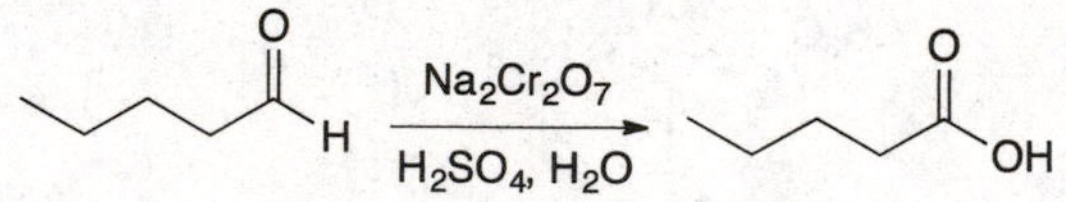

e. $Na_2Cr_2O_7$, H_2SO_4, H_2O

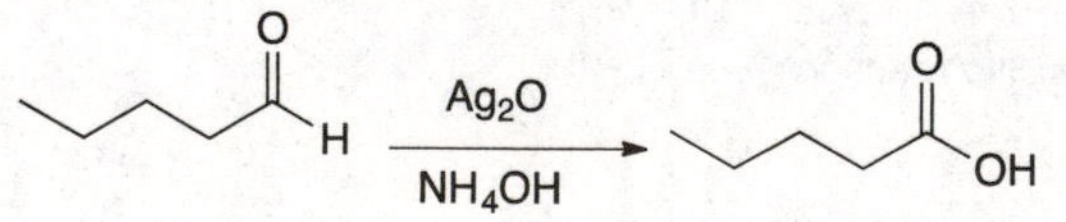

f. Ag_2O, NH_4OH

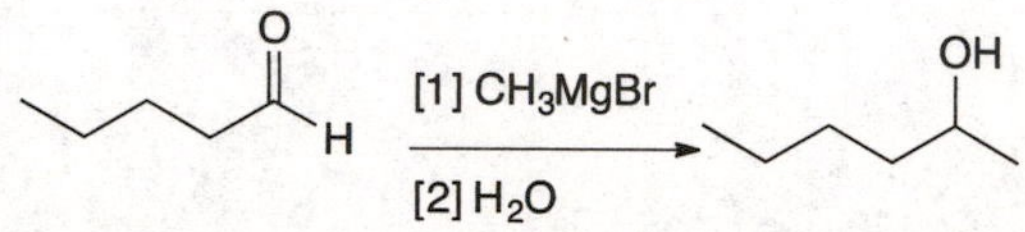

g. [1] CH_3MgBr, [2] H_2O

h. [1] C_6H_5Li, [2] H_2O

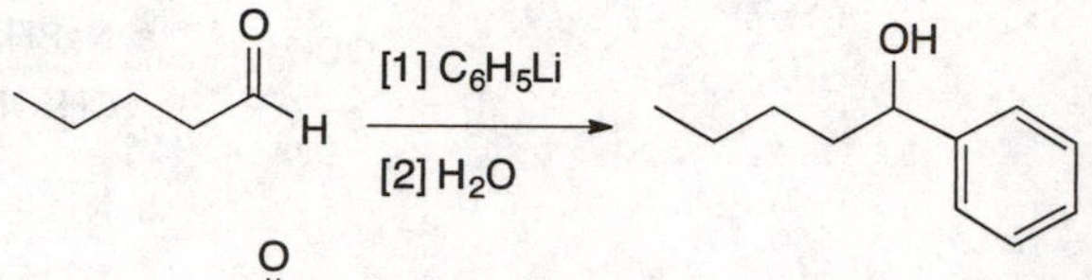

i. [1] $(CH_3)_2CuLi$, [2] H_2O → **No reaction**

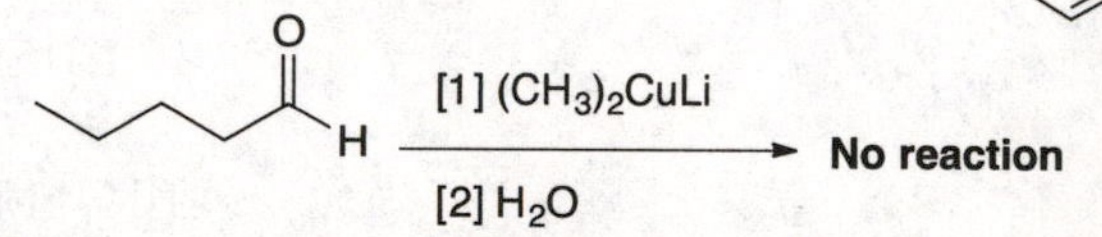

j. [1] HC≡CNa, [2] H_2O

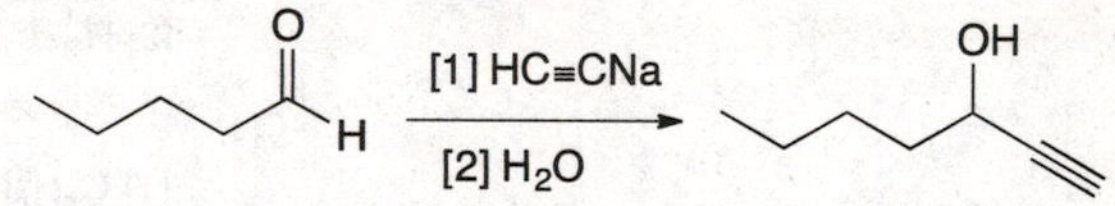

k. [1] $CH_3C{\equiv}CLi$, [2] H_2O

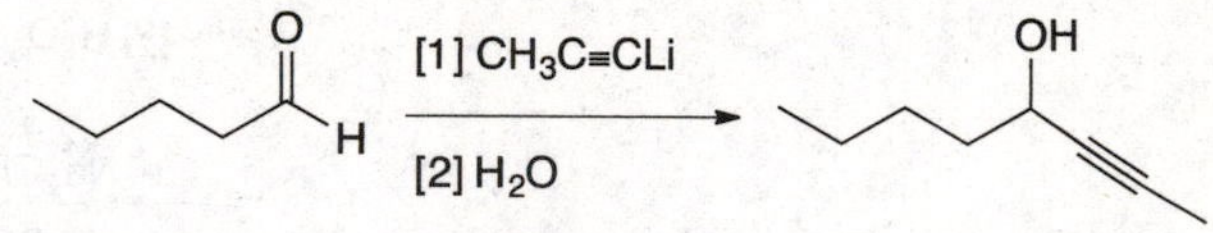

l. TBDMSCl, imidazole → O–TBDMS

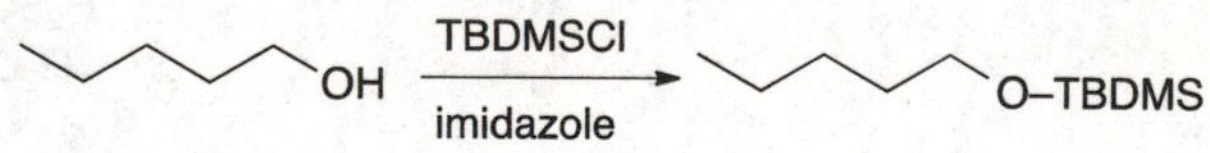

20.39

a. Cl; $(CH_3CH_2CH_2CH_2)_2CuLi$

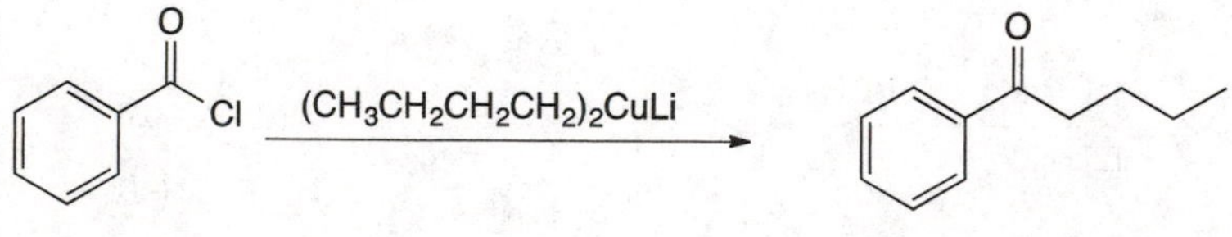

b. OCH_3; $(CH_3CH_2CH_2CH_2)_2CuLi$ → **No reaction**

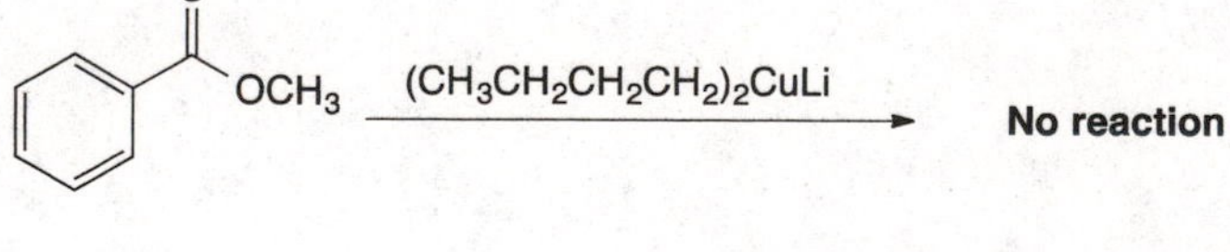

c. 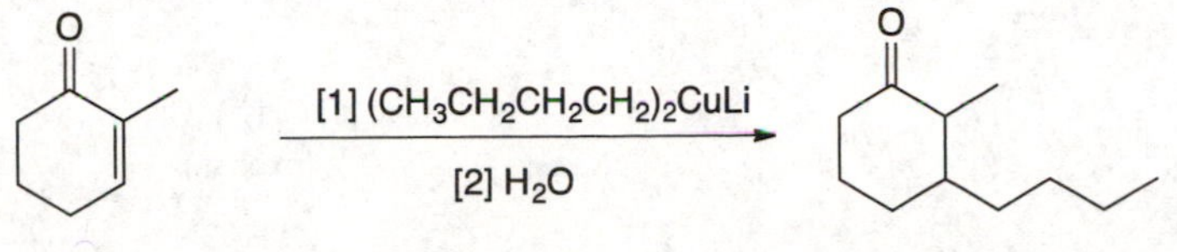 [1] $(CH_3CH_2CH_2CH_2)_2CuLi$ / [2] H_2O

d. 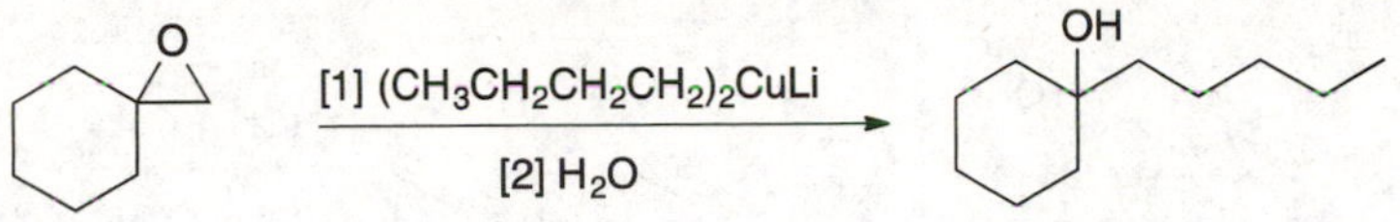 [1] $(CH_3CH_2CH_2CH_2)_2CuLi$ / [2] H_2O → OH

20.40 Arrange the larger group [$(CH_3)_3C–$] on the left side of the carbonyl.

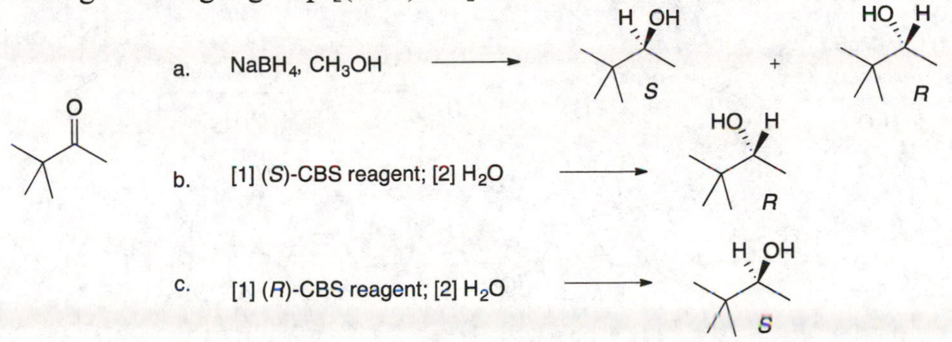

a. $NaBH_4$, CH_3OH → H OH *S* + HO H *R*

b. [1] (*S*)-CBS reagent; [2] H_2O → HO H *R*

c. [1] (*R*)-CBS reagent; [2] H_2O → H OH *S*

20.41

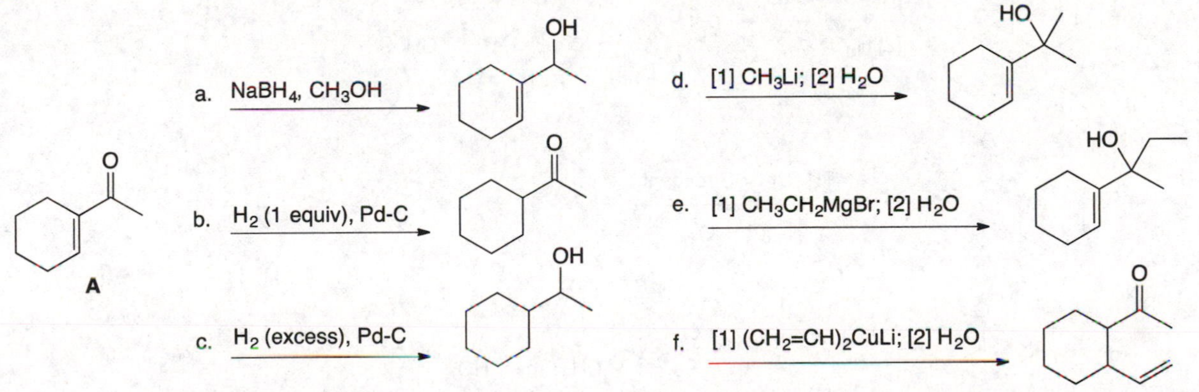

A

a. $NaBH_4$, CH_3OH → OH

b. H_2 (1 equiv), Pd-C →

c. H_2 (excess), Pd-C → OH

d. [1] CH_3Li; [2] H_2O → HO

e. [1] CH_3CH_2MgBr; [2] H_2O → HO

f. [1] $(CH_2{=}CH)_2CuLi$; [2] H_2O →

20.42

a. 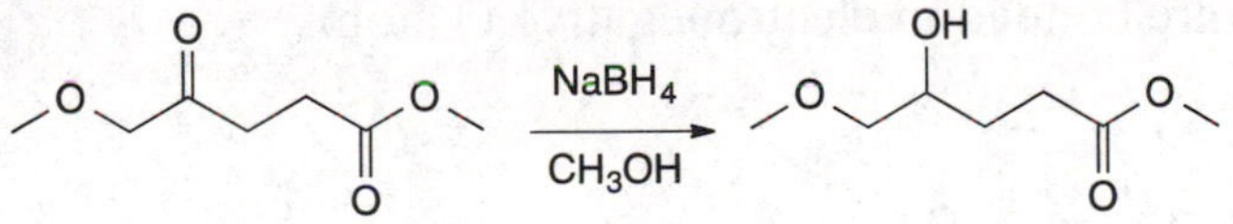 $NaBH_4$ / CH_3OH → OH

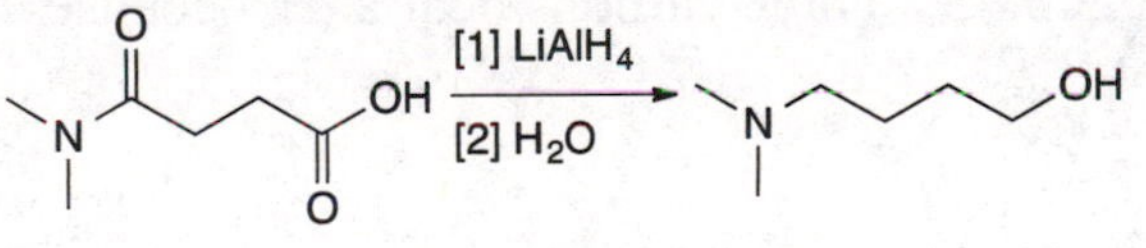

b. 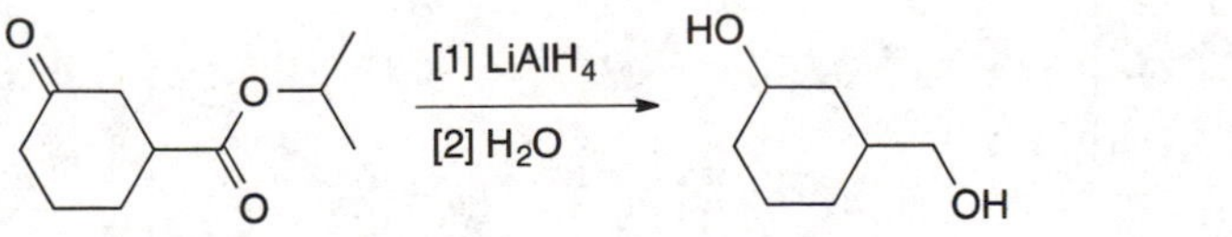 [1] $LiAlH_4$ / [2] H_2O → HO, OH

c. OH; [1] $LiAlH_4$ / [2] H_2O → N, OH

d. 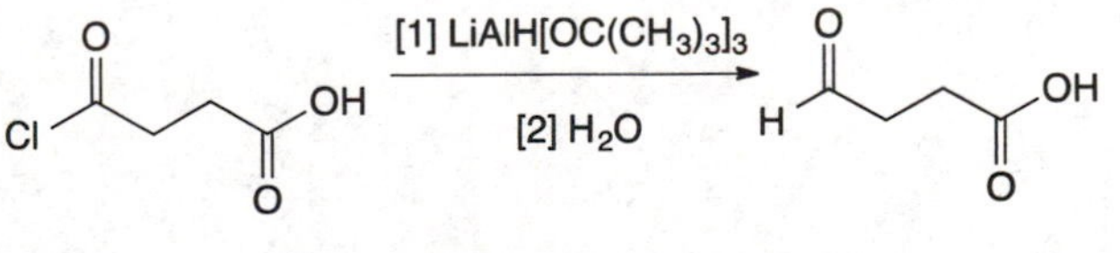 Cl, OH; [1] $LiAlH[OC(CH_3)_3]_3$ / [2] H_2O → H, OH

20.43

a. MgBr; [1] CO_2; [2] H_3O^+; O; OH

d. O; O; [1] CH_3MgCl (excess); [2] H_2O; OH

b. O; [1] CH_3CH_2MgBr; [2] H_2O; OH

e. O; [1] $(CH_3)_2CuLi$; [2] H_2O; O

c. O; Cl; [1] C_6H_5MgBr (excess); [2] H_2O; HO

f. C_6H_5; O; [1] $(CH_3)_2CuLi$; [2] H_2O; OH

20.44

a. O; [1] CH_3Li; [2] H_2O; OH + OH

b. O; [1] $(CH_2{=}CH)_2CuLi$; [2] H_2O; OH + OH

c. O; [1] (*S*)-CBS reagent; [2] H_2O; H OH

d. O; O; [1] $LiAlH_4$; [2] H_2O; OH + HO

20.45 Three carbons bear a δ+ because they are bonded to electronegative O atoms.

C3; O; C2; O; O; C1

C1 is an sp^3 hybridized C bonded to an O so it bears a δ+. There are no additional resonance structures that affect C1. C2 is part of a carbonyl that has three resonance structures, only one of which places a (+) charge on C. The O atom of the ester donates its electron pair, making the carbonyl C less electrophilic than C3. C3 is most electrophilic. Its carbonyl is stabilized by two resonance structures, one of which places a (+) charge on carbon.

C2

O donates electron density.

C3

full (+) charge
most electrophilic site

20.46 The organolithium reagent is a nucleophile and a base. As a base it can remove the most acidic proton (between the benzene ring and C=O) to form an enolate that is protonated by D_3O^+.

OCH_3 H
HC≡C–Li
OCH_3
A
OCH_3 + HC≡CH
+ Li^+
$D-OD_2$
OCH_3 D
+ D_2O
B
$C_{12}H_{13}DO_3$

20.47 Both ketones are chiral molecules with carbonyl groups that have one side more sterically hindered than the other. In both reductions, hydride approaches from the less hindered side.

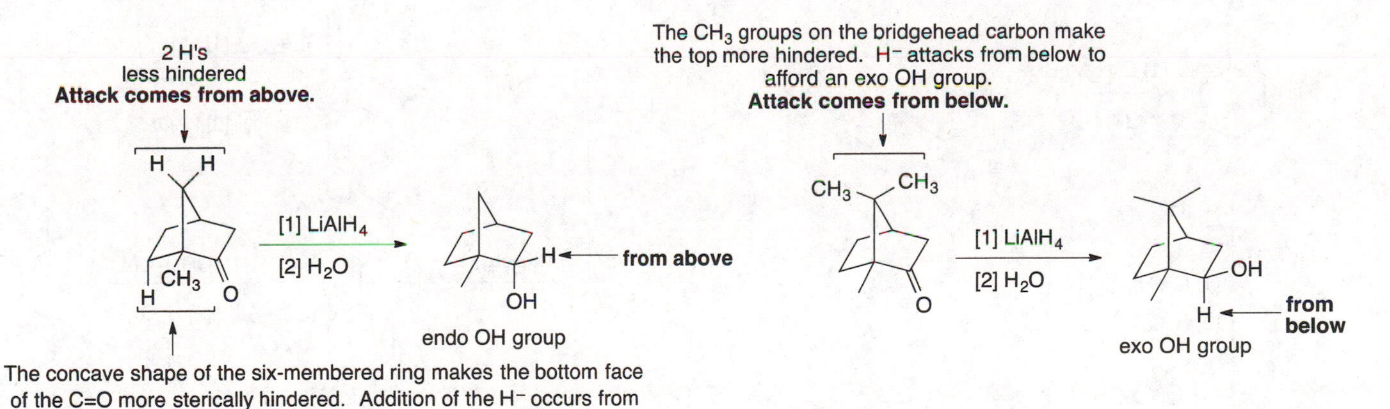

2 H's
less hindered
Attack comes from above.
H H
CH3
H
O
[1] LiAlH4
[2] H2O
H
from above
OH
endo OH group
The concave shape of the six-membered ring makes the bottom face of the C=O more sterically hindered. Addition of the H⁻ occurs from above to place the new C–H bond exo, making the OH endo.
The CH3 groups on the bridgehead carbon make the top more hindered. H⁻ attacks from below to afford an exo OH group.
Attack comes from below.
CH3
CH3
O
[1] LiAlH4
[2] H2O
OH
H
from below
exo OH group

20.48 A Grignard reagent contains a carbon atom with a partial negative charge, so it acts as a base and reacts with the OH of the starting halide, $BrCH_2CH_2CH_2CH_2OH$. This acid–base reaction destroys the Grignard reagent, so that addition cannot occur. To get around this problem, the OH group can be protected as a *tert*-butyldimethylsilyl ether, from which a Grignard reagent can be made.

basic site
acidic site
δ–
Br ... OH —Mg→ BrMg ... OH —proton transfer→ ... O⁻ + HOMgBr
These will react.

INSTEAD: Use a protecting group.

Br ... OH —TBDMS–Cl, imidazole→ Br ... OTBDMS —Mg, ether→ BrMg ... OTBDMS
protected OH
[1] cyclohexanone
[2] H_2O
(Bu)$_4$NF
OH, OH, OTBDMS
A

20.49 Compounds **F, G,** and **K** are all alcohols with aromatic rings so there will be many similarities in their proton NMR spectra. These compounds, however, will show differences in absorptions due to the CH protons on the carbon bearing the OH group. **F** has a CH_2OH group, which will give a singlet in the 3–4 ppm region of the spectrum. **G** is a 3° alcohol that has no protons on the C bonded to the OH group so it will have no peak in the 3–4 ppm region of the spectrum. **K** is a 2° alcohol that will give a doublet in the 3–4 ppm region of the spectrum for the CH proton on the carbon with the OH group.

[1] C_6H_5MgBr
[2] H_2O
OH
A
H_2SO_4
B
[1] O_3
[2] CH_3SCH_3
O **C**
CHO
HCl
Cl **D**
[1] Mg
[2] CH_2=O
[3] H_2O
OH
F
mCPBA
O
E
[1] $(CH_3)_2CuLi$
[2] H_2O
OH
G
[1] $LiAlH_4$
[2] H_2O
OH
H
PBr_3
Br
I
Mg
MgBr
J
[1] C_6H_5CHO
[2] H_2O
OH
K

20.50

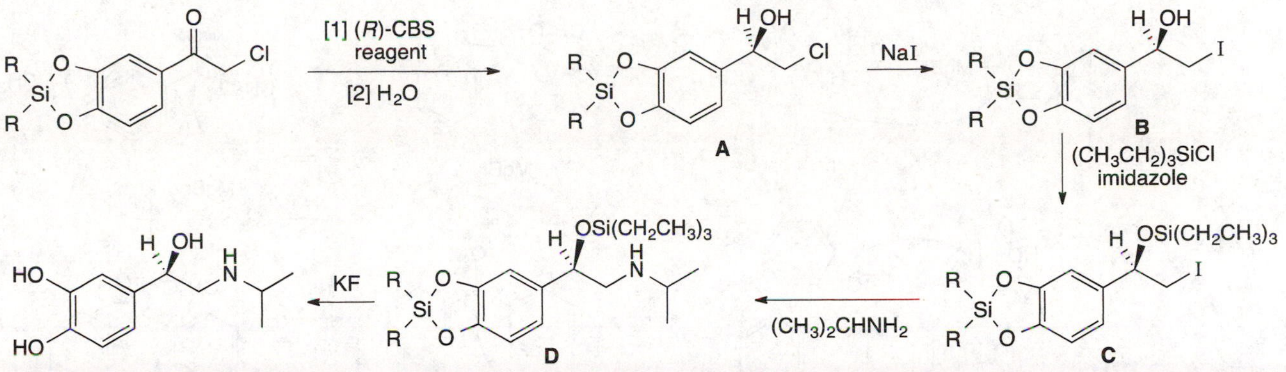

[1] (*R*)-CBS reagent
[2] H_2O

A

NaI

B

$(CH_3CH_2)_3SiCl$
imidazole

C

$(CH_3)_2CHNH_2$

D

KF

20.51

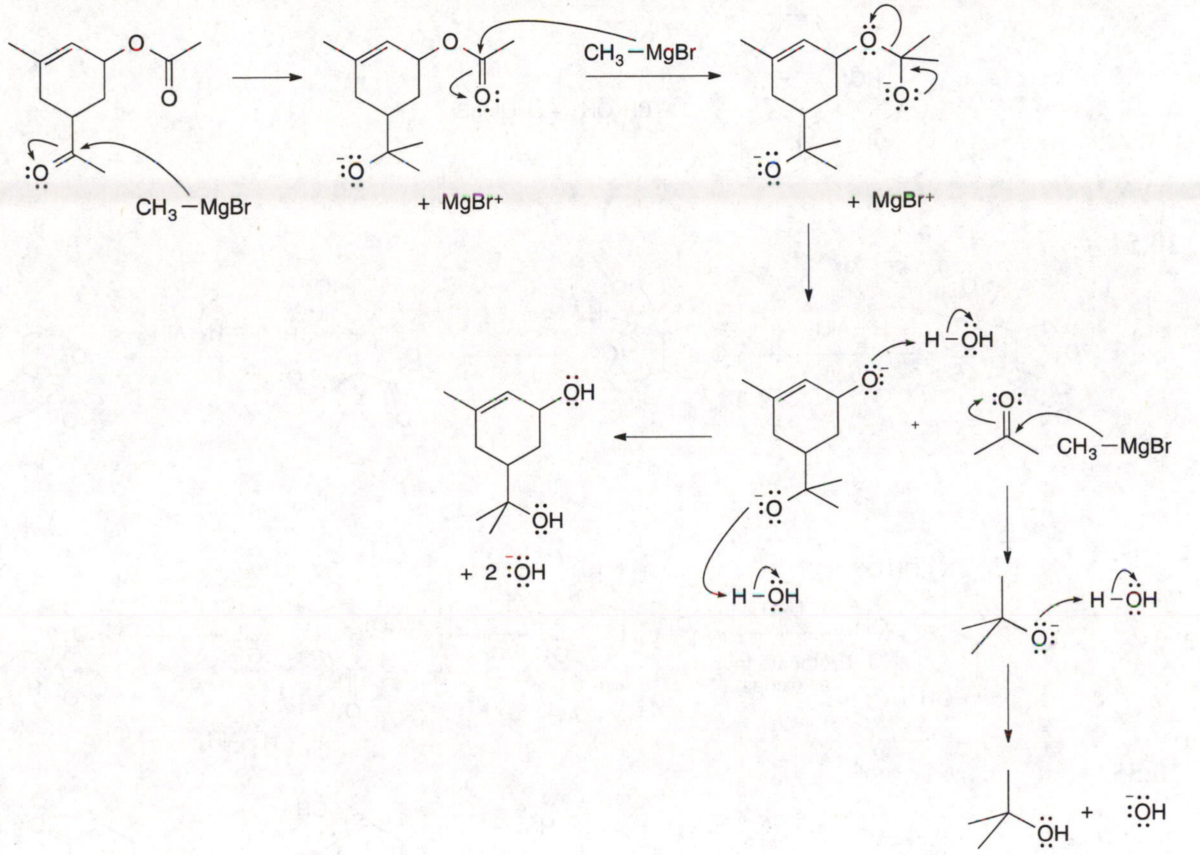

CH_3-MgBr

+ $MgBr^+$

CH_3-MgBr

+ $MgBr^+$

H–OH

CH_3-MgBr

H–OH

+ 2 $:\ddot{O}H^-$

H–OH

20.52

BrMg MgBr

+ $\overset{+}{M}gBr$

BrMg

H–OH H–OH

+ $\overset{+}{M}gBr$

HOMgBr +

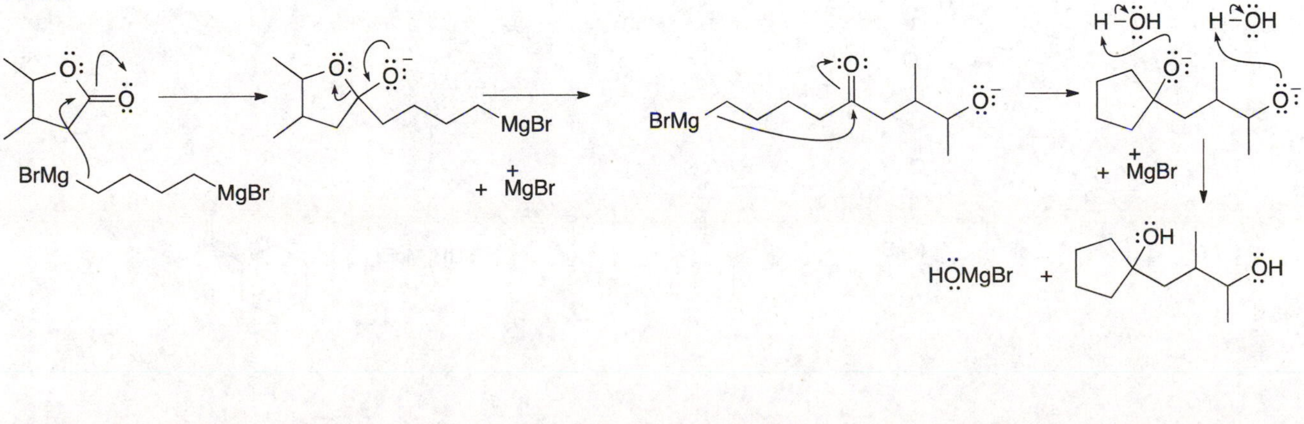

20.53

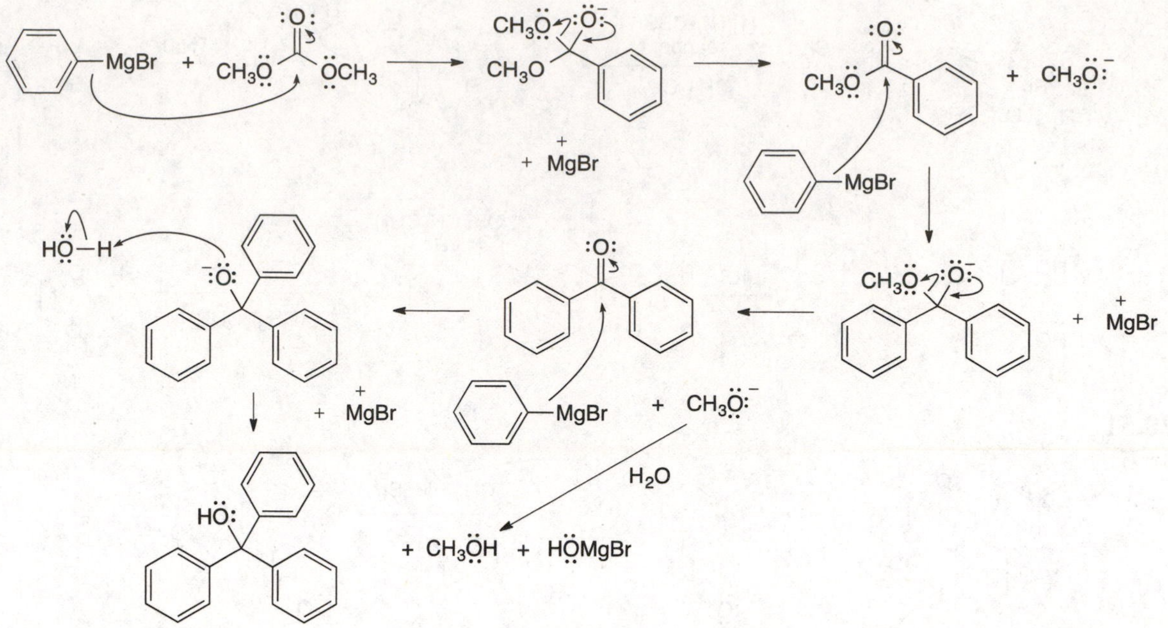

MgBr + CH_3O ... OCH_3 → CH_3O ... CH_3O + $\overset{+}{M}gBr$ → CH_3O + CH_3O^-

MgBr → CH_3O + $\overset{+}{M}gBr$

HO–H ... $^-$O: + $\overset{+}{M}gBr$ ... MgBr + CH_3O^-

H_2O

HO: + CH_3OH + HOMgBr

20.54

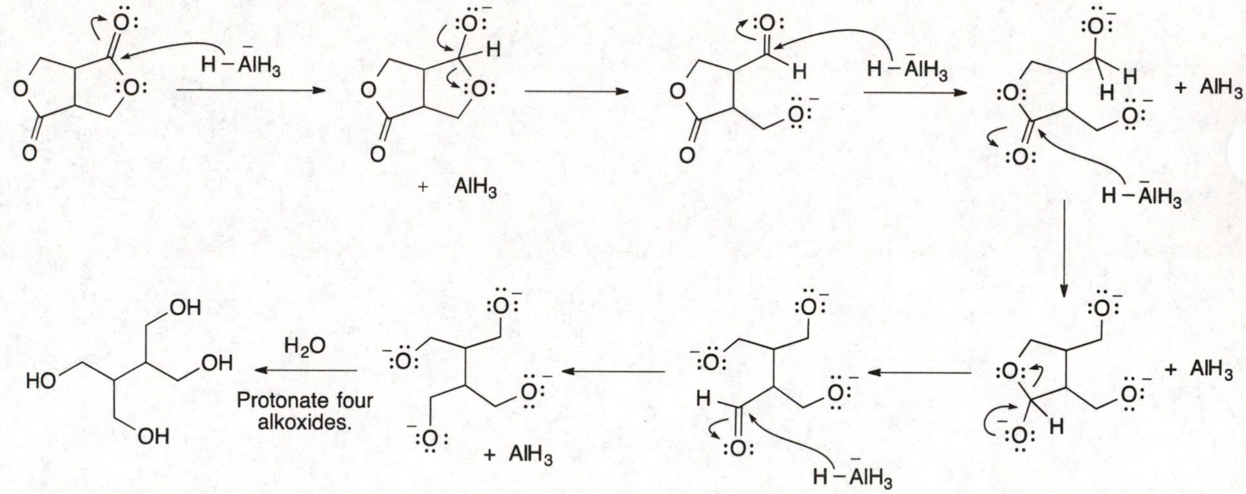

H – $\bar{A}lH_3$ → + AlH_3 → H – $\bar{A}lH_3$ → + AlH_3

H – $\bar{A}lH_3$

OH, OH, HO, OH ← H_2O Protonate four alkoxides. ← + AlH_3 ← H – $\bar{A}lH_3$ ← + AlH_3

20.55

a. OH ⟹ MgBr + H, O

or

H, O + MgBr

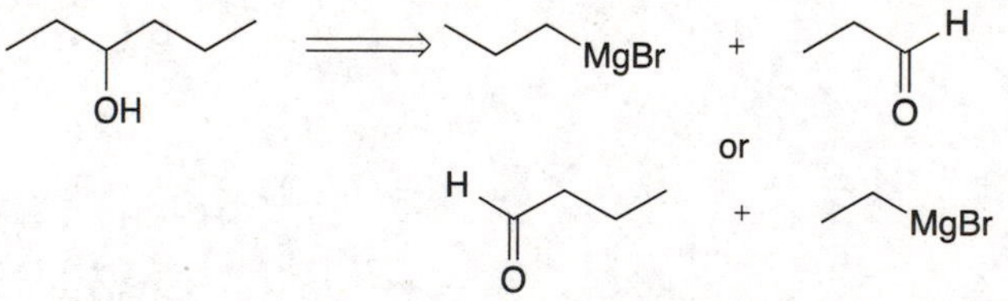

c. OH ⟹ O + MgBr

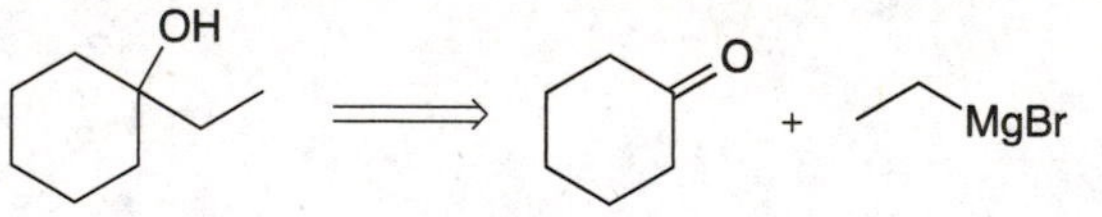

b. HO ⟹ O + BrMg

or MgBr + O or CH_3MgBr + O

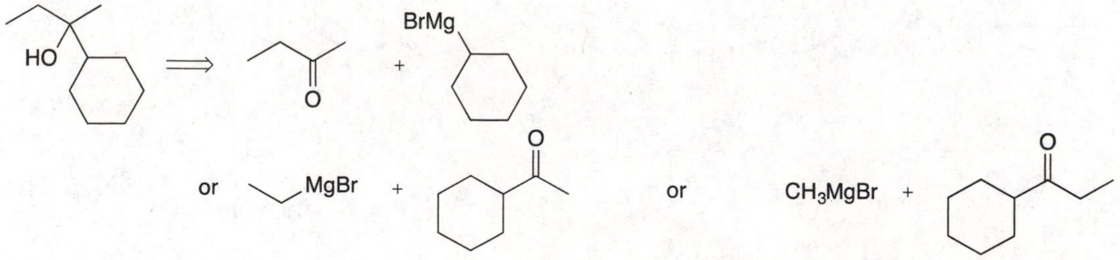

20.56

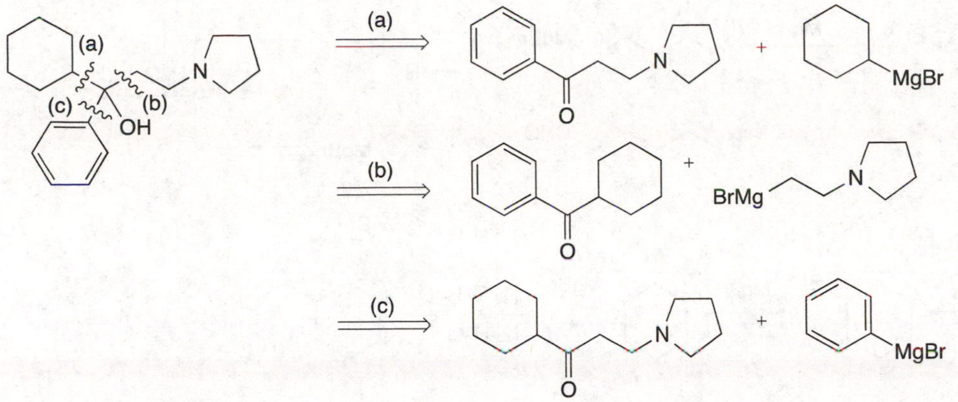

(a) (b) (c) OH N

(a) ⟹ O N + MgBr

(b) ⟹ O + BrMg N

(c) ⟹ O N + MgBr

20.57

a. 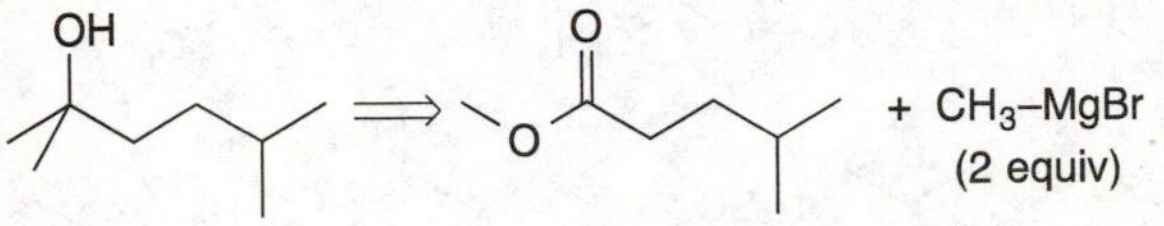 OH ⟹ O O + CH_3–MgBr (2 equiv)

b. 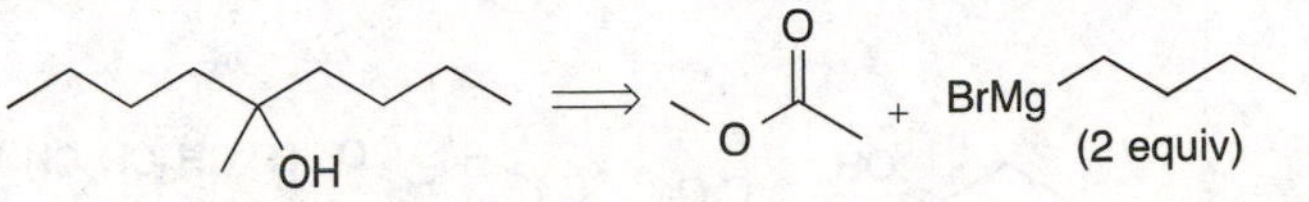 OH ⟹ O O + BrMg (2 equiv)

20.58

a. 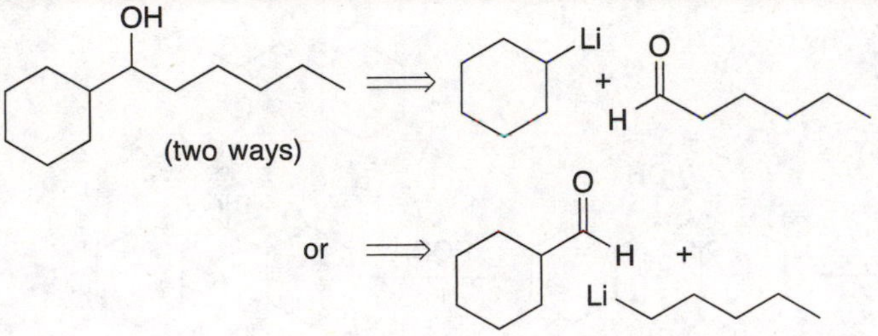 OH (two ways) ⟹ Li + O H

or ⟹ O H + Li

b. 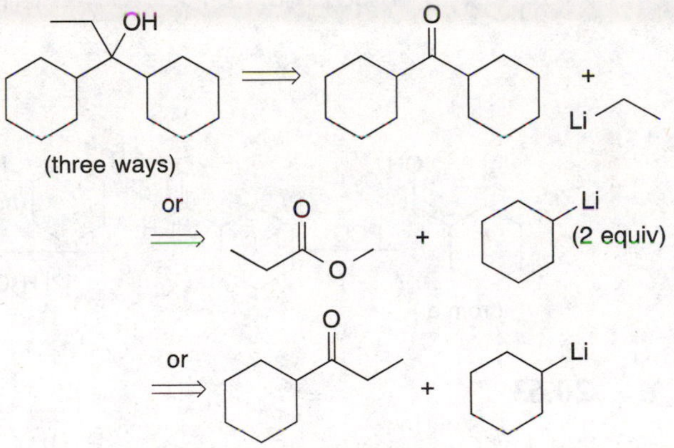 OH (three ways) ⟹ O + Li

or ⟹ O O + Li (2 equiv)

or ⟹ O + Li

20.59

a. 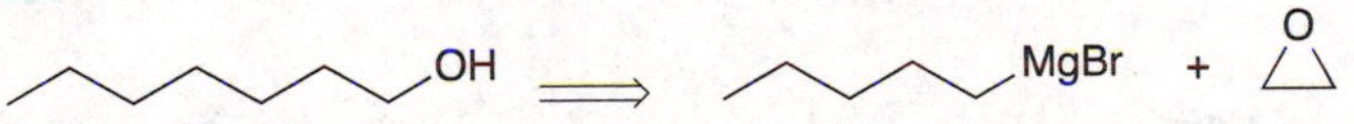 OH ⟹ MgBr + O

b. 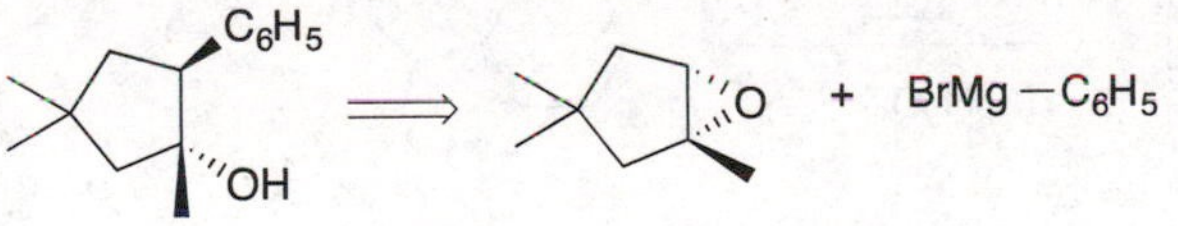 C_6H_5 OH ⟹ O + BrMg — C_6H_5

c. 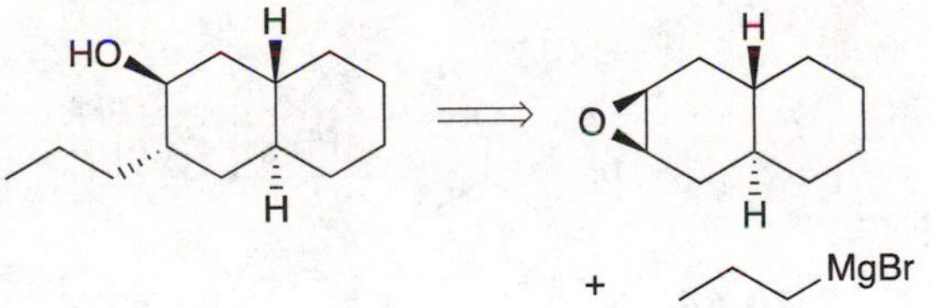 HO H H ⟹ O H H + MgBr

20.60

C_6H_5 Br —Mg→ C_6H_5 MgBr —H_2O→ C_6H_5

C_6H_5 Br —$KOC(CH_3)_3$→ C_6H_5 —H_2, Pd-C→ C_6H_5

C_6H_5 Br —$LiAlH_4$→ C_6H_5

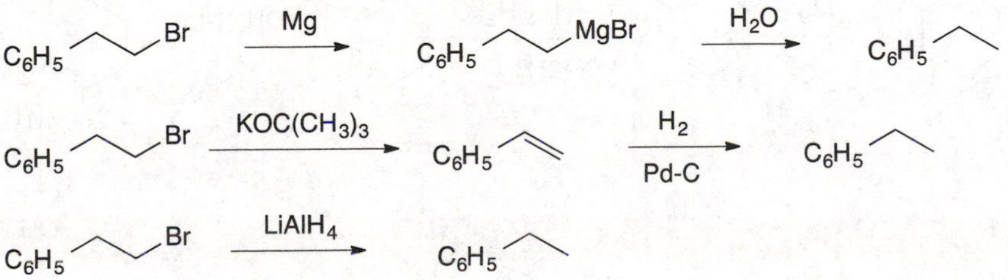

20.61

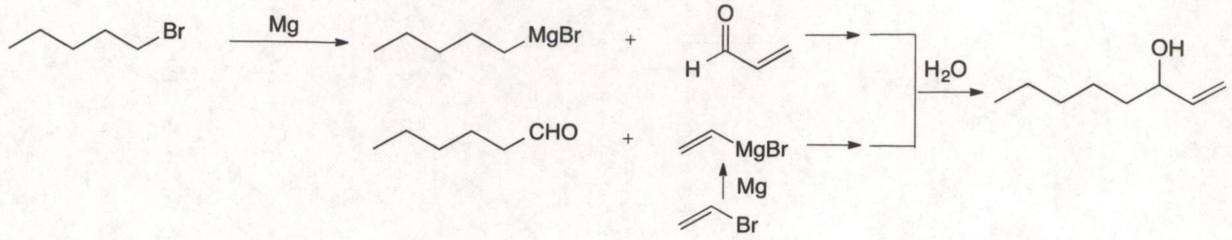

Br —Mg→ MgBr + H(C=O)CH=CH$_2$ →
CHO + MgBr (from Br, Mg) →
H_2O → OH

20.62

a. OH —PBr_3→ Br —Mg→ MgBr —[1] CH_2=O, [2] H_2O→ OH

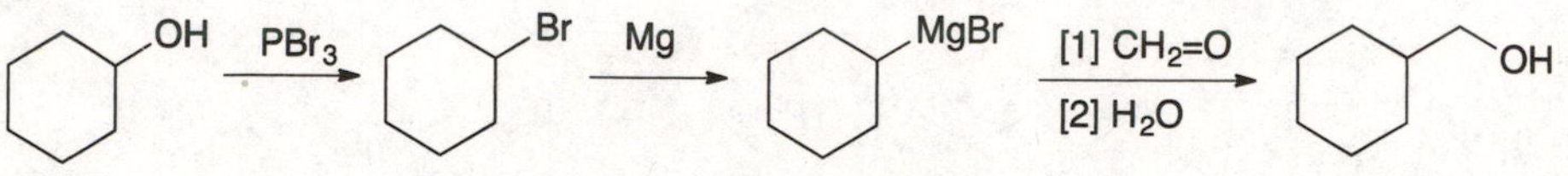

b. OH —H_2SO_4→ —mCPBA→ O —[1] C_6H_5MgBr, [2] H_2O→ OH, C_6H_5 —PCC→ O, C_6H_5

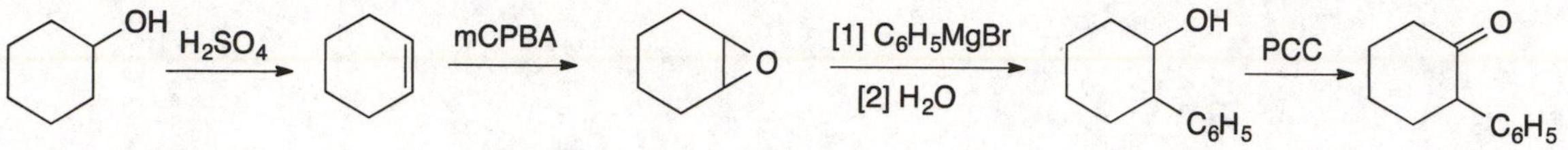

c. OH —CrO_3, H_2SO_4, H_2O→ O —[1] $CH_3CH_2CH_2MgBr$, [2] H_2O→ OH —H_2SO_4→ + —H_2, Pd-C→

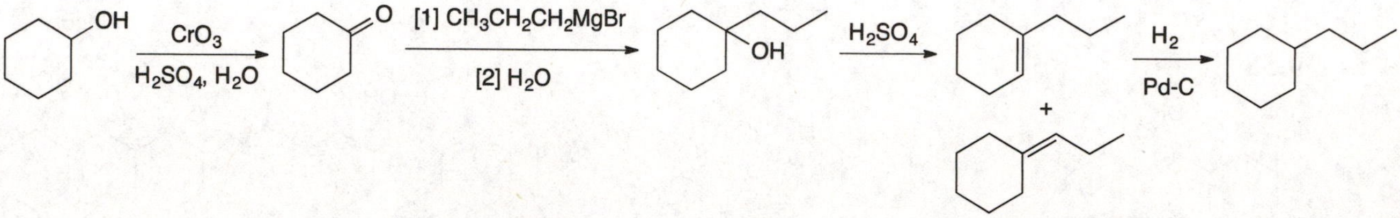

d. OH (from a.) —PCC→ O, H —[1] MgBr (from a.), [2] H_2O→ OH —PCC→ O

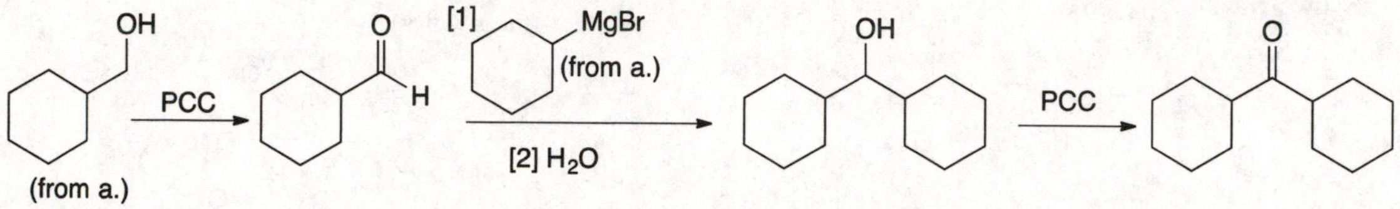

20.63

a. OH —PCC→ O —[1] MgBr, [2] H_2O→ OH

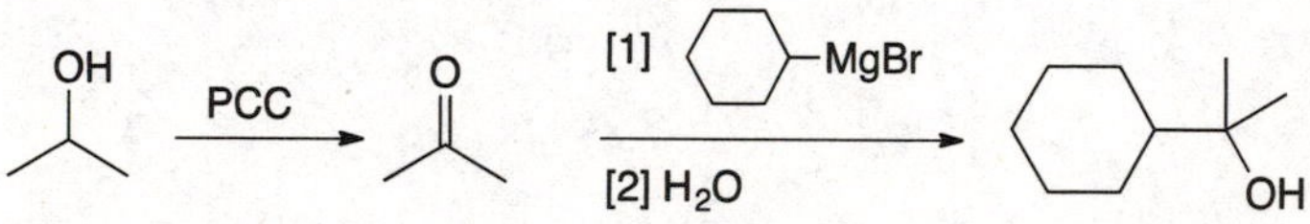

b. OH —PBr_3→ Br —Mg→ MgBr —[1] =O, [2] H_2O→ OH

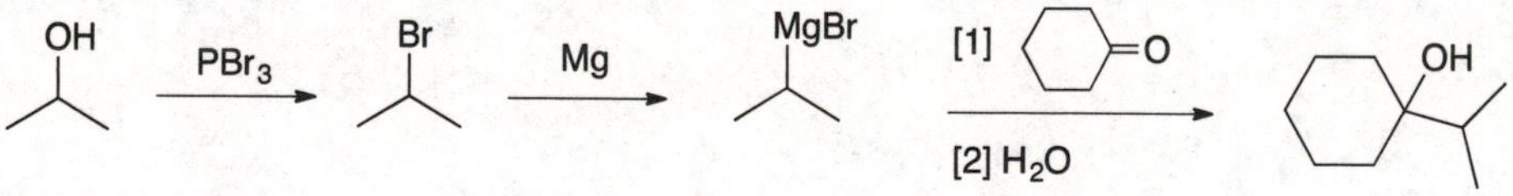

20.64

a. —Br_2, $FeBr_3$→ Br —Mg→ MgBr —[1] CH_2=O (CH_3OH, PCC), [2] H_2O→ OH —PCC→ O, H —[1] PhMgBr, [2] H_2O→ OH —PCC→ O

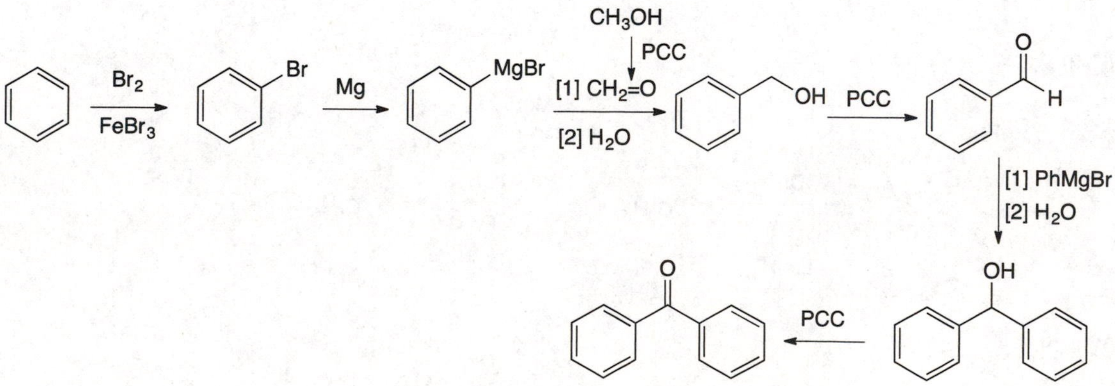

b. OH → PCC → H, O + MgBr (from a.) → H_2O → OH → PCC → O → Br_2, $FeBr_3$ → Br, O

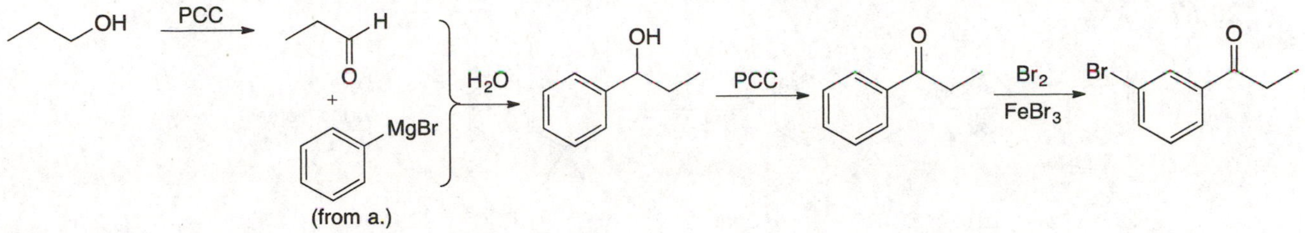

20.65

a. OH → PCC → H, O → [1] MgBr [2] H_2O → OH → PCC → O

OH → PBr_3 → Br → Mg → MgBr

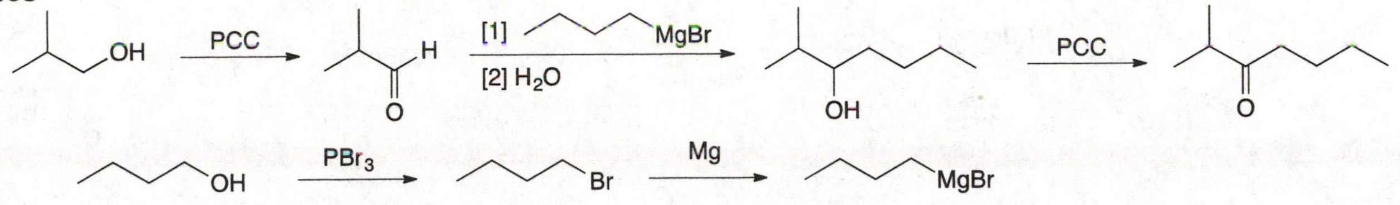

b. H, O (from a.) → [1] MgBr [2] H_2O → OH → PBr_3 → Br → Mg → MgBr → [1] CO_2 [2] H_3O^+ → CO_2H

OH → PBr_3 → Br → Mg → MgBr

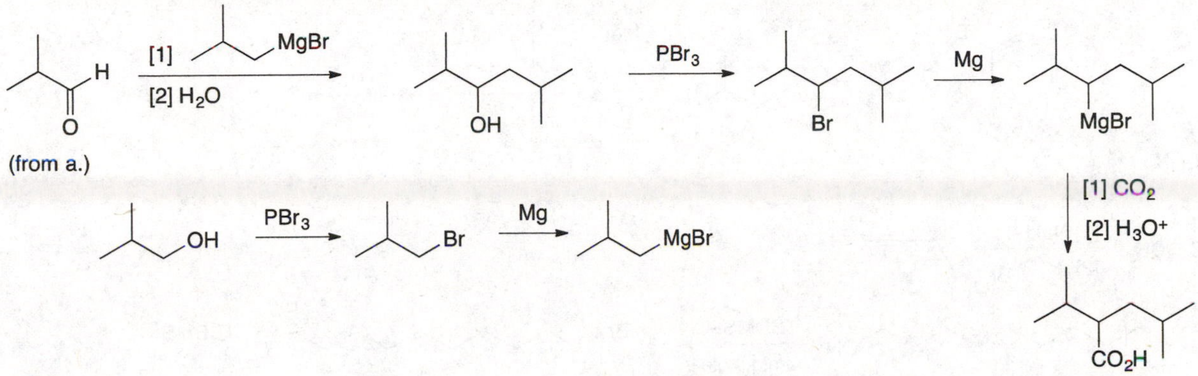

c. OH → HBr → Br → Mg → MgBr + H, O (from a.) → H_2O → OH → PBr_3 → Br → K^+ $^-OC(CH_3)_3$

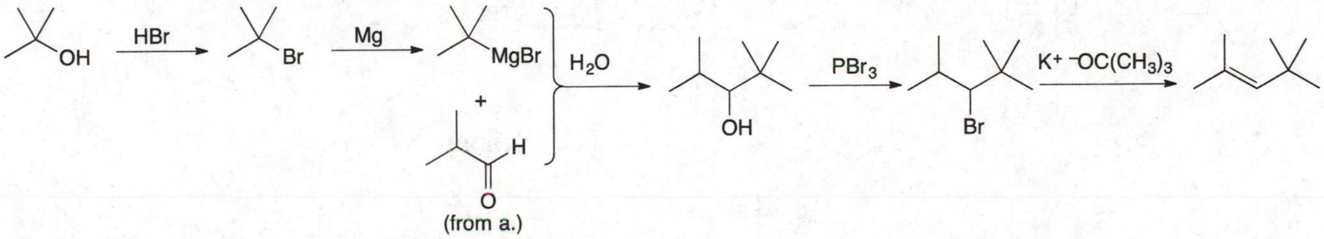

20.66

a. O → [1] $(CH_3)_2CuLi$ [2] H_2O → O → [1] CH_3MgBr [2] H_2O → OH → H_2SO_4 → + → H_2, Pd-C

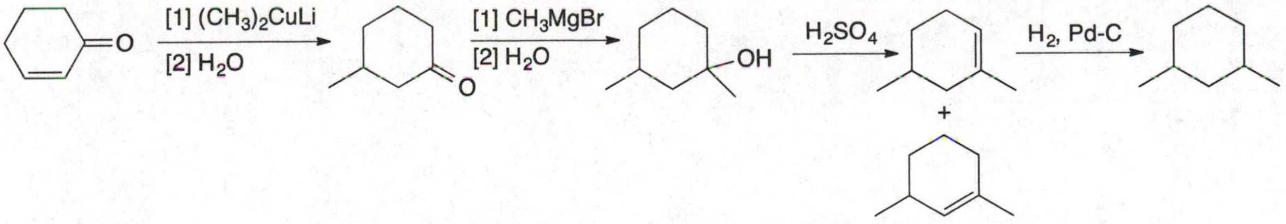

b. HC≡CH → [1] NaH [2] $CH_3CH=O$ [3] H_2O → HO → TBDMS–Cl, imidazole → TBDMSO → NaH → TBDMSO → [1] O (epoxide) [2] H_2O → TBDMSO, OH → $(Bu)_4NF$ → HO, OH

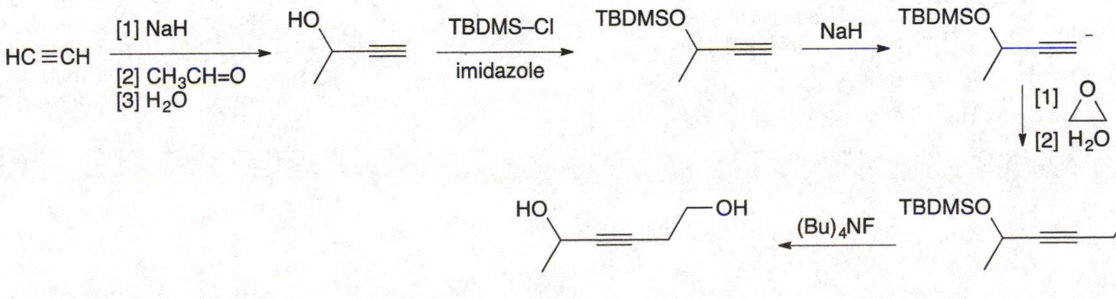

20.67

OH
H_2SO_4
Br_2
$FeBr_3$
Br
+ ortho isomer

OH
PCC
O
H
OH
PBr$_3$
Br
Mg
MgBr
H_2O
OH
PCC
O

Br
Mg
MgBr
O
H_2O
OH

20.68

HC≡CH
[1] NaH
[2] Br

Br
OH
TBDMS–Cl
imidazole
Br
O
TBDMS
A

H
[1] NaH
[2] **A**
O
TBDMS

Na, NH_3

O
TBDMS

Bu_4NF

OH
PCC
O
H

(*E*)-tetradec-11-enal

20.69

H_b H_b
O
CH_3 CH_3
H_a H H H_a
CH_3 CH_3
H_b H_b
$C_7H_{14}O$
A

IR peak: 1716 cm^{-1} (C=O)
^{1}H NMR: 2 signals (ppm)
doublet 1.2 (H_b)
septet 2.7 (H_a)

$NaBH_4$
CH_3OH

H_d H_b H_d
H
OH_a
CH_3 CH_3
H_c H H H_c
CH_3 CH_3
H_d H_d
$C_7H_{16}O$
B

IR peak: 3600–3200 cm^{-1} (OH)
^{1}H NMR: 4 signals (ppm)
doublet 0.9 (H_d)
singlet 1.5 (H_a)
multiplet 1.7 (H_c)
triplet 3.0 (H_b)

20.70

Hb → H, Hb → H, O, CH_3 ← Ha, CH_3 ← Ha

C_4H_8O

C

[1] C_6H_5MgBr

[2] H_2O

Hc Hc, H H, OH_b, CH_3 CH_3, Ha Ha

$C_{10}H_{14}O$

D

^1H NMR: 2 signals (ppm)
singlet (6 H) 1.3 (H_a)
singlet (2 H) 2.4 (H_b)

IR peak 3600–3200 cm^{-1} (OH)
^1H NMR: 4 signals (ppm)
singlet (6 H) 1.2 (H_a)
singlet (1 H) 1.6 (H_b)
singlet (2 H) 2.7 (H_c)
multiplet (5 H) 7.2 (benzene ring)

20.71

$CH_3C(=O)OCH_2CH_3$ (Ha, Hb, Hc)

$C_4H_8O_2$

E

IR peak 1743 cm^{-1} (C=O)
^1H NMR: 3 signals (ppm)
triplet (3 H) 1.2 (H_c)
singlet (3 H) 2.0 (H_a)
quartet (2 H) 4.1 (H_b)

[1] CH_3CH_2MgBr (excess)

[2] H_2O

$CH_3CH_2-C(OH_d)(CH_3)-CH_2CH_3$ (Ha, Hb, Hc)

$C_6H_{14}O$

F

IR peak 3600–3200 cm^{-1} (OH)
^1H NMR: 4 signals (ppm)
triplet (6 H) 0.9 (H_a)
singlet (3 H) 1.1 (H_c)
quartet (4 H) 1.5 (H_b)
singlet (1 H) 1.55 (H_d)

20.72 Molecular ion at $m/z = 86$: $C_5H_{10}O$ (possible molecular formula).

[1] CH_3MgBr

[2] H_3O^+

O, H H (Hb Hb), CH_3 (Hc), H H (Hd Hd), CH_3 (Ha)

G

IR peak 1721 cm^{-1} (C=O)
^1H NMR: 4 signals (ppm)
triplet (3 H) 0.9 (H_a)
sextet (2 H) 1.6 (H_b)
singlet (3 H) 2.1 (H_c)
triplet (2 H) 2.4 (H_d)

20.73 Molecular ion at $m/z = 86$: $C_5H_{10}O$ (possible molecular formula).

[1] $(CH_3)_3CLi$

[2] $CH_2=O$

[3] H_2O

Hb Hd Hd, CH_3 H H, H (He), H (Hf), H H (Hc Hc), OH (Ha)

H

IR peaks: 3600–3200 cm^{-1} (OH)
1651 cm^{-1} (C=C)
^1H NMR: 6 signals (ppm)
singlet (1 H) 1.7 (H_a)
singlet (3 H) 1.8 (H_b)
triplet (2 H) 2.2 (H_c)
triplet (2 H) 3.8 (H_d)
two signals at 4.8 and 4.9 due to 2 H's: H_e and H_f

20.74

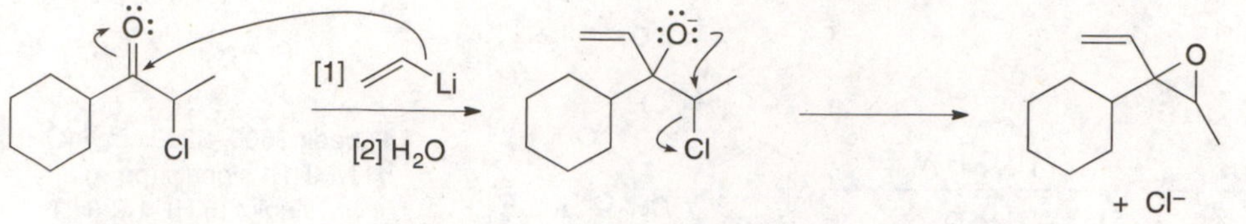

[1] Li (vinyllithium)
[2] H_2O

Cl

Cl

+ Cl^-

20.75

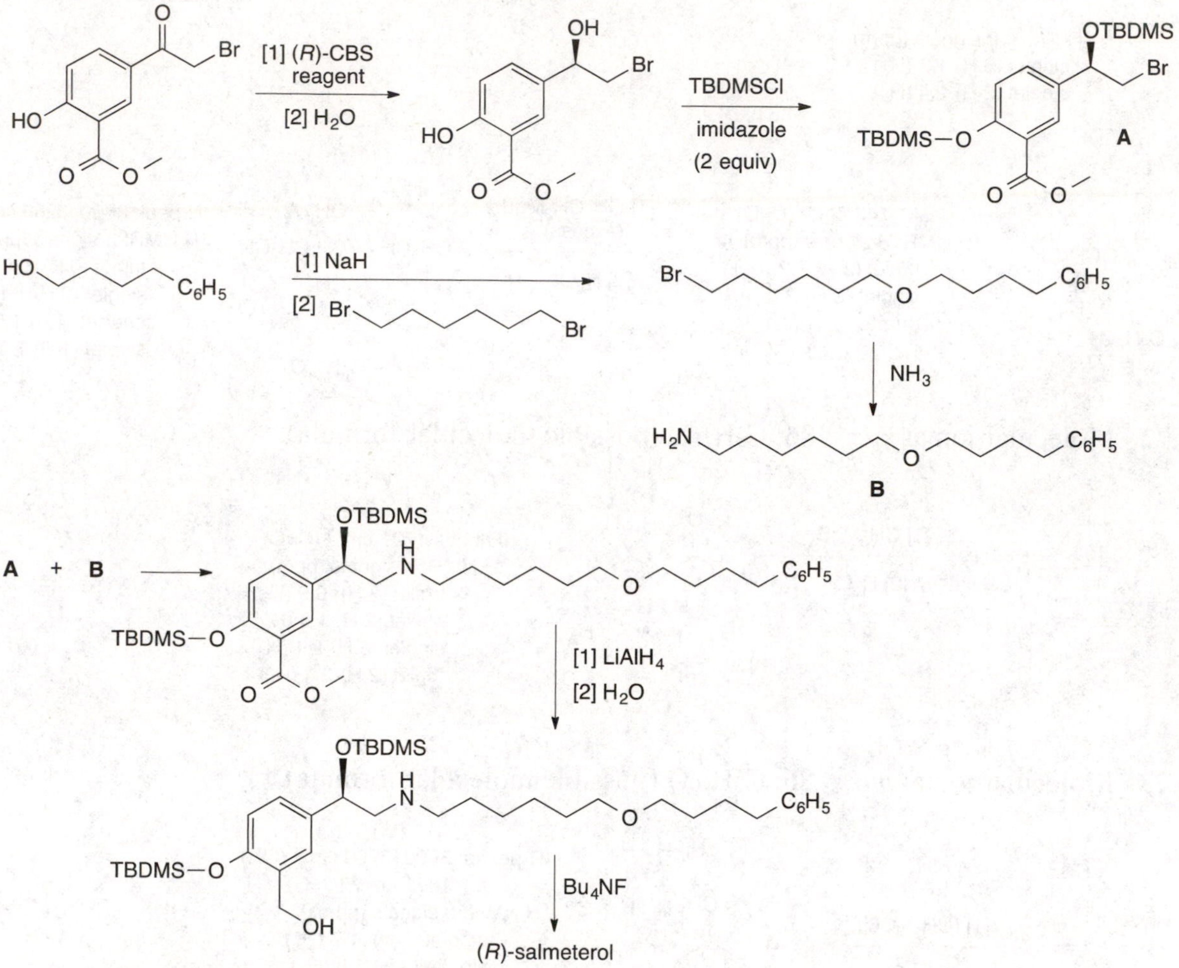

Br
[1] (*R*)-CBS reagent
[2] H_2O
HO
OH
Br
HO
TBDMSCl
imidazole
(2 equiv)
OTBDMS
Br
TBDMS—O
A

HO C_6H_5
[1] NaH
[2] Br Br
Br O C_6H_5
NH_3
H_2N O C_6H_5
B

A + **B** →
OTBDMS
H
N
O C_6H_5
TBDMS—O
[1] $LiAlH_4$
[2] H_2O
OTBDMS
H
N
O C_6H_5
TBDMS—O
OH
Bu_4NF
(*R*)-salmeterol

20.76

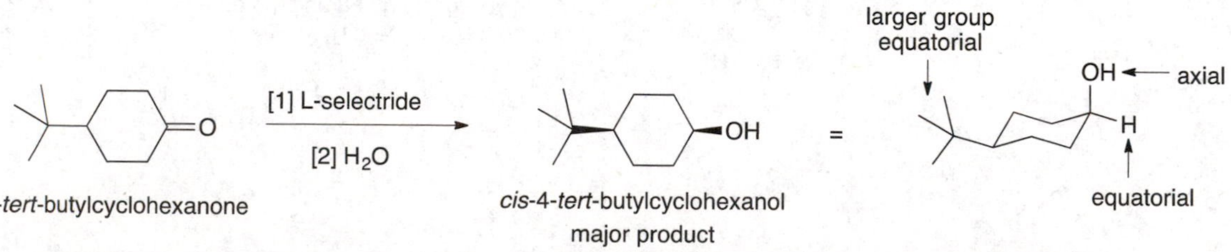

4-*tert*-butylcyclohexanone
[1] L-selectride
[2] H_2O
OH
cis-4-*tert*-butylcyclohexanol
major product
=
larger group equatorial
OH ← axial
H
equatorial

L-Selectride adds H^- to a C=O group. There are two possible reduction products—cis and trans isomers—but the cis isomer is favored. The key element is that the three *sec*-butyl groups make L-selectride a large, bulky reducing agent that attacks the carbonyl group from the less hindered direction.

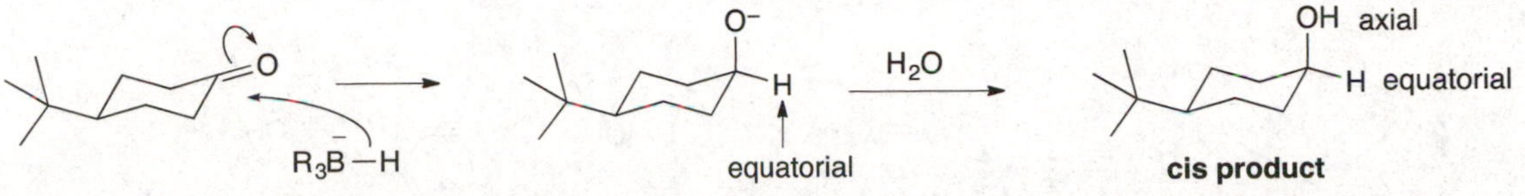

O– H H_2O OH axial H equatorial

R_3B—H equatorial **cis product**

When H⁻ adds from the equatorial direction, the product has an axial OH and a new equatorial H. Since the equatorial direction is less hindered, this mode of attack is favored with large bulky reducing agents like L-selectride. In this case, the product is cis.

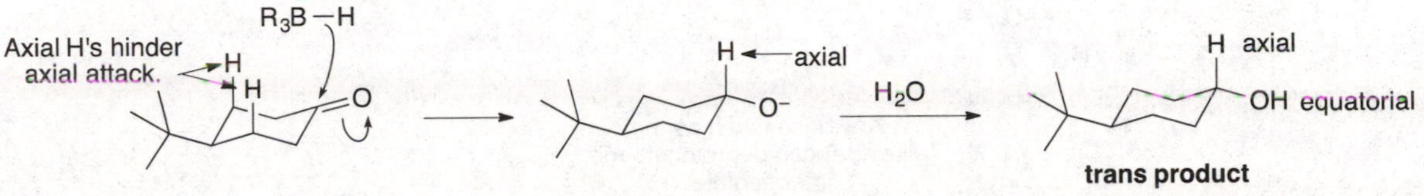

Axial H's hinder axial attack. $R_3\bar{B}$—H H ←axial O– H_2O H axial OH equatorial

trans product

The axial H's hinder H⁻ attack from the axial direction. As a result, this mode of attack is more difficult with larger reducing agents. In this case the product is trans. This product is not formed to any appreciable extent.

20.77 The β carbon of an α,β-unsaturated carbonyl compound absorbs farther downfield in the ^{13}C NMR spectrum than the α carbon, because the β carbon is deshielded and bears a partial positive charge as a result of resonance. Because three resonance structures can be drawn for an α,β-unsaturated carbonyl compound, one of which places a positive charge on the β carbon, the decrease of electron density at this carbon deshields it, shifting the ^{13}C absorption downfield. This is not the case for the α carbon.

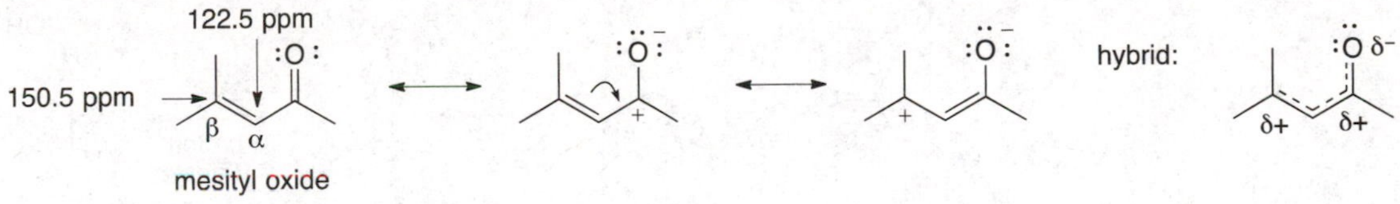

122.5 ppm

150.5 ppm β α

mesityl oxide

hybrid: δ– δ+ δ+

20.78

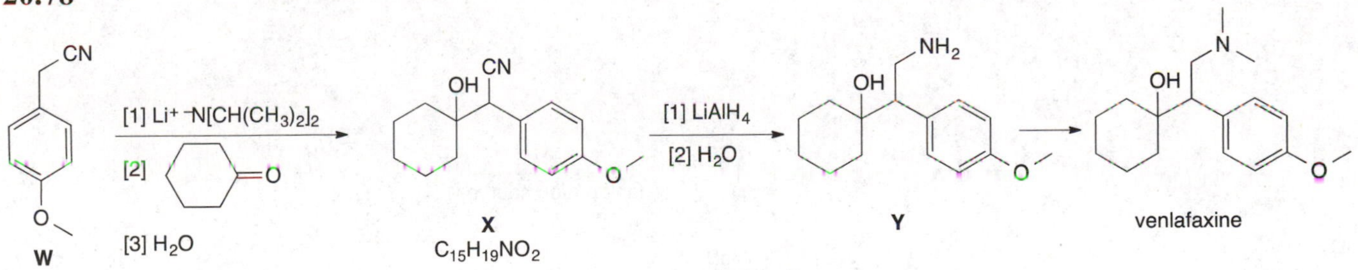

W: [1] Li⁺ ⁻N[CH(CH₃)₂]₂ [2] cyclohexanone [3] H_2O → X $C_{15}H_{19}NO_2$ → [1] $LiAlH_4$ [2] H_2O → Y → venlafaxine

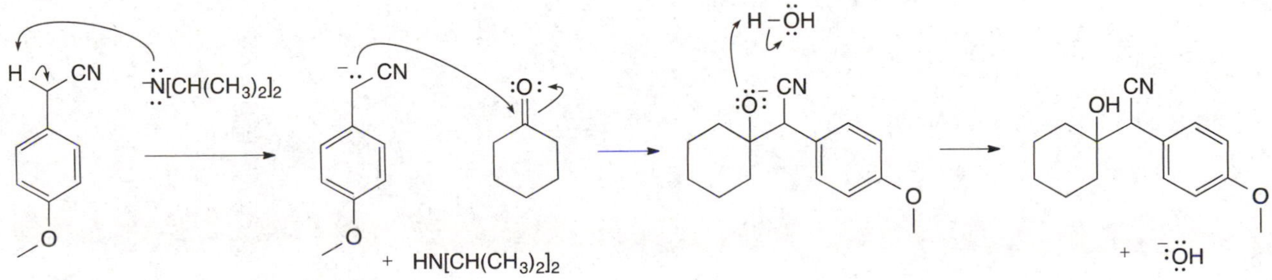

H CN ⁻N[CH(CH₃)₂]₂ → CN :O: + HN[CH(CH₃)₂]₂ → H–OH, CN → OH CN + ⁻:OH

20.79

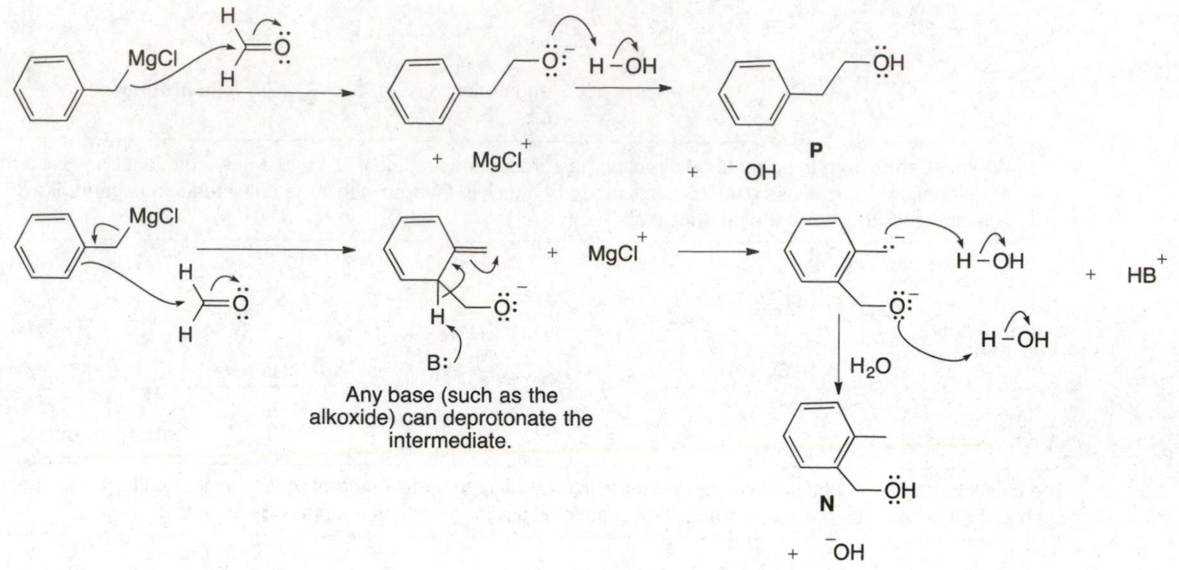

MgCl H H $\ddot{O}$: $\ddot{O}$:⁻ H–OH + $MgCl^+$ $\ddot{O}$H P + ⁻OH

MgCl H H $\ddot{O}$: H B: $\ddot{O}$:⁻ + $MgCl^+$ H–OH + HB^+ H–OH H_2O $\ddot{O}$H N + ⁻OH

Any base (such as the alkoxide) can deprotonate the intermediate.

20.80

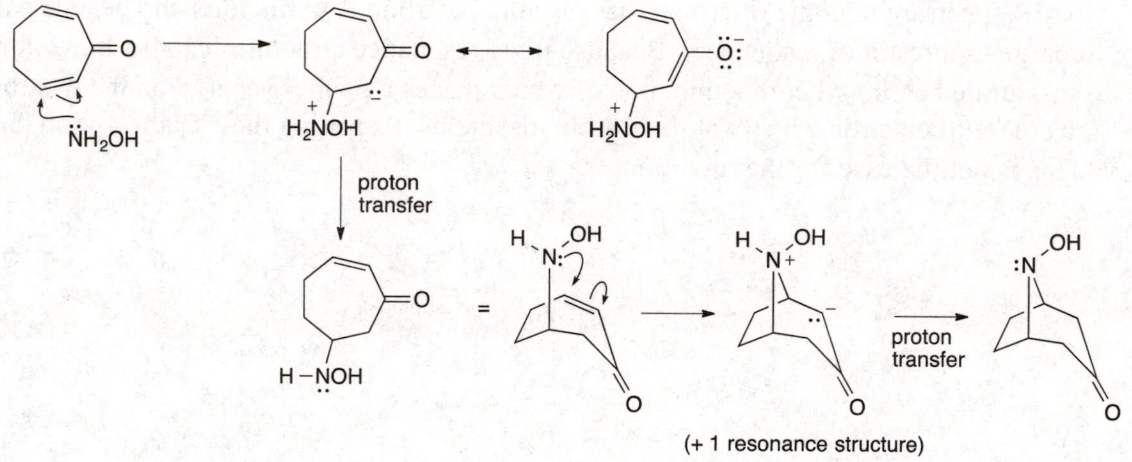

O $\ddot{N}H_2OH$ O $H_2\overset{+}{N}OH$ $\ddot{O}$:⁻ $H_2\overset{+}{N}OH$

proton transfer

O H–$\ddot{N}$OH = H N: OH O H N+ OH O proton transfer :N OH O

(+ 1 resonance structure)

20.81

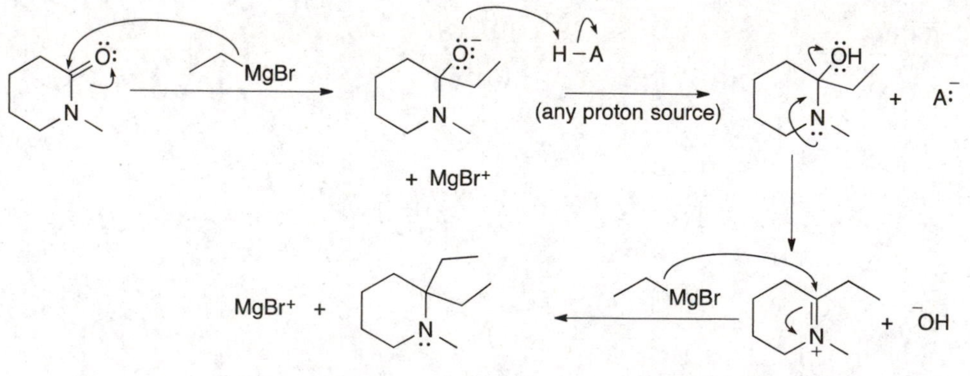

$\ddot{O}$: N MgBr $\ddot{O}$:⁻ N + $MgBr^+$ H–A (any proton source) $\ddot{O}$H N + A:⁻

$MgBr^+$ + N MgBr N+ + ⁻OH

Chapter 21 Aldehydes and Ketones—Nucleophilic Addition

Chapter Review

General facts

- Aldehydes and ketones contain a carbonyl group bonded to only H atoms or R groups. The carbonyl carbon is sp^2 hybridized and trigonal planar (21.1).
- Aldehydes are identified by the suffix *-al,* whereas ketones are identified by the suffix *-one* (21.2).
- Aldehydes and ketones are polar compounds that exhibit dipole–dipole interactions (21.3).

Summary of spectroscopic absorptions of RCHO and R_2CO (21.4)

IR absorptions	C=O	~1715 cm^{-1} for ketones • increasing frequency with decreasing ring size ~1730 cm^{-1} for aldehydes • For both RCHO and R_2CO, the frequency decreases with conjugation.
	C_{sp^2}–H of CHO	~2700–2830 cm^{-1} (one or two peaks)
^{1}H NMR absorptions	CHO	9–10 ppm (highly deshielded proton)
	C–H α to C=O	2–2.5 ppm (somewhat deshielded C_{sp^3}–H)
^{13}C NMR absorption	C=O	190–215 ppm

Nucleophilic addition reactions

[1] Addition of hydride (H^-) (21.8)

R(C=O)R′ (R′ = H or alkyl) —[NaBH$_4$, CH$_3$OH or [1] LiAlH$_4$ [2] H$_2$O]→ R(R′)C(OH)H — 1° or 2° alcohol

- The mechanism has two steps.
- H:$^-$ adds to the planar C=O from both sides.

[2] Addition of organometallic reagents (R^-) (21.8)

R(C=O)R′ (R′ = H or alkyl) —[[1] R″MgX or R″Li [2] H$_2$O]→ R(R′)C(OH)R″ — 1°, 2°, or 3° alcohol

- The mechanism has two steps.
- R:$^-$ adds to the planar C=O from both sides.

[3] Addition of cyanide ($^-$CN) (21.9)

R(C=O)R′ (R′ = H or alkyl) —[NaCN, HCl]→ R(R′)C(OH)CN — cyanohydrin

- The mechanism has two steps.
- $^-$CN adds to the planar C=O from both sides.

[4] Wittig reaction (21.10)

R(R')C=O + Ph_3P=C(R")(R") → R(R')C=C(R")(R") (alkene)

R' = H or alkyl; Wittig reagent

- The reaction forms a new C–C σ bond and a new C–C π bond.
- Ph_3P=O is formed as by-product.

[5] Addition of 1° amines (21.11)

RC(=O)R' —($R''NH_2$, mild acid)→ RC(=NR")R' (imine)

R' = H or alkyl

- The reaction is fastest at pH 4–5.
- The intermediate carbinolamine is unstable, and loses H_2O to form the C=N.

[6] Addition of 2° amines (21.12)

R'C(=O)CH₃ —(R_2NH, mild acid)→ R'C(NR_2)=CH₂ (enamine)

R' = H or alkyl

- The reaction is fastest at pH 4–5.
- The intermediate carbinolamine is unstable, and loses H_2O to form the C=C.

[7] Addition of H_2O—Hydration (21.13)

RC(=O)R' ⇌(H_2O, H^+ or ^-OH) RR'C(OH)₂ (*gem*-diol)

R' = H or alkyl

- The reaction is reversible. Equilibrium favors the product only with less stable carbonyl compounds (e.g., H_2CO and Cl_3CCHO).
- The reaction is catalyzed with either H^+ or ^-OH.

[8] Addition of alcohols (21.14)

RC(=O:)R' + R"OH (2 equiv) ⇌(H^+) RR'C(OR")₂ (acetal) + H_2O

R' = H or alkyl

- The reaction is reversible.
- The reaction is catalyzed with acid.
- Removal of H_2O drives the equilibrium to favor the products.

Other reactions

[1] Synthesis of Wittig reagents (21.10A)

RCH₂X —([1] Ph_3P: [2] Bu—Li)→ Ph_3P=CHR

- Step [1] is best with CH_3X and RCH_2X because the reaction follows an S_N2 mechanism.
- A strong base is needed for proton removal in Step [2].

[2] Conversion of cyanohydrins to aldehydes and ketones (21.9)

R(R')C(OH)CN —(^{-}OH)→ **RCOR' (aldehyde or ketone)** + H_2O + ^{-}CN

- This reaction is the reverse of cyanohydrin formation.

[3] Hydrolysis of nitriles (21.9)

R(R')C(HO)–C≡N —(H_2O; H^+ or ^{-}OH; Δ)→ **R(R')C(HO)COOH (α-hydroxy carboxylic acid)**

R' = H or alkyl

[4] Hydrolysis of imines and enamines (21.12)

imine or enamine —(H_2O, H^+)→ **aldehyde or ketone** + RNH_2 or R_2NH

[5] Hydrolysis of acetals (21.14)

R(R')C(OR'')$_2$ + H_2O ⇌(H^+) **RCOR' (aldehyde or ketone)** + R''OH (2 equiv)

R' = H or alkyl

- The reaction is acid catalyzed and is the reverse of acetal synthesis.
- A large excess of H_2O drives the equilibrium to favor the products.

Practice Test on Chapter Review

1. Give the IUPAC name for the following compounds.

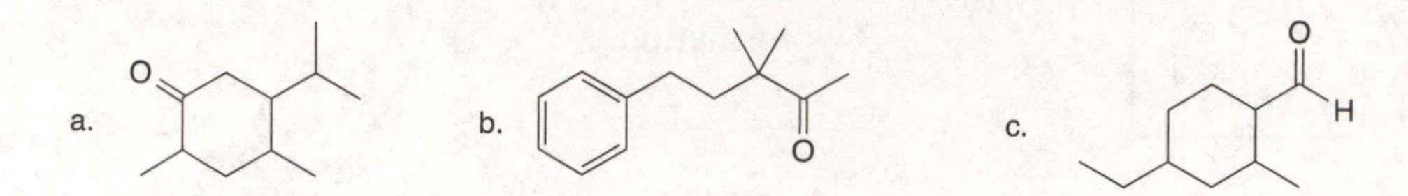

2. (a) Considering compounds **A–D,** which compound forms the smallest amount of hydrate? (b) Which compound forms the largest amount of hydrate?

CH_3O—C₆H₄—CHO (**A**) Cl—C₆H₄—C(O)CH₃ (**B**) O_2N—C₆H₄—CHO (**C**) CH_3O—C₆H₄—C(O)CH₃ (**D**)

3. (a) Considering compounds **A–D,** which compound absorbs at the ***lowest*** wavenumber in its IR spectrum? (b) Which compound absorbs at the ***highest*** wavenumber in its IR spectrum?

A **B** **C** **D**

4. Fill in the lettered reagents (**A–G**) in the following reaction scheme.

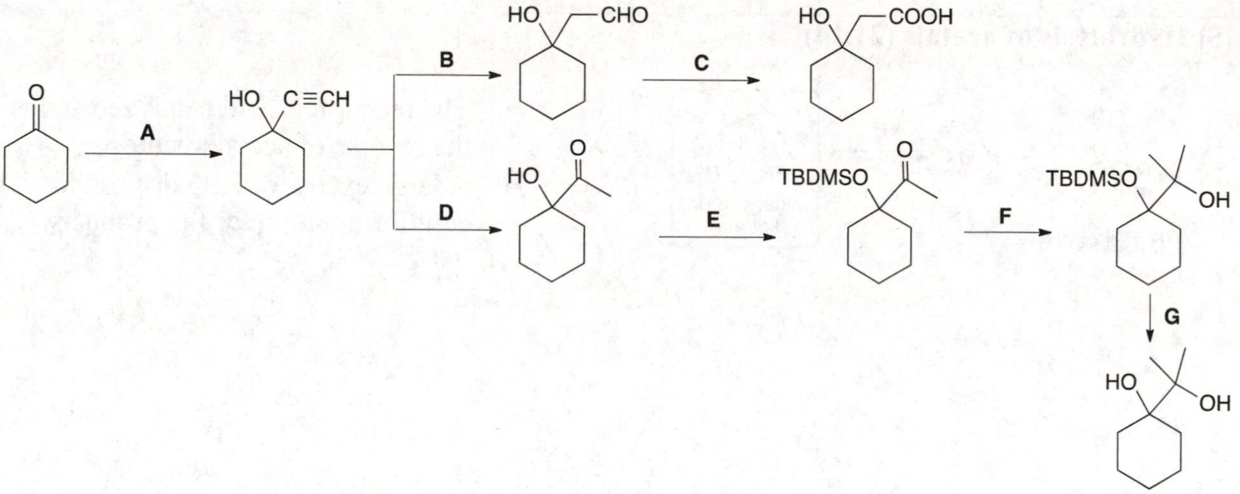

O → **A** → HO C≡CH → **B** → HO CHO → **C** → HO COOH

→ **D** → HO O → **E** → TBDMSO O → **F** → TBDMSO OH → **G** → HO OH

5. Draw the organic products formed in the following reactions.

a. 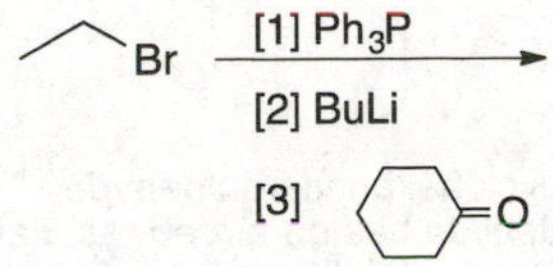 Br — [1] Ph_3P; [2] BuLi; [3] cyclohexanone

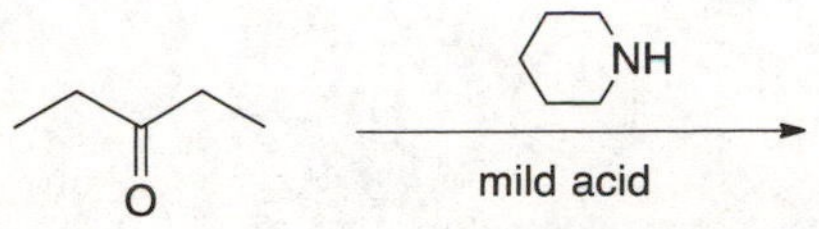

b. 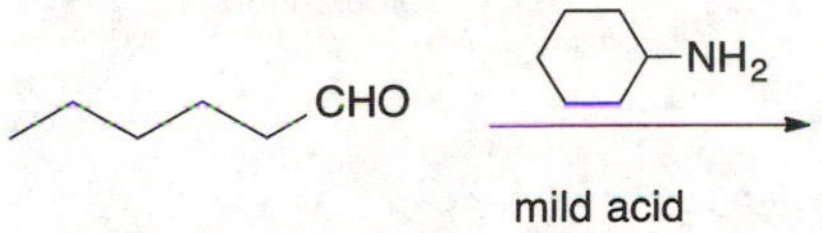 CHO — cyclohexyl-NH_2, mild acid

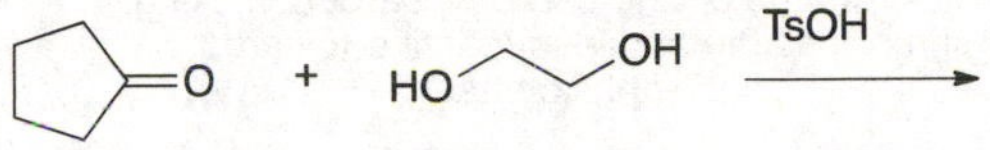

c. 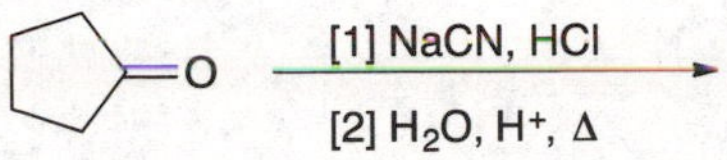 cyclopentanone — [1] NaCN, HCl; [2] H_2O, H^+, Δ

d. 3-pentanone — piperidine (NH), mild acid

e. cyclopentanone + HO–CH2CH2–OH — TsOH

f. 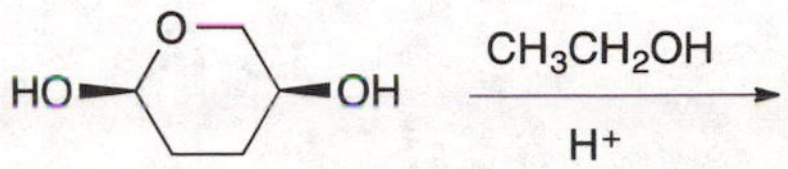 HO–(tetrahydropyran)–OH — CH_3CH_2OH, H^+

[Indicate stereochemistry.]

Answers to Practice Test

1.a. 5-isopropyl-2,4-dimethyl-cyclohexanone
 b. 3,3-dimethyl-5-phenylpentan-2-one
 c. 4-ethyl-2-methyl-cyclohexane-carbaldehyde

2.a. **D**
 b. **C**

3.a. **B**
 b. **C**

4.
A = [1] LiC≡CH; [2] H_2O
B = [1] R_2BH; [2] H_2O_2, ^-OH
C = Ag_2O, NH_4OH
D = H_2O, H_2SO_4, $HgSO_4$
E = TBDMS–Cl, imidazole
F = [1] CH_3Li; [2] H_2O
G = $Bu_4N^+F^-$

5.
a. 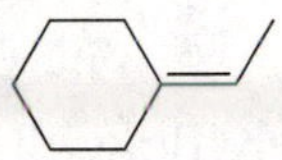
b. 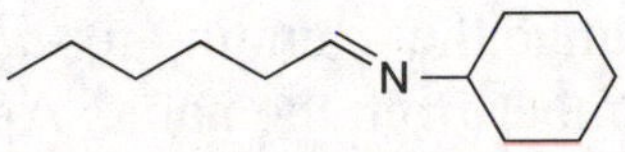 N
c. 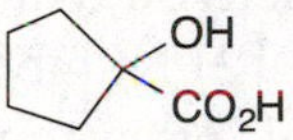 OH, CO_2H
d. 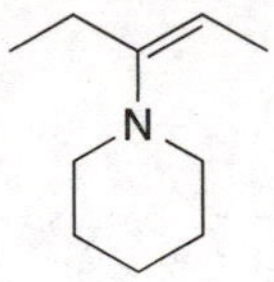 N (*E* + *Z*)
e. 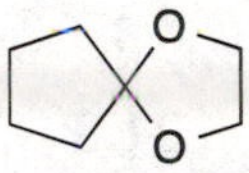 O, O
f. 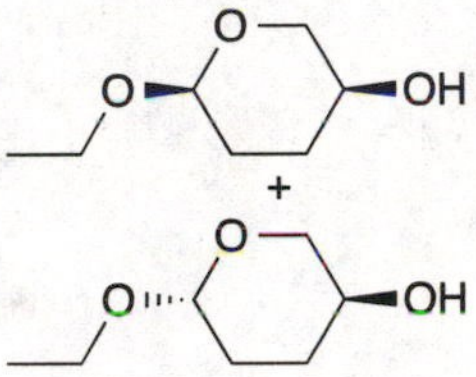 O, OH + O, OH

Answers to Problems

21.1 As the number of R groups bonded to the carbonyl C increases, reactivity towards nucleophilic attack decreases. Steric hindrance decreases reactivity as well.

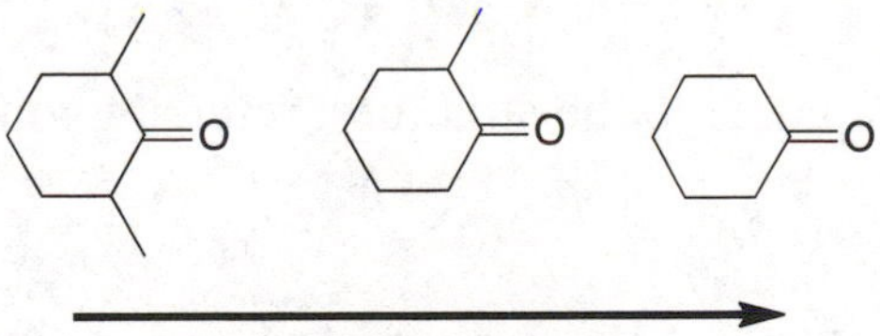

O O O

Increasing reactivity
decreasing steric hindrance

21.2 More stable aldehydes are less reactive towards nucleophilic attack.

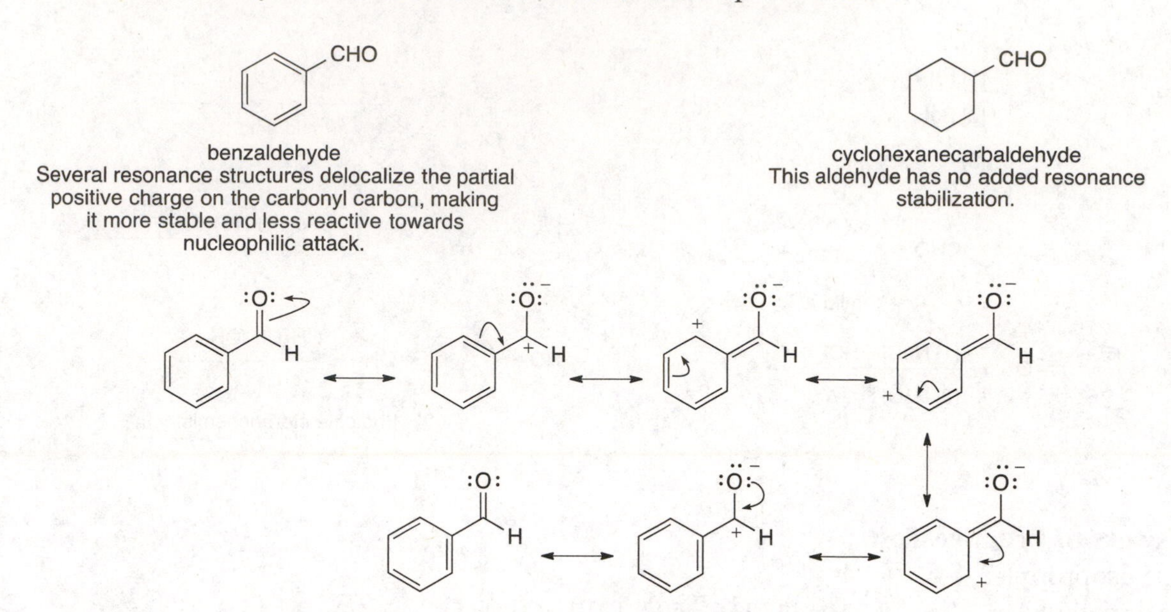

CHO
benzaldehyde
Several resonance structures delocalize the partial positive charge on the carbonyl carbon, making it more stable and less reactive towards nucleophilic attack.
CHO
cyclohexanecarbaldehyde
This aldehyde has no added resonance stabilization.
:O:
H
+

21.3 • To name an aldehyde with a chain of atoms: [1] Find the longest chain with the CHO group and change the *-e* ending to *-al*. [2] Number the carbon chain to put the CHO at C1, but omit this number from the name. Apply all other nomenclature rules.

• To name an aldehyde with the CHO bonded to a ring: [1] Name the ring and add the suffix *-carbaldehyde*. [2] Number the ring to put the CHO group at C1, but omit this number from the name. Apply all other nomenclature rules.

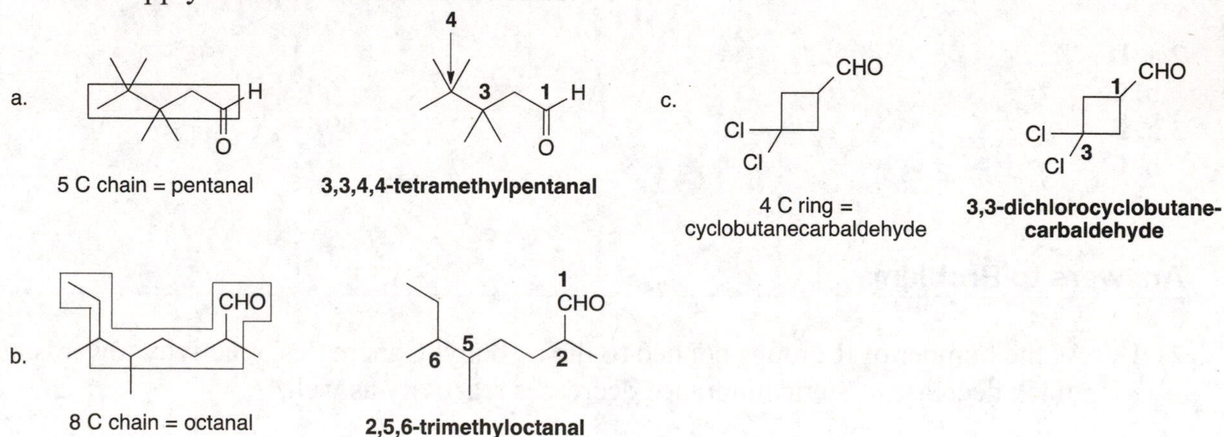

21.4 Work backwards from the name to the structure, referring to the nomenclature rules in Answer 21.3.

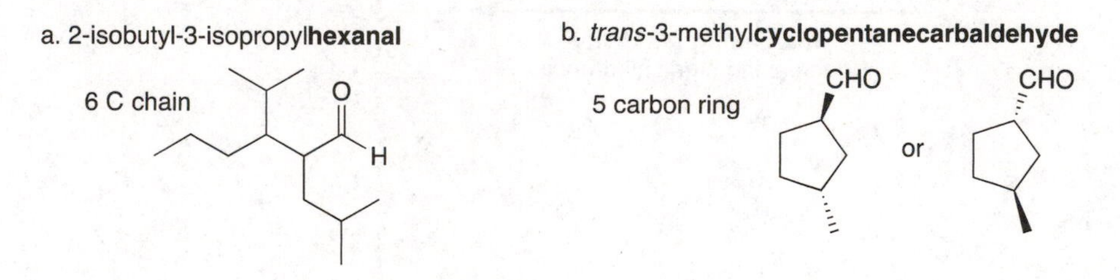

c. 1-methyl**cyclopropanecarbaldehyde**

3 carbon ring

CHO

CH_3

d. 3,6-diethyl**nonanal**

9 C chain

H

O

21.5 • To name an acyclic ketone: [1] Find the longest chain with the carbonyl group and change the *-e* ending to *-one*. [2] Number the carbon chain to give the carbonyl C the lower number. Apply all other nomenclature rules.

• To name a cyclic ketone: [1] Name the ring and change the *-e* ending to *-one*. [2] Number the C's to put the carbonyl C at C1 and give the next substituent the lower number. Apply all other nomenclature rules.

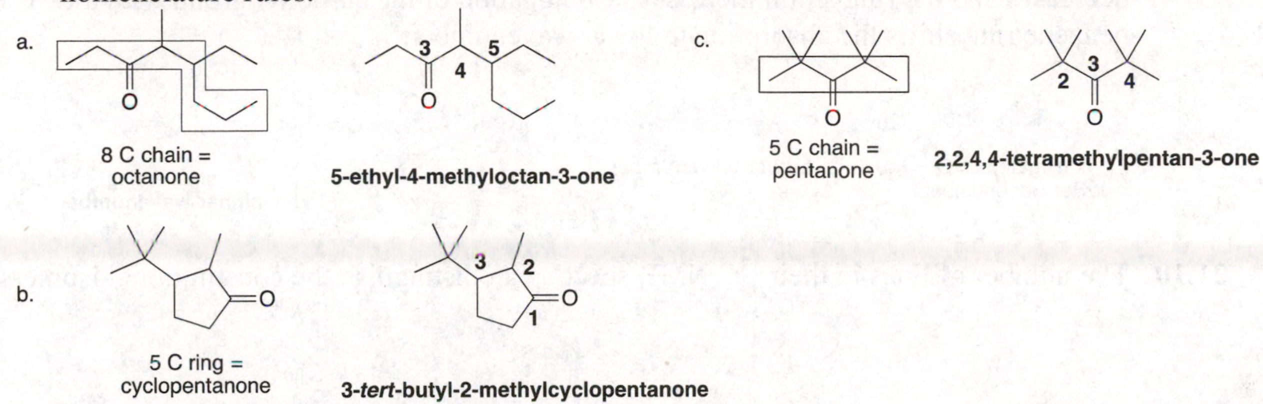

21.6 Most common names are formed by naming both alkyl groups on the carbonyl C, arranging them alphabetically, and adding the word ketone.

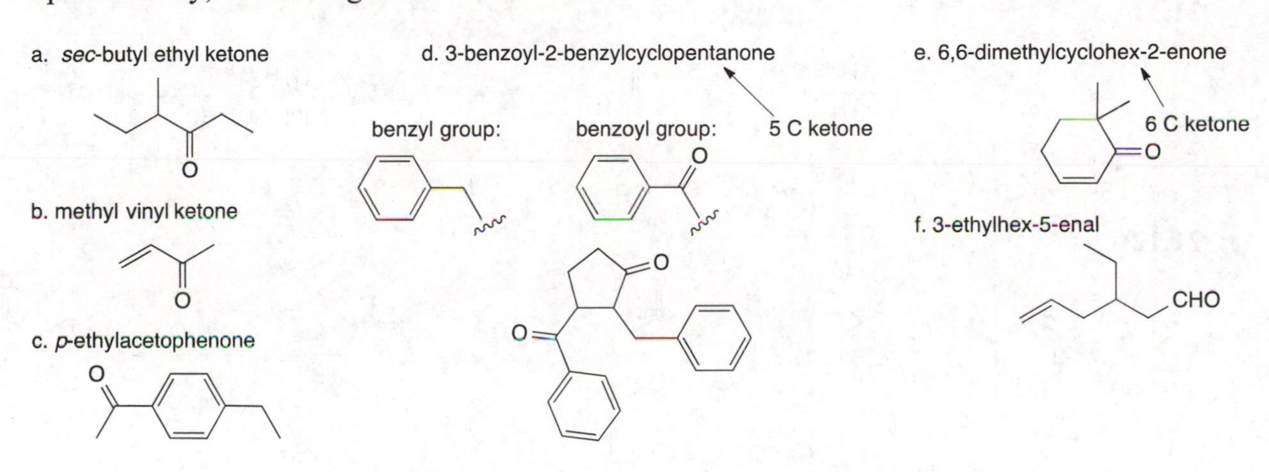

21.7 Compounds with both a C–C double bond and an aldehyde are named as enals.

a. (*Z*)-3,7-dimethylocta-2,6-dienal

7 6 3 2 1 CHO

neral

b. (2*E*,6*Z*)-nona-2,6-dienal

7 6 3 2 1 CHO

cucumber aldehyde

21.8 Even though both compounds have polar C–O bonds, the electron pairs around the sp^3 hybridized O atom of diethyl ether are more crowded and less able to interact with electron-deficient sites in other diethyl ether molecules. The O atom of the carbonyl group of butan-2-one extends out from the carbon chain, making it less crowded. The lone pairs of electrons on the O atom can more readily interact with the electron-deficient sites in the other molecules, resulting in stronger forces and a higher boiling point.

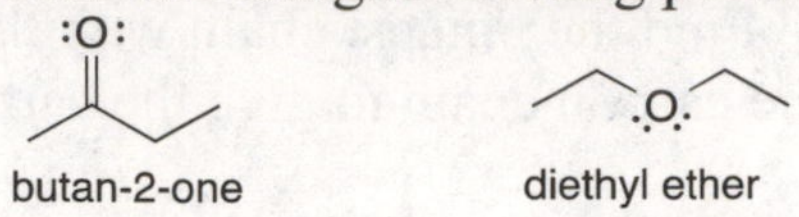

:O:

butan-2-one　　diethyl ether

21.9 For cyclic ketones, the carbonyl absorption shifts to higher wavenumber as the size of the ring decreases and the ring strain increases. Conjugation of the carbonyl group with a C=C or a benzene ring shifts the absorption to lower wavenumber.

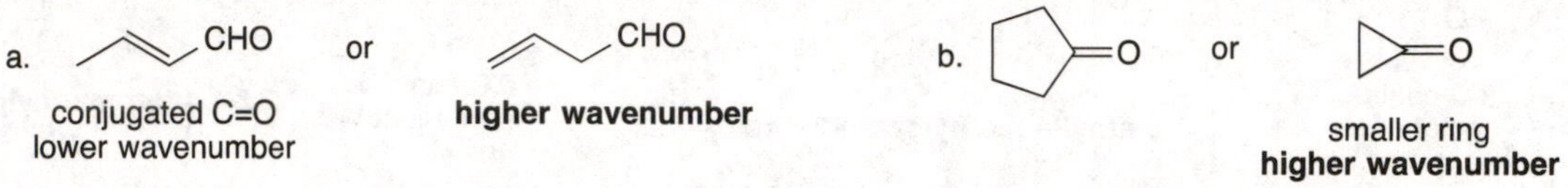

a. CHO or CHO

conjugated C=O
lower wavenumber

higher wavenumber

b. =O or =O

smaller ring
higher wavenumber

21.10 The number of lines in their ^{13}C NMR spectra can distinguish the constitutional isomers.

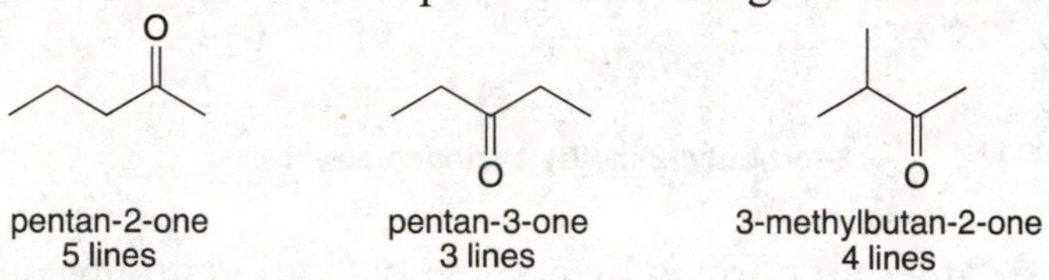

pentan-2-one
5 lines

pentan-3-one
3 lines

3-methylbutan-2-one
4 lines

21.11

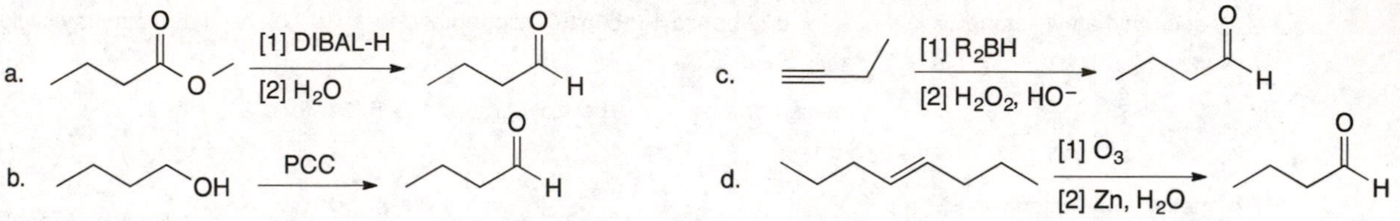

a. [1] DIBAL-H, [2] H_2O → aldehyde (H)

b. OH, PCC → aldehyde (H)

c. [1] R_2BH, [2] H_2O_2, HO− → aldehyde (H)

d. [1] O_3, [2] Zn, H_2O → aldehyde (H)

21.12

a. Cl (acetyl chloride), $AlCl_3$ → acetophenone

b. Cl (benzoyl chloride), [1] $(CH_3)_2CuLi$, [2] H_2O → acetophenone

c. H_2O, H_2SO_4, $HgSO_4$ → acetophenone

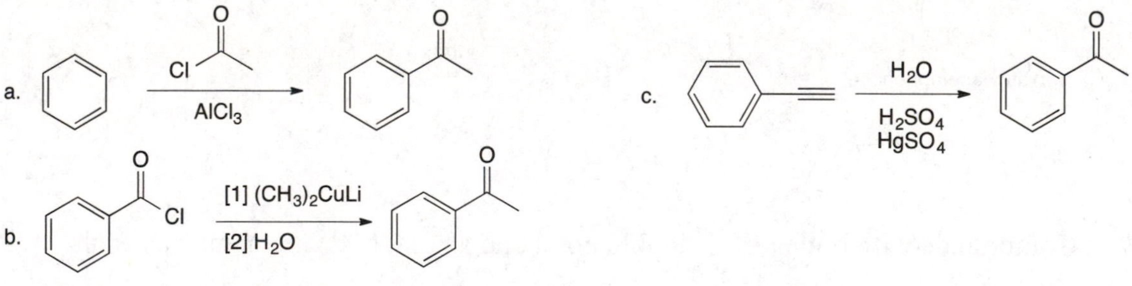

21.13

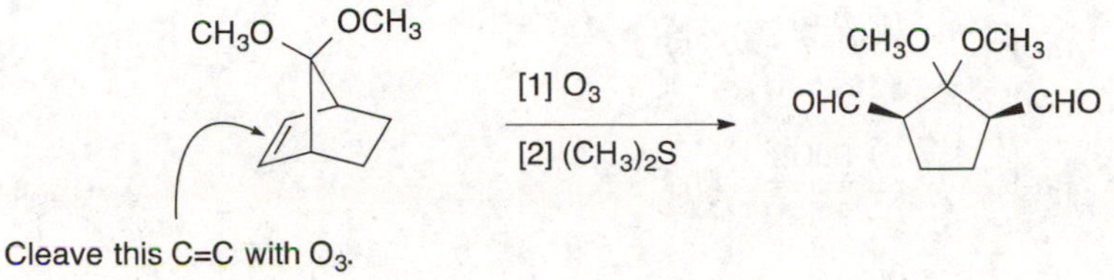

CH_3O OCH_3 — [1] O_3, [2] $(CH_3)_2S$ → OHC, CHO, CH_3O OCH_3

Cleave this C=C with O_3.

21.14 Addition of hydride or R–M occurs at a planar carbonyl C, so two different configurations at a new stereogenic center are possible.

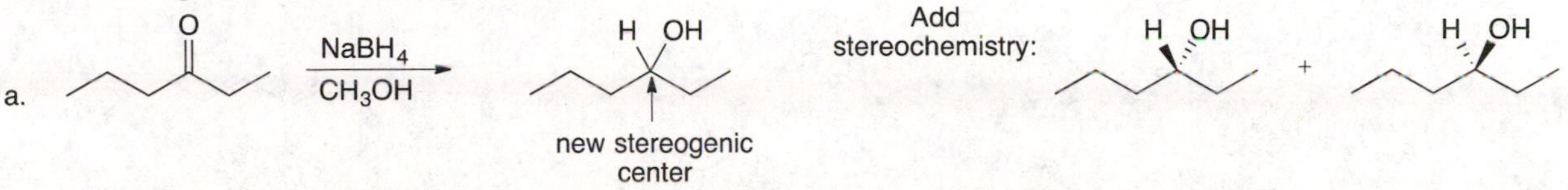

a. NaBH₄ / CH_3OH → H OH; new stereogenic center. Add stereochemistry: H OH + H OH

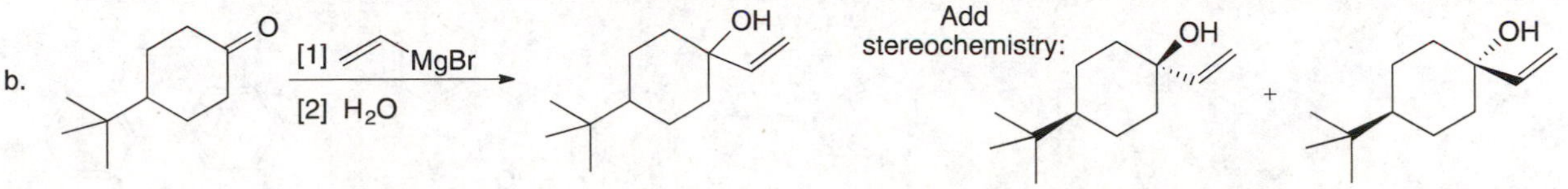

b. [1] MgBr, [2] H_2O → OH. Add stereochemistry: OH + OH

21.15 Treatment of an aldehyde or ketone with NaCN, HCl adds HCN across the double bond. Cyano groups are hydrolyzed by H_3O^+ to replace the three C–N bonds with three C–O bonds.

a. CHO — NaCN, HCl → OH, CN

b. OH, CN — H_3O^+, Δ → OH, COOH

21.16

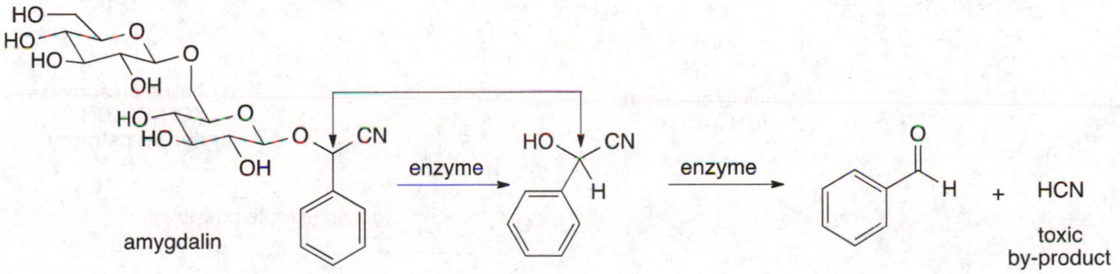

amygdalin (HO, HO, HO, O, OH, HO, HO, O, OH, O, CN) — enzyme → HO, CN, H — enzyme → O, H + HCN

toxic by-product

21.17

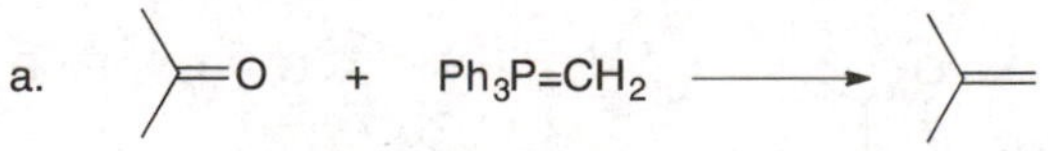

a. =O + $Ph_3P{=}CH_2$ →

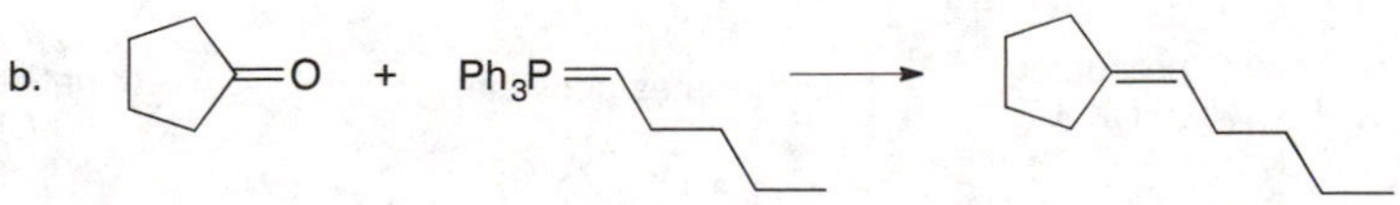

b. =O + Ph_3P= →

21.18

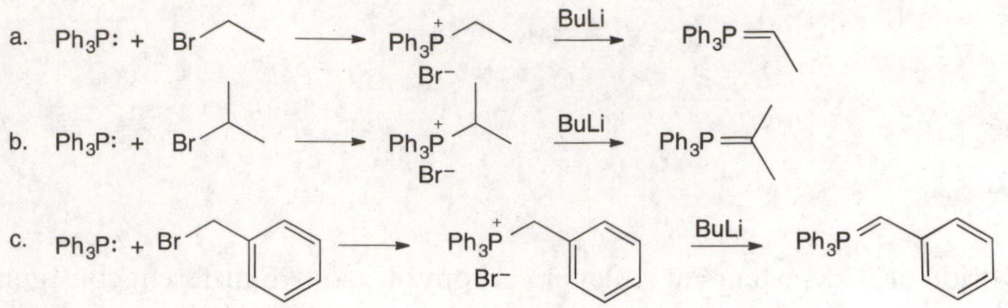

a. $Ph_3P:$ + Br⁀ → $Ph_3\overset{+}{P}$ Br^- —BuLi→ $Ph_3P=$

b. $Ph_3P:$ + Br → $Ph_3\overset{+}{P}$ Br^- —BuLi→ $Ph_3P=$

c. $Ph_3P:$ + Br → $Ph_3\overset{+}{P}$ Br^- —BuLi→ Ph_3P

21.19

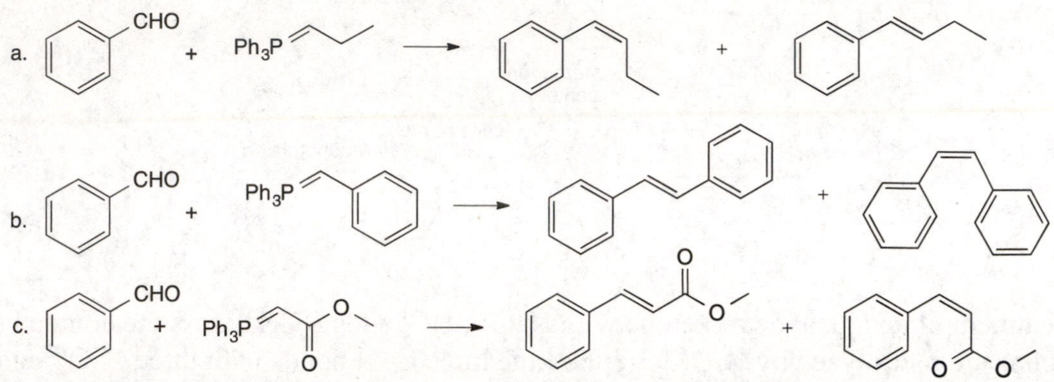

a. CHO + Ph_3P → +

b. CHO + Ph_3P → +

c. CHO + Ph_3P O O → O O + O O

21.20 To draw the starting materials of the Wittig reactions, find the C=C and cleave it. Replace it with a C=O in one half of the molecule and a C=PPh3 in the other half. The preferred pathway uses a Wittig reagent derived from a less hindered alkyl halide.

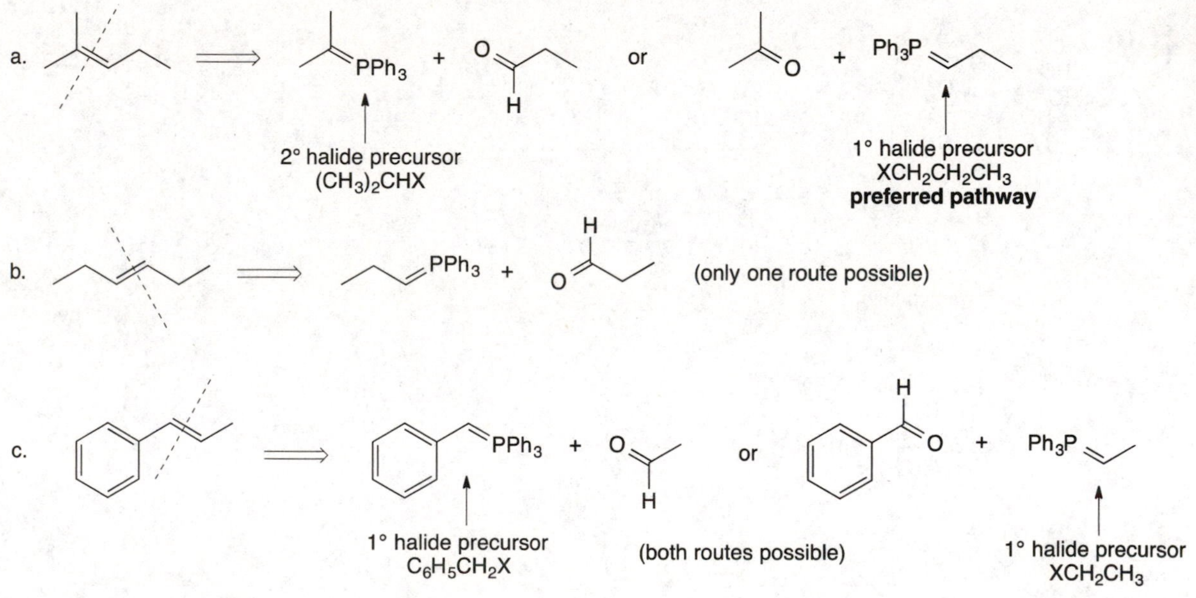

a. ⟹ PPh_3 + O H or O + Ph_3P

2° halide precursor $(CH_3)_2CHX$

1° halide precursor $XCH_2CH_2CH_3$ **preferred pathway**

b. ⟹ PPh_3 + H O (only one route possible)

c. ⟹ PPh_3 + O H or H O + Ph_3P

1° halide precursor $C_6H_5CH_2X$

(both routes possible)

1° halide precursor XCH_2CH_3

21.21

a. Two-step sequence:

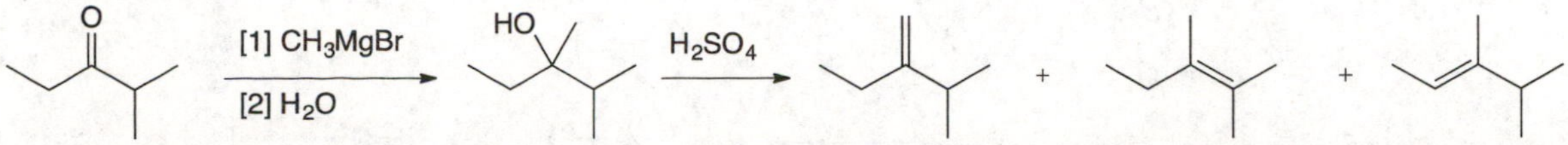

[1] CH_3MgBr
[2] H_2O

HO

H_2SO_4

minor product

tetrasubstituted
major product

(*E* and *Z* isomers)

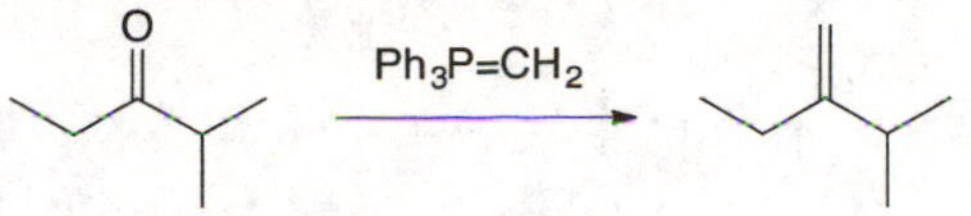

One-step sequence:

$Ph_3P{=}CH_2$

only product

b. Two-step sequence:

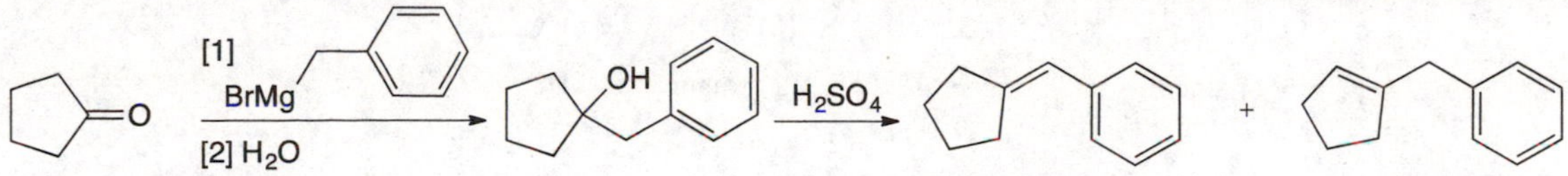

[1] BrMg
[2] H_2O

OH

H_2SO_4

trisubstituted
conjugated C=C

trisubstituted

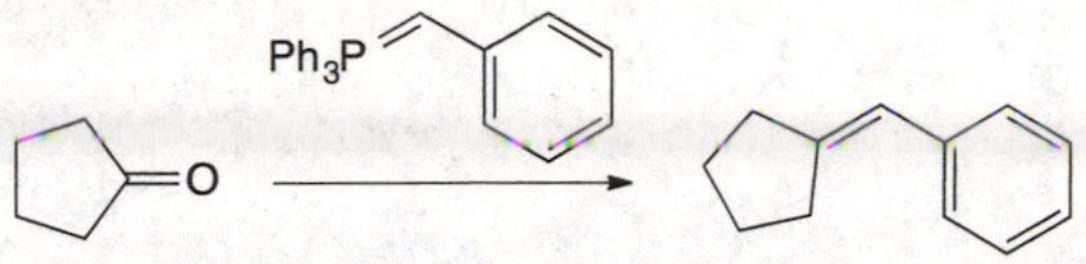

One-step sequence:

Ph_3P

only product

21.22 When a 1° amine reacts with an aldehyde or ketone, the C=O is replaced by C=NR.

a. CHO NH_2 N

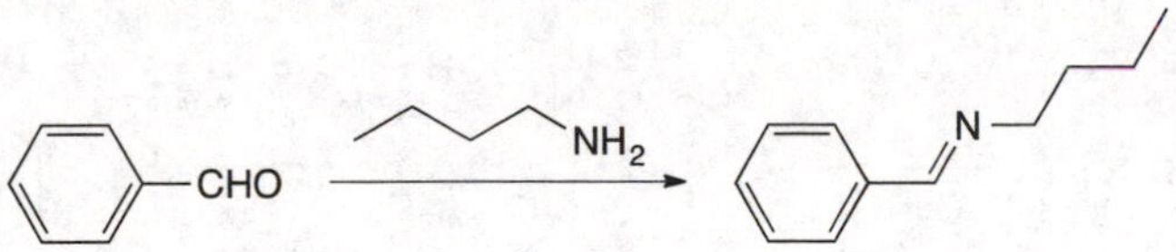

b. O NH_2 N

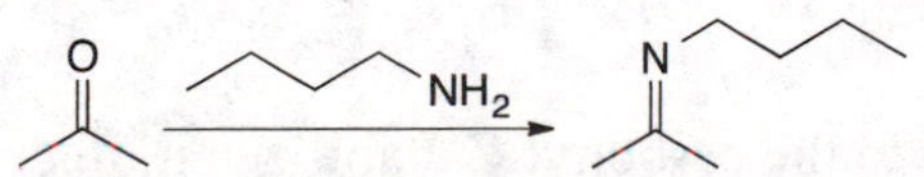

c. O NH_2 N

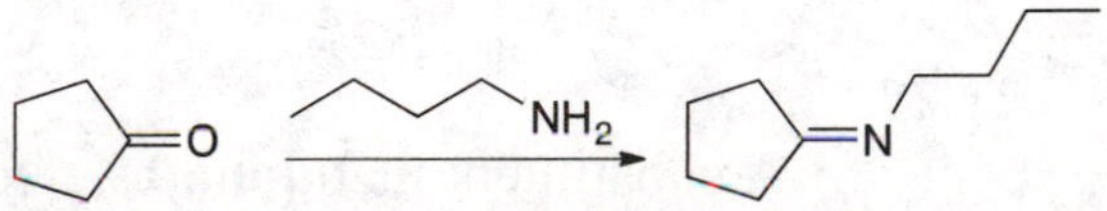

21.23 Remember that the C=NR is formed from a C=O and an NH_2 group of a 1° amine.

a. N ⟹ H O + H_2N

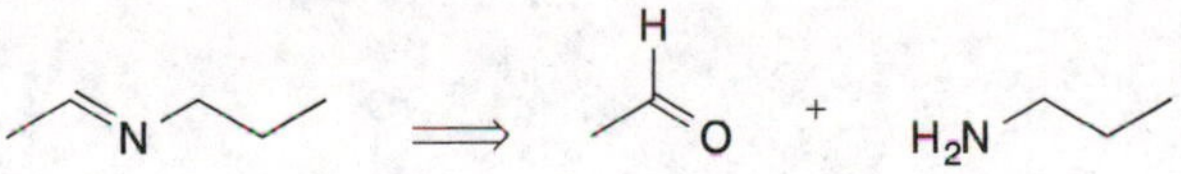

b. N ⟹ O + NH_2

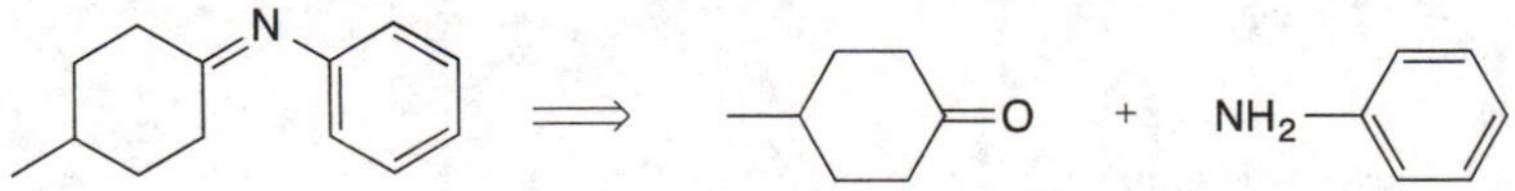

21.24

O + CH_3 N—H CH_3 ⟶ $N(CH_3)_2$ + $N(CH_3)_2$

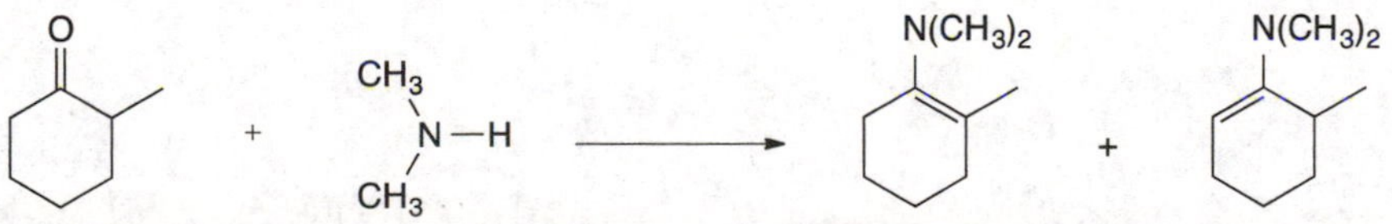

21.25 • Imines are hydrolyzed to 1° amines and a carbonyl compound.
• Enamines are hydrolyzed to 2° amines and a carbonyl compound.

a. imine $\xrightarrow[H^+]{H_2O}$ benzaldehyde + H_2N–cyclopentyl (1° amine)

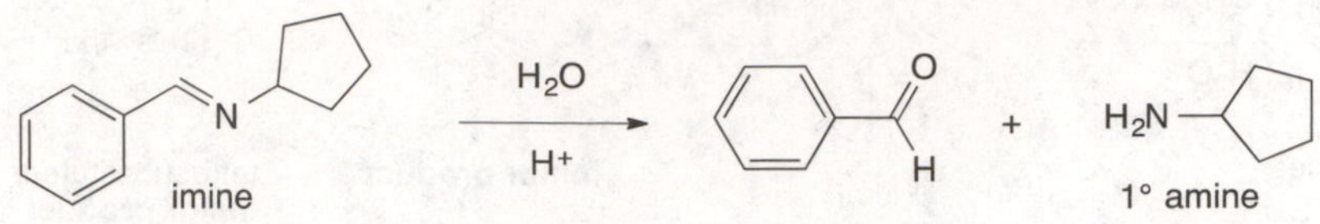

b. enamine $\xrightarrow[H^+]{H_2O}$ 2° amine + cyclopentanone

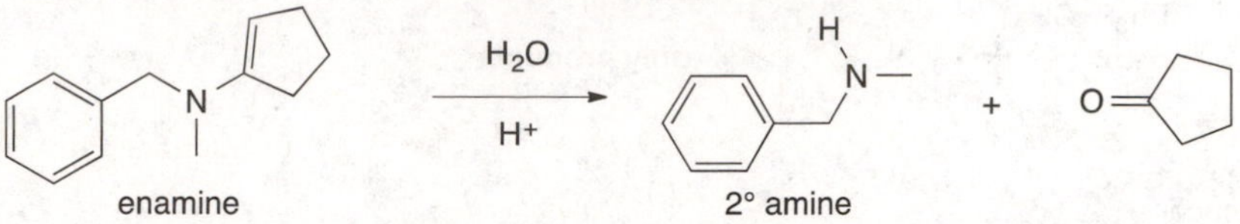

c. enamine $\xrightarrow[H^+]{H_2O}$ $(CH_3)_2NH$ (2° amine) + aldehyde

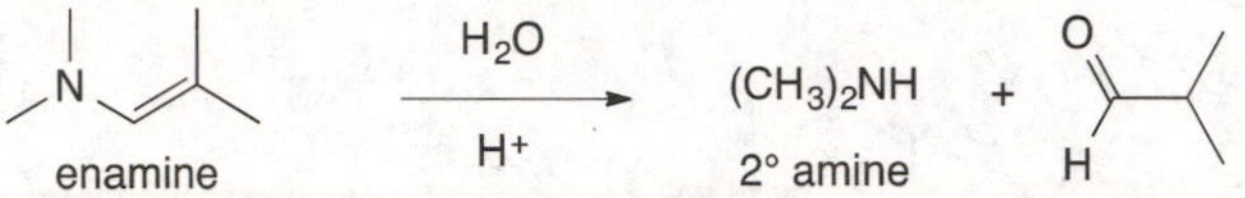

21.26

$H-\ddot{O}H_2^+$, $H_2\ddot{O}:$, $H_2\ddot{O}:$, $+ H_3O^+$

proton transfer

$H-\ddot{O}H_2^+$, $\ddot{N}H_2$, $\overset{+}{N}H_3$, $H_2\ddot{O}:$ +

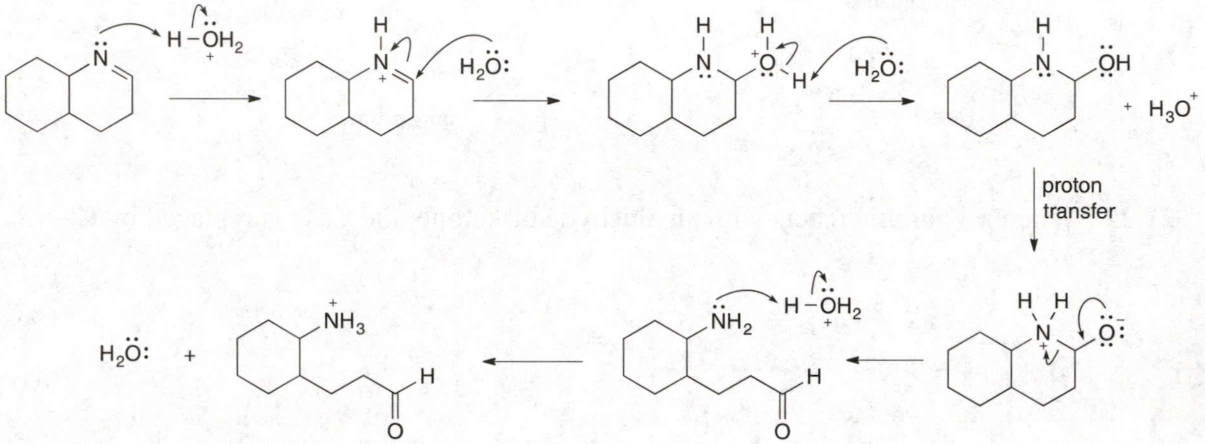

21.27

• A substituent that **donates** electron density to the carbonyl C stabilizes it, **decreasing** the percentage of hydrate at equilibrium.
• A substituent that **withdraws** electron density from the carbonyl C destabilizes it, **increasing** the percentage of hydrate at equilibrium.

a. one R group on C=O higher percentage of hydrate or 2 R groups on C=O

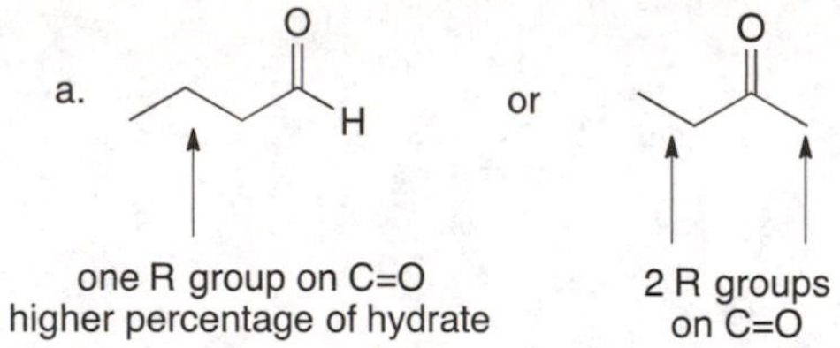

b. F atoms are electron withdrawing. higher percentage of hydrate or

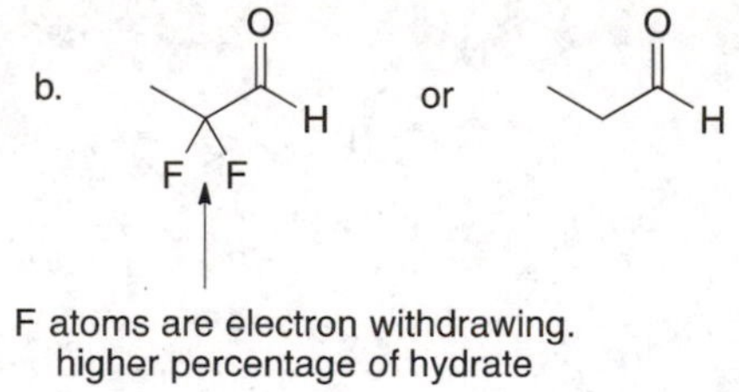

21.28

$H-\ddot{O}H_2^+$ → + $H_2\ddot{O}$ → (+ 1 resonance structure) + $H_2\ddot{O}$ → cyclohexanone + $H_3\ddot{O}^+$

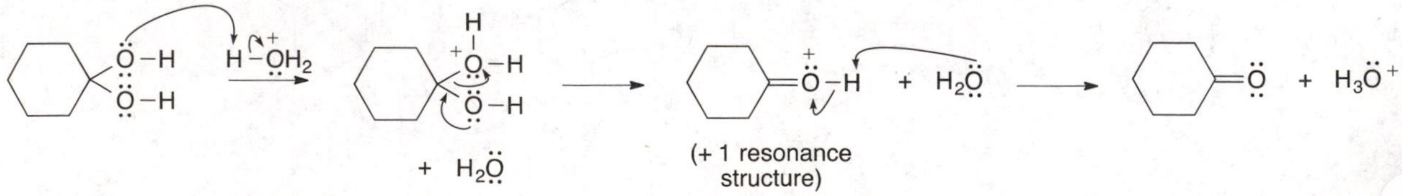

21.29 Treatment of an aldehyde or ketone with two equivalents of alcohol results in the formation of an acetal (a C bonded to two OR groups).

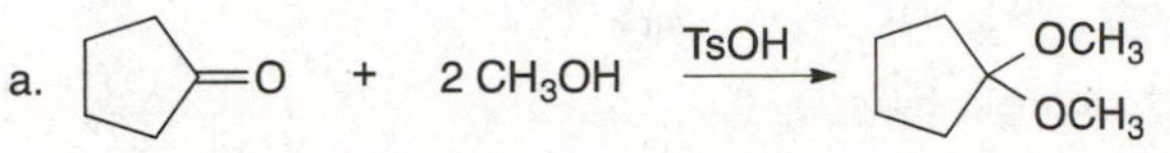

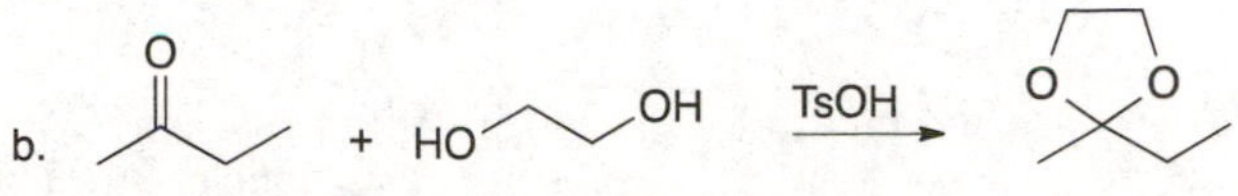

a. =O + 2 CH_3OH —TsOH→ OCH_3 OCH_3

b. O + HO OH —TsOH→ O O

21.30

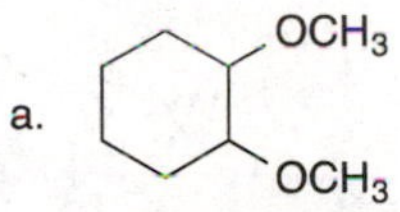

a. OCH_3 OCH_3
2 OR groups
on different C's
2 ethers

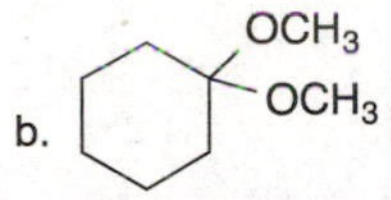

b. OCH_3 OCH_3
2 OR groups
on same C
acetal

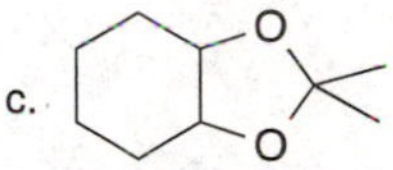

c. O O
2 OR groups
on same C
acetal

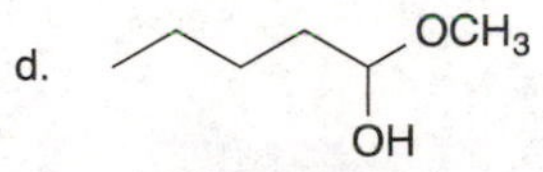

d. OCH_3 OH
1 OR group and
1 OH group
on same C
hemiacetal

21.31 The mechanism has two parts: [1] nucleophilic addition of ROH to form a hemiacetal; [2] conversion of the hemiacetal to an acetal.

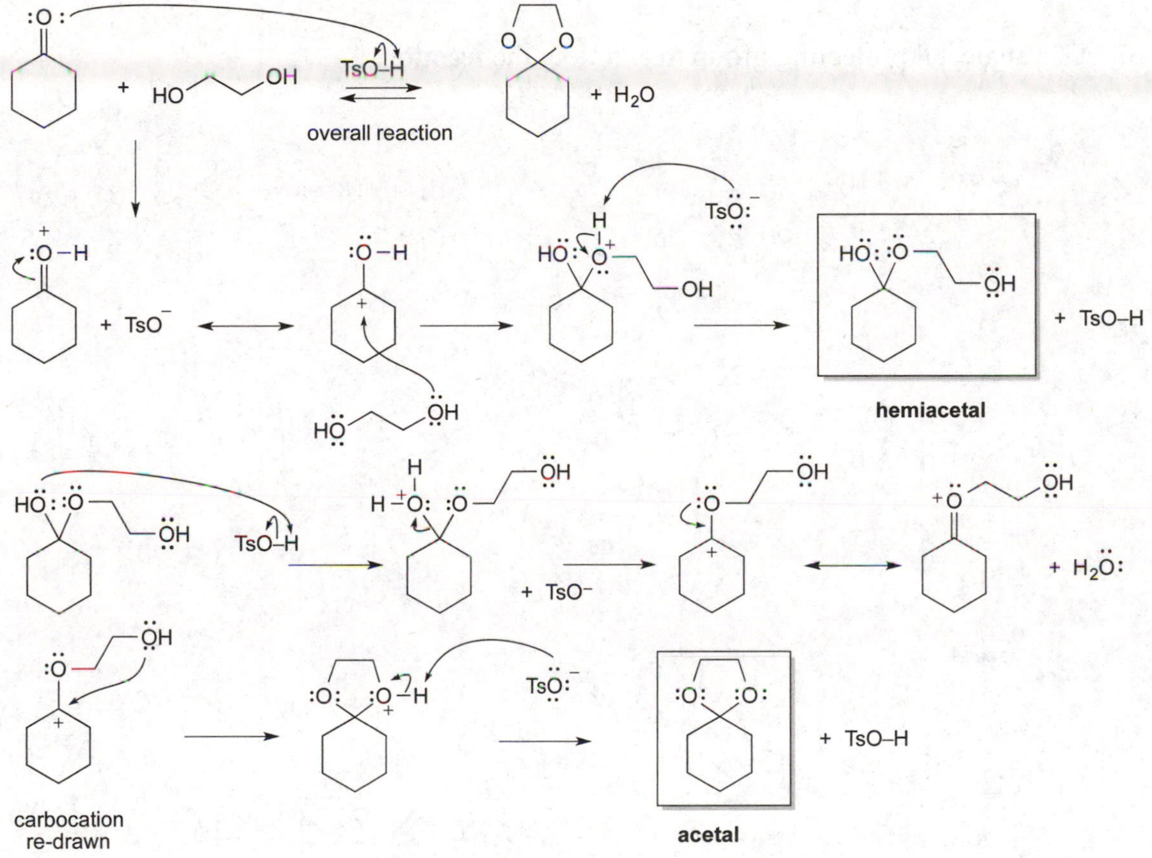

:O: + HO OH —TsO–H→ O O + H_2O

overall reaction

$^+$:O–H + TsO^- ⟷ :O–H + ⟶ HÖ: H O^+ OH ⟶ HÖ: :Ö ÖH + TsO–H

HÖ ÖH

TsÖ:⁻

hemiacetal

HÖ: :Ö ÖH TsO–H ⟶ H–O^+: :Ö ÖH + TsO^- ⟶ :Ö ÖH + ⟷ $^+$Ö ÖH + $H_2\ddot{O}$:

:Ö ÖH ⟶ :O: :O^+–H TsÖ:⁻ ⟶ :O: :O: + TsO–H

carbocation
re-drawn

acetal

21.32

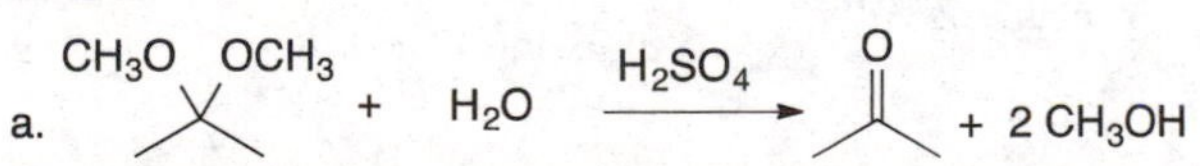

a. CH_3O OCH_3 + H_2O —H_2SO_4→ O + 2 CH_3OH

b. O O + H_2O —H_2SO_4→ O + OH OH

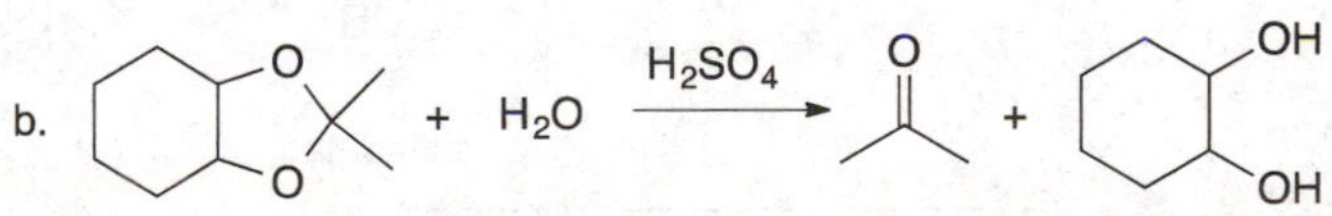

21.33

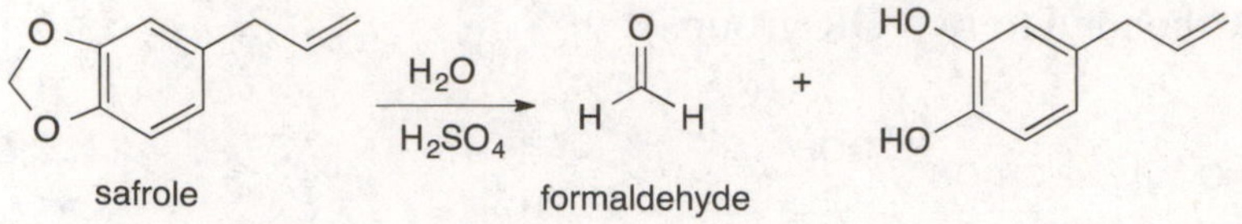

safrole $\xrightarrow[H_2SO_4]{H_2O}$ formaldehyde +

21.34

acetal

oleandrin $\xrightarrow{H_3O^+}$ +

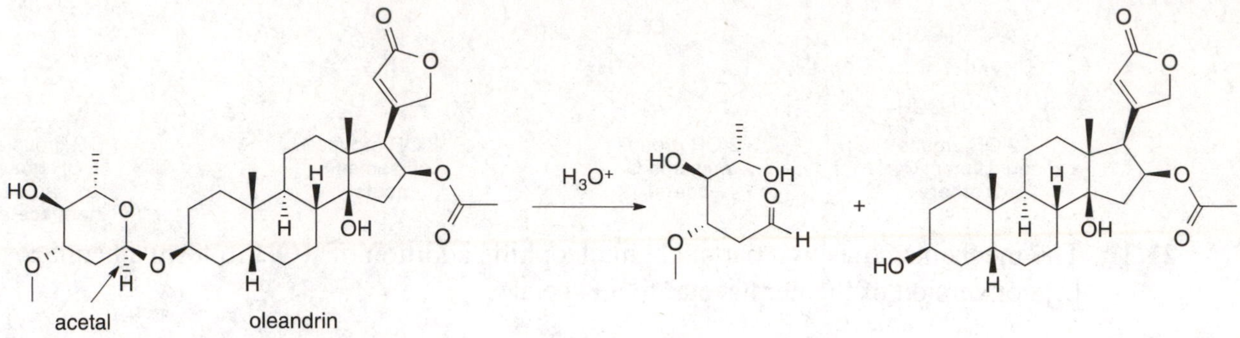

21.35 Use an acetal protecting group to carry out the reaction.

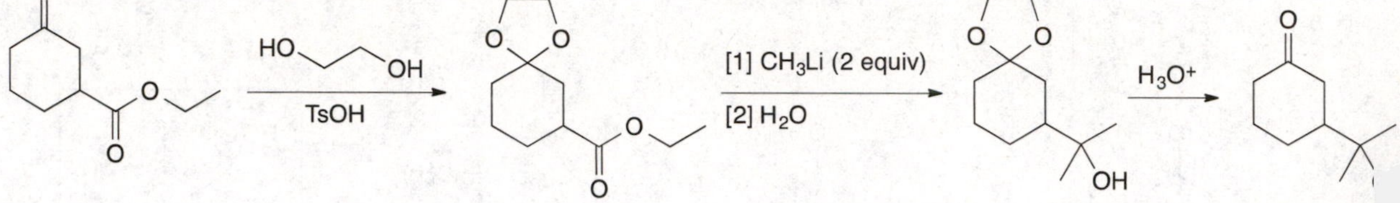

$\xrightarrow[TsOH]{HOCH_2CH_2OH}$ $\xrightarrow[\text{[2] } H_2O]{\text{[1] } CH_3Li \text{ (2 equiv)}}$ $\xrightarrow{H_3O^+}$

21.36

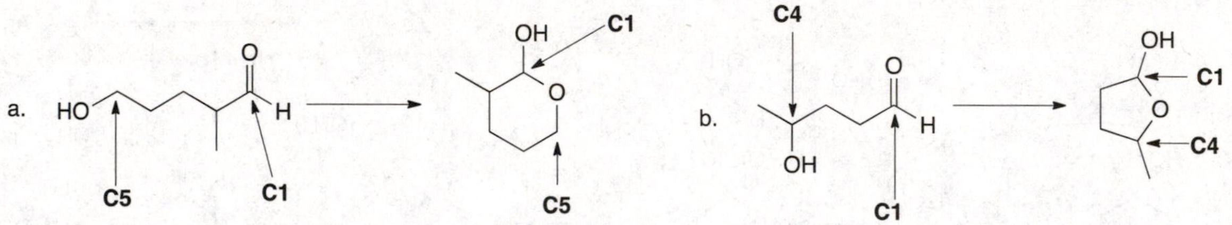

a. C5, C1 → C1, C5

b. C4, C1 → C1, C4

21.37

acetal

hemiacetal

monensin

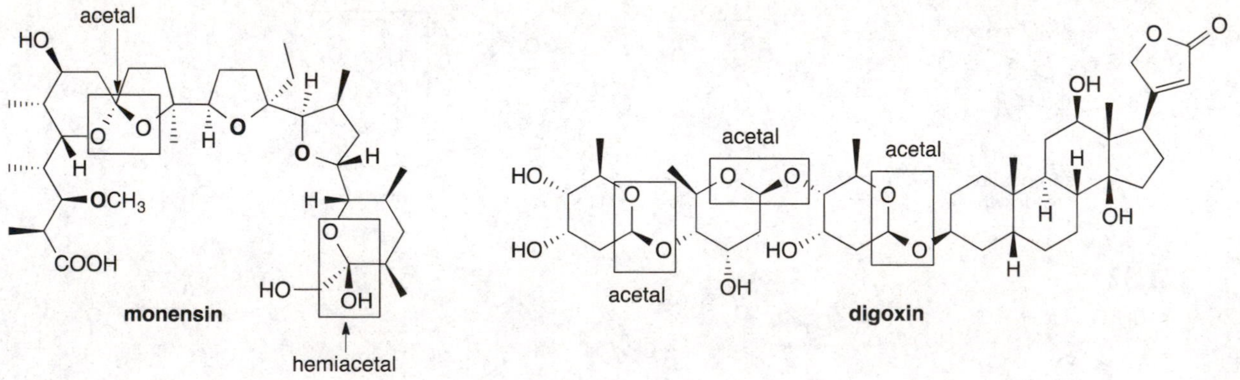

Ether **O** atoms are indicated in **bold.**

acetal

acetal

acetal

digoxin

21.38 The hemiacetal OH is replaced by an OR group to form an acetal.

a. 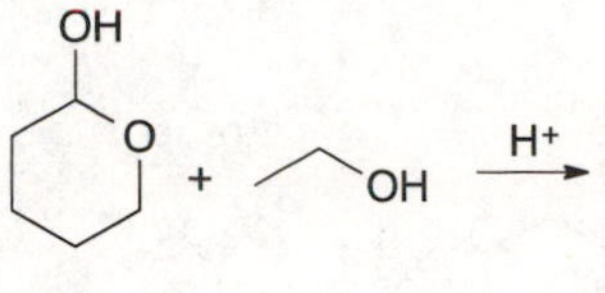 OH, O + OH $\xrightarrow{H^+}$ O, O

b. 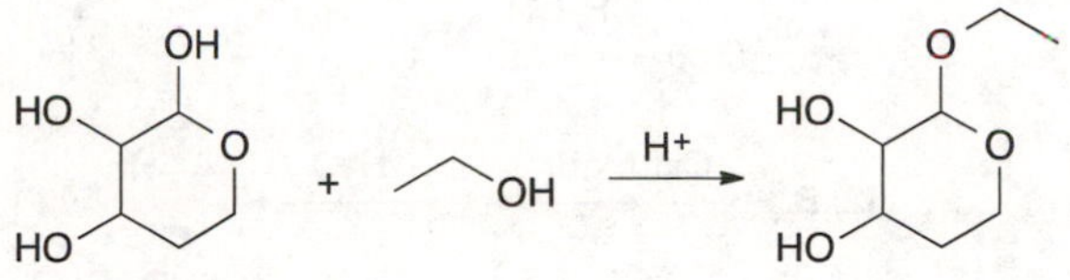 OH, HO, O, HO + OH $\xrightarrow{H^+}$ O, HO, O, HO

21.39

a. 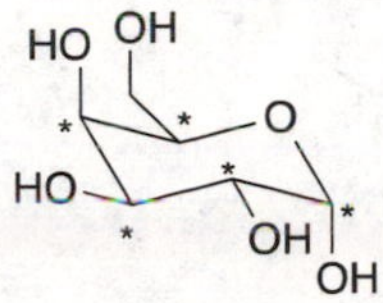 HO OH, O, HO, OH, OH — 5 stereogenic centers (labeled with *)

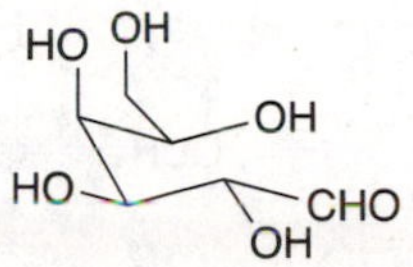

b. 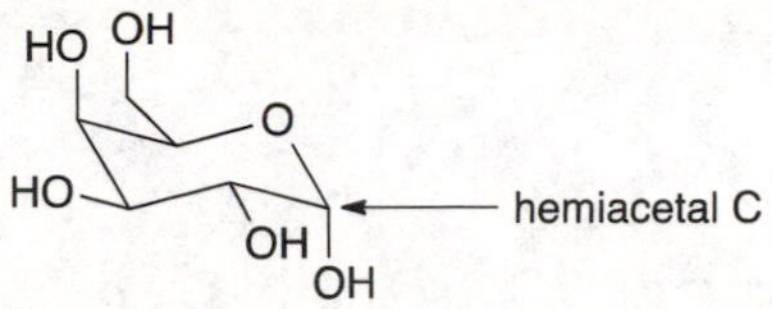 HO OH, O, HO, OH, OH ← hemiacetal C

α-D-galactose

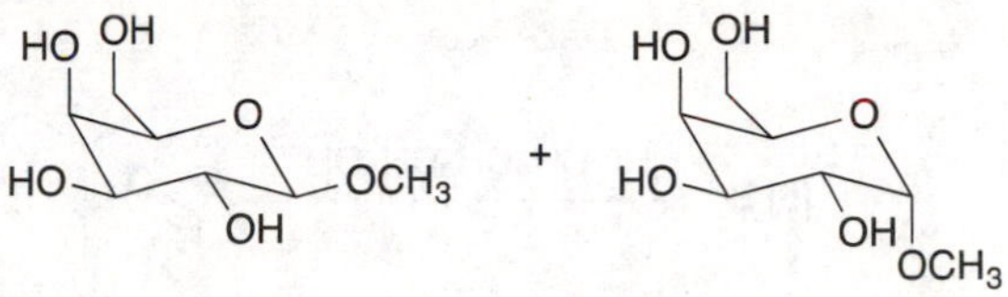

c. 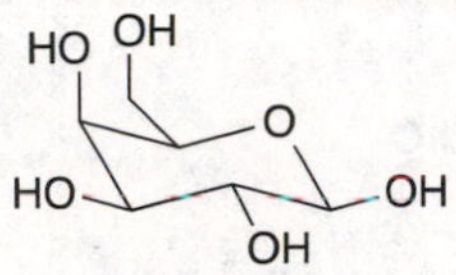 HO OH, O, HO, OH, OH

β-D-galactose

d. HO OH, OH, HO, CHO, OH

e. HO OH, O, HO, OCH_3, OH + HO OH, O, HO, OH, OCH_3

21.40

a. 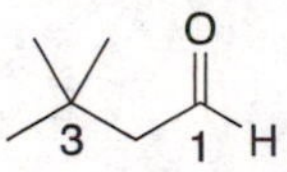 O, 3, 1, H

3,3-dimethylbutanal
A

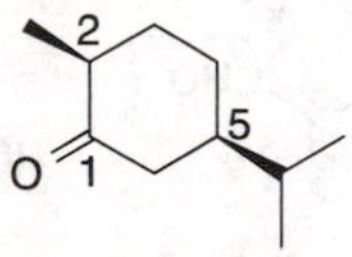

2, 5, O, 1

cis-5-isopropyl-2-methylcyclohexanone
B

b. [1] **A** $\xrightarrow[CH_3OH]{NaBH_4}$ OH

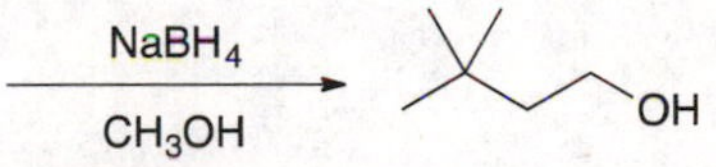

[2] **A** $\xrightarrow{CH_3MgBr}$ $\xrightarrow{H_2O}$ H, OH + H, OH

[3] **A** $\xrightarrow{Ph_3P=CHOCH_3}$ OCH_3 (+ cis isomer)

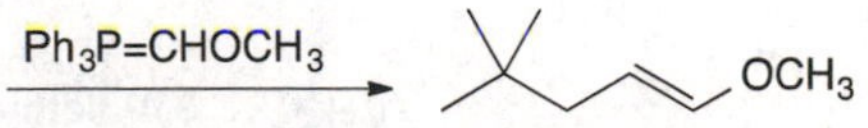

[4] **A** $\xrightarrow[\text{mild } H^+]{CH_3CH_2CH_2NH_2}$ N

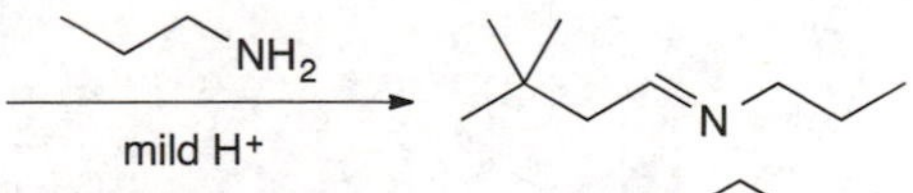

[5] **A** $\xrightarrow[H^+]{HOCH_2CH_2CH_2OH}$ O, O

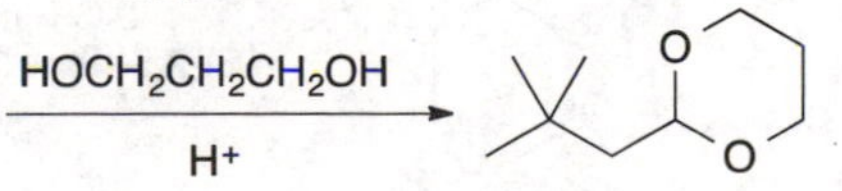

[1] **B** $\xrightarrow[CH_3OH]{NaBH_4}$ HO + HO

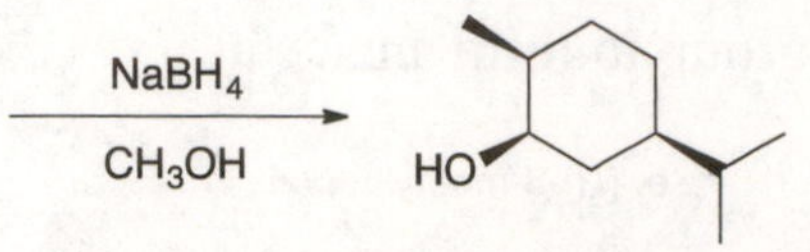

[2] **B** $\xrightarrow{CH_3MgBr}$ $\xrightarrow{H_2O}$ HO + HO

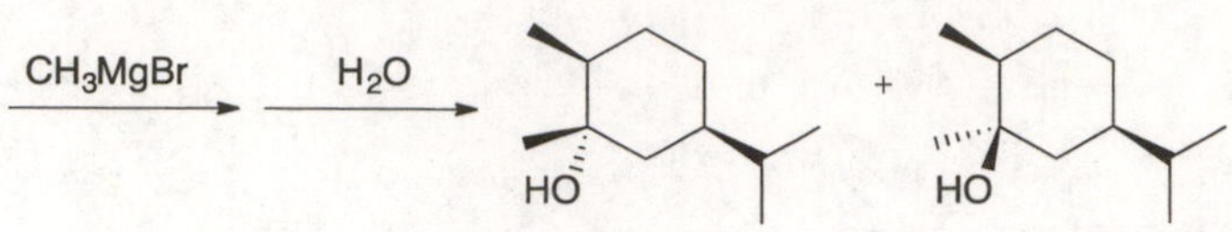

[3] **B** $\xrightarrow{Ph_3P=CHOCH_3}$ OCH_3 + CH_3O

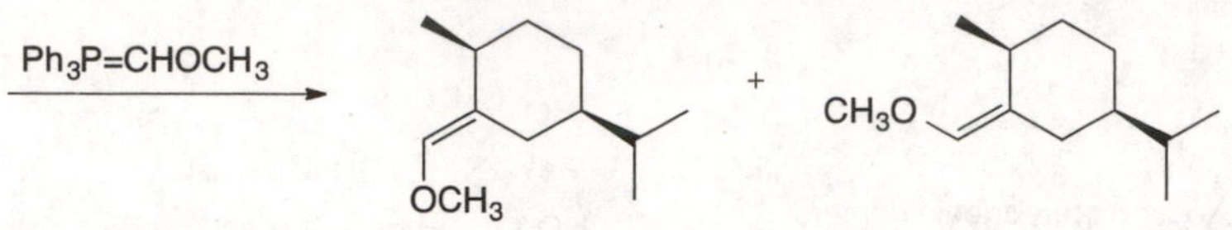

[4] **B** $\xrightarrow[\text{mild } H^+]{\text{NH}_2}$ N

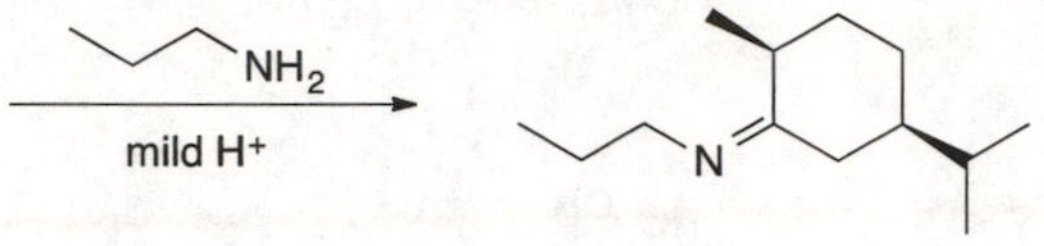

[5] **B** $\xrightarrow[H^+]{HOCH_2CH_2CH_2OH}$ O O

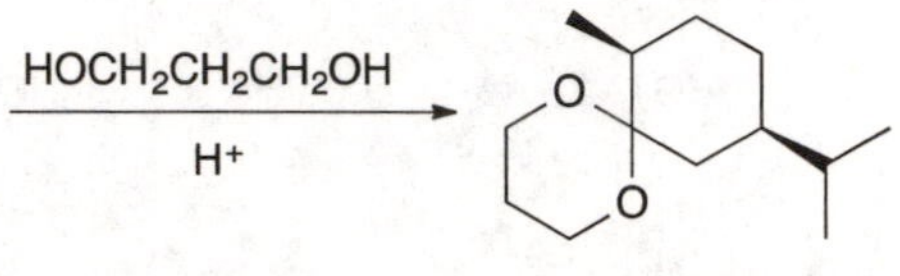

21.41 The least hindered carbonyl group is the most reactive.

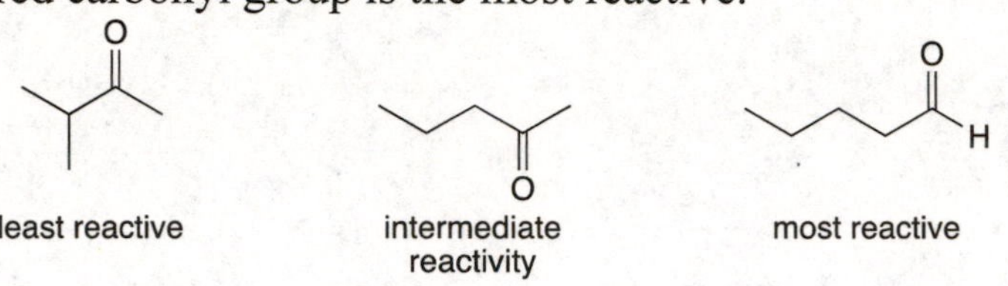

O O O H

least reactive | intermediate reactivity | most reactive

21.42

a. O O ⟹ CHO + HO OH

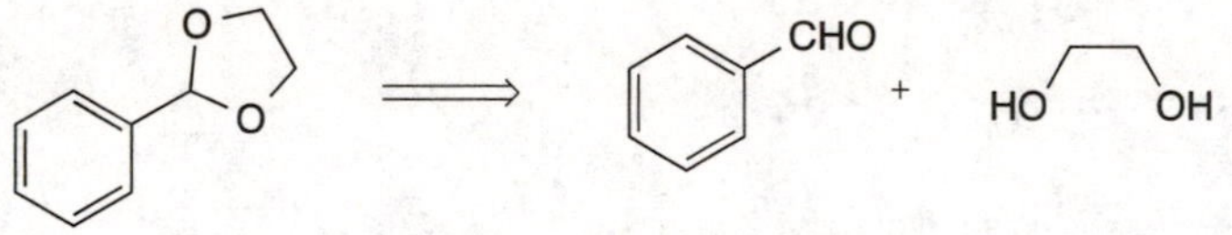

b. O O ⟹ OH OH + O

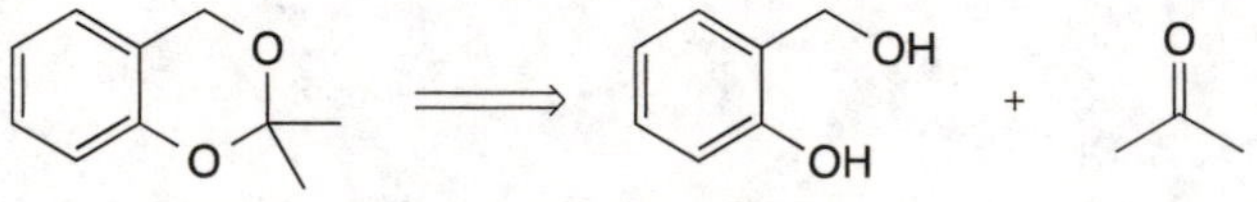

21.43 Use the rules from Answers 21.3 and 21.5 to name the aldehydes and ketones.

a. O — 6 C ring = cyclohexanone **5-ethyl-2-methyl-cyclohexanone**

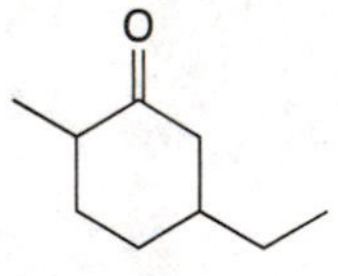

b. CHO — ***trans*-2-benzylcyclohexane-carbaldehyde**

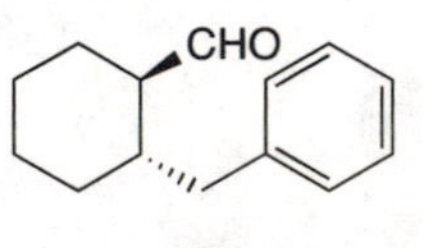

c. O NO$_2$ — ***o*-nitroacetophenone**

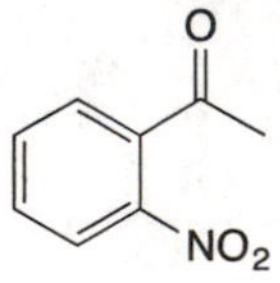

d. CHO — 6 C = hexanal **3,4-diethylhexanal**

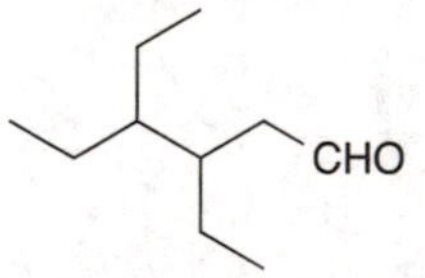

e. O *E* — 8 C = octenone **(*E*)-2,5-dimethyloct-5-en-4-one**

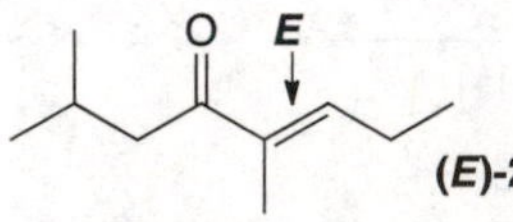

f. CHO — 6 C = hexenal **3,4-diethyl-2-methylhex-3-enal**

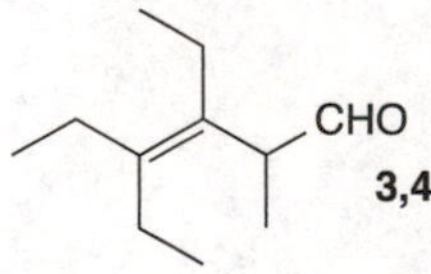

21.44

a. 2-methyl-3-phenylbutanal

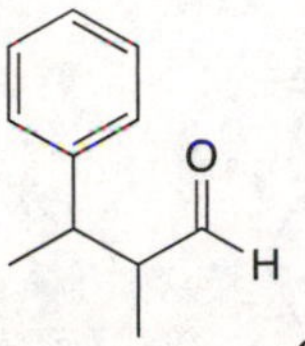

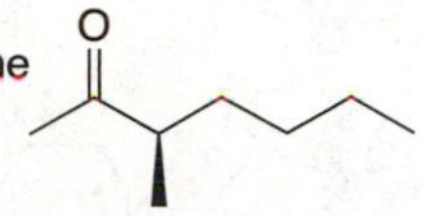

b. 3,3-dimethylcyclohexanecarbaldehyde

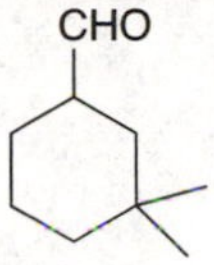

CHO

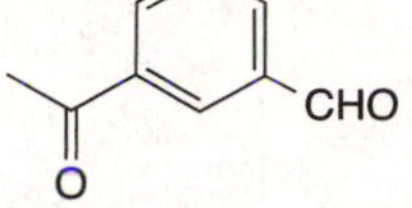

c. 3-benzoylcyclopentanone

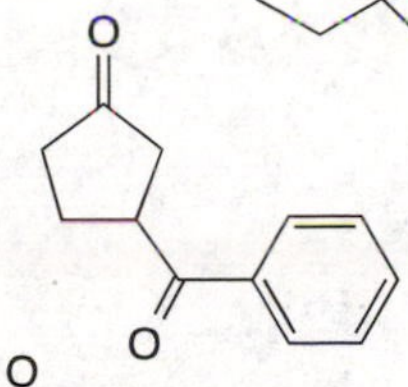

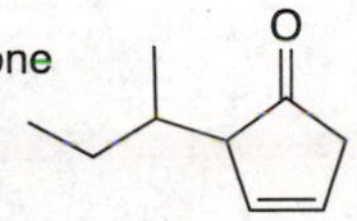

d. 2-formylcyclopentanone

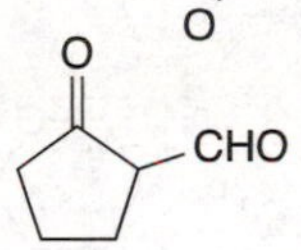

CHO

e. (*R*)-3-methylheptan-2-one

f. *m*-acetylbenzaldehyde

CHO

g. 2-*sec*-butylcyclopent-3-enone

h. 5,6-dimethylcyclohex-1-enecarbaldehyde

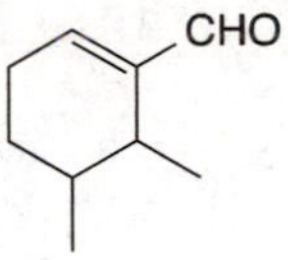

CHO

21.45

a. Br; [1] Ph_3P; [2] BuLi; [3] $C_6H_5CH_2CH_2CHO$ → (+ *Z* isomer)

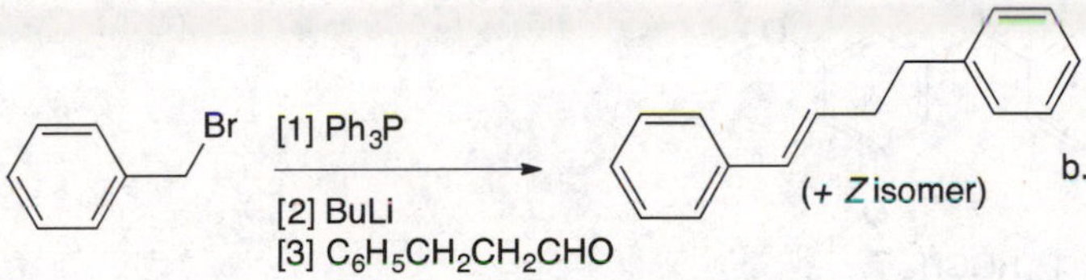

b. Cl; [1] Ph_3P; [2] BuLi; [3] $CH_3CH_2CH_2CHO$ → (+ *Z* isomer)

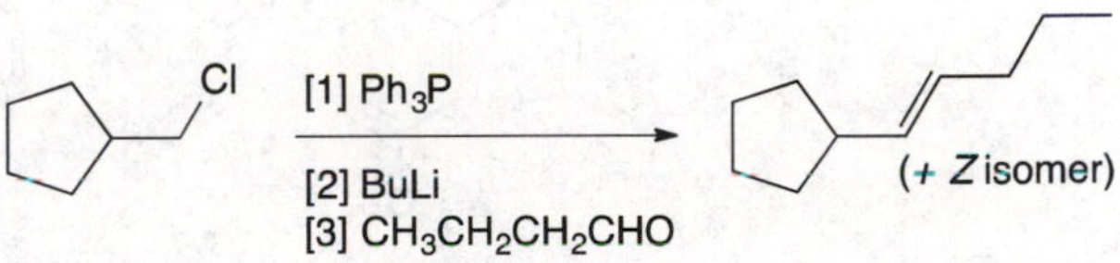

21.46

a. + H_2N— ; mild acid → =N—

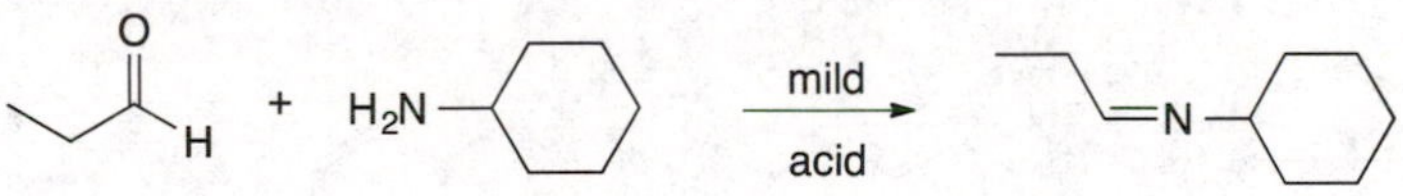

b. HO, OH; H^+ →

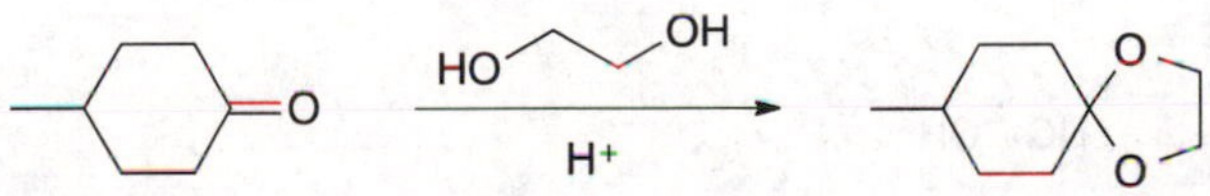

c. H_3O^+ → + H_2N

d. C_6H_5 + N H; mild acid → N, C_6H_5 (*E* and *Z* isomers)

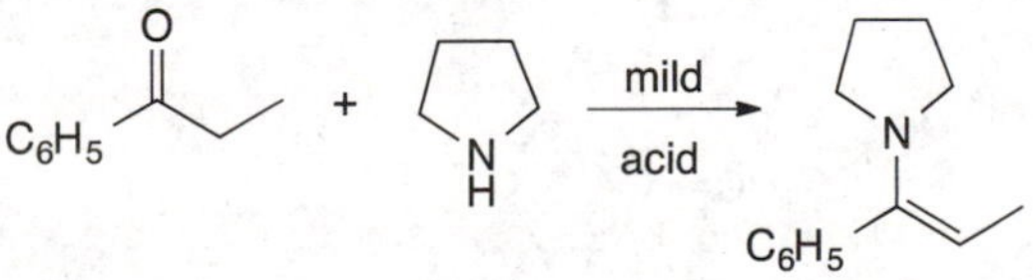

e. HO CN; H_3O^+, Δ → HO, OH

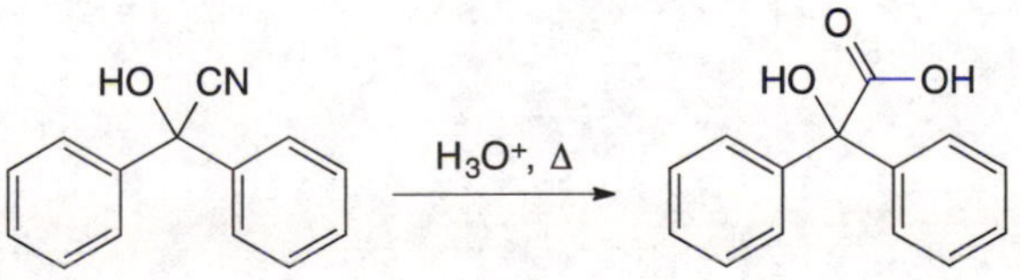

f. 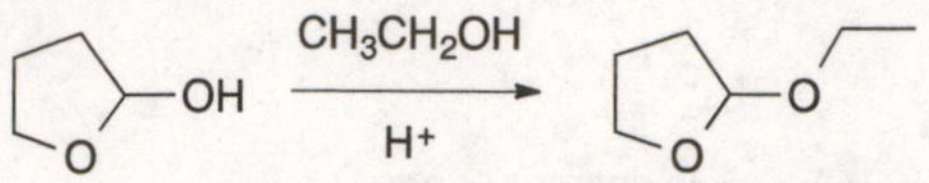 OH, CH_3CH_2OH, H^+, O, O, O

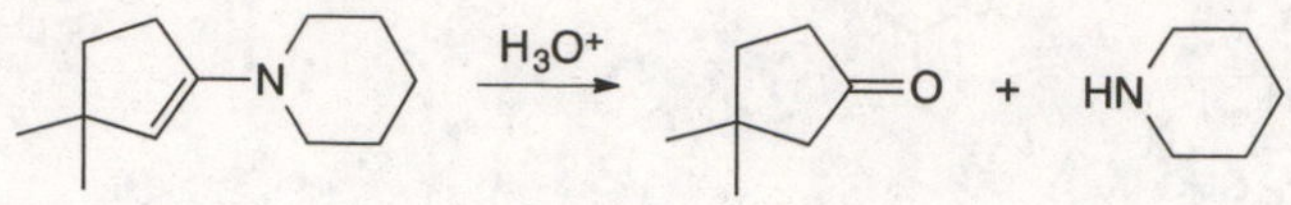

g. N, H_3O^+, O + HN

h. O, Ph_3P, O, OCH_3, O, OCH_3

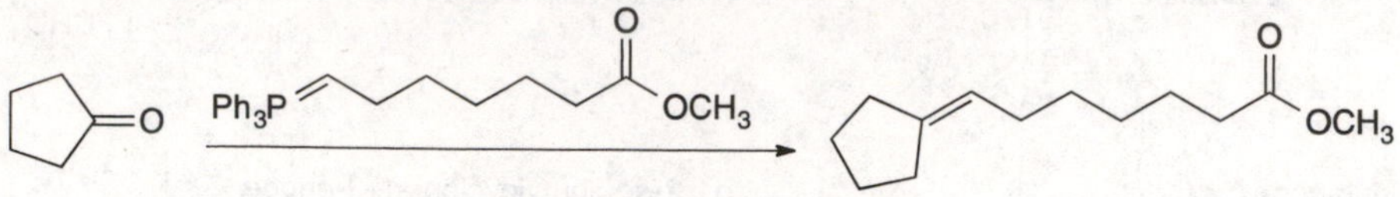

21.47

a. 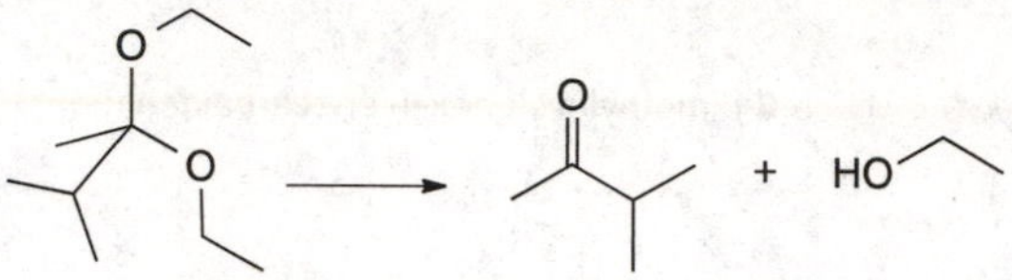 O, O → O + HO

c. 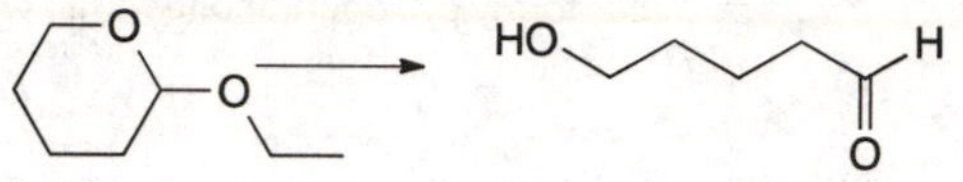 O, O → HO, H, O

+ $HOCH_2CH_3$

b. 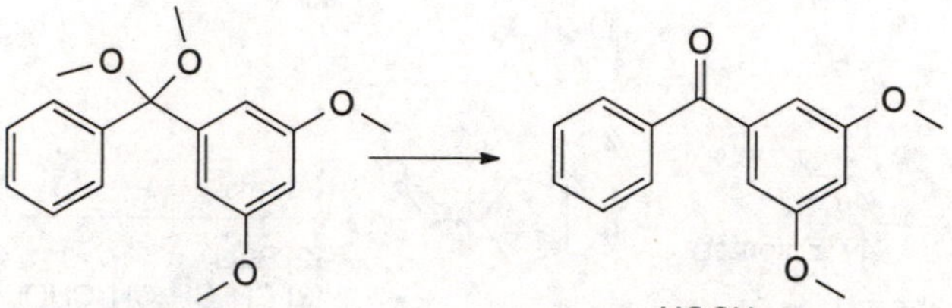 O, O, O, O → O, O, O

+ $HOCH_3$

21.48

a. 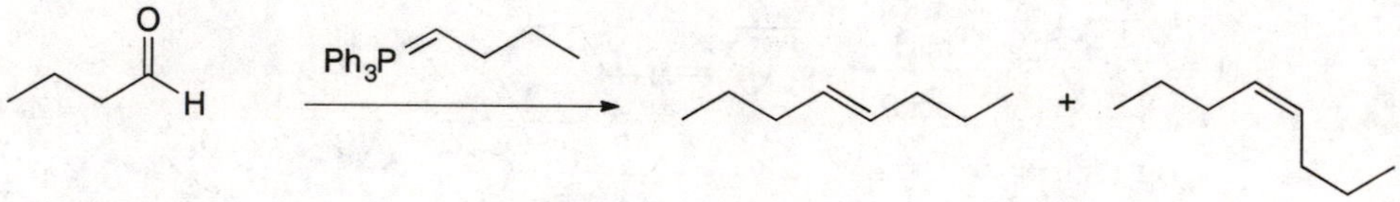 O, H, Ph_3P → +

b. 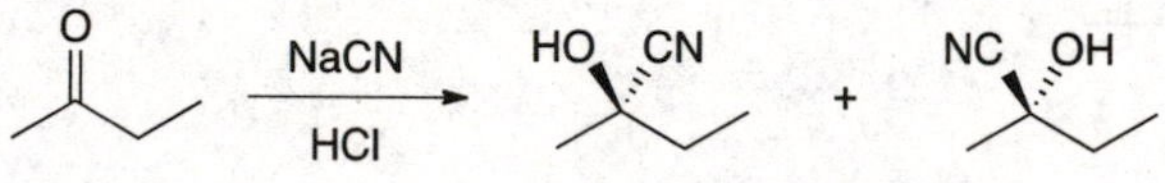 O, NaCN, HCl → HO CN + NC OH

c. 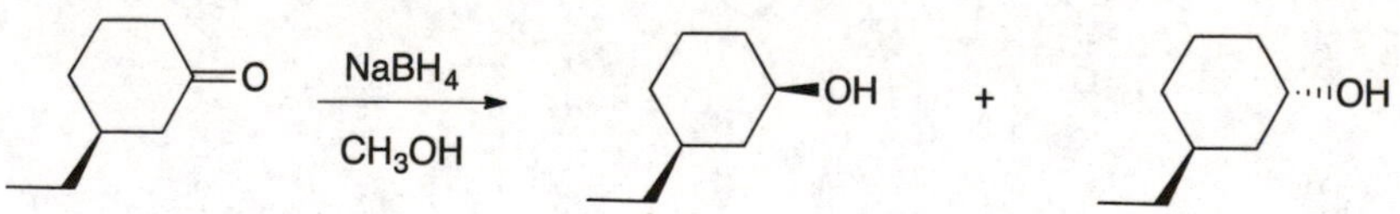 O, $NaBH_4$, CH_3OH → OH + OH

d. 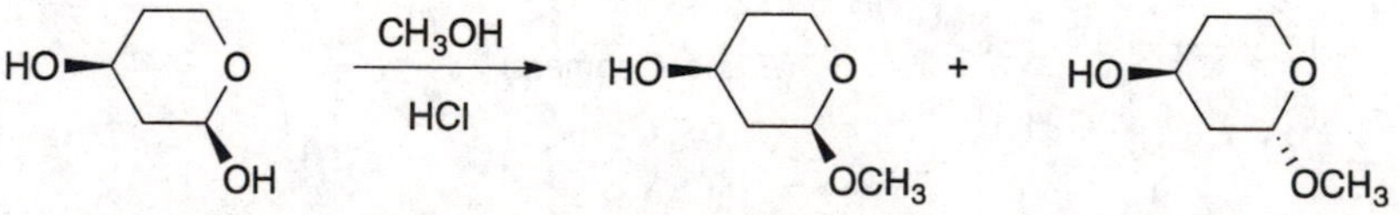 HO, O, OH, CH_3OH, HCl → HO, O, OCH_3 + HO, O, OCH_3

21.49

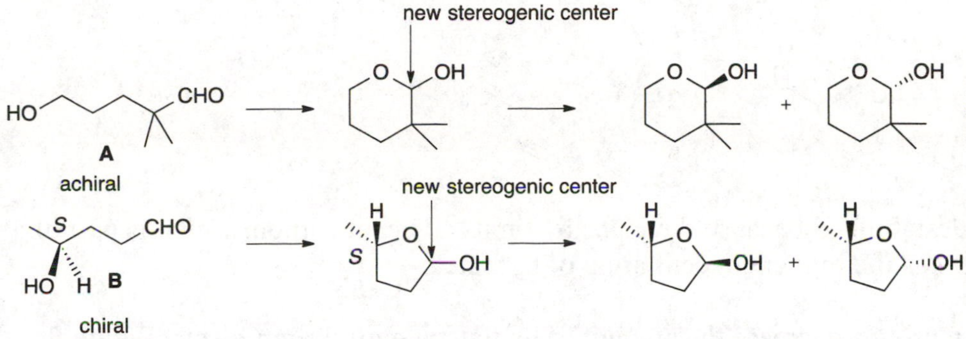

new stereogenic center

HO–CH₂CH₂CH₂C(CH₃)₂–CHO → (cyclic hemiacetal) → mixture

A

achiral

An equal mixture of enantiomers results, so the product is optically inactive.

new stereogenic center

B

chiral

A mixture of diastereomers results. Both compounds are chiral and they are not enantiomers, so the mixture is optically active.

21.50

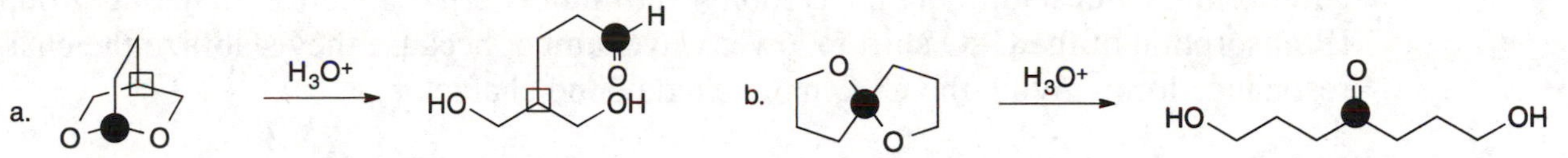

a. $\xrightarrow{H_3O^+}$

b. $\xrightarrow{H_3O^+}$

21.51

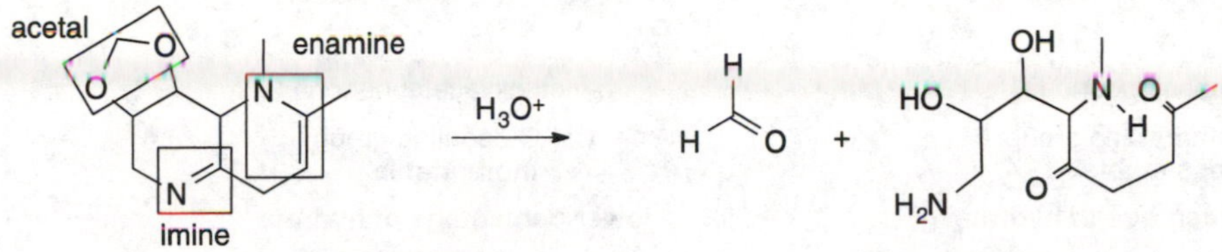

acetal

enamine

imine

$\xrightarrow{H_3O^+}$ H₂C=O +

21.52

a.

acetal

acetal

acetal

etoposide

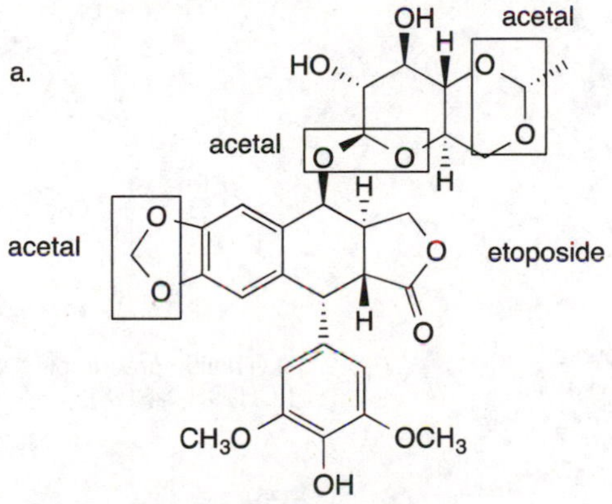

b. Lines of cleavage are drawn in.

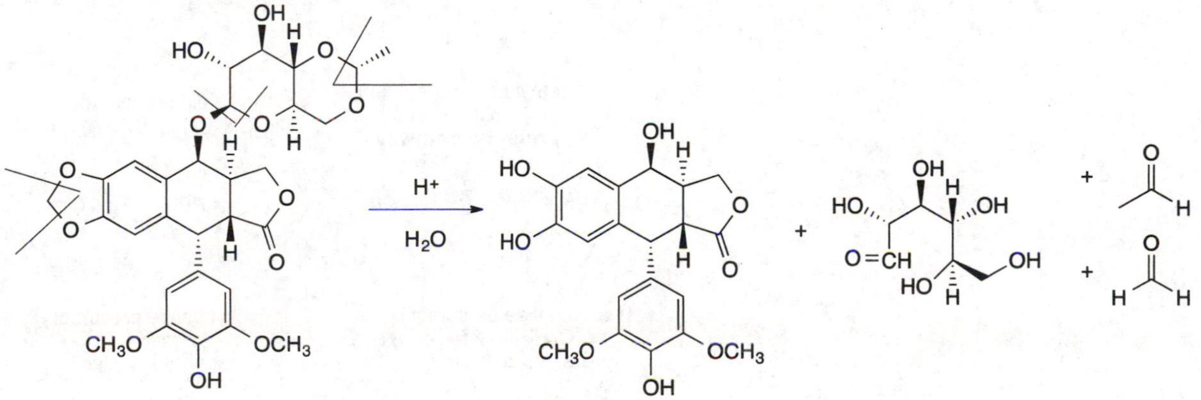

$\xrightarrow[H_2O]{H^+}$

21.53 The less stable carbonyl compound forms the higher percentage of hydrate.

δ+

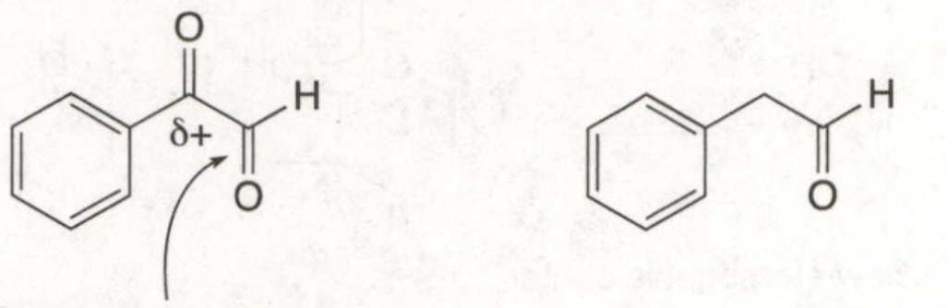

The aldehyde is destabilized because there is a δ+ on its adjacent carbon, which is part of a C=O. Thus, PhCOCHO has the higher concentration of hydrate.

21.54 Electron-donating groups decrease the amount of hydrate at equilibrium by stabilizing the carbonyl starting material. Electron-withdrawing groups increase the amount of hydrate at equilibrium by destabilizing the carbonyl starting material. Electron-donating groups make the IR absorption of the C=O shift to lower wavenumber because they stabilize the charge-separated resonance form, giving the C=O more single bond character.

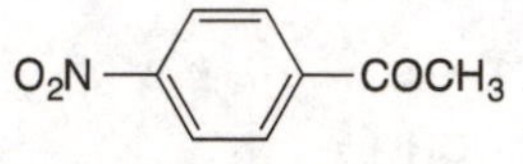

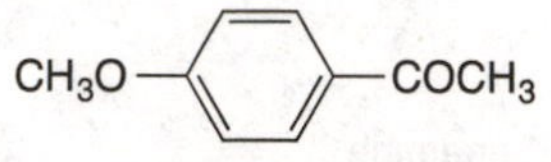

O_2N–C_6H_4–$COCH_3$ | CH_3O–C_6H_4–$COCH_3$

	p-nitroacetophenone	*p*-methoxyacetophenone
a.	NO_2 withdrawing group **less stable**	CH_3O donating group **more stable**
b.	**higher percentage of hydrate**	**lower percentage of hydrate**
c.	**higher wavenumber**	**lower wavenumber**

21.55 Use the principles from Answer 21.20.

a. ⟹ (ketone) + Ph_3P=CHCH₂CH₂CH₃ or $(CH_3CH_2)_2C$=PPh_3 + O=CH(CH₂CH₂CH₃)

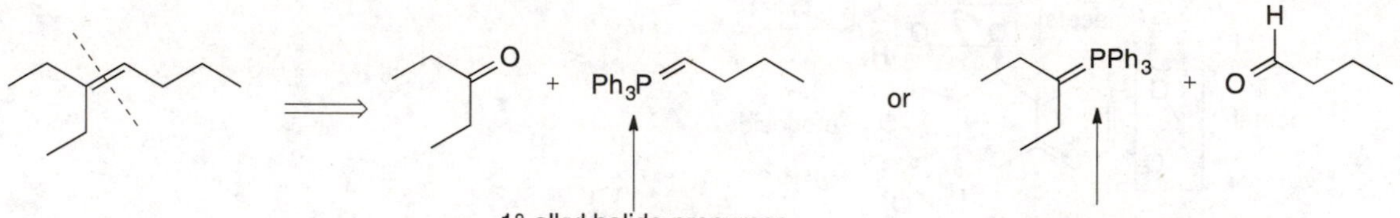

1° alkyl halide precursor
$(XCH_2CH_2CH_2CH_3)$
preferred pathway

2° alkyl halide precursor
$[(CH_3CH_2)_2CHX]$

b. ⟹ (ketone) + $Ph_3P{=}CH_2$ or (ylide with PPh_3) + $CH_2{=}O$

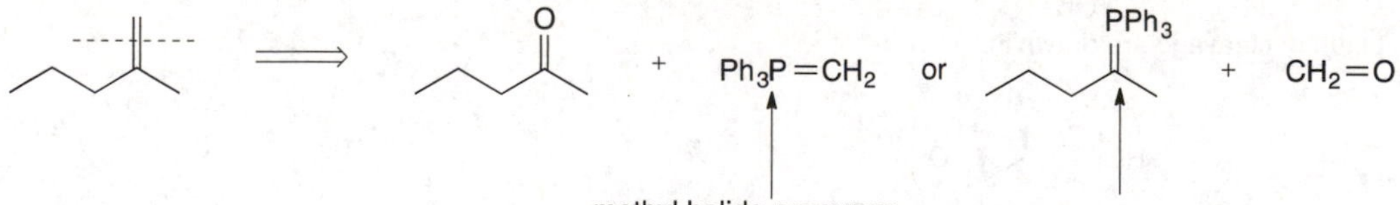

methyl halide precursor
(CH_3X)
preferred pathway

2° alkyl halide precursor
$(CH_3CH_2CH_2CHXCH_3)$

c. (cyclohexylidene with C_6H_5) ⟹ cyclohexanone + Ph_3P=CHC_6H_5 or cyclohexylidene=PPh_3 + O=CH(C_6H_5)

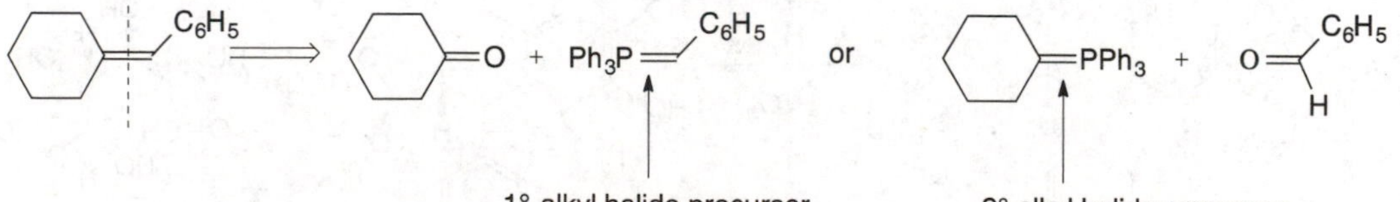

1° alkyl halide precursor
$(C_6H_5CH_2X)$
preferred pathway

2° alkyl halide precursor
(cyclohexyl–X)

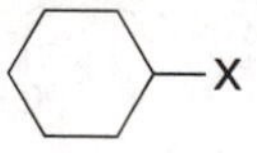

21.56

a. ⟹ + NH_2

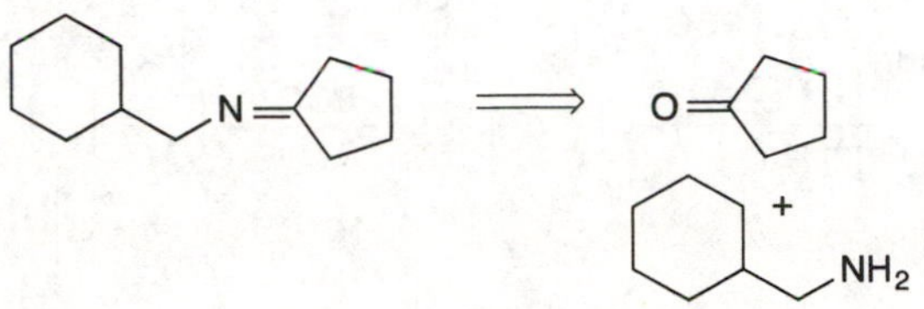

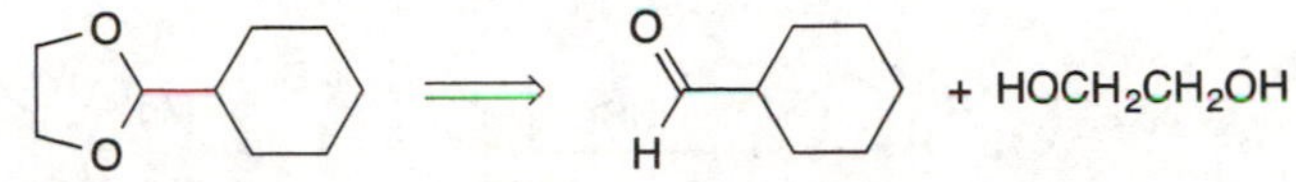

b. ⟹ =O + HN

c. ⟹ O H + $HOCH_2CH_2OH$

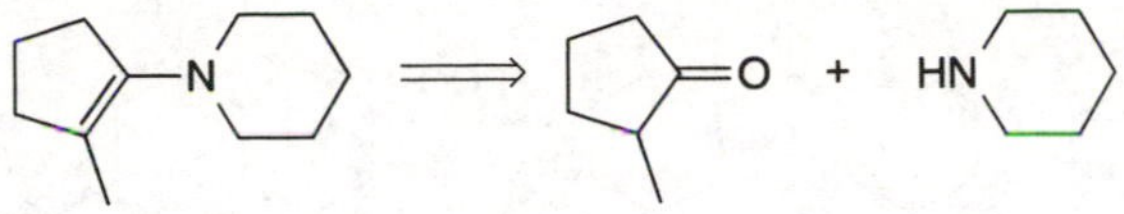

d. CH_3O OCH_3 OCH_3 ⟹ CH_3O H O + $HOCH_3$

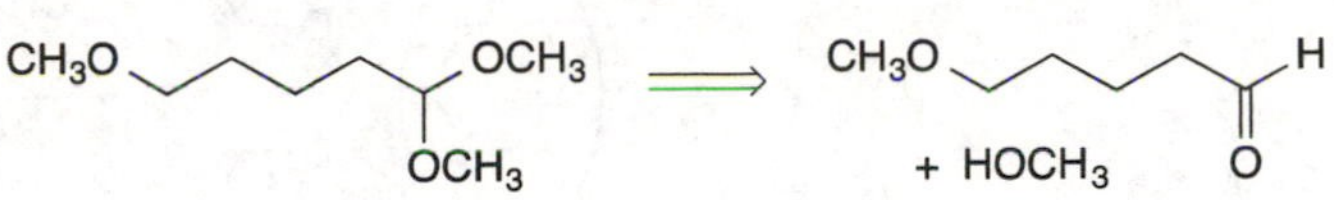

21.57

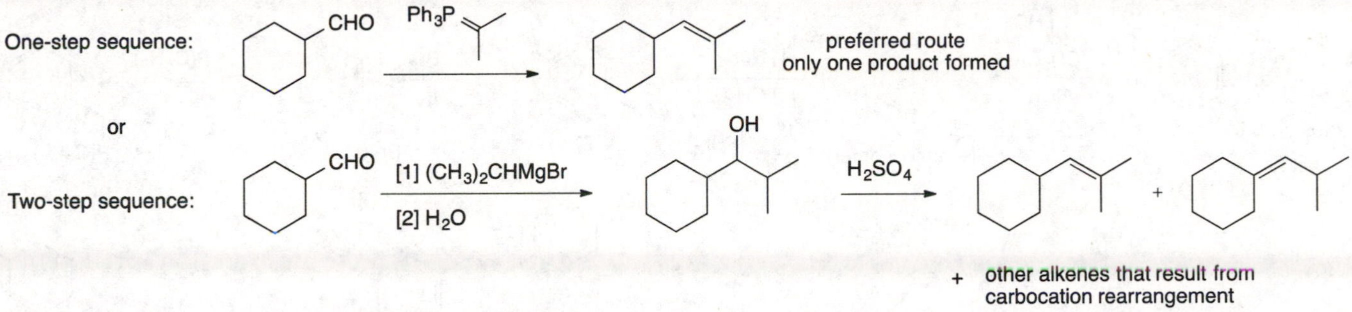

One-step sequence: CHO Ph_3P preferred route only one product formed

or

Two-step sequence: CHO [1] $(CH_3)_2CHMgBr$ [2] H_2O OH H_2SO_4 +

+ other alkenes that result from carbocation rearrangement

21.58

a.

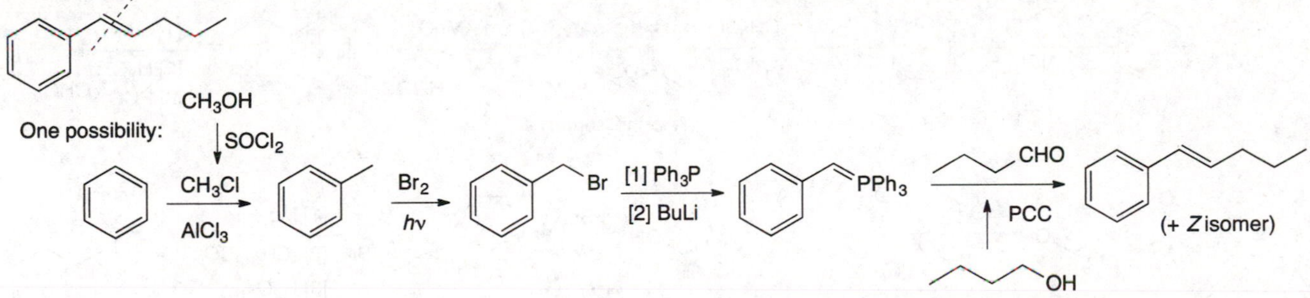

One possibility: CH_3OH $SOCl_2$ CH_3Cl $AlCl_3$ Br_2 $h\nu$ Br [1] Ph_3P [2] BuLi PPh_3 CHO PCC OH (+ *Z* isomer)

b.

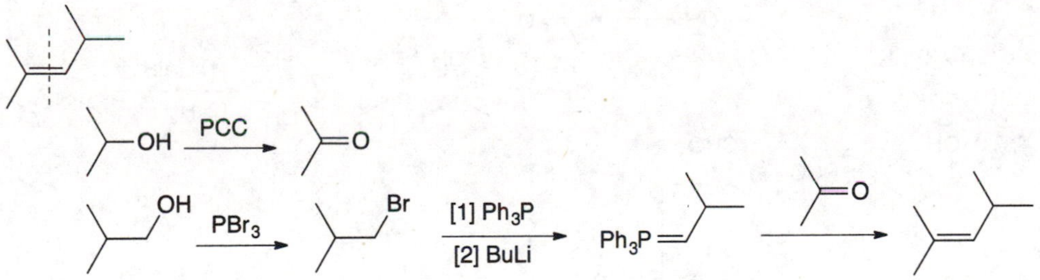

OH PCC =O

OH PBr_3 Br [1] Ph_3P [2] BuLi Ph_3P= =O

21.59

a. H_2O H_2SO_4 OH PCC =O H_2N mild acid =N

NH_3(excess) Br PBr_3 HO

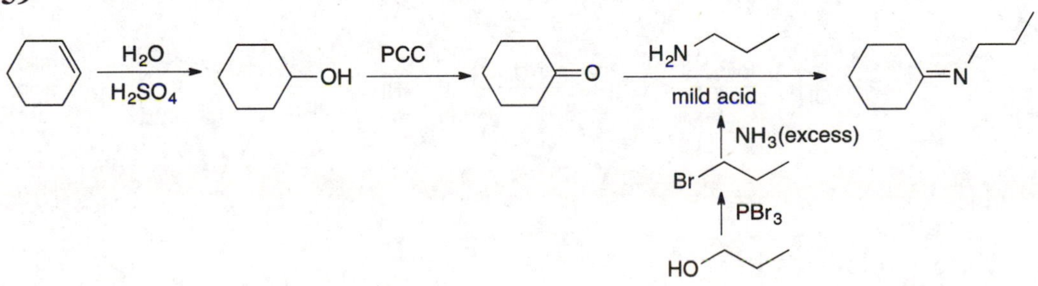

b. OH (from a.) HBr → Br; Mg → MgBr; CH_3OH / PCC; [1] $H_2C{=}O$ [2] H_2O → OH; PCC → O, H; [1] MgBr [2] H_2O → OH; PCC → O; CH_3CH_2OH (2 equiv) TsOH → O, O, 2

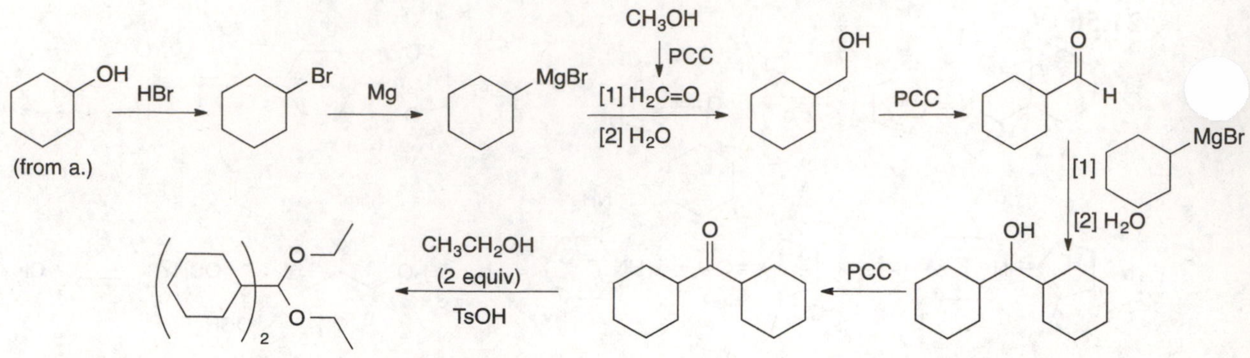

21.60

a. OH; CH_3OH / HCl; CH_3Cl, $AlCl_3$ → OH (+ ortho isomer); [1] NaH [2] CH_3Br; PBr_3 / CH_3OH → OCH_3; Br_2, $h\nu$ → OCH_3, Br; $P(C_6H_5)_3$ → OCH_3, $^{+}P(C_6H_5)_3$, Br^{-}; BuLi → OCH_3, $P(C_6H_5)_3$ +

OH; HCl → Cl; $AlCl_3$; CH_3OH / HCl; CH_3Cl, $AlCl_3$ → (+ ortho isomer); $KMnO_4$ → HOOC; [1] $LiAlH_4$ [2] H_2O [3] PCC → OHC

O (+ *Z* isomer)

b. O, OH; HO, OH, H^+ → O, O, OH; PBr_3 → O, O, Br; [1] Mg [2] $(CH_3)_2CO$ [3] H_2O → O, O, OH; OH, PCC → O; H_2O, H^+ → O, OH; [1] $LiAlH_4$ [2] H_2O → OH, OH

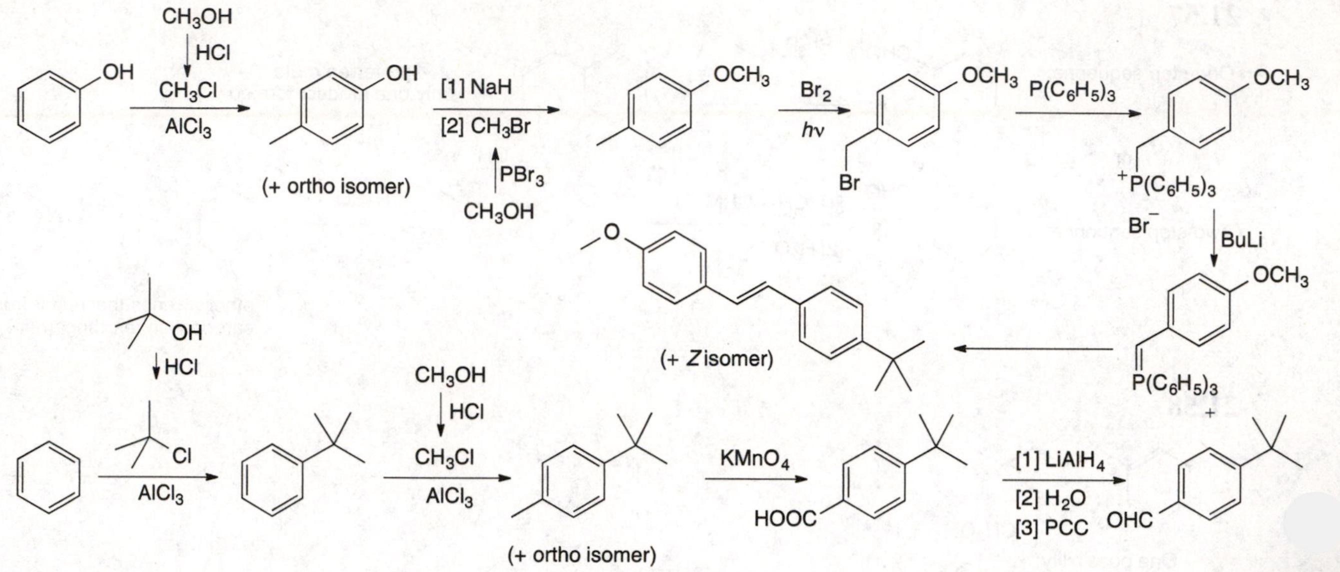

c. OH, PCC → CHO; Br_2, $FeBr_3$ → Br; Mg → MgBr; H_2O → OH; PCC → O

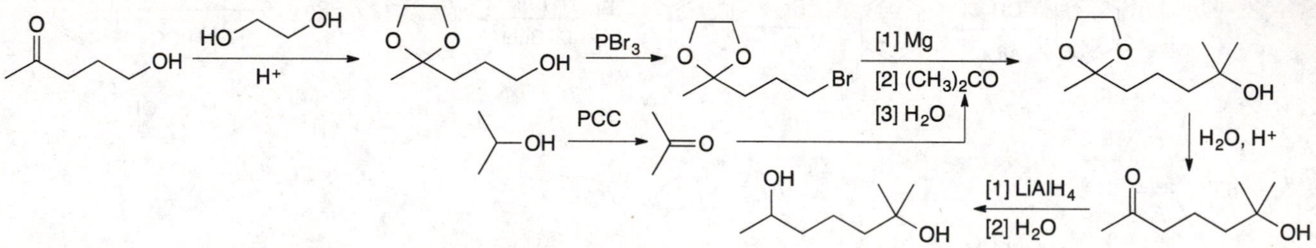

OH; PBr_3 → Br; NH_3 → NH_2; Br → N, H; mild acid → N

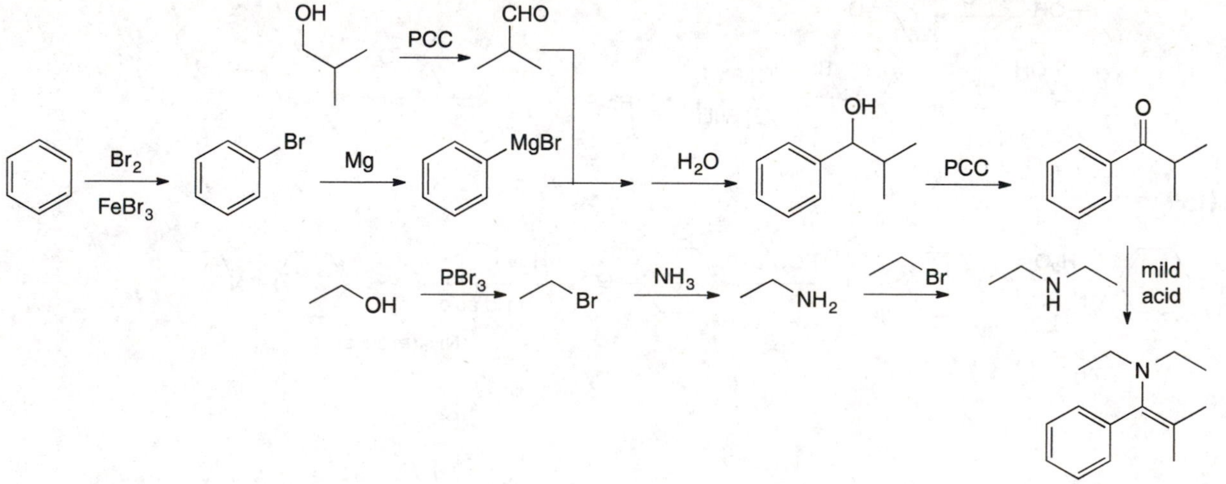

21.61

a. OH — PCC → O, H

PBr_3

Br — Mg → MgBr — [1] O, H [2] H_2O → OH — PCC → O — OH, TsOH → O, O

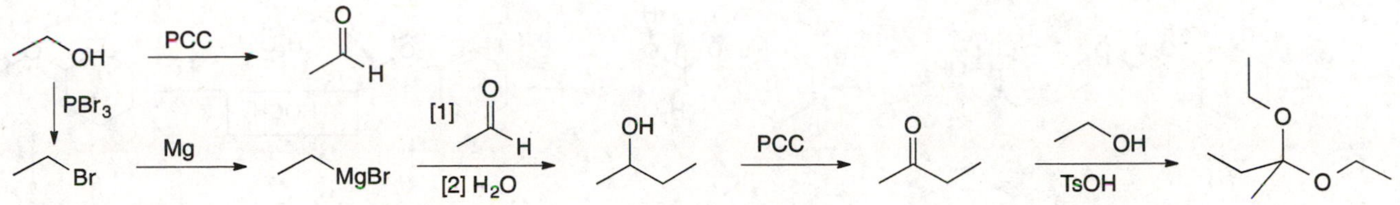

b. OH — H_2SO_4 → $CH_2{=}CH_2$ — Br_2 → Br, Br — 2 $NaNH_2$ → HC≡CH — NaH → $HC{\equiv}C^{-}$ Na^{+}

[1] OsO_4 [2] $NaHSO_3$, H_2O

[1] O, H (from a.) [2] H_2O

O, O ← HO, OH, TsOH ← O ← PCC ← OH

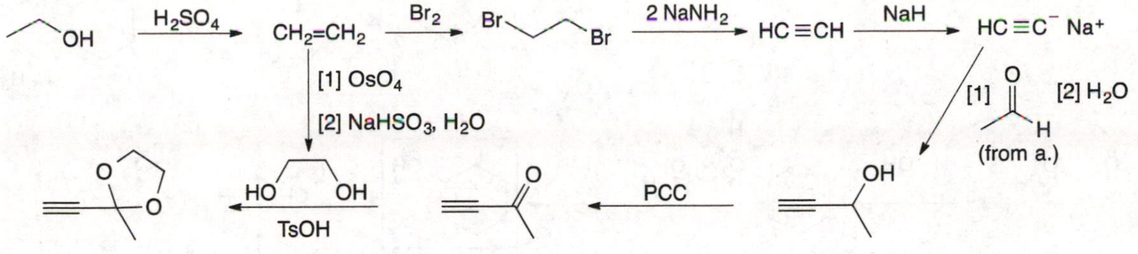

21.62

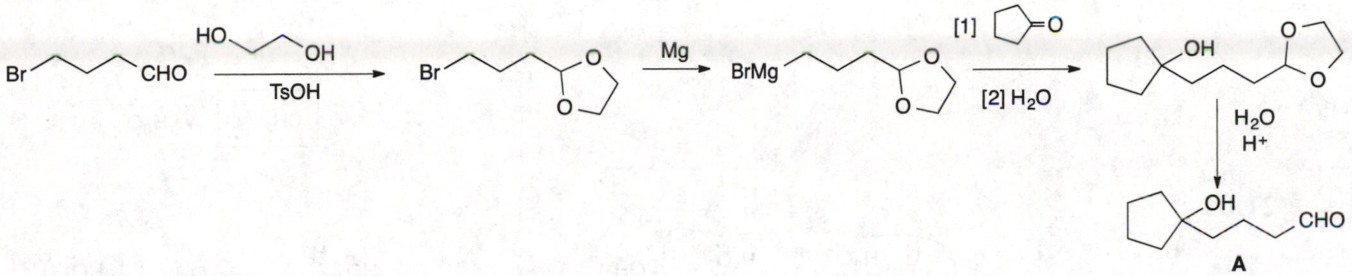

Br, CHO — HO, OH, TsOH → Br, O, O — Mg → BrMg, O, O — [1] =O [2] H_2O → OH, O, O

H_2O H^{+}

OH, CHO

A

21.63

a. Ö–H + $Na^{+}H^{-}$ → Ö: Cl, O, CH_3 → O, O + Cl^{-}

+ H_2 + Na^{+}

methoxy methyl ether

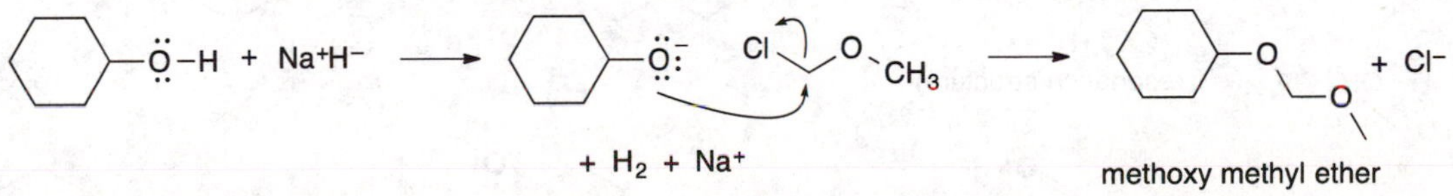

b. O, O acetal

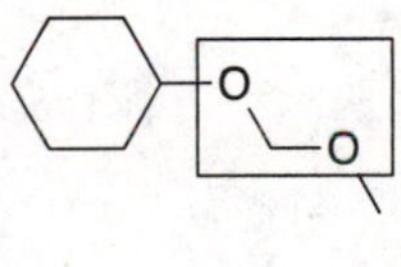

c.

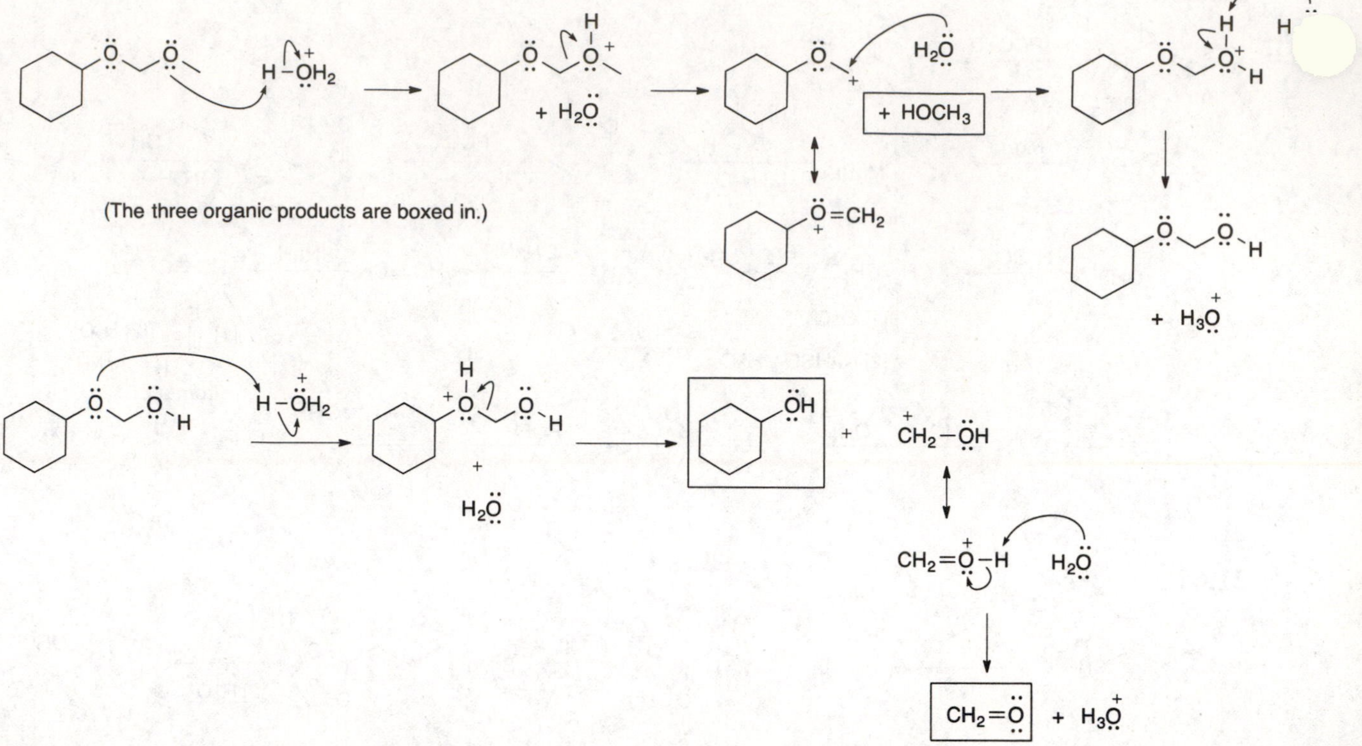

$H-\overset{+}{O}H_2$ → $+ H_2O$ → H_2O $+ HOCH_3$ → $+ H_3\overset{+}{O}$

$\overset{+}{O}=CH_2$

(The three organic products are boxed in.)

$H-\overset{+}{O}H_2$ → $+ H_2O$ → OH $+ \overset{+}{C}H_2-OH$ ↔ $CH_2=\overset{+}{O}-H$ H_2O → $CH_2=O$ $+ H_3\overset{+}{O}$

21.64

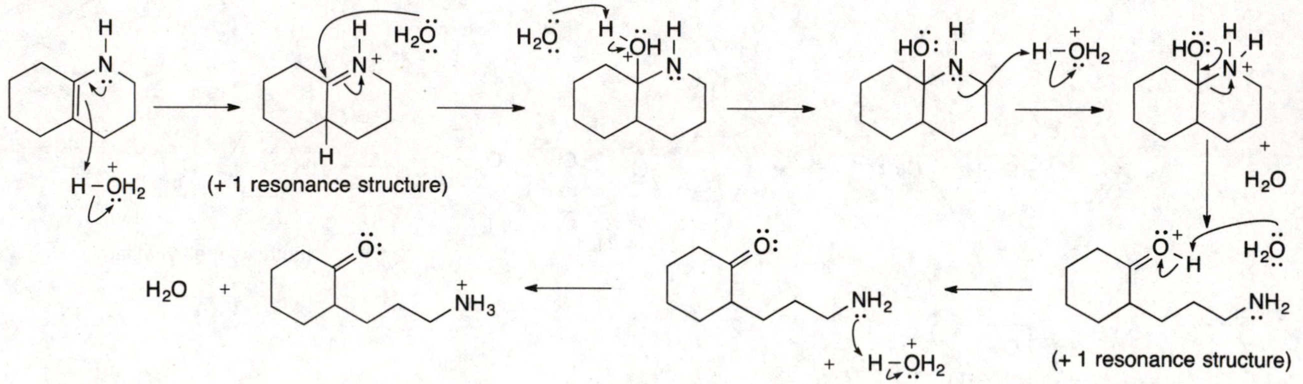

$H-\overset{+}{O}H_2$ → (+ 1 resonance structure) H_2O → H_2O → HO: → $H-\overset{+}{O}H_2$ → HO: $+ H_2O$ → H_2O (+ 1 resonance structure) → NH_2 $+ H-\overset{+}{O}H_2$ → H_2O $+$ $\overset{+}{N}H_3$

21.65

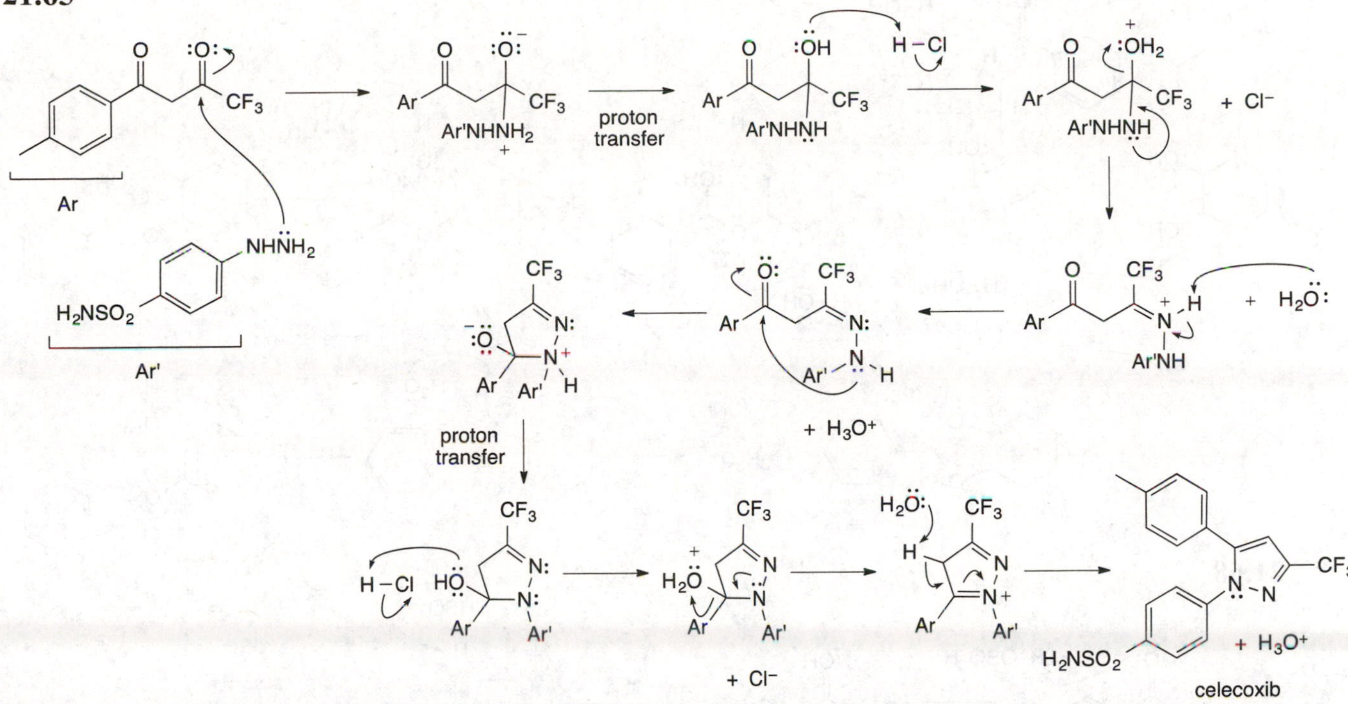

Ar

Ar'

proton transfer

$+ Cl^-$

$+ H_3O^+$

proton transfer

$+ Cl^-$

$+ H_3O^+$

celecoxib

21.66 The OH groups react with the C=O in an intramolecular reaction, first to form a hemiacetal, and then to form an acetal.

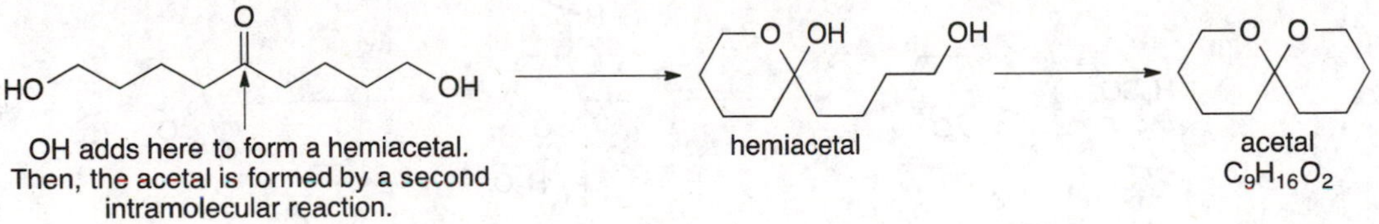

OH adds here to form a hemiacetal. Then, the acetal is formed by a second intramolecular reaction.

hemiacetal

acetal
$C_9H_{16}O_2$

21.67

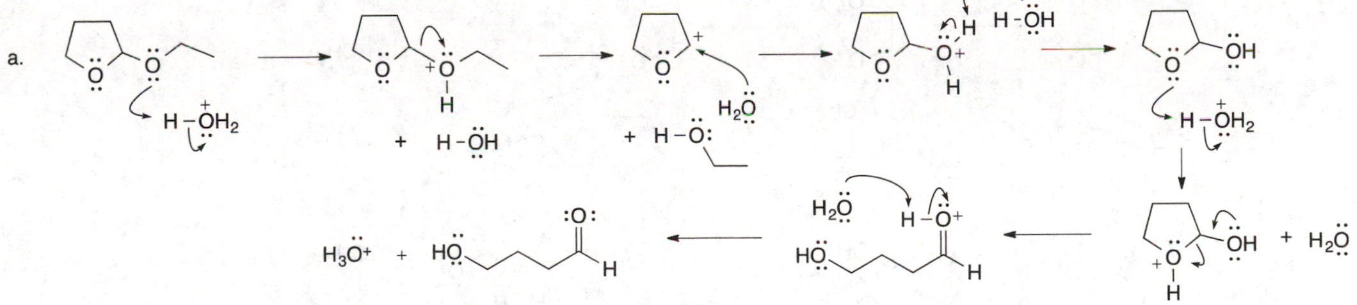

a.

b.

$:O:$ $H-\ddot{O}H_2^+$ H

$\ddot{O}H$ + $:\overset{+}{O}H$ H → $:\ddot{O}H$ → $:\ddot{O}H$ $H-\ddot{O}H_2^+$ → $:\overset{+}{O}H_2$ + $H-\ddot{O}H$

+ $H-\ddot{O}H$ $H-\ddot{O}H$

$H_2\ddot{O}$ + $H_3\ddot{O}^+$ + $H-\ddot{O}H$

21.68

$:O:$ $H-OSO_3H$ → $:\overset{+}{O}H$ $:O:$ + HSO_4^- → HSO_4^- H $H\ddot{O}:$ → $H\ddot{O}:$ $H-OSO_3H$

H_2SO_4 + ← HSO_4^- H O + H_2O ← $H_2\overset{+}{O}$ + HSO_4^-

21.69

$H-Cl$ → $+\overset{H}{O}$ + Cl^- → $\ddot{O}H$ $\ddot{O}-\overset{+}{C}H_2$ ↔ $\ddot{O}H$ $\overset{+}{O}=CH_2$

S $\ddot{N}H_2$

S N H OH $H-Cl$ ← S $\overset{+}{N}$ H H OH $:\ddot{Cl}:^-$

S N H $\overset{+}{O}$ H OH → S $\overset{+}{N}=CH_2$ H + $H\ddot{O}$ OH → $:\ddot{Cl}:^-$ H S + $:NH$ → S NH + $H-\ddot{Cl}:$

21.70

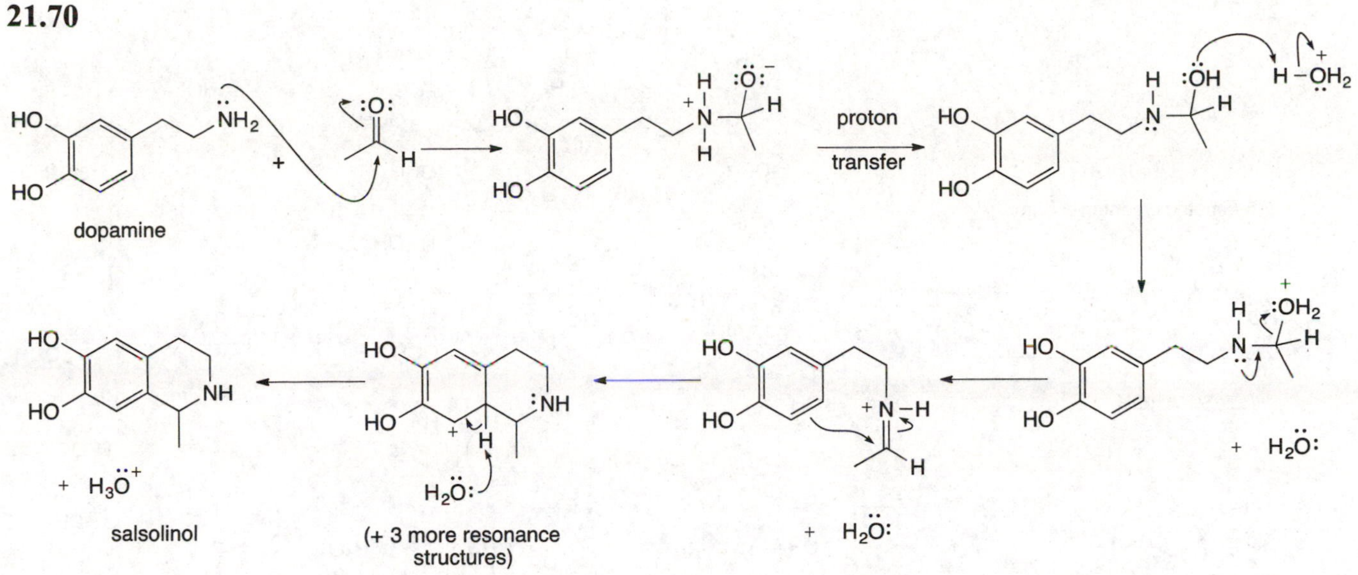

dopamine + acetaldehyde → (addition) → proton transfer → … + H_2O:

(+ 3 more resonance structures)

salsolinol + H_3O^+

21.71 Hemiacetal **A** is in equilibrium with its acyclic hydroxy aldehyde. The aldehyde can undergo hydride reduction to form butane-1,4-diol and a Wittig reaction to form an alkene.

a. **A** ⇌ hydroxy aldehyde → butane-1,4-diol

This can now be reduced with $NaBH_4$.

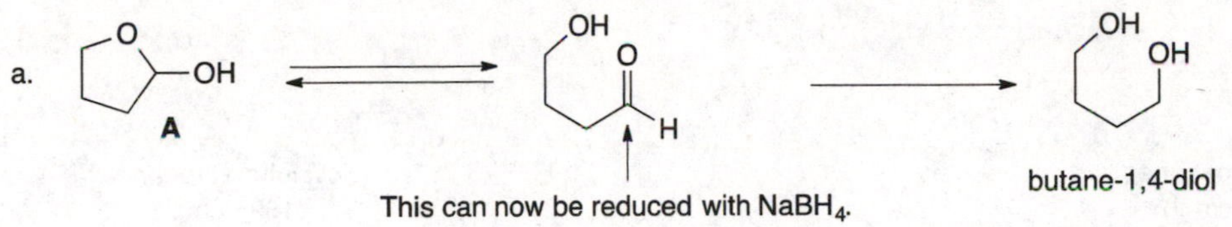

b. **A** ⇌ hydroxy aldehyde —(PPh_3 ylide)→ alkenol (+ *Z* isomer)

reacts with the Wittig reagent

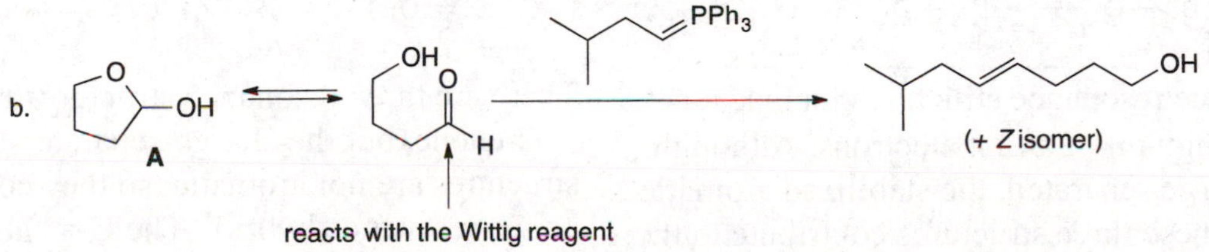

c. **Y** ($\overset{+}{P}Ph_3$) —$NaOCH_2CH_3$→ ylide (PPh_3)

X ⇌ aldehyde acid + ylide → isotretinoin

aldehyde for Wittig reaction

New C=C could be cis or trans. The more stable trans C=C is drawn.

isotretinoin

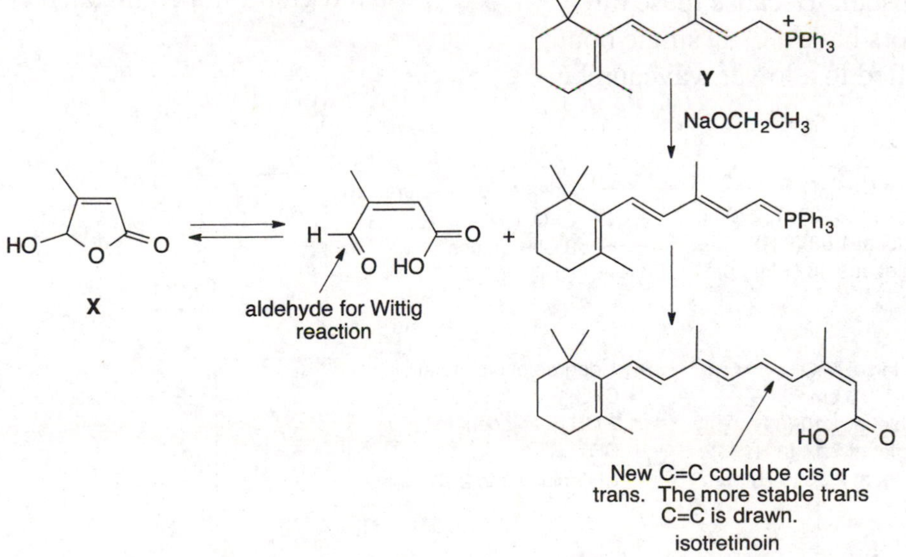

21.72

5,5-dimethoxypentan-2-one

(+ 1 resonance structure)

(+ 1 resonance structure)

Y

21.73

cyclopropenone ($1640\ cm^{-1}$)

cyclohex-2-enone ($1685\ cm^{-1}$)

These three resonance structures include an aromatic ring; $4n + 2 = 2\ \pi$ electrons. Although they are charge separated, the stabilized aromatic ring makes these three structures contribute to the hybrid more than usual. Because these three resonance contributors have a C–O single bond, the absorption is shifted to a lower wavenumber.

There are three resonance structures for cyclohex-2-enone, but the charge-separated resonance structures are not aromatic, so they contribute less to the resonance hybrid. The C=O absorbs in the usual region for a conjugated carbonyl.

21.74

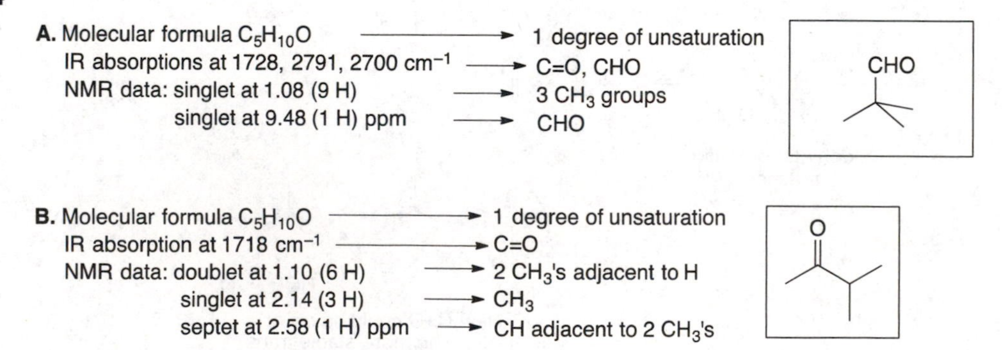

A. Molecular formula C5H10O → 1 degree of unsaturation
IR absorptions at 1728, 2791, 2700 cm−1 → C=O, CHO
NMR data: singlet at 1.08 (9 H) → 3 CH3 groups
singlet at 9.48 (1 H) ppm → CHO
CHO
B. Molecular formula C5H10O → 1 degree of unsaturation
IR absorption at 1718 cm−1 → C=O
NMR data: doublet at 1.10 (6 H) → 2 CH3's adjacent to H
singlet at 2.14 (3 H) → CH3
septet at 2.58 (1 H) ppm → CH adjacent to 2 CH3's
O

C. Molecular formula $C_{10}H_{12}O$ → 5 degrees of unsaturation (4 due to a benzene ring)
IR absorption at 1686 cm^{-1} → C=O
NMR data:

- triplet at 1.21 (3 H) → CH_3 adjacent to 2 H's
- singlet at 2.39 (3 H) → CH_3
- quartet at 2.95 (2 H) → CH_2 adjacent to 3 H's
- doublet at 7.24 (2 H) → 2 H's on benzene ring
- doublet at 7.85 (2 H) ppm → 2 H's on benzene ring

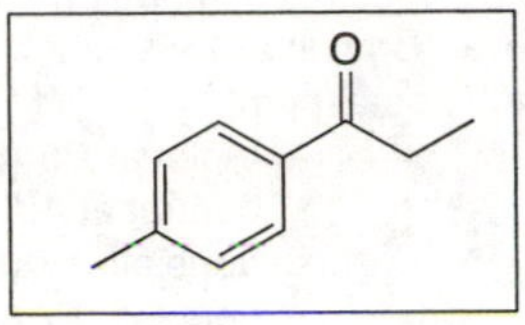

D. Molecular formula $C_{10}H_{12}O$ → 5 degrees of unsaturation (4 due to a benzene ring)
IR absorption at 1719 cm^{-1} → C=O
NMR data:

- triplet at 1.02 (3 H) → CH_3 adjacent to 2 H's
- quartet at 2.45 (2 H) → 2 H's adjacent to 3 H's
- singlet at 3.67 (2 H) → CH_2
- multiplet at 7.06–7.48 (5 H) ppm → a monosubstituted benzene ring

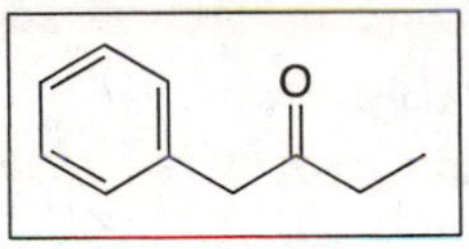

21.75

$C_7H_{16}O_2$: 0 degrees of unsaturation
IR: 3000 cm^{-1}: **C–H bonds**
NMR data (ppm):
H_a: quartet at 3.8 (**4 H**), split by 3 H's
H_b: singlet at 1.5 (**6 H**)
H_c: triplet at 1.2 (**6 H**), split by 2 H's

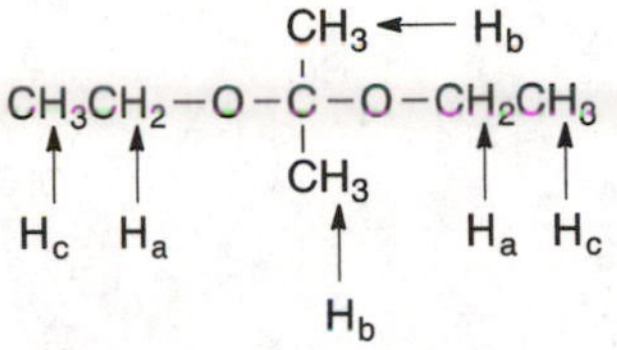

$CH_3CH_2–O–C(CH_3)_2–O–CH_2CH_3$
(CH_3 ← H_b; CH_3 of ethyl ← H_c; CH_2 ← H_a)

21.76

A. Molecular formula $C_9H_{10}O$
5 degrees of unsaturation
IR absorption at 1700 cm^{-1} → C=O
IR absorption at ~2700 cm^{-1} → CH of RCHO
NMR data (ppm):
- triplet at 1.2 (2 H's adjacent)
- quartet at 2.7 (3 H's adjacent)
- doublet at 7.3 (2 H's on benzene)
- doublet at 7.7 (2 H's on benzene)
- singlet at 9.9 (CHO)

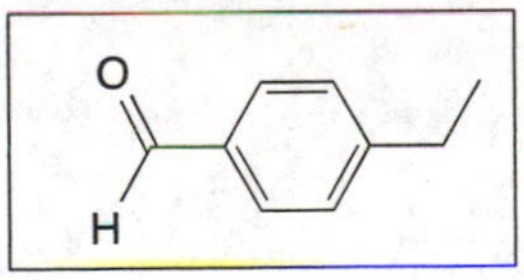

B. Molecular formula $C_9H_{10}O$
5 degrees of unsaturation
IR absorption at 1720 cm^{-1} → C=O
IR absorption at ~2700 cm^{-1} → CH of RCHO
NMR data (ppm):
- 2 triplets at 2.85 and 2.95 (suggests $–CH_2CH_2–$)
- multiplet at 7.2 (benzene H's)
- signal at 9.8 (CHO)

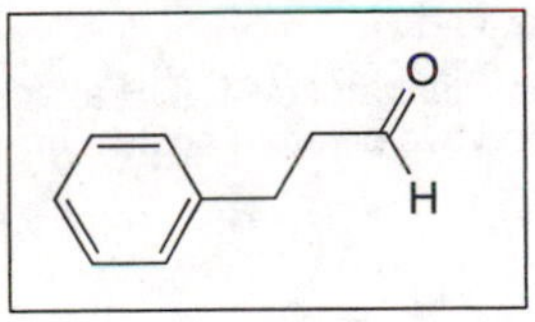

21.77

C. Molecular formula $C_6H_{12}O_3$
1 degree of unsaturation
IR absorption at 1718 cm^{-1} → C=O
NMR data (ppm):
- singlet at 2.1 (3 H's)
- doublet at 2.7 (2 H's)
- singlet at 3.3 (6 H's – 2 OCH_3 groups)
- triplet at 4.8 (1 H)

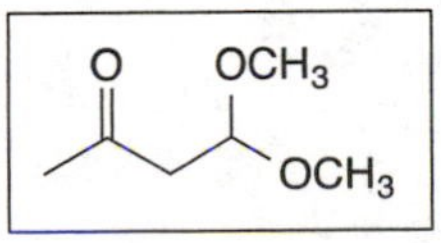

O OCH_3 OCH_3

21.78

D. Molecular ion at $m/z = 150$: $C_9H_{10}O_2$ (possible molecular formula)
5 degrees of unsaturation
IR absorption at 1692 cm^{-1} → C=O
NMR data (ppm):
- triplet at 1.5 (3 H's – CH_3CH_2)
- quartet at 4.1 (2 H's – CH_3CH_2)
- doublet at 7.0 (2 H's – on benzene ring)
- doublet at 7.8 (2 H's – on benzene ring)
- singlet at 9.9 (1 H – on aldehyde)

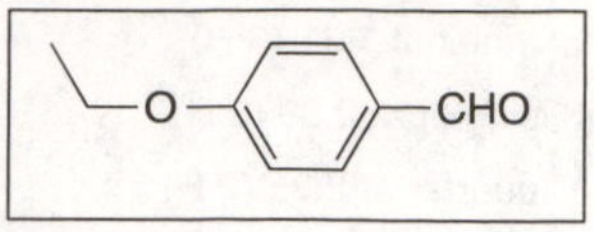

O CHO

21.79

a. OH HO OH OH HO CHO

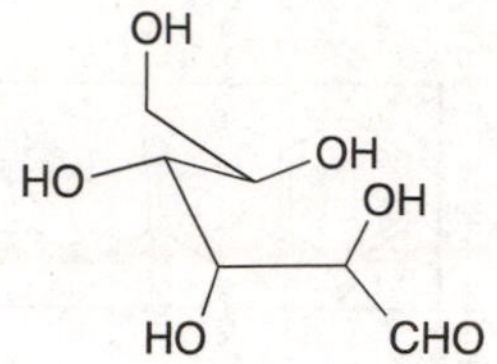

b. HO OH HO CHO OH HO

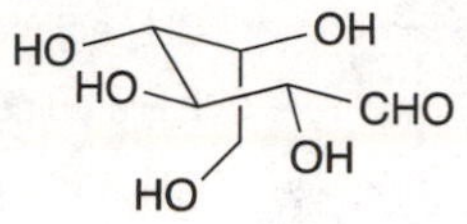

21.80

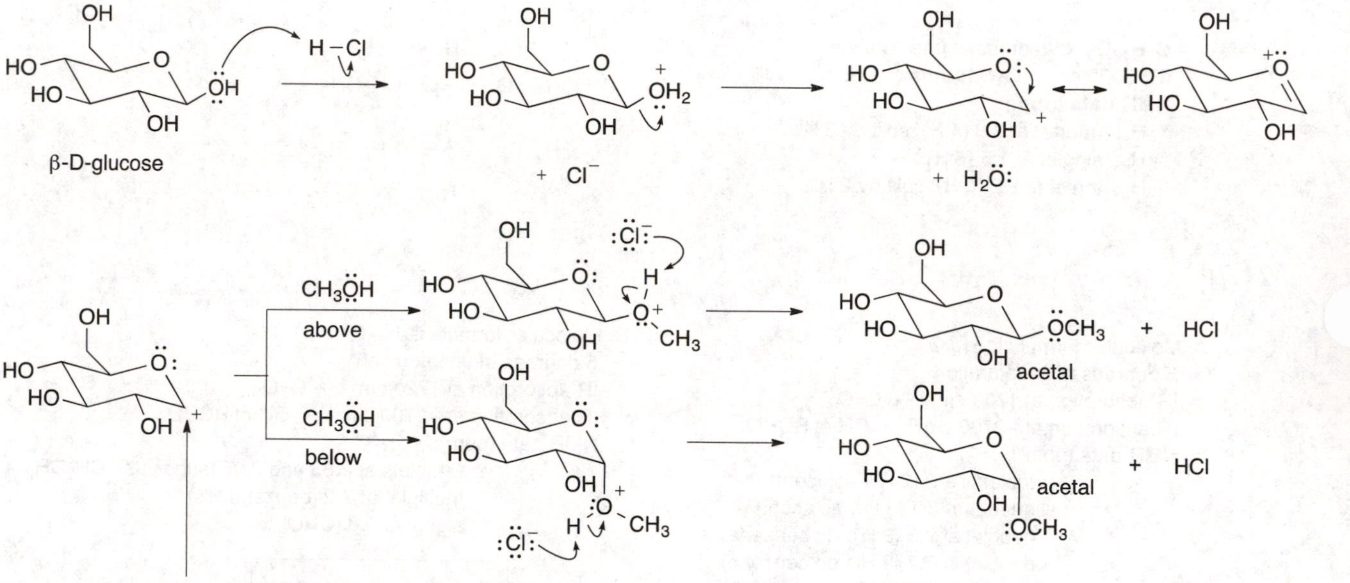

β-D-glucose

H–Cl

$+ Cl^-$

$+ H_2\ddot{O}$:

$CH_3\ddot{O}H$ above

$CH_3\ddot{O}H$ below

$\ddot{O}CH_3$ + HCl acetal

acetal + HCl $:\ddot{O}CH_3$

The carbocation is trigonal planar, so CH_3OH attacks from two different directions, and two different acetals are formed.

21.81

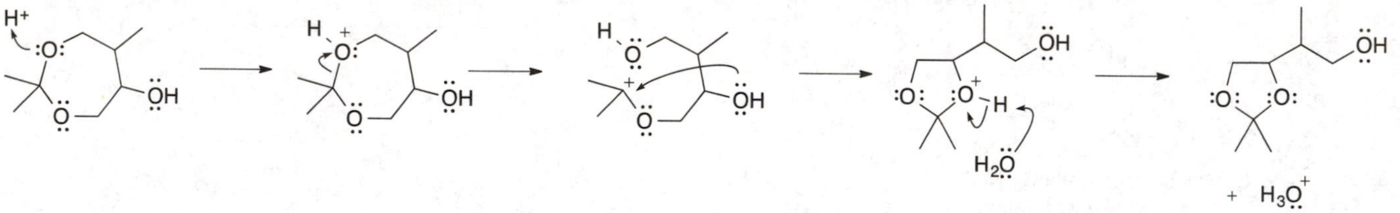

H^+

$H_2\ddot{O}$

$+ H_3O^+$

21.82

a. brevicomin

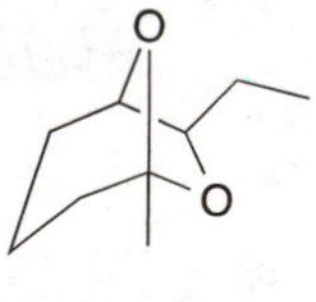

b.

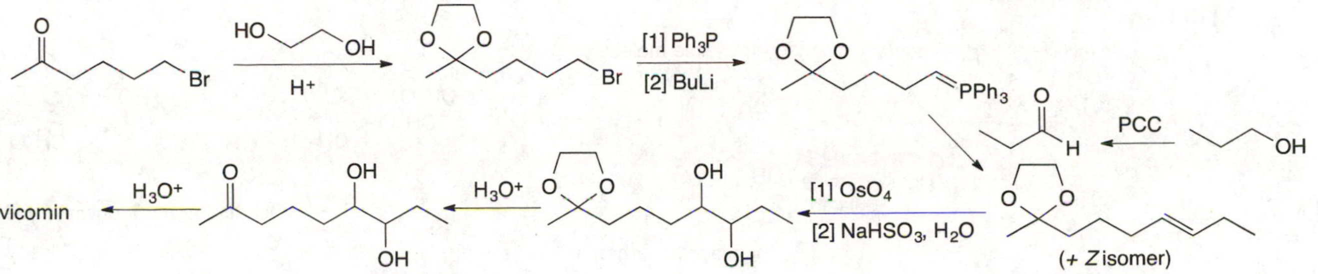

HO–CH2CH2–OH, H^+

[1] Ph_3P
[2] BuLi

PPh_3

PCC

OH

H

(+ *Z* isomer)

[1] OsO_4
[2] $NaHSO_3$, H_2O

H_3O^+

OH

OH

H_3O^+

brevicomin

21.83

H^+

:$\ddot{O}CH_3$

CH_3O

H

OCH_3

$CH_3\ddot{O}H$

$H_2\ddot{O}$

$H_3\ddot{O}^+$

H^+

:$\ddot{O}H$

CH_3

O

OCH_3

$CH_3\ddot{O}H$

H_3O^+

H^+

$\ddot{O}H_2$

$H_2\ddot{O}$:

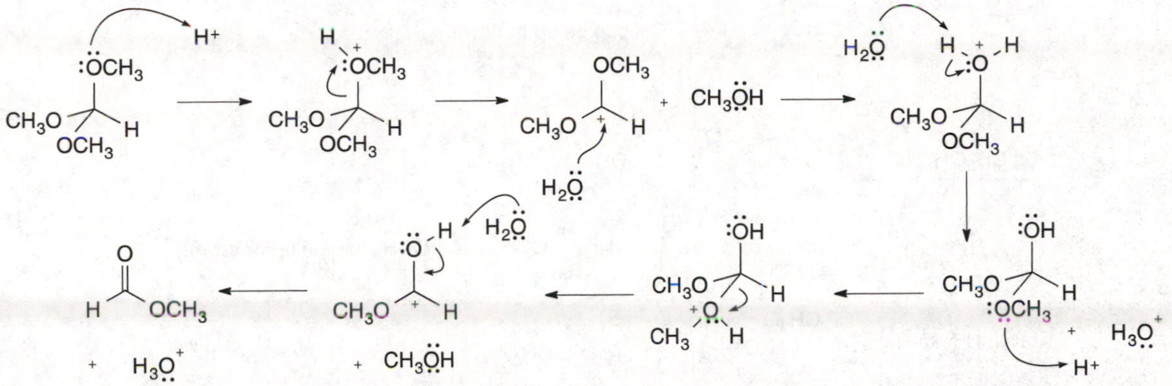

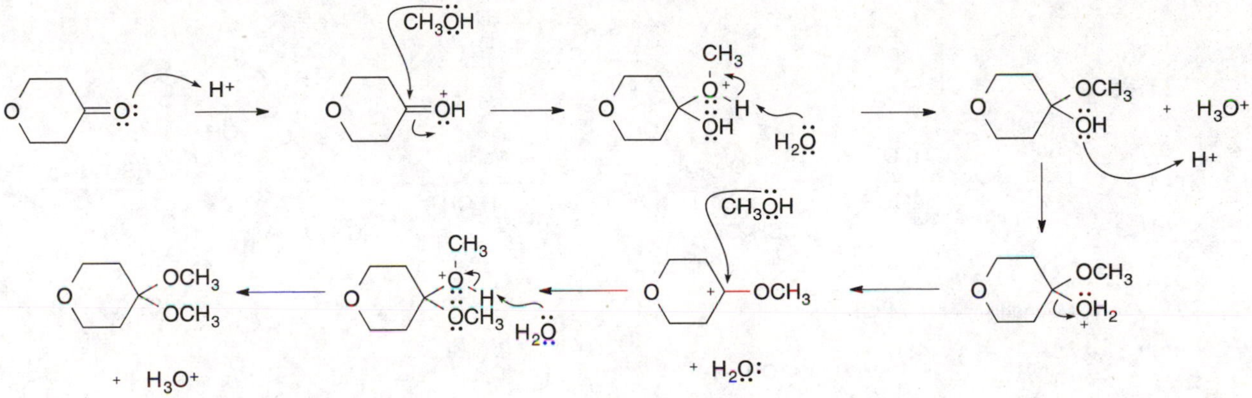

21.84

a. OH, HO, HO, HO, O, O, acetal carbon, OH, O, HO, HO, OH, hemiacetal carbon

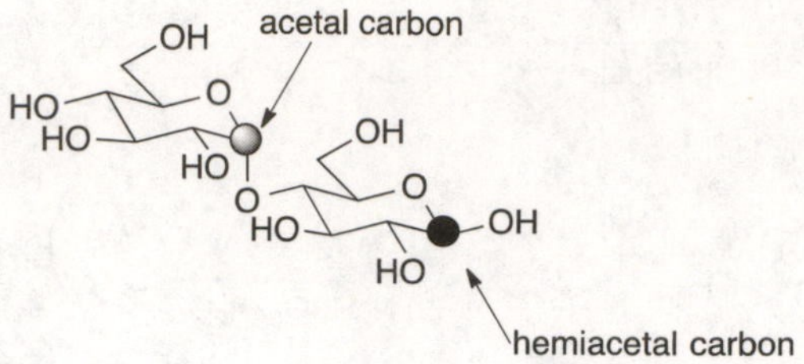

c. OCH_3, CH_3O, CH_3O, O, CH_3O, OH + OCH_3, HO, CH_3O, O, CH_3O, OH

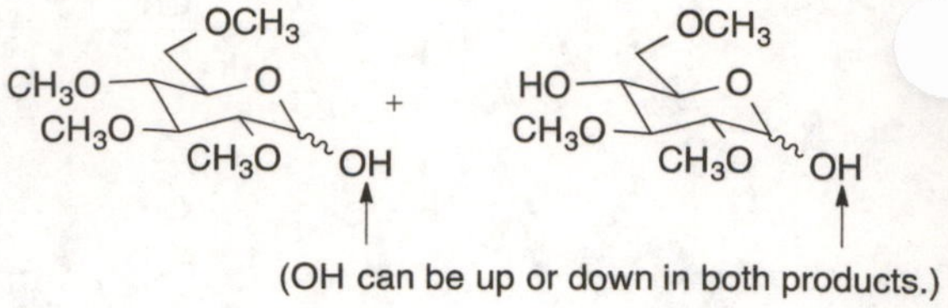

(OH can be up or down in both products.)

b. [1] H_3O^+ → OH, HO, HO, O, HO, OH ← (OH can be up or down.)

[2] CH_3OH, HCl → OH, HO, HO, O, HO, O, OH, HO, O, HO, OCH_3 ← (OCH_3 can be up or down.)

[3] NaH (excess), CH_3I (excess) → OCH_3, CH_3O, CH_3O, O, CH_3O, O, OCH_3, CH_3O, O, OCH_3, OCH_3

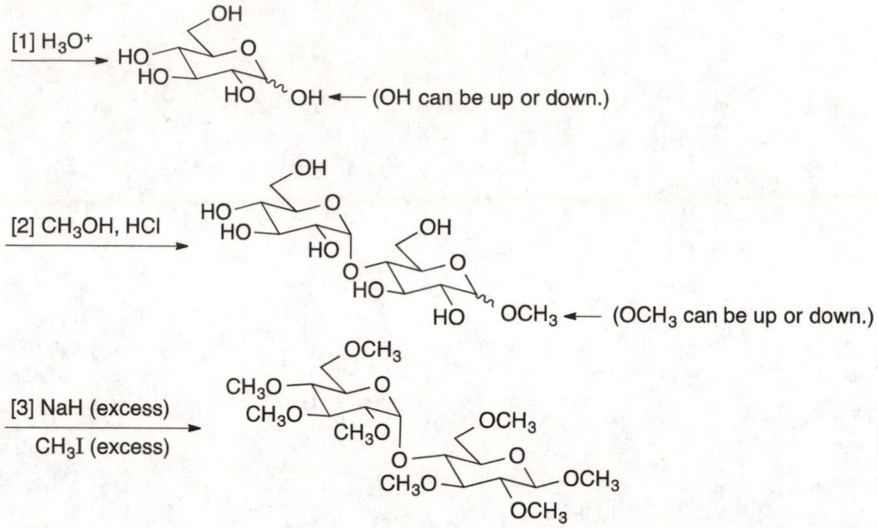

21.85

O, OH, CH_3O → [1] O_3, [2] $(CH_3)_2S$ → O, OH, CH_3O, CHO → $NaBH_4$, CH_3OH → O, OH, CH_3O, OH, R → RSO_3H → O, OH, H, H, O, S, $C_6H_{10}O_3$

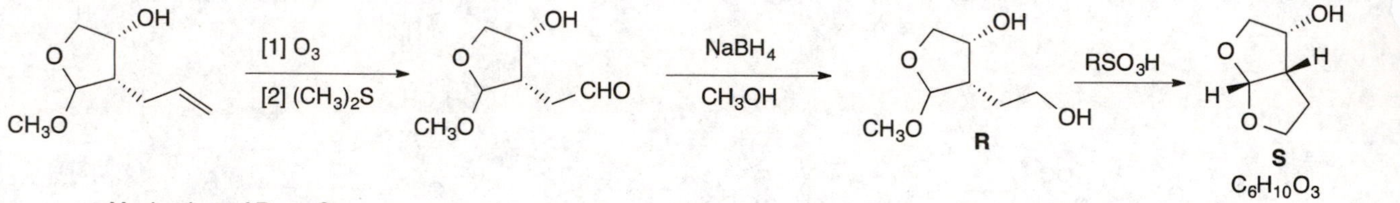

Mechanism of R → S:

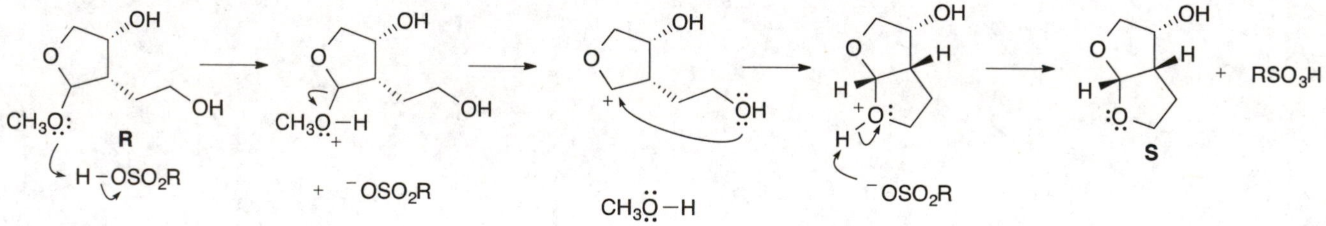

O, OH, $CH_3\ddot{O}$:, OH, R, H–OSO_2R → O, OH, $CH_3\ddot{O}$–H (+), OH, + $^-OSO_2R$ → O, OH, +, $\ddot{O}$H, $CH_3\ddot{O}$–H → O, OH, H, H, $^+$O, H, $^-OSO_2R$ → O, OH, H, H, :$\ddot{O}$, S + RSO_3H

21.86 The mechanism involves S_N2 displacement of Br, followed by intramolecular enamine formation.

H_2N N H H CO_3^{2-} → H_2N N H + HCO_3^- Br O N R S_N2 → H_2N N H O N R + Br^- CO_3^{2-} → H_2N N: + HCO_3^- :O N R

H H N+ N $^-$:O N R proton transfer → H–N N HO H–A N R (any acid) → H–N N H H_2O CO_3^{2-} N R

HN N N O O NH + HCO_3^- + H_2O

conivaptan

Chapter 22 Carboxylic Acids and Their Derivatives—Nucleophilic Acyl Substitution

Chapter Review

Summary of spectroscopic absorptions of RCOZ (22.5)

IR absorptions
- All RCOZ compounds have a C=O absorption in the region 1600–1850 cm^{-1}.
  - RCOCl: 1800 cm^{-1}
  - $(RCO)_2O$: 1820 and 1760 cm^{-1} (two peaks)
  - RCOOR': 1735–1745 cm^{-1}
  - $RCONR'_2$: 1630–1680 cm^{-1}
- Additional amide absorptions occur at 3200–3400 cm^{-1} (N–H stretch) and 1640 cm^{-1} (N–H bending).
- Decreasing the ring size of a cyclic lactone, lactam, or anhydride increases the frequency of the C=O absorption.
- Conjugation shifts the C=O to lower wavenumber.

^{1}H NMR absorptions
- C–H α to the C=O absorbs at 2–2.5 ppm.
- N–H of an amide absorbs at 7.5–8.5 ppm.

^{13}C NMR absorption
- C=O absorbs at 160–180 ppm.

Summary of spectroscopic absorptions of RCN (22.5)

IR absorption
- C≡N absorption at 2250 cm^{-1}

^{13}C NMR absorption
- C≡N absorbs at 115–120 ppm.

Summary: The relationship between the basicity of Z^- and the properties of RCOZ

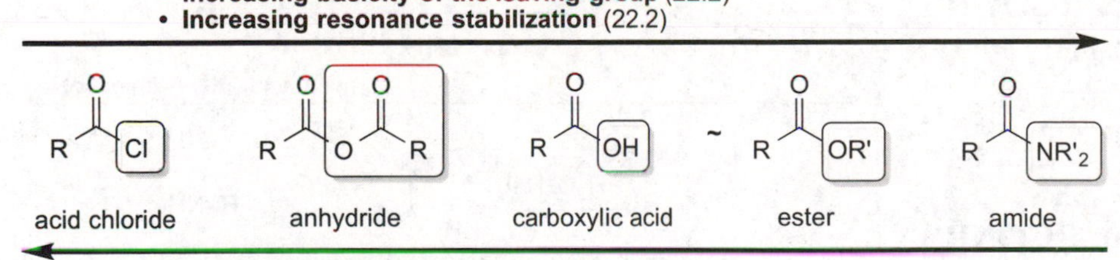

General features of nucleophilic acyl substitution

- The characteristic reaction of compounds having the general structure RCOZ is nucleophilic acyl substitution (22.1).
- The mechanism consists of two steps (22.7A):
  [1] Addition of a nucleophile to form a tetrahedral intermediate
  [2] Elimination of a leaving group
- More reactive acyl compounds can be used to prepare less reactive acyl compounds. The reverse is not necessarily true (22.7B).

Nucleophilic acyl substitution reactions

[1] Reaction that synthesizes acid chlorides (RCOCl)

From RCOOH (22.10A):

$$RCOOH + SOCl_2 \longrightarrow RCOCl + SO_2 + HCl$$

[2] Reactions that synthesize anhydrides [(RCO)₂O]

a. From RCOCl (22.8):

$$RCOCl + {}^{-}OCOR' \longrightarrow RCOOCOR' + Cl^-$$

b. From dicarboxylic acids (22.10B):

$$HOOC\text{-}CH_2CH_2\text{-}COOH \xrightarrow{\Delta} \text{cyclic anhydride} + H_2O$$

cyclic anhydride

[3] Reactions that synthesize carboxylic acids (RCOOH)

a. From RCOCl (22.8):

$$RCOCl + H_2O \xrightarrow{\text{pyridine}} RCOOH + C_5H_5NH^+\ Cl^-$$

b. From (RCO)₂O (22.9):

$$RCOOCOR + H_2O \longrightarrow 2\ RCOOH$$

c. From RCOOR' (22.11):

$$RCOOR' + H_2O \xrightarrow{(H^+ \text{ or } {}^-OH)} RCOOH \text{ (with acid) or } RCOO^- \text{ (with base)} + R'OH$$

d. From RCONR'₂ (R' = H or alkyl, 22.13):

R' = H or alkyl

$$RCONR'_2 \xrightarrow{H_2O,\ H^+} RCOOH + R'_2\overset{+}{N}H_2$$

$$RCONR'_2 \xrightarrow{H_2O,\ {}^-OH} RCOO^- + R'_2NH$$

[4] Reactions that synthesize esters (RCOOR')

a. From RCOCl (22.8):

$$RCOCl + R'OH \xrightarrow{\text{pyridine}} RCOOR' + C_5H_5NH^+\ Cl^-$$

b. From $(RCO)_2O$ (22.9):

$$RCOOCOR + R'OH \longrightarrow RCOOR' + RCOOH$$

c. From RCOOH (22.10C):

$$RCOOH + R'OH \xrightarrow{H_2SO_4} RCOOR' + H_2O$$

[5] Reactions that synthesize amides ($RCONH_2$) [The reactions are written with NH_3 as the nucleophile to form $RCONH_2$. Similar reactions occur with $R'NH_2$ to form RCONHR', and with R'_2NH to form $RCONR'_2$.]

a. From RCOCl (22.8):

$$RCOCl + \underset{\text{(2 equiv)}}{NH_3} \longrightarrow RCONH_2 + NH_4^+Cl^-$$

b. From $(RCO)_2O$ (22.9):

$$RCOOCOR + \underset{\text{(2 equiv)}}{NH_3} \longrightarrow RCONH_2 + RCOO^-NH_4^+$$

c. From RCOOH (22.10D):

$$RCOOH \xrightarrow[\text{[2] }\Delta]{\text{[1] }NH_3} RCONH_2 + H_2O$$

$$RCOOH + R'NH_2 \xrightarrow[\text{DCC}]{} RCONHR' + H_2O$$

d. From RCOOR' (22.11):

$$RCOOR' + NH_3 \longrightarrow RCONH_2 + R'OH$$

Nitrile synthesis (22.18)

Nitriles are prepared by S_N2 substitution using unhindered alkyl halides as starting materials.

$$R{-}X + {}^-CN \xrightarrow[S_N2]{} R{-}C{\equiv}N + X^-$$

$R = CH_3$, 1°

Reactions of nitriles

[1] Hydrolysis (22.18A)

$$R{-}C{\equiv}N \xrightarrow[\text{(H}^+\text{ or }^-\text{OH)}]{H_2O} \underset{\text{(with acid)}}{RCOOH} \text{ or } \underset{\text{(with base)}}{RCOO^-}$$

[2] Reduction (22.18B)

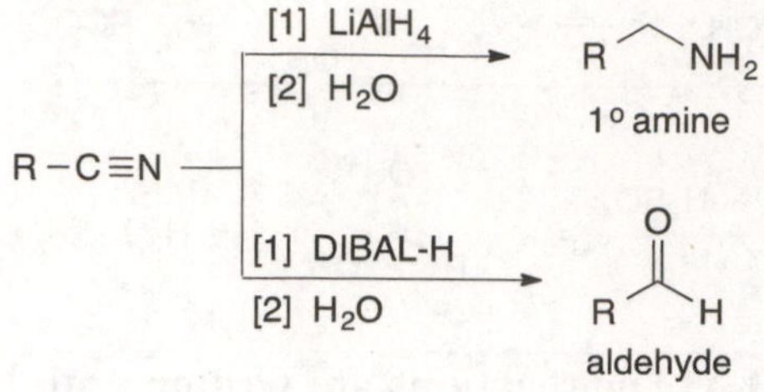

R−C≡N — [1] $LiAlH_4$ / [2] H_2O → RCH_2NH_2 (1° amine)

R−C≡N — [1] DIBAL-H / [2] H_2O → RCHO (aldehyde)

[3] Reaction with organometallic reagents (22.18C)

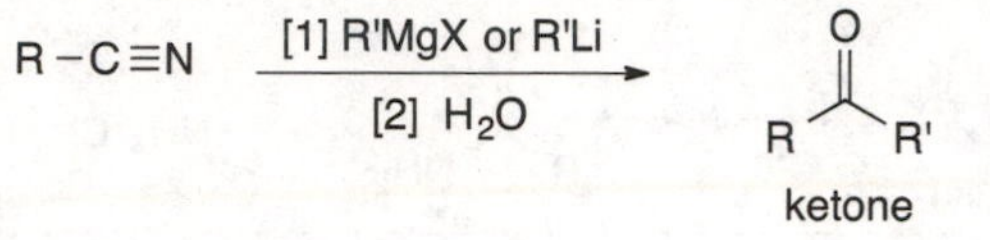

R−C≡N — [1] R'MgX or R'Li / [2] H_2O → RCOR' (ketone)

Practice Test on Chapter Review

1. Give the IUPAC name for each of the following compounds.

a. 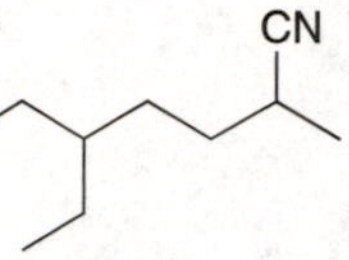 b. 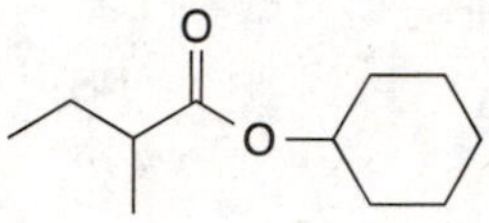 c.

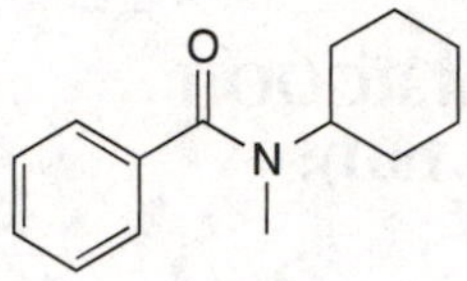

2. (a) Which compound absorbs at the *lowest* wavenumber in the IR? (b) Which compound absorbs at the *highest* wavenumber?

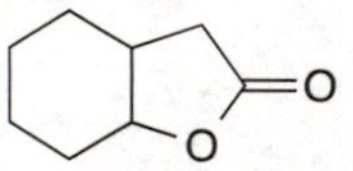

A 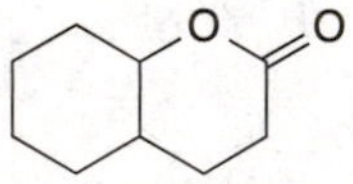 B 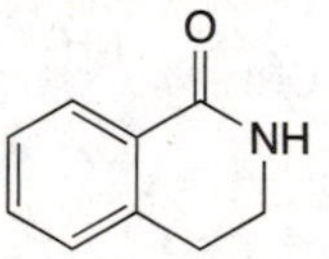 C 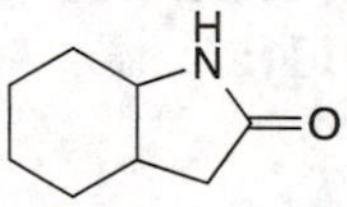 D

3.a. Which of the following reaction conditions can be used to synthesize an ester?

1. RCOCl + R'OH + pyridine
2. RCOOH + R'OH + H_2SO_4
3. RCOOH + R'OH + NaOH
4. Both methods (1) and (2) can be used to synthesize an ester.
5. Methods (1), (2), and (3) can all be used to synthesize an ester.

b. Which of the following compounds is most reactive in nucleophilic acyl substitution?

1. CH_3COCl
2. CH_3COOCH_3
3. $CH_3CON(CH_3)_2$
4. $(CH_3CO)_2O$
5. CH_3COOH

4. What reagent is needed to convert $(CH_3CH_2)_2CHCOOH$ into each compound?
   a. $(CH_3CH_2)_2CHCOO^-Na^+$
   b. $(CH_3CH_2)_2CHCOCl$
   c. $(CH_3CH_2)_2CHCON(CH_3)_2$
   d. $(CH_3CH_2)_2CHCO_2CH_2CH_3$
   e. $[(CH_3CH_2)_2CHCO]_2O$

5. What reagent is needed to convert $CH_3CH_2CH_2CN$ to each compound?
   a. $CH_3CH_2CH_2COOH$
   b. $CH_3CH_2CH_2CH_2NH_2$
   c. $CH_3CH_2CH_2COCH_2CH_3$
   d. $CH_3CH_2CH_2CHO$

6. Draw the organic products formed in the following reactions.

a. [1] NaCN; [2] $LiAlH_4$; [3] H_2O [Indicate stereochemistry.]

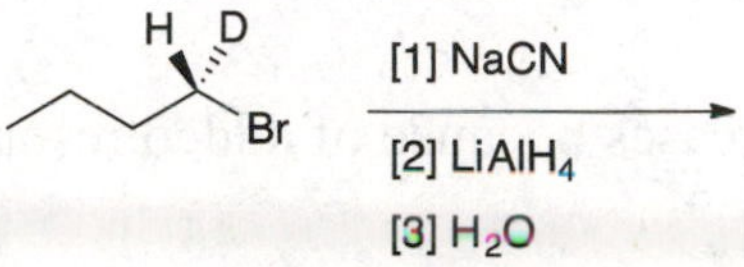

d. [1] $SOCl_2$; [2] $(CH_3CH_2)_2NH$

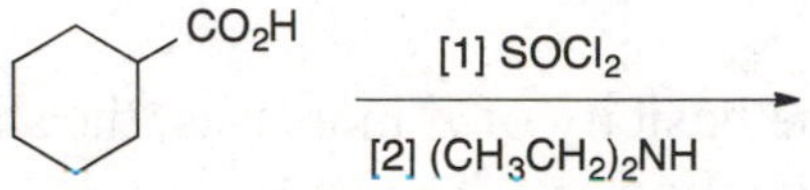

b. NaOH, H_2O

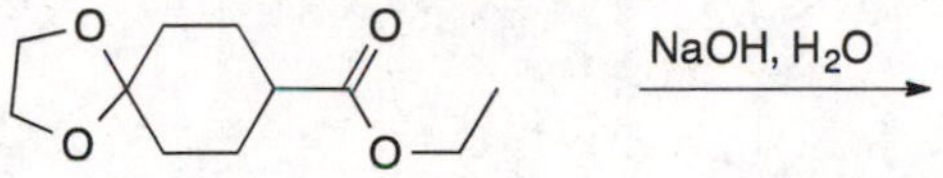

e. + NH_2 (excess)

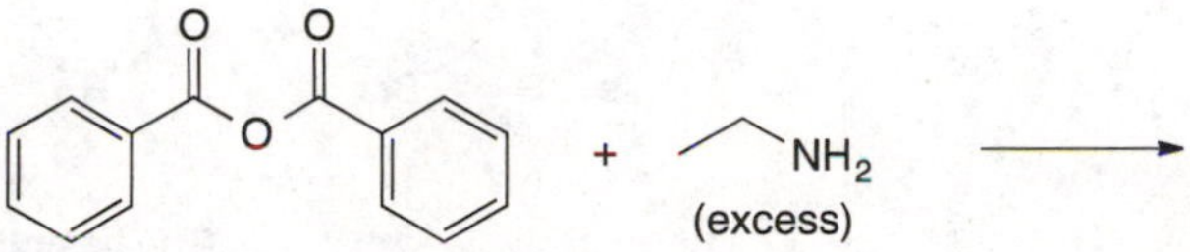

c. +

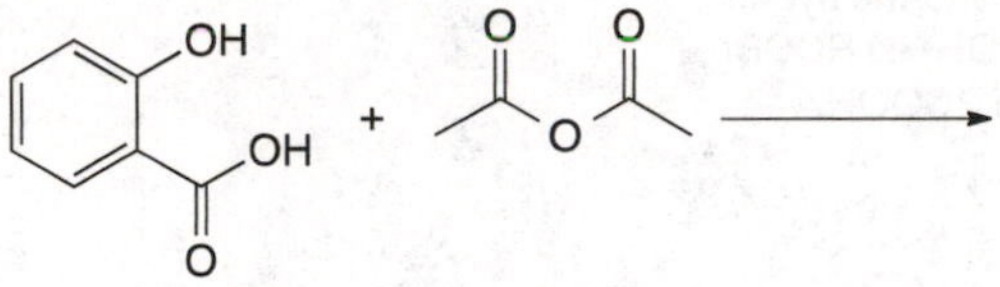

f. [1] NaCN; [2] CH_3CH_2MgBr; [3] H_2O

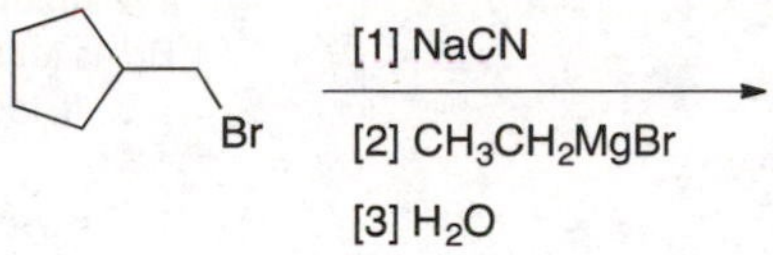

Answers to Practice Test

1.a. 5-ethyl-2-methylheptanenitrile
  b. cyclohexyl 2-methylbutanoate
  c. *N*-cyclohexyl-*N*-methylbenzamide

2.a. **C**
  b. **A**

3.a. 4
  b. 1

4.a. NaOH
  b. $SOCl_2$
  c. $HN(CH_3)_2$, DCC
  d. CH_3CH_2OH, H_2SO_4
  e. heat

5.a. H_3O^+
  b. [1] $LiAlH_4$; [2] H_2O
  c. [1] CH_3CH_2Li; [2] H_2O
  d. [1] DIBAL-H; [2] H_2O

6.
a. NH_2
b. O^- + HO
c. OH + OH
d. 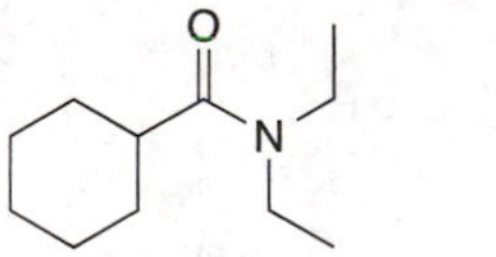
e. + O^- $H_3\overset{+}{N}$
f.

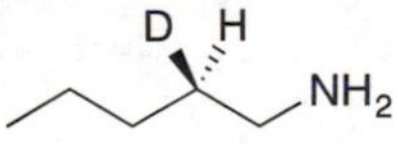

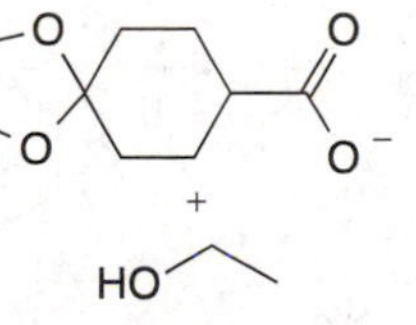

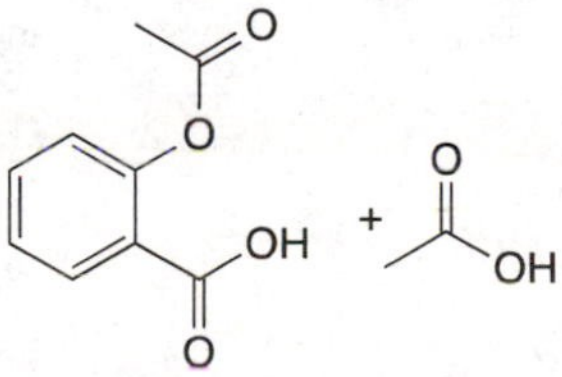

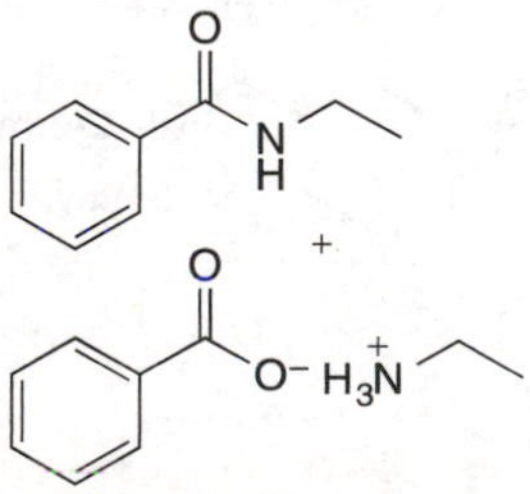

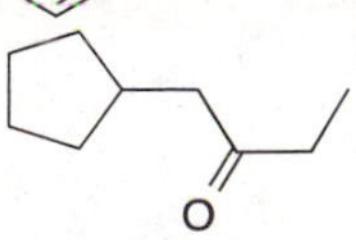

Answers to Problems

22.1 The number of C–N bonds determines the classification as a 1°, 2°, or 3° amide.

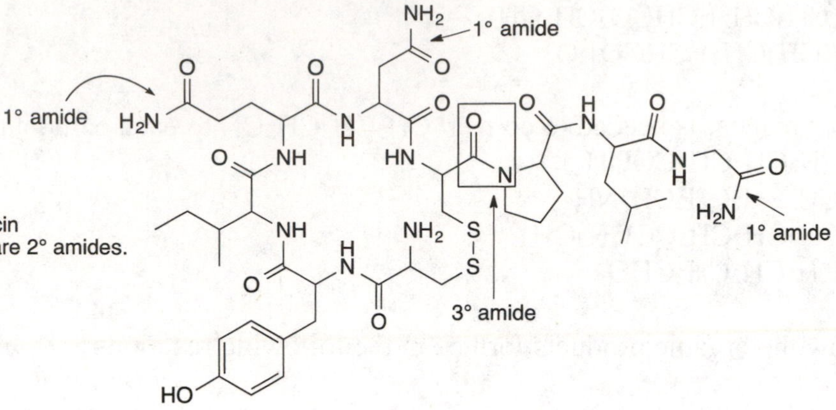

1° amide

1° amide

1° amide

oxytocin
All seven others are 2° amides.

3° amide

22.2 As the basicity of Z increases, the stability of RCOZ increases because of added resonance stabilization.

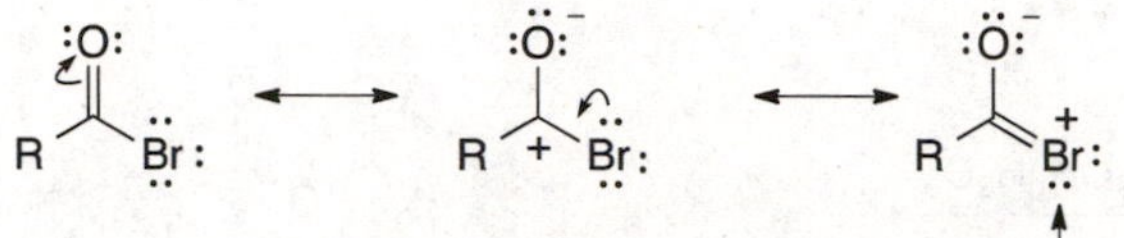

The **basicity of Z** determines how much this structure contributes to the hybrid. Br^- is less basic than ^-OH, so RCOBr is less stable than RCOOH.

22.3

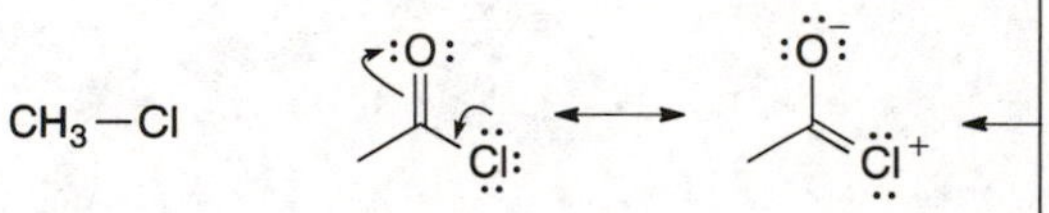

CH_3-Cl

This resonance structure contributes little to the hybrid because Cl^- is a weak base. Thus, the C–Cl bond has little double bond character, making it similar in length to the C–Cl bond in CH_3Cl.

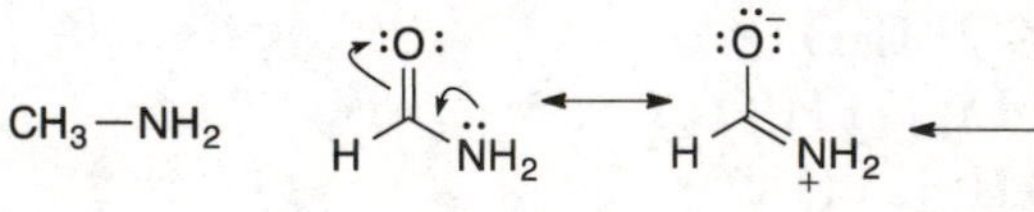

CH_3-NH_2

This resonance structure contributes more to the hybrid because $^-NH_2$ is more basic. Thus, the C–N bond in $HCONH_2$ has more double bond character, making it shorter than the C–N bond in CH_3NH_2.

22.4

a. 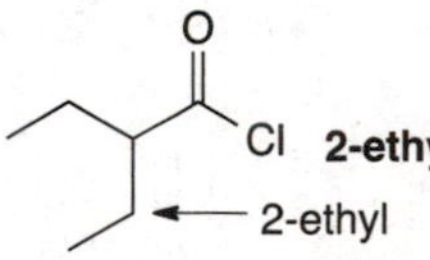 **2-ethylbutanoyl chloride**

2-ethyl

b. 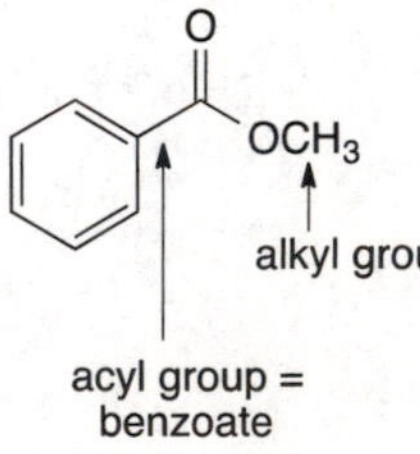 **methyl benzoate**

alkyl group = methyl

acyl group = benzoate

c. acyl group = propanamide; N-ethyl-N-methyl

N-ethyl-N-methylpropanamide

d. acyl group = formate; alkyl group = ethyl

ethyl formate

e. acyl group = propanoic; acyl group = benzoic

benzoic propanoic anhydride

f. 6 carbon chain = hexanenitrile

3-ethylhexanenitrile

22.5

a. 5-methylheptanoyl chloride

b. isopropyl propanoate

c. acetic formic anhydride

d. *N*-isobutyl-*N*-methylbutanamide

e. 3-methylpentanenitrile

f. *o*-cyanobenzoic acid

g. *sec*-butyl 2-methylhexanoate

h. *N*-ethylhexanamide

22.6 CH_3CONH_2 has a higher boiling point than CH_3CO_2H because it has more opportunities for hydrogen bonding. Each CH_3CONH_2 can hydrogen bond to three other molecules (at its O atom and two N–H bonds), whereas CH_3CO_2H has only two intermolecular hydrogen bonds possible (at its carbonyl O atom and OH bond).

22.7

a. and

amide: C=O at lower wavenumber

b. and

smaller ring: C=O at a higher wavenumber

c. and

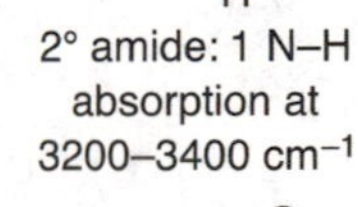

2° amide: 1 N–H absorption at 3200–3400 cm^{-1}

1° amide: 2 N–H absorptions

d. and

anhydride: 2 C=O peaks

22.8

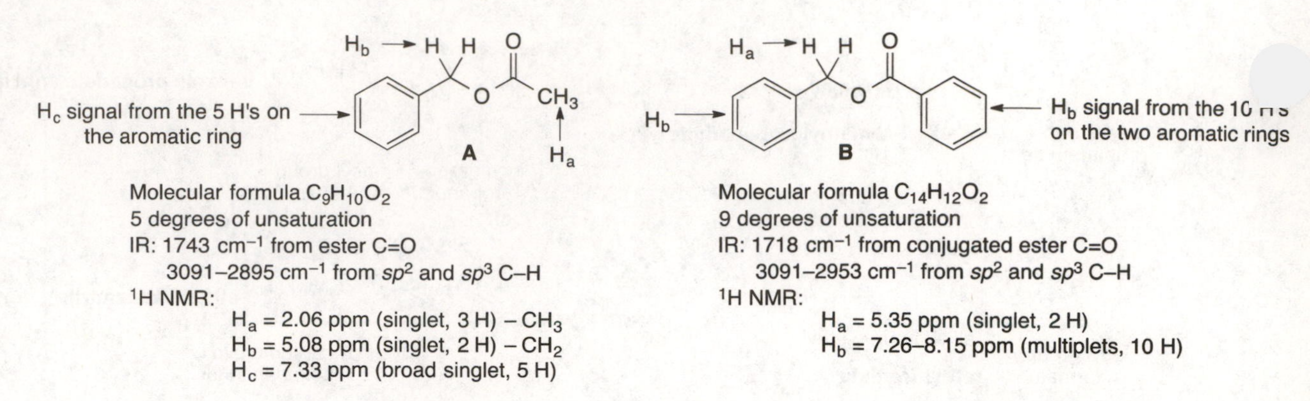

H_b → H H O
H_c signal from the 5 H's on the aromatic ring
CH_3
A
H_a
H_a → H H O
H_b
H_b signal from the 10 H's on the two aromatic rings
B
Molecular formula $C_9H_{10}O_2$
5 degrees of unsaturation
IR: 1743 cm^{-1} from ester C=O
3091–2895 cm^{-1} from sp^2 and sp^3 C–H
^1H NMR:
H_a = 2.06 ppm (singlet, 3 H) – CH_3
H_b = 5.08 ppm (singlet, 2 H) – CH_2
H_c = 7.33 ppm (broad singlet, 5 H)
Molecular formula $C_{14}H_{12}O_2$
9 degrees of unsaturation
IR: 1718 cm^{-1} from conjugated ester C=O
3091–2953 cm^{-1} from sp^2 and sp^3 C–H
^1H NMR:
H_a = 5.35 ppm (singlet, 2 H)
H_b = 7.26–8.15 ppm (multiplets, 10 H)

22.9 More reactive acyl compounds can be converted to less reactive acyl compounds.

a. CH_3COCl (more reactive) → **YES** → CH_3COOH (less reactive)

b. $CH_3CONHCH_3$ (less reactive) → **NO** → CH_3COOCH_3 (more reactive)

c. CH_3COOCH_3 (less reactive) → **NO** → CH_3COCl (more reactive)

d. $(CH_3CO)_2O$ (more reactive) → **YES** → CH_3CONH_2 (less reactive)

22.10 The better the leaving group is, the more reactive the carboxylic acid derivative. The weakest base is the best leaving group.

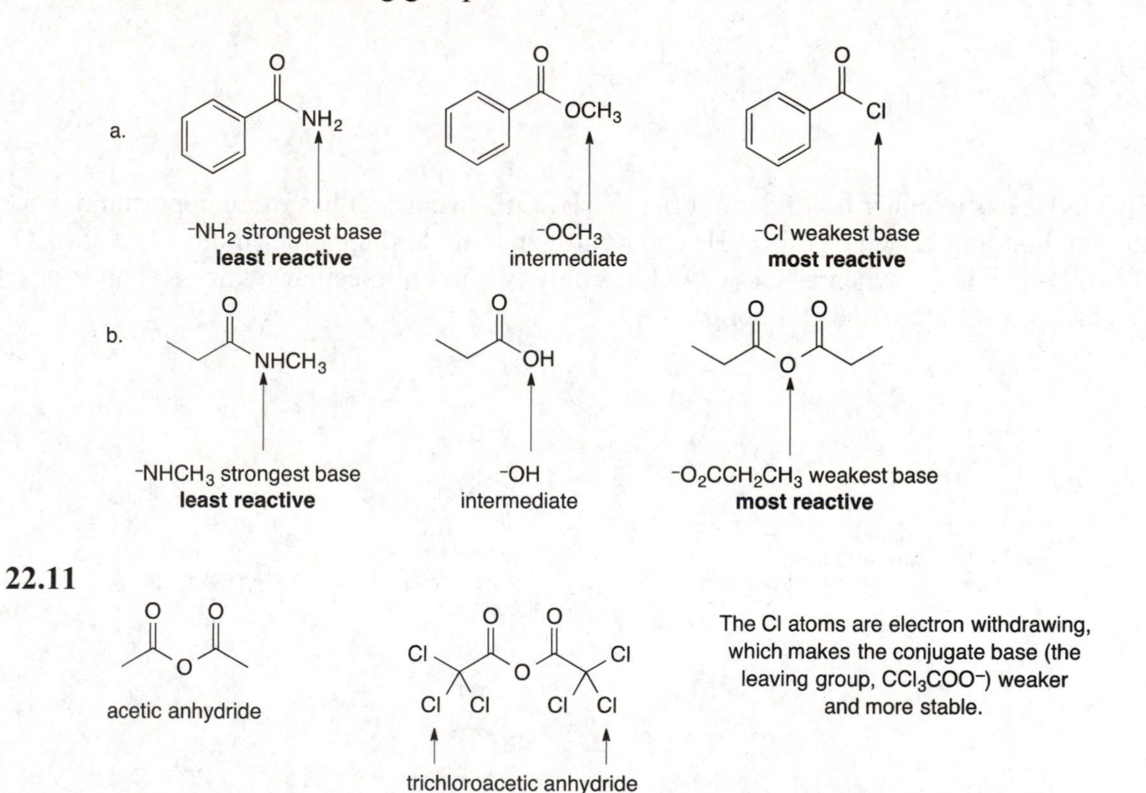

22.12

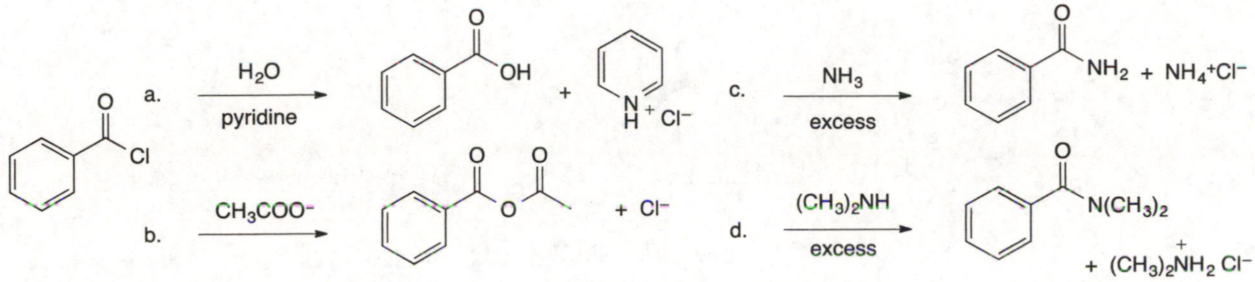

a. $\xrightarrow[\text{pyridine}]{H_2O}$ + Cl^-

b. $\xrightarrow{CH_3COO^-}$ + Cl^-

c. $\xrightarrow[\text{excess}]{NH_3}$ NH_2 + $NH_4^+Cl^-$

d. $\xrightarrow[\text{excess}]{(CH_3)_2NH}$ $N(CH_3)_2$ + $(CH_3)_2\overset{+}{N}H_2\ Cl^-$

22.13 The mechanism has three steps: [1] nucleophilic attack by O; [2] proton transfer; and [3] elimination of the Cl^- leaving group to form the product.

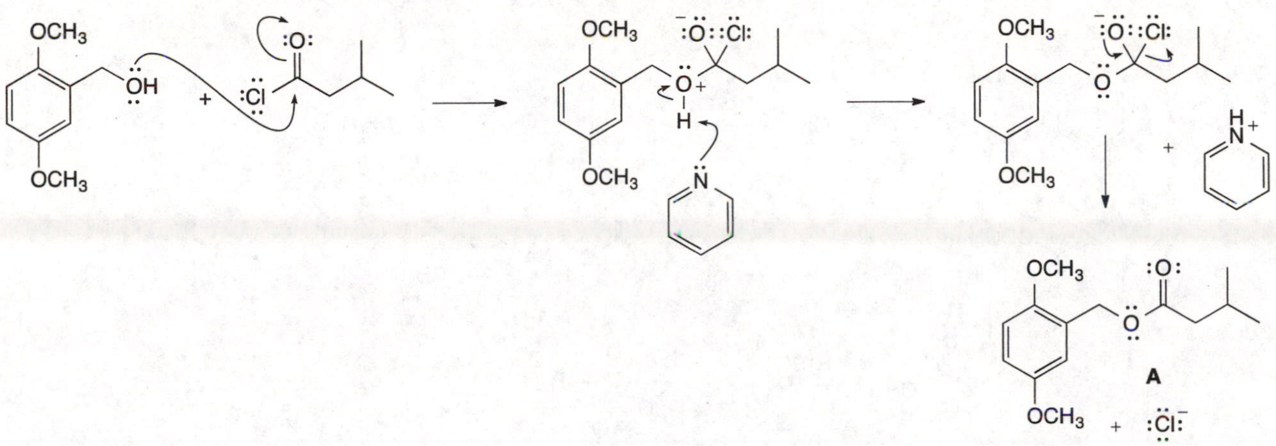

OCH_3 … A + $:\ddot{Cl}:^-$

22.14

a. $\xrightarrow{H_2O}$ OH + HO

b. $\xrightarrow{CH_3OH}$ OCH_3 + HO

c. $\xrightarrow[\text{excess}]{NH_3}$ NH_2 + $\overset{+}{N}H_4$ ^-O

d. $\xrightarrow[\text{excess}]{(CH_3)_2NH}$ $N(CH_3)_2$ + $(CH_3)_2\overset{+}{N}H_2$ ^-O

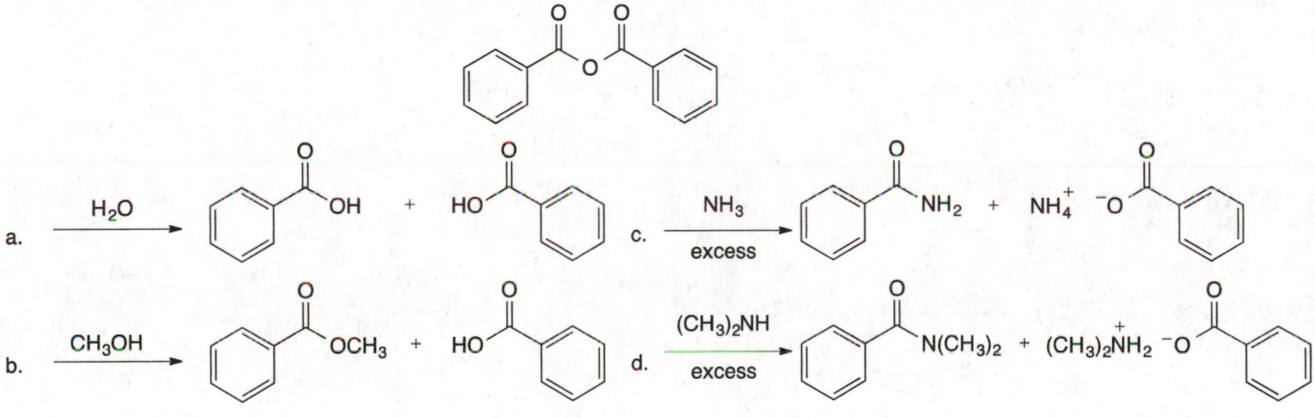

22.15 Reaction of a carboxylic acid with thionyl chloride converts it to an acid chloride.

a. O, OH → $SOCl_2$ → O, Cl

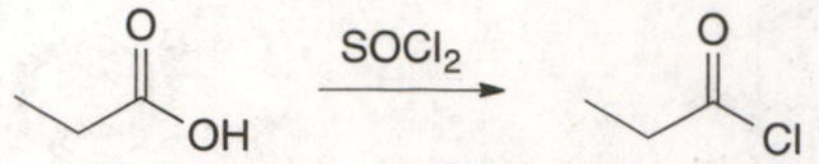

b. O, OH → [1] $SOCl_2$ → O, Cl → [2] $(CH_3CH_2)_2NH$ (excess) → O, N + $\overset{+}{N}H_2$ Cl^-

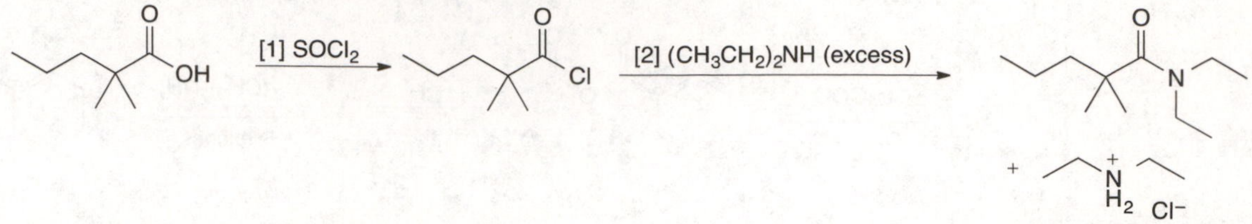

22.16

a. COOH + OH → H_2SO_4 → O, O + H_2O

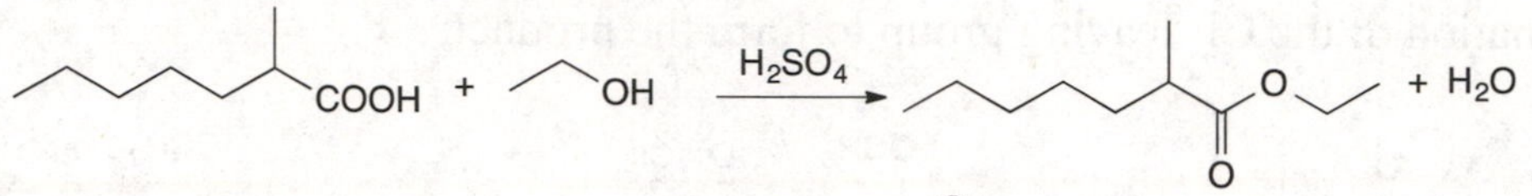

b. COOH + OH → H_2SO_4 → O, O + H_2O

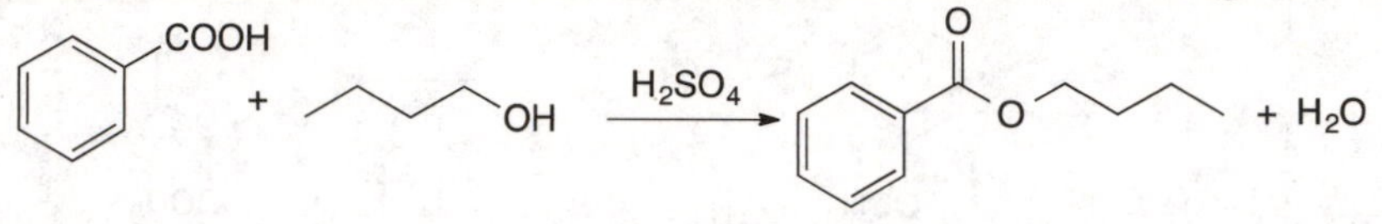

c. COOH + $NaOCH_3$ → O, O^- Na^+ + CH_3OH

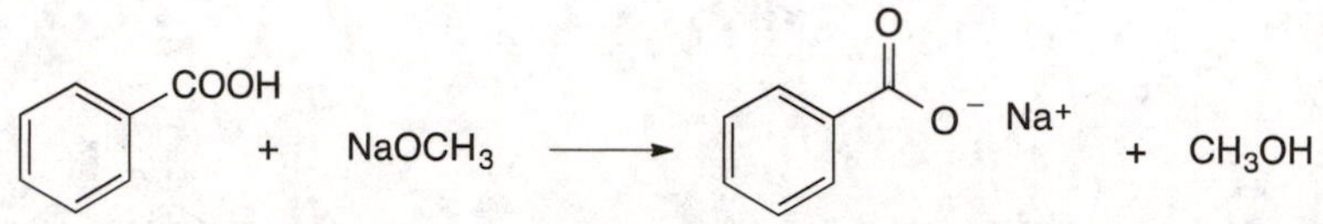

d. HO, O, OH → H_2SO_4 → O, O + H_2O

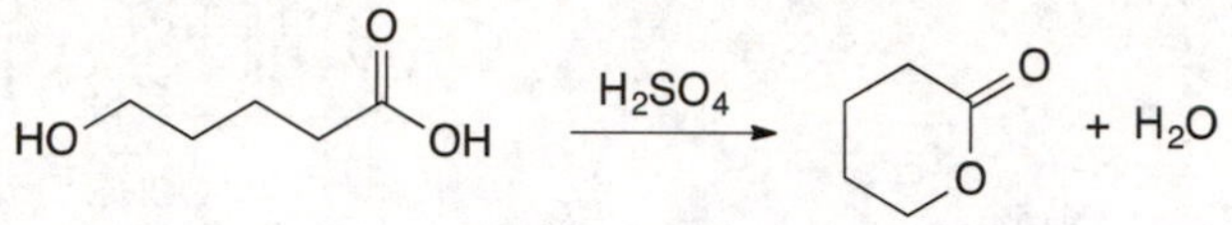

22.17

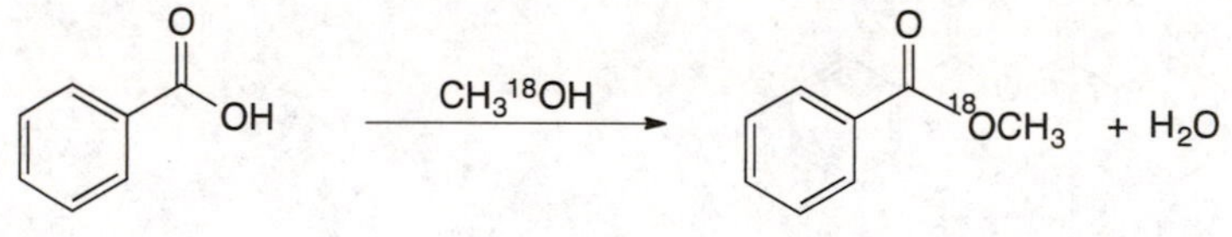

O, OH → $CH_3{}^{18}OH$ → O, $^{18}OCH_3$ + H_2O

22.18

HÖ, :O:, ÖH + H–OSO_3H → HÖ, $\overset{+}{O}H$, ÖH + HSO_4^- → HÖ: :ÖH, $\overset{+}{O}$–H, $^-OSO_3H$ → HÖ: :ÖH, :O: + H–ÖSO_3H → HÖ: :$\overset{+}{O}H_2$, :O: + HSO_4^- → :$\overset{+}{O}$–H, HSO_4^-, :O: → $H_2Ö$ + :O:, Ö: + H_2SO_4

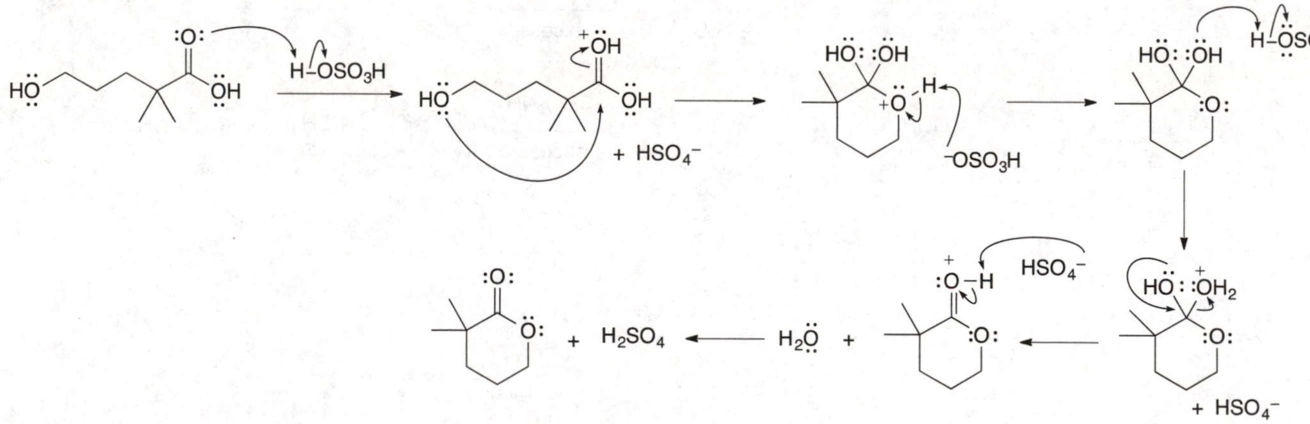

22.19

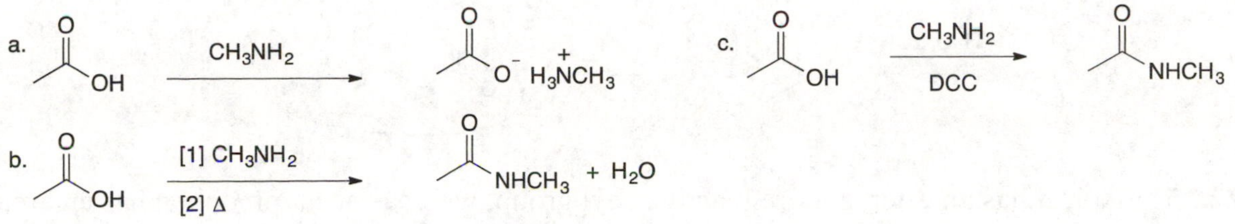

a. CH_3COOH —(CH_3NH_2)→ CH_3COO^- $H_3\overset{+}{N}CH_3$

b. CH_3COOH —([1] CH_3NH_2; [2] Δ)→ $CH_3CONHCH_3$ + H_2O

c. CH_3COOH —(CH_3NH_2, DCC)→ $CH_3CONHCH_3$

22.20

fenofibrate —hydrolysis→ fenofibric acid + HO (isopropyl alcohol)

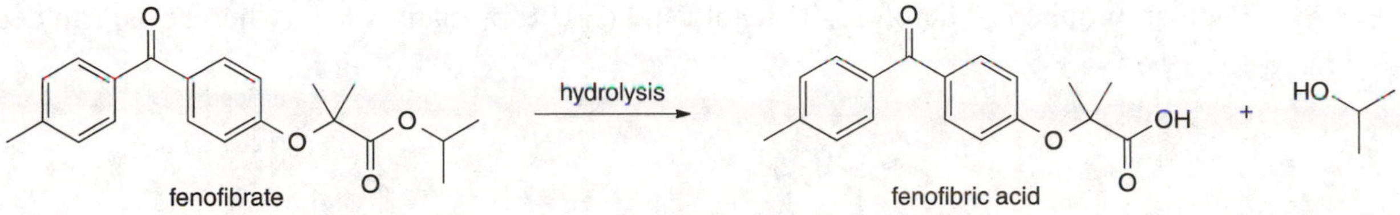

22.21 The three bonds broken during hydrolysis are indicated.

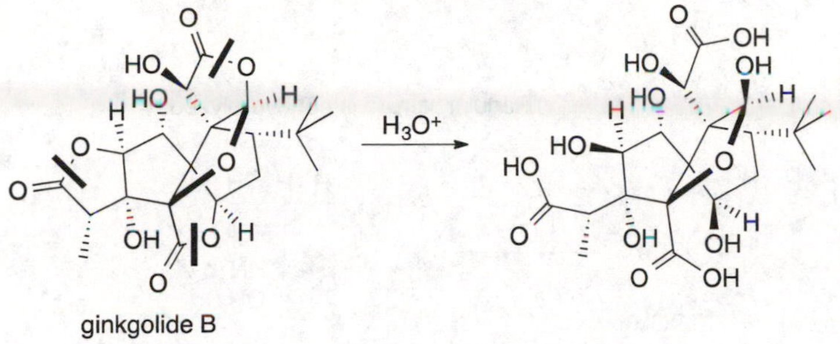

ginkgolide B —H_3O^+→

22.22

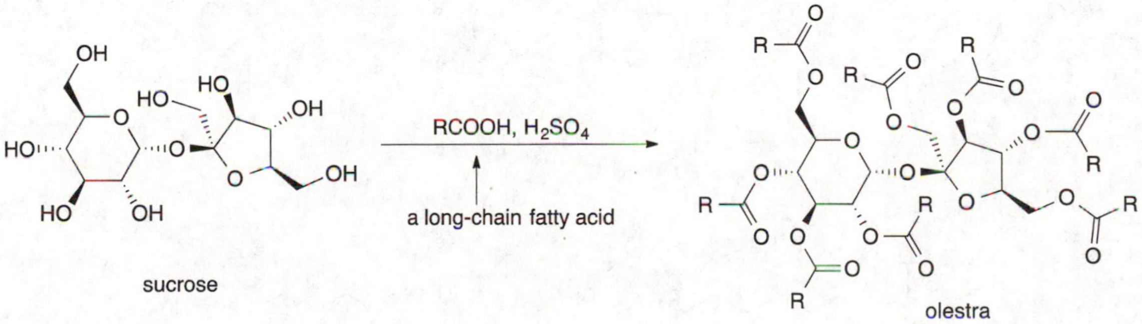

sucrose —RCOOH, H_2SO_4→ olestra

a long-chain fatty acid

22.23

—hydrolysis→ glycerol + soap

cis

Na$^+$ $^-$O

cis

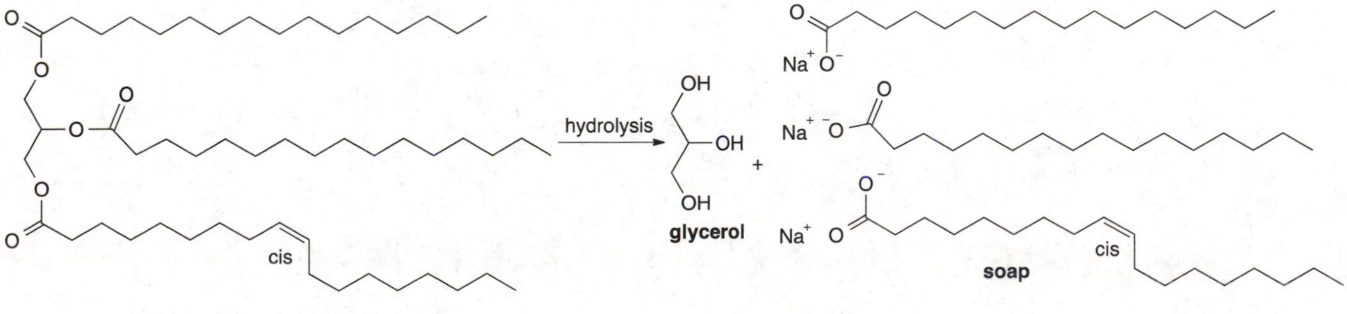

22.24

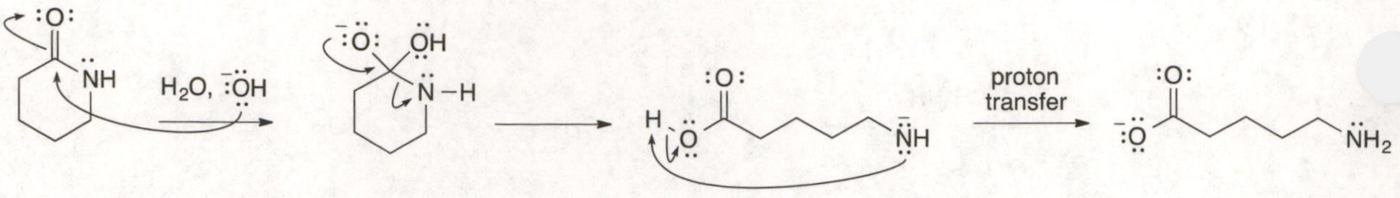

H_2O, $^{-}$:OH

proton transfer

22.25 Aspirin has an ester, a more reactive acyl group, whereas acetaminophen has an amide, a less reactive acyl group. The ester makes aspirin more easily hydrolyzed with water from the air than acetaminophen. Therefore, Tylenol can be kept for many years, whereas aspirin decomposes.

ester

acetylsalicylic acid

amide

acetaminophen

22.26

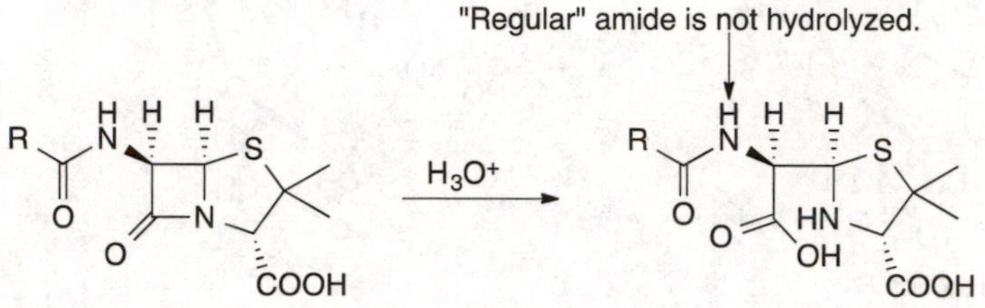

"Regular" amide is not hydrolyzed.

H_3O^+

22.27

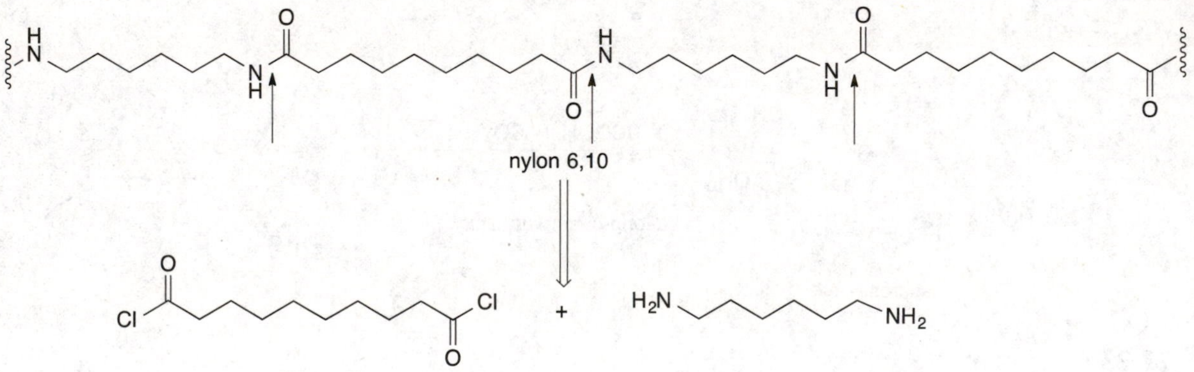

nylon 6,10

22.28

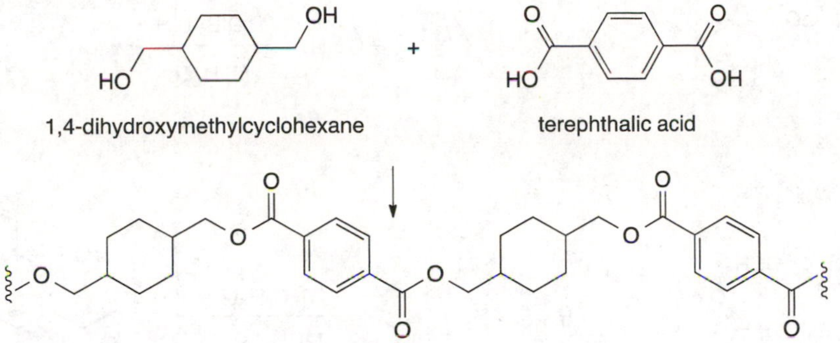

HO — 1,4-dihydroxymethylcyclohexane — OH + HO — terephthalic acid — OH

Kodel

In the polyester Kodel, most of the bonds in the polymer backbone are part of a ring, so there are fewer degrees of freedom. Fabrics made from Kodel are stiff and crease resistant, due to these less flexible polyester fibers.

22.29

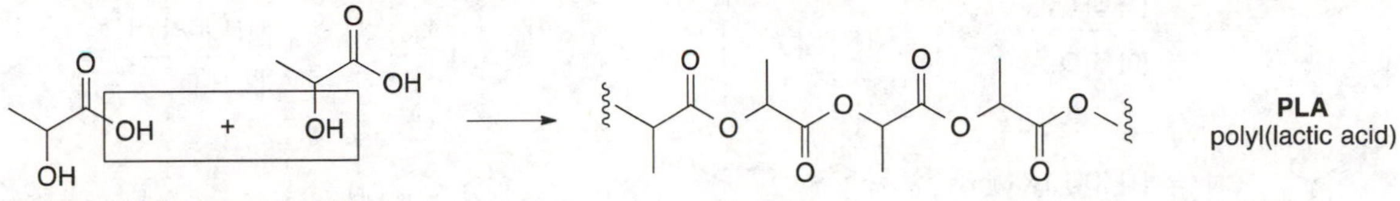

Reaction occurs here ($-H_2O$).

PLA
polyl(lactic acid)

22.30 Acetyl CoA acetylates the NH_2 group of glucosamine, because the NH_2 group is the most nucleophilic site.

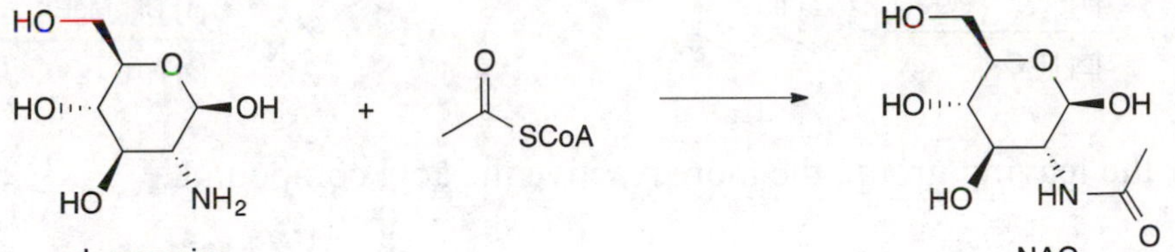

glucosamine + SCoA → NAG

22.31

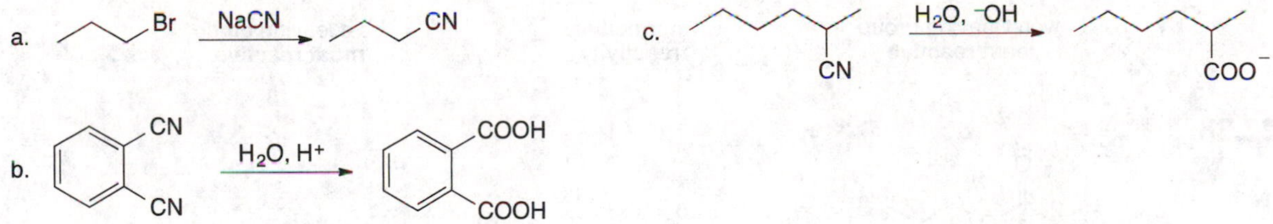

a. Br —NaCN→ CN

b. CN, CN —H_2O, H^+→ COOH, COOH

c. CN —H_2O, ^-OH→ COO^-

22.32

a. H_2N ⇌ HN, OH

b. OH O NHCH$_3$ ⇌ OH OH NCH$_3$

c. NH OH ⇌ NH$_2$ O

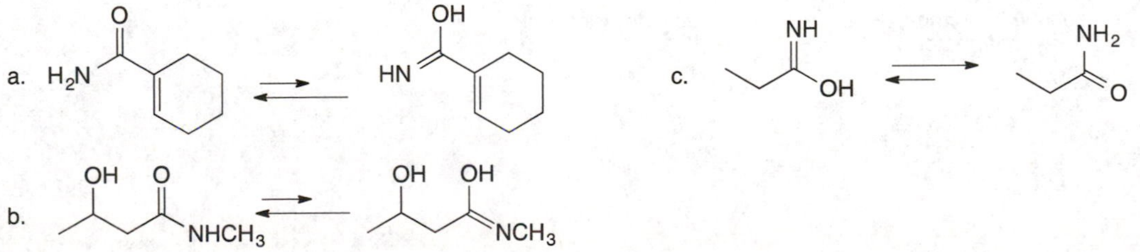

22.33

a. O Br [1] NaCN [2] $LiAlH_4$ [3] H_2O → O NH_2

b. $-C{\equiv}N$ [1] DIBAL-H [2] H_2O → O H

22.34

a. OCH_3 CN [1] MgCl [2] H_2O → OCH_3 O

b. CN [1] C_6H_5Li [2] H_2O → O

22.35

a. CN [1] CH_3MgBr [2] H_2O → O

b. CN [1] $(CH_3)_3CMgBr$ [2] H_2O → O

c. CN [1] DIBAL-H [2] H_2O → O H

d. CN H_3O^+ → O OH

22.36

$CH_3-C{\equiv}N$ [1] MgBr [2] H_2O → O

N [1] CH_3MgBr [2] H_2O → O

22.37 The better the leaving group, the more reactive the acyl compound.

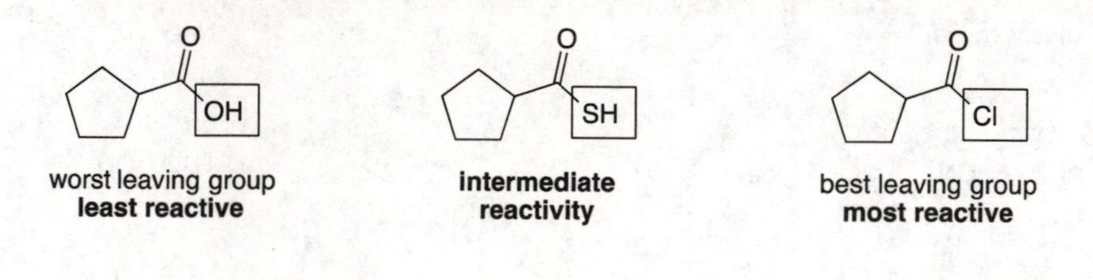

22.38

a. O 3 2 1 O

1 CN 6 5 4 3 2

A
isobutyl 2,2-dimethylpropanoate

B
2-ethylhexanenitrile

b. A [1] H_3O^+ → (CH3)3C–COOH + HO–CH2CH(CH3)2

A [2] ^-OH, H_2O → (CH3)3C–COO^- + HO–CH2CH(CH3)2

A [3] CH3CH2CH2MgBr → H_2O → (CH3)3C–C(OH)(CH2CH2CH3)2 + HO–CH2CH(CH3)2

A [4] $LiAlH_4$ → H_2O → (CH3)3C–CH2OH + HO–CH2CH(CH3)2

B [1] H_3O^+ → COOH (2-ethylhexanoic acid)

B [2] ^-OH, H_2O → COO^-

B [3] CH3CH2CH2MgBr → H_2O → ketone (O, propyl)

B [4] $LiAlH_4$ → H_2O → NH_2

22.39 Better leaving groups make acyl compounds more reactive. **C** has an electron-withdrawing NO_2 group, which stabilizes the negative charge of the leaving group, whereas **D** has an electron-donating OCH_3 group, which destabilizes the leaving group.

C: NO_2 — an electron-withdrawing substituent

D: OCH_3 — an electron-donating substituent

leaving group from **C** ⟷ one possible resonance structure

Delocalizing the negative charge on the NO_2 stabilizes the leaving group, making **C** more reactive than **D**.

leaving group from **D** ⟷ one possible resonance structure

Adjacent negative charges destabilize the leaving group.

22.40

a. **cyclohexanecarboxylic anhydride**

b. **phenyl phenylacetate**

c. ***N*-cyclohexylbenzamide**

d. ***m*-chlorobenzonitrile**

e. ***cis*-2-bromocyclohexane-carbonyl chloride**

f. ***N,N*-diethylcyclohexanecarboxamide**

22.41

a. cyclohexyl propanoate

b. cyclohexanecarboxamide

c. 4-methylheptanenitrile

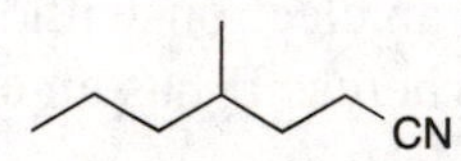

d. vinyl acetate

e. benzoic propanoic anhydride

f. 3-methylhexanoyl chloride

g. octyl butanoate

h. *N,N*-dibenzylformamide

22.42

imidazolide

resonance structures for the leaving group

The leaving group is both resonance stabilized and aromatic (6 π electrons), making it a much better leaving group than exists in a regular amide.

22.43

Reaction as an acid:

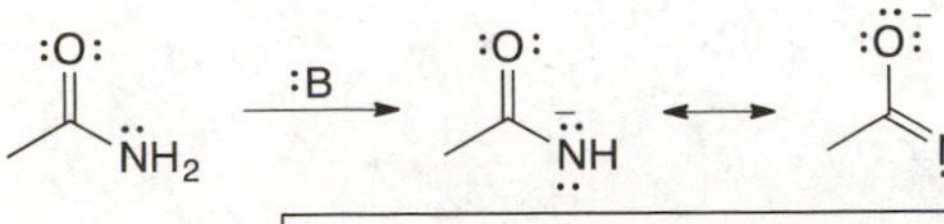

:B

These two resonance structures make the conjugate base more stable, and therefore CH_3CONH_2 a stronger acid.

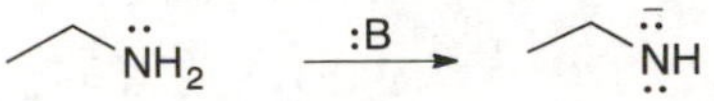

:B

no resonance stabilization of the conjugate base

Reaction as a base:

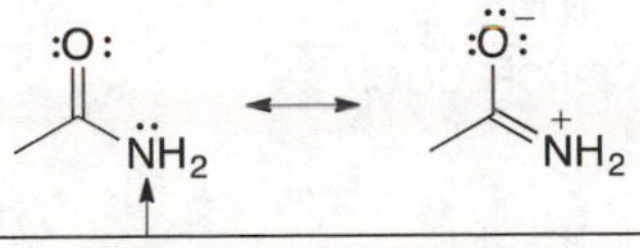

This electron pair is delocalized by resonance, making it less available for electron donation. Thus, CH_3CONH_2 is a much weaker base.

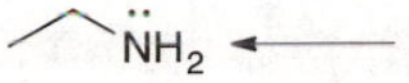

This electron pair is localized on N.

22.44 a. The electron pair on the O atom in phenyl acetate is delocalized on both the carbonyl group and the benzene ring. Thus, the C=O has less single bond character than the C=O of cyclohexyl acetate, so the absorption occurs at higher wavenumbers.

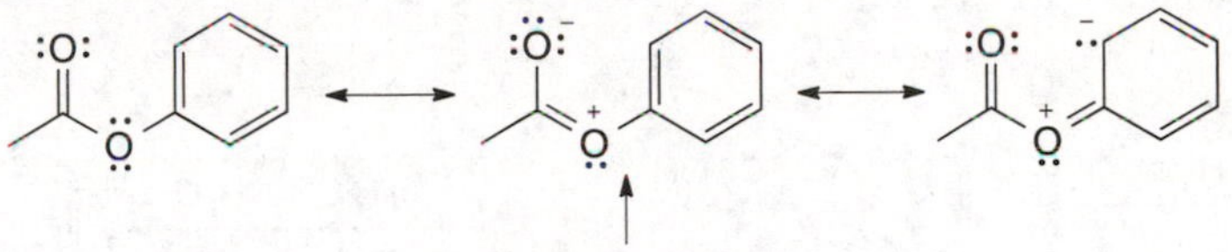

+ other forms with a C=O

This form contributes less to the hybrid.

b. Because the lone pair on the O atom in cyclohexyl acetate can only be delocalized on the C=O, the C=O of cyclohexyl acetate is more effectively stabilized by resonance.

only two resonance forms

c. Phenyl acetate has a better leaving group and is more reactive.

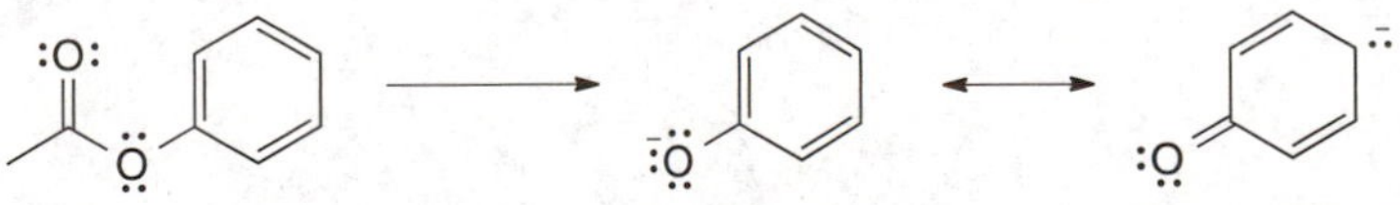

weaker base
better leaving group

+ 3 more resonance structures

stronger base

22.45

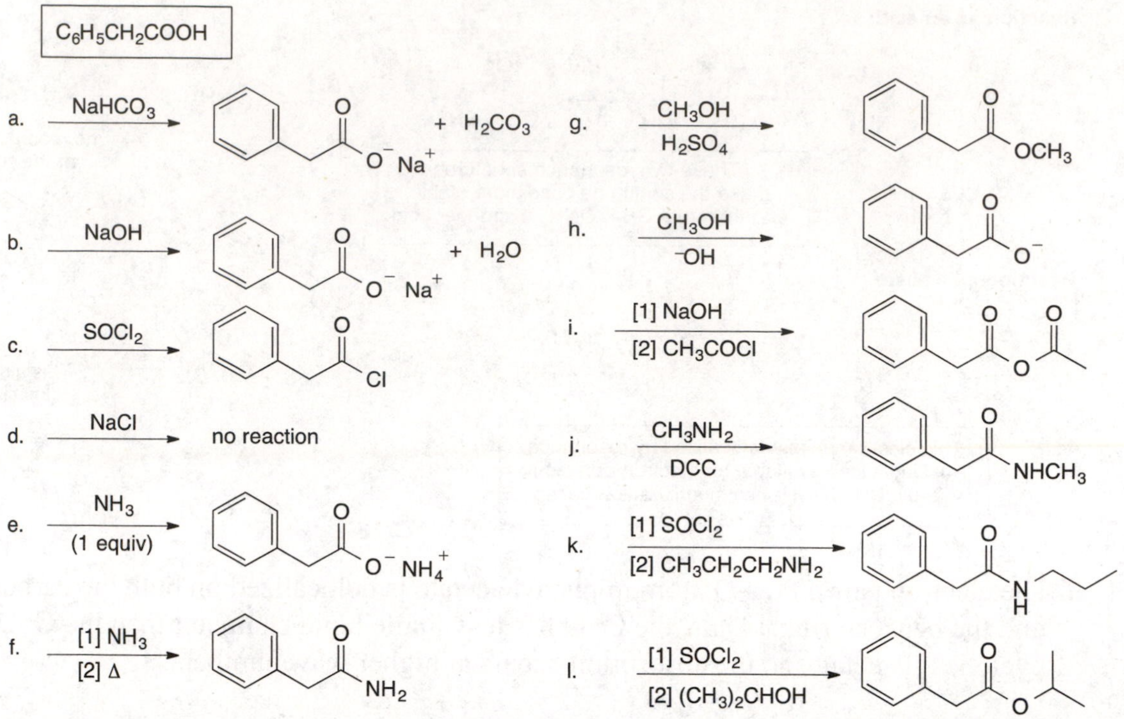

$C_6H_5CH_2COOH$

a. $NaHCO_3$ → O, O^- Na^+ + H_2CO_3

b. NaOH → O, O^- Na^+ + H_2O

c. $SOCl_2$ → O, Cl

d. NaCl → no reaction

e. NH_3 (1 equiv) → O, O^- NH_4^+

f. [1] NH_3 [2] Δ → O, NH_2

g. CH_3OH, H_2SO_4 → O, OCH_3

h. CH_3OH, ^-OH → O, O^-

i. [1] NaOH [2] CH_3COCl → O, O, O

j. CH_3NH_2, DCC → O, $NHCH_3$

k. [1] $SOCl_2$ [2] $CH_3CH_2CH_2NH_2$ → O, N, H

l. [1] $SOCl_2$ [2] $(CH_3)_2CHOH$ → O, O

22.46

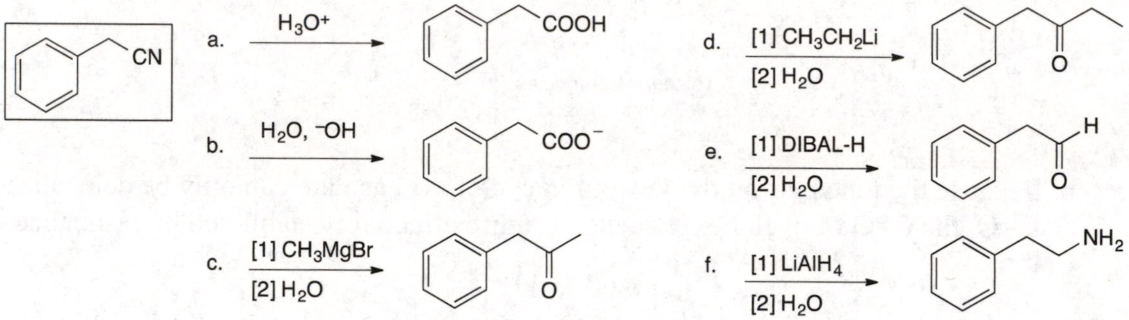

CN

a. H_3O^+ → COOH

b. H_2O, ^-OH → COO^-

c. [1] CH_3MgBr [2] H_2O → O

d. [1] CH_3CH_2Li [2] H_2O → O

e. [1] DIBAL-H [2] H_2O → H, O

f. [1] $LiAlH_4$ [2] H_2O → NH_2

22.47

a. 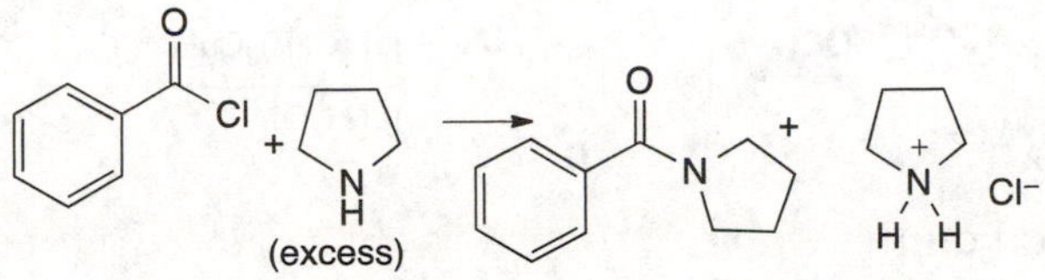 + (excess) → + Cl^-

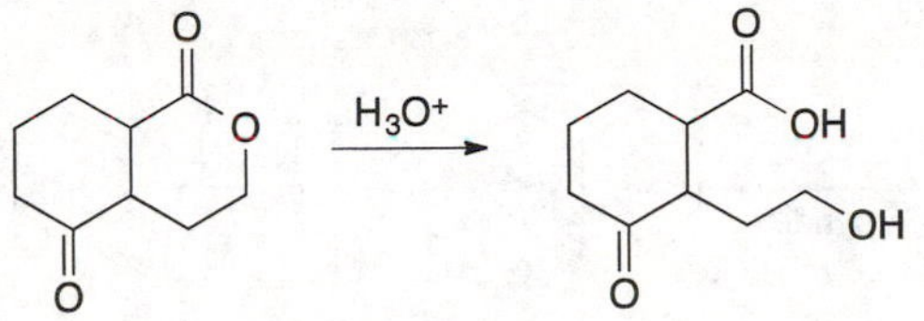

e. 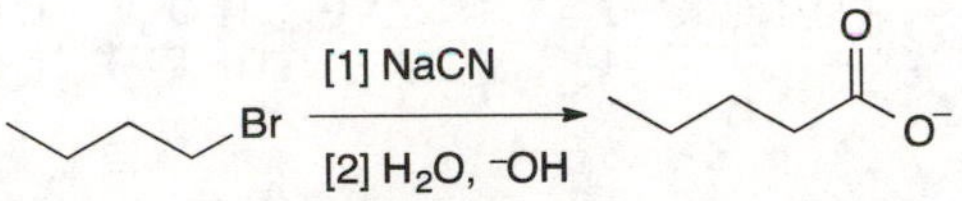 Br [1] NaCN [2] H_2O, ^-OH →

b. CN [1] MgBr [2] H_2O →

f. OH [1] $SOCl_2$ [2] $CH_3CH_2CH_2CH_2NH_2$ [3] $LiAlH_4$ [4] H_2O → N H

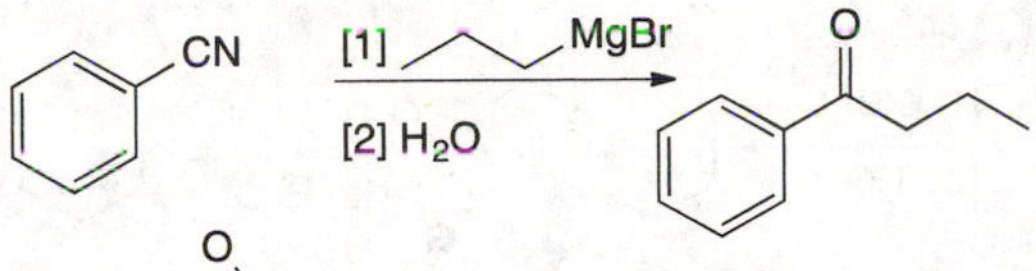

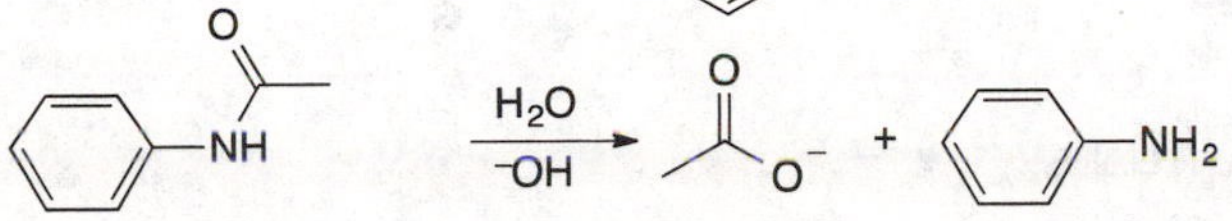

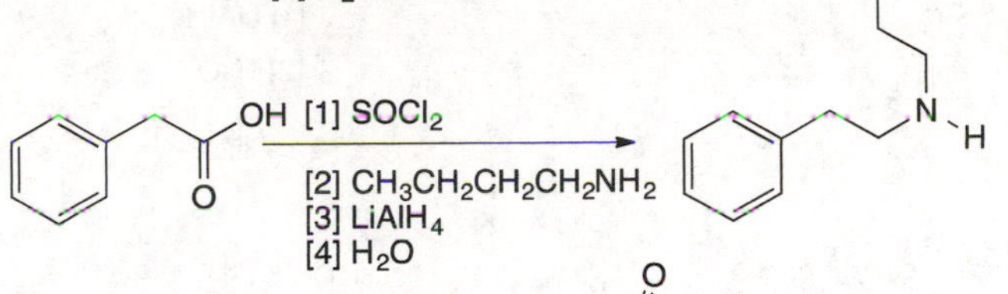

c. NH H_2O ^-OH → + NH_2

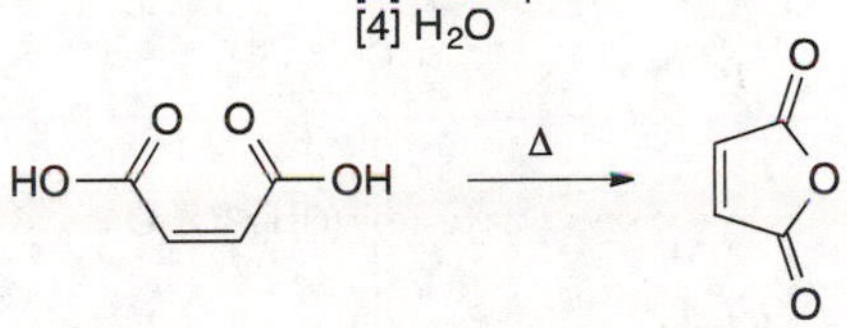

g. HO OH Δ →

d. H_3O^+ → OH OH

h. + NH_2 (excess) → NH

+ CH_3COO^- $H_3\overset{+}{N}$

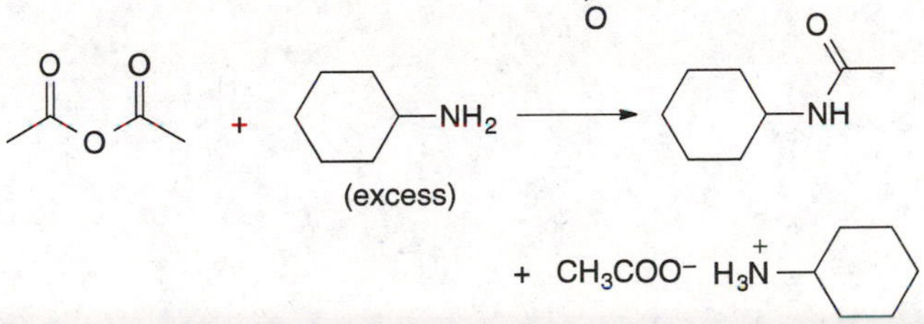

22.48

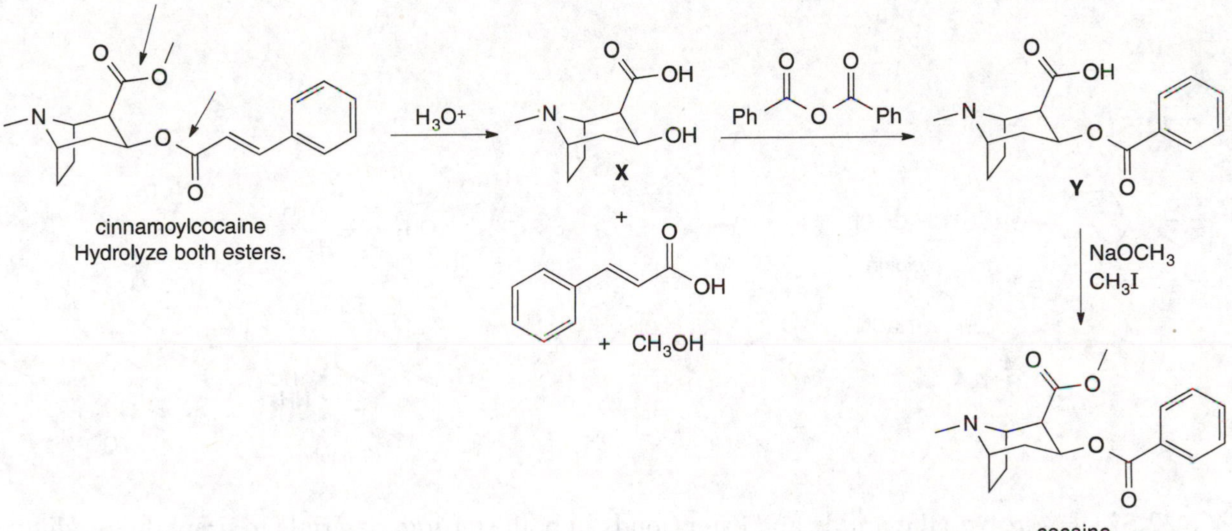

cinnamoylcocaine
Hydrolyze both esters.

H_3O^+ → X + OH + CH_3OH

Ph O Ph → Y

NaOCH₃, CH_3I →

cocaine

22.49

a. 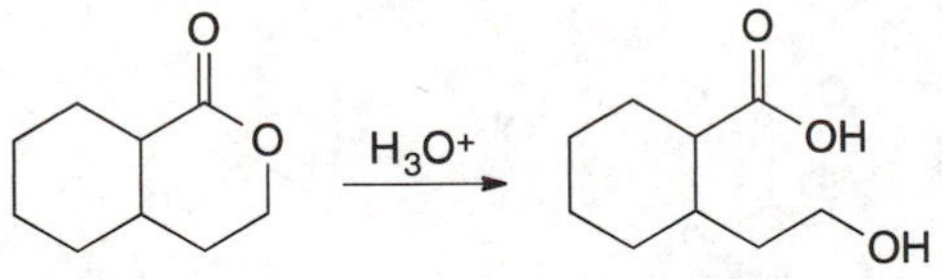 H_3O^+ → OH OH

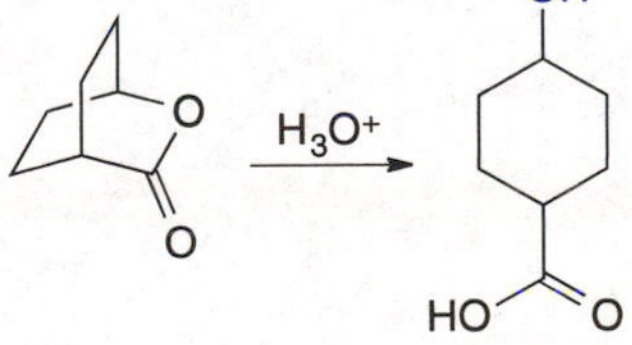

c. 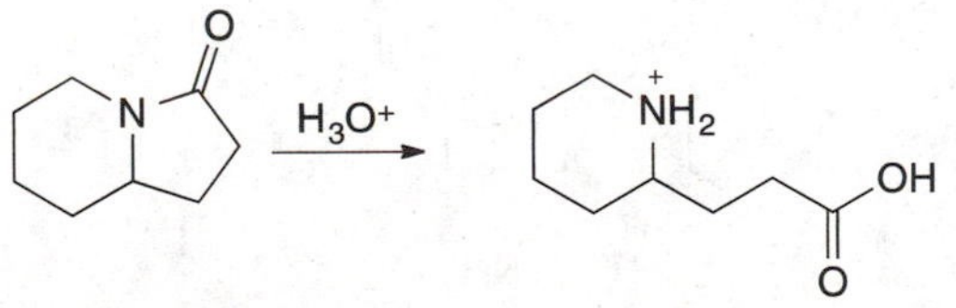 N H_3O^+ → $\overset{+}{N}H_2$ OH

b. H_3O^+ → OH HO

d. 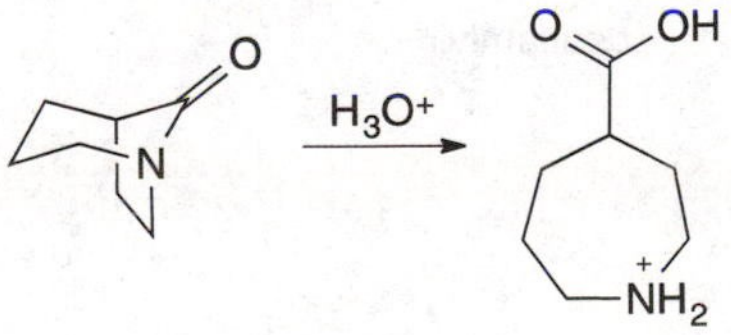 N H_3O^+ → OH $\overset{+}{N}H_2$

22.50

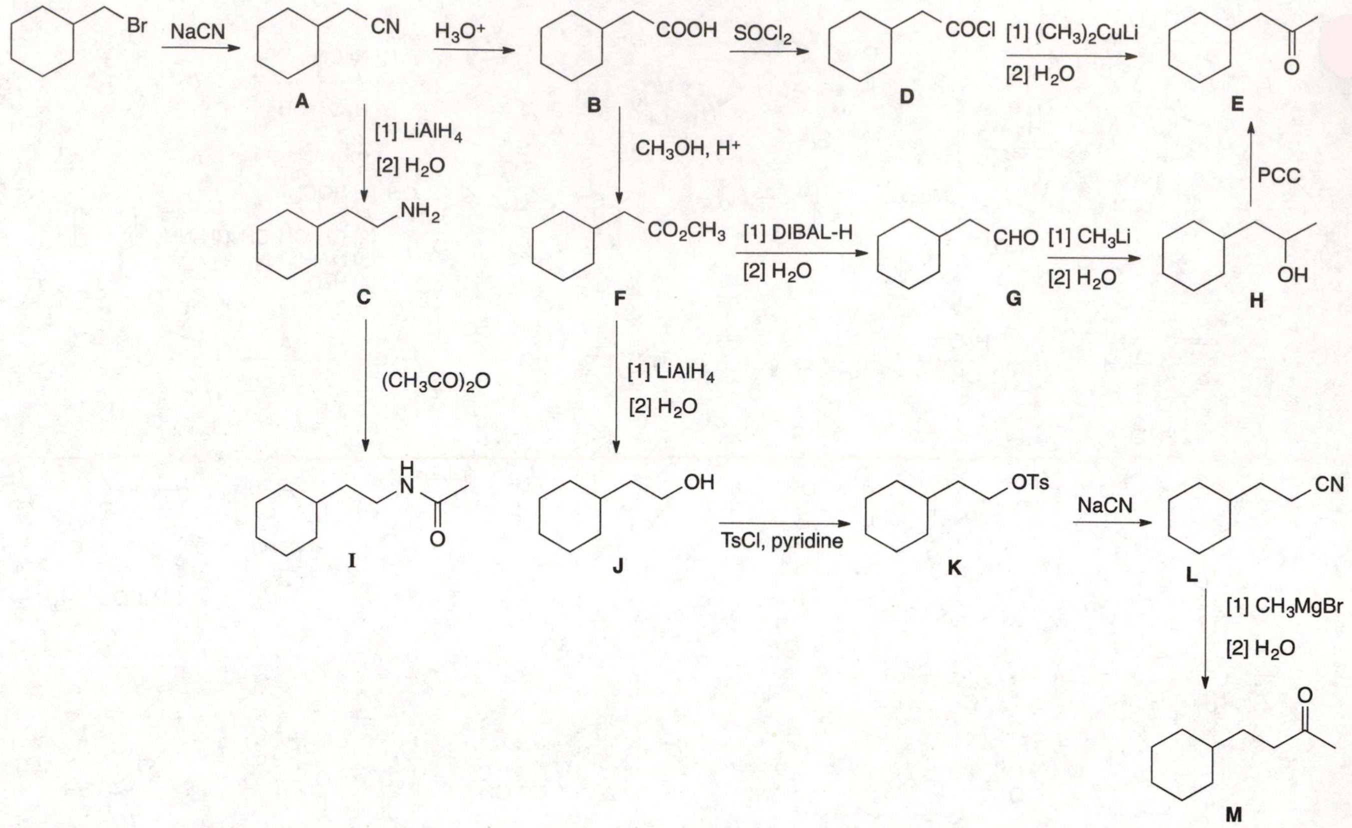

Br NaCN → A (CN) H_3O^+ → B (COOH) $SOCl_2$ → D (COCl) [1] $(CH_3)_2CuLi$ [2] H_2O → E (O)

A → [1] $LiAlH_4$ [2] H_2O → C (NH_2)

B → CH_3OH, H^+ → F (CO_2CH_3) [1] DIBAL-H [2] H_2O → G (CHO) [1] CH_3Li [2] H_2O → H (OH) PCC → E

C → $(CH_3CO)_2O$ → I (H N, O)

F → [1] $LiAlH_4$ [2] H_2O → J (OH) TsCl, pyridine → K (OTs) NaCN → L (CN) [1] CH_3MgBr [2] H_2O → M (O)

22.51

a. OH CH_3COCl, pyridine → O, O

b. Br, H D NaCN → CN, D H

c. H D, OH, O; OH, H^+ → H D, O, O

d. O, Cl + NH_2, H (2 equiv) → NH, H, O + NH_3, H, +, Cl^-

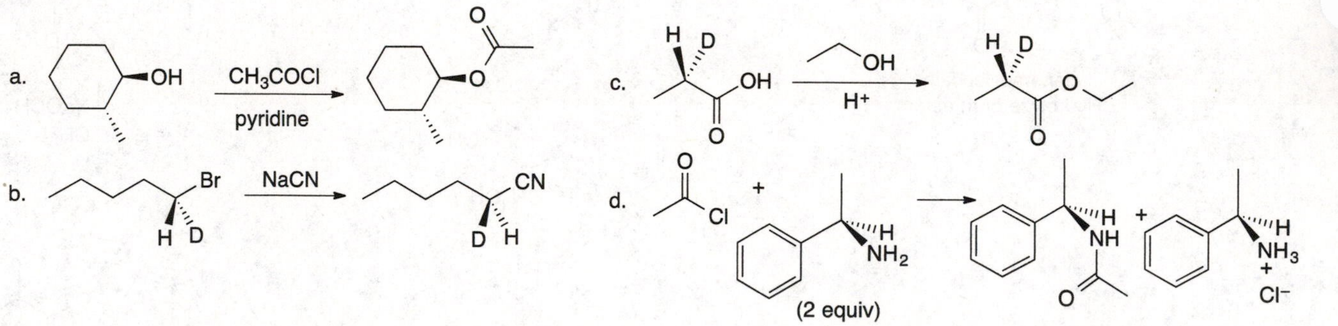

22.52 Hydrolyze the amide and ester bonds in both starting materials to draw the products.

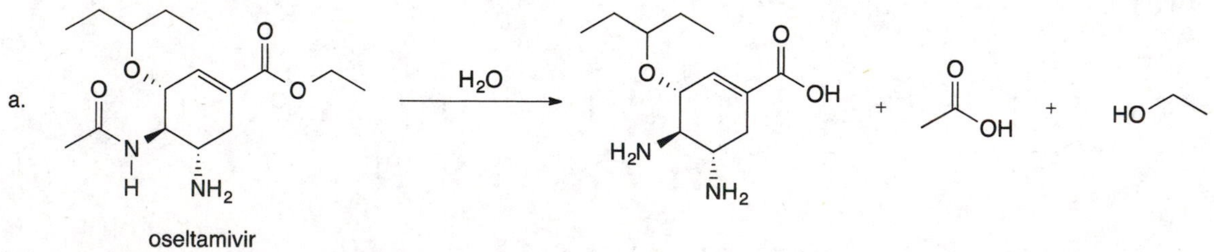

a. O, O, O, N, H, NH_2 (oseltamivir) H_2O → O, O, OH, H_2N, NH_2 + O, OH + HO

b. H_2N ... aspartame $\xrightarrow{H_2O}$ HO ... NH_2 ... OH + H_2N ... OH phenylalanine + CH_3OH

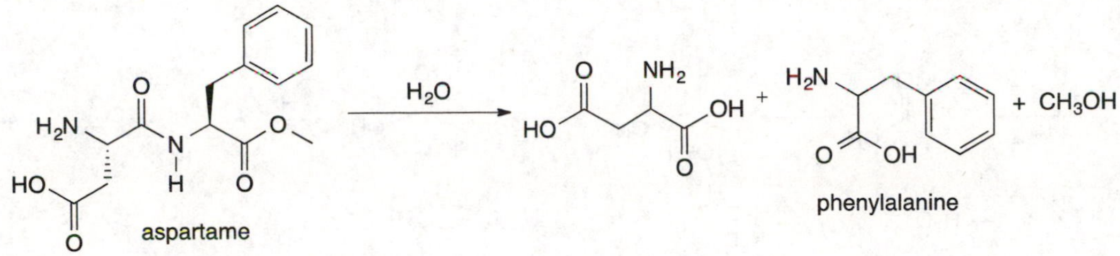

aspartame

phenylalanine

22.53

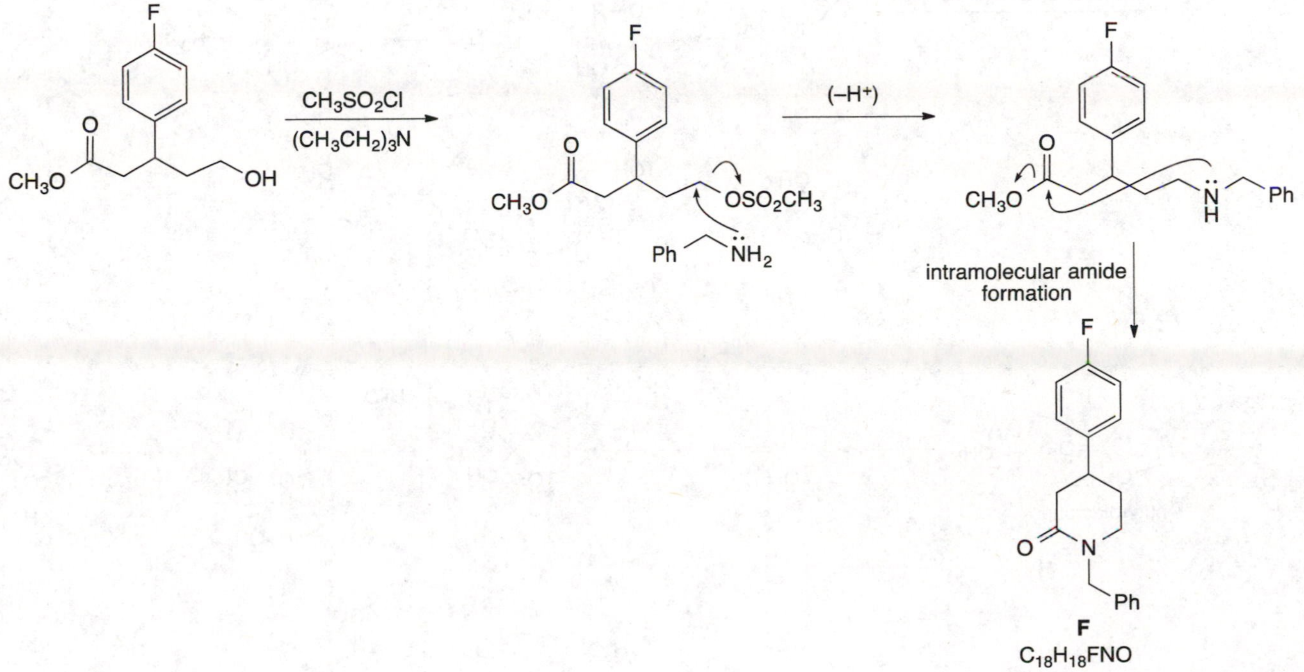

CH_3O ... OH (F) $\xrightarrow[(CH_3CH_2)_3N]{CH_3SO_2Cl}$ CH_3O ... OSO_2CH_3 (F) + Ph $\ddot{N}H_2$ $\xrightarrow{(-H^+)}$ CH_3O ... $\ddot{N}$(H) Ph (F)

intramolecular amide formation

F

$C_{18}H_{18}FNO$

22.54

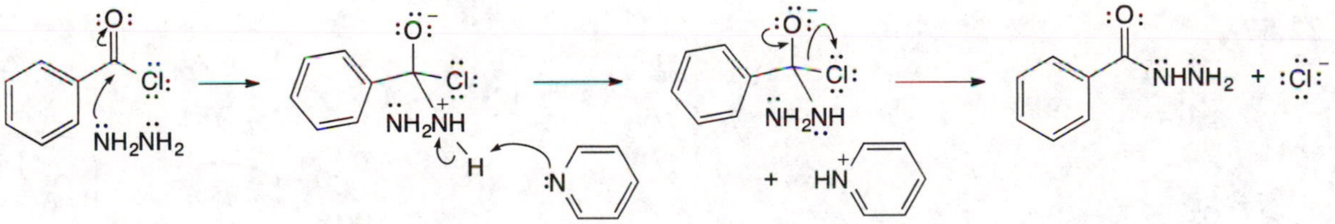

:O: ... Cl: + $\ddot{N}H_2\ddot{N}H_2$ → :O:⁻ ... Cl: $\ddot{N}H_2\overset{+}{N}H$ ... H + :N(pyridine) → :O:⁻ ... Cl: $\ddot{N}H_2\ddot{N}H$ + $H\overset{+}{N}$(pyridine) → :O: ... $\ddot{N}H\ddot{N}H_2$ + :Cl:⁻

22.55

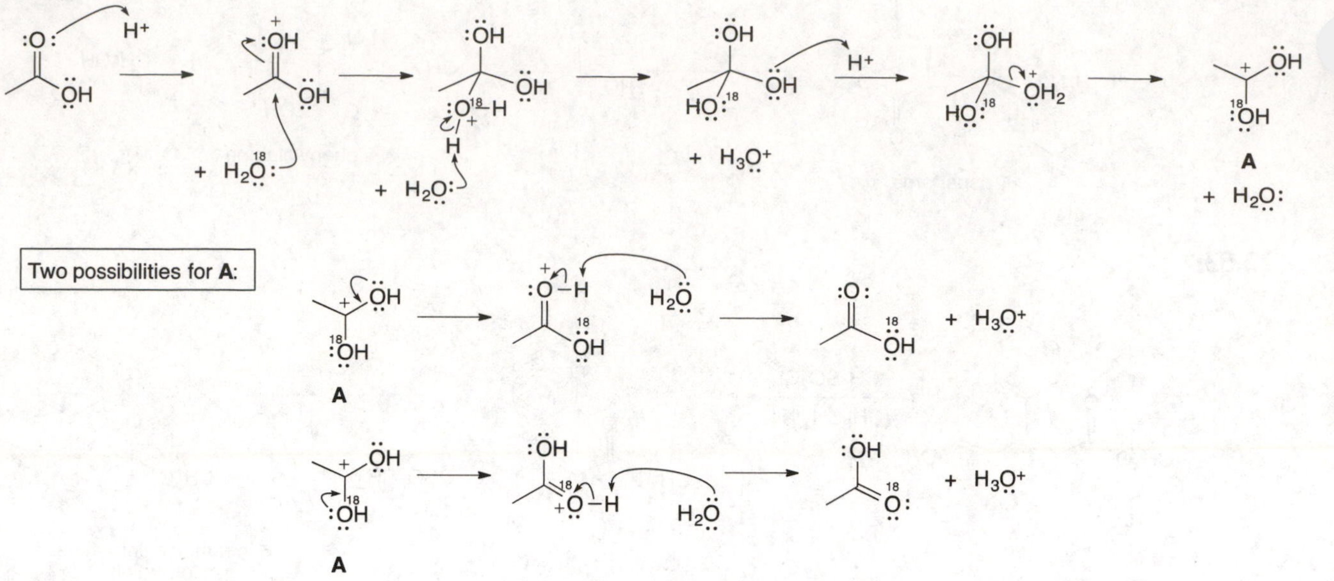

+ $H_2{}^{18}O$ + H_2O + H_3O^+ A + H_2O

Two possibilities for **A**:

A + H_3O^+

A + H_3O^+

22.56

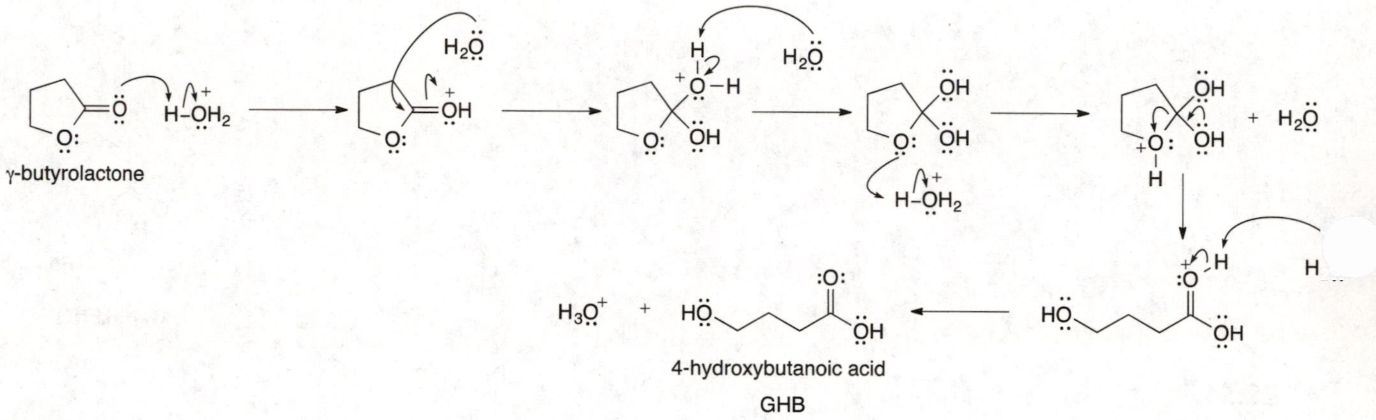

γ-butyrolactone

H_3O^+ + 4-hydroxybutanoic acid GHB

22.57

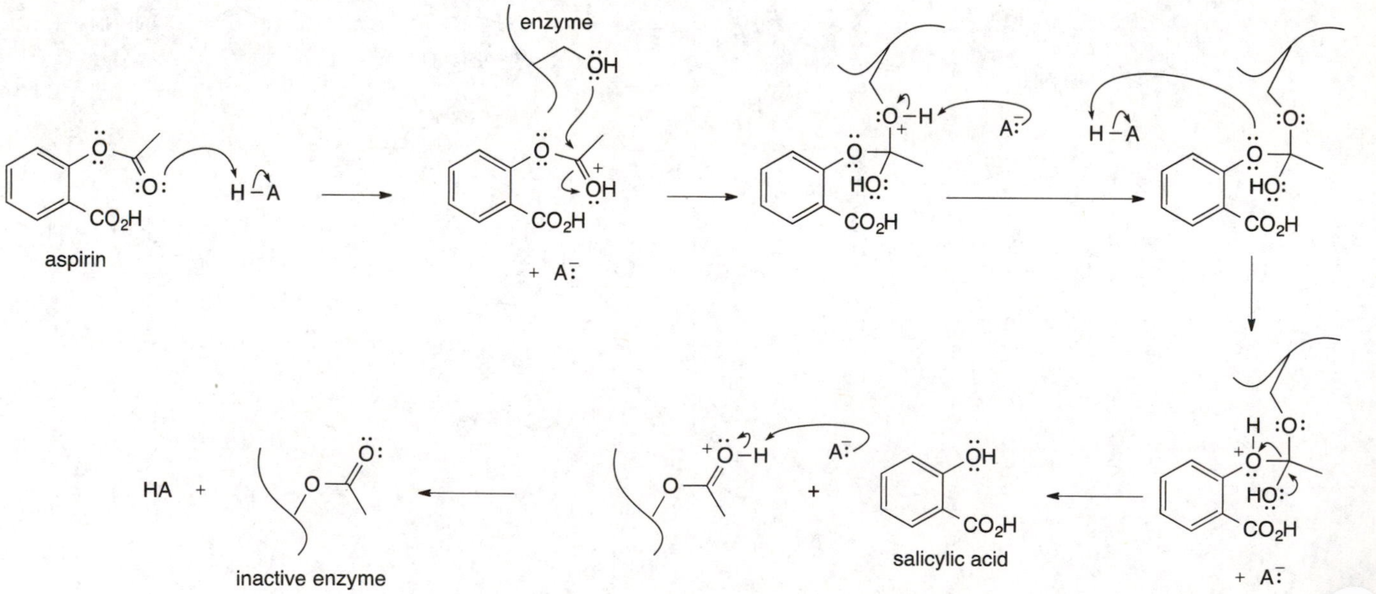

aspirin

enzyme

+ A:⁻

salicylic acid

HA + inactive enzyme

22.58

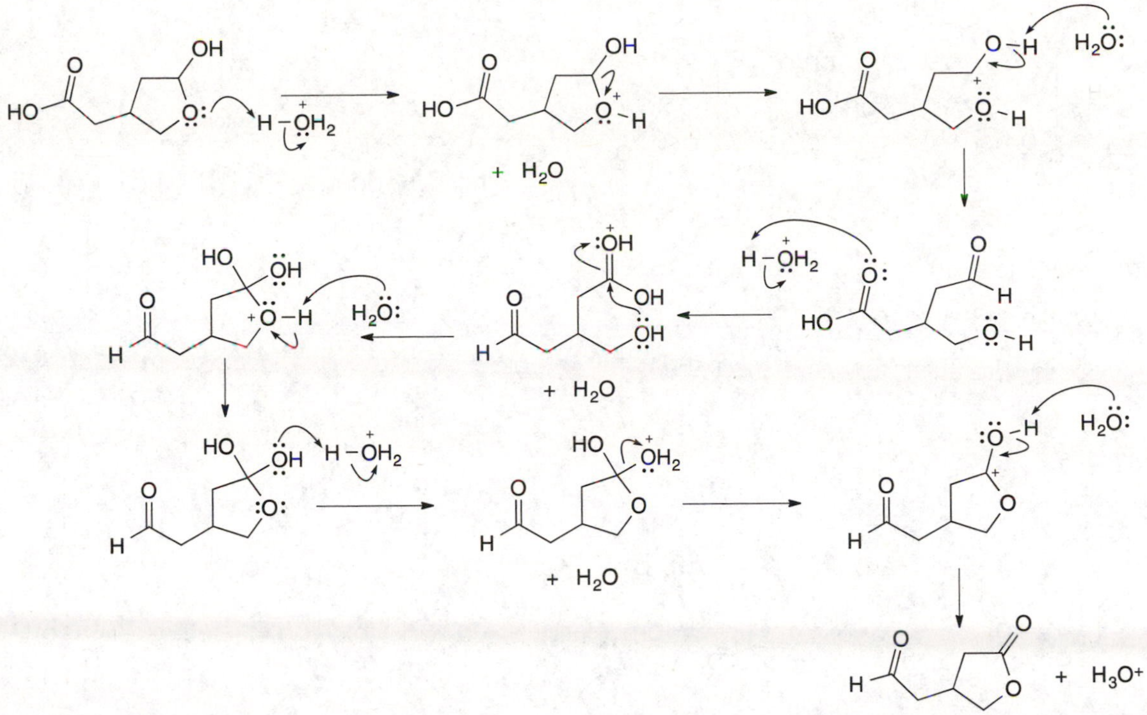

O OH HO O: H–OH2 +
O OH HO O H + + H2O
O O–H H2Ö: HO + Ö–H
HO OH O +O–H H2Ö: H
:OH + O OH ÖH H + H2O
H–OH2 + Ö: O H HO Ö–H
HO ÖH H–OH2 + O :O: H
HO OH2 + O O H + H2O
:Ö–H H2Ö: O + O H
O O H O + H3O+

22.59

O Ar O F Ar' RO N H CH3CH2–MgBr
O Ar RO N Ar' + MgBr+ + CH3CH3
Ar –:Ö N: Ar' OR
O O N F + RO–

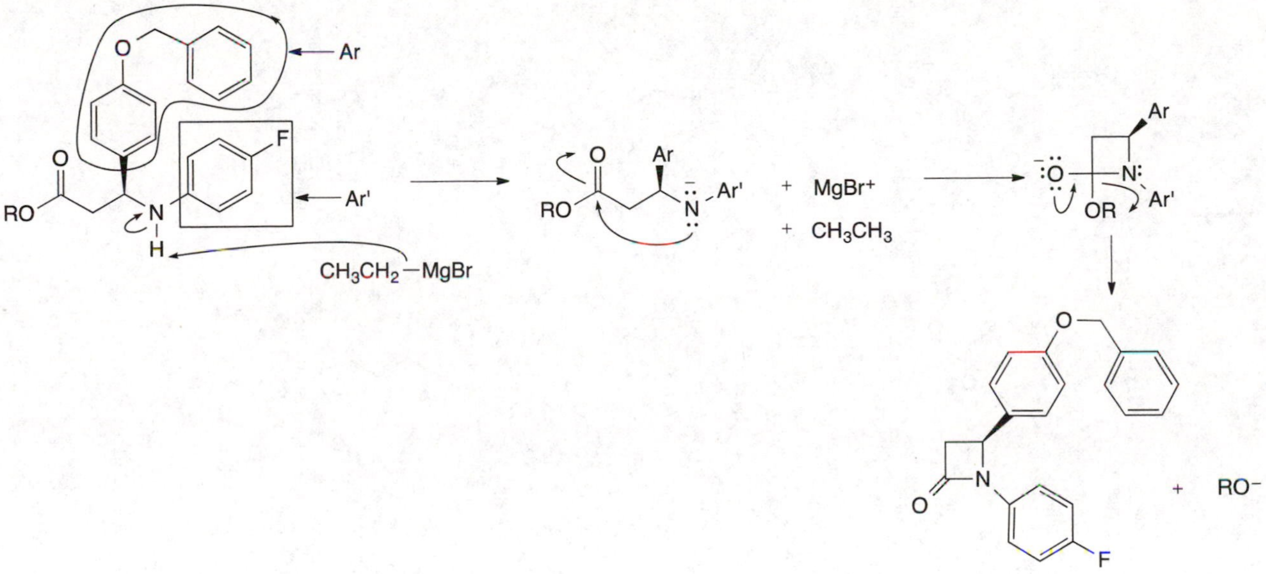

22.60

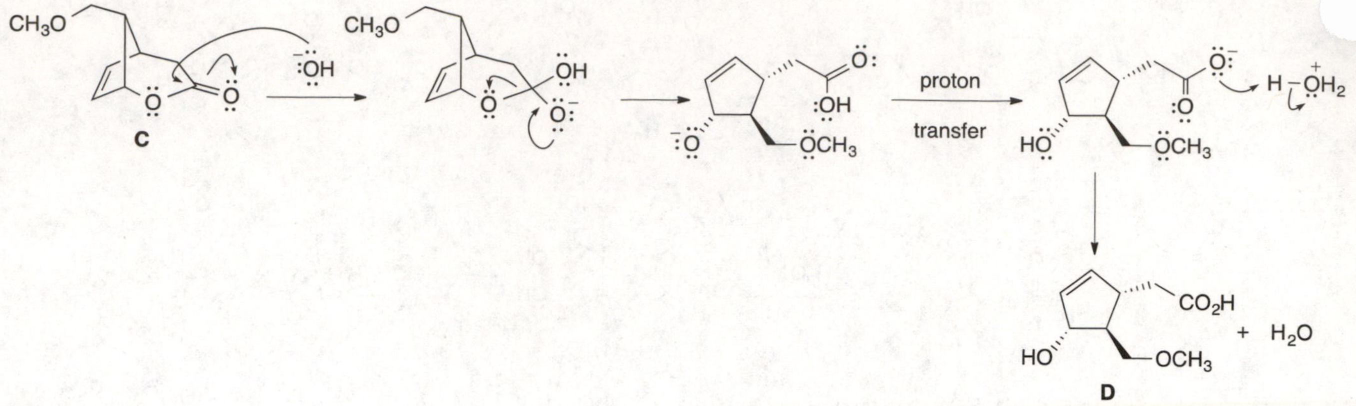

CH_3O C :ÖH CH_3O ÖH Ö: ÖH $ÖCH_3$ proton transfer HO $ÖCH_3$ $H-ÖH_2$

HO CO_2H OCH_3 + H_2O D

22.61

A H–Cl H–Ö + :Cl:⁻ :Cl:⁻ :ÖH H–Cl

:ÖH $:ÖH_2$ + :Cl:⁻ :ÖH H–Cl :ÖH :Cl:⁻ :ÖH + :Cl:⁻

:Cl:⁻ + H_2O Cl = Cl B

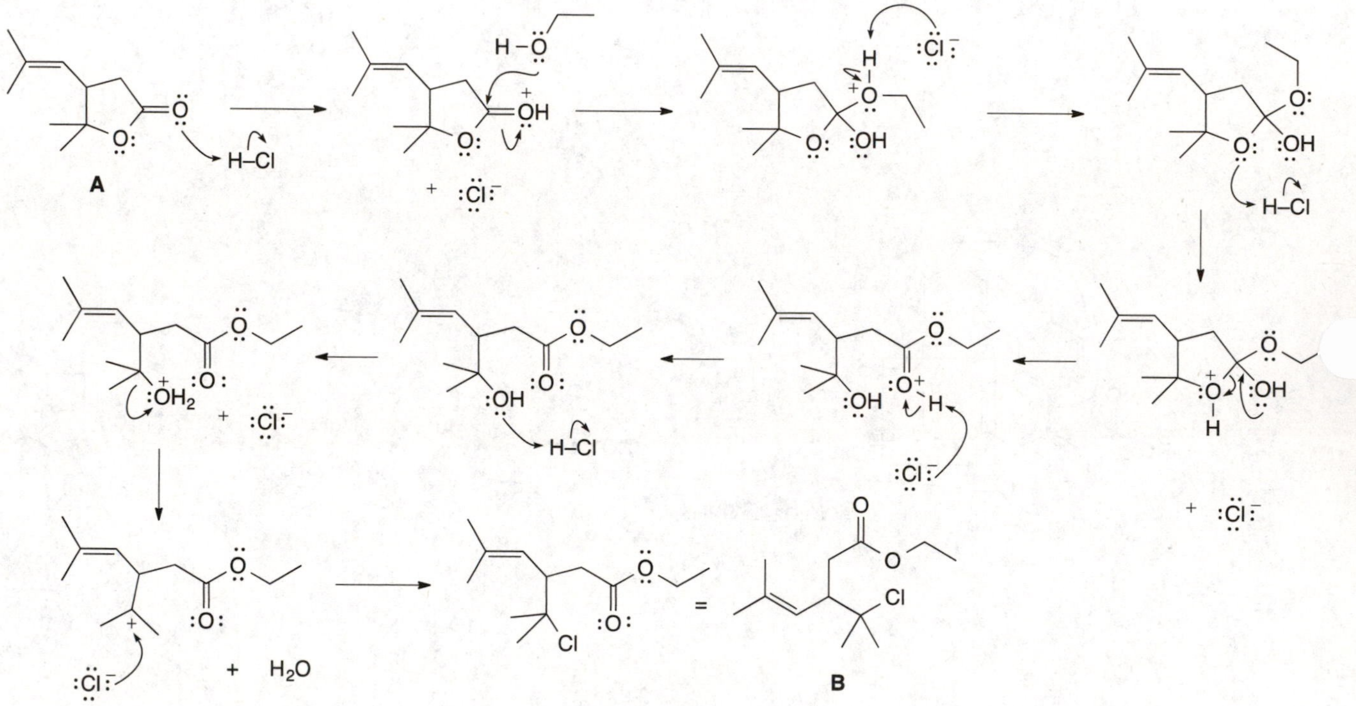

22.62 The mechanism is composed of two parts: hydrolysis of the acetal and intramolecular Fischer esterification of the hydroxy carboxylic acid.

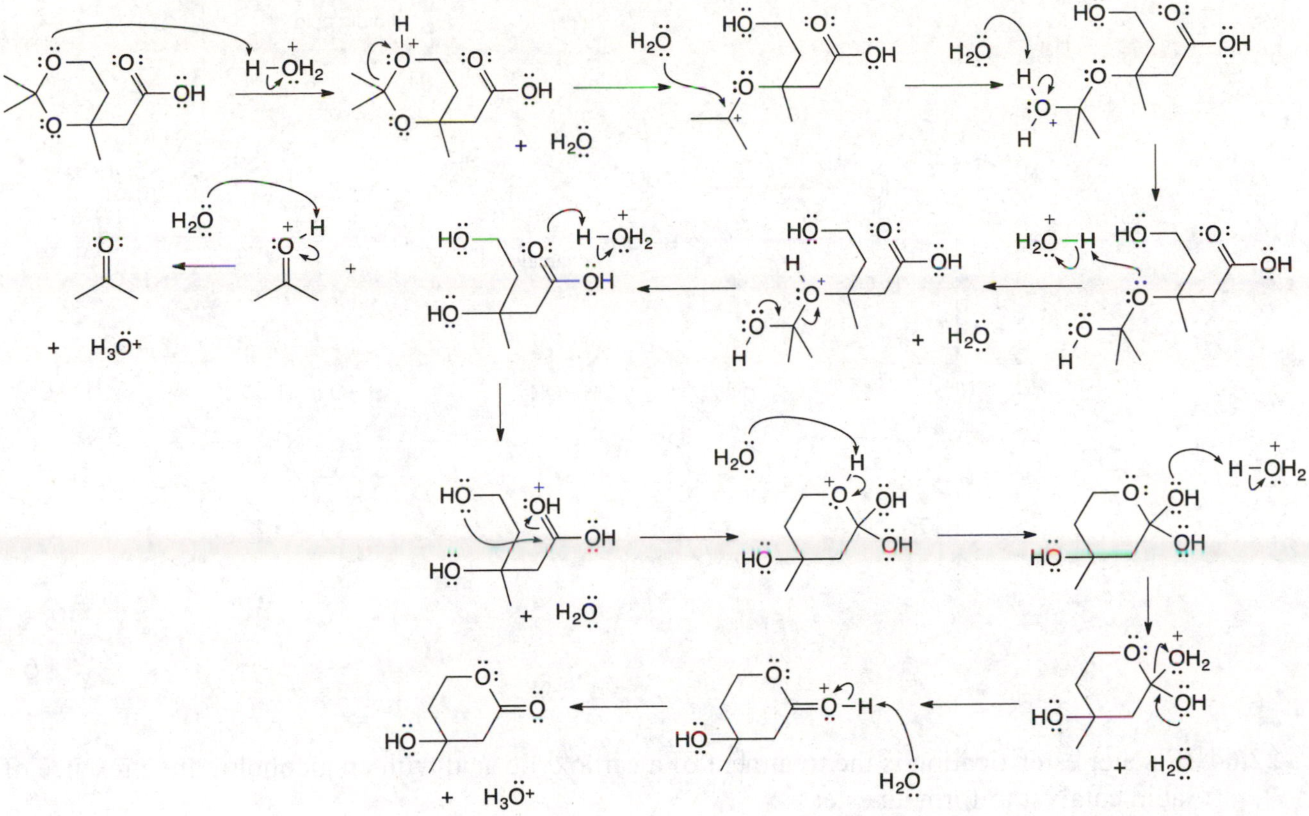

22.63

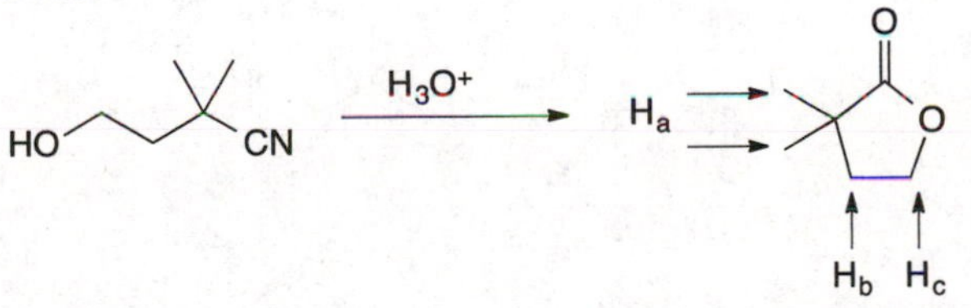

H_3O^+

A

$C_6H_{10}O_2$: 2 degrees of unsaturation

IR: 1770 cm^{-1} from ester C=O in a five-membered ring

^{1}H NMR:

H_a = 1.27 ppm (singlet, 6 H) – 2 CH_3 groups
H_b = 2.12 ppm (triplet, 2 H) – CH_2 bonded to CH_2
H_c = 4.26 ppm (triplet, 2 H) – CH_2 bonded to CH_2

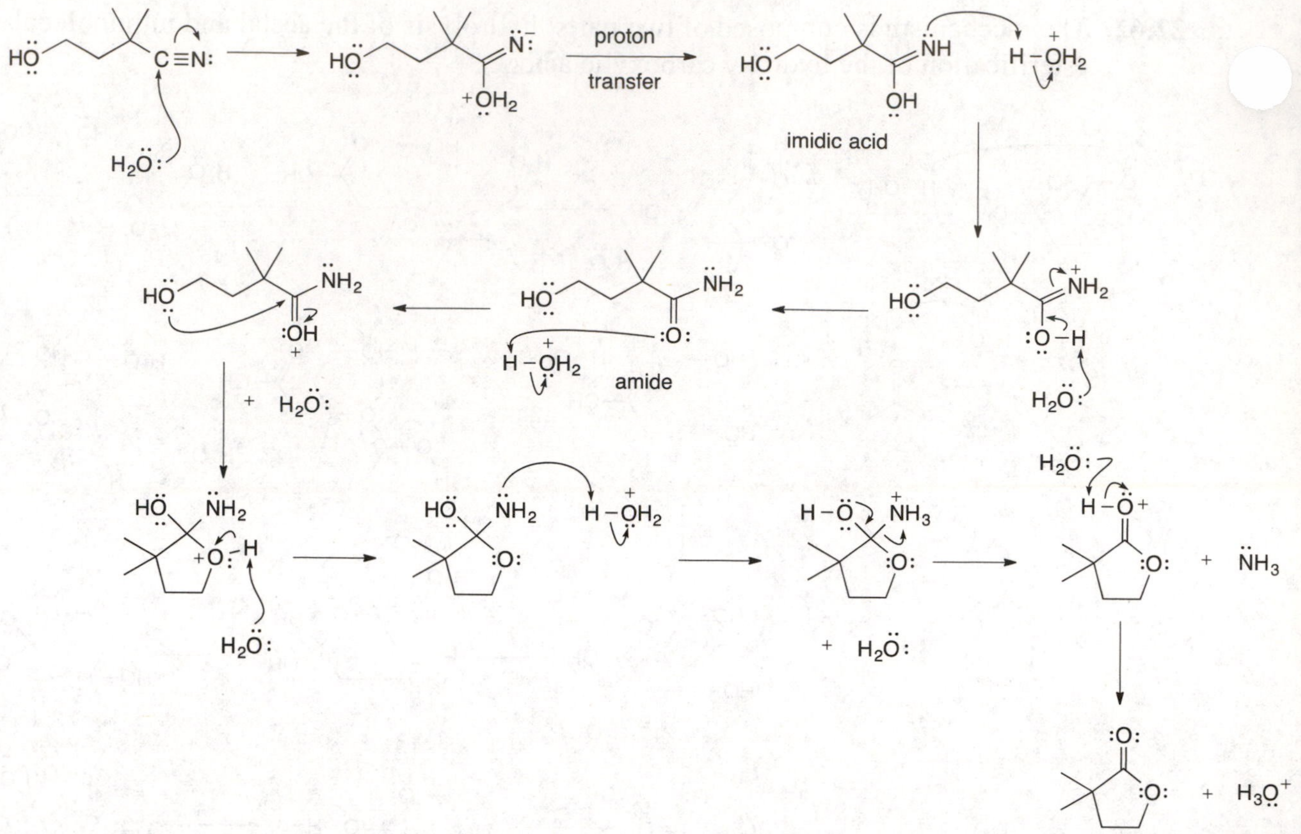

HO C≡N: → HO N: +OH2 → proton transfer → HO NH OH H–OH2+ imidic acid

H2O:

HO NH2 :OH+ ← HO NH2 :O: H–OH2+ amide ← HO NH2+ :O–H H2O:

+ H2O:

HO NH2 +O–H → HO NH2 :O: H–OH2+ → H–O NH3+ :O: → H–O+ :O: + NH3

H2O: + H2O: H2O:

:O: :O: + H_3O^+

22.64 Fischer esterification is the treatment of a carboxylic acid with an alcohol in the presence of an acid catalyst to form an ester.

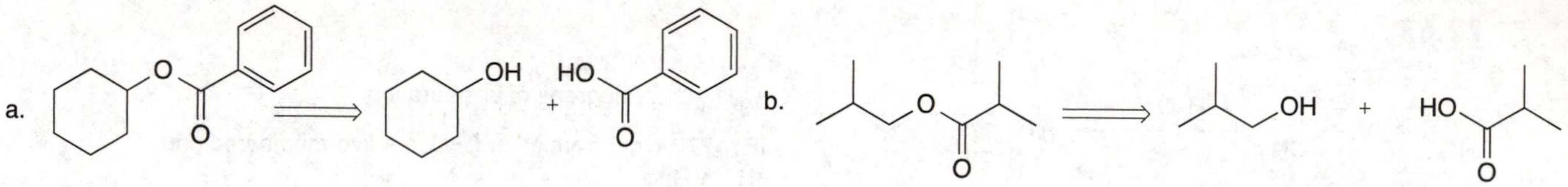

a. O O ⟹ OH + HO O   b. O O ⟹ OH + HO O

22.65

a. Br —NaCN→ CN —[1] MgBr, [2] H_2O→ O   Br —Mg→ MgBr

b. Br —NaOH→ OH —CrO_3, H_2SO_4, H_2O→ O OH —H_2SO_4, OH→ O O

[1] MgBr (from a.) (2 equiv)
[2] H_2O

OH

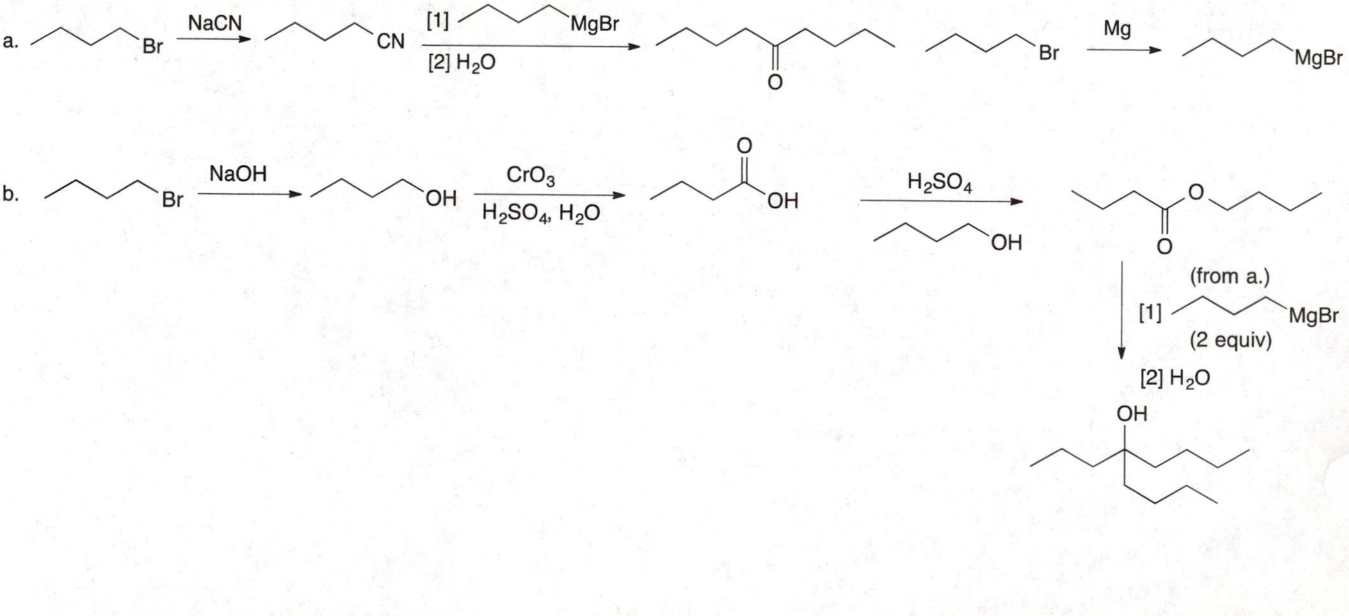

22.66

a. CH_3Cl + NaCN ⟶ CH_3-CN $\xrightarrow{H_3O^+}$ CH_3-COOH

CH_3-Cl + Mg ⟶ CH_3-MgCl $\xrightarrow[\text{[2] } H_3O^+]{\text{[1] } CO_2}$ CH_3-COOH

b. Br (on benzene, sp^2) + NaCN ⟶ This method can't be used because an S_N2 reaction can't be done on an sp^2 hybridized C.

Br + Mg ⟶ MgBr $\xrightarrow[\text{[2] } H_3O^+]{\text{[1] } CO_2}$ COOH

c. HO~~~~Br + NaCN ⟶ HO~~~~CN $\xrightarrow{H_3O^+}$ HO~~~~C(=O)OH

HO~~~~Br + Mg ⟶ This method can't be used because you can't make a Grignard reagent with an acidic OH group.

22.67

CH_3OH $\xrightarrow{SOCl_2}$ CH_3Cl

benzene $\xrightarrow[AlCl_3]{CH_3Cl}$ toluene $\xrightarrow[H_2SO_4]{HNO_3}$ O_2N–C₆H₄–CH₃ (+ ortho isomer) $\xrightarrow{KMnO_4}$ O_2N–C₆H₄–COOH $\xrightarrow[H_2SO_4]{CH_3CH_2OH}$ O_2N–C₆H₄–C(=O)OCH₂CH₃ $\xrightarrow{H_2,\ Pd\text{-}C}$ H_2N–C₆H₄–C(=O)OCH₂CH₃

22.68

a. phenol (OH) $\xrightarrow[AlCl_3]{CH_3CH_2Cl}$ o-ethylphenol (OH) (+ para isomer) $\xrightarrow{KMnO_4}$ C(=O)OH, OH $\xrightarrow{SOCl_2}$ C(=O)Cl, OH $\xrightarrow[\text{(2 equiv)}]{NH_3}$ C(=O)NH₂, OH

salicylamide

b. phenol (OH) $\xrightarrow[H_2SO_4]{HNO_3}$ HO–C₆H₄–NO_2 (+ ortho isomer) $\xrightarrow{H_2,\ Pd\text{-}C}$ HO–C₆H₄–NH_2 $\xrightarrow{CH_3COCl}$ HO–C₆H₄–NHC(=O)CH₃

(More nucleophilic NH_2 reacts first.)

acetaminophen

c. HO–C₆H₄–NHC(=O)CH₃ $\xrightarrow{NaH}$ ⁻O–C₆H₄–NHC(=O)CH₃ $\xrightarrow{CH_3CH_2Br}$ CH₃CH₂O–C₆H₄–NHC(=O)CH₃

acetaminophen (from b.)

p-acetophenetidin

22.69

a. Benzene → (CH_3Cl, $AlCl_3$; CH_3Cl from CH_3OH, $SOCl_2$) → toluene → (HNO_3, H_2SO_4) → p-nitrotoluene (O_2N) (+ ortho isomer) → ($KMnO_4$) → p-nitrobenzoic acid (COOH) → (CH_3OH, H^+) → methyl p-nitrobenzoate → (H_2, Pd-C) → methyl p-aminobenzoate (H_2N) → (propanoyl chloride, from 1-propanol: [1] CrO_3, H_2SO_4, H_2O [2] $SOCl_2$) → methyl 4-(propanamido)benzoate (N–H)

b. Benzene → (2-chloro-2-methylpropane, $AlCl_3$; from tert-butanol + HCl) → tert-butylbenzene → (CH_3Cl, $AlCl_3$; CH_3Cl from CH_3OH, $SOCl_2$) → 4-tert-butyltoluene (+ ortho isomer) → ($KMnO_4$) → 4-tert-butylbenzoic acid (HO) → ($SOCl_2$) → 4-tert-butylbenzoyl chloride (Cl)

\+

Benzene → (Br_2, $FeBr_3$) → bromobenzene → (CH_3Cl, $AlCl_3$) → 4-bromotoluene (+ ortho isomer) → ($KMnO_4$) → 4-bromobenzoic acid (HO) → (NaOH) → 4-bromobenzoate (^-O)

→ mixed anhydride: 4-tert-butylbenzoic 4-bromobenzoic anhydride (Br)

22.70

a. HO–cyclohexane–OH and HO(O)C–CH2CH2–C(O)OH → polyester (–C(O)CH2CH2C(O)O–cyclohexyl–O–)n

b. Terephthaloyl chloride (Cl, Cl) and H_2N–(CH2)5–NH_2 → polyamide (–C(O)–C6H4–C(O)–NH–(CH2)5–NH–)n

22.71

a.

b.

22.72

a. Docetaxel has fewer C's and one more OH group than paclitaxel. This makes docetaxel more water soluble than paclitaxel.

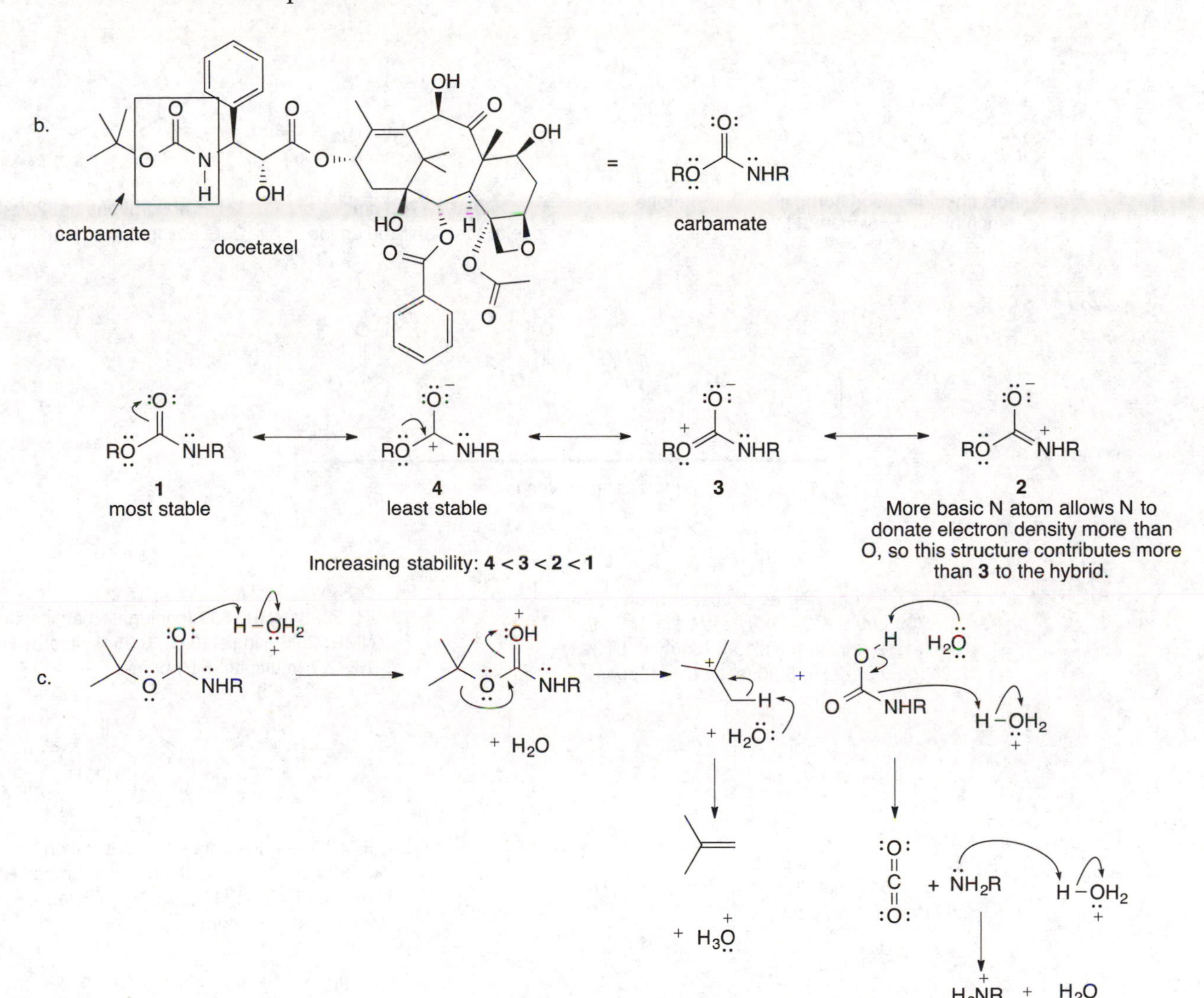
b.
carbamate
docetaxel
carbamate
1
most stable
4
least stable
3
2
More basic N atom allows N to donate electron density more than O, so this structure contributes more than 3 to the hybrid.
Increasing stability: 4 < 3 < 2 < 1
c.

d.

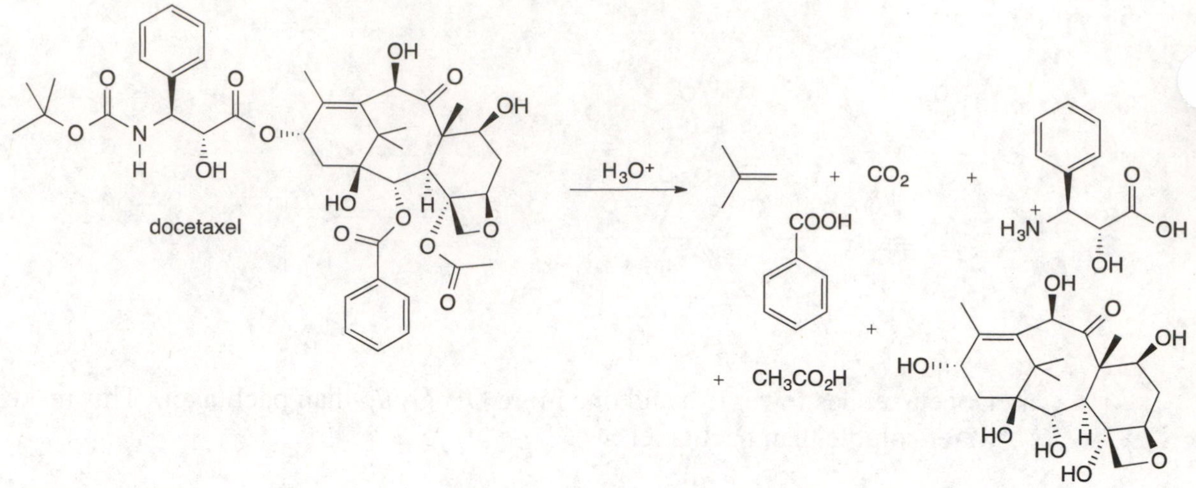

docetaxel

H_3O^+

+ CO_2 +

COOH

+

+ CH_3CO_2H

22.73

a. and

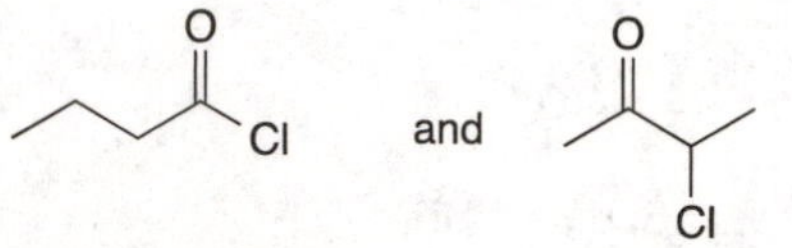

Acid chloride CO absorbs at much higher wavenumber.

ketone

b. $N(CH_3)_2$ and NH_2

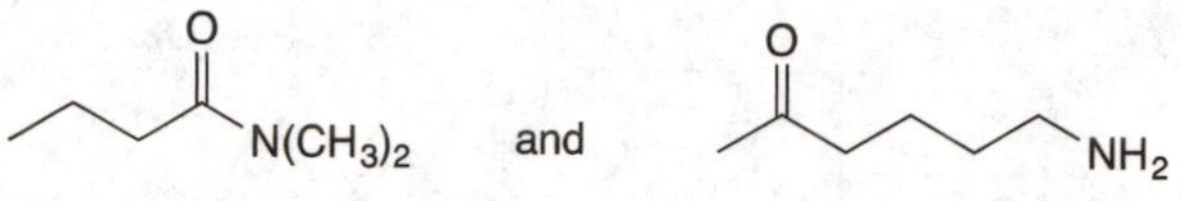

C=O at < 1700 cm^{-1} due to the stabilized amide

2 NH absorptions at 3200–3400 cm^{-1}
C=O absorption at higher wavenumber

22.74

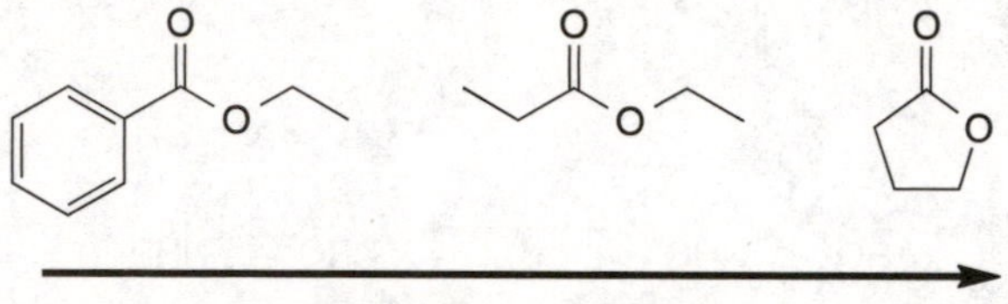

Increasing wavenumber

22.75

a. $C_6H_{12}O_2$ → one degree of unsaturation
IR: 1738 cm^{-1} → C=O
NMR: 1.12 (triplet, 3 H), 1.23 (doublet, 6 H), 2.28 (quartet, 2 H), 5.00 (septet, 1 H) ppm

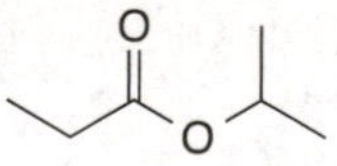

b. C_4H_7N
IR: 2250 cm^{-1} → triple bond
NMR: 1.08 (triplet, 3 H), 1.70 (multiplet, 2 H), 2.34 (triplet, 2 H) ppm

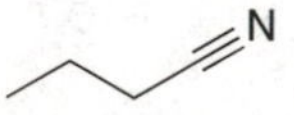

c. C_8H_9NO
IR: 3328 (NH), 1639 (conjugated amide C=O) cm^{-1}
NMR: 2.95 (singlet, 3 H), 6.95 (singlet, 1 H), 7.3–7.7 (multiplet, 5 H) ppm

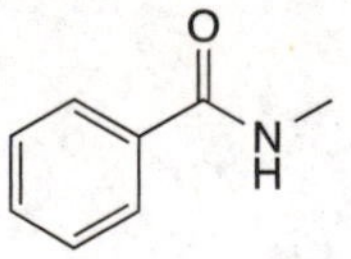

d. C_4H_7ClO → one degree of unsaturation
IR: 1802 cm^{-1} → C=O (high wavenumber, RCOCl)
NMR: 0.95 (triplet, 3 H), 1.07 (multiplet, 2 H), 2.90 (triplet, 2 H) ppm

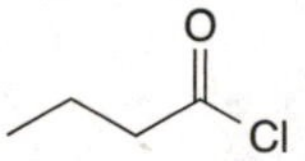

e. $C_{10}H_{12}O_2$ → five degrees of unsaturation
IR: 1740 cm^{-1} → C=O
NMR: 1.2 (triplet, 3 H), 2.4 (quartet, 2 H),
5.1 (singlet, 2 H), 7.1–7.5 (multiplet, 5 H) ppm

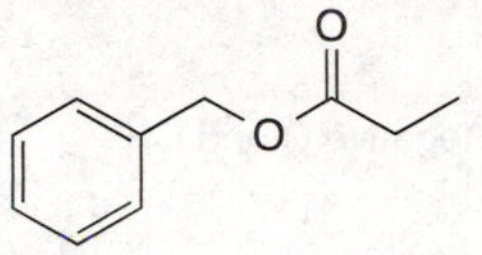

22.76

A. Molecular formula $C_{10}H_{12}O_2$ → five degrees of unsaturation
IR absorption at 1718 cm^{-1} → C=O
NMR data (ppm):
- triplet at 1.4 (CH_3 adjacent to 2 H's)
- singlet at 2.4 (CH_3)
- quartet at 4.4 (CH_2 adjacent to CH_3)
- doublet at 7.2 (2 H's on benzene ring)
- doublet at 7.9 (2 H's on benzene ring)

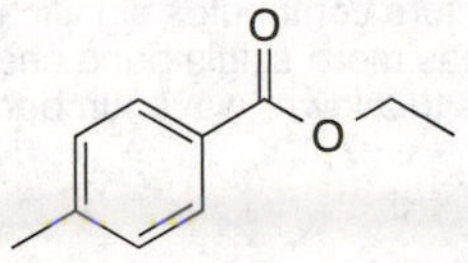

B. IR absorption at 1740 cm^{-1} → C=O
NMR data (ppm):
- singlet at 2.1 (CH_3)
- triplet at 2.9 (CH_2 adjacent to CH_2)
- triplet at 4.3 (CH_2 adjacent to CH_2)
- multiplet at 7.3 (5 H's, monosubstituted benzene)

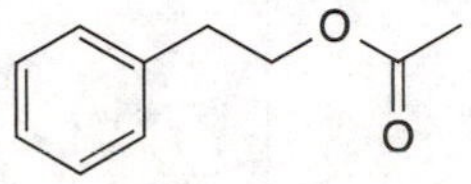

22.77

Molecular formula $C_{10}H_{13}NO_2$ → five degrees of unsaturation
IR absorptions at 3300 (NH) and 1680 (C=O, amide or conjugated) cm^{-1}
NMR data (ppm):
- triplet at 1.4 (CH_3 adjacent to CH_2)
- singlet at 2.2 (CH_3C=O)
- quartet at 3.9 (CH_2 adjacent to CH_3)
- doublet at 6.8 (2 H's on benzene ring)
- singlet at 7.1 (NH)
- doublet at 7.3 (2 H's on benzene ring)

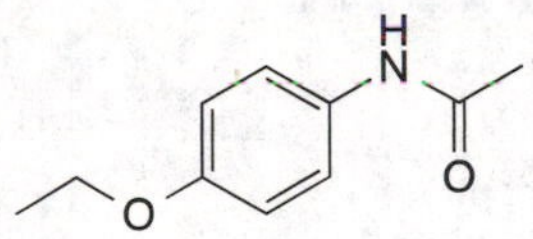

phenacetin

22.78

Molecular formula $C_{11}H_{15}NO_2$ → five degrees of unsaturation
IR absorption 1699 (C=O, amide or conjugated) cm^{-1}
NMR data (ppm):
- triplet at 1.3 (3 H) (CH_3 adjacent to CH_2)
- singlet at 3.0 (6 H) (2 CH_3 groups on N)
- quartet at 4.3 (2 H) (CH_2 adjacent to CH_3)
- doublet at 6.6 (2 H) (2 H's on benzene ring)
- doublet at 7.9 (2 H) (2 H's on benzene ring)

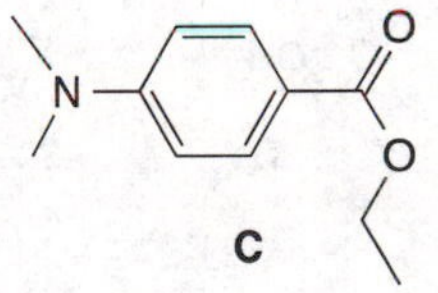

C

22.79

a. Molecular formula $C_6H_{12}O_2$ → one degree of unsaturation
IR absorption at 1743 cm–1 → C=O
^1H NMR data (ppm):
- triplet at 0.9 (3 H) – CH_3 adjacent to CH_2
- multiplet at 1.35 (2 H) – CH_2
- multiplet at 1.60 (2 H) – CH_2
- singlet at 2.1 (3 H – from CH_3 bonded to C=O)
- triplet at 4.1 (2 H) – CH_2 adjacent to the electronegative O atom and another CH_2

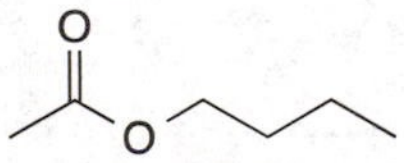

D

b. Molecular formula $C_6H_{12}O_2$ → one degree of unsaturation
IR absorption at 1746 cm–1 → C=O
^1H NMR data (ppm):
doublet at 0.9 (6 H) – 2 CH_3's adjacent to CH
multiplet at 1.9 (1 H)
singlet at 2.1 (3 H) – CH_3 bonded to C=O
doublet at 3.85 (2 H) – CH_2 bonded to electronegative O and CH

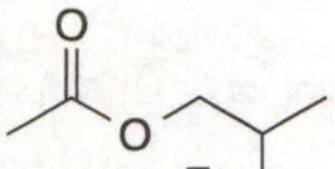

E

22.80 The extent of resonance stabilization affects the position of the C=O absorption in the IR of an amide.

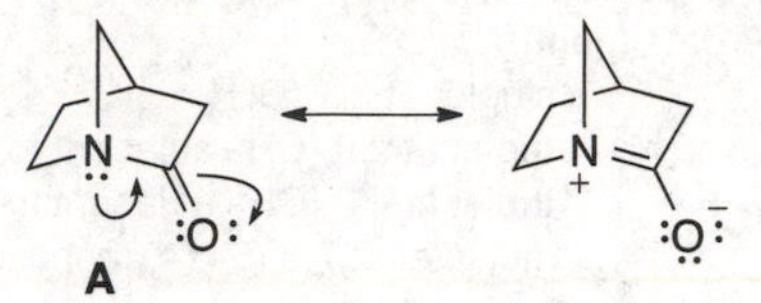

A

This resonance structure does not contribute significantly to the hybrid because it places a double bond at the bridgehead N, an impossible geometry in small rings. The C=O has more double bond character, so the absorption is at a higher wavenumber.

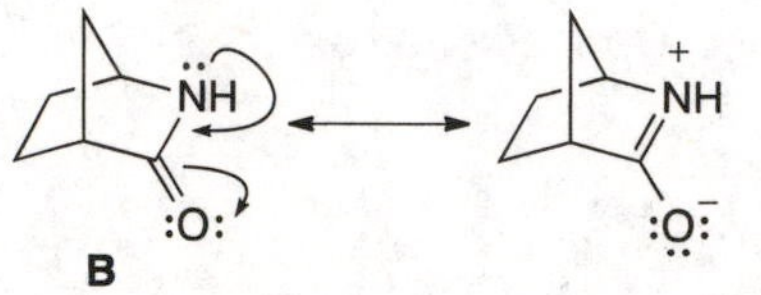

B

This resonance structure contributes significantly to the hybrid, so the C=O has more single bond character. The absorption shifts to a lower wavenumber.

22.81

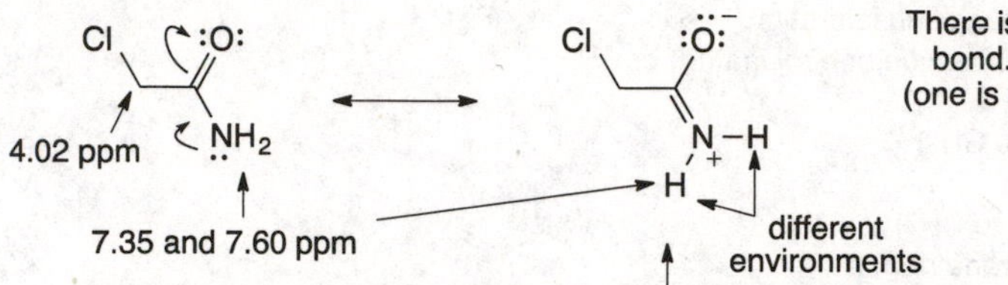

4.02 ppm

7.35 and 7.60 ppm

different environments

There is restricted rotation around the amide C–N bond. The 2 H's are in different environments (one is cis to an O atom, and one is cis to CH_2Cl), so they give different NMR signals.

This resonance structure makes a significant contribution to the resonance hybrid.

22.82

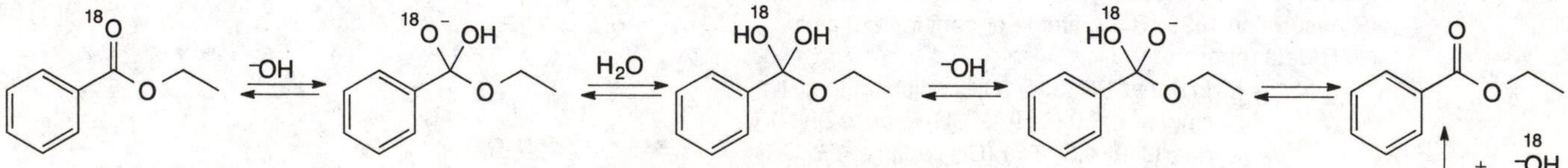

ethyl benzoate

Two OH groups are now equivalent and either can lose H_2O to form labeled or unlabeled ethyl benzoate.

Unlabeled starting material was recovered.

22.83

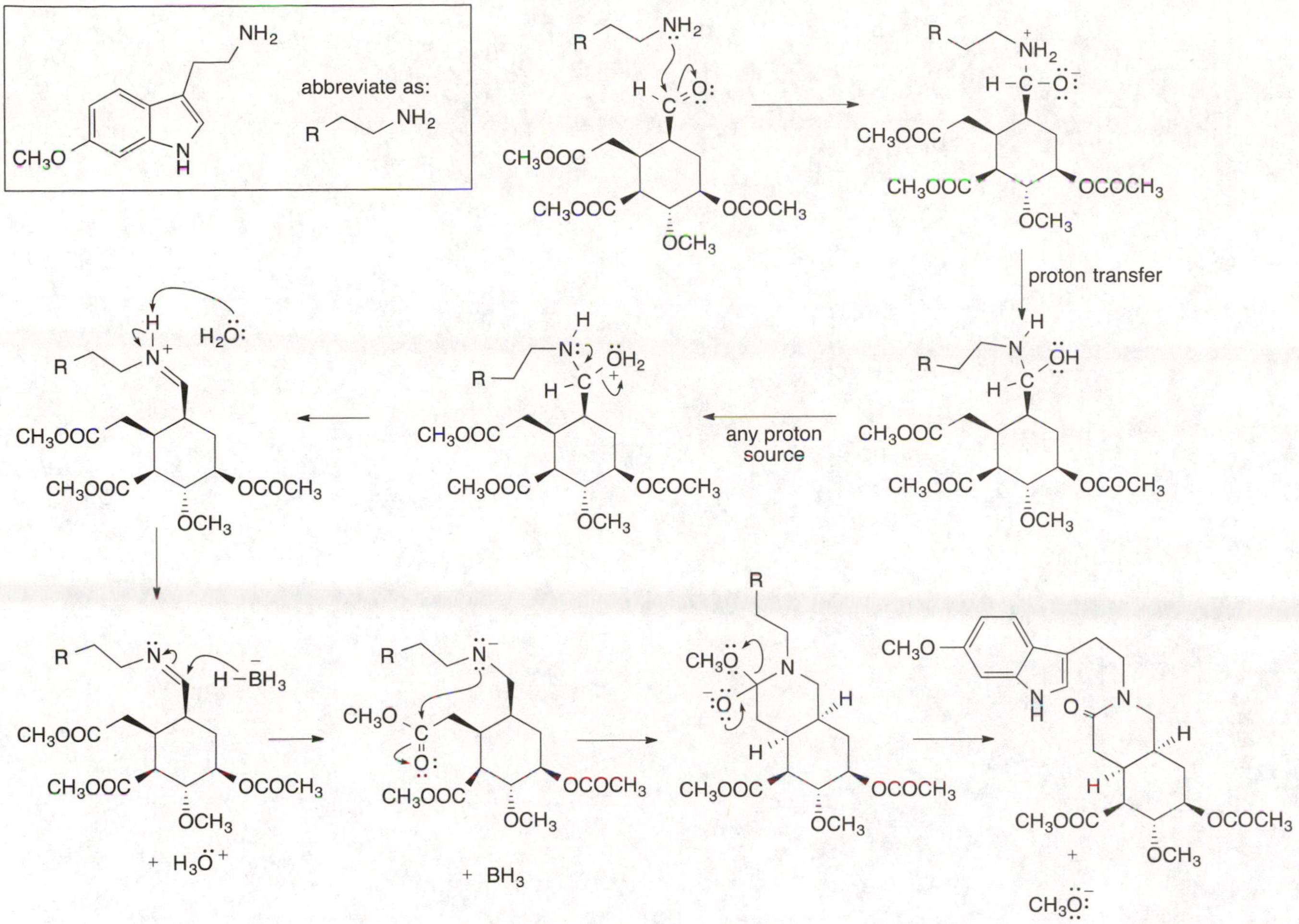
NH2
abbreviate as:
R NH2
CH3O
N
H
proton transfer
any proton source
H2Ö:
+ H3Ö+
H–BH3
+ BH3
CH3OOC
OCOCH3
OCH3
CH3O
+
CH3Ö:

22.84 Both acetals are hydrolyzed by the usual mechanism for acetal hydrolysis (Steps [1 –[6]), forming four new OH groups. Intramolecular esterification forms a lactone (Steps [10]–[15]), followed by conversion of a carbonyl tautomer to an enol (Steps [16]–[17]). Both acetals are hydrolyzed at once in the given mechanism. For ease in drawing this long mechanism, the stereochemistry of intermediates is omitted.

[1] + 2 Cl⁻ [2] [3] [4] [5] + 2 Cl⁻ [6] + (2 equiv) [7] + Cl⁻ [8] [9] + 2 HCl (2 equiv) [10] + Cl⁻ [11] [12] [13] + Cl⁻ [14] + H_2O [15] [16] + Cl⁻ [17] + HCl

vitamin C

22.85

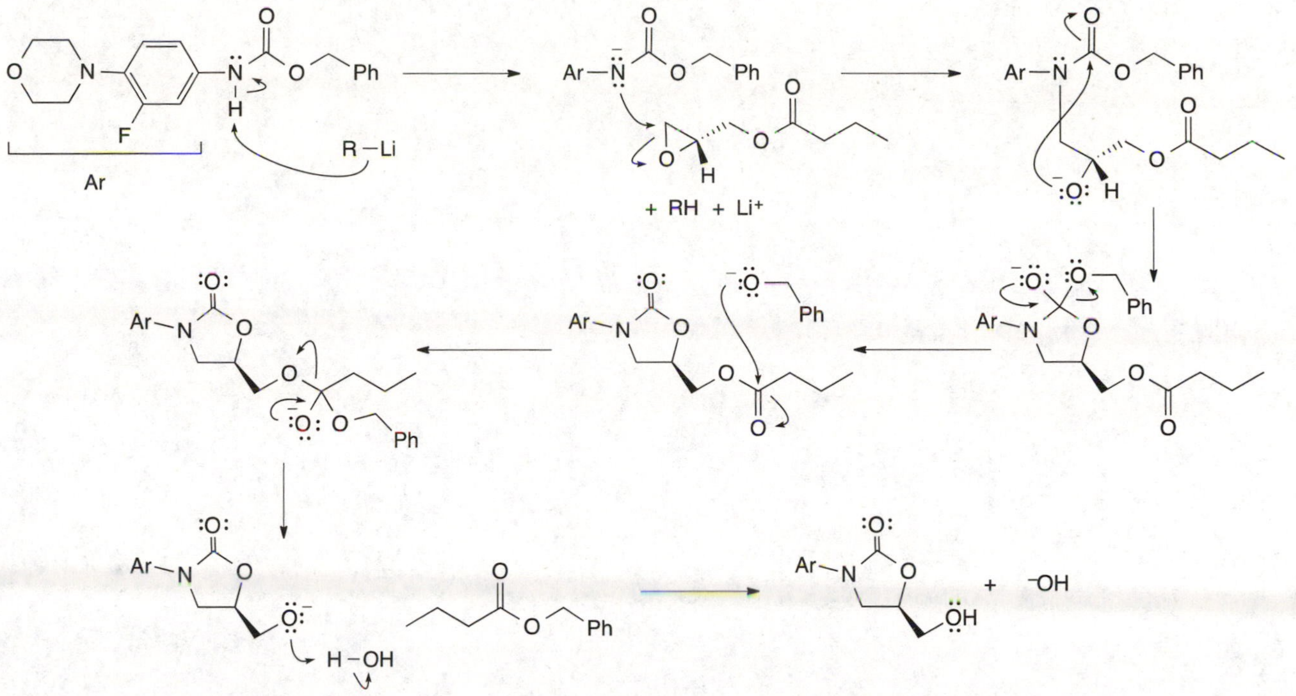

Ar

R–Li

+ RH + Li+

Ph

$^{-}$OH

Chapter 23 Substitution Reactions of Carbonyl Compounds at the α Carbon

Chapter Review

Kinetic versus thermodynamic enolates (23.4)

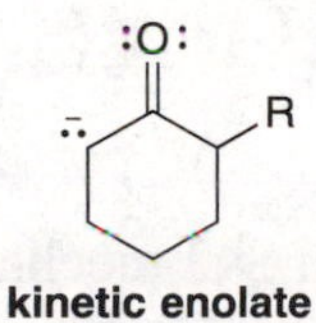

kinetic enolate

Kinetic enolate

- The less substituted enolate
- Favored by strong base, polar aprotic solvent, low temperature: LDA, THF, –78 °C

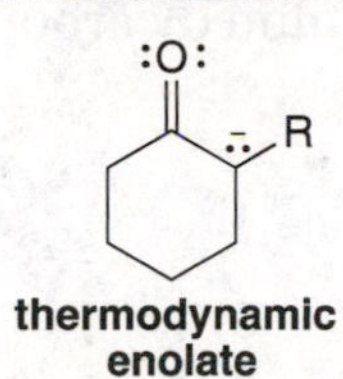

thermodynamic enolate

Thermodynamic enolate

- The more substituted enolate
- Favored by strong base, protic solvent, higher temperature: $NaOCH_2CH_3$, CH_3CH_2OH, room temperature

Halogenation at the α carbon

[1] Halogenation in acid (23.7A)

X_2, CH_3COOH

$X_2 = Cl_2, Br_2$, or I_2

α-halo aldehyde or ketone

- The reaction occurs via enol intermediates.
- Monosubstitution of X for H occurs on the α carbon.

[2] Halogenation in base (23.7B)

X_2 (excess), ^-OH

$X_2 = Cl_2, Br_2$, or I_2

- The reaction occurs via enolate intermediates.
- Polysubstitution of X for H occurs on the α carbon.

[3] Halogenation of *methyl* ketones in base—The haloform reaction (23.7B)

X_2 (excess), ^-OH

$X_2 = Cl_2, Br_2$, or I_2

+ HCX_3 haloform

- The reaction occurs with methyl ketones and results in cleavage of a carbon–carbon σ bond.

Reactions of α-halo carbonyl compounds (23.7C)

[1] Elimination to form α,β-unsaturated carbonyl compounds

R–CO–CH(Br)–CH$_3$ (α) → [Li_2CO_3, LiBr, DMF] → R–CO–CH=CH$_2$ (α, β)

- Elimination of the elements of Br and H forms a new π bond, giving an α,β-unsaturated carbonyl compound.

[2] Nucleophilic substitution

R–CO–CH$_2$–Br (α) → [:Nu$^-$] → R–CO–CH$_2$–Nu (α)

- The reaction follows an SN2 mechanism, generating an α-substituted carbonyl compound.

Alkylation reactions at the α carbon

[1] Direct alkylation at the α carbon (23.8)

α-H carbonyl compound → [1] Base; [2] RX → α-R carbonyl compound + X–

- The reaction forms a new C–C bond to the α carbon.
- LDA is a common base used to form an intermediate enolate.
- The alkylation in Step [2] follows an SN2 mechanism.

[2] Malonic ester synthesis (23.9)

EtO–CO–CH$_2$(α)–CO–OEt (diethyl malonate)

→ [1] NaOEt; [2] RX; [3] H_3O^+, Δ → R–CH$_2$(α)–COOH

→ [1] NaOEt; [2] RX → [1] NaOEt; [2] R'X; [3] H_3O^+, Δ → R(R')CH(α)–COOH

- The reaction is used to prepare carboxylic acids with one or two alkyl groups on the α carbon.
- The alkylation in Step [2] follows an SN2 mechanism.

[3] Acetoacetic ester synthesis (23.10)

EtO–CO–CH$_2$(α)–CO–CH$_3$ (ethyl acetoacetate)

→ [1] NaOEt; [2] RX; [3] H_3O^+, Δ → R–CH$_2$(α)–CO–CH$_3$

→ [1] NaOEt; [2] RX → [1] NaOEt; [2] R'X; [3] H_3O^+, Δ → R(R')CH(α)–CO–CH$_3$

- The reaction is used to prepare ketones with one or two alkyl groups on the α carbon.
- The alkylation in Step [2] follows an SN2 mechanism.

Practice Test on Chapter Review

1.a. Which of the following compounds can be prepared using either the acetoacetic ester synthesis or the malonic ester synthesis?

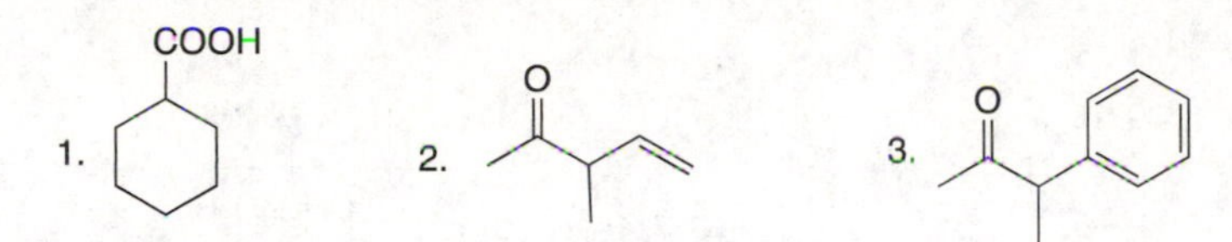

4. Both (1) and (2) can be prepared by one of these routes.
5. Compounds (1), (2), and (3) can all be prepared by these routes.

b. Which of the following compounds is ***not*** an enol form of dicarbonyl compound **A?**

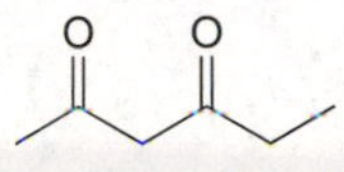

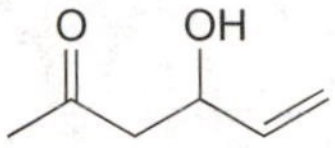

5. Compounds (1)–(4) are all enols of **A.**

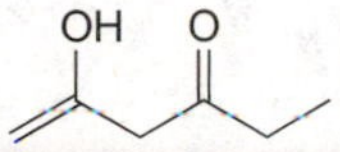

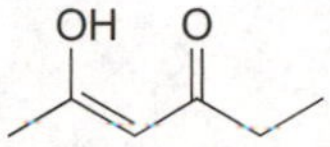

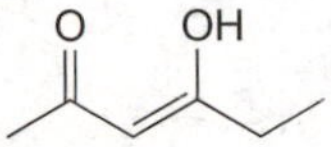

2.a. Which proton in compound **A** has the *lowest* p*K*a?
b. Which proton in compound **A** has the *highest* p*K*a?

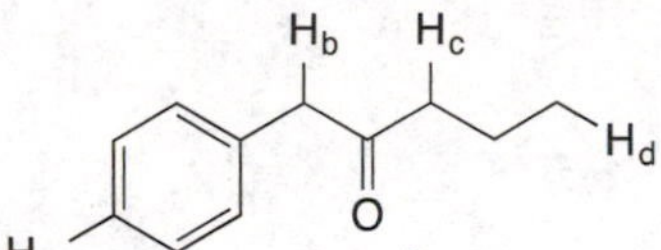

c. Which proton in compound **B** is the least acidic?
d. Which proton in compound **B** is the most acidic?

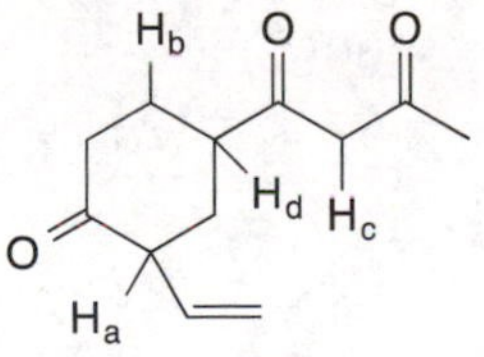

3. What reagents are needed to convert heptan-4-one to each compound?
   a. 3-bromoheptan-4-one
   b. 3-methylheptan-4-one
   c. 3,5-dimethylheptan-4-one
   d. hept-2-en-4-one
   e. 2-methylheptan-4-one

4. Draw the organic products formed in each of the following reactions.

a. 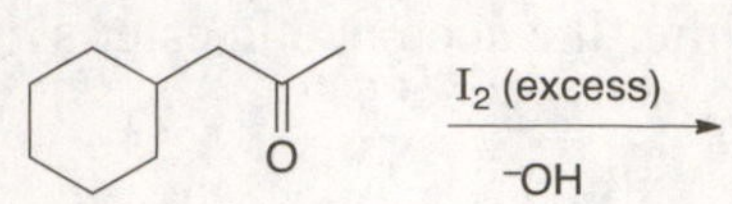 I_2 (excess), ^{-}OH

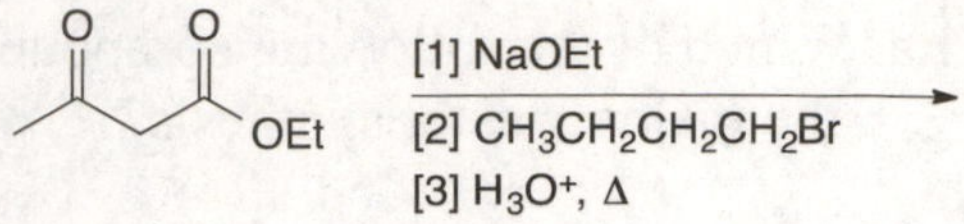

b. 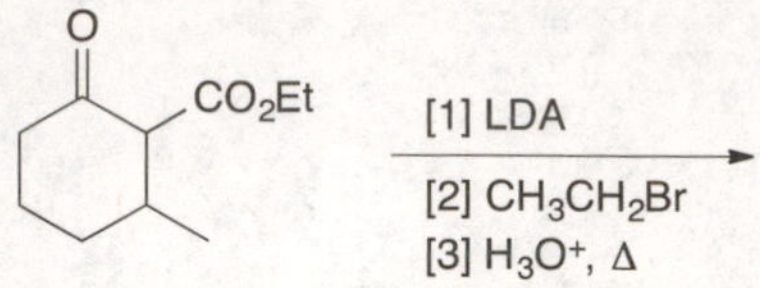 [1] LDA; [2] CH_3CH_2Br; [3] H_3O^+, Δ (CO_2Et)

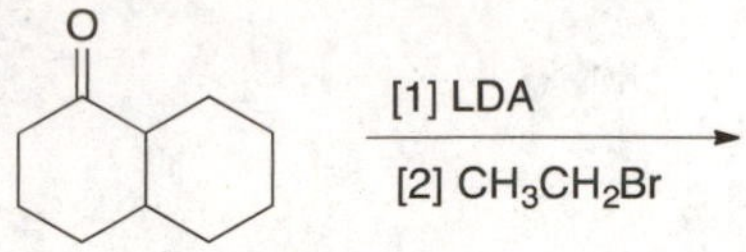

c. $CH_2(CO_2Et)_2$ — [1] NaOEt; [2] CH_3CH_2Br — [1] NaOEt; [2] $C_6H_5CH_2Cl$; [3] H_3O^+, Δ

d. (OEt) [1] NaOEt; [2] $CH_3CH_2CH_2CH_2Br$; [3] H_3O^+, Δ

e. [1] LDA; [2] CH_3CH_2Br

5.a. What starting materials are needed to synthesize carboxylic acid **A** by a malonic ester synthesis?

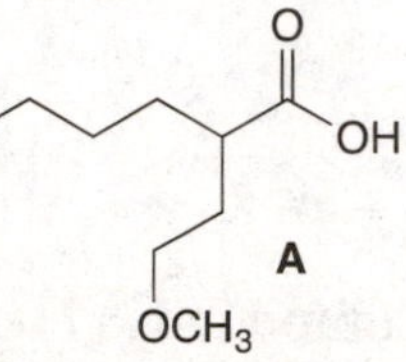

OH, OCH_3

A

b. What starting materials are needed to synthesize ketone **B** by the acetoacetic ester synthesis?

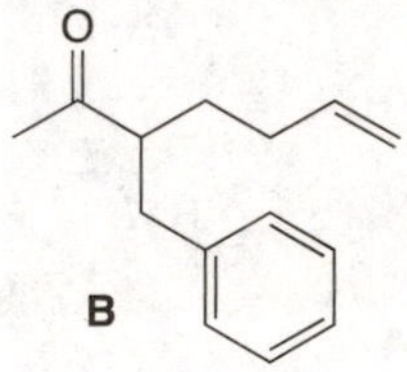

B

Answers to Practice Test

1.a. 1
b. 4

2.a. H_b
b. H_d
c. H_b
d. H_c

3.a. Br_2, CH_3CO_2H
b. [1] LDA; [2] CH_3I
c. []1 LDA; [2] CH_3I; [3] LDA; [4] CH_3I
d. [1] Br_2, CH_3CO_2H; [2] Li_2CO_3, LiBr, DMF
e. [1] Br_2, CH_3CO_2H; [2] Li_2CO_3, LiBr, DMF; [3] $(CH_3)_2CuLi$; [4] H_2O

4.
a. 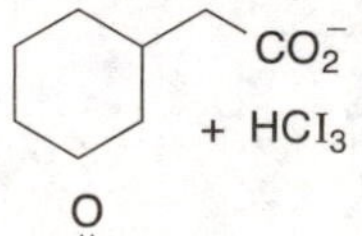 CO_2^- + HCI_3
b. 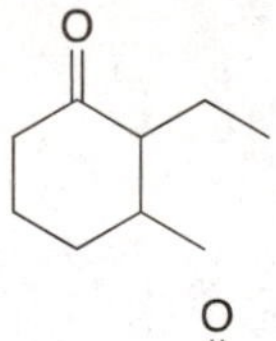
c. 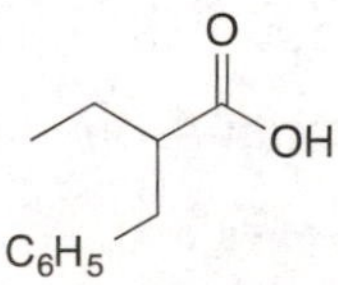 OH, C_6H_5
d. 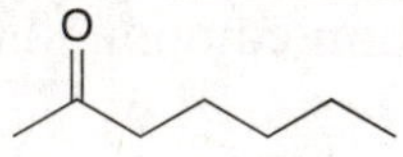
e.

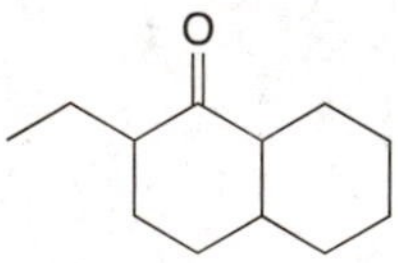

5.
a. 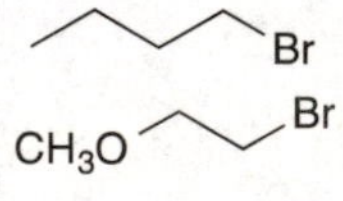 Br; CH_3O Br; $CH_2(CO_2Et)_2$
b. 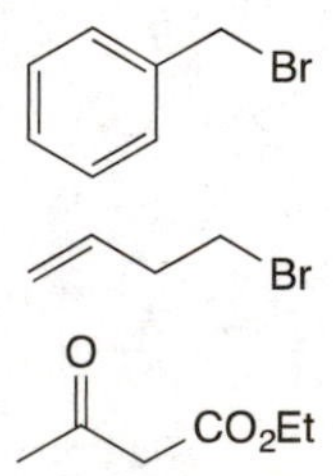 Br; Br; CO_2Et

Answers to Problems

23.1 • To convert a ketone to its enol tautomer, change the C=O to C–OH, make a new double bond to an α carbon, and remove a proton at the other end of the C=C.
• To convert an enol to its keto form, find the C=C bonded to the OH. Change the C–OH to a C=O, add a proton to the other end of the C=C, and delete the double bond.

[In cases where *E* and *Z* isomers are possible, only one stereoisomer is drawn.]

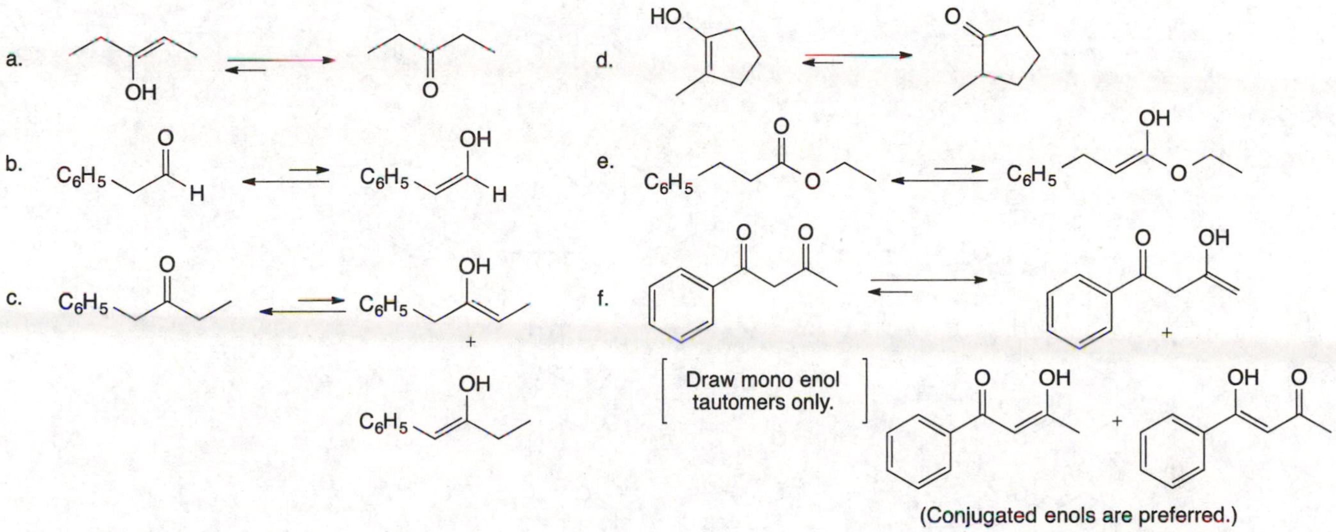

a. OH ⇌ O

b. C_6H_5, O, H ⇌ C_6H_5, OH, H

c. C_6H_5, O ⇌ C_6H_5, OH + C_6H_5, OH

d. HO ⇌ O

e. C_6H_5, O, O ⇌ C_6H_5, OH, O

f. O, O ⇌ O, OH + O, OH + OH, O

[Draw mono enol tautomers only.]

(Conjugated enols are preferred.)

23.2 The mechanism has two steps for each part: protonation followed by deprotonation.

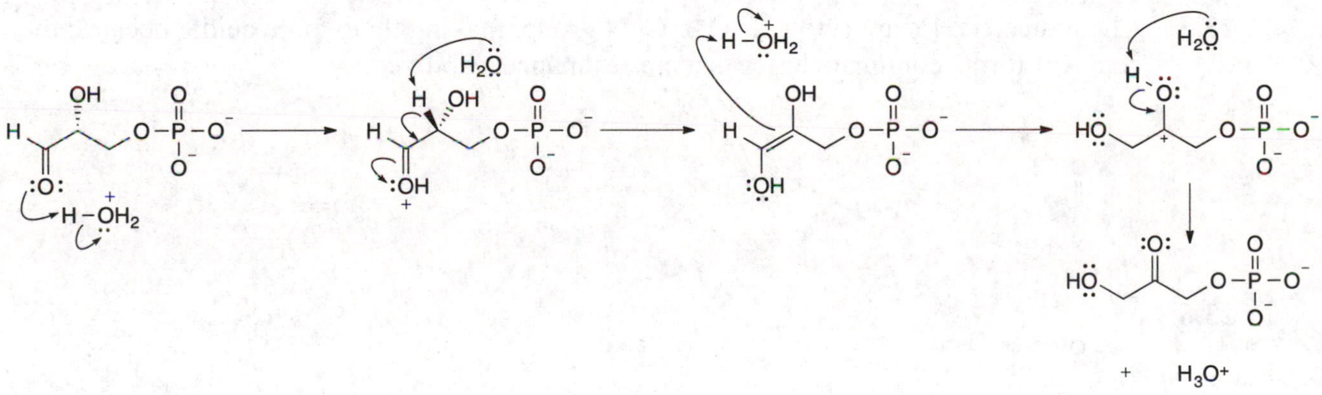

H, OH, O, O–P–O⁻, O⁻, :O:, H–ÖH₂⁺ → H, H OH, H₂Ö, O–P–O⁻, :ÖH⁺ → H–ÖH₂⁺, H, OH, O–P–O⁻, :ÖH → H₂Ö, H, Ö:, HÖ, O–P–O⁻ → :O:, HÖ, O–P–O⁻ + H_3O^+

23.3

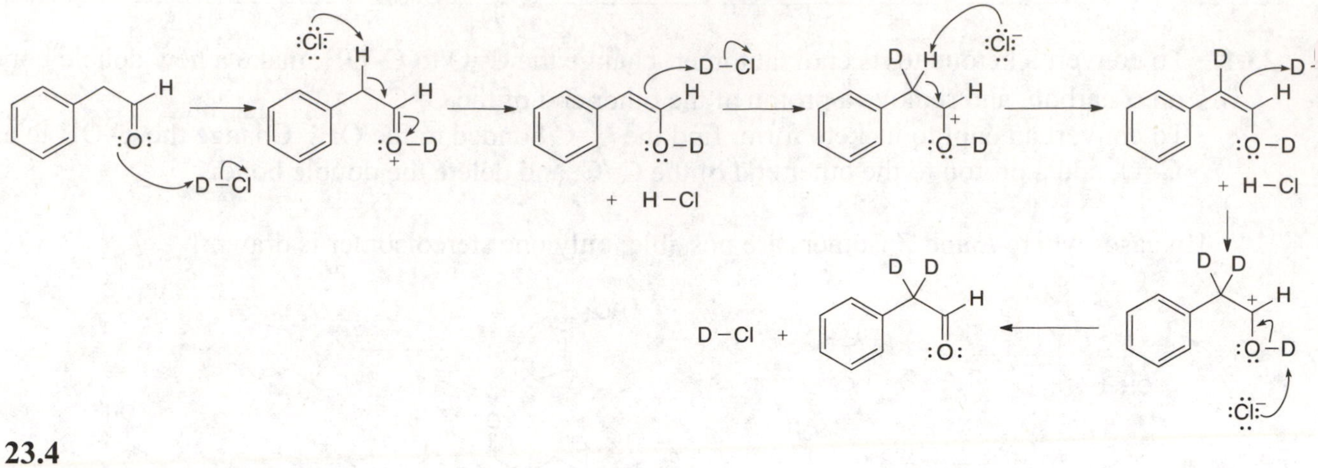

23.4

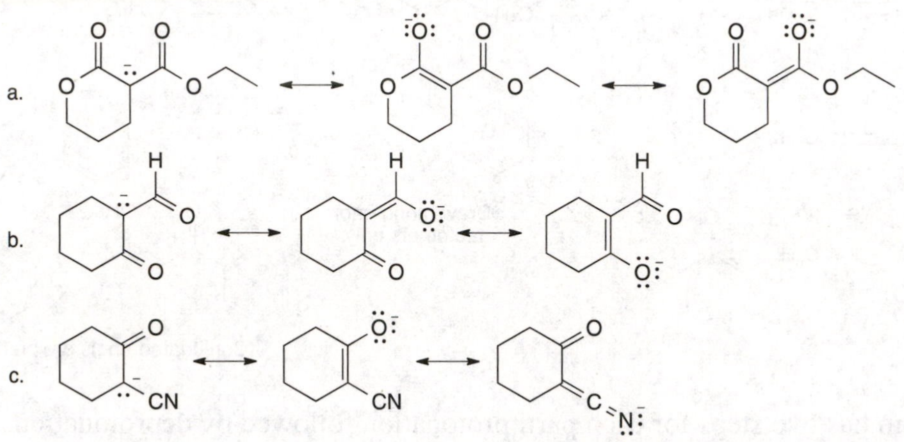

a.

b.

c.

23.5 The indicated H's are α to a C=O or C≡N group, making them more acidic because their removal forms conjugate bases that are resonance stabilized.

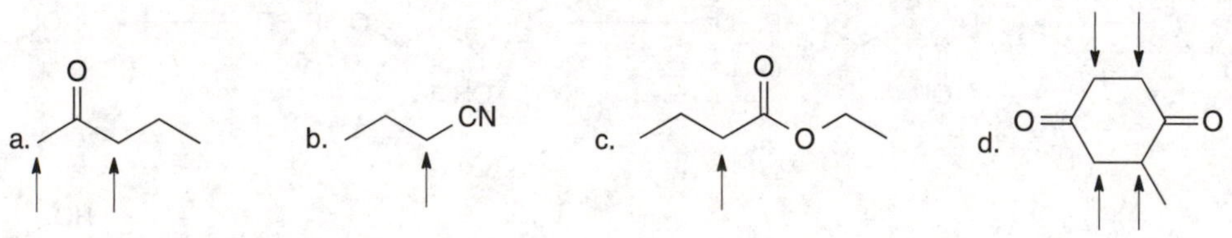

a. b. c. d.

23.6

no resonance stabilization
least acidic

Two resonance structures stabilize the conjugate base.
intermediate acidity

Three resonance structures stabilize the conjugate base.
most acidic

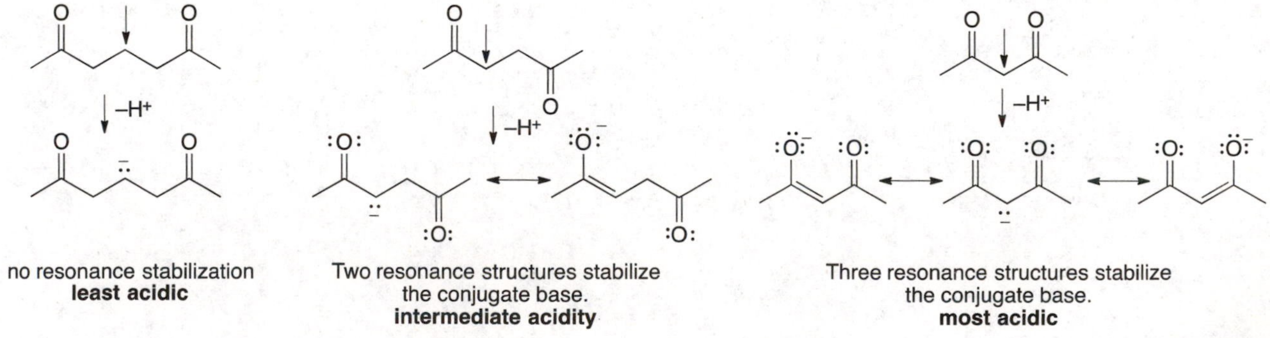

23.7 In each of the reactions, the LDA pulls off the most acidic proton.

a. LDA, THF

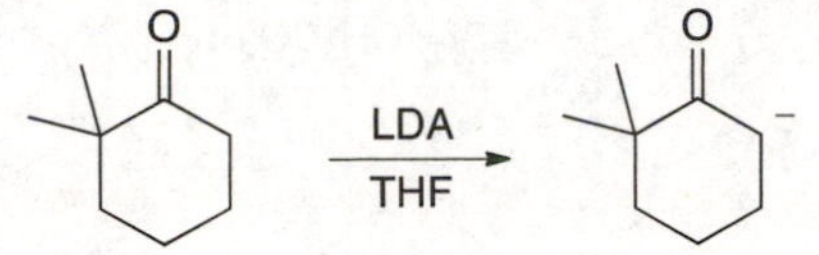

c. LDA, THF

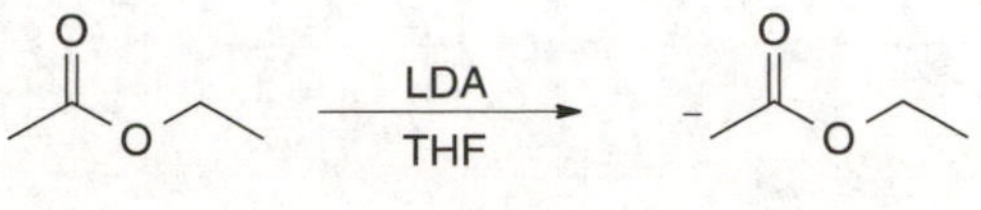

b. CHO LDA, THF CHO

d. CN LDA, THF CN

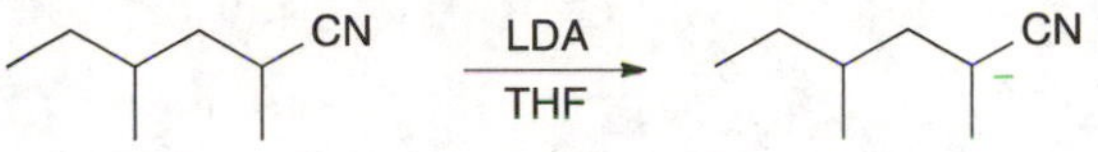

23.8 In addition to being strong bases, organolithiums are good nucleophiles that can add to a carbonyl group instead of pulling off a proton to generate an enolate.

23.9
- LDA, THF forms the kinetic enolate by removing a proton from the less substituted C.
- Treatment with $NaOCH_3$, CH_3OH forms the thermodynamic enolate by removing a proton from the more substituted C.

a. LDA, THF; $NaOCH_3$, CH_3OH

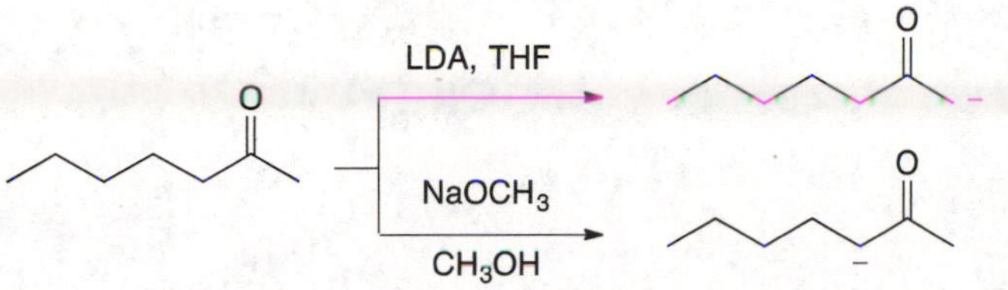

c. LDA, THF; $NaOCH_3$, CH_3OH

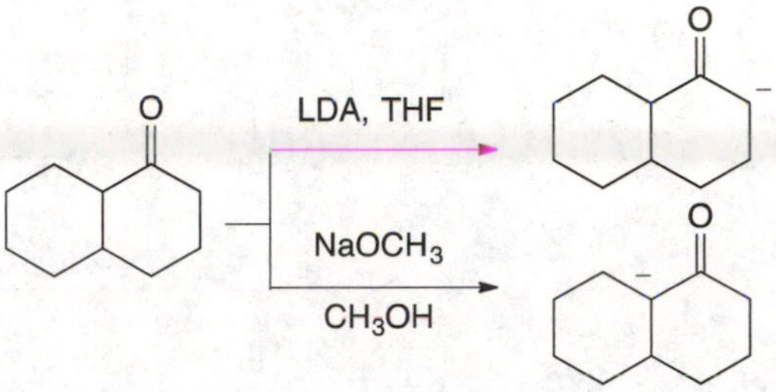

b. LDA, THF; $NaOCH_3$, CH_3OH

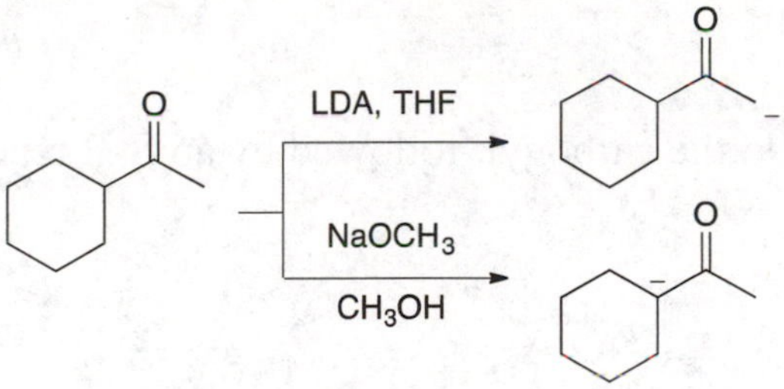

23.10

a. This acidic H is removed with base to form an achiral enolate.

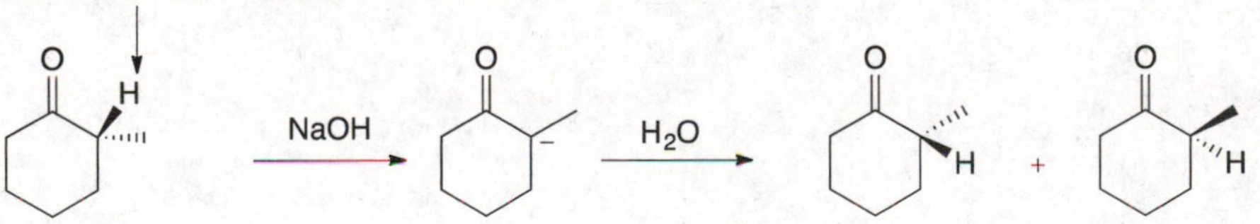

(*R*)-2-methylcyclohexanone → NaOH → achiral → H_2O → +

Protonation of the planar achiral enolate occurs with equal probability from two sides, so a racemic mixture is formed. The racemic mixture is optically inactive.

b. (*R*)-3-methylcyclohexanone → NaOH → or → H_2O →

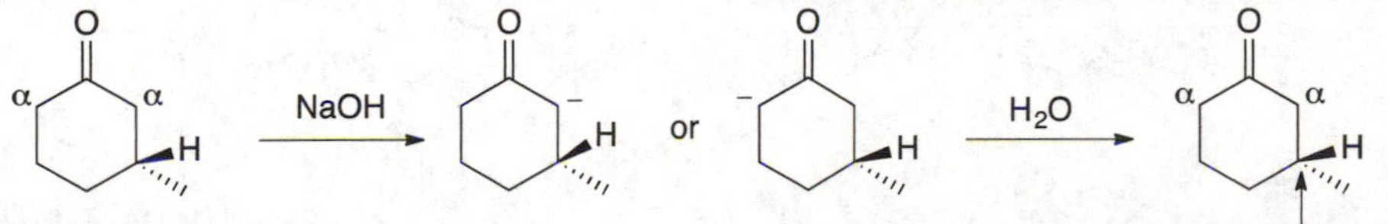

This stereogenic center is not located at the α carbon, so it is not deprotonated with base. Its configuration is retained in the product, and the product remains optically active.

23.11

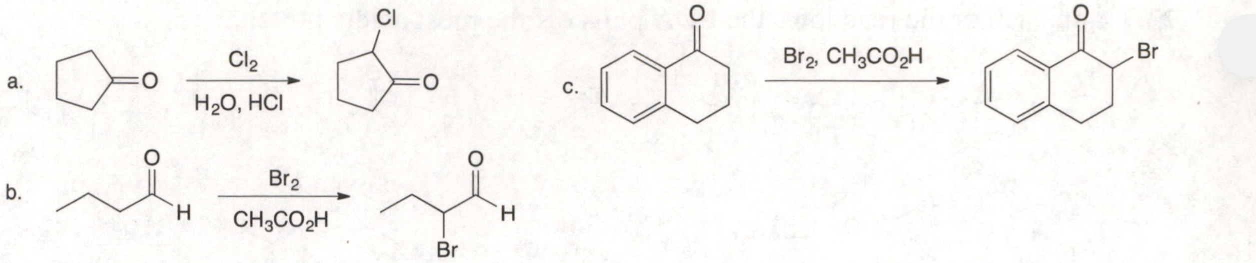

a. O → Cl$_2$, H$_2$O, HCl → Cl, O

b. O, H → Br$_2$, CH$_3$CO$_2$H → O, H, Br

c. O → Br$_2$, CH$_3$CO$_2$H → O, Br

23.12

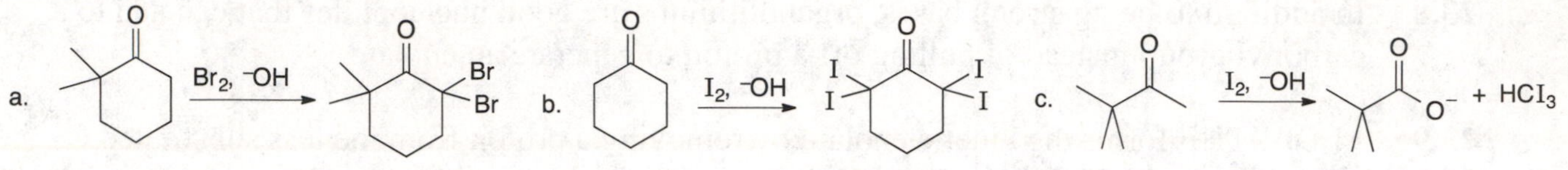

a. O → Br$_2$, $^-$OH → O, Br, Br

b. O → I$_2$, $^-$OH → O, I, I, I, I

c. O → I$_2$, $^-$OH → O, O$^-$ + HCI$_3$

23.13

a. O, Br → Li$_2$CO$_3$, LiBr, DMF → O

b. O, Br → CH$_3$CH$_2$NH$_2$ → O, NHCH$_2$CH$_3$

c. O, Br → CH$_3$SH → O, SCH$_3$

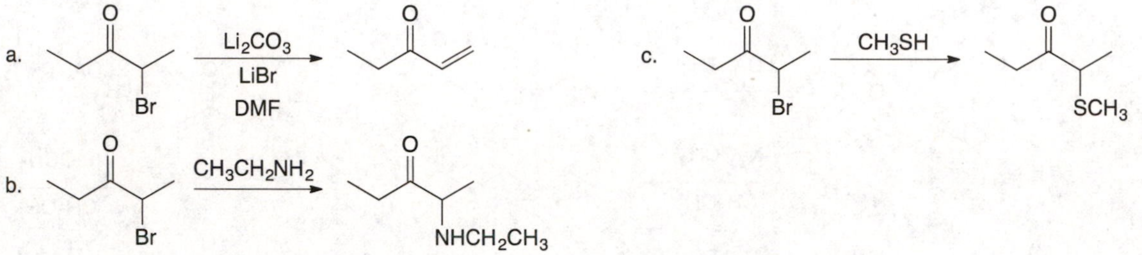

23.14 Bromination takes place on the α carbon to the carbonyl, followed by an S$_N$2 reaction with the nitrogen nucleophile.

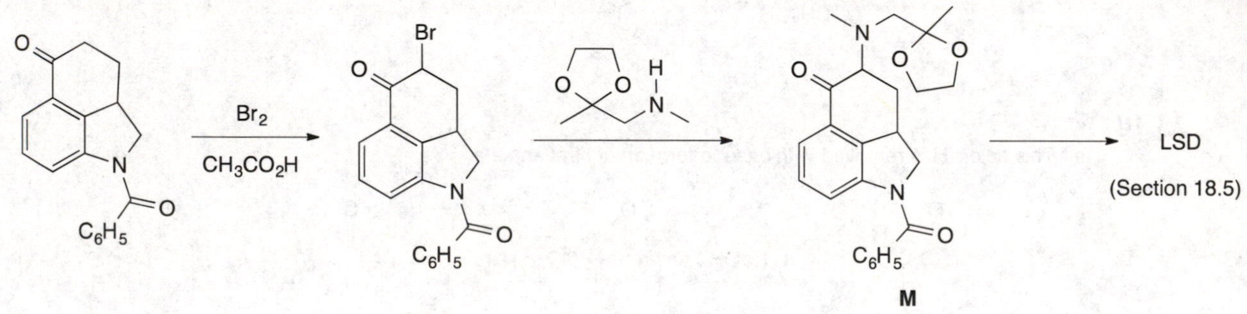

O, N, O, C$_6$H$_5$ → Br$_2$, CH$_3$CO$_2$H → Br, O, N, O, C$_6$H$_5$ → O, O, H, N → N, O, O, O, N, O, C$_6$H$_5$ **M** → LSD (Section 18.5)

23.15

a. O → [1] LDA, THF [2] CH$_3$CH$_2$I → O

b. O → [1] LDA, THF [2] CH$_3$CH$_2$I → O

c. O, O → [1] LDA, THF [2] CH$_3$CH$_2$I → O, O

d. CN → [1] LDA, THF [2] CH$_3$CH$_2$I → CN

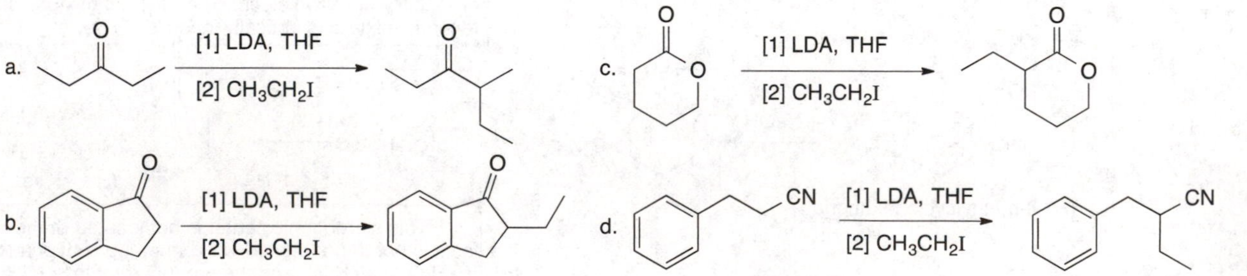

23.16

a. 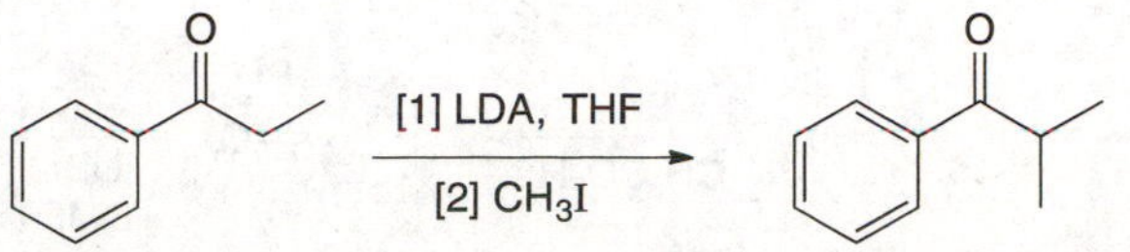 [1] LDA, THF; [2] CH_3I

b. 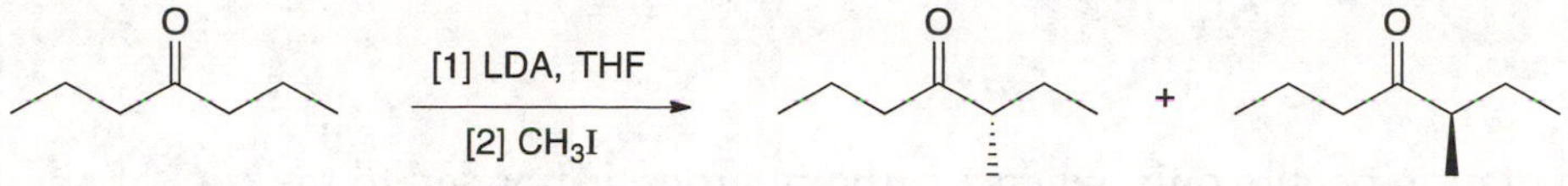 [1] LDA, THF; [2] CH_3I +

c. 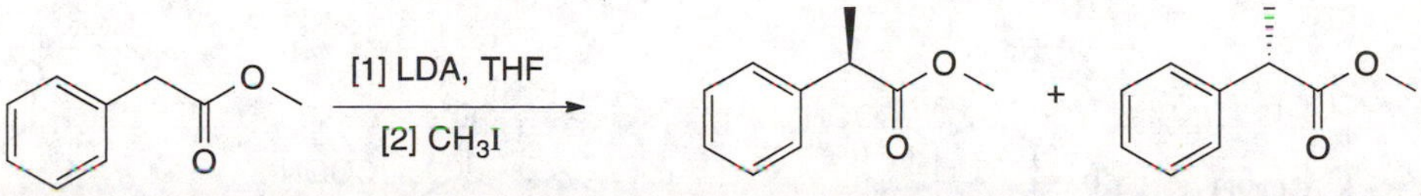 [1] LDA, THF; [2] CH_3I +

23.17 Three steps are needed: [1] formation of an enolate; [2] alkylation; [3] hydrolysis of the ester.

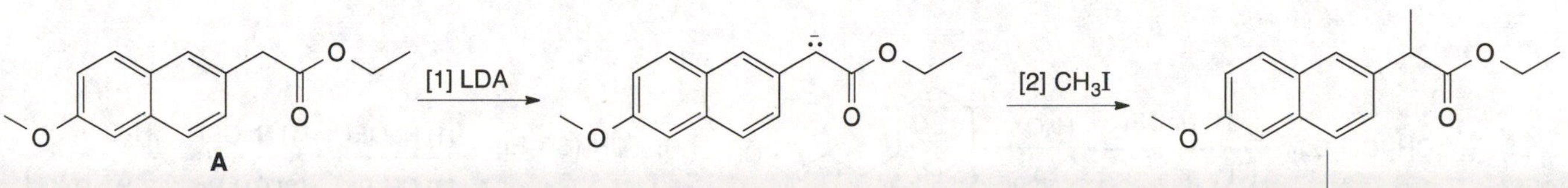

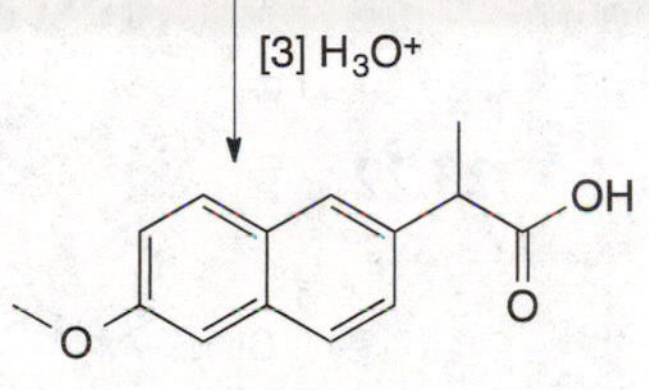

A [1] LDA → [2] CH_3I → [3] H_3O^+ → OH

The product is racemic because the new stereogenic center is formed by alkylation of a planar enolate with equal probability from above and below.

naproxen

23.18

a. 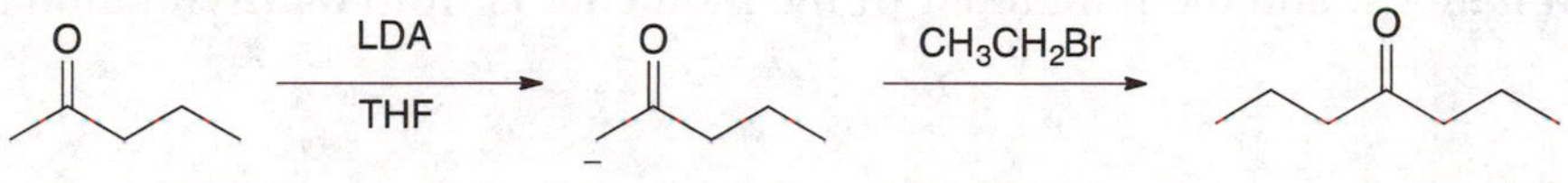 LDA / THF → CH_3CH_2Br →

b. 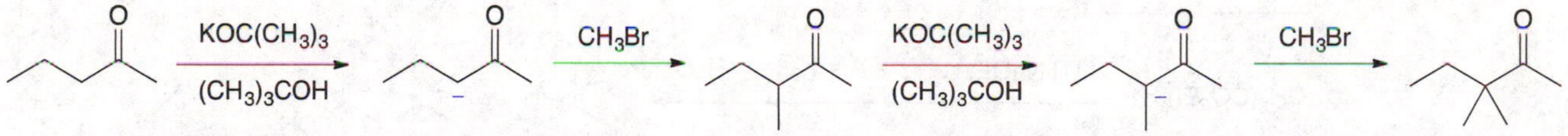 $KOC(CH_3)_3$ / $(CH_3)_3COH$ → CH_3Br → $KOC(CH_3)_3$ / $(CH_3)_3COH$ → CH_3Br →

c. 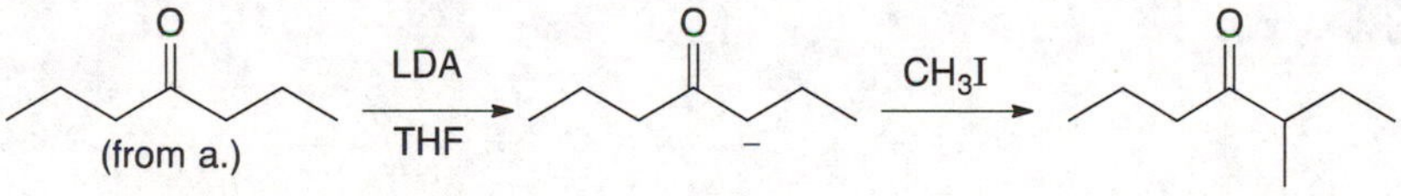 (from a.) LDA / THF → CH_3I →

d. 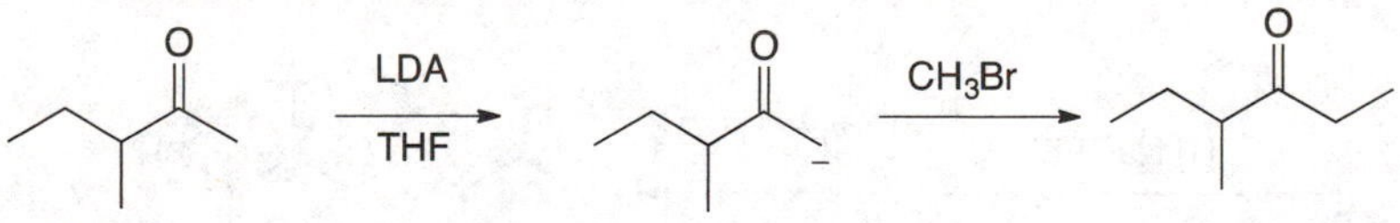 (from b.) LDA / THF → CH_3Br →

23.19

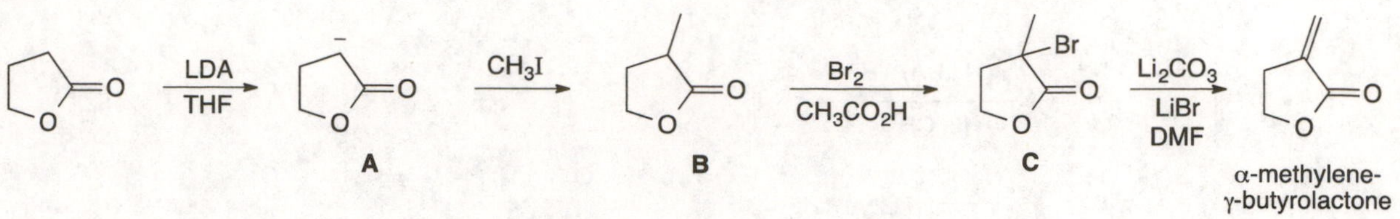

LDA, THF → A → CH_3I → B → Br_2, CH_3CO_2H → C → Li_2CO_3, LiBr, DMF → α-methylene-γ-butyrolactone

23.20 Decarboxylation occurs only when a carboxy group is bonded to the α C of another carbonyl group.

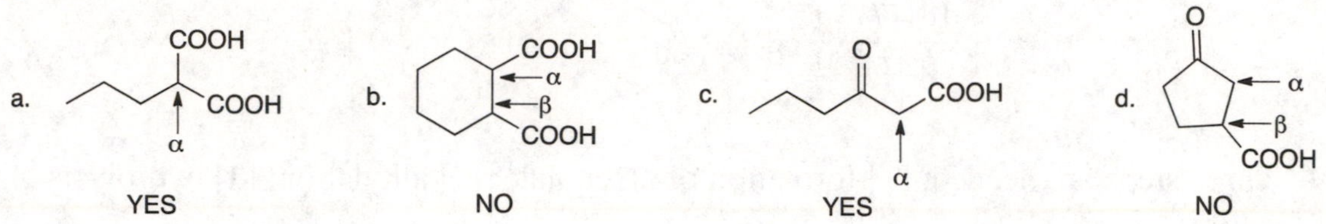

a. COOH, COOH, α — YES

b. COOH, α, β, COOH — NO

c. O, COOH, α — YES

d. O, α, β, COOH — NO

23.21

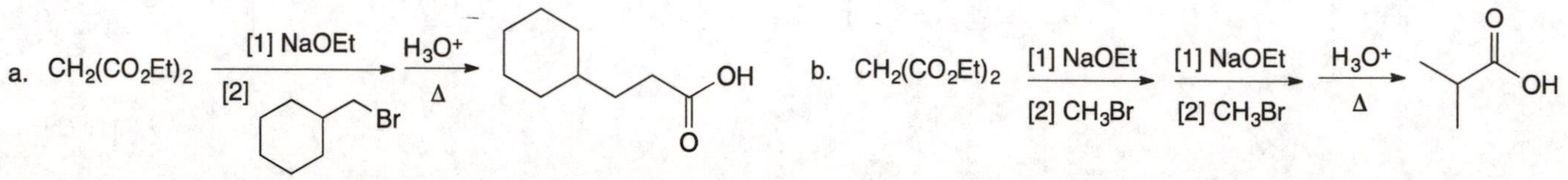

a. $CH_2(CO_2Et)_2$ → [1] NaOEt, [2] (cyclohexylmethyl bromide, Br) → H_3O^+, Δ → OH, O

b. $CH_2(CO_2Et)_2$ → [1] NaOEt, [2] CH_3Br → [1] NaOEt, [2] CH_3Br → H_3O^+, Δ → O, OH

23.22

a. Cl, Cl → COOH

b. Br, O, Br → O, COOH

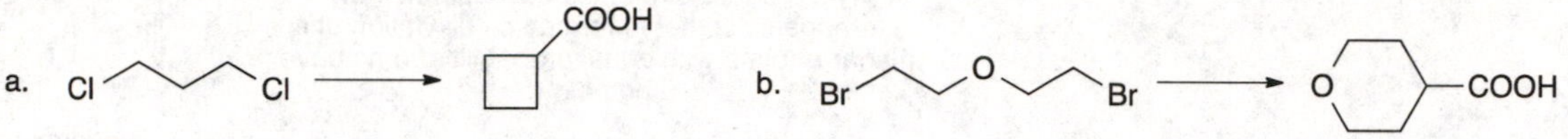

23.23 Locate the α C to the COOH group, and identify all of the alkyl groups bonded to it. These groups are from alkyl halides, and the remainder of the molecule is from diethyl malonate.

a. O, OH, α

$CH_2(CO_2Et)_2$ → [1] NaOEt, [2] Br → H_3O^+, Δ → O, OH

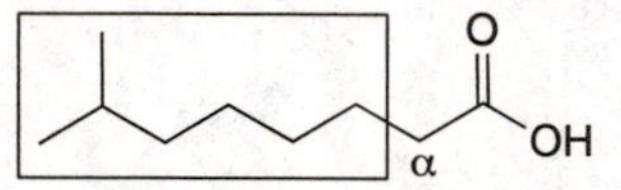

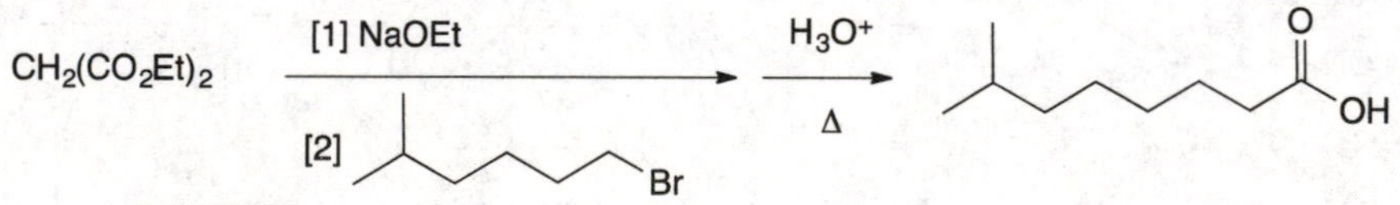

b. OH, α, O

$CH_2(CO_2Et)_2$ → [1] NaOEt, [2] Br → [1] NaOEt, [2] Br → H_3O^+, Δ → OH, O

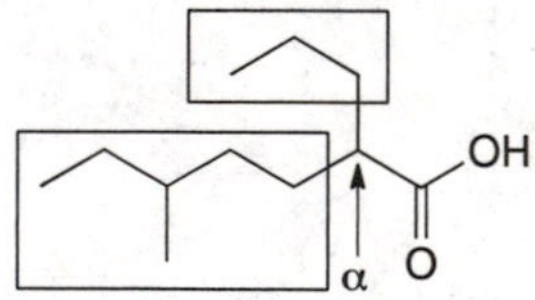

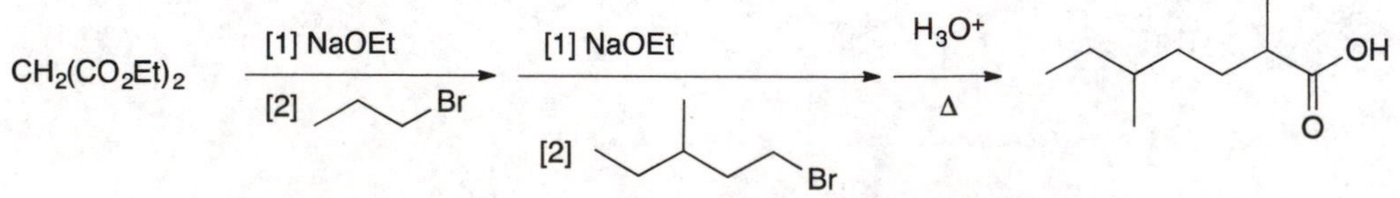

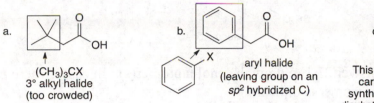

c.

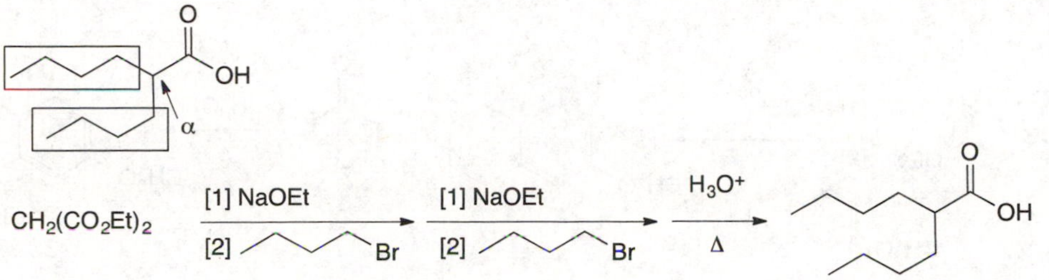
OH
α
$CH_2(CO_2Et)_2$
[1] NaOEt
[2]
Br
[1] NaOEt
[2]
Br
H_3O^+
Δ
O
OH

23.24 The reaction works best when the alkyl halide is 1° or CH_3X, because this is an S_N2 reaction.

a.

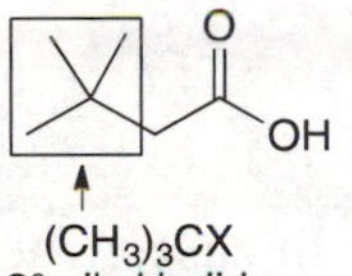
O
OH

$(CH_3)_3CX$
3° alkyl halide
(too crowded)

b.

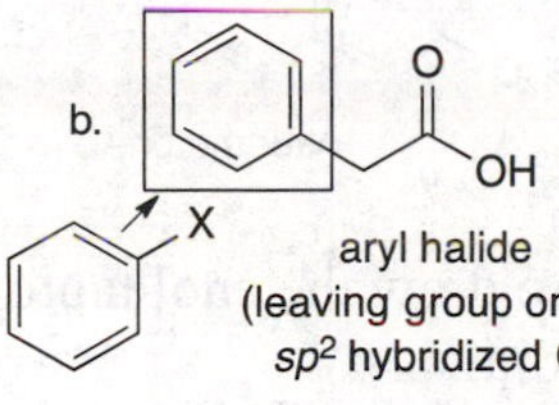
O
OH
X

aryl halide
(leaving group on an *sp*[2] hybridized C)

Aryl halides are unreactive in S_N2 reactions.

c.

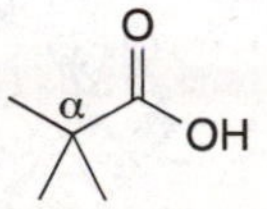
O
α
OH

This compound has 3 CH_3 groups on the α carbon to the COOH. The malonic ester synthesis can be used to prepare mono- and disubstituted carboxylic acids only: RCH_2COOH and $R_2CHCOOH$, but not R_3CCOOH.

23.25

a.

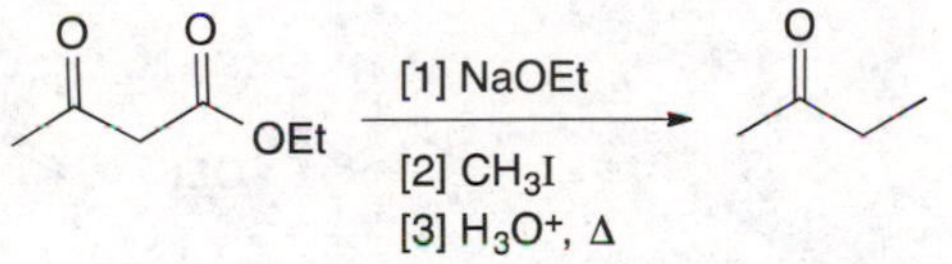
O
O
OEt
[1] NaOEt
[2] CH_3I
[3] H_3O^+, Δ
O

b.

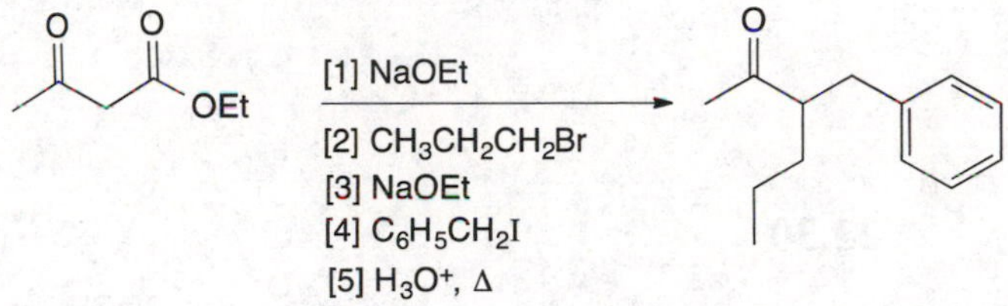
O
O
OEt
[1] NaOEt
[2] $CH_3CH_2CH_2Br$
[3] NaOEt
[4] $C_6H_5CH_2I$
[5] H_3O^+, Δ
O

23.26 Locate the α C. All alkyl groups on the α C come from alkyl halides, and the remainder of the molecule comes from ethyl acetoacetate.

a.

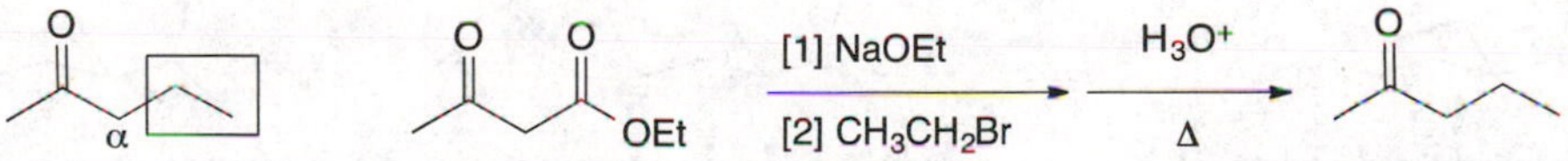
O
α
O
O
OEt
[1] NaOEt
[2] CH_3CH_2Br
H_3O^+
Δ
O

b.

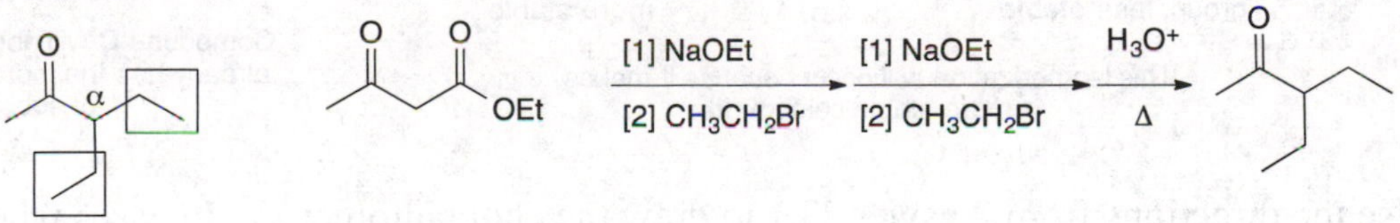
O
α
O
O
OEt
[1] NaOEt
[2] CH_3CH_2Br
[1] NaOEt
[2] CH_3CH_2Br
H_3O^+
Δ
O

c.

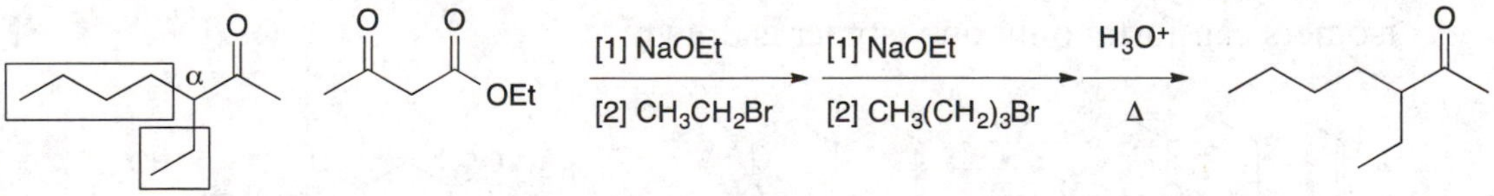
α
O
O
O
OEt
[1] NaOEt
[2] CH_3CH_2Br
[1] NaOEt
[2] $CH_3(CH_2)_3Br$
H_3O^+
Δ
O

23.27

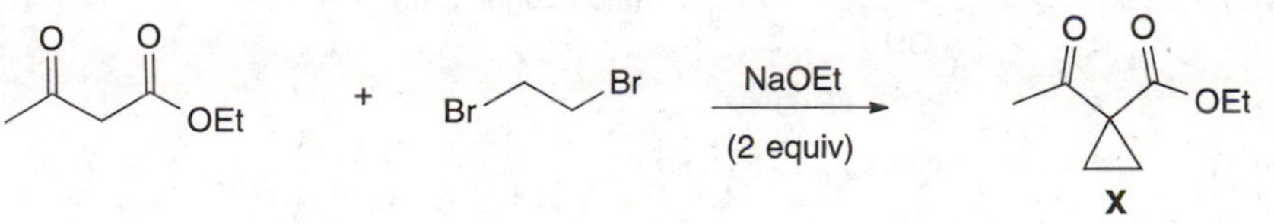
O
O
OEt
+
Br
Br
NaOEt
(2 equiv)
O
O
OEt
X

23.28

a. [1] NaOEt, [2] CH_3O-naphthyl-CH_2Br → (CO$_2$Et intermediate) → H_3O^+, Δ → nabumetone

OEt

Br

CH_3O

CO_2Et

nabumetone

b. [1] LDA, THF; [2] CH_3O-naphthyl-CH_2Br → nabumetone

Br

CH_3O

nabumetone

23.29 Use the directions from Answer 23.1 to draw the enol tautomer(s). In cases where *E* and *Z* isomers can form, only one isomer is drawn.

a. O, H ⇌ OH, H

b. O, O, OEt ⇌ OH O, OEt + OH O, OEt

unconjugated enol (less stable)

+

O OH, OEt + OH OH, OEt

23.30

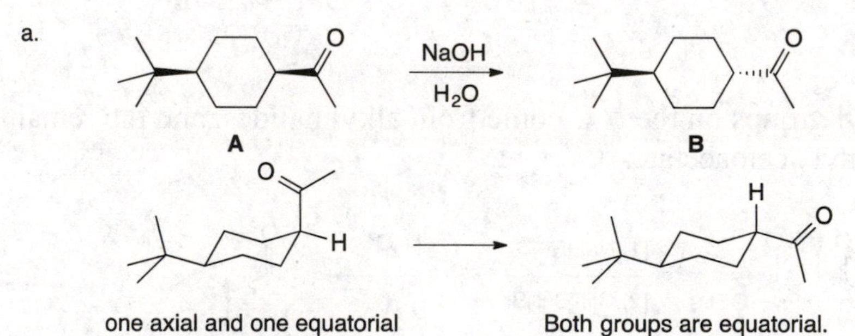
a.
NaOH
H_2O
A
B
H
H
one axial and one equatorial group, **less stable**
Both groups are equatorial. **more stable**
This isomerization will occur because it makes a more stable compound.

b. C

Both groups are equatorial = cis.

Compound **C** will not isomerize because it already has the more stable arrangement of substituents.

23.31 Use the directions from Answer 23.1 to draw the enol tautomer(s). In cases where *E* and *Z* isomers can form, only one isomer is drawn.

a. O ⇌ OH + OH

conjugated enol (more stable)

b. O, O ⇌ OH, O

c. O, O ⇌ HO, O + HO, O

(mono enol form)

conjugated enol (more stable)

23.32

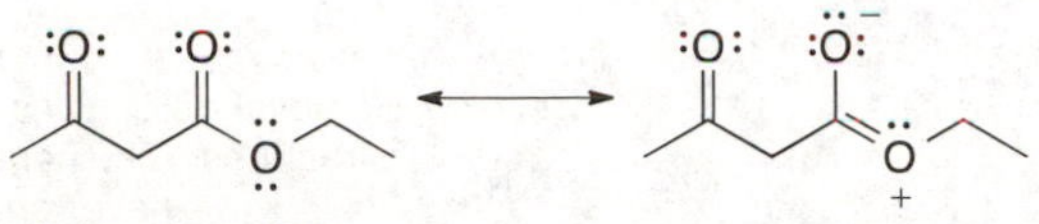

:O: :O: ↔ :O: :Ö:⁻ O⁺

ethyl acetoacetate

The ester C=O is resonance stabilized, and is therefore less available for tautomerization. Since the carbonyl form of the ester group is stabilized by electron delocalization, less enol is present at equilibrium.

23.33

a.

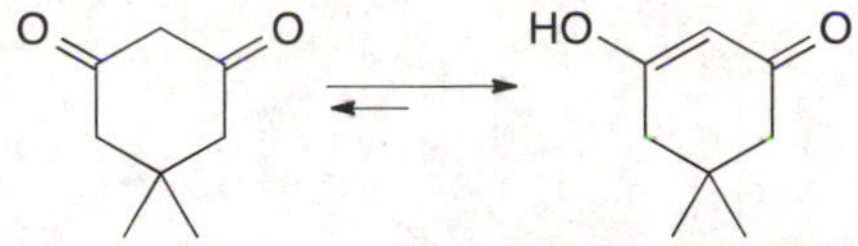

Conjugation stabilizes this enol.
higher percentage of enol

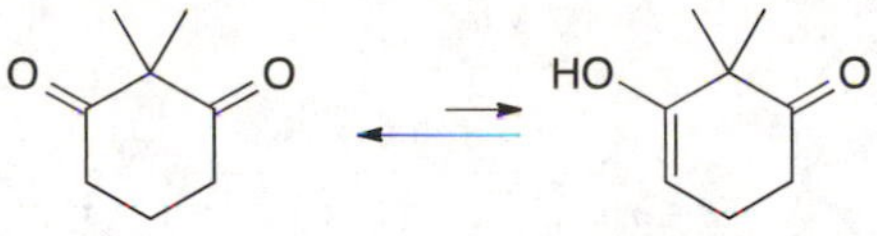

not conjugated

b.

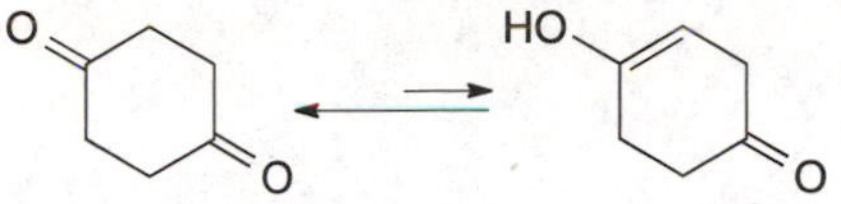

not conjugated

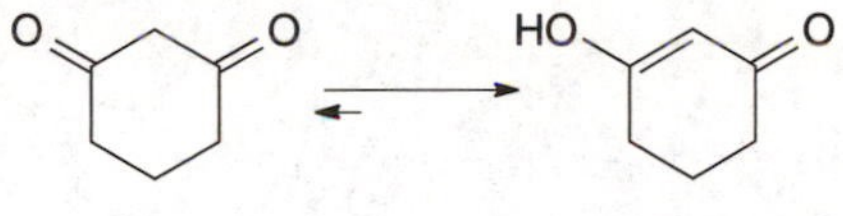

conjugated C=C
higher percentage of enol

23.34

a. b. 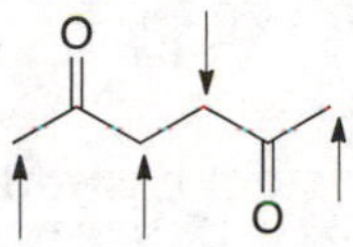 c.

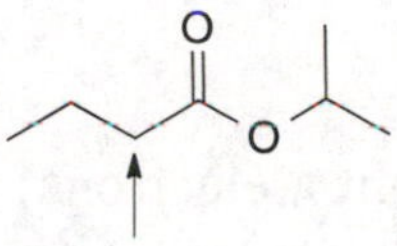

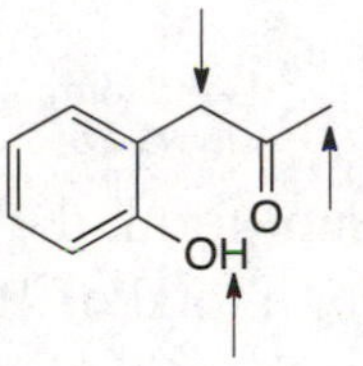

23.35

a. H_b O H_a H_c H

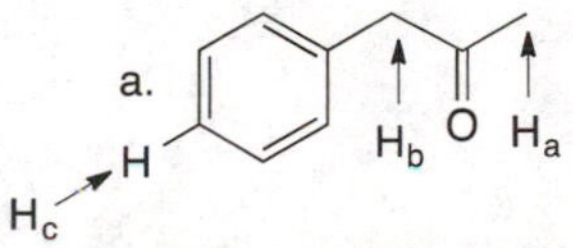

H_c is bonded to an *sp*^{2} hybridized C = **least acidic.**
H_a is bonded to an α C = **intermediate acidity.**
H_b is bonded to an α C, and is adjacent to a benzene ring = **most acidic.**

b. HO H_a H H H_b H_c O

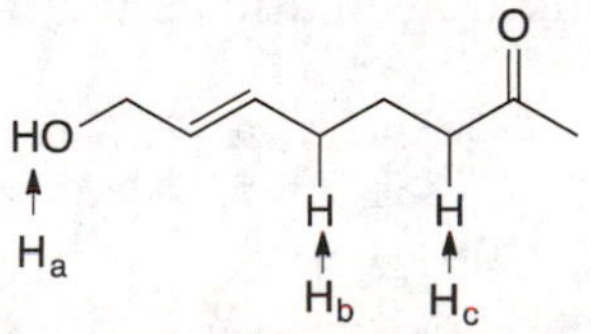

H_b is bonded to an *sp*^{3} hybridized C = **least acidic.**
H_c is bonded to an α C = **intermediate acidity.**
H_a is bonded to O = **most acidic.**

23.36

a. LDA / THF

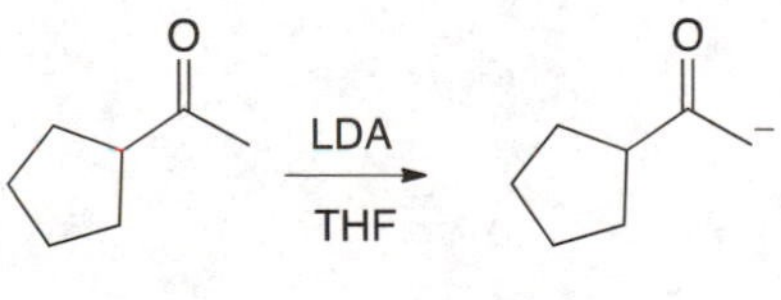

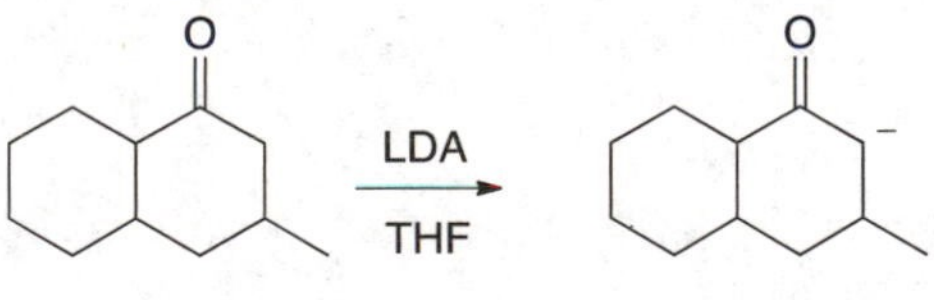

b. CN LDA / THF CN

c. LDA / THF

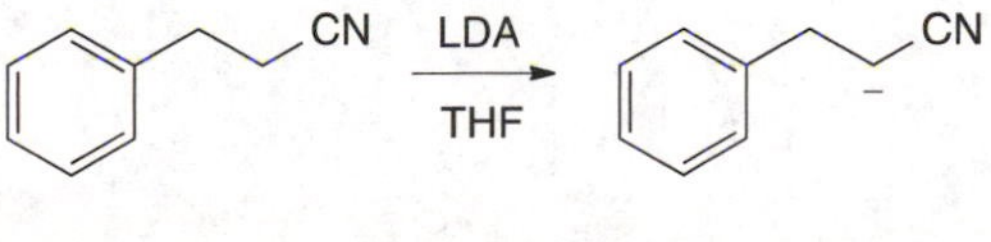

d. LDA / THF

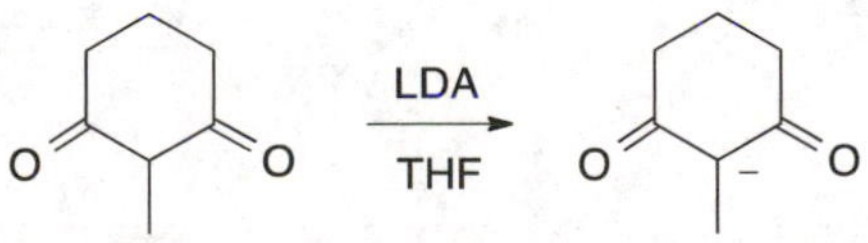

23.37

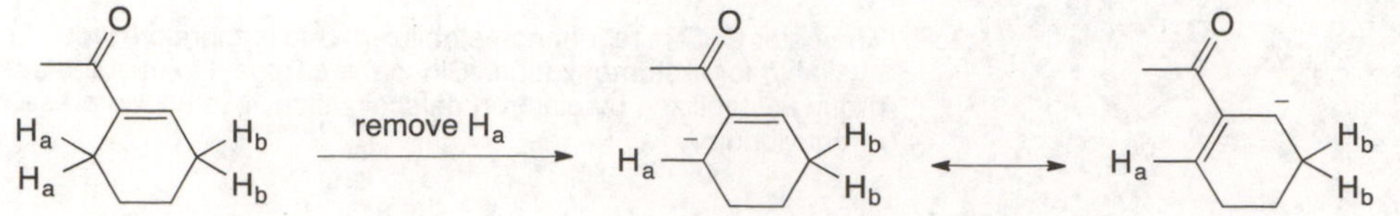

remove H_a

Removal of H_a gives two resonance structures. The negative charge is never on O.

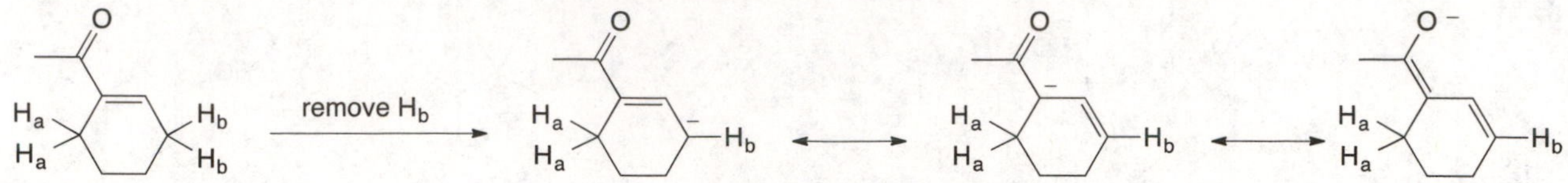

remove H_b

Removal of H_b gives three resonance structures. The negative charge is on O in one resonance structure, making the conjugate base more stable and H_b more acidic (lower pK_a).

23.38 a.

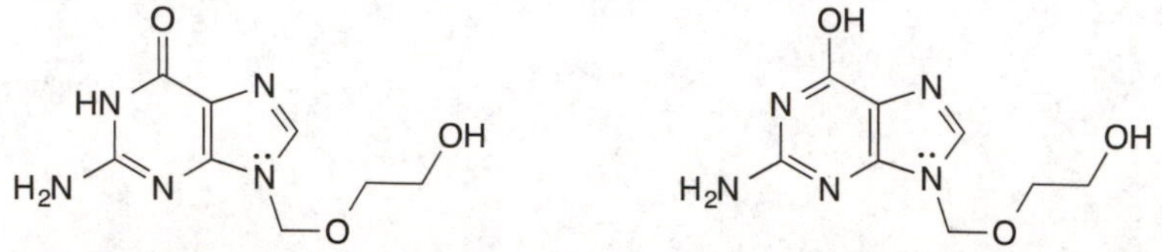

keto form
acyclovir

enol form

In the enol form, the bicyclic ring system has four π bonds (eight π electrons) and a lone pair on N, for a total of 10 π electrons. This makes it aromatic by Hückel's rule (Section 17.7).

b. The keto form of acyclovir can also be drawn in a resonance form that gives it 10 π electrons making it aromatic as well.

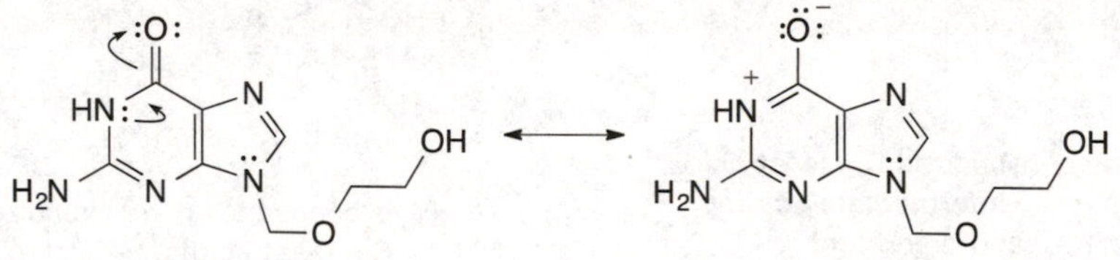

aromatic
completely conjugated
10 π electrons

23.39

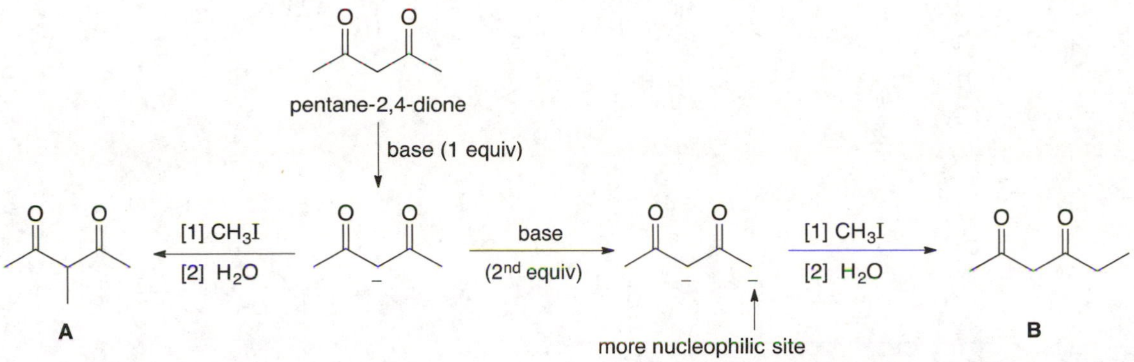

pentane-2,4-dione

base (1 equiv)

[1] CH_3I
[2] H_2O

A

base
(2nd equiv)

more nucleophilic site

[1] CH_3I
[2] H_2O

B

One equivalent of base removes the most acidic proton between the two C=O's, to form **A** on alkylation with CH_3I.

With a second equivalent of base a dianion is formed. Since the second enolate is less resonance stabilized, it is more nucleophilic and reacts first in an alkylation with CH_3I, forming **B** after protonation with H_2O.

23.40

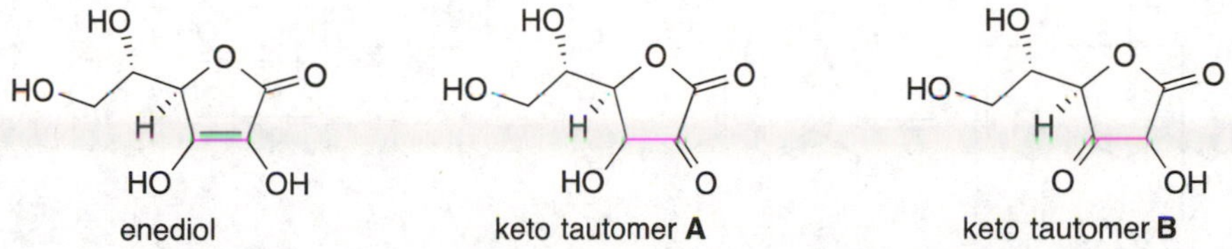

enediol

keto tautomer **A**

keto tautomer **B**

The enediol is more stable than either of the two keto tautomers because it is more highly conjugated. More resonance structures can be drawn, which delocalize the lone pairs of the two OH groups bonded to the C=C. Such delocalization is not possible with either keto tautomer.

23.41 The mechanism of acid-catalyzed halogenation consists of two parts: **tautomerization** of the carbonyl compound to the enol form and **reaction of the enol with halogen.**

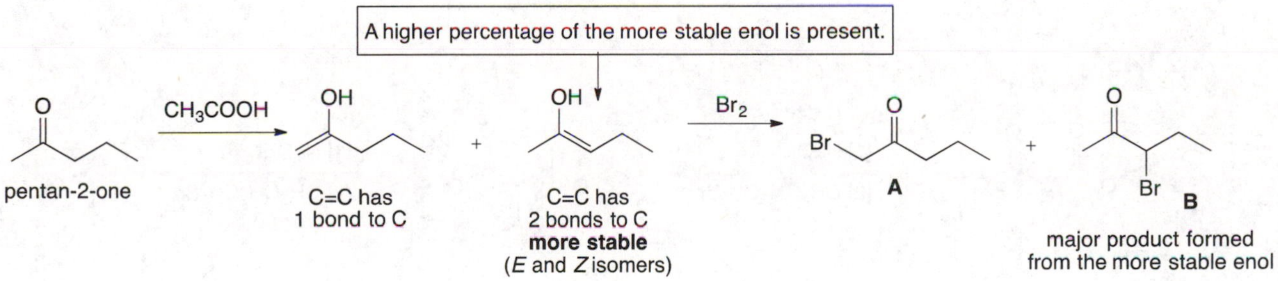

A higher percentage of the more stable enol is present.

pentan-2-one

CH_3COOH

C=C has
1 bond to C

C=C has
2 bonds to C
more stable
(*E* and *Z* isomers)

Br_2

A

B

major product formed
from the more stable enol

23.42 In the haloform reaction, the three H's of the CH_3 group are successively replaced by X, to form an intermediate that is oxidatively cleaved with base.

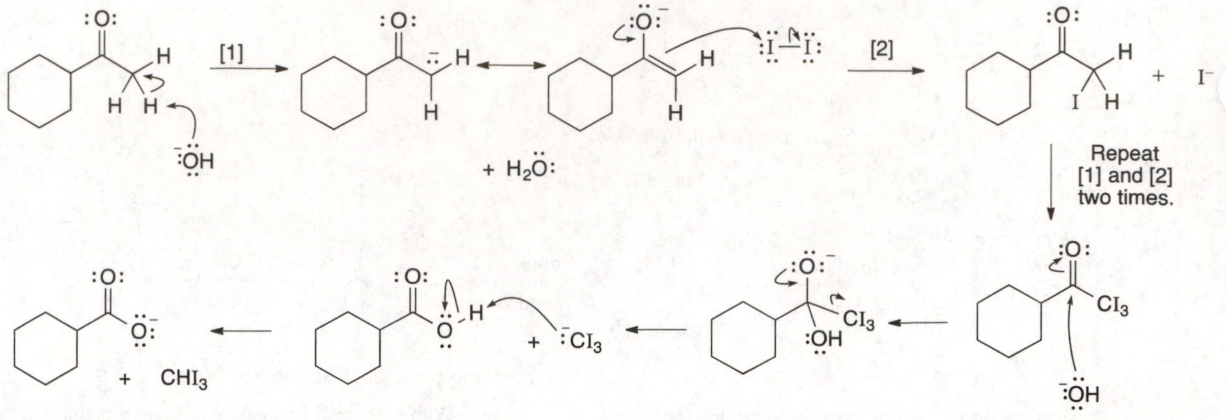

:O: H H H :ÖH [1] :O: H H + $H_2\ddot{O}$: :Ö:⁻ H H :I–I: [2] :O: H I H + I^-

Repeat [1] and [2] two times.

:O: CI_3 :ÖH → :Ö:⁻ CI_3 :ÖH → :O: Ö H + $:CI_3$ → :O: Ö:⁻ + CHI_3

23.43 Use the directions from Answer 23.23.

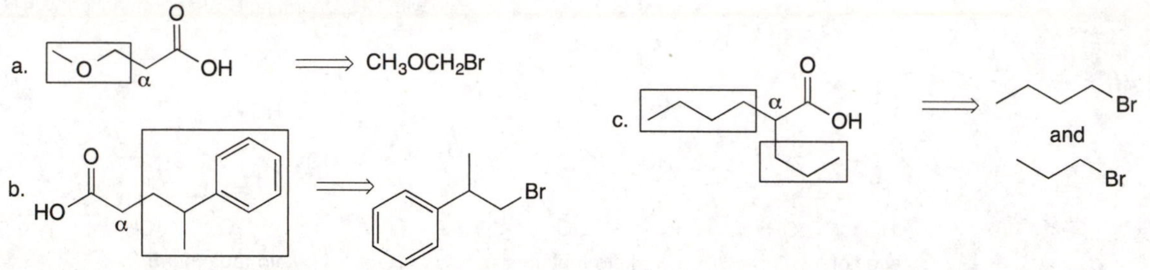

a. O α OH $\Longrightarrow CH_3OCH_2Br$

b. HO O α $\Longrightarrow$ Br

c. α O OH $\Longrightarrow$ Br and Br

23.44

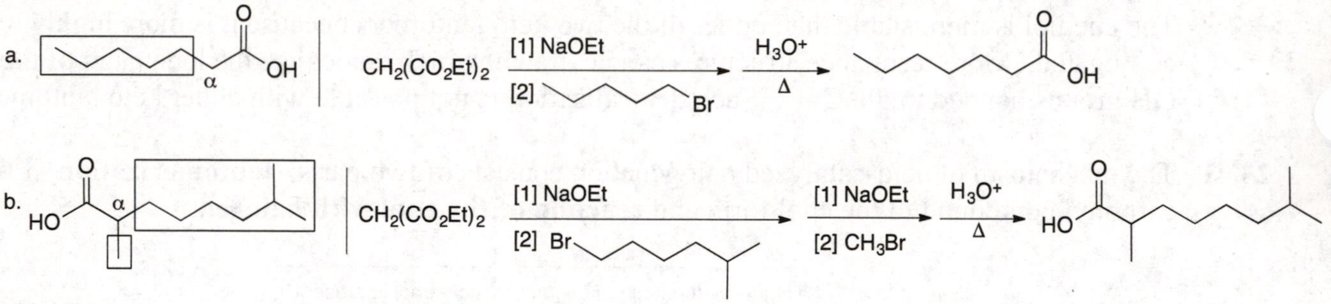

a. α O OH | $CH_2(CO_2Et)_2$ [1] NaOEt [2] Br $\rightarrow$ H_3O^+ Δ $\rightarrow$ O OH

b. O HO α | $CH_2(CO_2Et)_2$ [1] NaOEt [2] Br $\rightarrow$ [1] NaOEt [2] CH_3Br $\rightarrow$ H_3O^+ Δ $\rightarrow$ O HO

23.45

O O O O [1] NaOEt [2] $CH_3CH_2CH_2Br$ $\rightarrow$ O O O O [1] NaOEt [2] $CH_3CH_2CH_2Br$ $\rightarrow$ O O O O

H_3O^+ Δ ↓

O H O

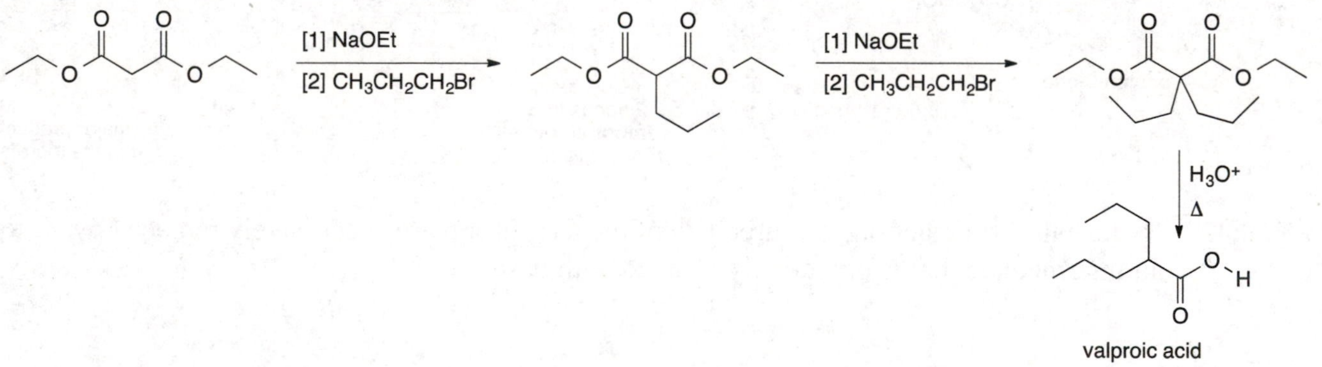

valproic acid

23.46

a. $CH_2(CO_2Et)_2$ —[1] NaOEt; [2] $BrCH_2CH_2CH_2CH_2CH_2Br$; [3] NaOEt→ —H_3O^+, Δ→ cyclohexyl–COOH —[1] $LiAlH_4$; [2] H_2O→ cyclohexyl–CH_2OH

b. cyclohexyl–COOH (from a.) —CH_3OH, H_2SO_4→ cyclohexyl–CO_2CH_3 —[1] CH_3MgBr (2 equiv); [2] H_2O→ OH

23.47

a. (epoxide; nucleophilic attack here) —[1] $Na^+ {}^-CH(COOEt)_2$; [2] H_2O→ EtO, OEt, OH

b. H, H (formaldehyde) —[1] $Na^+ {}^-CH(COOEt)_2$; [2] H_2O→ EtO, OEt, OH

c. (acetyl chloride, Cl) —[1] $Na^+ {}^-CH(COOEt)_2$; [2] H_2O→ EtO, OEt

d. (acetic anhydride) —[1] $Na^+ {}^-CH(COOEt)_2$; [2] H_2O→ EtO, OEt + CH_3COOH

23.48 Use the directions from Answer 23.26.

a. (α) ethyl acetoacetate (OEt) —[1] NaOEt; [2] Br–(alkyl bromide)→ —H_3O^+, Δ→ ketone product

b. (α) ethyl acetoacetate (OEt) —[1] NaOEt; [2] CH_3CH_2Br→ —[1] NaOEt; [2] Br–CH_2–cyclohexyl→ —H_3O^+, Δ→ ketone product

c. (α) ethyl acetoacetate (OEt) —[1] NaOEt; [2] Br–(dibromide)–Br; [3] NaOEt→ —H_3O^+, Δ→ ketone product

23.49

a. ethyl acetoacetate (OEt) —[1] NaOEt; [2] CH_3Br→ —[1] NaOEt; [2] CH_3Br→ —H_3O^+, Δ→ ketone product

b. ketone (from a.) —LDA, THF→ enolate —CH_3I→ ketone product

23.50

a. Δ

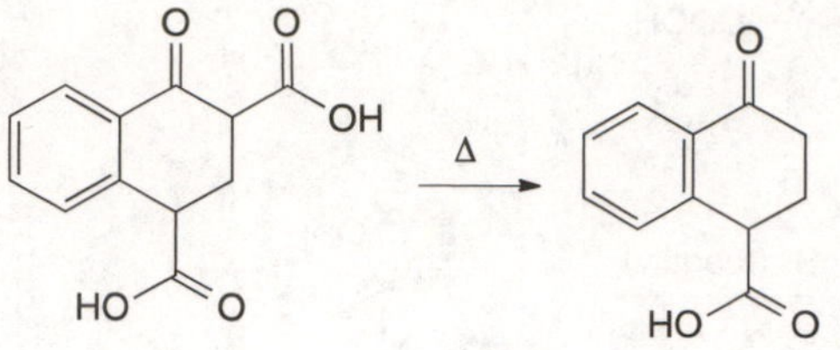

b. [1] LDA [2] CH_3CH_2I OEt OEt

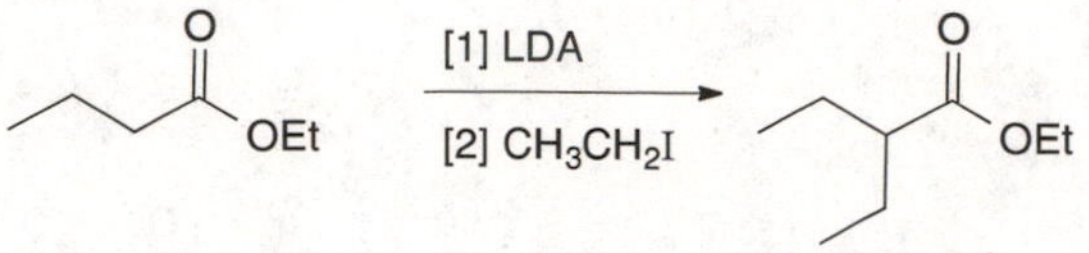

c. Br NH_2 NH

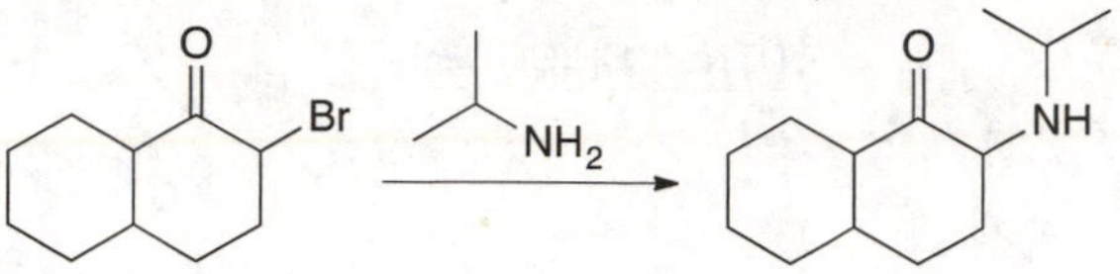

d. [1] LDA [2] CH_3CH_2I

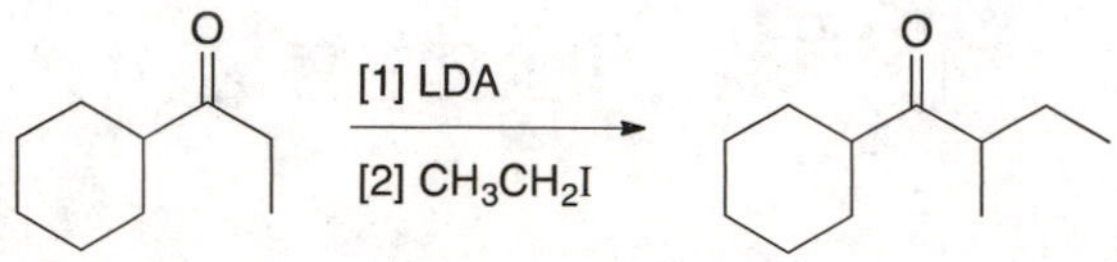

e. [1] Br_2, CH_3CO_2H [2] Li_2CO_3, LiBr, DMF

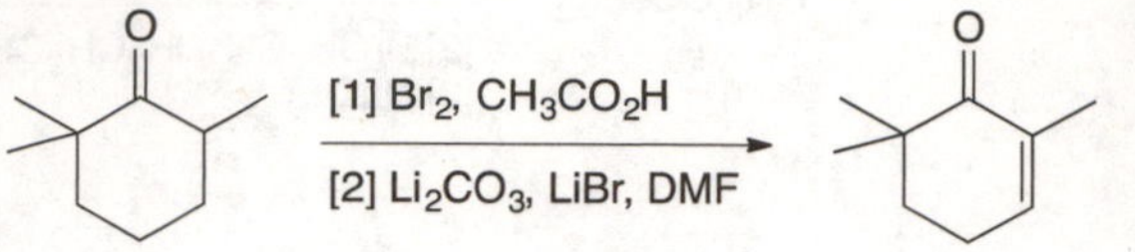

f. I_2 (excess) ^-OH + CHI_3

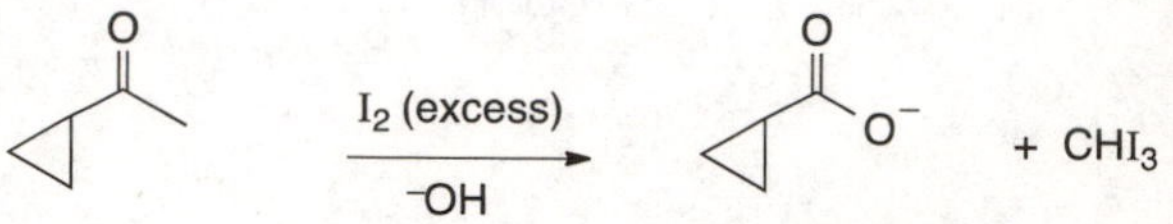

g. Cl CN NaH C≡N

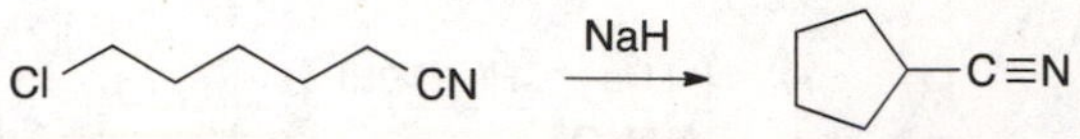

h. Br_2 (excess) ^-OH Br Br

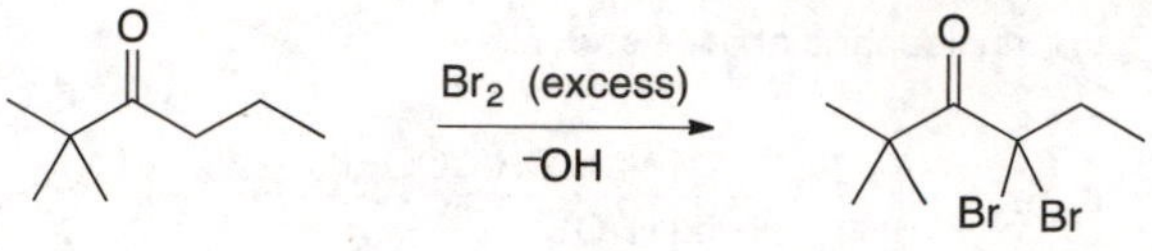

23.51

a. [1] LDA [2] Cl +

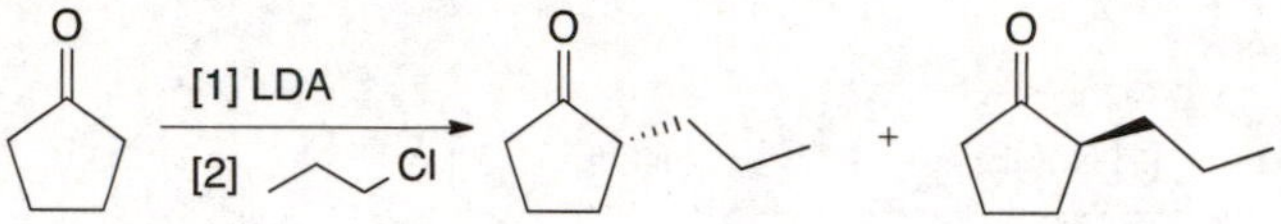

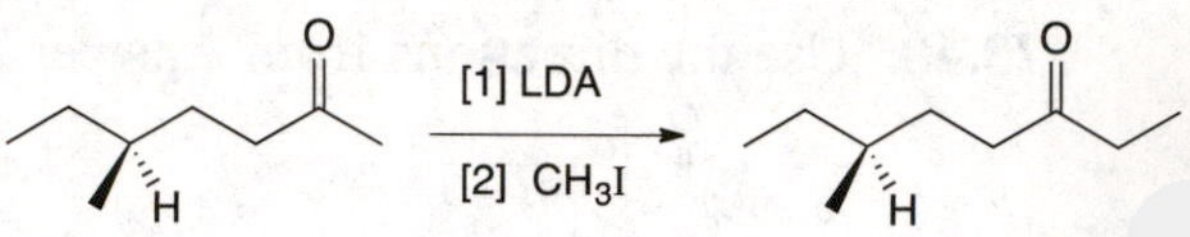

b. [1] LDA [2] H D I D H

c. H [1] LDA [2] CH_3I H

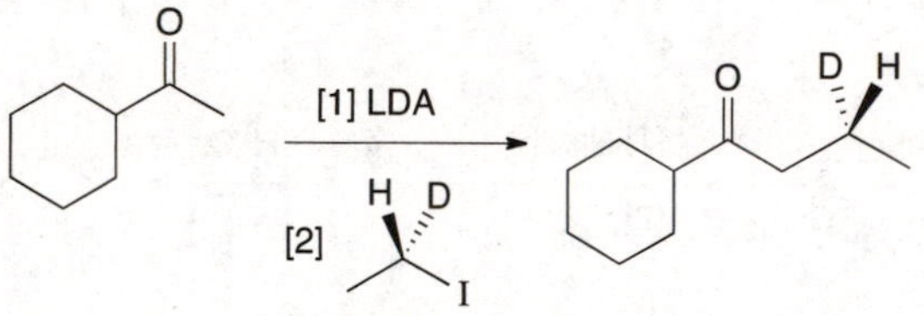

23.52

a. CHO $NaBH_4$ CH_3OH OH [1] PBr_3 [2] NaCN CN [1] LDA [2] CH_3I CN

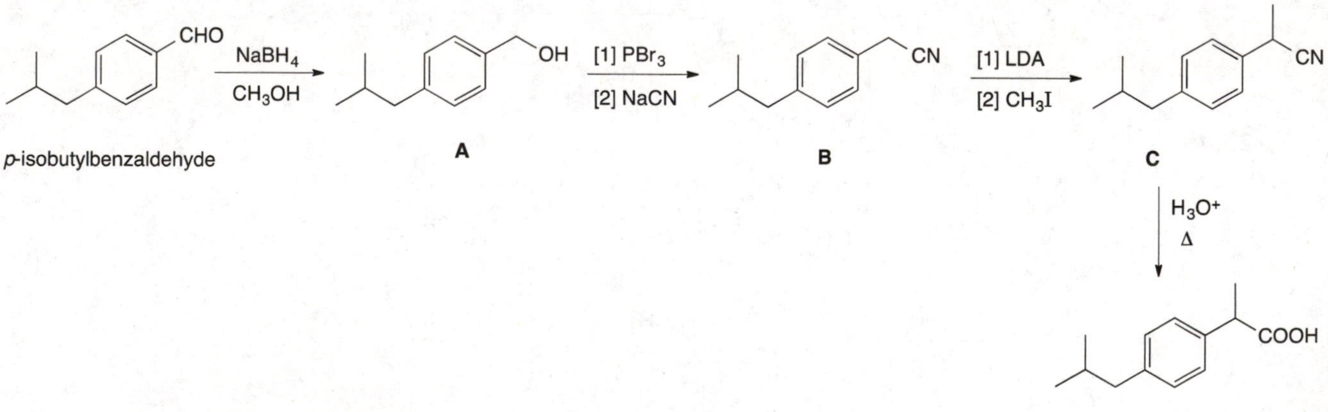

p-isobutylbenzaldehyde A B C

H_3O^+ Δ

COOH

ibuprofen

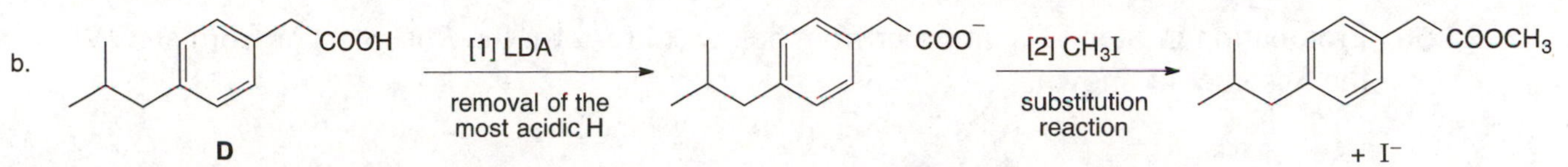

b. [1] LDA — removal of the most acidic H; [2] CH_3I — substitution reaction

D

$+ I^-$

Removal of the most acidic proton with LDA forms a carboxylate anion that reacts as a nucleophile with CH_3I to form an ester as substitution product.

23.53

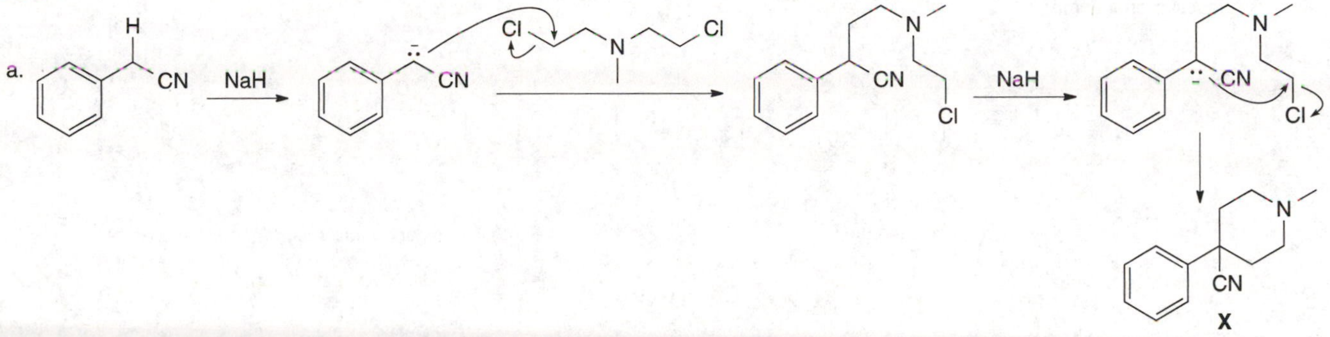

a. NaH → … → NaH → X

b. **X** can be converted to meperidine by hydrolysis of the nitrile and esterification.

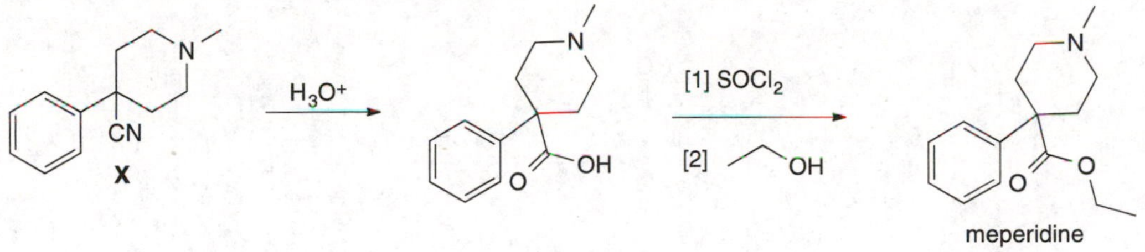

X — H_3O^+ → — [1] $SOCl_2$ [2] ethanol → meperidine

23.54

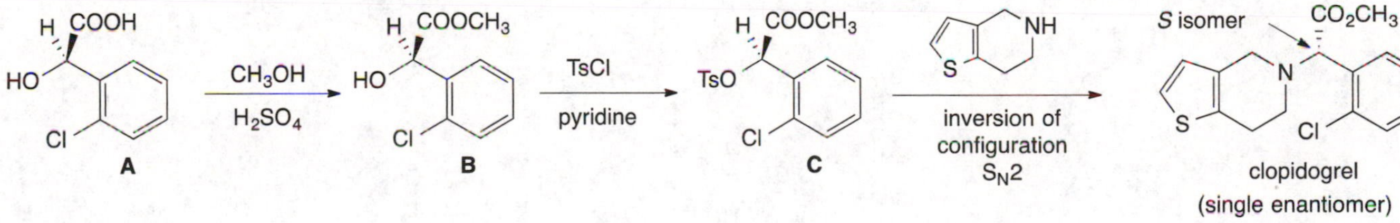

A — CH_3OH, H_2SO_4 → B — TsCl, pyridine → C — inversion of configuration S_N2 → clopidogrel (single enantiomer)

S isomer

23.55

A — LDA, THF, –78 °C → B

A — $KOC(CH_3)_3$, $(CH_3)_3COH$, room temperature → C

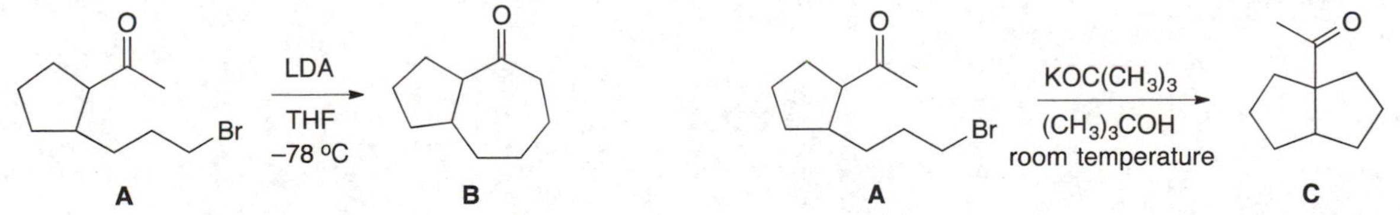

23.56 Protonation in Step [3] can occur from below (to re-form the *R* isomer) or from above to form the *S* isomer as shown.

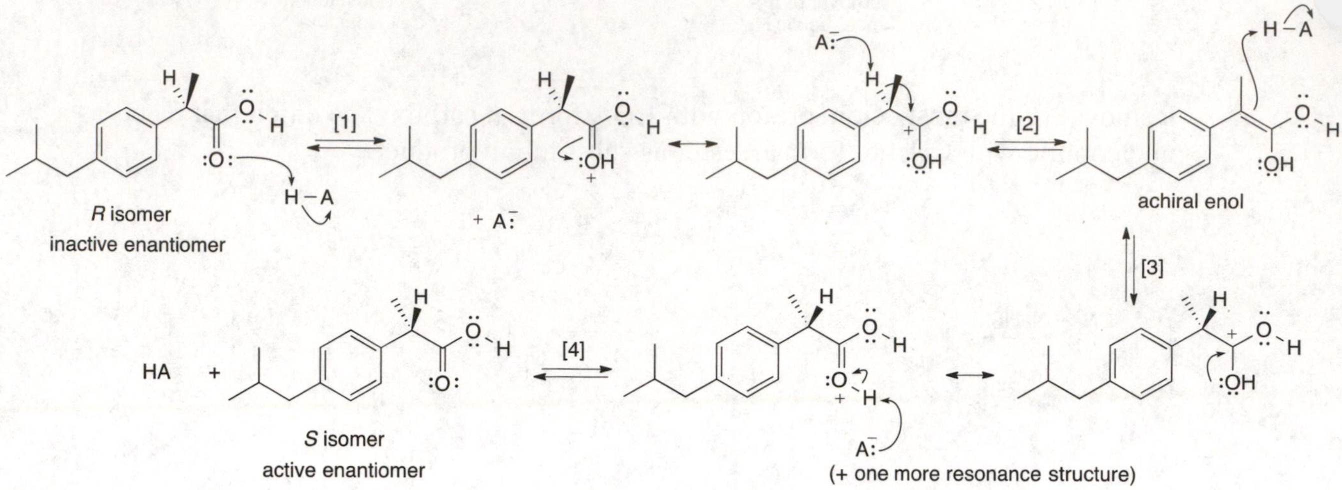

R isomer
inactive enantiomer
[1]
+ A:⁻
[2]
achiral enol
[3]
HA +
[4]
S isomer
active enantiomer
(+ one more resonance structure)

23.57

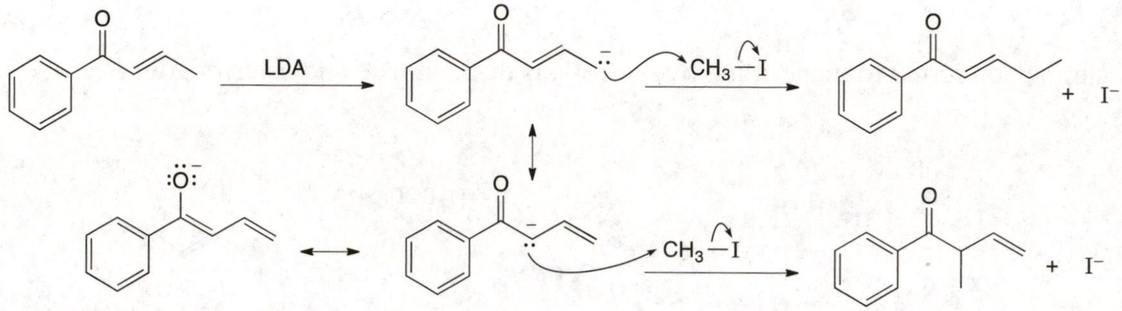

LDA
CH_3—I
+ I^-

23.58

EtO OEt
:ÖEt
+ HÖEt
CO_2Et
H_3O^+

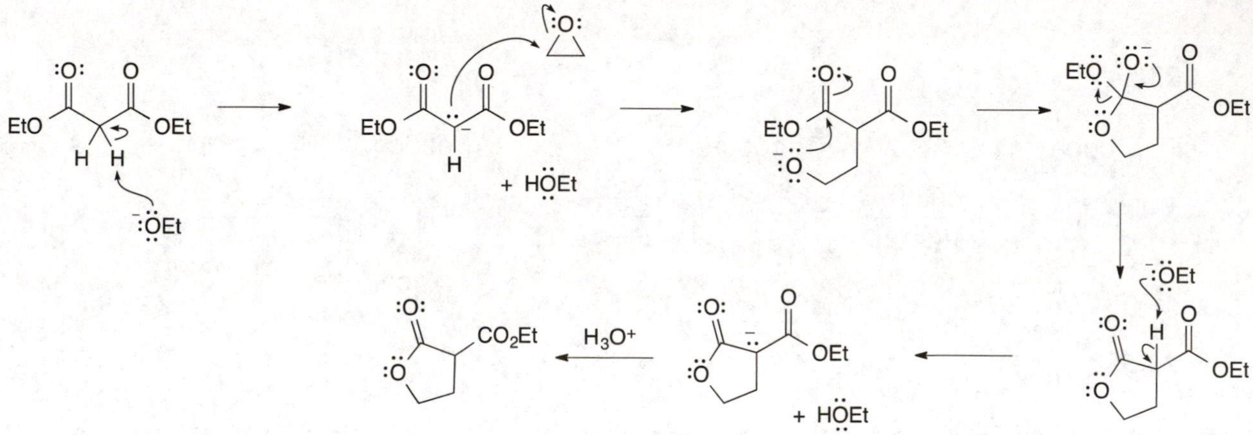

23.59 Protons on the γ carbon of an α,β-unsaturated carbonyl compound are acidic because of resonance.

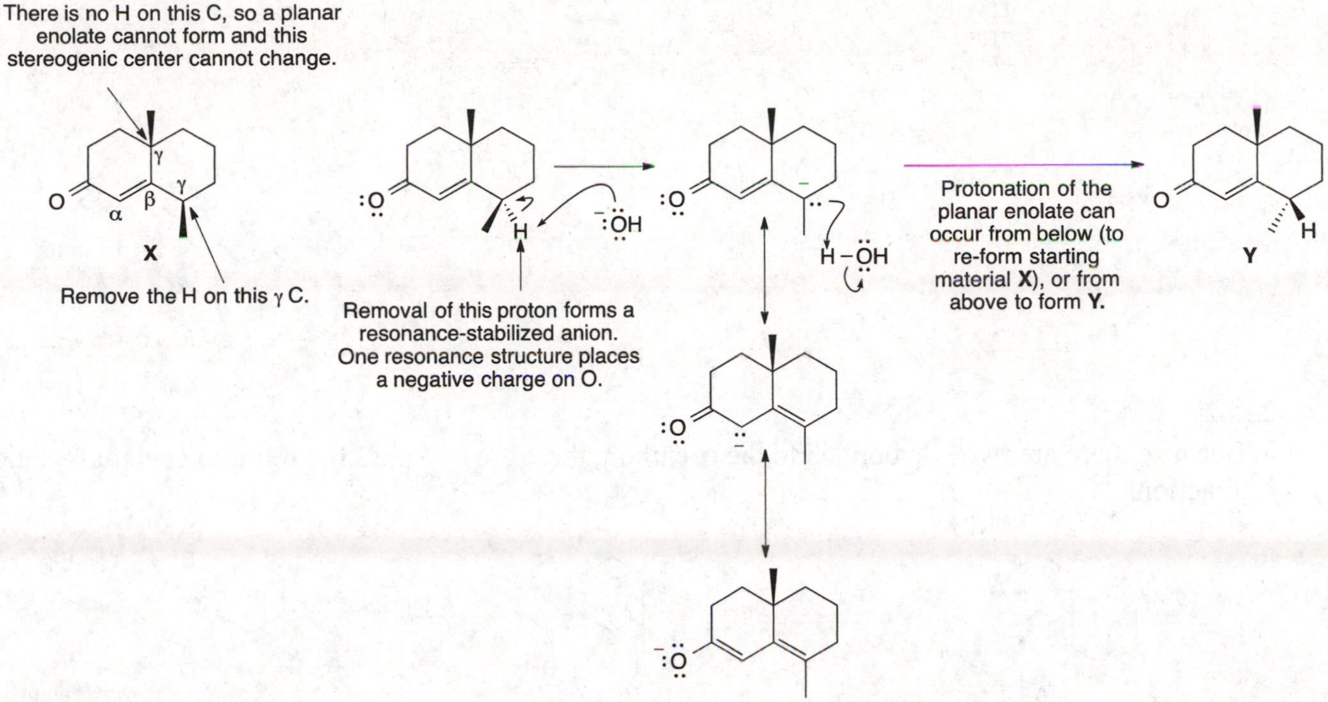

There is no H on this C, so a planar enolate cannot form and this stereogenic center cannot change.

γ γ α β X

Remove the H on this γ C.

Removal of this proton forms a resonance-stabilized anion. One resonance structure places a negative charge on O.

$^{-}$:OH H–OH

Protonation of the planar enolate can occur from below (to re-form starting material **X**), or from above to form **Y**.

Y

23.60

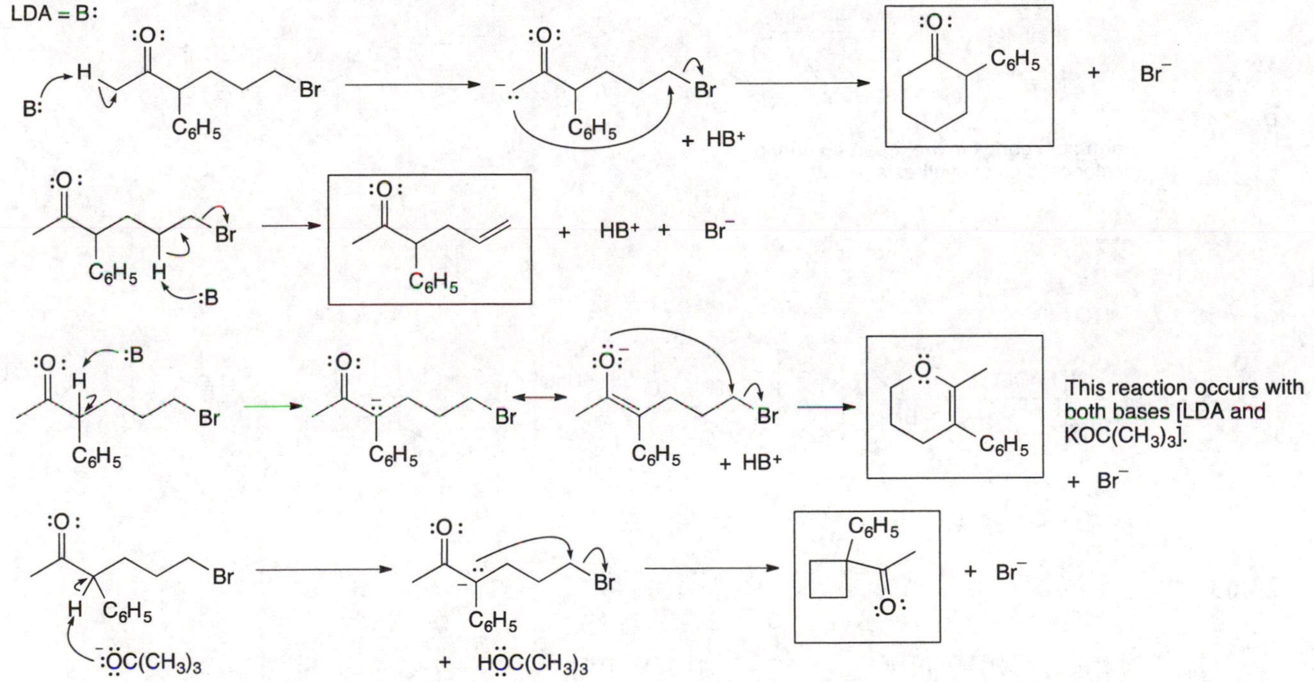

LDA = B:

B: H Br C_6H_5 → Br C_6H_5 + HB^+ → C_6H_5 + Br^-

Br C_6H_5 H :B → C_6H_5 + HB^+ + Br^-

:B H Br C_6H_5 → Br C_6H_5 ↔ Br C_6H_5 + HB^+ → C_6H_5 + Br^-

This reaction occurs with both bases [LDA and $KOC(CH_3)_3$].

Br H C_6H_5 $:\ddot{O}C(CH_3)_3$ → Br C_6H_5 + $H\ddot{O}C(CH_3)_3$ → C_6H_5 + Br^-

23.61

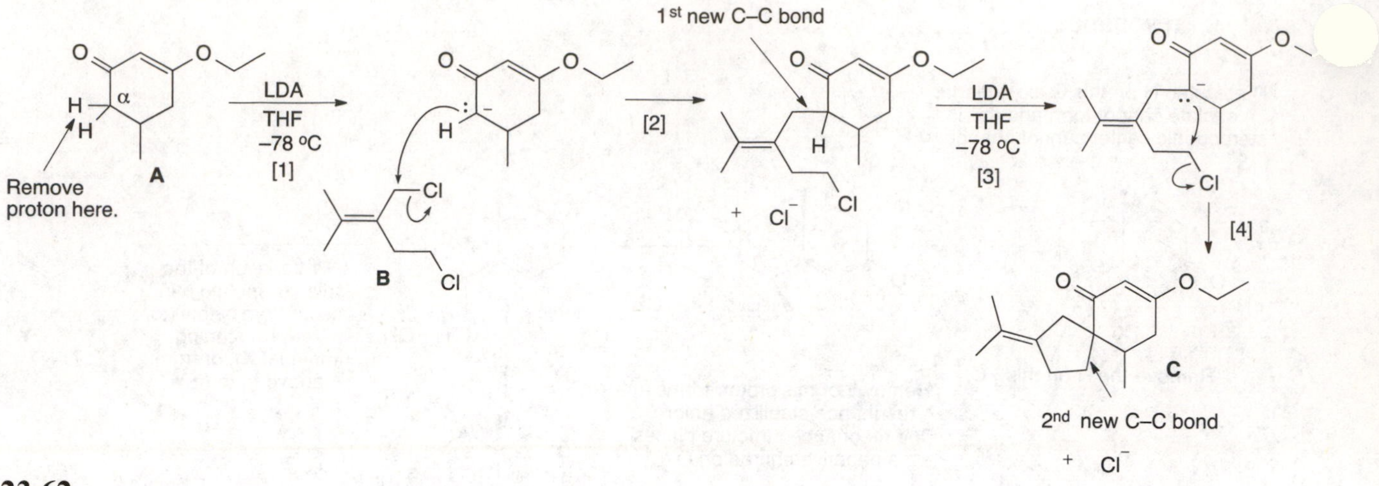

H, α, H; Remove proton here. A —LDA, THF, −78 °C [1]→ (enolate) + B (Cl, Cl) —[2]→ 1st new C–C bond; + Cl^- —LDA, THF, −78 °C [3]→ —[4]→ C; 2nd new C–C bond; + Cl^-

23.62

a. Because there are two C's bonded to the α carbon, there are two possible intramolecular alkylation reactions.

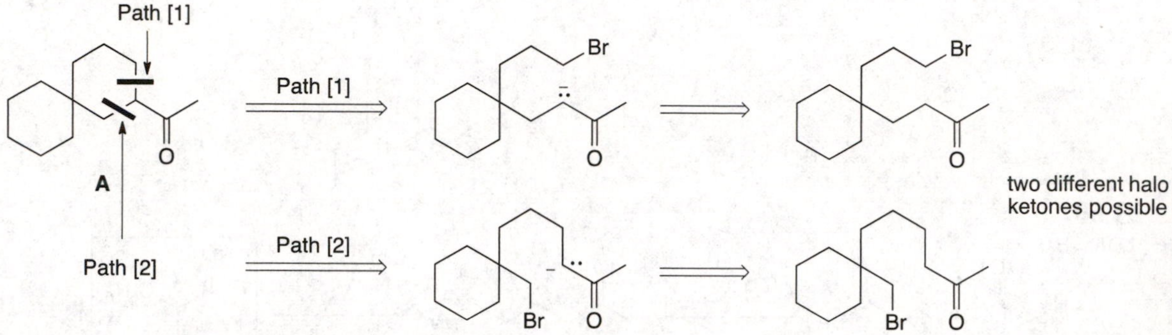

A: Path [1] ⟹ (carbanion, Br) ⟹ (bromo ketone)

Path [2] ⟹ (carbanion, Br) ⟹ (bromo ketone)

two different halo ketones possible

b.

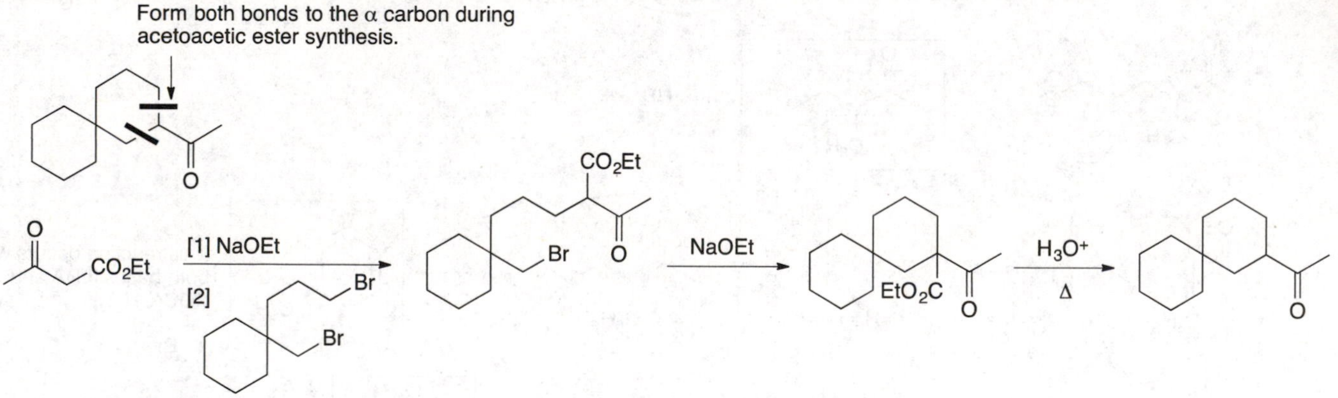

Form both bonds to the α carbon during acetoacetic ester synthesis.

$CH_3COCH_2CO_2Et$ —[1] NaOEt; [2] (dibromide, Br, Br)→ (CO_2Et, Br, O) —NaOEt→ (EtO_2C, O) —H_3O^+, Δ→ (spiro ketone)

23.63

a. cyclohexanone —[1] LDA, THF; [2] CH_3Br→ 2-methylcyclohexanone —[1] LDA, THF; [2] CH_3CH_2Br→ 2-ethyl-6-methylcyclohexanone

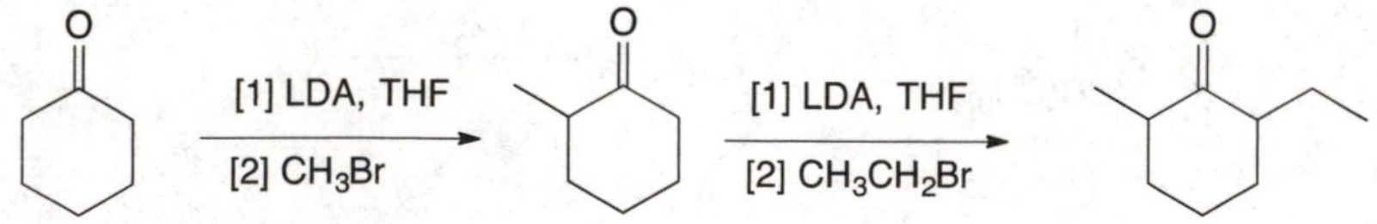

b.

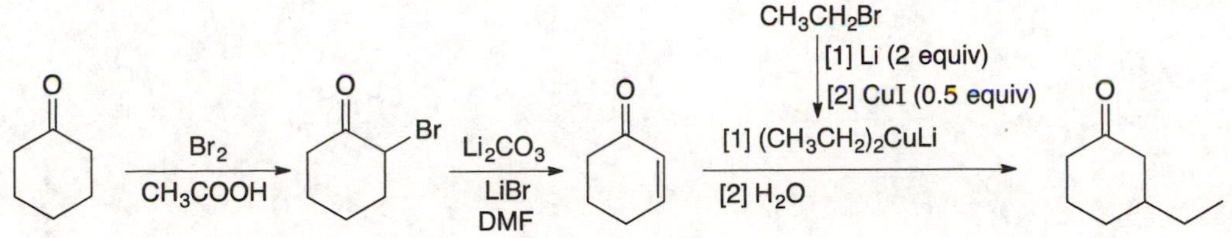

Br_2 / CH_3COOH

Li_2CO_3 / LiBr / DMF

CH_3CH_2Br → [1] Li (2 equiv) [2] CuI (0.5 equiv)

[1] $(CH_3CH_2)_2CuLi$ [2] H_2O

c.

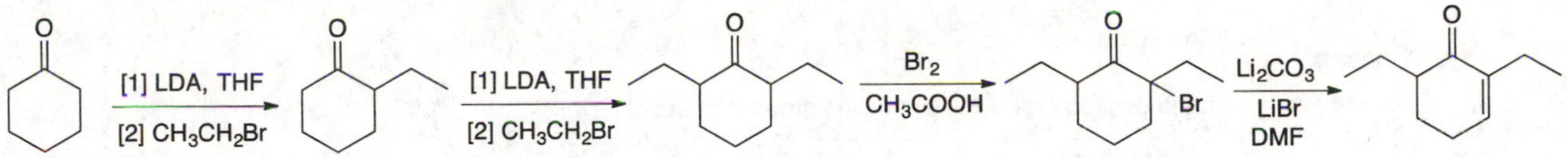

[1] LDA, THF [2] CH_3CH_2Br

[1] LDA, THF [2] CH_3CH_2Br

Br_2 / CH_3COOH

Li_2CO_3 / LiBr / DMF

23.64

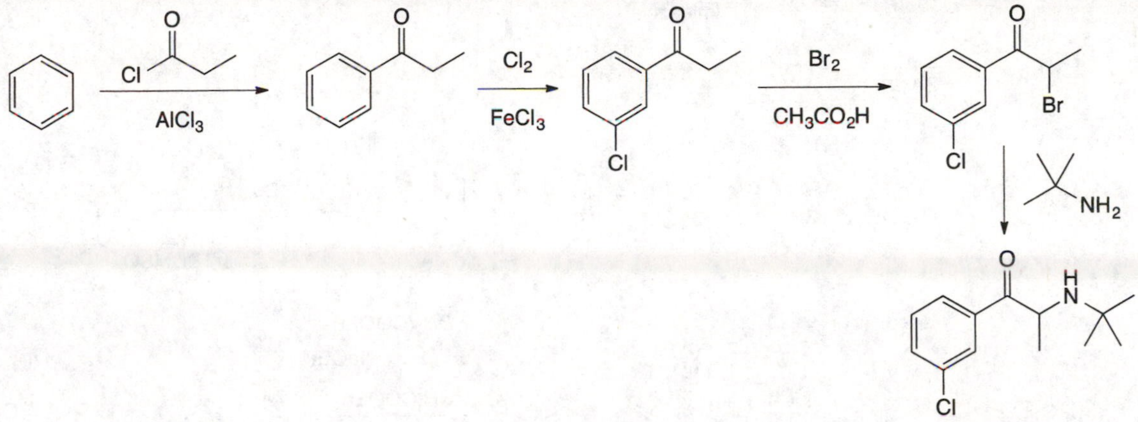

Cl (propanoyl chloride) / $AlCl_3$

Cl_2 / $FeCl_3$

Br_2 / CH_3CO_2H

NH_2 (tert-butylamine)

bupropion

23.65

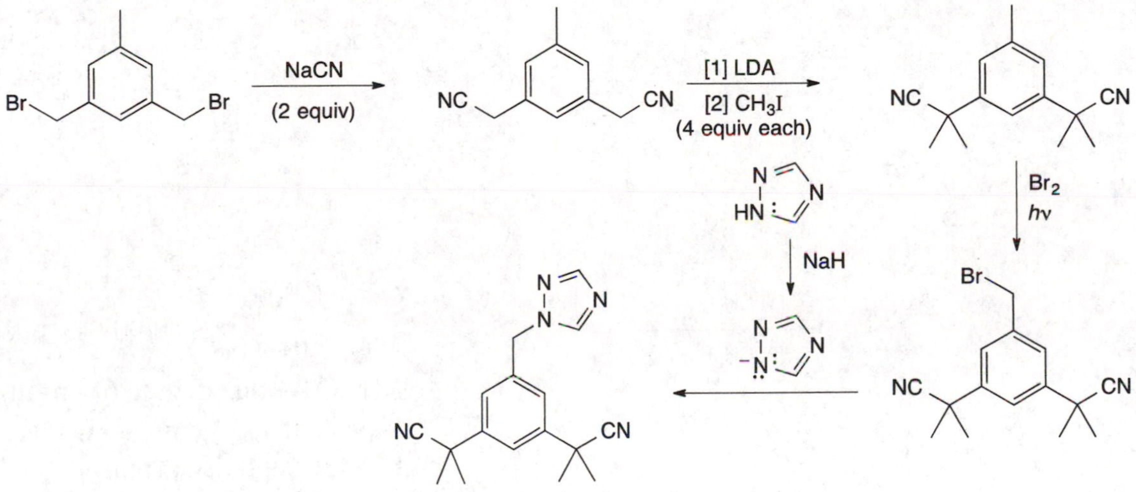

NaCN (2 equiv)

[1] LDA [2] CH_3I (4 equiv each)

Br_2 / $h\nu$

NaH

anastrozole

23.66

[1] NaOEt [2] $HC\equiv CCH_2Br$

COOEt

H_3O^+ / Δ

$HOCH_2CH_2OH$ / TsOH

[1] NaH [2] CH_3I

H_2 / Lindlar catalyst

H_3O^+

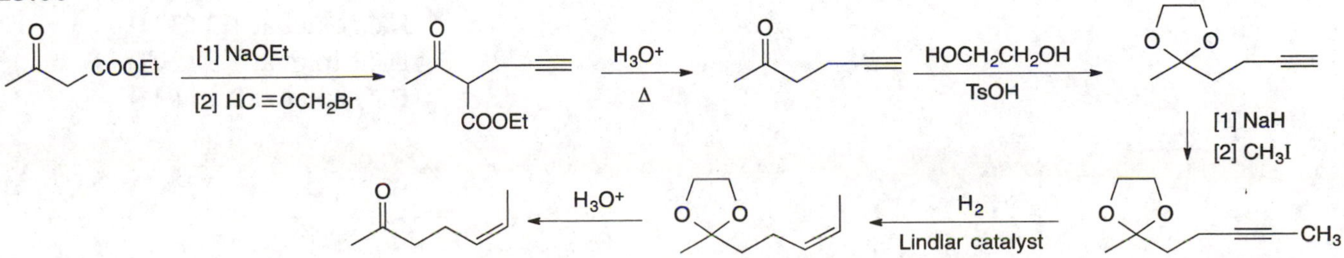

23.67

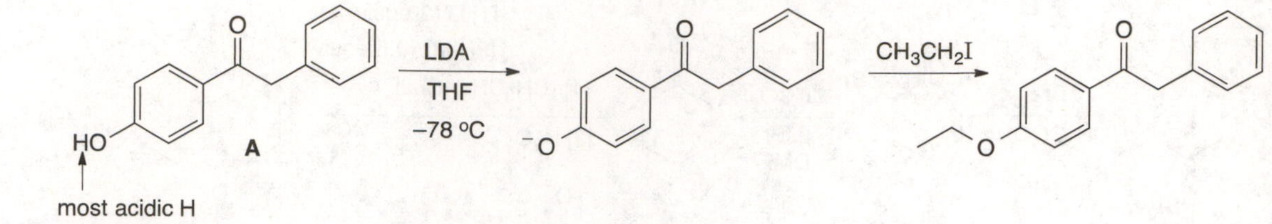

LDA
THF
–78 °C

CH_3CH_2I

HO
A
most acidic H

To synthesize the desired product, a protecting group is needed:

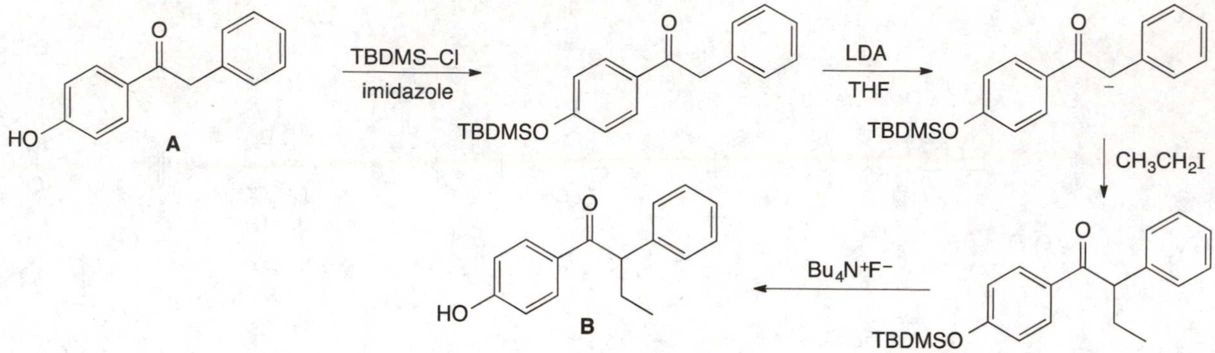

TBDMS–Cl
imidazole

LDA
THF

CH_3CH_2I

$Bu_4N^+F^-$

HO
A
TBDMSO
TBDMSO
TBDMSO
HO
B

23.68

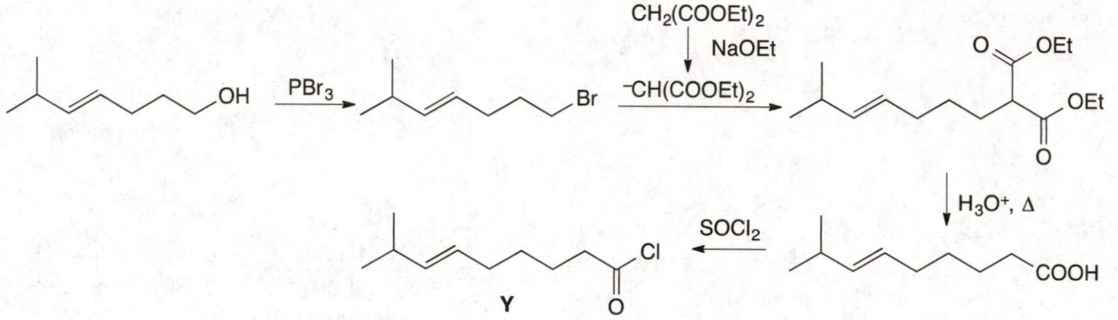

PBr$_3$

$CH_2(COOEt)_2$
NaOEt
$^-CH(COOEt)_2$

H_3O^+, Δ

SOCl$_2$

OH
Br
OEt
OEt
COOH
Cl
Y

23.69

a.

[1] CH_3Li
[2] CH_3I

W
Y

Y = $(CH_3)_2CH-C(=O)-CH_2CH_2CH_3$ (H$_a$: $(CH_3)_2$; H$_b$: CH; H$_c$, H$_d$: CH_2CH_2; H$_e$: CH_3)

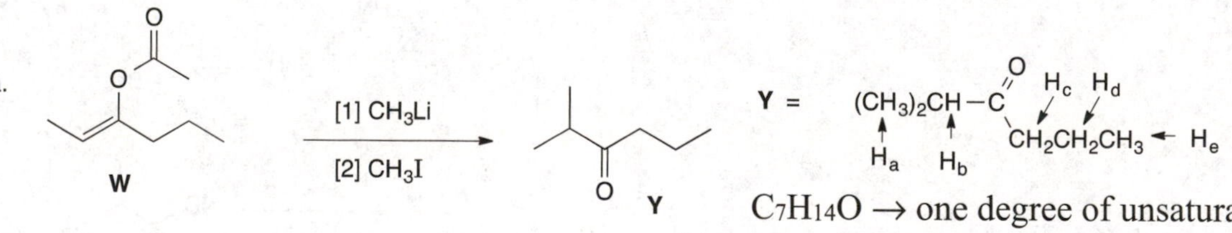

$C_7H_{14}O$ → one degree of unsaturation
IR peak at 1713 cm^{-1} → C=O
^{1}H NMR signals at (ppm)
H$_e$: triplet at 0.7 (3 H)
H$_a$: doublet at 0.9 (6 H)
H$_d$: sextet at 1.3 (2 H)
H$_c$: triplet at 1.9 (2 H)
H$_b$: septet at 2.1 (1 H)

b.

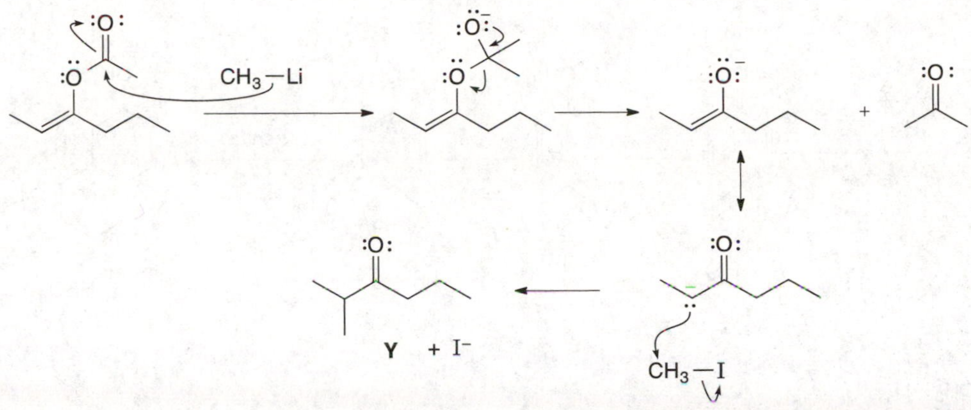

:O:

:Ö

CH_3—Li

:Ö:

:Ö

:Ö:

:O:

:O:

:O:

Y + I^-

CH_3—I

23.70 Removal of Ha with base does not generate an anion that can delocalize onto the carbonyl O atom, whereas removal of Hb generates an enolate that is delocalized on O.

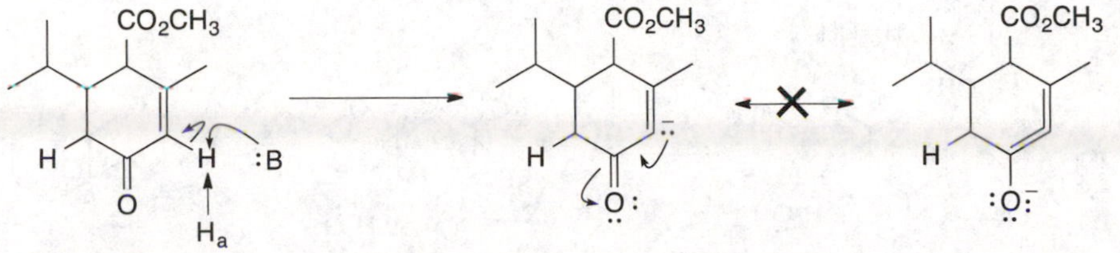

CO_2CH_3 CO_2CH_3 CO_2CH_3

H O H :B

H_a

H :Ö:

H :Ö:

Delocalization of this sort can't occur by removal of H_a, making H_a less acidic.

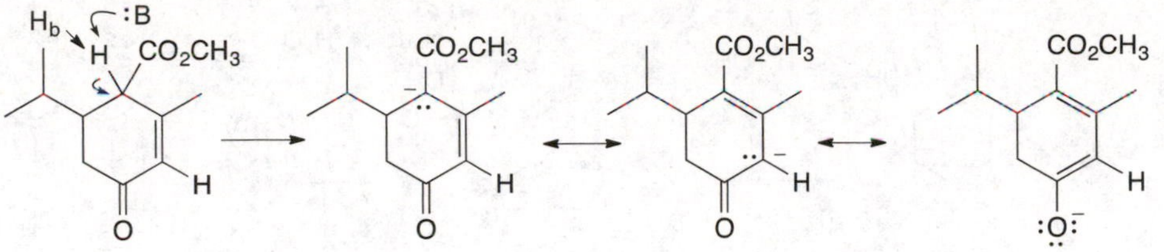

H_b :B H CO_2CH_3 CO_2CH_3 CO_2CH_3 CO_2CH_3

H O H O H O H :Ö:

Removal of H_b gives an anion that is resonance stabilized, so H_b is more acidic.

Mechanism:

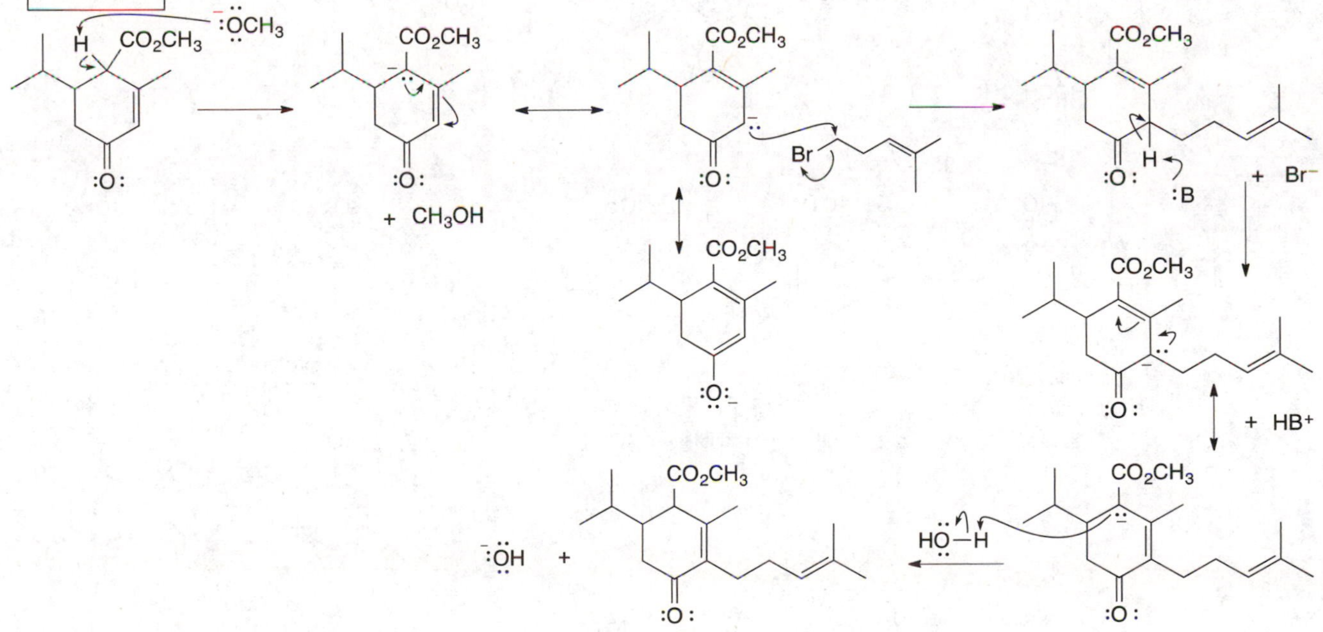

H CO_2CH_3 $:\ddot{O}CH_3$

:O: CO_2CH_3 :O: + CH_3OH

CO_2CH_3 :O: Br

CO_2CH_3 :Ö:

CO_2CH_3 :O: H :B + Br^-

CO_2CH_3 :O: + HB^+

CO_2CH_3 :O:

HÖ—H

$:\ddot{O}H$ + CO_2CH_3 :O:

23.71

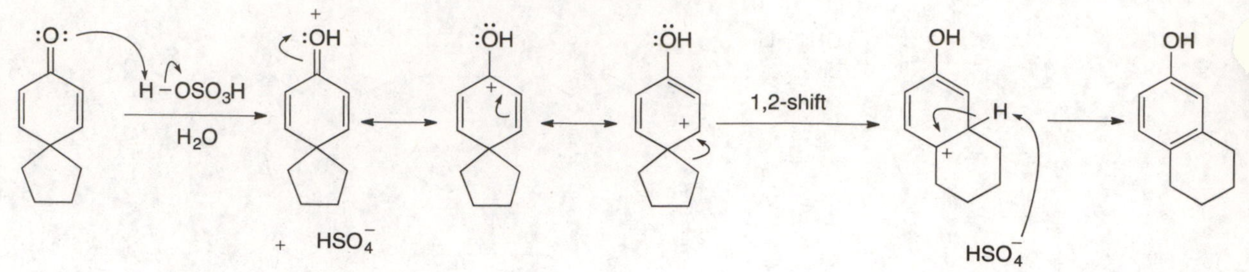

:O: H–OSO$_3$H H$_2$O :OH + HSO$_4^-$:ÖH :ÖH 1,2-shift OH H HSO$_4^-$ OH

23.72

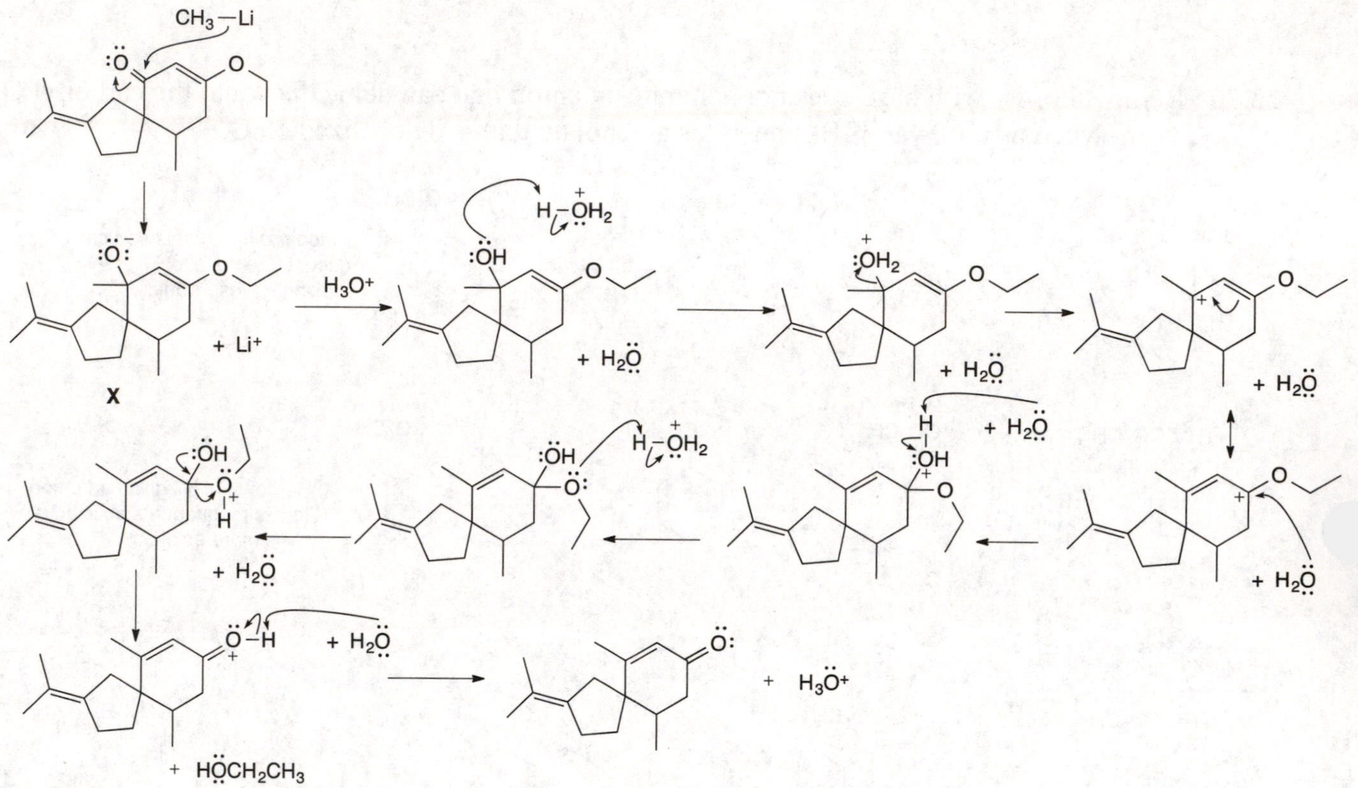

CH$_3$–Li :Ö O

:Ö:$^-$ O + Li$^+$ X H_3O^+ :ÖH H–ÖH$_2$ O + H$_2$Ö :ÖH$_2$ O + H$_2$Ö O + H$_2$Ö

:ÖH Ö H + H$_2$Ö :ÖH Ö: H–ÖH$_2$ H :ÖH O + H$_2$Ö O + H$_2$Ö

Ö–H + H$_2$Ö + HÖCH$_2$CH$_3$ Ö: + $H_3Ö^+$

23.73

:O: e$^-$:O: :Ö:$^-$ e$^-$:Ö:$^-$ H–NH$_2$:Ö:$^-$ H :O: H R–X

or :Ö:$^-$ H–NH$_2$:ÖH e$^-$ + $^-$NH$_2$:ÖH proton transfer :O: R + X$^-$

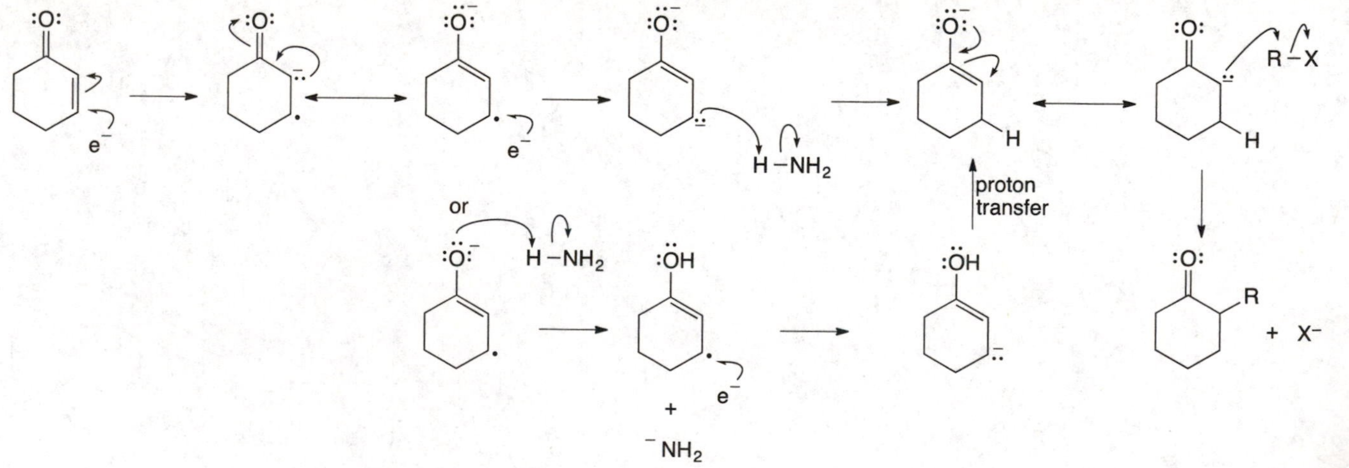

23.74 a. In the presence of base an achiral enolate that can be protonated from both sides is formed.

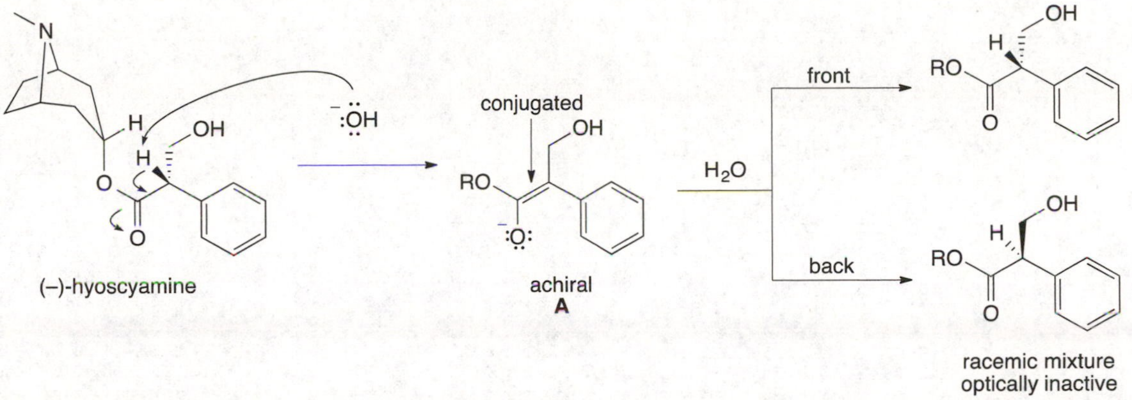

(–)-hyoscyamine

conjugated

achiral
A

H_2O

front

back

racemic mixture
optically inactive

b. The enolate **A** formed from (–)-hyoscyamine is conjugated with the benzene ring, making it easier to form. The enolate **B** formed from (–)-littorine is not conjugated, so it is less readily formed.

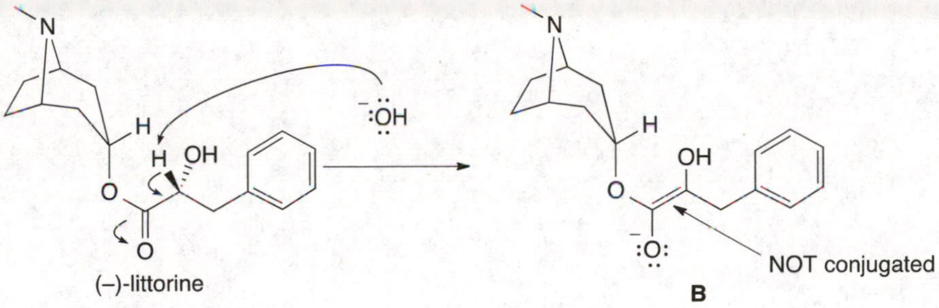

(–)-littorine

NOT conjugated

B

Chapter 24 Carbonyl Condensation Reactions

Chapter Review

The four major carbonyl condensation reactions

Reaction type	Reaction
[1] Aldol reaction (24.1)	2 aldehyde (or ketone) ⇌ ($^{-}$OH, H_2O) β-hydroxy carbonyl compound → ($^{-}$OH or H_3O^+) (*E* and *Z*) α,β-unsaturated carbonyl compound
[2] Claisen reaction (24.5)	2 ester → ([1] NaOR'; [2] H_3O^+) β-keto ester
[3] Michael reaction (24.8)	α,β-unsaturated carbonyl compound + carbonyl compound → ($^{-}$OR' or $^{-}$OH; H_2O) 1,5-dicarbonyl compound
[4] Robinson annulation (24.9)	α,β-unsaturated carbonyl compound + carbonyl compound → ($^{-}$OH, H_2O) cyclohex-2-enone

Useful variations

[1] Directed aldol reaction (24.3)

R" = H or alkyl → ([1] LDA; [2] RCHO; [3] H_2O) β-hydroxy carbonyl compound → ($^{-}$OH or H_3O^+) (*E* and *Z*) α,β-unsaturated carbonyl compound

[2] Intramolecular aldol reaction (24.4)

a. With 1,4-dicarbonyl compounds:

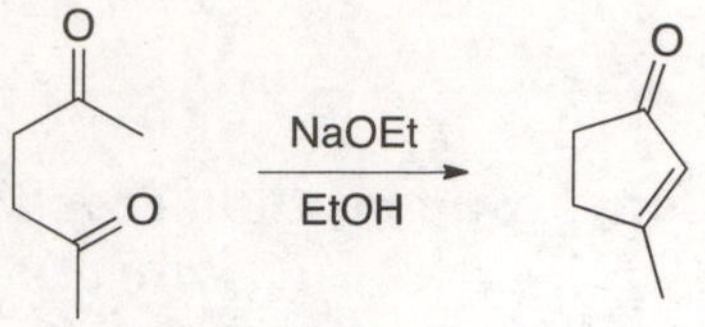

NaOEt
EtOH

b. With 1,5-dicarbonyl compounds:

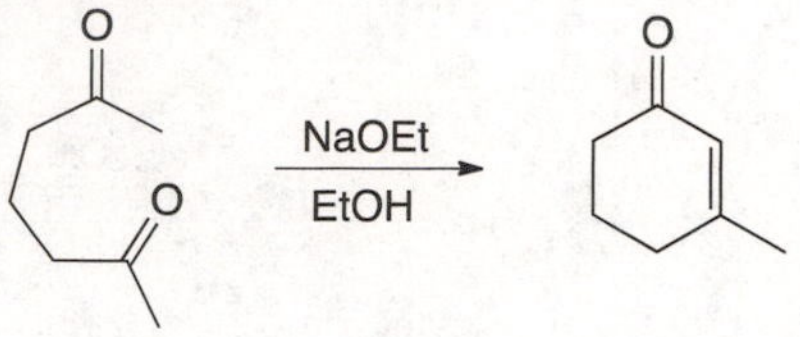

NaOEt
EtOH

[3] Dieckmann reaction (24.7)

a. With 1,6-diesters:

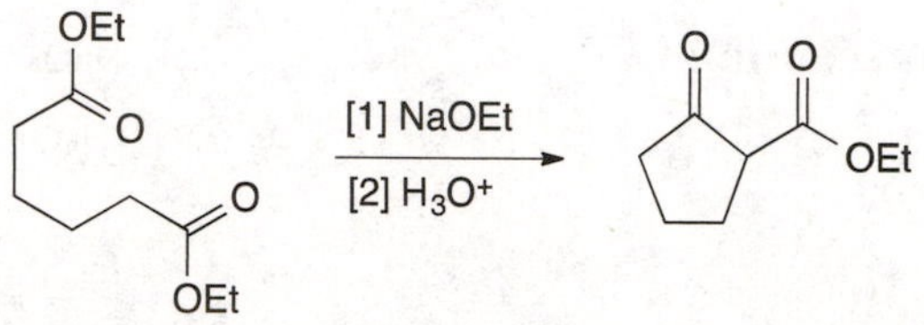

OEt O O OEt

[1] NaOEt
[2] H_3O^+

O O OEt

b. With 1,7-diesters:

OEt O O OEt

[1] NaOEt
[2] H_3O^+

O O OEt

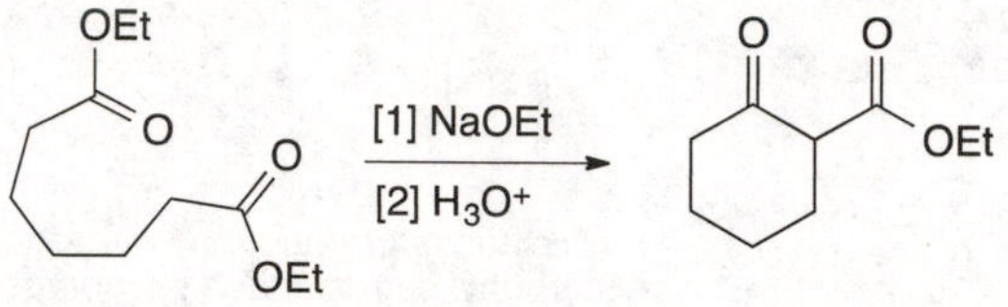

Practice Test on Chapter Review

1.a. Which compounds are possible Michael acceptors?

1. 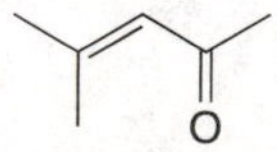 O

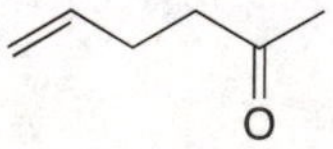

2. 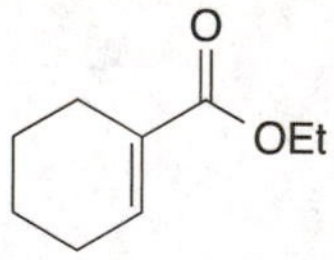 O OEt

3. O

4. Both compounds (1) and (2) are Michael acceptors.

5. Compounds (1), (2), and (3) are all Michael acceptors.

b. Which of the following compounds can be formed by an aldol reaction?

1. 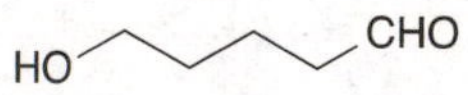 HO CHO

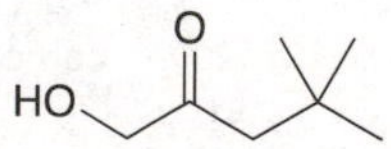

2. 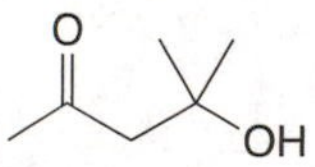 O OH

3. HO O

4. Both (1) and (2) can be formed by aldol reactions.

5. Compounds (1), (2), and (3) can all be formed by aldol reactions.

c. Which compounds can be formed in a Robinson annulation?

1. 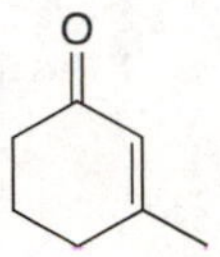

3. 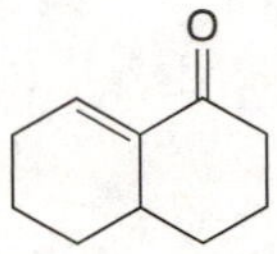

2. 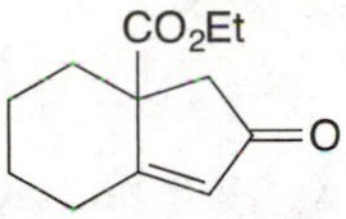

4. Compounds (1) and (2) can be formed by Robinson annulations.

5. Compounds (1), (2), and (3) can all be formed by Robinson annulations.

d. What compounds can be used to form **A** by a condensation reaction?

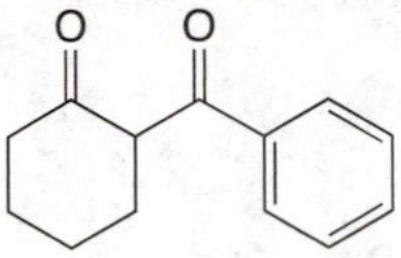

A

1. 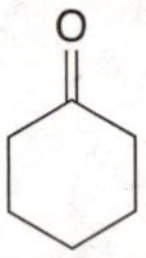 and 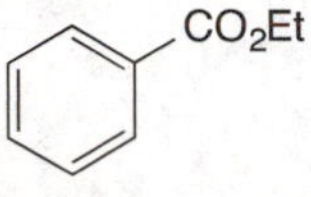

3. 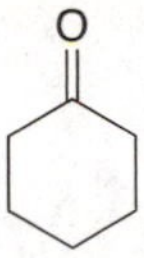 and 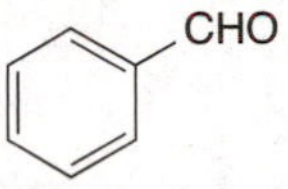

2. 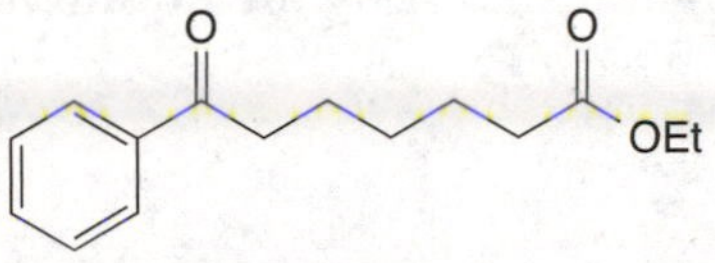

4. Compounds (1) and (2) can be used to form **A**.

5. Compounds (1), (2), and (3) can all be used to form **A**.

2. Give the reagents required for each step.

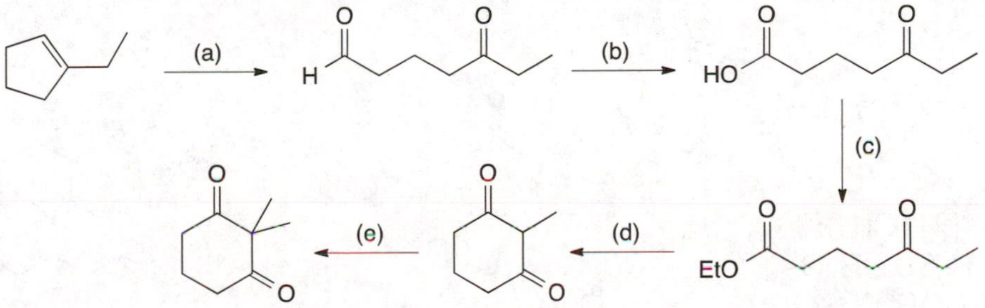

3. Draw the organic products formed in the following reactions.

a. 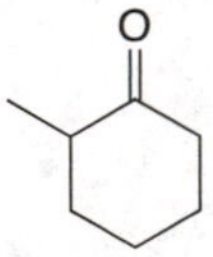 [1] LDA; [2] CH_3CH_2CHO; [3] H_2O

c. 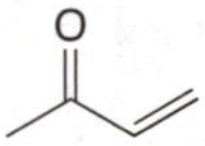 + 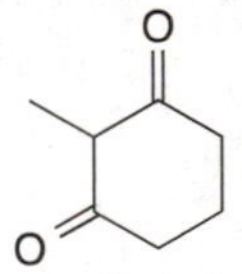 NaOEt / EtOH →

b. 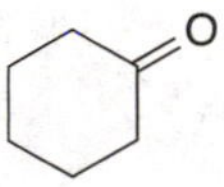 + HCO_2Et [1] NaOEt, EtOH; [2] H_3O^+ →

d. 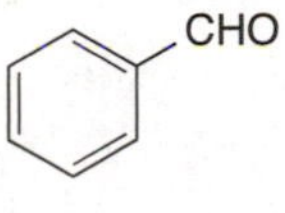 + 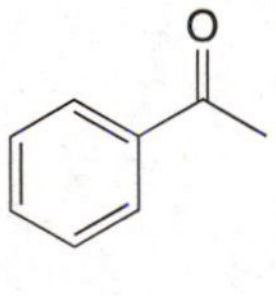 ^-OH, H_2O →

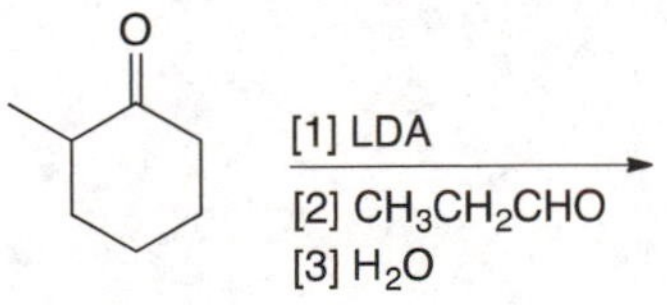

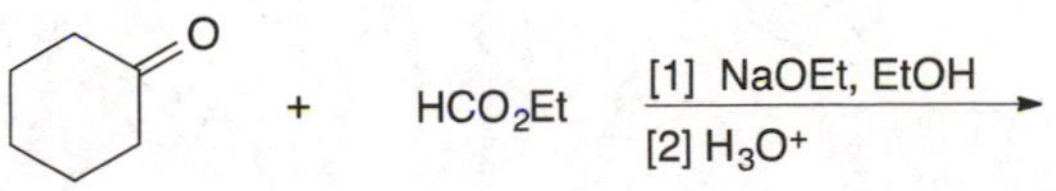

4.a. What organic starting materials are needed to synthesize **D** by a Robinson annulation reaction?

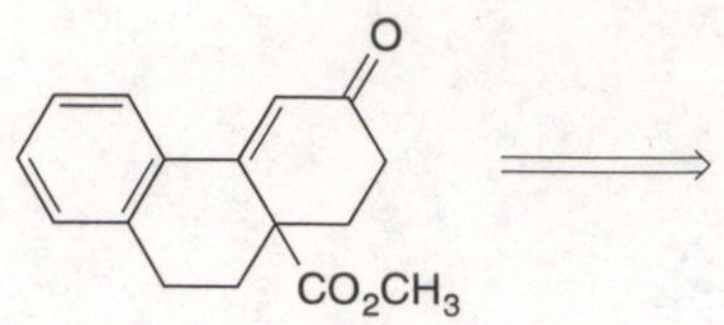

O

CO_2CH_3

D

b. What organic starting materials are needed to synthesize β-keto ester **B** by a Dieckmann reaction?

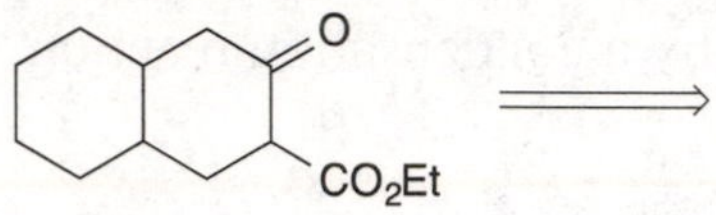

O

CO_2Et

B

c. What starting materials are needed to synthesize **A** by an aldol reaction?

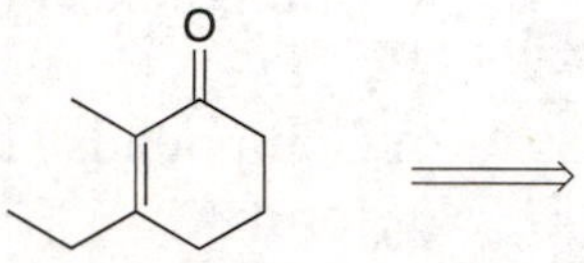

O

A

Answers to Practice Test

1.a. 4
   b. 2
   c. 1
   d. 4

2.a. [1] O_3; [2] $(CH_3)_2S$
   b. CrO_3, H_2SO_4, H_2O
   c. CH_3CH_2OH, H_2SO_4
   d. [1] NaOEt, EtOH; [2] H_3O^+
   e. [1] LDA; [2] CH_3I

3.

a. 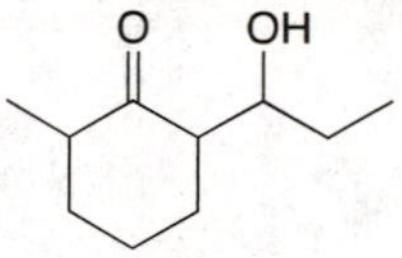 O OH

b. 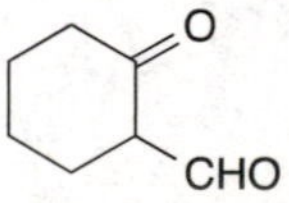 O CHO

c. 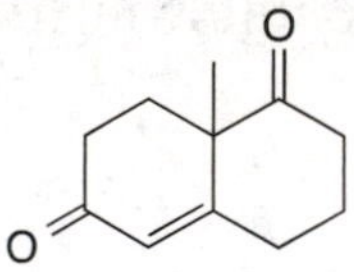 O O

d. 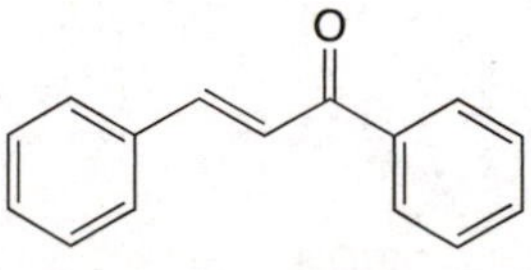 O

(*E* and *Z* isomers)

4.

a. 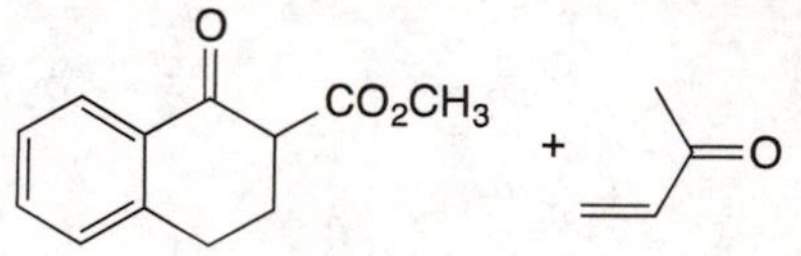 O CO_2CH_3 + O

b. 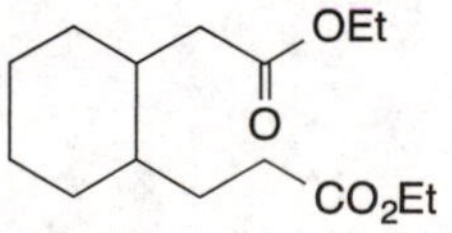 OEt O CO_2Et

c. 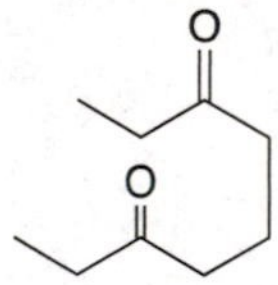 O O

Chapter 24: Answers to Problems

24.1

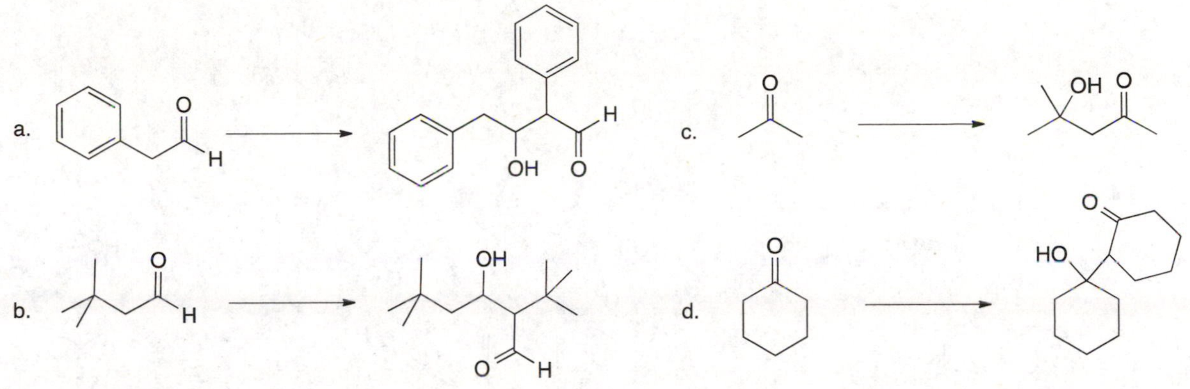

a. O, H → H, OH, O

b. O, H → OH, O, H

c. O → OH O

d. O → O, HO

24.2

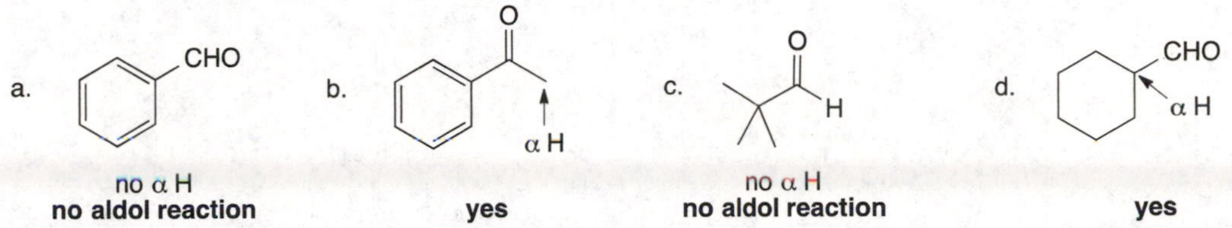

a. CHO — no α H — **no aldol reaction**

b. O, α H — **yes**

c. O, H — no α H — **no aldol reaction**

d. CHO, α H — **yes**

24.3

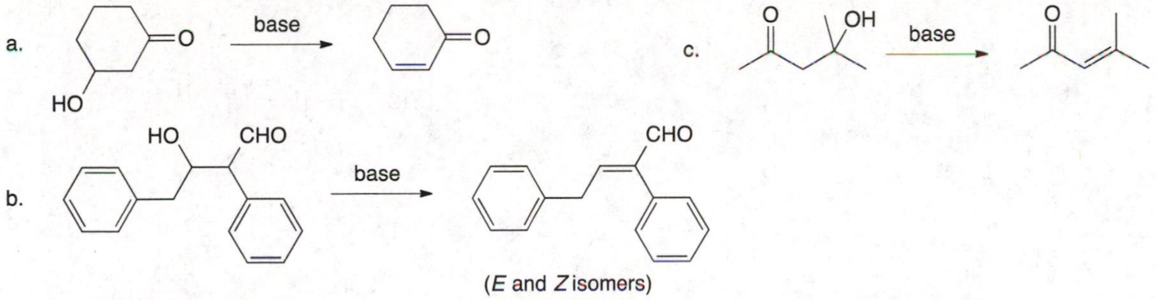

a. O, HO —base→ O

b. HO, CHO —base→ CHO (*E* and *Z* isomers)

c. O, OH —base→ O

24.4

:O: :ÖH, H, H + H–OSO_3H → :O: $:OH_2^+$, H, H → :O:, H, H, + + $H_2\ddot{O}$: + HSO_4^- → :O:, H + H_2SO_4

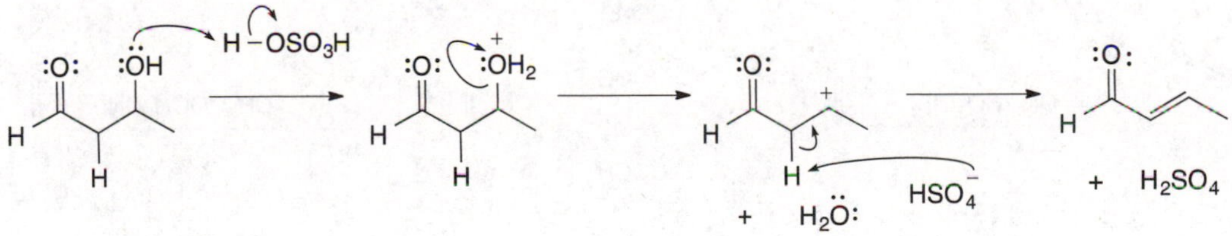

24.5 Locate the α and β C's to the carbonyl group, and break the molecule into two halves at this bond. The α C and all of the atoms bonded to it belong to one carbonyl component. The β C and all of the atoms bonded to it belong to the other carbonyl component.

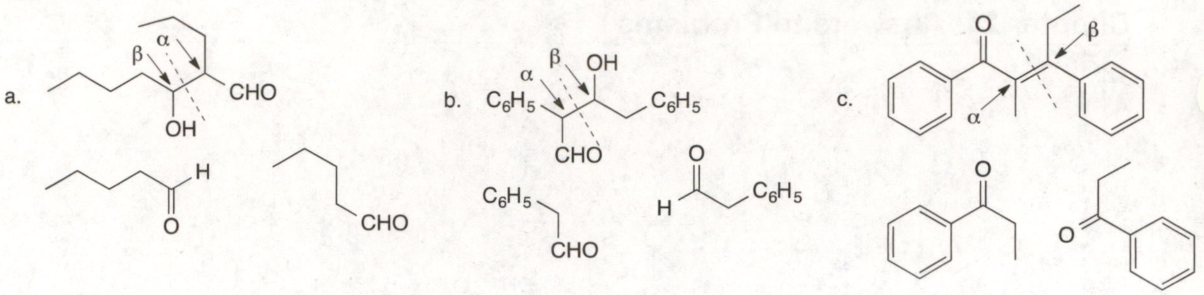

a. β α CHO OH H O CHO

b. α β OH C_6H_5 C_6H_5 CHO C_6H_5 CHO O H C_6H_5

c. O β α O O

24.6

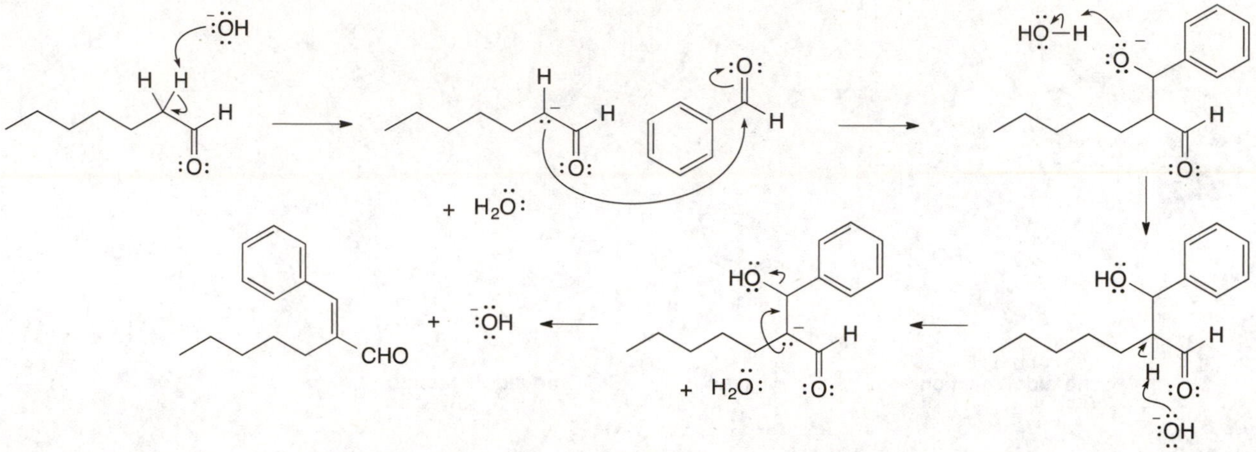

$^-$:ÖH H H H :O: → H − H :O: + $H_2\ddot{O}$: :O: H → HÖ—H :Ö:$^-$ H :O: ↓ HÖ H H :O: $^-$:ÖH ← HÖ − H + $H_2\ddot{O}$: :O: ← + $^-$:ÖH CHO

24.7

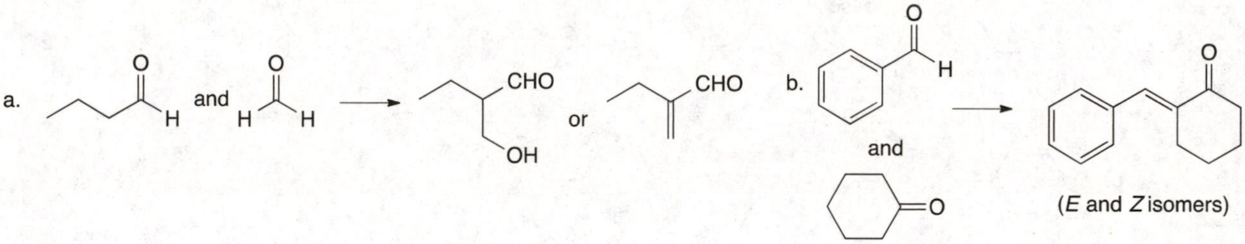

a. O H and O H H → CHO OH or CHO

b. O H and O → O (*E* and *Z* isomers)

24.8

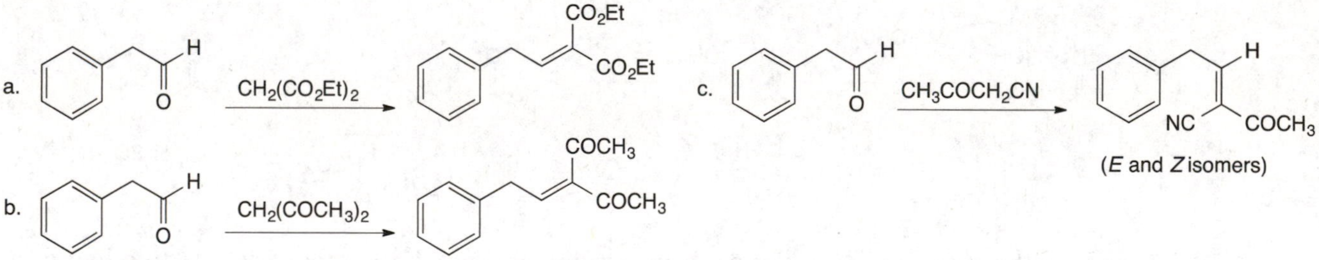

a. H O $CH_2(CO_2Et)_2$ → CO_2Et CO_2Et

b. H O $CH_2(COCH_3)_2$ → $COCH_3$ $COCH_3$

c. H O CH_3COCH_2CN → H NC $COCH_3$ (*E* and *Z* isomers)

24.9

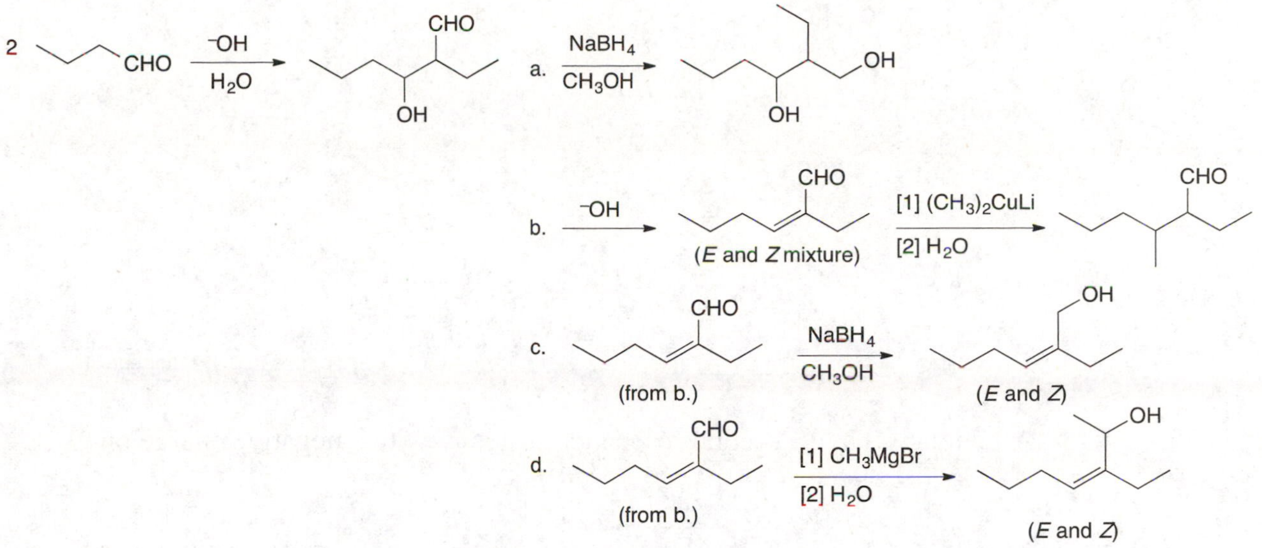

2 butanal (CHO) $\xrightarrow[H_2O]{^-OH}$ (CHO, OH)

a. $\xrightarrow[CH_3OH]{NaBH_4}$ (OH, OH)

b. $\xrightarrow{^-OH}$ CHO (*E* and *Z* mixture) $\xrightarrow[[2]\ H_2O]{[1]\ (CH_3)_2CuLi}$ CHO

c. CHO (from b.) $\xrightarrow[CH_3OH]{NaBH_4}$ OH (*E* and *Z*)

d. CHO (from b.) $\xrightarrow[[2]\ H_2O]{[1]\ CH_3MgBr}$ OH (*E* and *Z*)

24.10 Find the α and β C's to the carbonyl group and break the bond between them.

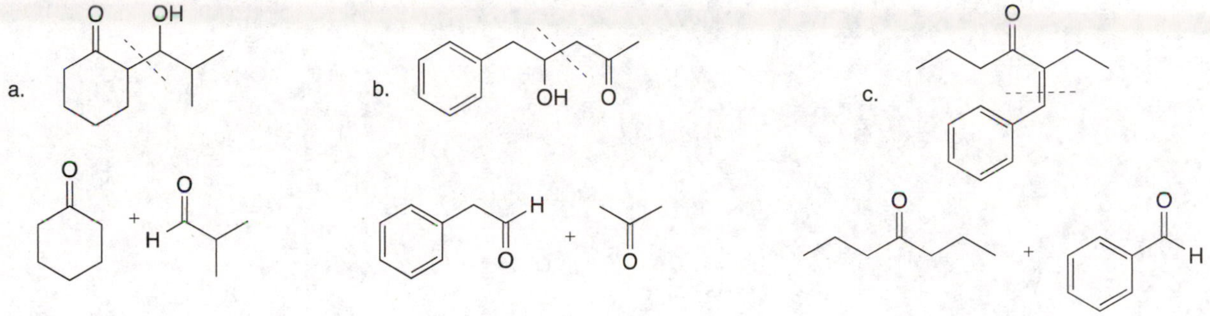

a. (O, OH) → cyclohexanone (O) + H–C(=O)–isopropyl (O, H)

b. (OH, O) → phenylacetaldehyde (H, O) + acetone (O)

c. (O) → ketone (O) + benzaldehyde (O, H)

24.11

TBDMS—O (aryl ketone, O) **A** $\xrightarrow{LDA}$ TBDMS—O (enolate, O)

↓ H–C(=O)–pentyl (O, H)

TBDMS—O (O, OH) $\xrightarrow{Bu_4NF}$ HO (O, OH)

gingerol

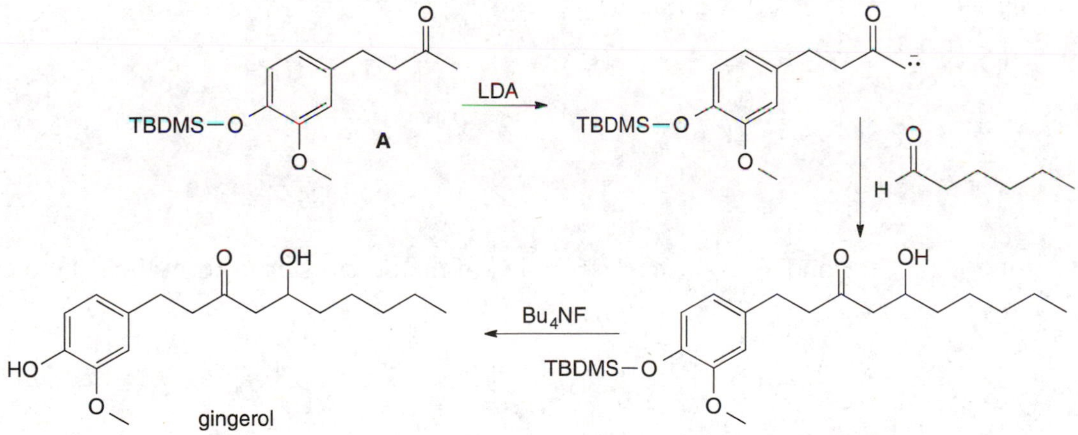

24.12

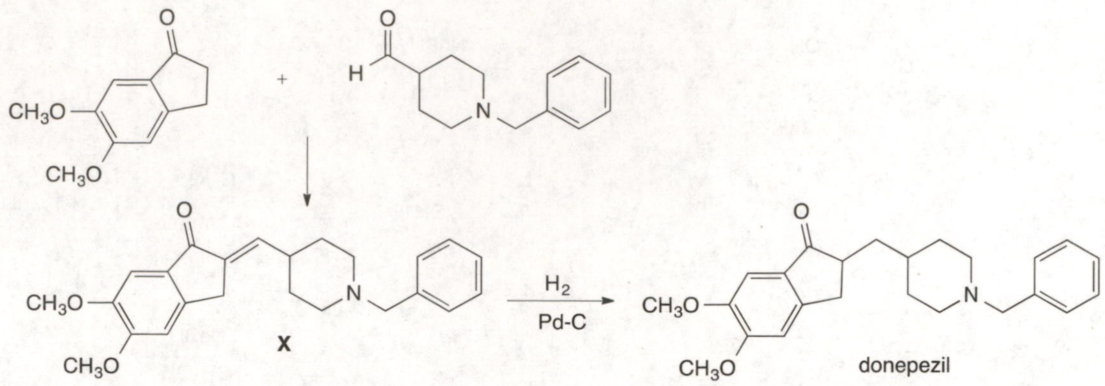

CH₃O, CH₃O, H, N, X, H_2, Pd-C, donepezil

24.13 All enolates have a second resonance structure with a negative charge on O.

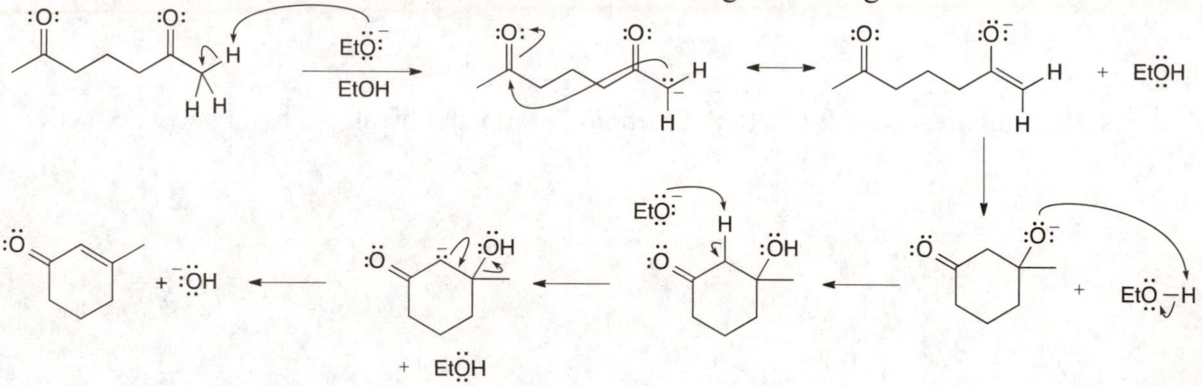

EtO:⁻, EtOH, + EtOH, + :OH⁻, + EtOH

24.14

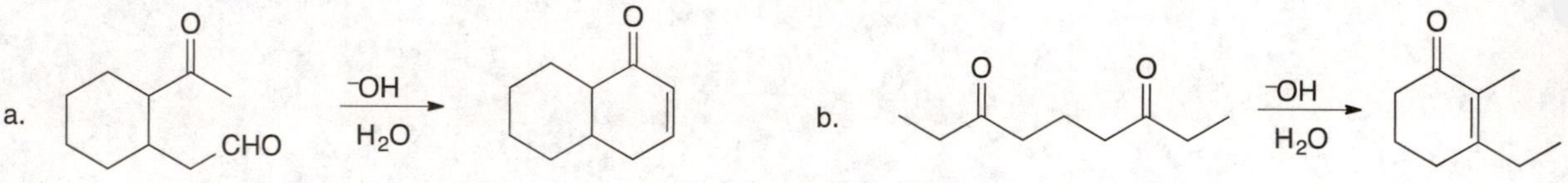

a. CHO, ⁻OH, H_2O

b. ⁻OH, H_2O

24.15

A, [1] O_3, [2] $(CH_3)_2S$, CHO, ⁻OH, H_2O, B

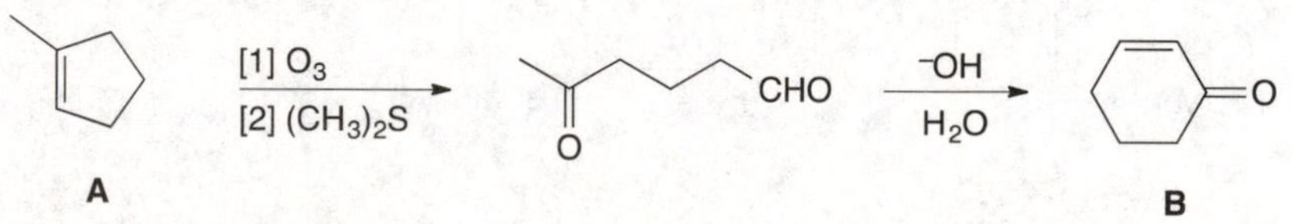

24.16 Join the α C of one ester to the carbonyl C of the other ester to form the β-keto ester.

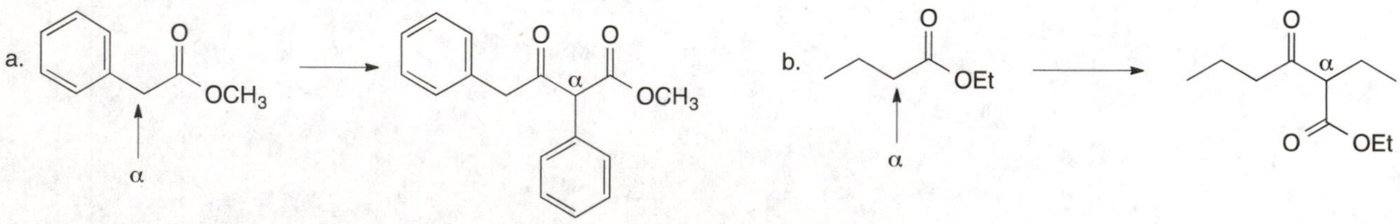

a. OCH_3, α, OCH_3

b. OEt, α, OEt

24.17 In a crossed Claisen reaction between an ester and a ketone, the enolate is formed from the ketone, and the product is a β-dicarbonyl compound.

a. $CH_3(CH_2)_5CO_2Et$ and HCO_2Et → (product)

Only this compound can form an enolate.

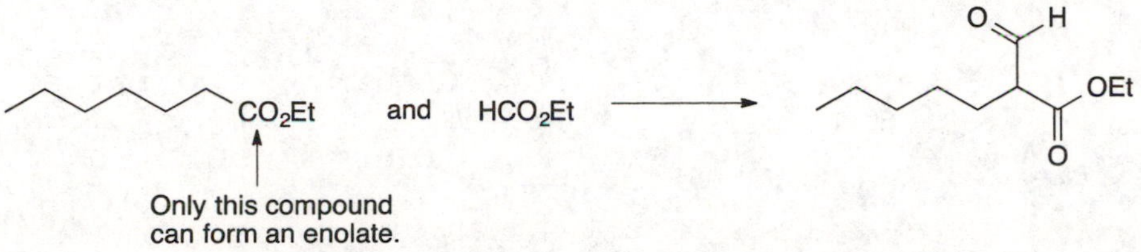

b. (cyclohexanone) and (ethyl benzoate) → (product)

The ketone forms the enolate.

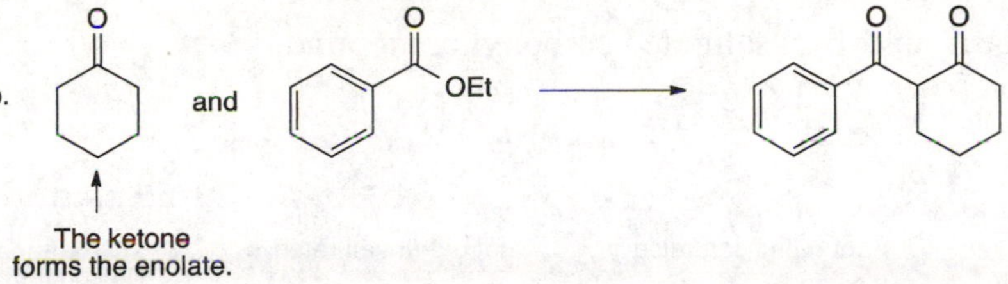

24.18 A β-dicarbonyl compound like avobenzone is prepared by a crossed Claisen reaction between a ketone and an ester.

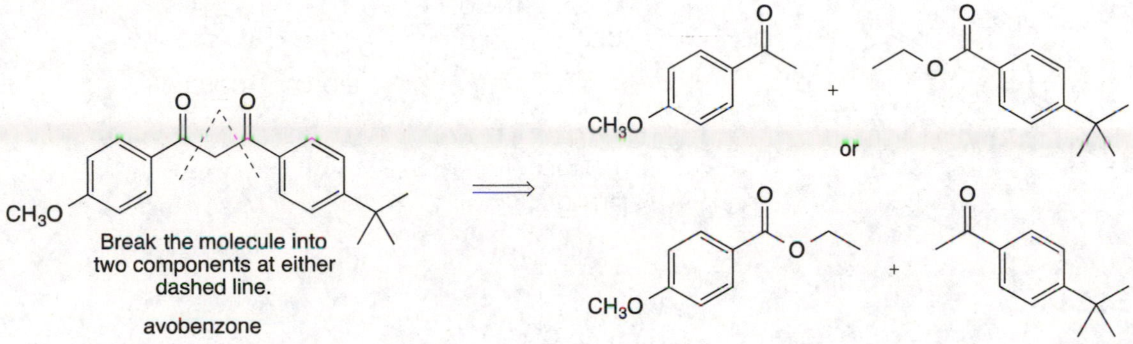

Break the molecule into two components at either dashed line.

avobenzone

CH_3O ... + ... or CH_3O ... + ...

24.19

a. [1] NaOEt, [2] $(EtO)_2C{=}O$ → OEt

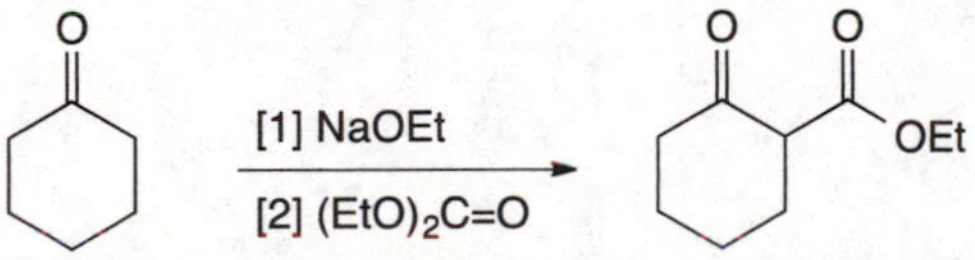

b. OEt [1] NaOEt, [2] $ClCO_2Et$ → OEt, EtO

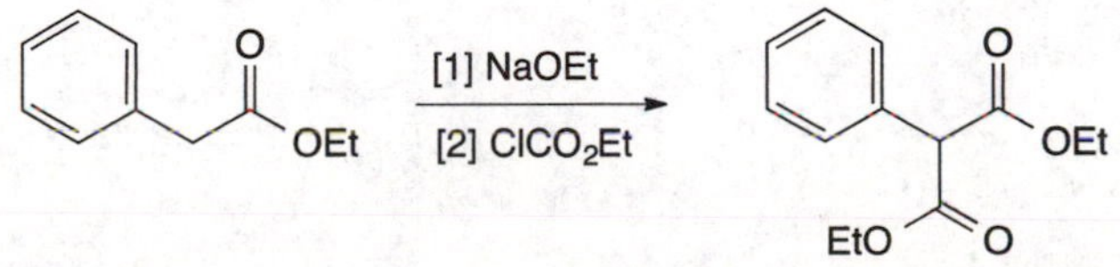

24.20

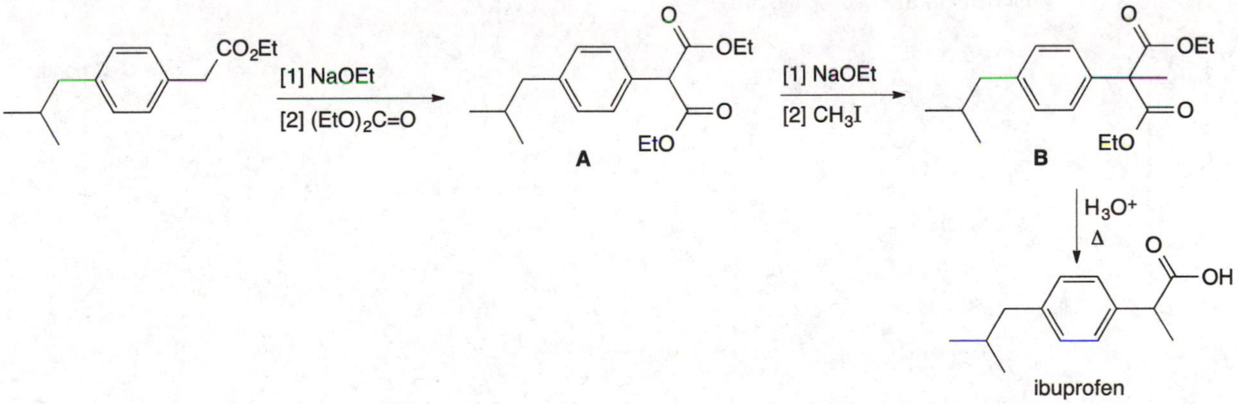

CO_2Et [1] NaOEt, [2] $(EtO)_2C{=}O$ → **A** (OEt, EtO) [1] NaOEt, [2] CH_3I → **B** (OEt, EtO)

H_3O^+, Δ → OH

ibuprofen

24.21

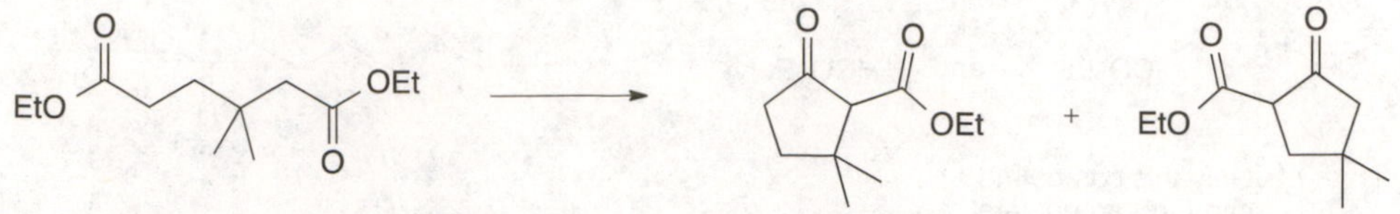

EtO OEt O O → O O OEt + EtO O O

24.22 A Michael acceptor is an α,β-unsaturated carbonyl compound.

a. 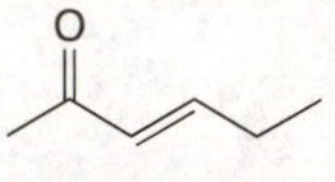 α,β-unsaturated **yes—Michael acceptor**

b. 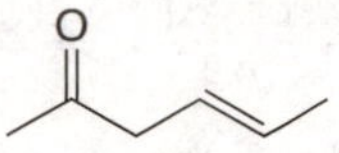 not α,β-unsaturated

c. 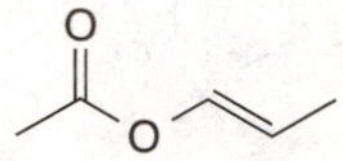 not α,β-unsaturated

d. 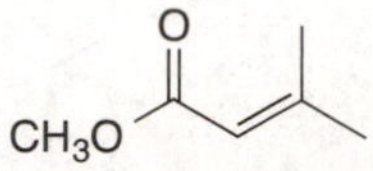 CH_3O α,β-unsaturated **yes—Michael acceptor**

24.23

a. 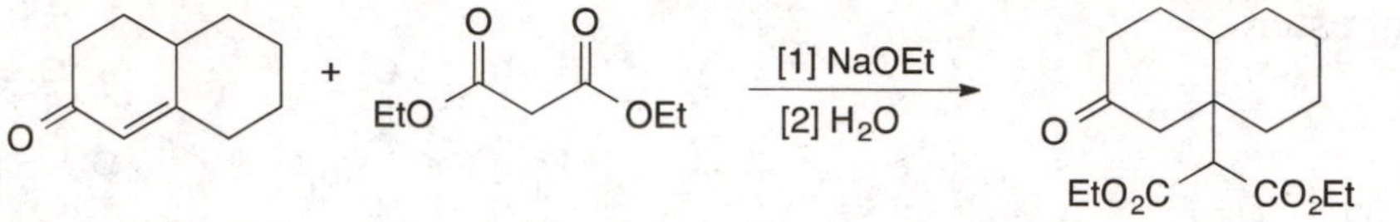 + EtO OEt → [1] NaOEt [2] H_2O → EtO_2C CO_2Et

b. 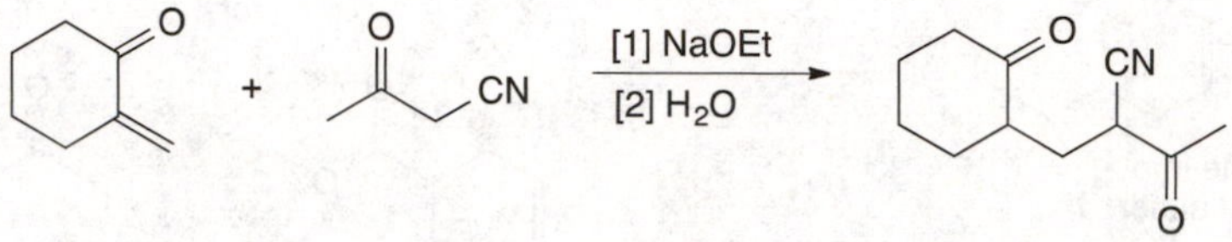 + CN → [1] NaOEt [2] H_2O → CN

24.24

a. 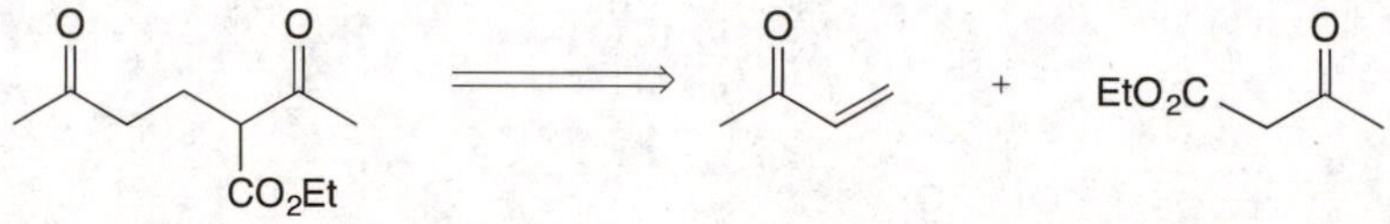 CO_2Et ⟹ + EtO_2C

b. 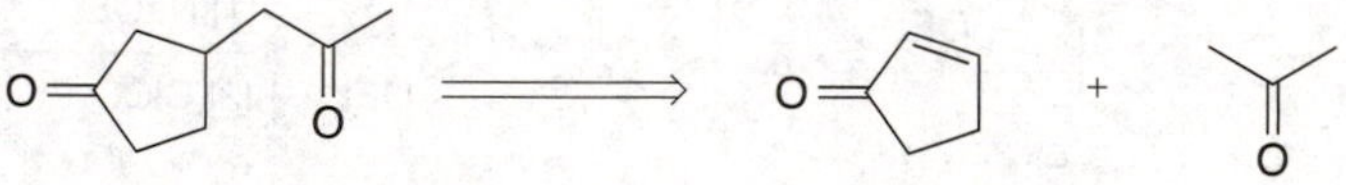 ⟹ +

24.25 The Robinson annulation forms a six-membered ring and three new carbon–carbon bonds: two σ bonds and one π bond.

a. 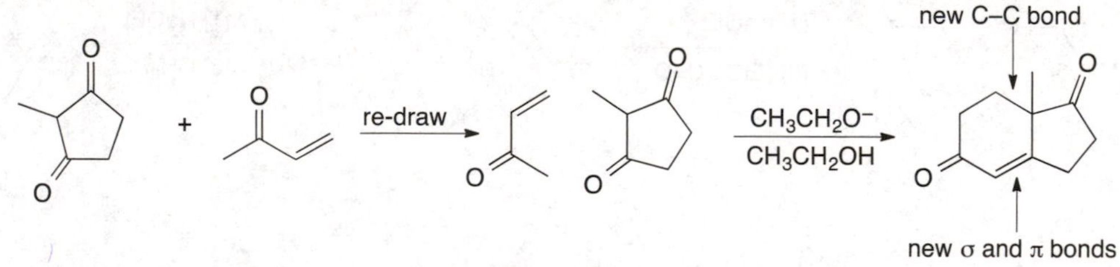 + → re-draw → → $CH_3CH_2O^-$ / CH_3CH_2OH →

new C–C bond

new σ and π bonds

b. 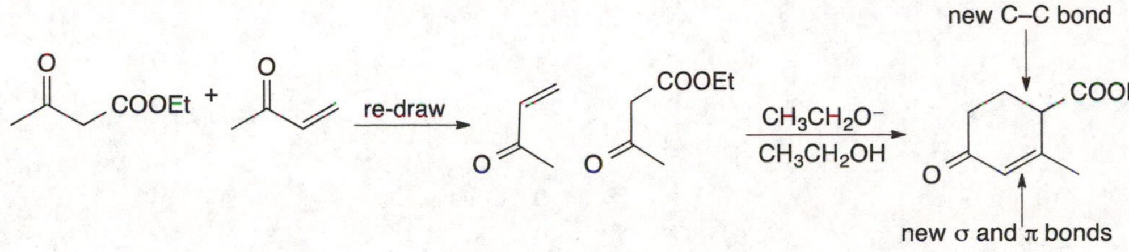 COOEt + re-draw COOEt $CH_3CH_2O^-$ / CH_3CH_2OH → new C–C bond, COOEt, new σ and π bonds

c. + re-draw $CH_3CH_2O^-$ / CH_3CH_2OH → new C–C bond, new σ and π bonds

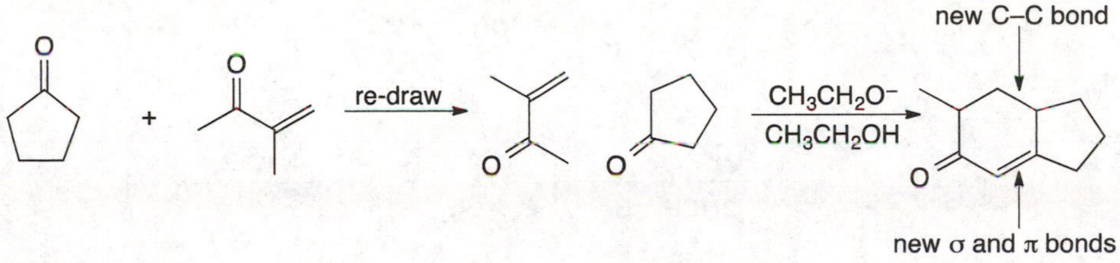

d. 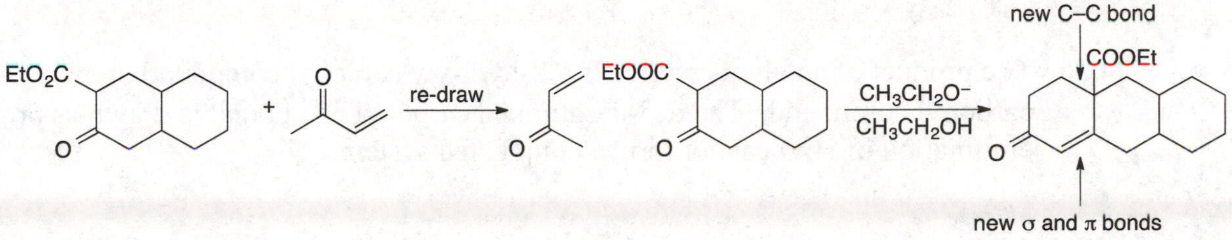 EtO_2C + re-draw EtOOC $CH_3CH_2O^-$ / CH_3CH_2OH → new C–C bond, COOEt, new σ and π bonds

24.26 A Robinson annulation forms a conjugated cyclohex-2-enone. In a bicyclic product, one carbon of the C=C must be shared by both rings.

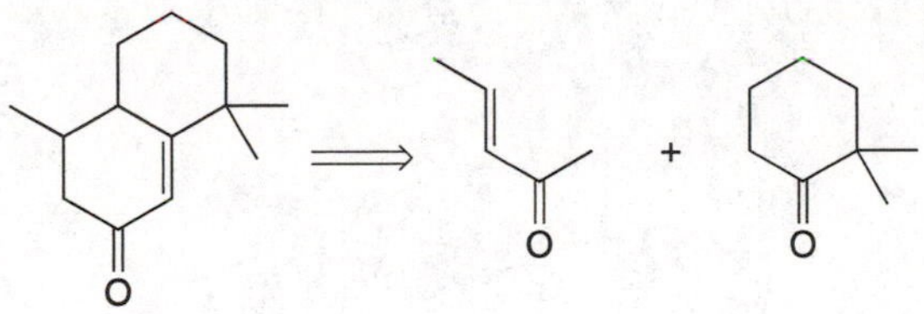

A yes ⟹ +

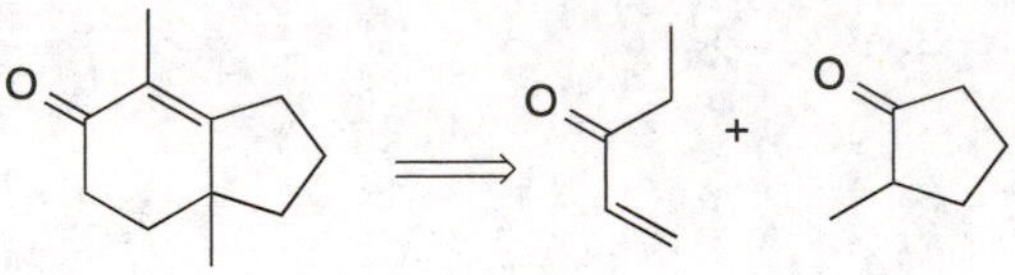

C yes ⟹ +

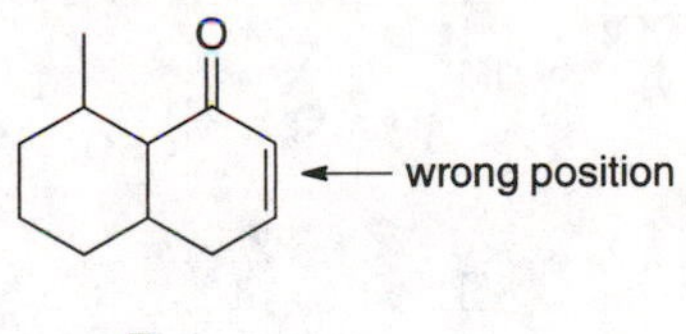

B no ← wrong position

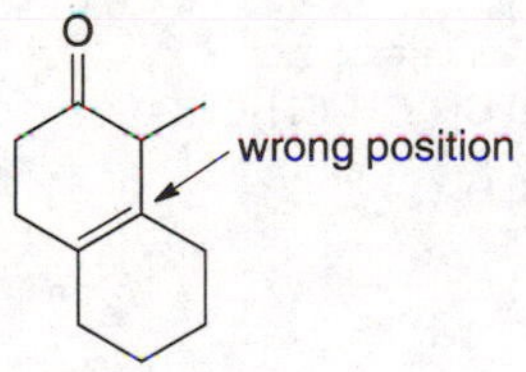

D no ← wrong position

24.27

a. 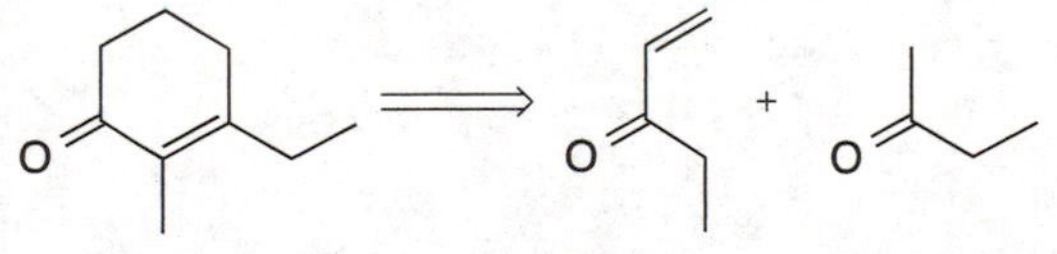 ⟹ +

c. 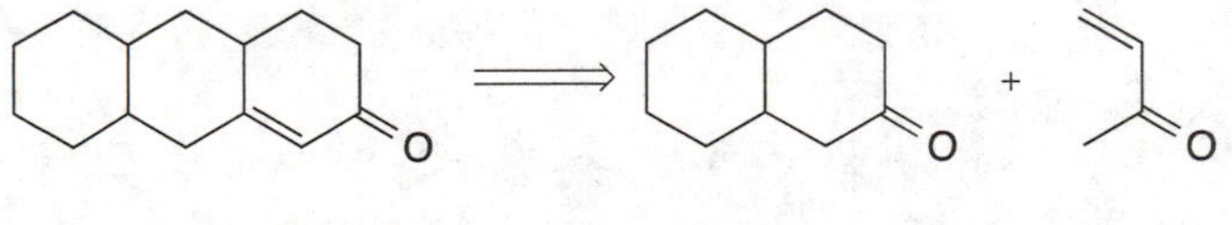 ⟹ +

b. 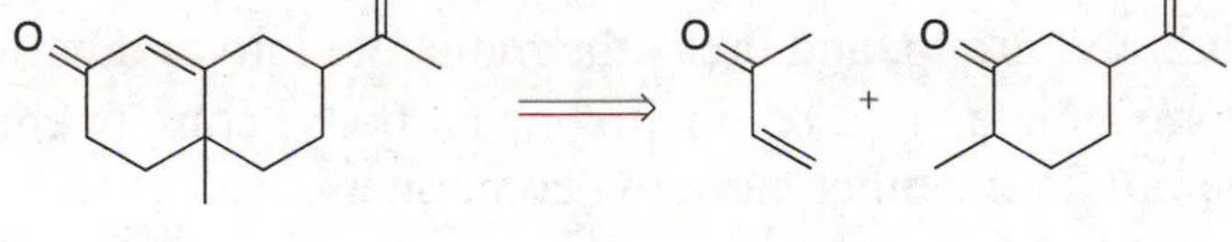 ⟹ +

24.28

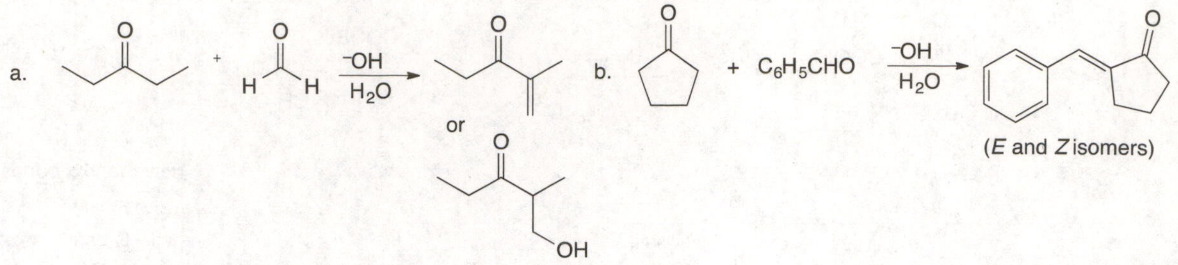

a. + ^{-}OH, H_2O → or OH

b. + C_6H_5CHO $\xrightarrow[H_2O]{^{-}OH}$ (*E* and *Z* isomers)

24.29

A $\xrightarrow[\text{[2] }(CH_3)_2S]{\text{[1] }O_3}$ CHO CHO $\xrightarrow[EtOH]{NaOEt}$ CHO B

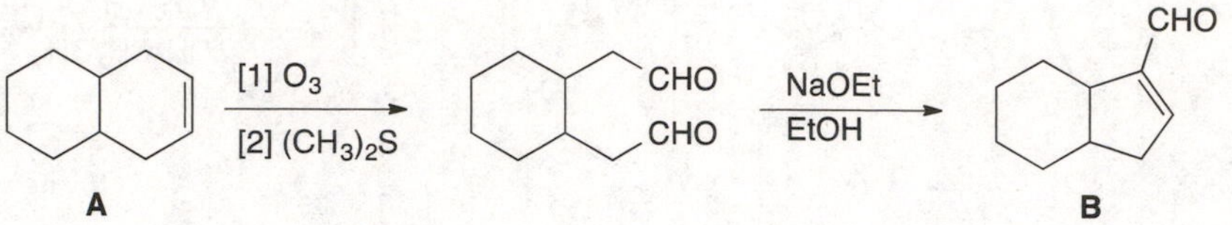

24.30 The product of an aldol reaction is a β-hydroxy carbonyl compound or an α,β-unsaturated carbonyl compound. The α,β-unsaturated carbonyl compound is drawn as product unless elimination of H_2O cannot form a conjugated system.

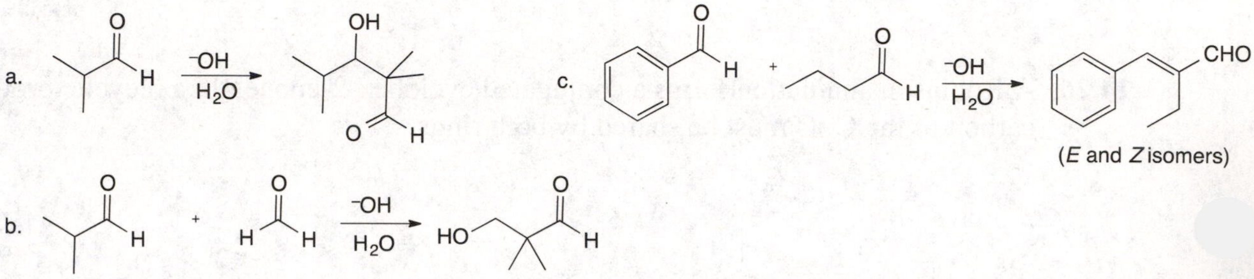

a. $\xrightarrow[H_2O]{^{-}OH}$ OH

b. + $\xrightarrow[H_2O]{^{-}OH}$ HO

c. + $\xrightarrow[H_2O]{^{-}OH}$ CHO (*E* and *Z* isomers)

24.31

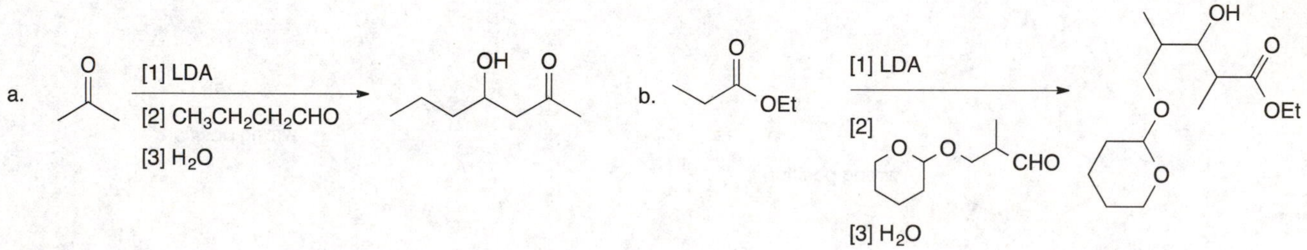

a. $\xrightarrow{\text{[1] LDA; [2] }CH_3CH_2CH_2CHO\text{; [3] }H_2O}$ OH O

b. OEt $\xrightarrow{\text{[1] LDA; [2] ... CHO; [3] }H_2O}$ OH O OEt

24.32

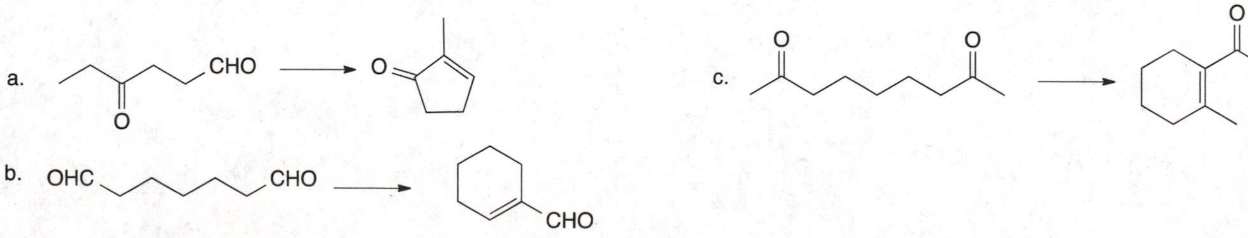

a. CHO → O

b. OHC CHO → CHO

c. O O → O

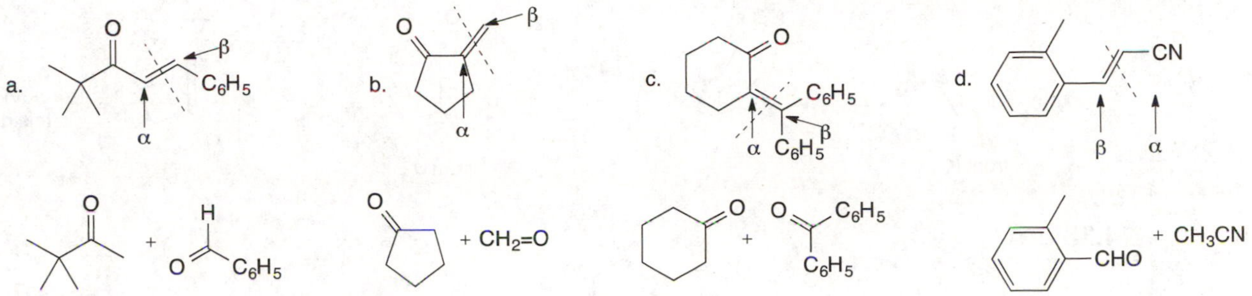

a. α β C_6H_5 + C_6H_5

b. β α + CH_2=O

c. C_6H_5 α C_6H_5 β + C_6H_5 C_6H_5

d. CN β α CHO + CH_3CN

24.34 Base removes the most acidic proton between the two C=O's in **B.** This enolate reacts with the aldehyde in **A** to form a product that loses H_2O.

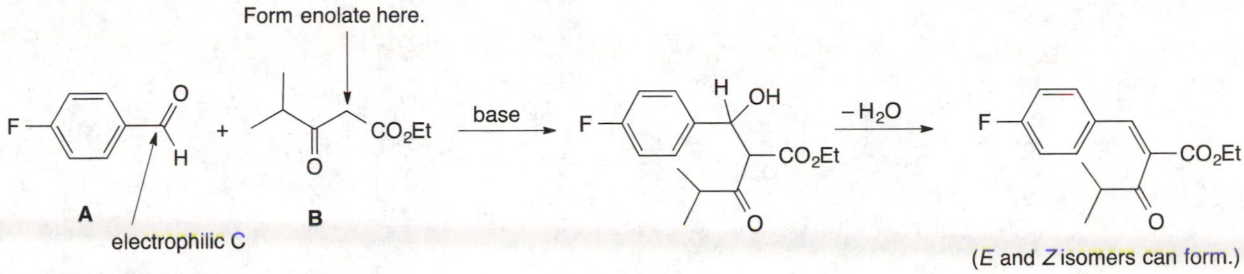

Form enolate here.

F **A** electrophilic C + **B** CO_2Et base → F H OH CO_2Et $-H_2O$ → F CO_2Et

(*E* and *Z* isomers can form.)

24.35

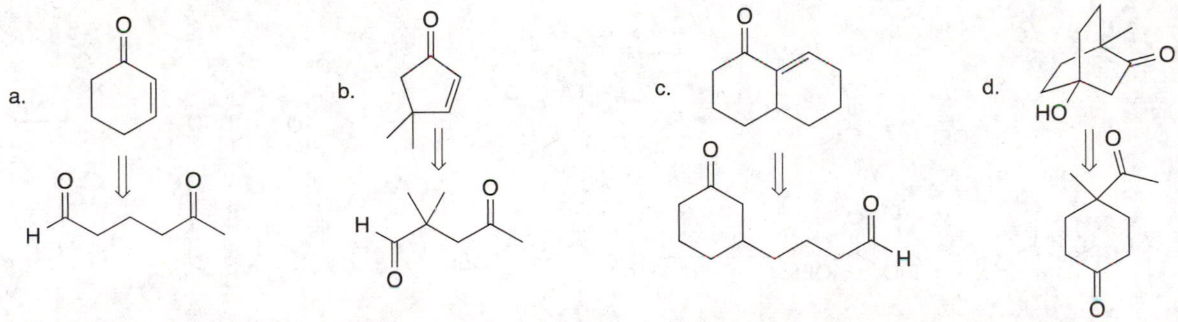

a. b. c. d. HO

24.36 Ozonolysis cleaves the C=C, and base catalyzes an intramolecular aldol reaction.

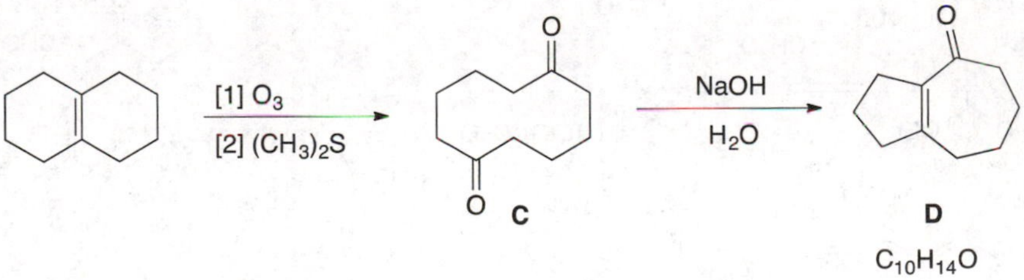

[1] O_3 [2] $(CH_3)_2S$ → **C** NaOH H_2O → **D** $C_{10}H_{14}O$

24.37 Aldol reactions proceed via resonance-stabilized enolates. **K** can form an enolate that allows for delocalization of the negative charge on O. Delocalization is not possible in **J,** because a double bond would be placed at a bridgehead carbon, which is geometrically impossible.

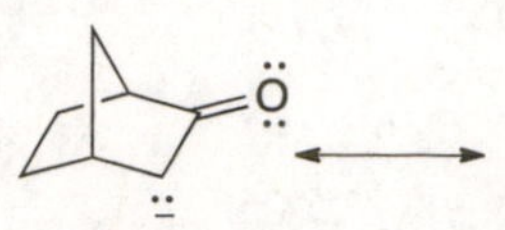

from **K**

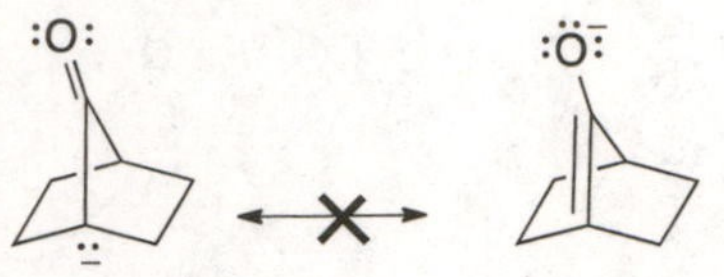

from **J**

Bond angles don't allow this double bond.

24.38

a.

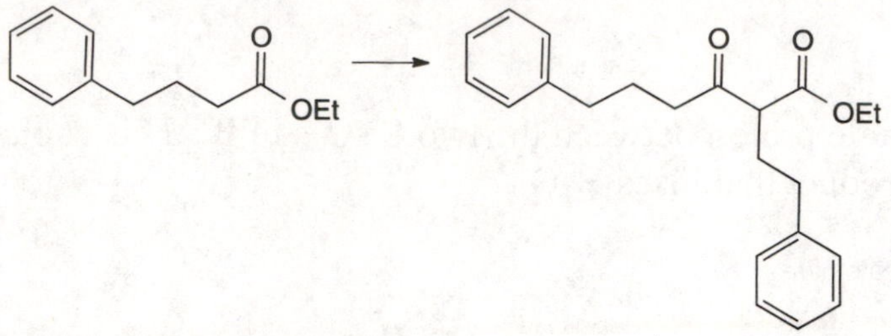

OEt

OEt

b.

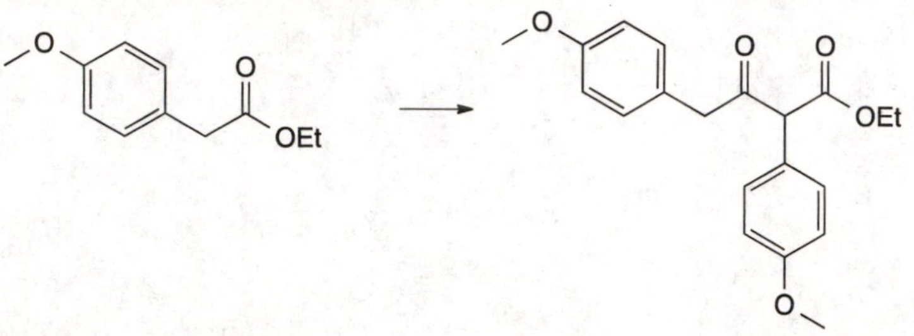

OEt

OEt

24.39

a. 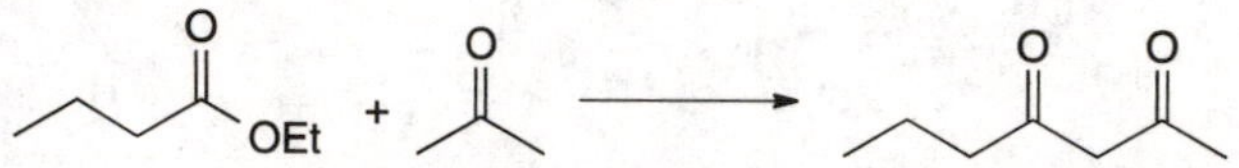 OEt + →

c. 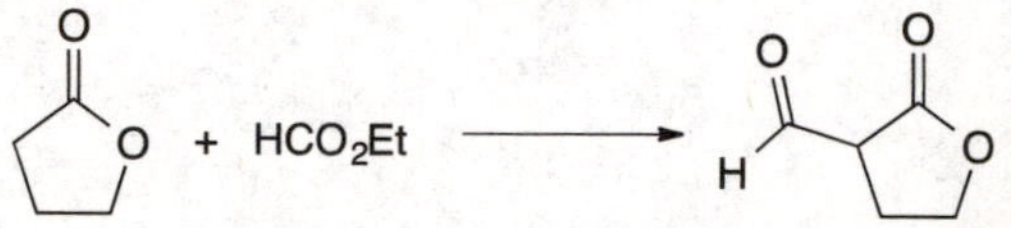 + HCO_2Et → H

b.  OEt + EtO OEt → EtO OEt

d. 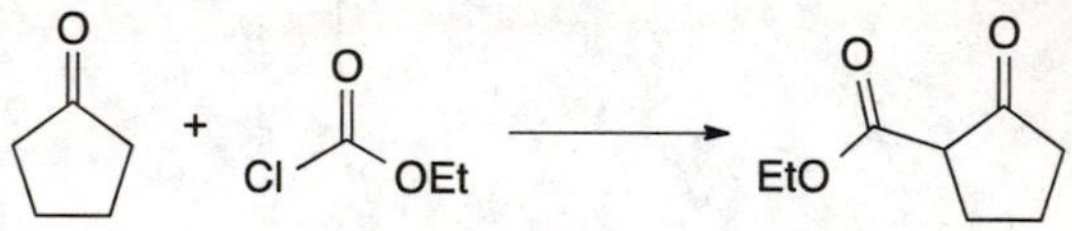 + Cl OEt → EtO

24.40

a. 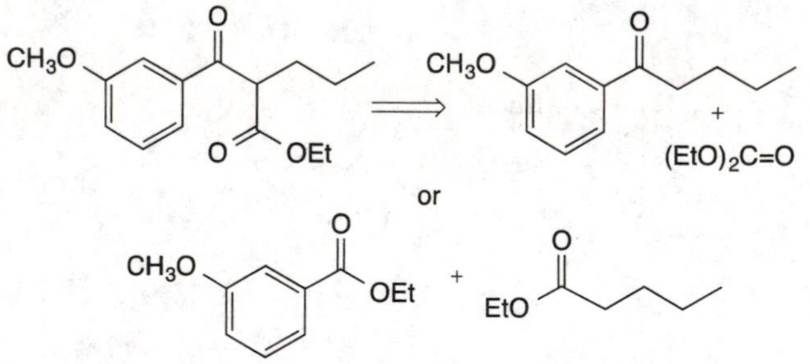 CH_3O ⟹ CH_3O + $(EtO)_2C{=}O$

OEt

or

CH_3O OEt + EtO

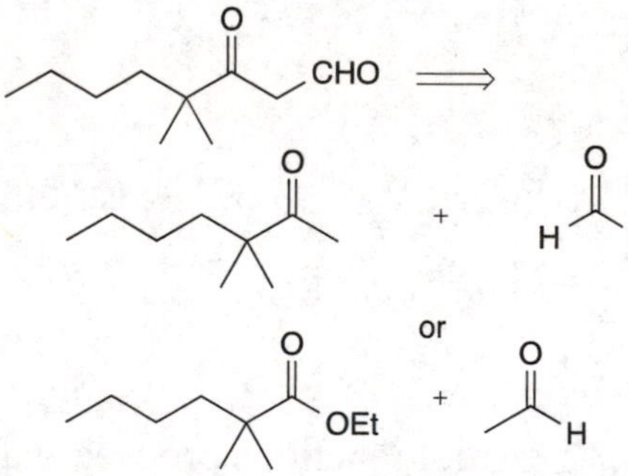

b. 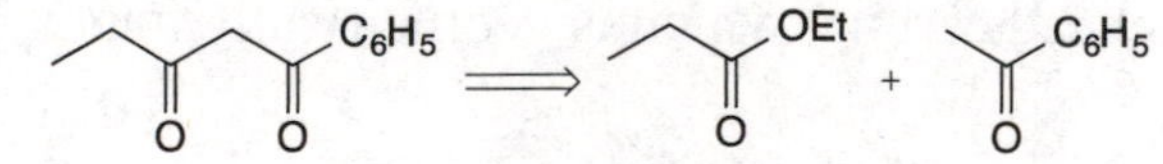 C_6H_5 ⟹ OEt + C_6H_5

c. CHO ⟹ + H OEt

or

OEt + H

d. 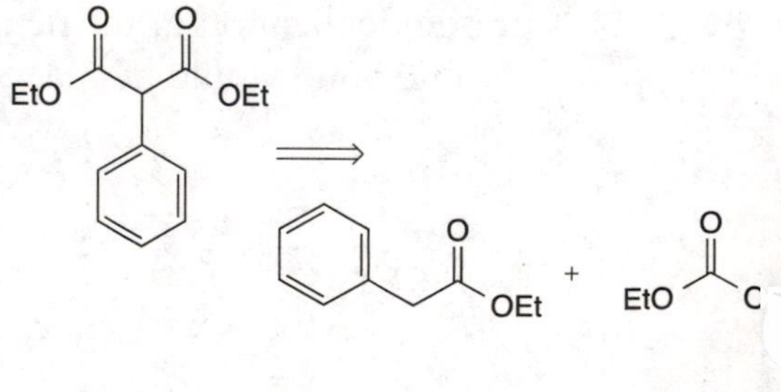 EtO OEt ⟹ OEt + EtO C

24.41 Only esters with two H's or three H's on the α carbon form enolates that undergo Claisen reaction to form resonance-stabilized enolates of the product β-keto ester. Thus, the enolate forms on the CH_2 α to one ester carbonyl, and cyclization yields a five-membered ring.

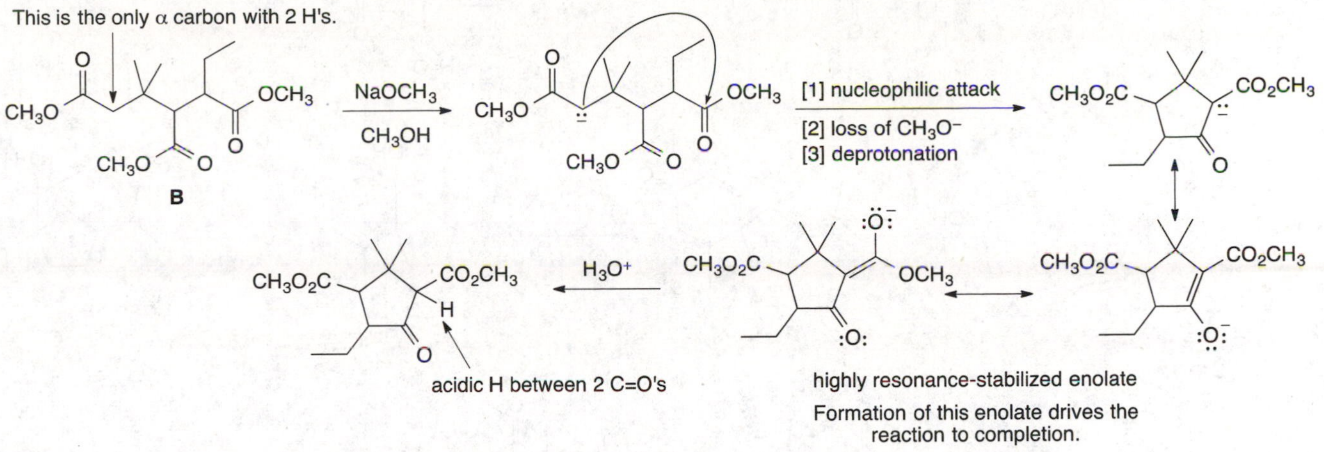

This is the only α carbon with 2 H's.

O, CH_3O, OCH_3, O, CH_3O, O

B

$NaOCH_3$ / CH_3OH →

O, CH_3O, OCH_3, O, CH_3O, O

[1] nucleophilic attack
[2] loss of CH_3O^-
[3] deprotonation

CH_3O_2C, CO_2CH_3, O

CH_3O_2C, CO_2CH_3, :Ö:$^-$

CH_3O_2C, :Ö:$^-$, OCH_3, :O:

highly resonance-stabilized enolate

Formation of this enolate drives the reaction to completion.

H_3O^+ ←

CH_3O_2C, CO_2CH_3, H, O

acidic H between 2 C=O's

24.42

a. O + C_6H_5 O C_6H_5 —^-OEt, EtOH→ O O C_6H_5 C_6H_5

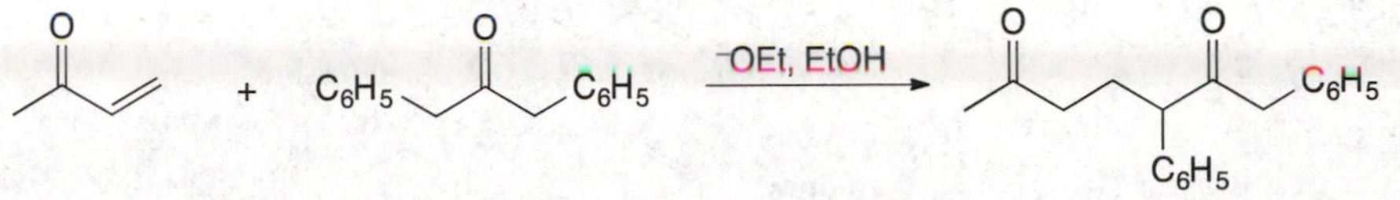

b. O + NC CN —^-OEt, EtOH→ O CN CN

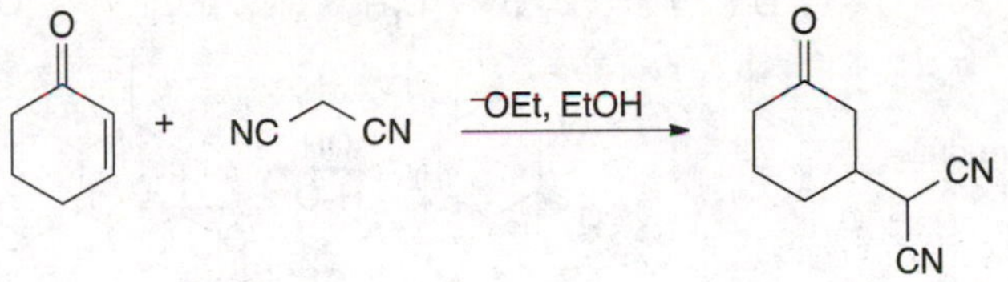

24.43

a. O O O ⇓ O + O O

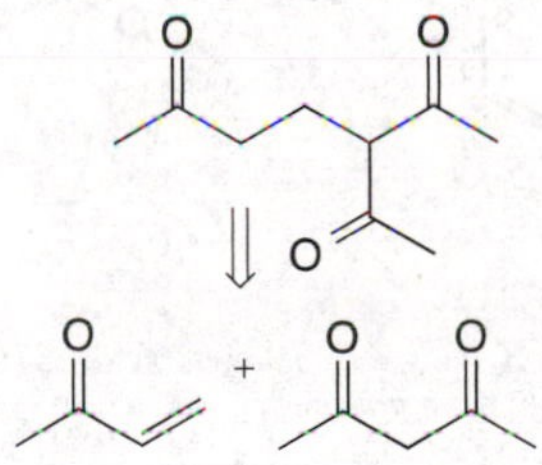

b. O O CO_2Et ⇓ O + EtO_2C O

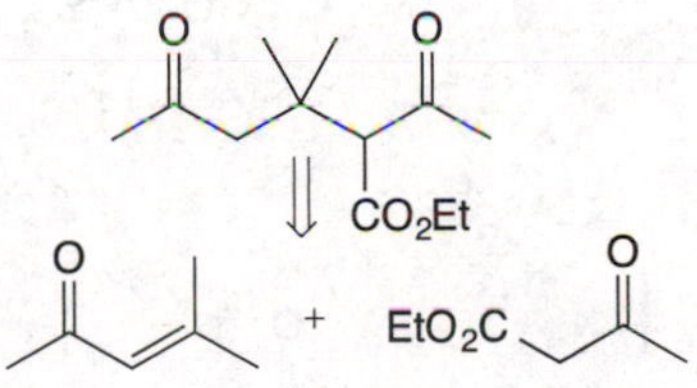

c. O CO_2Et C_6H_5 ⇓ O + C_6H_5 CO_2Et

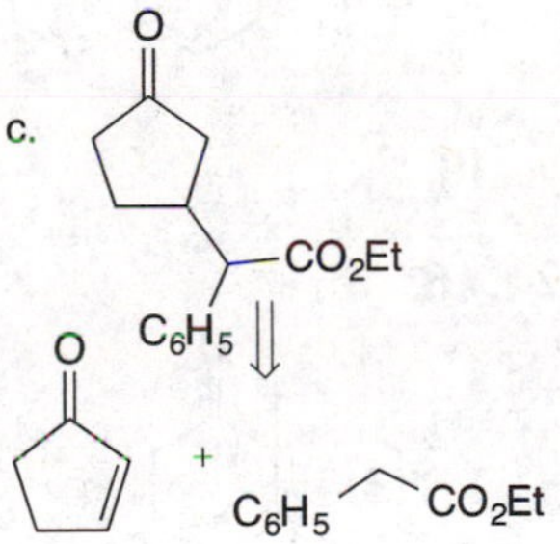

24.44

a. O A + H O (E or Z isomer can be used.) —Michael reaction→ O O

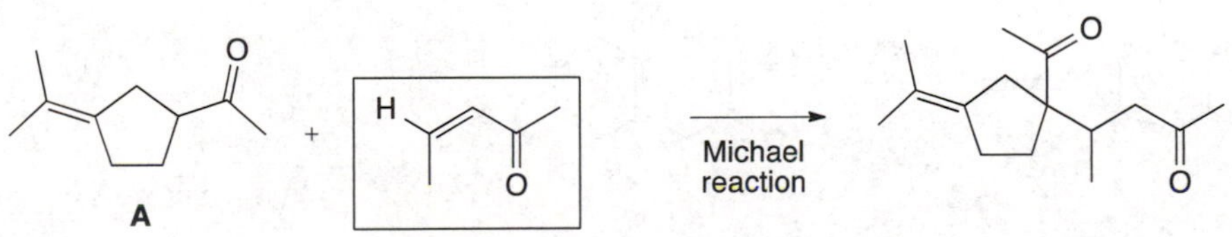

b. 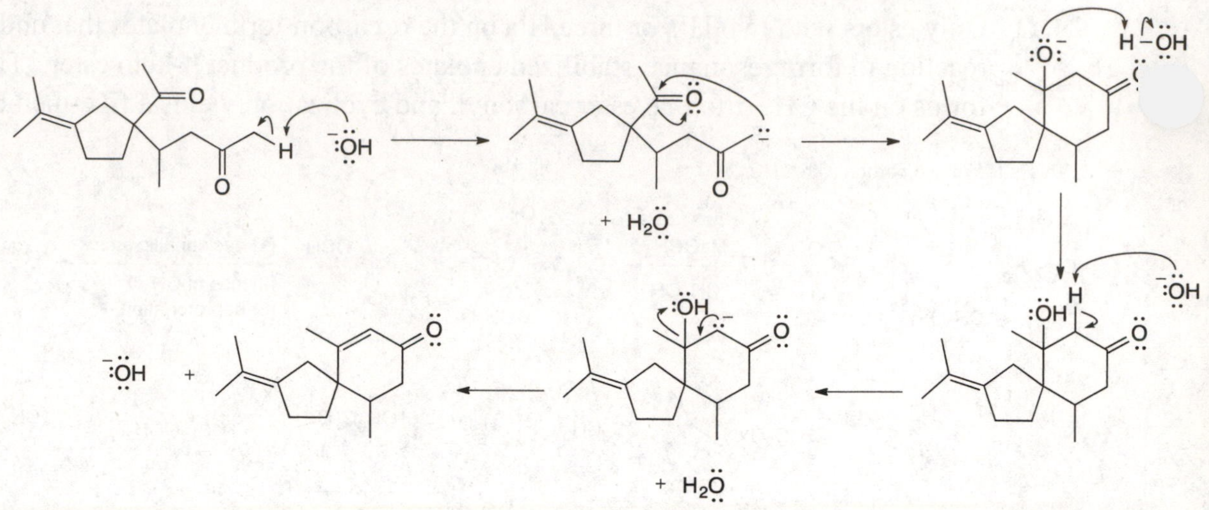 :ÖH → + $H_2\ddot{O}$ → $H-\ddot{O}H$ → :ÖH → :ÖH + + $H_2\ddot{O}$

24.45

a. 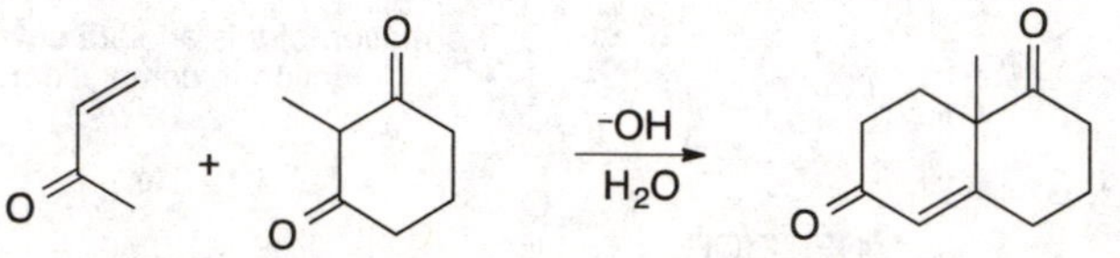 + $\xrightarrow[H_2O]{^{-}OH}$

b. 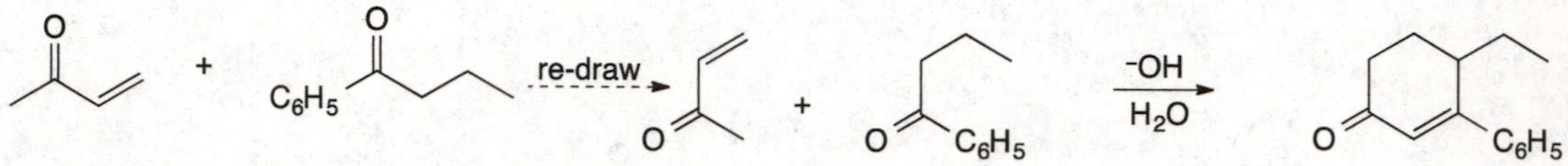 + C_6H_5 re-draw → + C_6H_5 $\xrightarrow[H_2O]{^{-}OH}$ C_6H_5

c. + re-draw → + $\xrightarrow[H_2O]{^{-}OH}$

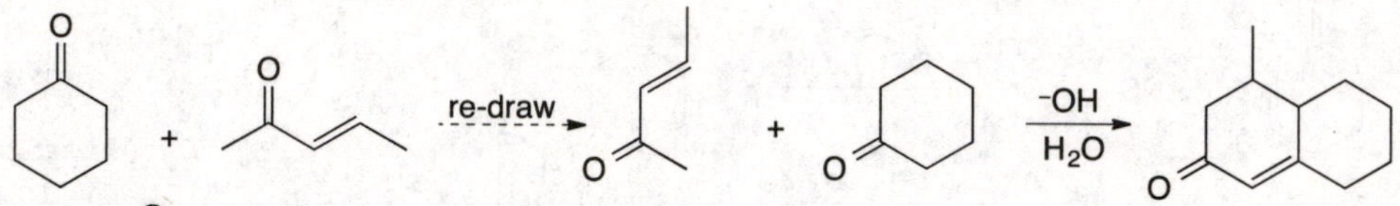

d. 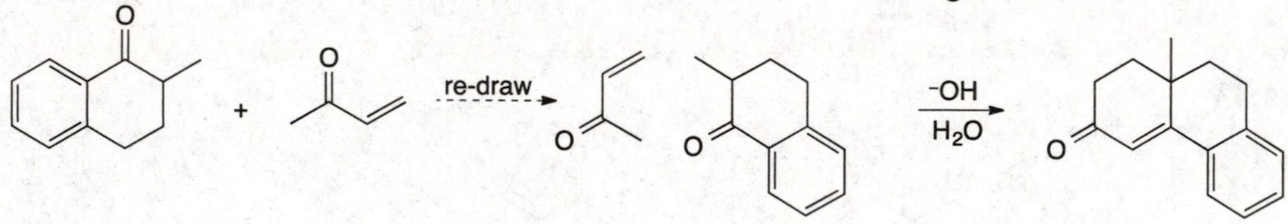 + re-draw → $\xrightarrow[H_2O]{^{-}OH}$

24.46

a. 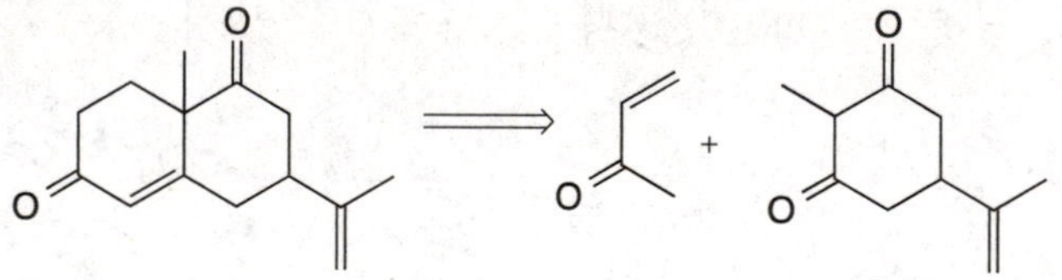 ⟹ +

c. 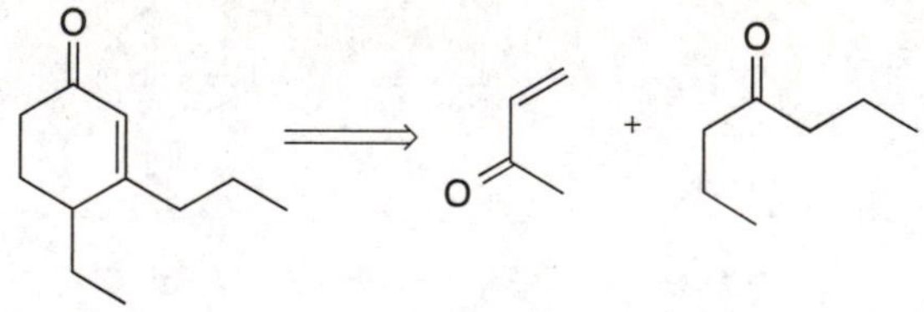 ⟹ +

b. 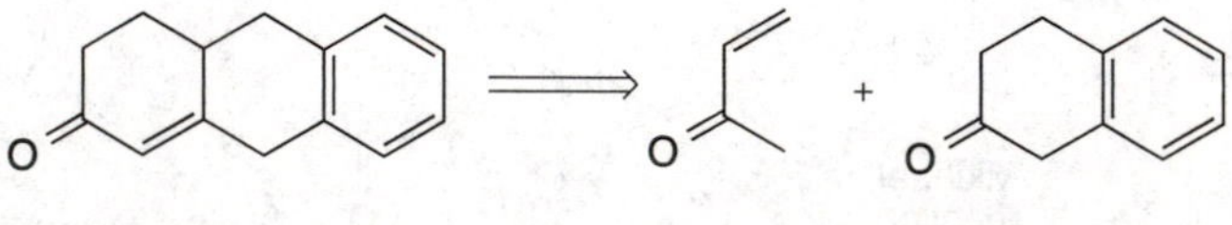 ⟹ +

24.47

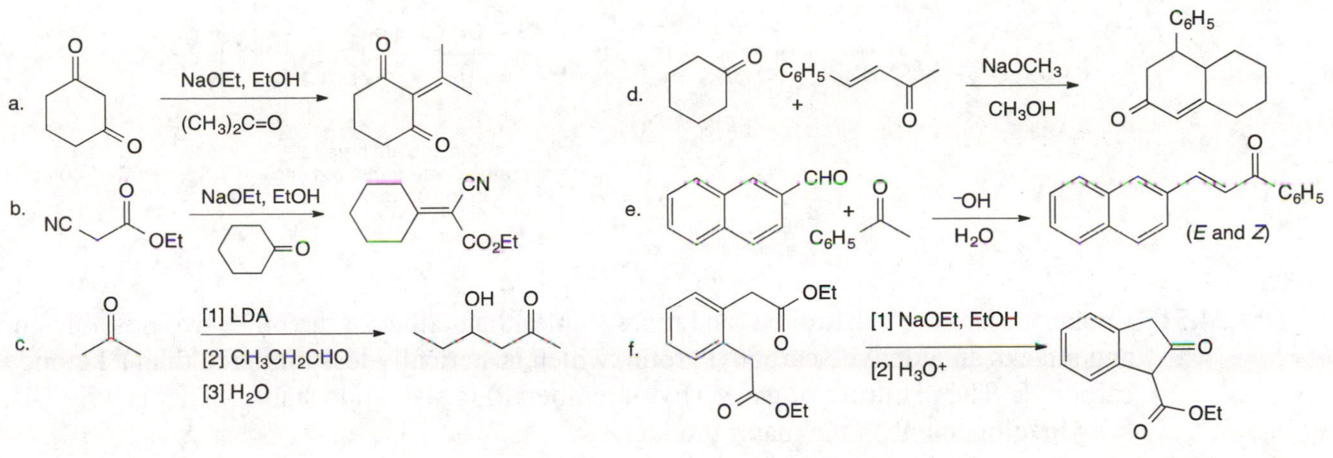

a. NaOEt, EtOH; $(CH_3)_2C{=}O$

b. NaOEt, EtOH; NC, OEt, CN, CO_2Et

c. [1] LDA; [2] CH_3CH_2CHO; [3] H_2O; OH

d. C_6H_5 + ; $NaOCH_3$, CH_3OH; C_6H_5

e. CHO + C_6H_5; ^-OH, H_2O; C_6H_5 (*E* and *Z*)

f. OEt, OEt; [1] NaOEt, EtOH; [2] H_3O^+; OEt

24.48

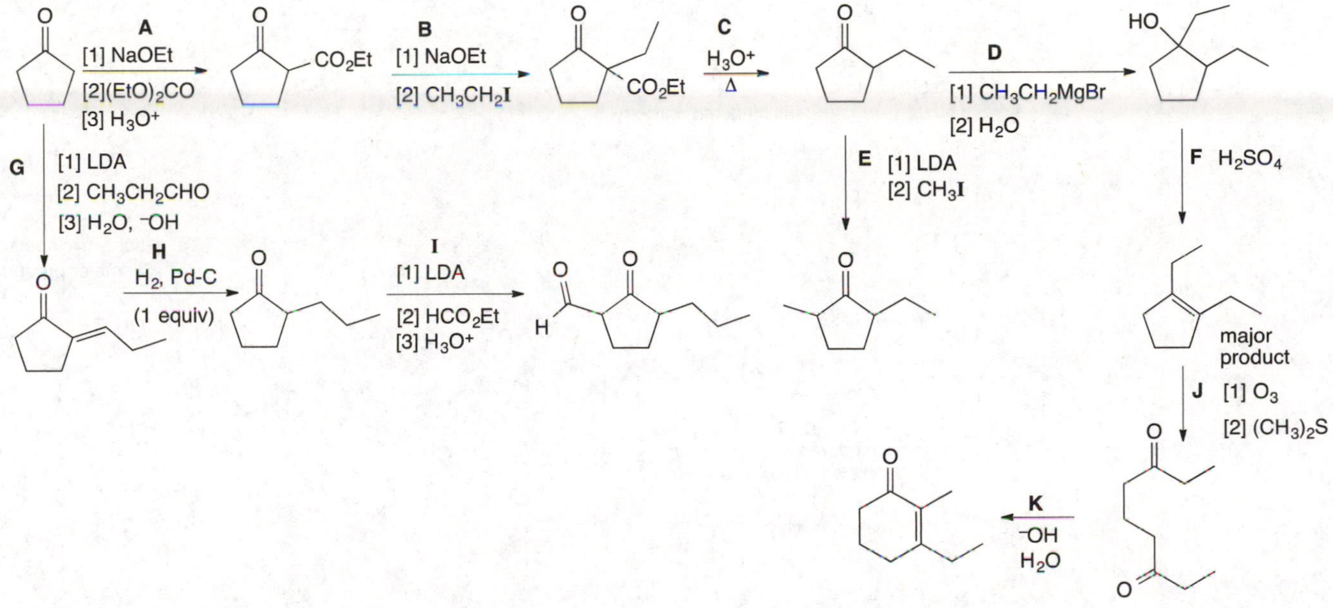

A [1] NaOEt; [2] $(EtO)_2CO$; [3] H_3O^+; CO_2Et

B [1] NaOEt; [2] CH_3CH_2I; CO_2Et

C H_3O^+, Δ

D [1] CH_3CH_2MgBr; [2] H_2O; HO

E [1] LDA; [2] CH_3I

F H_2SO_4; major product

G [1] LDA; [2] CH_3CH_2CHO; [3] H_2O, ^-OH

H H_2, Pd-C (1 equiv)

I [1] LDA; [2] HCO_2Et; [3] H_3O^+; H

J [1] O_3; [2] $(CH_3)_2S$

K ^-OH, H_2O

24.49

a. This reaction is a crossed Claisen between an ester and ketone. Cyclization forms the more stable six-membered ring.

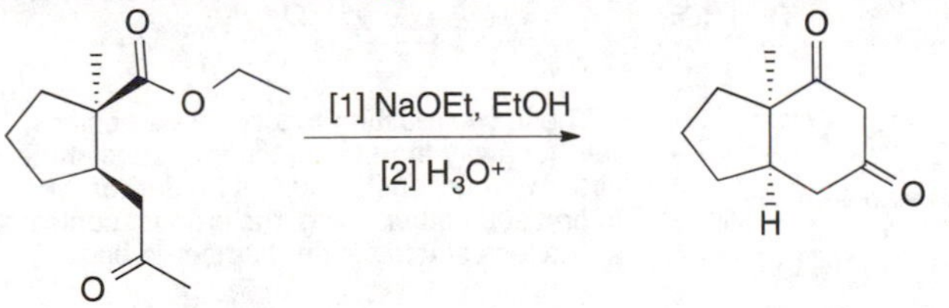

[1] NaOEt, EtOH; [2] H_3O^+; H

b.

Br; NaOEt, EtOH; Br

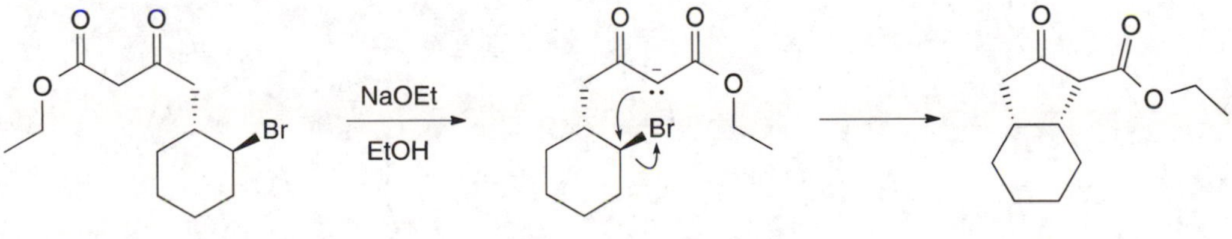

24.50

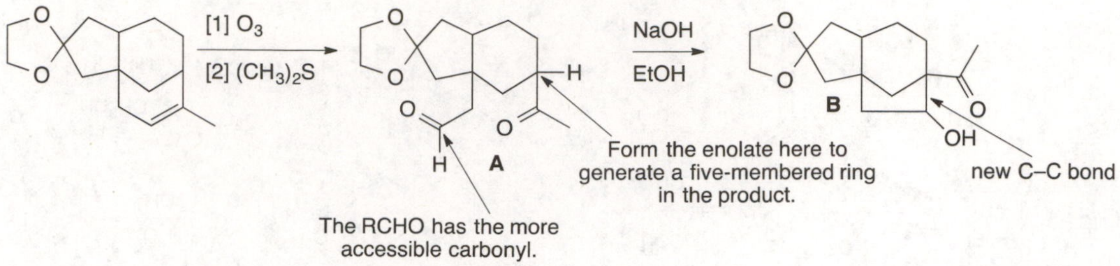

[1] O_3
[2] $(CH_3)_2S$

NaOH
EtOH

A

B

Form the enolate here to generate a five-membered ring in the product.

The RCHO has the more accessible carbonyl.

new C–C bond

24.51 Enolate **A** is more substituted (and more stable) than either of the other two possible enolates and attacks an aldehyde carbonyl group, which is sterically less hindered than a ketone carbonyl. The resulting ring size (five-membered) is also quite stable. That is why 1-acetylcyclopentene is the major product.

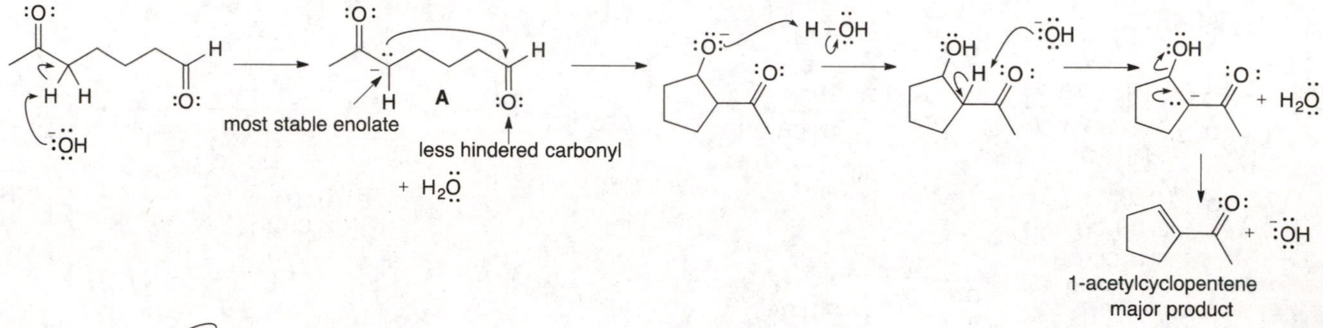

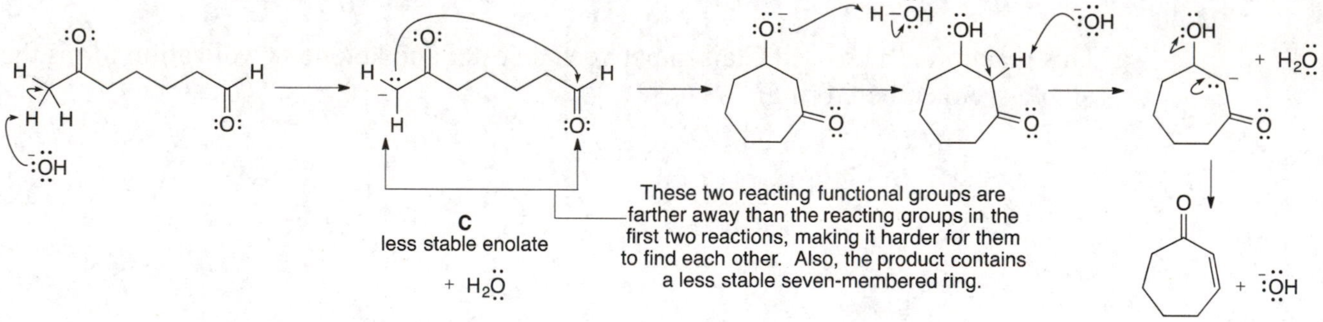

most stable enolate

A

less hindered carbonyl

+ H_2O

1-acetylcyclopentene
major product

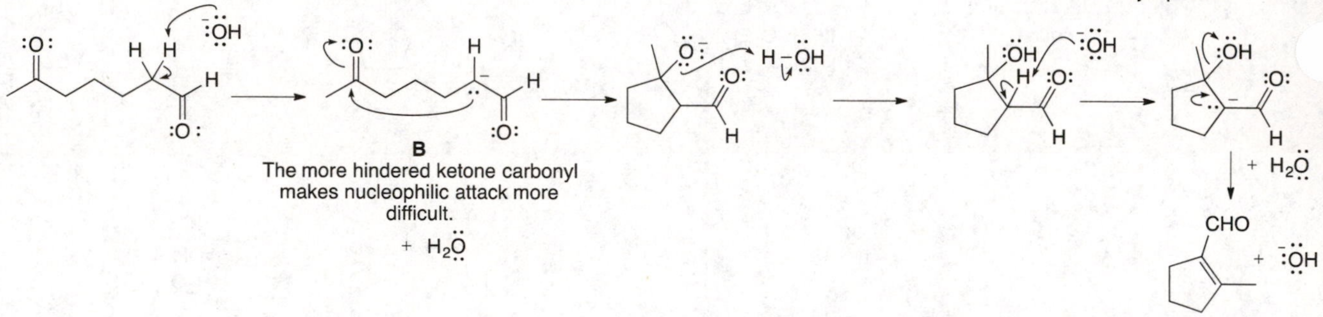

B

The more hindered ketone carbonyl makes nucleophilic attack more difficult.

+ H_2O

CHO

C

less stable enolate

+ H_2O

These two reacting functional groups are farther away than the reacting groups in the first two reactions, making it harder for them to find each other. Also, the product contains a less stable seven-membered ring.

24.52 Removal of a proton from CH_3NO_2 forms an anion for which three resonance structures can be drawn.

$H-CH_2-N^+(=O)O^-$ + ^-OH → $^-CH_2-N^+(=O)O^-$ ⟷ $CH_2=N^+(O^-)O^-$ ⟷ $^-CH_2-N^+(O^-)=O$

→ + H_2O → + ^-OH

24.53 All enolates have a second resonance structure with a negative charge on O.

+ H_2O → + $H-OH$ → + $H-OH$ → + $H-OH$ → + ^-OH

24.54

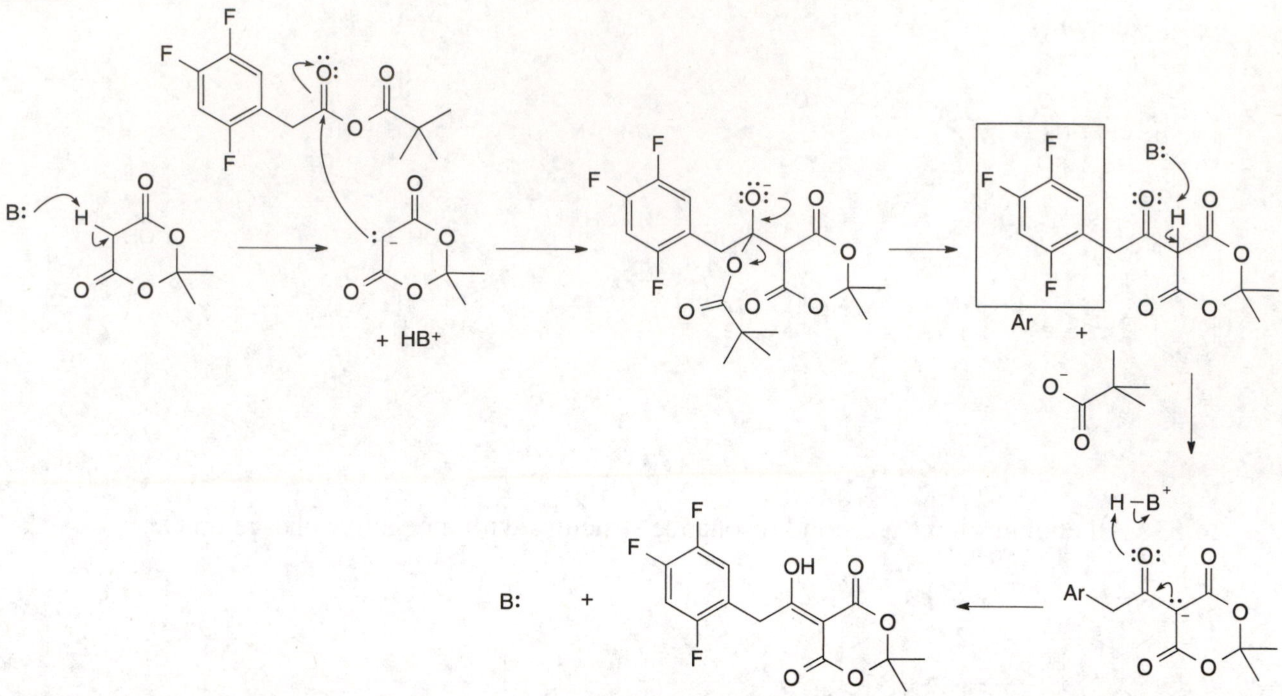

B: H F F F O O O O O O O + HB^+ Ar + B: H O O O O O O F F F OH O O O O B: + $H{-}B^+$:O: Ar

24.55 Et_3N reacts with phenylacetic acid to form a carboxylate anion that acts as a nucleophile to displace Br, forming **Y.** Then an intramolecular crossed Claisen reaction yields rofecoxib.

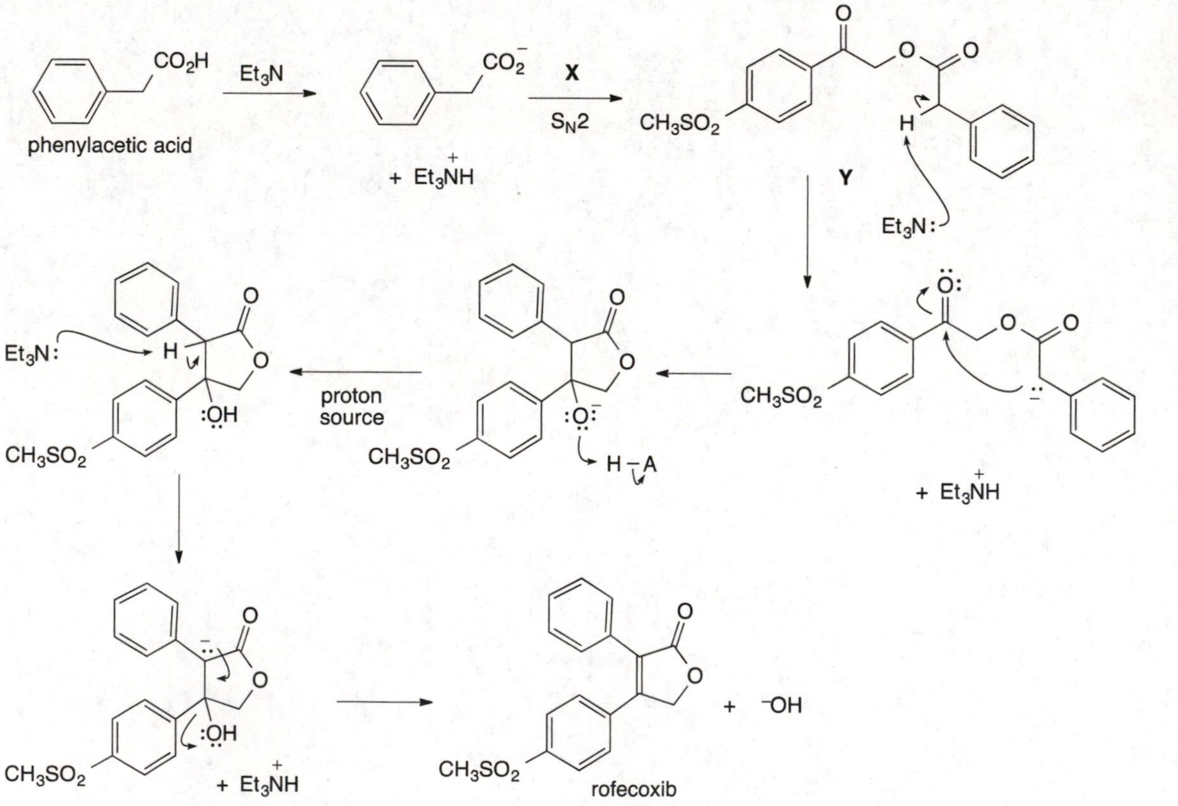

CO_2H phenylacetic acid Et_3N CO_2^- + $Et_3\overset{+}{N}H$ **X** S_N2 CH_3SO_2 O O O H **Y** Et_3N:

CH_3SO_2 :O: O O + $Et_3\overset{+}{N}H$

O O CH_3SO_2 :O:$^-$ H–A proton source

Et_3N: H O O :OH CH_3SO_2

O O :OH CH_3SO_2 + $Et_3\overset{+}{N}H$

O O CH_3SO_2 rofecoxib + ^-OH

24.56

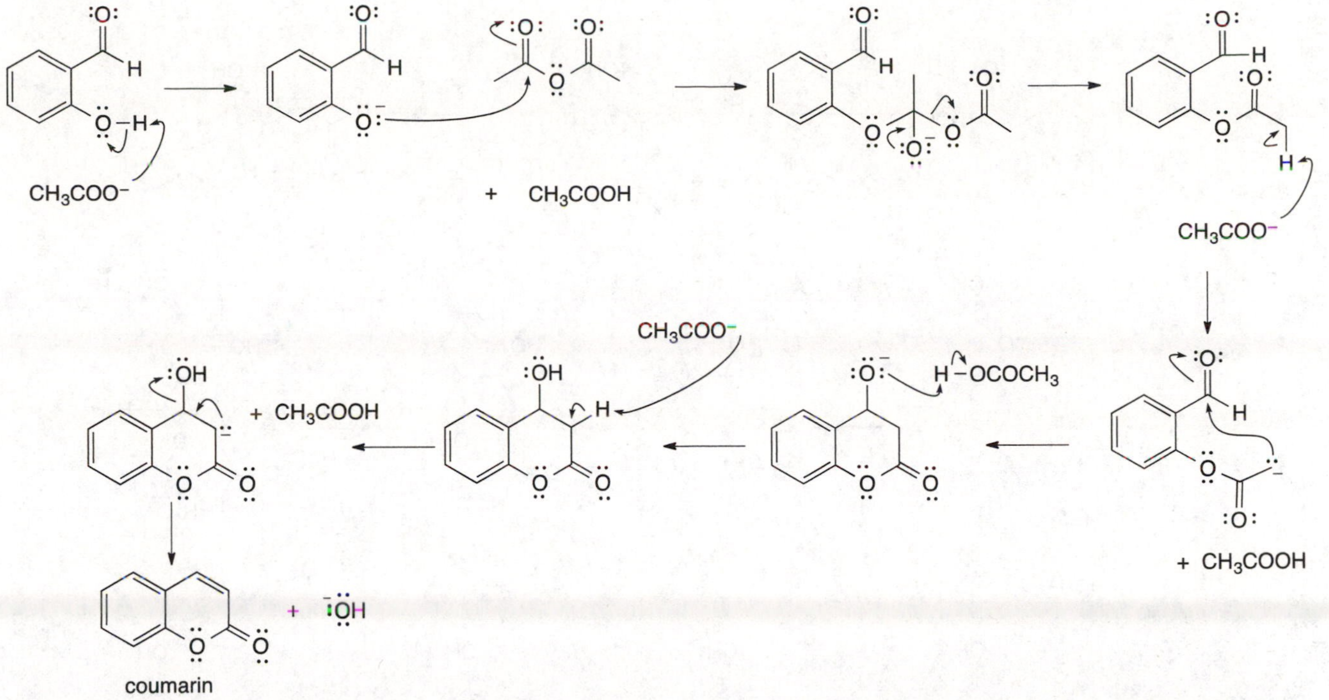

CH_3COO^-

+ CH_3COOH

CH_3COO^-

+ CH_3COOH

CH_3COO^-

$H-OCOCH_3$

+ CH_3COOH

+ :ÖH

coumarin

24.57 The mechanism consists of an intramolecular aldol reaction using an enolate formed by removal of a γ proton of an α,β-unsaturated carbonyl. The β-hydroxy carbonyl compound loses H_2O by the two-step E1cB mechanism to generate the final product.

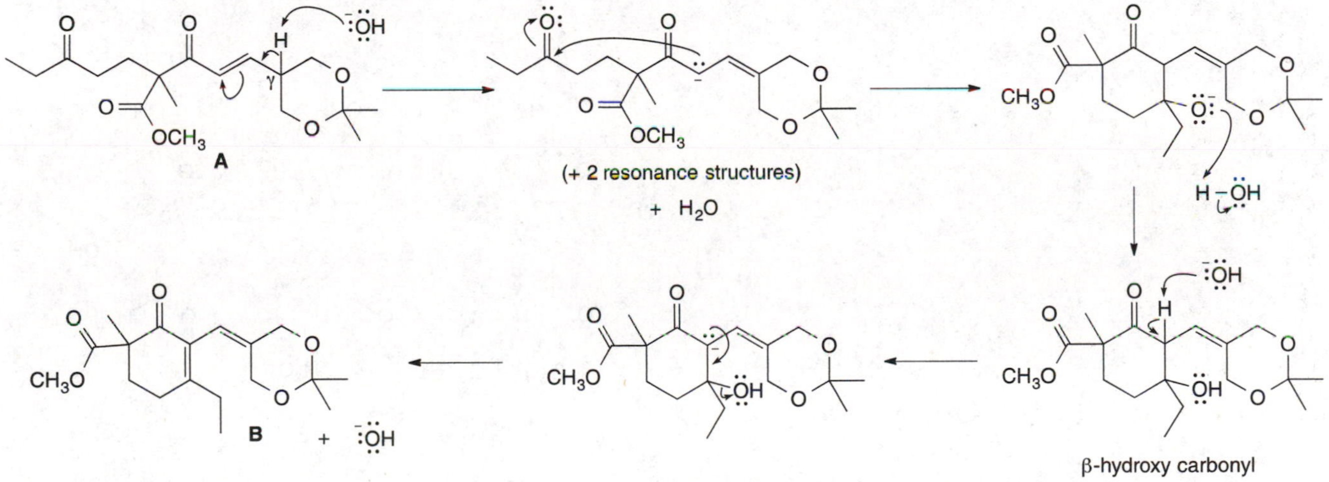

A

(+ 2 resonance structures)

+ H_2O

β-hydroxy carbonyl

B + :ÖH

24.58 Because the reaction is carried out in acid, enols (not enolates) must be intermediates.

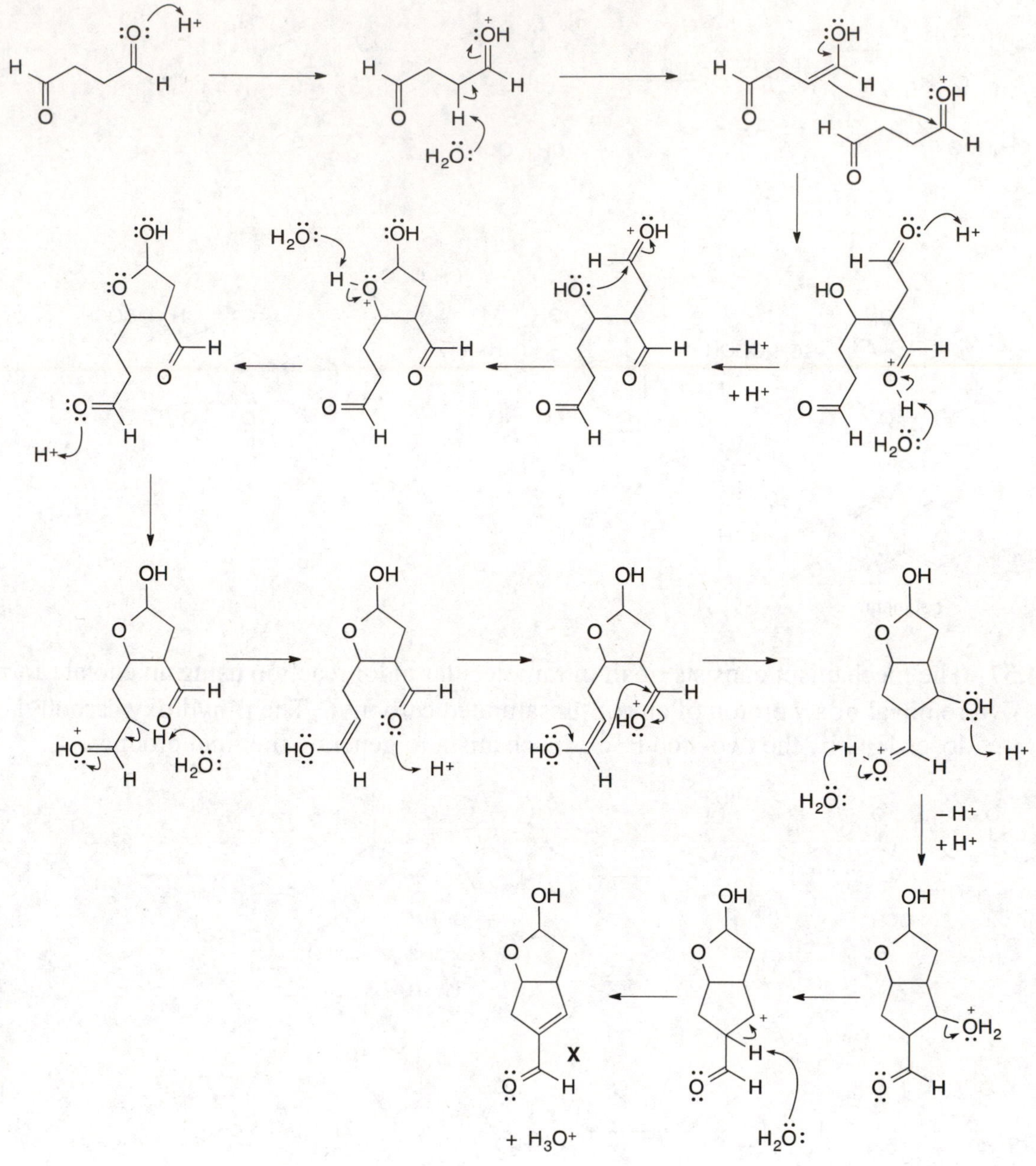

$:\ddot{O}H$ H^+ $H_2\ddot{O}:$ $-H^+$ $+H^+$ X $+ H_3O^+$

24.59

a.

C6

ethyl hexa-2,4-dienoate

$+$ EtOH

diethyl oxalate

$^-:\ddot{O}Et$

CO_2Et EtO_2C OEt

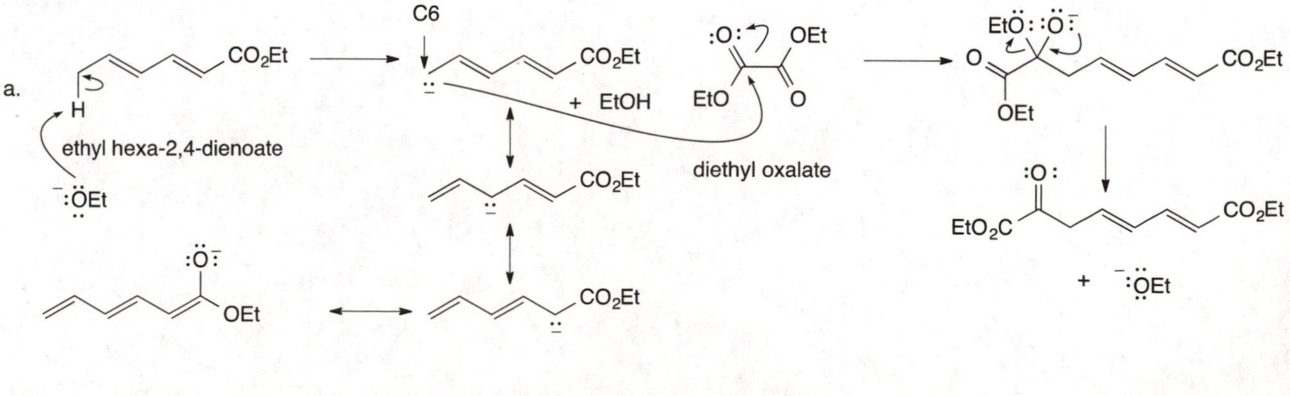

b. The protons on C6 are more acidic than other sp^3 hybridized C–H bonds because a highly resonance-stabilized carbanion is formed when a proton is removed. One resonance structure places a negative charge on the carbonyl O atom. This makes the protons on C6 similar in acidity to the α H's to a carbonyl.

c. This is a crossed Claisen because it involves the enolate of a conjugated ester reacting with the carbonyl group of a second ester.

24.60

a.

^{-}OH, H_2O

H_2 (excess), Pd-C

OH

b.

+

^{-}OH, H_2O

[1] $(CH_3)_2CuLi$

[2] H_2O

24.61

a.

[1] LDA

[2] CH_3CO_2Et

[1] NaOEt

[2] CH_3Br

OH, H_2SO_4

PBr_3

CH_3OH

OH

CrO_3, H_2SO_4, H_2O

OH

b.

CH_3OH

$SOCl_2$

CH_3Cl, $AlCl_3$

[1] Br_2, $h\nu$

[2] ^{-}OH

OH

PCC

[1] LDA

[2] C_6H_5CHO

OH

C_6H_5

unstable

$-H_2O$

C_6H_5

H_2 (excess), Pd-C

HO

C_6H_5

c.

[1] CH_3MgBr

[2] H_2O

Mg

CH_3OH $\xrightarrow{PBr_3}$ CH_3Br

OH

H_2SO_4

major product

[1] O_3

[2] $(CH_3)_2S$

CHO

^{-}OH, H_2O

24.62

a. OEt, O → [1] NaOEt, [2] H_3O^+ → CO_2Et, O → [1] NaOEt, [2] C_6H_5 Br → C_6H_5, CO_2Et, O → [1] $LiAlH_4$, [2] H_2O → H_5, OH, OH

$CH_3OH + SOCl_2$ → [1] CH_3Cl, $AlCl_3$, [2] Br_2, $h\nu$ → Br

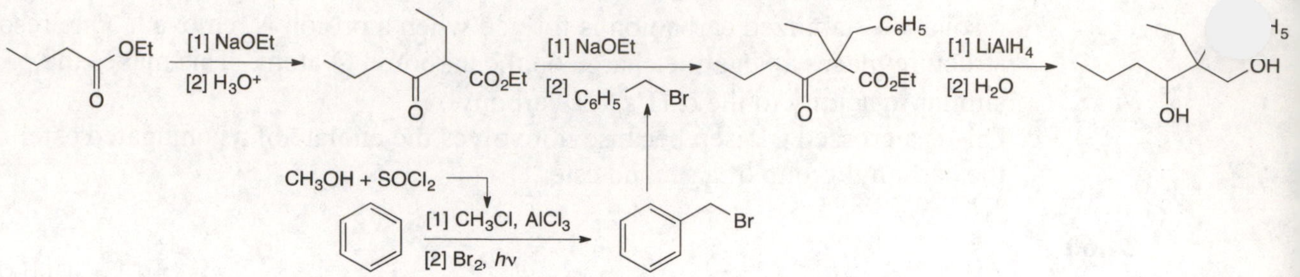

b. OEt, O + C_6H_5, O → [1] NaOEt, [2] H_3O^+ → C_6H_5, O, O

$AlCl_3$

+ O, Cl

$SOCl_2$

O, OH ← CrO_3, H_2SO_4, H_2O ← OH

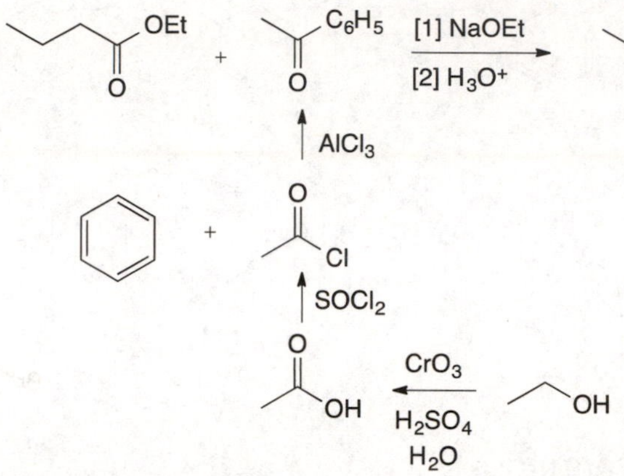

24.63

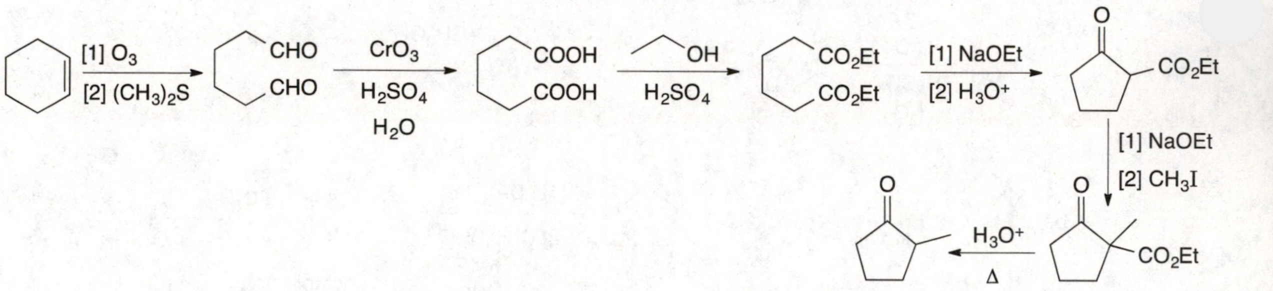

[1] O_3, [2] $(CH_3)_2S$ → CHO, CHO → CrO_3, H_2SO_4, H_2O → COOH, COOH → OH, H_2SO_4 → CO_2Et, CO_2Et → [1] NaOEt, [2] H_3O^+ → O, CO_2Et → [1] NaOEt, [2] CH_3I → O, CO_2Et → H_3O^+, Δ → O

24.64

a. O, O, CH_3O — octinoxate ⟹ CHO, CH_3O + O, O

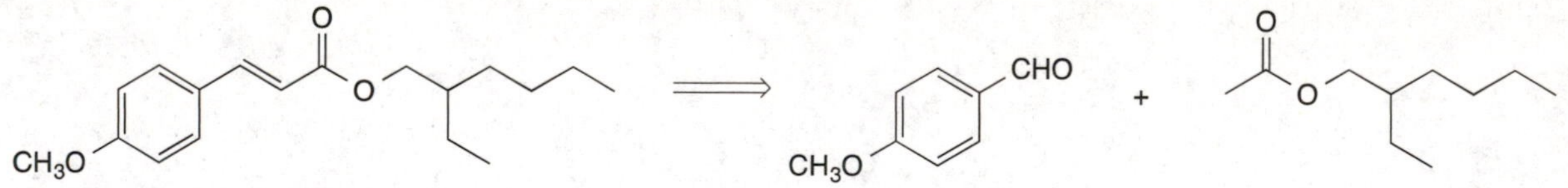

b.

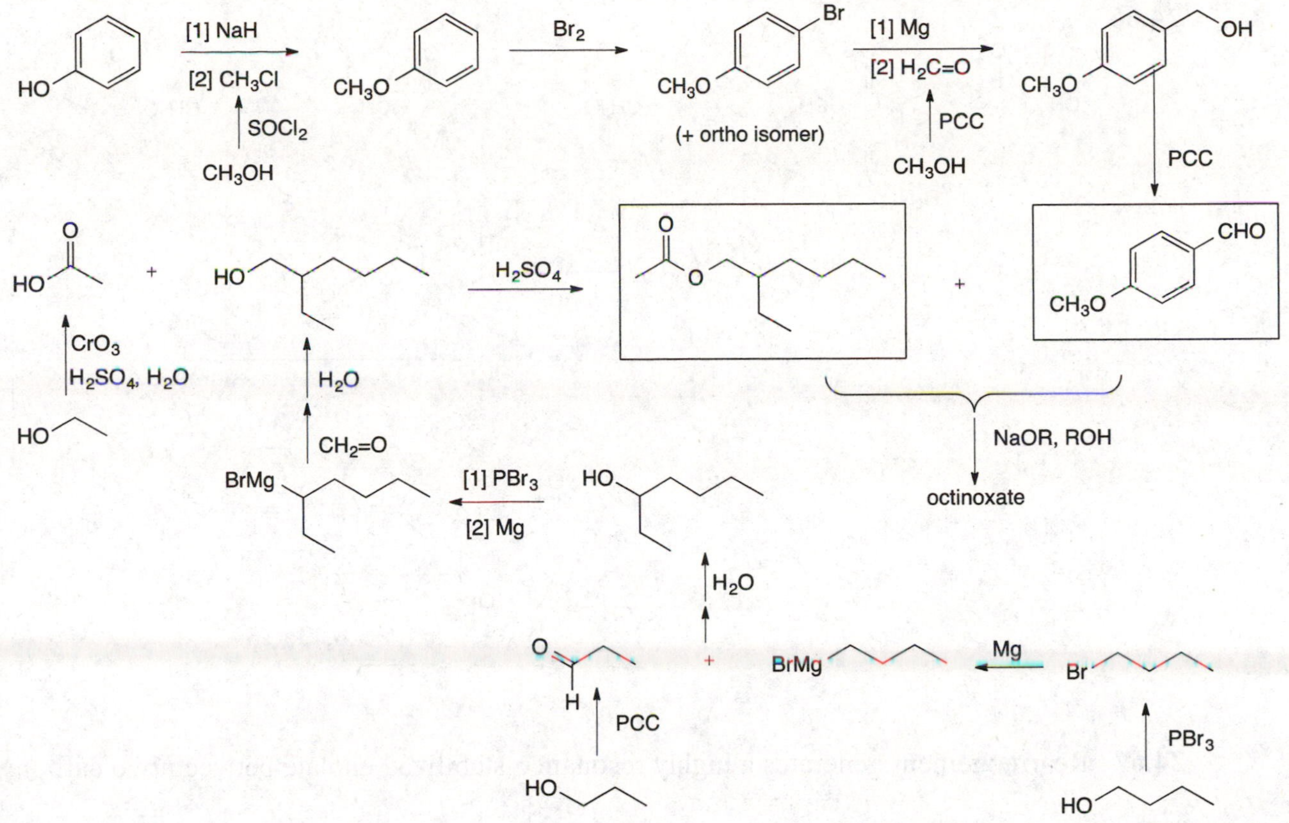

HO (phenol) —[1] NaH, [2] CH_3Cl→ CH_3O (anisole) —Br_2→ CH_3O–C₆H₄–Br (+ ortho isomer) —[1] Mg, [2] $H_2C{=}O$→ CH_3O–C₆H₄–CH_2OH —PCC→ CH_3O–C₆H₄–CHO

CH_3OH —$SOCl_2$→ CH_3Cl

CH_3OH —PCC→ $H_2C{=}O$

HO–C(=O)–CH_3 + HO–CH_2CH(Et)Bu —H_2SO_4→ ester + aldehyde —NaOR, ROH→ octinoxate

HO–CH_2CH_3 —CrO_3, H_2SO_4, H_2O→ acetic acid

BrMg (2-ethylhexyl) —$CH_2{=}O$, then H_2O→ alcohol

HO (3-heptanol) —[1] PBr_3, [2] Mg→ BrMg

propanal (O, H) + BrMg (butyl) —H_2O→ 3-heptanol

HO (propanol) —PCC→ propanal

HO (butanol) —PBr_3→ Br (butyl) —Mg→ BrMg

24.65

a. EtO

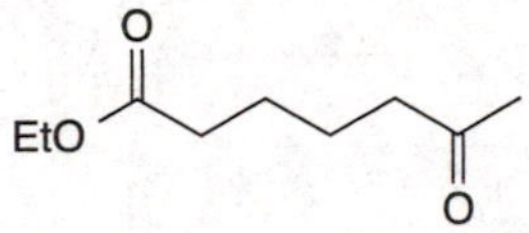

b. (cyclopentanone) + EtO–C(=O)–CH_3

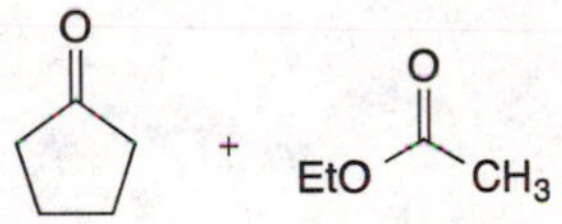

c. 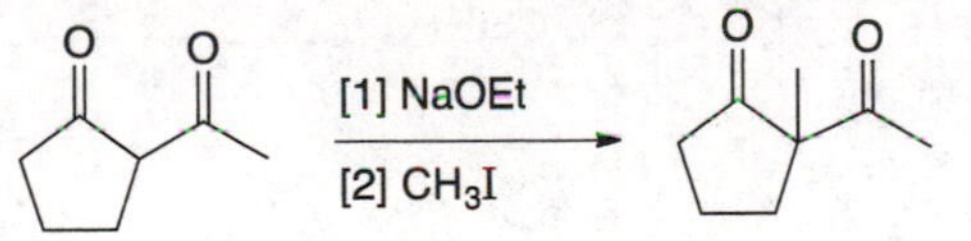 —[1] NaOEt, [2] CH_3I→

d. base → ; base →

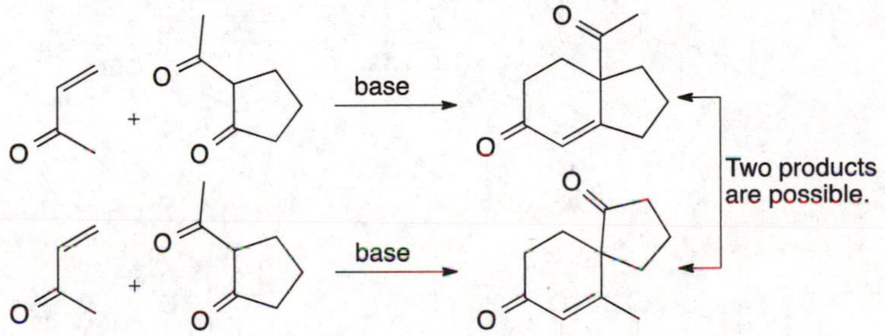

Two products are possible.

e. ⇌ OH O or O OH

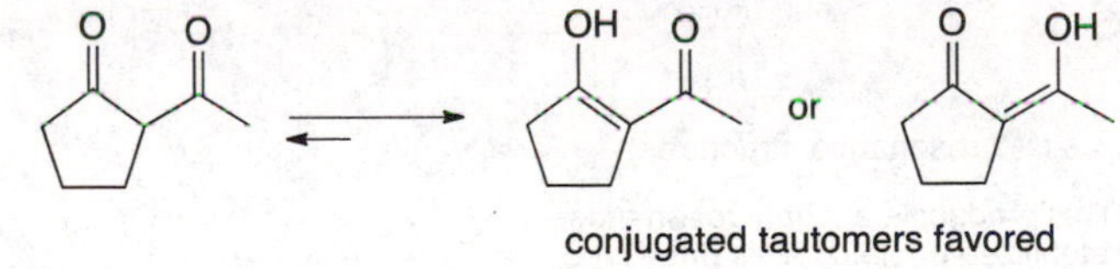

conjugated tautomers favored

24.66

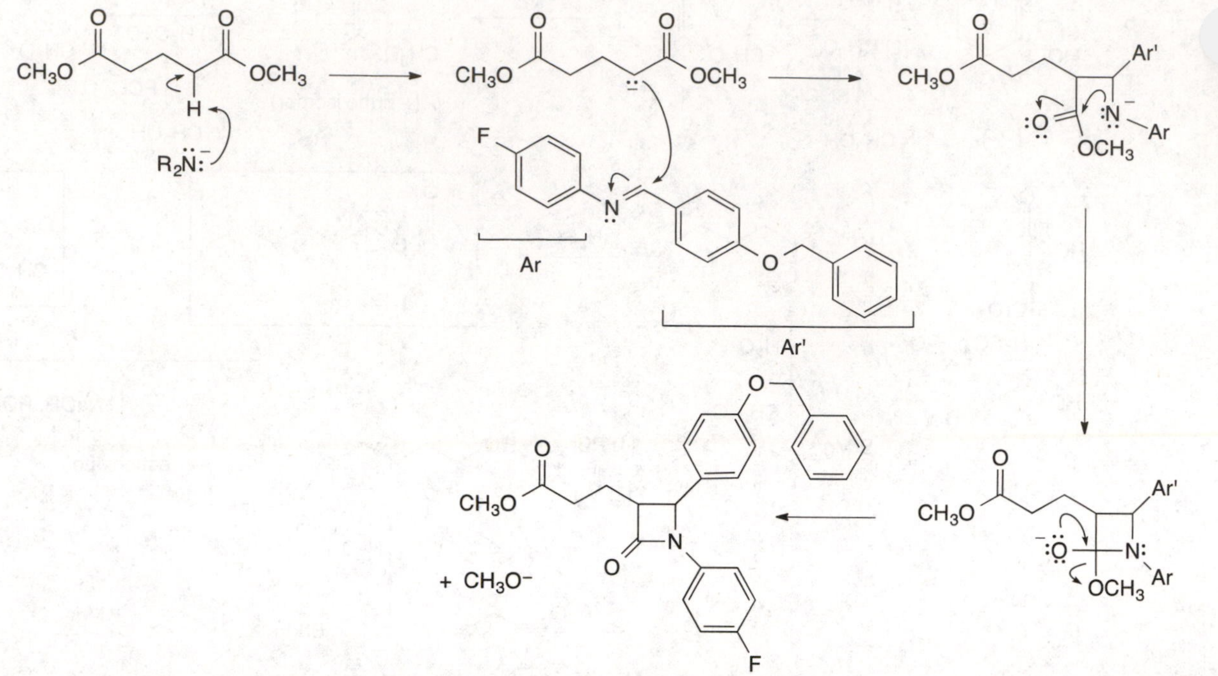

O O CH_3O OCH_3 H $R_2\ddot{N}:^-$ → O O CH_3O OCH_3 F N Ar O Ar' → O CH_3O Ar' :O: N Ar OCH_3 → O CH_3O Ar' :Ö: N: Ar OCH_3 → O CH_3O O N F O + CH_3O^-

24.67 Rearrangement generates a highly resonance-stabilized enolate between two carbonyl groups.

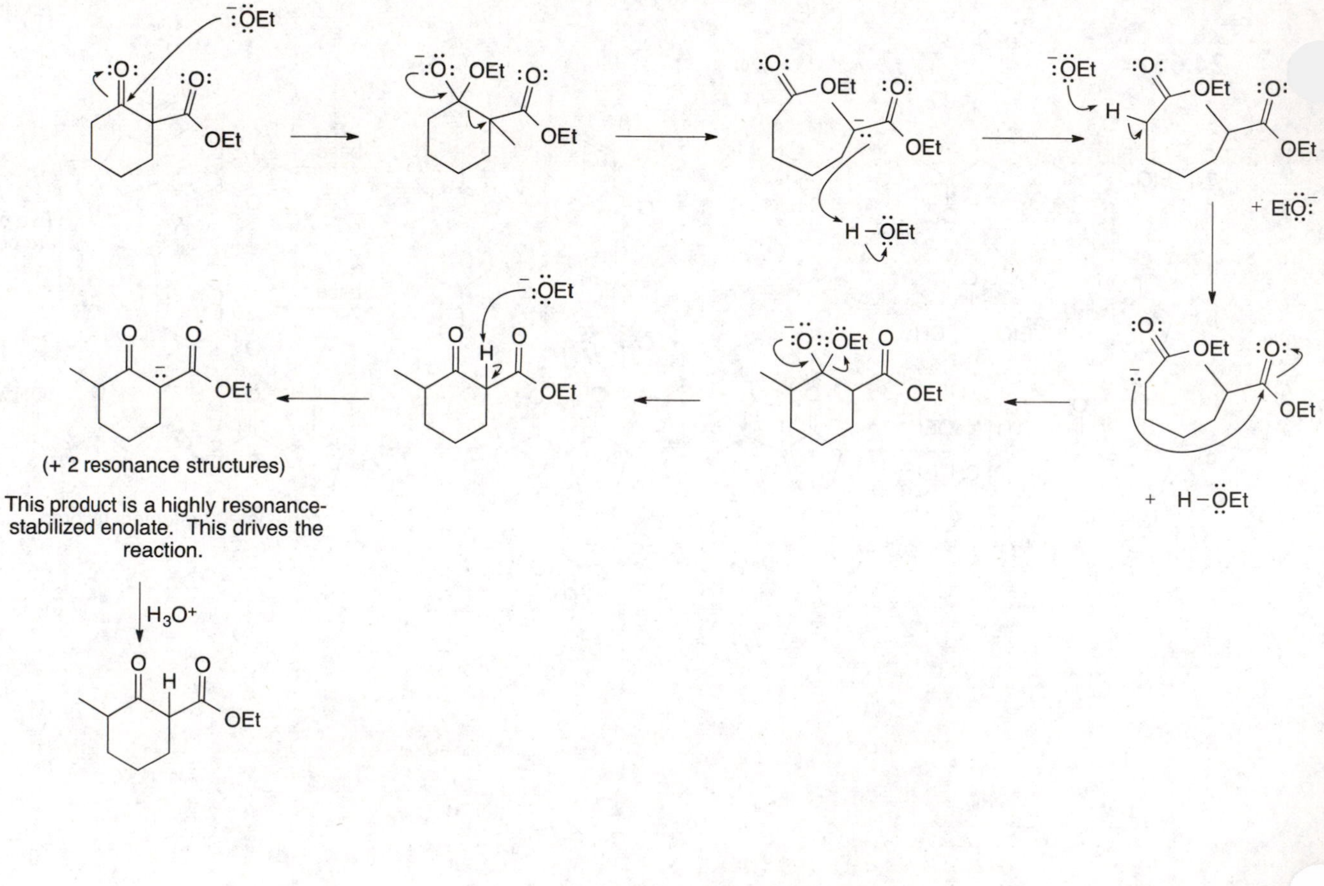

:ÖEt :O: :O: OEt → :Ö: OEt :O: OEt → :O: OEt :O: OEt H–ÖEt → :ÖEt :O: H OEt :O: OEt + EtÖ: → :O: OEt :O: OEt + H–ÖEt → :O::ÖEt O OEt → :ÖEt O H O OEt → O O OEt

(+ 2 resonance structures)

This product is a highly resonance-stabilized enolate. This drives the reaction.

H_3O^+

O H O OEt

24.68 All enolates have a second resonance structure with a negative charge on O.

B: [1] + HB^+ [2] [3] + $^-$:OH B: [4] + HB^+ [5] + $^-$:OH

Repeat steps [1]–[5] by deprotonating the indicated CH_3.

+ HB^+ (+ 2 resonance structures)

$^-$:OH + isophorone

24.69

1,4-addition

proton transfer

β-alkoxy carbonyl

1,4-dicarbonyl compound

intramolecular aldol

+ $^-$:OH

24.70 All enolates have a second resonance structure with a negative charge on O.

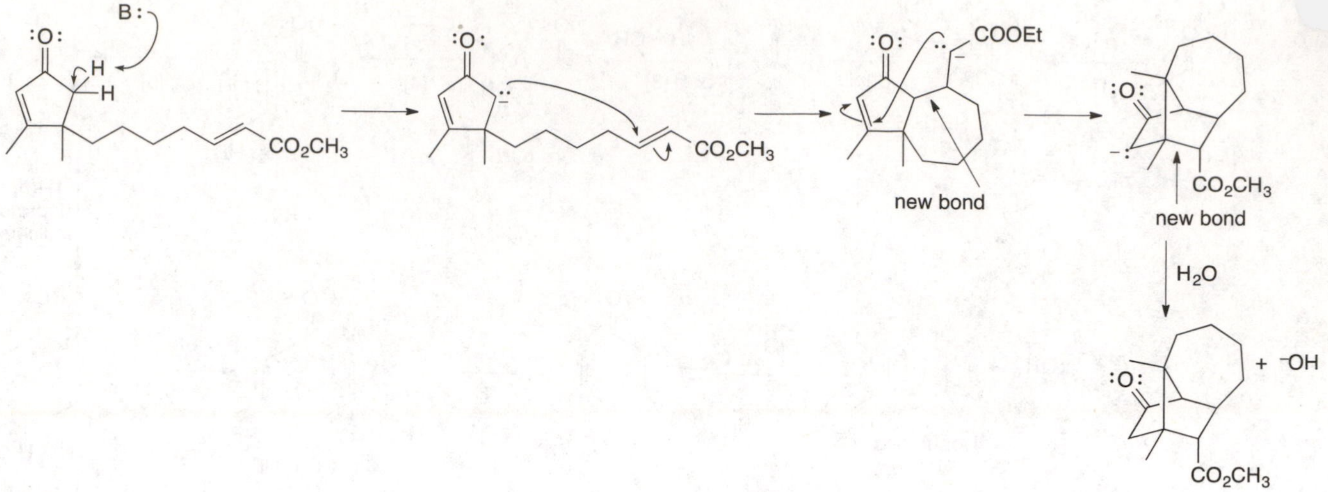

B:
:O:
H
H
CO_2CH_3
:O:
CO_2CH_3
:O:
COOEt
new bond
:O:
CO_2CH_3
new bond
H_2O
:O:
+ ^-OH
CO_2CH_3

24.71

a. $:\ddot{O}H$ H N: → N: ↔ :N: (+ other resonance structures)

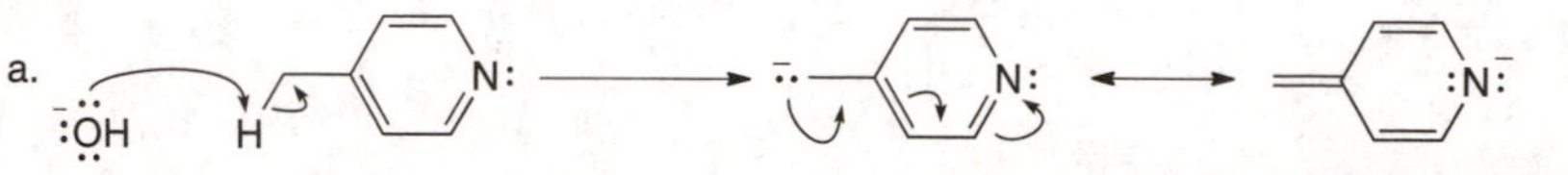

one possible resonance structure
The negative charge is delocalized on the electronegative N atom. This factor is what makes the CH_3 group bonded to the pyridine ring more acidic, and allows the condensation to occur.

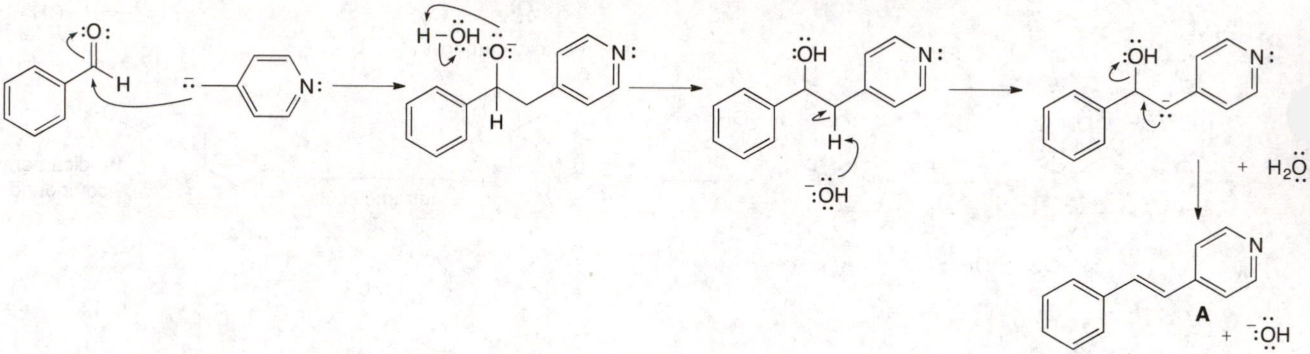

:O: H N: → H–ÖH :O: N: H → :OH N: H :OH → :OH N: + H_2O → N A + $:\ddot{O}H$

b. The condensation reaction can occur only if the CH_3 group bonded to the pyridine ring has acidic hydrogens that can be removed with ^-OH.

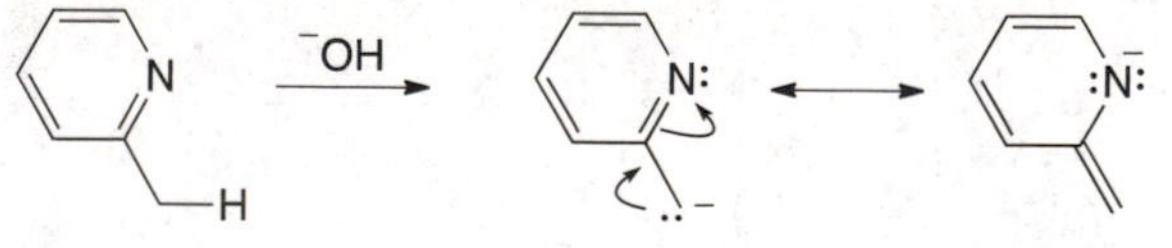

2-methylpyridine N H ^-OH → N: ↔ :N: Since the negative charge is delocalized on the N, the CH_3 contains acidic H's and reaction will occur.

(+ other resonance structures)

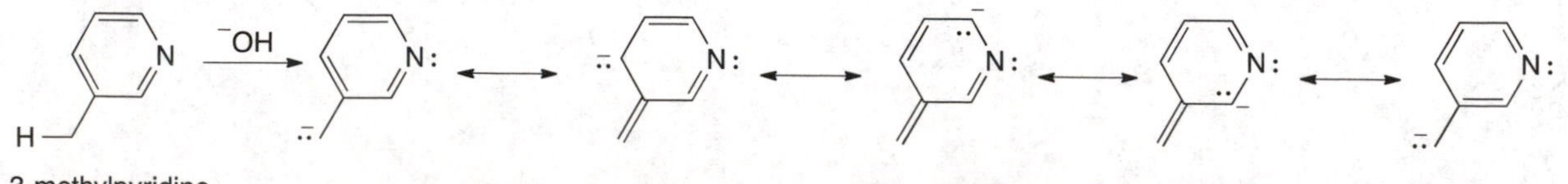

3-methylpyridine N H ^-OH → N: ↔ N: ↔ N: ↔ N: ↔ N:

No resonance structure places the negative charge on the N, so the CH_3 is not acidic and condensation does not occur.

24.72 The reaction takes place in acid, so enols are involved. After the initial condensation reaction, the NH_2 and C=O groups form an imine by an intramolecular reaction.

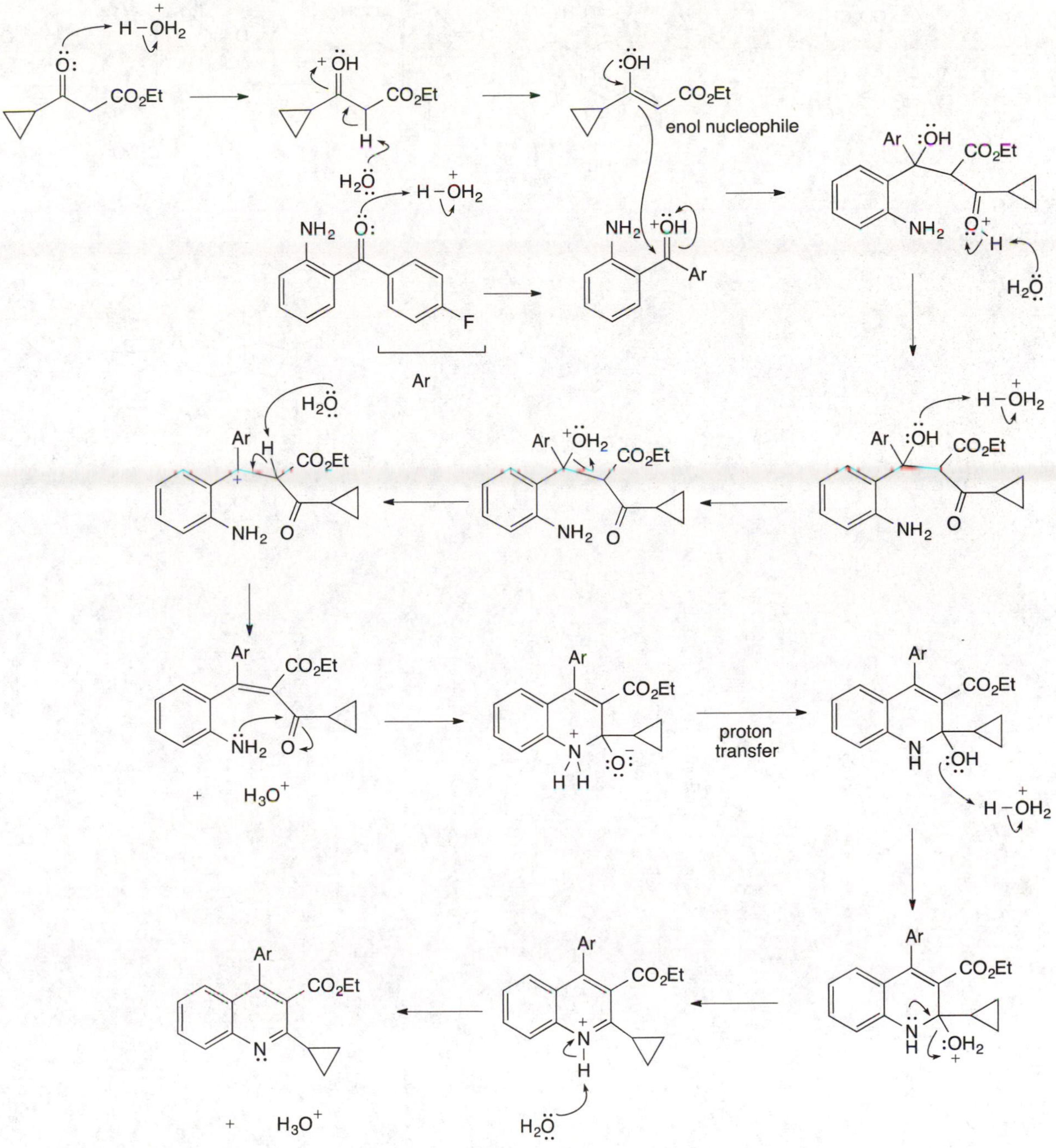

H–OH2
CO2Et
enol nucleophile
NH2
F
Ar
Ar
proton
transfer
H3O+

24.73

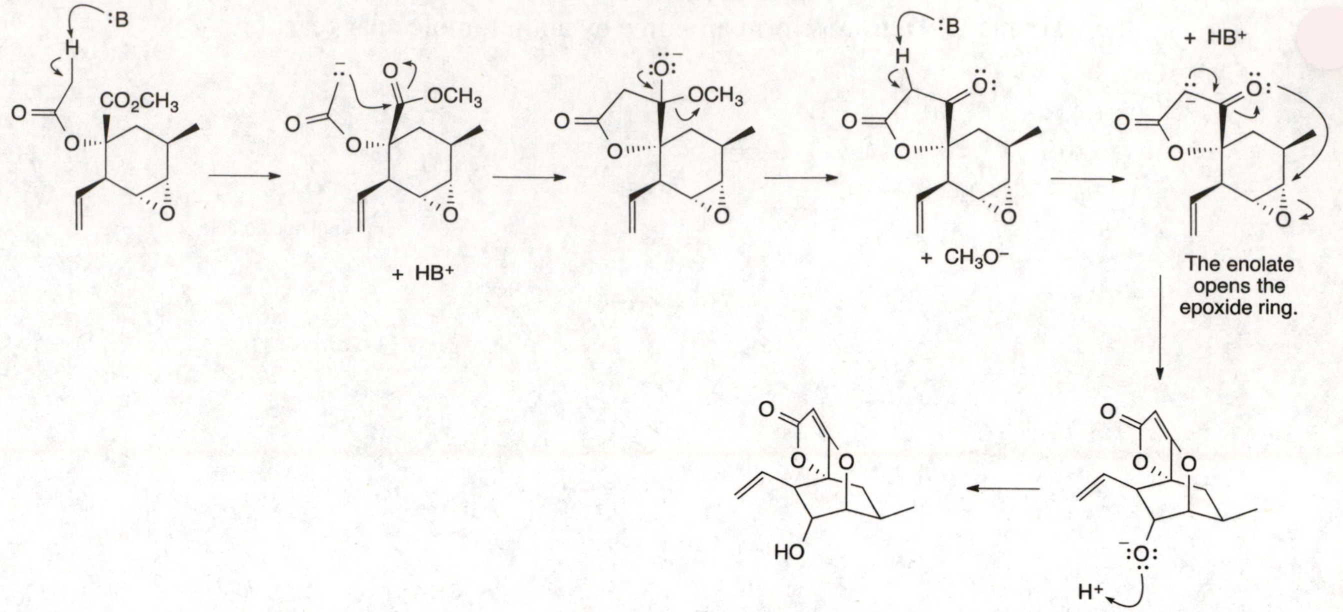

:B

H

CO_2CH_3

O

+ HB^+

OCH_3

+ CH_3O^-

+ HB^+

The enolate opens the epoxide ring.

H^+

HO

Chapter 25 Amines

Chapter Review

General facts

- Amines are organic nitrogen compounds having the general structure RNH_2, R_2NH, or R_3N, with a lone pair of electrons on N (25.1).
- Amines are named using the suffix *-amine* (25.3).
- All amines have polar C–N bonds. Primary (1°) and 2° amines have polar N–H bonds and are capable of intermolecular hydrogen bonding (25.4).
- The lone pair on N makes amines strong organic bases and nucleophiles (25.8).

Summary of spectroscopic absorptions (25.5)

Mass spectra	Molecular ion	Amines with an odd number of N atoms give an odd molecular ion.
IR absorptions	N–H	3300–3500 cm^{-1} (two peaks for RNH_2, one peak for R_2NH)
^{1}H NMR absorptions	NH	0.5–5 ppm (no splitting with adjacent protons)
	CH–N	2.3–3.0 ppm (deshielded Csp^3–H)
^{13}C NMR absorption	C–N	30–50 ppm

Comparing the basicity of amines and other compounds (25.10)

- Alkylamines (RNH_2, R_2NH, and R_3N) are more basic than NH_3 because of the electron-donating R groups (25.10A).
- Alkylamines (RNH_2) are more basic than arylamines ($C_6H_5NH_2$), which have a delocalized lone pair from the N atom (25.10B).
- Arylamines with electron-donor groups are more basic than arylamines with electron-withdrawing groups (25.10B).
- Alkylamines (RNH_2) are more basic than amides ($RCONH_2$), which have a delocalized lone pair from the N atom (25.10C).
- Aromatic heterocycles with a localized electron pair on N are more basic than those with a delocalized lone pair from the N atom (25.10D).
- Alkylamines with a lone pair in an sp^3 hybrid orbital are more basic than those with a lone pair in an sp^2 hybrid orbital (25.10E).

Preparation of amines (25.7)

[1] Direct nucleophilic substitution with NH_3 and amines (25.7A)

$$R{-}X + \underset{\text{excess}}{\ddot{N}H_3} \longrightarrow \underset{1^\circ\ \text{amine}}{\boxed{R{-}\ddot{N}H_2}} + NH_4^+\ X^-$$

$$\underset{\text{excess}}{R{-}X} + R'{-}\ddot{N}H_2 \longrightarrow \underset{\text{ammonium salt}}{\boxed{R_3\overset{+}{N}R'\ X^-}}$$

- The mechanism is S_N2.
- The reaction works best for CH_3X or RCH_2X.
- The reaction works best to prepare 1° amines and ammonium salts.

[2] Gabriel synthesis (25.7A)

H–N: (phthalimide) —[1] KOH, [2] RX→ $R-NH_2$ (1° amine) + (benzene-1,2-dicarboxylate, CO_2^- / CO_2^-)

- The mechanism is S_N2.
- The reaction works best for CH_3X or RCH_2X.
- Only 1° amines can be prepared.

[3] Reduction methods (25.7B)

a. From nitro compounds

$R-NO_2$ —H_2, Pd-C or Fe, HCl or Sn, HCl→ $R-NH_2$ (1° amine)

b. From nitriles

$R-C\equiv N$ —[1] $LiAlH_4$, [2] H_2O→ RCH_2NH_2 (1° amine)

c. From amides

$RCONR'_2$ (R' = H or alkyl) —[1] $LiAlH_4$, [2] H_2O→ $RCH_2NR'_2$ (1°, 2°, and 3° amines)

[4] Reductive amination (25.7C)

$RCOR'$ + $R_2''NH$ (R', R" = H or alkyl) —$NaBH_3CN$→ $RCH(NR''_2)R'$ (1°, 2°, and 3° amines)

- Reductive amination adds one alkyl group (from an aldehyde or ketone) to a nitrogen nucleophile.
- Primary (1°), 2°, and 3° amines can be prepared.

Reactions of amines

[1] Reaction as a base (25.9)

$R-\ddot{N}H_2$ + $H-A$ ⇌ $R-\overset{+}{N}H_3$ + $:A^-$

[2] Nucleophilic addition to aldehydes and ketones (25.11)

With 1° amines:

$RCOCH_3$ (R = H or alkyl) —$R'NH_2$→ $RC(=NR')CH_3$ (imine)

With 2° amines:

$RCOCH_3$ (R = H or alkyl) —R'_2NH→ $RC(NR'_2)=CH_2$ (enamine)

[3] Nucleophilic substitution with acid chlorides and anhydrides (25.11)

$RCOZ$ + R'_2NH (2 equiv) ⟶ $RCONR'_2$

Z = Cl or OCOR
R' = H or alkyl

1°, 2°, and 3° amides

---

[4] Hofmann elimination (25.12)

[1] CH_3I (excess)
[2] Ag_2O
[3] Δ

alkene

- The less substituted alkene is the major product.

---

[5] Reaction with nitrous acid (25.13)

With 1° amines:

$R{-}NH_2$ $\xrightarrow[HCl]{NaNO_2}$ $R{-}\overset{+}{N}{\equiv}N{:}\ Cl^-$

alkyl diazonium salt

With 2° amines:

R_2NH $\xrightarrow[HCl]{NaNO_2}$ $R_2N{-}N{=}O$

N-nitrosamine

Reactions of diazonium salts

[1] Substitution reactions (25.14)

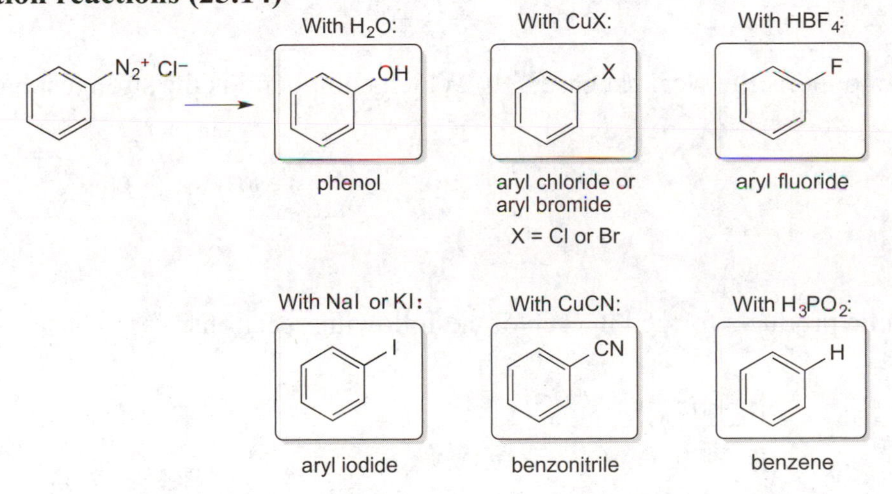

[2] Coupling to form azo compounds (25.15)

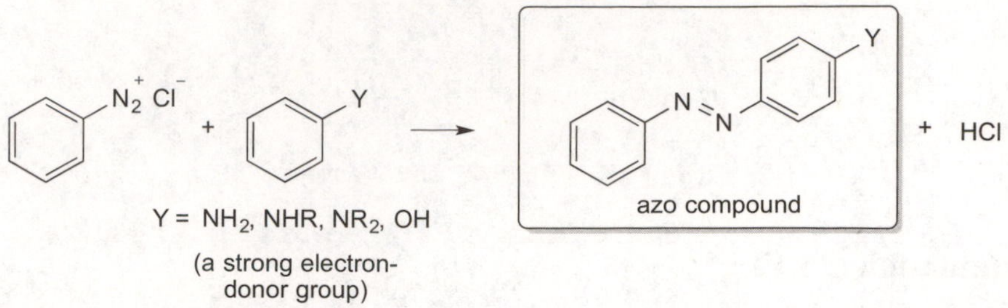

$\overset{+}{N_2}\ Cl^-$
+
Y
Y = NH_2, NHR, NR_2, OH
(a strong electron-donor group)
N=N
Y
azo compound
+ HCl

Practice Test on Chapter Review

1. Give a systematic name for each of the following compounds.

a.

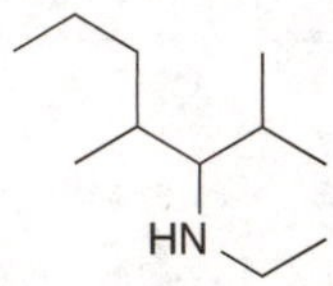

HN

b.

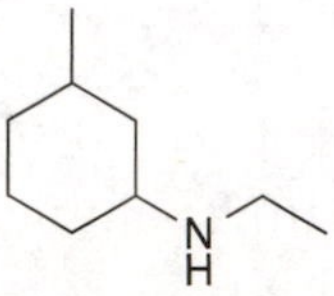

N
H

2. (a) Which compound is the weakest base? (b) Which compound is the strongest base?

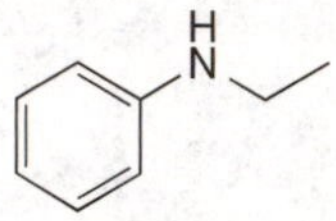

H
N

A

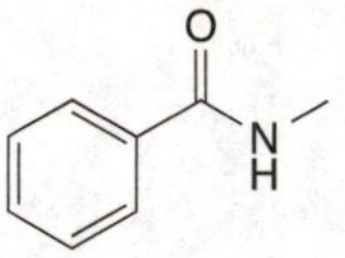

O
N
H

B

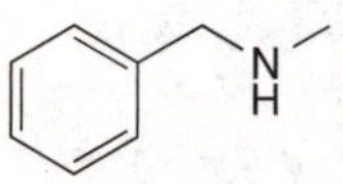

N
H

C

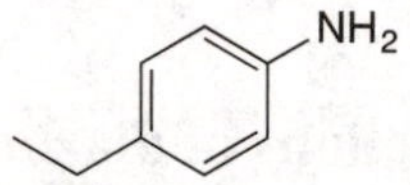

NH_2

D

3. (a) Which compound is the weakest base? (b) Which compound is the strongest base?

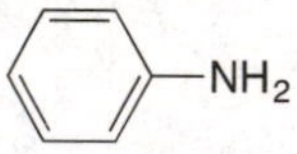

NH_2

A

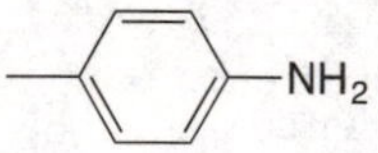

NH_2

B

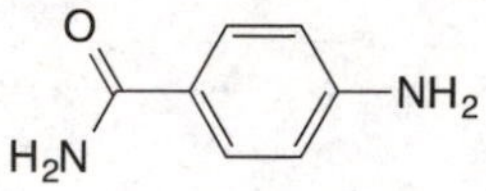

O
H_2N
NH_2

C

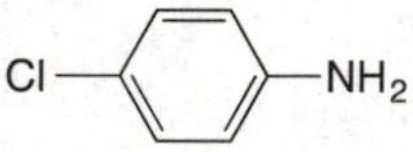

Cl
NH_2

D

4. Draw the organic products formed in each of the following reactions.

a.

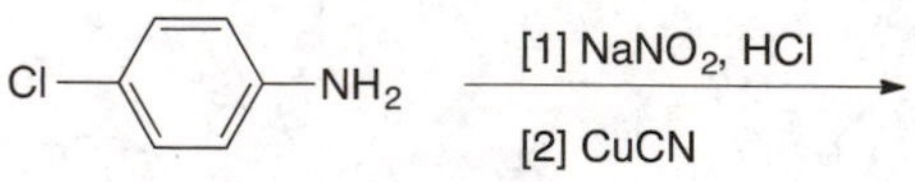

Cl
NH_2
[1] $NaNO_2$, HCl
[2] CuCN

b.

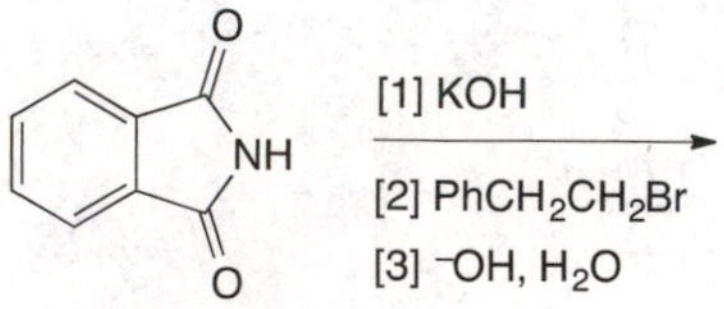

O
NH
O
[1] KOH
[2] $PhCH_2CH_2Br$
[3] ^-OH, H_2O

c.

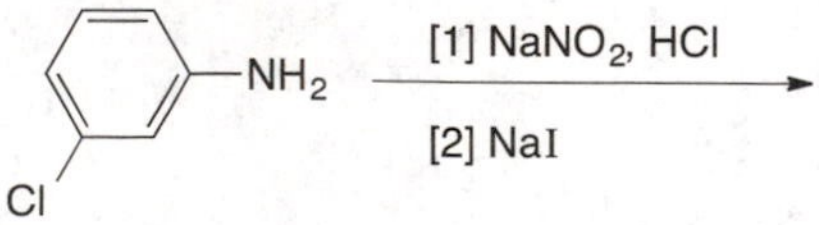

NH_2
Cl
[1] $NaNO_2$, HCl
[2] NaI

d.

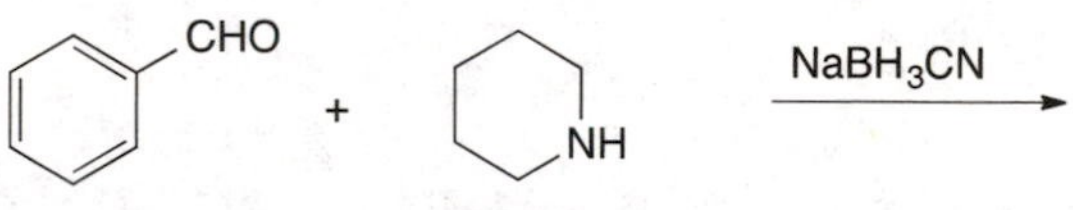

CHO
+
NH
$NaBH_3CN$

e.

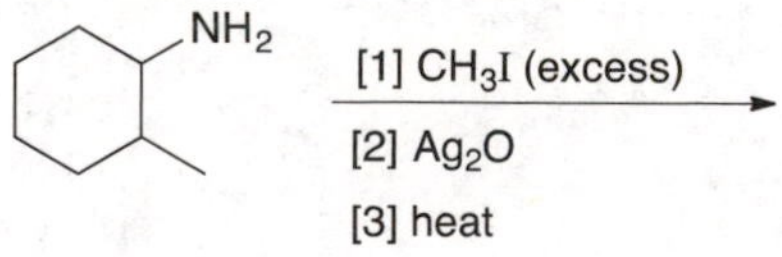

NH_2
[1] CH_3I (excess)
[2] Ag_2O
[3] heat

5. Draw the products formed when the given amine is treated with [1] CH_3I (excess); [2] Ag_2O; [3] Δ, and indicate the major product. You need not consider any stereoisomers formed in the reaction.

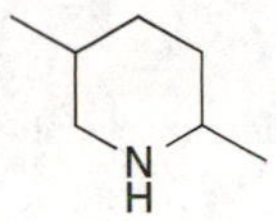

N
H

6. What organic starting materials are needed to synthesize **B** by reductive amination?

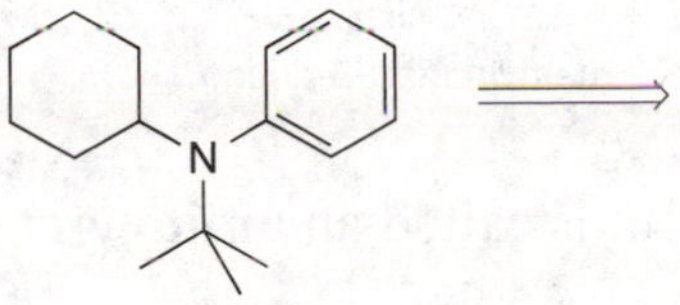

N

B

Answers to Practice Test

1.a. *N*-ethyl-2,4-dimethyl-heptan-3-amine
b. *N*-ethyl-3-methylcyclohexanamine

2.a. **B**
b. **C**

3.a. **C**
b. **B**

4.
a.  Cl —CN
b. 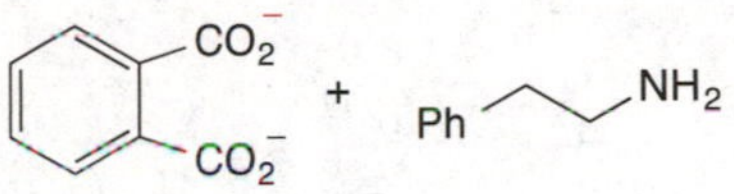 CO_2^- CO_2^- + Ph NH2
c. 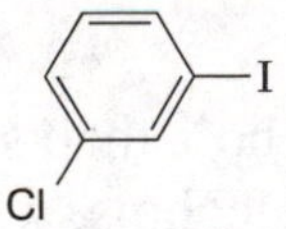 I Cl
d. 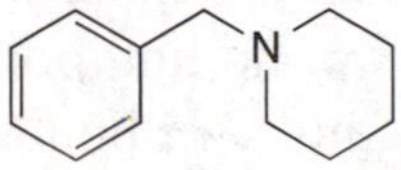 N
e. 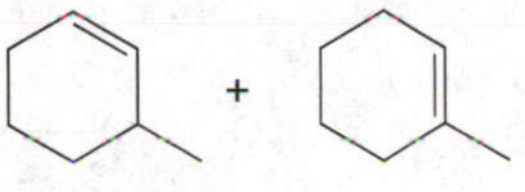 +

5. N + N + N
major

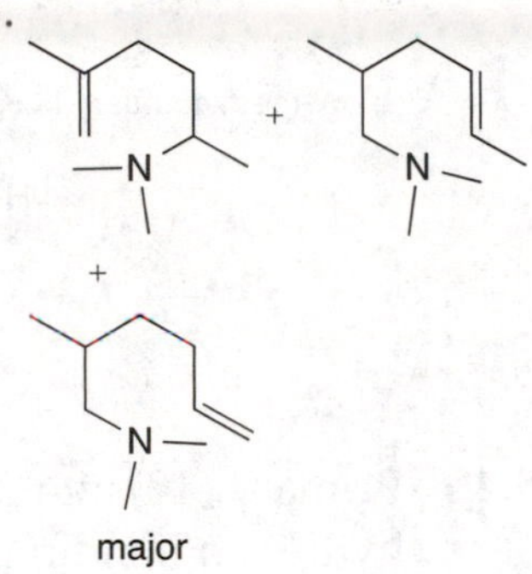

6. O + H N

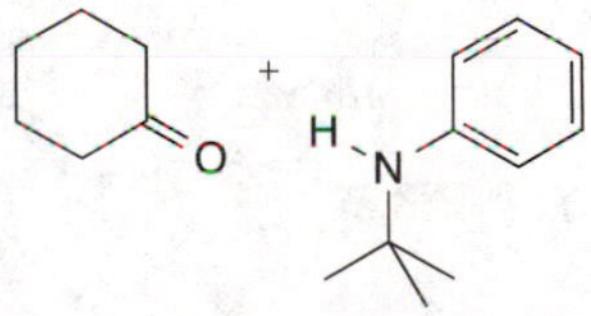

Answers to Problems

25.1 The N atom of an ammonium salt is a stereogenic center when the N is surrounded by four different groups. All stereogenic centers are circled.

a. 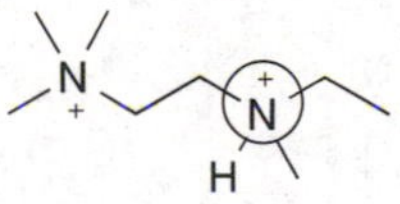 N+ N+ H

b. 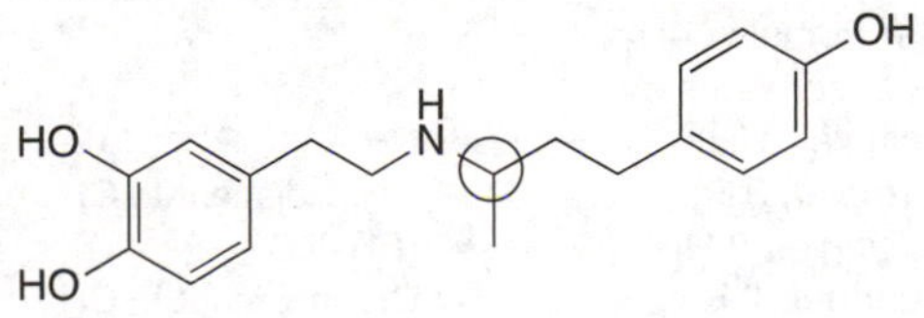 HO HO H N OH

25.2

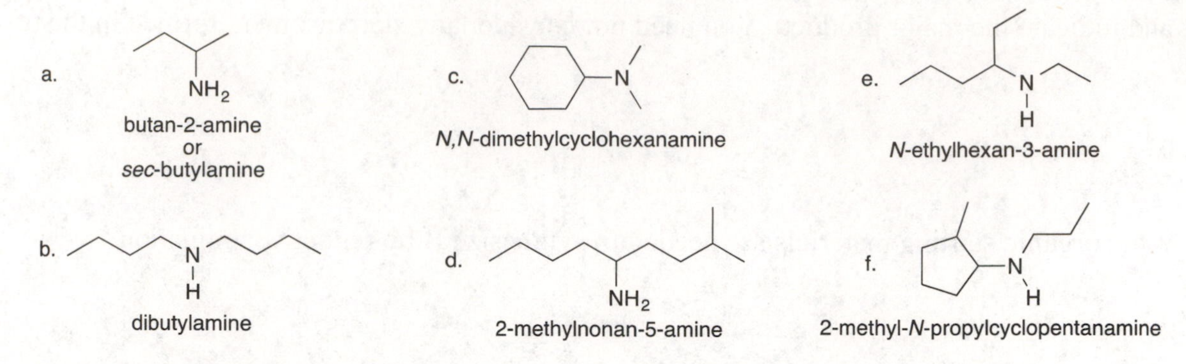

25.3 An **NH_2** group named as a substituent is called an **amino group.**

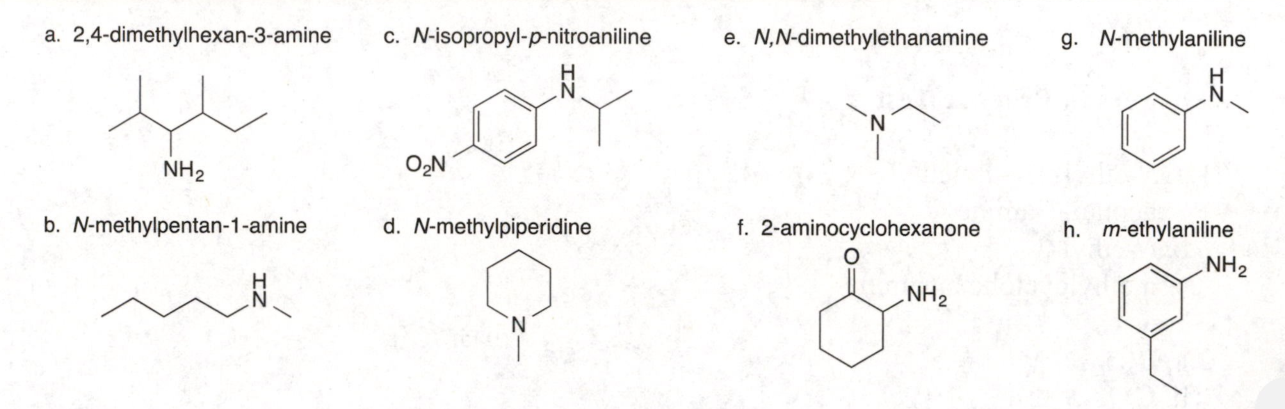

25.4 Primary (1°) and 2° amines have higher bp's than similar compounds (like ethers) incapable of hydrogen bonding, but lower bp's than alcohols, which have stronger intermolecular hydrogen bonds. Tertiary amines (3°) have lower boiling points than 1° and 2° amines of comparable molecular weight because they have no N–H bonds with which to form hydrogen bonds.

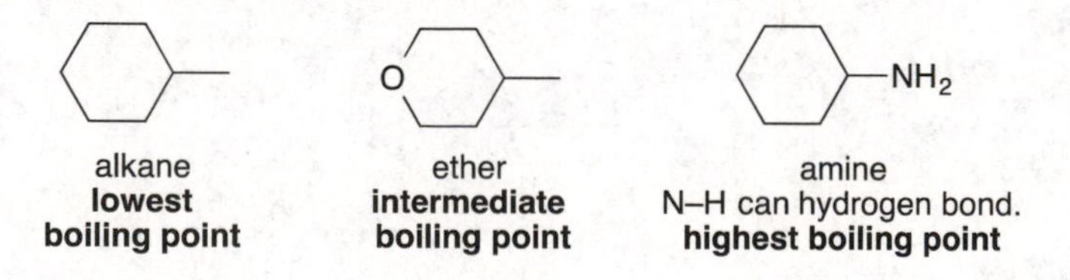

25.5 **The NH signal occurs between 0.5 and 5.0 ppm.** The protons on the carbon bonded to the amine nitrogen are deshielded and typically absorb at 2.3–3.0 ppm. The NH protons are not split.

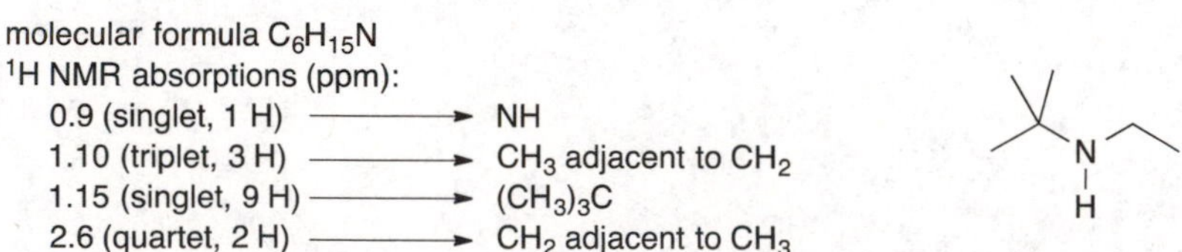

25.6 The atoms of 2-phenylethanamine are in bold.

a. LSD
lysergic acid diethyl amide

b. codeine

25.7 S_N2 reaction of an alkyl halide with NH_3 or an amine forms an amine or an ammonium salt.

a. NH_3 excess

b. Br, excess, Br^-

25.8 The Gabriel synthesis converts an alkyl halide into a 1° amine by a two-step process: nucleophilic substitution followed by hydrolysis.

a. b. c. CH_3O ... NH_2 ⟹ CH_3O ... Br

25.9 The Gabriel synthesis prepares 1° amines from alkyl halides. Because the reaction proceeds by an S_N2 mechanism, the halide must be CH_3 or 1°, and X can't be bonded to an sp^2 hybridized C.

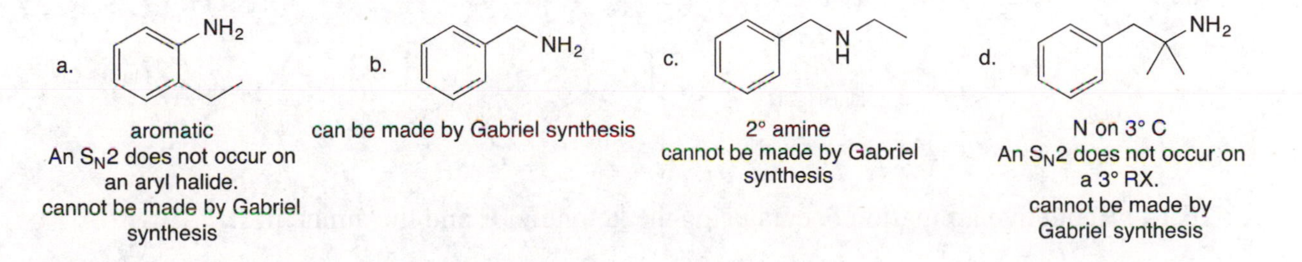

25.10 Nitriles are reduced to 1° amines with $LiAlH_4$. Nitro groups are reduced to 1° amines using a variety of reducing agents. **Primary (1°), 2°, and 3° amides are reduced to 1°, 2°, and 3° amines** respectively, using $LiAlH_4$.

a. NH_2 ⟹ NO_2 C≡N NH_2

b. NH_2 ⟹ NO_2 C≡N NH_2

c. 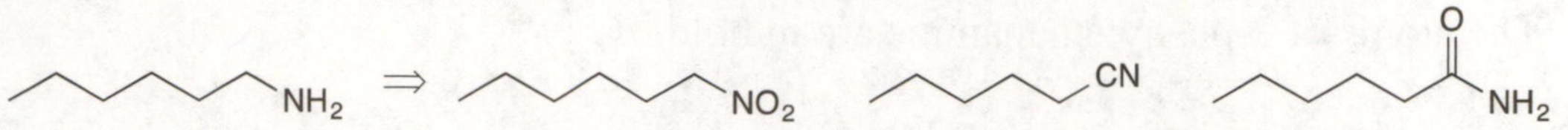 NH_2 ⇒ NO_2 CN O NH_2

25.11 Primary (1°), 2°, and 3° amides are reduced to 1°, 2°, and 3° amines respectively, using $LiAlH_4$.

a. 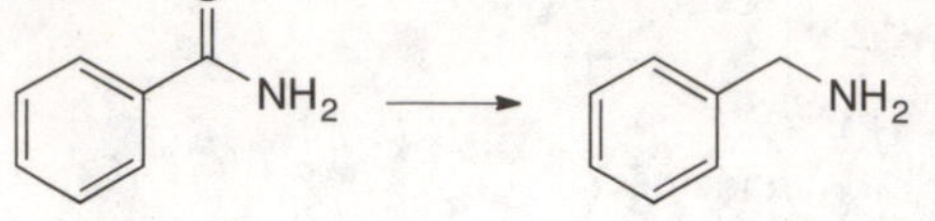 O NH_2 → NH_2

c. 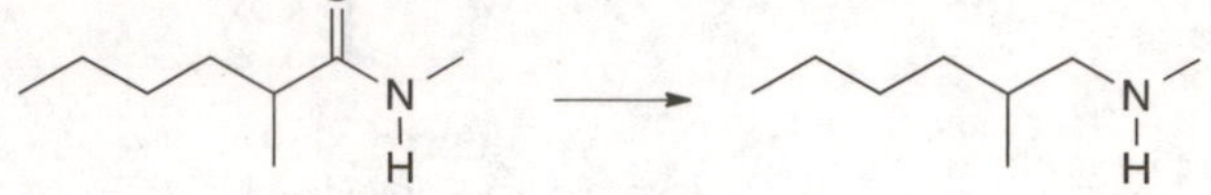 O N H → N H

b. 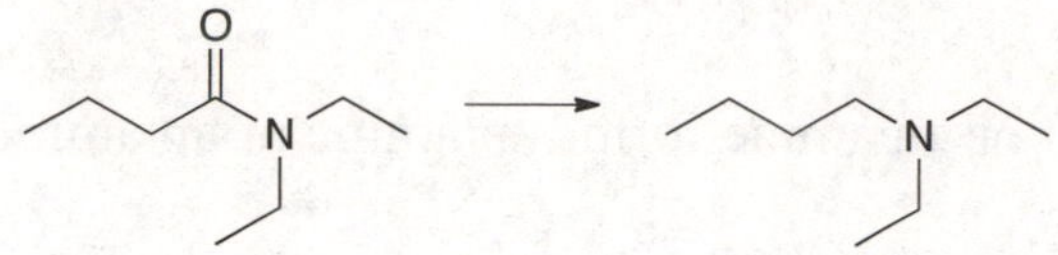 O N → N

25.12 Only amines with a CH_2 or CH_3 bonded to the N can be made by reduction of an amide.

a. 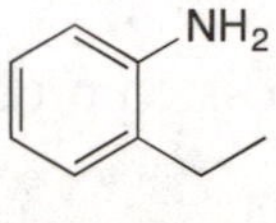 NH_2
N bonded to benzene
cannot be made by reduction of an amide

b. 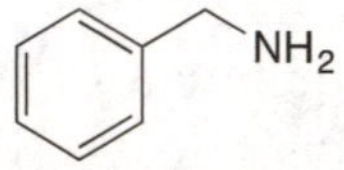 NH_2
N bonded to CH_2
can be made by reduction of an amide

c. 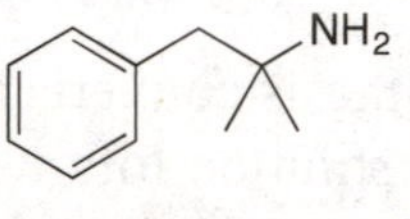 NH_2
N bonded to a 3° C
cannot be made by reduction of an amide

d. 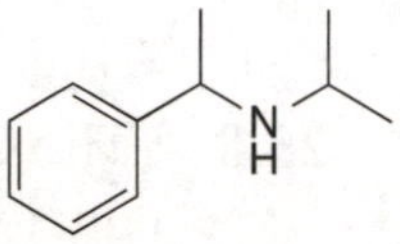 N H
N with 2° C on both sides
cannot be made by reduction of an amide

25.13 Reductive amination is a two-step method that converts aldehydes and ketones into 1°, 2°, and 3° amines. Reductive amination replaces a C=O by a C–H and C–N bond.

a. 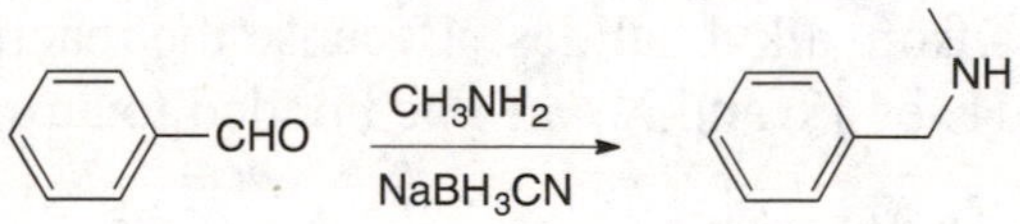 CHO $\xrightarrow[NaBH_3CN]{CH_3NH_2}$ NH

c. 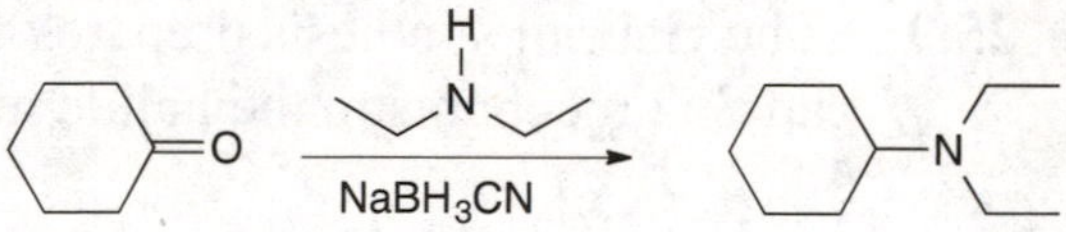 O + H–N $\xrightarrow{NaBH_3CN}$ N

b. 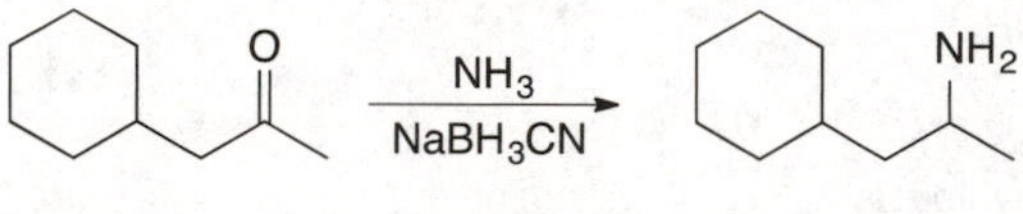 O $\xrightarrow[NaBH_3CN]{NH_3}$ NH_2

d. 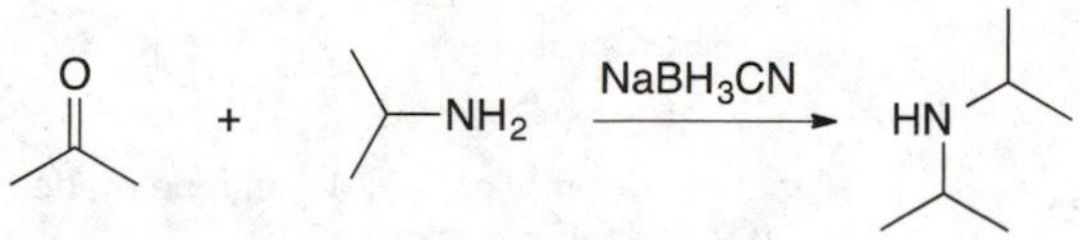 O + NH_2 $\xrightarrow{NaBH_3CN}$ HN

25.14 Reductive amination occurs using the ketone in **E** and the amine in **D**.

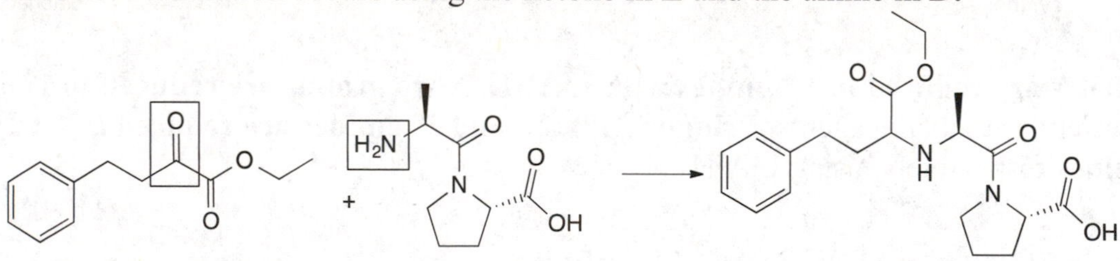

E + D (H_2N, O, N, O, OH) → enalapril (O, O, N H, O, N, O, OH)

E **D** enalapril

25.15

a. NH_2 ⇒ O + NH_3

b.

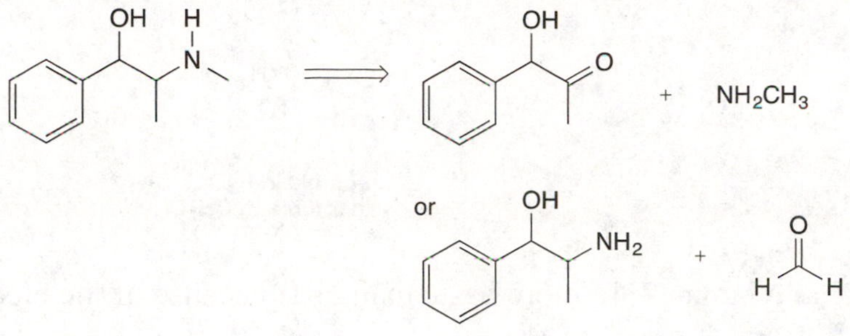

OH H N ⟹ OH O + NH_2CH_3

or OH NH_2 + O H H

25.16

a.

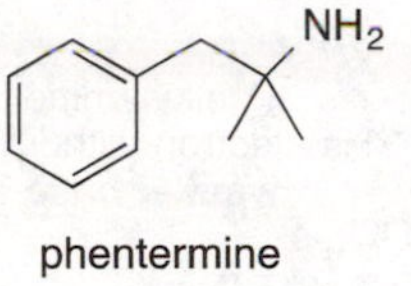

NH_2

phentermine

Only amines that have a C bonded to a H and N atom can be made by reductive amination; that is, an amine must have the following structural feature:

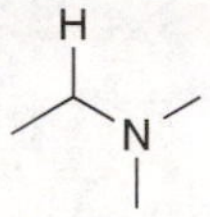

H N

In phentermine, the C bonded to N is not bonded to a H, so it cannot be made by reductive amination.

b. systematic name: 2-methyl-1-phenylpropan-2-amine

25.17

The pK_a of many protonated amines is 10–11, so the pK_a of the starting acid must be **less than 10** for equilibrium to favor the products. Amines are thus readily protonated by strong inorganic acids (e.g., HCl and H_2SO_4) and by carboxylic acids.

a. NH_2 + HCl ⇌ $\overset{+}{N}H_3$ + Cl^-

pK_a = −7 · pK_a ≈ 10 weaker acid products favored

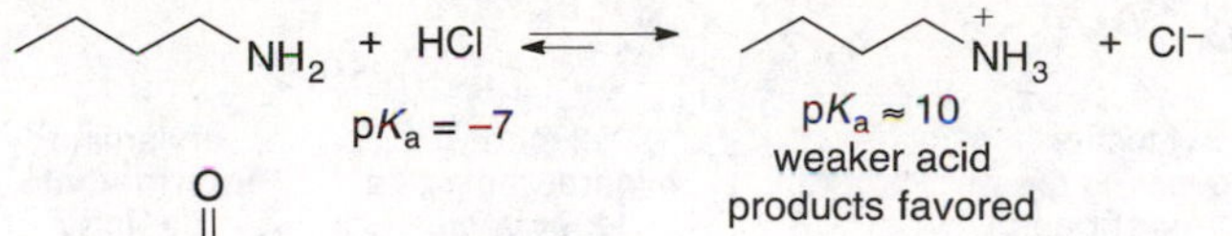

b. OH (O) + N–H ⇌ $\overset{+}{N}$ H H + O⁻ (O)

pK_a = 4.2 · pK_a = 10.7 weaker acid products favored

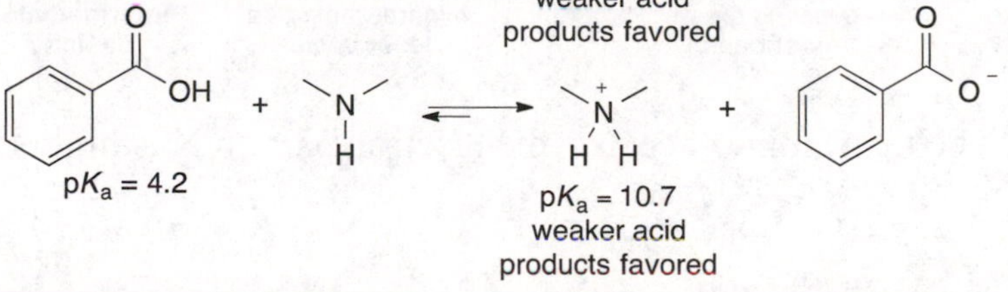

c. N–H + H_2O ⇌ $\overset{+}{N}$ H H + HO^-

pK_a = 15.7 weaker acid reactants favored · pK_a ≈ 10

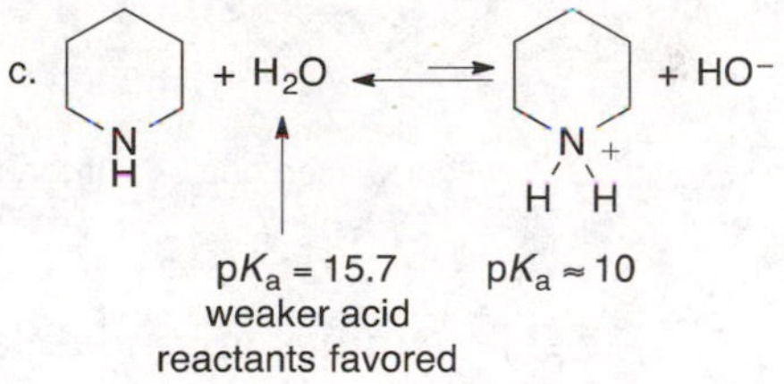

25.18

An amine can be separated from other organic compounds by converting it to a water-soluble ammonium salt by an acid–base reaction. In each case, the extraction procedure would employ the following steps:

- Dissolve the amine and either **X** or **Y** in CH_2Cl_2.
- Add a solution of 10% HCl. The amine will be protonated and dissolve in the aqueous layer, while **X** or **Y** will remain in the organic layer as a neutral compound.
- Separate the layers.

a. NH_2 and **X** —H–Cl→ $\overset{+}{N}H_3$ Cl^- + **X**

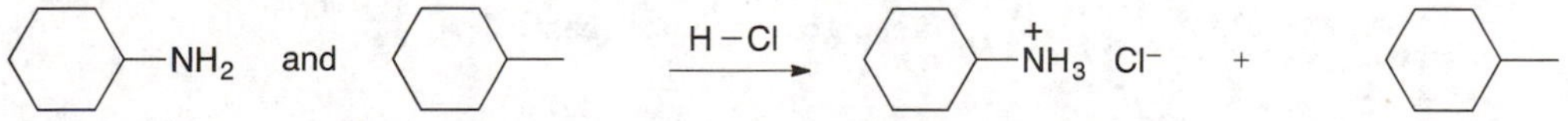

- **soluble in H_2O**
- **insoluble in CH_2Cl_2**

X

- insoluble in H_2O
- soluble in CH_2Cl_2

b. and Y $\xrightarrow{\text{H–Cl}}$ Cl^- + Y

- **soluble in H_2O**
- **insoluble in CH_2Cl_2**

- insoluble in H_2O
- soluble in CH_2Cl_2

25.19 Primary (1°), 2°, and 3° alkylamines are more basic than NH_3 because of the electron-donating inductive effect of the R groups.

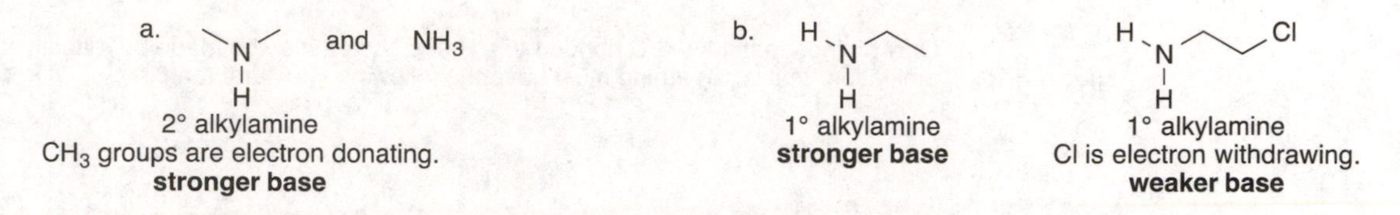

25.20 Arylamines are less basic than alkylamines because the electron pair on N is delocalized. Electron-donor groups add electron density to the benzene ring making the arylamine more basic than aniline. Electron-withdrawing groups remove electron density from the benzene ring, making the arylamine less basic than aniline.

a.

electron-withdrawing group **least basic**	arylamine **intermediate basicity**	electron-donating group **most basic**

b.

electron-withdrawing group **least basic**	arylamine **intermediate basicity**	alkylamine **most bas**

25.21 Amides are much less basic than amines because the electron pair on N is highly delocalized.

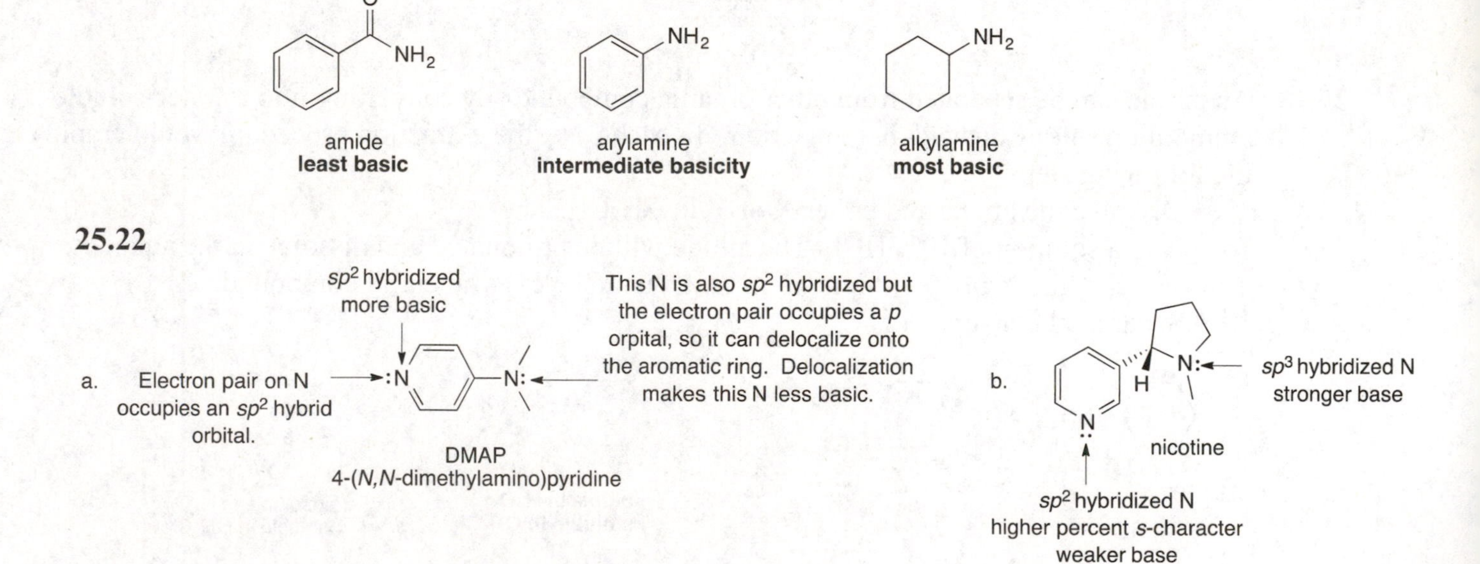

25.23 HCl protonates the more basic N atom.

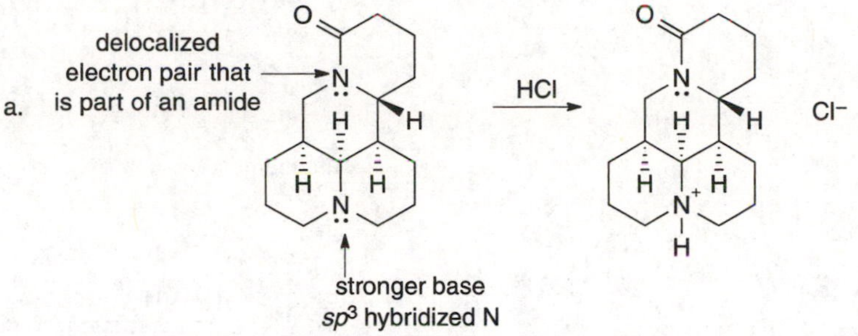

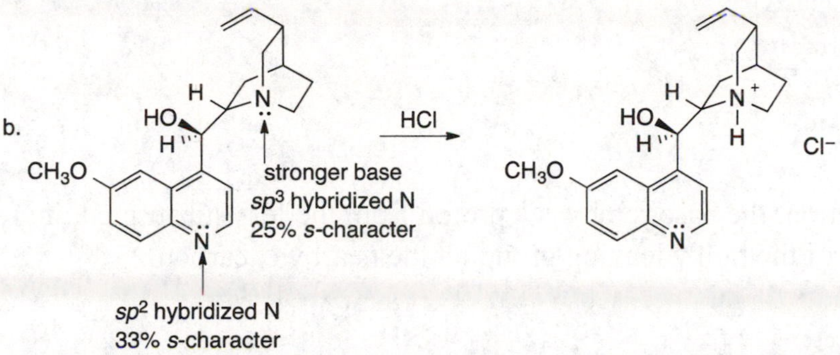

a. delocalized electron pair that is part of an amide → N

O N H H H H N

stronger base
sp^3 hybridized N

HCl

O N H H H H N^+ H Cl^-

b. H HO H N CH_3O N

stronger base
sp^3 hybridized N
25% *s*-character

sp^2 hybridized N
33% *s*-character

HCl

H HO H N^+ H CH_3O N Cl^-

25.24 Amines attack carbonyl groups to form products of nucleophilic addition or substitution.

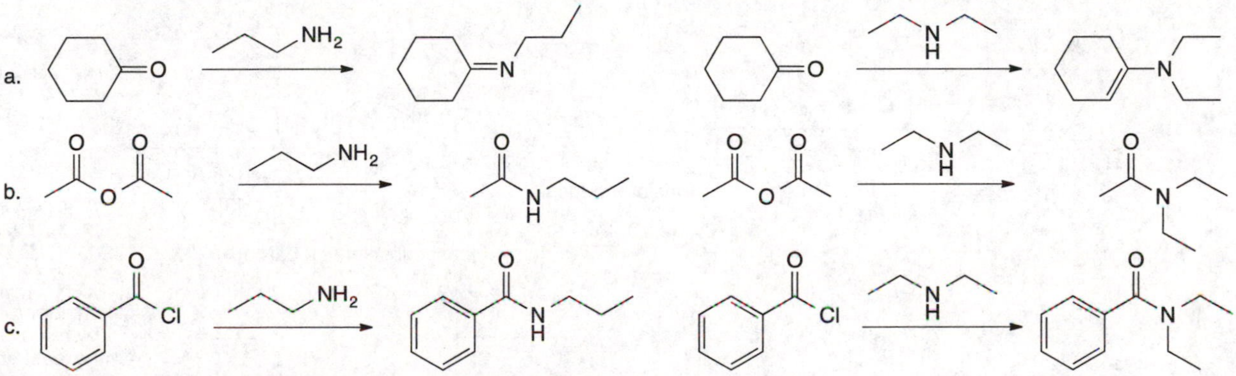

a. O NH_2 N O N H N

b. O O O NH_2 O N H O O O N H O N

c. O Cl NH_2 O N H O Cl N H O N

25.25 [1] Convert the amine (aniline) into an amide (acetanilide).
[2] **Carry out the Friedel–Crafts reaction.**
[3] **Hydrolyze the amide** to generate the free amino group.

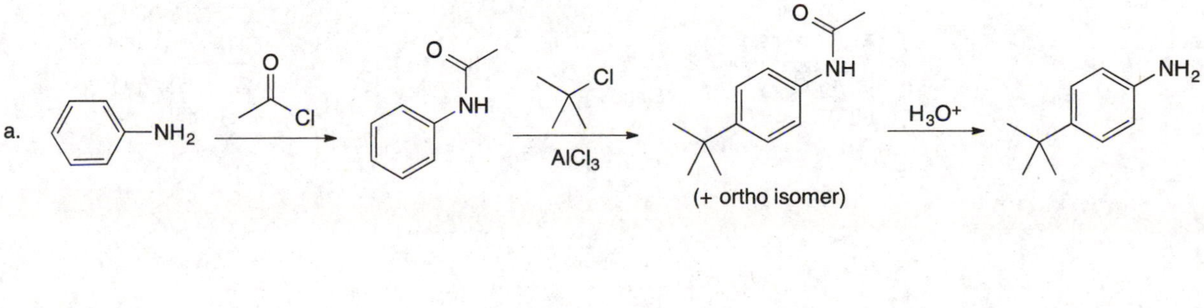

a. NH_2 O Cl O NH Cl $AlCl_3$ O NH H_3O^+ NH_2

(+ ortho isomer)

b. 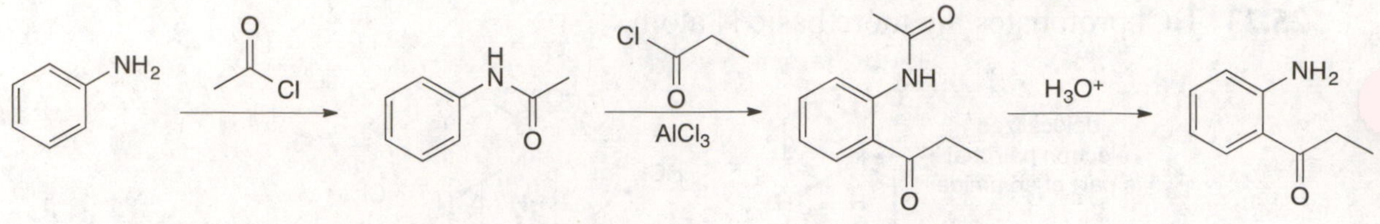 NH$_2$ —(O, Cl)→ H N, O —(Cl, O; AlCl$_3$)→ NH, O, O (+ para isomer) —H$_3$O$^+$→ NH$_2$, O

25.26

a. 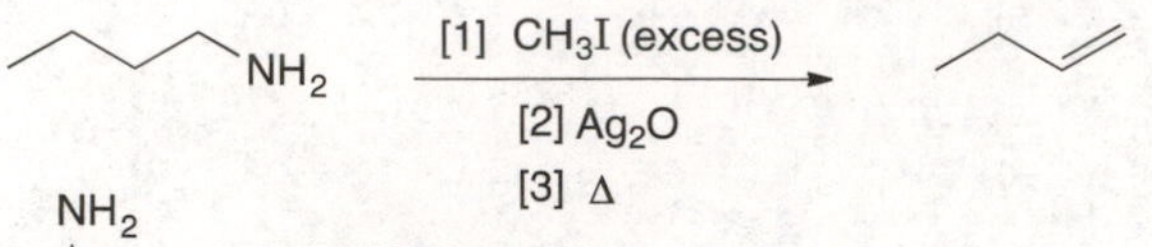 NH$_2$ —[1] CH_3I (excess) [2] Ag_2O [3] Δ→

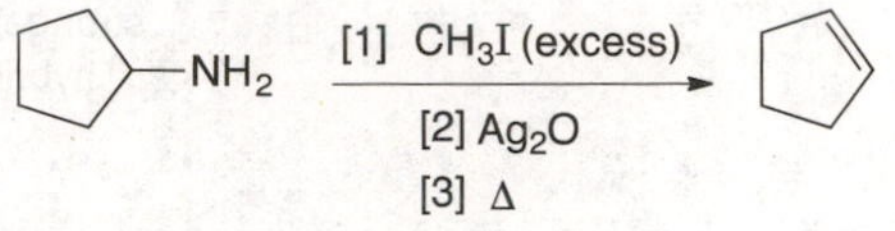

b. 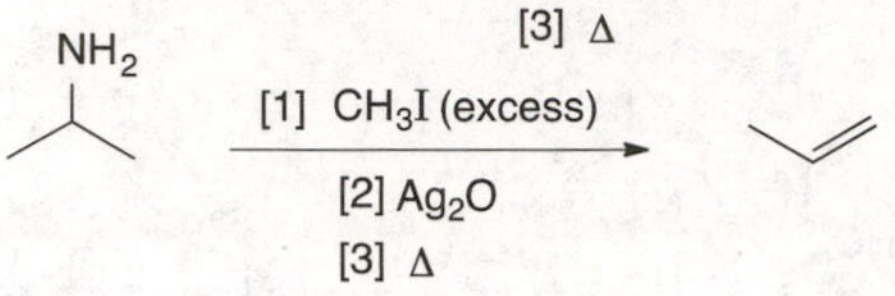 NH$_2$ —[1] CH_3I (excess) [2] Ag_2O [3] Δ→

c. NH$_2$ —[1] CH_3I (excess) [2] Ag_2O [3] Δ→

25.27 In a Hofmann elimination, the base removes a proton from the less substituted, more accessible β carbon atom, because of the bulky leaving group on the nearby α carbon.

a. 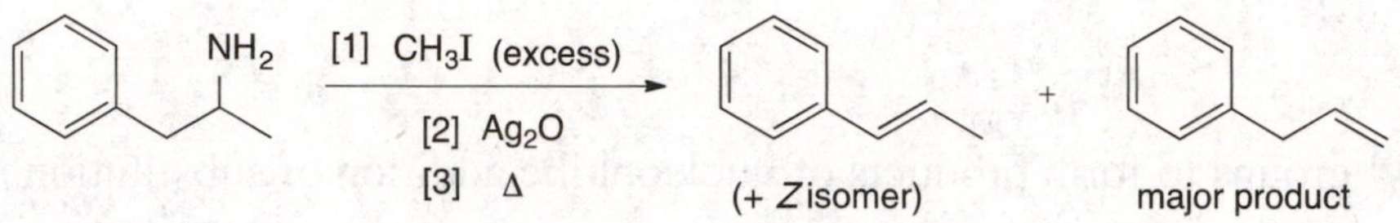 NH$_2$ —[1] CH_3I (excess) [2] Ag_2O [3] Δ→ (+ *Z* isomer) + major product

b. 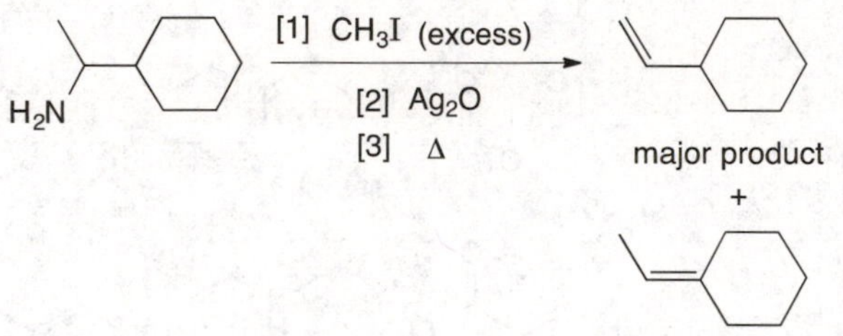 H$_2$N —[1] CH_3I (excess) [2] Ag_2O [3] Δ→ major product +

c. β β β N H (3 β C's) least substituted β carbon —[1] CH_3I (excess) [2] Ag_2O [3] Δ→ N (+ *Z* isomer) + N + N

major product, formed by removal of a H from the least substituted β C

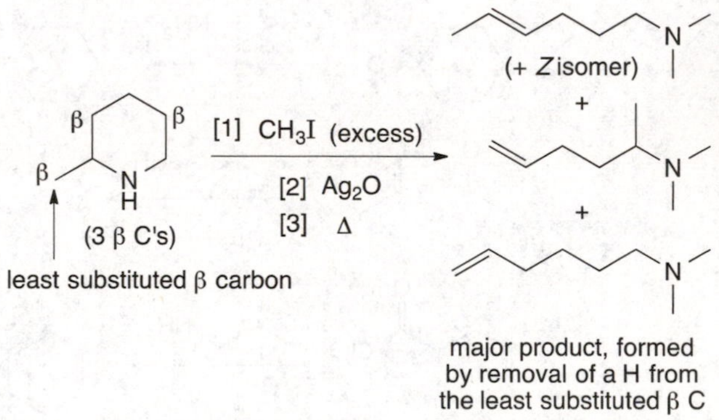

25.28

a. 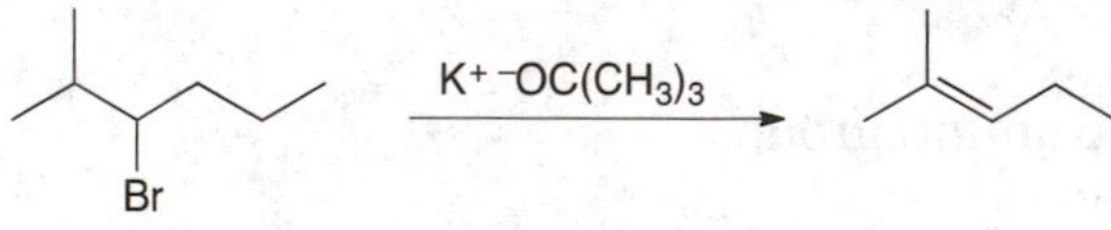 Br —K$^+$ $^-$OC(CH$_3$)$_3$→

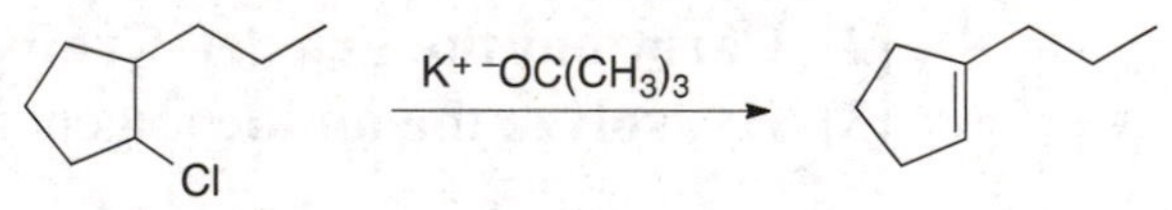

b. 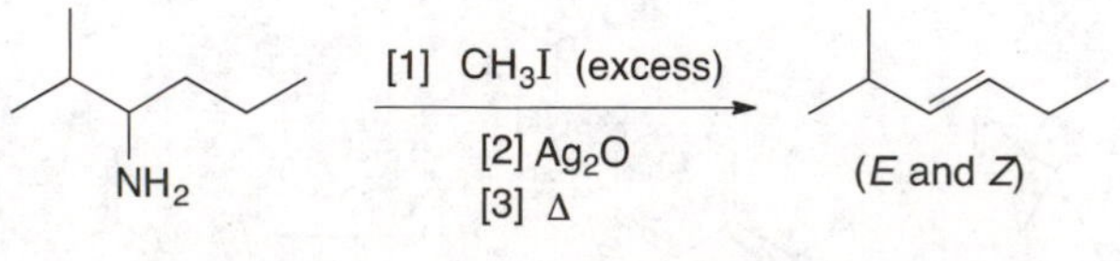 NH$_2$ —[1] CH_3I (excess) [2] Ag_2O [3] Δ→ (*E* and *Z*)

c. Cl —K$^+$ $^-$OC(CH$_3$)$_3$→

d. 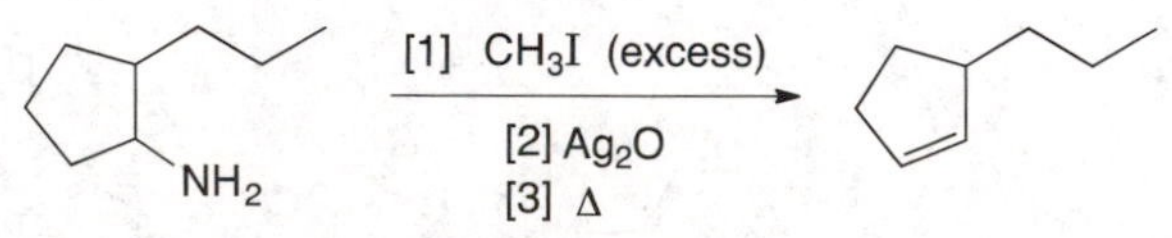 NH$_2$ —[1] CH_3I (excess) [2] Ag_2O [3] Δ→

25.29

a. 2-methylaniline $\xrightarrow[HCl]{NaNO_2}$ 2-methylbenzenediazonium $N_2^+\ Cl^-$

b. $CH_3CH_2NHCH_3$ $\xrightarrow[HCl]{NaNO_2}$ $CH_3CH_2N(NO)CH_3$

c. piperidine (N–H) $\xrightarrow[HCl]{NaNO_2}$ N-nitrosopiperidine (N–NO)

d. 2-butanamine (NH_2) $\xrightarrow[HCl]{NaNO_2}$ $\overset{+}{N_2}\ Cl^-$ (2-butyl diazonium chloride)

25.30

a. aniline ($-NH_2$) $\xrightarrow[\text{[2] CuBr}]{\text{[1] NaNO}_2\text{, HCl}}$ bromobenzene ($-Br$)

b. m-nitroaniline (O_2N, $-NH_2$) $\xrightarrow[\text{[2] H}_2\text{O}]{\text{[1] NaNO}_2\text{, HCl}}$ m-nitrophenol (O_2N, $-OH$)

c. $CH_3O-C_6H_4-NH_2$ $\xrightarrow[\text{[2] HBF}_4]{\text{[1] NaNO}_2\text{, HCl}}$ $CH_3O-C_6H_4-F$

d. 2-chlorobenzenediazonium ($-N_2^+\ Cl^-$, Cl) $\xrightarrow[\text{[2] LiAlH}_4\text{ [3] H}_2\text{O}]{\text{[1] CuCN}}$ 2-chlorobenzylamine (NH_2, Cl)

25.31

a. benzene $\xrightarrow[H_2SO_4]{HNO_3}$ nitrobenzene (NO_2) $\xrightarrow[\text{Pd-C}]{H_2}$ aniline (NH_2) $\xrightarrow[\text{[2] HBF}_4]{\text{[1] NaNO}_2\text{, HCl}}$ fluorobenzene (F)

b. nitrobenzene (NO_2) (from a.) $\xrightarrow[H_2SO_4]{HNO_3}$ m-dinitrobenzene (NO_2, NO_2) $\xrightarrow[\text{Pd-C}]{H_2}$ m-phenylenediamine (NH_2, NH_2) $\xrightarrow[\text{[2] H}_2\text{O}]{\text{[1] NaNO}_2\text{, HCl}}$ resorcinol (OH, OH)

c. aniline (NH_2) (from a.) $\xrightarrow[\text{[2] NaI}]{\text{[1] NaNO}_2\text{, HCl}}$ iodobenzene (I) $\xrightarrow[AlCl_3]{CH_3Cl}$ o-iodotoluene (I) (+ para isomer)

d. aniline (NH_2) (from a.) $\xrightarrow[FeCl_3]{Cl_2\text{ (excess)}}$ 2,4,6-trichloroaniline (NH_2; Cl, Cl, Cl) $\xrightarrow[\text{[2] H}_3\text{PO}_2]{\text{[1] NaNO}_2\text{, HCl}}$ 1,3,5-trichlorobenzene (Cl, Cl, Cl)

25.32

a. aniline (NH_2) $\longrightarrow$ $C_6H_5-N{=}N-C_6H_4-NH_2$

b. $HO-C_6H_5$ $\longrightarrow$ $C_6H_5-N{=}N-C_6H_4-OH$

c. $HO-C_6H_4-OH$ $\longrightarrow$ $C_6H_5-N{=}N-$(2,5-dihydroxyphenyl) (OH, HO)

25.33 To determine what starting materials are needed to synthesize a particular azo compound, always divide the molecule into two components: **one has a benzene ring with a diazonium ion, and one has a benzene ring with a very strong electron-donor group.**

a. H_2N–C_6H_4–N=N–C_6H_4–NO_2 (ortho) ⟹ H_2N–C_6H_5 + Cl^- $^+N_2$–C_6H_4–O_2N

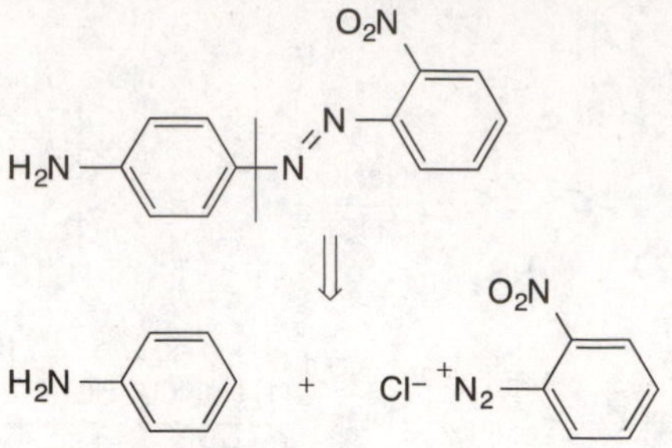

b. HO–C_6H_4–N=N–(2-chloro-4-methylphenyl) ⟹ HO–C_6H_5 + Cl^- $^+N_2$–(2-chloro-4-methylphenyl)

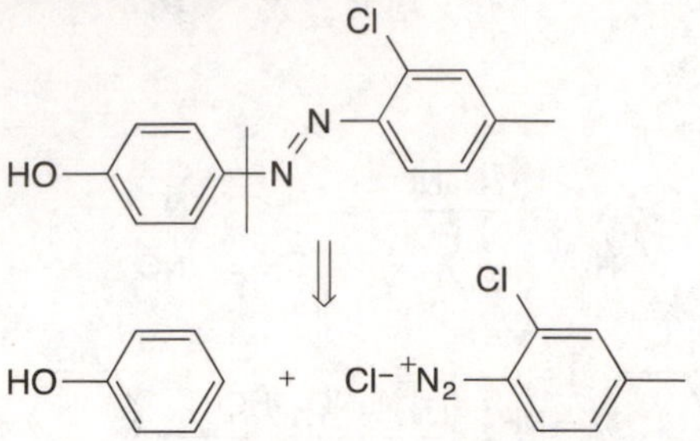

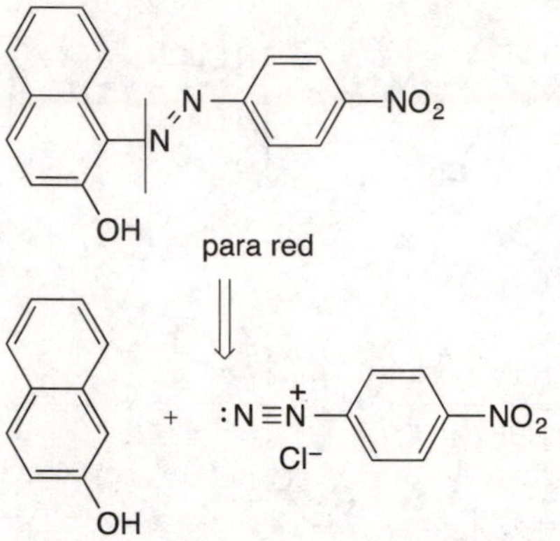

25.34

a. para red ⟹ 2-naphthol (OH) + $:N{\equiv}\overset{+}{N}$–C_6H_4–NO_2 Cl^-

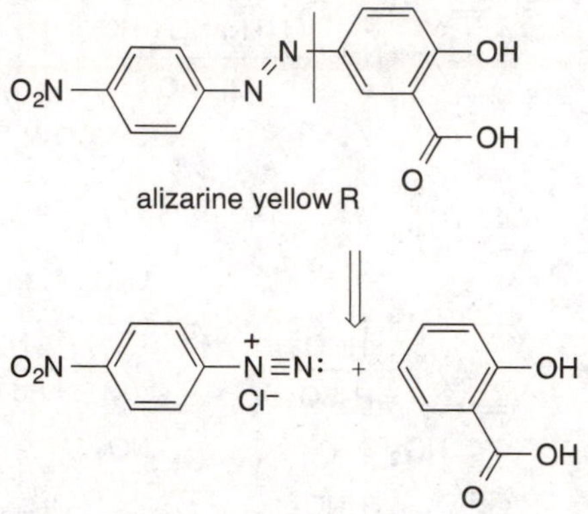

b. alizarine yellow R ⟹ O_2N–C_6H_4–$\overset{+}{N}{\equiv}N:$ Cl^- + salicylic acid (OH, COOH)

25.35

a. 4,6-dimethylheptan-1-amine

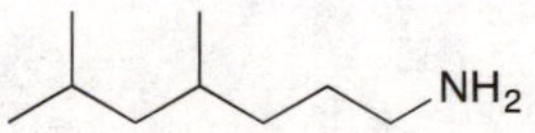

b. *N,N*-diethylcycloheptanamine

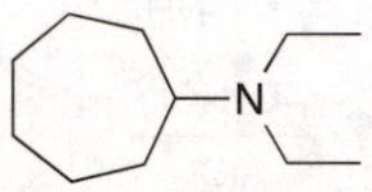

25.36

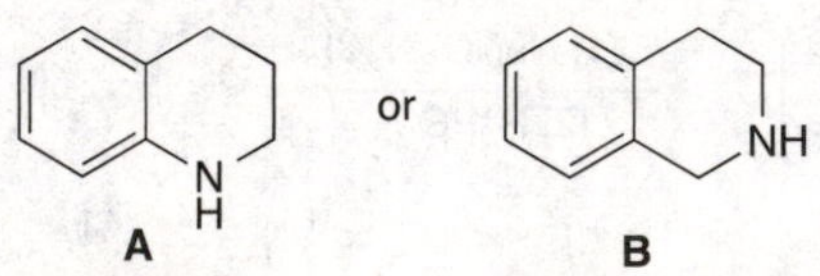

A or B

A
weaker base
(delocalized electron pair on N)

B
stronger base

25.37

a, b. varenicline (NH) $\xrightarrow{HCl}$ $^+NH_2$ + Cl^-

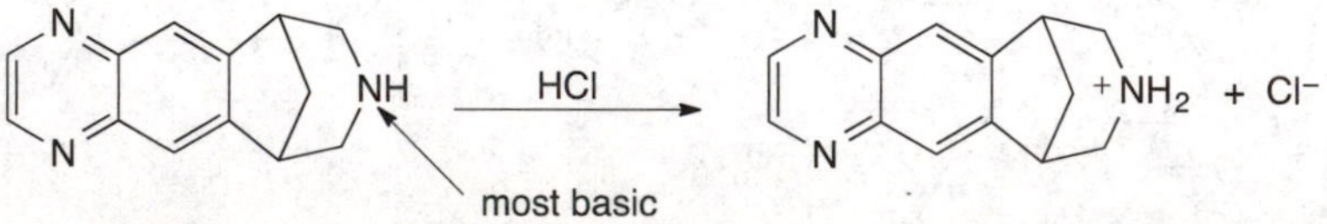

most basic
only sp^3 hybridized N

25.38

a. *N*-ethyl-2-methylbutan-1-amine

b. 4-ethyl-2-methyloctan-1-amine

c. *N*-ethyl-*N*-methylcyclohexanamine

d. *N*-*tert*-butyl-*N*-ethylaniline

e. 4-aminocyclohexanone

f. 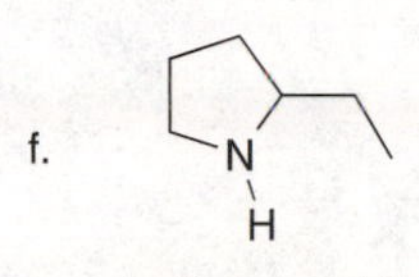

2-ethylpyrrolidine

g. 2-methylhexan-3-amine

h. 3-ethyl-2-methylcyclohexanamine

25.39

a. *N*-isobutylcyclopentanamine

b. tri-*tert*-butylamine

$N[C(CH_3)_3]_3$

c. *N*,*N*-diisopropylaniline

d. *N*-methylpyrrole

e. *N*-methylcyclopentanamine

f. 3-methylhexan-2-amine

g. 2-*sec*-butylpiperidine

h. (*S*)-heptan-2-amine

25.40

Cl^-

2 stereogenic centers
4 stereoisomers

Cl^- Cl^- Cl^- Cl^-

25.41

a.

delocalized electron pair on N
least basic

sp^2 hybridized N
intermediate basicity

sp^3 hybridized N
most basic

b.

electron-withdrawing group
least basic

intermediate basicity

electron-donating group
most basic

25.42 The most basic N atom is protonated on treatment with acid.

a.

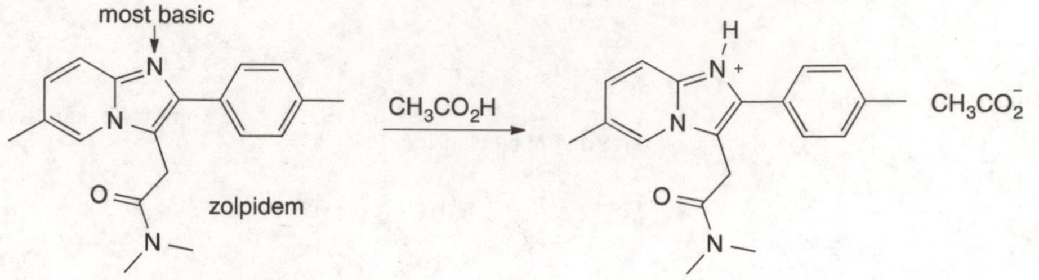

most basic

zolpidem $\xrightarrow{CH_3CO_2H}$ $CH_3CO_2^-$

a 3° alkylamine with an *sp*^{3} hybridized N
most basic

b.

aripiprazole $\xrightarrow{CH_3CO_2H}$ $CH_3CO_2^-$

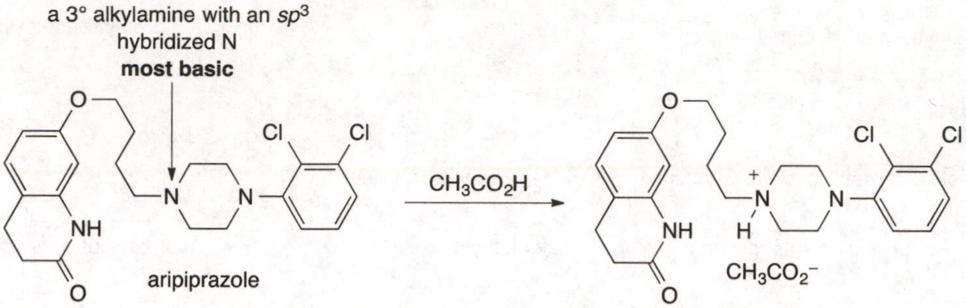

25.43 Pyrimidine has a second electronegative N atom that withdraws electron density from the other N, making the electron pair less available for electron donation.

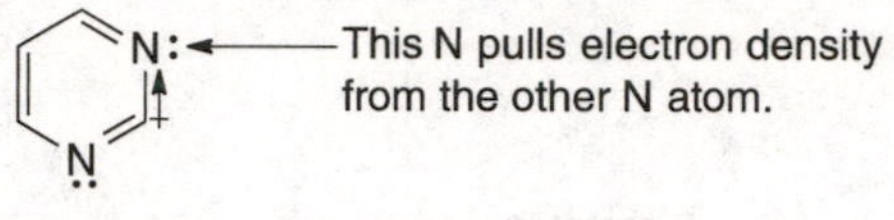

This N pulls electron density from the other N atom.

pyrimidine

25.44

a. $N_b < N_a < N_c$

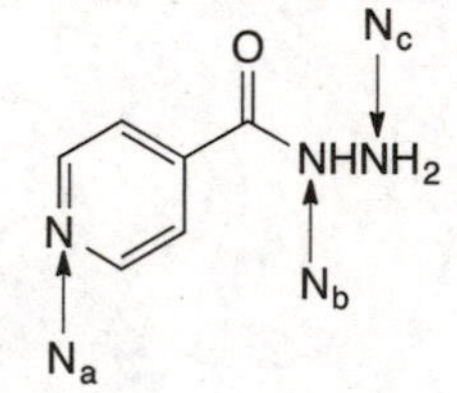

Order of basicity: $N_b < N_a < N_c$
N_b – The electron pair on this N atom is delocalized on the O atom; least basic.
N_a – The electron pair on this N atom is not delocalized, but is on an *sp*^{2} hybridized atom.
N_c – The electron pair on this N atom is on an *sp*^{3} hybridized N; most basic.

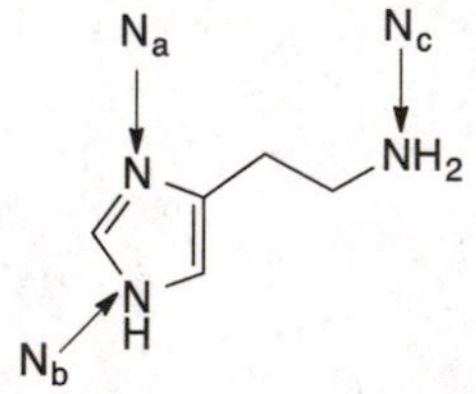

b. $N_b < N_a < N_c$

Order of basicity: $N_b < N_a < N_c$
N_b – The electron pair on this N atom is delocalized on the aromatic five-membered ring; least basic.
N_a – The electron pair on this N atom is not delocalized, but is on an *sp*^{2} hybridized atom.
N_c – The electron pair on this N atom is on an *sp*^{3} hybridized N; most basic.

25.45 The para isomer is the weaker base because the electron pair on its NH_2 group can be delocalized onto the NO_2 group. In the meta isomer, no resonance structure places the electron pair on the NO_2 group, and fewer resonance structures can be drawn:

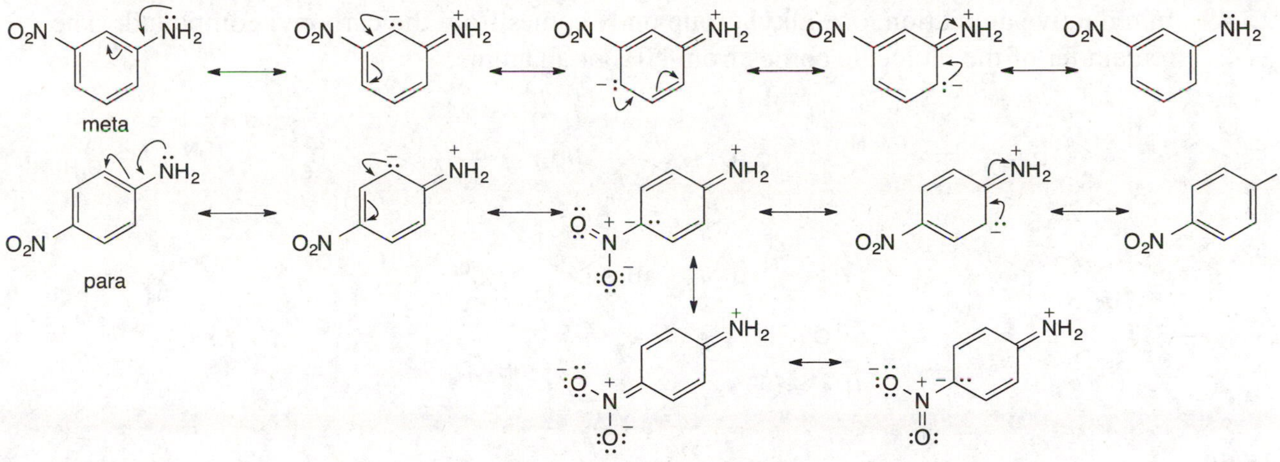

O_2N $\ddot{N}H_2$ meta ⟷ ... $\overset{+}{N}H_2$

para

25.46

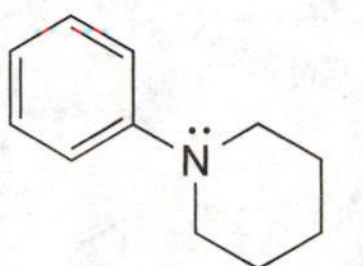

A

pK_a of the conjugate acid = 5.2
stronger conjugate acid
weaker base
The electron pair of this arylamine is delocalized on the benzene ring, decreasing its basicity.

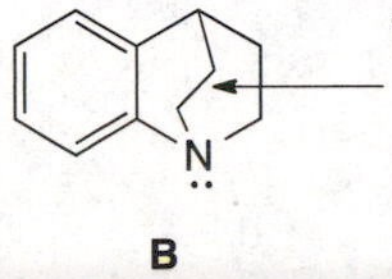

B

This two-carbon bridge makes it difficult for the lone pair on N to delocalize on the aromatic ring.

pK_a of the conjugate acid = 7.29
weaker conjugate acid
stronger base

Resonance structures that place a double bond between the N atom and the benzene ring are destabilized. Since the electron pair is more localized on N, compound **B** is more basic.

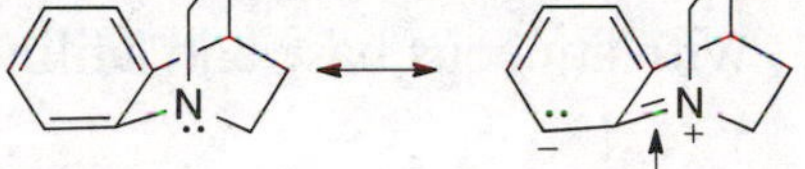

B

Geometry makes it difficult to have a double bond here.

25.47

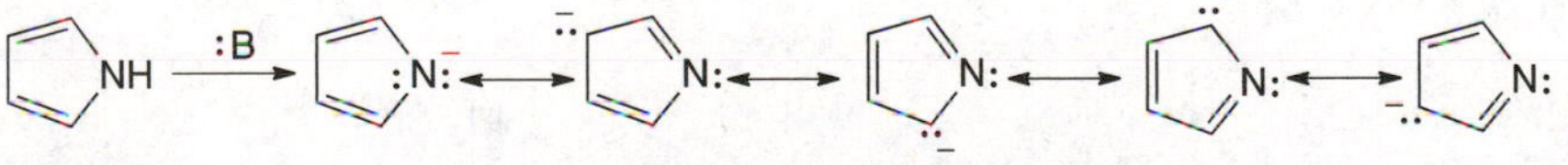

pyrrole
pK_a = 23
stronger acid

:B

weaker conjugate base
The electron pair is delocalized, decreasing the basicity. The N atom is sp^2 hybridized.

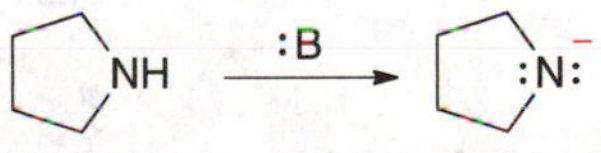

pyrrolidine
pK_a = 44
weaker acid

:B

stronger conjugate base
The electron pair is not delocalized on the ring. The N atom is sp^3 hybridized.

25.48

a. 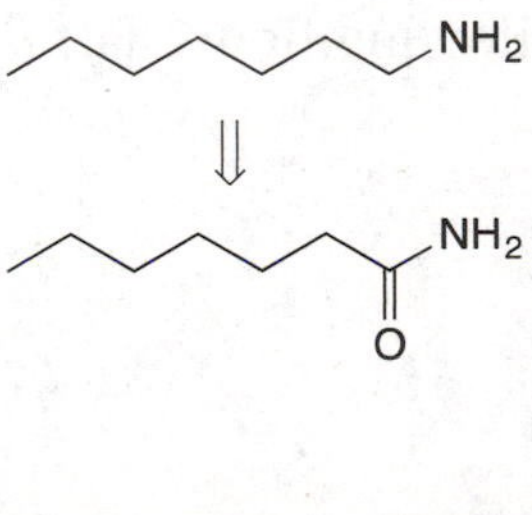 NH_2 ⇓ NH_2 O

b. 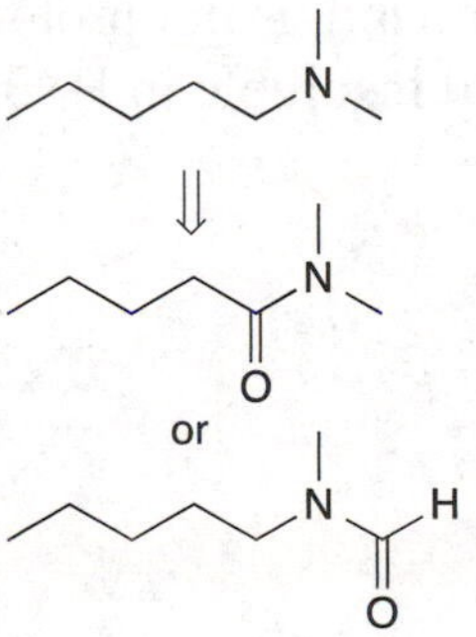 N ⇓ N O or N H O

c. 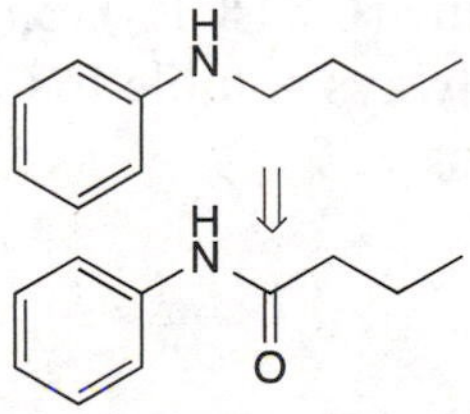 H N ⇓ H N O

25.49 In reductive amination, one alkyl group on N comes from the carbonyl compound. The remainder of the molecule comes from NH_3 or an amine.

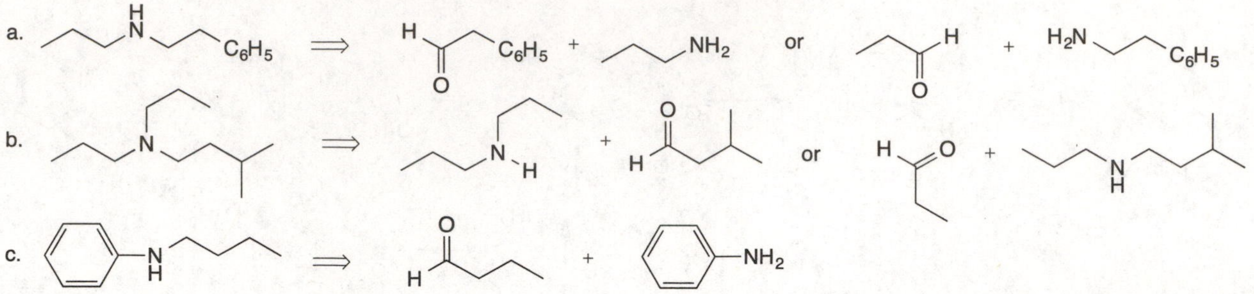

a. C_6H_5 ⟹ C_6H_5 + NH_2 or + H_2N C_6H_5

b. ⟹ + or +

c. ⟹ + NH_2

25.50

a. C_6H_5 —(NH_2, $NaBH_3CN$)→ C_6H_5 NH

b. =O —($(CH_3)_2NH$, $NaBH_3CN$)→ $N(CH_3)_2$

c. C_6H_5 CHO —(NH_3, $NaBH_3CN$)→ C_6H_5 NH_2

d. —(NH_2, $NaBH_3CN$)→ NH

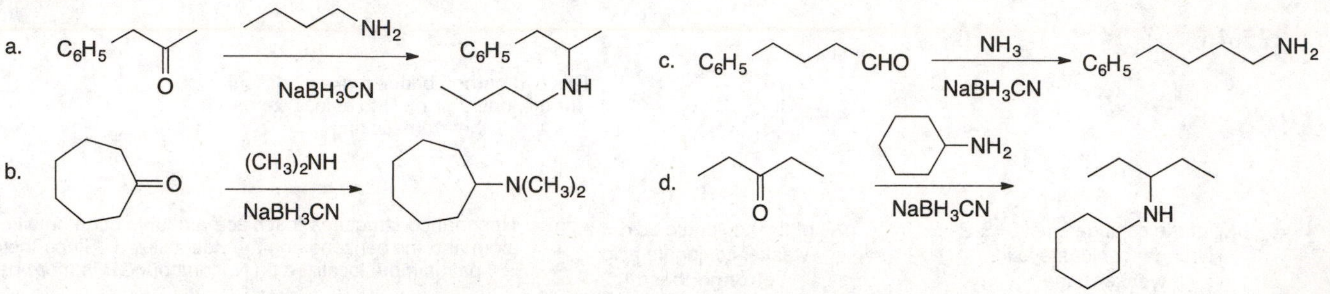

25.51 Use the directions from Answer 25.18. Separation can be achieved because benzoic acid reacts with aqueous base and aniline reacts with aqueous acid according to the following equations:

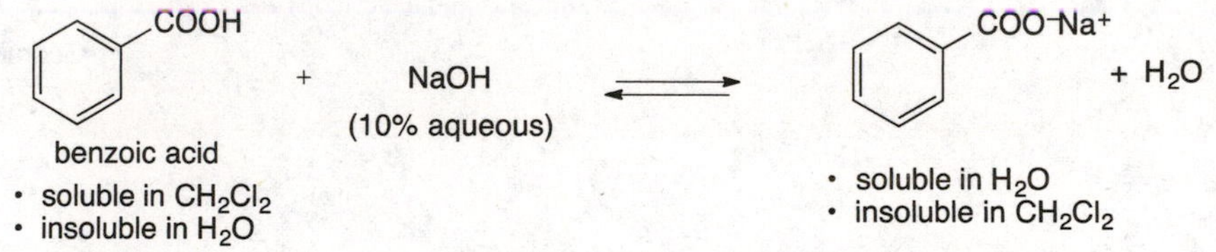

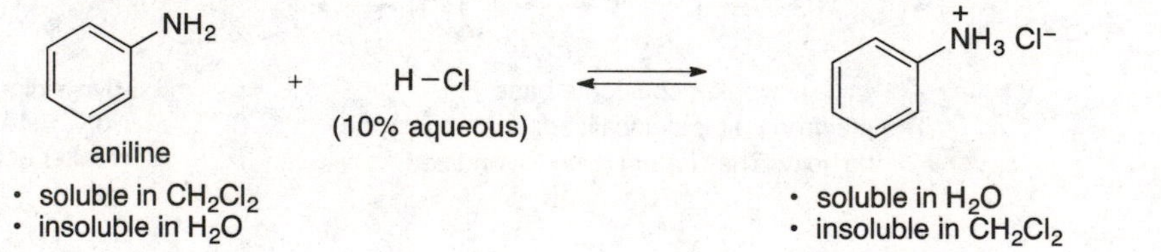

COOH + NaOH (10% aqueous) ⇌ COO^-Na^+ + H_2O

benzoic acid
- soluble in CH_2Cl_2
- insoluble in H_2O

- soluble in H_2O
- insoluble in CH_2Cl_2

NH_2 + H−Cl (10% aqueous) ⇌ $\overset{+}{N}H_3$ Cl^-

aniline
- soluble in CH_2Cl_2
- insoluble in H_2O

- soluble in H_2O
- insoluble in CH_2Cl_2

Toluene ($C_6H_5CH_3$), on the other hand, is not protonated or deprotonated in aqueous solution, so it is always soluble in CH_2Cl_2 and insoluble in H_2O. The following flow chart illustrates the process.

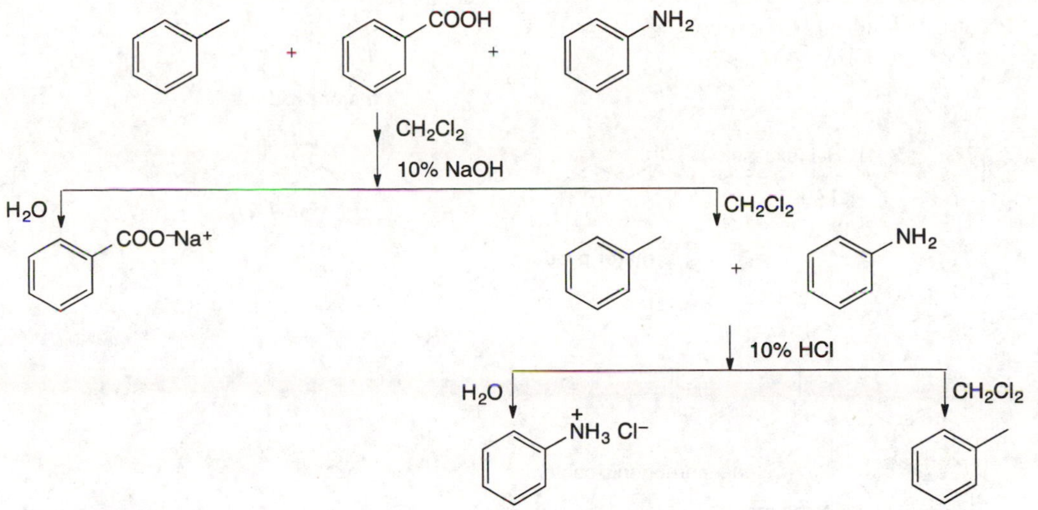

+ $COOH$ + NH_2

CH_2Cl_2
10% NaOH

H_2O COO^-Na^+

CH_2Cl_2 + NH_2

10% HCl

H_2O $\overset{+}{N}H_3$ Cl^-

CH_2Cl_2

25.52

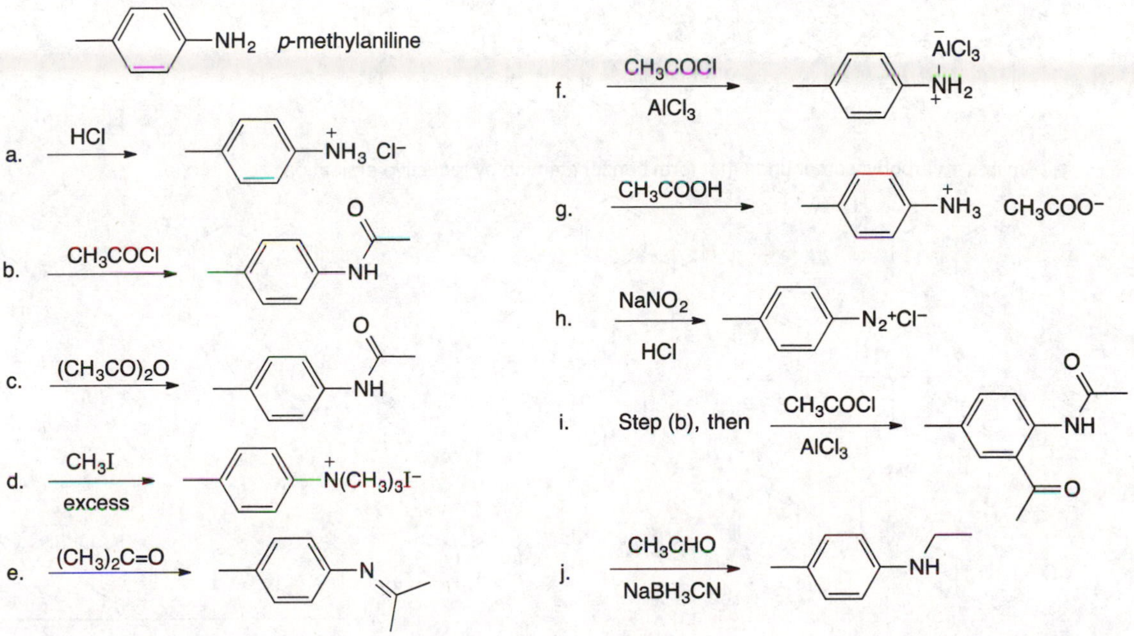

NH_2 *p*-methylaniline

a. HCl → $\overset{+}{N}H_3$ Cl^-

b. CH_3COCl → NH, O

c. $(CH_3CO)_2O$ → NH, O

d. CH_3I excess → $\overset{+}{N}(CH_3)_3I^-$

e. $(CH_3)_2C{=}O$ → N

f. CH_3COCl, $AlCl_3$ → $\bar{A}lCl_3$, $\overset{+}{N}H_2$

g. CH_3COOH → $\overset{+}{N}H_3$ CH_3COO^-

h. $NaNO_2$, HCl → $N_2^+Cl^-$

i. Step (b), then CH_3COCl, $AlCl_3$ → NH, O, O

j. CH_3CHO, $NaBH_3CN$ → NH

25.53

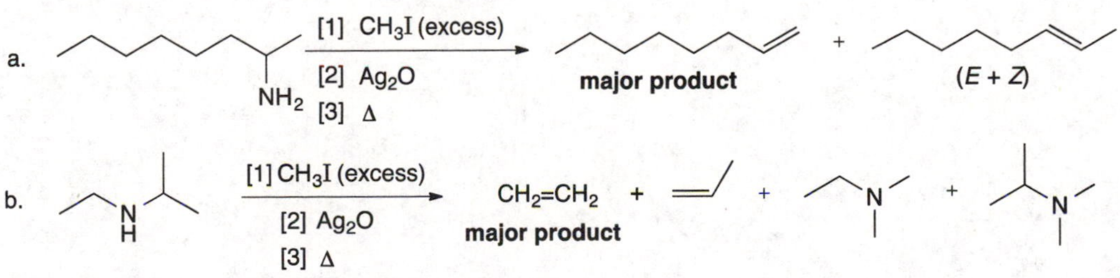

a. NH_2 → [1] CH_3I (excess) [2] Ag_2O [3] Δ → major product + (*E* + *Z*)

b. N H → [1] CH_3I (excess) [2] Ag_2O [3] Δ → $CH_2{=}CH_2$ major product + + N + N

c. [1] CH_3I (excess), [2] Ag_2O, [3] Δ

NH_2

major product

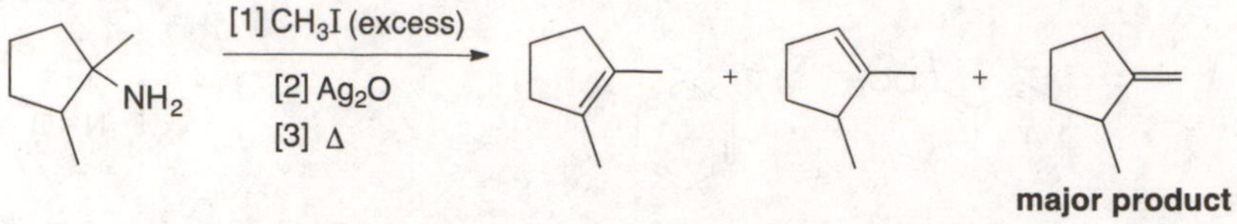

d. β_1 β_2 N H β_3 [1] CH_3I (excess), [2] Ag_2O, [3] Δ

β_1 N major product + N β_2 (*E* + *Z*) + N β_3 (*E* + *Z*)

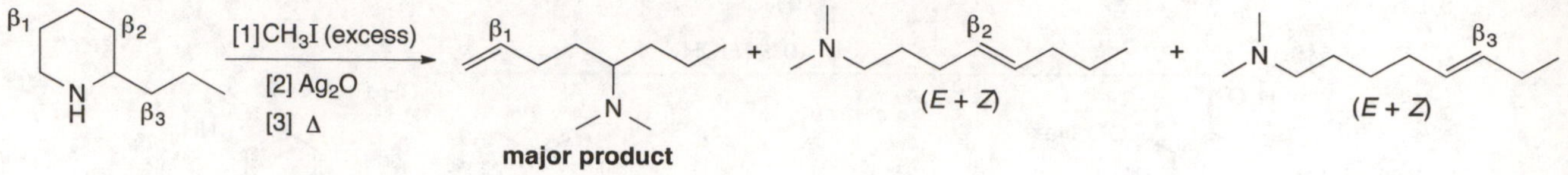

25.54

a. N

one stereogenic center

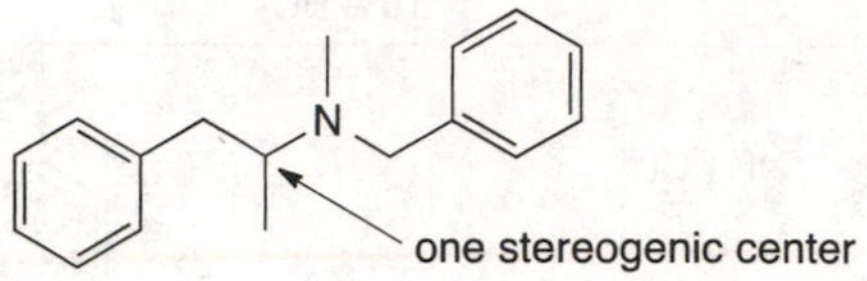

benzphetamine

b. Amides that can be reduced to benzphetamine:

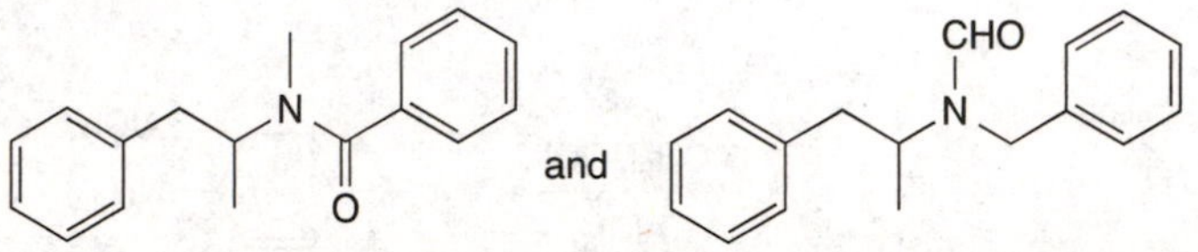

N O and CHO N

c. Amines + carbonyl compounds that form benzphetamine by reductive amination:

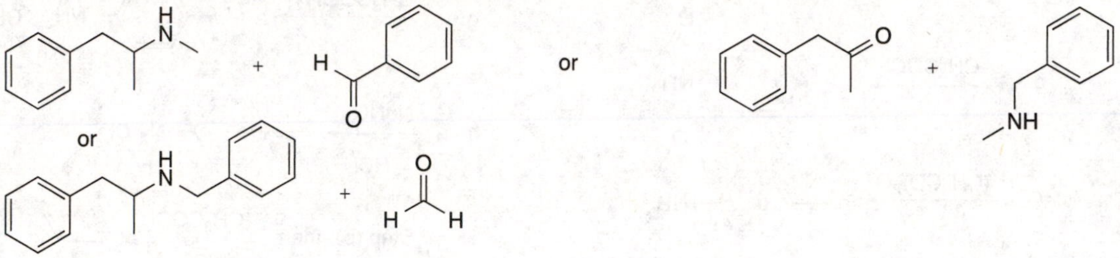

H N + H O or O + NH

or

H N + O H H

d. β_1 α N β_2 [1] CH_3I [2] Ag_2O [3] Δ

major product (elimination across α, β_2) + + elimination across α, β_1

+ N

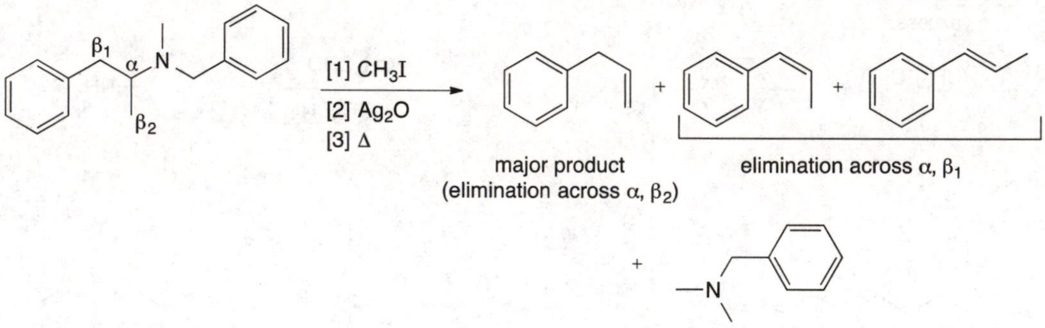

25.55

a. cyclohexyl–CH_2CH_2Cl $\xrightarrow[\text{excess}]{NH_3}$ cyclohexyl–$CH_2CH_2NH_2$

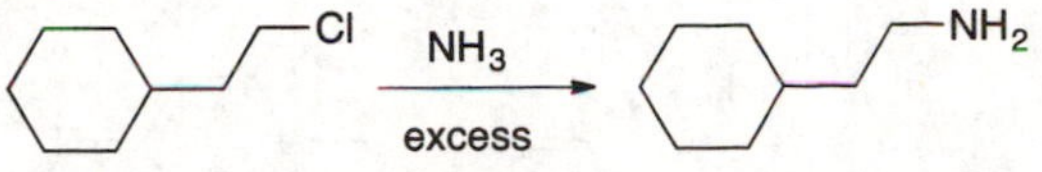

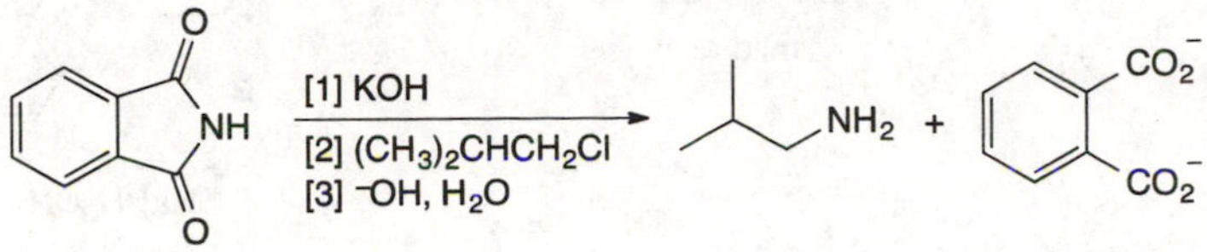

b. phthalimide (NH) $\xrightarrow{[1]\ KOH;\ [2]\ (CH_3)_2CHCH_2Cl;\ [3]\ ^-OH,\ H_2O}$ $(CH_3)_2CHCH_2NH_2$ + benzene-1,2-$(CO_2^-)_2$

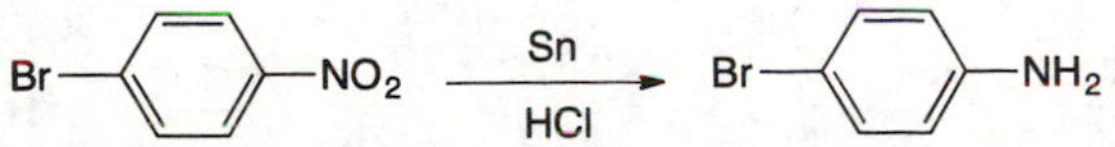

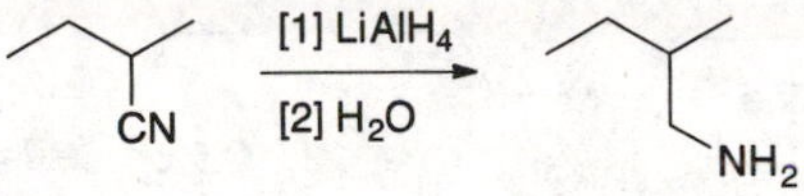

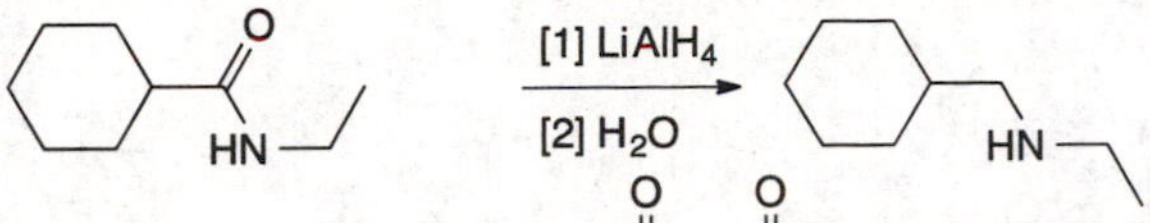

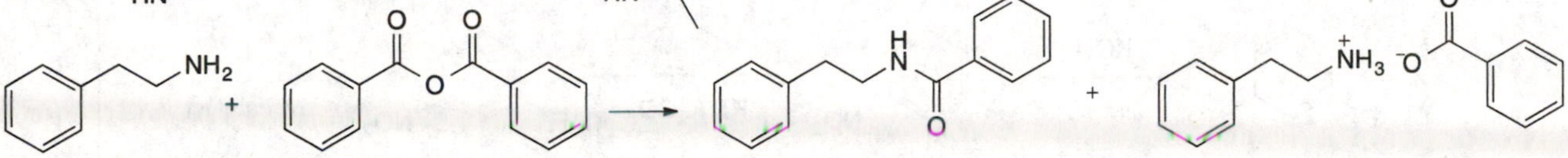

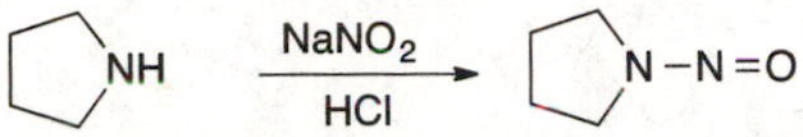

c. Br–C_6H_4–NO_2 $\xrightarrow[HCl]{Sn}$ Br–C_6H_4–NH_2

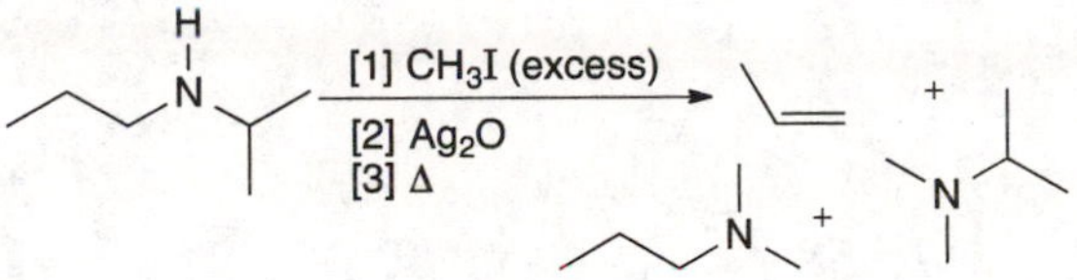

d. nitrile (CN) $\xrightarrow{[1]\ LiAlH_4;\ [2]\ H_2O}$ CH_2NH_2 product

e. cyclohexyl–C(=O)–NH–CH_2CH_3 $\xrightarrow{[1]\ LiAlH_4;\ [2]\ H_2O}$ cyclohexyl–CH_2–NH–CH_2CH_3

f. $C_6H_5CH_2CH_2NH_2$ + $(C_6H_5CO)_2O$ ⟶ $C_6H_5CH_2CH_2$–NH–C(=O)–C_6H_5 + $C_6H_5CH_2CH_2\overset{+}{N}H_3$ ^-O–C(=O)–C_6H_5

g. pyrrolidine (NH) $\xrightarrow[HCl]{NaNO_2}$ pyrrolidine N–N=O

h. pyrrolidine (NH) + C_6H_5CHO $\xrightarrow{NaBH_3CN}$ N-benzylpyrrolidine

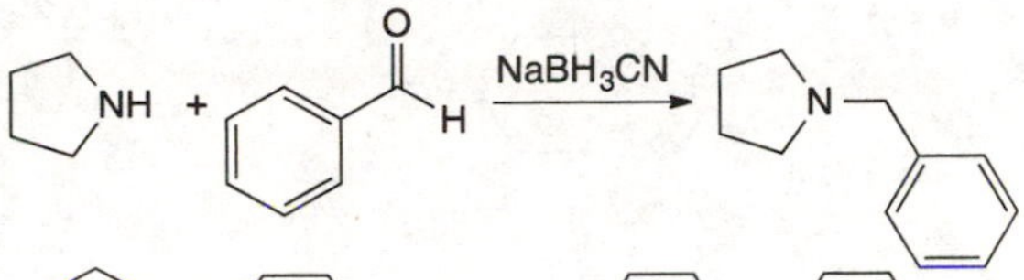

i. piperidine (N–H) + cyclohexanone (=O) ⟶ cyclohexenyl–N(piperidine)

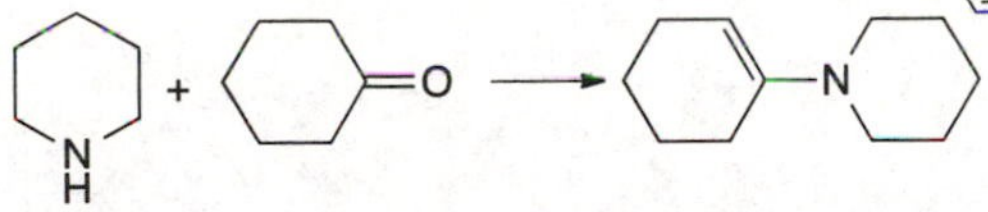

j. $CH_3CH_2CH_2$–NH–$CH(CH_3)_2$ $\xrightarrow{[1]\ CH_3I\ (excess);\ [2]\ Ag_2O;\ [3]\ \Delta}$ $CH_2=CHCH_3$ + $(CH_3)_2N$–$CH(CH_3)_2$ + $CH_3CH_2CH_2$–$N(CH_3)_2$ + $CH_2=CHCH_3$...

25.56 NH_2 and H must be anti for the Hofmann elimination. Rotate around the C–C bond so the NH_2 and H are anti.

a. Newman projection: front: CH_3, NH_2, C_6H_5; back: H, CH_2CH_3, $C(CH_3)_3$

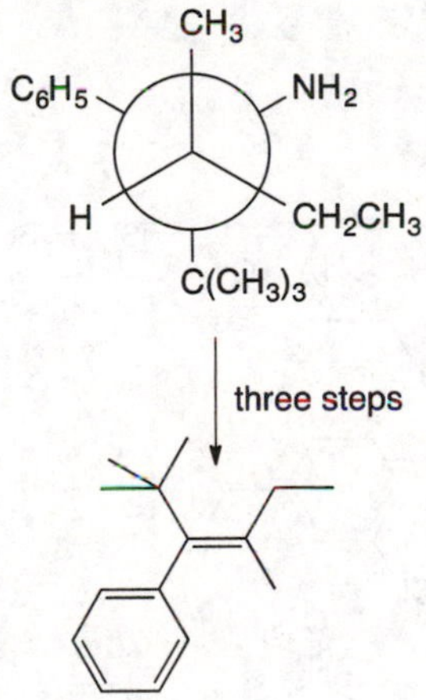

three steps ⟶ alkene product

b. Newman projection: NH_2, CH_3, H; C_6H_5, $C(CH_3)_3$, CH_2CH_3

rotate 120° counterclockwise

NH_2, CH_3, CH_2CH_3; C_6H_5, $C(CH_3)_3$, H

three steps ⟶ alkene product

c. Newman projection: C_6H_5, CH_3, H; $(CH_3)_3C$, NH_2, CH_2CH_3

rotate 120° counterclockwise

C_6H_5, H, CH_2CH_3; $(CH_3)_3C$, NH_2, CH_3

three steps ⟶ alkene product

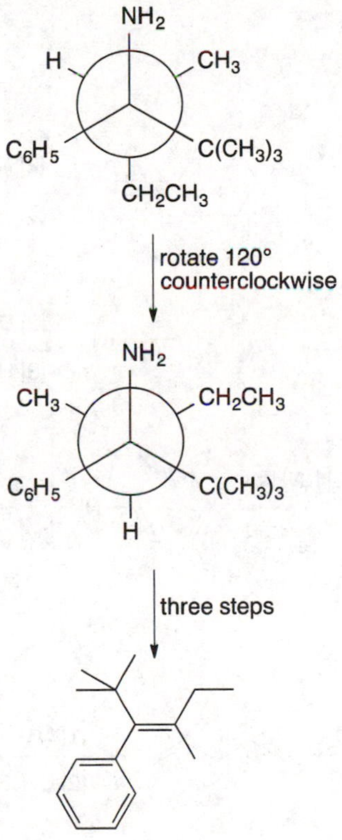

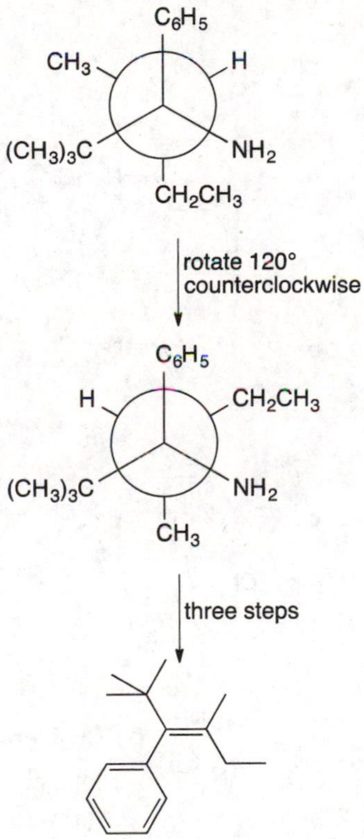

25.57

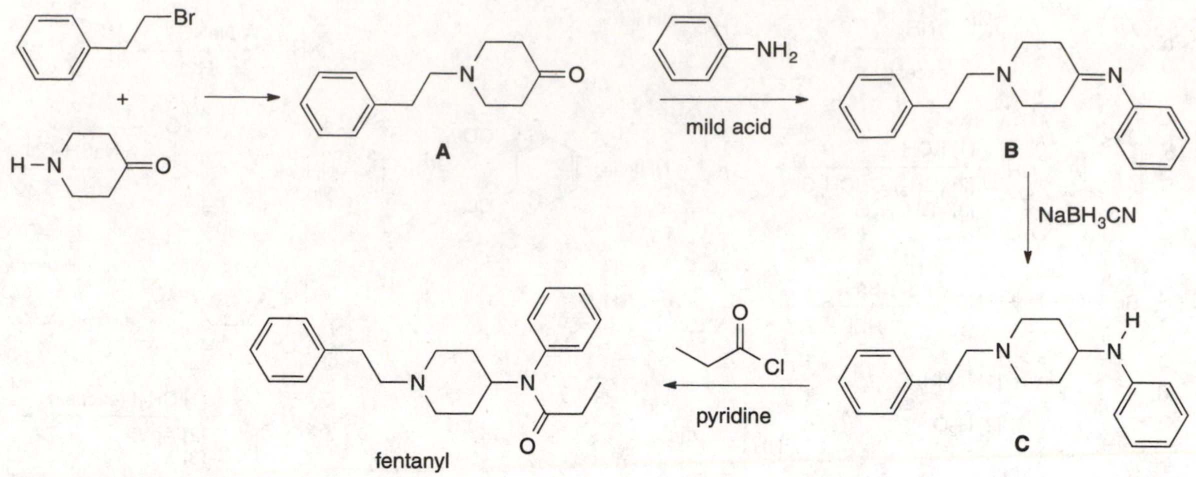

Br (2-phenylethyl bromide) + H–N (4-piperidone) ⟶ **A** —[$C_6H_5NH_2$, mild acid]⟶ **B** —[$NaBH_3CN$]⟶ **C** —[propanoyl chloride (Cl), pyridine]⟶ fentanyl

25.58

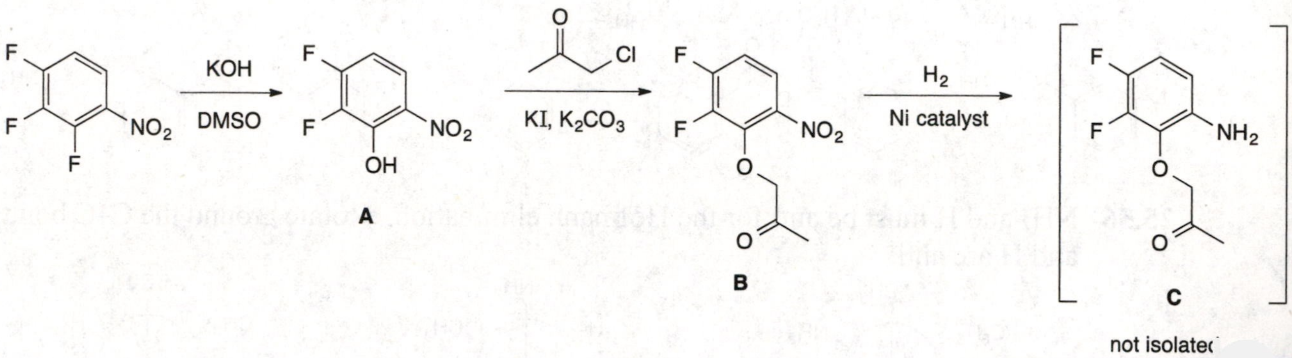

(trifluoronitrobenzene: F, F, F, NO_2) —[KOH, DMSO]⟶ **A** (F, F, OH, NO_2) —[chloroacetone (O, Cl), KI, K_2CO_3]⟶ **B** (F, F, O, NO_2) —[H_2, Ni catalyst]⟶ [**C** (F, F, O, NH_2)] not isolated

25.59

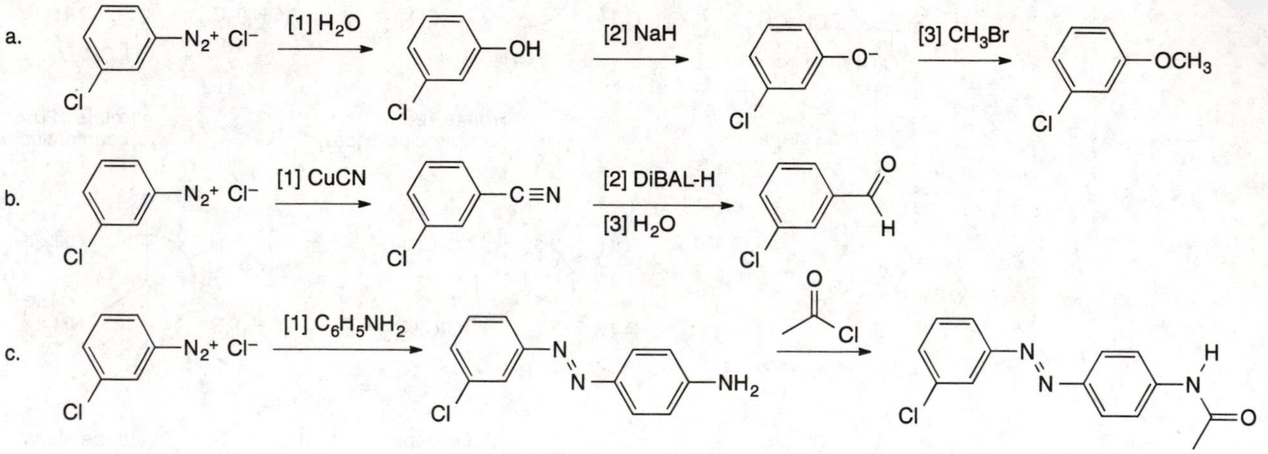

a. (Cl, $-N_2^+$ Cl^-) —[1] H_2O⟶ (Cl, OH) —[2] NaH⟶ (Cl, O^-) —[3] CH_3Br⟶ (Cl, OCH_3)

b. (Cl, $-N_2^+$ Cl^-) —[1] CuCN⟶ (Cl, $C\equiv N$) —[2] DiBAL-H, [3] H_2O⟶ (Cl, CHO: O, H)

c. (Cl, $-N_2^+$ Cl^-) —[1] $C_6H_5NH_2$⟶ (Cl, N=N, NH_2) —[acetyl chloride (O, Cl)]⟶ (Cl, N=N, N–H, C=O)

25.60

A (NH_2, *R*, H) —[1] CH_3I (excess), [2] Ag_2O, [3] Δ⟶ **B** —[1] O_3, [2] CH_3SCH_3⟶ H(C=O)H + H(C=O)CH₂CH₂CH₃

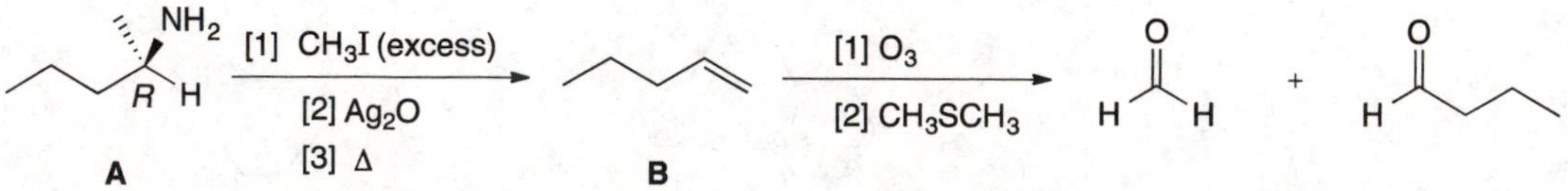

25.61

a. Br(CH$_2$)$_4$Br + $CH_3CH_2\ddot{N}H_2$ → Br(CH$_2$)$_4$$\overset{+}{N}H_2CH_2CH_3$ + Br^- (Na+ :ÖH:⁻) → Br(CH$_2$)$_4$N̈HCH$_2$CH$_3$ + Na+ + $H_2\ddot{O}$ → N-ethylpyrrolidinium (H) + Br^- (Na+ :ÖH⁻) → N-ethylpyrrolidine + $H_2\ddot{O}$ + Na+

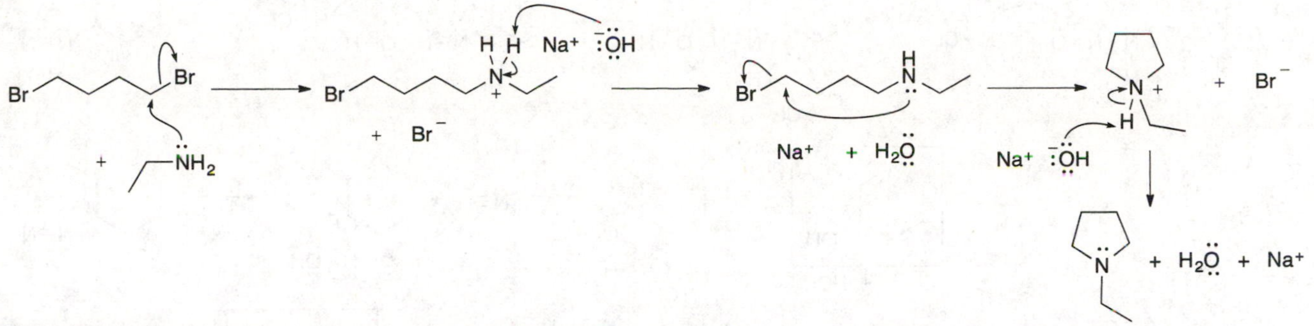

b. 2-(2-aminoethyl)cyclohexanone → (:Ö:⁻, $\overset{+}{N}H_2$) → proton transfer → (OH, N̈H) → proton source (H–A) → ($\overset{+}{O}H_2$, N̈H) + A:⁻ → iminium ($H_3\bar{B}$–H) + $H_2\ddot{O}$ → octahydroindole + BH_3

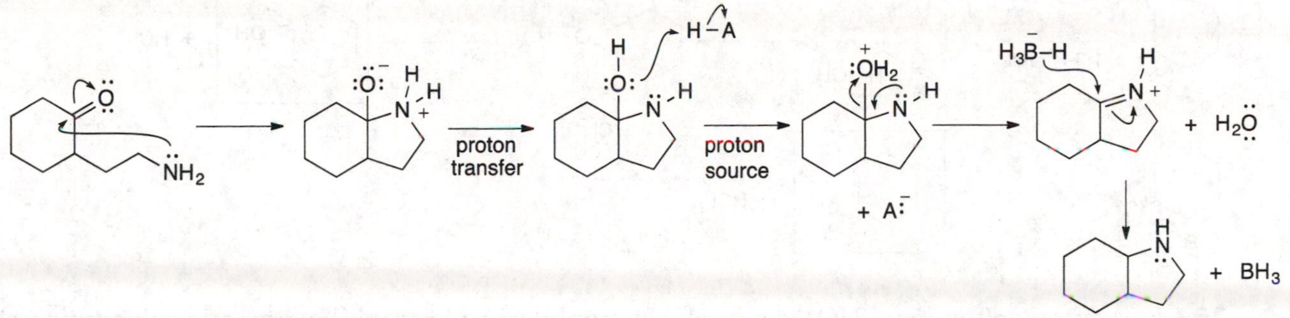

25.62

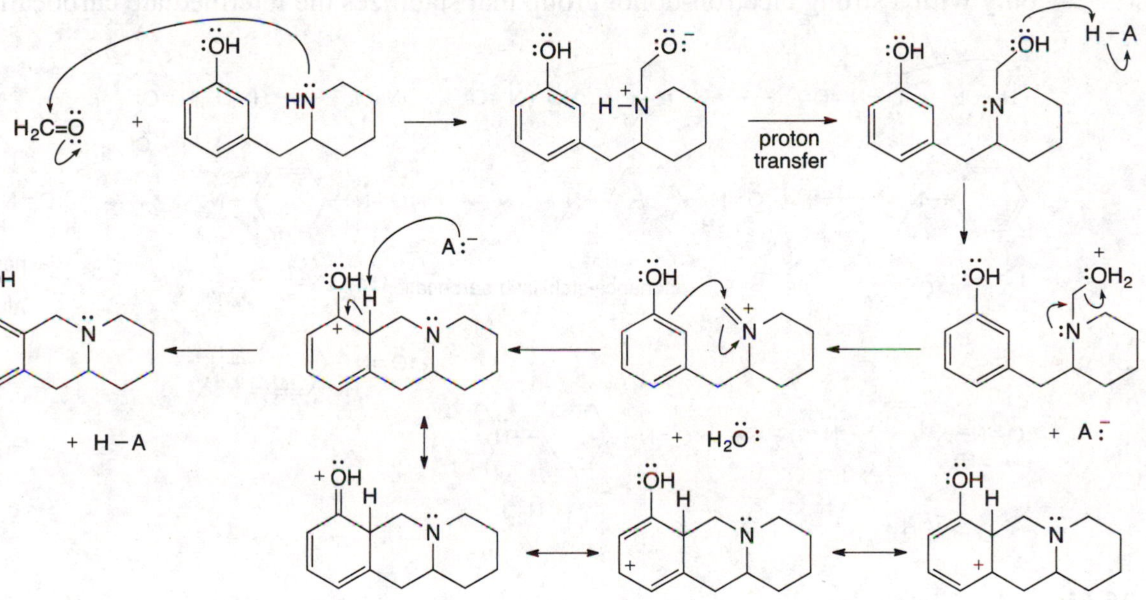

$H_2C{=}\ddot{O}$ + phenol–piperidine (HN̈) → (ÖH, H–$\overset{+}{N}$, :Ö:⁻) → proton transfer → (ÖH, :N, :ÖH; H–A) → (ÖH, :N, $\overset{+}{:O}H_2$) + A:⁻ → iminium (=$\overset{+}{N}$) + $H_2\ddot{O}$: → (ÖH, H, +; A:⁻) → tricyclic product (ÖH, N̈) + H–A

Resonance structures: $\overset{+}{\ddot{O}}H$ ↔ ÖH (+) ↔ ÖH (+)

25.63

Overall reaction: 2-methylcyclohexylamine (NH$_2$) —$NaNO_2$; HCl, H_2O→ 2-methylcyclohexanol (OH) + 1-methylcyclohexanol (OH) + 1-methylcyclohexene

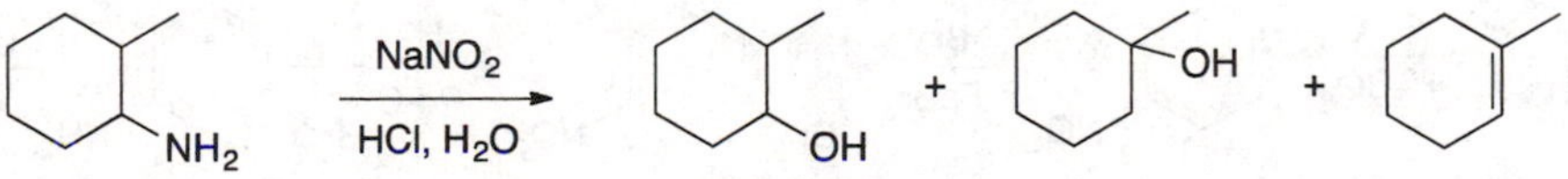

The steps:

$NaNO_2$ + HCl → $\overset{+}{N}{=}\ddot{O}$:

2-methylcyclohexylamine (N̈H$_2$) + $\overset{+}{N}{=}\ddot{O}$: → (H, $\overset{+}{N}$–N̈=Ö:, H) + :C̈l:⁻ → (N̈–N̈=Ö:, H) + HCl

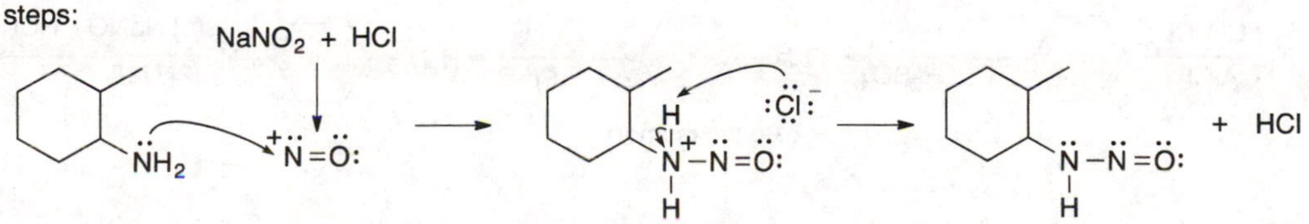

N–N=O: H–Cl → N–N=O⁺–H (H, :Cl:⁻) → N=N–O–H H–Cl → N=N–OH₂⁺ +

↓

N≡N + H₂O ← (A) + N≡N ← H–OH → O⁺–H, H, :Cl:⁻ → OH (boxed) + HCl

A —(1,2-H shift)→ B; B + H–OH → O⁺–H, H, :Cl:⁻ → OH (boxed) + HCl

B + H, :Cl:⁻ → (boxed alkene) + HCl

25.64 A nitrosonium ion (^{+}NO) is a weak electrophile, so electrophilic aromatic substitution occurs only with a strong electron-donor group that stabilizes the intermediate carbocation.

H–Cl :O–N=O: Na⁺ → H–Cl HO–N=O: + NaCl → H–O⁺(H)–N=O: + Cl⁻ → ⁺N=O: + H_2O:

(N,N-dimethylaniline) + ⁺N=O: → :O=N(H)–C₆H₅⁺–N ↔ :O=N(H)–⁺–N ↔ :O=N(H)–C₆H₄=N⁺ ↕ :O=N(H)–⁺–N

resonance-stabilized carbocation

especially good resonance structure
All atoms have octets.

:O=N(H)–⁺–N + :OH⁻ → :O=N–C₆H₄–N + H_2O:

25.65

a. benzene —(HNO_3, H_2SO_4)→ NO_2 —(Br_2, $FeBr_3$)→ Br, NO_2 —(H_2, Pd-C)→ Br, NH_2 —(Cl–C(=O)–CH₃)→ Br, N(H)–C(=O)–CH₃

b. benzene —(CH_3Cl, $AlCl_3$)→ toluene —(HNO_3, H_2SO_4)→ O_2N–C₆H₄–CH₃ (+ ortho isomer) —(H_2, Pd-C)→ H_2N–C₆H₄–CH₃ —([1] $NaNO_2$, HCl; [2] NaI)→ I–C₆H₄–CH₃

c. (from a.) O_2N–benzene → (H_2, Pd-C) → H_2N–benzene → ([1] $NaNO_2$, HCl; [2] CuCN) → NC–benzene → (Br_2, $FeBr_3$) → m-Br–C6H4–CN

d. benzene → (Br_2, $FeBr_3$) → bromobenzene → (HNO_3, H_2SO_4) → o-bromonitrobenzene (+ para isomer) → (H_2, Pd-C) → o-bromoaniline → ([1] $NaNO_2$, HCl; [2] H_2O) → o-bromophenol

e. (from b.) O_2N–C6H4–CH3 → ($KMnO_4$) → O_2N–C6H4–COOH → (H_2, Pd-C) → H_2N–C6H4–COOH → ([1] $NaNO_2$, HCl; [2] H_2O) → HO–C6H4–COOH

f. (from c.) C6H5–NH_2 → ([1] $NaNO_2$, HCl; [2] H_2O) → C6H5–OH → ($C_6H_5N_2{}^+Cl^-$ (from c.)) → C6H5–N=N–C6H4–OH

25.66

a. C6H5–NH_2 → ([1] $NaNO_2$, HCl; [2] NaCN) → C6H5–CN → (H_3O^+) → C6H5–COOH → ([1] $SOCl_2$; [2] NH_2CH_3) → C6H5–C(=O)NHCH3

b. C6H5–NH_2 → (CH_3COCl) → C6H5–NHC(=O)CH3 → (Br_2, $FeBr_3$) → p-Br–C6H4–NHC(=O)CH3 (+ ortho isomer) → (CH_3Cl, $AlCl_3$) → 4-Br-2-CH3–C6H3–NHC(=O)CH3 → (^-OH, H_2O) → 4-Br-2-CH3–C6H3–NH_2 → ([1] $NaNO_2$, HCl; [2] H_3PO_2) → m-Br–C6H4–CH3

c. H_2N–C6H5 → ([1] $NaNO_2$, HCl; [2] H_2O) → HO–C6H5 → (CH_3Cl, $AlCl_3$) → p-HO–C6H4–CH3 (+ ortho isomer) → ($C_6H_5N_2{}^+Cl^-$ (from a., Step [1])) → C6H5–N=N–(2-HO-5-CH3–C6H3)

25.67

[1] benzene → (CH_3Cl, $AlCl_3$) → toluene → (Br_2, $h\nu$) → C6H5CH2Br → (NH_3 (excess)) → C6H5CH2NH_2 → ($H_2C{=}O$, $NaBH_3CN$) → C6H5CH2NHCH3

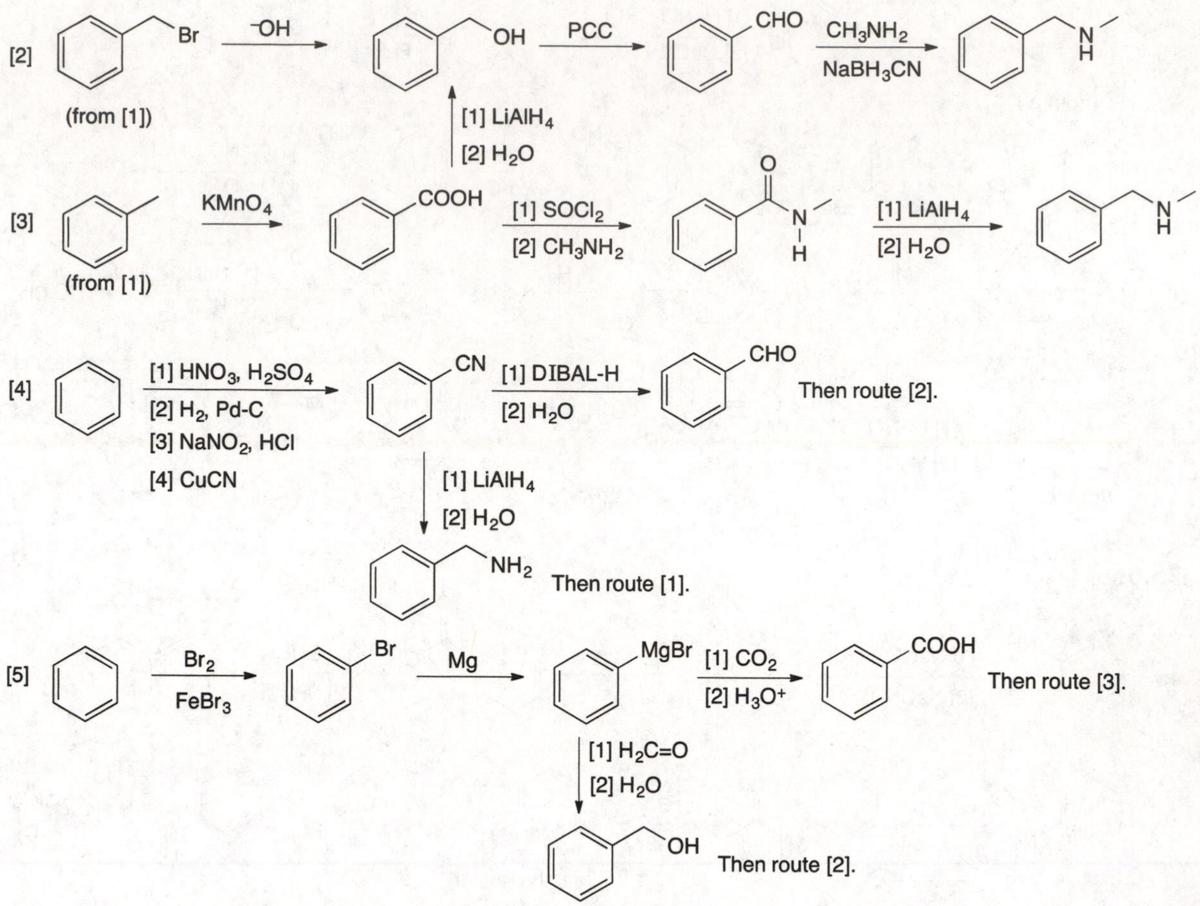

[2] (from [1]) Br → ^{-}OH → OH → PCC → CHO → CH_3NH_2, $NaBH_3CN$ → N–H

[1] $LiAlH_4$ [2] H_2O

[3] (from [1]) → $KMnO_4$ → COOH → [1] $SOCl_2$ [2] CH_3NH_2 → O, N–H → [1] $LiAlH_4$ [2] H_2O → N–H

[4] → [1] HNO_3, H_2SO_4 [2] H_2, Pd-C [3] $NaNO_2$, HCl [4] CuCN → CN → [1] DIBAL-H [2] H_2O → CHO Then route [2].

[1] $LiAlH_4$ [2] H_2O → NH_2 Then route [1].

[5] → Br_2 $FeBr_3$ → Br → Mg → MgBr → [1] CO_2 [2] H_3O^+ → COOH Then route [3].

[1] $H_2C{=}O$ [2] H_2O → OH Then route [2].

25.68

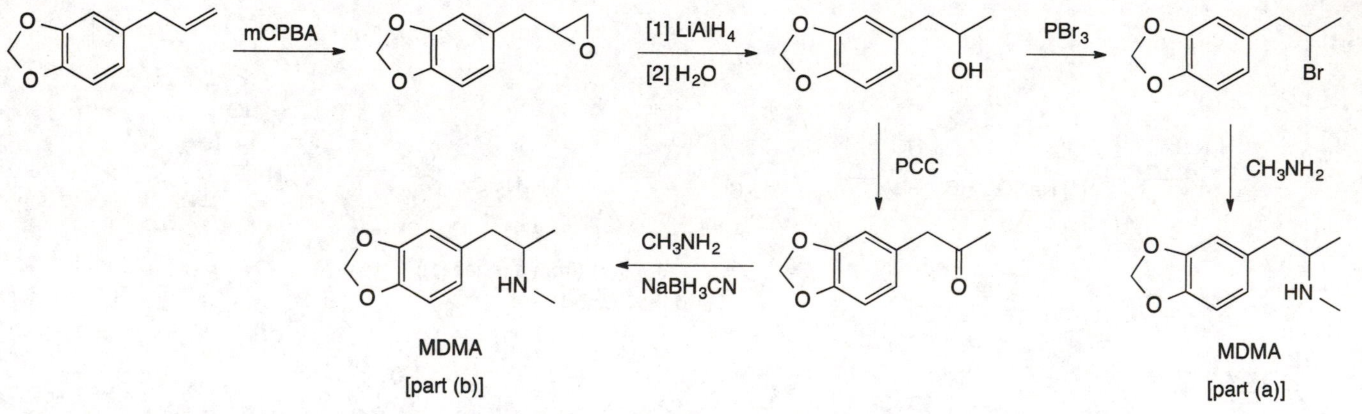

mCPBA → O → [1] $LiAlH_4$ [2] H_2O → OH → PBr_3 → Br → CH_3NH_2 → HN

PCC → O → CH_3NH_2 $NaBH_3CN$ → HN

MDMA [part (b)]

MDMA [part (a)]

25.69

a. → CH_3Cl $AlCl_3$ → HNO_3 H_2SO_4 → NO_2 (+ ortho isomer) → H_2 Pd-C → NH_2 → [1] $NaNO_2$, HCl [2] NaCN → CN → [1] $LiAlH_4$ [2] H_2O → NH_2

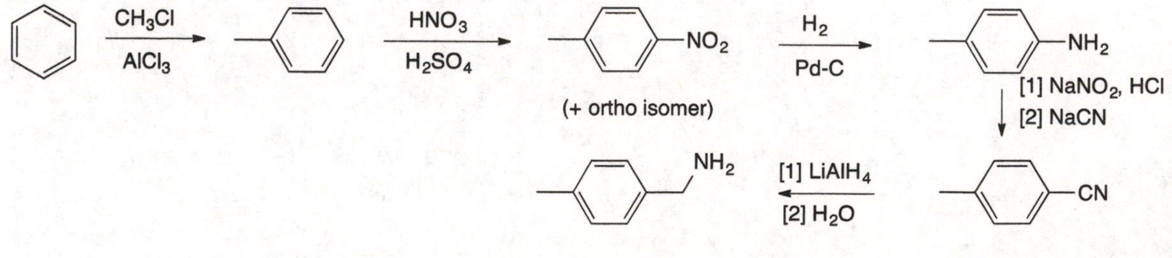

b. benzene $\xrightarrow[AlCl_3]{CH_3CH_2Cl}$ ethylbenzene $\xrightarrow[H_2SO_4]{HNO_3}$ p-ethylnitrobenzene (–NO_2) (+ ortho isomer) $\xrightarrow[Pd\text{-}C]{H_2}$ p-ethylaniline (–NH_2) $\xrightarrow{[1]\ NaNO_2,\ HCl;\ [2]\ CuCN}$ p-ethylbenzonitrile (–CN) $\xrightarrow{H_3O^+}$ p-ethylbenzoic acid (–COOH)

c. benzene $\xrightarrow[H_2SO_4]{HNO_3}$ nitrobenzene (NO_2) $\xrightarrow[Pd\text{-}C]{H_2}$ aniline (NH_2) $\xrightarrow{CH_3COCl}$ acetanilide ($NHCOCH_3$) $\xrightarrow[H_2SO_4]{HNO_3}$ p-nitroacetanilide ($NHCOCH_3$, NO_2) $\xrightarrow[FeCl_3\ (2X)]{Cl_2}$ 2,6-dichloro-4-nitroacetanilide (Cl, Cl, $NHCOCH_3$, NO_2) $\xrightarrow[H_2O]{^-OH}$ 2,6-dichloro-4-nitroaniline (Cl, Cl, NH_2, NO_2) $\xrightarrow{[1]\ NaNO_2,\ HCl;\ [2]\ H_3PO_2}$ 1,3-dichloro-5-nitrobenzene (Cl, Cl, NO_2) $\xrightarrow[Pd\text{-}C]{H_2}$ 3,5-dichloroaniline (Cl, Cl, NH_2) $\xrightarrow{[1]\ NaNO_2,\ HCl;\ [2]\ \text{aniline} (–NH_2)\ \text{(from above)}}$ 3,5-dichlorophenyl–N=N–phenyl–NH_2 (Cl, Cl)

25.70

a. benzene $\xrightarrow[H_2SO_4]{HNO_3}$ O_2N–benzene $\xrightarrow[Pd\text{-}C]{H_2}$ H_2N–benzene $\xrightarrow{[1]\ NaNO_2,\ HCl;\ [2]\ H_2O}$ HO–benzene $\xrightarrow[H_2SO_4]{HNO_3}$ p-nitrophenol (HO, NO_2) (+ ortho isomer) $\xrightarrow[Pd\text{-}C]{H_2}$ p-aminophenol (HO, NH_2) $\xrightarrow{\text{acetic anhydride}}$ HO–C_6H_4–NHC(=O)CH_3

b. benzene $\xrightarrow[AlCl_3]{\text{Cl–C(=O)CH}_2\text{CH}_3}$ propiophenone $\xrightarrow[H_2O,\ HCl]{Cl_2}$ 2-chloro-1-phenyl-1-propanone (Cl) $\xrightarrow{NH_2CH_3}$ PhC(=O)CH(CH_3)NHCH$_3$ $\xrightarrow[CH_3OH]{NaBH_4}$ PhCH(OH)CH(CH_3)NHCH$_3$

25.71

OH

HCl

Cl

a. AlCl$_3$

Br$_2$ FeBr$_3$

Br

(+ ortho isomer)

[1] Mg [2] CH$_2$=O

PCC CH$_3$OH

H$_2$O

OH

PCC

O H

A

HNO$_3$ H$_2$SO$_4$

NO$_2$

H$_2$ Pd-C

NH$_2$

[1] NaNO$_2$, HCl [2] H$_2$O

OH

O OH

CrO$_3$

OH

O O

HNO$_3$ H$_2$SO$_4$

O$_2$N O O

H$_2$ Pd-C

H$_2$N O O

B

O H

A

+

O O H$_2$N

B

mild H$^+$

N O O

CH$_3$OH

SOCl$_2$

b. CH$_3$Cl AlCl$_3$

HNO$_3$ H$_2$SO$_4$

O$_2$N

(+ ortho)

Br$_2$ $h\nu$

O$_2$N Br

NaOH

O$_2$N OH

H$_2$ Pd-C

H$_2$N OH

[1] NaNO$_2$, HCl [2] NaCN

NC OH

[1] LiAlH$_4$ [2] H$_2$O

H$_2$N OH

CH$_3$OH PCC CH$_2$=O

NaBH$_3$CN

H N OH

CrO$_3$

H N O OH

A

HNO_3, H_2SO_4 → O_2N-benzene; Cl_2, $FeCl_3$ → Cl; H_2, Pd-C → H_2N … Cl; [1] $NaNO_2$, HCl [2] H_2O → HO … Cl

B

O, OH, H, N, **A** + HO, Cl, **B** → H^+ → O, O, Cl, H, N

c. Probably a strong enough activator that the Friedel–Crafts reaction will still occur.

Make two parts:

H, N, O, CN + O, Cl, I → $AlCl_3$ → O, NH, O, I, CN

CH_3CH_2OH → CrO_3, H_2SO_4, H_2O → CH_3COOH → $SOCl_2$ → CH_3COCl

[1] HNO_3, H_2SO_4 [2] H_2, Pd-C → NH_2 → H, N, O → [1] HNO_3, H_2SO_4 [2] H_2, Pd-C → H, N, O, NH_2 (+ ortho isomer) → [1] $NaNO_2$, HCl [2] CuCN → H, N, O, CN

CH_3OH → $SOCl_2$ → CH_3Cl, $AlCl_3$ → HNO_3, H_2SO_4 → O_2N → [1] H_2, Pd-C [2] $NaNO_2$, HCl [3] NaI → I → [1] $KMnO_4$ [2] $SOCl_2$ → O, Cl, I

25.72

NH | HN— | H_2N—

Compound A: $C_8H_{11}N$
IR absorption at 3400 cm^{-1} → 2° amine
^1H NMR signals at (ppm):
- 1.3 (triplet, 3 H) CH_3 adjacent to 2 H's
- 3.2 (quartet, 2 H) CH_2 adjacent to 3 H's
- 3.6 (singlet, 1 H) amine H
- 6.8–7.2 (multiplet, 5 H) benzene ring

Compound B: $C_8H_{11}N$
IR absorption at 3310 cm^{-1} → 2° amine
^1H NMR signals at (ppm):
- 1.4 (singlet, 1 H) amine H
- 2.4 (singlet, 3 H) CH_3
- 3.8 (singlet, 2 H) CH_2
- 7.3 (multiplet, 5 H) benzene ring

Compound C: $C_8H_{11}N$
IR absorption at 3430 and 3350 cm^{-1} → 1° amine
^1H NMR signals at (ppm):
- 1.3 (triplet, 3 H) CH_3 near CH_2
- 2.5 (quartet, 2 H) CH_2 near CH_3
- 3.6 (singlet, 2 H) amine H's
- 6.7 (doublet, 2 H) | para disubstituted
- 7.0 (doublet, 2 H) | benzene ring

25.73

HO–CH₂CH₂–C≡N → written as: $HOCH_2CH_2C\equiv N$ | $HOCH_2CH_2CH_2NH_2$

Compound **D**:
Molecular ion at $m/z = 71$: C_3H_5NO (possible formula)
IR absorption at 3600–3200 cm^{-1} → OH
2263 cm^{-1} → CN
^1H NMR signals at (ppm):
2.5 (triplet, 2 H) CH_2 adjacent to 2 H's
3.1 (singlet, 1 H) OH
3.8 (triplet, 2 H) CH_2 adjacent to 2 H's

Compound **E**:
Molecular ion at $m/z = 75$: C_3H_9NO (possible formula)
IR absorption at 3600–3200 cm^{-1} → OH
3636 cm^{-1} → N–H of amine
^1H NMR signals at (ppm):
1.6 (quintet, 2 H) CH_2 split by 2 CH_2's
2.5 (singlet, 3 H) NH_2 and OH
2.8 (triplet, 2 H) CH_2 split by CH_2
3.7 (triplet, 2 H) CH_2 split by CH_2

25.74 Guanidine is a strong base because its conjugate acid is stabilized by resonance. This resonance delocalization makes guanidine easily donate its electron pair; thus it's a strong base.

guanidine —HA→ conjugate acid ($pK_a = 13.6$) ↔ resonance structures

25.75 The compound with the most available electron pair or the compound with the highest electron density on an atom (N in this case) is the strongest base. Pyrrole is the weakest base because its lone pair is delocalized on the five-membered ring to make it aromatic. Both imidazole and thiazole contain sp^2 hybridized N atoms with electron pairs that are localized on N. Imidazole is a stronger base than thiazole, because its second N atom is more basic than thiazole's S atom, so it places more electron density on N by a resonance effect.

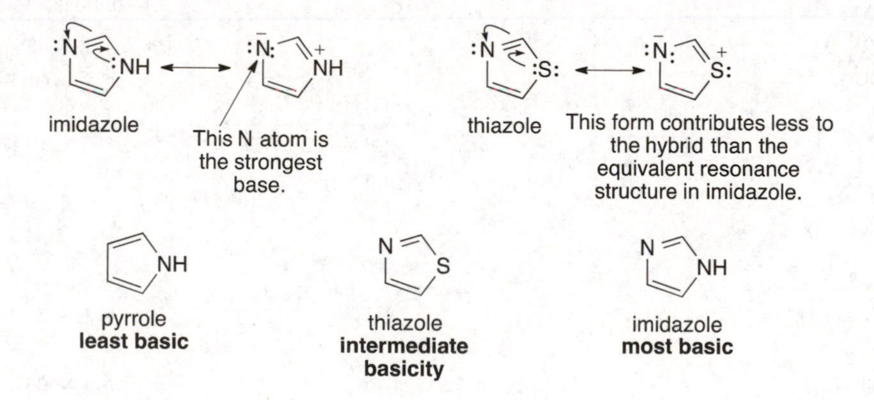

imidazole
This N atom is the strongest base.
thiazole
This form contributes less to the hybrid than the equivalent resonance structure in imidazole.
pyrrole
least basic
thiazole
intermediate basicity
imidazole
most basic

25.76

[1] CH_3I (excess) → [2] Ag_2O, [3] Δ → $N(CH_3)_2$ → [1] CH_3I (excess) → I^- $^+N(CH_3)_3$ → [2] Ag_2O, [3] Δ → C_8H_{10} **Y**

25.77

One possibility:

a. $\xrightarrow{CH_3COCl,\ AlCl_3}$ (+ isomer) $\xrightarrow{Br_2,\ CH_3CO_2H}$ $\xrightarrow{H_2NC(CH_3)_3}$ $\xrightarrow{NaBH_4,\ CH_3OH}$ $\xrightarrow{H_3O^+}$ albuterol

b. epoxide + diethylamine → A

benzene $\xrightarrow{HNO_3,\ H_2SO_4}$ $\xrightarrow{H_2,\ Pd\text{-}C}$ $\xrightarrow{[1]\ NaNO_2,\ HCl;\ [2]\ H_2O}$ $\xrightarrow{[1]\ NaH;\ [2]\ \text{propyl-Cl}}$ $\xrightarrow{Br_2}$ (+ ortho) $\xrightarrow{Mg}$ $\xrightarrow{CO_2}$ $\xrightarrow{H_3O^+}$ $\xrightarrow{H_2SO_4,\ HNO_3}$ $\xrightarrow{SOCl_2}$ $\xrightarrow{A}$ $\xrightarrow{H_2,\ Pd\text{-}C}$

25.78 $CH_2{=}O$ reacts with the amine to form an intermediate imine, which undergoes an intramolecular Diels–Alder reaction.

X $\xrightarrow{CH_2{=}O,\ H_2O}$ → Y ($C_{17}H_{23}NO$) $\xrightarrow{\text{one step}}$ lupinine + epilupinine

Chapter 26 Carbon–Carbon Bond-Forming Reactions in Organic Synthesis

Chapter Review

Coupling reactions

[1] Coupling reactions of organocuprate reagents (26.1)

R'—X + R_2CuLi ⟶ [R'—R] + RCu + LiX

X = Cl, Br, I

- R'X can be CH_3X, RCH_2X, 2° cyclic halides, vinyl halides, and aryl halides.
- X may be Cl, Br, or I.
- With vinyl halides, coupling is stereospecific.

[2] Suzuki reaction (26.2)

R'—X + R–B(Y_2) $\xrightarrow[\text{NaOH}]{Pd(PPh_3)_4}$ [R'—R] + HO–BY_2 + NaX

X = Br, I

- R'X is most often a vinyl halide or aryl halide.
- With vinyl halides, coupling is stereospecific.

[3] Heck reaction (26.3)

R'—X + CH_2=CH–Z $\xrightarrow[\text{P(o-tolyl)}_3,\ \text{Et}_3\text{N}]{Pd(OAc)_2}$ [R'–CH=CH–Z] + $Et_3\overset{+}{N}H$ X^-

X = Br or I

- R'X is a vinyl halide or aryl halide.
- Z = H, Ph, COOR, or CN
- With vinyl halides, coupling is stereospecific.
- The reaction forms trans alkenes.

Cyclopropane synthesis

[1] Addition of dihalocarbenes to alkenes (26.4)

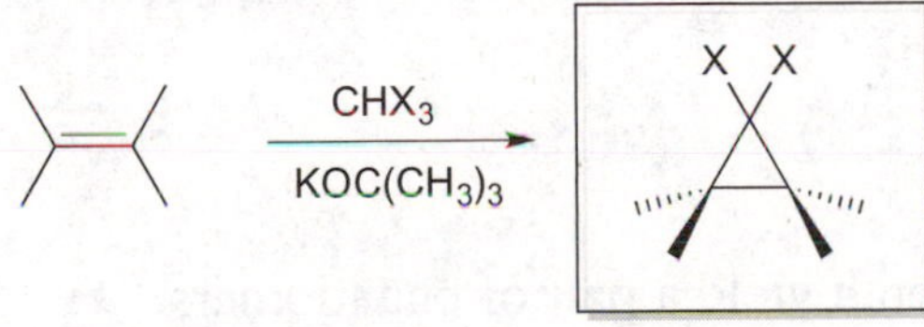

- The reaction occurs with syn addition.
- The position of substituents in the alkene is retained in the cyclopropane.

[2] Simmons–Smith reaction (26.5)

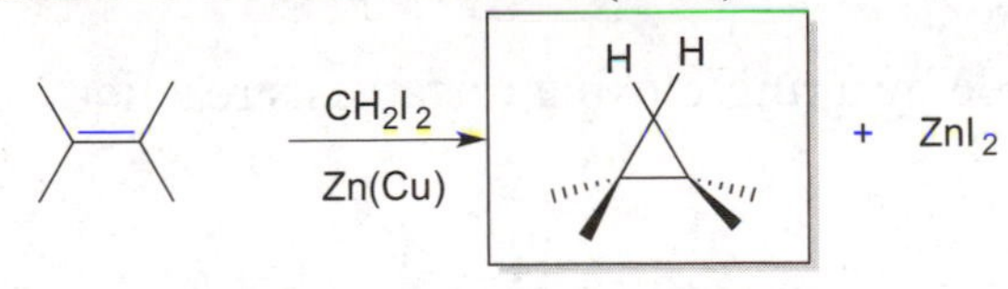

- The reaction occurs with syn addition.
- The position of substituents in the alkene is retained in the cyclopropane.

Metathesis (26.6)

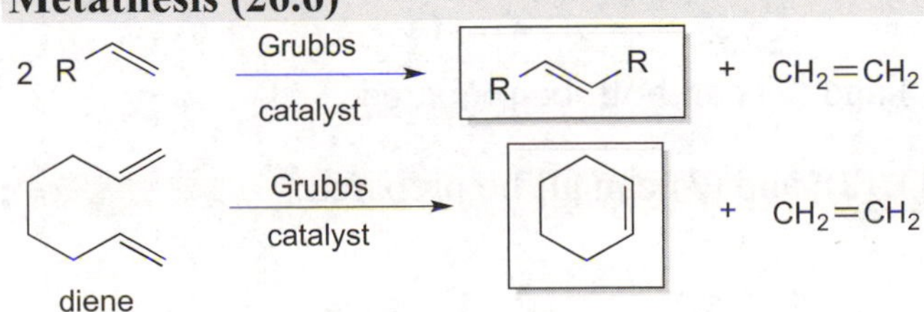

- Metathesis works best when CH_2=CH_2, a gas that escapes from the reaction mixture, is formed as one product.

Practice Test on Chapter Review

1.a. Which functional groups react with lithium dialkyl cuprates?
1. epoxides
2. vinyl halides
3. acid chlorides
4. Compounds (1) and (2) both react with R_2CuLi.
5. Compounds (1), (2), and (3) all react with R_2CuLi.

b. Which of the following statements is (are) true for the Suzuki reaction?
1. Arylboranes can serve as one reactant.
2. The reaction is stereospecific.
3. The reaction occurs between a vinyl or aryl halide and an alkene in the presence of a palladium catalyst.
4. Statements (1) and (2) are both true.
5. Statements (1), (2), and (3) are all true.

c. Which of the following compounds can react with $CH_2{=}CHCN$ in a Heck reaction?

1.  Br
2. 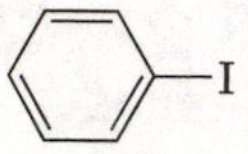 —I
3. $CH_2{=}CH_2$
4. Both (1) and (2) can react.
5. Compounds (1), (2), and (3) can all react.

d. Which of the following compounds yield(s) a pair of enantiomers on reaction with the Simmons–Smith reagent?

1. 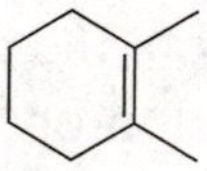
2. 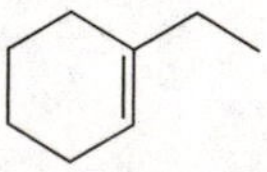
3. 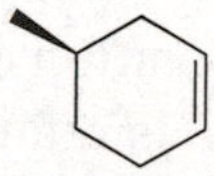
4. Compounds (1) and (2) both yield a pair of enantiomers.
5. Compounds (1), (2), and (3) all yield a pair of enantiomers.

e. Which of the following compounds can be made by a ring-closing metathesis reaction?

1. 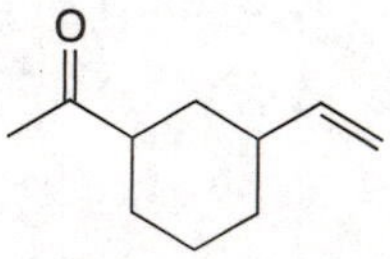 O
2. 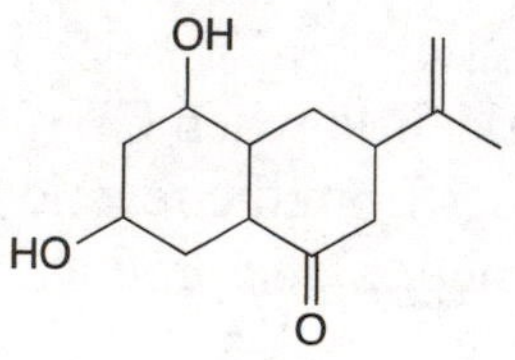 OH HO O
3. 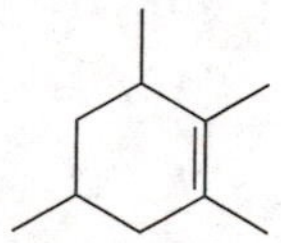
4. Compounds (1) and (2) can both be prepared.
5. Compounds (1), (2), and (3) can all be prepared.

f. Which of the following compounds can be prepared from $CH_3–C≡C–H$ by a Suzuki reaction? You may use other organic compounds or inorganic reagents.

1. 

3.

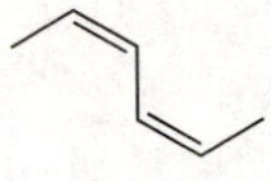

2.

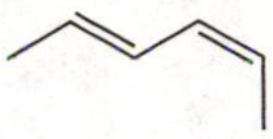

4. Compounds (1) and (2) can both be prepared.

5. Compounds (1), (2), and (3) can all be prepared.

2. Draw the product formed in each reaction. Indicate the stereochemistry around double bonds and stereogenic centers when necessary.

a. 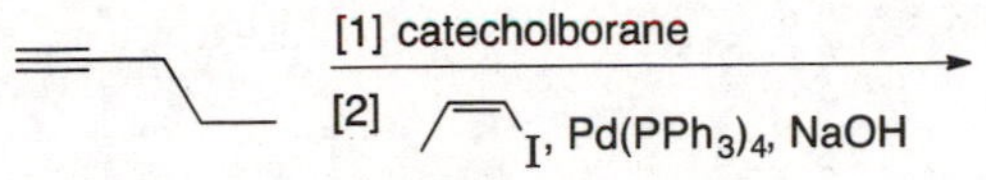 [1] catecholborane; [2] I, $Pd(PPh_3)_4$, NaOH

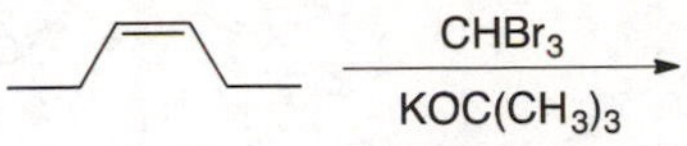

b. 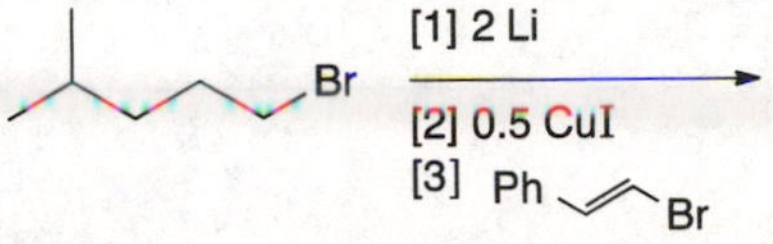 Br; [1] 2 Li; [2] 0.5 CuI; [3] Ph Br

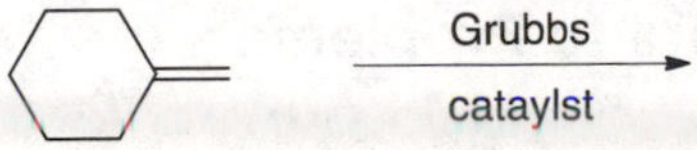

c. 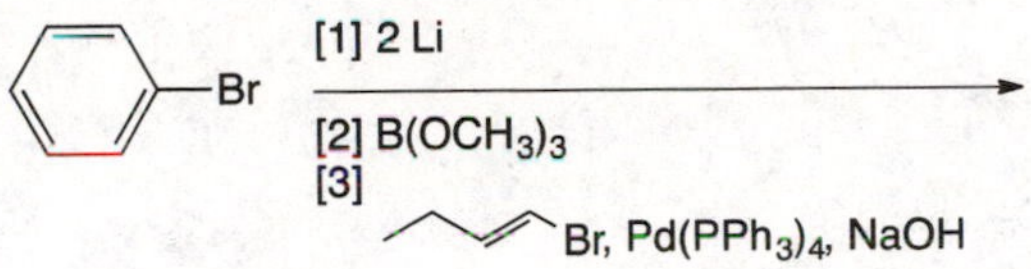 Br; [1] 2 Li; [2] $B(OCH_3)_3$; [3] Br, $Pd(PPh_3)_4$, NaOH

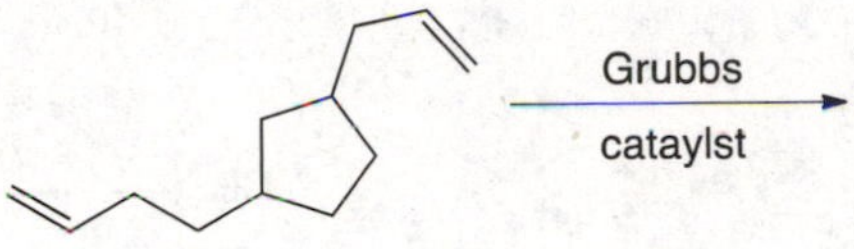

d. 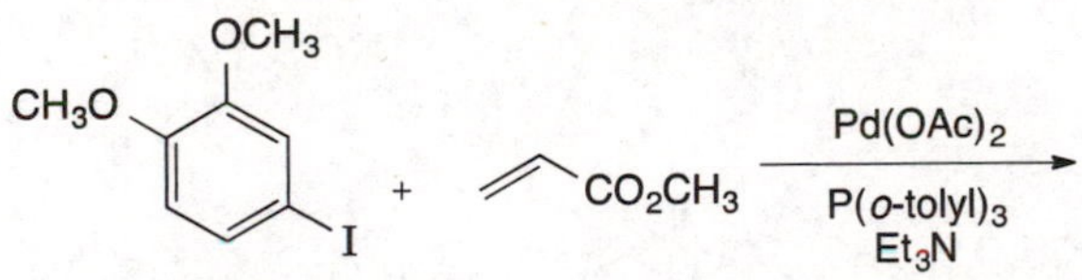 OCH_3, CH_3O, I + CO_2CH_3; $Pd(OAc)_2$; $P(o\text{-tolyl})_3$; Et_3N

e. $CHBr_3$; $KOC(CH_3)_3$

f. Grubbs catalyst

g. Grubbs catalyst

3. What starting material is needed to synthesize each compound by ring-closure metathesis?

a. 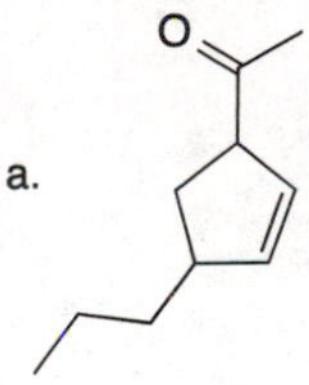 O

b. 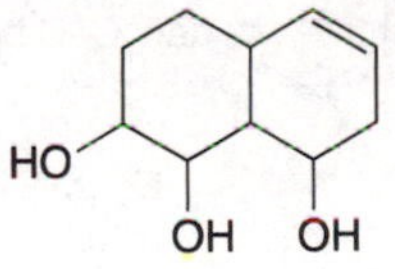 HO, OH, OH

c. 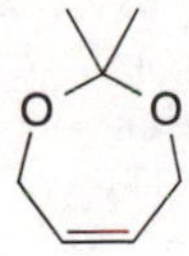 O, O

Answers to Practice Test

1.a. 5
b. 4
c. 4
d. 2
e. 3
f. 4

2.
a. 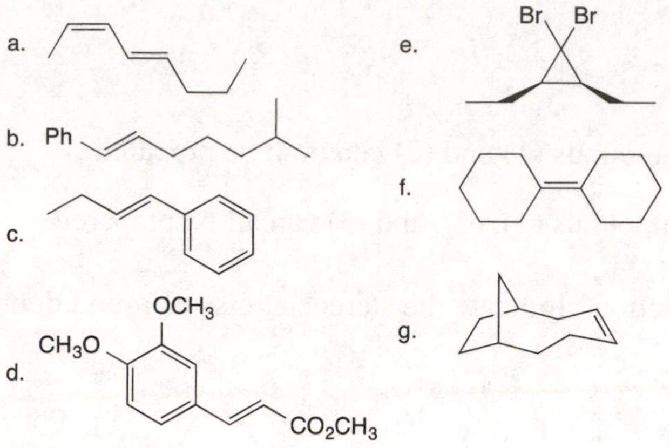
b. Ph
c.
d. CH_3O, OCH_3, CO_2CH_3
e. Br Br
f.
g.

3.
a.
b. HO, OH, OH
c.

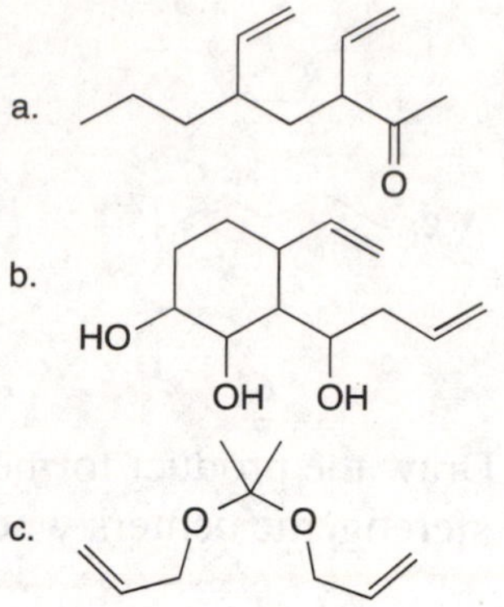

Answers to Problems

26.1 A new C–C bond is formed in each coupling reaction.

a. Cl → $(\text{CH}_2{=}\text{CH})_2CuLi$ → new C–C bond

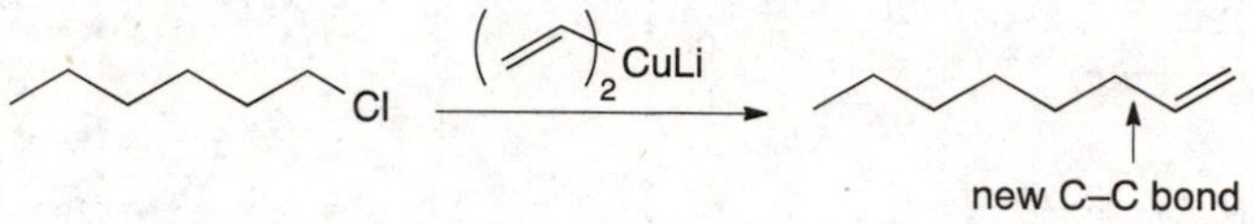

b. Br → $(CH_3)_2CuLi$ → new C–C bond

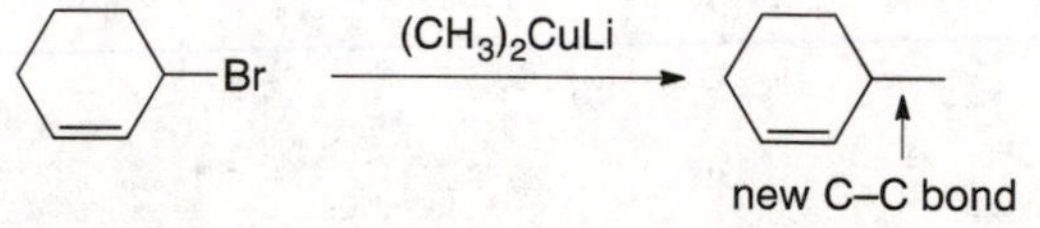

c. CH_3O, CH_3O, I → $(CH_3CH_2CH_2CH_2)_2CuLi$ → CH_3O, CH_3O, new C–C bond

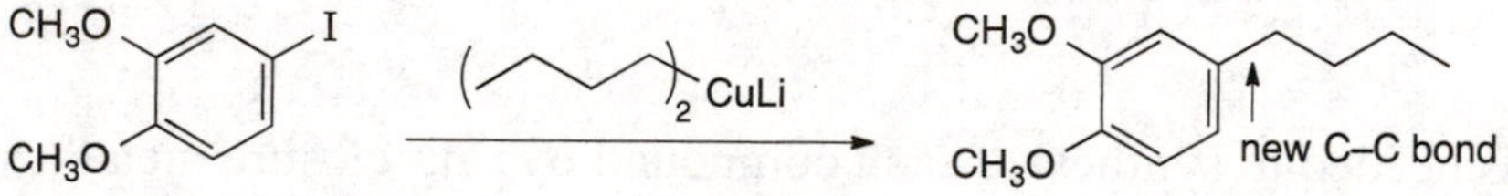

d. Br → $((CH_3)_2CHCH_2)_2CuLi$ → new C–C bond

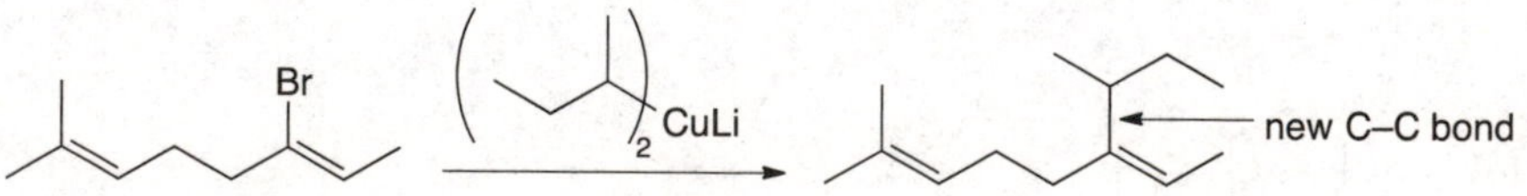

26.2

I, OH → A = $(CH_3CH_2)_2CuLi$ → OH → several steps → I, OH → B = $(CH_3)_2CuLi$ → OH → several steps → O, CO_2CH_3

C_{18} juvenile hormone

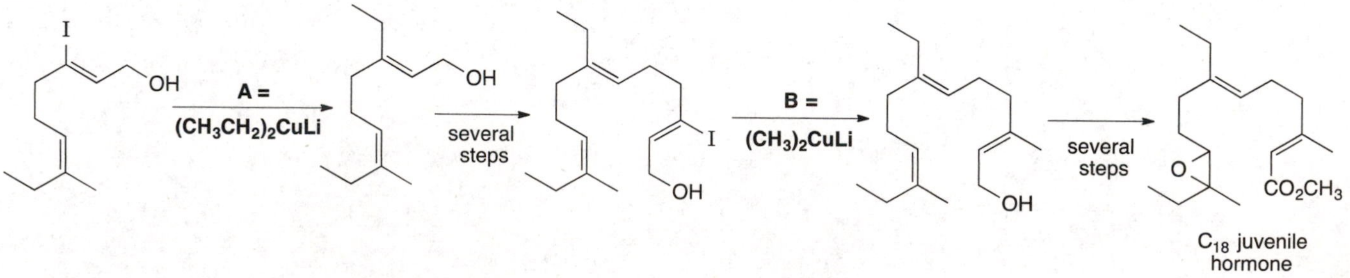

26.3

a. 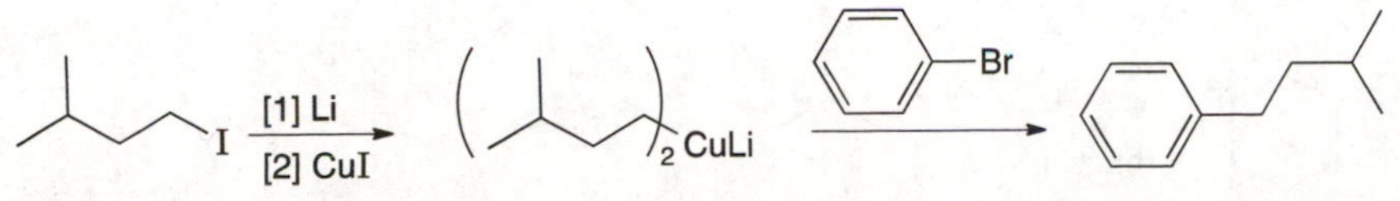 (1-iodo-3-methylbutane) $\xrightarrow[\text{[2] CuI}]{\text{[1] Li}}$ $(\text{R})_2CuLi$ $\xrightarrow{\text{PhBr}}$ product

or

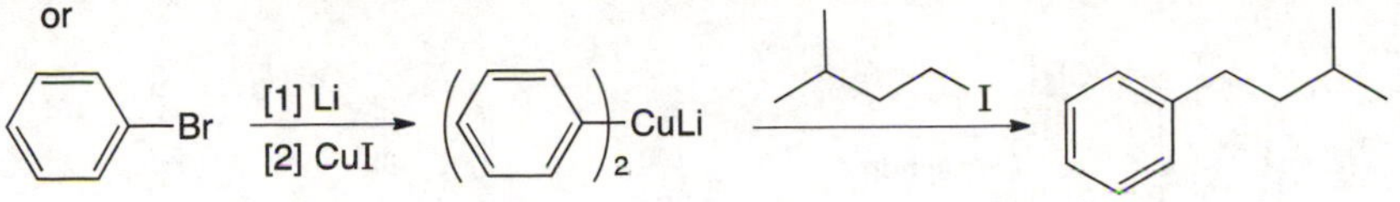

PhBr $\xrightarrow[\text{[2] CuI}]{\text{[1] Li}}$ $(Ph)_2CuLi$ $\xrightarrow{\text{R–I}}$ product

b. 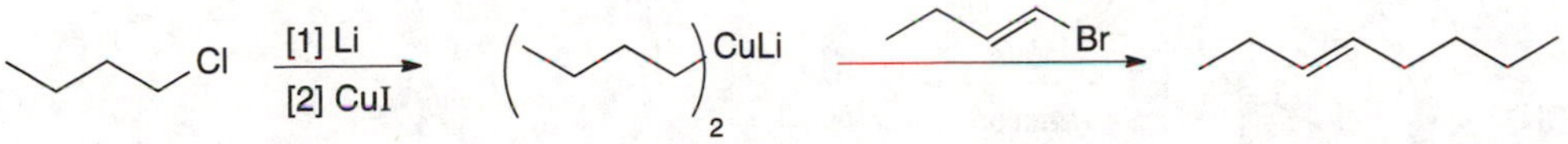 Cl $\xrightarrow[\text{[2] CuI}]{\text{[1] Li}}$ $(\ldots)_2CuLi$ $\xrightarrow{\text{Br}}$ product

c. 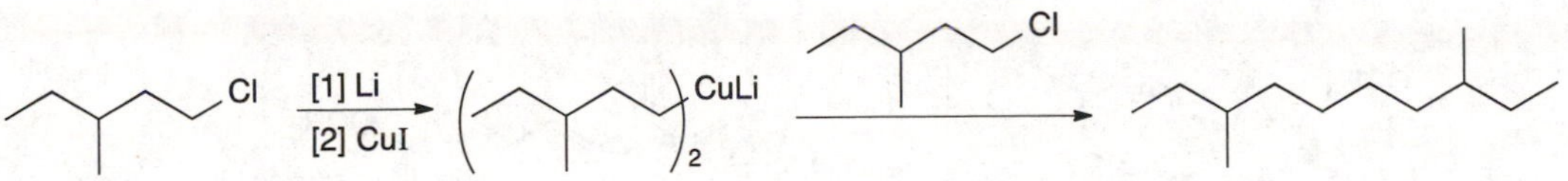 Cl $\xrightarrow[\text{[2] CuI}]{\text{[1] Li}}$ $(\ldots)_2CuLi$ $\xrightarrow{\text{Cl}}$ product

26.4

a. 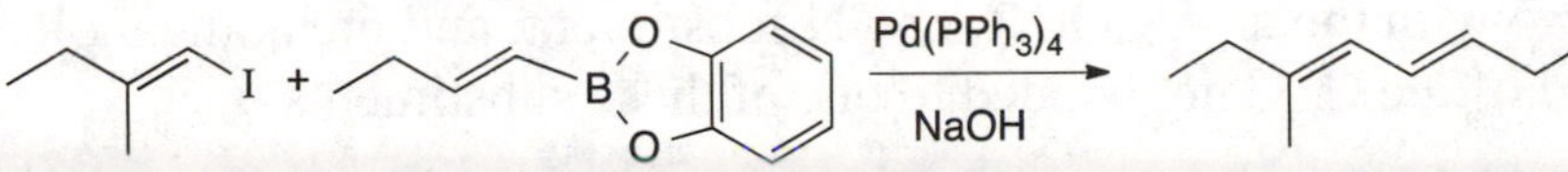 I + B(catechol) $\xrightarrow[\text{NaOH}]{Pd(PPh_3)_4}$ product

b. $\text{Ph–}B(OCH_3)_2$ + Br $\xrightarrow[\text{NaOH}]{Pd(PPh_3)_4}$ product

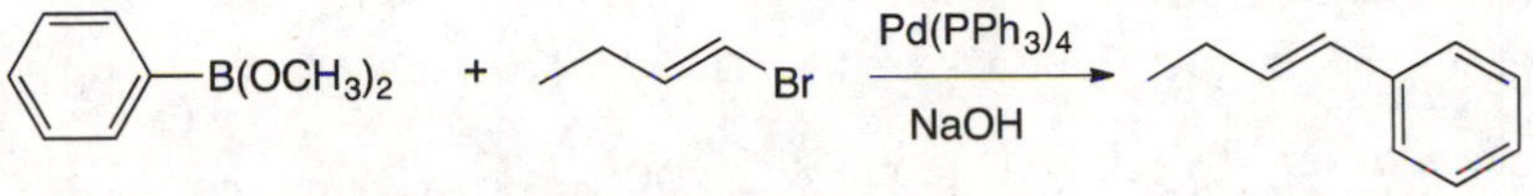

c. 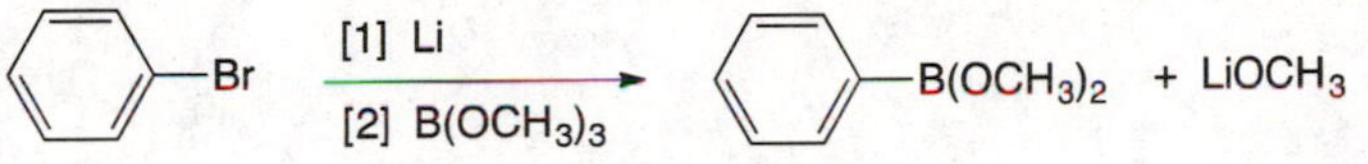 PhBr $\xrightarrow[\text{[2] }B(OCH_3)_3]{\text{[1] Li}}$ $\text{Ph–}B(OCH_3)_2$ + $LiOCH_3$

d. 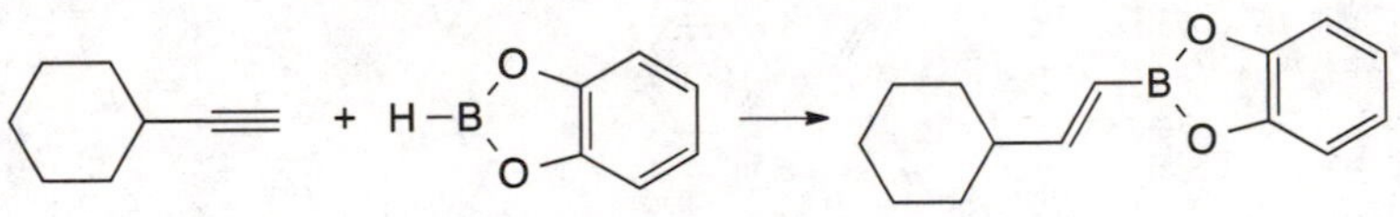 cyclohexylacetylene + H–B(catechol) ⟶ product

26.5 The Suzuki reaction forms a new carbon–carbon bond between a vinyl halide and an arylborane.

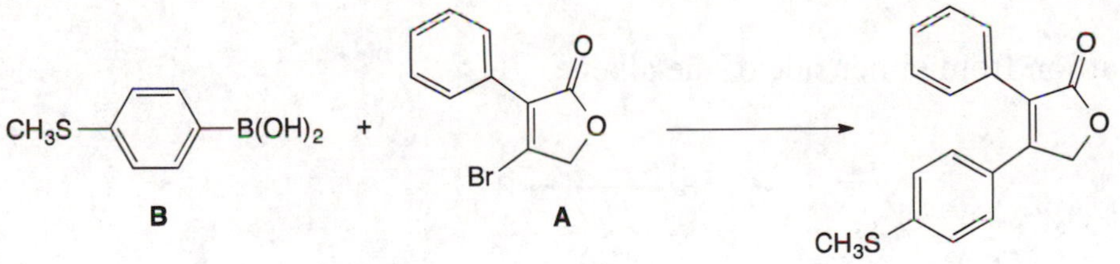

$CH_3S\text{–C}_6H_4\text{–}B(OH)_2$ (**B**) + Br (**A**) ⟶ product (CH_3S)

26.6

a. 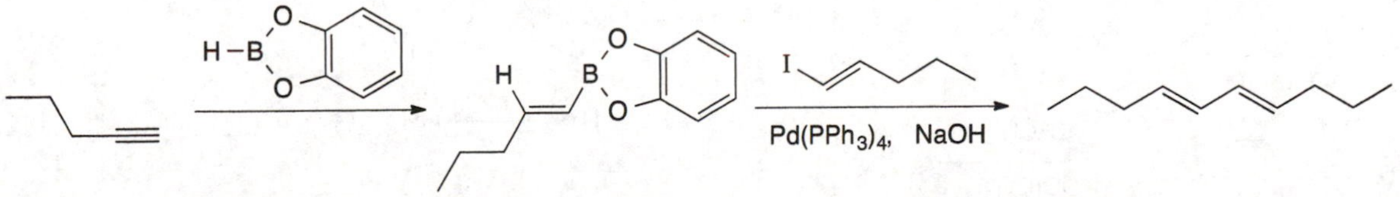 alkyne $\xrightarrow{\text{H–B(catechol)}}$ H, B $\xrightarrow[Pd(PPh_3)_4,\ NaOH]{\text{I}}$ product

b. PhI $\xrightarrow{\text{Li}}$ PhLi $\xrightarrow{B(OCH_3)_3}$ $\text{Ph–}B(OCH_3)_2$ + PhI $\xrightarrow[\text{NaOH}]{Pd(PPh_3)_4}$ biphenyl

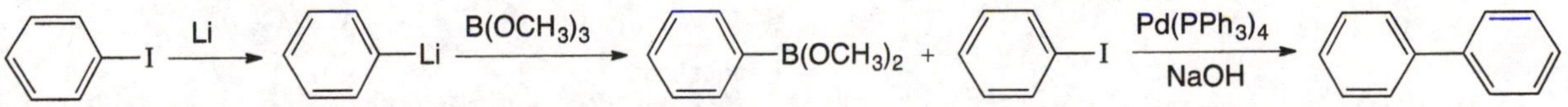

c. 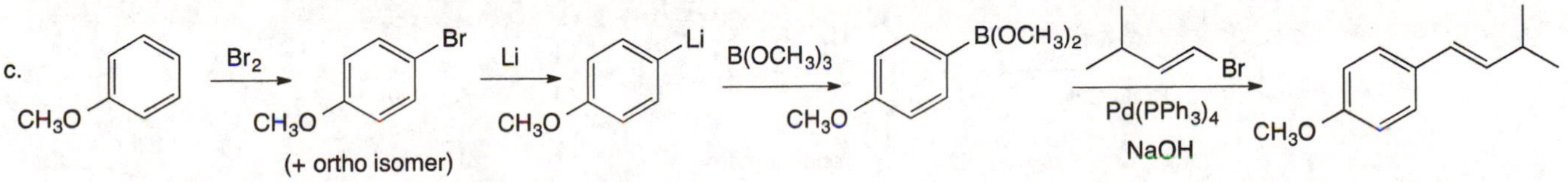 CH_3O–Ph $\xrightarrow{Br_2}$ CH_3O, Br (+ ortho isomer) $\xrightarrow{\text{Li}}$ CH_3O, Li $\xrightarrow{B(OCH_3)_3}$ CH_3O, $B(OCH_3)_2$ $\xrightarrow[\text{NaOH}]{Pd(PPh_3)_4,\ \text{Br}}$ CH_3O product

26.7

a. 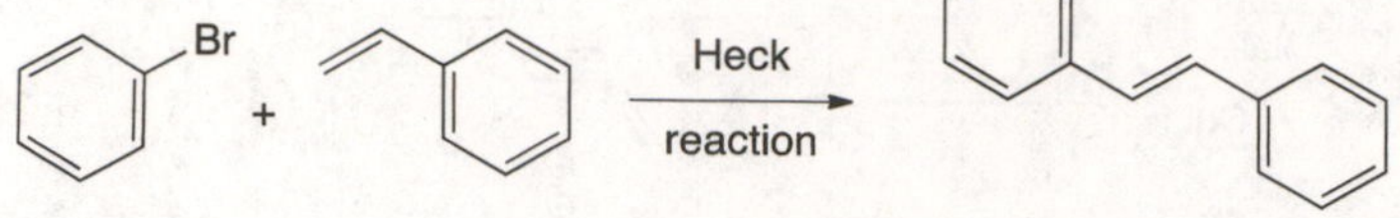 Bromobenzene + styrene —Heck reaction→ stilbene

b. 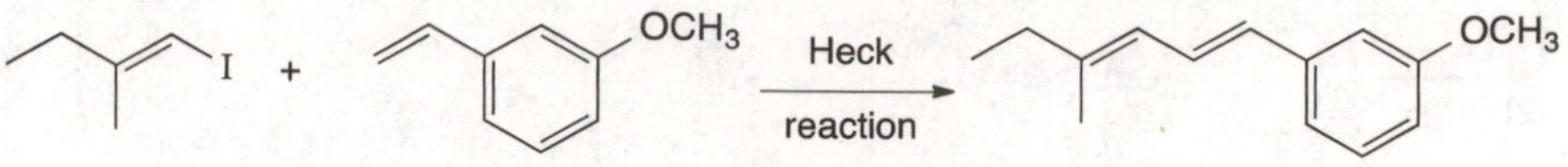 (vinyl iodide) + 3-vinylanisole (OCH_3) —Heck reaction→ diene product with OCH_3

c. 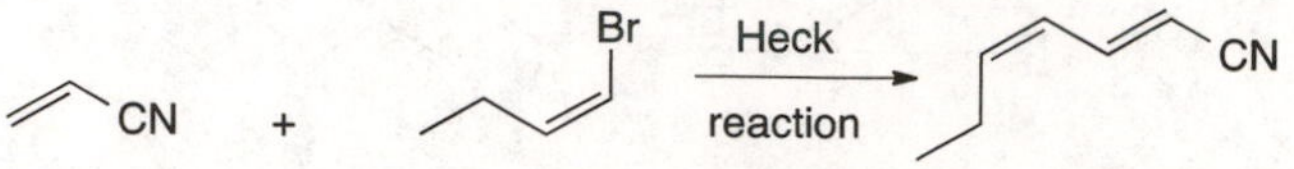 Acrylonitrile (CN) + vinyl bromide (Br) —Heck reaction→ diene nitrile (CN)

d. 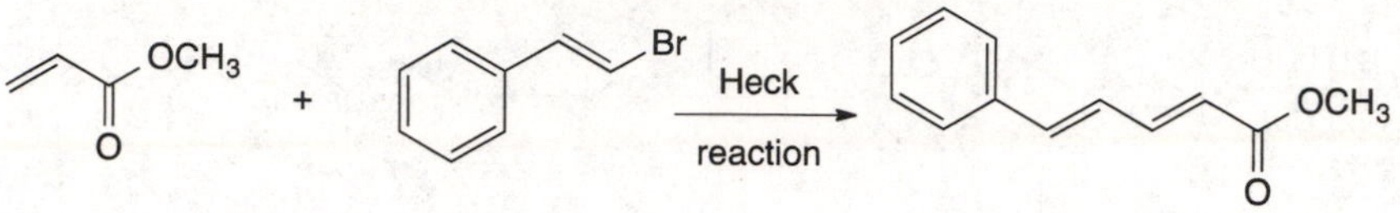 Methyl acrylate (OCH_3, O) + (2-bromovinyl)benzene (Br) —Heck reaction→ dienoate (OCH_3, O)

26.8 Locate the double bond with the aryl, COOR, or CN substituent, and break the molecule into two components at the end of the C=C not bonded to one of these substituents.

a. 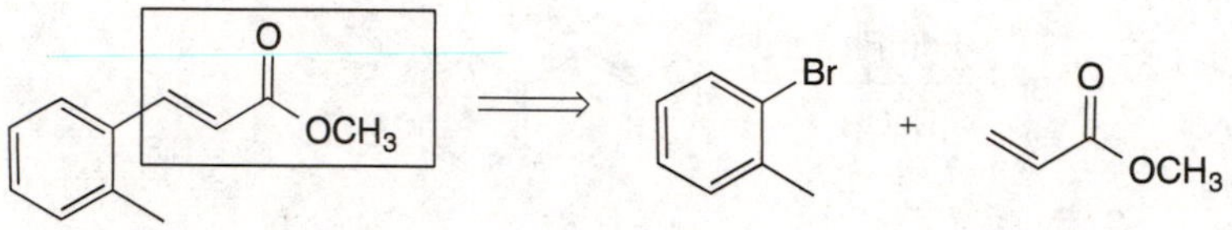 ⟹ Br (2-bromotoluene) + methyl acrylate (OCH_3)

b. 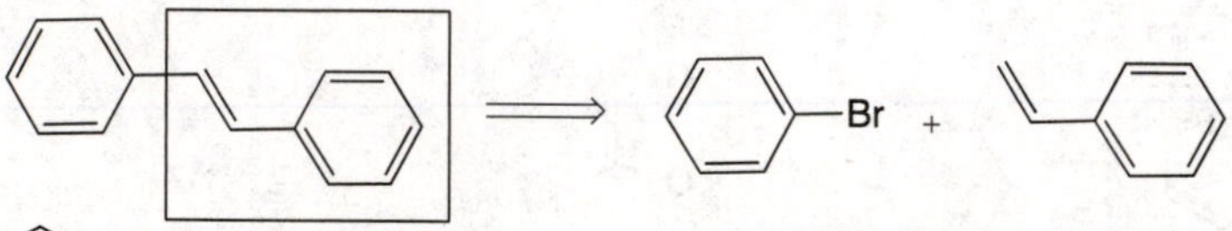 ⟹ Bromobenzene (Br) + styrene

c. 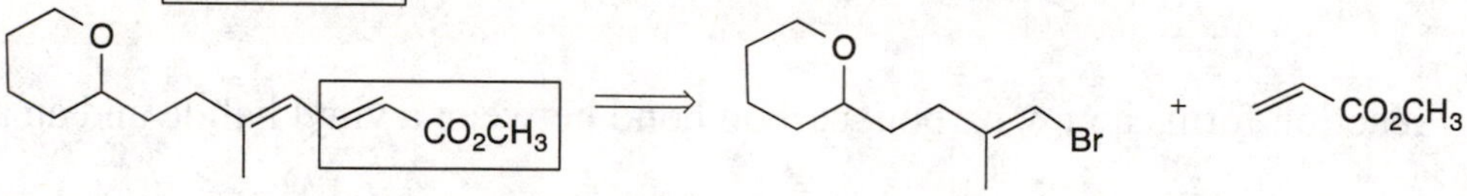 ⟹ vinyl bromide (Br) + CO_2CH_3 (methyl acrylate)

26.9 Add the carbene carbon from either side of the alkene.

a. 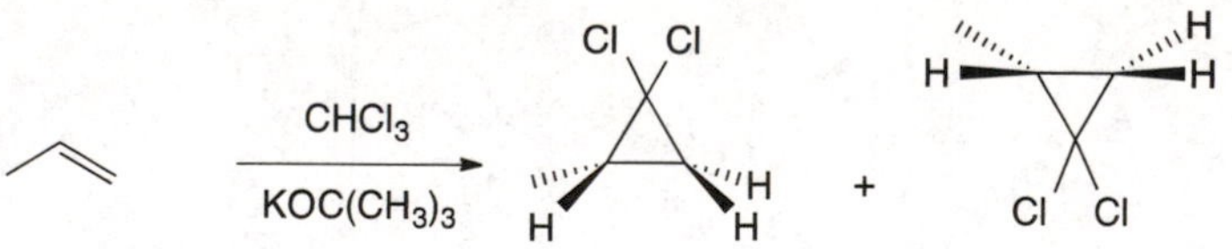 $\xrightarrow[KOC(CH_3)_3]{CHCl_3}$ (Cl, Cl, H, H, H) + (H, H, H, Cl, Cl)

enantiomers

b. 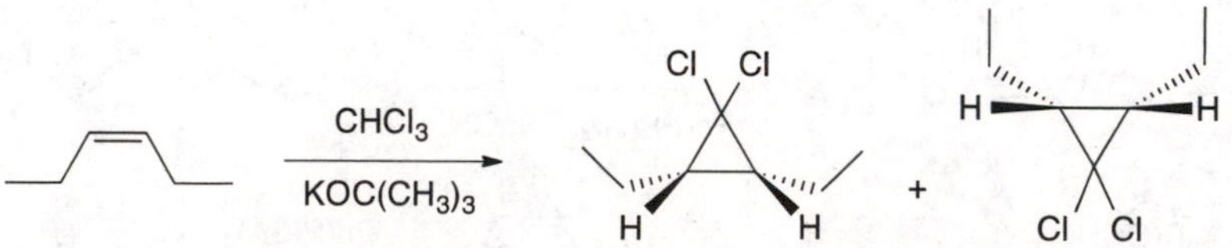 $\xrightarrow[KOC(CH_3)_3]{CHCl_3}$ (Cl, Cl, H, H) + (H, H, Cl, Cl)

identical

c. 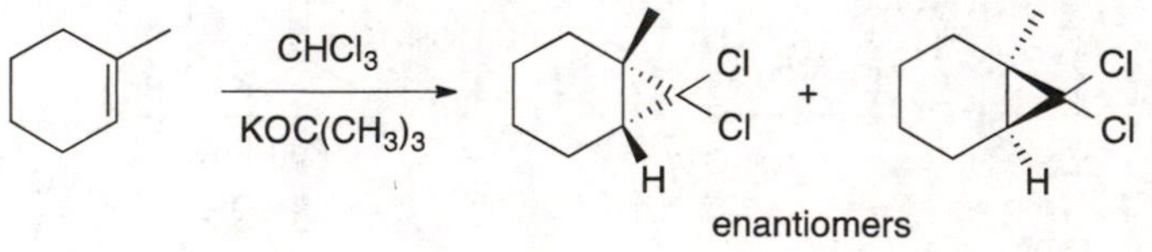 $\xrightarrow[KOC(CH_3)_3]{CHCl_3}$ (Cl, Cl, H) + (Cl, Cl, H)

enantiomers

26.10

a. $CHCl_3$ / $KOC(CH_3)_3$ → Cl, Cl

b. $CHBr_3$ / $KOC(CH_3)_3$ → Br, Br

c. Br, Br (from b.) $LiCu(CH_3)_2$ →

26.11

a. CH_2I_2 / Zn(Cu) → + ZnI_2

b. CH_2I_2 / Zn(Cu) → H, H + ZnI_2

c. CH_2I_2 / Zn(Cu) → + ZnI_2

26.12 The relative position of substituents in the reactant is retained in the product.

trans-hex-3-ene (H, H) CH_2I_2 / Zn(Cu) → (H H, H, H) + (H, H, H H)

two enantiomers of *trans*-1,2-diethylcyclopropane

26.13

a. Grubbs catalyst → (*E* and *Z*)

b. OCH_3 Grubbs catalyst → OCH_3, OCH_3 (*E* and *Z*)

c. Grubbs catalyst →

($CH_2{=}CH_2$ is also formed in each reaction.)

26.14

cis-pent-2-ene → + + +

There are four products formed in this reaction including stereoisomers, and therefore, it is not a practical method to synthesize 1,2-disubstituted alkenes.

26.15

a. Ph O (O) → Ph O (O)

b. CO_2CH_3, CO_2CH_3 → CO_2CH_3, CO_2CH_3

26.16

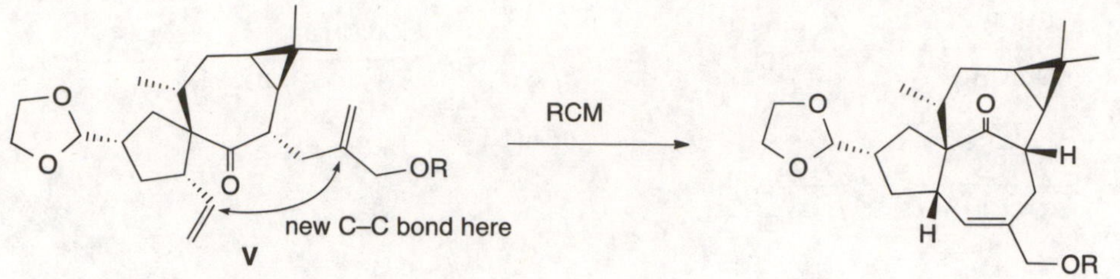

RCM

OR

new C–C bond here

V

H

H

OR

26.17 Cleave the C=C bond in the product, and then bond each carbon of the original alkene to a CH_2 group using a double bond.

a.

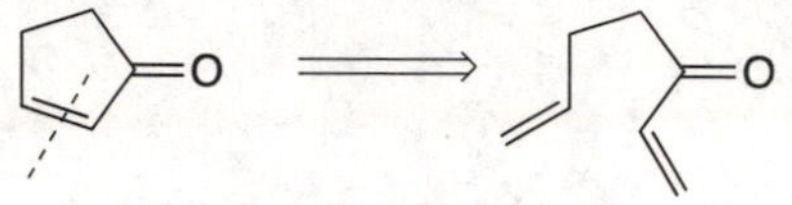

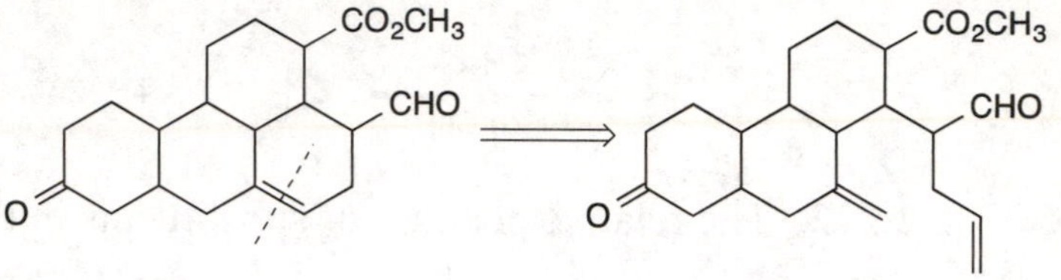

b.

CH_3O

OH

c.

CO_2CH_3

CHO

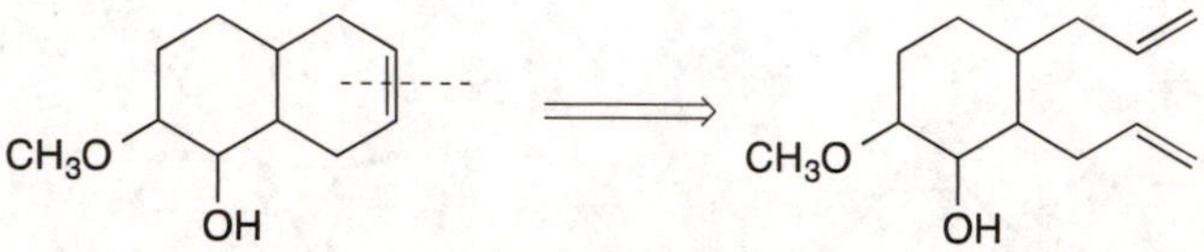

26.18 Inversion of configuration occurs with the substitution of the methyl group for the tosylate.

a.

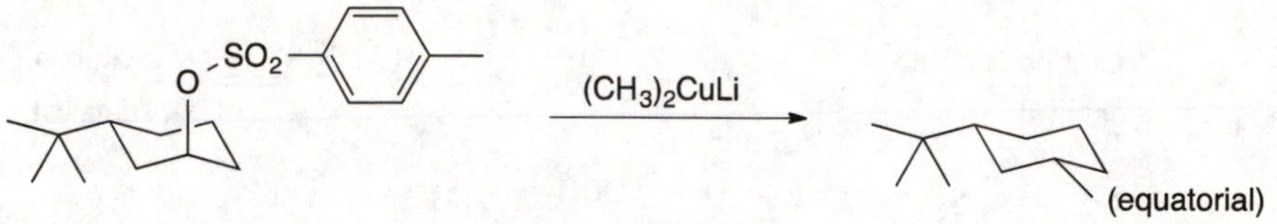

SO_2

O

$(CH_3)_2CuLi$

(equatorial)

b.

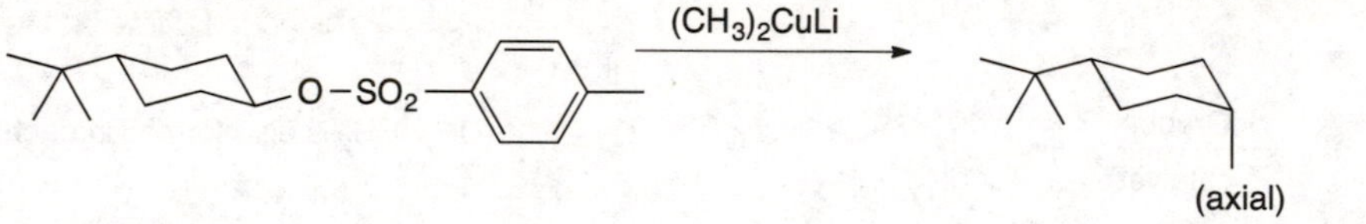

$O-SO_2$

$(CH_3)_2CuLi$

(axial)

26.19

a.

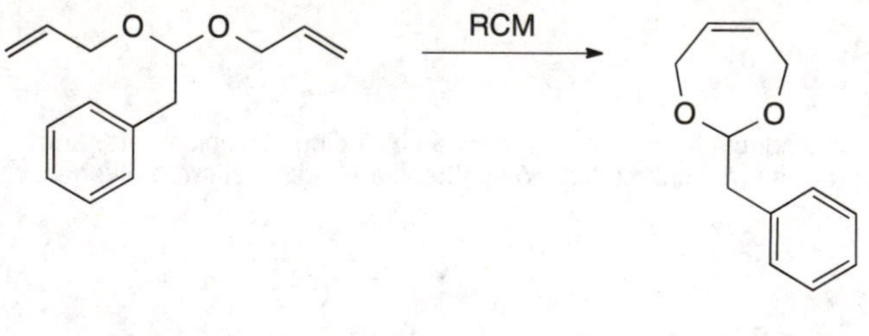

RCM

b.

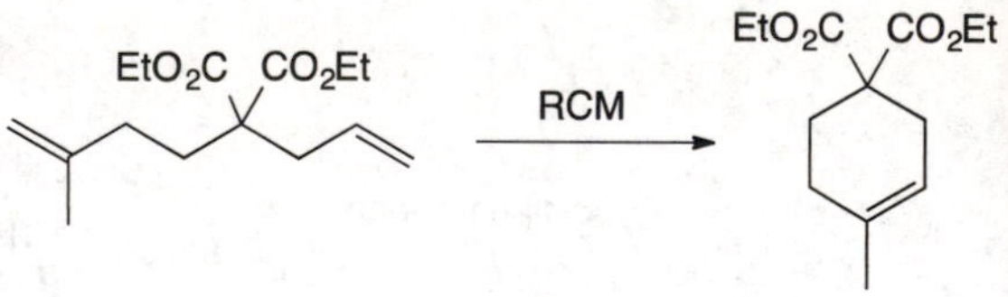

EtO_2C CO_2Et

RCM

EtO_2C CO_2Et

26.20

a. 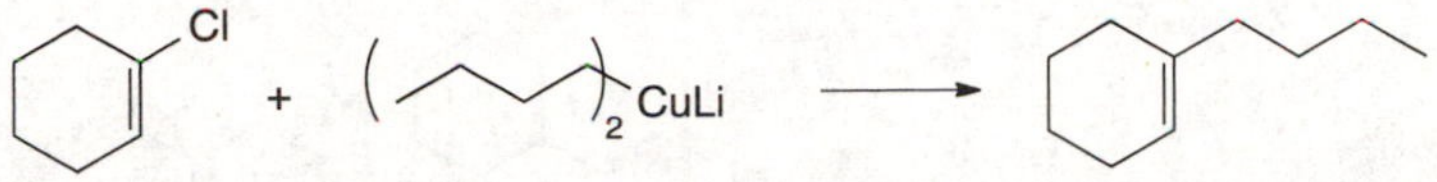 Cl + ( )$_2$CuLi →

b. 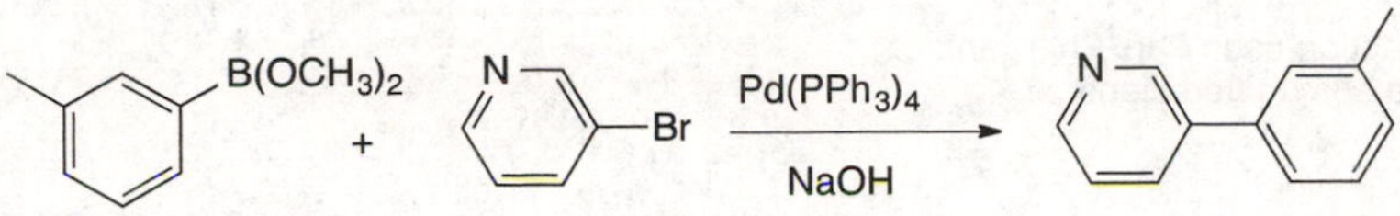 $B(OCH_3)_2$ + N Br —$Pd(PPh_3)_4$, NaOH→ N

c. 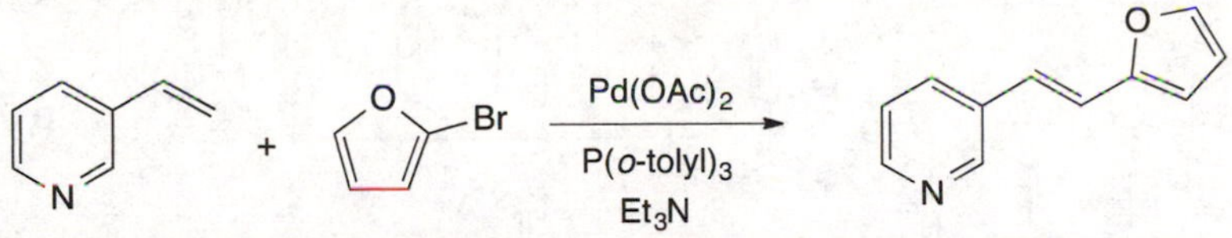 N + O Br —$Pd(OAc)_2$, $P(o\text{-tolyl})_3$, Et_3N→ O N

d. 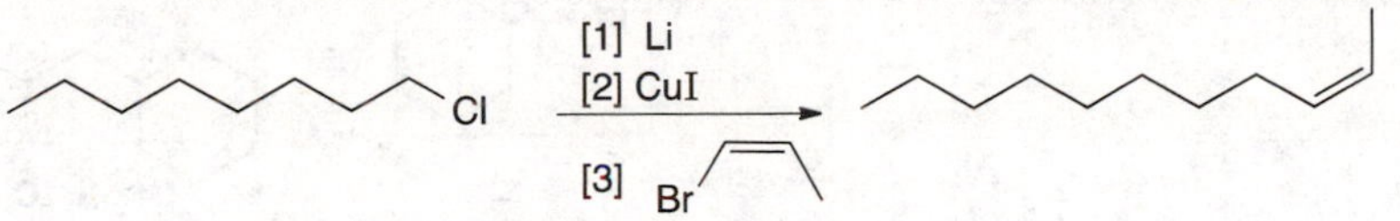 Cl —[1] Li, [2] CuI, [3] Br→

e. 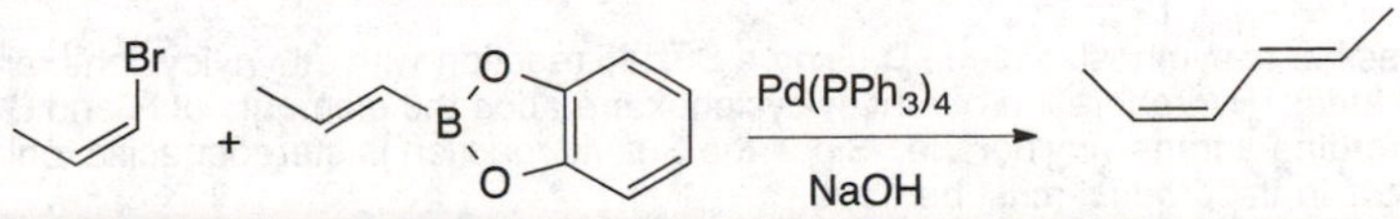 Br + B O O —$Pd(PPh_3)_4$, NaOH→

f. 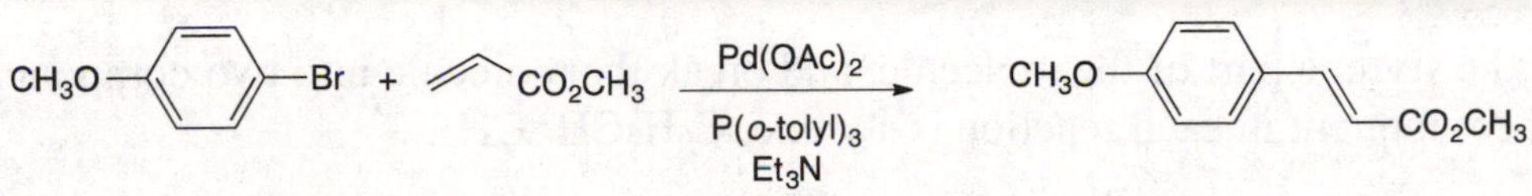 CH_3O—Br + CO_2CH_3 —$Pd(OAc)_2$, $P(o\text{-tolyl})_3$, Et_3N→ CH_3O CO_2CH_3

g. 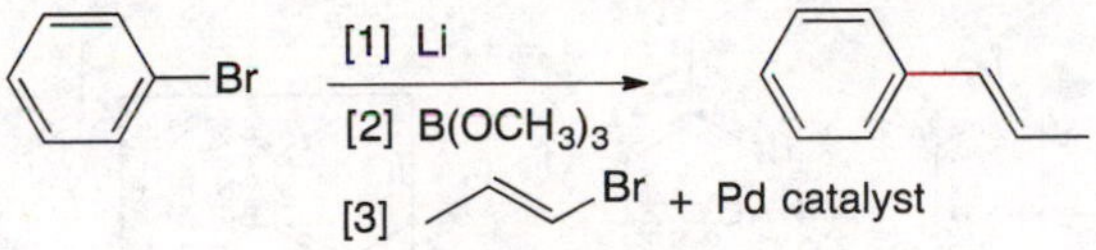 Br —[1] Li, [2] $B(OCH_3)_3$, [3] Br + Pd catalyst→

h. 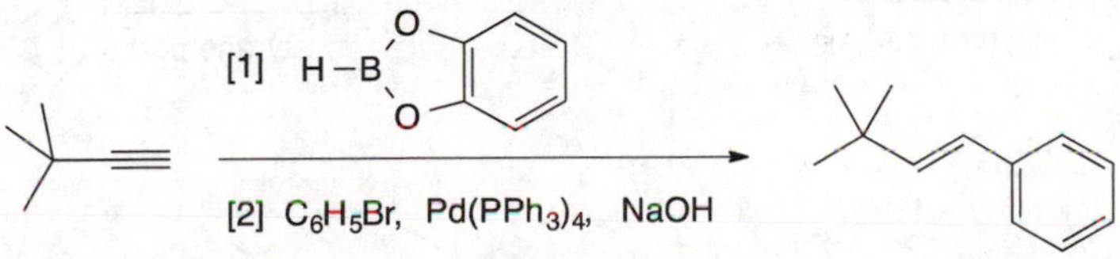 —[1] H–B(O)(O), [2] C_6H_5Br, $Pd(PPh_3)_4$, NaOH→

26.21

a. 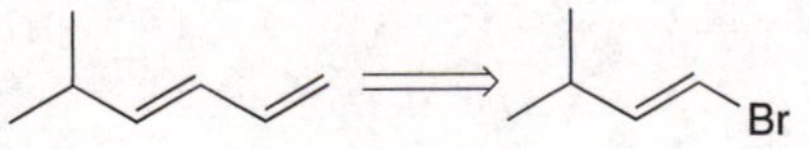 ⟹ Br

b. 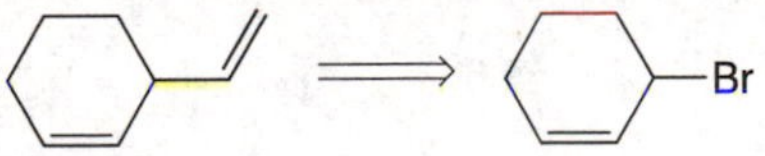 ⟹ Br

c. 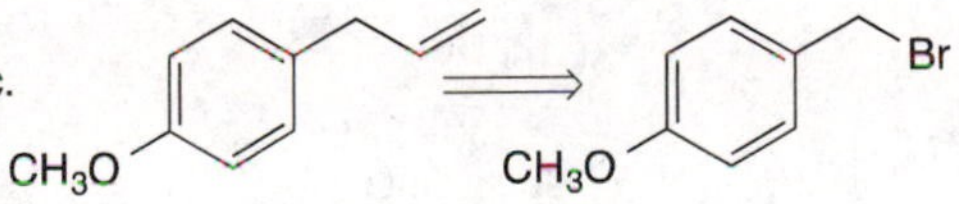 CH_3O ⟹ CH_3O Br

26.22

Each coupling reaction uses $Pd(PPh_3)_4$ and NaOH to form the conjugated diene.

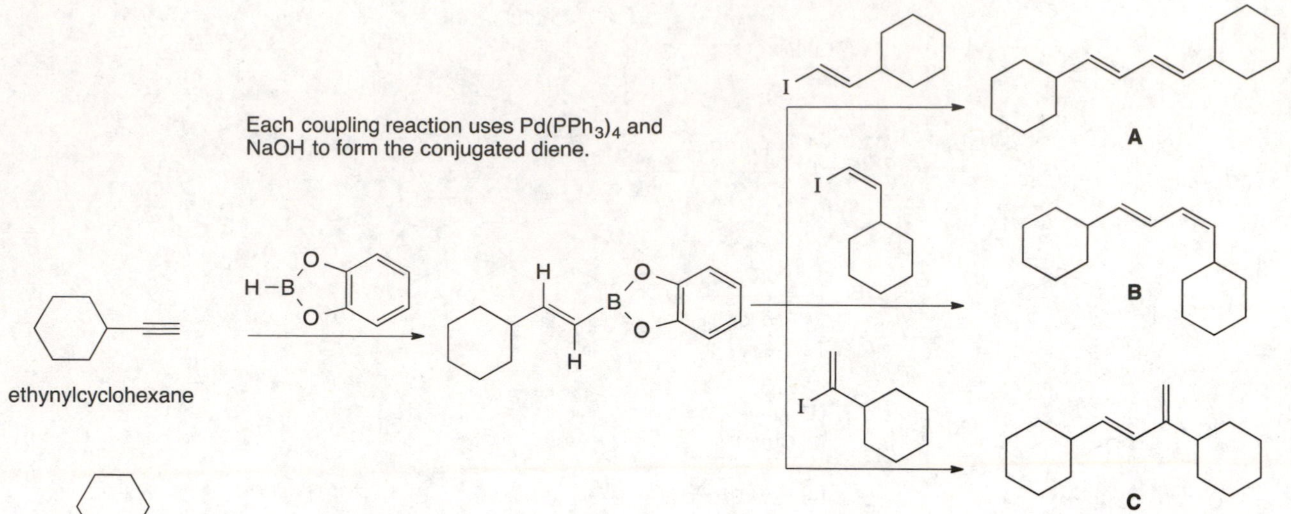

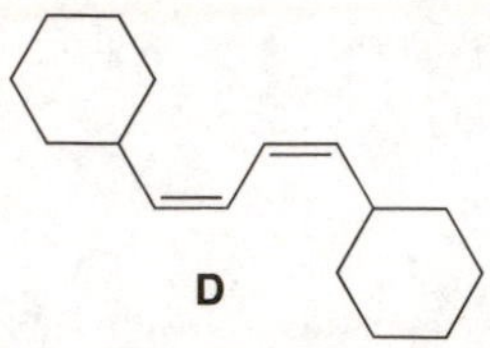

ethynylcyclohexane

H–B

A

B

C

D

It is not possible to synthesize diene **D** using a Suzuki reaction with ethynylcyclohexane as starting material. Hydroboration of ethynylcyclohexane adds the elements of H and B in a syn fashion, affording a trans vinylborane. Since the Suzuki reaction is stereospecific, one of the double bonds in the product must be trans.

26.23 Locate the styrene part of the molecule, and break the molecule into two components. The second component in each reaction is styrene, $C_6H_5CH{=}CH_2$.

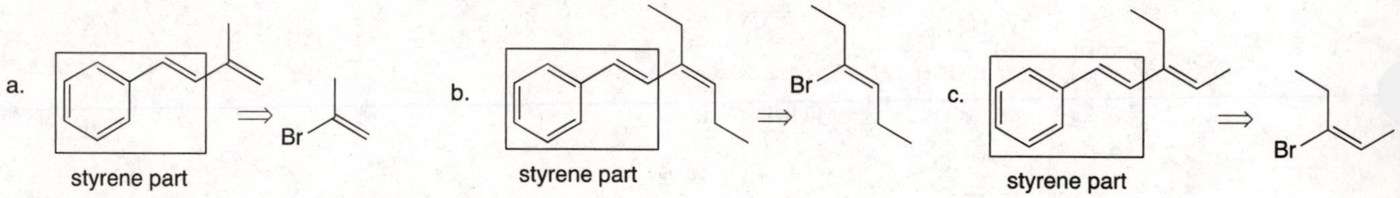

a. styrene part ⇒ Br

b. styrene part ⇒ Br

c. styrene part ⇒ Br

26.24

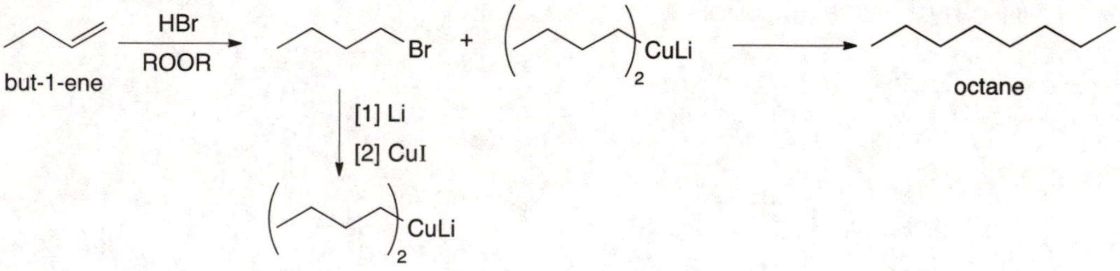

but-1-ene —HBr, ROOR→ Br + $(\ldots)_2$CuLi → octane

[1] Li
[2] CuI

$(\ldots)_2$CuLi

26.25

HO, Cl, N, N, Br — **A**

\+

Ph, Ph, Ph, N–N, N, N, B(OH)$_2$ — **B**

Suzuki reaction →

HO, Cl, N, N, Ph Ph, Ph, N–N, N, N

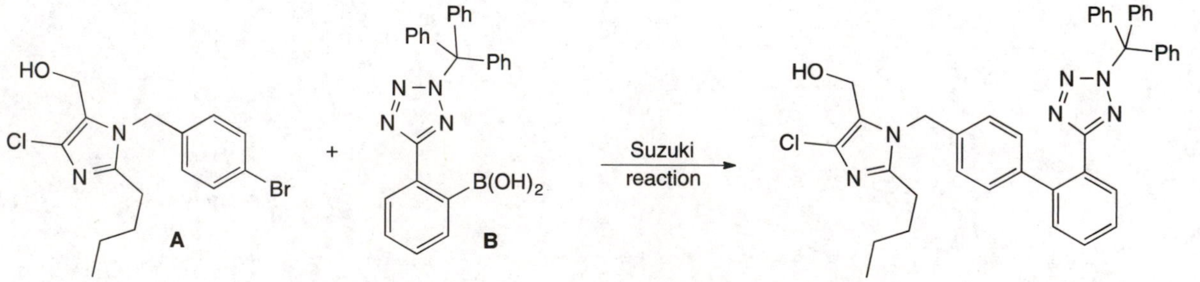

26.26

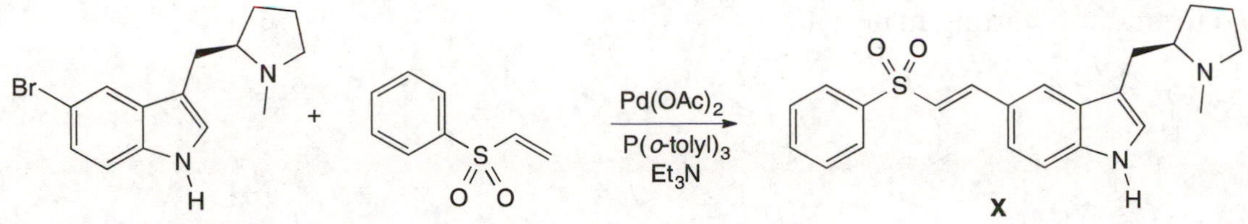

Br, N, H, +, S, O O, $Pd(OAc)_2$, $P(o\text{-tolyl})_3$, Et_3N, O O, S, N, N, H, **X**

26.27 Add the carbene carbon from either side of the alkene.

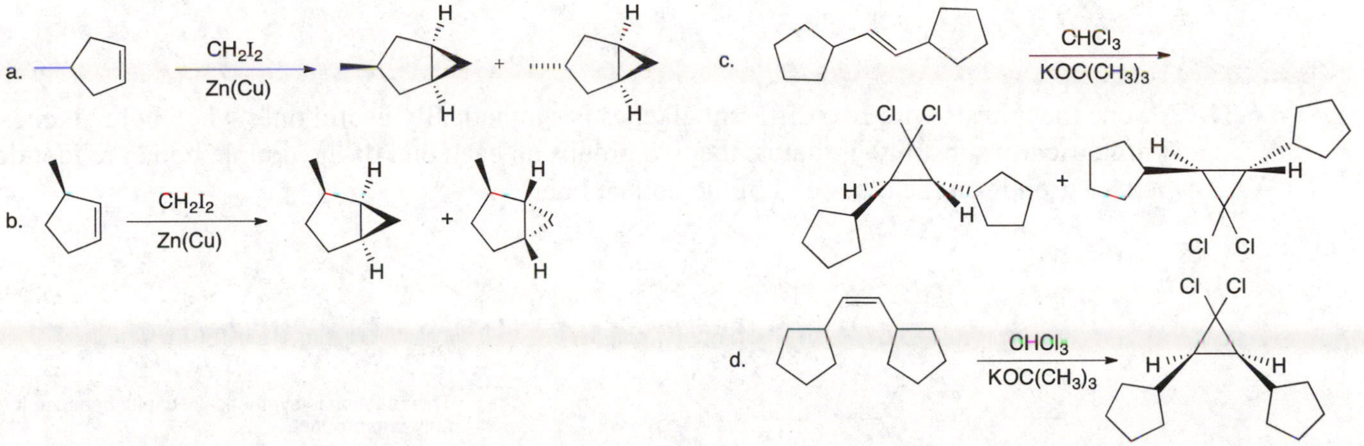

a. CH_2I_2 / Zn(Cu) → H, H + H, H

b. CH_2I_2 / Zn(Cu) → H, H + H, H

c. $CHCl_3$ / $KOC(CH_3)_3$ → Cl Cl, H, H + H, H, Cl Cl

d. $CHCl_3$ / $KOC(CH_3)_3$ → Cl Cl, H, H

26.28 The new three-membered ring has a stereogenic center on the C bonded to the phenyl group, so the phenyl group can be oriented in two different ways to afford two stereoisomers. These products are diastereomers of each other.

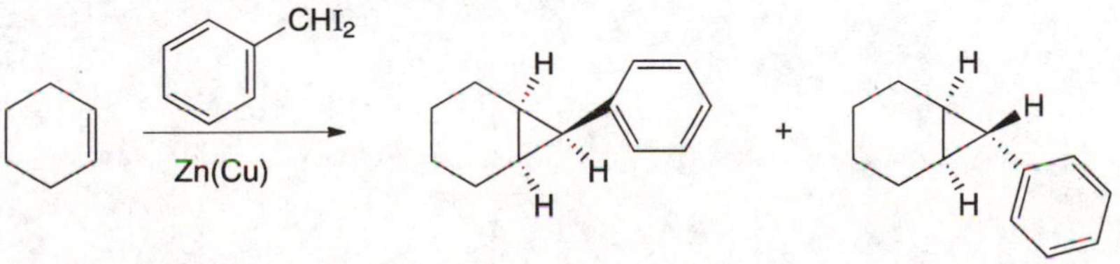

CHI_2, Zn(Cu) → H, H, H + H, H, H

26.29 High-dilution conditions favor intramolecular metathesis.

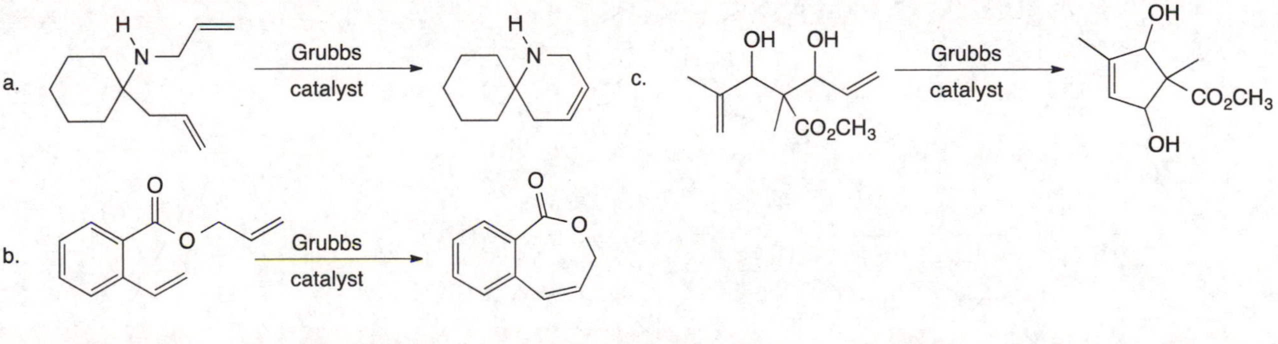

a. H, N, Grubbs catalyst → H, N

b. O, O, Grubbs catalyst → O, O

c. OH, OH, CO_2CH_3, Grubbs catalyst → OH, CO_2CH_3, OH

26.30 Retrosynthetically break the double bond in the cyclic compound and add a new =CH2 at each end to find the starting material.

a.

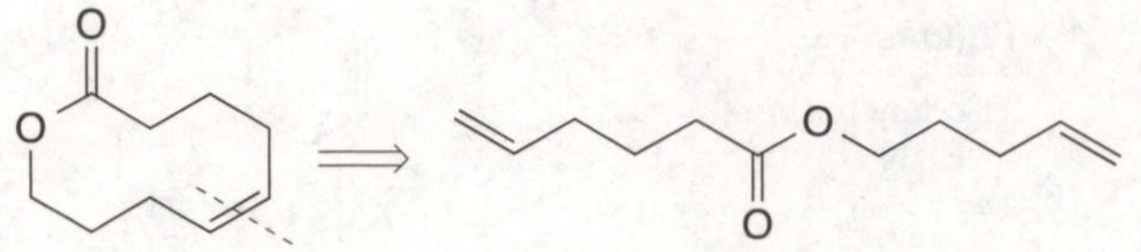

b. 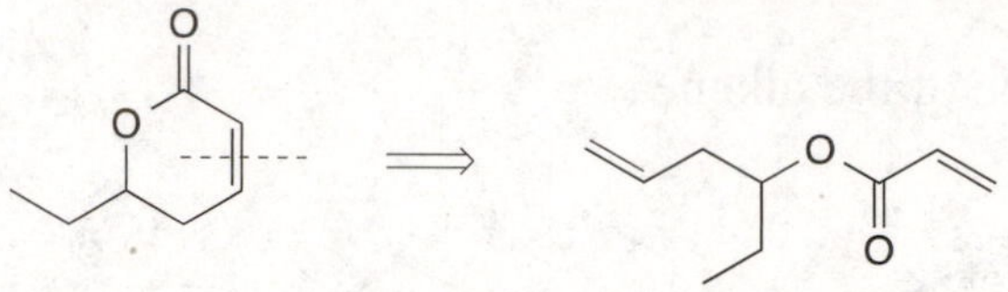 c.

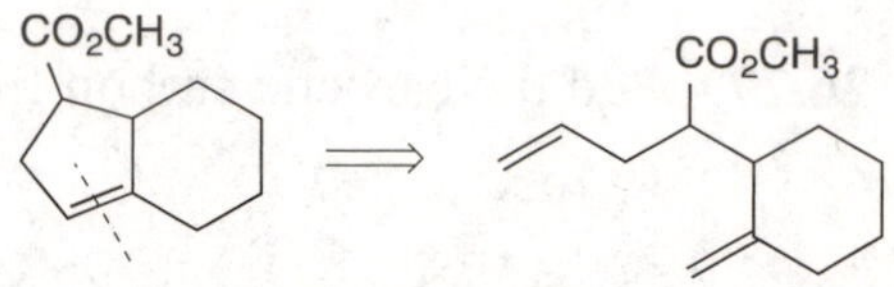

CO_2CH_3 CO_2CH_3

26.31 Alkene metathesis with two different alkenes is synthetically useful only when both alkenes are symmetrically substituted; that is, the two groups on each end of the double bond are identical to the two groups on the other end of the double bond.

a. + → (Z + E) + (Z + E) + (Z + E) +

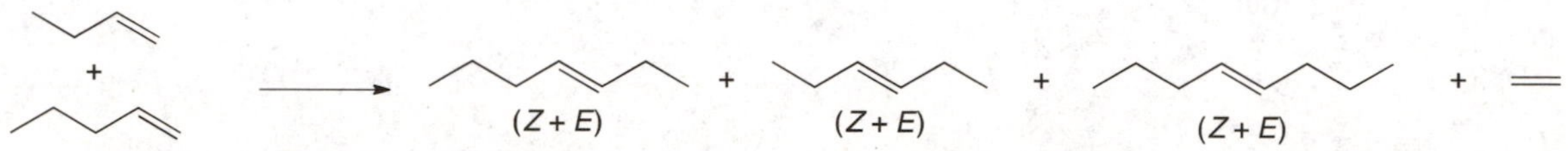

b. 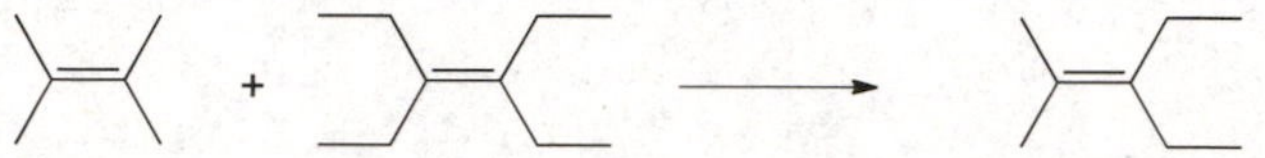 + → This reaction is synthetically useful because it yields only one product.

c. + → + + + + + + + (Z + E) (Z + E)

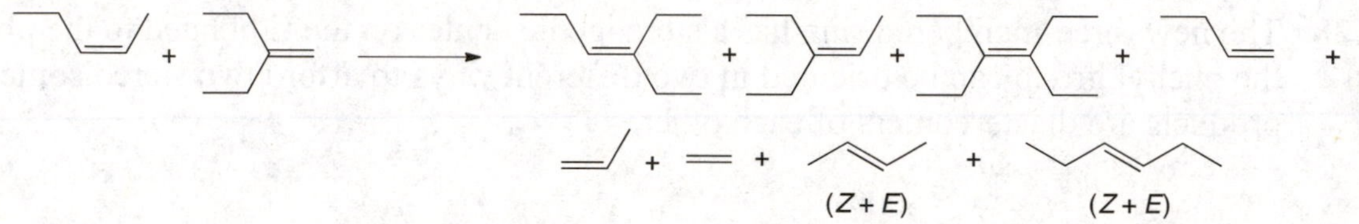

26.32

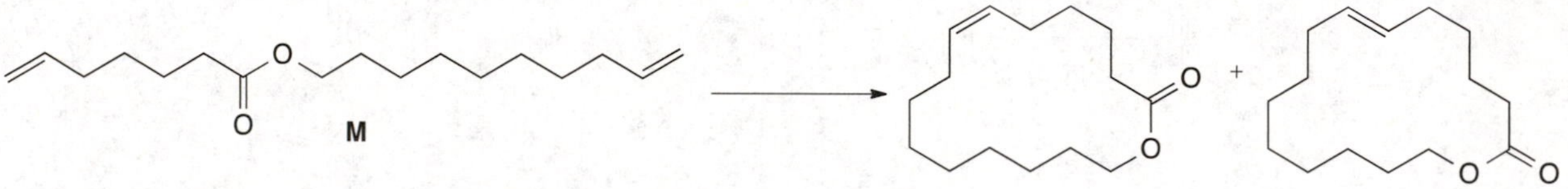

M

26.33 All double bonds can have either the *E* or *Z* configuration.

a.

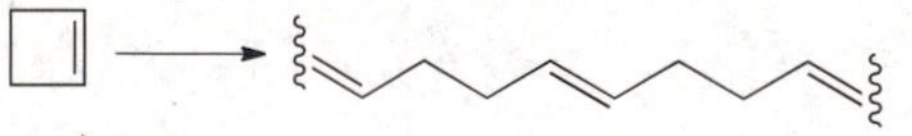

b.

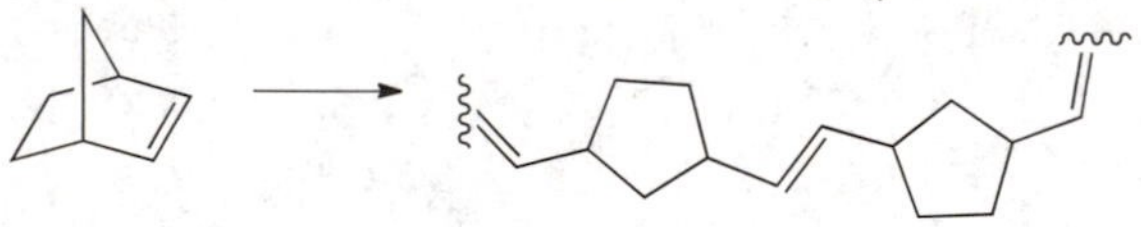

c.

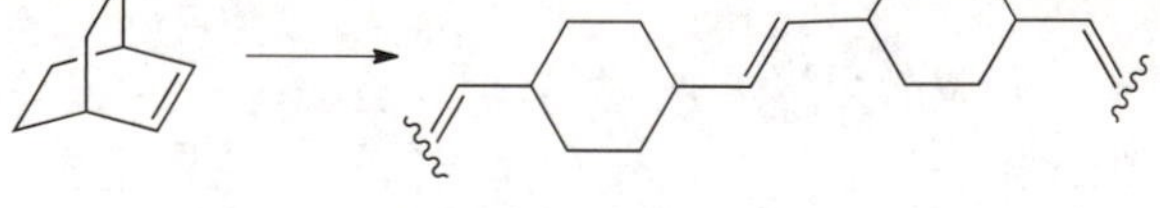

26.34

a. [1] $CHBr_3$ / $KOC(CH_3)_3$ → Br Br → [2] $(CH_3)_2CuLi$ →

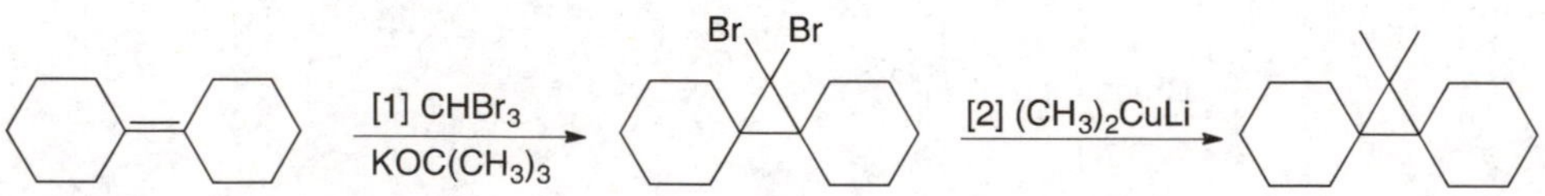

b. Br COOH + CO_2CH_3 → $Pd(OAc)_2$ / $P(o\text{-tolyl})_3$ / Et_3N → CH_3O_2C COOH

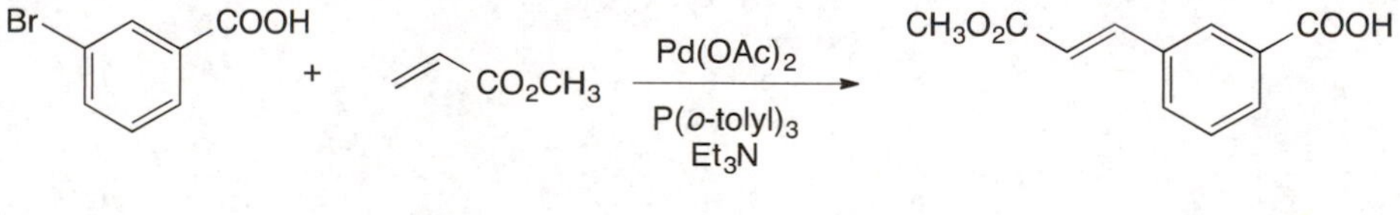

c. 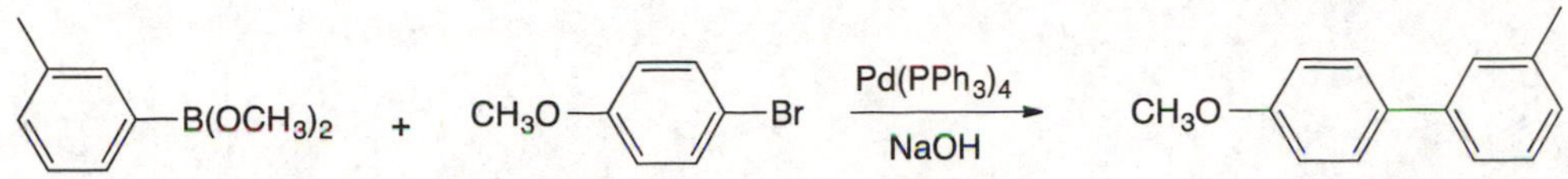 $CH_3C_6H_4B(OCH_3)_2$ + CH_3O–C_6H_4–Br $\xrightarrow[\text{NaOH}]{\text{Pd(PPh}_3)_4}$ CH_3O–C_6H_4–$C_6H_4CH_3$

d. 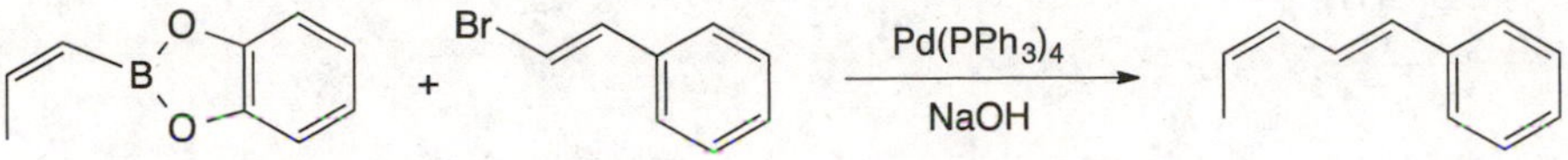 B (catecholborane) + Br–CH=CH–Ph $\xrightarrow[\text{NaOH}]{\text{Pd(PPh}_3)_4}$

e. 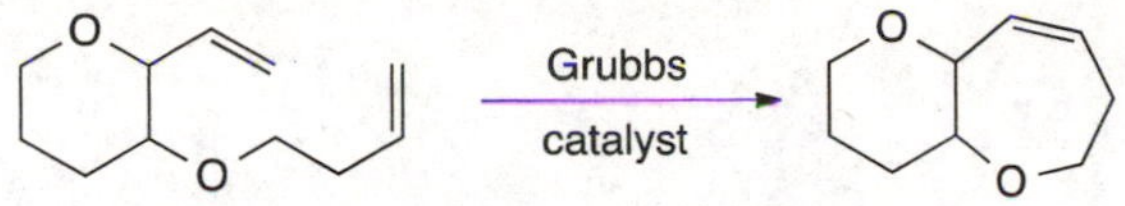 Grubbs catalyst

f. 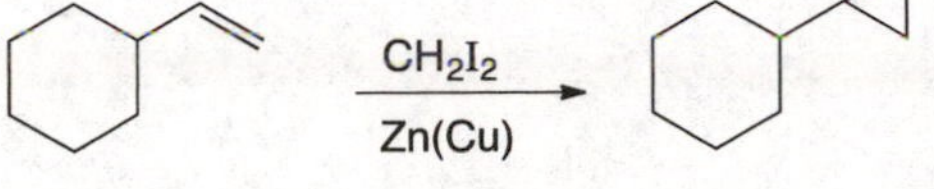 $\xrightarrow[\text{Zn(Cu)}]{CH_2I_2}$

g. 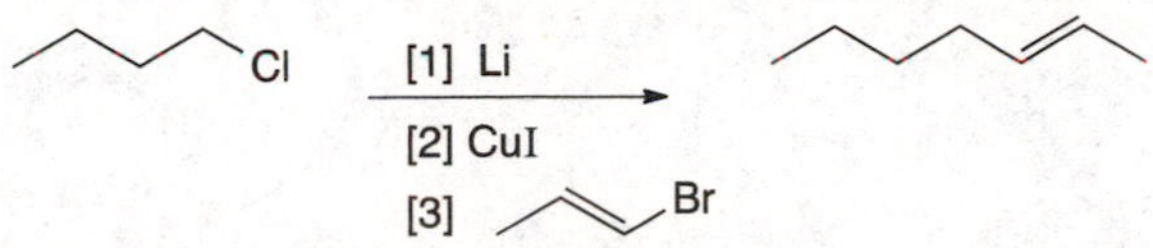 Cl $\xrightarrow{\text{[1] Li; [2] CuI; [3] Br}}$

h. [1] H–B (catecholborane) [2] Br, Pd(PPh₃)₄, NaOH → H, H

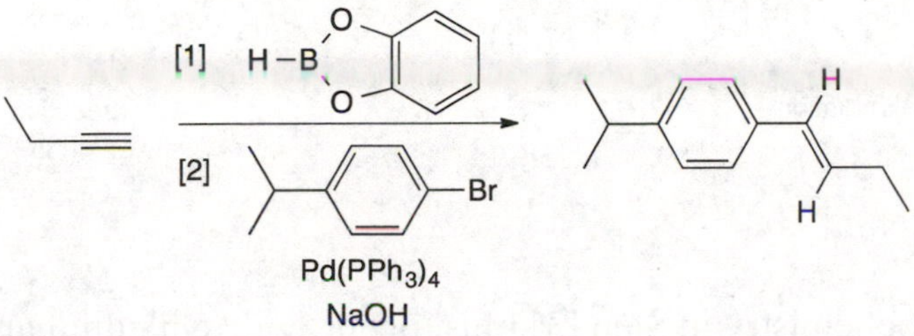

26.35

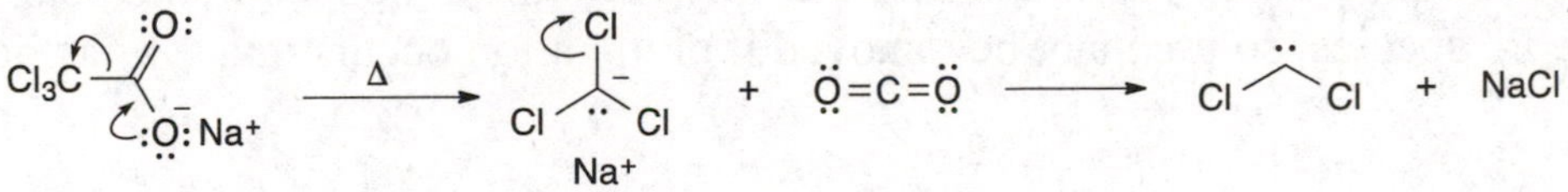

Cl_3C–C(=:O:)–:Ö:⁻ Na⁺ $\xrightarrow{\Delta}$ $^-CCl_3$ Na⁺ + Ö=C=Ö ⟶ Cl–C–Cl (dichlorocarbene) + NaCl

26.36 This reaction follows the Simmons–Smith reaction mechanism illustrated in Mechanism 26.5.

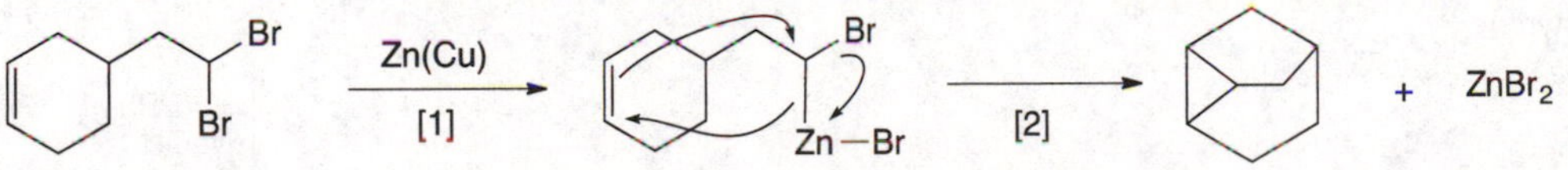

Br, Br $\xrightarrow[\text{[1]}]{\text{Zn(Cu)}}$ Br, Zn–Br $\xrightarrow{\text{[2]}}$ + $ZnBr_2$

26.37

:O: ÖCH₃ + Ph₂S⁺ (a sulfur ylide) → 1,4-addition of the sulfur ylide to the β carbon → Ph₂S⁺, :O:, ÖCH₃ (a resonance-stabilized enolate) ⟶ OCH₃, O (methyl *trans*-chrysanthemate) + Ph₂S

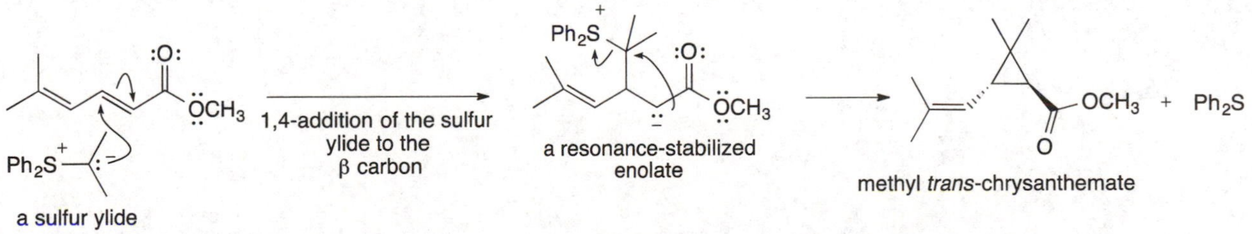

26.38

A

diazo compound

CO_2CH_3

CO_2CH_3

$+ N_2$

CO_2CH_3

B

CO_2CH_3

26.39

a.

[1]

$Pd(PAr_3)_2$
oxidative addition

PAr_3
$Ar_3P-Pd-I$

[2]
syn addition

PAr_3
$Ar_3P-Pd-I$

[3]
syn elimination

E

PAr_3
$Ar_3P-Pd-I$
H

[4]
reductive elimination

$Pd(PAr_3)_2 + HI$

The H cis to Pd is eliminated.

This H is trans to Pd.

b. This suggests that the stereochemistry in Step [3] must occur with syn elimination of H and Pd to form **E.** Product **F** cannot form because the only H on the C bonded to the benzene ring is trans to the Pd species, so it cannot be removed if elimination occurs in a syn fashion.

26.40

(Z)-2-bromostyrene

$NaNH_2$
(– HBr)

$Pd(PPh_3)_4$
NaOH

A

26.41

Br_2
$FeBr_3$

$Pd(PPh_3)_4$, NaOH

26.42

HO~~~~OH $\xrightarrow{SOCl_2}$ HO~~~~Cl

$\downarrow$ TBDMS–Cl, imidazole

HC≡CH $\xrightarrow{NaH}$ $HC\equiv C^-$ + TBDMSO~~~~Cl

$\downarrow$

≡~~~~OTBDMS $\xrightarrow{\text{catecholborane (BH)}}$ catechol–B~~~~OTBDMS

$\uparrow$ Bu_4NF

catechol–B~~~~OH

26.43

a. CH_3O–C6H4–CH=CH–C6H5 $\Longrightarrow$ CH_3O–C6H4–Br + styrene

Synthesize these two components, and then use a Heck reaction to synthesize the product.

benzene $\xrightarrow[H_2SO_4]{HNO_3}$ C6H5–NO_2 $\xrightarrow[Pd\text{-}C]{H_2}$ C6H5–NH_2 $\xrightarrow[HCl]{NaNO_2}$ C6H5–$N_2^+Cl^-$ $\xrightarrow{H_2O}$ C6H5–OH

C6H5–OH $\xrightarrow[\text{[2] } CH_3Br]{\text{[1] NaH}}$ CH_3O–C6H5 $\xrightarrow{Br_2}$ CH_3O–C6H4–Br (boxed)

(CH_3Br from CH_3OH via PBr_3)

(CH_3CH_2Cl from CH_3CH_2OH via $SOCl_2$)

benzene $\xrightarrow[AlCl_3]{CH_3CH_2Cl}$ ethylbenzene $\xrightarrow[h\nu]{Br_2}$ C6H5–CH(Br)CH3 $\xrightarrow{KOC(CH_3)_3}$ styrene (boxed)

CH_3O–C6H4–Br + styrene $\xrightarrow[\text{P(}o\text{-tolyl)}_3,\ Et_3N]{Pd(OAc)_2}$ CH_3O–C6H4–CH=CH–C6H5

b. CH_3O–C6H4–Br (from a.) $\xrightarrow[AlCl_3]{CH_3CH_2Cl}$ 4-bromo-2-ethylanisole (CH_3O, Br)

(CH_3CH_2Cl from CH_3CH_2OH via $SOCl_2$)

4-bromo-2-ethylanisole + $CH_2{=}CHCO_2Et$ $\xrightarrow[\text{P(}o\text{-tolyl)}_3,\ Et_3N]{Pd(OAc)_2}$ Ar–CH=CH–CO_2Et $\xrightarrow[Pd\text{-}C]{H_2}$ Ar–CH_2CH_2–CO_2Et

benzene $\xrightarrow[FeBr_3]{Br_2}$ C6H5–Br $\xrightarrow{Mg}$ C6H5–MgBr

$\downarrow$ [1] (2 equiv) [2] H_2O

Ar–CH_2CH_2–C(OH)(C6H5)2 (CH_3O, OH)

26.44

a. Br Br, OH ⟹ OH ⟹

NBS, $h\nu$ → Br; NaOH → OH; $CHBr_3$, $KOC(CH_3)_3$ → Br Br, OH

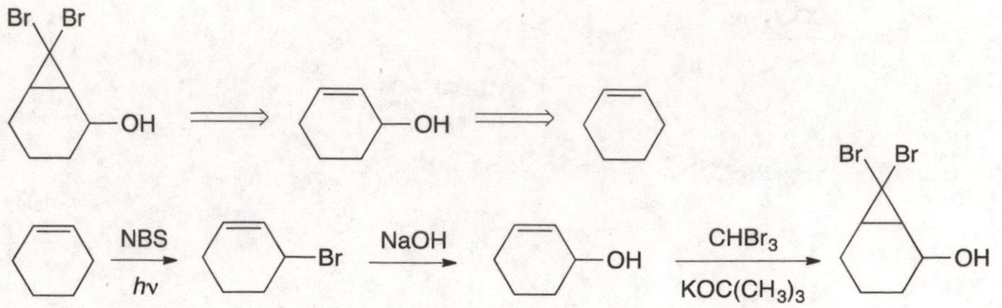

b. ⟹ ⟹ Br ⟹

Br (from a.); $(C_6H_5)_2CuLi$ → ; CH_2I_2, Zn(Cu) →

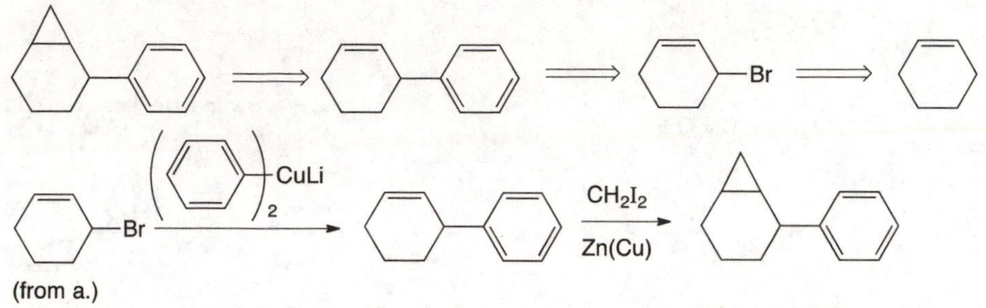

26.45

a. Br_2, $FeBr_3$ → Br; $CH_2{=}CH_2$, $Pd(OAc)_2$, $P(o\text{-tolyl})_3$, Et_3N → styrene; $CHCl_3$, $KOC(CH_3)_3$ → Cl Cl

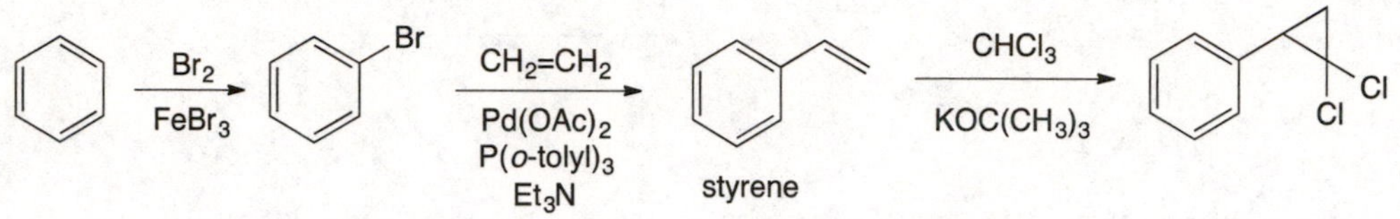

b. Br_2, $FeBr_3$ → Br; $CH_2{=}CHCO_2CH_3$, $Pd(OAc)_2$, $P(o\text{-tolyl})_3$, Et_3N → CO_2CH_3

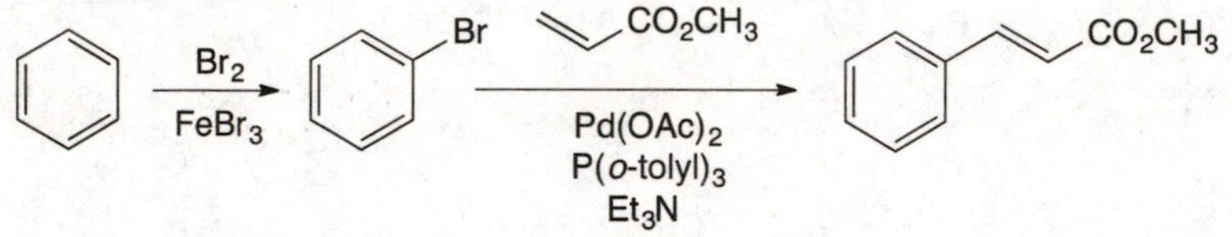

26.46

a. $HC{\equiv}CH$ — [1] NaH, [2] $CH_3CH_2CH_2Cl$ → $C{\equiv}CH$ — [1] NaH, [2] $(CH_3)_2CHCH_2Cl$ → ; H_2, Lindlar catalyst → ; CH_2I_2, Zn(Cu) → + enantiomer

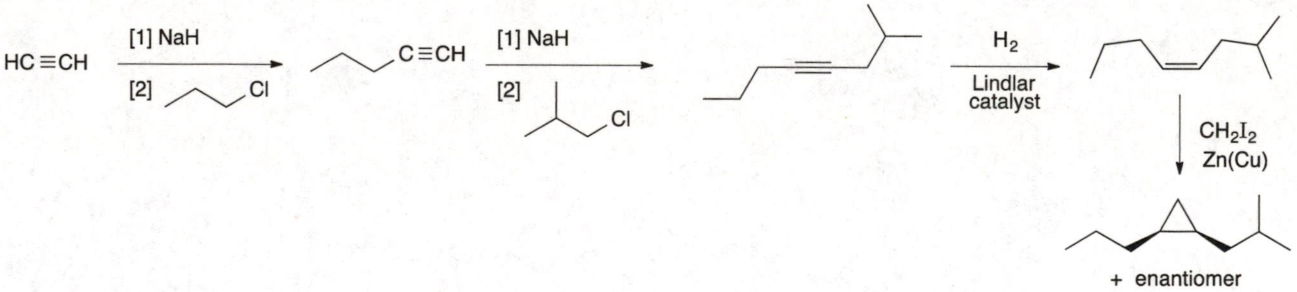

b. =O — Br_2, CH_3CO_2H → Br, =O — LiBr, Li_2CO_3, DMF → =O — [1] $(CH_3)_2CuLi$, [2] H_2O → =O — $Ph_3P{=}CH_2$ → — $CHCl_3$, $KOC(CH_3)_3$ → Cl Cl — [1] $(CH_3)_2CuLi$, [2] H_2O →

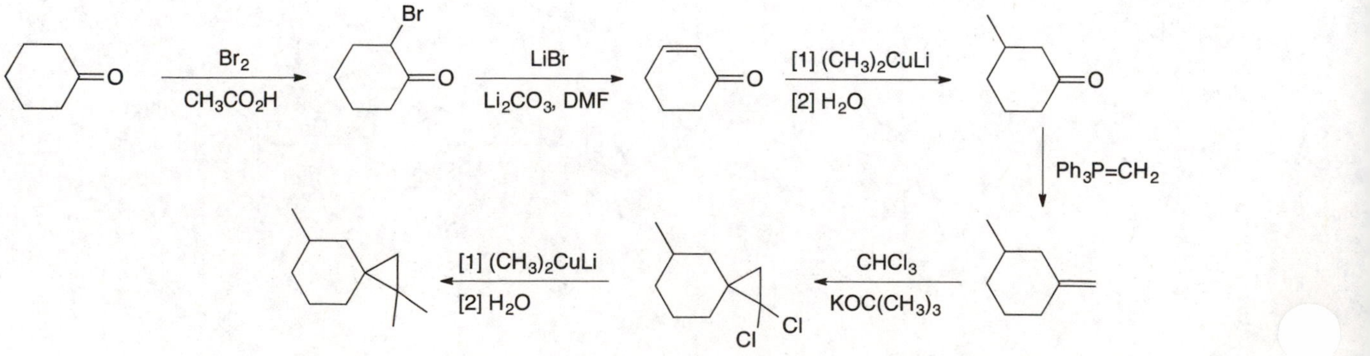

26.47

a. 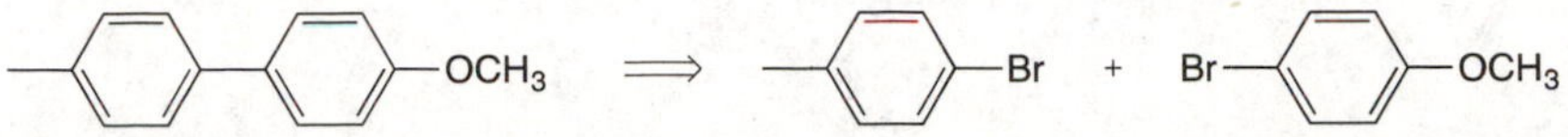 $CH_3C_6H_4C_6H_4OCH_3$ ⟹ $CH_3C_6H_4Br$ + $BrC_6H_4OCH_3$

Either compound can be used to synthesize the organoborane, so two routes are possible.

Possibility [1]:

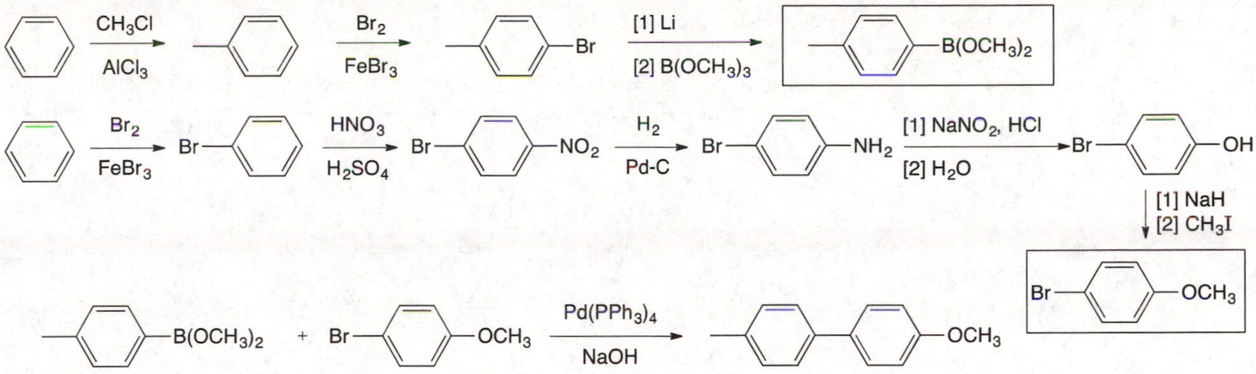

Benzene $\xrightarrow[AlCl_3]{CH_3Cl}$ toluene $\xrightarrow[FeBr_3]{Br_2}$ $CH_3C_6H_4Br$ $\xrightarrow[[2]\ B(OCH_3)_3]{[1]\ Li}$ **$CH_3C_6H_4B(OCH_3)_2$**

Benzene $\xrightarrow[FeBr_3]{Br_2}$ BrC_6H_5 $\xrightarrow[H_2SO_4]{HNO_3}$ $BrC_6H_4NO_2$ $\xrightarrow[Pd\text{-}C]{H_2}$ $BrC_6H_4NH_2$ $\xrightarrow[[2]\ H_2O]{[1]\ NaNO_2,\ HCl}$ BrC_6H_4OH $\xrightarrow{[1]\ NaH;\ [2]\ CH_3I}$ **$BrC_6H_4OCH_3$**

$CH_3C_6H_4B(OCH_3)_2$ + $BrC_6H_4OCH_3$ $\xrightarrow[NaOH]{Pd(PPh_3)_4}$ $CH_3C_6H_4C_6H_4OCH_3$

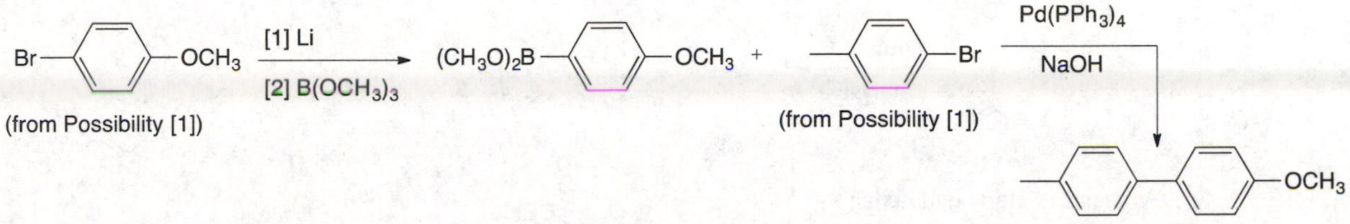

Possibility [2]:

$BrC_6H_4OCH_3$ (from Possibility [1]) $\xrightarrow[[2]\ B(OCH_3)_3]{[1]\ Li}$ $(CH_3O)_2BC_6H_4OCH_3$ + $CH_3C_6H_4Br$ (from Possibility [1]) $\xrightarrow[NaOH]{Pd(PPh_3)_4}$ $CH_3C_6H_4C_6H_4OCH_3$

b. 3-HOC$_6$H$_4$C$_6$H$_5$ ⟹ 3-HOC$_6$H$_4$Br + BrC_6H_5

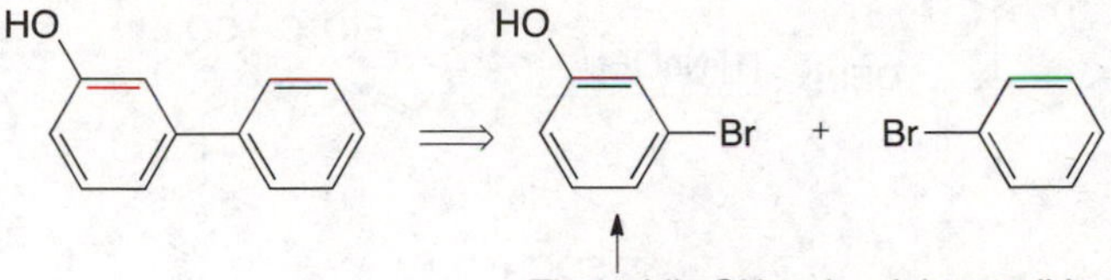

The acidic OH makes it impossible to prepare an organolithium reagent from this aryl halide, so this compound must be used as the aryl halide that couples with the organoborane from bromobenzene.

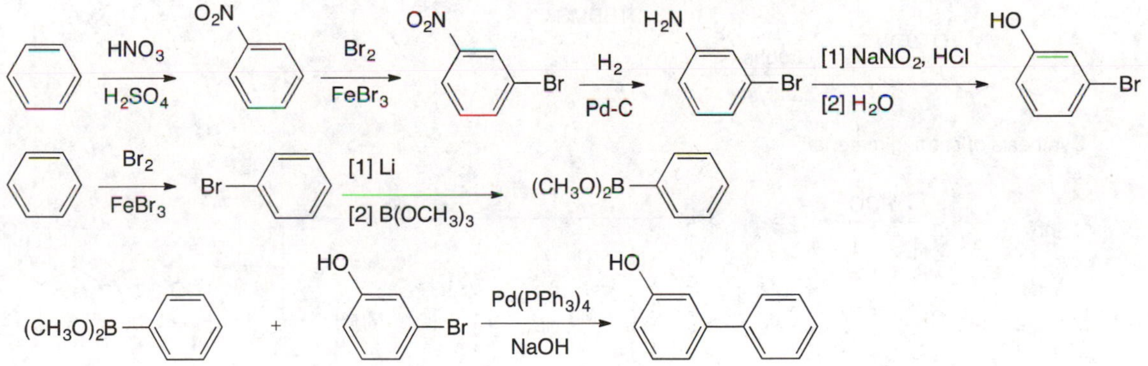

Benzene $\xrightarrow[H_2SO_4]{HNO_3}$ $O_2NC_6H_5$ $\xrightarrow[FeBr_3]{Br_2}$ 3-$O_2NC_6H_4Br$ $\xrightarrow[Pd\text{-}C]{H_2}$ 3-$H_2NC_6H_4Br$ $\xrightarrow[[2]\ H_2O]{[1]\ NaNO_2,\ HCl}$ 3-HOC$_6$H$_4$Br

Benzene $\xrightarrow[FeBr_3]{Br_2}$ BrC_6H_5 $\xrightarrow[[2]\ B(OCH_3)_3]{[1]\ Li}$ $(CH_3O)_2BC_6H_5$

$(CH_3O)_2BC_6H_5$ + 3-HOC$_6$H$_4$Br $\xrightarrow[NaOH]{Pd(PPh_3)_4}$ 3-HOC$_6$H$_4$C$_6$H$_5$

c.

This can't be converted to an organoborane reagent via an organolithium reagent.

three steps → $B(OCH_3)_2$ (as in 26.47b)

Br_2 / $FeBr_3$ → Br; HNO_3 / H_2SO_4 → Br, NO_2; H_2 / Pd-C → Br, NH_2; [1] $NaNO_2$, HCl [2] H_2O → Br, OH; CH_3COCl pyridine → Br, O, O

$B(OCH_3)_2$ + Br, O, O; $Pd(PPh_3)_4$ / NaOH → O, O

26.48

a. EtO_2C CO_2Et → Grubbs catalyst → EtO_2C CO_2Et

Synthesis of starting material:

EtO, OEt, O, O → [1] NaOEt [2] Cl (from OH, $SOCl_2$) → OEt, O, OEt → [1] NaOEt [2] Cl (from OH, $SOCl_2$) → EtO_2C CO_2Et

b. OTBDMS → Grubbs catalyst → OTBDMS

Synthesis of starting material:

OH → PCC → H, O

OH → PBr_3 → Br → Mg → MgBr

→ H_2O → OH → TBDMS–Cl imidazole → OTBDMS

26.49

a. O → Grubbs catalyst → O

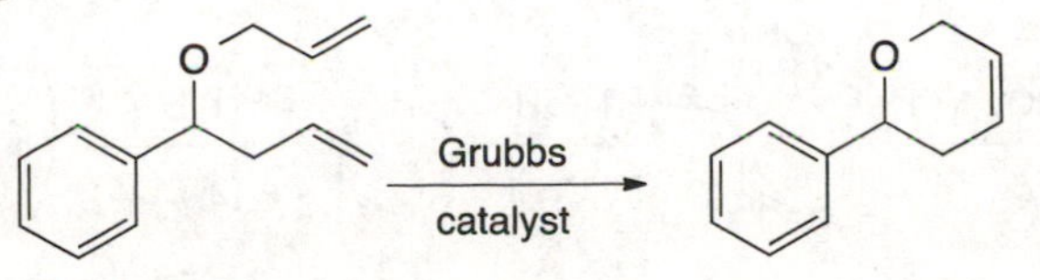

Synthesis of starting material:

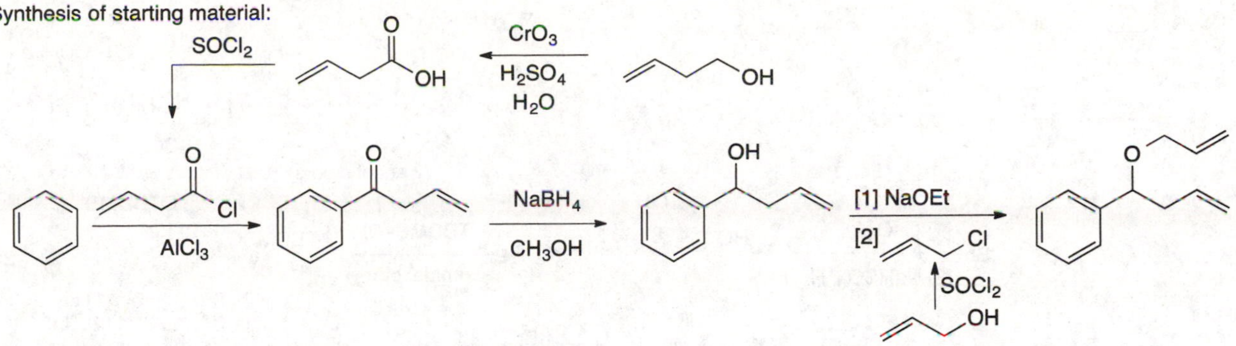

$SOCl_2$ ← O, OH ← CrO_3, H_2SO_4, H_2O ← OH

O, Cl, $AlCl_3$ → O → $NaBH_4$, CH_3OH → OH → [1] NaOEt, [2] Cl (↑ $SOCl_2$, OH) → O

b. O, O → Grubbs catalyst → O, O

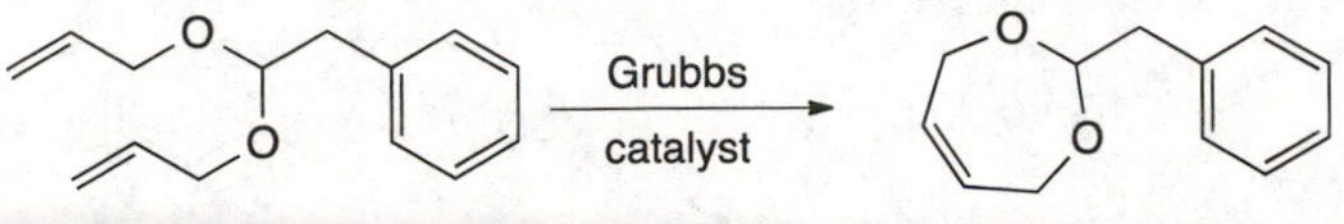

Synthesis of starting material:

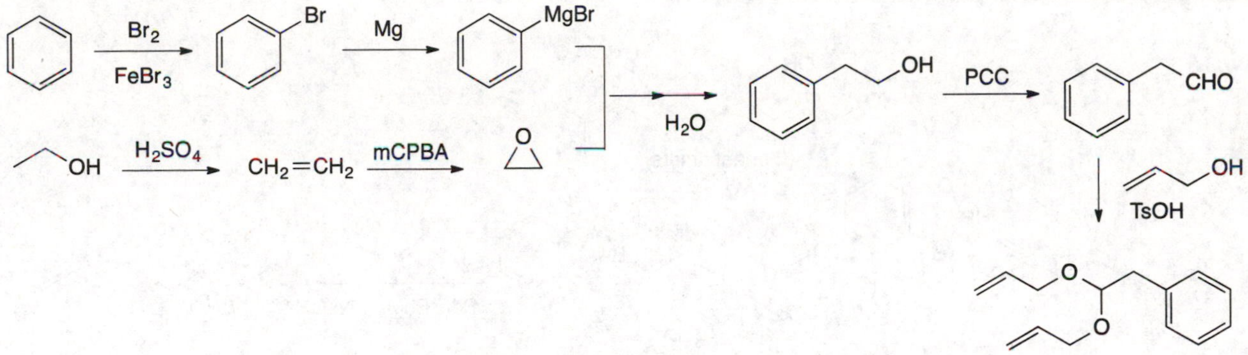

Br_2, $FeBr_3$ → Br → Mg → MgBr

OH → H_2SO_4 → $CH_2{=}CH_2$ → mCPBA → O

→ H_2O → OH → PCC → CHO → OH, TsOH → O, O

26.50

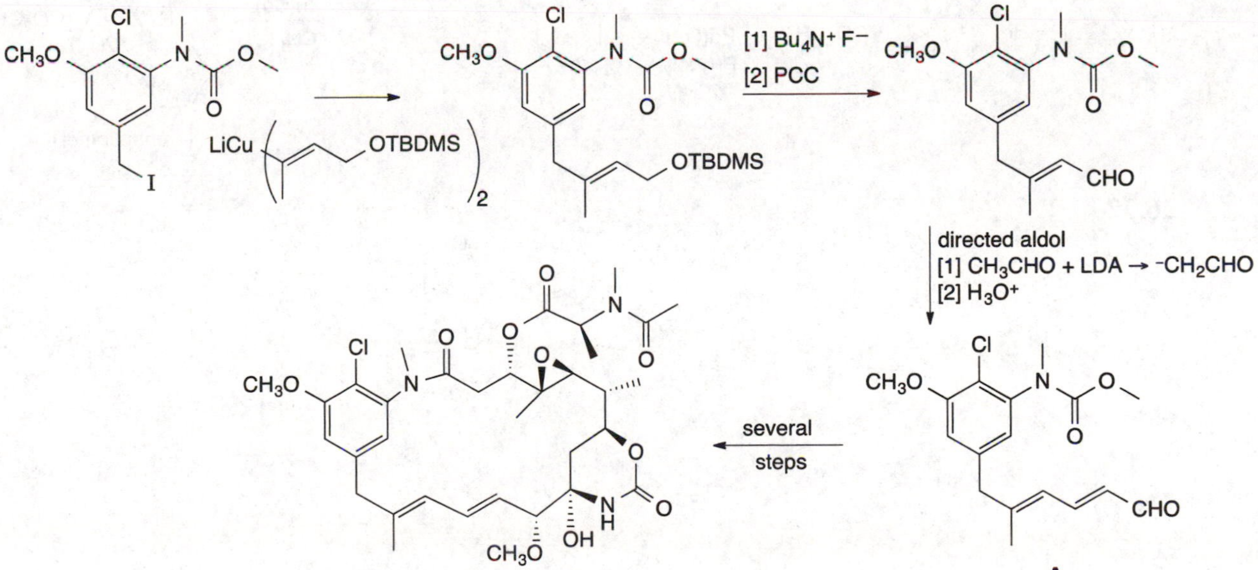

Cl, CH_3O, N, O, O, I

LiCu (OTBDMS)$_2$ →

Cl, CH_3O, N, O, O, OTBDMS

[1] $Bu_4N^+\ F^-$
[2] PCC

Cl, CH_3O, N, O, O, CHO

directed aldol
[1] CH_3CHO + LDA → $^-CH_2CHO$
[2] H_3O^+

Cl, CH_3O, N, O, O, CHO

A

several steps

O, N, O, Cl, CH_3O, N, O, O, O, O, N, H, O, OH, CH_3O

maytansine

26.51

a. Br-benzene $\xrightarrow{HNO_3,\ H_2SO_4}$ o-bromonitrobenzene $\xrightarrow{H_2,\ Pd\text{-}C}$ o-bromoaniline $\xrightarrow{[1]\ NaNO_2,\ HCl;\ [2]\ H_2O}$ o-bromophenol $\xrightarrow{[1]\ NaH;\ [2]\ CH_3I}$ o-bromoanisole $\xrightarrow{[1]\ Li;\ [2]\ CuI}$ (OCH_3-aryl)$_2$Cu… ; + Br(CH$_2$)$_4$CH=CH$_2$ → o-(hex-5-enyl)anisole $\xrightarrow{H_2O,\ H_2SO_4}$ OCH$_3$, OH product (2 enantiomers)

b. CH$_2$=CH(CH$_2$)$_4$Br $\xrightarrow{[1]\ OsO_4;\ [2]\ NaHSO_3,\ H_2O}$ HO, OH, Br diol $\xrightarrow{TBDMS\text{–}Cl,\ imidazole}$ TBDMSO, OTBDMS, Br $\xrightarrow{[1]\ Li;\ [2]\ CuI}$ (TBDMSO, OTBDMS)$_2$CuLi; + Br (prenyl bromide) → TBDMSO, OTBDMS alkene $\xrightarrow{Bu_4N^+F^-}$ HO, OH alkene (2 enantiomers)

c. hex-1-yne + HB(catecholborane) → alkenyl-B(catechol) (H) ; + Br-alkene, Pd(PPh$_3$)$_4$, NaOH → diene

d. Br-benzene + CH$_2$=CHCO$_2$CH$_3$ $\xrightarrow{Pd(OAc)_2,\ P(o\text{-tolyl})_3,\ Et_3N}$ PhCH=CHCO$_2$CH$_3$ $\xrightarrow{Zn(Cu),\ CH_2I_2}$ H, CO$_2$CH$_3$, H cyclopropane (+ enantiomer)

26.52

a. O, O diallyl ether of cyclobutene → bis(dihydrofuran) product

b.

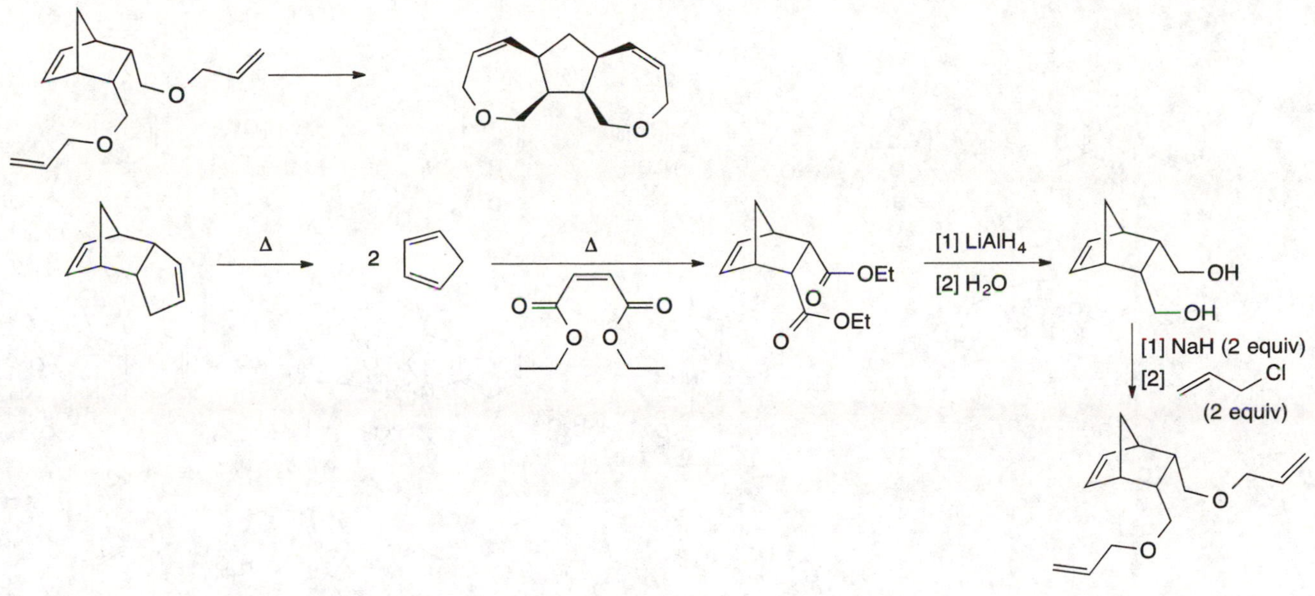

26.53 There is more than one way to form **Z** by metathesis reactions. One possibility involves ring opening of the bicyclic alkene followed by successive ring closures to generate the five- and seven-membered rings.

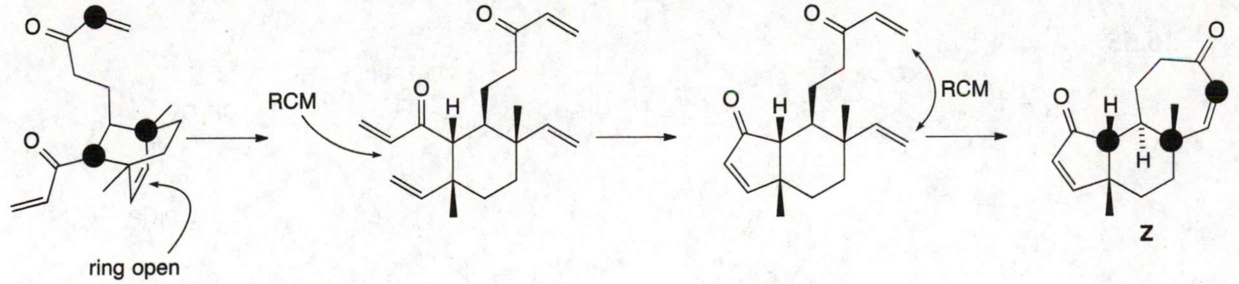

26.54

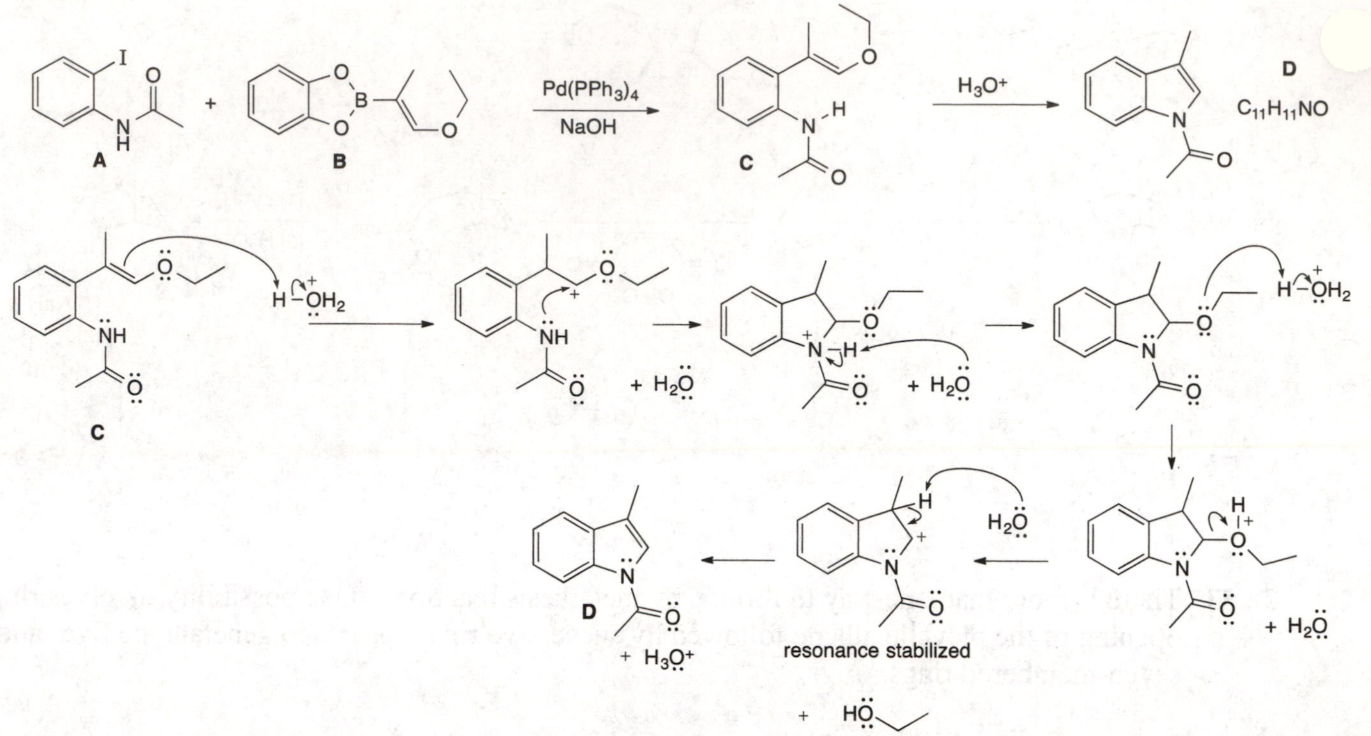

A + B → (Pd(PPh3)4, NaOH) → C → (H_3O^+) → D ($C_{11}H_{11}NO$)

C + $H-\ddot{O}H_2^+$ → + $H_2\ddot{O}$ → + $H_2\ddot{O}$ → $H-\ddot{O}H_2^+$ → + $H_2\ddot{O}$ → $H_2\ddot{O}$ resonance stabilized + $H\ddot{O}$ → D + H_3O^+

26.55

X + Y (PPh_3) → CH_3O ... $O-PPh_3$ → CH_3O ... + $Ph_3P{=}O$ → Ph_3P^+ → CH_3O ... Ph_3P^+ → CH_3O ... + PPh_3

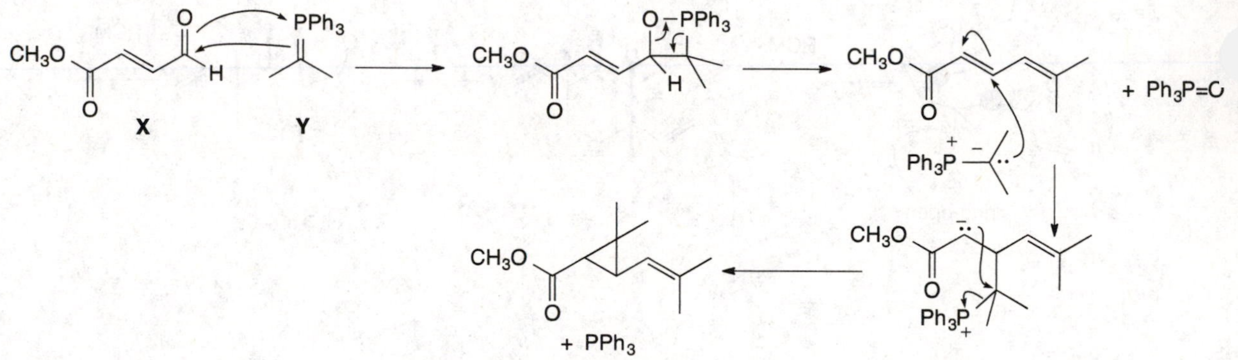

26.56

a. Reaction of a terminal alkene with the catalyst forms a metal–carbene that undergoes an intramolecular reaction with the triple bond, generating a new metal–carbene. A second intramolecular reaction forms the bicyclic product.

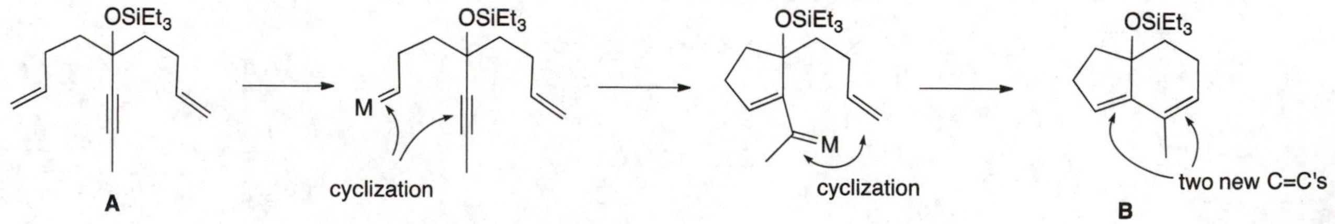

A ($OSiEt_3$) → (M, cyclization) → (M, cyclization) → B ($OSiEt_3$) two new C=C's

b. Two products are possible because the cascade of reactions can begin at two different double bonds.

OSiEt$_3$ OSiEt$_3$ OSiEt$_3$ OSiEt$_3$

M M

C

Begin here.

OSiEt$_3$ OSiEt$_3$ OSiEt$_3$ OSiEt$_3$

M M

C Begin here.

Chapter 27 Pericyclic Reactions

Chapter Review

Electrocyclic reactions (27.3)

Woodward–Hoffmann rules for electrocyclic reactions

Number of π bonds	Thermal reaction	Photochemical reaction
Even	Conrotatory	Disrotatory
Odd	Disrotatory	Conrotatory

Examples

The stereochemistry of a thermal electrocyclic reaction is opposite to that of a photochemical electrocyclic reaction.

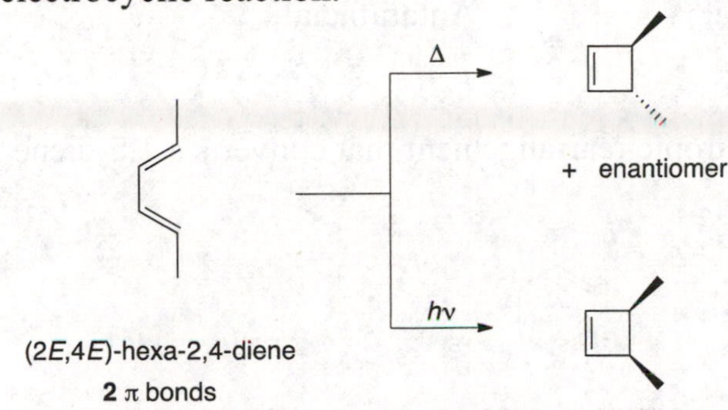

- A thermal electrocyclic reaction with an even number of π bonds occurs in a conrotatory fashion.
- A photochemical electrocyclic reaction with an even number of π bonds occurs in a disrotatory fashion.

Cycloaddition reactions (27.4)

Woodward–Hoffmann rules for cycloaddition reactions

Number of π bonds	Thermal reaction	Photochemical reaction
Even	Antarafacial	Suprafacial
Odd	Suprafacial	Antarafacial

Examples

[1] A thermal [4 + 2] cycloaddition takes place in a suprafacial fashion with an odd number of π bonds. An antarafacial photochemical [4 + 2] cycloaddition to form a six-membered ring cannot occur, because of the geometrical constraints of forming a six-membered ring.

+ $CH_2{=}CH_2$ —Δ, suprafacial→ **cis** product only

[2] A photochemical [2 + 2] cycloaddition takes place in a suprafacial fashion with an even number of π bonds. An antarafacial thermal [2 + 2] cycloaddition to form a four-membered ring cannot occur, because of the geometrical constraints of forming a four-membered ring.

C_6H_5 C_6H_5 + $CH_2=CH_2$ $\xrightarrow[\text{suprafacial}]{h\nu}$ C_6H_5 C_6H_5

cis product only

Sigmatropic rearrangements (27.5)

Woodward–Hoffmann rules for sigmatropic rearrangements

Number of electron pairs	Thermal reaction	Photochemical reaction
Even	Antarafacial	Suprafacial
Odd	Suprafacial	Antarafacial

Examples

[1] A **Cope rearrangement** is a thermal [3,3] sigmatropic rearrangement that converts a 1,5-diene into an isomeric 1,5-diene.

Δ

1,5-diene isomeric 1,5-diene

[2] An **oxy-Cope rearrangement** is a thermal [3,3] sigmatropic rearrangement that converts a 1,5-dien-3-ol into a δ,ε-unsaturated carbonyl compound, after tautomerization of an intermediate enol.

HO O Δ [3,3]

1,5-dien-3-ol δ,ε-unsaturated carbonyl compound

[3] A **Claisen rearrangement** is a thermal [3,3] sigmatropic rearrangement that converts an unsaturated ether into a γ,δ-unsaturated carbonyl compound.

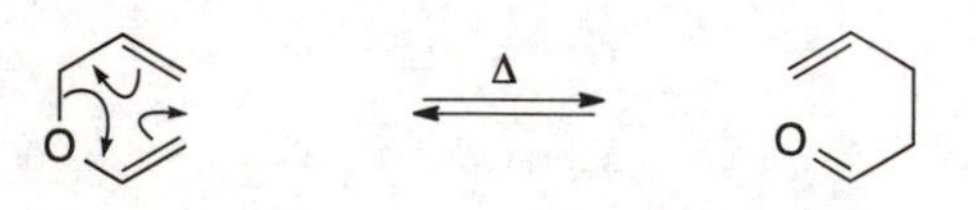

unsaturated ether γ,δ-unsaturated carbonyl compound

Practice Test on Chapter Review

1. a. Which of the following pericyclic reactions is symmetry allowed and readily occurs?
   1. a photochemical conrotatory electrocyclic ring closure of a conjugated triene
   2. a disrotatory thermal electrocyclic ring opening of a substituted cyclohexadiene
   3. a thermal [2 + 2] cycloaddition
   4. Reactions (1) and (2) will both occur.
   5. Reactions (1), (2), and (3) will all occur.

   b. Which of the following reactions requires suprafacial stereochemistry to be symmetry allowed?
   1. a photochemical [1,5] sigmatropic rearrangement
   2. a thermal [8 + 2] cycloaddition
   3. a photochemical [4 + 2] cycloaddition
   4. Reactions (1) and (2) are both suprafacial.
   5. Reactions (1), (2), and (3) are all suprafacial.

   c. What product(s) are formed from the photochemical [2 + 2] cycloaddition of (*E*)-hex-3-ene?

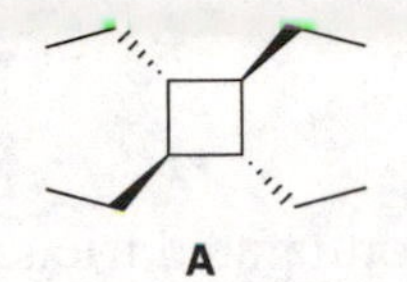

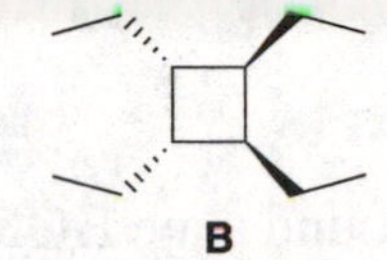

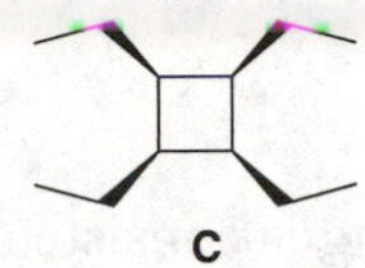

   1. **A** only
   2. **B** only
   3. **C** only
   4. **A** and **B**
   5. **A, B,** and **C**

   d. What product(s) are formed from the photochemical electrocyclic ring opening of *cis*-3,4-dimethylcyclobutene?

   1. (2*E*,4*E*)-hexa-2,4-diene
   2. (2*E*,4*Z*)-hexa-2,4-diene
   3. (*Z*)-hexa-1,3,5-triene
   4. Compounds (1) and (2) are both formed.
   5. Compounds (1), (2), and (3) are all formed.

e. What product(s) are formed by the [3,3] sigmatropic rearrangement of cyclodeca-1,5-dien-3-ol?

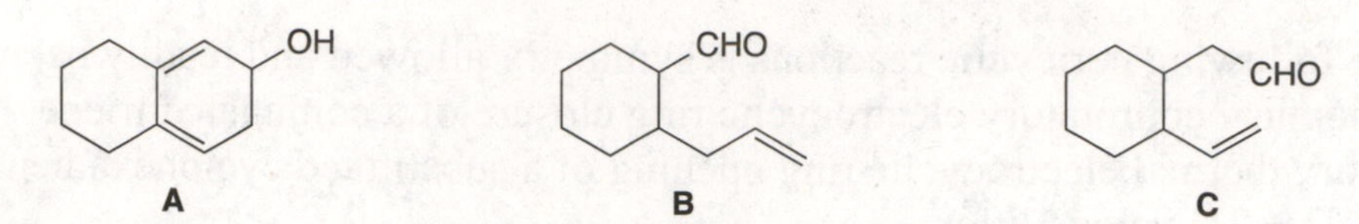

1. **A** only
2. **B** only
3. **C** only
4. **A** and **B**
5. **A, B,** and **C**

2. Consider the *p* orbitals of the terminal carbons of a conjugated polyene with like phases on the same side of the molecule (as in **A**) or opposite sides of the molecule (as in **B**), and answer each question.

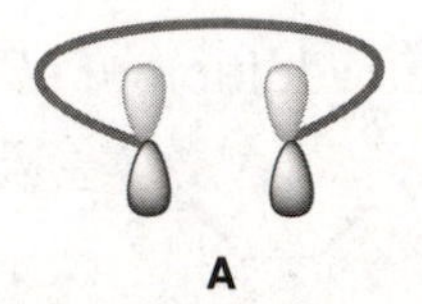

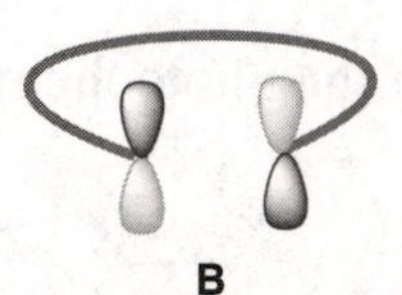

a. Which drawing is consistent with the ground state HOMO of a conjugated triene?
b. Which drawing is consistent with the excited state LUMO of a conjugated diene?
c. Which drawing is consistent with the ground state LUMO for a conjugated tetraene?

3. What type of sigmatropic rearrangement is depicted in each reaction?

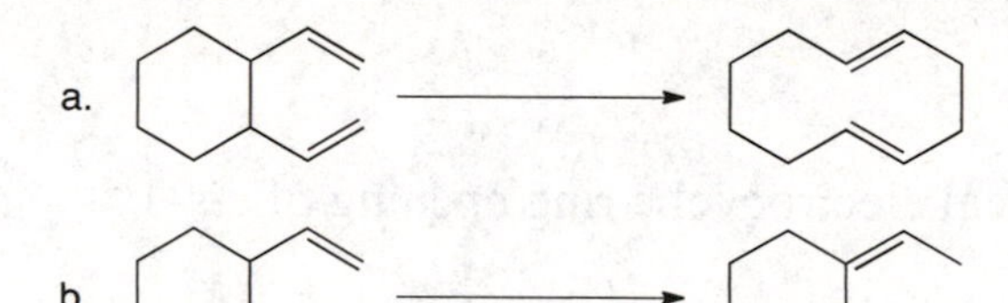

Answers to Practice Test

1. a. 4	2. a. **A**	3. a. [3,3]
b. 2	b. **B**	b. [1,3]
c. 4	c. **A**	
d. 1		
e. 3		

Answers to Problems

27.1 Use the following definitions:

- An **electrocyclic ring closure** is an intramolecular reaction that forms a cyclic product containing one more σ bond and one fewer π bond than the reactant. An **electrocyclic ring opening** is a reaction in which a σ bond of a cyclic reactant is cleaved to form a conjugated product with one more π bond.
- A **cycloaddition** is a reaction between two compounds with π bonds that forms a cyclic product with two new σ bonds.
- A **sigmatropic rearrangement** is a reaction in which a σ bond is broken in the reactant, the π bonds rearrange, and a σ bond is formed in the product.

a. 2 π bonds; σ bond broken → 3 π bonds — **electrocyclic reaction**

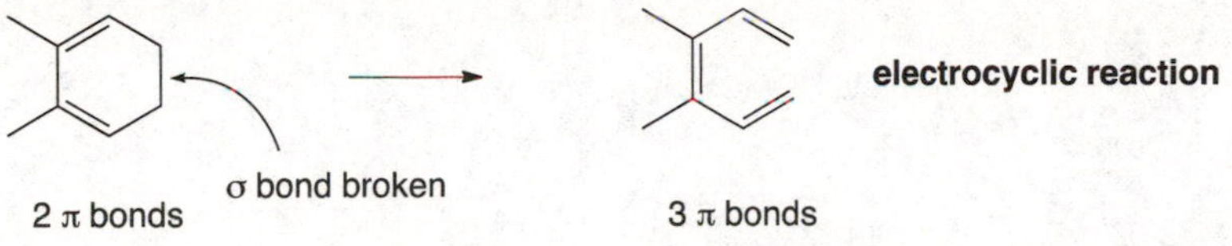

b. + $H_2C{=}CH_2$ → σ bond formed; σ bond formed — **cycloaddition**

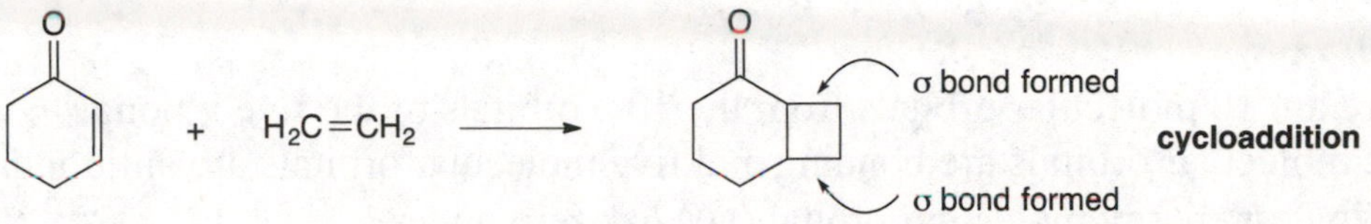

c. 4 π bonds → 3 π bonds; σ bond formed — **electrocyclic reaction**

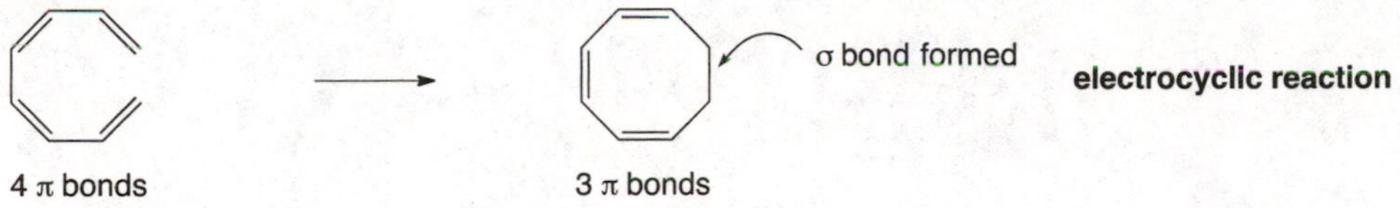

d. 1 π bond; σ bond broken → 1 π bond; σ bond formed — **sigmatropic rearrangement**

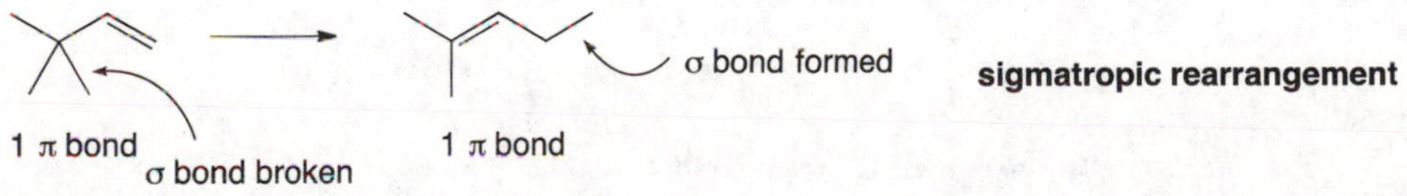

27.2

a. For a bonding molecular orbital, the number of bonding interactions is greater than the number of nodes.

b. For an antibonding molecular orbital, the number of bonding interactions is less than the number of nodes.

For buta-1,3-diene:

	Bonding	Nodes	Type of MO
ψ_1	3	0	bonding MO
ψ_2	2	1	bonding MO
ψ_3*	1	2	antibonding MO
ψ_4*	0	3	antibonding MO

27.3 The molecular orbitals of all conjugated dienes look similar.

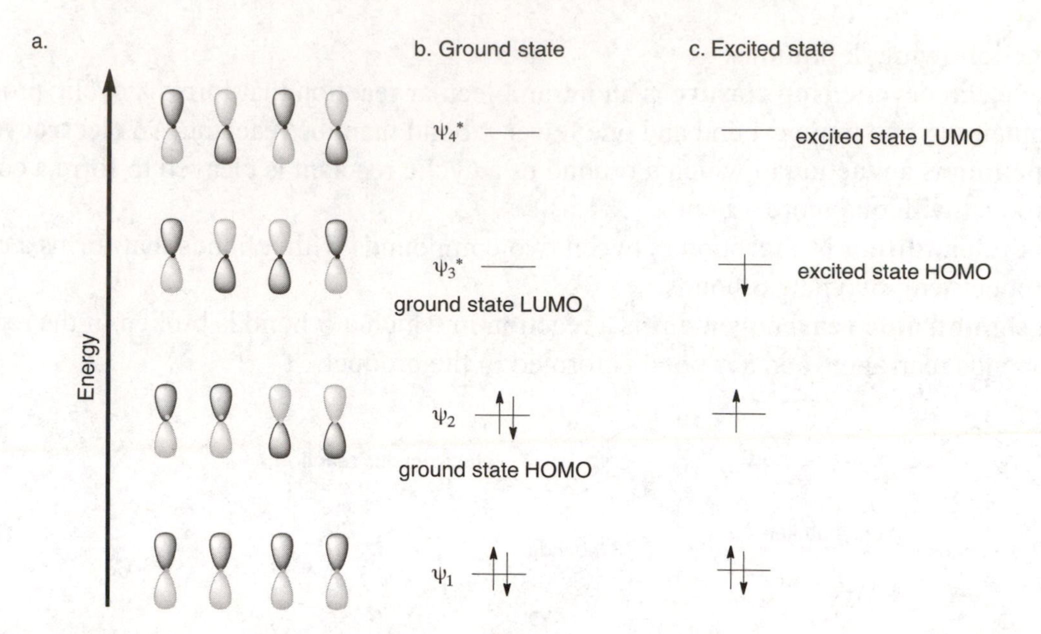

27.4 a. There are 10 molecular orbitals from the 10 *p* orbitals of the five π bonds.
b. Five molecular orbitals are bonding and five molecular orbitals are antibonding.
c. The lowest energy molecular orbital (ψ_1) has zero nodes.

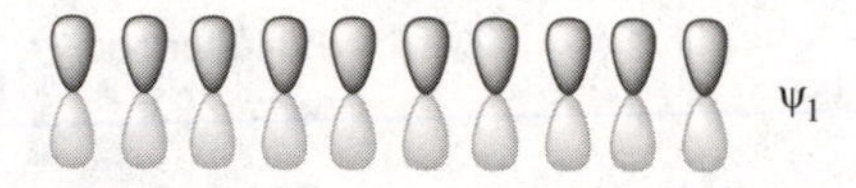

d. The highest energy molecular orbital (ψ_{10}*) has nine nodes.

ψ_{10}^*

27.5 To draw the product of an electrocyclic reaction, use curved arrows and begin at a π bond. Move the π electrons to an adjacent carbon–carbon bond, and continue in a cyclic fashion.

a. Δ

b. Δ

c. re-draw Δ

27.6 Thermal electrocyclic reactions occur in a *disrotatory* fashion for a conjugated polyene with an *odd* number of π bonds, and in a *conrotatory* fashion for a conjugated polyene with an *even* number of π bonds.

a. C_6H_5 C_6H_5 3 π bonds → Δ disrotatory → C_6H_5 C_6H_5

b. → Δ conrotatory → 2 π bonds

27.7 For an *even* number of π bonds, thermal electrocyclic reactions occur in a *conrotatory* fashion.

a. re-draw → Δ → + enantiomer

b. re-draw → Δ →

27.8 Photochemical electrocyclic reactions occur in a *conrotatory* fashion for a conjugated polyene with an *odd* number of π bonds, and in a *disrotatory* fashion for a conjugated polyene with an *even* number of π bonds.

C_6H_5 C_6H_5 → *hν* conrotatory → C_6H_5 C_6H_5 + enantiomer

b. → *hν* disrotatory →

[The (*E*,*E*) diene is favored over the (*Z*,*Z*) diene.]

27.9 The photochemical electrocyclic reaction cleaves a six-membered ring to form a hexatriene.

HO H H → *hν* → HO H

7-dehydrocholesterol

provitamin D_3

27.10 Use the rules for electrocyclic reactions found in Answers 27.6 and 27.8.

a. re-draw → 3 π bonds → Δ disrotatory →

hν conrotatory → + enantiomer

b. re-draw

3 π bonds

Δ

disrotatory

+ enantiomer

hν

conrotatory

27.11 Use the rules for electrocyclic reactions found in Answer 27.6. A reaction with three π bonds and a disrotatory cyclization is thermal.

H

H

27.12 Count the number of π electrons in each reactant to classify the cycloaddition.

a. O + CH_2 = CH_2 → O (one possibility) [2 + 2] cycloaddition

2 π electrons 2 π electrons

b. O CH_2 = CH_2 → O [4 + 2] cycloaddition

4 π electrons 2 π electrons

c. O CH_2 = CH_2 → O [6 + 2] cycloaddition

6 π electrons 2 π electrons

27.13 A thermal suprafacial addition is symmetry allowed in a [4 + 2] cycloaddition because like phases interact.

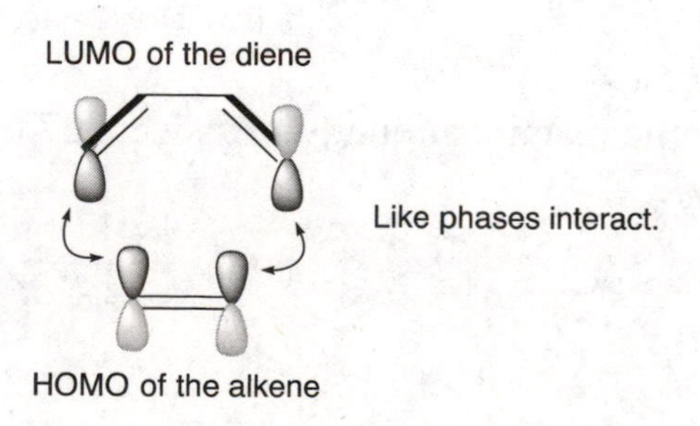

27.14 A thermal [4 + 2] cycloaddition is suprafacial.

a.

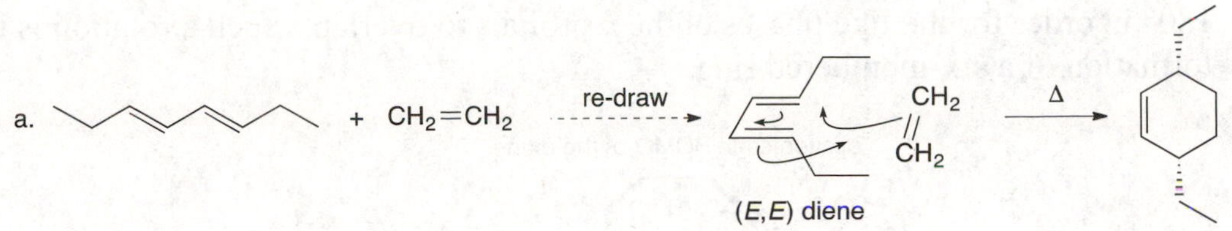

a. + $CH_2{=}CH_2$ re-draw CH_2 CH_2 Δ

(*E,E*) diene

b. + CN CN re-draw NC CN Δ CN CN + enantiomer

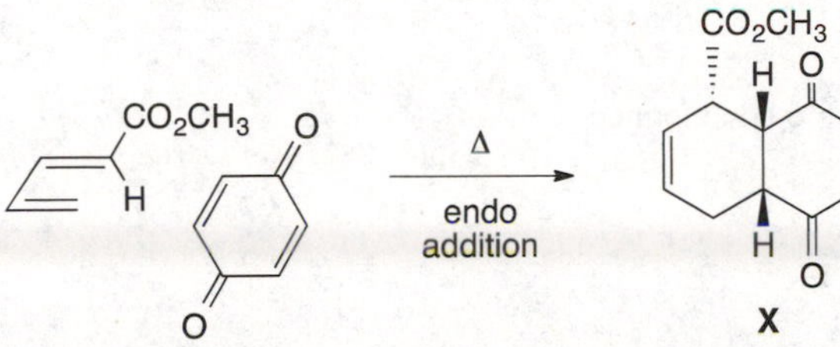

(*E,Z*) diene

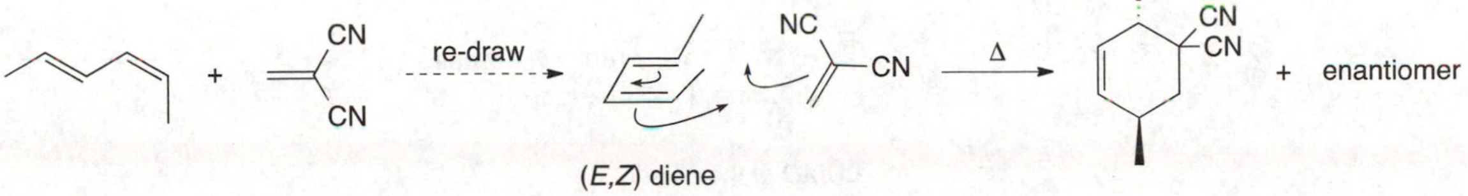

27.15

a. CO_2CH_3 H O Δ endo addition O CO_2CH_3 H O H O **X**

b. CO_2CH_3 H diene HOMO O H dienophile LUMO O H CO_2CH_3 H H O H O **X**

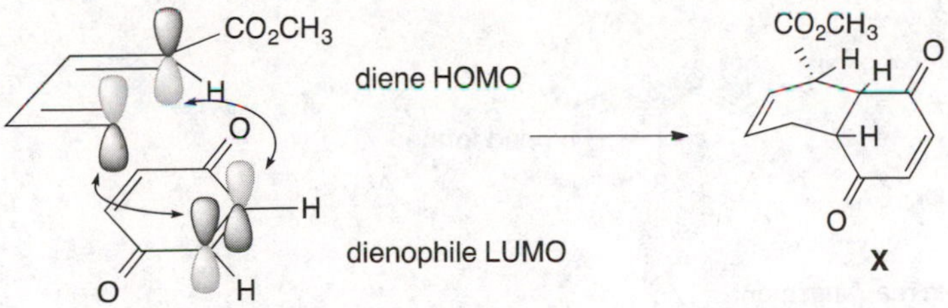

The dienophile is under the diene, by the rule of endo addition (Section 16.13). The H's at the ring fusion are cis to each other, but trans to the CO_2CH_3 group.

27.16 A photochemical [2 + 2] cycloaddition is suprafacial.

a. C_6H_5 + C_6H_5 $h\nu$ C_6H_5 C_6H_5 + enantiomer

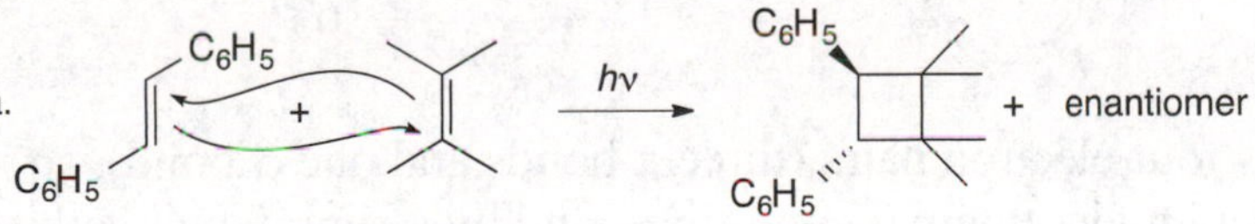

b. O H + CH_2 CH_2 CN O H CN + enantiomer

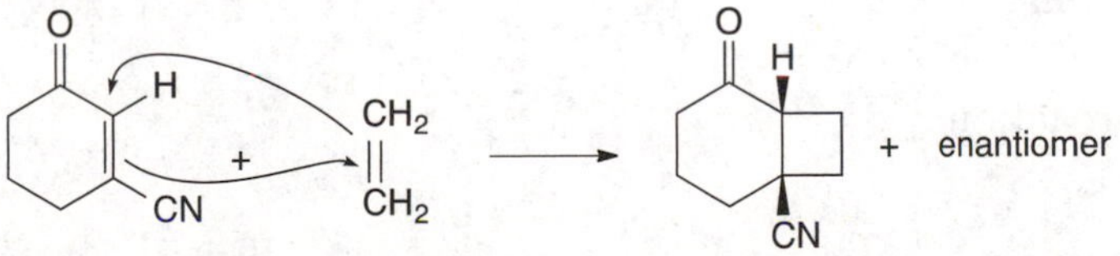

27.17

a. The photochemical [6 + 4] cycloaddition involves five π bonds (the total number of π electrons divided by two) and is antarafacial.

b. A thermal [8 + 2] cycloaddition involves five π bonds and is suprafacial.

27.18 A photochemical [4 + 2] cycloaddition like the Diels–Alder reaction must proceed by an antarafacial pathway. This would require either the 1,3-diene or the alkene component to twist 180° in order for the like phases of the *p* orbitals to overlap. Such a rotation is not possible in the formation of a six-membered ring.

excited state HOMO of the diene

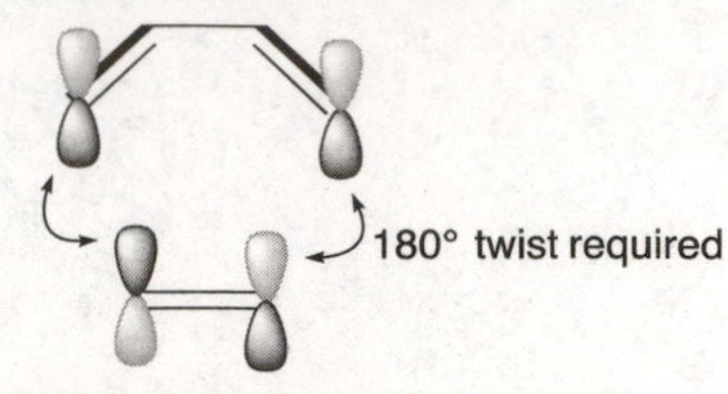

LUMO of the alkene

27.19 Locate the σ bonds broken and formed, and count the number of atoms that connect them.

a. σ bond broken → σ bond formed

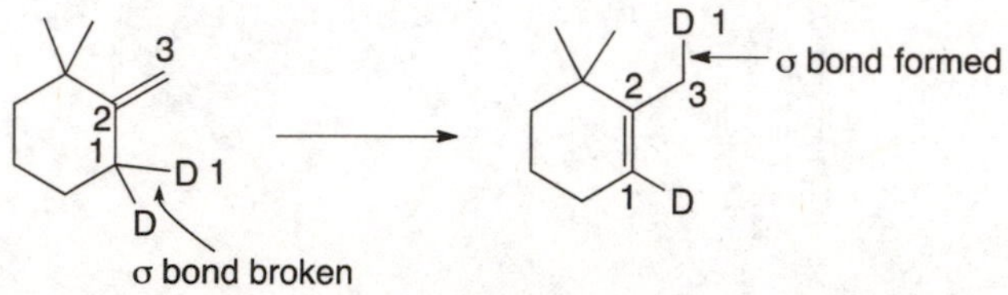

[1,3] sigmatropic rearrangement

b. σ bond broken → σ bond formed

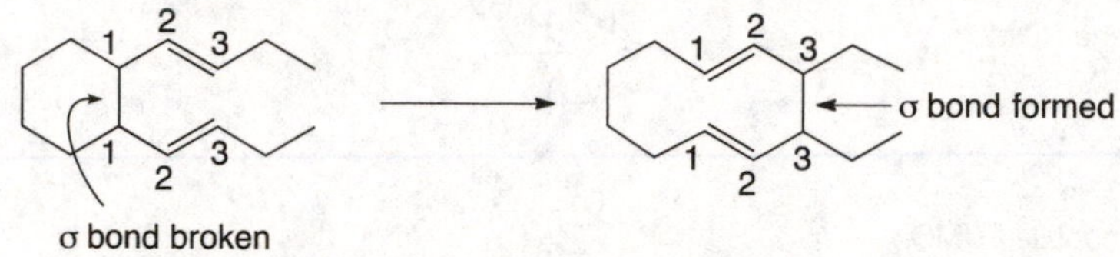

[3,3] sigmatropic rearrangement

27.20

a. re-draw → [1,7]

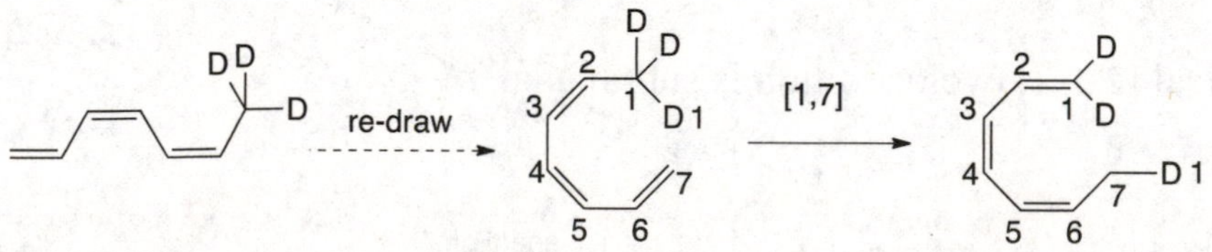

b, c. The reaction involves four electron pairs (three π bonds and one σ bond), so it proceeds by an antarafacial pathway under thermal conditions, and by a suprafacial pathway under photochemical conditions.

27.21 Draw the products of each reaction.

a. re-draw → [3,3]

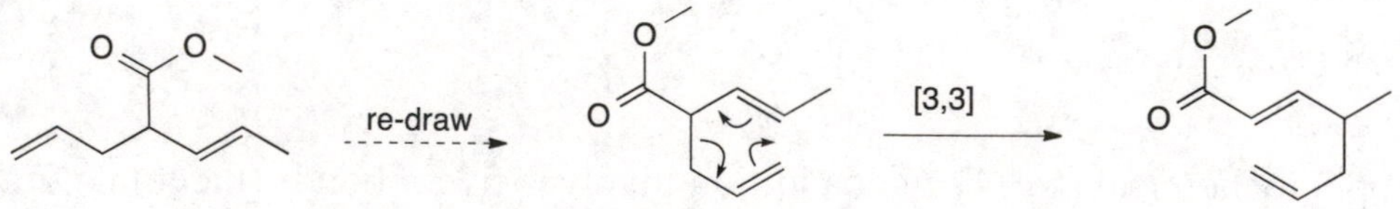

b. 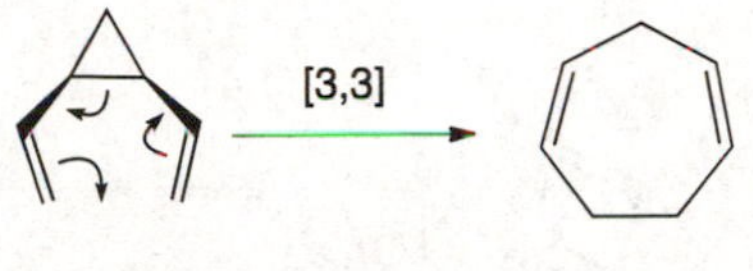 [3,3]

c. 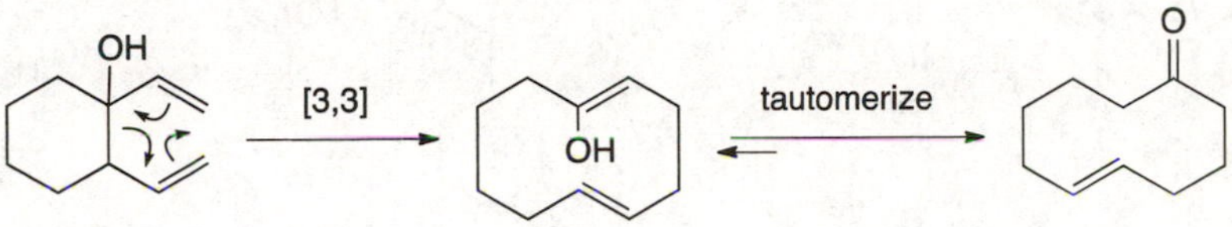 OH [3,3] OH tautomerize O

27.22 Draw the product after protonation.

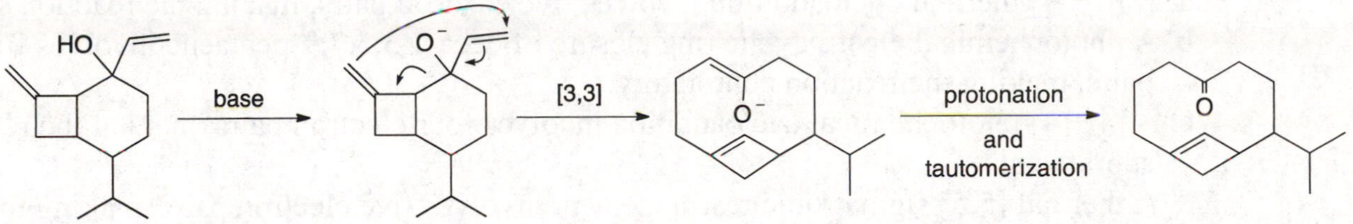

HO base O⁻ [3,3] O⁻ protonation and tautomerization O

27.23 Re-draw geranial to put the ends of the 1,5-diene close together. Then draw three curved arrows, beginning at a π bond.

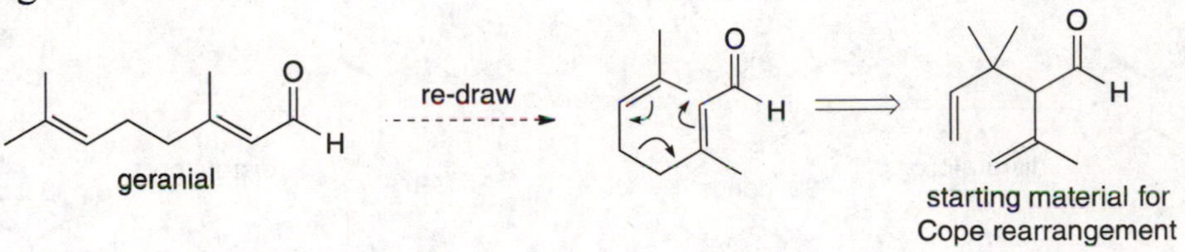

O H geranial re-draw O H ⟹ O H starting material for Cope rearrangement

27.24 Draw the product of Claisen rearrangement.

a. 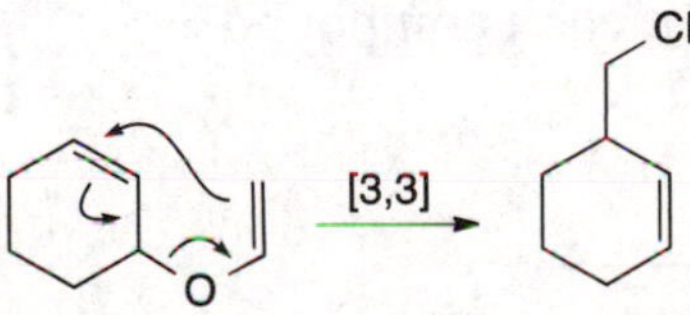 O [3,3] CHO

c. 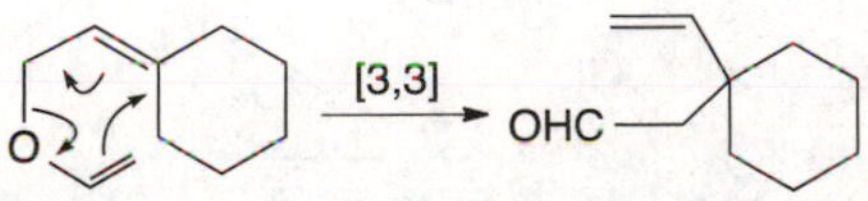 O [3,3] OHC

b. 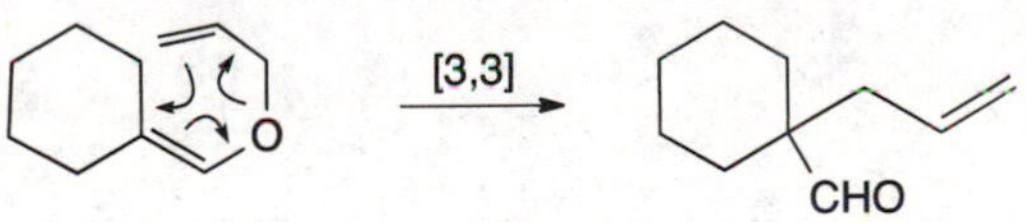 O [3,3] CHO

27.25

a. 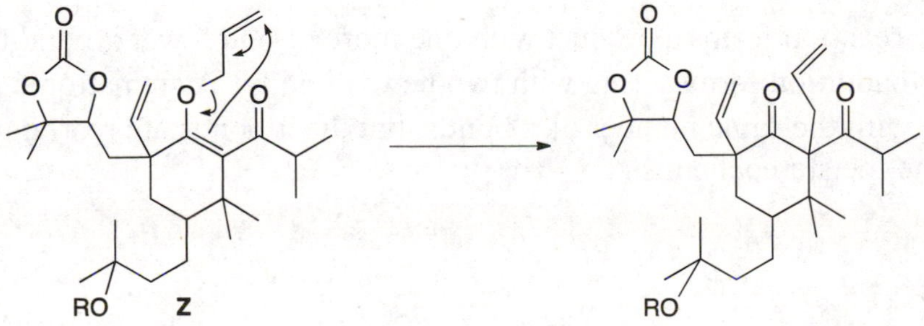 O O O O O RO Z O O O O O RO

b.

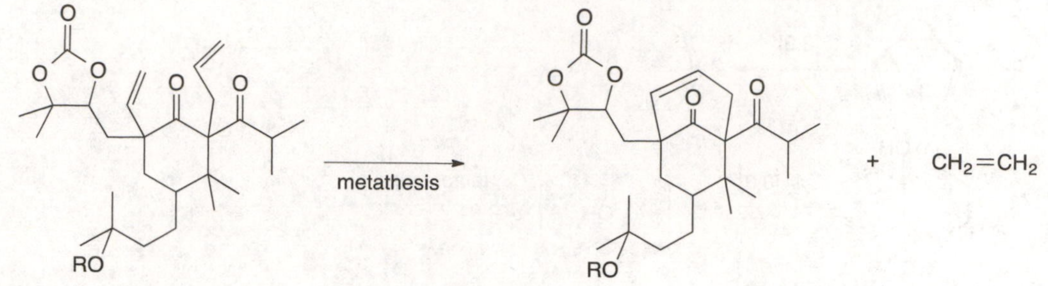

metathesis

RO RO

+ $CH_2{=}CH_2$

27.26 Predict the stereochemistry of each reaction using Table 27.4.

a. A [6 + 4] thermal cycloaddition involves five electron pairs, making the reaction suprafacial.

b. A photochemical electrocyclic ring closure of deca-1,3,5,7,9-pentaene involves five electron pairs, making the reaction conrotatory.

c. A [4 + 4] photochemical cycloaddition involves four electron pairs, making the reaction suprafacial.

d. A thermal [5,5] sigmatropic rearrangement involves five electron pairs, making the reaction suprafacial.

27.27 Use the rules found in Answers 27.6 and 27.8.

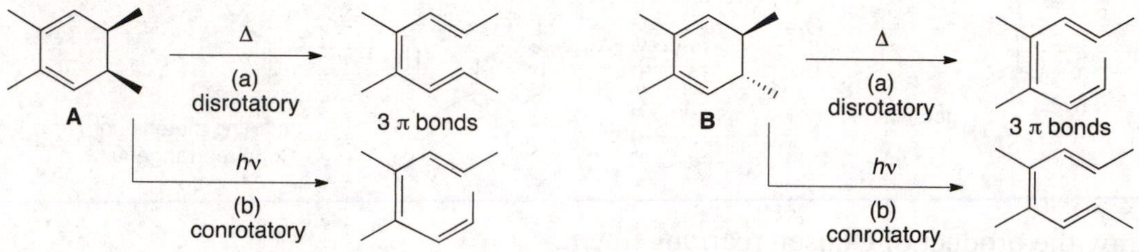

A

Δ

(a) disrotatory

3 π bonds

hν

(b) conrotatory

B

Δ

(a) disrotatory

3 π bonds

hν

(b) conrotatory

27.28 Draw the product of [3,3] sigmatropic rearrangement of each compound.

a.

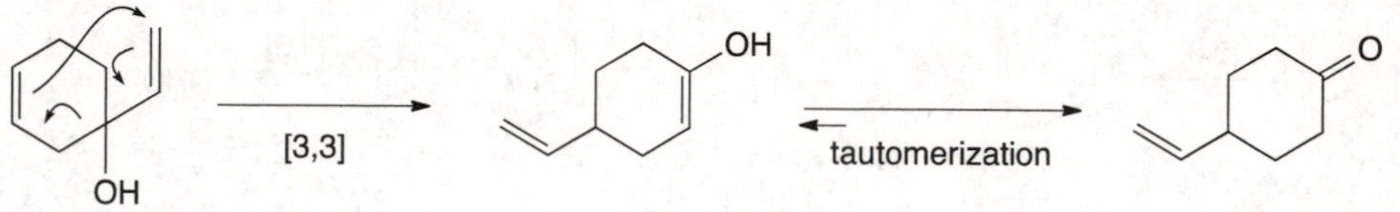

OH

[3,3]

OH

tautomerization

O

b.

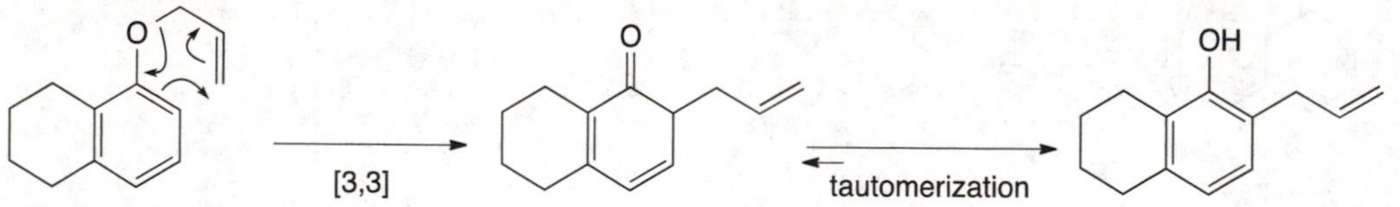

O

[3,3]

O

tautomerization

OH

27.29 An electrocyclic reaction forms a product with one more or one fewer π bond than the starting material. A cycloaddition forms a ring with two new σ bonds. A sigmatropic rearrangement forms a product with the same number of π bonds, but the π bonds are rearranged. Use Table 27.4 to determine the stereochemistry.

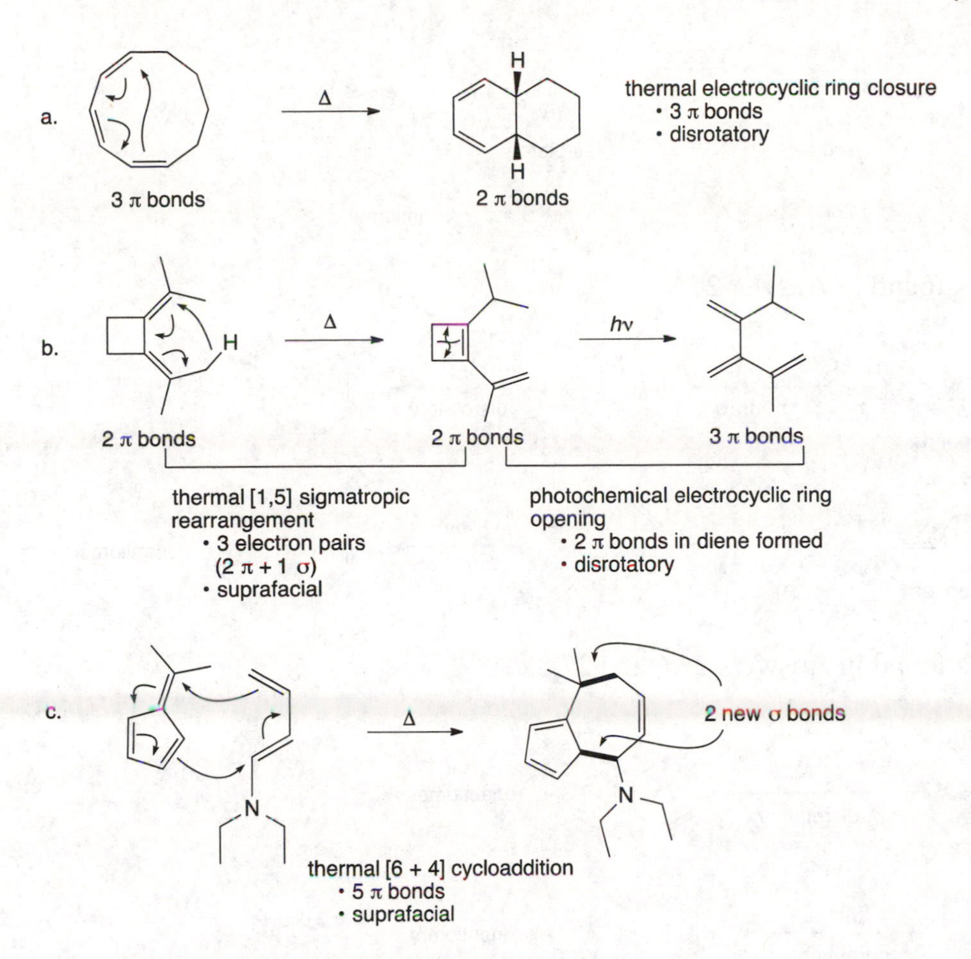

27.30 Use the rules for thermal electrocyclic reactions found in Answer 27.6.

a. Δ disrotatory — *E*, *Z*, *Z* — 3 π bonds

b. 2 π bonds — re-draw — Δ conrotatory

27.31 Use the rules for photochemical electrocyclic reactions found in Answer 27.8.

a. hν conrotatory — *E*, *Z*, *E* — 3 π bonds

Although conrotatory ring opening could also form, at least in theory, an all-(*Z*) triene, steric hindrance during ring opening would cause the terminal CH_3's to crash into one another, making this process unlikely.

b. re-draw; $h\nu$ disrotatory; + enantiomer

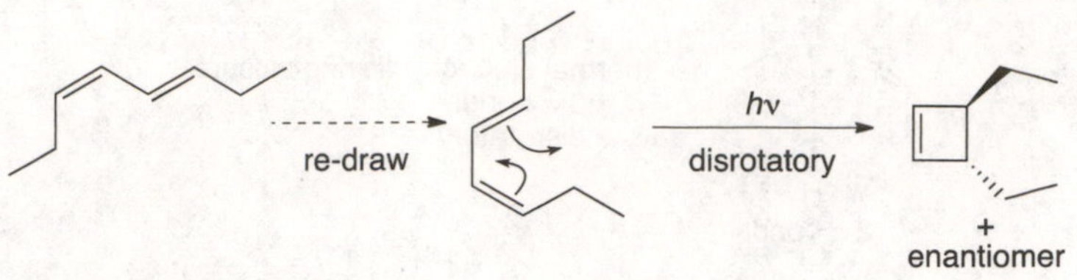

27.32 Use the rules found in Answer 27.8.

a. 4 π bonds; re-draw; $h\nu$ disrotatory

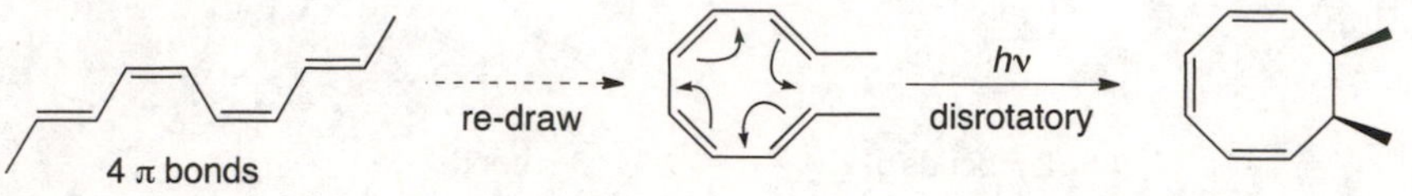

b. 4 π bonds; re-draw; $h\nu$ disrotatory; + enantiomer

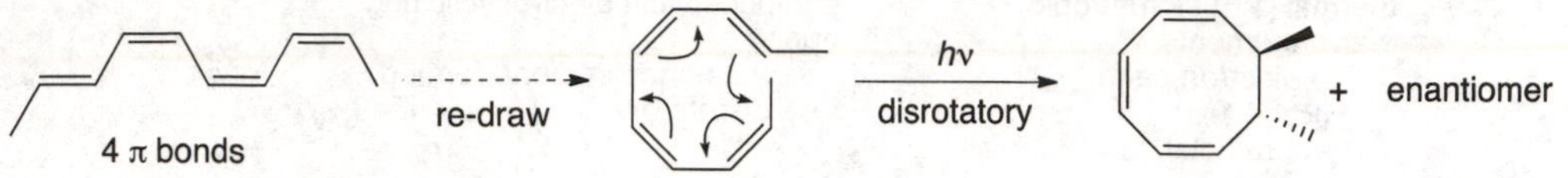

27.33 Use the rules found in Answers 27.6 and 27.8.

a, b. 2*E*, 4*Z*, 6*Z*; 3 π bonds; Δ disrotatory; + enantiomer; $h\nu$ conrotatory; + enantiomer

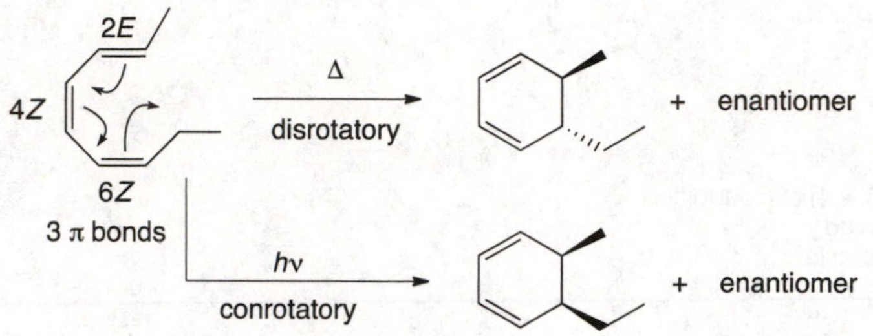

c. (one enantiomer); Δ disrotatory; *E*, *E*; 3 π bonds

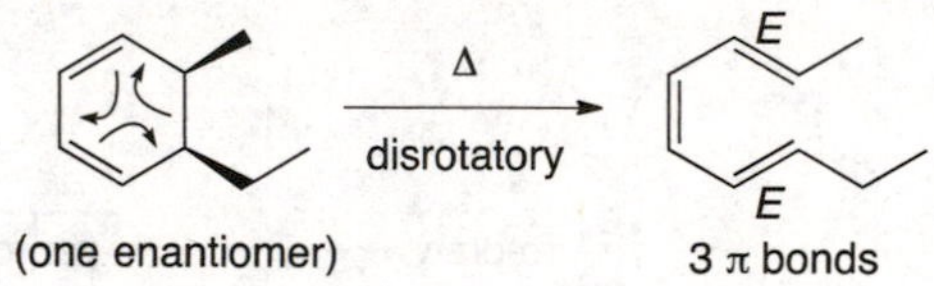

d. (one enantiomer); $h\nu$ conrotatory; *E*, *E*; 3 π bonds

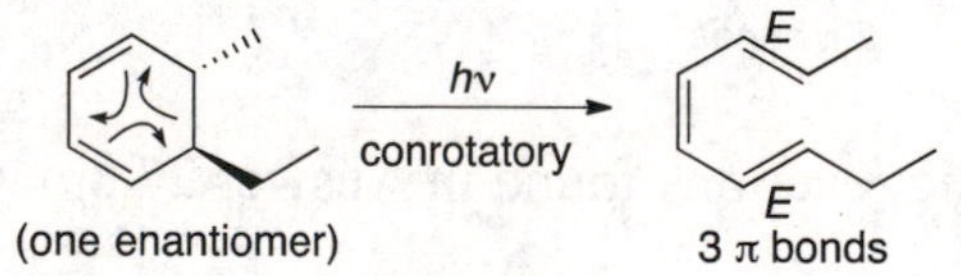

27.34 The trans product is indicative of a disrotatory ring closure from the cyclic triene with the given stereochemistry at the double bonds. A disrotatory ring closure with a polyene having three π bonds must occur under thermal conditions.

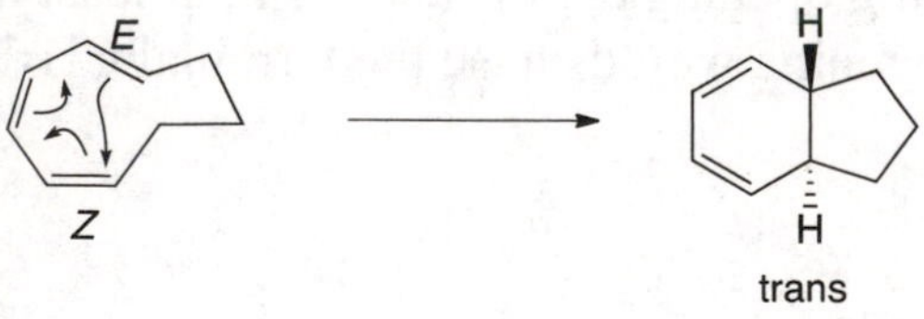

E, *Z*; H, H; trans

27.35 A disrotatory cyclization of a reactant with an even number of π bonds must occur under photochemical conditions.

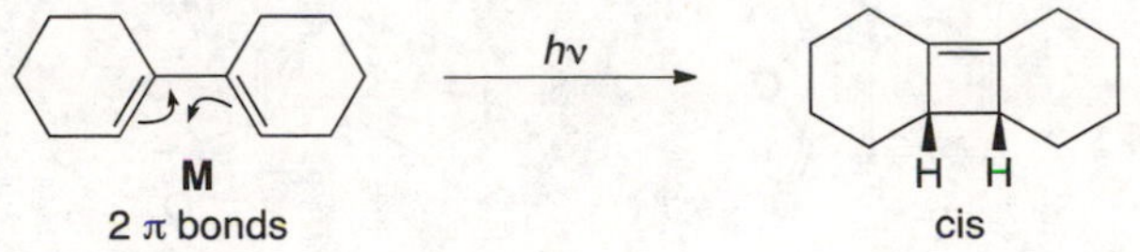

M
2 π bonds
$h\nu$
H H
cis

27.36 Use the rules found in Answers 27.6 and 27.8.

a, c.
N
3 π bonds
Δ
disrotatory
H
H
cis

b, c.
N
3 π bonds
$h\nu$
conrotatory
H
H
trans
+ enantiomer

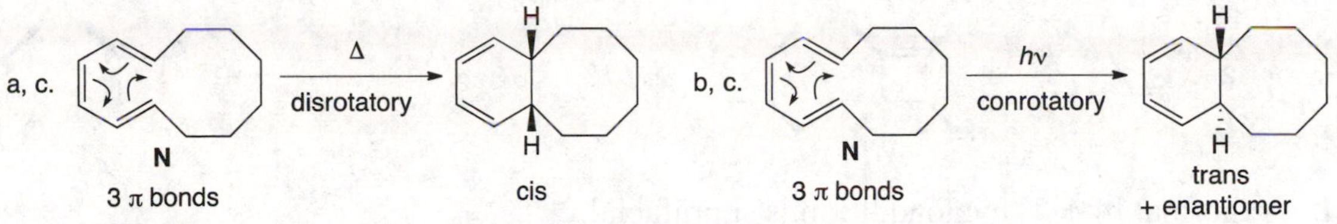

27.37 Use the rules found in Answers 27.6 and 27.8.

H
H
P
Δ
conrotatory
E
Z
Q
2 π bonds
$h\nu$
disrotatory
Z
Z
R

(A 10-membered ring cannot contain two *E* double bonds.)

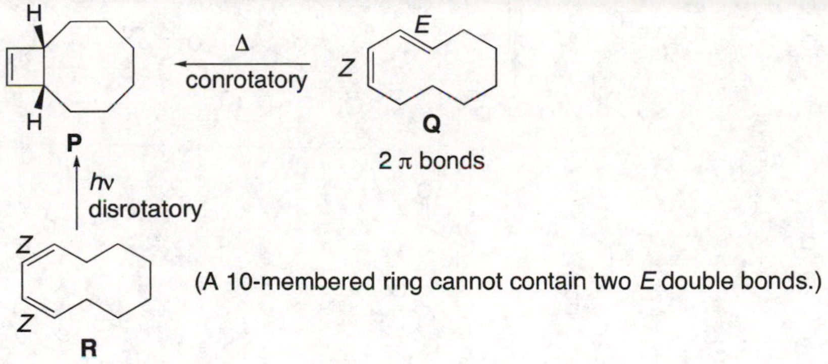

27.38 The reaction involves three π bonds in one reactant and two π bonds in the second reactant, so the reaction is a [6 + 4] cycloaddition. A suprafacial cycloaddition with five π bonds must proceed under thermal conditions.

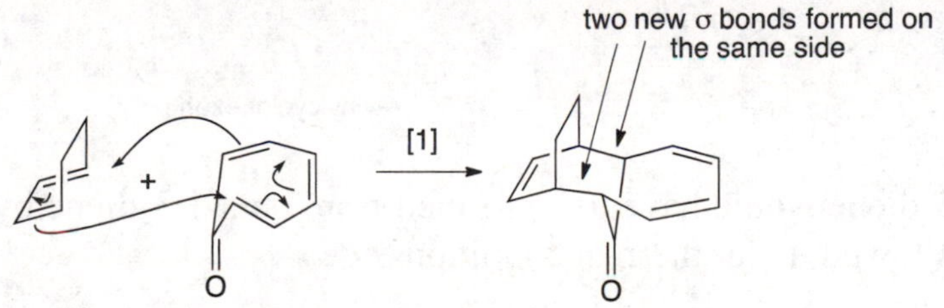

two new σ bonds formed on the same side

+
[1]
O
O

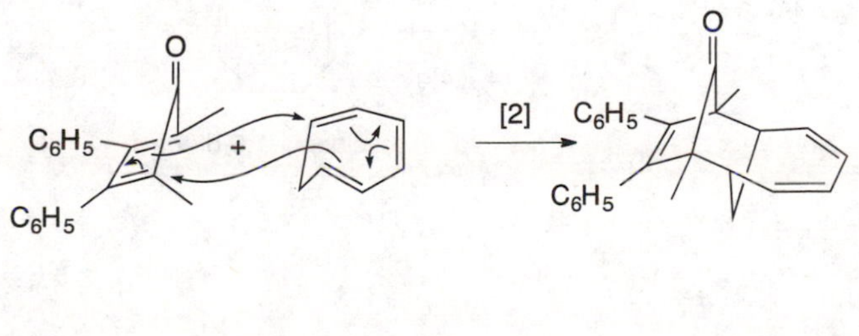

O
C_6H_5
C_6H_5
+
[2]
O
C_6H_5
C_6H_5

27.39 The Diels–Alder reaction is a thermal, suprafacial [4 + 2] cycloaddition.

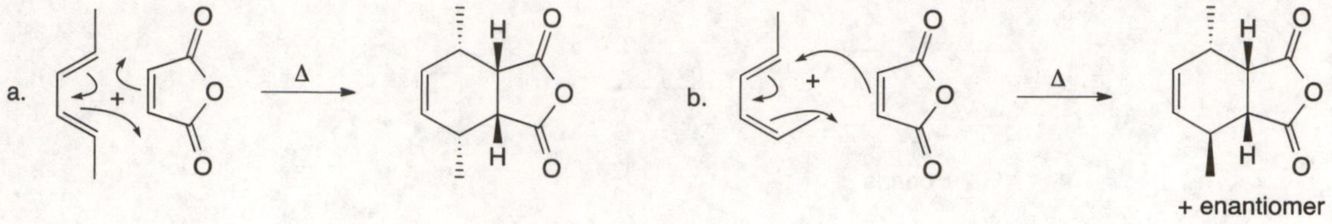

a. Δ

b. Δ

+ enantiomer

27.40 A photochemical [2 + 2] cycloaddition is suprafacial.

a. 2 $h\nu$ +

b. 2 $h\nu$ +

27.41 A thermal [4 + 2] cycloaddition is suprafacial.

a. Δ +

c. Δ +

b. Δ +

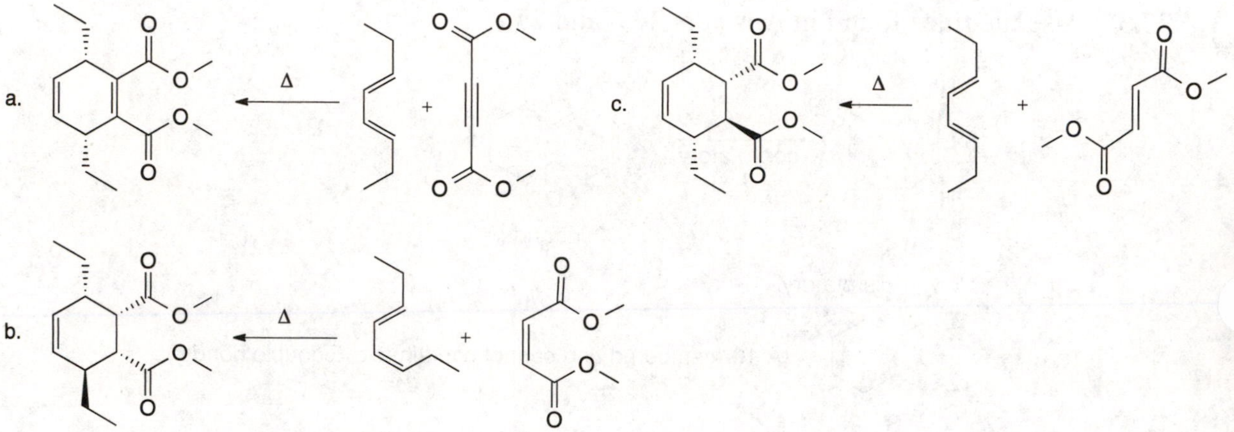

27.42 Buta-1,3-diene can react with itself in a symmetry-allowed thermal [4 + 2] cycloaddition to form 4-vinylcyclohexene.

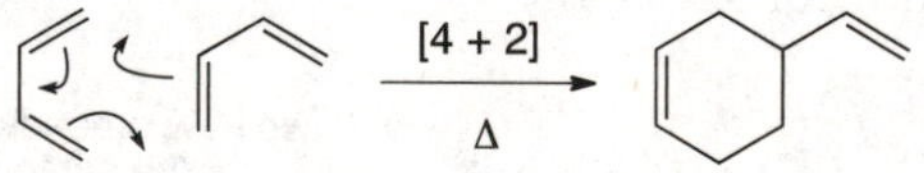

[4 + 2]
Δ

4-vinylcyclohexene

Cycloocta-1,5-diene would have to be formed from buta-1,3-diene by a [4 + 4] cycloaddition, which is not allowed under thermal conditions.

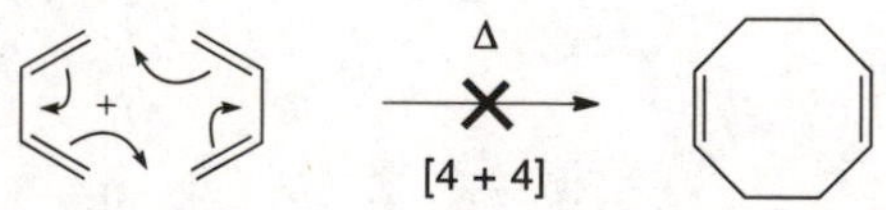

Δ
[4 + 4]

cycloocta-1,5-diene

27.43 A series of three [2 + 2] cycloadditions with *E* alkenes forms **X**.

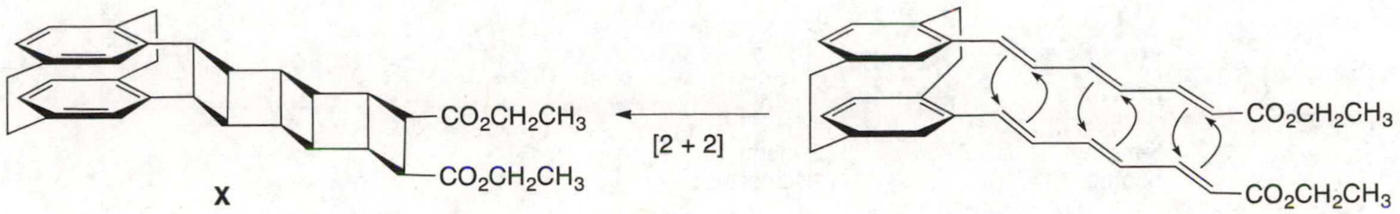

$-CO_2CH_2CH_3$ ← [2 + 2] $-CO_2CH_2CH_3$

$-CO_2CH_2CH_3$ $-CO_2CH_2CH_3$

X

27.44 Re-draw the reactant and product to more clearly show the relative location of the bonds broken and formed.

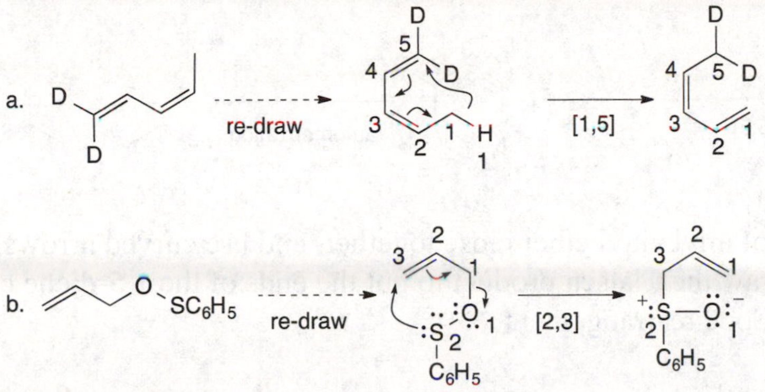

a. D, D — re-draw → (D, 5, 4, D, 3, 2, 1, H, 1) — [1,5] → (D, 4, 5, D, 3, 2, 1)

b. O, SC_6H_5 — re-draw → (2, 3, 1, O, 1, S, 2, C_6H_5) — [2,3] → (2, 3, 1, $^+$S—O$^-$, 2, 1, C_6H_5)

27.45 Draw the products of each reaction.

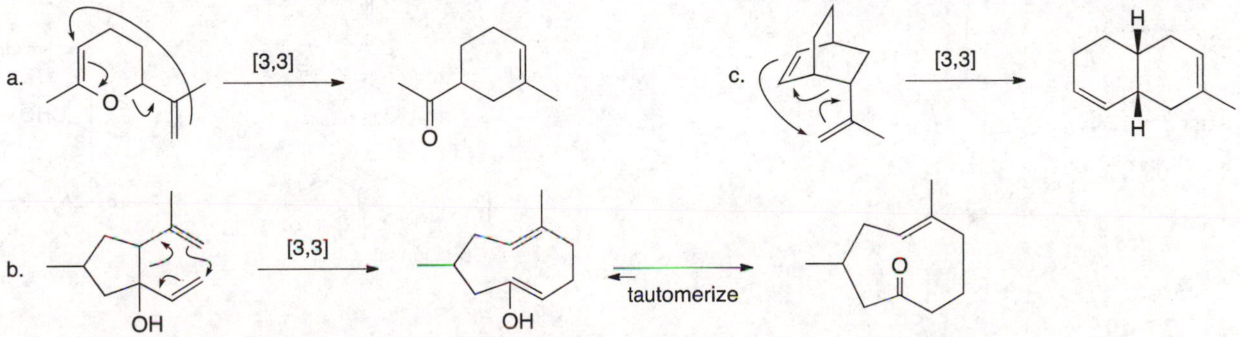

a. [3,3] → O

c. [3,3] → H, H

b. OH — [3,3] → OH ⇌ tautomerize — O

27.46 a. Two [1,5] sigmatropic rearrangements occur.

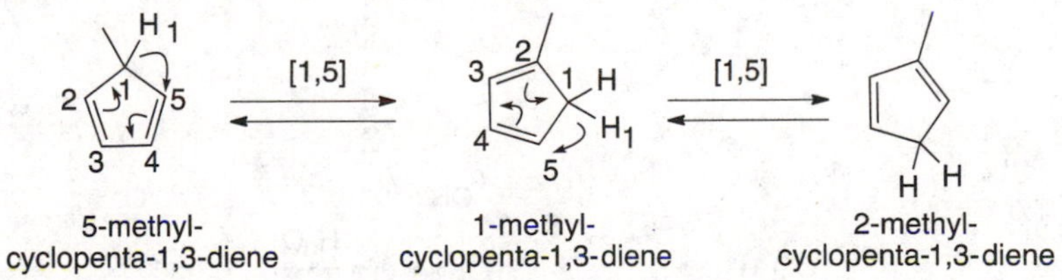

(H 1, 1, 2, 3, 4, 5) ⇌ [1,5] ⇌ (2, 3, 1, H, H1, 4, 5) ⇌ [1,5] ⇌ (H, H)

5-methyl-cyclopenta-1,3-diene 1-methyl-cyclopenta-1,3-diene 2-methyl-cyclopenta-1,3-diene

b.

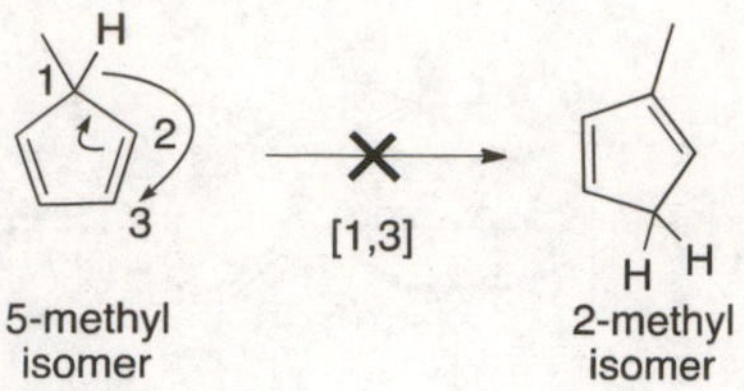

5-methyl isomer → [1,3] (X) → 2-methyl isomer

A [1,3] sigmatropic rearrangement requires photochemical conditions not thermal conditions, so 5-methylcyclopenta-1,3-diene cannot rearrange directly to its 2-methyl isomer by a [1,3] shift.

27.47

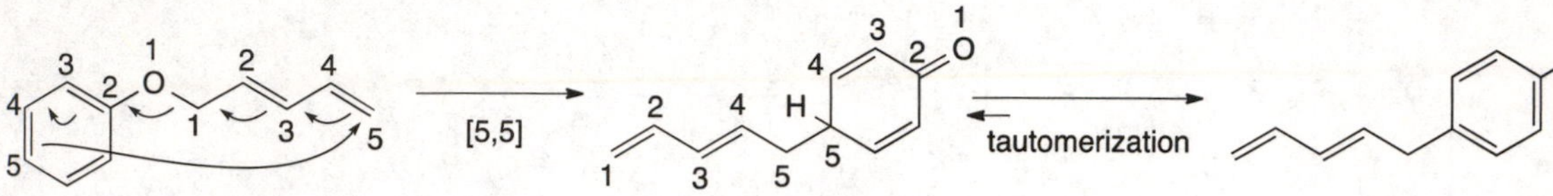

[5,5] → tautomerization

27.48 Re-draw **A** to put the ends of allyl vinyl ether close together, and use curved arrows to draw the Claisen product. Then re-draw the Claisen product to put the ends of the 1,5-diene close together to draw the product of the Cope rearrangement.

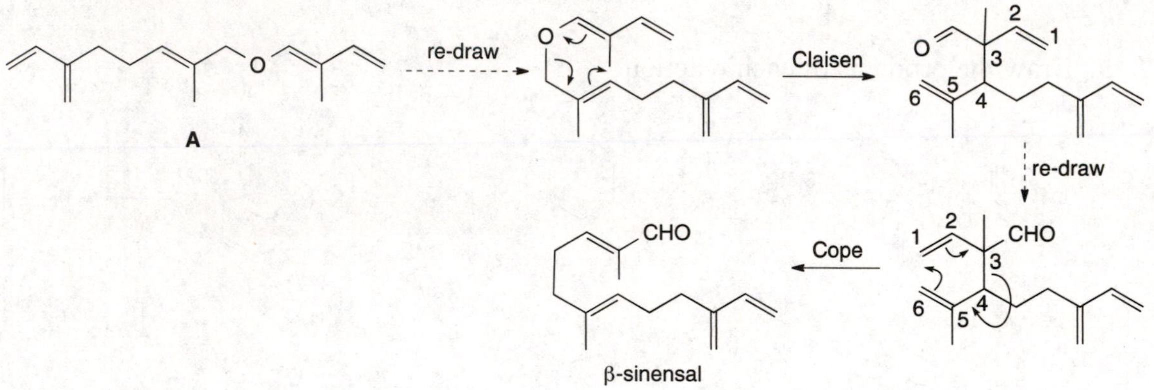

A → re-draw → Claisen → re-draw → Cope → β-sinensal

27.49

+ pyridine → **A** → LDA → ↔ **B** → re-draw → Δ [3,3] → **C** → H_3O^+ → re-draw → CO_2H

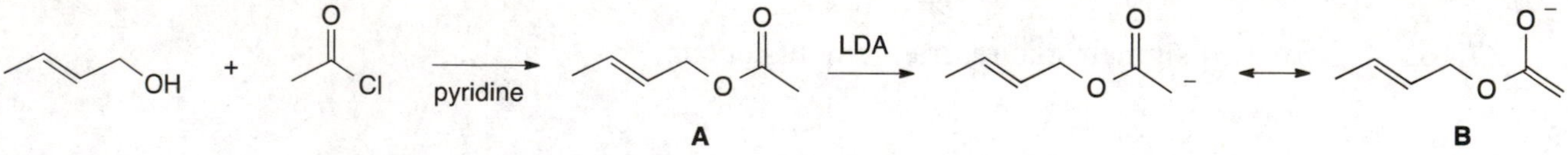

The conversion of **B** to **C** is a [3,3] sigmatropic rearrangement of an intermediate enolate.

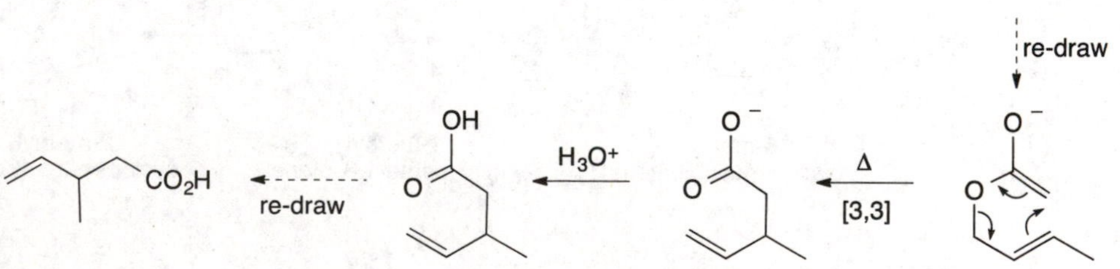

27.50 Use the definitions found in Answer 27.1.

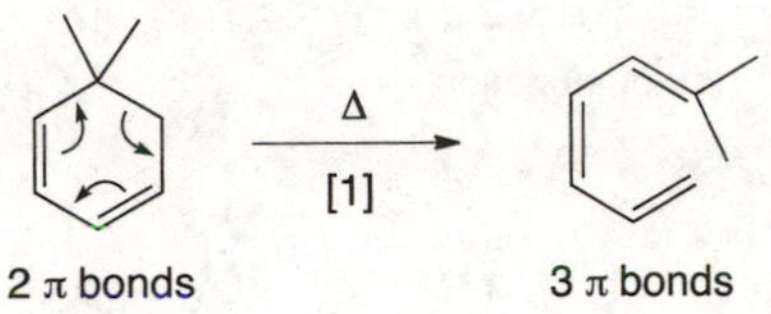

Δ [1]

2 π bonds → 3 π bonds: an electrocyclic ring opening

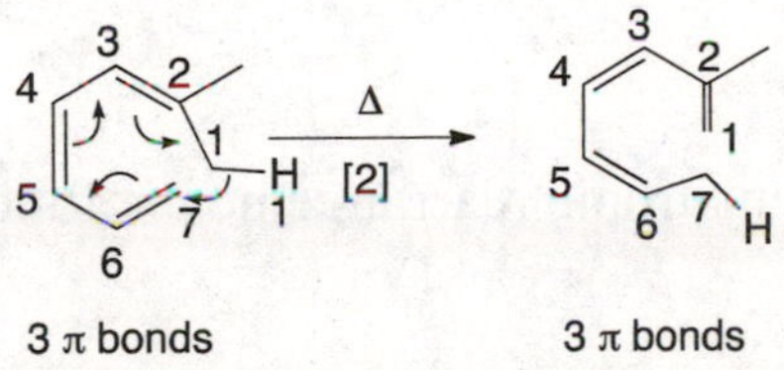

3 2 4 1 H 5 7 1 6 —Δ [2]→ 3 4 2 1 5 6 7 H

3 π bonds → 3 π bonds: a [1,7] sigmatropic rearrangement

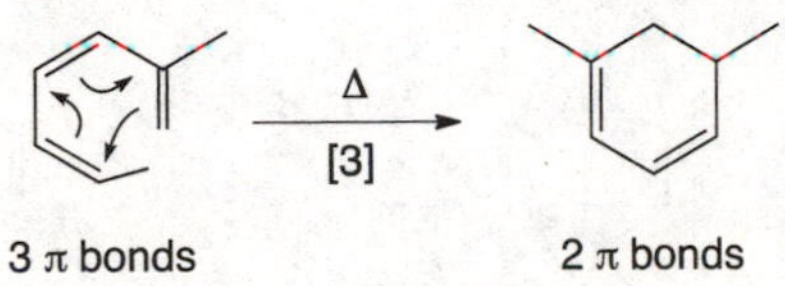

Δ [3]

3 π bonds → 2 π bonds: an electrocyclic ring closure

27.51

a. 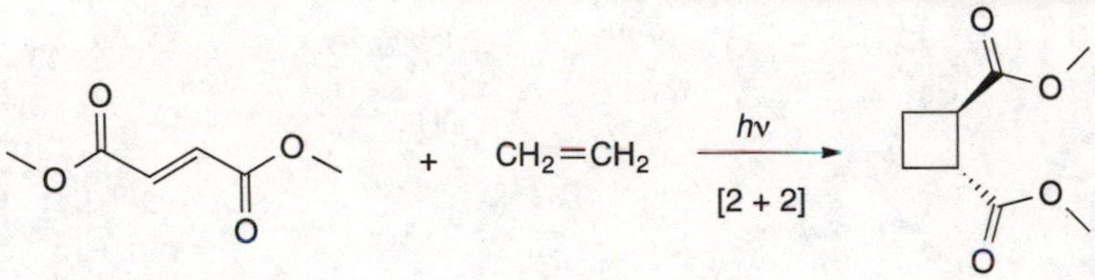 O O O O + $CH_2{=}CH_2$ —hν, [2 + 2]→ O O O O + enantiomer

b. 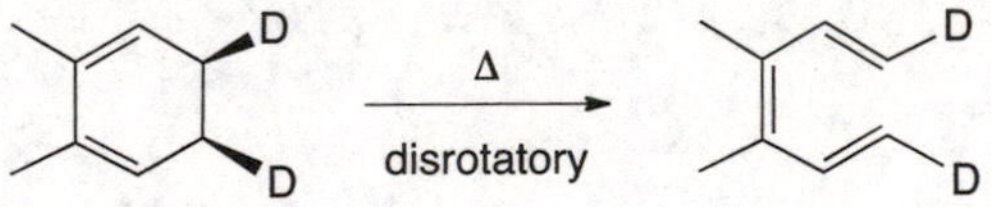 D D —Δ, disrotatory→ D D

c. 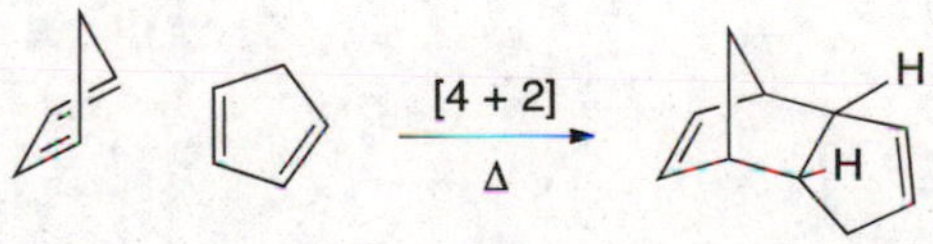 —[4 + 2], Δ→ H H

d. D D —hν, conrotatory→ D D

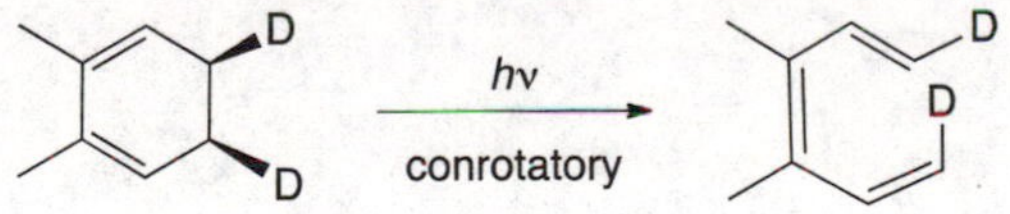

27.52

a. 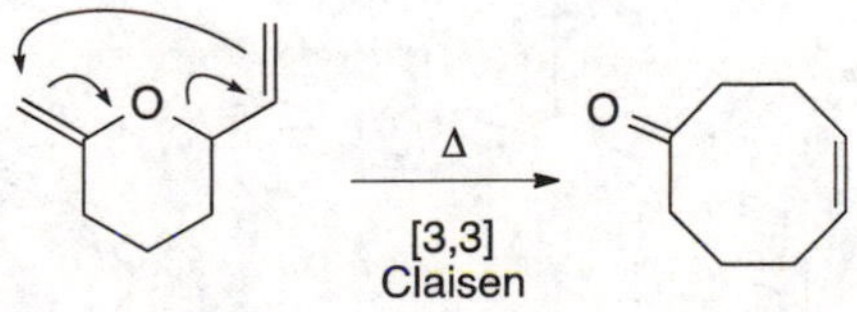 O —Δ, [3,3] Claisen→ O

b. 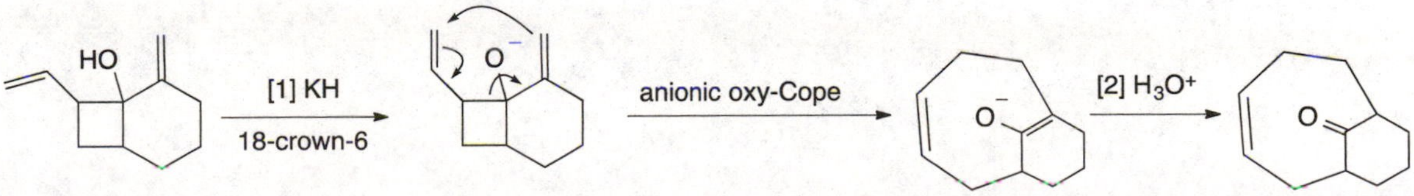 HO —[1] KH, 18-crown-6→ O⁻ —anionic oxy-Cope→ O⁻ —[2] H_3O^+→ O

c. Δ Claisen → tautomerization

d. $h\nu$ [2 + 2] → C_7H_8

27.53 The mechanism consists of sequential [3,3] sigmatropic rearrangements, followed by tautomerization.

[3,3] → [3,3] → H⁺ source →

27.54

[3,3] ← allyl vinyl ether + HB⁺

27.55

H, H, Z, E

Δ — conrotatory electrocyclic ring opening forming 4 π bonds

Δ — disrotatory ring closure involving only 3 of the 4 π bonds

27.56

H, O, **E**

[1,5] sigmatropic rearrangement

re-draw

[4 + 2] cycloaddition

27.57

SCH_3, OCH_3 — Δ — electrocyclic ring opening forming two π bonds

CH_3S, CH_3O — Δ — [4 + 2] cycloaddition suprafacial

27.58 The mechanism consists of a [4 + 2] cycloaddition, followed by intramolecular imine formation.

NH_2 — [4 + 2] — proton transfer

+ H_3O^+

27.59

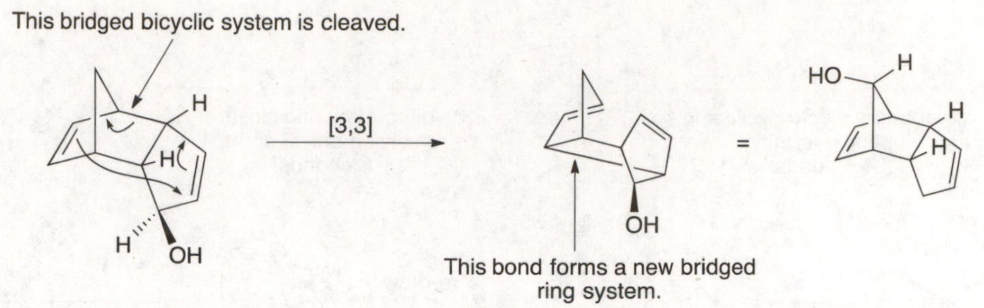

This bridged bicyclic system is cleaved.

H

H

H

OH

[3,3]

This bond forms a new bridged ring system.

OH

=

HO

H

H

H

27.60

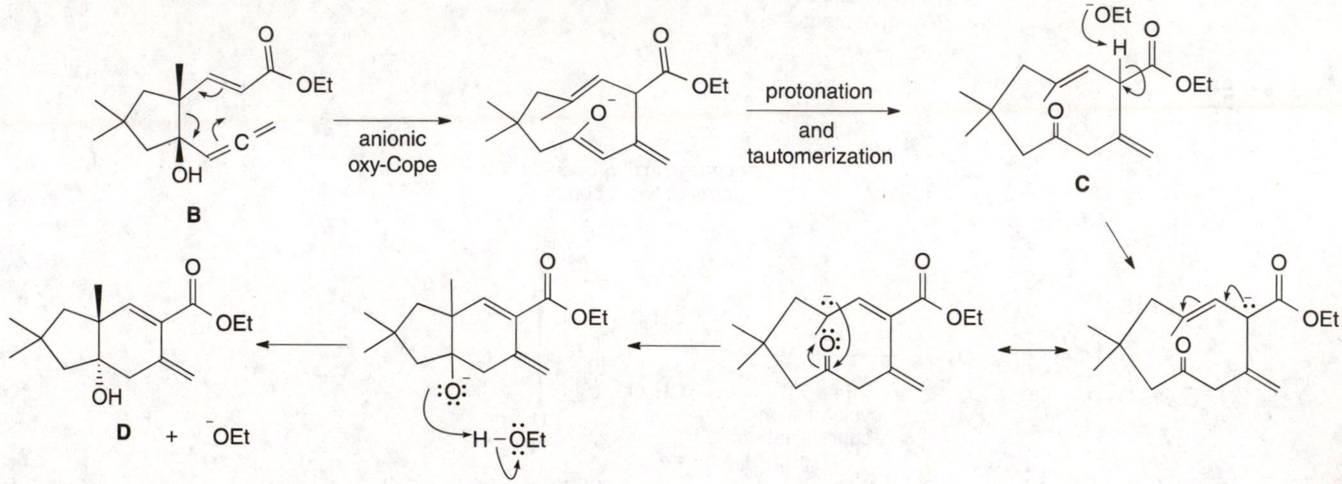

O

OEt

C

OH

B

anionic oxy-Cope

O

OEt

O^-

protonation and tautomerization

^-OEt

H

O

OEt

O

C

O

OEt

O

O

OEt

O:

O

OEt

:O:

H–ÖEt

O

OEt

OH

D + ^-OEt

27.61

:B

Ö–H

+ HB^+

$Ö^-$

O O

O

O :Ö:

O

O

O

B:

H

+

$Ö^-$

H_2O

ÖH

O :Ö:

O

O :Ö:

O

H–ÖH

O H O

O

+ $^-$:ÖH

:Ö: H

:B

+ CO_2

:Ö:

:O:

H–ÖH

:O:

H

+ HO^-

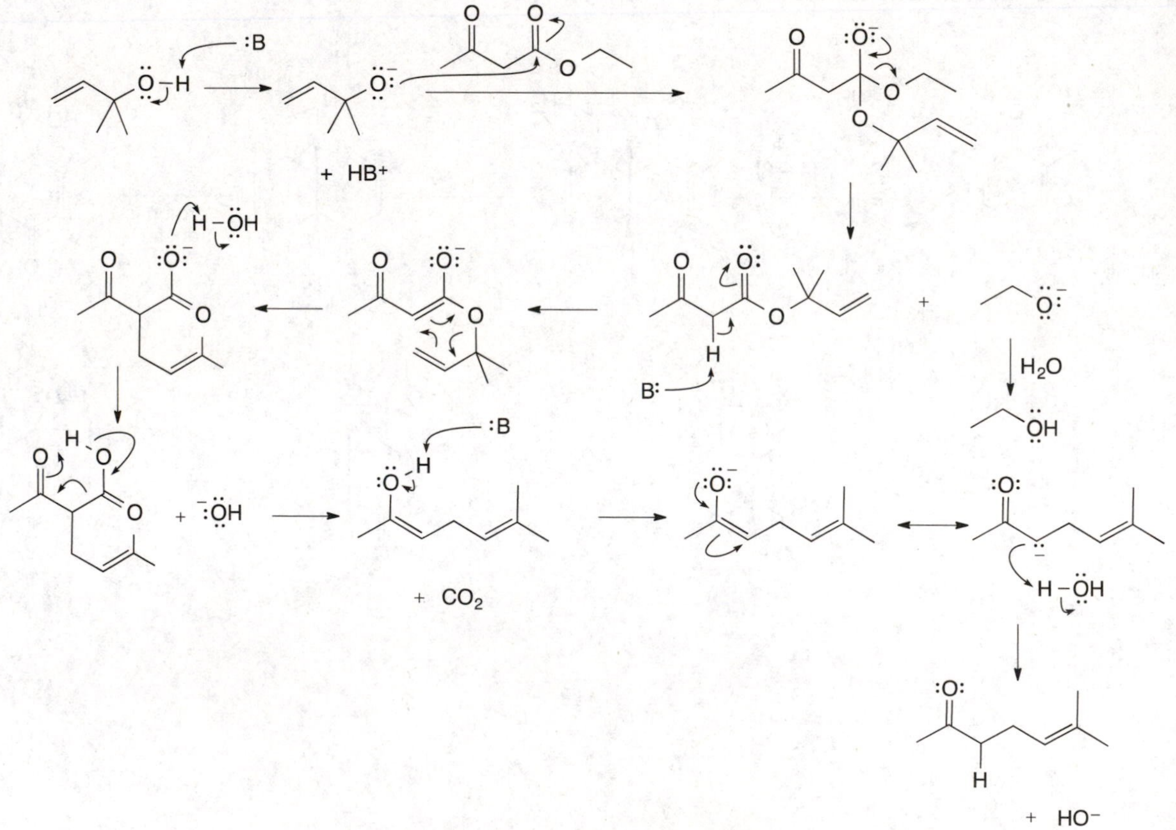

27.62 Conrotatory cyclization of **Y** using four π bonds forms **X.** Disrotatory ring closure of **X** can occur in two ways—on the top face or the bottom face of the eight-membered ring to form diastereomers.

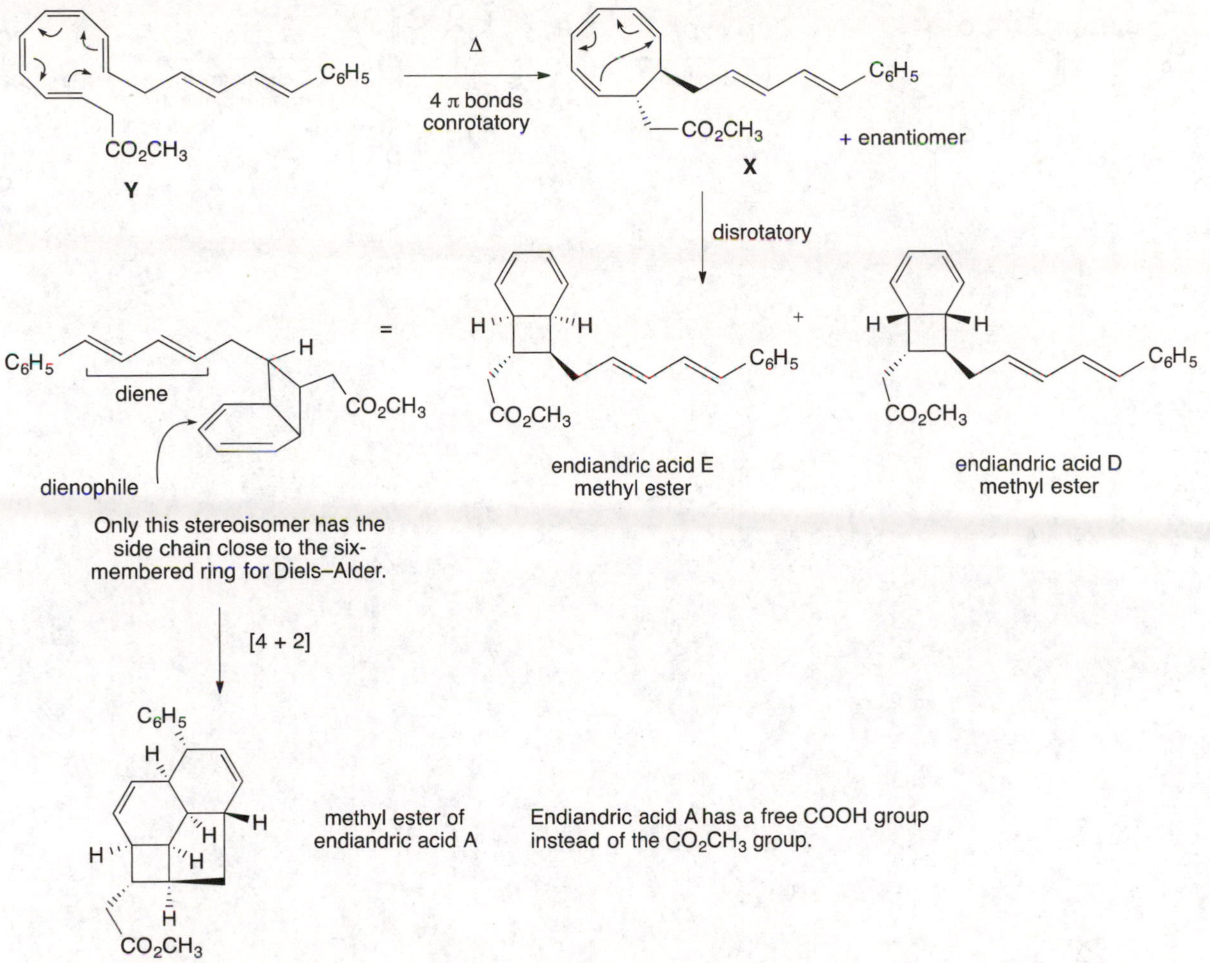

C_6H_5 CO_2CH_3 **Y** Δ 4 π bonds conrotatory C_6H_5 CO_2CH_3 **X** + enantiomer

disrotatory

C_6H_5 diene H CO_2CH_3 dienophile =

Only this stereoisomer has the side chain close to the six-membered ring for Diels–Alder.

H H C_6H_5 CO_2CH_3 endiandric acid E methyl ester + H H C_6H_5 CO_2CH_3 endiandric acid D methyl ester

[4 + 2]

C_6H_5 H H H H H H CO_2CH_3 methyl ester of endiandric acid A

Endiandric acid A has a free COOH group instead of the CO_2CH_3 group.

27.63

a. O O C_6H_5 diene dienophile → O O C_6H_5 or O O C_6H_5 diene dienophile → O O C_6H_5

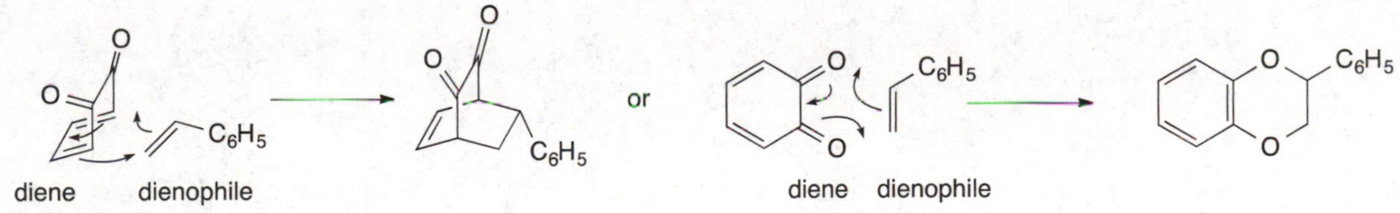

b. O O diene dienophile → O O or O O dienophile diene → O O

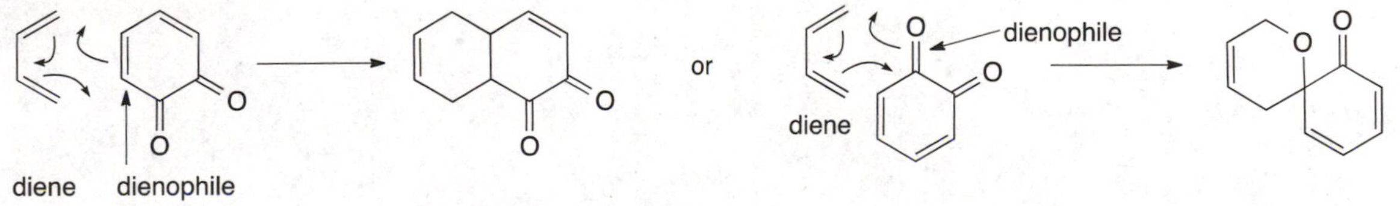

c.

C_4H_9 O O O O C_4H_9 O O O

[4 + 2]

cycloaddition

C_4H_9 O O O O O C_4H_9 O O

[3,3]

sigmatropic rearrangement

O O O O C_4H_9 O O O C_4H_9

Chapter 28 Carbohydrates

Chapter Review

Important terms

- **Aldose** A monosaccharide containing an aldehyde (28.2)
- **Ketose** A monosaccharide containing a ketone (28.2)
- **D-Sugar** A monosaccharide with the O bonded to the stereogenic center farthest from the carbonyl group drawn on the right in the Fischer projection (28.2C)
- **Epimers** Two diastereomers that differ in configuration around one stereogenic center only (28.3)
- **Anomers** Monosaccharides that differ in configuration at only the hemiacetal OH group (28.6)
- **Glycoside** An acetal derived from a monosaccharide hemiacetal (28.7)

Acyclic, Haworth, and 3-D representations for D-glucose (28.6)

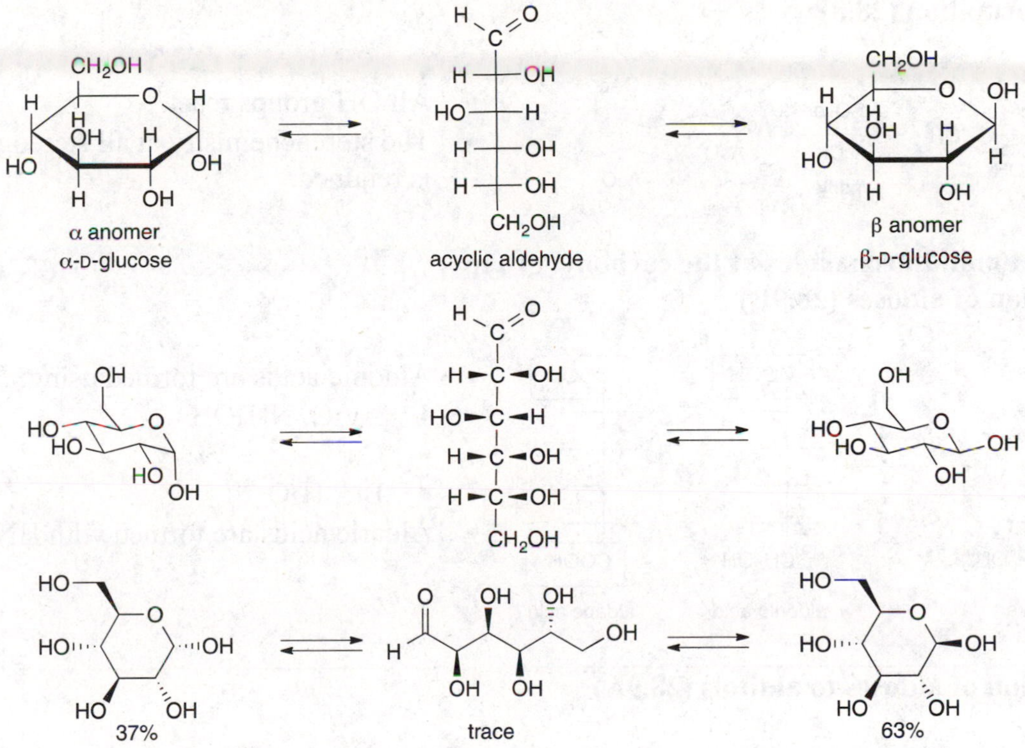

α anomer
α-D-glucose

acyclic aldehyde

β anomer
β-D-glucose

37%

trace

63%

Reactions of monosaccharides involving the hemiacetal

[1] Glycoside formation (28.7A)

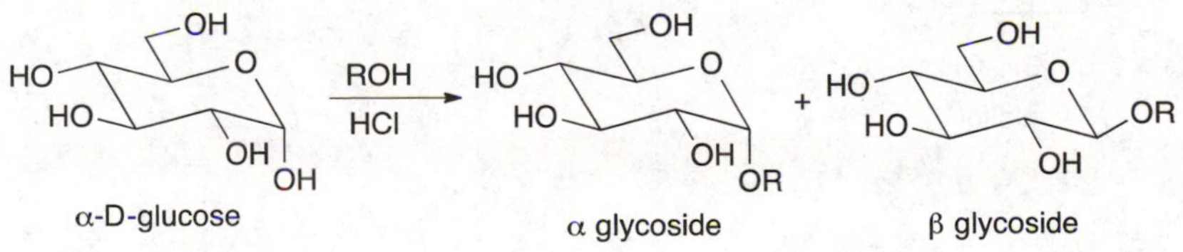

α-D-glucose —ROH, HCl→ α glycoside + β glycoside

- Only the hemiacetal OH reacts.
- A mixture of α and β glycosides forms.

[2] Glycoside hydrolysis (28.7B)

H$_3$O$^+$ → α anomer + β anomer + ROH

- A mixture of α and β anomers forms.

Reactions of monosaccharides at the OH groups

[1] Ether formation (28.8)

Ag$_2$O, RX →

- All OH groups react.
- The stereochemistry at all stereogenic centers is retained.

[2] Ester formation (28.8)

Ac$_2$O or AcCl, pyridine →

- All OH groups react.
- The stereochemistry at all stereogenic centers is retained.

Reactions of monosaccharides at the carbonyl group

[1] Oxidation of aldoses (28.9B)

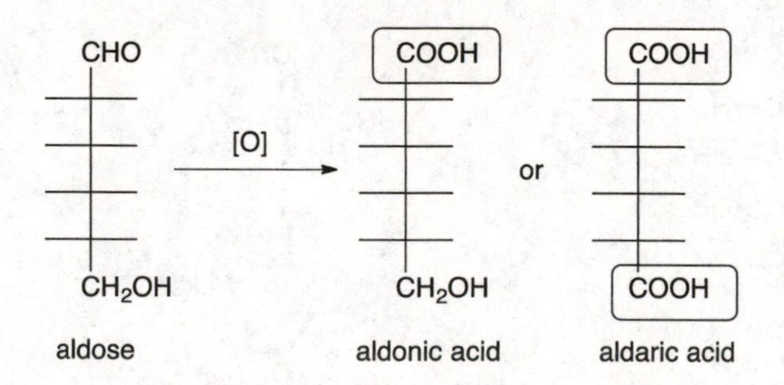

- Aldonic acids are formed using:
  - Ag$_2$O, NH$_4$OH
  - Cu^{2+}
  - Br$_2$, H$_2$O
- Aldaric acids are formed with HNO$_3$, H$_2$O.

[2] Reduction of aldoses to alditols (28.9A)

CHO ... CH$_2$OH (aldose) → NaBH$_4$, CH$_3$OH → CH$_2$OH ... CH$_2$OH (alditol)

[3] Wohl degradation (28.10A)

This C–C bond is cleaved.

CHO, H–C2–OH, ..., CH_2OH → [1] NH_2OH [2] Ac_2O, NaOAc [3] $NaOCH_3$ → CHO, ..., CH_2OH

- The C1–C2 bond is cleaved to shorten an aldose chain by one carbon.
- The stereochemistry at all other stereogenic centers is retained.
- Two epimers at C2 form the same product.

[4] Kiliani–Fischer synthesis (28.10B)

CHO, ..., CH_2OH → [1] NaCN, HCl [2] H_2, Pd-$BaSO_4$ [3] H_3O^+ → CHO, H–C2–OH, ..., CH_2OH + CHO, HO–C2–H, ..., CH_2OH

- One carbon is added to the aldehyde end of an aldose.
- Two epimers at C2 are formed.

Other reactions

[1] Hydrolysis of disaccharides (28.11)

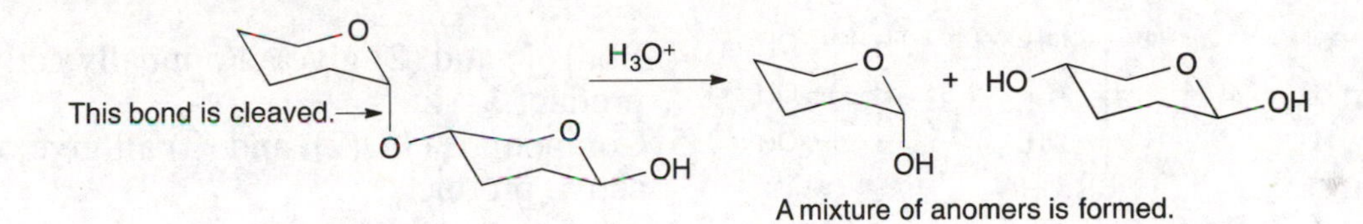

A mixture of anomers is formed.

[2] Formation of *N*-glycosides (28.13B)

CH_2OH furanose (H, OH at C1) → RNH_2, mild H^+ → CH_2OH furanose (H, NHR at C1) + CH_2OH furanose (NHR, H at C1)

- Two anomers are formed.

Practice Test on Chapter Review

1.a. How are the following two representations related to each other?

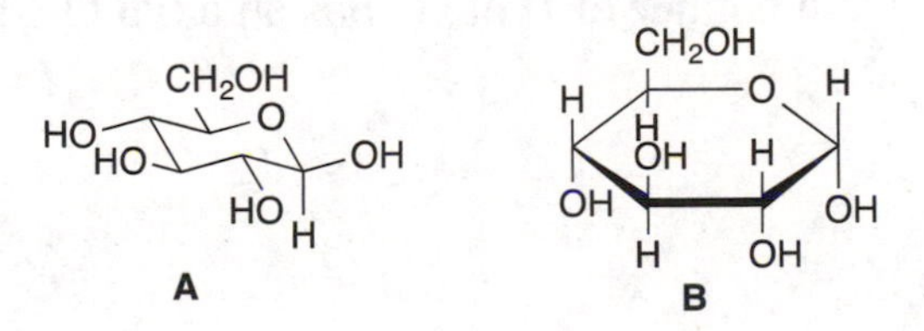

1. **A** and **B** are anomers of each other.
2. **A** and **B** are epimers of each other.
3. **A** and **B** are diastereomers of each other.
4. Statements (1) and (2) are both true.
5. Statements (1), (2), and (3) are all true.

b. Which of the following statements is (are) true about monosaccharide **C**?

OH, HO, O, HO, OH, HO **C**

1. **C** is a D-sugar.
2. The β anomer is drawn.
3. **C** is an aldohexose.
4. Statements (1) and (2) are both true.
5. Statements (1), (2), and (3) are all true.

c. Which of the following are different representations for monosaccharide **D**?

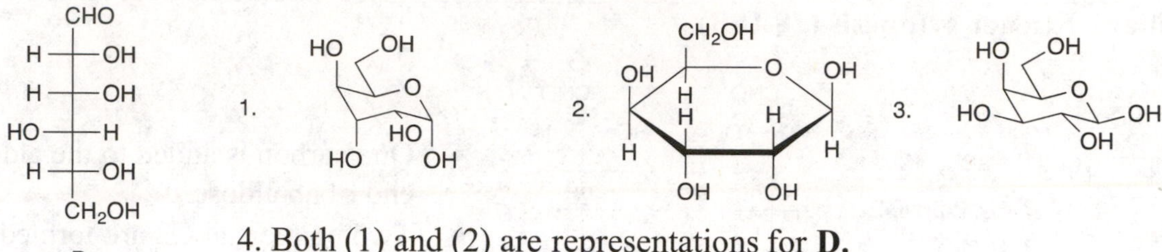

4. Both (1) and (2) are representations for **D**.
5. Compounds (1), (2), and (3) all represent **D**.

d. Which aldoses give an optically active compound upon reaction with $NaBH_4$ in CH_3OH?

1. CHO; H—OH; HO—H; HO—H; CH_2OH
2. CHO; H—OH; HO—H; H—OH; CH_2OH
3. CHO; H—OH; H—OH; H—OH; CH_2OH
4. Both (1) and (2) give an optically active product.
5. Compounds (1), (2), and (3) all give optically active products.

2. Answer each question about monosaccharide **D** as True (T) or False (F).

OH, HO, O, HO, HO, OH **D**

a. **D** is a D-sugar.

b. **D** is drawn as an α anomer.

c. **D** is an aldohexose.

d. Reduction of **D** with $NaBH_4$ in CH_3OH forms an optically inactive alditol.

e. Oxidation of **D** with Br_2, H_2O forms an optically active aldonic acid.

f. Oxidation of **D** with HNO_3 forms an optically active aldaric acid.

g. C2 has the *R* configuration.

h. Treatment of **D** with CH_3OH, HCl forms two products.

i. Treatment of **D** with Ag_2O, and CH_3I (excess) forms two products.

j. An epimer of **D** at C3 has an axial OH group.

3. Answer the following questions about the three monosaccharides (**A–C**) drawn below.

a. Draw the α anomer of **A** in a Haworth projection.
b. Draw the β anomer of **B** in a three-dimensional representation using a chair conformation.
c. Convert **C** into the acyclic form of the monosaccharide using a Fischer projection.
d. Which two aldoses yield **A** in a Wohl degradation?

4. Draw the product of each reaction with the starting material D-xylose.

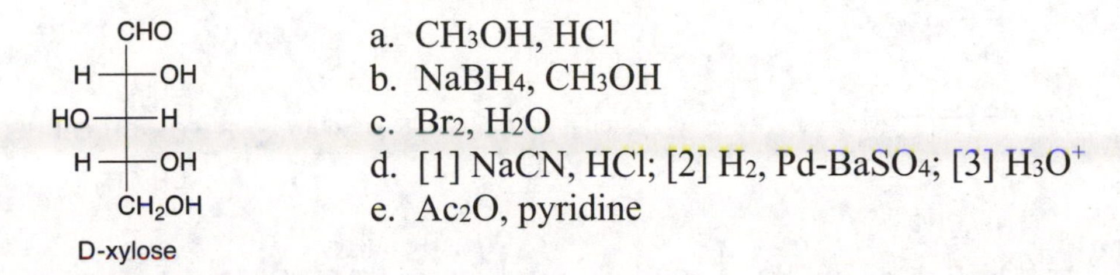

a. CH_3OH, HCl
b. $NaBH_4$, CH_3OH
c. Br_2, H_2O
d. [1] NaCN, HCl; [2] H_2, Pd-$BaSO_4$; [3] H_3O^+
e. Ac_2O, pyridine

Answers to Practice Test

1.a. 5
b. 5
c. 4
d. 1

2. a. T
b. F
c. T
d. F
e. T
f. T
g. F
h. T
i. F
j. T

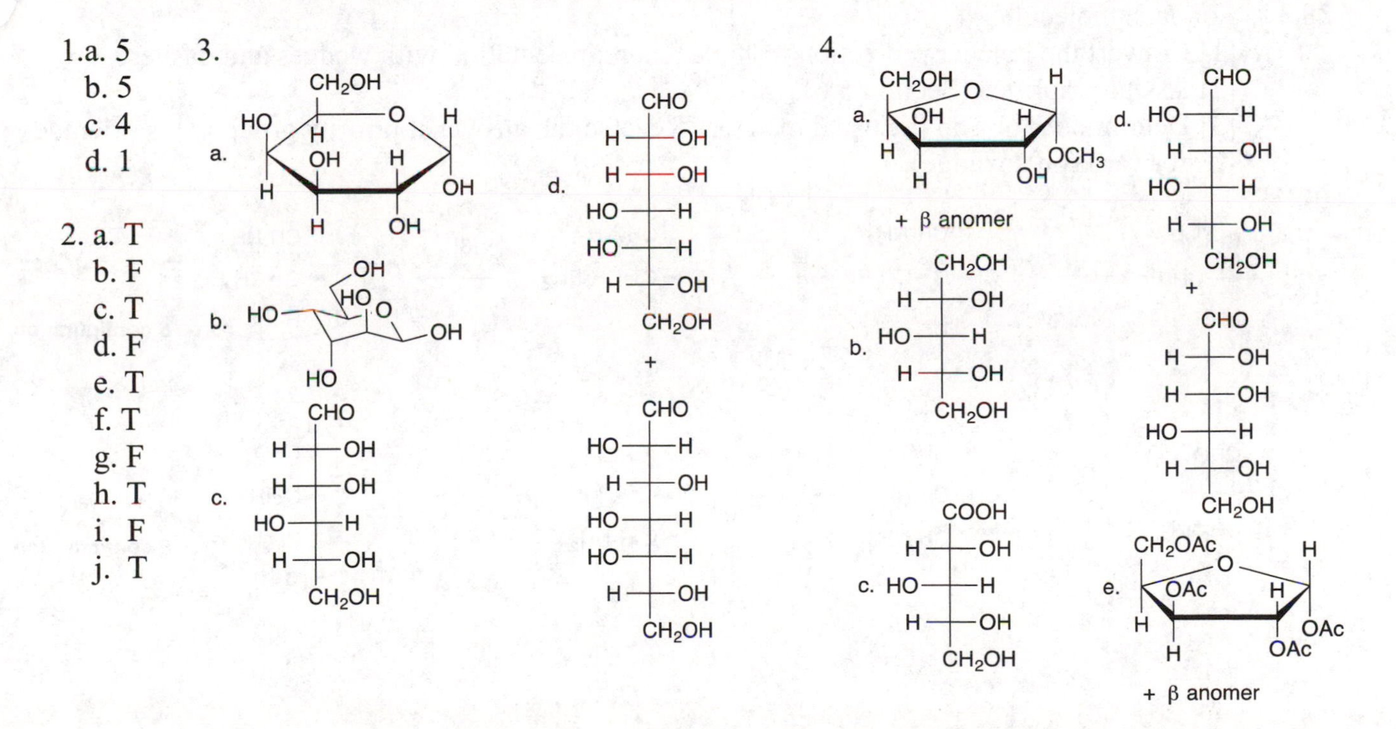

Answers to Problems

28.1 A *ketose* is a monosaccharide containing a ketone. An *aldose* is a monosaccharide containing a aldehyde. A monosaccharide is called: a *triose* if it has three C's, a *tetrose* if it has four C's, a *pentose* if it has five C's, a *hexose* if it has six C's, and so forth.

a. a ketotetrose

CH_2OH
C=O
H–C–OH
CH_2OH

b. an aldopentose

CHO
H–C–OH
H–C–OH
H–C–OH
CH_2OH

c. an aldotetrose

CHO
H–C–OH
H–C–OH
CH_2OH

28.2 Rotate and re-draw each molecule to place the horizontal bonds in front of the plane and the vertical bonds behind the plane. Then use a cross to represent the stereogenic center in a Fischer projection formula.

a. HO–C=O
CH_3–C–OH
CH_2CH_2OH
=
HO–C=O
CH_3–+–OH
CH_2CH_2OH

b. O=C–H
HO–C–H
CH_3
re-draw →
CHO
HO–C–CH_3
H
=
CHO
HO–+–CH_3
H

c. O
H
HO
H
re-draw →
CHO
H–C–CH_2OH
CH_2CH_3
=
CHO
H–+–CH_2OH
CH_2CH_3

d. OH H
H–C–C=O
CH_3
re-draw →
CHO
HO–C–CH_3
H
=
CHO
HO–+–CH_3
H

28.3 For each molecule:
[1] Convert the Fischer projection formula to a representation with wedges and dashes.
[2] Assign priorities (Section 5.6).
[3] Determine *R* or *S* in the usual manner. Reverse the answer if priority group [4] is oriented forward (on a wedge).

a. CH_2NH_2
Cl–+–CH_2Br
H
[1] →
CH_2NH_2
Cl–C–CH_2Br
H
[2] →
3
CH_2NH_2
1 Cl–C–CH_2Br 2
H
4
[3] →
3
CH_2NH_2
1 Cl–C–CH_2Br 2
H
S configuration

b. CHO
Cl–+–H
CH_2NH_2
[1] →
CHO
Cl–C–H
CH_2NH_2
[2] →
2
CHO
1 Cl–C–H 4
CH_2NH_2
3
[3] →
2
CHO
1 Cl–C–H ← H forward
CH_2NH_2
3
S configuration

c. CHO, Cl—C—H, CH₂OH [1]→ Cl◄C◄H (CHO, CH₂OH) [2]→ 2 CHO; 1 Cl—C—H 4; CH₂OH 3 [3]→ 1 Cl, 2 CHO, 3 CH₂OH, H forward

S configuration

d. COOH, Cl—C—CH₂Br, H [1]→ Cl—C—CH₂Br (COOH, H) [2]→ 3 COOH; 1 Cl—C—CH₂Br 2; H 4 [3]→ 1 Cl, 2 CH₂Br, 3 COOH, H

S configuration

28.4

HO H H OH, HO, CHO, HO H HO H — rotate → OHC, H, OH, HO, H, H, OH, H, HO, HO — Convert staggered to eclipsed. → OHC, OH, H, OH, H, OH, H, H, OH, OH — re-draw → CHO; HO—C—H; HO—C—H; HO—C—H; H—C—OH; CH₂OH = CHO; HO—H; HO—H; HO—H; H—OH; CH₂OH

28.5

CHO *R*; H—C—OH *S*; HO—C—H *R*; H—C—OH *R*; H—C—OH; CH₂OH

D-glucose

28.6 A D sugar has the OH group on the stereogenic center farthest from the carbonyl on the right. An L sugar has the OH group on the stereogenic center farthest from the carbonyl on the left.

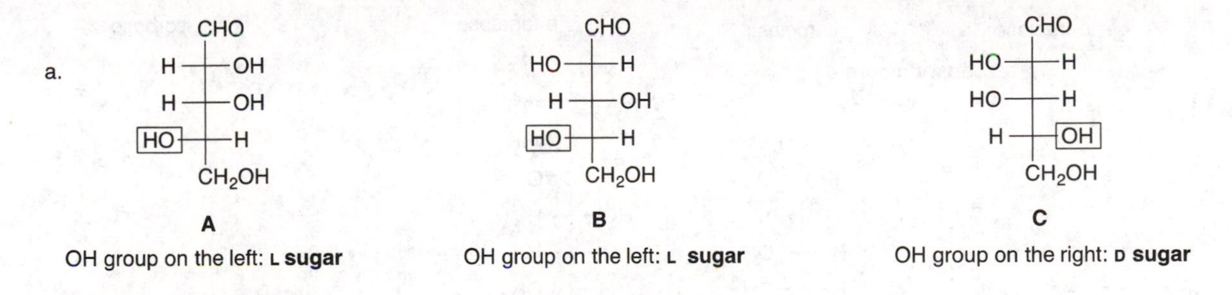

b. **A** and **B** are diastereomers.
A and **C** are enantiomers.
B and **C** are diastereomers.

28.7 There are 32 aldoheptoses; 16 are D sugars.

28.8 *Epimers* are two diastereomers that differ in the configuration around only one stereogenic center.

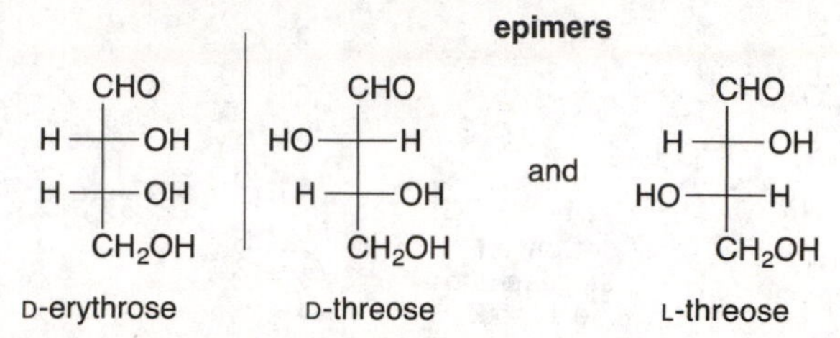

28.9
a. D-allose and L-allose: **enantiomers**
b. D-altrose and D-gulose: **diastereomers** but not epimers
c. D-galactose and D-talose: **epimers**
d. D-mannose and D-fructose: **constitutional isomers**
e. D-fructose and D-sorbose: **diastereomers** but not epimers
f. L-sorbose and L-tagatose: **epimers**

28.10

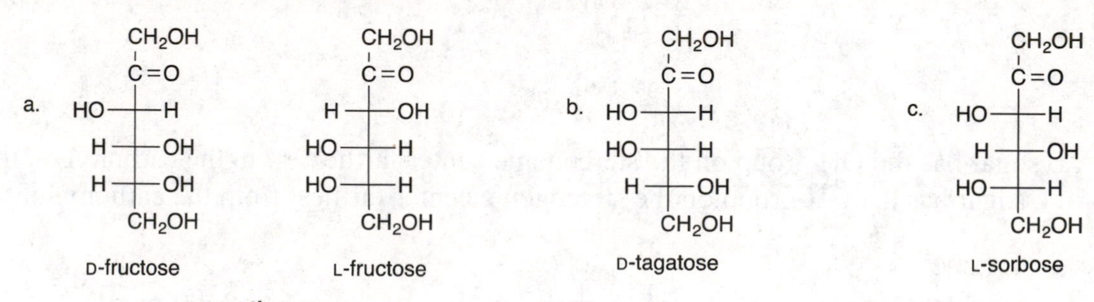

28.11

D-fructose

D-tagatose

28.12

a. α anomer — OH trans to CH_2OH

b. β anomer — OH cis to CH_2OH

28.13 Step [1]: Place the O atom in the upper right corner of a hexagon, and add the CH_2OH group on the first carbon counterclockwise from the O atom.

Step [2]: Place the anomeric carbon on the first carbon clockwise from the O atom.

Step [3]: Add the substituents at the three remaining stereogenic centers, clockwise around the ring.

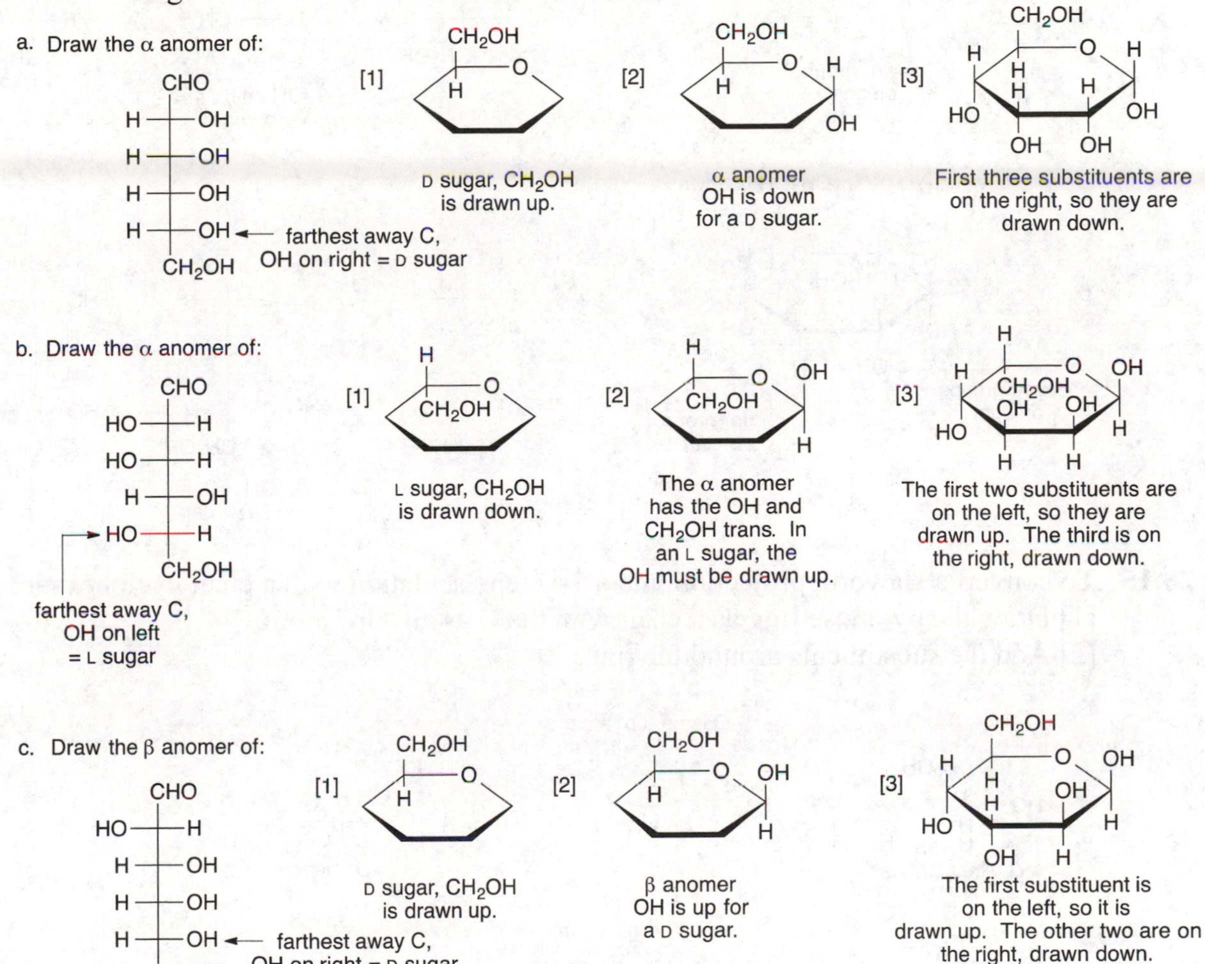

a. Draw the α anomer of:
farthest away C, OH on right = D sugar
D sugar, CH2OH is drawn up.
α anomer OH is down for a D sugar.
First three substituents are on the right, so they are drawn down.
b. Draw the α anomer of:
farthest away C, OH on left = L sugar
L sugar, CH2OH is drawn down.
The α anomer has the OH and CH2OH trans. In an L sugar, the OH must be drawn up.
The first two substituents are on the left, so they are drawn up. The third is on the right, drawn down.
c. Draw the β anomer of:
farthest away C, OH on right = D sugar
D sugar, CH2OH is drawn up.
β anomer OH is up for a D sugar.
The first substituent is on the left, so it is drawn up. The other two are on the right, drawn down.

28.14 To convert each Haworth projection into its acyclic form:

[1] Draw the C skeleton with the CHO on the top and the CH_2OH on the bottom.

[2] Draw in the OH group farthest from the C=O.

A CH_2OH group drawn up means a D sugar; a CH_2OH group drawn down means an L sugar.

[3] Add the three other stereogenic centers.

"Up" groups go on the left, and "down" groups go on the right.

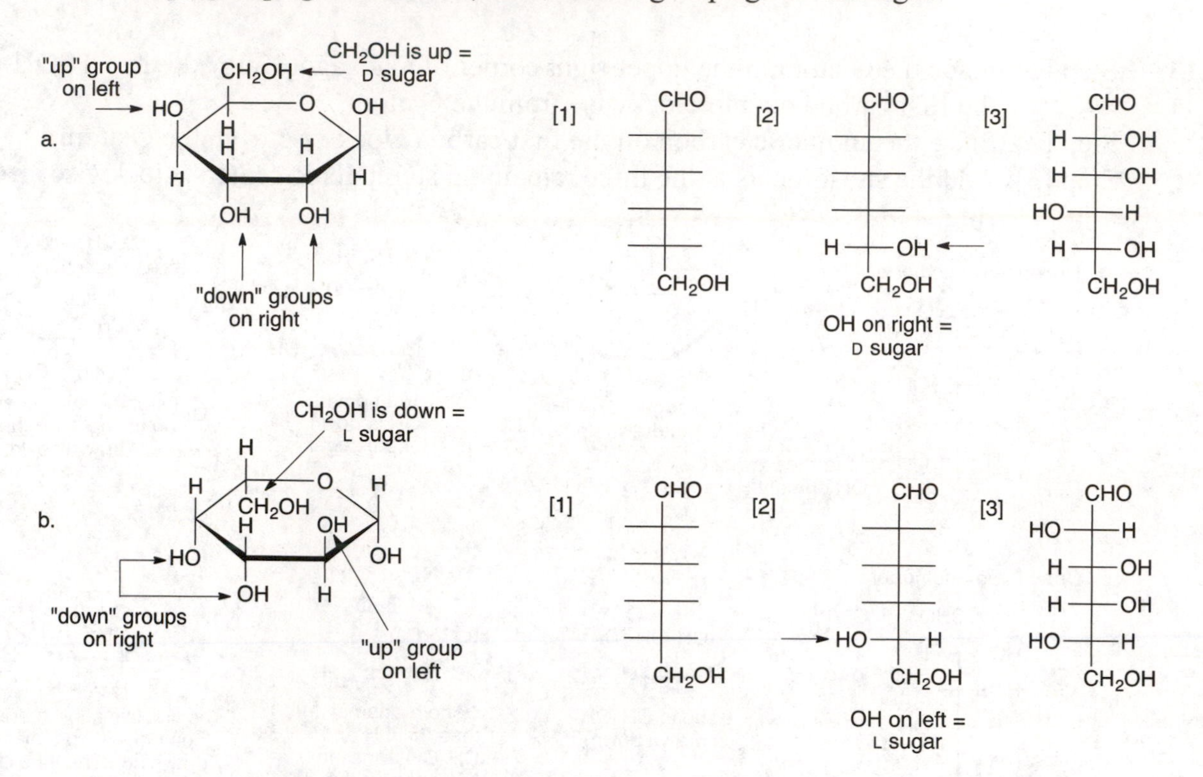

28.15 To convert a Haworth projection into a 3-D representation with a chair cyclohexane:

[1] Draw the pyranose ring as a chair with the O as an "up" atom.

[2] Add the substituents around the ring.

a. O is an "up" atom.

b. O is an "up" atom.

With so many axial groups, this is not the more stable conformation of this sugar.

28.16 Cyclization always forms a new stereogenic center at the anomeric carbon, so two different anomers are possible.

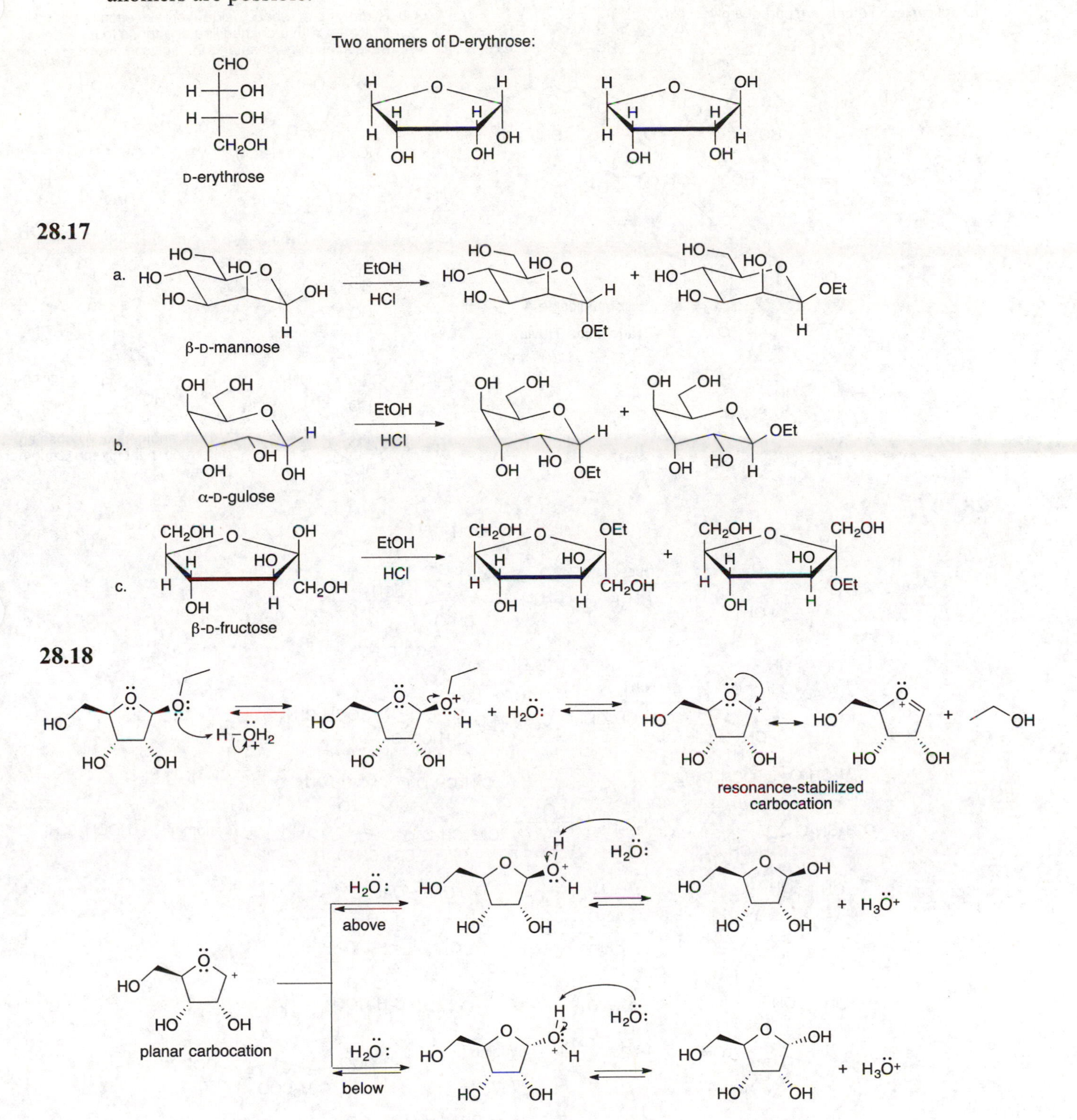

28.19

a. All circled O atoms are part of a glycoside.

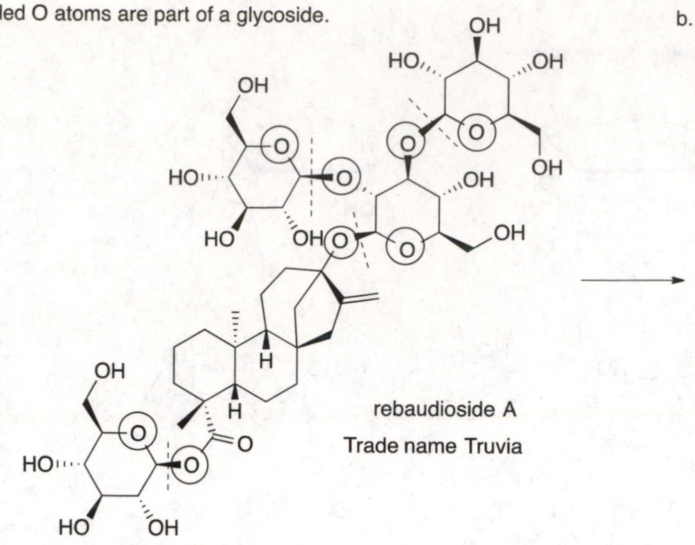

b. Hydrolysis of rebaudioside A breaks each bond indicated with a dashed line and forms four molecules of glucose and the aglycon drawn.

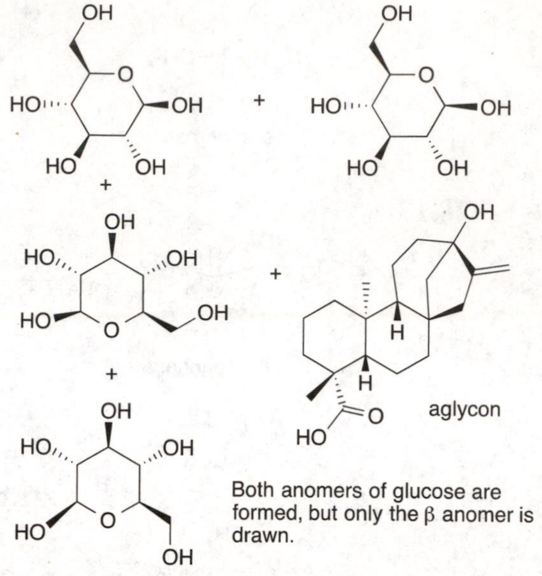

Both anomers of glucose are formed, but only the β anomer is drawn.

28.20

a.

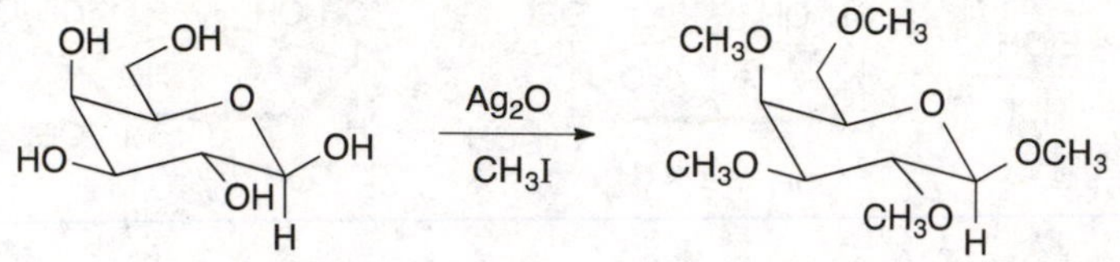

b.

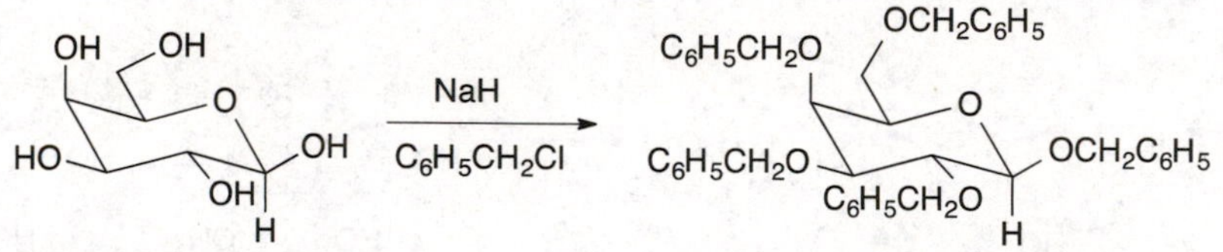

c.

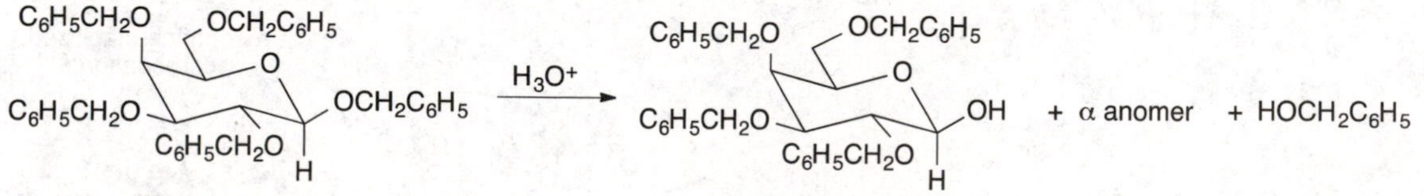

d.

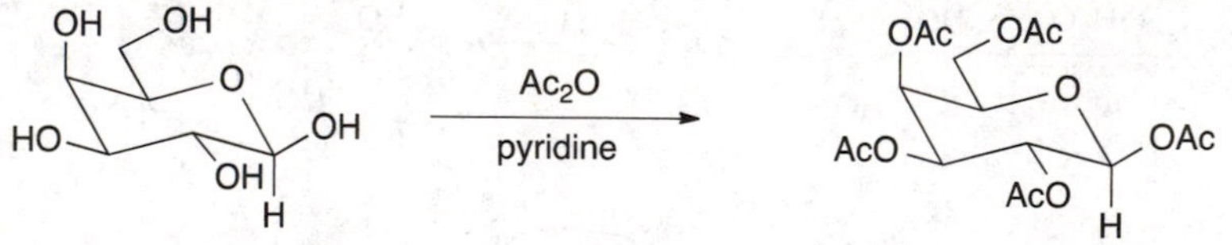

e.

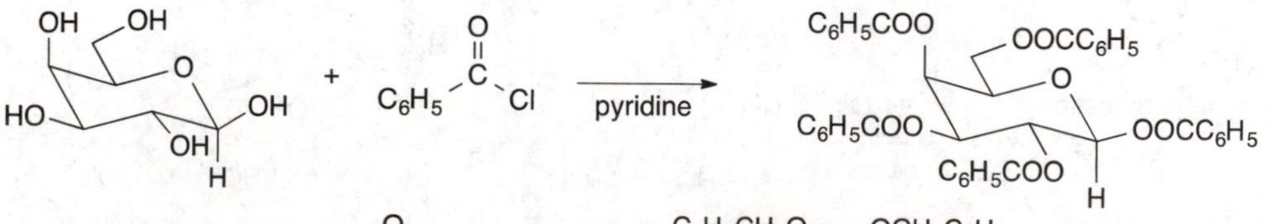

f. product in (c)

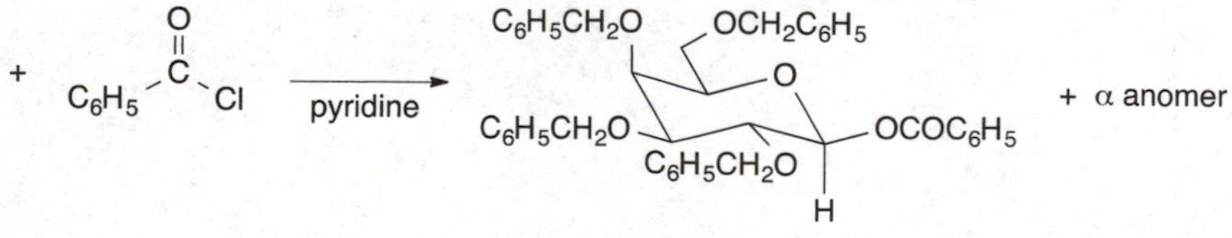

28.21

CH₂OH
=O
HO—H
HO—H
H—OH
CH₂OH
D-tagatose

$\xrightarrow[CH_3OH]{NaBH_4}$

CH₂OH
H—OH
HO—H
HO—H
H—OH
CH₂OH
D-galactitol

+

CH₂OH
HO—H
HO—H
HO—H
H—OH
CH₂OH
D-talitol

28.22 Carbohydrates containing a hemiacetal are in equilibrium with an acyclic aldehyde, making them reducing sugars. Glycosides are acetals, so they are not in equilibrium with any acyclic aldehyde, making them nonreducing sugars.

a. hemiacetal
CH₂OH O OH HO H OH H H H H OH
reducing sugar

b. acetal
O O HO HO OH
nonreducing sugar

c. hemiacetal
HO HO O O O OH HO HO OH HO OH
lactose
reducing sugar

28.23

a.
CHO
HO—H
H—OH
H—OH
CH₂OH

$\xrightarrow[NH_4OH]{Ag_2O}$

COOH
HO—H
H—OH
H—OH
CH₂OH

b.
CHO
HO—H
H—OH
H—OH
CH₂OH

$\xrightarrow[H_2O]{Br_2}$

COOH
HO—H
H—OH
H—OH
CH₂OH

c.
CHO
HO—H
H—OH
H—OH
CH₂OH

$\xrightarrow[H_2O]{HNO_3}$

COOH
HO—H
H—OH
H—OH
COOH

28.24 Molecules with a plane of symmetry are optically inactive.

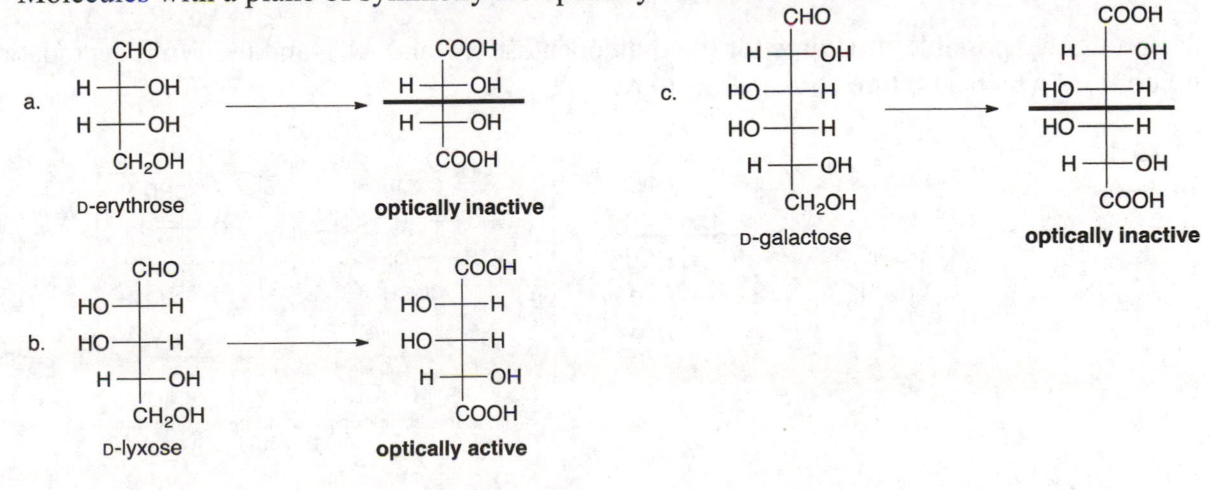

28.25

CHO / HO—H / H—OH / HO—H / H—OH / CH_2OH
D-idose

or

CHO / H—OH / H—OH / HO—H / H—OH / CH_2OH
D-gulose

→

CHO / H—OH / HO—H / H—OH / CH_2OH
D-xylose

28.26

a. CHO / HO—H / H—OH / CH_2OH (D-threose) → CHO / HO—H / HO—H / H—OH / CH_2OH + CHO / H—OH / HO—H / H—OH / CH_2OH

b. CHO / H—OH / H—OH / H—OH / CH_2OH (D-ribose) → CHO / HO—H / H—OH / H—OH / H—OH / CH_2OH + CHO / H—OH / H—OH / H—OH / H—OH / CH_2OH

c. CHO / H—OH / HO—H / HO—H / H—OH / CH_2OH (D-galactose) → CHO / HO—H / H—OH / HO—H / HO—H / H—OH / CH_2OH + CHO / H—OH / H—OH / HO—H / HO—H / H—OH / CH_2OH

28.27

Possible optically inactive D-aldaric acids:

CHO / H—OH / H—OH / H—OH / CH_2OH
A'

⇐ COOH / H—OH / H—OH / H—OH / COOH ← plane of symmetry → COOH / H—OH / HO—H / H—OH / COOH ⇒

CHO / H—OH / HO—H / H—OH / CH_2OH
A"

This OH is on the **right** for a D sugar.

There are two possible structures for the D-aldopentose (**A'** and **A"**), and the Wohl degradation determines which structure corresponds to **A**.

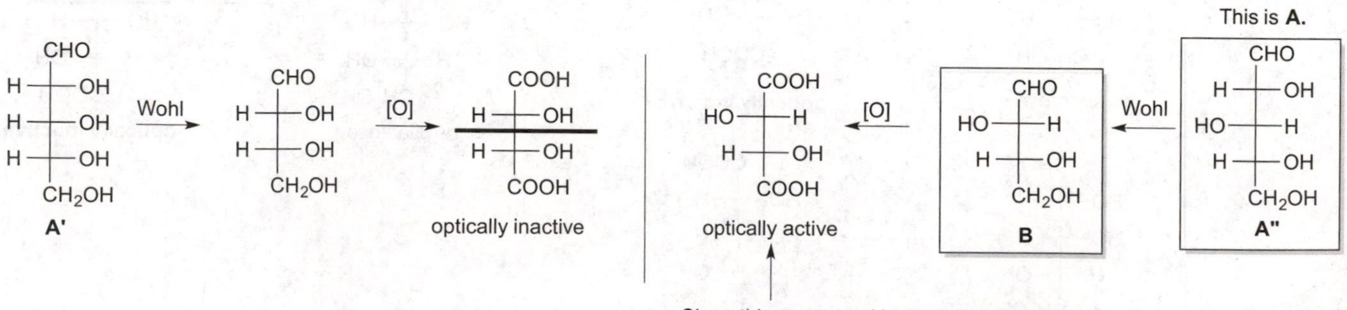

Since this compound has no plane of symmetry, its precursor is **B**, and thus **A" = A.**

28.28

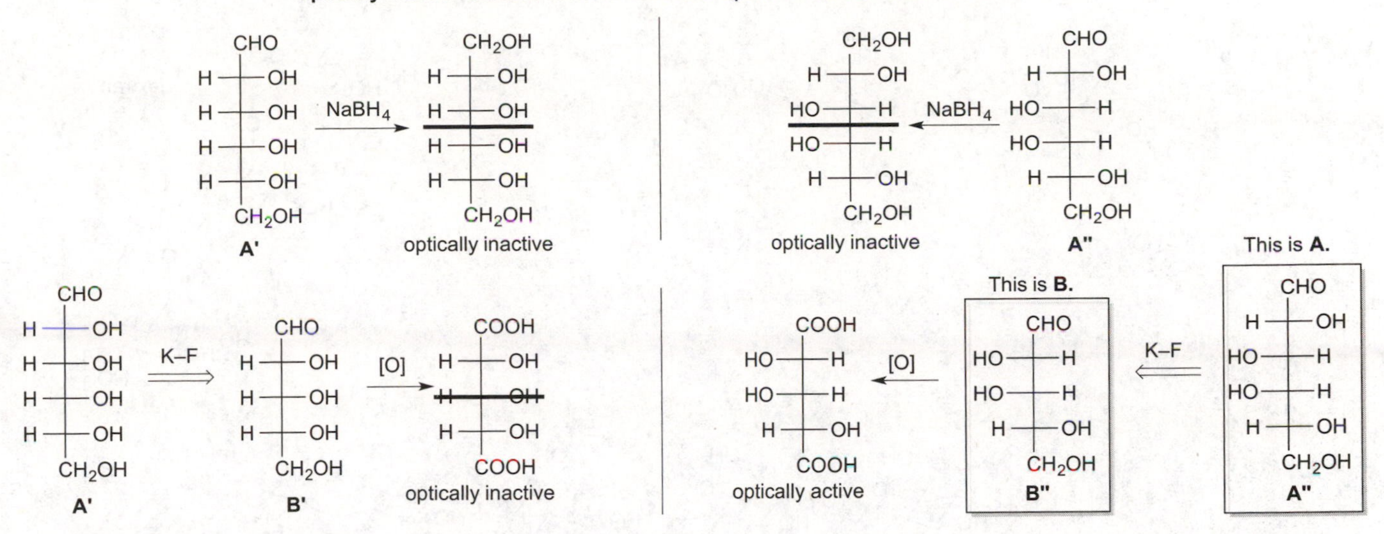

Two D-aldohexoses (**A'** and **A"**) give optically inactive alditols on reduction. **A"** is formed from **B"** by Kiliani–Fischer synthesis. Because **B"** affords an optically active aldaric acid on oxidation, **B"** is **B** and **A"** is **A.** The alternate possibility (**A'**) is formed from an aldopentose **B'** that gives an optically inactive aldaric acid on oxidation.

28.29

H_3O^+

α anomer

α-D-glucose + β-D-glucose

The same products are formed on hydrolysis of the α and β anomers of maltose.

28.30

β glycoside bond

cellobiose

Two possible anomers here. β OH is drawn.

28.31

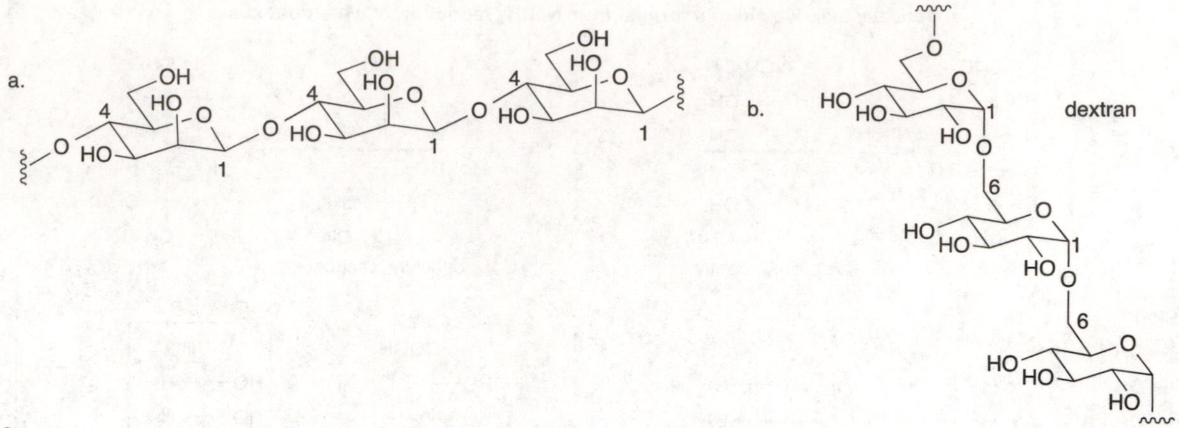

a.

OH HO O 4 O HO 1 O OH HO O 4 HO 1 O OH HO O 4 HO 1

b.

O HO HO O 1 HO O 6 HO HO O 1 HO O 6 HO HO O HO

dextran

28.32

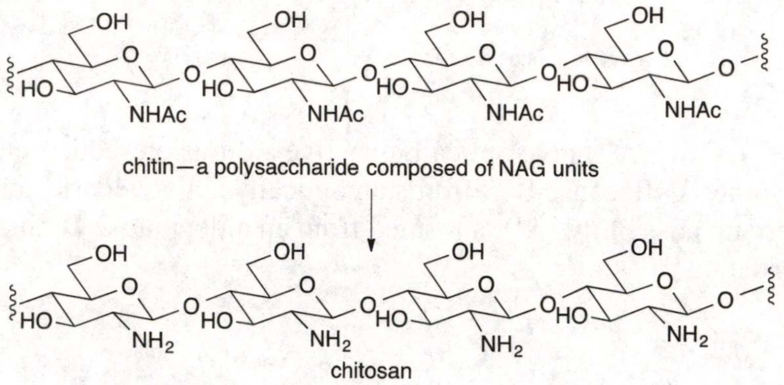

OH O HO O NHAc OH O HO O NHAc OH O HO O NHAc OH O HO O NHAc

chitin—a polysaccharide composed of NAG units

OH O HO O NH_2 OH O HO O NH_2 OH O HO O NH_2 OH O HO O NH_2

chitosan

28.33

a. HO O HO OH HO OH $\xrightarrow[\text{mild } H^+]{CH_3NH_2}$ HO O HO $NHCH_3$ HO OH + HO O HO $NHCH_3$ HO OH + H_2O

b. OH HO O OH OH H OH $\xrightarrow[\text{mild } H^+]{C_6H_5NH_2}$ OH HO O NHC_6H_5 OH H OH + OH HO O H OH NHC_6H_5 OH + H_2O

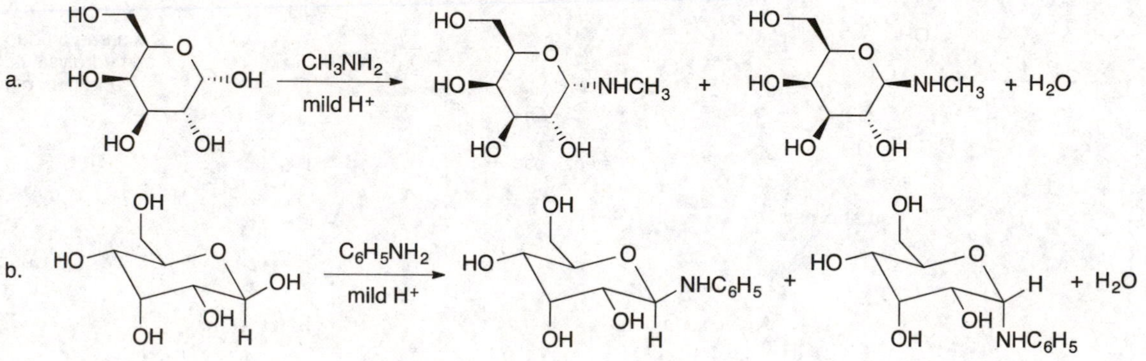

28.34

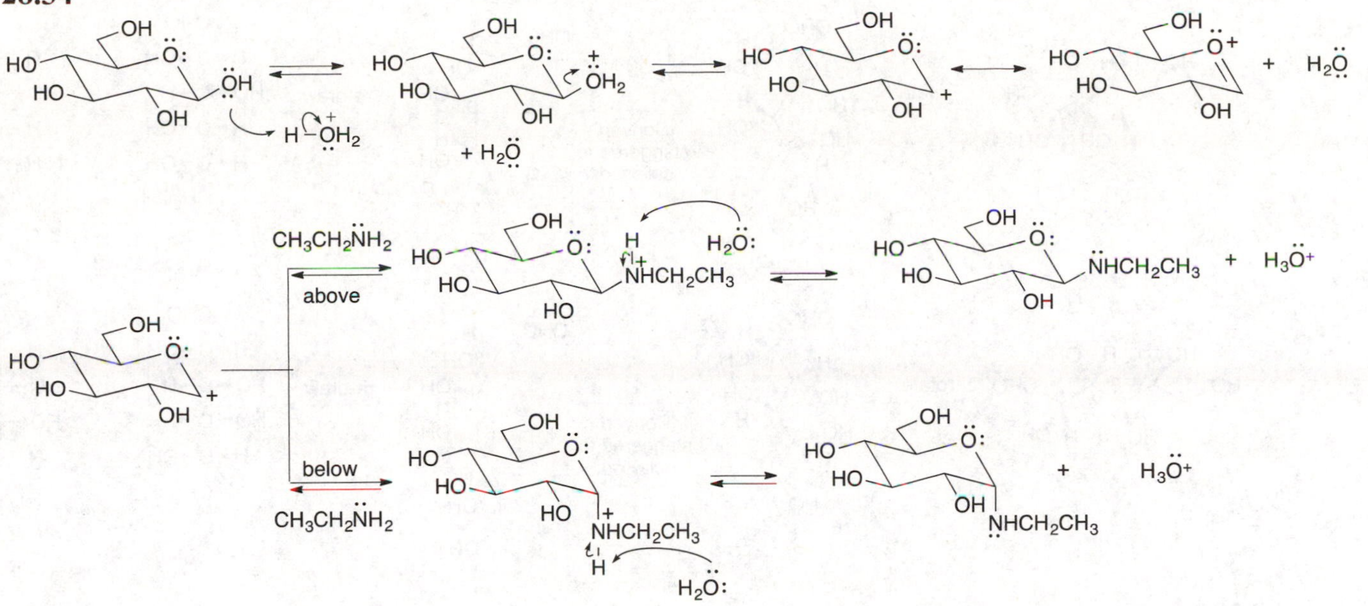

28.35

a.

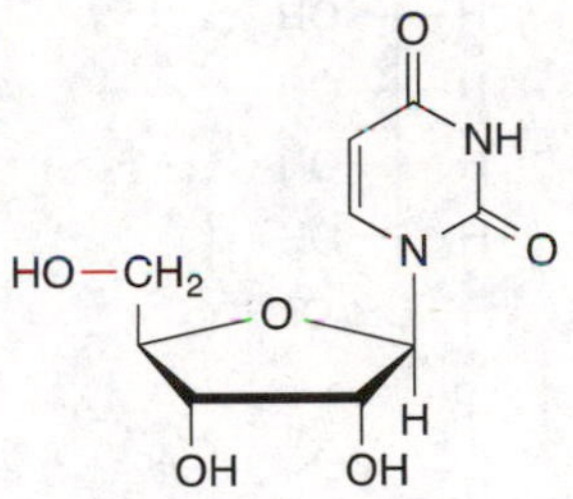

b.

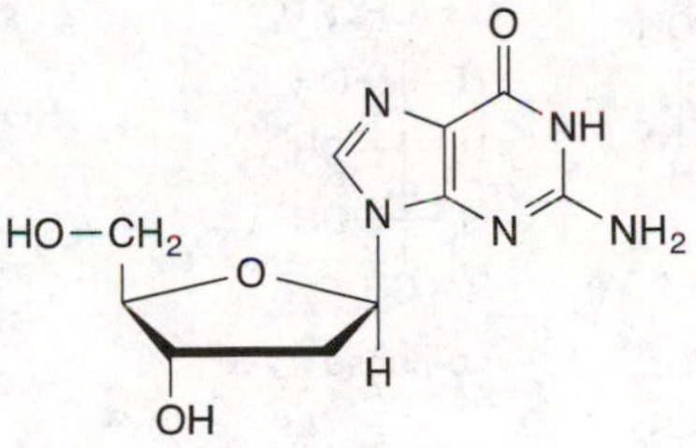

28.36

a. Two purine bases (A and G) are both bicyclic bases. Therefore they are too big to hydrogen bond to each other on the inside of the DNA double helix.

b. Hydrogen bonding between guanine and cytosine has three hydrogen bonds, whereas between guanine and thymine there are only two. This makes hydrogen bonding between guanine and cytosine more favorable.

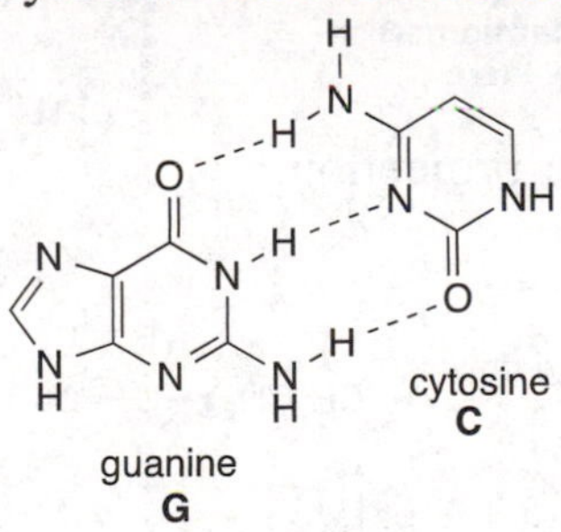

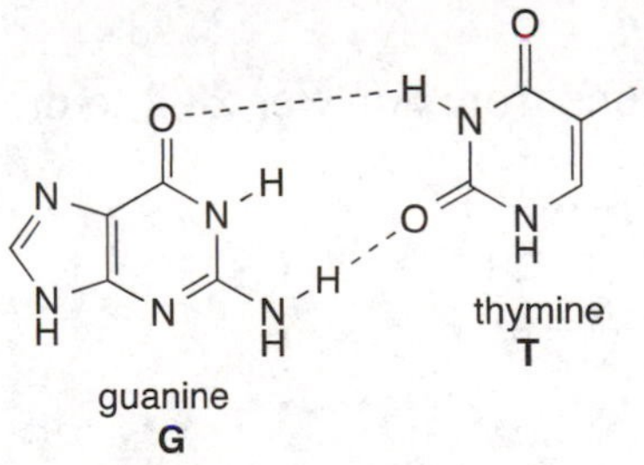

guanine
G

cytosine
C

guanine
G

thymine
T

28.37

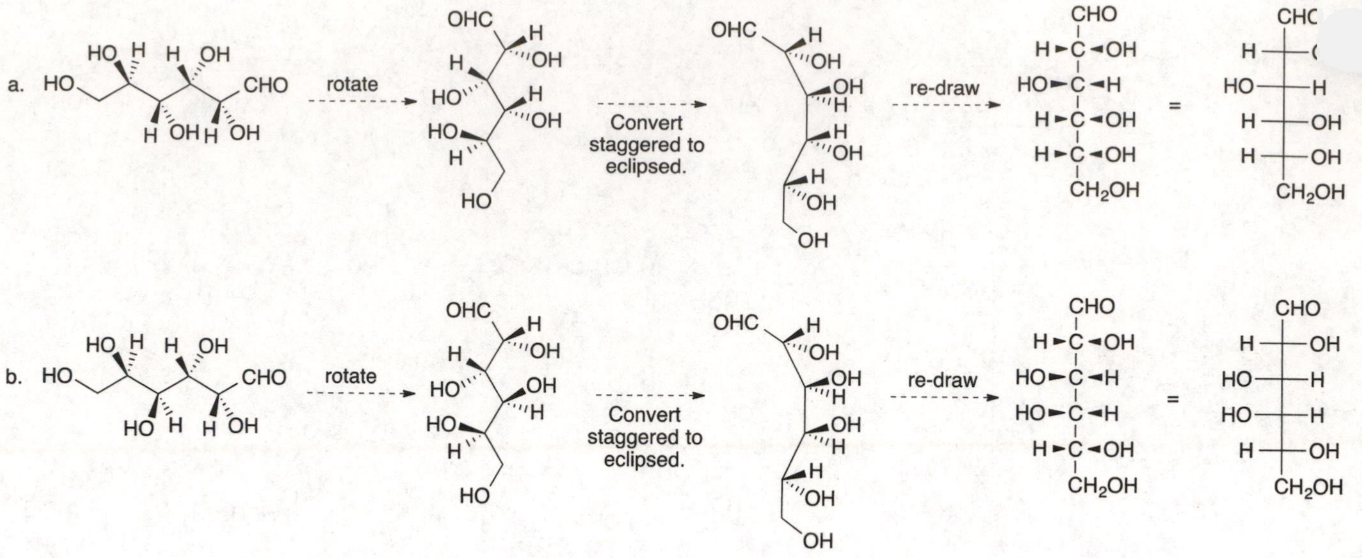

a. rotate → Convert staggered to eclipsed. → re-draw

	CHO	
H		OH
HO		H
H		OH
H		OH
	CH_2OH	

b. rotate → Convert staggered to eclipsed. → re-draw

	CHO	
H		OH
HO		H
HO		H
H		OH
	CH_2OH	

28.38

A β anomer = D-ribose

B α anomer = D-allose

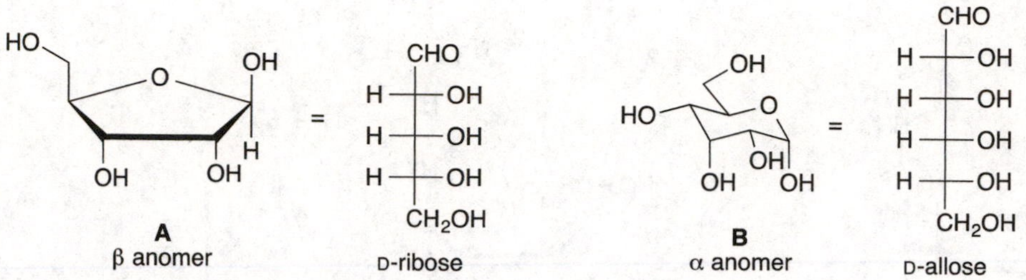

28.39 Label the compounds with *R* or *S* and then classify.

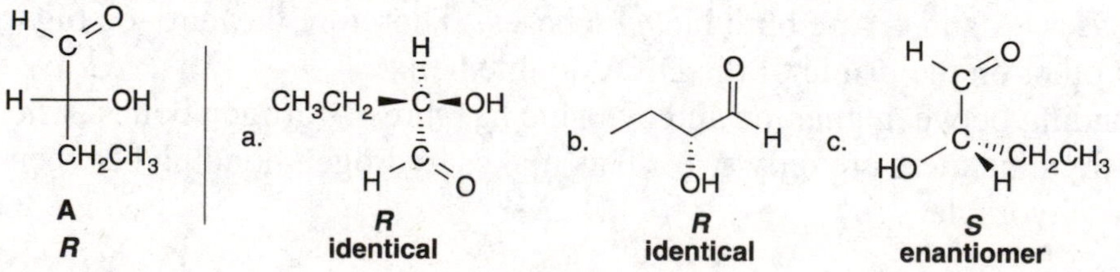

A *R*

a. *R* identical

b. *R* identical

c. *S* enantiomer

28.40 Use the directions from Answer 28.2 to draw each Fischer projection.

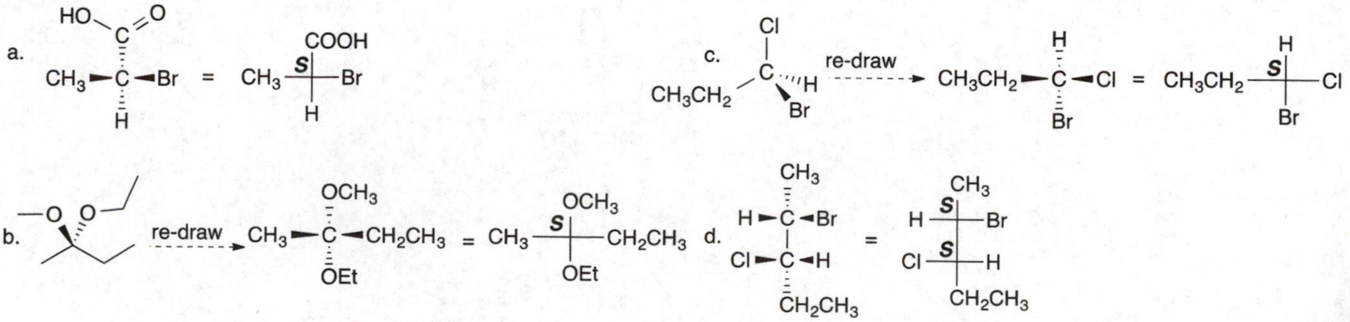

a. CH_3–C(COOH)(Br)(H) = Fischer projection, *S*

b. re-draw → CH_3–C(OCH_3)(OEt)–CH_2CH_3 = Fischer projection, *S*

c. re-draw → CH_3CH_2–C(H)(Br)–Cl = Fischer projection, *S*

d. = Fischer projection, *S*, *S*

e. re-draw = S S

f. re-draw = S S R

28.41

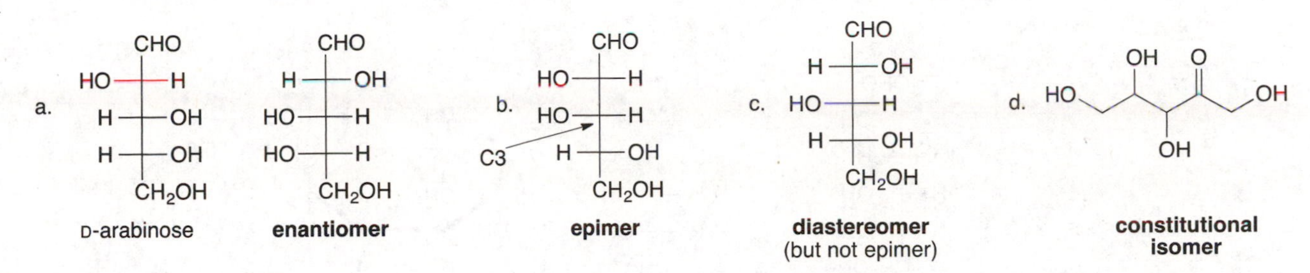

28.42

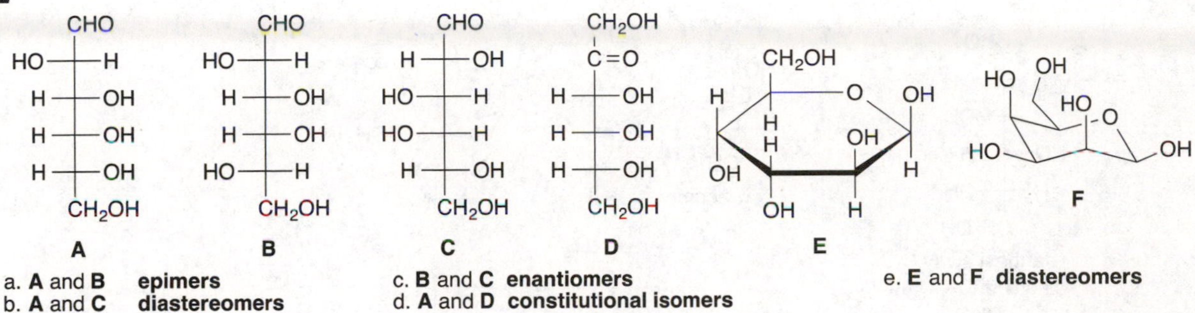

a. **A** and **B** **epimers**
b. **A** and **C** **diastereomers**
c. **B** and **C** **enantiomers**
d. **A** and **D** **constitutional isomers**
e. **E** and **F** **diastereomers**

28.43 Use the directions from Answer 28.13.

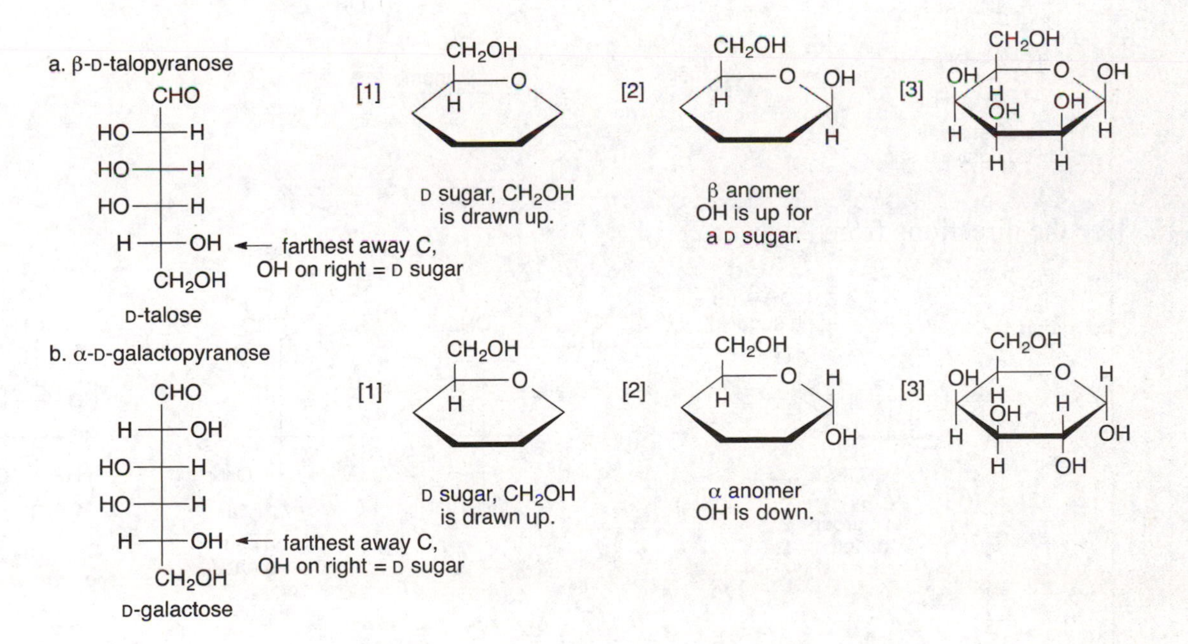

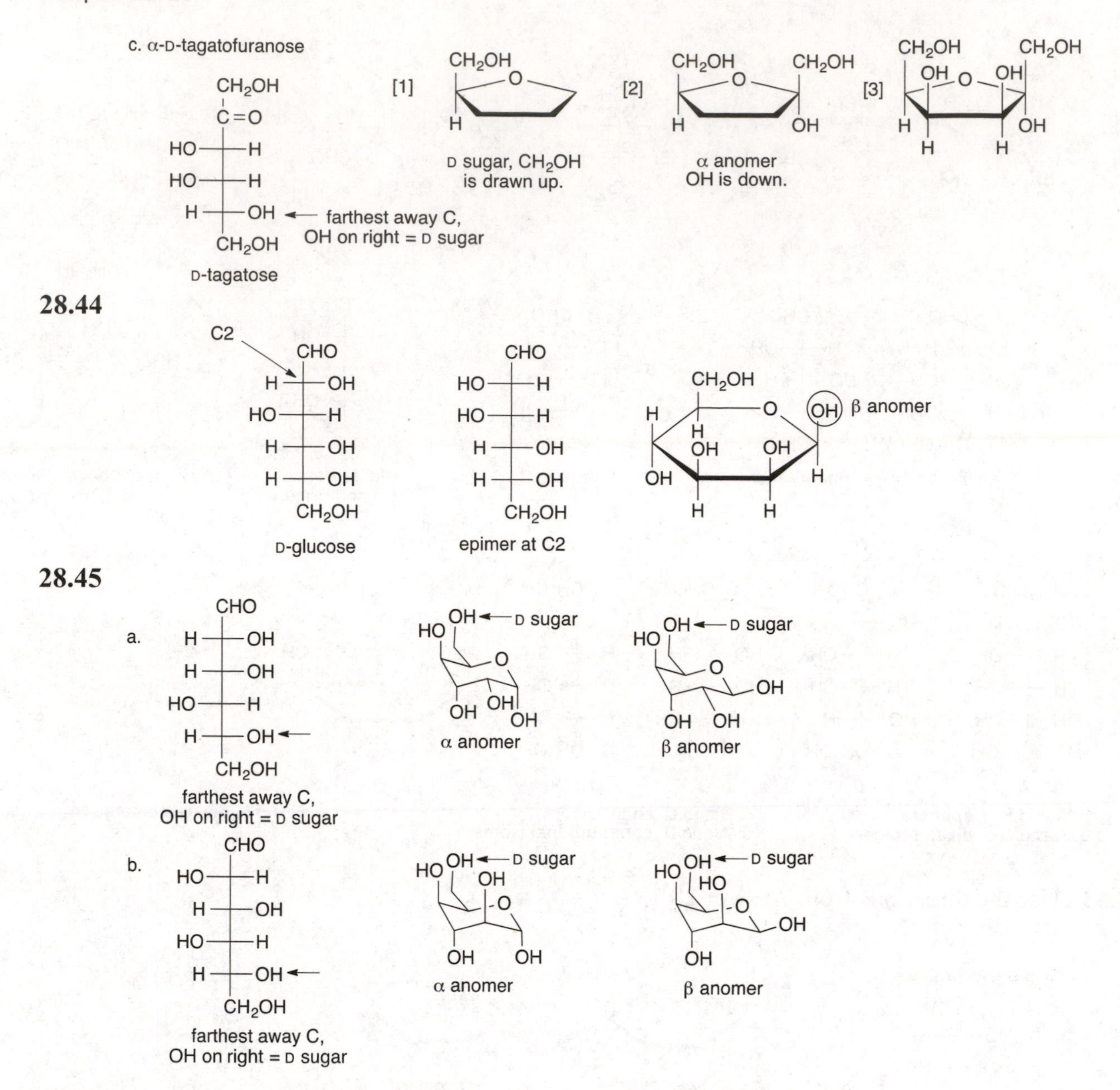

28.46 Use the directions from Answer 28.14.

a.

"up" group on left

CH_2OH is up = D sugar

"down" group on right

"up" group on left

[1]

[2]

OH on right = D sugar

[3]

b. "up" group on left → OH; CH_2OH is down = L sugar; "down" group on right; "up" group on left

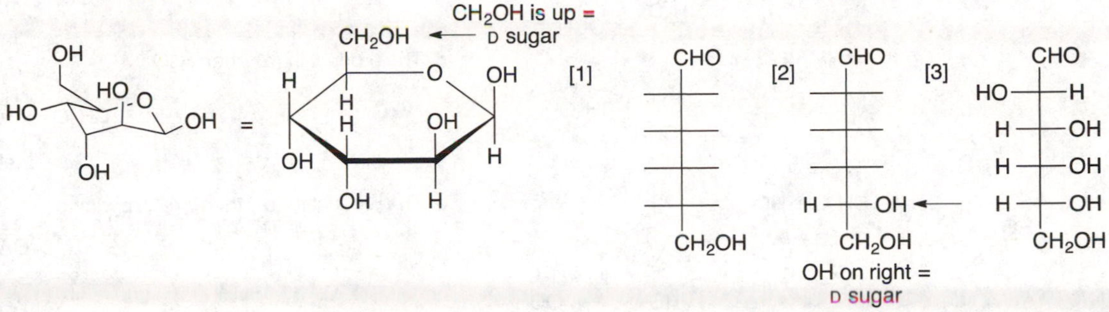

[1] CHO ... CH_2OH [2] CHO ... HO—H, CH_2OH — OH on left = L sugar [3]

CHO
HO—H
H—OH
HO—H
HO—H
CH_2OH

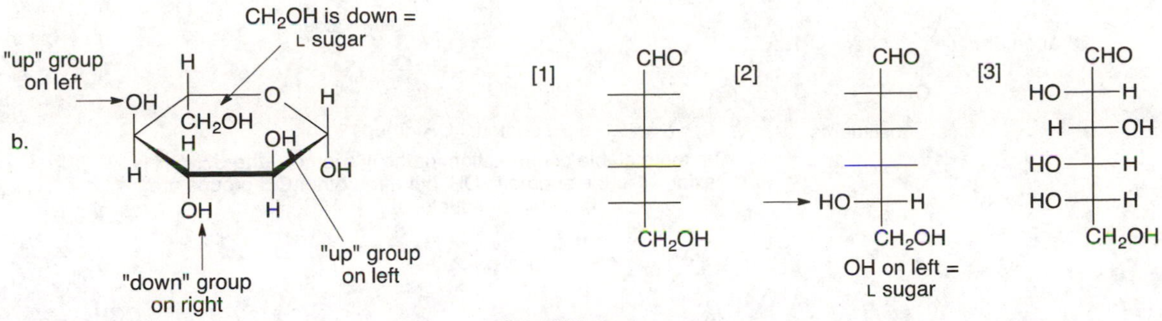

c. CH_2OH is up = D sugar

[1] CHO ... CH_2OH [2] CHO ... H—OH, CH_2OH — OH on right = D sugar [3]

CHO
HO—H
H—OH
H—OH
H—OH
CH_2OH

d. D sugar

CH_2OH
C=O
H—OH
H—OH
H—OH
CH_2OH

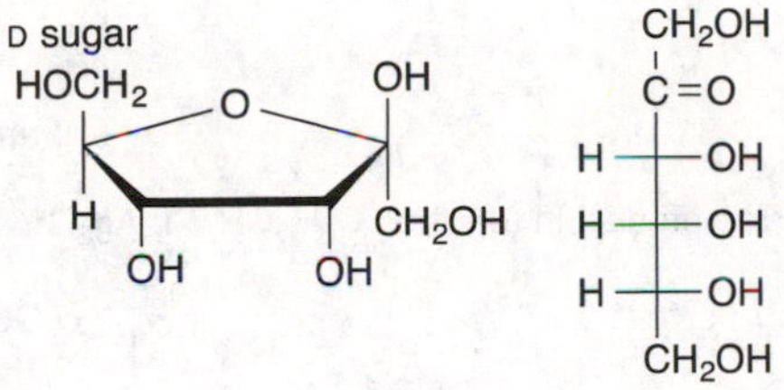

28.47

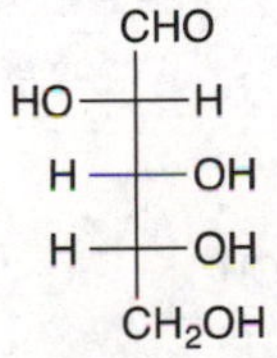

CHO
HO—H
H—OH
H—OH
CH_2OH

D-arabinose

a. β anomer α anomer

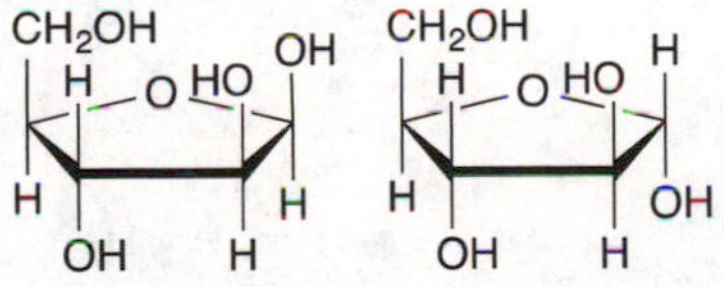

b. two anomers in the pyranose form

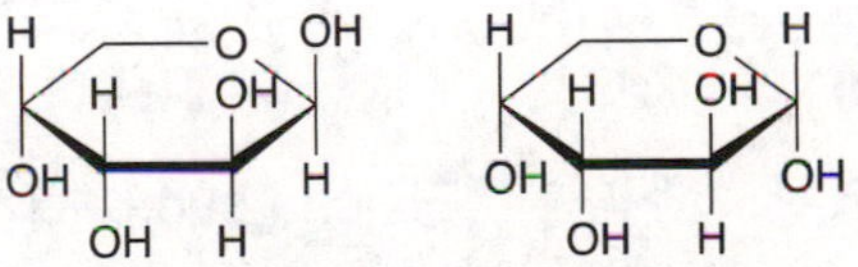

28.48

Two anomers of D-idose, as well as two conformations of each anomer:

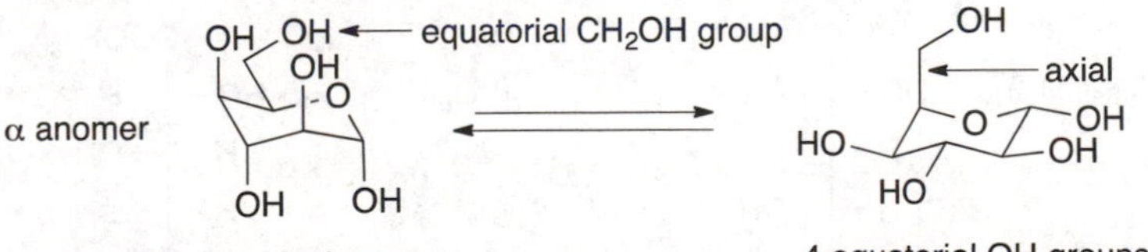

α anomer — equatorial CH_2OH group — 4 axial substituents ⇌ axial — 4 equatorial OH groups

More stable conformation for the α anomer—the CH_2OH is axial, but all other groups are equatorial.

β anomer

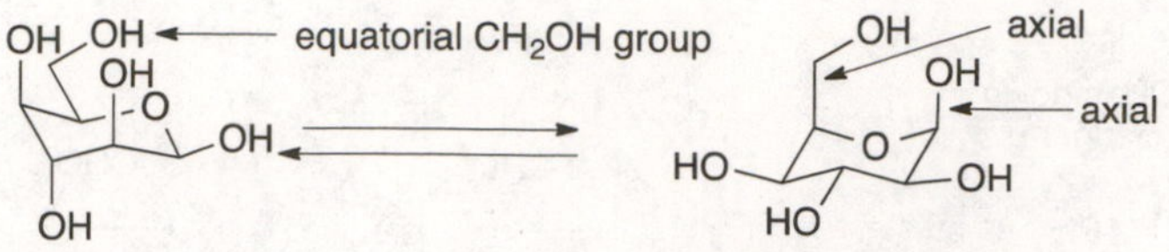

equatorial CH_2OH group

axial

axial

3 axial substituents

3 equatorial OH groups

The more stable conformation for the β anomer—the CH_2OH is axial, as is the anomeric OH, but three other OH groups are equatorial.

28.49

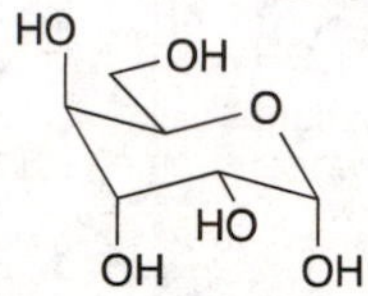

a. CH_3I, Ag_2O

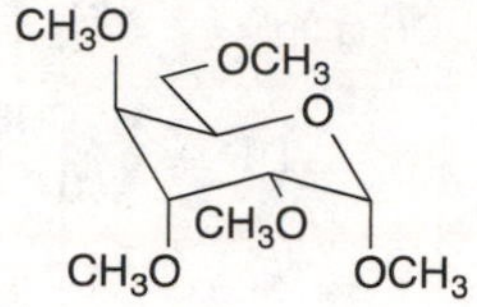

b. CH_3OH, HCl

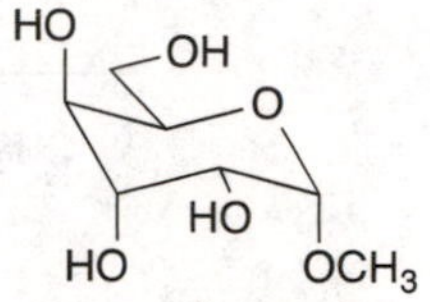

+ β anomer

D-gulose

c. Ac_2O, pyridine

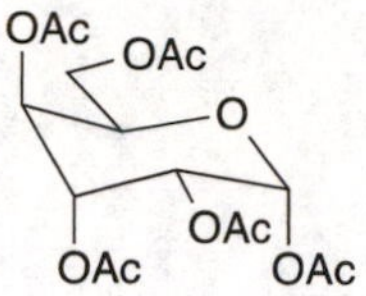

d. The product in (a), then H_3O^+

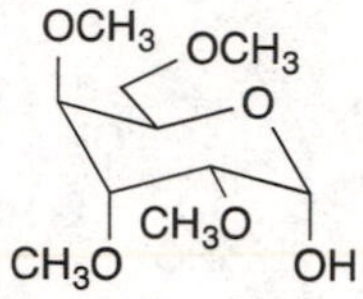

+ β anomer

e. The product in (b), then Ac_2O, pyridine

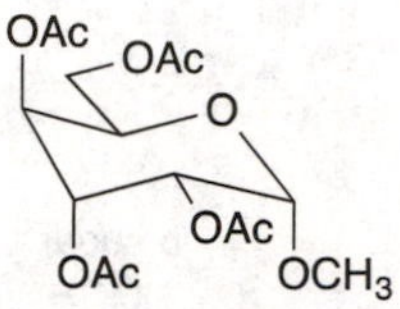

+ β anomer

f. The product in (d), then $C_6H_5CH_2Cl$, Ag_2O

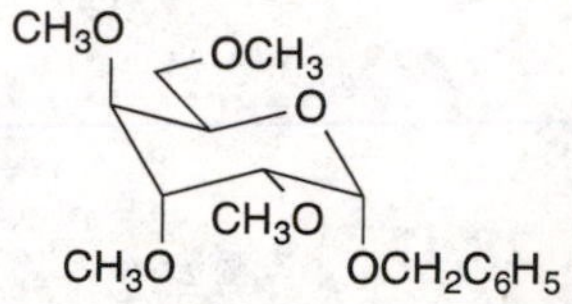

+ β anomer

28.50

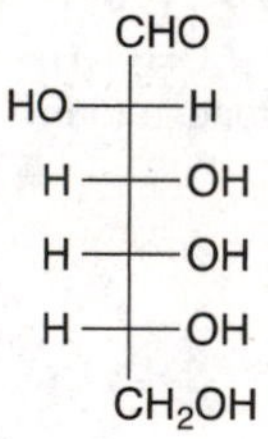

D-altrose

a. $(CH_3)_2CHOH$, HCl

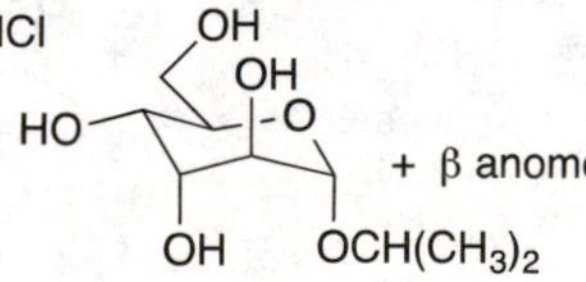

+ β anomer

b. $NaBH_4$, CH_3OH

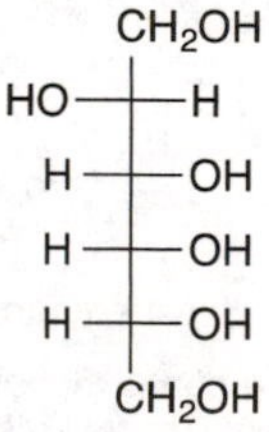

c. Br_2, H_2O

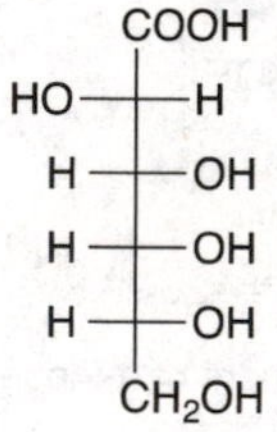

d. HNO_3, H_2O

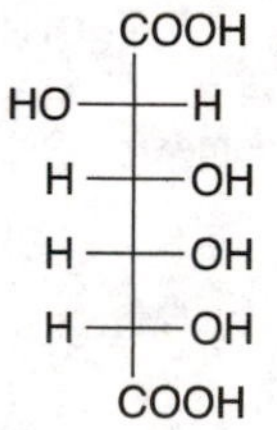

e. [1] NH_2OH
[2] $(CH_3CO)_2O$, $NaOCOCH_3$
[3] $NaOCH_3$

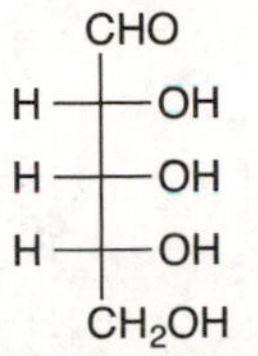

CHO
H—OH
H—OH
H—OH
CH_2OH

g. CH_3I, Ag_2O

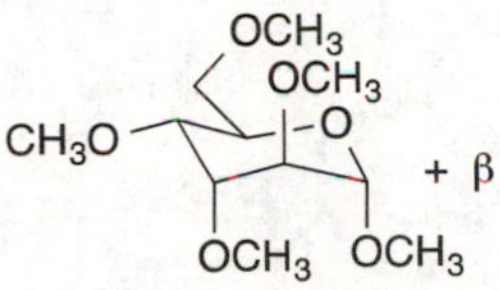

OCH_3, OCH_3, CH_3O, OCH_3, OCH_3 + β anomer

f. [1] NaCN, HCl
[2] H_2, Pd-$BaSO_4$
[3] H_3O^+

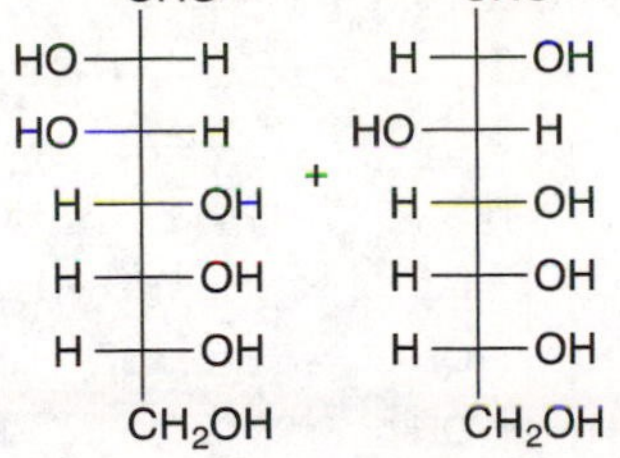

CHO, HO—H, HO—H, H—OH, H—OH, H—OH, CH_2OH + CHO, H—OH, HO—H, H—OH, H—OH, H—OH, CH_2OH

h. $C_6H_5CH_2NH_2$, mild H^+

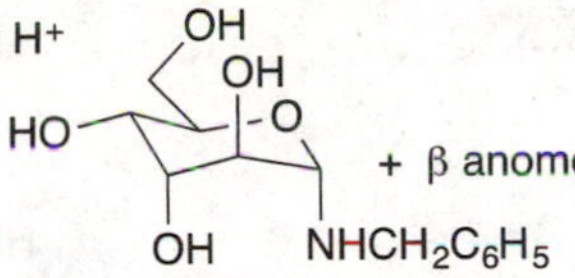

+ β anomer, $NHCH_2C_6H_5$

28.51

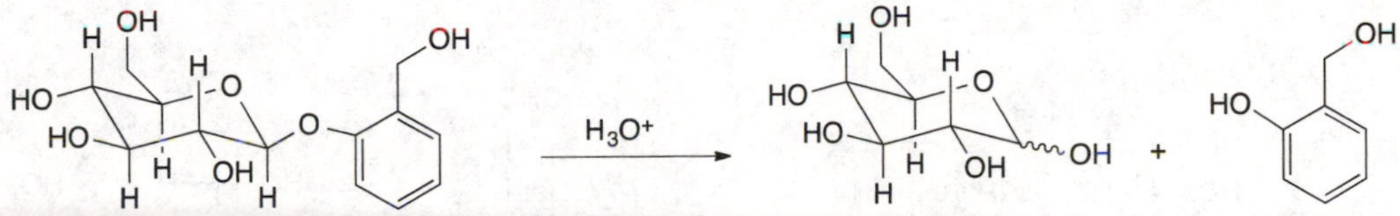

salicin —H_3O^+→ monosaccharide (both anomers) + aglycon

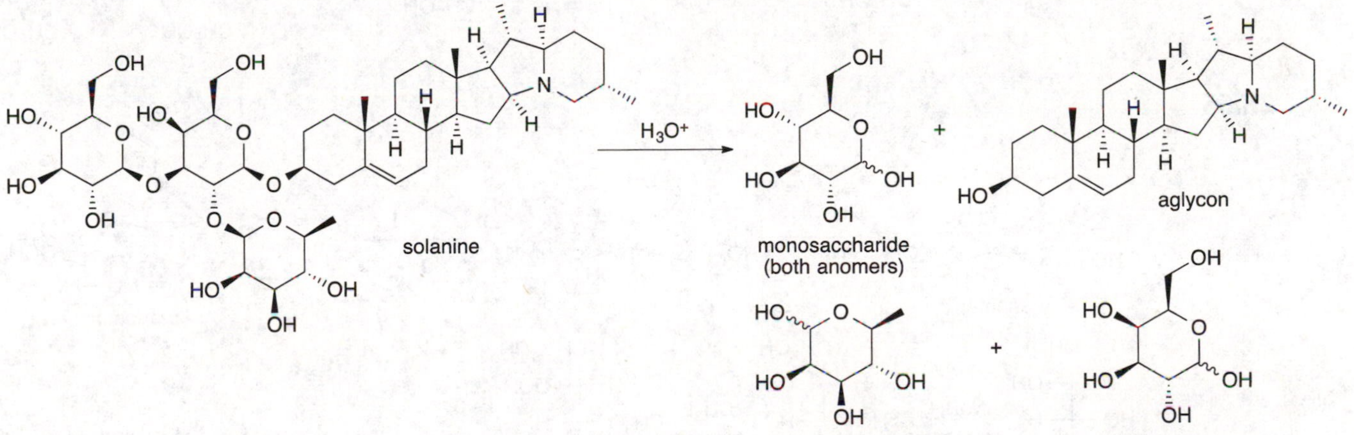

solanine —H_3O^+→ monosaccharide (both anomers) + aglycon

monosaccharide (both anomers) + monosaccharide (both anomers)

28.52

a.

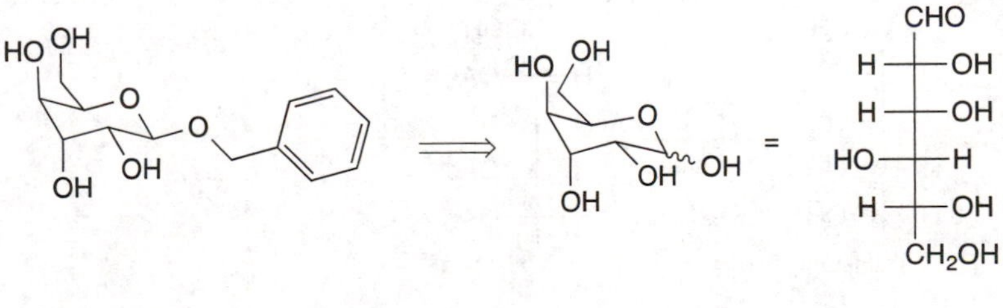

CHO
H—OH
H—OH
HO—H
H—OH
CH_2OH

b.

HO, O, OH, OH, OH ⟹ HO, OH, O, OH, OH, OH ($\alpha + \beta$) =

CH_2OH	
C=O	
H	OH
H	OH
H	OH
CH_2OH	

28.53

D-glucose: CHO; H–OH; HO–H; H–OH; H–OH; CH_2OH → D-arabinose: CHO; HO–H; H–OH; H–OH; CH_2OH ← D-mannose: CHO; HO–H; HO–H; H–OH; H–OH; CH_2OH

28.54

a. CHO; HO–H; HO–H; H–OH; CH_2OH → CHO; HO–H; HO–H; HO–H; H–OH; CH_2OH + CHO; H–OH; HO–H; HO–H; H–OH; CH_2OH

b. CHO; HO–H; HO–H; HO–H; H–OH; CH_2OH → CHO; HO–H; HO–H; HO–H; HO–H; H–OH; CH_2OH + CHO; H–OH; HO–H; HO–H; HO–H; H–OH; CH_2OH

28.55

a. HO, HO, OH, O, OH, OH + α anomer $\xrightarrow[HCl]{CH_3OH}$ HO, HO, OH, O, OCH_3, OH + α anomer $\xrightarrow[Ag_2O]{CH_3CH_2I}$ O, O, O, O, OCH_3, O + α anomer

b. CHO; H–OH; HO–H; H–OH; H–OH; CH_2OH $\xrightarrow[CH_3OH]{NaBH_4}$ CH_2OH; H–OH; HO–H; H–OH; H–OH; CH_2OH $\xrightarrow[Ag_2O]{CH_3I}$ CH_2OCH_3; H–OCH_3; CH_3O–H; H–OCH_3; H–OCH_3; CH_2OCH_3

c. CHO; H–OH; HO–H; H–OH; H–OH; CH_2OH $\xrightarrow[H_2O]{Br_2}$ COOH; H–OH; HO–H; H–OH; H–OH; CH_2OH $\xrightarrow[pyridine]{Ac_2O}$ COOH; H–OAc; AcO–H; H–OAc; H–OAc; CH_2OAc

28.56 Molecules with a plane of symmetry are optically inactive.

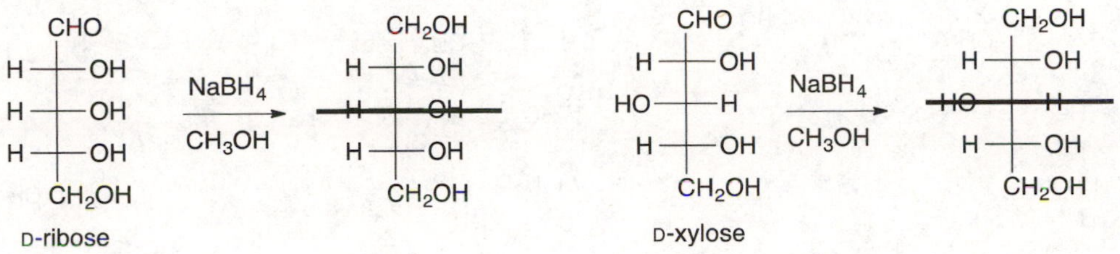

D-ribose $\xrightarrow[CH_3OH]{NaBH_4}$

D-xylose $\xrightarrow[CH_3OH]{NaBH_4}$

28.57

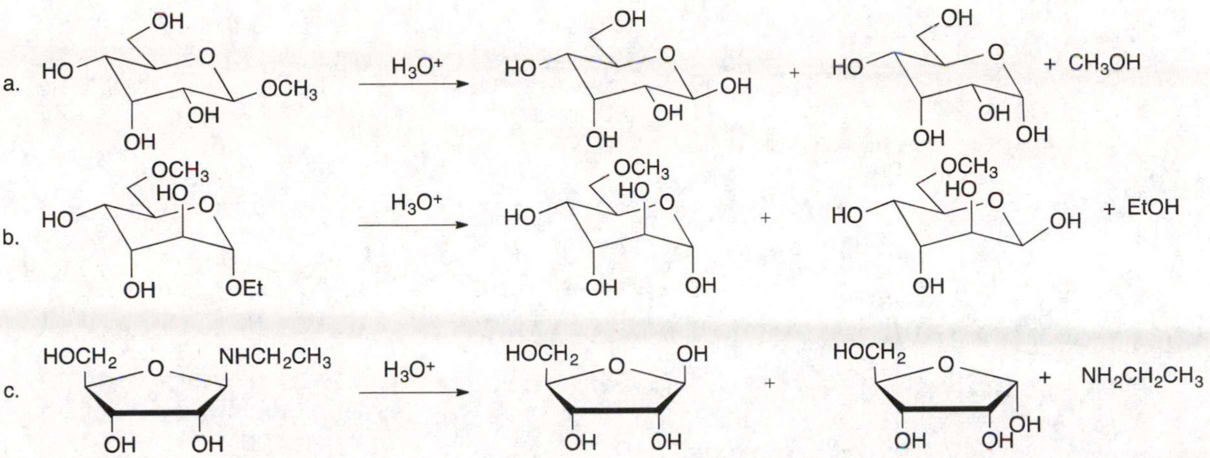

a. $\xrightarrow{H_3O^+}$ + CH_3OH

b. $\xrightarrow{H_3O^+}$ + EtOH

c. $\xrightarrow{H_3O^+}$ + $NH_2CH_2CH_3$

28.58

:B (any base)

+ HB^+

:B (any base)

+ H_2O:

+ HB^+

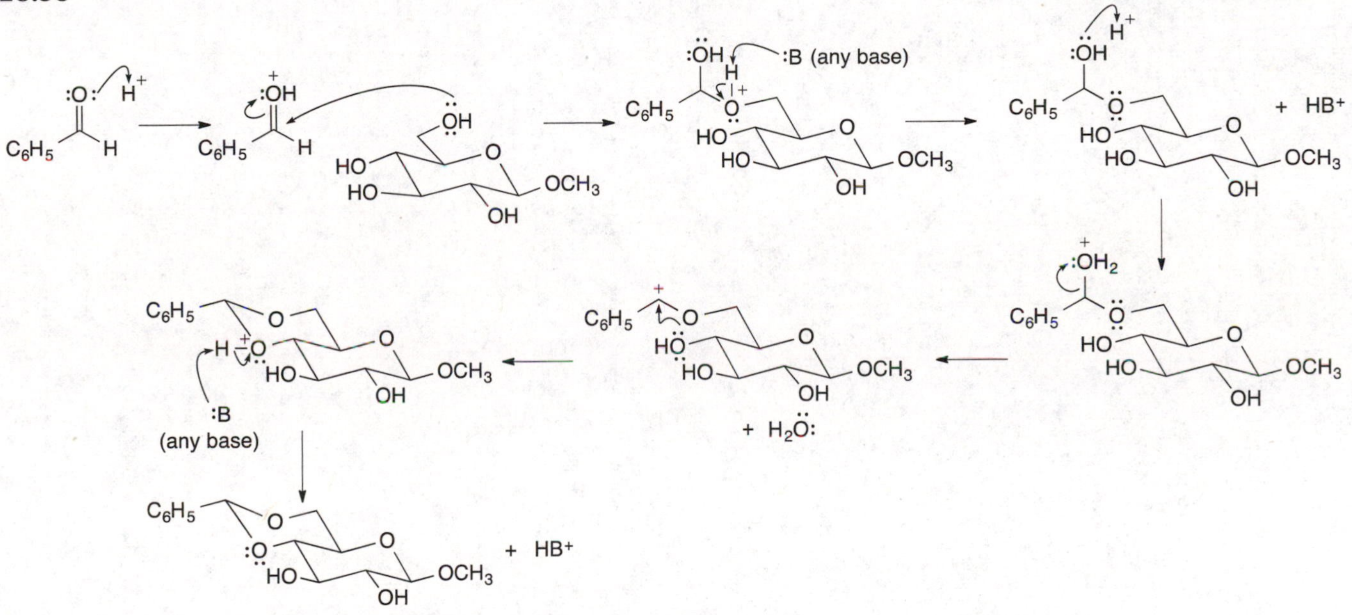

28.59

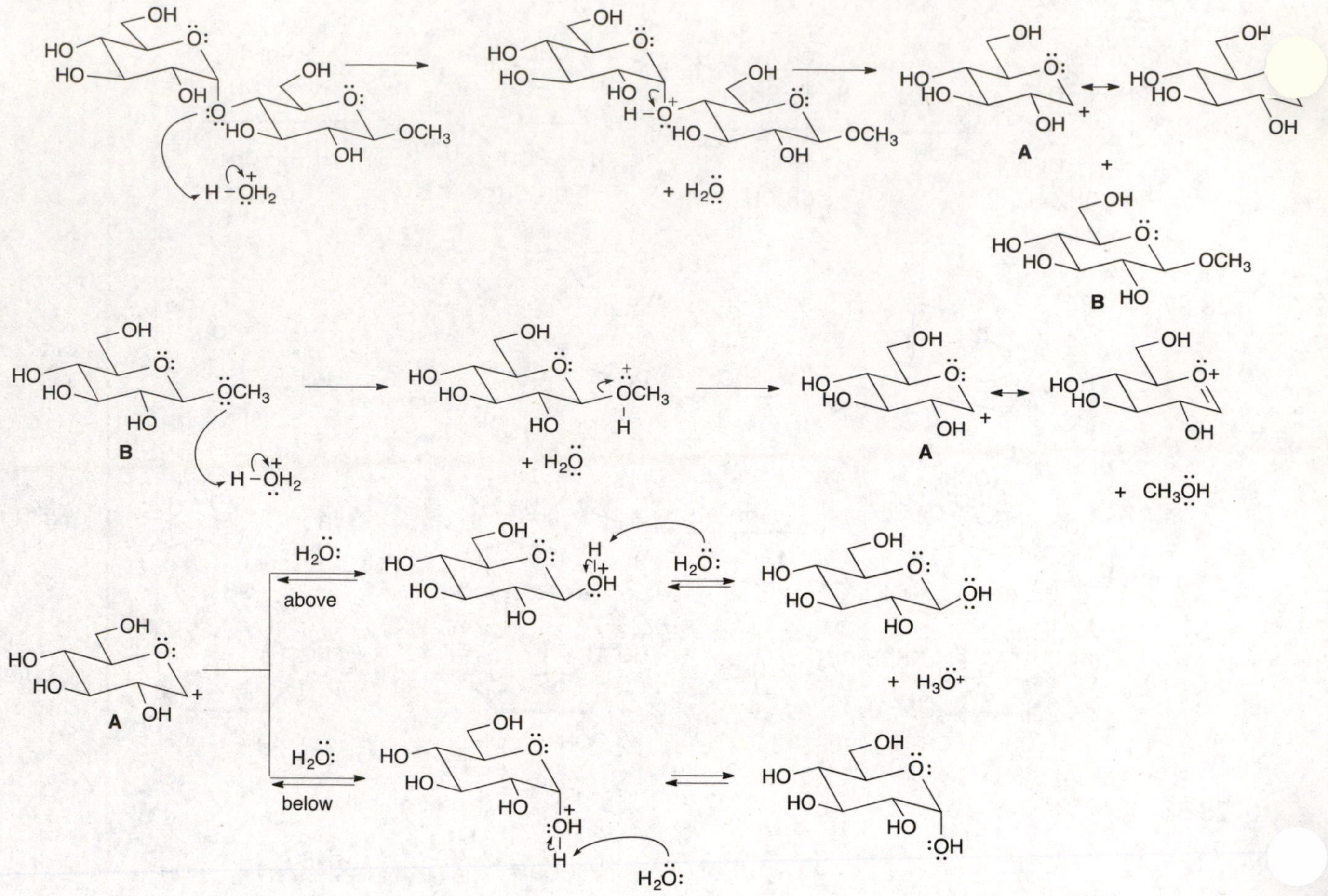

HO, HO, OH, OH, OH, HO, OCH$_3$, OH

H–OH_2^+

+ H_2O

A

B

CH_3OH

above

below

H_2O:

+ H_3O^+

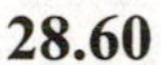

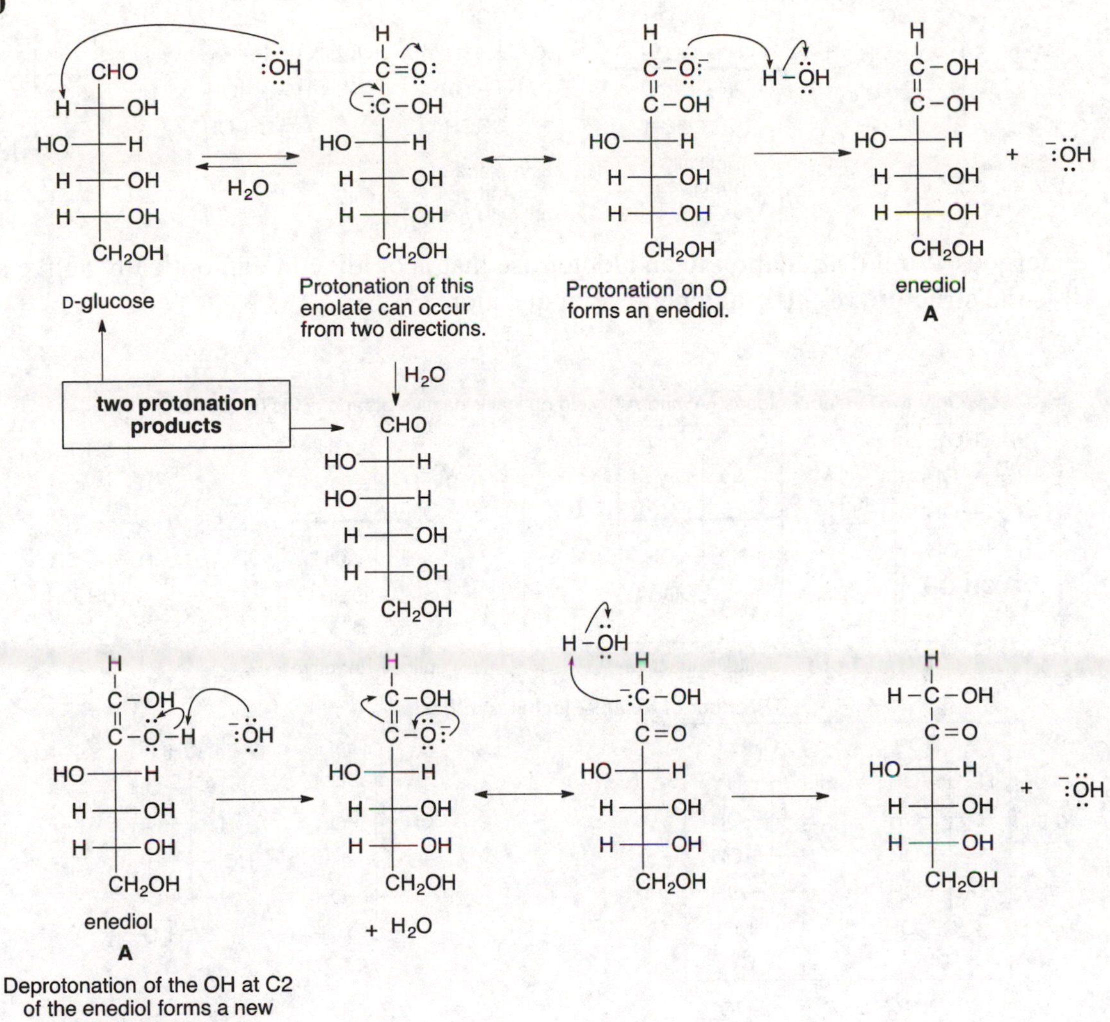
D-glucose
Protonation of this enolate can occur from two directions.
Protonation on O forms an enediol.
enediol
A
two protonation products
enediol
A
Deprotonation of the OH at C2 of the enediol forms a new enolate that goes on to form the ketohexose.

28.61

Two D-aldopentoses (**A'** and **A"**) yield optically active aldaric acids when oxidized.

Optically active D-aldaric acids:

A' (CHO; HO–H; H–OH; H–OH; CH_2OH) —[O]→ COOH; HO–H; H–OH; H–OH; COOH — optically active

A" D-lyxose (CHO; HO–H; HO–H; H–OH; CH_2OH) —[O]→ COOH; HO–H; HO–H; H–OH; COOH — optically active

A' —Wohl→ —[O]→ optically inactive

no plane of symmetry optically active ←[O]— ←Wohl— A"

Only **A"** undergoes Wohl degradation to an aldotetrose that is oxidized to an optically active aldaric acid, so **A"** is the structure of the D-aldopentose in question.

28.62

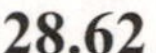

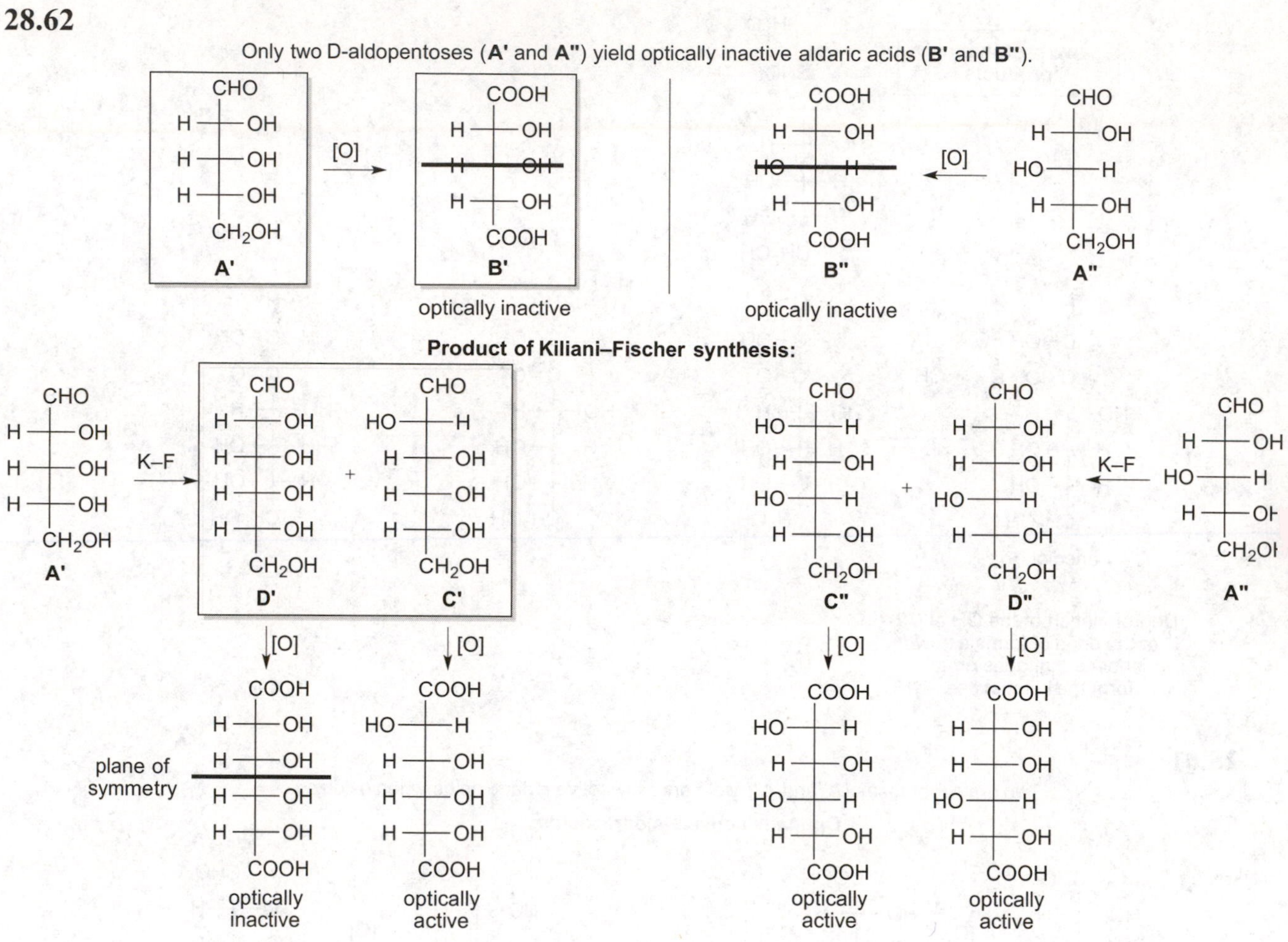

Only **A'** fits the criteria. Kiliani–Fischer synthesis of **A'** forms **C'** and **D'**, which are oxidized to one optically active and one optically inactive aldaric acid. A similar procedure with **A"** forms two optically active aldaric acids. Thus, the structures of **A–D** correspond to the structures of **A'–D'**.

28.63

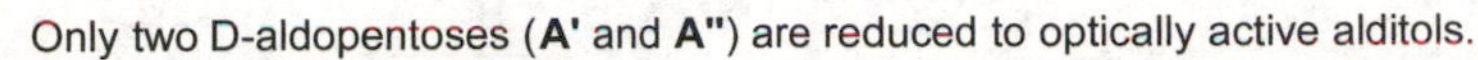
Only two D-aldopentoses (**A'** and **A"**) are reduced to optically active alditols.

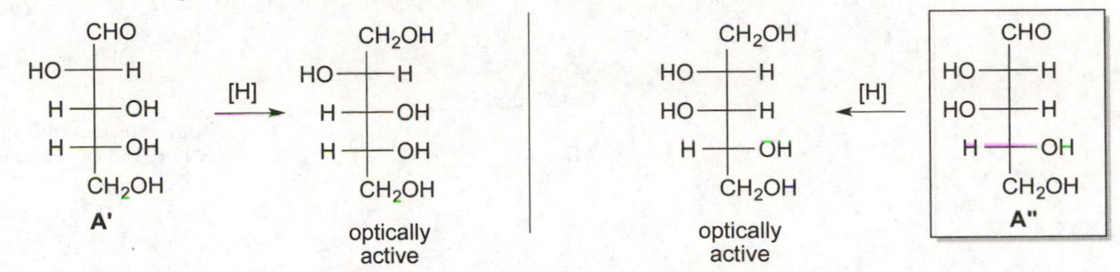

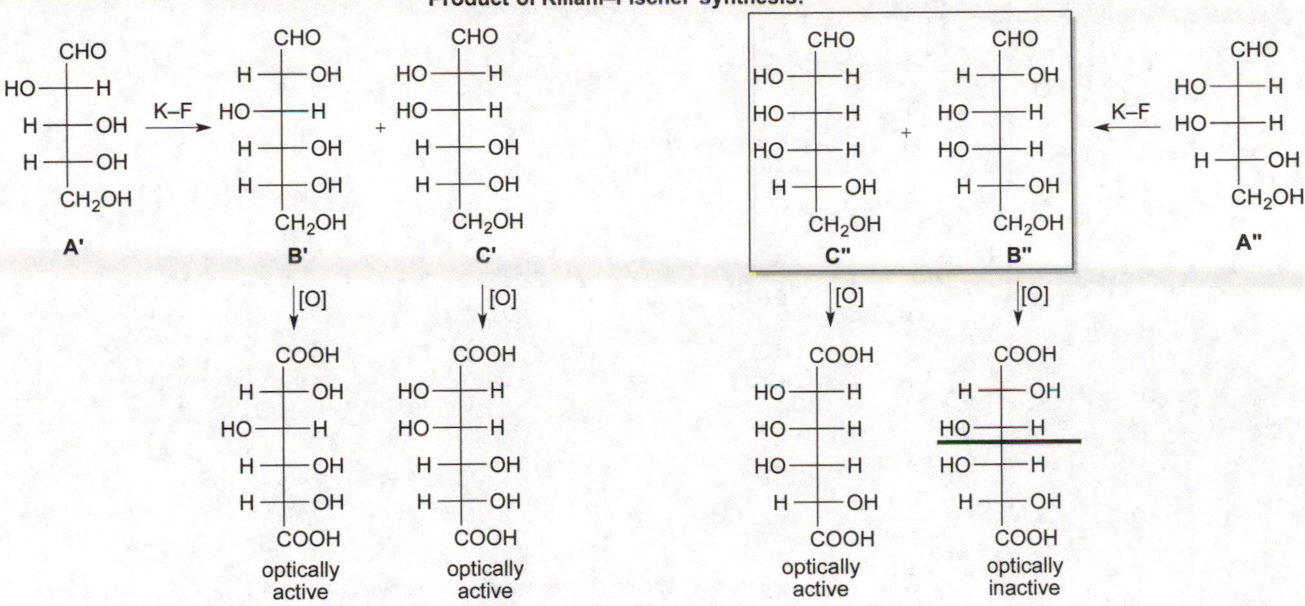

Only **A"** fits the criteria. Kiliani–Fischer synthesis of **A"** forms **B"** and **C"**, which are oxidized to one optically inactive and one optically active diacid. A similar procedure with **A'** forms two optically active diacids. Thus, the structures of **A–C** correspond to **A"–C"**.

28.64 A disaccharide formed from two mannose units in a 1→4-α-glycosidic linkage:

α glycoside bond

28.65

a.

α glycoside bond

1→6-α–glycoside bond

reducing sugar (hemiacetal)

b.

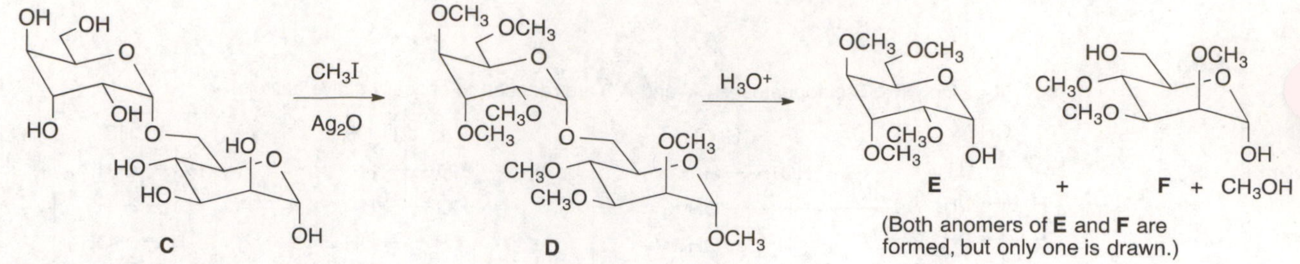

C → (CH_3I, Ag_2O) → D → (H_3O^+) → E + F + CH_3OH

(Both anomers of **E** and **F** are formed, but only one is drawn.)

28.66

a, b.

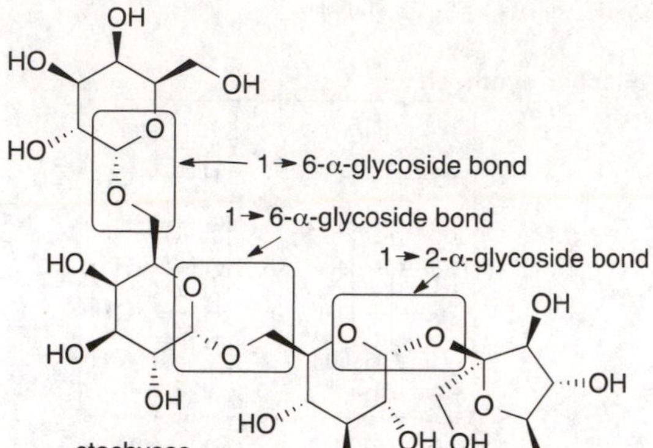

$1 \rightarrow 6$-α-glycoside bond

$1 \rightarrow 6$-α-glycoside bond

$1 \rightarrow 2$-α-glycoside bond

stachyose

c. H_3O^+ →

identical

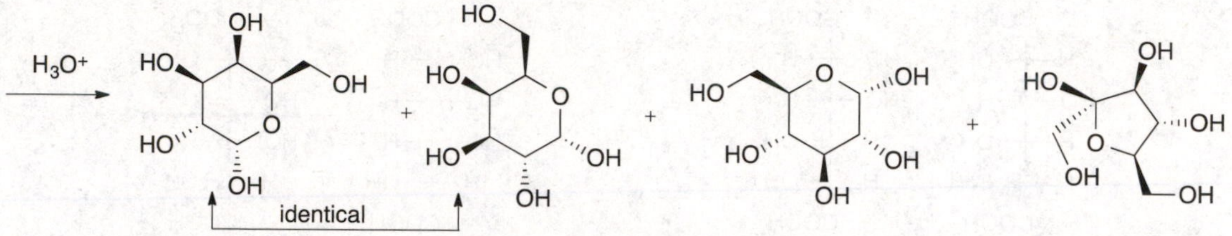

Two anomers of each monosaccharide are formed, but only one anomer is drawn.

d. Stachyose is not a reducing sugar because it contains no hemiacetal.

e. CH_3I, Ag_2O →

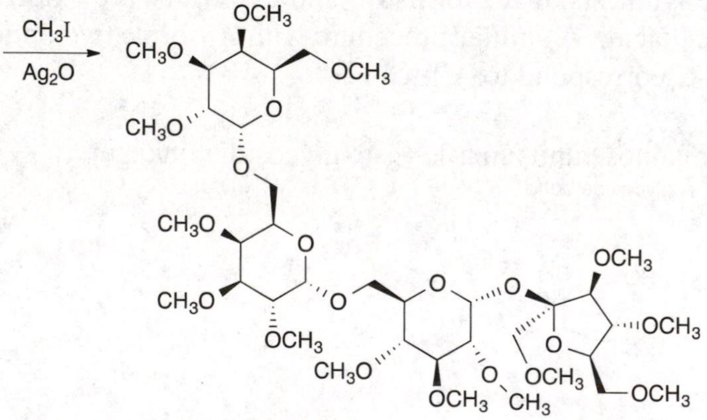

f. product in (e) → (H_3O^+)

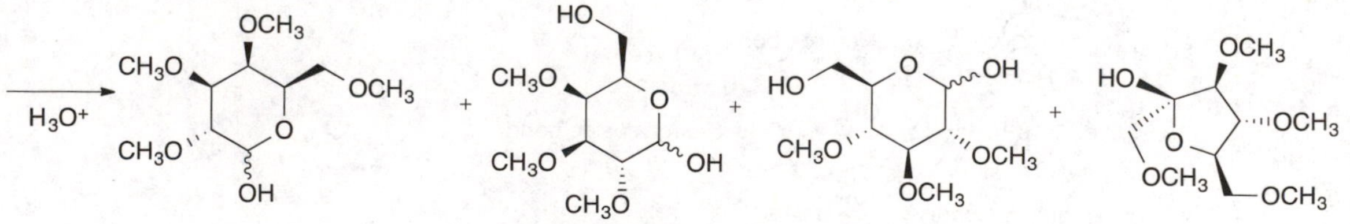

Two anomers of each monosaccharide are formed.

28.67

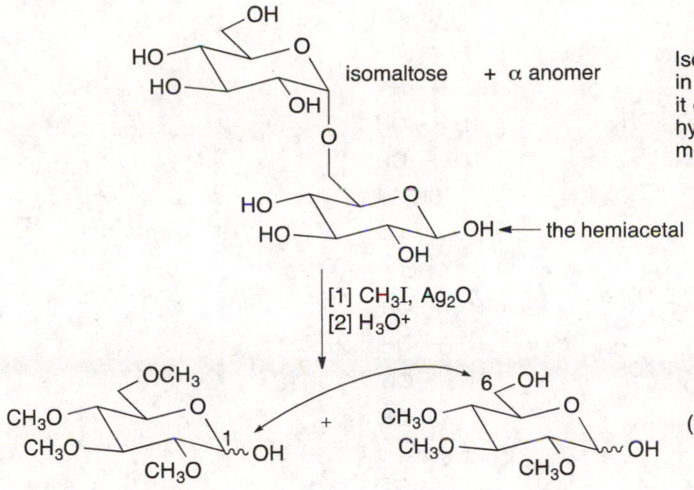

isomaltose + α anomer

the hemiacetal

[1] CH_3I, Ag_2O
[2] H_3O^+

(Both anomers are present.)

Isomaltose must be composed of two glucose units in an α-glycosidic linkage. Since it is a reducing sugar it contains a hemiacetal. The free OH groups in the hydrolysis products show where the two monosaccharides are joined.

28.68

a.

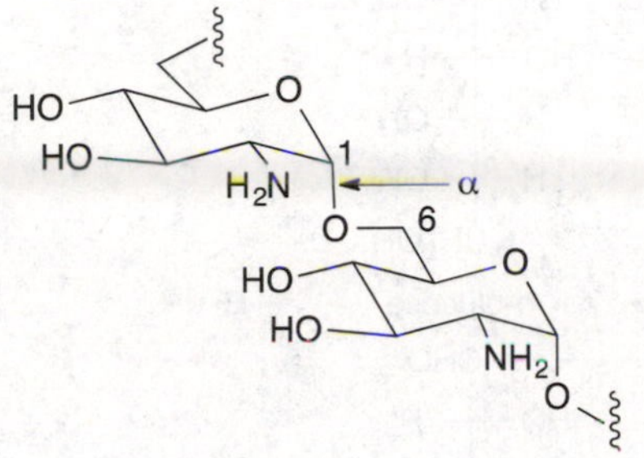

c.

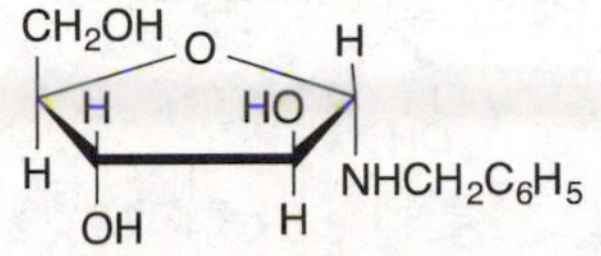

b.

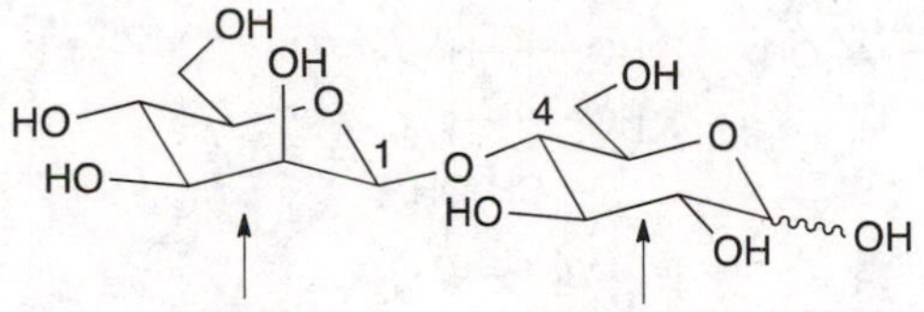

mannose glucose

d.

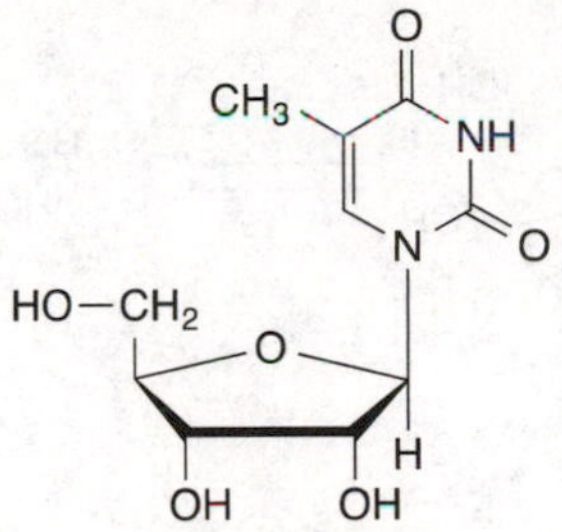

28.69

a.

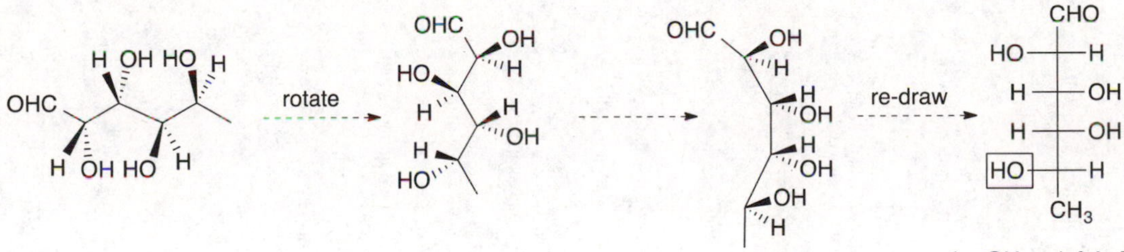

rotate

re-draw

b. OH on left in Fischer L-monosaccharide

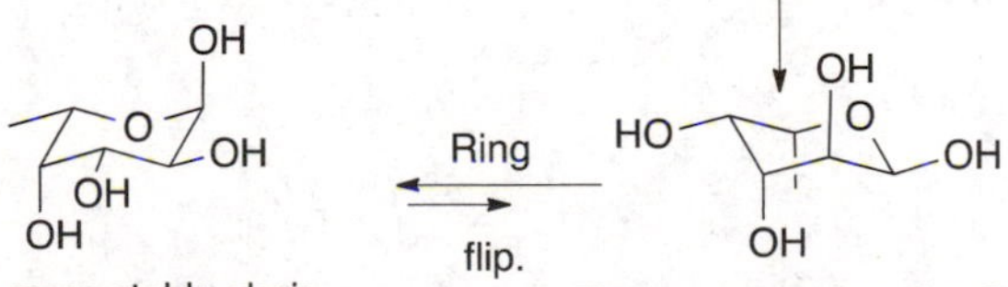

more stable chair

Ring flip.

The α anomer has the CH_3 on C5 and the anomeric OH trans.

c. Fucose is unusual because it is an L-monosaccharide and it contains a CH_3 group rather than a CH_2OH group on its terminal carbon.

28.70

a. D-fructose

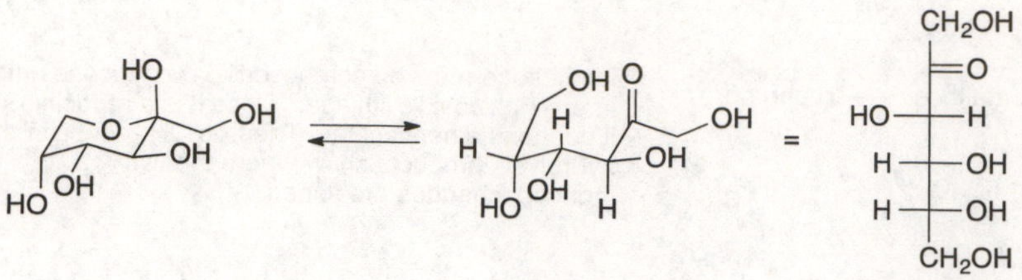

b. D-ribose

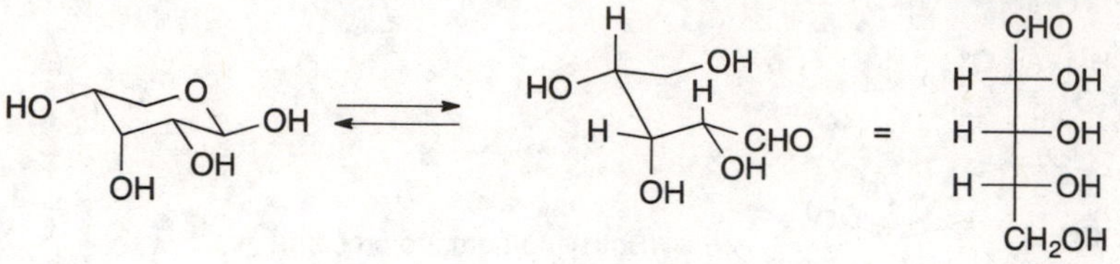

c. D-glucose

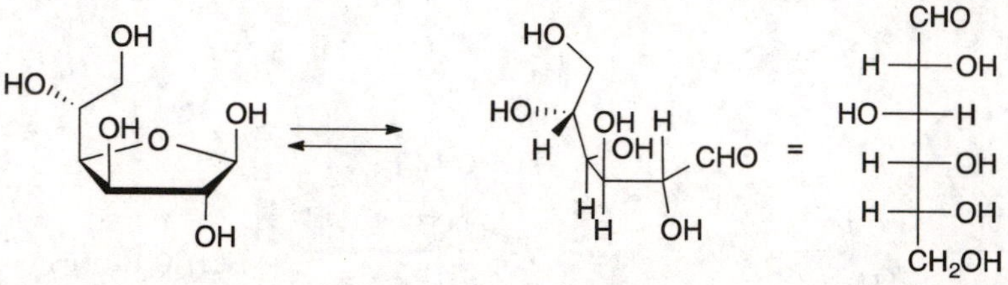

d. L-glucose

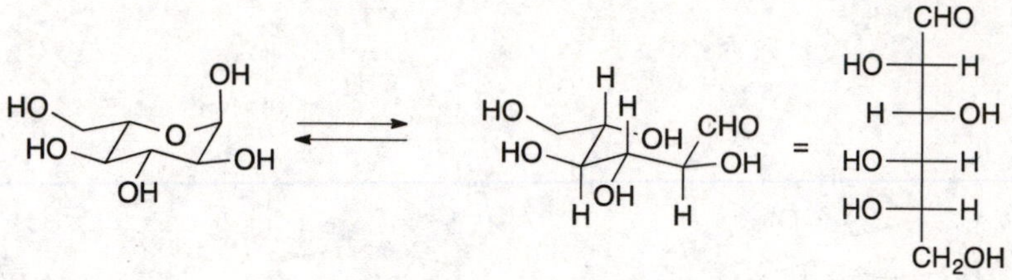

28.71

Ignoring stereochemistry along the way:

[1] [2] [3] [4] [5] [6] + [7] [8] $-H_2O$: $+ H_2O$ [9] [10] [11] [12] [13] [14] $-H_2O$: [15] [16] + HA

Chapter 29 Amino Acids and Proteins

Chapter Review

Synthesis of amino acids (29.2)

[1] From α-halo carboxylic acids by S_N2 reaction

R–CH(Br)–COOH → (NH_3 (large excess), S_N2) → R–CH(NH_2)–COO^- NH_4^+ + NH_4^+ Br^-

[2] By alkylation of diethyl acetamidomalonate

$CH_3CONHCH(COOEt)_2$ → ([1] NaOEt; [2] RX; [3] H_3O^+, Δ) → H_3N^+–CH(R)–COOH

- Alkylation works best with unhindered alkyl halides—that is, with CH_3X and RCH_2X.

[3] Strecker synthesis

RCHO → (NH_4Cl, NaCN) → RCH(NH_2)CN (α-amino nitrile) → (H_3O^+) → RCH(NH_3^+)COOH

Preparation of optically active amino acids

[1] Resolution of enantiomers by forming diastereomers (29.3A)

- Convert a racemic mixture of amino acids into a racemic mixture of *N*-acetyl amino acids [(*S*)- and (*R*)-$CH_3CONHCH(R)COOH$].
- React the enantiomers with a chiral amine to form a mixture of diastereomers.
- Separate the diastereomers.
- Regenerate the amino acids by protonation of the carboxylate salt and hydrolysis of the *N*-acetyl group.

[2] Kinetic resolution using enzymes (29.3B)

(S) isomer
(R) isomer
enantiomers
Ac_2O
enantiomers
acylase
(S) isomer
recovered amide
No reaction
separate

[3] By enantioselective hydrogenation (29.4)

H_2 / Rh*
S enantiomer
H_2O, ^-OH
S amino acid
Rh* = chiral Rh hydrogenation catalyst

Summary of methods used for peptide sequencing (29.6)

- Complete hydrolysis of all amide bonds in a peptide gives the identity and amount of the individual amino acids.
- Edman degradation identifies the N-terminal amino acid. Repeated Edman degradations can be used to sequence a peptide from the N-terminal end.
- Cleavage with carboxypeptidase identifies the C-terminal amino acid.
- Partial hydrolysis of a peptide forms smaller fragments that can be sequenced. Amino acid sequences common to smaller fragments can be used to determine the sequence of the complete peptide.
- Selective cleavage of a peptide occurs with trypsin and chymotrypsin to identify the location of specific amino acids (Table 29.2).

Adding and removing protecting groups for amino acids (29.7)

[1] Protection of an amino group as a Boc derivative

$(Boc)_2O$ / Et_3N

[2] Deprotection of a Boc-protected amino acid

Boc—NH–CH(R)–COOH → (CF_3CO_2H or HCl or HBr) → $H_3\overset{+}{N}$–CH(R)–COO^-

[3] Protection of an amino group as an Fmoc derivative

$H_3\overset{+}{N}$–CH(R)–COO^- + Fmoc–Cl → (Na_2CO_3, H_2O) → Fmoc—NH–CH(R)–COOH

[4] Deprotection of an Fmoc-protected amino acid

Fmoc—NH–CH(R)–COOH → (piperidine) → $H_3\overset{+}{N}$–CH(R)–COO^-

[5] Protection of a carboxy group as an ester

$H_3\overset{+}{N}$–CH(R)–COO^- → (CH_3OH, H^+) → H_2N–CH(R)–$COOCH_3$

methyl ester

$H_3\overset{+}{N}$–CH(R)–COO^- → ($PhCH_2OH$, H^+) → H_2N–CH(R)–$COOCH_2Ph$

benzyl ester

[6] Deprotection of an ester group

H_2N–CH(R)–$COOCH_3$ (methyl ester) → (^-OH, H_2O) → H_2N–CH(R)–COO^-

H_2N–CH(R)–$COOCH_2Ph$ (benzyl ester) → (H_2O, ^-OH or H_2, Pd-C) → $H_3\overset{+}{N}$–CH(R)–COO^-

Synthesis of dipeptides (29.7)

[1] Amide formation with DCC

Boc—NH–CH(R)–COOH + H_2N–CH(R)–$COOCH_2Ph$ → (DCC) → Boc—NH–CH–CO–NH–CH(R)–$COOCH_2Ph$

[2] Four steps are needed to synthesize a dipeptide:

a. **Protect** the amino group of one amino acid using a Boc or Fmoc group.
b. **Protect** the carboxy group of the second amino acid using an ester.
c. Form the amide bond with **DCC.**
d. **Remove both protecting groups** in one or two reactions.

Summary of the Merrifield method of peptide synthesis (29.8)

[1] Attach an Fmoc-protected amino acid to a polymer derived from polystyrene.
[2] Remove the Fmoc protecting group.
[3] Form the amide bond with a second Fmoc-protected amino acid using DCC.
[4] Repeat steps [2] and [3].
[5] Remove the protecting group and detach the peptide from the polymer.

Practice Test on Chapter Review

1.a. Which statement is true about the peptide Ala–Gly–Tyr–Phe?

1. The N-terminal amino acid is Ala.
2. The N-terminal amino acid is Phe.
3. The peptide contains four peptide bonds.
4. Statements (1) and (3) are true.
5. Statements (2) and (3) are true.

b. Which of the following peptides is hydrolyzed by trypsin?

1. Glu–Ser–Gly–Arg
2. Arg–Gln–Trp–Asp
3. Glu–Val–Leu–Lys
4. Peptides (1) and (2) are hydrolyzed.
5. Peptides (1), (2), and (3) are all hydrolyzed.

c. In which types of protein structure is hydrogen bonding observed?

1. α-helix
2. β-pleated sheet
3. 3° structure
4. Hydrogen bonding is present in (1) and (2).
5. Hydrogen bonding is present in (1), (2), and (3).

2. Answer the following questions about peptides.

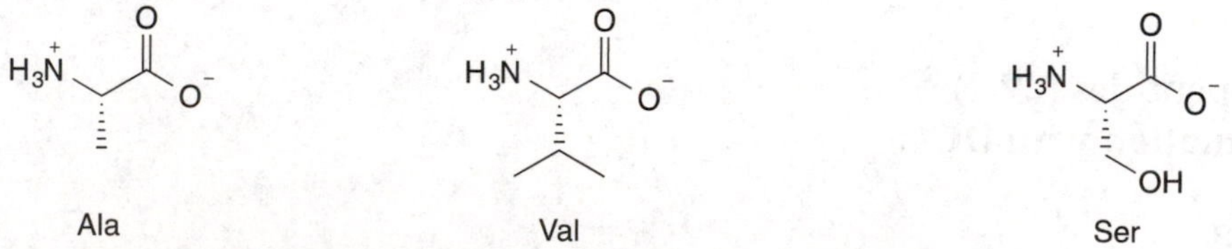

Ala Val Ser

a. Draw the structure of the following tripeptide: Val–Ser–Ala.
b. Give the three-letter abbreviation for the N-terminal amino acid.
c. Give the three-letter abbreviation for the C-terminal amino acid.

3. Answer the following questions about the amino acid leucine (2-amino-4-methylpentanoic acid), which has p*K*a's of 2.33 and 9.74 for its ionizable functional groups.
   a. Draw a Fischer projection for L-leucine and label the stereogenic center as *R* or *S*.
   b. What is the p*I* of leucine?
   c. Draw the structure of the predominant form of leucine at its isoelectric point.
   d. Draw the structure of the predominant form of leucine at pH 10.
   e. Is leucine an acidic, basic, or neutral amino acid?

4. What product is formed when the amino acid phenylalanine is treated with each reagent?
   a. $PhCH_2OH$, H^+
   b. Ac_2O, pyridine
   c. PhCOCl, pyridine
   d. $(Boc)_2O$
   e. $C_6H_5N{=}C{=}S$

5. Draw the amino acids and peptide fragments formed when the octapeptide Tyr–Gly–Ala–Lys–Val–Ser–Phe–Met is treated with each reagent or enzyme:
   a. chymotrypsin
   b. trypsin
   c. carboxypeptidase
   d. $C_6H_5N{=}C{=}S$

Answers to Practice Test

1.a. 1
  b. 2
  c. 5

2.a

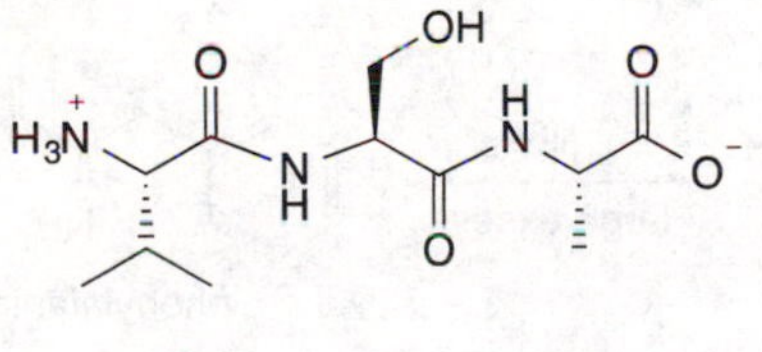

$H_3\overset{+}{N}$ ... N–H ... N–H ... O^-

Val Ser Ala

  b. Val
  c. Ala

3.

a. 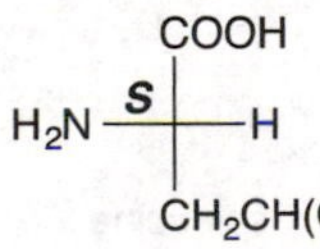 COOH; H_2N — *S* — H; $CH_2CH(CH_3)_2$

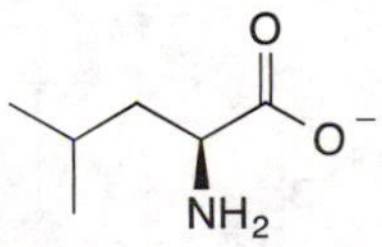

b. 6.04

c. NH_3^+ ... O^-

d. NH_2 ... O^-

e. neutral

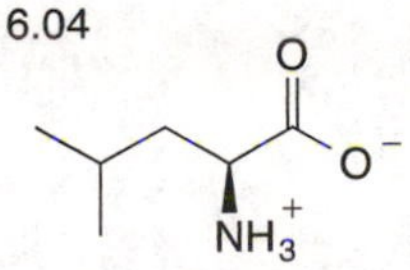

4.

a. 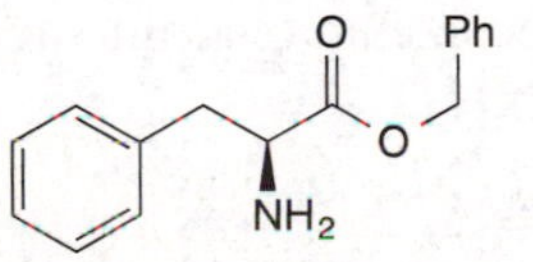 O, Ph, O, NH_2

b. 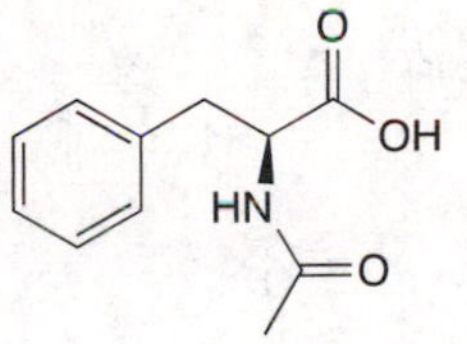 O, OH, HN, O

c. 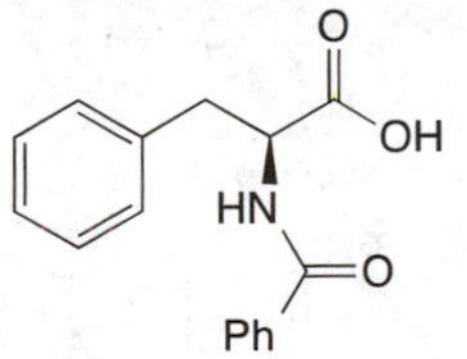 O, OH, HN, O, Ph

d. 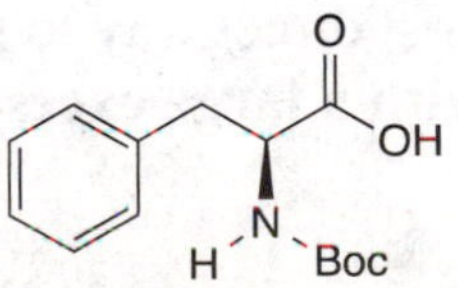 O, OH, N, H, Boc

e. 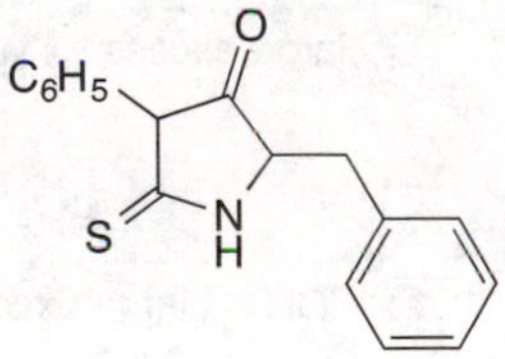 O, C_6H_5, S, N, H

5.a. Tyr, Gly–Ala–Lys–Val–Ser–Phe, Met
  b. Tyr–Gly–Ala–Lys, Val–Ser–Phe–Met
  c. Tyr–Gly–Ala–Lys–Val–Ser–Phe, Met
  d. Tyr, Gly–Ala–Lys–Val–Ser–Phe–Met

Answers to Problems

29.1

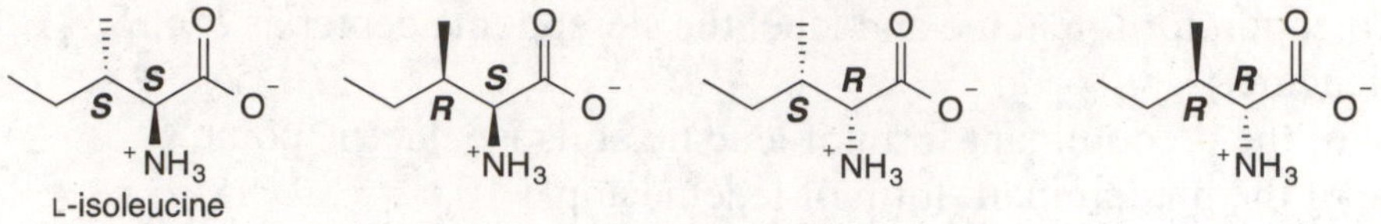

S S R S S R R R

$^+NH_3$ $^+NH_3$ $^+NH_3$ $^+NH_3$

L-isoleucine

29.2

a. $^+NH_3$ b. $^+NH_3$ c. NH_2^+ d. HO $^+NH_3$

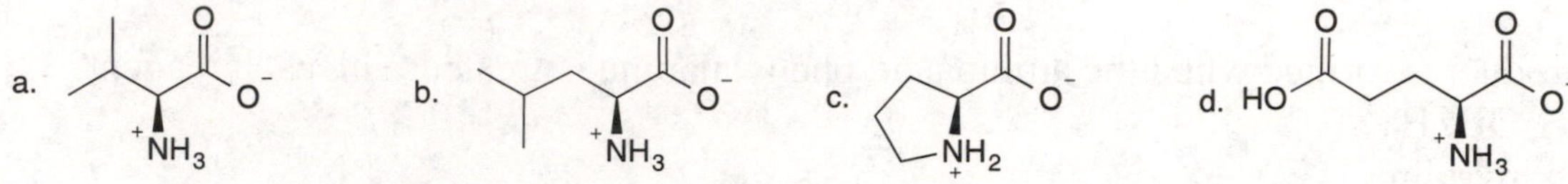

29.3 In an amino acid, the electron-withdrawing carboxy group destabilizes the ammonium ion ($-NH_3^+$), making it more readily donate a proton; that is, it makes it a stronger acid. Also, the electron-withdrawing carboxy group removes electron density from the amino group ($-NH_2$) of the conjugate base, making it a weaker base than a 1° amine, which has no electron-withdrawing group.

29.4

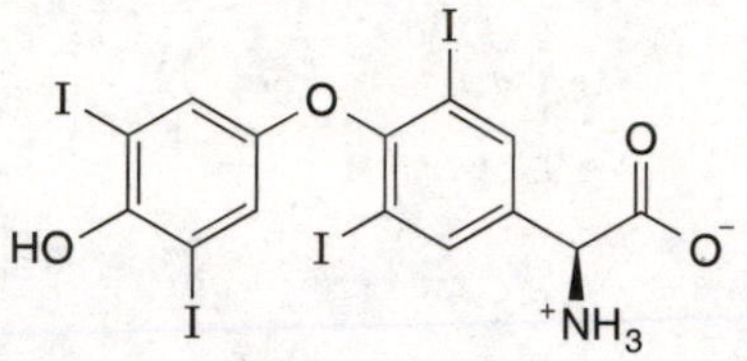

I I HO I I O $^+NH_3$

zwitterionic form

29.5 The most direct way to synthesize an α-amino acid is by **S_N2 reaction of an α-halo carboxylic acid with a large excess of NH_3.**

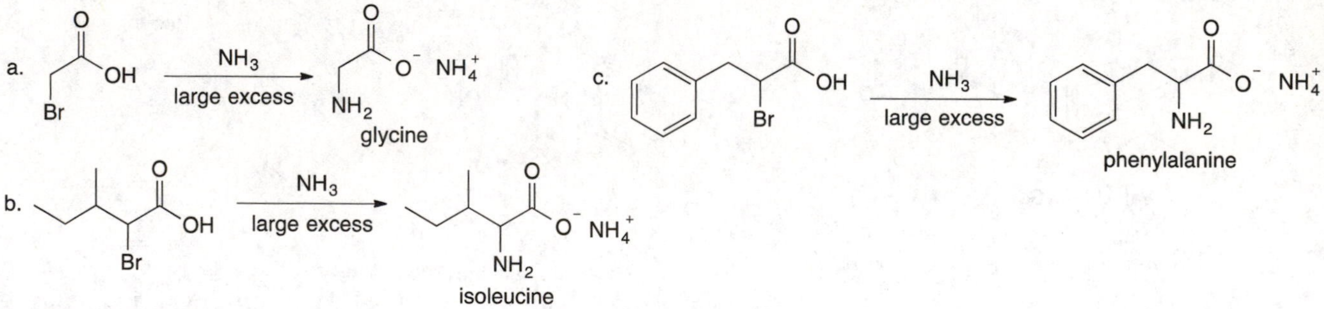

a. Br OH → NH_3 large excess → NH_2 O^- NH_4^+ glycine

b. Br OH → NH_3 large excess → NH_2 O^- NH_4^+ isoleucine

c. Br OH → NH_3 large excess → NH_2 O^- NH_4^+ phenylalanine

29.6

a. CH_3I → $^+NH_3$ alanine

b. Cl → $^+NH_3$ leucine

c. Br → $^+NH_3$ isoleucine

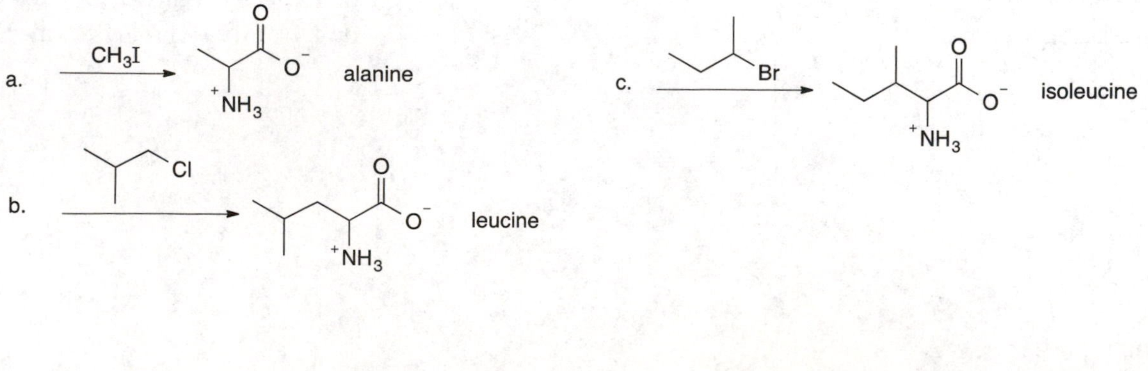

29.7

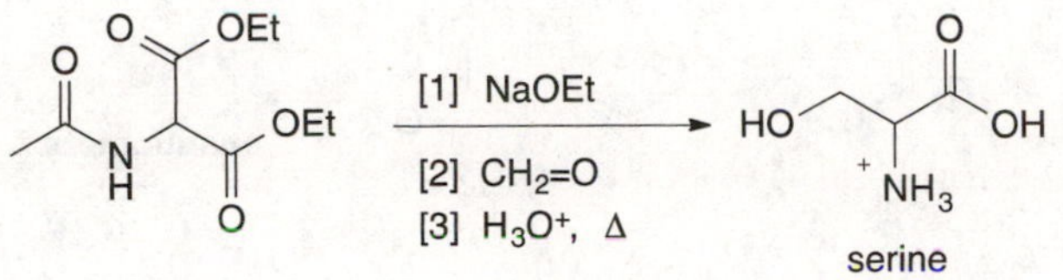

O OEt O OEt N H O

[1] NaOEt
[2] $CH_2{=}O$
[3] H_3O^+, Δ

HO O OH $\overset{+}{N}H_3$

serine

29.8

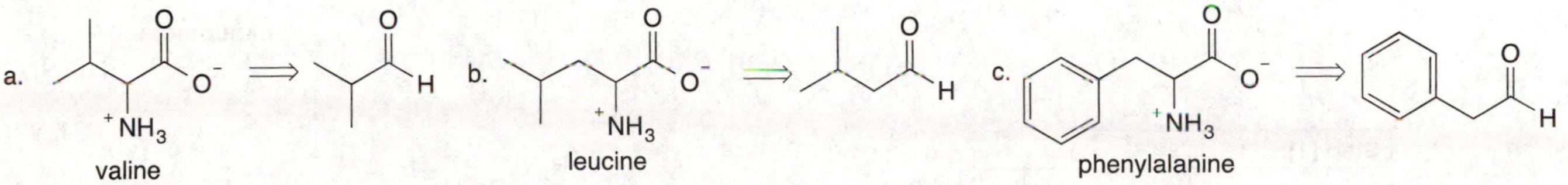

a. O O^- $^+NH_3$ valine ⟹ O H

b. O O^- $^+NH_3$ leucine ⟹ O H

c. O O^- $^+NH_3$ phenylalanine ⟹ O H

29.9

a. O OH Br → NH_3 large excess → O O^- NH_4^+ NH_2

b. H N O OEt O O OEt

[1] NaOEt
[2] $(CH_3)_2CHCl$
[3] H_3O^+, Δ

O OH $^+NH_3$

c. O H

[1] NH_4Cl, NaCN
[2] H_3O^+

O OH $^+NH_3$

d. H N O OEt O O OEt

[1] NaOEt
[2] $BrCH_2CO_2Et$
[3] H_3O^+, Δ

HO O O OH $^+NH_3$

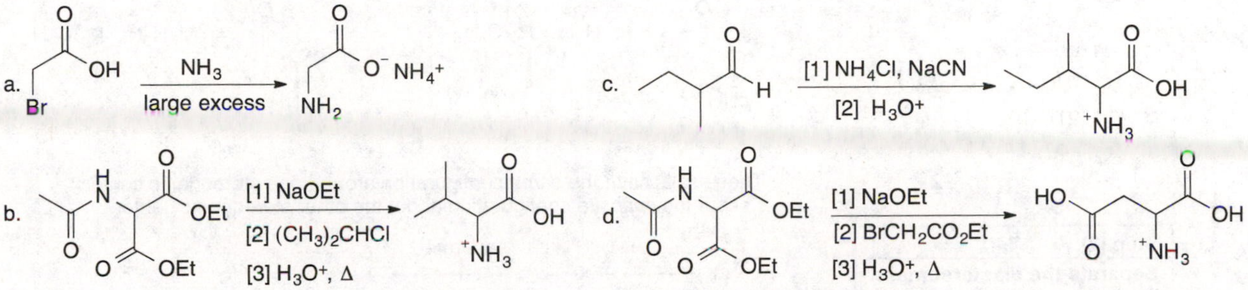

29.10 A chiral amine must be used to resolve a racemic mixture of amino acids.

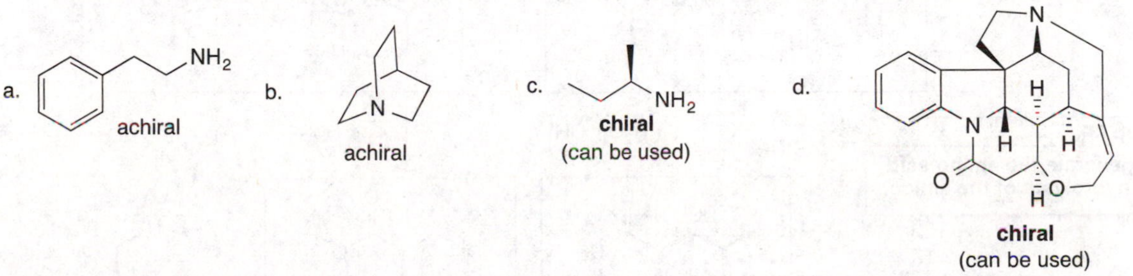

a. NH_2 achiral

b. N achiral

c. NH_2 **chiral** (can be used)

d. N N H H H O O H **chiral** (can be used)

29.11

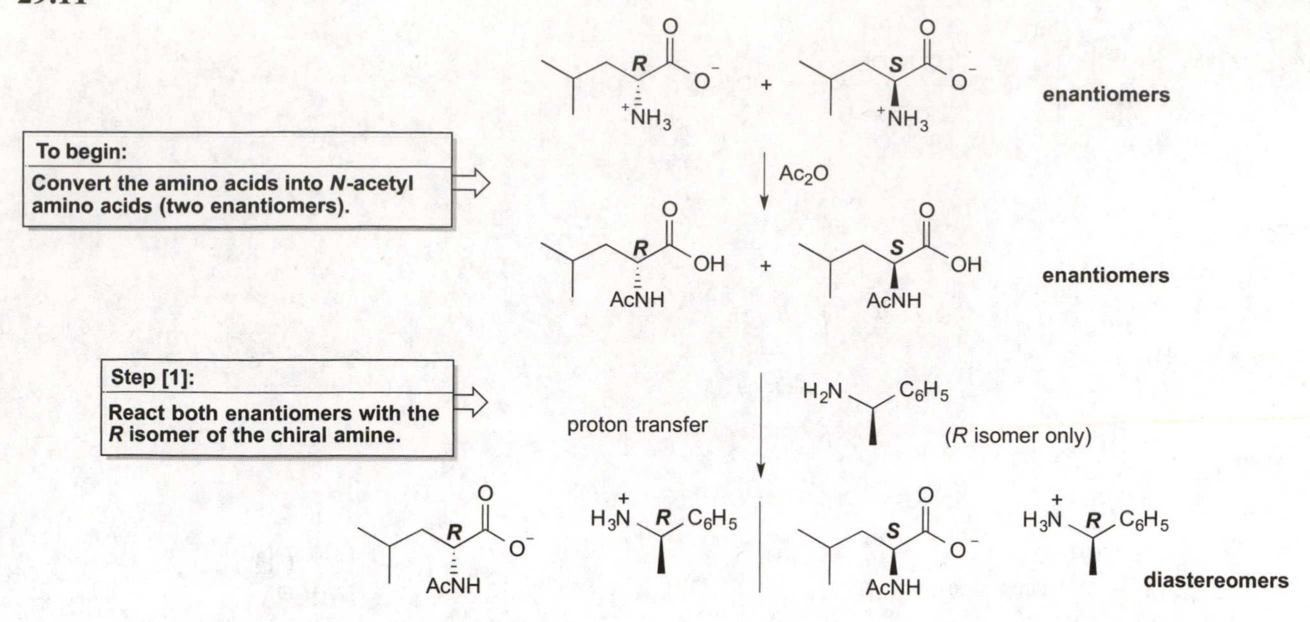

To begin:
Convert the amino acids into *N*-acetyl amino acids (two enantiomers).
R
S
$^+NH_3$
enantiomers
Ac_2O
OH
AcNH
enantiomers
Step [1]:
React both enantiomers with the *R* isomer of the chiral amine.
proton transfer
H_2N
C_6H_5
(*R* isomer only)
$H_3\overset{+}{N}$
diastereomers

These salts have the *same* configuration around one stereogenic center, but the *opposite* configuration about the other stereogenic center.

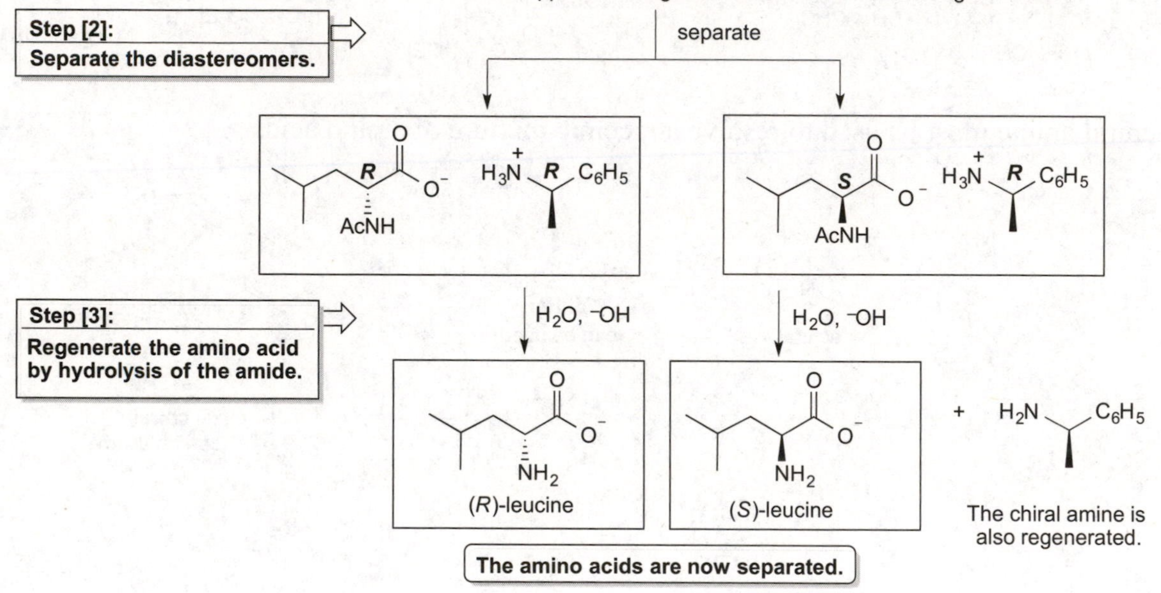

Step [2]:
Separate the diastereomers.
separate
AcNH
$H_3\overset{+}{N}$
R
S
C_6H_5
Step [3]:
Regenerate the amino acid by hydrolysis of the amide.
H_2O, ^-OH
NH_2
(*R*)-leucine
(*S*)-leucine
H_2N
The chiral amine is also regenerated.
The amino acids are now separated.

29.12

(mixture of enantiomers) — [1] $(CH_3CO)_2O$; [2] acylase → (*S*)-leucine + *N*-acetyl-(*R*)-leucine

29.13

a. 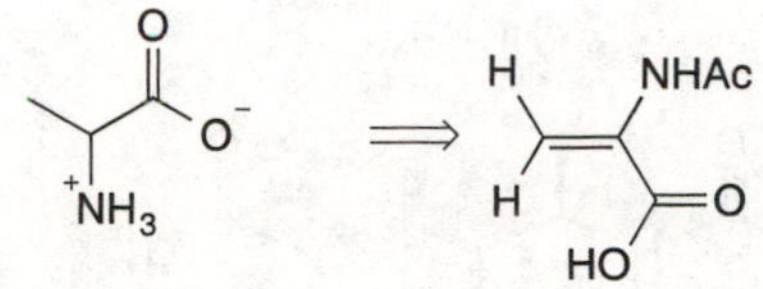 O, O^-, $^+NH_3$ $\Longrightarrow$ H, NHAc, H, O, HO

c. 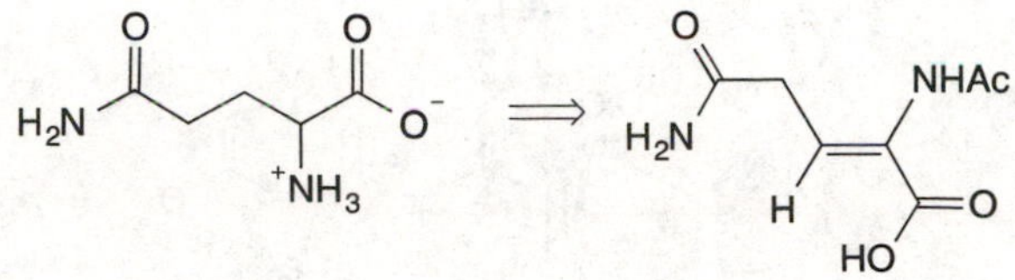 O, O, H_2N, O^-, $^+NH_3$ $\Longrightarrow$ O, NHAc, H_2N, H, O, HO

b. 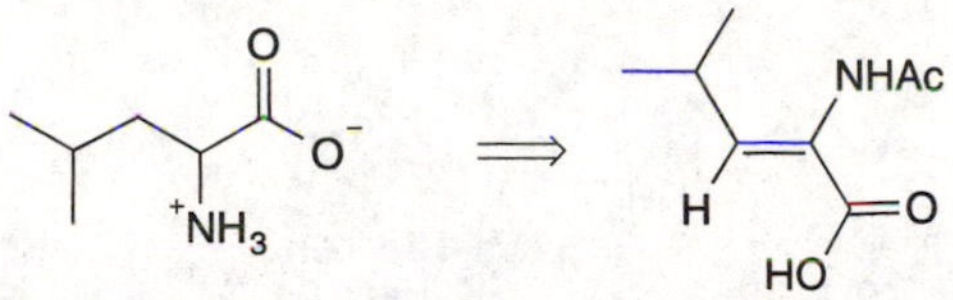 O, O^-, $^+NH_3$ $\Longrightarrow$ NHAc, H, O, HO

29.14 Draw the peptide by joining adjacent COOH and NH_2 groups in amide bonds.

a. 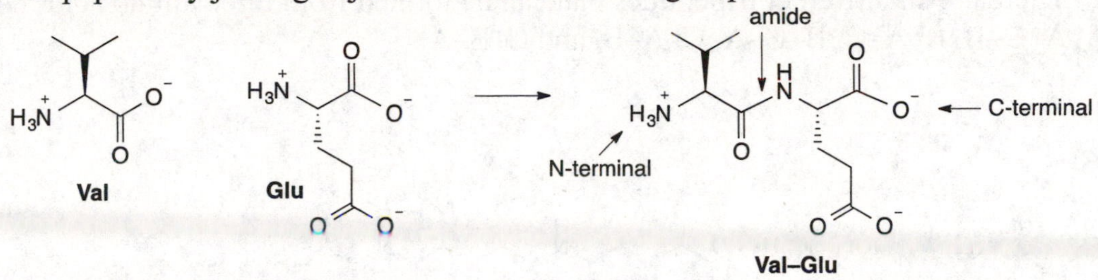 $H_3\overset{+}{N}$, O^-, O — **Val**

$H_3\overset{+}{N}$, O, O^-, O, O^- — **Glu**

→ amide, H, N, O, $H_3\overset{+}{N}$, O, O^- ← C-terminal; N-terminal; O, O, O^-

Val–Glu

b. 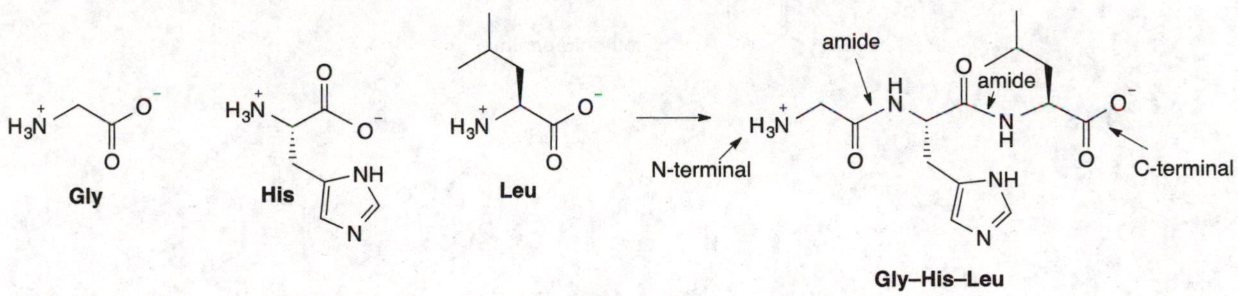 $H_3\overset{+}{N}$, O^-, O — **Gly**

$H_3\overset{+}{N}$, O, O^-, NH, N — **His**

$H_3\overset{+}{N}$, O^-, O — **Leu**

→ amide, H, N, O, amide, O^-, $H_3\overset{+}{N}$, N, H, N-terminal, O, O, C-terminal, NH, N

Gly–His–Leu

c. S, $H_3\overset{+}{N}$, O^-, O — **M**

$H_3\overset{+}{N}$, O, O^- — **A**

OH, $H_3\overset{+}{N}$, O^-, O — **T**

$H_3\overset{+}{N}$, O, O^-, OH — **T**

→

amide, amide, OH, S, H, N, O, H, N, O, N-terminal → $H_3\overset{+}{N}$, N, H, O, O, O^- ← C-terminal, OH, amide

M–A–T–T

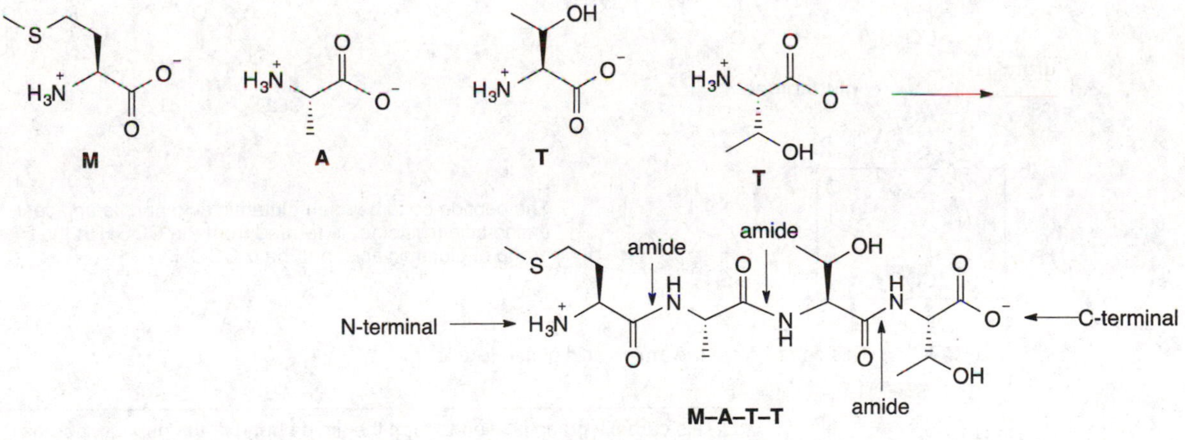

29.15

a. 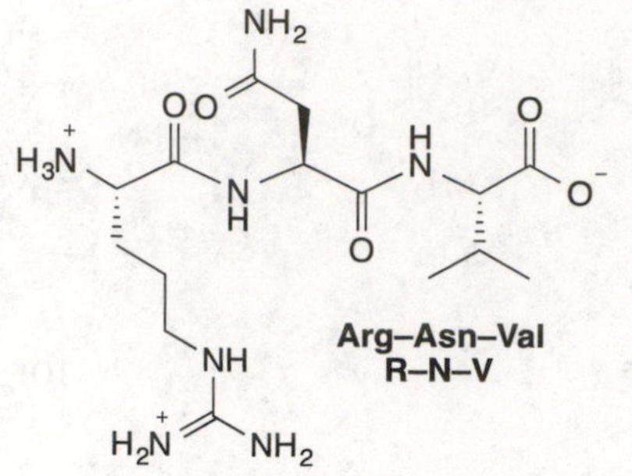 Arg–Asn–Val
R–N–V

b. 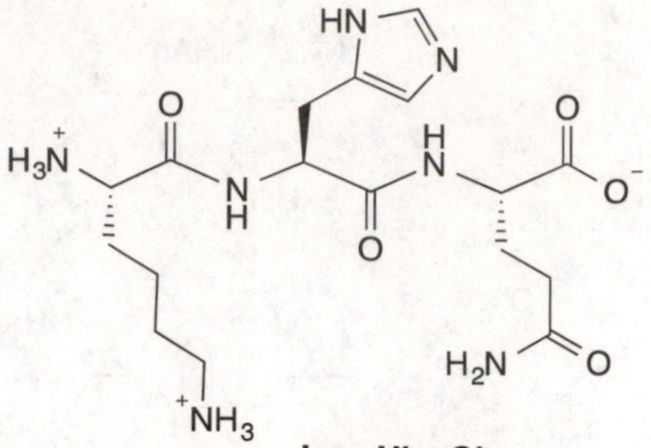 Lys–His–Gln
K–H–Q

29.16 There are six different tripeptides that can be formed from three amino acids (A, B, C): A–B–C, A–C–B, B–A–C, B–C–A, C–A–B, and C–B–A.

29.17

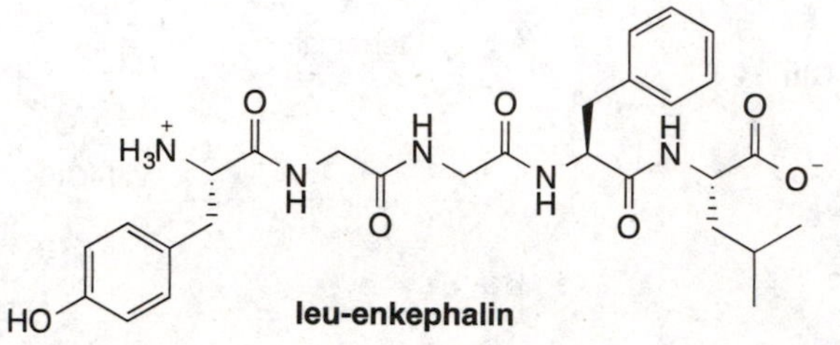

leu-enkephalin

29.18

a. 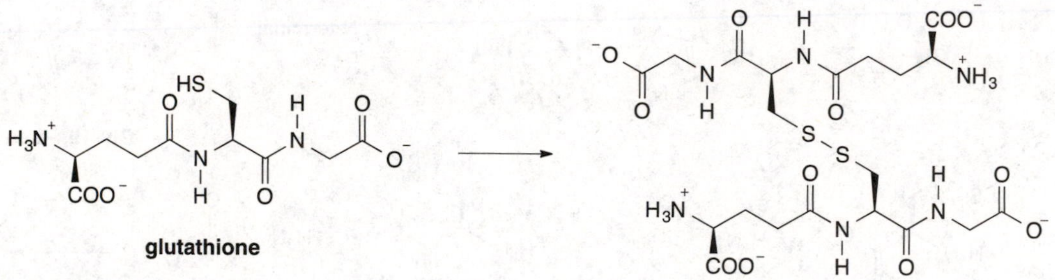 glutathione

b. 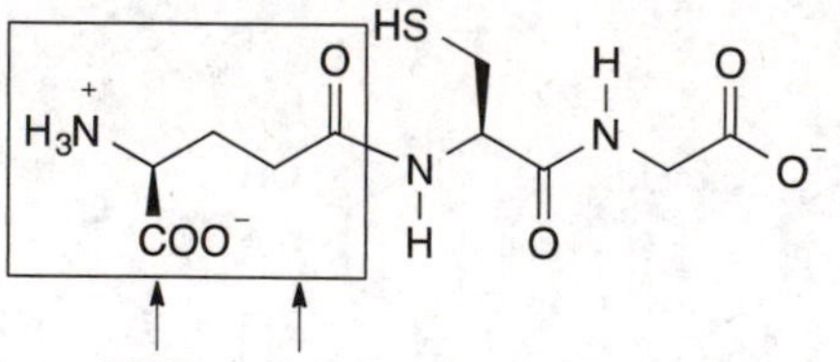 The peptide bond beween glutamic acid and its adjacent amino acid (cysteine) is formed from the COOH in the R group of glutamic acid, not the α COOH.

α COO– This comes from the amino acid glutamic acid.

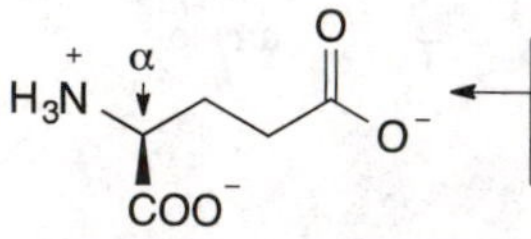

This carboxy group is used to form the amide bond in the peptide, not the α COOH, as is usual. That's what makes glutathione's structure unusual.

glutamic acid

29.19

a. C_6H_5–N, O, S, N–H

from Ala

b. C_6H_5–N, O, S, N–H

from Val

29.20 Determine the sequence of the octapeptide as in Sample Problem 29.2. Look for overlapping sequences in the fragments.

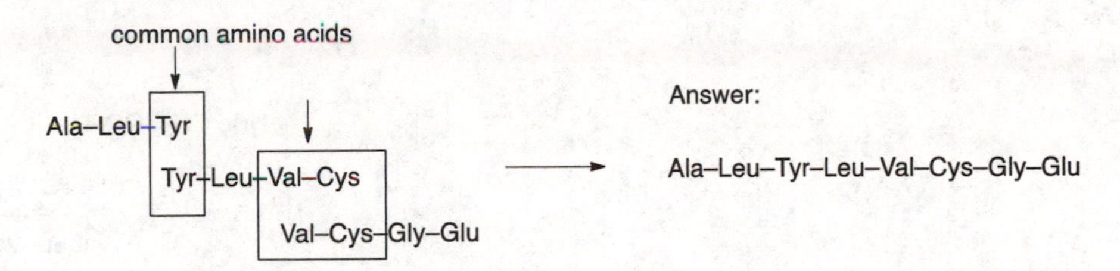
common amino acids
Ala–Leu–Tyr
Tyr–Leu–Val–Cys
Val–Cys–Gly–Glu
Answer:
Ala–Leu–Tyr–Leu–Val–Cys–Gly–Glu

29.21 Trypsin cleaves peptides at amide bonds with a carbonyl group from Arg and Lys. Chymotrypsin cleaves at amide bonds with a carbonyl group from Phe, Tyr, and Trp.

a. [1] Gly–Ala–Phe–Leu–Lys + Ala
[2] Phe–Tyr–Gly–Cys–Arg + Ser
[3] Thr–Pro–Lys + Glu–His–Gly–Phe–Cys–Trp–Val–Val–Phe

b. [1] Gly–Ala–Phe + Leu–Lys–Ala
[2] Phe + Tyr + Gly–Cys–Arg–Ser
[3] Thr–Pro–Lys–Glu–His–Gly–Phe + Cys–Trp + Val–Val–Phe

29.22

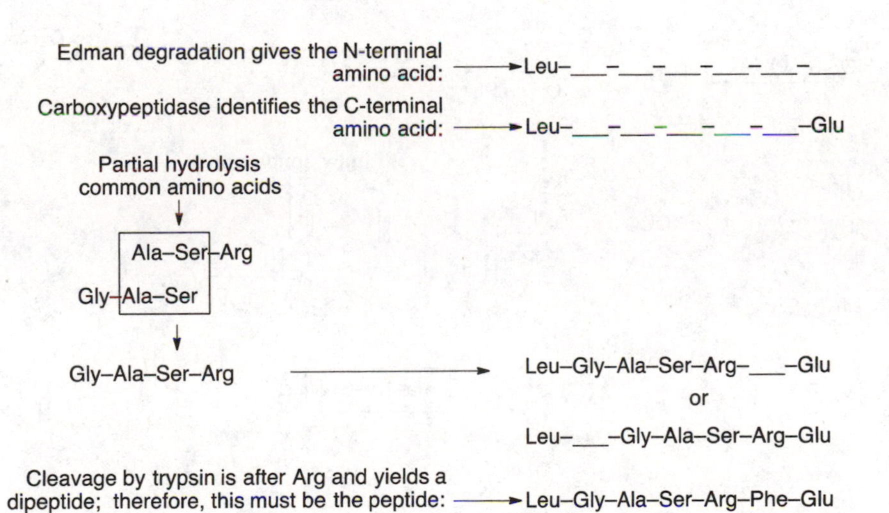
Edman degradation gives the N-terminal amino acid:
Leu–___–___–___–___–___–___
Carboxypeptidase identifies the C-terminal amino acid:
Leu–___–___–___–___–___–Glu
Partial hydrolysis
common amino acids
Ala–Ser–Arg
Gly–Ala–Ser
Gly–Ala–Ser–Arg
Leu–Gly–Ala–Ser–Arg–___–Glu
or
Leu–___–Gly–Ala–Ser–Arg–Glu
Cleavage by trypsin is after Arg and yields a dipeptide; therefore, this must be the peptide:
Leu–Gly–Ala–Ser–Arg–Phe–Glu

29.23

a. $H_3\overset{+}{N}$–Leu–O^- $\xrightarrow[Et_3N]{Boc_2O}$ Boc–NH–Leu–OH | $H_3\overset{+}{N}$–Val–O^- $\xrightarrow[H^+]{HOCH_2Ph}$ H_2N–Val–OCH_2Ph

Leu

Val

Boc–NH–Leu–OH + H_2N–Val–OCH_2Ph $\xrightarrow{DCC}$ Boc–NH–Leu–NH–Val–OCH_2Ph

new amide bond

HBr, CH_3COOH

$H_3\overset{+}{N}$–Leu–NH–Val–O^-

Leu–Val

b. $H_3\overset{+}{N}$–Ala–O^- $\xrightarrow[Et_3N]{Boc_2O}$ Boc–NH–Ala–OH (A) | $H_3\overset{+}{N}$–Ile–O^- $\xrightarrow[H^+]{HOCH_2Ph}$ H_2N–Ile–OCH_2Ph (B)

Ala

Ile

A + B $\xrightarrow{DCC}$ Boc–NH–Ala–NH–Ile–OCH_2Ph $\xrightarrow[Pd\text{-}C]{H_2}$ Boc–NH–Ala–NH–Ile–OH (C)

new amide bond

$H_3\overset{+}{N}$–Gly–O^- $\xrightarrow[H^+]{HOCH_2Ph}$ H_2N–Gly–OCH_2Ph

Gly

C + H_2N–Gly–OCH_2Ph $\xrightarrow{DCC}$ Boc–NH–Ala–NH–Ile–NH–Gly–OCH_2Ph $\xrightarrow[CH_3COOH]{HBr}$

new amide bond

Ala–Ile–Gly $H_3\overset{+}{N}$–Ala–NH–Ile–NH–Gly–O^-

29.24 The dipeptide depicted in the 3-D model has alanine as the N-terminal amino acid and cysteine as the C-terminal amino acid.

Ala $\xrightarrow[Et_3N]{Boc_2O}$ A

Cys $\xrightarrow[H^+]{HOCH_2Ph}$ B

A + B $\xrightarrow{DCC}$ Boc-protected dipeptide benzyl ester $\xrightarrow{HBr,\ CH_3COOH}$ Ala–Cys

29.25

All Fmoc-protected amino acids are made by the following general reaction:

H_3N^+–CH(R)–COO^- $\xrightarrow[Na_2CO_3,\ H_2O]{Fmoc\text{–}Cl}$ Fmoc–NH–CH(R)–COOH

The steps:

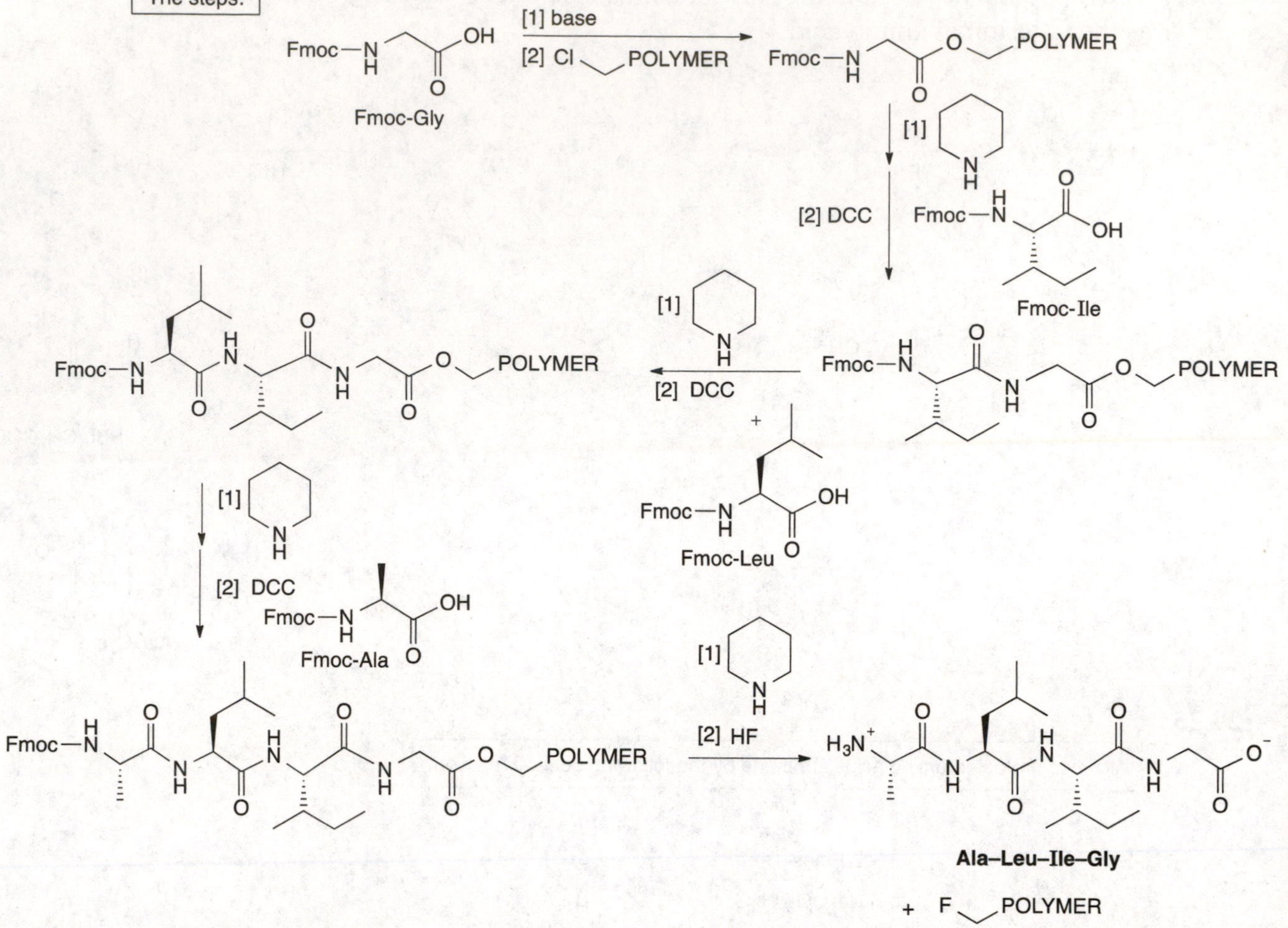

Fmoc-N(H)-CH₂-C(=O)OH (Fmoc-Gly) → [1] base, [2] Cl⌄POLYMER → Fmoc-N(H)-CH₂-C(=O)-O⌄POLYMER

[1] piperidine; [2] DCC, Fmoc-Ile → Fmoc-Ile-Gly-O⌄POLYMER

[1] piperidine; [2] DCC + Fmoc-Leu → Fmoc-Leu-Ile-Gly-O⌄POLYMER

[1] piperidine; [2] DCC, Fmoc-Ala → Fmoc-Ala-Leu-Ile-Gly-O⌄POLYMER

[1] piperidine; [2] HF → H_3N^+-Ala-Leu-Ile-Gly-O^-

Ala–Leu–Ile–Gly

+ F⌄POLYMER

29.26 In a *parallel* β-pleated sheet, the strands run in the *same* direction from the N- to C-terminal amino acid. In an *antiparallel* β-pleated sheet, the strands run in the *opposite* direction.

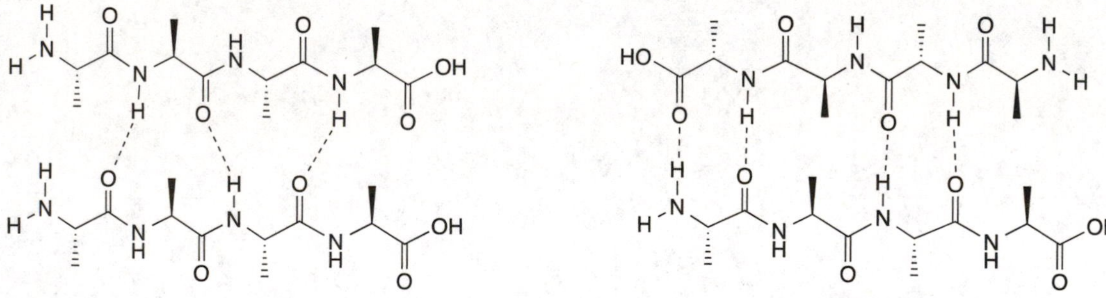

parallel

antiparallel

29.27

a. Ser and Tyr

side chains with OH groups
hydrogen bonding

b. Val and Leu

side chains with only C–C and C–H bonds
van der Waals forces

c. 2 Phe residues

van der Waals forces

29.28 a. The R group for glycine is a hydrogen. The R groups must be small to allow the β-pleated sheets to stack on top of each other. With large R groups, steric hindrance prevents stacking.

b. Silk fibers are water insoluble because most of the polar functional groups are in the interior of the stacked sheets. The β-pleated sheets are stacked one on top of another, so few polar functional groups are available for hydrogen bonding to water.

29.29

phenylalanine

a. CH_3OH, H^+

b. CH_3COCl, pyridine

c. HCl (1 equiv)

d. NaOH (1 equiv)

e. $C_6H_5N{=}C{=}S$

29.30

a. N-terminal amino acid: alanine
C-terminal amino acid: serine

b. A–Q–C–S

c. Amide bonds are **bold** (not wedges).

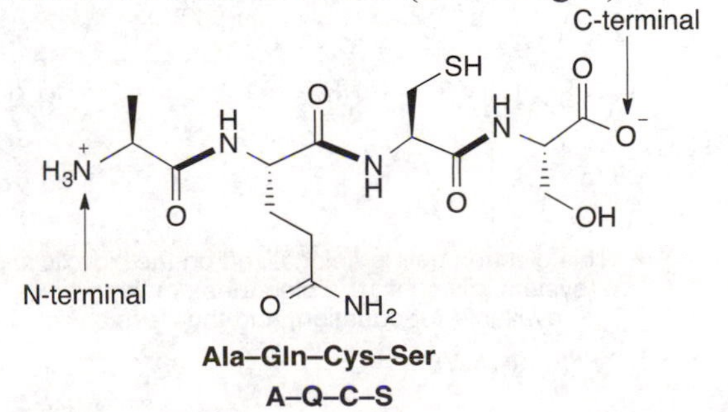

29.31 The dipeptide is composed of phenylalanine and leucine.

Boc_2O, Et_3N

Phe

A

HO–Ph, H^+

Leu

B

A + B → DCC

HBr, CH_3COOH

Phe–Leu

29.32

a. (*R*)-penicillamine (*S*)-penicillamine

b.

29.33 The electron pair on the N atom not part of a double bond is delocalized on the five-membered ring, making it less basic.

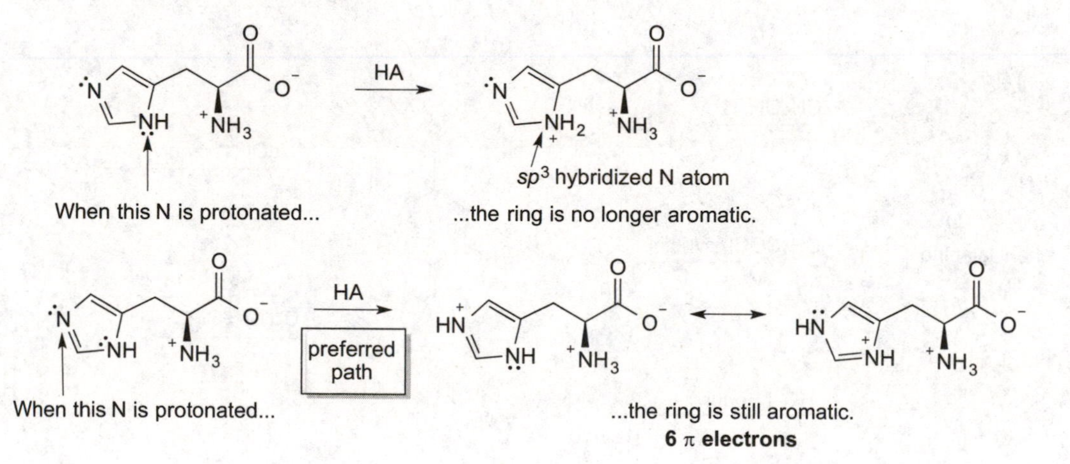
HA
sp^3 hybridized N atom
When this N is protonated...
...the ring is no longer aromatic.
HA
preferred path
When this N is protonated...
...the ring is still aromatic.
6 π electrons

29.34

no *p* orbital on N

This electron pair is delocalized on the bicyclic ring system (giving it 10 π electrons), making it less available for donation, and thus less basic.

The ring structure on tryptophan is aromatic because each atom contains a *p* orbital. Protonation of the N atom would disrupt the aromaticity, making this a less favorable reaction.

29.35 At its isoelectric point, each amino acid is neutral.

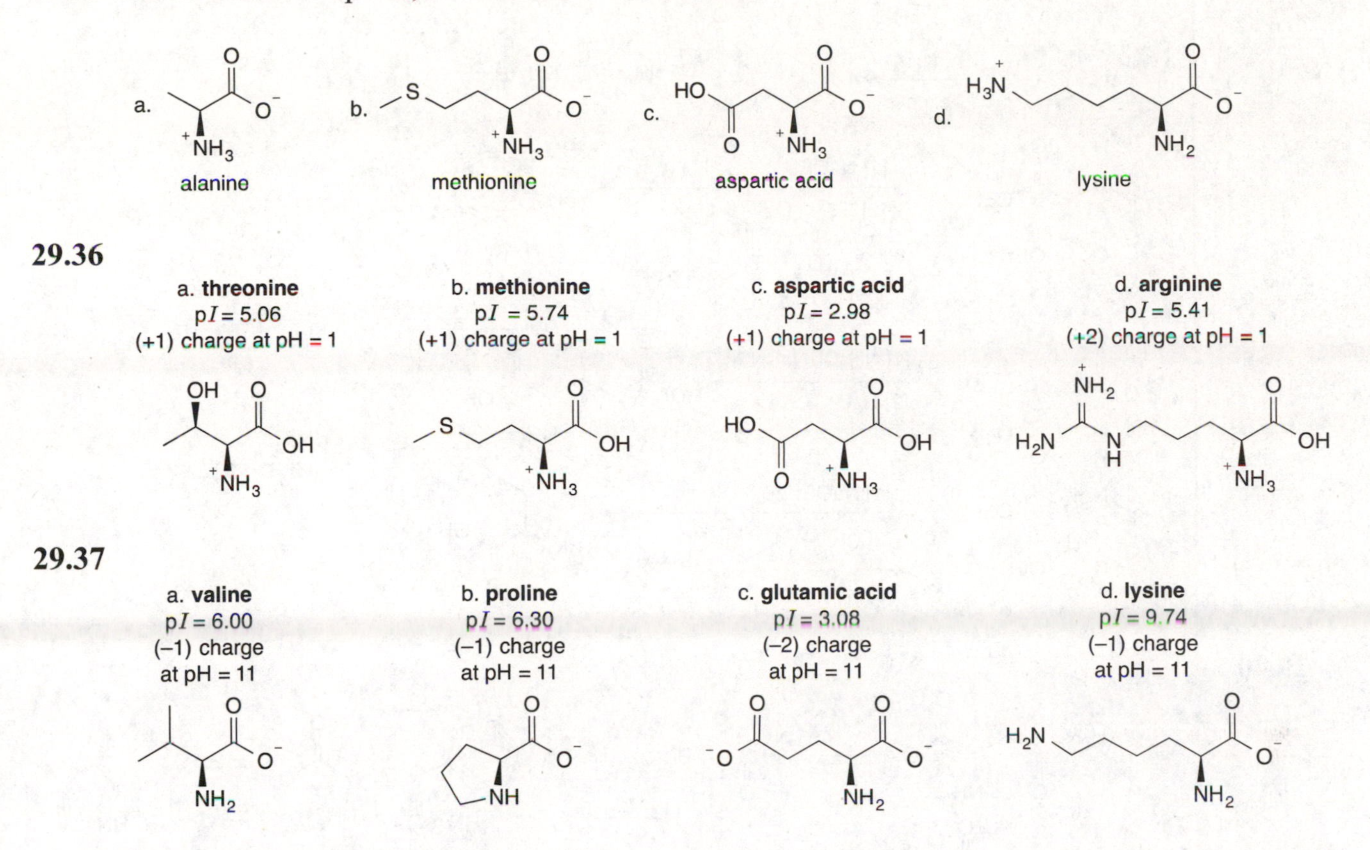

29.38 The terminal NH_2 and COOH groups are ionizable functional groups, so they can gain or lose protons in aqueous solution.

a. Ala → **A–A–A**

(drawn with all uncharged atoms)

b. At pH = 1

c. The pK_a of the COOH of the tripeptide is higher than the pK_a of the COOH group of alanine, making it less acidic. This occurs because the COOH group in the tripeptide is farther away from the $–NH_3^+$ group. The positively charged $–NH_3^+$ group stabilizes the negatively charged carboxylate anion of alanine more than the carboxylate anion of the tripeptide because it is so much closer in alanine. The opposite effect is observed with the ionization of the $–NH_3^+$ group. In alanine, the $–NH_3^+$ is closer to the $COO^–$ group, so it is more difficult to lose a proton, resulting in a higher pK_a. In the tripeptide, the $–NH_3^+$ is farther away from the $COO^–$, so it is less affected by its presence.

29.39

a. 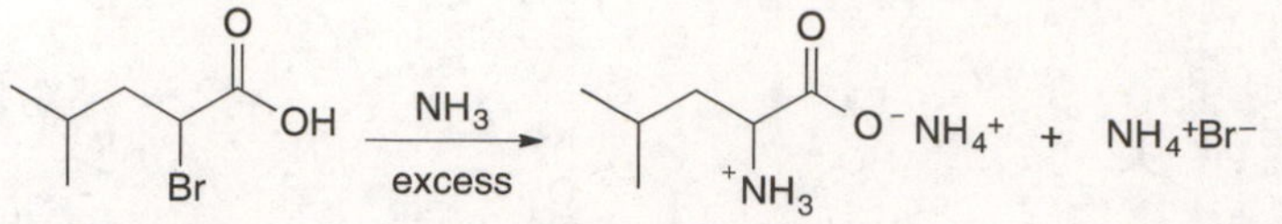 OH, Br → NH_3 / excess → $O^- NH_4^+$ + $NH_4^+Br^-$; $^+NH_3$

b. 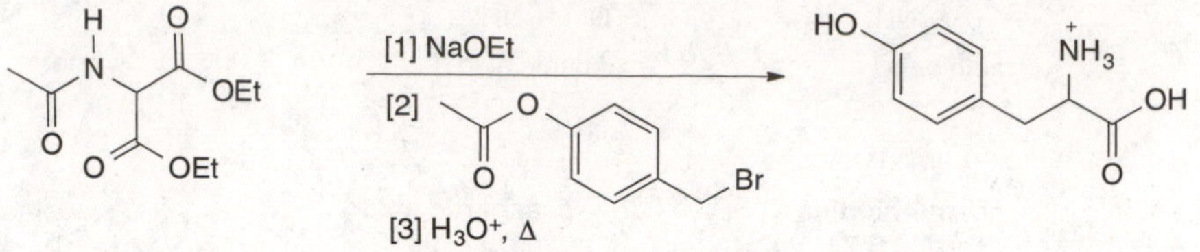 H, N, O, OEt, O, OEt → [1] NaOEt [2] O, O, Br [3] H_3O^+, Δ → HO, $\overset{+}{NH_3}$, OH, O

c. 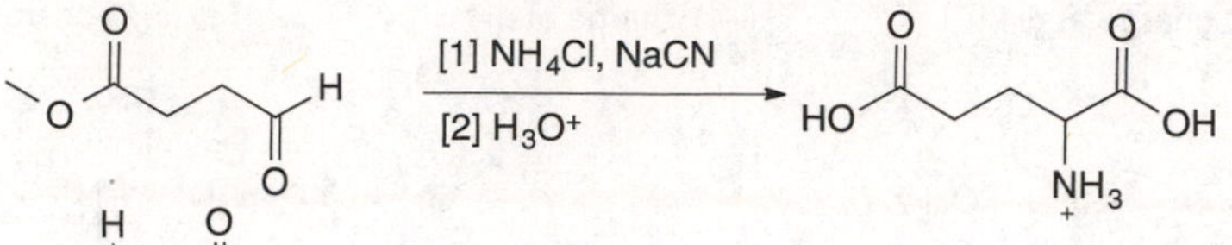 O, O, H, O → [1] NH_4Cl, NaCN [2] H_3O^+ → O, O, HO, OH, $\overset{+}{NH_3}$

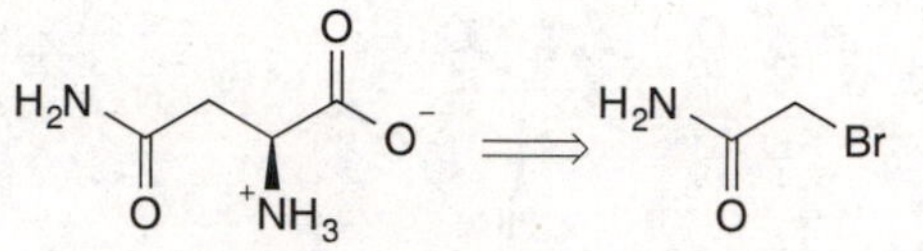

d. 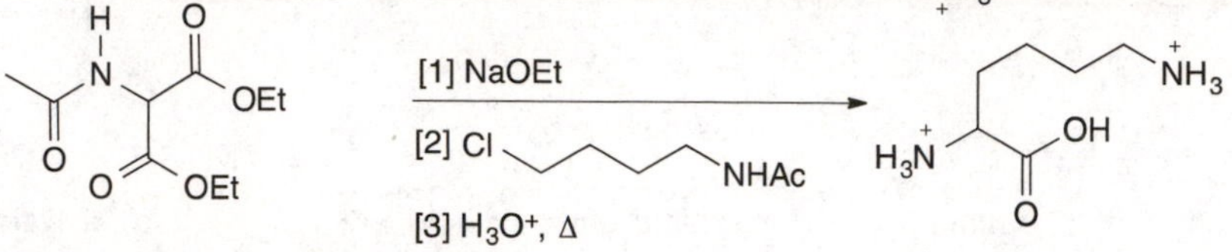 H, N, O, OEt, O, O, OEt → [1] NaOEt [2] Cl, NHAc [3] H_3O^+, Δ → $\overset{+}{NH_3}$, $H_3\overset{+}{N}$, OH, O

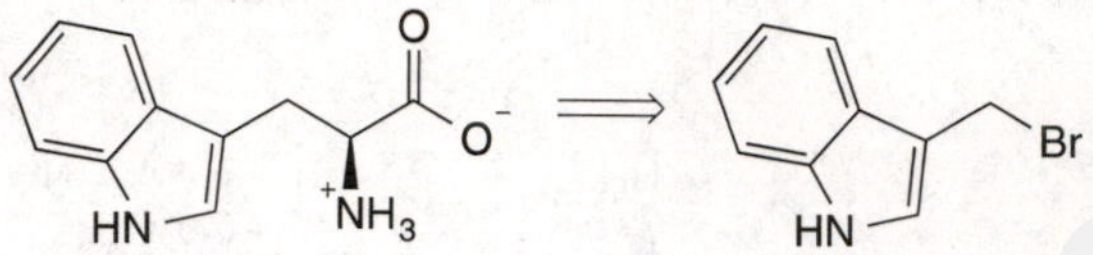

29.40

a. Asn H_2N, O, O, O^-, $^+NH_3$ ⟹ H_2N, O, Br

b. His N, NH, O, O^-, $^+NH_3$ ⟹ N, NH, Br

c. Trp HN, O, O^-, $^+NH_3$ ⟹ HN, Br

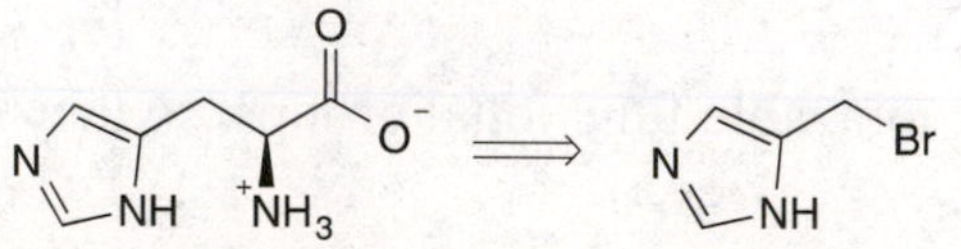

29.41

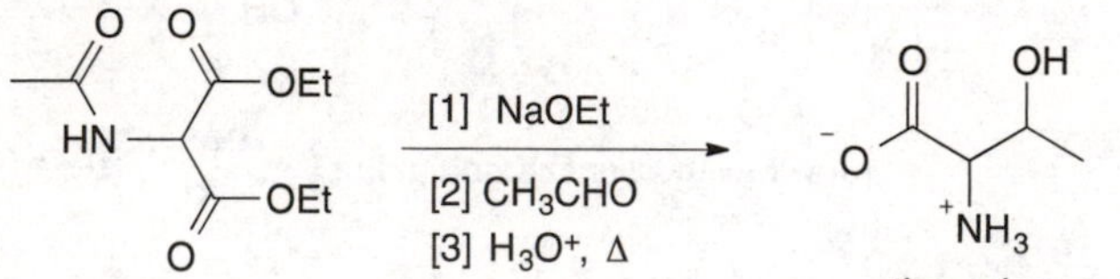

O, O, OEt, HN, OEt, O → [1] NaOEt [2] CH_3CHO [3] H_3O^+, Δ → O, OH, ^-O, $^+NH_3$

threonine

29.42

a. 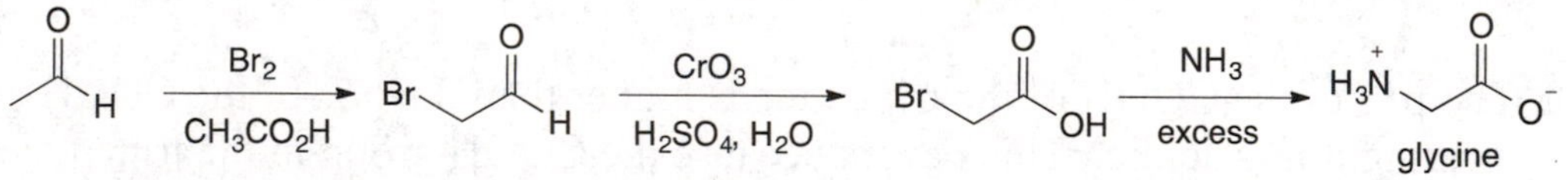 O, H → Br_2 / CH_3CO_2H → Br, O, H → CrO_3 / H_2SO_4, H_2O → Br, O, OH → NH_3 / excess → $H_3\overset{+}{N}$, O, O^-

glycine

b. 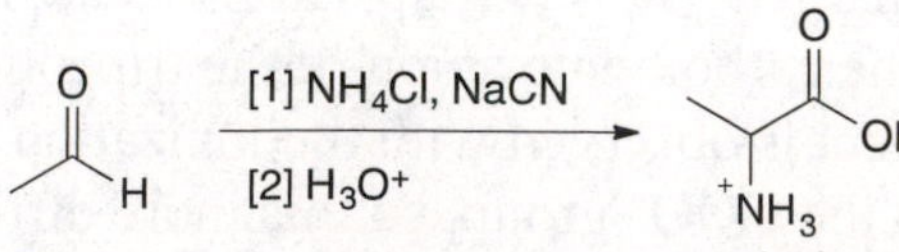 O, H → [1] NH_4Cl, NaCN [2] H_3O^+ → O, OH, $^+NH_3$

alanine

29.43

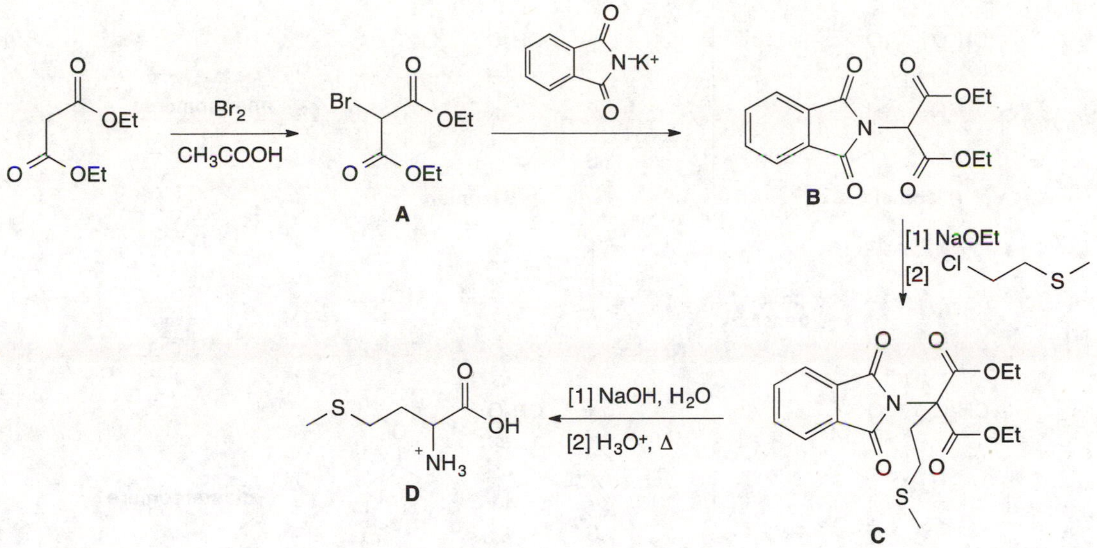

O OEt O OEt → Br$_2$, CH$_3$COOH → Br O OEt O OEt **A**

N$^-$K$^+$ O O →

O O OEt N O OEt **B** O O

[1] NaOEt
[2] Cl S

O O OEt N O OEt O O S **C**

[1] NaOH, H$_2$O
[2] H$_3$O$^+$, Δ

S O OH $^+$NH$_3$ **D**

29.44

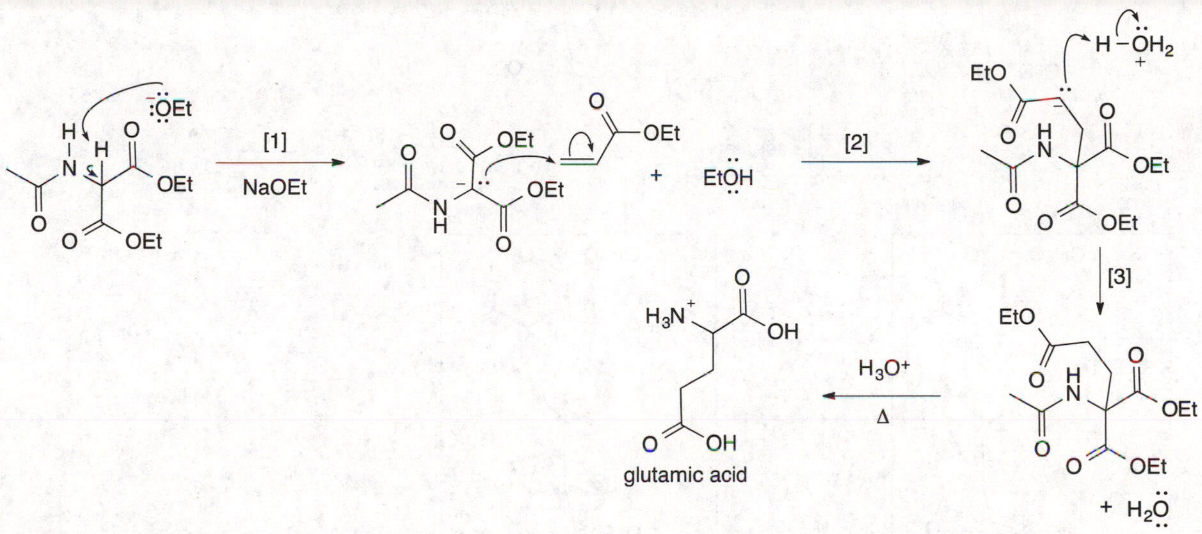

H H N H O :ÖEt$^-$ O OEt O OEt

[1] NaOEt →

O O OEt N H O OEt O O OEt + EtÖH

[2] →

EtO H–ÖH$_2^+$ O O H N O OEt O OEt

[3]

EtO O O H N O OEt O OEt + H$_2$Ö

H$_3$O$^+$, Δ

H$_3$N$^+$ O OH O OH

glutamic acid

29.45

CN N → [1] DIBAL-H [2] H$_2$O → CHO N **A**

Ph$_3$P=CHCO$_2$Et → CO$_2$Et N (+ cis isomer) **B**

H$_2$, Pd-C → CO$_2$Et N **C**

KOH, H$_2$O → CO$_2$H N **D**

H$_2$N O OCH$_3$ H N Boc

DCC →

O NH O OCH$_3$ N **E** Boc N H

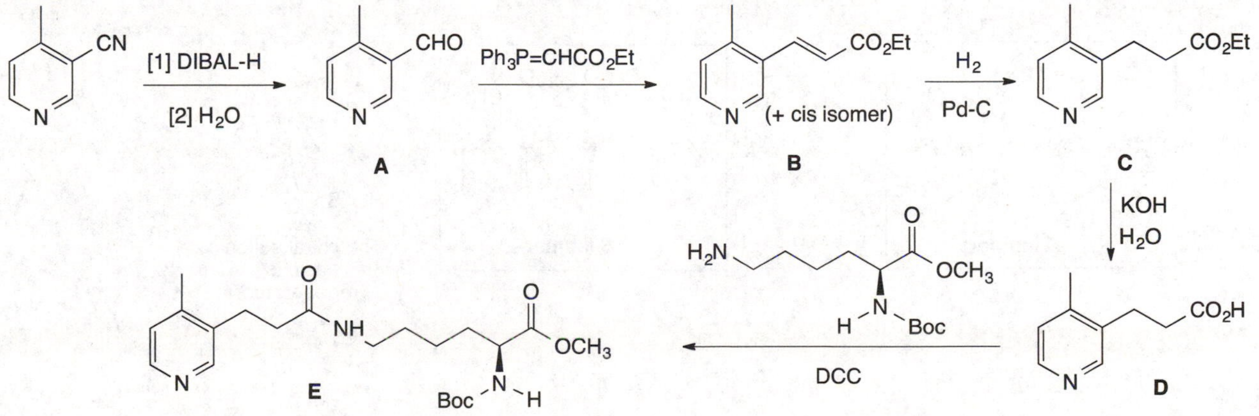

29.46

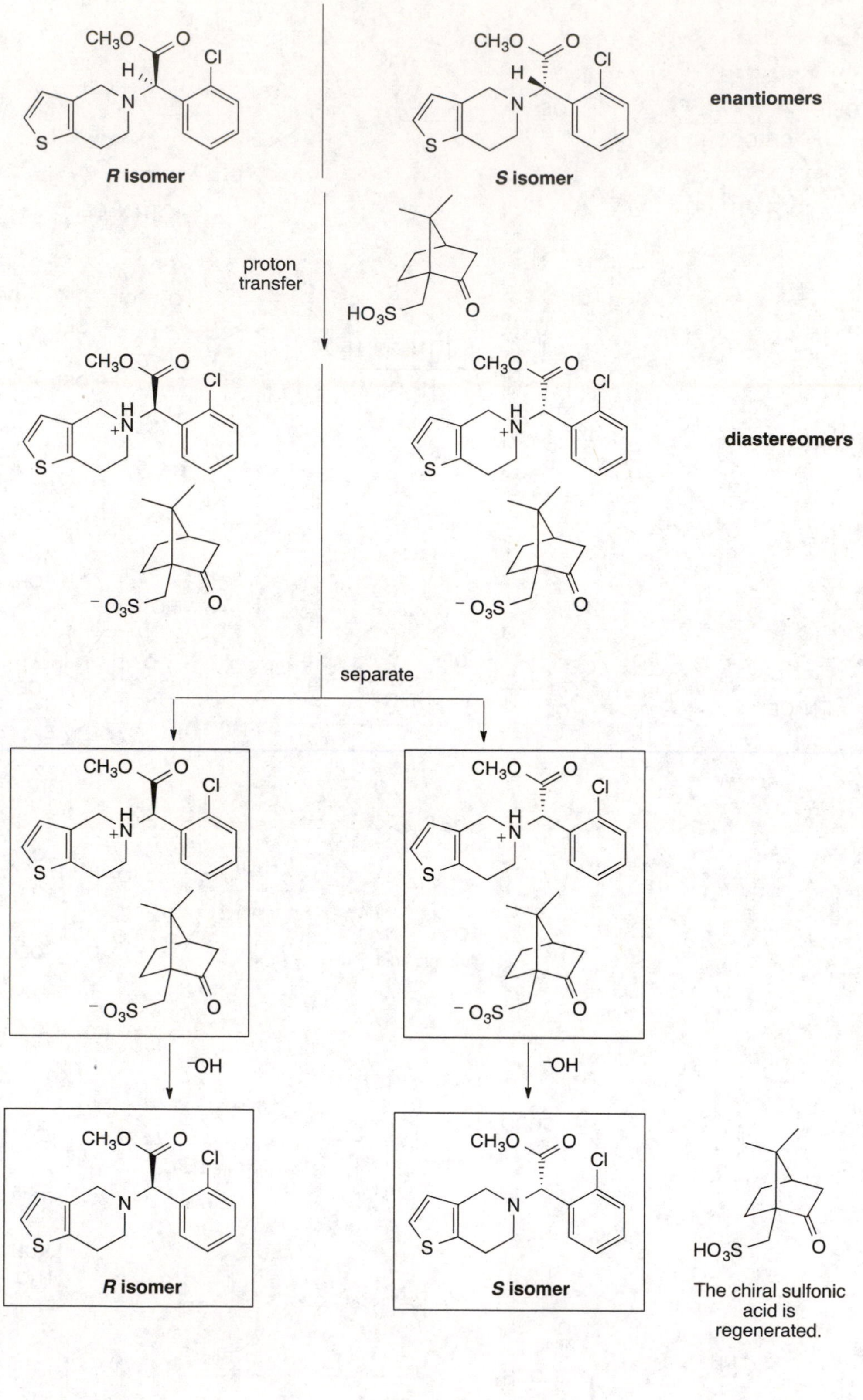

CH_3O O Cl H N S
***R* isomer**

CH_3O O Cl H N S
***S* isomer**

enantiomers

proton transfer

HO_3S O

CH_3O O Cl H +N S
^-O_3S O

CH_3O O Cl H +N S
^-O_3S O

diastereomers

separate

CH_3O O Cl H +N S
^-O_3S O

CH_3O O Cl H +N S
^-O_3S O

^-OH

^-OH

CH_3O O Cl N S
***R* isomer**

CH_3O O Cl N S
***S* isomer**

HO_3S O

The chiral sulfonic acid is regenerated.

29.47

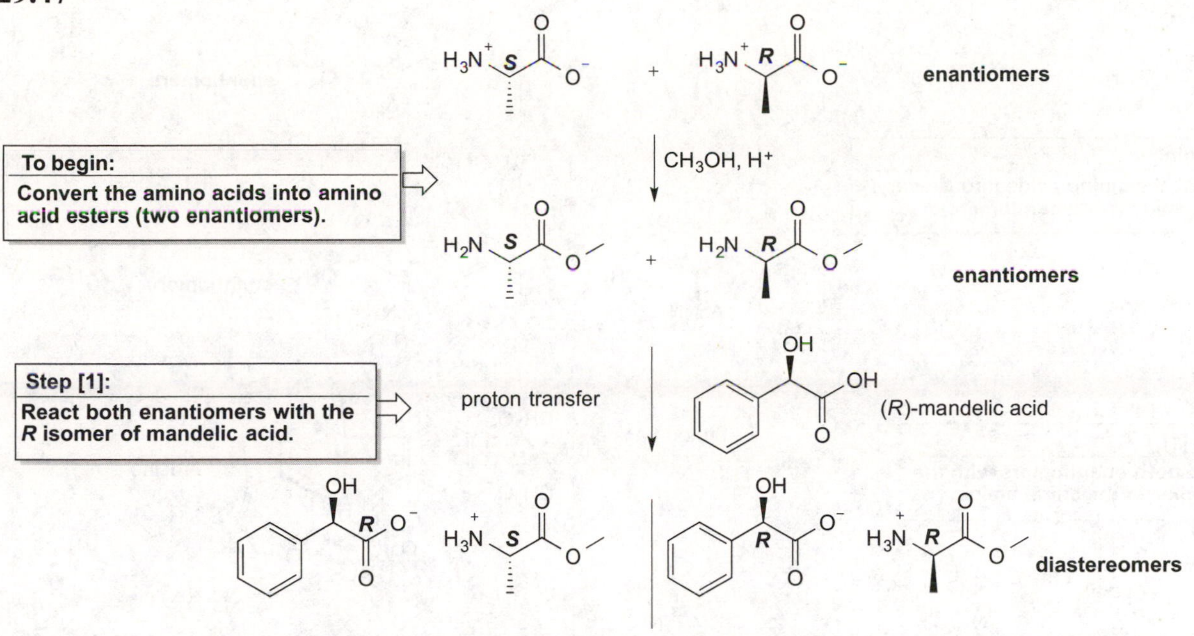
enantiomers
To begin:
Convert the amino acids into amino acid esters (two enantiomers).
CH3OH, H+
enantiomers
Step [1]:
React both enantiomers with the R isomer of mandelic acid.
proton transfer
(R)-mandelic acid
diastereomers

These salts have the *same* configuration around one stereogenic center, but the *opposite* configuration about the other stereogenic center.

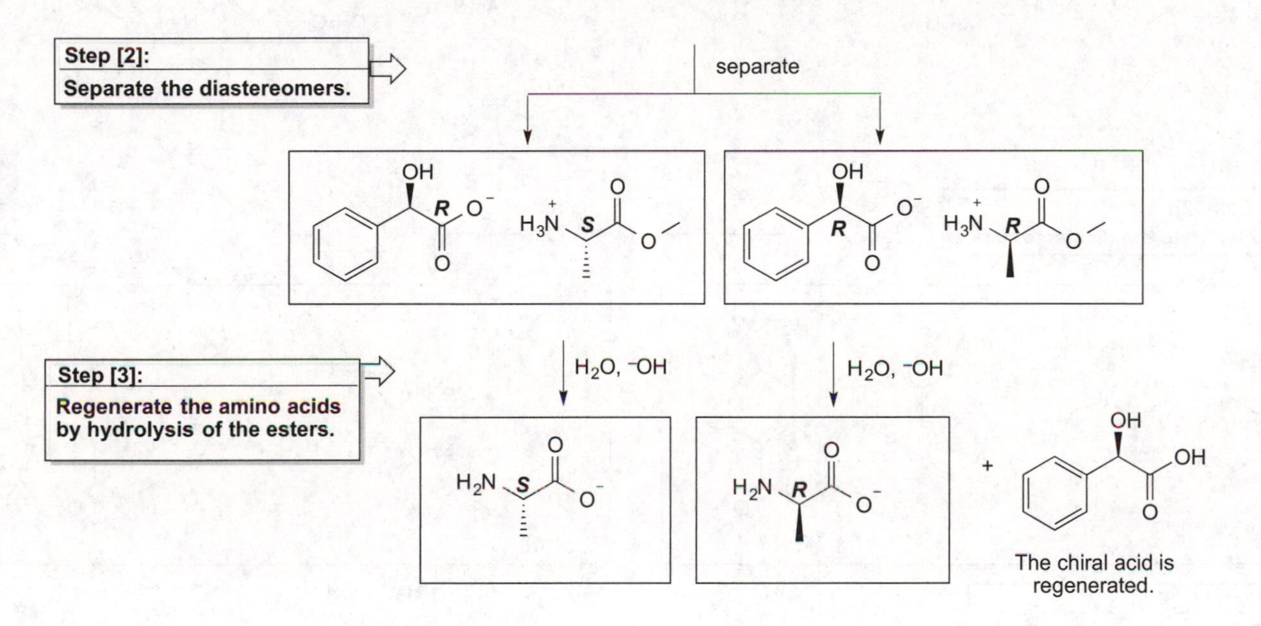
Step [2]:
Separate the diastereomers.
separate
Step [3]:
Regenerate the amino acids by hydrolysis of the esters.
H2O, −OH
H2O, −OH
The chiral acid is regenerated.

29.48

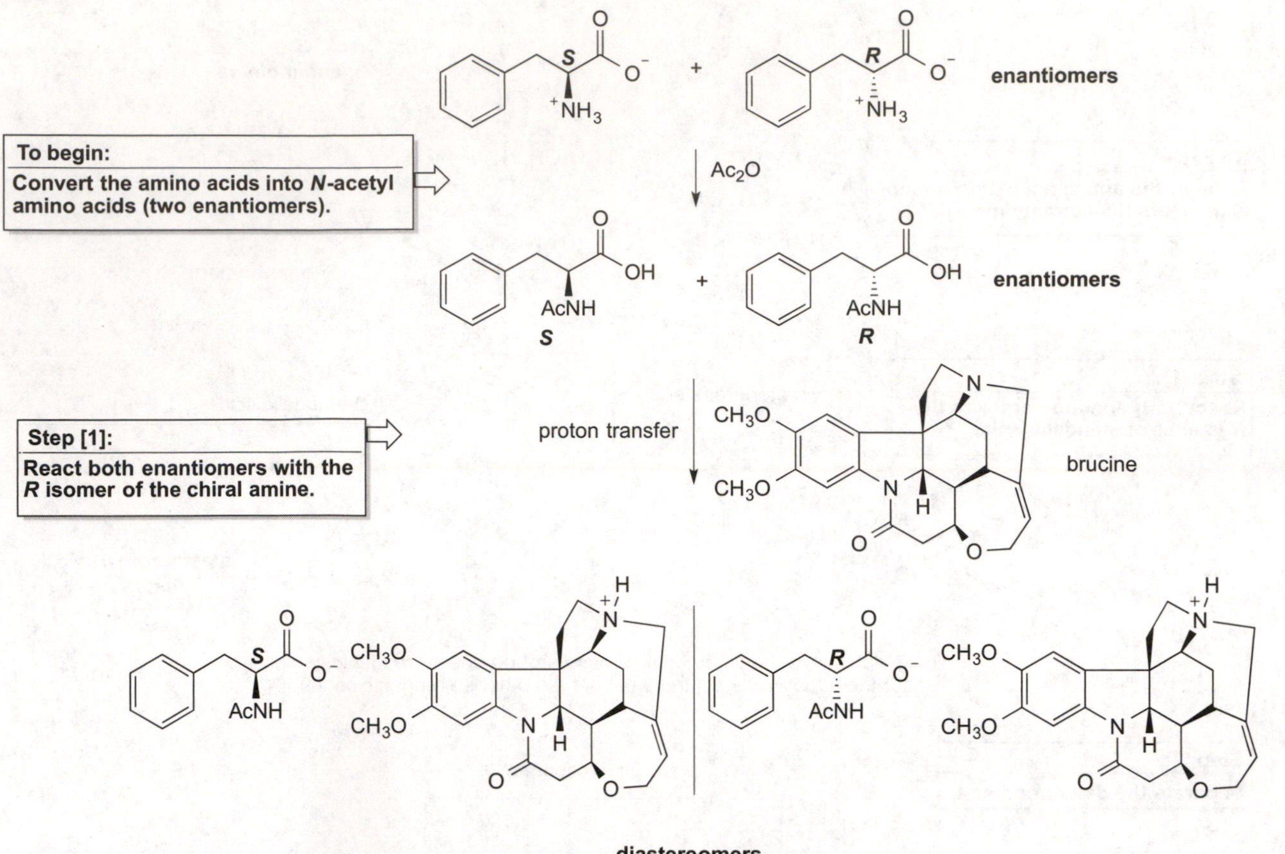
enantiomers
To begin:
Convert the amino acids into N-acetyl amino acids (two enantiomers).
Ac2O
AcNH
OH
enantiomers
Step [1]:
React both enantiomers with the R isomer of the chiral amine.
proton transfer
CH3O
brucine

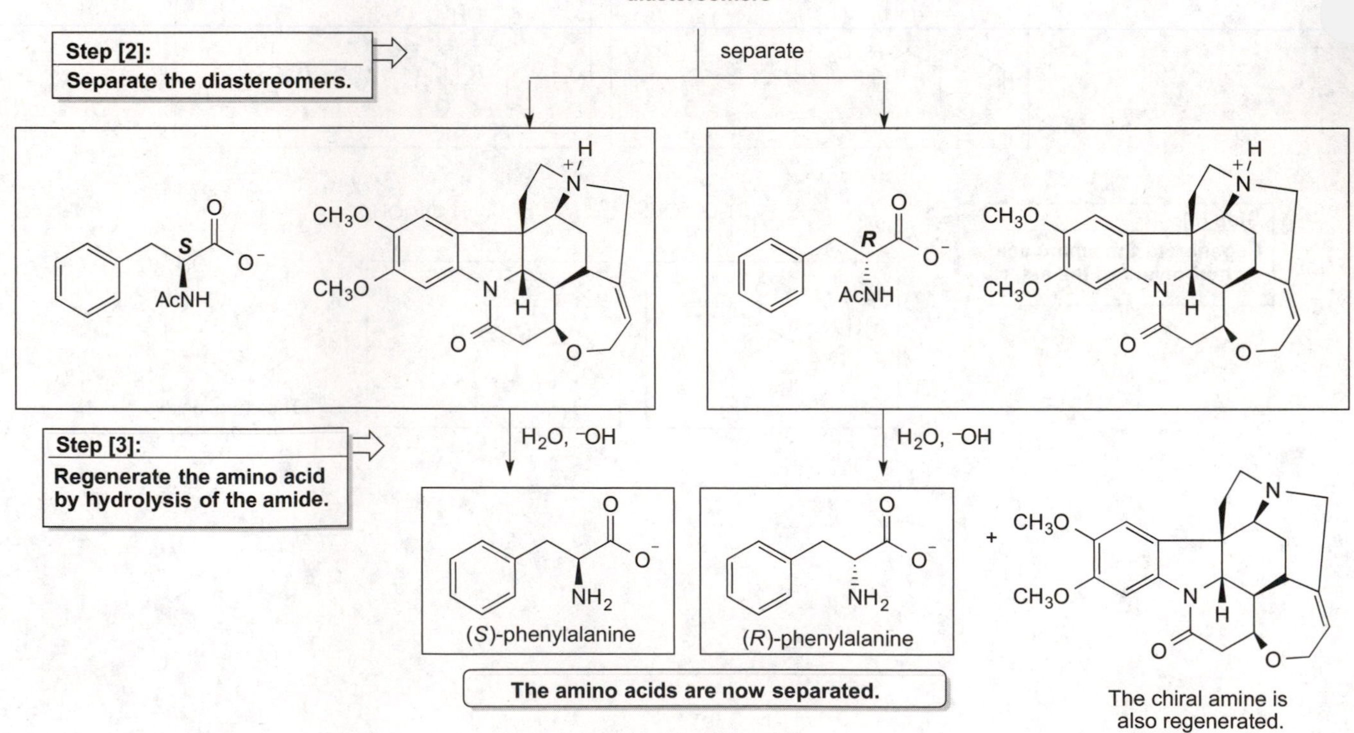
diastereomers
Step [2]:
Separate the diastereomers.
separate
Step [3]:
Regenerate the amino acid by hydrolysis of the amide.
H2O, −OH
(S)-phenylalanine
(R)-phenylalanine
The amino acids are now separated.
The chiral amine is also regenerated.

29.49

a. $H_3\overset{+}{N}$ ... O^- —Ac_2O→ —acylase→ $H_3\overset{+}{N}$ ... O^- + AcNH ... OH

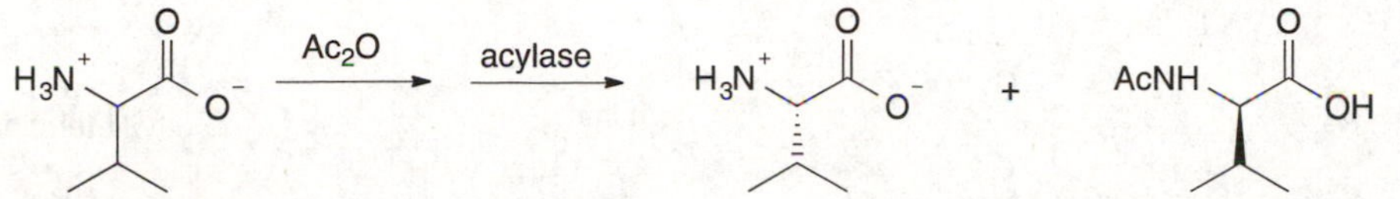

racemic mixture

b. OH, NHAc, N H —H_2, chiral Rh catalyst→ —^-OH, H_2O→ H_2N ... O^- (*S*)-isomer, N H

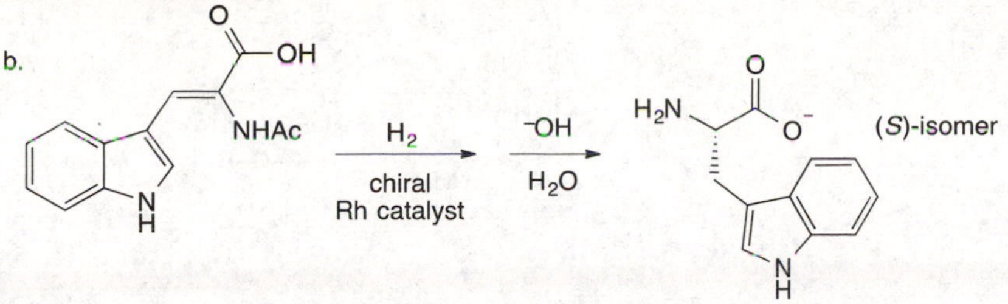

29.50

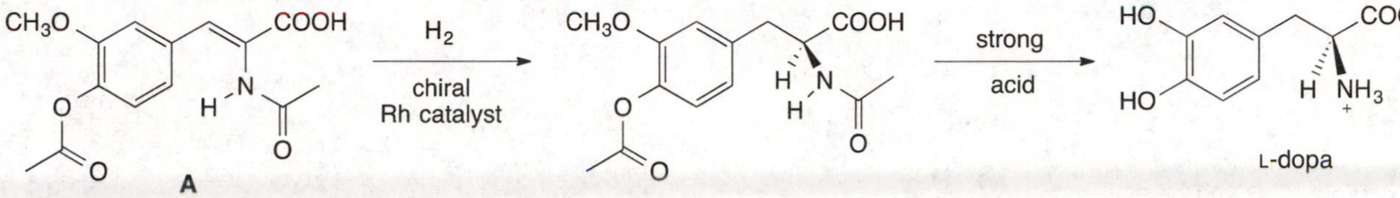

CH_3O ... COOH, H N, O, O, **A** —H_2, chiral Rh catalyst→ CH_3O ... COOH, H N H, O, O —strong acid→ HO ... COOH, HO, H $\overset{+}{N}H_3$

L-dopa

29.51

a. $H_3\overset{+}{N}$ ... H N ... O^-, O

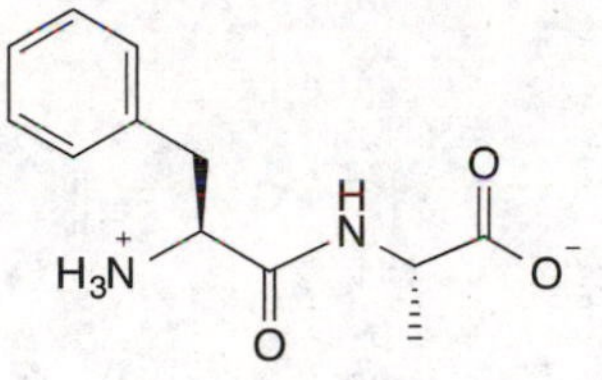

Phe–Ala

c. $H_3\overset{+}{N}$ ... $H_3\overset{+}{N}$ ... H N ... O^-, O

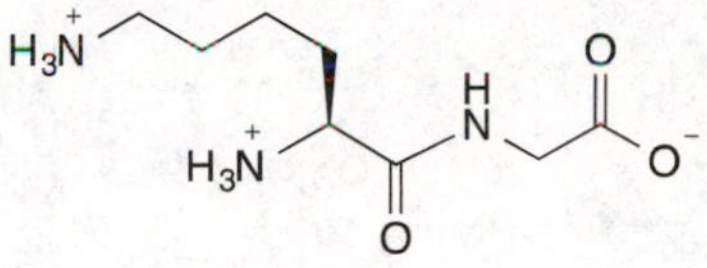

Lys–Gly

b. $H_3\overset{+}{N}$ ... H N ... O^-, O, H_2N O

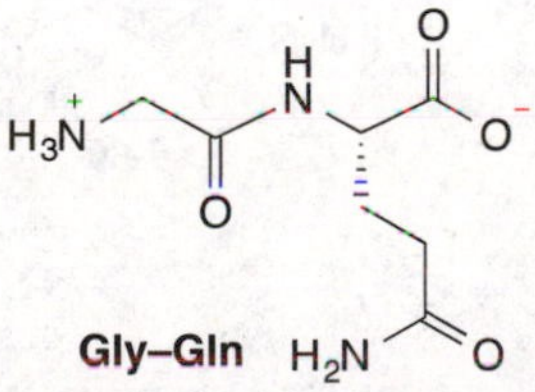

Gly–Gln

d. H_2N ... H N, $\overset{+}{N}H_2$, $H_3\overset{+}{N}$ ... H N ... O^-, O, NH, N

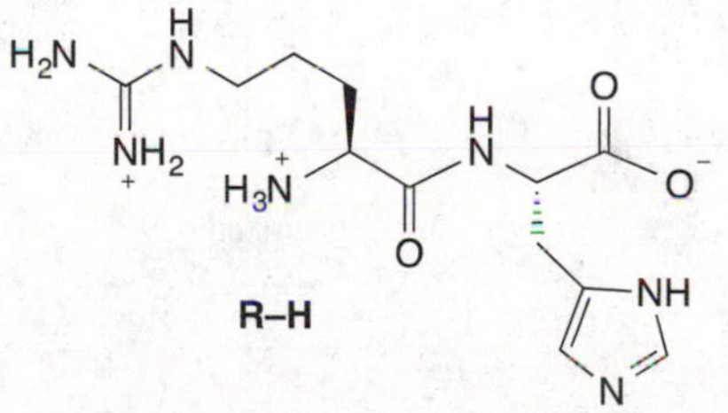

R–H

29.52 Amide bonds are bold lines (not wedges). The structure is drawn with all uncharged atoms.

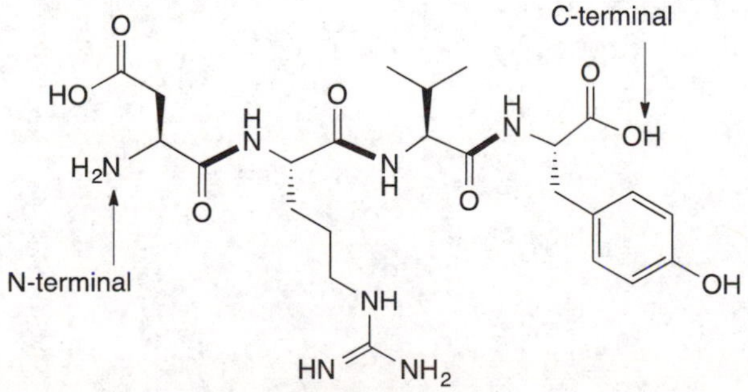

C-terminal

HO, O, H_2N, H N, O, N H, H N, O, OH, NH, HN, NH_2, OH

N-terminal

Asp–Arg–Val–Tyr

D–R–V–Y

29.53 Name a peptide from the N-terminal to the C-terminal end.

a.

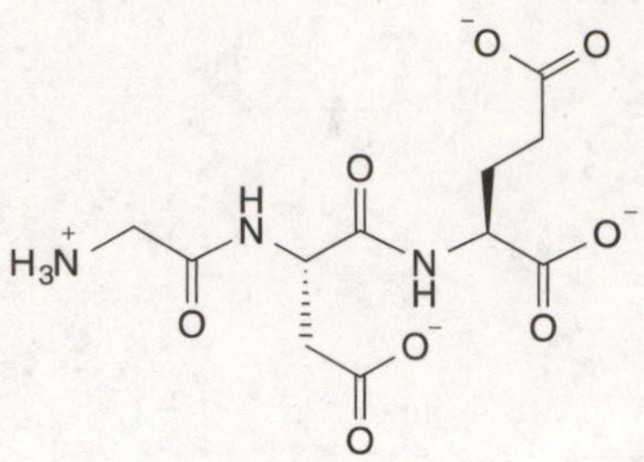

Gly–Asp–Glu

G–D–E

b.

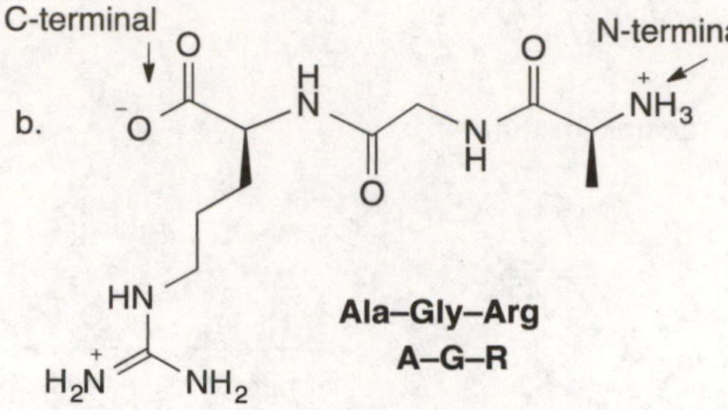

C-terminal

N-terminal

Ala–Gly–Arg

A–G–R

29.54 The unusual features of gramicidin S are the presence of the uncommon enantiomer of phenylalanine (D-Phe) in the molecule, as well as the presence of an uncommon amino acid, labeled **X**.

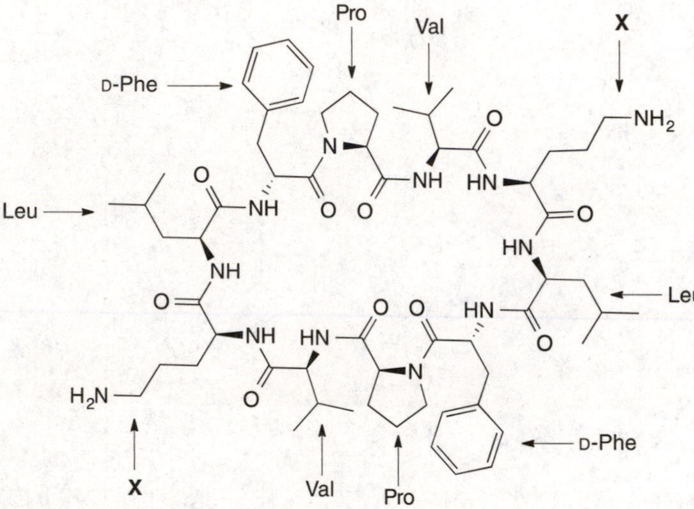

D-Phe, Pro, Val, X, Leu, Leu, X, Val, Pro, D-Phe

gramicidin S

Amino acids:

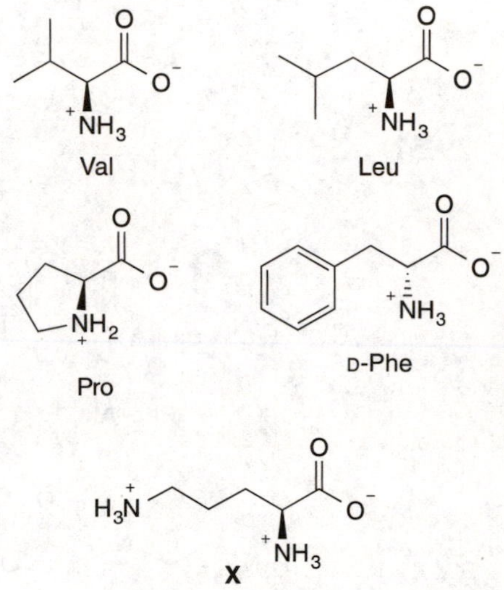

Val

Leu

Pro

D-Phe

X

29.55

a. A–P–F + L–K–W + S–G–R–G
b. A–P–F–L–K + W–S–G–R + G
c. A–P–F–L–K–W–S–G–R + G
d. A + P–F–L–K–W–S–G–R–G

29.56

angiotensin I

Three-letter abbreviation: Asp–Arg–Val–Tyr–Ile–His–Pro–Phe–His–Leu

a. Trypsin cleavage products: Asp–Arg + Val–Tyr–Ile–His–Pro–Phe–His–Leu

b. Chymotrypsin cleavage products: Asp–Arg–Val–Tyr + Ile–His–Pro–Phe + His–Leu

c. ACE cleavage products: Asp–Arg–Val–Tyr–Ile–His–Pro–Phe (angiotensin II) + His–Leu

29.57

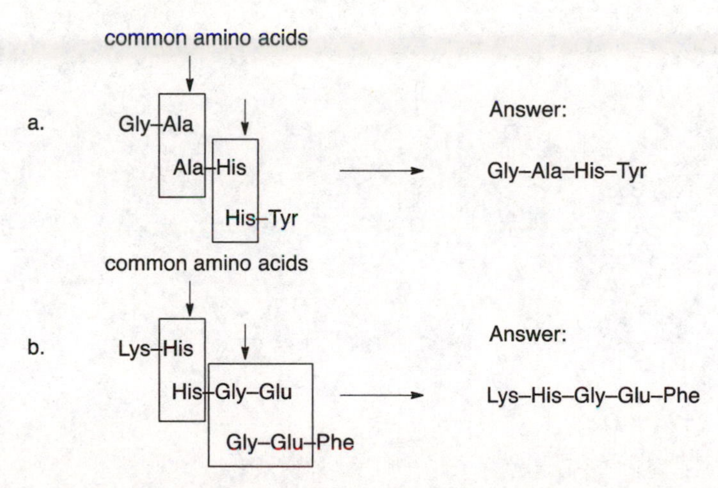

29.58 Gly is the N-terminal amino acid (from Edman degradation), and Leu is the C-terminal amino acid (from treatment with carboxypeptidase). Partial hydrolysis gives the rest of the sequence.

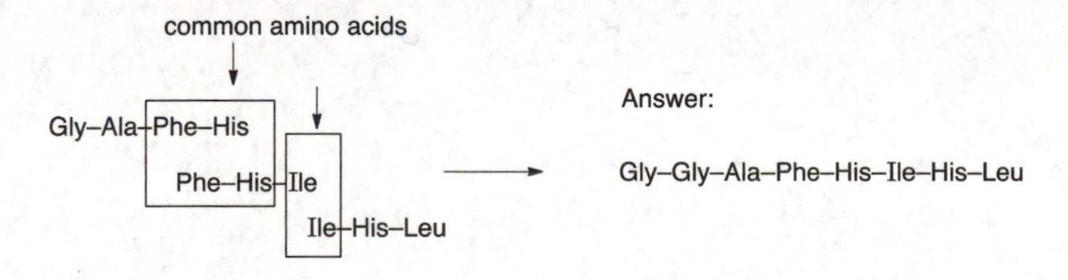

29.59 Edman degradation data give the N-terminal amino acid for the octapeptide and all smaller peptides.

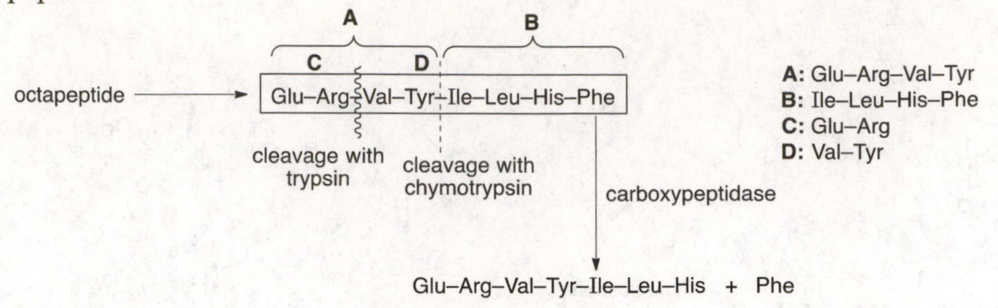

Glu–Arg–Val–Tyr–Ile–Leu–His + Phe

29.60

a. 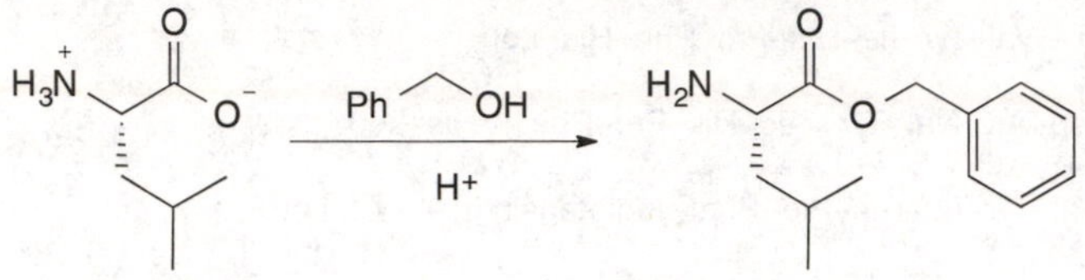 $H_3\overset{+}{N}$ … O^- — Ph⁀OH, H^+ → H_2N … O … (benzyl ester)

b. 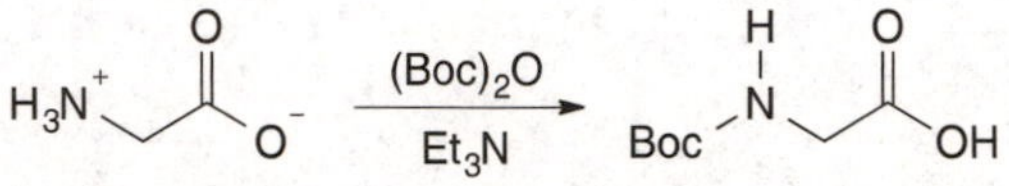 $H_3\overset{+}{N}$ … O^- — $(Boc)_2O$, Et_3N → Boc–NH … OH

c. product in (a) + product in (b) 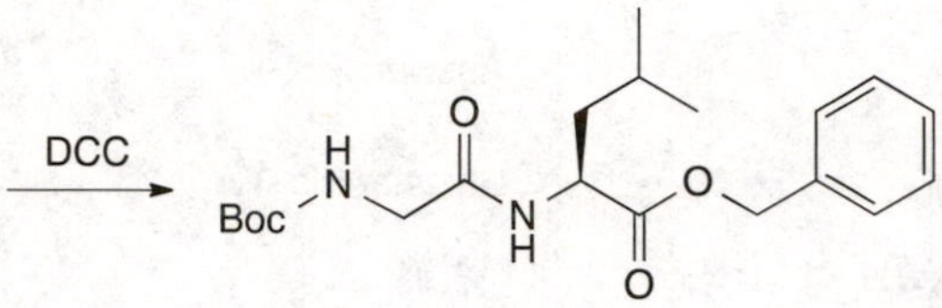 — DCC → Boc–NH … NH … O (benzyl ester)

d. 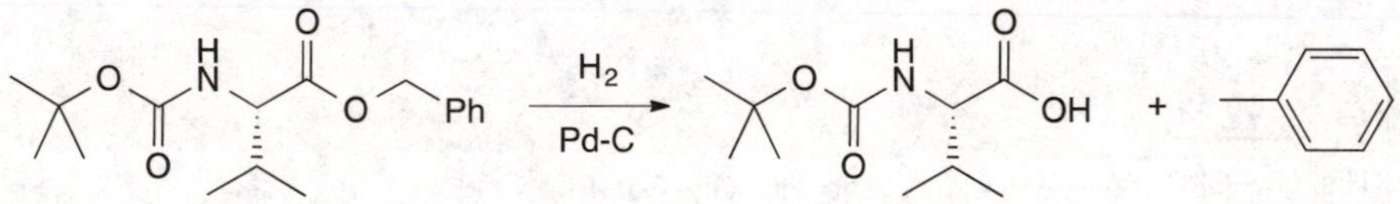 (Boc-Val benzyl ester) … O–Ph — H_2, Pd-C → (Boc-Val) … OH + (toluene)

e. product in (d) 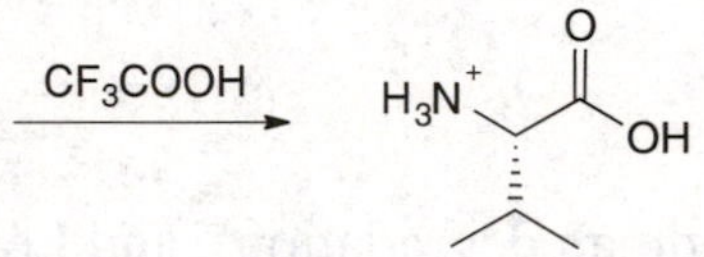 — CF_3COOH → $H_3\overset{+}{N}$ … OH

f. 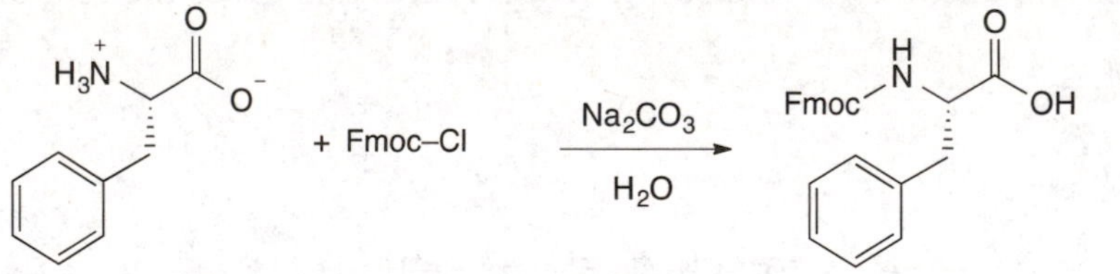 $H_3\overset{+}{N}$ … O^- + Fmoc–Cl — Na_2CO_3, H_2O → Fmoc–NH … OH

29.61

a. $H_3\overset{+}{N}$ O^- O Gly $\xrightarrow[Et_3N]{Boc_2O}$ Boc N H OH O **A**

$H_3\overset{+}{N}$ O O^- Ala $\xrightarrow[H^+]{HO \quad Ph}$ H_2N O O Ph **B**

A + **B** $\xrightarrow{DCC}$ Boc N H O H N O O Ph $\xrightarrow[CH_3COOH]{HBr}$ $H_3\overset{+}{N}$ O H N O O^-

Gly–Ala

b. $H_3\overset{+}{N}$ O^- O Ile $\xrightarrow[Et_3N]{Boc_2O}$ Boc N H OH O **A**

$H_3\overset{+}{N}$ O O^- Ala $\xrightarrow[H^+]{HO \quad Ph}$ H_2N O O Ph **B**

A + **B** $\xrightarrow{DCC}$ Boc N H O H N O O Ph $\xrightarrow[Pd\text{-}C]{H_2}$ Boc — N H O H N O OH **C**

$H_3\overset{+}{N}$ O^- O Phe $\xrightarrow[H^+]{HO \quad Ph}$ H_2N O O Ph $\xrightarrow[DCC]{C}$ Boc N H O H N O N H O O Ph

$\downarrow$ HBr CH_3COOH

$H_3\overset{+}{N}$ O H N O N H O^- O

Ile–Ala–Phe

29.62 Make all the Fmoc derivatives as described in Problem 29.25.

a.

(Fmoc-Phe)

[1] base

[2] Cl POLYMER

[1]

[2] DCC

(Fmoc-Phe)

[1]

[2] DCC +

(Fmoc-Leu)

[1]

[2] DCC

(Fmoc-Ala)

[1]

[2] HF

Ala–Leu–Phe–Phe

+ F POLYMER

b.

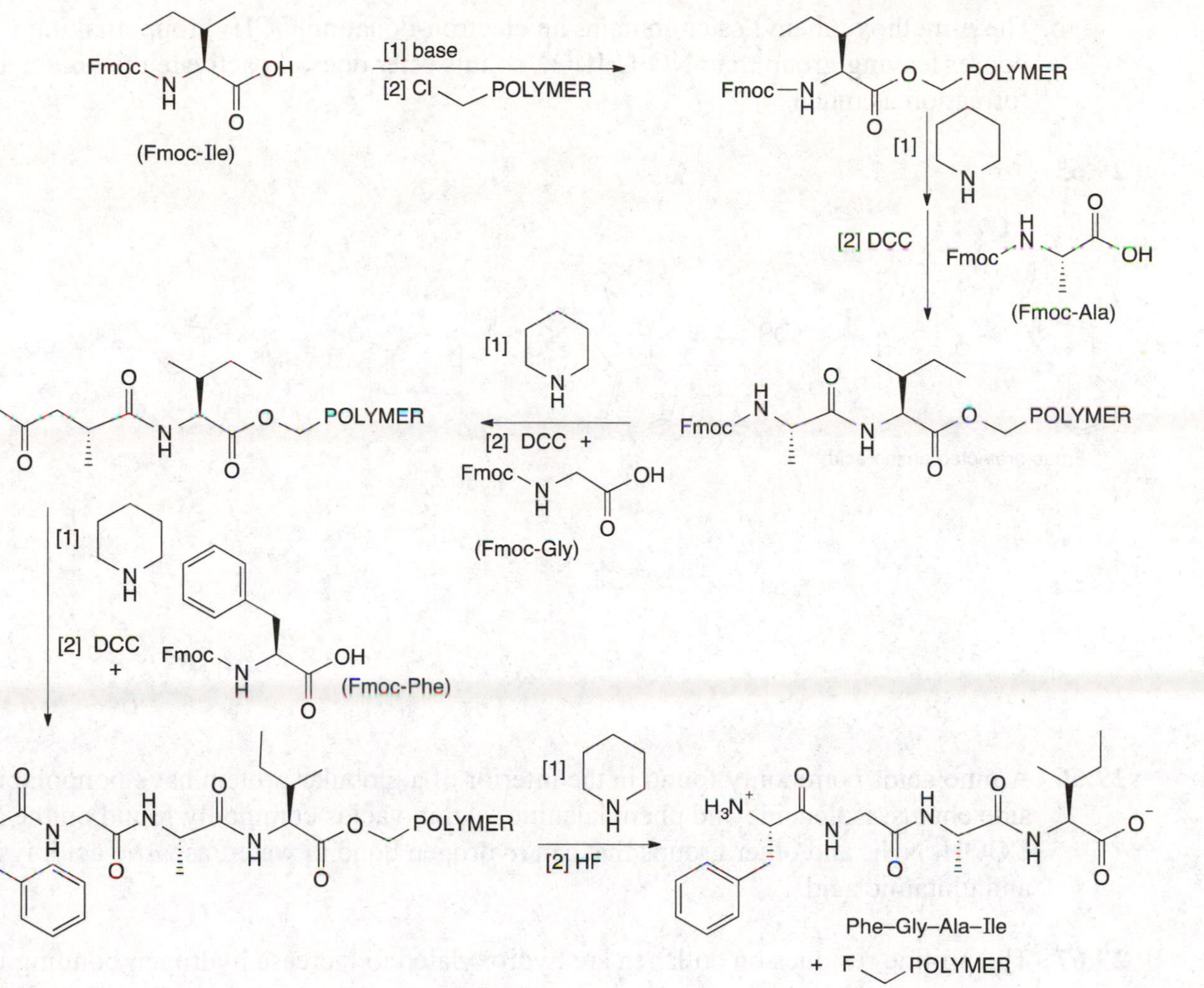

Fmoc-Ile: [1] base, [2] Cl–CH2–POLYMER → Fmoc–NH–Ile–O–CH2–POLYMER

(Fmoc-Ile)

[1] piperidine

[2] DCC + Fmoc-Ala

(Fmoc-Ala)

[1] piperidine

[2] DCC + Fmoc-Gly

(Fmoc-Gly)

[1] piperidine

[2] DCC + Fmoc-Phe

(Fmoc-Phe)

[1] piperidine

[2] HF

Phe–Gly–Ala–Ile

+ F–CH2–POLYMER

29.63

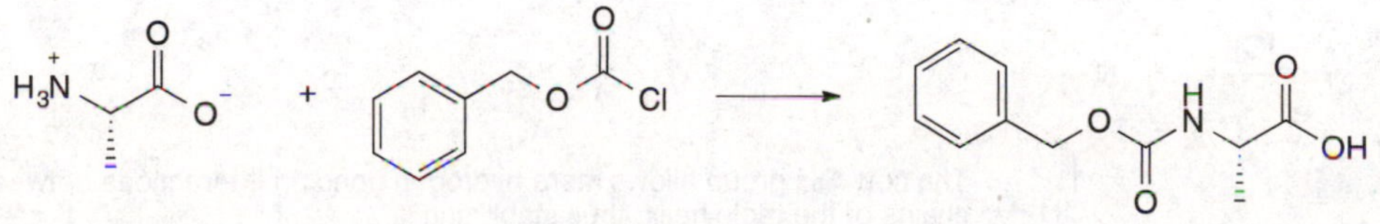

29.64

a. A *p*-nitrophenyl ester activates the carboxy group of the first amino acid to amide formation by converting the OH group into a good leaving group, the *p*-nitrophenoxide group, which is highly resonance stabilized. In this case the electron-withdrawing NO_2 group further stabilizes the leaving group.

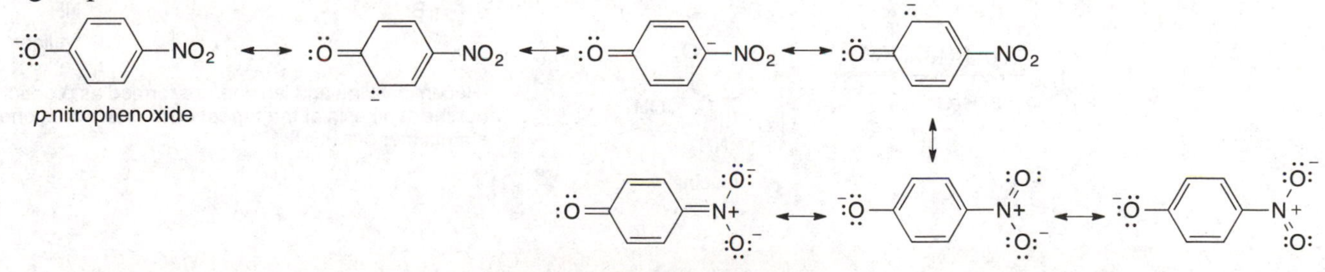

p-nitrophenoxide

The negative charge is delocalized on the O atom of the NO_2 group.

b. The *p*-methoxyphenyl ester contains an electron-donating OCH_3 group, making $CH_3OC_6H_4O^-$ a poorer leaving group than $NO_2C_6H_4O^-$, so this ester does *not* activate the amino acid to amide formation as much.

29.65

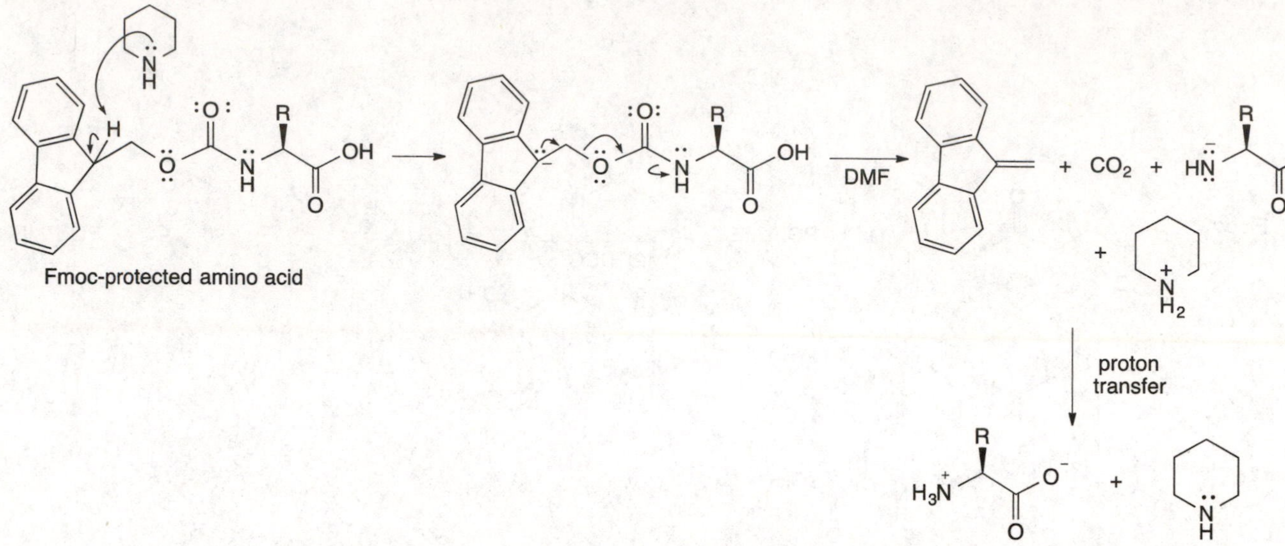

Fmoc-protected amino acid

DMF

+ CO_2 +

proton transfer

29.66 Amino acids commonly found in the interior of a globular protein have nonpolar or weakly polar side chains: isoleucine and phenylalanine. Amino acids commonly found on the surface have COOH, NH_2, and other groups that can hydrogen bond to water: aspartic acid, lysine, arginine, and glutamic acid.

29.67 The proline residues on collagen are hydroxylated to increase hydrogen bonding interactions.

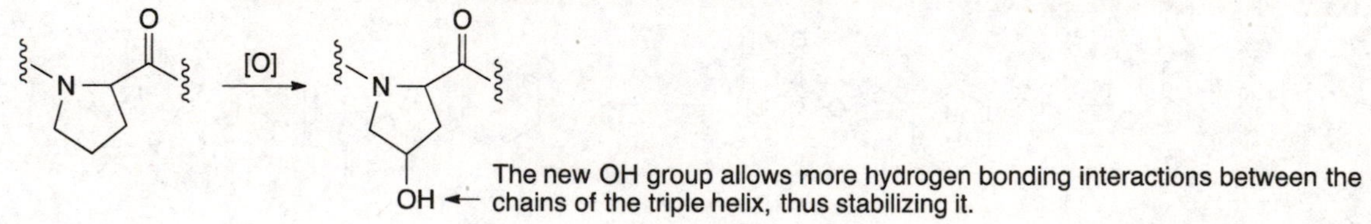

[O]

The new OH group allows more hydrogen bonding interactions between the chains of the triple helix, thus stabilizing it.

29.68

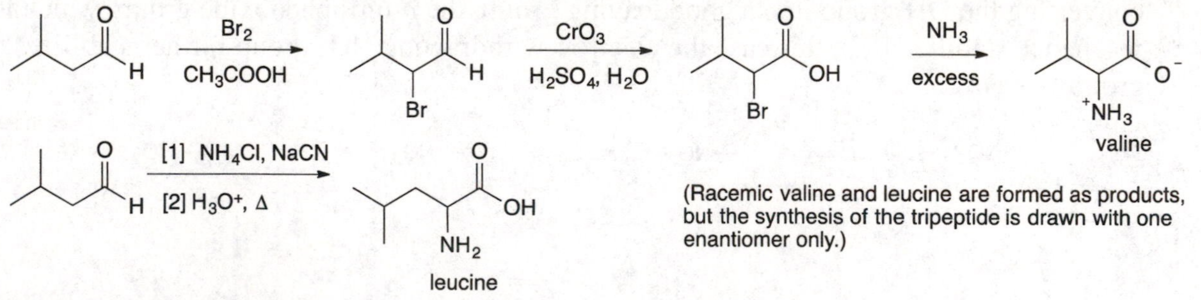

Br_2 / CH_3COOH → CrO_3 / H_2SO_4, H_2O → NH_3 excess → valine

[1] NH_4Cl, NaCN / [2] H_3O^+, Δ → leucine

(Racemic valine and leucine are formed as products, but the synthesis of the tripeptide is drawn with one enantiomer only.)

valine; Boc_2O, Et_3N → A; $C_6H_5CH_2OH$, H^+ → C

leucine; HO–CH$_2$–Ph, H^+ → B

A + B, DCC → Boc–NH–Val–Leu–O–CH$_2$Ph; H_2, Pd-C → Boc–NH–Val–Leu–OH

Boc–NH–Val–Leu–OH + C, DCC → Boc–Val–Leu–Val–OCH$_2$Ph; HBr, CH_3COOH → Val–Leu–Val

29.69 Perhaps using a chiral amine R*NH2 (or related chiral nitrogen-containing compound) to make a chiral imine, will now favor formation of one of the amino nitriles in the Strecker synthesis. Hydrolysis of the CN group and removal of R* would then form the amino acid.

RCHO + NH_2R^* (chiral amine) → chiral imine; ^-CN → amino nitrile (Perhaps a large excess of one stereoisomer will be formed.); Hydrolyze nitrile → R–CH(NHR*)–COOH; Remove R* → amino acid enantiomerically enriched (?)

29.70 This reaction is similar to the reaction of penicillin with the glycopeptide transpeptidase enzyme discussed in Section 22.14. Serine has a nucleophilic OH, which can open the strained β-lactone to form a covalently bound, inactive enzyme.

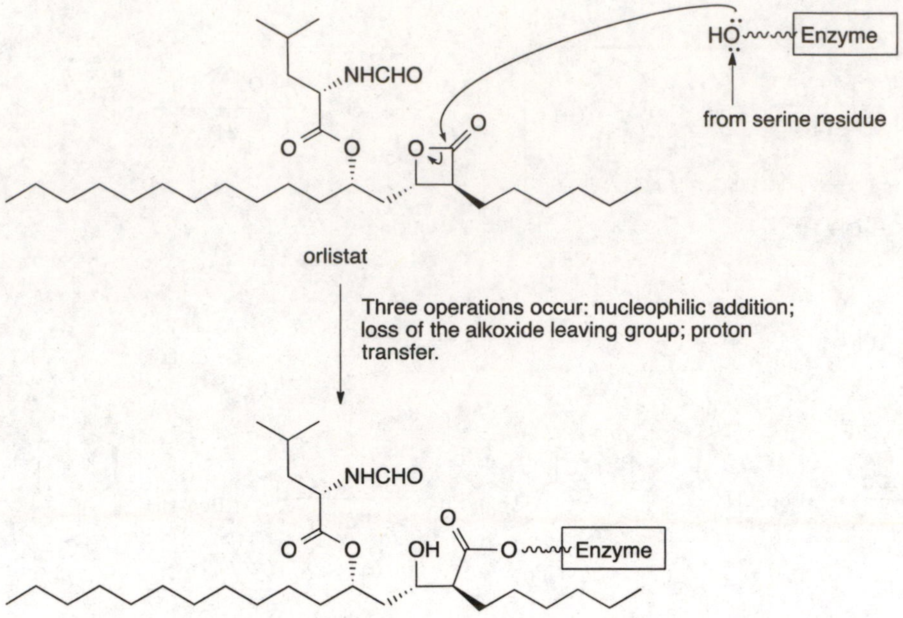

HO: Enzyme

from serine residue

orlistat

Three operations occur: nucleophilic addition; loss of the alkoxide leaving group; proton transfer.

NHCHO

OH

O–Enzyme

29.71

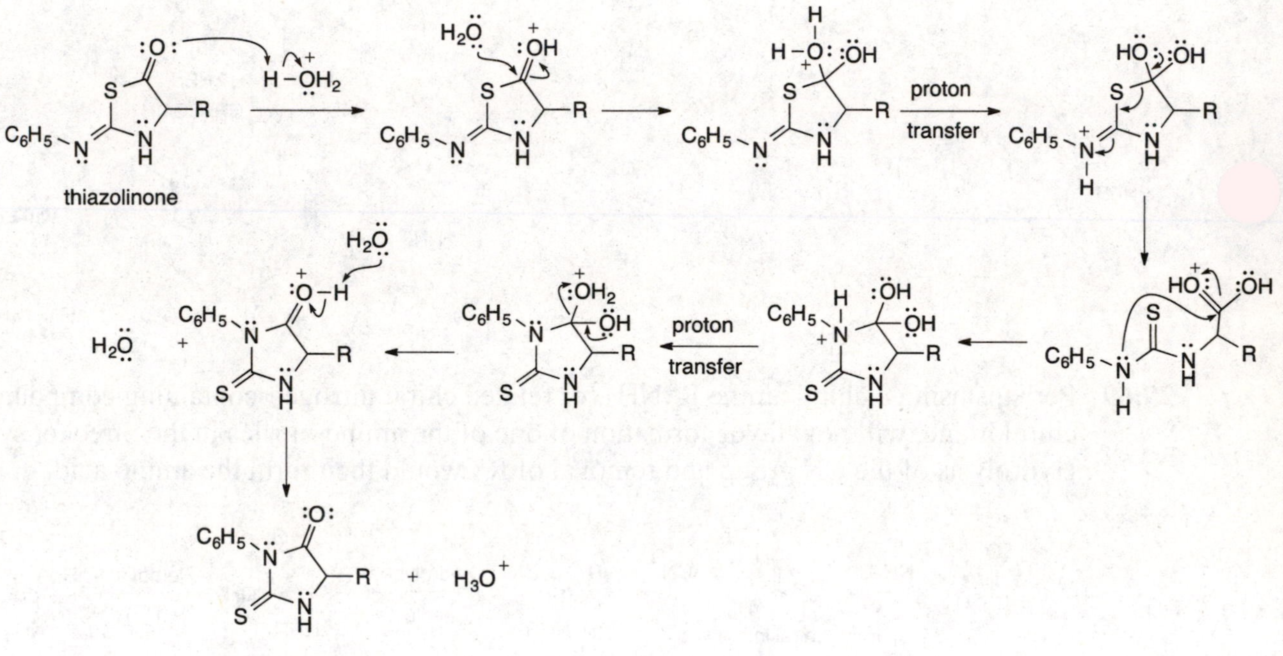

thiazolinone

proton transfer

proton transfer

N-phenylthiohydantoin

Chapter 30 Synthetic Polymers

Chapter Review

Chain-growth polymers—Addition polymers

[1] Chain-growth polymers with alkene starting materials (30.2)

- General reaction:

initiator

- Mechanism—three possibilities, depending on the identity of Z:

Type	Identity of Z	Initiator	Comments
[1] radical polymerization	Z stabilizes a radical. Z = R, Ph, Cl, etc.	A source of radicals (ROOR)	Termination occurs by radical coupling or disproportionation. Chain branching occurs.
[2] cationic polymerization	Z stabilizes a carbocation. Z = R, Ph, OR, etc.	H–A or a Lewis acid (BF_3 + H_2O)	Termination occurs by loss of a proton.
[3] anionic polymerization	Z stabilizes a carbanion. Z = Ph, COOR, COR, CN, etc.	An organolithium reagent (R–Li)	Termination occurs only when an acid or other electrophile is added.

[2] Chain-growth polymers with epoxide starting materials (30.3)

$^{-}$OH

- The mechanism is S_N2.
- Ring opening occurs at the less substituted carbon of the epoxide.

Examples of step-growth polymers—Condensation polymers (30.6)

Polyamides

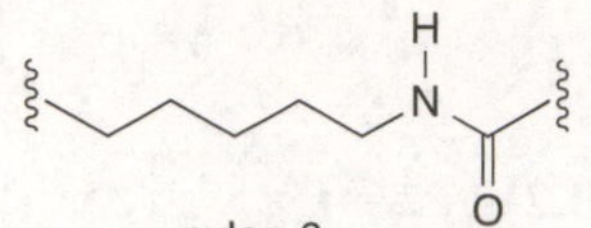

nylon 6

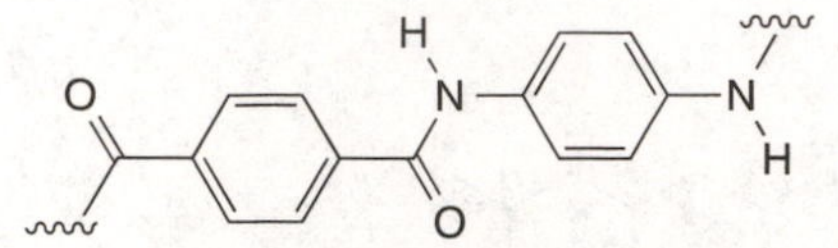

Kevlar

Polyesters

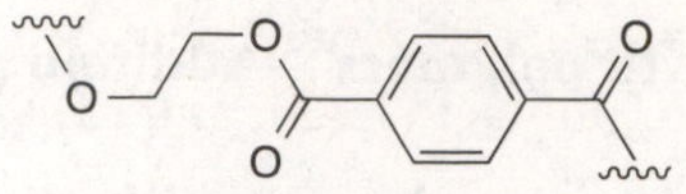

polyethylene terephthalate

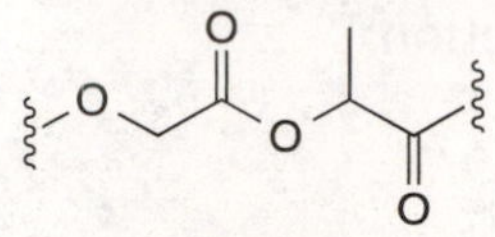

copolymer of glycolic and lactic acids

Polyurethanes

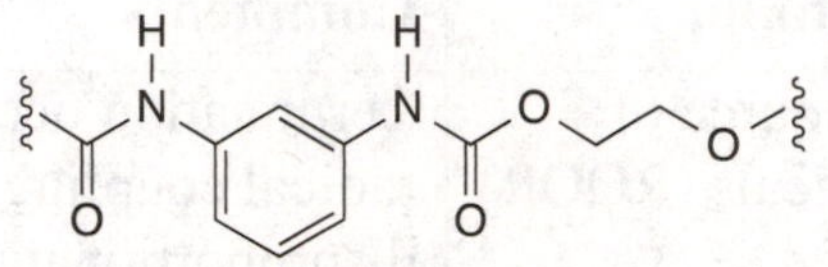

a polyurethane

Polycarbonates

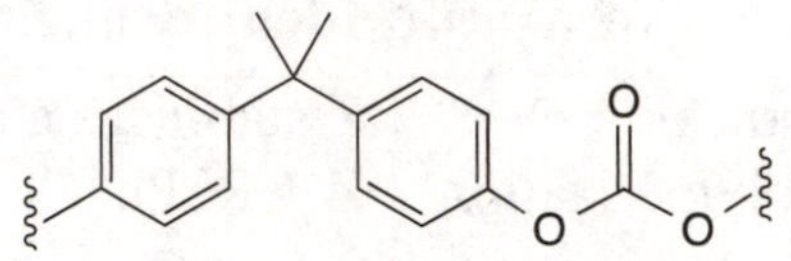

Lexan

Structure and properties

- Polymers prepared from monomers having the general structure CH_2=CHZ can be **isotactic, syndiotactic,** or **atactic,** depending on the identity of Z and the method of preparation (30.4).
- **Ziegler–Natta catalysts** form polymers without significant branching. Polymers can be isotactic, syndiotactic, or atactic, depending on the catalyst. Polymers prepared from 1,3-dienes have the *E* or *Z* configuration, depending on the monomer (30.4, 30.5).
- Most polymers contain ordered crystalline regions and less ordered amorphous regions (30.7). The greater the crystallinity, the harder the polymer.
- **Elastomers** are polymers that stretch and can return to their original shape (30.5).
- **Thermoplastics** are polymers that can be molded, shaped, and cooled such that the new form is preserved (30.7).
- **Thermosetting polymers** are composed of complex networks of covalent bonds, so they cannot be melted to form a liquid phase (30.7).

Practice Test on Chapter Review

1.a. Which of the following statements is (are) true about chain-growth polymers?
   1. The reaction mechanism involves initiation, propagation, and termination.
   2. The reaction may occur with anionic, cationic, or radical intermediates.
   3. Epoxides can serve as monomers.
   4. Statements (1) and (2) are both true.
   5. Statements (1), (2), and (3) are all true.

b. Which of the following alkenes is likely to undergo anionic polymerization?
   1. $CH_2{=}CHCO_2CH_3$
   2. $CH_2{=}CHOCH_3$
   3. $CH_2{=}CHCH_2CO_2CH_3$
   4. Both (1) and (2) will react.
   5. Compounds (1), (2), and (3) will all react.

c. Which of the following compounds can serve as an initiator in cationic polymerization?
   1. butyllithium
   2. $(CH_3)_3COOC(CH_3)_3$
   3. BF_3
   4. Both (1) and (2) can serve as initiators.
   5. Compounds (1), (2), and (3) can all serve as initiators.

d. Which of the following statements is (are) true about step-growth polymers?
   1. A small molecule such as H_2O or HCl is extruded during synthesis.
   2. Polycarbonates are an example of a step-growth polymer.
   3. Step-growth polymers are also called addition polymers.
   4. Statements (1) and (2) are both true.
   5. Statements (1), (2), and (3) are all true.

2. Label each statement as True (T) or False (F).
   a. A polyester is the most easily recycled polymer.
   b. Natural rubber is a polymer of repeating isoprene units in which all double bonds have the *E* configuration.
   c. A syndiotactic polymer has all Z groups bonded to the polymer chain on the same side.
   d. A polyether can be formed by anionic polymerization of an epoxide.
   e. An epoxy resin is a chain-growth polymer.
   f. A branched polymer is more amorphous, giving it a higher T_m.
   g. Polystyrene is a thermoplastic that can be melted and molded in shapes that are retained when the polymer is cooled.
   h. A polyurethane is a condensation polymer.
   i. Ziegler–Natta catalysts are used to form highly branched chain-growth polymers.
   j. Using a feedstock from a renewable source is one method of green polymer synthesis.

3. What monomer(s) are needed to synthesize each polymer?

a. 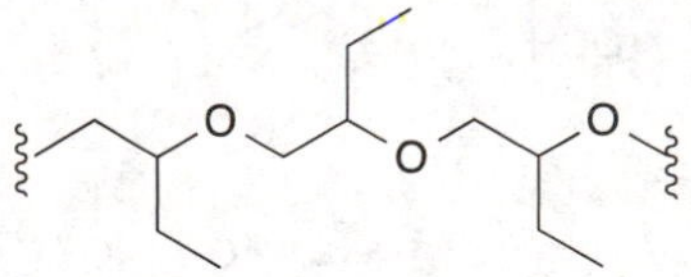 O O O

b. 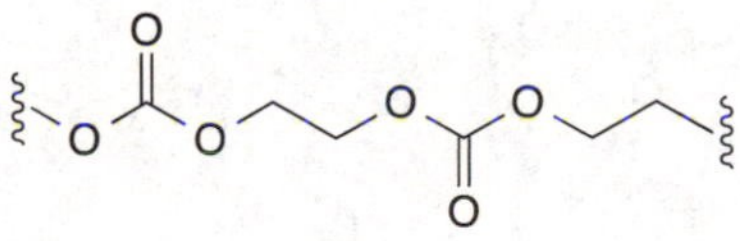 O O O O O O

c. 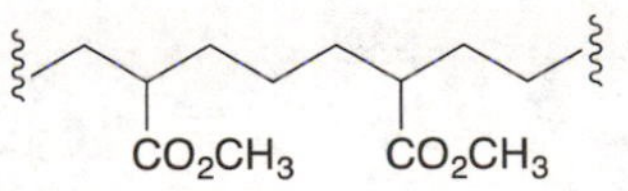 CO_2CH_3 CO_2CH_3

d. 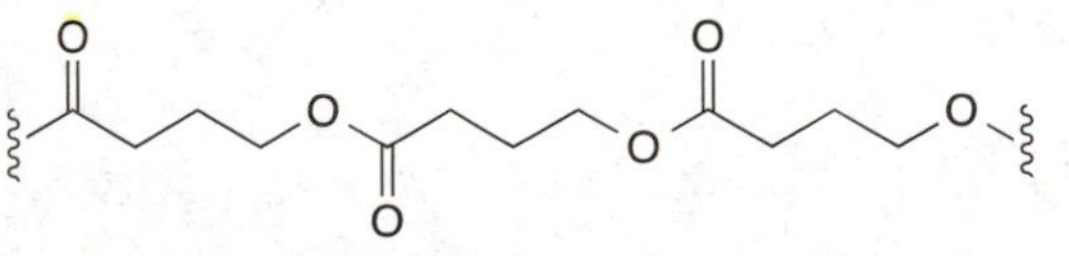 O O O O O O

e. 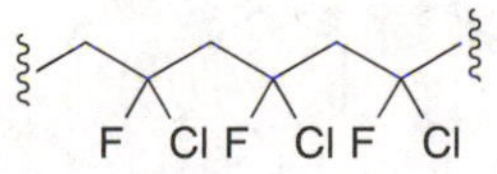 F Cl F Cl F Cl

Answers to Practice Test

1.a. 5
b. 1
c. 3
d. 4

2.a. T
b. F
c. F
d. T
e. F
f. F
g. T
h. T
i. F
j. T

3.
a. 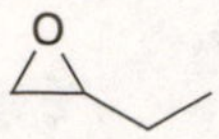
b. 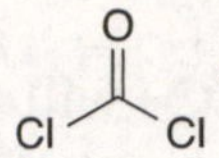 and HO OH
c. 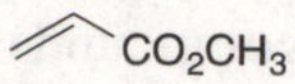 CO_2CH_3 and $CH_2{=}CH_2$
d. 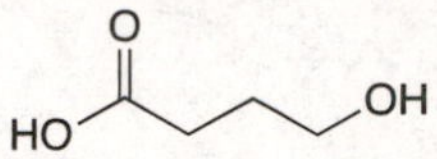 HO OH
e. 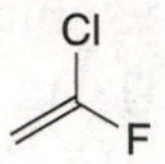 Cl F

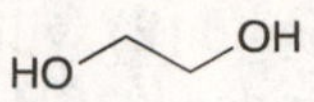

Answers to Problems

30.1 Place brackets around the repeating unit that creates the polymer.

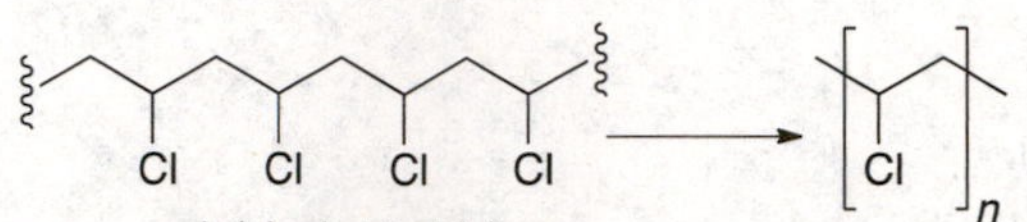

poly(vinyl chloride)

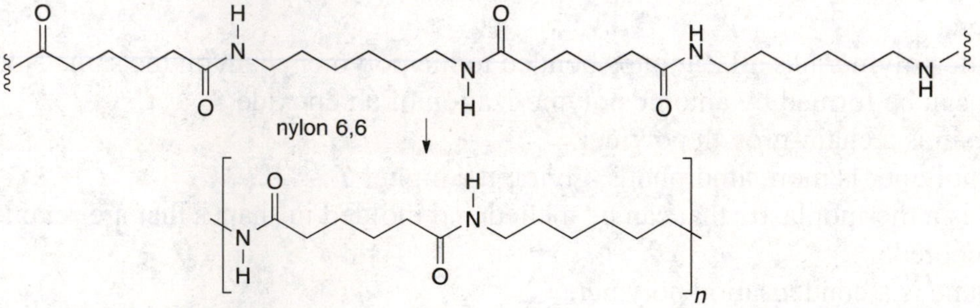

nylon 6,6

30.2 Draw each polymer formed by chain-growth polymerization.

a. 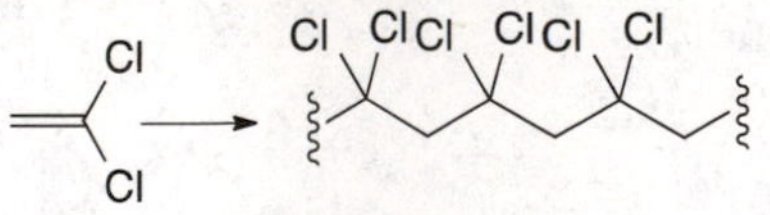
b. 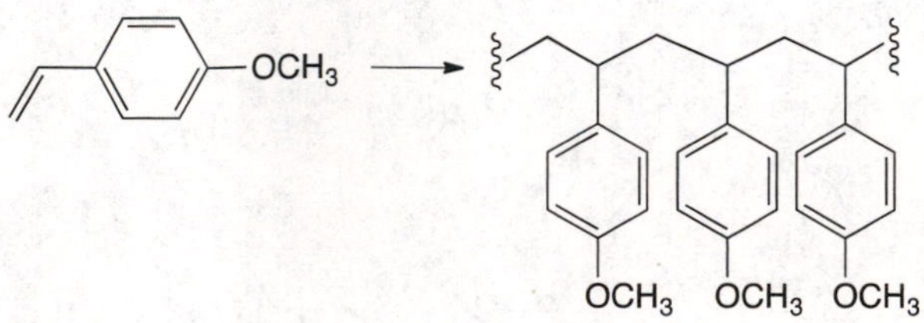 OCH_3
c. OCH_3
d.

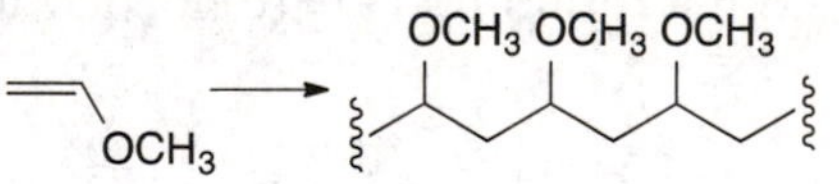

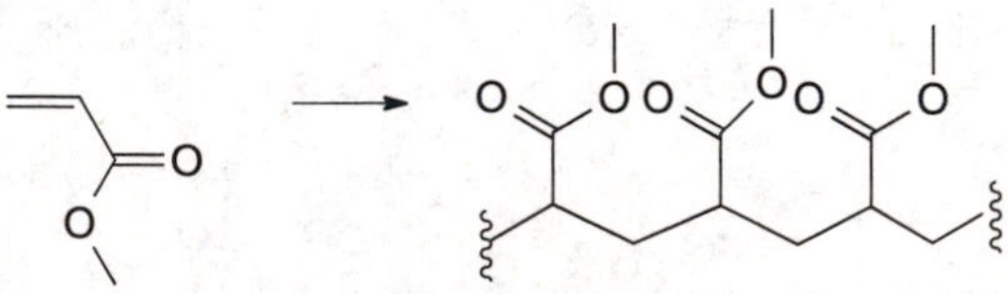

30.3 Draw each polymer formed by radical polymerization.

a.

b.

CN CN CN

CN

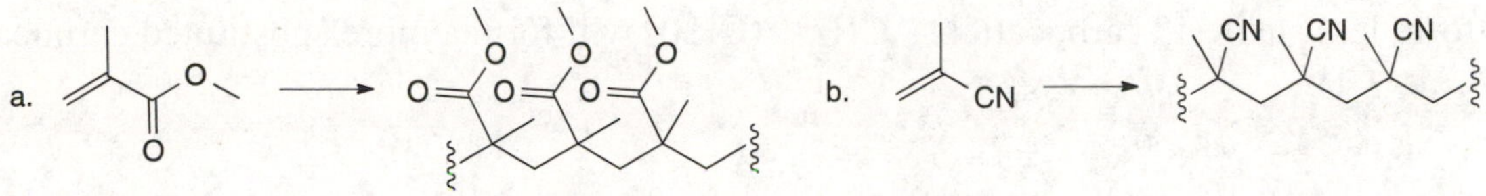

30.4 Use Mechanism 30.1 as a model of radical polymerization.

Initiation:

$(CH_3)_3CO–OC(CH_3)_3$ [1] → 2 $(CH_3)_3CO\cdot$ [2] → $(CH_3)_3CO$

Propagation:

$(CH_3)_3CO$ + [3] → $(CH_3)_3CO$ Repeat Step [3] over and over.

Termination:

[4] → OAc OAc

(Ac = CH_3CO–)

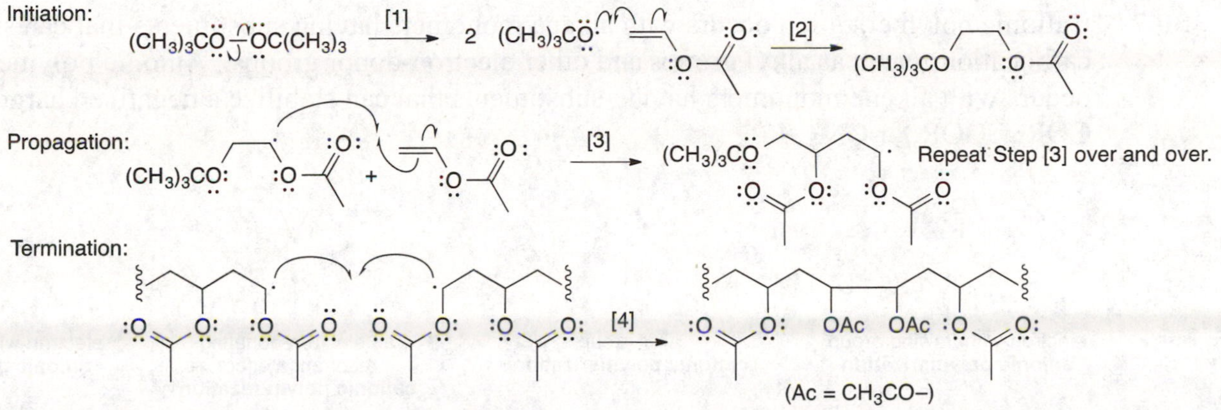

30.5 Radical polymerization forms a long chain of polystyrene with phenyl groups bonded to every other carbon. To form branches on this polystyrene chain, a radical on a second polymer chain abstracts a H atom. Abstraction of H_a forms a resonance-stabilized radical **A'**. The 2° radical **B'** (without added resonance stabilization) is formed by abstraction of H_b. Abstraction of H_a is favored, therefore, and this radical goes on to form products with 4° C's (**A**).

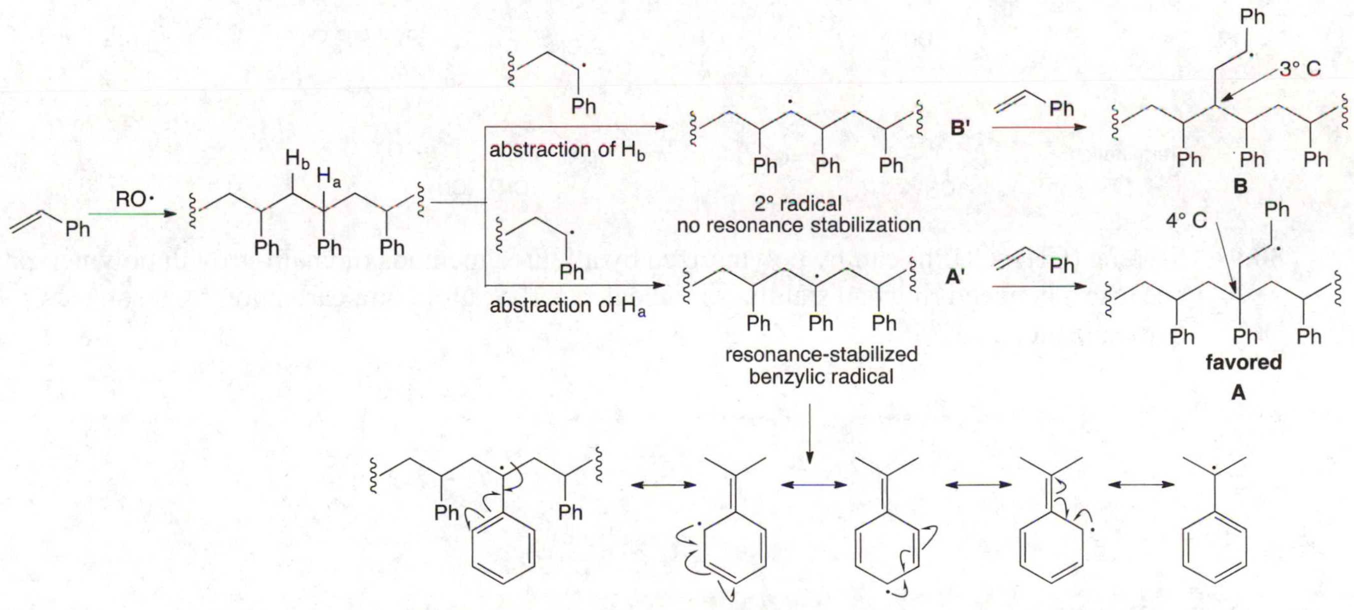

Ph RO· H_b H_a Ph Ph Ph

Ph abstraction of H_b Ph Ph Ph

Ph abstraction of H_a Ph Ph Ph

2° radical
no resonance stabilization

B' Ph → Ph 3° C Ph Ph Ph **B**

resonance-stabilized
benzylic radical

A' Ph → 4° C Ph Ph Ph Ph **favored** **A**

Ph Ph

30.6 Cationic polymerization proceeds via a carbocation intermediate. Substrates that form more stable 3° carbocations react more readily in these polymerization reactions than substrates that form less stable 1° carbocations. $CH_2=C(CH_3)_2$ will form a more substituted carbocation than $CH_2=CH_2$.

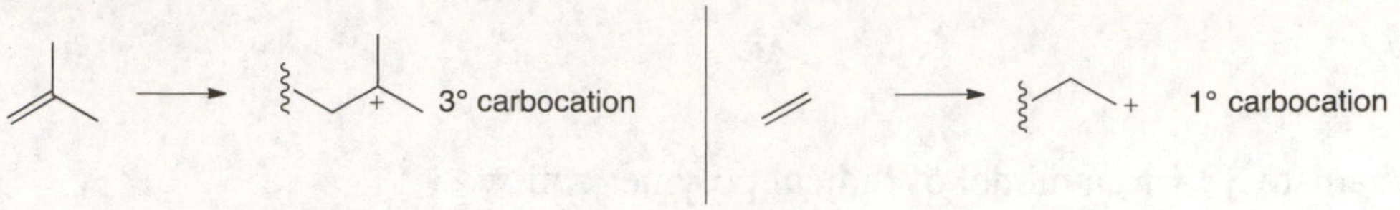

3° carbocation

1° carbocation

30.7 Cationic polymerization occurs with alkene monomers having substituents that can stabilize carbocations, such as alkyl groups and other electron-donor groups. Anionic polymerization occurs with alkene monomers having substituents that can stabilize a negative charge, such as COR, COOR, or CN.

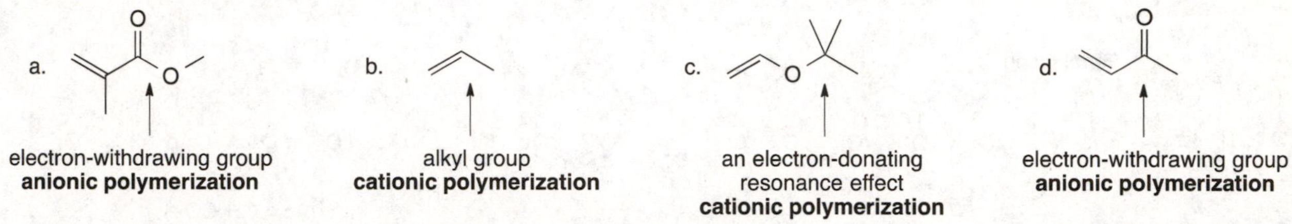

a. electron-withdrawing group **anionic polymerization**

b. alkyl group **cationic polymerization**

c. an electron-donating resonance effect **cationic polymerization**

d. electron-withdrawing group **anionic polymerization**

30.8 Use Mechanism 30.4 as a model of anionic polymerization.

Initiation: Bu—Li + CH2=CH–CN [1] → Bu–CH2–CH(–)(CN) Li+

Propagation: [2] → Repeat Step [2] over and over.

Termination: O=C=O [3] →

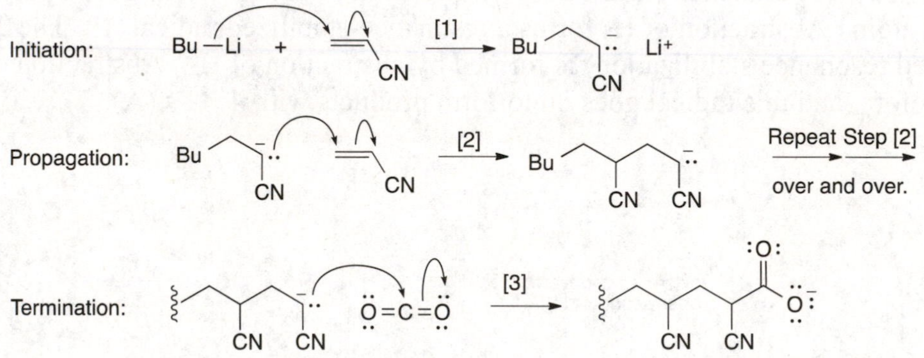

30.9 Styrene (CH_2=CHPh) can by polymerized by all three methods of chain-growth polymerization because a benzene ring can stabilize a radical, a carbocation, or a carbanion by resonance delocalization.

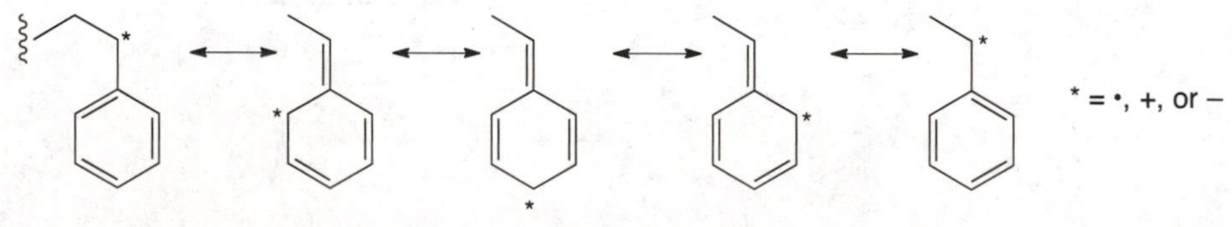

* = •, +, or –

30.10 Draw the copolymers formed in each reaction.

a. 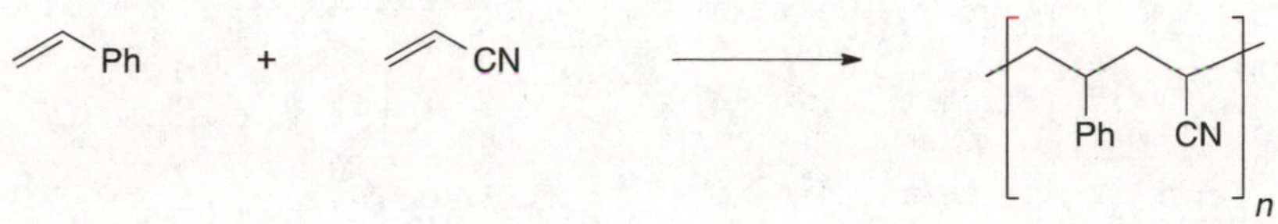 Ph + CN ⟶ Ph CN *n*

b. F F CF_3 F + F F ⟶ F F F CF_3 F F *n*

30.11

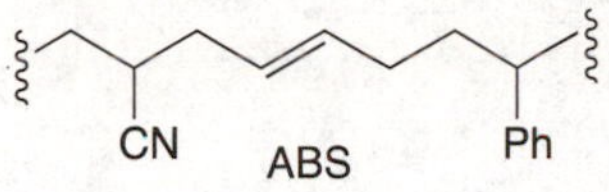

CN ABS Ph

30.12

a. 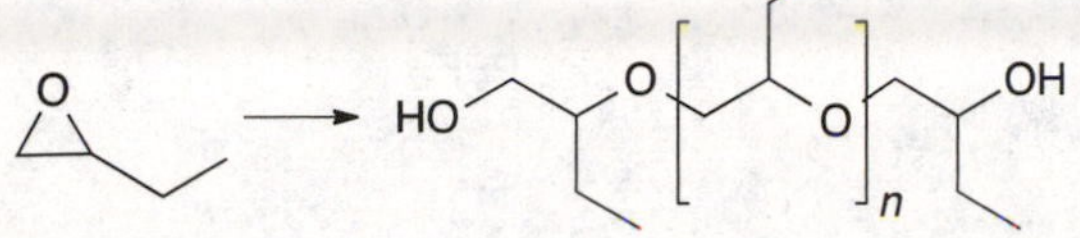 O ⟶ HO O O *n* OH

b. 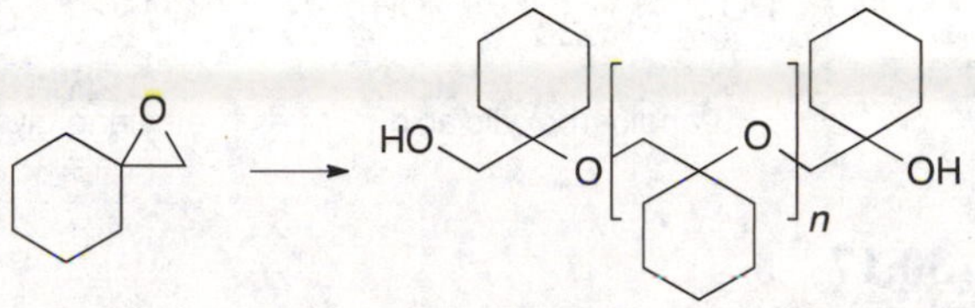 O ⟶ HO O O *n* OH

30.13

Cl Cl Cl

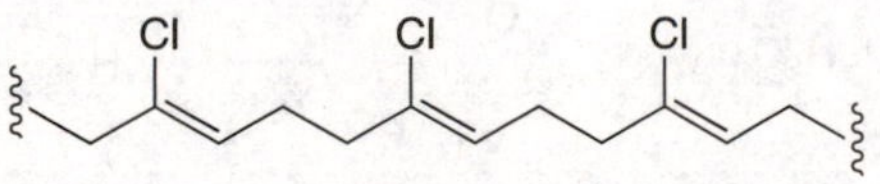

neoprene

All double bonds have the *Z* configuration.

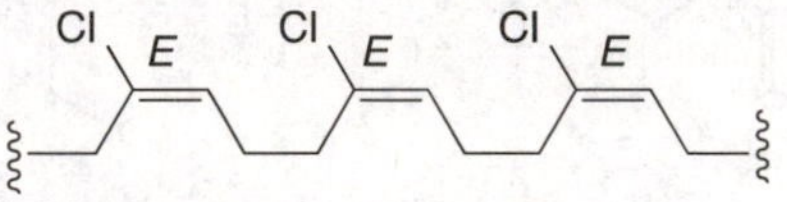

1 Cl 1 CH_2 CH_2 2 H 2

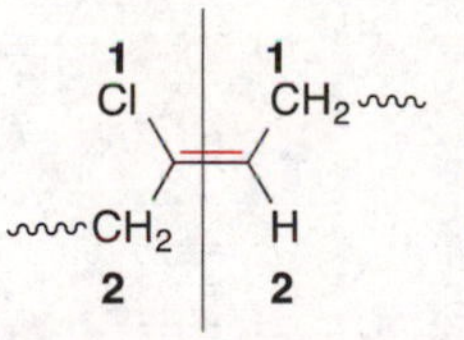

Two higher priority groups (1's) are on the same side of the double bond - - - > *Z* configuration.

Cl *E* Cl *E* Cl *E*

E configuration of each double bond

30.14

RO–OR ⟶ 2 RO· + ⟶ RO ⟷ RO

RO ⟷ RO

A

The resonance-stabilized radical can react at two carbons.

RO

B

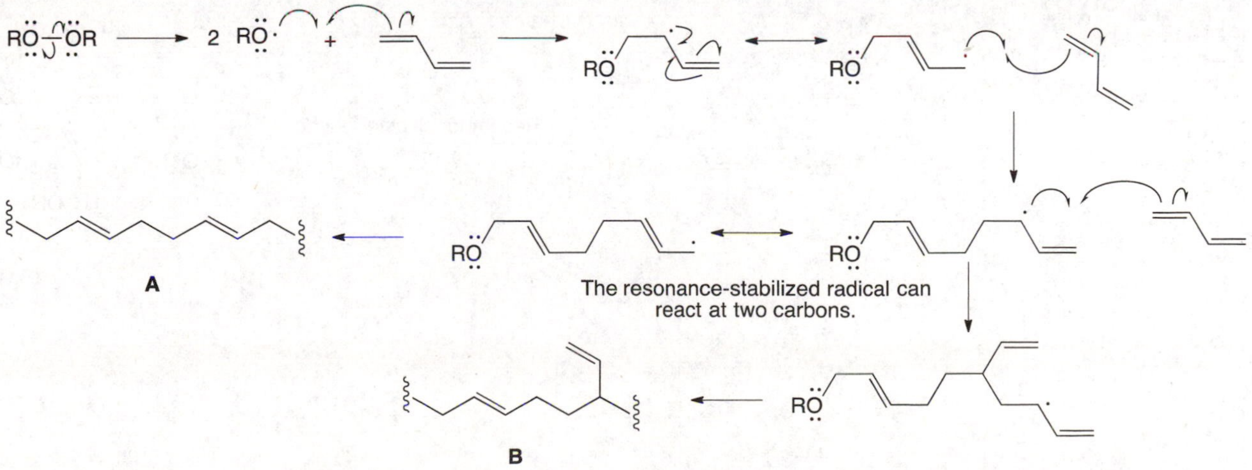

30.15

a. HO ... OH + H_2N ... NH_2 →

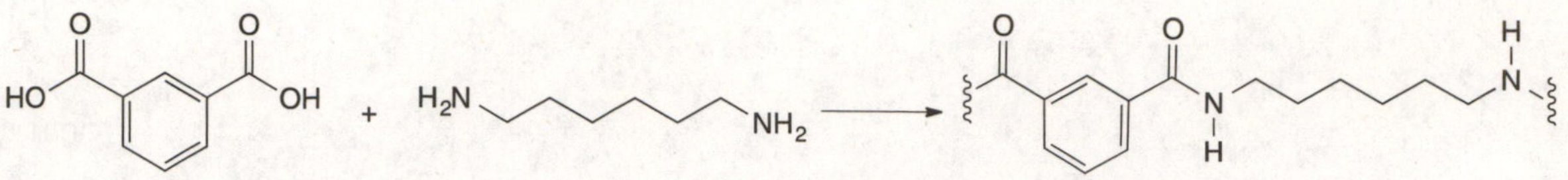

b. NH, H_2O → HO ... NH_2 →

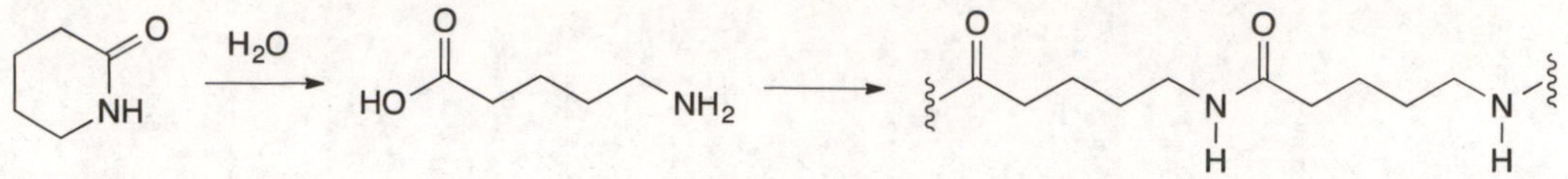

c. H_2N ... NH_2 + HO ... OH →

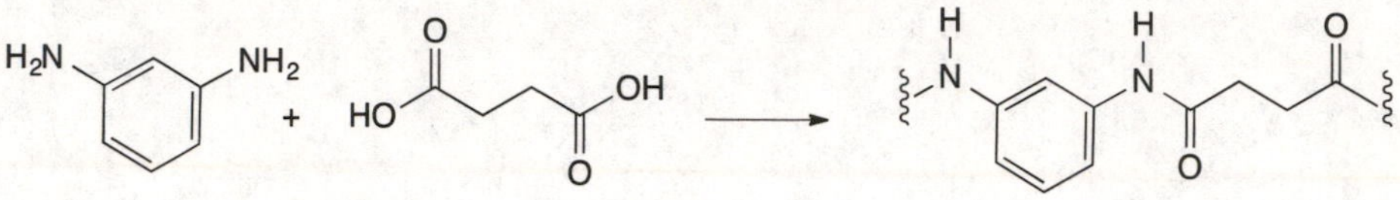

30.16

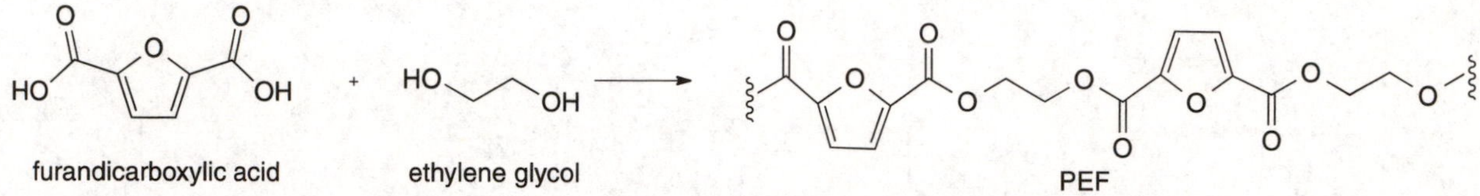

furandicarboxylic acid + ethylene glycol → PEF

30.17

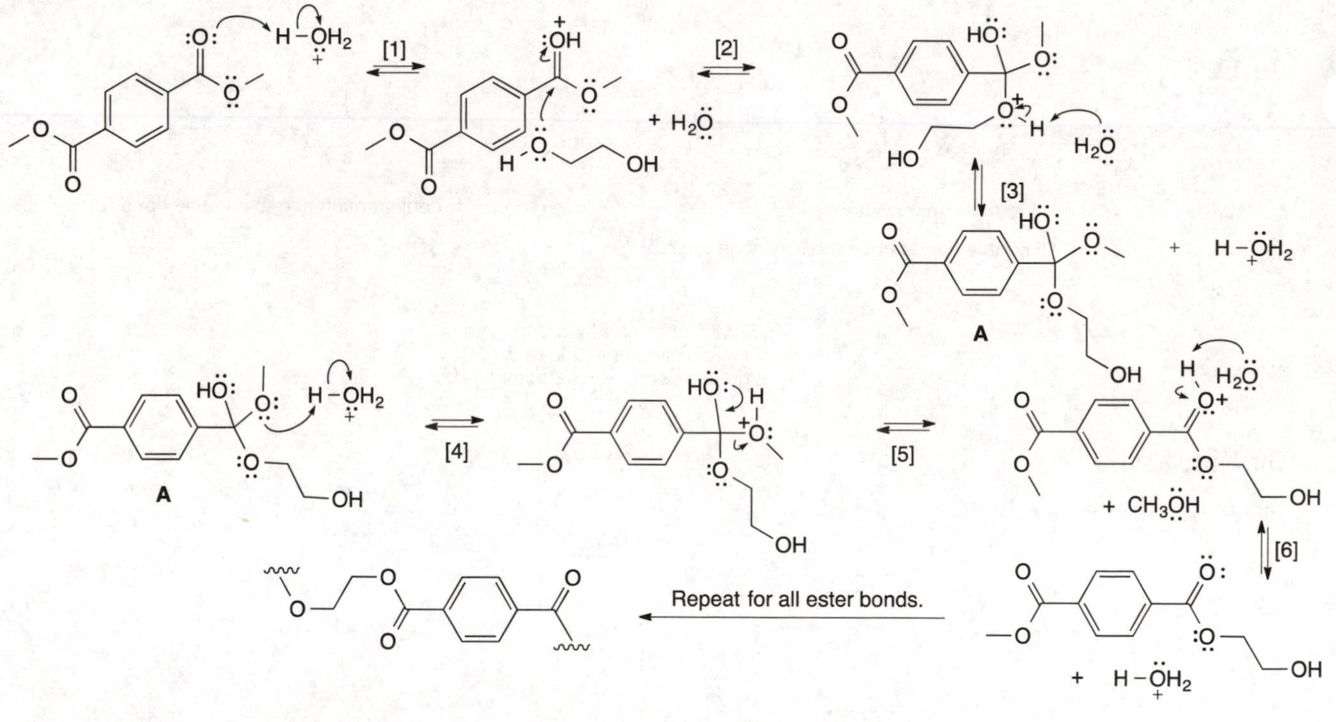

[1] [2] [3] A

A [4] [5] + $CH_3\ddot{O}H$ [6] + H–$\ddot{O}H_2$

Repeat for all ester bonds.

30.18

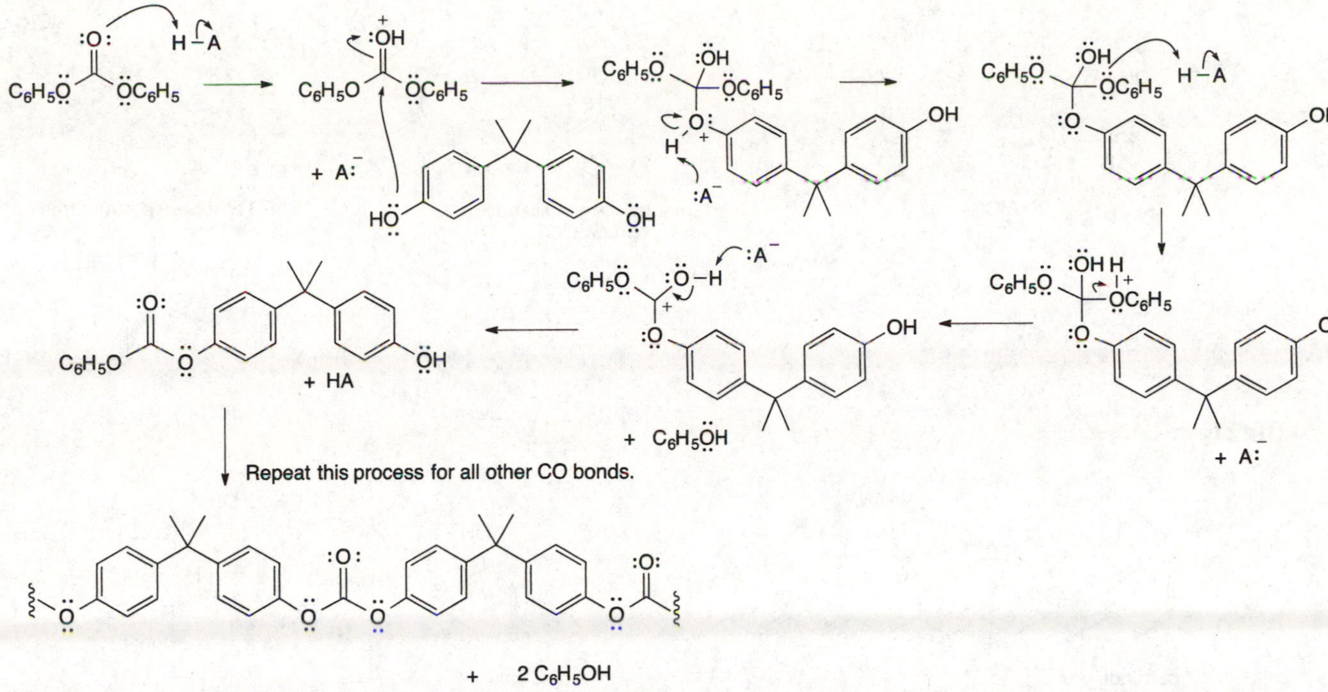

:O: H–A
$C_6H_5\ddot{O}$ $\ddot{O}C_6H_5$ → :ÖH+ C_6H_5O $\ddot{O}C_6H_5$ + A:⁻ HÖ ÖH → $C_6H_5\ddot{O}$:ÖH $\ddot{O}C_6H_5$ H O+ OH :A⁻ → $C_6H_5\ddot{O}$:ÖH $\ddot{O}C_6H_5$ H–A :O OH

$C_6H_5\ddot{O}$:ÖH H OC_6H_5 :O OH + A:⁻ → $C_6H_5\ddot{O}$:Ö–H :A⁻ :O OH + $C_6H_5\ddot{O}H$ → :O: C_6H_5O O ÖH + HA

Repeat this process for all other CO bonds.

:O: :O:

+ 2 C_6H_5OH

30.19

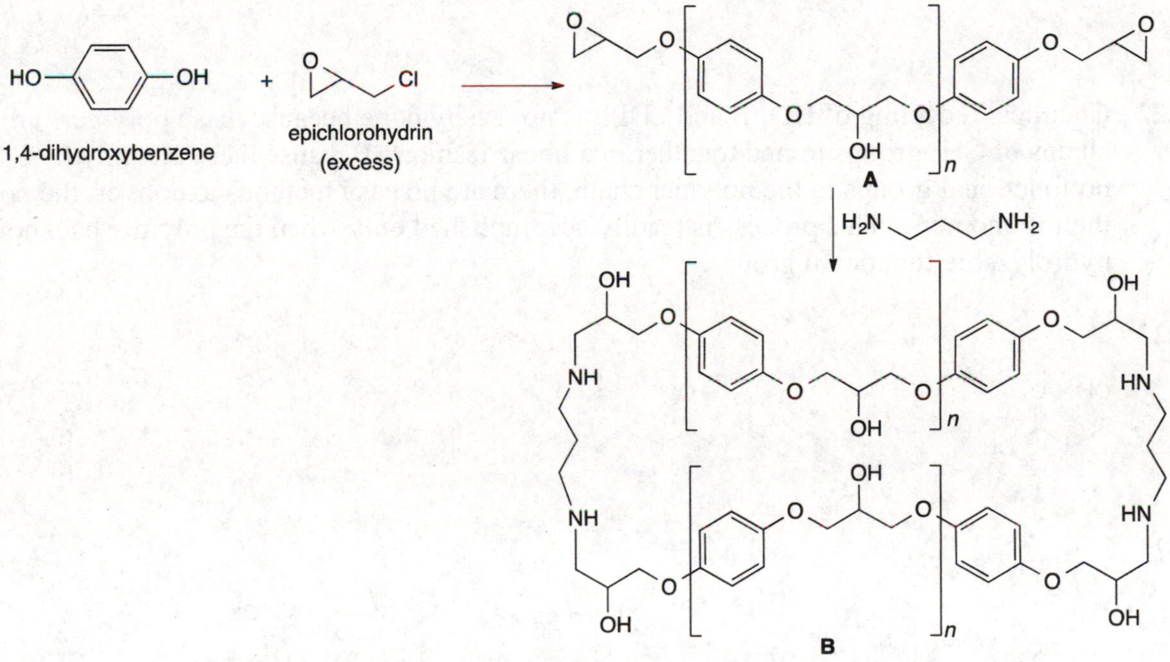

HO OH + O Cl →
1,4-dihydroxybenzene
epichlorohydrin (excess)

O O O O OH n O O
A

H_2N NH_2

OH O O O OH n O OH NH HN
NH OH O O HN OH n O O OH
B

30.20

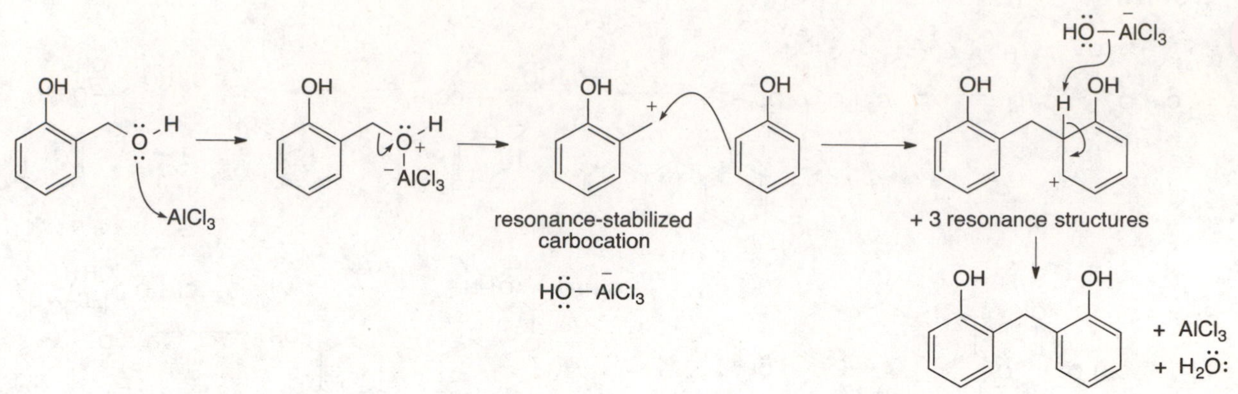

OH, $\ddot{O}$–H, $AlCl_3$ → OH, $\ddot{O}^+$–H, $^-AlCl_3$ → OH, + (resonance-stabilized carbocation), $H\ddot{O}-\bar{A}lCl_3$, OH → OH, H, OH, +, $H\ddot{O}-\bar{A}lCl_3$ (+ 3 resonance structures) → OH, OH + $AlCl_3$ + $H_2\ddot{O}$:

30.21

cardinol (OH) = OH, R $\xrightarrow[H^+]{H_2C=O}$ HO, OH, R, R, OH, R

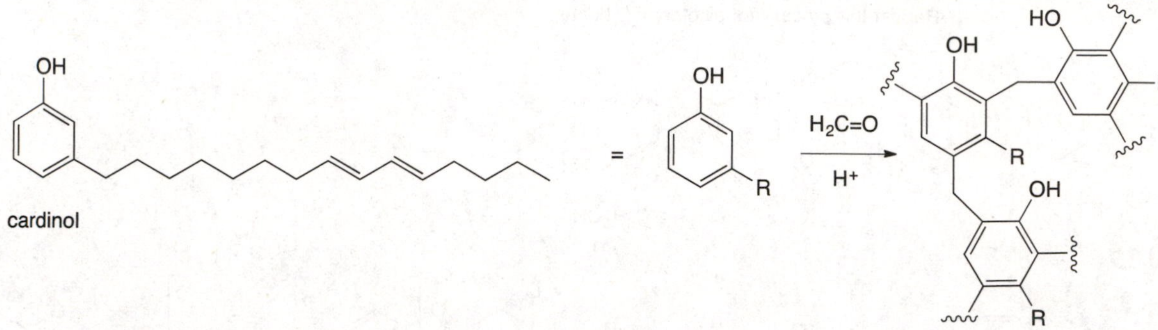

30.22 Chemical recycling of HDPE and LDPE is not easily done because these polymers are both long chains of CH_2 groups joined together in a linear fashion. Because there are only C–C bonds and no functional groups in the polymer chain, there are no easy methods to convert the polymers to their monomers. This process is readily accomplished only when the polymer backbone contains hydrolyzable functional groups.

30.23

a. 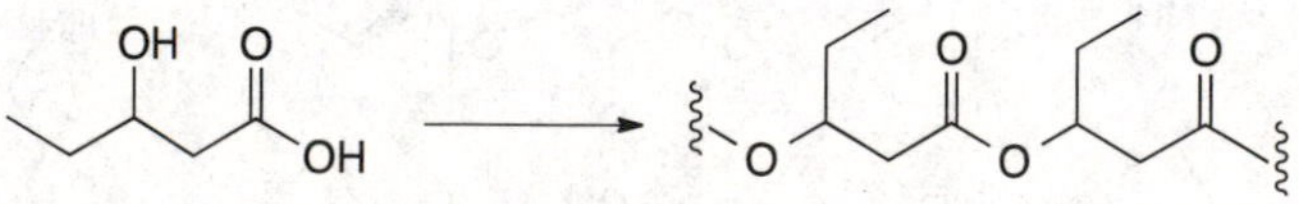 OH O OH → O O O O

b. 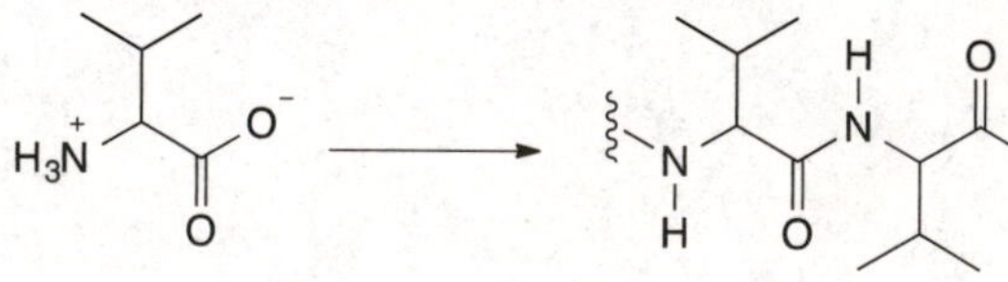 $H_3\overset{+}{N}$ O^- O → N–H, O, H–N, O

30.24

a. 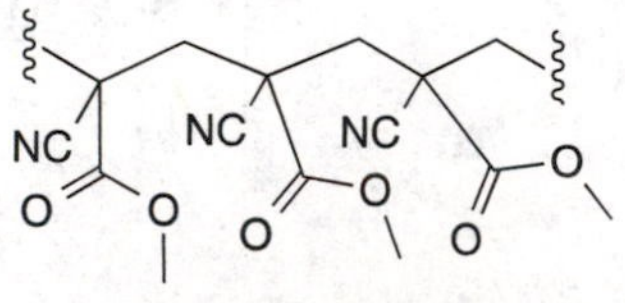 NC, NC, NC, O, O, O, O, O, O

b. 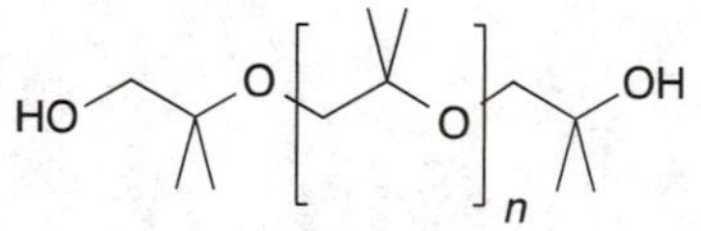 HO, O, O, n, OH

30.25

a. 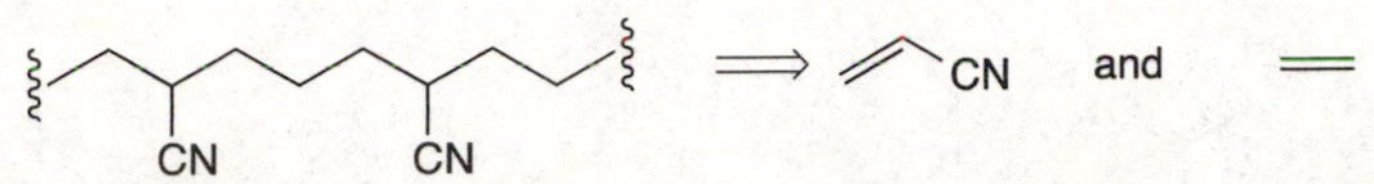 CN CN ⟹ CN and

b. 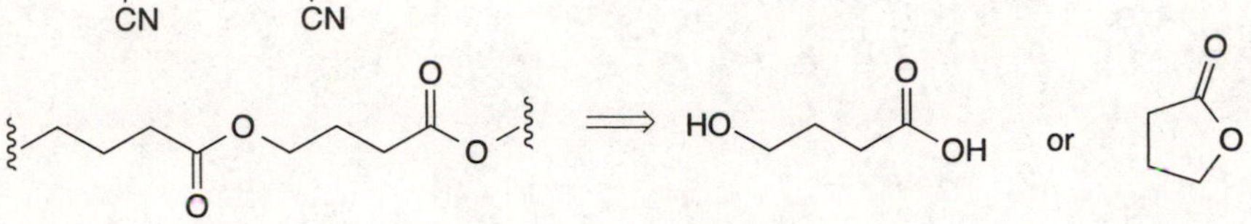 O O O O ⟹ HO OH O or O O

30.26

a. 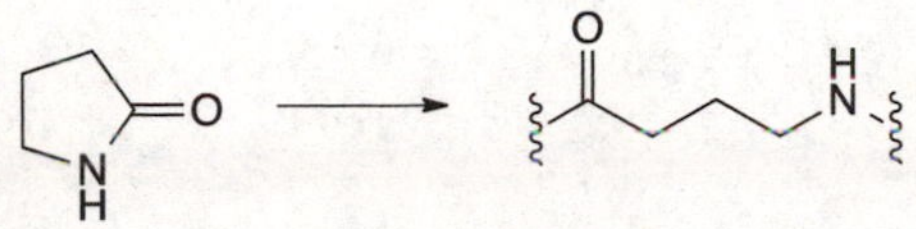 N H =O → O H N b. 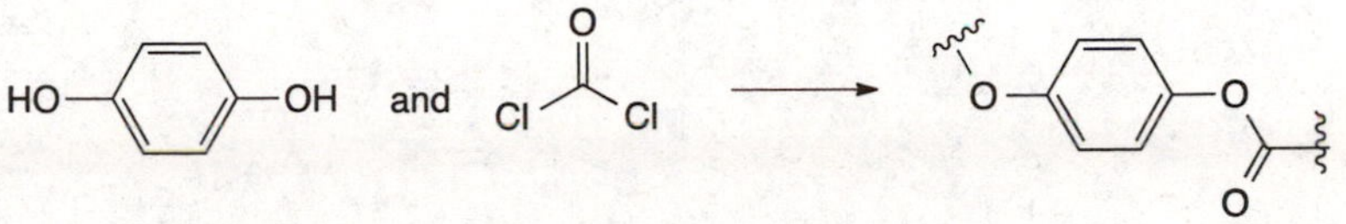 HO OH and Cl O Cl → O O O

30.27 Draw the polymer formed by chain-growth polymerization as in Answer 30.2.

a. 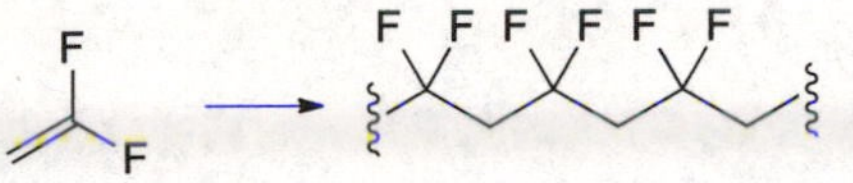 F F → F F F F F F

c. 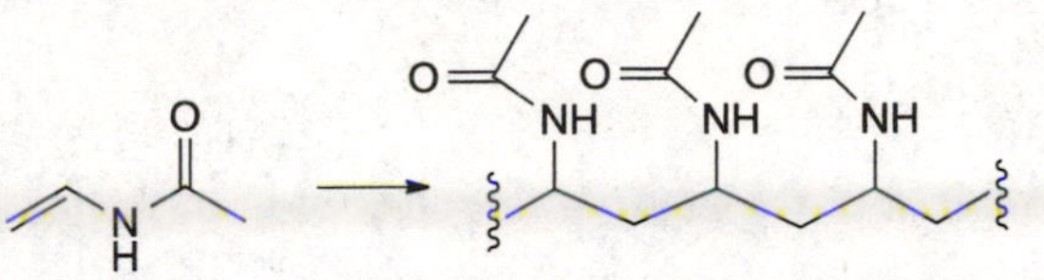 N H O → O NH O NH O NH

b. 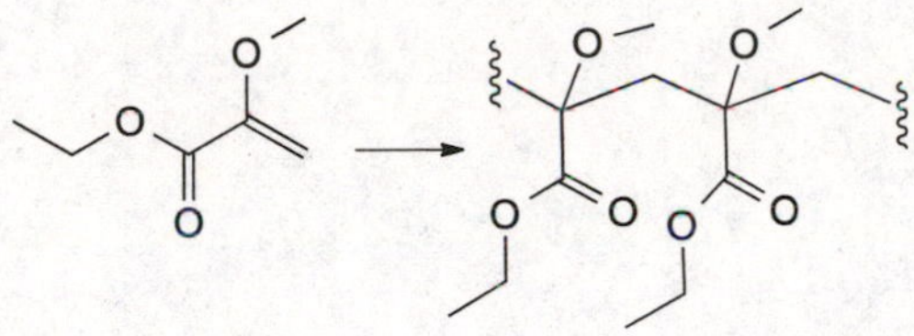 O O O → O O O O O O

d. 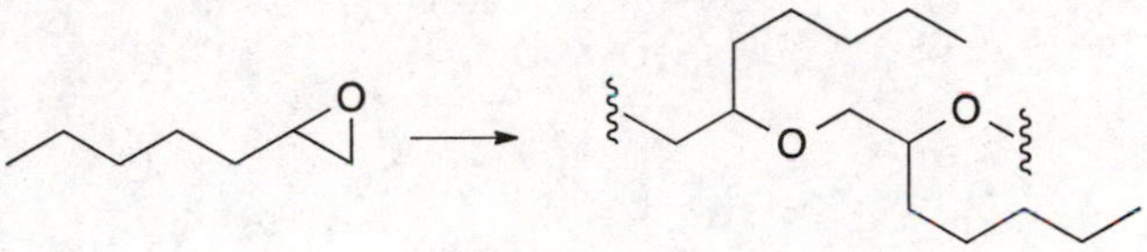 O → O O

30.28 Draw the copolymers.

a. 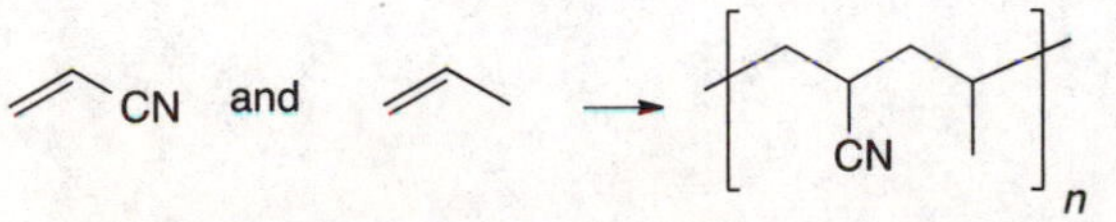 CN and → [CN]$_n$

c. 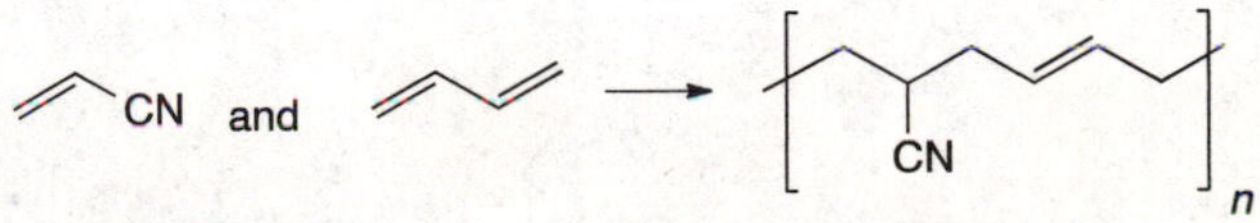 CN and → [CN]$_n$

b. 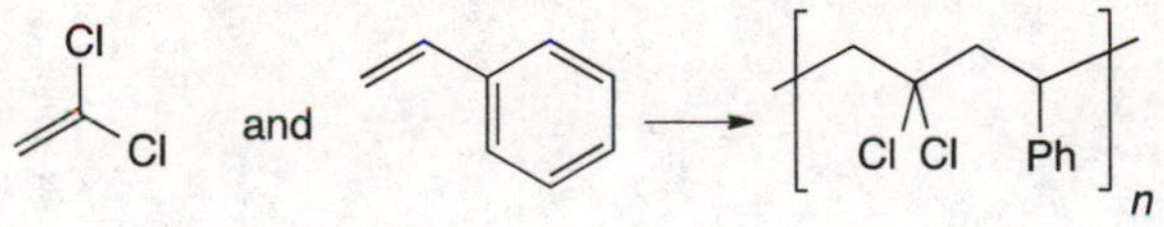 Cl Cl and → [Cl Cl Ph]$_n$

d. 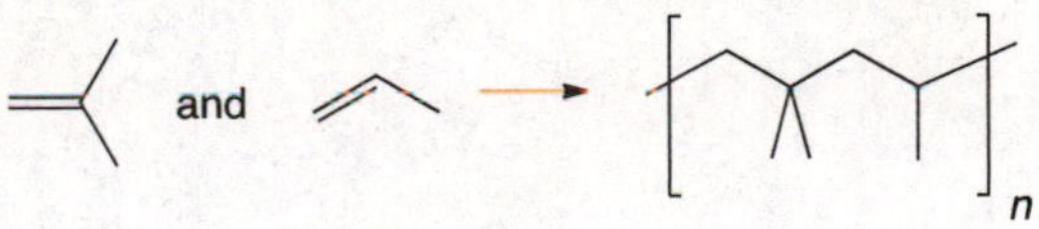 and → [ ]$_n$

30.29

a. 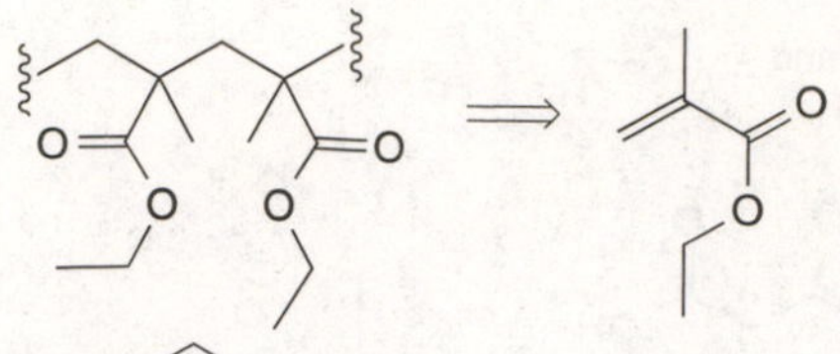 $\Rightarrow$

b. 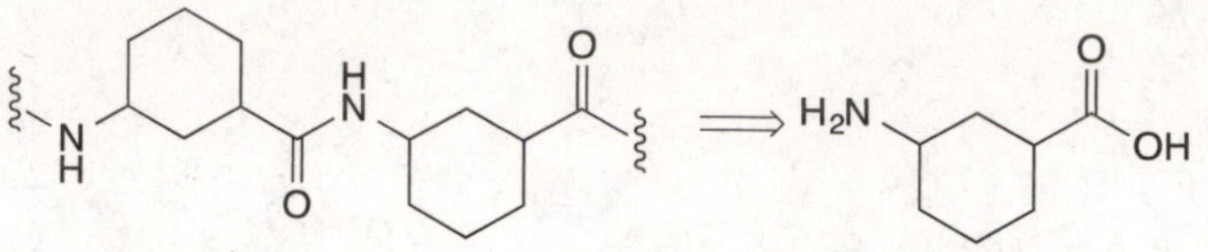 $\Rightarrow$ H_2N … OH

c. 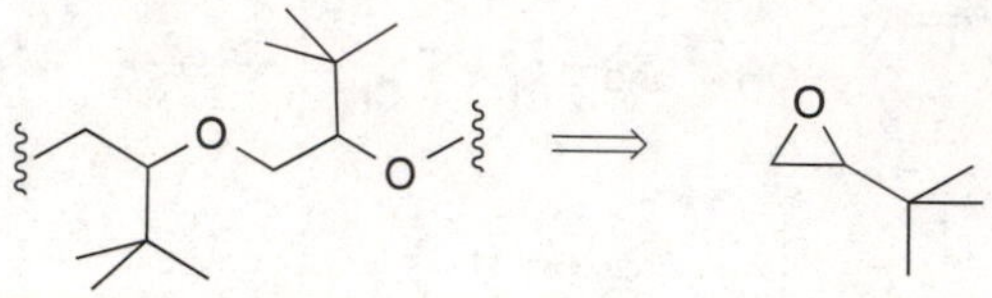 $\Rightarrow$

d. 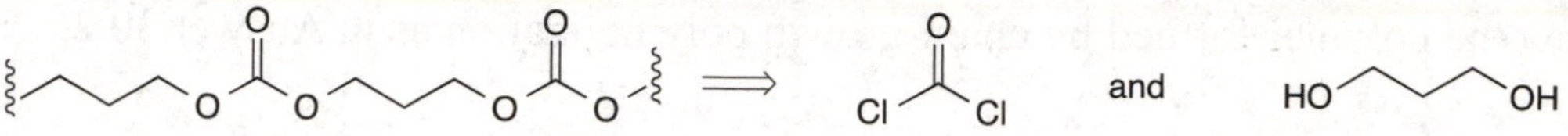 $\Rightarrow$ Cl Cl and HO OH

30.30

a. 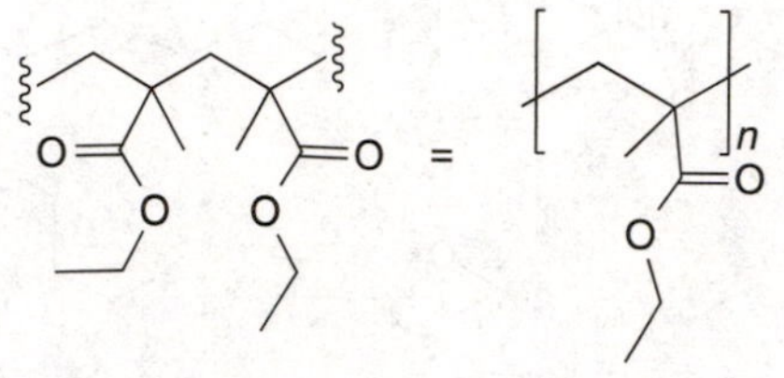 =

b. 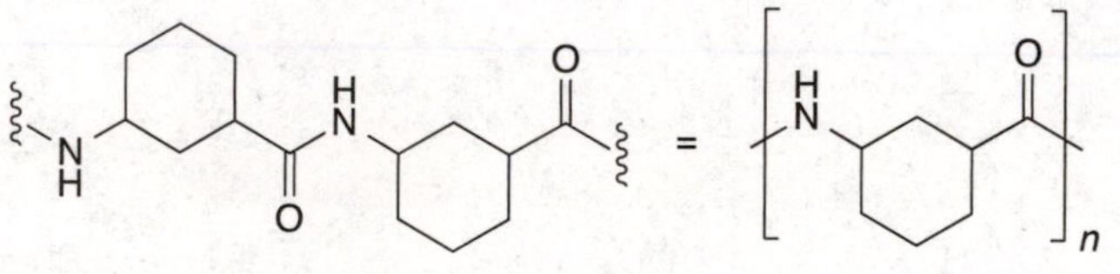 =

c. 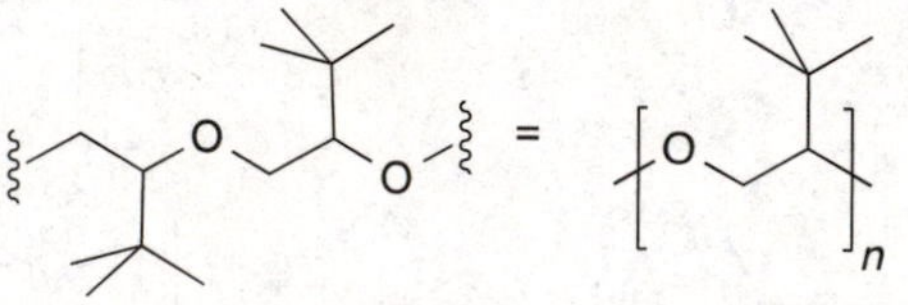 =

d. 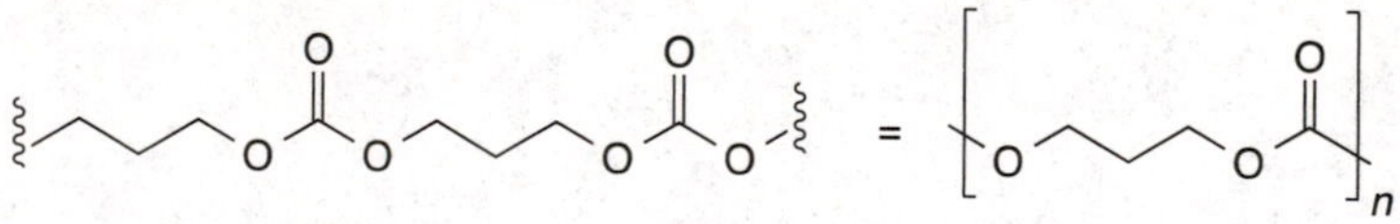 =

30.31 An **isotactic polymer** has all Z groups on the same side of the carbon backbone. A **syndiotactic polymer** has the Z groups alternating from one side of the carbon chain to the other. An **atactic polymer** has the Z groups oriented randomly along the polymer chain.

a. 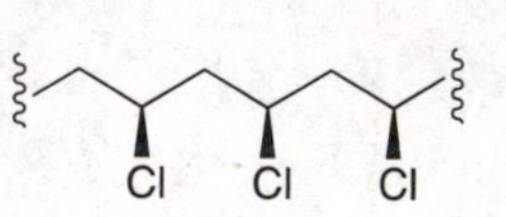 Cl Cl Cl

b. 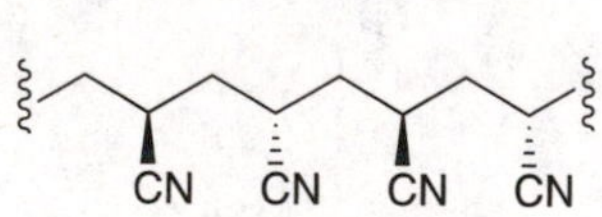 CN CN CN CN

c. 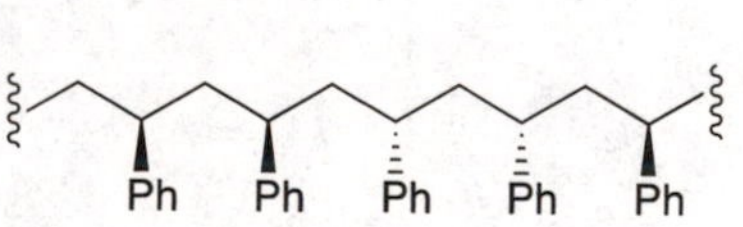 Ph Ph Ph Ph Ph

30.32

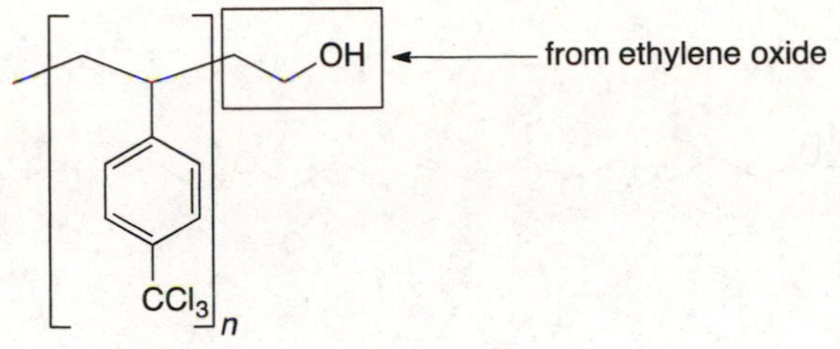

OH ← from ethylene oxide

CCl_3 *n*

30.33

a. H_2N NH_2 and HO O O OH → O H N H N O

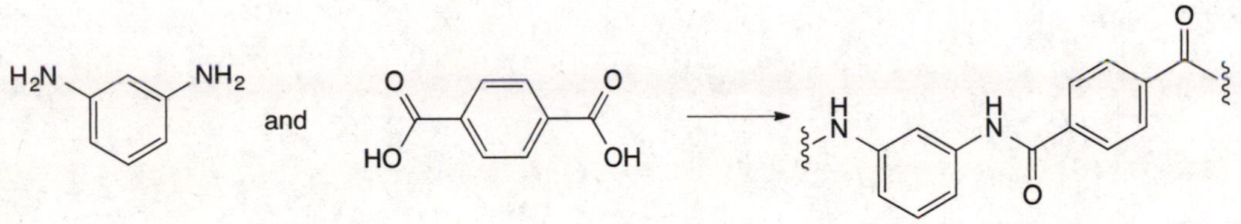

b. O=C=N N=C=O and HO OH → O H N O O NH O

c. O Cl Cl O and HO OH → O O O O

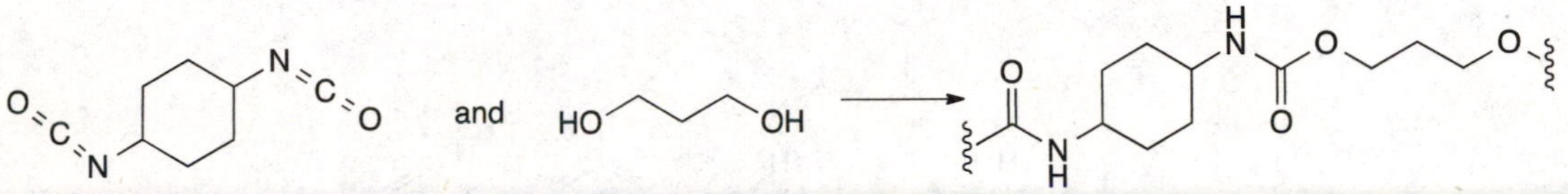

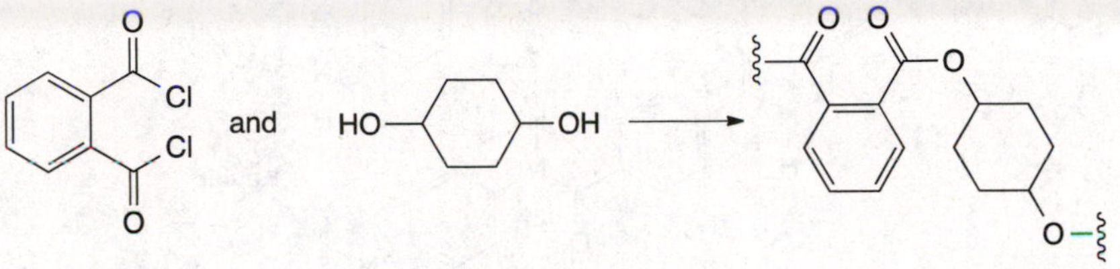

d. HO O OH → O O

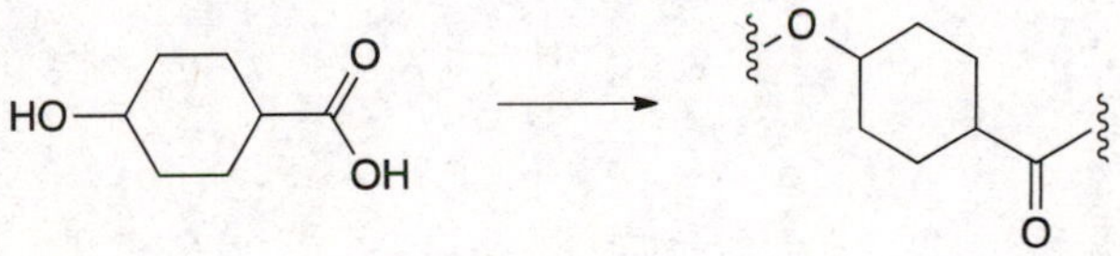

30.34

a. H_2N NH_2 + HO O O OH → NH O N H O

Quiana

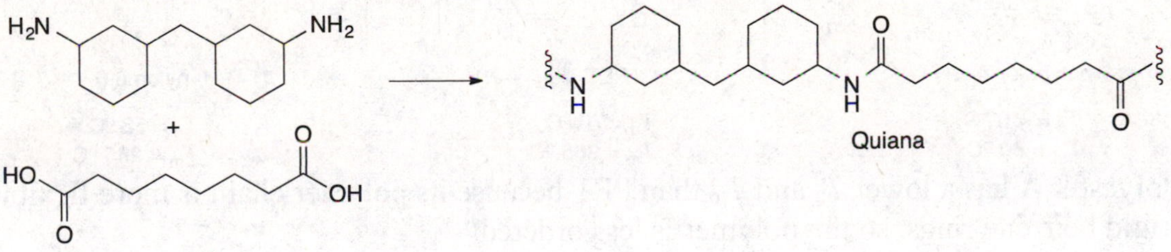

b. O Cl O Cl + H_2N NH_2 → O O N H N H

Nomex

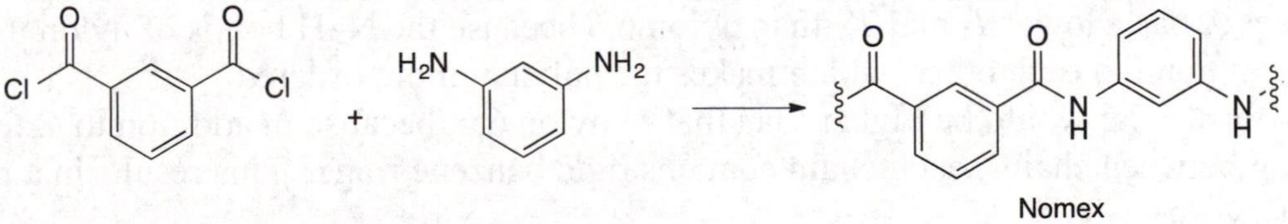

30.35

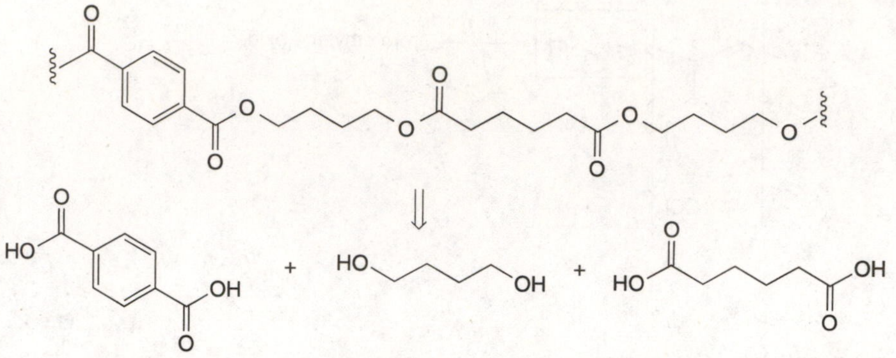

HO OH + HO OH + HO OH

30.36

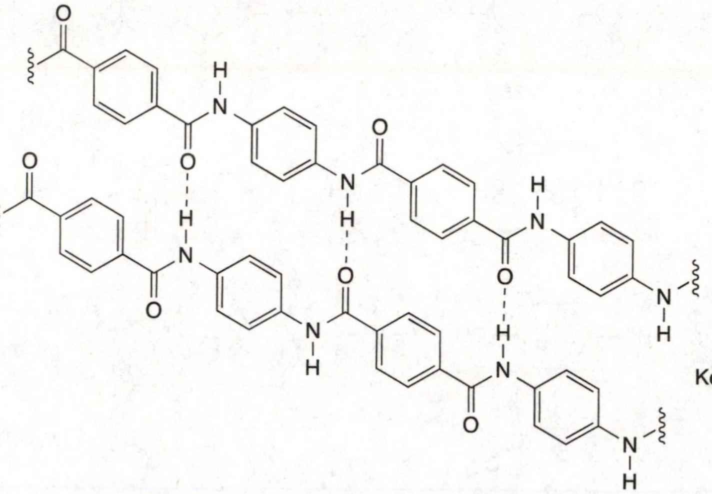

Kevlar

30.37

polyester **A**

T_g = < 0 °C
T_m = 50 °C

PET

T_g = 70 °C
T_m = 265 °C

nylon 6,6

T_g = 53 °C
T_m = 265 °C

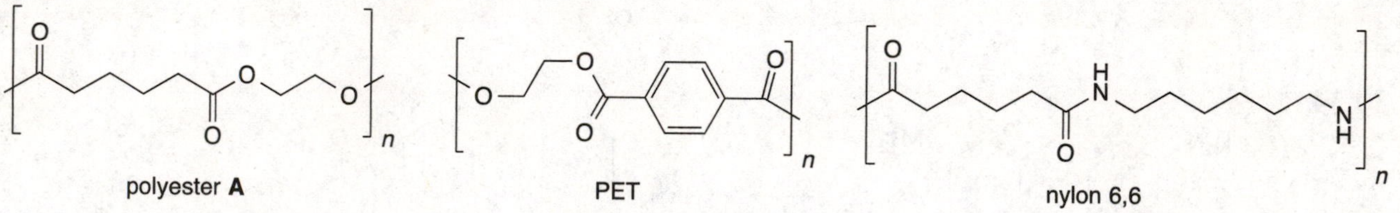

a. Polyester **A** has a lower T_g and T_m than PET because its polymer chain is more flexible. There are no rigid benzene rings, so the polymer is less ordered.

b. Polyester **A** has a lower T_g and T_m than nylon 6,6 because the N–H bonds of nylon 6,6 allow chains to hydrogen bond to each other, which makes the polymer more ordered.

c. The T_m for Kevlar would be higher than that of nylon 6,6, because in addition to extensive hydrogen bonding between chains, each chain contains rigid benzene rings. This results in a more ordered polymer.

30.38

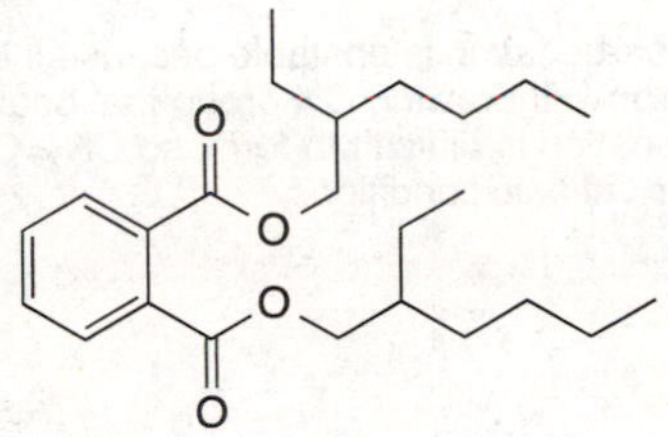

A

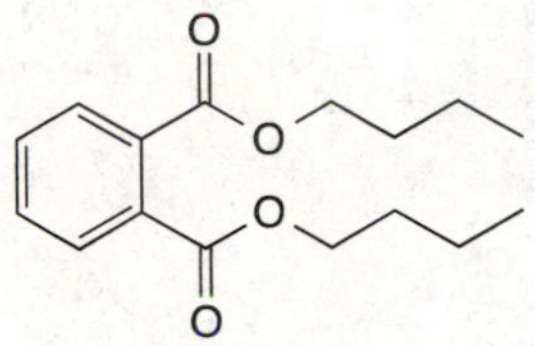

dibutyl phthalate

Diester **A** is often used as a plasticizer in place of dibutyl phthalate because it has a higher molecular weight, giving it a higher boiling point. **A** should therefore be less volatile than dibutyl phthalate, so it should evaporate from a polymer less readily.

30.39

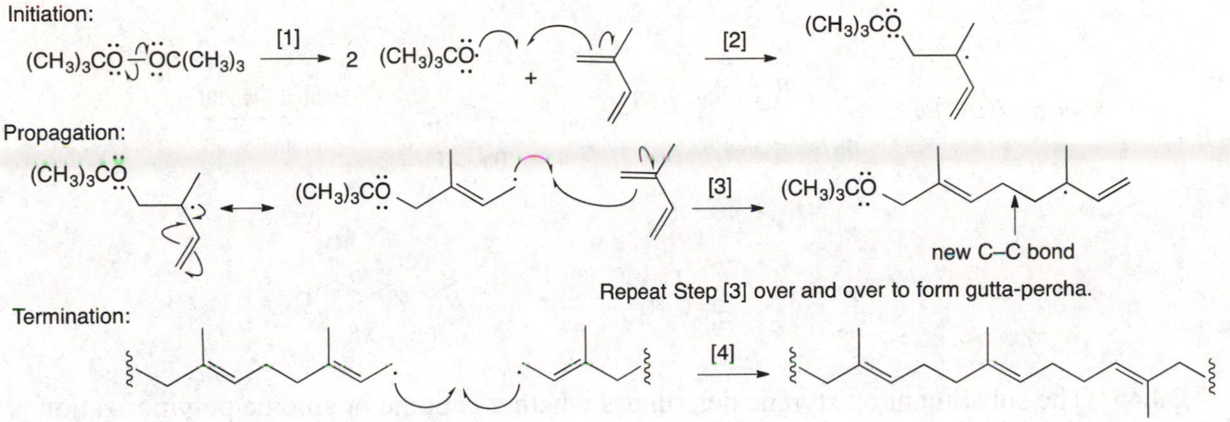

Initiation:

$(CH_3)_3CO{-}OC(CH_3)_3 \xrightarrow{[1]} 2\ (CH_3)_3CO\cdot$ + $\xrightarrow{[2]}$ $(CH_3)_3CO$

Propagation:

$(CH_3)_3CO$ ⟷ $(CH_3)_3CO$ $\xrightarrow{[3]}$ $(CH_3)_3CO$

new C–C bond

Repeat Step [3] over and over to form gutta-percha.

Termination:

$\xrightarrow{[4]}$

30.40

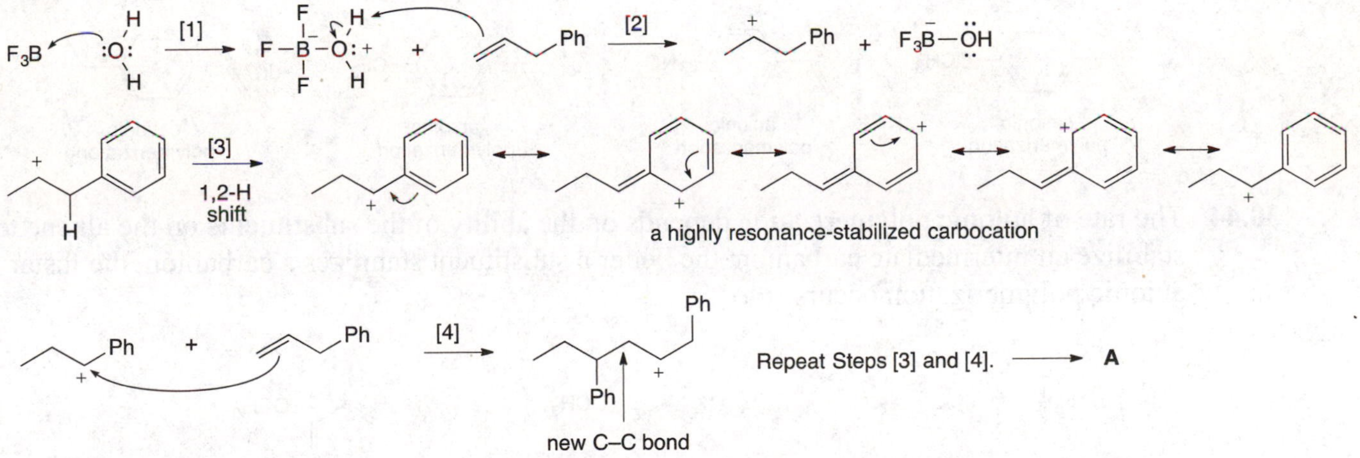

F_3B :O: (H, H) $\xrightarrow{[1]}$ $F{-}B(F)(F){-}O(H)(H)$ + Ph $\xrightarrow{[2]}$ Ph + $F_3\bar{B}{-}OH$

$\xrightarrow[\text{shift}]{[3]\ 1,2\text{-H}}$ ⟷ ⟷ ⟷ ⟷

a highly resonance-stabilized carbocation

Ph + Ph $\xrightarrow{[4]}$ Ph, Ph

new C–C bond

Repeat Steps [3] and [4]. ⟶ **A**

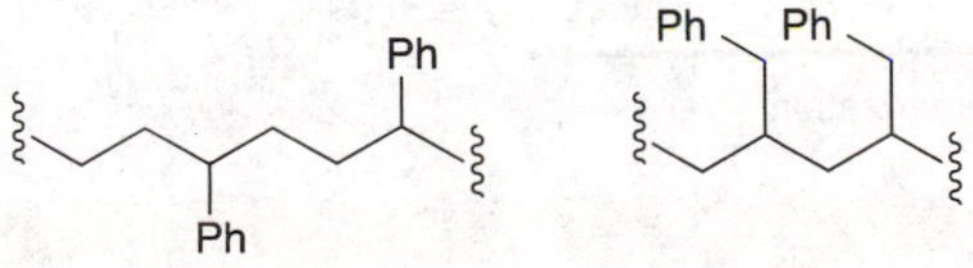

Ph, Ph

A

major product

Ph, Ph

B

A is the major product formed due to the 1,2-H shift (Step [3]) that occurs to form a resonance-stabilized carbocation.
B is the product that would form without this shift.

30.41

This carbocation is unstable because it is located next to an electron-withdrawing CN group that bears a δ+ on its C atom. This carbocation is difficult to form, so CH_2=CHCN is only slowly polymerized under cationic conditions.

This 2° carbocation is more stable because it is not directly bonded to the electron-withdrawing CN group. As a result, it is more readily formed. Thus, cationic polymerization can occur more readily.

30.42

Initiation: Bu—Li + styrene —[1]→ carbanion, Li^+

Propagation: —[2]→ Repeat Step [2] over and over.

Termination: + O=C=O —[3]→ carboxylate

30.43 The substituent on styrene determines whether cationic or anionic polymerization is preferred. When the substituent stabilizes a carbocation, cationic polymerization will occur. When the substituent stabilizes a carbanion, anionic polymerization will occur.

a. OCH_3 — cationic polymerization

b. NO_2 — anionic polymerization

c. CF_3 — anionic polymerization

d. ethyl — cationic polymerization

30.44 The rate of anionic polymerization depends on the ability of the substituents on the alkene to stabilize an intermediate carbanion: the better a substituent stabilizes a carbanion, the faster anionic polymerization occurs.

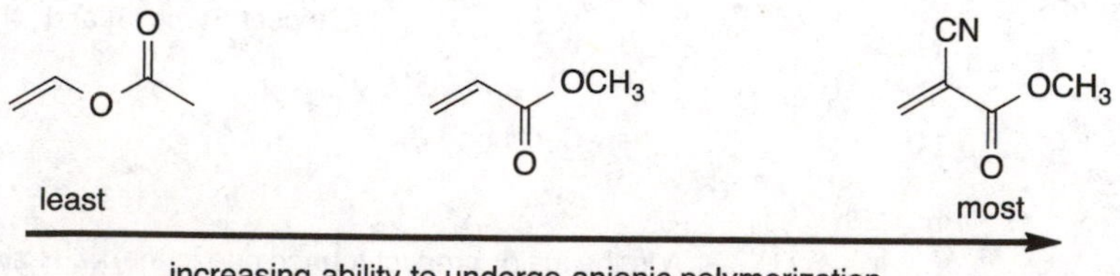

30.45 The reason for this selectivity is explained in Figure 9.9. In the ring opening of an unsymmetrical epoxide under acidic conditions, nucleophilic attack occurs at the carbon atom that is more able to accept a δ+ in the transition state; that is, nucleophilic attack occurs at the more substituted carbon. The transition state having a δ+ on a C with an electron-donating CH_3 group is more stabilized (lower in energy), permitting a faster reaction.

[1] [2] [3] [4] [5]

Repeat Steps [4] and [5] over and over.

30.46

A

resonance-stabilized carbocation

+ $AlCl_4^-$

cyclohexane-1,4-diol

+ $AlCl_3$ + HCl

Repeat.

B

30.47

resonance-stabilized cation

(+ 3 additional resonance structures)

bisphenol A

+ H_2SO_4

(+ 3 additional resonance structures)

+ H_2O

(+ 5 additional resonance structures)

+ HSO_4^-

30.48

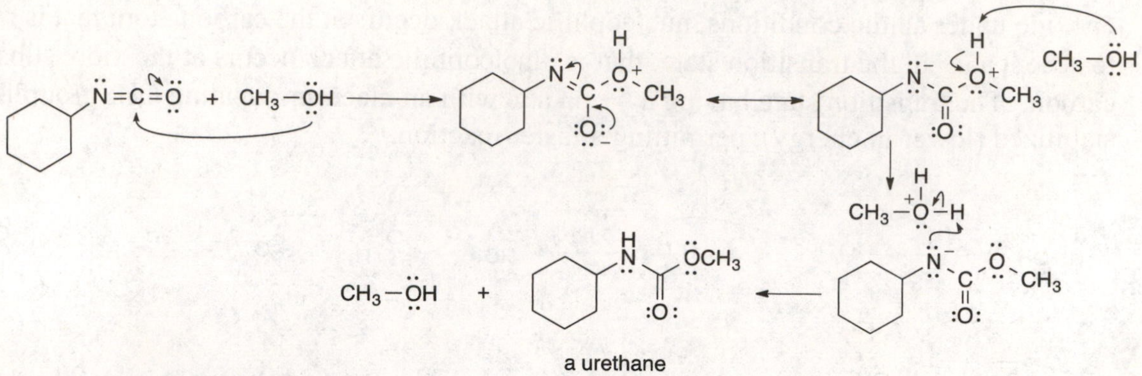

$\ddot{N}=C=\ddot{O}$ + $CH_3-\ddot{O}H$ → $\ddot{N}$ H O^+ CH_3 C $:\ddot{O}:^-$ ↔ N^- H O^+ CH_3 C $:O:$; $CH_3-\ddot{O}H$

H $CH_3-\overset{+}{O}-H$ N^- C O CH_3 $:O:$

$CH_3-\ddot{O}H$ + H N $\ddot{O}CH_3$ $:O:$ ←

a urethane

30.49

a. 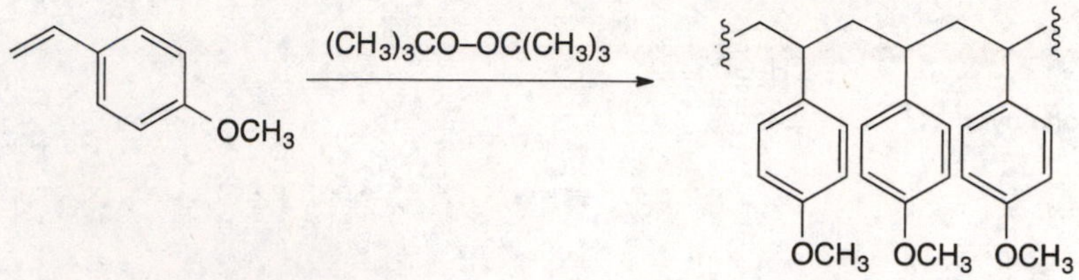 $(CH_3)_3CO-OC(CH_3)_3$ → OCH_3 ; OCH_3 OCH_3 OCH_3

b. 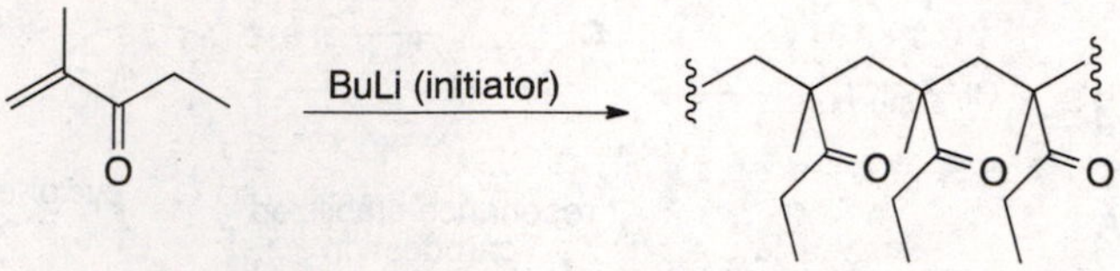 BuLi (initiator) → O O O O

c. 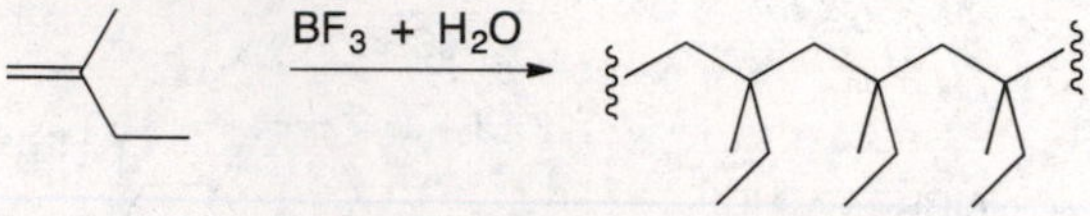 BF_3 + H_2O →

d. O ; ^{-}OH → O O

e. OH OH + O Cl (excess) → O O O OH O O O O n

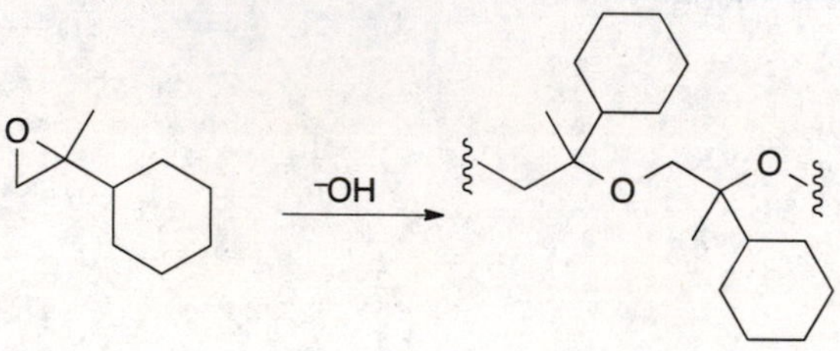

f. O O O O ; ^{-}OH, H_2O → HO OH + ^{-}O O O O^-

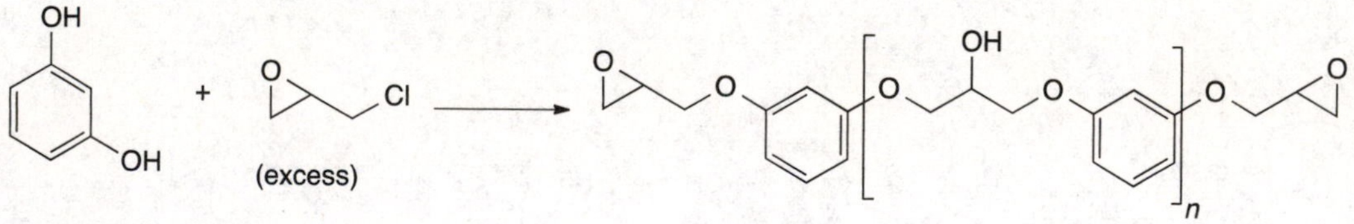

g. OCN NCO + HO OH → H N O O N H O O

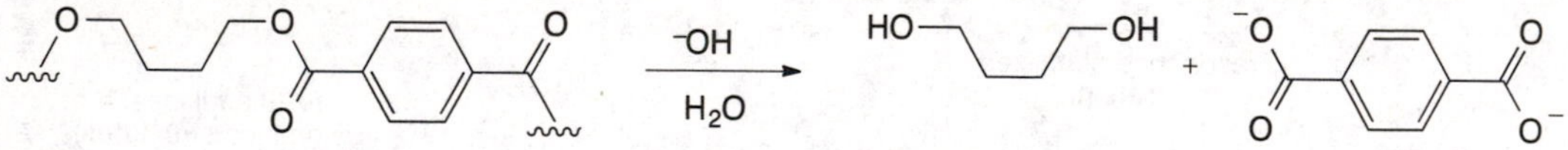

h. HO OH ; $Cl_2C=O$ → O O O

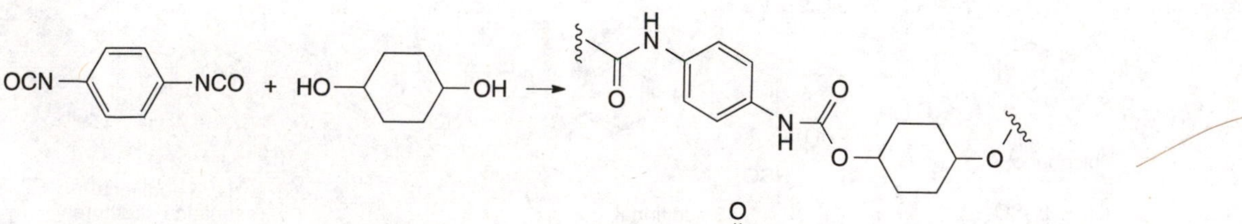

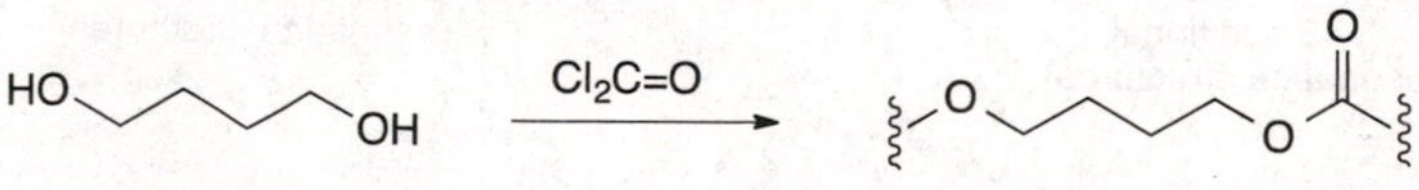

30.50 Polyethylene bottles are resistant to NaOH because they are hydrocarbons with no reactive sites. Polyester shirts and nylon stockings both contain functional groups. Nylon contains amides and polyester contains esters, two functional groups that are susceptible to hydrolysis with aqueous NaOH. Thus, the polymers are converted to their monomer starting materials, creating a hole in the garment.

30.51

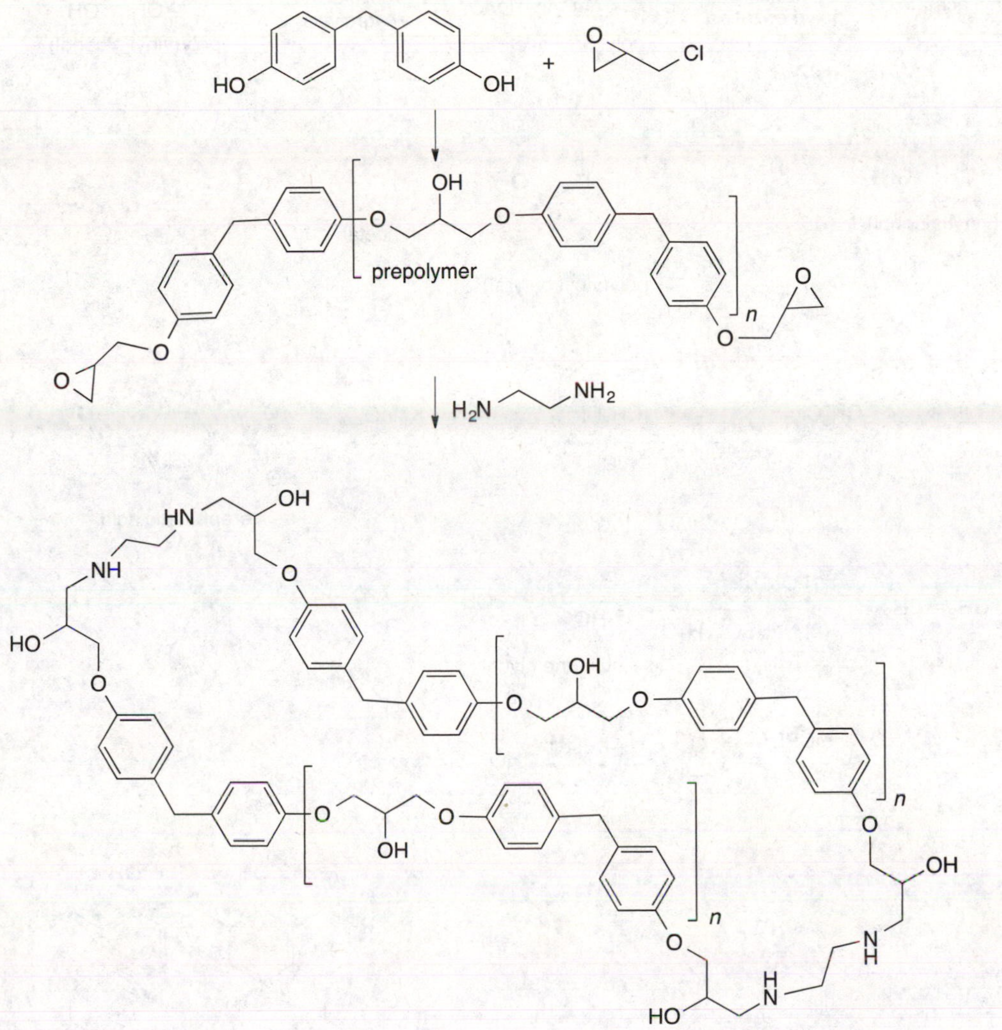

HO OH + O Cl

OH O O prepolymer O n O O

H_2N NH_2

OH HN NH HO O O OH O O n OH O O OH O n O H N HO N H

30.52

a. vinyl alcohol → (X) poly(vinyl alcohol)

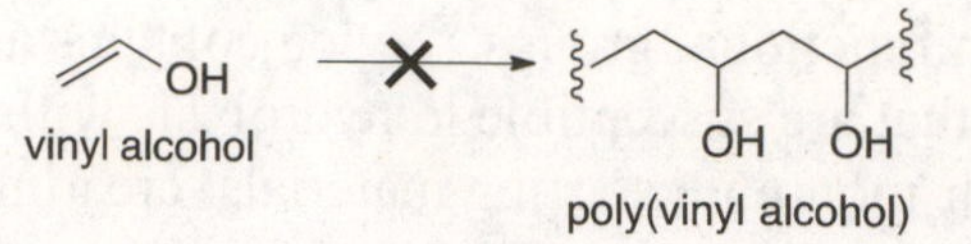

Poly(vinyl alcohol) cannot be prepared from vinyl alcohol because vinyl alcohol is not a stable monomer. It is the enol of acetaldehyde (CH_3CHO), and thus it can't be converted to poly(vinyl alcohol).

b. vinyl acetate —ROOR, radical polymerization→ poly(vinyl acetate) —^{-}OH, H_2O, hydrolysis→ poly(vinyl alcohol) + $CH_3CO_2^-$

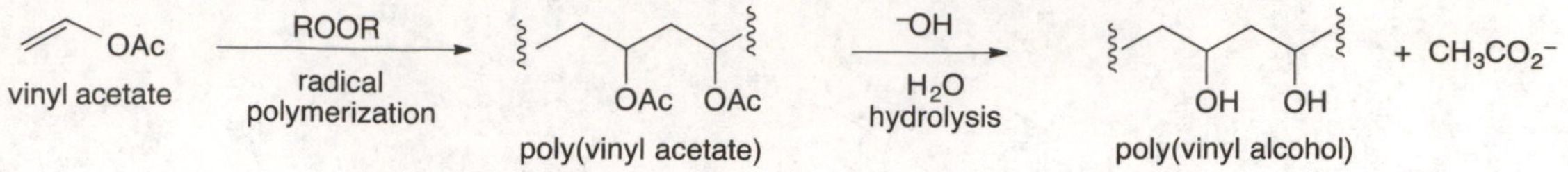

c. poly(vinyl alcohol) —butanal, H^+→ poly(vinyl butyral) (an acetal)

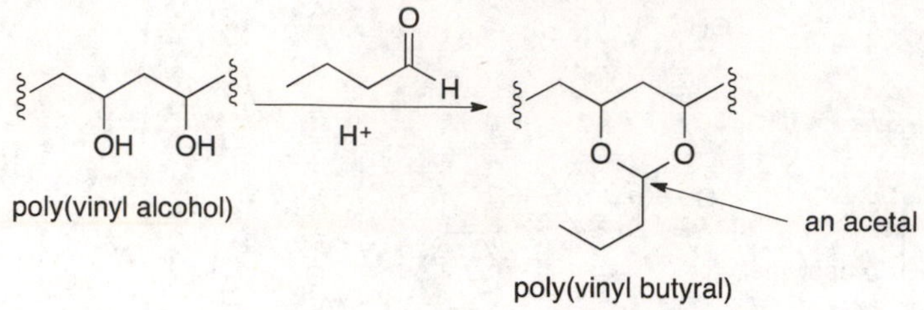

30.53

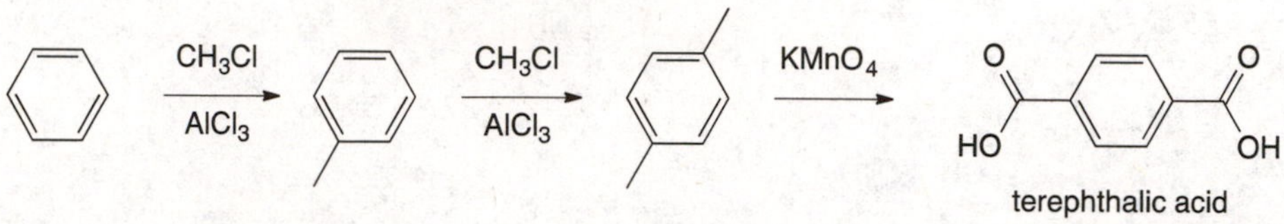

benzene —CH_3Cl, $AlCl_3$→ toluene —CH_3Cl, $AlCl_3$→ p-xylene —$KMnO_4$→ terephthalic acid

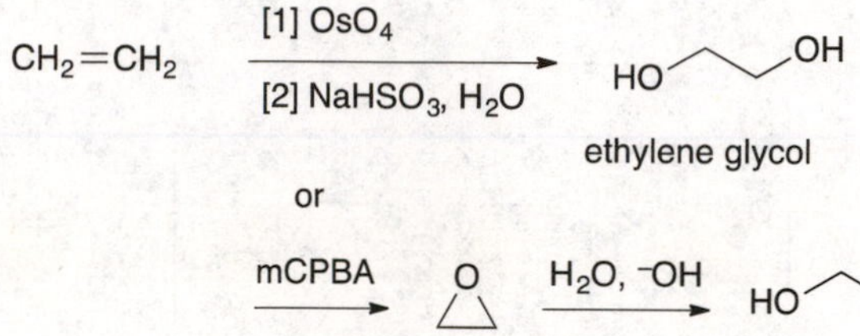

$CH_2{=}CH_2$ —[1] OsO_4; [2] $NaHSO_3$, H_2O→ HOCH₂CH₂OH (ethylene glycol)

or

—mCPBA→ ethylene oxide —H_2O, ^{-}OH→ $HOCH_2CH_2OH$

30.54

phenol + $CH_2{=}O$ → Bakelite

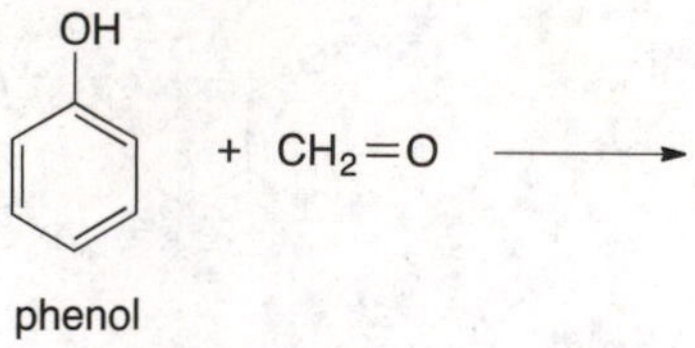

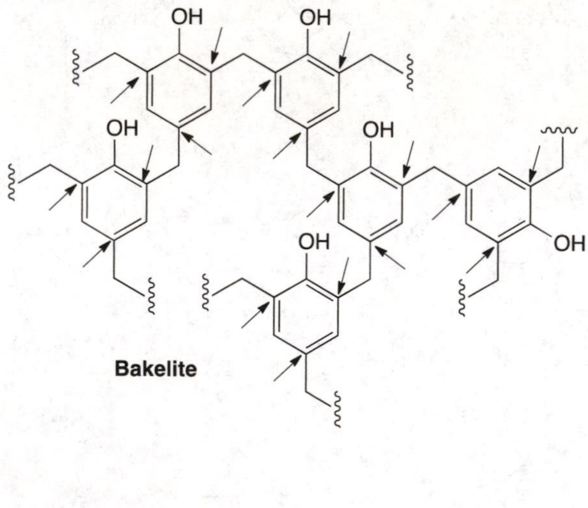

Since phenol has no substituents at any ortho or para position, an extensive network of covalent bonds can join the benzene rings together at all ortho and para positions to the OH groups.

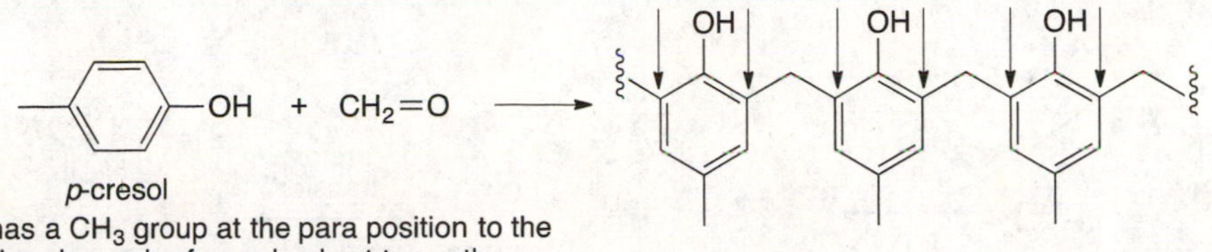

p-cresol + $CH_2=O$ ⟶ [polymer with OH groups]

p-cresol

Since *p*-cresol has a CH_3 group at the para position to the OH group, new bonds can be formed only at two ortho positions, so that a less extensive three-dimensional network can form.

30.55

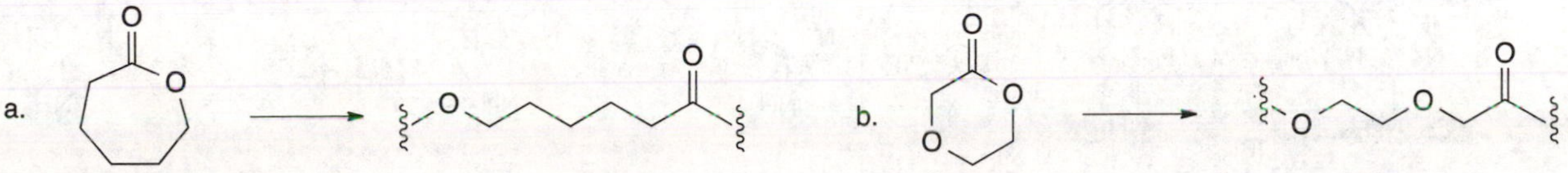

a. ε-caprolactone ⟶ polycaprolactone

b. *p*-dioxanone ⟶ polydioxanone

30.56

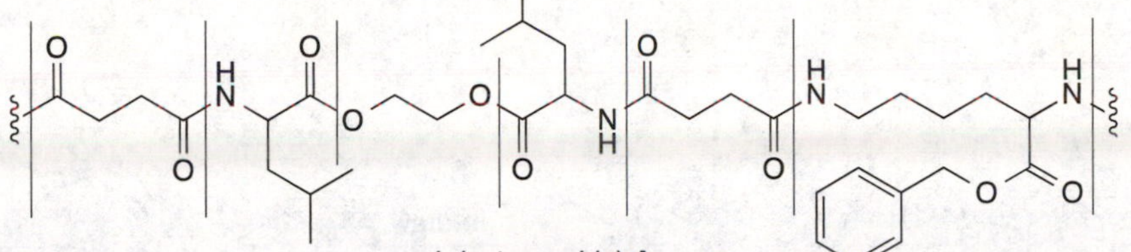

poly(ester amide) **A**

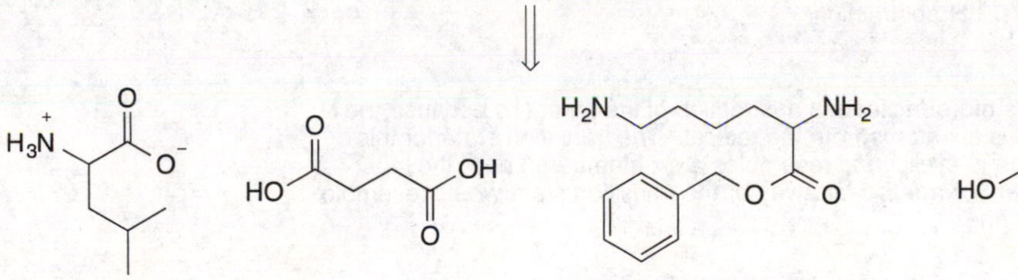

⇓

$H_3\overset{+}{N}$ … O^- (leucine); HO…OH; H_2N … NH_2 (the benzyl ester of lysine); HO…OH

leucine

the benzyl ester of lysine

30.57

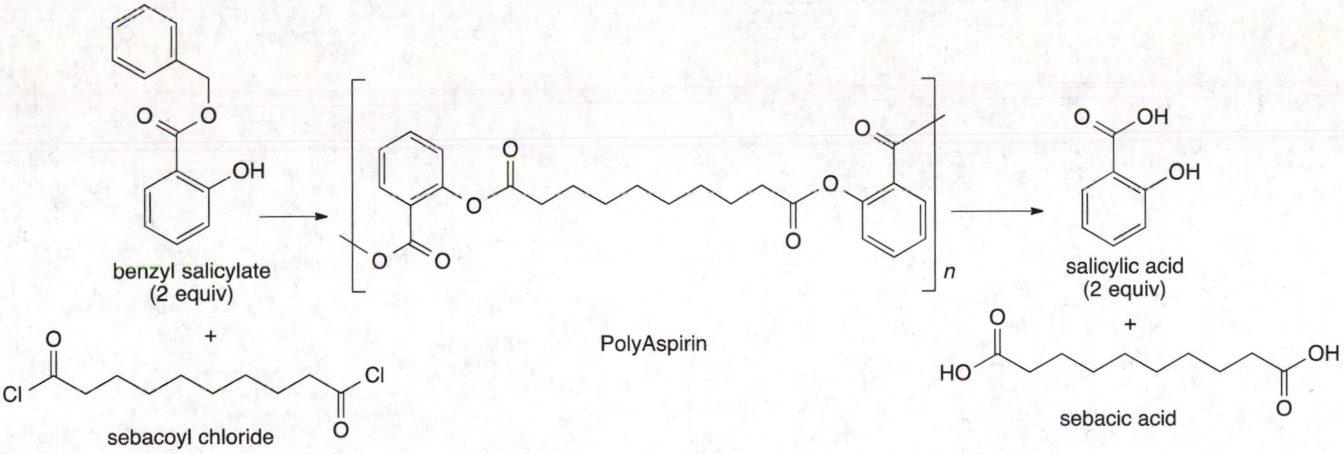

benzyl salicylate (2 equiv) + sebacoyl chloride ⟶ PolyAspirin ⟶ salicylic acid (2 equiv) + sebacic acid

benzyl salicylate
(2 equiv)

+

sebacoyl chloride

PolyAspirin

salicylic acid
(2 equiv)

+

sebacic acid

30.58

melamine

[1]

proton transfer

[2]

[3]

[4]

+ H_2O

+ A:

[5]

[6]

+ H–A

30.59

a.

[1]

intramolecular H abstraction

re-draw

[2]

[3] Repeat Step [2].

n

butyl substituent

b. Abstraction of the H is more facile than abstraction of the other H's because the H atom that is removed is six atoms from the radical. The transition state for this intramolecular reaction is cyclic, and resembles a six-membered ring, the most stable ring size. Other H's are too far away or the transition state would resemble a smaller, less stable ring.

30.60

urea + formaldehyde

repeat

repeat

ISBN-13: 978-1-259-96892-1
ISBN-10: 1-259-96892-8